ENCYCLOPEDIA OF PHYSICS

EDITED BY

S. FLÜGGE

VOLUME XLVIII

GEOPHYSICS II

GROUP EDITOR

J. BARTELS

WITH 497 FIGURES

Springer-Verlag Berlin Heidelberg GmbH
1957

HANDBUCH DER PHYSIK

HERAUSGEGEBEN VON

S. FLÜGGE

BAND XLVIII

GEOPHYSIK II

REDAKTION

J. BARTELS

MIT 497 FIGUREN

Springer-Verlag Berlin Heidelberg GmbH
1957

ISBN 978-3-642-45883-5 ISBN 978-3-642-45881-1 (eBook)
DOI 10.1007/978-3-642-45881-1

Ursprünglich erschienen bei Springer-Verlag oHG. Berlin · Göttingen · Heidelberg 1957
Softcover reprint of the hardcover 1st edition 1957

Inhaltsverzeichnis.

Dynamic Meteorology.

By

A. ELIASSEN and E. KLEINSCHMIDT jr.[1]

With 41 Figures.

I. Basic theory.

a) Physical properties of atmospheric air.

1. Fluid models. All motions in the atmosphere connected with weather are formed as a result of heating from the sun. The atmosphere acts as a kind of heat engine, continually converting heat into mechanical energy; and mechanical energy is in turn converted back to heat by frictional dissipation. This indicates that neither pure hydrodynamics, nor pure thermodynamics suffices to explain the motions in the atmosphere; the appropriate science is a combination of both, which is called "physical hydrodynamics".

In order that a fluid shall be able to serve as the working substance of a heat engine, which converts heat into mechanical energy by circular processes without any progressive change of state of the fluid, it is necessary that the possible thermodynamic states of each fluid particle form a two-dimensional manifold, at least. This is the case for atmospheric air, where two functions of state, e.g. pressure and density, are free to vary independently under the influence of heat sources.

The theory of the motion of such a general fluid is rather complicated, and very few results have been obtained except some general theorems. Most theories in dynamic meteorology have been based on a simpler fluid model where the possible thermodynamic states of each fluid particle form a one-dimensional manifold. Such fluids were named *piezotropic* by V. BJERKNES and coll. [*2*].

A piezotropic fluid cannot convert heat into mechanical energy by circular processes, and the assumption that the air is piezotropic is therefore inadequate to explain the maintenance of atmospheric motions against frictional dissipation. Nevertheless, the concept of a piezotropic fluid has proven useful in the study of motion phenomena of relatively short time scale which utilize energy stored as potential, internal or kinetic energy. The hydrodynamics of piezotropic fluids has therefore played an important role in dynamic meteorology. In a piezotropic fluid, there is for each particle a relation between density ϱ and pressure p. In the most general case, this *equation of piezotropy* may be different for all particles; it may then be written

$$\varrho = f(p, A, B, C), \tag{1.1}$$

where A, B and C are a set of LAGRANGian coordinates, which define individual particles. Letting the symbol D denote the differential derived from the *individual* change of a quantity, i.e. the change experienced by a fluid particle, one has for the LAGRANGian coordinates

$$\frac{DA}{Dt} = \frac{DB}{Dt} = \frac{DC}{Dt} = 0. \tag{1.2}$$

[1] Chap. I to III are written mainly by ELIASSEN, Chap. IV and V by KLEINSCHMIDT.

The differential quotient

$$\gamma = \frac{D\varrho}{Dp} = \frac{\partial f}{\partial p} \tag{1.3}$$

is a measure of the compressibility and is called the *coefficient of piezotropy*.

It is not necessary that as many as three LANGRANGian coordinates appear in the equation of piezotropy. In the case most commonly considered, the adiabatic motion, this equation contains just one such parameter, viz. the entropy s. The equation of piezotropy then assumes the form

$$\varrho = f(p, s) \tag{1.4}$$

with the condition

$$Ds/Dt = 0. \tag{1.5}$$

In this case, the relationship between pressure and density will be the same for particles with the same entropy, but will be different for particles with different entropy.

Another example of piezotropy is the *incompressible* fluid (provided its density is independent of the temperature), for which pressure drops out in the equation of piezotropy.

An important special case occurs if the relationship between pressure and density is the same for all particles within the fluid. In this case, no parameters appear in the equation of piezotropy, so that this equation becomes

$$\varrho = f(p). \tag{1.6}$$

The surfaces of constant density must then at all times coincide with the surfaces of constant pressure (barotropy), and the fluid is said to be *autobarotropic*. In an autobarotropic fluid the interchange of two particles will not cause any change of the field of any function of state. An adiabatic fluid in which all particles have the same entropy, and a homogeneous and incompressible fluid are both examples of autobarotropic fluids.

The "classical" hydrodynamics may be defined as the hydrodynamics of autobarotropic fluids, in particular of the homogeneous and incompressible fluid. In contrast, the "physical" hydrodynamics is the hydrodynamics of piezotropic fluids and more general fluid models of two ore more independent variables of state.

The theoretical basis of physical hydrodynamics was laid by V. BJERKNES, who systematized this science, and introduced the classification of fluid models outlined above, and by MARGULES, who studied the energy conversions in general fluid models.

2. Thermodynamic properties of dry air. Atmospheric air is a mixture of a number of perfect gases and water vapor, sometimes also containing suspended condensation products of water. Whereas the amount of water vapor is highly variable, the mixing ratio of the perfect gases is constant up to great heights. Dry air is therefore a well-defined gaseous mixture which with great accuracy behaves like an ideal gas.

Thus the pressure p, density ϱ and absolute temperature T of dry air fulfill the *gas equation*

$$p/\varrho = R\,T, \tag{2.1}$$

where $R = 2.8704 \times 10^6$ cm^2 sec^{-2} per degree absolute, is the gas constant of dry air.

The *internal energy* per unit mass of dry air is

$$e = c_v\,T + \text{constant}, \tag{2.2}$$

where $c_v = 7.17 \times 10^6$ cm^2 sec^{-2} per degree is the specific heat at constant volume.

The *enthalpy* per unit mass is

$$h = e + p\,\alpha = c_p\,T + \text{constant}, \tag{2.3}$$

where $\alpha = \varrho^{-1}$ is the specific volume, and

$$c_p = c_v + R. \tag{2.4}$$

The ratio

$$\varkappa = \frac{c_p}{c_v} \tag{2.5}$$

has for dry air the value 1.4.

For any gas or liquid, or for any gaseous mixture of constant mixing ratio, the *entropy* per unit mass is defined by

$$T\,ds = de + p\,d\alpha = dh - \alpha\,dp. \tag{2.6}$$

It should be noted that the differentials in this equation are quite general; the equation holds for individual changes as well as for changes from one particle to another.

In the case of dry air, one has

$$T\,ds = c_v\,dT + p\,d\alpha = c_p\,dT - \alpha\,dp. \tag{2.7}$$

Eliminating α by means of (2.1) and integrating, one obtains

$$s = c_p \log T - R \log p + \text{constant} = c_v \log p - c_p \log \varrho + \text{constant}. \tag{2.8}$$

A convenient measure of the entropy of dry air is the *potential temperature*, θ, which may be defined by

$$s = c_p \log \theta + \text{constant}. \tag{2.9}$$

Comparison between (2.8) and (2.9) shows that

$$\theta = \left(\frac{p_0}{p}\right)^{1-\frac{1}{\varkappa}} T = \frac{p_0^{1-\frac{1}{\varkappa}}\, p^{\frac{1}{\varkappa}}}{R\varrho}, \tag{2.10}$$

where p_0 is a constant which is given the value 1000 mbar = 1 bar[1].

It is often convenient to introduce the "EXNER function"

$$\Pi = c_p \left(\frac{p}{p_0}\right)^{1-\frac{1}{\varkappa}}. \tag{2.11}$$

One has

$$\alpha\,dp = \theta\,d\Pi, \tag{2.12}$$

$$T\,ds = \Pi\,d\theta \tag{2.13}$$

and

$$h = \Pi\theta + \text{constant}. \tag{2.14}$$

These formulas will be referred to later.

Dry air is a two-parameter system, the state of which is determined by the two independent functions of state, e.g. p and ϱ. If in a mass of dry air the changes of state of the particles are *isentropic*, then the air is piezotropic, and the equation of piezotropy is obtained from (2.10), where θ is now a constant for each particle. Isentropic changes of state of dry air are referred to as

[1] The potential temperature is the temperature that the air would assume when brought adiabatically from its actual pressure to the pressure of 1000 mb.

dry-adiabatic. The corresponding coefficient of piezotropy is

$$\gamma_d = \left(\frac{D\varrho}{Dp}\right)_d = \frac{1}{\varkappa R T} \tag{2.15}$$

and from (2.7) one obtains

$$\left(\frac{DT}{Dp}\right)_d = \frac{\alpha}{c_p}. \tag{2.16}$$

3. Thermodynamic properties of moist air. α) *Thermodynamics of the water component.* Water vapor does not follow the laws of an ideal gas nearly as accurately as does dry air. Nevertheless, it suffices for most purposes to consider also water vapor as an ideal gas. The errors introduced by this assumption will be small as far as the meteorological properties of moist air are concerned, since the amount of water in atmospheric air constitutes only a small fraction of the total mass.

Denoting the partial pressure and density of water vapor by p' and ϱ', its equation of state may therefore be written

$$\varrho' = \sigma \frac{p'}{RT}. \tag{3.1}$$

Here R is the gas constant of dry air, and $\sigma = 0.622$ is the relative density of water vapor with respect to dry air. Similarly, the enthalpy of water vapor may be written

$$h' = c_p' T + \text{constant}, \tag{3.2}$$

where c_p' is the specific heat of water vapor at constant pressure.

Liquid water may be treated as incompressible. Denoting its specific heat by c'', the enthalpy per unit mass may be written.

$$h'' = c'' T + \text{constant}. \tag{3.3}$$

By definition, the latent heat of evaporation is

$$L = h' - h'', \tag{3.4}$$

where h' and h'' refer to states of vapor and liquid water which are in equilibrium with each other. However, since both h' and h'' are assumed to be functions of T only, (3.4) may be considered to be true also for vapor and liquid water which are not in equilibrium, provided only that their temperature is the same. The enthalpy of water vapor may therefore be written

$$h' = c'' T + L + \text{constant}, \tag{3.5}$$

where the arbitrary constant is identic with the constant in (3.3). The heat of condensation varies with the temperature, but the variation is slow, and within the range of temperatures occuring in the atmosphere one may for most purposes consider L as a constant.

Let $P(T)$ denote the *saturation pressure* of water vapor with respect to liquid water with a plane interface. The details of the condensation processes in the atmosphere are beyond the scope of this article[1]; it suffices here to assume that: 1. if the water component is present as vapor only, then $p' \lesseqgtr P(T)$; and 2. if the air contains drops of liquid water, then $p' = P(T)$. Case 1. will be referred to as unsaturated (or clear) air, and case 2. as saturated (or cloudy) air.

[1] See the chapter by MASON and LUDLAM in this volume.

What is said above about the liquid phase is true for the solid phase also, and the same formulas apply, provided only c'' and L are interpreted as specific heat of ice and latent heat of sublimation, respectively.

β) Unsaturated air. Consider a mass of moist air (of pressure p and density ϱ) which is a mixture of dry air (of partial pressure p_d and partial density ϱ_d) and water vapor (of partial pressure p' and partial density ϱ'). Applying the laws for ideal gaseous mixtures, one has

$$p = p_d + p', \qquad \varrho = \varrho_d + \varrho'. \tag{3.6}$$

The equation of state for the dry air component is

$$\varrho_d = \frac{p_d}{R\,T} \tag{3.7}$$

and for the water vapor component (3.1). The mixing ratio is defined as

$$\mu' = \frac{\varrho'}{\varrho_d} = \sigma \frac{p'}{p - p'}. \tag{3.8}$$

In unsaturated air, this quantity will change only as a result of diffusion and turbulent mixing. This is usually a slow process, and in many cases one may consider μ' as an individual constant so that

$$D\mu'/Dt = 0. \tag{3.9}$$

From (3.1), (3.6), (3.7) and (3.8) one derives

$$\frac{p}{\varrho} = \left(1 + \frac{1-\sigma}{\sigma}\,\frac{\mu'}{1+\mu'}\right) R\,T. \tag{3.10}$$

Hence, if the mixing ratio is constant, moist air satisfies the gas equation, and its gas constant is slightly larger than the gas constant of dry air. It is customary in meteorology to write this equation in the form

$$p/\varrho = R\,T_v, \tag{3.11}$$

where

$$T_v = \left(1 + \frac{1-\sigma}{\sigma}\,\frac{\mu'}{1+\mu'}\right) T \tag{3.12}$$

is called the *virtual temperature*. The difference $(T_v - T)$ may amount to a few degrees centigrade. If no great accuracy is needed, one may neglect the effect of humidity upon the equation of state and use the gas equation for dry air (2.1).

From (2.3) and (3.5), one obtains for the enthalpy per unit mass of moist air

$$h = \frac{c_p + \mu' c''}{1+\mu'}\,T + \frac{\mu'}{1+\mu'}\,L + \text{constant}. \tag{3.13}$$

Since μ' is a small quantity, of the order 1% in the lower troposphere and still smaller at greater heights, a suitable approximation for most purposes is

$$h \approx c_p\,T + \mu' L + \text{constant}. \tag{3.14}$$

In the same approximation, the internal energy is

$$e \approx c_v\,T + \mu' L + \text{constant}. \tag{3.15}$$

One may therefore write, for any process which satisfies (3.9),

$$T\,Ds = Dh - \alpha\,Dp \approx c_p\,D\,T - \alpha\,Dp. \tag{3.16}$$

Thus, nonsaturated air behaves in this approximation in the same way as dry air. In particular, the isentropic change of state will be very nearly unaffected by the humidity.

γ) Saturated air. Let a mass of moist air contain liquid water, so that one mass unit of dry air is connected with μ' mass units of water vapor and μ'' mass units of liquid water. The vapor is assumed to be in equilibrium with the liquid phase, so that

$$p' = P(T). \tag{3.17}$$

In this case, (3.8) yields

$$\mu' = \sigma \frac{P(T)}{p - P(T)} \approx \sigma \frac{P(T)}{p} \tag{3.18}$$

i.e. the mixing ratio in saturated air is a function of p and T. The quantity will change as a result of condensation and evaporation, so that (3.9) is no longer applicable. From (3.18) one obtains

$$D\mu' \approx \frac{\sigma}{p}\frac{dP}{dT} DT - \frac{\sigma P}{p^2} Dp. \tag{3.19}$$

The total mass of the water component per unit mass of dry air is

$$\mu = \mu' + \mu''. \tag{3.20}$$

In the absence of diffusion, turbulent mixing and precipitation, one has

$$D\mu/Dt = 0 \tag{3.21}$$

which now replaces (3.9).

The water drops will slightly increase the mean density of the air, and this effect can easily be expressed in terms of μ''. Such a formula is of little use, because μ'' is not known. The effect is very small, however, since μ'' is always a small quantity, and one may therefore as in the case of nonsaturated air in most applications use the gas equation for dry air (2.1).

The enthalpy per unit mass of the two-phase system is

$$h = \frac{c_p + \mu c''}{1 + \mu} T + \frac{\mu'}{1 + \mu} L + \text{constant}. \tag{3.22}$$

As in the case of nonsaturated air, this formula may be replaced by the approximate formula

$$h \approx c_p T + \mu' L + \text{constant} \tag{3.23}$$

which may be used whether the air is saturated or not. However, in the saturated case, μ' is no longer an individual constant. From (3.23) and (3.19) one finds

$$T Ds = Dh - \alpha Dp \approx \left(c_p + \frac{\sigma L}{p}\frac{dP}{dT}\right) DT - \left(\alpha + \frac{\sigma L P}{p^2}\right) Dp. \tag{3.24}$$

The difference between the nonsaturated and the saturated case is apparent from a comparison between (3.16) and (3.24).

Isentropic changes of state of saturated air are referred to as *moist-adiabatic*. For such processes, we have according to (3.24)

$$\left(\frac{DT}{Dp}\right)_m \approx \frac{\alpha + \dfrac{\sigma L P}{p^2}}{c_p + \dfrac{\sigma L}{p}\dfrac{dP}{dT}}. \tag{3.25}$$

When its magnitude is compared with the dry-adiabatic change of state (2.16) one finds

$$\left(\frac{DT}{Dp}\right)_m < \left(\frac{DT}{Dp}\right)_d. \tag{3.26}$$

The difference between these quantities is appreciable for relatively high temperatures. However, since P and dP/dT tend rapidly towards zero as T decreases, it follows that $(DT/Dp)_m$ approaches $(DT/Dp)_d$ for low temperatures; and their difference is for most purposes negligible at temperatures below $-10°$ Centigrade.

As a conclusion, we may say that atmospheric air may, as far as its thermodynamic properties are concerned, behave in two different ways: 1. Non-saturated (or clear) air behaves nearly like dry air, regardless of its humidity; and 2. saturated (or cloudy) air has properties which differ from those of non-saturated air, and which are nearly independent of the content of liquid water or ice. Therefore, the total content of the water constituent is of little importance as long as we know whether the air is saturated or not, and we may then consider the thermodynamic state of the air as given by two functions of state, e.g. pressure and density. However, knowledge of the total water content is essential if we want to determine when saturation will begin and when it will end.

In many theories in dynamic meteorology, the condensation processes are disregarded, and the air is considered as non-saturated, or dry.

For a more complete treatment of atmospheric thermodynamics, the reader is referred to the textbooks [*1*], [*4*], [*9*], [*11*], [*12*] and [*16*, I].

b) Basic equations of physical hydrodynamics.

4. The hydrodynamic equations. *α) The* EULER*ian representation.* The EULERian representation, with three space coordinates and time as independent variables, is most commonly used in dynamic meteorology. The space coordinates may be CARTESian, spherical or any other system; however, the vector notation permits us to write equations without specifying the coordinate system.

The EULERian description of a hydrodynamic process involves the knowledge of the velocity vector $\boldsymbol{v}$ and the thermodynamic state in all points and at all times. In most cases, the thermodynamic state will be considered as determined by two thermodynamic functions, e.g. pressure p and density ϱ. If we want to take into account the effect of humidity, another variable, e.g. the mixing ratio μ must be added. Therefore the dependent variables, which are to be considered as functions of the space coordinates and time, are $\boldsymbol{v}$, p, ϱ and possibly μ.

The operator D/Dt expressing individual differentiation with respect to time may be expanded in the following way

$$\frac{D}{Dt} = \frac{\partial}{\partial t} + \boldsymbol{v} \cdot \text{grad}, \tag{4.1}$$

where the first term on the right is referred to as the local rate of change, and the second term as the convective rate of change. This formula is applicable to vector fields as well as scalar fields; and it is valid in any frame of reference.

β) The equation of continuity. The vector $\varrho\boldsymbol{v}$ represents the flux of mass; and the net flow of mass out of a resting volume element $d\tau$ per unit time is therefore $\operatorname{div}(\varrho\boldsymbol{v})\,d\tau$. According to the principle of conservation of mass, this expression must be equal to the rate of decrease of the mass contained within the same volume element, which is $-\frac{\partial \varrho}{\partial t}\, d\tau$. Hence

$$\frac{\partial \varrho}{\partial t} + \operatorname{div}(\varrho \boldsymbol{v}) = 0. \tag{4.2}$$

This equation is called the *equation of continuity*. By means of (4.1) and the identity

$$\operatorname{div}(\varrho \boldsymbol{v}) = \varrho \operatorname{div} \boldsymbol{v} + \boldsymbol{v} \cdot \operatorname{grad} \varrho, \tag{4.3}$$

the equation of continuity may be written in the alternative form

$$\operatorname{div} \boldsymbol{v} = -\frac{1}{\varrho}\frac{D\varrho}{Dt} = \frac{1}{\alpha}\frac{D\alpha}{Dt}. \tag{4.4}$$

If the humidity is taken into account, we will need a corresponding continuity equation for the water component. However, no simple expression is obtained for the flux of water, since it will depend upon such processes as precipitation, evaporation and diffusion.

From the equation of continuity (4.2) and the differentiation rule (4.1), we may derive the following relation

$$\varrho \frac{D\psi}{Dt} = \frac{\partial(\varrho \psi)}{\partial t} + \operatorname{div}(\varrho \boldsymbol{v} \psi), \tag{4.5}$$

where ψ denotes any scalar or vector function of the space coordinates and time.

γ) The equation of motion referred to an inertial frame. The forces acting upon an air particle referred to unit mass, are the following: 1. The force of pure gravity $\boldsymbol{g}_a = -\operatorname{grad}\varphi_a$, where φ_a is the potential of the force of gravitational attraction; 2. the pressure force $-\alpha \operatorname{grad} p$; and 3. the force of molecular friction $\alpha \operatorname{div} \mathcal{F}$, where $\mathcal{F}$ is the NAVIER-STOKES *frictional stress tensor*.

Denoting by $\boldsymbol{v}_a$ the velocity relative to an inertial frame of reference, and by D_a the individual change relative to an inertial frame of reference, the acceleration is $D_a \boldsymbol{v}_a/Dt$. NEWTON's second law, written for a mass unit of air, therefore gives

$$D_a \boldsymbol{v}_a/Dt = -\operatorname{grad}\varphi_a - \alpha \operatorname{grad} p + \alpha \operatorname{div} \mathcal{F}. \tag{4.6}$$

This is the equation of motion referred to an inertial frame. According to (4.1), the acceleration may be written

$$\frac{D_a \boldsymbol{v}_a}{Dt} = \left(\frac{\partial \boldsymbol{v}_a}{\partial t}\right)_a + \boldsymbol{v}_a \cdot \operatorname{grad} \boldsymbol{v}_a, \tag{4.7}$$

where $(\partial \boldsymbol{v}_a/\partial t)_a$ is the local acceleration in the inertial frame, and $\boldsymbol{v}_a \cdot \operatorname{grad} \boldsymbol{v}_a$ the convective acceleration. An alternative expression for the convective acceleration which is often useful is

$$\boldsymbol{v}_a \cdot \operatorname{grad} \boldsymbol{v}_a = \operatorname{curl} \boldsymbol{v}_a \times \boldsymbol{v}_a + \operatorname{grad}\left(\tfrac{1}{2} \boldsymbol{v}_a^2\right). \tag{4.8}$$

If we let the frame of reference follow the earth in its orbital motion without participating in its rotation, Eq. (4.6) will hold provided the inertial force due to the acceleration in the orbital motion is added to the forces. This inertial force will nearly balance the gravitational attraction from the moon and the sun, however, their resultant being the small tidal forces. Therefore, if we neglect the tidal force, Eq. (4.6) will hold in a non-rotating frame centered at the center of the earth, provided φ_a is the potential of gravitational attraction from the earth only. We shall refer to motions relative to this coordinate system as "*absolute*".

δ) The equation of motion referred to a frame rotating with the solid earth. The earth rotates around its axis with an angular velocity $\Omega = 0.729 \times 10^{-4}\,\text{sec}^{-1}$. Applying the right-handedscrew convention of correspondence between positive direction of rotation and positive direction of translation along the axis, the angular velocity may be represented as a vector $\boldsymbol{\Omega}$ pointing toward the celestial North Pole.

Let $\boldsymbol{v}$ denote velocity of the air relative to a frame of reference rotating with the earth (which will be referred to as "*relative velocity*"); let $\boldsymbol{r}$ denote position vector from the center of the earth, and $\boldsymbol{r}_e$ the equator projection of $\boldsymbol{r}$. In the rotating frame, NEWTON's second law still holds, provided two fictitious inertial forces are added, viz. the centrifugal force and the CORIOLIS force. The centrifugal force per unit mass is

$$\Omega^2 \boldsymbol{r}_e = \operatorname{grad} \tfrac{1}{2} (\boldsymbol{\Omega} \times \boldsymbol{r})^2$$

and the CORIOLIS force $-2\boldsymbol{\Omega} \times \boldsymbol{v}$.

The centrifugal force and the force of the earth's gravitational attraction must be considered as a unity since only their sum is observable. This sum

$$\boldsymbol{g} = \boldsymbol{g}_a + \Omega^2 \boldsymbol{r}_e = -\operatorname{grad} \varphi \tag{4.9}$$

may be called the force of apparent gravity, but is usually referred to simply as the force of gravity. Its potential

$$\varphi = \varphi_a - \tfrac{1}{2} (\boldsymbol{\Omega} \times \boldsymbol{r})^2 \tag{4.10}$$

is called the *geopotential*.

When the centrifugal force and the CORIOLIS force are added on the right-hand side of (4.6), we obtain the equation of motion in the rotating frame,

$$\frac{D\boldsymbol{v}}{Dt} = -\operatorname{grad} \varphi - 2\boldsymbol{\Omega} \times \boldsymbol{v} - \alpha \operatorname{grad} p + \alpha \operatorname{div} \mathscr{F}, \tag{4.11}$$

where D denotes individual change relative to the rotating frame.

By means of (4.1), the acceleration may be written as

$$\frac{D\boldsymbol{v}}{Dt} = \frac{\partial \boldsymbol{v}}{\partial t} + \boldsymbol{v} \cdot \operatorname{grad} \boldsymbol{v} \tag{4.12}$$

where the terms on the right are the local and convective accelerations in the frame rotating with the earth.

Eq. (4.11) is referred to a mass unit. The equation of motion referred to a unit volume is obtained by multiplication with ϱ. Making use of (4.5), we obtain the equation of motion in the equivalent form

$$\frac{\partial (\varrho \boldsymbol{v})}{\partial t} + \operatorname{div} (\varrho \boldsymbol{v} \boldsymbol{v} + p \mathscr{E} - \mathscr{F}) = -\varrho \operatorname{grad} \varphi - 2\boldsymbol{\Omega} \times \varrho \boldsymbol{v} \tag{4.13}$$

where $\mathscr{E}$ denotes the unit tensor. The first term on the left is the rate of increase of impulse per unit volume; the second term is the divergence of a tensor which may be interpreted as an impulse flux, and the right-hand side represents the rate of production of impulse.

ε) Boundary conditions. On a rigid boundary surface, $F(x, y, z) = 0$, the normal velocity component must vanish; i.e.

$$\boldsymbol{v} \cdot \operatorname{grad} F = 0 \quad \text{when} \quad F(x, y, z) = 0. \tag{4.14}$$

This is the only condition to be fulfilled at a rigid boundary in a nonviscid fluid.

In a viscous fluid, also the tangential velocity component must vanish, so that

$$\boldsymbol{v} = 0 \quad \text{when} \quad F = 0. \tag{4.15}$$

These conditions are easily generalized to the case of a boundary surface of specified motion.

There occur in the atmosphere abrupt transition zones (fronts and inversions) which are most conveniently treated as internal *surfaces of discontinuity*. On such surfaces, the equations of motion and continuity lose their meaning, because the derivatives involved do not exist; therefore they must be replaced by interface conditions.

Suppose that $F(x, y, z, t) = 0$ is the equation of the interface (which is usually unknown); and suppose further that the interface is a material surface. Letting subscripts 1 and 2 indicate the dependent variables on either side of the interface, we have

$$\left.\begin{aligned} \frac{\partial F}{\partial t} + \boldsymbol{v}_1 \cdot \operatorname{grad} F = 0 \\ \frac{\partial F}{\partial t} + \boldsymbol{v}_2 \cdot \operatorname{grad} F = 0 \end{aligned}\right\} \quad \text{when } F = 0. \tag{4.16}$$

These equations are called the *kinematic interface conditions*; they replace the continuity equation at the interface. When these equations are subtracted, one obtains

$$(\boldsymbol{v}_1 - \boldsymbol{v}_2) \cdot \operatorname{grad} F = 0 \quad \text{when} \quad F = 0 \tag{4.17}$$

showing that the normal velocity components are the same on both sides. This equation is not sufficient, however, since F is not known, and (4.17) must be supplemented by one of the Eqs. (4.16); this would suffice in the non-viscid case. In the case of viscous fluids, one has

$$\left.\begin{aligned} \boldsymbol{v}_1 = \boldsymbol{v}_2 \\ \frac{\partial F}{\partial t} + \boldsymbol{v} \cdot \operatorname{grad} F = 0 \end{aligned}\right\} \quad \text{when} \quad F = 0. \tag{4.18}$$

The equations of motion must be replaced by the dynamic interface condition. This is, in the case of a nonviscous fluid,

$$p_1 - p_2 = 0 \quad \text{when} \quad F = 0. \tag{4.19}$$

For a viscous fluid, one must write instead

$$\operatorname{grad} F \cdot \left[-(p_1 - p_2)\,\mathscr{E} + \mathscr{F}_1 - \mathscr{F}_2\right] = 0 \quad \text{when} \quad F = 0. \tag{4.20}$$

5. Circulation and vorticity. α) *Circulation theorem of* V. BJERKNES *in absolute motion.* Consider an arbitrary closed curve L within the fluid, with a chosen positive direction along it. Neglecting frictional forces and taking the line integral of (4.6) in positive direction along L, the force of gravity drops out and one obtains

$$\oint_L \frac{D_a \boldsymbol{v}_a}{Dt} \cdot d\boldsymbol{r} = -\oint_L \alpha \operatorname{grad} p \cdot d\boldsymbol{r}. \tag{5.1}$$

If the closed curve is defined as a material curve, which follows the air in its motion, then (5.1) may be written

$$\frac{dC_a}{dt} = \oint_L (-\alpha \operatorname{grad} p) \cdot d\boldsymbol{r} = -\oint_L \alpha\, dp, \tag{5.2}$$

where

$$C_a = \oint_L \boldsymbol{v}_a \cdot d\boldsymbol{r} \tag{5.3}$$

is called the *absolute circulation.* Eq. (5.2) is the *first circulation theorem* of V. BJERKNES[1]. By means of the theorem of STOKES, (5.2) may also be written

$$\frac{dC_a}{dt} = \int_\sigma \boldsymbol{S} \cdot d\boldsymbol{\sigma}, \tag{5.4}$$

where $d\boldsymbol{\sigma}$ is the vectorial surface element of any surface σ which has L as boundary curve, and where

$$\boldsymbol{S} = \operatorname{curl}(-\alpha \operatorname{grad} p) = -\operatorname{grad}\alpha \times \operatorname{grad} p \tag{5.5}$$

will be referred to as the "*baroclinicity vector*".

The field of $\boldsymbol{S}$ is non-divergent or "solenoidal". A convenient representation of any vector field with this property is obtained by dividing the space into tubes or "solenoids" of unit vector flux. A set of such solenoids for the field $\boldsymbol{S}$ is formed by the family of isosteric surfaces $\alpha =$ constant (spaced one volume unit apart) and the family of isobaric surfaces $p =$ constant (spaced one pressure unit apart). These solenoids will have parallellogrammatic cross-section; they may be given positive direction in the direction of $\boldsymbol{S}$.

The right-hand side of (5.5) is seen to represent the number of such isosteric—sobaric solenoids embraced by the curve L; the number is to be counted algebraically, so that a solenoid counts $+1$ or -1 according to whether it runs through the curve L (through the surface σ) in positive or negative direction.

By means of (2.6) and (2.12), the baroclinicity vector $\boldsymbol{S}$ may also be written

$$\boldsymbol{S} = -\operatorname{grad} s \times \operatorname{grad} T = -\operatorname{grad}\theta \times \operatorname{grad}\Pi \tag{5.6}$$

and alternative solenoids of $\boldsymbol{S}$ are therefore obtained from the isentropic surfaces $s =$ constant and the isothermal surfaces $T =$ constant, or from the surfaces $\theta =$ constant and $\Pi =$ constant.

According to V. BJERKNES[2], the density field is said to be *barotropic* if $\boldsymbol{S} = 0$ in every point within a region. This implies that the isosteric surfaces $\alpha =$ constant coincide with the isobaric surfaces $p =$ constant, or that grad $\alpha = 0$ (the case grad $p = 0$ is uninteresting). In other words, the barotropic state is characterized by a constant density in each isobaric surface. The *coefficient of barotropy* Γ is defined by

$$\operatorname{grad}\varrho = \Gamma \operatorname{grad} p. \tag{5.7}$$

Within a barotropic region, the circulation theorem takes the form

$$dC_a/dt = 0. \tag{5.8}$$

On the other hand, the density field is said to be *baroclinic* if $\boldsymbol{S} \neq 0$. This implies that grad α and grad p are both different from zero and that the isosteric surfaces do not coincide with the isobaric surfaces. Hence, baroclinicity means that the density distribution in each isobaric surface is non-uniform.

If the fluid remains in the state of rest, then the circulation of every material closed curve remains zero, and it follows that the state must be barotropic. Such a state of rest can be disturbed by means of heat and cold sources, provided the fluid is not piezotropic. The density in an isobaric surface will decrease when the fluid is heated and increase where the fluid is cooled, so that the initially uniform

[1] V. BJERKNES: Vidensk. selsk. Skr., Christiania (Oslo), Math.-naturv. Kl. **1898**, No. 5. — Sv. Vetensk. Akad. Handl., Stockh. **31**, Nr. 4 (1898).

[2] V. BJERKNES: Geof. Publ. **2**, No. 4 (1921).

density distribution in the isobaric surface will become non-uniform, and the state will become baroclinic. As a result, material closed curves will acquire a circulation in accordance with (5.4). Roughly speaking, the motion thus created is directed upwards towards lower pressure where the fluid was heated, and downwards towards higher pressure where the fluid was cooled.

In the atmosphere, heat sources and sinks are continually producing circulation changes in this manner. However, the resulting motions are strongly modified by friction and by the earth's rotation. This latter effect is relatively weak in the case of such local circulations as land and sea breeze, and mountain and valley breeze, whereas the earth's rotation is a dominating factor for large-scale phenomena such as the monsoon circulation and the general circulation between low and high latitudes.

Whether a fluid is piezotropic or not, the density field is normally baroclinic when the fluid is in the state of motion (although barotropy may occur accidentally at certain instants). However, if in a piezotropic fluid the piezotropic relation between pressure and density is the same for all particles, then the density must necessarily be constant on each isobaric surface and the density field is always barotropic, no matter how the fluid moves. A fluid with this physical property is called *autobarotropic* (compare Sect. 1). In an autobarotropic fluid, Eq. (5.8) holds at all times, expressing KELVIN's theorem of the conservation of circulation; and the vortex-theorems by HELMHOLTZ hold for autobarotropic fluids only.

β) The second circulation theorem of V. BJERKNES. The circulation theorem (5.4) eliminates the force of gravity from the equation of motion (4.6). We may instead eliminate the pressure force if we multiply by ϱ before taking the line integral. This gives, when friction is neglected,

$$\oint_L \varrho \frac{D_a \boldsymbol{v}_a}{Dt} \cdot d\boldsymbol{r} = -\oint_L \varrho \operatorname{grad} \varphi_a \cdot d\boldsymbol{r} = -\oint_L \varrho \, d\varphi_a, \tag{5.9}$$

which is the *second circulation theorem* of V. BJERKNES (ref. 1, p. 11). Here the operator D_a/Dt cannot be taken outside the integral sign on the left-hand side, as in the case of Eq. (5.1).

The right-hand side of (5.9) vanishes for all closed curves if the density is constant in equipotential surfaces, i.e. if the density field is *potentiotropic*. If the density field is *potentioclinic*, i.e. if the density has a non-vanishing gradient in equipotential surfaces, then the right-hand side of (5.9) is different from zero for most closed curves.

γ) Vorticity theorem in absolute motion. In the absence of friction, the equation of motion (4.6) may be written

$$\frac{\partial \boldsymbol{v}_a}{\partial t} + \operatorname{curl} \boldsymbol{v}_a \times \boldsymbol{v}_a + \operatorname{grad}\left(\frac{1}{2} \boldsymbol{v}_a^2\right) = -\operatorname{grad} \varphi_a - \alpha \operatorname{grad} p, \tag{5.10}$$

where the acceleration has been expanded according to (4.7) and (4.8). Performing curl operation on both sides one obtains

$$\frac{D_a}{Dt} \operatorname{curl} \boldsymbol{v}_a = -\operatorname{curl} \boldsymbol{v}_a \operatorname{div} \boldsymbol{v}_a + (\operatorname{curl} \boldsymbol{v}_a) \cdot \operatorname{grad} \boldsymbol{v}_a + \boldsymbol{S}, \tag{5.11}$$

which is the vorticity equation. The two first terms on the right express the kinematic vorticity generation (rotation and stretching of the vortex tubes); and the last term is the vorticity generation due to baroclinicity. This latter effect was discovered by SILBERSTEIN[1].

[1] L. SILBERSTEIN: Bull. Internat. Acad. Sci. Cracow, Juin **1896**.

The component ζ_a of the absolute vorticity along the local vertical plays a particularly important role in dynamic meteorology. Let r denote distance from the center of the earth, and let us consider the geopotential surfaces as spheres $r=$ constant; then $\boldsymbol{k}=\operatorname{grad} r$ is a unit vector pointing towards the local zenith. Let $\boldsymbol{v}_{ah}$ denote the horizontal velocity component and $w=\boldsymbol{k}\cdot\boldsymbol{v}_a$ the vertical velocity; similarly, let curl $\boldsymbol{v}_a$ be split up into its horizontal component $(\operatorname{curl}\boldsymbol{v}_a)_h$ and its vertical component

$$\zeta_a=\boldsymbol{k}\cdot\operatorname{curl}\boldsymbol{v}_a=\operatorname{grad} r\cdot\operatorname{curl}\boldsymbol{v}_a. \tag{5.12}$$

Applying gradient operation on the identity

$$w=\frac{Dr}{Dt}=\frac{D_a r}{Dt}=\left(\left(\frac{\partial}{\partial t}\right)_a+\boldsymbol{v}_a\cdot\operatorname{grad}\right)r$$

we obtain

$$\frac{D_a\boldsymbol{k}}{Dt}=\frac{D_a}{Dt}\operatorname{grad} r=\operatorname{grad} w-(\operatorname{grad}\boldsymbol{v}_a)\cdot\boldsymbol{k}. \tag{5.13}$$

Individual time differentiation of (5.12) and application of (5.11) and (5.13) gives

$$\frac{D\zeta_a}{Dt}=-\zeta_a\operatorname{div}_h\boldsymbol{v}_a+(\operatorname{curl}\boldsymbol{v}_a)_h\cdot\operatorname{grad} w+\boldsymbol{k}\cdot\boldsymbol{S}, \tag{5.14}$$

where

$$\operatorname{div}_h\boldsymbol{v}_a=\operatorname{div}\boldsymbol{v}_a-\frac{\partial w}{\partial r} \tag{5.15}$$

is the horizontal divergence.

Eq. (5.14) expresses that the vertical component of the absolute vorticity may change individually as a result of three different effects: (1) horizontal divergence, (2) conversion of horizontal vorticity into vertical vorticity through the action of horizontal gradients of vertical velocity, and (3) solenoids intersecting the horizontal surfaces. In addition the force of friction, which has been left out in the derivation of (5.14), may also cause ζ_a to change individually.

δ) Potential vorticity. Consider the quantity

$$\chi=\operatorname{curl}\boldsymbol{v}_a\cdot\alpha\operatorname{grad}\psi \tag{5.16}$$

where ψ is an arbitrary scalar function. Individual differentiation gives

$$\frac{D\chi}{Dt}=\frac{D_a}{Dt}(\operatorname{curl}\boldsymbol{v}_a)\cdot\alpha\operatorname{grad}\psi+\operatorname{curl}\boldsymbol{v}_a\cdot\left(\frac{D\alpha}{Dt}\operatorname{grad}\psi+\alpha\frac{D_a}{Dt}(\operatorname{grad}\psi)\right).$$

On the right-hand side we substitute for $\frac{D_a}{Dt}\operatorname{curl}\boldsymbol{v}_a$ from (5.11), for $\frac{D\alpha}{Dt}$ from (4.4), and for $\frac{D_a}{Dt}\operatorname{grad}\psi$ from the identity

$$\frac{D_a}{Dt}\operatorname{grad}\psi=\operatorname{grad}\frac{D\psi}{Dt}-\operatorname{grad}\boldsymbol{v}_a\cdot\operatorname{grad}\psi.$$

This gives

$$\frac{D\chi}{Dt}=\operatorname{curl}\boldsymbol{v}_a\cdot\alpha\operatorname{grad}\frac{D\psi}{Dt}+\boldsymbol{S}\cdot\alpha\operatorname{grad}\psi. \tag{5.17}$$

This theorem was derived by ERTEL[1]. If in particular ψ is taken to be a thermodynamic function of state (i.e. a function of p and α), the second term on the right vanishes, because $\boldsymbol{S}$ is normal to both grad α and grad p. If ψ is taken to be the potential temperature (2.10), χ is termed the *potential vorticity*, which will be denoted by Z,

$$Z=\operatorname{curl}\boldsymbol{v}_a\cdot\alpha\operatorname{grad}\theta. \tag{5.18}$$

---

[1] H. ERTEL: Meteor. Z. **59**, 277 (1942).

Hence

$$\frac{DZ}{Dt} = \operatorname{curl} \boldsymbol{v}_a \cdot \alpha \operatorname{grad} \frac{D\theta}{Dt} \tag{5.19}$$

so that Z can change individually only as a result of non-adiabatic heating or cooling of the air. It should be noted, however, that this result was obtained by ignoring the vorticity-producing effect of friction; if this effect is taken into account, we shall find that potential vorticity may change individually also as a result of friction. Thus, if the force of turbulent friction per unit volume (compare Sects. 15 to 17) is denoted by $\boldsymbol{F}$, we obtain instead of (5.19)

$$\frac{DZ}{Dt} = \operatorname{curl} \boldsymbol{v}_a \cdot \alpha \operatorname{grad} \frac{D\theta}{Dt} + \operatorname{curl}(\alpha \boldsymbol{F}) \cdot \alpha \operatorname{grad} \theta. \tag{5.20}$$

If the motion is frictionless and adiabatic, (5.20) gives

$$\frac{DZ}{Dt} = 0 \tag{5.21}$$

expressing individual conservation of potential vorticity. This theorem was first derived, in a somewhat simplified form, by ROSSBY[1].

ε) Circulation and vorticity theorems in relative motion. Between the absolute velocity $\boldsymbol{v}_a$ and the relative velocity $\boldsymbol{v}$ one has the relation

$$\boldsymbol{v}_a = \boldsymbol{\Omega} \times \boldsymbol{r} + \boldsymbol{v}, \tag{5.22}$$

where $\boldsymbol{r}$ denotes the position vector from a reference point on the earth's axis. The velocity field $\boldsymbol{\Omega} \times \boldsymbol{r}$, being a field of solid rotation about the earth's axis, has a constant vorticity $2\,\boldsymbol{\Omega}$ and no divergence. Hence,

$$\operatorname{curl} \boldsymbol{v}_a = 2\,\boldsymbol{\Omega} + \operatorname{curl} \boldsymbol{v} \tag{5.23}$$

$$\operatorname{div} \boldsymbol{v}_a = \operatorname{div} \boldsymbol{v}. \tag{5.24}$$

By means of these relationships we may express the circulation and vorticity theorems developed in this section in terms of relative motion.

Using STOKES' theorem and (5.23), the absolute circulation (5.3) of a closed curve L (which is the boundary curve of a surface σ) may be written

$$C_a = \int_\sigma \operatorname{curl} \boldsymbol{v}_a \cdot d\boldsymbol{\sigma} = 2\,\boldsymbol{\Omega} \cdot \int_\sigma d\boldsymbol{\sigma} + \int_\sigma \operatorname{curl} \boldsymbol{v} \cdot d\boldsymbol{\sigma}. \tag{5.25}$$

Introducing the relative circulation

$$C = \oint_L \boldsymbol{v} \cdot d\boldsymbol{r} = \int_\sigma \operatorname{curl} \boldsymbol{v} \cdot d\boldsymbol{\sigma} \tag{5.26}$$

and the vector area $\boldsymbol{\sigma}$ of the closed curve L,

$$\boldsymbol{\sigma} = \int_\sigma d\boldsymbol{\sigma} = \oint_L \tfrac{1}{2} \boldsymbol{r} \times d\boldsymbol{r}, \tag{5.27}$$

Eq. (5.25) becomes

$$C_a = 2\,\boldsymbol{\Omega} \cdot \boldsymbol{\sigma} + C. \tag{5.28}$$

When this expression is substituted for C_a in the circulation theorem (5.4), we obtain the *circulation theorem in relative motion*,

$$\frac{dC}{dt} = -2\,\boldsymbol{\Omega} \cdot \frac{d\boldsymbol{\sigma}}{dt} + \int_\sigma \boldsymbol{S} \cdot d\boldsymbol{\sigma}. \tag{5.29}$$

This theorem was derived by V. BJERKNES[2].

[1] C. G. ROSSBY: Quart. J. Roy. Met. Soc. **66**, Suppl., 68 (1940).
[2] V. BJERKNES: Meteor. Z. **19**, 97 (1902).

The first term on the right may be called the induced circulation change, owing to its formal correspondence with the electromotive force induced in a closed conductor which moves in a constant magnetic field $2\boldsymbol{\Omega}$. The induced circulation change is seen to depend upon the change of the area of the equator projection of the curve L. The second term on the right is the effect of baroclinicity.

Next let us express Eq. (5.14) for the rate of change of vertical absolute vorticity ζ_a in terms of relative motion. Denoting by ζ the vertical component of the relative vorticity, we obtain from (5.23) by taking the vertical component,

$$\zeta_a = \lambda + \zeta, \tag{5.30}$$

where

$$\lambda = 2\Omega \sin\varphi \quad (\varphi \text{ latitude}) \tag{5.31}$$

is the vertical component of $2\boldsymbol{\Omega}$ or the CORIOLIS *parameter*. Using also (5.24), Eq. (5.14) may be written

$$\frac{D}{Dt}(\lambda + \zeta) = -(\lambda + \zeta)\,\mathrm{div}_h\,\boldsymbol{v} + \left(2\boldsymbol{\Omega}_h + (\mathrm{curl}\,\boldsymbol{v})_h\right)\cdot \mathrm{grad}\,w + \boldsymbol{k}\cdot\boldsymbol{S}, \tag{5.32}$$

where subscript h denotes horizontal component. This equation holds for a spherical earth and expresses the *vorticity theorem in relative motion*.

The CORIOLIS *parameter* λ is a function of latitude only, and therefore

$$D\lambda/Dt = \boldsymbol{v}_h \cdot \mathrm{grad}\,\lambda = \beta v, \tag{5.33}$$

where v is the component of $\boldsymbol{v}_h$ toward the north, and, for a spherical earth of radius a,

$$\beta = \frac{1}{a}\frac{d\lambda}{d\varphi} = \frac{2\Omega\cos\varphi}{a} \tag{5.34}$$

is the gradient of the CORIOLIS parameter. The quantity β is zero at the poles and has its maximum value ($2.29 \times 10^{-13}\,\mathrm{cm}^{-1}\,\mathrm{sec}^{-1}$) on the equator.

6. Energy and entropy. *α) Generalization of the first law of thermodynamics.* The first law of thermodynamics is derived from the principle of the impossibility of a perpetuum mobile of the first kind. The law expresses that the energy fed into a system through the boundaries by work and heat flux from the surroundings is stored within the system, the level of stored energy depending only upon the state of the system. In pure thermodynamics, the only energy store is the internal energy, which is a function of the thermodynamical state of the system. In a system such as a part of the atmosphere, on the other hand, energy may be stored not only as internal energy, but as kinetic and potential energy as well; and the amount of stored energy will depend upon the thermodynamic state, the state of motion and the distribution of mass. The generalization of the first law of thermodynamics, applicable to a mass of air moving in the field of gravity, is therefore

$$d(K + \Phi + E) = dA + dQ, \tag{6.1}$$

where K is the kinetic, Φ the potential and E the internal energy of the system, dA the work done on the system by the surroundings and dQ the amount of heat entering the system by radiation and conduction. It should be noticed that K and Φ as well as dA will depend upon the frame of reference applied.

It is assumed here that the system is closed in the sense that matter is not allowed to enter or leave the system. A body of air in the atmosphere, whether large or small, is strictly speaking not a closed system in this sense, because mass is exchanged through its boundaries by molecular diffusion, turbulence and precipitation. The effect upon the energy equations can be accounted for by making use of the theory of thermodynamics of open systems (compare VAN MIEGHEM[1]). In the following, such exchange processes will be ignored, and every air particle will be considered as a closed system.

Let $d\sigma$ designate a surface element (normal pointing outwards) of the surface σ enclosing the system. The force exerted by the surroundings on the system over the surface $d\sigma$ is $-p\,d\sigma + \mathcal{F}\cdot d\sigma$, and thus

$$\frac{dA}{dt} = -\int_\sigma p\,\boldsymbol{v}\cdot d\sigma + \int_\sigma \boldsymbol{v}\cdot\mathcal{F}\cdot d\sigma. \tag{6.2}$$

Moreover, let $\boldsymbol{W}$ denote the flux of heat by radiation and conduction; then we have

$$\frac{dQ}{dt} = -\int_\sigma \boldsymbol{W}\cdot d\sigma \tag{6.3}$$

and the first law may be written:

$$\frac{d}{dt}(K + \Phi + E) = \int_\sigma [\boldsymbol{v}\cdot(-p\mathcal{E} + \mathcal{F}) - \boldsymbol{W}]\cdot d\sigma. \tag{6.4}$$

In the limit for an infinitely small system, Eq. (6.4) becomes

$$\frac{D}{Dt}\left(\frac{1}{2}\boldsymbol{v}^2 + \varphi + e\right) = -\alpha\,\mathrm{div}\,(p\,\boldsymbol{v}) + \alpha\,\mathrm{div}\,(\mathcal{F}\cdot\boldsymbol{v}) - \alpha\,\mathrm{div}\,\boldsymbol{W}, \tag{6.5}$$

which expresses the first law in differential form. The left-hand side represents the individual rate of change of total energy per unit mass. On the right-hand side, the first term is the work done by the pressure force per unit mass; it may be written by means of (4.4),

$$\left.\begin{aligned} -\alpha\,\mathrm{div}\,(p\boldsymbol{v}) &= -\alpha\boldsymbol{v}\cdot\mathrm{grad}\,p - \alpha\,p\,\mathrm{div}\,\boldsymbol{v} \\ &= -\alpha\boldsymbol{v}\cdot\mathrm{grad}\,p - p\frac{D\alpha}{Dt} \end{aligned}\right\} \tag{6.6}$$

showing that the total work of the pressure forces may be separated into a "translatory" work and a work by expansion.

The second term on the right of (6.5) is the total work done by friction per unit mass. It may be partitioned into a "translatory" work and a work due to deformation:

$$\alpha\,\mathrm{div}\,(\mathcal{F}\cdot\boldsymbol{v}) = \alpha\boldsymbol{v}\cdot\mathrm{div}\cdot\mathcal{F} + \delta, \tag{6.7}$$

where

$$\delta = \alpha\mathcal{F}\cdot\mathrm{grad}\cdot\boldsymbol{v} \tag{6.8}$$

is STOKES' dissipation function. It has the important property that it is always positive, except in the case of rigid body motion, when it is zero.

The last term on the right of (6.5)

$$Dq/Dt = -\alpha\,\mathrm{div}\,\boldsymbol{W} \tag{6.9}$$

is the heat received per unit mass and unit time[2].

By means of (4.5), Eq. (6.5) may be written in the form

$$\frac{\partial(\varrho\,e_{\mathrm{tot}})}{\partial t} + \mathrm{div}\,\boldsymbol{F} = 0, \tag{6.10}$$

where

$$e_{\mathrm{tot}} = \tfrac{1}{2}\boldsymbol{v}^2 + \varphi + e \tag{6.11}$$

[1] J. VAN MIEGHEM: Comp. of Meteor., Amer. Meteor. Soc. **1951**, 531.

[2] Note that q is not a function of state.

is the total energy per unit mass, and

$$\boldsymbol{F} = (\tfrac{1}{2}\,\boldsymbol{v}^2 + \varphi + h)\,\varrho\boldsymbol{v} - \mathscr{F}\cdot\boldsymbol{v} + \boldsymbol{W} \tag{6.12}$$

is the energy flux.

β) The mechanical energy equation. Multiplying (4.11) scalarly by $\boldsymbol{v}$, we obtain, since $\partial\varphi/\partial t = 0$,

$$\frac{D}{Dt}\left(\frac{1}{2}\,\boldsymbol{v}^2 + \varphi\right) = -\alpha\,\boldsymbol{v}\cdot\operatorname{grad} p + \alpha\,\boldsymbol{v}\cdot\operatorname{div}\mathscr{F}, \tag{6.13}$$

which is the mechanical energy equation in differential form. The left-hand side is the individual rate of change of the kinetic plus potential energy per unit mass, and the right-hand side represents the translatory work of the pressure force and friction.

According to (6.6) and (6.7), Eq. (6.13) may be written

$$\frac{D}{Dt}\left(\frac{1}{2}\,\boldsymbol{v}^2 + \varphi\right) = -\alpha\operatorname{div}(p\boldsymbol{v}) + \alpha\operatorname{div}(\mathscr{F}\cdot\boldsymbol{v}) + p\,\frac{D\alpha}{Dt} - \delta. \tag{6.14}$$

By means of (4.5) it may also be written in the form

$$\frac{\partial}{\partial t}\left[\varrho\left(\frac{1}{2}\,\boldsymbol{v}^2 + \varphi\right)\right] + \operatorname{div}\left[\varrho\left(\frac{1}{2}\,\boldsymbol{v}^2 + \varphi\right)\boldsymbol{v} + p\boldsymbol{v} - \mathscr{F}\cdot\boldsymbol{v}\right] = \varrho\,p\,\frac{D\alpha}{Dt} - \varrho\,\delta. \tag{6.15}$$

We may interpret the vector

$$(\tfrac{1}{2}\boldsymbol{v}^2 + \varphi)\,\varrho\boldsymbol{v} + p\boldsymbol{v} - \mathscr{F}\cdot\boldsymbol{v}$$

as the flux of mechanical energy, and the right-hand side $\varrho p\,\frac{D\alpha}{Dt} - \varrho\delta$ as the rate of production of mechanical energy[1]. Whereas the expansion work $\varrho p\,\frac{D\alpha}{Dt}$ may act as a source or a sink of mechanical energy, the dissipation $-\varrho\delta$ will always represent a sink of mechanical energy, since δ is positive.

When the mechanical energy Eq. (6.13) or (6.14) is integrated over a finite mass of air of volume τ, enclosed by a closed surface σ, one obtains

$$\left.\begin{aligned} \frac{d}{dt}(K + \Phi) &= -\int_\tau \boldsymbol{v}\cdot\operatorname{grad} p\,d\tau + \int_\tau \boldsymbol{v}\cdot\operatorname{div}\mathscr{F}\,d\tau \\ &= -\int_\sigma p\,\boldsymbol{v}\cdot d\sigma + \int_\sigma \boldsymbol{v}\cdot\mathscr{F}\cdot d\sigma + \int_\tau \varrho\left(p\,\frac{D\alpha}{Dt} - \delta\right)d\tau. \end{aligned}\right\} \tag{6.16}$$

Therefore, the rate of increase of mechanical energy of the system does not depend only on the work done by the surroundings on the boundaries, but also on the processes occuring inside the system itself.

γ) The thermodynamic energy equation. When (6.14) is subtracted from (6.5), there results

$$\frac{De}{Dt} = -p\,\frac{D\alpha}{Dt} + \delta + \frac{Dq}{Dt}\,. \tag{6.17}$$

This is the thermodynamic energy equation, referred to a unit mass; it expresses that the internal energy is increased by compression, dissipation and heat supply.

[1] It should be noted, however, that the interpretation of flux and rate of production is ambiguous; we may, formally, add any vector field to the flux, provided its divergence is added to the production rate. For instance, (6.13) shows that we may, alternatively, consider $(\frac{1}{2}\boldsymbol{v}^2 + \varphi)\,\varrho\boldsymbol{v}$ as the flux of mechanical energy, and $(-\boldsymbol{v}\cdot\operatorname{grad} p + \boldsymbol{v}\cdot\operatorname{div}\mathscr{F})$ as the rate of production.

By means of (4.5), it may be written in the form

$$\frac{\partial(\varrho e)}{\partial t} + \operatorname{div}(\varrho e \boldsymbol{v} + \boldsymbol{W}) = -\varrho p \frac{D\alpha}{Dt} + \varrho \delta. \tag{6.18}$$

Thus, when $\varrho e \boldsymbol{v} + \boldsymbol{W}$ is considered the flux of internal energy, the rate of production is $-\varrho p \frac{D\alpha}{Dt} + \varrho\delta$. Comparing with (6.15), it is seen that the rate of production of internal energy represents at the same time a sink of mechanical energy, i.e. this expression is the rate of transformation of mechanical energy into internal energy. Whereas $-\varrho p \frac{D\alpha}{Dt}$ represents a reversible energy transformation, which may work both ways, the dissipation function $\varrho\delta$ is irreversible, always converting mechanical energy into internal energy.

When (6.17) is integrated over a finite system of volume τ, we obtain

$$\frac{dE}{dt} = \int_\tau \left(-p\frac{D\alpha}{Dt} + \delta\right) \varrho\, d\tau - \int_\sigma \boldsymbol{W} \cdot d\boldsymbol{\sigma}. \tag{6.19}$$

This equation shows that the change of internal energy does not result only from exchange of energy between the system and the surroundings, but will depend also upon the processes going on inside the system.

δ) The entropy equation. By combining (2.6) and (6.17), we obtain the following equation for the individual rate of change of entropy per unit mass

$$\frac{Ds}{Dt} = \frac{\frac{Dq}{Dt} + \delta}{T}. \tag{6.20}$$

Integrating over a finite system, we obtain

$$\frac{dS}{dt} = \int_\tau \frac{\frac{Dq}{Dt} + \delta}{T} \varrho\, d\tau, \tag{6.21}$$

where S is the entropy of the system. It is seen that frictional dissipation represents an irreversible effect, which will tend to increase the entropy. Also exchange of heat inside the system by radiation and conduction, which tends to make the temperature uniform, will represent an irreversible entropy increase. In reality, the entropy will be increased irreversibly also by diffusion and possibly also by other effects. These effects are not contained in the above equations, since they have been disregarded in deriving the energy equations.

ε) The energy equations applied to the entire atmosphere. The energy of the entire atmosphere is fairly constant. There will be certain yearly and diurnal periodic variations, and also aperiodic changes; but there is no progressive change, so that on the average, one may put

$$\frac{dK}{dt} = \frac{d\Phi}{dt} = \frac{dE}{dt} = 0, \qquad \frac{dS}{dt} = 0 \tag{6.22}$$

for the entire atmosphere.

The boundary surface σ of the atmosphere will consist of a closed surface σ_1 outside the atmosphere, and the surface σ_2 of the earth. Clearly, p and $\mathscr{F}$ vanish over σ_1, and the total energy Eq. (6.4) may therefore be written

$$-\int_{\sigma_2} p\boldsymbol{v} \cdot d\boldsymbol{\sigma} + \int_{\sigma_2} \boldsymbol{v} \cdot \mathscr{F} \cdot d\boldsymbol{\sigma} - \int_{\sigma_1+\sigma_2} \boldsymbol{W} \cdot d\boldsymbol{\sigma} = 0. \tag{6.23}$$

It is known that the atmosphere does work on the oceans, thereby maintaining ocean currents and waves, and also on the solid earth (wind erosion etc.). Therefore, the two first terms in (6.23) are both negative, so that the net heat received by the atmosphere must be positive.

However, the work done by the atmosphere over the surface σ_2, although all-important for the energy balance of the oceans, is of minor importance for the energy budget of the atmosphere. If this boundary work is neglected, i.e. if the surface σ_2 is considered as the surface of a rigid body, the wind velocity $\boldsymbol{v}$ at σ_2 must vanish, owing to viscosity. In this approximation, the two first integrals of (6.23) vanish, and there results

$$\frac{dQ}{dt} = -\int\limits_{\sigma_1+\sigma_2} \boldsymbol{W} \cdot d\boldsymbol{\sigma} = 0, \tag{6.24}$$

i.e. the atmosphere receives as much heat as it loses.

From (6.6), (6.7), (6.19) and (6.24) one finds

$$\int\limits_\tau p \frac{D\alpha}{Dt} \varrho\, d\tau = -\int\limits_\tau \boldsymbol{v} \cdot \operatorname{grad} p\, d\tau = -\int\limits_\tau \boldsymbol{v} \cdot \operatorname{div} \mathscr{F} d\tau = \int\limits_\tau \varrho\,\delta\, d\tau > 0. \tag{6.25}$$

In order to compensate the dissipation of kinetic energy, the atmosphere must do a positive expansion work $\int\limits_\tau p \frac{D\alpha}{Dt} \varrho\, d\tau$; one may therefore say that expansion takes place on the average at higher pressure than contraction.

Substituting for $p D\alpha$ from (2.6) and using (6.22), the first of the inequalities (6.25) may also be written

$$\int\limits_\tau T \frac{Ds}{Dt} \varrho\, d\tau > 0. \tag{6.26}$$

Thus, on the average, entropy increase takes place at higher temperature than entropy decrease.

This inequality may again by (2.12) be transformed into

$$\int\limits_\tau \Pi \frac{D\theta}{Dt} \varrho\, d\tau > 0. \tag{6.27}$$

which says that increase of potential temperature takes place, on the average, at higher pressure than decrease of potential temperature. Eq. (6.27) is the mathematical expression for SANDSTRÖM's *theorem*[1], (also found by WENGER[2]) which is usually stated in the following way: *heat sources are on the average situated at higher pressure than heat sinks.* This interpretation is correct, provided the word "heat sources" is taken to include frictional dissipation, as well as released latent heat.

From (6.21) and (6.22) follows

$$\int\limits_\tau \frac{1}{T} \frac{Dq}{Dt} \varrho\, d\tau = -\int\limits_\tau \frac{\delta}{T} \varrho\, d\tau < 0. \tag{6.28}$$

This inequality shows that in order to compensate the entropy increase due to dissipation, *the heat sources in the atmosphere must on the average be situated at higher temperature than the heat sinks.*

---

[1] J. W. SANDSTRÖM: Göteborgs Kgl. Vetensk. o. Vitterh. Samh. Handl., 4 F, 17, 1916.
[2] R. WENGER: Physikal. Z. **17**, 547 (1916).

In view of (6.25), it appears that the mean total dissipation in the atmosphere is a basic characteristic of the atmospheric heat engine. SVERDRUP[1] estimated the mean dissipation to about 4 watts per atmospheric column of 1 m^2 horizontal cross-section. This agrees well with the figure 5 watts/m^2 given by BRUNT[2] [*4*]. A more recent estimate by LETTAU[3] is 2 watts/m^2. These figures may be compared with the rate at which solar radiation is absorbed by the earth and the atmosphere; on the average for the whole earth, this amounts to about 200 watts/m^2.

The first thorough study of the energy transformations is due to MARGULES [*14*]. Derivations of the energy equations given by V. BJERKNES and coll. [*2*] and by BRUNT [*4*] suffer from the defect that the effect of the dissipation in the thermodynamic energy equation has been left out. A recent treatment of the energy equations has been given by MILLER[4].

7. The complete system of equations. As long as we ignore the effects of water vapour, the thermodynamic-hydrodynamic state of the atmosphere is characterized by the velocity field $\boldsymbol{v}$ and the two scalar fields of p and ϱ. The change with time of these fields is described by the equation of motion (4.11), the equation of continuity (4.2) and the thermodynamic energy Eq. (6.17). If the heat sources are considered as known, these equations form a closed system of governing equations, which together with the boundary conditions determine the future (or past) state if the initial state is known.

The effect of water vapour may be taken into account by introducing the specific humidity as another dependent variable; we must then also introduce another prognostic equation, viz. the continuity equation for the water component.

The system of equations referred to above is very general and may be applied to virtually all types of motion which take place in the atmosphere, from the large-scale currents which are shown on weather maps to the smallest scale turbulent motions.

In practice, however, we are not able to deal with the smallest details in the fields, partly because these details are not defined by the observations, and partly because we are unable to handle mathematically fields of such a complicated structure. We must therefore instead deal with simpler fields where the smallest details have been smoothed out. Following the technique developed by O. REYNOLDS [*18*], one finds that such smoothed fields satisfy differential equations which are similar to the system referred to above, except for certain additional terms of which the REYNOLDS *stresses* and the *eddy-flux of heat* are the most important. These effects will be briefly dealt with in Sect. 15 sequ. The molecular viscosity and heat conduction is important only for the smallest scale motions; their effects upon the smoothed fields are negligible compared with the REYNOLDS stresses and the eddy-flux of heat, and these terms can therefore be neglected in the equations used in dynamic meteorolgy.

c) The quasi-static equations.

8. The quasi-static approximations. In treatments of large-scale motions in the atmosphere, of horizontal dimensions of the order 500 km or more, it is customary to apply the following two *approximations*:

1. The vertical acceleration is ignored in the equation of motion.

2. The horizontal component of $\boldsymbol{\Omega}$ is ignored in the expression for the CORIOLIS force.

---

[1] H. U. SVERDRUP: Veröff. Geophys. Inst. Univ. Lpz., Ser. II **2**, 173 (1918).
[2] D. BRUNT: Phil. Mag., Ser. VII **1**, 523 (1926).
[3] H. LETTAU: Arch. Meteor., Geophys. u. Biokl., Ser. A **7**, 133 (1954).
[4] J. E. MILLER: Comp. of Meteor., Amer. Meteor. Soc. **1951**, 483.

Although these approximations are mutually independent, they are usually applied simultaneously and are often referred to in common as the quasi-static approximation. No rigorous justification of these approximations have been given; all that can be offered, therefore, is a demonstration of their plausibility.

Any vector field may be considered as the sum of a *solenoidal (non-divergent)* field and an *irrotational* field. Thus we may write the momentum vector as

$$\varrho \boldsymbol{v} = (\varrho \boldsymbol{v})_1 + (\varrho \boldsymbol{v})_2, \tag{8.1}$$

where

$$\operatorname{div} (\varrho \boldsymbol{v})_1 = 0, \qquad \operatorname{curl} (\varrho \boldsymbol{v})_2 = 0. \tag{8.2}$$

One important property of the motions in which we are interested is that *the solenoidal field $(\varrho \boldsymbol{v})_1$ is the dominating part of the total momentum field, whereas $(\varrho \boldsymbol{v})_2$ is very much weaker* (the opposite is true for sound waves). We may therefore consider $(\varrho \boldsymbol{v})_1$ as an approximation to $\varrho \boldsymbol{v}$.

The solenoidal field $(\varrho \boldsymbol{v})_1$ has closed streamlines, from which we may form closed tubes or solenoids of unit mass flux. In large-scale motion systems, a typical streamline will have horizontal dimensions of several hundred km or more, whereas the vertical distance between its highest and lowest point is limited because the motion is confined to the troposphere and lower stratosphere. The streamlines will therefore in most points have nearly horizontal tangent, showing that the vertical component of $(\varrho \boldsymbol{v})_1$ in these points is very small[1]. There may also be regions where the streamlines run almost vertically; but in such regions the streamlines will normally spread out, and the solenoids widen so much that the vertical component of $(\varrho \boldsymbol{v})_1$ is small in these places also.

Since we have assumed that $\varrho \boldsymbol{v}$ is approximately equal to $(\varrho \boldsymbol{v})_1$, it follows that the field of vertical velocity is much weaker than the horizontal velocity field in large-scale systems. In fact, estimates show that typical vertical velocities in large-scale motion systems are of the order 0.05 m/sec, whereas the horizontal velocities are 100 or 1000 times stronger. For such motion systems, the kinetic energy is well approximated by the kinetic energy of the *horizontal* motion; and likewise, the relative vorticity is well approximated by the relative vorticity of the horizontal motion. Therefore the effects of vertical accelerations upon the mechanical energy Eq. (6.13) and the vorticity Eq. (5.11) are negligible compared with the effects of horizontal accelerations. This makes it plausible that the vertical accelerations may be neglected altogether, in accordance with approximation 1.

The vertical accelerations are in large-scale motions extremely small compared with either of the two large vertical forces, viz. the force of gravity and the vertical pressure force. It should be noted, however, that this in itself is not enough to justify approximation 1. Consider for instance internal gravity oscillations in a stable fluid (compare Sect. 22). The vertical accelerations can clearly be made arbitrarily small by reducing the static stability of the fluid; but the horizontal accelerations will then become small also, and approximation 1 is not applicable, even in the limit for vanishing stability, as long as the horizontal wavelength is comparable with the vertical wavelength. In order for approximation 1 to be legitimate, it is essential that the horizontal extent of the motion system is large compared with its vertical extent, so that the vertical accelerations are negligible compared with the horizontal accelerations.

The justification for approximation 2 above is that the horizontal component of $\boldsymbol{\Omega}$ is usually negligible compared with the horizontal component of the relative vorticity, so that the horizontal component of the absolute vorticity may with

[1] This conclusion need not be true if the streamlines had small wiggles; these, however, should be considered as due to superimposed small-scale motions which should be smoothed out in dealing with large-scale motions.

sufficient accuracy be identified with the horizontal component of the relative vorticity.

In addition to approximations 1 and 2 considered above, we shall apply the following approximations:

3. $\left(\frac{D\boldsymbol{v}}{Dt}\right)_h$ is replaced by $\left(\frac{D\boldsymbol{v}_h}{Dt}\right)_h$ $\left(\text{ignoring}\ \frac{w\boldsymbol{v}_h}{a}\right)$,

4. $\operatorname{div}_h(\varrho\boldsymbol{v})$ is replaced by $\operatorname{div}_h(\varrho\boldsymbol{v}_h)$ $\left(\text{ignoring}\ \frac{2\varrho w}{a}\right)$,

where the subscript h means horizontal component, and a is the radius of the earth. The ignored terms are small effects of the earth's curvature.

9. The quasi-static equations for frictionless motions, with height as the vertical coordinate. *α) The equations of motion.* In the quasi-static theory it is convenient to use a notation which brings out the distinction between vertical and horizontal directions. We shall therefore express space differentiation in vertical direction as $\partial/\partial z$, where z is height above mean sea level, whereas horizontal space differentiation will be expressed vector-analytically by means of a horizontal gradient-operator grad_h without specification of the horizontal coordinates. Similarly, the velocity will be subdivided into a horizontal component $\boldsymbol{v}_h$ and a vertical component $w\boldsymbol{k}$. In this notation, the expansion (4.1) of the individual time differentiation is expressed as

$$\frac{D}{Dt} = \frac{\partial}{\partial t} + \boldsymbol{v}_h \cdot \operatorname{grad}_h + w\frac{\partial}{\partial z}. \tag{9.1}$$

With approximations 2 and 3 of Sect. 8, the horizontal part of the equation of motion (4.11) without the friction term becomes

$$\left(\frac{D\boldsymbol{v}_h}{Dt}\right)_h = -\alpha\operatorname{grad}_h p - \lambda\boldsymbol{k}\times\boldsymbol{v}_h. \tag{9.2}$$

The vertical part of (4.11) becomes, on account of approximations 1 and 2,

$$0 = -\alpha\frac{\partial p}{\partial z} - g \quad \text{or} \quad \varrho = -\frac{1}{g}\frac{\partial p}{\partial z}, \tag{9.3}$$

where g is the acceleration of gravity. This equation, which expresses that the force of gravity is balanced by the vertical pressure force, is referred to as the *hydrostatic equation*. Ignoring the horizontal variation of g, horizontal differentiation of (9.3) yields

$$\frac{\partial p}{\partial z}\operatorname{grad}_h\alpha + \alpha\operatorname{grad}_h\left(\frac{\partial p}{\partial z}\right) = 0. \tag{9.4}$$

By means of this relation, the horizontal component $\boldsymbol{S}_h$ of the baroclinicity vector $\boldsymbol{S}$ (5.5) may be written as

$$\boldsymbol{S}_h = -\boldsymbol{k}\times\left(\frac{\partial\alpha}{\partial z}\operatorname{grad}_h p - \frac{\partial p}{\partial z}\operatorname{grad}_h\alpha\right) = -\boldsymbol{k}\times\frac{\partial}{\partial z}(\alpha\operatorname{grad}_h p). \tag{9.5}$$

The vertical component of the baroclinicity vector is

$$S_z = -\boldsymbol{k}\cdot\operatorname{grad}_h\alpha\times\operatorname{grad}_h p. \tag{9.6}$$

If the circulation theorem is derived from (9.2) and (9.3), we obtain instead of (5.29),

$$\frac{d}{dt}\oint_L \boldsymbol{v}_h\cdot d\boldsymbol{r}_h = \frac{dC_h}{dt} = 2\boldsymbol{\Omega}\cdot\frac{d}{dt}\int_\sigma \boldsymbol{k}\,d\sigma_z + \int_\sigma(\boldsymbol{S}_h\cdot d\boldsymbol{\sigma}_h + S_z\,d\sigma_z). \tag{9.7}$$

In the quasi-static approximation, therefore, the circulation is approximated by the circulation of the horizontal motion; and the induced circulation change is expressed by the change of the equator projection of the horizontal projection of the closed curve.

Similarly, we may derive the vorticity theorem on the basis of (9.2) and (9.3). Applying the operation $\boldsymbol{k}\times\frac{\partial}{\partial z}$ on (9.2), we obtain the horizontal part of the quasi-static vorticity equation,

$$\boldsymbol{k}\times\frac{\partial}{\partial z}\frac{D\boldsymbol{v}_h}{Dt}=\boldsymbol{S}_h+\lambda\frac{\partial\boldsymbol{v}_h}{\partial z}. \tag{9.8}$$

The vertical vorticity equation is obtained by applying the operator $\boldsymbol{k}\cdot$ curl on (9.2). This gives

$$\frac{D}{Dt}(\lambda+\zeta)=-(\lambda+\zeta)\operatorname{div}_h\boldsymbol{v}_h+\boldsymbol{k}\times\frac{\partial\boldsymbol{v}_h}{\partial z}\cdot\operatorname{grad}_h w+S_z, \tag{9.9}$$

where ζ is the vertical component of the relative vorticity. Comparing with the exact Eq. (5.32), it will be noticed that the only difference is that the horizontal component of the absolute vorticity is approximated quasi-statically by $\boldsymbol{k}\times\frac{\partial\boldsymbol{v}_h}{\partial z}$.

Another equation derived from the equations of motion is the mechanical energy Eq. (6.13). In the quasi-static approximation, this equation breaks up into two equations. Multiplying (9.2) by $\boldsymbol{v}_h$, we obtain

$$\frac{D}{Dt}\left(\frac{1}{2}\boldsymbol{v}_h^2\right)=-\alpha\,\boldsymbol{v}_h\cdot\operatorname{grad}_h p. \tag{9.10}$$

Multiplication of (9.3) by w gives

$$\frac{D\varphi}{Dt}=-\alpha\,w\frac{\partial p}{\partial z}. \tag{9.11}$$

Thus, in the quasi-static approximation, the kinetic energy is approximated by the kinetic energy of the horizontal motion, $\frac{1}{2}\boldsymbol{v}_h^2$. The work done by the horizontal pressure force increases the kinetic energy, and the work done by the vertical pressure force increases the potential energy.

β) The continuity equation and the tendency equation. The hydrostatic Eq. (9.3) is assumed to hold for all times; this equation may therefore be differentiated with respect to time, with the result

$$\frac{\partial\varrho}{\partial t}=-\frac{1}{g}\frac{\partial}{\partial z}\frac{\partial p}{\partial t}. \tag{9.12}$$

An independent expression for $\partial\varrho/\partial t$ is obtained from the continuity Eq. (4.2), which in the notation of the present section becomes

$$\frac{\partial\varrho}{\partial t}=-\operatorname{div}_h(\varrho\,\boldsymbol{v}_h)-\frac{\partial(\varrho\,w)}{\partial z}. \tag{9.13}$$

Elimination of $\partial\varrho/\partial t$ between these equations gives, when g is considered as a constant,

$$\frac{\partial}{\partial z}\left(\frac{\partial p}{\partial t}-g\,\varrho\,w\right)=g\operatorname{div}_h(\varrho\,\boldsymbol{v}_h). \tag{9.14}$$

This equation may be solved for the pressure tendency $\partial p/\partial t$ by integrating from a level z to a level ($z=\infty$) high enough so that p and ϱw can be neglected. Thus we obtain

$$\frac{\partial p}{\partial t}=g\,\varrho\,w-g\int\limits_z^\infty\operatorname{div}_h(\varrho\,\boldsymbol{v}_h)\,dz. \tag{9.15}$$

This is the "*tendency equation*", which was derived by MARGULES[1]. The right-hand side measures the rate of increase in the weight of the air contained within a vertical column of unit cross-section, extending from the height z to the outer limit of the atmosphere, due to inflow of mass through the bottom and the sides.

Assuming that the earth's surface is the level surface $z=0$, the kinematic boundary condition gives $w=0$ at $z=0$, and the tendency equation applied at the surface becomes

$$\left(\frac{\partial p}{\partial t}\right)_{z=0} = -g\int_0^\infty \operatorname{div}_h(\varrho\,\boldsymbol{v}_h)\,dz. \tag{9.16}$$

γ) RICHARDSON'S *equation*. The thermodynamic energy Eq. (6.17) becomes, when friction is ignored, and e and α are expressed in terms of p and ϱ, assuming the air to be non-saturated,

$$\frac{D\varrho}{Dt} - \frac{1}{\varkappa RT}\frac{Dp}{Dt} = -\frac{\varrho}{c_p T}\frac{Dq}{Dt}. \tag{9.17}$$

By eliminating $\partial\varrho/\partial t$ and $\partial p/\partial t$ between (9.13), (9.15) and (9.17), we find

$$\frac{\partial w}{\partial z} = -\frac{1}{\varrho}\operatorname{div}_h(\varrho\,\boldsymbol{v}_h) + \frac{g}{\varkappa p}\int_z^\infty \operatorname{div}_h(\varrho\,\boldsymbol{v}_h)\,dz - \boldsymbol{v}_h\cdot\frac{1}{\theta}\operatorname{grad}_h\theta + \frac{1}{c_p T}\frac{Dq}{Dt} \tag{9.18}$$

which does not contain any time derivatives of the dependent variables[2]. This equation was derived by RICHARDSON [*19*]. It may be solved for w by integrating vertically from the earth's surface where w is known from the kinematic boundary condition.

Inspecting the quasi-static equations, we notice that only the fields of $\boldsymbol{v}_h$ and p can be assigned independently; when these fields are given at a certain instant, the instantaneous distributions of ϱ and w follow from (9.3) and (9.18). The evolution in time of the fields of $\boldsymbol{v}_h$ and p is governed by (9.2) and (9.15).

10. Other coordinate systems. *α) Pressure as vertical coordinate.* The quasi-static equations assume a simpler form when pressure is used as a vertical coordinate instead of height; this transformation was introduced by SUTCLIFFE[3], VAN MIEGHEM[4] and ELIASSEN[5]. The independent variables then are p, t, and two horizontal coordinates. Partial differentiation with respect to the horizontal coordinates or time are then taken at constant pressure and will be denoted by the subscript p. Thus $(\partial/\partial t)_p$ means differentiation at constant pressure and constant geographical coordinates, and grad_p is the horizontal gradient operator in a horizontal projection of an isobaric surface. Similarly we define the operations div_p and curl_p by differentiating with respect to horizontal coordinates at constant pressure. These operations can all be evaluated from constant pressure maps.

In these coordinates, the expansion of the individual time derivative is

$$\frac{D}{Dt} = \left(\frac{\partial}{\partial t_p}\right) + \boldsymbol{v}_h\cdot\operatorname{grad}_p + \omega\frac{\partial}{\partial p}, \tag{10.1}$$

where

$$\omega = Dp/Dt \tag{10.2}$$

formally takes the role of the vertical velocity.

---

[1] M. MARGULES: Über die Beziehung zwischen Barometerschwankungen und Kontinuitätsgleichung. Boltzmann-Festschrift, p. 585, Leipzig 1904.

[2] The rate of heating Dq/Dt is usually considered as a given function.

[3] R. C. SUTCLIFFE: Quart. J. Roy. Meteor. Soc. **73**, 370 (1947).

[4] J. VAN MIEGHEM: Inst. Roy. Meteor. Belg. Misc. **1947**, Nr. 17.

[5] A. ELIASSEN: Geof. Publ. **17**, No. 3 (1949).

Height, or geopotential, appears as a dependent variable, and the hydrostatic equation reads

$$\frac{\partial \varphi}{\partial p} = -\alpha. \tag{10.3}$$

The expression for the horizontal pressure force per unit mass is $-\operatorname{grad}_p \varphi$, and the horizontal equation of motion (9.2) therefore becomes

$$\left(\frac{D\boldsymbol{v}_h}{Dt}\right)_h = -\operatorname{grad}_p \varphi - \lambda \boldsymbol{k} \times \boldsymbol{v}_h. \tag{10.4}$$

The horizontal component of the baroclinicity vector, which is given by (9.5), may now be expressed as

$$\boldsymbol{S}_h = -g\,\boldsymbol{k} \times \frac{1}{\alpha} \operatorname{grad}_p \alpha = -g\,\boldsymbol{k} \times \frac{1}{T} \operatorname{grad}_p T = -g\,\boldsymbol{k} \times \frac{1}{\theta} \operatorname{grad}_p \theta. \tag{10.5}$$

Thus $\boldsymbol{S}_h$ is directed along the isotherms in a constant pressure map (which coincide with the isosteric and isentropic lines), with warmer air to the left. The vertical component of the baroclinicity vector (9.6) may be written

$$S_z = -g\,\boldsymbol{k} \cdot \frac{1}{\alpha} \operatorname{grad}_p \alpha \times \operatorname{grad}_p z = \boldsymbol{S}_h \cdot \operatorname{grad}_p z. \tag{10.6}$$

Hence it follows that S_z vanishes when the isosteric lines (or isotherms) in the isobaric surface coincide with the contour lines of the same surface.

The equivalent to the vertical vorticity equation is obtained by applying the operation curl_p to (10.4). The result is

$$\frac{D}{Dt}(\lambda + \zeta_p) = -(\lambda + \zeta_p) \operatorname{div}_p \boldsymbol{v}_h + \boldsymbol{k} \times \frac{\partial \boldsymbol{v}_h}{\partial p} \cdot \operatorname{grad}_p \omega, \tag{10.7}$$

where

$$\zeta_p = \operatorname{curl}_p \boldsymbol{v}_h \left(= \left(\frac{\partial v}{\partial x}\right)_p - \left(\frac{\partial u}{\partial y}\right)_p \right) \tag{10.8}$$

is the vorticity measured on a constant pressure map. Eq. (10.7) is formally identic with (9.9), except that the solenoid term is absent, although its effect is implicitely included.

The continuity equation assumes the simple form

$$\operatorname{div}_p \boldsymbol{v}_h + \frac{\partial \omega}{\partial p} = 0 \tag{10.9}$$

showing that the air is "incompressible" when considered in a space where pressure is the vertical coordinate. Eq. (10.9) does not contain any time derivatives; in this respect it takes over the role of RICHARDSON's Eq. (9.18).

The thermodynamic Eq. (6.17) is conveniently written as

$$\frac{1}{\alpha}\left(\frac{\partial \alpha}{\partial t}\right)_p + \frac{\boldsymbol{v}_h}{\alpha} \cdot \operatorname{grad}_p \alpha + \frac{\omega}{\theta} \frac{\partial \theta}{\partial p} = \frac{1}{c_p T} \frac{Dq}{Dt}. \tag{10.10}$$

By means of (10.3), this equation may also be written

$$\frac{\partial}{\partial p}\left(\frac{\partial \varphi}{\partial t}\right)_p = \boldsymbol{v}_h \cdot \operatorname{grad}_p \alpha + \alpha\,\omega \frac{\partial \log \theta}{\partial p} - \frac{R}{c_p p} \frac{Dq}{Dt}. \tag{10.11}$$

Integrating between two pressure values p_1 and p_2, we obtain

$$\frac{\partial}{\partial t}(\varphi_1 - \varphi_2) = -\int_{p_1}^{p_2} \boldsymbol{v}_h \cdot \operatorname{grad}_p \alpha\, dp - \int_{p_1}^{p_2} \alpha\,\omega \frac{\partial \log \theta}{\partial p}\, dp + \frac{R}{c_p} \int_{p_1}^{p_2} \frac{Dq}{Dt} \frac{dp}{p}. \tag{10.12}$$

The left-hand side represents the change with time of the thickness of the isobaric layer, measured in geopotential. The three integrals on the right-hand side represent three different processes which produce such thickness changes, viz. (1) isobaric advection of air of different density, (2) density changes produced by adiabatic contraction or expansion due to individual pressure changes, and (3) heating or cooling.

The boundary conditions are

$$\omega = 0 \qquad \text{when} \qquad p = 0 \tag{10.13}$$

and, when the earth's surface is level,

$$\frac{D\varphi}{Dt} = \left(\frac{\partial\varphi}{\partial t}\right)_p + \boldsymbol{v}_h \cdot \mathrm{grad}_p\, \varphi + \omega \frac{\partial\varphi}{\partial p} = 0 \qquad \text{when} \qquad \varphi = 0. \tag{10.14}$$

Whereas all the previous equations of this section are formally simpler than the corresponding equations written with z as vertical coordinate, the boundary condition at the ground (10.14) is difficult to apply, particularly because it applies at a pressure level which is not known explicitely, but is dependent upon the motion itself. However, since the surface pressure varies within only about $\pm 5\%$, we may to a good approximation apply (10.14) at a standard pressure value (usually 1000 mb) and thus avoid the difficulty.

Eqs. (10.3), (10.4), (10.9) and (10.11), together with the boundary conditions (10.13) and (10.14), form a closed system which define the future evolution if the heat sources are known and the distributions of $\boldsymbol{v}_h$ and φ are known at an initial instant.

β) Potential temperature as vertical coordinate. In the cases when the changes of state are considered as dry-adiabatic, so that

$$D\theta/Dt = 0, \tag{10.15}$$

it will often prove convenient to use θ as the vertical coordinate. Differentiation with respect to time and horizontal coordinates will then have to be carried out at constant entropy, and will be denoted by the subscript θ. Such derivatives can be directly evaluated from isentropic maps.

With this notation, individual differentiation is expressed as

$$\frac{D}{Dt} = \left(\frac{\partial}{\partial t}\right)_\theta + \boldsymbol{v}_h \cdot \mathrm{grad}_\theta. \tag{10.16}$$

As a consequence of (10.15), Eq. (10.16) does not contain any term involving $\partial/\partial\theta$.

Instead of the geopotential it is convenient to use as dependent variable the quantity

$$\Psi = \varphi + h = \varphi + c_p T \tag{10.17}$$

which was introduced by MONTGOMERY[1]. Making use of the relations (2.12 to 2.14), the hydrostatic equation now becomes

$$\frac{\partial\Psi}{\partial\theta} = \Pi \tag{10.18}$$

and the equation of motion

$$\left(\frac{D\boldsymbol{v}_h}{Dt}\right)_h = -\,\mathrm{grad}_\theta \Psi - \lambda\, \boldsymbol{k} \times \boldsymbol{v}_h. \tag{10.19}$$

[1] R. B. MONTGOMERY: Bull. Amer. Meteor. Soc. **18**, 210 (1937).

The corresponding vorticity equation is obtained by performing the operation curl_θ on this equation; this operation means curl differentiation with respect to the horizontal coordinates at constant θ, i.e. curl differentiation in an *isentropic map*. With the notation $\zeta_\theta = \text{curl}_\theta \, \boldsymbol{v}_h$ we obtain

$$\frac{D}{Dt}(\lambda + \zeta_\theta) + (\lambda + \zeta_\theta)\,\text{div}_\theta\, \boldsymbol{v}_h = 0. \tag{10.20}$$

The equation of continuity assumes the form

$$\frac{D}{Dt}\frac{\partial p}{\partial \theta} + \frac{\partial p}{\partial \theta}\,\text{div}_\theta\, \boldsymbol{v}_h = 0\,. \tag{10.21}$$

Elimination of $\text{div}_\theta\, \boldsymbol{v}_h$ between (10.20) and (10.21) gives

$$\frac{D}{Dt}\left(\frac{\lambda + \zeta_\theta}{\left(\frac{\partial p}{\partial \theta}\right)}\right) = 0. \tag{10.22}$$

This is the *theorem of the conservation of potential vorticity* (5.21). With the approximation 2 of Sect. 8, it is easily verified that the potential vorticity (5.18) assumes the simple form

$$Z = \frac{\lambda + \zeta_\theta}{\frac{\partial p}{\partial \theta}}\,. \tag{10.23}$$

This was pointed out by ROSSBY[1].

The quasi-static equations for adiabatic motion assume a particularly simple and attractive form when written with θ as vertical coordinate. This is because the vertical motion is accounted for implicitely and does not appear explicitely, owing to (10.15). The two conservation theorems (10.15) and (10.22) show that θ and Z represent two LAGRANGian coordinates; and in regions where turbulent diffusion of water vapour and condensation phenomena can be ignored, a third LAGRANGian coordinate is represented by the specific humidity. The quasi-conservative nature of these three quantities can be utilized synoptically in connection with the isentropic analysis technique (STARR and NEIBURGER[2], KLEINSCHMIDT[3]).

It should be noted, however, that the simplicity of the differential equations in the θ-system is counteracted by the complex form assumed by the boundary condition at the earth's surface in this system of coordinates. The potential temperature at the earth's surface varies within wide limits, and cannot even approximately be replaced by a standard value, as in the case of the surface pressure. Moreover, the attractive simplicity of the differential equations in the θ-system is strictly dependent upon the assumption of adiabatic changes of state; if heat sources are taken into consideration, nothing is gained by using θ as the vertical coordinate.

d) Equilibrium states.

11. Statics. Within a mass of air which remains in the state of rest, we have $\boldsymbol{v} = 0$, $\frac{D\boldsymbol{v}}{Dt} = 0$, and $\mathscr{F} = 0$, and the equation of motion (4.11) gives

$$0 = -\,\text{grad}\,\varphi - \alpha\,\text{grad}\,p. \tag{11.1}$$

[1] C. G. ROSSBY: Quart. J. Roy. Meteor. Soc. **66**, Suppl., 68 (1940).
[2] V. STARR and M. NEIBURGER: J. Mar. Res. **3**, 202 (1940).
[3] E. KLEINSCHMIDT: Tellus **7**, 96 (1955).

Curl differentiation of this equation gives $\boldsymbol{S}=0$ so that the state of rest is necessarily barotropic. Pressure and density are therefore functions of φ (or z) only, and (11.1) may be written, by means of (2.12) and (2.1),

$$d\varphi = -\alpha\, dp = -\theta\, d\Pi = -RT\, d\log p. \tag{11.2}$$

By means of (11.2) and the gas Eq. (2.1) we find the following expression for the coefficient of barotropy (5.7) in resting air,

$$\Gamma = \frac{d\varrho}{dp} = -\frac{1}{\varrho}\frac{d\varrho}{d\varphi} = \frac{1}{gT}\left(\frac{dT}{dz} + \frac{g}{R}\right). \tag{11.3}$$

If α, θ or T is known as a function of p, the thickness of a layer between two specified pressure values p_1 and p_2 can be obtained from (11.2) by integration:

$$\varphi_1 - \varphi_2 = \int_{p_1}^{p_2} \alpha\, dp = \int_{\Pi_1}^{\Pi_2} \theta\, d\Pi = \int_{\log p_1}^{\log p_2} RT\, d\log p. \tag{11.4}$$

If, on the other hand, α, θ and T are known as functions of height, the following integrals can be evaluated,

$$p_2 - p_1 = \int_{z_2}^{z_1} g\varrho\, dz, \qquad \Pi_2 - \Pi_1 = \int_{z_2}^{z_1} \frac{g\, dz}{\theta}, \qquad \log\frac{p_2}{p_1} = \int_{z_2}^{z_1} \frac{g\, dz}{RT}. \tag{11.5}$$

In the quasi-static theory, (11.2) is assumed to hold along any vertical line at every instant, although p and ϱ are permitted to vary horizontally as well as with time. Therefore (11.4) and (11.5) are also assumed to apply locally even though the air is not at rest. In particular, the last integral of (11.4) is used in evaluating the relationship between φ and p from radiosonde observations of the relationship between T and p [correction for humidity is made by replacing T by the virtual temperature in (11.4)].

In the first two integrals of (11.5) we may put $z_1 = \infty$, $p_1 = 0$; this shows that p_2 is determined if ϱ or θ is given at all levels above z_2. It is important to note, however, that we cannot put $p_1 = 0$ in the third integral of (11.5); in fact, knowledge of T at all levels above z_2 is not sufficient to determine p_2. Therefore, a temperature rise (fall) at all levels is not necessarily connected with a fall (rise) of surface pressure, as has sometimes been assumed in meteorology.

Certain theoretical distributions of pressure, temperature and density, compatible with the hydrostatic equation and the gas equation, have been used extensively as model atmospheres in theoretical work. The simplest is the *isothermal atmosphere* (T = constant). Assuming g constant, we obtain in this case

$$p = p_0\, e^{-\frac{gz}{RT}}, \qquad \varrho = \varrho_0\, e^{-\frac{gz}{RT}}, \qquad \frac{p_0}{\varrho_0} = RT, \tag{11.6}$$

where p_0 is the pressure and ϱ_0 the density at $z=0$. We note that such an atmosphere has an infinite vertical extent, and the coefficient of barotropy (Γ) and the dry-adiabatic coefficient of piezotropy [Eq. (2.15)] are both constants.

Another simple model atmosphere is the *polytropic atmosphere*, which is characterized by a constant lapse rate m of the temperature. Denoting by T_0 the temperature at $z=0$, the temperature distribution is therefore

$$T = T_0 - mz. \tag{11.7}$$

The corresponding distributions of p and ϱ, compatible with (11.2) and (2.1), are given by

$$\frac{p}{p_0}=\left(\frac{\varrho}{\varrho_0}\right)^{\frac{g}{g-Rm}}=\left(\frac{T}{T_0}\right)^{\frac{g}{Rm}},\qquad \frac{p_0}{\varrho_0}=RT_0, \tag{11.8}$$

which is a polytropic distribution of the class $g/(g\text{-}Rm)$. The isothermal atmosphere considered above is obtained as a special case by taking $m=0$.

With $m>0$ (T decreasing with height) we obtain an atmosphere with a finite vertical extent; its height is

$$H=T_0/m. \tag{11.9}$$

At this level p and T are zero, and ϱ is also zero provided $m<g/R$.

If $m=g/c_p$, the potential temperature is constant, as will be seen by comparing (11.8) and (2.10). The height of such an *"isentropic" atmosphere* is $c_p T_0/g$. The coefficient of barotropy becomes in this case $\Gamma=(\varkappa RT)^{-1}$, which is equal to the coefficient of piezotropy for dry-adiabatic changes of state [Eq. (2.15)]; this means that the atmosphere is autobarotropic. The quantity g/c_p is called the *dry-adiabatic lapse rate.*

In the real atmosphere, the lapse rate of temperature cannot appreciably exceed the dry-adiabatic lapse rate (except in a very thin layer near the ground) because such an equilibrium is unstable. Such static instability (see Sect. 21) will manifest itself in the occurance of convection currents, which will quickly change the vertical structure of the atmosphere until ultimately a new equilibrium state is reached where the lapse rate is at every level smaller than the dry-adiabatic lapse rate.

If the atmosphere is saturated with moisture, the equilibrium is statically unstable when the lapse rate exceeds the moist-adiabatic lapse rate, which is smaller than the dry-adiabatic lapse rate (comp. Sect. 21).

Whether saturated or not, a polytropic model atmosphere with $m>g/c_p$, (θ decreasing with height) is certainly unstable. One such unrealistic model is the *"homogeneous"* or rather *equidense atmosphere* obtained when $m=g/R$; its height is RT_0/g. See also Eq. (32.10).

For a more detailed treatment of atmospheric statics, the reader is referred to the textbooks [*1*], [*2*], [*3*], [*7*], [*11*] and [*12*].

12. Geostrophic flow. If the motion is frictionless and adiabatic, and if the earth's curvature is neglected and the CORIOLIS-parameter is regarded as constant, then the quasi-static equations will permit a particularly simple solution which represents steady flow along parallel, straight horizontal streamlines. In this case the acceleration vanishes, and the equation of motion (9.2) expresses that the horizontal pressure force is balanced by the CORIOLIS force. Choosing the x-axis in the direction of the flow, we obtain from (9.2) and (9.3),

$$\frac{\partial p}{\partial x}=0,\quad \frac{\partial p}{\partial y}=-\lambda\varrho u,\quad \frac{\partial p}{\partial z}=-g\varrho. \tag{12.1}$$

It follows that $\frac{\partial\varrho}{\partial x}=0$ and $\frac{\partial u}{\partial x}=0$, so that u, ϱ and p are functions of y and z only; the continuity equation and the adiabatic equation are then satisfied also. The wind velocity is under these conditions determined by the horizontal pressure force; this wind velocity is called the *geostrophic wind,* and the flow is referred to as *geostrophic flow.*

Eqs. (12.1) show that the geostrophic wind blows in a direction normal to the horizontal pressure gradient with lower pressure to the left in the northern hemisphere (to the right in the southern hemisphere). The geostrophic wind speed may be written as

$$u_g = -\frac{\alpha}{\lambda}\frac{\partial p}{\partial y} = -\frac{1}{\lambda}\left(\frac{\partial \varphi}{\partial y}\right)_p = -\frac{1}{\lambda}\left(\frac{\partial \Psi}{\partial y}\right)_\theta . \tag{12.2}$$

The two last expressions are obtained from the equation of motion in the forms (10.4) and (10.19), respectively. We note that

$$\left(\frac{\partial z}{\partial y}\right)_p = \frac{1}{g}\left(\frac{\partial \varphi}{\partial y}\right)_p = -\frac{\lambda}{g} u_g \tag{12.3}$$

is the slope of the isobaric surfaces. For geostrophic velocities of the order 10 m/sec the isobaric slope is therefore of the order 10^{-4}.

It follows from (12.3) that the isobaric surfaces are parallel sloping planes if the distribution of u_g is uniform. However, nothing prevents that u_g varies from one streamline to the next; and we may also allow the CORIOLIS parameter λ to vary horizontally with y. For the variation of u_g along the vertical, we obtain from (12.2) in virtue of (9.5),

$$\frac{\partial u_g}{\partial z} = -\frac{1}{\lambda} S_x, \tag{12.4}$$

where S_x is the x-component of the baroclinicity vector. From (10.5) we obtain the equivalent equations

$$\frac{\partial u_g}{\partial z} = -\frac{g}{\lambda}\frac{1}{\alpha}\left(\frac{\partial \alpha}{\partial y}\right)_p = -\frac{g}{\lambda}\frac{1}{T}\left(\frac{\partial T}{\partial y}\right)_p = -\frac{g}{\lambda}\frac{1}{\theta}\left(\frac{\partial \theta}{\partial y}\right)_p . \tag{12.5}$$

Since in the present case $S_y = S_z = 0$, it will be seen that $\partial u_g/\partial z$ is a measure of the baroclinicity; in particular, the flow is barotropic if $\partial u_g/\partial z = 0$.

In a geostrophic flow the fields of velocity and density are interrelated; elimination of p between the two last Eqs. (12.1) gives

$$\lambda \frac{\partial(\varrho u_g)}{\partial z} = g \frac{\partial \varrho}{\partial y} . \tag{12.6}$$

In particular, if the density field is potentiotropic, there is no change of momentum with height.

We may further generalize the concept of geostrophic flow by noting that u_g and ϱ need not even be continuous functions of y and z, as long as p is continuous. The Eqs. (12.4) and (12.5) are in this case inapplicable in the points of discontinuity, because the derivatives appearing in these equations do not exist. However, we may still eliminate p, and thus obtain the relationship between the fields of u_g and ϱ by applying V. BJERKNES' second circulation theorem (5.9) to a closed curve in the yz-plane. This gives

$$-\oint dp = \oint (\lambda \varrho u_g\, dy + g \varrho\, dz) = 0 \tag{12.7}$$

for any closed curve. This equation applies whether u_g and ϱ are continuous or not, and is therefore more general than (12.6), to which it reduces if u_g and ϱ are differentiable.

Of particular importance in meteorology is the case of a *surface of discontinuity* $F(y,z) = 0$, with continuous fields on both sides. The atmospheric frontal zones, which are transition layers with strong gradients of wind velocity and potential temperature, may for many purposes be treated as abrupt discontinuity surfaces of this kind. Let u_{g1}, ϱ_1 and u_{g2}, ϱ_2 denote the velocity and density below and above the discontinuity surface, respectively. When (12.7) is applied to an infinitesimal rectangular curve with sides dy and dz, whose diagonal is part of the discontinuity surface, (Fig. 1), we obtain

$$\frac{dz}{dy} = -\frac{\lambda}{g}\frac{\varrho_1 u_{g1} - \varrho_2 u_{g2}}{\varrho_1 - \varrho_2} = -\frac{\lambda}{g}\frac{T_2 u_{g1} - T_1 u_{g2}}{T_2 - T_1}. \tag{12.8}$$

This is MARGULES' *formula*[1,2] which relates the slope (dz/dy) of the discontinuity surface to the values of u_g and ϱ (or T) on both sides. This formula replaces (12.6) in points on the surface of discontinuity.

Formula (12.8) shows that the discontinuity surface is horizontal if the momentum ϱu_g is continuous, and vertical if the density is continuous; if the velocity is continuous, then the discontinuity surface will coincide with an isobaric surface.

The slope of atmospheric frontal layers is generally of the order 10^{-2}, which is compatible with temperature and wind discontinuities of the order 5° C and 20 m/sec respectively.

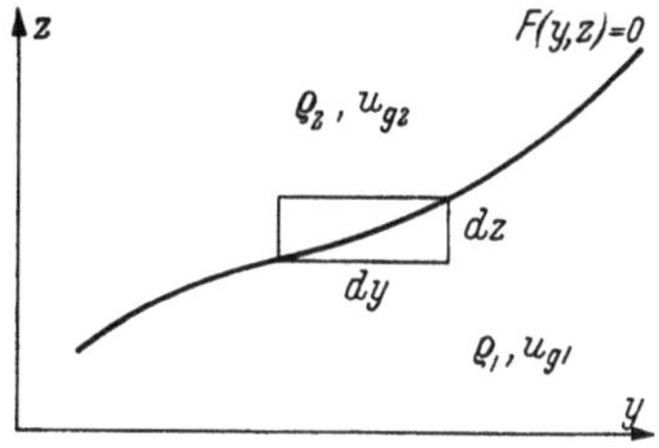

Fig. 1. MARGULES' formula.

In order for a sloping frontal surface to be statically stable so that it will not be destroyed by convection currents, it is necessary that the density *decreases* as we pass upwards through the frontal discontinuity surface. In other words, when subscript 1 refers to the air below the frontal surface, we must have $\varrho_1 - \varrho_2 > 0$ (excluding the case of vertical discontinuity surfaces). Furthermore, we may choose the positive direction along the y-axis such that the frontal slope $dz/dy > 0$; this will also determine the positive x-direction since our coordinate system is assumed right-handed. MARGULES' formula (12.8) then gives

$$\lambda(\varrho_2 u_{g2} - \varrho_1 u_{g1}) = \left(\frac{\partial p}{\partial y}\right)_1 - \left(\frac{\partial p}{\partial y}\right)_2 > 0. \tag{12.9}$$

This shows that the horizontal shear of momentum is *cyclonic*, and that the isobaric surfaces are refracted upwards in a frontal surface.

13. The geostrophic approximation. Although the geostrophic flow considered in the previous section requires straight and parallel isobars, the formula (12.2) permits us to define a geostrophic wind velocity in every point of an arbitrary pressure field. The geostrophic wind $\boldsymbol{v}_g$ thus defined is just another expression for the horizontal pressure force and differs in general from the true wind velocity.

In vector form, (12.2) may be written

$$\boldsymbol{v}_g = \boldsymbol{k} \times \frac{\alpha}{\lambda}\,\mathrm{grad}_h\, p = \boldsymbol{k} \times \frac{1}{\lambda}\,\mathrm{grad}_p\, \varphi = \boldsymbol{k} \times \frac{1}{\lambda}\,\mathrm{grad}_\theta\, \Psi \tag{13.1}$$

which can also be obtained from the equation of motion (9.2), (10.4) or (10.19) by dropping the acceleration and replacing $\boldsymbol{v}_h$ by $\boldsymbol{v}_g$. The geostrophic wind vector is directed along the isobars in a horizontal surface, or what is equivalent, along the contour lines of an isobaric surface, or along the curves $\Psi = $ constant in an isentropic surface.

The pressure force per unit volume can be expressed as $\lambda \boldsymbol{k} \times \boldsymbol{v}_g$, and the equation of motion (9.2) may therefore be written in the form

$$\left(\frac{D\boldsymbol{v}_h}{Dt}\right)_h = \lambda \boldsymbol{k} \times (\boldsymbol{v}_g - \boldsymbol{v}_h) \tag{13.2}$$

[1] M. MARGULES: Meteor. Z. (Hann.-Bd.) **1906**, 243.

[2] See also V. BJERKNES: Geof. Publ. **2**, No. 4 (1921).

or

$$\boldsymbol{v}_h - \boldsymbol{v}_g = \boldsymbol{k} \times \frac{1}{\lambda} \left(\frac{D \boldsymbol{v}_h}{D t} \right)_h. \tag{13.3}$$

Thus, in the absence of friction, the *geostrophic departure* $\boldsymbol{v}_h - \boldsymbol{v}_g$ is just an expression for the acceleration.

It is an empirical fact that in large-scale motion systems in middle and high latitudes, the accelerations are usually so small that the geostrophic departure is generally considerably smaller than the wind itself. Consequently, *the geostrophic wind can be used as a first approximation for the wind velocity.* This is clearly brought out by weather maps over areas where the network of observations is dense enough to permit independent analyses of pressure and wind.

The geostrophic wind approximation is not a very accurate one, however. The geostrophic departures are generally of the order 1 to 10 m/sec and amount to about 10 to 20% of the wind, on the average. Nevertheless, the geostrophic wind approximation is extensively used in synoptic meteorolgy and is also applied in theory (comp. Chap. III). We shall therefore consider some of the properties of the geostrophic wind field.

We note first that the change of $\boldsymbol{v}_g$ with height (or the horizontal vorticity of the geostrophic wind) is related to the baroclinicity. Differentiation of (13.1) with respect to height gives, on account of (9.5),

$$\frac{\partial \boldsymbol{v}_g}{\partial z} = - \frac{1}{\lambda} \boldsymbol{S}_h. \tag{13.4}$$

This equation, or rather its integral over a definite vertical distance, is referred to as the *thermal wind equation.*

From (13.4) and (13.1) we derive, on account of (9.3) and (9.6)

$$\boldsymbol{k} \cdot \boldsymbol{v}_g \times \frac{\partial \boldsymbol{v}_g}{\partial z} = \frac{g}{\lambda} \varrho \, \boldsymbol{v}_g \cdot \operatorname{grad}_h \alpha = \frac{1}{\lambda^2} S_z. \tag{13.5}$$

Therefore, if the geostrophic wind does not change with height, the density field is barotropic; and if the geostrophic wind changes with height in magnitude but not in direction, then S_z vanishes.

Eq. (13.4), in connection with the geostrophic approximation, may be used to determine the variation of wind with height when $\boldsymbol{S}_h$ is known; or vice versa, if the change of wind with height is measured, we may find the isobaric temperature gradient (or $\boldsymbol{S}_h$).

From (13.1) we derive the following expression for the vertical geostrophic vorticity

$$\boldsymbol{k} \cdot \operatorname{curl} \boldsymbol{v}_g = \frac{\alpha}{\lambda} \nabla^2 p + \operatorname{grad}_h \left(\frac{\alpha}{\lambda} \right) \cdot \operatorname{grad}_h p. \tag{13.6}$$

The second term on the right is generally small and is often ignored. However, it is usually more convenient to consider the *"isobaric" vorticity* of $\boldsymbol{v}_g$, which is obtained by performing the operation curl_p on (13.1). This gives

$$\boldsymbol{k} \cdot \operatorname{curl}_p \boldsymbol{v}_g = \frac{1}{\lambda} \nabla_p^2 \varphi + \operatorname{grad}_h \left(\frac{1}{\lambda} \right) \cdot \operatorname{grad}_p \varphi. \tag{13.7}$$

Here we are even more justified in ignoring the second term on the right. In similar manner we may find $\operatorname{curl}_\theta \boldsymbol{v}_g$.

Generally, the geostrophic vorticity is found to be a fair approximation to the true vertical vorticity. The two may differ significantly in certain areas; this is particularly true in regions where the vorticity is large, or where the horizontal velocity field has strong deformation, as pointed out by PETTERSSEN[1]. The geostrophic vorticity is usually algebraically greater than the true vertical vorticity in the vicinity of cyclonic or anticyclonic centers, and smaller in the vicinity of saddle points in the pressure field.

It follows from (13.1) that $\operatorname{div}_h(\lambda\,\varrho\,\boldsymbol{v}_g)=0$. Hence we obtain, when v_g denotes the component of $\boldsymbol{v}_g$ towards north, and β the gradient of λ (5.34),

$$\operatorname{div}_h(\varrho\,\boldsymbol{v}_g) = -\,\varrho\,\boldsymbol{v}_g\cdot\frac{\operatorname{grad}_h\lambda}{\lambda} = -\,\frac{\beta}{\lambda}\,\varrho\,v_g \tag{13.8}$$

so that the horizontal divergence of geostrophic momentum is due only to the variation of the CORIOLIS parameter. This effect is quantitatively small; the horizontal divergence of the true momentum is mainly due to the nongeostrophic part of the wind field and cannot be approximated geostrophically. Similarly, we find

$$\operatorname{div}_p\boldsymbol{v}_g = -\boldsymbol{v}_g\cdot\frac{\operatorname{grad}_h\lambda}{\lambda} = -\frac{\beta}{\lambda}\,v_g. \tag{13.9}$$

The isobaric divergence of the geostrophic wind field is generally considerably smaller than $\operatorname{div}_p\boldsymbol{v}_h$ and cannot be used as an approximation for the latter expression.

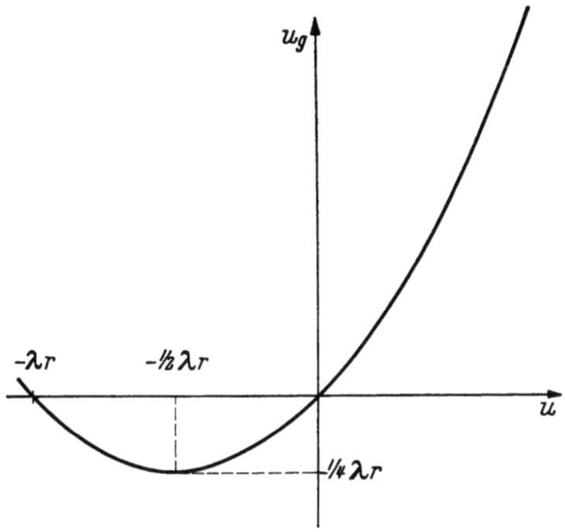

Fig. 2. Relation between gradient wind u and geostrophic wind u_g.

14. Gradient wind. When the CORIOLIS parameter is considered constant and the earth's curvature is neglected, the equations of frictionless, adiabatic motion permit a solution which represents a *steady circular vortex motion* around a vertical axis.

Since the centripetal acceleration and the CORIOLIS force are both directed normal to the streamlines, it follows that the horizontal pressure force is also normal to the streamlines, i.e. the isobars in a level surface coincide with the streamlines. Let r denote distance from the center of the streamlines, and u the velocity, so that the acceleration is u^2/r. The equation of motion in the form (13.2) then yields

$$u^2/r = \lambda(u_g - u), \tag{14.1}$$

where u is reckoned positive for cyclonic rotation (counter-clockwise in the northern hemisphere) and negative for anticyclonic rotation. Similarly, u_g is reckoned positive when the horizontal pressure force is directed toward the center, so that the horizontal pressure field represents a low, and u_g is negative for a pressure high.

The relation (14.1) between u and u_g is represented graphically in Fig. 2. It is seen that u_g has a minimum $(-\tfrac{1}{4}\lambda r)$ when $u = -\tfrac{1}{2}\lambda r$; this gives the strongest outward directed pressure force, i.e. the strongest pressure high that can exist under balanced conditions. We have $|u| > |u_g|$ when $u<0$ (except for very large negative values of u which are without meteorological interest), and $|u|<|u_g|$ when $u>0$. The wind will be approximately geostrophic when $|u|\ll\lambda r$.

e) Atmospheric turbulence.

15. Averaging the hydrodynamic-thermodynamic equations. If one could perform a complete analysis of the motions in the atmosphere into motion systems of different scale, one would obtain an unbroken spectrum of motions ranging

[1] S. PETTERSSEN: Tellus 5, 231 (1953).

from the large-scale currents encircling the globe and down to systems of microscopic scale. The phenomenon could conceivably be dealt with by means of the methods of statistical turbulence, using such concepts as the spectral distribution of kinetic energy and the transfer of such energy from one part of the spectrum to another. In some parts of the spectrum there will be energy input, and in other parts energy output, by processes which transform potential and internal energy into kinetic energy and vice versa. We must assume, however, that there is no energy input or output of this kind when we go far enough down toward the short-wave end of the spectrum. Here the motion obeys the statistical laws for isotropic turbulence which have recently been formulated (theories of KOLMOGOROFF[1] and HEISENBERG[2]), i.e. the kinetic energy is here transferred to motions of ever smaller scale, until in the very smallest scale motions the energy leaves the spectrum, being dissipated into heat by molecular friction.

Passing from the short-wave end of the spectrum towards systems of larger scale, we shall find that such effects as static stability or instability, the decrease of density with height and the kinematic constraint imposed upon the motion by the earth's surface will cause the motion systems to become increasingly anisotropic as their scale increases; and the statistical behaviour of these larger scale motions is entirely different from isotropic turbulence. Some knowledge about the energy transfer processes in the long-wave end of the spectrum has recently been gained; reference is made to Chap. V about the general circulation and to Sect. 39 (energy transfer in two-dimensional, non-divergent flow). The part of the spectrum which we know least about is the intermediate section, where the motion systems are too small to be studied from the synoptic network of observations, and at the same time large enough so that their statistical behaviour differs radically from isotropic turbulence. Here we find in particular such phenomena as convection currents.

The principal aim of dynamic meteorology is to study the motion itself, and not only its statistical characteristics. However, in an attempt to extract useful information about the motion from the governing differential equations, it is virtually impossible to deal with the entire spectrum of motions. In order to obtain fields which we are able to handle at all, it is necessary to disregard the short-wave end of the spectrum and consider only motion systems of dimensions larger than a certain selected smallest scale; this is true whether we seek analytic solutions of the differential equations, or we try to integrate them numerically. The smallest scale to be considered should be chosen according to the nature of the motions to be studied. It should be noted that in the problem of weather prediction, where the initial condition is to be determined from observations, this smallest scale will depend upon the density of observing stations; the initial state is defined only for motion systems of scale larger than the distance between the observing stations, whereas all smaller scale motions are lost between the meshes in the observational network.

In order to deal with a truncated spectrum of motions, it is necessary to derive governing equations which apply to that part of the motion which we want to study. Such equations have been derived by O. REYNOLDS [*18*] for a homogeneous and incompressible fluid; and his method has been generalized to compressible fluids by CALDER[3], MILLER[4] and VAN MIEGHEM[5]. The following outline follows, in its broad features, the derivation given by VAN MIEGHEM[5].

[1] A. N. KOLMOGOROFF: C. R. [Doklady] Acad. Sci. URSS. **30**, 301 (1941); **32**, 16 (1941).
[2] W. HEISENBERG: Z. Physik **124**, 628 (1948).
[3] K. L. CALDER: Quart. J. Roy. Meteor. Soc. **75**, 71 (1949).
[4] J. E. MILLER: Comp. of Meteor., Amer. Meteor. Soc. **1951**, 483.
[5] J. VAN MIEGHEM: Tellus **4**, 334 (1952).

We first define a linear averaging operator (denoted by a bar) which, when applied to a field, effectively filters away the small-scale features which we want to disregard. REYNOLDS applied a time mean, but a space mean, e.g. over a horizontal surface of suitable size, may also be used. We assume that with sufficient accuracy

$$\overline{\overline{A}} = \overline{A}, \qquad \overline{\overline{A}\,B} = \overline{A}\,\overline{B}. \tag{15.1}$$

Hence, we may write

$$\varrho = \overline{\varrho} + \varrho', \qquad \overline{\varrho'} = 0. \tag{15.2}$$

We shall also make use of a weighted mean (denoted by ˆ), which was introduced by HESSELBERG[1]; it is defined by

$$\overline{\varrho \boldsymbol{v}} = \overline{\varrho}\,\hat{\boldsymbol{v}}, \qquad \boldsymbol{v} = \hat{\boldsymbol{v}} + \boldsymbol{v}'', \qquad \overline{\varrho \boldsymbol{v}''} = 0. \tag{15.3}$$

The mean value of the kinetic energy then becomes

$$\tfrac{1}{2}\,\overline{\varrho \boldsymbol{v}^2} = \tfrac{1}{2}\,\overline{\varrho}\,\hat{\boldsymbol{v}}^2 + \tfrac{1}{2}\,\overline{\varrho \boldsymbol{v}''^2} \tag{15.4}$$

showing that the mean value of the kinetic energy may be partitioned into the kinetic energy of the mean motion and the "eddy energy" $\frac{1}{2}\overline{\varrho \boldsymbol{v}''^2}$.

We shall now apply the averaging operator to the hydrodynamic-thermodynamic equations. The equation of continuity (4.2) becomes

$$\frac{\partial \overline{\varrho}}{\partial t} + \operatorname{div}(\overline{\varrho}\,\hat{\boldsymbol{v}}) = 0 \tag{15.5}$$

i.e. the mean density field $\overline{\varrho}$ and the weighted mean velocity $\hat{\boldsymbol{v}}$ satisfy the equation of continuity.

Averaging of the equation of motion in the form (4.13) gives

$$\frac{\partial(\overline{\varrho}\,\hat{\boldsymbol{v}})}{\partial t} + \operatorname{div}\left(\overline{\varrho}\,\hat{\boldsymbol{v}}\,\hat{\boldsymbol{v}} + \overline{\varrho \boldsymbol{v}''\boldsymbol{v}''} + \mathscr{E}\overline{p} - \overline{\mathscr{F}}\right) = -\overline{\varrho}\operatorname{grad}\varphi - 2\boldsymbol{\Omega}\times\overline{\varrho}\,\hat{\boldsymbol{v}}. \tag{15.6}$$

Using (15.5) and the notation

$$\frac{\hat{D}}{Dt} = \frac{\partial}{\partial t} + \hat{\boldsymbol{v}}\cdot\operatorname{grad} \tag{15.7}$$

equation (15.6) may also be written

$$\overline{\varrho}\,\frac{\hat{D}\hat{\boldsymbol{v}}}{Dt} = -\,\overline{\varrho}\operatorname{grad}\varphi - 2\boldsymbol{\Omega}\times\overline{\varrho}\,\hat{\boldsymbol{v}} - \operatorname{grad}\overline{p} + \operatorname{div}\left(\overline{\mathscr{F}} - \overline{\varrho \boldsymbol{v}''\,\boldsymbol{v}''}\right). \tag{15.8}$$

Eq. (15.6) or (15.8) represents the equation of motion for a fluid with velocity $\hat{\boldsymbol{v}}$, density $\overline{\varrho}$, pressure $\overline{p}$ and a virtual friction stress tensor $\overline{\mathscr{F}} - \overline{\varrho \boldsymbol{v}''\boldsymbol{v}''}$. The components of $-\overline{\varrho \boldsymbol{v}''\boldsymbol{v}''}$ are called the REYNOLDS *stresses* or *eddy stresses*.

Scalar multiplication of (15.8) by $\hat{\boldsymbol{v}}$ and application of (15.5) gives

$$\left.\begin{aligned} &\frac{\partial}{\partial t}\left(\frac{1}{2}\,\overline{\varrho}\,\hat{\boldsymbol{v}}^2 + \overline{\varrho}\,\varphi\right) + \operatorname{div}\left[\left(\frac{1}{2}\,\overline{\varrho}\,\hat{\boldsymbol{v}}^2 + \overline{\varrho}\,\varphi\right)\hat{\boldsymbol{v}} + \overline{p}\,\hat{\boldsymbol{v}} + \overline{\varrho \boldsymbol{v}''\,\boldsymbol{v}''}\cdot\hat{\boldsymbol{v}} - \overline{\mathscr{F}}\cdot\hat{\boldsymbol{v}}\right] \\ &\qquad = \overline{p}\operatorname{div}\hat{\boldsymbol{v}} + \overline{\varrho \boldsymbol{v}''\,\boldsymbol{v}''}\cdot\operatorname{grad}\cdot\hat{\boldsymbol{v}} - \overline{\mathscr{F}}\cdot\operatorname{grad}\cdot\hat{\boldsymbol{v}}. \end{aligned}\right\} \tag{15.9}$$

Here the first term on the left represents the local rate of change of mechanical energy of the mean motion per unit volume. The vector expression within the brackets on the left-hand side may be interpreted as the flux, and the right-hand side as the rate of production of mechanical energy of the mean motion.

[1] TH. HESSELBERG: Beitr. Phys. fr. Atmosph. **12**, 141 (1926).

A corresponding equation for the eddy energy is obtained by averaging the mechanical energy Eq. (6.15) and subtracting (15.9). On account of (15.4), we obtain

$$\left.\begin{aligned}\frac{\partial}{\partial t}\left(\frac{1}{2}\overline{\varrho \boldsymbol{v}''^2}\right) + \operatorname{div}\left[\frac{1}{2}\overline{\varrho \boldsymbol{v}''^2}\,\hat{\boldsymbol{v}} + \frac{1}{2}\overline{\varrho \boldsymbol{v}''^2 \boldsymbol{v}''} + \overline{p \boldsymbol{v}''} - \overline{\mathscr{F}\cdot \boldsymbol{v}''}\right] \\ = \overline{p \operatorname{div} \boldsymbol{v}''} - \overline{\varrho \boldsymbol{v}'' \boldsymbol{v}''}\cdot \operatorname{grad}\cdot \hat{\boldsymbol{v}} - \overline{\mathscr{F}\cdot \operatorname{grad}\cdot \boldsymbol{v}''}.\end{aligned}\right\} \quad (15.10)$$

This equation defines the flux of eddy energy (within the brackets) and the rate of production of eddy energy (the right-hand side).

Finally we perform the averaging operation on the thermodynamic energy Eq. (6.18), and obtain

$$\left.\begin{aligned}\frac{\partial(\bar{\varrho}\,\hat{e})}{\partial t} + \operatorname{div}\left[\bar{\varrho}\,\hat{e}\,\hat{\boldsymbol{v}} + \overline{\varrho\, e'' \boldsymbol{v}''} + \overline{\boldsymbol{W}}\right] \\ = -\bar{p}\operatorname{div}\hat{\boldsymbol{v}} - \overline{p \operatorname{div} \boldsymbol{v}''} + \overline{\mathscr{F}}\cdot \operatorname{grad}\cdot \hat{\boldsymbol{v}} + \overline{\mathscr{F}\cdot \operatorname{grad}\cdot \boldsymbol{v}''}.\end{aligned}\right\} \quad (15.11)$$

Here the vector expression within the brackets signifies the flux of mean internal energy, and the right-hand side is the rate of production of mean internal energy[1].

Comparing the right-hand sides of (15.9), (15.10) and (15.11) we notice that their sum vanishes, i.e. the rate of production of the total energy is zero. The various right-hand terms represent energy transformation functions. The mean expansion work $\overline{p \operatorname{div} \boldsymbol{v}}$ is separated into one part $\bar{p} \operatorname{div} \hat{\boldsymbol{v}}$ which represents transformation of internal energy into mechanical energy of the mean motion, and another part $\overline{p \operatorname{div} \boldsymbol{v}''}$ representing transformation of internal energy into eddy energy. Similarly, the mean dissipation function $\overline{\mathscr{F}\cdot \operatorname{grad}\cdot \boldsymbol{v}}$ (≥ 0) is broken up into $\overline{\mathscr{F}}\cdot \operatorname{grad}\cdot \hat{\boldsymbol{v}}$ which represents transformation of mechanical energy of the mean motion into internal energy, and $\overline{\mathscr{F}\cdot \operatorname{grad}\cdot \boldsymbol{v}''}$ representing transformation of eddy energy into internal energy. Besides, the term $-\overline{\varrho \boldsymbol{v}''\boldsymbol{v}''}\cdot \operatorname{grad}\cdot \hat{\boldsymbol{v}}$ on the right of (15.9) and (15.10) represents transformation of mechanical energy of the mean motion into eddy energy. Each of these energy transformation functions may be positive or negative, so that the energy can be converted both ways; however, the sum of the two dissipation functions must be non-negative.

The viscous stresses ($\mathscr{F}$) depend upon the velocity gradients and will be noticeable only when these gradients are very strong, i.e. for motions of very small scale. Generally, the averaging process will eliminate such exceedingly strong gradients, with the result that the averaged viscous stress tensor $\overline{\mathscr{F}}$ will be negligible in comparison with the eddy stresses. An exception occurs in the immediate neighbourhood of the ground where the large vertical velocity gradients will not be smoothed out if time averaging or horizontal space averaging is applied. Except in this shallow layer, we may safely ignore $\overline{\mathscr{F}}$ in (15.6) and (15.8), and also the terms containing $\overline{\mathscr{F}}$ in (15.9) and (15.11). This implies that $\overline{\mathscr{F}\cdot \operatorname{grad}\cdot \boldsymbol{v}''}$ is considered the entire dissipation and is therefore non-negative, always converting eddy-energy into internal energy. The rate of transformation between mechanical energy of the mean motion $(\frac{1}{2}\bar{\varrho}\,\hat{\boldsymbol{v}}^2 + \bar{\varrho}\,\varphi)$, mean eddy energy $(\frac{1}{2}\overline{\varrho\, \boldsymbol{v}''^2})$ and mean internal energy $(\bar{\varrho}\,\hat{e})$ is illustrated schematically in Fig. 3.

The subdivision into "mean motion" and "eddy motion" depends upon our choice of averaging operation. By taking the average over a sufficiently short period, or over a sufficiently small area, we may conceivably attain that the eddy motion consists of pure isotropic turbulence. In this case, the first term on the

[1] It should be stressed that the definitions of flux and rate of production are ambiguous. Compare footnote p. 17.

right of (15.10) would be negligible, and the eddy motion would receive energy at the rate $-\overline{\varrho \boldsymbol{v}'' \boldsymbol{v}''} \cdot \operatorname{grad} \cdot \hat{\boldsymbol{v}}$ from the mechanical energy of the mean motion, and lose energy at the same rate by dissipation.

In synoptic meteorology, it is customary to define the *wind* as an *average velocity over 10 min.* The eddy motions defined by this kind of average will generally contain motions which cannot be characterized as isotropic turbulence. This case is more complicated because the transformation function $\overline{p \operatorname{div} \boldsymbol{v}''}$ of (15.10) is generally different from zero so that both the mechanical energy of the mean motion and the mean internal energy may supply energy to the eddy motion; or the eddy motion may receive energy from one of these sources and lose energy to the other. One may assume that the mean internal energy will constitute an energy *sink* when the air is statically *stable*, and an energy *source* when the air is

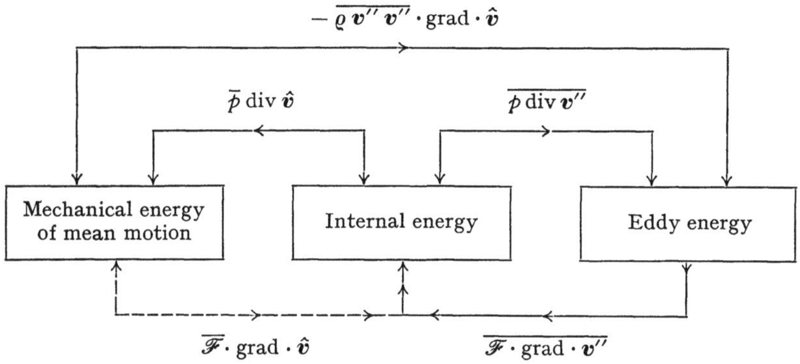

Fig. 3. Scheme of energy transformations.

statically *unstable* (compare Sect. 19); in the latter case, the eddy motion will at least partly have the character of *thermal convection*. Eddy motions maintained mainly by the mechanical energy of the mean motion is sometimes labelled "*mechanical turbulence*", and similarly, eddy motions which feed upon the mean internal energy may be called "*thermal turbulence*".

The *numerical forecasting techniques* use a representation of the variables in a grid of points spaced about 300 km apart. This corresponds roughly to a space averaging over horizontal areas of linear dimensions of the order 300 km. Very little is known so far about the structure of the eddy motions defined by such averages, and their effect upon the mean motion. Clearly, such eddy motions will contain convective currents of several kinds.

There are also other kinds of averaging operations used in meteorology, such as 5 day means, monthly means, annual means etc. A method which is particularly useful in the study of the general circulation is *averaging along latitude circles*; this kind of average has the advantage that Eqs. (15.1) are rigorously fulfilled. The averaging procedure may also be applied in steps, first smoothing out small-scale turbulence (thereby introducing REYNOLDS-stresses), and afterwards averaging along latitude circles, thus introducing another kind of eddy stresses (VAN MIEGHEM, ref. p. 34).

The equations for the mean motion (15.5), (15.6) and (15.11), with suitable boundary conditions, constitute a closed system provided the four eddy terms

$$-\overline{\varrho \boldsymbol{v}'' \boldsymbol{v}''}, \qquad \overline{\varrho e'' \boldsymbol{v}''}, \qquad \overline{p \operatorname{div} \boldsymbol{v}''}, \qquad \overline{\mathscr{F} \cdot \operatorname{grad} \cdot \boldsymbol{v}''}$$

can be expressed in terms of the mean fields $\hat{\boldsymbol{v}}, \bar{p}$ and $\bar{\varrho}$. Unfortunately, our present knowledge about these statistical properties of the eddy motions (for various amounts of smoothing) is not sufficient to establish such relationships.

One is therefore often forced to neglect these eddy terms altogether. There exists, however, crude hypotheses which bring the eddy stress and the eddy flux of internal energy due to small scale turbulence into relation with the mean fields of wind and temperature. These relations are dealt with briefly in the following sections.

For a more complete treatment of atmospheric turbulence, reference is made to LETTAU [*13*], PRIESTLEY and SHEPPARD [*15*] and SUTTON [*22*] and [*23*].

16. The effect of small-scale turbulence. The wind profile within the friction layer near the ground. *α) Approximations for the friction layer.* The state of small-scale turbulence in the atmosphere is highly variable, depending upon the mean wind field, the static stability, and near the ground also upon the roughness of the earth's surface. The turbulence is particularly pronounced in a layer near the ground, where the strongest vertical gradients of the mean wind are also found. This "*friction layer*" near the ground is also the layer where the turbulence has been most thoroughly studied. There is evidence that the motion is often turbulent also at higher levels; but this "*free-air turbulence*" is generally assumed to be of less importance for the dynamics of the mean motion. Although this is questionable, we shall here deal only with the friction layer near the ground.

It follows from (15.8) that the force of virtual friction is $\operatorname{div}(\overline{\mathcal{F}-\varrho\boldsymbol{v}''\boldsymbol{v}''})$. Because the friction layer is relatively shallow, its depth being of the order 500 to 1000 m, the vertical stress gradients implicite in this expression will dominate against the horizontal stress gradients. We may therefore ignore the horizontal gradients and write the horizontal force of virtual friction per unit volume as

$$\boldsymbol{F}_h = \frac{\partial}{\partial z}\left(\mu\frac{\partial\bar{\boldsymbol{v}}_h}{\partial z} - \overline{\varrho w''\boldsymbol{v}_h''}\right) = \frac{\partial\boldsymbol{\tau}_h}{\partial z}, \tag{16.1}$$

where $\boldsymbol{\tau}_h$ is the horizontal stress vector on a horizontal surface. Moreover, it is customary to neglect the vertical force of virtual friction altogether.

The only stress which is considered is therefore the horizontal stress vector on a horizontal surface, which is the sum of the mean viscous stress and the REYNOLDS stress. The viscous stress is generally negligible compared to the REYNOLDS stress, except in a very shallow layer near the ground, of thickness of the order 1 mm, where the REYNOLDS stresses are reduced as a result of the kinematical constraint of the earth's surface. Within this layer, the mean velocity increases extremely rapidly with height and vanishes at the ground itself. Its direction does not vary with height and coincides with the direction of the horizontal stress.

Above this layer, viscosity may be neglected, and the horizontal stress on a horizontal surface is therefore

$$\boldsymbol{\tau}_h = -\overline{\varrho w''\boldsymbol{v}_h''}. \tag{16.2}$$

β) The layer up to about 10 m above the ground. The mixing length hypothesis. The variation of mean wind with height is particularly strong in the lowest part of the friction layer. The stress, on the other hand, varies rather slowly with height, and in the study of the wind profile we may here consider the stress as a constant, i.e.

$$\boldsymbol{\tau}_h = \boldsymbol{\tau}_{hs} \tag{16.3}$$

where $\boldsymbol{\tau}_{hs}$ is the surface stress. In the expression (16.2) for the REYNOLDS stress we shall furthermore neglect the turbulent fluctuations of density, which are of little importance compared with the velocity fluctuations; and since the layer considered is very shallow, we may also ignore the variation of density with height, so that ϱ may be considered as a constant.

In order to establish a relation between the REYNOLDS stress and the mean velocity distribution, PRANDTL[1] introduced the concept of *mixing length* (l) as a statistical measure of the linear dimensions of the turbulent motions. He assumed that $\overline{u''^2}$, $\overline{v''^2}$, $\overline{w''^2}$ and $l^2\left(\frac{d\hat{\boldsymbol{v}}_h}{dz}\right)^2$ are all of nearly the same magnitude, and that the correlations between the velocity components are such that

$$\boldsymbol{\tau}_h = \boldsymbol{\tau}_{hs} = -\overline{\varrho\, w''\boldsymbol{v}_h''} = \varrho\, l^2 \left|\frac{d\hat{\boldsymbol{v}}_h}{dz}\right| \frac{d\hat{\boldsymbol{v}}_h}{dz}. \tag{16.4}$$

It follows that $\hat{\boldsymbol{v}}_h$ varies with height in magnitude only and has everywhere the same direction as $\boldsymbol{\tau}_{hs}$. When the x-axis is chosen in the direction of $\boldsymbol{\tau}_{hs}$, Eq. (16.4) may be written

$$\frac{d\hat{u}}{dz} = \frac{1}{l}\sqrt{\frac{\tau}{\varrho}}, \tag{16.5}$$

where τ is the numerical value of $\boldsymbol{\tau}_h$. Furthermore, PRANDTL introduced the hypothesis

$$l = k\,z, \tag{16.6}$$

where k is a universal constant (KÁRMÁN's constant) whose value is about 0.4. The two last equations give

$$u = \frac{1}{k}\sqrt{\frac{\tau}{\varrho}}\log\frac{z}{z_0}, \tag{16.7}$$

where z_0 is a constant of integration which depends upon the roughness of the underlying surface. Eq. (16.7) shows that the distribution of mean wind with height for any given stress obeys a logarithmic law; but it shows also that the stress is proportional to the square of the mean velocity at any given level.

The logarithmic velocity profile is in good agreement with observations up to about 10 meters above the ground when the air is statically indifferent. If the air is statically stable or unstable, the velocity distribution will be modified; this effect has been studied by ROSSBY and MONTGOMERY[2], SVERDRUP[3] and DEACON[4].

γ) The layer from about 10 m to about 1000 m. The EKMAN *spiral.* The assumptions underlying the logarithmic velocity profile are no longer valid at heights greater than about 10 m; thus $\boldsymbol{\tau}_h$ cannot be treated as a constant, neither is assumption (16.6) valid. Still it may be assumed that $\boldsymbol{\tau}_h$ is directed along $d\hat{\boldsymbol{v}}_h/dz$; and considerations based upon the exchange of momentum brought about by the turbulent motions (SCHMIDT[5]) lead to the formula

$$\boldsymbol{\tau}_h = A\frac{d\hat{\boldsymbol{v}}_h}{dz} = \bar{\varrho}\,K\frac{d\hat{\boldsymbol{v}}_h}{dz}. \tag{16.8}$$

A is called the *Austausch coefficient*, and K the *eddy viscosity*. The value of K is highly variable, ranging between 10^3 and 10^5 cm² sec⁻¹. In comparison, the kinematic viscosity due to molecular friction is of the order 10^{-1} cm² sec⁻¹.

EKMAN[6] calculated the variation with depth of an ocean current driven by a given wind stress at the surface, under the assumption of horizontal stationary

[1] L. PRANDTL: Z. angew. Math. Mech. **5** (2), 136 (1925). — Beitr. Phys. d. fr. Atmosph. **19**, 188 (1932).

[2] C. G. ROSSBY and R. B. MONTGOMERY: Papers phys. ocean. meteor., Mass. Inst. Techn., Woods Hole ocean. Inst. **3**, No. 3 (1935).

[3] H. U. SVERDRUP: Geofys. Publ. **11**, No. 7 (1936).

[4] E. L. DEACON: Quart. J. Roy. Meteor. Soc. **75**, 89 (1949).

[5] W. SCHMIDT: Der Massenaustausch in freier Luft etc. In: Probleme der kosmischen Physik, Bd. 7. Hamburg: H. Grand 1925.

[6] V. W. EKMAN: Nyt Mag. Naturv. **40**, 1 (1902). See also the chapter Oceanography by H. U. SVERDRUP in this Volume, Sect. 38.

flow and constant eddy viscosity. The corresponding calculation for the atmosphere was first made by ÅKERBLOM[1]; the following derivation follows mainly that of TAYLOR[2].

When $\hat{\boldsymbol{v}}_h$ is a function of z only, the acceleration vanishes, and the equation of motion in the form (13.2) becomes, when the force of friction $d\boldsymbol{\tau}_h/dz$ is added on the right-hand side

$$0 = -\lambda \boldsymbol{k} \times (\hat{\boldsymbol{v}}_h - \hat{\boldsymbol{v}}_g) + K \frac{d^2 \hat{\boldsymbol{v}}_h}{dz^2}. \tag{16.9}$$

We have here neglected the variation of $\bar{\varrho}$ with height.

In order to integrate (16.9), let the two horizontal vectors $\hat{\boldsymbol{v}}_h$ and $\hat{\boldsymbol{v}}_g$ be represented as complex numbers V and V_g in the complex plane (u, iv); Eq. (16.9) may then be written

$$\frac{d^2(V - V_g)}{dz^2} - i\frac{\lambda}{K}(V - V_g) = 0 \tag{16.10}$$

since V_g is considered as a constant. The general solution of (16.10) is

$$V - V_g = C_1 e^{m(1+i)z} + C_2 e^{-m(1+i)z},$$

where C_1 and C_2 are arbitrary complex constants, and

$$m = \sqrt{\lambda/2K}. \tag{16.11}$$

We are interested in a solution which approaches V_g as z increases; consequently we put $C_1 = 0$ and write the solution in the form

$$V = V_g + C e^{-m(1+i)(z - z_1)}, \tag{16.12}$$

where C is a complex constant, and z_1 a constant height.

The corresponding value of the stress, represented as a complex number, is

$$\tau = \varrho K \frac{dV}{dz} = \varrho K m \sqrt{2}\, e^{-\frac{3\pi}{4} i} \cdot C e^{-m(1+i)(z - z_1)}. \tag{16.13}$$

The stress forms an angle $-\frac{3}{4}\pi$ with the geostrophic deviation $(V - V_g)$.

This solution does not apply to the bottom layer, where K is highly variable. It may be assumed, however, that the solution is valid down to a level which is sufficiently close to the ground so that the stress and the wind velocity have practically the same direction; let z_1 denote this level. Then, when $z = z_1$, we have $\arg(\tau) = \arg(V)$, or

$$\arg\left(C e^{-\frac{3\pi}{4} i}\right) = \arg(V_g + C). \tag{16.14}$$

Denoting by α the angle between the geostrophic wind and the surface stress (or wind), this condition gives:

$$C = \sqrt{2} \sin\alpha\, V_g e^{i\left(\frac{3\pi}{4} + \alpha\right)}. \tag{16.15}$$

Thus the arbitrary complex constant C is reduced to a real constant α. With this value of C the wind and stress at $z = z_1$, become

$$\left.\begin{aligned} V_1 &= V_g + C = (\cos\alpha - \sin\alpha) V_g e^{i\alpha}, \\ \tau_1 &= 2\varrho K m \sin\alpha\, V_g e^{i\alpha}. \end{aligned}\right\} \tag{16.16}$$

[1] F. A. ÅKERBLOM: Nova Acta Reg. Soc. Sci. Uppsala, Ser. IV, **2**, 2 (1908).

[2] G I. TAYLOR: Phil. Trans. Roy. Soc. Lond., Ser. A **215**, 1 (1915).

The constant α can be related to $|V_g|$ if the solution can be linked to the solution for $z < z_1$, so that the boundary conditions at the ground can be utilized. If the logarithmic profile (16.7) is assumed to hold for $z < z_1$, we have

$$|\tau_1| = \varkappa \varrho |V_1|^2, \tag{16.17}$$

where $\varkappa$ is a constant. Substituting Eq. (16.16), this gives

$$\frac{\sin \alpha}{(\cos \alpha - \sin \alpha)^2} = \frac{\varkappa |V_g|}{\sqrt{2K\lambda}}. \tag{16.18}$$

This equation gives values of α around 20° for moderate and strong winds, in fair agreement with observed values. Owing to the many crude assumptions involved, however, one generally prefers to leave α as an empirical constant in the formulas.

The solution (16.12) is shown in Fig. 4 in the form of a hodograph, i.e. a curve connecting the end points of the velocity vector as z varies. This curve is a logarithmic spiral around the point $\hat{\boldsymbol{v}}_h = \hat{\boldsymbol{v}}_g$. The geostrophic deviation $\hat{\boldsymbol{v}}_h - \hat{\boldsymbol{v}}_g$ rotates clockwise and decreases exponentially as z increases.

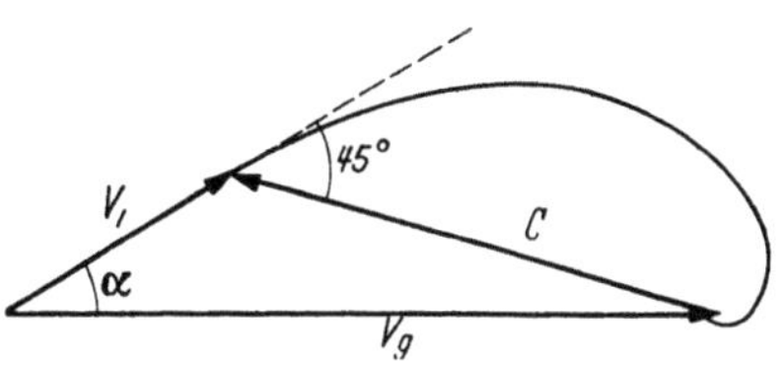

Fig. 4. Vertical variation of the wind in the friction layer (EKMAN spiral).

Through a layer of depth π/m, $\hat{\boldsymbol{v}}_h - \hat{\boldsymbol{v}}_g$ is rotated 180° and reduced by a factor of $e^{-\pi} = 0.0432$. To find the order of magnitude of π/m we may put $K = 5 \times 10^4$ cm² sec⁻¹ and $\lambda = 10^{-4}$ sec⁻¹ and get $\pi/m \approx 1000$ m. On the other hand, TAYLOR (ref. p. 40) estimated π/m from observed wind curves, and used this estimate to determine K.

Owing to the crudeness of the underlying assumptions, in particular the constancy of K with height, the EKMAN spiral should not be considered as more than a rough approximation. Nevertheless, the EKMAN spiral explains several features typical of observed wind distributions. Thus, the surface wind is normally weaker than the geostrophic wind, and deflected toward the left (toward lower pressure). The angle of deflection decreases with height, whereas the wind speed increases, so that the wind tends toward the geostrophic wind, which is attained with fair approximation at about 1000 m above the ground. Above this level, friction does not seem to cause any significant deviation from the geostrophic wind.

17. Mass transport and dissipation in the friction layer. In the case of steady horizontal mean motion along straight streamlines, the equation of motion for the friction layer becomes

$$0 = -\lambda \boldsymbol{k} \times \bar{\varrho} (\hat{\boldsymbol{v}}_h - \hat{\boldsymbol{v}}_g) + \frac{\partial \boldsymbol{\tau}_h}{\partial z}. \tag{17.1}$$

This equation may be used also if the above assumptions are not fulfilled, for the accelerations connected with large-scale motion systems are generally small enough so that (17.1) may still be considered as a useful approximation.

When integrated from $z = 0$ to a level $z = H$, high enough so that the frictional stress at that level may be ignored (i.e. $z = H$ is the top of the friction layer), (17.1) yields

$$\int_0^H \bar{\varrho} \hat{\boldsymbol{v}}_h \, dz = \int_0^H \bar{\varrho} \hat{\boldsymbol{v}}_g \, dz + \frac{1}{\lambda} \boldsymbol{k} \times \boldsymbol{\tau}_{hs} \tag{17.2}$$

showing that the surface stress $\boldsymbol{\tau}_{hs}$ is a measure of the non-geostrophic mass transport within the friction layer. This equation does not depend upon any specific assumptions concerning the variation of wind and stress with height in the friction layer.

Observations show that the surface wind, and hence $\boldsymbol{\tau}_{hs}$ is directed to the left of $\hat{\boldsymbol{v}}_g$ (Northern hemisphere), forming an acute angle with $\hat{\boldsymbol{v}}_g$. The mass transport in the friction layer will therefore have a component across the isobars towards lower pressure, carrying mass out of high pressure areas into low pressure areas. Taking the divergence of (17.2) we obtain, since $\bar{\varrho}\,\hat{\boldsymbol{v}}_g$ is approximately solenoidal,

$$\operatorname{div}\int_0^H \bar{\varrho}\,\hat{\boldsymbol{v}}_h\,dz = -\frac{1}{\lambda}\,\boldsymbol{k}\cdot\operatorname{curl}\boldsymbol{\tau}_{hs}. \tag{17.3}$$

This is the vorticity equation (somewhat simplified) integrated over the friction layer, expressing that the vorticity change brought about by friction is compensated by a mass divergence in the friction layer.

Since there is no significant accumulation or depletion of mass in the friction layer, the divergence of the horizontal mass transport must be balanced by a vertical transport of mass at the top of the friction layer. Thus we obtain

$$(\bar{\varrho}\,\hat{w})_{z=H} = \frac{1}{\lambda}\,\boldsymbol{k}\cdot\operatorname{curl}\boldsymbol{\tau}_{hs}. \tag{17.4}$$

This gives ascending motion in cyclones and descending motion in anticyclones. These frictionally controlled vertical motions undoubtedly contribute to the maintenance of clouds and precipitation in cyclonic areas and dry weather conditions in anticyclones, although other dynamic factors are much more important in this respect. The vertical motions at the top of the friction layer represent a downward energy flux from the free atmosphere to the friction layer, because $(\bar{p}\,\hat{w})$, averaged over large horizontal areas, is negative. As a consequence, the dissipative effect of friction will be felt not only in the friction layer itself, but also at higher elevations where the direct effect of friction is ignored. The mass transport towards lower pressure in the friction layer must from continuity reasons be compensated by an equally strong mass transport in the opposite direction at higher levels. Although the pressure gradient changes with height, the compensating mass transport above will generally have a cross-isobaric component towards higher pressure, causing a decay of the kinetic energy in the free atmosphere in accordance with Eq. (9.10).

The positive work done by the horizontal pressure gradient force in the friction layer does not lead to an increase in the kinetic energy of the mean motion there, because it is compensated by the negative work done by the force of turbulent friction. A mathematical expression for this is obtained by multiplication of (17.1) by $\hat{\boldsymbol{v}}_h$ giving

$$\lambda\,\boldsymbol{k}\cdot\hat{\boldsymbol{v}}_g\times\bar{\varrho}\,\hat{\boldsymbol{v}}_h + \hat{\boldsymbol{v}}_h\cdot\frac{\partial\boldsymbol{\tau}_h}{\partial z} = 0. \tag{17.5}$$

Integration of this equation from the surface to the top of the friction layer gives, when $\hat{\boldsymbol{v}}_g$ is considered constant,

$$\lambda\,\boldsymbol{k}\cdot\hat{\boldsymbol{v}}_g\times\int_0^H \bar{\varrho}\,\hat{\boldsymbol{v}}_h\,dz + \int_0^H \hat{\boldsymbol{v}}_h\cdot\frac{\partial\boldsymbol{\tau}_h}{\partial z}\,dz = 0. \tag{17.6}$$

Here we substitute expression (17.2) for the mass transport, and integrate by parts in the last term. Since $\hat{\boldsymbol{v}}_h = 0$ when $z=0$ and $\boldsymbol{\tau}_h = 0$ when $z=H$, we obtain

$$\hat{\boldsymbol{v}}_g\cdot\boldsymbol{\tau}_{hs} = \int_0^H \boldsymbol{\tau}_h\cdot\frac{\partial\hat{\boldsymbol{v}}_h}{\partial z}\,dz. \tag{17.7}$$

Thus the total work of the horizontal pressure gradient force within the friction layer, per unit time and unit horizontal area, is simply $\hat{\boldsymbol{v}}_g\cdot\boldsymbol{\tau}_{hs}$. The mechanical energy produced by this work is immediately transformed into eddy energy; for if we make use of the approximations for the friction layer (Sect. 16α), we find

$$-\overline{\varrho \boldsymbol{v}'' \boldsymbol{v}''} \cdot \operatorname{grad} \cdot \hat{\boldsymbol{v}} = \boldsymbol{\tau}_h \cdot \frac{\partial \hat{\boldsymbol{v}}_h}{\partial z} \tag{17.8}$$

showing that the integral on the right of (17.7) is precisely the total transformation of mean mechanical energy into eddy kinetic energy [compare Eqs. (15.9) and (15.10)].

This energy transformation is the only source of eddy energy if the air in the friction layer is autobarotropic; and if the eddy energy is constant in time, this production of eddy energy must equal the annihilation of eddy energy by frictional dissipation. Therefore, when the friction layer is statically indifferent, $\hat{\boldsymbol{v}}_g \cdot \boldsymbol{\tau}_{hs}$ is also an expression for the frictional dissipation, which must be positive.

If the friction layer is statically stable, the energy transformation function $\overline{p \operatorname{div} \boldsymbol{v}''}$ of Eq. (15.10) will be negative, and eddy energy will be transformed into internal energy. The work of the pressure gradient force $\hat{\boldsymbol{v}}_g \cdot \boldsymbol{\tau}_{hs}$ must then be larger than the dissipation. On the other hand, if the friction layer is statically unstable, then the eddy motion will in part have the character of convection currents with a positive value of the transformation function $\overline{p \operatorname{div} \boldsymbol{v}''}$. In this case, $\hat{\boldsymbol{v}}_g \cdot \boldsymbol{\tau}_{hs}$ is smaller than the dissipation.

So far, the deductions of this section are independent upon how $\hat{\boldsymbol{v}}_h$ and $\boldsymbol{\tau}_h$ vary with height; the only important assumption is the existence of a relatively shallow friction layer, above which $\boldsymbol{\tau}_h$ is negligible.

The above formulas can be given in a more explicit (but less general) form by making use of the theory of the EKMAN spiral. In vector notation, Eq. (16.16) reads

$$\boldsymbol{\tau}_{hs} = \bar{\varrho} \sqrt{2K\lambda} \sin\alpha \left(\cos\alpha\, \hat{\boldsymbol{v}}_g + \sin\alpha\, \boldsymbol{k} \times \hat{\boldsymbol{v}}_g\right). \tag{17.9}$$

With this expression for the surface stress, (17.2) gives, when variations in $\bar{\varrho}$ within the friction layer are ignored,

$$\int_0^H \bar{\varrho}\, \hat{\boldsymbol{v}}_h\, dz = \left(H - \sqrt{\frac{2K}{\lambda}} \sin^2\alpha\right) \bar{\bar{\varrho}}\, \hat{\boldsymbol{v}}_g + \sqrt{\frac{2K}{\lambda}} \sin\alpha \cos\alpha\, \boldsymbol{k} \times \bar{\varrho}\, \hat{\boldsymbol{v}}_g. \tag{17.10}$$

Similarly, the work of the horizontal pressure gradient force within the friction layer per unit time and unit horizontal area becomes

$$\hat{\boldsymbol{v}}_g \cdot \boldsymbol{\tau}_{hs} = \sqrt{2K\lambda} \sin\alpha \cos\alpha\, \bar{\varrho}\, \hat{\boldsymbol{v}}_g^2. \tag{17.11}$$

Eqs. (17.10) and (17.11) are given by BRUNT [2] who has used the latter to estimate the total dissipation in the friction layer.

Combination of (17.9) and (17.4) gives

$$\hat{w}_{z=H} = \sqrt{\frac{2K}{\lambda}} \sin\alpha \cos\alpha\, \zeta_g, \tag{17.12}$$

where ζ_g is the vertical vorticity of the geostrophic wind. This formula was derived by CHARNEY and ELIASSEN[1].

The concept of a friction layer, above which the eddy stress resulting from small-scale turbulence is negligible, has recently been subject to strong criticism by SHEPPARD, CHARNOCK and FRANCIS[2]. They point out that the variation of

[1] J. G. CHARNEY and A. ELIASSEN: Tellus 1, 38 (1949).

[2] P. A. SHEPPARD, H. CHARNOCK and J. R D. FRANCIS: Quart. J. Roy. Meteor. Soc. 78, 563 (1952).

geostrophic wind with height is not to be neglected in comparison with the frictional variation of wind with height, and find from observations over the eastern North Atlantic that, in cases where the geostrophic wind increases with height, there is no substantial decrease of the stress with height as required by the friction layer theory. This indicates that the "friction layer" may under certain conditions constitute the entire troposphere, so that we can no longer distinguish between the "friction layer" and the "free atmosphere" above. Compare also PRIESTLEY and SHEPPARD [*15*].

18. The eddy-flux of internal energy. The convective eddy-flux of internal energy is $\overline{\varrho e'' \boldsymbol{v}''}$ [see (15.11)], or what is equivalent, $\overline{\varrho e \boldsymbol{v}''}$. In many cases it proves advantageous to consider instead the eddy-flux of enthalpy

$$\overline{\varrho h \boldsymbol{v}''} = \overline{\varrho e \boldsymbol{v}''} + \overline{p \boldsymbol{v}''}. \tag{18.1}$$

With reference to (3.14) and (3.15), we have

$$\overline{\varrho e \boldsymbol{v}''} = c_v \overline{\varrho T \boldsymbol{v}''} + L \overline{\varrho \mu \boldsymbol{v}''}, \tag{18.2}$$

where μ is the mixing ratio of water vapor.

Small-scale turbulence in the friction layer causes a mixing of the air and tends to smooth out the gradients of quasi-conservative properties such as potential temperature and mixing ratio (when the air is unsaturated). One should therefore expect the eddy-flux of such properties to be directed along their mean gradient, and for a given intensity of the turbulent motions, to be proportional to this gradient. From such considerations, TAYLOR[1], SCHMIDT[2], BRUNT [*4*] and others have arrived at the formulas

$$\overline{\varrho T w''} = -K_c \bar{\varrho} \frac{\partial \bar{\theta}}{\partial z}, \tag{18.3}$$

$$\overline{\varrho \mu w''} = -K_d \bar{\varrho} \frac{\partial \bar{\mu}}{\partial z}. \tag{18.4}$$

Here K_c is called the *eddy-conductivity*, and K_d, (which may be set equal to K_c) the *eddy-diffusivity*; they differ from the eddy-viscosity, but are of the same order of magnitude. The usefulness of (18.3) and (18.4) is limited because K_c and K_d are highly variable with height as well as from case to case. In particular, these coefficients depend strongly upon the static stability, i.e. upon $\frac{\partial \bar{\theta}}{\partial z}$, being large when $\frac{\partial \bar{\theta}}{\partial z}$ is algebraically small. From this circumstance, it is possible to explain qualitatively the frequent formation of a strongly mixed layer next to the ground, where $\bar{\theta}$ and $\bar{\mu}$ are nearly constant with height, bounded above by a "*turbulence inversion*" where $\bar{\theta}$ increases and $\bar{\mu}$ decreases rapidly with height. The inversion represents a sharp upper boundary of the turbulent layer.

Studies of the heat balance of the atmosphere (LONDON[3], HOUGHTON[4]) indicate that the release of latent heat in the atmosphere is insufficient to compensate for the radiational cooling of the air. The energy balance requires that the mean eddy-flux of "sensible heat" ($c_p \overline{\varrho T w''}$) near the ground is directed upwards. It was previously assumed that this flux was on the average directed

[1] G. I. TAYLOR: Phil. Trans. Roy. Soc. Lond., Ser. A **215**, 1 (1915).
[2] W. SCHMIDT: Der Massenaustausch usw. In Probleme der kosmischen Physik, Bd. 7. Hamburg: H. Grand 1925.
[3] J. LONDON: J. Meteor. **9**, 145 (1952).
[4] H. G. HOUGHTON: J. Meteor. **11**, 1 (1954).

downwards because $\overline{\theta}$ on the average increases with height; but this conclusion does not follow from (18.3) when the dependency of K_c upon $\frac{\partial\overline{\theta}}{\partial z}$ is taken into consideration.

ERTEL[1] and PRIESTLEY and SWINBANK[2] have shown that the upward eddy-flux of sensible heat must be larger than indicated by (18.3), because this formula does not account for the fact that rising eddies are systematically warmer than sinking eddies because of the effect of buoyancy.

The reader is referred to the reviews by SUTTON [*22*], [*23*] and PRIESTLEY and SHEPPARD [*15*] for further details concerning eddy-flux of heat and turbulent diffusion.

19. RICHARDSON's criterion. The right-hand side of (15.10) represents the rate of production of eddy energy. The last term represents energy loss by dissipation; in order that the eddy energy shall be maintained, it is therefore necessary that

$$\overline{p \operatorname{div} \boldsymbol{v}''} - \overline{\varrho \boldsymbol{v}'' \boldsymbol{v}''} \cdot \operatorname{grad} \cdot \hat{\boldsymbol{v}} > 0. \tag{19.1}$$

In the friction layer we may assume that $|\operatorname{div} \hat{\boldsymbol{v}}| \ll |\operatorname{div} \boldsymbol{v}''|$; using the continuity equation, we may write

$$\operatorname{div} \boldsymbol{v}'' \approx \operatorname{div} \boldsymbol{v} = -\frac{1}{\varrho}\frac{D\varrho}{Dt} \approx -\frac{1}{\overline{\varrho}}\frac{\partial\overline{\varrho}}{\partial z} w'' \approx \frac{g}{R\overline{\theta}} w'', \tag{19.2}$$

and hence

$$\overline{p \operatorname{div} \boldsymbol{v}''} \approx \frac{g}{R\overline{\theta}} \overline{p w''} = \frac{g}{\overline{\theta}} \overline{\varrho T w''}, \tag{19.3}$$

Using (18.3), this may be written

$$\overline{p \operatorname{div} \boldsymbol{v}''} \approx -\frac{g}{\overline{\theta}} \overline{\varrho} K_c \frac{\partial\overline{\theta}}{\partial z}. \tag{19.4}$$

On the other hand we obtain from (17.8) and (16.8)

$$-\overline{\varrho \boldsymbol{v}'' \boldsymbol{v}''} \cdot \operatorname{grad} \cdot \hat{\boldsymbol{v}} = \boldsymbol{\tau}_h \cdot \frac{\partial \hat{\boldsymbol{v}}_h}{\partial z} = \overline{\varrho} K \left(\frac{\partial \hat{\boldsymbol{v}}_h}{\partial z}\right)^2. \tag{19.5}$$

From (19.1), (19.4) and (19.5) we obtain the following criterion for maintenance of turbulence, which is due to RICHARDSON[3],

$$Ri = \frac{\frac{g}{\overline{\theta}} \frac{\partial\overline{\theta}}{\partial z}}{\left(\frac{\partial \hat{\boldsymbol{v}}_h}{\partial z}\right)^2} < \frac{K}{K_c}. \tag{19.6}$$

Ri is called the RICHARDSON *number*, and the criterion states that this number must be smaller than a critical value (K/K_c) in order that turbulence shall persist. The criterion shows that turbulence is favored by vertical shear and suppressed by static stability. The actual value of the critical RICHARDSON number in the atmosphere is very uncertain.

II. Adiabatic and frictionless motions. Stability properties and the theory of small-amplitude oscillations and waves.

20. Introduction. It was shown in Sect. 6 that heat sources and sinks are necessary in order to maintain the motions in the atmosphere against frictional dissipation. It appears, however, that heat sources and friction are comparatively slow effects which do not *directly* lead to such rapid changes as are usually seen

[1] H. ERTEL: Meteor. Z. **59**, 250 (1942).
[2] C. H. B. PRIESTLEY and W. C. SWINBANK: Proc. Roy. Soc. Lond., Ser. A **189**, 543 (1947).
[3] L. F. RICHARDSON: Proc. Roy. Soc. Lond., Ser. A **67**, 354 (1920).

on weather maps in disturbed regions. Thus for instance, an extratropical cyclone may grow from a weak system to its full strength in 24 hours; in such cases one is forced to assume that the rapid change is either a result of a redistribution of pre-existing kinetic energy, or of a release of potential and internal energy; and estimates show that heat sources and friction must be secondary effects which can have only modifying influence during such a short span of time. One may therefore assume that the changes during a limited time interval from a given initial state will in many cases take place almost as if the motion were adiabatic and frictionless. On the other hand, heat sources and friction are of vital importance for explanation of how the initial state with its supply of releasable energy was created.

The theory of adiabatic and frictionless motion, although inadequate to explain changes over long periods of time and the general circulation in the atmosphere, is very useful in the study of changes of relatively short time scale, such as the development or decay of individual cyclones. In fact, most of what has been done in dynamic meteorology has been made on the basis of this theory, which has thus played a dominating role within the science of meteorology.

There is also another important assumption made in the application of the adiabatic and frictionless motion to meteorological phenomena, namely, that the important energy transformations are taking place within the portion of the atmosphere under consideration, while exchange of energy between this system and remote regions as a result of energy flux is irrelevant during the relatively rapid changes to which the theory is supposed to apply. This means that the processes can be studied in a system which is surrounded by rigid walls, so that the system is energetically closed. Such an assumption may seem highly artificial, since there are no such walls in the atmosphere. It should be remembered, however, that unless we want to solve the motion problem of the entire atmosphere as a whole, some such artificial boundary condition is inevitable.

With the assumptions made above, the energy equation states that the sum of kinetic, potential and internal energy of the fluid mass under consideration is constant in time, so that the kinetic energy can grow only at the expense of the potential and internal energy of the system.

A great deal of work on piezotropic and frictionless systems has been concentrated on the study of certain steady states and their stability properties. A steady state is termed *stable* if any small disturbance of the state will remain small indefinitely, and *unstable* if growing disturbances exist such that, no matter how small these disturbances are initially, they will ultimately develop into a finite deviation from the steady state.

The simplest steady state is the state of rest. In this case, the stability criterion can be derived directly from the energy equation, in much the same way as for mechanical systems. This method can also be used to derive the stability properties of a stationary baroclinic vortex motion with respect to axially symmetric disturbances. In the case of arbitrary disturbances of a steady current, the stability problem is more complicated.

When the system is in the vicinity of a steady state, the quantities which define the deviations from the steady state will approximately satisfy perturbation equations, which have the important property that they are linear and therefore tractable by standard mathematical methods. Although the use of perturbation theory in hydrodynamics is quite old, it was first systematized and adopted to meteorological problems by V. BJERKNES[1, 2]. The perturbation equations lead to

[1] V. BJERKNES: Beitr. Phys. fr. Atmosph. **13**, 1 (1927).
[2] V. BJERKNES: Geof. Publ. **5**, No. 11 (1929).

an eigenvalue problem, the solution of which may be expressed in terms of eigensolutions or fundamental modes of motion. For certain simple systems, the *eigenvalues*, or fundamental frequencies, form a *discrete spectrum*, and the general solution is represented as an infinite sum of eigensolutions. In other cases, the spectrum of eigenvalues will be partly or totally continuous, and the general solution will contain integrals over the continuous part of the spectrum (see ELIASSEN, HÖILAND and RIIS[1]).

The perturbation theory can be used to derive the stability properties of the steady state. The existence of one discrete eigenfrequency which is imaginary or complex so that the corresponding eigensolution will grow exponentially with time suffices to prove that the state is unstable. Instability may also result from a *continuous eigenvalue spectrum*, in which case the growth need not be of exponential character. In order to prove the stability of a system, on the other hand, it is necessary to show that no such growing eigenvalues exist. A stability proof will thus require a certain amount of knowledge of all eigenvalues, and will also require a demonstration of the completeness of the set of eigenfunctions.

In any kind of oscillation or wave motion, the energy will fluctuate between kinetic energy and some other energy form. It is natural to classify the various types of oscillations and waves according to the energy forms involved. In this respect, there are three fundamental types of hydrodynamic oscillations and waves:

1. Pure *gravitational* oscillations and waves, where the energy is stored as potential energy when it is not in kinetic form.

2. Pure *compressibility* or *acoustic* oscillations and waves, where only kinetic and internal energy is involved.

3. In the case of perturbations of steady currents, the kinetic energy of the current represents a supply of energy from which the kinetic energy of the perturbation may be fed. We may then have pure *inertial* oscillations and waves where only kinetic energy is involved.

Besides these three fundamental types there are mixed types, where, apart from the kinetic perturbation energy, two or three forms of energy are involved. All wave motions in the atmosphere are of such mixed character; but in many cases, one energy form predominates so that we may speak of waves of predominantly gravitational, acoustic or inertial character.

It should be noted that many authors refer to mixed gravitational—sound waves simply as gravitational, and also that the term "inertial oscillations" are often restricted to oscillations in a circular vortex.

a) Motion under the influence of gravity on a resting earth.

21. Static stability. As shown by FJÖRTOFT[2] [8], the criterion of static stability can be derived rigorously from the energy equation.

We consider an adiabatic and frictionless mass of air enclosed by rigid boundaries, or alternatively, bounded laterally and below by rigid surfaces and extending upwards to a level where the pressure vanishes. In any case, the boundary work will vanish so that the energy equation becomes

$$K + \Phi + E = \text{constant}. \tag{21.1}$$

Consequently, an increase of kinetic energy must be supplied from the energy store

$$J = \Phi + E = \int_\tau (\varphi + e)\, \varrho\, d\tau. \tag{21.2}$$

[1] A. ELIASSEN, E. HÖILAND, E. RIIS: Inst. Weather and Climate Res., Norw. Acad. of Science and Letters **1953**, Publ. 1.

[2] R. FJÖRTOFT: Geof. Publ. **16**, No. 5 (1946).

Now consider a velocity field $\boldsymbol{v}(x, y, z, t)$ which satisfies the kinematic boundary conditions, but which is otherwise arbitrary. The displacements of the air particles during a time δt,

$$\delta \boldsymbol{r} = \boldsymbol{v}\,\delta t \tag{21.3}$$

may then be considered as a set of virtual displacements of the system.

The energy store J has the important property that its variation δJ is uniquely determined by the virtual displacements. We find

$$\frac{dJ}{dt} = \int_\tau \left(\frac{D\varphi}{Dt} + \frac{De}{Dt}\right) \varrho\, d\tau. \tag{21.4}$$

By means of

$$\frac{D\varphi}{Dt} = \boldsymbol{v} \cdot \operatorname{grad} \varphi, \tag{21.5}$$

$$\frac{De}{Dt} = -p\frac{D\alpha}{Dt} = -p\,\alpha \operatorname{div} \boldsymbol{v} \quad \text{(isentropic change, continuity equation)}, \tag{21.6}$$

(21.4) becomes

$$\left.\begin{aligned} \frac{dJ}{dt} &= \int_\tau (\boldsymbol{v} \cdot \varrho \operatorname{grad} \varphi - p \operatorname{div} \boldsymbol{v})\, d\tau \\ &= \int_\tau \boldsymbol{v} \cdot (\varrho \operatorname{grad} \varphi + \operatorname{grad} p)\, d\tau \end{aligned}\right\} \tag{21.7}$$

and therefore

$$\delta J = \int_\tau \delta \boldsymbol{r} \cdot (\varrho \operatorname{grad} \varphi + \operatorname{grad} p)\, d\tau. \tag{21.8}$$

An equilibrium state is characterized by the energy store J being an extremum, i.e. $\delta J = 0$ for any set of virtual displacements. Since $\delta \boldsymbol{r}$ is unrestricted inside the fluid, this gives

$$\varrho_0 \operatorname{grad} \varphi + \operatorname{grad} p_0 = 0 \tag{21.9}$$

where index 0 indicates equilibrium values. Thus we refind the condition of equilibrium (11.1).

In the case of an incompressible fluid, we have $J = \Phi$, $\delta J = \int_\tau \delta \boldsymbol{r} \cdot \varrho \operatorname{grad} \varphi\, d\tau$. In this case, the equilibrium condition $\delta J = 0$ does not require $\varrho \operatorname{grad} \varphi = 0$, because $\delta \boldsymbol{r}$ are restricted: $\operatorname{div} \delta \boldsymbol{r} = 0$. As shown by FJÖRTOFT, it suffices that ϱ grad φ be solenoidal, i.e. $\varrho_0 \operatorname{grad} \varphi + \operatorname{grad} p_0 = 0$, where p_0 is the pressure, which as shown by LAGRANGE, plays the role of a LAGRANGian multiplier introduced by the side condition of incompressibility.

Now let us calculate the second variation of J. From (21.7), we find

$$\begin{aligned} \frac{d^2 J}{dt^2} &= \left(\frac{d}{dt} \int_\tau \boldsymbol{v} \cdot (\varrho \operatorname{grad} \varphi + \operatorname{grad} p)\, d\tau\right)_0 \\ &= \int_\tau \frac{\partial \boldsymbol{v}}{\partial t} \cdot (\varrho_0 \operatorname{grad} \varphi + \operatorname{grad} p_0)\, d\tau + \int_\tau \boldsymbol{v} \cdot \left[\left(\frac{\partial \varrho}{\partial t}\right)_0 \operatorname{grad} \varphi + \operatorname{grad} \left(\frac{\partial p}{\partial t}\right)_0\right] d\tau. \end{aligned}$$

Here the first integral on the right vanishes by (21.9). Using the integral theorem of GAUSS, the second integral becomes

$$\frac{d^2 J}{dt^2} = \int_\tau \left[\boldsymbol{v} \cdot \operatorname{grad} \varphi \left(\frac{\partial \varrho}{\partial t}\right)_0 - \operatorname{div} \boldsymbol{v} \left(\frac{\partial p}{\partial t}\right)_0\right] d\tau. \tag{21.10}$$

The density may be considered as a function of pressure and entropy, so that we may write

$$d\varrho = \gamma\, dp - \mu\, ds, \tag{21.11}$$

where γ is the coefficient of piezotropy for isentropic changes (compare Sect. 1). In the case of dry or nonsaturated air we have

$$\gamma = \frac{1}{\varkappa R T}, \qquad \mu = \frac{\varrho}{c_p}. \tag{21.12}$$

From the continuity Eq. (4.2) and (21.11), we find

$$\frac{\partial \varrho}{\partial t} = -\varrho \operatorname{div} \boldsymbol{v} - \boldsymbol{v} \cdot \operatorname{grad} \varrho = -\varrho \operatorname{div} \boldsymbol{v} - \gamma \boldsymbol{v} \cdot \operatorname{grad} p + \mu \boldsymbol{v} \cdot \operatorname{grad} s. \tag{21.13}$$

Similarly, we have

$$\frac{\partial p}{\partial t} = \frac{D p}{D t} - \boldsymbol{v} \cdot \operatorname{grad} p = \frac{1}{\gamma} \frac{D \varrho}{D t} - \boldsymbol{v} \cdot \operatorname{grad} p = -\frac{\varrho}{\gamma} \operatorname{div} \boldsymbol{v} - \boldsymbol{v} \cdot \operatorname{grad} p. \tag{21.14}$$

When (21.13) and (21.14) are introduced into (21.10), we finally obtain

$$\frac{d^2 J}{d t^2} = \int\limits_\tau \left[\boldsymbol{v} \cdot \operatorname{grad} \varphi\, \mu \operatorname{grad} s_0 \cdot \boldsymbol{v} + \frac{\varrho_0}{\gamma} (\operatorname{div} \boldsymbol{v} - \gamma \boldsymbol{v} \cdot \operatorname{grad} \varphi)^2\right] d\tau$$

and the second variation of J is therefore

$$\delta^2 J = \frac{1}{2} \int\limits_\tau \left[\delta \boldsymbol{r} \cdot (\operatorname{grad} \varphi\, \mu \operatorname{grad} s_0) \cdot \delta \boldsymbol{r} + \frac{\varrho_0}{\gamma} (\operatorname{div} \delta \boldsymbol{r} - \gamma\, \delta \boldsymbol{r} \cdot \operatorname{grad} \varphi)^2\right] d\tau. \tag{21.15}$$

The integral on the right cannot attain negative values if $\operatorname{grad} \varphi$ and $\operatorname{grad} s_0$ have everywhere the same direction, i.e. if

$$\mu \frac{d s_0}{d \varphi} > 0 \quad \text{at all levels.} \tag{21.16}$$

The integral is then zero if all displacements are horizontal and non-divergent; for all other displacements it is positive. Therefore, the energy store J is a minimum in the equilibrium state, and there is no energy available for the growth of disturbances. Consequently, the equilibrium is stable, and (21.16) is the criterion of static stability.

If on the other hand,

$$\mu \frac{d s_0}{d \varphi} < 0 \quad \text{at any level,}$$

then $\delta^2 J$ will be negative for certain displacements. This entails that J can be reduced, so that the energy equation does not prohibit unstable growth of certain disturbances. A strict proof of the instability would require a demonstration that such growing motions actually exist.

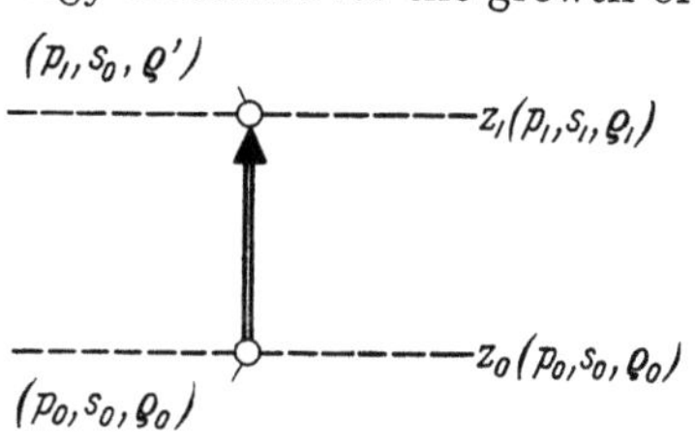

Fig. 5. Static stability: Vertically displaced particle.

The criterion of static stability may be arrived at more simply in the following way: Consider an air mass in equilibrium, and let a small air particle at the level z_0 where the pressure is p_0, the density ϱ_0 and the entropy s_0 be displaced to a level z_1 where the equilibrium state is characterized by p_1, ϱ_1 and s_1, as shown in Fig. 5. It is assumed that the displacement of the particle will not disturb the equilibrium pressure field, so that the particle will arrive at the level z_1 in a state characterized by pressure p_1, density ϱ' and entropy s_0. As a result the particle will be exposed to a vertical buoyancy force per unit mass

$$F = -g \frac{\varrho' - \varrho_1}{\varrho'}.$$

Since the particle and the environment have the same pressure, but different entropy, we may write, in accordance with (21.11)

$$\varrho' - \varrho_1 = -\mu(s_0 - s_1) = \mu \frac{ds}{dz}(z_1 - z_0)$$

and hence

$$F = -\frac{g\mu}{\varrho}\frac{ds}{dz}(z_1 - z_0). \tag{21.17}$$

Here F is reckoned positive upward. It will be seen that this buoyancy force will accelerate the particle towards the equilibrium position if $\mu \frac{ds}{dz} > 0$, and away from the equilibrium position if $\mu \frac{ds}{dz} < 0$, thus indicating stability in the first case and instability in the latter, in agreement with the result found by the energy method.

It should be noticed, however, that the "particle method" is far from rigorous. In the first place, it is based on the displacement of a single particle, which is unrealistic; and secondly, the method rests upon the assumption that the pressure field is unaffected by the displacement of the particle. Therefore, the particle method cannot be considered as more than a demonstration of the plausibility of the stability criterion.

As a *measure of the static stability*, we may use the restoring buoyancy force per unit length of vertical displacement. This quantity has the dimension of a frequency squared, and will be denoted by ν_s^2. From (21.17), we find

$$\nu_s^2 = -\frac{F}{z_1 - z_0} = \frac{g\mu}{\varrho}\frac{\partial s}{\partial z}. \tag{21.18}$$

By means of (21.11), the definition of the coefficient of barotropy (5.7) and the hydrostatic Eq. (9.3), this expression may also be written

$$\nu_s^2 = \frac{g}{\varrho}\left(\gamma \frac{\partial p}{\partial z} - \frac{\partial \varrho}{\partial z}\right) = (\Gamma - \gamma)\, g^2. \tag{21.19}$$

Using the gas equation, we may also write

$$\nu_s^2 = \frac{g^2 \varrho}{T}\left(\frac{DT}{Dp} - \frac{\partial T/\partial z}{\partial p/\partial z}\right). \tag{21.20}$$

These expressions hold whether the air is saturated or not. For dry or non-saturated air, we may use Eqs. (2.7) and (2.9), and so we find

$$\nu_s^2 = \frac{g}{\theta}\frac{\partial \theta}{\partial z} = \frac{g}{T}\left(\frac{\partial T}{\partial z} + \frac{g}{c_p}\right). \tag{21.21}$$

Hence it follows that dry or non-saturated air is statically stable when the potential temperature increases with height, or when the lapse-rate of temperature $(-\partial T/\partial z)$ does not exceed $g/c_p = 0{,}01°$ centrigrade per meter. If θ decreases with height, or if T decreases with height at a rate exceeding g/c_p, then such air is statically unstable, and convection currents will occur.

On the other hand, if the air is saturated, then Eq. (3.25) for moist-adiabatic changes must be used. The result may be written in the form

$$\nu_s^2 = \frac{g}{T}\left(\frac{\partial T}{\partial z} + \delta_m\right), \tag{21.22}$$

where the moist-adiabatic lapse-rate δ_m is a function of p and T. At very low temperatures, we have $\delta_m \approx g/c_p$; therefore, the effect of the condensation phenomena upon the static stability is negligible in the upper troposphere and in the stratosphere. At temperatures corresponding to conditions in the lower troposphere, we have $\delta_m < g/c_p$. Thus at 1000 mb, $\delta_m = 0.0065°$/m at 0° and $\delta_m = 0.0044°$/m at 20° centigrade, showing that the static stability of cloudy air is considerably less than that of clear air with the same lapse-rate. If $\delta_m < -\frac{\partial T}{\partial z} < \frac{g}{c_p}$, the air mass is statically stable as long as it is non-saturated; but if saturation occurs, the mass will be unstable. Such cases are usually referred to as *conditional instability*.

While the role of condensation processes upon the static stability is of vital importance for such smaller scale phenomena as convection currents and tropical cyclones, its importance for larger scale phenomena such as extratropical cyclones is uncertain (compare Chap. IV).

The principal difficulty in dealing with systems where condensation occurs lies in the circumstance that the location of the regions of saturated air will depend upon the motion itself. Thus, condensation will normally take place within the ascending currents, whereas clear air prevails in regions of subsidence.

22. Combined gravitational-compressibility oscillations and waves. *α) Perturbation equations.* Consider a mass of air in the state of rest in a uniform field of gravity on a non-rotating, flat earth. Let $p_0(z)$ and $\varrho_0(z)$ denote equilibrium pressure and density. In a CARTESian coordinate system with the z-axis pointing vertically upwards, the equilibrium equations are

$$\frac{\partial p_0}{\partial x} = \frac{\partial p_0}{\partial y} = 0, \quad \frac{\partial p_0}{\partial z} = -g\varrho_0, \tag{22.1}$$

$$\frac{\partial \varrho_0}{\partial x} = \frac{\partial \varrho_0}{\partial y} = 0, \quad \frac{\partial \varrho_0}{\partial z} = -\Gamma g \varrho_0 \tag{22.2}$$

where Γ is the coefficient of barotropy.

Now suppose that the equilibrium state is slightly disturbed. For simplicity, only two-dimensional disturbances in the xz-plane will be considered; but the theory can be extended to three dimensions. Let ξ and ζ denote small displacements of the particles in the x- and z- direction respectively. These displacements are considered as functions of x, z and t.

As a result of the displacements, individual air particles will be subject to an expansion $\partial\xi/\partial x + \partial\zeta/\partial z$, correct to the first order in the displacements. The equation of continuity may therefore be written

$$\frac{\partial \xi}{\partial x} + \frac{\partial \zeta}{\partial z} = -\frac{\varrho_i'}{\varrho_0}, \tag{22.3}$$

where ϱ_i' denotes the individual density perturbation. The piezotropic character of the changes of state is expressed by

$$\varrho_i' = \gamma p_i', \tag{22.4}$$

where p_i' is the individual pressure perturbation. The individual perturbations of pressure and density are related to the corresponding local perturbations p' and ϱ' by the formulas, correct to first order,

$$p_i' = p' + \frac{\partial p_0}{\partial z}\zeta = p' - g\varrho_0\zeta, \tag{22.5}$$

$$\varrho_i' = \varrho' + \frac{\partial \varrho_0}{\partial z}\zeta = \varrho' - \Gamma g \varrho_0 \zeta. \tag{22.6}$$

From (22.4), (22.5) and (22.6) we find

$$\varrho' = (\Gamma - \gamma)\, g\varrho_0\zeta + \gamma p'. \tag{22.7}$$

The components of the acceleration are $\partial^2\xi/\partial t^2$ and $\partial^2\zeta/\partial t^2$, and the equations of motion are, correct to the first order,

$$\varrho_0 \frac{\partial^2 \xi}{\partial t^2} = -\frac{\partial(p_0 + p')}{\partial x}, \tag{22.8}$$

$$\varrho_0 \frac{\partial^2 \zeta}{\partial t^2} = -\frac{\partial(p_0 + p')}{\partial z} - g(\varrho_0 + \varrho'). \tag{22.9}$$

Eliminating the zero order terms from these equations by means of the equilibrium Eqs. (22.1), and expressing ϱ' in terms of p' and ζ by means of (22.7), we obtain the perturbation equations

$$\varrho_0 \frac{\partial^2 \xi}{\partial t^2} = -\frac{\partial p'}{\partial x}, \tag{22.10}$$

$$\varrho_0 \frac{\partial^2 \zeta}{\partial t^2} = -\frac{\partial p'}{\partial z} - \gamma p' g - \nu_s^2 \varrho_0 \zeta \tag{22.11}$$

where ν_s is the frequency of vertical oscillations defined by (21.19). On account of (22.4) and (22.5), the equation of continuity (22.3) becomes

$$\frac{\partial \xi}{\partial x} + \frac{\partial \zeta}{\partial z} - \gamma g \zeta = -\frac{\gamma}{\varrho_0} p'. \tag{22.12}$$

The three Eqs. (22.10), (22.11) and (22.12) form a complete system of perturbation equations in the three dependent variables ξ, ζ and p'. These equations are in LAGRANGian form, since the displacements occur as dependent variables. An equivalent set of equations can be derived in EULERian representation, but the LAGRANGian system is preferred because the potential energy is related to the displacements. The derivation above suffers from the formal defect that the coordinates x, z are used partly as LAGRANGian coordinates referring to the equilibrium positions of the particles, and partly as EULERian coordinates. It is noteworthy, however, that in perturbations of equilibrium states, the distinction between LAGRANGian and EULERian coordinates is unnecessary in the first-order theory, since it will be represented by second order terms. We may therefore regard x and z as LAGRANGian or EULERian coordinates as we please.

With the substitution

$$\xi = \varrho_0^{-\frac{1}{2}} \xi'', \qquad \zeta = \varrho_0^{-\frac{1}{2}} \zeta'', \qquad p' = \varrho_0^{\frac{1}{2}} p'', \tag{22.13}$$

the perturbation Eqs. (22.10), (22.11) and (22.12) assume the form

$$\frac{\partial^2 \xi''}{\partial t^2} = -\frac{\partial p''}{\partial x}, \tag{22.14}$$

$$\frac{\partial^2 \zeta''}{\partial t^2} + \nu_s^2 \zeta'' = -\frac{\partial p''}{\partial z} - l p'', \tag{22.15}$$

$$\frac{\partial \xi''}{\partial x} + \frac{\partial \zeta''}{\partial z} - l \zeta'' = -\gamma p'', \tag{22.16}$$

where

$$l = (\gamma - \tfrac{1}{2}\Gamma) g. \tag{22.17}$$

β) Wave energy and wave energy flux. When Eq. (22.10) is multiplied by $\partial\xi/\partial t$, Eq. (22.11) multiplied by $\partial\zeta/\partial t$, Eq. (22.12) differentiated with respect to t and multiplied by p', and the three equations thus derived are added, there results

$$\varrho_0 \frac{\partial q}{\partial t} = -\frac{\partial F_x}{\partial x} - \frac{\partial F_z}{\partial z}, \tag{22.18}$$

where

$$\left.\begin{aligned}\varrho_0 q &= \frac{1}{2}\left[\varrho_0\left(\left(\frac{\partial\xi}{\partial t}\right)^2+\left(\frac{\partial\zeta}{\partial t}\right)^2\right)+\nu_s^2\,\varrho_0\,\zeta^2+\frac{\gamma}{\varrho_0}\,p'^2\right]\\ &= \frac{1}{2}\left[\left(\frac{\partial\xi''}{\partial t}\right)^2+\left(\frac{\partial\zeta''}{\partial t}\right)^2+\nu_s^2\,\zeta''^2+\gamma\,p''^2\right]\end{aligned}\right\} \tag{22.19}$$

is the "wave energy" per unit volume, and

$$F_x = p'\,\frac{\partial\xi}{\partial t} = p''\,\frac{\partial\xi''}{\partial t}\,,\qquad F_z = p'\,\frac{\partial\zeta}{\partial t} = p''\,\frac{\partial\zeta''}{\partial t} \tag{22.20}$$

may be interpreted as the components of a vector representing the flux of wave energy.

If the mass of air considered is supposed to be confined within rigid boundaries, the kinematic boundary condition requires the flux of wave energy through the boundary contour to vanish. In many wave problems, however, the system is regarded to extend over the infinite half-plane $z>0$, with the earth's surface $z=0$ as the only boundary. In order that the problem shall have a definite physical meaning, one must then require that the total wave energy of the system, $Q=\iint q\varrho_0\,dx\,dz$ be finite; this condition implies that the perturbations tend to zero at great distances fast enough so that the wave energy flux through a semi-circle centred at the origin approaches zero when its radius increases indefinitely. (The compressibility is important in this respect, because it secures a finite signal velocity.) Therefore, the system is energetically closed, whether it is bounded or infinite, and Eq. (22.18) gives, when integrated over the system

$$Q = \iint q\,\varrho_0\,dx\,dz = \text{constant}. \tag{22.21}$$

The total wave energy can be shown to represent the difference between the total energy (as kinetic, potential and internal) in the disturbed state and the total energy in the equilibrium state, correct to the second order (as shown in Sect. 21, the first order effect vanishes on account of the equilibrium condition). The physical content of (22.21) is therefore identic with that of the energy Eq. (21.1).

The wave energy per unit volume consists of kinetic energy

$$\varrho_0\,q_k = \frac{\varrho_0}{2}\left[\left(\frac{\partial\xi}{\partial t}\right)^2+\left(\frac{\partial\zeta}{\partial t}\right)^2\right] \tag{22.22}$$

and potential + internal wave energy

$$\varrho_0\,(q_p+q_i) = \frac{1}{2}\,\varrho_0\,\nu_s^2\,\zeta^2+\frac{1}{2}\,\frac{\gamma}{\varrho_0}\,p'^2. \tag{22.23}$$

It is tempting to regard the first term on the right as potential and the second term as internal wave energy, but this interpretation is not strictly correct.

Multiplying Eq. (22.10) by ξ, Eq. (22.11) by ζ, Eq. (22.12) by p' and adding, we obtain

$$\varrho_0\left(q_k-(q_p+q_i)\right) = \frac{\varrho_0}{4}\,\frac{\partial^2}{\partial t^2}\,(\xi^2+\zeta^2)+\frac{1}{2}\,\frac{\partial}{\partial x}\,(p'\xi)+\frac{1}{2}\,\frac{\partial}{\partial z}\,(p'\zeta). \tag{22.24}$$

If this equation is integrated over the entire system, the two last terms will give an integral over the boundary surface which will vanish in the same manner as the wave energy flux. If the equilibrium state is stable, then $(\xi^2+\zeta^2)$ will have a limited variation, so that the first term on the right may be eliminated by averaging the equation over a sufficiently long time. Denoting this average by a bar, we thus have

$$\overline{Q_k} = \overline{Q_p+Q_i} = \tfrac{1}{2}\,\overline{Q} \tag{22.25}$$

showing that there is on the average equipartition of wave energy between kinetic and potential + internal form.

γ) Fundamental solutions for an isothermal atmosphere. It is assumed that for the infinite half-plane $z>0$, the general solution can be obtained by FOURIER analysis from fundamental solutions of the type $\zeta''=A(z)\,e^{i(kx+\nu t)}$ and similar expressions for ξ'' and p''. The differential equation for the z-dependency of these fundamental solutions will in general have variable coefficients, since Γ and γ vary with z, and the determination of the fundamental solutions is therefore quite complicated. The case of a constant lapse rate of temperature is treated in Phys. Hydr. [2]. Most work has been made under the assumption that Γ and γ are constants, independent of height. Inspection of Eqs. (21.12) and (11.3) shows that this assumption is equivalent to assuming that the atmosphere has constant temperature, and therefore implies a static stability (ν_s^2) which is considerably higher than the average value in the troposphere. This difficulty may be overcome by assuming that the thermodynamic properties of the isothermal model atmosphere differ somewhat from those of the real atmosphere; thus by varying the value of the specific heat ratio $\varkappa$, any desired value of ν_s^2 can be attained. This method will not give correct values of l and γ; but these coefficients are in most wave problems considered to be of much less importance.

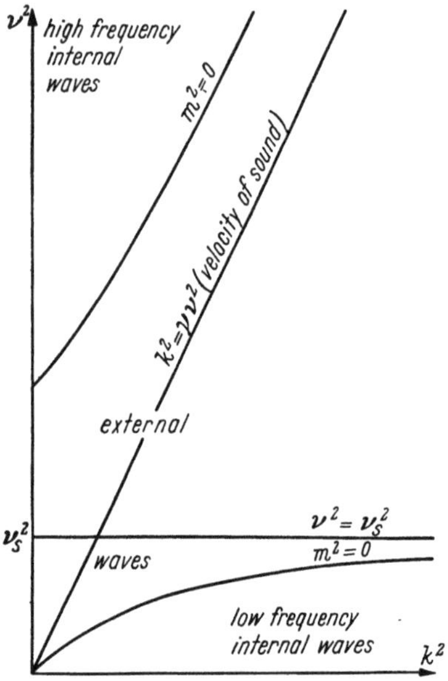

Fig. 6. Wave classification in the $k^2\nu^2$-diagram.

For an isothermal atmosphere, the coefficients ν_s^2, l and γ are constants, and the system of perturbation Eqs. (22.14), (22.15), (22.16) are satisfied by fundamental solutions of the form

$$\left.\begin{aligned} \xi'' &= -A\,\frac{i k}{\nu^2}\,\frac{\nu_s^2-\nu^2}{l+i m}\,e^{i(kx+mz+\nu t)}, \\ \zeta'' &= A\,e^{i(kx+mz+\nu t)}, \\ p'' &= -A\,\frac{\nu_s^2-\nu^2}{l+i m}\,e^{i(kx+mz+\nu t)}, \end{aligned}\right\} \tag{22.26}$$

where A is an arbitrary amplitude factor, and k, m and ν are three constants satisfying the relation

$$(l^2+m^2)\,\nu^2=(k^2-\gamma\nu^2)\,(\nu_s^2-\nu^2). \tag{22.27}$$

The constant k will be referred to as the *horizontal wave number*, and the constant ν as the *frequency*. We shall in the following assume a statically stable equilibrium state ($\nu_s^2>0$), and shall restrict the consideration to the case that k and ν are real. The different types of fundamental solutions may then be conveniently represented in a $k^2\nu^2$-diagram as shown in Fig. 6. For a complete discussion of the various wave types, reference is made to Phys. Hydr. [2].

The horizontal line $\nu^2=\nu_s^2$ on this diagram represents purely vertical oscillations in an undisturbed pressure field (thus satisfying the assumption underlying the particle method described in Sect. 21). The line $k^2=\gamma\nu^2$ represents waves whose horizontal phase velocity $-\nu/k$ equals the velocity of sound ($c_s=\gamma^{-\frac{1}{2}}$).

For given values of k^2 and ν^2, Eq. (22.27) gives a corresponding value of m^2. The curves $m^2=$ constant are a family of hyperbolas. The hyperbola $m^2=0$ separates the domain of external (or non-cellular) waves ($m^2<0$), between the two branches of the hyperbola, from the domain of internal (or cellular) waves ($m^2 > 0$) on both sides. The internal waves may therefore be subdivided into low-frequency internal waves (lower region) and high-frequency internal waves (upper region).

If k^2, m^2 and ν^2 satisfy (22.27), then any of the eight combinations $\pm k$, $\pm m$, $\pm \nu$ will satisfy (22.27). Therefore, each point in the diagram represents eight fundamental solutions of the type (22.26). Linear combinations of these eight solutions give the various types of real solutions which are possible. These are the following[1]:

A. External waves. $-m^2 = m_1^2 > 0$

1. Standing oscillations, represented by

$$\zeta'' = A \sin k x \, e^{\pm m_1 z} \sin \nu t. \tag{22.28}$$

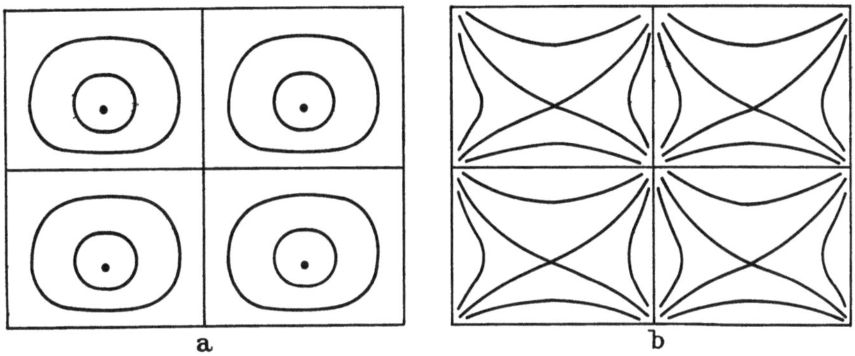

Fig. 7a and b. Streamline patterns pertaining to internal oscillations of low frequency (a) and high frequency (b).

2. Waves with horizontal propagation

$$\zeta'' = A \, e^{\pm m_1 z} \sin (k x + \nu t). \tag{22.29}$$

In these formulas, $e^{\pm m_1 z}$ may be replaced by hyperbolic functions of z. Since these can have only one zero, it follows that external waves cannot occur in a layer of air confined between two rigid horizontal planes. Neither can external waves exist as free waves over a level ground in an infinite atmosphere of uniform stability, since this would imply that the total wave energy is infinite. External waves can exist as forced waves, produced by motions of the boundaries, as in the problem of mountain waves. Besides, external waves may occur as free waves in an atmospheric layer which is bounded by other layers with different properties; thus for instance, waves in an inversion layer of high static stability may produce external wave motions in the adjacent layers of lower static stability.

It should be noticed that propagating external waves will transport wave energy horizontally in the direction of propagation. External waves cannot transport wave energy in vertical direction.

B. Internal waves, $m^2 > 0$.

3. Standing oscillations

$$\zeta'' = A \sin k x \sin m z \sin \nu t. \tag{22.30}$$

The wave energy flux vanishes when averaged over one period. Fig. 7 shows the streamline pattern for such oscillations. Diagram a refers to low-frequency

[1] For the sake of brevity, only the solutions for ζ'' are given; the other variables may be obtained on the basis of (22.26). Arbitrary phase constants are omitted.

internal oscillations and diagram b to high-frequency internal oscillations. It is seen that the low frequency type represents a circulatory motion, similar to the streamline pattern in pure gravitational waves in an incompressible fluid, whereas the high-frequency oscillation represents a quite different kinematics. It is therefore customary to characterize low-frequency internal oscillations as *gravitational oscillations*, modified by compressibility, and high-frequency internal oscillations as *compressibility oscillations*, modified by gravity.

4. Waves with horizontal propagation.

$$\zeta'' = A \sin m z \sin (k x + \nu t). \tag{22.31}$$

The wave energy is easily calculated from (22.19). Its mean value, averaged with respect to z and x (or t) is

$$\bar{Q} = \frac{A^2}{4} \frac{\nu_s^2 k^2 - \gamma \nu^4}{k^2 - \gamma \nu^2}. \tag{22.32}$$

The mean wave energy flux is found from (22.20) to be

$$\bar{F}_x = -\frac{A^2}{4} \frac{k}{\nu} \frac{(\nu_s^2 - \nu^2)^2}{l^2 + m^2}, \qquad \bar{F}_z = 0 \tag{22.33}$$

i.e. the wave transports energy in the direction of propagation. Analysis of the kinematics of progressive waves reveals a typical difference between low-frequency and high-frequency waves, as in the case of standing oscillations discussed above.

5. Waves with vertical propagation.

$$\zeta'' = A \sin k x \sin (m z + \nu t). \tag{22.34}$$

The mean wave energy (averaged with respect to x and z, or t) equals that of horizontally propagated waves, and is represented by (22.32). However, the mean wave energy flux is now

$$\bar{F}_x = 0, \quad \bar{F}_z = \frac{A^2}{4} \frac{m}{\nu} \frac{\nu^2 (\nu_s^2 - \nu^2)}{l^2 + m^2}. \tag{22.35}$$

When we compare the sign of the vertical energy flux with that of the vertical phase velocity $(-\nu/m)$, we find that the signs are equal for high-frequency internal waves $(\nu^2 > \nu_s^2)$ and opposite for low-frequency internal waves $(\nu^2 < \nu_s^2)$. Thus, low-frequency internal waves with vertical propagation possess the unusual property that they transport energy in the direction opposite to the direction of propagation. This was pointed out by ELIASSEN and PALM[1].

6. Waves with tilting phase-lines.

$$\zeta'' = A \sin (k x + m z + \nu t). \tag{22.36}$$

These waves have a horizontal as well as a vertical propagation. The mean wave energy is twice the value (22.32). The mean horizontal wave energy flux is twice the value (22.33), and the mean vertical energy flux is twice the value (22.35). We have again the abnormal vertical energy propagation when $\nu^2 < \nu_s^2$. The waves with tilting phase-lines are important in the theory of mountain waves (Sect. 24).

The different solutions considered above are not independent of each other; thus for instance, progressive waves can be represented as a sum of standing oscillations, and vice versa. It should be noted, however, that the mean wave

[1] A. ELIASSEN and E. PALM: Inst. Weather and Climate Res., Norw. Acad. of Science and Letters, Publ. 1, 1954.

energy and the mean wave energy flux are quadratic functions of the perturbation quantities and are therefore not additive.

The initial value problem will generally require the knowledge of the initial distribution of four quantities, e.g. $\xi, \frac{\partial \xi}{\partial t}, \zeta, \frac{\partial \zeta}{\partial t}$. In most meteorological applications, only waves with velocities of propagation comparable to typical wind velocities are of interest, whereas waves which propagate at speeds comparable to the velocity of sound are unimportant for weather, although of interest for tidal oscillations (see chapter by KERTZ in this volume). In these cases it is customary to simplify the perturbation equations by putting $\gamma=0$ in Eq. (22.16), whereas l and ν_s^2 are retained unchanged. This is equivalent to assuming the velocity of sound infinite. In the diagram Fig. 6, the line $k^2=\gamma\nu^2$ then becomes vertical, and the entire regime of high-frequency internal waves is lost. The initial value problem now requires only two quantities to be assigned initially, e.g. ζ and $\frac{\partial \zeta}{\partial t}$ (not ζ and ξ!). The approximation is very good for waves with moderate speed of propagation.

23. HÖILAND's frequency formula. HÖILAND [10] has applied the second circulation theorem by V. BJERKNES to small oscillations of a fluid in a gravity field. His method is most easily demonstrated in the case of pure gravitational oscillations of an inhomogeneous, incompressible fluid. Putting $\gamma=0$ in the perturbation Eqs. (22.10) and (22.11), we obtain

$$\varrho_0 \frac{\partial^2 \xi}{\partial t^2} = -\frac{\partial p'}{\partial x}, \tag{23.1}$$

$$\varrho_0 \frac{\partial^2 \zeta}{\partial t^2} = -\frac{\partial p'}{\partial z} - \nu_s^2 \varrho_0 \zeta. \tag{23.2}$$

Multiplying the first equation by dx, the second by dz, adding and integrating along a closed curve, the perturbation pressure drops out, and we obtain

$$\oint \varrho_0 \left(\frac{\partial^2 \xi}{\partial t^2} dx + \frac{\partial^2 \zeta}{\partial t^2} dz\right) = -\oint \nu_s^2 \varrho_0 \zeta \, dz. \tag{23.3}$$

The left-hand side of this equation is the circulation of density times acceleration, and the equation expresses that this circulation equals the circulation of the vertical restoring force $-\nu_s^2 \varrho_0 \zeta$. If the fluid is enclosed by rigid boundaries, any set of vertical displacements, which produce restoring forces, will for continuity reasons be connected with a set of horizontal displacements, which do not produce such forces. According to (23.3), the restoring forces produced by the vertical displacements must drive the vertical as well as the horizontal accelerations. The magnitude of the resulting accelerations will therefore depend upon the kinematics of the motion and will be quite small if the motion is predominantly horizontal.

Let us consider in particular a pure eigen-oscillation of the fluid, assuming that the form of the boundary is such that eigensolutions exist. The motion will then have the character of small harmonic oscillations along streamlines which form stationary, closed curves. When ν is the eigen-frequency, we have

$$\frac{\partial^2 \xi}{\partial t^2} = -\nu^2 \xi, \qquad \frac{\partial^2 \zeta}{\partial t^2} = -\nu^2 \zeta \tag{23.4}$$

and Eq. (23.3) gives in this case

$$\nu^2 = \frac{\oint \nu_s^2 \varrho_0 \zeta \, dz}{\oint \varrho_0 (\xi \, dx + \zeta \, dz)}. \tag{23.5}$$

This is HÖILAND'S frequency formula, which enables us to compute the frequency when the kinematics is known. The curve of integration may conveniently be chosen as one of the streamlines; the displacements are then directed tangentically to the curve in the same direction along the curve in all points. When the integration along the curve is performed in the direction of the displacements, we have $\xi\, dx \geqq 0$, $\zeta\, dz \geqq 0$ in all points, so that

$$\oint \varrho_0 \xi\, dx > 0, \qquad \oint \varrho_0 \zeta\, dz > 0. \tag{23.6}$$

As a consequence we have, when ν_s^2 is constant (independent of z),

$$0 < \nu^2 < \nu_s^2 \tag{23.7}$$

showing that ν_s is the highest frequency that can be produced by the static stability alone, i.e. the maximum frequency of internal wave motion.

If the fluid consists of several layers with different static stability, then ν^2 may exceed the value of ν_s^2 in the layers where ν_s^2 is smallest. It follows from (23.5), however, that no closed streamlines can exist within these layers; all streamlines must intersect layers where $\nu_s^2 > \nu^2$. In layers where $\nu_s^2 < \nu^2$, the oscillations will be of external type. A necessary condition for the existence of external oscillations in a layer is therefore that the layer is bounded by other layers with higher static stability; in this respect, a stable discontinuity surface or a free surface will act as a shallow layer of very high stability which may produce external oscillations in the adjacent layers.

Returning to the case of internal oscillations in a fluid of uniform stability, it follows from (23.5) that ν^2 will approach ν_s^2 if the kinematics is such that $\oint \varrho_0 \xi\, dx \ll \oint \varrho_0 \zeta\, dz$, i.e. the motion is predominantly vertical, whereas ν^2 approaches zero when the motion is predominantly horizontal, so that $\oint \varrho_0 \xi\, dx \gg \oint \varrho_0 \zeta\, dz$. While the horizontal accelerations are produced entirely by the horizontal pressure force, as is shown by (23.1), the vertical perturbation pressure force in (23.2) will in internal oscillations act to reduce the restoring force which produces the vertical acceleration so that the resulting frequency will be less than ν_s^2. We notice that the horizontal pressure force is in phase with the horizontal acceleration, while the vertical pressure force and the vertical acceleration have opposite phase. On the basis of this circumstance, which is easily seen from the perturbation equations, we may deduce the abnormal energy flux for gravitational waves with vertical propagation, discussed in the previous section.

HÖILAND'S frequency formula (23.5) may also be used to find a criterion for the applicability of the quasi-static method. The quasi-static approximation consists in neglecting the vertical acceleration in (23.2); and when HÖILAND'S frequency formula is derived quasi-statically, we therefore obtain

$$\nu^2 = \frac{\oint \nu_s^2 \varrho_0 \zeta\, dz}{\oint \varrho_0 \xi\, dx}. \tag{23.8}$$

In order that this quasi-static frequency shall represent an approximation to the exact frequency (23.5), it is necessary that

$$\oint \varrho_0 \xi\, dx \gg \oint \varrho_0 \zeta\, dz \tag{23.9}$$

i.e. the motion is predominantly horizontal. The quasi-static theory can be used only when this criterion is satisfied for all normal modes of importance in the problem considered.

Although HÖILAND'S frequency formula applies to standing oscillations, the results apply equally well to progressive waves, because these may be obtained by superposition of standing oscillations of different phase.

HÖILAND'S frequency formula may also be used in the compressible case. We must then multiply the perturbation equation by the integrating factor $e^{\int \gamma g\,dz}$ before taking the line integral, in order to eliminate the pressure. In this case, it does not follow from the continuity equation that the streamlines are closed curves, nor does it follow that a displacement along a closed streamline has the same direction along the curve in all points. Therefore, (23.6) and (23.7) are no longer necessarily true. Although HÖILAND'S frequency formula (23.5) still holds in all cases, its usefulness is restricted to the low-frequency domain; in the high-frequency domain, the numerator and the denominator in (23.5) both tend to become small.

24. Application to mountain waves. When the air blows over hilly country, the horizontal flow will be disturbed by the unevenness of the surface. The character of the resulting motion depends primarily on the static stability of the air, the vertical wind profile and the size of the topographic features of the surface. In this section we shall be concerned with the disturbing influence of hills or mountains whose horizontal extent in the wind direction is comparatively small; in this case the effect of the earth's rotation upon the motion is unimportant. When the static stability is relatively high, the motion is observed to be almost stationary, having the character of a steady wave, through which the air is blowing. With sufficiently high humidity, stationary clouds (normally *altocumulus lenticularis*) will form in the wave crests and so make these visible. The upwind fields in such wave motions have been utilized and explored by soaring planes.

The most remarkable feature of the stationary wave motion is a system of almost regular stationary sine waves formed on the lee side of mountain ranges *(lee waves)*. They are usually most pronounced in the mid-troposphere, and often visible as a number of equally spaced lenticularis clouds.

The perturbation theory of gravitational-compressibility waves (Sect. 22) has been successfully applied to mountain waves (LYRA[1], QUENEY[2], PALM[3], WURTELE[4]) and has been able to explain several features of the motion.

The underlying assumptions then are: The wind velocity U is considered as independent of height; the static stability is constant with height, so that the isothermal model atmosphere can be used; and the mountains are low and their slopes gentle, so that the perturbation theory applies.

The theory of the mountain waves has so far been satisfactorily developed for two-dimensional motion only, i.e. one considers an infinite mountain range parallel to the y-axis which is crossed at right angle by air moving at the speed U in the positive x-direction. Let $x'z$ denote a coordinate system at rest relative to the earth, and xz a coordinate system moving with the air, i.e. at the speed U. Then we have

$$x' = x + Ut. \tag{24.1}$$

Assuming two-dimensional motion, we seek a solution of the form

$$\zeta = \zeta(x', z) = \zeta(x + Ut, z) \tag{24.2}$$

which satisfies the boundary condition at the ground

$$\zeta(x', 0) = h(x') = h(x + Ut), \tag{24.3}$$

[1] G. LYRA: Z. angew. Math. Mech. **23**, 1 (1943).
[2] P. QUENEY: Univ. of Chicago, Dept. of Meteor., Misc. Rep. 23, 1947.
[3] E. PALM: Astrophys. Norv. **5**, 3 (1953).
[4] M. WURTELE: Univ. of Calif., Dept. of Meteor., Sci. Rep. No. 3, Contract No AF 19 122)—263, 1953.

$h(x')$ representing the profile of the mountain. The solution (24.2) represents a steady motion relative to the ground (in the $x'z$-system). Relative to the moving xz-system, the motion is a disturbance of the state of rest, produced by a mountain moving with the velocity $-U$; and the theory developed in Sect. 22 applies in this coordinate system. The motion has the character of a composite wave, moving upstream with the velocity $-U$ without change in shape.

By means of FOURIER'S integral theorem, applied to the x-coordinate, the problem may be reduced to that of determining the sine wave with phase velocity $-U$, produced by a sine-shaped ground-profile with arbitrary wave number k. Let $\zeta_k''=\zeta_k''(x+Ut, z)$ denote the sine-wave solution for $\zeta''=\varrho_0^{\frac{1}{2}}\zeta$ (compare Eq. (22.13)] and let the boundary condition be

$$\zeta_k''(x+Ut, 0) = a_k \cos k(x+Ut) \tag{24.4}$$

where a_k and k are given. It follows that the frequency of the wave is

$$\nu = Uk \tag{24.5}$$

so that the waves which have to be considered are represented by a straight line $\nu^2 = U^2k^2$ in the ν^2k^2-diagram Fig. 6. In the cases which have meteorological interest, $U^2 \ll c_s^2 = \gamma^{-1}$, so that the line $\nu^2 = U^2k^2$ has small inclination and cuts through the domain of low-frequency internal waves. For these waves, it is permissible to simplify the mathematics by putting $\gamma = 0$, which means that the velocity of sound is considered as infinite (compare Sect. 22). Eq. (22.27) then gives together with (24.5)

$$m^2 = k_s^2 - k^2, \tag{24.6}$$

where

$$k_s^2 = \frac{\nu_s^2}{U^2} - l^2. \tag{24.7}$$

Eq. (24.6) shows that the solution is an external wave when $k > k_s$ and an internal wave where $k < k_s$. To each value of k correspond two values of m of equal magnitude but opposite sign.

For external waves, $k > k_s$, the two solutions are of the form

$$\zeta_k'' \sim e^{\pm m_1 z} \cos k(x+Ut), \quad \text{with} \quad m_1^2 = -m^2.$$

When m_1 denotes the positive root, we must reject the solution with the positive exponent, since a FOURIER integral containing such solutions will represent a motion whose wave energy per unit volume will increase indefinitely with height. The solution is therefore in this case

$$\zeta_k'' = a_k e^{-m_1 z} \cos k(x+Ut) \tag{24.8}$$

and is thus unique.

Consider next the case of internal waves ($k < k_s$). The solution may then be written

$$\left.\begin{aligned} \zeta_k'' &= a_k \cos mz \cos k(x+Ut) + \\ &+ \sin mz\,[A_k \cos k(x+Ut) + B_k \sin k(x+Ut)] \end{aligned}\right\} \tag{24.9}$$

which satisfies the boundary condition (24.4) for any values of the constants A_k and B_k. There are no further boundary conditions which can serve to determine A_k and B_k, and the problem is therefore indeterminate, when presented in this form.

There is a corresponding ambiguity in the solution for a solitary mountain ridge. As shown by LYRA[1] there is generally one solution which is a monotoneous function of x' on the upwind side ($x' < 0$) and an oscillatory function of x' on the

[1] G. LYRA: Z. angew. Math. Mech. 23, 1 (1943).

lee side. Let $\zeta^*(x', z)$ denote this solution. In the case of a symmetric mountain ridge, centered at the origin $(h(x') = h(-x'))$, $\zeta^*(-x', z)$ must also be a solution of the problem; for from a frictionless, stationary motion we may always derive another stationary motion by reversing the velocity field. Therefore, if the equations permit a stationary solution with waves on the lee side only, the same equations will permit a solution with waves only on the upwind side. Moreover, since the problem is linearized,

$$\zeta(x', z) = a\zeta^*(x', z) + (1 - a)\zeta^*(-x', z) \tag{24.10}$$

is a stationary solution for any real value of the constant a.

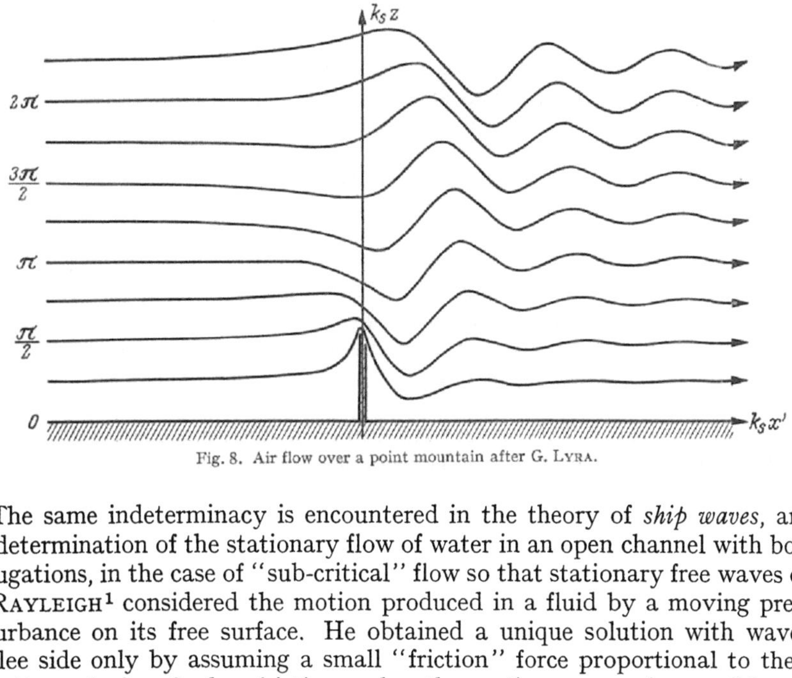

Fig. 8. Air flow over a point mountain after G. LYRA.

The same indeterminacy is encountered in the theory of *ship waves*, and in the determination of the stationary flow of water in an open channel with bottom corrugations, in the case of "sub-critical" flow so that stationary free waves exist.

RAYLEIGH[1] considered the motion produced in a fluid by a moving pressure disturbance on its free surface. He obtained a unique solution with waves on the lee side only by assuming a small "friction" force proportional to the perturbation velocity. Such a friction makes the motion process irreversible, which is necessary in order to remove the ambiguity; but it does not increase the order of the differential equation, and therefore does not require any additional boundary conditions.

LYRA applied the RAYLEIGH friction to the mountain wave problem, and was thus able to obtain a unique solution with waves on the lee side only. In polar coordinates r, θ in the xz-plane, LYRA's solution for a "point-mountain" with unit cross-section located at the origin is as follows:

$$\left.\begin{aligned} \zeta = Z(x', z) \equiv -2k_s \sqrt{\frac{\varrho_0(0)}{\varrho_0(z)}} \times \\ \times \left[\frac{1}{4} N_1(k_s r) \sin\theta + \frac{1}{\pi} \sum_{\nu=1}^{\infty} \frac{2\nu}{(2\nu)^2 - 1} J_{2\nu}(k_s r) \sin 2\nu\theta\right] \end{aligned}\right\} \tag{24.11}$$

where $\varrho_0(z)$ is undisturbed density, and J and N the BESSEL functions of the first and second kind. The corresponding streamline pattern is shown in Fig. 8.

[1] Lord RAYLEIGH: Sci. Papers 2, 258 (1883).

The horizontal wave length of the lee waves is

$$L_s = \frac{2\pi}{k_s} \approx \frac{2\pi U}{\nu_s}. \qquad (24.12)$$

When the lapse rate is known, ν_s may be evaluated from Eq. (21.21). In typical cases $\left(\frac{\partial T}{\partial z} = -0.006°\,\mathrm{C/m},\ U = 10-20\ \mathrm{m/sec}\right)$ Eq. (24.12) gives $L_s = 5$ to 10 km, in good agreement with observations.

The expression (24.11) has a singularity at $r=0$, and is therefore not applicable near the origin. The solution for an arbitrary mountain profile $h(x')$ is

$$\zeta(x', z) = \int_{-\infty}^{+\infty} Z(x'-\xi, z)\, h(\xi)\, d\xi. \qquad (24.13)$$

QUENEY[1] applied the RAYLEIGH-friction to the solution for a sine-profile of the form (24.4) and found that in the solution for internal waves (24.9), we must now put $A_k = 0$, $B_k = -a_k$, so that the solution becomes

$$\zeta_k'' = a_k \cos(k\,x + m\,z + k\,U\,t), \qquad (24.14)$$

where we have assumed $k > 0$, $m > 0$. This is a sine-wave with the phase-lines tilting towards the wind; since the wave moves against the wind relative to the air, it has also a downward propagation. By means of the fundamental solutions (24.8) (when $k > k_s$) and (24.14) (when $k < k_s$), we may build up a unique solution for an arbitrary mountain by FOURIER-analysis. As shown by QUENEY, this composite solution has waves on the lee side only, in agreement with the result of LYRA.

So far, the theory is able to give a solution in fair agreement with observations. One may, however, raise objections against the method which has been used to make the solution unique. The RAYLEIGH-friction is an artifice; and from the rapidness with which the lee-waves form when the current starts, it follows that friction is not the factor which makes nature select this solution. HÖILAND[2] has proposed to treat the mountain-wave problem as an initial value problem. If the mountain is initially at rest in a resting atmosphere, and then starts moving with the velocity $-U$ relative to the resting air, a non-stationary motion will be set up in the air, which ultimately approaches a stationary state relative to the mountain. Because this is an initial value problem, the solution must in this case be unique, and the artifice of the RAYLEIGH-friction becomes unnecessary. PALM[3] and WURTELE[4] have carried out the calculation of the ultimate stationary motion, and have found that the solution thus obtained shows lee-waves and is in agreement with the solution obtained by LYRA and QUENEY. Moreover, it is found that the ultimate stationary motion is independent upon the way in which the motion of the mountain is started, and also independent upon the initial state of motion in the air as long as the total wave energy at $t=0$ is finite.

The solution of the initial value problem is a laboreous procedure. ELIASSEN and PALM[5] have proposed to consider the stationary mountain wave as a motion

[1] P. QUENEY: Univ. of Chicago, Dept. of Meteor., Misc. Rep. 23, 1947.

[2] E. HÖILAND: Univ. of Calif., Dept. of Meteor., 5th progr. Rep., Appendix A, Contract No. AF 19 (122)—263, 1951.

[3] E. PALM: Astrophys. Norv. 5, No. 3 (1953).

[4] M. WURTELE: Univ. of Calif., Dept. of Meteor., Sci. Rep. No. 3, Contract No. AF 19 (122)—263, 1953.

[5] A. ELIASSEN and E. PALM: Inst. Weather and Climate Res., Norw. Acad. of Science and Letters, Publ. 1, 1954.

where wave energy is "radiated" by the mountain. In the moving xz-system, the motion is entirely produced by the moving mountain, which represents the only energy source.

Suppose that the air is initially at rest in the moving xz-system, so that the total wave energy is zero. When the mountain starts moving, a wave-motion will be set up in the air; the total wave energy of the air will increase as a result of the work done on the air by the mountain. The wave energy flux will convey wave energy to an ever increasing region. After a sufficiently long time, the motion is nearly stationary (relative to the mountain) in the vicinity of the mountain; at greater distance the motion is transitory, and at still greater distance, there is no disturbance at all (provided the velocity of sound is considered finite). The domain of transitory motion represents a region of increasing wave energy, and the only source of energy is the mountain. Hence it follows:

1. The work done by the mountain on the air must be positive, i.e.

$$W = \int_{-\infty}^{+\infty} p' \frac{\partial \zeta}{\partial t} dx = U \int_{-\infty}^{+\infty} p' \frac{\partial \zeta}{\partial x} dx > 0 \qquad \text{when } z = 0. \tag{24.15}$$

The pressure is therefore higher on the upwind side of the mountain than on the lee side. The drag exerted by the air on the mountain, or the downward momentum transport, is

$$F = W/U. \tag{24.16}$$

Therefore the drag on the mountain is always in the direction of the wind, so that the mountain acts as a brake on the air current.

2. The stationary flow must be such that the wave energy flux is directed away from the mountain. The definition (22.20) of the wave energy flux applies in the moving xz-system. It is convenient here to introduce the flux relative to the $x'z$-system, in which the motion is stationary. This energy flux has the components

$$\left.\begin{aligned} F'_x &= p' \frac{\partial \xi}{\partial t} + U \varrho_0 q = U\left(p' \frac{\partial \xi}{\partial x} + \varrho_0 q\right), \\ F'_z &= p' \frac{\partial \zeta}{\partial t} \qquad\quad\;\; = U p' \frac{\partial \zeta}{\partial x}. \end{aligned}\right\} \tag{24.17}$$

It follows from (22.18) that this flux is solenoidal. Its streamlines start at the mountain and cannot stop anywhere within the region of stationary motion. It follows that the stationary motion must be such that

$$\lim_{z \to \infty} U p' \frac{\partial \zeta}{\partial x} \geq 0 \qquad \text{for any value of } x'. \tag{24.18}$$

It is shown by ELIASSEN and PALM that this condition suffices to determine the motion uniquely; and the solution thus obtained is identic with QUENEY's solution and has waves on the lee side. It is noteworthy that QUENEY's solution (24.14) for a bottom with sine-profile is characterized by an upward directed wave energy flux in agreement with the argument above.

It should be noted that QUENEY's solution (24.14) for internal waves holds in an infinite atmosphere of uniform stability. The result must be modified if layers exist which are able to reflect part of the "radiation" of wave energy from below, as for instance a discontinuity surface separating layers of different stability.

SCORER[1] has pointed out that the solutions of QUENEY and LYRA give an increase with height of the wave amplitude [due to the factor $\varrho_0^{-\frac{1}{2}}$ of (22.13)] and a strong dampening of the waves downstream, whereas in reality the maximum amplitude is often observed to be in the middle or lower troposphere, with less dampening downstream. SCORER has shown that these discrepancies can be explained as effects of the variation of static stability and wind with height (see also ZIEREP[2], WURTELE[3] and PALM[4]); these effects must therefore be taken into consideraton in order to arrive at a completely satisfactory theory of mountain waves. The wind variation with height brings in another possible energy source or sink, viz. the kinetic energy of the basic wind field. Waves in a non-uniform wind field are not of the pure gravitational-compressibility type, and the energy considerations of the present section are no longer applicable.

The stratospheric mother of pearl clouds at heights about 25 km, which are sometimes observed on the lee side of mountain ranges, notably in Scandinavia, are generally believed to be manifestations of particularly high-reaching lee waves.

b) Axially symmetric motions in a circular vortex.

25. Pure inertial stability. A revolving mass of fluid possesses a particular kind of stability, usually referred to as inertial stability, which was first found by RAYLEIGH[5], and later treated by V. BJERKNES and SOLBERG[6].

In order to study the phenomenon in pure form, let us consider a homogenous and incompressible fluid rotating around an axis (no external force). Let R denote distance from the axis and z distance from a plane normal to the axis. The velocity field may be characterized by the angular velocity ω, or by the circulation c of a circle $R=\text{const}$, $z=\text{const}$. We have

$$c = 2\pi R^2 \omega . \tag{25.1}$$

Suppose that the vortex is bounded by a rigid surface of revolution, symmetric with respect to the axis, enclosing a volume τ. The kinetic energy of the vortex is

$$K_v = \int_\tau \varphi_v \varrho c^2 d\tau , \tag{25.2}$$

where

$$\varphi_v = 1/8\pi^2 R^2 . \tag{25.3}$$

In expression (25.2), c may be any function of R and z.

Suppose now that an axially symmetric meridional motion, characterized by a meridional velocity field $\boldsymbol{v}_m(R, z)$ is superimposed upon the vortex motion. This will not destroy the axial symmetry of the vortex. The total kinetic energy of the fluid is then $K_m + K_v$, where

$$K_m = \int_\tau \tfrac{1}{2} \boldsymbol{v}_m^2 \varrho \, d\tau \tag{25.4}$$

is the kinetic energy of the meridional motion. The energy equation of the system is

$$K_m + K_v = \text{constant} . \tag{25.5}$$

[1] R. S. SCORER: Quart. J. Roy. Meteor. Soc. **75**, 41 (1949); **79**, 70 (1953); **80**, 417 (1954).
[2] J. ZIEREP: Z. Flugwiss. **1**, 9 (1953).
[3] M. WURTELE: Univ. of Calif., Dept. of Meteor., Sci. Rep. No. 4, Contract No. AF 19 (122)—263, 1953.
[4] E. PALM: Univ. of Calif., Dept. of Meteor., Sci. Rep. No. 3, Contract No. 19 (604)—728, 1955.
[5] Lord RAYLEIGH: Sci. Papers **6**, 447 (1916).
[6] V. BJERKNES and H. SOLBERG: Avh. Norske Vidensk. Akad. Oslo (I) **1929**, 7.

Let us use the term "symmetric circle" to denote a circle $R = \text{const}$, $z = \text{const}$. It follows from the assumption of axial symmetry that a fluid line which at one time forms a symmetric circle will always remain a symmetric circle; and its circulation c will be conserved by KELVIN's theorem. Therefore the change in the field $c(R, z)$, and hence, the change in K_v, depends only upon the meridional displacements. It follows that K_v plays the role of potential energy for the meridional motion. This circumstance was pointed out by FJÖRTOFT [8] who showed that the stability criteria for a vortex with respect to axially symmetric motions can be derived from its energy equation in a manner which is formally identic with the derivation of the criterion for static stability given in Sect. 21. Referring to the method developed in that section, the stability criterion of the vortex is obtained as follows:

A field $c_0(R, z)$ which corresponds to an extremum of K_v is a stationary or balanced vortex with $\boldsymbol{v}_m \equiv 0$. The condition of balance is

$$\varrho\, c_0^2 \operatorname{grad} \varphi_v + \operatorname{grad} p_0 = 0 \tag{25.6}$$

which may be obtained by requiring the variation δK_v to vanish for any set of virtual meridional displacements compatible with the kinematic boundary condition and the condition of incompressibility.

The equation shows that the pressure increases with R and is independent of z. It follows also that c_0 depends on R only and is thus independent of z.

The energy Eq. (25.5) shows that the balanced vortex is stable with respect to axially symmetric motions when K_v is a minimum. As is easily seen, this is the case when c_0^2 increases monotoneously with R. The criterion of inertial stability may therefore be written

$$d c_0^2/dR > 0 \qquad \text{for all values of } R. \tag{25.7}$$

On the other hand, if

$$d c_0^2/dR < 0 \qquad \text{for some values of } R, \tag{25.8}$$

then K_v is not a minimum. In this case, K_v can be reduced by suitable meridional displacements. The energy thus released can be used to speed up the meridional motion, and the vortex is unstable.

The stability criterion can also be derived, in a less rigorous manner, by means of the particle method. The "particle" is then chosen as a symmetric circle which is assumed to expand (or contract) without change in circulation, through an undisturbed environment. If c_0^2 increases with R, then a symmetric circle which has been brought to expand will arrive to an environment which rotates faster than the circle itself. The pressure gradient which balances the centrifugal force of the environment [Eq. (25.6)] will exceed the centrifugal force of the displaced circle and thus give a net restoring force, which may be termed a *dynamic buoyancy*. The vortex is thus inertially stable. The restoring force per unit mass and per unit length displacement normal to the axis is found to be

$$\nu_i^2 = \operatorname{grad} c_0^2 \cdot (-\operatorname{grad} \varphi_v) = \frac{c_0}{2\pi^2 R^3} \frac{d c_0}{dR}\,. \tag{25.9}$$

This quantity has the dimension of a frequency squared and may be taken as a measure of the inertial stability. Introducing the vorticity $\operatorname{curl} \boldsymbol{v}_0$ in the undisturbed vortex which is parallel the axis, and is defined by $2\pi R\, dR (\operatorname{curl} \boldsymbol{v}_0)_z = d c_0$, the formula may also be written

$$\nu_i^2 = 2\omega_0 (\operatorname{curl} \boldsymbol{v}_0)_z = 2\boldsymbol{\omega}_0 \cdot \operatorname{curl} \boldsymbol{v}_0. \tag{25.10}$$

If c_0^2 decreases with R, the dynamic buoyancy acting on an expanding or contracting circle will be directed away from its equilibrium position, thus indicating inertial instability. In this case ν_i^2 is negative. A vortex where $c_0^2 = \text{constant}$ (i.e. $\omega_0 \sim R^{-2}$ or $\omega_0 R \sim R^{-1}$) is indifferent and gives $\nu_i^2 = 0$. A vortex rotating as a rigid body ($\omega_0 = \text{const}$) is inertially stable ($\nu_i^2 = 4\omega_0^2$).

In an inertially stable vortex, an axially symmetric disturbance of the balanced state will result in inertial oscillations and waves. In these motions, the meridional displacements satisfy differential equations analogeous to the perturbation equations for two-dimensional gravitational waves, with ν_i^2 instead of ν_s^2. For inertial oscillations, the displacements normal to the axis are the ones which give rise to restoring forces, while the displacements parallel to the axis are inactive and must be driven by the pressure force alone. HÖILAND [*10*] has applied his frequency formula to inertial oscillations, and has thus found that if the displacements are predominantly normal to the axis, the frequency of a fundamental mode approaches ν_i^2 (which is the largest frequency possible); but when the motion is predominantly parallel to the axis, then the frequency is very small. We may also have inertial waves travelling in the meridional plane, in the direction parallel to the axis as well as normal to it. The theory of these waves is very similar to the theory of gravitational waves.

26. Stability of a compressible baroclinic vortex in a gravity field, with respect to axially symmetric motions. In the preceding sections, we have studied separately two types of stability, viz. *static stability* and *inertial stability*. We shall now proceed to study their combined effect. The theory goes back to HELMHOLTZ[1]. The explicite stability criterion was derived from a parcel method by SOLBERG[2]. It was later derived independently by KLEINSCHMIDT[3] and ERTEL, JAW and LI[4] who used a similar method, and by HÖILAND[5], who derived it from the circulation theorem. Also MÖLLER[6], VAN MIEGHEM[7,8] and HOLMBOE[9] have contributed to the theory. FJÖRTOFT[10], [*8*] derived the stability criterion on the basis of the energy equation. The derivation below follows in its broad features the method given by FJÖRTOFT.

We consider a compressible fluid performing a vortex motion under the influence of a gravity field. The equipotential surfaces are supposed to be surfaces of revolution around the axis of the vortex. The fluid is bounded by a rigid surface of revolution, whose axis coincides with the axis of the vortex. Superimposed upon the vortex motion is a meridional velocity field $\boldsymbol{v}_m$ which is supposed to be axially symmetric.

The energy equation of this system is

$$K_m + K_v + \Phi + E = \text{constant}, \tag{26.1}$$

where the total kinetic energy has been subdivided in a part K_m due to the meridional motion, and another part K_v due to the vortex motion, as was done in Sect. 25.

The motion is such that a fluid line which at one time forms a "symmetric circle" around the axis will always remain a symmetric circle, and its circulation c as well as its specific entropy s will remain unchanged. As a consequence, the change of the quantity $J = K_v + \Phi + E$ will depend only upon the meridional displacements $\delta \boldsymbol{r}_m$. Therefore, J may be considered as an "effective potential

[1] H. v. HELMHOLTZ: Meteor. Z. **5**, 329 (1888).
[2] H. SOLBERG: U.G.G.I., VIième Assemblée, Edinbourgh 1936.
[3] E. KLEINSCHMIDT: Ann. Hydrogr., Berlin **69**, 305 (1941).
[4] H. ERTEL, J. JAW and S. LI: Meteor. Z. **58**, 389 (1941).
[5] E. HÖILAND: Avh. Norske Vidensk. Akad. Oslo (I) **1941**, No. 11.
[6] F. MÖLLER: Meteor. Z. **60**, 269 (1943).
[7] J. VAN MIEGHEM: Bull. Acad. Belg., Cl. Sci. **5**, 30 (1944).
[8] J. VAN MIEGHEM: Comp. of Meteor., Amer. Meteor. Soc. **1951**, 434.
[9] J. HOLMBOE: J. Mar. Res. **7**, 163 (1948).
[10] R. FJÖRTOFT: Geof. Publ. **16**, No. 5 (1946).

energy" for the meridional motion. From (21.2) and (25.2) we find

$$J = \int_\tau (\varphi + e)\,\varrho\, d\tau + \int_\tau \varphi_v\, \varrho\, c^2\, d\tau, \tag{26.2}$$

where φ_v is the function defined by (25.3). Time differentiation of this equation gives, since $\varrho\, d\tau$ and c^2 are conserved individually,

$$\frac{dJ}{dt} = \int_\tau \left(\frac{D\varphi}{Dt} + \frac{De}{Dt}\right) \varrho\, d\tau + \int_\tau \frac{D\varphi_v}{Dt}\, \varrho\, c^2\, d\tau.$$

By means of (21.3), (21.5), (21.6) and the equations

$$\frac{D\varphi_v}{Dt} = \boldsymbol{v} \cdot \operatorname{grad} \varphi_v = \boldsymbol{v}_m \cdot \operatorname{grad} \varphi_v,$$

$$\operatorname{div} \boldsymbol{v} = \operatorname{div} \boldsymbol{v}_m,$$

we find the first variation of J,

$$\delta J = \int_\tau \delta \boldsymbol{r}_m \cdot (\varrho \operatorname{grad} \varphi + \varrho\, c^2 \operatorname{grad} \varphi_v + \operatorname{grad} p)\, d\tau. \tag{26.3}$$

Therefore, J is stationary with respect to any set of axially symmetric meridional displacements $\delta \boldsymbol{r}_m$ if

$$\operatorname{grad} \varphi + c_0^2 \operatorname{grad} \varphi_v + \alpha_0 \operatorname{grad} p_0 = 0, \tag{26.4}$$

where the subscript 0 denotes equilibrium distribution. This is the equation of motion for a stationary vortex where the accelerations are purely centripetal. It will be seen that the stationary state may be barotropic or baroclinic. Curl differentiation of (26.4) gives

$$\operatorname{grad} \alpha_0 \times (-\operatorname{grad} p_0) + \operatorname{grad} c_0^2 \times (-\operatorname{grad} \varphi_v) = 0, \tag{26.5}$$

whence it is seen that in a barotropic stationary vortex, c_0 is independent of z (in cylinder coordinates R, z), whereas in a baroclinic vortex c_0 varies with z. Eq. (26.5) expresses nearly the same as the thermal wind Eq. (12.4), the difference being due to somewhat different assumptions.

By means of (21.11), Eq. (26.5) may also be written

$$\alpha_0 \mu_0 \operatorname{grad} s_0 \times (-\alpha_0 \operatorname{grad} p_0) + \operatorname{grad} c_0^2 \times (-\operatorname{grad} \varphi_v) = 0. \tag{26.6}$$

For dry (or non-saturated) air, we have $s = c_p \log \theta$ and $\alpha\mu = 1/c_p$, and the equation becomes

$$\operatorname{grad} \log \theta_0 \times (-\alpha_0 \operatorname{grad} p_0) + \operatorname{grad} c_0^2 \times (-\operatorname{grad} \varphi_v) = 0. \tag{26.7}$$

The second variation of J may be derived in the same manner as shown in Sect. 21 by means of (21.13) and (21.14) and the equation

$$\frac{\partial c^2}{\partial t} = -\boldsymbol{v}_m \cdot \operatorname{grad} c^2. \tag{26.8}$$

The result is

$$\delta^2 J = \frac{1}{2} \int_\tau \left[\delta \boldsymbol{r}_m \cdot \mathcal{M} \cdot \delta \boldsymbol{r}_m + \frac{1}{\gamma} (\operatorname{div} \delta \boldsymbol{r}_m + \gamma\, \delta \boldsymbol{r}_m \cdot \alpha_0 \operatorname{grad} p_0)^2\right] \varrho_0\, d\tau, \tag{26.9}$$

where $\mathcal{M}$ is a tensor defined by

$$\mathcal{M} = \alpha_0 \mu_0 \operatorname{grad} s_0 (-\alpha_0 \operatorname{grad} p_0) + \operatorname{grad} c_0^2 (-\operatorname{grad} \varphi_v). \tag{26.10}$$

The balance Eq. (26.6) expresses that this tensor is symmetric.

It will be seen that if the quadratic form $\delta \boldsymbol{r}_m \cdot \mathcal{M} \cdot \delta \boldsymbol{r}_m$ is positive definite, then $\delta^2 J$ is positive for all virtual axially symmetric meridional displacements, so that the stationary value J_0 of J is a minimum. The energy Eq. (26.1) then shows that K_m is bounded, so that the vortex is stable with respect to axially symmetric disturbances.

If, on the other hand, the quadratic form $\delta \boldsymbol{r}_m \cdot \mathcal{M} \cdot \delta \boldsymbol{r}_m$ is capable of negative values, then there will be virtual displacements which make $\delta^2 J$ negative, indicating that the stationary value J_0 is not a minimum. In this case there exists therefore releasable energy which may conceivably be used to speed up meridional motions, which is a necessary condition for instability. A demonstration that this condition is also sufficient would require a closer investigation, but there seems to be little doubt that such a demonstration can be given.

The stability of the vortex with respect to axially symmetric disturbances depends thus on the character of the quadratic form $\delta \boldsymbol{r}_m \cdot \mathcal{M} \cdot \delta \boldsymbol{r}_m$, which again depends on the sign of the determinant of the matrix $\mathcal{M}$,

$$\left.\begin{aligned} \operatorname{Det}(\mathcal{M}) &= \alpha_0 \mu_0 \operatorname{grad} s_0 \times \operatorname{grad} c_0^2 \cdot (-\alpha_0 \operatorname{grad} p_0) \times (-\operatorname{grad} \varphi_v) \\ &= \alpha_0 \mu_0 \operatorname{grad} s_0 \times \operatorname{grad} c_0^2 \cdot \operatorname{grad} \varphi \times (-\operatorname{grad} \varphi_v) \end{aligned}\right\} \quad (26.11)$$

and the sign of its trace

$$\operatorname{Trace}(\mathcal{M}) = \alpha_0 \mu_0 \operatorname{grad} s_0 \cdot (-\alpha_0 \operatorname{grad} p_0) + \operatorname{grad} c_0^2 \cdot (-\operatorname{grad} \varphi_v). \quad (26.12)$$

The vortex is stable ($\delta^2 J > 0$ for all $\delta \boldsymbol{r}_m$) when

$$\operatorname{Det}(\mathcal{M}) > 0, \qquad \operatorname{Trace}(\mathcal{M}) > 0 \quad (26.13)$$

which may therefore be considered as the stability criteria of the vortex. On the other hand, if

$$\operatorname{Det}(\mathcal{M}) > 0, \qquad \operatorname{Trace}(\mathcal{M}) < 0 \quad (26.14)$$

then the quadratic form $\delta \boldsymbol{r}_m \cdot \mathcal{M} \cdot \delta \boldsymbol{r}_m$ in expression (26.9) is negative definite, and $\delta^2 J < 0$ for all displacements which make $|\operatorname{div} \delta \boldsymbol{r}_m + \gamma\, \delta \boldsymbol{r}_m \cdot \alpha_0 \operatorname{grad} p_0|$ sufficiently small. Finally, if

$$\operatorname{Det}(\mathcal{M}) < 0 \quad (26.15)$$

then the sign of $\delta \boldsymbol{r}_m \cdot \mathcal{M} \cdot \delta \boldsymbol{r}_m$ will depend upon the direction of $\delta \boldsymbol{r}_m$; in this case, we may talk about "stable" (positive) sectors and "unstable" (negative) sectors.

We shall not discuss all cases which may conceivably occur, but shall restrict the discussion to the case that $\operatorname{Trace}(\mathcal{M}) > 0$. Suppose we know the directions of the four vectors $\alpha_0 \mu_0 \operatorname{grad} s_0$, $-\alpha_0 \operatorname{grad} p_0$, $\operatorname{grad} c_0^2$ and $-\operatorname{grad} \varphi_v$ in a certain point in the meridional plane (Fig. 9). From (26.10) we have

$$\left.\begin{aligned} \delta \boldsymbol{r}_m \cdot \mathcal{M} \cdot \delta \boldsymbol{r}_m = \delta \boldsymbol{r}_m \cdot \alpha_0 \mu_0 \operatorname{grad} s_0 (-\alpha_0 \operatorname{grad} p_0) \cdot \delta \boldsymbol{r}_m + \\ + \delta \boldsymbol{r}_m \cdot \operatorname{grad} c_0^2 (-\operatorname{grad} \varphi_v) \cdot \delta \boldsymbol{r}_m . \end{aligned}\right\} \quad (26.16)$$

The signs of each of the two terms on the right depend upon the direction of $\delta \boldsymbol{r}_m$. In Fig. 9, the negative sector for each of these terms is indicated by hatching. As shown by HÖILAND[1], the vortex is stable $(\operatorname{Det}(\mathcal{M}) > 0)$ if these negative sectors do not overlap (diagram a), but unstable $(\operatorname{Det}(\mathcal{M}) < 0)$ if such overlapping occurs (diagram b). In the latter case, the sector where both right-hand terms are negative (doubly hatched) is of course contained within the "unstable" sector where the left-hand side is negative.

[1] E. HÖILAND: Avh. Norske Vidensk. Akad., Oslo (I) **1941**, No. 11.

The instability occuring in the latter case discussed above is closely related to the baroclinicity of the vortex; for in a barotropic vortex, neither of the right-hand terms of (26.16) would have any negative sector. It is noteworthy that this kind of baroclinic instability may occur even if the entropy increases with height (static stability) and the circulation increases with increasing distance from the axis (inertial stability), so that static stability and inertial stability cannot simply be added.

The tensor $\mathcal{M}$ is symmetric and therefore possesses two orthogonal principal axes. We may write

$$\delta \boldsymbol{r}_m \cdot \mathcal{M} \cdot \delta \boldsymbol{r}_m = \nu_y^2 (\delta y')^2 + \nu_z^2 (\delta z')^2, \tag{26.17}$$

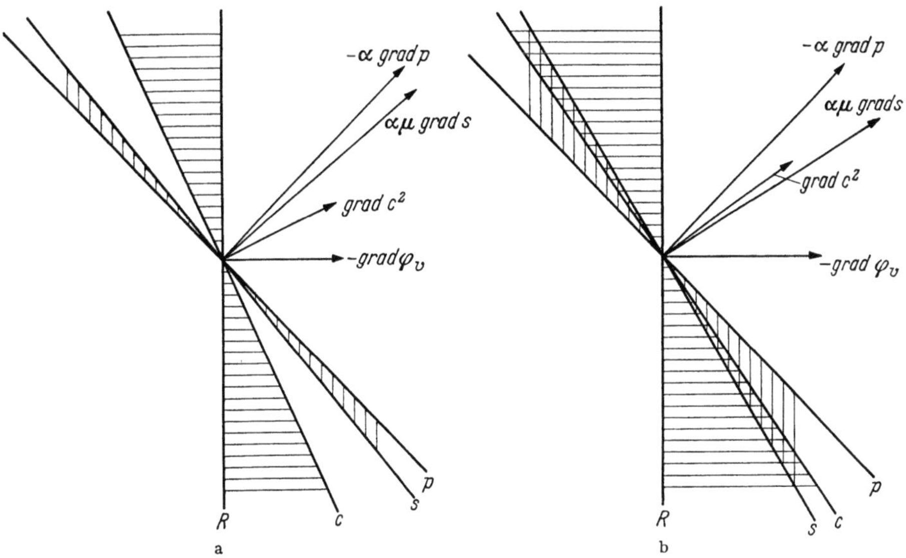

Fig. 9a and b. Stability conditions for a vortex with respect to axially symmetric disturbances; a stable, b unstable.

where $\delta y'$ and $\delta z'$ are the components of $\delta \boldsymbol{r}_m$ along the principal axes (y', z'). Since the determinant and the trace are invariant with respect to rotation of the coordinate system, we have

$$\text{Det}\,(\mathcal{M}) = \nu_y^2 \nu_z^2, \qquad \text{Trace}\,(\mathcal{M}) = \nu_y^2 + \nu_z^2. \tag{26.18}$$

The stability criteria (26.13) may therefore be written

$$\nu_y^2 > 0, \qquad \nu_z^2 > 0. \tag{26.19}$$

For axially symmetric motion in the vicinity of a balanced state, we may derive linear perturbation equations, similar to those which hold for gravitational waves. Denoting by $\boldsymbol{r}_m$ the displacement vector in the meridional plane, and by p' the perturbation pressure, the linearized equations of motion may be expressed as the vector equation

$$\frac{\partial^2 \boldsymbol{r}_m}{\partial t^2} = -\mathcal{M} \cdot \boldsymbol{r}_m - \alpha_0 \nabla p' + \gamma \alpha_0^2 (\nabla p_0)\, p'. \tag{26.20}$$

The first term on the right represents the restoring force, which in a stable vortex ($\boldsymbol{r}_m \cdot \mathcal{M} \cdot \boldsymbol{r}_m$ positive definite) has a component along $-\boldsymbol{r}_m$. We notice that a displacement in any direction will produce a restoring force; there are no inactive directions as in the case of pure static stability, or pure inertial stability. An exception occurs for a vortex where grad φ and grad φ_v are parallel (equatorial conditions); in this case the static stability is simply added to the inertial stability, and horizontal displacements will be inactive.

The component equations of (26.20) are particularly simple when referred to the principal axes of the tensor $\mathcal{M}$.

The stability criteria for the vortex may be derived in a less rigorous manner by means of a particle method where the particle is a symmetric circle which is displaced along the axis of the vortex, and at the same time expanded (or contracted) so that its intersection point in a meridional plane is displaced a distance $\boldsymbol{r}_m$. The environment is supposed to be undisturbed, so that there is no pressure perturbation ($p' = 0$). Under these assumptions, the displaced particle is subject to a force

$$\boldsymbol{F} = -\mathcal{M} \cdot \boldsymbol{r}_m \tag{26.21}$$

which will tend to restore or destroy the equilibrium according as $\boldsymbol{r}_m \cdot \mathcal{M} \cdot \boldsymbol{r}_m \gtrless 0$. The reason why this crude method gives the correct stability criteria lies in the circumstance that for an indifferent motion, which is transitional between stability and instability, the accelerations are zero, and the pressure perturbation vanishes, so that the assumption $p' = 0$ is justified in this case.

It should be stressed that the stability criteria developed in this section are derived for *axially symmetric* motions and do not apply for disturbances which vary with longitude.

27. Stability criteria for the baroclinic vortex applied to the atmosphere. In an absolute frame of reference, the motion of the atmosphere is nearly a circular vortex motion around the earth's axis. Although the disturbances or deviations from a pure vortex motion are not predominantly axially symmetric, it is nevertheless of interest to consider the stability of the atmosphere with respect to axially symmetric disturbances. To this purpose, we shall consider a baroclinic atmosphere performing a strict balanced vortex motion.

We shall consider the earth as a sphere with radius a. In a point with latitude Φ and height z above the earth's surface, the distance from the axis is

$$R = (a + z) \cos \Phi. \tag{27.1}$$

Here z will be considered as negligible compared with a. Let Ω denote the angular speed of rotation of the earth, and u the relative zonal wind speed (assumed independent of longitude); then

$$c_a = 2\pi (\Omega R^2 + R u) \tag{27.2}$$

is the absolute circulation of a symmetric circle.

According to observation, the atmosphere is mostly statically stable, so that the entropy increases with height. Moreover, the pressure gradient is always very nearly vertical, and the first term on the right of (26.12) is therefore positive. The second term on the right of (26.12) is also positive, since c_a^2 is found to increase with the distance from the axis in the direction parallel to the equator plane. We may therefore assume that Trace $\mathcal{M} > 0$ in the atmosphere, and the stability conditions will consequently depend only on the sign of Det $\mathcal{M}$.

Let $\boldsymbol{i}$ denote a unit vector pointing toward the East. In the Northern hemisphere we then have, on account of (25.3) and (26.4)

$$-\alpha \operatorname{grad} p \times (-\operatorname{grad} \varphi_v) = \operatorname{grad} \varphi \times (-\operatorname{grad} \varphi_v) = \frac{g}{4\pi^2 R^3} \boldsymbol{i}, \tag{27.3}$$

and

$$\operatorname{curl} \boldsymbol{v}_a = \frac{1}{2\pi R} \operatorname{grad} c_a \times \boldsymbol{i}. \tag{27.4}$$

Consequently, (26.11) gives

$$\operatorname{Det} \mathcal{M} = \frac{g \alpha \mu c_a}{2 \pi^2 R^3} (\operatorname{grad} s \times \operatorname{grad} c_a \cdot \boldsymbol{i}) = \frac{g \alpha \mu c_a}{\pi R^2} \operatorname{grad} s \cdot \operatorname{curl} \boldsymbol{v}_a. \quad (27.5)$$

Here the scalar factors are positive so that the stability depends on the sign of the vector products in expression (27.5). The stability criterion may therefore be written

$$\left(\frac{\partial c_a}{\partial \Phi}\right)_s \quad \begin{matrix} < 0 & \text{stable} \\ > 0 & \text{unstable} \end{matrix} \quad (27.6)$$

where the index s means differentiation at constant s, i.e. in an isentropic surface. By means of (27.1) and (27.2), (27.6) may also be written

$$\left(2\Omega + \frac{u}{R}\right)\left[\left(1 + \frac{z}{a}\right)\sin\Phi - \frac{1}{a}\left(\frac{\partial z}{\partial \Phi}\right)_s \cos\Phi\right] - \frac{1}{a}\left(\frac{\partial u}{\partial \Phi}\right)_s \quad \begin{matrix} > 0 & \text{stable} \\ < 0 & \text{unstable.} \end{matrix}$$

Here we neglect z versus a. Also, since the slope of the isentropic surfaces $\frac{1}{a}\left(\frac{\partial z}{\partial \Phi}\right)_s$ seldom exceeds $^1/_{100}$ (except by almost vanishing static stability), we neglect the term containing $\cos\Phi$. The criterion then becomes

$$\lambda + \frac{u}{a} \operatorname{tg} \Phi - \left(\frac{\partial u}{\partial y}\right)_s \quad \begin{matrix} > 0 & \text{stable} \\ < 0 & \text{unstable,} \end{matrix} \quad (27.7)$$

where y indicates horizontal length toward the North. If u is not too great, we may also neglect the second term, and write

$$\lambda - \left(\frac{\partial u}{\partial y}\right)_s \quad \begin{matrix} > 0 & \text{stable} \\ < 0 & \text{unstable.} \end{matrix} \quad (27.8)$$

In this form, the criterion applies strictly to a straight, parallel current of arbitrary direction on a flat earth, rotating with the angular velocity $\lambda/2$.

It is noteworthy that the direction of least stability is very nearly along the isentropic slope. In the case of weak instability, the unstable directions will be confined to a narrow sector around the isentropic surfaces; this will lead to the growth of such meridional circulations where the meridional motion takes place predominantly in the unstable directions. The result is therefore superimposed meridional circulations in flat cells orientated roughly along the isentropic slope.

It is reasonable to assume that the stability criterion (27.8) is applicable not only to strictly zonal currents which encircle the globe, but also to a limited portion of a current of arbitrary direction (provided the variation of λ along the current can be neglected).

Application of the above criterion to atmospheric currents has indicated two kinds of flow patterns where the stability with respect to meridional (or transverse) circulations is normally very small, and where the current may even be unstable in pronounced cases.

The first type is the frontal zone (Sect. 46) in the lower troposphere. Since the air is here normally saturated, the isentropic surfaces must be defined as the surfaces of constant potential wet-bulb (or equivalent) temperature; these surfaces are roughly parallel to the frontal layer and thus have a rather steep slope. The wind shear in these surfaces is normally found to be strongly anticyclonic so that the left-hand side of (27.8) is only slightly positive and may even be negative, as has been pointed out by VAN MIEGHEM[1]. The most unstable direction coincides with the frontal slope, approximately. The significance of the weak

[1] J. VAN MIEGHEM: Inst. Roy. Meteor. Belg., Mem. 23 (1946).

stability, or instability of frontal zones is that the air will make very little resistance, if any, against an upgliding motion in the direction of the frontal slope.

The other type of flow where the stability is very small is found in the upper troposphere to the south of strong jets (Sect. 46). Here the isentropic surfaces are almost horizontal; but the horizontal wind shear is strongly anticyclonic so that the left-hand side of (27.8) or (27.7) is nearly zero, as has been pointed out by PALMÉN and NAGLER[1] and PALMÉN and NEWTON[2]. We may therefore conclude that there is very little resistance just south of a jet stream maximum against meridional or transverse circulations in flat horizontal cells, superimposed upon the current.

In all other parts of the atmosphere a pronounced stability with respect to meridional motions seems to prevail.

c) Two-dimensional disturbances of linear currents and vortices.

28. Two-dimensional wave disturbances of linear currents. The stability of a linear flow $U(z)$ of an invisquid, homogeneous and incompressible fluid bounded

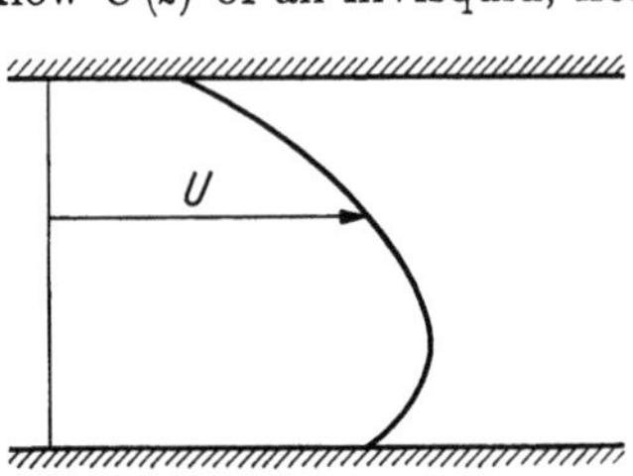

Fig. 10.

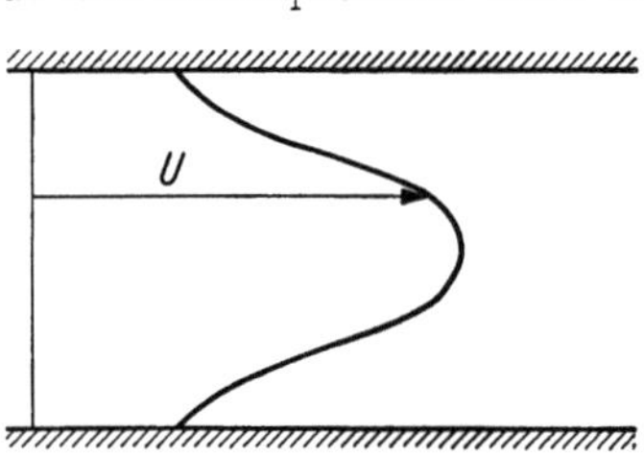

Fig. 11.

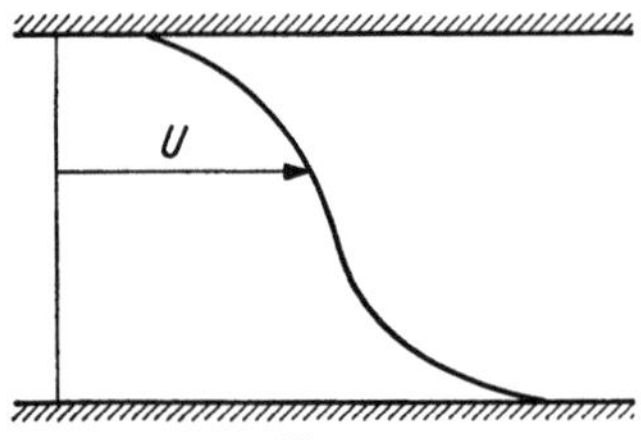

Fig. 12.

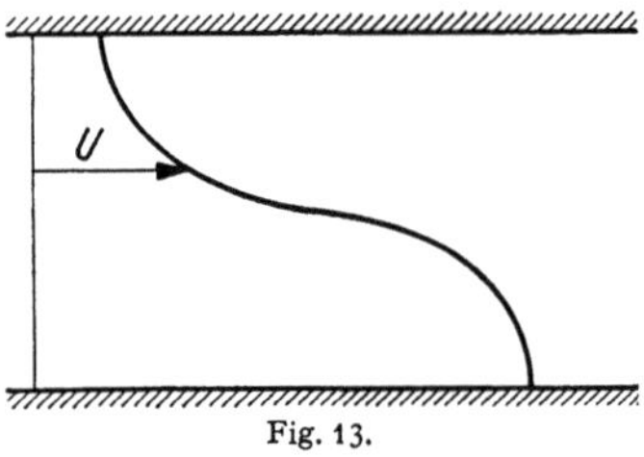

Fig. 13.

Fig. 10—13. Velocity profiles of linear flow between two rigid planes.

by rigid planes $z = z_1$ and $z = z_2$ with respect to two-dimensional wave disturbances in the xz-plane is a classical problem in hydrodynamics. RAYLEIGH[3] and TAYLOR[4] have proved that the linear flow is stable when the velocity profile $U(z)$ is such that $d^2 U/dz^2$ has the same sign for all values of z between z_1 and z_2. In this case the vorticity is a monotonous function of z as shown in Fig. 10. TOLLMIEN[5] has shown that a flow with a velocity profile symmetrical to the axis of the channel is unstable, if the profile exhibits in each half at least one inflection point, as is the case in Fig. 11. FJÖRTOFT [8] has proved that a flow with a velocity profile of the type shown in Fig. 12, with a minimum of vorticity somewhere in the middle of the layer, is stable. On the other hand, a profile of the kind shown in Fig. 13, with a maximum of vorticity somewhere in the layer may under circumstances be unstable. This is not a sufficient criterion of instability, however;

[1] E. PALMÉN and K. M. NAGLER: J. Meteor. **5**, 58 (1948).
[2] E. PALMÉN and C. W. NEWTON: J. Meteor. **5**, 220 (1948).
[3] Lord RAYLEIGH: Sci. Papers **6**, 197 (1913).
[4] G. I. TAYLOR: Phil. Trans. Roy. Soc. Lond., Ser. A **215**, 1 (1915).
[5] W. TOLLMIEN: Nachr. Ges. Wiss. Göttingen, N. F. **1**, Nr. 5 (1935).

HÖILAND[1] has shown that in the case of a sine profile, the current is stable as long as the wave length of the profile exceeds twice the thickness of the layer.

It was noticed by FJÖRTOFT [8] that the stability proofs referred to above must be based on a definition of the concept of stability that differs somewhat from the definition used in the case of static stability and inertial stability. What can be proved in the case of a linear current is that the displacements of the particles in the z-direction will remain small provided the perturbation *vorticities* are small initially; it is not sufficient that the perturbation *energy* is small.

If we define a mean current $\bar{u}(z, t)$ by averaging with respect to x, and write

$$u = \bar{u} + u', \qquad w = w' \qquad (\overline{w} = 0),$$

we may write the transport of momentum through a line $z =$ constant as

$$W_m = \overline{u' w'} \tag{28.1}$$

(apart from an irrelevant density factor).

For the changes in the mean velocity, we then have

$$\frac{\partial \bar{u}}{\partial t} = -\frac{\partial}{\partial z} \overline{u' w'}. \tag{28.2}$$

The rate of change of kinetic energy of the mean flow is accordingly

$$\frac{\partial}{\partial t} \frac{1}{2} \bar{u}^2 = -\bar{u} \frac{\partial}{\partial z} \overline{u' w'}. \tag{28.3}$$

Integration through the layer gives, when K_m denotes the total kinetic energy of the mean motion,

$$\frac{d K_m}{d t} = -\int_{z_1}^{z_2} \bar{u} \frac{\partial \overline{u' w'}}{\partial z} d z = \int_{z_1}^{z_2} \overline{u' w'} \frac{\partial \bar{u}}{\partial z} d z. \tag{28.4}$$

For amplifying perturbations, this expression must be negative, i.e. the momentum transport must be directed from layers with large $\bar{u}$-values to layers with small $\bar{u}$-values. This means that the tilt of the perturbation patterns must be opposite to the tilt of the velocity profile curve.

29. Stability of zonal currents in the case of horizontal non-divergent motion. FJÖRTOFT [8] has generalized the theory of stability of linear currents so that it applies to two-dimensional non-divergent motion on a sphere. This case is of considerable interest in meteorology, because it has been found that at a certain level, which seems to be somewhere near 500 mb, the large-scale motions may to a certain approximation be considered as horizontal and non-divergent (compare Sect. 38). The following outline of the stability of zonal currents with respect to horizontal, non-divergent disturbances follows FJÖRTOFT's derivation in its broad features.

We consider the earth as a sphere with radius a, and shall use geographic coordinates, latitude Φ and East longitude Λ. The distance from the axis to a point on the sphere with latitude Φ is

$$R = a \cos \Phi. \tag{29.1}$$

The fluid is supposed to be confined within a zone bounded by the latitude circles Φ_1 and Φ_2; the area of the zone is denoted by σ. Let

$$F(\Phi) = 2\pi a^2 \sin \Phi \tag{29.2}$$

[1] E. HÖILAND: Geof. Publ. **18**, No. 9 (1953).

denote the area between the equator and the latitude circle Φ. It will be convenient in the following to use occasionally $F(\Phi)$ as a coordinate instead of Φ. In particular we write $F_1 = F(\Phi_1)$ and $F_2 = F(\Phi_2)$, so that $\sigma = F_2 - F_1$. The area of the ring between Φ and $\Phi + d\Phi$ then is

$$dF = 2\pi a^2 \cos\Phi \, d\Phi = 2\pi a R \, d\Phi. \tag{29.3}$$

Let the absolute velocity components be u_a (toward the East) and v (toward the North). The absolute circulation along a latitude circle is

$$c_a(\Phi) = \int_0^{2\pi} R u_a \, d\Lambda = 2\pi R \bar{u}_a, \tag{29.4}$$

where a bar is used to designate mean values over a latitude circle. Since the velocity field is assumed to be solenoidal, we have $\bar{v} = 0$. The total velocity field may therefore be expressed as

$$u_a = \frac{c_a}{2\pi R} + u', \qquad v = v', \tag{29.5}$$

where u' and v' are the deviations from the mean values, satisfying the condition $\overline{u'} = \overline{v'} = 0$.

The two-dimensional velocity field defines a scalar vorticity, viz. the absolute vorticity component ζ_a normal to the sphere. Under the assumptions made, ζ_a is individually conserved. We may write

$$\zeta_a = \bar{\zeta}_a + \zeta', \qquad \bar{\zeta}' = 0. \tag{29.6}$$

From the theorem of STOKES, we have

$$c_a(\Phi) - c_a(\Phi_1) = -\int_{F_1}^{F} \bar{\zeta}_a \, dF \tag{29.7}$$

and hence

$$\frac{\partial c_a}{\partial F} = -\bar{\zeta}_a. \tag{29.8}$$

The kinematic boundary condition requires that the latitude circles $\Phi = \Phi_1$ and $\Phi = \Phi_2$ will remain material curves. Therefore, KELVIN's theorem shows that $c_a(\Phi_1)$ and $c_a(\Phi_2)$ are independent of time. Hence, we obtain, by putting $\Phi = \Phi_2$ in (29.7),

$$\int_{F_1}^{F_2} \bar{\zeta}_a \, dF = \int_\sigma \zeta_a \, d\sigma = c_a(\Phi_1) - c_a(\Phi_2) = \text{constant}. \tag{29.9}$$

Another quantity which does not vary with time is, under the assumptions made, the total absolute angular momentum M_a of the system, which is given by

$$M_a = \int_\sigma R u_a \, d\sigma = \int_{F_1}^{F_2} R \bar{u}_a \, dF = \frac{1}{2\pi} \int_{F_1}^{F_2} c_a \, dF. \tag{29.10}$$

Integrating by parts and using (29.8), we find

$$\int_{F_1}^{F_2} F \bar{\zeta}_a \, dF = \int_\sigma F \zeta_a \, d\sigma = \text{constant}. \tag{29.11}$$

The possible distributions of $\bar{\zeta}_a$ which can develop from a given initial distribution are restricted to such distributions which leave the two integrals (29.9) and (29.11) unchanged. For instance, it follows that $\bar{\zeta}_a$ cannot increase at high latitudes and decrease at low latitudes, or vice versa.

From (29.9) and (29.11) follows also the stability of a zonal current where $\bar{\zeta}_a$ varies monotoneously with latitude. The proof can be given by referring to an analogy with the potential energy of a non-homogeneous incompressible fluid in a gravity field. Consider the integral

$$Q = \int\limits_0 F \zeta_a \, d\sigma. \tag{29.12}$$

If ζ_a is interpreted as the individually conserved density of a hypothetic fluid, and F as the potential of a hypothetic gravity field, then Q represents the potential energy of the fluid, and (29.9) expresses that the total mass is constant.

With reference to the method developed in Sect. 21, we may calculate the change in potential energy resulting from an arbitrary set of virtual displacements of the fluid [without regard to the restriction imposed by (29.11)]. From this analogy it follows immediately that Q is a maximum when $\zeta' = 0$ (equidense surfaces horizontal) and $\bar{\zeta}_a$ increases monotoneously with latitude (density increasing with height). Similarly, Q is a minimum when $\zeta' = 0$ and $\bar{\zeta}_a$ decreases monotoneously with φ.

Suppose that the motion is initially characterized by $\bar{\zeta}_a$ varying monotoneously with latitude, and in addition, small values of ζ' (compared with the total variation of $\bar{\zeta}_a$) so that Q is near its maximum (or minimum) value. The condition (29.11) then states that Q will always remain near its maximum (or minimum) value. This, however, implies that the north-south displacements must remain small almost everywhere. A zonal current is therefore stable if its vorticity has a monotoneous variation with latitude. If c_a for such a current is plotted as a function of F, the curvature of the profile curve will be of one sign in all latitudes.

The stability can be proved in certain cases, even if the curvature of the profile $c_a(F)$ changes sign. Let

$$\bar{\Omega}_a = c_a / 2\pi R^2 \tag{29.13}$$

denote the mean absolute angular speed of rotation along a latitude circle. Then we have:

A zonal current is stable if the values of $\bar{\Omega}_a$ in all latitudes where $\partial\bar{\zeta}_a/\partial\Phi < 0$ are larger than any value of $\bar{\Omega}_a$ in latitudes where $\partial\bar{\zeta}_a/\partial\Phi > 0$. This theorem was proved by Fjörtoft from (29.9), (29.11) and the energy equation.

The above stability criteria are sufficient but not necessary. By reversing them, we therefore obtain necessary, but insufficient criteria of instability.

The stability properties of various profiles may also be studied by investigating the solutions of the corresponding perturbation equations. It should be noted, however, that continuous eigenvalue spectra will occur in this case, and we cannot get full account of the stability properties by considering only the ordinary eigensolutions which correspond to discrete eigenvalues.

In order to apply the above criteria to the atmosphere, we introduce the *relative circulation*

$$c = 2\pi R \bar{u} \tag{29.14}$$

and have

$$c_a = 2\pi R^2 \Omega + c = \frac{\Omega}{2\pi a^2}(4\pi^2 a^4 - F^2) + c, \tag{29.15}$$

$$\bar{\zeta}_a = -\frac{\partial c_a}{\partial F} = \frac{\Omega}{\pi a^2} F - \frac{\partial c}{\partial F} = 2\Omega \sin\Phi - \frac{\partial c}{\partial F}, \tag{29.16}$$

$$\frac{\partial \bar{\zeta}_a}{\partial F} = \frac{\Omega}{\pi a^2} - \frac{\partial^2 c}{\partial F^2}, \tag{29.17}$$

$$\bar{\Omega}_a = \Omega + \frac{c}{2\pi R^2}. \tag{29.18}$$

There are no latitude walls in the atmosphere which represent the rigid boundaries assumed in the theory. However, observations show that the disturbances of the westerlies have maximum amplitude in middle latitudes and are small in high and low latitudes. It is therefore not quite unrealistic to assume that the zonal flow is undisturbed south of 20° and north of 70° N, say. This assumption is equivalent to assuming rigid boundaries in these latitudes.

Fig. 14 shows the quantities $\bar{u}$, c_a, $\bar{\zeta}_a$ and $\bar{\Omega}_a$ plotted as a function of F for three different hypothetic distributions of the mean zonal wind. The solid curves refer to the case $c=0$, so that the undisturbed flow is a rotation as a rigid body. This current is stable, since the vorticity $\bar{\zeta}_a$ increases monotonously with latitude. The dotted curves also correspond to a stable current, for the same reason. The dashed

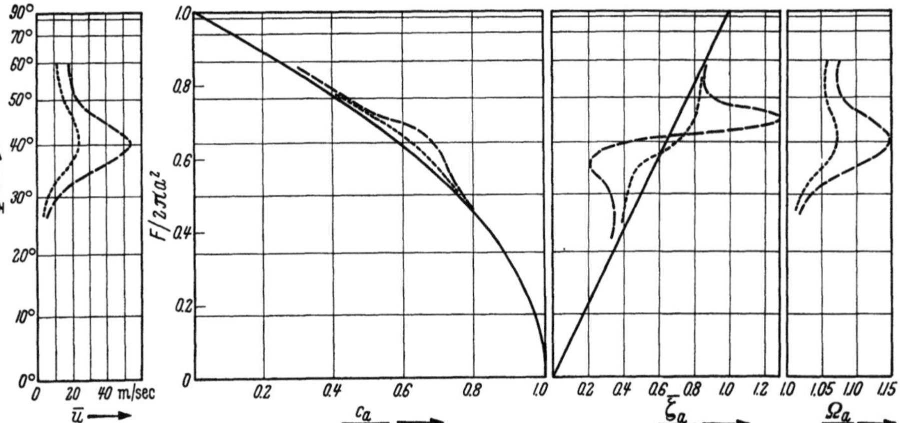

Fig. 14. Stability of zonal currents in case of horizontal, non-divergent motion. Zonal velocity, and normalized absolute circulation, absolute vorticity and absolute angular velocity for 3 different zonal wind distributions.

curves correspond to a concentrated zonal current or jet stream of the type often observed in the upper troposphere (Sect. 46). In this case, $\bar{\zeta}_a$ is a maximum in the vicinity of 45° N, so that a necessary condition for instability is fulfilled. We also note that $\bar{\Omega}_a$ in a large part of the zone where $\partial\bar{\zeta}_a/\partial\Phi > 0$ is larger than anywhere North of 45° N, where $\partial\bar{\zeta}_a/\partial\Phi < 0$. Although this is not a sufficient criterion of instability, it serves to make it plausible that this current is unstable.

Since there is a possibility that narrow, concentrated zonal currents are unstable with respect to horizontal, non-divergent motion, one might think that the observed wave disturbances in the upper westerlies may develop as a result of this kind of instability. Such a possibility would require conversion of kinetic energy of the mean flow into kinetic energy of the perturbations.

The kinetic energy of the mean flow is

$$K_m = \int_{F_1}^{F_2} \frac{1}{2} \bar{u}_a^2 \, dF = \int_{F_1}^{F_2} \frac{c_a^2}{8\pi^2 R^2} \, dF \tag{29.19}$$

and hence

$$\frac{dK_m}{dt} = \int_{F_1}^{F_2} \frac{\bar{\Omega}_a}{2\pi} \frac{\partial c_a}{\partial t} \, dF. \tag{29.20}$$

The rate of change of c_a can be expressed as

$$\frac{\partial c_a}{\partial t} = -2\pi \frac{\partial W_M}{\partial F}, \tag{29.21}$$

where

$$W_M = 2\pi R^2 \overline{u' v'} \tag{29.22}$$

is the northward eddy-flux of angular momentum. Hence

$$\frac{dK_m}{dt} = -\int_{F_1}^{F_2} \bar{\Omega}_a \frac{\partial W_M}{\partial F} dF = \int_{F_1}^{F_2} W_M \frac{\partial \bar{\Omega}_a}{\partial F} dF. \tag{29.23}$$

This equation corresponds to (28.4) in the linear case.

Measurements of the mean meridional eddy-flux of angular momentum (compare Sect. 53) indicate a pronounced northward flux south of the latitude of maximum westerlies, in agreement with the fact that a northeast-southwest tilt of troughs and ridges, corresponding to a positive correlation between u' and v', is predominant in these latitudes. North of the maximum westerlies, on the other hand, only small values of W_M are observed, which are partly directed southward. Therefore, the right-hand side of (29.23) is on the average positive in the atmosphere, indicating that kinetic energy of the perturbations is continually being converted into kinetic energy of the mean current. The perturbations consequently must take their energy from other sources than the kinetic energy of the mean zonal current, and instability of the type discussed in this section cannot be responsible for their growth in general, as has been pointed out by KUO[1] and STARR[2]. This conclusion must only be true in the mean and does not exclude the possibility that *some* growing disturbances (perhaps the fastest growing ones) develop as a result of instability of the type discussed above, taking part of their energy from the kinetic energy of the mean zonal current.

30. ROSSBY-waves. The application of the theory of horizontal, non-divergent motion to large-scale currents in the atmosphere is due to ROSSBY and coll. [*20*].

As a consequence of this theory, the vertical component of the absolute vorticity $\zeta_a = \lambda + \zeta$ must be conserved, so that

$$D(\lambda + \zeta)/Dt = 0. \tag{30.1}$$

With the notation

$$\beta = \frac{1}{a}\frac{\partial \lambda}{\partial \Phi} = \frac{2\Omega \cos \Phi}{a}, \tag{30.2}$$

Eq. (30.1) may also be written

$$\frac{\partial \zeta}{\partial t} = -\boldsymbol{v} \cdot \operatorname{grad} \zeta - \beta v. \tag{30.3}$$

The first term on the right represents the horizontal advection of relative vorticity; the second term is an effect of the earth's curvature which was first discussed by EKMAN[3], who termed it "*the planetary vorticity effect*". Owing to this effect, particles moving northward will lose relative vorticity, and particles moving southward will gain relative vorticity.

ROSSBY and coll. [*20*] used Eq. (30.3) to derive the velocity of propagation of waves of a particularly simple kinematics. Since the most important effect of the earth's curvature is the appearance of the term βv in (30.3), we may now neglect any further effects of the curvature and consider a wave motion in a horizontal plane, satisfying (30.3) with a constant value of β. Let us choose the

[1] H.-L. KUO: Tellus **3**, 268 (1951).
[2] V. STARR: Tellus **5**, 494 (1953).
[3] V. W. EKMAN: Ark. Mat., Astronom. Fys. **2**, No. 11 (1905).

x-axis toward the East and the y-axis toward the North, and consider a motion independent of $y\left(\frac{\partial u}{\partial y}=\frac{\partial v}{\partial y}=0\right)$. Since the divergence is assumed to vanish, we have

$$\frac{\partial u}{\partial x}=0, \quad \text{or} \quad u=u_0=\text{constant}. \tag{30.4}$$

Furthermore,

$$\zeta=\partial v/\partial x, \tag{30.5}$$

and (30.3) becomes

$$\frac{\partial^2 v}{\partial x\,\partial t}+u_0\frac{\partial^2 v}{\partial x^2}+\beta v=0. \tag{30.6}$$

The equation is linear in v and it is not necessary to neglect second order terms. The equation permits solutions of the form

$$v=B\cdot\sin k(x-ct), \tag{30.7}$$

where the velocity of propagation is

$$c=c_R=u_0-\frac{\beta}{k^2}. \tag{30.8}$$

This is ROSSBY's *formula*, which expresses that the waves propagate westward relative to the air at a speed proportional to the square of its wavelength. In particular, the waves are seen to be stationary for a certain wave number k_s defined by

$$k_s=\sqrt{\beta/u_0}. \tag{30.9}$$

Waves with wave lengths greater than the stationary wave length $L_s=2\pi/k_s$ move westward, and shorter waves eastward.

J. BJERKNES[1] discussed in 1937 the distribution of horizontal divergence in horizontal waves in a westerly current as a function of the speed of propagation. His discussion contains in implicite form the effect of ROSSBY's β-term and the concept of the stationary wavelength.

HAURWITZ[2] and ERTEL[3] have generalized ROSSBY's formula (30.8) to the case where the velocities vary with y. Since the velocity field is solenoidal, we may introduce a stream function ψ, defined by

$$u=-\partial\psi/\partial y, \qquad v=\partial\psi/\partial x, \qquad \zeta=\nabla^2\psi. \tag{30.10}$$

Eq. (30.3) then becomes

$$\frac{\partial}{\partial t}\nabla^2\psi-\frac{\partial\psi}{\partial y}\frac{\partial}{\partial x}\nabla^2\psi+\frac{\partial\psi}{\partial x}\frac{\partial}{\partial y}\nabla^2\psi+\beta\frac{\partial\psi}{\partial x}=0. \tag{30.11}$$

This equation has solutions of the form

$$\psi=-Uy+A\cos my\sin k(x-ct), \tag{30.12}$$

where

$$c=U-\frac{\beta}{k^2+m^2}. \tag{30.13}$$

If the solution (30.12) is substituted into (30.11), the second order terms cancel out, so that linearization of the equation is unnecessary. However, if we want to superimpose solutions with different wave lengths, their sum will no longer

[1] J. BJERKNES: Meteor. Z. **54**, 462 (1937).
[2] B. HAURWITZ: J. Mar. Res. **3**, 35 (1940).
[3] H. ERTEL: Ann. Hydrogr., Berlin **68**, 421 (1940).

satisfy the non-linear Eq. (30.11), but will still be a solution of the linearized equation

$$\frac{\partial}{\partial t}\nabla^2\psi' + U_0\frac{\partial}{\partial x}\nabla^2\psi' + \beta\frac{\partial\psi'}{\partial x} = 0, \tag{30.14}$$

with

$$\psi = -U_0 y + \psi'. \tag{30.15}$$

With appropriate choice of length unit, x may be taken to measure longitude in radians along a chosen representative latitude circle. The solution must then be periodic in x with period 2π. A solution which contains the factor $\cos my$, but has an arbitrary variation in the x-direction may be represented as a FOURIER series in x,

$$\psi'(x, y, t) = \cos m y \sum_{k=-\infty}^{+\infty} A_k e^{ik(x-c_k t)}. \tag{30.16}$$

Suppose the initial distribution $\psi'(x, y, 0)$ is known. Then it follows that

$$\psi'(x + U_0 t, y, t) = \psi'(x, y, 0) + \int_0^{2\pi} \psi'(\xi, y, 0)\, I_{m^2}(x-\xi, t)\, d\xi, \tag{30.17}$$

where

$$I_{m^2}(x, t) = \frac{1}{2\pi}\sum_{k=-\infty}^{+\infty}\left(e^{\frac{ik\beta t}{k^2+m^2}} - 1\right) e^{ikx} \tag{30.18}$$

may be called the *influence function*. In (30.17), the stream function may be replaced geostrophically by the height of an isobaric surface. In this manner, (30.17) was used by CHARNEY and ELIASSEN[1] to predict the height changes of the 500 mb surface along the latitude circle 45° N. The formula was found to give best results with $m^2 \approx 10$. This one-dimensional method is very crude, and should only be considered as a very rough first approximation.

31. Shearing instability. It was mentioned in Sect. 28 that a linear current of a homogeneous and incompressible fluid is possibly unstable if the shear of the velocity profile has a maximum somewhere within the layer. Instability of this kind is particularly pronounced if the shear of the current is concentrated in a shear line, which represents a discontinuity in the velocity field. The instability of such a current was discovered by HELMHOLTZ[2].

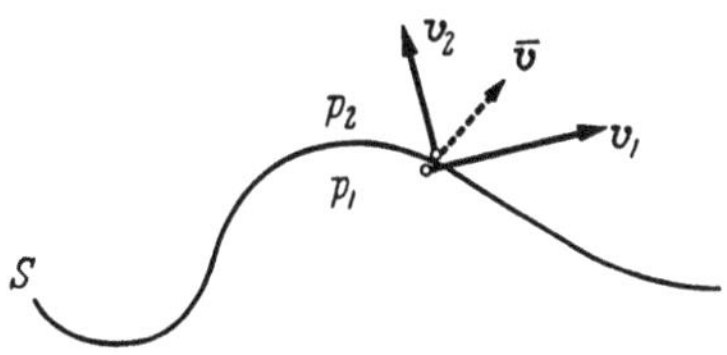

Fig. 15. Velocities and pressure values on both sides of a shear line S.

HÖILAND[3] has derived the shearing instability by adapting the vorticity equation so as to apply to the shearing vorticity on a shear line. Let us consider two-dimensional motion of a homogenous and incompressible fluid in a plane (no CORIOLIS force). Let a curve S denote the shear line, and let $\boldsymbol{v}_1$ and $\boldsymbol{v}_2$ denote the velocity fields and p_1 and p_2 the pressure fields on both sides of S (Fig. 15). The equation of motion applies on both sides, so that

$$\frac{\partial \boldsymbol{v}_1}{\partial t} + \boldsymbol{v}_1\cdot\operatorname{grad}\boldsymbol{v}_1 = -\operatorname{grad}\frac{p_1}{\varrho}, \qquad \frac{\partial \boldsymbol{v}_2}{\partial t} + \boldsymbol{v}_2\cdot\operatorname{grad}\boldsymbol{v}_2 = -\operatorname{grad}\frac{p_2}{\varrho} \tag{31.1}$$

[1] J. G. CHARNEY and A. ELIASSEN: Tellus **1**, No. 2 (1949).
[2] H. v. HELMHOLTZ: Berl. Mber., Apr. **1868**. — Phil. Mag. **36**, 337 (1868). — Wissensch. Abh. I, Leipzig **1882**, 146.
[3] E. HÖILAND: Arch. Math. Naturvidensk. **46**, No. 2 (1942).

hold in any points of S. Subtracting these equations, we obtain

$$\frac{\partial}{\partial t}(\boldsymbol{v}_1-\boldsymbol{v}_2)+\frac{\boldsymbol{v}_1+\boldsymbol{v}_2}{2}\cdot\operatorname{grad}(\boldsymbol{v}_1-\boldsymbol{v}_2)+(\boldsymbol{v}_1-\boldsymbol{v}_2)\cdot\operatorname{grad}\frac{\boldsymbol{v}_1+\boldsymbol{v}_2}{2}=-\operatorname{grad}\frac{p_1-p_2}{\varrho}. \quad (3$$

Let us write

$$\frac{\boldsymbol{v}_1+\boldsymbol{v}_2}{2}=\overline{\boldsymbol{v}}, \tag{31.3}$$

where the field $\overline{\boldsymbol{v}}$ need only be defined in points on S. Moreover, we write

$$\frac{\delta}{\delta t}=\frac{\partial}{\partial t}+\overline{\boldsymbol{v}}\cdot\operatorname{grad} \tag{31.4}$$

so that (31.2) becomes

$$\frac{\delta(\boldsymbol{v}_1-\boldsymbol{v}_2)}{\delta t}+(\boldsymbol{v}_1-\boldsymbol{v}_2)\cdot\operatorname{grad}\overline{\boldsymbol{v}}=-\operatorname{grad}\frac{p_1-p_2}{\varrho}. \tag{31.5}$$

Let us introduce a unit vector $\boldsymbol{t}$ tangential to S, and let the sliding vorticity ζ in a point on S be defined by

$$\zeta=(\boldsymbol{v}_1-\boldsymbol{v}_2)\cdot\boldsymbol{t}, \quad \text{or} \quad \boldsymbol{v}_1-\boldsymbol{v}_2=\zeta\boldsymbol{t}. \tag{31.6}$$

The second equation follows as result of the kinematic boundary condition, which expresses that $\boldsymbol{v}_1-\boldsymbol{v}_2$ is parallel with $\boldsymbol{t}$. From (31.6) we obtain

$$\frac{\delta\zeta}{\delta t}=\boldsymbol{t}\cdot\frac{\delta(\boldsymbol{v}_1-\boldsymbol{v}_2)}{\delta t}+\zeta\boldsymbol{t}\cdot\frac{\delta\boldsymbol{t}}{\delta t}. \tag{31.7}$$

The second term on the right is zero because $\boldsymbol{t}$ is a unit vector. In the first term, we introduce expression (31.5). Noting that

$$\boldsymbol{t}\cdot\operatorname{grad}\frac{p_1-p_2}{\varrho}=0 \tag{31.8}$$

on account of the dynamic boundary condition, we find

$$\frac{\delta\zeta}{\delta t}=-(\boldsymbol{v}_1-\boldsymbol{v}_2)\cdot\operatorname{grad}\overline{\boldsymbol{v}}\cdot\boldsymbol{t}=-\zeta\boldsymbol{t}\cdot\frac{\partial\overline{\boldsymbol{v}}}{\partial s}. \tag{31.9}$$

Let ds denote a line element on S, whose end-points move with the velocity $\overline{\boldsymbol{v}}$; as a consequence, the end-points will not leave the curve. It is easily seen that

$$\frac{1}{ds}\frac{\delta}{\delta t}ds=\boldsymbol{t}\cdot\frac{\partial\overline{\boldsymbol{v}}}{\partial s}. \tag{31.10}$$

Thus it is seen that $\boldsymbol{t}\cdot\frac{\partial\overline{\boldsymbol{v}}}{\partial s}$ is an expression for the one-dimensional divergence in the curve S of the velocity field $\overline{\boldsymbol{v}}$. From the two last equations, we find

$$\frac{\delta}{\delta t}(\zeta\,ds)=0. \tag{31.11}$$

This equation expresses that the total sliding vorticity contained between two points on S which move with the velocity $\overline{\boldsymbol{v}}$ is constant in time. Hence, if the distribution of $\overline{\boldsymbol{v}}$ is such that the two points approach each other, then the mean sliding vorticity between them will increase with time, and vice versa.

Let us assume that there is potential flow on both sides of S, and that the fluid extends to infinity so that there are no vorticities at finite distances except the sliding vorticity on S. Furthermore, let us choose the coordinate system so that $\boldsymbol{v}_1=-\boldsymbol{v}_2$ at great distances from S. Then the vorticities at infinity will give zero velocity everywhere, and it follows from the theory of vortex sheets that the velocity in any point A may be obtained as an integral along S,

$$\boldsymbol{v}_A = \int\limits_S \frac{\boldsymbol{r} \times \boldsymbol{k}}{2\pi r^2} \zeta \, ds, \tag{31.12}$$

where $\boldsymbol{r}$ is the vector from the point A to a point on S. The integral diverges when A is on S, but its principal value (P.V.) exists and is equal to $\overline{\boldsymbol{v}}$,

$$\overline{\boldsymbol{v}} = \text{P. V.} \int\limits_S \frac{\boldsymbol{r} \times \boldsymbol{k}}{2\pi r^2} \zeta \, ds \tag{31.13}$$

so that $\overline{\boldsymbol{v}}$ is determined by the form of the shear line and the distribution of shearing vorticity along it.

In particular, if we consider a shear line which has initially the shape of a sine-curve with a uniform distribution of shearing vorticity ($\zeta = \text{constant}$), it follows from (31.13) that $\overline{\boldsymbol{v}}$ is horizontal in the wave crests and troughs (points A and C in Fig. 16) and that $\overline{\boldsymbol{v}}$ vanishes in the inflection points B and D. Hence it follows that there will be convergence in the vector field $\overline{\boldsymbol{v}}$ along the shear line between A and C, and divergence between C and the next trough point A'.

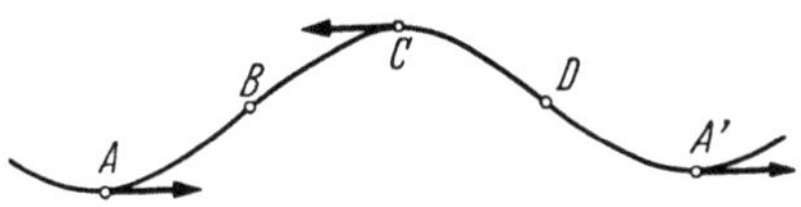

Fig. 16. Mean velocities on a sine-shaped shear line with uniform shearing vorticity.

From this circumstance and Eq. (31.9) or (31.11) it follows that ζ will increase in the vicinity of the inflection point B and decrease in the vicinity of the next inflection point D. As a result of this redistribution of ζ, the initially symmetric waves on the shear line will start to grow and assume an asymmetric form. Eqs. (31.11) and (31.13) show that this development will continue and result in a rolling up of the shear line around the point B, so that the vorticity will ultimately be collected in the surroundings of the point B. The end result is therefore a number of vortices.

If the two fluid layers are bounded by two rigid planes parallel with the undisturbed shear line, the method outlined above may still be used in combination with the method of images. In this manner it is easily seen that the effect of the bounding planes will be to increase the speed with which the shearing vorticity is concentrated near the inflection point B. Therefore, the growth-rate of the waves will be the greater the smaller the distance between the two rigid planes.

Eq. (31.9) may be generalized to the case when the two fluid layers have different densities, and there is a uniform gravity field directed normal to the undisturbed interface. The shearing vorticity will then change not only as a result of one-dimensional convergence in the interface, but also as a result of the solenoids which are concentrated in the interface. If the lower fluid is the heavier one, the latter effect will act to stabilize the system. As a result, waves longer than a certain critical wavelength will be stable, and shorter waves unstable. The perturbation theory gives in this case the wellknown formula for the velocity of propagation:

$$c = \frac{\varrho_1 U_1 + \varrho_2 U_2}{\varrho_1 + \varrho_2} \pm \sqrt{\frac{g}{k} \frac{\varrho_1 - \varrho_2}{\varrho_1 + \varrho_2} - \frac{\varrho_1 \varrho_2}{(\varrho_1 + \varrho_2)^2} (U_1 - U_2)^2}. \tag{31.14}$$

The system considered above is somewhat unrealistic because strict velocity discontinuities do not exist in nature. If the velocity discontinuity is replaced by a transition layer of continuous velocity variation, this will not have any noticeable effect upon waves whose wavelength is considerably longer than the depth of the transition layer; however, waves comparable with, or shorter than the depth of the transition layer will now be stable, as shown by RAYLEIGH[1]. In this case, the amplifying waves will therefore be confined to an intermediate band of wavelengths, whose width will depend upon the differences in velocity and density in the upper and lower layer and the depth of the transition layer.

Billow clouds, which are formed in a layer of high static stability (inversion) and strong vertical wind shear, are generally assumed to be a result of instability of this kind. Contributions to their theory have been given by HAURWITZ[2] and SEKERA[3]. The wavelengths calculated from the theory by HAURWITZ are in good agreement with the observed spacing of the clouds.

d) Wave motions in zonal currents.

32. External waves in a barotropic current. Disregarding the curvature of the earth except for the northward variation of the CORIOLIS parameter, we may operate in a CARTESian system x (eastward) y (northward), and use pressure as the vertical coordinate. We consider a steady, linear current in the x-direction with velocity u_0. Let $z_0(y, p)$ denote the height; then, from (10.3) and (12.3),

$$\frac{\partial z_0}{\partial x} = 0, \quad \frac{\partial z_0}{\partial y} = -\frac{\lambda}{g} u_0, \quad \frac{\partial z_0}{\partial p} = -\frac{\alpha_0}{g}. \tag{32.1}$$

Assuming the current to be barotropic, it follows from the thermal wind equation that u_0 does not vary with p; and we shall also assume u_0 to be horizontally uniform.

Superimposed upon this current, we may clearly have internal inertial waves, and, if the stratification is stable, also internal gravitational waves. However, we shall in the following disregard these wave types and consider a special type of wave motion in which the particles move horizontally and the horizontal pressure force per unit mass is independent of height. Strictly speaking, this requires an autobarotropic fluid; however, the motion will be only slightly modified by an internal static stability.

Let us consider a disturbed motion defined by the velocity field u_0+u, v, and the height z_0+z; since the air is autobarotropic, the specific volume will remain unchanged in any isobaric surface, so that the perturbation of α_0 will vanish.

Writing the equations of motion for the disturbed motion, simplifying by means of (32.1), and neglecting second order terms in the perturbations, we obtain

$$\frac{\partial u}{\partial t} + u_0 \frac{\partial u}{\partial x} = -g \frac{\partial z}{\partial x} + \lambda v, \tag{32.2}$$

$$\frac{\partial v}{\partial t} + u_0 \frac{\partial v}{\partial x} = -g \frac{\partial z}{\partial y} - \lambda u. \tag{32.3}$$

Likewise, the quasistatic equation gives

$$\partial z / \partial p = 0, \tag{32.4}$$

showing that all isobaric surfaces have the same shape. If (32.2) and (32.3) are differentiated with respect to p and $\partial z/\partial p$ is eliminated by means of (32.4), we find that u and v must either be independent of height, or all particles must during the wave motion oscillate with the inertial frequency λ. We are here

[1] Lord RAYLEIGH: Sci. Papers **1**, 474 (1880).
[2] B. HAURWITZ: Gerlands Beitr. Geophys. **37**, 16 (1932).
[3] Z. ZEKERA: J. Meteor. **5**, 93 (1948).

interested in the first possibility and put

$$\partial u/\partial p = \partial v/\partial p = 0. \tag{32.5}$$

The equation of continuity (10.9), with the boundary condition $\omega = 0$ when $p = 0$, can then be integrated to

$$\omega = -p \operatorname{div} \boldsymbol{v}. \tag{32.6}$$

Finally we have to consider the boundary condition at the ground, which is assumed to be level. Thus we have, at $p = p_s$ (surface pressure),

$$w = \frac{\partial (z_0 + z)}{\partial t} + (u_0 + u)\frac{\partial (z_0 + z)}{\partial x} + v\frac{\partial (z_0 + z)}{\partial y} + \omega\frac{\partial (z_0 + z)}{\partial p} = 0. \tag{32.7}$$

Neglecting second order terms and applying (32.1) and (32.4), we obtain

$$\frac{\partial z}{\partial t} + u_0\frac{\partial z}{\partial x} - \frac{\lambda u_0}{g}v - \frac{\alpha_0}{g}\omega = 0 \quad \text{at} \quad p = p_s. \tag{32.8}$$

Elimination of ω between (32.6) and (32.8) gives

$$\frac{\partial z}{\partial t} + u_0\frac{\partial z}{\partial x} - \frac{\lambda u_0}{g}v = -H \operatorname{div} \boldsymbol{v} = -H\left(\frac{\partial u}{\partial x} + \frac{\partial v}{\partial y}\right), \tag{32.9}$$

where we have introduced

$$H = \frac{p_s \alpha_{0s}}{g} = \frac{R T_{0s}}{g}, \tag{32.10}$$

subscript s denoting surface values. It will be seen that H represents the height of a homogeneous fluid layer of density α_{0s}^{-1} and surface pressure p_s. Since the percentage variation of surface temperature is small, H may be treated as a constant[1]. If we insert average atmospheric values in (32.10), we find $H \approx 8000$ m.

The derivation of (32.9) shows that this equation holds at the surface; however, since z, u and v do not vary with height, (32.9) must be valid at any level. In an isothermal (and adiabatic) atmosphere, this would imply that the motion is strictly horizontal at all levels, whereas if the temperature varies with height, small vertical motions will take place at higher levels. Eq. (32.9) is the *linearized tendency equation for barotropic flow*.

Eqs. (32.2), (32.3) and (32.9) form a complete system in the three unknowns u, v and z. By eliminating z from the first two equations, we obtain the *linearized vorticity equation*

$$\left(\frac{\partial}{\partial t} + u_0\frac{\partial}{\partial x}\right)\left(\frac{\partial v}{\partial x} - \frac{\partial u}{\partial y}\right) = -\lambda\left(\frac{\partial u}{\partial x} + \frac{\partial v}{\partial y}\right) - \beta v. \tag{32.11}$$

We shall consider sine waves of wave number k propagated at the velocity c in the x-direction. Then we may write

$$\frac{\partial}{\partial x} = i k, \quad \frac{\partial}{\partial t} = -i k c. \tag{32.12}$$

Furthermore, we shall for simplicity make the assumption that

$$\partial u/\partial y = \partial v/\partial y = 0, \tag{32.13}$$

whereas z may vary with y. This assumption is admittedly inconsistent, but the error will be small. The vorticity and the divergence may then be written

$$\zeta = \frac{\partial v}{\partial x} - \frac{\partial u}{\partial y} = i k v, \quad \operatorname{div} \boldsymbol{v} = \frac{\partial u}{\partial x} + \frac{\partial v}{\partial y} = i k u \tag{32.14}$$

[1] In physics of the upper atmosphere, H is called *scale height*, as proposed by S. CHAPMAN.

and the vorticity Eq. (32.11) assumes the form

$$\left(u_0 - c - \frac{\beta}{k^2}\right) k^2 v = \lambda \operatorname{div} \boldsymbol{v} = \lambda\, i\, k\, u. \tag{32.15}$$

We notice that if we assume the horizontal flow to be non-divergent, then we refind ROSSBY's formula (30.8). However, if we do not make this assumption, we may eliminate z and u between the three Eqs. (32.2), (32.9) and (32.15). This gives

$$[g H - (u_0 - c)^2]\left(u_0 - c - \frac{\beta}{k^2}\right) = \frac{\lambda^2 c}{k^2} \tag{32.16}$$

which is an equation of the third degree in c. This equation has two large roots which are approximately given by

$$c \approx u_0 \pm \sqrt{g H}\,. \tag{32.17}$$

Recalling the definition of H (32.10), $\sqrt{g H}$ is seen to be comparable with the velocity of sound. In fact, these fast-moving waves are predominantly sound waves, slightly modified by gravity. Waves of this type do exist in the atmosphere; thus for instance, the small-amplitude tidal waves which are clearly shown on barograms from tropical or subtropical stations, are forced waves of this kind. On the other hand, it is quite apparent that these fast-moving waves do not carry any weather phenomena, and that their pressure amplitude is insignificant compared with the amplitude of the slow-moving pressure systems shown on weather maps. From the point of view of weather forecasting, we may therefore say that the fast moving waves are uninteresting solutions; they will be discussed in the chapter on atmospheric tides by W. KERTZ in this Volume.

An approximation for the third root of (32.16), which is much smaller, is found by neglecting $(u_0 - c)^2$ versus $g H$; this gives

$$c \approx \frac{u_0 - \dfrac{\beta}{k^2}}{1 + \dfrac{\lambda^2}{g H k^2}}\,. \tag{32.18}$$

This root agrees with the observed speed of waves in the westerlies in order of magnitude and is therefore the only solution which is of significance for weather forecasting. However, the solution can only serve as a rough approximation owing to the many crude assumptions made.

Eq. (32.18) was developed by ROSSBY and coll. [*20*] for an incompressible atmosphere with a free surface, and by HOLMBOE, FORSYTHE and GUSTIN [*11*] for a compressible, barotropic atmosphere. It may be considered as an improved form of the ROSSBY-formula (30.8), but the difference is very slight for short and moderate wavelengths (shorter than about 6000 km). The two formulae give the same value for the stationary wavelength.

It is noteworthy that (32.18) could be derived by replacing (32.2) by the corresponding geostrôphic relationship

$$- g \frac{\partial z}{\partial x} + \lambda v = 0 \tag{32.19}$$

and then eliminating z and u between (32.19), (32.9) and (32.15). This procedure gives (32.18) as the only solution; i.e. by introducing the geostrophic approximation at this point we automatically eliminate the solutions which are meteorologically uninteresting, whereas the significant solution is kept. This circumstance, which has been pointed out by CHARNEY [*5*], will be referred to later.

33. Waves in a baroclinic current. A next step towards reality is to consider the zonal current to be baroclinic, as the atmospheric westerlies are, with the temperature in isobaric surfaces decreasing with y (northward), and hence the wind speed increasing with height according to the thermal wind Eq. (12.4). Again, we assume that there is no horizontal (or alternatively, isobaric) wind shear.

Considering a small-amplitude sine wave perturbation of wave number k, which at all levels moves with the same speed c, and assuming again that the perturbation velocities do not vary with y, we find that the linearized vorticity equation has the same form (32.15) as in the barotropic case. The only difference is that both sides of this equation will now vary with height, whereas they were independent of height for a barotropic current. This equation was used by STARR[1] and by BJERKNES and HOLMBOE[2] to discuss the distribution of the horizontal divergence. One finds that ahead of a pressure trough, where v is ordinarily positive at all levels (provided the phase does not shift too much with height), div $\boldsymbol{v}$ will increase with height in the troposhere because u_0 increases with height.

CHARNEY [5] obtained a generalization of the ROSSBY-formula (30.8) and the formula (32.18) to the case of a baroclinic current. From the continuity equation, one has

$$\omega_s = -\int_0^{p_s} \operatorname{div} \boldsymbol{v}\, dp, \tag{33.1}$$

where subscript s denotes surface values; and from the boundary condition at the ground in the form (32.8), one obtains by using the geostrophic approximation (32.19),

$$\omega_s = -\frac{\lambda}{\alpha_{0s}} c\, v_s = -\frac{\lambda p_s}{g H} c\, v_s, \tag{33.2}$$

where H is defined by (32.10). Elimination of div $\boldsymbol{v}$ and ω_s between these two equations and the vorticity Eq. (32.15) gives

$$\overline{u_0 v} - c\,\bar{v} - \frac{\beta}{k^2}\bar{v} = \frac{\lambda^2}{g H k^2} c\, v_s, \tag{33.3}$$

where a bar denotes average with respect to pressure. This formula can be used to determine the wave speed c if the distribution of v with height is known. This is therefore not a complete solution of the problem, in which the distribution of v would follow as one of the results. We may, however, utilize our empirical knowledge of the distribution of v with height. For instance, if to a first, rough approximation we put $v = \text{const}$, (33.3) gives

$$\bar{u}_0 - c - \frac{\beta}{k^2} = \frac{\lambda^2}{g H k^2} c \tag{33.4}$$

which is identic with (32.18) except that the constant velocity u_0 of the barotropic current is now replaced by the pressure-average of u_0 in the baroclinic current. If we neglect the right-hand term, which is a small correction, we obtain a corresponding generalization of the ROSSBY-formula to baroclinic currents.

According to observations, the meridional velocity component will change with height both in amplitude and phase; and ordinarily one will observe an increase of amplitude with height in the troposphere. This will further reduce the importance of the right-hand member of (33.3), because $v_s < \bar{v}$. The change in phase of v with height implies that the wave amplitude will change with time; in particular, waves whose phase-lines slope westward will amplify (compare Sect. 28).

[1] V. STARR: Basic Principles of Weather Forecasting. New York and London: Harper Brs. 1942.

[2] J. BJERKNES and J. HOLMBOE: J. Meteor. **1**, 1 (1944).

The relatively small importance of the right-hand member of (33.3) expresses a highly significant fact, namely, that the divergence at different levels will have a tendency to compensate each other, so that *the pressure-averaged divergence is small compared with the values which the divergence may have at any particular level.* Therefore, if div $\boldsymbol{v}$ is not unusually small at all levels, as in the barotropic wave, there must be at least one level on every vertical line where div $\boldsymbol{v}=0$.

Let us consider for instance the condition along a vertical line to the east of a pressure trough. Assuming that v is positive at all levels and does not vary too fast with height, the vorticity Eq. (32.15) shows that div $\boldsymbol{v}$ will increase with height throughout the troposphere, where u_0 increases with height. Therefore if the pressure-average of div $\boldsymbol{v}$ shall be nearly zero, c must be such that div $\boldsymbol{v}>0$ in the upper troposphere and div $\boldsymbol{v}<0$ in the lower troposphere, whereas div $\boldsymbol{v}=0$ at a certain level in the mid-troposphere. With this distribution of div $\boldsymbol{v}$ we may infer from the continuity equation that $\omega<0$ (ascending motion) at most levels, with a negative maximum at the level of non-divergence. For a point in the rear (West) of the trough, one finds the opposite distribution of div $\boldsymbol{v}$, and a subsiding motion. This distribution of vertical motion agrees with the observed fact that there are more clouds and precipitation ahead of a trough than in its rear. In an amplifying wave, whose phase-lines tilt westward with height, the region of low-level convergence and ascending motion will be displaced somewhat towards the West so that it also contains the central part of the trough; this gives even better agreement with the observed distribution of cloudiness and precipitation.

The above considerations are incomplete insofar as the thermal energy equation, (conservation of entropy) which represents a linkage between the motion at different levels, has not been used. A very simple but crude theory for waves in a baroclinic current is obtained by introducing the advective assumption that the entropy changes are a result of horizontal advection only. When p is used as vertical coordinate, this implies that the local change of temperature, or specific volume, is caused solely by advection. Thus we may write

$$\frac{\partial(\alpha_0+\alpha)}{\partial t}+(u_0+u)\frac{\partial(\alpha_0+\alpha)}{\partial x}+v\frac{\partial(\alpha_0+\alpha)}{\partial y}=0. \tag{33.5}$$

Omitting second order terms and using the hydrostatic and geostrophic relations, this equation may be written

$$(u_0-c)\frac{\partial v}{\partial p}-\frac{\partial(u_0-c)}{\partial p}v=0, \tag{33.6}$$

showing that v varies with height in proportion to (u_0-c). Therefore, if the right-hand member of (33.3) is disregarded, we may on the left-hand side replace v by u_0-c, and obtain

$$\overline{\left(u_0-c-\frac{\beta}{k^2}\right)(u_0-c)}=0 \tag{33.7}$$

which is a quadratic equation in c, with the solution

$$c=\bar{u}_0-\frac{1}{2}\frac{\beta}{k^2}\pm\sqrt{\frac{\beta^2}{4k^4}-\left(\overline{u_0^2}-\bar{u}_0^2\right)}\,. \tag{33.8}$$

This formula was derived by FJÖRTOFT[1] and SUTCLIFFE[2]. The radicand vanishes for waves whose wavelength is

$$L_1=2\pi\sqrt{\frac{2}{\beta}}\sqrt[4]{\overline{u_0^2}-\bar{u}_0^2}\,. \tag{33.9}$$

[1] R. FJÖRTOFT: Comp. of Meteor., Amer. Meteor. Soc. **1951**, 454.

[2] R. C. SUTCLIFFE: Quart. J. Roy. Meteor. Soc. **77**, 226 (1951).

This quantity can be calculated if u_0 is known at all levels; clearly, L_1 will increase with increasing vertical wind shear. For vertical wind profiles of the type observed in the westerlies we find that L_1 is of the order of several thousand kilometers, even for a moderate vertical wind shear.

For waves whose wave length is smaller than L_1, (33.8) gives two complex roots of c, one of which gives an amplifying and the other a dampening wave. For waves longer than L_1, (33.8) gives two real roots, corresponding to stable waves. For large wavelengths, one of the roots approaches the value given by the ROSSBY formula, whereas the other is closer to u_0.

The theory outlined above, although resting on the crude advective assumption of temperature changes, is nevertheless capable of disclosing the main features of the mechanism which causes certain waves in baroclinic currents to grow. This instability is different from the types of instability which have been previously considered. Analysis of the fields of divergence and vertical motion by means of the vorticity equation shows that in the amplifying wave, density and vertical velocity are negatively correlated, so that on the average, cold air sinks and warm air ascends. The corresponding decrease of potential energy is the main energy source for the growth of perturbations; a small portion of the perturbation energy is also taken from the kinetic energy of the zonal current, in accordance with the fact that the phase lines of the amplifying waves tilt somewhat Westward with height [compare Eq. (28.4)].

In the advective theory, the effect of the static stability is lost; as a consequence, the theory over-emphasizes the instability for short wavelengths, for which the static stability has the greatest effect. Thus, according to (33.8), the growth-rate of the waves increases indefinitely as the wavelength tends toward zero, what is clearly not realistic. This difficulty is removed when the effect of vertical motion upon the temperature changes is taken into account. Such a non-advective theory was first developed by CHARNEY [*5*]; another version is given by EADY[1]; and FJÖRTOFT [*8*] has derived the instability from energy considerations. Other contributions to the theory have been given by BERSON[2] and KUO[3]. The theory is rather lengthy and will not be presented here; there are also a number of details which are as yet unsettled.

The perturbed motion consists primarily in a quasi-horizontal overturning in planes sloping upward toward the North. In the case of amplifying waves, the slope of these planes is less than the slope of the isentropic surfaces, so that the air particles which are displaced northward (southward) as a result of the wave motion are warmer (colder) than the undisturbed air in the same latitude. Since the warm air moving northward is ascending slightly, and the cold air moving southward is descending slightly, the potential (and internal) energy must decrease during the development of the wave; and the energy thus released is used to speed pu the perturbation velocities.

According to CHARNEY [*5*], amplification will take place provided the wave length is shorter than a certain critical value L_1, which increases with vertical wind shear as shown in Fig. 17. So far, the result is in accordance with the results of the advective theory; even the quantitative dependency of L_1 upon vertical wind shear agrees well with Eq. (33.9).

According to EADY and FJÖRTOFT, sufficiently short waves are again stable, so that amplification is confined to an intermediate band of wavelengths. Fig. 18 shows the critical curve according to FJÖRTOFT [*8*]. The correct extent of the unstable area in this diagram is still somewhat in doubt, but it seems quite certain that the growth-rate, for a given vertical shear, has a maximum for waves of a

[1] E. T. EADY: Tellus **1**, No. 3 (1949).
[2] F. A. BERSON: Ark. Geofys., Stockh. **1**, 187 (1950).
[3] H.-L. KUO: J. Meteor. **9**, 260 (1952); **10**, 236 (1953).

certain wavelength $L_{\max}$ of the order 3000 km. For perturbations formed as a result of baroclinic instability of this kind, one would expect the bulk of the energy

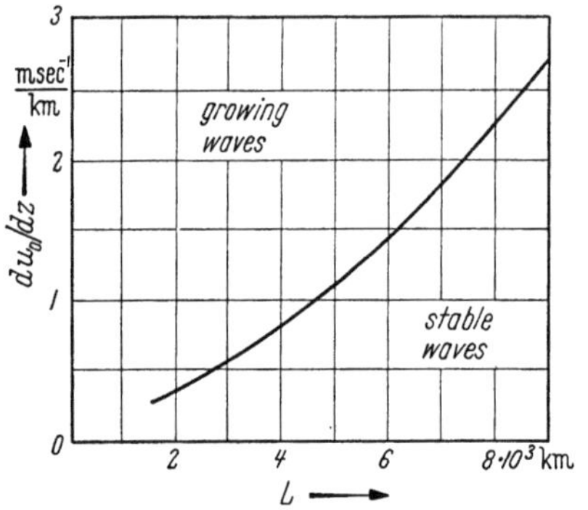

Fig. 17. The critical stability curve after J. G. CHARNEY [5].

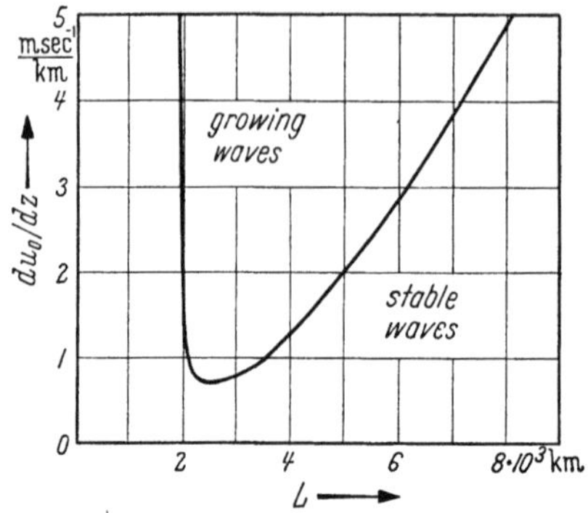

Fig. 18. The critical stability curve according to R. FJÖRTOFT [8].

to appear in wavelengths in the vicinity of $L_{\max}$. As pointed out by FJÖRTOFT [8], $L_{\max}$ depends upon latitude and static stability according to the formula

$$L_{\max} \sim \nu_s/\lambda. \tag{33.10}$$

BREISTEIN and PARRY[1] measured the wavelengths of 97 wave disturbances in the westerlies, and found that they had a positive correlation of about 0.82 with the

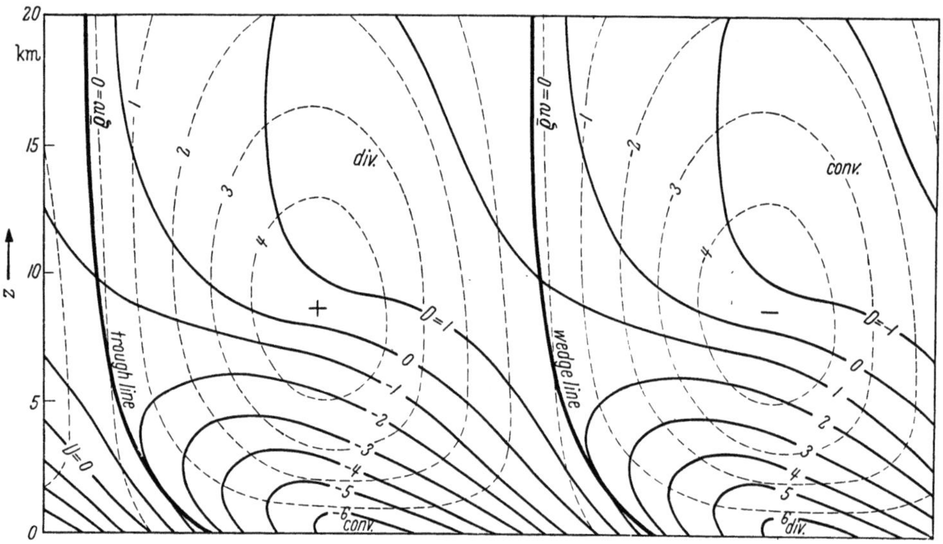

Fig. 19. Theoretical distribution of horizontal mass divergence (light solid lines, unit 10^{-8} gr cm^{-3} sec^{-1}) and vertical momentum (dashed lines, unit 10^{-3} gr cm^{-2} sec^{-1}) in the unstable baroclinic wave (after J. G. CHARNEY).

corresponding values of ν_s/λ, suitably averaged[2]. This result seems to support the hypothesis that baroclinic instability is a factor of importance for the formation of disturbances in the westerlies.

[1] P. M. BREISTEIN and H. D. PARRY: Univ. of Chicago, Techn. Rep. No. 2, Contract No. AF 19 (604)—390, 1954.

[2] A second set of data comprising 51 cases gave a correlation coefficient of only 0.45; however, these data had very little spread in latitude.

The computed growth-rate depends of course on the wave length and the vertical wind shear. For typical atmospheric values one obtains a doubling time for the amplitude which is of the order 24 hours, in reasonable agreement with observed doubling times for atmospheric disturbances. Also the theoretically determined kinematics of growing waves agrees in its broad features with the observed field of motion in growing wave-disturbances. Fig. 19, which is taken from CHARNEY [5], shows in a xz-plane the westward tilt of the trough and wedge lines and the distribution of horizontal mass divergence and vertical momentum.

The most important restriction in the theory of waves in zonal currents outlined above is probably the assumption that the velocity of the current is horizontally uniform. In the atmospheric westerlies, there is normally a considerable variation of wind with latitude, particularly in the upper troposphere where a pronounced wind maximum (the "jet stream") is found. A theory which takes into account also this property of the current has not yet been developed systematically, but its stability properties have been discussed by CHARNEY[1].

It was stated in Sect. 29 that the observed eddy flux of angular momentum is such that perturbation energy is converted into kinetic energy of the mean zonal current. It is clear therefore, that on the average, the effect of the horizontal shear of the mean zonal current is to reduce the wave growth caused by baroclinic instability.

34. Waves on a frontal surface. When the polar front had been discovered by the Norwegian school about 1920, V. BJERKNES and his collaborators put forward the hypothesis that the extratropical cyclones are amplifying wave motions which develop spontaneously on the frontal surface as a result of some kind of instability connected with it. H. SOLBERG[2] investigated theoretically the possible wave types in a system consisting of two statically stable, barotropic layers of different density moving zonally at different speeds and separated by a sloping interface (compare MARGULES' formula (12.8)). It would have been desirable in this system to introduce a horizontal rigid boundary to simulate the surface of the earth; but so far, this has not been possible due to the mathematical difficulties which such a boundary causes. Instead, SOLBERG introduced two rigid bounding planes parallel with the undisturbed frontal surface, or alternatively, considered the two fluid layers to be of infinite extent.

SOLBERG found that two types of amplifying waves are possible in this system. The first type is of very short wavelength, up to a few kilometers, and has clearly nothing to do with cyclones. The second type is found in a band of wavelengths of the order 1000 to 2000 km which is in agreement with the dimensions of cyclones; also, these waves have certain kinematical features in common with growing cyclones. This wave type therefore represents one possible explanation for the growth of cyclones.

J. BJERKNES and GODSKE[3,4] have given a physical interpretation of some of the wave types found by SOLBERG, which may be expressed briefly as follows. The growth of very short unstable waves is due to the shearing instability connected with the wind discontinuity at the frontal surface. The shearing instability will for sufficiently short wavelengths dominate against the static stability, so that the disturbances will amplify. However, as the wavelength increases, the static stability will become relatively more important, and from a certain wavelength on (of a few kilometers magnitude), the waves will be stable. So far, the earth's rotation has no noticeable effect; the fluid particles will move in orbits whose planes are nearly vertical, and the system behaves in agreement with Eq. (31.14). As the wavelength is increased further, the earth's rotation will have the effect to turn the

[1] J. G. CHARNEY: Int. Union Geodesy and Geophysics, 9ième Assemblée, Bruxelles 1951.
[2] H. SOLBERG: Geof. Publ. **5**, No. 9 (1928).
[3] J. BJERKNES and C. L. GODSKE: Astrophys. Norvegica **1**, No. 6 (1936).
[4] C. L. GODSKE: Meteor. Z. **53**, 445 (1936).

orbital planes towards the horizontal, so that the static stability becomes less effective; and from a certain wave length on, the role of the static stability has been thus reduced so much that the shearing instability again dominates, and the waves become unstable. These are the cyclone-like long unstable waves, with quasi-horizontal particle orbits. For still longer wavelengths, the inertial stability due to the earth's rotation will dominate, and the waves are again stable.

Thus, according to J. BJERKNES and GODSKE, the growth of short and long unstable waves are both attributed to the shearing instability. As for the long unstable waves, this interpretation has been questioned by HÖILAND[1] who has shown that long unstable waves of the same nature as those found by SOLBERG may appear as a result of an interaction between the shear and the internal static (or inertial) stability of the two layers. Thus, even on a resting earth, long unstable waves, with vertical orbital planes will appear in a system consisting of two layers separated by a horizontal statically stable shearing interface, provided the two layers possess internal static stability. If the two layers are autobarotropic, then no such long unstable waves exist. This indicates that the long unstable waves are not a simple manifestation of the shearing instability, but the result of a process of a more complex nature.

SOLBERG's long unstable waves are found to take their energy from the kinetic energy of the mean zonal motion whereas the potential energy of the system is increased during their growth. Since it is known that the perturbations in the westerlies on the average lose energy to the kinetic energy of the mean zonal flow, it does not seem possible that SOLBERG's type of instability is responsible for the growth of *all* kinds of disturbances in the westerlies. This does not exclude the possibility, however, that certain types of frontal waves of relatively short wavelength may develop in accordance with SOLBERG's theory.

III. Non-linear prognostic equations for the large-scale motions.

a) Filtering approximations.

35. Properties of the hydrodynamic equations in "primitive" form. The small-amplitude solutions considered in Chap. II are very useful in the study of certain stable wave motions such as gravity waves, and also for the study of stability properties and energy transformations. If it is assumed, however, that atmospheric disturbances such as cyclones develop as a result of baroclinic instability of zonal currents, then it follows that the waves must continue to amplify as long as the linear theory is valid, i.e. as long as the amplitude is sufficiently small. Not before the amplitude has grown so large that the linear theory does not hold any more is it possible for the growth to come to an end. Thus it follows from the linearized theory that this theory is itself inapplicable to disturbances in a *mature* stage. If the instability of the current is maintained, then the current will remain in a fully disturbed state with finite amplitude disturbances. This is actually what we observe in the atmospheric westerlies; and the fact that the perturbation theory does not apply very well to mature disturbances such as cyclones is therefore in agreement with what is to be expected from the instability theory.

Unfortunately, the general solution of the non-linear equations cannot be expressed as a superposition of eigensolutions, as in the linear case, and we are therefore unable to discuss the properties of the general solution analytically. What we can do, however, is to solve the non-linear initial value problem by numerical or graphical methods. A weather forecasting based on such a procedure

[1] E. HÖILAND: Arch. Math. Naturvidensk. **47**, No. 3 (1943); **50**, No. 3 (1948).

was first proposed by V. BJERKNES[1]. This method has in recent years become known as "*numerical forecasting*", a name which tends to emphasize the great numerical work involved. A more adequate name would be "*physical forecasting*", since the method is based upon the governing physical equations rather than upon extrapolation or statistics.

The first attempt on a physical forecast was made by L. F. RICHARDSON [*19*]. He devised a method by which the equations could be integrated numerically by replacing the space and time differentials by finite differences. He used the quasi-static system of equations (compare Sect. 9) which may be written in the form

$$\frac{\partial u}{\partial t} = -u\frac{\partial u}{\partial x} - v\frac{\partial u}{\partial y} - w\frac{\partial u}{\partial z} - \alpha\frac{\partial p}{\partial x} + \lambda v, \tag{35.1}$$

$$\frac{\partial v}{\partial t} = -u\frac{\partial v}{\partial x} - v\frac{\partial v}{\partial y} - w\frac{\partial v}{\partial z} - \alpha\frac{\partial p}{\partial y} - \lambda u, \tag{35.2}$$

$$\frac{\partial p}{\partial t} = g\varrho w - g\int_z^\infty \mathrm{div}_h(\varrho \boldsymbol{v})\,dz, \tag{35.3}$$

where ϱ is given by the hydro-static equation and w by RICHARDSON'S Eq. (9.18). Strictly speaking, the system used by RICHARDSON was more complicated; he took into account the earth's curvature, heat sources due to radiation, and turbulent friction, etc. However, the main properties of his equations were the same as those of the simplified system written above.

In principle, RICHARDSON'S procedure was the following: If the dependent variables u, v and p are known at a certain instant in a three-dimensional grid of points, ϱ and w can be calculated from (9.3) and (9.18) by replacing the differentials by the finite differences between the grid points and using a trapezoidal rule for evaluating the integrals. Using similar finite difference approximations in the three equations above, these equations will determine the time increments of u, v and p in the gridpoints during a short interval of time. By iterating the procedure, a step-wise integration of the equations in time is attained.

When RICHARDSON carried out the calculations with initial conditions taken from an actual weather situation, he found that the computed tendencies were utterly misleading; thus he obtained calculated surface pressure tendencies of the order 70 mb/3 hours, whereas only little change in the pressure field was actually observed.

The poor result obtained by RICHARDSON is due to an oversensitivity of the primitive Eqs. (35.1) to (35.3). Already MARGULES[2] pointed out that the tendency Eq. (35.3) is unsuitable for calculating the pressure tendency, because this would require an accuracy of at least $10^{-7}\,\mathrm{sec}^{-1}$ in the determination of $\mathrm{div}_h\,\boldsymbol{v}$, which is not attainable from wind observations. A similar difficulty, although not quite as marked, is met with when the equations of motion (35.1) and (35.2) are used for the determination of $\partial u/\partial t$ and $\partial v/\partial t$; the resultant force on the right-hand side is a small difference between two large forces, viz. the pressure force and the CORIOLIS force, and is therefore difficult to measure accurately.

It was pointed out by CHARNEY[3] that the over-sensitivity of the primitive equations is related to the fact that these equations are very general; they do not only possess solutions representing large-scale motions of the type observed on weather maps, but have also other solutions of different character, notably internal

[1] V. BJERKNES: Meteor. Z. **21**, 1 (1904).
[2] M. MARGULES: Boltzmann-Festschrift, 585, Leipzig 1904.
[3] J. G. CHARNEY: Geof. Publ. **17**, No. 2 (1948).

and external gravitational-sound-inertial waves and oscillations. These latter types of motion do not correspond to the motion patterns observed on weather maps and are therefore uninteresting from the point of view of weather forecasting; they were termed "*meteorological noise*" by CHARNEY. In general, the solution of the primitive equations will be a mixed type where both meteorologically significant motions and meteorological noise will be present. However, the oversensitivity of the equation applies to the noise component only; in other words, a small error in the initial conditions will affect the meteorologically significant part of the solution only slightly, but will give a large error in amplitude and phase of the computed noise-waves.

Although noise-waves may exist in the atmosphere, their amplitude must be very small, since they do not show up on weather maps or barograms[1]. This requires a rather delicate balance between the pressure and velocity fields. As a result of unavoidable errors in observation and analysis, this balance is not present in our initial data, and the computed solution will therefore contain fictitious noise-waves of a considerable amplitude. These noise-waves will give rapid oscillations of the calculated pressure and velocity changes and thus cause large fictitious tendencies which will completely blur the slower changes connected with the meteorologically significant motions. The poor result obtained by RICHARDSON was due to the occurrence of fictitious noise-waves of this kind.

In the case of wave motions in a barotropic atmosphere (Sect. 32), the noise-waves are represented by the two large roots (32.17) of the frequency Eq. (32.16), whereas the small root (32.18) represents the "meteorologically significant motion". A study of the role of the noise waves in this case has been made by HINKELMANN[2].

In order to overcome the difficulty caused by the presence of the "meteorological noise", so that the behaviour of motions which carry weather phenomena can be calculated, one of the following three methods can conceivably be used:

1. One may calculate the motion with the fictitious noise waves included, and afterwards perform some kind of smoothing upon the solution in order to eliminate the noise waves. This will require a very accurate calculation of the unsmoothed motion in all its details, and will therefore require much labor. The method is not considered as feasible, partly because some of the noise waves have too low frequencies to be eliminated by smoothing, and partly because the large-amplitude fictitious noise-waves will interact with the smoothed motion as a result of the non-linearity of the equations.

2. Before the time-integration is performed, the initial values of pressure and wind may be adjusted to each other in such a manner that no noise waves of appreciable amplitude will occur in the solution. A test of this method applied to a barotropic model atmosphere carried out in Princeton indicates that the method is feasible[3].

3. The third method consists in changing the prognostic equations so that they are no longer satisfied by noise-waves, whereas the meteorologically significant solutions are being kept with sufficient accuracy. The approximations which will have to be made in the equations are referred to as "*filtering approximations*", since their effect is to filter out the noise from the solution. The prognostic equations arrived at in this manner will have as their only solutions the meteorologically significant motions in pure form. This method has played an important role in dynamic meteorology in recent years.

[1] Except forced tidal waves.

[2] K. HINKELMANN: Tellus 3, 285 (1951).

[3] Personal communication from J. G. CHARNEY.

36. The quasi-geostrophic equations. *α) Derivation of the differential equations.* It is noteworthy that the large-scale motions in the atmosphere, of the type which is brought out by the weather maps, is characterized by:

1. The wind field is approximately geostrophic, and

2. the horizontal (or isobaric) divergence at any level is considerably smaller (about 10 times) than the relative vertical vorticity.

These two conditions are interrelated, because the geostrophic wind field itself is almost solenoidal [compare Eq. (13.8)].

The quasi-geostrophic method is designed to take advantage of the quasi-balance between horizontal pressure gradient and wind field, in much the same way as the quasi-static method utilizes the quasi-balance between the vertical pressure gradient and gravity.

Assuming the wind field to be nearly geostrophic, we would expect to obtain an approximate value of the geostrophic departure $(\boldsymbol{v}_h - \boldsymbol{v}_g)$ by replacing the wind by the geostrophic wind in the acceleration term of the equation of motion (13.3). Thus we obtain the quasi-geostrophic wind formula

$$\boldsymbol{v}_h = \boldsymbol{v}_g + \frac{1}{\lambda}\boldsymbol{k} \times \left(\left(\frac{\partial \boldsymbol{v}_g}{\partial t}\right)_p + \boldsymbol{v}_g \cdot \operatorname{grad}_p \boldsymbol{v}_g\right). \tag{36.1}$$

This method was introduced by HESSELBERG[1] and has been used by BRUNT and DOUGLAS[2], H. PHILIPPS[3], ERTEL[4] and VAN MIEGHEM[5].

Eq. (36.1) permits an approximate calculation of the work of the horizontal pressure force, and of the horizontal (or isobaric) divergence, both of which depend upon the geostrophic deviation. Thus we obtain

$$-\boldsymbol{v}_h \cdot \alpha \operatorname{grad}_h p = \frac{\partial}{\partial t}\frac{1}{2}\boldsymbol{v}_g^2 + \boldsymbol{v}_g \cdot \operatorname{grad}\frac{1}{2}\boldsymbol{v}_g^2, \tag{36.2}$$

showing that the work of the horizontal pressure force equals the geostrophically calculated increase in kinetic energy. This is a reasonable approximation of the more exact energy Eq. (9.10).

In taking the isobaric divergence of (36.1), we shall neglect the variation of λ with latitude except in the first term on the right. The result is

$$-\lambda \operatorname{div}_p \boldsymbol{v}_h = \left(\frac{\partial \zeta_g}{\partial t}\right)_p + \boldsymbol{v}_g \cdot \operatorname{grad}_p (\lambda + \zeta_g), \tag{36.3}$$

where ζ_g is the geostrophic vorticity

$$\zeta_g = \boldsymbol{k} \cdot \operatorname{curl}_p \boldsymbol{v}_g = \operatorname{div}_p\left(\frac{1}{\lambda}\operatorname{grad}_p \varphi\right) \approx \frac{1}{\lambda}\nabla_p^2 \varphi. \tag{36.4}$$

It was pointed out by ELIASSEN[6] that when (36.3) is introduced into the equation of continuity

$$\operatorname{div}_p \boldsymbol{v}_h + \frac{\partial \omega}{\partial p} = 0 \tag{36.5}$$

and formula (36.1) for $\boldsymbol{v}_h$ into the thermodynamic energy equation,

$$\frac{\partial}{\partial p}\left(\frac{\partial \varphi}{\partial t}\right)_p + \boldsymbol{v}_h \cdot \operatorname{grad}_p \frac{\partial \varphi}{\partial p} - \frac{\alpha}{\theta}\frac{\partial \theta}{\partial p}\omega = 0, \tag{36.6}$$

---

[1] TH. HESSELBERG: Veröff. Geophys. Inst. Leipzig **1** (1915).
[2] D. BRUNT and C. K. M. DOUGLAS: Mem. Roy. Soc. **3**, 22 (1928).
[3] H. PHILIPPS: Meteor. Z. **56**, 12 (1939).
[4] H. ERTEL: Meteor. Z. **58**, 3 (1941).
[5] J VAN MIEGHEM: Acad. roy. Belg. Sci., Mém. **19**, Fasc. 3 (1941).
[6] A. ELIASSEN: Geof. Publ. **17**, No. 3 (1949).

(assuming no heat sources), the resulting equations will, together with the boundary conditions, suffice to determine the future motion when the pressure field [or $\varphi(x, y, p)$] is known initially. These quasi-geostrophic equations are not encumbered with the over-sensitivity which was found to be such a disagreeable property of the primitive equations.

CHARNEY[1] derived essentially the same quasi-geostrophic equations by introducing the geostrophic approximation into the thermodynamic energy equation and into the equation expressing the conservation of potential vorticity (compare Sect. 5 δ), approximated by $(\lambda+\zeta)\,\partial\theta/\partial p$. This is equivalent to using the vorticity Eq. (10.7) in the simplified form

$$\left(\frac{\partial \zeta_p}{\partial t}\right)_p + \boldsymbol{v}_h \cdot \operatorname{grad}_p(\lambda + \zeta_p) = -(\lambda + \zeta_p)\operatorname{div}_p \boldsymbol{v}_h. \tag{36.7}$$

When the geostrophic approximation is introduced everywhere in this equation except in $\operatorname{div}_p \boldsymbol{v}_h$ (where it is not permitted), the resulting equation is identic with (36.3), except for the term $\zeta_g \operatorname{div}_p \boldsymbol{v}_h$ which was lost in (36.3) because the geostrophic approximation was introduced there before taking the curl operation. In this respect CHARNEY's derivation is more correct. It should be noted, however, that there seems to be very little justification for keeping the term $\zeta_g \operatorname{div}_p \boldsymbol{v}_h$ as long as the advection of vorticity is computed geostrophically, because the neglected advection effect $(\boldsymbol{v}_h - \boldsymbol{v}_g)\cdot \operatorname{grad}_p \zeta_g$ will be equally important. A similar reasoning applies also to the terms involving ω in the complete vorticity equation; we cannot expect to improve the result by keeping these terms, as long as the advection is computed geostrophically. A similar view has been expressed by SUTCLIFFE[2].

In accordance with the considerations above, we shall use the vorticity equation in the form (36.3). When the geostrophic values are introduced and the divergence eliminated by means of (36.5), there results

$$\nabla_p^2 \left(\frac{\partial \varphi}{\partial t}\right)_p - \lambda^2 \frac{\partial \omega}{\partial p} = -\lambda\, \boldsymbol{v}_g \cdot \operatorname{grad}_p(\lambda + \zeta_g). \tag{36.8}$$

Introduction of the geostrophic wind into (36.6) gives

$$\frac{\partial}{\partial p}\left(\frac{\partial \varphi}{\partial t}\right)_p + \sigma\omega = \boldsymbol{v}_g \cdot \operatorname{grad}_p \alpha, \tag{36.9}$$

where

$$\sigma = -\frac{\alpha}{\theta}\frac{\partial \theta}{\partial p} \tag{36.10}$$

is a measure of the static stability.

These are the *quasi-geostrophic equations*. Since the geostrophic approximation breaks down near the equator, they apply in middle and high latitudes only. The region where these equations are used must therefore necessarily have lateral boundaries, as well as an upper and lower boundary. The lateral boundary surfaces are conceived surfaces through which the air can flow freely.

If ω is eliminated between (36.8) and (36.9), a single non-linear differential equation of the third order in φ with x, y, p and t as independent variables is obtained. The third order terms are, when σ is considered as a standard function of the space coordinates,

$$\left(\left(\frac{\partial}{\partial t}\right)_p + \boldsymbol{v}_g \cdot \operatorname{grad}_p\right)\left(\sigma \nabla_p^2 \varphi + \lambda^2 \frac{\partial^2 \varphi}{\partial p^2}\right), \tag{36.11}$$

---

[1] J. G. CHARNEY: Geof. Publ. **17**, Nr. 2 (1948).

[2] R. C. SUTCLIFFE: Quart. J. Roy. Meteor. Soc. **72**, 370 (1947).

whence it is seen that the characteristics of the quasi-geostrophic equations in the four-dimensional $xypt$-continuum are:

1. the "hyperplanes" $t = \text{const}$,
2. the particle trajectories $d\boldsymbol{r} = \boldsymbol{v}_g\, dt$.

If we want to determine φ in a certain region R of the four-dimensional continuum we shall have to know φ or its normal derivative[1] in all points where the characteristics $t = \text{const}$ intersect the boundary of the region R. These are the boundary conditions, which will thus have to be specified at all times, because the system has an infinite signal velocity. Besides, we will have to know the value of the quantity $M = \sigma \nabla_p^2 \varphi + \lambda^2 \frac{\partial^2 \varphi}{\partial p^2}$ (or equivalent information) in one point of each characteristic line $d\boldsymbol{r} = \boldsymbol{v}_g\, dt$ that crosses through the region R. For a closed system, where mass does not enter through the boundaries, this information is furnished by the initial conditions at $t = 0$. However, if the system is open, then the initial conditions must be supplemented with information about the value of M for air which enters the region R at later instants. For two-dimensional motion, these requirements for initial and boundary value information were derived in a different manner by CHARNEY, FJÖRTOFT, VON NEUMANN[2]; the derivation of these requirements from the properties of the characteristics of the equation is due to HINKELMANN and SCHWARZENBERGER[3,4].

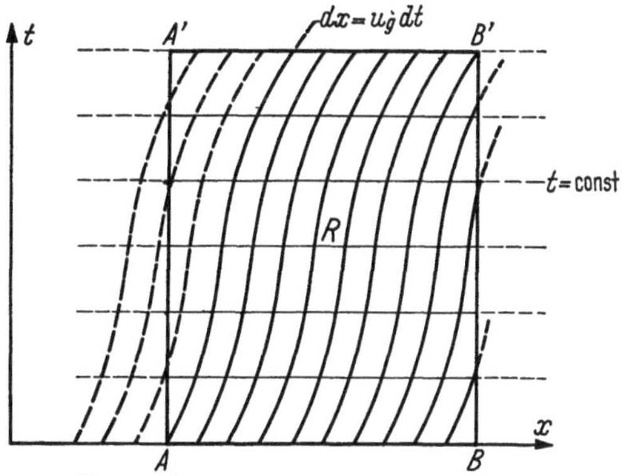

Fig. 20. Characteristic lines in an x, t-plane.

The matter is illustrated in Fig. 20, where for simplicity only one spatial dimension is considered. The boundary values must be given along the lines AA' and BB', where characteristic lines $t = \text{const}$ enter the region R. The family of curves indicate the trajectory characteristics $dx = u_g dt$. For the solid curves, M is specified at $t = 0$ from the initial conditions, which are given on the line AB. For the dashed curves, this is not so, and M must be specified whenever such a curve enters the region, i.e. along AA'.

β) The boundary conditions.

At the "top" of the atmosphere, we have

$$\omega = 0 \quad \text{when} \quad p = 0. \tag{36.12}$$

When the earth's surface is level, the kinematic boundary condition at the ground is given by (10.14). The condition becomes, when the geostrophic wind is introduced,

$$\left(\frac{\partial \varphi}{\partial t}\right)_p - \alpha\,\omega = 0 \quad \text{when} \quad p = p_s. \tag{36.13}$$

This equation implies that ω is nearly zero at the ground, since at higher levels $\alpha\omega$ is normally greater than $(\partial\varphi/\partial t)_p$ in order of magnitude. Eq. (36.13) may therefore be approximated by

$$\omega = 0 \quad \text{when} \quad p = p_s. \tag{36.14}$$

[1] Or equivalent information.

[2] J. G. CHARNEY, R. FJÖRTOFT and J. v. NEUMANN: Tellus 2, 237 (1950).

[3] K. HINKELMANN: Lecture given at Univ. of Stockholm, Inst. of Met. 1951.

[4] K. HINKELMANN and R. SCHWARZENBERGER: Ber. dtsch. Wetterdienst 2, Nr. 14 (1954).

There is actually not much justification for keeping the term $(\partial\varphi/\partial t)_p$ in (36.13), because the neglected term $\boldsymbol{v}_h \cdot \text{grad}_p\, \varphi$ is about equally important.

The condition (36.14) may easily be generalized so as to take account of the topography of the earth's surface. Let the geopotential of the ground be

$$\varphi_s = \varphi_s(x, y). \tag{36.15}$$

The boundary condition then becomes

$$\frac{D\varphi}{Dt} = \left(\frac{\partial\varphi}{\partial t}\right)_p + \boldsymbol{v}_h \cdot \text{grad}_p\, \varphi - \alpha\omega = \boldsymbol{v}_h \cdot \text{grad}\, \varphi_s \quad \text{when} \quad p = p_s. \tag{36.16}$$

With the same approximations as before, we obtain

$$\omega = -\varrho\, \boldsymbol{v}_g \cdot \text{grad}_h\, \varphi_s \quad \text{when} \quad p = p_s. \tag{36.17}$$

Since the force of turbulent friction has not been accounted for in the quasi-geostrophic vorticity Eq. (36.8), the accuracy of this equation is particularly poor within the "friction layer" near the ground. This difficulty can be avoided by applying the Eqs. (36.8) and (36.9) above a level p_F only, which corresponds to the height of the friction layer (500 m above the ground, say). We will then need a boundary condition to be applied at the level p_F. It was shown in Sect. 17 that the motion in the friction layer is such that there is horizontal convergence of mass in areas of cyclonic vorticity and vice versa, so that at the level p_F we will have frictionally controlled vertical motions which are approximately given by Eq. (17.12). These vertical velocities will have to be added to the vertical motion which is caused by the slope of the ground. The boundary condition thus becomes (since $\omega \approx -g\varrho w$),

$$\omega = -\varrho\, \boldsymbol{v}_g \cdot \text{grad}\, \varphi_s - g\sqrt{\frac{K}{2\lambda}}\sin(2\delta)\, \varrho\, \zeta_g, \quad \text{when} \quad p = p_F, \tag{36.18}$$

where K is the eddy viscosity and δ the angle between the surface wind and the geostrophic wind. Once this relation is established, it may without much error be applied at the level $p_F = 1000$ mb (although 950 or 900 mb would perhaps be preferable). We may also use constant values for K, δ and ϱ. It is convenient to write

$$F = \frac{g\varrho}{p_0}\sqrt{\frac{\lambda K}{2}}\sin 2\delta. \tag{36.19}$$

Eq. (36.18) then becomes

$$\omega = -\varrho\, \boldsymbol{v}_g \cdot \text{grad}\, \varphi_s - F\frac{p_0}{\lambda}\zeta_g \quad \text{when} \quad p = p_F, \tag{36.20}$$

where $p_0 = 1000$ mb and F^{-1} is of the order 5 to 10 days in middle latitudes.

The effect of mountains and surface friction upon the motion, as represented by the lower boundary condition (36.20) is best illustrated by taking the pressure average (designated by a bar) of the vorticity Eq. (36.3) from $p=0$ to $p=p_F$. This gives, on account of (36.5),

$$\overline{\left(\left(\frac{\partial}{\partial t}\right)_p + \boldsymbol{v}_g \cdot \text{grad}_p\right)(\lambda + \zeta_g)} = \frac{\lambda}{p_F}\omega_F, \tag{36.21}$$

where ω_F is the value of ω at $p = p_F$. Eliminating ω_F by means of (36.20) we obtain, with $p_F = p_0 = 1000$ mb,

$$\overline{\left(\left(\frac{\partial}{\partial t}\right)_p + \boldsymbol{v}_g \cdot \text{grad}_p\right)(\lambda + \zeta_g)} = -\frac{\lambda}{gH}\boldsymbol{v}_{g0} \cdot \text{grad}\, \varphi_s - F\zeta_{g0}. \tag{36.22}$$

The left-hand side represents the pressure-average of the individual rate of change of the absolute vorticity. The effect of the first term on the right is to give a mean vorticity decrease on the upwind slope of a mountain due to mean horizontal divergence, and a mean vorticity increase on the lee slope, where there is a mean horizontal convergence. This mountain effect will therefore tend to establish a stationary vorticity pattern such that there is anticyclonic relative vorticity over the mountain crests; this feature shows up distinctly on normal upper air maps.

The second term on the right of (36.22) is seen to cause a mean vorticity decrease in areas of low-level cyclonic vorticity ($\zeta_{g0} > 0$) and vice versa. In cases where the vorticity does not change too much in vertical direction, this effect will bring about a dampening of the relative vorticity, with a logarithmic decrement of the order F. The dampening will be ineffective in the case of high-level vorticity patterns which do not reach the ground (upper warm anticyclones and cold cyclones).

Along the lateral boundary, which may be chosen as a vertical surface enclosing a certain region of the map, we must know φ at all times, and furthermore, $M = \sigma \nabla_p^2 \varphi + \lambda^2 \frac{\partial^2 \varphi}{\partial p^2}$ must be known wherever the air is entering the region through this surface. Since $\partial^2 \varphi / \partial p^2$ is known when φ is known at all levels, it suffices to know the geostrophic vorticity $\frac{1}{\lambda} \nabla_p^2 \varphi$ for air entering the region.

In an actual forecast problem, we do not know the future values of φ and ζ_g on the lateral boundary, so that their values must be assigned more or less arbitrarily. This introduces boundary errors, which are felt most strongly in the vicinity of boundaries, in particular where there is strong inflow. The boundary error is probably minimized if the lateral boundary is chosen along the subtropical high pressure belt where the inflow is small and where also the pressure changes are usually small so that the assumption that φ is stationary along the boundary will not introduce too large errors.

γ) Properties of the quasi-geostrophic equations. If the pressure field, or the field of $\varphi(x, y, p)$ is known at a given instant, then the quasi-geostrophic Eqs. (36.8) and (36.9) form a linear system in the two unknowns $(\partial\varphi/\partial t)_p$ and ω. The system is of the elliptic type when $\sigma > 0$, and together with the boundary conditions, it determines these two unknowns uniquely.

If we eliminate $(\partial\varphi/\partial t)_p$ between the two equations we obtain, when σ for the sake of simplicity is regarded as horizontally uniform,

$$\sigma \nabla_p^2 \omega + \lambda^2 \frac{\partial^2 \omega}{\partial p^2} = \lambda \frac{\partial}{\partial p} [\boldsymbol{v}_g \cdot \mathrm{grad}_p (\lambda + \zeta_g)] + \nabla_p^2 [\boldsymbol{v}_g \cdot \mathrm{grad}_p \alpha]. \tag{36.23}$$

This is a Poisson-type equation for the vertical motion. The boundary conditions at $p = 0$ and $p = p_s$ are already expressed in terms of ω. On the lateral boundary surface, we have assumed that φ is known at all times; hence we also know $(\partial\varphi/\partial t)_p$, and ω can be obtained from (36.9). It is noteworthy, however, that the effect of the lateral boundary conditions on the solution of (36.23) will be felt only close to the boundary; this may be inferred from potential theory and is analogous to the fact that the field of electric potential between the plates of a condenser is almost unaffected by disturbing influences from the edges.

The solution of (36.23) may be expressed in terms of the Green's function, which is the solution for the case when the right-hand side vanishes everywhere except in one point. The effect of such a point-source upon the vertical motion will decrease with increasing distance from the point-source, but will also depend upon the direction, because the equation is anisotropic. It can be made isotropic, however, by stretching all vertical lengths in the ratio ν_s/λ where ν_s is the static stability; this ratio is normally of the order 100 in the troposphere, and about two or three times as large in the stratosphere. Therefore, the influence of a point-source is felt at much longer distances in horizontal direction, where the effective stability (the inertial stability λ) is relatively small, than in vertical direction where the stability (ν_s) is much larger.

In order to discuss qualitatively the content of (36.23), we notice that its left-hand side has a strong negative correlation with ω and may in a first approximation be replaced by a negative scale factor times ω.

Let us consider first the effect of the first right-hand term. In regions where this term is positive, i.e. where the vorticity advection increases with height, we must expect negative values of ω, corresponding to ascending motion; and vice versa, we must also expect subsiding motion when the vorticity advection decreases with height. Normally the vorticity advection increases with height in absolute magnitude, as a result of the increase of wind with height in the westerlies. Consequently, there will be a tendency of ascending motion ahead of vorticity maxima and descending motion in their rear.

The effect of the second right-hand term of (36.23) is to cause ascending motion (negative ω) where $\nabla_p^2(\boldsymbol{v}_g \cdot \operatorname{grad}_p \alpha) > 0$, i.e. where the advection of specific volume has a maximum (warm air advection). Conversely, a maximum of cold air advection will produce descending motion. It is significant that it is not the volume advection itself, but its horizontal LAPLACian that produces the effect; therefore, smaller scale advection patterns are more efficient in producing vertical motions than larger scale patterns of the same amplitude.

The above qualitative discussion of (36.23) is due to FJÖRTOFT[1].

Next let us assume that the boundary-value problem discussed above has been solved so that the distribution of ω is known. The field of $(\partial\varphi/\partial t)_p$ is then determined by (36.8) together with the lateral boundary condition. This gives for every level a two-dimensional POISSON equation, the solution of which is highly dependent upon the lateral boundary condition. Since the value of $(\partial\varphi/\partial t)_p$ along the lateral boundary is not known but must be assigned arbitrarily, we should expect the boundary errors to be very important throughout the region. The reason why this is not the case is that when the boundary curve is sufficiently long, positive and negative errors in the assumed distribution of $(\partial\varphi/\partial t)_p$ along the boundary will tend to cancel each other statistically, so that the boundary error in the tendencies will be annihilated by interference when the distance from the boundary is considerably larger than the linear dimension of the motion patterns. A similar statistical cancellation effect applies to the influence of the non-linear term of the POISSON equation over distances considerably larger than the linear dimension of the motion patterns; in this sense the solution in a given point will depend only upon the distribution of the right-hand side within a limited influence region surrounding the point.

We may also determine $(\partial\varphi/\partial t)_p$ directly by eliminating ω between (36.8) and (36.9); this gives

$$\left.\begin{aligned} &\sigma \nabla_p^2\left(\frac{\partial\varphi}{\partial t}\right)_p + \lambda^2\left(\frac{\partial^2}{\partial p^2} - \frac{\partial\log\sigma}{\partial p}\frac{\partial}{\partial p}\right)\left(\frac{\partial\varphi}{\partial t}\right)_p \\ &\qquad = -\lambda\sigma\boldsymbol{v}_g\cdot\operatorname{grad}_p(\lambda+\zeta_g) + \lambda^2\boldsymbol{v}_g\cdot\left(\frac{\partial}{\partial p} - \frac{\partial\log\sigma}{\partial p}\right)\operatorname{grad}_p\alpha. \end{aligned}\right\} \tag{36.24}$$

This is a POISSON-type differential equation for $(\partial\varphi/\partial t)_p$. The second-order terms are the same as in (36.23).

[1] R. FJÖRTOFT: Lecture, Norw. Weather Service, 1955.

The solution of (36.24) is complicated by the circumstance that the boundary conditions at $p=0$ and $p=p_s$ are not expressed in terms of $(\partial\varphi/\partial t)_p$ and thus need to be reformulated. Numerical calculations of the GREEN's function for Eq. (36.24) with reformulated boundary conditions have been carried out by HINKELMANN, ESSENWANGER, REYMANN and WIPPERMANN[1]. Their results show the tendency field produced by a single point source of vorticity advection (or stability advection), computed for various heights of the point source. The calculations show that the influence of the point source extends about 100 times as far in horizontal direction as in vertical direction.

The differential equations for ω (36.23) and for $(\partial\varphi/\partial t)_p$ (36.24) are very similar. However, the boundary conditions are quite different; as a consequence, ω depends primarily upon the distribution of the right-hand side of (36.23) in the nearest surroundings and is almost independent upon influences from distances greater than 1000 km. The tendency $(\partial\varphi/\partial t)_p$ on the other hand, will depend upon the distribution of the right-hand side of (36.24) over the entire region and also upon the lateral boundary conditions. The long distance influences are reduced only as a result of the statistical interference previously mentioned. Therefore, a qualitative discussion of the solution of (36.24) cannot be given as easily as for Eq. (36.23).

If the field of $\varphi(x, y, p)$ is known initially, the quasi-geostrophic equations determine the instantaneous tendency $(\partial\varphi/\partial t)_p$. The solution can be carried out numerically by replacing the continuous fields by the discrete values in a rectangular grid of points in space, and replacing the spatial derivatives by their finite difference analogues. Having thus determined $(\partial\varphi/\partial t)_p$, we may obtain the field of φ at a later instant by extrapolating the computed tendencies over a finite time interval Δt. By iterating this procedure, an integration of the equations in steps of time is attained.

In order for such a procedure to converge, it is necessary that the time step is short enough so that we do not extrapolate the value in any grid-point across the characteristic surface (whose generatrices are the space-time trajectories) which at the beginning of the time step runs through the adjacent grid-points. In other words, if we at the beginning of a time-step define a horizontal square whose corners are the four adjacent grid-points, and assume that this square moves with the geostrophic wind, then the time-step must be short enough so that the central grid-point remains inside the moving square during the time-step. This is secured provided

$$\Delta t < \frac{\Delta x}{\sqrt{2}\, V_{\max}} \tag{36.25}$$

where Δx is the horizontal grid-size and $V_{\max}$ the maximum geostrophic wind velocity which occurs. This is the COURANT-FRIEDRICHS-LEWY criterion, which was adapted to the quasi-geostrophic equations by CHARNEY, FJÖRTOFT and VON NEUMANN[2]. If this criterion is not satisfied, computational instability will occur; the computed solution will amplify quickly and will not converge toward the solution of the differential equations, however small Δx and Δt are chosen.

If we put $V_{\max} = 60$ m/sec and choose $\Delta x = 300$ km, (36.25) gives $\Delta t < 1$ hour (approx). This comparatively long time-step is permitted because the characteristic velocity in the quasi-geostrophic equations is slow. If the primitive equations were to be integrated numerically, $V_{\max}$ in (36.25) would have to be

[1] K. HINKELMANN, O. ESSENWANGER, G. REYMANN and F. WIPPERMANN: Ber. dtsch. Wetterdienst. US-Zone **38** (1952).

[2] J. G. CHARNEY, R. FJÖRTOFT and J. v. NEUMANN: Tellus **2**, 237 (1950).

replaced by the highest characteristic speed which is for these equations comparable with the velocity of sound; consequently the time step would have to be less than about ten minutes in that case.

37. The quasi-solenoidal approximation. The quasi-geostrophic equations form a consistent system of prognostic equations which is free from the oversensitivity of the primitive equations. However, the geostrophic approximation is a rather crude one, and the quasi-geostrophic equations do not give satisfactory accuracy. A recent trend in dynamic meteorology is therefore to look for better filtering approximations.

Performing divergence operation upon the equation of motion (10.4), we obtain the divergence theorem

$$\frac{D}{Dt}\operatorname{div}_p \boldsymbol{v}+\left[\left(\frac{\partial u}{\partial x}\right)^2+2\frac{\partial u}{\partial y}\frac{\partial v}{\partial x}+\left(\frac{\partial v}{\partial y}\right)^2\right]+\operatorname{grad}_p\omega\cdot\frac{\partial \boldsymbol{v}}{\partial p}+\beta u=\lambda\left(\zeta_p-\frac{1}{\lambda}\nabla_p^2\varphi\right). \quad (37.1)$$

The expression within the brackets can also be written

$$\left(\frac{\partial u}{\partial x}\right)^2+2\frac{\partial u}{\partial y}\frac{\partial v}{\partial x}+\left(\frac{\partial v}{\partial y}\right)^2=\frac{1}{2}\left[(\operatorname{div}_p \boldsymbol{v})^2+(\operatorname{def}_p \boldsymbol{v})^2-\zeta_p^2\right], \quad (37.2)$$

where

$$\operatorname{def}_p \boldsymbol{v}=\left[\left(\frac{\partial u}{\partial x}-\frac{\partial v}{\partial y}\right)^2+\left(\frac{\partial u}{\partial y}+\frac{\partial v}{\partial x}\right)\right]^{\frac{1}{2}} \quad (37.3)$$

is the horizontal (or isobaric) *deformation*.

The right-hand side of (37.1) measures the difference between the true vorticity and the geostrophic vorticity. By estimating the value of the left-hand side of the equation, PETTERSSEN[1] concluded that the geostrophic vorticity approximation is often very poor.

On the basis of the divergence theorem, we may arrive at a relationship between the wind and pressure field which differs from the geostrophic relation. Let the horizontal velocity field be partitioned into two parts,

$$\boldsymbol{v}_h=\boldsymbol{v}_\psi+\boldsymbol{v}_\varepsilon \quad (37.4)$$

where

$$\operatorname{div}_p \boldsymbol{v}_\psi=0, \qquad \operatorname{curl}_p \boldsymbol{v}_\varepsilon=0. \quad (37.5)$$

We may then write

$$\boldsymbol{v}_\psi=\boldsymbol{k}\times\operatorname{grad}_p\psi, \qquad \zeta_p=\nabla_p^2\psi, \quad (37.6)$$

$$\boldsymbol{v}_\varepsilon=\operatorname{grad}_p\frac{\partial\varepsilon}{\partial p}, \qquad \operatorname{div}_p \boldsymbol{v}_h=\nabla_p^2\frac{\partial\varepsilon}{\partial p}. \quad (37.7)$$

The equation of continuity is satisfied if we set

$$\omega=-\nabla_p^2\varepsilon. \quad (37.8)$$

Since $\operatorname{div}_p \boldsymbol{v}_h$ for large scale motion systems is generally considerably smaller than ζ_p, we may use $\boldsymbol{v}_\psi$ as an approximation for $\boldsymbol{v}_h$. This may be referred to as the *quasi-solenoidal approximation*. If this approximation is introduced in the divergence equation, and the term depending on ω is neglected, we obtain

$$\lambda\nabla_p^2\psi+2\left[\frac{\partial^2\psi}{\partial x^2}\frac{\partial^2\psi}{\partial y^2}-\left(\frac{\partial^2\psi}{\partial x\,\partial y}\right)^2\right]+\beta\frac{\partial\psi}{\partial y}=\nabla_p^2\varphi. \quad (37.9)$$

This equation expresses a balance between the fields of ψ and φ such that divergence is not being produced. If ψ is known, we have a POISSON equation for the

[1] S. PETTERSSEN: Tellus 5, 231 (1953).

determination of the corresponding field of φ. If on the other hand φ is known, the equation is a MONGE-AMPERE equation in ψ. In the particular case of circular concentric isobars, the equation reduces to the gradient wind Eq. (14.1).

Eq. (37.9) in ψ is of the elliptic, parabolic or hyperbolic type according as $\zeta_g + \lambda/2 \gtreqless 0$. Evaluation of ζ_g from constant pressure maps shows that although $\zeta_g + \lambda/2$ over most regions is positive, negative values do occur in anticyclones and upper ridges. We are thus faced with an equation which is hyperbolic in certain limited regions and elliptic outside these regions. This circumstance is a basic difficulty in the application of (37.9), because the numerical methods which have been designed to solve the equation in ψ can be shown to converge only in the elliptic case. Moreover, it has been shown by RELLICH[1] that Eq. (37.9), with ψ given on the boundary, has in the elliptic case only two solutions (one of which is meteorologically insignificant); but to the authors' knowledge, no such uniqueness theorem has been proven in the hyperbolic or mixed case.

The quasi-solenoidal approximation and the use of the balance Eq. (37.9) have been treated by CHARNEY[2], BOLIN[3] and THOMPSON[4].

b) Simplified atmospheric models.

38. The barotropic model. Since it is difficult to work with three-dimensional fields, it is natural to break the fields of the variables up into a number of two-dimensional fields, in accordance with the customary representation by horizontal (or isobaric) maps. It is convenient also to transform the dynamic equations so that they contain derivatives with respect to horizontal coordinates and time only. This can be done in various ways on the basis of assumptions or models of the vertical structure of the fields of motion and pressure. The simplest model of this kind is due to ROSSBY [*20*] who proposed to treat the large-scale motion as horizontal and non-divergent, so that the vorticity equation becomes

$$\left(\frac{\partial}{\partial t} + \boldsymbol{v}_h \cdot \operatorname{grad}\right)(\lambda + \zeta) = 0. \tag{38.1}$$

This equation together with $\operatorname{div}_h \boldsymbol{v} = 0$ suffices to determine the motion, and represents at the same time a filtering approximation (without using the geostrophic approximation), since its solution does not contain "noise" waves. With a slight and insignificant change of assumptions, we may assume $\omega = 0$ and $\operatorname{div}_p \boldsymbol{v} = 0$, in which case the derivatives of (38.1) are to be taken at constant pressure.

In order to give a justification for Eq. (38.1), we take the pressure average of the equation of continuity, and obtain

$$\overline{\operatorname{div}_p \boldsymbol{v}_h} = -\,\omega_s/p_s, \tag{38.2}$$

where subscript s means surface values. On each vertical line there must be at least one level p^* such that

$$\operatorname{div}_p \boldsymbol{v}_h = -\,\omega_s/p_s \quad \text{at} \quad p = p^*. \tag{38.3}$$

The barotropic model is based on the assumption that $(\operatorname{div}_p \boldsymbol{v} - \overline{\operatorname{div}_p \boldsymbol{v}})$ has the simplest possible variation with height, being of opposite sign in the upper and lower part of the atmosphere and zero at a constant level p^* in the middle.

---

[1] See R. COURANT and D. HILBERT: Methoden der Mathematischen Physik, II, p. 278. Berlin: Springer 1937.

[2] J. G. CHARNEY: Tellus **7**, 22 (1955).

[3] B. BOLIN: Tellus **7**, 27 (1955); **8**, 61 (1956).

[4] P. D. THOMPSON: A Theory of Large-Scale Disturbances in Nongeostrophic Flow. Report, Joint Numerical Weather Prediction Unit, Wash. (1955); J. Meteor. **13**, 251 (1956).

Over level ground, ω_s is very small and may be set equal to zero [compare (36.14)], so that p^* is a level of non-divergence. At this level the vorticity equation assumes the form (38.1), when the terms involving ω are neglected. Therefore, Eq. (38.1) does not apply at every level, but only at the level p^*. The best value of p^* is somewhere near 500 mb, and (38.1) is usually applied to 500 mb maps (see CHARNEY[1]; CHARNEY, FJÖRTOFT and VON NEUMANN[2]).

CHARNEY and PHILLIPS[3] have given another more formal derivation of (38.1), replacing the vertical derivative in the equation of continuity by a corresponding finite difference quotient. Thus one obtains

$$(\mathrm{div}_p\, \boldsymbol{v}_h)_{500\,\mathrm{mb}} = -\left(\frac{\partial\omega}{\partial p}\right)_{500\,\mathrm{mb}} \approx \frac{\omega_{p=0} - \omega_{1000\,\mathrm{mb}}}{1000\ \mathrm{mb}} = 0 \tag{38.4}$$

since ω vanishes at $p=0$ and $p=p_s \approx 1000$ mb according to the boundary conditions. Assuming that the individual change of vorticity is caused by $\mathrm{div}_p\, \boldsymbol{v}_h$ alone, Eq. (38.1) is obtained at the 500 mb level.

The barotropic model can be used either in connection with the geostrophic approximation, in which case (38.1) will contain the geopotential φ^* of the isobaric surface $p=p^*$ as the only dependent variable; or, alternatively, $\boldsymbol{v}^*$ can be expressed in terms of a streamfunction ψ^*, related to φ^* through the balance Eq. (37.9). In the latter case, the model could conceivably be applied also in the tropics.

Since the barotropic model involves one horizontal field as the only dependent variable, it may be referred to as the *one-parameter* (or one-level) *model*. The model avoids the use of the thermodynamic energy equation and does not contain any representation of the temperature field. This clearly restricts the types of processes which the model can describe. In the absence of boundary work on the lateral boundaries, Eq. (38.1) implies that the total kinetic energy is constant; this is true for the absolute motion as well as for the motion relative to the rotating earth. Growth of wave disturbances in a zonal current is possible in the barotropic model, but this must be a manifestation of the type of instability discussed in Sect. 29, and the perturbation energy must be supplied by the kinetic energy of the mean zonal current. Since intensifications of atmospheric disturbances in general appear to be connected with liberation of potential and internal energy, we must conclude that the barotropic model is not an adequate one for describing the growth of disturbances, such as cyclones, in the atmosphere.

Several numerical test forecasts have been computed from the barotropic model[2–8]. The observed and computed 24 hour changes are usually found to have about equal variance and a correlation coefficient of about 0.7—0.8. This indicates that the barotropic model is a serviceable first approximation, although it leaves a considerable portion of the observed changes unexplained.

39. Further theorems on two-dimensional, non-divergent flow. The theory of the barotropic model is based on the assumption that the horizontal motion is non-divergent at a certain level. Therefore, the properties of two-dimensional, non-divergent velocity fields are of considerable meteorological interest.

[1] J. G. CHARNEY: J. Meteor. **6**, 371 (1949).
[2] J. G. CHARNEY, R. FJÖRTOFT and J. v. NEUMANN: Tellus **2**, 237 (1950).
[3] J. G. CHARNEY and N. A. PHILLIPS: J. Meteor. **10**, 71 (1953).
[4] Staff Members, Inst. of Meteor., Univ. of Stockholm: Tellus **6**, 139 (1954).
[5] B. BOLIN: Tellus **7**, 27 (1955).
[6] K. GAMBO, K. KATOW, S. MANABE and M. AIHARA: J. Meteor. Soc. Japan **33**, No. 3 (1955).
[7] Staff Members, Geophys. Inst., Tokyo Univ.: J. Meteor. Soc. Japan **33**, No. 4 (1955).
[8] C. L. JORDAN: J. Meteor. **13**, 223 (1956).

A non-divergent, two-dimensional velocity field $\boldsymbol{v}_a$ (absolute motion) on a sphere may be written as

$$\boldsymbol{v}_a = \boldsymbol{k} \times \operatorname{grad} \psi_a, \tag{39.1}$$

where $\boldsymbol{k}$ is a unit vector pointing out of and normal to the sphere, ψ_a is a stream function on the sphere and grad denotes the gradient operator on the surface of the sphere. Denoting by ∇^2 the two-dimensional LAPLACIAN operator on the sphere, the absolute vorticity is

$$\zeta_a = \nabla^2 \psi_a. \tag{39.2}$$

Under the assumptions made, ζ_a is individually conserved; this may be expressed in the form

$$\partial\zeta_a/\partial t = -\boldsymbol{v}_a \cdot \operatorname{grad} \zeta_a = J(\zeta_a, \psi_a), \tag{39.3}$$

where J indicates the JACOBIan determinant on the sphere. Note that

$$J(\alpha,\beta) = -J(\beta,\alpha), \quad J(\alpha,\alpha) = 0 \quad \text{and} \quad J(\alpha, k\beta) = k\,J(\alpha,\beta).$$

Two quantities are constant during the motion. The first is the total kinetic energy

$$K = \int \tfrac{1}{2} (\operatorname{grad} \psi_a)^2 \, d\sigma = \tfrac{1}{2} \int \operatorname{div} (\psi_a \operatorname{grad} \psi_a)\, d\sigma - \tfrac{1}{2} \int \psi_a \nabla^2 \psi_a \, d\sigma.$$

The first integral on the right must vanish by GAUSS' theorem, and therefore

$$K = -\tfrac{1}{2} \int \psi_a \zeta_a \, d\sigma = \text{constant}. \tag{39.4}$$

The second invariant is the variance of the vorticity

$$L = \tfrac{1}{2} \int \zeta_a^2 \, d\sigma = \text{constant}; \tag{39.5}$$

this follows from the conservation of vorticity.

We shall now expand ψ_a in a series of harmonic functions, which are the eigenfunctions of the equation

$$\nabla^2 f = -q\,f, \quad q \text{ constant} \tag{39.6}$$

with the additional condition that f is analytic over the entire sphere[1]. This defines an infinite series of eigenfunctions f_k, $k=1, 2 \ldots$ with corresponding eigenvalues $q_1 \leq q_2 \leq q_3 \leq \cdots$. The eigenfunctions may be assumed to be orthogonal and normalized, so that

$$\int f_j f_k \, d\sigma = \begin{cases} 0 & \text{when } j \neq k \\ 1 & \text{when } j = k. \end{cases} \tag{39.7}$$

Since f_k form a complete set, we may write

$$\psi_a = \sum_{k=1}^{\infty} a_k(t)\, f_k, \tag{39.8}$$

where a_k are amplitude functions which vary with time.

This formula gives a resolution of the field of motion into harmonic fields of different scale. The scale of any one of the eigenfunctions f_k is given by the eigenvalue q_k, such that the smallest eigenvalue q_1 corresponds to the largest scale motion possible, and larger values of q_k give smaller scale fields which are broken up into a number of circulation cells.

[1] Alternatively, we may choose to restrict the motion to a certain region on the sphere, bounded by a closed curve S. In this case ,we must have $\psi_a = 0$ on S, and Eq. (39.6) must be supplemented by the boundary condition $f = 0$ on S. The theory applies not only to a sphere but to any smooth surface.

From (39.2), (39.6) and (39.8), we find

$$\zeta_a = \sum_{k=1}^{\infty} a_k(t)\, q_k\, f_k. \tag{39.9}$$

Suppose that at a given instant ψ_a coincides with one of the eigenfunctions,

$$\psi_a = a_j\, f_j. \tag{39.10}$$

Then, from (39.9),

$$\zeta_a = a_j\, q_j\, f_j \tag{39.11}$$

and (39.3) gives

$$\frac{\partial \zeta_a}{\partial t} = J(a_j\, q_j\, f_j,\, a_j\, f_j) = a_j^2\, q_j\, J(f_j,\, f_j) = 0 \tag{39.12}$$

showing that the motion does not change with time. Any one of the eigenfunctions therefore corresponds to a possible stationary motion (in an absolute frame).

In a similar manner, it is easy to prove that if $q_i = q_j$, then

$$\psi_a = a_i\, f_i + a_j\, f_j, \tag{39.13}$$

where a_i and a_j are arbitrary constants, represents a stationary motion.

For a general motion (39.8), the kinetic energy follows from (39.4) and (39.9) by using PARSEVAL's theorem; thus we obtain

$$K = \sum_{k=1}^{\infty} \tfrac{1}{2}\, a_k^2\, q_k = \sum_{k=1}^{\infty} e_k = \text{constant}, \tag{39.14}$$

where e_k denotes the kinetic energy associated with the k-th harmonic component.

Similarly, we obtain from (39.5)

$$L = \sum_{k=1}^{\infty} \tfrac{1}{2}\, a_k^2\, q_k^2 = \sum_{k=1}^{\infty} e_k\, q_k = \text{constant}. \tag{39.15}$$

Eqs. (39.14) and (39.15) show that if a certain amount of energy during the motion is transferred from one harmonic component to another which is of smaller scale (larger q), then another amount of energy must necessarily be transferred towards larger scale. In the case of two-dimensional motion, therefore, the energy will not be continually transferred towards the short-wave end of the spectrum, as in the case of three-dimensional turbulence. This result is due to FJÖRTOFT[1] and LORENZ[2].

In particular, it follows that if initially $e_1\, q_1 \gg \sum_{k=2}^{\infty} e_k\, q_k$ and therefore also $e_1 \gg \sum_{k=2}^{\infty} e_k$, then the energy is trapped in the largest scale harmonic component and cannot get out. In this sense, the largest scale motion is stable.

It is interesting to note that the theory outlined above is formally analogous to the theory of motion of a rigid body without external forces, the only difference being the dimension number which is infinite in the former case and three in the latter. For instance, the stability of the harmonic f_1 with the smallest eigenvalue corresponds to the stability of the principal axis with smallest moment of inertia.

So far, the theory outlined holds for two-dimensional motion in a closed region in general. We shall now consider briefly the particular eigenfunctions obtained for a sphere. In this case, (39.6) is solvable when

$$q_n = n\,(n+1)/a^2, \quad n = 1, 2, \ldots \tag{39.16}$$

where a is the radius of the sphere (earth).

[1] R. FJÖRTOFT: Tellus 5, 225 (1953).

[2] E. LORENZ: Unpublished manuscript.

To any particular eigenvalue q_n correspond $2n+1$ eigenfunctions:

$$\left.\begin{aligned} f_n &= P_n(\sin\Phi)\,, \\ f_n^m &= P_n^m(\sin\Phi)\cos m\Lambda \\ g_n^m &= P_n^m(\sin\Phi)\sin m\Lambda \end{aligned}\right\} \quad m=1,2,\ldots,n\,, \tag{39.17}$$

A general field of motion may therefore be expressed as

$$\psi_a = \sum_{n=1}^{\infty}\left[A_n f_n + \sum_{m=1}^{n}\left(A_n^m f_n^m + B_n^m g_n^m\right)\right] \tag{39.18}$$

where the amplitudes A_n, A_n^m and B_n^m are functions of time.

The harmonics with $n=1$ correspond to particularly simple motions. Thus $\psi_a = A_1 f_1$ represents a solid rotation around the axis with an angular velocity

$$\Omega_a = -A_1/a^2. \tag{39.19}$$

Similarly, f_1^1 and g_1^1 represent solid rotations about two axes in the equator plane. It is easily seen that the total angular momentum vector of the motion represented by (39.18) is given by the three amplitudes A_1, A_1^1 and B_1^1, which must therefore be constants. If we assume that the total angular momentum has no component in the equator plane, we obtain

$$A_1 = \text{constant}, \quad A_1^1 = B_1^1 = 0. \tag{39.20}$$

Next let us consider a motion where

$$\psi_a = A_1 f_1 + A_n f_n + \sum_{m=1}^{n}\left(A_n^m f_n^m + B_n^m g_n^m\right), \tag{39.21}$$

where n is an integer larger than 1. According to (39.9), the corresponding vorticity is

$$\zeta_a = A_1 q_1 f_1 + q_n\left[A_n f_n + \sum_{m=1}^{n}\left(A_n^m f_n^m + B_n^m g_n^m\right)\right]. \tag{39.22}$$

Introducing (39.21) and (39.22) into (39.3) we obtain

$$\left.\begin{aligned} \frac{dA_n}{dt} &= 0\,, \\ \frac{dA_n^m}{dt} &= \frac{q_n - q_1}{q_n}\frac{m}{a^2}A_1 B_n^m\,, \\ \frac{dB_n^m}{dt} &= -\frac{q_n - q_1}{q_n}\frac{m}{a^2}A_1 A_n^m \end{aligned}\right\} \tag{39.23}$$

with the solution

$$A_n^m(t) = -C_n^m \sin m\omega_a t, \quad B_n^m(t) = C_n^m\cos m\omega_a t, \tag{39.24}$$

where C_n^m is a constant, and

$$\omega_a = -\frac{q_n - q_1}{q_n}\frac{A_1}{a^2} = \frac{n(n+1)-2}{n(n+1)}\Omega_a \tag{39.25}$$

[compare Eqs. (39.16) and (39.19)]. The motion (39.21) is therefore of the form

$$\psi_a = A_1 f_1 + A_n f_n + \sum_{m=1}^{n} C_n^m P_n^m(\sin\Phi)\sin m(\Lambda - \omega_a t). \tag{39.26}$$

This result shows that whereas there is no interaction between f_1 and f_n, f_1 does interact with f_n^m and g_n^m, causing them to combine into a permanent wave which is propagated zonally with the absolute angular velocity ω_a.

If we want to consider the relative motion, we may write

$$\psi_a = -\Omega a^2 f_1 + \psi, \tag{39.27}$$

where ψ is the stream function of the relative velocity field. Putting

$$\Omega_a = \Omega + \Omega', \quad \omega_a = \Omega + \omega', \tag{39.28}$$

where Ω' is the relative angular velocity corresponding to the first harmonic and ω' the relative speed of propagation of the wave (39.26), we obtain from (39.25)

$$\omega' = \frac{n(n+1)-2}{n(n+1)}\Omega' - \frac{2}{n(n+1)}\Omega. \tag{39.29}$$

This formula was derived by HAURWITZ[1] from the linearized form of the vorticity equation; it may be considered as a generalization of HAURWITZ' formula (30.13) for the plane case. It is noteworthy, however, that the solution (39.26) and Eq. (39.29) follow rigorously from the non-linear vorticity equation; this was shown by NEAMTAN[2]. See also publications by ERTEL[3], CRAIG[4], HÖILAND[5], HAURWITZ and CRAIG[6].

If the stream function contains harmonics with more than one value of n (except $n=1$), they will interact with each other in a more complicated manner. These interactiones have been studied by SILBERMAN[7].

40. Stationary disturbances produced by the earth's topography, determined by means of the barotropic model. As shown by CHARNEY and ELIASSEN[8], the barotropic model may be generalized so as to account for the topography of the earth's surface and surface friction. This is attained by using the boundary condition (36.20) instead of $\omega_s = 0$ in the expression (38.3) for the divergence at the level p^*. When this is introduced into the simplified vorticity Eq. (36.3), there results

$$\left(\left(\frac{\partial}{\partial t}\right)_p + \boldsymbol{v}^* \cdot \mathrm{grad}_p\right)(\lambda + \zeta_p^*) = -\frac{\lambda}{gH}\boldsymbol{v}_s \cdot \mathrm{grad}\,\varphi_s - F\zeta_s, \tag{40.1}$$

where H is the height of the homogeneous atmosphere (32.10). In order to obtain an equation which contains the velocity field at the level p^* only, the additional assumption must be made that $\boldsymbol{v}_s$ and ζ_s can be expressed with sufficient accuracy in terms of $\boldsymbol{v}^*$ and ζ^*, e.g. by setting

$$\boldsymbol{v}_s = \varkappa \boldsymbol{v}^*, \quad \zeta_s = \varkappa \zeta^* \tag{40.2}$$

where $\varkappa$ is a constant factor less than one. Eqs. (40.1) and (40.2) were applied by CHARNEY and ELIASSEN in linearized form to compute the stationary disturbances in the westerlies produced by the mountain ranges of the northern hemisphere. Neglecting the earth's curvature except for the variation of the CORIOLIS parameter, introducing the geostrophic approximation and putting $\partial/\partial t = 0$, Eq. (40.1) becomes in linearized form

$$U\left(\frac{\partial^2}{\partial x^2} + \frac{\partial^2}{\partial y^2}\right)\frac{\partial\varphi}{\partial x} + \beta\frac{\partial\varphi}{\partial x} = -\frac{\lambda^2}{gH}\varkappa U\frac{\partial\varphi_s}{\partial x} - \varkappa F\left(\frac{\partial^2\varphi}{\partial x^2} + \frac{\partial^2\varphi}{\partial y^2}\right), \tag{40.3}$$

[1] B. HAURWITZ: J. Mar. Res. **3**, 254 (1940).
[2] S. M. NEAMTAN: J. Meteor. **3**, 53 (1946).
[3] H. ERTEL: Meteor. Z. **60**, 332 (1943).
[4] R. A. CRAIG: J. Meteor. **2**, 175 (1945).
[5] E. HÖILAND: Geof. Publ. **17**, No. 10 (1950).
[6] B. HAURWITZ and R. A. CRAIG: A. F. Cambridge Res. Center, Geophys. Res Papers 14 (1952).
[7] I. SILBERMAN: J. Meteor. **11**, 27 (1954).
[8] J. G. CHARNEY and A. ELIASSEN: Tellus **1**, No. 2 (1949).

where U is a constant zonal basic current in the x-direction, and φ the perturbation of the geopotential of the isobaric surface $p = p^*$. Assuming further that the disturbance has a simple harmonic variation in the y-direction so that $\partial^2/\partial y^2 = -m^2$, the equation becomes

$$\frac{\partial^3 \varphi}{\partial x^3} + \left(\frac{\beta}{U} - m^2\right)\frac{\partial \varphi}{\partial x} + \frac{\varkappa F}{U}\left(\frac{\partial^2 \varphi}{\partial x^2} - m^2 \varphi\right) = -\varkappa \frac{\lambda^2}{g H}\frac{\partial \varphi_s}{\partial x}. \tag{40.4}$$

This non-homogeneous equation was solved for the latitude circle 45° N, using for $\varphi_s(x)$ the actual profile of the earth's surface in this latitude, and with $U = 17°$ longitude per day and $\varkappa = 0.4$. The result was compared with the normal height profile of the 500 mb surface for January; and the agreement was quite good when m was chosen so that the lateral half-wavelength was 33° latitude, and the eddy-viscosity in the surface layer was taken to be 40 m²/sec (Fig. 21). This is perhaps an unreasonably high value of surface friction, and the reason why such a high value gives the best agreement may be that the important eddy-flux of vorticity, which acts as an additional internal friction, has been neglected.

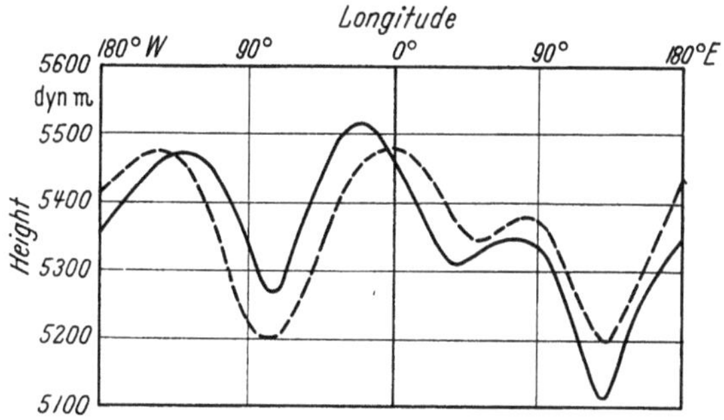

Fig. 21. Computed stationary profile around the hemisphere of the 500 mb surface (dashed curve) according to CHARNEY and ELIASSEN and observed mean conditions (solid curve) after SCHERHAG[2].

The two-dimensional stationary disturbance produced by an isolated circular mountain has been calculated by BOLIN[1].

The *geographically fixed disturbances in the westerlies* could also conceivably be caused by geographically fixed heat and cold sources, or by the difference in surface friction over continents and oceans. Estimates of the friction effect indicate that it is too small and cannot explain the magnitude of the observed disturbances. It was pointed out by CHARNEY and ELIASSEN[3] that if the stationary perturbations were due to heat and cold sources varying with longitude, one should expect a phase shift of the resulting disturbances from summer to winter, whereas the stationary troughs and ridges remain nearly in the same position throughout the year. This seems to indicate that heating and cooling is not the primary cause of the stationary disturbances in the westerlies; thus we are left with the mountain effect as the only explanation. This argument has been questioned by SMAGORINSKY[4] who has investigated the perturbations produced by stationary heat sources and demonstrated that the phase of the stationary waves can be accounted for. It is therefore quite possible that the observed stationary ridges and troughs in the westerlies are due to a combined effect of mountains and heat sources.

41. The two-parameter model. From the barotropic model with only one horizontal field as dependent variable we may proceed to consider the two-parameter model involving two horizontal fields. This model gives a representation of the horizontal distribution of temperature, and thus enables us in a crude manner to deal with processes connected with the baroclinicity. It is also possible to account for heat and cold sources; however, we shall here consider adiabatic motion only.

There is no unique way to set up a two-parameter model, and several versions have been designed. The first consistent model is due to PHILLIPS[5]; it consists of two barotropic layers separated by an interface. This model has the advantage

[1] B. BOLIN: Tellus **2**, 184 (1950).
[2] R. SCHERHAG: Wetteranalyse und Wetterprognose. Berlin: Springer 1948.
[3] J. G. CHARNEY and A. ELIASSEN: Tellus **1**, No. 2 (1949).
[4] J. SMAGORINSKY: Quart. J. Roy. Meteor. Soc. **79**, 342 (1953).
[5] N. A. PHILLIPS: J. Meteor. **8**, 381 (1951).

that it represents a real physical system; on the other hand, it is not easily identifiable with the atmosphere. Other two-parameter models have been designed by EADY[1], ELIASSEN[2], SAWYER and BUSHBY[3] and THOMPSON[4]; these are rather similar to each other. Still another version is due to CHARNEY and PHILLIPS[5].

So far, the two parameter model has been derived only in connection with the geostrophic approximation. The model may be based on the following assumption about the variation of geostrophic wind with height,

$$\boldsymbol{v}_g(x, y, p, t) = \overline{\boldsymbol{v}}_g(x, y, t) + A(p)\,\boldsymbol{v}_T(x, y, t), \qquad \boldsymbol{v}_T = \overline{\boldsymbol{v}}_g - \boldsymbol{v}_{gs}, \tag{41.1}$$

where the bar denotes average with respect to pressure from $p=0$ to $p=p_s \approx$ 1000 mb, which may be used as an approximation for the surface pressure. Eq. (41.1) means that the hodograph curve is considered as a straight line. Such an assumption was made by SUTCLIFFE[6] in a paper which contains much of the theory of the two-parameter model.

The function $A(p)$ should be determined so as to give the best fit in a statistical sense. Clearly we must have

$$\overline{A} = 0, \quad A(p_s) = -1. \tag{41.2}$$

$A(p)$ must have at least one zero p^*; at this level $\boldsymbol{v}_g = \overline{\boldsymbol{v}}_g$. The geostrophic wind formula gives

$$\overline{\boldsymbol{v}}_g = \frac{1}{\lambda}\,\boldsymbol{k} \times \operatorname{grad} \overline{\varphi}, \qquad \boldsymbol{v}_T = \frac{1}{\lambda} \times \operatorname{grad} \varphi_T \tag{41.3}$$

where

$$\varphi_T = \overline{\varphi} - \varphi_s = -\overline{p\frac{\partial \varphi}{\partial p}} = R\overline{T}. \tag{41.4}$$

The last expression is obtained by using the static Eq. (9.3) and the equation of state. In order to express the thermodynamic energy equation in terms of φ_T, we write it in the form

$$R\left(\left(\frac{\partial T}{\partial t}\right)_p + \boldsymbol{v}_g \cdot \operatorname{grad}_p T\right) + \frac{R\,T}{\theta}\frac{\partial \theta}{\partial p}\,\omega = 0 \tag{41.5}$$

assuming adiabatic changes of state. Moreover, we introduce an assumption about the distribution of ω with height,

$$\omega(x, y, p, t) = B(p) \cdot \overline{\omega}(x, y, t). \tag{41.6}$$

The boundary conditions (36.12) and (36.14) are satisfied when $B(p)$ is chosen so that

$$B(0) = B(p_s) = 0 \tag{41.7}$$

and furthermore, we must have

$$\overline{B} = 1. \tag{41.8}$$

When (41.6) and (41.1) are inserted into (41.5) and the average with respect to pressure is taken, we obtain since $\boldsymbol{v}_T \cdot \operatorname{grad}_p T = 0$,

$$\frac{\partial \varphi_T}{\partial t} + \overline{\boldsymbol{v}}_g \cdot \operatorname{grad} \varphi_T = \gamma\,\overline{\omega}, \tag{41.9}$$

---

[1] E. T. EADY: Tellus 4, 157 (1952).
[2] A. ELIASSEN: Tellus 4, 145 (1952).
[3] J. S. SAWYER and F. H. BUSHBY: J. Meteor. 10, 54 (1953).
[4] PH. D. THOMPSON: Quart. J. Roy. Meteor. Soc. 79, 51 (1953); see also PH. D. THOMPSON and W. L. GATES: J. Meteor. 13, 127 (1956).
[5] J. G. CHARNEY and N. A. PHILLIPS: J. Meteor. 10, 71 (1953).
[6] R. C. SUTCLIFFE: Quart. J. Roy. Meteor. Soc. 72, 370 (1947).

where

$$\gamma = -\overline{\frac{RT}{\theta}\frac{\partial\theta}{\partial p}B} \tag{41.10}$$

is a constant stability factor. Its value will depend upon our choice of the function $B(p)$, but it will be of the order $gH/4p_s$, where H is the height of the homogeneous atmosphere.

From (41.1) we find the corresponding vorticity field

$$\zeta_g = \overline{\zeta}_g + A(p)\,\zeta_T, \qquad \zeta_T = \overline{\zeta}_g - \zeta_{gs}, \tag{41.11}$$

where

$$\overline{\zeta}_g = \frac{1}{\lambda}\nabla^2\overline{\varphi}, \qquad \zeta_T = \frac{1}{\lambda}\nabla^2\varphi_T. \tag{41.12}$$

The vorticity equation in the simplified form (36.8) then becomes

$$\left.\begin{aligned}\nabla^2\left(\frac{\partial\overline{\varphi}}{\partial t} + A(p)\frac{\partial\varphi_T}{\partial t}\right) - \lambda^2 B'(p)\,\overline{\omega}\\ = -\lambda\left(\overline{\boldsymbol{v}}_g + A(p)\,\boldsymbol{v}_T\right)\cdot\operatorname{grad}\left(\lambda + \overline{\zeta}_g + A(p)\,\zeta_T\right).\end{aligned}\right\} \tag{41.13}$$

Eq. (41.13) will not be satisfied at all levels; in fact, since there are only two horizontal fields $\overline{\varphi}$ and φ_T as dependent variables, all that can be demanded is that the equation is satisfied at two separate levels, or that two equivalent conditions are fulfilled. Following ELIASSEN[1] we shall choose these conditions so as to minimize the error at all levels in the sense of least squares. This means that the equation shall be satisfied when we take its mean and its first moment in vertical direction. By forming the mean of (41.13), we obtain on account of (41.2) and (41.7),

$$\Delta^2\frac{\partial\overline{\varphi}}{\partial t} = -\lambda\,\overline{\boldsymbol{v}}_g\cdot\operatorname{grad}(\lambda + \overline{\zeta}_g) - \lambda\,\overline{A^2}\,\boldsymbol{v}_T\cdot\operatorname{grad}\zeta_T. \tag{41.14}$$

The first moment is obtained by multiplying with $(p - \frac{1}{2}p_s)$ and averaging. This gives, when $\overline{\omega}$ is eliminated by (41.9),

$$\left.\begin{aligned}(\nabla^2 - k_0^2)\,\frac{\partial\varphi_T}{\partial t}\\ = -\lambda\left[\overline{\boldsymbol{v}}_g\cdot\operatorname{grad}\zeta_T + \boldsymbol{v}_T\cdot\operatorname{grad}(\lambda + \overline{\zeta}_g) + m\,\boldsymbol{v}_T\cdot\operatorname{grad}\zeta_T\right] + k_0^2\,\overline{\boldsymbol{v}}_g\cdot\operatorname{grad}\varphi_T\end{aligned}\right\} \tag{41.15}$$

where

$$k_0^2 = \frac{\lambda^2}{\overline{(\frac{1}{2}p_s - p)\,A}\cdot\gamma}, \qquad m = \frac{\overline{(\frac{1}{2}p_s - p)\,A^2}}{\overline{(\frac{1}{2}p_s - p)\,A}}. \tag{41.16}$$

Eq. (41.14) and (41.15) are governing equations in the two-parameter model.

Eq. (41.14) which was derived by CHARNEY, FJÖRTOFT and VON NEUMANN[2], expresses that the local change of mean vorticity is caused not only by advection of mean absolute vorticity, but also by advection of thermal vorticity by the thermal wind. The latter effect may cause an intensification of a center of maximum mean vorticity when the thermal field (thickness field) is such that the thermal wind advects thermal vorticity into the mean vorticity maximum. As shown in Fig. 22, this effect will cause a baroclinic wave to amplify by increasing the mean vorticity amplitude, provided the thermal wave pattern remains behind the mean

[1] A. ELIASSEN: Tellus 4, 145 (1952).

[2] J. G. CHARNEY, R. FJÖRTOFT and J. v. NEUMANN: Tellus 2, 237 (1950).

wave. This effect may be investigated by linearizing (41.14) and (41.15), assuming a basic current with a mean wind $\bar{U}$ and a thermal wind U_T. The following formula is obtained for the velocity of waves with wave number k,

$$c = \bar{U} - \frac{\beta}{k^2} + \frac{\beta k_0^2}{2k^2(k_0^2 + k^2)} \pm \sqrt{\frac{\beta^2 k_0^4}{4k^4(k_0^2 + k^2)^2} - \overline{A^2}\, U_T^2 \frac{k_0^2 - k^2}{k_0^2 + k^2}}. \quad (41.17)$$

For sufficiently strong U_T, this formula gives complex values of c in a certain intermediate band of wave lengths, in qualitative agreement with the result obtained by FJÖRTOFT from energy considerations (compare Fig. 18). The type of

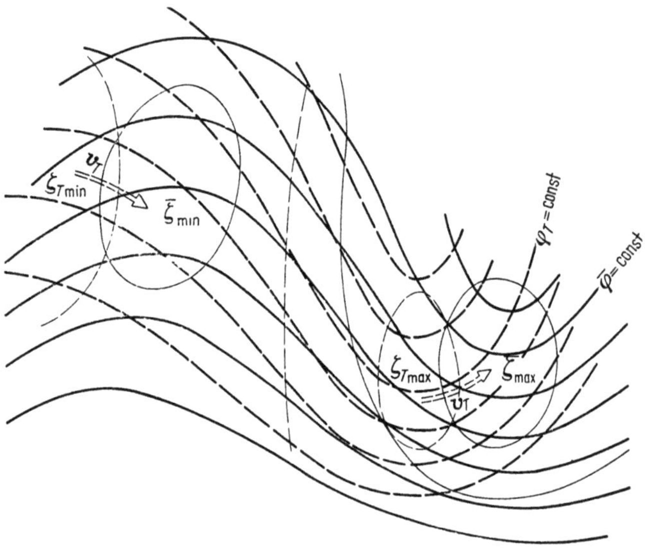

Fig. 22. Amplification of a baroclinic wave. Heavy full lines are equipotential lines for the mean wind and heavy broken lines, for the thermal wind $\mathfrak{v}_T$. Thin solid lines are isopleths of mean vorticity. Thin broken lines are isopleths of thermal vorticity.

instability thus described by the two-parameter model is the baroclinic instability discussed in Sect. 33. The growth rate found from the two-parameter model depends very much on the coefficient $\overline{A^2}$, and vanishes if $\overline{A^2} = 0$.

The instability thus obtained is an indication that the two-parameter model, in contrast to the barotropic model, is capable of describing processes where potential and internal energy is converted into kinetic energy; since such energy conversion must be an important factor in the atmospheric motion processes, we have reason to assume that the two-parameter model is a rather realistic one.

The various versions of the two-parameter model (see ref. p. 108) lead to essentially the same governing Eqs. (41.14) and (41.15), the only important difference between them being the somewhat different numerical values obtained for the three coefficients $\overline{A^2}$, k_0^2 and m which appear in these equations.

The models of EADY, SAWYER and BUSHBY, and THOMPSON assume a linear variation of wind with pressure, giving $\overline{A^2} = \frac{1}{3}$. The derivation by CHARNEY and PHILLIPS (see next section) gives $\overline{A^2} = \frac{1}{4}$. On the other hand, if $A(p)$ is chosen in accordance with observed wind profiles as proposed by ELIASSEN, a considerably larger value of $\overline{A^2}$ is obtained. The best value is still uncertain; the choice of this coefficient will be important, particularly for the intensity changes (deepenings and fillings) of pressure systems predicted from the two-parameter model.

The coefficient k_0 is found to be of the order $(500\text{ km})^{-1}$ in middle latitudes, although this value will also be somewhat different for the different versions of the two-parameter model. The third coefficient m can probably be set equal to zero without much error; it will be exactly zero if A is an antisymmetric function.

The equations of the two-parameter model can be integrated numerically in steps of time, in much the same way as the equation obtained in the barotropic model. Thus a forecast may be obtained of the fields of $\bar{\varphi}$, φ_T and $\bar{\omega}$. It is convenient to express the mean wind $\bar{\boldsymbol{v}}_g$ in terms of the wind at a particular level (e.g. 500 mb) by means of (41.1), so that the forecast is obtained for the geopotential at two levels.

Experimental forecasts of this kind have been calculated on electronic computers by BUSHBY and HINDS[1] and CHARNEY[2]. Their results are not remarkably good. This may be taken as an indication that the two-parameter model is still too crude, so that more parameters are needed; but it may also be due to improper coefficient values in the governing equations, or to geostrophic errors.

42. Models involving more than two parameters. CHARNEY and PHILLIPS[3] have derived equations for models involving an arbitrary number of horizontal fields. Somewhat simplified, their method may be described as follows.

Let the pressure range from $p=0$ to $p=p_s$ $(=1000\text{ mb})$ be divided into $2n$ equal parts by the $2n+1$ levels

$$p_i = \frac{p_s}{2n}\, i, \qquad i = 0, 1, \ldots, 2n. \tag{42.1}$$

Furthermore let φ_i (and hence, $\boldsymbol{v}_{g\,i}$ and $\zeta_{g\,i}$) at the level p_i be defined for all odd values of i, and let ω_i be defined for all even values of i; in particular, $\omega_0 = \omega_{2n} = 0$ owing to the boundary conditions.

We now write the vorticity Eq. (36.8) for any level p_i where i is odd $(i = 2j-1)$, and replace the derivative with respect to p by a finite difference quotient. This gives

$$\left.\begin{aligned} \nabla^2 \frac{\partial \varphi_{2j-1}}{\partial t} - \lambda^2 \frac{\omega_{2j} - \omega_{2j-2}}{p_s/n} &= -\lambda\, \boldsymbol{v}_{g(2j-1)} \cdot \operatorname{grad}\left(\lambda + \zeta_{g(2j-1)}\right) \\ &(j = 1, 2, \ldots, n). \end{aligned}\right\} \tag{42.2}$$

In the same manner, Eq. (36.9) becomes when written for any even number $i = 2j$,

$$\frac{\dfrac{\partial \varphi_{2j+1}}{\partial t} - \dfrac{\partial \varphi_{2j-1}}{\partial t}}{p_s/n} + \sigma_{2j}\, \omega_{2j} = -\frac{\boldsymbol{v}_{g(2j+1)} + \boldsymbol{v}_{g(2j-1)}}{2} \cdot \operatorname{grad} \frac{\varphi_{2j+1} - \varphi_{2j-1}}{p_s/n}. \tag{42.3}$$

Here σ_{2j} may be expressed in terms of φ at several levels. Eqs. (42.2) and (42.3) form a system of $(2n-1)$ equations with the same number of dependent variables

$$\frac{\partial \varphi_1}{\partial t}, \ \frac{\partial \varphi_3}{\partial t}, \ \ldots, \ \frac{\partial \varphi_{2n-1}}{\partial t}, \qquad \omega_2, \ \omega_4, \ \ldots, \ \omega_{2n-2}.$$

In particular, with $n=1$ we obtain the barotropic model; with $n=2$ we obtain another formulation of the two-parameter model, which can be shown to be equivalent with the formulation given in the previous section, (with $\overline{A^2} = \frac{1}{4}$). With $n=3$ we obtain a three-parameter (or three-level) model. This model

[1] F. H. BUSHBY and M. K. HINDS: Quart. J. Roy. Meteor. Soc. **80**, 16, 165 (1954).
[2] J. G. CHARNEY: Proc. Nat. Acad. Sci. U.S.A. **40**, 2 (1954).
[3] J. G. CHARNEY and N. A. PHILLIPS: J. Meteor. **10**, 71 (1953).

gives more detail in the vertical pressure distribution than the two-parameter model, and is capable of describing and take into account changes of static stability with space and time, whereas in the two-parameter model the static stability must be given a constant value.

An example of a relatively successful three-parameter numerical forecast has been published by CHARNEY[1].

IV. Cyclones and anticyclones.

43. Introduction. Cyclones and anticyclones are horizontal vortices with a diameter varying from 500 to a few thousand kilometres. With respect to the absolute frame of reference the air in a cyclone rotates faster than the earth's surface underneath, and more slowly in an anticyclone. Since the motion is approximately geostrophic, the pressure within a cyclone is relatively low and in an anticyclone relatively high. Apart from these general properties, the extent, structure and behaviour of the different vortices vary greatly. Large cyclones and anticyclones often persist, practically unchanged, over long periods of time. The quasi-permanent anticyclones of the subtropical belt are of this type. The smaller atmospheric vortices are mostly characterized by a clear-cut life cycle featuring growth and decay. Their life duration ranges from a few days to a few weeks.

The *decay* of a vortex may always plausibly be ascribed to the friction at the earth's surface, whereas the *origin* of vortices and their maintenance against friction are still in many respect an unsolved problem. One of the most striking processes in the atmophere is the development of a new strong cyclone in relatively short time, as is often observed in the surface maps. Over a large area the absolute vorticity in the lower layers increases to twice the value, or even more in, say, 24 hours. The process always takes place immediately polewards of the polar front which, thereby, usually undergoes a characteristic deformation known as the occlusion of a frontal wave.

In spite of the fact that the process is well-known in its details, opinions on its nature still differ in essential points. It is, for instance, a dispute of long standing whether the condensation of water vapor, which occurs along the warm front in abundance, is merely a symptomatic phenomenon without appreciable influence on the whole process, or whether it is necessary for the formation of the cyclone. In other words: Could the cyclone also arise in completely dry air, or could it not.

The question is answered in Sect. 49 to the effect that *condensation* in fact plays a decisive role in the process. This view is based upon consideration of the potential vorticity. As will be shown in the next section, a close relationship exists between the potential vorticity distribution and the wind field which is particularly significant in the case of cyclones and anticyclones. From this relationship it follows that the changes in potential vorticity produced by condensation in the case in question, are really necessary for the formation of the cyclone.

Still another point—which seems to be important especially with respect to cyclones and anticyclones—may be mentioned. In Sect. 46 and 47 some investigations are discussed according to which the assumption that short-period processes in the free atmosphere may be treated as frictionless (Sect. 20), seems not to be justified throughout. Certain results lead to the conclusion that forces which are not contained in the equation of frictionless motion occasionally play a considerable role. In case of such findings it is in general supposed that the additional force is accounted for by turbulent friction. It must however be remarked that not only do the forces in question often seem to be stronger than what would be

[1] J. G. CHARNEY: Proc. Nat. Acad. Sci. U.S.A. **40**, 2 (1954).

expected from normal turbulent friction, but they also have an accelerating action in regions where a braking effect would be expected in view of the usual conception of friction, as for instance expressed by the NAVIER-STOKES term.

It must already be noted at this stage that a satisfactory explanation of the phenomenon so far does not exist. Investigations concerning this matter are still sparse and are partly based on statistical methods, so that no detailed knowledge exists. Since the phenomenon furthermore seems to occur only in the upper regions, many meteorologists are inclined to ascribe it to the inaccuracy of upper air observations or to incorrect analyses because of the sparsity of soundings[1].

There is however one established phenomenon which appears to confirm the observations mentioned above. This is the occurence of strong and continuous vortex motion near the ground. The problem as to how this motion is conserved against surface friction, inevitably leads to the supposition of additional, accelerating forces. The analytical treatment of this problem is dealt with in Sect. 54, in connection with the general circulation.

The study of cyclones and anticyclones clearly leads to some unsolved problems. Important knowledge still lacks. A comprehensive treatment of the whole domain is therefore still impossible.

44. The structure of a geostrophic vortex in dependance on the potential vorticity of the air masses. The relations between potential vorticity and the wind and pressure fields may be best illustrated by considering simple cases. In an atmosphere completely at rest on a flat earth's surface ($\lambda = \text{const}$) the potential vorticity Z [see Eq. (5.18)] is a function of the potential temperature only. If this function is known, it determines, together with the surface values of pressure and potential temperature, the whole pressure field. We now consider an atmosphere in which Z is likewise a function of θ only, with the exception of an air mass bounded on all sides in which Z differs considerably from this function. The whole system is taken to be in static and geostrophic equilibrium. A system of this nature located on 45° of latitude is shown in Fig. 23. The potential vorticity in the imbedded body is six times that in the neighbouring air masses. The system is a cyclone.

The mathematical basis of the model is the following. We use the potential temperature as independent variable along the vertical (Sect. 10β). This is justified, because $\partial\theta/\partial z$ is assumed to be positive. Then the condition $Z = f(\theta)$ takes the form:

$$Z \equiv \frac{\lambda + \frac{v}{r} + \frac{\partial v}{\partial r}}{\varrho \frac{\partial z}{\partial \theta}} = \frac{\lambda}{\varrho_u \frac{\partial z_u}{\partial \theta}}. \tag{44.1}$$

Here v denotes the circular wind, r the distance from the axis and the subscript u the undisturbed atmosphere which is approximated in the case of indefinitely increasing r.

The thermal wind equation in our coordinates is:

$$\lambda\left(\theta \frac{\partial v}{\partial \theta} - v\right) = -g \frac{\partial z}{\partial r}. \tag{44.2}$$

Neglecting the small differences between ϱ and ϱ_u at the same value of θ and eliminating z, we obtain

$$U^2(\theta) \frac{\partial^2 v}{\partial \theta^2} + \frac{\partial^2 v}{\partial r^2} + \frac{1}{r}\frac{\partial v}{\partial r} - \frac{v}{r^2} = 0 \tag{44.3}$$

[1] H. LETTAU observed that, if the stratification of the friction layer is extremely stable, a wind maximum, much stronger than the gradient wind, may occur at a height of some 300 m. (Lecture at the Meteorological Meeting at Frankfurt 1955.) An explanation of this phenomenon on the basis of turbulent friction seems to be hardly possible.

where we have introduced

$$U^2(\theta) = \frac{\lambda^2 \theta}{g \dfrac{\partial z_u}{\partial \theta}}. \tag{44.4}$$

Eq. (44.3) is the differential equation for v in the surrounding atmosphere.

As boundary condition in the upper half-plane we have

$$v \to 0 \tag{44.5}$$

when the distance from the origin increases indefinitely. The surface condition depends on the distribution of the potential temperature at the ground. In many cases this temperature may with good approximation be considered as being constant ($= \theta_0$). Then we have the surface condition

$$\lambda \left(\theta \frac{\partial v}{\partial \theta} - v \right) = 0 \quad \text{when } \theta = \theta_0. \tag{44.6}$$

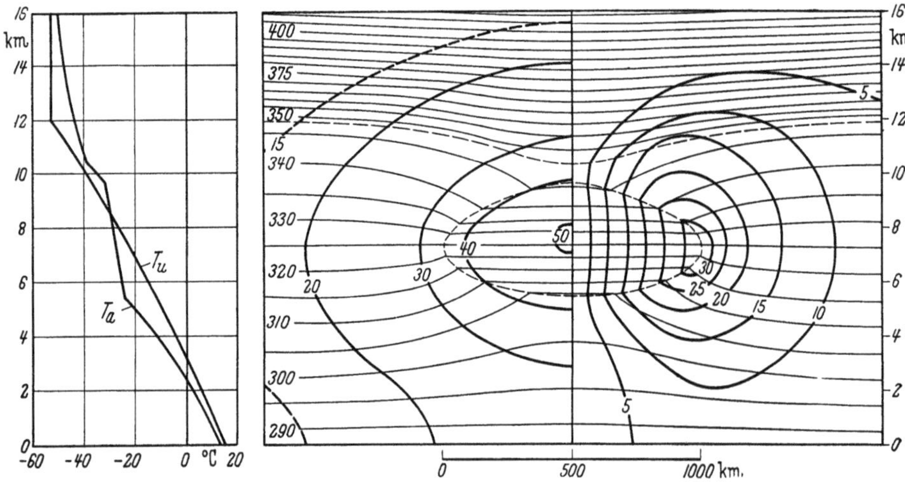

Fig. 23. Model of a cyclone produced by a body of 6 times the normal potential vorticity. Left-hand diagram: Temperature distribution on the axis (T_a) and in the undisturbed atmosphere (T_u). Cross-section: Thin lines are lines of equal potential temperature (degrees absolute). Heavy lines in the left half indicate the relative depression $(p_u - p)/p_u$ (per thousand). Heavy lines in the right half are isotachs (m/sec).

The general procedure for constructing the model is the following. A solution of the boundary value problem (44.3) to (44.6) is sought, which has a (finite or infinite) number of poles within the region of the r, θ-plane which is occupied by the body. Then, in the interior of the body, this solution is replaced by another function, which is regular and which assumes the same values as those of the outer solution on the surface of the body.

By inserting the combined solution into (44.2) and integrating over r, we obtain z as a function of r and θ, i.e. the topographies of the isentropic surfaces. Now, by means of (44.1) the potential vorticity within the body can be computed. By suitable choice of the poles this vorticity can be made to assume a nearly uniform, pre-given value.

Under certain assumptions simple analytical solutions may be obtained. The undisturbed troposphere may be approximated by putting $U = A = \text{const}$. Then Eq. (44.3) is satisfied by

$$v = \text{const} \cdot r \, \xi_\nu^{-3} \tag{44.7}$$

where

$$\xi_\nu = \sqrt{r^2 + (\theta - \theta_\nu)^2 / A^2}\,, \tag{44.8}$$

θ_ν being an arbitrary constant.

The undisturbed stratosphere may be represented up to great heights, by $U = (B\theta + C)^2$, where B and C are constants. With this assumption, Eq. (44.3) is satisfied by

$$v = \text{const} \cdot (B\theta + C) \, r \, \eta_\nu^{-3} \tag{44.9}$$

where

$$\eta_\nu = \sqrt{r^2 + \frac{(\theta - \theta_\nu)^2}{(B\theta + C)^2 (B\theta_\nu + C)^2}}\,, \tag{44.10}$$

The general solution for the outer troposphere and stratosphere consists of a linear combination of solutions (44.7) and (44.9) respectively. Condition (44.5) is fulfilled. The surface condition and the interface conditions at the tropopause may be satisfied by suitable choice of the single θ_ν-values.

Simple relations exist, when the body is located within the troposphere and its surface is of the form $\xi_i = \text{const}$, the subscript i henceforth referring to the centre of the body. This is assumed in the model of Fig. 23. Then the wind at a tropospheric point which is neither too near to the ground nor to the tropopause is mainly given by one single term, namely by

$$v = \frac{(Z_i - Z_u)\, M}{4\pi A} \cdot \frac{r}{\xi_i^3}\,, \tag{44.11}$$

M being the mass of the body. In the same region the isentropic surfaces are raised (or depressed) through approximately

$$z - z_u = \frac{\lambda}{g}\, \frac{(Z_i - Z_u)\, M}{4\pi A^3} \cdot \frac{\theta\,(\theta_i - \theta)}{\xi_i^3}\,. \tag{44.12}$$

When we approach the earth's surface or the tropopause, the wind becomes stronger, and the lifting of the isentropic surfaces weaker than the above formulae indicate. At the ground the wind is twice as strong, whereas $z - z_u$ vanishes, according to our assumption $\theta_0 = \text{const}$.

In general, it is found that in an isolated mass with abnormal potential vorticity the vertical stability as well as the absolute vorticity deviate from the normal in the same sense as the potential vorticity. (The normal value of the absolute vorticity is λ.) In our special case the relative deviation in absolute vorticity corresponds to about $\frac{2}{3}$, the relative stability deviation to about $\frac{1}{3}$ of the relative excess in potential vorticity.

In order to obtain the pressure field we have to express v as a function of r and z. Then integration of the geostrophic wind equation with respect to r produces the departure from the undisturbed pressure field. The central depression at the ground produced by a body of the type in question is

$$\Delta p_0 = -\frac{\varrho_0 \lambda\, (Z_i - Z_u)\, M}{2\pi\, (\theta_i - \theta_0)}\,. \tag{44.13}$$

In our model this value amounts to about 24 mb.

From the construction of such models the following general conclusions may be drawn. In geostrophic equilibrium an air-mass of relatively high potential vorticity ($Z_i > Z_u$) establishes a cyclone. The strongest cyclonic rotation is met with in the layer to which the air-mass belongs. Below the mass the isentropic surfaces are raised, above they are depressed. Thus the cyclone has a cold core below and a warm one above. If the "producing mass" is lying on the ground, we obtain a cyclone which is warm throughout.

An air-mass of relatively low potential vorticity ($Z_i < Z_u$) gives rise to an anticyclone. The deviations in pressure and temperature have the opposite directional sense to those in cyclones.

Since the boundary value problem (44.1) to (44.6) is linear and homogeneous in v and $(z - z_u)$, the effect of *several* bodies of abnormal potential vorticity is equal to the sum of the effects of the separate bodies. This is valid even if the bodies are not situated on one and the same vertical axis. The separate wind fields are added vectorially, the deviations in pressure and temperature algebraically.

From this it follows that the above statements may be reversed. For instance: a cyclone requires a mass of relatively high potential vorticity. Or: a cold air-mass only remains cold as long as there are masses of high potential vorticity above or masses of reduced potential vorticity below it. When this condition is no longer fulfilled, the air sinks down and loses the character of a cold mass by adiabatic heating.

All the statements just outlined must be modified if there is no uniform potential temperature at the ground. An isolated area or tongue of potentially low surface temperatures, *the other conditions being unchanged* shifts the whole wind field towards the anticyclonic side. By contrast, if the lowest isentropic layers are missing locally, then the wind field is shifted towards the cyclonic side. Both phenomena occur particularly in connection with cyclones. Their influence on wind and pressure, however, is appreciable only in the lower troposphere.

The wind systems of the middle and high latitudes have often their greatest intensity near the tropopause. Consequently, the strongest isentropic variations in potential vorticity occur here also. They are connected in a characteristic manner with the discontinuity in the temperature lapse rate which marks the tropopause. In the anticyclones the tropopause lies relatively high and hence at relatively high potential temperatures; in the cyclones both the elevation and potential temperature of the tropopause are relatively low. The tropopause is no coherent surface but rather consists of sections at different levels with discontinuous transitions between the sections. Thus a rather thick isentropic layer exists which at some places belongs to the stratosphere and at others to the troposphere.

The static stability in the stratosphere is 3 to 4 times as large as in the troposphere. According to what has been said above on the connection between stability and potential vorticity in producing masses, the potential vorticity varies by a factor of ten or more in the layer concerned. In the other layers of the atmosphere the range of variation is always appreciably smaller, if we disregard inversions and other small-scale phenomena. Thus the phenomenon of the tropopause appears to be closely connected with the upper-air flow pattern.

The tropopause cyclones, as they may be called, are accompanied by a cold troposphere and a warm stratosphere, according to Eq. (44.12). In the tropopause anticyclones these conditions are reversed.

The singular character of the tropopause level and the special form of the temperature field connected with it is illustrated by the well-known correlations between tropopause height, pressure and temperature at different heights as computed by DINES[1] and SCHEDLER[2].

45. Vortices in the general wind field. *α) Two-dimensional treatment.* For the purpose of studying the behaviour of vortices within the general wind field, two-dimensional non-divergent motions in a horizontal surface are often considered. The wind fields pertaining to vortices which extend through the greatest part of the atmosphere with nearly constant strength, can be approximately described in this way.

As is known from the kinematics of the two-dimensional motion, the non-divergent motion on a sphere is completely specified, the absolute vorticity being given in each point on the sphere. The vorticity field is subject to the single condition that its integral over the whole spherical surface vanishes[3]. If a vortex can be represented by a point—this is justified in the case of very small cyclones and anticyclones—, its displacement is, at any instant, determined by the distribution of absolute (or relative) vorticity over the earth.

In general the area covered by a vortex is of such extent that its representation by a point is not justified. In view of this, it is remarkable that most vortices change their shape only very slightly with time. The vortices are usually steady with respect to shape. This means that a frame of reference can be found with respect to which the motion in the region of the vortex is stationary. It is then possible to refer to the motion of the vortex as a whole which is characterised by the motion of that frame. This motion is in general composed of a translation and a rotation.

[1] W. H. DINES: Geophys. Mem. **1**, No. 2, 31 (1911).

[2] A. SCHEDLER: Beitr. Phys. fr. Atmosph. **7**, 88 (1915).

[3] See H. LAMB, Hydrodynamics, 6th ed. Cambridge University Press 1932. § 160. German edition: Lehrbuch der Hydrodynamik. Leipzig u. Berlin: 1931. § 160.

From the individual conservation of absolute vorticity in frictionless motion it follows that the streamlines in this frame must coincide with the lines of constant absolute vorticity[1]. Vortices which do not change their shape, may thus be distinguished by the property that the absolute vorticity is so distributed in their fields that the streamlines resulting in a certain frame of reference, coincide with the lines of equal absolute vorticity.

The only known stream patterns which satisfy this condition on the sphere, are the fields of motion given by Eq. (39.26). By this equation tesselated patterns of cyclones and anticyclones rotating uniformly about an axis of the sphere, may be represented. If the earth's surface is assumed to be flat ($\lambda = \text{const}$), a single circular vortex within a uniform basic flow may also fulfil the above condition[2]. In a CARTESian coordinate system moving in the x-direction with velocity a, the equations of frictionless motion take the form

$$\frac{\partial u}{\partial t} + u\frac{\partial u}{\partial x} + v\frac{\partial u}{\partial y} - \lambda v = -\frac{1}{\varrho}\frac{\partial p}{\partial x}, \tag{45.1}$$

$$\frac{\partial v}{\partial t} + u\frac{\partial v}{\partial x} + v\frac{\partial v}{\partial y} + \lambda(u + a) = -\frac{1}{\varrho}\frac{\partial p}{\partial y}. \tag{45.2}$$

They are satisfied if the stream function ψ is a time independent function of the distance r from the origin of the moving system:

$$\psi = \psi(r) \tag{45.3}$$

and if the pressure field is subject to the condition:

$$\frac{p}{\varrho} = \lambda\psi + \int \frac{(\psi')^2}{r}\,dr - \lambda a y. \tag{45.4}$$

The resulting field of flow obviously satisfies the above condition.

When referred to the solid earth, the motion is composed of the circular vortex (disturbance flow) and a uniform translation in the direction of x (basic flow). Fig. 24 shows the streamlines and pressure lines in a special case. For simplicity it is assumed that $\psi' \equiv 0$ beyond a radius of 750 km. Then the disturbance flow is spatially bounded, and the integral of the relative vorticity over the whole area of the vortex vanishes. The inner parts of the vortex of which the relative vorticity is positive, are surrounded by a ring with negative relative vorticity (Fig. 24b).

The velocity of the basic flow (20 mps) is taken to be little greater than the maximum velocity of the disturbance flow (18 mps). Then closed pressure lines still occur, whilst the streamlines are already open. With stronger basic flow, the closed pressure lines also vanish, leaving only a trough in the pressure field.

It may be shown that a stationary circular vortex on a flat earth can also be realised when the basic flow consists of a rigid rotation about a fixed point. In this case the centre of the vortex must likewise move with the basic flow, if absolute vorticity is to be individually conserved. As in the case discussed above, this field has the property that the basic flow is free of deformation.

If the basic flow involves deformation, a circular vortex—i.e. a circular area with circularly distributed vorticity—cannot maintain its shape[3]. However also in this case vortex shapes obviously exist which are stationary in some frame of

[1] Conversely the coincidence of streamlines and lines of equal absolute vorticity is a sufficient condition for the dynamic realisation of a stationary wind field.

[2] F. M. EXNER: Ann. Hydrogr., Berlin **48**, 414 (1920) or [7], p. 268.

[3] The practice, followed in some investigations, of regarding the vortex under these circumstances, as a circular rigid cylinder and applying the laws relevant to such a cylinder in shearing fluid, is therefore not justified.

reference. Such shapes are conceivably elliptic or oval. HESSELBERG[1] and MÖLLER[2] developed methods by means of which the wind field can be computed when the pressure field is given as a simple analytic function of time and space, or vice versa. Wind and pressure in the inner parts of an elliptic vortex can be represented approximately correctly by these methods.

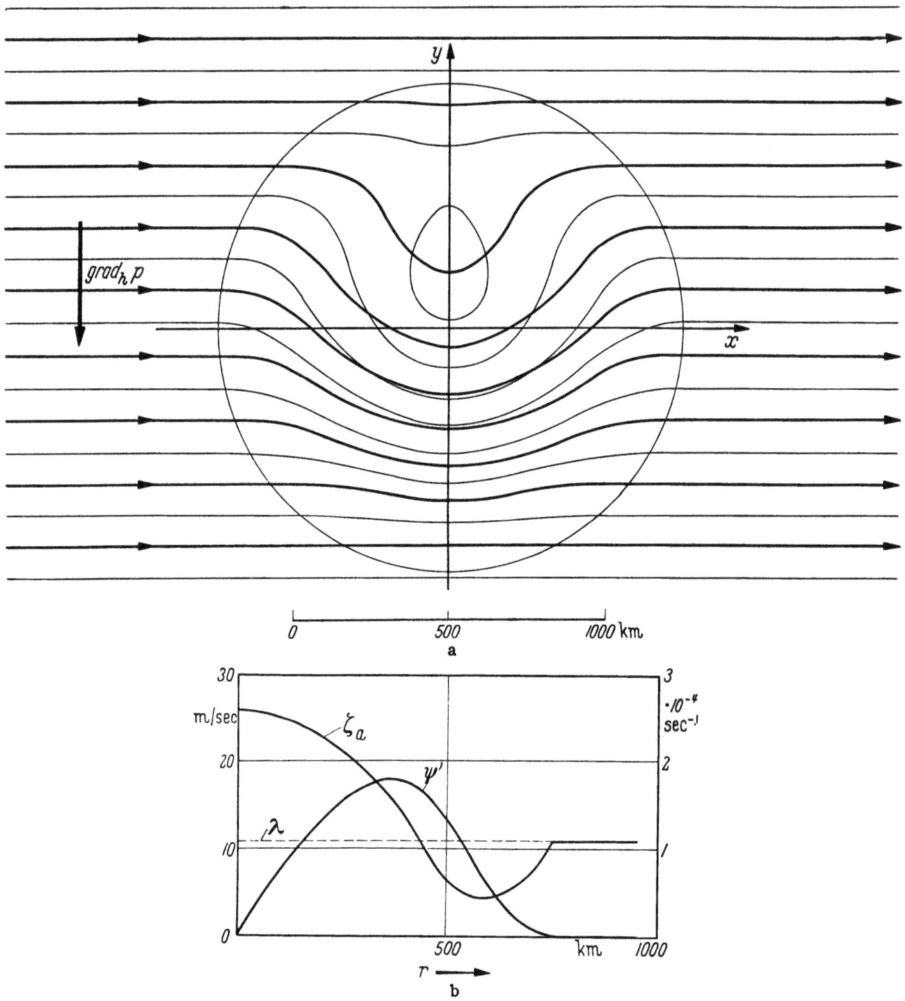

Fig. 24 a and b. Circular vortex in a uniform basic flow. (a) Heavy lines are streamlines, thin lines are pressure lines. (b) Circular velocity ψ' and absolute vorticity ζ_a in dependance on the distance from the cyclones centre.

The majority of atmospheric vortices are not exactly circular which corresponds to the fact that the basic flow as represented by the flow at larger distances from the vortex, is seldom free of deformation.

It has been maintained that the curvature of the earth's surface exerts an influence on the shape and motion of vortices. ROSSBY[3], taking account of the dependance of λ upon latitude, showed that a vortex is subject to a horizontal

[1] TH. HESSELBERG: Geophys. Publ. 9, No. 8 (1932).
[2] F. MÖLLER: Meteor. Z. 53, 284 (1936).
[3] C.-G. ROSSBY: J. Meteor. 6, 163 (1949).

force, the pressure along the outer edge of the vortex being assumed constant. Cyclones are driven polewards and anticyclones, towards the equator. It is however uncertain whether the assumption of constant pressure along the outer edge is justified[1].

β) Three-dimensional treatment. It is observed that, in many vortices, the masses with deviating potential vorticity are limited to a comparatively thin isentropic layer, the deviations themselves being accordingly pronounced. The cyclone of Fig. 23 may serve as an example although the structure of the potential vorticity field is mostly not as simple as is shown in the figure. In actual high-level cyclones the mass with high potential vorticity is usually surrounded by a ring of reduced potential vorticity.

Again, such complexes of masses with strongly deviating potential vorticity are observed to maintain their external shape nearly unchanged over long periods. In the three-dimensional case it is not the absolute vorticity which serves as individual constant but, if heat sources are neglected in addition to friction, potential vorticity and potential temperature. Consequently, steadiness of a vortex means now that the streamlines coincide with the intersection lines of the isentropic surfaces and surfaces of constant potential vorticity, the motion being referred to a frame moving with the vortex.

Once more the flow can only be exemplified by simple cases. One such example is the case in which a parallel basic flow without horizontal or vertical shear, is superimposed on a circular vortex with vertical axis. Since the isentropic surfaces are practically horizontal in the undisturbed basic flow, the superposition is not accompanied by a change in height of the isentropic surfaces. As is easily seen, the above condition is satisfied when the vortex moves with the basic flow. The motion is in all parts horizontal.

If a baroclinic basic flow is superimposed on a circular vortex with vertical axis, the condition for stationary motion can only be partly fulfilled. Let, e.g., a flow in the x-direction ($\boldsymbol{i}\, u_2$) be superimposed on the flow of the cyclone ($\boldsymbol{v}_1$), shown in Fig. 23, such that

$$u_2 = a\,\theta + b\,, \tag{45.5}$$

a and b being arbitrary constants. Since the thermal wind equation for a flow of this nature takes the form

$$\lambda\left(\theta\frac{\partial u_2}{\partial \theta} - u_2\right) = g\,\frac{\partial z_2}{\partial y}\,, \tag{45.6}$$

the isentropic surfaces are raised in the undisturbed basic flow over the standard height z_u by the amount

$$z_2 - z_u = -\frac{b\,\lambda}{g}\,y\,. \tag{45.7}$$

If this height increment is added to the height increment of the isentropic surfaces in the vortex at rest ($z_1 - z_u$), the resulting temperature field satisfies the thermal wind equation for the compound flow $\boldsymbol{v}_1 + \boldsymbol{i}\,u_2$.

When the vortex moves with the velocity $\boldsymbol{i}\,\bar{u}$, $\bar{u}$ being arbitrary, the condition that the streamlines lie in the isentropic surfaces in the moving frame, may be fulfilled by putting the vertical velocity,

$$w = \big(\boldsymbol{v}_1 + \boldsymbol{i}\,(u_2 - \bar{u})\big)\cdot \mathrm{grad}_\theta\,(z_1 + z_2)\,. \tag{45.8}$$

The other condition, that the streamlines lie on the surfaces of constant potential vorticity, is then also satisfied outside the imbedded body, since these surfaces here

[1] See R. W. JAMES: J. Meteor. 9, 447 (1952).

coincide with the isentropic surfaces. Within the body deviations occur[1]. These deviations are obviously smallest, when $\bar{u}$ is the speed of the basic flow at the body's centre; (furthermore, they decrease with the vertical extent of the body as compared to the vertical shear of the basic flow).

Fig. 25 shows the streamlines relative to the earth and the vertical velocities calculated from (45.8), both in the 300° surface, with $\bar{u}=u_2(325°)=20$ mps and $u_2(300°)=10$ mps. Since the first velocity is taken to be twice the latter, inversion of the figure shows the streamlines in the frame moving with the vortex. The

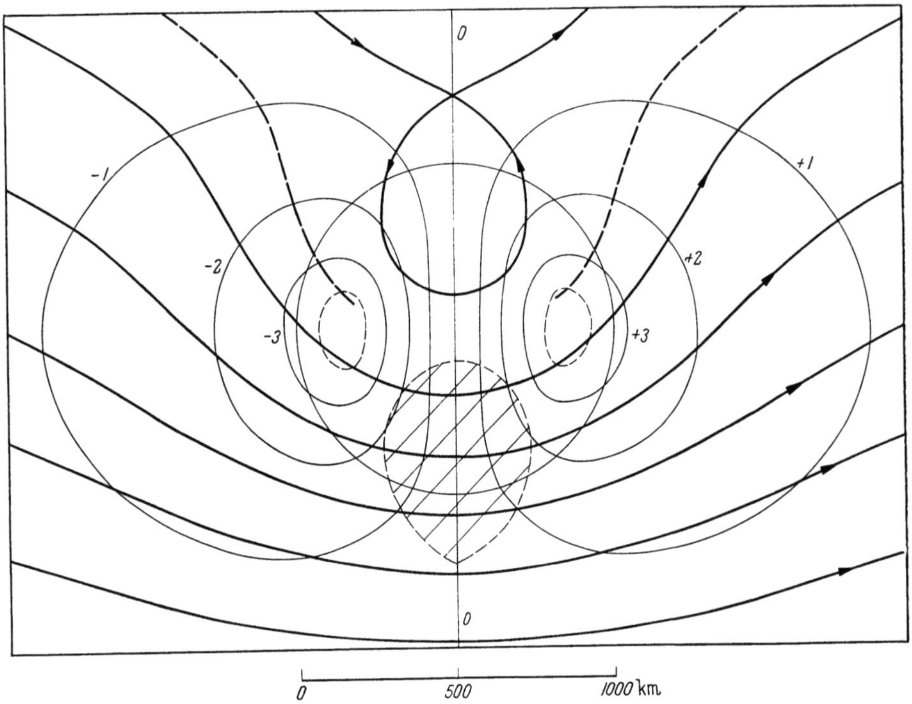

Fig. 25. The cyclone of Fig. 23 in a baroclinic basic flow. Streamlines (heavy lines) and lines of equal vertical velocity (thin lines, numbers indicate cm/sec) on the surface of 300° absolute potential temperature.

hatched area (the mirror image of the area of closed streamlines) presents the part of the 300° surface permanently accompanying the vortex. All other parts of the surface flow past the vortex. Also the air in the streamline loop is constantly being replaced.

It may, in general, be taken for granted that a vortex which is produced by a relatively shallow mass with deviating potential vorticity, moves with a mean velocity of the basic flow within the producing mass. It is likely that a set of useful prognostic rules can be explained by this fact, by means of which the shift of pressure patterns or pressure tendency areas in a certain level is traced back to the flow in some other level. One of the best known is e.g.: The surface pressure systems move with approximately half the speed of the 500 mb-wind over them. The validity of this rule is probably founded on the fact that the centre of the producing masses of the surface pressure systems is mostly located at a height between the 800 and the 700 mb-level where the basic flow is about half as strong

[1] Another defect is the fact that the equation of continuity and the kinematic boundary condition at the ground are only approximately satisfied.

as in the 500 mb-level. Further rules of this kind have been compiled in treatises on weather forecasting, e.g. by DUNN[1] and REUTER[2].

A study of models such as described above shows that, if a cyclone lies in a baroclinic basic flow, ascending motion takes place on that side of the cyclone, towards which the thermal wind vector of the basic flow points. Descending motion prevails on the opposite side. The descent and ascent take place below as well as above the producing masses. In an anticyclone within a baroclinic basic flow the vertical motions are reversed.

These vertical movements determine to a large extent the weather in the different sectors of a migrating vortex. Superimposed upon them are the vertical motions caused by surface friction. Thereby the regions of rising air in the lower layers are shifted towards the centre of the cyclones and the regions of subsidence towards that of the anticyclones.

The above statements were so formulated that their validity is independent of the motion of the vortex. In the following form this motion is taken into account: If the basic flow has the same direction in the producing mass as the thermal wind vector, the air ahead of a cyclone ascends, that in the rear descends. This case is predominantly realized in the temperate latitudes where the general west drift increases with height. It constitutes here the "*polar weather type*" after MÜGGE[3]. If, on the other hand, the basic flow in the producing masses has the opposite direction to the thermal wind vector, the converse is true. Then we have the "*subtropical weather type*", in which the bad weather is met with in the rear of the cyclones.

FLEAGLE[4] and MILLER[5] have determined the large-scale vertical motion in the neighbourhood of migrating vortices from aerological observations. As far as can be seen, the results obtained by these authors are in fairly good agreement with our statements.

46. General remarks on the origin of vortices. The jet stream and the polar front. From the relations outlined in Sect. 44 it follows that the genesis of a vortex is in all cases accompanied by a redistribution of potential vorticity. We can in principle distinguish two categories of vortex-development. The one is marked by the fact that heat sources—including the heat of condensation—and turbulent friction can be ignored. Then the redistribution is brought about only by advection. We can call this class of vortex-development the "*adiabatic*" *development*, where the word adiabatic is to mean dry-adiabatic and frictionless. In the second category heat sources and turbulent friction play a decisive role.

A tendency towards vortex development of the first kind always exists when a bounded air mass is meridionally displaced. The potential vorticity increases on the average from the equator to the poles. This holds true for the atmosphere as a whole as well as for the separate isentropic layers. Thus an air mass moving towards the equator usually comes into contact with masses of lower potential vorticity than its own. Consequently a cyclone arises. On the other hand a displacement towards the poles usually gives rise to anticyclogenesis.

Apart from this tendency, adiabatic vortex-developments are to be expected preferably, where there are pronounced local differences in potential vorticity within the isentropic layers and consequently also strong wind gradients. This is especially the case in the vicinity of the jet stream.

---

[1] G. E. DUNN: Short range weather forecasting, in Compendium of Meteorology. A.M.S., p. 747, Boston-Mass. 1951.

[2] H. REUTER: Methoden und Probleme der Wettervorhersage. Vienna 1954.

[3] R. MÜGGE: Meteor. Z. **48**, 1 (1931).

[4] R. G. FLEAGLE: J. Meteor. **4**, 165 (1947).

[5] J. E. MILLER: Meteor. Papers New York Univ. **1**, No. 1 (1948).

In view of this, it is remarkable that strong cyclonic or anticyclonic developments only occur in the neighbourhood of the jet stream. It is likely that some of them are practically adiabatic processes in the above sense. This has, for instance, been demonstrated by the fairly exact prediction of a strong cyclogenesis by means of a three-parameter model after Sect. 42, published by CHARNEY (ref. p. 112). Cyclogenesis from a frontal wave, on the other hand, depends essentially on the release of latent heat, as will be shown in Sect. 49. Again, as for other types of cyclogenesis, it has not yet been finally decided whether they are adiabatic processes or not (Sect. 47).

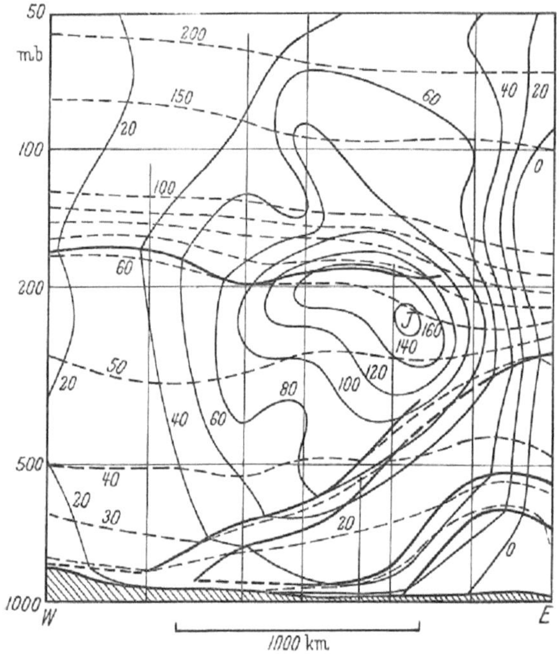

Fig. 26. Vertical cross-section through a north-south directed jet (after BOVILLE, CRESWICK and GILLIS[1], with minor alterations). Solid heavy lines: Discontinuities in temperature gradient. Solid thin lines: Isotachs (knots). Dashed lines: Lines of equal potential temperature (degrees centigrade).

Before we describe single types of cyclone formation, we discuss structure and properties of the jet stream.

The *jet stream* forms the core of the westerlies of temperate and subtropic latitudes in each hemisphere. It is a relatively narrow tube containing strong winds located at about the tropopause level and which more or less clearly circumscribes the whole hemisphere. The course of the jet stream varies strongly both in time and in space. Some times it flows nearly zonally over many degrees of longitude, at times its flow is wavelike or strongly meandering. The mean maximum velocity may reach some 40 mps. Sometimes, however, the velocity increases to appreciably higher values. In exceptional cases a maximum velocity of 150 mps may be reached. Fig. 26 gives a vertical cross section through such a high-speed jet. Its maximum velocity is some 170 Knots = 88 mps.

A characteristic feature of the temperature field of the jet stream is the discontinuity in height of the tropopause. On the equatorial side the tropopause is located appreciably higher than on the polar side, in agreement with what has been stated in Sect. 44 on the location of the tropopause in cyclones and anticyclones. The jet stream can be regarded as a large irregularly shaped high-level cyclone encircling the pole.

The region underneath the jet stream is marked by a strong horizontal temperature gradient. This region is termed the *frontal zone*. When the velocity in the jet reaches high values in temperate latitudes, the frontal zone is often intensified to front formation, as is the case in Fig. 26. The front usually lies a little steeper than the isentropic surfaces, and may take the form of a strong inversion of which the lower portion either ends near the ground in a high-pressure region or else intersects the ground in a trough of low pressure. A sharp front may be considered as a surface of discontinuity. Wind and temperature on both sides of the surface

[1] B. W. BOVILLE, W. S. CRESWICK and J. J. GILLIS: Tellus 7, 314 (1955).

approximately satisfy the formula of MARGULES (Sect. 12). The front underneath the jet stream is identical with the polar front discovered by the Norwegian meteorologists from the surface maps.

For further details regarding structure and behaviour of the jet stream the reader is referred to publications of PALMÉN[1], CRESSMAN[2] and to the comprehensive treatments of RIEHL[3] and NAMIAS and CLAPP[4].

As to the dynamics of the jet stream final clarity has not been reached. In Chap. V the jet stream is treated as a planetary phenomenon. It is shown there that, for the momentum balance of the zonally averaged jet, a meridional exchange of momentum plays a predominant role. This exchange is mainly conveyed by the large-scale disturbances of the westerlies such as cyclones, anticyclones etc. The motion in the separate current may for this purpose be regarded as frictionless. For the explanation of some details of the planetary jet stream, however, this assumption does not seem to be justified. This is particularly true of the high-speed jets.

The high-speed jets are some thousands of km in length. They are usually subject to a translation in their direction of flow but with a much smaller velocity than that of the air (order of magnitude 15 m/sec). They thus contain a stream flowing from rear to front. The air is accelerated in the entrance region, decelerated in the exit or "delta" region. As was already mentioned in Sect. 27, very small, possibly even negative values of absolute vorticity occur on the equatorial side of such jets. Since the isentropic surfaces lie horizontally in this region and are relatively widely spaced, the potential vorticity is also approximately zero. The region of nearly vanishing potential vorticity originates simultaneously with the high-speed jet. This shows that at least the formation of a high-speed jet is not an adiabatic process in the above sense. The low potential vorticity values could, according to Eq. (5.20), be produced by a suitable distribution of non-adiabatic heating and cooling. More probably they may be brought about by a locally bounded, additional field of force directed along the axis of the originating jet. This is suggested by the observation that simultaneously a region with very high potential vorticity originates on the polar side of the jet.

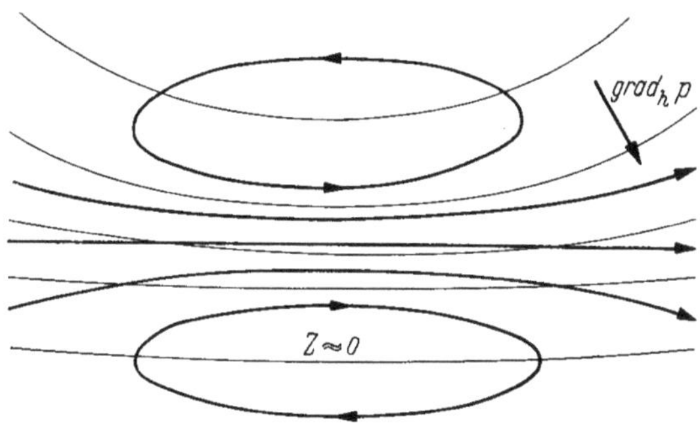

Fig. 27. Schematic representation of frictionless flow through a high-speed jet. Heavy lines: streamlines in a frame moving with the jet and lines of equal potential vorticity. Thin lines: pressure lines.

Still it is conceivable that at least the fully developed, stationary jet obeys the equations of frictionless motion. The stream lines would then have to coincide with the intersection lines of the isentropic surfaces and the surfaces of potential vorticity in a frame moving with the jet. The horizontal field of flow in this frame would, roughly, be given by Fig. 27. But even this does not seem to hold true.

In the absence of friction, the individual changes in speed require motions across the isobars. MURRAY and DANIELS[5] have proved that such motions actually

[1] E. PALMÉN: J. Meteor. 5, 20 (1948).
[2] G. P. CRESSMAN: J. Meteor. 7, 39 (1950).
[3] M. A. ALAKA, C. L. JORDAN and R. J. RENARD: The Jet Stream, ed. by H. RIEHL, U.S. Bureau of Aeronautics, 1953.
[4] J. NAMIAS and P. F. CLAPP: In: Comp. of Met., p. 551. Boston-Mass. 1951.
[5] R. MURRAY and S. M. DANIELS: Quart, J. Roy. Meteor. Soc. 79, 236 (1953).

exist in the entrance and exit regions and are, in the mean, of such nature as to account for the respective accelerations. On the average over all parts of the jet however, they found a definite component directed towards the higher pressure. This means that the total work done by the horizontal pressure forces on the unit mass during its flow through the high-speed jet (time t_1 to t_2) is negative

$$-\int_{t_1}^{t_2} \alpha\, \boldsymbol{v}_h \cdot \mathrm{grad}_h\, p\, dt < 0. \tag{46.1}$$

Since the kinetic energy before and after the jet is about the same, the mechanical energy Eq. (9.10) is obviously not satisfied. In order to satisfy it, an additional force $\boldsymbol{A}$ must be postulated such that

$$\int_{t_1}^{t_2} \alpha\, \boldsymbol{v}_h \cdot \boldsymbol{A}\, dt > 0. \tag{46.2}$$

FAUST[1] also found that in the high-speed jets a component of motion towards higher pressures exists in the mean. He furthermore ascertained that the wind component parallel to the isobars is on the average stronger than the gradient wind. This result may possibly be explained by the fact that the isobars from which the gradient wind is computed, are more cyclonically curved than the trajectories, as a consequence of the flow across the isobars in the entrance and exit regions (Fig. 27). Another explanation was given by HOLLMANN[2].

In this connection an investigation by REINEKE[3] must be mentioned according to which a mean wind component towards higher pressure in general exists in the upper troposphere. Such a flow seems also to be necessary to compensate the mass transport towards lower pressure in the friction layer (Sect. 17) that would smooth out the pressure differences at the ground in a few days, if no counter-flux existed in the upper layers. Again, in order to account for such a flux, the existence of additional forces must be postulated which have, in the mean, an accelerating action (see also Sect. 54).

The formation of the polar front is also to a large extent an open question. The two most obvious properties of fronts are the strong temperature gradient and the strong wind shear between the warm and cold air. A theory, suggested by BERGERON[4] and developed by PETTERSSEN[5], explains the intensification of the temperature contrast purely kinematically. According to this theory, the front originates as the result of a deformation field within the single horizontal planes. If a constant temperature gradient initially exists in these planes, the isotherms after some time congregate along the axis of dilatation. With indefinitely increasing time a discontinuity in temperature appears. BERGERON[6] was able to show that the polar front preferably takes on form in regions where the mean geostrophic wind field is subject to a strong, zonally directed dilatation, that is, in certain saddle points of the mean horizontal pressure field. The theory, however, is unsatisfactory inasmuch as it does not account for the creation of the wind shear.

The relevant equation for the intensification of the 3-dimensional wind shear is, if friction may be neglected, Eq. (5.11) and for the increase of the horizontal shear Eq. (5.14) or (5.32), both applied to a field of motion which does not vary in

[1] H. FAUST: Arch. Meteor., Geophys. u. Biokl. **8**, 45 (1955).
[2] G. HOLLMANN: Meteor. Rdsch. **8**, 105 (1955).
[3] I. REINEKE: Meteor. Abh. fr. Univ. Berlin **1**, H. 1 (1950).
[4] T. BERGERON: Geof. Publ. **5**, No. 6 (1928).
[5] S. PETTERSSEN: Geof. Publ. **11**, No. 6 (1936).
[6] T. BERGERON: Meteor. Z. **47**, 246 (1930).

the direction of the initial flow. REED and SANDERS[1] found in a single case of frontogenesis that the "tilting term", $(\mathrm{curl}\, \boldsymbol{v}_a)_h \cdot \mathrm{grad}\, w$, became decidedly positive in the inner parts of the frontal zone due to strong subsidence on its warm side. Other investigations, however, failed to demonstrate a subsidence of this nature[2,3].

It is certain that the formation of a front can not be completely explained by a process which is dry-adiabatic as well as friction-free. From wind and temperature observations one finds, that the potential vorticity in the interior of a sharp front often amounts to ten times that in the adjacent masses of air. This difference must be generated during the formation of the front, since here as well as in the circular vortex a one-to-one correspondence exists between the geostrophic wind field and the potential vorticity distribution.

In cases where the front is accompanied by clouds and precipitation, moist-adiabatic upgliding of the air perpendicular to the basic flow may serve as an explanation for its formation. Such upgliding is possible if dynamic instability exists in the surfaces of equal wet-bulb potential temperature. In fact, fronts with upslide clouds and precipitation often show such instability[4]. They end at the ground in a trough of low pressure, and the upgliding air is supplied, from either or both sides, by the surface friction flows. The nature of the whole process is probably similar to that of the motion in the tropical cyclones (Sect. 48). The field of deformation in the planes normal to the basic flow may account for the front formation.

For the fronts which originate without cloud formation and which often do not immerge into the friction layer a complete theory does not yet exist.

47. The cut-off cyclone. As long as the jet stream flows in a straight course or in waves of not too large an amplitude, the surfaces of equal potential vorticity are approximately parallel to the streamlines. A coherent belt with high potential vorticity is located on the polar side of the jet, a belt with reduced potential vorticity on the equatorial side. This is no longer the case when the amplitudes become too large or when the flow pattern becomes meandering. This "break-down" of the predominantly zonal flow is always accompanied by the formation of separated, more or less compact masses of relatively high or low potential vorticity.

A typical process of this kind is the "*cutting-off*" process. Fig. 28 shows on four maps at 12 hours intervals the change in flow in the potential temperature surface $\theta = 35$ centigrade during a case of cutting-off. From a predominantly zonal flow of the jet stream a trough developes which stretches far to the south. Out of the southern part of this trough the cyclone arises.

The vertical structure of newly formed cut-off cyclones was investigated particularly by PALMÉN[5,6]. The characteristics of this structure are seen from Fig. 29. It shows a west-east cross section along the dashed line in map 28d. The cyclone contains a mass of air (A) in the upper troposphere whose potential vorticity is about 6 times the normal tropospheric value. It is almost completely surrounded by air with strongly reduced potential vorticity. The mass A is essentially the producing mass of the cyclone. Its origin and development constitute the most important problem in the cutting-off action.

---

[1] R. J. REED and F. SANDERS: J. Meteor. **10**, 338 (1953).
[2] W. E. HUBERT: Tellus **5**, 359 (1953).
[3] L. A. VUORELA: Geophysica **4**, 105 (1953).
[4] Compare J. BJERKNES: Comp. of Meteor., Amer. Meteor. Soc. **1951**, 590.
[5] E. PALMÉN: Tellus **1**, Nr. 1, 22 (1949).
[6] Comp. of Meteor., Amer. Meteor. Soc. **1951**, 599.

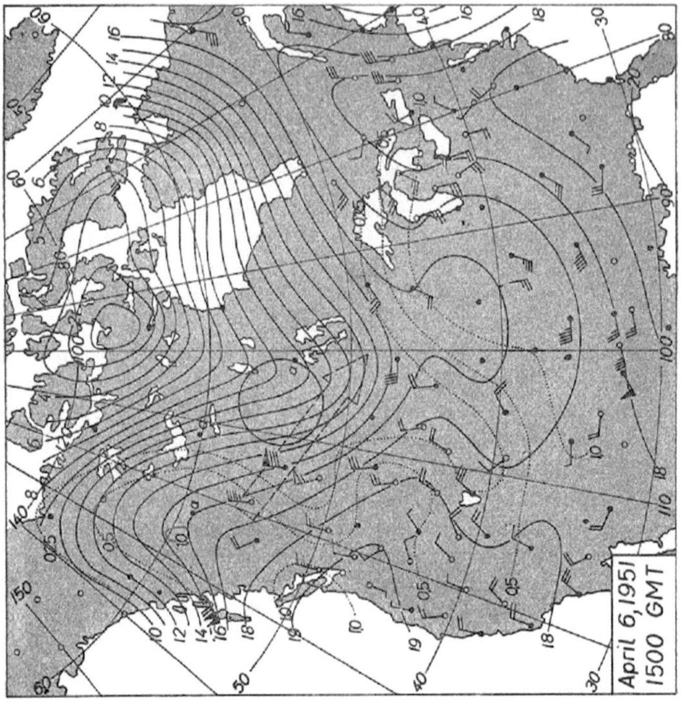

b

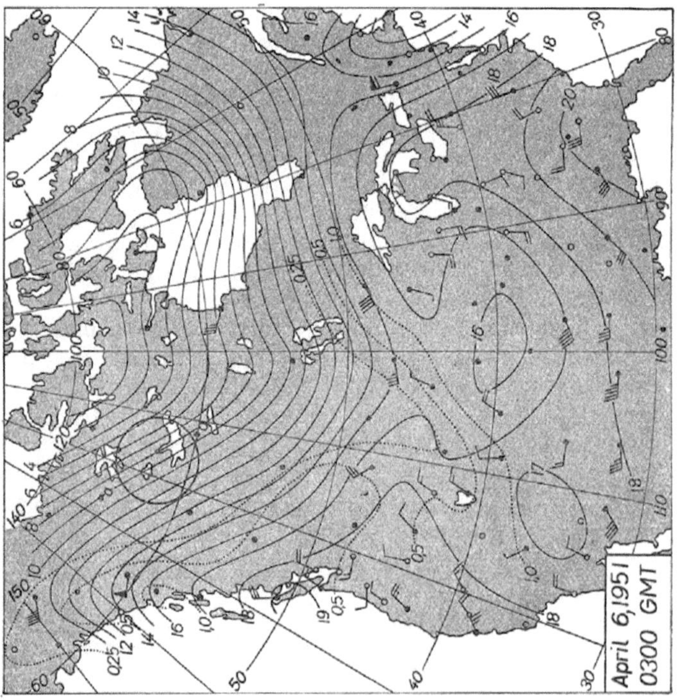

a

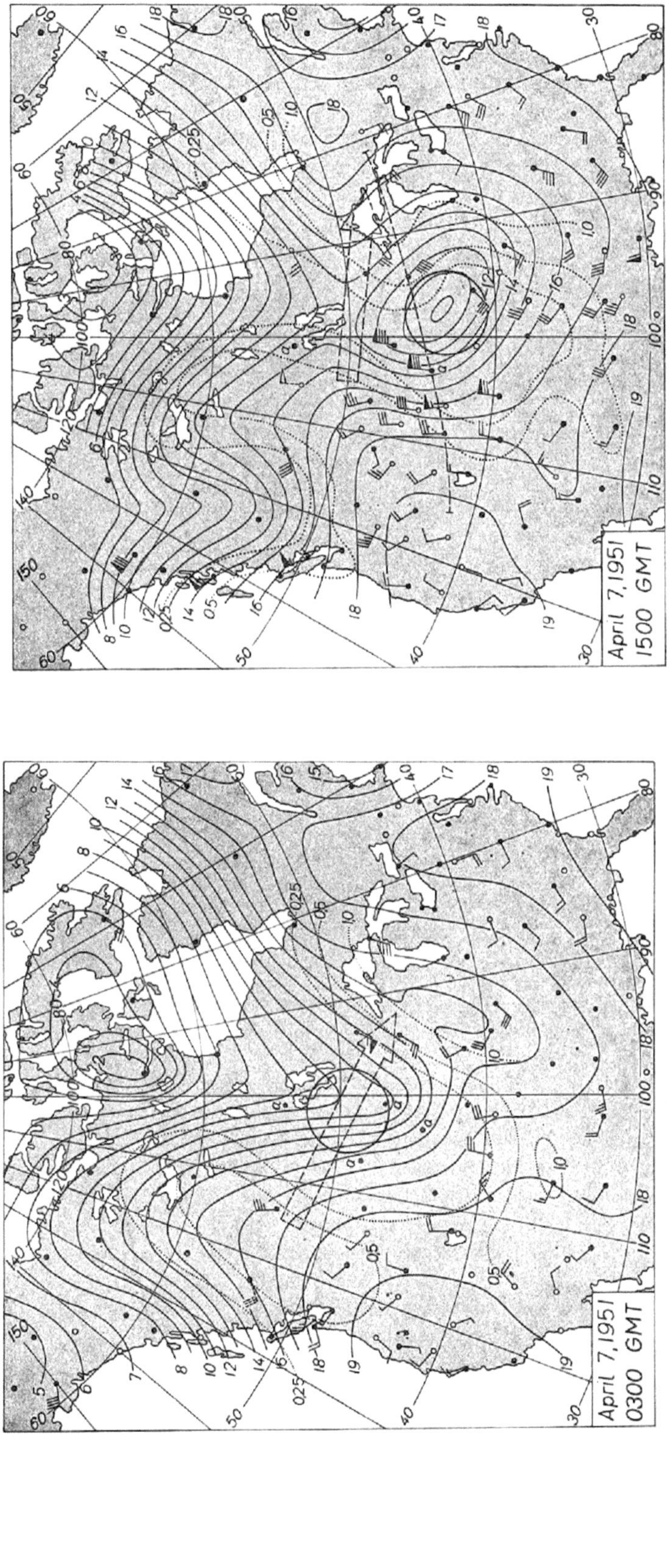

c d

Fig. 28a–d. Twelve-hourly sequence of four maps showing the flow in the isentropic surface $\theta = 35$ Centigrade. Full lines are streamlines of the geostrophic wind, that is lines of equal $(\varphi + h)$ (compare Sect. 10β, p. 26). The dotted lines are some lines of equal specific humidity.

The presumption nearest at hand is that this mass was originally part of the deep polar stratosphere north of the jet stream, and that it became detached through the action of some mechanism and was displaced towards lower latitudes.

In general the spontaneity of the process suggests that it is due to some instability of the jet stream. In addition to the strongly reduced dynamic stability on the equatorial side, the jet stream usually shows baroclinic instability. Even with moderate maximum velocity the vertical shear underneath the jet is of the order of magnitude of 3 mps/km so that the flow is unstable by the criteria given in Sect. 33 for a wave length region below, say 5000 km.

On the other hand, it is often observed that the jet stream maintains its rectilinear course for days, even with high velocities. Thus the impression is

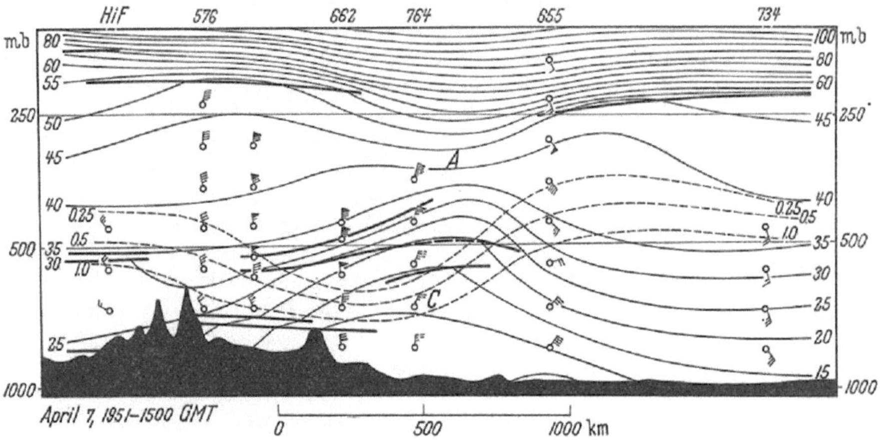

Fig. 29. Vertical cross section along the dashed line in map 28*d*, showing the producing mass *A* of the cut-off cyclone and the cold dome *C*. Thin solid lines are lines of equal potential temperature (degrees centigrade). Heavy lines indicate discontinuities in temperature gradient. Dotted lines are some lines of equal specific humidity.

created that it does not concern instability with respect to infinitely small perturbations, but rather that a disturbance of finite magnitude is necessary to initiate the process (compare PALMÉN in ref. [6] of page 125).

Unfortunately the origin of the producing mass *A* can only with difficulty be determined from upper air observations. KRUHL[1], studying a single case, found that the mass in question originated in the transition region between the polar stratosphere and the subtropical troposphere, that is more or less in the centre of the jet stream. Its potential vorticity was initially almost as great as later on in the fully developed cyclone.

In contrast, KLEINSCHMIDT[2] found in the case represented by Fig. 28, that the potential vorticity of the mass *A* has increased to more than the three-fold value in the 24 hours between map-time 1 and map-time 3, and subsequently decreased somewhat. The circles on the maps represent the probable positions of the mass. If this is true, cutting-off is decidedly not an adiabatic process in the sense of Sect. 46. A closer investigation revealed the probable existence of an additional field of force, which was approximately located as shown by the large arrows in Fig. 28.

NEWTON and coll.[3] while studying a similar development, viz. the formation of a pronounced high-level trough, determined that the deciding factor was the

[1] H. KRUHL: Ann. Meteor 6, 65, 321 (1953/54).

[2] E. KLEINSCHMIDT: Tellus 7, 96 (1955).

[3] C. W. NEWTON, N. A. PHILLIPS, J. E. CARSON and D. L. BRADBURY: Tellus 3, 154 (1951).

acceleration of the northerly flow on the western side of the trough. Using the BERNOULLI equation for frictionless motion they calculated a value of 15° for the angle of deviation from the geostrophic wind towards the low pressure. From the wind observations however no such deviation can be detected, again indicating the action of an additional accelerating force.

Final clarity on the character of the cutting-off process can only be achieved by the study of further cases.

In most cases of cutting-off action it is observed, that the jet stream is initially accelerated to high velocities on the west side of the growing trough. This high-speed jet under which a sharp front is usually formed, shifts gradually towards the southern and then the eastern side of the cyclone. Here a frontal cyclone almost regularly originates (Sect. 49).

Under the cut-off cyclone the isentropic levels are raised, thus forming a "*cold dome*". The cold dome becomes particularly clear-cut when polar air is transported to the south by the frontal cyclone originating in the east. The raising action by the cut-off cyclone produces very low temperatures in the lower troposphere. If the polar air is carried further south, it sinks down again. Investigations by CROCKER[1], HSIEH[2], PALMÉN and NEWTON[3] reveal details of these processes.

A cut-off cyclone, once formed, can persist for a week or even longer, because the frictional loss of angular momentum in the upper layers is small. If the vortex is imbedded in a baroclinic basic flow, it is often confined to the upper troposphere and the lower stratosphere. In the lower layers it can then only be recognized by the raised position of the isentropic surfaces. A detailed description of such a "drop of cold air" was e.g. given by BUSCHNER[4].

The high-level anticyclones, which originate at the jet stream are to a certain extent the counterpart of the cut-off cyclones. However, their origin is probably governed by other laws. This is born out by the conspicuous absence of developments such as shown in Fig. 28, but with changed sign. The formation of an anticyclone is in fact always preceded by the formation of a trough or a cut-off cyclone. The anticyclone then originates on the eastern side of the low pressure region. This process often repeats itself several times, each time somewhat further towards the west. This has the effect of destroying the zonal flow successively over many degrees of longitude and of splitting up the jet stream into several branches.

The newly-formed high pressure areas often fuse to form a powerful "*blocking high*", the centre of which lies somewhat further north than the original position of the jet stream. The separate cut-off cyclones lie on the south side of this high. Blocking actions have been studied e.g. by REX[5], BERGGREN, BOLIN and ROSSBY[6] and NAMIAS[7].

48. Tropical cyclones. This section precedes the section on frontal cyclones, because the tropical cyclones, or hurricanes or typhoons as they are called, involve *in pure form* the process which probably also leads to the formation of frontal cyclones.

With a life-time of many days, the motion in the hurricane may be considered to be stationary. The ideal hurricane is symmetrical with respect to a vertical

[1] A. M. CROCKER: Quart J. Roy. Meteor. Soc. **75**, 57 (1949).
[2] Y. HSIEH: J Meteor. **6**, 401 (1949).
[3] E. PALMÉN and C. W. NEWTON: J. Meteor. **8**, 25 (1951).
[4] W. BUSCHNER: Ber. dtsch. Wetterdienst US-Zone **1951**, Nr. 23.
[5] D. F. REX: Tellus **2**, 196, 275 (1950).
[6] R. BERGGREN, B. BOLIN and C.-G. ROSSBY: Tellus **1**, 2/14 (1949)
[7] J. NAMIAS: Extended Forecasting by mean Circulation Methods. U.S.Weather Bureau, Washington D.C., 1947.

axis. Fig. 30 shows a cross section running through the axis of a hurricane. The dimensions nearly correspond to that of the typhoon "Marge", one of the strongest typhoons investigated till now[1]. The innermost part, the "eye" of the hurricane is nearly calm. The motion of the air outside the eye is composed of a circular component and of a vertical circulation (1-2-3-1 in Fig. 30). The circular component increases strongly towards the inner parts, reaching velocities up to 100 mps near the eye. Further inwards it suddenly diminishes and becomes nearly zero at the edge of the eye. The total motion takes on the form of a double-spiral. The influx near the ground (1-2 in the figure) can be estimated from surface observa-

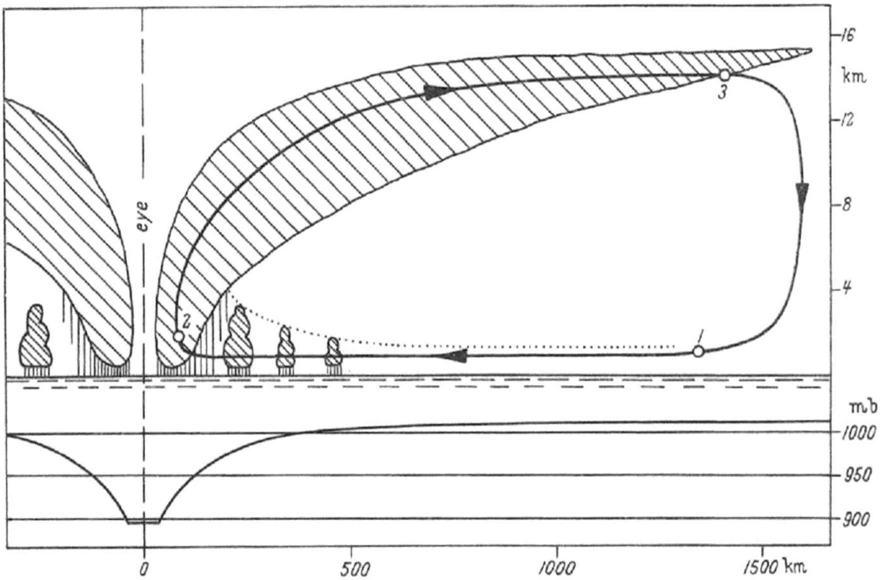

Fig. 30. Axial cross section and surface pressure distribution of a tropical cyclone.

tions. Existing alone it would fill up the typhoon in a few hours. The outflow (2-3) can be recognized by the widespread cloud canopy. In the part 3-1 of the circulation the air slowly subsides on a broad base. This subsidence is confirmed by a very low humidity under the outer edge of the cloud canopy[2].

For further details the reader is referred to the extensive literature. A recent survey over the problem of the tropical cyclone with a nearly complete list of references has been given by BERGERON[3].

In order to describe the motion in the storm zone we introduce cylinder coordinates around the axis. Then the equations of frictionless motion are

$$\frac{Du}{Dt} = -\frac{1}{\varrho}\frac{\partial p}{\partial r} + \lambda v + \frac{v^2}{r}, \tag{48.1}$$

$$\frac{Dv}{Dt} = -\lambda u - \frac{u v}{r}. \tag{48.2}$$

By introducing the absolute angular momentum of the unit mass with respect to the typhoon's axis,

$$m_i = r\left(v + \frac{\lambda}{2} r\right), \tag{48.3}$$

[1] R. H. SIMPSON: Bull. Amer. Meteor. Soc. **33**, 286 (1952).
[2] R. H. SIMPSON: Month. Weath. Rev. **75**, 53 (1947).
[3] T. BERGERON: Quart. J. Roy. Meteor. Soc. **80**, 131 (1954).

Eq. (48.1) becomes

$$\frac{Du}{Dt} = -\frac{1}{\varrho}\frac{\partial p}{\partial r} + \frac{m_t^2}{r^3} - \frac{\lambda^2}{4} r \tag{48.4}$$

and (48.2) becomes

$$\frac{Dm_t}{Dt} = 0, \tag{48.5}$$

which shows that the absolute angular momentum of a particle is constant in the absence of friction.

Observations now show that m_t decreases markedly towards the interior in the influx near the ground. This is e.g. seen from Fig. 31 which presents the circular wind component

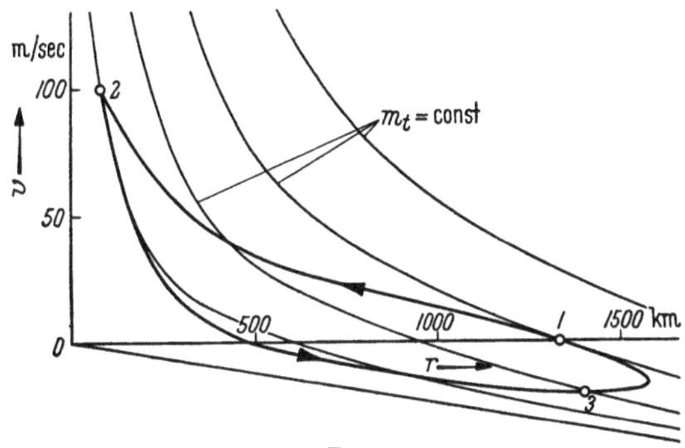

Fig. 31.

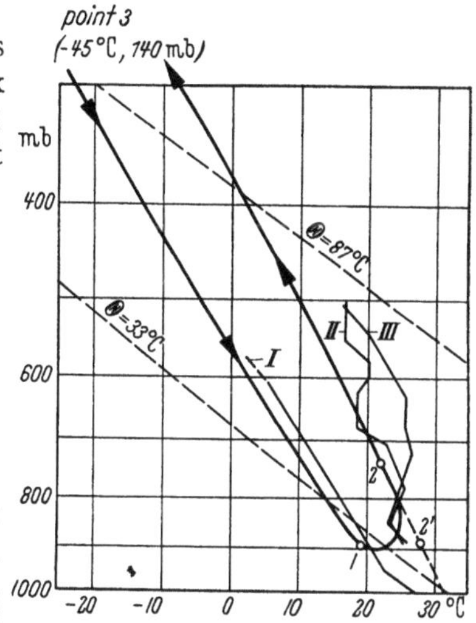

Fig. 32.

Fig. 31. Circular wind diagram for typhoon "Marge".

Fig. 32. The probable thermodynamic path of the circulating air in "Marge" (heavy curve). The two dashed lines denote dry-adiabates. Curve I: Sounding made at a distance of about 480 km from the centre. Curve II and III: Soundings made in the eye.

during one revolution as it approximately occurred in "Marge". From the decrease of m_t in section 1—2 it follows that the influx is a frictional flux and that above a certain frictional height no further influx occurs. The upper limit of the friction layer is approximately given by the dotted line in Fig. 30.

The behaviour of the circular wind component in the outflow as presented by the curve in Fig. 31 is based on the assumption that friction here also plays a role, although by far not to the same extent as in the influx because v rapidly decreases to small absolute values. Already at the distance of 500 km v becomes negative. The anticyclonic motion in the outer parts of the cloud canopy is confirmed by observation.

Fig. 32 shows part of the thermodynamic path described by the revolving air. The two branches crossing the upper edge of the diagram intersect in point 3 which is determined by the data quoted in the diagram. The ascending branch follows the moist-adiabate through point 2 which is defined by the lowest surface pressure and by the water temperature (point 2'). The descending branch approximately follows the vertical temperature distribution in the environs of the hurricane. The required decrease in potential temperature is effected by radiation.

Above the frictional layer the thermal wind equation is approximately valid. It takes the form:

$$\frac{1}{r^3}\frac{\partial}{\partial z}(\varrho\, m_t^2) = -g\frac{\partial \varrho}{\partial r}, \tag{48.6}$$

which we may deduce from (48.4), neglecting Du/Dt and eliminating p by means of the static equation. The density and wind fields within the circulation curve in Fig. 30 must satisfy Eq. (48.6). In addition the wind field must satisfy the boundary conditions on the circulation curve imposed by Fig. 31, and the density field, those imposed by Fig. 32. From these conditions it is probably possible to explain the essential features of the hurricane, especially the phenomenon of the eye and the shape of the outflow. This problem, however, has not yet been solved.

In the interior of the circulation curve the circular wind speed decreases upwards, the temperature increases inwards. Since above the inner part of the outflow, defined by positive values of v, the wind speed continues to decrease upwards, the temperature in this region likewise continues to rise inwards, as a consequence of Eq. (48.6). Thus the upper parts of the eye are warmer than the correspondent parts of the outflow. This has been verified by several temperature soundings made in the eyes of hurricanes. In Fig. 32 three temperature curves measured in "Marge" are entered.

The high temperatures in the eye of a tropical cyclone entail an upward weakening of the depression, turning it to a high in the upper layers. This is also required by the anticyclonic motion in the outer parts of the outflow.

In order to gain information on the energy conversions in tropical cyclones we use the Bernoulli equation

$$\Delta \frac{\boldsymbol{v}^2}{2} + g\,\Delta z + \int \frac{\partial p}{\varrho} + F = 0. \tag{48.7}$$

Here F denotes the work per gr done against friction. Applying the equation to a whole circuit 1-2-3-1 we get

$$F_{\text{total}} = -\oint \frac{\partial p}{\varrho}\,. \tag{48.8}$$

That means, the frictional work done during one revolution is equal to the thermodynamic energy gained during this revolution. In our example this energy amounts to 6600 m²/sec².

Again, applying (48.7) to the section 1—2 of the circuit and assuming that the entire frictional work is done in this section, we obtain for the velocity at point 2: $v = 96$ mps. If only $\frac{2}{3}$ of the total frictional dissipation takes place during the influx, we get $v = 117$ mps. These values well agree with the wind speeds estimated in strong typhoons.

The high value of the thermodynamic energy is only reached, if the air between point 1 and point 2 is heated up to the water temperature and becomes saturated with water vapour, so that it is immediately able to ascend along the moist-adiabatic from point 2. Both is possible only in virtue of the immense turbulence within the friction layer. Thus the hurricane itself procures the required energy. The ocean remains slightly cooled behind it. Strong hurricanes only occur over warm oceans. When they come over cooler water, they begin to wane, and likewise, when they move in over land, where the supply of latent heat is stopped.

As for the origin of tropical cyclones, a ring-shaped zone with dynamic instability is probably necessary to set the circulation in motion. The conditions for such an instability zone have been derived by Kleinschmidt[1]. The theoretical results agree with the conditions found empirically by Riehl[2].

49. Frontal cyclones. Since the frontal cyclones are by far the strongest cyclones appearing in the surface maps of the higher latitudes, they have for a long time been considered the prototype of cyclones. Now it has become evident that they are only a particular, although very interesting, species of cyclone.

[1] E. Kleinschmidt: Arch. Meteor., Geophys. u. Biokl., Ser. A **4**, 53 (1951).

[2] H. Riehl: Misc. Rep. No. 24, Dept. of Meteorol. Univ. Chicago, 1948.

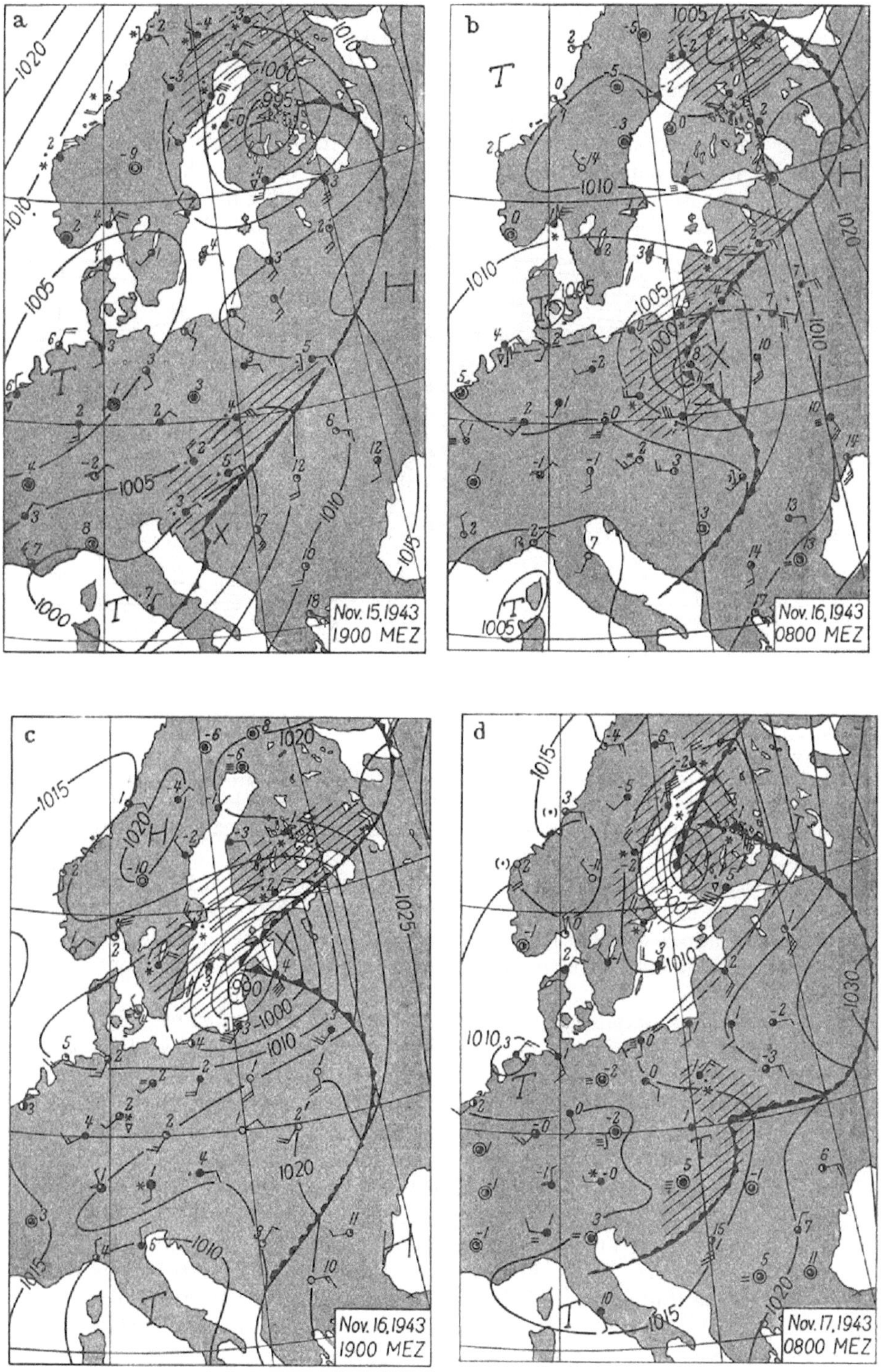

Fig. 33a—d. Twelve-hourly sequence of four surface maps, showing the origin of a frontal cyclone from a frontal wave (X).

V. BJERKNES[1] has inaugurated the theory that the frontal cyclone develops from a wave on the polar front. A survey of the theoretical work done in this connection has been given in Sect. 34. Indeed it is a well-known fact that frontal cyclones only originate below the jet stream, and that the initial stages of the development take on the form of a wave on a well-marked polar front. However, the pure wave motion, taken as a thoroughly dry-adiabatic process, cannot account for the formation of the cyclone. The latter requires the creation of an air mass of increased potential vorticity, and this producing mass is in fact always observed in newly formed frontal cyclones, lying immediately on the ground.

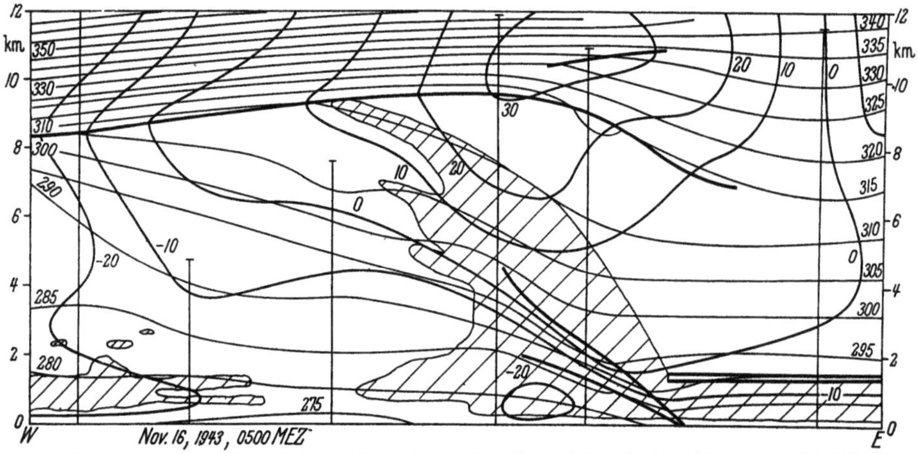

Fig. 34. Vertical cross section through the leading part of the frontal wave (along the dashed line in map 33b). Thin lines are lines of equal potential temperature (degrees absolute). Medium lines—normal component of the geostrophic wind relative to the wave (m/sec). Heavy lines—discontinuities in temperature gradient. Clouds are indicated by hatching.

KLEINSCHMIDT advanced the opinion that it is the difference between the thermodynamic behaviour of rising and subsiding air, which leads to the formation of the producing mass. The process may be described by means of a real case of frontal cyclogenesis[2].

In the weather situation represented by the surface map a of Fig. 33 the centre of a tropopause cyclone was situated over Denmark. On its east side the speed of the southerly upper winds reached 50 mps. A distinct front was lying below the jet stream (see Fig. 34). It intersected the ground along the surface front plotted in Fig. 33a.

On this front an amplifying wave runs towards NNE, marked by X in the four surface maps. Fig. 34 shows a cross section running through the leading part of the wave along the dashed line in Fig. 33b. In the cross section the normal component of the wind is plotted, referred to a frame moving with the wave. It is seen that the zero line nearly coincides with the upper surface of the front. This means, that the wave nearly moves with the mean velocity of the cold and the warm air on both sides of the front. Thus two opposite streams flow through the wave. The upper stream goes forward, the lower stream flows backwards.

Fig. 35 gives a picture of the three-dimensional flow, referred to a system moving with the wave. Since the flow is approximately stationary in this system —at least in the leading part of the wave—streamlines can be constructed from the observed wind and the slope of the dry- (or moist-) adiabatic surfaces. The five lines marked by $S1$, $S3$, $S5$, $S7$ and $S9$ are five streamlines which cross

[1] V. BJERKNES: Geof. Publ. **2**, No. 4 (1921).
[2] E. KLEINSCHMIDT: Meteor. Rdsch. **3**, 54 (1950).

station G in 1, 3, 5, 7, 9 km of height, respectively. Along these lines, values of height and potential temperature are plotted.

In the leading part of the wave the motion in all heights has a strong component towards the west. Hence the air flows upwards on both sides of the front, thus

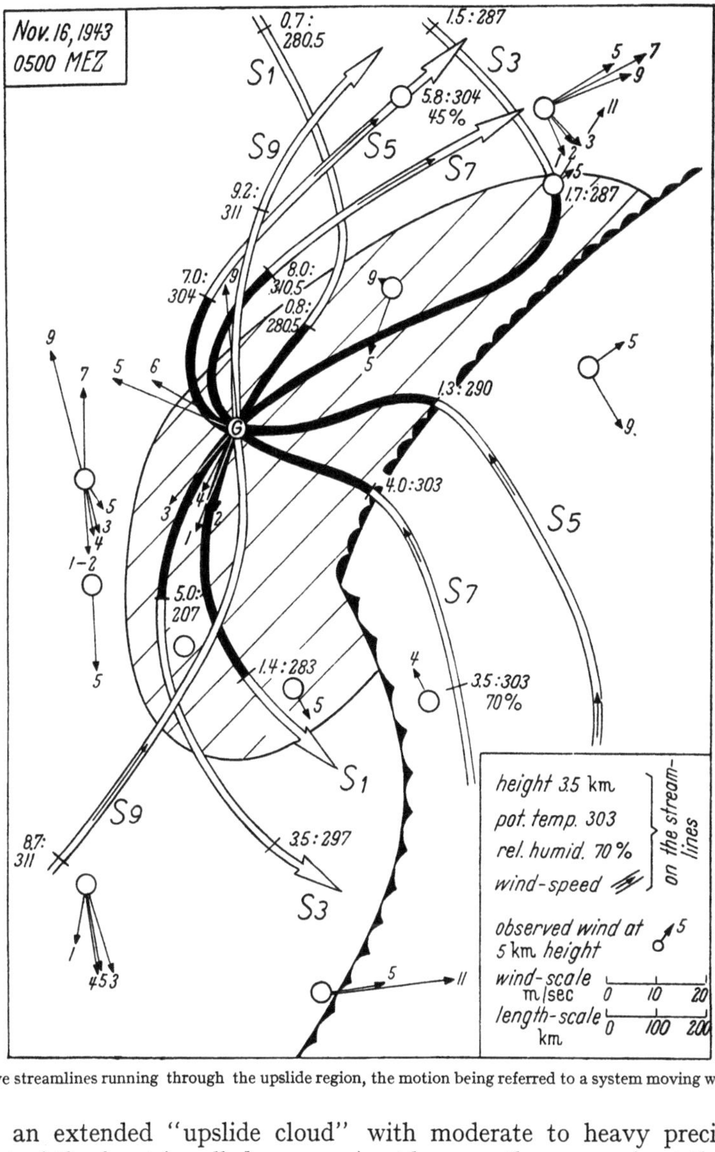

Fig. 35. Five streamlines running through the upslide region, the motion being referred to a system moving with the wave.

forming an extended "upslide cloud" with moderate to heavy precipitation. This part of the front is called a *warm front* because the warm air at the ground advances and the cold recedes.

To a certain extent the motion of the lower stream in this part is similar to that in the storm zone of a tropical cyclone. This leads to an interpretation of the warm-front motion, which differs somewhat from the conception of an unstable, increasing wave, namely: The release of latent heat is the moving agent of the whole process. The flow towards the front is governed by the surface friction. The field of motion tends towards a stationary state as far as the processes in the environs of the warm front admit. This state is determined by the amount of the released heat on the one hand, by the work done against friction and against the stabilizing forces in the surrounding masses on the other[1].

In Fig. 36 which corresponds to Fig. 34 the arrows indicate the vectors of the absolute vorticity as far as they depend on the wind component normal to the cross section. They represent the main part of curl $\boldsymbol{v}_a$. The closed lines are lines of equal vertical velocity w. As is seen from the figure, curl $\boldsymbol{v}_a \cdot$ grad w is predominantly positive in the lower stream, negative in the upper stream. This means, according to Eq. (5.11) that the vertical vorticity component increases individually in the lower current, whereas in the upper stream it decreases; for on the average, the values of curl $\boldsymbol{v}_a \cdot$ grad w exceed the vertical components of curl $\boldsymbol{v}_a$ div $\boldsymbol{v}_a$ and $\boldsymbol{S}$. The increase of the vorticity in the lower stream can immediately be seen from the surface maps, the decrease in the upper stream from the upper air maps or, for instance, from the cirrus tracks ahead of the upslide cloud, which are anticyclonically curved.

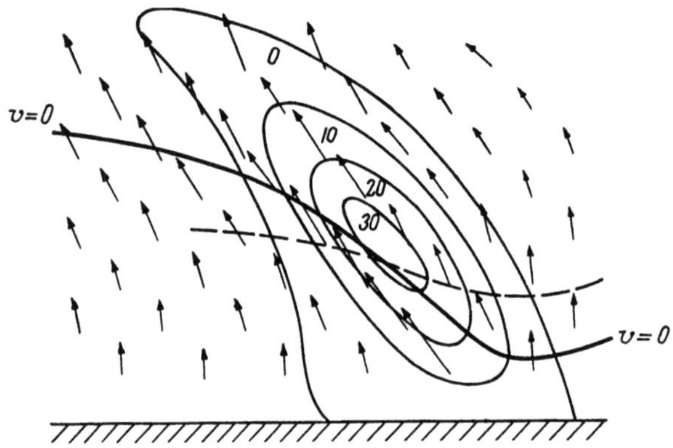

Fig. 36. Lines of equal vertical velocity (cm/sec) in the cross section of Fig. 34. The arrows indicate components of the absolute vorticity vector. The heavy line indicates the boundary between upper and lower stream.

Since the rate of change of potential temperature by condensation is proportional in rough approximation to the vertical velocity, it follows that for the term curl $\boldsymbol{v}_a \cdot \text{grad}\,\frac{D\theta}{Dt}$ almost the same holds as for curl $\boldsymbol{v}_a \cdot$ grad w. Thus, according to Eq. (5.19), the potential vorticity increases in the lower current and decreases in the upper current. In our case the average potential vorticity in the lower current rose to more than twice the initial value.

Behind the apex of the wave the air of the lower current sinks down. The cloud dissolves. The high potential vorticity, however, cannot be reduced, because now the air behaves dry-adiabatically again. Therefore the air does not subside to its former levels and for a great part retains the high absolute vorticity which it had upon leaving the cloud. Its trajectories remain cyclonically curved. In this part the front, as a "*cold front*", is pushed towards the warm side (Fig. 33d). The cold air which has flowed outside the cloud turns into its former direction.

The air with increased potential vorticity gathers in a continually growing body, which is the producing mass of the new cyclone. Thereby the warm air at the ground is being removed further and further from the cyclone's centre. The process is called the *occlusion* of the frontal cyclone. It leads to the termination of the warm-front process. Fig. 37 shows a cross section through the rear part of the newly-formed body, taken along the dashed line in the chart 33d. (Note that the cross-section time is 3 hours earlier than the map-time.) The dotted line denotes the approximate surface of the body.

If the jet stream is not too strong, the air masses leaving the cloud with the upper stream form a high-level anticyclone ahead of the cyclone. Its maximum

[1] In the tropical cyclones the "upper stream" does not exist, except in the initial stages of the cyclone. Here it causes a strong subsidence in the interior parts of the eye. Since thereby the eye becomes warmer and warmer, the upper stream rapidly dies away.

occurs at 5 to 8 km of height. In the surface maps it is only faintly distinguishable. Thus the warm-front process acts as a mechanism simultaneously producing a low cyclone and a high anticyclone. If the jet stream is strong, the upper masses are carried forward and may continue their path in the form of a ROSSBY wave (Sect. 30). In any case the jet stream slows down and broadens. Conversely, a broadening of the jet stream, such as occurs in the delta of a high-speed jet, is particularly favourable for the development of a cyclone from a frontal wave. This has been pointed out by SCHERHAG.

The strongest and most frequent warm-front cyclones arise below the jet streams directed towards the north-east. If the jet stream belongs to a high-level cyclone, the result is a large cyclonic system with an inclined axis, extending through the whole troposphere and far into the stratosphere. The two producing

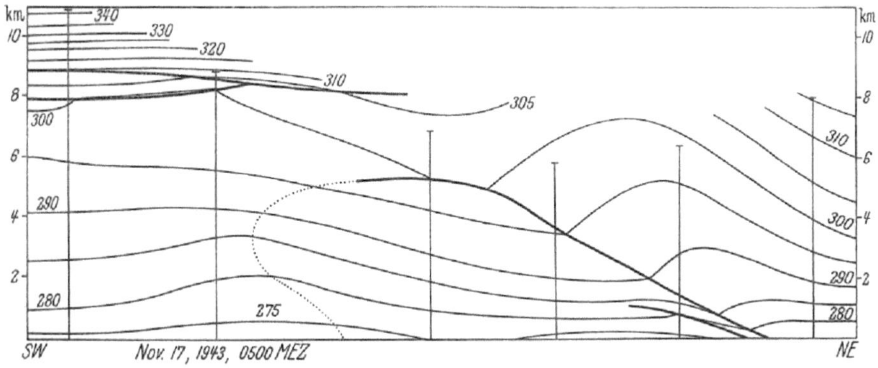

Fig. 37. Cross section through the rear part of the newly-formed cyclone (along the dashed line in map 33d).

masses, however, can always be distinguished clearly (compare Fig. 37). When the warm-front process is finished, the low cyclone is dissolved in a rather short time by the ground friction, and the cyclonic motion recedes to the upper levels. It appears as if an originally warm cyclone is transformed into a cold one.

V. The general circulation.

50. The zonally averaged circulation. The conception, "general circulation", comprises the largest wind systems in the spectrum of atmospheric motions down to systems of the magnitude of the cyclones and anticyclones. Averaging over sufficiently long periods of time does away with the smaller amongst these systems. The remaining ones are large streams encircling the entire globe which remain nearly unchanged when annual averages are taken. These are in the first instance implied by the term general circulation. Apart from certain irregularities which are caused by land and sea distribution and by mountains, they are symmetrically placed with respect to the earth's axis. The motion in these currents is predominantly zonally directed.

The principal problem pertaining to these streams is the question of their maintenance. Friction opposes their continuation, presumably seeking to reduce all velocities to that of a solid rotation with the earth, that is to produce a general calm. To maintain the general circulation against friction, heat sources and sinks are necessary (Sect. 6ε). A theory of the general circulation thus differs from the theories developed in the preceding chapters mainly inasmuch as both friction and heat sources are drawn into consideration a priori.

An analytic treatment of these two influences on the wind field is only possible in simple flow patterns e.g. in wind systems symmetric with respect to a vertical axis as is approximately the case in some cyclones and anticyclones. In order to investigate their influence on the global circulation, it is advantageous to approximate the latter with a wind field averaged in zonal direction over all longitudes. The zonally averaged circulation can easily be treated theoretically as it is strictly

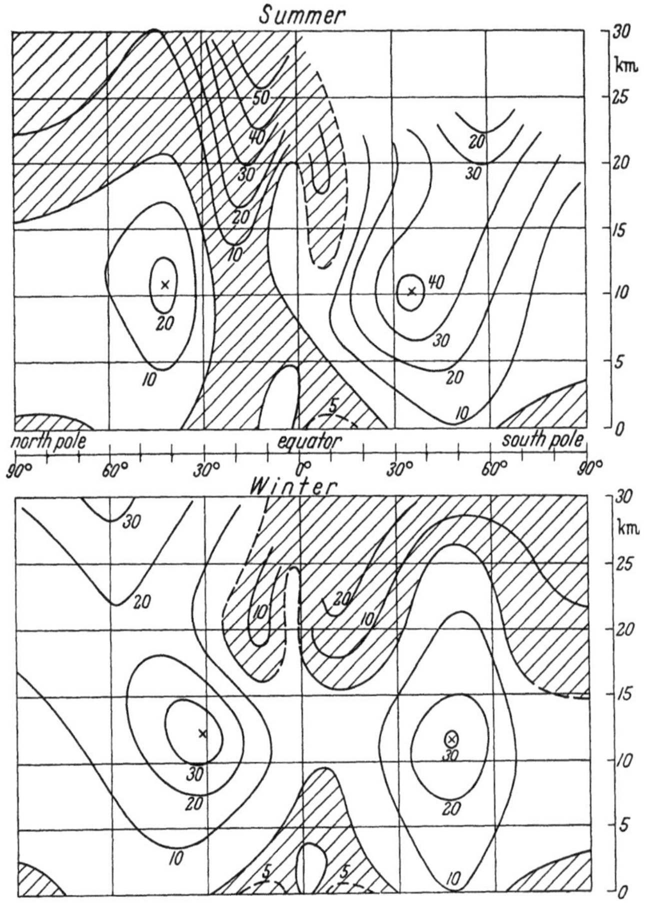

Fig. 38. The mean zonal wind distribution, half-annually averaged, after H. FLOHN[1]. Isotachs in m/sec. Easterly winds marked by hatching. Summer and winter denote seasons for the northern hemisphere.

symmetric with respect to the earth's axis. It can, moreover, be studied relatively easily empirically as the observed winds are referred to geographic coordinates. Most of the literature on general circulation does, in fact, treat the zonally averaged circulation.

Fig. 38 shows the *zonal* wind component after FLOHN[1], averaged half-annually over all longitudes, summer and winter being presented separately. The maxima of the westerly flow are found approximately in the tropopause level, slightly nearer to the poles than the subtropical high pressure belts on the surface. The latter are marked by the points in which the zero lines cut the surface. The two

[1] H. FLOHN: Ber. dtsch. Wetterdienst US-Zone **1950**, Nr. 12, 156.

maxima do not coincide with the average latitude of the jet streams, the latter flowing on the average nearer the poles. This discrepancy is explained by the fact that the meandering of the westerlies increases with increasing latitude. The jet streams often oscillate over 30 degrees of latitude and hence do not particularly make themselves felt in the zonal averages.

The easterly winds in tropical regions in surface proximity are brought about by the trade winds which form a rather steady flow over most longitudes.

The *meridional* components of the averaged circulation reach higher values only in the friction layer. The average friction wind follows from the mean pressure distribution in surface proximity. In the tropical east wind zones the mean friction flow is directed towards the equator, reaches a height of 1 to 2 km and has a velocity of about 1 mps. In the west wind regions, mean frictional flows are found of about the same strength which are directed towards the poles. In the polar regions the zonally averaged friction flow is again directed towards lower latitudes.

Hence it follows that on each hemisphere at least 3 closed mean meridional circulations must exist. The courses taken by these circulations in the free atmosphere, whether they extend to other zones aloft as has often been supposed etc., are well-nigh unknown. From the data published by STARR and WHITE[1] it becomes clear that the courses alternate strongly with time. In the annual mean, however, certain principal features seem to appear. Thus at 13° N the counterstream to the friction flow near the surface is for the greatest part found at a height of 10 to 14 km. Mostly the zonally averaged meridional velocities in the free atmosphere amount to only a few dm/sec, very seldom to more than 1 m/sec.

It is possible that in the westerlies of the low latitudes considerable systematic deviations from the geostrophic wind occur. This can be concluded for instance from an investigation by LOEWE and RADOK[2]. The flow in high latitudes, on the other hand, seems to be approximately geostrophic. For the zonally averaged flow this implies that the zonal wind component corresponds to an average meridional pressure gradient according to Eq. (12.2). In the same way the change of the mean zonal wind with height corresponds to an average meridional temperature gradient on the isobars according to Eq. (12.5). In the troposphere where an overall upwards increase of the westerly wind component occurs, an isobaric temperature rise directed from pole to equator exists. In the stratosphere the temperature rise is directed polewards, in summer generally, in winter partly.

The significance of heat sources is easily demonstrated in the case of a circulation assumed to be uniform along each circle of latitude. Friction continuously deprives the westerly and easterly streams of momentum. In order to balance this loss in the individual zones, meridional circulations effecting a transport of momentum are necessary. Meridional circulations, however, also involve a heat transport, bringing about local heating or cooling in the different zones. To preserve the initial state of equilibrium, heat sources and sinks must be distributed in such a way as to compensate for these temperature changes.

The following quasi-static theory of meridional motion, developed by ELIASSEN[3], contains the analytical description of this consideration.

51. Meridional circulations in a circular vortex generated by friction or heat sources. We consider the vortex treated in Sects. 25 to 27, referring the motion to a rotating earth. Introducing instead of absolute circulation, the absolute

[1] V. P. STARR and R. M. WHITE: Tellus **4**, 118, 332 (1952).
[2] F. LOEWE and U. RADOK: J. Meteor. **7**, 58 (1950).
[3] A. ELIASSEN: Astrophys. norvegica **5**, Nr. 2, 19 (1952).

angular momentum per unit mass

$$m_a = \Omega R^2 + R u, \tag{51.1}$$

the equation of balance (26.4) becomes

$$-\alpha \operatorname{grad} p + \frac{m_a^2}{R^3} \operatorname{grad} R - \operatorname{grad} \varphi = 0. \tag{51.2}$$

Using the distance from the axis R and the pressure p as coordinates in the meridional plane, we obtain from the vector Eq. (51.2) the two component equations for the directions R = const and p = const

$$-\alpha - \frac{\partial \varphi}{\partial p} = 0, \tag{51.3}$$

$$\frac{m_a^2}{R^3} - \frac{\partial \varphi}{\partial R} = 0. \tag{51.4}$$

"Equatorial" points, where the two directions coincide, have to be excluded from consideration. Elimination of φ yields

$$\frac{1}{R^3} \frac{\partial m_a^2}{\partial p} = -\frac{\partial \alpha}{\partial R}. \tag{51.5}$$

Now heat sources and frictional forces, distributed symmetrically with respect to the axis, are assumed to act on the vortex. Then the resulting meridional motions are likewise axially symmetrical and a symmetric circle of fluid always remains a symmetric circle. It is assumed that the vortex is stable throughout, that is, that everywhere and at all times,

$$\operatorname{Det} \mathcal{M} > 0, \qquad \operatorname{Trace} \mathcal{M} > 0 \tag{51.6}$$

where $\mathcal{M}$, the stability tensor, is

$$\mathcal{M} = -\frac{\alpha}{c_p} \operatorname{grad} s \operatorname{grad} p + \frac{1}{R^3} \operatorname{grad} m_a^2 \operatorname{grad} R. \tag{51.7}$$

It is further assumed that the heat sources and frictional forces are weak. This ensures that the velocities of the meridional motions are low and that Eqs. (51.2) to (51.5) approximately still hold true. We now assume that they are *exactly* valid at all times and attempt to determine the arising meridional motions from this condition. It will be seen that thereby the meridional circulation is uniquely determined.

By virtue of (51.3) we can put the equation of continuity in the following form

$$\frac{1}{R} \frac{\partial}{\partial R}\left(\frac{R}{g_a} \frac{DR}{Dt}\right) + \frac{\partial}{\partial p}\left(\frac{1}{g_a} \frac{Dp}{Dt}\right) = -\frac{\partial}{\partial t}\left(\frac{1}{g_a}\right). \tag{51.8}$$

Here g_a is the component of grad φ parallel to the axis. The term on the right hand side differs from zero when the isobaric levels move in the direction of the axis. In the atmosphere these movements are always so slow that the right hand side may be neglected. The rest of the equation is satisfied by putting

$$\frac{DR}{Dt} = \frac{g_a}{R} \frac{\partial \psi}{\partial p}, \qquad \frac{Dp}{Dt} = -\frac{g_a}{R} \frac{\partial \psi}{\partial R} \tag{51.9}$$

where ψ is a kind of streamfunction for the meridional motion.

The heat sources—which are assumed to include frictional dissipation—change the specific entropy of the individual air rings according to the formula

$$\frac{Ds}{Dt} = \frac{1}{T}\frac{Dq}{Dt}. \tag{51.10}$$

The torque χ of the frictional forces with respect to the axis changes the angular momentum:

$$\frac{D m_a^2}{Dt} = 2 m_a \chi. \tag{51.11}$$

Developing the individual derivatives on the left hand side of (51.10) and (51.11) according to the formula

$$\frac{D}{Dt} = \frac{\partial}{\partial t} + \frac{DR}{Dt}\frac{\partial}{\partial R} + \frac{Dp}{Dt}\frac{\partial}{\partial p} \tag{51.12}$$

while noting that, with p as one coordinate,

$$\frac{\partial s}{\partial t} = \frac{c_p}{\alpha}\frac{\partial \alpha}{\partial t} \tag{51.13}$$

and eliminating $\partial\alpha/\partial t$ and $\partial m_a^2/\partial t$ with the aid of the partial time derivatives of (51.3) and (51.4), we find,

$$-\frac{\partial}{\partial p}\frac{\partial \varphi}{\partial t} + A\frac{\partial \psi}{\partial R} + B\frac{\partial \psi}{\partial p} = E, \tag{51.14}$$

$$\frac{\partial}{\partial R}\frac{\partial \varphi}{\partial t} + B\frac{\partial \psi}{\partial R} + C\frac{\partial \psi}{\partial p} = F \tag{51.15}$$

where we have introduced

$$A = -\frac{g_a\alpha}{R c_p}\frac{\partial s}{\partial p}, \quad B = \frac{g_a\alpha}{R c_p}\frac{\partial s}{\partial R} = -\frac{g_a}{R^4}\frac{\partial m_a^2}{\partial p}, \quad C = \frac{g_a}{R^4}\frac{\partial m_a^2}{\partial p}, \tag{51.16}$$

$$E = \frac{\alpha}{c_p T}\frac{Dq}{Dt}, \quad F = \frac{2 m_a}{R^3}\chi. \tag{51.17}$$

Elimination of $\partial\varphi/\partial t$ from (51.14) and (51.15) yields

$$\frac{\partial}{\partial R}\left(A\frac{\partial \psi}{\partial R} + B\frac{\partial \psi}{\partial p}\right) + \frac{\partial}{\partial p}\left(B\frac{\partial \psi}{\partial R} + C\frac{\partial \psi}{\partial p}\right) = \frac{\partial E}{\partial R} + \frac{\partial F}{\partial p}. \tag{51.18}$$

In this self-adjoint second order differential equation for ψ no time derivatives appear. This means that the determination of the meridional circulation is no initial value problem in the quasistatic theory. ψ, in fact, only depends on the instantaneous distribution of heat sources and frictional moments and on the instantaneous structure of the vortex.

Using the notation of (51.16), the stability tensor (51.7) assumes the form

$$\mathcal{M} = \frac{R}{g_a}(A \operatorname{grad} p \operatorname{grad} p - B \operatorname{grad} p \operatorname{grad} R - B \operatorname{grad} R \operatorname{grad} p + C \operatorname{grad} R \operatorname{grad} R) \tag{51.19}$$

whence its determinant becomes

$$\operatorname{Det} \mathcal{M} = \frac{R^2}{g_a^2}(AC - B^2)(\operatorname{grad} R \times \operatorname{grad} p)^2. \tag{51.20}$$

This, together with (51.6), yields

$$AC - B^2 > 0 \tag{51.21}$$

which indicates that the differential Eq. (51.18) is of the elliptic type. Together with suitable boundary conditions, it determines the meridional circulation uniquely, which proves our statement.

The principal part of the GREEN's function of our boundary value problem in the neighbourhood of a point R_0, p_0 in the meridian plane is

$$G = \frac{1}{2\pi\sqrt{A_0 C_0 - B_0^2}} \ln\left[C_0 (R - R_0)^2 - 2 B_0 (R - R_0)(p - p_0) + A_0 (p - p_0)^2\right]^{\frac{1}{2}} \quad (51.22)$$

where the subscripts 0 refer to the values in the point R_0, p_0. With this the function ψ is known in the neighbourhood of an axially symmetric line source of heat or of angular momentum. In the case of a heat source we have

$$\psi = -\frac{\partial G}{\partial R_0} \varepsilon_0 \quad (51.23)$$

and in the case of a source of angular momentum

$$\psi = -\frac{\partial G}{\partial p_0} \eta_0 \quad (51.24)$$

where ε_0 and η_0 are the respective strengths of the line source in R_0, p_0.

Fig. 39 and 40 show examples of the meridional streamlines $\psi = \text{const}$ for each of the two cases. The streamlines are conformal ellipses going through the source point. In the case of a heat source they touch the curve $m_a = m_{a0}$; in the case of a source of angular momentum, the curve $s = s_0$. The direction of motion in the source point is, in the case of a heat source, towards lower pressure, in the case of a heat sink, towards high pressure; with positive frictional torque away from the earth's axis and with negative torque towards the earth's axis.

It may furthermore be proved that the true lengths of the diameters parallel to the lines $R = R_0$ (L_R) and $p = p_0$ (L_p) are related in each of the two cases as shown by the formula

$$\frac{L_R}{L_p} = \frac{\sqrt{C_0/A_0}}{|\operatorname{grad} p|_0}. \quad (51.25)$$

If $\boldsymbol{r}_R$ and $\boldsymbol{r}_p$ be two meridional unit vectors parallel to the abovementioned directions, we have from (51.19)

$$\boldsymbol{r}_R \cdot \mathcal{M} \cdot \boldsymbol{r}_R = \frac{R}{g_a} A (\operatorname{grad} p \cdot \boldsymbol{r}_R)^2, \quad (51.26)$$

$$\boldsymbol{r}_p \cdot \mathcal{M} \cdot \boldsymbol{r}_p = \frac{R}{g_a} C (\operatorname{grad} R \cdot \boldsymbol{r}_p)^2. \quad (51.27)$$

Denoting the left hand side respectively with ν_R^2 and ν_p^2, ν_R and ν_p are measures for the stability in the direction concerned (compare Sect. 26). From (51.26) and (51.27) it follows that

$$\frac{\nu_R^2}{\nu_p^2} = \frac{A}{C} \operatorname{grad}^2 p, \quad (51.28)$$

so that

$$\frac{L_R}{L_p} = \frac{\nu_p}{\nu_R}. \quad (51.29)$$

The ellipse diameters parallel to the two directions are thus related inversely as the two stabilities. This, however, only holds true for the mentioned directions.

In the atmosphere, the shape of the ellipses is in general approximately determined by L_R and L_p. For, as may easily be proved, the diameters parallel to the lines $R = R_0$ and $s = s_0$ are conjugated, and the angles between the isentropic surfaces and the isobaric surfaces are almost always very small becoming, in fact, equal to zero in the barotropic case.

Averaging over larger regions, we always have $\nu_R \gg \nu_p$, the horizontal extension of the ellipses being consequently much larger than the vertical extension (order of magnitude 100:1).

In the attempt to solve the boundary value problem for a vortex which resembles the actual atmosphere, the surface friction layer must be excepted in addition to the equatorial regions, as the forces of friction are so strong in this layer that (51.4) is no longer valid. We thus have to take the upper boundary of the friction layer as lower boundary for the range of integration, for the sake of simplicity about the 900 mb level. The boundary condition on this surface is given by the mass flux through the surface which is computed according to the Ekman theory. Using the notation of Sect. 17 we find

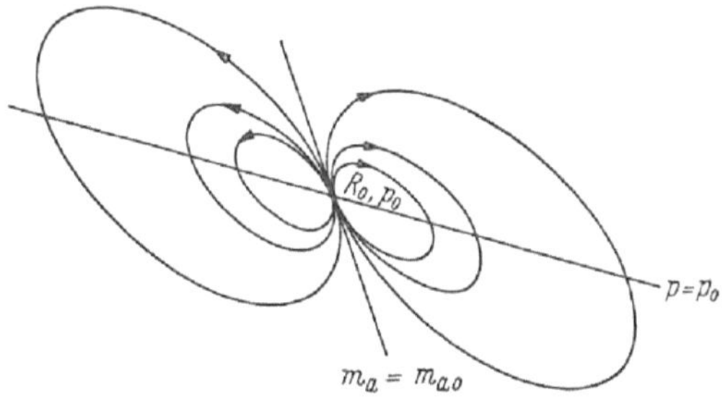

Fig. 39. Meridional streamlines in the vicinity of a circular line source of heat.

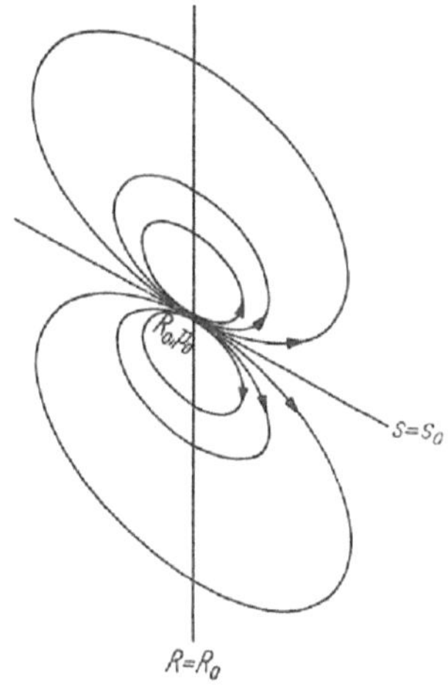

Fig. 40. Meridional streamlines in the vicinity of a circular line source of angular momentum.

$$\frac{\partial \psi}{\partial R} = \varrho \sqrt{K/2\lambda}\,\sin 2\alpha \left(\frac{\partial m_a}{\partial R} - 2R\Omega\right), \quad \text{when} \quad p = 900\,\text{mb}. \tag{51.30}$$

The upper boundary condition is

$$\frac{\partial \psi}{\partial R} = 0, \quad \text{when} \quad p = 0. \tag{51.31}$$

To every arbitrary distribution of heat sources and frictional torques one meridional circulation is thus assigned in which geostrophic equilibrium is everywhere preserved. The structure of the vortex itself, of course, is in general subject to changes due to the circulation.

If the total state is to remain stationary, as is the case with long period averages in the atmosphere, the distributions of friction and heat sources are no longer mutually independent. With $\partial \varphi / \partial t = 0$ we deduce from (51.14) and (51.15)

$$\frac{\partial \psi}{\partial R} = \frac{CE - BF}{AC - B^2}, \tag{51.32}$$

$$\frac{\partial \psi}{\partial p} = -\frac{BE - AF}{AC - B^2} \tag{51.33}$$

whence it follows

$$\frac{\partial}{\partial p}\frac{CE - BF}{AC - B^2} + \frac{\partial}{\partial R}\frac{BE - AF}{AC - B^2} = 0. \tag{51.34}$$

The boundary conditions take the form

$$\frac{CE - BF}{AC - B^2} = \varrho \sqrt{K/2\lambda}\, \sin 2\alpha \left(\frac{\partial m_a}{\partial R} - 2R\Omega\right), \quad \text{when} \quad p = 900\,\text{mb} \tag{51.35}$$

$$\frac{CE - BF}{AC - B^2} = 0, \quad \text{when} \quad p = 0. \tag{51.36}$$

The eventual aim of the theory is to explain the observed stationary state solely in terms of the external data. That is, in addition to the earth's rotation, particularly the incoming solar radiation. In order to solve this problem, the relationship between E and F and the structure of the vortex must be known. Let us assume that E and F are functions of m_a and s and their spatial derivatives and are as such known—with E in addition being a function of the incoming radiation. Substituting these functions in (51.34) to (51.36), Eqs. (51.5) and (51.34) form a system of differential equations for the two unknowns m_a and s and (51.35) and (51.36) are the appropriate boundary conditions.

The solution of the problem is provisionally rendered impossible especially by the fact that E is to a large extent based on an eddy flux of heat and F on an eddy flux of momentum. About the relationship between these fluxes and the structure of the zonally averaged state little is known. From recent investigations it can be concluded that the assumptions with respect to E and F upon which earlier computations of the general circulation were based[1], are hardly compatible with reality. One of the current aims of meteorology is, in fact, to gain greater clarity from observations on these relationships which are of basic importance for the general circulation.

ELIASSEN (ref. p. 139) pointed out that there is no *a priori* reason to assume that the boundary value problem for the stationary state is soluble, in other words that an ideally stationary circulation is possible. In the case of the non-existence of a solution one could, however, assume that the atmosphere oscillates about an average state. The variations in the strength of the zonal circulation, first described by ALLAN and others[2], may possibly be explained in this way. Periods with strong average transport of mass in the temperate latitudes (high *zonal index*) alternate with periods of weak mass transport (low zonal index). The fluctuations are in fact quite irregular, the period ranging from 3 to 8 weeks. For characteristic details of the course of a "*zonal index cycle*", the reader is referred to ROSSBY and WILLETT[3] and DEFANT[4].

52. The equations for zonally averaged angular momentum and specific entropy. For the application to the actual atmosphere, the frictional torque is to be replaced by the torque of virtual friction due mainly to the REYNOLDS stresses (Sect. 15). For this purpose we first derive the balance equation for the non-averaged angular momentum.

Let $\boldsymbol{i}$ denote a unit vector pointing towards the local east and R the distance from the axis of the earth. Then grad $R\boldsymbol{i}$ is an antisymmetric dyadic and consequently

$$\boldsymbol{v} \cdot \frac{D}{Dt} R\boldsymbol{i} = \boldsymbol{v} \cdot \operatorname{grad} R\boldsymbol{i} \cdot \boldsymbol{v} = 0. \tag{52.1}$$

Multiplying the equation of motion (4.11) scalarly by $R\boldsymbol{i}$ and adding (52.1), one obtains

$$\frac{D m_a}{Dt} = R\boldsymbol{i} \cdot (-\alpha \operatorname{grad} p + \alpha \operatorname{div} \mathscr{F}) \tag{52.2}$$

---

1 See for instance F. KROPATSCHECK: Beitr. Phys. fr. Atmosph. **22**, 272 (1935).

2 R. A. ALLAN and coll.: Papers in phys. Oceanogr. and Meteor. M.I.T. and Woods Hole Oceanogr. Institution **8**, No. 3 (1940).

3 C.-G. ROSSBY and H. C. WILLETT: Science, Lancaster, Pa. **108**, 643 (1948).

4 F. DEFANT: Arch. Meteor., Geophys. u. Biokl. **6**, 253 (1954).

which equation represents a special case of the general theorem that the rate of change of the absolute angular momentum of a particle is equal to the torque of the real forces.

Since $\mathcal{F}$ is a symmetric tensor, we have $\mathcal{F}\cdot\operatorname{grad}\cdot R\boldsymbol{i}=0$. Since, moreover, $\operatorname{div} R\boldsymbol{i}=0$, Eq. (52.2) can also be written

$$\frac{D m_a}{Dt} = -\alpha \operatorname{div}(R p\,\boldsymbol{i} - R\,\mathcal{F}\cdot\boldsymbol{i}). \tag{52.3}$$

Applying (4.5), one obtains

$$\frac{\partial(\varrho m_a)}{\partial t} + \operatorname{div}(\varrho m_a \boldsymbol{v} + R p\,\boldsymbol{i} - R\,\mathcal{F}\cdot\boldsymbol{i}) = 0. \tag{52.4}$$

This equation shows that the local rate of change of angular momentum can be expressed completely by the divergence of a flux $\varrho m_a\boldsymbol{v}+Rp\,\boldsymbol{i}-R\mathcal{F}\cdot\boldsymbol{i}$. With this interpretation there exists no production of angular momentum.

For zonal averaging we refer to a coordinate system symmetric with respect to the earth's axis. We take x, y, z to be the coordinates in zonal, meridional and vertical directions, u, v, w being the corresponding components of $\boldsymbol{v}$. Let us define the zonal average of a vector field $\boldsymbol{a}$ with components a_x, a_y, a_z, as the vector field $\bar{\boldsymbol{a}}$, the components of which are $\bar{a}_x$, $\bar{a}_y$, $\bar{a}_z$, the bar denoting zonal averaging along the zonal circle over all longitudes. Then it follows from the special metric of the coordinate system that

$$\operatorname{div}\bar{\boldsymbol{a}} = \frac{1}{R}\operatorname{div}_m R\overline{\boldsymbol{a}_m} \tag{52.5}$$

where div_m is the two-dimensional divergence in the meridional plane and $\overline{\boldsymbol{a}_m}$ the meridional component of $\bar{\boldsymbol{a}}$. Applying (52.5) to the vector field $Rp\,\boldsymbol{i}$, we obtain

$$\operatorname{div} R\bar{p}\,\boldsymbol{i} = 0, \tag{52.6}$$

which means that the torque of the pressure forces vanishes in the zonal mean. This, however, only holds, when the zonal circle does not intersect the earth's surface. For circles interrupted by a mountain, however narrow, a pressure difference between the western and eastern slopes renders (52.6) invalid. With (52.6), and since the torques of the molecular friction above the laminar boundary layer (Sect. 16α) may be neglected, Eq. (52.4) yields

$$\frac{\partial\bar{\varrho}\hat{m}_a}{\partial t} + \frac{1}{R}\operatorname{div}_m R\,\overline{\varrho m_a \boldsymbol{v}_m} = 0. \tag{52.7}$$

$\overline{\varrho m_a \boldsymbol{v}_m}$ is the mean meridional flux of angular momentum. It may be written as the sum of a *"circulation flux"* and an *eddy flux*:

$$\overline{\varrho m_a \boldsymbol{v}_m} = \bar{\varrho}\hat{m}_a\hat{\boldsymbol{v}}_m + \overline{\varrho m_a'' \boldsymbol{v}_m''}. \tag{52.8}$$

From the definition (51.1) of angular momentum, the eddy flux becomes

$$\overline{\varrho m_a'' \boldsymbol{v}_m''} = R\,\overline{\varrho u'' \boldsymbol{v}_m''} \tag{52.9}$$

and consequently, if we use the zonally averaged equation of continuity

$$\frac{\partial\bar{\varrho}}{\partial t} + \frac{1}{R}\operatorname{div}_m R\,\overline{\varrho\boldsymbol{v}_m} = 0. \tag{52.10}$$

(52.7) assumes the form

$$\frac{\hat{D}\hat{m}_a}{Dt} \equiv \frac{\partial\hat{m}_a}{\partial t} + \hat{v}\frac{\partial\hat{m}_a}{\partial y} + \hat{w}\frac{\partial\hat{m}_a}{\partial z} = -\frac{1}{\bar{\varrho}R}\operatorname{div}_m R^2\,\overline{\varrho u'' \boldsymbol{v}_m''}. \tag{52.11}$$

The expression on the right hand side of this equation assumes the role of frictional torque for the zonally averaged circulation as may be seen on comparing Eq. (51.11). As, in the identity

$$\frac{\hat{D}}{Dt}(\widehat{m_a^2}) = 2\hat{m}_a \frac{\hat{D}\hat{m}_a}{Dt} + \frac{\hat{D}}{Dt}(\widehat{m_m''^2}) \tag{52.12}$$

the second term on the right hand side is presumably of little importance compared to the first, we have

$$\frac{\hat{D}}{Dt}(\widehat{m_a^2}) \approx -\frac{2\hat{m}_a}{\bar{\varrho} R} \operatorname{div}_m R^2 \overline{\varrho u'' v_m''}. \tag{52.13}$$

Thus the expression

$$-\frac{2\hat{m}_a}{\bar{\varrho} R^4} \operatorname{div}_m R^2 \overline{\varrho u'' v_m''}$$

approximately corresponds to F in the quasistatic theory of meridional motion. The divergences of the eddy flux of angular momentum constitute the mechanical propulsion in the average meridional circulation.

In a similar way we deduce from the balance equation for specific entropy

$$\frac{\partial(\varrho s)}{\partial t} + \operatorname{div} \varrho s \boldsymbol{v} = \frac{\varrho}{T} \frac{Dq}{Dt} \tag{52.14}$$

the following equation:

$$\frac{\hat{D}\hat{s}}{Dt} = \frac{1}{\bar{\varrho}} \overline{\frac{\varrho}{T} \frac{Dq}{Dt}} - \frac{1}{\bar{\varrho} R} \operatorname{div}_m R^2 \overline{\varrho s'' v_m''}. \tag{52.15}$$

The first term on the right hand side contains, apart from the effects of radiation, heat conduction and dissipation, also heating due to phase transformation of water, since s is the entropy of dry air. The second term corresponds to the divergences of eddy flux of entropy. As the oscillations of α are relatively small along a parallel of latitude, the right hand side of (52.15), multiplied by $\bar{\alpha}/c_p$, corresponds to E in the quasistatic theory. The thermal impulse of the meridional circulation hence not only consists of heat sources, but also of the eddy flux of entropy.

The eddy fluxes in question are on principle carried by all magnitudinal classes of atmospheric motion, from the ordinary turbulence up to such flows as occur in cyclones and anticyclones. However the efficacy of the single classes is, of course, extremely different, that of the largest prevailing by far. The large irregular motions are sometimes termed *large-scale turbulence* and consequently the smaller ones, *small-scale turbulence* or *micro-turbulence*, the latter referring particularly to the normal turbulence of the wind.

It must, however, be stressed that the laws, valid for normal turbulence, cannot in general be extended to large-scale turbulence. The normal turbulence, for instance, gives rise to a flux of certain entities such as heat, momentum etc. which is proportional and equally-directed to the gradient of the quantity in question. Such laws do not necessarily hold good in large-scale turbulence, since the basic condition that the exchange motions are not correlated to the fluctuations of the quantity, is in general not fulfilled. For this reason and also in view of the different efficacies, it is advisable to separate the effects exerted by the different classes of eddy motion, by distinguishing REYNOLDS *stresses* (small scale, no appreciable correlation) and JEFFREYS *stresses* (large scale), as has been done by VAN MIEGHEM[1].

53. The mean meridional flux of angular momentum. The angular momentum flux from earth to atmosphere through an infinitesimal area is according to (52.4)

$$R p \boldsymbol{i} \cdot d\sigma - \boldsymbol{i} \cdot R \mathscr{F} \cdot d\sigma. \tag{53.1}$$

[1] J. VAN MIEGHEM: Tellus 4, 334 (1952).

Since the total angular momentum of the atmosphere is fairly constant, the integral of (53.1) over the whole surface of the earth must approximately vanish. To the integral over a latitudinal zone, the first term contributes a share, when the pressure values at the eastern and western slopes of mountains lying in the zone are unequal (mountain torque). Higher pressure on the western slope involves a net flux of angular momentum from atmosphere to earth and higher pressure on the eastern slope, a net flux in the opposite direction. The second term is the torque of the stress exerted by the infinitesimal surface on the air (friction torque). It may be estimated according to the EKMAN spiral theory from the geostrophic wind in surface proximity [Eq. (16.17)].

In the tropical easterlies, i.e. in the zone between about $\pm 32°$, the frictional flux is directed away from the earth, in the westerlies of the temperate latitudes, into the earth. The mountain torque in the northern hemisphere, as computed by WHITE[1] from pressure charts, shows a similar meridional distribution with values of the same order of magnitude as the frictional torque. The total flux of angular momentum entering the atmosphere in the tropics is according to estimates of PRIESTLEY[2] and others at all times so great, that it supplies the total relative angular momentum for the west wind regions in 2 to 4 weeks. Compared with this, the observed changes in angular momentum are small in the individual zones. Hence it follows that already in averages over a few days a nearly divergence-free flux of angular momentum flows permanently from the tropics to the temperate latitudes.

In the same way it follows from the predominantly easterly direction of the winds in the polar regions that also from here an angular momentum flux flows to the westerlies although considerably weaker.

With regard to the facts stated in the preceding section, it is important for the comprehension of general circulation structure, to know to what extent these fluxes exist as circulation fluxes and to what extent they exist as eddy fluxes. Since the total angular momentum flux is free of divergence, the divergences of eddy flux are convergences of circulation flux and vice versa.

As the frictional currents of the 3 zones belong to different cells of the mean meridional circulation, each of the two angular momentum fluxes must cross from one cell to another at least once. The normal component of $\hat{\boldsymbol{v}}_m$ now vanishes on the interfaces between the different cells. Consequently the cross-over takes on the form of a pure eddy flux. H. JEFFREYS[3] was the first to point out the necessity of eddy transports of angular momentum.

In the friction layers the angular momentum transport consists mainly of a vertical small-scale eddy flux which ends or begins at the upper boundary of the frictional layer, and a horizontal circulation flux. In the tropics where angular momentum flows out of the earth, the former is convergent and the latter, divergent. According to computations of PALMÉN and ALAKA[4], the main part of the angular momentum flowing from the earth is transported towards the equator by the circulation flux and does not rise to the upper troposphere before having crossed the 10° parallel.

The determination of the angular momentum transport in the free atmosphere and its separation in circulation transport and eddy transport is by no means easy. From aerological observations only the horizontal components of these fluxes can

[1] R. M. WHITE: J. Meteor. **6**, 353 (1949).
[2] C. H. B. PRIESTLEY: Austral. J. Sci. Res. A **4**, 315 (1951). — Quart. J. Roy. Meteor. Soc. **77**, 200 (1951).
[3] H. JEFFREYS: Quart. J. Roy. Meteor. Soc. **52**, 85 (1926).
[4] E. PALMÉN and M. A. ALAKA: Tellus **4**, 324 (1952).

be determined, the vertical components can only be uncertainly arrived at by means of indirect methods.

STARR and WHITE[1] showed from wind soundings at 29 stations around the 31° N parallel, that a steady northward angular momentum eddy flux occurs at all heights over this parallel with maximum between 300 and 200 mb and that its integral over all heights just about suffices to carry the angular momentum transport required by the surface observations[2]. Correspondingly, the vertically integrated circulation flux over the 31-st parallel of latitude was shown to be small, at least in annual averages.

PRIESTLEY (ref. p. 147), by contrast, came to the result, based on sparser data, that the eddy flux constitutes only about one half of the required angular momentum flux over the subtropical belt. He concluded from this result that the other half is carried by a circulation flux. Since the net mass flux over the parallel of latitude must vanish, and the vertical differences in absolute angular momentum here as in every other latitude, are relatively small, $\hat{v}$ must assume values of 1 to 2 mps in the different heights in order to produce such a strong flux in the integral over all heights. In view of the fact that the west wind freshens with height, $\hat{v}$ must be positive above and negative below.

Apart from the fact that other evaluations did not reveal such high values for $\hat{v}$ they are also improbable for another reason. In the subtropics, the zonal motion is, absolutely seen, approximately that of a rigid rotation, not very different from that of the solid earth. That means that

$$\frac{\partial \hat{m}_a}{\partial y} \approx \frac{\partial}{\partial y} R^2 \Omega = -\lambda R. \tag{53.2}$$

Upon substituting this in Eq. (52.11), it follows that a positive $\hat{v}$ of a few mps would require a strong divergence of the eddy flux of angular momentum in the long-period mean. This result may also be stated in the following way: a high value of $\hat{v}$ involves a strong convergence of circulation flux of angular momentum. Consequently the eddy flux must have an equally strong divergence.

This divergence must be produced predominantly by vertical eddy fluxes in view of the large horizontal extensions. It is true that the small scale eddy flux has a divergence in the upper troposphere of the subtropics, but this turns out to be so small when computed from the usual exchange coefficients, that it could counterbalance $\hat{v}$-values of only a few cm/sec. Thus a large scale eddy flux in the subtropics flowing from the upper to the lower troposphere must be postulated. There are however no indications for the existence of such a flux.

Because of the fact that the upper air wind observations are very sparse in some regions, the angular momentum flux as also the transport of other entities, are often computed from the geostrophic winds. The pressure charts from which the geostrophic wind is taken, can be drawn with good overall accuracy. In this method, however, a circulation flux is not recorded, since from

$$\varrho v_g = \frac{1}{\lambda} \frac{\partial p}{\partial x}, \tag{53.3}$$

it follows that

$$\bar{\varrho} \hat{v}_g = 0. \tag{53.4}$$

---

[1] V. P. STARR and R. M. WHITE: Tellus 4, 332 (1952). — Quart. J. Roy. Meteor. Soc. 78, 62 (1952).

[2] Hitherto, in averaging aerological material zonally the variance of ϱ in zonal direction has been left out of consideration so that the averages are unweighted. For the errors due to this omission (which are in general small) see VAN MIEGHEM, Tellus 4, 135 (1952).

The "geostrophic flux" of angular momentum thus only contains the eddy flux and even this only inaccurately. The deviations from the true eddy flux, however, do not seem to be very large[1].

Investigations of WIDGER jr.[2], BJERKNES[3], MINTZ[4], STARR and WHITE[5] have shown that the circulation flux of angular momentum which leaves the trades and rises up, begins to change into an eddy flux already in the tropics. This eddy flux flows mainly between 10 and 14 km of height towards the north and reaches its peak strength at about 32° N. Beyond this latitude it gradually changes back into a circulation flux. WHITE and COOLEY[6] have shown that part of the angular momentum flux in temperate latitudes sinks into the lower layers as large-scale eddy flux which is carried by vertical motions connected with migrating cyclones and anticyclones. An explanation of this fact was given by STARR[7].

The eddy flux of angular momentum directed towards the poles in the upper troposphere is principally carried by high-level cyclones in the tropics and by the meandering of the westerlies on the polar side of the horse latitudes. It often makes itself felt here through the distinct tilting of ridges and troughs into a south-west to north-east direction, with the currents flowing polewards having higher angular momentum than those flowing towards the equator.

The divergences and convergences of the eddy flux of angular momentum in the upper troposphere are presumably the primary moving agent for the mean meridional circulation. The sources in the tropics propel the air masses towards the poles and the sinks in the middle latitudes, towards the equator. This gives rise to the subtropical high pressure belt and consequently to the two friction streams in which the divergences of the eddy flux have the opposite sign to the respective divergences aloft.

The distribution of heat sources in the sense of Sect. 52 results from the course of the meridional circulation and from the field of entropy. Since the potential temperature increases with height, the sources are positive in the ascending branches, negative in the descending branches. Presumably the heat sources are of secondary nature, that is they are caused by the meridional circulation itself which, on its part, is driven by the sources of the eddy flux of angular momentum. This is suggested by the fact that the middle-latitude cell, and also the tropical cell in midsummer, rotate in an opposite sense to that dictated by the meridional distribution of insolation.

The causes of the eddy flux of angular momentum are uncertain. Doubtlessly it depends upon the structure and strength of the zonal circulation. LORENZ[8] detected a positive correlation between the zonal velocity and the simultaneous polward eddy flux in the lower layers.

The structure of the mean zonal circulation as shown in Fig. 38 has not yet been explained satisfactorily. It is characterized, amongst others, by the condition that the total transport of angular momentum be free of divergence. In this

[1] According to investigations by LORENZ and MACHTA, both quoted by WIDGER[2]. See also WHITE [Tellus **3**, 82 (1951)], who studied the same problem with regard to the eddy flux of sensible heat.

[2] W. K. WIDGER jr.: J. Meteor. **6**, 291 (1949).

[3] J. BJERKNES: Procès Verbeaux Séances de L'Assoc. de Météor. UGGI, Bruxelles, p. I–XXII, 1951.

[4] Y. MINTZ: Tellus **3**, 195 (1951).

[5] V. P. STARR and R. M. WHITE: Tellus **6**, 180 (1954).

[6] R. M. WHITE and D. S. COOLEY: Trans. Amer. Geophys. Un. **33**, 502 (1952).

[7] V. P. STARR: J. Meteor. **5**, 39 (1948).

[8] E. N. LORENZ: J. Meteor. **9**, 152 (1952).

connection it is of interest to ask for the distribution of zonal wind which involves no eddy transport of angular momentum. In zones where the eddy motion is intensive, the profile of the mean zonal wind will probably not deviate much from such a state of zero eddy transport, since otherwise the eddy transport of angular momentum would become excessive.

It is generally presumed that an isotropic and purely frictionally driven mixing produces a state of rigid rotation, for this corresponds to the state of least kinetic energy, the total amount of angular momentum being fixed. ROSSBY[1] advanced the view that a thermally driven exchange, on the other hand, causes an equalization of the absolute vorticity. RAETHJEN[2] suggested that, when the isentropic surfaces are inclined in meridional direction, the lateral mixing is not isotropic but intensified in the meridional direction. The degree of this intensification determines the state of vanishing angular momentum flux, lying between the two extremes, rigid rotation and motion with constant angular momentum. The occurence of the west wind maximum may be explained by assuming that the lateral exchange is approximately isotropic at high latitudes and more or less "heterotropic" at low latitudes (RAETHJEN[3]).

It is a highly significant fact that motions of the type observed in the atmosphere have been successfully reproduced in laboratory experiments, notably by FULTZ and his collaborators in Chicago. The best results have been obtained in a rotating cylindrical tank containing water, heated at the rim. The similarity theory for these experiments has been developed by FULTZ[4,5]. By varying the experimental conditions, different circulation regimes are obtained. Thus for slow rotation, and large thermal gradients, one obtains an axially symmetric motion with only "circulation flux" of angular momentum (this is called the HADLEY *regime* by FULTZ). For strong rotation and weak thermal gradients, the ROSSBY *regime* prevails, characterized by a pronounced large-scale eddy motion, which under suitable conditions is very similar to the motion in the atmosphere[5,6,7]. Also details such as frontal wave cyclones have been reproduced in the laboratory[8,9].

54. Application to a stationary cyclone or anticyclone. If the variation of the CORIOLIS parameter with latitude is disregarded, the results of Sect. 51 and 52 apply practically unchanged to a cyclonic or anticyclonic vortex. It is only necessary to modify the notions "zonal", "meridional" and "distance from the axis (R)" so as to refer to a vertical axis, and to replace Ω by $\lambda/2$ in the formulae. The "absolute angular momentum" thus becomes

$$m_a = \frac{\lambda}{2} R^2 + R u. \tag{54.1}$$

If the vortex reaches the surface, as actual vortices mostly do, considerations similar to those in Sect. 53 can be made regarding the friction flow near the ground.

[1] C.-G. ROSSBY: Bull. Amer. Meteor. Soc. **28**, 53 (1947).

[2] P. RAETHJEN: Arch. Meteor., Geophys. u. Biokl. Ser. A **3**, 173 (1951).

[3] P. RAETHJEN: Arch. Meteor., Geophys. u. Biokl., Ser. A **4**, 41 (1951).

[4] D. FULTZ: J. Meteor. **8**, 262 (1951).

[5] D. FULTZ and coll: Studies of thermal convection in a rotating cylinder with some implications for large-scale atmospheric motions. Hydrodyn. Lab., Dept. of Meteorology, Univ. of Chicago, Final Report I, Contr. AF 19 (122) — 160 (1956) (contains complete bibliography).

[6] J. CORN and D. FULTZ: Geophys. Res. Papers No. 34, U.S. Air Force Geophys. Research Directorate 1955.

[7] D. FULTZ: Proc. First Symposium Geophysical Models, John Hopkins Univ. 1956.

[8] D. FULTZ: J. Meteor. **9**, 379 (1952).

[9] A. J. FALLER: J. Meteor. **13**, 1 (1956).

Fig. 41 shows as an example schematically the meridional circulation in a cyclone, with the friction flow toward the axis and a counterflux somewhere aloft, as well as the distribution of the absolute angular momentum, which increases outward. Between L and H an angular momentum flux flows into the earth. This is often so strong that it would consume the total amount of relative angular momentum (Ru) stored in the cyclone in, say, 3 days. If the cyclone is to survive unchanged, an equally strong flux must enter the cyclone on the sides. This flux must cross the outer boundary of the circulation cell as pure eddy flux. The sources (+) and sinks (−) of eddy flux of angular momentum result, according to Eq. (52.11), from the course of the meridional circulation and the distribution of angular momentum.

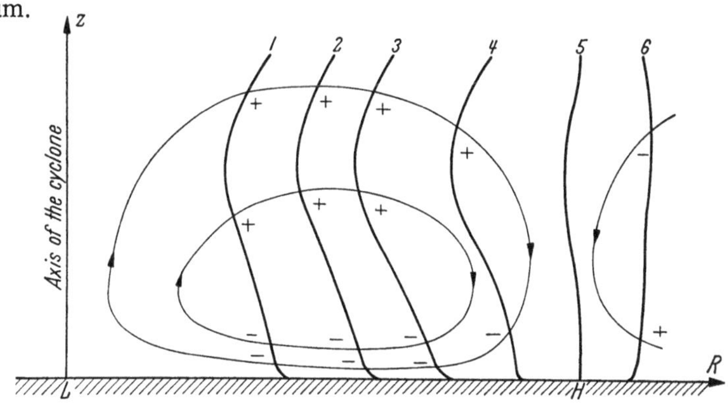

Fig. 41. Schematic meridional section through a stationary cyclone. Heavy lines: lines of equal absolute angular momentum Thin lines: meridional streamlines. Divergences (+) and convergences (−) of eddy flux of angular momentum.

In an anticyclone the sense of rotation of the meridional circulation is reversed; the angular momentum, however, still increases outward. Consequently divergence of eddy flux must exist aloft and convergence in the friction layer. Angular momentum steadily flows into the surroundings from the anticyclone.

The immediate effect of the eddy flux from and to the surroundings is a zonal acceleration and, in consequence, a motion towards the higher pressure in the upper layers of the vortex. As already mentioned in Sect. 46, the existence of such motions was proved by REINEKE (ref. p. 124).

The cause of the eddy flux of angular momentum is uncertain. RAETHJEN[1] suggested that, in a high-level cyclone with the isentropic surfaces ascending inwards in the lower parts, the inward angular momentum flux may be accounted for by a heterotropic lateral mixing (Sect. 53). In the case of an anticyclone, a corresponding explanation is not possible, since here the absolute angular velocity increases outward.

55. The balance of the kinetic energy of the mean zonal flow. To gain further insight into the nature of the general circulation, it is useful to divide the total kinetic energy into the kinetic energy of the zonally averaged motion and the supplementary eddy energy, and to study the conversions from the one form to the other. The relevant equations are (15.9) and (15.10), with the bar denoting zonal averaging.

Instead of (15.9) it is advantageous to consider the balance equation for the kinetic energy of the zonally averaged *zonal* motion. This energy ($\hat{u}$-energy) constitutes the main part of the kinetic energy of the averaged total motion.

[1] P. RAETHJEN: Arch. Meteor., Geophys. u. Biokl., Ser. A **1**, 295 (1949).

Multiplying Eq. (52.11) by $\hat{u}$ and using $\hat{m}_a = R^2\Omega + R\hat{u}$, we obtain

$$\left.\begin{aligned} R\frac{\partial}{\partial t}\frac{\hat{u}^2}{2} + R\hat{\boldsymbol{v}}_m\cdot\operatorname{grad}\frac{\hat{u}^2}{2} + (2R\Omega\hat{u}+\hat{u}^2)\,\hat{\boldsymbol{v}}_m\cdot\operatorname{grad}R \\ = -\frac{\hat{u}}{\bar{\varrho}R}\operatorname{div}_m R^2\,\overline{\varrho u''\boldsymbol{v}_m''}\,. \end{aligned}\right\} \tag{55.1}$$

Within the brackets $\hat{u}^2$ may be neglected as compared to $2R\Omega\hat{u}$. With the aid of the continuity Eq. (52.10) we obtain the wanted balance equation

$$\left.\begin{aligned} \frac{\partial}{\partial t}\left(\frac{1}{2}\,\bar{\varrho}\,\hat{u}^2\right) + \frac{1}{R}\operatorname{div}_m R\left(\frac{1}{2}\,\bar{\varrho}\,\hat{u}^2\,\hat{\boldsymbol{v}}_m + \hat{u}\,\overline{\varrho u''\boldsymbol{v}_m''}\right) \\ = R\,\overline{\varrho u''\boldsymbol{v}_m''}\cdot\operatorname{grad}\frac{\hat{u}}{R} - 2\bar{\varrho}\,\Omega\,\hat{u}\,\hat{\boldsymbol{v}}_m\cdot\operatorname{grad}R\,. \end{aligned}\right\} \tag{55.2}$$

It has the advantage over Eq. (15.9) that it does not contain terms involving pressure or geopotential. The terms within the bracket represent, by virtue of the smallness of $\hat{v}$ and $\hat{w}$, the main part of the meridional transport of the total kinetic energy. The first term on the right represents conversion of eddy energy into $\hat{u}$-energy (Arakawa[1]). It is positive whenever the meridional eddy flux of angular momentum is directed towards zones of higher mean angular velocity, and negative when the direction is inverted. The latter is the case on the polar side of the west wind maximum. On the equatorial side the term is positive, showing that eddy energy is here converted into kinetic energy of the mean motion. The second term represents the effect of the mean meridional circulation. Upon integrating over a vertical interface, erected on a latitude circle and extending through the whole atmosphere, the value of the integral depends upon the baroclinicity of the zonal circulation. Since in the troposphere $\hat{u}$ increases with height, the integral gives a positive contribution for a latitude lying within the tropical circulation cell. It contributes negatively in the middle latitudes, where the meridional circulation has the opposite sense of rotation.

Since the west wind maximum is located roughly on the boundary between the tropical and the middle latitude cell, the signs of the latitudinal interface integrals of the two terms change together from plus to minus. Thus a strong flux of kinetic energy must leave the tropics for the temperate latitudes. This flux is contributed in the subtropics mainly by the term $R\hat{u}\,\overline{\varrho u''\boldsymbol{v}_m''}$, its strength being thus proportional to that of the eddy flux of angular momentum.

Actual computations of production and consumption of $\hat{u}$-energy large-scale eddies in dependance of latitude have been made by Kuo[2], Starr[3] and Pisharoty[4]. The last of these publications also gives the poleward flux of kinetic energy and hence, by the aid of (55.2), the production of $\hat{u}$-energy by mean meridional circulation. Furthermore, the author studies the balance of the eddy energy, by computing, in addition, the meridional transport of this energy [the first two terms in the bracket of Eq. (15.10)]. He obtains the result that the divergences of this transport and the production of eddy energy from the mean motion, as given by $\overline{\varrho\boldsymbol{v}''\boldsymbol{v}''}\cdot\operatorname{grad}\cdot\hat{\boldsymbol{v}}$ do not balance each other, and that the balancing values of $\operatorname{div}\overline{p\boldsymbol{v}''} - \overline{p\operatorname{div}\boldsymbol{v}''} = \overline{\boldsymbol{v}''\cdot\operatorname{grad}p}$ have alternating signs in 4 subsequent zones between 20 and 75° N, beginning in the tropics with negative values. Negative values of $\overline{\boldsymbol{v}''\cdot\operatorname{grad}p}$ may be due to the preferred ascent of relatively warm air

[1] H. Arakawa: J. Meteor. Soc. Japan, II. Ser. **19**, 271 (1941).
[2] H.-L. Kuo: Tellus **3**, 205 (1951).
[3] V. P. Starr: Tellus **5**, 494 (1953).
[4] P. R. Pisharoty: Final Report Gen. Circ. Proj. Article XIV, U.C.L.A. 1955.

and descent of relatively cold air or to a preferred horizontal motion towards lower pressure. (The latter follows from the smallness of $\hat{v}$ as compared with the average amount of v). Positive values result under the reverse conditions. PISHAROTY attempts to connect the above result with characteristic features of the large-scale motions in the different zones.

56. The meridional energy flux. Investigations on the budget of radiation energy in the different latitudes, carried out by SIMPSON[1], ALBRECHT[2], BAUR and PHILIPPS[3] have shown that the lower latitudes exhibit a surplus of radiation energy, the higher latitudes, a deficit. Everywhere, the predominant part is ascribed to the atmosphere, the rest to the hydrosphere, where it is balanced by the ocean currents.

To balance the budget of total energy of the atmosphere in the individual zones, a poleward flowing energy flux must be postulated on each hemisphere. Ignoring the transport of energy by molecular friction and heat conduction which is quite insignificant, the energy flux over a parallel of latitude is according to (6.12)

$$2\pi \int_0^\infty R\,\overline{(k+\varphi+h)\,\varrho v}\,dz\,, \tag{56.1}$$

the bar denoting zonal averaging as before. This flux must, on long term averages, be equal to the loss of energy resulting from radiation in the part of the atmosphere on the polar side of the parallel of latitude.

We now have

$$\overline{(k+\varphi+h)\,\varrho\, v} = \bar{\varrho}\,(\hat{k}+\varphi+\hat{h})\,\hat{v} + \overline{\varrho\,(k''+h'')\,v''}. \tag{56.2}$$

The contributions of the various terms on the right-hand side to integral (56.1) have partly been estimated and partly computed more accurately. STARR[4] estimated that the kinetic energy flux—circulation and eddy flux taken together—is one order of magnitude smaller than the enthalpy flux. For the "geostrophic flux" (compare Sect. 53), this has been confirmed by PISHAROTY (ref. p. 152). Moreover, the circulation flux of enthalpy, kinetic and potential energy is doubtlessly of little importance. The quantity $\bar{\varrho}\,(\hat{k}+\varphi+\hat{h})$ increases in general monotonically with height. If the circulation flux is to balance the energy budget of the individual zones, $\hat{v}$ would have to reach values of 1 mps and more in temperate latitudes. Observations and the reasons outlined in Sect. 53 contradict this. The energy balance between the individual zones must thus, for the greatest part, be effected by an eddy flux of enthalpy.

The flux of enthalpy consists of two parts according to the definition (3.14). The constant in this definition falls away in integrating the enthalpy flux over the vertical interface erected on a parallel of latitude, as no net flux of mass crosses the parallel. The two components are then a flux of sensible heat corresponding to the term $c_p T$ and a flux of latent heat corresponding to $\mu' L$.

Calculations of the eddy flux of sensible heat in dependance upon height and geographic latitude, have been carried out by WHITE[5], BJERKNES[6], MINTZ[7] and

[1] G. C. SIMPSON: Mem. Roy. Meteor. Soc. **3**, No. 23 (1929).
[2] F. ALBRECHT: Meteor. Z. **48**, 57 (1931).
[3] F. BAUR u. H. PHILIPPS: Gerlands Beitr. Geophys. **45**, 82 (1935).
[4] V. P. STARR: J. Meteor. **6**, 160 (1949).
[5] R. M. WHITE: Tellus **3**, 82 (1951); **6**, 177 (1954). — Quart. J. Roy. Meteor. Soc. **77**, p. 188 (1951).
[6] J. BJERKNES: l. c. on page 149.
[7] Y. MINTZ: Final Report Gen. Circ. Proj. Article V, U.C.L.A. 1955.

PISHAROTY[1]. The flux is generally directed towards the poles. It is appreciably stronger in winter than in summer. According to investigations by PRIESTLEY[2], NYBERG and SCHMACKE[3] and NYBERG[4] the eddy flux of latent heat in middle and high latitudes is also directed polewards and of same order of magnitude as that of sensible heat. It seems that there are only small seasonal variations. Both fluxes together are about as strong as is required to balance the energy budget.

In contrast to the eddy flux of angular momentum which mainly flows in the upper half of the troposphere, the eddy flux of sensible heat is essentially carried within the lower layers of the troposphere. The reason is that in surface proximity the exchange currents are forced to flow horizontally. This, together with the north-south temperature gradient, produces a strong heat flux. Aloft the exchange currents seem rather to follow the isentropic surfaces.

General references.

[1] BERRY, F. A., E. BOLLAY and N. R. BEERS: Handbook of Meteorology. New York: McGraw-Hill 1945.
[2] BJERKNES, V., J. BJERKNES, H. SOLBERG u. T. BERGERON: Physikalische Hydrodynamik. Berlin: Springer 1933.
[3] BJERKNES, V., and J. W. SANDSTRÖM: Dynamic Meteorology and Hydrography. Part I. Statics. Washington: Carnegie Institution 1910.
[4] BRUNT, D.: Physical and Dynamical Meteorology. Cambridge University Press 1941.
[5] CHARNEY, J. G.: J. Meteor. **4**, 135 (1947).
[6] ERTEL, H.: Methoden und Probleme der dynamischen Meteorologie. Berlin: Springer 1938.
[7] EXNER, F. M.: Dynamische Meteorologie. Vienna: Springer 1925.
[8] FJÖRTOFT, R.: Geof. Publ. **17**, 6 (1950).
[9] HAURWITZ, B.: Dynamic Meteorology. New York and London: McGraw-Hill 1941.
[10] HÖILAND, E.: Arch. Math. Naturvidensk. **42**, No. 5 (1939).
[11] HOLMBOE, J., G. E. FORSYTHE and W. GUSTIN: Dynamic Meteorology. New York: Wiley 1945.
[12] KOSCHMIEDER, H.: Physik der Atmosphäre, Bd. 2. Dynamische Meteorologie. Leipzig: Akademische Verlagsgesellschaft 1951.
[13] LETTAU, H.: Atmosphärische Turbulenz. Leipzig: Akademische Verlagsgesellschaft 1939.
[14] MARGULES, M.: Über die Energie der Stürme. Jb. Zentr.-Anst. Meteor. Wien **1903**.
[15] PRIESTLEY, C. H. B., and P. A. SHEPPARD: Turbulence and Transfer Processes in the Atmosphere. Quart. J. Roy. Meteor. Soc. **78**, 488 (1952).
[16] RAETHJEN, P.: Einführung in die Physik der Atmosphäre. Bd. I u. II. Leipzig u. Berlin: Teubner 1942.
[17] RAETHJEN, P.: Dynamik der Zyklonen. Leipzig: Akademische Verlagsgesellschaft 1953.
[18] REYNOLDS, O.: On the Dynamical Theory of Incompressible Viscous Fluids and the Determination of the Criterion. Phil. Trans. Roy. Soc. Lond., Ser. A **186**, 123 (1895).
[19] RICHARDSON, L. F.: Weather Prediction by Numerical Process. Cambridge University Press 1922.
[20] ROSSBY, C.-G., and Coll.: J. Mar. Res. **2**, 1 (1939).
[21] SYONO, S.: Introduction into Dynamic Meteorology. Tokyo 1954 (in Japanese).
[22] SUTTON, O. G.: Atmospheric Turbulence. Methuen's Monographs. London 1949.
[23] SUTTON, O. G.: Atmospheric Turbulence and Diffusion. Comp. of Meteor., Amer. Meteor. Soc. **1951**, 492.

---

[1] P. R. PISHAROTY: Final Report Gen. Circ. Proj. Article XV, U.C.L.A. 1955.
[2] C. H. B. PRIESTLEY: Quart. J. Roy. Meteor. Soc. **75**, 28 (1949).
[3] A. NYBERG and E. SCHMACKE: Tellus **3**, 89 (1951).
[4] A. NYBERG: Arch. Meteor., Geophys. u. Biokl., Ser. A **7**, 99 (1945).

Strahlung in der unteren Atmosphäre.

Von

FRITZ MÖLLER.

Mit 38 Figuren.

Einleitung.

1. Die Strahlungsbilanz der Erde als Grundproblem. Der Planet Erde empfängt von der *Sonne* Strahlung, deren größte thermisch wirksame Energiedichte in dem Bereich von 0,2 bis etwa 3 μ liegt. Strahlung größerer Wellenlängen ist bis etwa 25 μ zwar nachweisbar, energetisch jedoch bedeutungslos. Auch die Strahlung unterhalb 3000 Å wird schon in sehr hohen Schichten der Atmosphäre durch Ozon, Sauerstoff und Stickstoff absorbiert, und dient zur Dissoziation von O_3 und O_2 oder zur Ionisation von O und N_2, so daß in die Troposphäre nur Wellenlängen $\lambda > \sim 3000$ Å eindringen. Dem steht gegenüber die von der *Erde* als Temperaturstrahler ausgehende Strahlung. Durch diese beiden, hereinkommenden und ausgehenden, Strahlungsströme steht die Erde im Energieaustausch mit dem Weltraum.

Die Bestrahlungsstärke aller Wellenlängen der Sonnenstrahlung im mittleren Abstand der Erde von der Sonne ist $J_0 = 2{,}0$ cal cm^{-2} min^{-1} oder 0,1396 Watt cm^{-2} (Ziff. 12). Beim Eindringen in die Atmosphäre unterliegt die Sonnenstrahlung vielfachen Streuungsvorgängen an Molekülen und größeren Schwebeteilchen; sie wird ferner durch Reflexion an der Erdoberfläche und an den Wolken teilweise zurückgeworden. Als Folge dieser Vorgänge wird der Anteil $A = 0{,}35$ der einfallenden Gesamtstrahlung ohne Änderung der Wellenlänge diffus wieder in den Weltraum reflektiert; A wird die *planetarische Albedo* genannt. Der von der Erde vereinnahmte Strahlungsstrom ist daher $J_0(1-A)R^2\pi$, wobei R der Erdradius ist. Der zum Ausgleich von der Erde abgestrahlte Strahlungsstrom muß der Einnahme gleich sein, daher

$$J_0(1 - A)\, R^2\pi = 4R^2\pi\,\sigma\, T_E^4. \tag{1.1}$$

Mit Einsetzen von Zahlenwerten ergibt sich daraus die *effektive Strahlungstemperatur der Erde* zu $T_E = 250$ °K $= -23°$ C. (Sie ist nur abhängig von der Sonnenentfernung und der Albedo, unabhängig von der Größe des Planeten.)

Die Temperatur von $-23°$ C wird in der Erdatmosphäre in etwa 6 km Höhe beobachtet, während die Temperatur an der Erdoberfläche im Mittel aller Breiten und Jahreszeiten $+14{,}3°$ beträgt. Die Höhe von 6 km kann man etwa als mittlere Obergrenze der Wolken annehmen. Wenn die Erde ganz mit Wolken bedeckt wäre, könnte so das Gleichgewicht durch schwarz ausstrahlende Wolken hergestellt sein. Da aber in wolkenlosen Gebieten die Erdoberfläche Strahlung breiter Wellenlängenbereiche direkt an den Weltraum abgibt, müssen auch Schichten der Atmosphäre an der Emission beteiligt sein, deren Temperatur unter 250 °K liegt. Dies ist eine Wirkung derjenigen Gase der Atmosphäre, H_2O, CO_2, O_3, N_2O u. a., die im Langwelligen so stark absorbieren, daß (wegen der geringen Reichweite der Strahlung) schon sehr hohe und daher kältere Luftschichten nach außen abstrahlen.

Die Sonnenstrahlung wird zum kleinen Teil direkt in der Atmosphäre zur Temperaturerhöhung verwendet. Von der zum Boden durchdringenden Strahlung dient dort auch nur etwa $\frac{1}{5}$ zur Temperaturerhöhung, während der größte Teil zur Verdampfung des Wassers an der Erdoberfläche verbraucht wird. Die latente Wärme wird mit dem Wasserdampf emporgetragen und bei der Kondensation in Wolken wieder frei. Ein weiterer Teil baut potentielle Energien auf, aus denen die Bewegungsvorgänge der Atmosphäre herrühren; deren Dissipation durch Reibung liefert schließlich wiederum thermische Energien, deren Abfuhr durch Strahlung nach außen möglich ist. Die Atmosphäre mit ihren Bewegungen kann deshalb als *Wärmekraftmaschine* zwischen Absorption und Emission der Strahlung aufgefaßt werden, für deren Funktionieren es von großer Bedeutung ist, daß die Absorption überwiegend am Erdboden stattfindet (die atmosphärische Wärmequelle also am Grunde des Luftmeeres liegt), während die Emission überwiegend aus hohen Schichten stammt (vgl. Ziff. 33).

Streuung, Absorptions- und Emissionsvorgänge sind Teilprobleme der Strahlungsvorgänge, die für alles meteorologische Geschehen die wichtige Rolle des Antriebsmotors darstellen.

Die kürzeste Wellenlänge der emittierten Strahlung ist, den Temperaturen entsprechend, etwa 5 μ. Die zugestrahlte Sonnenenergie ist hier aber schon so schwach, daß praktisch die eindringende und die ausgehende Strahlung sich nicht überlappen und beide zunächst getrennt betrachtet werden können[1].

A. Die Sonnenstrahlung.

2. Das Extinktionsgesetz. Die Luftmasse. Die Schwächung der Strahlung erfolgt in der Atmosphäre durch *Streuung* und *Absorption*. Beide Vorgänge sind streng zu unterscheiden; bei der — ohne Änderung der Phase oder Wellenlänge erfolgenden — Streuung tritt keine Umwandlung der Strahlungsenergie auf, während bei der Absorption eine Veränderung in Wärme-, Dissoziations- und andere Energieformen stattfindet. Wenn allgemein nur die Schwächung der einfallenden Strahlung betrachtet werden soll, werden Streuung und Absorption gemeinsam als *Extinktion* bezeichnet.

Für jede Extinktion monochromatischer Strahlung gilt das Gesetz von BOUGUER und LAMBERT.

$$d i_\lambda = - k_\lambda i_\lambda d l, \tag{2.1}$$

$$i_\lambda = i_{0\lambda} e^{-\int k_\lambda d l}. \tag{2.1a}$$

i_λ ist die Strahlungsstromdichte je Wellenlängeneinheit, meist als „*Strahlungsintensität*" bezeichnet, k_λ cm^{-1} der monochromatische *Extinktionskoeffizient.* Mit

$$k_\lambda d l = d\tau_\lambda$$

wird

$$d \ln i_\lambda = - d\tau_\lambda \qquad \text{und} \qquad i_\lambda / i_{0\lambda} = e^{-\tau_\lambda}. \tag{2.2}$$

τ_λ wird *optische Tiefe* genannt: $e^{-\tau_\lambda} = q_\lambda$ ist der *Transmissionsfaktor (transmissivity)*.

Bei reiner Streuung des Lichtes an den Molekülen gilt weiter das BEERsche Gesetz, nach dem

$$k_\lambda = k_{\lambda,n}\, \varrho / \varrho_n \tag{2.3}$$

ist, wo ϱ die Luftdichte ist und n den Normalzustand $p_n = 760$ Torr, $T_n = 273°$ bezeichnet.

[1] G. HOFMANN: Arch. Met., Geophys. u. Bioklim., Ser. B **6**, 274 (1955).

Für eine horizontal geschichtete und unendlich ausgedehnte Atmosphäre kann weiter $dl = -\sec\zeta\, dz$ geschrieben werden, wenn ζ die Zenitdistanz des von oben einfallenden Strahles ist. Damit wird

$$d \ln i_\lambda = + k_{\lambda,n} \frac{\varrho}{\varrho_n} \sec\zeta\, dz \tag{2.4}$$

und mit Verwendung der statischen Grundgleichung $-\varrho\, dz = dp/g$ (g = Fallbeschleunigung, p = Luftdruck, z = Höhe)

$$d \ln i_\lambda = - k_{\lambda,n} \sec\zeta\, dp/g\,\varrho_n. \tag{2.4a}$$

Die *Höhe der homogenen Atmosphäre (scale height)* ist $H = p_n/g\,\varrho_n = R_L \cdot T_n/g$, mit R_L = Gaskonstante der Luft, T = Temperatur. Wenn $a_\lambda = k_{\lambda,n} \cdot H$ der auf die homogene Atmosphäre als Einheit bezogene Extinktionskoeffizient ist, dann wird

$$d \ln i_\lambda = - a_\lambda \sec\zeta\, d(p/p_n)$$

oder

$$i_\lambda/i_{0\lambda} = \exp\left(-\int\limits_0^p a_\lambda \sec\zeta\, dp/p_n\right). \tag{2.5}$$

$i_{0\lambda}$ ist der Betrag der Strahlung am oberen Rande, d.h. außerhalb der Atmosphäre ($p = 0$).

Die wirkliche Erdatmosphäre ist keine ebene Schicht, sondern *gekrümmt*, so daß auch für einen horizontal einfallenden Strahl die durchstrahlte Luftmasse endlich bleibt. Außerdem tritt infolge der nach oben abnehmenden Luftdichte eine Strahlkrümmung nach unten ein, durch die die *astronomische Refraktion* hervorgerufen und zugleich der Lichtweg gegenüber der homogenen Atmosphäre verlängert wird. In (2.4) und (2.5) ist daher $\sec\zeta$ zu ersetzen durch die *relative Luftmasse*, die eine Funktion $m_r(\zeta_B)$ der am Boden beobachteten scheinbaren Zenitdistanz ζ_B ist. Allgemein sind die von BEMPORAD[1] berechneten Tabellen (Tabelle 1) im Gebrauch, die für $\zeta_B \leqq 60°$ fast keine, und auch bis 85° nur geringe Abweichungen von $\sec\zeta_B$ zeigen. Für $0 < \zeta_B < 88°$ ist die empirische Näherung $m_r = \sec\zeta_B - 40\,(90 - \zeta_B)^{-\frac{8}{3}}$ (ζ_B in Grad) brauchbar.

Streng ist m_r etwas mit λ und der Dichteschichtung veränderlich und daher auch mit der Höhe des Meßortes über Meeresniveau. Unter Vernachlässigung der Strahlkrümmung, aber Berücksichtigung einer exponentiell mit der Höhe abnehmenden Dichte hat HULBURT[2] die relativen Luftmassen durch Γ-Funktionen berechnet. LINK[3] hat die heutigen Kenntnisse der Höhenverteilung der Luftdichte berücksichtigt und neue Tabellen für m_r aufgestellt, die jedoch erst bei $\zeta_B > 83°$ kleine Aweichungen von BEMPORAD zeigen (Tabelle 1).

Die *absolute Luftmasse* ist gegeben durch

$$m = m_r(\zeta_B)\, p_B/p_n. \tag{2.6}$$

Wenn nicht die Luft oder eines der permanenten Gase, deren Konzentration unabhängig von p ist, extingierendes Medium ist, sondern eine Beimengung mit rascherer Höhenabnahme, z.B. Wasserdampf oder Staub, so gelten etwas andere relative Luftmassen, die sich um so mehr $\sec\zeta_B$ nähern, je stärker die Substanz in Bodennähe zusammengedrängt ist. Meist kann der Unterschied vernachlässigt werden. Tabelle 1 gibt in Zeile 3 für Wasserdampf gültige Werte $m_{r,w}$ nach SCHNAIDT[4].

[1] A. BEMPORAD: Meteor. Z. **24**, 306 (1907). — Mitt. Großh. Sternwarte Heidelberg **4** (1904). — In WINKELMANN, Handbuch der Physik, Bd. 6, Optik, S. 551. Leipzig 1906. — Ausführliche Wiedergabe der Tabellen in [*2*] und [*2a*].

[2] E. O. HULBURT: Phys. Rev. **55**, 639 (1939).

[3] F. LINK: J. des Observ. **17**, 41 (1934).

[4] F. SCHNAIDT: Meteor. Z. **55**, 296 (1938).

Ist dagegen die extingierende Substanz nur in einer hochgelegenen Schicht vorhanden, wie z.B. das absorbierende Ozon, so muß auf die Ableitung der Formel für die relative Luftmasse zurückgegangen werden[1]. Nach der Strahlgleichung für Refraktion gilt $\mu_\lambda(R+z)\sin\zeta=\mu_{\lambda B}(R+z_B)\sin\zeta_B$; μ_λ = Brechzahl der Luft, R = Erdradius. An die Stelle von $\sec\zeta$ in (2.4) und (2.5) tritt dann die von z (und damit von p) abhängige Größe

$$m_r' = (\cos^2\zeta_B + 2\varkappa\sin^2\zeta_B)^{-\frac{1}{2}} \tag{2.7}$$

mit

$$\varkappa = \frac{z-z_B}{R} + (\mu_{\lambda n}-1)\,\frac{\varrho_B-\varrho}{\varrho_n}.$$

Für ein Gas mit der Konzentration c gilt dann

$$i_\lambda/i_{0\lambda} = \exp\left(-\int k_{\lambda n}\, c\, m_r'\, H\, dp/p_n\right).$$

Tabelle 1, Z. 5 gibt als Beispiel m_r' für eine in 25 km Höhe bei $\varrho = 4{,}17\cdot 10^{-5}$ g/cm³ gelegene (Ozon-)Schicht.

Tabelle 1. *Durchstrahlte relative Luftmassen m_r.*

φ_B	0°	30°	60°	70°	80°	85°	88°	89°	90°
m_r (nach BEMPORAD)	1,0	1,154	1,995	2,904	5,60	10,40	19,8	27,0	39,7
m_r (nach LINK)	1,0	1,154	1,995	2,904	5,60	10,50	20,1	27,5	
$m_{r,w}$ (nach SCHNAIDT)	1,0	1,155	1,998	2,917	5,71	11,11	24,5	38,6	75,1
$\sec\zeta_B$	1,0	1,155	2,000	2,924	5,76	11,47	28,7	57,3	∞
m_r' (für z = 25 km)	1,0	1,153	1,975	2,835	5,11	7,92	10,20	10,72	10,91

3. Die Molekülstreuung nach RAYLEIGH. Die 3 Ursachen der Extinktion des Lichtes in der Atmosphäre sind:

1. Streuung an den Molekülen der atmosphärischen Gase;
2. Streuung an größeren Schwebteilchen, Dunst, Staub, Wassertröpfchen, Eiskristallen, usw.;
3. Absorption durch die atmosphärischen Gase.

Von diesen Vorgängen ist nur der erste keinen Schwankungen von Tag zu Tag unterworfen, wenn man von den Schwankungen des Luftdruckes absieht. Der Vorgang der *Molekülstreuung* ist zuerst durch Lord RAYLEIGH d. Ä., später in anderer Betrachtungsweise durch EINSTEIN und PLANCK theoretisch erfaßt worden [*1a*] (dieses Handbuch Bd. 26). Die von der einfallenden Lichtwelle getroffenen Moleküle werden selbst zu Schwingungen angeregt und strahlen die Energie aus, die dem primären Strahl verloren geht. Es wird dabei vorausgesetzt, daß die Teilchen kugelförmig sind, dielektrisch, klein gegen die Wellenlänge und ihre Anordnung im Raum so unregelmäßig, daß keine Phasenbeziehungen zwischen den von ihnen ausgesandten Strahlen bestehen.

Der Extinktionskoeffizient ist dann

$$k_{R\lambda} = \frac{8}{3}\pi^3(\mu_\lambda^2-1)^2 N^{-1}\frac{T_n}{T} f\lambda^{-4}. \tag{3.1}$$

μ_λ ist die Brechzahl der Luft, N die Molekülzahl je cm³ (AVOGADROsche Zahl), $N = 2{,}687\cdot 10^{19}$, f der Anisotropiekoeffizient [vgl. (3.4)]. Setzt man für die gesamte Atmosphäre den Wert von $k_{R\lambda}$ für $T_n = 273°$ an und bildet den auf die Luftmasse $m=1$ bezogenen Extinktionskoeffizienten $a_{R\lambda} = k_{R\lambda,\,T=273}\cdot H$, so ist der Fehler gegenüber einer exakten Berücksichtigung des Temperaturaufbaus

[1] E. SCHÖNBERG: In Handbuch der Astrophysik, Bd. 7, S. 23 u. 27. Berlin: Springer 1936.

der Atmosphäre nur —1,3% (PENNDORF[1]). Da $(\mu_\lambda - 1)$ sehr klein ist, kann man auch schreiben

$$a_{R\lambda} = \frac{32}{3}\pi^3 (\mu_\lambda - 1)^2 N^{-1} H f \lambda^{-4}. \tag{3.2}$$

Die Brechzahl μ_λ ist nur wenig von der Wellenlänge abhängig. Sie beträgt nach EDLÉN[2]

λ	0,2	0,3	0,4	0,5	0,6	0,8	1,0	1,2	1,5	2,0	4,0 μ
$(\mu_\lambda - 1)\, 10^6$	341,9	307,6	298,3	294,3	292,2	290,1	289,2	288,7	288,3	288,0	287,7

Man kann dieser Änderung nach LINKE [*1b*] Rechnung tragen durch eine Formel

$$a_{R\lambda} = 0{,}00879\, \lambda^{-4{,}09}; \qquad \lambda \text{ in } \mu. \tag{3.3}$$

Der Wasserdampfgehalt der Luft hat keinen Einfluß auf die Extinktion der staub- und dunstfreien Luft, da seine Moleküle keinen anderen Streuungseffekt als die Luft besitzen[1]. Das geht auch aus Messungen der Extinktion hervor, die SCHÜEPP während der Regenzeit in Belgisch-Kongo durchgeführt hat; diese ergeben trotz des hohen Betrages von 5 cm Niederschlagswasser fast reine RAYLEIGH-Extinktion (vgl. den Schluß der Ziff. 11). Messungen FOWLES (vgl. Ziff. 9), die eine größere Extinktion ergaben, sind entweder als Resteffekte einer Dunststreuung oder einer Wasserdampfabsorption, oder beider Vorgänge zu deuten.

Den Einfluß der Schichtung in der aus verschiedenen Gasen zusammengesetzten Atmosphäre hat DEIRMENDJIAN[3] untersucht. Er fand den auf die Masse bezogenen Extinktionskoeffizient bis 40 km Höhe unverändert, darüber bis 100 km nur geringe Zunahme um 2%.

Der Extinktionskoeffizient ist umgekehrt proportional λ^4, er ist also für das blaue Ende des sichtbaren Spektrums etwa 10mal größer als für das rote. Die daraus folgende große Schwächung des blauen und geringe Schwächung des roten Lichtes hat bei tiefem Stand der Sonne oder eines Gestirnes (bei großer Luftmasse m) zur Folge, daß das zum Boden durchdringende Licht mehr und mehr rot erscheint. Der optische Schwerpunkt der Sonnenstrahlung verlagert sich mit zunehmendem m nach längeren Wellen. Bei Betrachtung der Gesamtstrahlung aller Wellenlängen folgt daher auch der Mittelwert des Extinktionskoeffizienten dem Gang des optischen Schwerpunktes, d.h. er nimmt mit zunehmendem m ab.

Umgekehrt ist das aus der geraden Strahlrichtung abgelenkte Licht im Blau intensiver als im Rot, so daß die Farbe des Himmels blau wird. Auch die blaue Färbung des Meeres beruht zum Teil hierauf[4], ebenso das bläuliche Streulicht trüber Medien (TYNDALL-Licht).

Die Moleküle der atmosphärischen Gase sind nicht kugelförmig oder optisch isotrop, was nach BORN, GANS, CABANNES[5] u. a. in (3.1) den Faktor f, von CABANNES *Anisotropiekoeffizient* genannt, erforderlich macht. f ist aus der Depolarisation δ des senkrecht zum primären unpolarisierten Strahl abgebeugten Lichtes abzuleiten:

$$f = (6 + 3\delta)(6 - 7\delta)^{-1} \approx 1 + 1.617\,\delta. \tag{3.4}$$

Nach älteren Messungen wurde für Luft $\delta = 0{,}042$, $f = 1{,}074$ gefunden, nach neuen Messungen von GUCKER und BASU[6] ist $\delta = 0{,}035$ und damit $f = 1{,}061$.

---

[1] R. PENNDORF: Tables of the refractive index ..., AFCRC-TN-55-206, Cambridge (Mass.) 1955.
[2] B. EDLÉN: J. Opt. Soc. Amer. **43**, 339 (1953).
[3] D. DEIRMENDJIAN: Arch. Met., Geophys. u. Bioklim., Ser. B **6**, 452 (1955).
[4] F. LAUSCHER: In Handbuch der Geophysik, Bd. 8, Kap. 12. Berlin 1955.
[5] M. BORN: Verh. dtsch. phys. Ges. **19**, 43 (1917); **20**, 16 (1918). — R. GANS: Z. Physik **17**, 353 (1923). — J. CABANNES: Ann. de Phys. **15**, 5 (1921).
[6] F. T. GUCKER u. S. BASU: Sci. Rep. No. 1, AF 19 (122)—400, Univ. of Indiana (1952).

Eine *quantitative Nachprüfung* der Beziehung (3.1) für Luft ist im Labor wegen der äußerst geringen Intensität des Streulichtes kaum möglich. In der Atmosphäre stört die Absorption der permanenten Gase und des Ozons (für $\lambda < 0{,}33\,\mu$ und $0{,}5 < \lambda < 0{,}7\,\mu$) und die auch bei trockenster und reinster Luft noch vorhandenen Restabsorptionen und Streuungen besonders im Ultraroten. Götz[1] extrapolierte 459 Meßreihen, die er den sorgfältigen Sonnenstrahlungsmessungen der Smithsonian Institution in Montezuma (2700 m) entnahm, in ungestörten Wellenlängen auf den Wasserdampfgehalt $w = 0$, konnte jedoch den Wellenlängenexponenten nur zu $-3{,}7$ finden ($0{,}35 < \lambda < 0{,}80\,\mu$).

Tien Kiu und Dufay[2] verwandten das gleiche Beobachtungsmaterial wie Götz und konnten an den klarsten Tagen eine Beziehung $a_\lambda = A + B(\mu_\lambda - 1)^2 \lambda^{-x}$ nachweisen. A (wahrscheinlichster Wert 0,0034) und B zeigten eine Korrelation untereinander ($r = +0{,}78$) und mit dem gesamten Wasserdampfgehalt der Atmosphäre. Das bedeutet, daß auch zwischen den infraroten Absorptionsbanden des Wasserdampfes noch eine Restabsorption A zu finden ist. x ergab sich aus 14 besonders klaren Tagen im Mai 1923 zu $x = -3{,}99 \pm 0{,}08$ ($0{,}35 < \lambda < 1{,}6\,\mu$). Aus dem Zahlenwert von B konnte $N = (2{,}66 \pm 0{,}19)\,10^{19}$ abgeleitet werden, wahrscheinlich wegen einer Spur von Wasserdampfeinfluß etwas zu klein.

Auch das Himmelsstreulicht kann bei reiner trockener Luft zur Prüfung der λ-Abhängigkeit verwandt werden (vgl. Ziff. 16). J. Shaw, Adel und Anthony[3] haben Messungen im Infrarot durchgeführt, wobei letzterer mit zunehmender Winkeldistanz gegen die Sonne eine Annäherung an die Rayleigh-Streuung bis zum Exponent $x = 3{,}5$ fand (zwischen 0,56 und 1,0 μ). Ein Einfluß von apparativem Streulicht, aber auch von Mehrfachstreuung scheint nicht ausgeschlossen.

Die Gültigkeit des Rayleighschen Streuungsgesetzes für reine Luft wird heute allgemein anerkannt, auch wenn die experimentelle Bestätigung noch nicht sicher gelungen ist, jedoch wird die Berechtigung des Anisotropiekoeffizienten f von manchen Autoren bezweifelt.

Aus visuellen Extinktionsmessungen an Sternen konnte Paetzold[4] zeigen, daß in besonders klaren Nächten Extinktionen auftreten, die bis zu 20% kleiner sind als die nach (3.2) berechnete Rayleigh-Extinktion. Dieser Unterschied konnte bisher durch die bekannten Fehlermöglichkeiten nur zum Teil gedeutet werden. Auch in den Beobachtungen der Smithsonian Institution kommen an Tagen mit äußerster Luftklarheit Fälle mit größerer Durchlässigkeit der Atmosphäre im UV vor, als dem Rayleighschen Gesetz entspricht.

4. Die Dunstextinktion bei einheitlicher Teilchengröße. Subtrahiert man vom beobachteten Wert des Extinktionskoeffizienten a_λ die Molekülextinktion $a_{R\lambda}$ nach (3.2) oder (3.3) und die Absorption der atmosphärischen Gase, so verbleibt eine wellenlängenabhängige Extinktion $a_{D\lambda}$, die man der Streuung am Dunst zuschreibt. Unter „Dunst" faßt man alle atmosphärischen Schwebeteilchen zusammen, die optisch wirksam sind.

Diese sind zum größten Teil Partikel aus löslichen und unlöslichen Salzen (Kondensationskerne), Rauch- und Staubteilchen, allerkleinste Tröpfchen und Gemische aus allen diesen Bestandteilen. Nur weniger als 1 pro mille sind Pollen, Bakterien und dergleichen. Die nichtorganischen Teilchen entstehen entweder durch Kondensation und Sublimation von Flammengasen (Ionen, Aitkenkerne $r < 0{,}1\,\mu$, Rauchpartikel $r \approx 0{,}4\,\mu$) oder durch Zerteilung größerer Massen, Aufwirbelung von Staub oder Versprühen von Meerwasser ($r > 0{,}1\,\mu$). Die Teilchen wachsen durch Koagulation oder durch (reversible) Anlagerung von Wasserdampf-

---

[1] F. W. P. Götz: Meteor. Z. **57**, 414 (1940).

[2] Tien Kiu: J. de Phys. (7) **9**, 297 (1938). — J. Dufay u. T. Kiu: J. de Phys. (7) **7**, 198 (1936). — J. Dufay: J. de Phys. (8) **1**, 251 (1940).

[3] A. Adel: Phys. Rev. **74**, 446 (1949). — R. Anthony: J. Meteorology **10**, 60 (1953). — J. H. Shaw: Final Rep. Contr. AF 19 (122)—65, Columbus (Ohio) 1954.

[4] H. K. Paetzold: Astr. Nachr. **281**, 17 (1952). — Met. Rdsch. **5**, 56 (1952).

molekülen bei zunehmender Feuchtigkeit. Nach erfolgter Kondensation können sich Tröpfchen zusammenlagern und bei Wolkenauflösung wieder verdampfen, wodurch dann auch größere Kerne entstehen können. Die obere Grenzgröße ist durch das Ausfallen bedingt, das mit der Größe der Teilchen und deren Sinkgeschwindigkeit zunimmt; der Ausfallprozeß kann jedoch durch vertikale Durchmischung und Konvektion verzögert werden.

Daneben findet auch ein ständiger Verlust der in Niederschlagselementen gebundenen Kerne statt. Die untere Grenzgröße liegt nach Ionenmessungen etwa bei $r=10^{-6}$ cm. Optisch sind jedoch nur noch Partikel bis etwa 10^{-5} cm herab wirksam; dies ist durch die Bedingung der RAYLEIGH-Extinktion $r \ll \lambda$ gegeben [*1a*].

Messungen an Sonnen- oder Sternenlicht oder an irdischen Lichtquellen (Sichtweite, vgl. Beitrag MIDDLETON in diesem Band) haben meist eine Wellenlängenabhängigkeit der Dunstextinktion $a_{D\lambda} \sim \lambda^{-\delta}$ ergeben, wobei δ zwischen 1 und 1,5 liegt. Aus 141 aktinometrischen Messungen von SCHÜEPP [*6*] in Davos ergibt sich $\delta_m = 1{,}37$ mit einer Streuung $s_\delta = 0{,}86$; aus Photozellenmessungen in Frankfurt und Mainz (Juli 1952 bis März 1954) erhielt VOLZ[1] $\delta_m = 1{,}48$ mit $s_\delta = 0{,}42$ bei 426 Messungen. Neben solchen Potenzfunktionen sind auch wellenlängenunabhängige $a_{D\lambda}$ und solche mit Maximalwerten oder Minimalstellen im Sichtbaren beobachtet worden.

Zur Deutung der Messungen muß zurückgegriffen werden auf die MIEsche[2] *Theorie der Streuung* an kleinen dielektrischen Kugeln. Sie zeigt, daß der Extinktionskoeffizient *eines* Teilchens vom Halbmesser r ist

$$k_{D\,r\lambda} = \pi r^2 \sigma, \tag{4.1}$$

wo σ als optischer Wirkungsquerschnitt bezeichnet werden kann. σ ist nur abhängig von

$$\alpha = 2\pi r/\lambda \tag{4.2}$$

und von der Brechzahl m der Substanz des Teilchens. Bei sehr *kleinen* Teilchen, wenn $r \ll \lambda$ ist, ist $\sigma \sim \alpha^4$, also $k_{D\,r\lambda} \sim r^6 \lambda^{-4}$; das ist der Gültigkeitsbereich der RAYLEIGHschen Streuung. Bei sehr *großen* Teilchen, $r \gg \lambda$, dürfen die an der Kugel gebeugten, in ihr gebrochenen und an der Oberfläche reflektierten Strahlen näherungsweise getrennt, nach den Gesetzen der geometrischen Optik, betrachtet werden. BUCERIUS[3] hat gezeigt, daß dies direkt aus den Formeln MIEs durch Grenzübergang abgeleitet werden kann. Die Beugungsvorgänge sind nach der Theorie von FRAUNHOFER und AIRY zu behandeln, Brechung und Reflexion sind von WIENER berechnet worden (vgl. Beitrag BRICARD in diesem Band). Die untere Grenze des Gültigkeitsbereiches der geometrischen Optik kann nach MECKE[4] bis etwa $\alpha = 10$ herabgedrückt werden (bei $m = \frac{4}{3}$), wenn man die Interferenzen der einzelnen Strahlenbündel berücksichtigt.

Dagegen versagen alle Vereinfachungen in der Umgebung von $r = \lambda$, hier muß die strenge elektromagnetische Theorie nach MIE angewandt werden. Es liegen mehrere Berechnungen von $\sigma(\alpha)$ vor; am bekanntesten sind die älteren von STRATTON und HOUGHTON[5], neuere sind von LOWAN[6] und sehr detaillierte

[1] F. VOLZ: Geofis. pura e appl. **31**, 119 (1955).
[2] G. MIE: Ann. Phys. (4) **25**, 377 (1908).
[3] H. BUCERIUS: Optik **1**, 188 (1946).
[4] R. MECKE: Ann. Phys. **62**, 623 (1920).
[5] J. A. STRATTON u. H. G. HOUGHTON: Phys. Rev. **38**, 159 (1931). — H. G. HOUGHTON u. W. R. CHALKER: J. Opt. Soc. Amer. **39**, 955 (1949).
[6] A. N. LOWAN: Tables of Scatt. Funct. f. spherical particles (Nat. Bur. Stand., Wash. 1949) Appl. Math. Ser. 4.

für verschiedene m von PENNDORF und GOLDBERG[1] berechnet worden. Aus Fig. 1 ist zu erkennen, daß der Streuquerschnitt σ oberhalb des RAYLEIGH-Bereiches, in dem $\sigma \sim \alpha^4$ ist, bei $\alpha = 6$ (für Wasser, $m = 1{,}330$) ein Maximum erreicht, bei $\alpha = 12$ ein erstes Minimum, und sich für sehr große α in Wellen an den Wert 2 annähert. Es wird dann $k_{D_r\lambda} = 2\pi r^2$, während der geometrische Schatten nur πr^2 ist. Für sehr große α trifft fast das gesamte gebeugte Licht in einen sehr engen Winkelbereich nach vorn, so daß ein Meßinstrument auch bei kleinen Öffnungswinkel $\overline{\varphi}$ fast das gesamte gebeugte Licht mitmißt; für $\alpha = 400$ und $\overline{\varphi} = 1°$ wird deshalb $\sigma = 1{,}076$ gemessen[2].

Kleine Oszillationen sind dem Verlauf von σ überlagert, die vermutlich durch die Interferenz der ein- und mehrfach im Tropfen gespiegelten Strahlen untereinander und mit den gebrochenen oder gebeugten entstehen. Für größere Brechungszahlen m tritt das erste Maximum (und alle folgenden) bei kleineren α ein; die Lage der Extreme bleibt aber nahezu unverändert, wenn man σ als Funktion von

$$x = (m - 1)\,\alpha \qquad (4.3)$$

aufträgt, worauf V. D. HULST[3] zuerst hingewiesen hat. Das erste Maximum liegt dann etwa bei $x = 2{,}0$. Die Amplituden von σ sind bei größeren Werten m etwas größer. [Für $m < 1$, z.B. Luftblasen in Wasser, gilt $x = (1 - m)\,\alpha$.]

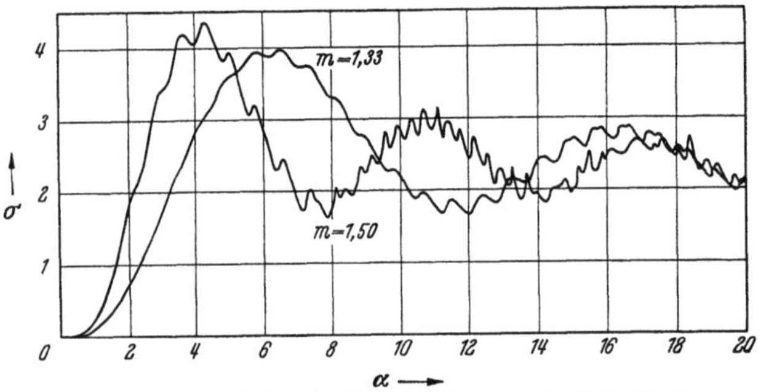

Fig. 1. Streuquerschnitt σ in Abhängigkeit von $\alpha = 2\pi r/\lambda$ für Brechungsindex m; nach PENNDORF u. GOLDBERG[1].

Die *Substanz* der Dunstteilchen ist nicht sicher bekannt, wohl auch nicht einheitlich. JUNGE [4] hat durch umfangreiche Untersuchungen nachgewiesen, daß Ammonium-Sulfat $(NH_4)_2SO_4$ einen Hauptbestandteil der kleineren Kerne bildet, während die Teilchen $r > 0{,}8\ \mu$ zumeist aus NaCl oder Meeressalz bestehen. Nur bei sehr trockener Luft können die Substanzen in Kristallform vorliegen, oder die Partikel bestehen aus Konglomeraten (Mischkerne), unter denen sich auch Kristalle befinden. Mit zunehmender Feuchte ist eine flüssige Hülle von gesättigten Lösungen, aber erst dicht vor 100% starke Verwässerung zu erwarten.

Bei Berücksichtigung dieser Erkenntnisse müssen Brechzahlen $m > 1{,}33$ und wahrscheinlich auch Absorption in der Partikelsubstanz erwartet werden.

Für komplexe Brechzahlen $m^* = m(1 - i\varkappa)$, d.h. für Medien mit Absorption, deren Absorptionskoeffizient in der Schicht $k = 4\pi m\varkappa\,\lambda^{-1}$ beträgt, hat der Streuquerschnitt σ einen anderen Verlauf[4,5], der Anstieg wird flacher — bei kleinem α nur mehr proportional α bis α^3 — und auch der Verlauf der Maxima und Minima wird stark abgeflacht (Fig. 2). Für undurchsichtige Teilchen ($m = \infty$) hat GÖTZ[6] den Verlauf von σ berechnet: nach dem steilen Anstieg proportional α^4 verbleibt σ für $\alpha > 2{,}0$ nahezu konstant bei $\sigma = 2$.

[1] R. PENNDORF u. B. GOLDBERG: Geophys. Res. Pap., Cambridge (Mass.) No. 45 (im Druck). — B. GOLDBERG: J. Opt. Soc. Amer. **43**, 1221 (1953). — R. PENNDORF: J. Meteorology **13**, 219 (1956).

[2] R. O. GUMPRECHT, C. M. SLIEPCEVICH: J. Phys. Chem. **57**, 90 (1953).

[3] H. C. V. D. HULST, in G. KUIPER: The Atmosph. of the Earth and Planets, S. 49—111. Chicago 1949.

[4] A. N. LOWAN: Tables of Scatt. Funct. f. spherical particles (Nat. Bur. Stand., Wash. 1949) Appl. Math. Ser. 4.

[5] J. C. JOHNSON u. J. R. TERRELL: J. Opt. Soc. Amer. **45**, 451 (1955).

[6] F. W. P. GÖTZ: Astr. Nachr. **255**, 63 (1935).

Experimentell konnte der wellenförmige Verlauf des Streuquerschnittes an Staubteilchen und Nebeln verschiedener, aber sehr einheitlicher Größe wiederholt bestätigt werden, insbesondere durch zahlreiche Arbeiten von SINCLAIR und LA MER[1]. Eine Zusammenstellung dieser Untersuchungen hat VOLZ in [*1e*] gegeben.

Es ist wiederholt versucht worden, die durch die Beobachtungen der atmosphärischen Transmission gefundene Proportionalität $a_{D\lambda} \sim \lambda^{-\delta}$ mit Hilfe des Verlaufs von σ zu deuten. Sind in der Volumeneinheit n Teilchen der Größe r enthalten, also in der ganzen Atmosphärensäule über der Flächeneinheit $H \cdot n$ solche Teilchen, so wird der Extinktionskoeffizient

$$\left.\begin{aligned} k_{D\lambda} &= n\, r^2 \pi\, \sigma(\alpha) \\ \text{und}\quad a_{D\lambda} &= H\, n\, \pi\, r^2 \sigma(\alpha). \end{aligned}\right\} \quad (4.4)$$

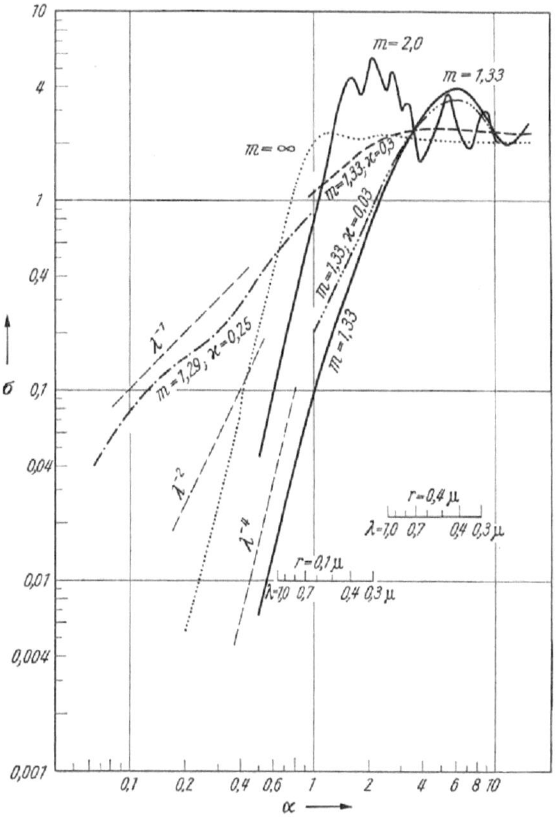

Fig. 2. Streuquerschnitt σ in Abhängigkeit von $\alpha = 2\pi r/\lambda$ für reelle Brechungsindices m und komplexe $m^* = m(1 - i\varkappa)$.

Bei einheitlicher Partikelgröße würde sich daher in dem kleinen Wellenlängenbereich des sichtbaren Lichtes, der in Fig. 3 auf der Wellenlängenskala angedeutet ist, ein Verlauf des Extinktionskoeffizienten mit λ ergeben, wie ihn Fig. 3 zeigt. Bei sehr kleinen Tröpfchen erhält man einen Verlauf mit großem negativem Exponenten δ, größere Tröpfchen ergeben flacheren Verlauf, und Tröpfchen, deren Radius $r > \lambda$, könnten auch ein Minimum im Extinktionsverlauf mit λ ergeben. Eine Stelle, bei der (4.3) $\sigma \sim x$ ist, was nach (4.1), (4.2), bei $r = \text{const}$ bedeutet $k_D \sim \lambda^{-1}$, findet sich bei $x = 1{,}5$. Beobachtet man also eine Dunstextinktion dieser Art im sichtbaren Licht, dann könnten Teilchen vorliegen mit einem Radius von $r = \frac{1{,}5\,\lambda}{(m-1)\,2\pi}$, also $r = 0{,}35\,\mu$ bei Wassertröpfchen, von $r = 0{,}24\,\mu$ bei $m = 1{,}5$. Dies ergibt einen ersten Anhaltspunkt für die Größe der extingierenden Teilchen. LINKE und V. D. BORNE[2] erhielten so einen effektiven Teilchenradius $r = 0{,}3\,\mu$, GÖTZ[3] $r = 0{,}29\,\mu$. Durch Hinzunahme von Messungen der Winkelverteilung des gestreuten Lichtes konnte SIEDENTOPF[4] genauere Angaben erlangen. Er erhielt einen Gehalt von 500 Wassertröpfchen mit 0,1 bis 0,5 μ Radius und von 20 Staubteilchen mit 0,2 bis 1 μ Radius im cm^3 bei einer Sichtweite von 20 km. DESSENS[5] glaubte insbesondere aus der Lage von Minimumstellen in seinen Spektralbeobachtungen der Extinktion schließen zu können, daß die gesamte Dunstextinktion durch Wassertröpfchen einheitlicher Größe zwischen 1 und 2 μ hervorgerufen sei.

5. Die Dunstextinktion bei nicht einheitlicher Teilchengröße. Daß die Deutungen SIEDENTOPFs und DESSENS' den wahren Sachverhalt nicht richtig wiedergeben, wissen wir vor allem aus den Untersuchungen des atmosphärischen

[1] D. SINCLAIR u. V. K. LA MER: Chem. Rev. **44**, 245 (1949).
[2] F. LINKE u. H. V. D. BORNE: Gerlands Beitr. Geophys. **37**, 64 (1932).
[3] F. W. P. GÖTZ: Verh. schweiz. naturforsch. Ges. Sils **1944**, 88.
[4] H. SIEDENTOPF: Z. Met. **1**, 417 (1947).
[5] H. DESSENS: Ann. Géophys. **2**, 68 (1946).

Aerosols durch Junge [4], der gezeigt hat, daß ein *kontinuierliches Größenspektrum* von Teilchen existiert, das von den Luftionen und Aitken-Kernen ($r < 10^{-5}$ cm) über die Dunsttröpfchen und -teilchen (große Kerne) bis zu den Staubpartikeln (Riesenkerne $r > 10^{-4}$ cm) reicht. Es sind jederzeit Teilchen aller Größenordnungen vorhanden, deren Verteilungsgesetz sich auch formelmäßig angeben läßt.

Wenn allgemein verschiedene Partikelgrößen vorkommen und demgemäß ein Größenverteilungsgesetz $dn/dr = f(r)$ existiert, dann ist der Extinktionskoeffizient der Volumeneinheit Luft:

$$k_{D\lambda} = \int_r \pi r^2 \sigma(\alpha) \frac{dn}{dr} dr. \tag{5.1}$$

Junge hat gezeigt, daß über einen großen Bereich von Teilchen, der von $r = 5 \cdot 10^{-6}$ cm bis 10^{-3} cm reicht, ein Verteilungsgesetz $f(r)$ gefunden wird, das einer Potenzfunktion mit dem Exponenten -4 gehorcht:

$$dn = M c_1 r^{-4} dr \tag{5.2}$$

M = Modul = $\log_{10} e$; die von Junge gewählte Definition der Konstanten c_1 ist hier beibehalten. Setzt man (5.2) in (5.1) ein, so ergibt sich mit (4.2)

$$k_{D\lambda} = M\, 2\pi^2 c_1 \lambda^{-1} \int_{\alpha_0}^{\alpha_1} \sigma(\alpha)\, \alpha^{-2} d\alpha = c_1 \lambda^{-1} K_4. \tag{5.3}$$

Der Wellenlängenexponent in der Dunstextinktion findet hierdurch seine Erklärung. Auch beim Bestehen einer Größenverteilung von streuenden Teilchen gemäß (5.2), und *nur* in diesem Falle, existiert also ein Dunst-Extinktionskoeffizient, der umgekehrt proportional λ ist. Die Größe K_4, die die Extinktion mit dem Exponenten (-4) der Aerosolverteilung verbindet, ist dabei nur von der Brechzahl m und von den Grenzgrößen α_0, α_1, innerhalb deren das Potenzgesetz (5.2) gilt, abhängig. Ohne wesentliche Einschränkung kann $\alpha_0 = 0$ gesetzt werden, weil für kleine α der Wert von $\sigma(\alpha) \sim \alpha^{-4}$, also sehr klein wird, bei großen α dagegen $\sigma(\alpha) \to 2{,}0$ geht und der Integrand ebenfalls verschwindet. Damit ist K_4 exakt von α unabhängig.

Fig. 3 veranschaulicht das Zustandekommen des λ^{-1}-Gesetzes durch Überlagerung der Extinktionen (Streuquerschnitte) einer Anzahl diskreter Teilchengrößen entsprechend der Häufigkeit ihres Vorkommens. Bricht die Jungesche Größenverteilung nach kleinen oder großen Werten bei r_0 bzw. r_1 ab, so ergibt sich bei kleinen λ eine Verflachung der λ-Abhängigkeit, an der oberen Grenze dagegen eine Versteilung. Ersteres ist mehrfach von Götz als anomale Extinktion im UV beobachtet worden (vgl. Ziff. 7). Da r_1 nach Junge bei etwa 10 μ anzunehmen ist, ist ein entsprechender Effekt in der Extinktion erst bei $\lambda > 10\,\mu$ zu erwarten. Messungen im Infrarot haben jedoch mehr auf Werte $\delta < 1{,}0$ anstatt auf größere δ geführt[1].

Verallgemeinert man (5.2) zu

$$dn = M c_1 r^{-\nu} dr, \tag{5.4}$$

so erhält man anstatt (5.3)

$$k_{D\lambda} = c_1 \lambda^{3-\nu} K_\nu, \quad \text{mit } K_\nu = M\, 2^{\nu-3} \pi^{\nu-2} \int \varkappa(\alpha)\, \alpha^{2-\nu} d\alpha. \tag{5.5}$$

Tabelle 2 gibt K_ν für verschiedene Exponenten ν und Brechzahlen m. Findet man also durch Extinktionsmessungen eine Wellenlängenabhängigkeit $\lambda^{-1,3}$, so so kann daraus geschlossen werden, daß die streuenden Teilchen (zumindest in

[1] A. Arnulf, J. Bricard u. C. Veret: C. R. Acad. Sci. Paris **238**, 503 (1954). — T. Elder u. J. Strong: J. Franklin Inst. **255**, 189 (1953).

dem optisch ausschlaggebenden Bereich $0{,}2\,\mu < r < 0{,}6\,\mu$) einem Größenspektrum nach (5.4) gehorchen mit einem Exponenten $\nu = 4{,}3$.

Bei direkten Kernzählungen wird der Exponent $\nu = 4$ vor allem im kontinentalen Aerosol gefunden, d.h. in Luftmassen, die über Land zahlreiche Verbrennungspartikel aufgenommen haben. Über dem Ozean scheint $\nu = 4$ (oder Werte

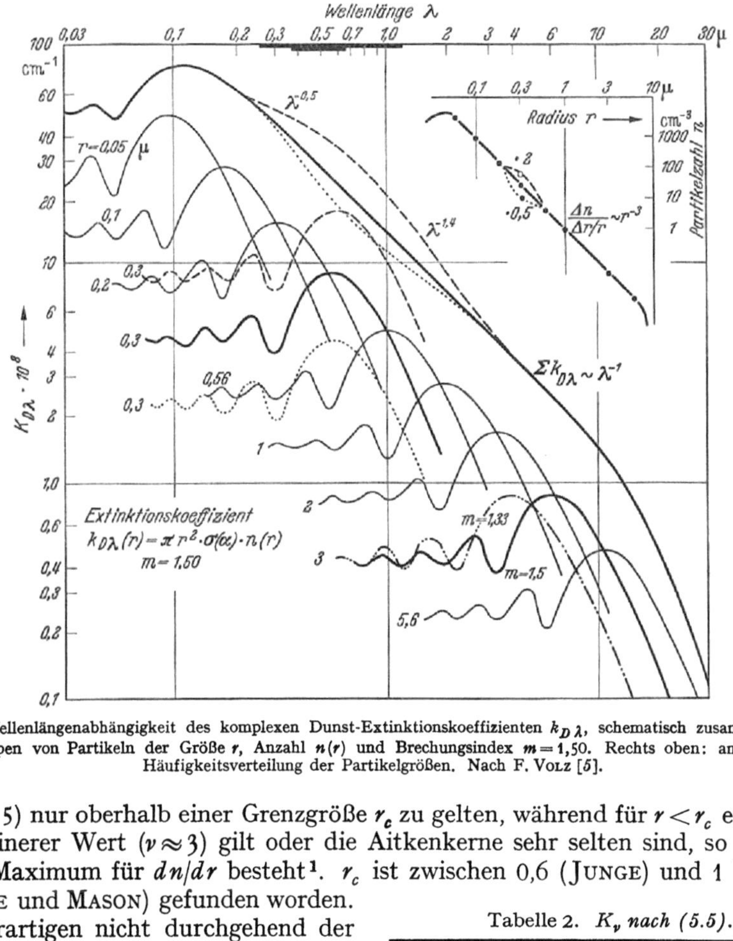

Fig. 3. Wellenlängenabhängigkeit des komplexen Dunst-Extinktionskoeffizienten $k_{D\lambda}$, schematisch zusammengesetzt aus Gruppen von Partikeln der Größe r, Anzahl $n(r)$ und Brechungsindex $m = 1{,}50$. Rechts oben: angenommene Häufigkeitsverteilung der Partikelgrößen. Nach F. Volz [5].

bis $\nu = 5$) nur oberhalb einer Grenzgröße r_c zu gelten, während für $r < r_c$ entweder ein kleinerer Wert ($\nu \approx 3$) gilt oder die Aitkenkerne sehr selten sind, so daß bei r_c ein Maximum für dn/dr besteht[1]. r_c ist zwischen 0,6 (Junge) und 1 bis 10 μ (Moore und Mason) gefunden worden. Bei derartigen nicht durchgehend der Form (5.4) gehorchenden Kernspektren folgt auch $a_{D\lambda}$ keinem einfachen Potenzgesetz mehr.

Tabelle 2. K_ν *nach (5.5)*.

$\nu =$	$m = 1{,}33$	1,4	1,5	2,0	∞
3,5	11,6	13,2	14,6	19,0	16,4
4,0	11,1	12,6	16,0	22,8	26,2
4,5	12,6	16,6	20,8	38,9	94,5
5,0	20,6	27,0	37,3	42,5	100,8

Besondere Aufmerksamkeit verlangen zeitliche Änderungen im Größenspektrum des Aerosols. Durch zunehmende Feuchtigkeit und Wasserdampfanlagerung an die Kerne wird eine Abnahme von ν hervorgerufen. Junge[2] hat gezeigt, daß dabei auch der Einfluß der oberen Grenzgröße r_1 zu berücksichtigen ist. Dann ergibt sich erst bei $\nu \to 2$ ein Wellenlängenexponent 0 anstatt bei $\nu \to 3$. Eine solche Umbildung des Tröpfchenspektrums erfolgt aber erst bei sehr großer Feuchtigkeit ($f > 97\%$).

[1] A. H. Woodcock: J. Meteorology 7, 397 (1950). — D. J. Moore u. B. J. Mason: Quart. J. Roy. Met. Soc. 80, 583 (1954).

[2] C. Junge: J. Meteorology 12, 13 (1955).

Da mit gewöhnlichen optischen Methoden nur der Bereich von etwa 0,3 bis 0,8 μ zugänglich ist (bei kleineren Wellenlängen stört die O_3-, bei größeren die H_2O-Absorption), können abweichende Werte des Wellenlängenexponenten auch durch wellenartige Überlagerungen in der Häufigkeitsverteilung der Partikeln bewirkt sein, wie es in Fig. 3 durch vergrößerte oder verkleinerte Anzahl der Teilchen mit $r = 0{,}3$ μ angedeutet ist. Die Wiederanpassung an das ungestörte Gesetz tritt hier erst jenseits des optischen Bereiches ein. — Vergrößerte Brechzahl kann gleiche Extinktion nur bei verkleinerter Teilchenzahl gleicher Halbmesser ergeben.

Bei Extinktionsmessungen des Sonnen -oder Sternenlichtes ist nicht der auf die Längeneinheit bezogene Extinktionskoeffizient $k_{D\lambda}$, sondern ein für die ganze Atmosphäre gültiger Wert

$$a_{D\lambda} = H_D\, k_{D\lambda} \tag{5.6}$$

zu verwenden [vgl. (4.4)]. Man darf annehmen, daß die Teilchenkonzentration im Mittel nach einer e-Funktion mit der Höhe abnimmt: $n = n_0 e^{-z/H_D}$. Dann ist $\int_0^\infty n\,dz = n_0 H_D$ die Gesamtzahl der über der Einheitsfläche befindlichen Teilchen in der Atmosphäre. H_D ist also die Mächtigkeit einer Schicht, die bei höhenunabhängiger Teilchendichte die gleiche Gesamtzahl enthalten würde wie die wirkliche Atmosphäre. Man bezeichnet H_D als *(äquivalente) Höhe der homogenen Dunstatmosphäre*; in erster Näherung kann sie zu etwa 1 bis 2 km angesetzt werden. In (5.6) ist vorausgesetzt, daß H_D unabhängig von λ und damit unabhängig von r ist, also daß große und kleine Teilchen nach dem gleichen Gesetz mit der Höhe abnehmen, oder daß der Exponent ν in (5.4) unabhängig von der Höhe ist. Kernmessungen von JUNGE im Tiefland und Hochgebirge (Zugspitze) haben dies bestätigt; eine weitere Nachprüfung ist wohl erforderlich.

6. Extinktionsmessungen bei Dunst. Bei Gültigkeit von (5.4) folgt aus (5.5) und (5.6):

$$a_{D\lambda} = c_1 H_D K_\nu \lambda^{3-\nu}. \tag{6.1}$$

Hat man Messungen von $a_{D\lambda}$ durchgeführt, so lassen sich aus der Wellenlängenabhängigkeit und den Absolutwerten allein die drei Konstanten in (6.1) noch nicht trennen. c_1 kann nur aus gleichzeitigen unabhängigen Messungen des Aerosolgehaltes der Luft, d.h. der Partikelzahlen bei verschiedenen Radien r gewonnen werden, während man H_D nach (5.6) durch eine gesonderte Bestimmung des Extinktionskoeffizienten $k_{D\lambda}$ auf horizontaler Meßstrecke, also z.B. durch eine Spektralmessung der Sichtweite (vgl. Beitrag MIDDLETON in diesem Band), ermitteln kann. Liegen diese beiden Messungen vor, dann könnte die Größe von K_ν bestimmt werden. Da K_ν gemäß (5.5) und Tabelle 2 außer vom Exponenten ν noch von der Brechzahl m abhängig ist, kann dann m für die Partikelsubstanz ermittelt werden.

Bolometrische Spektralmessungen der Extinktion der Sonnenstrahlung sind in großer Zahl von den Observatorien der Smithsonian-Institution in Washington und an einer Reihe von Bergstationen [*3*] ausgeführt worden, außerdem am Astrophysikalischen Observatorium in Potsdam. WEMPE[1] hat alle diese Messungen, sowie photometrische Messungen an Sternen einer sorgfältigen Kritik unterzogen sowie eigene Messungen beigesteuert. Die einzelnen Beobachtungen des Wellenlängenverlaufes der Extinktion zeigen meist keinen glatten Verlauf sondern manchmal sogar spitze Minima und Maxima. Es ist anzunehmen, daß diese durch die Schwierigkeit der Messungen besonders bei geringer Schwächung

[1] J. WEMPE: Astr. Nachr. **275**, 1 (1947).

Tabelle 3. *Parameter der Dunstextinktion* $a_{D\lambda} = B\lambda^{-\delta}$ *nach* J. WEMPE.

Ort und Zeit	Zahl der Beobachtungen	δ	B	$a_{D,\lambda=0,55}$	Bemerkungen
Jena (1941/42)	6	1,83	0,090	0,223	Sterne, phot.
	6	1,56	0,132	0,335	Sterne, phot.
	6	1,46	0,239	0,571	Sterne, phot.
Göttingen (1934/35)	6	1,02	0,058	0,107	Sterne, phot.
	7	0,84	0,178	0,294	Sterne, phot.
Kiel (1931/32)	8	1,58	0,067	0,172	Sterne, phot.
Babelsberg (1934)	6	1,52	0,034	0,085	Sterne, phot.
Potsdam (1913)	13	1,37	0,115	0,262	Sonne, bol.
Potsdam (1914)	13	1,13	0,092	0,180	Sonne, bol.
Washington (1903 bis 1907)	5	1,13	0,041	0,081	Sonne, bol.
	10	1,22	0,090	0,187	Sonne, bol.
	5	1,23	0,148	0,308	Sonne, bol.
Bassour, 1160 m (1912)	9 Tage	1,73	0,020	0,056	
	2 Tage	0,68	0,154	0,231	Katmai-Störung
Hump Mountain, 1500 m	8 Tage	1,24	0,025	0,053	
Harqua Hala, 1721 m	29 Tage	0,92	0,012	0,021	p.w. = 0,12 cm
	22 Tage	1,09	0,028	0,058	p.w. = 1,30 cm
Mt. Wilson, 1780 m	14 Tage	1,39	0,014	0,031	
	2 Tage	0,66	0,135	0,200	Katmai-Störung
Calama, 2250 m		0,48	0,019	0,026	gute Sicht
		1,33	0,049	0,109	schlechte Sicht
Table Mountain, 2286 m	39 Tage	0,93	0,015	0,028	p.w. = 0,11 cm
	31 Tage	1,48	0,023	0,055	p.w. = 1,62 cm
Montezuma, 2711 m	33 Tage	0,93	0,012	0,020	p.w. = 0,08 cm
	45 Tage	0,83	0,032	0,053	p.w. = 1,24 cm

p.w. = kondensierter Wasserdampfgehalt der Atmosphäre, precipitable water.

bedingt sind, weil jedes Aerosol, auch beim Vorhandensein diskreter Partikelgrößen, einen geglätteten Verlauf von k_λ besitzen muß, der durch den Verlauf der σ-Werte bedingt ist.

Zu Gruppen zusammengefaßte Mittelwerte WEMPEs gibt Tabelle 3 wieder, wobei ein Ansatz $a_{D\lambda} = B\lambda^{-\delta}$ zugrunde gelegt worden ist. Der Exponent ergibt sich bei allen Trübungsstärken B aus WEMPEs Jenaer Messungen zu $\delta = 1{,}5$ (unterhalb 450 mμ ist eine Verringerung von δ angedeutet), während die vom gleichen Beobachter nach der gleichen Methode gewonnenen Messungen in Göttingen deutlich kleinere und von der Trübung abhängige δ zeigen. Die Messungen auf Bergen führen oberhalb 2000 m auf geringere Werte von durchschnittlich $\delta = 0{,}9$, womit eine frühere Auswertung der gleichen Beobachtungen durch LINKE und v. D. BORNE [*1c*], die eine Zunahme von δ mit der Höhe fanden, nicht bestätigt werden konnte. Die kleineren Werte δ für hochgelegene Orte und bei geringerer Trübung bedeuten gröbere wirksame Teilchen, ein meteorologisch nicht verständlicher Befund. Weitere Messungen erscheinen erwünscht, vor allem auch gleichzeitig an mehreren Orten, um in denselben Luftmassen wirklich vergleichbare Werte zu erhalten.

Aus weitaus zahlreicheren Messungen in jeweils 2 bis 4 Spektralbereichen fanden SCHÜEPP und VOLZ Mittelwerte $\delta_m = 1{,}37$ bzw. 1,48 (vgl. Ziff. 4, S. 161) für normale Bedingungen im Hochgebirge bzw. im Flachland.

Im Infrarot kann nur bei hoher spektraler Dispersion zwischen den Absorptionsbanden oder Linien des Wasserdampfes gemessen werden. Für stark dunstige Luft können Messungen von GEBBIE u.a.[1] etwa durch eine Extinktion $a_{D\lambda}$ angenähert werden mit $B \approx 0{,}1\ \text{km}^{-1}$ und δ mit zunehmender Trübung von 0,4 auf 0,65 zunehmend. Bei Messungen zur Bestimmung

[1] Ausgewertet nach T. ELDER u. J. STRONG: J. Franklin Inst. **255**, 189 (1953).

der extraterrestrischen Sonnenstrahlung im Infrarot ($0{,}9 < \lambda < 2{,}3\,\mu$) fand PIERCE[1] auf Mount Wilson auch noch bei sehr klarer Luft Extinktionskoeffizienten a_λ, die entweder wellenlängenunabhängig in den Grenzen 0,005 bis 0,014 (für Atmosphärendicke $H = 6{,}44$ km) oder mit einer geringen Abnahme $\sim\lambda^{-\frac{1}{4}}$ bestimmt werden können.

Während der Trübung der Atmosphäre mit *Vulkanstaub* nach dem Ausbruch des Katmai 1912 wurden in Bassour (Algerien), Mt. Wilson (Calif.) und Uppsala übereinstimmend Exponenten $\delta = 0{,}68$ festgestellt (Tabelle 3). Allerdings zeigt der Spektralverlauf in doppelt logarithmischer Darstellung (Fig. 4) eine deutliche Krümmung[2], so daß eine Erfassung mit einem Potenzgesetz nur unvollkommen ist oder das Verteilungsgesetz nur über einen kurzen Größenbereich Gültigkeit hat.

Eine auch dem bloßen Auge sichtbare Störung in der Extinktion war die „blaue Sonne" von Ende September 1950, eine Erscheinung, die durch Transport von Rauchpartikeln in den Strömungen der oberen Troposphäre von ausgedehnten Waldbrandgebieten Westkanadas bis Europa hervorgerufen wurde. Die spektrale Verteilung der Extinktion wurde durch WILSON, WELLMANN und VOLZ mit einem Minimum im Blaugrün bestimmt, was nur durch das erste Minimum der Streuquerschnittskurve nach Fig. 1 gedeutet werden kann. Bei $m = \frac{4}{3}$ wäre daraus auf $r \approx 0{,}5\,\mu$ zu schließen, bei Voraussetzung eines größeren m auf etwas kleinere Halbmesser[3].

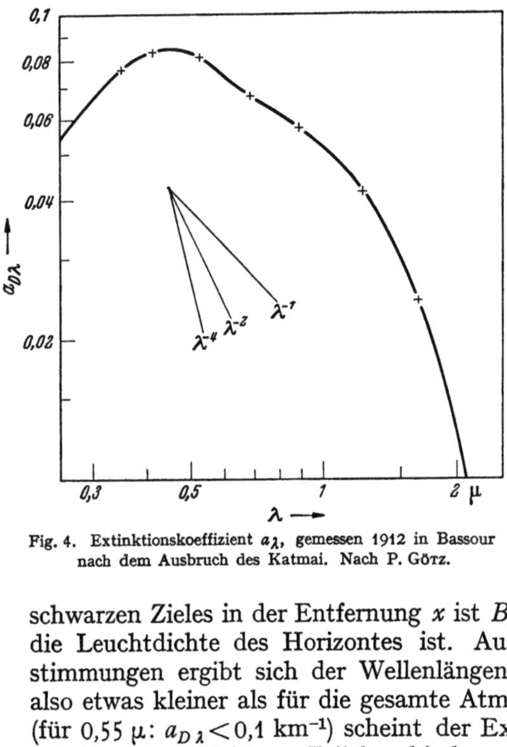

Fig. 4. Extinktionskoeffizient a_λ, gemessen 1912 in Bassour nach dem Ausbruch des Katmai. Nach P. GÖTZ.

Bestimmungen von $k_{D\lambda}$ bei horizontalem Strahlenverlauf sind vor allem durch Auswertungen von *Sichtmessungen* in den verschiedenen Wellenlängen gewonnen worden. Die spektrale Leuchtdichte $B_\lambda(x)$ eines schwarzen Zieles in der Entfernung x ist $B_\lambda(x) = B_\lambda(\infty)\,(1 - e^{-k_\lambda x})$, wenn $B_\lambda(\infty)$ die Leuchtdichte des Horizontes ist. Aus den in Tabelle 4 angegebenen Bestimmungen ergibt sich der Wellenlängenexponent δ bei oder knapp unter 1,0, also etwas kleiner als für die gesamte Atmosphäre. Bei sehr kleiner Extinktion (für $0{,}55\,\mu$: $a_{D\lambda} < 0{,}1$ km^{-1}) scheint der Exponent δ gegen 0,5 abzunehmen, was auf Mangel an kleineren Teilchen hindeuten würde. Nur die Messungen MIDDLETONs zeigen eine systematische Zunahme von δ mit abnehmender Extinktion bzw. zunehmender Sichtweite. Allerdings wird die Bestimmung um so unsicherer, je geringer die Dunstextinktion ist.

F. VOLZ[4] hat in Mainz zahlreiche Messungen durchgeführt, bei denen *gleichzeitig* Sonnenstrahlung und Sichtweite in je zwei Filterbereichen gemessen und $a_{D\lambda}$ durch ein $\lambda^{-\delta}$-Gesetz angenähert wurde. Der Mittelwert für δ_{Sicht} lag mit

[1] A. KEITH PIERCE: Astrophys. J. **119**, 312 (1954).

[2] F. W. P. GÖTZ: Procès Verbaux Assoc. Internat. Météorol. 8. Assemblée Gén. Oslo 1948, II, 89. Uccle 1950.

[3] R. WILSON: Month. Not. Roy. Astr. Soc. **111**, 478 (1951). — P. WELLMANN: Z. Astrophys. **28**, 310 (1951). — S. BAUMBACH u. F. VOLZ: Ber. dtsch. Wetterd. U.S.-Zone **35**, 228 (1952). — R. PENNDORF: Geophys. Res. Papers **1953**, No. 20. — F. W. P. GÖTZ u. F. VOLZ: Procès-Verbaux Assoc. Internat. Météorol., 9. Assemblée Gén. Brüssel 1951, 216. Brüssel 1953.

[4] F. VOLZ: Geofis. pura e appl. **31**, 119 (1955).

1,45 etwas höher als $\delta_{\odot}=1{,}35$. Bei etwas größerer, methodisch bedingter Streuung der Sichtwerte fand er eine Korrelation von $r=+0{,}55$ (132 Messungspaare). Die Kopplung ist erkennbar; der Unterschied zwischen dem Exponenten des bodennahen Aerosols und dem der gesamten Atmosphäre ist gering, so daß man $H_D=a_{D\lambda}:k_{D\lambda}$ nach (5.6) als unabhängig von der Wellenlänge annehmen kann. Damit wird dann H_D auch unabhängig von der Teilchengröße r, und der erwähnte Befund JUNGEs (Ziff. 5, S. 166) über die Unveränderlichkeit des Aerosolspektrums mit der Höhe wird auf optischem Wege bestätigt.

Tabelle 4. *Parameter der Dunstextinktion bei horizontalem Strahl.*

Ort und Autor	Zahl der Beobachtungen	δ	B [km⁻¹]	$k_{D\lambda=0{,}55}$ [km⁻¹]	Bemerkungen
Jena, SCHMOLINSKY[1]	5	0,82	0,062	0,116	Sichtmessungen bei Tage
	2	0,88	0,124	0,219	
	4	0,96	0,406	0,721	
Toronto, MIDDLETON[2]	7	2,1	0,0078	0,030[3]	
	40	1,7	0,0154	0,044[3]	
	57	1,4	0,054	0,129[3]	
	58	1,0	0,160	0,298[3]	
	8	0,8	0,275	0,461[3]	
Berlin, Wangeroog	—	0,7	?	∼1,3	Scheinwerfermessungen, nachts, Nebel nach Abb. 13 ausgewertet
Brocken,	—	0,5	?	∼3,9	
FOITZIK[4]	—	−0,1	?	> 5,0	
Saint Gaudens, DESSENS[5]	6	0,42	0,030	0,038	Sichtmessungen bei Tage nach den Einzelangaben des Verf. zu Gruppen zusammengefaßt
	9	0,33	0,045	0,055	
	10	0,58	0,051	0,068	
	5	0,85	0,070	0,114	
Lindenberg FOITZIK und ZSCHAECK[6]	4	0,66[7]	0,0385[7]	0,058[8]	Scheinwerfermessungen, nachts
	4	1,05[7]	0,063[7]	0,116[8]	
	4	1,05[7]	0,088[7]	0,161[8]	
	6	1,50[7]	0,097[7]	0,231[8]	
	4	1,35[7]	0,153[7]	0,335[8]	
	3	1,30[7]	0,278[7]	0,593[8]	

Durch gleichzeitige Messungen von $a_{D\lambda}$ gegen Sonne und von $k_{D\lambda}$ für die Sicht mit Rotfilter ($\lambda=0{,}63\,\mu$) erhält VOLZ an windschwachen Schönwettertagen folgende Daten für H_D:

Frankfurt Winter 1952/53 Mittagswert von 8 Tagen, $H_D=1{,}15$ km

Mainz April/August 1953 23 Vormittagswerte von 16 Tagen,
mittlere Zeit 9³⁰ Uhr $H_D=1{,}70$ km

Mainz April/August 1953 22 Nachmittagswerte von 16 Tagen,
mittlere Zeit 15⁰⁰ Uhr, $H_D=3{,}60$ km.

[1] F. SCHMOLINSKY: Meteor. Z. **61**, 199 (1944).
[2] W. E. K. MIDDLETON: Gerlands Beitr. Geophys. **44**, 358 (1935).
[3] Für $\lambda=0{,}528\,\mu$.
[4] L. FOITZIK: Wiss. Abh. Reichsamt Wetterd. **4**, Nr. 5 (1938).
[5] H. DESSENS: Ann. Géophys. **2**, 68 (1946).
[6] L. FOITZIK u. H. ZSCHAECK: Z. Met. **7**, 1 (1953).
[7] Neu ausgewertet nach Tabelle 7 der Verfasser.
[8] Für $\lambda=0{,}546\,\mu$.

Nicht unerwartet ist der große Tagesgang (Fig. 5), der sehr schön die Wirkung der nachts bis in Bodennähe abgesunkenen Dunstmassen (Industrie-Aerosol) und nachmittags ihre Verteilung auf große Höhen infolge der einsetzenden Konvektion erkennen läßt. Als Mittelwert von H_D war früher bereits von ÅNGSTRÖM 1,4 km, von P. GÖTZ 1,2 km angegeben worden[1].

7. Anomale Extinktion. P. GÖTZ[2] hat mehrfach darauf hingewiesen, daß er unterhalb einer Wellenlänge von 450 bis 350 mμ keine weitere Zunahme der Dunstextinktion entsprechend λ^{-1} sondern gleichbleibende Werte, unterhalb 300 mμ bis gegen 250 mμ hin sogar eine Abnahme gefunden hat (Fig. 4). Bei diesen kürzesten Wellenlängen können Messungen nur mit irdischen Lichtquellen durchgeführt werden, weil die Ozonabsorption das Sonnenlicht unterhalb 300 mμ abschneidet. Zwar fanden sich gleichbleibende Extinktionen auch schon zwischen Messungen mit UV-Filtern und mit Cadmiumzellen, die erst unterhalb 0,33 μ empfindlich werden, jedoch sind solche Messungen wegen der Wanderung des Schwerpunktes der Strahlung mit der Sonnenhöhe schwer auszuwerten [1b]. Messungen von DUNKELMAN[3] zeigen die anomale Extinktion, wenn man eine Absorption durch Ozon in Rechnung stellt; ebenso konnte MICHALOWSKY[4] mehrfach in dichtem Dunst den gleichen Effekt nachweisen. In Indien fanden RAMANATHAN und KARANDIKAR[5] bei Ozonmessungen häufig positive Werte des Exponenten für $0{,}30 < \lambda < 0{,}44$ μ.

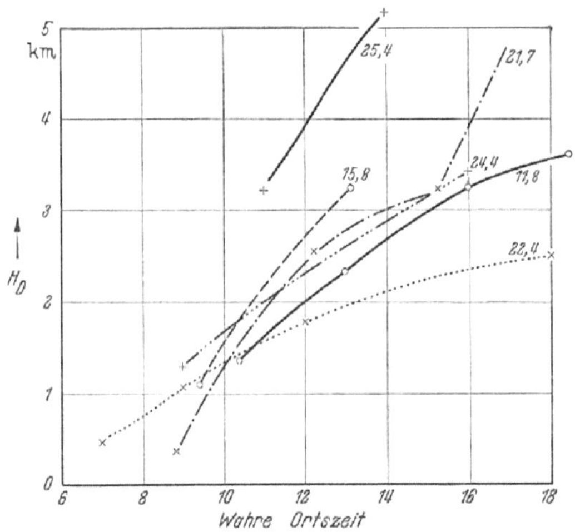

Fig. 5. Tageszeitlicher Gang der Höhe der Dunstschicht $H_D = a_{D\,\lambda}/k_{D\,\lambda}$, für $\lambda = 0{,}63$ μ. Nach F. VOLZ. Die angeschriebenen Zahlen geben die Beobachtungsdaten (25, 4 = 25. April, usw.).

Zur Erklärung dieser mit zunehmender Wellenlänge zunehmenden Extinktion versuchte GÖTZ die alleinige Wirksamkeit einer Partikelgröße von etwa 0,3 μ (Wassertröpfchen) verantwortlich zu machen, denn diese haben für $\alpha > 6$, also $\lambda < 0{,}3$ μ abnehmenden optischen Wirkungsquerschnitt σ (vgl. Fig. 1). Man muß allerdings voraussetzen, daß kleinere Partikel als $r = 0{,}3$ μ nicht vorhanden sind, während nach größeren Partikeln hin ein Verteilungsgesetz (5.2) oder (5.4) anschließen darf. Ein Abbruch sollte nach JUNGE im kontinentalen Aerosol erst unterhalb $r = 0{,}05$ μ erwartet werden, während in Entfernung von mehreren 100 km vom Land solche Verteilungen als Folge der Koagulation beobachtet sind. Die genannten Beobachtungen lassen jedoch eine andere Deutung kaum zu, wenngleich nicht verkannt werden darf, daß ein Abbrechen des kontinuierlichen Größenspektrums sich in ganz charakteristischen Färbungen im Streulicht am Himmel äußern müßte. Diese sind aber recht selten.

8. Die Absorption der atmosphärischen Gase (außer Wasserdampf). Die permanenten atmosphärischen Gase besitzen im Sonnenspektrum zahlreiche

[1] A. ÅNGSTRÖM: Geogr. Ann. **12**, 130 (1930). — F. W. P. GÖTZ u. E. SCHÖNMANN: Helv. phys. Acta **21**, 151 (1948).
[2] F. W. P. GÖTZ: Verh. schweiz. nat. Ges. Luzern **1924**. — Helv. phys. Acta **5**, 336 (1932).
[3] W. A. BAUM u. L. DUNKELMAN: J. Opt. Soc. Amer. **45**, 166 (1955).
[4] J. MICHALOWSKY: Dipl.-Arbeit Mainz 1955 (nicht veröffentlicht).
[5] K. R. RAMANATHAN u. R. V. KARANDIKAR: Quart. J. Roy. Met. Soc. **75**, 275 (1949).

einzelne Absorptionslinien, die in Spektralmessungen hoher Dispersion festgestellt werden können[1] und sich von den solaren FRAUNHOFERschen Linien durch die Zunahme ihrer Intensität mit der durchstrahlten absorbierenden Luftmasse m unterscheiden. Eine ganze Reihe von Spurengasen der Atmosphäre und Isotopen ist durch ihre Absorption im kurz- und langwelligen Infrarot des Sonnenspektrums gefunden oder bestätigt worden, z.B. CH_4, CO, N_2O, NO_2, NO (?), NH_3 (?), SO_2[2]. Für solche Untersuchungen sind Messungen auf hohen Berggipfeln, wie sie z.B. MIGEOTTE[3] auf dem Jungfraujoch durchgeführt hat oder in hochfliegenden Flugzeugen[4] besonders wertvoll. Die durch sie absorbierte Energie ist jedoch zu vernachlässigen gegenüber der Absorption in den großen Banden mehratomiger Gasmoleküle (O_2, H_2O, CO_2, O_3).

Der *Sauerstoff* absorbiert in den Banden A bei 759 mμ und B bei 687 mμ. Die Absorption von A hat nach SCHÜEPP [6] fast unabhängig von der durchstrahlten Luftmasse den Betrag von 8,9% der Strahlung im Bereich 760 bis 800 mμ; B ist vernachlässigbar klein. Die Absorptionsbanden unterhalb 2000 Å, deren Absorption schon in hohen Schichten der Atmosphäre erfolgt, werden im Beitrag PAETZOLD über Ozon behandelt. Durch Deutung der Banden ist auch der Nachweis der Modifikation O_4 und der Isotope $O^{16}O^{17}$ und $O^{16}O^{18}$ gelungen (GOLDBERG[5]).

Die *Kohlensäure* CO_2 ist mit dem Volumenanteil $3 \cdot 10^{-4}$ in der Atmosphäre vorhanden; zeitlich-räumliche Schwankungen treten nur in Bodennähe auf; die Abnahme mit der Höhe ist geringfügig und kann meist vernachlässigt werden. Der Gesamtgehalt beträgt daher 240 cm NTP. — Eine Reihe Absorptionsbanden liegen im nahen Infrarot bei den Wellenlängen 1,46, 1,60, 2,04, 2,75 und 4,27 μ. Weitere schwache Banden konnten im Sonnenspektrum nachgewiesen werden, ebenso fünf Banden der Isotope $C^{13}O_2^{16}$ und $C^{12}O^{16}O^{18}$[5]. Die Absorption fast aller dieser Banden ist gering; die Bande bei 2,75 μ ist intensiv, fällt aber mit der X-Bande des Wasserdampfes zusammen. Die letzte, sehr intensive Bande absorbiert im Bereich 4,255 bis 4,280 μ bei 0,2 cm CO_2 bereits 80%, jedoch enthält dieser Bereich nur 10^{-4} der Energie der extraterrestrischen Sonnenstrahlung[6].

Die Luftdruckabhängigkeit der CO_2-Absorption ist als Folge des linearen Aufbaues des Moleküls vom Partial-(Eigen-)Druck in der gleichen Weise abhängig wie vom Gesamtluftdruck. Der Effekt ist bei der 4,27 μ-Bande von WIMMER[7] sorgfältig untersucht worden; aus seinen Messungen konnte MÖLLER[8] ableiten, daß gleiche Absorption durch effektive Kohlensäuremengen $c_{\text{eff}} = c(p/p_0)^{\frac{3}{4}}$ bewirkt werden. Obwohl ELSASSER und KING[9] für die 15 μ-Bande Proportionalität mit p/p_0 gefordert haben, scheint das $p^{\frac{3}{4}}$-Gesetz in allen Wellenlängenbereichen den Beobachtungen besser gerecht zu werden.

---

[1] M. MINNAERT, G. F. W. MULDERS u. J. HOUTGAST: Photometric Atlas of the Solar Spectrum. Amsterdam 1940. — O. C. MOHLER, A. K. PIERCE, R. R. MCMATH u. L. GOLDBERG: Photometric Atlas of the near infrared solar spectrum, 8465 to 25242 A. Ann. Arbor Univ. of Michigan Press 1952. — J. H. SHAW, R. M. CHAPMAN, J. N. HOWARD u. M. L. OXHOLM: Astrophys. J. **113**, 268 (1951). — J. H. SHAW, M. L. OXHOLM, H. H. CLAASEN: Astrophys. J. **116**, 554 (1952). — M. V. MIGEOTTE, L. NEVEN u. J. SWENSSON: Mém. Soc. Roy. Sci. Liège, Coll. in 4°. — Eine Zusammenstellung der Ergebnisse gibt L. GOLDBERG in G. KUIPER: The Solar System, Vol. 2, S. 434. Chicago 1954.

[2] E. GLÜCKAUF: In Compendium of Meteor. Boston 1951.

[3] M. V. MIGEOTTE, L. NEVEN u. a.: Astrophys. J. **115**, 326 (1952). — Physica, Haag **18**, 982 (1952). — J. Opt. Soc. Amer. **43**, 1119 (1953). — Nuovo Cim. (9) **8**, Suppl. (1954).

[4] J. YARNELL u. R. M. GOODY: J. Sci. Instrum. **29**, 352 (1952).

[5] L. GOLDBERG, O. C. MOHLER u. R. R. MCMATH: Phys. Rev. **74**, 1881 (1948).

[6] CL. SCHAEFER u. PHILIPPS: Z. Physik **36**, 641 (1926). — F. SCHNAIDT: Gerlands Beitr. Geophys. **54**, 203 (1939). — G. S. CALLENDAR: Quart. J. Roy. Met. Soc. **67**, 263 (1941).

[7] M. WIMMER: Ann. Phys. **81**, 1091 (1926).

[8] F. MÖLLER: Meteor. Z. **61**, 37 (1944).

[9] W. M. ELSASSER u. J. I. KING: Contr. AF 19 (122)—392, Techn. Rep. No. 9, Univ. of Utah, 1953.

Das *Ozon*[1] ist mit der HARTLEY- und den vorgelagerten HUGGINS-Banden ($0{,}20 < \lambda < 0{,}35\,\mu$) für das Abbrechen des am Boden beobachteten Sonnenspektrums bei $0{,}3\,\mu$ verantwortlich; die CHAPPUIS-Bande ($0{,}44 < \lambda < 0{,}76\,\mu$) hat im Sichtbaren eine wesentlich schwächere aber nicht vernachlässigbare Absorption (Fig. 6). Beide Banden zeigen Temperaturabhängigkeit, jedoch keine Abweichung vom BEER-BOUGUERschen Gesetz, also keine Abhängigkeit vom Druck. Im Gegensatz dazu besitzt die $9{,}6\,\mu$-Bande eine Druckabhängigkeit, so daß J. STRONG[2] durch Vergleich beider Absorptionen eine interessante Methode der Höhenbestimmung des Ozons erhielt. Die Absorption im UV bewirkt Dissoziation des O_3 in der Schicht von 20 bis 60 kmHöhe. In der Troposphäre beträgt die Konzentration nur 1 bis $2 \cdot 10^{-3}$ cm NTP/km. Das besagt, daß bei 1 km Entfernung Licht von $\lambda = 2500$ Å bereits um 30 bis 50% geschwächt wird! Die Sonnenstrahlung verliert bei vertikaler Einstrahlung im UV bei einer Gesamtmenge von 0,3 cm O_3 2% der Energie der Solarkonstante, während unter den gleichen Bedingungen in der CHAPPUIS-Bande, die nahezu im maximalen Energiebereich der Sonnenstrahlung liegt, weniger als 1% absorbiert wird[*1b*]. Nur bei sehr tief stehender Sonne, wenn die durchstrahlte O_3-Schicht auf den zehnfachen Betrag steigt, erreicht die Einbuße im gelben Spektralbereich 40%; dies wird dann für die Farbe des vom

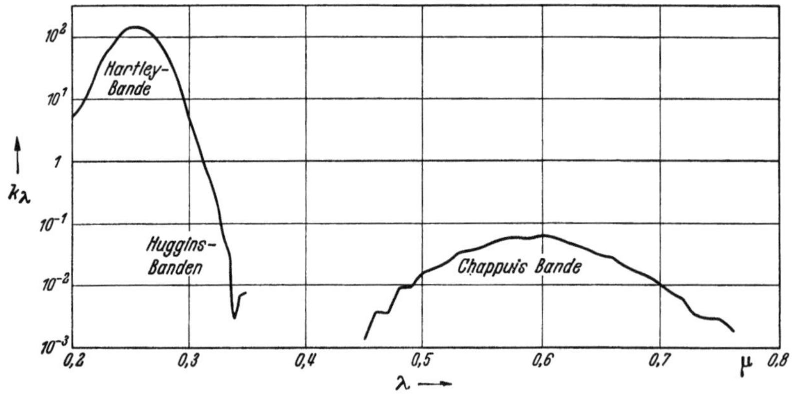

Fig. 6. Dekadische Absorptionskoeffizienten des Ozon, bezogen auf 1 cm Schichtdicke bei 1 Atm und 0° C.

Himmel kommenden Streulichtes entscheidend, aber auch für die Farbe des die Atmosphäre durchsetzenden Sonnenlichtes, was man speziell am Rande des Erdschattens bei Mondfinsternissen erkennen kann.

Näheres über Ozon im Beitrag PAETZOLD in diesem Band.

9. Die Absorption des Wasserdampfes. Der Wasserdampf ist weitaus der wichtigste Absorber für die Sonnenstrahlung. Man kennt insgesamt 19 Rotationsschwingungsbanden, deren Breiten und Intensitäten vom Sichtbaren bis zu der $6{,}3\,\mu$-Bande (Grundschwingung ν_2) zunehmen[3]. Die Absorptionen am roten Ende des sichtbaren Spektrums sind seit langem unter dem anschaulichen, aber etwas irreführenden Namen „*Regenbanden*" bekannt[4]. Zu ihnen gehören die Bereiche um 572 und 632 mμ, sowie von 588 bis 594, 649 bis 655 und 693 bis 703 mμ, in denen allein über 400 Absorptionslinien nachgewiesen sind[5].

Wirklich große Energiebeträge werden durch die breiten und tiefen Banden im Infrarot absorbiert; zwischen ihnen bleiben eine Reihe von „Fenstern" mit guter Transmission offen [*3*]. Von zahlreichen Autoren sind die absorbierten Energiebeträge zusammengestellt worden. Als am umfangreichsten und in sich am besten zusammenhängend müssen auch heute noch die Messungen von FOWLE[6]

1 R. CRAIG: Met. Monogr. Boston **1**, No. 2 (1950).
2 J. STRONG: J. Franklin Inst. **231**, 121 (1941).
3 T. G. COWLING: Rep. Progr. Phys. **9**, 29 (1942/43).
4 P. SMITH: Edinburgh Astr. Obs. **14**; Ref. in Öst. Z. Met. **16**, 22 (1881).
5 Smithson. Met. Tables, 4. ed., Washington 1918, Tab. 101, S. 229.
6 F. E. FOWLE: Astrophys. J. **35**, 149 (1912); **42**, 394 (1915). — Smithson. Miscell. Coll. **68**, No. 8 (1917). — Meteor. Z. **31**, 270 (1914).

zwischen 0,7 und 10 μ gelten, die bei geringen absorbierenden Dampfmengen (ausgedrückt in cm Niederschlagswasser, precipitable water oder g $H_2O \cdot cm^{-2}$) im Laboratorium durchgeführt und systematisch vor Sonne für zunehmende durchstrahlte Dampfmassen erweitert wurden. Die großen Banden im kurzwelligen Infrarot sind:

a	0,70 —0,74 μ,	Minimum 0,718 μ
0,8 μ	0,79 —0,84 μ,	Minimum 0,810 μ
$\varrho\sigma\tau$	0,926—0,978 μ,	Minimum 0,935 μ
Φ	1,095—1,165 μ,	Minimum 1,130 μ
Ψ	1,319—1,498 μ,	Minimum 1,395 μ
Ω	1,762—1,977 μ,	Minimum 1,870 μ
X	2,520—2,845 μ,	Minimum 2,68 μ.

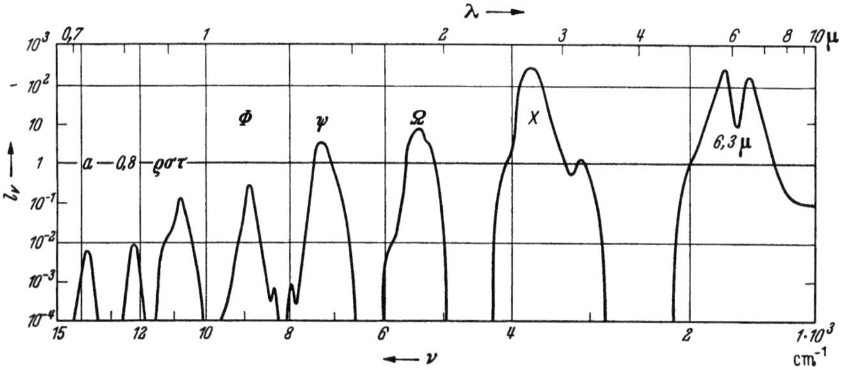

Fig. 7. Generalisierte Absorptionskoeffizienten nach ELSASSER, vgl. (26.7) und (26.8), für Wasserdampf. Nach YAMAMOTO und ONISHI.

Auch in schmalen Spektralbereichen folgt die Absorption wegen der *Linienstruktur* der Banden nicht dem BOUGUERschen Gesetz, sondern zeigt eine scheinbare Abnahme des nach (2.1) berechneten Absorptionskoeffizienten mit der Weglänge. YAMAMOTO und ONISHI[1] haben deshalb das Modell einer regelmäßigen Bande nach ELSASSER (26.5) auf FOWLEs Messungen angepaßt und die verallgemeinerten Absorptionskoeffizienten l_ν nach (26.8) berechnet (Fig. 7)[2].

Tabelle 5, Zeile 1, gibt die absorbierten Energiebeträge in Abhängigkeit von $m_r w$. Diese Zahlen dürfen besonders bei kleinen Dampfmengen ($m_r w < 10^{-1}$ cm) als einigermaßen zuverlässig angesehen werden. Auf den gleichen Messungen FOWLEs basierend haben MÖLLER und MÜGGE[3] eine einfache Interpolationsformel angegeben

$$\Delta J_{\mathrm{MM}} = 0{,}172\,(m_r\,w)^{0,3}\ \mathrm{cal\,cm^{-2}\,min^{-1}} \tag{9,1}$$

vgl. Tabelle 5, Zeile 2. Diese nimmt, ebenso wie YAMAMOTOs Auswertung, keine Rücksicht auf die Schwächung der Sonnenstrahlung infolge der Streuung auf dem Wege durch die Atmosphäre, gilt also gewissermaßen für eine reine Wasserdampf-Atmosphäre; allerdings ist die Streuung im infraroten Bandenbereich auch nur gering.

[1] G. YAMAMOTO u. G. ONISHI: J. Meteorology 9, 415 (1952).

[2] Hierbei wurde der Linienabstand $d = 2{,}3\ cm^{-1}$, die Halbwertsbreite $\delta = 0{,}1\ cm^{-1}$ angesetzt, die extraterrestrische Strahlung entsprechend derjenigen eines schwarzen Körpers von 6000° K bei einer Solarkonstante 1,94 cal cm^{-2} min^{-1} angenommen.

[3] R. MÜGGE u. F. MÖLLER: Z. Geophys. 8, 53 (1932).

Tabelle 5. *Absorption ΔJ (cal cm^{-2} min^{-1}) der Sonnenstrahlung durch den Wasserdampf w (ausgedrückt als „precipitable water" in cm).*

m_r = relative Luftmasse. YO nach YAMAMOTO-ONISHI[1], MM nach MÜGGE-MÖLLER[2] H nach HOELPER[3], ES nach ELDER-STRONG[4], MH nach MÖLLER-HOWARD.

Zeile	$m_r w =$	10^{-5}	10^{-4}	10^{-3}	$5 \cdot 10^{-3}$	10^{-2}	$3 \cdot 10^{-2}$	$6 \cdot 10^{-2}$	10^{-1}
1	ΔJ_{YO}	0,002	0,016	0,024	0,041	0,051	0,072	0,088	0,101
2	ΔJ_{MM}	0,005	0,011	0,021	0,035	0,043	0,060	0,073	0,086
3	ΔJ_{H}	—	—	—	—	—	—	—	0,057
4	ΔJ_{ES}	—	—	0,004	0,022	0,032	0,050	0,065	0,078
5		—	—	—	—	—	—	—	—
6	ΔJ_{MH}	—	—	—	—	—	—	—	0,090
7		—	0,00235	0,0074	0,017	0,025	0,044	0,060	0,075

Zeile	$m_r w =$	0,2	0,4	0,7	1,0	2	3	5	7	10
1	ΔJ_{YO}	0,122	0,145	0,167	0,183	0,217	0,240	0,268	0,288	0,312
2	ΔJ_{MM}	0,106	0,130	0,154	0,172	0,212	0,240	0,280	0,310	0,346
3	ΔJ_{H}	0,073	0,094	0,115	0,132	0,168	0,190	0,215	0,230	0,242
4	ΔJ_{ES}	0,101	0,138	0,176	0,200	0,247	0,276	0,309	0,332	0,356
5		—	—	***0,178***	***0,204***	***0,254***	***0,284***	***0,324***	***0,353***	***0,386***
6	ΔJ_{MH}	0,111	0,139	0,167	0,187	0,227	0,254	0,285	0,310	0,335
7		0,101	0,131	0,160	0,180	0,223	0,254	0,285	0,310	0,335

HOELPER[3] hat Messungen am Boden verwendet, wodurch die kontinuierliche Schwächung der Strahlung durch Molekül- und Dunststreuung auch im Bereich der Banden vorweg berücksichtigt ist und deshalb die verbleibende Bandenabsorption geringer ausfällt. Zahlenmäßig nahezu die gleichen Ergebnisse wie HOELPER hat KIMBALL[5] erhalten, jedoch hat er sich die — heute nicht mehr haltbare — Annahme FOWLEs [*3*] zu eigen gemacht, daß der Wasserdampf eine eigene, verhältnismäßig große Streuung bewirkt.

SCHÜEPP [*6*] hat aus sehr sorgfältigen Aktinometermessungen der gesamten Sonnenstrahlung nachzuweisen versucht, daß die Absorptionswerte von FOWLE unterhalb $w = 6$ cm zu groß, darüber aber zu klein sind. Schließlich haben ELDER und STRONG[4] neben FOWLEs auch neue, zum Teil noch unveröffentlichte Messungen der Durchlässigkeit in den verschiedenen Fenstern zwischen den Banden zusammengetragen und zunächst nach einer Faustformel interpoliert. Unter Verwendung ihrer Daten erhält man Absorptionszahlen, die zum Teil den Forderungen SCHÜEPPS entsprechen und in Tabelle 5, Zeile 4 wiedergegeben sind.

FOWLE [*3*] hat — wie erwähnt — bei der Reduktion seiner Extinktionsmessungen auf eine staub- und dunstfreie Atmosphäre eine kontinuierliche, vom Wasserdampfgehalt abhängige Extinktion $(e^{-a_{w\lambda} w})^{m_r}$ erhalten, die er einer eigenen Streuung der H_2O-Moleküle zuschreibt ([2a], Tabellen 145, 147). Die Größe dieser Koeffizienten übersteigt die der zu erwartenden molekularen Streuung des Wasserdampfes um das 50fache und würde daher auf größere Molekülaggregate $(H_2O)_n$ hindeuten, die im gasförmigen Wasserdampf mit Sicherheit nicht vorhanden sind[6]. FOWLEs Werte $a_{w\lambda}$ folgen im Kurzwelligen einem Gesetz $\lambda^{-\delta}$, wobei δ von 3,5 im UV auf 1,8 im Sichtbaren abnimmt; von $\lambda = 0{,}9\ \mu$ ab bleibt $a_{w\lambda}$ konstant. Man muß daher den Effekt im Sichtbaren einer nicht als solche erkannten, aber mit dem Wasserdampfgehalt gekoppelten Dunststreuung zuschreiben, während im Infrarot wahrscheinlich eine

[1] G. YAMAMOTO u. G. ONISHI: J. Meteorology **9**, 415 (1952).

[2] R. MÜGGE u. F. MÖLLER: Z. Geophys. **8**, 53 (1932).

[3] O. HOELPER: Dtsch. met. Jb. Aachen **1933**. — Wiss. Abh. Reichsamt Wetterd. **5**, No. 10 (1939). — Meteor. Z. **60**, 37 (1943).

[4] T. ELDER u. J. STRONG: J. Franklin Inst. **255**, 189 (1953). — J. N. HOWARD u. R. M. CHAPMAN: J. Opt. Soc. Amer. **42**, 423 (1952). — H. A. GEBBIE, W. R. HARDING, C. HILSUM, A. W. PRYCE u. V. ROBERTS: Proc. Roy. Soc. Lond., Ser. A **206**, 87 (1951).

[5] H. H. KIMBALL: Month. Weath. Rev. **55**, 155 (1927).

[6] T. KIU: J. de Phys. (7) **9**, 297 (1938).

schwache Absorption zwischen den Banden vorliegt, die entweder von vereinzelten Absorptionslinien oder von der kontinuierlichen Absorption in den Flügeln der stärksten Linien der Banden herrührt. Berücksichtigt man diese quasi-kontinuierliche Absorption im Infrarot, so erhöhen sich bei großen Wasserdampfbeträgen die nach STRONG berechneten Zahlen auf die in Tabelle 5, Zeile 5 kursiv gesetzten Werte.

Alle diese Absorptionen sind wegen der Druckverbreiterung der Linien vom Atmosphärendruck abhängig. YAMAMOTO[1] hat als erster diesen Effekt für die H_2O-Absorption im nahen Infrarot untersucht und seine Auswirkung bei der mittleren Verteilung des Wasserdampfes über den Luftdruck in der Atmosphäre berücksichtigt. ELDER und STRONG[2] geben für die nichtkontinuierliche Banden-Absorption eine Näherungsformel $T = T_0 (p_0/p)^{\frac{1}{2}}$, wobei T, T_0 die Transmissionen einer ganzen Bande bei dem auf dem Lichtwege herrschenden Luftdruck p, p_0 sind. In der Atmosphäre ist p auf dem Lichtwege variabel, was die Anwendung dieser Formel erschwert. Auch muß die nichtselektive Absorption zwischen den Banden einem anderen Gesetz gehorchen (vgl. Ziff. 28).

Wohl den wichtigsten Beitrag zu dieser Frage haben HOWARD u. a[3]. geleistet, die die Absorption des Wasserdampfes in den Infrarotbanden mit großen Absorptionsrohren spektral untersucht haben. Die Messungen wurden ausgeführt an Wasserdampfmengen von 0,02 bis 1,0 cm precipitable water unter verschiedenen Partialdrucken und unter Beifügung von reinem Stickstoff (Partialdrucke von 5 bis 750 mm) zur Linienverbreiterung. Die Absorption in diesen „synthetischen" Atmosphären wurde zunächst durch empirische Formeln wiedergegeben, und zwar bei geringer Absorption $A < A_c$ durch eine andere Formel als bei großer Absorption $A > A_c$. Ihre Ergebnisse und Formeln sind in Tabelle 6 wiedergegeben. (Entsprechende Zahlen sind für CO_2-Absorption mitgeteilt.) Der Luftdruck tritt hier in anderer Form als bei ELDER und STRONG auf. Aus der für langwellige Infrarotstrahlung entwickelten Theorie der Linienverbreiterung (Ziff. 28) würde ein Korrektionsfaktor bei w in der Größe p/p_n, oder bei starker Linienüberlappung $(p/p_n)^{0,5}$ zu fordern sein, während die Messungen meist die Potenz 0,6 ergeben; die doppelte Wirksamkeit des Wasserdampfes selbst durch $(p+\varepsilon)^n = (p_L + 2\varepsilon)^n$ konnte nach den Messungen bei den niedrigen atmosphärischen Dampfdrucken nur geschätzt werden. Für die Absorption in den Kohlensäurebanden finden die Verfasser analoge Exponenten n bzw. n' von durchschnittlich 0,83.

Tabelle 6. *Absorption A der Wasserdampfbanden nach* HOWARD *u. a.*[3]

$$\text{Für} \quad A < A_c \quad \text{ist} \quad A = c\sqrt{w\left(\frac{p+\varepsilon}{p_n}\right)^n}$$

$$\text{für} \quad A > A_c \quad \text{ist} \quad A = C + D\log\left[w\left(\frac{p+\varepsilon}{p_n}\right)^{n'}\right].$$

$p_n = 1000$ mb, Gesamtdruck $p = p_L + \varepsilon$, Partialdruck der trockenen Luft p_L und des Wasserdampfes ε; w in cm Niederschlagswasser (precipitable water).

Bande	Bereich $\Delta\nu\,[10^3\,\text{cm}^{-1}]$	c	n	A_c	C	D	n'
$\varrho\sigma\tau$	10,23—10,8	0,41	0,54	0,35	—	—	—
Φ	8,3 — 9,3	0,18	0,52	0,20	—	—	—
Ψ	6,5 — 8,0	0,82	0,60	0,23	0,515	0,307	0,43
Ω	4,8 — 5,9	1,14	0,60	0,25	0,492	0,211	0,62
X	3,34— 4,0	2,56	0,64	0,19	0,725	0,232	0,61
3,2 μ	2,8 — 3,34	0,56	0,60	0,93	—	—	—
6,3 μ	1,15— 2,05	2,97	0,60	0,18	0,837	0,242	0,72

[1] G. YAMAMOTO u. G. ONISHI: J. Meteorology **9**, 415 (1952).
[2] Siehe Fußnote 4, S. 174.
[3] J. N. HOWARD, D. E. BURCH u. D. WILLIAMS: Geophys. Res. Pap. Cambridge **1955**, No. 40

Die Güte der Messungen und damit die allgemeine Anwendbarkeit der empirischen Näherungsformeln wurde durch Vergleich der Messungen mit einer von Goody theoretisch abgeleiteten Transmissionsfunktion (26.12) erwiesen.

Eine Auswertung für die Atmosphäre ist in zweifacher Weise durchgeführt worden[1]. Nach Curtis und Godson[2] kann der Einfluß der Druckverbreiterung der Linien in der geschichteten Atmosphäre durch Verwendung einer mittleren Linienbreite $\delta_m = \int\limits_0^\infty \delta(p)\, dw : \int\limits_0^\infty dw$ berücksichtigt werden. Zeile 6 in Tabelle 5 gibt die damit bestimmte Absorption ΔJ für Messungen am Boden wieder, Zeile 7 für Messungen, die bei kleinen Werten w denjenigen (niedrigeren) Werten p des Luftdruckes entsprechen, unter denen solche Werte w vorkommen (z.B. kommt $w = 10^{-2}$ cm etwa in einem Niveau $p = 300$ mb vor). Man darf diese beiden letzten Zeilen der Tabelle wohl als die zur Zeit zuverlässigste Bestimmung der Gesamtabsorption ansehen.

Die Absorption in der Mitte der $\varrho\sigma\tau$-Bande, in der Totalabsorption nicht erreicht wird, ist erstmalig von Fowle [3], später von Herzing und anderen Autoren[3] als bestgeeignetes Maß für die gesamte Menge an Niederschlagswasser in der Atmosphäre benutzt worden. Bei der „Eichung" der Absorption gegen den auf andere Weise gemessenen Wasserdampfgehalt mußte man früher den Umweg über den Dampfdruck am Boden und eine mittlere angenommene Höhenverteilung der Dampfdichte $\varrho_w = \varrho_{w0}\, e^{-z/d}$ gehen, es war daher nur ein statistischer Vergleich möglich. Die regelmäßigen Radiosonde-Messungen gestatten heute den Vergleich von Einzelmessungen[4]. An Stelle der spektroskopischen Messungen sind auch solche mit Interferenzfiltern benutzt worden[5]. Das Verfahren ist jedoch auch zur Messung des lokalen Wasserdampfgehaltes, also zur Hygrometrie, ausgebildet worden[6].

10. Der Trübungsfaktor nach F. Linke. Für die beobachtende und vergleichende Meteorologie ist ganz allgemein die Gewinnung charakteristischer Maßzahlen mittels einfacher Meßverfahren ein wesentliches Anliegen. Zu einer solchen numerischen Beschreibung der Sonnenstrahlung einerseits, der gesamten Trübung der Atmosphäre andererseits sind spektrale Extinktionskoeffizienten ungeeignet, denn sie sind weder mit einfachen Feldmethoden zu messen, noch sind sie geeignet für eine klare und anschauliche Beschreibung der Strahlungsverhältnisse eines Ortes oder Klimas. Mittlere Extinktionskoeffizienten für die totale Strahlung aller Wellenlängen sind wegen der Verschiebung des optischen Schwerpunktes mit Veränderung der durchstrahlten Luftmasse stark von dieser, d.h. von der Sonnenhöhe und damit von der Jahreszeit und von der geographischen Breite abhängig.

Deshalb vergleicht F. Linke [*1b*] die Gesamtextinktion mit der Rayleigh-Extinktion und nennt den Quotienten *Trübungsfaktor* T. Er setzt also

$$J(m) = \int\limits_0^\infty J_{0\lambda}\, e^{-a_\lambda m}\, d\lambda = \int\limits_0^\infty J_{0\lambda}\, e^{-T a_{R\lambda} m}\, d\lambda. \tag{10.1}$$

m ist die absolute Luftmasse. Diese Definitionsgleichung könnte den gleichen Wert T für verschiedene Luftmassen m nur dann ergeben, wenn exakt $a_\lambda = T a_{R\lambda}$, d.h. wenn die gesamte Extinktion

$$a_\lambda = a_{R\lambda} + a_{D\lambda} + a_{A\lambda}, \tag{10.2}$$

[1] F. Möller: Noch unveröffentlicht.

[2] A. R. Curtis: Quart J. Roy. Met. Soc. **78**, 638 (1952). — W. L. Godson: J. Meteorology **12**, 533 (1955).

[3] F. Herzing: Gerlands Beitr. Geophys. **49**, 71 (1937). — E. Stenz: Meteor. Z. **50**, 219 (1933). — I. F. Hand: Month. Weath. Rev. **68**, 95 (1940). — A. u. E. Vassy: C. R. Acad. Sci. Paris **211**, 158 (1940); **219**, 327 (1944), — N. B. Foster u. L. W. Foskett: J. Opt. Soc. Amer. **35**, 601(1945).

[4] J. Goldschmidt: Z. Met. **5**, 152 (1951).

[5] J. Werner: Ber. dtsch. Wetterd. US-Zone **35**, 302 (1952).

[6] N. B. Foster u. L. W. Foskett: Bull. Amer. Met. Soc. **24**, 146 (1943).

die sich aus Molekül- und Dunststreuung und selektiver Absorption $a_{A\lambda}$ zusammensetzt, der gleichen Wellenlängenabhängigkeit wie $a_{R\lambda}$ gehorchen würde. Das ist nicht der Fall. Die Erfahrung hat aber gezeigt, daß zumindest bei nicht zu großen Luftmassen m der Trübungsfaktor in erster Näherung unabhängig von m ist, wenn die Extinktionsverhältnisse, also vor allem Dunst- und Wasserdampfgehalt, sich zeitlich nicht ändern (vgl. aber S. 179). Gerade dies ist die Forderung, die man an eine Maßzahl der genannten Art stellen muß.

Wir definieren zunächst den mittleren Transmissionsfaktor für RAYLEIGH-Streuung als

$$q_{R_m} = e^{-a_{Rm}} = [J_R(m) : J(0)]^{1/m} = [\int_0^\infty J_{0\lambda} e^{-a_{R\lambda} m} d\lambda : J(0)]^{1/m}, \tag{10.3}$$

wobei

$$J(0) = \int_0^\infty J_{0\lambda} d\lambda \tag{10.4}$$

die extraterrestrische Sonnenstrahlung ist. q_{R_m} ist nicht mehr unabhängig von der Luftmasse, denn nach Auslöschung der stärkst extingierten Wellenlängen verbleiben für weitere Schwächung nur noch die Spektralbereiche mit schwächeren Extinktionen verfügbar. q_{R_m} nimmt also mit m zu, a_{R_m} nimmt ab. Aus beobachteten Werten $J(m)$ wird dann T gewonnen nach

$$T = (m \log q_{R_m})^{-1} [\log J(m) - \log J(0)] = P_m [\log J(0) - \log J(m)], \tag{10.5}$$

wo

$$P_m = -(m \log q_{R_m})^{-1} = (\log_{10} e \cdot m \cdot a_{R_m})^{-1}. \tag{10.6}$$

In Tabelle 7 sind die Zahlenwerte für a_{R_m} und P_m nach den Berechnungen von FEUSSNER und DUBOIS[1] angegeben. Graphische Methoden zur Bestimmung von Trübungsfaktoren haben KAEMPFERT und LANGE[2] angegeben. Zugleich ist in Zeile 3 der Verlauf der Strahlung mit der durchstrahlten Luftmasse bei reiner, trockener Atmosphäre $J_R(m) = J(0) \exp(-a_{R_m} \cdot m)$ beigefügt. Der Trübungsfaktor kann nicht kleiner als 1 werden, denn $T=1$ bedeutet rein molekulare Streuung. Der große anschauliche Wert des Trübungsfaktors liegt dann darin, daß man ihn deuten kann als die Anzahl der Atmosphären reiner, trockener Luft, die die gleiche Gesamtextinktion wie die Beobachtungen ergeben würden.

Diese Anschaulichkeit hat dem Trübungsfaktor größte Beliebtheit verschafft, er ist für sehr viele Strahlungsmessungen in aller Welt berechnet worden [*1b*]. Auch ist es leicht, bei Kenntnis des Trübungsfaktors die Strahlungswerte oder Strahlungssummen längerer Zeitperioden, die von der Sonne auf die Erdoberfläche eingestrahlt werden, zu berechnen[3]. Eine umfangreiche Zusammenstellung von klimatischen Daten hat STEINHAUSER[4] gegeben. Hier sei nur eine Zusammenfassung nach der geographischen Breite und Jahreszeit wiedergegeben, die von BAUR und PHILIPPS[5] verwendet worden ist (Tabelle 8).

Tabelle 7. *Konstanten für die* RAYLEIGH-*Extinktion der Sonnenstrahlung.*

m	0	0,5	1	2	3	4	6	8	10
a_{R_m}	—	0.1016	0.0971	0.0890	0.0822	0.0766	0.0673	0.0598	0.0545
P_m	—	46.00	23.65	12.86	9.36	7.52	5.72	4.80	4.19
$J_R(m)$	1.940	1.841	1.757	1.623	1.517	1.428	1.297	1.200	1.120

[1] K. FEUSSNER u. P. DUBOIS: Gerlands Beitr. Geophys. **27**, 132 (1930).
[2] W. KAEMPFERT: Met. Rdsch. **1**, 215 (1947). — O. LANGE: Z. Met. **9**, 234 (1955).
[3] F. LINKE, in B. GUTENBERG: Lehrbuch der Geophysik. Berlin 1929.
[4] F. STEINHAUSER: Gerlands Beitr. Geophys. **42**, 110 (1934).
[5] F. BAUR u. H. PHILIPPS: Gerlands Beitr. Geophys. **42**, 160 (1943).

Tabelle 8. *Mittelwerte des Trübungsfaktors T nach* Baur *und* Philipps[1].

	Dezember und Januar	März	Juni und Juli	September
90–60° N	1,6	1,9	2,3	2,2
60–50	2,5	2,6	3,2	2,9
50–40	3,0	3,5	4,1	3,6
40–20	3,8	4,0	4,4	4,2
20–0	4,6	4,6	4,6	4,6

Die Trübungsfaktoren einzelner Orte unterliegen größeren Schwankungen als es in Mittelwerten zum Ausdruck kommt, z.B. zeigen Großstädte mit ihrer staub- und dunsthaltigen Atmosphäre deutlich größere Werte als Orte auf dem Lande. Jedoch scheinen die Mittelwerte häufig auch von der Auswahl, die der Beobachter trifft, bedingt; manche Beobachter lehnen es ab, bei dunstigem Himmel oder in Bewölkungslücken Messungen durchzuführen und erhalten dadurch eine schärfere Auswahl von Tagen geringer Bewölkung, was oft mit geringer Trübung gleichbedeutend ist. Eine bessere Kennzeichnung ergeben daher Trübungsfaktoren für einzelne Wetterlagen oder Luftmassen, die am gleichen Ort gewonnen sind. So wurden in Potsdam und am Taunus-Observatorium folgende Werte ermittelt:

Tabelle 9. *Abhängigkeit des Trübungsfaktors T von der synoptischen Luftmasse (Luftkörper).*

In Potsdam[2]		Am Taunus Observatorium 800 m NN [*1 b*].	
kontinentale Arktikluft	2,20	polare Kontinentalluft	1,85
maritime Arktikluft	2,10	Polarluft	2,17
kontinentale gemäßigte Luft	2,89	polar-maritime Luft	2,26
maritime gemäßigte Luft	2,85	maritime Luft	2,63
kontinentale Tropikluft	3,76	kontinentale und tropisch-kontinentale Luft	2,77
maritime Tropikluft	3,66	tropisch-maritime und Tropikluft	3,62

Die Unterschiede werden deutlich bestimmt durch den geringen Dunst- und Wasserdampfgehalt polarer Luftmassen und den hohen Gehalt in tropischen. Ähnliche Unterschiede sind auch für Moskau, Japan, USA und viele andere Orte nachgewiesen worden [*1 b*].

Die *Veränderung der Trübung mit der Höhe* kann durch die Änderung des Trübungsfaktors, der für die gesamte oberhalb liegende Atmosphäre gilt, beschrieben werden, wobei durch Verwendung des von Feussner und Dubois[3] angegebenen „projizierten" Trübungsfaktors die gesetzmäßige Veränderung von T mit der Höhe zu berücksichtigen ist. Besser bestimmt man die Trübung der einzelnen Schichten durch den „speziellen" Trübungsfaktor T^* [*1 b*], der angibt, um wieviel der Extinktionskoeffizient in einer Atmosphären*schicht* größer ist als der molekulare. Steinhauser[4] hat aus Messungen an Stationen in verschiedener Höhe der Ostalpen Mittelwerte bestimmt (Tabelle 10), die sehr deutlich die starke Abnahme mit der Höhe erkennen lassen.

Tabelle 10. *Spezielle Trübungsfaktoren verschiedener Höhenschichten im Alpengebiet, nach* Steinhauser[3].

Höhenschicht	Winter	Frühling	Sommer	Herbst
0,2–1,0	11,0	10,7	11,1	9,6
1,0–1,5	6,1	10,3	9,5	8,2
1,5–2,0	4,7	6,8	7,6	5,4
2,0–3,0	2,2	4,0	5,7	3,0

[1] Siehe Fußnote 5, S. 177.
[2] H. Hinzpeter: Z. Met. **4**, 1 (1950).
[3] Siehe Fußnote 1, S. 177.
[4] F. Steinhauser: Meteor. Z. **56**, 172 (1939).

Bis zu der Schicht 2 bis 3 km geht die Trübung im Winter auf $\frac{1}{5}$ des Wertes in Bodennähe, im Sommer dagegen nur auf die Hälfte zurück. Diese für Gesamtstrahlung gültigen Zahlen enthalten die Wirkung von Dunst- und Wasserdampfabsorption. Erstere allein wird mit dem Kurzstrahlungs-Trübungsfaktor (vgl. den vorletzten Absatz dieser Ziffer) erfaßt. Für diesen gibt es eine Auswertung LINKES[1] von Messungen auf einer Freiballonfahrt von Frankfurt am Main aus am 4. 3. 28 (Hochdrucklage), die die enorme Steigerung im bodennahen Aerosol erkennen läßt.

h	0,09 bis 0,55	0,55 bis 1,27	1,27 bis 1,66	1,66 bis 2,30 km
T_k^*	29	4,6	3,8	2,3

R. PENNDORF hat eine Serie von 8 Flugzeugaufstiegen von U. KRUG ausgewertet[2]:

h	0,57	1	2	3	4	5	6	7	8	9 km
a_{Dk}	0,258	0,084	0,031	0,017	0,006	0,008	0,007	0,013	0,006	km^{-1}
T_k^*	7,4	3,7	2,1	1,7	1,3	1,4	1,4	1,8	1,4	

Auch in diesen Messungen kommen die überaus rasche Abnahme der Dunsttrübung mit der Höhe und die hohen Werte in Bodennähe zur Geltung. Oberhalb 4 km bleibt die Trübung nahezu konstant. Messungen dieser Art sind leider sehr spärlich; sie könnten wesentlich mehr Aufschlüsse über die vertikale Verteilung des Aerosols geben als wir besitzen.

Für die Untersuchung des *Tagesganges der Trübung*, der sich unter dem Einfluß der nächtlichen Sedimentation des Dunstes und der tagsüber einsetzenden Konvektion einstellt, ist es ein Hindernis, daß der Trübungsfaktor T eine gesetzmäßige Veränderung mit der Luftmasse m zeigt, auch wenn die Trübung sich nicht ändert.

Diese ist hervorgerufen durch die ungleiche Wellenlängenabhängigkeit der Dunsttrübung ($\lambda^{-\delta}$) und der Wasserdampfabsorption (selektive Banden) einerseits, der RAYLEIGH-Streuung (λ^{-4}) als Vergleichsmaß andererseits. Die zusammengesetzte Strahlung weiter Spektralbereiche gehorcht keiner e-Funktion mehr, sondern einem für jede Trübungsart spezifischen, meist nur numerisch angebbaren Gesetz. Kennzeichnung der einen Trübungsart mit dem Gesetz der anderen (das geschieht beim Trübungsfaktor) führt dann zu einer virtuellen Veränderung des Vergleichsmaßes (T) mit der Sonnenhöhe, also mit der Tageszeit. Zur Veranschaulichung ist in Fig. 8 anstatt von Beobachtungen das Ergebnis von Rechnungen H. HINZPETERs mit definierter Trübung wiedergegeben. Zwischen Luftmasse 1 und 2 nimmt der Trübungsfaktor durchweg ab, man wird also im Sommer ein Mittagsmaximum von T in den Beobachtungen finden; nach höheren Luftmassen nimmt er, besonders bei stärkerer Trübung, erheblich zu und man wird daher morgens und abends zu hohe Werte von T beobachten. Beobachtungen solcher Tagesgänge sind also mit Vorsicht auf ihre Bedeutung zu prüfen.

In Erkennung dieses Mangels hat LINKE einen „neuen Trübungsfaktor Θ" definiert [*1b*], bei dem die beobachtete Trübung mit einer dunstfreien RAYLEIGH-Atmosphäre verglichen wird, die jedoch 1 cm Wasserdampf enthält. Das hat zunächst zur Folge, daß bei geringer Dunsttrübung und geringem Wasserdampfgehalt w die beobachteten Werte Θ kleiner werden können als 1. Der anschauliche Wert von T geht deshalb bei Θ verloren. Der Gang mit m ist in der Tat stark herabgesetzt, jedoch bei stärkeren Trübungen keineswegs verschwunden. — Bei diesen Verfahren ist statt $J_R(m)$ als „Normalwert" der Sonnenstrahlung in Gl. (10.3) einzusetzen:

$$J_N(m) = \int_0^\infty J_{0\lambda}\, q_{R\lambda}^{m}\, q_{w\lambda}^{m_r w}\, d\lambda,$$

[1] F. LINKE: Beitr. Phys. fr. Atmosph. **15**, 176 (1929).

[2] U. KRUG-PIELSTICKER: Ber. dtsch. Wetterd. US-Zone **1949**, Nr. 8. — R. PENNDORF: Geophys. Res. Papers **1954**, No. 25, A. F. Cambridge Res. Center.

wobei $q_{R\lambda}$ der RAYLEIGH-Transmissionsfaktor, $q_{w\lambda}$ der Transmissionsfaktor für Wasserdampfabsorption, m die durchstrahlte Luftmasse, w der Wasserdampfgehalt der Atmosphäre (in cm Niederschlagswasser) sind. Im Exponenten $m_r w$ sind streng die in Tabelle 1, Zeile 3 angegebenen Werte m_{r_w} zu verwenden. Es ist vorausgesetzt, daß $q_{w\lambda}$ unabhängig von der Höhe, d.h. von Druck und Temperatur ist. LINKE verwendete die von MÖLLER angegebenen Zahlenwerte nach (9.1), die sich auf Messungen FOWLEs stützen. Die Berechnungsgrundlagen sind im Meteorologischen Taschenbuch II [2] zu finden; graphische Auswerte-Tafeln hat DANČ[1] berechnet, während die in [1b] gegebenen Tafeln Fehler enthalten. Einfache Umrechnungsfaktoren von T in Θ, die verschiedentlich angegeben wurden, sind nur als Näherungen zu brauchen.

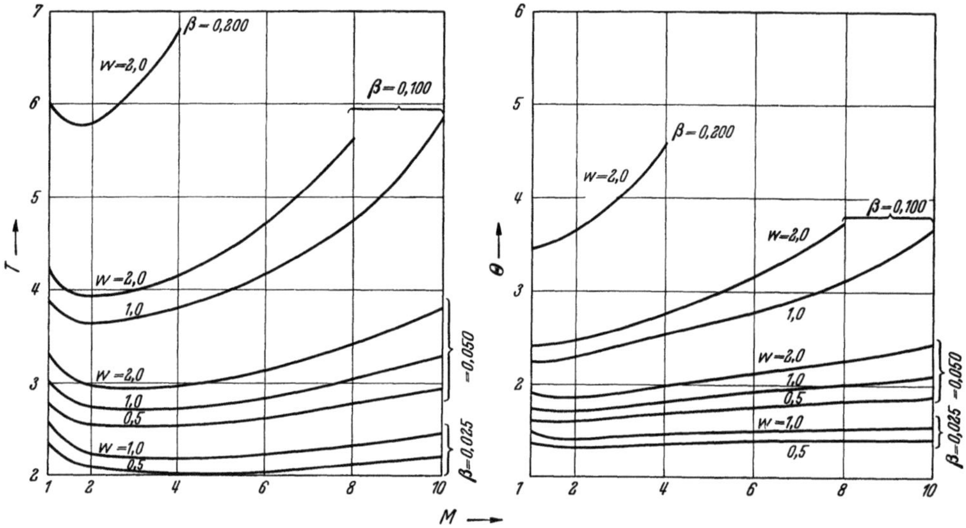

Fig. 8. Virtueller Gang des Trübungsfaktors T und neuen Trübungsfaktors Θ mit der Luftmasse M bei Anwesenheit einer Dunsttrübung $a_{D\lambda} = \beta\lambda^{-1,3}$ und absorbierendem Wasserdampf w (cm. precipitable water). Nach H. HINZPETER.

Ergiebiger ist eine *Unterteilung des Spektrums* in einen kurz- und einen langwelligen Teil (wenn man nicht zu spektralen Messungen übergehen will oder kann). Das geschieht durch das weitverbreitete Rotfilter RG 2 von SCHOTT (2,5 mm stark), das von 0,625 bis 2,5 μ fast ungeschwächt durchläßt, kürzere Wellenlängen vollkommen abfängt und damit das Spektrum in einen fast nur von RAYLEIGH- und Dunststreuung beeinflußten (für $\lambda < 625$ mμ ist $q_{w\lambda} \approx 1$) und einen fast nur von H_2O-Absorption beherrschten Teil (für $\lambda > 625$ mμ ist $q_{R\lambda} \approx 1$) trennt[2]. Noch besser wird dies durch das SCHOTT-Filter RG 8 bewirkt, dessen Kante bei 0,715 μ liegt. Auch OG 1, das das Spektrum im Gelben ($\lambda \gtrsim 525$ mμ) teilt, wird daneben gern verwendet. Bei beiden Filtern ist der Übergang von vollkommener Durchlässigkeit zu Undurchlässigkeit auf weniger als 10 mμ zusammengedrängt, das ideale Filter also gut angenähert.

Die hinter RG 2 gemessene Rotstrahlung R ist nur durch den doppelten Reflexionsverlust am Glas geschwächt, der ziemlich konstant 16% beträgt. Die „Kurzstrahlung" ist dann durch $K = J - 1{,}19\,R$ gegeben. Die *Trübungsfaktoren T_R für Rot- und T_K für Kurzstrahlung* werden analog Gl. (10.5) berechnet;

[1] J. DANČ: Meteorol. Zprávy **8**, 71 (1955).
[2] K. FEUSSNER: Meteor. Z. **49**, 242 (1932). — H. HINZPETER: Z. Met. **7**, 150 (1953). — [2] Tabelle 106.

die Berechnungskonstanten sind z.B. im Meteorologischen Taschenbuch [2], Tabelle 105a gegeben. Die erhaltenen Zahlenwerte T_R und T_K erweisen sich als wesentlich besser von der durchstrahlten Luftmasse unabhängig als T_G für die Gesamtstrahlung.

Aus den gemessenen Werten R oder aus T_R kann man unmittelbar den Betrag des Wasserdampfes bestimmen, wenn man den kleinen Betrag der molekularen und Dunststreuung vorher abzieht. Man hat nur die empirischen Werte der Absorptionsmessungen zugrunde zu legen, wobei wiederum angenommen wird, daß die $q_{w\lambda}$ druck- und temperaturunabhängig sind. Entsprechend kann T_K zu einer Angabe über den Aerosolgehalt der Atmosphäre verwendet werden, wie es von FEUSSNER und DUBOIS vorgeschlagen worden ist.

NEUWIRTH[1] hat versucht, den von ihm als Trübungsquotient bezeichneten Wert $\Theta_R : \Theta_K$ in Beziehung zum dynamischen Zustand der Atmosphäre zu bringen, und hat sogar Niederschlagsprognosen damit gegeben.

11. Der Trübungskoeffizient nach A. ÅNGSTRÖM. Da die RAYLEIGH-Streuung kein gutes Vergleichsmaß für die Streuwirkung des Dunstes ist, macht ÅNGSTRÖM[2] für die Dunstextinktion den Ansatz

$$a_{D\lambda} = \beta\,\lambda^{-1,3}; \qquad \lambda \text{ in } \mu \tag{11.1}$$

und bestimmt aus den Meßwerten $J(m)$ die Größe von β nach

$$J(m) = \int_0^\infty J_{0\lambda} \exp\left(-a_{R\lambda}\, m - \beta\, m_r\, \lambda^{-1,3} - a_{w\lambda}\, w\, m_r\right) d\lambda. \tag{11.2}$$

β wird Trübungskoeffizient genannt. Noch besser wird wiederum durch Verwendung des Rotfilters der langwellige Bereich mit fast alleiniger Wirksamkeit von Wasserdampfabsorption abgetrennt und β aus der Kurzstrahlung bestimmt. Berechnungsunterlagen sind von ÅNGSTRÖM[2], KIMBALL, HOELPER und FEUSSNER ausgearbeitet worden[3].

Der Exponent $\delta = -1{,}3$ war in Ziff. 6 als ein guter Mittelwert der Spektralbeobachtungen der Dunstextinktion gefunden worden, so daß β als ein gutes Maß für die atmosphärische Dunsttrübung angesehen werden darf. Die beobachteten Trübungskoeffizienten liegen für sehr reine Hochgebirgsluft unter 0,01, können aber bei großer Trübung Werte weit über 0,20 erreichen (vgl. die Messungen SCHÜEPPs unten). Allerdings ist der Trübungskoeffizient bei weitem nicht für so viele Messungen an verschiedenen Orten und Klimaten ausgewertet worden wie der Trübungsfaktor. HOELPER hat aus seinen Beobachtungen nachweisen können, daß der aus Kurzstrahlungsmessungen bestimmte Wert β während der Sommermonate April bis September kleiner ist als der aus der Gesamtstrahlung ermittelte, während der Monate Oktober bis März dagegen größer. Der Grund kann entweder in der Verwendung einer nicht ganz richtigen Wasserdampfabsorption (was rechnerisch, nach Tabelle 5 geschieht) liegen oder in einer jahreszeitlichen Veränderung des Wellenlängenexponenten δ, der durch $\delta = 1{,}3$ nicht exakt erfaßt wird.

Nur in diesem Falle, wenn nämlich

$$a_{D\lambda} = B\,\lambda^{-\delta}; \qquad \delta \neq 1{,}3 \tag{11.3}$$

[1] R. NEUWIRTH: Arch. Met., Geophys. u. Bioklim., Ser. B **6**, 431 (1955).

[2] A. ÅNGSTRÖM: Geogr. Annaler **11**, 156 (1929); **12**, 130 (1930).

[3] H. H. KIMBALL: Month. Weath. Rev. **61**, 80 (1933). — O. HOELPER: Dtsch. Met. Jb. Aachen **1933**; Wiss. Abh. Reichsamt. Wetterd. **5**, Nr. 10 (1939). — K. FEUSSNER: Tät. Ber. Preuß.-Met. Inst. 1931, 89 (1932). — [2], Tabelle 104.

ist, kann auch ein virtueller Tagesgang auftreten. Aus Tabelle 11 ist zu ersehen, daß β scheinbar mit der Luftmasse abnimmt.

Zugleich erkennt man, daß bei $\delta \gtrless 1{,}3$ die Zahlenwerte $\beta \gtrless B$ werden. Definitionsgemäß ist β der Extinktionskoeffizient des Dunstes für $\lambda = 1\,\mu$, eine Wellenlänge, die weit außerhalb des Meßbereiches liegt. Bei ungefähr gleicher Extinktion im optischen Bereich bedingt eine Vergrößerung von δ eine Verkleinerung der Extinktion bei $1\,\mu$ und umgekehrt. Es erscheint daher zweckmäßiger, einen Trübungskoeffizient β' oder verallgemeinert B' auf die Wellenlänge $0{,}5\,\mu$ (oder eine nahe dem Schwerpunkt des Meßbereiches gelegene) zu beziehen entsprechend $B\lambda^{-\delta} = 2^\delta B(\lambda/0{,}5)^{-\delta} = B'(2\lambda)^{-\delta}$ wo $B' = 2^\delta B$ ist, oder wenigstens neben β bzw. B auch $a_{D,\lambda=0{,}5}$ anzugeben (vgl. Tabelle 3 und 4). Es ist dann für

$\delta =$	1,0	1,3	1,6	2,0
$B'/B =$	2,00	2,46	3,03	4,00.

Tabelle 11. *Virtueller Gang des Trübungskoeffizienten β mit der Luftmasse m_r, wenn $a_{D\lambda} = 0{,}10\,\lambda^{-\delta}$ ist.*

	$m_r = 1$	2	4	6	10
$\delta = 1{,}0$	$\beta = 0{,}77$	0,75	0,73	0,72	0,70
$\delta = 1.6$	$\beta = 1{,}20$	1,14	1,08	1,05	1,00
$\delta = 2{,}0$	$\beta = 1{,}61$	1,52	1,45	1,40	1,36

SCHÜEPP [6] wählt als Basis der Exponentialfunktion in (11.2) nicht e sondern 10, wodurch die Vergrößerung von B' gegenüber B ungefähr aufgehoben wird; für $\delta = 1{,}3$ ergibt sich $2{,}46 \cdot \log e = 1{,}06$.

Die Größenverteilung des Aerosols kann durch Bestimmung des Exponenten δ oder der Wellenlängenabhängigkeit von $a_{D\lambda}$ erfaßt werden. W. SCHÜEPP [6] hat deshalb durch Vorschalten der Filter OG 1 oder RG 2 vor das Aktinometer eine doppelte Unterteilung des Spektrums bei 525 und 625 mμ erreicht und ein numerisch-graphisches Verfahren entwickelt, aus den Intensitäten der 3 Spektralbereiche den Wasserdampfgehalt, den Wellenlängenexponent δ und den verallgemeinerten Trübungskoeffizient B zu bestimmen. Da die Ausgangswerte für die Berechnung als Differenzen zwischen je 2 Meßgrößen gefunden werden müssen, ist das Verfahren sehr empfindlich gegen kleine Meß- oder Eichungsfehler oder Verschmutzungen des Filters, und die Bestimmung — insbesondere von δ — schwierig. Direkte Messungen schmaler Spektralbereiche unter Verwendung von Filterkombinationen oder Interferenzfiltern vor Aktinometern oder Photozellen ergeben zuverlässigere Werte. Zum Beispiel sind die Messungen von VOLZ (Ziff. 4, S. 161) mit Zellen und Filtern gewonnen. Eine gute Erfassung der spektralen Abhängigkeit ist zur Bestimmung der Trübung durch Extinktion in Zukunft anzustreben.

SCHÜEPP[1] hat für Stockholm je nach der Herkunft der Luftmassen Mittelwerte für δ zwischen 1,3 und 1,8 gefunden und Werte für B von 0,025 bis 0,150. Aus zahlreichen Messungen in Léopoldville (Belgisch-Kongo) in der Regenzeit ergab sich für B 0,03 bis 0,10 (bei bis zu 5 cm Niederschlagswasser!), in der Trockenzeit sehr hohe Werte von 0,30 bis 1,00; der Exponent δ ist während der Trocken- und Regenzeit $2{,}0 \pm 0{,}5$; nur an der nahe der Wüste gelegenen Station N'Délé wurden gelegentlich auch Werte δ unter 1,0 ($0{,}5 < \delta < 1{,}8$) festgestellt.

12. Die extraterrestrische Sonnenstrahlung. Als *Solarkonstante* bezeichnet man die Strahlungsstromdichte, die der Erde von der Sonne bei mittlerem Abstand, aber unbeeinflußt durch die Erdatmosphäre zugesandt wird. Die tatsächliche (extraterrestrische) Strahlung schwankt wegen der Elliptizität der Erdbahn um

[1] Briefliche Mitteilung, Auswertungen nach unveröffentlichten Messungen des Meteorologischen Instituts Stockholm und des Service Météorologique, Congo-Belge.

$\pm 3{,}35\%$, das Maximum tritt im Perihel der Erde Anfang Januar, das Minimum im Aphel Anfang Juni ein. Andere Schwankungen konnten in dem energetisch gemessenen Spektralbereich bisher nicht einwandfrei nachgewiesen werden, obwohl für das kurzwellige UV Schwankungen aus den Ionisierungswirkungen in der hohen Atmosphäre bekannt sind (vgl. Bd. IL dieses Handbuches) und auch im Bereich der elektromagnetischen cm- und m-Wellen erhebliche Variationen festgestellt wurden. Beide gehen ungefähr parallel mit der Sonnenaktivität, die durch die Sonnenfleckenzahl oder die photosphärische Eruptionshäufigkeit gekennzeichnet wird.

Die Methode der Bestimmung der Solarkonstante geht auf LANGLEY zurück. Das BOUGUERsche Gesetz $J_\lambda = J_{0\lambda}\, e^{-a_\lambda m}$ gilt monochromatisch streng, wenn a_λ unabhängig von der Höhe in der Atmosphäre, also von der Luftmasse m ist; das darf hier angenommen werden. Durch Messungen bei verschiedenen Luftmassen m kann man $J_{0\lambda}$ extrapolieren, wenn sich a_λ, also vor allem Dunst- und Wasserdampfgehalt der Atmosphäre, im Laufe des Tages nicht ändern. Sorgfältige spektrobolometrische Messungen mit hoher Auflösung sind seit Anfang des Jahrhunderts durch das Astrophysical Observatory der Smithsonian Institution Washington durchgeführt worden, vor allem auf deren Bergobservatorien Mt. Wilson und Mt. Whitney in Californien, ferner in Chile, New Mexico, Südwestafrika, auf der Sinai-Halbinsel u. a., heute routinemäßig noch auf dem *Table Mountain* (2250 m) in Californien [*3*]. Hochgelegene Stationen sind ausgewählt, weil man erwartete, in großer Höhe bei geringer Dunstextinktion und Wasserdampfabsorption auch nur geringe Tagesgänge dieser Größen anzutreffen.

Die Messungen selbst erfolgen nur im Bereich $0{,}346 < \lambda < 2{,}44\,\mu$, so daß für das ultraviolette und infrarote Ende Korrektionen angebracht werden müssen. Die absolute Genauigkeit des *Bolometers* ist wegen geringer zeitlicher Schwankungen nicht ganz ausreichend, deshalb werden gleichzeitig mit der Spektralaufnahme Messungen der Gesamtintensität mit *Absolut-Pyrheliometern* durchgeführt und das Integral über die Wellenlänge dieser gleichgesetzt. Die Frage nach der Solarkonstante ist daher eng verquickt mit der der Absolutmessung oder, da die Absolutinstrumente für den ständigen Gebrauch zu umständlich sind, mit der Frage der Skala, in der die Relativinstrumente geeicht sind.

Die sog. *Smithsonian revised Scale 1913*, die auf Messungen mit 2 verschiedenen Absolutinstrumenten, dem Waterflow- und dem Waterstir-Pyrheliometer beruhte, und in der die meisten Silverdisc-Aktinometer (als Subnormalinstrumente) geeicht sind, hat um etwa 2,5% zu hohe Meßwerte ergeben, wie durch einen Vergleich mit einem neuen elektrischen Kompensations-Waterflow-Pyrheliometer 1932 von ABBOT und ALDRICH[1] festgestellt wurde. Vergleiche mit dem ÅNGSTRÖMschen Kompensations-Pyrheliometer ergaben anfänglich noch größere Differenzen; nach Anbringung einer als notwendig erkannten Korrektur wegen der Wärmeableitung am Rande der schwarzen Streifen bei diesem Instrument verminderte sich die Differenz aber ebenfalls auf 2,3 bis 2,5%[2]. Auch Messungen mit anderen Absolutinstrumenten[3] führten auf dieselbe Korrektion, so daß heute die Smithsonian Scale 1913, vermindert um 2,4% als korrekt anzusehen ist[4].

---

[1] L. B. ALDRICH u. C. G. ABBOT: Smithson. Miscell. Coll. **92**, No. 13 (1934); **110**, No. 5 u. 11 (1948). — W. H. HOOVER u. A. G. FROILAND: Smithson. Miscell. Coll. **122**, No. 5 (1953).

[2] A. ÅNGSTRÖM: Meteor. Z. **31**, 369 (1914). — W. MARTEN: Meteor. Z. **39**, 342 (1922).

[3] K. FEUSSNER: Meteor. Z. **58**, 68 (1941); **60**, 73 (1943). — G. FALCKENBERG: Meteor. Z. **60**, 314 (1943).

[4] A. J. DRUMMOND: W. M. O. Bull. **5**, 75 (1956) macht den Vorschlag, eine Skala „Smithsonian minus 3%" als Vermittelung zwischen Smiths. rev. Scale 1913 und originaler ÅNGSTRÖM-Skala einzuführen, vgl. Anmerkung auf S. 253.

Jedoch hat (nach einer 1952 gegebenen Mitteilung von Aldrich und Hoover[1]) die Smithsonian Institution selbst die Rev. Scale 1913 in ihren Veröffentlichungen nie benutzt, sondern eine Skala, deren Angaben um 1,8 % vergrößert werden müssen, um auf die Skala von 1913 zu kommen, so daß tatsächlich alle früheren Bestimmungen der Solarkonstanten durch die Smithsonian Institution um 0,6% verringert werden müssen, um korrekt zu sein (vgl. Anmerkung auf S. 253).

Die UV-Korrektion für das nicht gemessene Gebiet $\lambda < 0{,}346\,\mu$ wurde nach verschiedenen Verfahren auf $+3{,}4\%$ geschätzt[2]. Heute liegen durch die Messungen mit Raketen in großen Höhen direkte Bestimmungen der Strahlung auch oberhalb der Ozonschicht vor[3]. Sie lassen bei hoher Auflösung im Bereich 220 bis 340 mμ eine außerordentlich hohe Zahl von solaren Fraunhofer-Linien erkennen und geben einen sehr detaillierten Verlauf der Strahlungsenergie. Mit vorläufigen Werten dieser UV-Strahlung wurden bereits zwei Neuberechnungen der Solarkonstanten durchgeführt[4]. Es sind aber auch im nahen UV und sichtbaren Spektrum neue Messungen der Spektralverteilung durchgeführt und auf Luftmasse $m=0$ extrapoliert worden. Aus allen diesen Messungen ergab sich eine Erhöhung der alten UV-Korrektion auf das 1,5fache[5].

Für das infrarote Ende des Sonnenspektrums bei $\lambda > 2{,}44\,\mu$ setzt man die Sonne als grauen Strahler von 6000° K an, was auch am besten der Spektralverteilung für $1\,\mu < \lambda < 2{,}4\,\mu$ entspricht und mit anderen Messungen in diesem Bereich[6] in Übereinstimmung steht. Johnson[5] hat jedoch einen Fehler in den Überlegungen Abbots bei der Bestimmung seiner Infrarotkorrektion nachgewiesen, wodurch sich diese auf das Doppelte vergrößert. Nach Anbringung der neuen Korrektionen stellt sich der beste zur Zeit angebbare Wert der Solarkonstanten auf

$$J_0 = 2{,}00 \text{ cal cm}^{-2} \text{ min}^{-1}.$$

Die Fehlergrenze wird auf $\pm 2\%$ geschätzt, wobei die größte Unsicherheit allerdings in der Frage der Berechtigung der 1,8%-Änderung aller früheren Messungen der Smithsonian-Institution liegt[7]. Bisher unveröffentlichte Messungen von H. R. Shackleton und A. L. Quirk in 27 km Höhe scheinen den Wert 2,0 cal cm^{-2} min^{-1} zu bestätigen.

Die extraterrestrische Intensitätsverteilung nach Johnson zeigt Fig. 9. Am langwelligen Ende des Spektrums ist der Verlauf sehr glatt; er ist zwar durch eine über die Fenster zwischen den großen Banden gelegte ausgeglichene Kurve entstanden, aber dadurch gesichert, daß auch bei hoher Auflösung im Infrarot viele atmosphärische, aber nur wenig solare Linien zu finden sind. Im Gegensatz dazu fällt der unruhige Verlauf im UV auf; er ist bedingt durch die außerordentliche Häufung von solaren Absorptionslinien. Schon Pettit, Hess und Götz-Schönmann[8] hatten auf die Einsenkungen bei 430 und 390 mμ aufmerksam

---

[1] L. B. Aldrich u. W. H. Hoover: Science, Lancaster, Pa. **116**, 12, 3 (1952).

[2] C. G. Abbot: Gerlands Beitr. Geophys. **16**, 344 (1927).

[3] F. S. Johnson, J. D. Purcell, R. Tousey u. N. Wilson: In Rocket exploration of the upper atmosphere, S. 279. London 1954.

[4] H. G. Houghton: J. Meteorology **8**, 270 (1951). — M. Nicolet: Arch. Met., Geophys. u. Bioklim. Ser. B **3**, 209 (1951). — Ann. Astrophys. **14**, 249 (1951).

[5] F. S. Johnson: J. Meteorology **11**, 431 (1954).

[6] R. Peyturaux: C. R. Acad. Sci. Paris **232**, 931, 1069 (1951). — Ann. d'Astrophys. **15**, 302 (1952). — A. Keith Pierce: Astrophys. J. **119**, 312 (1954).

[7] J. Georgi: Ann. Met. **5**, 83 (1952). — Met. Rdsch. **7**, 182 (1954). — W. Schüepp: Geofis. pura e appl. **30**, 179 (1955).

[8] E. Pettit: Astrophys. J. **75**, 219 (1932). — P. Hess: Untersuchungen über die spektrographische Energieverteilung im Sonnenspektrum. Diss. Frankfurt a. M. 1938. — F. W. P. Götz u. E. Schönmann: Helv. phys. Acta **21**, 151 (1948).

gemacht, die besonders auffallend im Spektrum des Himmelslichtes sind und dort zu einer ausgesprochenen Zweigipfeligkeit der Spektralkurve Anlaß geben.

Eine gute Stütze für die Bestimmung dieser Spektralverteilung ergibt sich aus den Abschätzungen von MULDERS[1], der Intensität und Anzahl der solaren FRAUNHOFER-Linien für enge Spektralbereiche auszählte und deren Gesamteffekt dem kontinuierlichen Photosphärenspektrum überlagerte. Die auf diese Weise erhaltene Spektralverteilung ist mehrfach (LINKE [*1b*], NICOLET) mit den Messungen der Smithsonian Institution verglichen worden und hat erstaunlich gute Übereinstimmung ergeben.

W. SCHÜEPP [*6*] hat aus Aktinometer-Messungen mit Rotfiltern auf einen höheren Betrag der extraterrestrischen Sonnenstrahlung im Infraroten geschlossen, als es der früheren Berechnung LINKEs entspricht. Ebenso haben GATES und Mitarbeiter[2] eine höhere Infrarotstrahlung im Bereich 1,2 bis 2,2 μ gemessen, die sich dem von ADEL für das 10 μ-Fenster angegebenen Verlauf einer 7000°-Sonne anpassen würde. JOHNSON ebenso wie NICOLET haben jedoch darauf hingewiesen, daß das nicht zu einer Erhöhung der Solarkonstante zu führen braucht, da es durch eine Erniedrigung im Sichtbaren ausgeglichen werden muß.

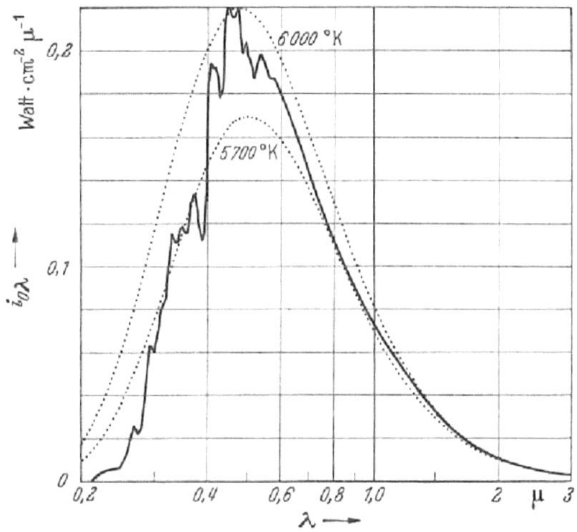

Fig. 9. Extraterrestrisches Sonnenspektrum bei mittlerer Entfernung Erde—Sonne. Nach F. S. JOHNSON.

Es ist mehrfach versucht worden, die *Temperatur T_S der Sonne* aus der Solarkonstante zu bestimmen. Es muß $4\pi r^2 \sigma T_s^4 = 4\pi R^2 J_0$ sein, wenn der Sonnenradius $r = 6{,}95 \cdot 10^{10}$ cm, der mittlere Erdabstand $R = 1{,}495 \cdot 10^{13}$ cm ist. Mit $J_0 = 2{,}00$ führt das auf $T_S = 5790°$ K. Die Überlegung läßt aber außer acht, daß das Spektrum der Sonne keineswegs als schwarz angesehen werden kann. Auch im IR sind zwischen den Wasserdampfbanden noch solare Absorptionslinien festgestellt (vgl. Fußnote 1, S. 171), so daß das infrarote Spektrum nach Fig. 9 nur 0,990 der schwarzen Strahlung einer 6000°-Sonne liefert. Im Sichtbaren erreicht (bei 10 mμ Intervallbreite) nur die Spitze bei 450 mμ die Schwarzstrahlung 6000°, jedoch kommen bei höherer Auflösung sowohl im Sichtbaren wie im UV mehrere linienarme Stellen vor, deren Intensität 6000° erreicht oder überschreitet.

Die laufende *Überwachung* der Solarkonstante ist das eigentliche Anliegen aller Untersuchungen der Smithsonian-Institution gewesen. Es haben sich dabei Schwankungen von Tag zu Tag ergeben, über deren Realität noch keine Einmütigkeit erzielt werden konnte. Seit 1919 wurde eine „kurze" Methode der Bestimmung[3] eingeführt: empirisch ist eine Beziehung zwischen den Transmissionsfaktoren in jeder der 40 zur Messung verwendeten Wellenlängen mit der Tiefe der $\varrho\sigma\tau$-Bande (Wasserdampfmaß) und der Intensität der zirkumsolaren Himmelsstrahlung (Streuung) festgestellt worden, die nun verwendet wird, um aus einer Spektralmessung bei nur einer Luftmasse die Solarkonstante zu

[1] G. F. W. MULDERS: Z. Astrophys. **11**, 132 (1936). — Publ. Astr. Soc. Pacif. **51**, 220 (1939). — [*2*], Tabelle 97, 98.

[2] D. M. GATES, R. F. CALFEE u. B. E. COHN: Phys. Rev. **88**, 157 (1952).

[3] C. G. ABBOT: Meteor. Z. **38**, 246 (1921).

extrapolieren (LINKE [1b] hat darauf hingewiesen, daß man so auf keinen Fall die extraterrestrische, sondern nur die „subozonale" Sonnenstrahlung erhalten kann). Bedenken gegen die so bestimmten Werte J_0 ergaben sich wegen der nicht sehr guten Korrelation zwischen gleichzeitigen Messungen von J_0 an weit entfernten Orten (z.B. Chile—Californien) und wegen der verhältnismäßig großen Korrelationen zwischen Trübungsfaktor und J_0[1]. Dies weist darauf hin, daß in den Schwankungen von J_0 Resteffekte von Wetterlagen-Einflüssen zu sehen sind. ABBOT[2] hat in mehreren Untersuchungen den gegenteiligen Standpunkt vorgetragen.

Durch Multiplikation der Strahlungsintensität mit der spektralen Augenempfindlichkeit und Umrechnung auf lux kann man die extraterrestrische *Beleuchtungsstärke* durch die Sonne ermitteln. Andererseits sind auch direkte photometrische Messungen durchgeführt und auf die Luftmasse $m=0$ reduziert worden. Die verschiedenen Bestimmungen hat KARANDIKAR[3] mit eigenen Messungen verglichen. Er erhält

$$E_0 = 131{,}000 \text{ lux}$$

als Mittelwert seiner eigenen Messungen, was einer mittleren Leuchtdichte von $\overline{B} = 193{,}00$ stilb und einer maximalen Leuchtdichte in der Mitte der Sonnenscheibe von $B_C = 243{,}00$ stilb entspricht. JOHNSONS extraterrestrische Energieverteilung führt auf $E_0 = 137{,}000$ lux.

B. Die Himmelsstrahlung.

Übersicht. Der nicht auf Absorption beruhende Anteil der Extinktion der Sonnenstrahlung wird durch die Streuung aus der geradlinigen Strahlrichtung abgelenkt und gelangt entweder als diffuse Himmelsstrahlung zum Boden oder wird in den Weltraum reflektiert. In reiner Luft erfolgt die Streuung proportional $(\mu_\lambda - 1)^2 \lambda^{-4}$, so daß das Spektrum des Himmelslichtes ebenso wie das des in den Weltraum zurückgestreuten Lichtes stark nach kurzen Wellenlängen verschoben ist; der Schwerpunkt liegt im UV. Die Mitwirkung von gröberen Partikeln bewirkt eine schwächere Abhängigkeit von λ (z.B. $\lambda^{-1,3}$), so daß die Farbe des Himmels der der Sonnenstrahlung — dem Weiß — näherkommt. — Die einmal gestreute Strahlung kann an den Luftteilchen zum zweitenmal oder auch mehrfach gestreut werden, oder sie kann vom Erdboden oder von Wolken reflektiert werden, bevor sie endgültig zum Boden oder in den Raum gelangt. Dieses Problem der *Vielfachstreuung* erweist sich — selbst in der wolkenlosen Atmosphäre — als sehr kompliziert.

13. Die Streuung des Lichtes. Die Streuung des Lichtes an den *Luftmolekülen* ist von Lord RAYLEIGH[4] als Streuung an kleinen kugelförmigen dielektrischen Teilchen, deren Radius $r \ll \lambda$ ist, behandelt worden. Wenn die Volumeinheit Luft mit der Strahlungsintensität J_λ bestrahlt wird, so ist die Intensität des gestreuten Lichtes im Raumwinkel $d\omega$

$$i_{\varphi\lambda}\, d\omega = J_\lambda \frac{3}{16\pi} k_{R\lambda} \frac{\varrho}{\varrho_0} (1 + \cos^2\varphi)\, d\omega. \qquad (13.1)$$

[1] F. LINKE: Meteor. Z. **41**, 74 (1924). — W. MILCH: Z. Geophys. **1**, 377 (1925). — M. M. PARANJPE: Quart. J. Roy. Met. Soc. **64**, 459 (1938).

[2] G. C. ABBOT: Quart. J. Roy. Met. Soc. **65**, 215 (1939). — Smithson. Miscell. Coll. **104**, No. 5 (1944).

[3] R. V. KARANDIKAR: J. Opt. Soc. Amer. **45**, 483 (1955).

[4] Lord RAYLEIGH: Phil. Mag. **41**, 107, 247 (1871;) **47**, 375 (1899). — Proc. Roy. Soc. Lond., Ser. A **84**, 25 (1911); **90**, 219 (1914).

Dabei ist $k_{R\lambda}$ der Extinktionskoeffizient nach RAYLEIGH (vgl. Ziff. 3), ϱ und ϱ_0 die vorhandene Luftdichte und diejenige unter Normalbedingungen, φ der Winkel zwischen einfallendem und gestreutem Lichtstrahl.

$$\int_{4\pi} i_{\varphi\lambda}\, d\omega = J_\lambda\, k_{R\lambda}\, \varrho/\varrho_0 \tag{13.2}$$

ist die gesamte Streustrahlung der Volumeinheit und dem Betrag nach gleich der extingierten Strahlung. Ist die einfallende Strahlung J_λ (teilweise) polarisiert, so treten kleine Zusatzglieder hinzu, die hier nicht berücksichtigt zu werden brauchen (vgl. Beitrag SEKERA in diesem Band).

Die Verteilung der Streustrahlung über den Winkel φ ist nach (13.1) unabhängig von λ und um $\varphi = \pi/2$ symmetrisch (vgl. Fig. 12, S. 191, dort allerdings wegen log φ-Skala unsymmetrisch erscheinend); es wird also der gleiche Betrag nach $\varphi = 0$ und $\varphi = \pi$ gestreut, während die Streustrahlung unter $\varphi = \pi/2$ den halben Wert der Vorwärts- und Rückwärtsstreuung besitzt. Die Abnahme des Streulichtes von $\varphi = 0$ bis 10° beträgt nur 3%, bis 20° 12%. Wenn nur die Luftmoleküle in der Atmosphäre Streulicht erzeugten, würde man also die Sonne vor einem Himmel sehen, dessen Leuchtdichte von großen Sonnenabständen gegen den Sonnenrand hin nur unmerklich heller wird. Das wird nur in der reinen Luft sehr großer Höhen über der Erdoberfläche beobachtet; in bodennahen Schichten bewirkt die Dunsttrübung eine starke Aufhellung der nächsten Sonnenumgebung.

Für den Grenzfall *sehr großer Teilchen* $r \gg \lambda$, wie sie im Dunst enthalten sind, ergibt sich eine andere Vereinfachung der allgemeinen Theorie, da man die gestreute Strahlung in drei Anteile aufspalten kann (vgl. Ziff. 4):

1. Das am Rande der Partikel *gebeugte* Licht i_B nach der Theorie von FRAUNHOFER und AIRY;
2. das an der Oberfläche *gespiegelte* Licht;
3. das durch den Tropfen hindurchgegangene und dabei 2mal *gebrochene* Licht, wenn es sich um eine transparente Substanz, z.B. einen Flüssigkeitstropfen handelt.

Die drei Anteile können getrennt behandelt und dann addiert werden.

Für das am Rande einer Kugel gestreute Licht gilt

$$i_B = \frac{\alpha^2}{4}\left(\frac{2J_1(z)}{z}\right)^2, \tag{13.3}$$

wobei $z = \alpha \sin \varphi/2$ und $\alpha = 2\pi r/\lambda$, J_1 die BESSEL-Funktion erster Ordnung ist. Dies liefert die bekannten *Beugungsringe* um die Lichtquelle; die Lage der Minima kann ziemlich genau mit $\sin \varphi_{\min} = (n + 0{,}22)\, \lambda/2r$ angegeben werden ([2], Tabelle 130). Je größer die Wellenlänge oder je kleiner der Radius, umso weiter sind die Ringe. Bei einfallendem weißen Licht erscheinen sie daher, bei einheitlicher Tropfengröße, farbig.

Der gebrochene und gespiegelte Anteil i_G wurde von CHR. WIENER[1] für Wasserkugeln ($n = 1{,}330$) nach den Gesetzen der geometrischen Optik berechnet (Tabelle 12). Die Winkelverteilung dieses Anteils zeigt einen schroffen Abfall mit φ, der vom Brechungsindex nur wenig abhängig ist. Berücksichtigt man auch innere Reflexionen im Tropfen, so liefert dieser Anteil bei den entsprechenden Winkeln auch die Regenbogen erster und höherer Ordnung. Messungen an großen Tropfen[2], die von BRICARD, SIEDENTOPF und BULLRICH an natürlichen, von

[1] CHR. WIENER: Abh. Kais. Leop. Carol. Dtsch. Akad. Naturf. Halle **73**, Nr. 1 (1900); **91**, Nr. 2 (1909).

[2] J. BRICARD: Ann. de Phys. **14**, 148 (1940). — C. R. Acad. Sci. Paris **213**, 136, 495 (1941). — E. REEGER u. H. SIEDENTOPF: Optik **1**, 15 (1946). — K. BULLRICH: Optik **2**, 301 (1947). — G. R. PARANJPE, Y. G. NAIK u. P. B. VAIDYE: Proc. Indian Acad. A **9**, 352 (1939). — M. S. ELSÄSSER u. K. WIRTZ: Z. angew. Phys. **5**, 133 (1953).

Tabelle 12. *Brechungs- und Spiegelungsanteil des Streulichtes an Tropfen ($n = 1,330$, Wasser, und $n = 1,50$) nach* WIENER[1] *und* VOLZ[2].

φ	0	10	20	30	45	60	70	80	90°
$n = 1,33$	4,086	3,372	2,107	1,165	0,408	0,112	0,031	0,008	0,007
$n = 1,50$	2,32	—	—		40°: 1,38	—	0,19	—	0,015

φ	100	110	120	135	150	160	170	180°
$n = 1,33$	0,006	0,011	0,020	0,059	0,084	0,094	0,101	0,102
$n = 1,50$	—	—	0,010	—	—	—	—	0,080

PARANJPE u. a. an künstlichen Nebeln, von ELSÄSSER und WIRTZ an einzelnen Tröpfchen in der WILSON-Kammer durchgeführt sind, liefern alle einen schwächeren Abfall des Streulichtes als die WIENERsche Funktion.

Unterhalb von etwa $\alpha = 40$ machen sich Interferenzen zwischen dem geometrischen Anteil i_G und dem Beugungsanteil i_B stärker bemerkbar, die von MECKE[3] und BRICARD[4] untersucht worden sind; sie äußern sich darin, daß die Farbfolge der Beugungsringe unregelmäßig wird, nur für das zentrale Feld konnte MECKE die Reihenfolge der Färbungen in Abhängigkeit von der Tropfengröße angeben.

Im Größenbereich $r \approx \lambda$ oder $\alpha \approx 10$, dessen Wichtigkeit bei atmosphärischem Dunst in Ziff. 4 und 5 festgestellt ist, kann die Streuung nur nach der exakten, 1908 von MIE[5] gegebenen Theorie berechnet werden. Die relative Streulichtintensität oder Streufunktion ist eine Funktion f von r, λ, φ und vom Brechungsindex m; da r und λ nur in der Verbindung $\alpha = 2\pi r/\lambda$ als unabhängige Variable eingehen, kann man setzen

$$f(\alpha, \varphi, m) = \frac{\lambda^2}{4\pi^2} \frac{i_1 + i_2}{2} = \frac{\lambda^2}{4\pi^2} i \tag{13.4}$$

wobei i_1 und i_2 die beiden senkrecht zueinander polarisierten Anteile der Streustrahlung sind (vgl. Beitrag SEKERA in diesem Bande). Es ist

$$i_1 = \left| \sum_{n=1}^{\infty} \frac{2n+1}{n(n+1)} (a_n \pi_n + b_n \tau_n) \right|^2; \qquad i_2 = \left| \sum_{n=1}^{\infty} \frac{2n+1}{n(n+1)} (a_n \tau_n + b_n \pi_n) \right|^2. \tag{13.5}$$

Dabei sind:

$$a_n = i^{n+\frac{1}{2}} \frac{S_n(\alpha)\, S_n'(\beta) - m\, S_n(\beta)\, S_n'(\alpha)}{S_n'(\beta)\, \Phi_n(\alpha) - m\, S_n(\beta)\, \Phi_n'(\alpha)},$$

$$b_n = i^{n+\frac{1}{2}} \frac{m\, S_n(\alpha)\, S_n'(\beta) - S_n(\beta)\, S_n'(\alpha)}{m\, S_n'(\beta)\, \Phi_n(\alpha) - S_n(\beta)\, \Phi_n'(\alpha)},$$

$$S(\alpha) = \sqrt{\pi\alpha/2}\, J_{n+\frac{1}{2}}(\alpha); \qquad \Phi_n(\alpha) = S_n(\alpha) + i\, C_n(\alpha); \qquad C_n(\alpha) = \sqrt{\pi\alpha/2}\, J_{-n-\frac{1}{2}}(\alpha),$$

$$\pi_n = \frac{dP_n(x)}{dx}; \qquad \pi_n' = \frac{d^2 P_n(x)}{dx^2}; \qquad \tau_n = x \cdot \pi_n - (1 - x^2)\, \pi_n'; \qquad \beta = m\alpha.$$

Umfangreiche Tabellen der Streulichtanteile i_1 und i_2 sind von HOLL, LOWAN, SLIEPCEVICH und Mitarbeitern und PENNDORF[6] gegeben worden, vornehmlich

[1] Siehe Fußnote 1, S. 187.

[2] F. VOLZ: Persönliche Mitteilung 1955.

[3] R. MECKE: Ann. Phys. **51**, 492 (1920); **52**, 623 (1920).

[4] J. BRICARD: Ann. Géophys. **1**, 231 (1946).

[5] G. MIE: Ann. Phys. (4) **25**, 377 (1908).

[6] H. HOLL: Optik **1**, 213 (1946); **4**, 173 (1948). — E. DE BARY: Optik **9**, 319 (1952). — A. N. LOWAN: Tab. of scatt. funct. f. spher. part., Nat. Bur. Stand., Appl. Math. Ser. 4, Wash. 1949. — R. O. GUMPRECHT, N.-L. SUNG, J. H. CHIN u. C. M. SLIEPCEVICH: J. Opt. Soc. Amer. **42**, 226 (1952). — R. PENNDORF: Briefl. Mitt. 1953. — M. KERKER: J. Opt. Soc. Amer. **45**, 1081 (1955); dort umfangreiche Literaturangaben, auch für andere Brechzahlen.

für $m = 1{,}33$, jedoch auch für größere Werte des Brechungsindex bis $m = 2{,}0$. LOWAN hat auch komplexe Brechungszahlen, also absorbierende Medien berücksichtigt, allerdings nur für $\alpha \leqq 6{,}0$. Fig. 10 gibt nach einer Zusammenstellung von VOLZ [5] die Streufunktionen für kugelförmige Wassertröpfchen ($m = 1{,}33$) in relativen Werten des Anfangswertes, d.h. $f(\varphi):f(0)$. Die Darstellung zeigt, daß in einem kleinen Winkelbereich um $\varphi = 0$ die Streulichtintensität für alle α groß ist gegenüber dem Wert bei größeren Winkelabständen; an einen scharfen Abfall schließen sich eine Reihe von sekundären Maxima und Minima an. Der Winkelbereich starker Vorwärtsstreuung ist um so enger, d.h., der Abfall erfolgt bei um so kleineren Winkelwerten, je größer α, d.h. je größer der Tropfen oder je kleiner die Wellenlänge ist. Der größte Teil des Streulichtes geht als Vorwärtsstreuung in diesen engen Kegel und liefert daher den weit überwiegenden Anteil des in die Kugel gehenden Streulichtes

$$\left.\begin{aligned} &\int_0^{\pi} f(\alpha,\varphi,m)\, 2\pi \sin\varphi\, d\varphi \\ &\quad = \pi r^2 \sigma(\alpha, m) \end{aligned}\right\} \quad (13.6)$$

vgl. (4.1), während der Anteil des seitwärts und rückwärts gestreuten Lichtes nur gering ist. Der Verlauf der Absolutwerte von $f(\varphi = 0)$ ist deshalb dem von $r^2\sigma(\alpha, m)$ ähnlich, zeigt also an den gleichen Stellen α (relative) Maxima und Minima wie σ (vgl. Fig. 1). Für sehr kleine Werte α nähert sich die Streufunktion der RAYLEIGH-Verteilung $(1 + \cos^2\varphi)$ mit nur einem Minimum bei $\varphi = \pi/2$ und wird wellenlängenunabhängig.

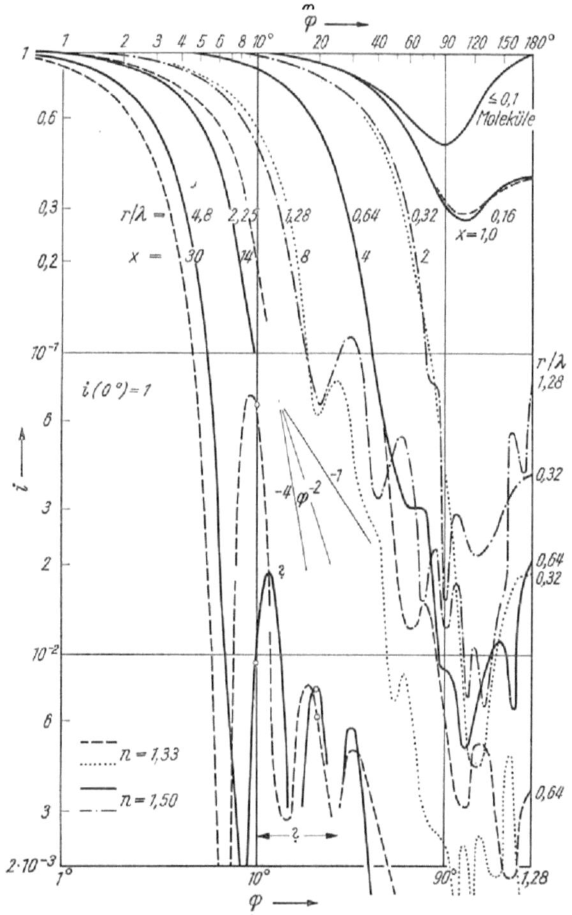

Fig. 10. Abhängigkeit des gestreuten Lichtes i vom Streuwinkel φ, $i(\varphi = 0) = 1$ gesetzt, für verschiedene Werte r/λ und Brechungsindices $n = 1{,}33$ und $1{,}50$. Nach VOLZ.

Für den atmosphärischen Dunst ist die Annahme homogener Wasserkugeln sicher nur angenähert zulässig (vgl. Ziff. 4). Je kleiner die Teilchen sind, um so mehr ist das Vorhandensein löslicher oder unlöslicher Kondensationskerne von Einfluß und macht die Berücksichtigung anderer Brechzahlen, vielleicht auch mit Absorption (komplexes m), oder fester Kerne in der Wasser- oder Lösungshülle erforderlich. Für größere m ist der Charakter der Streufunktionen der gleiche, der Abfall von $\varphi = 0$ bis $\pi/2$ wird größer, aber im übrigen sind die Differenzen unregelmäßig. KERKER und ADEN[1] haben die Rückstreuung für schmelzende Eiskörner, die mit einer Wasserhaut überzogen sind, im Mikrowellengebiet ($\alpha = 0{,}1$) untersucht

[1] A. L. ADEN u. M. KERKER: J. Appl. Phys. **22**, 1242 (1951). — M. KERKER, P. LANGLEBEN u. K. L. S. GUNN: J. Meteorology **8**, 424 (1951).

und finden schon bei einer Dicke der Wasserhaut von $^1/_{10}$ des Gesamtradius 91% des Streuquerschnittes einer homogenen Wasserkugel, was wohl auf den großen Unterschied in der Dielektrizitätskonstante für Wasser und Eis zurückzuführen ist. GÜTTLER[1] hat einen schalenförmigen Aufbau der Teilchen, mit absorbierenden Kernen aus Eisen, Quarz oder Kohle und Wasserhülle, untersucht. Er findet (nur für kleine Teilchen im RAYLEIGH-Bereich), daß die Wirkung gleich der einer homogenen Kugel mit dem mittleren Brechungsexponent ist.

Als Arbeitshypothese darf zunächst angenommen werden, daß die Dunstteilchen in der Atmosphäre aus kleinsten Wassertröpfchen bestehen. Dann gelten für sie die Streufunktionen nach Fig. 10. Allerdings sind nicht nur einzelne diskrete Tropfengrößen, sondern ein *kontinuierliches Größenspektrum* zu berücksichtigen, was aus den Untersuchungen zur Wellenlängenabhängigkeit der Extinktion und aus den direkten Zählungen bekannt ist (Ziff. 5). Es ist [vgl. (5.4)]

$$dN = M c r^{-\nu} dr, \tag{13.7}$$

wo dN die Teilchenzahl für $dr = 1$ cm in einer die Atmosphäre vertikal durchsetzenden Säule von 1 cm² Querschnitt, $M = 0{,}43 = \log_{10} e$, c eine Konstante ist. Diese Verteilung sei zwischen den Grenzgrößen r_0 und r_1 gültig; dann wird die komplexe Streufunktion des gesamten Aerosolspektrums[2]

$$f_{D\lambda}(\varphi) = \frac{M c \lambda^2}{4\pi^2} \int_{r_0}^{r_1} i(\alpha, \varphi, m)\, r^{-\nu} dr = M c \lambda^{3-\nu} (2\pi)^{\nu-3} \int_{\alpha_0}^{\alpha_1} i(\alpha, \varphi, m)\, \alpha^{-\nu} d\alpha. \tag{13.8}$$

Für monochromatisches Licht ergibt sich also eine ganz bestimmte Abhängigkeit der Streustrahlung $f_{D\lambda}(\varphi)$ vom Streuwinkel, deren Gestalt nur vom Größenspektrum der Dunstteilchen, hier also vom Exponenten ν abhängt. Die untere Grenze $\alpha_0 = 2\pi r_0/\lambda$ kann unbedenklich $\alpha_0 = 0$ gesetzt werden. Nach oben hin bricht das Gesetz (13.7) etwa bei $r_1 = 10\,\mu$ ab; das bedeutet, daß bis zu einem kleinen Winkel φ_c, der durch die Streufunktion i des Grenzwertes α_1 gegeben ist, das Streulicht konstant und gleich dem Wert für $\varphi = 0$ ist und erst bei größeren φ abfällt. Dieser Grenzwinkel ist beim gleichen Aerosolspektrum für rotes Licht etwas größer als für blaues. Gilt die Verteilung (13.7) bis zu einem größeren (oder kleineren) Wert r_1 und bricht dann ab, so reicht die Zunahme der Streulichtintensität bei Annäherung an die Lichtquelle von außen her noch bis zu kleineren Winkeln φ_c (oder nicht ganz so nahe) an die Lichtquelle heran. Im übrigen unterscheiden sich die Streufunktionen $f_{D\lambda}$ für verschiedene Wellenlängen nur durch einen von φ unabhängigen Faktor $\lambda^{3-\nu}$.

Fig. 11 veranschaulicht nach VOLZ [*5*] die Zusammensetzung einer komplexen Dunststreufunktion (13.8) aus den Streufunktionen der einzelnen Partikelgrößen, die den Dunst ausmachen. Es findet eine ganz charakteristische Überschneidung der einzelnen Kurven statt, so daß nur derjenige Teil der elementaren Streufunktion für die komplexe Streuung wirksam wird, der in der Nähe des „Knickwinkels" liegt. Für die Streuung unter kleineren Winkeln sind die größeren Partikel, trotz ihrer geringeren Zahl, wegen ihrer stärkeren Vorwärtsstreuung wichtiger, für größere Winkel sind die kleineren Tröpfchen wegen ihrer größeren Zahl mehr ausschlaggebend. Sind keine großen Teilchen vorhanden, dann ist die Streulichtabnahme mit zunehmendem φ sehr gering; je größere Teilchen in großer Zahl vorhanden sind, desto stärker wird die eng begrenzte Vorwärtsstreuung.

Die komplexe Streufunktion des Dunstes kann nach (13.8) berechnet werden. Mit $\nu = 4$ (dem Wert, der den Partikelzählungen JUNGEs [*4*] entspricht) findet

[1] A. GÜTTLER: Ann. Physik (6), **11**, 65 (1952).
[2] K. BULLRICH, E. DE BARY u. F. MÖLLER: Geofis. pura e appl. **23**, 69 (1952).

man die Zahlen der Tabelle 13, die eine außerordentlich große Vorwärtsstreuung zeigen, einen Abfall bis fast auf 1/1000 des Anfangswertes in $\varphi = 120°$ und einen schwachen Wiederanstieg bis $\varphi = 180°$. Die Streufunktion der dunsthaltigen Luft ergibt sich dann durch Addition zur Rayleigh-Streuung

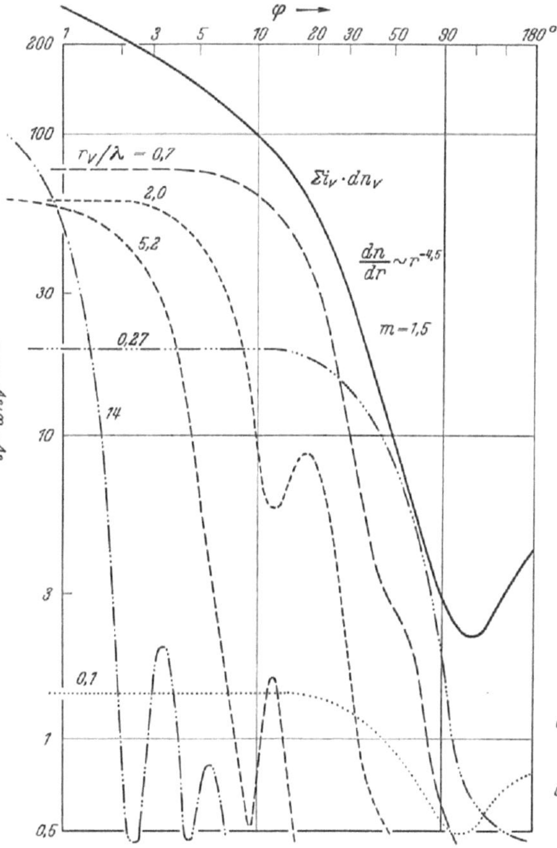

$$f_\lambda(\varphi) = f_{R\lambda}(\varphi) + f_{D\lambda}(\varphi). \quad (13.9)$$

Die Abhängigkeiten der beiden Anteile von der Wellenlänge, die mit λ^{-4} und λ^{-1} gehen, haben eine Veränderung der Streufunktion in der

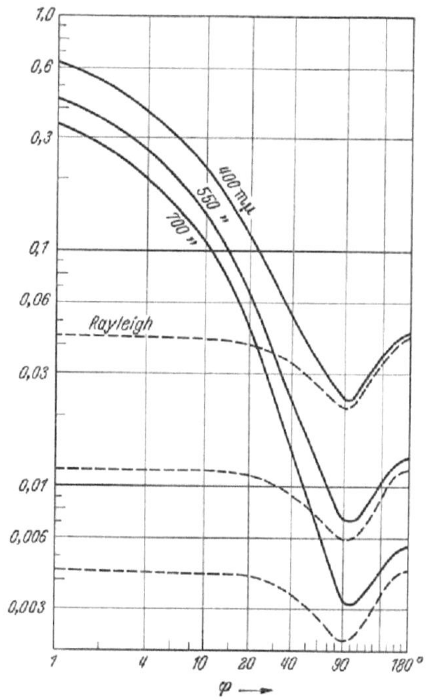

Fig. 11. Komplexe Streufunktion des Dunstes, schematisch zusammengesetzt für Gruppen von Partikeln der Größe r, Anzahl $n(r)$ und Brechungsindex $m = 1,50$. Nach F. Volz [5].

Fig. 12. Komplexe Streufunktion für dunsthaltige Atmosphäre bei verschiedenen Wellenlängen. Nach Bullrich, de Bary, Möller.

Weise zur Folge, daß für $\lambda = 400$, 550, 700 mμ das Verhältnis $f_\lambda(0) : f_\lambda(\pi) = 40$, 92, 157 wird (Fig. 12). Erst in der Nähe von $\varphi = \pi/2$ ist $f_{D\lambda}$ bis zur Größenordnung von $f_{R\lambda}$ herabgesunken. $f_D(\varphi = 0)$ ist jedoch im wesentlichen von der Größe des

Tabelle 13. *Streufunktion des Dunstes nach* (13.8) *für* $\lambda = 550$ mμ, $\nu = 4$, $c = 5 \cdot 10^{-7}\,\mathrm{cm}^{-2}$, $r_1 = 8\,\mu$, $m = 1,33$.

φ	0	2	4	10	20	30	45	60	90	120	150	180°
f_D	0,656	0,353	0,252	0,134	0,055	0,020	0,010	0,0046	0,0014	0,0010	0,0013	0,0015

Grenzradius r_1 abhängig wie oben gefunden wurde, und wächst rasch mit zunehmendem r_1. Die Veränderung der Streufunktion mit dem Exponent ν und dem Brechungsindex m der Tropfensubstanz hat Volz [5], [1e] eingehend studiert. Er findet, daß die Steilheit des Verlaufs der Streufunktion zwischen Sonnennähe

Tabelle 14. *Streufunktion des Dunstes, bei verschiedenen Exponenten ν des Partikelspektrums und verschiedenen Brechzahlen m der Partikelsubstanz.*

$\nu =$	4	4	5	5	5	beobachtet, $\lambda = 0{,}63\,\mu$	
$m =$	1,33	1,50	1,33	1,50	2,0	Typ A	Typ B
$f_D(10°):f_D(90°)$	82	45	31	20	6	9,2	23,5
$f_D(3°):f_D(90°)$	220	120	48	30	7	11,2	90

und 90° mit wachsendem ν und m stark abnimmt. Tabelle 14 stellt berechneten Zahlenwerten zwei typische Beobachtungen gegenüber.

Es ist einleuchtend, daß man diese Verhältnisse direkt am Himmel beim Blick zur Sonnenumgebung sehen kann, und daß die Analyse des Intensitätsabfalles und der Farbverteilung des sonnennahen Himmelslichtes ein wertvolles Hilfsmittel zur Erkennung der Größenverteilung des Aerosols ist[1].

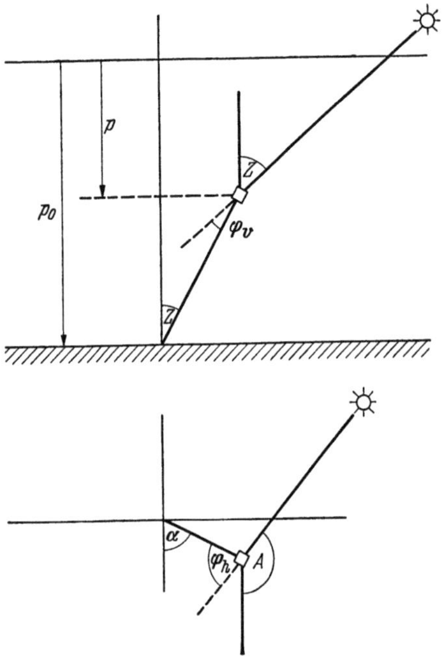

Fig. 13. Strahlengang des Lichtes bei einfacher (primärer) Streuung. Oben Seitenriß, unten Grundriß. φ_v, φ_h = Projektionen des Streuwinkels φ. (Der am Erdboden auftretende Winkel ist z, nicht Z.)

14. Einfache Theorie der Himmelsstrahlung; primäre Streuung. Die nach einmaliger Streuung des Sonnenlichtes von einer Stelle des Himmels ins Auge des Beobachters gelangende Strahlung ergibt sich durch Integration der von einem Volumenelement gestreuten Strahlung über den gesamten Sehstrahl. Es seien z, α und Z, A Zenitdistanz und Azimut des anvisierten Himmelspunktes und der Sonne (Fig. 13), m und M die z und Z entsprechenden relativen Luftmassen und φ der Winkel zwischen Sonnenstrahl und Blickstrahl:

$$\cos\varphi = \cos z \cos Z - \\ - \sin z \sin Z \cos(\alpha - A).$$

Wird weiterhin die (dunsterfüllte) Atmosphäre als homogen angenommen, dann kann der Luftdruck p als vertikale Koordinate verwendet werden. Die von einem sonnenbeleuchteten Massenelement der Atmosphäre ausgehende Streustrahlung ist dann nach [*1c*]:

$$d i_\lambda = J_{0\lambda}\, e^{-a_\lambda M p/p_n} \cdot e^{-a_\lambda m (p_0 - p)/p_n} f_\lambda(\varphi)\, m \frac{dp}{p_0}\, d\omega. \tag{14.1}$$

$d\omega$ ist der elementare Raumwinkel, p_n der Normaldruck, p_0 der Luftdruck am Boden.

Integration von $p=0$ bis $p=p_0$ ergibt

$$i_\lambda = J_{0\lambda} \frac{m}{m-M} \left(e^{-a_\lambda M} - e^{-a_\lambda m}\right) \frac{f_\lambda}{a_\lambda} \frac{p_n}{p_0}\, d\omega, \tag{14.2}$$

wobei für a_λ der Streuungskoeffizient dunsthaltiger Luft

$$a_\lambda = a_{R\lambda} + a_{D\lambda} \tag{14.3}$$

und für $f_\lambda(\varphi)$ die Streufunktion der Luft (13.9) einzusetzen ist. i_λ kann als primäre, d.h. durch eine einmalige Streuung entstandene Streustrahlung bezeichnet werden.

[1] F. Volz: Photogr. u. Wiss. **3**, 3 (1954).

Nur für dunstfreie Luft sind $a_\lambda = a_{R\lambda}$ und

$$f_{R\lambda} = a_{R\lambda} \frac{3}{16\pi} (1 + \cos^2\varphi) \tag{14.4}$$

durch analytische Ausdrücke angebbar. Die Verteilung der spektralen Leuchtdichte $i_{R\lambda}$ über den Himmel wird damit[1]

$$i_{R\lambda} = J_{0\lambda} \frac{m}{m - M} (e^{-a_{R\lambda} M} - e^{-a_{R\lambda} m}) \frac{3}{16\pi} (1 + \cos^2\varphi) \frac{p_n}{p_0} d\omega \tag{14.5}$$

und ist leicht zu berechnen. Fig. 14a zeigt das Ergebnis für $\lambda = 555$ mμ, $Z = 30°$. Man erkennt den dunkelsten Punkt in etwa 90° Abstand im Sonnenvertikal, der

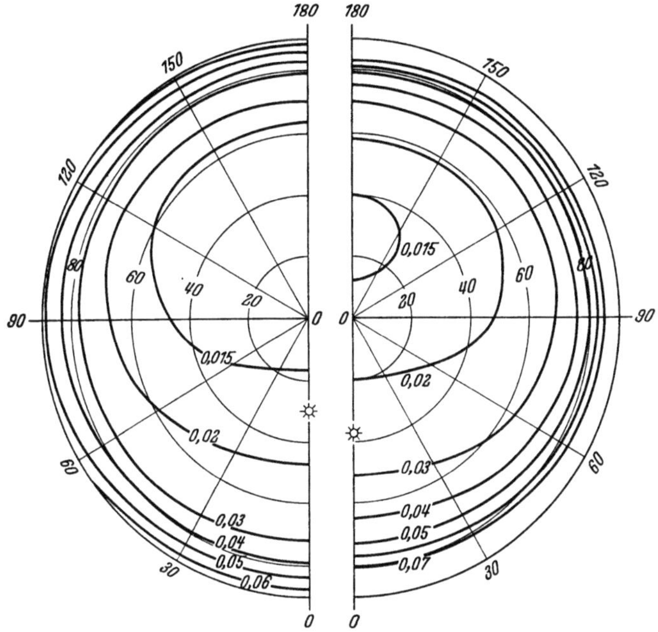

Fig. 14. Spektrale Leuchtdichte des Himmels bei reiner Molekülstreuung, $\lambda = 550$ μ. Links: Zenitdistanz der Sonne $Z = 30°$, primäre Streuung, nach MÖLLER u. a.; rechts: $Z = 37°$, Vielfachstreuung, Erdbodenalbedo $A = 0{,}10$, nach SEKERA.

dem Minimum der Streufunktion (13.1) entspricht. Eine Zone geringer Leuchtdichte zieht sich in 90° Abstand um die Sonne, der Horizont ist aufgehellt. Diese Aufhellung ist eine Folge der mit sec z zunehmenden optischen Masse und ist infolgedessen im Roten, wo die Extinktion innerhalb der strahlenden Masse gering ist, stärker als im Blauen, wo durch die starke Streuung eine gleichförmigere Verteilung über alle Zenitdistanzen z erreicht wird (vgl. Tabelle 15). Sonst ist die Verteilung über den Himmel in den verschiedenen Wellenlängen sehr ähnlich. Aus (14.5) erkennt man, daß in jedem Horizontalkreis m=const das Verhältnis der Leuchtdichte zweier beliebiger Wellenlängen λ_1 und λ_2 unabhängig von φ bzw. vom Azimut α ist, d.h., daß auch die Farbe des Himmelslichtes auf jedem Horizontalkreis unabhängig vom Azimut ist. Es ändert sich nur die Gesamtleuchtdichte.

Reine RAYLEIGH-Atmosphären ohne Dunst kommen nur in großen Höhen zur Beobachtung.

[1] F. LINKE u. P. HESS: Meteor. Z. 59, 313, 364, 400 (1942). — R. TOUSEY u. E. O. HULBURT: J. Opt. Soc. Amer. 37, 78 (1947).

In tieferen Lagen ist nach (14.3) und (13.9) $a_\lambda = a_{R\lambda} + a_{D\lambda}$ und $f_\lambda = f_{R\lambda} + f_{D\lambda}$ zu setzen. Damit sind die Streufunktionen zweier verschiedener Wellenlängen

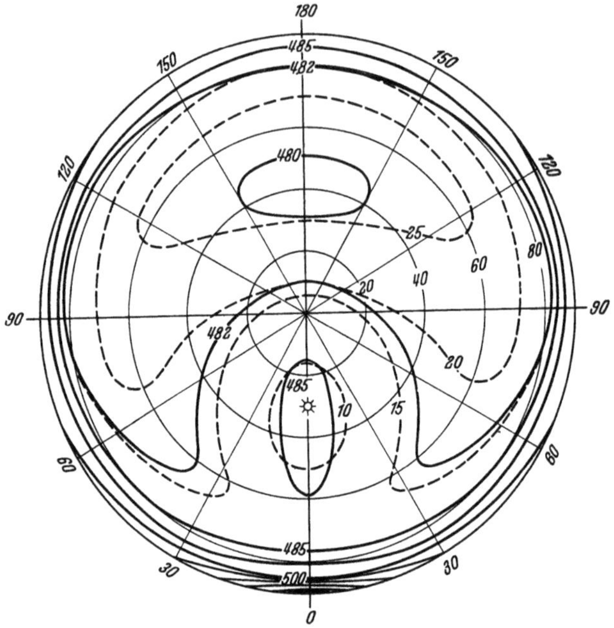

Fig. 15. Leuchtdichte des Himmels bei dunsthaltiger Atmosphäre. Nach Möller u. a.

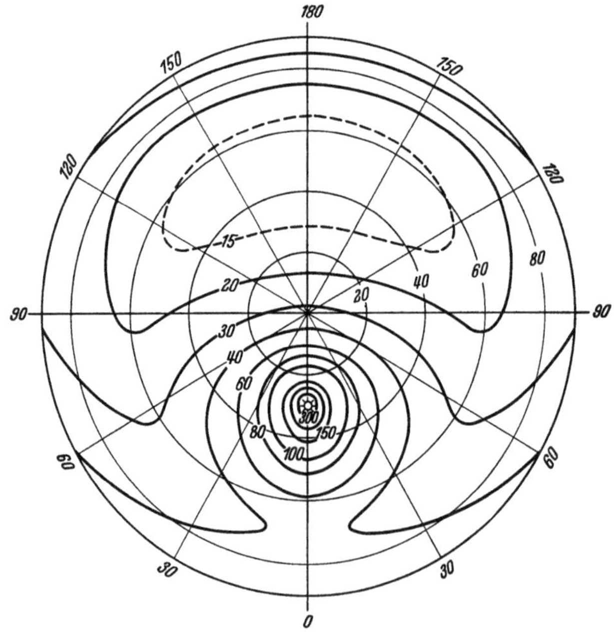

Fig. 16. Farbtongleiche Wellenlänge (mμ) und Farbsättigung (%) des Himmelslichtes bei dunsthaltiger Atmosphäre.

nicht mehr durch einen konstanten Faktor unterschieden, sondern haben verschiedenen Verlauf. Auch sind sie nicht mehr analytisch ausdrückbar. Eine mit der Streufunktion $f_{D\lambda}$ nach Tabelle 13 ermittelte spektrale Leuchtdichteverteilung über dem Himmel zeigt Fig. 15. Auch hier ist natürlich noch der dunkelste Punkt in 90° Sonnenabstand im Sonnenvertikal zu finden. Jedoch nimmt die Leuchtdichte zur Sonne hin stark zu, die nächste Sonnenumgebung ist stark aufgehellt. Die dunkle Zone zieht sich seitlich unter der Sonne hindurch, so daß auch zwischen Sonne und Horizont ein Minimum der Leuchtdichte auftritt.

Bei tiefstehender Sonne ist der dunkelste Punkt ziemlich genau in 90° Sonnenabstand zu finden. Je höher die Sonne steht, desto mehr gerät der 90°-Punkt in die Aufhellung des Horizontes hinein, und die dunkelste Stelle wird daher näher zur Sonne hin gedrängt, so daß sie bei Zenitsonne nur mehr 60 bis 70° Abstand hat. Mit zunehmender Trübung der Luft nimmt die Leuchtdichte an allen der Sonne nahegelegenen Himmelsstellen zu, dagegen in der Nähe des Horizontes ab. Möller und Mitarbeiter[1] haben aus den Spektralleuchtdichten nach der Farbtafel die Farbe des Himmelslichtes, also farbtongleiche Wellenlänge λ_F und Farbsättigung S bestimmt. Über einem weiten Teil des der Sonne gegenüberliegenden Himmels zeigt λ_F nur geringe Änderungen und liegt in der Nähe von 480 mμ, während die Farbsättigung dort mit zunehmendem Aerosolgehalt c [nach (13.7)] abnimmt (in reiner Luft $S \approx 35\%$, bei $c = 8 \cdot 10^{-12}$ Rückgang bis auf etwa 23%). Im übrigen ist die Verteilung der beiden Farbcharakteristiken über den Himmel ähnlich derjenigen der Leuchtdichte: zum Horizont hin geht der Farbton nach längeren Wellen (etwa 490 mμ) und die Sättigung nimmt ab, doch sind die Unterschiede so gering, daß das Auge sie wesentlich weniger empfindet als die Leuchtdichteunterschiede. Man sieht nicht eine „blaueste" Stelle am Himmel, sondern eine „dunkelste". In Sonnennähe und am Horizont unter der Sonne treten weiße ($S = 0\%$) oder sogar rötliche Farbtöne auf, wobei die Sättigung wieder etwas (im Gelbroten) zunimmt (Fig. 16).

15. Verfeinerung der Theorie, Mehrfachstreuung. Das einzelne Massenelement dunsthaltiger Luft wird nicht nur von der Sonne, sondern auch von der diffusen Strahlung des Himmels und der Reflexstrahlung des Erdbodens beleuchtet. Das von der Sonne kommende Licht erfährt also mehrfache Streuung. Wird eine zweimalige Streuung berücksichtigt, so kann die Rechnung noch durch Betrachtung des geometrischen Strahlenverlaufes bewältigt werden, was für den Fall der Rayleigh-Atmosphäre von Sato[2] und Bernhardt[3], für dunsthaltige Luft von Möller und Mitarbeitern[1] durchgeführt worden ist. Letztere konnten zeigen, daß in Sonnennähe die Intensität des sekundären Streulichtes nur wenige Prozent des primären ausmacht, das Verhältnis jedoch mit zunehmendem Winkelabstand des Himmelspunktes von der Sonne, mit zunehmender Trübung und mit abnehmender Wellenlänge zunimmt. In Horizontnähe und unter Azimutabständen von mehr als 90° gegen die Sonne erreicht das sekundäre Streulicht bei starker Trübung im Kurzwelligen (400 mμ) sogar fast die Intensität des primären (bei 700 mμ 70 bis 80% des primären).

Aus diesen Zahlen geht hervor, daß zumindest an sonnenfernen Punkten des Himmels die Mehrfachstreuung nicht vernachlässigt werden kann. Durch Abschätzungen mit Hilfe der Strahlungsstrommethode (Ziff. 20) haben Linke[4] und Tousey und Hulburt[5] für die Mehrfachstreuung 10 bis 30% der Primärstreuung angegeben. Bernhardt hat, wie früher schon Chr. Wiener[6], die

---

[1] K. Bullrich, E. de Bary u. F. Möller: Geofis. pura e appl. **23**, 69 (1952). — Z. Met. **8**, 303 (1954). — F. Möller, E. de Bary u. G. Krog: Geofis. pura e appl. **26**, 141 (1953).

[2] T. Sato: Sci. Rep. Tôhoku Univ. (5) Geophys. **2**, 1 (1950).

[3] F. Bernhardt: Z. Met. **6**, 257 (1952); **7**, 78 (1953).

[4] F. Linke u. P. Hess: Meteor. Z. **59**, 400 (1942). — [*1c*], S. 344.

[5] R. Tousey u. E. O. Hulburt: J. Opt. Soc. Amer. **37**, 78 (1947).

[6] Chr. Wiener: Vgl. Anm. 1, Ziff. 13 (S. 187).

Mehrfachstreuung als Summe einer geometrischen Reihe berechnet, indem er annahm, daß

$$i^{\mathrm{I}} : i^{\mathrm{II}} = i^{\mathrm{II}} : i^{\mathrm{III}} = q$$

eine unveränderliche Größe ist. HAMMAD[1] hat den Einfluß der Mehrfachstreuung auf die Polarisation des Himmelslichtes untersucht und gezeigt, daß das Problem durch Integralgleichungen gelöst werden kann. Eine exakte Lösung ist dann durch die umfassenden Arbeiten von CHANDRASEKHAR[2] ausgearbeitet worden. Sie führt auf zwei Systeme von je zwei nichtlinearen Integralgleichungen für vier Funktionen, die in die Lösung eingehen. Die numerische Durchrechnung für den einfachen Fall des RAYLEIGHschen Streuungsgesetzes haben SEKERA[3] sowie CHANDRASEKHAR und ELBERT[4] gegeben (vgl. den Beitrag SEKERAs in diesem Bande). Zugleich ist dabei eine diffuse Reflexionsfähigkeit des Erdbodens berücksichtigt worden, die für bewachsenen Boden bei 10% liegt, für frischen Schnee bis zu 80% betragen kann. Die Verteilung der Leuchtdichte über den Himmel nach SEKERAs Berechnungen ist in Fig. 14b wiedergegeben.

In dieser Darstellungsweise ist ein Unterschied gegen Fig. 14a kaum erkennbar, nur die Zahlenwerte an den Kurven sind höher. Klarer erkennt man aus Tabelle 15, daß die Berücksichtigung der Vielfachstreuung bei einer Bodenalbedo $A = 0{,}10$ eine Vermehrung der Himmelsleuchtdichte auf etwa das 1,4fache bei 495 mμ, aber auf etwa das 3fache bei 312 mμ bewirken kann. Die Sonnenumgebung fällt in Tabelle 15 und Abb. 14b in keiner Weise auf, wie es auch bei der reinen molekularen Streufunktion, die keine betonte Vorwärtsstreuung zeigt, nicht anders zu erwarten ist; allerdings entspricht dies nicht den in der Natur vorkommenden Bedingungen.

Der Einfluß der Albedo A oder des Reflexionsvermögens der Erdoberfläche ist selbst in diesem Falle der geringen Streufähigkeit der reinen Atmosphäre bemerkenswert groß (Tabelle 15). Bei einer Vermehrung der Albedo durch Schneefall von 10 auf 80% findet eine Zunahme der Himmelsstrahlung bei 495 (315) mμ um weitere 60 (110)% in Horizontnähe, um 85 (80)% in der Gegend des dunkelsten Punktes statt. Die Himmelsstrahlung beträgt dann im Sichtbaren das 2 bis 2,5-fache, im UV das 4- bis 7fache der primären; diese Zahlenwerte sind von VOLZ [5] durch Messungen bestätigt worden. Es ist anzunehmen, daß die Aufhellung an

Tabelle 15. *Relative spektrale Leuchtdichte des Himmels im Sonnenvertikal bei einer Sonnen-Zenitdistanz* $Z = 53°$ ($\cos Z = 0{,}60$).

RAYLEIGH-Streuung. P = nur primäre Streuung; $A = 0{,}1$ bzw. $0{,}8$ bei Berücksichtigung der Vielfachstreuung und Erdboden-Albedo A nach SEKERA. $J_{0\lambda}\, d\omega = 1$ gesetzt.

$Z =$		84	78,5	66,5 ⊙	45,5	0	26	45,5	66,5	78,5	84
		Sonnenseite			Gegenseite						
$a_{R\lambda} = 0{,}15$, $\lambda = 495$ mμ	P	0,0690	0,0498	0,0319	0,0200	0,0100	0,0084	0,0104	0,0204	0,0395	0,0612
	$A = 0{,}1$	0,0942	0,0675	0,0426	0,0264	0,0134	0,0118	0,0150	0,0290	0,0550	0,0853
	$A = 0{,}8$	0,148	0,104	0,0636	0,0395	0,0229	0,0220	0,0280	0,0496	0,0910	0,139
$a_{R\lambda} = 1{,}0$, $\lambda = 312$ mμ	P	0,0235	0,0296	0,0372	0,0361	0,0219	0,0174	0,0186	0,0237	0,0235	0,0208
	$A = 0{,}1$	0,0824	0,0961	0,1064	0,0942	0,0592	0,0535	0,0608	0,0795	0,0834	0,0767
	$A = 0{,}8$	0,1655	0,1738	0,1705	0,1440	0,0993	0,0961	0,1106	0,1440	0,1600	0,1630

[1] A. HAMMAD: Phil. Mag. **36**, 434 (1945); **38**, 515 (1947). — Astrophys. J. **108**, 338 (1948).

[2] S. CHANDRASEKHAR: Radiative Transfer, Oxford 1950.

[3] Z. SEKERA u. G. BLANCH: Sci. Rep. No. 3, Contract AF 19 (122)—239. Los Angeles 1952. — D. DEIRMENDJIAN u. Z. SEKERA: Tellus **6**, 382 (1954). — Nature, Lond. **175**, 459 (1955).

[4] S. CHANDRASEKHAR u. D. D. ELBERT: Trans. Amer. Phil. Soc. **44**, 643 (1954).

anderen Himmelspunkten, vor allem aber bei trüber Atmosphäre mit entsprechend gesteigertem Rückstreuvermögen noch stärker ist, jedoch liegen Berechnungen mit Dunststreuung bisher nur für die sekundäre — nicht für die Vielfachstreuung — vor. Eine ganz ähnliche Wirkung wie eine Schneedecke wird eine dicht unter Gipfelhöhe liegende geschlossene Wolkendecke in den Beobachtungen der Bergstationen hervorrufen[1].

Tabelle 16. *Himmelsstrahlung für* $Z = 0$, $a_{R\lambda} = 0{,}6$, $J_{0\lambda} d\omega = 1$, *nach Berechnungen* SEKERAS(S) *und* BERNHARDTS (B).

	$z = 0$	30	60	80	90°
S	0,064	0,065	0,070	0,079	0,085
B	0,054	0,057	0,073	0,098	0,099

Von Interesse ist auch die Frage, inwieweit die WIENER-BERNHARDTsche Approximation (Sekundärstreuung + Berücksichtigung höherer Streuungen durch eine geometrische Reihe) der exakten Lösung nahe kommt. Es wurden für Zenitstand der Sonne BERNHARDTS und SEKERAS Zahlen bei 355 mμ, also einem hohen Streuungskoeffizienten, verglichen. Tabelle 16 zeigt, daß in Sonnen- bzw. Zenitnähe BERNHARDTS Zahlen um etwa 15% kleiner, in Horizontnähe dagegen um 17% größer sind. Die geometrische Reihe ist also *keine* befriedigende Näherung der Vielfachstreuung.

16. Spektrale Messungen der Himmelsleuchtdichte. Messungen der spektralen Leuchtdichte Verteilung über den ganzen Himmel liegen nur wenige vor. HESS[2] hat an einzelnen Punkten des Sonnenvertikals das Spektrum der Himmelsstrahlung gemessen. Fig. 17 zeigt, wie stark der Schwerpunkt gegenüber der Sonnenstrahlung zum Violett hin verschoben ist. Relativ die größte Blauintensität zeigt der $\varphi = 90°$-Punkt, während das Spektrum in 10° Abstand von der Sonne und noch mehr das des horizontnahen Punktes am Gegenhimmel dem Sonnenspektrum wenigstens bei $\lambda < 500$ mμ ähnlicher ist.

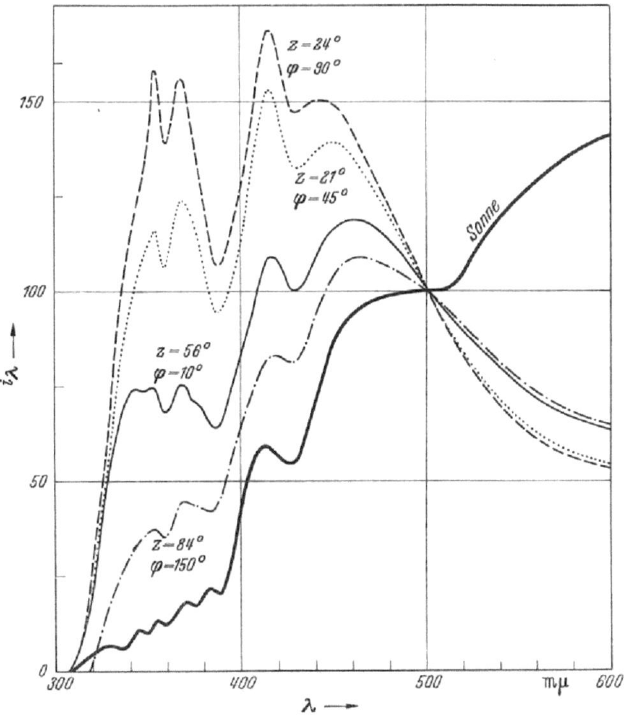

Fig. 17. Spektrale Energieverteilung an verschiedenen Himmelspunkten und für die Sonne, für $\lambda = 500$ mμ gleich 100 gesetzt. Nach P. HESS.

GÖTZ und SCHÖNMANN[3] haben gleichartige Messungen im Zenit durchgeführt und auch durch quantitative Berechnungen die Extinktion gedeutet. In einem Fall konnte sie zeigen, daß außer den Molekülen eine Gesamtzahl von $2{,}7 \cdot 10^6$ cm^{-2} Teilchen mit $r = 0{,}29\,\mu$ nötig war, um die beobachtete Intensitätsverteilung zu deuten.

[1] H. LIPP: Dtsch. met. Jb. Bayern **1928**.
[2] P. HESS: Gerlands Beitr. Geophys. **55**, 204 (1939).
[3] F. W. P. GÖTZ u. E. SCHÖNMANN: Helv. phys. Acta **21**, 151 (1948).

Meist ist nur die gesamte aktinometrisch wirksame Strahlung gemessen worden (vgl. Tabelle 17). Jedoch können Messungen der visuellen Leuchtdichte wegen des verhältnismäßig engen Spektralbereiches der Augenempfindlichkeit als spektrale Messungen für 550 mμ angesehen werden[1]. DORNO[2] hat für die reine Davoser Luft die in Fig. 18 wiedergegebene Isoplethenverteilung gemessen. Die Unterschiede gegen Fig. 15 fallen nicht sehr ins Auge. PACKER und LOCK[3] haben die Leuchtdichteverteilung am Himmel vom Flugzeug aus in Höhen von 5 bis 11 km über den Trockengebieten Arizonas und Südkaliforniens gemessen. In Winkelabständen von mehr als 40° von der Sonne läßt sie sich gut durch die von TOUSEY und HULBURT gegebene Theorie (primäre RAYLEIGH-Streuung und azimutunabhängige Mehrfachstreuung, nach der Strahlungsstrommethode berechnet) deuten, allerdings unter Verwendung eines visuellen Extinktionskoeffizienten $a_v = 0{,}0193$ km^{-1} (RAYLEIGH $a_{R\,v} = 0{,}0126$ km^{-1}) bzw. eines visuellen Trübungsfaktors $T_v = 1{,}5$. Bei Winkelabständen unter 40° müssen sie jedoch gegenüber der genannten Theorie Zuschläge anbringen, weil auch in diesen Höhen noch die Vorwärtsstreuung beträchtliche Werte besitzt.

Tabelle 17. *Himmelsstrahlung im Sonnenvertikal.*

a) Kalorisch wirksame Strahlung ($\lambda < 3\mu$) in 10^{-3} cal cm^{-2} min^{-1} bei $Z = 50$ bis 70°, extrapoliert auf Trübungsfaktor $T = 1$, nach KNEPPLE[4].

b) UV-Strahlung, gemessen mit Cd-Zellen, relative Werte, bezogen auf Zenit = 1. Nach ECKEL[5].

z	Sonnenseite		Gegenseite	
	a)	b)	a)	b)
0°	8,3	1,0	8,3	1,0
10	10,8		6,7	
20	14,4	1,19	6,4	0,92
30			6,7	
35		1,42		0,98
40			7,2	
50		1,88	9,1	1,15
60			12,8	
70		2,22		1,42
80				1,45
90		1,56		1,15

Die theoretisch für reine RAYLEIGH-Streuung gewonnenen Werte der Tabelle 15 zeigten, daß die Zunahme der Himmelsleuchtdichte vom Zenit zum Horizont im Langwelligen sehr viel stärker ist als im Kurzwelligen, wo sogar eine Wiederabnahme mit Horizontannäherung festzustellen ist. Messungen von DORNO und ECKEL[5] (vgl. Tabelle 17) mit Cd-Zellen bestätigen diese Erscheinung, die auf die geringe Reichweite der stark gestreuten UV-Strahlung zurückzuführen ist.

In vielen Fällen ist nur die Leuchtdichte im Zenit visuell oder spektral gemessen und mit den Veränderungen der Wetterlage oder des Trübungsfaktors in Zusammenhang gebracht worden[6]. Jedoch ist es kaum möglich aus diesen Messungen allein Schlüsse auf das atmosphärische Aerosol zu ziehen, da nicht nur der Extinktionskoeffizient, sondern auch die Streufunktion einen mit wechselnder Sonnenhöhe sehr wesentlichen Einfluß auf die Zenitstrahlung hat.

Für eine Meßreihe von SIEDENTOPF[7], die bei annähernd gleicher Zenitdistanz der Sonne in verschiedenen Höhen über dem Boden gewonnen worden ist, gilt dieser Einwand nicht. Tabelle 18 zeigt seine Mittelwerte der Änderung der Zenitleuchtdichte bei sommerlichen Hochdrucklagen. Zur Kennzeichnung der Farbänderung ist als Beispiel eine Meßserie des Verhältnisses der Intensität bei 400 mμ

[1] Zum Beispiel R. A. RICHARDSON u. E. O. HULBURT: J. Geophys. Res. **54**, 229 (1949).
[2] C. DORNO: Abh. preuß. met. Inst. **4** (1925).
[3] D. M. PACKER u. C. LOCK: J. Opt. Soc. Amer. **41**, 473 (1951).
[4] R. KNEPPLE: Gerlands Beitr. Geophys. **39**, 378 (1933).
[5] C. DORNO u. F. LINDHOLM: Meteor. Z. **46**, 287 (1929). — O. ECKEL: Meteor. Z. **51**, 180 280 (1934); **53**, 90 (1936).
[6] F. LINKE: Meteor. Z. **61**, 257 (1944). — H. LOSSNITZER: Arch. Met. Geophys. u. Bioklim., Ser. B **6**, 91 (1954).
[7] H. SIEDENTOPF: Zentr. wiss. Berichtswesen, Forschungsber. No. 1937/1 (1944).

zu der bei 600 mμ beigefügt, aus der das starke Überwiegen des Kurzwelligen gegenüber dem Langwelligen mit zunehmender Höhe deutlich wird. Ähnliche Messungen haben HARDING und LAMBERT[1] in 6 und 12 km durchgeführt, außerdem auch an 6 weiteren Punkten des Himmels gemessen.

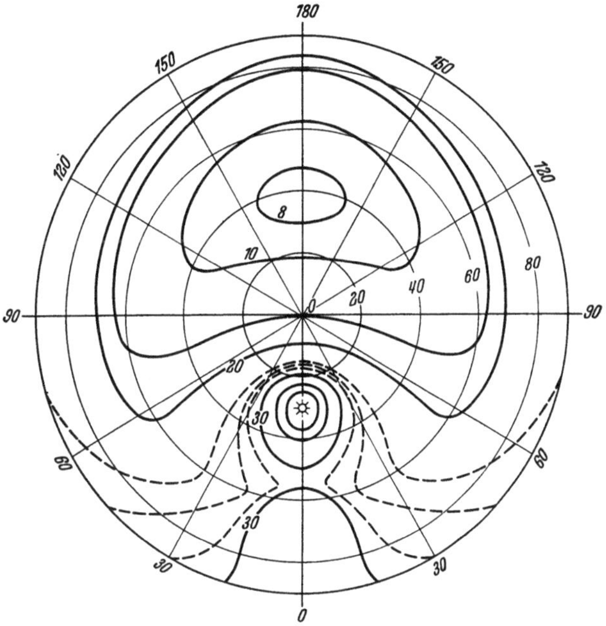

Fig. 18. Leuchtdichte des Himmels nach Messungen von C. DORNO in Davos.

In noch größeren Höhen sind von MILEY u. a. Messungen mittels Raketen durchgeführt worden[2]. Entsprechend der Abnahme des Luftdruckes geht auch das RAYLEIGH-Streulicht immer mehr zurück und nimmt bis 32 km exponentiell auf 0,14 des Wertes in 10 km für 559 mμ, auf 0,03 für 426 mμ ab. In dieser Höhe etwa wird das Streulicht vernachlässigbar gegenüber dem Eigenleuchten der Atmosphäre (*air-glow*), das am Boden nur bei Nacht als Nachthimmelslicht bekannt ist. Die Leuchtdichte bleibt dann bis zur nächsten Beobachtungshöhe (135 km) konstant. — Neuere Messungen von BERG[2] konnten so große Leuchtdichten nicht nachweisen, sondern lieferten zwischen 85 und 220 km Höhe Werte, die sicher kleiner als $^1/_{1000}$ der genannten sind und damit durch RAYLEIGH-Streulicht allein erklärt werden können.

Tabelle 18. *Änderung der Zenitleuchtdichte B_z mit der Höhe und Änderung des relativen Farbverhältnisses $B_{z,\lambda=400} : B_{z,\lambda=600}$ bei sommerlichen Hochdrucklagen.*

Höhe NN	0	0,5	1	2	3	4	6	8	10 km	
B_z	9,8	8,2	6,9	4,6	3,4	2,6	1,9	1,4	1,0	10^3 asb
$\frac{B_{z,\lambda=400}}{B_{z,\lambda=600}}$	—	1,0	—	1,3	—	1,6	1,9	3,2	—	

Theoretische Berechnungen der Zenitleuchtdichte für sehr tief oder im Horizont stehende Sonne ergeben merkwürdigerweise bei reiner Luft eine grünliche Färbung und nicht das bekannte Himmelsblau. Erst wenn man die Absorption der sichtbaren Sonnenstrahlung in der Chappuis-Bande des Ozons berücksichtigt,

[1] H. G. W. HARDING u. G. E. V. LAMBERT: Nature, Lond. **167**, 436 (1951).

[2] H. A. MILEY, E. H. CULLINGTON u. J. F. BEDINGER: Trans. Amer. Geophys. Union **34**, 680 (1953). — O. E. BERG: J. geophys. Res. **60**, 271 (1955).

die bei großen Luftmassen (für $Z=90°$ gilt für die Ozonschicht in 20 km Höhe etwa $m'=12$) bis 50% beträgt, wird die blaue Himmelsfarbe erklärlich[1]. Man kann also sagen, daß unter 10° Sonnenhöhe die Ozonabsorption an der blauen Himmelsfarbe beteiligt, unterhalb 5° dafür allein maßgebend ist, während sie bei größeren Sonnenhöhen unwesentlich ist. Die alte Ansicht von der „Farbe der Luft" kommt hier gegenüber der RAYLEIGHschen Theorie des Himmelsblau wenigstens zum Teil wieder zur Geltung. (Über den weiteren Verlauf der Farbe des Himmels s. Ziff. 22.)

17. Das Streulicht in Sonnenumgebung. Besonders wertvoll für Bestimmungen des atmosphärischen Aerosols sind Messungen des Himmelslichtes nahe an der Sonne. In Ziff. 14 ist gezeigt, daß für die Streulichtintensität in einem bestimmten Abstand von der Sonne nur ein enger Größenbereich von streuenden Partikeln maßgebend ist. In großem Abstand wird das Streulicht nur von den kleinsten Partikeln (Luftmolekülen) hervorgerufen, aber, besonders bei kurzen Wellen, auch durch die Mehrfachstreuung verstärkt; je kleiner der Winkel zwischen Sonne und Blickrichtung ist, um so größere Teilchen bewirken die Aufhellung des Himmels. Beobachtungen von i_λ am Himmel erlauben jedoch nicht unmittelbar Schlüsse auf die Streufunktion (13.9) zu ziehen, sondern i_λ enthält nach (14.2) einen von der Zenitdistanz der Sonne Z und des Blickpunktes z abhängigen Faktor. Für den Fall $z=Z$ bzw. $m=M$ wird dieser

$$\lim_{m\to M}(m-M)^{-1}(e^{-a_\lambda M}-e^{-a_\lambda m})=a_\lambda e^{-a_\lambda M}. \tag{17.1}$$

Da die Intensität der Sonnenstrahlung selbst $J_\lambda=J_{0\lambda}e^{-a_\lambda M}$ ist, fällt bei Beobachtung des Quotienten

$$(i_\lambda/J_\lambda)_{m=M}=M f_\lambda(\varphi)\,d\omega \tag{17.2}$$

der störende Luftmasseneinfluß heraus (M selbst ist bekannt). Messungen der Himmelsleuchtdichte relativ zur Sonnenleuchtdichte im Horizontalkreis durch die Sonne geben also unverfälscht die Streufunktion der Atmosphäre wieder [*5*], wenn man von der Vielfachstreuung absieht.

Beobachtungen im Sonnenhorizontal haben noch einen weiteren Vorteil. Die *Höhenverteilung* der gröberen Partikel gehorcht nicht dem gleichen Gesetz wie die der Luftmoleküle. Die Halbwertshöhe der Luftdichte ist etwa 6,5 km, die des Dunstes liegt im Mittel bei 1 bis 1,5 km, ist aber sehr großen Wetterlagen-Schwankungen unterworfen. In (14.2) ist angenommen worden, daß Luftmoleküle und Dunstteilchen dem gleichen Höhengesetz unterliegen. Man kann aber zeigen, daß eine geänderte Höhenverteilung des Dunstes die Leuchtdichte im Sonnenhorizontalkreis unbeeinflußt läßt, während bei einer Konzentration des Dunstes in tieferen Schichten der Atmosphäre die Himmelspunkte unterhalb des Sonnenhorizontals langwelligere Farbtöne und geringere Leuchtdichte, oberhalb davon eine kurzwelligere Farbe und größere Leuchtdichte annehmen[2].

Die Möglichkeit der Messung im Sonnenhorizontal ist von VOLZ zu eingehenden *Aerosoluntersuchungen* ausgenutzt worden. Da nach (13.9) die Streufunktion der dunsthaltigen Luft $f_\lambda(\varphi)=f_{D\lambda}(\varphi)+f_{R\lambda}(\varphi)$ aus Dunst- und Molekülstreulicht zusammengesetzt ist, könnte man versuchen, letzteres in ähnlicher Weise wie beim Extinktionskoeffizienten (Ziff. 4) rechnerisch von der Beobachtung gemäß (17.2) in Abzug zu bringen. Wegen der schwer abschätzbaren Störung durch

[1] E. O. HULBURT: J. Opt. Soc. Amer. **43**, 113 (1953). — F. MÖLLER u. a.: Z. Met. **8**, 303 (1954).

[2] [*5*] u. E. DE BARY, K. BULLRICH, F. MÖLLER: Z. Met. **8**, 303 (1954).

die Vielfachstreuung ist das bei großen Winkelabständen schwer durchzuführen, bei kleinen Winkeln und vor allem bei großen Wellenlängen ist aber der Einfluß der Dunststreuung so überwiegend[1] (vgl. Fig. 12), daß $f_{R\lambda}$ nicht berücksichtigt zu werden braucht. Nur bei sehr klarer Luft sind störende Einflüsse von Meßfehlern zu erwarten, die nur bei großer Trübung gering werden. Weiter ist bei Durchführung von Messungen in Sonnennähe sorgfältig auf die Vermeidung jedes störenden Streulichtes durch das Instrument zu achten. Die Meßanordnung muß also so gestaltet werden, daß die Eintrittsöffnung der Optik vor direktem Sonnenlicht abgeschirmt ist[2] [*5*].

Eine ähnliche Methode, die Größenverteilung der Teilchen eines Aerosols (oder Hydrosols) aus der Winkelabhängigkeit der Intensität des nach vorn gestreuten Lichtes bei sehr kleinen Winkeln abzuleiten, ist von SLIEPCEVICH u. a.[3] entwickelt und auch experimentell erprobt worden.

Für reine Molekülstreustrahlung ist nach (14.5) in jedem Horizontalkreis am Himmel, also auch im Sonnenhorizontal, das Verhältnis zweier Wellenlängen $i_{\lambda 1}:i_{\lambda 2}$ unabhängig vom Streuwinkel φ (gemessen im Großkreis) bzw. vom Azimut α. Wie aus (13.8) hervorging, gilt für den Dunstanteil der Streufunktion $f_{D\lambda}$, daß das Verhältnis zwischen zwei Wellenlängen, der Farbwert $F=f_{D\lambda 1}:f_{D\lambda 2}$ unabhängig vom Streuwinkel ist, sofern die Größenverteilung der Partikel einer Potenzfunktion gemäß (13.7) gehorcht. Dabei ist es gleichgültig, ob $\nu=4$ ist oder einen davon abweichenden Wert besitzt. Der Exponent ν ist in diesem Falle durch die Abhängigkeit der unter sich ähnlichen Streufunktionen von der Wellenlänge, $\lambda^{3-\nu}$, feststellbar. Hierin liegt ein wertvolles Mittel, das Vorhandensein von Potenzgesetzen nachzuprüfen. Zeigt der Farbwert Veränderungen mit dem Winkelabstand φ, dann liegen Abweichungen vom reinen Potenzverlauf des Aerosolspektrums vor, denn nur in dem Fall, daß ein Aerosolspektrum gemäß (13.7) vorliegt und daß dessen Gültigkeitsgrenzen $r_0=0$ und $r_1=\infty$ gesetzt werden dürfen, kann λ vor das Integral in (13.8) vorgezogen werden; bei beliebigem Größenspektrum ist keine Ausklammerung des λ vor das Integral möglich.

Bei seiner ersten Untersuchung [*5*] konnte VOLZ nach Streufunktion und Farbwert drei Typen der Streulichtverteilung unterscheiden (Fig. 19):

A. Gleichmäßig von etwa 1° Sonnenabstand bis 90° nach einer Potenzfunktion $f_{D\lambda}\sim\varphi^{-a}$ abnehmende Streufunktionswerte; die Intensität des Himmelslichtes nimmt von außen gegen die Sonne hin stetig zu. Der Exponent ist im roten Spektralbereich etwa $a=1$, im blauen ist er wesentlich kleiner. Auch der Farbwert, etwa als Quotient zwischen Rot und Grün oder zwischen Grün und Blau nimmt stetig von der Sonne nach außen hin ab.

B. Der Farbwert bleibt über einen großen Winkelbereich, mindestens von $\varphi\approx 5$ bis 50° konstant (nimmt jedoch oberhalb dieses Bereiches wegen Molekül- und Vielfachstreuung mit φ ab). Die Streufunktionen selbst nehmen unterhalb eines gewissen Grenzwinkels von etwa 5° einen fast konstanten Verlauf an, was bedeutet, daß die Sonne von einer gleichmäßig hellen, meist weißlichen Scheibe umgeben erscheint.

C. Die Streufunktion zeigt bei $\varphi=10$ bis 30° einen Buckel und der Farbwert Rot zu Grün nach einer Abnahme eine schwache Wiederzunahme, die Himmelsfarbe bekommt einen Anflug ins Bräunliche. Diese Erscheinung ist nach starken vulkanischen Ascheausbrüchen sehr deutlich beobachtet worden und wird

---

[1] F. MÖLLER: Met. Rdsch. **1**, 210 (1948).

[2] S. GÜNTHER: Optik **5**, 240 (1949).

[3] J. H. CHIN, C. M. SLIEPCEVICH u. M. TRIBUS: J. Phys. Chem. **59**, 841, 845 (1955).

Bishop-Ring genannt. Schwächer ausgebildet ist sie öfters zu sehen. Ein Vergleich von Fig. 19, Kurve a (Typ C) mit Fig. 11 zeigt sofort, daß es sich hier um ein besonders häufiges Vorkommen einer bestimmten Partikelgröße handeln muß.

Diese typischen Streufunktionen sind mit bestimmten Größenspektren der Dunstteilchen verbunden: bei B liegt eine ziemlich reine Potenzverteilung vor, bei A besteht ein Defizit von Teilchen um $r = 0{,}3\ \mu$, bei C ein Überschuß von etwas kleineren Teilchen. Die Krümmung der Partikelverteilungskurven im

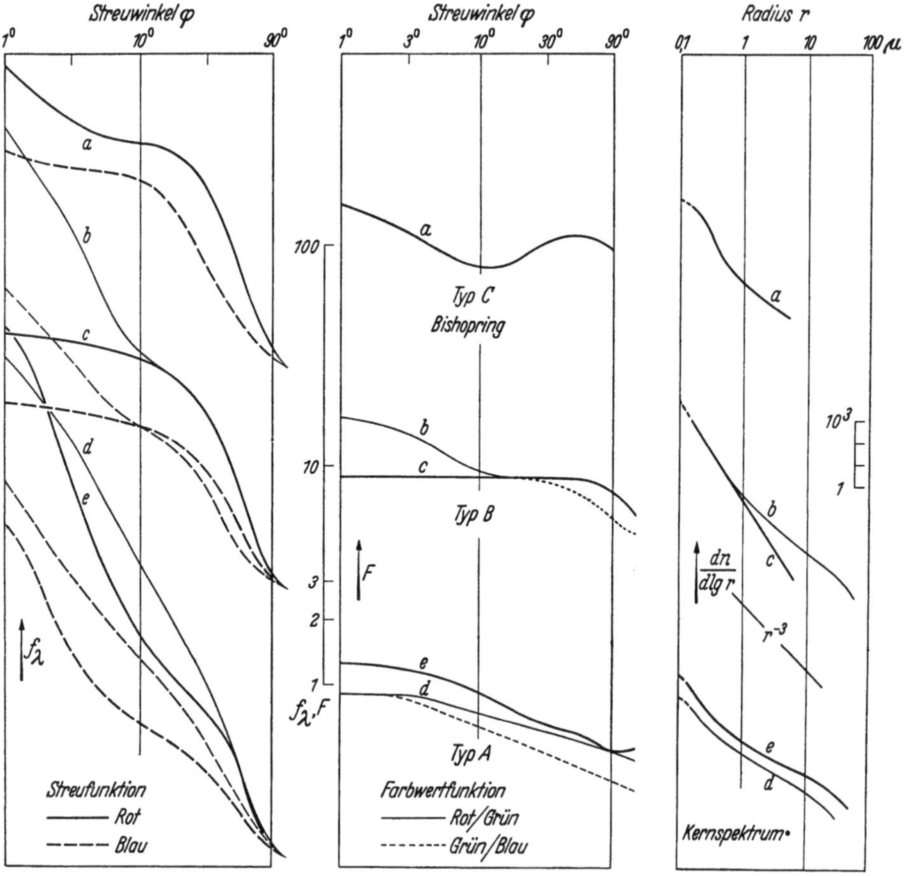

Fig. 19. Streulichttypen nach F. Volz [5]; links: Streufunktion $f_\lambda(\varphi)$, Mitte: Farbwert in Abhängigkeit vom Streuwinkel φ, rechts: Teilchenspektrum $dn(r)/d\lg r$. Typen A, B, C von unten nach oben.

doppeltlogarithmischen Diagramm, d.h. die Abweichung vom Potenzgesetz ist maßgebend für die Farbenverteilung im circumsolaren Licht. Es ist besonders wertvoll, daß diese Farbenverteilung mit bloßem oder nur durch eine dunkle Brille geschütztem Auge (ohne Spektralapparat) gesehen werden kann und damit schon eine erste Orientierung über das Aerosol gestattet. Bei A nimmt die Helligkeit und Weißverhüllung der Farbe stetig zur Sonne hin zu, so daß keine Farbabstufungen zu erkennen sind (Kaltlufthimmel). Bei B ist eine kräftige weiße Scheibe um die Sonne mit schroffem Übergang zum Himmelsblau in 30 bis 40° und bei C ein schwach bräunlicher Ring in 30° Abstand zu sehen. Natürlich sind Zwischenstufen und Übergangsformen vor allem zwischen A und B häufig. Eine über drei Jahre erstreckte Beobachtungsserie lieferte folgende

Zahlen von Beobachtungen der verschiedenen Typen je Jahr:

Typ	A	Mischtyp AB	B	C	Gesamtzahl je Jahr
Anzahl	34	61	50	6	151

Fig. 11 zeigt, wie man eine einzelne monochromatisch gemessene Streufunktion durch eine bestimmte Größenverteilung der streuenden Teilchen deuten kann. Nach Kenntnis der Verteilung kann man daraus die Streufunktion in anderen Wellenlängen eindeutig ableiten. Wenn man so verfährt, ergeben sich in den meisten Fällen Widersprüche zur Beobachtung. Eine Deutung gelingt aber wenn man nicht Wassertröpfchen, sondern Partikeln mit großen Brechungsindices voraussetzt. VOLZ [*5*] wurde auf diese Weise zur Annahme von Brechzahlen $n \approx 1{,}5$ geführt. Das würde bedeuten, daß die Teilchen sehr stark konzentrierte Salzlösungen oder Salzkristalle sind.

Der Einfluß der Mehrfachstreuung und der Reflexion vom Erdboden (Schneedecke) kann nach Berechnung der Sekundärstreuung (Ziff. 15) abgeschätzt werden und danach zwischen einem Faktor 1,2 für 0,4 μ und 3,0 für 0,7 μ liegen. Für RAYLEIGH-Streuung ist er aus der Tabelle 15 zu entnehmen.

Streufunktionen in dunsthaltiger Luft sind auch direkt an Scheinwerferstrahlen[1,2] und durch sorgfältige Auswertungen von Sichtbeobachtungen[3] bestimmt worden. Allerdings gelten solche Messungen nur für das *bodennahe Aerosol*, während die Himmelslichtuntersuchungen einen Mittelwert für die ganze Atmosphäre liefern. Scheinwerfermessungen zeigen meist Streufunktionen, die zwischen 10 und 90° Abstand einen bedeutend stärkeren Abfall zeigen als die des Himmelslichtes. Es läßt sich sogar eine Zunahme der Steilheit dieses Verlaufes mit dem Extinktionskoeffizienten klar feststellen[2], jedoch scheint die Deutung dieser Kurven schwierig, zumal nicht anzunehmen ist, daß sich das bodennahe Industrie-Aerosol ebenfalls mit dielektrischen Streuteilchen erklären läßt, ohne die Absorption der Partikelsubstanz und die Abweichung von der Kugelform zu berücksichtigen [*1e*].

18. Die diffuse Himmelsstrahlung bei wolkenlosem Himmel. Die aus der Halbkugel auf eine horizontale Flächeneinheit auffallende Strahlung aller Wellenlängen hat für den Energiehaushalt der Erde eine wesentlich größere Bedeutung als die Strahlung einzelner Himmelspunkte. Sie wird deshalb oft gemessen, da auch in den verschiedenen *Pyranometern* einfache Meßgeräte dafür zur Verfügung stehen, wenn man Sorge trägt, daß die direkte Sonnenstrahlung abgeschirmt wird. Dies wird entweder durch einen in 50 cm Abstand in parallaktischer Aufhängung der Sonne nachgeführten Schirm bewirkt oder durch einen ringförmigen Streifen, der die ganze Zone, die von der Sonne im Laufe des Tages durchwandert wird, ständig abschirmt. Der Fehler, der durch diese Abschirmung entsteht, wird von W. SCHÜEPP[4] auf 8 bis 10% bei klarem Himmel, auf 5% bei bedecktem Himmel angegeben.

Theoretisch ergibt sich die diffuse Himmelsstrahlung D_λ aus $i_\lambda(z, \alpha)$ durch Integration über die einzelnen Himmelszonen unter Berücksichtigung des Einfallswinkels z

$$D_\lambda = \int_0^{\pi/2} \sin z \cos z \, dz \int_0^{2\pi} i_\lambda(z, \alpha)\, d\alpha. \tag{18.1}$$

Für i_λ sind bei Berücksichtigung der primären Streuung die Werte nach (14.2) einzusetzen, besser die mit Hilfe der Mehrfachstreuung berechneten Zahlen (vgl.

---

[1] K. BULLRICH: Meteor. Z. **61**, 270 (1944). — Optik **2**, 301 (1947). — E. REEGER u. H. SIEDENTOPF: Optik **1**, 15 (1946). — H. SIEDENTOPF: Z. Met. **1**, 342 (1947).

[2] L. FOITZIK u. H. ZSCHAECK: Z. Met. **7**, 1 (1953).

[3] L. FOITZIK: Abh. met. Dienst D.D.R. **8** (1951).

[4] W. SCHÜEPP: Bull. Serv. Mét. Congo Belge **1952**, No. 8, 11.

S. 196). D_λ ist abhängig von der Zenitdistanz der Sonne Z und nimmt mit der Trübung der Atmosphäre zu. Es ist deshalb naheliegend, die gemessenen Werte der Himmelsstrahlung auf die Rayleigh-Atmosphäre zu beziehen[1] in der gleichen Weise, wie das für die direkte Sonnenstrahlung mit Hilfe des Trübungsfaktors oder Trübungskoeffizienten geschieht.

Nach Sekeras Berechnungen[2] (Ziff. 15) erhält man für die reine Molekülatmosphäre unter Berücksichtigung der Vielfachstreuung innerhalb der Atmosphäre die in Tabelle 19 angegebenen, auf die extraterrestrische Strahlung als Einheit bezogenen Werte. Die kurzwellige Strahlung (kleine Transmissionsfaktoren q_λ) wird viel stärker gestreut als die langwellige, und ist deshalb nicht nur prozentual wesentlich stärker als diese, sondern es liegt auch der optische Schwerpunkt der totalen Himmelsstrahlung im Blau. Für $\lambda > 1\,\mu$ kann die diffuse Himmelsstrahlung bei Rayleigh-Streuung ganz vernachlässigt werden.

Tabelle 19. *Diffuse Himmelsstrahlung $D_{R\lambda}$, bezogen auf die extraterrestrische Sonnenstrahlung $J_{0\lambda} = 100$ bei reiner Molekülstreuung; Z = Zenitdistanz der Sonne. Nach* Deirmendjian *und* Sekera[2].

λ	$a_{R\lambda}$	$Z = 0$	53,1	84,3	88,8°
312 mμ	1,0	29,2	21,2	3,0	0,5
495 mμ	0,15	6,9	6,6	3,7	0,9
618 mμ	0,06	2,9	2,8	2,2	0,9
gesamt, $A = 0$		4,10	3,66	1,71	0,56
gesamt, $A = 0,25$		5,64	4,52	1,84	0,58
gesamt, $A = 0,80$		9,69	6,80	2,08	0,62

Eine wesentlich einfachere Methode (Strahlungsströme, vgl. S. 195 und S. 210) hat H. Hinzpeter[3] zur Berechnung der diffusen Strahlung D angewendet und findet für vier verschiedene Zenitdistanzen der Sonne eine recht gute Übereinstimmung mit Deirmendjian und Sekera.

Im Integral über alle Wellenlängen zeigt die Himmelsstrahlung nur geringe Veränderung mit der Sonnenhöhe (Tabelle 20), so daß das Verhältnis zu der auf die horizontale Fläche auffallenden Sonnenstrahlung $S = J \cdot \cos Z$ mit abnehmender Sonnenhöhe stark zunimmt. Die symmetrische Form der molekularen Streufunktion läßt erwarten, daß die zur Erdoberfläche gelangende Himmelsstrahlung D etwa gleich der Hälfte des Strahlungsverlustes ΔS der direkten Sonnenstrahlung ist. Bernhardt hat dies an den Zahlen seiner Rechnung nachgeprüft und findet es für die drei Wellenlängen 0,356, 0,484 und 0,646 μ und für Zenitdistanzen bis 75° recht befriedigend erfüllt. Bei größeren Zenitdistanzen

Tabelle 20. *Diffuse Himmelsstrahlung D, Sonnenstrahlung S und Globalstrahlung $G = S + D$ auf die horizontale Fläche, bei reiner Molekülstreuung (R) und getrübter Atmosphäre (Trübungsfaktor T), nach* Deirmendjian *und* Sekera[2] *und* Nicolet *und* Dogniaux[4].

	$Z = 0$	40	60	70	75	80	85	88,8°
D_R	0,0816	0,0771	0,0695	0,0617	0,0554	0,0457	0,0318	0,0112 cal cm⁻² min⁻¹
S_R	1,824	1,365	0,845	0,535	0,387	0,240	0,104	0,0145 cal cm⁻² min⁻¹
$G_R = S_R + D_R$	1,906	1,442	0,914	0,597	0,442	0,286	0,136	0,0257 cal cm⁻² min⁻¹
$(D:S)_R$	0,045	0,056	0,082	0,115	0,143	0,190	0,306	0,78 cal cm⁻² min⁻¹
$(D:S)_{T=3}$	0,057	0,102	0,149	0,246	0,360			
$(D:S)_{T=6}$	0,175	0,234	0,295	0,418	0,570			

[1] W. Collmann: Ann. Met. **6**, 117 (1954).
[2] D. Deirmendjian u. Z. Sekera: Tellus **6**, 382 (1955). — Nature, Lond. **175**, 459 (1955).
[3] H. Hinzpeter: Z. Met. **9**, 308 (1955); **10**, 100 (1956).
[4] M. Nicolet u. R. Dogniaux: Inst. Roy. Mét. Belg., Mém. **47** (1951).

gelangt nach SEKERA ein sehr viel geringerer Teil der Sonnenstrahlung in Form von Himmelsstrahlung wieder zum Boden (bei $Z=0$; 53; 85° jeweils 49; 29; 4,5% der Gesamtstrahlung), obwohl keinerlei Absorption vorhanden ist. Natürlich kann dies nicht mehr für die unsymmetrische Streufunktion dunsthaltiger Luft mit ihrer weit überwiegenden Vorwärtsstreuung gelten; bei kleinen Werten des Wellenlängenexponenten $\delta=0{,}5$ beobachtet SCHÜEPP in Léopoldville $D>0{,}5\,\Delta S$ auch bei hochstehender Sonne, bei $\delta=1{,}7$ bis 2,0 dagegen $D=0{,}5\,\Delta S$.

Aus dem gleichen Grunde ist auch das Verhältnis zwischen D und S in dunsthaltiger Luft (vgl. Tabelle 20, die letzten beiden Zeilen) um ein Vielfaches größer als in reiner Luft. Bei starker Trübung und tiefstehender Sonne wird dann die Himmelsstrahlung sogar größer als die Sonnenstrahlung auf die horizontale Fläche. Bestimmt man das Verhältnis $D:S$ aus Tagessummen der Einstrahlung[1], so findet man für Stationen im Flachland einen Wert 0,30, in 1800 m Höhe in den Alpen 0,16 und in 3000 m Höhe 0,10. Die entsprechenden Zahlen für den Anteil der diffusen Himmelsstrahlung $D:(S+D)$ sind 0,23; 0,14; 0,09. Beide Zahlenreihen spiegeln deutlich die zunehmende Klarheit der Luft in der Höhe und das Dunklerwerden des Himmels und damit das Zurücktreten der Himmelsstrahlung gegenüber der Sonnenstrahlung S oder Globalstrahlung $S+D$ wieder. Fig. 20 zeigt nach REITZ[2] das Verhältnis der beobachteten Himmelsstrahlung zu der bei reiner Molekülstreuung, in Abhängigkeit vom Trübungsfaktor.

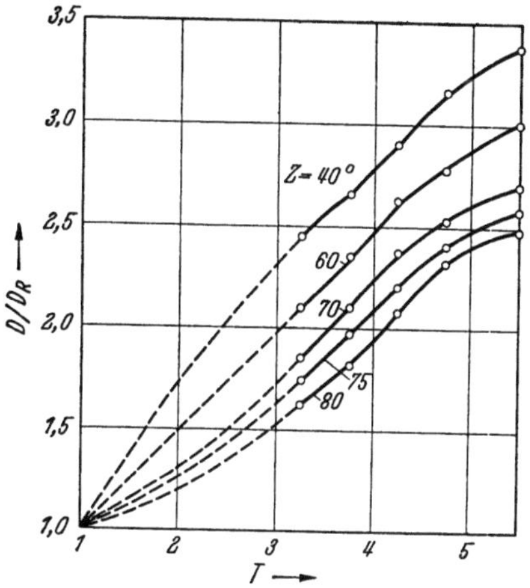

Fig. 20. Verhältnis des diffusen Himmelslichtes D zu demjenigen in dunstfreier Atmosphäre D_R, in Abhängigkeit von Trübungsfaktor T und Zenitdistanz der Sonne Z.

Verhältnismäßig zahlreiche Untersuchungen einzelner Spektralbereiche liegen nur für den visuellen und den ultravioletten Bereich vor. Das auf die horizontale Fläche einfallende sichtbare Licht wird als *Beleuchtungsstärke* bezeichnet und in lux angegeben. Messungen dieser Art wurden bereits von LEONHARD WEBER[3] mit seinem Milchglasphotometer durchgeführt und später an vielen Orten wiederholt [*1c*]. Die Daten schwanken unübersichtlich stark, zumal sie nur selten mit anders bestimmbaren Trübungsgrößen in Beziehung gesetzt sind. Einer von DORNO[4] für Davos gegebenen Zusammenstellung ist Tabelle 21, Zeile 1 bis 3 entnommen. Vergleicht man mit Tabelle 20 so erkennt man, daß das Verhältnis Himmelsstrahlung:Sonnenstrahlung, bei nicht zu starker Trübung oder Zenitdistanz, im Sichtbaren günstiger ist als für die Gesamtstrahlung.

Die laufende Messung und Überwachung der Ultraviolettstrahlung begegnet großen Schwierigkeiten, weil die Intensität der Strahlung von 400 nach 300 mμ sehr steil bis auf 0 zurückgeht. Ein jedes Filter oder eine UV-empfindliche Photozelle, z.B. eine Cd-Zelle, wirkt nie als ideales Filter, sondern besitzt eine vom Maximum nach beiden Seiten stark abfallende Empfindlichkeit. Verlagert sich nun mit sinkender Sonne und deshalb zunehmender Extinktion durch Streuung und O_3-Absorption die Grenze der Strahlung nach längeren Wellen, so kommt

[1] A. SCHEDLER: Arch. Met., Geophys. u. Bioklim., Ser. B **4**, 14 (1953).
[2] G. REITZ: Gerlands Beitr. Geophys. **55**, 253 (1939).
[3] L. WEBER: Meteor. Z. **2**, 163, 219, 451 (1885).
[4] C. DORNO: Abh. preuß. met. Inst. **6** (1919).

Tabelle 21. *Diffuse Himmelsstrahlung D und Sonnenstrahlung S auf die horizontale Fläche für Licht (v) und Ultraviolett (UVA und UVB), nach* DORNO[1], LEISTNER[2] *und* ECKEL[3].

Nr.		$Z=30$	40	42	45	50	60	70	80°
1	D_v	15,0	11,0	—	—	9,2	8,2	7,1	$6{,}0\cdot 10^3$ lux
2	S_v	126,0	109,3	—	—	89,3	63,6	36,6	$12{,}7\cdot 10^3$ lux
3	$(D:S)_v$	0,12	0,10	—	—	0,10	0,13	0,19	0,47
4	$(D:S)_{UVA}$	—	—	1,12	1,17	1,24	1,9	4,0	—
5	$(D:S)_{UVB}$	1,15	1,3	—	1,42	1,6	2,4	6,5	65

Strahlung nur noch in den wenig empfindlichen Wellenbereich des Empfängers und gibt daher zu geringe Meßwerte. Das gilt besonders für den Bereich des UV B, für den Meßwerte bei tiefstehender Sonne recht unsicher werden. (Man teilt ein: UV A 400 bis 320 mμ, UV B 320 bis 280 mμ, UV C kleiner 280 mμ. UV C kommt im natürlichen Licht nicht vor. UV B ist wegen seiner *biologischen Wirksamkeit* besonders wichtig.)

Die Zahlen von Tabelle 21, Zeile 4 für die Nordseeküste[2], Zeile 5 für die Kanzelhöhe (1500 m)[3] zeigen deutlich, daß die Himmelsstrahlung die von der Sonne kommende Strahlung im UV um so mehr überwiegt, je kürzer die Wellenlänge ist, eine verständliche Folge der mit λ^{-4} gehenden Streuung einerseits und Extinktion des Sonnenlichtes andererseits.

19. Die Globalstrahlung bei wolkenlosem Himmel. Die Summe von Sonnenstrahlung auf die horizontale Fläche + Himmelsstrahlung wird Globalstrahlung genannt:

$$G_\lambda = S_\lambda + D_\lambda \quad \text{mit } S_\lambda = J_\lambda \cdot \cos Z. \tag{19.1}$$

Ihr Integral über alle Wellenlängen ist diejenige Größe, die von *Pyranometern* ohne besondere Abschirmung angezeigt wird. Deshalb liegen zahlreiche Meßreihen vor[4]. Zur Beurteilung der Globalstrahlung wäre ein Vergleich mit der extraterrestrischen Sonnenstrahlung möglich, wie das von FUKUI[5] vorgeschlagen wurde, jedoch ist der Vergleich mit der über λ integrierten Globalstrahlung G_R der reinen Molekülatmosphäre zweckmäßiger. COLLMANN[6] hat für G_R Zahlenwerte in Abhängigkeit von der Luftmasse M und Tagessummen für jeden Tag des Jahres (Hamburg) angegeben, wobei auch der im Laufe des Jahres wechselnde Abstand der Sonne berücksichtigt worden ist. DEIRMENDJIAN und SEKERA haben ihre theoretischen Berechnungen mit Berücksichtigung der Vielfachstreuung in der RAYLEIGH-Atmosphäre und des Reflexionsvermögens der Erdoberfläche auch auf die Globalstrahlung ausgedehnt und konnten zeigen, daß die Rechnung sich sogar gegenüber der für einzelne Himmelspunkte vereinfacht. Ihre Zahlenwerte sind in Zeile 3 von Tabelle 20 wiedergegeben.

Die Ergebnisse von Meßwerten der Globalstrahlung sind von ALBRECHT[7] in einer einfachen Interpolationsformel zusammengefaßt worden, die für Strahlungshaushalts-Berech-

[1] Siehe Fußnote 4, S. 205.

[2] W. LEISTNER: Ber. dtsch. Wetterd. US-Zone **42**, 20 (1952).

[3] O. ECKEL: Wien. Sitzgsber. IIa **141**, 188 (1932). — Weitere Angaben bei K. BÜTTNER: Physikalische Bioklimatologie, S. 56. Leipzig 1938.

[4] J. GOLDSCHMIDT: Archiv. Met., Geophys. u. Bioklim., Ser. B **3**, 244 (1951). — H. HINZPETER: Veröff. met. hydr. Dienst. DDR **10** (1953). — I. DIRMHIRN: Arch. Met., Geophys. u. Bioklim., Ser. B **2**, 301 (1951). — F. PROHASKA: Gerlands Beitr. Geophys. **59**, 247 (1943). — [*1c*], S. 428. — LANDOLT-BÖRNSTEIN: Zahlenwerte und Funktionen, 6. Aufl., Bd. III, S. 689 u. 700. Berlin 1952.

[5] E. FUKUI: Geophys. Mag. Tokyo **23**, 429 (1952).

[6] W. COLLMANN: Ann. Met. **6**, 117 (1954).

[7] F. ALBRECHT: Meteor. Z. **52**, 454 (1935). — Forsch. u. Erf. Ber. Reichsamt Wetterd. Berlin A 18 (1943). — Arch. Met., Geophys. u. Bioklim., Ser. B **3**, 220 (1951).

nungen gut brauchbar ist:

$$G = (J_0 - J_w)\cos Z\,(1 - c\sqrt{M})\,(1 + 1{,}19\,A\,c\sqrt{p/1000})\,. \tag{19.2}$$

Hierbei ist J_0 die Solarkonstante, J_w der durch Wasserdampf in der Atmosphäre absorbierte Anteil der Sonnenstrahlung, Z und M die Zenitdistanz der Sonne und entsprechende absolute Luftmasse, p der Luftdruck am Boden in mb, A die Albedo des Erdbodens und c eine von ALBRECHT definierte „Rückstrahlungskonstante" der Atmosphäre, die von der Trübung abhängt und mit Werten zwischen 0,07 und 0,19 gefunden wurde.

COLLMANN[1] hat gemessene Tagessummen von G mit seinen Idealwerten G_R verglichen und findet als absolute Maximalwerte von G/G_R für Hamburg 83%, für Lissabon 97%. Im August wird in Hamburg der Wert 75% nicht überschritten, während im gleichen Monat in Lissabon 70% aller Werte über 80% liegen, ein Effekt von Bewölkungsgrad und Trübung.

Messungen der spektralen Globalstrahlung sind seltener durchgeführt worden. Aus kürzeren Meßreihen schlossen ALBRECHT[2] und später HERMANN[3], daß die *spektrale Verteilung der Globalstrahlung unabhängig von der Sonnenhöhe* sei. Das direkte Sonnenspektrum verschiebt seinen optischen Schwerpunkt mit zunehmender Luftmasse mehr und mehr nach langen Wellen infolge der starken Extinktion des kurzwelligen Endes. Die diffuse Himmelsstrahlung hat ihren Schwerpunkt im Blau und wird bei tiefstehender Sonne nur wenig zu längeren Wellen verschoben. Die Summe von beiden ergibt eine der extraterrestrischen Strahlung ähnliche Verteilung, die ihre Zusammensetzung vor allem deshalb nur wenig ändert, weil die stark rot werdende Sonnenstrahlung bei tief stehender Sonne mit dem $\cos Z$ nur schwach in die Globalstrahlung eingeht. SCHULZE[4] hat in einer längeren Meßreihe diese Frage geprüft und fand bei Sonnenhöhen über 20° zwar noch Veränderungen des Spektrums der Globalstrahlung mit der Sonnenhöhe, jedoch sind sie kleiner als 10%. Die Farbänderung ist jedenfalls wesentlich geringer als die der Sonnenstrahlung allein. Für die Erklärung ist der flache Einfallswinkel der Strahlung bei tiefstehender Sonne auf die horizontale Fläche wesentlich. Daraus folgt, daß das Gesetz nur für die Globalstrahlung Gültigkeit haben kann, und daß die Zusammensetzung der Strahlung oder die Beleuchtung einer senkrechten Wand durch Himmel und tiefstehende Sonne keineswegs dieselbe Farbe hat wie bei hochstehender Sonne, sondern wesentlich zum Rot verschoben ist[5], eine Tatsache, die den ganzen Reiz und die Stimmung abendlicher Beleuchtungen ausmacht.

Für die *visuelle Globalbeleuchtungsstärke* G_v bei wolkenlosem Himmel liegen Meßreihen von SIEDENTOPF und REEGER und BULLRICH[6] vor (Tabelle 22). Aus den Berechnungen der spektralen Globalstrahlung (mit Vielfachstreuung, nach

Tabelle 22. *Globalbeleuchtungsstärke G_v bei wolkenlosem Himmel nach* SIEDENTOPF *und* REEGER *(A) und* BULLRICH *(B).*

A	Z G_v	0 130,3	20 121,5	40 96,0	50 78,0	60 57,7	70 35,5	80 14,5	85° 6,5 · 10³ lux
B	Z G_v	85 5,0	86 3,6	87 2,5	88 1,6	89 0,98	90° 0,55 · 10³ lux		

[1] W. COLLMANN: Briefl. Mitteil. 1956.
[2] Siehe Fußnote 7, S. 206.
[3] H. HERMANN: Optik **3**, 394 (1947).
[4] R. SCHULZE: Ber. dtsch. Wetterd. US-Zone **12**, 210 (1950). — Ann. Met. **4**, 176 (1951).
[5] F. MÖLLER: Arch. Met., Geophys. u. Bioklim., Ser. B **5**, 1, 456 (1953).
[6] H. SIEDENTOPF u. E. REEGER: Meteor. Z. **61**, 114 (1944). — K. BULLRICH: Ber. dtsch. Wetterd. US-Zone **4** (1948).

Rayleigh, Albedo der Erdoberfläche $A=0$) von Deirmendjian und Sekera findet man, durch Multiplikation mit der spektralen Augenempfindlichkeit, für die Zenitdistanzen der Sonne $Z=0°$; 53°; 84,3° die Werte $G_v = 131$; 76,1; 9,5 × 10^3 lux, die beinahe zu gut in Siedentopfs Beobachtungsreihe (Dunsthimmel!) passen.

Weitere umfangreiche Meßreihen sind z.B. in Wien gewonnen worden, über die Steinhauser[1] berichtet. Im Winter kommen besonders hohe Werte, im Sommer besonders niedere Werte (positive bzw. negative Abweichungen vom Mittelwert) relativ selten vor, was eine Folge der relativen Seltenheit heiteren Himmels bzw. sehr dichter Bewölkung ist. Unter Gewitter- oder Schauerwolken sehr großer Mächtigkeit kann die Beleuchtungsstärke kurzzeitig auf $\frac{1}{100}$ oder sogar $\frac{1}{1000}$ des Wertes für heiteren Himmel zurückgehen, also auf den hundertsten Teil der normalen Helligkeit unter geschlossenen dichten Wolken[2].

Elvegard und Sjöstedt[3] haben eine bequeme Interpolationsformel für die Beleuchtungsstärke bei wolkenlosem Himmel gegeben

$$G_v = 123 \cos Z \cdot 10^{-0.1M} + 16{,}3 \sqrt{\cos Z}, \tag{19.3}$$

wobei Z und M wieder Zenitdistanz der Sonne und entsprechende Luftmasse sind. Das erste Glied der Formel gibt die von der Sonne, das zweite die vom Himmel herrührende Beleuchtung wieder. Die Formel ist ähnlich derjenigen Albrechts für die Globalstrahlung (19.2), wenn man $M = \cos^{-1} Z$ setzt.

Im ultravioletten Spektralgebiet (2900 bis 3150 Å) hat Coblentz langjährige Messungen mit Hilfe von Zirkonzellen durchgeführt und Mittelwerte der Tagessummen veröffentlicht[4].

Ein sehr handliches Meßgerät für die UVB-Strahlung ist das *UV-Dosimeter* der IG-Farbenindustrie. Es beruht auf der Rotverfärbung des Fuchsin-Leukosulfits unter dem Einfluß des UV, die gemessen wird. Die Empfindlichkeitskurve dieses Prozesses hat ein Maximum bei 291 mμ und ist damit der Empfindlichkeit der Erythembildung, d.h. der Rotverfärbung der menschlichen Haut bei UV-Bestrahlung (Sonnenbrand) sehr ähnlich. Die Messung liefert Relativwerte, die für biologische Untersuchungen wertvoll sind, jedoch keine sehr hohe Genauigkeit haben[5].

20. Strahlungsausbreitung in Wolken. Die am Boden gemessene kalorische Globalstrahlung bei vollkommen geschlossener Himmelsbedeckung beträgt im Durchschnitt nur $\eta = 0{,}21$ bis 0,25 derjenigen bei wolkenlosem Himmel[6]. Die Strahlungsdurchlässigkeit der einzelnen Wolkenarten ist jedoch sehr verschieden, sie wurde für Nebel zu 0,14, für tiefhängende dunkle Regenwolken zu 0,19, für stratus 0,25, sc 0,33, mittelhohe Wolken 0,40 bis 0,50 und Cirren 0,80 gefunden[7]. Ein auffallend starker Jahresgang des Monatsdurchschnitts von η, der sich aber durch den Anteil der Cirren an der Gesamtbewölkung erklären läßt, ist in den Tropen gefunden worden[8]. Entsprechend sind auch in hohen Breiten, wo der Wassergehalt der Wolken nur gering ist, höhere Werte (Helsinki 0,27, Sodankylä 0,35) und im Polargebiet $\eta = 0{,}54$ gemessen worden[9], ebenso auf Bergstationen (Sonnblick, 3000 m, im Mittel 0,45). Für diese hohen Zahlen sind

[1] F. Steinhauser: Arch. Met., Geophys. u. Bioklim., Ser. B **7**, 60 (1955).

[2] N. C. Helliwell u. M. J. Blackwell: Met. Mag., Lond. **84**, 342 (1955).

[3] E. Elvegard u. G. Sjöstedt: Gerlands Beitr. Geophys. **60**, 416 (1943).

[4] W. W. Coblentz: Bull. Amer. Met. Soc. **33**, 158 (1952).

[5] L. Schulz: Bioklim. Beibl. **6**, 137 (1939). — Beobachtungsergebnisse in der Zusammenstellung in [*1c*], S. 429; auch C. A. Pfeiffer: Bioklim. Beibl. **10**, 129 (1943). — E. Sydow: Met. Rdsch. **1**, 333 (1948). — J. Bleibaum: Met. Rdsch. **4**, 144 (1951).

[6] J. Dirmhirn: Arch. Met., Geophys. u. Bioklim., Ser. B **2**, 301 (1951).

[7] B. Haurwitz: J. Meteorology **2**, 154 (1945); **3**, 123 (1946); **5**, 110 (1948). — W. R. G. Atkins: Quart. J. Roy. Met. Soc. **77**, 659 (1951).

[8] F. H. W. Albrecht: Geof. pura e appl. **32**, 131 (1955).

[9] H. Mosby: Norw. North Polar Exp. Maud 1918—25, Sci. Res. **1**, Nr. 7 (1932). — A. Ångström u. O. Tryselius: Geogr. Annaler **16**, 53 (1934). — J. Dirmhirn: Wetter u. Leben **7**, 298 (1955). — [*1c*], S. 409.

jedoch nicht nur die geringere Wolkendichte und -mächtigkeit verantwortlich, sondern auch die Reflexwirkung einer geschlossenen Schneedecke im Verein mit der Vielfachreflexion zwischen Schneedecke (Albedo A) und Wolkendecke (Reflexionsvermögen γ), die mit einem Faktor $(1 - A\gamma)^{-1}$ zur Geltung kommt[1].

Es beruht ebenfalls auf der Wirkung der starken Vorwärtsstreuung an Wolkentropfen oder Eiskristallen in dünnen Wolken oder an Wolkenrändern, daß die diffuse Himmelsstrahlung D mit zunehmendem Bedeckungsgrad zunimmt, bei $\frac{8}{10}$ Bedeckung ein Maximum erreicht und dann bis $\frac{10}{10}$ rasch zurückgeht bis fast auf den gleichen Wert wie bei wolkenlosem Himmel. Dementsprechend nimmt die Globalstrahlung nicht linear mit dem Bewölkungsgrad N ab, wie ursprünglich von ÅNGSTRÖM angegeben wurde, sondern es sind Zusatzglieder zweiten oder dritten Grades in N erforderlich[2]. In größeren Höhen jedoch, wo die Wolkenmächtigkeiten meist geringer sind als man sie im Flachland antrifft, nimmt die Himmelsstrahlung bis $\frac{10}{10}$ zu, in 3000 m Höhe bis zum 7fachen des Wertes bei wolkenlosem Himmel.

Die *Verteilung der Strahlungsdichte über den Himmel* bei geschlossener Wolkendecke ist sehr gleichmäßig. E. J. MEYER[3] hat dies an der Leuchtdichte untersucht und gezeigt, daß diese bis in Horizontnähe ($z = 75°$) auf etwa die Hälfte der Zenitleuchtdichte abnimmt, unabhängig von der Sonnenhöhe; HOPKINSON[3] fand bei den gleichen Bedingungen ein Gesetz $B = B_0(1 + 2\cos z)/3$ gut erfüllt, und PETHERBRIDGE[3] über schneebedecktem Boden eine etwas schwächere Abnahme gemäß $B = B_0\,(1 + \cos z)/2$. Im Gegensatz dazu nimmt bei wolkenlosem Himmel die Strahlung von den dunklen Partien des Himmels zum Horizont hin wieder zu. Ein Vergleich mit der Leuchtdichte des Zenits ist in diesem Falle wertlos, weil dieser selbst mit zunehmender Sonnenhöhe in den Bereich des zirkumsolaren Streulichtes gerät und nicht als Bezugswert benutzt werden kann. Die entsprechenden Verhältnisse bei Dämmerung und bei Nacht sind von BULLRICH[3] eingehend untersucht worden. Er hat umfangreiche Tabellen der Globalbeleuchtungsstärke und der Leuchtdichteverteilung am Himmel in Abhängigkeit von der Sonnendepression, der Mondhöhe und -phase und der Bewölkung angegeben.

Der visuelle Extinktionskoeffizient a_v in Wolken ist sowohl auf Bergstationen wie vom Flugzeug aus gemessen worden. Neben einer direkten Bestimmung der Lichtschwächung wird hierfür gern die Methode der Messung der *Sichtweite* V angewendet, die nach KOSCHMIEDERs Formel eine Umrechnung auf den Extinktionskoeffizienten a_v für Licht gestattet (vgl. Beitrag MIDDLETON in diesem Band):

$$V \cdot a_v = 3{,}9. \tag{20.1}$$

Für die im Vergleich zur Wellenlänge sehr großen Wolkentropfen und den kleinen Sehwinkel[5] ist der Streuquerschnitt $\sigma \approx 2{,}0$ (vgl. Ziff. 4), daher $a_v = 2\pi N r^2$, wo N die Anzahl der Tropfen mit dem mittleren Radius r in der Volumeneinheit ist. Der Gehalt W flüssigen Wassers in der Volumeneinheit Wolkenluft ist $W = \frac{4}{3}\pi N r^3$. Daraus folgt $a_v = 3W/2r$ oder $V = 2{,}6\,r/W$. TRABERT[6] hat diese Formel zuerst angegeben, jedoch die Zahlenkonstante empirisch zu 6 ermittelt.

[1] W. PEPPLER: Z. wiss. Bäderkde. **4**, 170 (1929). — A. ÅNGSTRÖM: Gerlands Beitr. Geophys. **34**, 123 (1931).
[2] A. ÅNGSTRÖM: Medd. Statens Met. Hydr. Anstalt **4**, Nr. 3 (1928). — F. PROHASKA: Gerlands Beitr. Geophys. **59**, 247 (1943). — H. SCHIELDRUP-PAULSEN: Årbok Univ. Bergen 1948. — H. MATZKE: Abh. met. hydr. Dienst DDR **16** (1953). — I. DIRMHIRN: Sitzgsber. Österr. Akad. Wiss. **164**, 143 (1955).
[3] E. J. MEYER: Meteor. Z. **58**, 95 (1941). — R. G. HOPKINSON: J. Opt. Soc. Amer. **44**, 455 (1954). — P. PETHERBRIDGE: Quart. J. Roy. Met. Soc. **81**, 476 (1955).
[4] K. BULLRICH: Ber. dtsch. Wetterd. US-Zone **4** (1948).
[5] Für Meßinstrumente mit Öffnungswinkeln von einigen Grad wäre $\sigma \approx 1{,}0$ zu setzen.
[6] W. TRABERT: Meteor. Z. **18**, 518 (1901).

Der Unterschied rührt im wesentlichen daher, daß in der Formel genauer geschrieben werden muß

$$a_v = 1{,}5\, W\, \overline{r^2}/\overline{r^3} \tag{20.2}$$

und

$$V = 2{,}6\, W^{-1}\, \overline{r^3}\, (\overline{r^2})^{-1}, \tag{20.3}$$

und daß der Quotient zwischen dem kubischen und dem quadratischen Mittel nur bei sehr enger Tropfenverteilung gleich dem linearen Mittel der Tropfenradien ist[1]. In kräftig quellenden Cumulo-Nimbus-Wolken ist die Größenverteilung sehr breit (in den Grenzen $r = 2\,\mu$ bis $100\,\mu$) und stark unsymmetrisch, so daß $\overline{r^3}/\overline{r^2}$ leicht das doppelte oder gar vierfache von $\bar{r}$ betragen kann.

Für die großen Wolkentropfen ist σ unabhängig von $\alpha = 2\pi r/\lambda$ und dadurch bei gegebener Tropfenverteilung unabhängig von λ. Die visuelle Extinktion wird daher in Nebel und Wolken neutral (grau). Dies ist an Hand von Sichtmessungen besonders durch FOITZIK[2] wiederholt bestätigt worden; der Übergang von der wellenlängenabhängigen Dunstextinktion $a_{D\lambda} \sim \lambda^{-1}$ zur grauen Extinktion oder zu einem sehr kleinen positiven Exponenten $a_{\lambda,\,\mathrm{Nebel}} \sim \lambda^{0{,}1}$ erfolgt in seinen Darstellungen ziemlich schroff bei 800 m Sichtweite. LÖHLE und VOLZ [*1e*] fanden jedoch einen stetigen Übergang, was auch den Überlegungen JUNGEs über die Umbildung des Tropfenspektrums entspricht ([*5*], S. 14).

Die Ausbreitung des Lichtes in Wolken erfolgt nur auf ganz kurze Entfernungen geradlinig. Die nach allen Seiten hin erfolgende Streuung des Lichtes an den Wolkenelementen ($r = 5$ bis $20\,\mu$) ist ungleich stärker als die an Dunstteilchen ($r = 0{,}2$ bis $1\,\mu$), so daß nicht nur das auf eine Wolke auffallende diffuse Himmelslicht, sondern auch das eindringende parallele Sonnenlicht sehr rasch durch Streuung und vielfache Reflexionen vollständig diffus wird. Ein Beobachter im Flugzeug empfängt in der Wolke von allen Seiten gleichmäßige Strahlung. Die Veränderung dieser Strahlung kann daher nur durch Betrachtung der aufwärts- und abwärtsgehenden *Strahlungsströme* I_1 und I_2 erfolgen[3]. Setzt man den von einer Volumeneinheit Wolkenluft in den vorderen Halbraum gestreuten Anteil α, den nach hinten gestreuten β, den absorbierten γ und den gesamten Extinktionskoeffizienten a, so ist die Änderung der abwärtsgehenden Strahlung I_1

$$d I_1/dx = -\gamma a I_1 - \beta a (I_1 - I_2) \tag{20.4}$$

und die Änderung der aufwärtsgehenden Strahlung I_2

$$d I_2/dx = \gamma a I_2 - \beta a (I_1 - I_2)\,. \tag{20.5}$$

Die Lösung dieses Systems ergibt sich nach MECKE[4] zu:

$$\left.\begin{aligned} I_1 &= (1 + R_\infty)^{-1} (C R_\infty e^{\delta a x} + B e^{-\delta a x})\,;\\ I_2 &= (1 + R_\infty)^{-1} (C e^{\delta a x} + B R_\infty e^{-\delta a x}), \end{aligned}\right\} \tag{20.6}$$

wobei

$$R_\infty = \frac{\sqrt{1 - (\alpha - \beta)} - \sqrt{\gamma}}{\sqrt{1 - (\alpha - \beta)} + \sqrt{\gamma}} = \frac{\delta - \gamma}{\delta + \gamma} \tag{20.7}$$

und

$$\delta = \sqrt{\gamma\,(1 - (\alpha - \beta))} \tag{20.8}$$

---

[1] M. DIEM: Met. Rdsch. **1**, 261 (1948). — H. J. AUFM KAMPE u. H. WEICKMANN: J. Meteorology **7**, 54 (1950); **9**, 167 (1952); **10**, 204 (1953).

[2] L. FOITZIK: Wiss. Abh. Reichsamt Wetterd. **4**, Nr. 5 (1938).

[3] A. SCHUSTER: Astrophys. J. **21**, 1 (1950). — R. DIETZIUS: Beitr. Phys. fr. Atm. **10**, 202 (1922). — F. ALBRECHT: Meteor. Z. **50**, 478 (1933). — E. O. HULBURT u. R. TOUSEY: J. Opt. Soc. Amer. **37**, 78 (1947).

[4] R. MECKE: Ann. Phys. **65**, 257 (1925). — Meteor. Z. **61**, 195 (1944).

ist. Da $\alpha+\beta+\gamma=1$ ist, ist immer $\delta<1$. Für die Oberseite der Wolke gibt $I_1(0)=I_0$ den auftreffenden Strahlungsstrom und

$$(I_2/I_1)_{x=0}=A \tag{20.9}$$

die Albedo der Wolke. Für die Untergrenze $x=d$ gibt

$$(I_2/I_1)_{x=d}=R_0 \tag{20.10}$$

die Albedo der Unterlage, d.h. des Bodens und der darüberliegenden Luftschicht. Die durch die Wolke hindurchdringende Strahlung wird dann

$$I=I_0(1-R_\infty^2)\,e^{\delta a d}[(1-R_0R_\infty)+R_\infty(R_0-R_\infty)\,e^{-2\delta a d}]^{-1}. \tag{20.11}$$

Hier ist das Auftreten des Faktors δ im Exponenten im Zähler wichtig. Da das Exponentialglied im Nenner nur additiv auftritt, hat es geringeren Einfluß. Der diffuse Strahlungsstrom, der die Wolke durchdringt, wird daher nicht nach dem gleichen Gesetz wie die lineare Strahlung geschwächt, sondern langsamer, da $\delta<1$ ist. Dies ist die Wirkung der Vielfachstreuungsvorgänge innerhalb der Wolken. R_∞ ist die „*Grenzalbedo*" der unendlich dicken Wolke. Außerdem gilt

$$\beta=2\delta R_\infty(1-R_\infty^2)^{-1}. \tag{20.12}$$

Messungen der Extinktion innerhalb von Wolken sind meist im linearen Strahlengang durchgeführt worden. AUFM KAMPE[1] hat auf den Tragflächen eines Flugzeuges eine kleine Meßstrecke mit Scheinwerfer und Photozelle angebracht und den Extinktionskoeffizient a bestimmt. Tabelle 23, Zeile 1 bis 3 gibt für drei Wolkenarten charakteristische Zahlenwerte a. Dagegen hat FIEL[2] innerhalb von Wolken die von oben auf die horizontale Auffangfläche auftreffende diffuse Strahlung in verschiedenen Tiefen der Wolke gemessen. Die daraus abgeleiteten Extinktionskoeffizienten a' (Tabelle 23) sind um etwa eine Zehnerpotenz kleiner als a. Sie müssen nach Gl. (20.11) nahezu $a'=\delta a$ sein, da das von d abhängige Glied im Nenner nur wenig Einfluß hat. Durch Vergleich der Werte von FIEL und AUFM KAMPE erhält man δ und daraus nach (20.7) die Grenzalbedozahlen R_∞ der Tabelle, wenn man mit ALBRECHT den Anteil der Absorption zu $\gamma=0{,}015$ einsetzt; aus (20.12) und (20.8) findet BULLRICH[3] dann β und α. Für kurzwellige Strahlen (Licht) ist γ sicher kleiner einzusetzen. N und r sind die für die betreffende Wolkenart häufigsten Tropfenzahlen und -halbmesser. Nur bei Altostratus läßt sich ein starkes Überwiegen der Vorwärts- über die

Tabelle 23. *Strahlungsschwächung in Wolken.*

Extinktionskoeffizienten für lineare (a) und diffuse ($a'=\delta a$) Strahlung, Grenzalbedo R_∞, Anteil der Vorwärts- (α) und Rückwärts-Streuung (β) an der Extinktion. Typische Tropfenzahlen (N) und -halbmesser (r).

	a km^{-1}	a' km^{-1}	δ	R_∞	α	β	N cm^{-3}	r μ
Stratocumulus	38	4,75	0,12	0,78	0,50	0,48	770	4
Stratus	26	2,60	0,10	0,74	0,66	0,33	275	6
Altostratus	19,5	0,78	0,04	0,46	0,94	0,05	100	8
Stratus[4]	45	1,35	0,03	0,80	0,982	0,0147	75	10

[1] H. J. AUFM KAMPE: Zentr. wiss. Berichtsw., Bericht No. 3541 (1944).
[2] E. FIEL: Forsch. u. Erf. Ber. Reichsamt Wetterd., Berlin B **2** (1941).
[3] K. BULLRICH: Z. Met. **2**, 321 (1948).
[4] Nach M. NEIBURGER: J. Meteorology **6**, 98 (1949).

Tabelle 24. *Beleuchtungsstärke unter Wolken am Erdboden, relativ zur Beleuchtung bei wolkenlosem Himmel, in Abhängigkeit von Wolken-Art, -Mächtigkeit d und Erdbodenalbedo R_0, nach* BULLRICH.

Wolkenart	R_0	$d = 0{,}1$	0,2	0,6	1,0	2,0 km
Strato cumulus	0,1	0,48	0,28	0,11	0,07	0,03
	0,7	0,76	0,52	0,18	0,11	0,06
Stratus	0,1	0,72	0,48	0,20	0,13	0,07
	0,7	1,20	0,77	0,33	0,21	0,13
Altostratus	0,1	1,20	1,10	0,80	0,65	0,43
	0,7	1,60	1,40	1,10	0,85	0,57

Rückwärtsstreuung feststellen. An sich wäre dies bei allen Wolkenarten zu erwarten; das Auftreten anderer Zahlen rührt vielleicht von der unvermeidbaren Verwendung solcher Daten her, die zwar zu den gleichen Wolkenarten gehören, aber von verschiedenen Beobachtern und unter verschiedenen Bedingungen gewonnen sind. Messungen von *zusammengehörigen* Werten a und a' wären sehr wichtig; auch die Wellenlängenabhängigkeit ist nicht bekannt.

BULLRICH[1] hat unter Verwendung dieser Zahlen für die drei Wolkenarten die *Licht*durchlässigkeit bei verschiedenen Wolkendicken berechnet (Tabelle 24). Besonders auffallend an den Zahlen der Tabelle 24 sind die erhöhten Beleuchtungsstärken bei dünnen und wenig extingierenden Wolken. Sie rühren hier zum Teil von der Mehrfachreflexion zwischen Erdboden und Wolkenuntergrenze her, zum Teil auch von der starken Vorwärtsstreuung, die bei dünnen Wolken die Sonnenumgebung im Vergleich zum blauen Himmel sehr stark aufhellt, die direkte Strahlung aber nur wenig schwächt.

MIDDLETON[2] hat die *Farbe* des durch eine geschlossene Wolkendecke hindurchdringenden Tageslichtes berechnet. Er findet, daß die Absorption des Lichtes in den Wolkentröpfchen so wenig Einfluß hat, daß sie in den Gln. (20.4) und (20.5) vernachlässigt werden kann. Dagegen ist die spektrale Albedo der Erdoberfläche von Wichtigkeit, sie spiegelt sich in der Farbe des Wolkenlichtes stark abgeschwächt wieder; über Grasland und vor allem über städtischen Ansiedlungen wird die Farbtemperatur des einfallenden Lichtes (6350°) herabgesetzt auf 6200 bzw. 5700°, über Schneedecken dagegen erhöht auf 7500 bis 8150°. Auch der „*Eisblink*", die Aufhellung des Himmels über Eisfeldern im Polarmeer, und der gegenteilige Effekt, der blaue „*Wasserhimmel*" über offenen Wasserflächen innerhalb des Packeises, erhalten hierdurch ihre Erklärung[3].

Die Gln. (20.4) und (20.5) bieten auch die Möglichkeit, die *Absorption* der Strahlung in der Wolke zu untersuchen: der Verlust an Strahlung ist gegeben durch

$$\Delta I = I_1(0) - I_2(0) + I_2(d) - I_1(d) \tag{20.13}$$

oder

$$\Delta I = I_0(1 - A) - I_1(d)\,(1 - R_0).$$

Durch Einsetzen von (20.11) und (21.1) erhält man

$$\frac{\Delta I}{I_0} = (1 - R_\infty)\,\frac{(1 - R_0 R_\infty) + (R_\infty - R_0)\,e^{-2\delta a d} - (1 - R_0)\,(1 + R_\infty)e^{-\delta a d}}{(1 - R_0 R_\infty) + R_\infty\,(R_0 - R_\infty)\,e^{-2\delta a d}}. \tag{20.14}$$

Für unendlich dicke Wolken wird $\Delta I/I_0 = 1 - R_\infty$, denn in diesem Falle wird die durchgelassene Strahlung $I_1(d) = 0$ und $A = R_\infty$. Nach Tabelle 23 kann der absorbierte Betrag $\Delta I/I_0$ bei sehr dicken Schichten von Tröpfchenwolken (echter

[1] Siehe Fußnote 3, S. 211.
[2] W. E. K. MIDDLETON: J. Opt. Soc. Amer. **44**, 793 (1954).
[3] F. MÖLLER: Polarforschung **3**, 236 (1953).

Altostratus besteht immer aus Eis und kann eine gewisse Mächtigkeit nicht überschreiten) 20% erreichen. Das steht in Übereinstimmung mit Messungen von FRITZ und MACDONALD[1], die durch Messung der vier Strahlungsströme von (20.13) unter und über Wolken von etwa 5 km Mächtigkeit Absorptionswerte von 15%, in einem Extremfalle 27% erhielten. In den Berechnungen von MIDDLETON[2] und HEWSON[3] überschreitet die Absorption nicht 1 bzw. 7%, während das Reflexionsvermögen der Wolkenoberfläche den nur äußerst selten beobachteten Maximalwert von 80% weit überschreitet. Nebel, der weniger und größere Tropfen ($r = 20\,\mu$) als Wolken enthält, absorbiert nach HEWSON mehr und ist auch strahlungsdurchlässiger als Wolken gleicher Dicke, während sein Reflexionsvermögen geringer ist.

21. Die Albedo der Wolken und Bodenarten. Für den an der Wolkenoberfläche diffus zurückgeworfenen Teil der einfallenden Strahlung, die *Albedo der Wolke*, ergibt sich aus (20.9) und (20.6) nach MECKE

$$A = \frac{R_\infty(1 - R_0 R_\infty) + (R_0 - R_\infty)\,e^{-2\delta a d}}{(1 - R_0 R_\infty) + R_\infty(R_0 - R_\infty)\,e^{-2\delta a d}}. \tag{21.1}$$

Für verschwindende Wolkendicke $d = 0$ wird $A = R_0$ gleich der Rückstrahlung der Unterlage, für $d = \infty$ wird $A = R_\infty$ gleich der Grenzalbedo der Wolke. Für die vier in Tabelle 23 angegebenen Wolkenarten ergeben sich die Zahlenwerte von Tabelle 25, wenn für die Unterlage $R_0 = 0{,}10$ angenommen wird. Bereits eine sc-Wolke von 50 m oder ein st von 100 bis 150 m Dicke reflektieren 50% der auffallenden Strahlung, während für eine as-Wolke selbst bei sehr großer Mächtigkeit 50% nicht erreicht werden kann ($R_\infty = 0{,}46$). Leider ist die Zusammengehörigkeit der Zahlen in Tabelle 23 nicht ganz gesichert. Auch liegen für gewöhnlichen Bodennebel oder für Eiswolken keine Messungen von a und a' oder Albedomessungen vor, so daß eine entsprechende Bestimmung nicht möglich ist.

Tabelle 25. *Albedo verschiedener Wolkenarten nach Tabelle 23; berechnet nach (21.1). Wolkendicke = d.*

Wolkenart	$d = 0{,}02$	0,05	0,1	0,2	0,5	1,0	2,0 km
sc = Strato cumulus	$A = 0{,}33$	0,50	0,64	0,73	0,78	0,78	0,78
st = Stratus	0,25	0,36	0,48	0,61	0,72	0,74	0,74
as = Altostratus	0,12	0,13	0,16	0,21	0,31	0,40	0,45
st = Stratus	0,19	0,29	0,41	0,55	0,72	0,78	0,80

NEIBURGER[4] hat eine Reihe von Messungen der abwärts- und aufwärtsgerichteten Globalstrahlung im Küstenstratus vor der kalifornischen Küste mit Fesselballonen durchgeführt und daraus Albedo, Transmission und Absorption der Wolken in Abhängigkeit von ihrer Mächtigkeit d untersucht (Fig. 21). Sein Vergleich mit der Theorie liefert bei $d > 300$ m gute Übereinstimmung, dagegen sind für $d < 300$ m die beobachteten Albedowerte zu groß. Gibt man $R_\infty = 0{,}80$, $R_0 = 0{,}10$ und den gemessenen Wert A einer Höhenstufe vor, so kann nach (21.1) $a' = \delta a = 1{,}35$ bestimmt werden. Die wirksame Tropfengröße kann nach NEIBURGERs Angaben zu $r = 10\,\mu$, der Wassergehalt zu $W = 0{,}3\ \mathrm{g \cdot m^{-3}}$ angesetzt werden, woraus sich nach (20.2) $a = 45\ \mathrm{km^{-1}}$ ergibt (Tabelle 23, letzte Zeile).

[1] S. FRITZ u. T. MACDONALD: Bull. Amer. Met. Soc. **32**, 205 (1951).
[2] W. E. K. MIDDLETON: J. Opt. Soc. Amer. **44**, 793 (1954).
[3] E. W. HEWSON: Quart. J. Roy. Met. Soc. **69**, 47, 227 (1943).
[4] M. NEIBURGER: J. Meteorology **6**, 98 (1949).

Mit diesem Wert findet man eine recht gute Übereinstimmung zwischen Theorie und Beobachtung (Fig. 21). Aus $\delta = 0{,}03$ findet man nach (20.7) und (20.12) $\gamma = 0{,}0033$ für den Absorptions-, $\beta = 0{,}0147$ für den Rückstrahlungsanteil der Extinktion und daraus $\alpha = 0{,}982$, alles Werte (Tabelle 23, letzte Zeile), die glaubhafter erscheinen als die von BULLRICH für st und sc gefundenen (Tabelle 23, Zeile 1 bis 3), und auch mit den theoretischen Ergebnissen der MIE-Streuung für $2\pi r/\lambda > 40$ wie mit der Theorie FRAUNHOFERS und AIRYS und den Rechnungen WIENERS (Ziff. 13) besser im Einklang stehen. DIETZIUS hat β abhängig von der Zenitdistanz Z der Sonne gemäß $\beta = 0{,}073 + 0{,}642 \cdot 10^{-8} \cdot Z^4$ gefunden und im Mittel für diffuse Strahlung $\beta = 0{,}195$, was dem Wert für $Z = 66°$ entspricht; der Wert ist unwahrscheinlich groß.

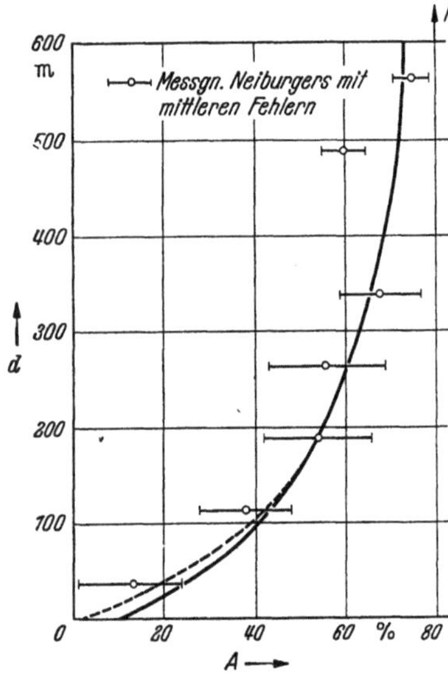

Fig. 21. Albedo von Stratuswolken verschiedener Mächtigkeit d nach Messungen NEIBURGERS. Mittelwerte mit mittlerem Fehler. Kurve nach (21.1).

FRITZ[1] hat unter Zugrundelegung der WIENERschen Streuung (Tabelle 12) eine neue Albedotheorie für großtropfige Wolken aufgestellt, in der er die Ausbreitung des Lichtes in der Wolke nach der Diffusionstheorie behandelt. Hierbei ergibt sich auch ein starker Einfluß der Zenitdistanz der Sonne auf die Albedo. Er findet z.B. für eine Wolke mit $ad = 2$ [a nach (20.2) bestimmt] und $R_0 = 0{,}10$ eine Zunahme der Albedo von 0,15 für $Z = 0°$ auf 0,24 bei $Z = 50°$ und 0,67 bei $Z = 80°$. Je mächtiger die Wolke, um so mehr verwischt sich der Einfluß der Zenitdistanz.

Zahlreiche Messungen der Albedo von Wolkenoberflächen[2] liegen vor, ohne genauere Daten der Wolkenstruktur und -mächtigkeit, jedoch sind Werte $A > 0{,}80$ nur selten gefunden worden. Im Mittel für alle Wolkenarten findet FRITZ[3] $A = 0{,}55$.

Für verschiedene *Bodenbedeckungen* sind pyranometrisch ($0{,}3 < \lambda < 3{,}0\ \mu$) folgende *Albedowerte* gemessen worden[2]:

sehr dunkler Mischwald	0,045	Wiesen und Felder	0,14
Nadelwald	0,07	heller Dünensand u. Brandg.	0,26 bis 0,63
heller Laub- und Mischwald	0,09	ältere Schneedecke	0,42 bis 0,70
Heide	0,10	frische Schneedecke	0,81 bis 0,85

Nur wenige Daten sind verfügbar über die spektrale Abhängigkeit der Bodenalbedo Von grünen Pflanzen ist das hohe Reflexionsvermögen (50%) bei etwa 800 mμ aus Infrarotphotographien von Landschaften bekannt, daneben besitzen sie ein schwaches Maximum der Albedo im Grünen (14% bei 550 mμ). Stadtsiedlungen zeigen einen stetigen Abfall von etwa 25% bei 800 mμ auf 5% bei 300 mμ, während die Albedo von Schnee bei 400 mμ einen Höchstwert von über 80% hat, von dem die Werte sowohl nach dem UV wie zum IR auf etwa 65% abfallen. Fig. 22 gibt Messungen von KRINOW wieder, die vom Erdboden aus gewonnen sind[4]. Die wenigen vorhandenen Messungen aus Flugzeugen deuten an, daß die Albedo von Wiesen die gleiche ist, die von Nadelwäldern aber bei gleicher spektraler Abhängigkeit auf $^1/_2$ bis $^1/_4$ des Wertes reduziert ist[5].

[1] S. FRITZ: J. Meteorology **11**, 291 (1954).

[2] Zusammenstellung von F. MÖLLER in [2], Tabelle 108, S. 522.

[3] S. FRITZ: J. Meteorology **6**, 277 (1949). — Bull. Amer. Met. Soc. **29**, 303 (1948); **31**, 25 (1950).

[4] E. L. KRINOV: Spectral reflective properties of natural surfaces, Akademia Nauk, Moskau 1947 [zitiert nach unveröff. Manuskript von R. PENNDORF, 1955]. — D. NICKERSON, K. L. KELLY u. K. F. STULTZ: J. Opt. Soc. Amer. **35**, 297 (1945). — E. V. ASHBURN u. R. G. WELDON: J. Opt. Soc. Amer. **46**, 583 (1956).

[5] R. SCHIMPF u. C. ASCHENBRENNER: Z. angew. Photogr. **2**, 41 (1940). — R. PENNDORF: Bull. Amer. Met. Soc. **37**, 142 (1956).

Von besonderem Interesse ist die Albedo von *Schneeflächen*, weil die Reflexion der direkten und diffusen Strahlung für den Strahlungshaushalt von Gletscheroberflächen und schneebedecktem Gelände von ausschlaggebender Wichtigkeit ist. SAUBERER und DIRMHIRN[1] haben aus zahlreichen Messungen für die Albedo von Totalstrahlung folgende Werte abgeleitet:

flockiger, feiner Neuschnee	0,90	Firn (Mittelwert)	0,51
grießliger Neuschnee	0,80	reines Eis	0,37
Altschnee je nach Alter und Verunreinigung	0,46 bis 0,75 (Mittel 0,56)	verschmutztes Eis	0,13

Schneedecken besitzen nicht ein ideal diffuses Reflexionsvermögen, weil die zahlreichen Kristalle, aus denen sie aufgebaut sind, zum Teil spiegelnd reflektieren. MIDDLETON und MUNGALL[2] haben Reflexionsmessungen des Lichtes an Schneeflächen unter verschiedenen Einfalls- und Reflexionswinkeln durchgeführt und fanden, daß bei Einfallswinkeln $< 45°$ die diffuse Reflexion alle Erscheinungen erklärt. Bei flacherem Einfall wird die Spiegelung beträchtlich, aber die Reflexion erfolgt flacher als der Einfall, was sich aus der verschiedenen Orientierung der Kristallflächen und der starken Zunahme der FRESNELschen Spiegelung bei Winkeln unter 45° erklärt. Eine theoretische Untersuchung der spektralen Abhängigkeit des Reflexionsvermögens haben DUNKLE und BEVANS[3] durchgeführt, wobei sie eine Veränderung mit der Korngröße des Schnees finden. Experimentelle Ergebnisse dieser Art stehen wohl noch aus.

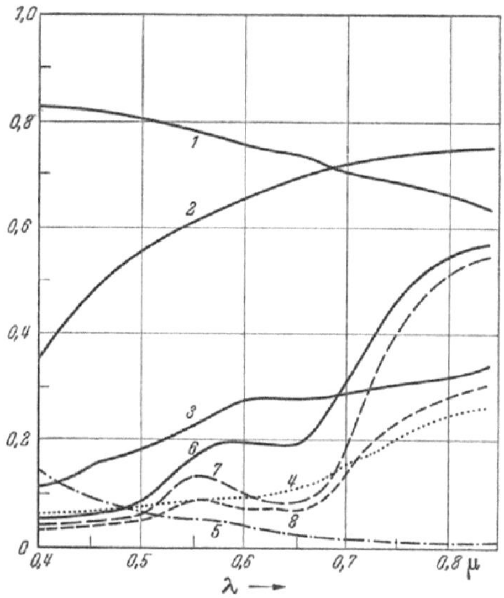

Fig. 22. Spektrales Reflexionsvermögen verschiedener Bodenbedeckungen nach E. L. KRINOW. *1* frischer Schnee *2* Kalkstein, heller Ton, *3* Sand, Wüste, nackter Fels, *4* Podsolböden, Lehm, Gebäude, gepflasterte Straßen, *5* Wasser *6* Herbstwälder, reife Feldfrüchte, *7* Laubwald im Sommer, saftige Wiesen, *8* Nadelwald im Sommer, trockene Wiesen.

SEKERA[4] hat aus seinen Untersuchungen über die Polarisation des Himmelslichtes unter dem Einfluß der Bodenalbedo den Schluß ziehen können, daß der Erdboden ganz allgemein nicht als diffus reflektierend angenommen werden darf, sondern immer ein gewisser Anteil spiegelnder Reflexion beteiligt ist.

Für die Albedo von *Wasseroberflächen*[5] könnten Zahlenwerte nach den FRESNELschen Formeln berechnet werden, jedoch sind solche höchstens in dem Fall ruhiger Binnengewässer brauchbar. Für die windbewegte Meeresoberfläche ist diese einfache Theorie nicht verwendbar. Messungen geben für hochstehende Sonne meist größere Werte als diese Theorie, für tiefstehende Sonne kleinere.

[1] F. SAUBERER u. J. DIRMHIRN: Arch. Met., Geophys. u. Bioklim., Ser. B **3**, 256 (1951). — I. DIRMHIRN u. E. TROJER: Arch. Met., Geophys. u. Bioklim., Ser. B **6**, 400 (1955).

[2] W. E. K. MIDDLETON u. A. G. MUNGALL: J. Opt. Soc. Amer. **42**, 572 (1952).

[3] R. V. DUNKLE u. J. T. BEVANS: J. Meteorology **13**, 212 (1956).

[4] Z. SEKERA: Final report, contract AF 19(122)—239, Univ. of California, Los Angeles 1955.

[5] F. LAUSCHER: In Handbuch der Geophysik, Bd. 8, Kap. 12, S. 723—740. Berlin 1955. — W. E. K. MIDDLETON: Quart. J. Roy. Met. Soc. **78**, 627 (1952). — H. GRIESSEIER: Z. Met. **6**, 53 (1952). — Acta Hydrophys. **1**, 107 (1953). — W. F. BURT: J. Meteorology **11**, 283 (1954). — M. NEIBURGER: Trans. Amer. Geophys. Union **29**, 647 (1948); **35**, 729 (1954).

Eine Reihe von verschiedenen Vorgängen spielen hierbei mit: die Reflexion der direkten und diffusen Strahlung bei den verschiedenen Neigungswinkeln der bewegten Meeresoberfläche gegen die Horizontale, diffuse Streuung des Lichtes an den Molekülen und trübenden kleinen Partikeln des Meereswassers und diffuse Streuung an den Luftblasen, die vor allem beim Brechen der Wellenkämme entstehen. Alle diese Effekte zusammen geben eine Albedo, die von 0,04 bei Zenitstand der Sonne auf 0,08 bei $Z=60°$ und auf 0,15 bis 0,25 (je nach Rauhigkeit der Meeresoberfläche und Trübungsgrad oder Bedeckung des Himmels) bei $Z=80°$ zunimmt.

Durch Summierung über die ganze Erde unter Berücksichtigung der Land- und Meerverteilung und vor allem der Wolkenbedeckung in verschiedenen Breitenzonen erhält man die *Albedo des Planeten Erde*, die nach den neuesten Berechnungen von FRITZ[1] mit 0,35 angenommen werden kann. Diese Größe ist einesteils sehr wesentlich für den gesamten Wärmehaushalt der Erde und ihrer Lufthülle, andererseits ist es erstaunlicherweise leichter sie zu messen, als die Albedowerte der verschiedenen Erdbodenbedeckungen: Das *aschgraue Mondlicht*, durch das die von der Sonne nicht beleuchtete Seite des Mondes vor und nach Neumond erhellt wird, stammt vom Widerschein der dann — vom Mond gesehen — voll sonnenbeschienenen Erde. Aus einer längeren Beobachtungsreihe von DANJON wird die visuelle Albedo der Erde wie folgt angegeben[2]:

λ	0,435	0,467	0,545	0,606 mμ
A_λ	0,59	0,49	0,39	0,29

Messungen dieses aschgrauen Mondlichtes an einander gegenüberliegenden Orten der Erde ergeben Albedowerte für verschiedene Erdgegenden und können in ihrer Veränderung mit dem wechselnden Bewölkungsgrad oder allgemein mit der Wetterlage in Beziehung gebracht werden.

22. Die Himmelsstrahlung in der Dämmerung und bei Nacht. Nach Sonnenuntergang oder vor Sonnenaufgang werden nur die hohen Schichten der Lufthülle beleuchtet, und zwar nur um so höhere, je tiefer die Depression der Sonne unter dem Horizont ist. Damit nimmt die Beleuchtungsstärke (und ihr nahezu proportional auch die Zenitleuchtdichte) stetig ab (Tabelle 26), bis bei etwa 16° Sonnendepression der letzte Widerschein der Sonne verschwunden, die astronomische Dämmerung beendet ist, und die Nacht beginnt. Das dann noch herrschende Nachthimmelslicht ist nur zu etwa $\frac{1}{4}$ Sternenlicht, während $\frac{3}{4}$ von einem Selbstleuchten hoher Schichten der Erdatmosphäre herrührt (vgl. Kapitel Aurora and Airglow in Bd. XLIX dieses Handbuchss). Der abendliche Abfall der Global-

Tabelle 26. *Globalbeleuchtungsstärke nach Sonnenuntergang in Abhängigkeit von der Sonnenhöhe, nach* BULLRICH.

Sonnenhöhe	wolkenlos	bedeckt	Sonnenhöhe	wolkenlos	bedeckt
0°	550 Lux	75 Lux	−10°	$3,1 \cdot 10^{-2}$	$5,5 \cdot 10^{-3}$
−2	120	11	−12	$6,5 \cdot 10^{-3}$	$1,5 \cdot 10^{-3}$
−4	15	1,4	−14	$2,3 \cdot 10^{-3}$	$6,0 \cdot 10^{-4}$
−6	1,6	$1,7 \cdot 10^{-1}$	−16	$1,8 \cdot 10^{-3}$	$5,0 \cdot 10^{-4}$
−8	$1,9 \cdot 10^{-1}$	$2,6 \cdot 10^{-2}$	−18	$1,8 \cdot 10^{-3}$	$5,0 \cdot 10^{-4}$

beleuchtungsstärke erfolgt in wolkenlosen, bewölkten und bedeckten Nächten ungefähr nach dem gleichen Gesetz, nämlich etwa um eine Zehnerpotenz bei

[1] S. FRITZ: J. Meteorology **6**, 277 (1949).
[2] R. PENNDORF: Meteor. Z. **54**, 349 (1937).

Zunahme der Sonnentiefe um 2,5°. Die Ausgangswerte bei Sonnenuntergang sind allerdings stark verschieden, bei bedecktem Himmel um etwa einen Faktor 10 niedriger als bei wolkenlosem, nur der Abfall erfolgt ähnlich. Der Vollmond gibt bei großer Höhe eine Vermehrung der Beleuchtungsstärke um etwa $\frac{1}{2}$ Lux [1].

Die Theorie der Himmelsbeleuchtung bei Dämmerung ist umständlich, wegen der über dem Beobachtungsort schräg ansteigenden Grenze des Erdschattens. Sie ist von GRUNER, SATO und HULBURT [2] gegeben worden. SATO hat auch den Einfluß der sekundären Streuung dabei berücksichtigt. Insbesondere für die Beleuchtung des Himmels über der untergehenden Sonne (bei der nicht nur die Extinktion des Lichtes in der tieferen dunsthaltigen Luft eine Rolle spielt, sondern auch die Streuung an dem bis zur Troposphärengrenze hinauf vorhandenen Staub und Dunst) kann die Dunststreuung nicht außer acht gelassen werden, wie es in allen Theorien bisher geschehen ist. Gerade die eigentlichen farbigen Dämmerungserscheinungen sind Zeugen für die Streuung an gröberem Aerosol (GRUNER) [2].

Die Änderung der Farbe der Globalstrahlung während der Dämmerung haben SIEDENTOPF und HOLL [1] gemessen. Sie wählen das für Farbkennzeichnungen sehr empfindliche Maß der *Farbtemperatur* und finden eine Änderung von $T_F = 6000°$ bei tiefstehender Sonne auf 15000° (blauer werdend) bei 10° Depression, danach einen raschen Abfall auf 4200° (gelber werdend) bei 18° Sonnentiefe. Diese letzte Änderung ist wohl schon der Einfluß der Emissionslinien des *Nachthimmelslichtes* (Air Glow, vgl. Bd. XLIX dieses Handbuches). Dagegen ist für die blaue Farbe des sonnenfernen Himmels in der Dämmerung die Absorption durch das Ozon in der Chappuis-Bande ausschlaggebend, wie MÖLLER [3] und HULBERT [2] gezeigt haben (vgl. den Schluß der Ziff. 16).

C. Die langwellige Wärmestrahlung in der Atmosphäre.

23. Übersicht. Die Probleme der langwelligen Strahlung des Bereiches von 4 bis über 100 μ sind wesentlich von denen der kurzwelligen unterschieden:

1. Die von außen auf die Erde auftreffende Strahlung ist geringfügig und kann fast bei allen Fragestellungen vernachlässigt werden. Sie ist nur in dem transparenten Bereich von 9 bis 12 μ beobachtbar.

2. Streuungsvorgänge an Gasen und Aerosolen spielen wegen der Kleinheit der Teilchen keine Rolle. In Wolken ist der überwiegende Teil der Tropfen wenig kleiner oder in der gleichen Größenordnung wie die Wellenlänge, trotzdem tritt die Streuung gegenüber dem Anteil der Absorption an der Gesamtextinktion zurück. Über beobachtbare Streuungsvorgänge ist nichts bekannt.

3. Die Absorption der Strahlung an den atmosphärischen Gasen und dementsprechend auch die *Emission von Temperaturstrahlung* sind die allein wesentlichen Vorgänge, auch die Emission des Erdbodens und der Wolkenflächen.

4. Es gibt in der Atmosphäre keine gerichtete, sondern nur *diffuse* Strahlung (was nicht ausschließt, daß die Verhältnisse im gerichteten Strahl untersucht werden müssen, schon zum Vergleich mit Laboratoriumsmessungen).

---

[1] K. BULLRICH: Ber. dtsch. Wetterd. US-Zone **4** (1948). — H. ISRAEL: Forsch. u. Erf. Ber. Reichsamt Wetterd. Berlin A Nr. 6 (1941). — H. SIEDENTOPF u. H. HOLL: Reichsber. Phys. **1**, 32 (1944). — M. J. KOOMEN, C. LOCK, D. M. PACKER, R. SCOLNIK, R. TOUSEY u. E. O. HULBURT: J. Opt. Soc. Amer. **42**, 353 (1952). — E. V. ASHBURN: J. Opt. Soc. Amer. **43**, 805 (1953); dort auch weitere Literaturangaben. — G. SCHEER: Met. Rdsch. **8**, 82 (1955).

[2] P. GRUNER: In Handbuch der Geophysik, Bd. 8, Kap. 8. Berlin 1943. — T. SATO: Sci. Rep. Tôhoku Univ. (5) **3**, 114 (1951). — E. O. HULBURT: J. Opt. Soc. Amer. **43**, 113 (1953).

[3] F. MÖLLER: Arch. Met., Geophys. u. Bioklim., Ser. B **5**, 1, 456 (1953).

Im Zusammenhang mit 2. steht, daß die untere Atmosphäre als im vollkommenen thermodynamischen Gleichgewicht befindlich angenommen werden kann. Absorption und Emission eines Massenelementes Luft stehen im Gleichgewicht mit der Energie der Temperaturbewegung der Atome und Moleküle; die Emission ist vollkommen gekennzeichnet durch die Angabe des Gases und seiner Temperatur. Dagegen sind im Ultraviolett die Absorption und die damit zusammenhängende Dissoziation oder Ionisation nicht im thermodynamischen Gleichgewicht, die übertragenen Energiequanten sind viel größer als die der spontanen (Temperatur-) Strahlung. Curtis[1] hat nachgewiesen, daß erst oberhalb 50 km Höhe kein thermodynamisches Gleichgewicht mehr herrscht und das Kirchhoffsche Gesetz nicht mehr gültig ist.

24. Die Ausbreitung monochromatischer Strahlung in der Atmosphäre[2]. Jedes Volumenelement absorbiert und emittiert Strahlung. Zur Vereinfachung nehmen wir an, daß die Atmosphäre aus horizontalen ebenen Schichten gleicher Temperatur und gleicher Dichte der strahlenden Materie aufgebaut ist, vernachlässigen also die Krümmung der Erdoberfläche und der ihr aufliegenden Lufthülle, sowie das Vorhandensein von horizontalen Unterschieden innerhalb der Atmosphäre, von Neigungen der Erdoberfläche oder vertikalen Begrenzungen von Wolken, die beide als Strahler und Absorber wirken können. Eine horizontale *Elementarschicht* der Dichte ϱ enthält über der Flächeneinheit die Masse $dm = \varrho\, dz$, eine endlich dicke Schicht die Masse $m = \int \varrho\, dz$. Durchdringt ein Strahl der Intensität I eine Elementarschicht unter dem Einfallswinkel ϑ, so wird der Betrag[3]

$$dI = -k I \sec\vartheta\, dm \tag{24.1}$$

absorbiert, wenn k der Absorptionskoeffizient ist. Sendet die Schicht auch selbst Strahlung aus, so ist diese nach dem Kirchhoffschen Gesetz gleich $k \cdot E_S$, wo E_S die Strahlung des schwarzen Körpers in den Raumwinkel 1 ist, und man erhält statt (24.1)

$$dI = -k(I - E_S) \sec\vartheta\, dm. \tag{24.2}$$

Wenn die einfallende Strahlung $I = 0$ ist, dann ist

$$dI = k E_S \sec\vartheta\, dm \tag{24.3}$$

die unter dem Winkel ϑ emittierte Strahlung.

Wir definieren den Strom der *diffusen Strahlung* f als das Integral der senkrechten Komponente aller Strahlen, die unter den Zenitdistanzen $0 < \vartheta < \pi/2$ einfallen. Setzen wir weiter voraus, daß zwischen der Elementarschicht dm und der Auffangfläche eine endlich dicke Schicht m liegt, welche selbst nicht strahlt, sondern nur absorbiert, dann ist der ankommende Strahlungsstrom der Elementarschicht

$$df = k\, dm \int_0^{2\pi} d\alpha \int_0^{\pi/2} E_S\, e^{-k m \sec\vartheta} \sin\vartheta\, d\vartheta. \tag{24.4}$$

Zur Ausführung der Integration über ϑ setzt man $\sec\vartheta = x$, dann wird

$$df = 2\pi k E_S\, dm \int_1^{\infty} e^{-k m x} x^{-2}\, dx,$$

---

[1] A. R. Curtis u. R. M. Goody: Scient. Proc., Internat. Assoc. Meteorol., 10. Gen. Assembly, Rome 1954. London 1955, 569. — Proc. Roy. Soc. Lond., A **236**, 193 (1956).

[2] E. Gold: Proc. Roy. Soc. Lond., Ser. A **82**, 43 (1909). — R. Emden: Sitzgsber. bayr. Akad. Wiss. **43**, 55 (1913). — O. F. T. Roberts: Proc. Roy. Soc. Edinburgh **50**, Nr. 19 (1930). — R. Mügge u. F. Möller: Z. Geophys. **8**, 53 (1932). — Außerdem [*1d*], [*7*], [*8*].

[3] Streng genommen müßte auf jeder Seite der Gleichung mit dem räumlichen Elementarwinkel $d\omega$ multipliziert werden.

wofür abgekürzt geschrieben wird

$$df = 2\pi E_S\, k\, dm\, Ei_2(km). \tag{24.5}$$

Die Funktion $Ei_2(u) = \int\limits_1^\infty e^{-ux} x^{-2} dx$ wird als das 2. Exponentialintegral bezeichnet. Allgemein ist definiert

$$Ei_n(u) = \int\limits_1^\infty e^{-ux} x^{-n} dx.$$

Durch Differentiation ergibt sich

$$d\,Ei_n(u)/du = -Ei_{n-1}(u) \tag{24.6}$$

und entsprechend

$$\int Ei_n(u)\, du = -Ei_{n+1}(u). \tag{24.7}$$

Durch partielle Integration findet man die Rekursionsformel

$$Ei_n(u) = \frac{e^{-u}}{n-1} - \frac{u}{n-1} Ei_{n-1}(u).$$

Das 1. Exponentialintegral ist tabuliert, ebenso das 2. und 3.[1].

Für die diffuse Strahlung, die von einer schwarzen *Fläche* durch die Schicht m hindurchgesandt wird, ergibt sich mit Verwendung der obigen Definition

$$f_a = E_S \int\limits_0^{2\pi} d\alpha \int\limits_0^{\pi/2} e^{-km\sec\vartheta} \sin\vartheta \cos\vartheta\, d\vartheta = 2\pi E_S\, Ei_3(km) \tag{24.8}$$

oder mit

$$f_S = \pi E_S, \tag{24.9}$$

$$f_a = f_S \cdot 2\, Ei_3(km) \text{ (vgl. Tabelle 27a)}. \tag{24.10}$$

ROBERTS hat gezeigt, daß man mit sehr guter Annäherung setzen kann

$$2\,Ei_3(km) \approx e^{-1{,}66\, km}. \tag{24.11}$$

Durch Integration von (24.5) über m erhält man die Strahlung einer endlich dicken Schicht $(m_1 - m_0)$, wobei zunächst angenommen werde, daß die Temperatur bzw. $f_S = \pi E_S$ in der ganzen Schicht unabhängig von m, die Schicht also isotherm ist:

$$f = -2 f_s \left[Ei_3(km_1) - Ei_3(km_0)\right].$$

Für $m_0 = 0$ gibt dies

$$f = 2 f_s \left[\tfrac{1}{2} - Ei_3(km_1)\right]. \tag{24.12}$$

Addition von (24.10) und (24.12) zeigt, daß die von einer schwarzen Fläche (f_a) und angrenzender absorbierender Schicht (f) der gleichen Temperatur ausgesandte Strahlung

$$f + f_a = f_s \tag{24.13}$$

gleich dem Strahlungsstrom (f_S) ist, der von der schwarzen Fläche allein in die Halbkugel geht. Gl. (24.13) ist nichts anderes als das KIRCHHOFFsche Gesetz für diffuse Strahlungsströme.

Im allgemeinen ist in der Atmosphäre die Temperatur und damit E_s bzw. f_s nicht unabhängig von der strahlenden Masse m, sondern eine Funktion von m bzw. von der Höhe über dem Boden, und zwar eine von Tag zu Tag wechselnde Funktion. Das Integral von (24.5) über m, das im Falle der isothermen Schicht auf (24.12) führt, ist also im allgemeinen nicht ausführbar. Man schreibt unter Verwendung von (24.6) zweckmäßiger

$$f = -\int\limits_{m_0}^{m_1} f_s(m) \cdot 2 \frac{dEi_3(km)}{dm}\, dm, \tag{24.14}$$

wo $f_s(m)$ eine beliebige, aber durch die Temperatur vorgegebene Funktion von m ist.

---

[1] E. JAHNKE u. F. EMDE: Tafeln höherer Funktionen, 4. Aufl. Leipzig: B. G. Teubner 1948. — F. LINKE: Meteorologisches Taschenbuch IV, Tabelle 144, S. 262. Leipzig 1939.

25. Die nicht-monochromatische Strahlung. Die in der langwelligen Wärmestrahlung als Absorber wirksamen atmosphärischen Gase besitzen *Bandenspektren*, die aus sehr zahlreichen Einzellinien zusammengesetzt sind. Um deren Gesamtwirkung zu erfassen, muß (24.14) über alle Frequenzen ν integriert werden, wobei der Absorptionskoeffizient k sowie die Schwarzstrahlung f_S Funktionen der Wellenlänge bzw. der Frequenz sind. Es ist unmöglich, in jedem einzelnen Fall einer bestimmten vorgegebenen Temperaturschichtung diese Integration gesondert durchzuführen. Man kann jedoch ein halb numerisches, halb graphisches Verfahren für die praktische Lösung durch Vorwegnahme der Integration über ν entwickeln [*1d*]. Man geht aus von dem Strahlungsstrom (24.12) einer isothermen Schicht mit der Masse m. Über ν integriert ist

$$F_{T_0}(m) = \int_0^\infty d\nu\, f_s(T_0)\,[1 - 2\,Ei_3(k_\nu m)]. \tag{25.1}$$

Die Strahlung einer Elementarschicht dm im Abstand m ist

$$d F_T(m) = -\int_0^\infty d\nu\, f_s(T)\, 2\,\frac{d\,Ei_3(k_\nu m)}{dm}\, dm. \tag{25.2}$$

Die Integrationen über ν in (25.1) und (25.2) müssen numerisch unter Vorgabe der Absorptionskoeffizienten $k(\nu)$ durchgeführt werden.

Bezeichnet man

$$F_{T_0}(m) = X(m)$$

und

$$d F_T(m) : d F_{T_0}(m) = Y(T, m)$$

als Koordinaten eines Diagrammes, dann kann der Strahlungsstrom einer beliebigen endlichen Schicht mit vorgegebener Temperaturverteilung $T(m)$ als Integral

$$F = \int_{X(m_0)}^{X(m_1)} Y\big(T(m), m\big)\, dX(m) \tag{25.3}$$

graphisch bestimmt werden (vgl. Ziff. 29).

Der Absorptionskoeffizient $k(\nu)$ kann im Bereich einer einzelnen Linie zwischen ihrem Zentrum und ihren Flügeln im Verhältnis 100:1 oder noch mehr variieren. Es ist deshalb eine wesentliche Vereinfachung für die Ausführung der Integrale in (25.1) und (25.2), nicht die *Transmissionsfunktionen* $Ei_3(km)$ und $d\,Ei_3(km)/dm$ für streng monochromatische Strahlung zu verwenden sondern Mittelwerte über den Bereich einer Linie oder eines Systems von gleichen Linien, und die Integrale zu ersetzen durch eine Summierung oder Integration über die ganzen Linienbereiche oder Liniengruppen. Auch die maximalen Absorptionskoeffizienten (im Zentrum der Linien) variieren im Bereich des gesamten Spektrums des Wasserdampfes noch wie 10^4:1 oder stärker (vgl. Ziff. 27).

Definieren wir nach [*7*] für einen schmalen Spektralbereich, in dem f_s noch keine nennenswerte, aber $k(\nu)$ beträchtliche Veränderung mit ν zeigt, die Transmission eines Strahles I als $\tau_I = [e^{-km}]_{\text{mittel}}$ und analog die Transmission einer Schicht $\tau_f = [f_a : f_s]_{\text{mittel}}$ oder nach (24.10)

$$\tau_f = [2\,Ei_3(k\,m)]_{\text{mittel}},$$

dann wird der Strahlungsstrom einer endlichen Schicht nach (25.2) zu:

$$F = -\int_0^\infty d\nu \int_{m_0}^{m_1} f_s \frac{d}{dm}\,\tau_f(m - m_0)\, dm.$$

In der gleichen Weise wird (25.1) vereinfacht. Durch partielle Integration wird

$$F = -\left| \int_0^\infty f_s \tau_f \, d\nu \right|_{m_0}^{m_1} + \int_0^\infty d\nu \int_{m_0}^{m_1} dm \frac{d f_s}{dm} \tau_f .$$

Mit Einführung von T als unabhängige Variable ergibt sich für den zweiten Term rechts

$$\int_0^\infty d\nu \int \frac{d f_s}{dm} \tau_f \, dm = \int_0^\infty d\nu \int \frac{d f_s}{dm} \cdot \frac{dm}{dT} \tau_f \, dT = \int_0^\infty d\nu \int \frac{d f_s}{dT} \tau_f \, dT .$$

Mit der Abkürzung

$$Q(m, T) = \int_0^\infty \frac{d f_s}{dT} \tau_f \, d\nu$$

wird

$$\left| \int_0^\infty f_s \tau_f \, d\nu \right|_{T_0(m_0)} = \int_0^{T_0} Q(m_0, T) \, dT .$$

Für $m_0 = 0$ wird dann

$$F = \int_0^{T_0} Q(0, T) \, dT - \int_0^{T_1} Q(m_1(T_1), T) \, dT + \int_{T_0}^{T_1} Q(m(T), T) \, dT , \tag{25.4}$$

wobei noch $\int_0^{T_0} Q(0, T) \, dT = \sigma T^4$, wegen $\tau_f(0) = 1$.

Es ist also auf diesem Wege F ebenfalls als graphisch zu bestimmendes Integral einer Funktion $Q(m, T)$ über T zu gewinnen, wie es oben mit dem Integral von $Y(m, T)$ über $X(m)$ gewonnen wurde. In der Tat sind die beiden Verfahren (25.3) und (25.4) durch eine flächentreue Transformation ineinander überführbar und daher prinzipiell gleichwertig. Einzelheiten über diese Integrationsdiagramme werden weiter unten behandelt (vgl. Ziff. 29).

26. Die Ausbreitung der Linien- und Bandenstrahlung. Ein isoliertes ruhendes Molekül würde nur eine durch einen Quantenübergang scharf definierte Frequenz absorbieren oder emittieren. Durch Bewegungen und Stöße wird ermöglicht, auch etwas abweichende Frequenzen mit bestimmter Wahrscheinlichkeit abzustrahlen. Die daraus resultierende Abhängigkeit des *Absorptionskoeffizienten* von der Schwingungszahl $\nu = 1/\lambda \, \mathrm{cm}^{-1}$ innerhalb einer Spektrallinie, kurz die *Linienform* genannt, ist im wesentlichen bestimmt durch die Störungen, die die Moleküle während des Strahlungsvorganges durch Zusammenstöße mit benachbarten Molekülen erleiden, und durch die Bewegungen der strahlenden Moleküle selbst. Die letzte Wirkung äußert sich in der DOPPLER-Verbreiterung der Linie; diese ist bei den atmosphärischen Temperaturen sehr viel kleiner als die *Stoßdämpfungs*-Verbreiterung, die daher allein die Linienform bestimmt. Erst in großen Höhen der Atmosphäre (über 40 km) kann die DOPPLER-Verbreiterung überwiegen. In den unteren Atmosphärenschichten gilt allgemein die Dispersions- oder LORENTZ-Form einer Linie:

$$k(\nu) = \frac{S}{\pi} \frac{\delta}{(\nu - \nu_0)^2 + \delta^2} \tag{26.1}$$

δ ist die Halbwertsbreite[1], denn für $\nu = \nu_0$ ist $k_{\max} = S/\pi\,\delta$, für $\nu = \nu_0 \pm \delta$ ist $k = S/2\pi\,\delta$. S wird die *Intensität* der Linie genannt, denn es ist $\int_{-\infty}^{+\infty} k(\nu)\, d\nu = \frac{S}{\pi} \left[\operatorname{arc\,tg} \frac{\nu - \nu_0}{\delta} \right]_{-\infty}^{+\infty} = S$. Eine von VAN VLECK und WEISSKOPF[2] angegebene

[1] In der Spektroskopie wird häufig 2δ als Halbwertsbreite bezeichnet.

[2] J. H. VAN VLECK u. V. F. WEISSKOPF: Rev. Mod. Phys. 17, 227 (1945).

exaktere Form zeigt im Ultraroten so wenig Unterschied gegen (26.1), daß sich ihre Verwendung erübrigt. Auch ist die Gültigkeit der Dispersionsform im Ultrarot experimentell schwer nachprüfbar, sie ist jedoch sowohl im Sichtbaren, wie im Mikrowellengebiet[1] mit solcher Exaktheit bestätigt worden, daß keine Zweifel möglich sind.

Plass und Warner[2] haben darauf hingewiesen, daß die Dispersionsform (26.1) nur in der Nähe der Linienmitte richtig ist, und daß nach der quantentheoretischen Begründung eine gewisse Unsymmetrie der Linienform besteht, die sich vor allem in den Linienflügeln bemerkbar macht; an der kurzwelligen Seite ist $k(\nu)$ geringer als im gleichen Abstand nach der langwelligen Seite. Diese Autoren haben auch die Auswirkung dieses Unterschiedes für die Grenzfälle starker und schwacher Absorption untersucht; vermutlich ist er jedoch zu geringfügig, um bei den Berechnungen der atmosphärischen Strahlungsvorgänge eine Rolle zu spielen. Zum gleichen Ergebnis kommen auch Curtis und Goody[2].

Die Transmissionsfunktion τ_I für einen geraden Strahl I ist in dem schmalen Spektralbereich $\Delta\nu$ gegeben durch

$$\tau_I = \frac{1}{\Delta\nu}\int\limits_{\Delta\nu} e^{-k(\nu)m}\,d\nu,$$

wenn vorausgesetzt wird, daß die einfallende Strahlungsintensität unabhängig von ν ist. Da $k(\nu)$ nach (26.1) für große $(\nu-\nu_0)$ sehr klein wird, darf das Integral bis in die Unendlichkeit erstreckt werden. Nach Ladenburg und Reiche[3] ist

$$\tau_I = 1-(2\pi\,\delta/\Delta\nu)\cdot e^{-x}\left(J_0(i\,x)-i\,J_1(i\,x)\right), \tag{26.2}$$

wo J_0 und J_1 die Bessel-Funktionen nullter und erster Ordnung, $x = S\,m/2\pi\delta$ ist. Für kleine x ist angenähert $\tau_I = 1-\frac{2\pi\,\delta\,x}{\Delta\nu} = 1-\frac{S\,m}{\Delta\nu}$. Für große x dagegen ergibt sich mit $J_n(i\,x)\to i^n e^x(2\pi x)^{-\frac{1}{2}}$

$$\tau_I = 1-\frac{2}{\Delta\nu}\sqrt{S\,\delta m}\,. \tag{26.3}$$

Dieses Wurzelgesetz für den — von einer oder mehreren gleichen Linien in einem schmalen Bereich — absorbierten Energiebetrag ist experimentell häufig bestätigt und seinerseits wiederum als Kennzeichen dafür angesehen worden, daß in einem untersuchten Spektralbereich Linienabsorption und nicht kontinuierliche Absorption vorliegt. Natürlich ist die Gültigkeit nach großen m hin beschränkt, denn τ kann nicht negativ werden.

Soll nicht lineare, sondern *diffuse* Strahlung behandelt werden, dann kann die von einer Spektrallinie in einer Schicht m absorbierte Strahlung numerisch bestimmt werden, indem man die e-Funktion durch $2Ei_3$ ersetzt. Mit Verwendung des Wurzelgesetzes (26.3) wird jedoch

$$\tau_f = 2\int\limits_0^{\pi/2}\left(1-\frac{2}{\Delta\nu}\sqrt{S\,\delta m\sec\vartheta}\right)\sin\vartheta\cos\vartheta\,d\vartheta = 1-\frac{2}{\Delta\nu}\sqrt{S\,\delta\cdot 1{,}78\,m}\,. \tag{26.4}$$

Die Absorption einer Schicht ist also numerisch gleich der Absorption eines Strahles bei der Masse $1{,}78\,m$, während dieser Faktor für monochromatische Strahlung nach (24.11) 1,66 betrug [8].

In *Bandenspektren* hat man es nicht mit einzelnen Linien zu tun, für die mit zunehmendem ν der Absorptionskoeffizient gegen 0 geht, denn in einem gewissen Abstand liegt die nächste Linie, und die Absorptionen benachbarter Linien überlappen einander.

---

[1] C. H. Townes u. F. R. Merritt: Phys. Rev. **70**, 558 (1946).

[2] G. N. Plass u. D. Warner: Phys. Rev. **86**, 138 (1952). — J. Meteorology **9**, 333 (1952). — A. R. Curtis u. R. M. Goody: Quart. J. Roy. Met. Soc. **80**, 58 (1954).

[3] R. Ladenburg u. F. Reiche: Ann. Phys. **42**, 181 (1911).

Schematisch kann man eine solche Bande, wenn sie sehr regelmäßig ist, durch eine Serie von Linien gleicher Intensität S mit untereinander gleichen Abständen d charakterisieren: $k(\nu) = \sum_n \frac{S\,\delta/\pi}{(\nu - n d)^2 + \delta^2}$. ELSASSER[1] hat gezeigt, daß dieser Ausdruck äquivalent ist dem Ansatz

$$k(\nu) = \frac{S}{d} \frac{\operatorname{Sin}\beta}{\operatorname{Cos}\beta - \cos(2\pi\nu/d)}, \tag{26.5}$$

wo $\beta = 2\pi\delta/d$ ein Maß für das Verhältnis von Linienbreite δ zu Linienabstand d ist. Auch für diese Funktion $k(\nu)$ kann die Transmission exakt bestimmt werden. Mit $y = Sm/d$ ergibt sich

$$\tau_I = \int_y^\infty e^{-y \operatorname{Cot}\beta} J_0(i\,y/\operatorname{Sin}\beta)\,dy. \tag{26.6}$$

Ausführliche numerische Tafeln sind von GODSON, FOWLES und KAPLAN veröffentlicht worden[2].

Zwei Grenzfälle seien genauer betrachtet: für große Werte β, also engbenachbarte Linien (etwa $\beta = 1$) zeigt τ_I nur sehr geringe Abweichungen von einer Exponentialfunktion. Das bedeutet, daß der Absorptionskoeffizient zwischen den Linien nur wenig absinkt oder ein fast konstanter Absorptionskoeffizient vorhanden ist. Der Fall kleiner Werte β (etwa $\beta = 0{,}1$) ist meist in der Atmosphäre gegeben. Führt man die Reihenentwicklung für kleine β durch, so ergibt sich als Näherungsform

$$\tau_I = 1 - \Phi\left(\sqrt{y\beta/2}\right) = 1 - \Phi\left(\sqrt{\pi S \delta m/d}\right), \tag{26.7}$$

wobei Φ das Wahrscheinlichkeitsintegral $\Phi(x) = 2\pi^{-\frac{1}{2}} \int_0^x e^{-x^2} dx$ ist. Die Absorption hängt also von der Linienintensität S, der Halbwertsbreite δ, dem reziproken Quadrat des Linienabstandes d^2 und der optischen Dicke m in der gleichen Weise ab. ELSASSER hat deshalb die Konstanten

$$2\pi S \delta/d^2 = l \tag{26.8}$$

zusammengefaßt und bezeichnet l als den *verallgemeinerten Absorptionskoeffizienten*. Diese Größe hat sich als ein sehr nützlicher Begriff erwiesen. Es wird damit

$$\tau_I = 1 - \Phi\left(\sqrt{l m/2}\right). \tag{26.9}$$

Wiederum kann aus dieser Form von τ_I durch Integration über alle Zenitdistanzen ϑ eine Transmissionsfunktion für diffuse Strahlungsströme abgeleitet werden. Man erhält mit $\tau_f = 2\int_0^{\pi/2} \tau_I(m \sec\vartheta) \sin\vartheta \cos\vartheta\, d\vartheta$ einen Ausdruck

$$\left.\begin{aligned} \tau_f &= \sqrt{\frac{y\beta}{2\pi}}\left[Ei_{\frac{1}{2}}(y\beta/2) - Ei_{\frac{3}{2}}(y\beta/2)\right] \\ &= \frac{2}{3}\sqrt{\frac{l m}{2\pi}}(1 - l m)\, e^{-l m/2} + \frac{1}{3}(3 + l^2 m^2)\left(1 - \Phi\left(\sqrt{l m/2}\right)\right). \end{aligned}\right\} \tag{26.10}$$

Jedoch kann an Stelle dieses etwas unbequemen Ausdruckes eine numerische Näherung verwendet werden: es ist mit geringen Abweichungen

$$\tau_f = 1 - \Phi\left(\sqrt{\frac{l}{2} \cdot 1.55\, m}\right). \tag{26.11}$$

Man kann also sowohl beim Linien- wie auch beim Bandenspektrum (vgl. [26.4)] die Bedingung der diffusen Strahlung formal durch Verwendung einer optischen Dicke m, die auf etwa das 5/3fache vergrößert ist, in den Transmissionsfunktionen für gerade Strahlen berücksichtigen [*8*].

Unter den atmosphärischen Absorbern kann nach KAPLAN nur die Kohlensäure (und nach den Untersuchungen von GOODY und WORMELL[3] auch das wenig bedeutsame N_2O) durch ein regelmäßig periodisches Spektrum nach (26.5) angenähert werden. Dagegen hat COWLING[4] gezeigt, daß Transmissionsmessungen

---

[1] W. M. ELSASSER: Phys. Rev. **54**, 126 (1938) und [*7*].

[2] W. L. GODSON: CIR-2099, TEC-113, Met. Div., Dept. of Transport, Toronto 1952. — G. R. FOWLES: Strat. Rad., Techn. Rep. No. 7, Contr. AF 19 (122—392), Dept. of Phys., Univ. of Utah 1952. — L. D. KAPLAN: J. Meteorology **9**, 1 (1952); **10**, 100 (1953). — W. LUCK: Z. Naturforsch. **6**a, 191 (1951).

[3] R. M. GOODY u. T. W. WORMELL: Proc. Roy. Soc. Lond., Ser. A **209**, 178 (1951).

[4] T. G. COWLING: Phil. Mag. **41**, 109 (1950).

am Wasserdampf in einer Reihe von Spektralbereichen nicht durch das ELSASSERsche Modell beschrieben werden können. Nach seiner Ansicht ist die Ursache

Tabelle 27. *Transmissionsfunktionen.*

a) $2\,Ei_3$ *für diffuse monochromatische Strahlung (24.10).*

km	0,01	0,02	0,05	0,1	0,2	0,5	1,	2	5
$2\,Ei_3(km)$	0,981	0,962	0,910	0,833	0,704	0,443	0,219	0,060	0,002

b) τ_f *nach (26.10) für diffuse Strahlung in* ELSASSERS *Bandenmodell (26.5).*

lm	0,002	0,01	0,04	0,08	0,16	0,2	0,3	0,4	0,6	1,0	2	4
$\beta = 0{,}1$	0,970	0,902	0,794	0,712	0,603	0,562	0,481	0,419	0,326	0,213	0,086	0,019
$\beta = 0{,}5$	0,992	0,963	0,874	0,785	0,660	0,614	0,519	0,450	0,349	0,226	0,092	0,021
$\beta = 1{,}0$	0,996	0,981	0,928	0,866	0,762	0,717	0,624	0,546	0,429	0,280	0,116	0,027

in dem *unregelmäßigen* Bau der *Banden* zu suchen, denn bereits die Anordnung von Dubletts oder Tripletts mit größeren Zwischenabständen führt zu wesentlich abweichenden Transmissionsverläufen. GOODY und GODSON haben nachgewiesen[1], daß das Wasserdampfspektrum besser durch Zufallsverteilungen der Linienintensitäten S und der Linienabstände d erfaßt werden kann — das gleiche gilt für das Ozonspektrum — und haben für diesen Fall Transmissionsfunktionen entwickelt. Für eine Zufallsverteilung der Linienstellung ν_0, und eine Wahrscheinlichkeitsverteilung der Linienintensitäten S entsprechend einer Exponentialfunktion erhält GOODY eine Transmissionsfunktion

$$\tau_I = \exp\left(-\frac{\bar{S}/\bar{d}}{\sqrt{1+\frac{\bar{S}\,m}{\pi\,\delta}}}\,m\right), \tag{26.12}$$

wobei $\bar{S}$, $\bar{d}$ die mittlere Linienintensität und der mittlere Linienabstand im Bereich $\Delta\nu$ sind. GODSON hat eine ähnliche Verteilung der Linienintensitäten zugrunde gelegt und dafür auf numerischem Wege den Faktor γ gesucht, mit dem m multipliziert werden muß, um

$$\tau_f(m) = \tau_I(\gamma\, m) \tag{26.13}$$

zu machen. Er ist von m bzw. τ_I abhängig, hält sich aber zwischen 1,4 und 1,8 und darf auch noch für das über ν integrierte Wasserdampfspektrum als $\gamma = 1{,}7$ beibehalten werden.

27. Die Absorption der atmosphärischen Gase im Ultrarot. Im fernen Ultrarot treten als wirksame Absorber nur die dreiatomigen Gase Wasserdampf und Kohlensäure, mit geringerer Bedeutung Ozon und — vom energetischen Standpunkt vernachlässigbar — N_2O auf. Vielleicht ist letzteres für die Strahlungsbedingungen der hohen Atmosphäre von Bedeutung, ähnlich wie es das Ozon bei etwa 50 km Höhe ist. Die permanenten zweiatomigen Gase haben Absorptionswirkungen nur im Sichtbaren und UV.

Bei weitem der wichtigste Absorber ist der *Wasserdampf*. Schon in den frühesten Untersuchungen PASCHENS und v. BAHRS[2] wurden seine starken Absorptionsbanden bei 6,3 μ und bei $\lambda > 20\,\mu$ erkannt und erstmalig von HETTNER[3] gründlich durchgemessen. Die Absorption ist so stark, daß bei Zimmertemperatur

[1] R. M. GOODY: Quart. J. Roy. Met. Soc. **78**, 165 (1952). — W. L. GODSON: Proc. Toronto Met. Conf. 1953, 35, London 1954. — J. Meteorology **12**, 272 (1955).

[2] F. PASCHEN: Ann. Phys. **51**, 1 (1894); **52**, 209 (1894); **53**, 335 (1894). — E. v. BAHR: Ann. Phys. **29**, 780 (1909); **33**, 585 (1910).

[3] G. HETTNER: Ann. Phys. **55**, 476, 545 (1918).

und -feuchtigkeit 5 m Lichtweg bei 6,3 μ bereits eine Absorption von 70% liefern, bei 50 μ geschieht dasselbe schon auf 50 cm. Dabei ist in dem Absorptions-„Fenster" von 8 bis 11 μ die Absorption so schwach, daß sie selbst bei der Schichtdicke der gesamten Atmosphäre nicht einwandfrei gemessen worden ist. Auch die gemittelten Werte der Absorptionskoeffizienten erfüllen noch einen Schwankungsbereich von 4 Zehnerpotenzen.

Quantitative Messungen der Absorption bei Zimmertemperatur[1] sind schwierig durchzuführen, weil — anders als im Sichtbaren — die vom absorbierenden Gas vermöge seiner Temperatur emittierte Strahlung in der gleichen Größenordnung liegt wie die absorbierte. Haben Strahlungsquelle und absorbierendes Gas die gleiche Temperatur, so ist nach dem KIRCHHOFFschen Gesetz überhaupt kein Meßeffekt vorhanden. Man kann sich so helfen, daß man dem Strahler die Temperatur der flüssigen Luft gibt, also praktisch nur die Eigenstrahlung des Gases mißt, die gleich der Absorption einer Schwarzstrahlung seiner Temperatur ist.

Messungen in heißem Dampf liefern Ergebnisse, die mit der Absorption von Dampf unter atmosphärischen Drucken und Temperaturen nicht vergleichbar sind.

Bei den geringen atmosphärischen Gasmengen braucht man sehr lange Lichtwege. Die spektrale Auflösung bei Intensitätsmessungen genügt meist nicht zur Erfassung der einzelnen Linien. Die gemessene Absorption eines breiteren Bereiches erfolgt aber nicht nach reinen Exponentialfunktionen, sondern etwa nach (26.6) oder (26.12). Deutet man die Absorption trotzdem durch ein Gesetz e^{-km}, so findet man eine starke Abnahme von k mit der Schichtdicke, was z.B. die Messungen FOWLEs[1] sehr deutlich gezeigt haben [*1d*].

Wesentlich genauer als die *Intensitäten* sind die *Lagen* der Absorptionslinien innerhalb der einzelnen Banden gemessen worden; besonders sorgfältige Messungen in weiten Spektralbereichen haben TAYLOR[2] u. a. an Wasserdampf von hoher Temperatur aber niedrigem Druck, SHAW u. a.[3] am Sonnenspektrum, d.h. im atmosphärischen Wasserdampf durchgeführt.

Solche Messungen liefern die Energieniveaus und die Trägheitsmomente der Moleküle und damit Aufschlüsse über deren Bau, was insbesondere durch die Arbeiten von MECKE für das *Wassermolekül* geschehen ist. Dieses ist ein gleichschenkliges Dreieck mit dem O-Atom an der Spitze, der Winkel an der Spitze beträgt 104° 31′, der Abstand O—H 0,958 Å. Die drei Hauptträgheitsmomente sind verschieden, das größte ist der Rotationsachse senkrecht zur Dreiecksebene zugehörig, das mittlere zu der zwischen den beiden H-Atomen verlaufenden Achse. In dieser Achse liegt auch der mit großem Moment versehene *Dipol.*

Die Bande um etwa 50 μ gehört dem reinen Rotationsspektrum an, das beim Wasserdampf wegen des geringen Trägheitsmomentes relativ weit nach kurzen Wellen vorgeschoben ist (bei schwereren Molekülen, z.B. O_3, liegt es im Mikrowellengebiet). Die 6,3 μ-Bande entspricht dem ersten Anregungszustand der Schwingungsform ν_2 mit zahlreichen überlagerten Rotationsübergängen, während die anderen Fundamentalbanden ν_1 und ν_2 nahe zusammen in der X-Bande bei 2,7 μ liegen und damit außerhalb des Wärmestrahlungsbereiches bei atmosphärischen Temperaturen. Die Absorptionsbanden des kurzwelligen Infrarot und des sichtbaren Spektrums gehören zu höheren Schwingungs-Anregungsstufen (vgl. Ziff. 9). Auch bei gewöhnlichen Temperaturen ist immer eine sehr große Zahl von Rotationsniveaus angeregt.

Die Deutung der Spektren und deren theoretische Analyse ist nach dem Vorgang von MECKE[4] nahezu vollständig durch RANDALL, DENNISON, GINSBURG und WEBER[5] für das Rotationsspektrum, durch NIELSEN für die 6,3 μ-Bande

[1] F. E. FOWLE: Smithson. Miscell. Coll. **68**, No. 8 (1917). — Astrophys. J. **42**, 409 (1915). — R. MECKE: Phys. Z. **30**, 907 (1924). — M. NEUNHOEFFER: Ann. Phys. **2**, 334 (1929); **4**, 352 (1930). — J. KÜHNE: Z. Physik **84**, 772 (1933).

[2] J. H. TAYLOR, W. S. BENEDICT u. J. STRONG: J. Chem. Phys. **20**, 1884 (1952).

[3] J. H. SHAW, R. M. CHAPMAN, J. N. HOWARD u. M. L. OXHOLM: Astrophys. J. **113**, 268 (1951).

[4] R. MECKE u. a.: Z. Physik **81**, 331, 445, 465 (1933).

[5] H. M. RANDALL, D. M. DENNISON, N. GINSBURG u. L. R. WEBER: Phys. Rev. **52**, 160 (1937). — H. H. NIELSEN: Phys. Rev. **59**, 565 (1941); **63**, 422 (1942). — W. S. BENEDICT, H. H. CLAASEN u. J. H. SHAW: J. Res. Nat. Bur. Stand. Wash. **49**, 91 (1952).

und von BENEDICT, CLAASSEN und SHAW für den weiten Bereich von 4,5 bis 13 μ gegeben worden, der auch den durchlässigen Bereich oberhalb 8,5 μ einschließt[1]. Kenntnis der Konstanten des Molekülbaues gestattet eine Bestimmung der Absolutwerte der Intensitäten der einzelnen Linien im Rotationsspektrum und wenigstens relativer Werte der Intensitäten im Rotations-Schwingungsspektrum, die dann an Absorptionsmessungen breiter Spektralbereiche gewissermaßen geeicht werden können. Eine solche vollständige Neuberechnung der Lagen ν_0 und Intensitäten S gemäß (26.1) von 456 Linien im Rotationsspektrum

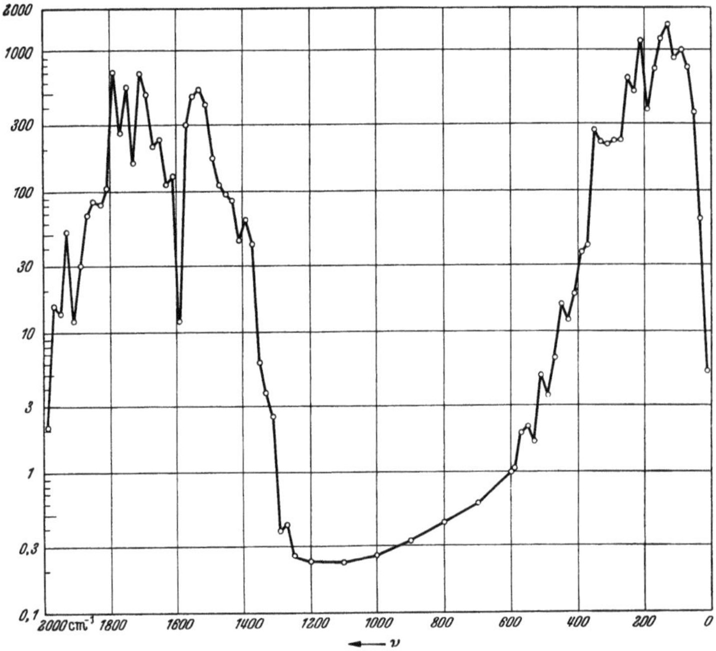

Fig. 23. Generalisierte Absorptionskoeffizienten l_ν nach ELSASSER in Abhängigkeit von der Schwingungszahl ν, für Wasserdampf, nach G. YAMAMOTO und G. ONISHI; für kurzwelliges Infrarot vgl. Fig. 7.

und 191 Linien in der 6,3 μ-Bande ist von YAMAMOTO und ONISHI[2] gegeben worden, ebenso von BENEDIKT u. a.[1] für denselben Bereich und das anschließende Fenster.

Zur Verwendung in (26.7) oder (26.12) wird noch die wichtige Größe der Halbwertsbreite δ benötigt. Nach Messungen von CORNELL[3] wurde früher $\delta = 0{,}25$ cm^{-1} angenommen. ADEL[4] fand dann an 2 Linien im kurzwelligen Ende des Rotationsspektrums $\delta = 0{,}105$. Sehr sorgfältige Ausmessungen der Linie $\lambda = 1{,}35$ cm im Mikrowellengebiet[5] lieferte bei 318° K für verschwindende Dampfdichte $\delta = 0{,}087$ und bei 50 g/m³ Dampfdichte $\delta = 0{,}107$ cm^{-1}. Nach der Theorie ist $\delta \sim \sigma N \sqrt{2kT}$, N ist die Molekülzahl in cm³, σ der Stoßdurchmesser des Wassermoleküls. Da $N \sim p/T$, wird

$$\delta = \frac{p}{p_0} \sqrt{\frac{T_0}{T}} \cdot \delta_0 . \tag{27.1}$$

[1] Siehe Fußnote 5, S. 225.
[2] G. YAMAMOTO u. G. ONISHI: Sci. Rep. Tôhoku Univ. (5) **1**, 5 (1949); (5) **3**, 73 (1951)
[3] S. D. CORNELL: Phys. Rev. **51**, 739 (1937).
[4] A. ADEL: Phys. Rev. **71**, 806 (1947).
[5] G. E. BECKER u. S. H. AUTLER: Phys. Rev. **70**, 300 (1946).

σ ist jedoch für Wasserdampf-Wasserdampf-Stöße (Eigendruckverbreiterung) verschieden von dem Wert für Wasserdampf-Luft-Stöße (Fremddruckverbreiterung). Aus diesen Beziehungen und durch Messungen konnte YAMAMOTO ableiten, daß bei $T=300°$ und 60% Feuchte $\delta = 0{,}098\, p/p_0$, bei $T=260$ (220) °K und verschwindendem Dampfgehalt $\delta = 0{,}096$ (0,105) $p/p_0\,\mathrm{cm}^{-1}$ ist (vgl. Ziff. 28). Die Wasserdampflinien haben nicht alle gleiche Intensitäten und Abstände, deshalb bestimmten YAMAMOTO und ONISHI den verallgemeinerten Absorptionskoeffizienten $l = 2\pi\,\delta\left(\frac{S}{d^2}\right)_m$ nach (26.8) für Intervalle von jeweils 20 cm^{-1} (Fig. 23). Eine Schwierigkeit, die bei diesen Bestimmungen noch keine Berücksichtigung

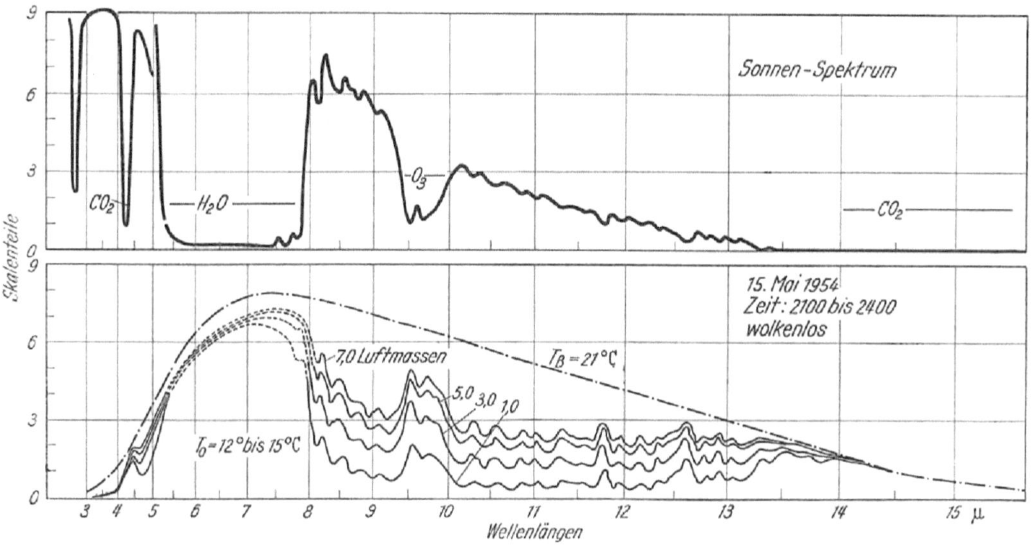

Fig. 24. Infrarotes Sonnenspektrum (oben). Spektrale Gegenstrahlung der Atmosphäre in einer wolkenlosen Nacht nach verschiedenen Zenitdistanzen (Luftmassen). Nach SLOAN, SHAW und WILLIAMS (Anm. 1, S. 239).

finden konnte, besteht darin, daß grundsätzlich — besonders bei Molekülen mit starken Dipolen wie beim Wasserdampf — zu erwarten ist, daß die Linienbreite von Linie zu Linie, von Bande zu Bande variiert. Hierüber liegen jedoch fast noch keine Messungen vor.

In dem Bereich des Fensters zwischen Rotations- und 6,3 μ-Bande ist die Absorption oder Emission der Atmosphäre quantitativ schwierig zu messen und deshalb teils bestätigt, teils bestritten worden. ELSASSER nahm deshalb an, daß die geringe Absorption als Wirkung der Flügel der weitentfernten, aber sehr starken Linien im Rotationsspektrum gedeutet werden könne, ein Effekt, der sich als zusätzliche kontinuierliche Absorption grundsätzlich auch in allen anderen Intervallen der Linienwirkung überlagert und berücksichtigt werden muß. Allerdings wird gerade im Fenster für diese Deutung der Absorption bedeutsam, daß die exakte Linienform in den fernen Flügeln eine Unsymmetrie zeigt mit geringerer Absorption am kurzwelligen Ende. Dadurch wird die berechnete Absorption zwischen 8 und 12 μ im Vergleich mit den Beobachtungen zu schwach, muß also ebenfalls eine Linienabsorption sein. ADEL und LAMPLAND[1] sowie SHAW und Mitarbeiter haben das Sonnenspektrum in diesem Bereich sorgfältig vermessen, BENEDICT u.a.[1] konnten die gefundenen Linien als dem Wasserdampf zugehörig nachweisen.

Mit seinen beiden intensiven Banden überdeckt der Wasserdampf den größten Teil der Schwarzstrahlung bei Atmosphärentemperaturen, nur zwischen 8 und 16 μ bleibt ein breites, transparentes Fenster (Fig. 24). Dieses wird in seiner

[1] A. ADEL u. C. O. LAMPLAND: Astrophys. J. **91**, 1, 481 (1940). — W. S. BENEDICT, H. H. CLAASEN u. J. H. SHAW: J. Res. Nat. Bur. Stand. Wash. **49**, 91 (1952).

langwelligen Hälfte durch das *Kohlendioxyd* verdeckt, welches um 15 μ eine ebenfalls außerordentlich stark absorbierende Bande besitzt. Das CO_2-Molekül ist geradlinig, das Kohlenstoffatom in der Mitte wird auf beiden Seiten flankiert von den O-Atomen. Infolgedessen fällt ein Schwingungszustand (ν_1) weg, die Rotationsbande liegt weit im Mikrowellengebiet, die Bande bei 15 μ ist das Fundamentalsystem der ν_2-Schwingung, während der ν_3-Schwingung eine Bande bei 4,3 μ entspricht, die wiederum in ein vom H_2O transparent gelassenes Gebiet fällt. Dort hört jedoch die Schwarzstrahlung bei Atmosphärentemperaturen schon auf.

Die älteren Beobachtungen der Absorption sind von CALLENDAR gesammelt[1] und durch eine empirische Absorptionsfunktion gut angenähert worden. KAPLAN[2] hat alle Einzellinien in dem Bereich 14 bis 16 μ hinsichtlich Lage und Intensität theoretisch berechnet und durch ältere Messungen bestätigt. Eingehende neuere Messungen sind in dem 100 ft langen Absorptionsrohr J. STRONGs in Baltimore mit einem Lichtweg, der durch mehrfache Spiegelung auf 190 m gebracht wurde, von PETERS und CLOUD durchgeführt worden, wobei mit verhältnismäßig weitem Spalt ($\frac{1}{2}$ μ) weniger auf die Linienstruktur als auf gute Absorptionsbestimmungen Wert gelegt wurde[3].

ELSASSER und KING [*8*] haben diese — bei verschiedenen Drucken und mit verschiedenen Fremdgasbeimengungen durchgeführten — Messungen einheitlich bearbeitet. Sie setzen voraus, daß der Absorptionsverlauf in jedem der sechs 1 μ-Intervalle zwischen 12 und 18 μ ein Teilstück einer universellen Absorptionsfunktion τ_c ist. Ein für jedes Wellenlängenintervall charakteristischer Faktor sagt dann aus, ob man bei den gemessenen Kohlesäuremengen c in dem Anfangstück der Funktion mit kleinen Absorptionen oder in dem Endstück mit großen Absorptionen liegt. Dieser Faktor ist eine Art verallgemeinerter Absorptionskoeffizient l_c, und die zugehörige Funktion τ_c wird empirisch festgelegt. l_c läßt sich nicht nach Art der Gl. (26.8) herleiten, weil es sich nicht auf die Funktion (26.7) bezieht. Das Verfahren setzt voraus, daß die Verteilung von Linienabständen und Intensitäten vom ELSASSERschen Bandenmodell (26.5) abweicht, ohne diesen Unterschied genau zu definieren, nimmt aber an, daß die Verteilung in jedem 1 μ-Intervall der 15 μ-Bande die gleiche ist. Der Abfall der Funktion erfolgt von $\tau_c = 0{,}9$ bis 0,1 in fast $3\frac{1}{2}$ Zehnerpotenzen von c, bei τ_f mit $\beta = 0{,}1$ nach Tabelle 27b bereits in 2 Zehnerpotenzen. Die Berechtigung dieser ELSASSERschen Annahme wäre nachzuprüfen.

Die so ermittelte Transmission der gesamten Bande in Teilen der Schwarzstrahlung des Bereiches 11 bis 19 μ ist in Tabelle 28 wiedergegeben. c ist die in cm gemessene Dicke der (auf Normaldruck und Normaltemperatur gebrachten) Gasschicht, $c_r = c p/p_0$ ist die effektiv absorbierende Kohlensäuremenge. — Der Kohlensäuregehalt der Luft ist mit nur geringen Schwankungen 0,03%, es befindet sich also zwischen Erdboden und 30 m Höhe bereits $c_r = 1{,}0$ cm NTP. In dieser Entfernung ist bereits ein Viertel des Spektralbereiches von 11 bis 19 μ absorbiert, also eine äquivalente Breite von 2 μ total ausgelöscht. Die Reichweite der Strahlung ist in der 15 μ-Bande sehr gering.

Tabelle 28. *Transmission T der 15 μ-Bande des CO_2 in Teilen der Schwarzstrahlung 11 bis 19 μ, in Abhängigkeit von* $c_r = c\,\dfrac{p}{p_0}$ *cm NTP.*

c_r	0,01	0,02	0,05	0,1	0,2	0,5	1	2	5	10	20	50	100	200	500	1000 cm
T	94,7	93,7	91,0	88,1	84,7	80,0	76,4	72,4	67,2	63,5	60,0	55,5	52,3	49,2	45,0	42,1%

[1] G. S. CALLENDAR: Quart. J. Roy. Met. Soc. **67**, 263 (1941). — H. RUBENS u. E. LADENBURG: Verh. phys. Ges. **7**, 170 (1905). — G. HERTZ: Verh. phys. Ges. **13**, 617 (1911). — C. SCHAEFER u. B. PHILIPPS: Z. Physik **36**, 641 (1926). — P. MARTIN u. E. BARKER: Phys. Rev. **41**, 176 (1932). — E. BARKER u. A. ADEL: Phys. Rev. **44**, 185 (1933).

[2] L. D. KAPLAN: J. Chem. Phys. **18**, 186 (1950).

[3] C. W. PETERS: Contr. N 5 ori-166, Task Order V, Johns Hopkins Univ. Baltimore Md. 1949. — W. H. CLOUD: Contr. N-onr 248-01, Hohns Hopkins Univ. Baltimore, Md. 1952.

Das *Ozon* besitzt im Langwelligen Absorptionsbanden bei 3,28 und 3,57 μ, die außerhalb des atmosphärischen Wärmestrahlungsbereiches liegen, eine Bande (ν_3) bei 9,6 μ (Fig. 25a), die sehr wirksam gerade im Wasserdampffenster liegt[1], und eine bei 14,2 μ (ν_2), die ziemlich genau mit der 15 μ-Bande des CO_2 zusammenfällt. Das Ozonmolekül ist ein Dreieck, bei dem der Winkel an der Spitze 117° beträgt, wie durch Messungen im Rotationsspektrum (Mikrowellengebiet) nachgewiesen wurde[2]. Die dem N_2O oder N_2O_5 angehörigen Banden bei 4,25, 6,7,

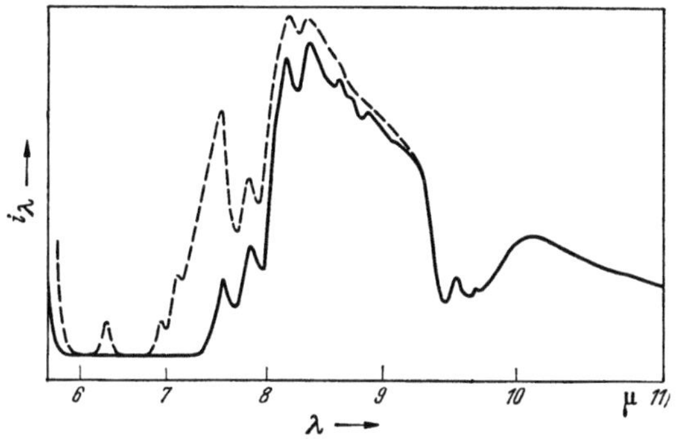

Fig. 25a. Sonnenspektrum im Infrarot, echte Veränderung der H_2O-Bande bei 6,3 μ, und an deren Flanke scheinbare Veränderungen der HDO- (7,1 μ) und N_2O-Banden (7,8 μ). O_3-Bande bei 9,6 μ. Nach A. ADEL[1].

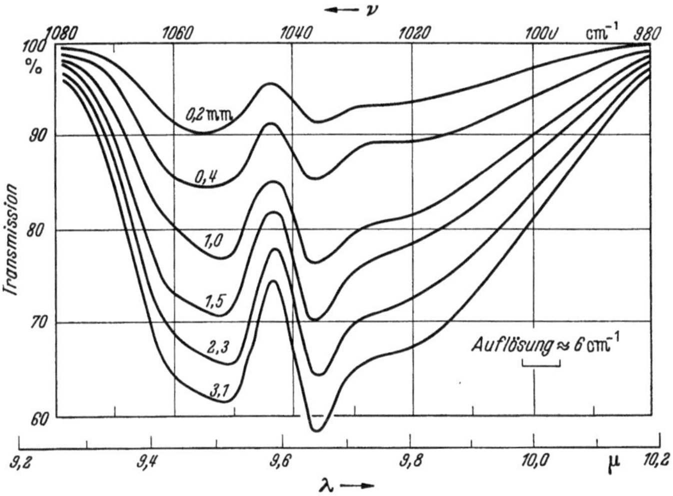

Fig. 25b. Ozonbande bei 9,6 μ für verschiedene absorbierende Massen (in mm N.T.P.) beim Gesamtdruck 60 mm Hg. Nach M. SUMMERFIELD und A. S. FILLER.

7,6 und 11,38 μ wurden früher dem Ozon zugeschrieben, weil diese Gase fast immer bei Ozonzerfall entstehen und mit ihm vergesellschaftet sind. Den besten Überblick über die Banden gab bisher eine Arbeit von HETTNER, POHLMANN und SCHUMACHER[3]; diese arbeiteten jedoch mit reinem Ozon unter hohem Druck und erhielten Absorptionskoeffizienten, die wesentlich kleiner sind als die des atmosphärischen Ozons.

[1] A. ADEL: in The atmospheres of the earth and planets, ed. by G. KUIPER, Kap. 10. Chicago 1949.

[2] R. H. HUGHES: Bull. Amer. Phys. Soc. **26**, No. 6, 21 (1951). — Phys. Rev. **85**, 717 (1952). — J. N. HOWARD u. J. H. SHAW: J. Chem. Phys. **21**, 558 (1953).

[3] G. HETTNER, R. POHLMANN u. H. J. SCHUMACHER: Z. Physik **91**, 372 (1934).

Die 9,6 μ-Bande ist von Strong[1] mit Ozonmengen und unter Fremdgasdrucken durchgemessen worden, die den Verhältnissen der Atmosphäre entsprechen. Dabei wurden etwa 10mal größere Absorptionen gefunden als von Hettner und Mitarbeitern. Die sorgfältigste Untersuchung ist offenbar in einer nicht veröffentlichten Arbeit von Summerfield[2] enthalten, die von Elsasser und King[3] bearbeitet worden ist (Fig. 25b). Diese wandten zur Analyse das gleiche Verfahren an, das oben bei der Kohlensäure beschrieben worden ist und definierten somit eine dem Ozon eigene Transmissionsfunktion τ_0 und einen darauf bezogenen verallgemeinerten Absorptionskoeffizienten l_0. Eine sorgfältige Analyse der einzelnen Linien der 9,6 μ-Bande hat Filler[4] durchgeführt.

Der Ozongehalt der Troposphäre und unteren Stratosphäre ist sehr gering, so daß sein Einfluß auf die Strahlungsvorgänge innerhalb dieser Schichten vermutlich zu vernachlässigen ist. Die 9,6 μ-Bande besitzt jedoch wegen ihrer Lage in der Lücke des Wasserdampfspektrums eine gewisse Abschirmwirkung auf die vom Boden ausgehende Effektivstrahlung. Insgesamt ist diese gering, auch die Schwankungen der Ozonmenge sind nicht so groß, daß sehr merkbare Schwankungen der effektiven Ausstrahlung dadurch bewirkt werden können. Im Sonnenspektrum, das von Wasserdampf und Kohlensäure um 10 μ durchgelassen wird, ist die Ozonbande deutlich zu erkennen.

Die Bedeutung der 14 μ-Bande ist umstritten. Für die Strahlung an der Erdoberfläche besitzt sie keine Wirkung, weil sie durch die CO_2-Bande vollständig überdeckt wird. Jedoch könnte sie bei der hohen Lage der Ozonschicht in der Atmosphäre eine Art Abschirmung für die Ausstrahlung der Kohlensäure in den Weltraum bewirken. Möller[5] konnte für eine solche Wirkung qualitative Argumente beibringen. Elsasser und King[3] weisen jedoch darauf hin, daß ihre Werte l_0, die sie durch Vergleich der Hettnerschen Messungen bei 14,1 und 9,6 μ mit ihren Werten abschätzen, so geringfügige Absorptionen ergeben, daß man keinen spürbaren Effekt erwarten darf. Auch ist die Absorption des Kohlendioxyds wahrscheinlich noch oberhalb der Ozonschicht sehr stark.

28. Die Abhängigkeit der Absorption von Druck und Temperatur. Auf die wichtige Rolle des Gesamtdruckes des Gasgemisches, in dem sich der Absorber befindet, wurde oben schon mehrfach hingewiesen. Die Dispersions-Linienbreite ist proportional der Stoßfrequenz der Moleküle und somit

$$\delta = \frac{1}{4\pi} \sum N_i \,(\sigma + \sigma_i)^2 \left[2\pi k T \left(\frac{1}{M} + \frac{1}{M_i}\right)\right]^{\frac{1}{2}} \tag{28.1}$$

mit den optischen Stoßdurchmessern σ, σ_i für das absorbierende und das i-te störende Molekül, N_i = Molekülzahl je cm³, M, M_i = Molekulargewichte, k = Boltzmann-Konstante. Damit ist die Halbwertsbreite proportional der Anzahl der Stöße, die das Molekül erleidet, und, wegen $N \sim p/T$, proportional dem Gasdruck. In erster Näherung (wenn $\sigma = \sigma_i$ gesetzt), sollte der Eigendruckeinfluß doppelt so groß sein wie der Einfluß eines Fremdgases. Jedoch sind die Stoßdurchmesser für verschiedene miteinander reagierende Gase durchaus verschieden, am größten sind sie für Moleküle mit großem Dipolmoment. Die absorbierenden Gase (H_2O, CO_2, O_3) haben in der Atmosphäre so geringe Partialdrucke, daß ihre Eigendruckverbreiterung meist vernachlässigt werden kann. Sinnfällig wird der Einfluß der Druckverbreiterung z. B. beim Vergleich der Hettnerschen

[1] J. Strong: J. Franklin Inst. **231**, 121 (1941).
[2] M. Summerfield: Thesis, Calif. Inst. Techn. 1941 (nicht veröffentlicht).
[3] W. M. Elsasser u. J. J. King: Stratosph. Rad. Techn. Rep. No. 9, Contract AF 19 (122)—392, Univ. of Utah 1953.
[4] A. S. Filler: Stratosph. Rad., Techn. Rep. No. 2, Contract AF 19 (122)—392, Univ. of Utah, 1951.
[5] F. Möller: Naturwiss. **31**, 148 (1943).

Absorptionskoeffizienten des Ozons, die bei 227 mm Hg Eigendruck um etwa eine Zehnerpotenz kleiner sind als beim gleichen Luftdruck, aber nur sehr kleinem Ozon-Partialdruck. Ebenso ist der Einfluß bedeutend in den Messungen von GOODY und WORMELL[1] bei der 7,8 μ-Bande des N_2O (Fig. 26), die eine Steigerung der Absorption auf das 8fache erkennen lassen bei Zunehmen des Gesamtdruckes von 2,6 auf 688 mm Hg aber gleicher absorbierender Gasmasse.

Nach (27.1) ist die Halbwertsbreite dem Druck direkt, der Wurzel aus der absoluten Temperatur umgekehrt proportional. Da T in der Atmosphäre nur zwischen etwa 200 und 300°, $\sqrt{T}$ also um einen Faktor 1,22 schwankt, ist dieser Einfluß wesentlich geringer als der des Luftdruckes mit 10:1 zwischen 0 und 16 km Höhe und wird deshalb neben dem Luftdruck meist vernachlässigt. Dagegen ist über die Veränderung des Stoßdurchmessers σ mit der Temperatur wenig bekannt. WEISSKOPF[2] erwartete $(\sigma+\sigma_i) \sim T^{-0,1}$ bei Stößen ungleicher, $(\sigma+\sigma_i) \sim T^{-0,25}$ bei Stößen zwischen gleichen Molekülen. Dagegen fanden GOODY und WORMELL[1] für N_2O den Exponenten zwischen $+0,25$ und $+0,75$. Weitere Messungen erscheinen wichtig.

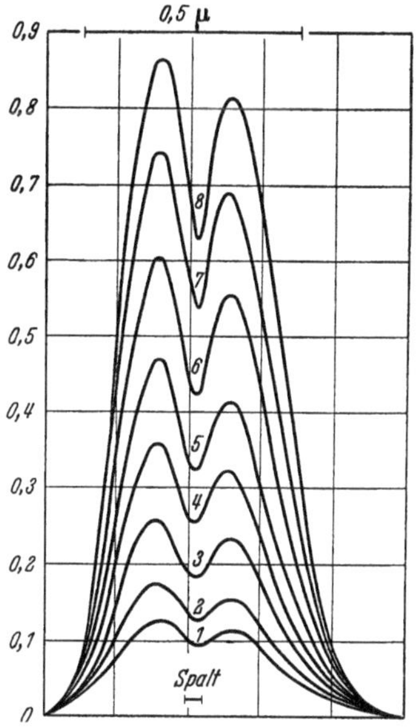

Fig. 26. Absorption der 7,8 μ-Bande des N_2O nach GOODY und WORMELL. Absorbierende Masse $m = 1{,}196 \cdot 10^{-3}$ g cm^{-2}, $T = 20{,}1°$ C, Weglänge 271 cm. Fremdgas trockene Luft. Gesamtdrucke: *1* 2,69, *2* 6,36, *3* 16,7, *4* 40,7, *5* 79,4, *6* 164,3, *7* 326,7, *8* 688,3 mm Hg.

Besonders übersichtlich ist der Einfluß der Linienbreite in der ELSASSERschen Bandenfunktion (26.7) zu erkennen, nach der die Halbwertsbreite δ den gleichen Einfluß hat wie m. Setzt man $\delta m = \delta_0 \frac{p}{p_0} m$, so kann der Faktor $\frac{p}{p_0}$ rechnerisch statt zu δ_0 auch zu m genommen werden. Das bedeutet, daß man dem Einfluß der veränderten Linienbreite auch durch eine veränderte absorbierende Masse

$$m_r = m\,p/p_0 \tag{28.2}$$

Rechnung tragen kann. Diese Reduktion der Massen auf gleiche Halbwertsbreite mit p/p_0 ist deshalb allgemein üblich (vgl. Kohlensäure, S. 228).

Durch sorgfältige *Messungen* der Linienbreite bei verändertem Druck konnte im Mikrowellengebiet einwandfrei festgestellt werden, daß die Verbreiterung wirklich proportional p verläuft[3]. Eine ausgezeichnete Bestätigung der genauen Proportionalität nach (27.1) erhielten GOODY und WORMELL bei ihren Untersuchungen der Absorptionsbanden von N_2O[2]. Dagegen schien sich nach Messungen FALCKENBERGS und deren rechnerischer Bearbeitung durch SCHNAIDT[4] herausgestellt zu haben, daß an m nicht ein Faktor p/p_0, sondern $\sqrt{p/p_0}$ anzubringen sei. Es ist deshalb lange Zeit so gerechnet worden. Jedoch kann der Wurzeleffekt in der Transmission vorgetäuscht sein durch die Verhältnisse, die bei dicht benachbarten Linien mit starker Überlappung (Dubletts) auftreten[5], wobei aber die Absorptions-Koeffizienten der einzelnen Linie noch streng die Formel (27.1) befolgen.

[1] R. M. GOODY u. T. W. WORMELL: Proc. Roy. Soc. Lond., Ser. A **209**, 178 (1951).
[2] W. WEISSKOPF: Z. Physik **75**, 287 (1932).
[3] C. H. TOWNES u. F. R. MERRITT: Phys. Rev. **70**, 558 (1946).
[4] G. FALCKENBERG: Meteor. Z. **55**, 174 (1938). — F. SCHNAIDT: Gerlands Beitr. Geophys. **54**, 203 (1939).
[5] M. SUMMERFIELD u. J. STRONG: Phys. Rev. **60**, 162 (1941). — F. MATOSSI u. E. RAUSCHER: Z. Physik **7**, 418 (1949). — L. D. KAPLAN: J. Meteorology **9**, 1 (1952).

HOWARD, BURCH und WILLIAMS[1] haben jeweils die Absorption einer ganzen Bande des Wasserdampfes in ihrer Abhängigkeit von der absorbierenden Masse m, vom Gesamtdruck p und Partialdruck ε des Dampfes und von der Temperatur gemessen. Sie verwendeten „synthetische Atmosphären" aus Wasserdampf mit Stickstoff als Fremdgas. Die ausführlich wiedergegebenen Meßresultate wurden zunächst durch empirische Näherungsformeln wiedergegeben (Ziff. 9, Tabelle 6), jedoch haben HOWARD und Mitarbeiter für schmale Frequenzbereiche, die jeweils nur $\frac{1}{10}$ oder $\frac{1}{20}$ einer Bande überdecken, die Formel GOODYS (26.12) in genauer Übereinstimmung mit ihren Meßresultaten gefunden. Bei Verwendung der Konstanten, die bei einem Gesamtdruck $p_0 = 750$ mm Hg gefunden wurden, aber Einsetzen einer nach (27.1) auf einen Druck p reduzierten Linienbreite ergab sich vorzügliche Übereinstimmung auch mit den Meßergebnissen, die bei p gefunden waren. Dies ist nicht nur eine Bestätigung für die Richtigkeit der Formel (26.12), sondern auch für die lineare Abhängigkeit $\delta \sim p$.

Sehr klar lassen sich die Verhältnisse auch bei einer einzelnen Linie überschauen. Hier tritt nach (26.1) in der Exponentialfunktion $k(\nu)\, m = S\,\delta\, m\, \pi^{-1}\left((\nu-\nu_0)^2+\delta^2\right)^{-1}$ auf, also im Zähler das Produkt $\delta m = \delta_0 m\, p/p_0$. Die Veränderung von δ^2 im Nenner mit $(p/p_0)^2$ spielt dann keine Rolle, wenn dieses Glied gegen $(\nu-\nu_0)^2$ vernachlässigt werden kann, also bei Abständen von der Linienmitte, die groß sind gegen die Halbwertsbreite. Die Verhältnisse in den Linienflügeln sind dann allein maßgebend, wenn durch starke Absorption die Linienmitte bereits vollständig ausgelöscht ist [7]. Dann ist mithin die Verwendung der proportional p reduzierten Masse m berechtigt [Formel (26.7) gilt auch nur für große Argumente]. Bei schwachen Absorptionen müssen andere Gesetze gelten, da das Glied δ^2 im Nenner nicht mehr gegen $(\nu-\nu_0)^2$ vernachlässigt werden kann.

PLASS und WARNER[2] haben außerdem darauf hingewiesen, daß die exakte Formulierung der Dispersionsform (26.1) lautet

$$k(\nu) = \frac{S}{\pi}\,\frac{\delta}{(\nu-\nu_0+A\,\delta)^2+\delta^2}, \tag{28.3}$$

wobei $A = 0{,}725$ gesetzt werden kann. Die Berücksichtigung des mit p veränderlichen δ im Nenner von (28.3) wird dann mit einer Druckverschiebung der Linie gleichbedeutend, die sehr wichtig sein kann. Sie konnten aber zeigen, daß dieser Einfluß bei atmosphärischen Bedingungen geringfügig ist.

Der Einfluß einer veränderten Halbwertsbreite bei anderen Modellen, z.B. dem Zufallsmodell GOODYS (26.12) ist komplizierter.

Der erwähnte Einfluß des Druckes oder der Temperatur auf die Linienbreite ändert nicht die Gesamtintensität der Linie, da $S = \int_{-\infty}^{+\infty} k(\nu)\,d\nu$ unabhängig von δ ist. Dagegen ändert sich hierbei die Absorption selbst, für die man mit (27.1) in der Näherung (26.3) setzen kann $A = (1-\tau) = \frac{2}{\Delta\nu}\sqrt{S\,\delta_0 m p/p_0}\,\sqrt[4]{T_0/T}$ oder A proportional mit der Wurzel aus p/p_0, aber umgekehrt proportional der vierten Wurzel aus T/T_0.

Neben dem darin enthaltenen *Temperatureinfluß* unterliegt jedoch auch der Absolutwert der Absorption einer Veränderung mit der Temperatur. Die Intensität einer Absorption ist proportional der Wahrscheinlichkeit, mit der der zugehörige Energieübergang $E_2 - E_1$ stattfindet. Diese ist $N_i = \exp(-E_i/kT)$, wobei k die BOLTZMANN-Konstante ist. Für kleine Energiestufen ist die Wahrscheinlichkeit etwa 1, für große wird sie selbst klein, aber ihre Temperatur-

[1] J. N. HOWARD, D. E. BURCH u. D. WILLIAMS: Geophys. Res. Papers, Cambridge (Mass.) **1955**, No. 40.

[2] G. N. PLASS u. D. WARNER: J. Meteorology **9**, 333 (1952).

abhängigkeit $dN_i/dT = N_i E_i/kT^2$ wird um so größer, je größer E_i ist. Die Linien mit großen Energieübergängen (Quantenzahlen) liegen an den Flügeln der Banden. Es ist also möglich, daß gerade die Absorptionen an den Grenzen des transparenten Bereiches oder in ihm besonders temperaturabhängig sind. Andeutungen dafür liegen in Beobachtungen ROBINSONS[1] vor, eine genauere Nachprüfung sowie eine quantitative Abschätzung für das ganze Spektrum fehlen noch.

Die Hauptschwierigkeit bei der Berücksichtigung der druckabhängigen Absorption liegt darin, daß die Strahlung in der Atmosphäre alle Schichten mit einem von etwa 1000 bis 0 mb variierenden Druck durchsetzt. Mit der einfachen Druckproportionalität (28.1) bereitet das keine Schwierigkeit; man hat nur zu setzen

$$m_r = \int \varrho_w(z) \frac{p(z)}{p_0} \, dz. \tag{28.4}$$

Jedoch ist dies nur eine Näherung, die auf den Näherungen (26.3) oder (26.7) der Absorptionsfunktion beruht und schon durch starke Überlappung gestört werden kann, auch wenn die Druckproportionalität (27.1) exakt gilt. Zum ersten Male hat 1942 PEDERSEN[2] auf die Wichtigkeit des Druckeffektes hingewiesen, in zahlreichen theoretischen Arbeiten haben sich später PLASS, KAPLAN, KING u. a.[3] mit der Frage der Transmissionsfunktion oder ihrer Annäherung durch leicht zu behandelnde Funktionen beschäftigt für den Fall einer Veränderung sowohl des Druckes wie der Konzentration des Absorbers auf dem Strahlenweg.

Diese komplizierten Zusammenhänge sind auch der Grund, daß MÖLLER[4] für den Wasserdampf auf halbempirischem Wege einen Druckfaktor μ fand, mit dem die Dichte der strahlenden Masse zu reduzieren ist

$$\varrho_{wr} = \mu_w(p)\, \varrho_w; \qquad \mu_w = 0{,}985 \left(\frac{p}{p_0}\right)^{\frac{4}{5}} + 0{,}015 \frac{p_0}{p}.$$

Diese Formel ist an atmosphärische Verhältnisse gebunden, wo eine starke Koppelung zwischen p, T und ϱ_w besteht. Der erste Term zeigt die Wirkung der Linienflügel in tiefen Schichten, der zweite die der Linienzentren in großen Höhen. Für Kohlensäure konnte MÖLLER aus Laboratoriumsmessungen von WIMMER[5] an der 4,27 μ-Bande ableiten: $\mu_C = (p/p_0)^{\frac{3}{4}}$. Ein geringerer Exponent als 1 wird auch durch Abweichungen wahrscheinlich gemacht, die die Messungen von CLOUD bei hohen Drucken gegen die verallgemeinerte Transmissionsfunktion für CO_2 in der Bearbeitung ELSASSERS und KINGS zeigen.

Der große Wert der in Ziff. 27 wiedergegebenen sorgfältigen Detailuntersuchungen der einzelnen Absorptionslinien innerhalb der Banden liegt darin, daß man bei Kenntnis der Frequenz ν_0, der Intensität S und der Halbwertsbreite δ die Messungen bei verschiedenen, aber in der absorbierenden Säule einheitlichen Bedingungen durch die Theorie deuten, und dann die Erweiterung auf natürliche Atmosphären mit ihren veränderlichen Drucken und Temperaturen entlang des Lichtweges vornehmen kann[6]. Dies erfordert immer noch viel experimentelle und theoretische Arbeit.

[1] G. D. ROBINSON: Quart. J. Roy. Met. Soc. **76**, 37 (1950). — F. MÖLLER: in Compendium of Met., S. 34—49. Boston 1951. — O. LÖNNQVIST: Ark. Geof. **2**, Nr. 12, 246 (1554).

[2] F. PEDERSEN: Met. Annaler **1**, 115 (1942).

[3] J. STRONG u. G. N. PLASS: Astrophys. J. **112**, 365 (1950). — G. N. PLASS: J. Meteorology **9**, 429 (1952); **11**, 163 (1954). — J. Opt. Soc. Amer. **42**, 677 (1952). — G. N. PLASS u. D. I. FIVEL: Astrophys. J. **117**, 225 (1953). — J. Meteorology **12**, 191 (1955). — Quart. J. Roy. Met. Soc. **81**, 48 (1955). — L. D. KAPLAN: J. Meteorology **9**, 1 (1952). — J. I. KING: J. Meteorology **9**, 311 (1952). — Astrophys. J. **116**, 491 (1952).

[4] F. MÖLLER: Meteor. Z. **61**, 37 (1944).

[5] M. WIMMER: Ann. Phys. **81**, 1091 (1926).

[6] A. R. CURTIS: Quart. J. Roy. Met. Soc. **78**, 638 (1952).

29. Die Strahlungsdiagramme. Die ersten Berechnungen der Strahlungsvorgänge sind von GOLD[1] durchgeführt worden und ergaben schon numerisch richtige Resultate. EMDEN[2] hat in einer weitgespannten Untersuchung das Problem erschöpfend behandelt, jedoch rechnete auch er noch mangels genauerer Kenntnisse in der damaligen Zeit mit grauer Absorption für den Wasserdampf. Das erste rohe, aber anschauliche Modell, das die Wasserdampfbanden berücksichtigte, wurde von SIMPSON[3] aufgestellt. Einzelfälle wurden exakt von ROBERTS[4] und ALBRECHT[5] behandelt. MÜGGE und MÖLLER[6] verwendeten erstmalig das Prinzip der Strahlungsdiagramme, das durch Einführung zahlreicher Verfeinerungen von ELSASSER[7] ausgebaut wurde. Die neuesten Modelle sind von MÖLLER, ELSASSER und KING und YAMAMOTO[8] berechnet worden.

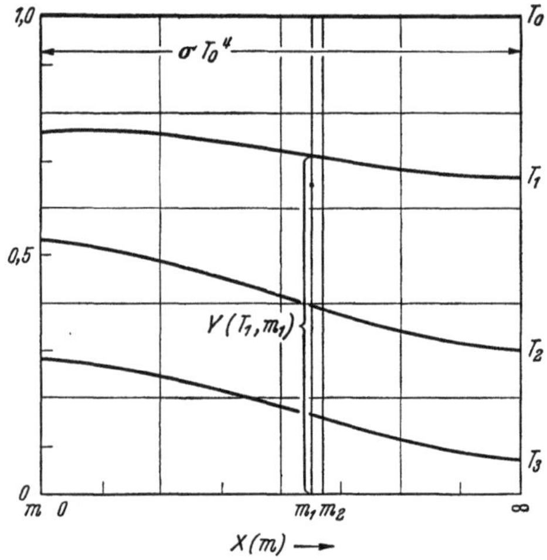

Fig. 27. Schematische Darstellung des Strahlungsdiagramms nach MÖLLER.

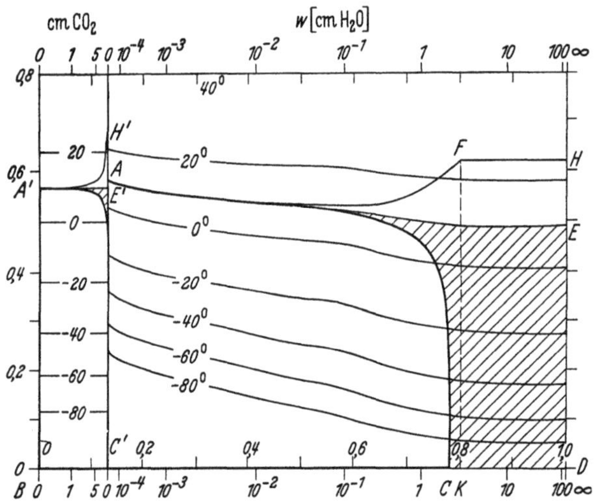

Fig. 28. Strahlungsdiagramm nach F. MÖLLER. Fläche $A'E'AEDBA'$ = Schwarzstrahlung fur 10° C; fur eine spezielle Atmosphärenschichtung gibt $A'C'BA' + ACC'A$ die von oben kommende (Gegen-) Strahlung G, die schraffierte Fläche die effektive Ausstrahlung $\sigma T^4 - G$, $A'H'AFHDBA'$ die von unten kommende Strahlung U, $FHDKF$ den von der Erdoberfläche stammenden Anteil von U, alles für das Niveau, wo die Temperatur $T = +10°$ herrscht.

Das Prinzip der Strahlungsdiagramme ist bereits in Ziff. 25 durch die Gln. (25.3) und (25.4) gegeben worden. Bei der numerischen Auswertung von (25.2) wird nicht mit der Transmissionsfunktion für monochromatische Strahlung Ei_3 gerechnet, sondern man verwendet auch hier τ_f nach einer der Gleichungen für Linien oder Banden (26.4). (26.10), (26.11) oder (26.12). Nach (25.3) ist $X(m_r)$ die Strahlung einer Wasserdampfschicht der Dicke m_r und der Temperatur T_0. Eine dünne Schicht $(m_2 - m_1)$ strahlt mit der Intensität $\Delta X = X(m_2) - X(m_1)$, was an der Funktionsskala X leicht abgelesen werden kann. Hat diese Schicht die Temperatur T, so ist

[1] E. GOLD: Proc. Roy. Soc. Lond., Ser. A **82**, 43 (1909).
[2] R. EMDEN: Sitzgsber. bayr. Akad. Wiss. **43**, 55 (1913).
[3] G. C. SIMPSON: Mem. Roy. Met. Soc. **3**, Nr. 21 (1928).
[4] O. F. T. ROBERTS: Proc. Roy. Soc. Edinburgh **50**, 225 (1929).
[5] F. ALBRECHT: Z. Geophys. **6**, 421 (1930).
[6] R. MÜGGE u. F. MÖLLER: Z. Geophys. **8**, 53 (1932).
[7] W. M. ELSASSER: Quart. J. Roy. Met. Soc. **66**, Suppl., 41 (1940). — Harvard Met. Stud. **6** (1942).
[8] F. MÖLLER: Das Strahlungsdiagramm. Reichsamt für Wetterdienst, Berlin 1943. — W. M. ELSASSER u. J. I. KING: Persönliche Mitteilung 1953. — G. YAMAMOTO: Sci. Rep. Tôhoku Univ. (5) **4**, 9, 98 (1952).

ΔX mit $Y(T, m)$ zu multiplizieren, um die Strahlung richtig zu erhalten. Definitions gemäß ist $Y(T_0, m) = 1$. In Fig. 27 ist, über $X(m)$ als Abszisse, $Y(T, m)$ aufgetragen und die Linien mit T beziffert. Die gesamte Fläche der Abbildung gibt σT_0^4, die Fläche unter einer T-Linie die Schwarzstrahlung dieser Temperatur, weil die Strahlung einer unendlich dicken Gasschicht gleich σT^4 ist, wenn nur in keiner Wellenlänge der Absorptionskoeffizient genau gleich 0 wird. Bei beliebig geschichteter Atmosphäre mit dem Gesamtgehalt M_r ist T als Funktion von m_r einzutragen und die erhaltene Fläche von 0 bis M_r zu integrieren. Sie gibt unmittelbar den Strahlungsstrom auf die Auffangfläche an.

In dem Diagramm Fig. 28 ist die Voraussetzung gemacht, daß die Kohlensäure im Bereich 13,5 bis 16,5 μ eine so hohe Absorption besitzt, daß der Wasserdampf dagegen vernachlässigt werden kann. Der abgetrennte Bereich am linken Rand enthält eine Wiederholung des ganzen Diagrammes für die Strahlung in diesem CO_2-Bereich.

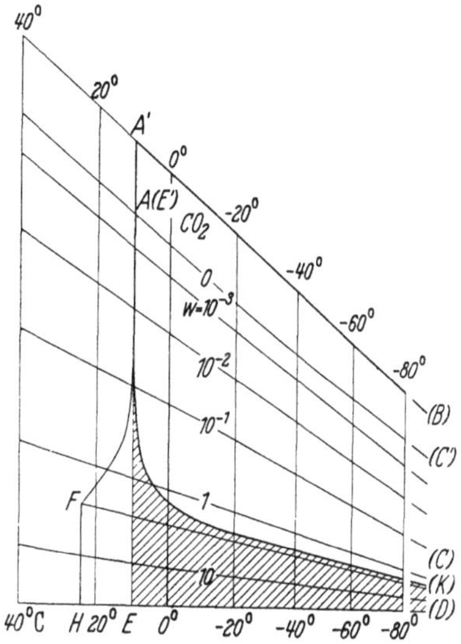

Fig. 29. Strahlungsdiagramm (radiation chart) nach W. M. ELSASSER. Buchstaben und Symbole haben die gleiche Bedeutung wie in Fig. 28. Fig. 28—30 sind ungefähr flächengleich.

ELSASSER und YAMAMOTO verwenden für ihre Diagramme die in (25.4) gegebene Umformung der Gl. (25.3), bei der die Temperatur als unabhängige Variable (Abszisse des Diagrammes) eingeführt wird. T ist im allgemeinen in der Atmosphäre mit m_r monoton verbunden, so daß diese Vertauschung keine wesentlichen Schwierigkeiten macht. In der Tat sind die Diagramme Fig. 29 und 30 flächentreue Transformationen von Fig. 28, bei denen die T-Linien gestreckt, dafür die m-Linien gekrümmt werden (wenn man von der Verschiedenheit der verwendeten Absorptionskonstanten absieht). Für den Entwurf eines Diagrammes hat das zweite Verfahren gewisse Vorzüge, für die anschauliche Durchdringung das erste.

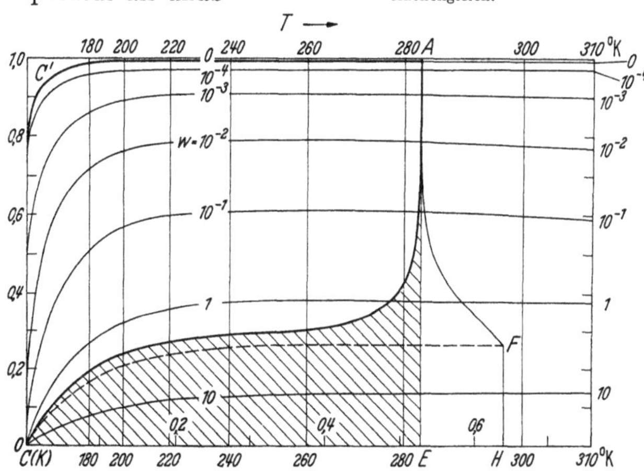

Fig. 30. Strahlungsdiagramm (radiation chart) nach G. YAMAMOTO. Buchstaben und Symbole haben die gleiche Bedeutung wie in Fig. 28. Fig. 28—30 sind ungefähr flächengleich.

Der Kohlensäurebereich um 15 μ wird von ELSASSER als schwarz absorbierend angenommen. In der Tat ist im Zentrum der Bande die Absorption schon bei sehr geringen Weglängen fast total; bei den niedrigen Drucken in der hohen Atmosphäre trifft dies jedoch nicht mehr zu. An den Flügeln der Banden kann man sicher nicht voraussetzen, daß die CO_2-Absorption diejenige des H_2O weit übertrifft. Deshalb berücksichtigt YAMAMOTO durch ein besonderes Verfahren die Überlagerung der beiden Absorptionen, für die am unteren Rande seines Diagrammes eine Eintragungsmöglichkeit vorgesehen ist.

Die in Fig. 28 dick umrandete Fläche $A'BC' + AC'C$ stellt die auf eine horizontale Flächeneinheit am Erdboden (Lufttemperatur $T_B = +10°$ C) von der darüber liegenden Atmosphäre (T mit m abnehmend) einfallende Strahlungsintensität G_B (Gegenstrahlung) dar. Die unter der Isotherme $T = +10°$ C (in Fig. 29 die rechts, in Fig. 30 links davon liegende Fläche) $A'BDEAE'A'$ ist die Schwarzstrahlung von 10° C, die der Erdboden selbst aussendet. Die Summe der beiden schraffierten Flächen ist deshalb gleich der *effektiven Ausstrahlung* der Erdoberfläche

$$E_B = \sigma T_B^4 - G_B. \tag{29.1}$$

Wenn die Stelle mit einer Temperatur von $T = +10°$ in dem Beispiel der Fig. 28 nicht an der Erdoberfläche liegt, sondern in der freien Atmosphäre, dann gibt der Betrag $E = \sigma T^4 - G$ die effektive Ausstrahlung einer in dieser Höhe befindlichen schwarzen Fläche oder eines auf der Temperatur T befindlichen Meßinstrumentes. Der wirklich von unten durch die Referenzfläche hindurchtretende Strahlungsstrom U stammt aus den zwischen Erdoberfläche und Referenzfläche liegenden Wasserdampfmassen und von der Erdoberfläche und ist in analoger Weise gegeben durch die Fläche $AC'KF$, die die Strahlung der Wasserdampfschicht M_r bedeutet und $FKDH$, die den Anteil der Schwarzstrahlung der Erdoberfläche wiedergibt, der durch M_r noch hindurchkommt. Dazu kommen die analog gelegenen Flächenstücke im Kohlensäurediagramm. Alle diese Anteile zusammen ergeben den von unten kommenden Strahlungsstrom U, dem von oben wiederum G entgegensteht. Der effektiv die Referenzfläche von unten nach oben durchsetzende Strahlungsstrom ist dann

$$E = U - G. \tag{29.2}$$

In den anderen Diagrammen sind die Strahlungsstromanteile in entsprechender Weise gegeben.

*Wolken*flächen dürfen als Schwarzstrahler angesehen werden, da alle Wellenlängen in den Wassertropfen so stark absorbiert werden, daß die Reichweite innerhalb der Wolke nur wenige Meter beträgt. Die Strahlung aller über oder unter den Wolken liegenden Luftschichten wird abgeschirmt und durch eine Strahlung der Temperatur der Wolkenfläche ersetzt. Bei Betrachtung von oben setzt also eine Wolkenschicht die Strahlung herab, bei Betrachtung von unten herauf.

F. A. BROOKS und ROBINSON haben auf rein empirischem Wege Strahlungsdiagramme ähnlicher Art entwickelt[1], indem sie Beobachtungen der Gegenstrahlung zur Bestimmung der Emission bzw. Absorption der gesamten Atmosphäre verwandten. Sie mußten allerdings voraussetzen, daß $(\sigma T^4)^{-1} \int f_s \tau_f d\nu$ unabhängig von der Temperatur ist. Dies widerspricht dem PLANCKschen Gesetz mit seiner Wellenlängenverschiebung des Maximums.

30. Die Gegenstrahlung an der Erdoberfläche. Diese ist der einzige Strahlungsstrom, für den zahlreiche Messungen vorliegen[2], während aus der freien Atmosphäre nur sehr wenig Strahlungsmessungen vorhanden sind.

Bei den Messungen in Bodennähe sind zwei verschiedene Prinzipien zu unterscheiden. Mit dem *Pyrgeometer* nach ÅNGSTRÖM, das aus einer blanken und einer schwarzen nach oben gerichteten Metallfläche besteht, wobei die schwarze elektrisch auf die Temperatur der blanken aufgeheizt wird, mißt man die effektive Ausstrahlung bei der Temperatur des Instrumentes und kann daraus die Gegenstrahlung bestimmen. Mit *Strahlungsbilanzmessern*, die nach oben und nach unten gerichtete schwarze Flächen besitzen und die Differenz der nach unten und nach oben gerichteten Strahlung messen, bestimmt man die Ausstrahlung der Erdoberfläche; diese kann wesentlich geringer sein, weil in klaren Nächten

[1] F. A. BROOKS: Pap. phys. Ocean. Met., Mass. Inst. Techn. and Woods Hole Ocean. Inst. **8**, No. 2 (1941). — G. D. ROBINSON: Quart. J. Roy. Met. Soc. **73**, 127 (1947); **76**, 37 (1950).

[2] C. DORNO: Meteor. Z. **39**, 316 (1922). — F. LAUSCHER: Meteor. Z. **45**, 371 (1928). — P. R. RAMAN: Proc. Ind. Acad. Sci. **1**, 815 (1935). — A. ÅNGSTRÖM: Medd. Stat. Met.-Hydr. Anstalt **1936**, Nr. 8. — F. SAUBERER: Meteor. Z. **53**, 296 (1936). — G. FALCKENBERG: Meteor. Z. **57**, 241 (1940). — G. YAMAMOTO u. T. SASAMORI: Sci. Rep. Tôhoku Univ. (5) Geoph. **6**, 19 (1954).

die Temperatur des Erdbodens oder die der abstrahlenden Pflanzenoberfläche oft erheblich unter der Lufttemperatur liegt. Beide Größen sind physikalisch sinnvoll, man muß deshalb die Begriffe *effektive Ausstrahlung* (bezogen auf Luft- oder Instrumententemperatur) und *nächtliche langwellige Strahlungsbilanz* (bezogen auf die wirklich strahlende Bodenoberfläche) unterscheiden. ALBRECHT[1] hat vom Flugzeug aus bolometrisch die Temperatur der strahlenden Oberfläche gemessen und in der Dämmerung oder frühmorgens bei klarem Himmel Temperaturunterschiede zwischen (warmem) Wasser und Land bis zu 14,5°, über verschiedenen Landflächen bis zu 10° gefunden. Bei Tage wurden sogar Unterschiede bis 40° C (Land wärmer) gefunden.

Die Beobachtungen der Gegenstrahlung G_B oder effektiven Ausstrahlung E_B sind meist durch *empirische Formeln* wie diejenige ÅNGSTRÖMS (hier mit den von LÖNNQVIST nach Messungen ROBINSONS gegebenen Zahlenkonstanten):

$$G_B = E_s(0{,}848 - 0{,}294 \cdot 10^{-0{,}030\,\varkappa\,\varrho_{wB}})$$

oder BRUNTS:

$$G_B = E_s\left(0{,}543 + 0{,}050\sqrt{\varkappa\,\varrho_{wB}}\right)$$

zusammengefaßt worden[2]. E_s ist die Schwarzstrahlung des Meßapparates, ϱ_{wB} g m^{-3} die Dampfdichte am Boden, $\varkappa = M_r/\varrho_{wB}$ cm p.w. $\times$ m^3 g^{-1} das (von der Schichtung der Atmosphäre abhängige) Verhältnis der gesamten reduzierten Wasserdampfmasse M_r[†] zur Bodendampfdichte. $\varkappa$ variiert im allgemeinen zwischen 1,1 und 3,6, Mittelwert 1,9, jedoch können in der Arktis auch wesentlich größere Werte vorkommen.

Theoretisch gut fundiert ist die Formel LÖNNQVISTS

$$E_B = \sigma T_B^4\left(A + B t_B + C(1+\varkappa)\gamma_m - D\log M_r\right);$$

hierin sind γ_m der mittlere Temperaturgradient in °C/km zwischen Erdboden und etwa 2 km Höhe $[\gamma_m = (T_B - T_2):2]$, t_B die Temperatur am Boden in °C. Als Zahlenwerte gibt LÖNNQVIST: $A = 0{,}514$, $B = 6\cdot 10^{-4}$, $C = 8\cdot 10^{-4}$, $D = 0{,}101$. Es ist ersichtlich, daß die einzelnen Konstanten nur bestimmt werden können, wenn gleichzeitig mit den Strahlungsmessungen auch Radiosondenaufstiege durchgeführt werden. An solchen Meßreihen[3] ist leider noch Mangel. — Die effektive Strahlung wird fast nur in dem schmalen Bereich von 8 bis 12 μ ausgesandt. Es wäre deshalb zweckmäßiger, sie nicht auf σT^4 zu beziehen, sondern auf die T-Abhängigkeit der Schwarzstrahlung in diesem Wellenlängenbereich, um das Glied t_B in der Klammer überflüssig zu machen. — Am wichtigsten ist der letzte, vom Wasserdampfgehalt abhängige Term, der die stärksten Schwankungen bewirkt. Nur bei starken Bodeninversionen, wenn $\gamma_m < 0$ ist und eine wesentlich wärmere Luftschicht dicht über dem Beobachter liegt, erreicht der dritte Term ähnliche Größe; die Ausstrahlung kann dann bis um 20% herabgesetzt werden [*1d*]. — Die Gegenstrahlung des Ozons in der 9,6 μ-Bande beträgt nur 2 bis 3% der gesamten Gegenstrahlung und ist wenig veränderlich, so daß sie leicht berücksichtigt werden kann[4].

---

[1] F. ALBRECHT: Berichte dtsch. Wetterd. U.S.-Zone **38**, 332 (1952).

[2] A. ÅNGSTRÖM: Astrophys. J. **37**, 305 (1918). — Meteor. Z. **33**, 529 (1916). — Gerlands Beitr. Geophys. **21**, 145 (1929). — D. BRUNT: Quart. J. Roy. Met. Soc. **58**, 389 (1932). — O. LÖNNQVIST: Ark. Geof. **1**, 79, 441 (1950/51); **2**, 151, 246 (1954). — Andere Zahlenwerte für die Konstanten sind angegeben worden von H. M. BOLZ u. G. FALCKENBERG: Z. Met. **3**, 97 (1949).

† Ausgedrückt in cm p.w. ≡ cm precipitable water ≡ cm Niederschlagswasser.

[3] K.-H. SCHMIDT: Z. Met. **5**, 331 (1951). — O. LÖNNQUIST: Ark. Geofys. **2**, 371 (1955).

[4] H. M. BOLZ: Z. Met. **2**, 225 (1948).

Nach verschiedenen *Zonen* des Himmels ist die effektive Ausstrahlung verschieden groß. Zum Zenit hat sie ein Maximum, während sie nach horizontnahen Stellen auf 0 zurückgeht. LINKE[1] zeigte nach Messungen von DUBOIS, daß die Abhängigkeit von der Zenitdistanz z einem Gesetz folgt

$$E^*(z) = E_0^* \cos^r z.$$

Der Exponent r wächst von 0,28 auf 0,46 mit Zunahme des Dampfdruckes von 3 bis 10 mm Hg. MÖLLER und MÜGGE[2] konnten auf theoretischem Wege nur eine sehr viel schwächere Abhängigkeit finden, aber auch einen Einfluß starker Bodeninversionen auf r nachweisen. LAUSCHER[3] hat die Veränderung der Ausstrahlung durch eine Einengung des Horizontes untersucht und Zahlenwerte für horizontal geneigte und vertikale Flächen bei verschiedener Horizonterhöhung angegeben.

Verschiedene Beobachter fanden unter gleichen Bedingungen sehr verschiedene Ausstrahlungswerte, aber auch rasche Veränderungen innerhalb weniger Minuten ohne erkennbare Änderung der meteorologischen Bedingungen. Eine solche Vergrößerung der Gegenstrahlung kann durch Dunst- oder Rauchschichten hervorgerufen werden, aber auch durch feinste Cirruswolken, die mitunter in großer Menge vorhanden, aber bei Nacht nicht sichtbar sind und auch bei Tage wegen der Dunstverhüllung vom Boden aus nicht gesehen werden.

Die stärkste Verminderung der effektiven Ausstrahlung wird durch *Wolkendecken* hervorgerufen[4]. ÅNGSTRÖM und ASKLÖF haben eine lineare Abhängigkeit des E vom Bewölkungsgrad gefunden und geben für bedeckten Himmel bei tiefen Wolken 14%, bei mittleren 23%, bei hohen 80% der effektiven Ausstrahlung bei wolkenlosem Himmel an. BOLZ, SAUBERER und KREITZ fanden quadratische Gesetze besser geeignet. Letzterer gibt im Mittel für alle Wolkenarten an:

$$G_N = G_0(1 + 0{,}27 N^2), \quad \text{wo} \quad 0 < N < 1 \text{ der Bedeckungsgrad ist.}$$

Aus sehr zahlreichen Messungen in Indien hat RAMAN[5] folgende Werte für das Verhältnis der effektiven Ausstrahlung bei bedecktem Himmel zu der bei wolkenlosem Himmel angegeben: tiefe Wolken, Mittelwerte verschiedener Wolkenarten zwischen 15 und 32% (Mittel 22%), mittelhohe Wolken 43%, hohe Wolken 56 bis 75% (Mittel 68%).

Die Wärmeabgabe des Bodens bei Nacht infolge seiner langwelligen Ausstrahlung bewirkt eine viel stärkere Abkühlung, als sie die Luft durch ihre eigene geringfügige Wärmestrahlung erfährt. Die Abkühlung breitet sich daher von der Oberfläche sowohl in den Boden hinein, wie durch turbulente Wärmeleitung der Luft nach oben hin aus, und ihr Betrag an der Oberfläche wird seinerseits durch die von unten und oben zufließenden Wärmeleitungsströme reguliert. Das wichtige Problem der Vorhersage der nächtlichen Tiefsttemperatur ist unter Berücksichtigung der Wärmeleitung im Boden und in der Luft, der Veränderung der Ausstrahlung mit der Zeit (sowohl infolge Temperaturänderungen des Bodens wie infolge Ausbildung einer Bodeninversion in der Luft) und unter verschiedenen Anfangsbedingungen mehrfach durchgerechnet worden[6]. Aus dem Vergleich der

---

[1] F. LINKE: Meteor. Z. **48**, 25 (1931). — P. DUBOIS: Gerlands Beitr. Geophys. **22**, 41 (1929).

[2] F. MÖLLER u. R. MÜGGE: Beitr. Phys. fr. Atm. **20**, 220 (1932). — Ö. LÖNNQVIST: Tellus **3**, 148 (1951).

[3] F. LAUSCHER: Sitzgsber. Akad. Wiss. Wien IIa **143**, 503 (1934).

[4] S. ASKLÖF: Geogr. Ann. **1920**. — F. SAUBERER: Arch. Met. Geophys. u. Bioklim., Ser. B **2**, 347 (1951). — H. M. BOLZ: Z. Met. **3**, 201 (1949). — E. KREITZ: Geof. pura e appl. **28**, 292 (1954).

[5] P. K. RAMAN: Ind. J. Met. a. Geoph. **4**, 261 (1953).

[6] D. BRUNT: Phys. and dyn. Meteorology. Cambridge 1939. — P. GROEN: Mededeel. en Verh. Ser. B, No. 9, De Bilt 1947. — J. Meteorology **4**, 63 (1947); **6**, 369 (1949). — H. PHILIPPS: Gerlands Beitr. Geophys. **56**, 229 (1940). — O. CZEPA: Z. Met. **4**, 359 (1950). — H. REUTER u. O. CZEPA: Arch. Met., Geophys. u. Bioklim., Ser. B **2**, 250 (1950). — W. STAAB: Diss. Mainz 1954.

Rechnungen mit der Beobachtung kann man den Windeinfluß und (in windschwachen Nächten) den Einfluß des Energietransports durch Strahlung abschätzen.

J. H. SHAW und Mitarbeiter[1] haben *spektrale* Messungen der Gegenstrahlung zum Zenit und unter verschiedenen Zenitdistanzen z ausgeführt (Fig. 24, S. 227). Diese zeigen sehr instruktiv, daß die Atmosphäre in dem Bereich der intensiven Banden (5.5 bis 8,0 μ, $> 14\,\mu$) nahezu schwarz, d.h. mit der Temperatur der bodennahen Luftschicht strahlt, und daß auch in dem durchlässigen Bereich die Strahlung an den Stellen stärkerer Linien des Wasserdampfes und der Ozonbande 9,6 μ heraufgesetzt ist. Mit zunehmender Luftmasse $m(z)$ wird auch die (kontinuierliche?) Strahlung im Fenster stärker (wahrscheinlich Dunstwirkung).

31. Die langwellige Strahlung in der freien Atmosphäre. Eine Wirkung der Strahlung auf die Atmosphäre ergibt sich aus der vertikalen Divergenz des Netto-Strahlungsstromes E. Diese verursacht eine Temperaturänderung

$$\frac{\partial T}{\partial t} = -\frac{1}{c_p \varrho_L}\frac{\partial E}{\partial z}, \tag{31.1}$$

wobei ϱ_L die Luftdichte und c_p die spezifische Wärme der Luft bei konstantem Druck ist. In bodennahen Schichten kann eine Strahlungswirkung auf die Atmosphäre durch eine stark überheizte oder eine sehr viel kältere Bodenoberfläche bewirkt werden, jedoch ist dieser Einfluß im wesentlichen auf die untersten 10 m beschränkt. In 60 m Höhe hat die Temperaturangleichung auch nach 6 Std nur 10% der anfänglichen Differenz erreicht[2].

In größeren Höhen kann die Veränderung von $E = U - G$ wie folgt abgeschätzt werden: der von unten kommende Strahlungsstrom U nimmt nach oben hin ab, weil mit zunehmender Höhe mehr und mehr warme, auf die Referenzfläche strahlende Schichten durch kältere ersetzt werden. Da der Erdboden im Hinblick auf die Strahlung durch eine unendlich mächtige Gasmasse ersetzt werden kann, ändert sich die Masse, von der U ausgeht, nicht. Die gesamte Abnahme von U bleibt daher kleiner als etwa 30%. Auch G nimmt wegen der Temperatur nach oben ab, aber außerdem verringert sich die strahlende Masse mit zunehmender Höhe z (und geht am „Rande" der Atmosphäre bis auf Null zurück), so daß G stärker als U abnimmt. E nimmt daher nach oben zu, es wird $\partial E/\partial z > 0$. Die *Atmosphäre kühlt sich* daher nach (31.1) *ab*.

Eine genauere Betrachtung zeigt, daß die Ursache nur zu einem kleinen Teil in der Wärmeabgabe an den Weltraum liegt, weil ein in tiefen oder mittleren Schichten gelegenes Massenelement gegen die Ausstrahlung doch noch gut abgeschirmt ist. Vielmehr ist wegen der raschen Konzentrationsabnahme des Wasserdampfes mit der Höhe die Reichweite der Strahlung nach oben größer als nach unten, die Wärmeabgabe durch Strahlung also nach oben gegen kältere Luftschichten gerichtet und damit größer als die Wärmeaufnahme von den wärmeren unteren Schichten.

Beispiele für die potentielle Abkühlung der Atmosphäre durch langwellige Strahlung sind oft gegeben worden. Fig. 31 zeigt die Wasserdampfwirkung für normale Temperaturverteilungen in verschiedenen Breiten[3]: die unteren Schichten einer tropischen Atmosphäre zeigen verhältnismäßig geringe Abkühlung, weil die höheren Schichten ziemlich feucht sind und damit gegen Strahlungsverluste schützen. In 30° Breite sehen wir den gegenteiligen Effekt: eine feuchte Grundströmung kann deshalb viel Wärme abstrahlen, weil sie in 2 km von einer sehr trockenen (absinkenden) Luftmasse überlagert wird. In der arktischen Atmosphäre

[1] R. SLOAN, J. H. SHAW u. D. WILLIAMS: J. Opt. Soc. Amer. **45**, 455 (1955).
[2] O. CZEPA: Z. Met. **5**, 292 (1951). — H. PANOFSKY: J. Meteorology **4**, 35 (1947).
[3] F. MÖLLER: Proc. Roy. Soc. Lond. A **236**, 148 (1956).

ergibt sich eine starke Abkühlung in 1,5 km Höhe als Folge eines Wärmeausgleichs durch Strahlung zwischen der warmen Luft in diesem Niveau und den kälteren Massen darüber und darunter. Die stärkste Abkühlung wird in der oberen Troposphäre erreicht. Sie verschiebt sich nach etwas tieferen Schichten mit abnehmender Temperatur und abnehmendem Dampfgehalt, von 10,5 km in den Tropen nach 7 km in mittleren Breiten und 4 km in der Arktis. Wenn man von den extremen arktischen Bedingungen absieht, zeigen die Kurven nicht gar zu viel Unterschiede. Ein von LONDON[1] gegebener Meridianschnitt läßt erkennen, daß die stärkste Abkühlung in den Tropen in etwa 10 km mit 2°/Tag erreicht wird und von da sowohl nach oben wie nach unten und polwärts abnimmt (Fig. 32).

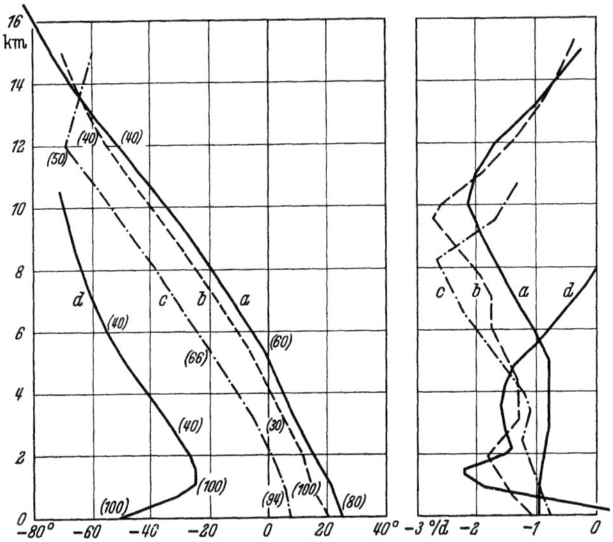

Fig. 31. Mittlere Abkühlung der Atmosphäre durch langwellige Ausstrahlung (rechts) und vertikale Temperaturverteilung (links); Zahlen in Klammern = relative Feuchtigkeit. a Tropen, b Roßbreiten, c mittlere Breiten, d Arktis, Winter.

Synoptische Betrachtungen der langwelligen Strahlung sind wegen der Umständlichkeit der Verfahren nur selten durchgeführt worden[2]. Deshalb hat MÖLLER[3] ein Kurzverfahren entwickelt, das nur die mittlere Strahlungsabkühlung dicker Atmosphärenschichten liefert. Er hat halbempirisch die Größe des Strahlungsstromes E in einem hohen Niveau der Troposphäre (400 oder 300 mb) als Funktion der dort herrschenden Temperatur und Feuchtigkeit angegeben (ähnlich wie es für E_B durch die Formeln von ÅNGSTRÖM oder BRUNT geschieht) und erhält die mittlere Abkühlung der ganzen Schicht als $-\frac{\partial T_m}{\partial t} = \frac{E_{400} - E_B}{c_p m}$, wo m die Masse zwischen den beiden Niveaus ist. Fig. 33 zeigt eine mittlere Verteilung für Januar 1954, die auf Grund von 31 täglichen Karten gewonnen ist. Die Größenunterschiede der Temperaturänderungen werden hervorgerufen durch Unterschiede der Feuchtigkeit in der Höhe, der Temperatur am Boden und der Bewölkungsverhältnisse.

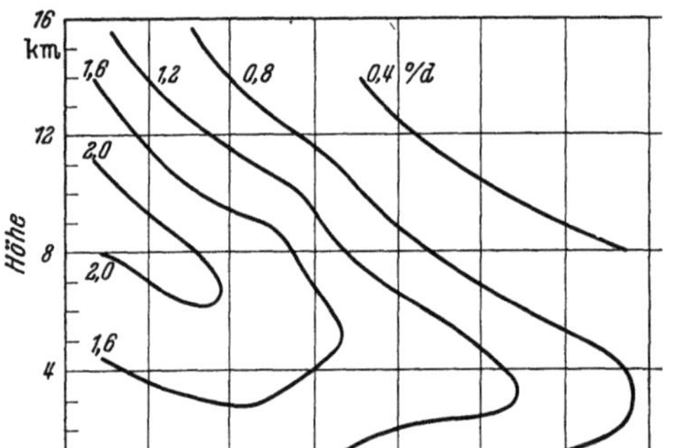

Fig. 32. Meridianschnitt der Abkühlung durch langwellige Ausstrahlung im März, wolkenlose Atmosphäre. Nach J. LONDON.

Fig. 31 und 32 gelten für wolkenlose Luft. *Wolken* wirken als schwarzstrahlende Körper. Das bedeutet, daß die Untergrenzen von Schichtwolken von den wärmeren, unter ihr gelegenen Luftmassen und der Erdoberfläche erwärmt werden, während ihre Oberflächen gegen den Weltraum und die kühleren Atmosphärenschichten Wärme abgeben; sie wirken wie in die Atmosphäre

[1] J. LONDON: J. Meteorology **9**, 145 (1952). — Geophys. Res. Papers, Cambridge (Mass) 1952 No. 18,

[2] A. H. THOMPSON u. M. NEIBURGER: J. Meteorology **10**, 167 (1953).

[3] F. MÖLLER: Arch. Met., Geophys. u. Bioklim., Ser. A **7**, 158 (1954).

gehobene Erdoberflächen. Insgesamt ist für tiefgelegene Wolkenschichten die Wärmeabgabe der Oberfläche größer als die Wärmeaufnahme der Untergrenze, so daß tiefe Wolken im ganzen die Wärmeabgabe der Lufthülle verstärken. Innerhalb der Wolken entwickelt sich allerdings am Grunde eine Art Wärmequelle und oben eine Kältequelle. Dies muß sehr rasch zu Veränderungen des Schichtungsgleichgewichtes in der Wolke und zu Zirkulationen führen, wenn nicht durch Sonneneinstrahlung an der Oberfläche und durch die thermodynamischen Vorgänge der Wolkenbildung selbst Gegenwirkungen hervorgebracht werden[1].

Es gibt nur sehr wenige *Messungen* der auf- und abwärtsgehenden Strahlungsströme U und G in der Atmosphäre. BREWER und HOUGHTON[2] haben im

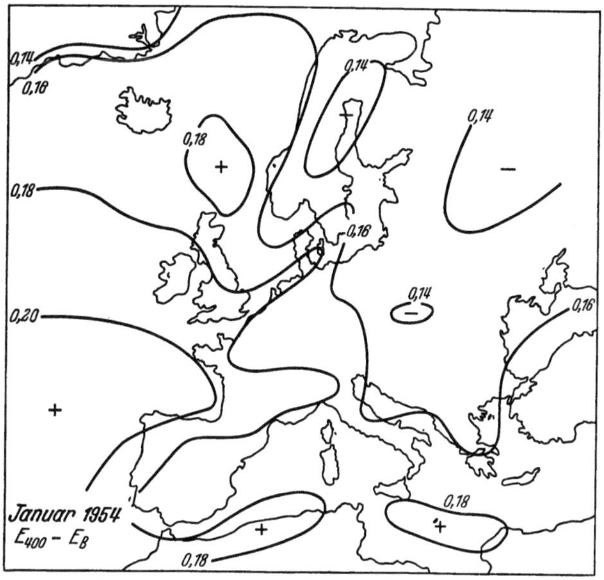

Fig. 33. Verteilung der mittleren Divergenz des langwelligen effektiven Strahlungsstromes zwischen 400 mb und Erdboden $E_{400} - E_B$ [cal cm^{-2} min^{-1}] im Monatsmittel des Januar 1954.

Flugzeug Messungen mit einem durch KRS5-Filter abgedeckten Radiometer durchgeführt. POHL[3] hat die Strahlungsbilanz $E = U - G$ in Radiosonden gemessen (Fig. 34). Er findet im Mittel eine Zunahme des Strahlungsstromes mit der Höhe, die derjenigen gleich ist, die sich aus Berechnungen mit Strahlungsdiagrammen ergibt. Jedoch liefern die Messungen von E in Bodennähe meist kleinere Werte, was sowohl auf einer niedrigen Strahlungstemperatur der Oberfläche beruhen, wie auf eine Strahlungswirkung des atmosphärischen Dunstes zurückgehen kann, die das Fenster bei 10 μ grau abschirmt. In der Troposphäre, aber auch in der Stratosphäre fallen starke Schwankungen von E in den Messungen sehr auf, die dünne Erwärmungs- und darüberliegende Abkühlungsschichten bedeuten. Auch diese könnten auf Dunstschichten, die wie schwache Wolkenschichten wirken, hinweisen. Weitere Messungen dieser Art werden wertvolle Ergebnisse bringen.

[1] F. MÖLLER: Meteor. Z. **60**, 212 (1943); **61**, 264 (1944).

[2] A. W. BREWER u. J. T. HOUGHTON: Proc. Toronto Met. Conf. 1953, London 1954, 68; u. Scient. Proc. Internat. Assoc. Meteorol., 10. Gen. Assembly, Rome 1954. London 1956, 571.

[3] W. POHL: Z. Geophys. **22**, 1 (1956). — H. G. MÜLLER u. W. POHL: Sitzgsber. bayr. Akad. Wiss. München **1955**, 303.

Die Kohlensäure hat in der Troposphäre wegen ihrer sehr geringen Reichweite keine nennenswerte Wirkung; erst in der *oberen Stratosphäre* wird durch die Verringerung der abschirmenden Masse und die Abnahme des Druckes und der Linienbreite die Wärmeabgabe nach außen ermöglicht. KAPLAN[1] hat die nachstehend wiedergegebene Tabelle berechnet, wonach sich bis zu der Schicht 100 bis 50 mb (16 bis 21 km) noch eine Zunahme der Abkühlung ergibt. MÖLLER[2] fand ein Maximum in 27 km Höhe. An der Tropopause befindet sich daher ein Minimum der Strahlungsabkühlung durch die gemeinsame Wirkung von Wasserdampf und Kohlensäure.

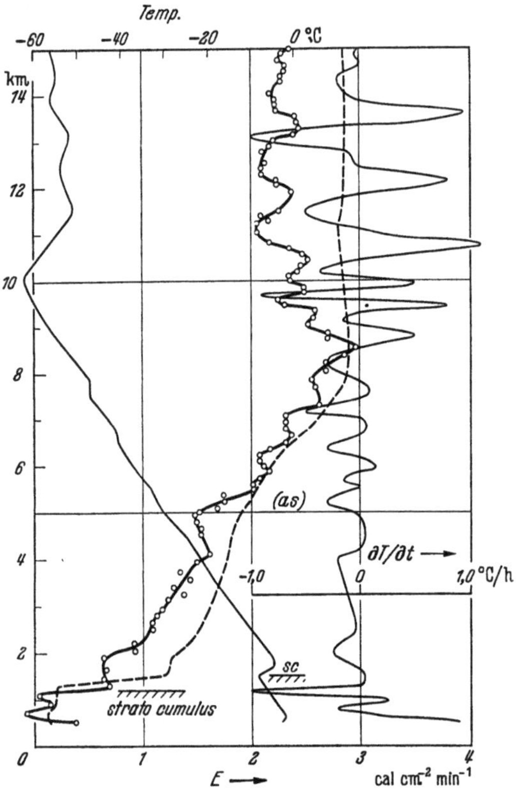

Fig. 34. Messung des langwelligen Strahlungsstromes E nach POHL, Aufstieg München 8. 3. 1955, 21.50 Uhr, Kreise und dicke Linie. Dasselbe berechnet (gestrichelt) nach der Temperaturverteilung (dünn) vom Aufstieg 9. 3. 1955, 03.00 Uhr. Durch den gemessenen Strahlungsstrom bedingte zeitliche Temperaturänderung, berechnet (dünn, rechts).

Die Wirkung der 9,6 μ-Bande des Ozons allein hat PLASS[3] untersucht und findet bis 20 km eine Heizwirkung von wenigen zehntel Grad, von 35 bis 60 km jedoch eine Abkühlung von etwa 2°/Tag. Dem steht die Aufheizung der Ozonschicht durch Absorption der Sonnenstrahlung im UV entgegen, die nach PRESSMAN[4] in 45 km Höhe ein Maximum von etwa +2°/Tag im Winter, +8°/Tag im Sommer hat (50° Breite). Darüber geht die Erwärmung zurück auf nahezu Null in 80 km Höhe und steigt darüber wegen der Absorption des Sauerstoffes an bis 100 km Höhe auf +2°/Tag im Winter, +15°/Tag im Sommer[5].

Luftdruck	234,5—200	200—150	150—100	100—50 mb
Höhe	10,8—11,8	11,8—13,6	13,6—16,1	16,1—20,4 km
Temperaturänderung durch CO_2	+0,023	−0,098	−0,245	−0,458°/Tag

Die numerische Differentiation (31.1) ist meist umständlich und nicht sehr genau. Eine Verbesserung der Genauigkeit ist folgendermaßen zu erreichen: betrachtet man eine einzelne Elementarschicht, so ist ihre Temperaturänderung bedingt durch den Strahlungsaustausch mit den darunter und darüber liegenden (wärmeren oder kälteren) Atmosphärenschichten. Die Intensität dieses Strahlungsaustausches ist abhängig von dem Wasserdampfgehalt, der zwischen den beiden Strahlungspartnern liegt. BRUINENBERG[6] findet durch Differentiation von

[1] L. D. KAPLAN: J. Meteorology 9, 1 (1952).
[2] F. MÖLLER: Meteor. Z. 61, 264 (1944).
[3] G. N. PLASS: Quart. J. Roy. Met. Soc. 82, 30 (1956).
[4] J. PRESSMAN: J. Meteorology 12, 87 (1955).
[5] R. M. GOODY (mitgeteilt durch R. J. MURGATROYD): Met. Mag. Lond. 85, 52 (1956).
[6] A. BRUINENBERG: Kon. Nederl. Met. Inst., Meded. en Verh. B 1, No. 1 (1946).

(29.2) nach der Höhe (31.1), wobei er aber die Differentiation unter den Integralausdrücken für E und G, vor Durchführung der Integration (25.3) ausführt. Er erhält

$$\frac{\partial T}{\partial t} = -\frac{q}{c_p}\left\{\int_0^{w_0}\left(\frac{\partial^2 F^{\downarrow}}{\partial T\,\partial w}\right)_w \frac{\gamma_w}{\varrho_w}\,dw + \frac{\partial F(T_B)}{\partial w} - \int_0^{w_u}\left(\frac{\partial^2 F^{\uparrow}}{\partial T\,\partial w}\right)_w \frac{\gamma_w}{\varrho_w}\,dw\right\};$$

dabei sind $q = \varrho_w/\varrho$ die spezifische Feuchtigkeit in der Referenzhöhe, für die $\partial T/\partial t$ bestimmt werden soll, ϱ_w die Dampfdichte, γ_u der vertikale Temperaturgradient an der Stelle w, $\partial F/\partial w$ die Integrale nach (25.2); die abwärts oder aufwärts gerichteten Pfeile beziehen sich auf die Integration über w_0 oder w_u, die oberhalb oder unterhalb der Referenzhöhe vorhandenen Wasserdampfmassen, T_B = Erdbodentemperatur. Durch Anwendung dieses Verfahrens können feinere Einzelheiten in der vertikalen Verteilung der Temperaturänderung heraus gearbeitet werden, als durch die grobe numerische Differentiation (31.1). Fig. 35 zeigt ein von BRUINENBERG gegebenes Beispiel. Jeder Knick in der Zustandskurve liefert eine Abkühlung, wenn geringe Gradienten nach oben in größere übergehen, im umgekehrten Falle eine Erwärmung. Diese können an den Unter- und Obergrenzen von Inversionen sehr hohe Beträge, im Beispiel bis zu $+1{,}5$ und $-3{,}1°$/Tag annehmen. Die Ähnlichkeit mit Wärmeleitungsvorgängen, welche in der gleichen Weise wirken würden, liegt auf der Hand. Es ist deshalb wiederholt versucht worden, die Strahlungsvorgänge formal durch die Wärmeleitungsgleichung zu erfassen und einen Koeffizienten der *Strahlungsscheinleitung* zu definieren[1]. Alle diese Versuche können aber den komplizierten Strahlungsvorgängen nicht voll gerecht werden[2]. — Ein vereinfachtes Modell des BRUINENBERGschen numerischen Verfahrens hat D. L. BROOKS[3] ausgearbeitet; HALES sowie YAMAMOTO und ONISHI haben graphische Methoden dafür angegeben[4].

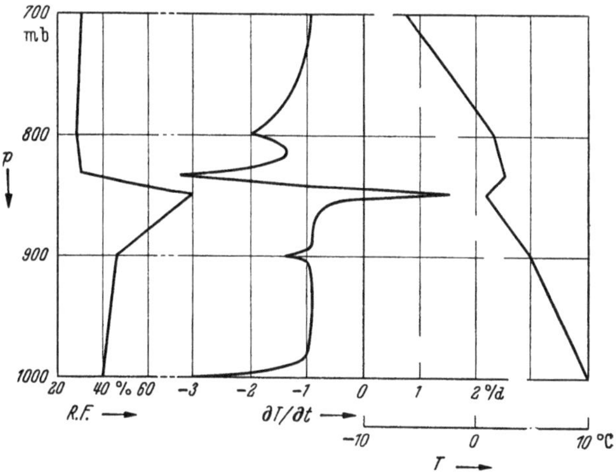

Fig. 35. Temperaturänderung durch langwellige Strahlung nach BRUINENBERG (Mitte), vertikale Temperaturverteilung (rechts), relative Feuchtigkeit (links).

32. Das Strahlungsgleichgewicht. Die langwellige Strahlung bedingt, — von den in Ziff. 31 erwähnten räumlich begrenzten Ausnahmen an den Knickstellen der Zustandskurve abgesehen, — durchweg Abkühlung in der freien Atmosphäre. Da die Erwärmung durch direkte Absorption der Sonnenstrahlung durchweg weniger als halb so groß ist, verbleibt eine Abkühlung, die in irgendeiner Form

[1] G. FALCKENBERG u. E. STÖCKER: Beitr. Phys. fr. Atm. **13**, 246 (1927). — D. BRUNT: Proc. Roy. Soc. Lond., Ser. A **124**, 201 (1929); A **130**, 98 (1930). — S. L. MALURKAR u. L. A. RAMDAS: Ind. J. Phys. **6**, 495 (1932).

[2] R. M. GOODY: Quart. J. Roy. Met. Soc. **79**, 282 (1953).

[3] D. L. BROOKS: J. Meteorology **7**, 313 (1950).

[4] J. V. HALES: An atm. rad. flux div. chart and merid. cross sections ... Dr. Thesis, Univ. Calif., Los Angeles 1951. — G. YAMAMOTO u. G. ONISHI: Sci. Rep. Tôhoku Univ. (5) **4**, 108 (1953).

durch andere Vorgänge kompensiert werden muß; denn die Atmosphäre kühlt sich nicht ständig ab. Als solche wärmezuführenden Vorgänge wirken die Wärmeleitung vom Erdboden und die Freimachung von latenter Wärme bei der Kondensation des Wasserdampfes. Offenbar ist keine Schicht der Troposphäre im Strahlungsgleichgewicht zwischen erwärmenden und abkühlenden Strahlungswirkungen[1]. Ein Gleichgewicht erscheint aber möglich, wenn der Differenz von Absorption der Sonnenstrahlung und Wärme-Abstrahlung an den Weltraum (die nach dem Gesagten negativ ist) eine Erwärmungswirkung durch die internen Wärmestrahlungsvorgänge zwischen den einzelnen Atmosphärenschichten gegenübersteht. Eine solche ist (bei angenäherter Betrachtung) gegeben, wenn $\partial^2 T/\partial m^2 > 0$ ist. Aus dieser nur qualitativen Überlegung folgt, daß eine Temperaturverteilung des *Strahlungsgleichgewichtes* nur existieren kann, wenn auch $\partial^2 T/\partial z^2$ sehr große Werte besitzt. In der Stratosphäre hat man lange Zeit das Vorhandensein eines Strahlungsgleichgewichtes dieser Art angenommen.

Die exakte Bedingung des Strahlungsgleichgewichtes ergibt sich aus dem Verschwinden der vertikalen Divergenz des Nettostrahlungsstromes E nach (31.1) leicht zu $\partial E/\partial z = 0$ oder mit (29.2) und unter Verwendung von $dm = \varrho_m\, dz$

$$\frac{\partial U}{\partial m} - \frac{\partial G}{\partial m} = 0. \qquad (32.1)$$

Die Berechnung der Temperaturverteilung des Strahlungsgleichgewichtes aus einer exakten Lösung dieser Beziehung, etwa unter Verwendung eines Strahlungsdiagrammes, ist noch nicht gelungen.

Geht man für monochromatische Strahlung von (24.2) aus, dann gilt für die auf- und abwärtsgehenden diffusen Ströme bei Weglassung des Index ν:

$$dU = -\gamma k (U - E_s)\, dm, \qquad (32.2)$$

$$dG = \gamma k (G - E_s)\, dm, \qquad (32.3)$$

wo γ nach (26.13) den Faktor zur Berücksichtigung diffuser Strahlung angibt. Macht man mit Emden[2] die Voraussetzung einer kontinuierlichen und wellenlängenunabhängigen „grauen" Absorption, dann gelten (32.2), (32.3) auch für die Gesamtstrahlung und es ist $E_s = \sigma T^4(m)$ zu setzen. Durch Integration von (32.1) und unter Hinzunahme der Sonnenstrahlung S ergibt sich

$$U(m) - G(m) - S(m) = \text{const} \qquad (32.4)$$

als Bedingung des Strahlungsgleichgewichtes. Das System von Differentialgleichungen (32.2) bis (32.4) ist lösbar und es kann die Temperaturverteilung des Strahlungsgleichgewichtes $T(m)$ angegeben werden.

Emden fand auf diesem Wege in der unteren Troposphäre einen weit überadiabatischen Temperaturgradient, in hohen Schichten konstante Temperatur von etwa $-60°$, was wenigstens mit der beobachteten Stratosphärentemperatur auffallend gut übereinstimmt. Jedoch ist dies ein von der Wahl des grauen Absorptionskoeffizienten abhängiges Zufallsergebnis. Es führte aber zu der jahrzehntelang anerkannten Ansicht, daß die Stratosphäre allein durch das Strahlungsgleichgewicht des Wasserdampfes erklärt werden könne. Schon eine (immer noch sehr schlechte) Approximation des Wasserdampfspektrums durch 2 Absorptionskoeffizienten zeigt, daß auch in hohen Schichten die Temperatur im Strahlungsgleichgewicht weiter nach oben hin abnimmt und bis auf weit unter $-100°$ C zurückgeht[3]. Die Absorption von Sonnenstrahlung $S(m)$ hat dabei nur geringen Einfluß auf das Ergebnis.

[1] F. Möller: Meteor. Z. **52**, 408 (1935).
[2] R. Emden: Sitzgsber. bayr. Akad. Wiss. München, Math.-phys. Kl. **43**, 55 (1913).
[3] F. Möller: Gerlands Beitr. Geophys. **58**, 11 (1941).

Eine brauchbarere Lösung kann erwartet werden, wenn es gelingt, die Linienstruktur der Absorption zu berücksichtigen. Eine Lösung des Systems (32.2), (32.3) für monochromatische Strahlung kann mit Einführung der Grenzwerte $G(M)$ und $U(0)$ für die von oben und unten eintretenden Strahlungsströme gegeben werden[1]:

$$U(m) = U(0)\exp\left(-\int_0^m \gamma\, k\, dm\right) + \int_0^m E_s(\mu)\,\frac{d}{d\mu}\exp\left(-\int_0^m \gamma\, k\, dm\right) d\mu, \tag{32.5}$$

$$G(m) = G(M)\exp\left(-\int_m^M \gamma\, k\, dm\right) - \int_m^M E_s(\mu)\,\frac{d}{d\mu}\exp\left(-\int_m^\mu \gamma\, k\, dm\right) d\mu. \tag{32.6}$$

Differentiation dieser Gleichungen nach m und Einsetzen in (32.1) führt zu einer Integralgleichung für $E_s(m)$, aus deren Lösung sich die Temperaturverteilung des Strahlungsgleichgewichtes ergeben muß. Zur Berücksichtigung der Linienstruktur ist für k die Absorption einer Linie nach (26.1) einzuführen mit der gegenüber (27.1) vereinfachten Druckabhängigkeit der Halbwertsbreite $\delta = b p$. Setzt man außerdem die strahlende Masse m proportional dem Luftdruck

$$dm = \varrho_m\, dz = -\frac{\varrho_m}{\varrho_L}\frac{dp}{g} = -c\,\frac{dp}{g}, \tag{32.7}$$

wobei c das von p unabhängige Mischungsverhältnis $c = \varrho_m/\varrho_L$ ist, dann können die in (32.5), (32.6) auftretenden Integrale folgendermaßen umgewandelt werden

$$\left.\begin{aligned} \exp\left(-\int_\mu^m \gamma\, k\, dm\right) &= \exp\left(+\int_q^p \gamma\, k\, c\, dp/g\right) = \exp\left(\frac{\gamma S c}{2\pi g b}\int_q^p \frac{2b^2 p\, dp}{(\nu-\nu_0)^2 + b^2 p^2}\right)\\ &= \exp\left(\eta \ln\frac{(\nu-\nu_0)^2 + b^2 p^2}{(\nu-\nu_0)^2 + b^2 q^2}\right) = \left[\frac{(\nu-\nu_0)^2 + b^2 p^2}{(\nu-\nu_0)^2 + b^2 q^2}\right]^\eta. \end{aligned}\right\} \tag{32.8}$$

Hierbei ist $\eta = \dfrac{S\gamma c}{2\pi g b}$ eine von p unabhängige Konstante. Durch diesen zuerst von STRONG und PLASS[2] angegebenen Kunstgriff können die Exponentialfunktionen in algebraische Funktionen umgewandelt und damit die Lösung der Integralgleichungen erleichtert werden. In ähnlicher Weise kann mit dem ELSASSERschen Bandenmodell (26.5) verfahren werden[3]. Die Annahme konstanter Konzentration c ist allerdings nur für die atmosphärische Kohlensäure brauchbar. Die mittlere Wasserdampfverteilung kann durch ein Gesetz $\varrho_w/\varrho_{w_0} = (p/p_0)^4$ recht gut approximiert werden. KAPLAN[4] hat auch für eine solche Konzentrationsverteilung eine Lösung angegeben.

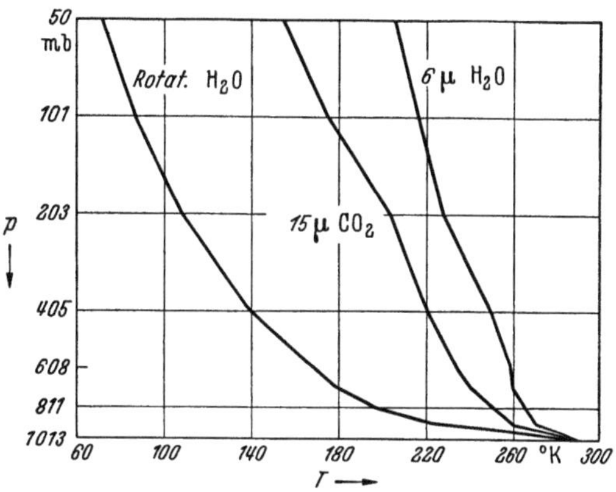

Fig. 36. Temperatur des Strahlungsgleichgewichtes bei alleiniger Wirkung eines Frequenzbereiches in der 6 μ- bzw. Rotationsbande des Wasserdampfes, bzw. der 15 μ-Bande des CO_2. Nach J. I. KING.

Fig. 36 zeigt die von KING gefundenen Lösungen für je einen Wellenlängenbereich in der 6,3 μ- und der Rotationsbande des Wasserdampfes und in der 15 μ-Bande der Kohlensäure. Eine für das ganze Wellenlängenspektrum gültige Lösung kann hieraus vielleicht abgeschätzt werden. Die Fig. 36 zeigt jedoch eindeutig, daß auch in der Stratosphäre (für $p < 200$ mb) die Gleichgewichtstemperatur weiter abnimmt und für die Wirkung der Rotationsbande sogar auf $-200°$ C in 20 km Höhe

[1] J. I. KING: Stratosph. Rad., Techn. Rep. No. 3, 4, 5. Contr. AF 19 (122)—392, Univ. of Utah 1951 u. 1952.

[2] J. STRONG u. G. N. PLASS: Astrophys. J. **112**, 365 (1950).

[3] J. I. KING: J. Meteorology **9**, 311 (1952). — Stratosph. Rad., Techn. Rep. No. 3 u. 4, Univ. of Utah 1951 u. 1952. — Ap. J. **116**, 491 (1952). — W. M. ELSASSER u. J. I. KING: Stratosph. Rad., Techn. Rep. No. 6, Univ. of Utah 1952. — L. D. KAPLAN: J. Meteorology **9**, 1, 139 (1952). — G. YAMAMOTO: Sci. Rep. Tôhoku Univ. (5) Geoph. **5**, 45 (1953); **6**, 127 (1955); **7**, 1 (1955).

[4] L. D. KAPLAN: J. Meteorology **11**, 16 (1954).

zurückgeht. In dieser Form ist das Strahlungsgleichgewicht eine mit den Beobachtungen nicht vereinbare Vorstellung.

Die Beobachtungen zeigen, daß die Tropopause und die untere Stratosphäre über dem Äquator bei 17 km und -80° C, in mittleren und hohen Breiten bei 10 km und -50° C liegt. Die örtlichen starken Schwankungen von Temperatur und Höhe sind mit den Veränderungen der Wetterlage und den dadurch bedingten vertikalen und horizontalen Bewegungen deutbar. Für die großräumige Meridionalverteilung sind solche Einflüsse unwahrscheinlich, wenn auch gewisse stationäre Zirkulationen zwischen niedrigen und hohen Breiten in der Stratosphäre anzunehmen sind[1]. Zieht man einen Vergleich mit der Troposphäre, so ist hier bekannt, daß die Temperaturverteilung stark beeinflußt wird von den synoptischen Vorgängen mit ihren Warmluftvorstößen und Kälteausbrüchen, daß aber die mittlere Temperaturverteilung: Tropen warm, Polargebiete kalt, allein auf die Strahlungsverhältnisse zurückzuführen ist. Man muß annehmen, daß in ähnlicher Weise die großräumige Anordnung der Tropopause und ihrer Temperatur ebenfalls durch die Strahlung bedingt ist, die täglichen Veränderungen dagegen durch die Wettervorgänge.

Der Mechanismus der Strahlungseinflüsse muß in einem Zusammenwirken der *verschiedenen Absorber* (oder auch von Strahlungs- und thermodynamischen Vorgängen gemeinsam) gesucht werden. MÖLLER[2] hat die Vorstellung entwickelt, daß die Strahlung in der 15 μ-Bande der Kohlensäure den wichtigsten Einfluß besitzt. Ihre Ausstrahlung wird teilweise abgeschirmt durch die an gleicher Stelle liegende Absorptionsbande des Ozons, das seinen Schwerpunkt in 20 km Höhe hat. Wo die Ozonmenge klein ist (Tropen, Polarwinter) ist die Wärmeabgabe der Kohlensäure groß und die Temperatur der unteren Stratosphäre niedrig, an anderen Stellen ist sie hoch. Eine quantitative Nachprüfung ist bisher kaum möglich, weil Absorptionsmessungen der 15 μ-Bande des Ozons unter geringen Drucken mit Fremdgasverbreiterung noch nicht existieren.

GOODY[3] hat die durch Wasserdampf, Kohlensäure und Ozon an der Tropopause hervorgerufenen Temperaturänderungen und ihre Variation mit Temperatur und Luftdruck untersucht. Er findet, daß ein Gleichgewicht zwischen der Abkühlung durch den Wasserdampf und der Erwärmung, die durch die Kohlensäure an dieser Stelle bewirkt wird, möglich ist. Es ist aber wahrscheinlich, daß die Erwärmung des CO_2 nur ein Spitzeneffekt wie in Fig. 35 ist, dagegen beim Wasserdampf die Ausstrahlung überwiegt, weil wegen des extrem geringen Feuchtegehaltes der Stratosphäre die Wärmeabgabe in diesen Höhen sehr stark ist. Das Gleichgewicht ist dann nur auf die Tropopausenfläche beschränkt, während darüber der Erwärmungseffekt des CO_2 und damit die Möglichkeit des Strahlungsgleichgewichtes entfällt. KAPLAN[4] hat in qualitativen Überlegungen darauf hingewiesen, daß es für ein Strahlungsgleichgewicht zwischen den drei Absorbern Ozon (O), Kohlendioxyd (C), Wasserdampf (W) von Wichtigkeit ist, daß 1. die Konzentrationsverteilungen der Ungleichung genügen $\frac{\partial \varrho_O}{\partial z} > \frac{\partial \varrho_C}{\partial z} > \frac{\partial \varrho_W}{\partial z}$ und daß 2. die wichtigen Absorptionsbanden so verteilt sind, daß die Schwarzstrahlung in den entsprechenden Frequenzbereichen $\frac{\partial \ln E_O}{\partial T} \gg \frac{\partial \ln E_C}{\partial T} \gg \frac{\partial \ln E_W}{\partial T}$ sind.

---

[1] K. R. RAMANATHAN: Scient. Proc. Internat. Assoc. Meteorol., 10. Gen. Assembly, Rome 1954. London 1956, 3.

[2] F. MÖLLER: Naturwiss. **31**, 148 (1943).

[3] R. M. GOODY: Proc. Roy. Soc. Lond., Ser. A **197**, 487 (1949). — G. M. B. DOBSON, A. W. BREWER u. B. M. CWILONG: Proc. Roy. Soc. Lond., Ser. A **185**, 144 (1946).

[4] L. D. KAPLAN: Scient. Proc. Internat. Assoc. Meteorol., 10. Gen. Assembly, Rome 1954. London 1956, 583.

Die Bedeutung der 2. Bedingung erkennt man schon in den Berechnungen KINGS in Fig. 36, wo die Temperaturverteilung des Gleichgewichtes um so größere Temperaturgradienten zeigt, je kleiner die Frequenz der betrachteten Bande ist.

Abschließend ist zu sagen, daß die Frage eines Strahlungsgleichgewichtes der drei beliebigen Absorber in der Stratosphäre noch nicht definitiv und quantitativ geklärt ist. Es ist jedoch wahrscheinlich, daß stationäre, vertikale und horizontale Strömungen und Wärmetransporte durch turbulenten Austausch dabei ebenso eine Rolle spielen wie die Strahlungsprozesse, zumal die Mitwirkung dieser Vorgänge bei der Entstehung der Ozonverteilung erwiesen ist.

D. Überblick.

33. Die Strahlungsbilanz.

Die Untersuchungen der Strahlung in der Atmosphäre kann man unter verschiedenen zusammenfassenden Gesichtspunkten betrachten: Das jahrzehntelang verfolgte Ziel des Astrophysical Observatory de Smithsonian Institution war, Schwankungen der *Solarkonstanten* nachzuweisen. Dies ist nicht gelungen, aber eine Fülle von Erkenntnissen über die Streuungs- und Absorptionsvorgänge der Sonnenstrahlung ging aus diesen Untersuchungen hervor. Die Untersuchung der *Streuung* und *Trübung* hat — auch wieder in neuerer Zeit — viel zur Erforschung der atmosphärischen Spurengase und des Aerosols beigetragen, und mit letzterem wiederum in die praktische Frage der *Sichtweite* (optisch und in anderen Spektralgebieten) eingegriffen. Die Wirkung der Strahlung auf, und ihre Beeinflussung durch Nebel und Wolken gibt wichtige Ausgangspunkte für die *Wolkenphysik* (Zahl, Größe und Wachstum der Tröpfchen, Energiebilanz der Wolke). Die Erklärung der natürlichen *Beleuchtung* und ihrer Farbe durch Sonnen- und Himmelsstrahlung oder die Untersuchung des *Ultraviolett*klimas sind für das tägliche Leben und die biologischen Einflüsse auf den Menschen von Wichtigkeit. Schließlich hat die Untersuchung der langwelligen Strahlung in mancher Hinsicht fördernd auf das Studium der *Bandenstrahlung* mehratomiger Moleküle eingewirkt und umgekehrt.

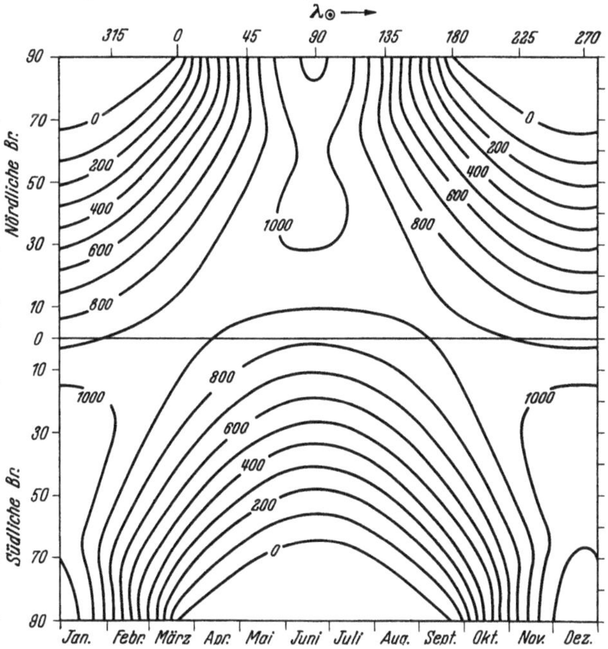

Fig. 37. Tagessummen (cal cm^{-2} d^{-1}) der extraterrestrischen Zustrahlung der Sonne in Abhängigkeit von geographischer Breite und Jahreszeit. Nach M. MILANKOVITCH.

Hier soll als zusammenfassender Gesichtspunkt die *Strahlungsbilanz* der Erdoberfläche und der Atmosphäre gewählt werden, d.h. die zeitliche und räumliche Verteilung der hereinkommenden und herausgehenden Strahlung. Die Verteilung der von außen auf die Erdoberfläche auftreffenden Strahlung über die geographische Breite in Abhängigkeit von der Jahreszeit ist von MILANKOVITCH[1] berechnet worden. Fig. 37 läßt erkennen, wie im Laufe des Jahres das tropische

[1] M. MILANKOVITCH: In KÖPPEN-GEIGER, Handbuch der Klimatologie, Bd. IA. Berlin 1930.

Maximum der Zustrahlung nordwärts und südwärts pendelt. Die Polargebiete erhalten (als Folge der 24 Std währenden Bestrahlung) im Sommer sogar größere Tagessummen als die Tropen. Würde man aus der Jahressumme der Strahlung in jeder Breite analog zu (1.1) eine Strahlungsgleichgewichts-Temperatur der Erdoberfläche (ohne Berücksichtigung der Albedo) berechnen, so würde man die Werte T_{St} erhalten, der in Tabelle 29 die wirklich gemessenen Temperaturen[1] T_m gegenübergestellt sind. Der Unterschied von durchschnittlich $+10°$ beruht, wie in Ziff. 1 bereits ausgeführt, zum Teil auf der Reflexion der Strahlung an den Wolken und der Rückstrahlung an der Atmosphäre, zum Teil auf der „Glashauswirkung" der atmosphärischen Absorber, die die Sonnenstrahlung bis zur Erdoberfläche durchlassen aber deren Ausstrahlung mit Ausnahme des Fensters bei 10 μ zurückhalten und dadurch einen Stau der Rückstrahlung bewirken.

Der von außen in die Atmosphäre eindringende Strahlungsstrom unterliegt der Streuung an Luftmolekülen (Ziff. 3) und Aerosolpartikeln (Ziff. 5), wodurch ein Teil nach außen reflektiert wird, der andere Teil als diffuse Himmelsstrahlung (Ziff. 18) den Boden erreicht; gleichzeitig erfolgt Absorption durch Gase (Ziff. 8 und 9), zum Teil auch an Partikeln. An Wolken erfolgt zunächst Reflexion an der Oberfläche (Ziff. 21), innerhalb der Wolke zusätzlich Absorption an Tröpfchen (Ziff. 20). Schließlich verbleibt an der Erdoberfläche, was nicht von ihr reflektiert wird (Ziff. 21). Alle diese Einflüsse sind quantitativ mehrfach abgeschätzt worden, so von SIMPSON[2], von BAUR und PHILIPPS[3] für Sommer und Winter, von LONDON[4] für den Monat März und von HOUGHTON[5] für das Jahresmittel, alle für die Nordhalbkugel. Jedoch bleiben in allen Arbeiten einige Größen schwer abschätzbar, weil sichere Angaben dafür fehlen, z.B. die Absorption am Wasserdampf der freien Atmosphäre (vgl. Tabelle 6) oder die Absorption in Wolken.

Tabelle 29. *Strahlungsgleichgewichts- und beobachtete Mittel-Temperaturen der Breitenkreise.*

$\varphi =$ 0	20	40	60	80	90°
$T_{St} = +20$	+16	+3	−18	−35	−38
$T_m = +26$	+24	+13	−2	−22	−28

In Fig. 38 sind die von HOUGHTON angegebenen Zahlenwerte dargestellt. Kurve 2 gibt die an der Erdoberfläche absorbierte Globalstrahlung, im Mittel aller Breiten 47% der extraterrestrischen Strahlung nach Kurve 1. Das schwache Maximum in $\varphi = 20°$ wird durch die geringe Bewölkung der Roßbreiten bewirkt. Die Absorption der Atmosphäre (einschließlich Wolken), Kurve 3, beträgt im Mittel 20% von J_0 und ist wahrscheinlich der unsicherste Wert in der Bilanzberechnung. Ausstrahlung der Erdoberfläche (Kurve 4) und der Atmosphäre (Kurve 5) sind mit Hilfe eines Strahlungsdiagrammes unter Berücksichtigung der Wolkenverteilung ermittelt.

Fig. 38 zeigt, daß die Bilanz der Erdoberfläche (Kurve 6) im Jahresdurchschnitt überall positiv ist, bis 30° fast konstant auf 0,3 cal cm^{-2} min^{-1}, dann polwärts bis auf 0 abnehmend, während die Bilanz der freien Atmosphäre (Kurve 7) durchweg negativ und mit der Breite wenig veränderlich ist. In „kurzen Cyclen", vertikalen Zirkulationen mit geringer horizontaler Verfrachtung, werden die Energiebeträge zwischen Boden und Atmosphäre ausgetauscht; in diesem Sinne wirken Wärmeübergang durch Konvektion, Verdampfung des Wassers am Boden und Kondensation in der Höhe bei Schauerniederschlägen,

[1] HANN-SÜRING: Lehrbuch der Meteorologie, 5. Aufl. Leipzig 1939.
[2] G. C. SIMPSON: Mem. Roy. Met. Soc. **3**, No. 21 (1928).
[3] F. BAUR u. H. PHILIPPS: Gerlands Beitr. Geophys. **42**, 160 (1934); **45**, 82 (1935).
[4] J. LONDON: J. Meteorology **9**, 145 (1952).
[5] H. G. HOUGHTON: J. Meteorology **11**, 1 (1954).

während Aufgleit-Regen kaum ohne größere horizontale Transporte vor sich gehen, aber im gleichen Sinne wirken.

Nach Ausgleich am Orte durch kurze Cyclen verbleibt die Gesamtbilanz des Systems Erdoberfläche + Atmosphäre, Kurve 8, mit positiven Werten bis 35° Breite und negativen in der Polarkalotte. Zu deren Ausgleich sind nun große horizontale Luftverfrachtungen nötig, wie sie in den Passaten und Monsunen

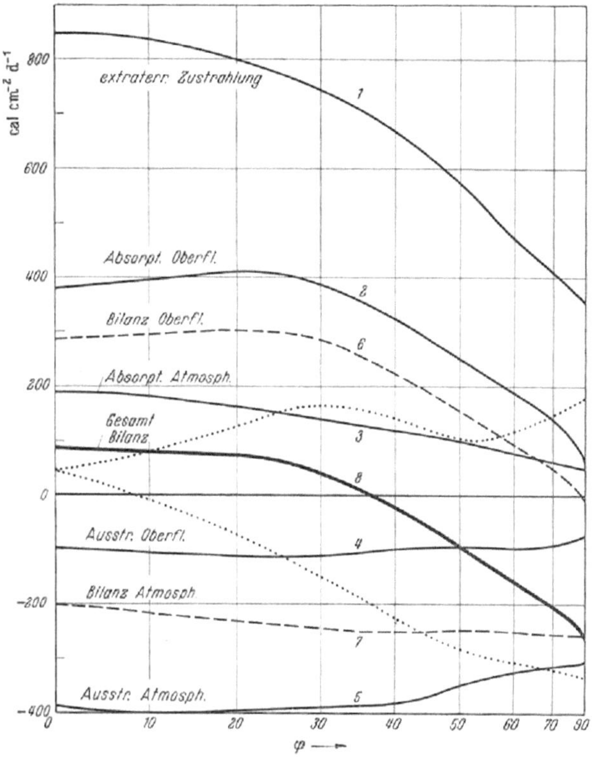

Fig. 38. Strahlungsbilanz der Atmosphäre und des Erdbodens, Jahresmittel, Nordhalbkugel, nach H. G. Houghton. Punktiert: Gesamtbilanz von Erde und Atmosphäre getrennt für Sommer- (obere) und Wintersolstitium (untere Kurve), nach F. Baur und H. Philipps.

niederer Breiten, in Zyklonen und Antizyklonen mittlerer und hoher Breiten vor sich gehen; unterstützt werden sie durch die gleichzeitigen Verdampfungs- und Kondensationsvorgänge.

Im Sommer ist die positive Strahlungsbilanz wesentlich weiter polwärts ausgedehnt, im Winter auf wenige Breitengrade in Äquatornähe beschränkt[1], und damit wird der gesamte Charakter des Wettergeschehens im Sommer mehr durch kurze, im Winter durch horizontal weit ausgreifende Cyclen beherrscht. Noch wesentlich differenzierter ergibt sich die Strahlungsbilanz-Verteilung in Abhängigkeit von der Unterlage[2] (Kontinent-Ozean) oder von der Wetterlage[3], wobei auch die durch atmosphärische Vorgänge geschaffenen Größen (warme und kalte Meeresströmung, Änderung der Albedo durch Schneedecke und Änderung der Zu- und

[1] Siehe Fußnote 3, S. 248.

[2] F. Albrecht: Z. Met. **1**, 97 (1947). — Ann. Met. **2**, 129 (1949). — Ber. dtsch. Wetterd. U.S.-Zone **17** (1950).

[3] F. Möller: Met. Abh., Inst. Met. Geoph. fr. Univ. Berlin **2**, H. 4, 125 (1955); Scient. Proc. Internat. Assoc. Meteorol., 10. Gen. Assembly, Rome 1954. London 1956, 561.

Abstrahlung wegen Bewölkung) wiederum bestimmend in die Strahlungsbilanz eingreifen. Damit beeinflußt also das Wetter wiederum die Strahlungsbilanz oder den Vorgang, der für die Weiterentwicklung die Impulse geben muß: der Kreislauf ist geschlossen und kann zu einer Abschwächung oder Aufschaukelung von eingeleiteten Prozessen führen.

34. Überblick über die Meßmethoden. Die zu messenden Strahlungsgrößen sind in allen Fällen *Energieströme* je Flächeneinheit (Strahlungsstrom-Dichten). Die Meßinstrumente müssen auf diese Eigenschaft des Energieflusses Rücksicht nehmen. Nach der Art und Weise der physikalischen Ausnutzung dieses Energiestromes zur Messung kann man eine systematische Klassifikation der Strahlungsmeßgeräte aufstellen[1].

Eine *direkte Umsetzung* der Energie der Photonen in die potentielle oder kinetische Energie der Elektronen arbeitet selektiv und kann daher nur für die Untersuchung spezieller Wellenlängen Verwendung finden. Daher wird in der Regel eine Umsetzung des Strahlungsenergie-Stromes in einen Strom von *Wärme*energie angewandt. Dies geschieht durch Vermittlung eines schwarzen Körpers oder einer schwarzen Fläche, wobei die Güte der Schwärzung den Wirkungsgrad der Transformation ausmacht.

Das häufigst angewandte Verfahren ist es, den Wärmestrom durch Wärmeleitung, Konvektion und langwelligen Strahlungsaustausch weiterzuleiten, ohne daß Phasenumänderungen in den Körpern des Meßsystems stattfinden. Hierbei kann stationär und instationär gearbeitet werden. Im *stationären* Zustand wird der gesamte Wärmefluß an die Umgebung abgeführt, die Temperaturen im Meßsystem ändern sich nicht; die Richtung, in die die Wärme abströmt, ist jedoch nur von der Güte der verschiedenen Wärmeleiter abhängig. Es läßt sich eine lehrreiche Analogie zu elektrischen Systemen aufstellen, indem die Wärmewiderstände durch elektrische Widerstände, die Wärmeströme durch elektrische Ströme und die Temperaturunterschiede als Spannungsdifferenzen gedeutet werden. Mit Hilfe solcher Ersatzschaltbilder haben Courvoisier und Wierzejewski[2] erfolgreich die Eigenschaften von Strahlungsmeßinstrumenten diskutiert und Vorschläge für konstruktive Maßnahmen geben können.

Wenn lediglich eine *Relativmessung* angestrebt wird, dann genügt es, die Wärmewiderstände konstant zu halten, um Proportionalität zwischen Wärmestrom und gemessener Temperaturdifferenz zu erhalten. Die Schwierigkeiten, Wärmeströme in kontrollierte Bahnen zu lenken und die Wärmeleitfähigkeit unabhängig von der Temperatur zu halten, sind jedoch meist sehr groß. — Nach diesem Prinzip arbeiten Instrumente mit mechanischer Temperaturmessung wie das primitive Schwarzkugel- oder Doppelthermometer nach Arago-Davy und das Bimetallaktinometer nach Michelson mit seinen verschiedenen weiter entwickelten Formen, aber auch solche, die die Temperaturen elektrisch messen, wie das Panzeraktinometer von Linke und Feussner und die verschiedenen Aktinometer und Pyranometer nach Moll-Gorczynski und nach Eppley[3].

---

[1] P. Courvoisier u. H. Wierzejewski: Arch. Met., Geophys. u. Bioklim., Ser. B **1**, 45 (1949).

[2] P. Courvoisier u. H. Wierzejewski: Arch. Met., Geophys. u. Bioklim., Ser. B **1**, 156 (1949); B **5**, 413 (1954).

[3] Beschreibung und Wirkungsweise der verschiedenen Strahlungsmesser findet man in: F. Albrecht u. K. Büttner: Gerlands Beitr. Geophys. **22**, 13 (1929); **26**, 241 (1930). — F. Albrecht: Die kalorischen Meßmethoden der atmosphärischen Strahlungsforschung in E. Kleinschmidt, Handbuch meteorologischer Instrumente Berlin 1935. — W. Mörikofer: Meteorologische Strahlungsmeßmethoden in E. Abderhalden, Handbuch biologischer Arbeitsmethoden, Abt. II, Teil 3, S. 4005. Wien 1939.

Bei *Absolutmessungen* erweist sich ein Kompensationsverfahren als zweckmäßige Maßnahme, d.h. es wird die Heizung durch die Strahlungsenergie ersetzt durch eine elektrische Heizung. Auch hier hat die Schwierigkeit, den Weg des Wärmestromes in beiden Fällen gleich zu gestalten, Unterschiede im Ergebnis der Absolutmessungen zur Folge gehabt; dies gilt besonders für die Unterschiede zwischen der *Smithsonian Scale 1913* und der ursprünglichen ÅNGSTRÖM-*Skala*, die auf dem Kompensations-Pyrheliometer nach ÅNGSTRÖM als dem typischen Vertreter der Kompensations-Instrumente beruht.

Bei der Verwendung eines Instruments im *instationären* Zustand wird der Erwärmungsvorgang selbst als Meßeffekt benutzt, wie es z.B. beim Silverdisk-Aktinometer geschieht. Dies Instrument kann auch als Absolutinstrument verwendet werden.

Ein gänzlich neues Element kommt in die Strahlungsmeßtechnik, wenn *Phasenumwandlungen* im Gerät zur Messung verwendet werden. Das geschieht z.B. bei den Eis-Pyrheliometern nach RÖNTGEN und EXNER[1] oder nach VOLOCHINE oder auch beim Kugelpyranometer nach BELLANI[2], in dem die Destillation von Alkohol zur Anzeige verwendet wird. Mit solchen Geräten können allerdings rationell nur Strahlungs*summen* gemessen werden, weil die Meßgenauigkeit mit der Dauer der Bestrahlung wächst.

Auf einer ganz anderen Einteilung beruht die *Benennung* der Strahlungs-Meßinstrumente[3]. Man bezeichnet als *Pyrheliometer* im allgemeinen nur Geräte — gleich welchen Prinzips — zur Absolutmessung der direkten Sonnenstrahlung, die auf eine normal zur Strahlungsrichtung stehende Fläche auffällt. Dagegen nennt man alle Relativinstrumente, die eine Eichung benötigen, *Aktinometer*. Beide Instrumententypen können natürlich auch zur Messung der Strahlung eines Himmelsausschnittes Verwendung finden. Die Verwendung des schwarzen Körpers als Strahlungsempfänger bereitet in diesem Falle keine grundsätzlichen Schwierigkeiten.

Für die Untersuchung der *spektralen* Verteilung der Sonnen- oder Himmelsstrahlung, aus der sich der Aerosolgehalt, Teilchengröße und Wasserdampfgehalt der Luft bestimmen lassen, liefern alle spektrographischen Verfahren natürlich die besten Unterlagen. Bei den in der Meteorologie für klimatische Vergleiche notwendigen Feldmeßmethoden, die auch an zahlreichen Stationen eines Beobachtungsnetzes von weniger geschulten Beobachtern durchgeführt werden können, sind Messungen hinter Interferenzfiltern vorzuziehen. Jedoch pflegen Aktinometer bisher nur mit einfachen Glasfiltern ausgerüstet zu werden, unter denen sich die Schott-Filter OG 1 und RG 2 (auch RG 8) der größten Beliebtheit erfreuen, weil sie das sichtbare oder totale Spektrum der Sonne etwa halbieren und besonders scharfe Abfallkanten besitzen.

Pyranometer nennt man Instrumente mit horizontaler (oder geneigter) Auffangfläche (Öffnungswinkel 180°), die die Globalstrahlung von Sonne und Himmel oder — bei Abschirmung der Sonne — die Strahlung des Himmels allein messen. In diesem Falle ist der schwarze Körper (Hohlraum) nicht als Strahlungsempfänger zu verwenden, denn das Instrument muß im Freien aufgestellt werden können und bei jeder Witterung zur Messung oder Registrierung verwendbar sein. Das besagt, daß auch Regen und starker Wind die Messung nicht beeinträchtigen

---

[1] W. C. RÖNTGEN u. F. EXNER: Sitzgsber. Akad. Wiss. Wien **60**, 228 (1874).

[2] P. COURVOISIER u. H. WIERZEJEWSKI: Arch. Met., Geophys. u. Bioklim., Ser. B **5**, 413 (1954). — F. BORREL: J. Sci. Mét. **5**, Nr. 20 (1953). — D. L. GUNN u. D. YEO: Quart. J. Roy. Met. Soc. **77**, 293 (1951). — H. MROSE: Ang. Meteorol. **2**, 147 (1955).

[3] W. MÖRIKOFER: Arch. Met., Geophys. u. Bioklim., Ser. B **1**, 32 (1949).

dürfen. Aus der Diskrepanz zwischen der notwendigen Wetterfestigkeit meteorologischer Instrumente und der physikalischen Definiertheit ergeben sich leider sehr viele konstruktive Schwierigkeiten, die bei Labormessungen unbekannt sind. Der Windeinfluß auf eine frei exponierte schwarze Fläche bewirkt einen ständig wechselnden Wärmeabfluß. Deshalb überdeckt man Pyranometer mit einer Glasglocke, verwendet aber dann als Empfänger Paare von schwarzen und weißen Flächen, deren Temperaturdifferenz gemessen wird. Das zur Weißung meist verwendete Magnesiumoxyd absorbiert oberhalb 3 μ nahezu schwarz, also den gleichen Betrag wie die schwarze Auffangfläche, so daß das System gegen die Wärmestrahlung der Glasglocke nahezu unempfindlich wird. Wiederholte Nullpunktskontrollen sind aber nötig, besonders bei Verwendung von Filtergläsern, die sich bei der Bestrahlung stark erwärmen.

Mit *Pyrgeometern* oder *Effektiv-Pyranometern* wird die langwellige Wärmestrahlung des Himmels oder die Summe aus Globalstrahlung und langwelliger Wärmestrahlung gemessen. Schließlich messen *Strahlungsbilanzmesser* die Differenz der aufwärts und abwärts gerichteten Strahlungsströme aller Wellenlängen. Alle derartigen Instrumente müssen frei der Strahlung exponierte horizontale Flächen besitzen; Bilanzmesser haben je eine nach oben und nach unten exponierte Fläche. An diesen kann der Wind ungehindert angreifen. Um die dadurch bedingte wechselnde Wärmeabfuhr unschädlich zu machen, sind verschiedene Wege eingeschlagen worden. Entweder wird (beim Bilanzmesser) die kältere Platte durch einen Regelmechanismus auf die Temperatur der wärmeren aufgeheizt[1]. Oder es wird ein zweites Plattenpaar daneben gesetzt und je eine nach oben und nach unten exponierte Platte mit konstanter Leistung elektrisch geheizt[2]; aus den Temperaturdifferenzen und der Heizleistung kann man dann die Strahlungsbilanz ermitteln. Schließlich kann auch das Plattenpaar kräftig durch einen horizontalen Luftstrom ventiliert werden, so daß einesteils die Temperaturdifferenzen klein aber noch meßbar sind, andernteils der natürliche Wind keinen Einfluß mehr hat[3]; die Grenzschichtbedingungen sind dabei sorgfältig zu beachten.

Eine weitere Möglichkeit hat sich durch die Entwicklung von Filtermaterialien ergeben, die bis über 30 μ hinaus transparent sind und damit vor allem das „Wasserdampffenster" von 8 bis 11 μ noch frei lassen, das für den langwelligen Strahlungsaustausch in Bodennähe fast allein maßgebend ist. Hier ist das KRS 5 zu nennen, ein gezüchteter Mischkristall aus TlJ und TlBr, das allerdings im Sichtbaren fast undurchlässig ist und sich deshalb bei Tagesbestrahlung sehr stark erwärmt. Vor allem hat sich aber das Polyäthylen (Handelsname Lupolen-H) als sehr günstig erwiesen, dessen Durchlässigkeit von 0,4 bis 100 μ sehr gleichmäßig ist und nur bei 3,5, 6,9 und 13,7 μ schmale Absorptionsbanden besitzt. Diese fallen gerade mit den starken Absorptionsbanden des Wasserdampfes und der Kohlensäure zusammen, in deren Bereich fast kein Strahlungsaustausch zwischen Erdboden und Atmosphäre stattfindet. Glocken aus Lupolen haben sich daher bei modernen Effektivpyranometern und Strahlungsbilanzmessern hervorragend bewährt[4].

---

[1] S. BAUMBACH u. a.: Sonderh. Techn. Mitteil., Met. Amt. f. NW-Deutschl., Hamburg 1950.

[2] G. HOFMANN: Forstwiss. Zbl. **71**, 330 (1952).

[3] M. FRANSSILA: Geophysica, Helsinki **4**, 131 (1953). — V. E. SUOMI, M. FRANSSILA u. N. F. ISLITZER: J. Meteorology **11**, 276 (1954). — J. T. GIER u. R. V. DUNKLE: Trans. Amer. Inst. Electr. Engrs. **70**, 339 (1951). — P. COURVOISIER: Verh. schweiz. naturforsch. Ges., Davos **130**, 152 (1950).

[4] R. SCHULZE: Geofis, pura e appl. **24**, 107 (1953). — Ann. Met. Hamburg **6**, 127 (1953/54). — Strahlentherapie **94**, 152 (1954).

Für die Messung schmälerer Spektralbereiche der Global- oder diffusen Himmelsstrahlung ergeben sich Schwierigkeiten bei der Verwendung von Filtern, weil die Durchlässigkeit jedes ebenen Filters mit der Durchstrahlungsrichtung variiert; Verwendung von Halbkugeln vermindert diese Schwierigkeit etwas, scheitert allerdings oftmals an der Kostenfrage. In vielen Fällen kann man sich durch Vorlegen einer ULBRICHTschen Kugel helfen, wie sie durch LARCHÉ und SCHULZE[1] eingeführt worden ist; der Strahlengang wird dadurch parallel gemacht und es können Filter, gegebenenfalls sogar Interferenzfilter, dazwischen geschaltet werden. Zur Aussonderung einzelner Spektralbereiche können sonst photoelektrische oder photochemische Verfahren angewendet werden. *Photozellen* haben jedoch für den Dauergebrauch den Nachteil, daß sie sich zeitlich unter dem Einfluß der Strahlung verändern und auch temperaturabhängig sind. *Photochemische Empfänger* absorbieren nicht flächenhaft, sondern im Volumen und können daher nicht das Cosinusgesetz befolgen; auch bei diesen stört zudem häufig eine Temperaturabhängigkeit.

Anmerkung zu S. 183 und 184: Die Internationale Strahlungskommission hat in Davos 1956 die Einführung einer „*Internationalen Pyrheliometer Skala 1956*" beschlossen, die zwischen der ursprünglichen ÅNGSTRÖM-Skala und der Smithsonian Skala 1913 liegt. Messungen auf Grund der ÅNGSTRÖM-Skala sind demnach um 1,5% zu vermehren, solche nach der Smithsonian Skala um 2% zu vermindern. Für diese Festlegung sind sowohl die Untersuchungen der beiden Standard-Instrumente selbst herangezogen worden sowie zahlreiche Vergleiche zwischen geeichten Sekundärinstrumenten bei Messungen der Sonnenstrahlung und bei Laboratoriumsmessungen.

Literatur.

[*1*] Handbuch der Geophysik, Bd. VIII, herausgeg. v. F. LINKE u. F. MÖLLER. Berlin: Gebr. Borntraeger 1942—1956. Insbesondere die Kapitel:
   [*1a*] Kap. 5. Theorie der Zerstreuung, von M. A. SCHIRMANN.
   [*1b*] Kap. 6. Die Sonnenstrahlung und ihre Schwächung in der Atmosphäre, von F. LINKE.
   [*1c*] Kap. 7. Die kurzwellige Himmelsstrahlung, von F. LINKE.
   [*1d*] Kap. 11. Langwellige Strahlungsströme in der Atmosphäre und die Strahlungsbilanz, von F. LINKE u. F. MÖLLER.
   [*1e*] Kap. 14. Optik der Tropfen, Abschn. 1: Optik des Dunstes, von F. VOLZ.
[*2*] LINKEs Meteorologisches Taschenbuch, neue Ausgabe, Bd. 2, herausgeg. v. F. BAUR. Leipzig 1953.
[*2a*] Smithsonian Meteorological Tables, 6. ed. (Smiths. Miscell. Coll., Vol. 114), Wash. 1951.
[*3*] Annals of the Astrophysical Observatory of the Smithsonian Inst., Vol. 1—7. Washington 1900—1955.
[*4*] JUNGE, CHR.: Ber. dtsch. Wetterd. U.S.-Zone **35**, 261 (1952). — Ann. Met. **5**, Beih. (1952). — Tellus **5**, 1 (1952). — J. Meteorology **11**, 323 (1954); **12**, 13 (1955).
[*5*] VOLZ, F.: Ber. dtsch. Wetterd. **2**, No. 13 (1954)
[*6*] SCHÜEPP, W.: Arch. Met., Geophys. u. Bioklim., Ser. B **1**, 257 (1949).
[*7*] ELSASSER, W. M.: Harvard Met. Studies, No. 6, 1942.
[*8*] ELSASSER, W. M., u. J. I. KING: Stratosph. Rad., Techn. Rep. No. 6, Contr. AF 19 (122)—392, Univ. of Utah 1952.

[1] K. LARCHÉ: Licht **12**, 110 (1942). — K. LARCHÉ u. R. SCHULZE: Z. techn. Phys. **23**, 114 (1942). — H. HERMANN: Optik **3**, 394 (1947). — R. SCHULZE: Ber. dtsch. Wetterd. US-Zone **12**, 210 (1950). — Ann. Met. Hamburg **4**, 176 (1951).

Vision through the Atmosphere.

By

W. E. K. MIDDLETON.

With 36 Figures.

1. Introduction. The distance at which a given object can be seen through the atmosphere is a function of three variables: (1) The optical properties of the atmosphere, (2) the properties of the object itself and of its background, and (3) the state of adaptation of the eyes of the observer. If the object of regard is a source of light, the luminous intensity (candlepower) of the light source is the property implied by the second variable.

It will be the purpose of this article to present a summary of our knowledge of the optical properties of the atmosphere as they affect vision, to summarize the pertinent properties of the eye, and to show how this knowledge may be used to predict the visual range of a given object or source of light. The instruments used to measure the appropriate properties of the atmosphere will be classified and briefly described; and in addition, a short account will be given of the effect of the atmosphere on the apparent colors of objects and lights.

It must be assumed throughout that the reader is familiar with the concepts of elementary photometry. Certain less usual relations will be developed as they are needed. The radiometric and photometric nomenclature will be that adopted by the Committee on Colorimetry of the Optical Society of America[1].

I. The extinction of light in the atmosphere.

a) Extinction by pure air.

2. RAYLEIGH scattering. If the air were a pure mixture of permanent gases the problems dealt with in this article would be of no practical importance, since in such an atmosphere one would be able to see more than 300 km. at sea level. Nevertheless the pure gaseous atmosphere is a limiting case and thus its properties are of interest.

Suppose a collimated beam of light incident on pure dry air. The volume scattering function was shown[2] by RAYLEIGH to be

$$\beta'(\varphi, \lambda) = A\,\lambda^{-4}\,(1 + \cos^2\varphi). \tag{2.1}$$

The units of β' are watts (or lumens) scattered in the direction φ per unit solid angle per unit volume for one watt (or lumen) incident on unit area. Note that the scattered intensity varies as the inverse fourth power of the wavelength λ, but not markedly with direction, the intensity at 90° to the incident beam[3] being half that at 0 or 180°.

The light scattered in all directions is abstracted from the incident beam. To express the depletion of the incident light we may define a *scattering coefficient* b by the relation

$$dF = -\,b\,F\,dr \tag{2.2}$$

---

[1] The Science of Color, p. 229. New York: Thomas Y. Crowell Co. 1953.

[2] Lord RAYLEIGH: Phil. Mag. **47**, 375 (1899).

[3] In this article 0° represents the direction of the incident light.

where dF is the change of flux in the beam as it crosses a layer of thickness dr. The dimensions of b are (1/length). Since all the light removed from the beam by scattering must go into directions between $\varphi=0$ and $\varphi=\pi$ (radians) we have the following relation between β' and b:

$$2\pi \int_0^\pi \beta'(\varphi) \sin\varphi \, d\varphi = b. \tag{2.3}$$

For pure air at S.T.P. the values of b in and near the visible spectrum are given in Table 1, from VAN DE HULST[1].

Table 1. *Scattering coefficient for pure air, pressure 1013.2 mb, temperature 0° C.*

Wavelength (microns)	0.3	0.4	0.5	0.6	0.7	0.8	0.9	1.0
$(m^{-1}) \times 10^5$	14.79	4.358	1.750	0.833	0.466	0.260	0.162	0.106

Although some writers have doubted the correctness of the RAYLEIGH theory, mainly on the evidence of exceptionally great visual ranges, the general opinion is in favor of its adequacy. It is necessary to interpret such observations with attention to the elevation of the light path above sea level. COLBERG[2] has given the theory necessary for this.

b) Extinction by spherical particles.

3. Nature of the particles in the atmosphere. Far from being a mixture of pure gases, the atmosphere may be thought of as an aerosol, which is a suspensoid with air as the medium. The disperse phase of this aerosol may consist of solid particles and liquid droplets of any size which will remain suspended in the air. In addition, the atmosphere is often made less transparent by rain, snow, sand, and dust, and even swarms of insects.

The most important of these particles are the small liquid droplets, resulting from the condensation of water on very small hygroscopic nuclei[3]. The next most important are the solid or liquid products of combustion other than those which act as nuclei; but almost everywhere except in snowstorms or dust storms the particles which limit the visual range are predominantly spherical droplets consisting mainly of water.

4. The theory of MIE. The results referred to in Sect. 2 above apply to particles of radii very much smaller than a wavelength (λ) of light. For particles of radius $a > 0.1\,\lambda$, we must go to the much more complex theory of MIE[4].

From a consideration of the electromagnetic waves inside and outside a small dielectric sphere of radius a, MIE derived differential equations which may be solved to give the electric and magnetic vectors at any point in space due to the light scattered by the sphere. The illuminance in any direction making an angle φ with the incident light can be expressed as the sum of two squares, each of which is a slowly converging infinite series in φ. The results are usually expressed in terms of a parameter $\alpha = 2\pi\, a/\lambda$.

---

[1] H. C. VAN DE HULST: In The atmosphere of the earth and planets, p. 55. Chicago: Univ. Press 1949.

[2] R. COLBERG: Meteor. Rdsch. **2**, 23 (1949).

[3] See the article "Cloud Physics".

[4] G. MIE: Ann. Phys. **25**, 377 (1908). See also the article on Scattering etc. in Vol. 26, and J. A. STRATTON, Electromagnetic theory, p. 563. New York: McGraw-Hill Book Co. 1941.

Tables of these functions are available for water drops ($\mu = 1.33$) for various values of α up to 40.0, and also for droplets of other indices of refraction[1-3]. In the first two of these references the argument is an angle $\gamma = 180° - \varphi$, so that light of which the direction is unchanged corresponds to $\gamma = 180°$. As α increases, the form of the scattering function becomes more and more complex, particularly near the direction $\varphi = 0$.

For our present purpose we are interested mainly in the scattering coefficient b rather than in the scattering function $S(\varphi)$. This may be expressed in the form

$$b = \sum_{i=1}^{n} N_i K_i \pi a_i^2 \tag{4.1}$$

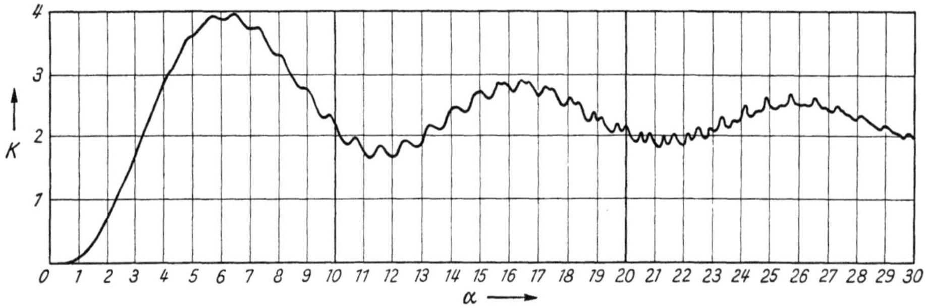

Fig. 1. Scattering area ratio K as a function of α (after GOLDBERG).

in which we are assuming an atmosphere including n different sizes of particles, the i-th having a radius a_i and occurring in a quantity N_i per unit of volume. The *scattering area ratio* K is found from the scattering function $S(\varphi)$ by integrating it with respect to φ:

$$K = \frac{\lambda^2}{2\pi^2 a^2} \int_0^\pi S(\varphi) \sin\varphi \, d\varphi = \frac{2}{\alpha^2} \int_0^\pi S(\varphi) \sin\varphi \, d\varphi. \tag{4.2}$$

This quantity K is a pure number, and is in fact the ratio of the area of wave front affected by the particle to the cross-sectional area of the particle itself. It turns out to be an extremely complicated function of α, as is shown by Fig. 1[4], approaching the value 2 for very large droplets. The reason for this value 2 is that the MIE theory includes the effect of diffraction. The larger the droplet, the narrower the cone into which this diffracted light is sent, so that the experimental verification of the theory for large droplets demands a small receiver and well-collimated light, as has been shown by SINCLAIR[5].

It is possible to extend the MIE theory to absorbing particles by the use of a complex index of refraction.

The theory has now been tested in so many ways that its adequacy may be considered established.

5. The treatment by geometrical and physical optics. For droplets of radius greater than two or three wavelengths the labor of computing results from the MIE theory is tremendous, and the method would not be feasible without modern

[1] A. N. LOWAN: Tables of Scattering functions for Spherical Particles. Nat. Bur. Stand., Washington D. C., Appl. Math. Ser. 4 **1949**.
[2] R. O. GUMPRECHT, N. L. SUNG, J. H. CHIN and C. M. SLIEPCEVICH: J. Opt. Soc. Amer. **42**, 226 (1952).
[3] H. HOLL: Optik **1**, 213 (1946).
[4] BERNICE GOLDBERG: J. Opt. Soc. Amer. **43**, 1221 (1953).
[5] D. SINCLAIR: J. Opt. Soc. Amer. **37**, 475 (1947).

electronic computers. It is possible to obtain a good approximation to the scattering function without resort to electromagnetic theory, by investigating separately the effects of *reflection* (external and internal) by the droplet, *refraction* by its surfaces, and *diffraction* by its outline.

The first two of these effects were investigated in the first decade of the twentieth century by WIENER[1], the third by many authors[2].

The whole theory has been assembled, with numerical examples, by BRICARD[3].

Suppose a spherical droplet of radius a in a beam of monochromatic light of wavelength λ, and let H [watts/m.²] be the irradiance at the drop. Let $\alpha = 2\pi a/\lambda$, $u = \alpha \sin \varphi$.

The radiant intensity I_1 at an angle φ due to diffraction by the droplet is

$$I_1 = \tfrac{1}{4}\alpha^2 [2 J_1(u)/u]^2 \cos^4(\varphi/2) \cdot H a^2 \tag{5.1}$$

in which $J_1(u)$ is the BESSEL function of the first order.

WIENER showed that the intensity due to the light refracted by the drop is

$$I_2 = \frac{(1-k_1)^2 H a^2}{2} \frac{\sin 2 i_1}{\sin\varphi} \cdot \frac{d i_1}{d\varphi}, \tag{5.2}$$

where $\varphi = 2(i_1 - r_1)$, as will be seen from Fig. 2, and

$$k_1 = \frac{1}{2}\left[\frac{\sin^2(i_1 - r_1)}{\sin^2(i_1 + r_1)} + \frac{\tan^2(i_1 - r_1)}{\tan^2(i_1 + r_1)}\right],$$

$$d i_1/d\varphi = (\sin i_1 \cos r_1)/[2\sin(\varphi/2)].$$

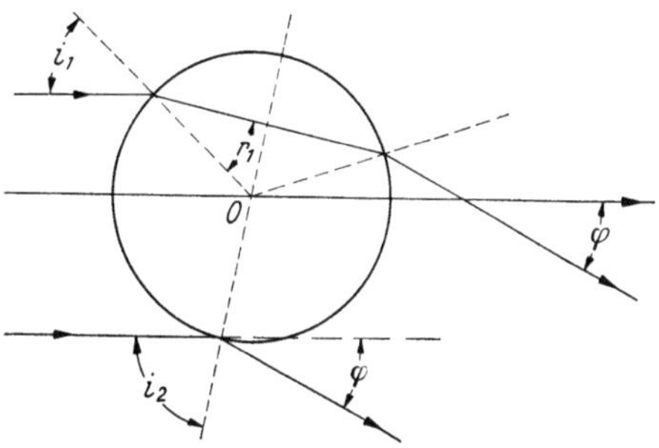

Fig. 2. Illustrating WIENER's calculations.

Also the intensity due to light reflected from the outside of the droplet is

$$I_3 = \tfrac{1}{4} K_2 H a^2, \tag{5.3}$$

where

$$K_2 = \frac{1}{2}\left[\frac{\sin^2(i_2 - r_2)}{\sin^2(i_2 + r_2)} + \frac{\tan^2(i_2 - r_2)}{\tan^2(i_2 + r_2)}\right],$$

$$\varphi = \pi - 2 i_2, \qquad \sin i_2 = \mu \sin r_2 .$$

The total scattered intensity would be $I = I_1 + I_2 + I_3$ if the radiation scattered by the three processes were completely incoherent. BRICARD[4] has shown how to calculate the effect of interference between the various beams, and has also pointed out that for a polydisperse aerosol it is not very important.

The intensity I_1, due to diffraction, falls off very rapidly as φ increases from zero. It is not generally measured at all by such instruments as polar scattering meters (see Sect. 49 below).

Compared to the scattering the effect of actual absorption is very small in the water droplets actually found in the atmosphere. ZANOTELLI[5] has demonstrated this for pure water, and MIDDLETON[6] has shown that even serious pollution of the cloud droplets has little effect on the spectral radiance of the

[1] C. WIENER: Nova Acta K.-Leopold dtsch. Akad. Naturf., Halle **73**, 3 (1907); **91**, 81 (1910).

[2] For a good summary see J. M. PERNTER and F. M. EXNER, Meteorologische Optik, 2nd ed., p. 476. Vienna 1922.

[3] J. BRICARD: J. Phys. Radium **4**, 57 (1943). — Ann. Géophys. **2**, 231 (1946). See also BRICARD's article in this Volume.

[4] J. BRICARD: Ann. Géophys. **2**, 231 (1946).

[5] G. ZANOTELLI: Atti, Acad. Ital., Rend., Cl. Sci. Fis. Mat. Nat. **2**, 42 (1940).

[6] W. E. K. MIDDLETON: J. Opt. Soc. Amer. **44**, 793 (1954).

overcast sky. Nevertheless some absorption is always present (especially in industrial regions), and we must therefore define an *extinction coefficient* σ:

$$\sigma = b + K \tag{5.4}$$

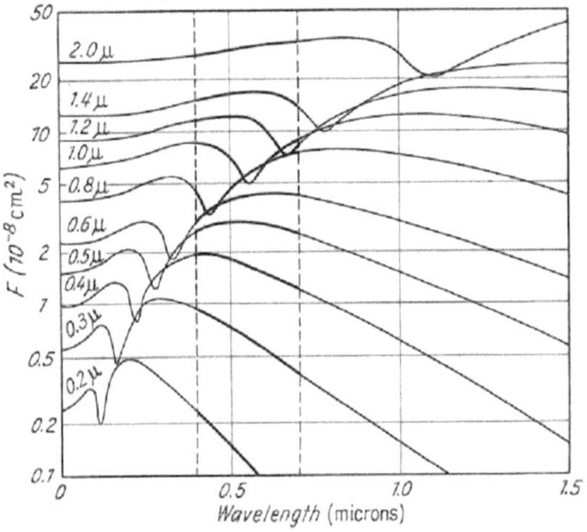

Fig. 3. Extinction cross-section vs. wavelength for drops of various sizes, after FOITZIK.

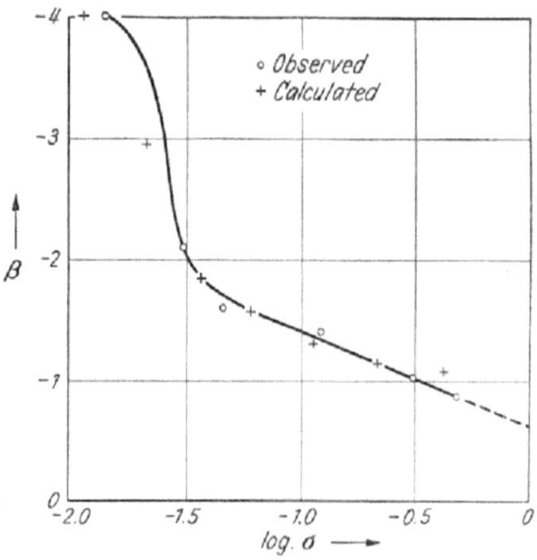

Fig. 4. Relation between log σ and n. Observations by MIDDLETON, calculations by JUNGE.

which is the sum of the scattering and absorption coefficients, each defined in the manner of Eq. (2.2). For nearly all our problems the mechanism of the extinction is not in question, and we shall generally make use of the quantity σ.

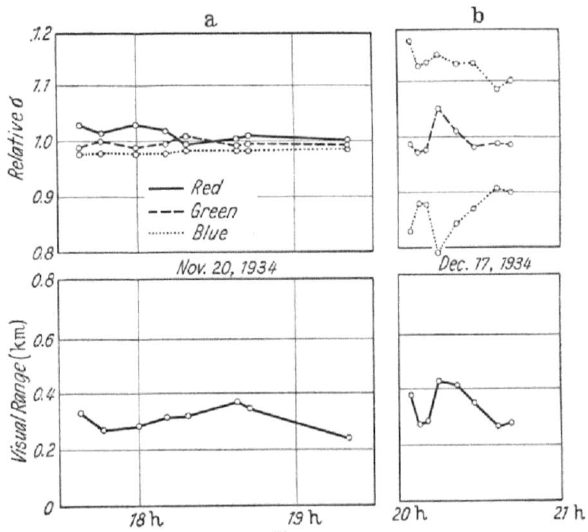

Fig. 5a and b. Relative extinction coefficients in fog (a) and in dense haze (b) of about the same obscurity, on two occasions at Berlin-Adlershof, after FOITZIK.

6. Empirical relations between extinction and wavelength. By analogy with the RAYLEIGH law, there have been many attempts[1] to find a relation of the

[1] e.g. Y. ROCARD and P. DE ROTHSCHILD: Rev. d'Opt. **6**, 353 (1927). — F. W. P. GÖTZ: Meteor. Z. **51**, 472 (1934). — W. E. K. MIDDLETON: Gerlands Beitr. Geophys. **44**, 358 (1935). — L. FOITZIK: Wiss. Abh. Reichsamt Wetterdienst, Berlin **4**, No. 5 (1938).

form

$$\sigma = c_1 \lambda^n \tag{6.1}$$

between extinction coefficient and wavelength.

In this, n was expected to vary between -4 and some small positive number. Since the permanent gases are always present, the expression

$$\sigma = c_1 \lambda^n + c_2 \lambda^{-4} \tag{6.2}$$

is a better one. It is now realized that such expressions are at best only rough approximations. That this is so is well shown by Fig. 3, after FOITZIK[1], derived from the MIE theory. Obviously no value of n is going to correspond to the behavior of droplets of radii $0.5 < a < 1.4$ microns.

JUNGE[2] has shown that the size distribution of MIE particles over the continent of Europe can be expressed by the equation

$$dN/d\log a = c\,a^{-3} \tag{6.3}$$

and has calculated values of n in the region $0.46\mu < \lambda < 0.64\mu$ corresponding to various values of σ, which agree closely with experimental values obtained in Canada by MIDDLETON[3], as is shown in Fig. 4. The success of this procedure is due to the polydisperse nature of the atmospheric aerosol.

Actual fog, which consists predominantly of droplets several microns in radius, is very nearly non-selective, that is to say n is nearly zero.

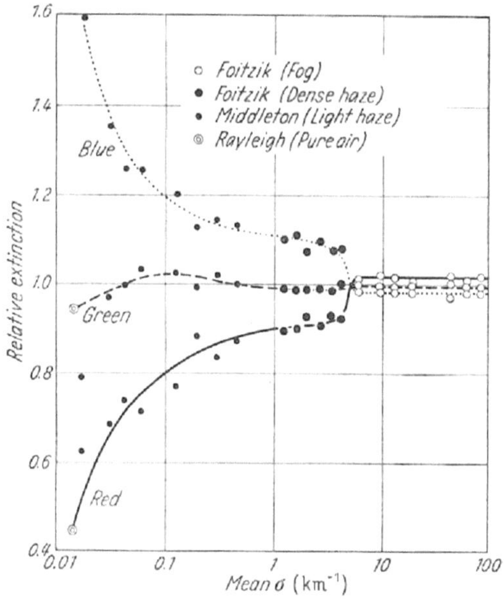

Fig. 6. Mean values of relative extinction coefficient as a function of the mean value for three colors. Adapted from FOITZIK.

7. "Anomalous" extinction. City "fog" or industrial haze is, on the other hand, frequently highly selective even when very dense, making lights appear orange at comparatively short distances. We present Fig. 5 to illustrate this[4]. The phenomenon has been discussed further by SCHOBER[5] and by MIDDLETON[6].

8. Experimental studies of the extinction coefficient. A very large number of investigations of natural haze have been made[7]. The general trend of the results is shown by the curve of Fig. 6, which summarizes observations by FOITZIK[4] and by MIDDLETON[3].

Note the sudden change of selectivity near $\sigma = 4$ km.$^{-1}$, which corresponds to the change from haze to fog. The results shown in this figure probably represent fairly closely the behavior of country air, not seriously polluted by the products of combustion.

[1] L. FOITZIK: Meteor. Z. **52**, 458 (1935). These curves are worked out from the earlier less detailed values of K given by J. A. STRATTON and H. G. HOUGHTON, Phys. Rev. **38**, 159 (1931).

[2] C. JUNGE: Ber. dtsch. Wetterdienst. US-Zone **35**, 261 (1952).

[3] W. E. K. MIDDLETON: Gerlands Beitr. Geophys. **44**, 358 (1935).

[4] L. FOITZIK: Wiss. Abh. Reichsamt Wetterdiesnt, Berlin **4**, No. 5 (1938).

[5] H. SCHOBER: Meteor. Z. **51**, 233 (1934).

[6] W. E. K. MIDDLETON: Q. J. Roy. Met. Soc. **62**, 473 (1936).

[7] See [2], p. 42, for numerous references.

9. Observations on the scattering function. The only really satisfactory method of measuring the scattering function is by means of a special apparatus such as those referred to in Sect. 49 below. WALDRAM[1] has made the most complete series of such measurements at altitudes up to 9200 m. Figs. 7 and 8 indicate the sort of results he obtained in "clear" air and in industrial haze. The lowest line in Fig. 7 is for pure air at sea level, from the RAYLEIGH theory.

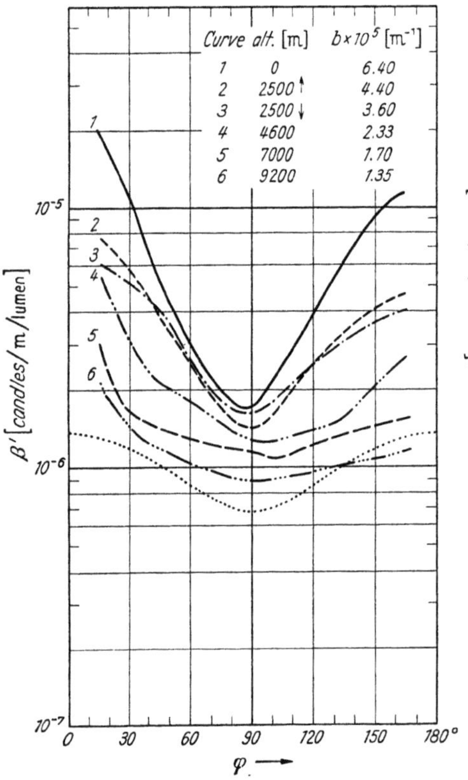

Fig. 7.

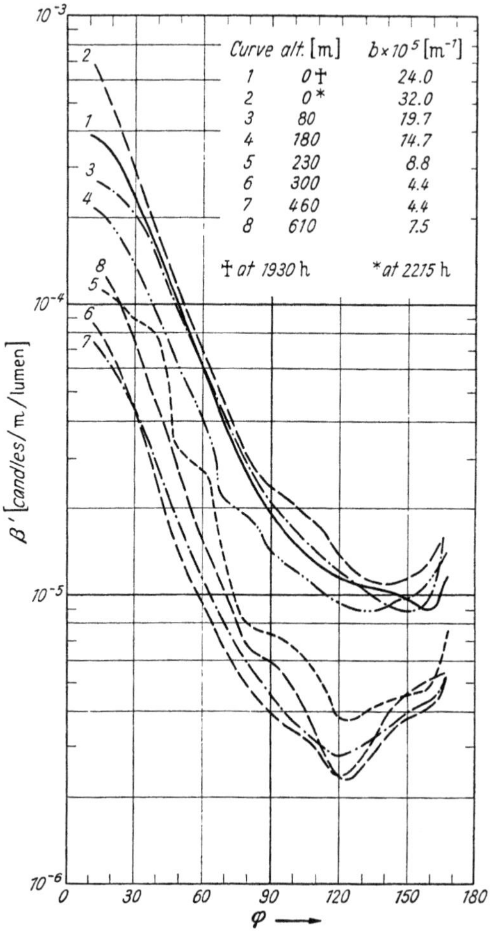

Fig. 8.

Fig. 7. Volume scattering functions $\beta'(\varphi)$ in clear air over England at various altitudes, 1 June 1942, after WALDRAM. The values of b are integrated from the curves.

Fig. 8. Volume scattering functions in industrial haze over England, 7 October 1942, after WALDRAM.

The curves of Fig. 8 are what would be expected from the MIE theory, but it is difficult to account for those of Fig. 7. Fig. 9 is of interest in showing the frequent stratification of the aerosols even in apparently clean air (air in which absorption is negligible).

Similar observations have been made by BULLRICH[2], SIEDENTOPF[3], and FOITZIK and ZSCHAECK[4], using a horizontal searchlight beam. FOITZIK and

[1] J. M. WALDRAM: Quart. J. Roy. Met. Soc. **71**, 319 (1945). — Trans. Illum. Engng. Soc. Lond. **10**, 147 (1945).

[2] K. BULLRICH: Optik **2**, 301 (1947).

[3] H. SIEDENTOPF: Z. Meteorol. **1**, 342 (1947).

[4] L. FOITZIK and H. ZSCHAECK: Z. Meteorol. **7**, 1 (1953).

ZSCHAECK calculated the ratio ν of the energy scattered in forward directions to that scattered in backward directions, and expressed it as a function of the extinction coefficient (Fig. 10).

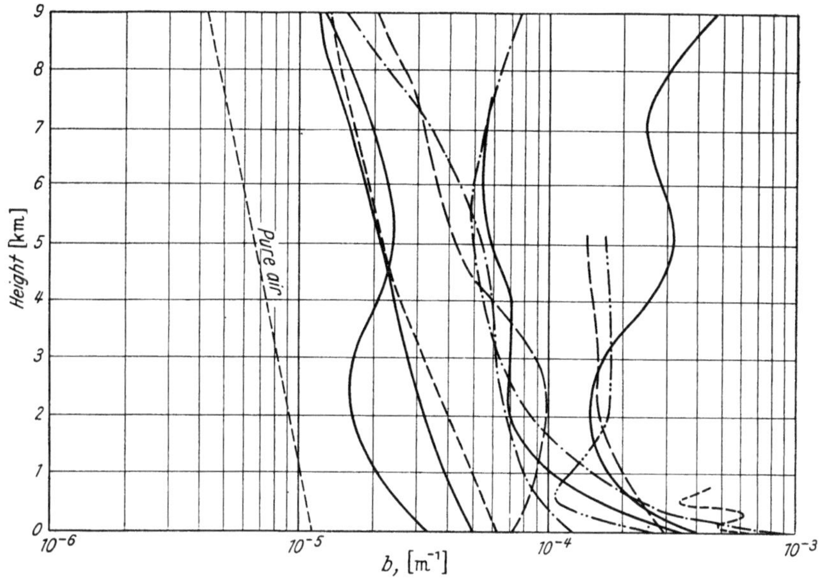

Fig. 9. Variation of scattering coefficient with height on various occasions in "clean air", after WALDRAM.

10. Direct observations of haze particles. It has recently become possible to make direct observations on haze particles, both by optical and by electron microscopy. It was discovered by DESSENS that certain small spiders spin fibres

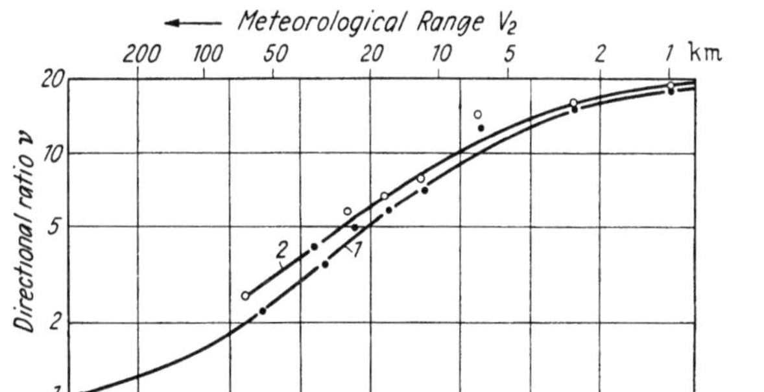

Fig. 10. Ratio ν of energy scattered in forward directions to that scattered in backward directions, as a function of σ (after FOITZIK and ZSCHAECK). Curve 1, entire atmosphere; curve 2, aerosol particles alone.

of the order of 10^{-6} cm. in diameter, to which minute droplets and other particles will adhere[1]. The size distribution can be obtained by actual counting, and the effect of changing relative humidity traced. Other workers have taken up this study[2]. DESSENS was able to calculate from his counts the dependence of σ on λ, and compare the results with direct measurements.

[1] H. DESSENS: C. R. Acad. Sci. Paris **223**, 915 (1946). — Ann. Géophys. **2**, 343 (1946); **3**, 68 (1947). — La Météorologie, p. 321. 1947. — Quart. J. Roy. Met. Soc. **75**, 23 (1949).

[2] See the article "Cloud Physics" in this volume.

II. The alteration of contrast by the atmosphere.

11. Definition of contrast. An object is recognizable because it differs in color or brightness, or both, from its surroundings. For the present we shall confine ourselves to the consideration of differences in brightness, which are measured by corresponding differences in luminance. Let the luminance of an object be B, that of its background B'; then the *contrast* between the two is defined by the expression

$$C = \frac{B - B'}{B'}. \tag{11.1}$$

Numerically, C can vary from -1 (ideally black object) to $+\infty$. Very large values of C arise when we consider lights at night, but in the daytime C is almost always less than about 10.

a) Theory of the air light.

12. KOSCHMIEDER's theory of the air light[1]. All the atmosphere between the observer and a distant object scatters light which comes from sun, sky, and ground, and some of this scattered light enters the eyes of the observer. The atmosphere between observer and object thus acquires a certain luminance which is greater the greater the intervening distance. The theory of this was first clearly stated by KOSCHMIEDER in 1924 for vision along the horizontal.

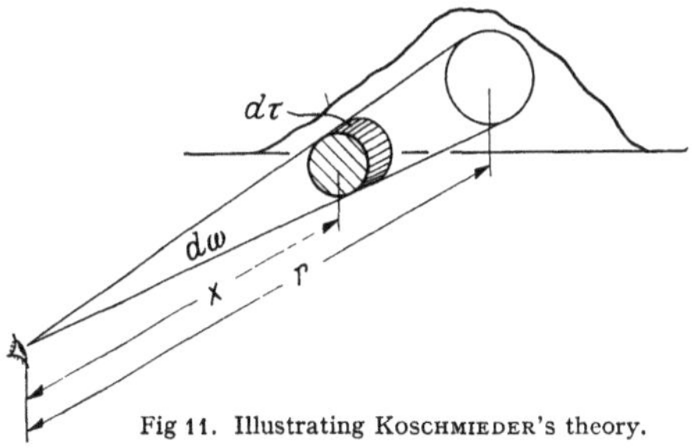

Fig 11. Illustrating KOSCHMIEDER's theory.

We must assume (i) that the light scattered from an element of volume may be considered as coming from a point source of which the intensity is proportional to the volume, (ii) that atmospheric refraction may be neglected, (iii) that all parts of the atmosphere in the horizontal plane are equally illuminated (this assumes that the surface of the earth be considered a uniform horizontal plane), (iv) that the scattering coefficient b is constant in the horizontal plane, (v) that the linear dimensions of the object are small in comparison to its distance from the observer.

Let us now derive an expression for the apparent luminance B_b of a black object at a distance r in the horizontal plane of the observer's eye.

Consider an element $d\tau = d\omega \cdot x^2\, dx$ of the cone of air (Fig. 11) having the eye at its apex and part of the object at its base. Since each element of this cone is illuminated in the same way, the intensity of $d\tau$ in the direction of the eye will be

$$dI = d\tau \cdot A\, b \tag{12.1}$$

where A is a constant to be determined.

The illuminance at the eye due to this light is

$$dE = dI \cdot x^{-2} \cdot e^{-b x} \tag{12.2}$$

and the luminance of $d\tau$ as seen from the eye is

$$dB = dE/d\omega = (1/d\omega)\, dI \cdot x^{-2}\, e^{-b x} = A\, b\, e^{-b x}\, dx \tag{12.3}$$

[1] H. KOSCHMIEDER: Beitr. Phys. freien Atm. **12**, 33, 171 (1924).

which can be integrated to give the apparent luminance of the black object

$$B_b = \int_0^r A\, b\, e^{-bx}\, dx = A(1 - e^{-br}). \tag{12.4}$$

To determine A, suppose the black object to be moved to infinity; then its luminance will equal B_h, the luminance of the horizon sky.

Therefore

$$B_h = \int_0^\infty A\, b\, e^{-bx}\, dx = A,$$

and finally

$$B_b = B_h(1 - e^{-br}). \tag{12.5}$$

Eq. (12.5) is known as KOSCHMIEDER'S *law*. We can extend it to objects which are not black; such an object of intrinsic luminance B_0 will have an apparent luminance

$$B = B_0\, e^{-br} + B_h(1 - e^{-br}). \tag{12.6}$$

This formula, and (12.5), apply only to horizontal vision.

13. The "two-constant" theory of the air light. If we are to apply the theory of the air light to non-horizontal vision, or to an absorbing atmosphere, we must adopt a slightly different point of view originally developed by DUNTLEY[1].

Consider Fig. 12. Let an object at O have an intrinsic luminance B_0 in the direction of the observer at R. At a distance r let its apparent luminance be $B(r)$. Let $B_a(r)$ be the apparent luminance of a layer of air 1 m. thick as seen in the direction RO. Then

$$dB(r)/dr = -\sigma(r)\, B(r) + B_a(r) \tag{13.1}$$

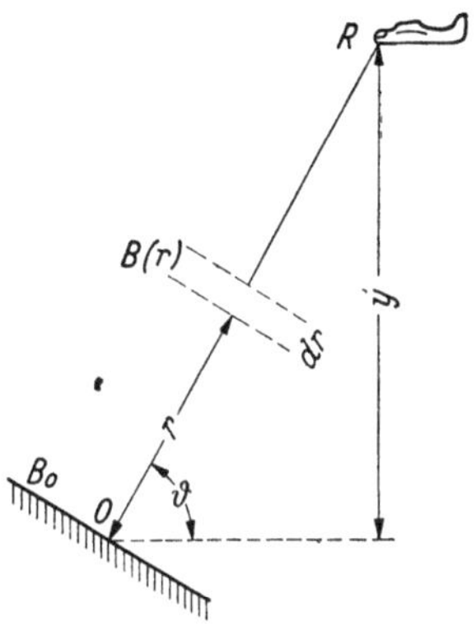

Fig. 12. Illustrating the two-constant theory (after DUNTLEY).

where σ (the extinction coefficient) and B_a may be functions of r. In practice it is essential to assume that σ and B_a are similar functions of r;

$$\sigma(r) = \sigma_0 f(r), \quad B_a(r) = B_a(0)\, f(r). \tag{13.2}$$

This limits the theory to atmospheres in which absorption is negligible if non-horizontal vision is concerned. For horizontal vision there is no limitation other than KOSCHMIEDER'S assumptions, since both σ and B_a are then constant.

Let us write (13.1)

$$dB = [-\sigma_0 B + B_a(0)]\, f(r)\, dr \tag{13.3}$$

and integrate. We have

$$\int_{B_0}^{B_R} \frac{dB}{\sigma_0 B - B_a(0)} = + \int_0^R f(r)\, dr \tag{13.4}$$

under the conditions of Fig. 12, or

$$\int_{B_0}^{B_r} \frac{dB}{\sigma_0 B - B_a'(0)} = - \int_0^R f(r)\, dr \tag{13.5}$$

[1] S. Q. DUNTLEY: J. Opt. Soc. Amer. **38**, 179 (1948). The treatment given here is a corrected one, condensed from [2], p. 64.

if the observer is looking upward. There is no difference between these equations except in the value of $B_a(0)$.

Write $\overline{R} = \int_0^R f(r)\,dr$. This quantity was called by DUNTLEY the *optical slant range*, and "represents the horizontal distance in a homogeneous atmosphere for which the attenuation is the same as that actually encountered along the true path of length R". From (13.4) we obtain

$$B_R = \frac{B_a(0)}{\sigma_0}\left[1 - e^{-\sigma_0 \overline{R}}\right] + B_0\, e^{-\sigma_0 \overline{R}}, \tag{13.6}$$

and from (13.5)

$$B_R = \frac{B'_a(0)}{\sigma_0}\left[1 - e^{-\sigma_0 \overline{R}}\right] + B_0\, e^{-\sigma_0 \overline{R}}. \tag{13.7}$$

These equations differ only in the value of $B_a(0)$, which depends on the relation of the direction of view to the distribution of luminance over the surroundings.

If now we consider horizontal vision, and compare Eqs. (12.8) and (13.6), we shall observe at once that they are identical if $B_h = B_a(0)/\sigma_0$, and if we substitute σ_0 for the scattering coefficient b. Thus (12.8) may legitimately be extended to a scattering and absorbing atmosphere, as long as only horizontal vision is concerned.

b) The reduction of contrast.

14. The general case. Let two objects, or an object and its background, have luminances B_0 and B'_0 when seen close at hand, B_R and B'_R when seen from a distance R. Then from (13.6) or (13.7) we can write

$$B_R - B'_R = (B_0 - B'_0)\, e^{-\sigma_0 \overline{R}}. \tag{14.1}$$

Referring to Eq. (11.1) we may define an *inherent contrast* $C_0 = (B_0 - B'_0)/B'_0$, and an *apparent contrast* $C_R = (B_R - B'_R)/B'_R$.

From (14.1) we immediately obtain[1]

$$C_R = C_0 (B'_0/B'_R)\, e^{-\sigma_0 \overline{R}} \tag{14.2}$$

which is the most general possible expression of the law of the reduction of contrast by the atmosphere.

15. The observer looking upward. It has been shown by DUNTLEY that the contrast of an object against the sky follows the law

$$C_R = C_0\, e^{-\sigma_0 \overline{R}} \cdot \frac{1 - \exp\left(-\sigma_0 \int_R^\infty f(r)\,dr\right)}{1 - \exp\left(-\sigma_0 \int_0^\infty f(r)\,dr\right)}. \tag{15.1}$$

The difficulty in applying such a formula lies in obtaining the necessary information about the function $f(r)$.

16. Horizontal vision. Let us refer to Eq. (14.2). Since we have agreed to consider the earth as flat, the luminance of the horizon sky, B_h, will be independent of r, and thus the apparent contrast of an object seen against the horizon sky will be

$$C_R = C_0\, e^{-\sigma R}. \tag{16.1}$$

---

[1] S. Q. DUNTLEY: J. Opt. Soc. Amer. **38**, 179 (1948).

17. Effect of absorption. It has been pointed out by DESSENS[1] that the addition of pure absorption in the path of horizontal vision may either increase or decrease the apparent contrast, depending on the values of R and C_0. It may be shown that $dC_R/d\sigma > 0$ if $R < (1/\sigma)(1 + 1/C_0)$. The increase in contrast results, of course, from the darkening of the horizon, not the brightening of the object.

18. "Optical equilibrium". From (13.1) it will be seen that $dB(r)/dr$ vanishes when $B(r) = B_a(r)/\sigma(r)$ or [Eq. (13.2)] when $B(r) = B_a(0)/\sigma_0$. A luminance of this magnitude will not change with distance. For horizontal vision this "equilibrium luminance" is that of the horizon sky in the direction of view. For oblique vision, it was pointed out by DUNTLEY[2] that there are usually two horizontal directions from which sunlight is scattered through the same angle as from the line of sight, and the luminance B_m of the horizon sky in these directions will be the equilibrium luminance.

19. The observer looking downward. From Eq. (13.6) and (14.2) we may derive the expression

$$C_R = C_0 \left[1 - \frac{B_m}{B'_0}\left(1 - e^{+\sigma_0 \bar{R}}\right)\right]^{-1} \tag{19.1}$$

in which B'_0 is the intrinsic luminance of the ground (as seen nearby in the direction of view). The quantity (B_m/B'_0) is called by DUNTLEY[3] the *sky-ground ratio*, and may vary from about 0.2 for fresh snow under a clear sky to about 2.5 for forest under an overcast. It is, however, not clear that the theory can be used unless we can assume that nearly all of the illumination is due to direct sunlight.

20. The optical slant range. The optical slant range $\bar{R} = \int_0^R f(r)\, dr$ is a function not only of the distance R and the angle of view ϑ but also of the optical state of the atmosphere. Unless it can be estimated the above theory cannot be used. DUNTLEY[3] has suggested that in many conditions we may assume the number of suspended particles to be proportional to that of the molecules of the permanent gases, and, proceeding from the well-known "standard atmosphere" used in aeronautics, define an "optical standard atmosphere". Under these assumptions the optical slant range between any two points R_1 and R_2 can be shown to be

$$\bar{R} = 9200 \operatorname{cosec} \vartheta \left[e^{-R_1 \sin\vartheta/9200} - e^{-R_2 \sin\vartheta/9200}\right] \tag{20.1}$$

if R is in meters. But there will be many occasions on which these assumptions will be so far from the truth that the theory cannot be applied (see Sect. 9 above).

21. Experimental tests. The theory of the reduction of contrast in the horizontal has been tested several times, usually by the telephotometry of similar dark screens (or pairs of screens, black and white) at various distances[4]. The most elaborate tests have been those of COLEMAN and his co-workers[5], in which all

---

[1] H. DESSENS: C. R. Acad. Sci. Paris **218**, 685 (1944).

[2] S. Q. DUNTLEY: J. Opt. Soc. Amer. **38**, 179 (1948). The principle of equilibrium luminance in horizontal vision was first enunciated by M. HUGON, Sci. et Ind. Phot. **1**, 161 and 201 (1930).

[3] S. Q. DUNTLEY: J. Opt. Soc. Amer. **38**, 179 (1948).

[4] F. LÖHLE: Meteor. Z. **46**, 49 (1929). — H. KOSCHMIEDER: Forsch.-Arb. staatl. Obs. Danzig **1930**, No. 2. — H. RÜHLE: Forsch.-Arb. staatl. Obs. Danzig **1930**, No. 3.

[5] H. S. COLEMAN, F. J. MORRIS, H. E. ROSENBERGER and M. J. WALKER: J. Opt. Soc. Amer. **39**, 515 (1949).

the targets subtended the same angle, and whose elaborate telephotometer was particularly free from stray light. There can be no doubt that contrast is actually attenuated exponentially in a uniform and uniformly illuminated atmosphere.

22. Other atmospheric effects on the appearance of distant objects. It is popularly believed that the edges of objects are made diffuse by a fog. There is, however, a theoretical basis[1] for doubting the correctness of this belief, and several experimental attempts[2] to find it by photographic and photoelectric photometry have given clear negative results; the phenomenon is therefore almost certainly subjective[3].

A more important phenomenon is that variously called "*shimmer*", "*boil*", and "*optical haze*" in the daytime, and "*scintillation*" or "*twinkling*" at night. This is caused by the bending of light as it passes through air of which the density is non-uniform because of small-scale convection currents.

The subject is difficult to handle theoretically, but an attempt has recently been made[4], based on the hypothesis that the path of sight may be divided into sections within which the optical effects of the shimmer are uncorrelated. The average apparent contrast of an object with its surroundings will be reduced by shimmer, and the reduction is calculable if enough information can be obtained about the small-scale turbulence.

There is a large experimental literature, mostly astronomical in emphasis[5] and dealing with the twinkling of stars. In recent years modern techniques have been applied with success both by day and by night. These experiments have been photographic[6] or photoelectric[7,8], the latter making use of wave analyzers to determine the frequency spectrum of the scintillation. The findings of GOLDSTEIN[8] are typical, the most important being:

1. the clearer the weather, the greater the scintillation,
2. the scintillation is greatest in warm weather,
3. large sources scintillate much less than small ones,
4. the longer the path, the greater the scintillation.

SIEDENTOPF and WISSHAK[7] investigated the diurnal variation, finding it to be similar to that of the turbulence due to smallscale convection in the lower atmosphere.

ZWICKY[9] has studied the dependence of stellar scintillation on wavelength by putting a prism or a grating in front of a telescope and letting the spectrum of the star drift across a photographic plate. He found that the color of the star, as well as its magnitude, appears to be constantly changing. He speaks of *aerial mollusks* (or *blobs*), volumes of air of locally altered density, ranging in size from millimeters to many meters.

---

[1] W. E. K. MIDDLETON: J. Opt. Soc. Amer. **32**, 139 (1942). — G. A. FRY, C. S. BRIDGMAN and V. J. ELLERBROCK: J. Opt. Soc. Amer. **37**, 635 (1947).

[2] S. Q. DUNTLEY: J. Opt. Soc. Amer. **38**, 179 (1948). — D. R. BARBER: Proc. Phys. Soc. Lond. B **63**, 364 (1950); also see [*2*], p. 79.

[3] It can be imitated without the presence of a fog by producing sharp-edged stimuli of very small contrast on a bright background.

[4] S. Q. DUNTLEY, W. H. CULVER, FRANCES R. CULVER and R. W. PREISENDORFER: J. Opt. Soc. Amer. **42**, 877 (1952) (abstract).

[5] For a survey see [*3*].

[6] L. A. RIGGS, C. G. MUELLER, C. H. GRAHAM, and F. A. MOTE: J. Opt. Soc. Amer. **37**, 415 (1947).

[7] H. SIEDENTOPF and F. WISSHAK: Optik **3**, 430 (1948).

[8] E. GOLDSTEIN: NRL (Washington) reports N-3462 (1949) and N-3710 (1950).

[9] F. ZWICKY: Publ. Astron. Soc. Pacific **62**, 150 (1950). — J. Amer. Rocket Soc. **23**, 370 (1953).

III. The relevant properties of the eye.

23. Stimulus and sensation. Definition of a threshold. Up to this point the discussion might as well have been in radiometric as in photometric terms; but as a preliminary to our next task, the discussion of the visual range, we must now inquire into those properties of the human eye which are relevant to our subject.

Radiation in the visible range of wavelengths, modified by the atmosphere in the ways discussed in the first two chapters, reaches the retina of the eye and acts as a physical *stimulus*. In ways which are not our immediate concern, each such stimulus or combination of stimuli produces a *sensation* of luminosity, shape, and color. While the magnitude of a stimulus can be measured, that of a sensation cannot. But sensations, juxtaposed in time or space, can be compared, and we measure the least difference in two stimuli which will produce a perceptible difference in sensation. Such a difference in stimuli is called a *threshold*.

24. The necessary information. For the purposes of the present study, two types of threshold are of interest: (a) the threshold of brightness contrast $\Delta B/B=\varepsilon$; (b) the threshold illuminance at the eye for the detection of point sources of light (E_t). Each of these thresholds is a function of the luminance of the visual field, and ε also depends on the angular subtense of the object of regard, and to a lesser extent upon its shape. The datum known as visual acuity is a special case of the threshold of brightness contrast.

25. Adaptation to changes in field luminance. As the luminance of the visual field changes, the eye changes with it. Apart from the well known expansion and contraction of the pupil, the eye can be in either of two conditions, the light-adapted state above about 2×10^{-3} candles/m.^{2}, and the dark-adapted state at lower values of luminance. In the former the cones of the retina play the major part in vision, while in the dark adapted state only the rods function. As the central part of the retina (fovea) is without rods, in the dark-adapted state an object or light is seen most easily if the view is directed somewhat away from it. Light signals emitting only red light (to which the rods are insensitive) form an exception to this rule. Except for red signals a signal can always be detected by this indirect vision (parafoveal vision) at a much lower illuminance than that required to appreciate its color[1]. Color vision is entirely confined to the cones.

26. The threshold of brightness contrast. The threshold of brightness contrast ε is to be measured by providing a field of view divided into two parts, the luminance of one of which can be varied. Different methods of experiment give somewhat different results. For our purposes we may confine our attention to researches in which free binocular vision is employed. Of these, those using the method of constant stimuli[2,3] seem to give the most stable results, with the great advantage that an exact definition of what is meant by "detection" is implicit in the data. In brief, the experimenter presents to the observer a limited set (4 to 7) of different stimuli a large number of times in completely random order. Calculations from the fraction of the observations in which each stimulus was perceived make it possible to state what contrast, for example, will be seen as often as not, i.e. 50% of the time, and this is a common definition of ε.

---

[1] For a very full and clear discussion of all such matters, see [*1*].

[2] See J. P. Guilford: Psychometric methods, pp. 166ff. New York: Mc-Graw-Hill Book Co. 1936.

[3] H. R. Blackwell: J. Opt. Soc. Amer. **42**, 606 (1952).

Of the researches of this kind, that carried on at the Tiffany Foundation in the United States and reported on by BLACKWELL is by far the most important[1]. In this monumental research about 450000 observations (out of a total of more than two million) were considered suitable for analysis. Circular stimuli of

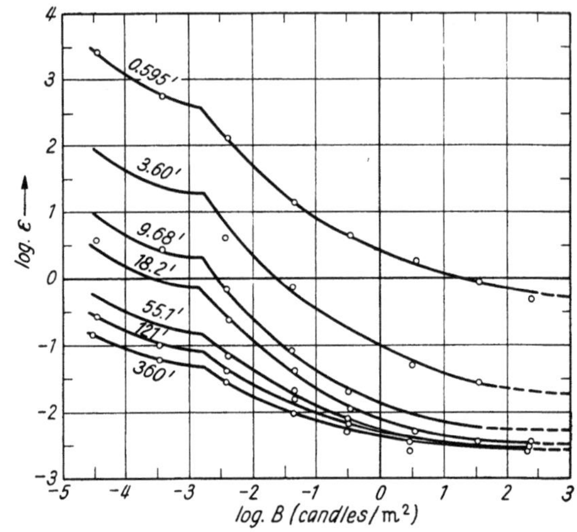

Fig. 13. Thresholds of brightness contrast for 50% detection for circular stimuli of seven diameters (minutes of arc), after BLACKWELL. Stimuli brighter than background. Unlimited times of exposure.

various sizes from 0.6 to 360 minutes in angular diameter were presented on a large uniform background. The military purpose of the research confined the observers to the age group 19 to 26.

One set of results is shown in Fig. 13. Note the break in the curves, which represents the transition from parafoveal to foveal vision. Note also the wide

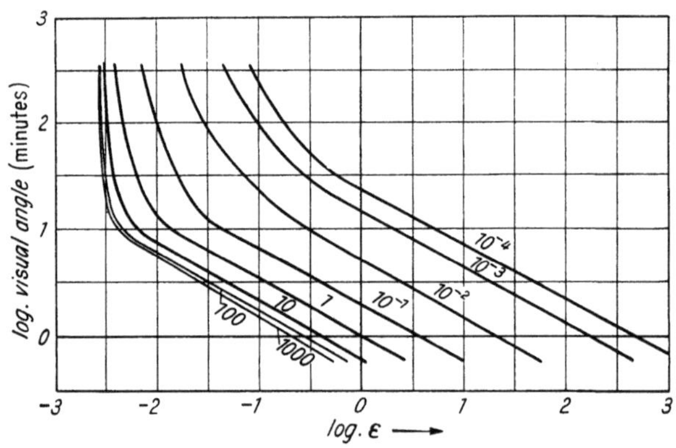

Fig. 14. Interpolations from Fig. 13, after BLACKWELL. Each curve refers to a given adaptation luminance in candles/m.².

range of background luminance B, and of ε. Another and perhaps more useful way of presenting these results is shown in Fig. 14, which represents interpolations from Fig. 13. Note that in the region of small visual angles, the curves are straight, indicating that the product of contrast and area of stimulus required for detection is constant (RICCO's law)[2].

[1] H. R. BLACKWELL: J. Opt. Soc. Amer. **36**, 624 (1946).
[2] RICCO: Ann. Ottalm. **6**, 373 (1877).

The angle at which this occurs is shown in Fig. 15. Any stimulus which plots in the region below this curve is effectively a "point source". It need not be very small against an almost dark background.

It is probable from other researches[1] that the Tiffany results for circular targets apply to targets of other shapes as long as the ratio of length to width does not exceed about 7:1.

It was found during the Tiffany researches that if the thresholds of contrast for 50% detection were multiplied by two, the probability of detection rose to about 95% in all cases. For practical purposes this is a more realistic value (see Sect. 35 below).

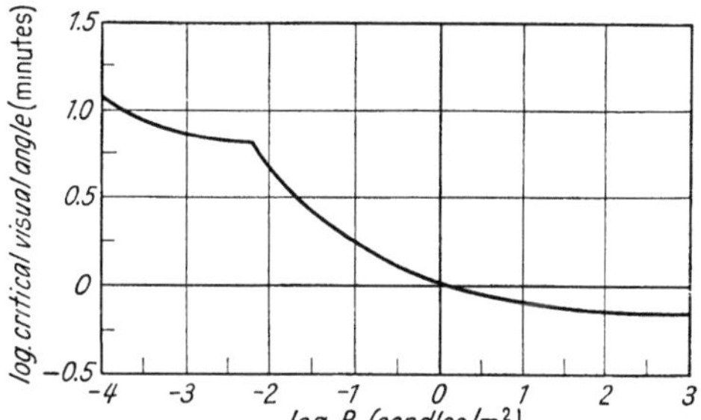

Fig. 15. Critical visual angle as a function of adaptation luminance. The region under the curve represents "point sources" (after BLACKWELL).

27. Effect of a diffuse boundary. As stated in Sect. 22 above, the atmosphere seldom or never causes sharp edges to appear diffuse. Nevertheless diffuse boundaries do occur, as for example the outlines of some kinds of trees when seen from a sufficient distance. It has been found by laboratory investigations[2,3] that a diffuse boundary has little effect on ε until it attains a width of several minutes. The question is therefore of little practical importance.

28. Effect of other stimuli in the field of view. Only in a minority of cases does the scene consist of an isolated object on a uniform field. A. T. CHUPRAKOV[4] experimented on the effect of a dark ring surrounding the object of regard, finding that the threshold is raised in all cases, more when the ring is closer, darker, or wider; except that if the image of the ring falls on a certain zone of the parafovea, the foveal threshold may be little affected. More work should be done along these lines.

The important case of a bright source of light in the field of view has been extensively studied[5]. It has been found that the effect of such a "*dazzle source*" is similar to that of a veiling luminance B_1 where

$$B_1 = K_1 E/\varphi^2. \tag{28.1}$$

In this K_1 is a constant, E the illuminance produced at the eye by the dazzle source, and φ the angle between the line of vision and the line from the eye the source. HOLLADAY found this relation to hold at least for $\varphi \leq 16°$. If B_1 is in cd/m.², E in lumens/m.², φ in degrees, HOLLADAY's value of K_1 is 13.7 ± 1.6 degrees^{-2}. The effect of a number of lights may be found by simple summation. One would suppose that the effect of an extended luminous field might be found by integration, but some results on the threshold illuminance for point sources[6] throw doubt on this possibility. The whole subject is of immense interest to illuminating engineers.

[1] E. S. LAMAR, S. HECHT, S. SHLAER and C. D. HENDLEY: J. Opt. Soc. Amer. **37**, 531 (1947).

[2] W. E. K. MIDDLETON: J. Opt. Soc. Amer. **27**, 112 (1937).

[3] A. M. KRUITHOF: Philips Techn. Rev. **11**, 333 (1950).

[4] A. T. CHUPRAKOV: Vestnik. Oftalm. **17** (5), 680 (1940).

[5] L. L. HOLLADAY: J. Opt. Soc. Amer. **12**, 271 (1926) and **14**, 1 (1927). See also Y. ROCARD [*4*]).

[6] H. A. KNOLL, R. TOUSEY and E. O. HULBURT: J. Opt. Soc. Amer. **36**, 480 (1946).

29. Practical interpretation of thresholds. The contrast thresholds indicated in Figs. 13 and 14 appear surprisingly low to many meteorologists, and indeed they have little relation to the contrasts usually reported to be at the visual range in meteorological observations[1]. Nevertheless it has been determined by direct experiment in the field that the Tiffany results can actually be used to predict the contrast at which actual outdoor targets can be detected.

30. Vision through telescopes. It has been found[2,3] that the contrast threshold for vision through binoculars (held in the hand) is not much different from that for unaided vision.

31. The threshold for point sources. By a simple calculation ([2], p. 96) the thresholds of contrast for small bright signals found in the Tiffany researches may be transformed into thresholds of illuminance from a point source. These are presented as a function of the background luminance B_0 in Fg. 16, together with comparable data from two other researches[4,5]. A scale of stellar magnitudes is included. In comparing these curves, it should be noted that we ought to add 0.3 log units to the Tiffany data in order to make them represent the practical certainty of seeing represented by the other curves. The Tiffany curve represents many more observations than the others, and clearly shows the change from foveal to parafoveal vision. In the opinion of the writer it may be used with some confidence for an isolated signal.

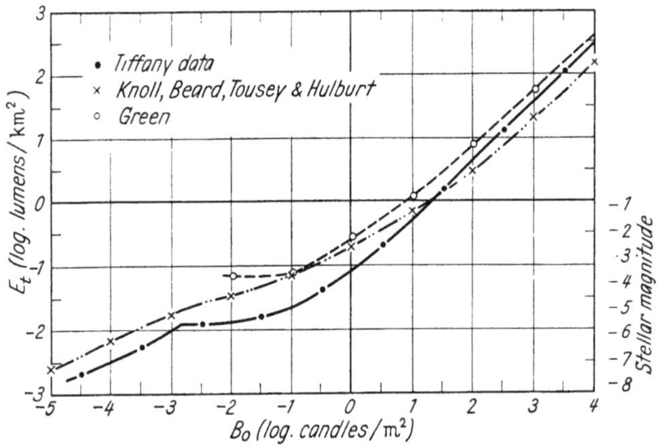

Fig. 16. Threshold illuminance from a fixed, achromatic point source, as a function of background luminance.

It is probable that observers in marine and air transport never become adapted to the lower values of background luminance, and never use parafoveal vision. At any rate, the practical values adopted for lighthouses and airway lights are in the region of 0.12 lumens/km.².

In order to be equally visible, a flashing light must provide more illumination than a steady light. HAMPTON[6] derived an expression

$$I/I_\infty = t\Big/\left[\left(\frac{0.0098}{E_c}\right)^{0.81} + t\right] \tag{31.1}$$

for the apparent intensity I of a flashing light shown t seconds which would have intensity I_∞ if allowed to appear without interruption. This is a function of the illuminance E_c of the fixed light. The constant in (31.1) assumes E_c in lumens/km.².

For an extended discussion of the threshold for point sources, see [5].

[1] W. E. K. MIDDLETON and A. G. MUNGALL: Trans. Amer. Geophys. Union **33**, 507 (1952).
[2] S. H. BARTLEY and ELOISE CHUTE: O.S.R.D. Rep. 4433, Washington (1944).
[3] H. S. COLEMAN and W. S. VERPLANCK: J. Opt. Soc. Amer. **38**, 250 (1948).
[4] See footnote 6, p. 269.
[5] H. N. GREEN: Illum. Engr. Lond. **28**, 146 (1935).
[6] W. M. HAMPTON: Illum. Engr. Lond. **27**, 46 (1934).

IV. Visual range of objects in natural light.

32. Definitions. We shall define the *visual range* V as the distance at which a given object can just be detected against its background, the proper value being assigned to the liminal contrast, as described in Chap. III.

KOSCHMIEDER[1] and others after him dealt with the subject as if the threshold of contrast were always 0.02. This, of course, is far from the truth. A distance calculated on this basis has, however, found its way into the literature under the name *meteorological range*, with the symbol V_2; and since about 1950 another "meteorological range" V_5 has been advocated in the United States of America, using the threshold 0.05. In what follows it will be clear from the notation which definition is being used. It should be emphasized that V_2 and V_5 are nothing more than convenient substitutes for the extinction coefficient, easily visualized because they are distances of the same order of magnitude as the "visibility" reported by the meteorological observer.

a) Visual range of objects seen against the sky.

33. Against the horizon sky. As we recede from an object which has an inherent contrast C_0 with the sky behind it, the apparent contrast decreases according to Eq. (14.2), until it reaches ε, the threshold contrast appropriate to the luminance of the field of view and the dimensions of the object (see Sect. 26). The distance at which this takes place is V, the visual range. At this distance

$$\varepsilon = C_0 e^{-\sigma_0 V} \quad [|C_0| > \varepsilon]$$

and therefore

$$V = (1/\sigma_0) \log_e |C_0/\varepsilon| . \tag{33.1}$$

The use of this equation needs a knowledge of σ_0 (see Chap. VIII) and of C_0. For many objects C_0 may be estimated; the simplest case is an ideal black object for which $C_0 = -1$, und for which Eq. (33.1) becomes

$$V = (1/\sigma_0) \log_e (1/\varepsilon) . \tag{33.2}$$

Similarly the meteorological range V_2 is

$$V_2 = (1/\sigma_0) \log_e 50 = 3.912/\sigma_0 . \tag{33.3}$$

For objects which are not black, calculation of C_0 is difficult unless the sky is completely overcast. An overcast sky has a distribution of luminance which may be approximated by[2]

$$B_\xi = B_{90} (1 + 2 \cos \xi) \tag{33.4}$$

in which B_ξ is the luminance at zenith angle ξ. The consequences of this have been worked out by MIDDLETON[3] for vertical white objects and grey objects of reflectance R, as a function of the reflectance R' of the ground. The general equation is

$$C_0 = \frac{R}{2} (1 + R') + \frac{2R}{3} \left(\frac{2}{\pi} + R'\right) - 1 \tag{33.5}$$

and is plotted in Fig. 17.

It is interesting to enquire what value of ε is actually implied in the estimates of meteorological observers. MIDDLETON and MUNGALL[4] measured with a photoelectric telephotometer the actual contrast between visibility marks

[1] H. KOSCHMIEDER: Beitr. Phys. freien Atm. **12**, 33, 171 (1924).
[2] P. MOON and DOMINA E. SPENCER: Illum. Engr. N. Y. **37**, 707 (1942).
[3] W. E. K. MIDDLETON: Quart. J. Roy. Met. Soc. **73**, 456 (1947).
[4] W. E. K. MIDDLETON and A. G. MUNGALL: Trans. Amer. Geophys. Union **33**, 507 (1952).

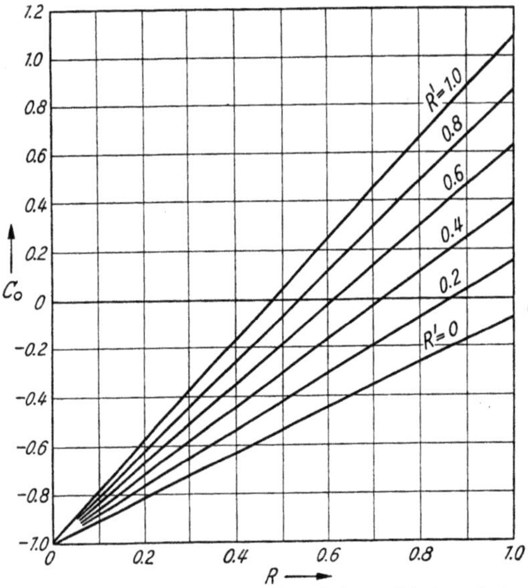

Fig. 17. Intrinsic contrast of vertical grey objects against the horizon sky, under overcast conditions.

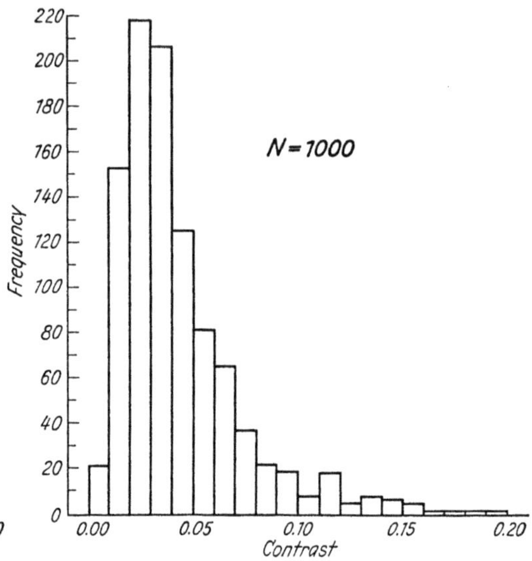

Fig. 18. Frequency distribution of values of ε from 1000 direct observations at Ottawa.

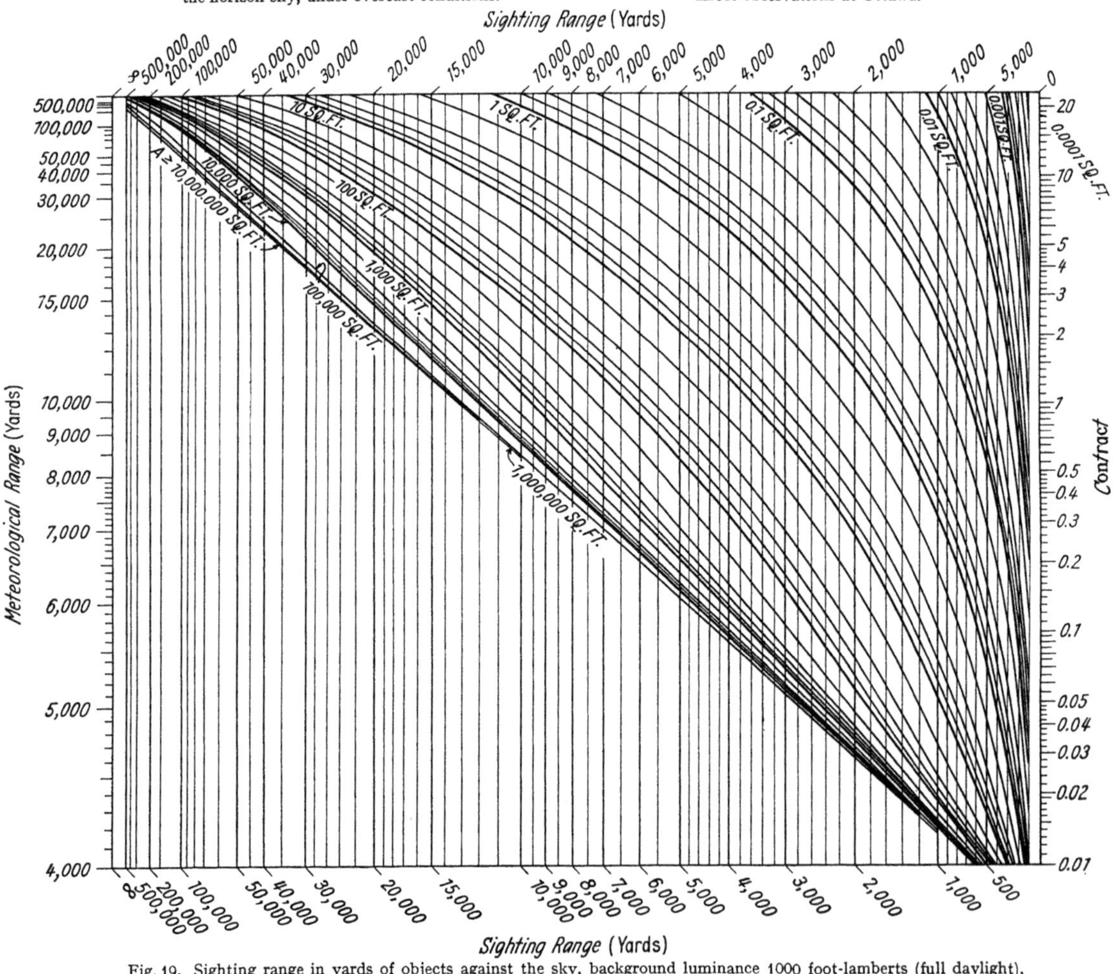

Fig. 19. Sighting range in yards of objects against the sky, background luminance 1000 foot-lamberts (full daylight), based on the Tiffany data for circular targets, at a probability of detection of 95%.

and the sky at the time when the observers picked them out as being at the distance of "visibility". The results of 1000 observations are plotted in Fig. 18. The median contrast is 0.031, but the great spread in the distribution is noteworthy.

34. Against the sky along slant paths. We may refer to Eq. (15.1), substituting ε for C_R and solving for $\bar{R}$, provided that we have enough information about the function $f(r)$. This application of the theory also assumes that B_a and σ [Eq. (13.5)] are similar functions of r, or in other words that absorption may be neglected. Particular solutions for certain forms of $f(r)$ have been presented by LÖHLE[1], FOITZIK[2], and BURKHART[3], the last two being fairly general nomographic methods based on the division of the atmosphere into layers. However, these papers do not tell us how the function $f(r)$ is to be measured, and at present there seems to be no possibility of doing this from the ground, at least to any considerable height.

35. Nomograms for the visual range. Eq. (33.1) is inconvenient for the actual calculation of the visual range of a given object, because ε depends on the angular subtense of the object (see Sect. 26) and this depends on the range. However, a series of nomograms was devised at the Tiffany foundation in the United States, which make possible an immediate graphical solution[4]. One of these nomograms, corresponding to full daylight, is reproduced here (Fig. 19). They are developed from the equation

$$\varepsilon = (B_0'/B_R')\, C_0\, e^{-3.912\,\bar{R}/V_2}, \tag{35.1}$$

the scale of V_2 appearing at the left-hand side, C_0 at the right. We must, of course, know V_2 *in the direction of the object*, which practically limits the use of these nomograms to horizontal vision (where $B_0' = B_R'$). To use the chart, lay a straight edge across it so as to join the existing values of V_2 and C_0. The sighting range V is read off the upper or lower margin, directly above or below the intersection of the straight edge with the line representing the appropriate target area. While these nomograms were designed for circular targets, they are approximately correct for targets of any not-too-extended shape.

For $V_2 < 4000$ yards, multiply the scales of both V_2 and V by the same factor, and the designations of area on each curved line by the square of this factor.

There is a special use for these nomograms in the calculation of air-to-air visual range at high altitudes, where it may be assumed that V_2 is practically infinite. In such circumstances a moderately precise guess at C_0 will yield useful, if not exact, information.

36. Extremely extensive objects. A very extensive object will shade an appreciable part of the light path from much of the sky. Thus its apparent luminance will be less than that predicted by Eq. (12.6), and its visual range greater than would be deduced from Eq. (33.1). This effect is particularly important in fog, for which FOITZIK[5] calculates, for example, that a semicircular black object 4° in radius will have twice the visual range that would be given by the usual formula.

[1] F. LÖHLE: Meteor. Z. **52**, 435 (1935).

[2] L. FOITZIK: Z. Meteorol. **1**, 161 (1947).

[3] K. BURKHART: Z. Meteorol. **2**, 106 (1948).

[4] S. Q. DUNTLEY: J. Opt. Soc. Amer. **38**, 237 (1948). Nine of these nomograms are reproduced in [2], altered to correspond to 95% probability of detection, as is Fig. 19.

[5] L. FOITZIK: Wiss. Abh. Reichsamt Wetterdienst, Berlin **4**, No. 5 (1938).

37. Visual range of colored objects. As the distance increases, even brightly colored objects become greyer in tone, almost always appearing neutral in color before the visual range is reached. (See also Sect. 46 below). For this reason no special theory of the visual range of colored objects is required.

38. Visual range in fog and its relation to water content. For a monodisperse fog, $\sigma = 2\pi N a^2$ [m.$^{-1}$] if there are N droplets per m.^{3}, each of radius a meters. From Eq. (33.2) the visual range of a black object in such a fog is

$$V = (2\pi N a^2)^{-1} \log_e (1/\varepsilon). \tag{38.1}$$

But the weight of water per unit volume of fog is

$$w = \tfrac{4}{3}\pi N a^3 \varrho \quad \text{(grams)}$$

where ϱ is the density of water (10^6 gm m.$^{-3}$). Therefore

$$V = (2a\varrho/3w) \log_e (1/\varepsilon). \tag{38.2}$$

Actual fogs are polydisperse, and information about droplet sizes is necessary[1].

b) Visual range of objects seen against terrestrial backgrounds.

39. Along the horizontal. Consider an object of luminous reflectance β, seen against a background of reflectance β', both being at the same distance r. For simplicity, assume that object and background are parallel planes, therefore similarly illuminated. To describe the illumination, imagine a white object in place of the object under consideration, and let its contrast with the horizon sky be C_W. By writing out expressions for the apparent luminances of the two surfaces [using Eq. (13.6)] and simplifying, we find that

$$V = \frac{1}{\sigma} \log_e \left\{\frac{1}{\varepsilon} \left[(\beta - \beta')(C_W + 1) - \beta' \varepsilon (C_W + 1) + \varepsilon\right]\right\}. \tag{39.1}$$

If β' is small compared to $(\beta - \beta')$, this becomes

$$V = \frac{1}{\sigma} \log_e \left\{\frac{1}{\varepsilon} \left[(\beta - \beta')(C_W + 1) - \varepsilon\right]\right\}. \tag{39.2}$$

This depends very greatly on C_W; from which we may deduce (for example) that a small black object against a snowfield will be much more easily visible opposite the sun than towards it.

SIEDENTOPF[2] dealt with this subject by the introduction of a "detail contrast", C_D, and showed that this is a function

$$C_D = C_D(0) \frac{1}{1 + (B_h/B_0')\,(e^{3.912\,r/V_2} - 1)} \tag{39.3}$$

of the intrinsic contrast $C_D(0)$, the ratio B_h/B_0', and the ratio r/V_2. The fraction may be denoted by $G(B_h/B_0', r/V_2)$, and a useful nomogram prepared (Fig. 20). It will be seen to what an extent detail in the shadows vanishes with distance[3].

Another interesting case is that of a dark object against another dark object at a considerably greater distance. Without going into details[4], we may state that for $\varepsilon = 0.02$, the visual range of a black object against another black object 1.5 times as far away is only 3.3% less than if it had been exposed against the horizon sky. This result permits the choice of many visibility marks that would otherwise be excluded.

---

[1] See H. J. AUFM KAMPE and H. K. WEICKMANN: J. Meteorol. **9**, 167 (1952).
[2] H. SIEDENTOPF: Z. Meteorol. **2**, 110 (1948).
[3] M. HUGON: Sci. et Ind. phot. **1**, 161, 201 (1930).
[4] See [2], p. 125.

40. Vision looking downwards. Nomographic methods have been devised by several authors[1,2]. DUNTLEY[3] has constructed nomograms somewhat similar to that of Fig. 19, but with a scale for the "sky-ground ratio" (see Sect. 19 above). These are not as easy to use, because the actual sighting range R is different from $\bar{R}$, on account of the non-uniformity of the atmosphere in the vertical. An interpolation is therefore necessary. The additional difficulty of estimating V_2 and the sky-ground ratio makes the practical value of these charts open to question, except as a general guide.

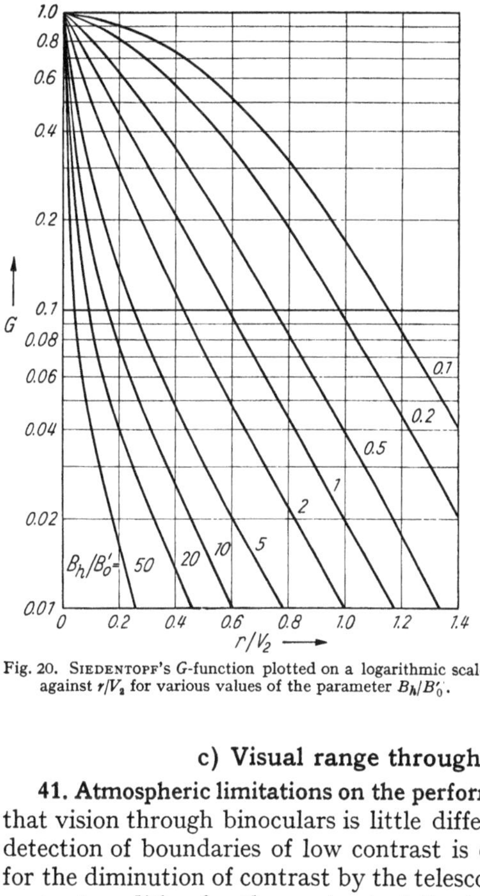

Fig. 20. SIEDENTOPF's G-function plotted on a logarithmic scale against r/V_2 for various values of the parameter B_h/B_0'.

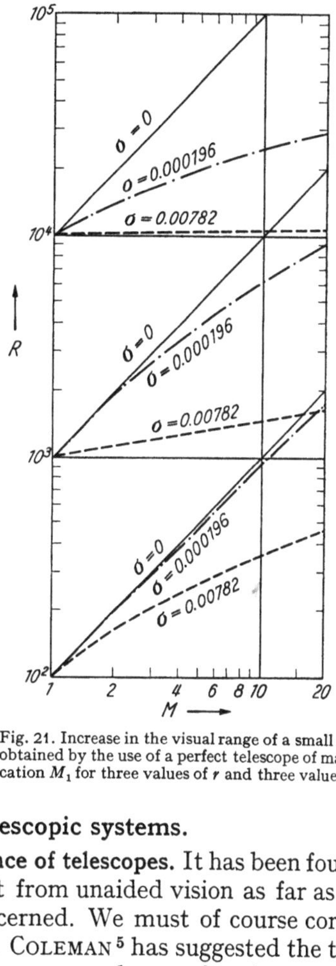

Fig. 21. Increase in the visual range of a small object obtained by the use of a perfect telescope of magnification M_1 for three values of r and three values of σ.

c) Visual range through telescopic systems.

41. Atmospheric limitations on the performance of telescopes. It has been found[4] that vision through binoculars is little different from unaided vision as far as the detection of boundaries of low contrast is concerned. We must of course correct for the diminution of contrast by the telescope. COLEMAN[5] has suggested the term *contrast rendition* for the contrast of the image expressed as a fraction of that of the object, and it is found that this quantity can be as high as 95%. Prism binoculars usually have a *transmittance* of from 45 to 75%. In general, binoculars may be thought of as magnifying devices with little effect on contrast, if (as is almost always the case) their exit pupil is larger than the pupil of the eye.

HARDY[6] has shown that the magnification M required to increase the visual range of a *small* target from r to R in an atmosphere of extinction coefficient σ is

[1] K. BURKHART: Z. Meteorol. **2**, 106 (1948).
[2] L. FOITZIK: Z. Meteorol. **1**, 161 (1947).
[3] S. Q. DUNTLEY: J. Opt. Soc. Amer. **38**, 237 (1948).
[4] S. B. BARTLEY and ELOISE CHUTE: O.S.R.D. report 4433, Washington 1944.
[5] H. S. COLEMAN: J. Opt. Soc. Amer. **37**, 434 (1947).
[6] A. C. HARDY: J. Opt. Soc. Amer. **36**, 283 (1946).

given by the equation
$$\log_{10} M = \log_{10}(R/r) + 0.2171\,(R-r)\,\sigma. \tag{41.1}$$
How this behaves is shown in Fig. 21, the three sets of curves corresponding to (1) no atmosphere, (2) $V_2 = 20$ km.; (3) $V_2 = 0.5$ km. respectively. It will be abundantly clear that no telescope will help much in a fog.

If the target is large to begin with, no increase in its apparent dimensions will decrease its threshold of contrast, and the telescope will not increase its visual range though it may enable details to be seen more clearly.

Nomograms such as Fig. 19 may be used for vision through telescopes by (1) multiplying the contrast by the contrast rendition of the telescope before entering the scale of contrast, and (2) reading the visual range of an area M^2 times as large as the actual object. In elaborate trials at sea, COLEMAN and VERPLANCK[1] showed that the sighting range of an object in the direction of the horizon could actually be predicted in this way.

d) Limitations of the theory.

42. Nature and results of the assumptions. The theory of the reduction of contrast, even along a horizontal path, is based on a number of assumptions, three of which are only approximately true, namely (1) that the earth is flat, (2) that the atmosphere is uniform in the horizontal, (3) that the illumination is uniform. The difficulty of making useful predictions from the theory arises chiefly from the large fluctuations in time and space of both light and atmosphere. Nevertheless, for horizontal vision, the theory is of considerable practical value, particularly for marine operations.

Its value for oblique vision is less, because of the difficulty of measuring the appropriate quantities at some distance from the surface of the earth. There is, however, no reasonable doubt of the correctness of the theory itself.

From a technical standpoint, reference should be made to a paper by DOUGLAS[2] which discusses several factors affecting visibility from aircraft.

V. Visual range of light sources.

43. ALLARD's law. For sources of light which are point sources according to the definition of Sect. 26, there is a very simple theory of the visual range. A distant light of intensity I produces at a distance r the illuminance
$$E = I\,r^{-2}\,e^{-\sigma r} \tag{43.1}$$
where σ is the extinction coefficient of the intervening atmosphere. This law was set down by ALLARD in 1876[3].

If now we substitute the threshold illuminance E_t (Sect. 31) for E, and put $r = V$, we obtain
$$V = \frac{1}{\sigma}\left[\log_e(I/E_t) - 2\log_e V\right], \tag{43.2}$$
an awkward expression for which useful tables are available[4]. This equation should be used only for point sources; signals of appreciable extent should be treated as areas, their contrast with the background computed, and the data put into Eq. (33.1).

A nomogram rather similar to Fig. 19 is available for the sighting range of point sources[5].

---

[1] H. S. COLEMAN and W. S. VERPLANCK: J. Opt. Soc. Amer. **38**, 250 (1948).
[2] C. A. DOUGLAS: Nat. Bur. Stand., Rep. 2715, Washington, Aug. 12, 1953.
[3] E. ALLARD: Mémoire sur l'intensité et la portée des phares. Paris: Dunod. 1876.
[4] Smithsonian Institution, Smithsonian Meteorological Tables, Washington 1951, Table 160.
[5] It is reproduced in [2], p. 139.

VI. Visual range of objects in a searchlight beam.

44. Searchlight directed horizontally. In this section we shall follow the work of HAMPTON[1] and of CHESTERMAN and STILES[2], but the reader should also consult the extensive paper by FOITZIK[3].

In Fig. 22, A is the searchlight, B the observer, T the target at a distance r. The observer looks at T in a direction making an angle ϑ with the axis of the beam. Let $AB = h$.

The target will appear to have luminance $B = B_1 + B_2$, where B_1 is from the light reaching it from the searchlight, B_2 from the air between the observer and the target. It will appear on a background of illuminated air of luminance B_3.

If I_0 is the axial intensity of the searchlight, I_ψ its intensity at an angle ψ to the axis, σ the extinction coefficient of the air, R the luminous reflectance of the target, it can easily be shown that for a target on the axis

$$B_1 = R I_0 e^{-2\sigma r}/\pi r^2. \quad (44.1)$$

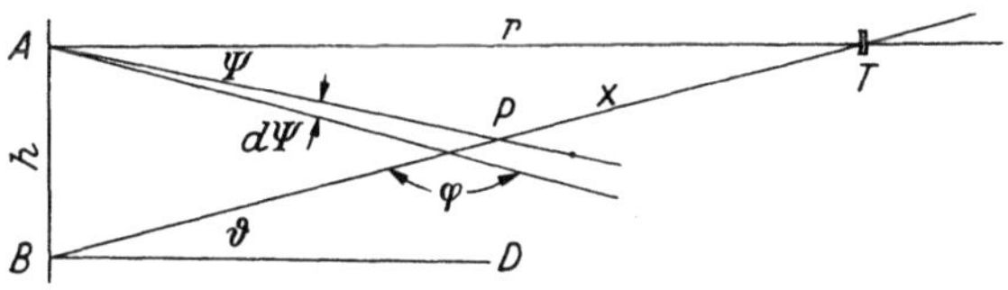

Fig. 22. Illustrating the theory of searchlight illumination.

B_2 and B_3 can be calculated as a function of ψ. It so happens that in a horizontal plane the intensity distribution of an arc searchlight closely follows the law

$$I_\psi = I_0 e^{-K\psi^2} \quad (44.2)$$

CHESTERMAN and STILES made use of this relation, and also showed that over the sea the scattering function at angles near $\varphi = 180°$ is nearly constant at a value $\beta' = \sigma/8\pi$. Under such conditions

$$B_2 = \frac{I_0}{8\pi h} \int_0^{\pi/2} \exp\left[-K\psi^2 - 2\sigma r h/(r\psi + h)\right] d\psi \quad (44.3)$$

and

$$B_3 = \frac{I_0}{8\pi h} \int_{-\vartheta}^{\pi/2} \exp\left[-K\psi^2 - 2\sigma r h/(r\psi + h)\right] d\psi. \quad (44.4)$$

CHESTERMAN and STILES further made a careful study of the contrast thresholds of small ship outlines in simulated searchlight beams. Their results turned out later to be in remarkably close agreement with the Tiffany data (see Sect. 26 above). Finally they made numerous calculations of the visual range, r_0. The results may be summarized as follows:

1. $\log r_0$ varies very approximately inversely as $\frac{1}{2}\log\sigma$.
2. A white target has only about six times the visual range of one with reflectance 0.01, three times the range of one with $R = 0.04$.
3. $\log r_0$ is approximately linear in $\log h$.
4. $\log r_0$ is approximately linear in the logarithm of the beam spread.
5. r_0 depends very little on the peak intensity I_0.
6. No general statement was possible about the effect of target size on r_0, the relationships being extremely complex.

[1] W. M. HAMPTON: Proc. Phys. Soc. Lond. **45**, 663 (1933).
[2] W. D. CHESTERMAN and W. S. STILES: In Symposium on searchlights, London, Illum. Eng. Soc. 1948, pp. 75–102.
[3] L. FOITZIK: Abh. meteorol. u. hydrol. Dienst. dtsch. dem. Republ. **2**, No. 9 (1952).

45. The searchlight directed upwards. The anti-aircraft searchlight presents a more complicated problem, because of (1) the non-uniformity of the atmosphere in the vertical, (2) the lack of axial symmetry in the beam, (3) the effects of perspective. The reader should be referred to papers by FOITZIK[1] and by STEVENS and WALDRAM[2]. The technical importance of this subject is not now very great, and a summary of these papers is not justified here.

VII. The effect of the atmosphere on the apparent colors of objects.

46. The apparent colors of objects and lights. It is well known to artists that distant scenery is generally bluer, and always paler, than that near at hand, while distant clouds appear a pale orange. An extension of Eq. (12.8) affords a basis for quantitative calculation of these effects. The theory of this has been set out in several papers by MIDDLETON[3].

The possibility of specifying such colors numerically rests on a large body of data which has been codified by the International Commission on Illumination (C.I.E.) in the form of a hypothetical standard observer. Given such a standard observer, a given set of spectroradiometric data will yield a definite numerical specification for the color of the light source or illuminated object to which these data apply. For reasons of space we cannot detail the C.I.E. color metric here, but must refer the reader to [2] and [4], or to books by BOUMA[4], WRIGHT[5], JUDD[6] and HARDY[7].

The necessary spectroradiometric data may be obtained by restricting Eq. (12.8) to one wavelength interval at a time, and writing

$$B_{r,\lambda} = B_{h,\lambda}(1 - e^{-\sigma_\lambda r}) + f_\lambda B_{0,\lambda} e^{-\sigma_\lambda r} \tag{46.1}$$

where f_λ is the spectral reflectance of the object at wavelength λ, $B_{0,\lambda}$ the (monochromatic) luminance of a white object in the position of the object of regard. If the horizon sky has the color of the prevailing illumination, this becomes

$$B_{r,\lambda} = B_{h,\lambda}\left[1 + (f_\lambda C_w + f_\lambda - 1)\, e^{-\sigma_\lambda r}\right] \tag{46.2}$$

where C_w is the inherent contrast of a white object, again in the position of the colored object in which we are interested. $B_{h,\lambda}$ refers to the prevailing illumination from sun and sky, which may be supposed known.

The clearer the air, the more saturated the colors produced. In very pure air, a black object becomes a fairly saturated blue-purple at a distance of a few kilometers, gradually becoming a paler blue at greater distances, until, before it vanishes, it is sensibly the same color as the horizon sky. A white object becomes pale orange, reaching its maximum saturation at 50 to 100 km. and then moving towards white. Green grass goes towards blue-purple and then towards white. In a non-selective atmosphere, such as a fog, any color will remain the same hue, merely becoming less saturated as the distance increases.

Lights at night generally become more orange in color; in some types of haze the color of the light from a full radiator merely passes along or near to the

---

[1] See footnote 3, p. 277.

[2] W. R. STEVENS and J. M. WALDRAM: Illum. Eng. Soc. (London), Proc. of Convention, London, May 14–16, 1946, p. 51.

[3] W. E. K. MIDDLETON: Trans. Roy. Soc. Canada, Sec. III **29**, 127 (1935); **37**, 39 (1943). — Quart. J. Roy. Meteorol. Soc. **62**, 473 (1936). — J. Opt. Soc. Amer. **40**, 373 (1950).

[4] P. J. BOUMA: Physical Aspects of Colour. Eindhoven: N. V. Philips 1948.

[5] W. D. WRIGHT: The measurement of colour. London: A. Hilger Ltd. 1944.

[6] D. B. JUDD: Color in business, science, and industry. New York: Wiley 1952.

[7] A. C. HARDY: Handbook of colorimetry. Cambridge, Mass.: Technology Press 1936.

PLANCKian locus in the direction of lower color-temperatures as the distance increases. Discharge lamps sometimes show striking changes of color[1]. In the case of light sources, or very small objects, the matter is complicated by certain properties of the eye which cannot be dealt with here[2].

VIII. Instrumentation.

47. Classification of Instruments. Instruments for the measurement of the visual range and associated quantities belong to one of three groups:

1. those which measure the extinction coefficient
2. those which measure the scattering function
3. instruments which aid in the estimation of the visual range without directly measuring the optical properties of the atmosphere. These can be further classified into

3a. instruments making use of an "artificial haze"

3b. miscellaneous "visibility meters".

The generic name for instruments of the first class is "*telephotometer*". These can be further classified into visual and photoelectric telephotometers. *Scattering meters* can be divided according to whether they measure the scattering function at one or several angles, or whether they integrate the scattering function. The observation of a searchlight beam also affords a means of measuring atmospheric scattering.

48. Telephotometers. *α) General remarks.* Telephotometers are usually telescopes fitted with some sort of measuring device. Like all optical systems, they suffer from internally scattered light, which falsifies the measurement unless reduced to a negligible amount or corrected for[3]. A much more serious error can occur with some types of telephotometer in the measurement of the illuminance from a distant light, because of multiple scattering in the atmosphere near the line of sight. The theory of this has been given by MIDDLETON[4] for a foggy atmosphere. The error depends on the angular aperture of both the telephotometer and the light source, on the distance between them, and on the particle sizes in the fog or haze. Under unfavorable conditions the secondary light can exceed that directly transmitted.

Experimental work in clearer atmospheres has been reported by STEWART and CURCIO[5], who derived the empirical formula

$$T_\vartheta = T + 0.5\,(1 - T)\,(1 - e^{-\vartheta}) \qquad (48.1)$$

where T is the transmittance for collimated light, T_ϑ the transmittance as measured with a telephotometer of angular aperture ϑ (radians). They also made the interesting suggestion that a measurement of T_θ/T might furnish a value of T itself, and found that this formula yielded good results in rather clear air ($V_2 > 11$ sea miles), in the region of Chesapeake Bay.

It is obvious that the angular aperture of a telephotometer to be used with a distant light source should be kept as small as possible. The folding of the beam by means of a mirror is one way to do this; but back-scattered light from

---

[1] W. E. K. MIDDLETON: Quart. J. Roy. Meteorol. Soc. **62**, 473 (1936).

[2] See [2], chap. 8.

[3] For one method of correction see H. RÜHLE, Forsch.-Arb. staatl. Obs. Danzig **1930**, No. 3.

[4] W. E. K. MIDDLETON: J. Opt. Soc. Amer. **39**, 576 (1949).

[5] H. S. STEWART and J. A. CURCIO: J. Opt. Soc. Amer. **42**, 801 (1952).

the air near the source is still not excluded. This is not important in the telephotometry of a large extended surface, nor in an *image forming* visual photometer, but it is the plague of photoelectric telephotometry.

β) Visual telephotometers[1]. Visual telephotometers are of three kinds: those for measuring extended sources; those using the so-called "MAXWELLian view" in the photometry of a distant lamp, and those using an "artificial star".

Fig. 23. The telephotometer of KOSCHMIEDER and RÜHLE. G Incandescent lamp; M Ground glass disc; D_1 Diaphragm; D_2 Diaphragm; N_1 Fixed nicols; N_2 Moveable nicol; L Lummer cube; E Eyepiece; O Objective.

The optical system of a typical instrument of the first class is shown in Fig. 23, which is self-explanatory. Such instruments are intended for the measurement of the apparent luminance of a dark object at a distance[2]. It should be noted that the polarizing apparatus is in the comparison beam; the polarization of skylight makes it impossible to have any polarizing optical elements in the path from the distant object. It is also essential to ensure that the exit pupils of the two optical systems coincide.

A number of "relative" telephotometers have been designed[3-5], which compare two neighbouring fields. The simplest of these is due to LÖHLE[7], and is shown in Fig. 24. There are two telescopes F_h and F_z, one having an adjustable diaphragm ("cat's-eye") in front of the objective. One is hinged at D.

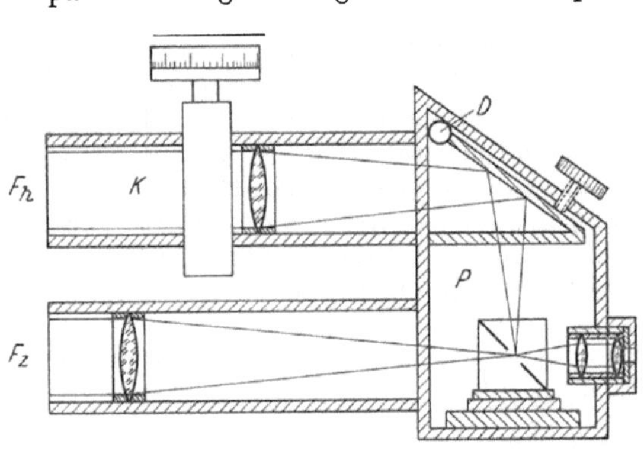

Fig. 24. Principle of LÖHLE's telephotometer.

The idea for the elaborate meter shown diagramatically in Fig. 25 has been ascribed to KOSCHMIEDER[6]. A beam of light from a source S is collimated and projected towards a distant mirror, whence it is reflected back to M_1'. Part of the beam is taken out by the inclined glass plate G, and directed on to M_2'. The rest of the instrument is the well-known PULFRICH photometer. Instead of a plane mirror, a corner reflector is used, simplifying adjustment.

If a real image of a source of light is formed at the pupil of the eye by a lens, the whole surface of the lens appears uniformly bright. This principle is known

[1] See L. J. COLLIER: Trans. Illum. Engng. Soc. Lond. **3**, 141 (1938) for an extended discussion..

[2] For a severe criticism of the use of such measurements for calculating the visual range, see L. FOITZIK, Abh. meteorol. u. hydrol. Dienst. dtsch. dem. Republ. **1951**, Nr. 8.

[3] M. GUREWITSCH and W. KASTROW: Zhurnal Geofis. **4**, 400 (1934).

[4] W. E. K. MIDDLETON: Gerlands Beitr. Geophys. **44**, 358 (1935).

[5] F. LÖHLE: Z. techn. Phys. **16**, 73 (1935).

[6] L. FOITZIK: Meteor. Z. **50**, 473 (1933). — Naturwiss. **22**, 384 (1934). — Wiss. Abh. Reichsamt. Wetterdienst, Berlin **4**, No. 5 (1938). — Z. Meteorol. **1**, 330 (1947). These papers describe successive models.

as the "MAXWELLian view", and may be used in visual telephotometers for the measurement of the transmittance of the atmosphere[1-3]. A diagram of the

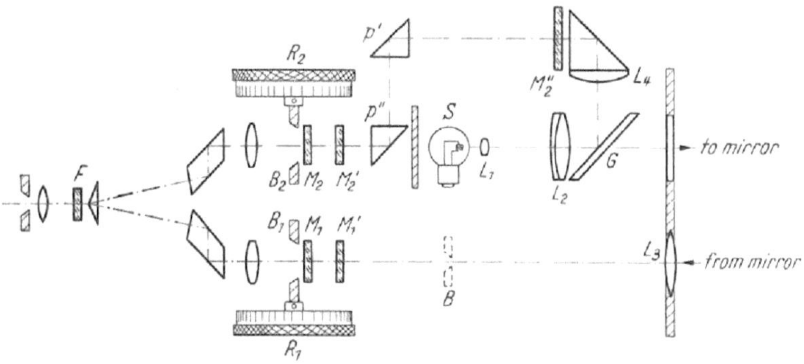

Fig. 25. The KOSCHMIEDER-ZEISS *Sichtmesser*.

instrument due to COLLIER and TAYLOR[3] is given in Fig. 26. The source of light to be measured is far away to the left, the comparison source being at S', with a neutral wedge W in front of it.

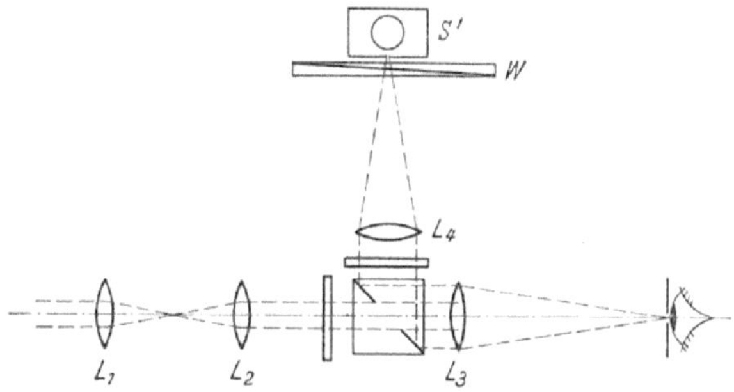

Fig. 26. Diagram of the telephotometer of COLLIER and TAYLOR.

All such telephotometers run foul of the STILES-CRAWFORD effect[4]. STILES and CRAWFORD showed that the luminosity produced by a narrow beam of light depends on the region of the pupil through which it passes, being greatest at or near the centre. Thus the apparent brightness of the lens depends on the size of the image of the distant lamp. Fig. 27 shows that this effect is by no means small. The PULFRICH photometer is also subject to this error.

The remaining type of visual telephotometer uses an "artificial star", and is exemplified by the instrument of MIDDLETON[5] in which a neutral wedge is used to attenuate the light from a distant lamp until it appears to have the same luminosity as a minute source contained in the instrument and projected to infinity by a positive lens. Such an instrument is very simple, but the precision of setting is low[6]. An important advantage is that

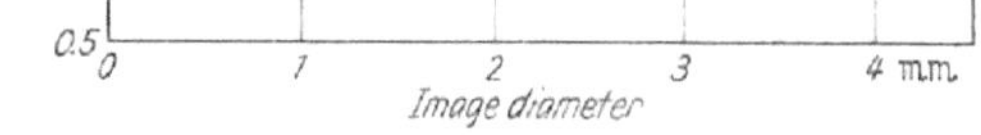

Fig. 27. Results of the STILES-CRAWFORD effect. Dependence of measured intensity on image diameter in a telephotometer using the MAXWELLian view.

[1] G. GEHLHOFF and H. SCHERING: Z. techn. Phys. **1**, 247 (1920).
[2] C. FABRY and H. BUISSON: J. de Phys. **1**, 25 (1920).
[3] L. J. COLLIER and W. G. A. TAYLOR: J. Sci. Instrum. **15**, 5 (1938).
[4] W. S. STILES and B. H. CRAWFORD: Proc. Roy. Soc. Lond., Ser. B **112**, 428 (1933).
[5] W. E. K. MIDDLETON: Trans. Roy. Soc. Canada, Sec. 3, **25**, 39 (1931); **26**, 25 (1932).
[6] For a discussion see L. J. COLLIER, Trans. Illum. Engng. Soc. Lond. **3**, 141 (1938).

the instrument is affected little by scattered light, since it is effectively a telephotometer with a very small angle of view.

γ) Photoelectric telephotometers. A photoelectric telephotometer uses as a receiver either a barrier-layer photocell or an emission phototube, more commonly the latter. In all such applications the spectral sensitivity of the receiver, always

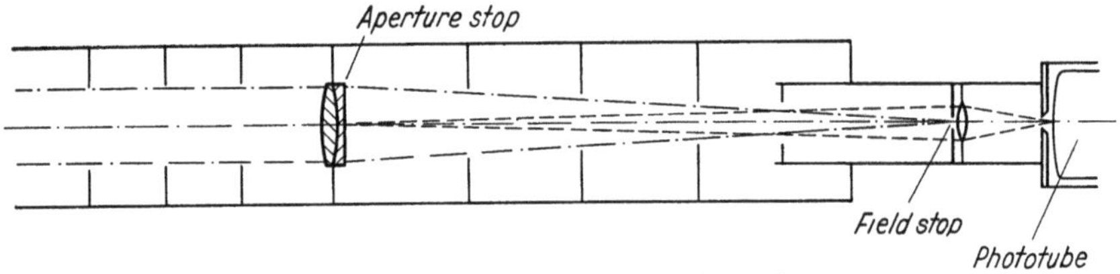

Fig. 28. Preferred arrangement of a photoelectric telephotometer.

very different from that of the eye, must be kept in mind, and for the purpose of predicting the visual range, suitable filters must be provided.

Photoelectric telephotometers may be used (1) to measure the luminance of a distant surface, or (2) to measure the illuminance from a distant light. A very advanced telephotometer of the first class was built by COLEMAN and his associates[1]. It is necessary to provide baffles to reduce stray light, preferably in front of the objective as well as behind it, and to this end the telescope tube may well be larger in diameter than the objective. It is advisable to limit the field of view by a diaphragm at the principal focus, and to provide a field lens immediately behind this, forming an image of the objective on the cathode of the phototube (Fig. 28). A second diaphragm near the phototube further reduces stray light.

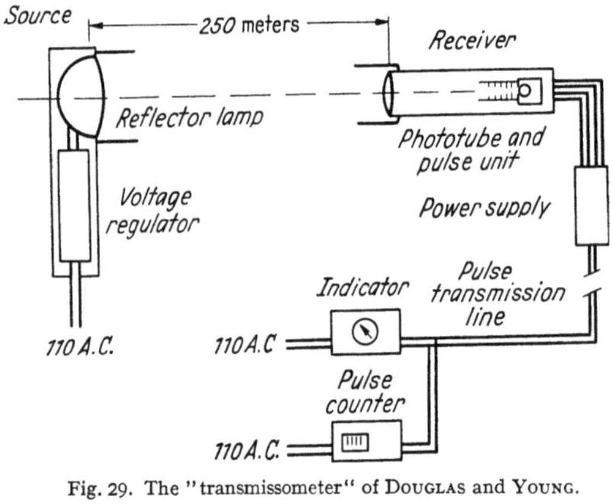

Fig. 29. The "transmissometer" of DOUGLAS and YOUNG.

As a measuring device a photomultiplier tube is frequently employed; by the use of comparatively simple circuits this can be made to record on a commercial recording potentiometer. For some purposes a logarithmic response is desirable, and circuits for achieving this have been described by SWEET[2] and by PLYMALE[3], which measure the voltage across the dynodes of the tube required to maintain a constant low anode current.

For routine measurements of atmospheric transmission the "transmissometer" of DOUGLAS and YOUNG[4] is much used in the United States. The general scheme of this is shown in Fig. 29; it should be noted that a pulse system of transmission is employed. The "background" due to daylight, which is small, can be measured by merely turning off the lamp. The calibration is done on clear days when the transmittance ot the light path is very nearly unity.

[1] H. S. COLEMAN, F. J. MORRIS, H. E. ROSENBERGER and M. J. WALKER: J. Opt. Soc. Amer. **39**, 515 (1949).
[2] M. H. SWEET: J. Opt. Soc. Amer. **37**, 432 (1947).
[3] W. S. PLYMALE: Electronics **26**, 143 (1953).
[4] C. A. DOUGLAS and L. L. YOUNG: U.S. Dept. of Commerce, C. A. A. Tech. Dev. Rep. 47, 1945.

BIBBY[1] in England has used a barrier-layer cell and galvanometer in a simple apparatus. One would expect the temperature coefficient of the photocell to be rather serious unless it were corrected for.

The effect of stray light in the field of such an instrument can be eliminated by modulating the beam. This was first done by BERGMANN[2], and later by BRADBURY and FRYER[3].

We shall describe a German instrument, due to JUNGINGER and reported on by SCHÖNWALD and MÜLLER[4] which seems to combine the best features of previous instruments.

This instrument has a triple mirror at one end of a base line and a projector and receiver at the other (Fig. 30). A small tungsten lamp L is enclosed in a

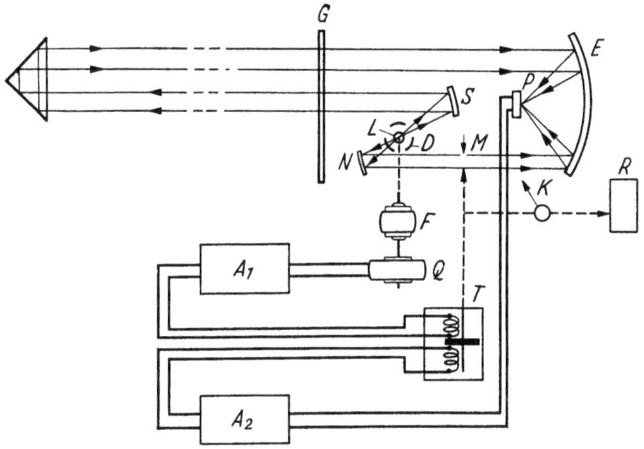

Fig. 30. The recording telephotometer of JUNGINGER (after SCHÖNWALD and MÜLLER).

rotating cylinder D with three slits, which allows light to fall alternately on the mirrors S and N. The mirror S sends light out towards the triple mirror, whence it returns to be collected by the paraboloid E and focussed on the photocell P. The smaller mirror N sends the comparison beam through the measuring diaphragm M to E, whence it is also reflected to the photocell. G is a clear glass window.

The cylinder D is driven by a motor, on the shaft of which is an A.C. generator which supplies one field of a small two-phase motor which drives the measuring diaphragm. The other field is fed by the amplified photocurrent, the relative phase of the two currents being such as to cause the measuring diaphragm to move towards the point of balance. At balance it indicates on a scale K, and records on a drum R.

The use of a triple mirror in this way greatly reduces the errors due to multiply-scattered light[5]. These can be almost entirely eliminated by modulating the light, not at the source, but at the distant mirror; and an elementary instrument on these lines was constructed by VAN LEAR[6] and used in an investigation on Mount Washington. The principle deserves further development.

[1] J. R. BIBBY: Gr. Britain, Air Ministry, Met. Res. Comm. M.R.P. 236, 1945.
[2] L. BERGMANN: Phys. Z. **35**, 177 (1934).
[3] N. E. BRADBURY and E. M. FRYER: Bull. Amer. Meteor. Soc. **21**, 391 (1940).
[4] B. SCHÖNWALD and TH. MÜLLER: Z. techn. Phys. **23**, 30 (1942).
[5] L. FOITZIK: Wiss. Abh. Reichsamt Wetterdienst **4**, No. 5 (1938).
[6] G. A. VAN LEAR jr.: O.S.R.D. Rep. 6201, Washington 1945.

δ) *Choice of base line for telephotometry.* From Eq. (33.3), $\sigma = 3.912/V_2$; and the transmittance T of an air path r_0 will therefore be

$$T = e^{-3.912\, r_0/V_2}.$$

Differentiating, it follows that

$$d\,V_2/V_2 = 0.26\,(V_2/r_0)\,(d\,T/T). \tag{48.2}$$

If $d\,V_2/V_2$ is to be less than 10%, and we can measure T to 1%, we shall find that V_2/r_0 must not be greater than about 40. This sort of discussion is suitable for visual photometry, or for logarithmic photomultiplier tube circuits, where $d\,T/T$ will be fairly constant.

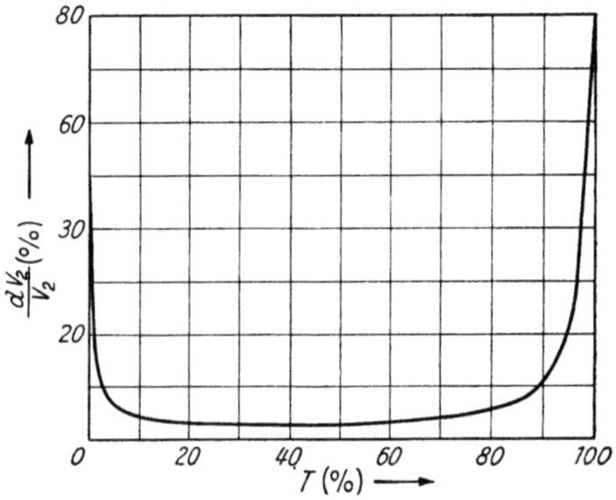

Fig. 31. Relative error in visual range as a function of measured transmittance, for a photoelectric telephotometer.

In photoelectric instruments with uniform scales $d\,T$ is likely to be constant over much of the range. In such a case we may note that $1/T = e^{\sigma r_0}$, so that

$$\sigma = (1/r_0)\,\log_e(1/T)$$

and

$$V_2 = 3.912/\sigma = 3.912\, r_0/\log_e(1/T).$$

Differentiating,

$$d\,V_2/V_2 = \left[(1/T)/\log_e(1/T)\right]\,d\,T. \tag{48.3}$$

A further differentiation shows a minimum at $T = e^{-1}$ ($= 0.37$). Fortunately the minimum is very flat, as shown in Fig. 31, after SCHÖNWALD and MÜLLER[1].

49. Instruments for measuring scattering. As long as industrial pollution is absent, a measurement of the scattering function may be substituted for a measurement of transmission. The technique is specially useful over the open ocean. The scattering coefficient may be calculated from a polar diagram of the scattering function, measured in many directions; but since this is a procedure suitable only for research, we shall not here describe apparatus for obtaining such a polar diagram, referring instead only to two easily-obtainable papers by J. M. WALDRAM[2]. If the form of the polar diagram were constant, a measurement at one angle would do; and WALDRAM and his co-workers decided that the most suitable angle was $\varphi = 30°$. A scattering meter[3] was designed and built which measured the luminance of a portion of a light beam at 30°; this instrument was enclosed, so that the air to be measured had to be brought into the instrument. This alone would preclude its use in fog, apart from the circumstance that the scattering function for fog is vrey different from that for clear air.

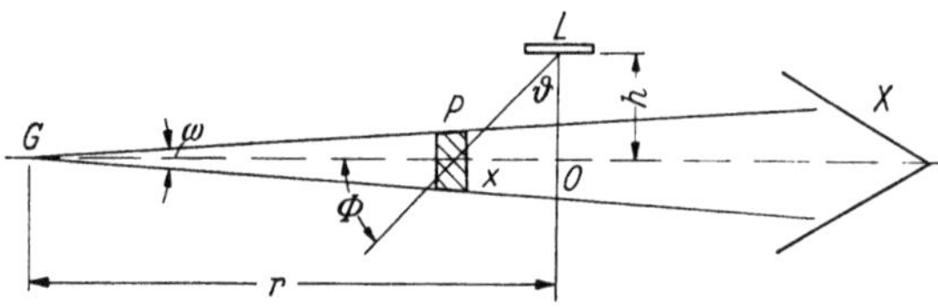

Fig. 32. Illustrating the theory of the "visibility meter" of BEUTTELL and BREWER.

[1] B. SCHÖNWALD and TH. MÜLLER: Z. techn. Phys. **23**, 30 (1942). For a fuller discussion of these matters see N. T. GRIDGEMAN, Anal. Chem. **24**, 445 (1952).

[2] J. M. WALDRAM: Quart. J. Roy. Meteorol. Soc. **71**, 319 (1945). — Trans. Illum. Engng. Soc. Lond. **10**, 147 (1945).

[3] Great Britain, Admiralty Res. Lab., Report A.R.L./R. 1/K. 904 (1949).

A more fruitful approach would seem to be that of BEUTTELL and BREWER[1]. Referring to Fig. 32, let L be a source of light of intensity I_0 in the direction LO, and intensity $I_0 \cos\vartheta$ in any direction ϑ. If we look from G, through the air which is illuminated by this source, into a dark cavity X, it can be shown that the coefficient of scattering b is given by

$$b = 2\pi h B/I_0 \tag{49.1}$$

a simple result which depends entirely on the source L being a diffuse radiator.

Fig. 33 shows the original design of BEUTTELL, in which the light beam is folded by the concave mirror M_1. Looking through the eyepiece E the observer sees an image of the light-trap B adjacent to one of the opal glass O_1, illuminated by light from the lamp L through a second opal glass O_2 and a measuring diaphragm W.

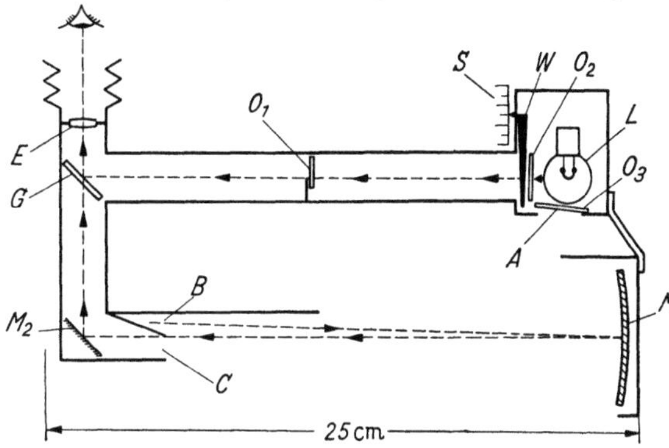

Fig. 33. Diagram of the "visibility meter" of BEUTTELL and BREWER (courtesy of the Journal of Scientific Instruments).

This meter may be calibrated without any absolute measurements of intensity or luminance by the use of two small disks coated with magnesium oxide[2].

Experience has shown that this meter is too compact in its present form, but a larger version might be a very good instrument.

An interesting application of a vertical *search-light* beam modulated[3] (e.g., by means of a rotating Venetian-blind type of shutter) has been made for the measurement of air densities[4] up to 62 km. A receiver tuned to the modulation frequency measures the intensity of the light scattered from the beam by the atmosphere; the polarization[5] of the scattered light can be used to separate it from the background air glow. Numerous other similar investigations have been reported[6-9].

Of more direct interest for our present subject would be the measurement from the ground of the transmittance of the first 100 or 200 m. of the atmosphere.

BEGGS and WALDRAM[10] devised a method in which a searchlight is directed over a photometer at an angle of about 15° from the horizontal. The luminance of the beam is measured at an angular elevation of about 35°. This method makes certain assumptions which are absent in a scheme proposed at the U.S. Naval Research Laboratory in 1949[11]. Referring to Fig. 34, a searchlight beam SR is directed upwards at an acute angle to the horizontal plane. Two points on this

[1] R. G. BEUTTELL and A. W. BREWER: J. Sci. Instrum. **26**, 357 (1949).
[2] See [2], p. 205.
[3] M. A. TUVE, E. A. JOHNSON and O. R. WULF: Terr. Magn. **40**, 452 (1935).
[4] L. ELTERMAN: J. Geophys. Res. **56**, 509 (1951).
[5] E. V. ASHBURN: J. Geophys. Res. **58**, 116 (1953).
[6] E. O. HULBURT: J. Opt. Soc. Amer. **27**, 377 (1937).
[7] E. A. JOHNSON, R. C. MEYER, R. E. HOPKINS and W. H. MOCK: J. Opt. Soc. Amer. **29**, 512 (1939).
[8] I. I. ROMANTZOV and I. A. KHVOSTIKOV: C. R. Acad. Sci. URSS. **53**, 703 (1946).
[9] L. M. MIKHAILIN and I. A. KHVOSTIKOV: C. R. Acad. Sci. URSS. **54**, 223 (1946).
[10] S. S. BEGGS and J. M. WALDRAM: G. E. Co. Ltd. Res. Labs. Rep. 8303, Wembley, England 1943.
[11] H. S. STEWART, L. F. DRUMMETER and C. A. PEARSON: NRL Rep. N-3484, Washington 1949.

plane, M_1 and N_1 are so chosen that if P is some point on the axis of the beam, the two angles RPM and RPN are equal, and also so that $PM \neq PN$. If the luminance of the beam at P is B_m as seen from M and B_n as seen from N, it can be shown that the vertical transmittance between P and the ground is

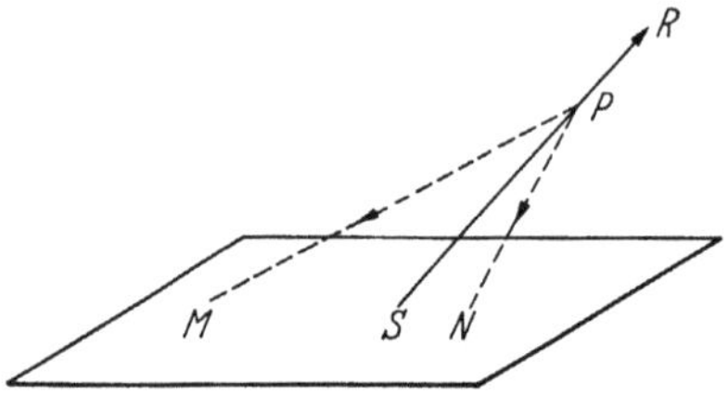

Fig. 34. The N.R.L. Searchlight method.

$$T_p = (B_m/B_n)^{\frac{1}{\operatorname{cosec}\vartheta_m - \operatorname{cosec}\vartheta_n}} \tag{49.2}$$

where ϑ_m and ϑ_n are the elevation angles of P as measured from M and N respectively[1].

This method is theoretically unexceptionable, but from a practical standpoint the trouble with it lies in the extremely small signals available in bad weather. It might, however, be developed for use at night. Since the searchlight beam is used merely as an elevated source of light, nothing need be known about it.

50. Meters using an "artificial haze". At the very beginning of the scientific study of the visual range it was felt that it might be possible to measure the desired quantity without taking the intermediate step of measuring the extinction coefficient. This approach to the problem resulted in a class of instrument which depends for its action on the addition of further obscuration to a field of view, until an object at a known distance can no longer be seen. The earliest of these were the *Sichtmesser* of WIGAND[2] and the visibility meter of JONES[3]. They were entirely empirical, and are now only of historical interest, as is the visibility meter of BENNETT[4], a variant of the WIGAND meter.

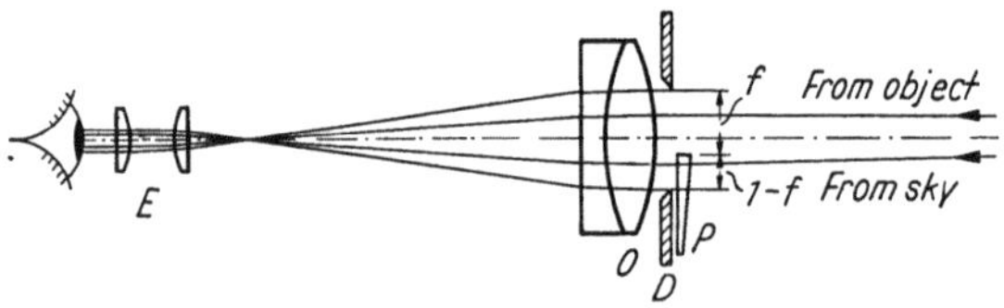

Fig. 35. Optical system of the meters of SHALLENBERGER and LITTLE and of WALDRAM.

A much sounder solution was found entirely independently by SHALLENBERGER and LITTLE[5] and by WALDRAM[6]. The theory of both these instruments is the same, and though the mechanical details differ greatly, each has the form of an attachment to a telescope. The form devised by WALDRAM is called the *Disappearance range gauge.*

The transmittance of the atmosphere up to the meteorological range, V_2 for example, is equal to the value of ε for which the range is calculated, for example 0.02. The transmittance for a given fraction of the range $aV_2 = D$ is also constant and independent of V_2. An object at a distance $D (< V_2)$ can be made to disappear if we can simulate in the field of view the remainder of the atmosphere between D and V_2. These instruments imitate this atmosphere both as to its transmittance and as to the air-light which it adds to the field of view.

In Fig. 35, O and E are the objective and eyepiece of a telescope; in front of the objective is placed a diaphragm of such a size that the exit pupil of the instrument is less than 2 mm. in its greatest dimension. Immediately in front

[1] The corresponding equation in the original paper is incorrect.
[2] A. WIGAND: Meteorol. Z. **36**, 342 (1919). — Phys Z. **22**, 484 (1921). — Gerlands Beitr. Geophys. **17**, 348 (1927) and other papers.
[3] L. A. JONES: Phil. Mag. **39**, 96 (1920).
[4] M. G. BENNETT: J. Sci. Instrum. **8**, 122 (1931).
[5] G. D. SHALLENBERGER and E. M. LITTLE: J. Opt. Soc. Amer. **30**, 168 (1940).
[6] J. M. WALDRAM: Wembley, England, G. E. Co. Res. Lab. Rep. 8672, May 23, 1945.

of this diaphragm slides a prism P of small angle so that in general two images of the scene are observed, separated by about half a degree. If a fraction f of the area of the aperture is left uncovered, the luminance of the undeviated image will be proportional to f and that of the added image to $(1-f)$.

If we decrease f to the point where the upper image (say of the horizon) just disappears, it can be shown that

$$V_2/D = \log 0.02/\log (0.02/f) \tag{50.1}$$

giving a ratio by which the distance D of the horizon must be multiplied to give V_2.

Because the aperture in front of the objective has to be small, a very precisely made mechanism has to be provided to move the prism. This makes the instrument somewhat expensive.

51. Miscellaneous "visibility meters". Several elementary devices for estimating the visual range have been described by PISKUN[1]. In general they consist of a board extending horizontally against the horizon sky at some distance from the observer, and painted in such a way as to appear to be graduated uniformly from black to white. Two black targets are set up parallel to this screen and at different distances behind it (see Fig. 36). The observer, moving his head as necessary, judges the positions along the board where it looks equal in brightness to each target. It will easily be seen how the extinction coefficient may be derived.

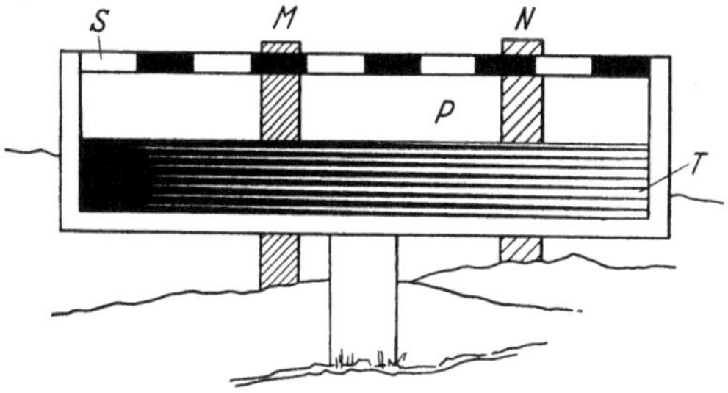

Fig. 36. One of the „photometric screens" of PISKUN.

GOLD[2] has given prominence to a method of measuring visual range at night by observing the setting of a simple optical wedge (variable-density filter) at which a calibrated lamp is no longer visible. HARCOMBE[3] and the present author[4] had earlier abandoned this method as being insufficiently precise under the practical conditions of observation.

General references.

[1] LEGRAND, Y.: Optique Physiologique, Tome II, Lumière et couleurs. Paris: Editions de la Revue d'Optique 1949.
[2] MIDDLETON, W. E. K.: Vision through the atmosphere. Toronto: University of Toronto Press 1952.
[3] PERNTER, J. M., and F. M. EXNER: Meteorologische Optik, 2nd ed. Vienna and Leipzig: W. Braumüller 1922.
[4] ROCARD, Y.: Etude sur la visibilité. Paris: Editions de la Revue d'Optique 1935.
[5] STILES, W. S., M. G. BENNETT and H. N. GREEN: Great Britain, Aer, Res. Comm. Rep. & Mem. no. 1793. 1937.
[6] WALSH, J. W. T.: Photometry, 2nd ed. London: Constable & Co. Ltd. 1953.

[1] V. F. PISKUN: Met. and Hydrol. (Moscow) April 1938, p. 21; June 1939, p. 114; Sept. 1939, p. 100.
[2] E. GOLD: Quart. J. Roy. Met. Soc. **65**, 139 (1939).
[3] S. HARCOMBE: Proc. Opt. Conv. Lond. **1**, 388 (1926).
[4] W. E. K. MIDDLETON: Trans. Roy. Soc. Canada, Sec. III **25**, 39 (1931).

Polarization of Skylight.

By

ZDENĚK SEKERA.

With 20 Figures.

A. Introduction.

1. Discovery and early development. Around 1811, ARAGO discovered the partial polarization of the light from a clear sky during his extensive studies of the polarization of light. He found the polarized component to be either perpendicular (positive polarization) or parallel (negative polarization) to the plane defined by the directions from the observer to sun and from the observer to point observed, and to reach its *maximum intensity* at about 90° from the sun. In the sun vertical on the antisolar side ARAGO found a point separating the region with positive polarization (toward the zenith) from the region with negative polarization (close to the horizon). The light from this point was neutral. The position of this point later called "*Arago neutral point*" varied with meteorological conditions, as did the maximum polarization. In 1840 two other neutral points were discovered in the sun vertical, above the sun by BABINET, and below the sun by BREWSTER.

The period of these discoveries was followed by years of rapid development of instruments for polarization measurements, a development which reached its highest point in SAVART'S polariscope and in CORNU'S photopolarimeter. Up to the present, these have been the standard instruments for visual measurement of the position of neutral points and of the degree of polarization. Numerous measurements of skylight polarization from different locations and at different times have built a detailed picture of the magnitude and distribution of skylight polarization, including all local, daily, seasonal and secular variations. Measurements made after violent volcanic eruptions (Krakatoa in 1883, Katmai in 1912) showed by the anomalies in position of neutral points and in the degree of polarization, of the order of 50 to 100% of normal values, the great sensitivity of skylight polarization to the amount of dust particles in the atmosphere.

The accumulated results of polarization measurements gave rise to several attempts at explanation of skylight polarization and its observed variations, ending with the classical theory of molecular scattering, presented by Lord RAYLEIGH in 1871. Under the assumption of primary scattering only (scattering occurring only once), the RAYLEIGH theory explains the linear polarization of skylight and the observed orientation of the polarized component. At the point of maximum polarization, however, the skylight should—according to the theory—be totally polarized and the neutral points should be situated in the directions toward the sun and toward the antisolar point.

This discrepancy between the theory and the observations is obviously the consequence of two main restrictive assumptions: 1. that of primary scattering only, 2. that of the molecular size of the scattering particles (negligible with respect to the wavelength of the incident light). In 1888 SORET estimated the effect of secondary scattering. Considering for simplicity reasons the scattering particles to be concentrated uniformly in a ring around the horizon, he was able to show that the secondary scattering introduces partial polarization 90° from the sun and is responsible for the shift of the neutral points to the observed positions, where the positive polarization of the primary scattering is compensated by the negative polarization due to the secondary scattering by particles at the horizon. SORET's original computations, done for the sun at the horizon, were extended later by AHLGRIMM for other sun elevations.

The first attempt to include scattering by particles of the size comparable to the wavelength or larger, was done by MILCH much later (1920). The numerical computations according to the MIE-DEBYE theory of scattering by large spherical particles, done by BLUMER and some experimental work done by SCHIRMANN led MILCH to attempt to obtain a better agreement between theory and observation by including primary scattering by large particles rather than higher order molecular scattering. BLICKHAN in 1934 found a dependence of skylight polarization upon LINKE's turbidity factor, as predicted by MILCH, and thus was able to estimate a value of polarization corresponding to the molecular atmosphere (by extrapolating to a turbidity factor of 1). The extrapolated values were much closer to the values computed by AHLGRIMM than to the values given by the RAYLEIGH original theory.

During recent years much progress has been made in efforts for more exact interpretation of measured skylight polarization. The use of *photoelectric methods* brought—among other advantages—more emphasis on the study of polarization in narrow spectral ranges and its potential usefulness in the separation of effects of large particle scattering from the effects of molecular scattering. The greatest credit for this progress, however, is to be attributed to the new methods developed by CHANDRASEKHAR. These methods lead to the exact solution of the problem of radiative transfer in the atmosphere with both multiple scattering and ground reflection included. Such a solution for a plane parallel atmosphere with RAYLEIGH scattering can now be numerically evaluated and the values of intensity and polarization of skylight are now available. From measurements the deviation of the actual polarization from the polarization of a molecular atmosphere can be directly obtained and thus the turbidity of the actual atmosphere can be accurately determined. Since the polarization can be expressed in terms of several independent parameters, it is reasonable to expect that polarization measurements may lead to a unique determination of the effective size distribution and total number of large scattering particles in the atmosphere.

B. Theory of skylight polarization.

2. STOKES' polarization parameters. As is well known, in the general state of polarization, the so called *partial elliptical polarization*, the radiation field can be decomposed into two parts:

1. a *neutral* radiation, in which the electric and magnetic field intensity (electric and magnetic vector) oscillates in such a random way that in the mean no preferable direction can be determined, and

2. an *elliptically polarized* radiation, in which, during one oscillation, the endpoint of the *electric* vector sweeps out an ellipse.

For a quantitative description of such a state of polarization four parameters are needed; the following quantities are most often used: 1. the *total intensity*, defining the flow of the radiant energy of both components, 2. the *degree of polarization*, giving the ratio of the intensity of the polarized component to the total intensity, 3. the *position of the plane of polarization*, containing the direction of the major axis of the ellipse described by the electric vector, 4. the *ellipticity*, defined by the ratio of the minor and major axis of the ellipse.

Such representation of the state of polarization is not suitable of course for an analytical treatment, since the four parameters, mentioned above, have different dimensions and are not additive for different incoherent radiation fields. On the other hand, the set of four polarization parameters introduced by Sir GEORGE STOKES in 1852 have the same dimensions and are additive for incoherent radiation fields. They are therefore most conveniently suited for analytical treatment, and CHANDRASEKHAR'S great success in the solution of the basic problems in skylight polarization may be attributed partly to his introduction of STOKES' parameters in the equations of radiative transfer.

A general definition of STOKES' parameters can be obtained in the following manner. Since the human eye and other photometric elements are unable to directly describe the character of the oscillations of the electric or magnetic vector it is necessary to use a special set-up for the detection or measurement of polarization. This special set up consists of the following optical systems:

1. an *analyzer*, which transmits electric or magnetic oscillations in only one direction in a plane normal to the direction of the propagation (to the high accuracy realized in the NICOL or GLAN-THOMPSON prism);

2. a *retardation plate*, which transmits electric or magnetic oscillations in two perpendicular directions called the fast and slow axes and introduces a known phase difference between the oscillations along these two directions (realized by different kinds of mica or quartz plates).

If a retardation plate of known phase difference (retardation) δ is placed normal to the direction of studied radiation, and the radiation leaving the plate is observed through an analyzer with the direction of transmission deviated from the direction of the fast axis of the plate by the angle ψ, then the intensity of the radiation leaving the analyzer varies with ψ and δ according to the relationship

$$I(\psi, \delta) = \tfrac{1}{2}\{I + Q\cos 2\psi + (U\cos\delta - V\sin\delta)\sin 2\psi\} \tag{2.1}$$

The quantities appearing in this relationship are STOKES' *polarization parameters*.

They have a simple physical meaning: if I_l, I_r denote the intensities for $\psi = 0°$ and $\psi = 90°$, respectively, then

$$I = I_l + I_r, \qquad Q = I_l - I_r. \tag{2.2}$$

The first parameter, the sum of the intensities in two perpendicular but otherwise quite arbitrary directions, measures the total flow of radiant energy and thus is identical with the total intensity defined above. The position of the plane of polarization can be determined from the maximum intensity $I(\psi, \delta)$ either without the retardation plate ($\delta = 0$), or with the use of a half-wave plate, ($\delta = \pi$), in which case the plane of polarization of the light leaving the plate is rotated by 90°. For $\delta = 0$ or π, the maximum intensity in (2.2) will appear for

$$U = \pm Q\tan 2\psi_0. \tag{2.3}$$

where ψ_0 is the direction of the plane of polarization. Hence U is the parameter defining the position of the plane of polarization. The last parameter V measures the ellipticity of polarization. It can be shown[1], that

$$\sin 2\beta = \frac{V}{\sqrt{Q^2+U^2+V^2}} \tag{2.4}$$

where $\beta = \arctan\,(b/a)$ with a and b denoting the lengths of the major and minor axis of the ellipse described by the end of the electric vector.

It is easy to see that for neutral (unpolarized) radiation $Q=U=V=0$; for linearly polarized radiation $V=0$; for circular polarization $Q=U=0$, $V=I$. In general the mixture of incoherent radiational fields may contain some part which is neutral. In such a case, because of the additiveness of these parameters, and of the above characteristics of the neutral radiation,

$$I > \sqrt{Q^2+U^2+V^2}.$$

The *degree of polarization*, mentioned above is then given by

$$P = \frac{\sqrt{Q^2+U^2+V^2}}{I}. \tag{2.5}$$

If there are no polarized components in the mixture, $P=0$; if there is no neutral component present (case of total polarization) $P=1$.

3. Scattering of light in the atmosphere, Mie-Debye and Rayleigh theory of scattering. In the visible range of the spectrum scattering is the most important manifestation of the interaction between radiation and matter (air molecules, dust and haze particles, etc.) suspended in the atmosphere. During this process the energy received by a scattering center is re-radiated in all directions at different rates, therefore the incident radiation appears to be attenuated by the amount of energy scattered from the center in all directions. The law of such attenuation can be expressed in the form (Beer's law)

$$d I_\nu = -\, k_\nu \varrho\, I_\nu\, d s \tag{3.1}$$

where I_ν is the intensity of incident radiation of the frequency ν, and dI_ν is its change after traversing the distance ds in the direction of propagation through the scattering medium of density ϱ. The proportionality factor k_ν is called the *mass scattering coefficient*, because the total amount of energy scattered from a pencil of radiation incident on the mass element dm of the scattering medium in a solid angle $d\omega$† is given by

$$k_\nu I_\nu\, dm\, d\omega. \tag{3.2}$$

[1] Cf. [*10*] pp. 25—35. The relationship between the components of the electric vector and Stokes' parameters can be established by means of Poynting's vector. If the electric vector of the studied radiation is written in components along the direction of the fast and slow axes of the plate, in the form

$$\boldsymbol{E} = (E_i \boldsymbol{i} + E_j \boldsymbol{j})\, e^{i\omega t} \quad \text{or} \quad = (a\,\boldsymbol{i} + b\, e^{-i\varepsilon} \boldsymbol{j})\, e^{i\omega t} \tag{2.5}$$

the electric vector of the radiation leaving the plate will contain in the second term the factor $e^{-i\delta}$. If the position of the analyzer is given by the unit vector $\boldsymbol{l}$, the intensity $I(\psi, \delta)$ is proportional to the expression $\mathrm{Re}\,\{(\boldsymbol{E}'\cdot\boldsymbol{l})(\boldsymbol{E}'^*\cdot\boldsymbol{l})\}$ where $'$ denotes the factor $e^{-i\delta}$ included and $*$ the complex conjugate. Substitution from (2.5) into this expression and comparison with (2.1) leads to the proportionalities

$$\left.\begin{aligned} I_l \sim \overline{E_i E_i^*} = \overline{a^2}, \quad I_r \sim \overline{E_j E_j^*} = \overline{b^2} \\ U \sim \overline{E_i E_j^* + E_j E_i^*} = \overline{2ab\cos\varepsilon}, \quad V \sim i\,(\overline{E_j E_i^* - E_i E_j^*}) = \overline{2ab\sin\varepsilon}. \end{aligned}\right\} \tag{2.6}$$

The bar denotes the time mean over eventual short period fluctuations of the corresponding quantities.

† Per unit time and per unit frequency interval.

For any particular direction deviated from the direction of the incident radiation by the angle Θ, the fraction of the total energy scattered into the solid angle $d\omega'$ can be written as follows

$$k_\nu I_\nu \, dm \, d\omega \, P(\cos\Theta) \frac{d\omega'}{4\pi} \tag{3.3}$$

where the *phase function* $P(\cos\Theta)$ specifies the angular distribution of the energy or the specific intensity of the scattered radiation. Comparison with the expression in (3.2) leads to the normalization condition for the phase function,

$$\int P(\cos\Theta) \frac{d\omega'}{4\pi} = 1, \tag{3.4}$$

where the integration is to be performed over all solids angles.

The explicit form for the phase function can be derived from the electromagnetic theory of scattering. A complete theory has been developed for the dielectric sphere. Its extension[1] to an asymmetric shape (prolate ellipsoid) has indicated that the effect of the shape of the scattering particle becomes noticeable only for a large refractive index. Furthermore the extension[2] of this theory to the case of two concentric spheres of different refractive indices has shown that with a thin coating the effect of the optically different interior disappears. Hence the refractive index of the atmospheric scatters may be assumed to be low, since these particles are almost always covered with a coating of liquid water or salt solution. Thus in the first approximation the atmospheric particles may be considered to be *dielectric spheres*.

In the theory of scattering by a dielectric sphere it is assumed that parallel radiation of the wavelength λ is incident on a single dielectric sphere of the radius a and of the dielectric constant ε. The incident radiation sets up electromagnetic waves both in the interior and in the exterior of the sphere. At a sufficiently large distance from the sphere, the solution of MAXWELL's equations which satisfies the boundary conditions on the surface of the sphere has the form of a diverging spherical wave from the center of the sphere; this spherical wave can be identified with the scattered radiation. The components of the electric vector of such a wave in the direction normal and parallel to the plane of scattering (containing the direction of the incident and scattered wave) substituted in (2.6), yield the expressions for the STOKES' parameters characterizing the scattered radiation. These consist of the specific intensities $I_{\parallel}^{(s)}$, $I_{\perp}^{(s)}$, parallel and normal to the plane of scattering and the parameters $U^{(s)}$ and $V^{(s)}$. They are related to the corresponding parameters of the net flux of the incident wave by the linear relationship

$$\boldsymbol{I}^{(s)} = \{P_{ij}\} \cdot \boldsymbol{F}^{(0)} \qquad (i, j = 1, 2, 3, 4) \tag{3.5}$$

where $\boldsymbol{I}^{(s)}$ stands for a one column matrix with the elements $I_{\perp}^{(s)}$, $I_{\parallel}^{(s)}$, $U^{(s)}$, $V^{(s)}$ and $\boldsymbol{F}^{(0)}$ stands for the corresponding matrix with the elements $F_i^{(0)}$ ($i = \perp, \parallel, u, v$). The matrix $\{P_{ij}\}$ if properly normalized, corresponds to the phase function $P(\cos\Theta)$. On the basis of general considerations[3] it follows that the elements

$$P_{11} = F_1, \quad P_{22} = F_2, \quad P_{33} = P_{44} = F_3, \quad P_{34} = -P_{43} = F_4$$

[1] R. GANS: Ann. Phys. **37**, 881 (1912). — D. ATLAS, M. KERKER and W. HITSCHFELD: J. Atmos. and Terres. Phys. **3**, 108 (1953).

[2] A. L. ADEN and M. KERKER: J. Appl. Phys. **22**, 1242 (1951). — A. GÜTTLER: Ann. Phys. **11**, 65 (1952).

[3] F. PERRIN: J. Chem Phys. **10**, 415 (1942). — F. PERRIN and A. ABRAGAM: J. Phys. Radium, Ser. VIII **12**, 69 (1951).

in the matrix $\{P_{ij}\}$ are the only ones that do not vanish. If the solution of MAXWELL's equations is expanded into the infinite series, derived originally by MIE and DEBYE, then

$$\left.\begin{aligned} F_1 &= S_R S_R^*, & F_3 &= \mathrm{Re}\,\{S_L S_R^*\},\\ F_2 &= S_L S_L^*, & F_4 &= \mathrm{Im}\,\{S_L S_R^*\} \end{aligned}\right\} \tag{3.6}$$

where S_R and S_L are the series appearing in the MIE-DEBYE theory[1]. For the special case of a very small sphere ($\alpha \to 0$), only the first term in the series S_R and S_L can be considered, yielding the expressions

$$S_R = \frac{\lambda}{2\pi}\,\alpha^3\,\frac{\varepsilon-1}{\varepsilon+2}, \qquad S_L = S_R \cos\Theta. \tag{3.8}$$

With (3.8) substituted in (3.6), the relation (3.5) reduces to the form

$$\boldsymbol{I}^{(s)} = \left(\frac{2\pi}{\lambda}\right)^4 a^6 \left(\frac{\varepsilon-1}{\varepsilon+2}\right)^2 \begin{pmatrix} 1 & 0 & 0 & 0\\ 0 & \cos^2\Theta & 0 & 0\\ 0 & 0 & \cos\Theta & 0\\ 0 & 0 & 0 & \cos\Theta \end{pmatrix} \cdot \boldsymbol{F}^{(0)}. \tag{3.9}$$

The intensity of the scattered radiation has the same form as the intensity of radiation from an oscillating dipole with the polarizability

$$\mathfrak{P} = a^3\,\frac{\varepsilon-1}{\varepsilon+2}. \tag{3.10}$$

Lord RAYLEIGH in 1871 arrived at the same results, assuming only that the dimension of the scattering center is very small compared to the wavelength and that the dielectric constant ε approaches 1, without specification of its shape. That the scattering by very small particles is independent of their size and shape, is implied in the fact that the radius a appears in the expression for their polarizability $\mathfrak{P}$. This constant can be related to the macroscopically measurable refractive index n of a specific gas, for example by the LORENTZ-LORENZ law

$$\mathfrak{P} = \frac{3}{4\pi N}\,\frac{n^2-1}{n^2+2}, \tag{3.11}$$

in which only the number of molecules of the gas in a unit volume N appears; hence $\mathfrak{P}$ is independent of the size and shape of the particles. The use of the infinite series S_L, S_R is impractical for very large values of α since they converge very slowly. Several approximations [*14f*] for the solution of the problem for large α have been recently suggested, but an explicit form for the STOKES' parameters has not yet been derived.

So far only the scattering by a single particle has been discussed. In the atmosphere, however, in addition to the large number of the molecules of the air constituents every unit volume contains a considerable number of particles of dimensions comparable to or larger than the wavelength. Hence a scattering

---

[1]
$$\left.\begin{aligned} S_L(\alpha,\varepsilon,\Theta) &= (\lambda/2\pi)\sum_{n=1}^{\infty}\left[a_n(\alpha,\varepsilon)\,p_n(\Theta) + b_n(\alpha,\varepsilon)\,t_n(\Theta)\right],\\ S_R(\alpha,\varepsilon,\Theta) &= (\lambda/2\pi)\sum_{n=1}^{\infty}\left[a_n(\alpha,\varepsilon)\,t_n(\Theta) + b_n(\alpha,\varepsilon)\,p_n(\Theta)\right], \end{aligned}\right\} \quad \left(\alpha = \frac{2\pi a}{\lambda}\right) \tag{3.7}$$

where a_n, b_n are complex coefficients, and with $x = \cos\Theta$,

$$p_n(\Theta) = dP_n(x)/dx, \qquad t_n(\Theta) = n(n+1)\,P_n(x) - x\,dP_n(x)/dx,$$

$P_n(x)$ denoting the LEGENDRE polynomial of n-th degree. (Cf. p. 188 and p. 333 of this volume.)

center should be considered illuminated not only by the incident (parallel) radiation, but also by the scattered radiation from other centers. Although it is not difficult to formulate the problem of scattering in such a case *(multiple scattering)*, the attempt [*14f*] for a solution has been made only for scattering centers situated so far apart that the scattered radiation can be considered as a pure radiational field (transverse wave). A complete discussion of such an effect on the STOKES' parameters of the scattered radiation is still not available.

However, it is possible to include the effect of multiple scattering under the assumption that the scattered radiations from different scattering centers are *incoherent*. This assumption is justified by the random distribution and by the irregular motion of individual molecules and other particles. These prevent the establishment of any phase relations between the components of the scattered radiation, corresponding to the illumination from other scattering centers. In such a case—as a consequence of the additiveness of the STOKES' parameters—the mass scattering coefficient can simply be determined from (3.2), by considering a unit volume of the scattering medium containing N scattering centers. If the net flux of the incident (neutral) parallel radiation is normalized to unity, the total amount of radiant energy scattered from the unit volume is

$$\beta_\nu = k_\nu \varrho = \int \sum_N (I_\perp^{(s)} + I_\parallel^{(s)})\, d\omega = \int \sum_N \tfrac{1}{2}(F_1 + F_2)\, d\omega \tag{3.12}$$

where β_ν, the volume scattering coefficient, has a meaning analagous to the mass scattering coefficient. Because a part of the scattered radiation from the particles around the element dm is scattered in the direction of the incident radiation, there will be an apparent increase of its intensity (or flux). The effect of multiple scattering leads therefore to a virtual emission from the element dm. With this emission included, the relation (3.1) becomes a fundamental equation for the variation of the intensity in a scattering medium, as discussed in the following section.

4. Equation of radiative transfer. Its approximate and exact solutions. If the radiation field, resulting from the scattering, is described by the specific intensity $I_\nu(\mu, \varphi)$ at every point of the scattering medium as a function of the directional parameters (for example μ the cosine of the zenith distance and φ the azimuth), then the amount of energy received by the mass element dm through the solid angle $d\omega'$ from the direction (μ', φ') and scattered into the solid angle $d\omega$ in the direction (μ, φ) can be written according to (3.3) in the form

$$\frac{1}{4\pi} k_\nu\, dm\, d\omega\, P(\mu, \varphi; \mu', \varphi')\, I_\nu(\mu', \varphi')\, d\omega', \tag{4.1}$$

where another notation is used for the phase function $P(\cos\Theta')$, Θ' being the angle between the directions (μ, φ) and (μ', φ'). From the net change of the radiant energy occuring during its passage through the element dm, given by the sum of the loss in (3.2) and of the gain, obtained by adding the contributions (4.1) from all solid angles $d\omega'$, the equation for the specific intensity $I_\nu(\mu, \varphi)$ can be easily derived in the form

$$dI_\nu(\mu, \varphi) = -k_\nu \varrho\, ds\, I_\nu(\mu, \varphi) + k_\nu \varrho\, ds \int P(\mu, \varphi; \mu', \varphi')\, I_\nu(\mu', \varphi') \frac{d\omega'}{4\pi}. \tag{4.2}$$

The *equation of radiative transfer* (4.2) enables one to solve the problem of radiative transfer in a medium with a known law of scattering so that the intensity $I_\nu(\mu, \varphi)$ at every point of the medium in any particular direction can be determined. In this equation the second term on the right hand side represents the virtual transmission of a unit mass due to the multiple scattering in the medium.

Eq. (4.2) refers to the total intensity as derived from energy considerations without taking account of the polarization produced by the scattering. With the method outlined in Sect. 1, corresponding equations for the STOKES' parameters can be derived. The equation of radiative transfer then assumes a form identical to (4.2), with the intensity $I_\nu(\mu, \varphi)$ replaced by the intensity vector $\boldsymbol{I}_\nu(\mu, \varphi)$, written in the form of a one column matrix with the elements $I_l(\mu, \varphi)$, $I_r(\mu, \varphi)$, $U(\mu, \varphi)$ and $V(\mu, \varphi)$, where the subscript l and r refer to the intensity parallel or normal respectively to the vertical plane in which the zenith distance $\vartheta = \arccos\mu$ is measured. Furthermore the phase function $P(\mu, \varphi; \mu', \varphi')$ is to be replaced by the *phase matrix* $\boldsymbol{P}(\mu, \varphi; \mu', \varphi') \equiv \{P_{ik}\}$ $(i, k = 1, 2, 3, 4)$, which can be obtained from the scattering matrix (3.6) by normalization to unity, in the sense that the scattering mass coefficient be defined as the energy scattered in all directions by the unit mass, illuminated by a neutral (unpolarized) radiation of unit specific intensity or of unit net flux. In other words the condition

$$\int (P_{11} + P_{22})\, d\omega / 8\pi = 1 \tag{4.3}$$

is to be satisfied with the integration performed over all solid angles.

In general it is quite difficult to solve the integro-differential eq. (4.2) and the solution for special cases has been discussed by CHANDRASEKHAR in his treatise [*10*]. In solving the problem of radiative transfer in the earth's atmosphere quite an essential simplification can be achieved by the assumption of a plane parallel atmosphere of infinite horizontal extent with a stratified density distribution. In such a case $ds = -dz/\mu$, and since ϱ and k_ν are functions of the vertical coordinate z only, it is convenient to express the vertical distances in terms of the *optical thickness* τ, defined by

$$\tau = \int_z^\infty k_\nu \varrho\, dz, \tag{4.4}$$

and measured positively downwards from the top of the atmosphere. (A plane layer of optical thickness τ reduces radiation at normal incidence to the fraction $e^{-\tau}$.) Furthermore, in the problems dealing with radiation transfer in the atmosphere illuminated by the sun, it is convenient to separate the parallel *sun radiation* from the radiation scattered in the atmosphere (called *diffuse sky radiation*). If $\pi\boldsymbol{F}$ denotes the matrix of STOKES' parameters of the net flux of solar radiation per unit area normal to its direction specified by $(-\mu_0, \varphi_0)$[1] incident on the top of the atmosphere, then this flux will be attenuated due to the scattering according to (3.1) to the value $\pi\boldsymbol{F}e^{-\tau/\mu_0}$ at the level of the optical thickness τ. The mass element dm at that level will be illuminated by this reduced flux, and the virtual emission in the eq. (4.2) has to be increased by the amount of the radiation scattered into the direction (μ, φ). The equation of radiative transfer in a plane parallel atmosphere illuminated by the sun radiation as defined above, assumes the form

$$\left.\begin{aligned} \mu \frac{d\boldsymbol{I}(\tau; \mu, \varphi)}{d\tau} &= \boldsymbol{I}(\tau; \mu, \varphi) - \frac{1}{4} e^{-\tau/\mu_0} \boldsymbol{P}(\mu, \varphi; -\mu_0, \varphi_0) \cdot \boldsymbol{F} - \\ &\quad - \frac{1}{4\pi} \int_{-1}^{1} \int_0^{2\pi} \boldsymbol{P}(\mu, \varphi; \mu', \varphi') \cdot \boldsymbol{I}(\tau; \mu', \varphi')\, d\mu'\, d\varphi'. \end{aligned}\right\} \tag{4.5}$$

[1] In the following μ is used as always *positive* so that the upward direction is specified b $+\mu$, downward direction by $-\mu$.

The subscript ν can be omitted in the eq. (4.5) since the frequency dependance is included implicitly in the optical thickness τ.

The form of the eq. (4.5) suggests two methods for the approximate solution: first, by successive iteration and second, by approximative quadrature. In the first method the first approximation obtained by putting $\boldsymbol{I}(\tau;\mu',\varphi')\equiv 0$ in the integrand of the third term on the right hand side has an obvious physical meaning, namely it gives the intensity and the polarization under the consideration of the primary scattering only. The second approximation obtained by substituting for $\boldsymbol{I}(\tau;\mu',\varphi')$ the solution for primary scattering gives the values of diffuse sky radiation with the primary and secondary scattering included. In both these approximations the eq. (4.5) reduces to a simple differential equation of the first order in τ and can be expressed analytically in terms of some exponential integrals[1].

In the second method the integration in the third term can be replaced by a summation over finite intervals or any approximate quadrature can be used which leads to the solution of a finite set of algebraic equations by standard methods (cf. CHANDRASEKHAR [*10*], ROBLEY[2]). The physical meaning of this approximation is that the radiation is considered only in discrete streams instead of a continuous distribution in different directions.

CHANDRASEKHAR, however, developed an ingenious method by which the integro-differential eq. (4.5) can be reduced to a few systems of simultaneous integral equations, with solutions accessible by successive iterations. The intensity of the radiation emerging from the atmosphere at its top, (as the sun radiation reflected diffusely by the atmosphere), and at its bottom (as the diffuse sky radiation), can be expressed in their dependance on the extraterrestrial flux $\pi\boldsymbol{F}$ of the sun radiation by means of two matrices, the reflection matrix[3] $\boldsymbol{S}$ and the transmission matrix $\boldsymbol{T}$ such that

$$\left.\begin{aligned} \boldsymbol{I}(0;+\mu,\varphi) &= \frac{1}{4\mu}\,\boldsymbol{S}(\tau_1;\mu,\varphi;\mu_0,\varphi_0)\cdot\boldsymbol{F}, \\ \boldsymbol{I}(\tau_1;-\mu,\varphi) &= \frac{1}{4\mu}\,\boldsymbol{T}(\tau_1;\mu,\varphi;\mu_0,\varphi_0)\cdot\boldsymbol{F}. \end{aligned}\right\} \tag{4.6}$$

If the intensities of the radiation leaving any atmospheric layer at the top and at the bottom are expressed in terms of the matrices defined in (4.6), a series of fundamental relationships can be derived, which amount to CHANDRASEKHAR'S *principles of invariance*[4]. From HELMHOLTZ's principle of reciprocity, it is pos-

[1] With the intensity vector and with the phase matrix replaced by the intensity and by the phase function this method leads to the results obtained by CHAPMAN and HAMMAD [Phil. Mag. (7) **28**, 99 (1939); **36**, 434 (1945); **39**, 956 (1948)] for the case of RAYLEIGH scattering. The use of the phase function cannot, however, lead to correct values of the skylight polarization since its use excludes the polarization of the scattered radiation.

[2] R. ROBLEY: Ann. Géophys. **8**, 1 (1952).

[3] CHANDRASEKHAR calls the matrix $\boldsymbol{S}$ in his treatment "scattering" matrix, which name in this discussion is used for the matrix in eq. (3.5), defining the general law of scattering.

[4] As an example, one of the invariance principles may be mentioned: the intensity of the downward diffuse radiation at an intermediate level can be considered composed of two parts: 1. diffuse sky radiation from the layer above (of the optical thickness τ), 2. diffusely reflected radiation emerging from the layer below, diffusely reflected downwards by the layer above. In mathematical form:

$$\boldsymbol{I}(\tau;-\mu,\varphi) = \frac{1}{4\mu}\,\boldsymbol{T}(\tau;\mu,\varphi;\mu_0,\varphi_0)\cdot\boldsymbol{F} + \\ + \frac{1}{4\pi\mu}\int\limits_0^1\int\limits_0^{2\pi}\boldsymbol{S}(\tau;\mu,\varphi;\mu',\varphi')\cdot\boldsymbol{I}(\tau;+\mu',\varphi')\,d\mu'\,d\varphi'.$$

sible to find the form of the matrices $\boldsymbol{S}$ and $\boldsymbol{T}$ and from the principles of invariance and their derivative with respect to τ it is possible to derive a sufficient number of equations which allow to separate the individual functions constituting the matrices and to find the set of integral equations they have to satisfy.

Since the elements of the phase matrix are functions of

$$\cos\Theta = \mu\,\mu' + (1-\mu^2)^{\frac{1}{2}}(1-\mu'^2)^{\frac{1}{2}}\cos(\varphi'-\varphi)$$

after transformation to the form $\boldsymbol{P}(\mu,\varphi;\mu',\varphi')$, the elements of the phase matrix can be developed into harmonic series in $(\varphi'-\varphi)$. Correspondingly, the intensities $\boldsymbol{I}$ and the other matrices can be written as series

$$\left.\begin{aligned} \boldsymbol{I}(\tau;\mu,\varphi) &= \sum_n \boldsymbol{I}^{(n)}(\tau;\mu,\varphi),\\ \boldsymbol{M}(\tau;\mu,\varphi;\mu_0,\varphi_0) &= \sum_n \boldsymbol{M}^{(n)}(\tau;\mu,\varphi;\mu_0,\varphi_0), \qquad (\boldsymbol{M}\equiv \boldsymbol{P},\boldsymbol{S},\boldsymbol{T}) \end{aligned}\right\} \tag{4.7}$$

where $\boldsymbol{I}^{(0)}$ and $\boldsymbol{M}^{(0)}$ are azimuth independent and $\boldsymbol{I}^{(n)}$ and $\boldsymbol{M}^{(n)}$ contain elements with $\cos n(\varphi_0-\varphi)$ and $\sin n(\varphi_0-\varphi)$. Substitution of the forms (4.7) into the eq. (4.5) and comparison of the corresponding terms in $(\varphi_0-\varphi)$ on both sides of the equation lead to separate equations for $\boldsymbol{I}^{(n)}(\tau,\mu,\varphi)$.

If the matrices $\boldsymbol{P}^{(n)}$ satisfy—as they do for the case of RAYLEIGH scattering—the relation

$$\left.\begin{aligned} &\frac{1}{4\pi}\int_0^{2\pi} \boldsymbol{P}^{(n)}(\mu,\varphi;\mu',\varphi')\cdot \boldsymbol{P}^{(n)}(\mu',\varphi';-\mu_0,\varphi_0)\,d\varphi'\\ &\qquad\qquad = \boldsymbol{P}^{(n)}(\mu,\varphi;-\mu_0,\varphi_0)\,f^{(n)}(\mu'), \end{aligned}\right\} \tag{4.8}$$

then the eq. (4.5) admits solutions of the form

$$\boldsymbol{I}^{(n)}(\tau;\mu,\varphi) = \boldsymbol{P}^{(n)}(\mu,\varphi;-\mu_0,\varphi_0)\cdot \boldsymbol{F}\,\Phi^{(n)}(\tau,\mu) \tag{4.9}$$

where $\Phi^{(n)}(\tau,\mu)$ are only scalar functions in τ and μ. For the standard solution, in which it is assumed that there is no flux of radiation entering the atmosphere from the bottom, the following boundary conditions are to be satisfied

$$\boldsymbol{I}(0;-\mu,\varphi)\equiv 0, \qquad \boldsymbol{I}(\tau_1;+\mu,\varphi)\equiv 0 \tag{4.10}$$

with the obvious physical meaning of non existence of downward diffuse radiation at the top and upward radiation at the bottom. In such a case the matrices $\boldsymbol{S}^{(n)}$ and $\boldsymbol{T}^{(n)}$ assume the following form[1]

$$\left.\begin{aligned} \boldsymbol{S}^{(n)}(\tau_1;\mu,\varphi;\mu_0,\varphi_0) &= \frac{\mu\,\mu_0}{\mu+\mu_0}\boldsymbol{P}^{(n)}(\mu,\varphi;-\mu_0,\varphi_0)\times\\ &\quad\times[X^{(n)}(\mu)X^{(n)}(\mu_0)-Y^{(n)}(\mu)Y^{(n)}(\mu_0)],\\ \boldsymbol{T}^{(n)}(\tau_1;\mu,\varphi;\mu_0,\varphi_0) &= \frac{\mu\,\mu_0}{\mu-\mu_0}\boldsymbol{P}^{(n)}(-\mu,\varphi;-\mu_0,\varphi_0)\times\\ &\quad\times[Y^{(n)}(\mu)X^{(n)}(\mu_0)-X^{(n)}(\mu)Y^{(n)}(\mu_0)]. \end{aligned}\right\} \tag{4.11}$$

The functions $X^{(n)}$ and $Y^{(n)}$ satisfy the simultaneous integral equations

$$\left.\begin{aligned} X(\mu) &= 1+\mu\int_0^1 \Psi(\mu')[X(\mu)X(\mu')-Y(\mu)Y(\mu')]\frac{d\mu'}{\mu+\mu'},\\ Y(\mu) &= e^{-\tau_1/\mu}+\mu\int_0^1 \Psi(\mu')[Y(\mu)X(\mu')-X(\mu)Y(\mu')]\frac{d\mu'}{\mu-\mu'}, \end{aligned}\right\} \tag{4.12}$$

[1] The reduction to the form (4.11), (4.12) has been done by CHANDRASEKHAR for the case of RAYLEIGH scattering only, but it can be shown that the same reduction is possible, whenever the relation (4.8) is valid.

where the form of the characteristic function $\Psi^{(n)}(\mu')$ is given by the form of the phase matrix, explicitely by the form of the function $f^{(n)}(\mu')$ in (4.8).

The corresponding reduction for the azimuth independent terms is in general more complicated and has to be done by means of the equations for the matrices $\boldsymbol{S}^{(0)}$, $\boldsymbol{T}^{(0)}$ obtained from (4.5) by elimination, based on the invariance principles.

For the case of RAYLEIGH scattering the phase matrix $\boldsymbol{P}(\mu, \varphi; \mu', \varphi')$ derived from the matrix in (3.9) by introducing the normalization factor $\frac{3}{8}$ and a transformation to the intensities I_l and I_r, parallel and normal to the vertical plane, has the following properties:

a) the series (4.7) contains only three terms, $n=0, 1, 2$ as a consequence of the element $\cos^2\Theta = \frac{1}{2}(1+\cos 2\Theta)$,

b) the first term in this series, the azimuth independent matrix $\boldsymbol{P}^{(0)}$ has only two columns and two rows,

c) the matrix is reducible with respect to the last row and column so that the STOKES parameter V can be determined independently, from (4.2) with $P(\cos\Theta) = \frac{3}{8}\cos\Theta$ and V substituted for $I_\nu(\mu, \varphi)$.

It follows from b) that the matrices $\boldsymbol{S}^{(0)}$ and $\boldsymbol{T}^{(0)}$ have only two columns and two rows and from the discussion of the equation for these matrices it can be seen that they are expressible in the form

$$\left.\begin{aligned} \boldsymbol{S}^{(0)}(\tau_1; \mu, \mu') &= \frac{3\mu'}{16(\mu+\mu')}\{\boldsymbol{K}(\mu)\cdot\tilde{\boldsymbol{K}}(\mu') - \boldsymbol{L}(\mu)\cdot\tilde{\boldsymbol{L}}(\mu')\}, \\ \boldsymbol{T}^{(0)}(\tau_1; \mu, \mu') &= \frac{3\mu'}{16(\mu-\mu')}\{\boldsymbol{L}(\mu)\cdot\tilde{\boldsymbol{K}}(\mu') - \boldsymbol{K}(\mu)\cdot\tilde{\boldsymbol{L}}(\mu')\}, \end{aligned}\right\} \tag{4.13}$$

where

$$\boldsymbol{K}(\mu) = \begin{pmatrix} \psi(\mu), & 2^{\frac{1}{2}}\varphi(\mu) \\ \chi(\mu), & 2^{\frac{1}{2}}\zeta(\mu), \end{pmatrix}, \quad \boldsymbol{L}(\mu) = \begin{pmatrix} \xi(\mu), & 2^{\frac{1}{2}}\eta(\mu) \\ \sigma(\mu), & 2^{\frac{1}{2}}\vartheta(\mu) \end{pmatrix} \tag{4.14}$$

and $\sim$ denotes the transposed matrix. The eight functions in the matrices (4.14) are linear functions of two pairs of functions $X_l(\mu)$, $Y_l(\mu)$, $X_r(\mu)$, $Y_r(\mu)$, which satisfy the integral equations (4.12) with the characteristic functions

$$\Psi_l(\mu) = \tfrac{3}{4}(1-\mu^2), \quad \Psi_r(\mu) = \tfrac{3}{8}(1-\mu^2). \tag{4.15}$$

All constants in the linear expressions for the functions in (4.14) in terms of X_i, Y_i $(i=l, r)$ can be uniquely determined by the moments of these X- and Y-functions.

From the explicit form of the phase matrix, the characteristic functions for the X- and Y-functions appearing in the second and third terms of the reflection and transmission matrices can be found,

$$\Psi^{(1)}(\mu) = \tfrac{3}{8}(1-\mu^2)(1+2\mu^2), \quad \Psi^{(2)}(\mu) = \tfrac{3}{16}(1+\mu^2)^2. \tag{4.16}$$

Finally, the elements of the reflection and transmission matrices corresponding to the STOKES parameter V can be expressed as linear functions of $X_r(\mu)$, $Y_r(\mu)$, $X_V(\mu)$, $Y_V(\mu)$; the characteristic functions for the last pair of functions has the form

$$\Psi_V(\mu) = \tfrac{3}{4}\mu^2. \tag{4.17}$$

The last expression concludes the reduction of the equations of radiative transfer in a plane parallel atmosphere with RAYLEIGH scattering to the solution of five pairs of integral equations (4.12) with the characteristic functions given in (4.15), (4.16), (4.17). Once the solution of these equations, the X- and Y-functions, are known, the intensities of the emergent radiation from the top and from the bottom of the atmosphere can be computed without great difficulty. Furthermore, with the use of invariance principles, the intensities of downward and upward diffuse radiation at any level within the atmosphere can be determined in terms of the reflection and transmission matrices of the layers above and below [*14g*].

The method of computation of the intensities of the emergent radiation from the atmosphere, mentioned above, has been extended by CHANDRASEKHAR to include the effect of the ground reflection. In the case of LAMBERT'S law of reflection, according to which the reflected radiation is unpolarized and isotropic in the outward hemisphere, independently of the polarization and the angular distribution of the incident radiation, and the outward normal flux of the reflected radiation is always a fraction (albedo λ_0) of the inward normal flux of the radiation incident on the ground, this can be done without great difficulty. If $\boldsymbol{I}_g \equiv (\frac{1}{2} I_g, \frac{1}{2} I_g, 0, 0)$ denotes the intensity of the reflected radiation, the intensity of the downward radiation is increased by

$$\boldsymbol{I}^* (\tau_1; -\mu) = \frac{1}{4\pi\mu} \int_0^1 \int_0^{2\pi} \boldsymbol{S}(\tau_1; \mu, \varphi; \mu', \varphi') \cdot \boldsymbol{I}_g \, d\mu' \, d\varphi' = \frac{1}{2\mu} I_g \begin{pmatrix} S_l(\mu) \\ S_r(\mu) \end{pmatrix},$$

where

$$S_i(\mu) = \int^1 (S_{il}^{(0)} + S_{ir}^{(0)}) \, d\mu', \qquad (i = l, r),$$

$S_{ik}^{(0)} (i, k = l, r)$ being the elements of the azimuth independent matrix $\boldsymbol{S}^{(0)}$ in (4.13). The constant I_g can be eliminated by using the condition for the flux of the reflected radiation. The resulting expression for the azimuth independent matrix $\boldsymbol{I}^* (\tau_1; -\mu)$ can be written in terms of the functions $\gamma_l(\mu)$, $\gamma_r(\mu)$ linearly dependent on the X- and Y-functions corresponding to the characteristic functions in (4.15). A similar result can be obtained for the diffusely reflected radiation at $\tau = 0$.

5. Polarization of skylight in a molecular atmosphere. If it is assumed that the atmosphere contains only scattering particles of a size negligible with respect to the wavelength, then the theory of RAYLEIGH scattering can be applied and the intensity and polarization of the diffuse sky radiation can be computed exactly by the method mentioned in the preceding section. The results would then refer to a plane parallel atmosphere where the presence of haze and dust particles or other particulate matter is disregarded, i.e. to a pure, molecular atmosphere.

The mass scattering coefficient for each air constituent can be obtained from (3.12) after substitution from (3.9), (3.10) and (3.11) in the form

$$k_\nu = \frac{24\pi^3}{\lambda^4 N \varrho} \left(\frac{n^2 - 1}{n^2 + 2}\right)^2 \simeq \frac{32\pi^3}{3\lambda^4} \frac{(n-1)^2}{N\varrho}. \tag{5.1}$$

When the variation of the refractivity $(n-1)$ with the density is considered and substituted in (5.1) and the contributions from the different air constituents are added, the value of the scattering coefficient and its variation with height can be found from the known composition and density distribution. The optical thickness τ can then be obtained by performing the integration in (4.4). In the following table a few values of the optical thickness are given, as obtained with the use of the most recent upper air data[1].

[1] D. DEIRMENDJIAN: Arch. Meteor., Geophys. u. Biokl., Ser. B **6**, 452 (1955).

Because the variation of the refractivity with the wavelength is almost negligible, the optical thickness τ varies primarily with the negative fourth power of the wavelength and thus it can be easily interpolated for other wavelengths. The optical thickness of the molecular atmosphere, of the same composition and density distribution as the actual atmosphere, is at sea level less than 1 for $\lambda \geq 308$ mμ.

The computation of the polarization of diffuse skylight in a molecular atmosphere can be simplified in one respect since the solar extraterrestrial radiation may be considered to be unpolarized, in which case $F_l = F_r = \frac{1}{2}F$, $F_U = F_V = 0$. Because of the property of the phase matrix, mentioned before eq. (4.13) sub c), it follows that the sky radiation is only linearly polarized ($V \equiv 0$), and thus only four pairs of X- and Y-functions are to be computed. The successive iterations of the eq. (4.12) for the characteristic functions defined in (4.15) and (4.16) have been performed independently at two different places[1] with the use of high speed computers, for several values of $\tau \leq 1$. The values obtained were used for the computation of all eight functions in (4.14) and of the functions $\gamma_i(\mu)$, $(i = l, r)$ (cf. [11], [12], [13]).

Table 1. *Optical thickness of the molecular atmosphere according to recent data on the composition and density distribution in the upper atmosphere.*

z (in km)	$\lambda = 375$ mμ	$\lambda = 520$ mμ	$\lambda = 835$ mμ
0	0.454	0.118	0.017
10	0.120	0.031	0.005
20	$0.025\frac{1}{2}$	0.007	0.001

The intensities I_l and I_r and the parameter U can be readily computed for any vertical plane of the azimuth φ from the relationships

$$\left.\begin{aligned} I_l(\mu,\varphi) &= I_l(\mu)\cos^2\tfrac{1}{2}(\varphi_0-\varphi) + I_{la}(\mu)\sin^2\tfrac{1}{2}(\varphi_0-\varphi) + \mu^2 Z(\mu)\sin^2(\varphi_0-\varphi)\,,\\ I_r(\mu,\varphi) &= I_r(\mu) - Z(\mu)\sin^2(\varphi_0-\varphi)\,,\\ U(\mu,\varphi) &= \frac{1}{2\mu}\left[I_l(\mu) - I_{la}(\mu)\right]\sin(\varphi_0-\varphi) - \mu Z(\mu)\sin 2(\varphi_0-\varphi)\,, \end{aligned}\right\} \tag{5.2}$$

where

$$I_l(\mu) \equiv I_l(\mu,\varphi_0)\,, \quad I_{la}(\mu) \equiv I_l(\mu,\varphi_0+\pi)\,, \quad I_r(\mu) \equiv I(\mu,\varphi_0)$$

are the intensities along the sun vertical and

$$Z(\mu) = \frac{3(1-\mu_0^2)}{16\mu_0(\mu-\mu_0)}\{X^{(2)}(\mu_0)\,Y^{(2)}(\mu) - X^{(2)}(\mu_0)\,X^{(2)}(\mu)\}\,.$$

In order to adhere to ARAGO's original definition of positive and negative polarization, it is preferable to write the expression in (2.7) for the sun vertical ($U = 0$) in the form

$$P = (I_r - I_l)/(I_r + I_l)\,. \tag{5.3}$$

The distribution of the degree of polarization computed [14c] from the formula (5.3), is represented in Fig. 1, for the sun close to the zenith ($\mu_0 = 0.98$) and close to the horizon ($\mu_0 = 0.20$) for $\tau = 0.15$. The curves show the basic properties of skylight polarization as found by the earliest discoveries: 1. maximum polarization about 90° from the sun, 2. region of negative polarization in the vicinity of the sun and antisolar point, 3. BABINET and BREWSTER neutral point for high sun and BABINET and ARAGO point for the sun close to the horizon.

The dependence of the degree of polarization on the optical thickness is illustrated in Fig. 2, where the distribution of polarization along the sun vertical

[1] At the WATSON Scientific Computing Laboratory, New York and at the Institute for Numerical Analysis (National Bureau of Standards), Los Angeles.

is shown for optical thickness $\tau = 0.15$, 0.25, 1.00, for the sun's zenith distance $\vartheta_0 = 53.1°$. With increasing optical thickness the maximum polarization derceases and is shifted slightly towards the sun. The distance of the neutral points from the sun or antisolar point increase with τ at a greater rate for the BREWSTER than for the BABINET point.

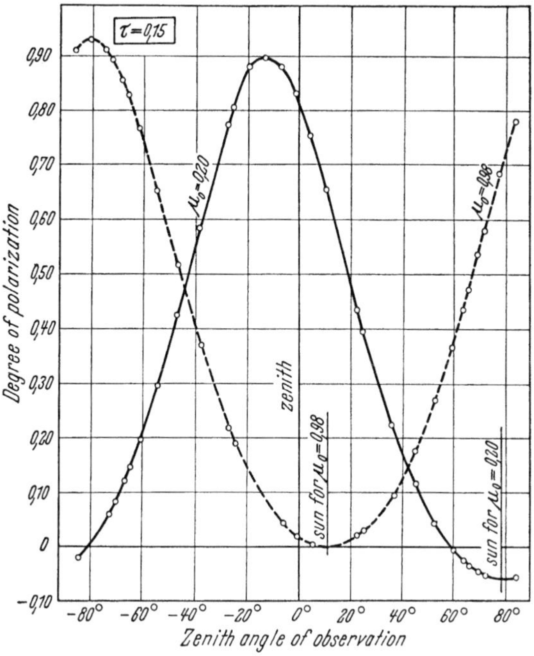

Fig. 1. Distribution of the degree of polarization along the sun vertical for two different positions of the sun and for the optical thickness $\tau = 0.15$.

The effect of the ground reflection (according to LAMBERT'S law) on the polarization in the sun vertical is illustrated in Fig. 3, where the distribution of the degree of polarization is shown for $\tau = 0.25$ and for different values of albedo (0, 0.25, 0.80), for the same zenith distance of the sun as in Fig. 2. The value of the maximum polaiization decreases with increasing albedo, its position is shifted very little and the positions of the neutral points, especially of the BABINET point, are not appreciably effected.

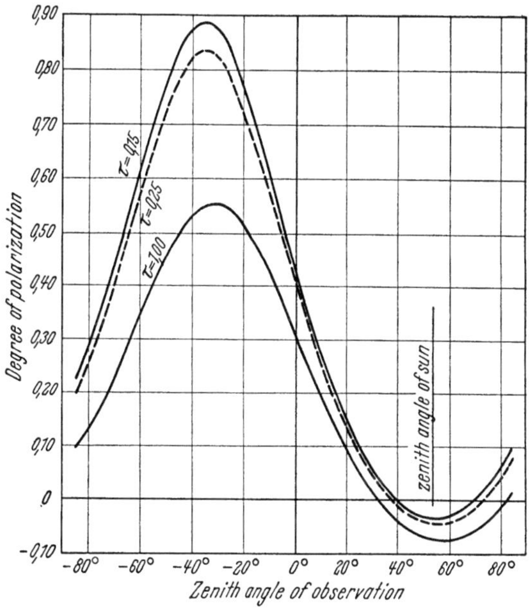

Fig. 2. Dependance of the degree of polarization along the sun vertical on the optical thickness ($\tau = 0.15$, 0.25, 1.00).

The positions of the neutral points in the sun vertical can be obtained by solving the equation $I_r(\mu) - I_l(\mu) = 0$, for the BREWSTER and BABINET point, $I_r(\mu) - I_{la}(\mu) = 0$ for the ARAGO point [*14c*]. The variation of the elevations of the neutral points above the horizon with the sun elevation for different optical thicknesses is illustrated in Fig. 4. This figure shows in greater detail the properties of the neutral points mentioned by CHANDRASEKHAR[1], namely that:

a) the neutral points coincide with the sun when the sun is in the zenith, for all optical thicknesses,

b) at other elevations of the sun the distances of the neutral points from the sun or from the antisolar point increase with increasing optical thickness,

[1] S. CHANDRASEKHAR and D. D. ELBERT: Nature, Lond. **167**, 51 (1951).

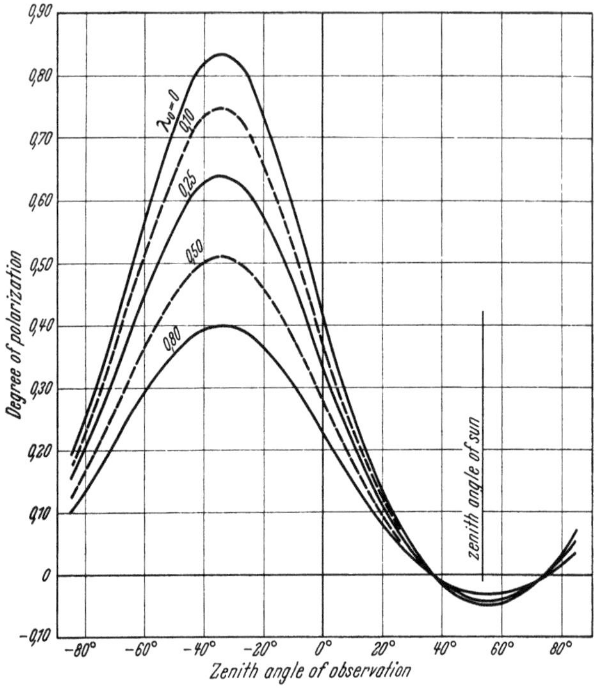

Fig 3. Distribution of the degree of polarization along the sun vertical for different values of the albedo ($\lambda_0 = 0, 0.25, 0.80$) for the optical thickness $\tau = 0.25$.

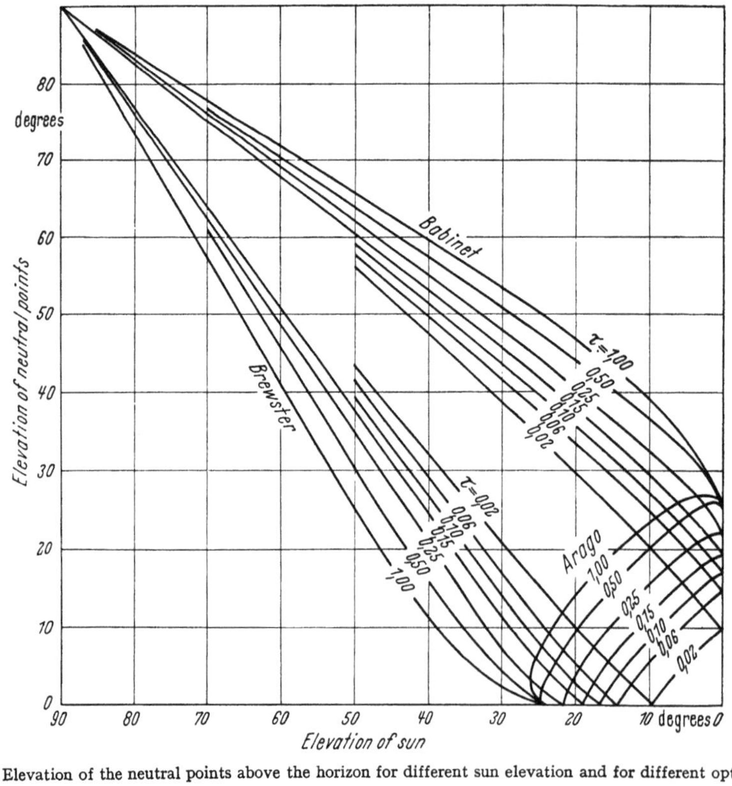

Fig. 4. Elevation of the neutral points above the horizon for different sun elevation and for different optical thickness.

c) the BREWSTER point rises (or sets) at the time when the ARAGO point sets (or rises) so that only two neutral points are always present in the sun's vertical,

d) for the sun at the horizon the BABINET and ARAGO points have the same elevation above the horizon.

The effect of the ground reflection is more pronounced for low sun elevations and for large optical thicknesses as is apparent from the enlarged portion of Fig. 4 reproduced in Fig. 5, with the dotted curve corresponding to the elevation of the neutral points for the albedo 0.80. This figure shows an interesting feature of the curve representing the ARAGO point for $\tau = 1.00$, suggesting the appearance of two ARAGO points for a sun elevation exceeding 24°. With increasing sun elevation the secondary neutral point rises and finally merges with the regular

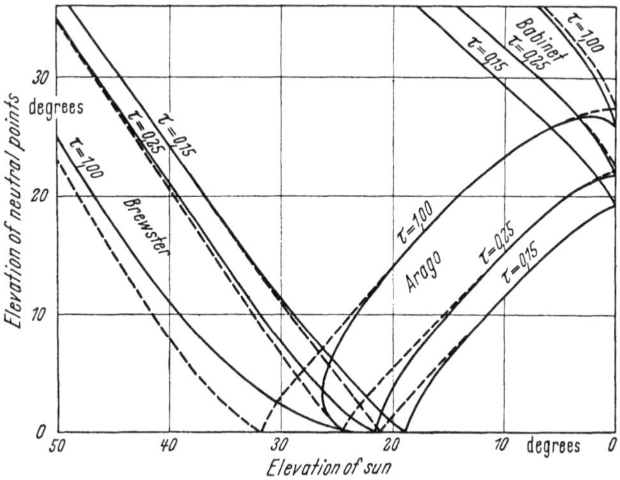

Fig. 5. Effect of the LAMBERT ground reflection of a high albedo ($\lambda_0 = 0.80$) on the position of the neutral points.

point descending toward the horizon. For higher albedos this feature disappears. The largest effect of the ground reflection can be noticed in the BREWSTER and ARAGO points: while the elevation of the BREWSTER point is decreased, the elevation of the ARAGO point is increased. The ARAGO point stays above the horizon longer and appears sooner, the BREWSTER point sets sooner and rises later when a ground reflection with a high albedo is present.

The diurnal variations of the distance of the neutral points from the sun or antisolar point are illustrated in Fig. 6. The cusps in the upper curves correspond to the position of the BREWSTER and ARAGO point at the horizon. The increase of the distance with increasing optical thickness is again visible.

Away from the sun vertical $U \neq 0$, except along the line where

$$U_1 \equiv I_l(\mu) - I_{la}(\mu) - 4\mu^2 Z(\mu) \cos(\varphi_0 - \varphi) = 0. \tag{5.4}$$

According to the definition of the degree of polarization in (2.7), the neutral points outside the sun vertical will be situated at the places of intersection of the line $U = 0$ and the so called *neutral lines* given by

$$I_r(\mu, \varphi) - I_l(\mu, \varphi) = 0.$$

At all other points where $U = 0$, but $I_r - I_l \neq 0$, the plane of polarization coincides with the vertical; where $U \neq 0$, but $I_r - I_l = 0$, the plane of polarization is inclined from the vertical by 45° or 135°, since the deviation χ of the plane of polarization form the vertical (i.e. from the direction of I_l) is given according to (2.3) by

$$\tan 2\chi = U(\mu, \varphi)/[I_l(\mu, \varphi) - I_r(\mu, \varphi)]. \tag{5.5}$$

When the line $U(\mu, \varphi) = 0$ is plotted, it can be seen that this line does not cross the neutral line except at the zenith, where the line $U = 0$ crosses the sun vertical at right angles, but the neutral lines under the angle of 45°[1]. From the distribution of the degree of polarization in the sun vertical, it is however evident, that the zenith is not a neutral point, and the crossing of the lines $U = 0$ and of the neutral lines at the zenith simply expresses the fact that the plane

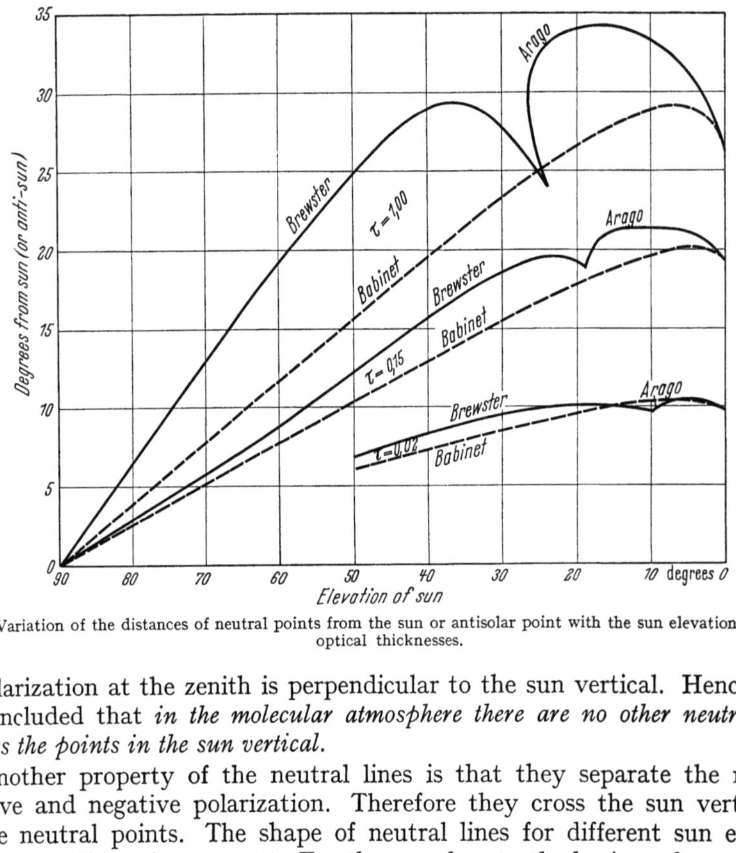

Fig. 6. Variation of the distances of neutral points from the sun or antisolar point with the sun elevation for different optical thicknesses.

of polarization at the zenith is perpendicular to the sun vertical. Hence it may be concluded that *in the molecular atmosphere there are no other neutral points besides the points in the sun vertical.*

Another property of the neutral lines is that they separate the region of positive and negative polarization. Therefore they cross the sun vertical also at the neutral points. The shape of neutral lines for different sun elevations is shown in Fig. 7, for $\tau = 0.15$. For the sun close to the horizon the neutral line has the shape of a lemniscate (called BUSCH's *lemniscate*) connecting the ARAGO point and the BABINET point and having a double point in the zenith. For higher sun elevations, when the BREWSTER point is above the horizon, the lemniscate changes into a closed curve of two separate branches, one passing through the BABINET, the other through the BREWSTER point, and crossing at the double point in the zenith.

Along the sun vertical the state of polarization can be expressed by one quantity only, the degree of polarization in (5.3) with the sign + or − indicating

[1] This is a consequence of the fact, that the denominator in (5.5) can be written in the form $A(\mu) + B(\mu)\cos(\varphi_0 - \varphi) + C(\mu)\cos 2(\varphi_0 - \varphi)$ and for $\mu = 1$, $A = B = C = 0$; $A = 0$ follows from the property of the functions in (4.14) that for $\mu = 1$; $\psi = \chi$, $\varphi = \zeta$, $\xi = \sigma$, $\eta = \vartheta$. This property can be used as a check of the computation of these functions.

the position of the plane of polarization. However, for all other azimuths the state of polarization is determined by the degree of polarization, always a positive quantity defined by (2.7), and by the deviation of the plane of polarization from the vertical plane given by (5.5). The distribution of the degree of polarization over the sky is illustrated in Fig. 8, where the only available theoretical values (2.7) for $\tau = 0.15$ are reproduced for the zenith angle of the sun 43.9°. The degree of polarization varies continously from the zero value at the BREWSTER

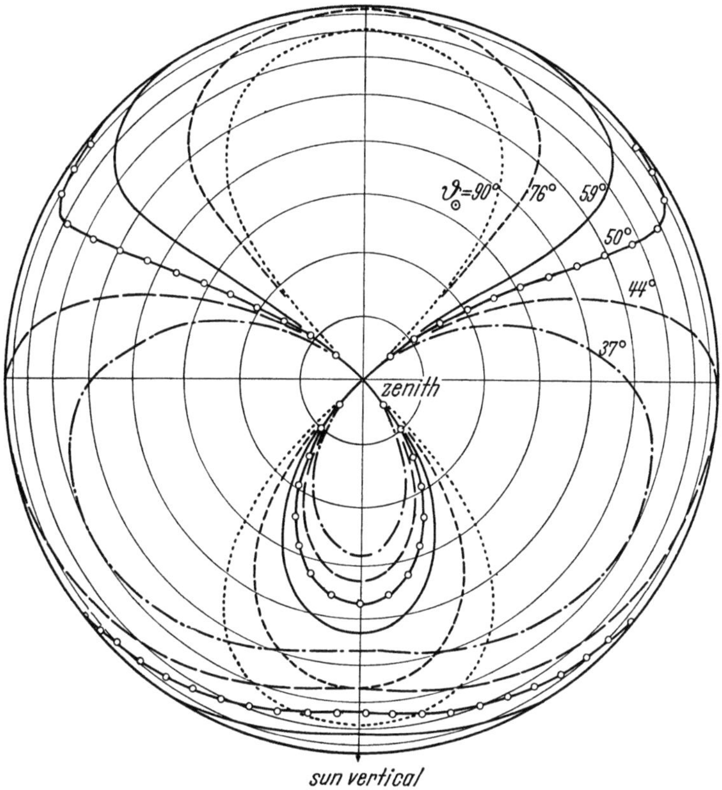

Fig. 7. Neutral lines for different sun zenith distances for the optical thickness $\tau = 0.15$ (after CHANDRASEKHAR).

and BABINET points to the maximum in the vicinity of the sun vertical about 90° from the sun. In determining the position of the plane of polarization the ambiguity inherent in the computation of χ from $\tan 2\chi$ in (5.5) should be kept in mind. The neutral lines and the lines where $U(\mu, \varphi) = 0$ indicate the lines along which the denominator and the numerator in (5.5) change their signs and hence delineate the regions of positive or negative values of $\tan 2\chi$.

The line $U(\mu, \varphi) = 0$ has the shape of a simple closed curve, crossing the sun vertical at right angles at the zenith and in the vicinity of the sun (above the sun). From the discussion of the expression for U_1 in (5.4), it follows that $U < 0$ in the interior of the curve $U = 0$, including the BABINET point, and $U > 0$ in the exterior of this curve, as long as $\varphi_0 - \varphi \leq 180°$. The neutral line through the BABINET point divides the region of negative U into two parts: a) a region containing the sun vertical between the BABINET point and the zenith, where the denominator of (5.5) is negative, so that $\tan 2\chi > 0$, b) the remaining part of the interior of the curve $U = 0$, where $\tan 2\chi < 0$. The exterior of the curve $U=0$ is divided by the neutral line through the BREWSTER point to the zenith into two parts again, c) the region of negative $\tan 2\chi$ inside of the neutral line, and d) the region of positive $\tan 2\chi$ outside of the neutral line towards the horizon. If the angle χ is considered positive and changing from 0 to 90° when the plane of polarization is rotated towards the right from the vertical plane, so that turning to the left corresponds to the change of χ from 180 to 90°, then the change of χ within the limits $0° \leq \chi \leq 45°$ in the region c),

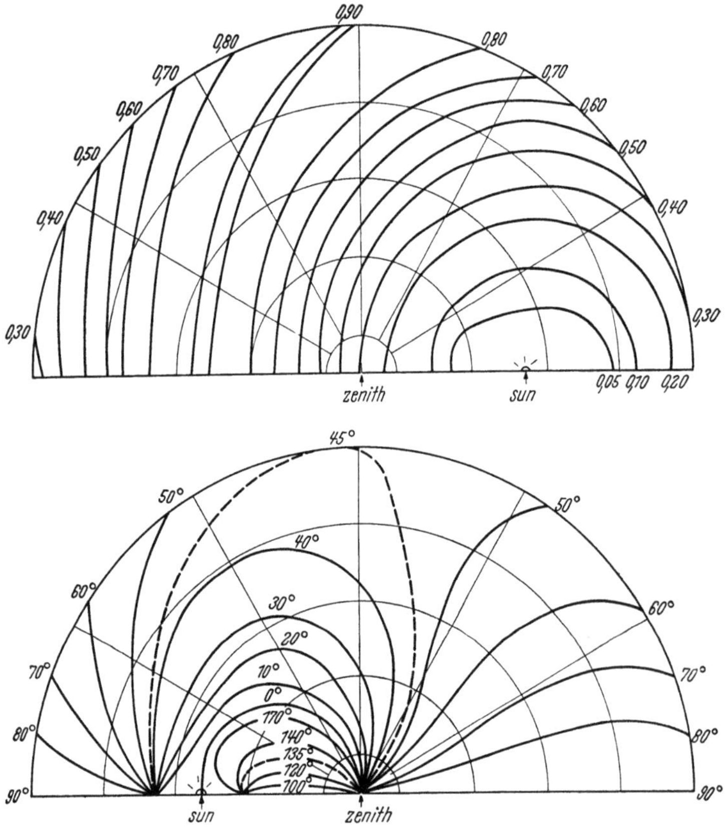

Fig. 8. Distribution of the degree of polarization (shown above) and of the position of the plane of polarization (shown below) over the sky for the zenith distance of the sun $\vartheta_0 = 43.9°$ and for the optical thickness $\tau = 0.15$.

$45° \leq \chi \leq 90°$ in the region d), $90° \leq \chi \leq 135°$ in the region a) and finally $135° \leq \chi \leq 180°$ in the region b) corresponds to the observed position of the plane of polarization in the vicinity of the sun (F. BUSCH[1]). The lines of equal deviation of the plane of polarization from the vertical plane are shown in Fig. 8 for $\tau = 0.15$ and for the zenith angle of the sun 43.9°.

6. Polarization of skylight in a turbid atmosphere. The actual atmosphere always contains a considerable number of particles, solid or liquid, of a varying nature, size etc. It is customary to treat the atmosphere as an aerosol and to

[1] Meteor. Z. **6**, 81–95 (1889). In this paper the earlier definition of the plane of polarization is used, as the plane normal to the direction of the maximum intensity, so that the agreement with the results above is obtained by applying a rotation of the plane of polarization by 90°.

denote all particulate matter other than air molecules by the term *aerosol particles*. They are responsible for the turbidity of the atmosphere, namely for the deviations of the optical properties of the atmosphere from those of a pure molecular atmosphere.

If the theory of skylight polarization for a RAYLEIGH atmosphere is to be extended to include the effect of the atmospheric aerosol, the first difficult task is to establish the appropriate law of scattering. According to Sect. 3, if the scattered radiation from each aerosol particle may be considered incoherent with respect to the others, the STOKES' parameters and the scattering coefficient can be obtained by a summation of these quantities over all the particles in a unit volume. For particles of a size comparable to the wavelength or larger, the quantities to be summed depend upon the radius of the particles and their refractive index or dielectric constant, so that it is desirable to know the *size distribution* of the particles present rather than their *total number*. The content of particles of different sizes varies so much from place and from time to time that it is almost impossible to estimate its actual value exactly at a given time. Recent measurements of the particle content, however, provide convincing evidence that, under normal conditions and in the average, special types of the size distribution of particles can be expected according to the location or the life history of the air mass under consideration so that the distribution can even be expressed by a simple mathematical law.

In an air mass which has travelled over the continent for a sufficiently long period, the number of particles per unit volume $n(a)$ having a radius contained in the interval $\langle a, a+da\rangle$ is proportional to $a^{-\beta}da$, where the average value of β is equal to 4[1]. The size distribution of particles close to or over the sea or large water surfaces can be expressed by a similar law, but with two different exponents; for radii smaller than a critical size a_c, at which the discontinuity in β appears, the index β is much smaller $(1 \leq \beta \leq 2.5)$ than for radii larger than a_c, in which range β approaches the value of 4 or is even larger[2]. Measurements of the particle content at higher levels indicate an exponential decrease in the number of particles, at the same rate for all sizes, at least up to a level of 4 to 5 km, with a slight increase at higher levels[3]. Although a theoretical explanation of these characteristics of size distribution of aerosol particles has not been given yet, the distribution laws mentioned above can be used as the first approximation.

If $n(a, m)$ denotes the number of particles of the refractive index m and of a radius within the interval $\langle a, a+da\rangle$, then the functions appearing in the scattering matrix (3.5) and the volume scattering coefficient for a unit volume of atmospheric aerosol are given by

$$\left.\begin{aligned} \bar{F}_i(\Theta) &= \sum_m \int_{a_0}^{\infty} F_i(\alpha, m; \Theta)\, n(a, m)\, da \qquad (i = 1, 2, 3, 4), \\ \bar{\beta}_\nu &= \sum_m \int_{a_0}^{\infty} K(\alpha, m)\, \pi a^2 n(a, m)\, da \qquad \left(\alpha = \frac{2\pi a}{\lambda}\right) \end{aligned}\right\} \tag{6.1}$$

[1] CH. JUNGE: Ber. dtsch. Wetterdienst, US-Zone 35, 261 (1952). — Tellus 5, 1 (1953). — J. Meteorology 12, 13 (1955).

[2] H. DESSENS: Ann. Géophys. 2, 276, 343 (1946); 3, 68 (1947). — A. H. WOODCOCK: J. Meteorology 7, 397 (1950). — J. Y. GILBERT: Unpubl. Report, 1954. — D. J. MOORE and B. J. MASON: Quart. J. Roy. Meteor. Soc. 80, 583 (1954). — See F. MÖLLER's discussion p. 163 ff. of this volume.

[3] R. PENNDORF: Geophys. Res. Papers, G.R.D. 25, 1 (1954). — J. Meteorology 11, 245 (1954).

where a_0 is the radius of the smallest particle, F_i are functions given in (3.6) and the total scattering cross section $K(\alpha, m)$ is introduced, defined by the relation

$$\int \tfrac{1}{2}(F_1 + F_2)\, d\omega = K(\alpha, m)\, \pi a^2 .$$

The summation in (6.1) is to be performed over all discrete values of m.

From the available tables of scattering functions for spherical particles it would be possible to compute the functions F_i and after performing the integration to arrive at the value of the functions $\bar{F}_i(\Theta)$ for different values of Θ. However, in eq. (4.5) the phase matrix appears in the form $\boldsymbol{P}(\mu, \varphi; \mu', \varphi')$. The transformation of the matrix in (3.5) into this form is possible only if the scattering angle Θ is separated from the other two variables α and m. This can be done conveniently by developing the functions F_i in series of LEGENDRE polynomials of the form

$$F_i(\alpha, m; \Theta) = \sum_{r=0}^{\infty} p_r^{(i)}(\alpha, m)\, P_r(\cos\Theta), \qquad (i = 1, 2, 3, 4). \tag{6.2}$$

The coefficients in the corresponding expansion for the functions $\bar{F}_i(\Theta)$ are then given by expressions similar to (6.1),

$$\bar{p}_r^{(i)} = \sum_m \int_{a_0}^{\infty} p_r^{(i)}(\alpha, m)\, n(a, m)\, da, \qquad (i = 1, 2, 3, 4). \tag{6.3}$$

In the transformation of the matrix in (3.5) to the form $P(\mu, \varphi; \mu', \varphi')$ the azimuth difference $\varphi' - \varphi$ and the parameters μ, μ' can be introduced simply by using the addition theorem for LEGENDRE polynomials. The phase matrix for the scattering in the turbid atmosphere can be decomposed into two parts; the matrix for RAYLEIGH scattering $\boldsymbol{P}_R$, and the matrix $\boldsymbol{P}_L$ which evidently contains the terms responsible for the asymmetry of scattering by large particles. When this separation of the phase matrix is incorporated in eq. (6.3) and proper normalization factors are introduced, the equation of radiative transfer in a turbid atmosphere assumes the form

$$\left.\begin{aligned} \mu \frac{d\boldsymbol{I}(\tau;\mu,\varphi)}{d\tau} &= \boldsymbol{I}(\tau;\mu,\varphi) - \frac{1}{4} e^{-\tau/\mu_0} \{\boldsymbol{P}_R(\mu,\varphi;-\mu_0,\varphi_0) + \\ &\quad + T(\tau)\boldsymbol{P}_L(\mu,\varphi;-\mu_0,\varphi_0)\}\cdot \boldsymbol{F} - \frac{1}{4\pi}\int_{-1}^{+1}\int_0^{2\pi} \{\boldsymbol{P}_R(\mu,\varphi;\mu',\varphi') + \\ &\quad + T(\tau)\boldsymbol{P}_L(\mu,\varphi;\mu',\varphi')\}\cdot \boldsymbol{I}(\tau,\mu',\varphi')\, d\mu'\, d\varphi', \end{aligned}\right\} \tag{6.4}$$

where τ is the optical thickness of the *turbid* atmosphere, namely $\tau = \int_z^{\infty} (k_\nu \varrho + \bar{\beta}_\nu)\, dz$ and $T(\tau) = \bar{\beta}_\nu / (k_\nu \varrho + \bar{\beta}_\nu)$, a function of the height z, thus expressible in terms of the optical thickness τ. This quantity is always ≤ 1 and has the meaning of some kind of a turbidity factor, being zero for the pure molecular atmosphere, and reaching the maximum value of 1 if the turbidity is so high that the molecular scattering coefficient can be neglected with respect to the scattering coefficient of the atmospheric aerosol. If $T(\tau)$ is small, the solution of the eq. (6.4) can be obtained by putting

$$\boldsymbol{I}(\tau;\mu,\varphi) = \boldsymbol{I}_R(\tau;\mu,\varphi) + \boldsymbol{I}_L(\tau;\mu,\varphi) \tag{6.5}$$

where $\boldsymbol{I}_R$ is the solution for RAYLEIGH scattering for the optical thickness τ, and $\boldsymbol{I}_L$ the deviation of the intensity of diffuse sky radiation in the turbid atmosphere from that of the molecular atmosphere of the same optical thickness. When

For oscillatory motion the direct flow has the time factor $e^{i\nu\tau}$; the reverse flow, $e^{-i\nu\tau}$. For this case the theorem has been given by FLAX [38], the general case for non-steady flow by HEASLET and SPREITER [39]. The same relations hold in supersonic main flow.

References.

[1] THEODORSEN, TH.: General theory of aerodynamic instability and the mechanism of flutter. NACA Techn. Rep. No. 496, 1935.

[2] KÜSSNER, H. G.: Zusammenfassender Bericht über den instationären Auftrieb von Flügeln. Luftf.-Forsch. **13**, 410–424 (1936).

[3] KÜSSNER, H. G.: Allgemeine Tragflächentheorie. Luftf.-Forschg. **17**, 370–378 (1940).

[4] KÜSSNER, H. G.: A review of the two-dimensional problem of unsteady lifting surface theory during the last thirty years. University of Maryland Lecture Series, No. 23, 1953.

[5] CIGALA, P.: Lo stato attuale delle ricerche sul moto instazionario di una superficie portante. L'Aerotecnica **21**, 671–685, 759–777 (1941).

[6] TIMMAN, R.: La théorie des profils minces en écoulement non stationnaire en fluide incompressible ou compressible. N.L.L. (Nat. Luchtv. Lab.) Publ. No. MP 69.

[7] HADAMARD, J.: Lectures on CAUCHY's problem. Yale 1923.

[8] RIESZ, M.: Acta math. **81**, 1–223 (1949).

[9] KÜSSNER, H. G.: Das zweidimensionale Problem der beliebig bewegten Tragfläche unter Berücksichtigung von Partialbewegungen der Flüssigkeit. Luftf.-Forschg. **17**, 355–363 (1940).

[10] SEARS, W. R.: Operational methods in the theory of airfoils in non-uniform motion. J. Franklin Inst. **230**, 95–111 (1940).

[11] V. D. VOOREN, A. I.: Generalization of the Theodorsen function to stable oscillations. J. Aeronaut. Sci. **19**, 209–211 (1952).

[12] WAGNER, H.: Über die Entstehung des dynamischen Auftriebs von Tragflügeln. Z. angew. Math. Mech. **5**, 17 (1925).

[13] TIMMAN, R.: The oscillating airfoil between two parallel walls. Appl. Sci. Res. A **3**, 31–57 (1951).

[14] MORSE and FESHBACH: Methods of theoretical physics. New York, 1953.

[15] SCHADE, TH.: Theorie der schwingenden kreisförmigen Tragfläche auf potentialtheoretischer Grundlage. Luftf.-Forschg. **17**, 387–400 (1940).

[16] VAN SPIEGEL, E.: Theory of the circular wing in steady incompressible flow. NLL Report F. 189, 1956.

[16a] VAN SPIEGEL: Boundary value problems in lifting surface theory. Thesis: Delft, 1959.

[17] KÜSSNER, H. G.: A general method for solving problems of the unsteady lifting surface theory in the subsonic range. J. Aeronaut. Sci. **21**, 17–27 (1954).

[18] HASKIND, M. D.: Oscillations of a wing in a subsonic gas flow. Prikl. Mat. i Mehk. (Moscow) **11**, 129–146 (1947). Also available as Translation No. A 9-T-22. Air Mat. Command and Brown Univ.

[19] REISSNER, E.: On the application of Mathieu functions in the theory of subsonic compressible flow past oscillating airfoils. NACA Techn. Note 2363, 1951.

[20] BILLINGTON, A. E.: Harmonic oscillations of an aerofoil in subsonic flow. ARL Report A 65, 1949.

[21] TIMMAN, R.: Beschouwingen over de luchtkrachten op trillende vliegtuigvleugels. Thesis, Technological University Delft 1946.

[22] TIMMAN, R., and A. I. V. D. VOOREN: Theory of the oscillating wing with aerodynamically balanced control surface in a two-dimensional subsonic compressible flow. NLL Report F. 54, 1949.

[23] MCLACHLAN, N. W.: Theory and application of Mathieu functions. Oxford: Clarendon Press 1947.

[24] MEIXNER, J., u. F. W. SCHÄFKE: Mathieusche Funktionen und Sphäroidfunktionen. Springer, 1953.

[25] BURGER, A. P.: On the asymptotic solution of wave propagation and oscillation problems. NLL Report F. 157, 1955.

[26] BAKER, B. B., and E. T. COPSON: The Mathematical theory of HUYGENS' principle. 2nd edition, Oxford: Clarendon Press 1939.

[27] WATKINS, CH. E., H. L. RUNYAN and D. S. WOOLSON: On the kernel function of the integral equation relating the lift and downwash distributions of oscillating finite wings in subsonic flow. NACA Techn. Note 3131, 1954.

[28] POSSIO, C.: L'azione aerodinamica sul profilo oscillante in un fluido compressibile a velocità iposonora. L'Aerotecnica **18**, 441–458 (1938).

[29] SCHADE, W.: The numerical solution of POSSIO's integral equation for an oscillating aerofoil in a two-dimensional subsonic stream. A.R.C. Report 9506, 1946.
[30] DIETZE, F.: Die Luftkräfte des harmonisch schwingenden Flügels im kompressibelen Medium bei Unterschallgeschwindigkeit. D.V.L. Forschungsber. 1733, 1943.
[31] FETTIS, H. E.: An approximate method for the calculation of non-stationary air forces at subsonic speeds. WADC Techn. Rep. 52—56, 1952.
[32] HEASLET, M., and H. LOMAX: Supersonic and transonic small perturbation theory in High Speed Aerodynamics and Jet Propulsion, Vol. VI: General theory of high speed aerodynamics (W. R. SEARS, ed.). Princeton: University Press 1954.
[33] GARDNER, C.: Time dependent linearized supersonic flow past planar wings. Comm. Pure Appl. Math. **3**, 33—38 (1950).
[34] GOODMAN, TH. R.: The quarter infinite wing oscillating at supersonic speeds. Quart. Appl. Math. **10**, 189—192 (1952).
[35] PINES, S., J. DUGUNDJI and J. NEURINGER: Aerodynamic flutter derivatives for a flexible wing with supersonic and subsonic edges. J. Aeronaut. Sci. **22**, 693—700 (1955).
[36] WATKINS, CH. E., and J. H. HERMAN: Velocity potential and air forces associated with a triangular wing in supersonic flow with subsonic leading edges, and deforming harmonically according to a general quadratic equation. NACA Techn. Note No. 3009, 1953.
[37] RAYLEIGH, Lord.: The theory of sound. Dover Publications 2nd ed. 1945.
[38] FLAX, A. H.: Reverse flow and variational theorems for lifting surfaces in non-stationary compressible flow. Cornell Aero. Lab. Cal-42, 1952.
[39] HEASLET, M. A., and J. R. SPREITER: Reciprocity relations in aerodynamics. NACA Techn. Note No. 2700, 1952.

Jets and Cavities.

By

DAVID GILBARG.

With 60 Figures.

1. Introduction. The problems of free streamline theory have long attracted both hydrodynamicists and mathematicians, the former because of the many applications to jet, cavity, and wake phenomena, and the latter because of the unusual and exciting mathematical features of this branch of potential theory. The different interests of these two groups have resulted in a parallel development of the field, emphasizing on the one hand particular solutions and technical applications, and on the other hand general theory and results of a more qualitative nature. These two trends are reflected in the following exposition in that Chap. II and V are concerned with special flows and numerical methods, and Chap. III und IV with the general theory. While this is not an altogether natural division from the physical point of view, we have adopted it as a convenient means of presenting the various methods and of tracing the development of the ideas. In devoting a separate chapter to qualitative features (Chap. III), we hope to direct attention to an elegant part of the general theory which should be useful to those with theoretical and applied interests alike.

The emphasis in this survey is on the basic models of cavity and jet flow and the mathematical methods that have been devised for their study. The discussion centers almost entirely on the exact theory, which is now in a relatively mature (but by no means final) state. From limitation of space and the essentially different nature of the material, a number of allied topics which might reasonably be included here have been touched on only perfunctorily or not at all. Among the most important are compressible fluid jets, and wakes in real fluids. For these and other subjects related to this work we refer to the literature. In particular, the reader is directed to BIRKHOFF and ZARANTONELLO [*1*][1] for topics not covered here, for references to numerous applications, and also for a more detailed account of many parts of this article.

In recent years a large portion of the literature on free surface phenomena has appeared in the form of laboratory reports with only limited circulation. We have attempted where possible to present these results, although a complete exposition is of course impossible.

I. Physical and mathematical foundations.

2. Experimental background. We review here some of the observations and experimental data concerning water entry, cavitation, wakes, and jet flow, that seem most pertinent in motivating and appraising the theory. For discussion of equipment, accuracy of measurement, etc., we refer the reader to the original sources.

[1] I am grateful to these authors for having provided me with a pre-publication copy of their valuable work.

α) Cavity flows[1]. Although one of the first applications of the concept of streamline of discontinuity was to the theory of wakes, the assumptions of the theory are better realized in the cavity phenomena associated with high speed underwater motion. These occur, for example, when a missile enters water from the atmosphere at moderate or high speed, its inertia creating behind it an air filled cavity; or, if the body is completely immersed and moves with such high speed that the local pressures descend to the vapor pressure of the liquid, the fluid "boils" and forms a vapor cavity behind the body. The practical importance of underwater motion at high speeds has led in recent years to extensive research on the properties of these cavity flows.

The phenomena accompanying water entry[2] include a surface impact phase, in which the fluid has not yet detached from the body; the splash; cavity formation and growth; subsurface and surface closure of the cavity; jet formation at each closure; and decay of the cavity. Of particular interest for the later discussion are the jets which form at the rear of the closed cavity and which can be observed (in high speed photographs) traveling down its length, often overtaking the missile and sometimes affecting its motion. The scaling of water entry phenomena is a problem complicated by several factors, including the influence of atmospheric density and, sometimes, surface tension and viscosity[3].

Water entry cavities, because of their non-stationary gravity-dependent character, are yet too complex for quantitative theoretical treatment. However, water tunnel experiments make it possible to observe steady-state cavities behind obstacles under conditions very close to those assumed in the theory.

Letting p_∞ denote the pressure in the undisturbed stream, and p_c the cavity pressure, we define the *cavitation parameter* σ by

$$\sigma = \frac{p_\infty - p_c}{\frac{1}{2}\varrho U^2}, \tag{2.1}$$

where U is the velocity of the stream and ϱ the fluid density. Assuming ideal fluid theory in the absence of gravity, one sees that σ has the significance of being the only non-dimensional similarity parameter associated with the flow. That is, two cavity flows under geometrically similar conditions will be geometrically and dynamically similar if and only if they have equal values of σ, and any non-dimensional quantity depending only on p_∞, p_c, ϱ and U (such as the drag coefficient of a fixed obstacle), is a function of σ alone. The free stream values p_∞ and U are experimentally assigned, while p_c is the (known) vapor pressure of the liquid[4]. The cavitation number is therefore an experimentally controlled variable and can be considered a datum in any flow problem.

A number of investigators[5] have carried out water tunnel studies of forces and cavity dimensions as functions of σ. The results of drag measurements on cones and discs in three different tunnels are shown in Fig. 1. At least for small

[1] For a survey of the experimental work on cavitation, see P. EISENBERG: David Taylor Model Basin Reports 712 (1950) and 842 (1953); also Forschungsh. Schiffstechn. **3**, 111–124 (1953); **4**, 155–168 (1953); **5**, 201–212 (1954). An early report on cavitation phenomena is contained in J. ACKERET, Techn. Mech. Thermodyn. **1** (1930).

[2] D. GILBARG and R. ANDERSON: J. Appl. Phys. **19**, 127–139 (1948). — A. WORTHINGTON: A study of splashes. New York: Longmans Green 1908. These contain numerous photographs of the phenomena involved.

[3] Cf. GILBARG and ANDERSON: Preceding footnote; also H. WAYLAND: Naval Ord. Test Station, NAVORD Rep. 978, 1947.

[4] In water tunnel experiments with air-filled cavities the cavity pressure must be determined by direct measurement.

[5] H. REICHARDT: Min. Aircraft Prod. Rep. and Transl. 766 (1946). — P. EISENBERG: Footnote 1. — M. PLESSET and B. PERRY: Mémoire sur la mécanique des fluides offerts à

collimator axes with respect to the sun is recorded by means of electrical contacts corresponding to certain directions and activating a system of special pens at each recorder.

The recording unit contains a power supply for the photomultiplier tubes, a tuned amplifier with a built in, 60 cycle rejecting circuit, a double pen BROWN recording potentiometer, an oscilloscope and a phasemeter, another BROWN potentiometer for recording the phase of 40 cycle components, amplifier and electronic phase shifter for the reference signal. The power supply provides a variable voltage, regulated to keep the output of the photomultiplier at a suitably chosen constant value. This arrangement was found to be necessary to prevent the shift of the working range from the linear characteristics and the fatigue effects, that were caused by the large variation of the intensity of the cathode illumination during the sweep from one horizon to the other. If $s(e)$ denotes the sensitivity of the photomultiplier as a function of the voltage e applied to its anode and dynods, then

$$s(e)\,C\left[I + Q\cos^2\frac{\delta}{2}\cos 2\chi\right] = s(e)\,C\,I = b = \text{constant} \qquad (7.2)$$

if δ or χ are chosen so that the last term in the bracket vanishes. According to (7.2) the voltage e is dependent on the intensity I and can be used as its measure. It is recorded by one pen of the BROWN potentiometer. Furthermore, with respect to (7.2) the amplitude of the 40 cycle component

$$A_{40} \sim s(e)\,C\,I\,P\sin^2\frac{\delta}{2} = b\,P\sin^2\frac{\delta}{2}$$

measures directly the degree of polarization, independently of the intensity, and is recorded by the second pen of the BROWN potentiometer.

After all instrumental constants are determined by a proper calibration, the intensity I, the degree of polarization P and the position of the plane of polarization χ can be computed from the continuous recording of the voltage e, of the amplitude A_{40} and of the phase angle of the 40 cycle component, at any zenith distance along the sun vertical or along any other vertical of a given azimuth. The neutral points can be easily recognized by a sharp minimum in the recording of P, after a proper increase of the gain of the amplifier, associated with a sudden change of the plane of polarization by 90° (either visible on the scope or in the recording of the phase). Their positions can be measured with an accuracy of $\pm 0.5°$. All neutral points, except when they are too close to the sun ($\leq 5°$), can be measured with the same ease; the BREWSTER point, because of its larger distance from the sun, can be recognized much oftener than the BABINET point. In this way it is possible to measure the position of the neutral points continuously from sunrise to sunset; for a few filters it is even possible to extend measurements into the twilight (up to 5 or 6° of sun depression). With the speed of the sweep of 24° per minute, it is possible to obtain one position of the neutral point every one half minute if only the neutral points are measured and the distribution of the degree of polarization along the entire sun vertical every 8 min. With the use of an automatic switch, switching the two channels alternately it is possible to obtain simultaneous measurements at two different wavelengths.

8. Measured magnitude and distribution of skylight polarization. Before the photoelectric methods were introduced, the measurements of skylight polarization had been made visually and almost entirely without any filters defining

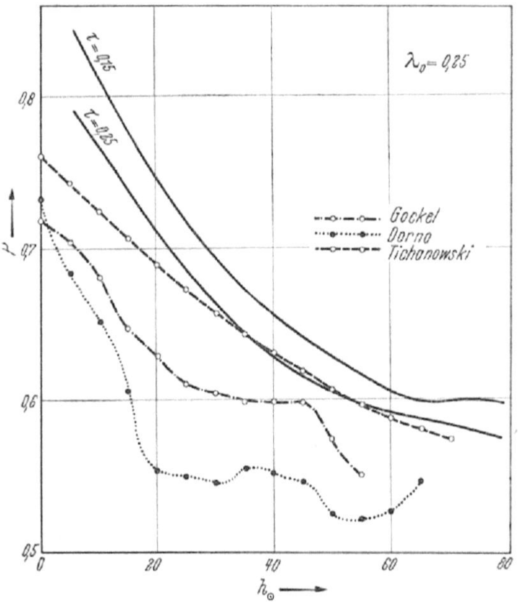

Fig. 11. Degree of polarization 90° from the sun in the sun vertical as measured by several authors, compared with the theoretical values for a molecular atmosphere.

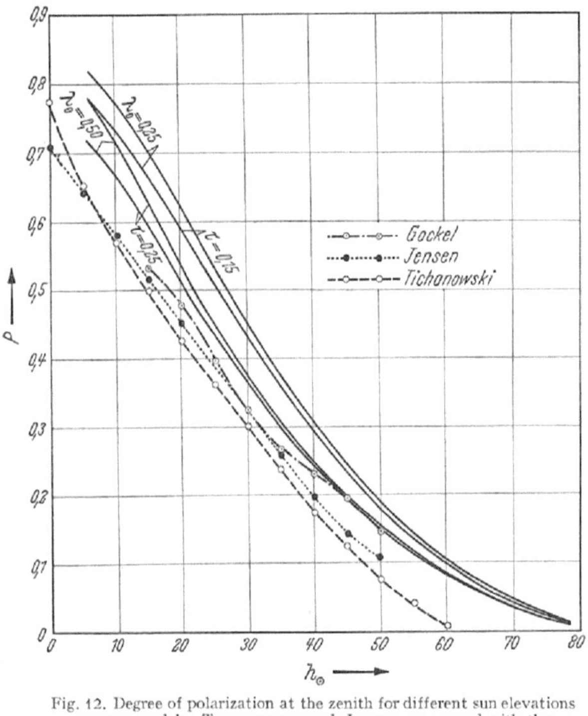

Fig. 12. Degree of polarization at the zenith for different sun elevations as measured by TICHANOWSKI and JENSEN, compared with the theoretical values for a molecular atmosphere.

a narrow spectral region, in so called "*white*" *light*. Obviously the results of such measurements cannot be compared with the theoretical values, which are valid for monochromatic light only, since the measured quantities are of the form

$$I_i = \int_0^\infty s(\lambda)\, I_i(\lambda)\, d\lambda \quad (i = l, r) \qquad (8.1)$$

where $I_i(\lambda)$ are the monochromatic intensities and $s(\lambda)$ the sensitivity of the observer's eye, a strongly varying quantity, dependent on the intensity of the measured light.

In spite of this difficulty the degree of polarization measured at the point of maximum polarization (90° from the sun in the sun vertical) and at the zenith shows the same general character as the theoretical values for the optical thickness corresponding to the dominant wavelength of the color of the sky, as can be seen in Fig. 11 and 12. The slight irregularities in the curve of measured values can be explained by the fact that they represent the mean of measurements obtained under quite different atmospheric conditions. Similarly the measured positions of the neutral points (BABINET and ARAGO) follow closely the theoretical curves, except for very low sun elevations, where the theory, because of the assumption of a plane parallel atmosphere, fails to represent the actual conditions.

The distribution of the polarization and of the position of the plane of polarization has been systematically studied at DAVOS by DORNO[1], as part of his excellent work on skylight polarization. His measurements of the position

[1] C. DORNO: Abh. Preuß. Meteor. Inst. Berlin **5** (1917). — Meteor. Z. **34**, 153, 246 (1917).

of the neutral lines, shown in Fig. 13, agree extremely well with the theory, reproduced in Fig. 8. The only persistent deviation can be found on the *SE* horizon, and is probably due to the effect of the terrain configuration.

The ground reflection has been found to affect the measured values in the same way as the theory predicts, namely in lowering the values of the degree of polarization 90° from the sun in the presence of snow cover on the ground. The position of the ARAGO point was found unaffected by the increased albedo of the ground when covered with snow[1].

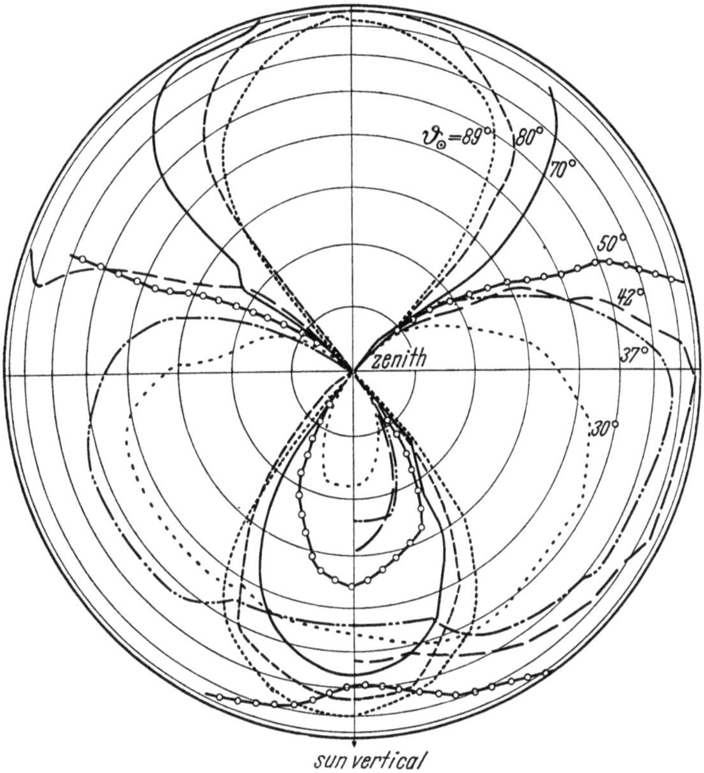

Fig. 13. Neutral lines for different sun zenith distances as measured by DORNO.

When the degree of polarization is measured by a comparison of the intensities normal and parallel to the plane of polarization, as is done in the MARTENS polarimeter, so called *polychroism* may be noticed, namely a difference in the color of the two halves of the field visible in the eye piece. This phenomenon, noticed as early as 1873 by PICKERING[2], was studied in detail by TICHANOWSKI[3]. The color of the field illuminated by the normal component to the polarization plane showed a more reddish tone when the turbidity of the atmosphere was low. With increasing turbidity the difference disappeared, and changed into the reverse conditions, the parallel component was richer in the red tone than the normal component, when the turbidity was very large. This effect can be easily explained[4] by the variation of the degree of polarization

[1] H. NEUBERGER: Bull. Amer. Meteor. Soc. **31**, 119 (1950).
[2] E. C. PICKERING: Proc. Amer. Acad. Boston **9**, 20 (1873).
[3] J. J. TICHANOWSKI: Meteor. Z. **41**, 173 (1924).
[4] Z. SEKERA: Čas. Pěst. Math. **67**, 278 (1938).

with wavelength, the so called *dispersion of the skylight polarization*. The polychroism disappears when the degree of polarization is the same for all wavelengths, the parallel component is richer in the red (blue), if the polarization in red is larger than the polarization in blue (or vice versa).

The variation of the dispersion of polarization with the turbidity, as discussed in the next section, corresponds exactly to the observations made by TICHANOWSKI. Already the first measurements of skylight polarization revealed quite a large daily, local and seasonal variation of much higher order of magnitude than can be expected in a molecular atmosphere, where the only variation of the polarization can be caused by a change of optical thickness which is almost independent of the weather and local conditions, and its variations never exceed 1%. The observed variations have been attributed to the strongly varying content of the aerosol particles. BLICKHAN[1] measured simultaneously the *turbidity factor*[2] T_k for shorter wavelengths and the maximum polarization (90° from the sun) P and found a very close correlation between these two quantities, leading to the empirical relationship $P = A/(B + T_k)$, where A, B are constant. Similar investigations were carried out by WOERNER[3], who compared the degree of polarization measured at the zenith with the simultaneous measurement of the turbidity factor T. For the turbidity factor T in the interval 2.0 to 2.5, the mean value of measured P was about 14% higher than the normal obtained by JENSEN[4] from a large series of measurements; for $T \geqq 5$, the mean was lower by 31.7%.

9. Dispersion of the skylight polarization. Measurements of skylight polarization in narrow spectral ranges were carried out as far back as 1884, when CORNU found quite a difference between the degree of polarization in the red and blue part of the spectrum. The results of measurements made by several authors do not agree with each other as well as the results of the measurements in "white" light, some of them are even quite contradictory They show much greater variability with location, weather, etc., with elements responsible for different degrees of the turbidity of the atmosphere. Since the turbidity during these measurements was neither actually measured (except in the latest studies) nor estimated from other phenomena (like visibility, color of the sky), it is very difficult to compare the results obtained by different authors. As an illustration, in Fig. 14 the values of the degree of polarization as measured by different authors[5] 90° from the sun in the sun vertical for different sun elevations are compared with the theoretical values of the molecular atmosphere, for three

---

[1] F. BLICKHAN: Gerlands Beitr. Geophys. **42**, 208 (1934).

[2] If the mean optical thickness for the RAYLEIGH atmosphere $\bar{\tau}$ is defined by the equation

$$F_0 e^{-\bar{\tau}/\mu_0} = \int_0^\infty F_{0\lambda} e^{-\tau/\mu_0} \, d\lambda$$

where F_0 is the solar constant and $F_{0\lambda}$ the monochromatic flux of extraterrestrial sun radiation, then the *turbidity factor*, as introduced by F. LINKE is given by

$$T = \frac{\mu_0}{\bar{\tau}} \log \frac{F_0}{F}$$

where F is the measured flux of sun radiation. If $F_k = F - F_R$ where F_R is the flux of sun radiation measured through a (red) filter transmitting at $\lambda > 0.625\,\mu$, is substituted above instead of F, the turbidity for shorter wavelengths T_k is obtained. See also F. MÖLLER's discussion on p. 177.

[3] H. WOERNER: Z. Meteor. **3**, 166 (1949).

[4] These normal values are plotted in Fig. 12.

[5] J. J. TICHANOWSKI: Meteor. Z. **43**, 288 (1926). — N. N. KALITIN: Meteor. Z. **43**, 132 (1926). — G. DIETZE: Z. Meteor. **5**, 86 (1951).

different values of the albedo a. The wide scatter of values can be attributed to the great variety of conditions, especially as to the turbidity.

Nevertheless, once it is realized that the degree of polarization itself can be considered as a measure of turbidity (cf. the high correlation found by BLICKHAN[1]), then it is possible to compare the difference between the degree of polarization in red and blue $(P_r - P_b)$ with the degree of polarization in red (P_r), as shown in Fig. 15. Although the points are still considerably scattered, the correlation of the difference $(P_r - P_b)$ with P_r is quite evident (correlation coefficient $r = 0.80$). For very low turbidity (TICHANOWSKI's values from the

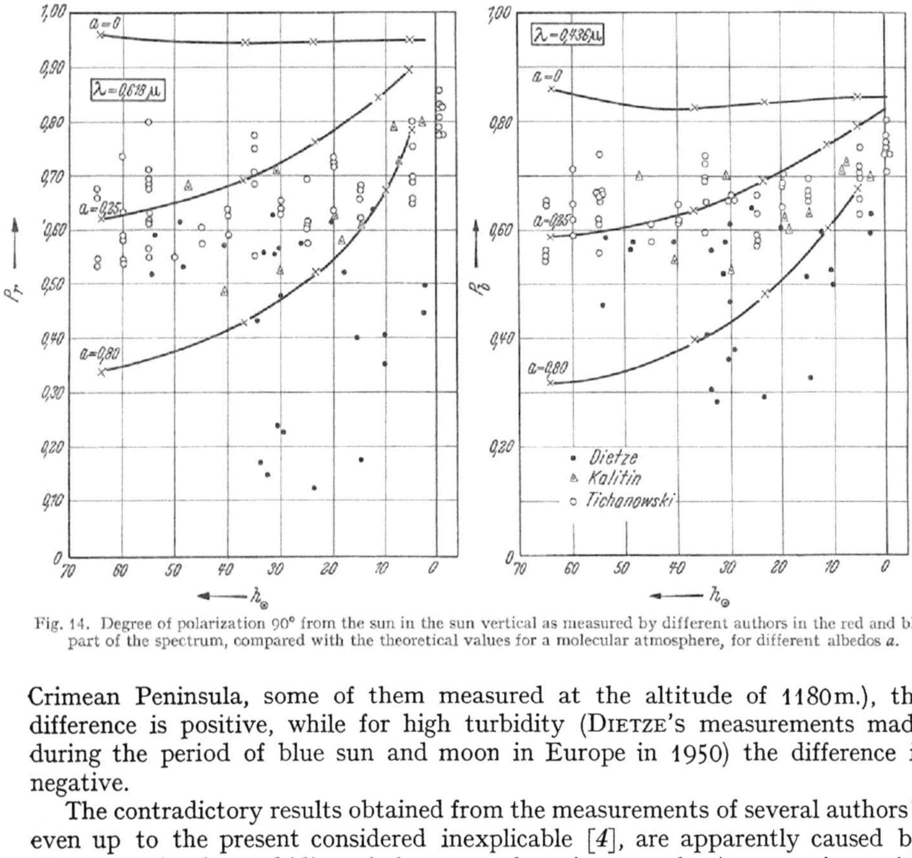

Fig. 14. Degree of polarization 90° from the sun in the sun vertical as measured by different authors in the red and blue part of the spectrum, compared with the theoretical values for a molecular atmosphere, for different albedos a.

Crimean Peninsula, some of them measured at the altitude of 1180m.), the difference is positive, while for high turbidity (DIETZE's measurements made during the period of blue sun and moon in Europe in 1950) the difference is negative.

The contradictory results obtained from the measurements of several authors[2], even up to the present considered inexplicable [4], are apparently caused by differences in the turbidity of the atmosphere in general. A comparison with the theoretical values for the molecular atmosphere suggests that the change of the sign of $(P_r - P_b)$ cannot be explained by RAYLEIGH scattering and can, therefore be attributed to the presence of particles of a size comparable to or larger than the wavelength. In this case the degree of polarization, as well as all the intensities and STOKES' parameters, depend on the size distribution, the total amount and the refractive index of the aerosol particles, all three factors of great variability in the atmosphere, causing the scatter of the points in the diagram.

[1] L. c. p. 316.

[2] C. DORNO: Meteor. Z. **36**, 109, 181 (1919). — A. GOCKEL: Ann. Phys. (4) **62**, 283 (1920). N. N. KALITIN: Meteor. Z. **43**, 132 (1926). — N. PILTSCHIKOFF: C. R. Acad. Sci. Paris **115**, 555 (1892). — J. J. TICHANOWSKI: Meteor. Z. **43**, 288 (1926).

Similar variations can be noticed in the position of the neutral points. As an example, in Fig. 16, the positions of the ARAGO point (for the sun elevation

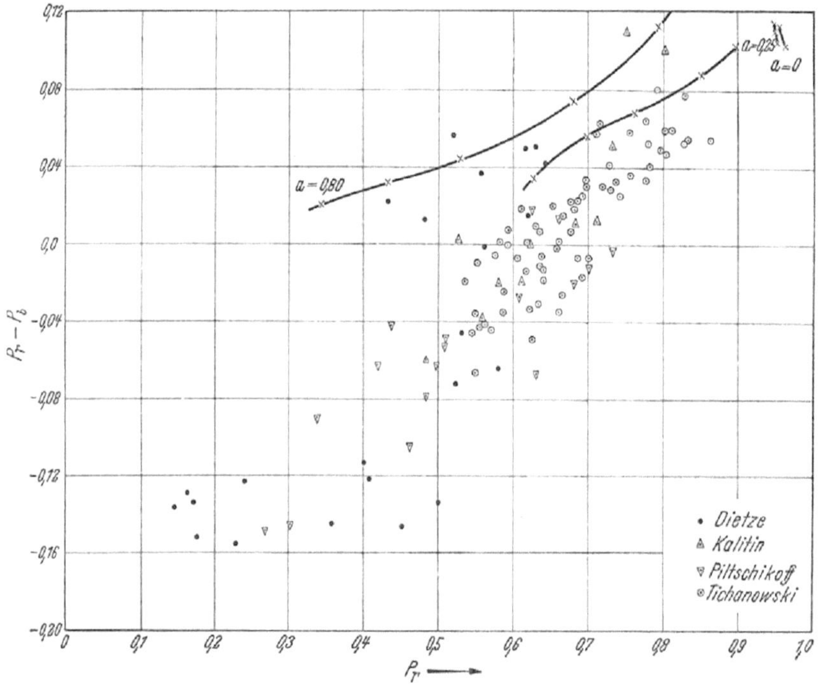

Fig. 15. Measured dispersion of the skylight polarization ($P_r - P_b$) as function of the degree of polarization in the red (P_r) (at 90° from the sun in the sun vertical) compared with the corresponding values for a molecular atmosphere.

10°5), for two different spectral regions are shown as a function of the turbidity, measured by the decrease of solar radiation[1]. Again, as in the previous case, the variation with the turbidity for longer wavelengths is more pronounced.

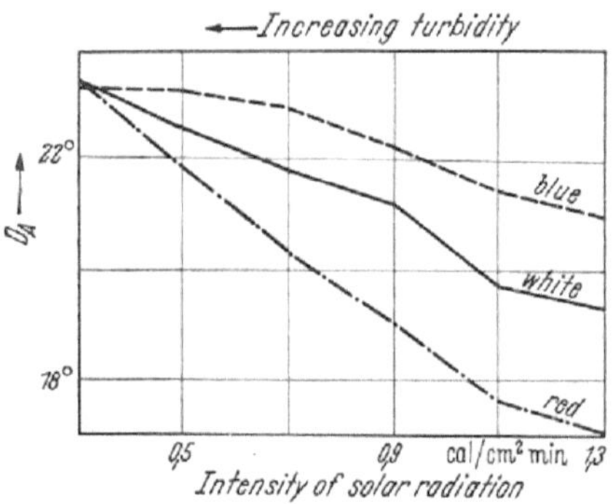

Fig. 16. Variation of the position of the ARAGO neutral point in the red and blue part of the spectrum with the turbidity (after NEUBERGER).

The recent measurements, made by the photoelectric polarimeter [*14h*] at different locations and under different weather conditions, give more quantitative data about the dependance of the dispersion of skylight polarization on the variability of the aerosol particles in the atmosphere. These measurements are superior to all those done previously because of a greater continuity with respect to the diurnal variation and to the spectral distribution. The diurnal variation due to the changing sun elevation was eliminated for the first time by comparison with the results of the theory for molecular scattering, corresponding to the local optical thickness for the central wavelength of the narrow spectral range in which the measurements were made (with the assumption of albedo $\lambda_0 = 0$). Hence the computed deviations from the theory (given by the difference: observed

[1] H. NEUBERGER: l. c. p. 309, 315.

values minus theoretical) include the effect of aerosol particles of the size comparable to or larger than the wavelength, as well as of the ground reflection. (In the values for small sun elevations another effect, namely, the effect of the sphericity of the earth, also appears.)

The magnitude of the theoretical and measured positions of the neutral points can be estimated from Fig. 17, showing the results of measurements made on November 12, 1953, on Table Mountain, California at 2280 m. above sea level[1].

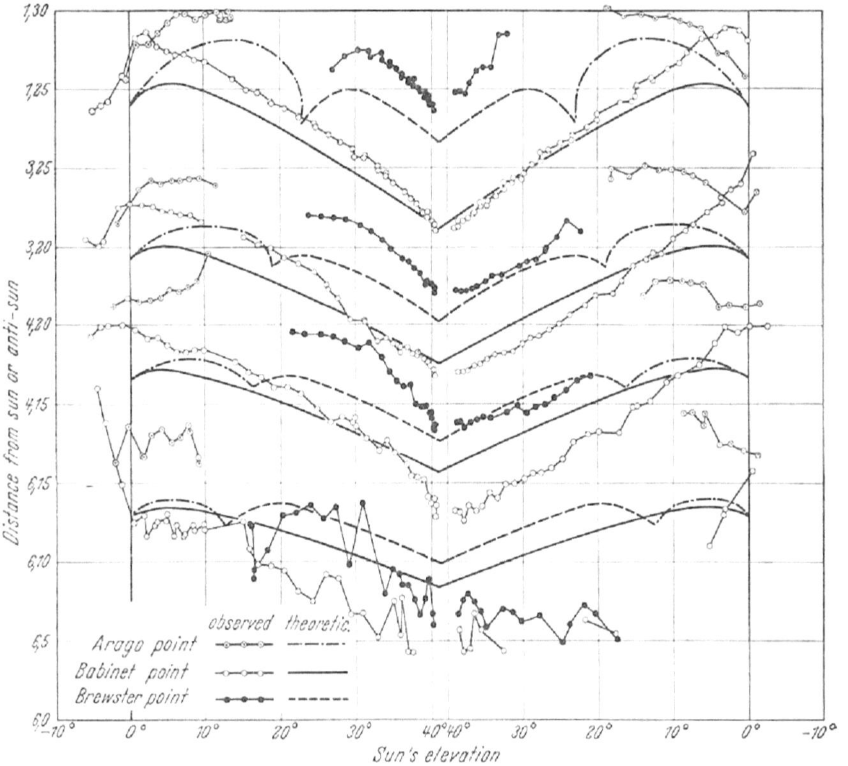

Fig. 17. Theoretical and measured position of the neutral points at Table Mountain, California (Nov. 12, 1953) for different wavelengths.

This diagram shows clearly the typical trend in the deviations, namely that the magnitude of the deviations, short periodical fluctuations and the asymmetry in the diurnal variations increase with the wavelength. Fig. 18a and b shows the results of measurements made at Los Angeles, California (UCLA campus) during a few days (in September, 1953) with similar conditions (intense haze, varying visibility below 15 miles). The deviations from the theory as derived from the observations at Los Angeles are in general of greater magnitude and the asymmetry with respect to noon is more pronounced in the curves corresponding to the Babinet and Brewster points. This asymmetry is also visible in the respective positions of the curves for the Brewster and Arago points. These two points should appear at the horizon at equal distances from the sun and from the antisolar point. However, the positive deviation for the Arago and the

[1] The figures to the left of the numbers indicating distances from the sun in Fig. 17 refer to the filters used, and thus to the wavelength (Filter 1: 365 mμ, 3: 460 mμ, 4: 515 mμ, 6: 625 mμ).

negative deviation for the BREWSTER point indicate that the ARAGO point is shifted further away from the antisolar point, while the BREWSTER point is shifted toward the sun This shift of the ARAGO point is almost the same in the morning and in the evening, while the shift of the BREWSTER point in the evening is more than twice as big as in the morning (for $\lambda = 625$ mμ).

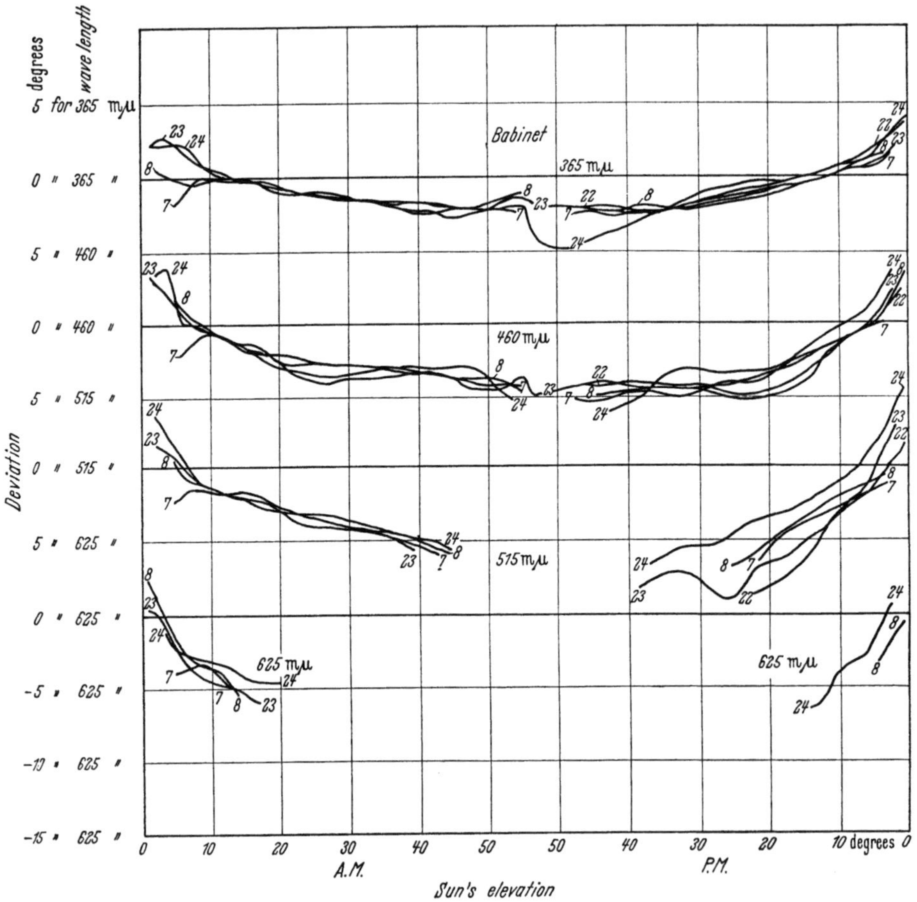

Fig. 18a. Deviations of the observed positions of the BABINET neutral points from the theoretical ones, for a series of hazy days at Los Angeles, California (Sept. 1953).

Quite interesting are the *day to day variations* in these curves. The deviations for the BABINET point in the ultraviolet ($\lambda = 365$ mμ) have almost the same magnitude so that the curves corresponding to different days are very close together. For longer wavelengths the deviations show a much greater variability and the shift of the BABINET point towards the sun is so large that the BABINET point is no more observable because of the shield[1]. On some occasions, however, the BABINET point disappeared far enough from the shield so that a flat minimum in the degree of polarization could be observed, but no turning of the plane of polarization by 90°.

On the other hand the day to day variation in the deviations of the BREWSTER point are much greater for shorter wavelengths. The unusual variations

[1] Cf. the description of the instrument on p. 312.

observed on Sept. 24 indicate that some part of the variations in the deviation of the BREWSTER point is due to a change in ground reflection, since during this day a stratus deck over the ocean (5 miles distant) became gradually thicker while rising until the late afternoon when it started to drift over the place of observation. A similar change in the deviation is apparent in the curves for longer

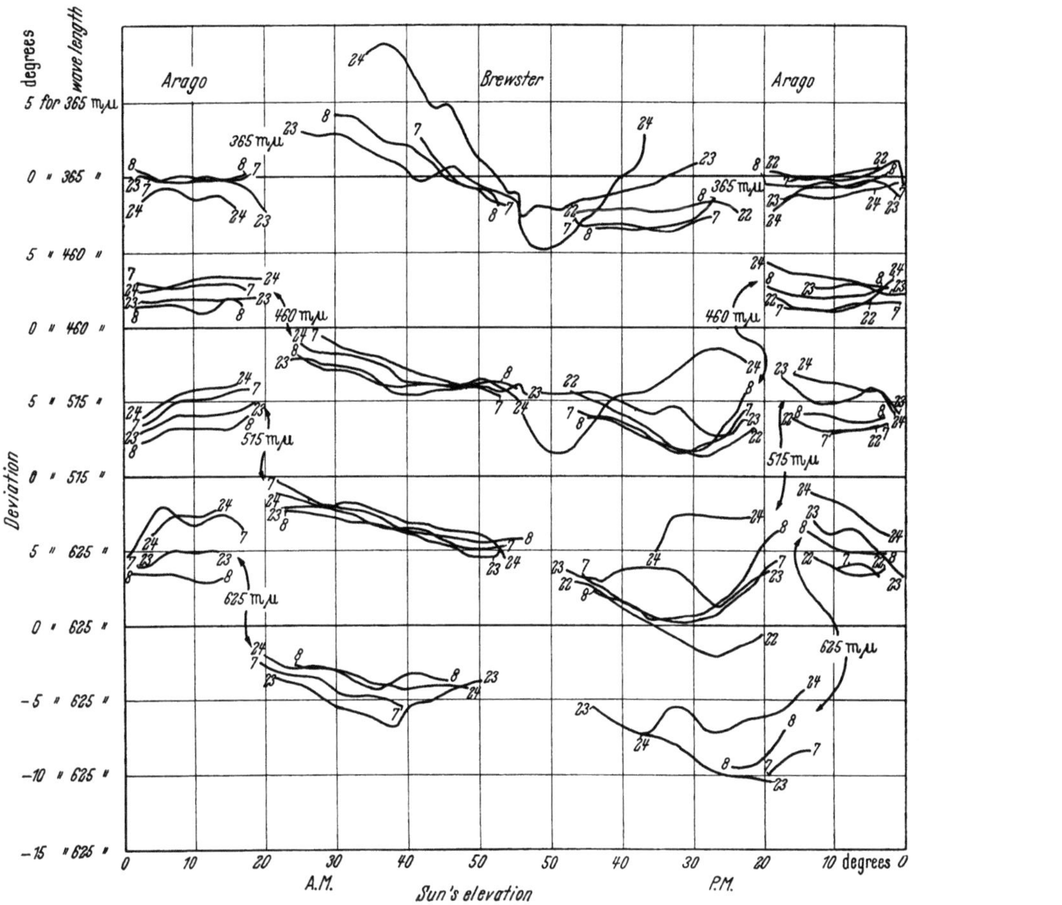

Fig. 18b. Deviations of the observed positions of the BREWSTER and ARAGO neutral points from the theoretical ones, for a series of hazy days at Los Angeles, California (Sept. 1953). (Numbers next to the lines refer to the date in September 1953. For example *24* refers to Sept. 24, 1953.)

wavelengths, but less pronounced. The effect of ground reflection on the position of the BREWSTER neutral point seems to decrease with the wavelength as it does in a molecular atmosphere. For this reason measurements of the position of the BREWSTER point in the shorter wavelengths are quite important and can be used to distinguish the effects due to the ground reflection from those due to the large particle scattering.

The results of the few available measurements of the degree of polarization made with the photoelectric polarimeter, agree well with the results of previous visual measurements and show a tendency in the deviation from the theory similar to that found in the position of the neutral points. The negative deviations from the theoretical values again increase in magnitude with the wavelength. This increase of negative deviations with the wavelength compensates for the increase of the theoretical degree of polarization so that the observed maximum degree of polarization is largest in the central part of the visible range (green), and the degree of polarization in the red P_r occasionally falls below the values of P_b[1].

Since the degree of polarization along the sun vertical has been measured continuously for the first time, the position of the maximum in the degree of polarization has been easily determined for different sun elevations and different wavelengths. For the shorter wavelengths the maximum was found to be shifted from the point 90° from the sun by the amount predicted by the theory for molecular scattering (one or two degrees towards the sun). For longer wavelengths the maximum degree of polarization is shifted in the opposite direction, i.e., from 90° to 93 to 95°, with an apparent daily variation.

The measured deviations from the theory for the molecular atmosphere give an altogether consistent picture of the effect of large aerosol particles on the skylight polarization, in agreement with the theory of scattering in a turbid atmosphere. The effect of large particles is more pronounced for the wavelengths at which the components of $\boldsymbol{I}_L$ in eq. (6.5) are comparable to or larger than $\boldsymbol{I}_R$. Since the magnitude of the intensities in $\boldsymbol{I}_R$ are in the first approximation proportional to the optical thickness, they decrease with λ^{-4} for RAYLEIGH scattering. The optical thickness corresponding to the atmospheric aerosol, however, decreases with λ^{-1} for the continental type (cf. p. 164) and is almost constant or even increases with λ in the coastal type. For the short wavelengths the optical thickness of the molecular atmosphere, τ_R, is much larger than τ_L, the optical thickness of the atmospheric aerosol, so that RAYLEIGH scattering is predominant and the observed values are close to the theoretical ones. With increasing wave length, $\boldsymbol{I}_R$ decreases more rapidly than $\boldsymbol{I}_L$ and the effect of the atmospheric aerosol is more pronounced.

Furthermore, because of the part $\boldsymbol{P}_R$ in the expression for the phase matrix of the atmospheric aerosol, the observed polarization should correspond to the theoretical values for a molecular atmosphere of the larger thickness $\tau = \tau_R + \tau_L$. The distance of the neutral points from the sun or antisolar point should be larger than distances in the molecular atmosphere, as is observed for the ARAGO point (positive deviations). The asymmetric part $\boldsymbol{P}_L$, because it includes the very large forward scattering, will change the distribution of polarization mostly in the vicinity of the sun, where it introduces an additional neutral radiation in the forward directions. The intensity of the scattered light by large particles increases by several orders of magnitude in the directions close to the forward scattering and therefore the effect of large particles overcompensates the increase due to the increased optical thickness from τ_R to $\tau = \tau_R + \tau_L$. The resulting deviations from the molecular theory change from positive to negative values, in agreement with the measured deviations for the BABINET and BREWSTER points, and with the measured position of the maximum degree of polarization in the sun vertical.

In order to support this hypothetical explanation of the observed effect of the aerosol particles a quantitative analysis of this effect, outlined in Sect 6, is obviously needed. From the discussion above it is also evident that for a proper evaluation of the effect of the aerosol particles, and thus for a correct measure of the turbidity of the atmosphere, the measurement of the position of the neutral points, and of the distribution of the degree of

[1] Similar characteristics of the dispersion of the polarization at the zenith was found by R. DUMOND, C. R. Acad. Sci. Paris **238** (26), 2512 (1954).

polarization, at least along the sun vertical, is necessary. Any conclusion made from partial measurements, say of the ARAGO point only, may lead to an erroneous result.

10. Anomalies in the observed skylight polarization. Polarization of the skylight during the twilight. As can be seen from the discussions above the observed polarization of skylight follows closely the pattern predicted by the theory of polarization due to the RAYLEIGH scattering in a molecular atmosphere. The observed deviations from the theoretical values show a systematic trend indicating that they are primarily caused by aerosol particles of larger sizes. These deviations, however, are usually of such a magnitude that the characteristic feature of the polarization in a molecular atmosphere is not substantially changed.

Much larger deviations from the theory can be noticed if the *ground reflection* is governed by a law quite different from LAMBERT'S law. Such anomalies have been observed for example in the vicinity of large *water surfaces* where at least a considerable portion of the light is reflected according to FRESNEL'S[1] law. The maximum distance of the ARAGO neutral point from the antisolar point has been observed to occur at much lower sun elevations over or close to the sea than over continents (cf. [4] p. 594). Furthermore new neutral points were occasionally observed in the vicinity of large water surfaces (sea, lakes) either in the sun vertical below the BABINET point and below the ARAGO point[2], or on both sides of the sun at the same elevation[3] [3]. The observations made by BREWSTER, however, indicate that a haze layer over the water surface may be responsible for such anomalies rather than the specular reflection of the direct sun radiation. The quantitative evaluation of such effects will be possible only after the theoretical studies of the effect of different types of ground reflections are complete and more observational evidence is available.

The largest *anomalies* were observed after such violent volcanic eruptions as those of Krakatoa, Mont Pelé, and Katmai during the periods 1883 to 1885, 1902 to 1903, 1912 to 1914. The effect was noticeable in the decrease of the maximum polarization; in 1884 CORNU[4] observed a decrease from 0.75 to 0.48, with the degree of polarization much smaller in the red than in the blue. DORNO'S mean value of the zenith polarization was 0.557 for 1913, 0.739 for 1915. A much larger effect was observed in the position of the neutral points. In Fig. 19 the mean values of the position of the BABINET and ARAGO points before and after the volcanic eruption of Katmai are compared. In addition to the regular neutral points (BABINET and ARAGO) CORNU[4] in 1884 observed four additional neutral points, symmetrically situated on both sides of the sun and the antisolar point, at the same elevation as the sun or the antisolar point.

A quantitative explanation of these anomalies is definitely one of the tasks of the theory of radiative transfer in a turbid atmosphere.

Another phenomenon which the theory of molecular scattering in a plane parallel atmosphere is unable to explain involves the variations of the skylight polarization during the twilight. As can be seen in Fig. 19, a minimum in the distance of the neutral points from the sun or antisolar point appears for small negative sun elevations and is followed by a steep increase. The minimum shifts towards larger sun depressions with increasing turbidity. In the degree of

[1] E. O. HULBURT: J. Opt. Soc. Amer. **24**, 35 (1934).
[2] H. NEUBERGER: Arch. dtsch. Seewarte **56** (1936).
[3] J. L. SORET: C. R. Acad. Sci. Paris **107**, 867 (1888).
[4] A. CORNU: C. R. Acad. Sci. Paris **99**, 488 (1884).

polarization at the zenith, the maximum does not appear when the sun is at the horizon, as it should, but is also shifted towards small sun depressions The increase of the degree of polarization P during the twilight is larger the lower

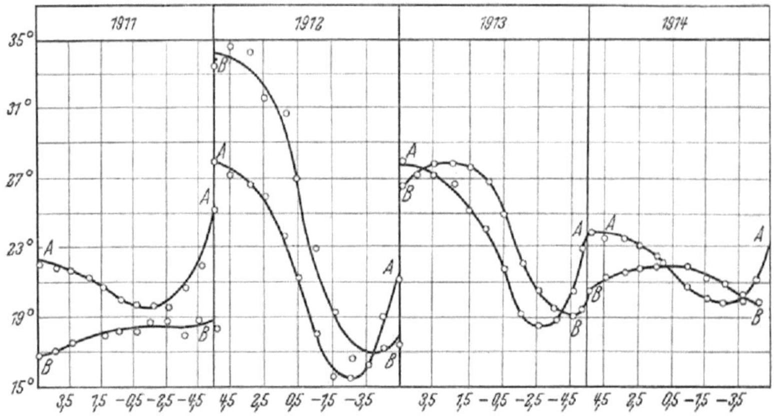

Fig. 19. Positions of BABINET and ARAGO neutral points during the period of volcanic eruption at Katmai 1912 (after JENSEN [3]). Numbers along the horizontal axis refer to the sun's elevation, in degrees.

the value of P with the sun at the horizon. There is no doubt that this increase of the degree of polarization as the earth shadow rises is due to the elimination of the lower levels of the atmosphere with the largest concentration of aerosol particles, which are responsible for the decrease of the

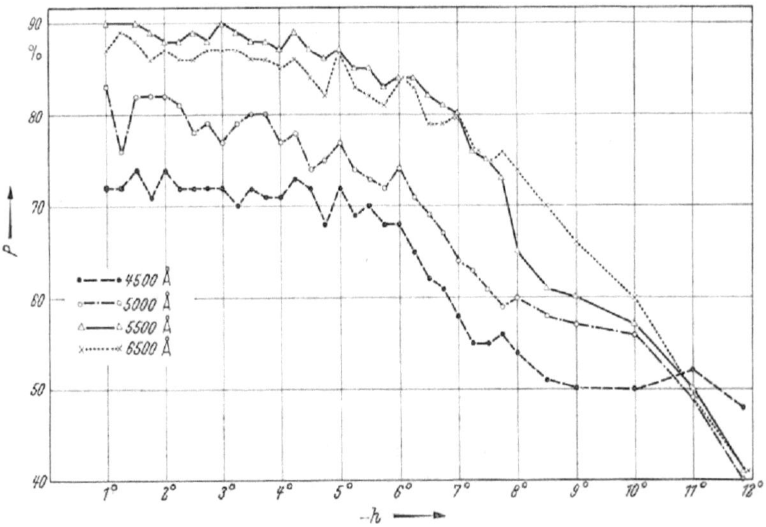

Fig. 20. Degree of polarization at the zenith for negative sun elevations (after ROBLEY).

degree of polarization. However, the observed variation of the polarization at the zenith for larger sun depressions, shown in Fig. 20, cannot be explained by such simple reasoning[1].

[1] Similar results were obtained from the measurements of the zenithal polarization at Mt. Abu, J. V. DAVE and K. R. RAMANATHAN, Proc. Ind. Acad. Sci. **43**, 67 (1956).

The decrease of the intensity and polarization at the zenith during the twilight has been studied intensively in the belief that valuable information about the upper levels of the atmosphere could be obtained from such measurements. Unfortunately the lower levels of the atmosphere are also illuminated, not by the direct sun light, but by the skylight from the illuminated part of the atmosphere. Since the intensity of the radiation scattered once only decreases rapidly with the height, it can be expected that the intensity of the skylight from the zenith due to the illumination by the skylight from the sunlit part (i.e. the effect of secondary and multiple scattering) may become even larger than that due to the primary scattering of the direct sun radiation According to HULBURT[1], this may be expected to occur when the sun depressions exceed 8°. His estimate is based on direct measurements of the intensity of skylight from the horizon close to the sun vertical, and thus includes also the effect of aerosol particles at lower levels. For this reason HULBURT's estimate of the ratio of the intensity due to the multiple scattering in lower levels to the intensity due to the primary scattering of the sunlit upper levels is much higher than the estimate, made by LINK[2] for the molecular atmosphere, and more recent results, obtained by ROBLEY[3], who derived the values due to the multiple scattering from the equations of radiative transfer for RAYLEIGH scattering by a simplified treatment of the illumination of the atmosphere during the twilight.

The exact evaluation of these anomalies will be possible only after the solution of the problem of radiative transfer is available for a spherical atmosphere. It is expected that it will be possible to incorporate in the treatment the effect of aerosol particles and from a comparison of the measurements with the theory it will be possible to derive the distribution of aerosol particles with height. However, it is doubtful whether any information about the upper, directly illuminated levels of the earth's atmosphere can be obtained by this approach to the problem.

D. Practical application of the investigation of skylight polarization.

11. The polarization of skylight and its distribution over the sky offer a few possibilities for use in practical problems. In the region of maximum polarization, the polarization of skylight can be used to eliminate skylight as the background thus increasing the visibility of faint objects or week light sources (such as stars[4]) during the day or during the twilight. The small variability of the position of the plane of polarization has been used as a means of orientation over the sky, i.e. air navigation, when the sun is obstructed by an object or by a cloud[5]. In the latter case however, the method may not give the correct result since individual clouds, asymmetrically distributed over the sky, can introduce a large deviation in the plane of polarization from its theoretical position.

---

[1] E. O. HULBURT: J. Opt. Soc. Amer. **28**, 227 (1938).

[2] F. LINK: Meteor. Z. **59**, 7 (1942).

[3] R. ROBLEY: Ann. Géophys. **8**, 1 (1952).

[4] R. TOUSEY u. M. J. KOOMEN: J. Opt. Soc. Amer. **43**, 177 (1953).

[5] While man in general needs an instrument to observe the polarization of the blue sky-light, many animals (insects) appear to be capable of using the plane of polarization as a means of orientation. See K. v. FRISCH: Experientia (Basel) **5**, 142 (1949); **6**, 210 (1950). Naturwiss. **38**, 165 (1951). — M. KERZ: Experientia (Basel) **6**, 429 (1950). — H. AUTRUM u. H. STUMPF: Z. Naturforsch **5**b, 116 (1930).

Another possible application may be based on the great sensitivity of skylight polarization to the varying content of aerosol particles. Since the direct measurement of the amount of smaller particles (for example, of a diameter less than 1 μ) still involves several major difficulties, the question arises whether the measurement of skylight polarization might be used for the determination of the size distribution of aerosol particles. From the preceding discussion it will be apparent that the determination of the particle size distribution from the deviations of the observed skylight polarization from the theoretical values for a molecular atmosphere is not a simple matter. It is not difficult to obtain a qualitative picture of the main characteristics of scattering by the atmospheric aerosol from an appropriate analysis of the deviations in several polarization parameters for different wavelengths, as indicated in Sect. 4. The greatest difficulties appear when a quantitative determination is to be made. The theory of the skylight polarization in a turbid atmosphere is so complex that a direct computation of the size distribution [like the quantity $n(a, m)$ in eq. (6.1) and (6.3)] is practically impossible. The determination of the size distribution can be done only by an indirect method by computing the polarization parameters for a given model or type of the size distribution of aerosol particles.

An attempt at a simplified treatment of this problem has been made by DIETZE[1]. The atmosphere is considered to be composed of two layers. In the lower turbid layer the scattering and the polarization are governed by empirical laws, derived from measurements from a searchlight beam and with an artificial aerosol. In the upper layer the scattering is assumed to follow the RAYLEIGH theory. From a consideration of primary scattering only the degree of polarization is computed as a function of different constants and parameters, defining the scattering and the turbidity in the lower layer and the optical thickness of the upper layer. The computation shows a great variability in the degree of polarization with varying constants and parameters. The actual measurement (only the degree of polarization 90° from the sun is used) on a particular day gives a curve in the range of the computed curves and similar to one of them, but the determination of the parameters defining the turbidity by interpolation is a little doubtful since the very important effect of the secondary and multiple scattering is not included.

Since the multiple scattering ought to be included, the computation of the polarization parameters for different models of size distribution should be done by the method outlined in Sect. 6. It may very well be that the same type of variation in a given polarization parameter with sun elevation or with wavelength may correspond to quite different models of the size distribution of aerosol particles. To avoid such ambiguities, the greatest possible number of polarization parameters should be used and the comparison with observation should be made in several wavelengths. In addition, the polarization measurements should be made simultaneously with measurements of the atmospheric transmission of solar radiation, of the distribution of the intensity of skylight in the vicinity of the sun (aureole), etc., since all these quantities also depend on the size distribution of aerosol particles although in a different way. The question of which of these quantities together with the skylight polarization gives the best result, being most sensitive to small variations in the type of the aerosol distribution, can be decided only after the theory of scattering and of the skylight polarization in a turbid atmosphere is available.

---

[1] G. DIETZE: Abh. Met. Dienstes Dtsch. Dem. Rep. **4**, 44 (1954).

Bibliography.

A detailed discussion of early study of skylight polarization and of its development with an extensive bibliography can be found in the following references:

[1] BUSCH, F., u. C. JENSEN: Tatsachen und Theorien der atmosphärischen Polarisation. Jb. hamburg. wiss. Anst. **28** (1910). Berlin: Duemmler 1912.

[2] PERTNER, J. M., u. F. M. EXNER: Meteorologische Optik, 2. Aufl., S. 644—725. Wien u. Leipzig: Braunmüller 1922.

[3] JENSEN, C.: Die Himmelsstrahlung. In Handbuch der Physik, Bd. 19, S. 70—152. Berlin: Springer 1928.

[4] JENSEN, C.: Die Polarisation des Himmelslichtes. In Handbuch der Geophysik, Bd. 8, S. 527—620. Berlin: Gebrüder Bornträger 1942.

[5] LINKE, F.: Die Theorie der Zerstreuung, Extinktion und Polarisation des Lichtes in der Atmosphäre. In Handbuch der Geophysik, Bd. 8, S. 120—238. Berlin: Gebrüder Bornträger 1942.

[6] SEKERA, Z.: Polarization of Skylight, Compendium of Meteorology, pp. 79—89, Boston, Amer. Met. Soc., 1951.

[7] JENSEN, C.: Die Apparate zur Untersuchung der atmosphärischen Polarisationserscheinungen. In E. KLEINSCHMIDT, Handbuch der meteorologischen Instrumente, S. 666—693. Berlin: Springer 1935.

In the reference [7] a detailed description of instruments used for polarization measurements can be found.

The exact theory of the radiation transfer in a RAYLEIGH atmosphere has been developed in the series of the following papers:

[8] CHANDRASEKHAR, S.: On the Radiative Equilibrium of a Stellar Atmosphere. XXI. Astrophys. J. **106**, 152 (1947). — XXII. Astrophys. J. **107**, 48, 188 (1948).

[9] CHANDRASEKHAR, S.: The Transfer of Radiation in Stellar Atmospheres. Bull. Amer. Math. Soc. **53**, 641 (1947)

and has been summarized in the treatise:

[10] CHANDRASEKHAR, S.: Radiative Transfer. Oxford: Clarendon Press 1950.

The basic functions for the computation of the intensity and polarization of the skylight in a RAYLEIGH atmosphere have been published in the following tables:

[11] CHANDRASEKHAR, S., and D. D. ELBERT: The Illumination and Polarization of the Sunlit Sky on Rayleigh Scattering. Trans Amer. Phil Soc. **44** (6), 643 (1954)

contains a) the functions X_i, $Y_i (i = 1, 2, l, r)$, the eight functions in eq. (4.14) on p. 298 and $\gamma_l(\mu)$, $\gamma_r(\mu)$ with the constants needed for their computations, tabulated for $\tau = 0.05$ (0.05) 0.25, 0.50, 1.00 for $\mu = 0.0$ (0.02) 1.00, b) the total intensity $I_l + I_r$, the degree of polarization P in eq. (2.7) and the deviation of the plane of polarization from the vertical χ for $\tau = 0.15$, for $\mu = 0.0$ (0.04) 1.00, $\varphi - \varphi_0 = 0°$ (10°) 90° and for $\mu_0 = 0$, 0.08, 0.24, 0.52, 0.64, 0.72, 0.80, 0.94, 1.00, furthermore the same quantity for $\tau = 0.10$, 0.20, for less values of $\varphi - \varphi_0$, but with the effect of ground reflection with the albedo $\lambda_0 = 0.10$, 0.20, c) the values of the same functions as in a) but computed from the corrected second approximation for $\tau = 0.01$ (0.01) 0.20 (0.05) 0.50 (0.10) 1.0 for $\mu = 0$, 0.05, 0.75, 0.10 (0.05) 0.3 (0.1) 1.0.

[12] Tables Relating to Rayleigh Scattering of Light in the Atmosphere, Sci. Report No. 3, Contract No. AF 19 (122)—239, Dept. of Meteorology, Univ. of Calif. at Los Angeles, (1952),

contains the same functions as in [11] a) for $\tau = 0.15$, 0.25, 1.00, for $\mu = 0.0$ (0.02) 0.20 (0.04) 0.60 (0.08) 1.00, the intensities I_l, I_r and their correction I_l^*, I_r^* along the sun vertical for the albedo $\lambda_0 = 0$, 0.10, 0.25, 0.50, 0.80, for $\tau = 0.15$, 0.25, 1.00, for $\mu_0 = 0.10$, 0.20 (0.20) 0.80, 0.90 (0.04) 0.98, for $\mu = 0.1$ (0.1) 1.0 and a few additional values.

[13] SEKERA, Z., and E. V. ASHBURN: Tables Relating to Rayleigh Scattering of Light in the Atmosphere, NAVORD Report 2061, U.S. Naval Ordnance Test Station, Inyokern, Calif. (1953),

contains the values of the same functions for almost the same ranges of the parameters as in [12] for $\tau = 0.001$, 0.003, 0.01, 0.02, 0.05, 0.10, based on the corrected second approximation only.

The partial results of the research on skylight polarization, conducted under the Contract No. AF 19 (122)—239 with U.S. Air Force Cambridge Research Center at the Dept. of Meteorology University of California at Los Angeles, have been published so far in the following reports:

[14a] Progress Report No. 2 (Feb. 28, 1951), Appendix A: SEKERA, Z.: The Theory of Polarization Measurement Suitable for the Investigation of Skylight Polarization.

Scientific Reports:

[*14b*] No. 2; Additional Scattering Functions for Spherical Particles. 1952.

[*14c*] No. 4; COULSON, K. L.: Polarization of Light in the Sun Vertical for a Rayleigh Atmosphere. 1952.

[*14d*] No. 5; SEKERA, Z.: Legendre Series for the Scattering Functions for Spherical Particles. 1952.

[*14e*] No. 7; COULSON, K. L.: Neutral Points of Skylight Polarization in a Rayleigh Atmosphere. 1954.

[*14f*] No. 9; SAXON, D. S.: Lectures on the Scattering of Light. 1955.

[*14g*] No. 10; SEKERA, Z.: Diffuse Radiation in Upper Levels of the Atmosphere. In preparation.

Final Report (June 1955):

[*14h*] A: SEAMAN, C. L., and Z. SEKERA: Photoelectric Polarimeter for Measurements of Skylight Polarization.

[*14i*] B: COULSON, K. L., and Z. SEKERA: Distribution of Polarization and the Orientation of the Plane of Polarization of Sky Radiation Over the Entire Sky in a RAYLEIGH Atmosphere.

[*14j*] C: FRASER, R. S.: Theoretical Positions of Maximum Degree of Polarization.

[*14k*] D: SEKERA, Z.: Scattering Matrix for Spherical Particles and Its Transformation.

[*14l*] E: FRASER, R. S., and Z. SEKERA: The Effect of Specular Reflection on a RAYLEIGH Atmosphere.

Diffusion des radiations par les gouttes d'eau en suspension dans l'atmosphère.

Par

J. Bricard.

Avec 23 Figures.

Introduction.

1. Développement de nos connaissances. Il n'est pas inutile de rappeler l'état de nos connaissances sur la diffusion de la lumière par les particules en suspension dans l'atmosphère lors de la rédaction de la précédente édition (Vol. XX, 1926) de cet ouvrage.

Les traités d'Optique classique de Pernter et Exner [1] ou de Mascart [2] contenaient depuis longtemps, et de facon très détaillée, l'exposé de l'action des particules de *glace* sur la lumière solaire; nos connaissances dans ce domaine, encore à l'heure actuelle, n'ont pas beaucoup évolué. On y trouvait également l'explication des franges de diffraction produites par les gouttes d'eau, soit en arrière (Arcs en Ciel, les phénomènes d'Anticouronnes, ou Spectre du Brocken, y étant simplement mentionnés sans autres commentaires), soit en avant (Couronnes), l'étude de ce dernier phénomène ayant été largement développée par Mecke [3].

Lorsqu'une particule transparente est éclairée par un faisceau parallèle, elle diffuse dans toutes les directions le flux qu'elle a emprunté au faisceau incident. Le calcul de la répartition de ce flux diffusé autour de la direction incidente avait été abordé au point de vue théorique par Mie [4], en se basant sur la théorie électromagnétique de la lumière, et quelques applications en avaient été données dans le cas de particules de diamètre inférieur à la longueur d'onde, mais les calculs numériques paraissaient d'une telle compléxité qu'on n'avait pas réussi à les étendre à des particules de dimensions supérieures, et correspondant aux gouttes nuageuses par exemple, de sorte que le calcul de Mie n'avait pas atteint le domaine de l'Optique atmosphérique.

Le calcul de Wiener [5] permettant d'évaluer cette répartition en se basant sur les lois de l'Optique géométrique avait été introduit par Mecke [6] dans le domaine des gouttes de brouillard, ainsi que la remarque fondamentale suivante: on peut admettre que la lumière diffusée par une goutte d'eau se compose exclusivement de deux parties: le flux lumineux diffusé par réfléxion et réfraction d'une part, et celui diffusé par diffraction d'autre part. Or ces deux flux sont égaux; en effet, d'après le théorème de Babinet, le flux diffracté par les bords d'un écran est égal au flux diffracté par le trou complémentaire. Le flux total diffusé par un obstacle transparent, dont la section peut d'ailleurs être quelconque, est ainsi toujours égal au double du flux intercepté par son contour géométrique.

Depuis lors, les techniques ont profondément évolué: techniques des récepteurs énergétiques vers l'Infra Rouge et des monochromateurs, et techniques des Ondes centimétriques, qui se sont développées rapidement, et ont fait leur apparition dans le domaine de la Météorologie, entrainant des recherches théoriques dans le même domaine.

En même temps, de nouvelles méthodes d'observation météorologiques ont fait leur apparition, ou se sont perfectionnées: étude directe des particules en suspension dans l'air par captation sur des fils fins, captations par chocs ou par l'action d'un champ électrique, mesures de masse d'eau condensée dans les nuages, de sorte qu'il devint possible d'établir des granulométries des milieux diffusants, en même temps que l'on étudiait leurs propriétés optiques.

Même dans le domaine théorique, les techniques nouvelles ont eu leur intervention grâce à l'apparition des machines à calculer électroniques, qui ont permis l'extension du calcul de MIE au domaine des grosses particules, permettant ainsi de faire la liaison Théorie de MIE et Optique géométrique.

On passa de l'étude d'une particule isolée, à celle d'un milieu diffusant indéfini, et on étudia théoriquement, puis expérimentalement, la transmission de la lumière par les atmosphères brumeuses, par les nuages, et dans le domaine des Ondes Radio, par la pluie.

Mais ceci n'était qu'une partie du problème et en fait, lorsqu'on étudie la transmission de la lumière par un milieu trouble, il faut tenir compte non seulement de la lumière transmise, mais également de la lumière diffusée par le milieu lui même. Cette question, qui vient de faire son apparition, est déja sérieusement entamée au point de vue théorique, et les résultats acquis dès maintenant demandent des vérifications expèrimentales qui sont d'ailleurs dans les possibilités de la technique actuelle.

A. Diffusion des radiations par une goutte d'eau.

I. Théorie électromagnétique.

2. Généralités. Les méthodes de calcul de la théorie électromagnétique permettent de donner une solution générale au problème de la diffusion des radiations par des particules de dimensions quelconques, dont l'indice est nettement différent de celui du milieu environnant. Cette solution a été développée par G. MIE [4] en 1908, puis par P. DEBYE[1] en 1909, à propos d'un problème connexe (pression de radiation exercée sur les particules par l'onde incidente). Lord RAYLEIGH[2], qui parait ignorer le travail de MIE, a étudié le problème en 1910, et a appliqué la solution aux particules d'indice relatif $n=1,5$ par rapport au milieu ambiant. Nous nous bornerons ici à indiquer schématiquement la marche des calculs, en suivant l'exposé de J. A. STRATTON [7].

Une onde périodique qui vient frapper un corps matériel donne naissance à des oscillations forcées des charges libres ou liées, synchrones avec le champ appliqué. Ces mouvements imposés aux charges produisent un champ secondaire, à la fois à l'intérieur et à l'extérieur du corps, le champ résultant en un point quelconque étant le vecteur somme du champ primaire et du champ secondaire.

Le problème de cette nature le plus simple, et en même temps d'un grand intérêt pratique, est celui d'une onde plane qui vient frapper une sphère. Soit a le rayon de la sphère. Posons:

$$k = \sqrt{\mu\,\varepsilon\,\omega^2 - i\,\mu\,\sigma\,\omega} \tag{2.1}$$

où $\omega = 2\pi c/\lambda$, c étant la vitesse de la lumière, λ la longueur d'onde; μ représente la perméabilité magnétique, ε est la constante diélectrique et σ la conductibilité

[1] P. DEBYE: Ann. d. Phys. **30**, 67 (1909).
[2] Lord RAYLEIGH: Proc. Roy. Soc. Lond., Ser. A **84**, 25 (1910).

électrique du milieu dans lequel on se place. On désignera par k_1 et k_2 les valeurs qui se rapportent à l'intérieur de la sphère et au milieu extérieur; si celui-ci est de l'air, $k_2 = 2\pi/\lambda$.

Considérons une onde plane de composante électrique et magnétique E_i et H_i, polarisée suivant la direction Ox, et qui se propage dans la direction Oz positive. On écrira, a_x et a_y représentant les vecteurs unités suivant Ox et Oy:

$$E_i = a_x E_0 e^{-ik_2 z + i\omega t}, \qquad H_i = a_y \cdot \frac{k_2}{\mu_2 \omega} E_0 e^{-ik_2 z + i\omega t}. \tag{2.2}$$

Pour traiter le problème, il est nécessaire d'exprimer le vecteur incident $a_x e^{-ik_2 z}$ et $a_y e^{-ik_2 z}$ en coordonnées sphériques soit (Fig. 1)

$$\left.\begin{aligned} x = r\sin\vartheta\cos\Phi, \quad y = r\sin\vartheta\sin\Phi, \quad z = r\cos\vartheta. \\ 0 < \vartheta < \pi, \quad 0 < \Phi < 2\pi. \end{aligned}\right\} \tag{2.3}$$

Posons:

$$\left.\begin{aligned} m_{\substack{e\\0}n}^{(p)} &= \pm \frac{1}{\sin\vartheta} z_n^{(p)}(kr)\, P_n^1(\cos\vartheta) \frac{\cos}{\sin} \Phi\, i_2 - z_n^{(p)}(kr) \frac{\partial P_n^1}{\partial\vartheta} \frac{\sin}{\cos} \Phi\, i_3, \\ n_{\substack{e\\0}n}^{(p)} &= \frac{n(n+1)}{kr} z_n^{(p)}(kr)\, P_n^1(\cos\vartheta) \frac{\sin}{\cos} \Phi\, i_1 + \frac{1}{kr}\frac{\partial}{\partial r}\left[r z_n^{(p)}(kr)\right] \frac{\partial P_n^1}{\partial\vartheta} \frac{\sin}{\cos} \Phi\, i_2 \pm \\ &\pm \frac{1}{kr\sin\vartheta}\frac{\partial}{\partial r}\left[r z_n^{(p)}(kr)\right] P_n^1(\cos\vartheta) \frac{\cos}{\sin} \Phi\, i_3. \end{aligned}\right\} \tag{2.4}$$

P_n^1 est la polynôme de LEGENDRE associè de 1ère espèce, i_1, i_2, i_3 les vecteurs unités tracés dans les directions croissantes de r, ϑ et Φ (Fig. 1) sur la sphère de rayon r (les vecteurs i_3 et i_2 sont donc tangents à la sphère le long d'un cercle parallèle et d'un cercle méridien). L'index p peut prendre deux valeurs différentes. S'il s'agit de l'énergie incidente, ou de l'énergie transmise, il a la valeur 1. S'il s'agit de l'énergie diffusée, il a la valeur 3.

$$z_n^1(u) = \left(\frac{\pi}{2u}\right)^{\frac{1}{2}} J_{n+\frac{1}{2}}(u), \quad z_n^3(u) = \left(\frac{\pi}{2u}\right)^{\frac{1}{2}} H_{n+\frac{1}{2}}^{(2)}(u) \tag{2.5}$$

ou $J_{n+\frac{1}{2}}(u)$ est une fonction de BESSEL de 1ère espèce et $H_{n+\frac{1}{2}}^{(2)}(u)$ une fonction de HANKEL de 2ème espèce.

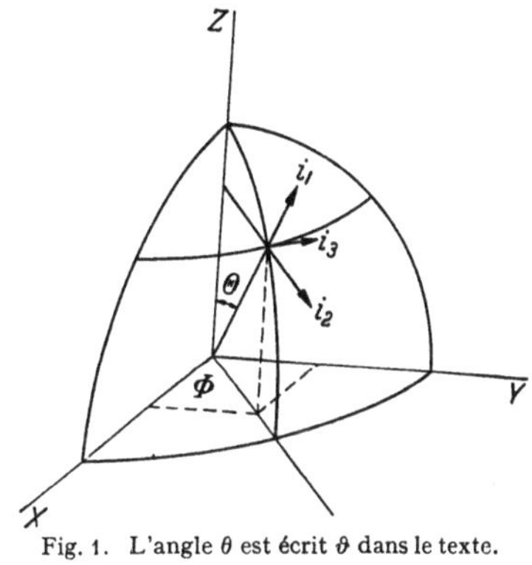

Fig. 1. L'angle θ est écrit ϑ dans le texte.

Dans ces conditions, les champs électrique et magnétique incidents peuvent se mettre sous la forme:

$$\left.\begin{aligned} E_i &= E_0 \sum_{n=1}^{\infty} (-i)^n \frac{2n+1}{n(n+1)} \left(m_{0n}^{(1)} + i\, n_{en}^{(1)}\right) e^{i\omega t}, \\ H_i &= -E_0 \frac{k_2}{\mu_2\omega} \sum_{n=1}^{\infty} (-i)^n \frac{2n+1}{n(n+1)} \left(m_{en}^{(1)} - i\, n_{0n}^{(1)}\right) e^{i\omega t}. \end{aligned}\right\} \tag{2.6}$$

Admettons que le champ secondaire induit à l'intérieur de la sphère ($r < a$) soit de la forme:

$$\left.\begin{aligned} E_t &= E_0 e^{i\omega t} \sum_{n=1}^{\infty} (-i)^n \frac{2n+1}{n(n+1)} \left(a_n^t m_{0n}^{(1)} + i\, b_n^t n_{en}^{(1)}\right), \\ H_t &= -\frac{k_1}{\mu_1\omega} E_0 e^{i\omega t} \sum_{n=1}^{\infty} (-i)^n \frac{2n+1}{n(n+1)} \left(b_n^t m_{en}^{(1)} - i\, a_n^t n_{0n}^{(1)}\right) \end{aligned}\right\} \tag{2.7}$$

et que le champ diffusé à l'extérieur de la sphère ($r>a$) soit de la forme:

$$\left.\begin{aligned} E_d &= E_0\, e^{i\omega t} \sum_{n=1}^{\infty} (-i)^n \frac{2n+1}{n(n+1)} \left(a_n^d\, m_{0n}^{(3)} + i\, b_n^d\, n_{en}^{(3)}\right), \\ H_d &= -\frac{k_2}{\mu_2\omega} E_0\, e^{i\omega t} \sum_{n=1}^{\infty} (-i)^n \frac{2n+1}{n(n+1)} \left(b_n^d\, m_{en}^{(3)} - i\, a_n^d\, n_{0n}^{(3)}\right). \end{aligned}\right\} \tag{2.8}$$

Les conditions limites pour $r=a$ sont:

$$i_1(E_i+E_d) = i_1 E_t, \qquad i_1(H_i+H_d) = i_1 H_t. \tag{2.9}$$

En introduisant ces conditions dans (2.7) et (2.8) on peut calculer les coefficients a et b et on obtient en particulier:

$$a_n^d = -\frac{\mu_1 z_n^{(1)}(N\varrho)\,[\varrho\, z_n^{(1)}(\varrho)]' - \mu_2 z_n^{(1)}(\varrho)\,[N\varrho\, z_n^{(1)}(N\varrho)]'}{\mu_1 z_n^{(1)}(N\varrho)\,[\varrho\, z_n^{(3)}(\varrho)]' - \mu_2 z_n^{(3)}(\varrho)\,[N\varrho\, z_n^{(1)}(N\varrho)]'}, \tag{2.10}$$

$$b_n^d = -\frac{\mu_1 z_n^{(1)}(\varrho)\,[N\varrho\, z_n^{(1)}(N\varrho)]' - \mu_2 N^2 z_n^{(1)}(N\varrho)\,[\varrho\, z_n^{(1)}(\varrho)]'}{\mu_1 z_n^{(3)}(\varrho)\,[N\varrho\, z_n^{(1)}(N\varrho)]' - \mu_2 N^2 z_n^{(1)}(N\varrho)\,[\varrho\, z_n^{(3)}(\varrho)]'} \tag{2.11}$$

avec $N=k_1/k_2$ et $\varrho=k_2 a$.

Ces relations permettent d'évaluer E_d et H_d, vibrations diffusées à l'extérieur de la sphère. En raisonnant de la même façon, on calculerait a_n^t et b_n^t, correspondant aux vibrations à l'intérieur de la sphère. Les primes des parenthèses carrées indiquent la différentiation par rapport à l'argument de la fonction de BESSEL à l'intérieur de la parenthèse.

Dans la pratique, pour calculer la vibration diffusée par la sphère ou la fraction d'énergie absorbée par elle, on se place à des distances du centre $r \gg a$ ou $k_2 r \gg k_2 a$. Or les coefficients a_n et b_n demeurent petits lorsque $n>k_2 a$, et on peut limiter la somme à $n \sim k_2 a$. D'autre part, aux grandes distances, $k_2 r>n$, ce qui signifie que l'ordre n des termes dont il faut tenir compte est inférieur à l'árgument $k_2 r$ des fonctions sphériques de BESSEL. Dans ces conditions on peut utiliser les expressions asymptotiques (STRATTON [7]):

$$\left.\begin{aligned} z_n^{(1)}(kr) &\sim \frac{1}{kr}\cos\left(kr - \frac{n+1}{2}\pi\right), \\ z_n^{(3)}(kr) &\sim \frac{1}{kr}\, e^{-i\left(kr-\frac{n+1}{2}\pi\right)}. \end{aligned}\right\} \tag{2.12}$$

Il résulte de ceci que la composante radiale du champ diffusé peut être négligée, car elle diminue avec r comme $1/r^2$, alors que les composantes en ϑ et Φ diminuent comme $1/r$. Pour les grandes valeurs de r, le champ est donc transversal et:

$$H_r^d = E_r^d = 0; \qquad r \gg a \tag{2.12 bis}$$

et avec $\sigma_2=0$

$$\left.\begin{aligned} E_\vartheta^d &= \left(\frac{i}{kr}\right) E_0\, e^{-ikr}\, e^{i\omega t} \sum_{n=1}^{\infty} \frac{2n+1}{n(n+1)} \left(a_n^d \frac{P_n^1}{\sin\vartheta} + b_n^d \frac{\partial P_n^1}{\partial\vartheta}\right)\cos\Phi, \\ E_\Phi^d &= -\left(\frac{i}{kr}\right) E_0\, e^{-ikr}\, e^{i\omega t} \sum_{n=1}^{\infty} \frac{2n+1}{n(n+1)} \left(a_n^d \frac{\partial P}{\partial\vartheta} + b_n^d \frac{P_n^1}{\sin\vartheta}\right)\sin\Phi. \end{aligned}\right\} \tag{2.13}$$

3. Fonctions de diffusion. Plaçons nous dans le cas des Ondes Optiques. C'est en effet dans ce domaine que paraissent avoir été effectuées des recherches expérimentales directes en ce qui concerne la répartition de la lumière diffusée par des particules éclairées par un faisceau parallèle.

En utilisant les notations du National Bureau of Standards[1], on peut, à partir des relations (2.13), obtenir les relations suivantes, correspondant à des particules non conductrices en suspension dans l'air:

$$\left.\begin{aligned} i_1 &= \left|\sum_{n=1}^{\infty}\{A_n\pi_n + P_n[x\pi_n - (1-x^2)\pi_n']\}\right|^2,\\ i_2 &= \left|\sum_{n=1}^{\infty}\{A_n[x\pi_n - (1-x^2)\pi_n'] + P_n\pi_n\}\right|^2 \end{aligned}\right\} \tag{3.1}$$

où

$$\left.\begin{aligned} A_n &= \left[\frac{(-1)^{n+\frac12}(2n+1)}{n(n+1)}\right]\left[\frac{S_n'(\beta)\,S_n(\alpha) - m\,S_n'(\alpha)\,S_n(\beta)}{S_n'(\beta)\,\Phi_n(\alpha) - m\,\Phi_n'(\alpha)\,S_n(\beta)}\right],\\ P_n &= \left[\frac{(-1)^{n+\frac12}(2n+1)}{n(n+1)}\right]\left[\frac{m\,S_n(\alpha)\,S_n'(\beta) - S_n(\beta)\,S_n'(\alpha)}{m\,\Phi_n(\alpha)\,S_n'(\beta) - S_n(\beta)\,\Phi_n'(\alpha)}\right] \end{aligned}\right\} \tag{3.2}$$

avec

$$\left.\begin{aligned} S_n(\alpha) &= \left(\frac{\pi\alpha}{2}\right)^{\frac12} J_{n+\frac12}(\alpha),\\ \Phi_n(\alpha) &= S_n(\alpha) + i\,C_n(\alpha),\\ C_n(\alpha) &= (-1)^n\left(\frac{\pi\alpha}{2}\right)^{\frac12} J_{-n-\frac12}(\alpha), \end{aligned}\right\} \tag{3.3}$$

$$S_n'(\alpha) = \frac{\partial S_n(\alpha)}{\partial\alpha},\qquad \Phi_n'(\alpha) = \frac{\partial\Phi_n(\alpha)}{\partial\alpha} \tag{3.4}$$

en posant $\beta = m\alpha$. Les J sont des fonctions de BESSEL, m est l'indice de réfraction de la goutte par rapport au milieu environnant et λ la longueur de la lumière monochromatique incidente.

$$\pi_n = \frac{\partial P_n}{\partial x},\qquad \pi_n' = \frac{\partial^2 P_n}{\partial x^2} \tag{3.5}$$

où P_n est le polynome de LEGENDRE de degré n. $x = \cos\gamma$ (γ angle entre la direction d'observation et la direction inverse de propagation du faisceau éclairant la goutte) et $\alpha = 2\pi a/\lambda$.

Quand la particule est éclairée par un faisceau de lumière parallèle d'éclairement unité, $\left(\frac{\lambda^2}{8\pi^2}\right)i_1$ est l'intensité diffusée par elle par unité d'angle solide pour la composante dont le vecteur champ électrique est perpendiculaire au plan d'observation [plan contenant la direction d'observation et la direction de propagation du faisceau incident]; $\left(\frac{\lambda^2}{8\pi^2}\right)i_2$ est l'intensité diffusée dont le vecteur champ électrique est parallèle au plan d'observation.

L'intensité totale par unité d'angle solide dans une direction quelconque a pour valeur

$$i = \frac{\lambda^2}{8\pi^2}(i_1 + i_2) = f(\alpha, \gamma, \lambda, m) \tag{3.6}$$

la quantité f représentant la fonction de diffusion de la goutte.

Quand la particule est éclairée par de la lumière polarisée rectilignement, d'éclairement unité, $\left(\frac{\lambda^2}{4\pi^2}\right)i_1$ est l'intensité de la composante diffusée dans une direction perpendiculaire au vecteur lumineux de l'onde incidente; $\left(\frac{\lambda^2}{4\pi^2}\right)i_2$ est l'intensité de la composante diffusée dans un plan contenant le champ électrique incident et sa direction de propagation; pour les autres directions, on trouve

[1] Tables of Scattering Functions of Spherical Particles. Nat. Bur. of Stand. Appl. Math. Ser. IV. Washington DC: US Government Printing Office 1948. — Voir Note A, p. 369.

l'intensité diffusée en considérant les deux composantes du vecteur lumineux incident parallèles et perpendiculaires au plan d'observation. Cette lumière est polarisée elliptiquement.

On remarquera que si la vibration incidente polarisée est normale au plan d'observation, la vibration diffusée ne possède aucune composante dans ce plan. Quelque soient les dimensions de la particule, si elle est sphérique et isotrope, la lumière diffusée est complètement polarisée, et vibre alors parallèlement à la lumière polarisée incidente.

On voit que la vibration diffusée se présente comme la résultante d'un nombre théoriquement infini de vibrations partielles. En fait, ces expressions forment des séries convergentes, dont il suffit de calculer un nombre fini de termes, les autres devenant négligeables. Mais le nombre de vibrations partielles dont il faut tenir compte augmente au fur et à mesure que le diamètre des particules augmente, et le calcul de i_1 et i_2 se complique à mesure que α est plus élevé. Il a pu étre mené à bien pour des valeurs de α supérieures à 10 ces dernières années seulement par GUMPRECHT et ses collaborateurs, grace à l'emploi de machines à calculer électroniques.

La quantité de lumière polarisée aura pour expression dans la direction correspondante:

$$p = \frac{i_1 - i_2}{i_1 + i_2}. \tag{3.7}$$

Tableau 1. *Répartition de la lumière diffusée par un gouttelette d'eau sphérique parfaitement transparente.* a = rayon de la sphère, λ = longueur d'onde, $\alpha = 2\pi a/\lambda$.

γ	$\alpha = 6{,}0$		$\alpha = 8{,}0$		$\alpha = 15$		$\alpha = 20$	
	i_1	i_2	i_1	i_2	i_1	i_2	i_1	i_2
0°	3,506	3,506	3,810	3,810	214,1	214,1	239,9	239,9
10°	2,082	3,677	0,971	6,891	63,06	112,1	23,14	483,3
20°	3,011	3,899	19,43	10,02	35,47	12,37	19,23	85,81
30°	5,973	3,320	8,999	3,151	37,02	2,972	67,71	0,058
40°	3,098	2,013	7,678	3,052	6,768	17,56	39,79	51,26
50°	0,415	1,310	12,60	6,294	8,971	11,00	52,25	72,45
60°	4,104	1,356	0,236	5,285	4,620	18,34	29,00	16,56
70°	4,303	1,801	9,993	4,093	12,75	24,18	3,964	5,373
80°	0,218	2,884	1,514	4,987	12,97	22,85	2,714	49,34
90°	3,453	3,694	6,134	8,239	13,28	30,82	3,575	72,05
100°	7,692	4,136	8,220	7,366	7,618	47,90	1,437	44,63
110^5	2,184	5,543	5,083	18,37	5,774	57,91	6,469	34,81
120°	7,044	11,76	25,76	25,84	43,38	68,21	15,69	104,5
130°	30,04	26,27	21,53	31,23	139,3	109,1	24,09	212,6
140°	23,86	31,37	70,93	87,52	243,5	208,1	44,60	265,5
150°	43,11	70,05	156,0	154,9	131,1	203,8	111,1	267,0
160°	340,1	357,9	158,9	216,4	232,1	133,0	405,3	742,9
170°	926,1	928,2	1536	1589	4218	3416	832,8	1596
171°	981,9	983,4	1747	1792	5920	5038	25,42	527,5
172°	1034	1035	1955	1994	7968	7066	624,4	510,2
173°	1083	1083	2157	2188	10310	9454	3283	2384
174°	1126	1126	2345	2369	12860	12100	8310	6721
175°	1164	1164	2515	2532	15480	14860	15510	13570
176°	1196	1195	2662	2673	17980	17540	24160	22300
177°	1221	1221	2781	2788	20190	19920	33070	31670
178°	1239	1239	2869	2872	21930	21800	40830	40080
179°	1250	1250	2923	2924	23040	23000	46130	45920
180°	1254	1254	2941	2941	23420	23420	48010	48010

Tableau 2. *Répartition de la lumière diffusée par une gouttelette d'eau sphérique parfaitement transparente (Suite).* $\alpha = 40$.

γ	0°	10°	20°	30°	40°	50°	60°	70°	80°	90°
i_1	183,9	52,96	103,6	166,9	223,8	132,3	10,37	41,73	18,73	9,926
i_2	183,9	305,9	285,5	109,3	130,7	44,78	63,57	22,58	41,44	58,42
γ	100°	110°	120°	130°	140°	150°	160°	170°	171°	172°
i_1	51,36	94,80	147,8	97,57	263,2	1979	4000	3279	1566	827,6
i_2	73,37	138,8	291,3	457,3	745,6	1682	3460	2318	843,8	656,4
γ	173°	174°	175°	176°	177°	178°	179°	180°		
i_1	268,7	5823	34310	106100	227200	374200	497400	545700		
i_2	129,4	4248	29980	99320	220200	369500	495900	545700		

Les Tableaux 1 et 2 représentent les résultats obtenus par GUMPRECHT[1] pour des gouttes parfaitement transparentes d'indice $m = 1{,}33$. On voit sur ces tableaux que l'indicatrice de diffusion dépend essentiellement des dimensions des particules (rayon a) par rapport à la longueur d'onde incidente λ, ($\alpha = 2\pi a/\lambda$). En particulier, l'intensité diffusée par la goutte dans la direction même de propagation du faisceau incident augmente par rapport à l'intensité diffusée dans la direction inverse à mesure que le rapport $\alpha = 2\pi\, a/\lambda$ augmente[2].

L'état de polarisation de la lumière diffusée dans une direction donnée par rapport à celle du faisceau incident dépend également de α ainsi que sa répartition lorsque la direction varie.

4. Approximation de BUCERIUS[3]. Les relations (3.3) et (3.4) peuvent étre simplifiées dans le cas ou le quantité $\alpha = 2\pi a/\lambda$ est très grande, ce qui correspond aux gouttes nuageuses dans la visible; on écrit alors les relations asymptotiques suivantes:

$$S_n(\alpha) = \cos\left(\alpha f_0 - \frac{\pi}{4}\right), \qquad S'_n(\alpha) = \sqrt{\sin\tau_0}\cos\left(\alpha f_0 + \frac{\pi}{4}\right), \tag{4.1}$$

$$\Phi_n(\alpha) = e^{-i\left(\alpha f_0 - \frac{\pi}{4}\right)}, \qquad \Phi'_n(\alpha) = \sqrt{\sin\tau_0}\, e^{-i\left(\alpha f_0 + \frac{\pi}{4}\right)} \tag{4.2}$$

en posant

$$\cos\tau_0 = \frac{n+\frac{1}{2}}{\alpha}, \qquad \cos\tau_1 = \frac{n+\frac{1}{2}}{m\alpha}, \qquad \frac{\pi}{2} > \tau > 0; \tag{4.3}$$

$$f_0 = \sin\tau_0 - \tau_0\cos\tau_0, \qquad f_1 = \sin\tau_1 - \tau_1\cos\tau_1. \tag{4.4}$$

Cette simplification[3] peut étre introduite dans les relations (3.1) ce qui permet d'exprimer i_1 et i_2 en fonction des coefficients de FRESNEL:

$$\left.\begin{aligned} r_1 &= \frac{m\sin\tau_0 - \sin\tau_1}{m\sin\tau_0 + \sin\tau_1} = \frac{\operatorname{tg}(\tau_0 - \tau_1)}{\operatorname{tg}(\tau_0 + \tau_1)}, \\ r_2 &= \frac{\sin\tau_0 - m\sin\tau_1}{\sin\tau_0 + m\sin\tau_1} = \frac{\sin(\tau_0 - \tau_1)}{\sin(\tau_0 + \tau_1)} \end{aligned}\right\} \tag{4.5}$$

en posant

$$\cos\tau_0 = m\cos\tau_1. \tag{4.6}$$

On retrouve alors les résultats de WIENER [5], que nous exposerons plus loin, ce qui permet de lier la théorie de MIE et les calculs basés sur les lois de

[1] R. O. GUMPRECHT, NEN LIN DING, JIN H. CHAN et C. M. SLIEPCEVICH: J. Opt. Soc. Amer. **42**, 226 (1952). γ est expliqué après (3.5), p. 333; voir aussi p. 336, avant (4.8), et la fin de la section 5, p. 337.

[2] Voir Note B, p. 369.

[3] H. BUCERIUS: Optik **1**, 188 (1946).

l'optique géométrique que nous allons maintenant étudier. On trouve notamment que cette théorie contient comme approximation la théorie de l'Arc en Ciel, indiquée au cours de la Sect. 11.

Le calcul de BUCERIUS est particulièrement important pour l'interprétation des couronnes et anticouronnes (Sect. 8, 9 et 11) c'est à dire pour les deux cas distincts ou les angles $\vartheta = \pi - \gamma$ et γ sont petits.

Dans le cas des couronnes (Sect. 8 et 9) on trouve que i_1 et i_2 correspondant à (3.1) sont égaux entre eux et proportionnels à la même quantité:

$$\left(\frac{\pi a^2}{\lambda}\right)^2 \left[\frac{2 J_1\left(\frac{4\pi a}{\lambda} \sin\frac{\vartheta}{2}\right)}{\frac{4\pi a}{\lambda} \sin\frac{\vartheta}{2}}\right]^2 \tag{4.7}$$

ou J_1 représente la première fonction de BESSEL.

Il se produit ainsi, au voisinage de la direction de propagation du faisceau incident, une série de maxima et de minima d'intensité qui sont ceux de $J_1\left(\frac{4\pi a}{\lambda}\sin\frac{\vartheta}{2}\right)$. Ce sont les couronnes de diffraction, que nous examinerons en détails au cours du chapitre suivant. Notons toutefois que la relation (4.7) diffère de (8.12) et (8.20) en ce sens que la quantité $\frac{2\pi a}{\lambda}\sin\vartheta$ que prévoit la théorie classique de la diffraction est remplacée par $\frac{4\pi a}{\lambda}\sin\frac{\vartheta}{2}$; mais ces quantités sont pratiquement les mêmes pour les petites valeurs de ϑ.

Si on examine la lumière diffusée au voisinage de la direction inverse de propagation du faisceau incident (observateur dont la tête se trouve dans le faisceau d'un projecteur placé dans le brouillard, et tournant le dos au projecteur), on observe les *anticouronnes*. Ceci correspond à un angle γ très petit. On trouve alors dans l'approximation de BUCERIUS, que les quantités i_1 et i_2 sont égales entre elles, et proportionnelles à:

$$\frac{a^2}{\pi^2} J_2^2\left(\frac{4\pi a}{\lambda}\sin\frac{\gamma}{2}\right) \tag{4.8}$$

ou J_2 représente la seconde fonction de BESSEL.

Il en résulte une série d'anneaux concentriques à centre noir ($J_2=0$), les emplacements des maxima d'intensité étant déterminés par:

$$\frac{4\pi a}{\lambda}\sin\frac{\gamma}{2} = 3{,}05;\quad 6{,}7;\quad 10{,}0;\quad 13{,}2 \tag{4.9}$$

leurs valeurs étant proportionnelles aux carrés des valeurs de la fonction de BESSEL d'indice 2, soit 0,24; 0,09; 0,06; 0,05 .... En lumière monochromatique, ces maxima sont séparés par des anneaux sombres concentriques, dont les emplacements sont déterminés par:

$$\frac{4\pi a}{\lambda}\sin\frac{\gamma}{2} = 5{,}7;\quad 8{,}4;\quad 11{,}6;\quad 14{,}8. \tag{4.9 bis}$$

5. Cas d'un milieu diffusant. Mais en fait, surtout dans la nature, on n'observe pas une goutte isolée, mais un élément de volume contenant un certain nombre de gouttes d'eau, éclairées par un faisceau monochromatique par exemple, et qui diffuse dans toutes les directions. C'est donc la superposition des effets de toutes les gouttelettes contenues dans cet élément de volume que l'on observe.

D'après les calculs de TRINKS[1], il suffit d'ajouter purement et simplement ces effets, que l'on calcule à partir de la théorie de MIE, à condition que les particules diffusantes ne soient pas trop rapprochées, et que leur distance soit égale au moins au double de leur diamètre[2]. De même, le principe de BABINET, dont nous parlerons plus loin, ne s'applique plus lorsque les écrans ne sont pas suffisamment éloignés les uns des autres. Or les atmosphères brumeuses et les nuages comportent quelques centaines de gouttes par cm^3, de rayon inférieur à une dizaine de microns ; les gouttes de pluie sont également trés peu nombreuses, de sorte que la condition requise ci dessus se trouve toujours verifiée dans les cas que nous examinerons. Mais il s'agit d'une condition théorique qui se trouve vérifiée par ses conséquences, et il ne semble pas exister de travaux expérimentaux pour la vérifier directement.

Lorsqu'on se place à distance assez faible de la source lumineuse pour que l'éclairement dû à la diffusion des autres particules soit négligeable par rapport à l'éclairement directement transmis E, on peut admettre que l'intensité diffusée latéralement par l'élément dv est proportionnelle à l'éclairement direct, et a pour expression, $f(\gamma)$ étant définie en (3.6):

$$dI = E \sum N_i f(a_i, \gamma, m, \lambda)\, dv, \tag{5.1}$$

N_i et a_i représentant la granulométrie des particules (N_i = nombre de particules de rayon a_i par unité de volume). Ceci peut s'écrire, si l'on ne précise pas la constitution du milieu:

$$dI = E f_1(\gamma)\, dv. \tag{5.2}$$

La fonction f_1 s'exprime en lumen par unité d'angle solide par unité de volume par lumen par unité de surface; c'est la fonction de diffusion en volume.

Le flux diffusé dans une direction faisant un angle γ avec la direction inverse de propagation du faisceau incident et dans un angle solide $d\omega$ a pour expression:

$$d\Phi = E\, dv\, f_1(\gamma)\, d\omega \tag{5.3}$$

et le flux total diffusé dans toutes les directions sera, par unité de volume diffusant:

$$\Phi = 2\pi E \int_0^{\pi} f_1(\gamma) \sin\gamma\, d\gamma \tag{5.4}$$

le flux diffusé en avant étant:

$$\Phi_1 = 2\pi E \int_{\pi/2}^{\pi} f_1(\gamma) \sin\gamma\, d\gamma, \tag{5.5}$$

tandisque le flux diffusé en arrière s'écrit:

$$\Phi_2 = 2\pi E \int_0^{\pi/2} f_1(\gamma) \sin\gamma\, d\gamma. \tag{5.6}$$

6. Résultats expérimentaux. Les travaux de LA MER[3] et de ses collaborateurs, effectués au laboratoire sur des solutions colloidales, dont l'exposé nous entrainerait en dehors du cadre de cet article, permettent de vérifier les considérations

[1] R. TRINKS: Ann. d. Phys. **22**, 561 (1935).

[2] En fait, ces calculs n'ont été poussés jusqu'au bout que dans le cas de l'approximation de RAYLEIGH.

[3] V. K. LA MER et MARION D. BARNES: J. Colloid, Sci. **1**, 71 (1946). — V. K. LA MER et S. HOCHBERG: O.R.S.D. Rep. No. 940 part 2, sect. 1 (PBL 8080) Washington 1942. — V. K. LA MER et D. SINCLAIR: O.R.S.D. Rep. No. 1857. (Dept. of Commerce, Office of Publications Board No. 944) Washington 1943.

théoriques développées ci dessus, et peuvent être considérés comme des bases solides à la théorie de MIE [4].

Les travaux de DESSENS[1] ont confirmé l'existence dans l'atmosphère, de particules liquides très fines, de diamètres compris entre quelques dixièmes de μ à quelques μ (captation sur des fils d'araignées de diamètre égal à quelques centièmes de μ). Le rôle diffusant de ces particules, indiqué par WRIGHT[2], se

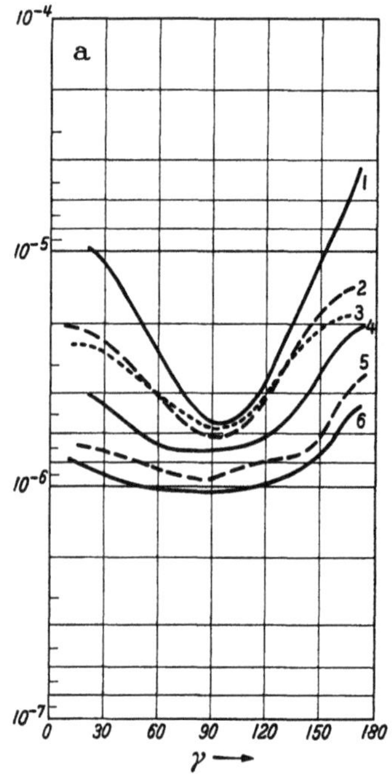

Courbe $n°$	Altitude en mètres	Densité optique par mètre d'èpaisseur $\times 10^5$
1	0	6,40
2	2500 (montée)	4,40
3	2500 (descente)	3,60
4	4600	2,33
5	7000	1,70
6	9200	1,35

Fig. 2a. Fonction de diffusion en volume $f_1(\gamma)$ en air clair au dessus de la Grande Bretagne d'après WALDRAM. Convertie en unités métriques par MIDDLETON. (Voir texte.)

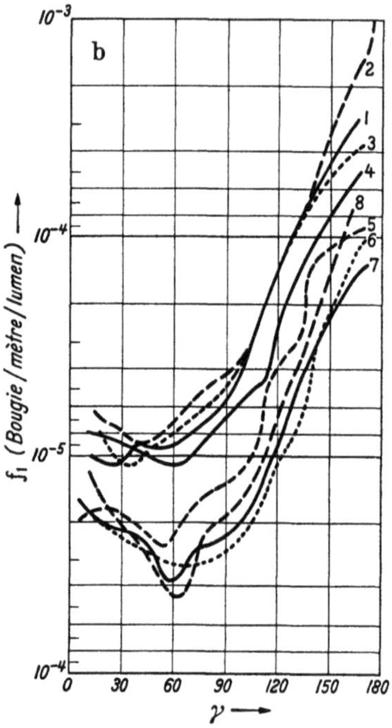

Courbe $n°$	Altitude en mètres	Densité optique par mètre d'épaisseur $\times 10^6$
1	0	24,0
2	0	32,0
3	80	19,7
4	180	14,7
5	230	8,8
6	300	4,4
7	460	4,4
8	610	7,5

Fig. 2b. Fonction de diffusion en volume $f_1(\gamma)$ en air brumeux au dessus de la Grande Bretagne d'après WALDRAM, convertie en unités métriques par MIDDLETON. (Voir texte.)

trouve confirmé quantitativement par les mesures de DESSENS[1] et d'ARNULF, BRICARD et VERET[3], sur lesquelles nous reviendrons plus loin. On peut ainsi leur attribuer les propriétés diffusantes des atmosphères brumeuses.

Il n'existe pas d'étude systématique des variations expérimentales de la fonction f_1 effectuées en plein air, en fonction à la fois de la direction et de la granulométrie, mais un certain nombre d'auteurs ont étudié expérimentalement le diagramme de diffusion d'un élément dv dans des conditions de transparence

[1] H. DESSENS: Ann. de Géophys. **2**, 343 (1946); **3**, 68 (1947).

[2] W. D. WRIGHT: Quart. J. Roy. Met. Soc. **61**, 71 (1935).

[3] A. ARNULF, J. BRICARD et C. VERET: C. R. Acad. Sci. Paris **238**, 503 (1954). — Rev. Opt. **33**, 658 (1954).

déterminées (Fessenkov[1], Grisollet[2], Bullrich[3], Siedentopf[4], Hulburt[5]). Le travail le plus récent et le plus complet dans cet ordre d'idées parait étre celui de Waldram[6], effectué avec une source de lumière artificielle, que nous examinerons rapidement.

Les mesures de Waldram[6] ont permis de déterminer directement la fonction f_1; la densité optique du milieu étant calculée par intégration. Elles étaient faites à différentes altitudes (avion), en atmosphère pure, par temps brumeux, et dans des régions industrielles à forte pollution.

On distingue deux types différents de fonctions de diffusion, l'un montrant un minimum à 90°, l'autre, un minimum entre 120° et 140° (On considère la direction de la lumière incidente comme 180°) tandisqu'il y a moins de symétrie que dans le premier cas.

Le premier type correspond aux atmosphères les plus pures, et sur quelques exemples, on a trouvé des diagrammes de formes très voisines de la diffusion moléculaire. La Fig. 2a en représente quelques exemples; l'altitude et la densité optique y sont représentées. La Fig. 2b correspond aux atmosphères du second type, et les répartitions se rapprochent grossièrement de celles que nous avons déduites du calcul de Mie [*4*]. Leur examen permet de déduire les dimensions des gouttelettes diffusantes; elles indiquent des particules de brume compatibles avec celles que l'on mesure directement sur les fils d'araignées. Des mesures analogues, mais utilisant le soleil comme source lumineuse ont été faites par Möller et Voltz[7], et conduisent à de semblables conclusions.

II. Théorie basée sur les lois de la réflexion, de la réfraction et de la diffraction.

7. Généralités. On sait que les gouttes de brouillard, ou les gouttes nageuses, ont un rayon compris entre quelques μ et 10 à 15 μ suivant les circonstances. Ceci correspond dans le visible ($\lambda = 0{,}5\ \mu$) à des valeurs de $\alpha = 2\pi a/\lambda$ supérieures à 100.

Lorsque les dimensions des particules se rapprochent de la longueur d'onde, ou lui sont supérieures, les calculs de Mie deviennent de plus en plus compliqués, les termes de série qui en résultent convergeant alors très lentement. Les calculs de Mie [*4*] ont été effectués par Gumprecht[8] et ses collaborateurs pour des gouttes d'eau en suspension dans l'air ($n = 1{,}33$) jusqu'à $\alpha = 40$; ceci correspond à des particules de rayon voisin de $3\,\mu$ (Tableau 2 et Fig. 3). Mais il existe une autre méthode de calcul, moins rigoureuse mais plus simple, valable dans le cas ou le diamètre de la particule est au moins égal à 10 fois la longueur d'onde, et qui consiste à leur appliquer les lois de l'optique géométrique. Cette méthode est basée sur la remarque suivante: On peut admettre que la lumière diffusée se compose exclusivement de deux parties: le flux lumineux diffusé par réflexion et réfraction d'une part et celui diffusé par diffraction d'autre part.

Mecke[9] a montré que le flux lumineux Φ diffusé est le même pour les deux processus. Le flux lumineux réfracté et réfléchi est en effet celui qui tombe

[1] B. Fessenkov: Zhurnal Geofis. (Moscou) **4**, 143 (1934).
[2] H. Grisollet: C. R. Acad. Sci. Paris **206**, 1912 (1938).
[3] K. Bullrich: Optik **2**, 301 (1947).
[4] H. Siedentopf: Z. Meteor. **1**, 342 (1947).
[5] E. O. Hulburt: J. Opt. Soc. Amer. **31**, 467 (1941).
[6] J. M. Waldram: Quart J. Roy. Meteor. Soc. **71**, 319 (1945).
[7] F. Voltz: Inst. für Meteor. und Geophys. der freien Univ. Berlin. Meteor. Abh. **2**, Fasc. 4, 99 (1955).
[8] Gumprecht et al., voir Sect. 3.
[9] R. Mecke: Ann. d. Phys. **65**, 257 (1921).

sur la goutte:

$$\Phi_1 = E\pi a^2. \tag{7.1}$$

E étant l'éclairement produit par le faisceau incident, et πa^2 la section de la goutte.

D'autre part d'après le théoréme de BABINET, le flux lumineux Φ_2 diffracté par les bords d'un écran est égal au flux diffracté par le trou complémentaire; il est par conséquent égal au lux qui tombe sur l'obstacle, soit:

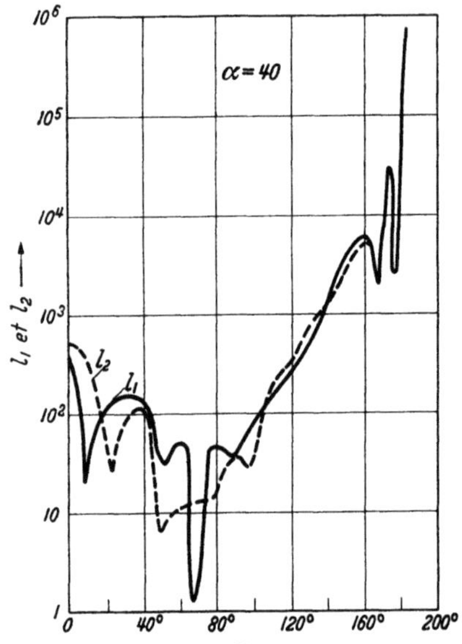

Fig. 3. Répartition de la lumière diffusée par une goutte d'eau d'après GUMPRECHT pour $\alpha = 2\pi a/\lambda = 40$, a = rayon des gouttes; λ = longueur d'onde de la lumière.

$$\Phi_1 = \Phi_2. \tag{7.2}$$

Le flux total diffusé par un obstacle transparent, dont la section peut d'ailleurs être quelconque, est ainsi toujours égal au double du flux intercepté par son contour géométrique. Ce flux 2Φ, enlevé à l'onde incidente, détermine, comme nous le verrons plus loin, le coefficient d'absorption d'un brouillard de gouttelettes. Toutefois il convient de noter que la plus grande partie de celui-ci se trouve rediffusée au voisinage de la direction de propagation du faisceau parallèle incident.

8. Lumière diffusée par la goutte par réflexion et réfraction et diffraction. WIENER [5] a fait le calcul complet de l'intensité lumineuse diffusée par une goutte par réflexion et réfraction en lui appliquant les lios de l'optique géométrique, et en tenant compte des faisceaux ayant subi jusqu'à trois réflexions intérieures. Rappelons brièvement ce calcul pour les termes correspondant à la lumière réfléchie à la surface extérieure de la goutte et transmise directement après deux réfractions:

α) Faisceau directement réfléchi. Soit E l'éclairement produit par le faisceau incident au niveau de la goutte, k le pouvoir réflecteur et I_1 l'intensité lumineuse diffusée par ce processus. Le flux lumineux $d\Phi$ compris entre les deux cylindres de rayon r et $r + dr$ donne naissance au flux réfléchi $d\Phi_1$ compris entre les cônes d'angle $\vartheta = \pi - \gamma$ et $\vartheta + d\vartheta$ (Fig. 4a); on a:

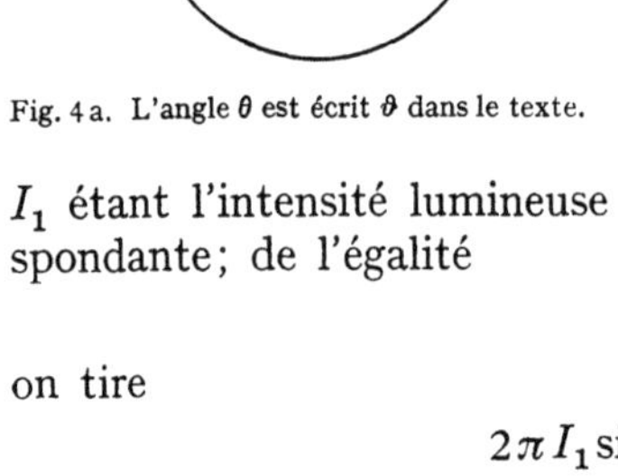

Fig. 4a. L'angle θ est écrit ϑ dans le texte.

$$\vartheta = \pi - 2i, \quad r = a \sin i, \tag{8.1}$$

d'où

$$\left.\begin{aligned} d\Phi &= 2\pi E r\, dr = \pi a^2 E \sin 2i\, di \\ &= \tfrac{1}{2}\pi E a^2 \sin\vartheta\, d\vartheta, \end{aligned}\right\} \tag{8.2}$$

d'autre part

$$d\Phi_1 = I_1\, d\omega = I_1\, 2\pi \sin\vartheta\, d\vartheta, \tag{8.3}$$

I_1 étant l'intensité lumineuse diffusée par la goutte dans la direction correspondante; de l'égalité

$$d\Phi_1 = k_1\, d\Phi, \tag{8.4}$$

on tire

$$2\pi I_1 \sin\vartheta\, d\vartheta = \frac{k_1}{2}\pi a^2 E \sin\vartheta\, d\vartheta, \tag{8.5}$$

d'où

$$I_1 = \tfrac{1}{4} k_1 E a^2. \tag{8.6}$$

L'intensité ainsi déterminée serait uniforme dans toutes les directions si le pouvoir réflecteur de la goutte était constant; celui-ci dépend de l'angle d'incidence, et a pour expression

$$k_1 = \frac{1}{2}\left[\frac{\sin^2(i-r)}{\sin^2(i+r)} + \frac{\operatorname{tg}^2(i-r)}{\operatorname{tg}^2(i+r)}\right] \tag{8.7}$$

avec

$$\sin i = m \sin r. \tag{8.8}$$

Il est égal à l'unité pour $i = 90°$, soit $\vartheta = 0$, c'est pourquoi, observée dans une direction voisine de celle de la source, la goutte éclairée parait très brillante. L'intensité correspondante a pour valeur $I_{1\,0} = 0{,}250\,E\,a^2$, elle tombe à $0{,}066\,E\,a^2$ pour $\vartheta = 25°$.

Le résultat de ce calcul pour des angles égaux au plus à 25° avec la direction du faisceau incident, se trouve dans le Tableau 3, extrait de la Table 4 de WIENER.

Tableau 3. *Intensité lumineuse diffusée par une goutte d'eau par réfléxion et réfraction I_1 = lumière réfléchie, I_2 = lumière refractée, $I = I_1 + I_2$. Ces indications correspondent aux valeurs de l'intensité divisée par la facteur $E a^2$; E = eclairement des gouttes, a = rayon des gouttes (voir Texte).*

ϑ	0°	2°	4°	6°	8°	10°	12°
I_1	0,250	0,223	0,201	0,181	0,163	0,146	0,132
I_2	3,836	3,802	3,726	3,602	3,435	3,226	2,987
I	4,086	4,025	3,927	3,783	3,598	3,372	3,119

ϑ	14°	16°	18°	20°	22°	24°
I_1	0,118	0,106	0,095	0,087	0,078	0,070
I_2	2,731	2,476	2,240	2,020	1,819	1,630
I	2,849	2,582	2,335	2,107	1,897	1,700

β) Lumière réfractée ou transmise par la goutte. La plus grande partie de la lumière diffusée au voisinage de la direction d'incidence est réfractée par la goutte, qui fonctionne alors comme une lentille sphérique (Fig. 4b). Dans ce cas, un

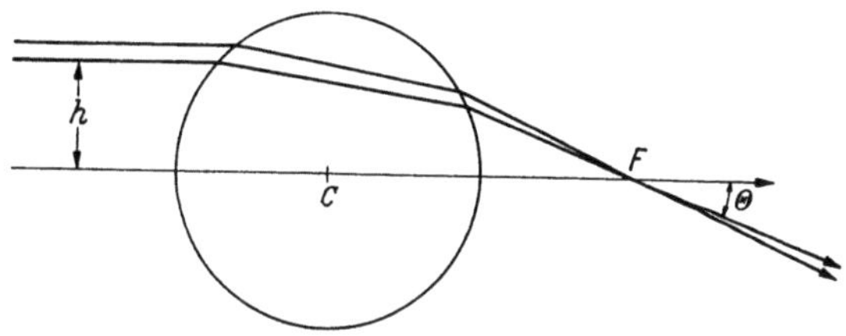

Fig. 4b. L'angle θ est écrit ϑ dans le texte.

raisonnement analogue au précédent conduit, pour l'intensité lumineuse de celle-ci examinée sous un angle $\vartheta = \pi - \gamma$ avec la direction du faisceau parallèle incident, à l'expression suivante:

$$I_2 = \frac{1}{2}\, t\, E\, a^2 \frac{\sin 2i}{\sin\vartheta}\frac{di}{d\vartheta}, \qquad I_2 = \frac{1}{2}(1-k_2)^2 E\, a^2 \frac{\sin 2i}{\sin\vartheta}\frac{di}{d\vartheta} \tag{8.9}$$

où t représente le coefficient de transmission de la goutte, k_2 ayant pour expression:

$$k_2 = \frac{1}{2}\left[\frac{\sin^2\frac{\vartheta}{2}}{\sin^2(i+r)} + \frac{\operatorname{tg}^2\frac{\vartheta}{2}}{\operatorname{tg}^2(i+r)}\right] \quad \text{avec} \quad \vartheta = 2(i-r), \quad \sin i = m \sin r \tag{8.10}$$

et

$$\frac{di}{d\vartheta} = \frac{\sin i \cos r}{2\sin\frac{\vartheta}{2}}. \tag{8.11}$$

Les résultats de ce calcul se trouvent représentés dans le Tableau 3 indiqué ci dessus, dans lequel on a pris la quantité $Ea^2 = 1$. On voit que l'intensité décroit rapidement lorsque ϑ augmente. Le Tableau 3 représente les variations de l'intensité totale $I = I_1 + I_2$ pour les valeurs de ϑ comprises entre 0 et 25°. Ces intensités ont été nommées en divisant par le facteur Ea^2. La Fig. 5 en représente l'ensemble, exprimé en coordonnées polaires en divisant par le facteur $E\pi a^2$. Contrairement au cas où les gouttes sont très petites cette répartition est indépendante des dimensions des gouttes.

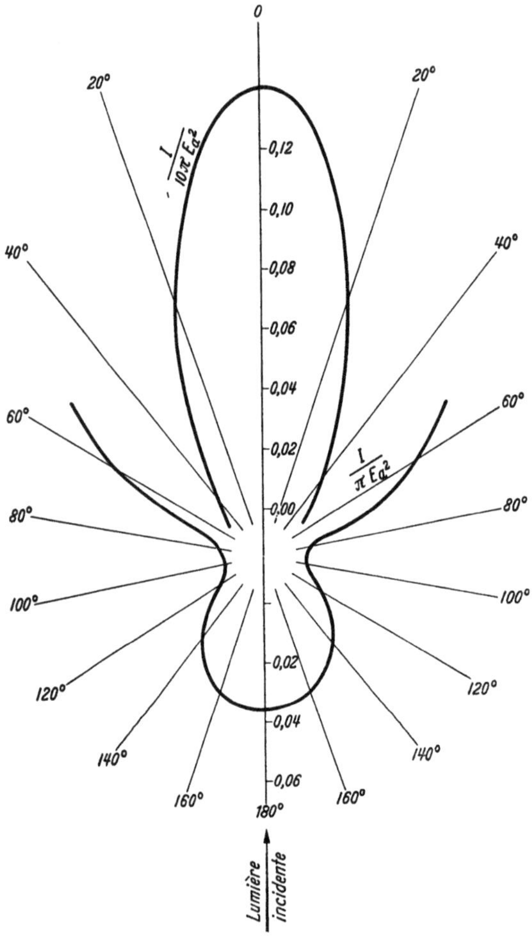

Fig. 5. Répartition de l'intensité lumineuse I diffusée par une goutte d'eau de rayon a par réflexion et réfraction.

γ) Lumiere diffractée par la goutte. La lumière diffractée par le contour de la goutte donne naissance aux *couronnes de diffraction*. Soit J l'intensité lumineuse diffractée par la goutte dans la direction ϑ. Elle est donnée par l'expression:

$$J = J_0 \left[\frac{2 J_1(u)}{u}\right]^2 \tag{8.12}$$

ou $J_1(u)$ représente la première fonction de BESSEL et u une variable auxiliaire:

$$u = \frac{2\pi a}{\lambda} \sin\vartheta. \tag{8.13}$$

J_0 est l'intensité du maximum de la tache centrale de diffraction. Si on veut comparer l'intensité de la lumière diffractée à celle de la lumière réfractée et réfléchie, il faut connaître J_0 en fonction de l'éclairement incident et du rayon a de la goutte; la théorie de la diffraction montre que J_0 est proportionnel à a^4/λ^2, mais ne permet pas la calcul de J_0 en valeur absolue. Ce calcul a été fait par BLUMER[1] par interpolation graphique pour des valeurs de $\alpha = 2\pi a/\lambda$ comprises entre 0,1 et 12; toutefois d'après KASTLER[2], il est plus facile de raisonner de la façon suivante: d'après le théoréme de BABINET le flux total diffracté:

$$\Phi = \int_{\vartheta} J(\vartheta)\, d\omega \tag{8.14}$$

est égal au flux lumineux tombant sur la goutte:

$$E\pi a^2 = \int_{\vartheta} J(\vartheta)\, 2\pi \sin\vartheta\, d\vartheta = 2\pi J_0 \int_{\vartheta} \left[\frac{2 J_1(u)}{u}\right]^2 \sin\vartheta\, d\vartheta. \tag{8.15}$$

---

[1] H. BLUMER: Z. Physik **32**, 119 (1925); **38**, 308, 942 (1926); **39**, 193 (1926).

[2] A. KASTLER: La diffusion de la lumière par les milieux troubles. Paris 1943.

Pour de petits angles ϑ, cette expression se simplifie. On peut écrire

$$\vartheta = \frac{\lambda}{2\pi a} u \quad \text{et} \quad \sin\vartheta\, d\vartheta = \vartheta\, d\vartheta = \left(\frac{\lambda}{2\pi a}\right)^2 u\, du, \tag{8.16}$$

$$\frac{E a^2}{J_0} = \frac{2\lambda^2}{\pi^2 a^2} \int_0^\infty \frac{J_1^2(u)}{u} du = \frac{2\lambda^2}{\pi^2 a^2} \frac{1}{\mu} \tag{8.17}$$

en posant

$$\int_0^\infty \frac{J_1^2(u)}{u} du = \frac{1}{\mu}, \qquad J_0 = E\frac{\pi^2 a^4}{\lambda^2}\mu. \tag{8.18}$$

D'après WATSON[1]

$$\int_0^\infty \frac{J_1^2(u)}{u} = \frac{1}{\mu} = \frac{1}{2} \tag{8.19}$$

d'où

$$J_0 = \frac{E\pi^2 a^4}{\lambda^2} = \frac{1}{4}\left(\frac{2\pi a}{\lambda}\right)^2 E a^2 = \frac{1}{4}\alpha^2 E a^2. \tag{8.20}$$

Il est facile de comparer l'intensité correspondant à $\vartheta = 0$: ($\gamma = 180°$) pour la lumière réfléchie et réfractée I_0 et la lumière diffractée J_0. La première a pour expression, d'après le Tableau 3, $I_0 = 4{,}086\,E a^2$ et la seconde $\frac{1}{4}\alpha^2 E a^2$; le rapport des deux est égal à $J_0/I_0 = \alpha^2/4 \cdot 4{,}086 = 6{,}12 \cdot 10^{-2}\alpha^2$. On voit qu'il est inférieur à l'unité pour les valeurs de $\alpha < 4$ et devient rapidement beaucoup plus grand pour les valeurs élevées, soit 882 pour $\alpha = 120$.

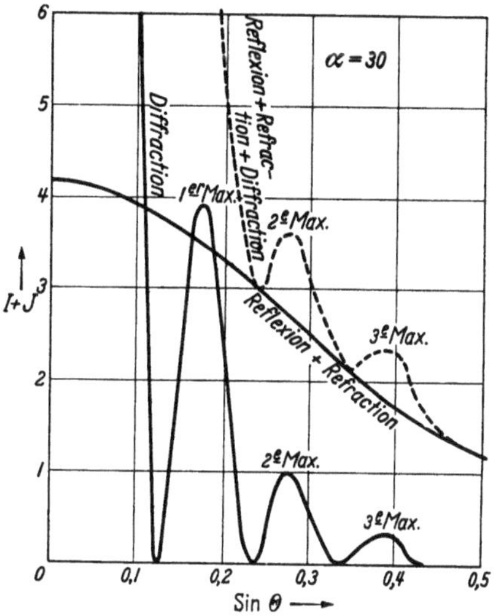

Fig. 6. Superposition de l'intensité lumineuse réfléchie, réfractée et diffractée par une goutte d'eau pour $\alpha = 30$. Champ des couronnes de diffraction. $\alpha = 2\pi a/\lambda$, a rayon des gouttes; λ = longueur d'onde de la lumière.

Les intensités des premières couronnes entourant la tache centrale de diffraction sont:

$$J_1 = 0{,}0175\, J_0, \quad J_2 = 0{,}0041\, J_0, \quad J_3 = 0{,}00016\, J_0; \tag{8.21}$$

soit en fonction α et de l'intensité I_0 réfléchie et réfractée dans la direction de propagation

$$J_1 = 1{,}071 \cdot 10^{-3} I_0\alpha^2; \quad J_2 = 2{,}509 \cdot 10^{-4} I_0\alpha^2; \quad J_3 = 9{,}792 \cdot 10^{-5} I_0\alpha^2 \tag{8.22}$$

pour $\alpha = 2\pi a/\lambda = 30$ on obtient

$$J_0 = 55{,}980\, I_0; \quad J_1 = 0{,}964\, I_0; \quad J_2 = 0{,}226\, I_0; \quad J_3 = 0{,}088\, I_0. \tag{8.23}$$

Dans ce cas, l'intensité des couronnes de diffraction est du même ordre de grandeur que celle de la lumière diffusée par la goutte par réflexion et réfraction: les couronnes sont noyées dans la lumière diffusée. Cet exemple montre que la superposition de la lumière réfractée et réfléchie à la lumière diffractée diminue le contraste des couronnes.

D'une façon générale, l'intensité lumineuse totale diffusée dans une direction faisant un angle ϑ quelconque avec celle du faisceau incident ($\vartheta = \pi - \gamma$) est égale à la somme $I + J$. La Fig. 6 empruntée a BRICARD[2] illustre les résultats

[1] WATSON: Theory of BESSEL Functions, p. 104.

[2] J. BRICARD: J. de Phys., Ser. VIII 4, 57 (1943).

de ce calcul, et indique les variations de I, J, $I+J$ dans le cas examiné ci dessus. On voit qu'entre le troisième minimum et le troisième maximum, l'intensité remonte à peine; à partir de la quatrième couronne, les maxima auront disparu.

Bien entendu, le calcul que nous venons d'exposer est loin d'être rigoureux. D'une part, la relation (8.12) doit étre corrigée, et remplacée par:

$$J = k J_0 \left[\frac{2 J_1(u)}{u}\right]^2 \tag{8.24}$$

le coefficient k étant égal à l'unité dans la direction correspondant au centre de la figure de diffraction, et diminuant lorsque ϑ augmente; on pose $k=\cos^4\vartheta/2$.

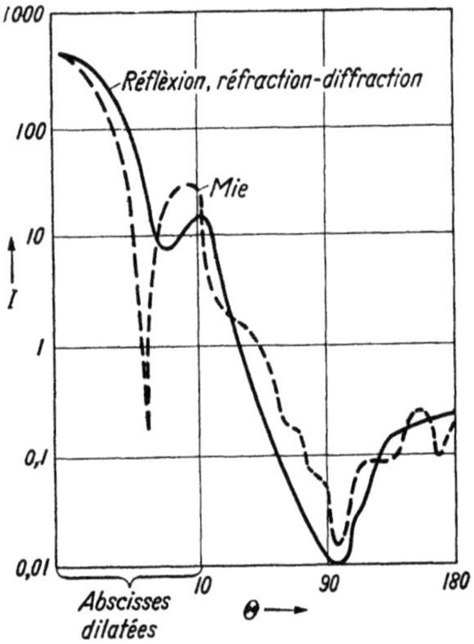

Fig. 7. Comparaison des indicatrices de diffusion d'une goutte calculées à partir de la théorie de MIE et de la théorie de la réfléxion, de la réfraction et de la diffraction.

D'autre part, dans la nature, il faut tenir compte de l'hétérogénéité du brouillard, et introduire dans les calculs la granulométrie du brouillard, ce qui a pour effet d'étaler les maxima et minima de diffraction, et de diminuer le contraste couronnes fond continu.

La Fig. 7, empruntée à OLIVIER[1], représente la comparaison entre les résultats ci dessus et ceux de GUMPRECHT pour $\alpha=30$; on voit qu'ils sont en bon accord.

En lumière blanche, pour chaque point du champ des couronnes, la couleur correspondante est obtenue en enlevant de la lumière blanche la teinte qui disparait au point considéré.

9. Cas où on ne peut pas considérer les vibrations réfléchies, réfractées et diffractées comme incohérentes. Nous venons de calculer l'intensité résultante dans une direction donnée comme étant la somme de trois quantités I_1, I_2, J. Mais lorsque les particules sont suffisamment petites, on ne peut plus considérer comme incohérentes les trois vibrations calculées ci-dessus (MECKE, KASTLER). If faut tenir compte des phases, et comparer les amplitudes. D'après (8.6), (8.9) et (8.20), ces dernières ont pour expression, en supprimant le facteur constant $\frac{1}{2}$ pour chacune des composantes pour simplifier les calculs:

$$\left.\begin{aligned} a_0 &= \alpha \cos^2\frac{\vartheta}{2}\left[\frac{2J_1(u)}{u}\right] a\sqrt{E}\,, \\ a_1 &= \sqrt{k_1}\, a\sqrt{E}\,, \qquad a_2 = (1-k_2)\sqrt{E}\sqrt{2}\,\sqrt{\frac{\sin 2i}{m\vartheta}\,\frac{di}{d\vartheta}}\,a\,, \end{aligned}\right\} \tag{9.1}$$

et leur résultante s'écrit:

$$a_0\cos(\omega t+\varphi_0)+a_1\cos(\omega t+\varphi_1)+a_2\cos(\omega t+\varphi_2)\,. \tag{9.2}$$

En ce qui concerne la diffraction, un trou circulaire de diamètre égal à celui de la particule diffracterait une vibration de forme $a_0\cos(\omega t+\pi/2)$ l'avance de phase $\pi/2$ étant caractéristique des phénomènes de diffraction; mais la particule diffracte comme l'écran complémentaire du trou et fournira, d'après le théorème de BABINET, une vibration de même amplitude et de phase opposée, soit

$$a_0\cos\left(\omega t-\frac{\pi}{2}\right).$$

[1] J. OLIVIER: Bull. Soc. franç. Electr., Ser. VII 5 (1955).

En ce qui concerne la réfraction il est facile d'évaluer sur la Fig. 8 la différence de marche entre le rayon $CABD$ et le rayon diffracté EOF, qui, en prenant comme origine le plan d'onde de trace DF, sera

$$\Delta = CABD - EOF \tag{9.3}$$

ce qui correspond, à un retard entre les deux rayons

$$\Delta = 2a(m\cos r - \cos i) \tag{9.4}$$

ou à un retard de phase:

$$\frac{4\pi a}{\lambda}(m\cos r - \cos i) = 2\alpha(m\cos r - \cos i), \tag{9.5}$$

et en tenant compte du fait que les rayons convergents passent par un foyer, puis divergent ensuite, ce qui correspond à une avance de phase de π, on trouve que l'avance de phase du rayon réfracté a pour valeur:

$$\varphi_2 = \pi - 2\alpha(m\cos r - \cos i). \tag{9.6}$$

En ce qui concerne la réflexion, le rayon réfléchi en I gagne par rapport au rayon passant par 0 une avance de marche $2a\sin\frac{\vartheta}{2}$, soit en tenant compte du renversement de phase du à la réflexion, une avance de phase:

$$\varphi_1 = \pi + \frac{4\pi a}{\lambda}\sin\frac{\vartheta}{2} = \pi + 2\alpha\sin\frac{\vartheta}{2}. \tag{9.7}$$

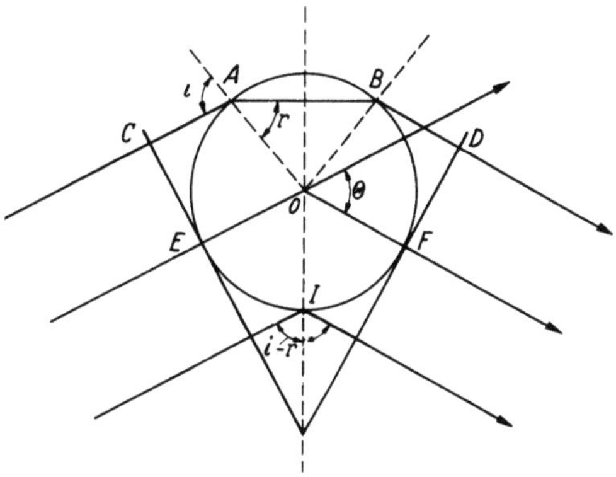

Fig. 8. Voir Fig. 1, p. 331.

En prenant pour origine des phases celle de la vibration diffractée, on a pour résultante de ces trois vibrations:

$$A = a_0\cos\omega t + a_1\sin(\omega t - \Phi_1) + a_2\sin(\omega t - \Phi_2) \tag{9.8}$$

avec

$$\Phi_2 = 2\alpha(m\cos r - \cos i) = 2\alpha\frac{\sin\frac{\vartheta}{2}}{\sin i}, \qquad \Phi_1 = 2\alpha\sin\frac{\vartheta}{2}. \tag{9.9}$$

La vibration résultante est de la forme:

$$A = M\cos\omega t + N\sin\omega t \tag{9.10}$$

et l'intensité lumineuse diffusée a pour expression:

$$I = M^2 + N^2 \tag{9.11}$$

soit

$$I = a_0^2 + a_1^2 + a_2^2 - 2a_0 a_2\sin\Phi_2 + 2a_0 a_1\sin\Phi_1 + 2a_1 a_2\cos(\Phi_1 + \Phi_2) \tag{9.12}$$

qui définit l'intensité dans une direction quelconque à partir des calculs de WIENER et de BESSEL.

Le Tableau 4 représente d'aprés BRICARD[1] les résultats de la relation (9.12) multipliés par le rapport $\frac{1}{E}\left(\frac{4\pi}{\lambda}\right)$, E représentant toujours l'éclairement de la goutte.

[1] J. BRICARD: Ann. de Géophys. 2, 231 (1946).

Tableau 4. *Répartition de la lumière diffusée par une goutelette d'eau sphérique dans le champ des couronnes de diffraction pour $\alpha = 2\pi a/\lambda = 10$ et 20. a = rayon des gouttes, λ = longueur d'onde de la lumière. Pour la définition de u, voir Equ. (8.13)*

u	0	1	2	3	4	5	6	7	8	9	10
$\alpha = 10$	1,310	1,861	0,592	0,012	0,286	0,230	0,028	0,062	0,019		
$\alpha = 20$	27,671	24,591	8,676	0,160	2,215	3,952	1,726	0,590	1,512	1,247	0,316

On voit que la lumière est répartie de façon différente dans ce cas, et par exemple, le maximum central de diffraction qui se manifestait encore pour $\alpha = 20$, fait place à un minimum pour $\alpha = 10$.

10. Vérifications experimentales. D'une façon générale, il n'existe pas de vérifications de ces travaux effectuées au laboratoire comme dans le cas des petites particules. Les seules mesures que l'on connaisse ont été effectuées en plein air, en observant latéralement la lumière diffusée par un faisceau parallèle dans le brouillard. En dehors du champ des couronnes, la répartition de la lumière diffusée parait en accord avec le calcul de WIENER [*5*], et ne dépend pas des dimensions des gouttes (BRICARD[1]). L'accord semble seulement qualitatif pour des valeurs de la déviation supérieures à 40°.

Des mesures plus récentes paraissent avoir pour résultat de contester cet accord (UNO et YOSIDA[2]). LANDER et NIELSEN[3] trouvent notamment un accord satisfaisant avec la théorie de WIENER pour a (rayon des gouttes) $= 4{,}8\,\mu$, et contrairement à ce qu'on pourrait attendre en se plaçant au point de vue théorique ils trouvent que la proportion de lumière diffusée vers l'arrière augmente lorsque le rayon atteint la valeur $a = 14\,\mu$.

L'accord entre la position angulaire des maxima et minima de diffraction a été longuement étudié par MECKE en milieu artificiel. Les plus récentes mesures, effectuées par BRICARD, confirment les résultats précédemment obtenus, et confirment également l'accord avec la théorie de la diffraction.

Dans le cas du Spectre Infra Rouge, qui correspondrait dans les nuages ou les brouillards naturels à des valeurs de α inférieures à 10, le problème est encore compliqué par l'absorption propre des gouttes (dans les régions de grande absorption comme dans la bande des $6\,\mu$, une couche d'eau d'épaisseur $10\,\mu$ transmet 10% du rayonnement incident); d'autre part, le pouvoir réflecteur de l'eau est très faible (quelques centièmes seulement du rayonnement incident sont réfléchis entre 5 et $15\,\mu$) de sorte que l'intensité diffusée dans de telles conditions doit être très faible. Enfin, il faut tenir compte de l'absorption propre de la vapeur d'eau, et vers les grandes longueurs d'onde, du rayonnement propre. On ne dispose pas, à l'heure actuelle, de renseignements concernant ce domaine spectral.

11. Étude particulière de la lumière diffusée en sens inverse de propagation du faisceau. *α) Arcs en ciel.* Nous venons d'étudier la lumière diffusée vers l'avant, qui correspond soit à une réflexion sur la surface extérieure de la goutte, soit à la lumière transmise après deux réfractions. Mais ceci ne constitue qu'une partie de l'énergie lumineuse incidente, le reste correspondant à la lumière qui a subi une ou plusieurs réflexions à l'intérieur de la goutte. Sur la Fig. 9, on voit que le rayon AB est partiellement réfléchi en B et émerge en E, ou il subit encore une réflexion partielle, le reste émerge en F, avec une seconde réflexion partielle etc.

Il est facile de montrer que la déviation subie par un rayon qui a subi k réflexions successives a pour expression:

$$\vartheta = k\pi + 2i - 2(k+1)\,r \tag{11.1}$$

[1] J. BRICARD: J. de Phys., Ser. VIII **4**, 57 (1943).
[2] K. UNO et Z. YOSIDA: Low Temp. Sci. Hokkaido Univ. **2**, 130 (1949).
[3] R. L. LANDER et C. E. NIELSEN: Rev. Sci. Instrum. **24**, 20 (1953).

soit pour une seule réflexion interne:

$$\vartheta = \pi - 2(i - 2r). \tag{11.2}$$

La déviation passe par un minimum, correspondant aux rayons qu'on appelle *rayons efficaces*, quand $d\vartheta/di = 0$ (il s'agit bien d'un minimum car on montre facilement que la dérivée seconde $d^2\vartheta/di^2$ est toujours positive).

On montre facilement que la valeur I de i correspondant au minimum de déviation est donnée par

$$\sin I = \sqrt{\frac{(k+1)^2 - m^2}{(k+1)^2 - 1}}. \tag{11.2 bis}$$

On en déduit la déviation ϑ correspondante.

On trouve que pour l'eau ($m = 1{,}33$) et $k = 1$, la direction des rayons efficaces correspond à $\vartheta = 138°$; les rayons qui leur donnent naissance frappent la goutte sous l'incidence efficace. On calculerait de même la déviation des rayons efficaces ayant subi deux, trois réflexions etc.

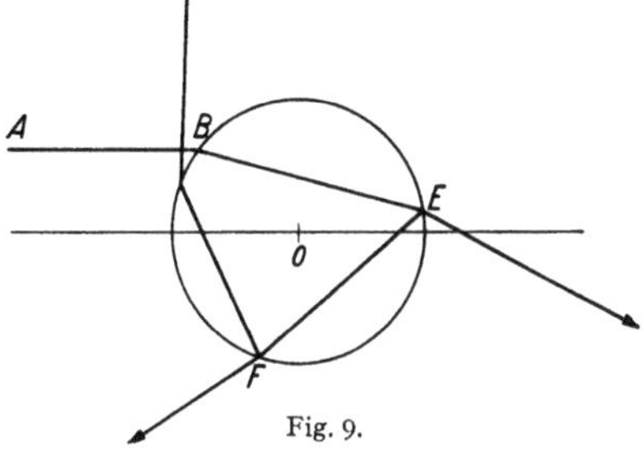

Fig. 9.

Au voisinage du minimum de déviation, les rayons conservent sensiblement la même direction, et peuvent être confondus pour des incidences voisines de celle qui correspond au minimum; la théorie de DESCARTES consiste à supposer que dans l'arc en Ciel naturel, ces rayons seuls ont pratiquement une intensité suffisante pour être perçus. On peut ainsi déterminer la position des Arcs de différents ordres, qui correspondent à un nombre croissant de réflexions intérieures. Tous ces Arcs paraissent évidemment circulaires, puisqu'ils sont déterminés par des droites qui font un angle constant avec la direction de la source.

En fait, les phénomènes naturels s'observent en lumière blanche, dans le cas de gouttes de pluie éclairées par le soleil. En tenant compte de la dispersion de l'eau, il est facile de montrer que le rayon angulaire du premier Arc en Ciel (correspondant à une seule réflexion interne), qui se produit à l'opposé du soleil, est compris entre $\gamma = \pi - \vartheta = 42°18$ pour le rouge et 40°23 pour le violet, le rouge paraissant ainsi à l'extérieur. L'Arc du second ordre correspond à une déviation supérieure à 180° tout en se produisant encore à l'opposé du soleil, avec un diamètre angulaire de 50°34 pour le rouge et 53°56 pour le violet, son éclat étant beaucoup moindre que celui du premier ordre à cause des deux réflexions successives et de l'étalement beaucoup plus grand des couleurs, dont la disposition est inversée par rapport à celle du premier.

Les mesures faites sur des Arcs en Ciel Naturels ont donné des résultats sensiblement en accord avec ceux ci, mais il n'est pas inutile d'insister sur le caractère approximatif de ces mesures, les limites du phénomène étant elles-mêmes très mal définies. D'autre part, à la simple observation, il est facile de s'apercevoir qu'il est en réalité plus complexe que nous ne l'indiquons ci dessus. Outre la bande des couleurs de spectre, il apparait à l'intérieur du premier Arc, et à l'extérieur du second, une série d'arcs colorés qui sont les *Arcs surnuméraires*, qui paraissent être la répétition de l'Arc principal sur une moindre largeur, en rappelant par leur aspect les franges de diffraction d'un écran à bords rectilignes.

On peut montrer que les rayons qui sortent de la goutte sont tangents à une surface caustique de révolution autour de la droite parallèle a la lumière primitive, qui passe par le centre de la sphère. La méridienne de cette caustique, représentée par la Fig. 10, possède une asymptote, dont la direction est celle des rayons efficaces.

AIRY a étudié théoriquement les phénomènes de diffraction qui se produisent au voisinage de cette caustique (on trouvera à sujet des explications détaillées dans les traités classiques de MASCART [2] ou de PERNTER et EXER [1]) et a conclu qu'il se forme un système de franges alternativement sombres et brillantes, dont les emplacements angulaires sont déterminés de la facon suivante: Soit α l'angle que fait une certaine direction, située dans le plan de la figure, avec celle des rayons efficaces, cet angle étant compté positivement dans le même sens que ϑ (Fig. 10).

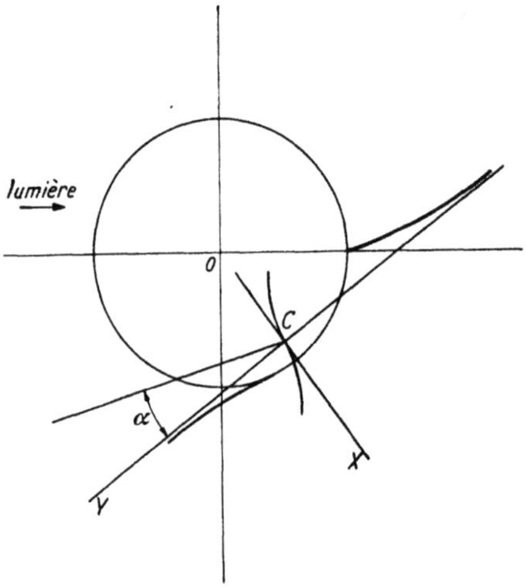

Fig. 10 La lumière vient de la gauche.

Posons:

$$\alpha = \frac{Z}{2}\left(\frac{h\,\lambda^2}{6a^2}\right)^{\frac{1}{3}} \tag{11.3}$$

ou a représente le rayon des gouttes, λ la longueur d'onde et h une constante qui, pour l'eau a pour valeur $h = 4{,}896$. Les maxima et minima du système de franges ont caractérisés par les valeurs suivantes de Z:

	Maxima	Minima
1	$Z = 2{,}4955$	$Z = 1{,}0845$
2	$Z = 4{,}3631$	$Z = 3{,}4669$
3	$Z = 5{,}8922$	$Z = 5{,}1446$
4	$Z = 7{,}2436$	$Z = 6{,}5782$

Les mesures de rayons angulaires effectuées en lumière monochromatique vérifient approximativement ce calcul.

Pour établir ces résultats, on raisonne de la façon suivante:

Les surfaces d'Onde relatives à différents instants sont elles mêmes de révolution et leurs courbes méridiennes sont les développantes de la caustique. Comme on peut, pour les calculs, considérer cette surface pour une quelconque de ses positions, nous prendrons celle qui passe par le point C de la perpendiculaire OC abaissée du centre sur l'asymptote (voir MASCART, HUMPHREYS, PERNTNER-EXNER). Le point C correspond à un point d'inflèxion de la méridienne de la surface d'onde.

Prenons la tangente CO au point d'inflèxion comme axe des X et comme axe OY le rayon le moins dévié (ou l'asymptote elle même, Fig. 10); cette méridienne, qui possède à l'origine un point d'inflèxion a pour équation:

$$y = b\,x^3. \tag{11.4}$$

Pour calculer b, remarquons qu'au voisinage du minimum, le formule de TAYLOR donne:

$$D = (\vartheta + \pi) = D_m + \frac{1}{2}\frac{d^2 D}{d i^2}\,d i^2 + \cdots. \tag{11.5}$$

Un calcul simple permet d'évaluer $d^2\vartheta/di^2$ en se basant sur les relations (11.1) et (11.2). On obtient pour l'Arc:

$$D = D_m + \frac{(k+1)^2 - 1}{(k+1)^2}\sqrt{\frac{(k+1)^2 - m^2}{m^2 - 1}}\,d i^2. \tag{11.6}$$

Le rayon ϑ_m est le moins dévié; sa distance au centre de la goutte est $\delta = a \sin I$, I étant la valeur de i correspondant à la direction des rayons efficaces. Le rayon ϑ voisin n'est pas tout à fait perpendiculaire à l'axe des x, mais il fait avec lui un angle très petit, dont le cosinus peut être confondu avec l'unité; il coupe donc l'axe Ox à une distance du premier égale à:

$$x = d\delta = a \cos I\, d i = a\sqrt{\frac{m^2 - 1}{(k+1)^2 - 1}}\,d i. \tag{11.7}$$

Ces deux rayons, qui font entre eux l'angle $D - D_m = \vartheta - \vartheta_m$ sont tous les deux normaux à une courbe dont l'équation est (11.4). Il est facile de voir que:

$$\frac{dy}{dx} = \operatorname{tg}(\vartheta - \vartheta_m) = \vartheta - \vartheta_m = 3b x^2. \tag{11.8}$$

On trouve ainsi:

$$b = \frac{3}{4a^2}\sqrt{\frac{4 - m^2}{(m^2-1)^3}}. \tag{11.9}$$

Pour l'eau $\left(m = \frac{4}{3}\right)\sqrt{\frac{4-m^2}{(m^2-1)^3}} = 2{,}176.$

Le retard de la vibration envoyée par le point A (Fig. 10 bis) de la section droite par rapport à la vibration envoyée par le point O dans la même direction est en première approximation

$$\varphi = (DC - AC) = y\cos\beta + x\sin\beta = bx^3\cos\beta + x\sin\beta. \tag{11.10}$$

Nous admettons que nous nous plaçons toujours dans une direction voisine de OY, c'est à dire des rayons efficaces. β est toujours petit. On peut écrire:

$$\varphi = bx^3 + x\beta. \tag{11.11}$$

Le mouvement envoyé par la portion de l'onde utile est:

$$\int \sin\left(\omega t - \frac{2\pi\varphi}{\lambda}\right)dx = \sin\omega t\int\cos\frac{2\pi\varphi}{\lambda}\,dx - \cos\omega t\int\sin\frac{2\pi\varphi}{\lambda}\,dx. \tag{11.12}$$

φ étant une fonction impaire de x, la seconde intégrale s'annule pour les mêmes limites de part et d'autre du point O.

L'intensité cherchée est donc définie, à un facteur constant près, par le carré de l'intégrale:

$$\int\cos\frac{2\pi\varphi}{\lambda}\,dx = \int\cos\frac{2\pi}{\lambda}(b x^3 + \beta x)\,dx. \tag{11.13}$$

En réalité, les limites de cette intégrale sont très voisines l'une de l'autre, mais le terme en x^3 croissant très vite, le cosinus prend toutes les valeurs comprises entre -1 et $+1$ pour une variation très faible de x: la valeur moyenne de l'intégrale pour cette variation est nulle. On commet donc une erreur négligeable en calculant entre $-\infty$ et $+\infty$, ou entre 0 et $+\infty$ puisqu'on calcule l'intensité à un facteur constant près. Posons

$$u^3 = \frac{4b}{\lambda}x^3, \qquad -zu = \frac{4x\beta}{\lambda}. \tag{11.14}$$

L'intensité est définie par le carré de l'intégrale:

$$f = \int_0^\infty \cos\frac{\pi}{2}(u^3 - zu)\,du. \tag{11.15}$$

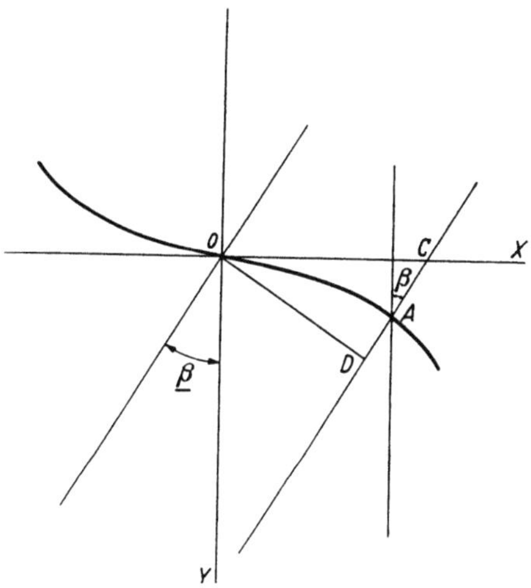

Fig. 10 bis.

Les figures précédentes montrent que les angles β qui correspondent aux valeurs négatives de z, définissent des directions pour lesquelles il n'existe aucun rayon lumineux. Quand on se trouve à l'intérieur de la caustique, dans l'ombre géométrique, la fonction f^2 ne présente ni maximum ni minimum. Il n'y a pas de franges dans l'ombre géométrique.

Au contraire, pour des valeurs de z positives, c'est à dire pour des valeurs de β correspondant à des rayons réels, il y a des franges. La fonction f^2 présente alternativement des maxima et des minima, qui satisfont approximativement à la condition

$$z = 3\sqrt[3]{\left(\frac{2m+1}{4}\right)^2}. \tag{11.16}$$

L'intensité est nulle pour m entier impair, maxima pour m entier pair.

Les angles α correspondants avec la direction des rayons efficaces seront ainsi en valeur absolue:

$$\alpha = \frac{3}{2}\sqrt[3]{\frac{b}{2}}\sqrt[3]{\left(\frac{2m+1}{4}\right)^2\lambda^2} = \frac{3}{2}\sqrt[3]{\frac{4{,}896}{6a^2}}\sqrt[3]{\left(\frac{2m+1}{4}\right)^2\lambda^2}. \tag{11.17}$$

Les résultats numériques que l'on obtient à partir de cette relation approchée sont légèrement différents de ceux du Tableau 5, qui proviennent de l'intégration de la relation (11.15) effectuée par Stokes (voir Mascart [*2*]).

En lumière blanche, on voit des arcs colorés, conformément à ce que prévoit la formule (11.3), la séparation des couleurs étant d'autant plus nette que le rayon des gouttes est plus grand (gouttes de pluie, rayon compris en 0,1 et 1 mm).

Si on est en présence de brouillard (rayon des gouttes compris entre quelques μ à quelques dizaines de μ), on aperçoit, même avec des sources de lumière artificielle un *arc en Ciel blanc*. Une particularité de cet arc blanc est que son rayon angulaire varie entre des limites très écartées, comprises entre une trentaine de degrés(correspondant à des gouttes de quelques μ de rayon) et une quarantaine de degrés.

La diminution du rayon de l'arc s'explique facilement d'après la formule (11.3) par la petitesse des gouttes, puisque le déplacement des maxima dépasse 6° pour un diamètre de 10 μ. Quant à la superposition des couleurs, elle tient en partie à l'existence de gouttes de diamètres très différents, mais la cause principale en est due à l'extension progressive des franges à mesure que le diamètre des gouttes diminue. Dans ce cas, les intensités relatives des diffèrentes couleurs conservent assez longtemps des valeurs égales sur l'Arc principal. On trouvera des discussions très complétes à ce sujet dans les ouvrages classiques cités plus haut [*1*], [*2*].

On constate également, dans tous les cas, que l'état de polarisation des franges confirme les prévisions que l'on peut faire en se basant sur le mécanisme décrit ci dessus (vecteur lumineux tangent aux arcs). Mais il existe toute fois un second systèmes de franges polarisées radialement, qui n'a été signalé que dans le cas des Arcs blancs (Bricard[1]) et que cette théorie ci dessus ne permet pas d'expliquer; par contre, on voit sur la Fig. 3 que la théorie de Mie prévoit deux systèmes différents de franges, caractérisés l'un par un vecteur lumineux tangent aux franges, l'autre par un vecteur lumineux radial; aux environs d'un diamètre angulaire $\gamma = 20°$, pour $2\pi a/\lambda = 40$, un maximum d'intensité du premier correspond à un minimum du second, ce qui parait en accord avec les résuls expérimentaux cités ci dessus.

β) Anticouronnes. Ainsi que nous l'avons déja vu, les anticouronnes se voient naturellement pour un observateur qui se trouve placé entre le soleil et un nuage, de telle facon que son ombre se projette sur celui-ci. On distingue en général toutes les parties de l'ombre, et autour de la tête, une auréole constituée par trois ou quatre couronnes concentriques de couleurs très vives. C'est autour de ces anticouronnes que l'on aperçoit concentriquement l'Arc en Ciel blanc. Les mêmes phénomènes sont d'ailleurs visibles pour un observateur qui se trouve dans le faisceau d'un projecteur, et qui regarde dans le sens de propagation de la lumière.

Considérons la méridienne de la surface d'onde se rapportant aux rayons émergents après une réflexion intérieure, et passant par le point d'intersection de l'asymptote de la caustique avec la surface sphérique de la goutte (Fig. 11). D'après le calcul de Wiener [*5*], la partie LD correspond à tous les rayons dont l'incidence est inférieure à l'incidence efficace et lorsqu'on dépasse cette incidence efficace, la méridienne dépasse l'asymptote au point d'inflèxion D, jusqu'en M ou elle rencontre la méridienne de la caustique, puis revient en arrière. Si on confond au point D la méridienne de la caustique avec son asymptote, la méridienne de la surface d'onde se réduit à l'arc LD; sauf au voisinage du point D,

[1] J. Bricard: Ann. de Phys. **14**, 148 (1940).

ou sa courbure est infinie, l'arc LD peut étre confondu avec un arc de cercle de centre B, le long duquel l'amplitude de la vibration qui subit deux réfractions et une réflexion interne reste constante.

D'après RAY[1], on pourrait assimiler cette surface d'onde à une simple calotte sphérique de rayon convenable. En première approximation, on supposera que la centre de cette onde sphérique est l'image d'un point placé à l'infini sur l'axe, après deux réfractions et une réflexion; on expliquera les anticouronnes comme des franges de diffraction produites par une petite ouverture circulaire, dont le diamètre est facile à déterminer, et un point lumineux placé à faible distance de cette ouverture sur son axe. Ce problème a été traité théoriquement par LOMMEL[2] et GRAY et MATHEWS[3].

Les diamètres angulaires des anticouronnes que l'on observe vérifient approximativement cette façon de voir; notons qu'en général, les anticouronnes qui se produisent dans les nuages sont assez mal définies.

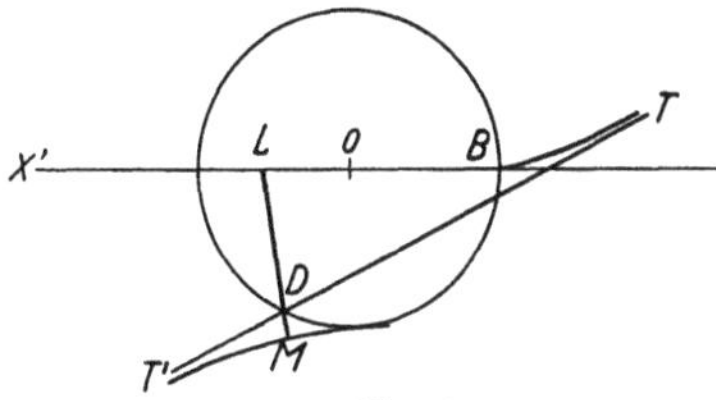

Fig. 11.

VAN DE HULST[4] considère le raisonnement ci dessus comme une première approximation. La solution proposée par lui, basée également sur les lois de la réfraction et de la réflexion, conduisant à une impossibilité en ce qui concerne les gouttes d'eau ($m = 1{,}33$), il faut en conclure que les lois de l'optique géométrique ne permettent pas d'expliquer de façon rigoureuse les anticouronnes que l'on observe dans la nature; il faut avoir recours à la théorie de MIE [*4*]. L'approximation de BUCERIUS (Sect. 4) en indique une solution[5].

B. Transmission et diffusion de la lumière par un nuage de gouttelettes.

Au cours de l'exposé ci dessus, on a étudié la répartition de la lumière diffusée par une particule isolée, éclairée par un faisceau de lumière parallèle, puis la répartition de l'énergie diffusée par un élément de volume contenant un certain nombre de particules de dimensions identiques, ou ayant une granulomètrie connue.

Nous nous proposons maintenant d'étendre ces résultats au cas d'un nuage diffusant, en tenant compte non seulement de l'éclairement direct que recoivent les particules, en provenance du soleil ou d'une source artificielle, mais aussi de la lumière reçue par une particule en provenance des autres.

Nous calculerons d'abord l'éclairement direct provenant de la source, reçu par un élément de surface, ou ce qui revient au même l'énergie transmise directement par une couche diffusante d'épaisseur finie.

I. Transmission des radiations par les atmosphères diffusantes.

12. Définitions. Lorsqu'elles se propagent dans l'atmosphère, les radiations sont absorbées, c'est à dire transformées en chaleur, et diffusées, chaque particule matérielle (molécules d'air, noyaux de condensation, gouttelettes d'eau) diffusant

[1] B. RAY: Proc. Ind. Assoc. Cultiv. Sci. **8**, 36, 42, 29 (1923).
[2] A. LOMMEL: Abh. bayer. Akad. Wiss. **15** (1886).
[3] GRAY et MATHEWS: Treatise on BESSEL Functions. London 1931.
[4] C. VAN DE HULST: J. opt. Soc. Amer. **37**, 16 (1947).
[5] Il serait intéressant de vérifier, comme le veut le calcul de BUCERIUS, que le centre des Anticouronnes est toujours noir, ce qui ne parait pas très solidement établi par l'expérience.

dans toutes les directions une partie de l'énergie incidente. Sauf dans le domaine visible, où l'absorption vraie est pratiquement négligeable, il faut donc tenir compte des bandes d'absorption des gaz et vapeurs, dont nous ne nous occuperons pas ici, et dont l'effet se superpose à celui des gouttelettes d'eau.

Considérons un faisceau cylindrique et monochromatique, se propageant à travers une couche nuageuse ou brumeuse d'épaisseur x, et une tranche élémentaire dx perpendiculaire à l'axe du faisceau. Soit Φ le flux venant frapper la tranche. Une partie $d\Phi$ est diffusée ou absorbée par les gouttes; nous admettrons qu'elle est proportionnelle à Φ et nous écrirons:

$$d\Phi = -K\Phi\, d x, \tag{12.1}$$

K représentant le *coefficient d'absorption totale*; si les gouttes sont parfaitement transparentes, ce qui est pratiquement le cas pour l'eau dans le visible, il y a absorption apparente, par opposition à l'absorption vraie, et toute l'énergie empruntée au faisceau incident se retrouve sous forme d'énergie diffusée. On peut écrire dans ce cas d'après (5.1), (5.2) et (5.4)

$$K = 2\pi \int_0^\pi \sum N_i f_i(a_i, \gamma, m, \lambda) \sin\gamma\, d\gamma = 2\pi \int_0^\pi f_1(\gamma) \sin\gamma\, d\gamma. \tag{12.2}$$

En intégrant pour une épaisseur x:

$$\Phi = \Phi_0 e^{-Kx} \tag{12.3}$$

Φ_0 représentant le flux incident et Φ le flux transmis.

On désigne par *facteur de transmission* le rapport $\tau = \Phi/\Phi_0$, et la *densité optique* décimale du milieu est égale à $\frac{1}{x}\log_{10}\frac{1}{\tau}$; la quantité $\frac{10}{x}\log_{10}\frac{1}{\tau}$ s'exprime en décibels.

Introduisons dans ces relations les propriétés spécifiques du milieu diffusant, et exprimons K en fonction des grandeurs qui le caractérisent. Considérons d'abord une sphère unique.

Le champ résultant en un point quelconque extérieur à la sphére est la somme du champ incident et du champ diffusé, soit:

$$E = E_i + E_d, \qquad H = H_i + H_d. \tag{12.4}$$

Le vecteur complexe de Poynting associé avec ce champ résultant (2.12 bis) est radial et s'écrit:

$$S_c = \tfrac{1}{2}(E_\vartheta H_\Phi^* - E_\Phi H_\vartheta^*) \tag{12.5}$$

l'astérisque caractérisant le conjugué complexe.

En utilisant la relation (12.4), on écrira:

$$\left.\begin{aligned} S_c = {}& \tfrac{1}{2}(E_\vartheta^i H_\Phi^{i*} - E_\Phi^i H_\vartheta^{i*}) + \tfrac{1}{2}(E_\vartheta^d H_\Phi^{d*} - E_\Phi^d H_\vartheta^{d*}) + \\ & + \tfrac{1}{2}(E_\vartheta^i H_\Phi^{d*} + E_\vartheta^d H_\Phi^{i*} - E_\Phi^i H_\vartheta^{d*} - E_\Phi^d H_\vartheta^{i*}). \end{aligned}\right\} \tag{12.6}$$

Considérons une sphère concentrique à la sphère diffusante, et de rayon R. La partie réelle de S_c intégrée sur cette sphère est égale au flux total à travers sa surface. Pour simplifier, admettons que le milieu extérieur [2] soit un diélectrique parfait, de sorte que $k_2 = \omega\sqrt{\varepsilon_2 \mu_2}$ (voir 2.1). Si la sphère diffusante est également non conductrice, le flux total résultant qui traverse la surface sphérique entourant la sphère est nul, mais si une partie de l'énergie incidente est transformée en chaleur, ce flux résultant est égal à la quantité d'énergie absorbée, et dirigé vers l'intérieur. Appelons W_a l'énergie absorbée par la sphère, soit:

$$W_a = \operatorname{Re} \int_0^{\pi} \int_0^{2\pi} (-S_c)\, R^2 \sin\vartheta\, d\Phi\, d\vartheta. \tag{12.7}$$

Le premier terme de (12.6) représente le flux d'énergie dans l'onde incidente; il s'annule lorsque $\sigma_2 = 0$ quand on l'intègre à travers une surface fermée. Le second représente l'énergie qui sort, c'est à dire l'énergie diffusée par la sphère, qui a pour expression:

$$W_d = \tfrac{1}{2} \operatorname{Re} \int_0^{\pi} \int_0^{2\pi} (E_\vartheta^d H_\Phi^{d*} - E_\Phi^d H_\vartheta^{d*})\, R^2 \sin\vartheta\, d\Phi\, d\vartheta \tag{12.8}$$

ce qui peut s'écrire d'après (2.12 bis) et (2.13):

$$W_d = \frac{1}{2} \frac{k_2}{\mu_2 \omega} \int_0^{\pi} \int_0^{2\pi} (|E_\vartheta^d|^2 + |E_\Phi^d|^2)\, R^2 \sin\vartheta\, d\Phi\, d\vartheta. \tag{12.9}$$

Pour maintenir l'équilibre, le 3ème terme de (12.6) intégré sur la sphère, doit être égal à la somme des énergies absorbées et diffusées, soit:

$$\left.\begin{aligned} W_t &= W_a + W_d \\ &= -\tfrac{1}{2} \operatorname{Re} \int_0^{\pi} \int_0^{2\pi} (E_\vartheta^i H_\Phi^{d*} + E_\vartheta^d H_\Phi^{i*} - E_\Phi^i H_\vartheta^{d*} - E_\Phi^d H_\vartheta^{i*})\, R^2 \sin\vartheta\, d\vartheta\, d\Phi. \end{aligned}\right\} \tag{12.10}$$

W_t représentant la quantité totale d'énergie soustraite par la sphère au flux incident.

On trouve ainsi, en se basant sur les relations (2.13)

$$W_t = \pi \frac{E_0^2}{k_2} \sqrt{\frac{\varepsilon_2}{\mu_2}} \operatorname{Re} \sum_{n=1}^{\infty} (2n+1)(a_n^d + b_n^d), \tag{12.11}$$

$$W_d = \pi \frac{E_0^2}{k_2} \sqrt{\frac{\varepsilon_2}{\mu_2}} \sum_{n=1}^{\infty} (2n+1)(|a_n^d|^2 + |b_n^d|^2), \tag{12.12}$$

avec $k_2 = 2\pi/\lambda$.

L'énergie moyenne incidente par unité de surface est:

$$W = \frac{1}{2} E_0^2 \sqrt{\frac{\varepsilon_2}{\mu_2}}. \tag{12.13}$$

La section droite d'absorption et la section droite de diffusion de la sphère sont définies comme le rapport de l'énergie absorbée ou diffusée à la densité d'énergie de l'onde incidente, soit:

$$Q_t = \frac{W_t}{W} = \frac{\lambda^2}{2\pi} \operatorname{Re} \sum_{n=1}^{\infty} (2n+1)(a_n^d + b_n^d), \tag{12.14}$$

$$Q_d = \frac{W_d}{W} = \frac{\lambda^2}{2\pi} \sum_{n=1}^{\infty} (2n+1)(|a_n^d|^2 + |b_n^d|^2). \tag{12.15}$$

Nous avons vu (Sect. 5) que l'on peut étendre ces résultats à un élément de volume contenant un certain nombre de particules de granulométrie connue, en ajoutant purement et simplement les effets de chacune. En se reportant à la définition (12.1), on exprimera le coefficient d'absorption totale K_t et le coefficient d'absorption par diffusion K_d par:

$$K_t = \sum N_i Q_{it}, \quad K_d = \sum N_i Q_{id} \qquad (12.15\text{ bis})$$

Ces deux quantités étant égales lorsque les gouttes sont constituées par un milieu transparent.

13. Cas des ondes radioélectriques courtes. Si λ est beaucoup plus grand que le diamètre des particules $\left(\frac{2\pi a}{\lambda} \ll 1\right)$, les relations (12.14) et (12.15) se simplifient, du fait que les relations (2.5) peuvent s'écrire dans ce cas

$$\left.\begin{aligned} z_n^{(1)}\left(\frac{2\pi a}{\lambda}\right) &= 2^n\left(\frac{2\pi a}{\lambda}\right)^n \sum_{m=0}^{\infty} \frac{(-1)^m (n+m)!}{m!(2n+2m+1)!}\left(\frac{2\pi a}{\lambda}\right)^{2m}, \\ z_n^{(3)}\left(\frac{2\pi a}{\lambda}\right) &= 2^n\left(\frac{2\pi a}{\lambda}\right)^n \sum_{m=0}^{\infty} \frac{(-1)^m (n+m)!}{m!(2n+2m+1)!}\left(\frac{2\pi a}{\lambda}\right)^{2m} + \\ &\quad + i\,\frac{1}{2^2\left(\frac{2\pi a}{\lambda}\right)^{n+1}} \sum_{m=0}^{\infty} \frac{(2n-2m)!}{m!(n-m)!}\left(\frac{2\pi a}{\lambda}\right)^{2m}. \end{aligned}\right\} \qquad (13.1)$$

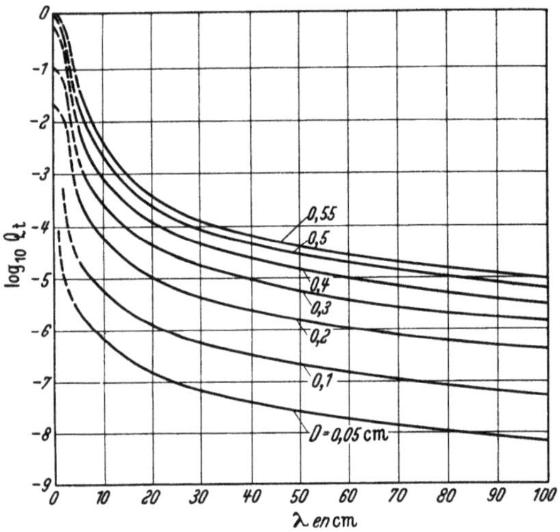

Fig. 12. Section droite d'absorption Q_t (= rapport de l'énergie absorbée à la densité d'énergie de l'onde incidente) en fonction de la longueur d'onde λ et du diamètre de la goutte $D = 2a$.

Les coefficients a_n^d et b_n^d des relations (2.10) et (2.11) peuvent être simplifiés en tenant compte de ceci, et la section droite d'absorption Q_t peut être mise sous la forme

$$Q_t = \frac{3\pi V}{\lambda}(C_1 + C_2\alpha^2 + C_3\alpha^3 + \cdots) \qquad (13.2)$$

ou V représente le volume de la goutte; les C sont des coefficients que l'on calcule en fonction de $m = k_1/k_2$; ils dépendent de la longueur d'onde et de la température. On écrit:

$$m^2 = \varepsilon_r - i\,\varepsilon_i = \varepsilon. \qquad (13.3)$$

Le Tableau 5, emprunté à Burrows et Atwood [8], indique les valeurs de Q_t, exprimées en cm^2 en fonction du diamètre de la goutte (exprimé en cm) et de la longueur d'onde (exprimée en cm) obtenues en tentant compte des coefficients C_1, C_2, C_3. Voir Fig. 12.

Dans le cas d'un milieu à granulométrie hétérogène (N_i gouttes par unité de volume de rayon R_i), le coefficient d'absorption totale, défini en (12.15 bis) aura pour expression:

$$K_t = \sum N_i Q_{it} = \frac{3\pi}{\lambda} \sum N_i V_i (C_1 + C_2\alpha^2 + C_3\alpha^3 + \cdots). \qquad (13.4)$$

L'absorption des ondes radioélectriques dépend ainsi des circonstances météorologiques (forte pluie, averses etc.). Il est donc fort difficile de vérifier quantitativement les résultats ci dessus, mais les ordres de grandeur qu'il permettent de prévoir sont en bon accord avec l'expérience.

Tableau 5. *Section droite d'absorption Q_t en cm^2 de gouttes d'eau de diamètre $D = 2a$ exprimé en cm pour différentes valeurs de la longueur d'onde λ exprimée en cm.*

λ cm	D/cm					
	0,05	0,10	0,15	0,20	0,25	0,30
1,25	$6{,}19 \cdot 10^{-5}$	$9{,}6 \cdot 10^{-4}$	$5{,}66 \cdot 10^{-3}$	$1{,}89 \cdot 10^{-2}$	$5{,}04 \cdot 10^{-2}$	$1{,}13 \cdot 10^{-1}$
3	$9{,}15 \cdot 10^{-6}$	$1{,}52 \cdot 10^{-4}$	$1{,}30 \cdot 10^{-3}$	$5{,}53 \cdot 10^{-3}$	$1{,}64 \cdot 10^{-2}$	$3{,}73 \cdot 10^{-1}$
5	$2{,}84 \cdot 10^{-6}$	$2{,}75 \cdot 10^{-5}$	$1{,}20 \cdot 10^{-4}$	$3{,}79 \cdot 10^{-4}$	$9{,}85 \cdot 10^{-4}$	$2{,}24 \cdot 10^{-3}$
8	$1{,}09 \cdot 10^{-6}$	$9{,}49 \cdot 10^{-6}$	$3{,}65 \cdot 10^{-5}$	$1{,}02 \cdot 10^{-4}$	$2{,}40 \cdot 10^{-4}$	$4{,}98 \cdot 10^{-4}$
10	$6{,}90 \cdot 10^{-7}$	$5{,}85 \cdot 10^{-6}$	$2{,}16 \cdot 10^{-5}$	$5{,}76 \cdot 10^{-5}$	$1{,}46 \cdot 10^{-4}$	$2{,}59 \cdot 10^{-4}$
15	$2{,}98 \cdot 10^{-7}$	$2{,}45 \cdot 10^{-6}$	$8{,}16 \cdot 10^{-6}$	$2{,}18 \cdot 10^{-5}$	$4{,}59 \cdot 10^{-5}$	$8{,}65 \cdot 10^{-5}$
20	$1{,}67 \cdot 10^{-7}$	$1{,}36 \cdot 10^{-6}$	$1{,}71 \cdot 10^{-6}$	$1{,}15 \cdot 10^{-5}$	$2{,}36 \cdot 10^{-5}$	$4{,}31 \cdot 10^{-5}$
30	$7{,}36 \cdot 10^{-8}$	$5{,}93 \cdot 10^{-7}$	$2{,}02 \cdot 10^{-6}$	$4{,}88 \cdot 10^{-6}$	$9{,}74 \cdot 10^{-6}$	$1{,}73 \cdot 10^{-5}$
50	$2{,}67 \cdot 10^{-8}$	$2{,}14 \cdot 10^{-7}$	$7{,}27 \cdot 10^{-7}$	$1{,}36 \cdot 10^{-6}$	$3{,}41 \cdot 10^{-6}$	$5{,}96 \cdot 10^{-6}$
75	$1{,}19 \cdot 10^{-8}$	$9{,}44 \cdot 10^{-8}$	$3{,}21 \cdot 10^{-7}$	$7{,}63 \cdot 10^{-7}$	$1{,}49 \cdot 10^{-6}$	$2{,}60 \cdot 10^{-6}$
100	$6{,}77 \cdot 10^{-9}$	$5{,}41 \cdot 10^{-8}$	$1{,}83 \cdot 10^{-7}$	$4{,}34 \cdot 10^{-7}$	$8{,}50 \cdot 10^{-7}$	$1{,}47 \cdot 10^{-6}$

λ cm	D/cm				
	0,35	0,40	0,45	0,50	0,55
1,25	$2{,}15 \cdot 10^{-1}$	$3{,}66 \cdot 10^{-1}$	$5{,}66 \cdot 10^{-1}$	$7{,}62 \cdot 10^{-1}$	1,01
3	$6{,}65 \cdot 10^{-2}$	$1{,}08 \cdot 10^{-1}$	$1{,}52 \cdot 10^{-1}$	$2{,}15 \cdot 10^{-1}$	$2{,}72 \cdot 10^{-1}$
5	$4{,}49 \cdot 10^{-3}$	$8{,}68 \cdot 10^{-3}$	$1{,}54 \cdot 10^{-2}$	$2{,}59 \cdot 10^{-2}$	$4{,}18 \cdot 10^{-2}$
8	$9{,}63 \cdot 10^{-4}$	$1{,}74 \cdot 10^{-3}$	$2{,}97 \cdot 10^{-3}$	$4{,}85 \cdot 10^{-3}$	$7{,}63 \cdot 10^{-3}$
10	$4{,}81 \cdot 10^{-4}$	$8{,}44 \cdot 10^{-4}$	$1{,}40 \cdot 10^{-3}$	$2{,}25 \cdot 10^{-3}$	$3{,}47 \cdot 10^{-3}$
15	$1{,}51 \cdot 10^{-4}$	$2{,}51 \cdot 10^{-4}$	$3{,}98 \cdot 10^{-4}$	$6{,}10 \cdot 10^{-4}$	$9{,}06 \cdot 10^{-4}$
20	$7{,}29 \cdot 10^{-5}$	$1{,}17 \cdot 10^{-4}$	$1{,}78 \cdot 10^{-4}$	$2{,}66 \cdot 10^{-4}$	$3{,}85 \cdot 10^{-4}$
30	$2{,}83 \cdot 10^{-5}$	$4{,}38 \cdot 10^{-5}$	$6{,}48 \cdot 10^{-5}$	$9{,}29 \cdot 10^{-5}$	$1{,}30 \cdot 10^{-4}$
50	$9{,}57 \cdot 10^{-6}$	$1{,}45 \cdot 10^{-5}$	$2{,}09 \cdot 10^{-5}$	$2{,}93 \cdot 10^{-5}$	$3{,}97 \cdot 10^{-5}$
75	$4{,}15 \cdot 10^{-6}$	$6{,}23 \cdot 10^{-6}$	$8{,}94 \cdot 10^{-6}$	$1{,}24 \cdot 10^{-5}$	$1{,}66 \cdot 10^{-5}$
100	$2{,}35 \cdot 10^{-6}$	$3{,}51 \cdot 10^{-6}$	$5{,}02 \cdot 10^{-6}$	$6{,}92 \cdot 10^{-6}$	$9{,}27 \cdot 10^{-6}$

Tableau 6. *Absorption en décibels par km de chutes de pluie d'intensité donnée exprimée en mm par heure pour différentes valeurs de la longueur d'onde λ exprimée en cm.*

mm/heure	λ cm					
	1,25	3	5	8	10	15
2,46	$1{,}93 \cdot 10^{-1}$	$4{,}92 \cdot 10^{-2}$	$4{,}24 \cdot 10^{-3}$	$1{,}23 \cdot 10^{-3}$	$7{,}34 \cdot 10^{-4}$	$2{,}80 \cdot 10^{-4}$
6,0	$6{,}15 \cdot 10^{-1}$	$1{,}92 \cdot 10^{-1}$	$1{,}25 \cdot 10^{-2}$	$3{,}02 \cdot 10^{-3}$	$1{,}67 \cdot 10^{-3}$	$5{,}84 \cdot 10^{-4}$
22,6	2,40	$7{,}28 \cdot 10^{-1}$	$5{,}29 \cdot 10^{-2}$	$1{,}21 \cdot 10^{-2}$	$6{,}93 \cdot 10^{-3}$	$2{,}27 \cdot 10^{-3}$
34,3	4,51	1,28	$1{,}12 \cdot 10^{-1}$	$2{,}32 \cdot 10^{-2}$	$1{,}17 \cdot 10^{-2}$	$3{,}64 \cdot 10^{-3}$

mm/heure	λ cm				
	20	30	50	75	100
2,46	$1{,}52 \cdot 10^{-4}$	$6{,}49 \cdot 10^{-5}$	$2{,}33 \cdot 10^{-5}$	$1{,}03 \cdot 10^{-5}$	$5{,}85 \cdot 10^{-6}$
6,0	$3{,}02 \cdot 10^{-4}$	$1{,}25 \cdot 10^{-4}$	$4{,}34 \cdot 10^{-5}$	$1{,}93 \cdot 10^{-5}$	$1{,}09 \cdot 10^{-5}$
22,6	$1{,}17 \cdot 10^{-3}$	$4{,}81 \cdot 10^{-4}$	$1{,}66 \cdot 10^{-4}$	$7{,}4 \cdot 10^{-5}$	$4{,}19 \cdot 10^{-5}$
34,3	$1{,}75 \cdot 10^{-3}$	$6{,}83 \cdot 10^{-3}$	$2{,}24 \cdot 10^{-4}$	$9{,}95 \cdot 10^{-5}$	$5{,}63 \cdot 10^{-5}$

Il existe une relation, d'ailleurs assez mal établie, entre le taux de précipitation d'une chute de pluie donnée, et la granulométrie des gouttes qui la constituent; ceci permet, grace à la relation (13.4) de calculer le coefficient d'absorption correspondant. Le Tableau 6 en représente les résultats, exprimés en décibels par kilomètre, pour différents taux de précipitation et différentes longueurs d'onde, et donne ainsi un ordre de grandeur des absorptions que l'on doit attendre (Burrows et Atwood [*8*]).

Lorsque, toujours dans les conditions d'approximation correspondant au Tableau 5 (on tient compte seulement des trois premiers termes $C_1 + C_2\alpha^2 + C_3\alpha^3$) on se trouve dans des conditions de transparence de l'eau ($\varepsilon_i = 0$) dans (13.3), les expressions (13.2) se simplifient et on trouve

$$Q_t = Q_d = \frac{128\pi^5 a^6}{3\lambda^4}\left(\frac{\varepsilon_r - 1}{\varepsilon_r + 2}\right)^2 = \frac{128\pi^5 a^6}{3\lambda^4}\left(\frac{m^2 - 1}{m^2 + 2}\right)^2. \tag{13.5}$$

m est l'indice de réfraction. C'est l'approximation bien connue de RAYLEIGH, correspondant à des particules d'indice pas trop éloigné de l'unité, très petites par rapport à la longueur d'onde.

Enfin, lorsque $2\pi a/\lambda$ est suffisamment faible, on peut écrire d'après (13.2)

$$Q_t = \frac{3\pi V}{\lambda} C_1 = \frac{3\pi V}{\lambda} \frac{6\,\varepsilon_i}{(\varepsilon_r + 2)^2 + \varepsilon_i^2} \tag{13.6}$$

en négligeant les termes en $2\pi a/\lambda$. D'après (13.4), le coefficient d'absorption totale correspondant s'écrira:

$$K_t = \sum N_i Q_{it} = \frac{3\pi}{\lambda} \sum N_i V_i \frac{6\,\varepsilon_i}{(\varepsilon_r + 2)^2 + \varepsilon_i^2}. \tag{13.7}$$

La densité de l'eau étant prise égale à l'unité, ceci peut s'écrire, M représentant la masse d'eau condensée par unité de volume:

$$K_t = \frac{3\pi}{\lambda} M \frac{6\,\varepsilon_i}{(\varepsilon_r + 2)^2 + \varepsilon_i^2}. \tag{13.8}$$

Dans ces conditions (petites gouttes, grandes longueur d'onde), l'absorption ne dépend pas des dimensions des gouttes, mais seulement de la masse d'eau condensée par unité de volume. C'est en particulier le cas des particules nuageuses (diamètre de quelques μ à quelques dizaines de μ) en présence des ondes centimètriques. En fait, le calcul montre que dans ces conditions, l'absorption correspondant à $\lambda = 1{,}25$ cm est seulement de 0,28 décibels par km et par gramme par mètre cube d'eau condensée: il faut donc une grande épaisseur nuageuse pour perturber les observations au radar. Pour des longueurs d'onde supérieures à 3 cm, l'absorption par les nuages est inexistante.

14. Cas des ondes lumineuses. Désignons par $f(\alpha)$ la quantité $Q_t/\pi a^2$. α) Le calcul de $f(\alpha)$ pour différentes valeurs du paramètre $\alpha = 2\pi a/\lambda$ dans le cas de sphères parfaitement transparentes (soit $Q_t = Q_d$) d'indice égal à l'indice moyen

Tableau 7. *Valeurs de $f(\alpha)$ en fonction de $\alpha = 2\pi a/\lambda$ pour $m = 1{,}33$ dans le cas de sphères transparentes. a = rayon des gouttes, λ = longueur d'onde.*

α	$f(\alpha)$	α	$f(\alpha)$	α	$f(\alpha)$	α	$f(\alpha)$	α	$f(\alpha)$
0,5	0,00676	5,0	3,592	11,125	1,728	13,5	2,192	18,0	2,598
0,6	0,0138	6,0	3,888	11,333	1,734	14,0	2,474	18,5	2,432
1,0	0,0938	7,0	3,722	11,4	1,768	14,5	2,528	19,0	2,218
1,2	0,171	8,0	3,282	11,5	1,858	15,0	2,744	19,25	2,090
1,5	0,322	9,0	2,738	11,75	1,758	15,5	2,740	19,50	1,998
1,8	0,523	9,5	2,394	12,0	1,670	16,0	2,870	19,75	1,976
2,0	0,710	10,0	2,512	12,125	1,776	16,25	2,872	20,0	2,092
2,4	1,126	10,25	2,052	12,250	1,1892	16,5	2,850	20,25	2,180
2,5	1,212	10,50	1,886	12,333	1,936	16,75	2,810	20,5	2,078
3,0	1,754	10,625	1,918	12,5	1,938	17,0	2,632	21,0	1,834
3,6	2,376	10,75	1,930	12,6	1,822	17,125	2,704	22,0	1,922
4,0	2,826	10,875	1,832	12,75	1,850	17,375	2,820	24,0	2,438
4,8	3,490	11,0	1,740	13,0	2,012	17,5	2,738		

de l'eau dans le visible ($m = 1{,}33$) a été fait à partir de la relation (12.15) par STRATTON et HOUGHTON[1], LA MER et SINCLAIR[2], puis précisé par HOUGHTON et CHALKER[3] pour des valeurs de α comprises entre 0,5 et 24 (Tableau 7 et Fig. 13).

Etant donné l'hypothèse faite (sphères diélectriques parfaitement transparentes[4]) Q_d correspond également à la totalité de l'énergie empruntée au faisceau incident et diffusée dans toutes les directions. On voit sur la Fig. 13 que pour des valeurs de α suffisamment élevées ($\alpha = 30$ ou 40) $f(\alpha)$ tend vers la valeur 2. Ainsi, le flux emprunté par la goutte à l'onde incidente et diffusé par elle, soit $2\pi a^2 E$, E désignant l'éclairement incident auquel est soumise la goutte, est égal au double du flux intercepté par son contour géométrique. On retrouve ainsi les résultats que l'obtient en se basant sur le principe de BABINET (Sect. 7). — Voir Note C, p. 369.

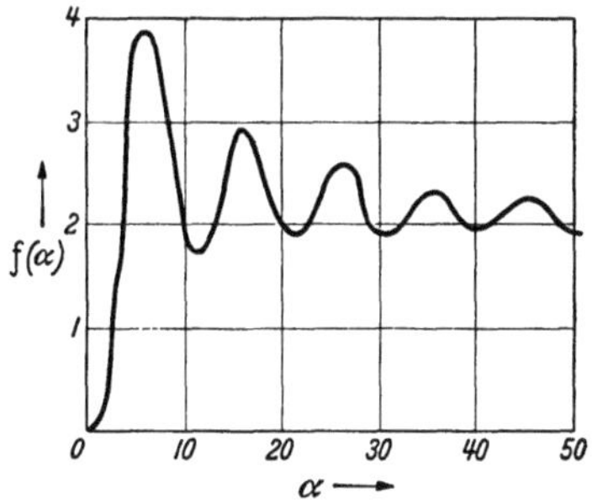

Fig. 13. Variations de $f(\alpha)$ en fonction de $\alpha = 2\pi a/\lambda$ (a = rayon des gouttes, λ = longueur d'onde de la lumière) pour des gouttes d'eau transparentes d'indice $m = 1{,}33$.

β) Il n'est pas indifférent d'évaluer l'importance de l'effet d'absorption propre dans les gouttes, et de cet effet dans le visible (MIDDLETON [9]).

En admettant que le coefficient d'absorption propre k de l'eau soit suffisamment faible, on peut écrire:

$$1 - e^{-2kx} = 2kx \tag{14.1}$$

k étant de l'ordre de $2 \cdot 10^{-2}\,\mathrm{cm}^{-1}$ pour l'eau dans le visible, ceci est évidemment valable pour les gouttes que nous envisageons.

Pour évaluer l'effet d'absorption, MIDDLETON se base sur le fait que le flux F_a absorbé par une particule de rayon a soumise à un éclairement unité a pour valeur (ZANOTELLI[5]):

$$F_a = \frac{4\pi k a^3}{3m} \tag{14.2}$$

ou m est l'indice de l'eau ($m = \frac{4}{3}$). Prenons une goutte de rayon $3\,\mu$; on peut admettre la validité du principe de BABINET, de sorte que le flux diffusé par une telle particule dans ces conditions sera:

$$F_d = 2\pi a^2 \tag{14.3}$$

tandis que le flux absorbé aura pour valeur d'après (14.2)

$$F_a = \pi k a^3 \tag{14.4}$$

le rapport $F_a/F_d = ka/2$ est de l'ordre de 10^{-5}, ce qui justifie l'approximation. Pour des gouttes de pluie de 1 mm de diamètre, le flux absorbé serait de l'ordre du millième du flux incident.

Ceci n'est pas valable en toute rigueur dans le cas du spectre Infra Rouge, où l'absorption propre est beaucoup plus élevée. Il faut alors admettre un indice complèxe $m^* = m(1 - ik)$ et tenir compte de la dispersion, c'est à dire des variations de m et k en fonction de la longueur d'onde. La Fig. 14 représente les variations de $f(\alpha)$ calculées par JOHNSON[6] et ses collaborateurs en admettant

[1] J. A. STRATTON et H. G. HOUGHTON: Phys. Rev. **38**, 159 (1931).

[2] V. K. LA MER et D. SINCLAIR: ORSD Rep. No. 1857 Dept. of Commerce, Office of Publications Board No. 944, Washington 1943.

[3] H. G. HOUGHTON et W. R. CHALKER: J. Opt. Soc. Amer. **39**, 955 (1949).

[4] En fait, en ce qui concerne les gouttelettes de brume notamment, on se trouve en présence de solutions plus ou moins concentrées.

[5] G. ZANOTELLI: Atti. Accad. Ital. Rend. Cl. Sci. Fis. Mat. Nat. **2**, 42 (1940).

[6] J. C. JOHNSON, R. G. ELDRIDGE et J. R. TERREL: Mass. Inst. of Techn. Sci. Rep. No. FA 19 (122)—245, Juillet 1954.

$m = 1{,}29$, k variant entre $k = 0$ (cas de HOUGHTON) et $k = 1{,}732$. On voit que l'absorption a pour effet d'atténuer, ou même de faire disparaitre les fluctuations de la fonction $f(\alpha)$ représentée par la Fig. 14, sans en changer la valeur moyenne.

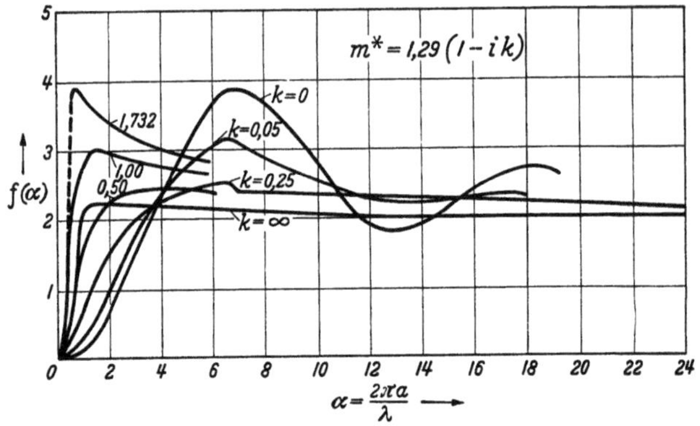

Fig. 14. Variations de $f(\alpha)$ en fonction de $\alpha = 2\pi a/\lambda$ pour des sphères absorbantes d'indice complèxe m^*. Indice réel = 1,29. k = Indice d'absorption.

Etant donné l'hétérogénéité des milieux absorbants naturels (BRUME ou BROUILLARD), dans lesquels α varie dans de larges limites, le fait de tenir compte de l'absorption propre des gouttes dans les calculs ne doit donc pas en changer sensiblement les résultats.

γ) Ces résultats ont été vérifiés dans le spectre visible au laboratoire par LA MER et ses collaborateurs[1], tous ces travaux pouvant être considérés comme autant de bases solides à des recherches atmosphèriques.

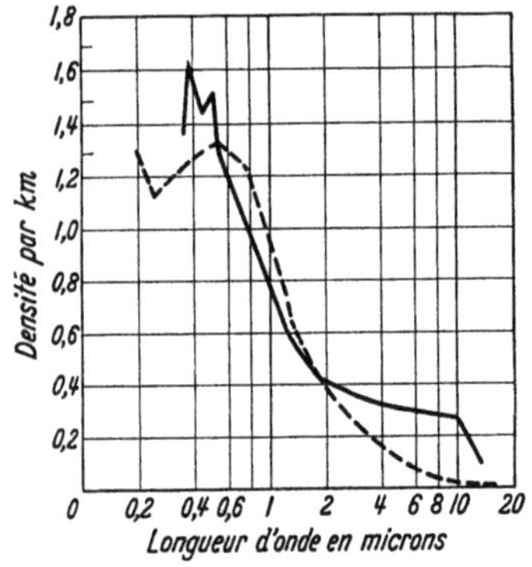

Fig. 15. Variations de la densité optique dans une atmosphère brumeuse mesurées (traits pleins) et calculées (traits pointillés) à partir de la granulométrie du milieu déterminée directement.

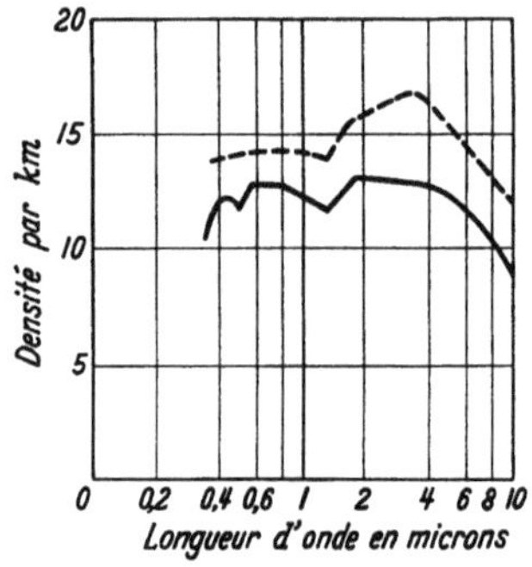

Fig. 16. Variations de la densité optique dans un brouillard mesurées (traits pleins) et calculées (traits pointillés) à partir de la granulométrie du milieu déterminée directement.

Pour en faire une vérification dans l'atmosphère, il faut pouvoir déterminer à la fois le coefficient d'absorption pour une longueur d'onde donnée par une mesure directe, et simultanément la granulométrie des particules responsables de l'absorption. Ce travail a été récemment entrepris par ARNULF, BRICARD et VERET[2], dans un domaine de radiations compris entre 0,4 et 10 μ, en se plaçant,

[1] V. K. LA MER et D. SINCLAIR: ORSD Rep. No. 1857 Dept. of Commerce Off. Publ. Board. Rep. No. 944, Washington.

[2] ARNULF, BRICARD et VERET: C. R. Acad. Sci. Paris **238**, 603 (1954). — Rev. Opt. **33**, 658 (1954).

pour le domaine infra rouge, dans les fenêtres de transmission des constituants de l'atmosphère, ou n'intervient sensiblement que l'action des particules en suspension dans l'air sur les diverses radiations.

Les granulométries (nombre de particules de dimensions déterminées par unité de volume) étaient effectuées par captations des particules diffusantes avec des fils d'araignées, complétées le cas échéant par une mesure de masse d'eau condensée par unité de volume. A partir de ces observations, il était possible de calculer en valeur absolue la densité optique pour une longueur d'onde donnée, ainsi que ses variations en fonction de la longueur d'onde, en se basant sur les considérations ci dessus.

Les Fig. 15 et 16 représentent les densités optiques par km calculées à partir de ces indications (traits pointillés) et mesurées pour différentes longueurs d'onde dans deux cas typiques. On voit ainsi que les brumes (densité optique inférieure à 2 par km pour 0,4 μ, soit quelques centaines de particules par cm^3 de diamètre en majorité inférieur à 1 ou 2 μ) sont très sélectives, tandis que dans les brouillards (densité optique de l'ordre de 10 à 20 par km, soit quelques centaines de gouttes par cm^3 de rayon compris entre quelques μ et 10 ou 20 μ), la sélectivité est inexistante dans le visible, et faible dans l'infra rouge. Ces mesures constituent une vérification des calculs ci dessus.

II. Diffusion des radiations par les nuages de gouttelettes (brume, nuages, pluie).

Lorsqu'on observe une source lumineuse à travers un milieu diffusant, une partie seulement de la lumière de la source parvient directement à l'observateur; c'est cette partie que nous venons d'étudier. Le reste est diffusé par le milieu. Nous nous proposons d'étudier la partie diffusée, en procédant par approximations successives: d'abord diffusion primaire, en introduisant dans les calculs uniquement l'éclairement direct d'un élément de volume du milieu par une source lumineuse intérieure ou extérieure au milieu, puis diffusion secondaire, en tenant compte au point considèré de l'éclairement dû aux autres parties du milieu diffusant.

15. Diffusion primaire. On se bornera à exposer les cas qui se présentent le plus souvent dans la pratique.

α) Ondes radioélectriques. Diffusion vers l'arrière. Posons dans (12.12) $d\omega = \sin\vartheta\, d\vartheta\, d\Phi$. A partir de cette relation, on trouve que la puissance diffusée vers l'arrière ($\vartheta=\pi$) par une sphère par unité d'angle solide a pour expression:

$$\left(\frac{d W_d}{d\omega}\right)_{\vartheta=\pi} = \frac{E_0^2}{8\mu_2^2\omega^2}\,\mathrm{Re}\sum_{n=1}^{\infty}\sum_{m=1}^{\infty}(-1)^{n+m}(2n+1)(2m+1)(a_n^d-b_n^d)(a_m^{d*}-b_m^{d*}), \quad (15.1)$$

(l'astérisque caractérisant toujours le conjugué complexe).

En raisonnant comme pour établir les relations (12.14) et (12.15), on trouve que la section droite de diffusion vers l'arrière, c'est à dire le rapport de la puissance diffusée par la sphère dans l'unité d'angle solide et dans la direction inverse de propagation de l'énergie incidente, à la densité de l'énergie incidente, a pour expression:

$$\left.\begin{aligned}\left(\frac{d Q_d}{d\omega}\right)_{\vartheta=\pi} &= \sigma(\pi)\\ &= \left(\frac{\lambda}{4\pi}\right)^2\mathrm{Re}\sum_{n=1}^{\infty}\sum_{m=1}^{\infty}(-1)^{n+m}(2n+1)(2m+1)(a_n^d a_m^{d*}+b_n^d b_m^{d*}-2a_n^d b_m^{d*}).\end{aligned}\right\} \quad (15.2)$$

Tableau 8. *Section droite de diffusion vers l'arrière $\sigma(\pi)$ exprimée en cm^2 pour des gouttes de diamètre $D = 2a$ exprimé en cm et différentes valeurs de la longueur d'onde λ exprimée en cm.*

D cm	λ cm				
	3	5	8	10	15
0,05	$4,25 \cdot 10^{-9}$	$5,55 \cdot 10^{-10}$	$8,63 \cdot 10^{-11}$	$3,50 \cdot 10^{-11}$	$6,96 \cdot 10^{-12}$
0,10	$2,64 \cdot 10^{-7}$	$3,52 \cdot 10^{-8}$	$5,47 \cdot 10^{-9}$	$2,24 \cdot 10^{-9}$	$4,44 \cdot 10^{-10}$
0,15	$2,88 \cdot 10^{-6}$	$3,97 \cdot 10^{-6}$	$6,28 \cdot 10^{-8}$	$2,54 \cdot 10^{-8}$	$5,10 \cdot 10^{-9}$
0,20	$1,48 \cdot 10^{-5}$	$2,15 \cdot 10^{-6}$	$3,45 \cdot 10^{-7}$	$1,42 \cdot 10^{-7}$	$2,84 \cdot 10^{-8}$
0,25	$5,02 \cdot 10^{-5}$	$7,42 \cdot 10^{-6}$	$1,30 \cdot 10^{-6}$	$5,34 \cdot 10^{-7}$	$1,07 \cdot 10^{-7}$
0,30	$1,34 \cdot 10^{-4}$	$2,25 \cdot 10^{-5}$	$3,80 \cdot 10^{-6}$	$1,57 \cdot 10^{-6}$	$3,19 \cdot 10^{-7}$
0,35	$2,48 \cdot 10^{-4}$	$5,40 \cdot 10^{-5}$	$9,37 \cdot 10^{-6}$	$3,91 \cdot 10^{-6}$	$8,01 \cdot 10^{-7}$
0,40	$5,04 \cdot 10^{-4}$	$1,12 \cdot 10^{-4}$	$2,03 \cdot 10^{-5}$	$8,55 \cdot 10^{-6}$	$1,77 \cdot 10^{-6}$
0,45	$7,76 \cdot 10^{-4}$	$2,12 \cdot 10^{-4}$	$3,99 \cdot 10^{-5}$	$1,70 \cdot 10^{-5}$	$3,55 \cdot 10^{-6}$
0,50	$9,91 \cdot 10^{-4}$	$3,65 \cdot 10^{-5}$	$7,30 \cdot 10^{-5}$	$3,14 \cdot 10^{-5}$	$6,63 \cdot 10^{-6}$

D cm	λ cm				
	20	30	50	75	100
0,05	$2,18 \cdot 10^{-12}$	$4,32 \cdot 10^{-11}$	$5,60 \cdot 10^{-14}$	$1,11 \cdot 10^{-14}$	$3,50 \cdot 10^{-15}$
0,10	$1,40 \cdot 10^{-10}$	$2,77 \cdot 10^{-11}$	$3,59 \cdot 10^{-11}$	$7,09 \cdot 10^{-13}$	$2,24 \cdot 10^{-13}$
0,15	$1,60 \cdot 10^{-9}$	$3,18 \cdot 10^{-10}$	$4,12 \cdot 10^{-11}$	$8,14 \cdot 10^{-12}$	$2,57 \cdot 10^{-12}$
0,20	$8,94 \cdot 10^{-9}$	$1,77 \cdot 10^{-9}$	$2,29 \cdot 10^{-10}$	$4,52 \cdot 10^{-11}$	$1,43 \cdot 10^{-11}$
0,25	$3,42 \cdot 10^{-8}$	$6,73 \cdot 10^{-9}$	$8,72 \cdot 10^{-10}$	$1,72 \cdot 10^{-10}$	$5,45 \cdot 10^{-11}$
0,30	$1,20 \cdot 10^{-7}$	$5,04 \cdot 10^{-8}$	$2,62 \cdot 10^{-9}$	$5,17 \cdot 10^{-10}$	$1,64 \cdot 10^{-10}$
0,35	$2,58 \cdot 10^{-7}$	$2,02 \cdot 10^{-7}$	$6,53 \cdot 10^{-9}$	$1,29 \cdot 10^{-9}$	$4,08 \cdot 10^{-10}$
0,40	$5,75 \cdot 10^{-7}$	$1,13 \cdot 10^{-7}$	$1,46 \cdot 10^{-8}$	$2,88 \cdot 10^{-9}$	$9,13 \cdot 10^{-10}$
0,45	$1,16 \cdot 10^{-6}$	$2,32 \cdot 10^{-7}$	$3,00 \cdot 10^{-8}$	$5,92 \cdot 10^{-9}$	$1,87 \cdot 10^{-9}$
0,50	$2,18 \cdot 10^{-6}$	$4,32 \cdot 10^{-7}$	$5,60 \cdot 10^{-8}$	$1,11 \cdot 10^{-8}$	$3,50 \cdot 10^{-9}$

Lorsque $2\pi a/\lambda$ est faible par rapport à l'unité, nous avons vu que l'on pouvait introduire les simplifications (13.1) dans les expressions (2.10) et (2.11) de a_n^d et b_n^d; $\sigma(\pi)$ peut être mis sous la forme

$$\sigma(\pi) = \left(\frac{\lambda}{4\pi}\right)^2 \alpha^6 \left[A_0 + A_2 \alpha^2 + A_3 \alpha^3 + \cdots\right] \quad (15.3)$$

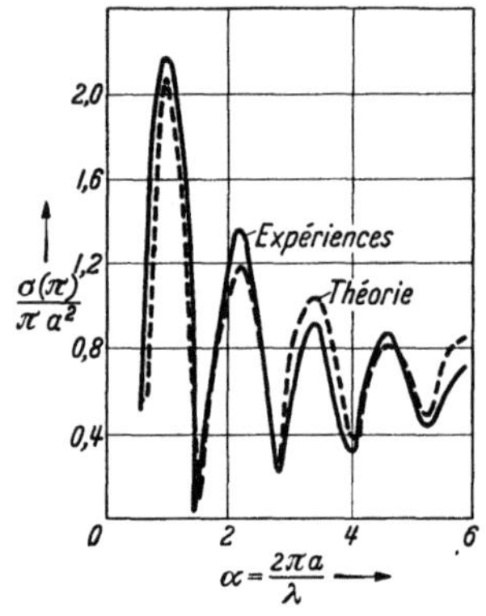

Fig. 17. Variations de $\sigma(\pi)/\pi a^2$ en fonction de $\alpha = 2\pi a/\lambda$ mesurées (traits pleins) et calculées (traits pointillés). a = rayon des gouttes, λ = longueur d'onde de la radiation.

les A étant des fonctions des constantes diélectriques ε_i et ε_r de l'eau, qui dépendent de la longueur d'onde. Le Tableau 8 représente d'après BURROWS et ATWOOD [*8*] les valeurs de la section droite de diffusion vers l'arrière exprimée en cm^2 pour des gouttes d'eau sphériques de diamètres compris entre 0,05 et 0,50 cm et des valeurs de la longueur d'onde comprises entre 3 et 100 cm. Lorsque l'absorption est négligeable ($\varepsilon_i = 0$) on obtient une expression de même forme que (13.5) et correspondant à l'approximation de RAYLEIGH [*1*].

Ces considérations théoriques ont été vérifiées par les mesures d'ADEN, effectuées au laboratoire (détermination de la fraction d'énergie diffusée vers l'arrière par des particules sphériques de différentes dimensions et pour différentes longueurs d'onde). La Fig. 17 représente les résultats obtenus, comparés aux résultats du calcul; on voit que l'accord est satisfaisant.

Dans le Tableau 6, nous avions indiqué l'absorption exprimée en décibels par km correspondant à une chute de pluie d'intensité donnée en fonction de la

Tableau 9. *Valeurs de K_π pour des couches de pluie de 1 km d'épaisseur de différentes intensités exprimées en mm par heure pour différentes valeurs de la longueur d'onde λ exprimée en cm.*

mm/heure	λ cm				
	3	5	8	10	15
2,46	$2,94 \cdot 10^{-5}$	$4,21 \cdot 10^{-6}$	$7,01 \cdot 10^{-7}$	$2,86 \cdot 10^{-7}$	$5,74 \cdot 10^{-8}$
4,00	$5,06 \cdot 10^{-5}$	$7,31 \cdot 10^{-6}$	$1,17 \cdot 10^{-6}$	$4,81 \cdot 10^{-7}$	$9,63 \cdot 10^{-8}$
6,00	$1,44 \cdot 10^{-6}$	$2,32 \cdot 10^{-5}$	$3,90 \cdot 10^{-6}$	$1,61 \cdot 10^{-6}$	$3,25 \cdot 10^{-7}$
15,2	$6,12 \cdot 10^{-4}$	$1,73 \cdot 10^{-4}$	$3,30 \cdot 10^{-5}$	$1,41 \cdot 10^{-5}$	$2,94 \cdot 10^{-6}$
34,2	$1,28 \cdot 10^{-3}$	$3,02 \cdot 10^{-4}$	$5,58 \cdot 10^{-5}$	$2,36 \cdot 10^{-5}$	$4,90 \cdot 10^{-6}$
mm/heure	λ cm				
	20	30	50	75	100
2,46	$1,82 \cdot 10^{-8}$	$3,60 \cdot 10^{-9}$	$4,66 \cdot 10^{-10}$	$9,20 \cdot 10^{-11}$	$2,91 \cdot 10^{-11}$
4,00	$3,03 \cdot 10^{-8}$	$5,99 \cdot 10^{-9}$	$7,66 \cdot 10^{-10}$	$1,53 \cdot 10^{-10}$	$4,85 \cdot 10^{-11}$
6,00	$1,17 \cdot 10^{-7}$	$2,31 \cdot 10^{-8}$	$2,99 \cdot 10^{-9}$	$5,91 \cdot 10^{-10}$	$1,87 \cdot 10^{-10}$
15,2	$7,62 \cdot 10^{-7}$	$1,50 \cdot 10^{-7}$	$1,94 \cdot 10^{-8}$	$3,83 \cdot 10^{-9}$	$1,21 \cdot 10^{-9}$
34,2	$1,63 \cdot 10^{-6}$	$3,22 \cdot 10^{-7}$	$4,17 \cdot 10^{-8}$	$8,24 \cdot 10^{-9}$	$2,61 \cdot 10^{-9}$

longueur d'onde. En raisonnant de la même façon, on peut évaluer le coefficient d'absorption K_π correspondant uniquement à la portion d'énergie diffusée en arrière par unité d'angle solide pour un taux de précipitation et une longueur d'onde donnés; ceci correspond aussi à la fraction d'énergie diffusée vers l'arrière par unité d'angle solide par unité d'épaisseur du milieu diffusant. En négligeant l'absorption de l'énergie diffusée en arrière par le milieu lui même dans son épaisseur, on obtient alors les résultats du Tableau 9 (BURROWS et ATWOOD [*8*]).

Il est intéressant de déduire du Tableau 9 la fraction de la puissance incidente P_i diffusée vers l'arrière pour différentes chutes de pluie de 1 km d'épaisseur. Pour une épaisseur Δx de la couche diffusante, la puissance diffusée vers l'arrière est:

$$\Delta P_\pi = -K_\pi P_i \Delta x. \tag{15.4}$$

La fraction de la puissance P_i cherchée pour une couche $\Delta x = 1$ km est alors $10 \log \frac{\Delta P_\pi}{P_i}$ ou $10 \log_{10} K_\pi$ décibels, K_π étant donné par le Tableau 9.

Le faisceau d'émission limite généralement l'étendue de la couche pluvieuse diffusante. La section droite de diffusion S_π de la couche est constituée par toutes les gouttes comprises dans le volume intervenant dans l'écho. Ce volume peut être considéré comme limité par la calotte sphérique d'épaisseur Δd dont la base est un segment sphérique de surface $2\pi d^2(1-\cos\vartheta)$, ϑ désignant approximativement la demi ouverture du faisceau. La section droite d'écho est ainsi, N_i et σ_i étant caractéristiques de la granulométrie:

$$S_\pi = 2\pi d^2 (1-\cos\vartheta)\, \Delta d \sum [N_i \sigma_i(\pi)], \tag{15.5}$$

à partir de laquelle il est facile d'évaluer la puissance diffusée vers l'arrière dans un cas concrétisé.

Comme dans le cas de la transmission des ondes centimétriques par les nuages pluvieux, une vérification précise de ces relations n'est pas possible étant donné les difficultés que présente l'établissement d'une granulométrie de pluie sur une aussi vaste étendue, ainsi que la détermination même de cette étendue. On peut dire seulement que les ordres de grandeur expérimentaux les vérifient qualitativement.

β) Lumière diffusée en avant. Source lumineuse placée dans le brouillard. Considérons une source ponctuelle monochromatique placée en A et un observateur en B à la distance r_0 de A. La source rayonne dans toutes les directions, et la lumière qui arrive en B peut être la lumière directe, la lumière diffusée par le brouillard directement éclairé par la source, et la lumière qui a subi plusieurs diffusion successives, processus que nous négligerons en première approximation. On se propose de calculer l'éclairement en B en tenant compte des deux premiers processus seulement (MIDDLETON [9])[1].

La lumière en provenance directe de la source produit un éclairement:

$$E = \frac{I_0}{r_0^2} e^{-K r_0} \tag{15.6}$$

K représentant le coefficient de transmission défini par la relation (12.1).

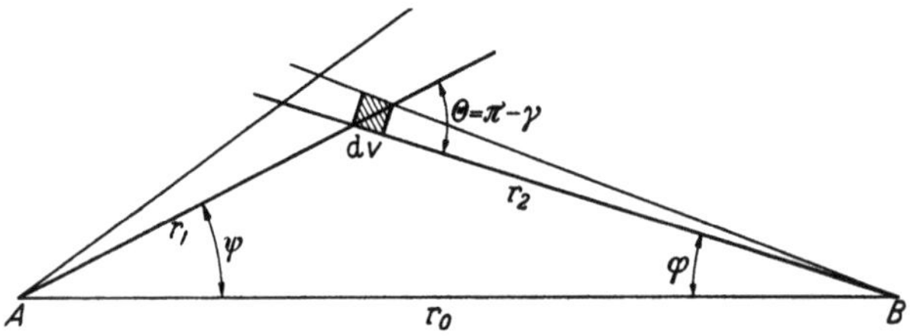

Fig. 18

La lumière diffusée une fois peut être calculée en considérant un élément de volume torique (Fig. 18):

$$dv = 2\pi r_2^2 \sin\varphi \, dr_2 \, d\varphi. \tag{15.7}$$

L'intensité dI_1 diffusée vers B est le produit de ce volume dv par la fonction $f_1(\gamma)$ définie par la relation (5.2), et par l'éclairement direct de dv par la source (γ représente l'angle entre les rayons diffusés vers l'observateur et la direction de la source). Ce dernier a pour expression:

$$E_1 = \frac{I_0}{r_1^2} e^{-K r_1}. \tag{15.8}$$

Il en résulte que dI_1 a pour valeur:

$$dI_1 = 2\pi r_2^2 I_0 \sin\varphi \, f(\gamma) \frac{e^{-K r_1}}{r_1^2} dr_2 \, d\varphi. \tag{15.9}$$

L'éclairement en B provenant de cet élément, compte tenu de l'absorption sur le trajet $B\,dv$ s'écrira:

$$dE_2 = 2\pi I_0 e^{-K(r_1+r_2)} \frac{\sin\varphi}{r_1^2} f_1(\gamma) \, dr_2 \, d\varphi, \tag{15.10}$$

soit

$$E_2 = 2\pi I_0 \int_0^\infty \int_0^\pi \frac{e^{-K(r_1+r_2)} \sin\varphi \, f_1(\gamma)}{(r_0 - r_2\cos\varphi)^2 + r_2^2 \sin^2\varphi} dr_2 \, d\varphi. \tag{15.11}$$

Cette quantité que l'on peut calculer numériquement en se basant sur les considérations développées Sect. 9 a été évaluée par MIDDLETON en faisant les hypothèses simplificatrices suivantes:

D'une part, pour des gouttes nuageuses de diamètre supérieur à plusieurs microns, la forme de l'indicatrice de diffusion est telle que le flux diffusé vers l'arrière de la goutte par rapport à la direction du faisceau incident, est négligeable comparé au flux diffusé vers l'avant. Il en résulte qu'on peut négliger tout ce qui correspond à $r_2 > r_0$.

[1] Voir également F. VOLZ (référence p. 253, [1e]), p. 892.

D'autre part, et ceci est encore légitimé dans une certaine mesure par la forme des diagrammes de diffusion, on admet que le rayonnement de la source n'est pas réparti suivant un angle solide 4π, mais dans un certain angle ψ (2 cas sont envisagés par MIDDLETON dans son calcul, $\psi = 1$ rad. et $\psi = 0{,}02$ rad.). Enfin, on place en B un récepteur de champ également limité à l'angle Φ par un système de diaphragmes convenables. C'est le problème des *téléphotomètres* pour la mesure de la *densité optique du brouillard*.

Dans ces conditions, on peut facilement confondre r_0 avec la somme $r_1 + r_2$; il en résulte que:

$$E_2 = 2\pi I_0 e^{-K r_0} \int\limits_0^{\Phi} \int\limits_0^{\frac{r_0}{1+\Phi\cot\psi}} \frac{\varphi f_1(\gamma)}{(r_0 - r)^2 + r^2 \varphi^2}\, d\varphi\, dr. \quad (15.12)$$

La Fig. 19 représente les résultats que l'on obtient à partir de ce calcul. On voit que le rapport E_2/E, E étant défini par la relation (15.6) peut avoir une valeur importante, et les résultats numériques auxquels on est conduit sont en bon accord avec l'expérience.

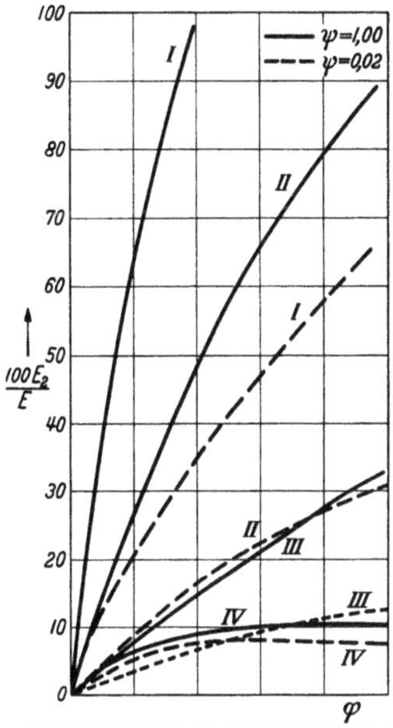

Courbe n°	Rayon des gouttes	Nombre par m³	Distance r_0 en mètres
I	2,5 μ	10^8	1000
II	2,5 μ	10^8	500
III	2,5 μ	10^8	200
IV	10 μ	$6{,}25 \cdot 10^6$	500

Fig. 19. Calcul à partir de la relation (15.12) du rapport 100. E_2/E pour deux atmosphères hypothétiques et deux valeurs de ψ.

16. Théorie des deux flux. L'absorption de la lumière solaire par les nuages a été étudiée de façon approchée en admettant l'existence d'un flux montant et d'un flux descendant d'énergie, chacun d'eux entrainant diffusion et absorption (SCHUSTER[1], SCHWARZSCHILD[2]). En se basant sur la théorie de WIENER [5] (gouttes de rayon supérieur à la longueur d'onde de la lumière) MECKE [6] a montré que la fraction $\beta = 0{,}0725$ du flux lumineux qui subit la diffusion, en ne tenant compte que de la réflexion et de la réfraction, est diffusée vers l'arrière (entre $\pi/2$ et $3\pi/2$ avec la direction incidente), de sorte que le reste, soit 92,75 % se retrouve dans la composante diffusée en avant. (En fait, ceci ne fait intervenir que la moitié du flux total diffusé $\Phi = 2\pi a^2$ et emprunté au faisceau incident, l'autre moitié, uniquement diffusée vers l'avant, étant constituée par le flux diffracté par le contour de la goutte.) D'autre part, dans un faisceau parallèle, une fraction $(1 - \alpha)\, dx$ passera directement sans subir aucune diffusion à travers une couche d'épaisseur dx, α étant la surface totale $\sum N_i \pi a_i^2$ occultée par les $N = \sum N_i$ gouttes présentes dans l'unité de volume. Les fractions $\beta\alpha$ et $\alpha(1 - \beta)$ sont diffusés respectivement vers l'arrière et vers l'avant.

Soient A et B les densités de flux dirigées vers le haut et vers le bas. Leurs variations après traversée de la couche seront:

$$dA = -\alpha\beta(A - B)\, dx, \qquad dB = -\alpha\beta(A - B)\, dx. \quad (16.1)$$

Bien entendu, les gouttes sont supposées parfaitement transparentes. Les conditions limites sont déterminées par le fait qu'au sommet de la couche, $A = A_0$,

[1] A. SCHUSTER: Astrophys. J. **21**, 1 (1905).

[2] K. SCHWARZSCHILD: Göttinger Nachr. **1906**, 41.

densité de flux venant du soleil et du Ciel, et qu'en bas de la couche d'épaisseur H, $B_H = A_H a_s$, a_s désignant l'albédo du sol.

Ceci n'est valable que si la lumière tombe perpendiculairement sur la couche nuageuse. En fonction de la distance zénithale z du soleil, les variations de β s'expriment par (ALBRECHT[1]):

$$\beta = 0{,}073 + 0{,}6421 \cdot 10^{-3} z^4 \tag{16.2}$$

où z est exprimé en degrés; β est maximum ($\beta = 0{,}5$) pour $z = 90°$. Pour la lumière diffuse

$$\beta = \frac{2}{\pi} \int_0^{\pi} \beta(z) \sin z \, dz = 0{,}195\,, \tag{16.3}$$

Les deux dernières relations donnent des valeurs de β sensiblement égales pour $z = 67°$, et si la distance zénithale est différente de celle-ci, β varie à mesure que l'on pénètre dans le nuage: la solution devient alors très compliquée.

NEIBURGER[2] a étudié l'albédo et la transmission des nuages en tenant compte de l'épaisseur et de la constitution des couches. Il se trouve que l'albédo augmente avec l'épaisseur de la couche, conformément aux résultats auxquels conduit la théorie ci dessus qui, du point de vue quantitatif, ne représente qu'imparfaitement les résultats expérimentaux.

17. Théorie de FRITZ[3]. LANGMUIR et WESTENDORP[4] avaient eu l'idée d'analogie entre la diffusion de la lumière dans le brouillard, et l'interdiffusion des gaz, ou encore la diffusion de la chaleur dans un conducteur. Une idée semblable a été développée par FRITZ.

Considérons un nuage de gouttelettes limité à sa partie supérieure par un plan horizontal indéfini. Supposons le faisceau parallèle provenant du soleil perpendiculaire à ce plan, et imaginons une couche d'épaisseur Δz assez mince pour que la lumière provenant du faisceau ne soit diffusée qu'une seule fois. En toute rigueur, une telle couche devrait être infiniment mince; mais pour faire les calculs, nous admettrons pour Δz la longueur $L/4$ où L, libre parcours moyen de la lumière dans le nuage a pour valeur:

$$L = \frac{1}{\sum \pi N_i a_i^2}\,, \tag{17.1}$$

N_i et a_i désignant comme plus haut la granulométrie du nuage.

En admettant le répartition de WIENER [5] et en admettant également que toute l'énergie diffusée dans la zone 60—180° avec la verticale descendante est isotrope, on peut calculer les énergies directe et diffusée en avant, ainsi que leur répartition vers le bas. A partir du faisceau direct, et de l'énergie diffusée vers l'avant, on peut obtenir la distribution de l'énergie diffuse engendrée dans le nuage. Pour traiter mathématiquement le problème de la diffusion de cette énergie engendrée, il est nécessaire d'en trouver une expression analytique en fonction de la profondeur et du libre parcours; en se basant sur des études de laboratoire, FRITZ admet la relation empirique:

$$n_g = 0{,}02 \left(\frac{4}{L}\right)^2 z \exp\left(-0{,}5656 \frac{z}{L}\right) \tag{17.2}$$

ou n_g représente l'énergie diffuse engendrée par unité de volume et par unité de temps et z l'épaisseur nuageuse, mesurée à partir du sommet de la couche.

[1] F. ALBRECHT: Meteor. Z. **50**, 478 (1933).
[2] M. NEIBURGER: J. Meteorology **6**, 98 (1949).
[3] S. FRITZ: J. Meteorology **11**, 291 (1954).
[4] I. LANGMUIR et W. F. WESTENDORP: Physics **1**, 273 (1931).

Pour un état stationnaire, en l'absence d'absorption et en admettant L constant, l'équation de diffusion est:

$$-D\frac{d^2 n}{d z^2} = n_g, \tag{17.3}$$

ou n représente le nombre de photons par unité de volume et D un coefficient de diffusion dont la valeur est prévue par la théorie. En introduisant les conditions limites, on peut intégrer la relation (17.3).

18. Solution générale. *α) Equation de transfert de* CHANDRASEKHAR[1]. Considérons une atmosphère diffusante plane, homogène, d'étendue horizontale indéfinie. Une direction sera définie par les angles ϑ avec la verticale ascendante

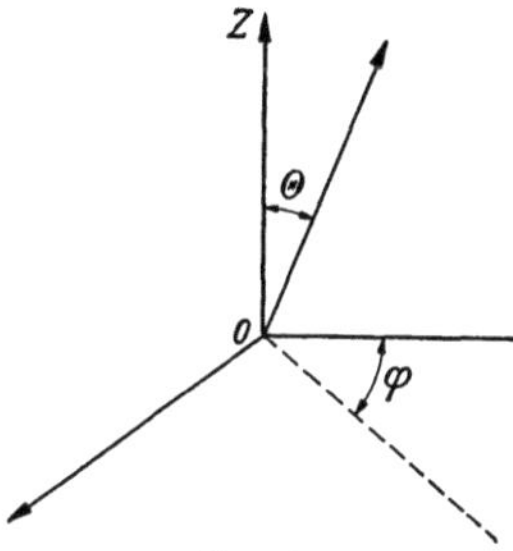

Fig. 20.

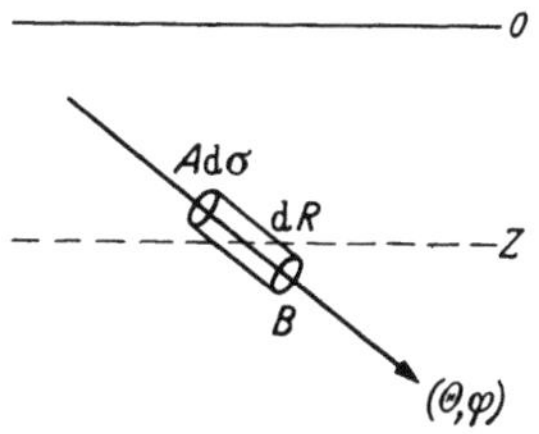

Fig. 21.

(compris entre 0 et π) et φ avec le plan vertical origine (Fig. 20). Soit z l'altitude, comptée négativement vers le bas, à partir de la surface ($z=0$) recevant le rayonnement incident (Fig. 21).

Désignons par K le coefficient d'absorption totale défini par la relation (12.1). Soit $E(z)$ l'éclairement en un point situé à l'intérieur du nuage diffusant. Considérons un élément de volume cylindrique de surface droite $d\sigma$ et d'épaisseur dR. Le flux traversant la surface A est $B(z, \vartheta, \varphi)\, d\sigma\, d\omega$ dans un pinceau d'angle solide $d\omega$ autour de la direction ϑ, φ; B désignant la luminance de l'atmosphère due aux diffusions au point considéré. Le flux traversant la surface B est:

$$\left[B(z, \vartheta, \varphi) + \frac{d}{dR} B(z, \vartheta, \varphi)\, dR\right] d\sigma\, d\omega. \tag{18.1}$$

Le flux absorbé à l'intérieur du cylindre est:

$$B(z, \vartheta, \varphi)\, d\sigma\, d\omega\, K\, dR. \tag{18.2}$$

On met le coefficient de diffusion dans la direction ϑ, φ faisant l'angle α avec la direction d'incidence ϑ', φ', sous la forme

$$\frac{K}{4\pi} p(\cos\alpha). \tag{18.3}$$

La fonction $p(\cos\alpha)$ s'appelle *fonction de phase.* Pour un milieu donné, on peut exprimer cette fonction de phase à partir des relations (3.6) et (5.2)

$$f_1(\gamma) = \frac{K}{4\pi} p[\cos(\pi-\gamma)]. \tag{18.4}$$

Notons que l'éclairement dE envoyé par un élément dv' sur un élément dv peut s'écrire $B(z, \vartheta', \varphi')\, d\omega'$. D'après la définition du coefficient de diffusion et de la fonction de phase, l'intensité diffusée dans la direction ϑ, φ par un élé-

[1] S. CHANDRASEKHAR: Radiative Transfer. Oxford 1950.

ment dv recevant l'éclairement dE dans la direction ϑ', φ' a pour expression:

$$\frac{K}{4\pi} p(\vartheta, \varphi; \vartheta', \varphi')\, dE\, dR\, d\sigma, \tag{18.5}$$

soit

$$\frac{K}{4\pi} p(\vartheta, \varphi; \vartheta', \varphi')\, B(z, \vartheta', \varphi')\, dR\, d\sigma\, d\omega'. \tag{18.6}$$

Le flux émis par l'élément $dv = d\sigma\, dR$ éclairé par le reste du nuage aura ainsi, pour expression:

$$\frac{K}{4\pi} dR\, d\sigma\, d\omega \int_0^{2\pi}\int_0^{\pi} p(\vartheta, \varphi; \vartheta', \varphi')\, B(z, \vartheta', \varphi') \sin\vartheta'\, d\vartheta'\, d\varphi'. \tag{18.7}$$

Finalement, l'équation de transfert s'établit facilement en écrivant le principe de conservation de l'énergie d'après (18.1), (18.2) et (18.7). On trouve la relation:

$$-\frac{d}{dR} B(z, \vartheta, \varphi) = K\, B(z, \vartheta, \varphi) + \frac{K}{4\pi} \int_0^{2\pi}\int_0^{\pi} p(\vartheta, \varphi; \vartheta', \varphi')\, B(z, \vartheta', \varphi') \sin\vartheta'\, d\varphi'\, d\vartheta'. \quad (18$$

Posons

$$\tau = -Kz, \qquad \mu = \cos\vartheta. \tag{18.9}$$

On trouve:

$$\mu \frac{dB(\tau, \mu, \varphi)}{d\tau} = B(\tau, \mu, \varphi) - \frac{1}{4\pi} \int_0^{2\pi}\int_{-1}^{+1} p(\mu, \varphi; \mu', \varphi')\, B(\tau, \mu', \varphi')\, d\varphi'\, d\mu'. \tag{18.10}$$

Pour établir ces relations, on ne fait intervenir que la lumière diffusée par le milieu. Mais il faut également tenir compte de la lumière directement transmise.

Soit πF l'éclairement envoyé directement par le soleil à la surface supérieure de la couche sur un plan perpendiculaire à sa direction de propagation, ϑ_0, φ_0 les coordonnées du soleil. Il faudra ajouter un flux diffusé dans la direction ϑ, φ du à cet éclairement direct qui, en conservant les mêmes notations que dans les relations (18.5), (18.6), (18.7), aura pour expression:

$$\tfrac{1}{4} K\, p(\mu, \varphi; \mu_0\, \varphi_0)\, F\, e^{\tau/\mu_0}\, d\omega\, dR\, d\sigma \tag{18.11}$$

et la relation (18.10) devient:

$$\left.\begin{aligned} \mu \frac{dB(\tau, \mu, \varphi)}{d\tau} &= B(\tau, \mu, \varphi) - \frac{1}{4\pi} \int_0^{2\pi}\int_{-1}^{+1} p(\mu, \varphi; \mu', \varphi')\, B(\tau, \mu', \varphi')\, d\varphi'\, d\mu' - \\ &\quad - \frac{1}{4} p(\mu, \varphi; \mu_0, \varphi_0)\, F\, e^{\tau/\mu_0}. \end{aligned}\right\} \tag{18.12}$$

Il faut associer à cette équation les conditions limites au sol et au dessus de la couche.

β) Résultats. Le premier problème consiste à représenter les fonctions de diffusion (ou la fonction de phase), sous une forme mathématique. On utilise pour cela le développement en série de polynomes de LEGENDRE:

$$p(\mu, \varphi; \mu', \varphi') = \sum_{\varrho=0}^{N} \overline{\omega}_\varrho\, p_\varrho [\cos(\mu, \varphi; \mu', \varphi')]. \tag{18.13}$$

Si on développe la luminance sous la forme:

$$B(\tau, \mu, \varphi) = \sum_{m=0}^{N} B^m(\tau, \mu) \cos m(\varphi' - \varphi_0), \tag{18.14}$$

l'équation de transfert se sépare en N équations intégro-différentielles en τ et μ.

La méthode des ordonnées discrètes de CHANDRASEKHAR consiste, en utilisant la formule de quadrature de GAUSS, à remplacer chacune de ces équations par un système de $2n$ équations différentielles linéaires en τ, dont la résolution ne présente plus de difficultés mathématiques, et qui donne une valeur approchée de $B^m(\tau, \mu)$ pour $2n$ directions μ_i; ce systéme s'écrit:

$$\left.\begin{aligned} &\mu_i \frac{dB^m(\tau_i \mu_i)}{d\tau} - B^m(\tau_i \mu_i) + \frac{1}{2}\sum_{\varrho=m}^{N} \overline{\omega}_\varrho \frac{(\varrho - m)!}{(\varrho + m)!} P_l^m(\mu_i) \sum_{j=-n}^{+n} a_j P_l^m(\mu_j)\, B^m(\tau, \mu_j) \\ &\quad = -\frac{F}{4} e^{\tau/\mu_0}(2 - \delta_{0m}) \sum_{\varrho=m}^{N} \overline{\omega}_\varrho \frac{(\varrho - m)!}{(\varrho + m)!} P_l^m(\mu_i)\, P_l^m(\mu_0) \end{aligned}\right\} \quad (18.15)$$

où

$$i = \pm 1, 2, \ldots n.$$

Les μ_i sont les zéros des polynomes de LEGENDRE d'ordre $2n$ et les a_i les nombres de CHRISTOFFEL correspondants; $P_l^m(\mu_i)$ sont des fonctions de LEGENDRE; δ_{0m} est égal à 1 si $m = 0$ et à zéro dans les autres cas.

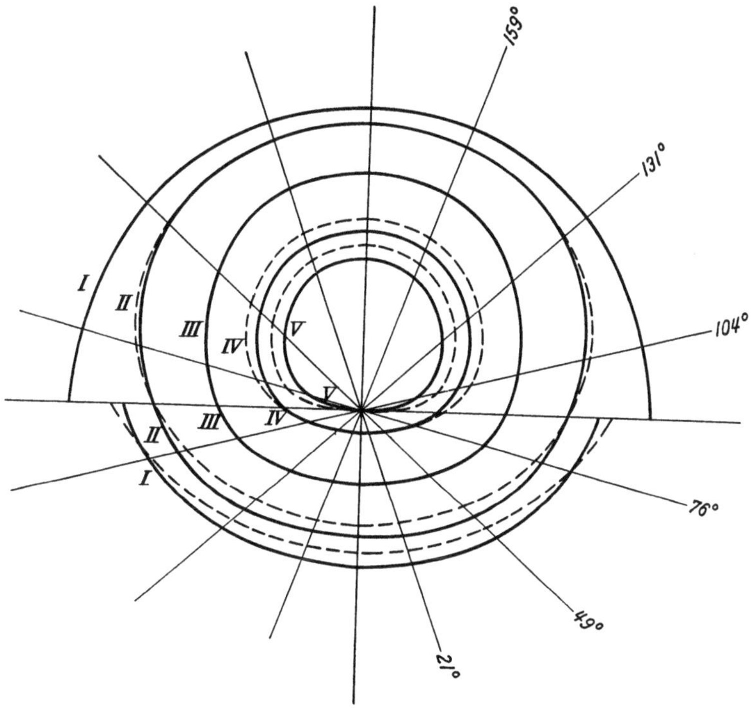

Fig. 22. Diagramme polaire de la luminance pour un brouillard de gouttelettes sans absorption propre d'épaisseur 200 mètres et de densité optique décimale 20 par km éclairé par un Ciel de luminance uniforme au dessus d'un sol noir. Traits pleins, calculs de J. LENOBLE. Traits pointillés, calculs de J. OLIVIER. *I* Sommet de la couche. *II* 30 m au dessous. *III* Milieu de la couche. *IV* 30 m au dessus du sol. *V* Sol.

La forme de la relation (18.15) sans second membre est à rapprocher de celle de la relation (16.1) ,qui peut en étre considérée comme une approximation. L'idée fondamentale est la même, mais alors que (16.1) a été établie à partir de deux flux seulement, on fait intervenir dans (18.15) un nombre de flux élevé.

Cette méthode a été appliquée par J. LENOBLE[1] en utilisant un développement d'ordre $N = 2$. Les seuls cas étudiés numériquement (dans l'approximation $n = 3$)

[1] J. LENOBLE: Rev. Opt. **35**, 1 (1956). — J. Sci. Météorol. **8**, 23 (1956).

correspondent à une couche d'épaisseur optique $\tau = 4{,}61$, au dessus d'un sol noir, et éclairée par un ciel de luminance uniforme d'une part, et par le soleil à 30° au dessus de l'horizon d'autre part. Les valeurs B/B_0 en fonction de ϑ, B_0 représentant la luminance uniforme du ciel, sont portées en coordonnées polaires pour diverses altitudes optiques τ sur la Fig. 22 dans le premier cas. Sur les Fig. 23a

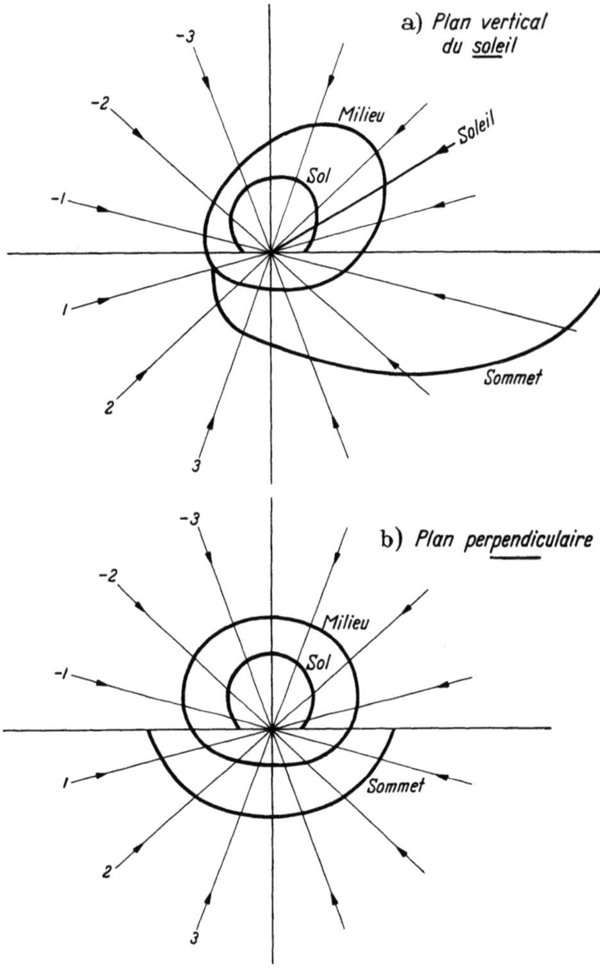

Fig. 23a et b. Diagramme polaire pour un brouillard d'épaisseur 200 mètres et de densité optique décimale 20 par km au dessus d'un sol noir éclairé par le soleil. a) Dans un plan vertical contenant le soleil. b) Dans un plan vertical perpendiculaire au précédent.

et 23b, on a porté les valeurs B/F correspondant au second cas. (πF est l'éclairement d'un écran perpendiculaire aux rayons solaires.) L'indicatrice de diffusion choisie est très voisine de celle de WIENER, et on a utilisé la 3ème approximation, ou ce qui revient au même, les calculs on été faits dans six directions qui sont indiquées sur les figures.

Les courbes en pointillé sur la Fig. 22 représentent les valeurs obtenues par OLIVIER et KUNTZMAN[1] dans le même cas par un calcul numérique direct, en utilisant l'indicatrice de diffusion de GUMPRECHT.

[1] J. OLIVIER: Bull. Soc. franç. Electr., Ser. VIII 5 (1955).

Une difficulté, dans ces calculs, se manifeste par les complications qu'entraine l'interpolation de l'indicatrice de diffusion, qui comporte une pointe extrêmement aigue dans le sens de propagation de la lumière incidente, et correspondant à la lumière diffractée. Cette question a été étudiée par OLIVIER, qui a cherché l'erreur que l'on commet en utilisant une courbe sans pointe et en même temps une densité optique réduite de la fraction correspondant à l'énergie ainsi perdue, et considérée comme directement transmise. D'après OLIVIER[1], l'erreur serait importante au voisinage de la direction de propagation du faisceau incident et au sommet de la couche.

Quoiqu'il en soit, on voit qu'il est possible, par des calculs relativement rapides, d'obtenir la répartition de la luminance d'une couche diffusante. Il reste donc à étendre ces calculs à d'autres exemples, et surtout à les vérifier expérimentalement.

Notes supplémentaires.

A. (p. 333). Voir, dans la même série de publications: GUMPRECHT et SLIEPCEVICH: Tables of light scattering functions for spherical particles. T. 16. Sept. 1951. — Tables of RICCATI BESSEL Functions for spherical particles. T. 14. Sept. 1951. — Tables of first and second partial derivatives of LEGENDRE Polynomials. T. 12. Sept. 1951. — Voir également HOLL, Optik **1**, p. 213 (1946).

B. (p. 335). Notons que dans la nature, et en particulier lorsqu'il s'agit de particules de brume qui sont constituées par des solutions plus ou moins concentrées de sels, l'indice m est certainement supérieur à celui de l'eau pure, et la goutte n'est pas diélectrique; la fonction f définie par (3.6) ne représente qu'une approximation.

C. (p. 357). La valeur limite des sections droites de diffusion pour les grosses particules diélectriques a fait l'objet d'une étude de L. BRILLOUIN (App., Math. Panel N.D.R.C. No. 871, No. 1, 1943; No. 2, 1944) qui a critiqué l'interprétation donnée par STRATTON aux divers termes du vecteur de POYNTING. Il a pu montrer sur les équations de STRATTON le phénomène bien connu de la localisation de l'énergie diffractée au voisinage de la direction de propagation.

Bibliographie générale.

[1] PERNTER, J. M., et F. M. EXNER: Meteorologische Optik. Wien u. Leipzig 1922.
[2] MASCART, E.: Traité d'Optique. Paris 1889.
[3] MECKE, R.: Handbuch der Physik, Bd. 20, S. 67. 1926.
[4] MIE, G.: Ann. d. Phys. **25**, 377 (1908).
[5] WIENER, C.: Abh. Leopold. Karol. dtsch. Akad. Naturforscher **73**, 1 (1907)
[6] MECKE, R.: Ann. d. Phys. **65**, 257 (1921).
[7] STRATTON, J. A.: Electromagnetic Theory. New York et Londres 1941.
[8] BURROWS, R., et S. ATWOOD: Radiowave Propagation. New York 1949.
[9] MIDDLETON, W. E. K.: Vision through the Atmosphere. Toronto 1952 (voir aussi la contribution de MIDDLETON dans ce volume).

[1] Travaux en cours de publication.

Ozon in der Erdatmosphäre.

Von

H. K. PAETZOLD und E. REGENER†[1].

Mit 47 Figuren.

A. Übersicht.

1. Geschichte. Die drei-atomige Form des Sauerstoffs, das Ozon, war schon 1785 von VAN MARUM durch einen charakteristischen Geruch bei elektrischen Entladungen festgestellt und später (1839, SCHÖNBEIN) auch bei chemischen Prozessen, wie der langsamen Oxydation des Phosphors an der Luft, gefunden worden, ohne daß allerdings die chemische Natur dieses Gases schon bekannt gewesen war. Diese wurde erst 1864 durch SORET geklärt, der die richtige Formel O_3 aufstellte[2].

Bereits 1858 wurde das Ozon von HOUZEAU[3] in der Luft als Spurengas auf chemischem Wege nachgewiesen. Die Existenz dieses atmosphärischen Ozons mußte überraschen, da das Ozon durch Abgabe eines O-Atomes außerordentlich stark oxydierend wirkt und deshalb durch die an der Erdoberfläche und in niederen Luftschichten immer vorhandenen oxydablen Substanzen rasch aufgezehrt wird. Wie man heute weiß, erfolgt die notwendige Nachbildung des Ozons photochemisch aus dem Luftsauerstoff durch kurzwelligeres ultraviolettes Sonnenlicht.

Die weitere Erforschung des atmosphärischen Ozons beruhte auf seinen spektralen Eigenschaften und ist aufs engste mit der Tatsache verknüpft, daß das vom Erdboden aus beobachtete Sonnenspektrum bei etwa 3000 Å ziemlich abrupt abbricht. Nach der Strahlungstemperatur der Sonne von etwa 6000° K im sichtbaren Spektralbereich sollte extrapoliert die spektrale Intensität bei 2200 Å noch mit 18% der bei 4000 Å herrschenden zu beobachten sein — bei Raketenaufstiegen wurden in 80 km Höhe 2% gemessen —, während es nur mit großem experimentellen Aufwand gelingt, vom Boden aus das Sonnenspektrum bis knapp unterhalb 2900 Å zu verfolgen[4].

CORNU[5] konnte nachweisen, daß die Verkürzung des ultravioletten Sonnenspektrums um so stärker ist, je größer die Zenitdistanz der Sonne ist. Er schreibt anschaulich: es sei, wie wenn ein beweglicher Schirm je nach Sonnenhöhe ein Spektrum zudeckte oder wieder freigäbe, welches sich so gut wie ungestört fortsetze. Hieraus schloß CORNU, daß der Abbruch des Sonnenspektrums durch Absorption in der Erdatmosphäre selbst verursacht wird. Im Jahre 1881 konnte HARTLEY[6] dann nach seinen Untersuchungen der ultravioletten Absorption des Ozons unterhalb von 3000 Å vermuten, daß die ultraviolette Begrenzung des Sonnenspektrums durch eine in der Atmosphäre in größeren Höhen eingelagerte

---

[1] Ausgearbeitet vom erstgenannten Verfasser; E. REGENER hat noch bei dem Entwurf mitwirken können (Herausgeber).

[2] Eine Übersicht und Literatur gibt FRONROBERT [*13*].

[3] HOUZEAU: C. R. Acad. Sci. Paris **46** (1858).

[4] F. W. P. GÖTZ: Strahlenther. **40**, 690 (1931).

[5] A. CORNU: C. R. Acad. Sci. Paris **88**, 1101, 1285 (1879).

[6] HARTLEY: Chem. News **42**, 268 (1880).

Ozonschicht verursacht wird. Eine quantitative Bestätigung konnten erst FABRY und BUISSON [*11*] im Jahre 1921 erbringen, als sie die Schwächung der ultravioletten Intensität des Sonnenlichtes bei verschiedenen Zenitabständen der Sonne, nach Abzug der allgemeinen atmosphärischen Extinktion, mit der Absorption einer Ozonschicht von 0,3 cm reduzierter Schichtdicke identifizierten. Danach beträgt die gesamte Ozonmenge nur 3 bis $4 \cdot 10^{-7}$ Volumenanteile der ganzen Atmosphäre, die eine homogene Höhe von 8 km aufweist. Durch optische Messungen wurde ferner gefunden[1], daß in den bodennahen Luftschichten das Ozon noch geringer, zu etwa $2 \cdot 10^{-8}$ Volumenanteilen vertreten ist, so daß das atmosphärische Ozon in größeren Höhen konzentriert sein muß. Dies ging auch aus Beobachtungen von MIETHE und LEHMANN[2] im Flachland und Gebirge sowie von WIGAND[3] bei Freiballonfahrten hervor, nach denen sich bis zu Höhen von 9 km keine merkliche Abnahme der atmosphärischen Ultraviolettabsorption ergab.

Der Betrag dieser hohen Ozonschicht zeigte aber schon nach den ersten Messungen [*11*] erhebliche *Schwankungen*, die eine dauernde Überwachung nahelegten. Die ersten Jahresreihen des Ozons von GÖTZ[4] in Arosa und von DOBSON und anderen [*6*] in Oxford zeigen einen ausgesprochenen *Ozonjahresgang* mit dem Maximum im Frühjahr. Vor allem hat DOBSON durch die Konstruktion und den Einsatz eines geeigneten Spektralphotometers an verschiedenen Orten für die homogene Erfassung der zeitlichen und geographischen Schwankungen des Ozons bahnbrechend gewirkt, so daß eine besondere internationale, der Internationalen Meteorologischen Assoziation angegliederte *Ozonkommission* gebildet wurde. Durch den Aufbau eines Ozon-Beobachtungsnetzes mit DOBSON-Instrumenten in Europa, Amerika, Australien und Asien, das in Europa mit über 10 Stationen am dichtesten ist, konnten enge Beziehungen zwischen dem meteorologischen Geschehen und den Ozonvariationen aufgefunden werden.

Während bis dahin in bezug auf die *Höhenverteilung* des Ozons nur geschlossen werden konnte, daß dieses sich in größeren Höhen befinden muß, bedeutete es einen großen Fortschritt, als es GÖTZ gelang, mit Hilfe des sog. *Umkehreffektes* die Höhe des Ozonmaximums zu etwa 25 km zu bestimmen. Auf dieselbe Weise wurde auch von GÖTZ, MEETHAM und G. M. B. DOBSON[5] 1934 die vertikale Ozonverteilung in Arosa während eines vollen Jahresturnus bestimmt. Im gleichen Jahr gelang es E. und V. H. REGENER in Stuttgart durch Registrier-*Ballonaufstiege* mit Quarzspektrographen bis 30 km Höhe die vertikale Ozonverteilung direkt zu messen und die Höhe des Ozonmaximums von 25 km zu bestätigen. Nach diesem Vorgang sind in den letzten Jahren zahlreiche Bestimmungen der Ozonverteilung durch V. H. REGENER [*39*] in Albuquerque (Neu Mexico) und PAETZOLD [*24*] in Weißenau durchgeführt worden, wobei sich sehr erhebliche Schwankungen ergaben.

Durch *Raketenaufstiege* in Amerika konnte in den letzten Jahren die Erforschung der Ozonschicht bis hinauf zu 70 km Höhe vorangetrieben werden. Als wertvolle Ergänzung zu den Messungen mit Ultraviolett-Spektrographen hat sich die von PAETZOLD [*25*] und BARBIER und anderen [*1*] entwickelte Methode der *Mondfinsternisse* erwiesen, besonders für geographische Breiten, in denen bislang noch keine Ballonaufstiege durchgeführt werden konnten.

---

[1] R. J. STRUTT: Proc. Roy. Soc. Lond., Ser. A **94**, 260 (1918).

[2] A. MIETHE u. E. LEHMANN: Ber. Berl. Akad. **1909**, 268.

[3] A. WIGAND: Verh. dtsch. phys. Ges. **15**, 1090 (1913).

[4] F. W. P. GÖTZ: Beitr. Phys. fr. Atmosph. **13**, 15 (1926).

[5] F. W. P. GÖTZ, A. R. MEETHAM u. G. M. B. DOBSON: Proc. Roy. Soc. Lond., Ser. A **145**, 416 (1934).

Für den Ozongehalt in bodennahen Luftschichten und in der Troposphäre wurde in neuerer Zeit wieder auf die *chemischen* Bestimmungen zurückgegriffen, die durch GLÜCKAUF und andere, V. H. REGENER und EHMERT zu hoher Genauigkeit entwickelt worden sind. Die damit beobachteten Schwankungen des *bodennahen Ozons* erscheinen vor allem von bioklimatischer Bedeutung.

Die für das ganze Ozonproblem zentrale Theorie der *Bildung* des atmosphärischen Ozons konnte von CHAPMAN 1930 [3] in der noch heute gültigen Form aufgestellt werden, nachdem die Quantentheorie die Molekülspektren und die photolytischen Prozesse verstehen ließ. Danach wird der Luftsauerstoff durch das kurzwelligere ultraviolette Sonnenlicht dissoziiert, und die dabei entstehenden Sauerstoffatome verbinden sich mit den zweiatomigen Sauerstoffmolekülen zu Ozon. Dieses wird aber durch die Sonnenstrahlung ebenfalls wieder dissoziiert, so daß beim Ozon ein typisches *photochemisches Gleichgewicht* vorliegt. Seine qualitative Fassung konnte soweit vorangetrieben werden, daß nach dem Vergleich mit den Beobachtungen kein Zweifel an der Richtigkeit dieser Vorstellungen mehr bestehen kann.

Nach dem heutigen Stand ist das Ozonproblem nach verschiedenen Richtungen hin von Bedeutung und weist eine innige Verflechtung mit anderen Problemen der Atmosphäre auf. Neben seiner Rolle als Schirm gegen die ultraviolette Sonnenstrahlung unterhalb von 3000 Å ist das Ozon durch seine ultrarote Absorption bei 9,6 μ, wo sich eine Lücke in den Absorptionsspektren des Kohlendioxyd und des Wassers befindet, am Strahlungshaushalt und Wärmegleichgewicht der Stratosphäre beteiligt. Seine örtlichen und zeitlichen Schwankungen sind ein wertvoller Indikator für Transportvorgänge in der Atmosphäre, der vor allem für das Studium weltweiter Zirkulationsvorgänge fruchtbar zu werden verspricht. Das Ozonproblem erscheint von besonderem Reiz, da bei ihm verschiedene Gebiete: Physik und Astrophysik, Geophysik und Meteorologie sowie Bioklimatik ineinander greifen.

2. Die Maßgrößen für das Ozon. Für die Bezeichnung der beim Ozon benutzten Größen hat sich bisher keine Einheitlichkeit durchgesetzt. Im folgenden werden daher die gebräuchlichsten Symbole verwandt:

$[O_3]$ Anzahl der O_3-Moleküle im cm^3.

$[M]$ Anzahl der Luftmoleküle im cm^3.

$x(h)$ Der über dem Meßort mit der Höhe h senkrecht befindliche Ozonbetrag, *reduziert* auf Normalbedingungen (0° C, 760 mm Hg). Dimension: cm O_3.

x_0 Der totale Ozonbetrag x_0 (cm O_3) für $h=0$ wird kurz als *Ozonwert* bezeichnet.

$\varepsilon(h)$ Die *Ozonkonzentration* in der Höhe h, ausgedrückt als die bei 1 km horizontaler Weglänge durchlaufene Ozonmenge, reduziert auf Normalbedingungen. Dimension: cm O_3/km.

Die Bezeichnung Ozonkonzentration für den in der Höhe h befindlichen *absoluten* Betrag des Ozons entspricht nicht ganz der sonst üblichen Bedeutung des Begriffs Konzentration, doch hat sie sich weitgehend eingebürgert. Es ist

$$x(h) = \int_h^\infty \varepsilon(h)\, dh; \qquad (h \text{ in km}). \tag{2.1}$$

Beim bodennahen Ozon wird auch öfter der *absolute* Ozongehalt in γ O_3/m^3 angegeben. 1 γ O_3/m^3 entspricht einem $\varepsilon = 4{,}67 \cdot 10^{-5}$ cm O_3/km.

$[O_3]/[M]$ Das Verhältnis: Ozon/Luft, das den *relativen* Gehalt an Ozon angibt.

Es gelten die folgenden Umrechnungsgleichungen:

$$[O_3] = \varepsilon(h) \cdot N_A \cdot 10^{-5}, \tag{2.2}$$

$$\frac{[O_3]}{[M]} = \frac{\varepsilon(h) \cdot 10^{-5}}{\varrho(h)}. \tag{2.3}$$

$\varrho(h)$ Luftdichte, bezogen auf die Luftdichte unter Normalbedingungen als Einheit.

N_A AVOGADROsche Zahl.

Es sind noch mehr Symbole vorgeschlagen worden, um das Ozon auf weitere Größen, wie das Luftgewicht usw. zu beziehen[1]. Doch erscheinen die obigen Größen hier als ausreichend, da in Einzelfällen eine erforderliche Umrechnung leicht möglich ist.

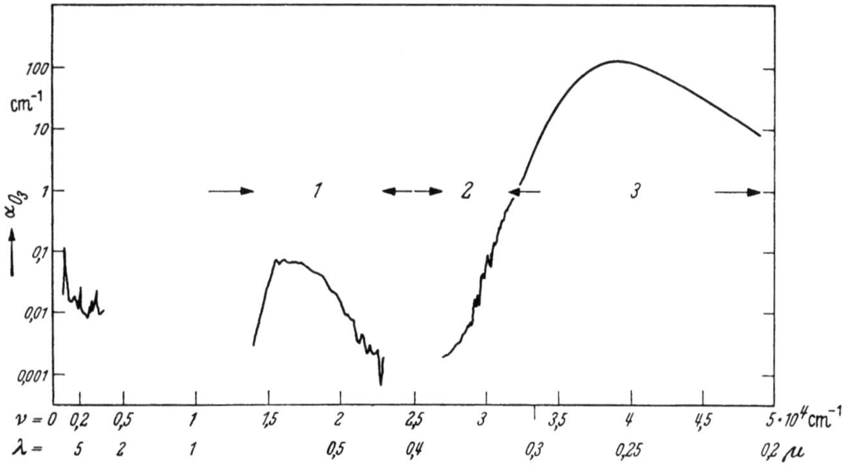

Fig. 1. Absorptionsspektrum des Ozons nach PENNDORF. *1* CHAPPUIS-Banden; *2* HUGGINS-Banden; *3* HARTLEY-Bande.

B. Das Absorptionsspektrum des Ozons.

Das Absorptionsspektrum des Ozons besteht aus der ultravioletten HARTLEY-*Bande* von 2000 bis 3200 Å und den bis etwa 3600 Å anschließenden HUGGINS-*Banden*, der im sichtbaren Spektralbereich liegenden CHAPPUIS-*Bande* (4300 bis 7500 Å) und den *Ultrarot-Banden* bei 4,7; 9,6 und 14,2 μ. Bei der Bestimmung der Absorptionskoeffizienten ist von den verschiedenen Beobachtern die Ozonmenge meistens chemisch gemessen worden. Eine Übersicht über des Ozon-Absorptionsspektrum gibt Fig. 1[2], während in Tabelle 1 die Absorptionskoeffizienten auszugsweise zusammengestellt sind; dabei wurde über die Welligkeit der Banden gemittelt.

3. Die HARTLEY-Bande. Sie stellt das Hauptabsorptionsgebiet des Ozons mit einem maximalen dekadischen Absorptionskoeffizienten von 145 cm^{-1} bei 2553 Å dar. Licht von dieser Wellenlänge wird demnach durch eine Ozonschicht von der reduzierten Dicke von nur $(1/145) = 0{,}007$ cm bis auf 1/10 seiner ursprünglichen Intensität geschwächt. (Diese starke Absorption des Ozons wird noch übertroffen von der Sauerstoffabsorption im SCHUMANN-RUNGE-Kontinuum, dessen maximale Absorption 213 cm^{-1} bei 1450 Å beträgt.)

[1] R. PENNDORF: J. Geophys. Res. **54**, 169 (1949).
[2] R. PENNDORF: Veröff. Geophys. Inst. Leipzig **8**, H. 4 (1936).

Die Absorption in der HARTLEY-Bande ist von verschiedener Seite gemessen worden, wobei sich zum Teil eine gute Übereinstimmung, aber auch zum Teil größere Differenzen ergaben[1-4]. Wegen des etwas welligen Charakters der Bande haben NY-TSI-ZE und CHOONG SHIN-PIAW[5] die Messungen mit dem kontinuierlichen Licht eines Wasserstoffkontinuums bis herab zu 2140 Å wiederholt. Ihre Ergebnisse passen gut zu den Werten von FABRY und BUISSON. Bis herab zu 2020 Å sind die Absorptionsmessungen von A. VASSY [*45*] ergänzt worden. In neuerer Zeit hat VIGROUX [*46*] das Absorptionsspektrum des Ozons über den gesamten Spektralbereich von 2300 bis 10000 Å neu gemessen. Dabei ergaben sich in der HARTLEY-Bande gegenüber den Messungen von NY-TSI-ZE und CHOONG SHIN-PIAW im Durchschnitt um 20% geringere Werte.

Tabelle 1. *Spektrale Ozonabsorption.*

Å	α_{O_3} (λ) cm^{-1}	Name
2000	5,0	
2200	30	
2400	100	HARTLEY-Bande
2600	138	
2800	60	
3000	4,9	
3200	0,4	
3400	0,02	HUGGINS-Bande
3500	0,01	
4500	0,0015	
5000	0,01	
5500	0,045	
6000	0,060	CHAPPUIS-Bande (290° K)
6500	0,030	
7000	0,011	
7500	0,003	

Von 2400 bis 2900 Å ist die HARTLEY-Bande von den sog. SHAVER-*Banden*[6] überlagert, die, wenigstens formal, die Fortsetzung der an die HARTLEY-Bande nach längeren Wellen anschließenden HUGGINS-Banden darstellen. Das kurzwellige Ende der HARTLEY-Bande zeigt keine merkliche Bandenstruktur mehr. Für die Bestimmung des Betrages des atmosphärischen Ozons aus optischen Messungen ist ausschlaggebend, daß im Bereich der HARTLEY-Bande keine Abweichung vom BEERschen Gesetz[7] (in einem Druckbereich von 10:1) und kein Fremdgaseinfluß festgestellt werden konnte[8]. Auch ein merklicher Temperatureinfluß ist nicht vorhanden[9] oder nur gering [*46*].

4. Die HUGGINS-Banden (3200 bis 3600 Å). Die HUGGINS-Banden wurden von LADENBURG und LEHMANN 1906 im Absorptionsspektrum des Ozons entdeckt[10]. Aber erst 1917 wurden sie mit Banden identifiziert[11], die HUGGINS[12] schon 1890 im Spektrum des Sirius gefunden hatte. Diese Banden zeigen einen starken *Temperatureffekt* in dem Sinne, daß die Absorptionsmaxima mit sinkender Temperatur kontrastreicher[9,10] [*44*] erscheinen (Fig. 2). Sie zeigen aber auch bei großen Dispersionen von einigen Å pro mm keine Anzeichen von Auflösung.

1 E. MEYER: Ann. Phys. **12**, 849 (1903).
2 F. KRÜGER u. M. MOELLER: Phys. Z. **13**, 729 (1912).
3 CH. FABRY u. H. BUISSON: J. Phys. (5) **3**, 196 (1913).
4 A. LÄUCHLI: Helv. phys. Acta **1**, 208 (1928).
5 NY-TSI-ZE u. CHOONG SHIN-PIAW: Chin. J. Phys. **1**, 38 (1938).
6 PR. SHAVER: Proc. Roy. Soc. Canada **15**, 5 (1921).
7 Absorption nur abhängig vom Produkt Druck mal durchlaufenem Weg. Vergl. Beitrag Z. SEKERA, Ziff. 3, S. 291.
8 B. MAHMONDIAN u. E. VASSY: C. R. Acad. Sci. Paris **234**, 2350 (1952).
9 O. R. WULF u. E. H. MELVIN: Phys. Rev. **38**, 330 (1931).
10 E. LADENBURG u. E. LEHMANN: Ann. Phys. (4) **21**, 305 (1906).
11 A. FOWLER u. R. J. STRUTT: Proc. Roy. Soc. Lond., Ser. A **93**, 577 (1917).
12 A. HUGGINS: Proc. Roy. Soc. Lond. **48**, 216 (1890).

Dementsprechend ist auch bei Variationen des Druckes im Verhältnis 40:1 kein Druckeinfluß auf den Absorptionskoeffizienten feststellbar [*44*].

Nach neueren Messungen[1] [*46*] zeigen sowohl die Absorptionsmaxima wie -minima einen parabolischen Temperaturgang; bei ersteren liegt aber das Minimum bei etwa $-40°$ C, bei den letzteren bei $-100°$ C, so daß in dem Temperaturintervall zwischen 0° C und $-100°$ C die Absorptionsstärke in den Absorptionsmaxima nur wenig variiert. Für optische Temperaturmessungen des atmosphärischen Ozons ist der Temperaturgang des *Kontrastes* der Banden maßgebend; dieser ist gut reproduzierbar und in dem Temperaturintervall von 0 bis $-100°$ C praktisch linear[2].

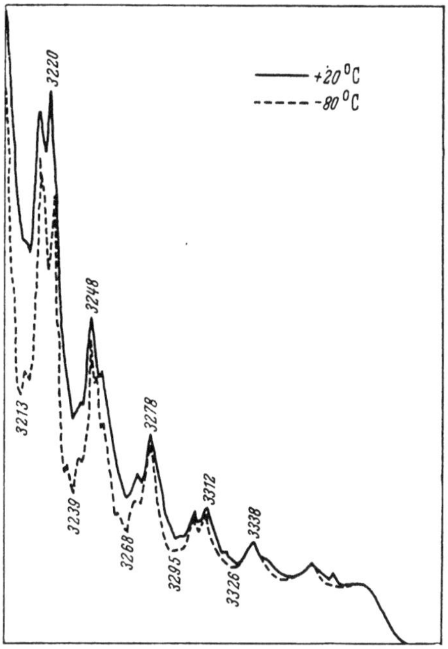

Fig. 2. Temperatureffekt der HUGGINS-Banden nach VASSY.

5. Die CHAPPUIS-Banden (4300 bis 7500 Å). Die von CHAPPUIS[3] 1882 gefundenen Banden im sichtbaren Spektralbereich sind nur schwach, mit einem maximalen Absorptionskoeffizienten von 0,06 cm^{-1} bei 6000 Å[4]. Bei einer reduzierten Schichtdicke von 10 cm bewirken sie die *blaue Farbe des Ozons*. Die in Fig. 3 nach VASSY[4] dargestellten Banden konnten ebenfalls noch nicht weiter aufgelöst werden. Sie zeigen nach E. VASSY [*44*] ebenfalls einen beträchtlichen Temperatureffekt, indem beim Absinken der Temperatur von 20 auf $-110°$ C die Absorption um 30% ansteigt. Im Gegensatz dazu findet LEFEBRVE[5] keinen Temperatureffekt. Auch nach den VIGROUXschen Messungen ist die Temperatur fast ohne Einfluß, so daß die Frage noch nicht ganz geklärt ist.

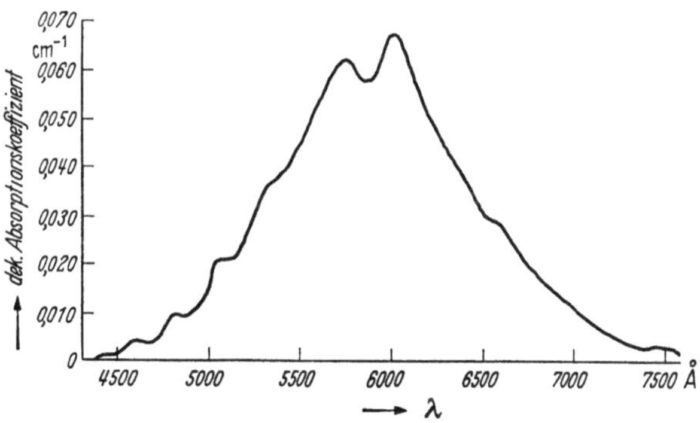

Fig. 3. CHAPPUIS-Banden nach VASSY.

VIGROUX [*46*] hat die CHAPPUIS-Banden über 4500 Å nach kürzeren Wellenlängen hin verfolgt. Bei 4100 Å ergibt sich ein Absorptionskoeffizient von nur 0,00035 cm^{-1}.

Wegen ihrer Schwäche sind die CHAPPUIS-Banden nur bei genauen quantitativen spektralen Beobachtungen des Sonnen- und Fixsternlichtes merklich[6].

[1] D. BARBIER u. D. CHALONGE: Ann. Phys. **17**, 280 (1942).
[2] H. ZSCHÖRNER u. H. K. PAETZOLD: Mitt. Max-Planck-Inst. Weißenau, Nr. 4, 1955.
[3] J. CHAPPUIS: J. Phys. (2) **1**, 484 (1882).
[4] A. VASSY: C. R. Acad. Sci. Paris **206**, 1638 (1938).
[5] L. LEFEBRVE: C. R. Acad. Sci. Paris **200**, 653 (1935).
[6] J. CABANNES u. J. DUFAY: J. Phys. et le Radium (6) **7**, 257 (1926).

In den letzten Jahren sind die CHAPPUIS-Banden für die Bestimmung der vertikalen Ozonverteilung aus photometrischen Beobachtungen bei Mondfinsternissen von Bedeutung geworden.

6. Die Ultrarot-Banden. Von 7500 bis 9000 Å zeigen sich nur sehr schwache Banden. Dagegen treten nach HETTNER und anderen[1] bei 4,7; 9,6 und 14,1 μ starke Rotationsschwingungsbanden auf (Fig. 4), die für das Strahlungsgleichgewicht der Stratosphäre von Bedeutung sind. Schwache Banden liegen bei 3,3 und 5,8 μ. Bei diesen Banden gilt *nicht* das BEERsche Gesetz; die Absorption hängt also nicht nur von dem Produkt Druck × durchlaufenem Weg in dem betreffenden Gas ab; ebenso ist auch der Fremdgaseinfluß nicht zu vernachlässigen.

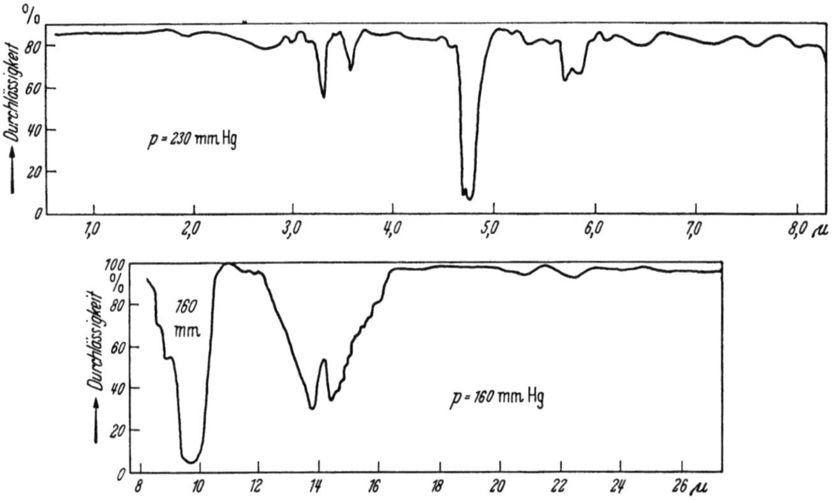

Fig. 4. Ultrarotbanden des Ozons nach HETTNER und anderen.

Nach neueren Messungen mit hochauflösenden Ultrarotspektrometern erscheint als das beste Modell für die 9,6 μ-Bande eine Anordnung von Triplets in gleichen Abständen[2]. VIGROUX [46] hat zahlreiche Linien des ultraroten Ozonspektrums neu vermessen. Die Theorie solcher sich aus vielen Einzellinien kompliziert aufbauenden Banden ist leider noch zu wenig entwickelt, als daß die Abhängigkeit der Absorption von der Schichtdicke und von dem Fremdgaseinfluß für die Anwendung rechnerisch genügend genau erfaßt werden könnte (vgl. Ziff. 25).

Nach eingehenden Messungen an der 9,6 μ-Bande des Ozons von STRONG[3] wird die durch eine Ozonschicht von der reduzierten Dicke x absorbierte Energie A (in Bruchteilen der Anfangsenergie) gegeben durch:

$$A = K\sqrt{x}, \tag{6.1}$$

wobei der Faktor K von dem Verhältnis $[O_2]/[O_3]$ abhängt. Bei Gültigkeit des BEERschen Gesetzes müßte gelten:

$$A = 1 - 10^{-\alpha x}, \tag{6.2}$$

wobei α wieder der Absorptionskoeffizient ist. Den Fremdgaseinfluß zeigt anschaulich Fig. 5. Aus ihr folgt ein Gang der Absorption mit $p^{\frac{1}{2}}$. GOODY[4] hat für

[1] G. HETTNER, R. POHLMANN u. H. J. SCHUHMACHER: Z. Physik **91**, 372 (1934).
[2] J. H. SHAW, H. L. OXHOLM u. H. H. CLAASEN: Astrophys. J. **116**, 554 (1952).
[3] J. STRONG: J. Franklin Inst. **231**, 121 (1941).
[4] R. M. GOODY: Quart. J. Roy. Met. Soc. **78**, 165 (1952).

ein statistisches Modell einer Rotationsschwingungsbande für die mittlere Transmission T^* die Beziehung abgeleitet:

$$T^* = \exp - \left[\frac{\pi \cdot z \cdot y}{\sqrt{1+z}}\right]; \quad z = \frac{\sigma \cdot m}{\pi \cdot \alpha}; \quad y = \frac{\alpha}{\delta}, \tag{6.3}$$

wobei σ die mittlere Linienintensität, δ der mittlere Linienabstand, α die Linienbreite und m der Betrag des absorbierenden Gases ist. Die experimentelle Prüfung bei der 9,6 μ-Ozonbande zeigte in einzelnen Bereichen gute, in anderen wieder schlechte Übereinstimmung mit der Theorie[1].

Die starke Druckabhängigkeit der Absorption in den ultraroten Ozonbanden kann dazu dienen, einen rohen Anhalt über die Höhenverteilung des atmosphärischen Ozons zu gewinnen[2,3]. Denn kennt man durch simultane Messungen sowohl den totalen Betrag der Ozonschicht wie die Absorption in der 9,6 μ-Bande, so erhält man aus Fig. 5 den mittleren Druck der Ozonschicht (vgl. Ziff. 9α).

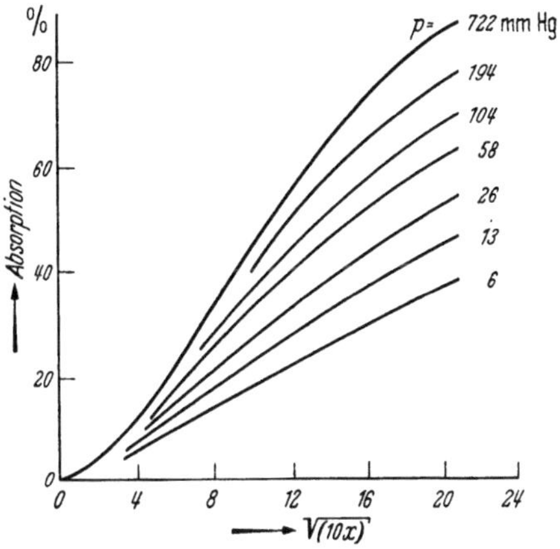

Fig. 5. Abhängigkeit der Absorption der 9,6 μ-Bande von der Schichtdicke x des Ozons und vom Druck p (nach STRONG).

C. Die Meßmethoden für das atmosphärische Ozon.

I. Die optischen Methoden.

7. Allgemeines. Für die Bestimmung des *Ozonwertes* x_0 vom Boden aus kommt nur die optische Methode in Frage, bei der die ultraviolette spektrale Intensitätsverteilung des Sonnen- oder Fixsternlichtes gemessen wird. Sie steht auch für die Erfassung der *vertikalen Ozonverteilung* gegenüber der chemischen Ozonmessung im Vordergrund.

Für die optische Ozonmessung im Bereiche der HARTLEY-Bande gilt die folgende Grundgleichung, aus der alle Spezialanwendungen abzuleiten sind:

$$I(h_0, \lambda) = I_0(\lambda) \cdot 10^{-f(z, h_0) \cdot x(h_0) \cdot \alpha_{O_3}} \cdot 10^{-M(z, h_0) \cdot k_R(\lambda) - D(z, h_0) \cdot k_D(\lambda)}. \tag{7.1}$$

Dabei bedeuten (vgl. auch die allgemeinen Symbole in Ziff. 2):

$I(h_0, \lambda)$ die spektrale Intensität der Sonnenstrahlung in der Höhe h_0;

$I_0(\lambda)$ die extraterrestrische Intensität der Sonne;

z die scheinbare Zenitdistanz der Sonne;

$M(z, h_0)$ die durchstrahlte Luftmasse, mit der zenitalen Luftmasse in Meereshöhe als Einheit;

$D(z, h_0)$ analog die durchstrahlte Dunstmasse;

$k_R(\lambda)$, $k_D(\lambda)$ die spektralen dekadischen Extinktionskoeffizienten der RAYLEIGH-Streuung und des atmosphärischen Dunstes.

---

[1] C. D. WALSHAW u. R. M. GOODY: Proc. of the Toronto Met. Conf., S. 49, 1953.
[2] K. WATANABE: J. Franklin Inst. **236**, 241 (1943).
[3] R. M. GOODY u. C. D. WALSHAW: Proc. of the Toronto Met. Conf., S. 27, 1953.

Die Funktion $f(z)$ wird unter Berücksichtigung der Strahlenbrechung gegeben durch[1]

$$f(z, h_0) = \frac{\int\limits_{h_0}^{\infty} \frac{\varepsilon(h) \cdot dh}{\sqrt{\cos^2 z + 2k \sin^2 z}}}{\int\limits_{h_0}^{\infty} \varepsilon(h) \cdot dh} \tag{7.2}$$

mit:

$$k = \frac{h - h_0}{a} - (n-1)\left(\varrho(h_0) - \varrho(h)\right) \tag{7.3}$$

n Brechungsindex der Luft; a Erdradius in km.

Da es nur auf den relativen Intensitätsverlauf ankommt, wird in logarithmischer Schreibweise aus (7.1) für 2 Wellenlängen λ_1 und λ_2:

$$\log \frac{I(h_0, \lambda_1)}{I(h_0, \lambda_2)} = \log \frac{I_0(\lambda_1)}{I_0(\lambda_2)} - f(z) \cdot x(h_0) \cdot \Delta\alpha_{O_3} - \left(M(z)\, \Delta k_R + D(z, h_0)\, \Delta k_D\right), \tag{7.4}$$

wobei $\Delta\alpha_{O_3}$ usw. Abkürzungen für die Differenzen der Absorptionen und Extinktionen für λ_1 und λ_2 sind.

8. Messung des Ozonwertes x_0. Die meßtechnischen Schwierigkeiten in der Anwendung von (7.4) bestehen in dem rapiden Intensitätsabfall am Ende des ultravioletten Sonnenspektrums, vor allem bei größerer Zenitdistanz z der Sonne, wie er durch Tabelle 2 [*11*] veranschaulicht wird. Deshalb ist auf die Vermeidung von störendem, langwelligerem *Streulicht* in der Apparatur besonders Wert zu legen.

Tabelle 2.
Spektrale Intensität des Sonnenlichtes am 7. Juni 1920 in Marseille.

λ (Å)	$z = 0°$	$z = 60°$
3143	22400	3230
3052	10200	410
3022	2700	46
2997	1320	8
2963	132	$1 \cdot 10^{-1}$
2946	25	$5 \cdot 10^{-3}$

FABRY und BUISSON benutzten einen Spektrographen mit gekreuzten Prismen. Das von dem ersten Spektralapparat erzeugte Spektrum dient als Eintrittsspalt für den zweiten, durch den das Streulicht unschädlich weggebrochen wird. Mit einem ähnlichen Spektrographen gelang es GÖTZ[2] in Arosa in 1600 m Höhe, das ultraviolette Sonnenspektrum bis herunter zu 2863 Å aufzunehmen.

Für eine laufende Messung des Ozonwertes ist jedoch die photographische Methode zu umständlich. Deshalb bedeutete die Konstruktion eines *lichtelektrischen Spektralphotometers* durch DOBSON[3] einen großen Fortschritt. Es besteht nach Fig. 6 aus einem Doppelmonochromator vom Typus VAN CITTERT, bei dem durch Spalte einzelne feste Wellenlängenpaare λ_1, λ_2 (7.4) ausgeblendet werden. Die Intensitäten der Strahlung je zweier Wellenlängen, die auf einen Multiplier (früher Photozelle) fällt, werden miteinander verglichen durch eine Nullmethode, mittels Wechsellichtes und Verstärker und einem optischen Absorptionskeil. Das Instrument ist so empfindlich, daß nicht nur im direkten Sonnenlicht, sondern auch bei bedecktem Himmel oder mit Mondlicht oder mit dem zenitalen Himmelsstreulicht bis in die Dämmerung hinein gemessen werden kann. Die letzte Möglichkeit erwies sich für die Erschließung der vertikalen Ozonverteilung als besonders fruchtbar (s. Ziff. 9α).

Die mit dem DOBSON-Gerät erzielte Genauigkeit ist beachtlich, besonders im Hochgebirgsklima, wo die Staub- und Dunstschicht nur schwach ist [7]. Fig. 7

[1] F. LINK: J. des Observ. **17**, 41 (1934).
[2] F. W. P. GÖTZ: Strahlenther. **40**, 690 (1931).
[3] G. M. B. DOBSON: Proc. Phys. Soc. Lond. **43**, 324 (1931).

zeigt Messungen in Arosa während eines Tages, an dem der Ozonwert nur wenig schwankte. Obwohl die wirksame Schichtdicke des Ozons wegen der verschiedenen

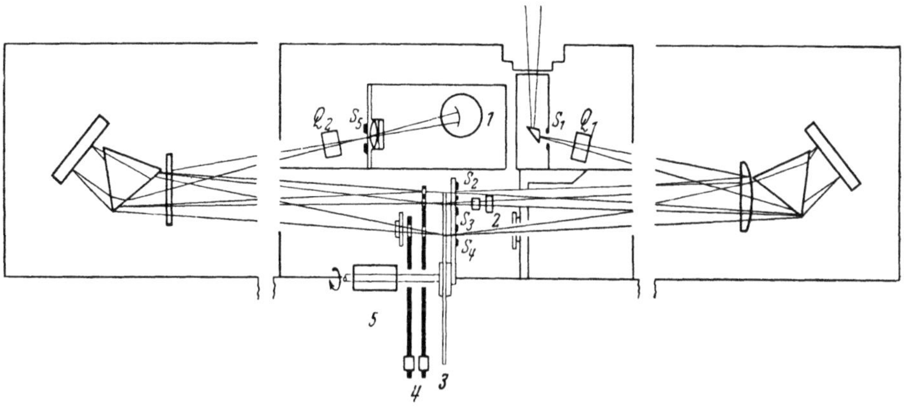

Fig. 6. DOBSON-Spektralphotometer. *1* Multiplier; *2* Absorptionskeile; *3* rotierende Sektorscheibe, die die Spalten S_2, S_3, S_4 freigibt; S_1 Eintritts-; S_5 Austrittsspalt des Monochromators; *4* Blendenschieber zur Auswahl der Spalten S_2, S_3, S_4; *5* Kommutator zur Gleichrichtung des Wechselstromes.

Sonnenhöhe stark variiert (Kreise im oberen Teil der Figur), treten bei dem Ozonwert x_0 für alle drei Wellenlängenpaare nur Schwankungen von $\pm 0{,}5\%$ auf.

Eine etwas einfachere Registrieranordnung haben COBLENTZ und STAIR im Anschluß an ihre *Ozonradiosonde* entwickelt. Sie benutzen eine Titanphotozelle

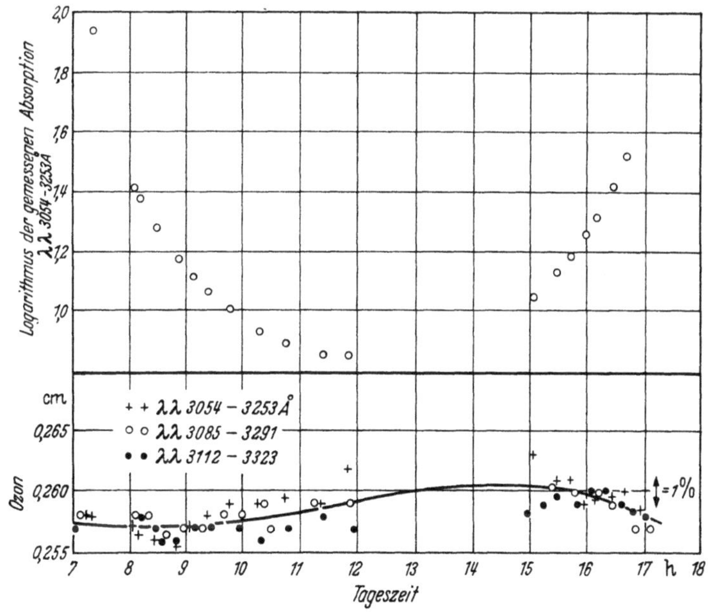

Fig. 7. Tagesmeßreihe mit dem DOBSON-Spektralphotometer.

als Empfänger, die nur unterhalb von 3400 Å empfindlich ist. Zur spektralen Aussonderung dienen vier verschiedene Filter.

Gleichgültig, welches apparative Verfahren im einzelnen angewandt wird, so ist in (7.4) doch neben dem Ozonbetrag noch die extraterrestrische spektrale Intensitätsverteilung $I_0(\lambda)$ der Sonne unbekannt. Sie wird für die betreffende

Apparatur bestimmt durch Messungen bei mehreren Zenitdistanzen z, wobei hier im allgemeinen

$$f(z) = \sec z$$

gesetzt werden kann *(lange Methode)*[1].

Das Dunstglied, das ja auch etwas variieren wird, wirkt sich nicht sehr kritisch aus, da der in Frage kommende Spektralbereich nur etwa 200 Å umfaßt und die Dunstextinktion nur mit λ^{-1} variiert. Es kann außerdem noch durch Messungen bei Wellenlängen, wo die Ozonabsorption verschwindet, berücksichtigt werden.

Ist das Glied $\log I_0(\lambda_1)/I_0(\lambda_2)$ mit Einschluß der Selektivität der Apparatur durch die lange Methode einmal bestimmt, so genügt jeweils eine einzelne Messung von $I(\lambda_1)/I(\lambda_2)$ zur Bestimmung des Ozonwertes x_0 *(kurze Methode)* [*14*].

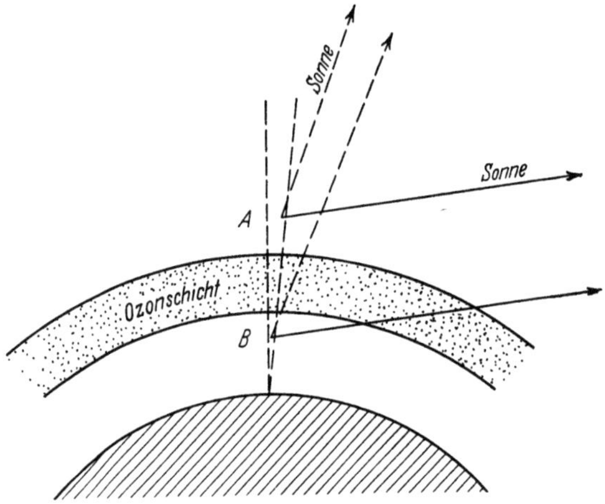

Fig. 8. Zum „Umkehreffekt".

9. Bestimmung der vertikalen Ozonverteilung $\varepsilon(h)$ (hohes Ozon). *α) Der Umkehreffekt* (GÖTZ-*Effekt*). Prinzipiell muß nach Gl. (7.1), durch Messungen bei sehr tiefen Sonnenständen, die Höhe des Schwerpunktes einer atmosphärischen Schicht zu bestimmen sein, wenn der Betrag der Schicht, hier der Ozonwert x_0, bekannt ist. Doch ergaben diesbezügliche Messungen beim Ozon nur unsichere Resultate, die, mit einer Schichthöhe von 40 bis 50 km, durch die nachfolgenden, genaueren Bestimmungen als viel zu hoch erkannt wurden. Ein wesentlicher Fortschritt in der Erschließung der vertikalen Ozonverteilung vom Erdboden aus wurde mit der Methode des *Umkehreffektes* durch GÖTZ[2] [*15*] erzielt.

Unter dem Umkehreffekt wird die folgende Erscheinung verstanden: Mißt man das *vom Zenit* herkommende *Himmelsstreulicht* bei tiefen Sonnenständen, so zeigt sich, daß bei Sonnenhöhen unterhalb von 9°, im Gegensatz zu den größeren Sonnenhöhen, mit sinkender Sonne die kurzwelligen Strahlungsanteile relativ zu den langwelligeren wieder zunehmen. Seien λ_1, λ_2 zwei Wellenlängen $(\lambda_1 > \lambda_2)$, und z die Zenitdistanz der Sonne, so durchläuft das Verhältnis $\frac{I(\lambda_2, z)}{I(\lambda_1, z)}$ mit wachsendem z ein Minimum und steigt dann wieder an.

Dieser zunächst überraschende Effekt erklärt sich anschaulich aus Fig. 8: Bei hochstehender Sonne wird das ultraviolette zenitale Streulicht vornehmlich aus Schichten mit größerer Luftdichte unterhalb des Hauptteils der Ozonschicht, etwa vom Punkt B her, gestreut. Bei tiefstehender Sonne dagegen gelangt das ultraviolette Licht, wegen der großen wirksamen Weglänge in der

[1] Streng müßte der Zenitwinkel z der Sonne am Erdboden durch den Zenitwinkel z' in der Höhe H der Ozonschicht ersetzt werden, wobei gilt: $\sin z' = \frac{a}{a+H} \cdot \sin z$, mit a = Erdradius. Doch ist die Korrektur nur für große Zenitwinkel merklich; Ozonmessungen bei niedriger Sonnenhöhe konnten deshalb zur Auswertung der Höhe H dienen. Diese anfänglichen Höhenbestimmungen erwiesen sich aber als ungenau, vgl. Ziff. 9α.

[2] F. W. P. GÖTZ, A. R. MEETHAM u. G. M. B. DOBSON: Proc. Roy. Soc. Lond., Ser. A **145**, 416 (1934).

Ozonschicht, nur mehr geschwächt nach B, so daß das Himmelsstreulicht vornehmlich von Punkten oberhalb der Ozonschicht, etwa von A, herabgestreut wird; dieses zenitale Streulicht durchsetzt die Ozonschicht vertikal, also auf kurzem Wege. Danach wird die genauere Form der Umkehrkurven von der jeweiligen vertikalen Ozonverteilung abhängen, wie Fig. 9 nach Messungen in Arosa und Tromsö zeigt [*15*].

Es gelingt, die vertikale O_3-Verteilung aus den Umkehrkurven mittels zweier verschiedener Auswerteverfahren abzuleiten: Bei dem synthetischen Verfahren (Methode B) werden die in den einzelnen Höhen versuchsweise angenommenen Ozonmengen solange variiert, bis die daraus abgeleitete Umkehrkurve mit der gemessenen übereinstimmt. Bei dem mehr analytischen Vorgehen (Methode A)

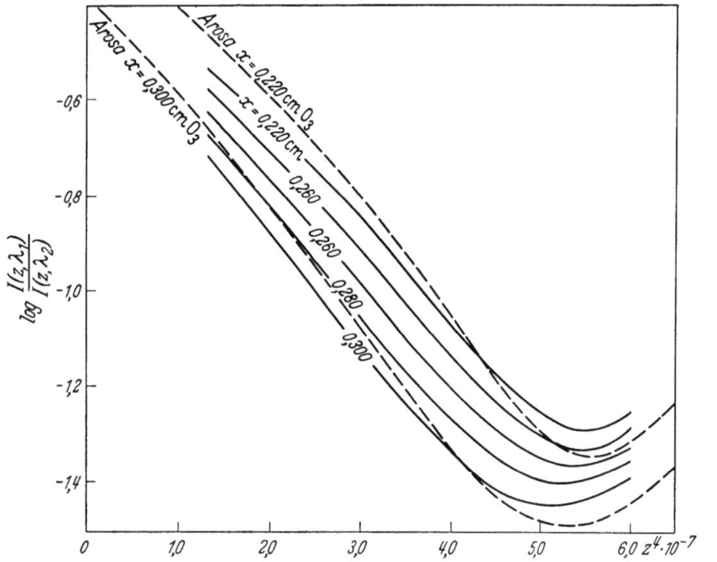

Fig. 9. Umkehrkurven in Tromsö (70° N) und in Arosa (47° N) (gestrichelt). z ist die scheinbare Zenitdistanz der Sonne, in Grad. Als Abszisse ist $z^4 \cdot 10^{-7}$ aufgetragen, um die Gegend des Umkehrens der Kurven auseinander zu ziehen.

werden die einzelnen Ozonbeträge in der (wieder in Schichten von etwa 10 km Dicke eingeteilten) Atmosphäre aus einem Gleichungssystem bestimmt, in dem die durchstrahlten Ozondicken und Luftmassen maßgeblich eingehen. Beide Verfahren A und B geben leider nicht ganz übereinstimmende Ergebnisse.

Ferner besteht in der schwer zu erfassenden Sekundärstreuung des Lichtes in der Erdatmosphäre eine weitere Fehlerquelle, die in jüngerer Zeit von WALTON[1] näher diskutiert worden ist. Der sekundäre Himmelsstreulichtanteil beträgt bei klarem Himmel immerhin rund 14% des primären. Und schließlich kann die Methode naturgemäß die Ozonverteilung nur im groben und keine feineren Einzelheiten wiedergeben, da sie bei zu feiner Schichteinteilung keine eindeutigen Resultate mehr liefert.

Ein Vergleich zwischen der indirekten Umkehrmethode und der direkten Aufstiegsmethode (Ziff. 9β) gibt Fig. 10 [*18*], wonach das spitze Ozonmaximum durch die erstere nur sehr eingeebnet wiedergegeben wird.

Trotz dieser Mängel hat die Umkehrmethode, als über die vertikale Ozonverteilung noch nichts bekannt war, die ersten wertvollen Aufschlüsse gegeben, z. B. die Höhe des Ozonmaximums bei 24 km; und auch heute ist sie unentbehrlich in Gegenden, wo bislang noch keine direkten Messungen mittels Ballonaufstiegen

[1] G. F. WALTON: UGGI, Association of Meteor., Mémoires and Discussions, Brüssel 1951.

möglich sind (polare und äquatoriale Gegenden). In Zukunft ist ein systematischer Vergleich mit direkten Messungen anzustreben, um das bisherige zahlreiche weltweite Beobachtungsmaterial auf eine sicherere Basis zu stellen.

Die Umkehrmessungen haben in neuester Zeit durch Messungen im ultraroten Spektralbereich eine willkommene Ergänzung erhalten. Bei ihnen wird aus *Absorptions-* und *Emissionsmessungen* vom Boden aus der mittlere *Druck* $\overline{p}$ und die mittlere *Temperatur* $\overline{\overline{T}}$ der Ozonschicht bestimmt. Aus diesen beiden Größen läßt sich zusammen mit dem Ozonwert x_0 die jeweilige Ozonverteilung ableiten, natürlich wieder nur in gröberen Zügen[1,2]. Im Gegensatz zum Umkehreffekt, der die Ozonverteilung nur in größeren Höhen, höchstens herab bis zu 10 km wiedergibt, gestatten die *Ultrarotmessungen* gerade eine Erfassung des troposphärischen Ozons[3,4].

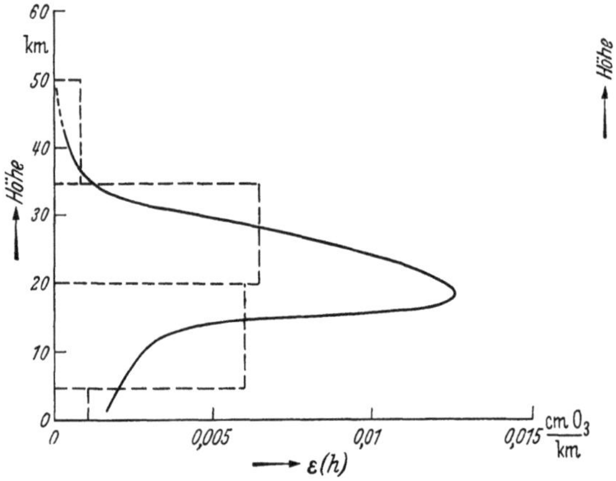

Fig. 10. Ozonverteilung über New Mexico am 2. 4. 1948 nach Umkehrmessungen (gestrichelt) und Raketenaufstieg (ausgezogen), nach Johnson und anderen.

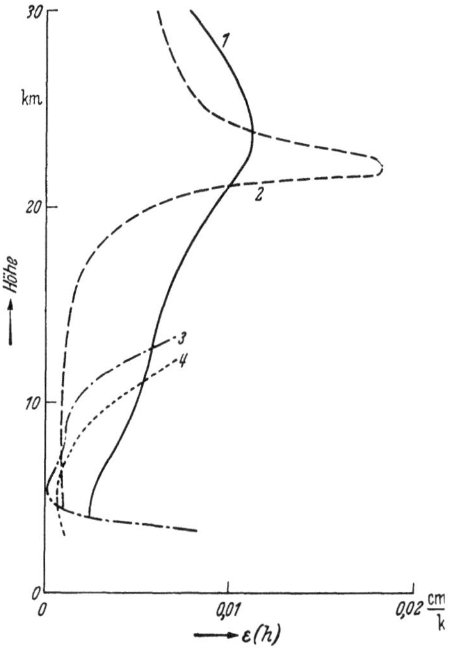

Fig. 11. Die ersten direkten Bestimmungen der vertikalen Ozonverteilung bei Ballonaufstiegen. *1* Stuttgart 31. 7. 1934; *2* Explorer 11. 11. 1935; *3* Stuttgart 30. 10. 1937; *4* Stuttgart 11. 12. 1937.

β) Ballon- und Raketenaufstiege. Es bedeutete einen großen Fortschritt, als mittels Quarzspektrographen bei Ballonaufstiegen in verschiedenen Höhen h ultraviolette Sonnenspektren aufgenommen wurden und daraus die jeweilige über der Apparatur befindliche Ozonmenge $x(h)$ bestimmt werden konnte. Die Ozonkonzentration $\varepsilon(h)$ (cm O_3/km) folgt dann direkt aus der gemessenen Kurve $x(h)$ durch Differentiation. Die ersten Messungen dieser Art wurden von E. und V. H. Regener[5] in Stuttgart 1934 und 1937 vervollkcmmnet [*38*] mit unbemannten Registrierballonen durchgeführt, die bis zu 31 km Höhe erreichten. Gleichzeitig wurden ähnliche Messungen in Amerika bei den bis 20 km Höhe reichenden bemannten *Explorer*-Aufstiegen 1934 und 1935 unternommen[6].

---

[1] A. Adel, E. S. Epstein u. Ch. Osterberg: Scientific Report HA 7, Contr. No. AF 19, 122—198 (1955).

[2] R. M. Goody u. W. T. Roach: Quart. J. Roy. Meteorol. Soc. **82**, 217 (1956).

[3] H. K. Paetzold: Z. Naturforsch. **11** a, 128 (1956).

[4] C. D. Walshaw u. R. M. Goody: Quart. J. Roy. Meteorol. Soc. **82**, 177 (1956).

[5] E. u. V. H. Regener: Phys. Z. **35**, 788 (1934).

[6] B. O'Brien, F. L. Mohler u. H. S. Stewart: Nat. Geogr. Soc. Contrib. Tech. Papers, Stratosphere Ser. **2**, 49 (1936).

Die Anforderungen an den Spektrographen sind verhältnismäßig gering, es genügt eine lineare Dispersion von 50 bis 100 Å/mm bei 3000 Å. Selbstverständlich müssen alle Regeln der photographischen Photometrie streng eingehalten werden. In Fig. 11 sind die ersten vier bis 1938 auf diese Weise bestimmten vertikalen Ozonverteilungen zusammengestellt; sie weisen schon auf beträchtliche Variationen derselben hin. Deshalb wurden nach 1950 sowohl in Amerika [*39*], [*40*] wie in Deutschland [*24*] weitere Ballonaufstiege durchgeführt, wobei vor allem auf die Konstruktion eines für Reihenaufstiege genügend stabilen Spektrographen Wert gelegt wurde. Fig. 12 gibt die in Höhen bis zu 30 km über Stuttgart aufgenommenen ultravioletten Sonnenspektren aus dem Jahre 1934 wieder. Als Lichtquelle für den Spektrographen diente eine von der Sonne beschienene MgO-Scheibe. Bei den späteren Weißenauer Aufstiegen wurde durch eine Ringblende das auf die MgO-Scheibe fallende Himmelsstreulicht um den Faktor 3 verringert. Bei einigen früheren Aufstiegen [*38*] war dies durch eine mittels einer Magnetnadel zur Sonne orientierten Blende erreicht worden. Die Reduzierung des Himmelsstreulichtes ist erforderlich, da für dieses die genaue Korrektur für die RAYLEIGH-Streuung und auch die exakte effektive Ozondicke (vgl. Ziff. 9α) nicht bekannt ist, wodurch in Höhen unterhalb von 10 km die Ergebnisse etwas gefälscht werden können.

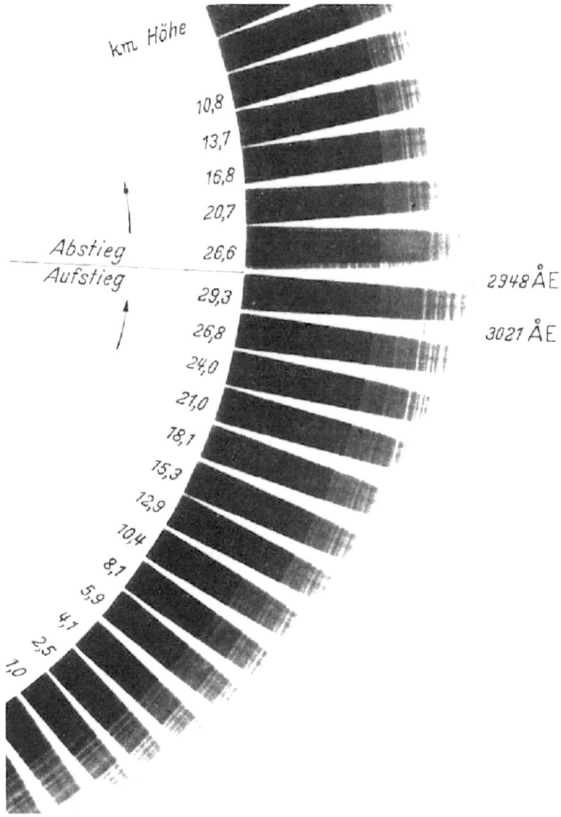

Fig. 12. Photographische Registrierung des ultravioletten Sonnenspektrums in der Stratosphare am 31. Juli 1934 nach E. und V. H. REGENER. Man beachte das Fortschreiten zu kürzeren Wellenlängen mit zunehmender Höhe.

Bei der Auswertung derartiger Aufstiegsspektrogramme nach (7.4) ist zu bedenken, daß hier die oben erwähnte „lange Methode" zur Eliminierung des Gliedes $\log\left(I_0(\lambda_1)/I_0(\lambda_2)\right)$ nicht angewendet werden kann. Doch ist durch die Messung der ultravioletten Sonnenintensität oberhalb der Ozonschicht durch die amerikanischen Raketenstarts jetzt die Unsicherheit in bezug auf dieses Glied weitgehend verringert worden. Allerdings ist zu beachten, daß wegen der starken Häufung der FRAUNHOFER-Linien im ultravioletten Sonnenspektrum die gemessene spektrale Intensitätsverteilung noch etwas von den Abbildungseigenschaften des benutzten Spektrographen abhängen wird. Deshalb ist zur Kontrolle der aus der gemessenen Ozonverteilung $\varepsilon(h)$ folgende Ozonwert x_0 an den mittels eines DOBSON-Instrumentes gemessenen anzuschließen. Ferner muß zur Erzielung einer hohen Genauigkeit ein genügend großes Wellenlängenintervall erfaßt werden, was durch ein vor den Spalt gesetztes Stufenfilter erreicht wird, das auch der photographisch-photometrischen Kalibrierung dient.

Bei den Weißenauer Aufstiegen konnte so eine Genauigkeit in $\varepsilon(h)$ von ± 1; 0,5 und $0,3 \cdot 10^{-3}$ cm O_3/km in 5; 25 und 30 km Höhe erzielt werden, die ausreichte, gröbere Schwankungen der Ozonverteilung auch in der Troposphäre zu erfassen, wo die Verhältnisse wegen der nur geringen Abnahme von $x(h)$ mit der Höhe besonders kritisch sind[1].

Die Aufstiegstechnik selbst ist durch die neuen amerikanischen in einem Stück im Tauchverfahren gefertigten Ballone[2] gegenüber früher sehr vereinfacht worden. Für eine Aufstiegsapparatur von 3 kg Gewicht genügen 2 Ballone mit je 2 kg Eigengewicht, um eine durchschnittliche Gipfelhöhe von 33 km zu erreichen, wobei die maximal erreichten Höhen knapp 40 km betragen. In jüngster Zeit konnte in Weißenau mit größeren 7,5 kg schweren Ballonen eine Höhe von 44 km erreicht werden. Mit der benutzten Apparatur (UV-Spektrograph und MgO-Scheibe) reichte in dieser Höhe das ultraviolette Sonnenspektrum auf den Spektralaufnahmen bis zu 2700 Å, d.h. bis zu einem Absorptionskoeffizienten des Ozons von 100 cm^{-1}.

Durch die Fortschritte der *Raketentechnik* konnte die vertikale Ozonverteilung zu noch größeren Höhen bis herauf zu 70 km bestimmt werden[3,4] [*18*], was besonders im Hinblick auf die Prüfung der photochemischen Theorie des Ozons wichtig ist (Ziff. 14). Allerdings geben die Raketenaufstiege die Ozonverteilung nur oberhalb von etwa 20 km Höhe genügend genau.

Die Methode der Aufstiege mit Quarzspektrographen ist bislang für das hohe Ozon die am genauesten arbeitende. Ein Nachteil ist, daß die Spektrographen nach der Landung für die Auswertung wieder aufgefunden werden müssen. Dies wird durch die *Ozonradiosonde* von COBLENTZ und STAIR [*5*] vermieden. Bei ihr werden mittels einer Cadmiumphotozelle und Filtergläsern mit verschiedener UV-Durchlässigkeit die Intensitäten verschiedener ultravioletter Spektralbereiche über einen Sender an eine Bodenempfangsstation heruntergegeben. Die Methode arbeitet naturgemäß nicht so genau; doch konnten mit ihr schon eine Reihe von Ozonbestimmungen erfolgreich durchgeführt werden. Für einen breiteren meteorologischen Einsatz ist jedoch die Konstruktion einer noch einfacheren und billigeren Sonde erforderlich, wie sie jetzt vorliegt[5].

γ) Methode der Mondfinsternisse. Die bisherigen Methoden geben die Ozonverteilung nur über dem jeweiligen Beobachtungsort. Hier bietet die in den letzten Jahren entwickelte Methode der Mondfinsternisse eine willkommene Ergänzung[6,7,8] [*1*] und [*25*]. Bei ihr wird die Ozonverteilung bestimmt aus der

[1] Diesen Nachteil der optischen Methode im ultravioletten Sonnenlicht, bei der aus der integralen Kurve $x(h)$ durch Differentiation die Ozonkonzentration $\varepsilon(h)$ bestimmt wird, vermeidet eine Anordnung, bei der $\varepsilon(h)$ direkt differentiell aus der Absorption der UV-Strahlung einer künstlichen Lichtquelle auf einem genügend langen Weg in verschiedenen Höhen gemessen wird. Eine solche Anordnung, bei der ein Lichtweg von mehreren hundert Metern durch wiederholte Reflexion an zwei Spiegeln verwirklicht wurde, ergab nach noch unveröffentlichten Messungen in Amerika (Signal Engineering Corps) eine Ozonverteilung mit einer „Feinschichtung", wie sie auch bei Messungen des Wasserdampfgehaltes in der Stratosphäre festgestellt wurde. Allerdings wird der Einsatz einer solchen etwa 100 kg schweren Apparatur bei Ballonaufstiegen nur in einzelnen seltenen Fällen möglich sein.

[2] Lieferfirma: Dewalco, Cambridge, Mass., USA.

[3] J. A. VAN ALLEN u. J. D. HOPFIELD: Mém. Soc. Roy. Sc., Liège **12**, Fasc. 4, 179 (1952).

[4] F. S. JOHNSON, J. D. PURCELL, R. TONSEY u. K. WATANABE: J. Geophys. Res. **57**, 157 (1952).

[5] W. KULCKE: Diss. Stuttgart 1956.

[6] H. K. PAETZOLD: Z. Naturforsch. **5**a, 661 (1950).

[7] H. K. PAETZOLD: J. Atmosph. a. Terres. Phys. **2**, 183 (1952).

[8] E. VIGROUX: Ann. d'Astrophys. X, **17**, 399 (1954).

im Erdschatten auf dem Mond senkrecht zur Schattengrenze gemessenen Ozonabsorption im Spektralbereich der CHAPPUIS-Bande. Je nach der Bahn des Mondes durch den Erdschatten können eventuell schon bei einer Finsternis verschiedene geographische Breiten erfaßt werden.

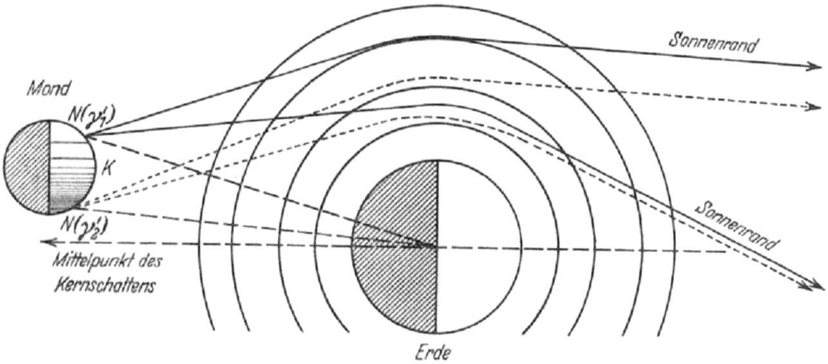

Fig. 13. Strahlengang bei einer Mondfinsternis; schematisch. γ' bedeutet die Entfernung von der Kernschattengrenze K. Positive Werte von γ' beziehen sich auf den Halbschatten-, negative auf den Kernschattenraum.

Die Wirksamkeit der Methode beruht darauf, daß einmal, wegen des tangentialen Strahlenganges, das Licht große, wirksame Schichtdicken Ozon (bis zu 15 cm O_3), aber im Verhältnis dazu nur kleine Luftmassen durchsetzt, und zum anderen, daß wegen der atmosphärischen Refraktion der Höhenbereich der Ozonschicht stark vergrößert auf dem Mond „abgebildet" wird. Beiläufig lassen sich auf diese Weise noch weitere sehr schwach absorbierende atmosphärische Schichten in größeren Höhen nachweisen.

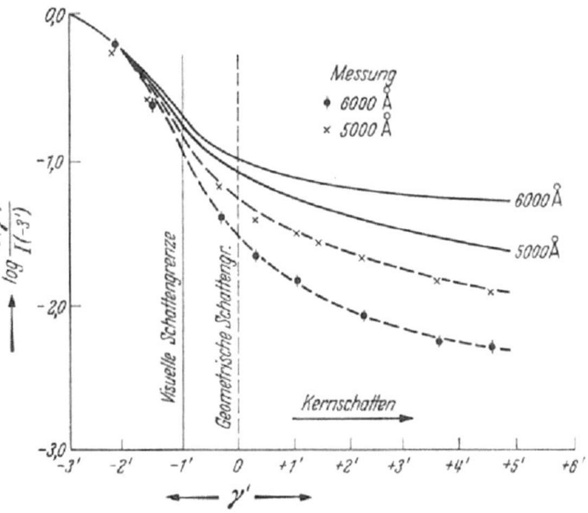

Fig. 14. Helligkeitsverlauf auf dem verfinsterten Mond in der Umgebung der Schattengrenze am 6. 10. 1949 nach PAETZOLD. Ausgezogen: berechnet ohne Ozon, gestrichelt: beobachtet. Die Werte sind auf die Helligkeit für $\gamma = -3'$ normiert. Die irdische Ozonschicht bewirkt bei 6000 Å eine zusätzliche Schwächung des den Mond beleuchtenden Lichtes auf $^1/_{10}$.

Bei den Messungen kommt es im Prinzip darauf an, die jeweilige wirksame Ozonmasse $O^*(\gamma')$ für ein Lichtbündel abzuleiten, das einen Punkt des Mondes, $N\gamma'$, der in der Entfernung γ' von der geometrischen Kernschattengrenze entfernt liegt, beleuchtet (Fig. 13). Dazu kann einmal mit einem Fernrohr mittels Filtern oder eines Spektrographen der spektrale Helligkeitsverlauf in der Umgebung der Schattengrenze gemessen und mit dem zu erwartenden Verlauf bei einer ozonfreien Atmosphäre verglichen werden, wie er nach einer von LINK[1] gegebenen Theorie folgt (Fig. 14). Oder aber man nimmt mit einem lichtstarken Spektrographen das Spektrum des verfinsterten Mondes zu verschiedenen Finsternisphasen auf und bestimmt die wirksame Ozonmasse $O^*(\gamma')$

[1] F. LINK: Bull. Astronomique 8, 77 (1932).

aus der Tiefe der im Spektrum beobachteten CHAPPUIS-Bande [1] (Fig. 15). Das erstere Verfahren gibt die bessere geographische Zuordnung.

Für die Auswertung bedingt die Tatsache, daß die Sonne kein punktförmiger Strahler ist, sondern in Erdentfernung einen scheinbaren Durchmesser von 30′ aufweist, eine Komplikation. Aus den gemessenen Ozonmassen $O^*(\gamma')$ muß die Ozonmasse $O(h_0)$, die von einem Einzelstrahl durchlaufen würde, abgeleitet werden, der die Erdatmosphäre in der Minimalhöhe h_0 passiert (Fig. 13). Es gilt, mit a = Erdradius,

$$O(h_0) = \sqrt{2a} \cdot \int_{h_0}^{\infty} \frac{\varepsilon(h)}{h - h_0} \cdot dh, \quad (9.5)$$

woraus das gesuchte $\varepsilon(h)$ als Lösung dieser ABELschen Integralgleichung folgt:

$$\varepsilon(h) = -\frac{1}{\pi\sqrt{2a}} \cdot \int_{h}^{\infty} \frac{\frac{dO(h_0)}{dh_0}}{h_0 - h} \cdot dh_0. \quad (9.6)$$

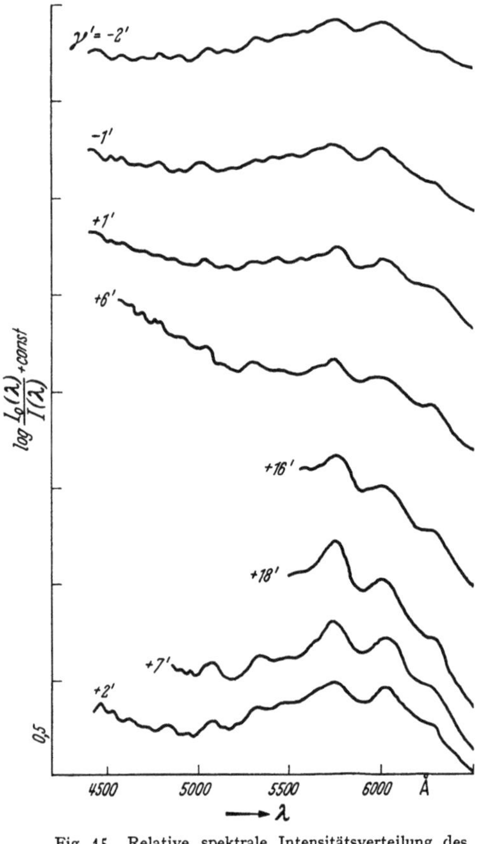

Fig. 15. Relative spektrale Intensitätsverteilung des den verfinsterten Mond beleuchtenden Lichtes für verschiedene Schattenpunkte nach VIGROUX. $I_0(\lambda)$ = Helligkeit auf dem unverfinsterten, $I(\lambda)$ auf dem verfinsterten Mond. Man beachte wieder die tiefe Einsattlung des Intensitätsverlaufs, die durch die CHAPPUIS-Bande der irdischen Ozonschicht bewirkt wird.

Die Umformung von $O^*(\gamma')$ in $O(h_0)$ gelingt in einem Iterationsverfahren, wobei die Atmosphäre in Schichten von mindestens 5 km Dicke einzuteilen ist. Damit gestattet die Methode der Mondfinsternisse, die Ozonverteilung etwas feiner zu erfassen als der Umkehreffekt. Fig. 16 zeigt vier aus früheren spektralen Helligkeitsmessungen im Erdschatten abgeleitete Ozonverteilungen, zwei äquatoriale und zwei polare, die die großen jeweiligen Unterschiede demonstrieren[1].

Es sei noch eine weitere optische Erscheinung bei Mondfinsternissen erwähnt, die durch die irdische Ozonschicht verursacht wird: Bei Beobachtungen der Farben des verfinsterten Mondes fällt auf, daß an der Schattengrenze in einem schmalen, nur wenige Minuten breiten Bereich häufig eine grünliche bis bläuliche Färbung auftritt, die über graue Tönungen in das bekannte Rostrot des mittleren und inneren Kernschattengebietes übergeht. Diese Erscheinung läßt sich auf die durch die CHAPPUIS-Bande des Ozons bewirkte starke Extinktion des den Mond beleuchtenden Lichtes im gelbroten Spektralbereich zurückführen (Fig. 15), durch die in der Nähe der Schattengrenze ein Intensitätsmaximum bei 5000 Å bewirkt wird[2,3]. Die grüne Schattengrenze stellt damit ein sichtbares optisches Abbild der irdischen Ozonschicht dar. Auf der Erde läßt sich eine analoge Erscheinung nicht beobachten, auch nicht bei sehr tief stehender Sonne, da hier immer die nach

[1] H. K. PAETZOLD: Z. Naturforsch. 7a, 325 (1952).
[2] H. K. PAETZOLD: Naturwiss. 38, 544 (1951).
[3] H. K. PAETZOLD: Die Sterne 30, 13 (1954).

kürzeren Wellen hin stark ansteigende RAYLEIGH-Extinktion den Ausschlag gibt, wie es auch bei Mondfinsternissen im inneren, rot erscheinenden Kernschattenbereich der Fall ist.

10. Das bodennahe Ozon. In der Nähe des Erdbodens und in der Troposphäre treten im allgemeinen nur kleine Ozonkonzentrationen zwischen 1 bis $4 \cdot 10^{-3}$ cm O_3/km auf. Sie werden optisch daher mit einer künstlichen ultravioletten Lichtquelle im Gebiet der maximalen Ozonabsorption in der Mitte der HARTLEY-Bande bei 2540 Å gemessen.

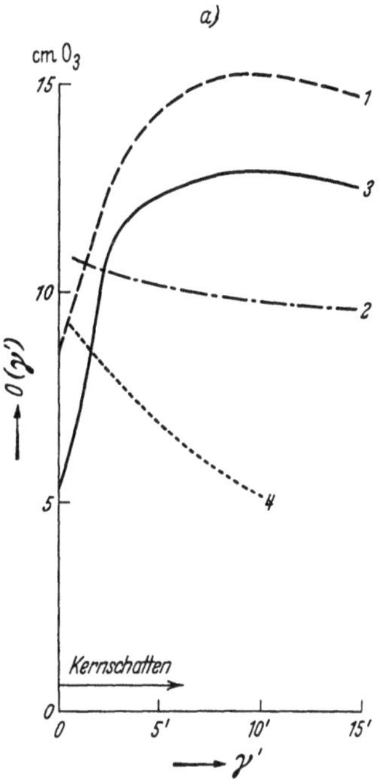

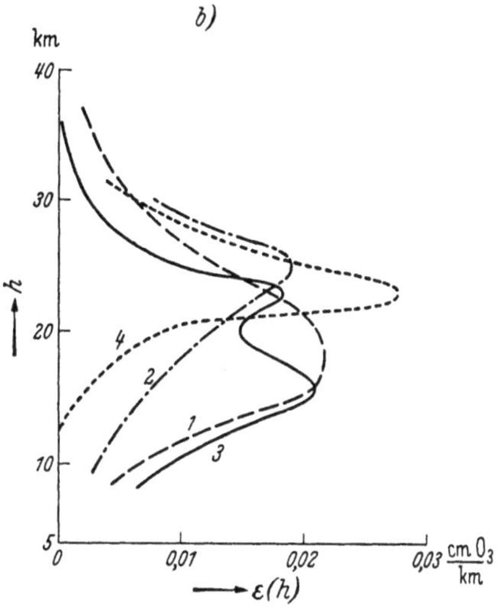

Fig. 16a u. b. Bestimmung der vertikalen Ozonverteilung durch Mondfinsternisse, nach PAETZOLD. a Mittlere effektive Ozonmassen $O^*(\gamma')$; b die daraus abgeleiteten Ozonverteilungen; *1* 14. 9. 1932 (60° S); *2* 14. 9. 1932 (0°); *3* 20. 2. 1943 (50° N); *4* 15. 8. 1943 (20° S).

Wegen des geringen Ozonbetrages muß die vom Licht durchlaufene Luftstrecke genügend groß sein. Da man das bodennahe Ozon zu Vergleichsmessungen nicht ausschalten kann, werden die Messungen jeweils mit zwei Entfernungen Lichtquelle — Spektrograph, einmal einige hundert Meter und dann 1 bis 2 km, durchgeführt. Die Ergebnisse sind sorgfältig nach den verschiedenen Extinktionen, Ozonabsorptionen, Sauerstoffabsorption, Dunst usw. zu diskutieren[1]. Bei 1 km Lichtweg betragen z.B. die durch die RAYLEIGH-Streuung, Sauerstoff und Ozon ($\varepsilon = 2 \cdot 10^{-3}$ cm O_3/km) verursachten Teilabsorptionen 26, 22 bzw. 52% der Gesamtabsorption.

Messungen in verschiedenen Höhen wurden im Berner Oberland[2, 3], in Lauterbrunnen (800 m) und auf dem Jungfraujoch (3450 m) und auf dem M. Ventouse [*45*] (1912 m) durchgeführt. Sie ergaben eine Konzentration des bodennahen Ozons um $2 \cdot 10^{-3}$ cm O_3/km. In neuerer Zeit haben STAIR und andere[4] die Methode photoelektrisch ausgebaut und automatisiert. Als Lichtquelle dient eine Hg-Niederdrucklampe, die die Hg-Linie 2537 mit überwiegender Intensität aussendet, und als Empfänger ein Multiplier.

[1] H. BUISSON, C. JAUSSERAN u. P. RONARD: Rev. Opt. **12**, 70 (1933).
[2] D. CHALONGE u. E. VASSY: J. Phys. Radium **5**, 309 (1934).
[3] F. W. P. GOETZ u. H. MAIER-LEIBNITZ: Z. Geophys. **9**, 253 (1933).
[4] R. STAIR, TH. C. BAGG u. R. G. JOHNSTON: J. Res. Nat. Bur. Stand. **52**, 133 (1954).

II. Die chemische Ozonmessung.

11. Angesichts der Schwierigkeit der optischen Methoden bei kleinen Ozonkonzentrationen erscheint für das bodennahe und troposphärische Ozon die chemische Methode am geeignetsten. Das am häufigsten angewendete Verfahren beruht auf der intensiven Oxydation von Kaliumjodid durch Ozon beim Durchsaugen der Luft durch eine neutrale 2%ige KJ-Lösung:

$$O_3 + 2\,KJ + H_2O = O_2 + 2\,KOH + J_2.$$

Der Lösung ist Natriumthiosulfat zugesetzt, das das freigewordene Jod sofort bindet:

$$2\,Na_2S_2O_3 + J_2 = 2\,NaJ + Na_2S_4O_6.$$

Damit ist die Ozonmessung auf die genaue Titration des verbrauchten $Na_2S_2O_3$ zurückgeführt. Eine mikroanalytische Ausarbeitung[1] derselben, wobei in bekannter Weise (durch Zugabe von Jodstärkelösung und Beobachten des Einsetzens der Blaufärbung) die Titration erfolgte, konnte auch bei einzelnen Flugzeugaufstiegen eingesetzt werden[2].

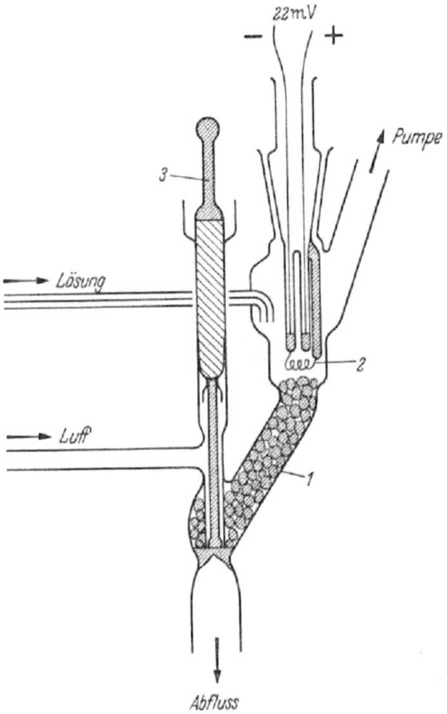

Fig. 17. Automatische chemische Ozonmessung nach V. H. REGENER.

Für laufende Ozonmessungen bedeutete es einen erheblichen Fortschritt, als die Titration des $Na_2S_2O_3$ durch GLÜCKAUF und andere zu einem elektrolytischen und automatisierten Verfahren ausgebildet wurde[3]. Die elektrolytische Titration beruht auf dem Umstand, daß in KJ-Lösung bei Anwesenheit von freiem Jod zwischen zwei Elektroden mit einer angelegten Spannung unterhalb von 0,5 V ein der Jodkonzentration proportionaler Strom fließt. Er wird vermutlich durch die Depolarisation an der Platinkathode bedingt, die durch das freie Jod verursacht wird.

Eine vereinfachte automatische Ausgestaltung des Verfahrens haben BOWEN und REGENER[4] angegeben (Fig. 17): Luft wird mit einer Geschwindigkeit von 1,5 Liter/min durch die Reaktionskammer 1 gesaugt, die mit 8 cm³ $KJ-Na_2S_2O_3$-Lösung gefüllt ist. Wenn alles $Na_2S_2O_3$ verbraucht ist und die Jodkonzentration ansteigt, fließt zwischen den Elektroden 2 ein Strom. Bei einer Stärke von $2 \cdot 10^{-9}$ Amp wird mittels einer Verstärkungsanordnung der Verschluß 3 betätigt; die verbrauchte Lösung fließt ab und neue ein. Die Zeit, in der ein Cyclus stattfindet, ist ein Maß für die Ozonkonzentration. Bei den durchschnittlich vorkommenden Werten des bodennahen Ozons werden etwa 30 min benötigt.

Etwas modifiziert arbeitet das halbautomatische Verfahren von EHMERT [9], das die höchste Genauigkeit aufweist. Bei ihm wird zur Titration durch ein

[1] V. H. REGENER: Meteor. Z. **12**, 459 (1938).
[2] A. EHMERT: Forschgs.- u. Erfahrungsber. dtsch. Wetterd., Reihe A **1941**, Nr. 13.
[3] E. GLÜCKAUF, H. G. HEAL, G. R. MARTIN u. F. A. PANETH: J. Chem. Soc. Trans. **1944**, 1.
[4] I. G. BOWEN u. V. H. REGENER: J. Geophys. Res. **56**, 307 (1951).

zweites Hilfselektrodenpaar mit bekannter Stromstärke in der Reaktionslösung elektrolytisch Jod erzeugt und die Zeit gemessen, die benötigt wird, bis zwischen dem Meßelektrodenpaar (Spannung 0,18 V) der Strom einsetzt. Mit 3 cm^3 Reaktionslösung und 10 Liter durchgesaugter Luft kann eine Genauigkeit von 0,5 γ O_3/m^3 erzielt werden, d. h. etwa 2% des mittleren Betrages des bodennahen Ozons.

Die gute Brauchbarkeit der chemischen Ozonmessung beruht auf dem Umstand, daß andere Gase wie NO_2 und H_2O_2, die ebenfalls Jod freisetzen, nur langsam mit KJ-Lösung reagieren. NO_2, das etwa in derselben Menge wie das Ozon in der unteren Atmosphäre vorhanden ist, wirkt etwa 20mal träger als das Ozon, das sehr schnell und vollständig zur Reaktion kommt. Den störenden Einfluß von Fremdgasen sucht CAUER[1] durch Pufferung der Kaliumjodidlösung mit Natriumazetat auszuschalten. Frei von derartigen Bedenken ist die von PANETH[2,3] benutzte Methode, die auf der unterhalb von $-120°$ C quantitativen Adsorption des Ozons durch Silicagel beruht. Leider ist das Verfahren für laufende Messungen zu umständlich. Nach einem eingehenden Vergleich der verschiedenen Verfahren durch TEICHERT[4] ist das Arbeiten mit neutraler Kaliumjodidlösung bei reiner Luft unbedenklich, während in der Nähe größerer Siedlungen, Industriezentren usw. Vorsicht geboten ist.

D. Photochemische Theorie des Ozons.

12. Allgemeines. Die Bildung des atmosphärischen Ozons muß zurückgeführt werden auf nur durch die ultraviolette Sonnenstrahlung ausgelöste photochemische Prozesse. Seit den ersten Arbeiten von CHAPMAN [*3*], MECKE[5], WULF und DENNING[6] ist das Problem in neuerer Zeit von SCHRÖER [*41*], DÜTSCH [*8*], BATES und NICOLET [*2*], PAETZOLD [*26*] und CRAIG [*4*] theoretisch behandelt worden. Es sei zuerst eine kurze Übersicht über die photochemischen Prozesse gegeben:

1. Die *primären* Prozesse: Sie werden durch Absorption von Lichtquanten bewirkt und sind temperaturunabhängig. Im einfachsten Fall wird ein Molekül XY zerlegt gemäß

$$XY + h\nu \rightarrow X^* + Y^*. \tag{12.1}$$

Die Bildungsgeschwindigkeit der Komponente X:

$$\frac{d[X]}{dt} = \varphi \cdot I_{\text{abs}, xy} \tag{12.2}$$

φ = Quantenausbeute, $I_{\text{abs}, xy}$ = Zahl der von XY im cm^3 absorbierten dissoziierenden Lichtquanten.

2. Die den primären folgenden *sekundären* Prozesse sind normale Dunkelreaktionen und temperaturabhängig. Es sind zu unterscheiden:

a) Der *bimolekulare* Prozeß

$$X + YZW \rightarrow XY + WZ \tag{12.3}$$

mit der Bildungsgeschwindigkeit:

$$\frac{d[XY]}{dt} = K_{\text{Bi}}[X][YZW]. \tag{12.4}$$

[1] H. CAUER: Z. anal. Chem. **103**, 321, 385 (1935).
[2] F. A. PANETH: Nature, Lond. **142**, 112, 571 (1938).
[3] F. A. PANETH: J. Chem. Soc. **1941**, 511.
[4] F. TEICHERT: Z. Meteor. **6**, 132 (1952).
[5] R. MECKE: Trans. Faraday Soc. **27**, 325 (1931).
[6] O. R. WULF u. L. S. DENNING: Terr. Magn. **41**, 299 (1936).

Die Reaktionskonstante K_{Bi} wird durch die ARRHENIUSsche Gleichung gegeben:

$$K_{\mathrm{Bi}} = \eta\, P\, e^{-q/R\,T} \tag{12.5}$$

η Stoßzahl bei Normalbedingungen, P ein sterischer Faktor und q die Aktivierungsenergie.

b) Die *Dreistoßreaktion* mit dem Stoßpartner M

$$X + Y + M \rightarrow X\,Y + M\,, \tag{12.6}$$

$$\frac{d[X\,Y]}{d\,t} = K_{\mathrm{Dr}}\,[X]\,[Y]\,[M]\,. \tag{12.7}$$

Die Reaktionskonstante K_{Dr} ist proportional der mittleren Stoßzahl τ und damit $\sqrt{T}$. Die bimolekularen Prozesse sind nach (12.5) stark temperatur- und die Dreistoßreaktionen (12.6) stark druckabhängig. Die Größenordnungen von K_{Bi} und K_{Dr} sind bei 273° K etwa $1 \cdot 10^{-16}\,\mathrm{cm^3/sec}$ bzw. $10^{-35}\,\mathrm{cm^3/sec}$.

13. Die ozonbildenden und ozonzerstörenden Reaktionen. Im folgenden benutzte Größen sind:

N_A AVOGADROsche Konstante;

C_{O_2} Volumenkonzentration des Sauerstoffs;

h Höhe in der Atmosphäre;

$p(h)$ Luftdruck, $\varrho(h)$ Luftdichte — mit den entsprechenden Größen unter Normalbedingungen als Einheit;

$\alpha_{O_2}(\lambda)$ der spektrale dekadische Absorptionskoeffizient des Sauerstoffs ($\mathrm{cm^{-1}}$) für $p = 1$;

$\alpha_{O_3}(\lambda)$ der spektrale dekadische Absorptionskoeffizient des Ozons ($\mathrm{cm^{-1}}$) für $p = 1$;

φ_1 Quantenausbeute der Sauerstoffdissoziation;

φ_2 Quantenausbeute der Ozondissoziation;

$I_0(\lambda)$ die extraterrestrische spektrale Intensität der Sonnenstrahlung $\left(\frac{\text{Quanten}}{\mathrm{cm^2\,sec\,Å}}\right)$;

$I(\lambda, h)$ die spektrale Intensität des Sonnenlichtes in der *Höhe* h $\left(\frac{\text{Quanten}}{\mathrm{cm^2\,sec\,Å}}\right)$;

$I_{\mathrm{abs},O_2}(h)$, $I_{\mathrm{abs},O_3}(h)$ — die vom O_2 bzw. O_3 absorbierten photochemisch wirksamen Quanten $\left(\frac{\text{Quanten}}{\mathrm{cm^3\,sec}}\right)$.

Allgemein ist die Photochemie des Ozons in komplizierterer Weise mit der Photochemie anderer atmosphärischer Bestandteile wie der des Wassers und des Stickstoffs verknüpft.

Die *primären* ozonisierenden und desozonisierenden Prozesse:

		kennzeichnende Größen		
$O_2 + h\nu = O + O$	$(\lambda < 2424\,Å)$	$I_{\mathrm{abs},O_2}(h)$	φ_1	(13.1)
$O_3 + h\nu = O_2 + O$	$(\lambda < 11340\,Å)$	$I_{\mathrm{abs},O_3}(h)$	φ_2,	(13.2)

wobei die Schreibweise die zahlreichen möglichen, verschieden angeregten Produkte nicht berücksichtigt.

Als *Sekundär*prozesse sind die wichtigsten:

	Reaktionskonstante	
$O_2 + O + M = O_3 + M$	K_1	(13.3)
$O_3 + O = O_2^* + O_2^*$	K_2	(13.4)
$O + O + M = O_2 + M$	K_3.	(13.5)

Ferner die sog. *thermische Ozonzersetzung*:

$O_3 + O_3 = 3\,O_2^*$	K_4	(13.6)
$O_3 + O_2 = 2\,O_2 + O$	K_5.	(13.7)

Außerdem sind Reaktionen mit N_2 möglich, wie:

$$O_3 + N_2 = N_2O + O_2 \tag{13.8}$$

$$O + N_2 + M = N_2O + M. \tag{13.9}$$

Noch verwickelter werden die Verhältnisse, wenn durch photolytische Dissoziation des Wasserdampfes H- und OH-Partner vorhanden sind, wobei unter anderen die folgenden Reaktionen möglich sind:

$$H + O_3 \rightarrow OH + O_2, \tag{13.10}$$

$$H + O_3 \rightarrow H_2O + O, \tag{13.11}$$

$$OH + O_3 \rightarrow H_2O + O_2. \tag{13.12}$$

Bei den meisten dieser Sekundärreaktionen sind die Reaktionskonstanten nicht genau bekannt und können nur abgeschätzt werden. Für eine quantitative Fassung der Theorie des Ozons ist es daher entscheidend, daß eine Reihe der erwähnten Reaktionen in dem für das Ozon hauptsächlich in Frage kommenden Höhenbereich unterhalb von 60 km vernachlässigt werden können. So wirkt sich die photolytische Dissoziation des H_2O-Dampfes merklich erst in Höhen oberhalb von 60 km aus. Hier ist allerdings mit einer erheblichen Verminderung (bis um den Faktor 2) des — absolut allerdings schon sehr geringen — Ozongehaltes bei Anwesenheit von Wasserdampf zu rechnen [*2*], [*28*].

Die Reaktionskonstanten sind in Tabelle 3 zusammengestellt. K_1, K_2 folgen aus Messungen von EUCKEN und PATAT[1], allerdings mit einiger Unsicherheit, da nur das Verhältnis K_1/K_2 gemessen wurde, was allerdings später allein maßgeblich eingeht. Die Reaktionskonstante K_3 der Rekombination der O-Atome zu normalen O_2-Molekülen ist ebenfalls nur größenordnungsmäßig bekannt. Die Konstanten K_6 und K_7 sind obere mögliche Werte, die aus der Annahme folgen, daß das gesamte in der Atmosphäre gemessene N_2O im Betrage von 0,4 cm durch die Reaktionen (13.8) und (13.9) erzeugt wird[2]. Diese Reaktionen

Tabelle 3. *Reaktionskonstanten* K_i.

T° K	K_1 cm⁶/sec	K_2 cm³/sec	K_3 cm⁶/sec	K_4 cm³/sec	K_5 cm³/sec	K_6 cm³/sec	K_7 cm⁶/sec
323	$9 \cdot 10^{-35}$	$1{,}4 \cdot 10^{-14}$		$3{,}8 \cdot 10^{-25}$	$6{,}6 \cdot 10^{-26}$	$1 \cdot 10^{-25}$	
273	$8 \cdot 10^{-35}$	$2{,}2 \cdot 10^{-15}$	$1 \cdot 10^{-34}$	$3{,}4 \cdot 10^{-28}$	$1 \cdot 10^{-29}$	$1 \cdot 10^{-27}$	$1 \cdot 10^{-38}$
223	$7 \cdot 10^{-35}$	$1{,}5 \cdot 10^{-16}$		$1{,}2 \cdot 10^{-32}$	$3{,}8 \cdot 10^{-35}$	$1 \cdot 10^{-31}$	

[1] A. EUCKEN u. F. PATAT: Z. phys. Chem. **33**, 459 (1952).
[2] A. P. MITRA: J. Atmos. a. Terres. Phys. **5**, 28 (1953).

sind für das Ozon zu vernachlässigen, was auch anschaulich aus den unterschiedlichen Lebenszeiten hervorgeht, bis ein N_2O- bzw. O_3-Molekül photolytisch zersetzt wird: Diese betragen am Boden und in 60 km Höhe beim N_2O 3 Jahre bzw. 2 Monate, und beim O_3 1 Std bzw. 3 min. Schließlich ist unterhalb von 60 km die Rekombination von O-Atomen zu O_2-Molekülen zu vernachlässigen; unterhalb dieser Höhe ist die O-Konzentration so gering, daß die O-Atome praktisch immer mit O_2-Molekülen reagieren. In Höhen zwischen 60 und 100 km ist dagegen Reaktion (13.5) von Einfluß, da hier die O-Konzentration gegenüber der O_2-Konzentration nicht mehr vernachlässigbar ist.

Die thermische Ozonzersetzung nach den Konstanten K_4 und K_5 [1] erfolgt in dem gesamten Höhenbereich so langsam, daß auch sie nicht berücksichtigt zu werden braucht.

14. Der photochemische Gleichgewichtszustand. Mit den obigen Vereinfachungen ergeben sich für die Bildungsgeschwindigkeit des Ozons bzw. des atomaren Sauerstoffs übersichtliche Ausdrücke:

$$d\,[O_3]/dt = K_1 \cdot 0{,}6 \cdot [O_2]\,[O]\,[M] - K_2 \cdot [O_3]\,[O] - \varphi_2 \cdot I_{\text{abs},\,O_3}, \tag{14.1}$$

$$d\,[O]/dt = \varphi_2 \cdot I_{\text{abs},\,O_3} + \varphi_1 \cdot I_{\text{abs},\,O_2} - K_1 \cdot 0{,}6 \cdot [O]\,[O_2]\,[M] - K_2 \cdot [O]\,[O_3]. \tag{14.2}$$

Der Faktor 0,6 berücksichtigt den Umstand, daß die Stöße bei der Dreistoßreaktion (13.3) mit Stickstoffmolekülen nur etwa halb so wirksam sind wie mit Sauerstoffmolekülen [2].

Im photochemischen Gleichgewicht erhält man für die im cm³ befindlichen O-Atome und O_3-Moleküle:

$$[O] = \frac{2\varphi_1 \cdot I_{\text{abs},\,O_2} + \varphi_2 \cdot I_{\text{abs},\,O_3}}{K_1 \cdot [O_2]\,[M] + K_2 \cdot [O_3]} \left(\frac{\text{O-Atome}}{\text{cm}^3}\right), \tag{14.3}$$

$$\left.\begin{aligned} [O_3] &= \frac{K_1}{K_2} \cdot 0{,}6 \cdot [O_2]\,[M] \cdot \frac{\varphi_1 \cdot I_{\text{abs},\,O_2}}{2\varphi_2 \cdot I_{\text{abs},\,O_3}} \left(\frac{O_3\text{-Moleküle}}{\text{cm}^3}\right) \\ &\text{mit } I_{\text{abs},\,O_2} \ll I_{\text{abs},\,O_3}. \end{aligned}\right\} \tag{14.4}$$

Für die Rechnung und Diskussion wird (14.4) bequemer auf die Ozonkonzentration $\varepsilon(h)$ bezogen:

$$\varepsilon(h) = C_{O_2} \cdot \sqrt{N_A} \cdot \sqrt{\varrho(h)^3} \cdot \sqrt{\frac{K_1}{K_2} \cdot 0{,}6 \cdot \frac{\varphi_1}{\varphi_2} \cdot \frac{i_{1,\,\text{abs}}(h)}{i_{2,\,\text{abs}}(h)}} \cdot 10^5 \left(\frac{\text{cm } O_3}{\text{km}}\right) \tag{14.5}$$

mit

$$i_{1,\,\text{abs}}(h) = 2{,}3 \int I(\lambda, h) \cdot \alpha_{O_2} \cdot f(p)\, dh, \tag{14.6}$$

$$i_{2,\,\text{abs}}(h) = 2{,}3 \int I(\lambda, h) \cdot \alpha_{O_3} \cdot dh. \tag{14.7}$$

Die Ozonkonzentration ist nach (14.5) bestimmt von dem Verhältnis: ozonisierende zur desozonisierenden Sonnenstrahlung, das stark mit der Höhe variiert. Die Funktion $f(p)$ berücksichtigt die Druckabhängigkeit des Absorptionskoeffizienten des Sauerstoffs. Die Form der Gl. (14.5) ist besonders bequem, um den Einfluß der Luftdichte, der Temperatur und der Strahlungsintensität zu diskutieren. Zur Berechnung der Ozonverteilung muß die Atmosphäre in einzelne Schichten aufgeteilt und für sie die Ozonkonzentration durch Iteration berechnet werden, da die spektrale Sonnenintensität $I(\lambda, h)$ auch von der Stärke der vom Licht durchstrahlten Ozonschicht oberhalb der Höhe h abhängt.

---

[1] A. GLISSMACHER u. H. J. SCHUHMACHER: Z. phys. Chem., Abt. B **21**, 325 (1933).
[2] H. J. SCHUHMACHER u. U. BERETTA: Z. phys. Chem. **13**, 17 (1932).

In (14.5) ist der Wert von K_1/K_2 wegen der bimolekularen Reaktion stark temperaturabhängig[1]:

$$\frac{K_1}{K_2} = 3{,}5 \cdot 10^{-20} \cdot e^{-\left(\frac{6180}{RT} - \frac{6180}{RT_0}\right)} \,(\mathrm{cm}^3)\,. \tag{14.8}$$

In einer warmen Atmosphäre von 250° K würde sich nur die Hälfte des Ozons nach dem photochemischen Gleichgewicht einstellen als in einer kälteren von 210° K [*26*].

Die Quantenausbeute φ_2 ist konstant $= 1$, auch ist in dem gesamten Absorptionsbereich für das Ozon das BEERsche Gesetz als gültig anzunehmen. Unterhalb von 15 km Höhe wirkt die Absorption in den CHAPPUIS-Banden etwa 5mal stärker photolytisch dissoziierend auf das Ozon als die Ultraviolettabsorption.

Bei dem *Sauerstoff* hingegen sind die Verhältnisse etwas verwickelter. Das Hauptabsorptionsgebiet des Sauerstoffs im SCHUMANN-RUNGE-Kontinuum von 1300 bis 1759 Å kommt für die Ozonbildung nicht in Betracht, da die Sonnenstrahlung dieses Spektralbereiches schon in Höhen von 80 km praktisch vollständig absorbiert ist. Bei den (bis 30 km herab mit wirksamen) SCHUMANN-RUNGE-FÜCHTBAUER-Banden von 1759 bis 2026 Å ist der Dissoziationsprozeß noch nicht ganz geklärt. Im Falle von Prädissoziation ist $\varphi_1 = 2$. Falls ein angeregtes O_2-Molekül entsteht, so geht dies entweder durch Aussendung von Strahlung oder durch Stöße mit N_2- und O_2-Molekülen in den Grundzustand über, wobei nur die letzten ozonbildend sind gemäß:

$$O_2^* + O_2 = O_3 + O\,.$$

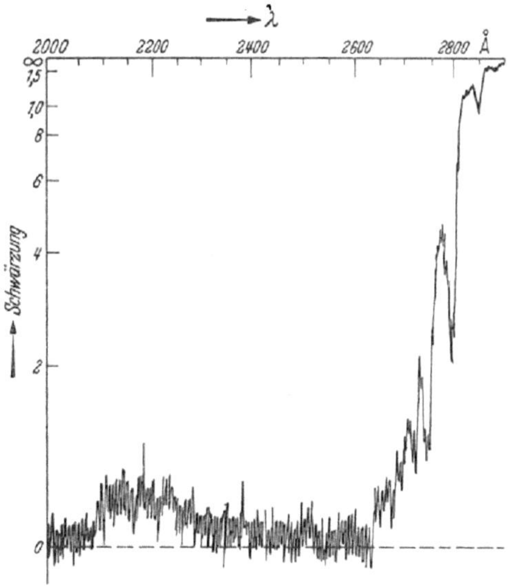

Fig. 18. Das Ozon-Sauerstoffenster bei 2150 Å in 35 km Höhe nach NEWELL jr.

In diesem zweiten Fall ist daher die Quantenausbeute φ_1 (14.9) druckabhängig, sie würde in dem hier in Frage kommenden Höhenbereich zwischen 0,04 und 0,4 liegen.

Diese Ungewißheit fällt jedoch nicht allzu stark ins Gewicht, da der Hauptteil des Ozons erzeugt wird durch die — in dem sog. atmosphärischen *Ozon-Sauerstoffenster* der Atmosphäre bei etwa 2150 Å — in tiefere Luftschichten dringende Sonnenstrahlung. Dieses Fenster wird durch das Zusammenwirken der Sauerstoff- und Ozonabsorption, die nach kürzeren Wellen ansteigen bzw. abfallen, bedingt[2,3]. Frühere Versuche, die Sonnenstrahlung in diesem Fenster zum Teil mit Lichtzählrohren[4] am Erdboden bzw. in der unteren Stratosphäre nachzuweisen, haben keinen eindeutigen Erfolg gehabt[5,6,7]. In Fig. 18 ist die photometrische Registrierung eines bei einem Raketenaufstieg in 35 km Höhe gewonnenen Spektrums wiedergegeben[8], die ein erstes Anzeichen dieser Strahlung erkennen läßt.

---

[1] A. EUCKEN u. F. PATAT: Z. phys. Chem. **33**, 459 (1952).
[2] F. W. P. GÖTZ: Vijschr. Astr. Ges. **70**, 343 (1935).
[3] E. VASSY: Rev. Opt. **15**, 81 (1936).
[4] F. W. P. GÖTZ, M. SCHEIN u. B. STOLL: Helv. phys. Acta **7**, 485 (1939).
[5] E. MEYER, M. SCHEIN u. B. STOLL: Nature, Lond. **134**, 535 (1934).
[6] K. O. KIEPENHEUER: Z. Astrophys. **14**, 348 (1937).
[7] V. H. REGENER: Naturwiss. **26**, 141 (1938).
[8] H. E. NEWELL jr.: Trans. Amer. Geophys. Un. **31**, 25 (1950).

Für die Sauerstoffabsorption in diesem Spektralbereich des Ozon-Sauerstoff-fensters, des sog. Tripletkontinuums ($\lambda < 2421$ Å) ist $\varphi_1 = 2$; allerdings gilt das BEERsche Gesetz nicht. Nachdem HEILPERN[1] zuerst eine exponentielle Druckabhängigkeit der Sauerstoffabsorption gemäß der Funktion $f(p) = p^{0,65}$ in (14.6) fand, ergaben seine letzten Messungen die folgende Beziehung[2] für $f(p)$:

$$f(p) = A + B \cdot C_{N_2} \cdot p + C \cdot \frac{C_{N_2}^2}{C_{O_2}} \cdot p \tag{14.10}$$

(mit C_{N_2} usw. Volumkonzentrationen), wonach auch ein beträchtlicher Fremdgaseinfluß bestehen würde. Doch folgt aus dem absoluten Wert der Konstanten für die hier in Frage kommenden Drucke ($p < 0.1$), daß die beiden letzten Glieder zu vernachlässigen sind, so daß das BEERsche Gesetz hier doch quasi gelten würde.

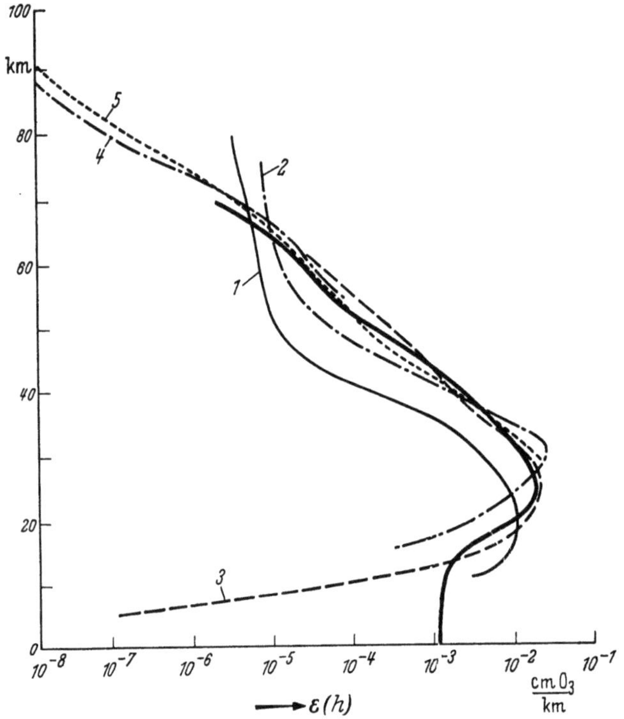

Fig. 19. Nach dem photochemischen Gleichgewicht berechnete Ozonverteilung. *1* DÜTSCH; *2* CRAIG; *3* PAETZOLD; *4* BATES und NICOLET; *5* JOHNSON und andere. Die in mittleren Breiten im Mittel beobachtete Ozonverteilung ist durch die starke Kurve gegeben.

Eine weitere Unsicherheit besteht noch in den Ozon- und Sauerstoff-Absorptionskoeffizienten in der für die Ozonbildung maßgeblichen Ozonsauerstofflücke, wenn auch durch die Arbeiten von VASSY [*45*] und HEILPERN[2] hier schon ein großer Fortschritt erzielt wurde. Schließlich ist die extraterrestrische spektrale Intensität der Sonnenstrahlung in diesem Spektralbereich trotz der Raketenaufstiege noch mit einer Unsicherheit etwa um den Faktor 2 behaftet[3]. Dagegen sind Temperatur und Luftdichte für Höhen über 30 km jetzt genügend bekannt. Nach Fig. 19 stimmen die neueren Berechnungen der vertikalen Ozonverteilung recht befriedigend überein und sind auch mit den Beobachtungen verträglich.

Eine ausführliche Diskussion [*26*] des Einflusses der noch bestehenden Unsicherheiten der einzelnen Parameter zeigt, daß dieselben von geringem Einfluß auf die Höhe des Ozonmaximums sind, die sich in Übereinstimmung mit den Beobachtungen immer zu 23 km ergibt. Das Ozonmaximum kommt durch das Zusammenwirken zweier Faktoren zustande: von oben nach unten nimmt die Luftdichte exponentiell und das Intensitätsverhältnis der ozonisierenden zur desozonisierenden Sonnenstrahlung stark ab. Nur der Ozonwert resultiert je nach den verwendeten Annahmen etwas verschieden.

---

[1] W. HEILPERN: Helv. phys. Acta **14**, 329 (1941).

[2] W. HEILPERN: Helv. phys. Acta **25**, 753 (1952).

[3] F. S. JOHNSON, J. D. PURCELL, R. TOUSEY u. W. WILSON: Rocket Exploration of the upper Atmosphere, S. 279. London 1954.

Doch zeigt sich beim Vergleich der berechneten und der beobachteten Abhängigkeit des Ozonwertes von der geographischen Breite eine wesentliche Diskrepanz: Die berechneten Werte ergeben sich am Äquator, mit 0,3 bis 0,4 cm O_3, immer am größten — entsprechend dem höheren Sonnenstand, bei dem die ozonisierende Sonnenstrahlung tiefer in die Atmosphäre hinabdringen kann — ; während in Wirklichkeit am Äquator mit 0,17 cm O_3 der geringste Ozonwert *gemessen* wird (Ziff. 23).

Nach [*8*] gelingt es nicht — auch nicht durch die Berücksichtigung der Streustrahlung des Himmels und des täglichen Ganges der Sonne —, die beobachtete Breitenabhängigkeit und die jahreszeitliche Variation des Ozons durch ein allein wirksames stationäres photochemisches Gleichgewicht zu deuten. Hier kommt der allgemeine Einfluß von *Luftbewegungen* zum Ausdruck, dessen Wirksamkeit von der Einstellzeit des photochemischen Gleichgewichtszustandes abhängen wird. Die Zeit t_{Gl}, in der die Abweichung von der Gleichgewichtsverteilung bei Einstrahlung bis auf $1/e$ des Anfangswertes abgesunken ist, wird in der oben benutzten Schreibweise mit guter Näherung gegeben durch [*8*], [*26*]:

Tabelle 4. *Einstellzeit t_{Gl} des photochemischen Gleichgewichts.*

h	WULF und DEMING	DÜTSCH	CRAIG	PAETZOLD
60 km	2^m	1^h	—	1^h
50	6^m	$1{,}5^h$	—	2^h
50	2^h	10^h	—	11^h
30	2^d	20^d	11^d	15^d
25	—	120^d	80^d	100^d
20	—	$1{,}3^a$	$1{,}4^a$	2^a

$$t_{Gl} = \frac{\sqrt{N_A^3}}{4} \cdot \sqrt{\frac{K_1}{K_2} \cdot \frac{2\varrho(h)}{i_{1,\,abs} \cdot i_{2,\,abs} \cdot \varphi_1 \cdot \varphi_2}}\,.$$

Wegen der Zunahme der Luftdichte und der starken Abnahme der ozonisierenden Strahlung bei Annäherung an den Erdboden wird t_{Gl} nach Tabelle 4 in tieferen Luftschichten sehr groß. Trotz der von den einzelnen Bearbeitern je nach den benutzten Voraussetzungen erhaltenen, etwas unterschiedlichen Werte ergibt sich ein einheitliches Bild: unterhalb von 25 bis 30 km kann sich der reine photochemische Gleichgewichtszustand überhaupt nicht einstellen; hier müssen die Transportvorgänge in der Atmosphäre von maßgeblichem Einfluß sein. Ferner wird die Ozonschicht kurzzeitigen Schwankungen der Sonnenstrahlung nur oberhalb von 50 km folgen; und der jährliche Gang der Sonnenhöhe wird sich wesentlich nur oberhalb von 25—30 km auswirken können[1] [*8*], [*26*].

15. Fehlende Einstrahlung (Nacht, Polarnacht). Für das Problem der zu erwartenden Tag- und Nachtvariationen des Ozons ist der Fall der fehlenden Einstrahlung zu diskutieren. Dann reduzieren sich (14.1) und (14.2) zu

$$d[O_3]/dt = K_1 \cdot 0{,}6\,[O_2]\,[O]\,[M]\,, \tag{15.1}$$

$$d[O]/dt = -K_1 \cdot [O_2]\,[O]\,[M] - 2K_3 \cdot [O]^2\,[M]\,. \tag{15.2}$$

Die thermische Ozonzersetzung nach (13.6) und (13.7) ist wiederum zu vernachlässigen. Denn die erste Reaktion, die sehr langsam verläuft, liefert freie O-Atome, die unter 70 km Höhe alle wieder O_3 bilden. Nach der 1000mal langsameren zweiten Reaktion nimmt aber (Tabelle 5) [*12*] bei den in der Atmosphäre vorkommenden Temperaturen eine Ozonkonzentration von $2{,}3 \cdot 10^{-1}$ cm O_3/km, wie sie nie

[1] H. U. DÜTSCH: Arch. Meteor., Geophys. u. Bioklim., Ser. A **9**, 87 (1956).

beobachtet wird, erst in mindestens $5 \cdot 10^4$ Jahren auf die Hälfte ab. Einmal gebildetes Ozon hält sich beim Fehlen von oxydablen und katalytisch wirksamen Substanzen in der freien Atmosphäre außerordentlich lange, so daß während der Nacht und der Polarnacht keine merkliche Menge Ozon thermisch zerfällt. Es ist vielmehr nach Sonnenuntergang eine Ozonzunahme nach (15.1) und (15.2) zu erwarten, da sich (15.2) unterhalb von etwa 100 km Höhe auf den ersten und oberhalb auf den zweiten Term reduziert. Allerdings wird dieser nächtliche Anstieg der Ozonkonzentration entsprechend der Reaktionskonstanten und der O_3- und O-Konzentration (Tabelle 6), nur in Höhen von 50 bis 70 km wesentlich sein; dort mag er aber zur Ausbildung eines nächtlichen zweiten oberen Ozonmaximums führen[1] [*28*]. Sein Beitrag zum Gesamtozon ist jedoch gering und sollte 3% nicht übersteigen. Dies ist in Übereinstimmung mit den langjährigen Ozonreihen in Europa. Im Gegensatz dazu ist in Indien[2] und Afghanistan [*19*] ein Ansteigen des Ozonwertes eine halbe Stunde nach Sonnenuntergang um etwa 10% beobachtet worden. Eine befriedigende Klärung steht noch aus.

Tabelle 5 [*12*].

Temperatur	Zeit, in der bei einem Ausgangswert von $2{,}3 \cdot 10^{-1}$ cm O_3/km das Ozon zur Halfte thermisch zerfallt
200°	$4 \cdot 10^{20}$ h
300°	$4 \cdot 10^{9}$ h
400°	$1 \cdot 10^{4}$ h
500°	6 h

Tabelle 6. *O- und O_3-Konzentration am Tag.*

h km	[O] cm⁻³	[O_3] cm⁻³
40	$1 \cdot 10^{10}$	$4 \cdot 10^{11}$
60	$3 \cdot 10^{10}$	$1 \cdot 10^{10}$
80	$1 \cdot 10^{11}$	$1 \cdot 10^{8}$
100	$1 \cdot 10^{12}$	$1 \cdot 10^{6}$

E. Schwankungen der vertikalen Ozonverteilung.

16. Umkehrmessungen und Mondfinsternisse. Obwohl der Umkehreffekt die Ozonverteilung nur verhältnismäßig grob wiedergibt, ließen sich mit ihm schon einige wesentliche charakteristische jahreszeitliche und geographische Unterschiede feststellen.

In Fig. 20 sind zwei Ozonverteilungen mit verschieden hohem Ozonwert x_0, wie sie von GÖTZ [*16*] in *mittleren* Breiten (in Arosa) gemessen wurden, dargestellt, die die Anordnung des Ozons in einer Schicht demonstrieren. Besonders fällt das zweite Maximum in 14 km Höhe auf, das auch bei Mondfinsternissen[3] (Fig. 16) und durch direkte Messungen bei Ballonaufstiegen bestätigt werden konnte. Dieser Typ tritt vornehmlich im Frühjahr auf, während das schlanke einzelne Maximum vor allem im Herbst beobachtet wird. Dies entspricht dem ausgeprägten Jahresgang des Ozonwertes, der in Arosa im langjährigen Mittel im Frühjahr 0,27 und im Herbst 0,20 cm O_3 beträgt (Ziff. 23).

In *höheren* Breiten, in Tromsö [*43*] und in Spitzbergen[4], wo sich die Extremwerte des Ozonbetrages (vgl. Fig. 40) noch stärker unterscheiden, zeigt auch die Ozonverteilung eine entsprechende jahreszeitliche Variation (Fig. 20).

In *äquatorialen* Breiten dagegen, wo der Ozonwert einen konstanten niedrigen Betrag von 0,17 cm O_3 aufweist, findet sich das Ozon immer im wesentlichen oberhalb von 20 km konzentriert, so daß in der Troposphäre im Mittel nur sehr geringe Ozonbeträge gemessen werden[5] (Fig. 21). Die Unterschiede zwischen den mittels Umkehrmessung und mittels Mondfinsternissen bestimmten Ozon-

[1] H. U. DÜTSCH: Arch. Meteor., Geophys. u. Bioklim., Ser. A **9**, 87 (1956).
[2] K. R. RAMANATHAN: [*32*].
[3] H. K. PAETZOLD: Z. Naturforsch. **7**a, 325 (1952).
[4] S. H. H. LARSEN: J. Atmosph. a. Terres. Phys. **6**, 46 (1955).
[5] R. V. KARANDIKAR: Proc. Ind. Akad. Soc. **35**, 290 (1952).

verteilungen sind vermutlich nicht reell, sondern methodisch bedingt. Im Gegensatz zu den mittleren und höheren Breiten scheinen am Äquator die (allerdings verhältnismäßig nur geringen) Ozonschwankungen vornehmlich in dem Höhenbereich zwischen 18 und 28 km lokalisiert zu sein[1].

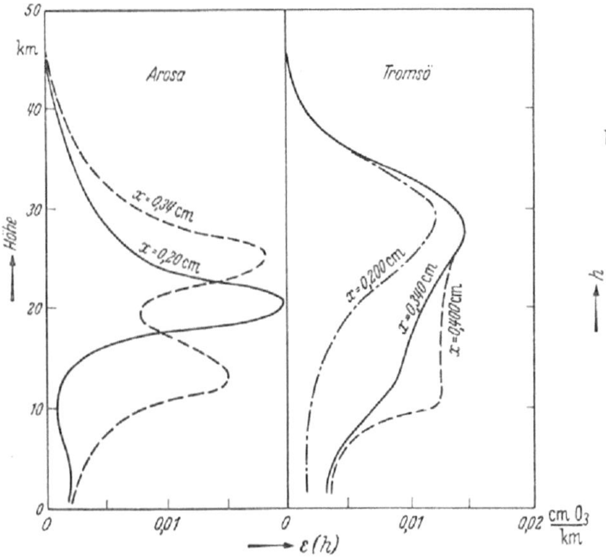

Fig. 20.

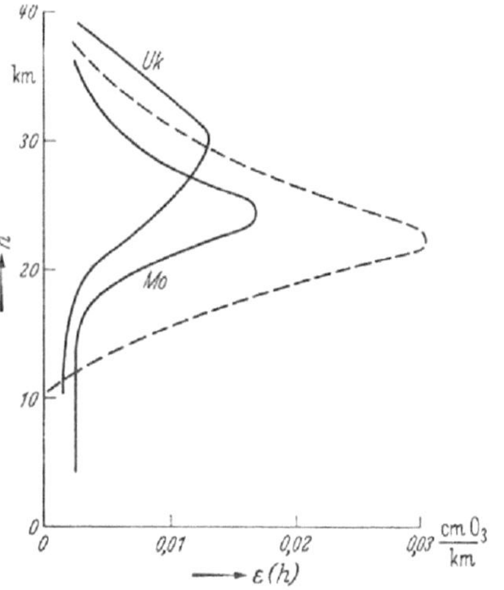

Fig. 21.

Fig. 20. Vertikale Ozonverteilungen nach Umkehrmessungen in Arosa (47° N) und Tromsö (70° N). Die hohen Ozonwerte werden vornehmlich im Frühjahr, die niedrigen im Herbst beobachtet.

Fig. 21. Äquatoriale Ozonverteilung nach Umkehrmessungen (U_K) und Mondfinsternissen (*Mo*). Die gestrichelte Kurve gibt die nach dem photochemischen Gleichgewicht zu erwartende Ozonverteilung.

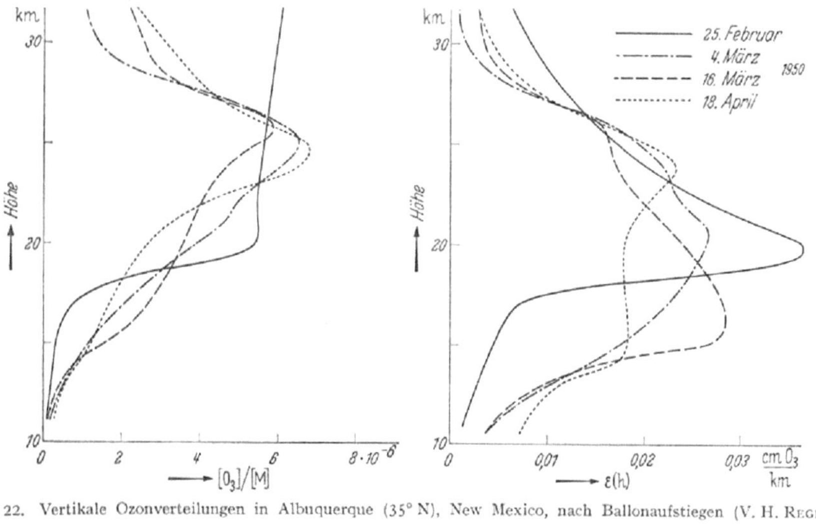

Fig. 22. Vertikale Ozonverteilungen in Albuquerque (35° N), New Mexico, nach Ballonaufstiegen (V. H. Regener).

17. Messungen bei Ballon- und Raketenaufstiegen. Die direkten Messungen, wie sie allerdings erst für die mittleren Breiten vorliegen, zeigen ein wesentlich detaillierteres Bild der beobachteten Variationen, wie es für Albuquerque [*3*] Fig. 22 und für Weißenau [*24*] Fig. 23 demonstrieren. Unterhalb von 20 km Höhe

[1] K. R. Ramanathan u. R. V. Karandikar: Proc. Ind. Acad. Soc. **29**, 330 (1949).

treten im Frühjahr und Sommer besonders starke Schwankungen auf. Ferner weist auch der Betrag des oberhalb von 30 km Höhe liegenden Ozons zu allen

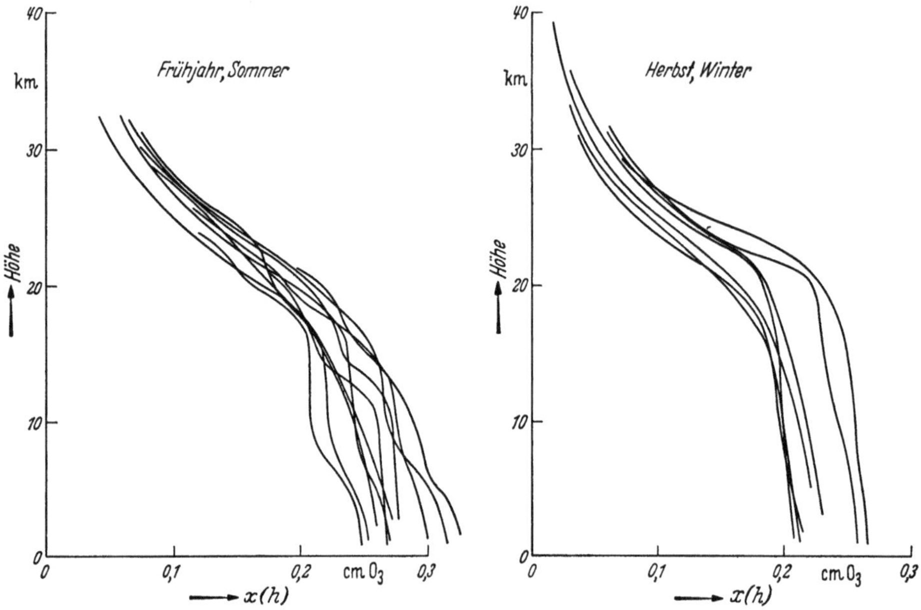

Fig. 23. Ozonmessungen bei Ballonaufstiegen in Weißenau nach Paetzold. Es ist die in den einzelnen Höhen über dem Spektrographen befindliche Ozonmenge $x(h)$ dargestellt.

Jahreszeiten beträchtliche Unterschiede auf, wie es auch die höher reichenden Ozonbestimmungen bei Raketenaufstiegen [*18*] (Fig. 24) und bei den Mondfinster-

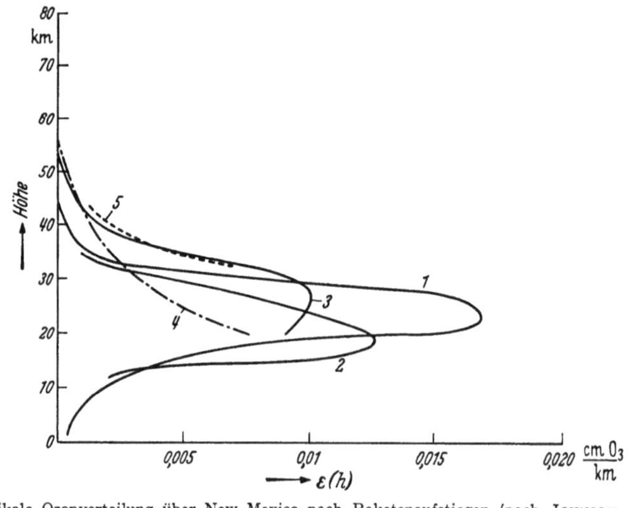

Fig. 24. Vertikale Ozonverteilung über New Mexico nach Raketenaufstiegen (nach Johnson und anderen). *1* 10. 10. 1946; *2* 2. 4. 1949; *3* 14. 6. 1949; *4* 25. 1. 1950; *5* 6. 6. 1951.

nissen (Fig. 16) zeigen. Dagegen scheinen zwischen 25 und 30 km Höhe die Variationen prozentual am geringsten zu sein.

Bei der Klassifikation der beobachteten Ozonverteilungen können mehrere in Fig. 25 dargestellte Haupttypen unterschieden werden, zwischen denen Übergangsformen bestehen: [*24*]

Type I: 1 Maximum in etwa 23 km Höhe (Kurve 1 und 2).
Type IIa: 2 Maxima in etwa 23 km und 15 km Höhe (Kurve 4).
Type IIb: 2 Maxima in etwa 23 km und 6 km Höhe (Kurve 3).
Type III: 3 Maxima in etwa 23, 15 und 6 km Höhe (Kurve 5).

Die beiden Maxima in 23 und 15 km Höhe sind nicht immer so diskret abgesetzt, sondern verlaufen oft miteinander zu einem breiten Buckel (Fig. 22). Bei dem primären Maximum in 20—25 km Höhe ist der Wechsel zwischen einer spitzen und einer flachen Form auffällig.

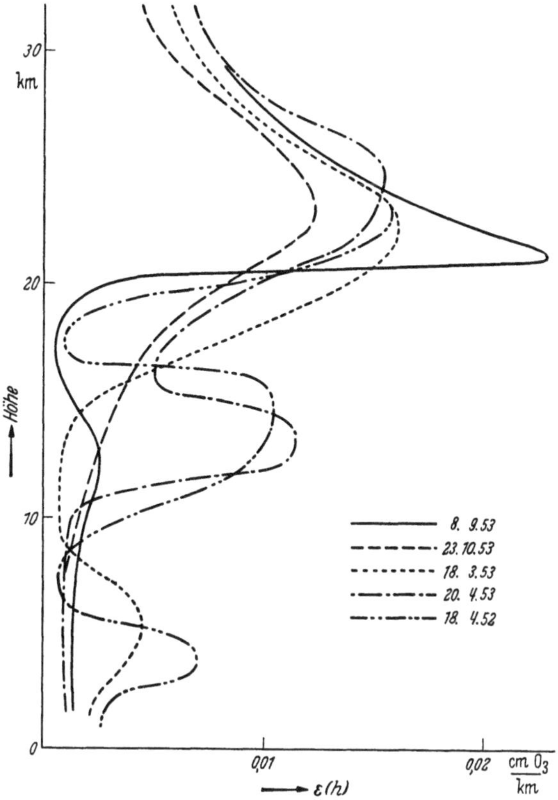

Fig. 25. Die verschiedenen Typen der vertikalen Ozonverteilung nach Ballonaufstiegen in Weißenau (48° N) (nach PAETZOLD).

18. Die die Ozonverteilung beeinflussenden Faktoren. Die Analyse der Ursachen für diese beträchtlichen Schwankungen des hohen Ozons steht wegen des doch noch verhältnismäßig geringen Beobachtungsmaterials erst in den Anfängen [*27*], [*34*]. Wegen der langen Einstellzeit des photochemischen Gleichgewichtes kommen als Ursachen unterhalb von etwa 35 km ausschließlich *Transportvorgänge* in der Atmosphäre in Frage, wenn von oxydablen Substanzen abgesehen wird, deren Wirkung aber oberhalb von 10 km Höhe im allgemeinen zu vernachlässigen sein wird.

Die durch Luftströmungen mit dem Vektor $\mathfrak{v}$ verursachte Änderung der Ozonkonzentration wird durch die Transportgleichung (18.1) gegeben:

$$\left(\frac{d[O_3]}{dt}\right)_{\text{Str}} = -(\mathfrak{v}\,\text{grad}\cdot[O_3]) - [O_3]\,\text{div}\,\mathfrak{v}. \tag{18.1}$$

Daneben ist noch der vertikale Massenaustausch wirksam

$$\left(\frac{d[O_3]}{dt}\right)_A = \frac{1}{\varrho_0}\cdot\frac{d}{dh}\cdot\left(A(h)\,\frac{d([O_3]/\varrho)}{dh}\right) \tag{18.2}$$

[$A(h)$ = *Austauschkoeffizient* (g/cm/sec)], der besonders bei der Frage des Ozonhaushaltes eine Rolle spielt (Ziff. 21).

Im allgemeinen überwiegen in der Atmosphäre die horizontalen Transportvorgänge die vertikalen bei weitem, so daß die Ozonbahnen sehr lang und sehr verwickelt sein werden. Da die Advektion auch in den einzelnen Höhenbereichen verschieden sein wird, kann es zur Ausbildung von mehrgipfligen Ozonverteilungskurven kommen; denn sowohl das sekundäre wie das troposphärische tertiäre

Ozonmaximum sind photochemisch nicht zu erklären, im Gegensatz zu dem primären Maximum in 23 km Höhe. Bei den Weißenauer Aufstiegen trat das

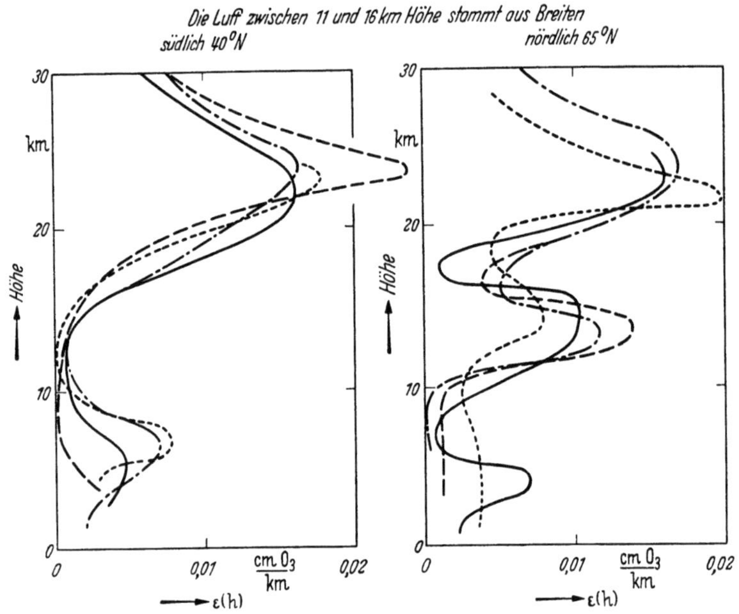

Fig. 26. Das sekundäre Ozonmaximum, nach PAETZOLD.

sekundäre Ozonmaximum immer dann auf (Fig. 26), wenn die Luft zwischen 11 und 16 km Höhen aus polaren Breiten stammte, in denen im Frühjahr der Ozongehalt ja unterhalb von 20 km hoch ist. Im Laufe des Sommers wird das sekundäre Maximum immer schwächer, entsprechend der Erschöpfung des tieferliegenden polaren Ozons (Fig. 20). Fig. 27 demonstriert die advektive Ursache des Ozonminimums in 15 km Höhe an dem in gleicher Höhe auftretenden Windsprung.

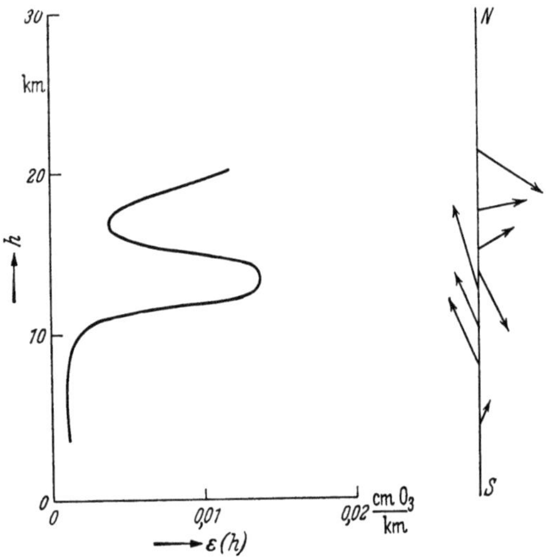

Fig. 27. Sekundäres Ozonmaximum und vertikale Windverteilung, nach PAETZOLD. Weißenau 26. 3. 1953.

Für die Wirkung *vertikaler* Luftströmungen scheinen einige Weißenauer Aufstiege im Herbst, wenn der meridionale Ozongradient besonders gering ist (Ziff. 23), einen Hinweis zu geben. In Fig. 28 ist neben der Ozonkonzentration auch das Verhältnis Ozon/Luft dargestellt [*27*], [*28*]. Für dieses deuten sich oberhalb von 20 km Höhe zwei Maxima an — wie sie bereits schon aus Umkehrmessungen folgten [*16*] —, die aber bei den einzelnen Aufstiegsmessungen in verschiedenen Höhen liegen, eine Erscheinung, die durch die Ergebnisse der Raketenaufstiege und der Mondfinsternisse gestützt wird[1]. Dieser Effekt ist plausibel nur durch

[1] H. K. PAETZOLD: J. Atmos. a. Terres. Phys. 7, 128 (1955).

die Annahme von Vertikalströmungen von einer Minimalstärke von 1 cm/sec zu erklären, die (nach der Kreuzung der Kurven $[O_3]/[M]$ zu schließen) ihr Vorzeichen in etwa 20 km Höhe umkehren würden. Ein ähnlicher Bewegungsvorgang in der Atmosphäre ist ja bei den Föhnwellen bekannt. Das obige Beispiel demonstriert die Möglichkeit, geringe Vertikalbewegungen der Luft durch die Variationen der Ozonverteilung direkt zu verfolgen.

Der Einfluß des vertikalen *Massenaustausches* auf die vertikale Ozonverteilung wird sich kaum in rascheren Änderungen der vertikalen Ozonverteilung auswirken. In Fig. 22 läßt der konstante Verlauf des Verhältnisses Ozon/Luft in

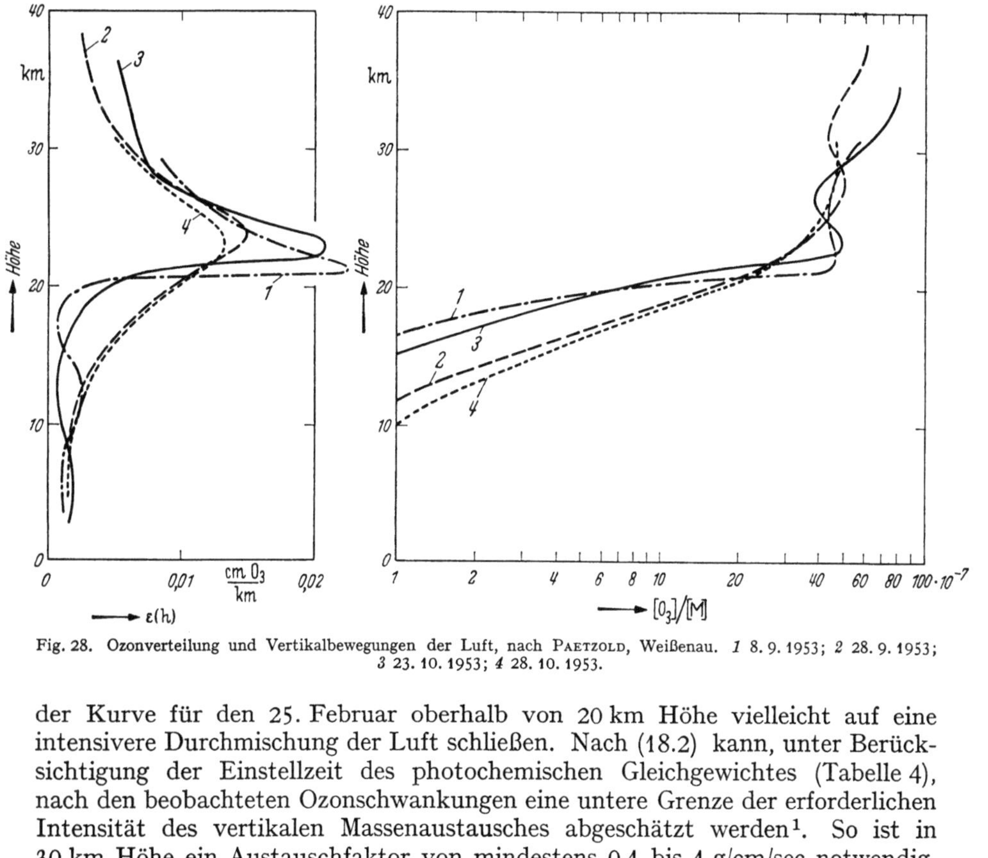

Fig. 28. Ozonverteilung und Vertikalbewegungen der Luft, nach PAETZOLD, Weißenau. *1* 8. 9. 1953; *2* 28. 9. 1953; *3* 23. 10. 1953; *4* 28. 10. 1953.

der Kurve für den 25. Februar oberhalb von 20 km Höhe vielleicht auf eine intensivere Durchmischung der Luft schließen. Nach (18.2) kann, unter Berücksichtigung der Einstellzeit des photochemischen Gleichgewichtes (Tabelle 4), nach den beobachteten Ozonschwankungen eine untere Grenze der erforderlichen Intensität des vertikalen Massenaustausches abgeschätzt werden[1]. So ist in 30 km Höhe ein Austauschfaktor von mindestens 0,1 bis 1 g/cm/sec notwendig, der allerdings nur temporär auftreten wird. Jedenfalls scheinen die Ozonschwankungen dafür zu sprechen, daß die Durchmischung in der Stratosphäre wesentlich intensiver ist als früher angenommen wurde — was auch mit den bei Raketenaufstiegen gemachten Beobachtungen übereinstimmt, nach denen bis hinauf zu 70 km Höhe keine merkliche Entmischung der atmosphärischen Gase im irdischen Schwerefeld festzustellen ist[2].

19. Ozonmessungen mit Flugzeugen. Die Schwankungen des troposphärischen Ozons lassen sich noch feiner mit der chemischen Methode bei Flugzeugaufstiegen verfolgen. Fig. 24 zeigt ein Beispiel derartiger Messungen von EHMERT [*10*], die ebenfalls öfters das tertiäre Ozonmaximum aufweisen. Nach neueren

[1] H. K. PAETZOLD: Ber. dtsch. Wetterdienst. US-Zone **1952**, Nr. 38, 292.
[2] F. A. PANETH: Rocket Exploration of the Upper Atm., p. 157. London 1954.

Flugzeugmessungen in England mit der Apparatur von EHMERT scheinen noch feinere Details vorzukommen. Ferner tritt im Mittel an der Tropopause in 11 km Höhe keine sprunghafte Änderung der Ozonkonzentration ein, die vielmehr vom Erdboden bis zu 12 km Höhe linear nur geringfügig um 30% zunimmt[1], in guter Übereinstimmung mit den spektrographischen Ozonmessungen, nach denen die Ozonkonzentration im Mittel erst oberhalb von etwa 15 km Höhe rascher anwächst. In einzelnen Fällen wurde aber bei den Flugzeug- und den Ballonaufstiegen ein abruptes Ansteigen der Ozonkonzentration direkt oberhalb der Tropopause beobachtet. Für die Dynamik der Atmosphäre scheint es von Bedeutung zu sein, daß der *Wasserdampfgehalt* der Luft das entgegengesetzte Verhalten zeigt wie das Ozon; er wird beim Durchgang durch die Tropopause in mittleren Breiten um etwa zwei Zehnerpotenzen kleiner (vgl. Ziff. 23).

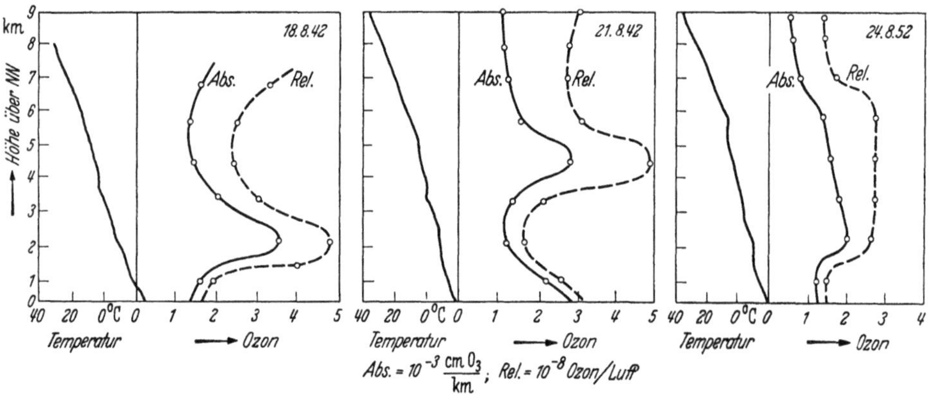

Fig. 29. Das troposphärische Ozon nach Flugzeugmessungen, nach A. EHMERT.

Nach dem Temperaturverlauf mit der Höhe, der ebenfalls in Fig. 29 wiedergegeben ist, lassen sich durch die Struktur der gemessenen vertikalen Ozonverteilung leicht verschiedene Luftkörper abgrenzen, die advektiv übereinandergeschoben worden sind. Dabei ist auch an vielleicht eingelagerte ozonzerstörende Schichten oxydabler oder katalytisch wirksamer Substanzen zu denken. Bekanntlich weist der atmosphärische Dunst bis herauf zu größeren Höhen oftmals eine ausgesprochen blättrige Struktur auf.

Instruktiv ist die Entwicklung des tertiären Ozonmaximums in Fig. 29, dessen Bereich durch Temperaturinversionen abgegrenzt ist. In der Zeit vom 19. 8. bis 21. 8. wird es gehoben, um bis zum 24. 8. stark eingeebnet zu werden; der zwischen 2 und 6 km Höhe konstante Verlauf des Partialdruckverhältnisses am 24. 8. läßt auf eine vorangehende intensive Durchmischung der Luft schließen.

Danach müssen für die Ozonverteilung in der *Troposphäre* der vertikale turbulente Massenaustausch, verbunden mit Abgrenzungen durch Inversionsschichten, die horizontale Advektion und schließlich Vertikalbewegungen die beherrschende Rolle spielen, wobei die dadurch bewirkten Ozonschwankungen wesentlich rascher als in der Stratosphäre verlaufen werden. Bemerkenswert ist die Beobachtung, daß während einer winterlichen Hochdruckwetterlage in einem durch eine Inversion in 1 bis 2 km Höhe nach oben abgeschlossenen *Kaltlufttropfen* niedrige Ozonkonzentrationen gemessen wurden[2]. Dies leitet über zu den Schwankungen des bodennahen Ozons.

---

[1] R. H. KAY: Rocket Exploration of the upper Atmosphere, S. 208. London 1954.
[2] A. EHMERT: Forschgs.- u. Erfahrungsber. dtsch. Wetterdienst, Reihe A **1941**, Nr. 13

20. Schwankungen des bodennahen Ozons. Obwohl zahlreiche Messungen des bodennahen Ozons schon in der zweiten Hälfte des 19. Jahrhunderts durchgeführt wurden, ist auch hier erst in den letzten Jahrzehnten ein repräsentatives Beobachtungsmaterial erhalten worden. Danach tritt, natürlich mit großen örtlichen Unterschieden, öfters ein ausgesprochener *Tagesgang* auf, wie ihn die Messungen von AUER[1] und EHMERT[2] in Friedrichshafen zeigen (Fig. 30). An windstillen, klaren Tagen werden in den Stunden nach Mittag Maximalwerte gemessen, während in der Nacht ein Minimum auftritt, in dem der Ozongehalt manchmal bis auf verschwindend kleine Werte absinkt. Neuere Messungen[3] in New Mexico auf dem Capillo Peak und in der in einem weiten Tal liegenden Station Aconita zeigen (Fig. 31), daß dieser Tagesgang nur im Tal, aber nicht auf dem Berg auftritt, und bestätigen damit frühere simul-

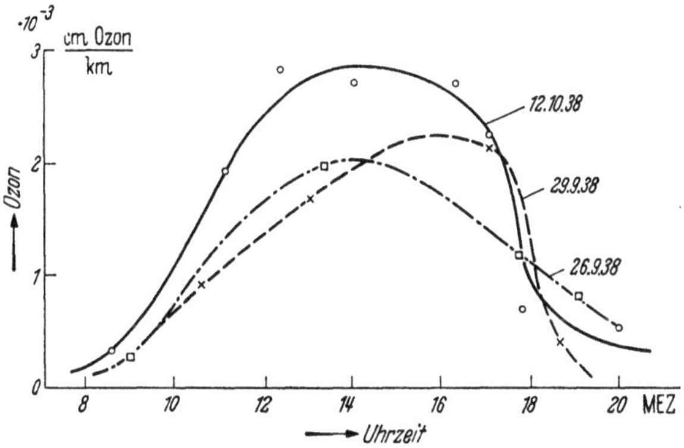

Fig. 30. Tagesgang des bodennahen Ozons an windstillen Tagen in Friedrichshafen, nach AUER.

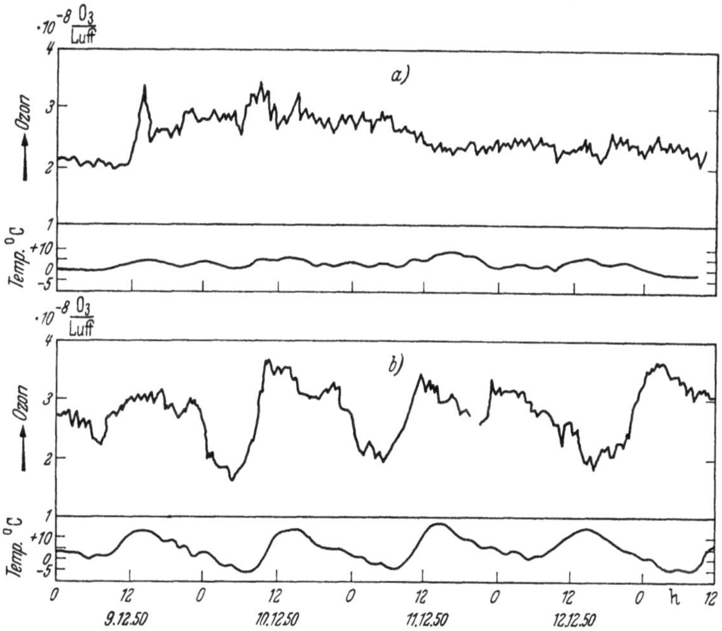

Fig. 31a. u. b. Tagesgang des bodennahen Ozons in New Mexico auf einem Berg (a) (Capillo Peak 2800 m) und in einem breiten Tal (b) (Aconita 2000 m), nach V. H. REGENER.

tane Beobachtungen auf dem Pfänder bei Bregenz und in Friedrichshafen[2]. Hier stieg auch während eines *Föhn*einbruches das bodennahe Ozon stark an[1].

[1] R. AUER: Gerlands Beitr. Geophys. **54**, 137 (1939).
[2] A. u. E. EHMERT: Forschgs.- u. Erfahrungsber. dtsch. Wetterdienst., Reihe A **1941**, Nr. 13.
[3] J. G. BOWEN u. V. H. REGENER: J. Geophys. Res. **56**, 307 (1951).

Besonders aufschlußreich sind Messungen in verschiedenen Höhen in der Nähe des Erdbodens. Dabei wurde z. B. in 80 m Höhe im Mittel etwa 15% mehr Ozon gemessen als in unmittelbarer Nähe des Erdbodens[1]. Sehr schön ist der Einfluß des Erdbodens an Beobachtungen von V. H. REGENER [*40*] an einem Mast von 12,5 m Höhe in verschiedenen Höhen angebrachten Meßstellen zu erkennen (Fig. 32).

Hieraus, sowie aus dem beschriebenen Tagesgang des bodennahen Ozons ist zu schließen, daß das Ozon in der Nähe vom Erdboden und in bodennahen Luftschichten durch Oxydation laufend zerstört wird und zum Ausgleich von oben dauernd nachgeliefert werden muß (Ziff. 21). Wird der turbulente Massenaustausch nach oben durch eine stabile Temperaturschichtung (Inversion) (Fig. 31) unterbunden, so wird das Ozon in der stagnierenden Luftschicht laufend

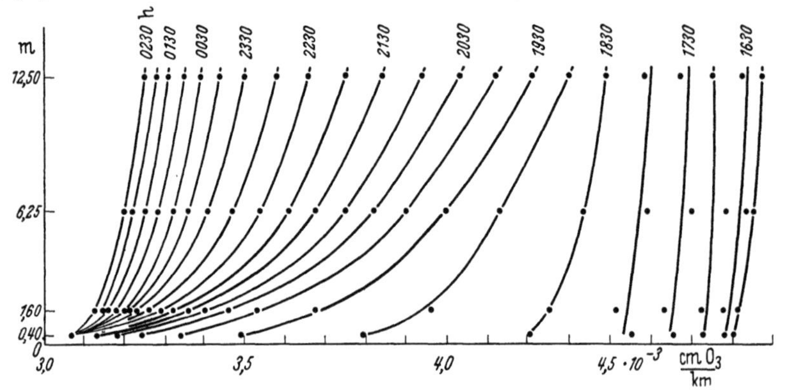

Fig. 32. Zeitliche Variation des bodennahen Ozons längs eines Mastes von 12,5 m Höhe am 23./24. 8. 1953, nach V. H. REGENER.

aufgezehrt. In Übereinstimmung damit wurden in Alaska[2] in den Monaten Januar bis März, wenn der Boden mit Schnee bedeckt ist, durchschnittliche Ozonkonzentrationen von 80 γ O_3/m^3 beobachtet, die etwa 2mal größer sind als in mittleren Breiten. Bei Schneebedeckung ist ja mit einer wesentlich herabgesetzten Ozonzerstörung zu rechnen.

Neben diesen — vornehmlich durch den örtlichen vertikalen Austausch bewirkten — Schwankungen des bodennahen Ozons wird dieses auch durch großräumige Luftbewegungen beeinflußt. Hohe Werte des bodennahen Ozons werden bei Einbruch einer Kaltfront beobachtet, während bei Zufuhr von subtropischer und tropischer Luft der Ozongehalt im allgemeinen niedrig ist[3]. Das gleiche Verhalten zeigte auch schon das hohe Ozon (Ziff. 17). Ferner besteht ein ausgesprochener *Jahresgang* des bodennahen Ozons mit einem Maximum im Sommer, das etwa doppelt so hoch ist wie die Winterwerte [*42*][3].

Im einzelnen werden die Schwankungen des bodennahen Ozons stark von den jeweiligen orographischen Verhältnissen abhängen und örtlich eigentümliche Züge aufweisen[4]. So kann z. B. an einem hochliegenden Ort wenig Ozon gemessen werden, wenn durch Talwind Luft heraufgeführt wird, deren Ozongehalt durch Stagnation oder durch Überstreichen stark ozonzerstörender Oberflächen (Wälder) schon erniedrigt ist. Für das allgemeine Ozonproblem repräsentative Messungen des bodennahen Ozons sind daher mit einiger Sicherheit nur auf isoliert stehenden Bergen mit geringer Pflanzendecke zu erwarten.

[1] F. TEICHERT: Z. Meteor. **9**, 21 (1955).
[2] W. S. WILSON: Trans. Amer. Geophys. Un. **32**, 552 (1951).
[3] F. W. P. GÖTZ u. F. VOLZ: Z. Naturforsch. **6**a, 634 (1951).
[4] R. NEUWIRTH: Abh. Bad Landeswetterdienst. **1952**, 152.

In das bisherige Bild fügen sich die in der Nähe von Los Angeles (Pazifik) manchmal am Tag beobachteten abnormal hohen Werte des bodennahen Ozons nicht ein, die nach Fig. 33 [*40*] von der Größenordnung der Ozonkonzentration im Ozonmaximum sein können. Sie sind als reell anzusehen, da die chemischen Messungen durch optische kontrolliert wurden. Durch Zufuhr von ozonreicher Luft ist eine Deutung nicht möglich. Sie ist vielleicht in der Richtung auf hygroskopische Aerosole zu suchen, bei denen temporär Wasserstoffsuperoxyd gebildet wird, das photochemisch dissoziiert zur Erzeugung von Ozon führt. Diese Möglichkeit erscheint jedoch an spezielle Bedingungen geknüpft zu sein, wie sie in Los Angeles herrschen. Über dem schmalen Streifen zwischen der Küste und der Gebirgskette der Sierra Nevada stagniert die Luft und wird durch die ausgedehnten Erdölraffinerien und einen starken Kraftwagen-Verkehr stark mit Kohlenwasserstoffen angereichert.

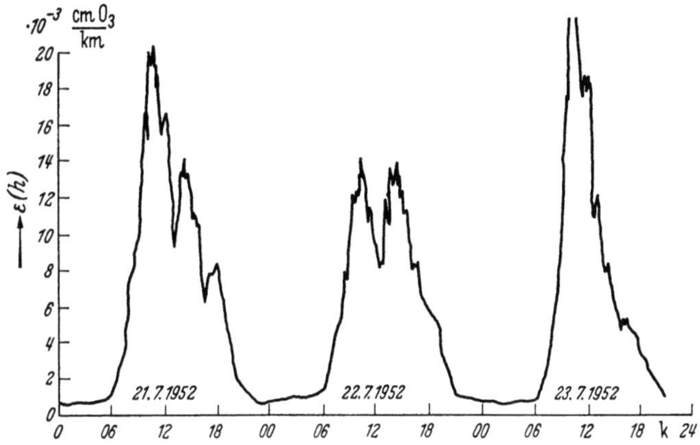

Fig. 33. Hohe Werte des bodennahen Ozons in Pasadena, nach V. H. Regener.

21. Ozonhaushalt und vertikale Turbulenz. Die Existenz des troposphärischen und bodennahen Ozons ist nicht selbstverständlich, da photochemisch nach Fig. 19 nur oberhalb von etwa 15 km Höhe merkliche Mengen von Ozon gebildet werden können. Unterhalb dieser Höhe ist die ozonisierende Sonnenstrahlung viel zu gering; in 5 km Höhe wird nur noch etwa 1 Lichtquant/cm³/sec des ozonbildenden Spektralbereiches absorbiert[1]. Andererseits erfolgt eine dauernde Zerstörung des bodennahen und troposphärischen Ozons, wofür nur Oxydationsprozesse und katalytische Reaktionen in Frage kommen. Denn die thermische Ozonzersetzung ist auch bei den höheren Temperaturen um 300° C in den bodennahen Luftschichten noch viel zu träge (Tabelle 5), und die Ozonzersetzung durch Licht, die in der Troposphäre hauptsächlich durch die Absorption in der Chappuis-Bande bewirkt wird, kann sich trotz der kurzen mittleren photolytischen Lebensdauer eines Ozonmoleküls von 1 Std kaum auswirken, da jedes O-Atom wieder ein O_3-Molekül zurückbildet. Von den oxydablen Substanzen werden organische besonders wirksam sein. Ferner scheinen Wasseroberflächen (z. B. Nebeltröpfchen) einen starken katalytischen Einfluß auszuüben[2].

Aus dem obigen Tagesgang des bodennahen Ozons läßt sich eine Ozonzerstörung von $1 \cdot 10^{11} \frac{O_3\text{-Moleküle}}{\text{cm}^2\text{-Säule} \cdot \text{sec}}$ abschätzen [*28*] und [*40*][3]. Für den Ersatz dieser laufend zersetzten Ozonmenge sind die Ultrastrahlung und atmosphärische

[1] H. K. Paetzold: J. Atmosph. a. Terres. Phys. **3**, 125 (1953).
[2] F. Volz: Ber. dtsch. Wetterdienst. US-Zone **1952**, Nr. 35, 257.
[3] H. K. Paetzold: J. Atmos. a. Terres. Phys. **7**, 128 (1955).

elektrische Entladungen (Gewitter) diskutiert worden. Nach der Anzahl der durch die Ultrastrahlung erzeugten Ionen können durch dieselbe im günstigsten Falle zwischen 0 und 10 km Höhe $1 \cdot 10^7 \frac{O_3\text{-Moleküle}}{\text{cm}^2\text{-Säule} \cdot \text{sec}}$ gebildet werden [28]. Da zwischen 0 und 10 km Höhe sich insgesamt etwa $6 \cdot 10^{17} \frac{O_3\text{-Moleküle}}{\text{cm}^2\text{-Säule}}$ befinden, würde durch die Ultrastrahlung allein das troposphärische Ozon erst in $2 \cdot 10^3$ Jahren erzeugt werden können, vorausgesetzt, daß kein Ozon zerstört wird.

Bei den elektrischen Entladungen liegen die Verhältnisse ganz ähnlich. Nach WARBURG[1] werden durch 100 Coulomb höchstens $1{,}2 \cdot 10^{24}$ O_3-Moleküle gebildet. In der Erdatmosphäre können durch die elektrischen Entladungen bei Gewittern lokal höhere Ozonkonzentrationen erzeugt werden. Doch ist dieser Effekt bei den 1800 Gewittern, die im Durchschnitt dauernd tätig sind, auf die ganze Erdoberfläche bezogen verschwindend klein. Denn nach ISRAEL[2] beträgt der Gewittervertikalstrom $3{,}3 \cdot 10^{-16} \frac{\text{Coulomb}}{\text{cm}^2\,\text{sec}}$, so daß dadurch höchstens $4 \cdot 10^6 \frac{O_3\text{-Moleküle}}{\text{cm}^2\text{-Säule} \cdot \text{sec}}$ erzeugt werden können. Sowohl die Ultrastrahlung wie die atmosphärischen elektrischen Entladungen erzeugen demnach viel zu wenig Ozon, um den Verlust zu decken.

Die Quelle für das troposphärische und bodennahe Ozon kann daher nur in größeren Höhen im Bereich der photochemischen Mutterschicht des Ozons zwischen 15 und 25 km Höhe gesucht werden [*36*], wo nach Fig. 19 und 21 ein zum Äquator hin ansteigendes Defizit der gemessenen Ozonverteilung gegenüber der photochemisch berechneten besteht, so daß dauernd Ozon photochemisch nachgebildet werden muß. Nach der Theorie folgt dafür in Übereinstimmung mit der obigen Ozonzerstörung ein durchschnittlicher Wert von $1 \cdot 10^{11}$ O_3-Molekül/cm²/sec [*28*]. Von dieser Ozonquelle fließt ein durch den *Massenaustausch* verursachter „Ozonstrom" $\vec{O}_3$ [*38*] nach unten, dessen Stärke gegeben ist durch

$$\vec{O}_3 = -\frac{A(h)}{\varrho_0}\,\frac{d([O_3]/\varrho)}{dh}\;\frac{O_3\text{-Moleküle}}{\text{cm}^2\,\text{sec}}. \tag{21.1}$$

Mit einem troposphärischen Austauschkoeffizienten $A(h)$ von 50 g/cm/sec folgt aus den gemessenen Ozonverteilungen ein abwärts gerichteter Strom von wiederum $1 \cdot 10^{11}$ O_3-Molekülen/cm²-Säule/sec im Mittel [*28*].

Die Übereinstimmung zwischen den obigen drei unabhängig folgenden Größen läßt es wahrscheinlich erscheinen, daß der Mechanismus:

— photochemische Ozonquelle in 15 bis 25 km Höhe — turbulenter Ozonstrom — Ozonsenke am Erdboden und in der Troposphäre —

eine der Ursachen für das äquatoriale Minimum des Ozonwertes ist, da alle drei Faktoren am Äquator am intensivsten sein werden. Nach Ziff. 23 werden aber an der meridionalen Abhängigkeit des Ozons noch weltweite Zirkulationsvorgänge beteiligt sein; doch sollten die obigen Größenordnungen für den Ozonhaushalt dadurch nicht wesentlich verschoben werden. Auf die ganze Erde bezogen ergibt sich, daß im Jahr 10^9 Tonnen Ozon an die Erdoberfläche verfrachtet und dort durch Oxydation zerstört werden, so daß der Luftsauerstoff in etwa 1 Million Jahren, d.h. in einer geologisch kurzen Frist, einmal das ozonisierte Stadium durchläuft. Ohne die dauernde photochemische Nachbildung würde das Ozon in 3 Jahren auf 1/10 seines jetzigen Gesamtbetrages abnehmen. Dies veranschaulicht, daß die vertikale Ozonverteilung unterhalb des Maximums als das Resultat

---

[1] E. WARBURG: Z. techn. Phys. **7**, 128—140 629 (1902).

[2] H. ISRAEL: Das Gewitter. Leipzig 1950.

eines dynamischen Gleichgewichts aufgefaßt werden muß, von dem einzelne Faktoren, wie der Luftaustausch usw., starken Schwankungen unterworfen sind. Diese Verhältnisse werden noch deutlicher bei den interdiurnen Schwankungen und dem Jahresgang des Ozonwertes.

F. Die Variationen des Ozonwertes x_0.

22. Die interdiurnen Ozonschwankungen und das Wetter. Schon bei den ersten Beobachtungen von FABRY und BUISSON [*11*] fielen die von Tag zu Tag auftretenden, manchmal beträchtlichen Variationen des totalen Ozonbetrages auf. Sie wurden dann durch die laufenden Messungen mit dem DOBSON-Instrument in

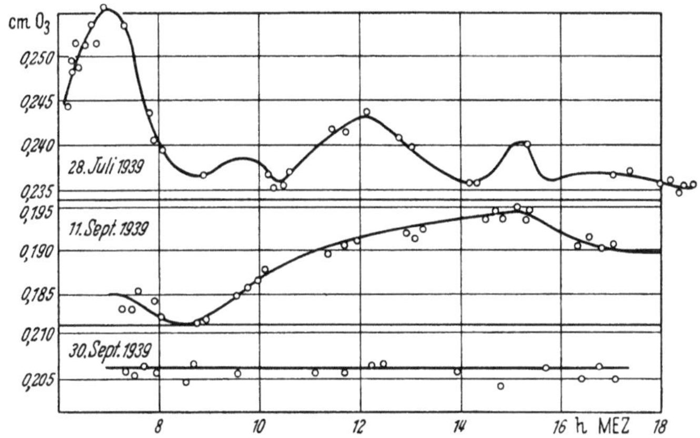

Fig. 34. Schwankungen des Ozonwertes in Arosa während eines Tages, nach GÖTZ.

dem Ozonbeobachtungsnetz näher erfaßt, ohne daß aber bis heute eine schon ganz befriedigende Deutung der maßgeblichen Faktoren gelungen wäre. Dies ist verständlich, da der totale Ozonbetrag sich aus den in den einzelnen Höhen befindlichen Ozonbeträgen zusammensetzt, die nach Abschnitt E in mannigfacher Weise variieren können[1].

Die Unterschiede des Ozonwertes von Tag zu Tag können in mittleren Breiten bis zu 30% betragen. Auch während eines Tages kommen beträchtliche Schwankungen vor, die nach dem Herbst zu abnehmen (Fig. 34)[2]. In höheren Breiten (Tromsö) sind die interdiurnen Schwankungen noch stärker ausgeprägt, während sie äquatorwärts abnehmen und dort nur noch geringe Amplituden aufweisen. Die mittleren interdiurnen Schwankungen zeigen nach Tabelle 7 [*16*] einen typisch jahreszeitlichen Gang; er ist im Frühjahr am größten und im Herbst am kleinsten. Im Sommerhalbjahr besteht keine große meridionale Abhängigkeit der interdiurnen Veränderlichkeit. Sie beträgt im Juli bis August in Italien 0,005, in Arosa 0,009, in Tromsö 0,010 und in Spitzbergen 0,007 cm O_3. Sie steigt dagegen im Dezember in Tromsö bis auf 0,056 cm O_3 an.

Tabelle 7. *Mittlere interdiurne Ozonschwankungen in Arosa* (in cm O_3).

Januar	0,017	Mai	0,011	September . .	0,008
Februar. . . .	0,017	Juni.	0,010	Oktober . . .	0,009
März	0,015	Juli	0,010	November . .	0,010
April	0,015	August . . .	0,009	Dezember . .	0,013

[1] H. K. PAETZOLD: J. Geophys. Res. **95**, 365 (1954).
[2] F. W. P. GÖTZ: Helv. phys. Acta **13**, 3 (1940).

Öfters werden Schwankungen des Ozonwertes während einer Messung beobachtet, die auf die Existenz von „*Ozonwolken*" schließen lassen, die über den Beobachtungsort hinwegziehen. In einem Fall konnte eine solche Ozonwolke, die der Vorläufer eines nachfolgenden starken Ozonanstiegs war, durch Beobachtungen von Sternen in verschiedenen Himmelsrichtungen direkt erfaßt werden[1].

In seinen klassischen Untersuchungen konnte DOBSON [*6*] zeigen, daß die interdiurnen Schwankungen eng mit der Luftdruckverteilung am Boden verknüpft sind, indem tiefe Ozonbeträge mit hohen Drucken und hohe Ozonbeträge mit niedrigen Drucken korreliert sind. In Westeuropa werden die höchsten bzw.

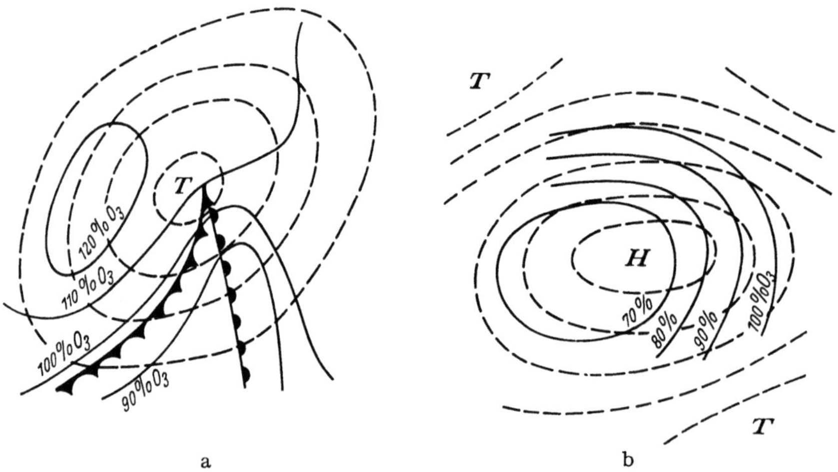

Fig. 35 a u. b. Ozonwert und Luftdruckverteilung nach DOBSON; a Tief; b Hoch. Der Ozonwert ist in Prozenten des Durchschnittswertes gegeben.

tiefsten Ozonwerte etwa 300 bis 400 km westlich bis südwestlich vom Kern einer Zyklone bzw. Antizyklone gefunden (Fig. 35). Die höchsten Ozonwerte treten bei einer Zyklone am Ort des Tropopausentrichters (PALMÉN[2] [*29*]) auf, also dort, wo die Tropopause mitunter bis auf 6 km Höhe und darunter hinabgesaugt wird. Niedrige Ozonwerte werden dagegen in dem Warmsektor einer Zyklone beobachtet. Dabei beginnt der Ozonbetrag nach Beobachtungen in Nordnorwegen [*43*] schon abzunehmen, wenn die Warmfront noch bis 1000 km entfernt ist. Wenn die Kaltfront passiert, steigt der totale Ozonbetrag jäh an; auch das bodennahe Ozon zeigt ja dann ein beträchtliches Emporschnellen (Ziff. 20). Diese Änderungen des Ozons mit der Hoch- und Tiefdrucktätigkeit sind am meisten ausgeprägt im Winter und Frühjahr, am schwächsten im Spätsommer und Herbst.

In einer neueren sorgfältigen Analyse dieser Zusammenhänge gibt NORMAND [*23*] einige schöne Beispiele, wie in Oxford in einem Tiefdrucktrog hohe Ozonwerte und in einem Hochdruckkeil niedrige Ozonwerte auftreten (Fig. 36). Der Wechsel von hohem zu hohem Ozon über ein Intervall niedrigen Ozons scheint in Westeuropa eine Quasiperiode von 6 bis 14 Tagen aufzuweisen. GÖTZ[3] glaubt in dem Arosaer Material Ozonwellen mit der Periode von 27 und 31 Tagen, weniger häufig auch 36,5, 10 oder 20 und 15,5 Tagen gefunden zu haben.

[1] D. BARBIER u. D. CHALONGE: J. Phys. Radium (7) **10**, 113 (1939).
[2] E. PALMÉN u. K. M. NAGLER: J. Meteorology **4**, 105 (1947).
[3] F. W. P. GÖTZ: Verh. schweiz. naturforsch. Ges. **1943**, 87.

Im Gegensatz zu den Verhältnissen in Westeuropa werden in Sibirien und Shanghai bei kalten Antizyklonen hohe Ozonwerte beobachtet, während in einem Trog negative Abweichungen von dem mittleren Ozonwert beobachtet werden[1]. Die Verhältnisse sind eben auch in derselben geographischen Breite unterschiedlich, je nach dem regionalen System der Zyklonen und Antizyklonen[2].

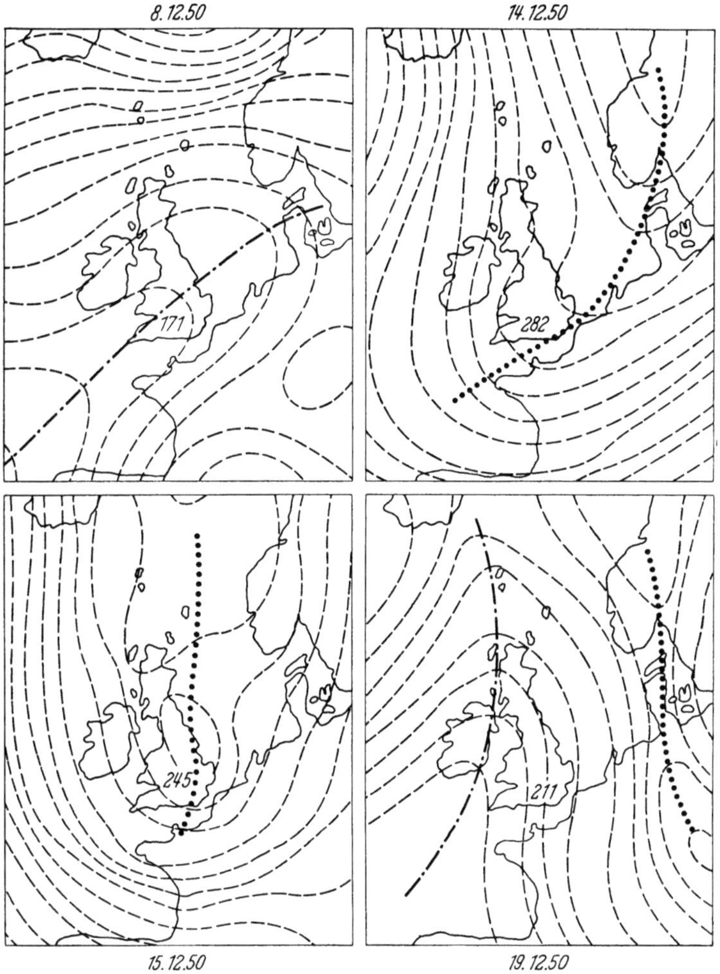

Fig. 36. Große und kleine Ozonwerte in Oxford in Tiefdrucktrögen (Kreise) bzw. Hochdruckkeilen (strichpunktiert), nach NORMAND. Die Ozonwerte über Oxford sind in der Einheit 0,001 cm gegeben.

Schon MEETHAM [*20*] fand eine enge Beziehung zwischen den Ozonvariationen und den Bedingungen in höheren Luftschichten: der Höhe der Tropopause und der potentiellen Temperatur. Ein Ozonanstieg um 0,01 cm O_3 ist danach verknüpft mit einer Senkung der Tropopause um 1 km und einer Erhöhung der potentiellen Temperatur um 3° C in 18 km Höhe, so daß eine hohe Lage der Tropopause korreliert mit einem relativ niedrigen Ozonwert und umgekehrt. Fig. 37 [*23*] zeigt an einem Beispiel die enge Korrelation. Doch ist bemerkenswert, daß zwar die Extrema der einzelnen Kurven zeitlich zusammenfallen, die relativen Größen

[1] P. LEJAY: Notes Méteor. Phys. de Zi-Ka-Wei **7** (1937).

[2] Y. MIYAKE u. K. KAWAMURA: Papers Meteor. a. Geophys. **5**, 178 (1954).

der Veränderungen aber keineswegs konstant sind. Als Faustregel für den Zusammenhang zwischen Ozon und meteorologischen Größen diene das folgende Schema:

klein	Ozonwert x_0	groß
Tropische Warmluft	Luftmasse	Arktische Kaltluft
warm	Temperatur der Troposphäre	kalt
hoch	Höhe der Tropopause	tief
kalt	Temperatur der Stratosphäre	warm

Die Trennung der für die interdiurnen Ozonschwankungen verantwortlichen Faktoren, Advektion und Vertikalbewegungen der Atmosphäre, stößt auf beträchtliche Schwierigkeiten und krankt an dem Mangel an Beobachtungsmaterial

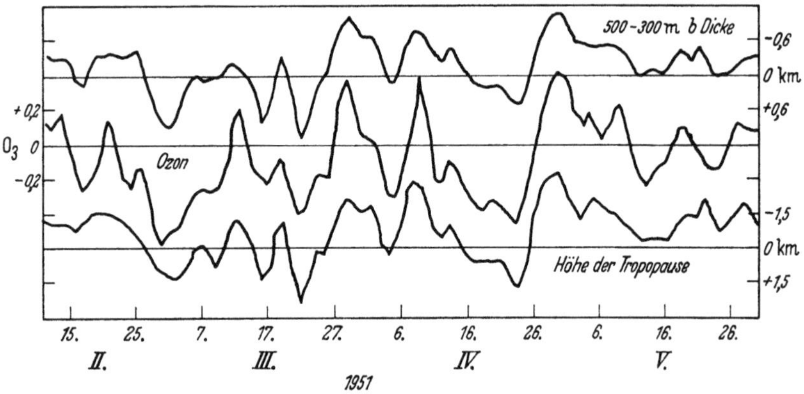

Fig. 37. Ozonschwankung, Tropopausenhöhe und potentielle Temperatur in 8 km Höhe, in Abweichungen vom Mittel. Die potentielle Temperatur ist durch den Abstand der Isobarenflächen 500 und 300 mb gegeben, nach NORMAND.

über die vertikale Ozonverteilung. Die Tatsache, daß bei Zustrom von polarer Luft der Ozonbetrag steigt, legt einen hauptsächlich advektiven Einfluß nahe, da der Ozongehalt der Troposphäre polwärts ja beträchtlich ansteigt. Auch der Umstand, daß im Herbst, wo der meridionale Gradient des Ozonwertes am geringsten ist (s. Ziff. 23), auch die interdiurnen Ozonschwankungen ein Minimum aufweisen, spricht für die Advektion.

MOSER [*22*] findet in 11 km die beste Korrelation zwischen Luftbahnen und Ozonbetrag. Nach ihm ist in Westeuropa der Ozonwert um so höher, je weiter die Luft aus Norden stammt, während Luftmassen aus den niederen Breiten niedrigere Ozonwerte aufweisen. Dies entspricht den direkten Messungen der vertikalen Ozonverteilung (Fig. 26) [*28*]. PENNDORF [*31*] findet die beste Korrelation zwischen dem Ozonwert und den Luftbahnen in 16 km Höhe. TÖNSBERG und OHLSEN [*43*] finden in Norwegen, daß die Ozonwerte enger mit der Bodenfront als mit dem Zentrum des Tiefdruckgebietes korreliert sind, was ebenfalls auf Advektionen schließen läßt, wobei die frontalen Einflüsse im Herbst verschwinden.

Trotzdem stehen der Annahme einer *alleinigen* Wirkung der Advektion wesentliche Argumente entgegen. Einmal sind auch im Herbst trotz des geringen meridionalen Ozongradienten die interdiurnen Ozonschwankungen noch beträchtlich, ferner sind bei polaren Luftmassen die beobachteten totalen Ozonwerte öfters erheblich größer als in dem Herkunftsbereich der Luft. Schließlich sind nach HAURWITZ[1] und REED [*33*] die geschlossenen Isoplethen westlich von Tief- oder

[1] B. HAURWITZ: Bull. Amer. Met. Soc. **19**, 417 (1938).

Hochdruckzentren advektiv kaum zu deuten (Fig. 35); sie weisen vielmehr auf *Vertikalbewegungen* der Atmosphäre.

Diese sind in zweifacher Hinsicht wirksam: Einmal führen absteigende Strömungen Ozon aus größeren Höhen, wo es photochemisch rasch genug nachgebildet wird, in tiefere, wo es sich photochemisch geschützt ansammelt — ein Effekt, der sich bei den interdiurnen Ozonschwankungen allerdings kaum auswirken wird. Andererseits führt eine vertikale Divergenz zu einer horizontalen Konvergenz, d.h. zu einer Kompression, die wegen des starken Anstiegs des Partialdruckverhältnisses Ozon/Luft mit der Höhe unterhalb von 25 km zu einem Anstieg des Ozonwertes führt. Umgekehrt müssen aufsteigende Luftströmungen eine Verminderung des Ozonbetrages bewirken. Nach den in Tief- und Hochdrucksystemen herrschenden Vertikalbewegungen[1] wird damit die Form der Isoplethen

Tabelle 8. *Ozonschwankungen durch Hebung und Senkung der Atmosphäre.*

Datum	Stratosphäre gehoben um 1 km zwischen		Stratosphäre gesenkt um 1 km zwischen	
	10 und 15 km cm O_3	10 und 20 km cm O_3	10 und 15 km cm O_3	10 und 20 km cm O_3
1. 1.	−0,004	−0,014	+0,007	+0,019
1. 4.	−0,008	−0,016	+0,011	+0,021
1. 7.	−0,007	−0,015	+0.009	+0,019
1. 10.	−0,002	−0,009	+0,003	+0,013

in Fig. 35 verständlich. Die Tabelle 8 gibt die Ozonänderung in mittleren Breiten, die durch die gleichmäßige Hebung bzw. Senkung der Stratosphäre um 1 km in den verschiedenen Jahreszeiten bewirkt werden [*28*].

Eine Abschätzung der Anteile des Einflusses von vertikalen und horizontalen Transportvorgängen ist von verschiedener Seite versucht worden. Nicolet[2] findet, daß über einer Antizyklone eine Ozonverminderung von je 0,015 cm O_3 durch den advektiven und den dynamischen Anteil bewirkt wird. Über einer Zyklone sollen, bei einem Ozonanstieg von 0,05 cm O_3, 40% der Advektion und je 30% dem dynamischen Anteil und den frontalen Vorgängen zuzuschreiben sein. Reed [*33*] findet, daß in New York ein starker Ozonanstieg immer mit einem markanten Absinken der Luft in 10, 13 und 16 km Niveau verbunden ist, wobei durch die dadurch bewirkte Kompression etwa die Hälfte des Gesamtanstieges des Ozons bewirkt wird.

Diese Aufteilung ist vermutlich nur im Winter und Frühjahr gültig, denn im Spätsommer und Herbst wird der vertikale Anteil stark überwiegen [*28*], [*23*]. Allgemein werden bei starken Ozonänderungen sowohl der advektive wie der vertikale Faktor zusammenwirken müssen. Detailliertere Aufschlüsse, die auch für das Problem der Wetterprognose von Bedeutung sein mögen, sind erst zu erwarten, wenn die Variationen der vertikalen Ozonverteilung in Verbindung mit den meteorologischen Bedingungen durch Schwarmaufstiege mit Ozonradiosonden genauer erfaßt werden können.

Bei den hohen Ozonwerten über einer sibirischen Antizyklone wird angenommen, daß entweder der Ozonanstieg durch die Zufuhr polarer Luft verursacht wird, oder daß das Hoch sich nur in geringe Höhen erstreckt und darüber ein Tief liegt, so daß der vertikale Anteil entscheidend wäre.

[1] R. G. Fleagle: J. Meteorology **4**, 165 (1947).
[2] M. Nicolet: Roy. Met. Inst. Belg. Misc. **1945**, Nr. 19.

23. Die meridionalen und jahreszeitlichen Variationen des Ozons. Die weltweiten Messungen des Ozonwertes mit DOBSON-Instrumenten ergaben zwei überraschende Züge: einmal die Abnahme des Ozonwertes nach niederen Breiten hin und zum andern die Existenz eines ausgesprochenen Jahresganges des Ozons [*6*]. In Fig. 38 [*16*] ist eine Isoplethendarstellung des Ozons gegeben, aus der die Existenz einer Zone maximalen Ozons (Ozongürtel) in 60° N zu erkennen ist. Der meridionale Ozongradient ist auf jeder Halbkugel der Erde im Frühjahr am größten und im Herbst am kleinsten, da die Amplitude des Ozonjahresganges stark von der Breite abhängt.

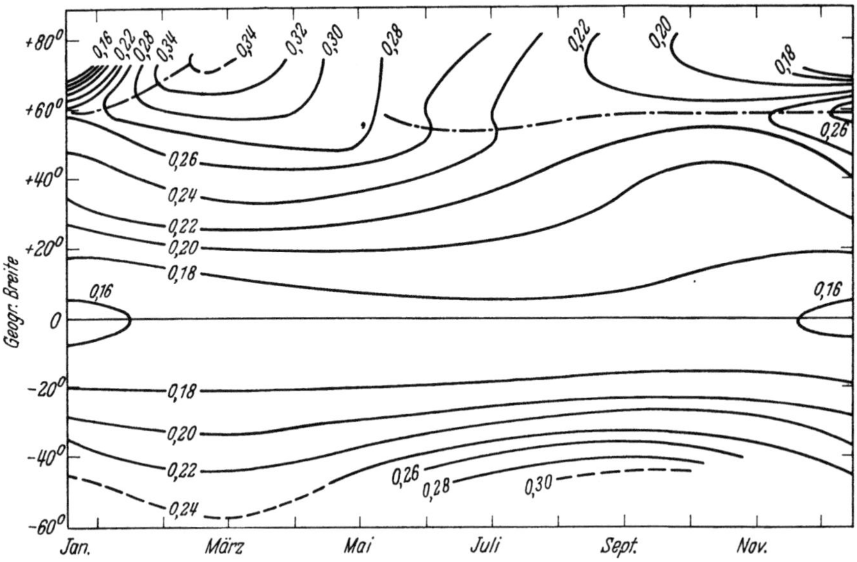

Fig. 38. Die meridionale und jahreszeitliche Variation des Ozonwertes, nach GÖTZ.

Am Äquator ist die Jahresamplitude verschwindend klein, um nach höheren Breiten zu beträchtlichen Werten anzusteigen. So beträgt sie in Tromsö (69,7°) 0,19 cm O_3, das sind 80% des Jahresmittelwertes, während in Arosa (48° N) sich 0,07 cm O_3 (= 30%) ergeben. Das Maximum des Ozonbetrages stellt sich auf beiden Erdhalbkugeln jeweils im Frühjahr ein, d.h. auf der Nordhalbkugel in den Monaten März bis Mai (Fig. 39) [*4*]. Das Minimum des Ozonbetrages wird in mittleren Breiten im Herbst erreicht (Nordhalbkugel Oktober). Im Spätherbst (Nordhalbkugel Ende November) setzt der Wiederanstieg des Ozonbetrages ein. Nach Fig. 40 [*16*] verläuft der Ozonjahresgang in Tromsö wesentlich unsymmetrischer als in Arosa.

Der Jahresgang wird ferner noch von den einzelnen regionalen klimatischen Verhältnissen abhängen. So scheint nach den Aufstiegen in Albuquerque (35° N) [*39*], mitten im nordamerikanischen Kontinent, das polare tiefere Ozon im Frühjahr wesentlich stärker aufzutreten und damit einen höheren Gesamtozonbetrag zu bewirken als es bei Orten in Europa auf derselben geographischen Breite der Fall ist. In der Tat reichen auf dem nordamerikanischen und asiatischen Kontinent polare Kaltlufteinbrüche wesentlich weiter nach Süden als in Europa mit seinem mehr maritimen Klima. Auch in Shanghai (31°) und New York (40° N), wo der kontinentale Klimatyp vorherrscht, werden vor allem im Frühjahr höhere Ozonwerte gemessen als auf den gleichen Breiten in Europa. In Zukunft sind daher auch die Ozonvariationen mit der geographischen Länge und dem Einfluß des Klimatyps zu beachten. Letzterer scheint sich auch in dem beobachteten zweiten schwächeren Ozonmaximum im Herbst zu manifestieren (Fig. 39).

Nach den Schwankungen der vertikalen Ozonverteilung (Ziff. 17) wird der Sitz der meridionalen und jahreszeitlichen Ozonvariationen vornehmlich in der Troposphäre und der unteren Stratosphäre bis 20 km Höhe lokalisiert sein; oberhalb davon weist das Ozon — ausgenommen vielleicht die extrem polaren Breiten — einen ziemlich konstanten Betrag von 0,16 bis 0,19 cm O_3 auf.

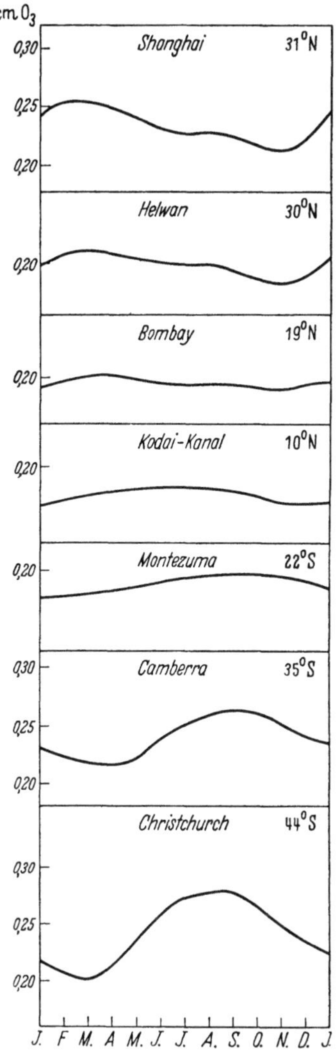

Fig. 39. Der Ozonjahresgang in verschiedenen geographischen Breiten, nach CRAIG.

Ein vermuteter sekundärer Zusammenhang zwischen Sonnentätigkeit (Fleckenrelativzahl) und Ozonwert konnte noch nicht gefunden werden [*16*]. Einige Messungen schienen anzudeuten, daß im Sonnenfleckenminimum der totale Ozonbetrag etwas größer ist als im Sonnenfleckenmaximum[1]. Auf jeden Fall werden die zu erwartenden Variationen nur klein sein, da eine erhebliche solare Änderung der ozonbildenden Sonnenstrahlung kaum zu erwarten ist.

Sowohl die meridionale Abhängigkeit des Ozons mit dem Äquatorminimum wie auch der Ozonjahresgang sind durch einen stationären photochemischen Gleichgewichtszustand nicht zu deuten. Nach diesem müßte am Äquator ja der Ozonwert am größten sein, und nach dem jahreszeitlichen Gang der Sonnenhöhe ist in 45° N nur eine Ozonschwankung von 0,01 cm O_3 zu erwarten [*8*], [*26*]. Ferner wirkt zwar der für den Ozonhaushalt ausschlaggebende vertikale Massenaustausch in der gewünschten Richtung, doch ergibt sich mit ihm allein eine zu kleine jahreszeitliche Amplitude des Ozons [*8*].

Sehr wahrscheinlich müssen daher zur Deutung großräumige Zirkulationsvorgänge mit herangezogen werden, die, wie es DOBSON schon 1929 vermutete [*6*], Ozon aus den äquatorialen Zonen im jahreszeitlichen Rhythmus nach höheren Breiten verfrachten[2]. Nach DÜTSCH [*8*] ist in 60° N eine absteigende Strömung von einer mittleren Stärke von $-0{,}4$ mm/sec zur Zeit des Ozonanstiegs im Frühjahr und im Spätsommer eine aufsteigende Strömung von $+0{,}4$ mm/sec erforderlich um den Jahresgang zu bewirken. Die zugehörige Horizontalkomponente dieser Zirkulation würde in der Größenordnung von $1 \cdot 10^1$ bis 10^2 cm/sec liegen. Eine solche Zirkulation erscheint nach dem meridionalen Gang des Luftdrucks, wie er aus der meridionalen Temperaturverteilung folgt, plausibel.

Diese Vorstellungen sind von RAMANATHAN [*32*] auf Grund der neueren Windmessungen in größeren Höhen näher ausgeführt worden, die in mittleren Breiten

[1] F. W. P. GÖTZ: Fundamenta radiol. **4**, 33 (1939).
[2] Y. MIYAKE u. K. SARUHASHI: IX. Congr. UGGI. Brüssel 1951.

westliche Winde mit einem Minimum bei 20 km ergaben[1]. Danach würde in Äquatornähe in einer Höhe bis zu 20 km ein Lufttransport polwärts existieren und darüber in umgekehrter Richtung (Fig. 41). Durch diese Zirkulation wird in äquatorialen Breiten Luft aus größeren Höhen, wo sich das Ozon photo-

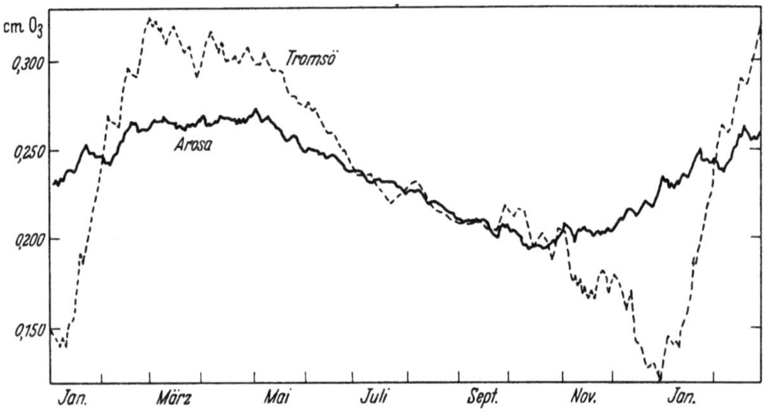

Fig. 40. Ozonjahresgang in Arosa (47° N) und Tromsö (70° N), nach Götz.

chemisch nachbildet, herabgeschafft, um in hohen Breiten das Ozonreservoir in geringeren Höhen unterhalb von 20 km Höhe aufzufüllen. Im Winter ist diese Zirkulationsbewegung am stärksten; wenn sich im Sommer die polare Stratosphäre erwärmt, schwächt sich die Zirkulation ab. Eine solche polwärts gerichtete Zirkulationsströmung entlang der Tropopause wird auch von Brewer[2] nach dem Verhalten des Wasserdampfgehaltes in der Umgebung der Tropopause angenommen.

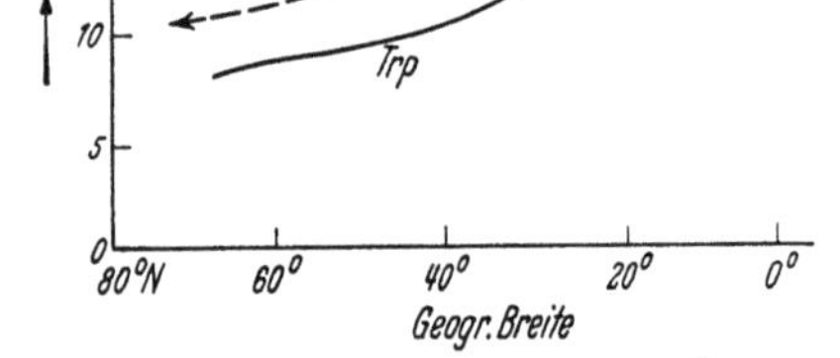

Fig. 41. Hypothetischer Ozontransport vom Äquator zum Pol (schematisch), nach Ramanathan. *Trp* Tropopause.

Gestützt werden diese Vorstellungen durch die Tatsache, daß auch für die meridionale Variation des Ozons die bei den interdiurnen Ozonschwankungen gefundene Beziehung gültig ist, daß eine hohe Tropopause mit einem niedrigen Ozonwert korreliert ist und umgekehrt. Je dicker die stratosphärische Schicht zwischen dem Ozonmaximum und der Tropopause ist, um so mehr Ozon enthält sie[3]. Nach neueren Messungen scheint auch der Ozonwert nach größeren Breiten zu nicht so stetig zuzunehmen wie in Fig. 38, sondern in der Zone des plötzlichen Überganges von der hohen tropischen Tropopause (der Zone der subtropischen Jet-Streams) zu der niedrigeren in mittleren Breiten mehr sprunghaft anzusteigen [*32*]. Dementsprechend sind in N-Indien im Sommer, wenn die Jet-Stream-Zone nördlich liegt, Ozonwert und tägliche Ozonschwankungen klein, um im Winter anzusteigen, wenn der Jet-Stream-Gürtel sich südwärts verschiebt und dadurch stratosphärische Luft mittlerer Breiten mit hohem Ozongehalt in die äquatoriale Troposphäre über Indien einströmen kann.

---

[1] F. J. Scrase: Quart. J. Roy. Met. Soc. **77**, 483 (1951).
[2] A. W. Brewer: Quart, J. Roy. Meteor. Soc. **75**, 351 (1949).
[3] K. R. Ramanathan u. R. N. Kulkarni: Proc. Ind. Acad. Soc. A **1953**, 37, 321.

SCHRÖER [*41*] und MOSER [*22*] nehmen an, daß sich an der Grenze der Polarnacht durch die Aufheizung der oberen warmen Schicht in 40 bis 60 km Höhe ein starkes meridionales Temperaturgefälle herausbildet. Dadurch muß eine kräftige zirkumpolare Strömung bedingt werden, die einen verstärkten vertikalen Massenaustausch zur Folge hat. Durch diesen wird wiederum Ozon aus größeren Höhen, wo es sich photochemisch nachbildet, herabgeschafft und in tieferen Schichten akkumuliert. Ergänzend dazu soll etwa ab Mitte November durch hochreichende Kaltlufttiefs Ozon aus der photochemischen Mutterschicht in die untere Stratosphäre verfrachtet werden.

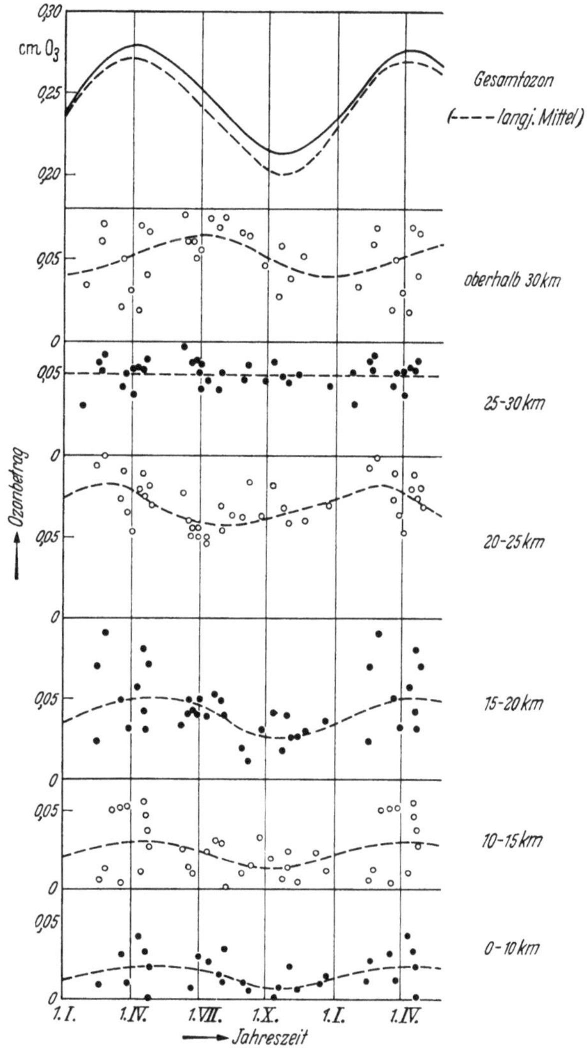

Fig. 42. Jahresgang des Ozons in verschiedenen Höhen in mittleren Breiten nach Ballonaufstiegen, nach PAETZOLD.

Für eine sichere Deutung des Ozonjahresganges ist die Kenntnis der jahreszeitlichen Variationen des in den verschiedenen Höhen befindlichen Ozons erforderlich, vor allem für polare Breiten. Für die mittleren Breiten ist mit den Ergebnissen der Ballonaufstiege ein erster Anfang von PAETZOLD[1] [*28*] versucht worden, den Fig. 42 zeigt. Danach besteht in verschiedenen Höhen ein etwas unterschiedlicher Jahresgang. Das Frühjahrsmaximum zwischen 0 und 20 km wird maßgeblich durch die Zufuhr polaren Ozons in einer mittleren Stärke von 0,04 cm O_3 bestimmt. Das Sommermaximum oberhalb von 30 km Höhe wird vermutlich hauptsächlich photochemisch durch die größere Sonnenhöhe bedingt. Daneben scheint aber auch noch parallel mit dem jahreszeitlichen Gang der Tropopausenhöhe eine im Herbst ab- und im Frühjahr aufsteigende Vertikalbewegung von einer Stärke von 0,1 mm/sec zu existieren, die durch Konvergenz bzw. Divergenz eine Ozonschwankung von etwa $\pm 0{,}02$ cm O_3 verursacht. Es ist bemerkenswert, daß zwischen 25 und 30 km Höhe keine jahreszeitlichen Ozonschwankungen zu erkennen sind. Dieses Bild wurde in jüngster Zeit von ADEL u. a.[2] (Flagstaff) durch Ultrarotmessungen bestätigt, nach denen sich ebenfalls der gegenläufige Jahresgang des Ozons ober- und unterhalb von 25 km Höhe ergab.

[1] H. K. PAETZOLD: Meteorologische Abhandlungen des Instituts für Geophysik, Freie Universität Berlin, Bd. II, S. 91. Berlin 1955.

[2] E. S. EPSTEIN: C. OSTERBERG u. A. ADEL: J. Atmosph. a. Terres. Phys. **8**, 347 (1956).

G. Ozonschicht und der Zustand der Stratosphäre.

24. Die Temperatur in der Stratosphäre. Die Temperaturempfindlichkeit der HUGGINS-Banden des Ozons (Ziff. 4) bietet die Möglichkeit zur Messung der Temperatur in der Stratosphäre (LADENBURG[1]). Derartige Beobachtungen sind vor allem von den französischen Ozonforschern[2] durchgeführt worden. Wegen der Komplikationen, die durch die Häufung der FRAUNHOFER-Linien im Sonnenspektrum in diesem Spektralbereich verursacht wird, wurden die HUGGINS-Banden des atmosphärischen Ozons vornehmlich im Licht von A- und B-Sternen mit ihrem ausgeglichenen UV-Spektrum beobachtet. BARBIER und CHALONGE[3] benutzten dazu einen Stern-Quarzspektrographen mit Objektivprisma, der sich durch einfache Handhabung und große Lichtstärke auszeichnet. Ein Nachteil ist die durch die atmosphärische Szintillation nicht genau definierte effektive „Spaltbreite" und die Störung durch Fremdlicht etwa bei Vollmond. In neuerer Zeit sind auch Messungen im Sonnenlicht durchgeführt worden, so von KAY[4] mit dem DOBSON-Instrument, KHALEK [*19*] in Kaboul und ZSCHÖRNER und PAETZOLD[5] mit UV-Spektrographen. Letztere diskutierten vor allem die Genauigkeit und die Fehlermöglichkeiten der Methode, auf die schon früher hingewiesen worden war[6].

Einen anderen Weg zur Messung der Ozontemperatur bietet die 9,6 μ-Ozonbande, deren Spektralbereich weitgehend von anderen atmosphärischen Absorptionen frei ist (Fig. 44). Außerdem liegt die Bande nahe dem Maximum der Strahlungskurve eines schwarzen Körpers von der Temperatur der Atmosphäre (Fig. 44). Nach dem KIRCHHOFFschen Gesetz ist:

$$E(\lambda) = A(\lambda) \cdot B(\lambda, T), \tag{24.1}$$

wenn $E(\lambda)$ die Emission, $A(\lambda)$ die Absorption der Ozonschicht im Wellenlängenintervall von λ bis $\lambda + d\lambda$, $B(\lambda, T)$ die PLANCKsche Funktion und T die Strahlungstemperatur sind. Die Messung von $A(\lambda)$ und $E(\lambda)$ erfolgt, indem einmal das Spektrometer auf die Sonne und dann auf eine Himmelsgegend mit demselben Zenitwinkel wie die Sonne gerichtet wird. Entsprechende Beobachtungen von ADEL[7,8] ergaben eine beträchtliche Genauigkeit von $\pm 5°$ K.

In Fig. 43 sind die Kontraste zwischen den Maxima und Minima der HUGGINS-Banden für Messungen bei Zimmertemperatur und bei Sternspektren zum Vergleich wiedergegeben[9]. Aus dem Unterschied folgt eine mittlere Ozontemperatur von $-30°$. Dieser Wert konnte auch für mittlere Breiten von anderen Beobachtern bestätigt werden[10]. Da die stratosphärischen Temperaturen bis herauf zu 30 km Höhe in mittleren Breiten etwa $-50°$ C betragen, muß der darüberliegende Teil der Ozonschicht erheblich wärmer sein als die gemessenen mittleren Temperaturen von $-30°$. In der Tat war ja schon aus der anomalen Schallfortpflanzung[11,12,13] und aus Meteorbeobachtungen[14] auf eine warme Schicht in etwa 50 km Höhe mit

[1] Rapport de la Réunion de l'ozone, Paris. Gerlands Beitr. Geophys. **24**, 1 (1929).
[2] G. DÉJARDIN: Cah. Phys. **1943**, Nr. 16.
[3] D. CHALONGE u. E. VASSY: Rev. Opt. **13**, 113 (1934).
[4] R. H. KAY: IX. Congress UGGI, Brüssel 1951.
[5] H. ZSCHÖRNER u. H. K. PAETZOLD: Mitt. Max-Planck-Inst. Weißenau, Nr. 4, 1955.
[6] D. BARBIER u. D. CHALONGE: Ann. Phys. **17**, 280 (1942).
[7] A. ADEL: Astrophys. J. **105**, 406 (1947).
[8] A. ADEL: Cent. Proc. Roy. Met. Soc. **5** (1950).
[9] D. BARBIER, D. CHALONGE u. E. VASSY: Rev. Opt. **14**, 425 (1935).
[10] ARNULF, DEJARDIN u. CAVASSILAS: C. R. Acad. Sci. Paris **205**, 809 (1937).
[11] F. J. W. WHIPPLE: Nature, Lond. **111**, 187 (1923).
[12] P. DUCKERT: Gerlands Beitr. Geophys. Suppl. **1**, 236 (1931).
[13] B. GUTENBERG: Gerlands Beitr. Geophys. **27**, 217 (1930).
[14] F. A. LINDEMANN u. G. M. B. DOBSON: Proc. Roy. Soc. Lond., Ser. A **102**, 411 (1923).

Temperaturen von 50° C geschlossen worden[1]. Die Raketenaufstiege über New Mexico haben zwar eine etwas niedrigere Temperatur von 0° C ergeben [*30*], doch ist die wichtige Tatsache eines stratosphärischen Temperaturmaximums bei 50 km voll bestätigt worden.

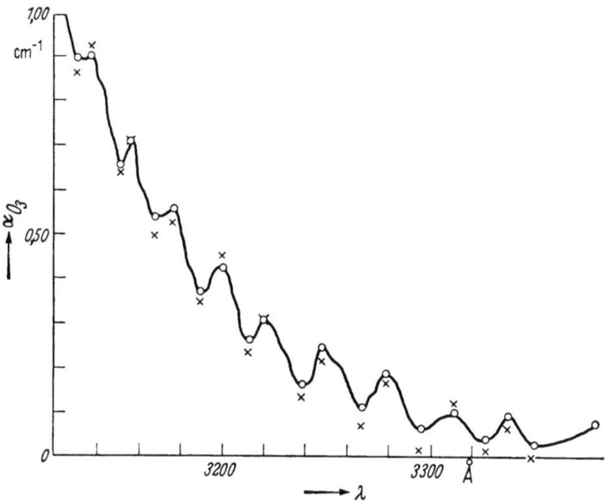

Fig. 43. Messung der Ozontemperatur mit Hilfe der HUGGINS-Banden, nach BARBIER und anderen. × HUGGINS-Banden bei Zimmertemperatur (15°); ○ HUGGINS-Banden im Siriusspektrum.

Es interessiert, in welcher Höhe die beobachtete mittlere Ozontemperatur $\overline{T}$ mit der herrschenden Lufttemperatur übereinstimmt. Man definiert den Absorptionskoeffizienten $\alpha(\lambda, \overline{T})$ durch [*44*]:

$$\alpha(\lambda, \overline{T}) = \frac{1}{x} \int_{h_0}^{\infty} \alpha(\lambda, T(h)) \cdot \varepsilon(h)\, dh, \tag{24.2}$$

wenn h_0 die Höhe des Meßortes ist. Man kann $\alpha(\lambda, T)$ genügend genau als lineare Funktion von T ansetzen und erhält

$$\overline{T}(h_0) = \frac{1}{x} \cdot \int_{h_0}^{\infty} T(h) \cdot \varepsilon(h)\, dh. \tag{24.3}$$

Mit einer mittleren Ozonverteilung stimmt nach (24.2) bei Messungen vom Erdboden aus die mittlere Ozontemperatur mit der in etwa 30 km Höhe herrschenden Temperatur überein. Bei Ballonaufstiegen oberhalb von 25 km Höhe läßt sich die Stratosphärentemperatur mit Hilfe der HUGGINS-Banden differentiell bis etwa 45 km Höhe messen[2].

Die mittlere Ozontemperatur zeigt ausgeprägte interdiurne Schwankungen[3], die mit den Ozonvariationen eine markante Korrelation aufweisen: Je größer der Ozonbetrag, um so höher ist auch die mittlere Ozontemperatur[4]. Dies muß als ein Anzeichen dafür gewertet werden, daß die Ozonvariationen maßgeblich von dem Lufttransport von einer geographischen Breite zur anderen verursacht worden sind[5]. Polare Luftmassen mit ihrem im allgemeinen gegenüber niederen

[1] Vgl. die Beiträge LOVELL und COX in diesem Bande, S. 439 und S. 475.
[2] H. ZSCHÖRNER u. H. K. PAETZOLD: Mitt. Max-Planck-Inst. Weißenau, Nr. 4, 1955.
[3] A. u. E. VASSY: C. R. Acad. Sci. Paris **207**, 338 (1938).
[4] D. BARBIER u. D. CHALONGE: J. Phys. Radium (7) **10**, 113 (1939).
[5] A. u. E. VASSY: Météorologie, S. 4. 1939.

Breiten größeren Ozongehalt weisen eine höhere Stratosphärentemperatur auf (um — 50° C), während diese am Äquator mit — 75° C erheblich niedriger liegt.

Während eines Tages können beträchtliche Schwankungen der mittleren Ozontemperatur auftreten. So wurde auf dem Jungfraujoch eine Temperaturänderung von 65° C in $2\frac{1}{4}$ Std beobachtet. In Kaboul [*19*] ergab sich ein typischer Tagesgang der Temperatur, indem diese um die Mittagszeit um 10 bis 20° höher liegt als nachts, wobei der Ozonbetrag tagsüber um etwa 15% erniedrigt ist (Ziff. 15). Der Versuch, diese Korrelation hauptsächlich durch den Temperatureinfluß auf die photochemischen Gleichgewichtsbedingungen zu deuten[1], erscheint wenig plausibel. Denn, wenn auch bei niederen Temperaturen sich ein höherer Ozonwert einstellt, sind doch die Einstellzeiten des photochemischen Gleichgewichts in den Höhen, wo die Hauptmasse des Ozons sich befindet, zu groß, um auf diese Weise die obigen Ozonvariationen erklären zu können (Ziff. 14). Dies gilt auch in bezug auf die jahreszeitlichen Ozonvariationen in höheren Breiten, bei denen ebenfalls im Verlauf des Frühjahrs und Sommers mit steigenden Stratosphärentemperaturen der Ozonwert absinkt.

Tabelle 9.
Mittlere Ozontemperatur in Abisko.

	°C
Ende Dezember . .	— 79
Januar	— 49
Februar	— 14
Anfang März . . .	+ 1
August	+ 20

In Abisko (63,3° N) haben VASSY und andere[2,3] nach dem Ende der Polarnacht einen sehr markanten Anstieg der mittleren Ozontemperatur festgestellt. Nach Tabelle 9 steigt sie bei dauernder Sonneneinstrahlung während eines Polartages bis auf 20° C im August. Dies leitet über zu dem Problem der Aufheizung der warmen Schicht und der Rolle, die das Ozon in dem Strahlungshaushalt der Stratosphäre spielt.

25. Der Strahlungshaushalt der Stratosphäre. Die warme Schicht in 50 km Höhe ist aus der Thermodynamik der Atmosphäre nicht zu deuten, sondern setzt eine Strahlungsenergie absorbierende eingelagerte Schicht voraus, für die nur das Ozon in Frage kommt, das in diesem Höhenbereich in der HARTLEY-Bande die Sonnenstrahlung von 2200 bis 2900 Å praktisch vollständig verschluckt. In diesem Spektralbereich beträgt die Strahlungsenergie etwa 1,5% der Solarkonstante. Die „Aufheizungstendenz“, d.h. die Erwärmung ΔT innerhalb 12 Std., die bei plötzlichem Einsetzen der Sonnenstrahlung (ohne Berücksichtigung der thermischen Strahlungsströme) eintritt, hat nach Tabelle 10 ein ausgesprochenes Maximum in 50 km Höhe in guter Übereinstimmung mit der Lage des beobachteten Temperaturmaximums [*28*]. Wegen ihrer Bedeutung für die Dynamik der Stratosphäre ist die von der Ozonschicht absorbierte Strahlungsenergie in Abhängigkeit von der Höhe, der Jahreszeit und der geographischen Breite von verschiedenen Seiten berechnet worden[4-6].

Tabelle 10.
„Aufheizungstendenz“ der Ozonschicht.

Höhe km	$\Delta T/12^h$ °K	Höhe km	$\Delta T/12^h$ °K
70	6	40	6
60	12	30	2
50	16		

Die sich im Strahlungsgleichgewicht einstellende Temperatur der Stratosphäre wird durch die Bilanz der Ein- und Ausstrahlung bestimmt. Wenn S, E

[1] A. u. E. VASSY: J. Phys. **2**, 82 (1941).
[2] D. BARBIER, D. CHALONGE u. E. VASSY: C. R. Acad. Sci. Paris **201**, 787 (1935).
[3] A. u. E. VASSY: Tellus **2**, 69 (1950).
[4] R. PENNDORF: Veröff. Geophys. Inst. Leipzig **8**, H. 4 (1936).
[5] R. V. KARANDIKAR: Proc. Indian Acad. Sci. **23**, 70 (1946).
[6] J. PRESSMAN: J. Geophys. Res. **59**, 485 (1954).

und D die Strahlung der Sonne, der Erde und der Troposphäre bzw. der Stratosphäre bedeuten und $K(\lambda)$ die Absorption und $B(\lambda, T)$ die Strahlungsfunktion eines Raumteiles sind, der als schwarzer Körper angenommen sei, dann ist im Strahlungsgleichgewicht:

$$\int K S\,dh + \int K E\,dh + \int K A\,dh = 2\int K B(\lambda, T)\,dh. \tag{25.1}$$

Einer numerisch befriedigenden Lösung dieser Gleichung stehen heute noch sehr große Schwierigkeiten gegenüber. Die grundlegende Voraussetzung der Gültigkeit des KIRCHHOFFschen Gesetzes kann bis zu etwa 50 km Höhe angenommen werden. Nach MILNE[1] muß die Geschwindigkeit der Desaktivierung der angeregten Zustände durch Stöße größer sein als die der spontanen Desaktivierung, damit Emissions- und Absorptionsniveaus sich im thermischen Gleichgewicht befinden. Bei den Schwingungsniveaus ist die Größenordnung der Lebensdauer zu 10^{-1} sec anzunehmen, während der Bruchteil der wirksamen Stöße zu $5 \cdot 10^{-6}$ abgeschätzt werden kann. In 60 km Höhe mit einer Stoßzahl zwischen den Luftmolekülen von $2 \cdot 10^{6}$ sec^{-1} ist damit die obige Gültigkeitsgrenze für Schwingungsniveaus erreicht. Bei den Rotationsniveaus liegen die Verhältnisse noch günstiger.

Wesentlich schwieriger ist das Problem der Absorptionsgrößen. Während diese im Ultravioletten und sichtbaren Spektralbereich bekannt sind, ist dies bei den ultraroten Molekülbanden, die für die globale Ausstrahlung allein maßgeblich sind, nicht der Fall. Wegen ihrer Zusammensetzung aus einzelnen Linien, die bei den atmosphärischen Gasen eine verschiedene Anordnung zeigen, hängt die Absorption in komplizierter Weise vom Druck ab. Die Breite einer einzelnen Linie setzt sich aus der Dopplerbreite und der durch Stoßdämpfung bedingten LORENTZ-Breite zusammen, die proportional dem Luftdruck ist. Bei den hier in Frage kommenden Gasen überwiegt die Dopplerbreite oberhalb von 30 bis 40 km Höhe [*17*]. Für ungeordnete Banden, wie sie bei H_2O und O_3 vorliegen, kann nach CURTIS[2] mit guter Näherung ein mittlerer effektiver Druck $\bar{p}$ eingeführt werden nach der Gleichung:

$$\bar{p} = \frac{\int p(h)\cdot\varrho(h)\,dh}{\int \varrho(h)\,dh}. \tag{25.2}$$

wo $\varrho(h)$ die Dichte des absorbierenden Gases ist.

Fig. 44 zeigt die Absorption in 11 km Höhe, die durch die Gase H_2O, CO_2 und O_3 bewirkt wird. Danach liegt die starke ultrarote Ozonbande bei 9,6 μ in einer Lücke des Absorptionsspektrums des Wasserdampfes und der Kohlensäure, was für den Strahlungshaushalt besonders der nächtlichen Atmosphäre von Belang ist. Das Ozon verstärkt damit, wie ANGSTRÖM[3] schon betont hat, die Treibhauswirkung der Atmosphäre, wobei sich auch die Ozonschwankungen auswirken mögen. Außerdem absorbiert im Bereich von 13 bis 16 μ sowohl die Kohlensäure wie das Ozon, so daß das letztere die Strahlung der ersteren in der unteren Stratosphäre gegen den Weltenraum abschirmt. Da in dem obigen Spektralbereich auch das Maximum der Temperaturstrahlung eines Strahlers von der Stratosphärentemperatur (250° K) liegt, muß die Ozonschicht sich wesentlich auf die Temperatur der unteren Stratosphäre auswirken[4]. Auf diese Weise ließe sich die besonders tiefe Temperatur der tropischen Tropopause (—85° C) vielleicht wenigstens zum Teil deuten durch den geringeren Ozonbetrag. Auch der

[1] E. A. MILNE: Handbuch der Astrophysik, Bd. III, S. 162. Berlin 1930.
[2] A. CURTIS: Quart. J. Roy. Met. Soc. **78**, 683 (1952).
[3] A. ANGSTRÖM: Gerlands Beitr. Geophys. **21**, 145 (1929).
[4] F. MÖLLER: Naturwiss. **31**, 148 (1943).

jährliche Temperaturgang der Tropopause kann vielleicht auf wechselnden Ozongehalt der unteren Stratosphäre zurückgeführt werden[1]. Im einzelnen sind jedoch die Zusammenhänge sehr verwickelt und noch ungeklärt[2].

Neuere Berechnungen des Strahlungshaushaltes sind von verschiedener Seite durchgeführt worden[3,4,5], nachdem die alten Ansätze von EMDEN zu niedrige Temperaturen ergeben hatten. Die eingehendste Berechnung der Temperatur nach (25.1) hat GOWAN[6] durchgeführt. Nach Fig. 45 ergibt sich das Temperaturmaximum in der beobachteten Höhe, allerdings etwas zu groß nach den Beobachtungen [*30*]. Wenn die Wasserdampfabsorption druckunabhängig angenommen wird, ist der Einfluß des Gehaltes an Wasserdampf sehr groß; dieser wurde, bei Annahme konstanter Durchmischung, zu 0, 10 und 40% der Sättigung in Tropopausenhöhe angesetzt, was einem Partialdruckverhältnis von etwa $0{,}1 \cdot 10^{-5}$ und

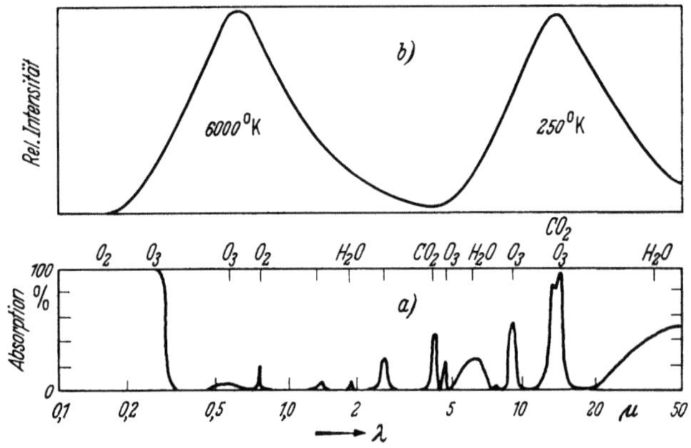

Fig. 44 a. u. b. Die Absorption a) von H_2O, CO_2 und O_3 in 11 km Höhe, nach GOODY. In b) sind die PLANCKsche Strahlungsfunktionen für 6000° und 250° K gegeben.

$4 \cdot 10^{-5}$ entspricht. Bei Ballonaufstiegen wurden Werte von $1 \cdot 10^{-4}$ bis $1 \cdot 10^{-5}$ gemessen[7]. Es ist noch zu bedenken, daß die Berücksichtigung des Druckeinflusses sich nur auf die LORENTZ-Breite der Linien bezieht, die aber oberhalb von 40 km Höhe gegenüber der DOPPLER-Breite zurücksteht. Dadurch ergibt sich für diesen Höhenbereich eine zu kleine thermische Emission und eine zu hohe Temperatur.

Von Interesse sind die Temperaturunterschiede zwischen Tag und Nacht, die durch die fehlende Strahlungsabsorption in der Ozonschicht nachts verursacht werden. Nach JOHNSON[8] sind nur geringe Schwankungen zu erwarten, die in 50 km Höhe mit 6° am größten sind. Dies ist in Übereinstimmung mit den Beobachtungen der Temperatur der Ozonschicht von ADEL[9], der keine merklichen Unterschiede zwischen Tag und Nacht feststellen konnte. Ferner müssen durch den jährlichen Sonnengang in der warmen Schicht jahreszeitliche Temperaturschwankungen auftreten, die am Pol am stärksten sein werden. Nach dem noch

---

[1] G. M. B. DOBSON, A. W. BREUER u. B. C. CWILONG: Proc. Roy. Soc. Lond., Ser. A **185**, 144 (1946).
[2] R. M. GOODY: Diss. Cambridge University.
[3] P. RAETHJEN: Meteor. Z. **57**, 317 (1940).
[4] R. PENNDORF: Veröff. Geophys. Inst. Leipzig **8** (1936).
[5] F. LINKE u. F. MÖLLER: Handbuch der Geophysik, Bd. VIII, S. 651. 1944.
[6] E. H. GOWAN: Proc. Roy. Soc. Lond., Ser. A **190**, 219, 227 (1947).
[7] E. W. BARRET, R. HERNDOU u. J. C. HOWARD: Tellus **2**, 302 (1950).
[8] F. S. JOHNSON: Bull. Amer. Met. Soc. **34**, 106 (1953).
[9] A. ADEL: Astrophys. J. **105**, 406 (1947). — Cent. Proc. Roy. Met. Soc. **5** (1950).

dürftigen Beobachtungsmaterial ist in Belmar (40° N)[1], Berlin (52° N)[2] und Abisko (70° N)[3] die Temperatur in 30 km Höhe im Sommer um 12, 20 und 70° C höher als im Winter. Durch die sommerliche Aufheizung und die winterliche Abkühlung der warmen Schicht in mittleren und polaren Breiten baut sich oberhalb von 20 km Höhe über den Polen eine sommerliche Antizyklone und eine winterliche Zyklone auf, die zwischen 20 und 80 km Höhe ein markantes jahreszeitliches zonales Windsystem zwischen 20 und 80 km Höhe in Gang setzen (Sommer: Ost-, Winter: Westwinde; Windgeschwindigkeit in 60 km Höhe 100 m pro sec)[4]. Dieser durchgreifende Einfluß des Spurengases Ozon auf die strato-

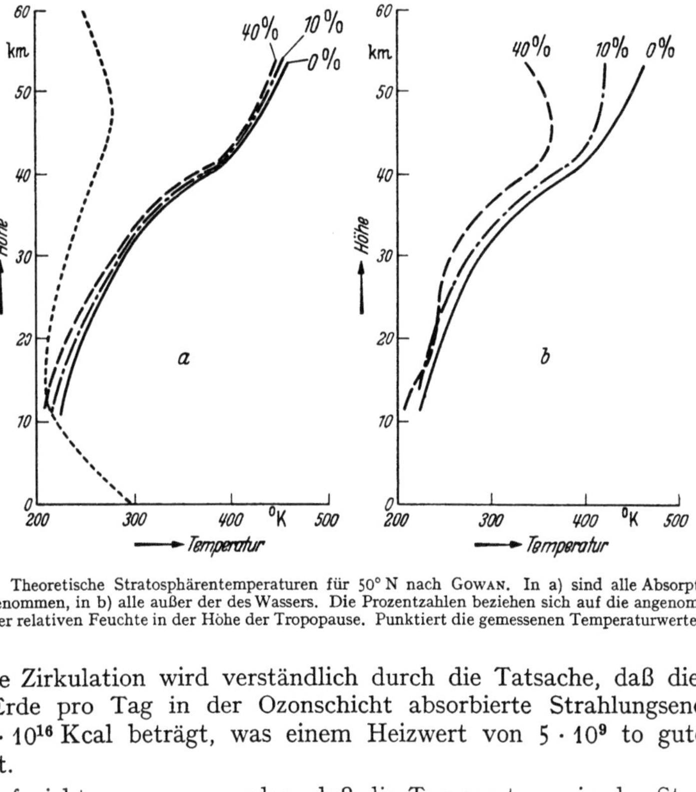

Fig. 45a u. b. Theoretische Stratosphärentemperaturen für 50° N nach GOWAN. In a) sind alle Absorptionen druckabhängig angenommen, in b) alle außer der des Wassers. Die Prozentzahlen beziehen sich auf die angenommenen Werte der relativen Feuchte in der Höhe der Tropopause. Punktiert die gemessenen Temperaturwerte.

sphärische Zirkulation wird verständlich durch die Tatsache, daß die auf der ganzen Erde pro Tag in der Ozonschicht absorbierte Strahlungsenergie der Sonne $4 \cdot 10^{16}$ Kcal beträgt, was einem Heizwert von $5 \cdot 10^9$ to guter Kohle entspricht.

Es darf nicht vergessen werden, daß die Temperaturen in der Stratosphäre noch von dynamischen Effekten beeinflußt werden, wenngleich die Strahlungsverhältnisse als primär anzusehen sind, da sie den Motor für die ersteren darstellen. Eine Synthese hat GOODY[5] zur Deutung des Phänomens der Tropopause versucht und die Höhe der Tropopause und die Diskontinuität des vertikalen Temperaturgradienten befriedigend ableiten können, wobei sich im einzelnen ergab, daß die Stratosphärentemperaturen steigen, wenn die Temperaturen am Erdboden sinken, was der Erfahrung entspricht. Es zeigte sich wieder die bedeutende Rolle, die die nur in kleinen Mengen vorkommenden Gase H_2O, O_3 und CO_2 spielen. Zum Beispiel ergibt die Annahme eines Strahlungsgleichgewichtes in einer grauen Atmosphäre — entgegen den Beobachtungen — immer einen

[1] W. C. CONNOVER u. C. J. WENTZIN: J. Meteor. **12**, 160 (1955).
[2] J. WIEHLER: Meteor. Abh. Fr. Univ. Berlin **3**, H. 1 (1955).
[3] A. u. E. VASSEY: Tellus **2**, 69 (1950).
[4] H. K. PAETZOLD: Mitt. dtsch. Wetterd. Nr. **22**, 43 (1956).
[5] R. M. GOODY: Proc. Roy. Soc. Lond., Ser. A **197**, 872 (1947).

negativen Temperaturgradienten. Diese Schwierigkeit verschwindet, wenn man annimmt, daß in der unteren Stratosphäre eine direkte Absorption der Sonnenstrahlung stattfindet, oder daß an dem Strahlungsgleichgewicht mehr als eine absorbierende Komponente beteiligt sind.

Für den ultravioletten Spektralbereich ist noch die starke Abschirmung zu erwähnen, die die Ozonschicht bei Sonnenaufgang oder -untergang auf die höhere Atmosphäre ausübt. Im Gebiet der HARTLEY-Bande ist die optische Obergrenze der Ozonschicht für die die Erdatmosphäre tangierenden Lichtstrahlen recht scharf: Lichtstrahlen mit einer Wellenlänge von 2550 Å erleiden noch keine wesentliche Absorption, wenn sie die Atmosphäre in einem Minimalabstand von 56 km passieren, sie werden aber schon bei einem nur um 3 km kleineren Abstand bis auf 1% geschwächt[1,2]. Diese Schirmwirkung des Ozons ist von Belang bei Dissoziations-, Ionisations- und Anregungsprozessen, die in der Atmosphäre durch Strahlung im Gebiet der HARTLEY-Bande erzeugt werden. So ergab sich bei der starken Emission der D-Linien des Natriums im Dämmerungsleuchten für die Obergrenze der die auslösende Strahlung abschirmenden Schicht eine Höhe von 56—60 km, die daher mit der Obergrenze der Ozonschicht identifiziert wurde[3,4]. Ferner scheinen die leuchtenden Nachtwolken nur aufzuleuchten, wenn die Sonnenstrahlung die Erdoberfläche in keinem kleineren Abstand als 30 bis 45 km Höhe passiert[5].

26. Die bioklimatische Bedeutung des Ozons. Die außerordentliche Bedeutung des Ozons für biologische Prozesse[6,7], die hier nur gestreift werden kann, beruht auf der Tatsache, daß die starke ultraviolette Ozonabsorption in einem Spektralbereich liegt, der eine Reihe wichtiger biologischer Reaktionen auslöst, da die Eiweiße eine starke spektrale Absorption mit dem Maximum bei etwa 2750 Å aufweisen. In Fig. 46 [*21*] sind einige spektrale Wirkungskurven für verschiedene biologische Vorgänge wiedergegeben. Besonders wichtig sind die Kurven für die Bildung des antirachitischen Vitamins D und für die Bakterientötung. Die lebensnotwendigen Vitamine D können von den Organismen nicht selbst aufgebaut werden, sondern werden durch UV-Strahlung unterhalb von 3100 Å aus den im Organismus vorhandenen Sterolen gebildet. Sowohl bei der Vitamin D-Bildung wie bei der Erythemerzeugung sind die Ausläufer der Wirkungskurven in das längere UV hinein ausschlaggebend, da dort die Intensität der Sonnenstrahlung noch verhältnismäßig groß, die Ozonabsorption aber schon schwach ist. Dadurch ist für das biologische Leben am Grund des irdischen Luftozeans eine ausreichende Vitaminbildung und eine fördernde Reizwirkung möglich. Umgekehrt reicht die Stärke der Ozonschicht aus, um die eiweißzerstörende Wirkung der ultravioletten Sonnenstrahlung abzuschirmen. Ohne Ozon wäre das „photochemische Klima" am Erdboden für das organische Leben viel zu hart, während es durch die spektralen Eigenschaften der atmosphärischen Gase einen ausgesprochen milden Charakter aufweist [*37*].

Zwischen der lebensfeindlichen und lebensfördernden Dosis der UV-Sonnenstrahlung am Erdboden besteht ein empfindliches Gleichgewicht, das maßgeblich vom Ozon abhängt. So wird die Intensität der Vitamin D-bildenden Strahlung

---

[1] R. PENNDORF: Z. Meteor. **1**, 345 (1943). — J. Meteorology **5**, 152 (1948).
[2] U. STILLE: Abh. Braunschweig. Wiss. Ges. **4**, 48 (1952).
[3] L. VEGARD u. E. TØNSBERG: Geofys. Publ. Oslo **13**, Nr. 1 (1940). — L. VEGARD u. G. KRIFTE: Geofys. Publ. Oslo **16**, Nr. 7 (1945).
[4] G. CARIO u. U. STILLE: Abh. Braunschweig. Wiss. Ges. **2**, 40 (1950).
[5] C. STÖRMER: Astrophys. norveg. **1**, 87 (1935).
[6] K. BÜTTNER: In Probleme der kosmischen Physik, Bd. 18. Leipzig 1938.
[7] R. LATARJET: Rev. Opt. **14**, 398 (1935).

allein durch die in unseren Breiten normalen Schwankungen des Ozonwertes zwischen 0,2 und 0,35 cm O_3 um den Faktor 5 variiert (berechnet für eine mittlere Sonnenhöhe von 45°). Ohne Ozon wäre die Erythemwirkung etwa 50mal stärker als jetzt, und es würden zumindest einfachste Lebewesen außerhalb des Wassers abgetötet werden (Fig. 46), was allein schon einen katastrophalen Eingriff in das heutige biologische Gleichgewicht bedeuten würde.

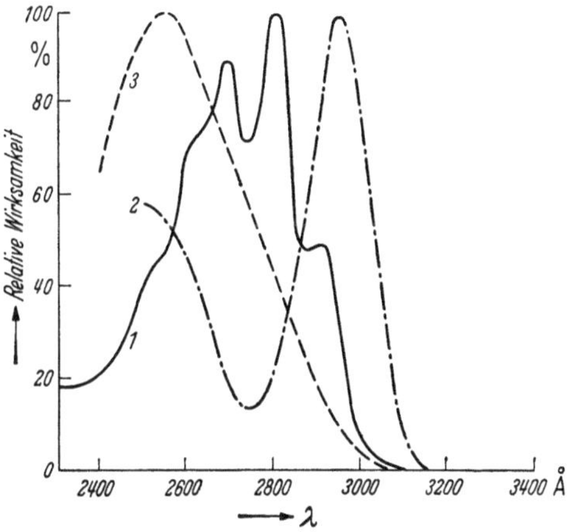

Fig. 46. Spektrale Wirkungskurve für *1* Vitamin-D-Bildung (Ergosterinabsorption); *2* Hauterythem und *3* Bakterientötung (Bact. coli), nach MEYER-SEITZ.

In diesem Zusammenhang ist es für die Entwicklung des irdischen Lebens von Bedeutung, daß die Stärke der sich nach dem photochemischen Gleichgewicht einstellenden Ozonschicht nur wenig von dem Sauerstoffgehalt der Atmosphäre abhängt. Noch bei einer Sauerstoffkonzentration von 1/100 ihres

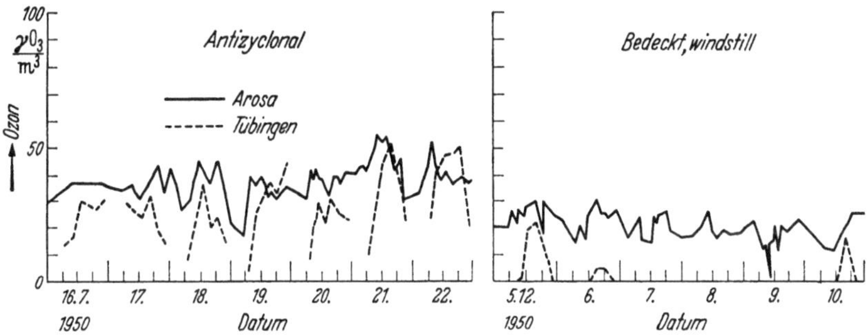

Fig. 47. Werte des bodennahen Ozons in Tübingen und Arosa bei verschiedener Unruhe der bodennahen Luftschichten, nach EHMERT.

heutigen Betrages würde sich eine Ozonschicht von 0,2 cm O_3 bilden, deren Maximum etwas tiefer, bei 15 km Höhe, liegen würde, so daß auch bei einem eventuell früher geringeren Sauerstoffgehalt schon ein ausreichender biologischer Strahlungsschutz gegeben war[1] [*28*].

Aber auch eine beträchtliche Vermehrung des Ozons würde sich verhängnisvoll auswirken, da sie für die obigen lebenswichtigen Prozesse eine „photochemische Finsternis" bedeuten würde. Nach (14.5) würde sich mehr Ozon bilden, wenn entweder die Strahlungsintensität der Sonne bei 2150 Å größer oder die Lufttemperatur niedriger wäre. So würde, wenn die Sonne im Ultra-

[1] H. K. PAETZOLD: Mitt. Astr. Ges. **1953**, 19.

violetten dieselbe Strahlungstemperatur wie im sichtbaren Spektralbereich (5700° K) aufweisen würde, sich ein doppelt so hoher Ozonwert einstellen wie jetzt, was eine Reduzierung der ultravioletten biologischen Wirkung auf $^1/_{10}$ bewirken würde. Auf dem Mars (tiefere Lufttemperatur und größere Sonnenentfernung) wäre dieselbe sogar bis auf $^1/_{100}$ des irdischen Wertes herabgesetzt.

Über direkte Wirkungen des Ozons in den am Erdboden vorkommenden geringen Konzentrationen auf Organismen, die schon öfters vermutet wurden, ist noch nichts bekannt. Doch haben die beobachteten Schwankungen des bodennahen Ozons (Ziff. 20) eine indirekte bioklimatische Bedeutung als Indikator für den *Reinheitsgrad* der Luft[1]. Hohe Ozonbeträge weisen auf Luft hin, die durch turbulente Mischungsvorgänge erneuert wird, während wenig Ozon bei stagnierender Luft mit eventuell großen Verunreinigungen auftritt. So ist nach Fig. 47[2] in dem durch sein Heilklima ausgezeichneten Höhenkurort Arosa (1900 m) der Ozongehalt, unabhängig von der Wetterlage, meistens höher als in dem tiefgelegenen Tübingen. Hier ist bei windstiller Wetterlage, bei der sich die Stagnation der Luft in einer über der Stadt liegenden Dunsthaube demonstriert, auch der Wert des bodennahen Ozons verschwindend klein.

Bibliographie.

[*1*] BARBIER, D., D. CHALONGE et E. VIGROUX: Utilisation des éclipses de lune à l'étude de la haute atmosphère. Ann. Astrophys. **5**, 1 (1942). — Genaue Darstellung der Beobachtung und Auswertung der Mondfinsternis am 2. 3. 1942.

[*2*] BATES, D. R., and M. NICOLET: The photochemistry of Atmospheric Water Vapour. J. Geophys. Res. **55**, 301 (1950). — Theorie der Photochemie in einer Sauerstoff-Wasserdampfatmosphäre oberhalb von 60 km Höhe.

[*3*] CHAPMAN, S.: On Ozone and Atomic Oxygen in the Upper Atmosphere. Phil. Mag. (7) **10**, 345 (1930). — Photochemische Theorie des Ozons.

[*4*] CRAIG, R. A.: The Observations and Photochemistry of Atmospheric Ozone and their Meteorological Significance. Meteor. Monogr. **1**, Nr. 2 (1950). — Ausführliche Darstellung und Diskussion des Beobachtungsmaterials über das Ozon sowie photochemische Berechnungen der Ozonverteilung.

[*5*] COBLENTZ, W. W., and R. STAIR: Distribution of Ozone in the Stratosphere. J. Res. Nat. Bur. Stand. **22**, 573 (1939). — Ausführlicher Bericht über Messungen der Ultraviolettstrahlung am Flagstaff Observatorium, des Ozonwertes und über die benutzte Ozonradiosonde.

[*6*] DOBSON, G. M. B., and D. N. HARRISON: Measurements of the amount of ozone in the earth's atmosphere and its relation to other geophysical conditions. Part 1. Proc. Roy. Soc. Lond. **110**, 660 (1926). — DOBSON, G. M. B., and LAWRENCE: Part 2. Proc. Roy. Soc. Lond. **114**, 521 (1927) und Part 3. Proc. Roy. Soc. Lond. **122**, 456 (1929). — DOBSON, G. M. B.: Part 4. Proc. Roy. Soc. Lond. **129**, 411 (1930). — Klassische eingehende Darstellung der geographischen und zeitlichen Ozonvariationen, insbesondere ihrer Zusammenhänge mit den meteorologischen Elementen.

[*7*] DOBSON, G. M. B.: Das Ozon in der Erdatmosphäre, Endeavour **11**, 215 (1952). — Kurze Zusammenfassung.

[*8*] DÜTSCH, H. U.: Photochemische Theorie des atmosphärischen Ozons unter Berücksichtigung von Nicht-Gleichgewichts-Zuständen und Luftbewegungen. Diss. Zürich 1946. — Ausführliche Berechnungen des Ozons mit besonderer Rücksicht auf Transportvorgänge in der Atmosphäre.

[*9*] EHMERT, A.: Ein einfaches Verfahren zur Messung kleinster Jodkonzentrationen, Jod- und Natriumthiosulfatmengen in Lösungen. Z. Naturforsch. **4**b, 321 (1949). — Eingehende Darstellung der elektrolytischen Titration.

[*10*] EHMERT, A.: Über das troposphärische Ozon. Mitt. dtsch. Wetterdienst. US-Zone **1949**, Nr. 11, 63. — Messungen des troposphärischen Ozons bei Flugzeugaufstiegen 1940.

[*11*] FABRY, CH., u. H. BUISSON: Étude de l'extrémité du spectre solaire. J. Phys. **2**, 197 (1921). — Klassische Untersuchung, die der Ausgangspunkt für die zahlreichen folgenden Arbeiten über das Ozonproblem geworden ist.

---

[1] E. REGENER: Naturwiss. Rdsch. **7**, 8 (1954).

[2] A. EHMERT: J. Atmos. a. Terres. Phys. **2**, 189 (1952).

[12] Fabry, Ch.: L'ozone atmosphérique. Éd. du Centre Nation. de la Recherche Scientifique. Paris 1950. — Umfassende Darstellung des ganzen Ozonproblems mit dem Schwerpunkt auf der Optik. Reiche Bibliographie.

[13] Fronrobert, E.: Das Ozon. Chemie in Einzeldarstellungen, Bd. 9. Stuttgart 1916. — Chemie des Ozons, seine Darstellung und geschichtliche Angaben.

[14] Götz, F. W. P.: Das atmosphärische Ozon. In Ergebnisse der kosmischen Physik, Bd. I, S. 180. Leipzig 1931. — Ausführliche Monographie, Geschichtliche Angaben, Meßmethoden und Ergebnisse, reiche Literatur.

[15] Götz, F. W. P.: Die vertikale Verteilung des atmosphärischen Ozons. In Ergebnisse der kosmischen Physik, Bd. III, S. 252. Leipzig 1938. — Ausführliche Monographie, Geschichtliche Angaben, Meßmethoden und Ergebnisse, reiche Literatur.

[16] Götz, F. W. P.: Ozone in the Atmosphere. Compend. of Meteorol. S. 275. Boston 1950. Knappe umfassende Darstellung mit vielen Zitaten.

[17] Goody, R. M.: The Physics of the Stratosphere. Cambridge 1954. — Übersichtliche zusammenfassende Darstellung der Probleme in der Stratosphäre. Literatur.

[18] Johnson, F. S., J. D. Purcell and R. Tousey: Studies of the Ozone Layer above Mexico. In Rocket Exploration of the Upper Atmosphere. London: Pergamon Press 1954. — Zusammenfassung der Ozonmessungen bei Raketenaufstiegen.

[19] Khalek, A.: Variations diurnes et annuelles de l'épaisseur réduite et de la température moyenne de l'ozone atmosphérique à Kaboul (Afghanistan). Thése Paris 1954. — Ozonmessungen mit Ultraviolettspektrographen.

20] Meetham, A. R.: The Correlations of the Amount of Ozone with other Characteristics of the Atmosphere. Quart. J. Roy. Met. Soc. **63**, 289 (1937). — Grundlegende Studie.

[21] Meyer, H., u. E. O. Seitz: Ultraviolette Strahlen. Berlin 1949. — Allgemeine Darstellung über UV-Strahlung, Erzeugung und Anwendung. Sehr ausführliches Literaturregister.

[22] Moser, H.: Ozon und Wetterlage. Ber. dtsch. Wetterdienst. US-Zone **1949**, Nr. 11 28. — Studie über die Zusammenhänge zwischen Ozon und Luftbahnen im Jahr, 1942.

[23] Normand, Ch.: Atmospheric Ozone and the Upper-Air Conditions. Quart. J. Roy. Met. Soc. **79**, 39 (1953). — Diskussion der neueren Ozonmeßreihen in Oxford.

[24] Paetzold, H. K.: Die vertikale Verteilung des atmosphärischen Ozons nach Ballonaufstiegen. Z. Naturforsch. **10**a, 33 (1955). — Zusammenfassende Darstellung der Weißenauer Ballonaufstiege.

[25] Paetzold, H. K.: Über die Erfassung der atmosphärischen Ozonverteilung durch die Benutzung von Mondfinsternissen. Z. Naturforsch. **6**a, 639 (1951). — Darstellung, Kritik und Anwendung der Methode.

[26] Paetzold, H. K.: Die vertikale Verteilung des atmosphärischen Ozons nach dem photochemischen Gleichgewicht. Geofisica Pura e Applicata **24**, 1 (1953) und Versuch einer Analyse der gemittelten vertikalen Ozonverteilung in verschiedenen geographischen Breiten. Geofisica Pura e Applicata **24**, 15 (1953). — Diskussion der photochemischen Theorie des Ozons in Verbindung mit den Beobachtungen.

[27] Paetzold, H. K.: Ozonschicht und Luftbewegungen in der Stratosphäre. Naturwiss. **41**, 320 (1954). — Kurze Übersicht.

[28] Paetzold, H. K.: Photochemische Prozesse in der Erdatmosphäre und Ozonschicht. Mitt. Max-Planck-Inst. Weißenau Nr. 8, 1957. — Eingehendere Darstellung von Theorie und Beobachtungen unter Berücksichtigung der Berührung mit der Meteorologie und Astrophysik.

[29] Palmén, E.: Über dreidimensionale Luftströmungen in einer Zyklone und die Ozonverteilung. UGGI, Assoc. Météor. Réunion Washington 1939, S. 31. — Grundlegende Diskussion des Zusammenhanges zwischen den Ozonvariationen und den zyklonal gesteuerten vertikalen Luftströmungen.

[30] The Rocket Panel, Pressures, Densities and Temperatures in the Upper Atmosphere. Phys. Rev. **88**, 1027 (1952). — Numerische Angaben nach den Ergebnissen der Raketenaufstiege.

[31] Penndorf, R.: Ozon und Wetter II. Meteor. Rdsch. **3**, 49 (1950). — Ausführliche Analyse zweier Wetterlagen im Jahre 1942.

[32] Ramanathan, K. B.: Atmospheric Ozone and the General Circulation of the Atmosphere. X. Congress UGGI, Scient. Proc. Assoc. Météorologique, Rom 1954; London: Butterworth 1956. — Zusammenfassende Diskussion.

[33] Reed, J. R.: The Effects of Atmospheric Motions on Ozone Distributions and Variations, Thesis, Mass. Inst. Techn. 1949. — Eingehende Analyse des Einflusses von vertikalen und horizontalen Transportvorgängen.

[34] Regener, E.: Über Schwankungen des Ozons in der Troposphäre und Stratosphäre. J. Atmos. a. Terres. Phys. **2**, 189 (1952). — Straffe Übersicht.

[35] REGENER, E.: Das atmosphärische Ozon. In Naturforschung und Medizin in Deutschland 1939—1946, S. 297. Wiesbaden 1948. — Kurze Zusammenfassung.

[36] REGENER, E.: Ozonschicht und atmosphärische Turbulenz. Meteor. Z. **60**, 253 (1943). — Grundlegende Studie über den Einfluß des vertikalen Massenaustausches auf das Ozon.

[37] REGENER, E.: Über das „photochemische" Klima der Erde. Naturwiss. **33**, 163 (1946). — Allgemeine Betrachtungen über den biologischen Strahlenschutz der atmosphärischen Komponenten.

[38] REGENER, V. H.: Neue Messungen der vertikalen Verteilung des Ozons in der Atmosphäre. Phys. Z. **109**, 642 (1938). — Ausführliche Darstellung der optischen Methode.

[39] REGENER, V. H.: Vertical Distribution of Atmospheric Ozone. Nature, Lond. **167**, 276 (1951). — Kurze Mitteilung über die Ballonaufstiege in Albuquerque.

[40] REGENER, V. H.: New Experimental Results on Atmospheric Ozone. X. Congress UGGI, Scient. Proc. Assoc. Meteorology Rom 1954; London: Butterworth 1956. — Knappe Übersicht über neue Messungen der vertikalen Ozonverteilung und des bodennahen Ozons.

[41] SCHRÖER, E.: Theorie der Entstehung, Zersetzung und Verteilung des atmosphärischen Ozons. Ber. dtsch. Wetterdienst. US-Zone **1949**, Nr. 11, 12. — Ausführliche Diskussion der einzelnen Prozesse.

[42] TEICHERT, FR., u. W. WARMBT: Ozonuntersuchungen am Meteorologischen Observatorium Wahnsdorf. Abh. Meteor. Hydrol. Dienst Dtsch. Dem. Repbl. **5**, 34 (1955). — Methodik, Diskussion der Meßergebnisse mit den meteorologischen Elementen. Ausführliche Bibliographie zum Problem.

[43] TØNSBERG u. K. OHLSEN: Investigations on Atmospheric Ozone at Nordlysobservatoriet, Tromsø. Geofys. Publ. **13**, Nr. 12 (1944). — Eingehende Darstellung der Meßreihen 1935—1942 mit Diskussion der meteorologischen Einflüsse.

[44] VASSY, E.: Sur quelques propriétées de l'ozone et leurs conséquences géophysiques. Ann. Phys. **8**, 679 (1937). — Umfassende Darstellung, vor allem der Einflüsse der Temperatur auf das Ozon und seine Absorption.

[45] VASSY, A.: Sur l'absorption atmosphérique dans l'ultra-violet. Ann. Phys. **16**, 145 (1941). — Eingehende Untersuchungen, vor allem der Absorption in der Ozon-Sauerstofflücke.

[46] VIGROUX, E.: Contribution à l'étude experimentale de l'absorption de l'ozone. Ann. Physique **8**, 709 (1953). — Umfassende Studie des Ozonabsorptionsspektrums von 2300 Å bis 10000 Å mit eingehender Darstellung der Versuche und der Meßergebnisse.

Geophysical Aspects of Meteors.

By

A. C. B. Lovell.

With 22 Figures.

1. Introduction. Although few meteors are visible to the unaided eye under normal clear sky conditions the total number entering the earth's atmosphere every day is very great. The majority of these meteors are minute pieces of stone or iron moving in closed orbits in the solar system. Occasionally the earth encounters concentrated groups of meteors known as *showers*, and these provide the most spectacular part of meteoric phenomena in which the visible hourly rate may rise to a hundred for a few nights. Even so these shower meteors contribute not more than 10 to 20% of the total meteoric influx; the remainder move in ungrouped orbits and are classed as *sporadic meteors*. There has, in the past, been a cleavage of opinion about the origin of the sporadic meteors. Some workers believed that they have an origin in interstellar space and move in hyperbolic orbits around the sun. However, the radio echo measurements during the past few years have not supported this belief and it is now generally agreed that these sporadic meteors move in short period orbits around the sun and essentially belong to the solar system. With regard to the shower meteors, many of these move in cometary orbits, although the relationship with particular comets is not clear in many cases.

The investigation of these astronomical aspects is an important and vigorous aspect of astronomy which cannot be considered within the scope of this chapter. The subject has been fully covered in a recent work[1], and here we are more concerned with the effect of meteors when they enter the atmosphere. The kinetic energy of the meteoric body is dissipated in heat, light and ionization. The major number of meteors are completely evaporated in the region of 80 to 110 km. above the earth's surface and the study of the light and ionized trail gives valuable information about the atmospheric conditions in this region. The following sections will first deal with the process by which the meteoric body is evaporated. The technique of measurement of the light and the ionized column will then be discussed and the subsequent sections will survey the conditions revealed in the high atmosphere from the study of these meteoric phenomena.

2. The evaporation of the meteoric body. The masses of the individual meteoric bodies in the visual range (say from -3 to $+5$ or $+6$ magnitude), and in the infra visual range down to $+12$ magnitude probably vary from about 1 gram to 10^{-6} gram. There is, at present, considerable uncertainty about the density of the meteoric material, but the diameters of the individual meteors before they enter the atmosphere must be very small. As an example of the orders of magnitude involved it is probable that a meteor on the limit of naked eye visibility (magnitude $+5$) has a mass of about 10^{-3} or 10^{-4} gram and a radius of about 10^{-2} cm. The mean free paths of the atmospheric atoms in the height range of

[1] A. C. B. Lovell: Meteor Astronomy. Oxford 1954.

80 to 110 km. in which these meteors evaporate are therefore large compared with the size of the meteor. Consequently the evaporation process may be considered to result from the impact of single air atoms on the meteor, which enters the atmosphere at velocities from 10 km./sec. to 72 km./sec. The original investigators considered that a compression cap would be formed in front of the meteor, but it now seems certain that except for very big meteors which appear as bright fireballs, or which reach the earth as meteorites, this view is untenable.

The modern theory of the evaporation process has been summarised by WHIPPLE[2] and HERLOFSON[3], and has been developed in considerable detail in the series of papers on *The Physical Theory of Meteors* from Harvard by WHIPPLE and his collaborators[4-11]. Qualitatively, the process is that the surface of the meteor is bombarded by single air molecules a great proportion of which are trapped in the meteor, and give up their kinetic energy as heat. This process brings the meteor to the temperature of evaporation and atoms distil off the meteor with thermal velocities relative to the meteor surface. The binding energies of the meteor atoms will be a few electron volts only, hence the heat supplied by each trapped molecule will be sufficient to evaporate a great number of atoms, and the retardation is small during the evaporation process. The evaporated meteor atoms move into the surrounding atmosphere with the velocity of the meteor and are slowed down by collision with the air molecules. The energy of these evaporated atoms will be determined by the meteor velocity and will vary from about 10^2 to 10^3 eV. In the collisions with the air molecules this kinetic energy is converted into *heat, light* and *ionization.*

Quantitatively the basic *evaporation equations* are obtained by writing down the conditions for the conservation of momentum, and the equation for the heat transfer between the impacting molecules and the latent heat carried away by the evaporated atoms. The rate of evaporation of meteor atoms n at any point is given by

$$n = \frac{1}{\mu}\frac{dm}{d} \tag{2.1}$$

and if p is the atmospheric pressure at this point it can be shown that

$$n = \frac{m\,v\cos\chi}{\mu H}\cdot\frac{p}{p_{\max}}\cdot\left(1-\frac{1}{3}\frac{p}{p_{\max}}\right)^2 \tag{2.2}$$

where m and v are the initial mass and velocity of the meteor before it enters the atmosphere at zenith angle χ. μ is the mass of an individual meteor atom, H the atmospheric scale height (expressing the decrease of atmospheric pressure p with height h, by $p = p_0 e^{-h/H}$). $p_{\max}$ is the atmospheric pressure at the point of maximum evaporation and is given by

$$p_{\max} = \frac{2l\,g}{\Lambda v^2 A}\cdot m^{\frac{1}{3}}\cos\chi \tag{2.3}$$

where l is the latent heat of evaporation of the meteor, g the gravitational acceleration, and A a shape factor defined by $A = \mathcal{A} m^{-\frac{2}{3}}$ where $\mathcal{A}$ is the cross

[2] F. L. WHIPPLE: Rev. Mod. Phys. **15**, 246 (1943).
[3] N. HERLOFSON: Phys. Soc. Rep. Progr. Phys. **11**, 444 (1948).
[4] M. A. COOK, H. EYRING and R. N. THOMAS: Astrophys. J. **113**, 475 (1951).
[5] F. L. WHIPPLE, and R. N. THOMAS: Astrophys. J. **114**, 448 (1951).
[6] R. N. THOMAS: Astrophys. J. **116**, 203 (1952).
[7] R. N. THOMAS, and W. C. WHITE: Astrophys. J. **118**, 555 (1953).
[8] H. J. SMITH: Astrophys. J. **119**, 438 (1954).
[9] A. F. COOK: Astrophys. J. **120**, 572 (1954).
[10] F. L. WHIPPLE: Astrophys. J. **121**, 241 (1955).
[11] L. JACCHIA: Astrophys. J. **121**, 521 (1955).

sectional area normal to the line of flight. Λ is the heat transfer coefficient. In the case where all impinging air molecules are trapped in the lattice $\Lambda = 1$. The actual value of Λ probably lies between 0.7 and 1.

If β is the probability that a single evaporated meteor atom produces a free electron in collision with the air molecules, then the line density of electrons produced by the meteor is, from (2.1)

$$\alpha = \frac{\beta}{\mu v} \cdot \frac{dm}{dt} \tag{2.4}$$

and from (2.2) and (2.3) at the point of maximum evaporation

$$\alpha_{\max} = \frac{\beta n_{\max}}{v} = \frac{4}{9} \frac{\beta}{\mu H} m \cos \chi \tag{2.5}$$

and at any point along the track

$$\alpha = \frac{\beta m \cos \chi}{\mu H} \cdot \frac{p}{p_{\max}} \left(1 - \frac{1}{3} \frac{p}{p_{\max}}\right)^2. \tag{2.6}$$

The energy emitted as light may be obtained by assuming that the luminosity is proportional to the kinetic energy of the mass lost. If τ is the luminous efficiency, the intensity of the visual radiation from the meteor is then given by

$$I = \frac{1}{2} \tau v^2 \frac{dm}{dt} \tag{2.7}$$

with forms corresponding to (2.5) and (2.6) in terms of p and $p_{\max}$.

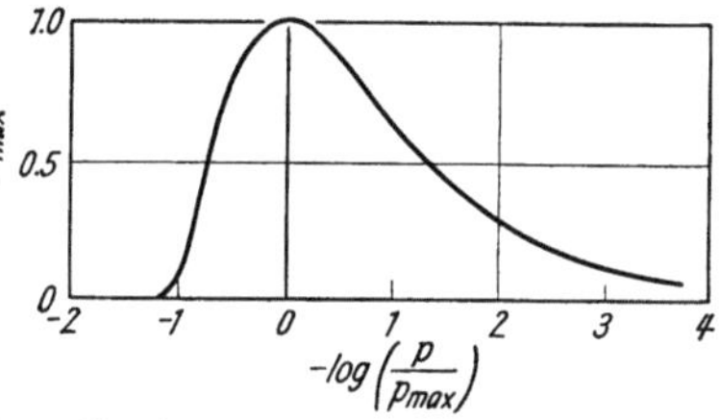

Fig. 1. The shape of the ionization curve produced by a meteor as it penetrates the atmosphere. α is the electron line density at atmospheric pressure p, and $\alpha_{\max}$ is the maximum electron line density produced at the height where the pressure is $p_{\max}$.

The general shape of the light and ionization curve predicted by the above theory is shown in Fig. 1. Considering the simplifying assumptions made, the experimental curves show reasonable general agreement. There is much uncertainty over the values of the ionizing efficiency β and the luminous efficiency τ, neither is the extent of the dependence of β and τ on the velocity yet agreed. By comparison of visual luminosity data and radio echo measurements of α, GREENHOW and HAWKINS[12] suggested that values of $\beta \simeq 0.2$ and of $\tau \simeq 10^{-3}$ gave satisfactory agreement with the observations. More recently MILLMAN and McKINLEY[13] have discussed the radio echo and visual data from over 3000 observations. Experimental work now in progress involving simultaneous photographic and radio echo studies of individual meteors should give more accurate information about β and τ and their dependence on velocity. Apart from the work on meteor ionization such results would be of considerable interest because of the difficulties of studying the transition probabilities in the collision of heavy atoms in the 100 to 1000 eV region in the laboratory.

The basic theory set out above will not apply to the extremes of meteor sizes. Meteors with masses of more than a gram are rather rare. They appear as brilliant *fireballs*, and if large enough penetrate the entire atmosphere before evaporation is complete. In such cases the fundamental assumption that the free path is large compared with the meteor becomes invalid, and a compression cap probably forms in front of the meteor. The appropriate theories are not well developed and will not be considered here. At the other extreme there exist

[12] J. S. GREENHOW, and G. S. HAWKINS: Nature, Lond. **170**, 355 (1952).
[13] P. M. MILLMAN and D. W. R. McKINLEY: Canad. J. Phys. **34**, 50 (1956).

large numbers of meteors with radii of 10^{-3} cm. or less which have a ratio of surface area to mass large enough to ensure that the energy of interaction with the atmosphere is radiated away so rapidly that evaporation does not occur. These so-called *micrometeorites* are stopped before appreciable ablation occurs, and eventually fall slowly through the atmosphere to reach the earth as very fine dust. The study of these micrometeorites is in an early stage. They have been investigated from the theoretical aspect by WHIPPLE[14], and will not be considered further here.

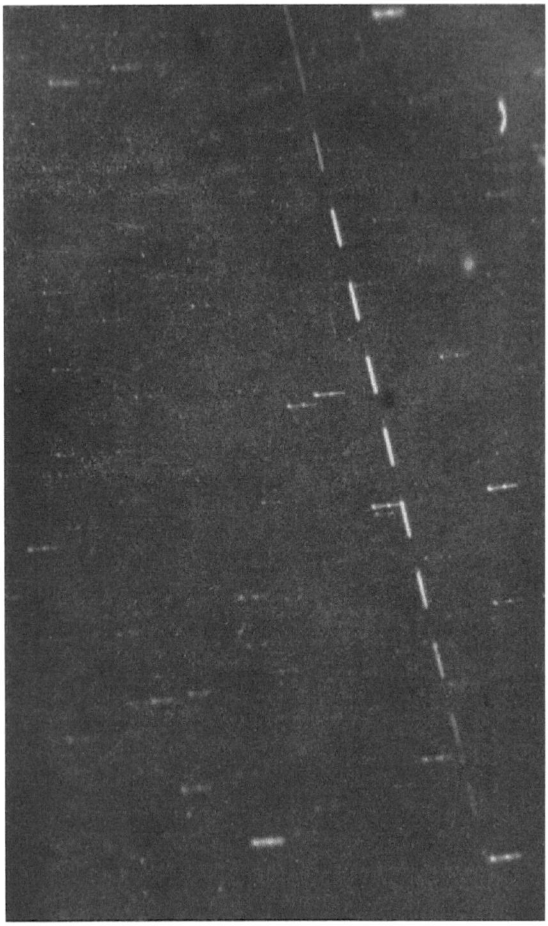

Fig. 2a. A Perseid meteor photographed with a Schmidt camera at Jodrell Bank on 1955 August 11. The camera shutter was rotating at 1500 r.p.m. and made 25 breaks per second in the trail. The effective aperture of the camera was $F = 0.78$ and the field of view 56°. Radio echo data were obtained simultaneously from the ionized trail.

3. Techniques of measurement. From the above estimates of β and τ it is clear that the major part of the kinetic energy of the meteor is probably dissipated in heat. However, no methods have yet been suggested for measuring the heat production, and the only methods at present available for the study of meteors are by the observation of either the luminous or the ionized trail. Much of the pioneer work on meteors was done purely by visual observation at night under clear sky conditions. Important results were obtained as regards the astronomy of meteors but the contribution of this type of visual observation to the geophysical aspects has been small compared with the recent photographic and radio echo studies. A full account of the contemporary photographic and radio techniques has been given by LOVELL[1], and here it is only possible to give a brief summary of the basic methods.

WHIPPLE at Harvard has been primarily responsible for the development of the *precision photographic methods* now in use. The meteor trail is photographed simultaneously from two cameras over a base line of about 40 km., so that its trajectory can be determined. The velocity and deceleration are determined by occulting the lens with a shutter rotating at about 1800 rpm so that the photographed trail is segmented. A typical photograph is shown in Fig. 2a. The attainment of sufficient sensitivity and field of view for an adequate photographic

[14] F. L. WHIPPLE: Proc. Nat. Acad. Sci. Wash. **36**, 687 (1950); **37**, 19 (1951).

rate is a major problem. For example the conventional cameras used in the earlier Harvard programme could only photograph meteors with magnitudes brighter than zero and the average yield was about 1 meteor per hundred hours exposure. Since 1950, however, the photographic techniques have been revolutionised by the development of *Super Schmidt meteor cameras*. The ones now in use in New Mexico for the Harvard programme have a field of view of 55°

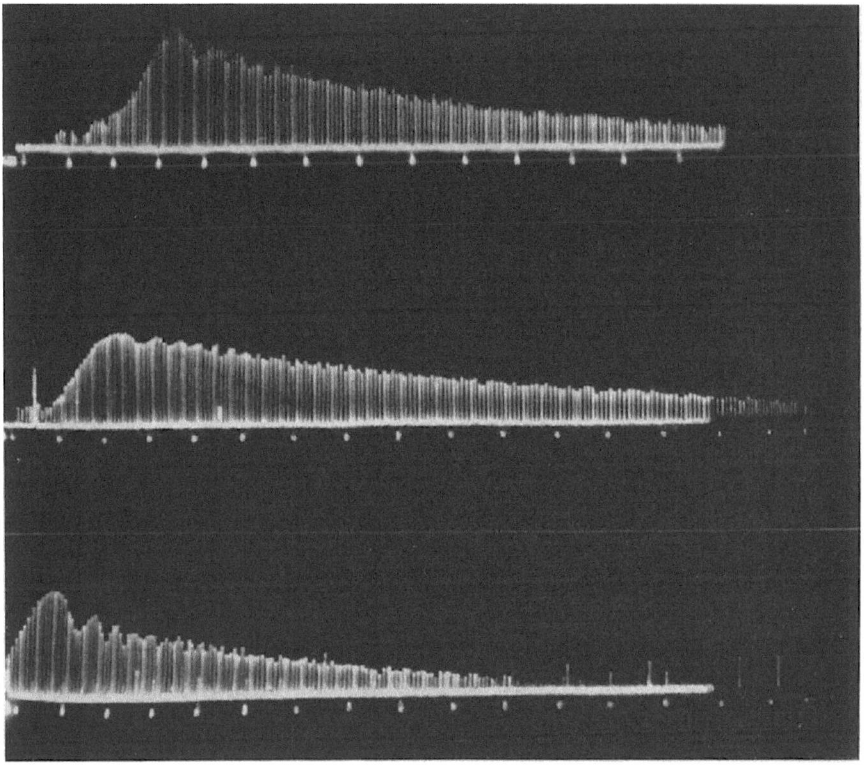

Fig. 2b. The radio echo from a meteor trail recorded simultaneously at three stations triangularly situated at spacings of 4 km. The vertical deflection is proportional to the amplitude of the echo and the time markers were at intervals of 1/50th second on the horizontal axis. The pulse recurrence frequency was 600 per second so that the individual pulses are separated by 1.66 milliseconds. The amplitude variations are the FRESNEL diffraction patterns from which the velocity of the meteor can be determined. The time separation of the three patterns enables the orbit of the meteor to be calculated. The range is measured on a separate display.

with a focal ratio of 0.65. The limiting magnitude is +3 or +4 and the yield of photographed meteors is some 50 or 100 times better than that obtained with the conventional cameras.

The *radio echo technique* for the study of meteors has been largely developed since 1945. The basis of the method consists of the transmission of pulses of radio waves through a suitable aerial system. The radio frequency is chosen to be well above the critical frequency of the ionosphere, and frequencies in the region 30 to 80 Mc/s are commonly used. The radio wave is scattered from the ionized meteor trail and the returned signal is received as a transient echo which may either be viewed on a cathode ray tube display or recorded photographically. A typical record from a single meteor is shown in Fig. 2b. This apparatus uses three receiving stations, triangularly spaced at 4 km. and the FRESNEL zone patterns as the meteor passes through each aerial beam are recorded on a single film. The velocity of the meteor is determined from the amplitude-time variations of the zone pattern and the displacement in time of the three patterns enables the trail to be positioned in space. This is possible because the reflection is specular, and the range is also measured. Thus complete orbital data can be determined for individual meteors and in the apparatus developed by DAVIES and GILL[15] at Jodrell Bank, between 200 and 300 such orbits could be determined in a 24 hour period. Some other developments of this radio echo technique for particular purposes will be mentioned later. Compared with the photographic techniques the radio echo method has the advantage of greater sensitivity—for example, it is readily possible to observe meteors down to magnitude +10 or fainter—and observations can be made under any sky conditions. It does not, however, possess the precision of the photographic techniques for studies of individual meteors.

4. The scattering of radio waves from the ionized column. The observation of the behaviour of the radio echo obtained from the ionized meteor trail is important in the study of the geophysical aspects of meteors. The theoretical ideas on which these studies are based will be discussed in this section.

It is desired to calculate the amount of energy which is scattered back to the radio receiver from an ionized meteor trail containing α electrons per cm. path. This will depend on the peak power of the transmitter P, the radio wavelength λ, the range of the meteor trail R and the power gain G over a half wave dipole of the transmitter and receiving aerials*. The calculation was first carried out by LOVELL and CLEGG[16] on the assumption that the density in the trail was at all times below the critical value for the wavelength in question, and that the electrons scattered independently and coherently. For a cluster of N electrons scattering under these conditions it can be shown that the amount of power scattered back to the receiver will be given by

$$\omega = \frac{8}{3}\pi\left(\frac{e^2}{m c^2}\right)^2 N^2 \frac{P\cdot G^2 \lambda^2}{16\pi^3 R^4} \tag{4.1}$$

where e and m are the electronic charge and mass and c is the velocity of light. LOVELL and CLEGG assumed that the diameter of the ionized trail was small compared with λ and calculated the number of electrons N which would be effective in the scattering process by applying optical diffraction theory. In terms of α it can be shown by using FRESNEL'S integral that

$$N = \alpha\sqrt{\lambda R/2} \tag{4.2}$$

and hence (4.1) reduces to

$$\omega = \frac{\alpha^2 P G^2 \lambda^3}{12\pi^2 R^3}\left(\frac{e^2}{m c^2}\right)^2. \tag{4.3}$$

The parameters P, G, λ are known for any particular equipment, and the received power ω can be measured from the amplitude of the radio echo. R is also measured for each individual meteor and hence the line density α can be determined. It is important to note that the reflection is specular and that the value of α determined in this way relates to an arbitrary point on the ionization curve of Fig. 1. This matter will be considered further in Sect. 7 and 8.

---

[15] J. G. DAVIES and J. C. GILL: Month. Not. Roy. Astronom. Soc. **116**, 105 (1956).

* For convenience the transmitting and receiving aerials are assumed to have the same power gain G. It is also convenient to express the gain in terms of a half wave dipole which itself has a gain of 1.64 over an isotropic radiator.

[16] A. C. B. LOVELL and J. A. CLEGG: Proc. Phys. Soc. Lond. **60**, 491 (1948).

More comprehensive scattering theories were subsequently developed by HERLOFSON[17], KAISER and CLOSS[18] and ESHELMAN[19]. It has been shown that the assumptions made in developing (4.3) were valid only when the line density α was less than about 10^{12} electrons/cm., and that for higher densities the trail could be treated as a metallic cylinder in which case (4.3) becomes

$$\omega = \frac{\alpha^{\frac{1}{2}} P G^2 \lambda^3}{20\pi^3 R^3} \left(\frac{e^2}{m c^2}\right)^{\frac{1}{2}}. \tag{4.4}$$

This is, in fact, now known to be the condition for all trails in the visual magnitude range, and that (4.3) applies only for meteors fainter than about +5 magnitude. Plasma resonance effects may also occur in trails with $\alpha < 10^{12}$ electrons/cm. when the electric vector is transverse to the ionized column. Experimental work has now fully justified the predictions of the above theories (see for example GREENHOW[20]), including the occurrence of the plasma resonance effects[21].

The main contribution to the reflection will come from the few FRESNEL zones in the vicinity of the specular reflection point. As the meteor crosses this point the formation of the ionized column through the successive zones gives the amplitude-time pattern illustrated in Fig. 2b, which enables the velocity of the meteor to be determined. Subsequently the trail will diffuse and the decay of echo amplitude is often superimposed on the zone pattern. It can be shown[17] that when formula (4.3) applies (that is for trails with $\alpha < 10^{12}$ electrons/cm.) the decay of echo amplitude is given by

$$V = V_0 \exp\left(\frac{-16\pi^2 D t}{\lambda^2}\right) \tag{4.5}$$

where V_0 is the initial amplitude, V the amplitude after time t and D the diffusion coefficient. The decay for this type of trail depends on λ^2 but is independent of α. On the other hand for meteors in the visual range for which $\alpha > 10^{12}$ electrons/cm. it can be shown[18, 20] that the duration of the echo ΔT varies with α and is given by

$$\Delta T = \frac{\alpha \lambda^2}{4\pi^2 D} \left(\frac{e^2}{m c^2}\right). \tag{4.6}$$

Further it is important to note that when $\alpha < 10^{12}$, the signal amplitude $V(::\sqrt{\omega})$ is directly related to α, but for $\alpha > 10^{12}$ the signal amplitude varies only as $\alpha^{\frac{1}{4}}$ and therefore provides only a very insensitive measure of this quantity. In these cases it is customary to use the duration ΔT which varies directly as α [Eq. (4.6)], in measurements of the line density.

5. The diurnal and seasonal variation of the meteor influx. In the earlier visual meteor observations it was evident that the sporadic meteor rate showed a diurnal and seasonal variation, and that the curve of variation was dominated by great increases in activity at irregular intervals due to the occurrence of major showers. It was generally believed that the sporadic meteor distribution was uniform so that for an earth at rest the number of meteors visible at any point would be constant. But the earth's orbital motion leads to an apparent concentration of radiants around the apex of the earth's way so that the observed meteor rate should show a diurnal variation with a maximum at 06 hours local time and a seasonal variation with a maximum in the autumn months (that is

[17] N. HERLOFSON: Ark. Fysik 3, 247 (1951).
[18] T. R. KAISER and R. L. CLOSS: Phil. Mag. 43, 1 (1952).
[19] V. R. ESHELMAN: Techn. Rep. 49, Electronics Res. Lab., Stanford 1952.
[20] J. S. GREENHOW: Proc. Phys. Soc. Lond. B 65, 169 (1952).
[21] I. C. BROWNE and E. R. BILLAM: Proc. Phys. Soc. Lond. B 69, 98 (1956).

in the northern hemisphere). The subject has been discussed in detail by LOVELL[1]. Here it is sufficient to summarise the recent radio echo results which, for the first time, have given the true 24 hour variation without the gaps in the visual observations occasioned by daylight. These contemporary results show that the old belief in the uniform distribution of sporadic meteors is incorrect, and that during the summer daytime in the northern hemisphere* the curve of variation is dominated to a remarkable extent by a concentration of shower and sporadic meteors not previously observable because of daylight.

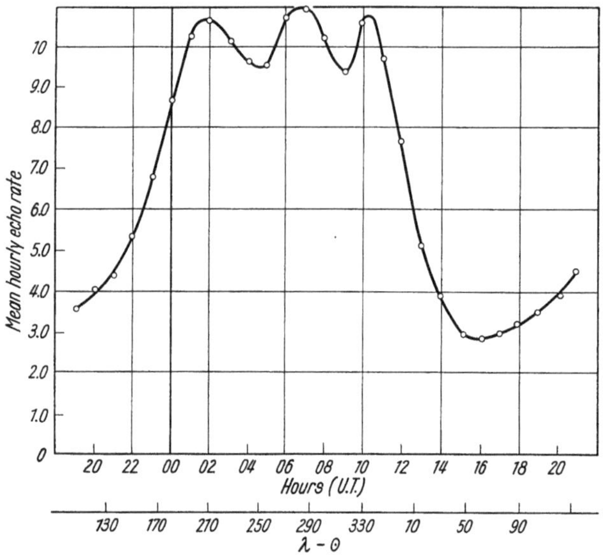

Fig. 3. The mean diurnal variation of the sporadic meteor rates, observed on a radio echo equipment. The abscissae show time in hours (*U.T.*) and in (longitude λ—sun's longitude ⊙).

The main results have been obtained by the continuous operation over a period of several years of two radio echo equipments at Jodrell Bank. One of these uses a rotating aerial which records the total meteor activity in the sky at the rate of two turns per minute. The other apparatus uses two fixed aerials beamed at azimuths 25° north and south of west respectively. The mean *diurnal rate curve* observed in this latter equipment, averaged over all days of the year, is shown in Fig. 3. This shows the concentration towards the apex at 06 h as predicted by the theory of uniform distribution, but in addition there are equal concentrations towards the helion and antihelion points. From this data it is possible to derive the true distribution in longitude of orbital directions of sporadic meteors in space by correcting for the earth's motion[22]. The individual daily hourly rate curves show considerable departures from the maximum at 06 h local time as demanded by the uniform distribution theory. For example, Fig. 4 shows the variation in time of the daily maximum in hourly rate throughout the year. In this analysis the effect of the major meteor showers has been excluded; the inclusion of the summer daytime showers would increase still further the variation of time of maximum shown in Fig. 4.

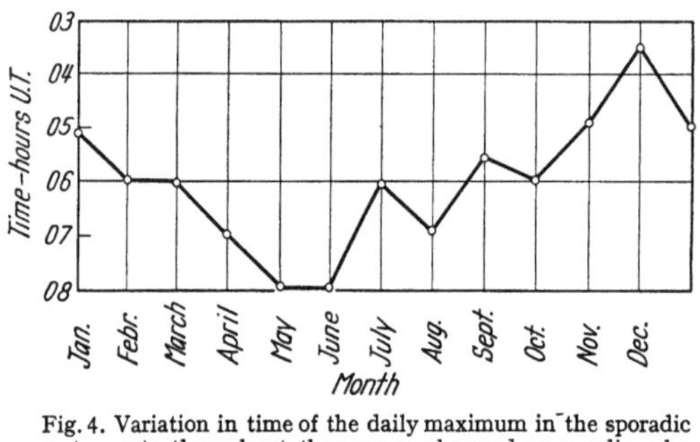

Fig. 4. Variation in time of the daily maximum in the sporadic meteor rate throughout the year as observed on a radio echo apparatus.

The *seasonal variation*, again excluding the major showers, is shown in Fig. 5. This reveals that in the northern hemisphere the seasonal maximum occurs

* The radio echo results from the southern hemisphere are not yet complete.

[22] G. S. HAWKINS: Month. Not. Roy. Astronom. Soc. **116**, 92 (1956).

in the summer months and not during the autumn. HAWKINS[22] has shown that this seasonal curve is well correlated with the variation of the cometary index. These radio echo surveys do not give the distribution in ecliptic latitude, but the visual results of PRENTICE[23] indicate that about 40% of the meteors are concentrated within $\pm 15°$ of the ecliptic plane. However, in the recent radio work by DAVIES[24] in which the orbits of individual sporadic meteors are measured, 50% of the meteors are concentrated within 15° of the ecliptic. An additional 15% move in rather circular orbits with inclinations grouped around 60° and 120°. This latter class appear to be quite absent in meteors brighter than magnitude +6. The cosmological reason for this type of distribution is not, at present, understood.

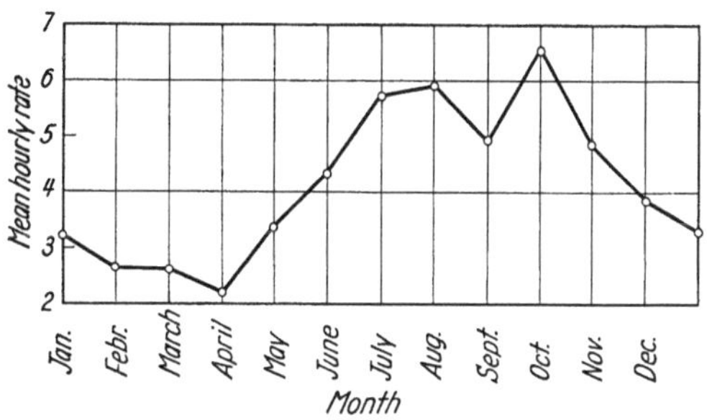

Fig. 5. The seasonal variation of sporadic meteor activity as recorded on a radio echo apparatus.

The occurrence of the *major meteor showers* gives high meteor rates for short periods of time and although they supply the most prominent visual phenomena their contribution to the total annual meteor influx is small compared with the sporadic contribution. The visual, photographic and radio echo data available at present on the major showers have been described by LOVELL[1] together with a discussion of their origin. For the purpose of this chapter the data are summarised in Table 1, and Fig. 6 for showers which are active at present. Table 1 gives

Table 1.

Shower	Date of maximum	Approximate hourly rate	Comments
Quadrantids	Jan. 3—4	50—100	
Lyrids	April 21	5—10	
η-Aquarids	May 4—6	15	
o-Cetids	May 14—23	10	Daytime shower
Arietids	May 29—June 18	70	Daytime shower
ζ-Perseids	June 1—16	40	Daytime shower
β-Taurids	June 26—July 5	20	Daytime shower
δ-Aquarids	July 28	30	
Perseids	Aug. 5—14	50	
Giacobinids	Oct. 10	Occasionally very great—see DAVIES and LOVELL[25]	
Orionids	Oct. 20—23	15	
Taurids	Nov. 3—10	10	
Leonids	Nov. 16—17	~10	Has produced great displays in the past
Geminids	Dec. 12—13	50—100	
Ursids	Dec. 22	10	

the date of maximum and apparent duration of each shower together with an indication of the hourly rate at maximum (for convenience this is taken as the rate seen by a single observer under clear sky conditions). Fig. 6 shows the influence of the showers on the overall day and night activity as recorded by the radio echo techniques.

[23] J. P. M. PRENTICE: Unpublished (quoted by LOVELL[1]).
[24] J. G. DAVIES: Unpublished.
[25] J. G. DAVIES and A. C. B. LOVELL: Month. Not. Roy. Astronom. Soc. **115**, 23 (1955).

The question of the distribution of meteor masses in the sporadic background and in the showers will be discussed in Sect. 6, and the total mass brought into the atmosphere will be discussed in Sect. 11.

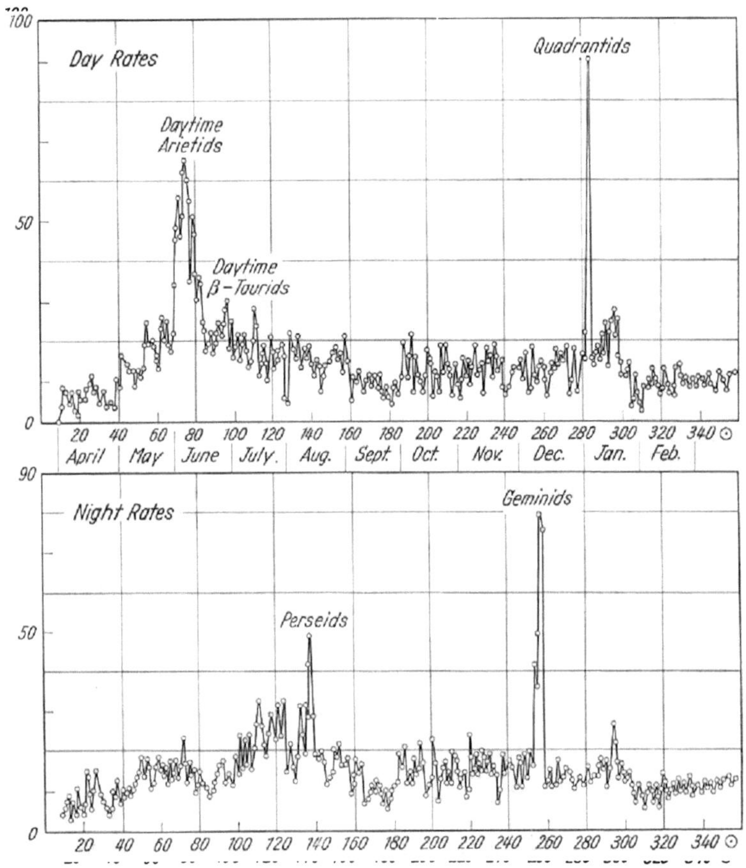

Fig. 6. The variation of meteor activity throughout the year for day and night periods as determined by the radio echo technique. Ordinates are relative hourly rates, abscissae show solar longitude ⊙.

6. The meteor mass distribution. The probable range of meteor masses has already been mentioned in Sect. 2. Here we shall be concerned with the distribution of masses in the shower and sporadic meteors. In the visual observations the basic observational data consist of a count of the number of meteors which appear in different magnitude groups. It was found that the change in number dN with magnitude M could be expressed in the form

$$dN = \chi^M \, dM. \tag{6.1}$$

Extensive visual and telescopic counts (see[1]) indicated that for sporadic meteors $\chi = 2.5$ over a magnitude range of -2 to $+9$. Since the magnitude M is proportional to the logarithm of the light intensity which from Sect. 2 depends directly on the meteor mass m, (6.1) becomes

$$dN = \frac{dm}{m^s} \tag{6.2}$$

Table 2. *The mass distribution (value of the exponent s in Eq. (6.2)) for shower meteors. Columns 1 to 4 are from the radio echo data. Column 5 is from the visual data.*

Shower	Magnitude range M				
	$+7$	$+7>M>+5$	$+3$	$+2>M>0$	$M<+3$
Quadrantids	—	1.78	—	1.8	2.5
Perseids	1.38	1.59	2.0	2.0	2.1
Arietids	—	2.7	—	1.8	—
ζ-Perseids	—	—	—	2.2	—
Geminids	1.45	1.62	2.24	2.3	2.3

for the mass frequency distribution. The relation between χ and s is given by (6.1) and (6.2). Thus for $\chi=2.5$, $s=2.0$, that is the mass distribution follows an inverse square law; and the total mass of meteoric material in each magnitude is roughly constant (see Sect. 11a).

Various methods for the determination of the mass distribution using the radio echo technique have been developed[26] and these agree with the visual observations in assigning a value of $s=2.0$ for the sporadic mass distribution in the magnitude range 0 to $+10$. Our knowledge of the mass distribution in the showers depends almost entirely on the radio methods, apart from some visual and photographic work on bright Perseid, Leonid and Giacobinid meteors[1]. The existing data are still only preliminary and almost our entire knowledge of the mass distributions of the shower meteors is summarised in Table 2 taken from Browne et al.[26].

It is apparent that compared with the sporadic meteors the value of s varies with magnitude and that with the exception of the daytime Arietids the showers appear to be deficient in faint meteors. Moreover, for the Quadrantids, Perseids and Geminids the greater part of the meteoric mass appears to be concentrated in the magnitude range $+2$ to $+5$ whereas in the case of the Arietids (and probably the ζ-Perseids) the main mass is contributed by meteors fainter than magnitude $+7$. Cosmologically these variations are probably connected with the age of the showers. From the geophysical aspect it is important to note that the estimates of the mass of meteors entering the atmosphere depend critically on the values of s, especially with regard to the relative influence of the shower meteors and sporadic background on the upper atmospheric and ionospheric phenomena.

7. Meteor heights. Eq. (2.3) predicts that the height at which a meteor evaporates should depend on its mass and velocity. The uncertainties and difficulties in the visual measurement of heights led to a widely different interpretation of the data, but the conclusions reached by Porter[27] from an analysis of the British meteor data are now known to be substantially correct. He concluded that the height of appearance and disappearance was determined mainly by the elongation and magnitude, quantities directly related to the velocity and mass respectively, and that shower and sporadic meteors behaved in an identical manner.

With the development of the radio echo techniques it became possible to measure meteor heights on a large scale, and at the same time to measure the velocity of the individual meteors. The apparatus in use at Jodrell Bank for this purpose has been described by Clegg and Davidson[28]. The reflection from the trail is specular and the height measured for individual meteors is at some arbitrary point on the trail. Thus, even for meteors of identical mass and velo-

[26] I. C. Browne, K. Bullough, S. Evans and T. R. Kaiser: Proc. Phys. Soc. Lond. B 69, 83 (1956).

city the heights measured by this apparatus would show a distribution about the mean. This will be discussed in Sect. 8. The effect of meteor velocity on the height can be seen from the analysis of the measurements on sporadic meteors, for which the mass distribution is known to be the same over the range of magnitudes involved. Some results are shown in Fig. 7 which gives the measured height distribution for different velocity groups of sporadic meteors.

The effect of meteor mass on the height distribution can be seen in the results from the shower meteors for which, as shown in Table 2, the mass distribution is often different from that of the sporadic meteors. Two extreme cases are illustrated in Fig. 8 which shows the height distribution for the

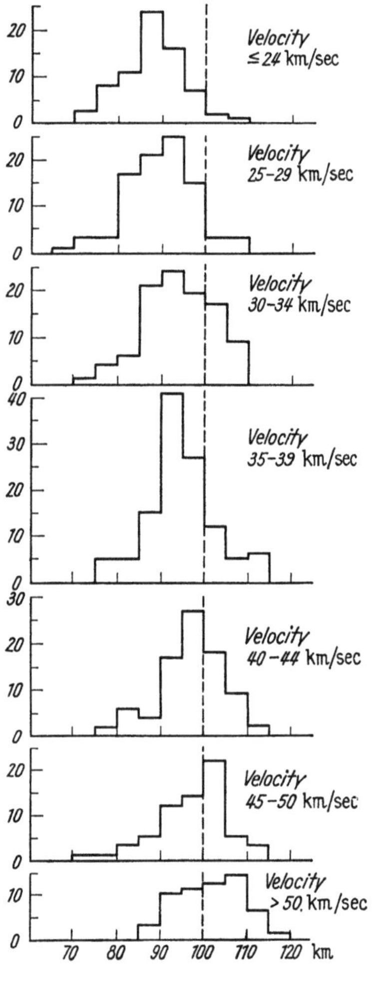

Fig. 7. Height distributions for different velocity groups of sporadic meteors as determined by the radio echo technique. Ordinates are relative numbers, abscissae are the heights in km.

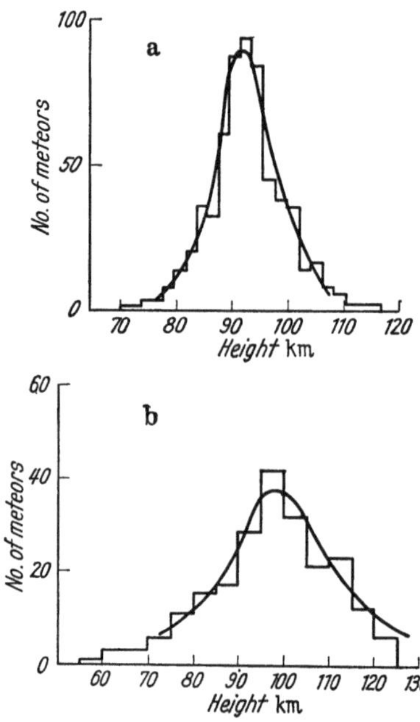

Fig. 8a and b. Radio echo measurements of the height distributions (a) for the daytime Arietid and (b) for the Perseid meteor showers, showing the effect of different mass distributions.

Arietids (velocity 38 km./sec.) in which faint meteors predominate, and for the Perseids (velocity 60 km./sec.) in which bright meteors are prominent. In each case the mean heights are in close agreement with the mean height for the appropriate velocity group of sporadic meteors in Fig. 7 but the spread in the distribution occasioned by the different mass distribution is very marked.

The use of such measurements in the determination of upper atmospheric scale heights and densities will be discussed below.

[27] J. G. Porter: Month. Not. Roy. Astronom. Soc. **103**, 134 (1943); **104**, 257 (1944).
[28] J. A. Clegg and I. A. Davidson: Phil. Mag. **41**, 77 (1950).

8. Determination of scale heights and densities. *α) Radio echo methods.* The full theory of the distribution of meteor heights observed on a radio echo apparatus has been given by KAISER[29]. The theory includes the effect of velocity and mass on the distributions referred to in Sect. 7, and also the effect of the aerial polar diagrams on the observations. This work is important because it leads to methods for the determination of scale heights and densities in the upper atmosphere. Firstly it is shown that for a given velocity group of meteors the width of the height distribution depends only on the mass distribution of the incident meteors and the atmospheric scale height. Qualitatively the reason for this can readily be seen since for a given velocity, large meteors require to intercept a greater air mass than small ones before evaporation is complete and can therefore be detected over a greater height range. Further this extension in height for a given mass depends on the atmospheric pressure gradient or scale height H. Experimentally it is most convenient to determine H from the observations on sporadic meteors for which the mass distribution has been well established [$s=2.0$ in Eq. (6.2)]. KAISER shows that in this case the r.m.s. deviation δh from the mean height distribution is given by

$$\delta h = (1.05 \pm 0.02)\, H \tag{8.1}$$

and hence H can be determined at various heights from the kind of data given in Fig. 7.

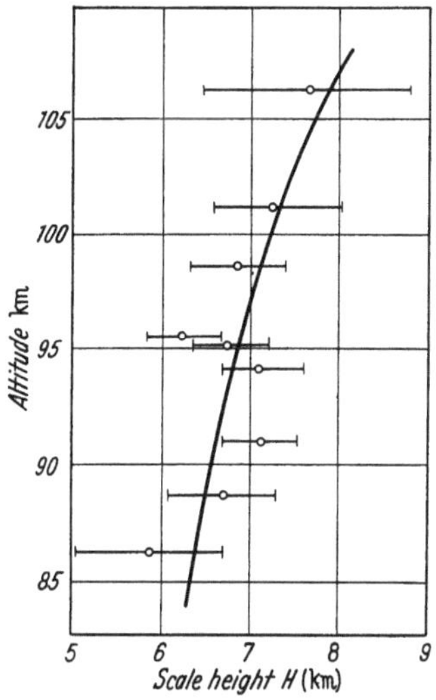

Fig. 9. The variation of scale height H with altitude as determined by the radio echo measurements on sporadic meteors. The full line is the variation determined by the rocket experiments.

The experimental investigation was made by EVANS[30], and the results for the variation of H with altitude are given in Fig. 9. The full line in this diagram is the profile adopted by the ROCKET PANEL[31] and it is seen that the two sets of results are in good agreement. A preliminary investigation of any seasonal variation has been made but none exceeding ± 0.5 km. ($\pm 17°$ K) has been found. An investigation of greater accuracy is quite possible. It would involve a greater number of observations and a detailed study to establish if any seasonal variation in the value of s for sporadic meteors exists.

Secondly, the investigations made by KAISER show that it is possible to use the radio echo measurements to determine the variation of pressure with height. Eq. (2.3) and (2.5) give

$$p_{\max} = \frac{2gl}{\Lambda A} \cdot \frac{\cos^{\frac{2}{3}}\chi}{v^2} \left(\frac{9H\mu\,\alpha_{\max}}{4\beta}\right)^{\frac{1}{3}} \tag{8.2}$$

and, in principle, this expression can be used to measure a series of values of p at different altitudes. In the experimental measurement of heights, however, the reflecting point does not necessarily lie at the point of maximum ionization $\alpha_{\max}$. KAISER has shown that for meteors moving with the same velocity the mean height is close to the height (h_m) at which the smallest detectable meteor produces its maximum ionization. The smallest detectable value of α can be calculated from the parameters of the apparatus using Eq. (4.3), and hence h_m can

[29] T. R. KAISER: Month. Not. Roy. Astronom. Soc. **114**, 39 (1954); **114**, 52 (1954).
[30] S. EVANS: Month, Not. Roy. Astronom. Soc. **114**, 63 (1954). — J. Atmos. a. Terres. Phys. Suppl. **2**, 86 (1955).
[31] ROCKET PANEL: Phys. Rev. **88**, 1027 (1952).

be determined for various values of v. Eq. (8.2) then gives the relation between p and h_m provided the other quantities are known.

The correction to the observed mean height to obtain h_m depends on the mass distribution (s) and the aerial polar diagram. This has been derived by EVANS[30] and the relationship between h_m and v for sporadic meteors is shown in Fig. 10. The minimum detectable line density for the apparatus in question, calculated from (4.3) was $\alpha = 2.3 \times 10^{11}$ electrons/cm. χ is known from the measured elevations and H from Fig. 9. Thus by using Fig. 10, the variation of p with h_m can be determined provided the remaining constants in (8.2)

$$\frac{l}{\Lambda A}\left(\frac{\mu}{\beta}\right)^{\frac{1}{3}} = Q \tag{8.3}$$

can be established. EVANS evaluated Q by using data obtained in the Harvard photographic meteor work and the radio echo-visual meteor correlations [Q

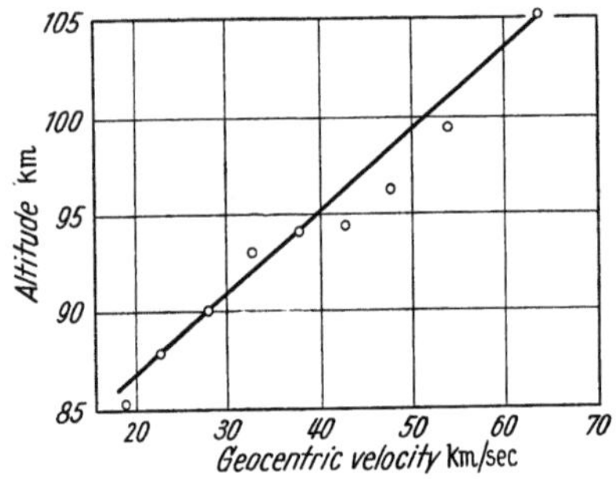

Fig. 10. The relation between the characteristic height h_m, at which the smallest detectable meteor produces its maximum ionization, and the meteor velocity, for the radio echo apparatus used in the atmospheric pressure measurements.

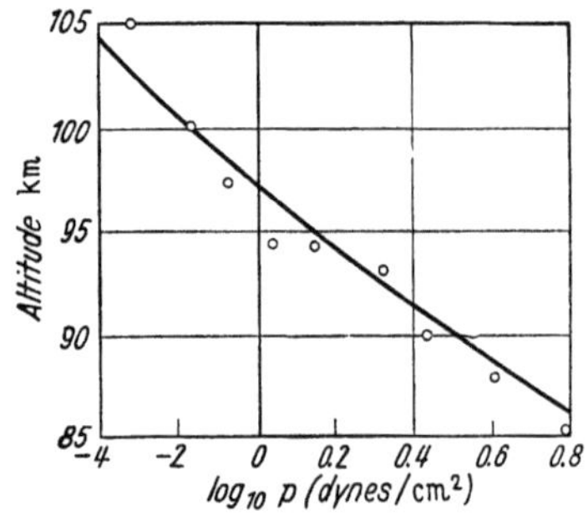

Fig. 11. The variation of atmospheric pressure p with height as determined in the radio echo measurements of meteors. The smooth curve is derived from the scale height data.

corresponds to the factor k in the photographic work—Eq. (8.7) below]. His final results for the variation of p with height are shown in Fig. 11. The smooth curve in Fig. 11 was obtained by integration of the scale height curve (Fig. 9) and was fitted to the experimental points for the minimum algebraic sum of the residuals in log p.

β) Photographic methods. The rotating shutter technique of meteor photography developed at Harvard provides a method for the measurement of the deceleration of the meteor at any point on its path. This has been used extensively by WHIPPLE and his collaborators for the measurement of upper atmospheric densities[32] and also by workers in Czechoslovakia[33].

The basic equation used in the determination of densities ϱ by this method is the *drag equation*

$$m\frac{dv}{dt} = -\gamma \mathcal{A} \varrho v^2. \tag{8.4}$$

The equation is based simply on the principle that the mass of air encountered in time dt is $\mathcal{A}\varrho v$, and the dimensionless drag coefficient γ measures the efficiency of the momentum transfer. (If $\gamma = 1$, and the trapped air molecules escape with negligible relative velocity (8.4) expresses the conservation of momentum). An expression for m is obtained from (2.7). It is assumed that the

---

[32] See for example, L. G. JACCHIA: Harv. Coll. Obs. Techn. Rep. No. 3, 1949; No. 10, 1952.

[33] See for example, Z. CEPLECHA: Bull. Astr. Czecho. 4, 55 (1953). — F. LINK: Bull. Astr. Czecho. 4, 168 (1953).

luminous efficiency τ is proportional to v so that $\tau = \tau_0 v$ where τ_0 is a constant. Then (2.7) becomes

$$I = \frac{1}{2}\,\tau_0 v^3 \frac{dm}{dt}$$

and at any instant t_0

$$m = \frac{2}{\tau_0}\int\limits_{t_0}^{\infty} \frac{I}{v^3}\,dt. \tag{8.5}$$

Since $\mathcal{A} = A\,m^{\frac{2}{3}}$, combination of (8.4) and (8.5) gives

$$\varrho = -\frac{2^{\frac{1}{3}}}{\gamma A \tau_0^{\frac{1}{3}} v^2}\cdot\frac{dv}{dt}\left[\int\limits_{t_0}^{\infty}\left(\frac{I}{v^3}\right)dt\right]^{\frac{1}{3}}. \tag{8.6}$$

All the quantities on the right hand side of (8.6) are observable except for the factor

$$\frac{2^{\frac{1}{3}}}{\gamma A \tau_0^{\frac{1}{3}}} = k. \tag{8.7}$$

The constants in k may be evaluated separately, from theoretical considerations, or k may be determined as a single constant by comparison with atmospheric densities measured by other methods in the 50 to 60 km. region. The agreement is good if values for γ near unity are taken.

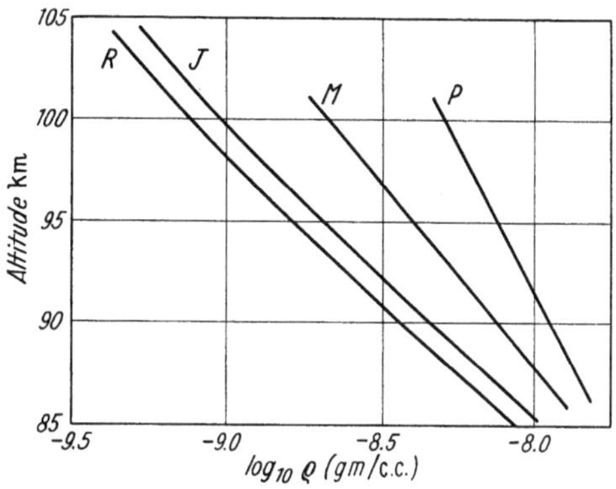

Fig. 12. The variation of upper atmospheric density ϱ with height as determined by various techniques. R rocket and photographic meteor data lat 32° N, J Jodrell Bank radio echo results lat 53° N, M photographic meteor data lat 42° N, P photographic meteor data from Czechoslovakia lat 50° N.

γ) Comparison of results. The photographic meteor results give a direct measure of ϱ, and the radio echo results measure p. For comparison the latter have been converted to densities by using $p = \varrho H g$. The results for the variation of ϱ with height over the range 85 to 105 km. are given in Fig. 12. R is the rocket and photographic meteor results from New Mexico at latitude 32° N, J is the Jodrell Bank radio echo results at latitude 53° N; M is the photographic meteor results at Massuchusetts at latitude 42° N, and P is the curve given by LINK[33] from photographic meteor observations near Prague at latitude 50° N. It is seen that R and J are in good agreement and also that rocket measurements at latitude 77° N[34] are similar to curves R and J. Hence it seems that the differences between M and R (carried out with identical techniques as far as the meteor work is concerned) cannot be due to a latitude effect as at one time believed. The Harvard workers now consider M to be unreliable since it was obtained from early results using an unfavourable selection of meteors. With regard to P, these results were obtained by a different method of reduction and may not be comparable with R. The greatest weight must be attached to R and J and it is particularly satisfactory that the radio echo results are in such reasonable agreement with the rocket and photographic meteor results.

The original Massachusetts meteor data showed a clear cut *seasonal effect*[35] in which the value of ϱ at a mean height of 78 km. followed the normal mean ground temperature variation with season. Later JACCHIA[36] concluded that the

[34] H. E. NEWELL: Report at Xth General Assembly of U.G.G.I. 1954.

[35] F. L. WHIPPLE, L. JACCHIA and Z. KOPAL: The Atmospheres of the Earth and the Planets, chap. 5. Chicago 1947.

[36] L. JACCHIA: Harv. Coll. Obs. Techn. Rep., No. 4, 1949.

amplitude of the seasonal variation decreased with height, and became very uncertain above about 85 km. The radio echo results have been inspected for a seasonal variation[30] but at present the probable error is of the same order as the indicated variation in the height of a given pressure level ($\sim \pm 1$ km.). The same remark applies to the *diurnal variation* where a *semidiurnal* component is indicated, but again the amplitude is little greater than the probable error. The phase, however, is consistent with that of the semidiurnal wind component discussed in Sect. 9.

9. Winds in the upper atmosphere. *α) Visual and photographic meteor observations.* The observation that the persistent trains left by some very bright meteors became distorted in a few minutes provided some of the earliest evidence for the existence of high winds in the upper atmosphere. OLIVIER[37] has made a comprehensive study of 1492 such trains which lasted for more than one minute. He divided the observations into day, twilight and night groups, and found that the mean height of the trains increased from 36 km. during the day, through 61 km. at twilight to 92 km at night. He concluded that the average wind velocity showed a similar increase from about 48 m./sec. by day to 56 m./sec. by night. From the zig-zag appearance of the trails OLIVIER also concluded that the winds were stratified and that a high degree of turbulence probably existed. The frequency of the night wind vectors over America derived by OLIVIER from these observations is shown in Fig. 13. Similar data have been presented by FEDYNSKY[38] for observations made in the U.S.S.R., although these results include some unusually high velocities in the neighbourhood of 3000 m./sec., and the average velocity of 110 m./sec.is also higher.

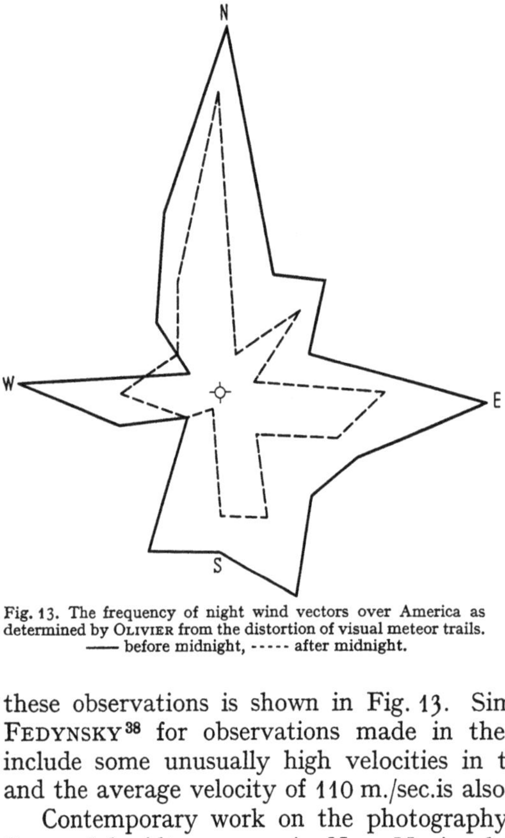

Fig. 13. The frequency of night wind vectors over America as determined by OLIVIER from the distortion of visual meteor trails. —— before midnight, ----- after midnight.

Contemporary work on the photography of persistent meteor trains using Super Schmidt cameras in New Mexico has been described by LILLER and WHIPPLE[39]. The authors describe the results from 5 trains photographed between August 1950 and May 1953. Fig. 14, which is a good example of the results obtained in this work, shows the variation of wind velocity with height for the meteor trains photographed on March 20, 1953. This shows that the velocity is very irregular in its variation with height and that the maximum wind component reaches over 100 m./sec. The mean speed for the 5 trains was 38 m./sec. which is lower than the value given by OLIVIER. However, when corrections are applied for projection effects the mean speed is increased to a value in reasonable agreement with the visual results.

[37] C. P. OLIVIER: Proc. Amer. Phil. Soc. **85**, 93 (1942); **91**, 315 (1948).
[38] V. V. FEDYNSKY: Astr. J. Sov. Un. **21**, 291 (1944).
[39] W. LILLER and F. L. WHIPPLE: J. Atmos. a. Terres. Phys. Suppl. **1**, 112 (1954)

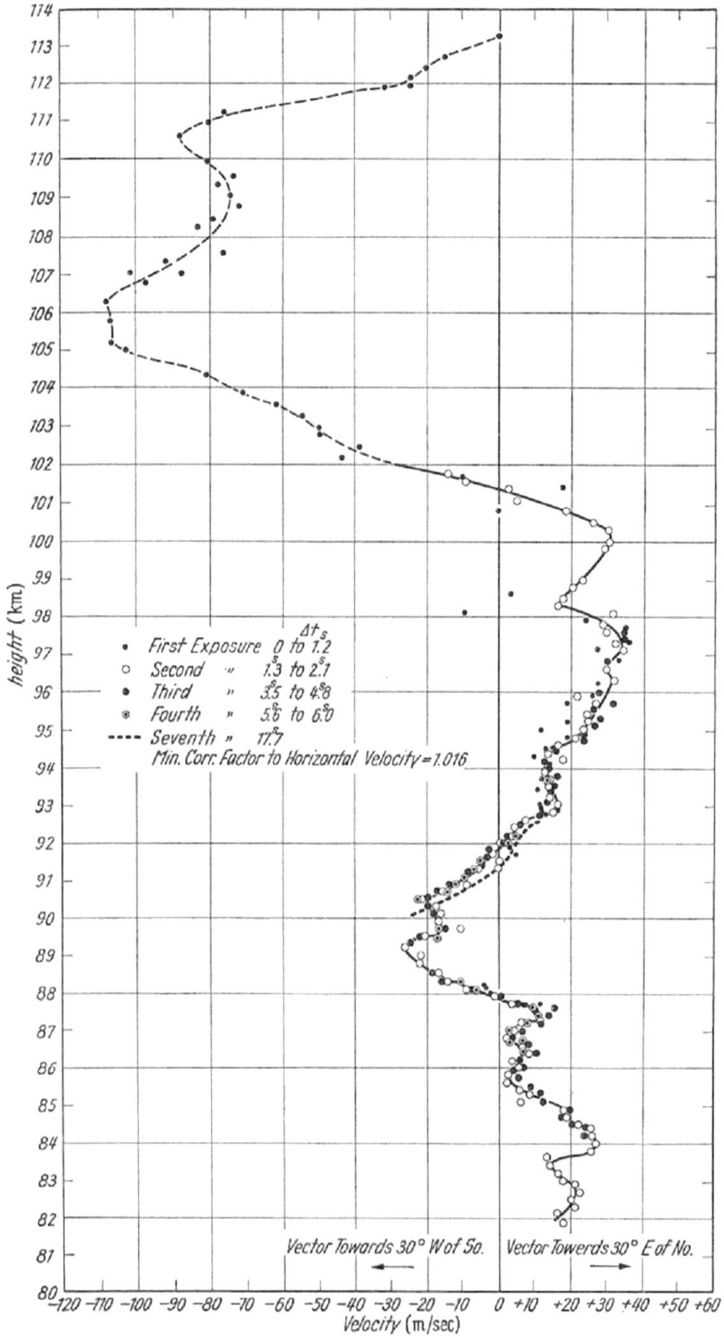

Fig. 14. The variation of wind velocity with height as determined by LILLER and WHIPPLE from a meteor trail photographed over New Mexico on 1953 March 20.

β) Radio echo results. LILLER and WHIPPLE point out that the meteor method of wind measurement possesses an inherent inaccuracy because of the local disturbances introduced by the meteor as it evaporates. Meteors which produce the type of trains discussed above are very bright and may weigh as much as 10 grams. The meteor is moving at, perhaps, 60 km./sec. along a 50 km. path, and hence about 2×10^{-6} gm./cm. is ablated where the air densities are only 10^{-9} gm./c.c. This type of disturbance is probably very small in the case of the radio echo measurements which refer to the very faint or infra visual meteors. The other advantage of the radio technique is, of course, the large amount of data which can be collected in a short time, which makes possible the determination of diurnal and seasonal effects.

As soon as the radio echo techniques began to be used extensively for the study of meteors it was observed that in the case of long duration echoes (i.e. greater than a few seconds), very marked fluctuations in the amplitude of the received echo were nearly always present. It was also found that many of the long duration echoes showed a change in range with time. Many of the workers in this field suggested that winds in the altitude range of 80 to 120 km. might be an important factor in these phenomena. At Jodrell Bank in 1949 to 1950 GREENHOW[40] studied the fluctuations on two wavelengths simultaneously and concluded that they were due to wind gradients in the high atmosphere which caused the trail of ionization to break up into separate reflecting centres. At the same time ELLYETT[41] reviewed the data on range movements observed at Jodrell Bank and elsewhere, and showed that the various types of drifts could be explained in terms of the movement of the ionized column under the influence of these high altitude winds.

The development of a specific radio technique using a continuous wave method was first made at Stanford by MANNING, VILLARD and PETERSON[42] in 1950. From 18 measurements they found velocities in the range from 13 to 58m./sec. with an average of 29 m./sec. The motions seemed to be concentrated along the north-south line and there were indications of differences between the summer and winter results. In Australia, ROBERTSON, LIDDY and ELFORD[43] developed a similar technique and obtained about 50 usable echoes daily. Even so it was necessary to integrate a month's continuous observations for a satisfactory wind determination. These results showed that at the 90 km. level both periodic and prevailing winds were present.

The problem of increasing the data rate to a level necessary for the study of the diurnal variations has been solved by GREENHOW[44], who has combined the advantages of the continuous wave and pulse methods. With this technique it is possible to record over 3000 usable echoes daily and his results now provide by far the most detailed study of these high altitude winds. A sample of the film record obtained is shown in Fig. 15. Four cathode ray tubes are photographed simultaneously for each meteor echo which trips the recording mechanism. *A* gives the range of the meteor on a conventional range-amplitude display. In *B* the speed of the time base is adjusted so that the individual pulses are separated; this gives the FRESNEL diffraction pattern as the meteor crosses the aerial beam and enables the velocity of the meteor to be measured (see Sect. 3). *C* and *D* record the phase variations as the ionized trails drift towards or away from the receiving station. A complete cycle in *C* and *D* represents a phase

[40] J. S. GREENHOW: Phil. Mag. **41**, 682 (1950). — Proc. Phys. Soc. Lond. B **65**, 169 (1952).
[41] C. D. ELLYETT: Phil. Mag. **41**, 694 (1950).
[42] L. A. MANNING, O. G. VILLARD and A. M. PETERSON: Proc. Inst. Radio Engrs. **38**, 877 (1950). — J. Geophys. Res. **57**, 387 (1952).
[43] D. S. ROBERTSON, D. T. LIDDY and W. G. ELFORD: J. Atmos. a. Terres. Phys. **4**, 255 (1953). — W. G. ELFORD and D. S. ROBERTSON: J. Atmos. a. Terres. Phys. **4**, 271 (1953).
[44] J. S. GREENHOW: Phil. Mag. **45**, 471 (1954). — J. S. GREENHOW and E. L. NEUFELD: Phil. Mag. **46**, 549 (1955).

change of 2π caused by a drift of the trail by $\lambda/2$. Since λ is 8 metres, and the separation between each pulse is 0.006 seconds it is clear that high drift speeds can be measured with considerable accuracy. This gives the radial component of

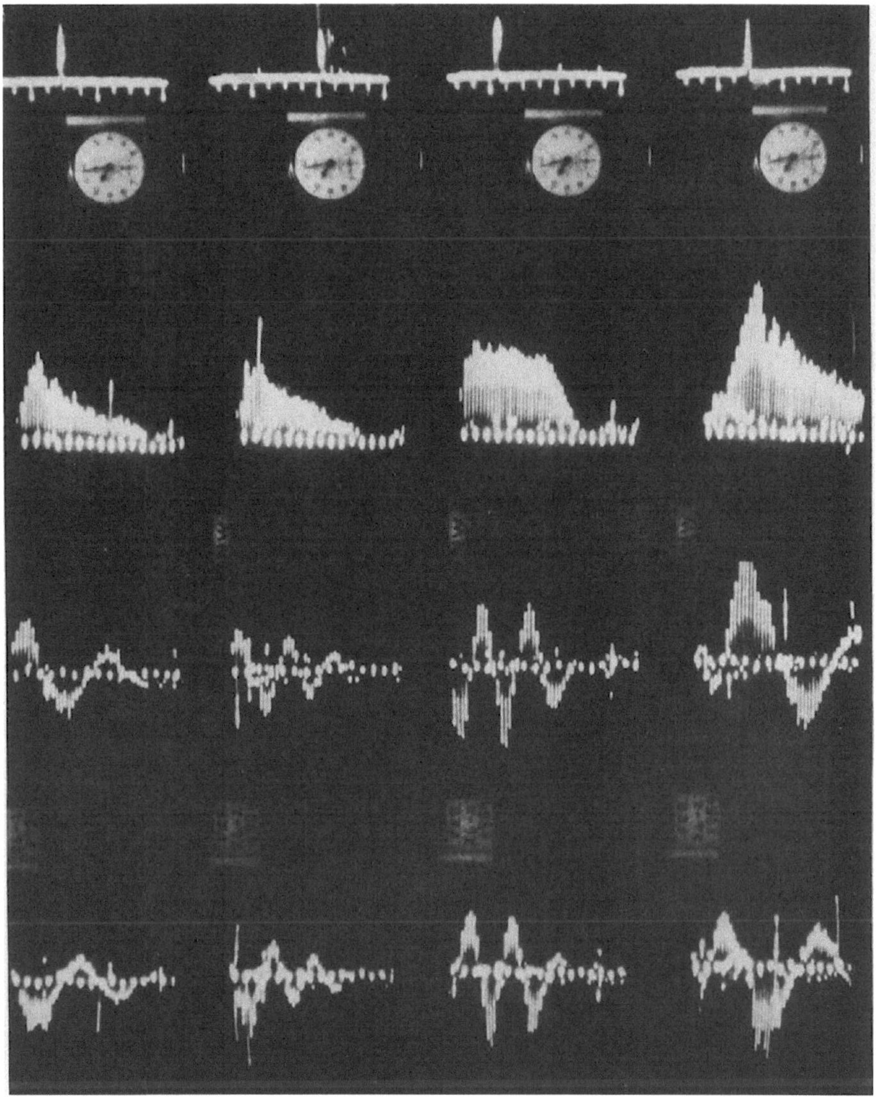
A
B
C
D

Fig. 15. Four successive records of radio echoes from meteor trails obtained in the radio echo technique for upper atmospheric wind measurements. In *A* the meteor echo is shown on a range time display, *B* shows the FRESNEL diffraction pattern and *C* and *D* the phase changes caused by the drift of the ionized trail.

the wind velocity, and in order to determine the absolute wind speed and direction a beamed aerial is pointed alternately North and East.

The original publications[44] give a large amount of detail obtained in the systematic measurements over a number of years. Some examples of the results are shown in Figs. 16 and 17. Fig. 16 is an example of the results for N–S: E–W components of the wind velocity for two days in September 1954. The wind reaches 40 m./sec. and has a semidiurnal variation with a phase difference of 3 hours between the two components, indicating a wind vector rotating in a

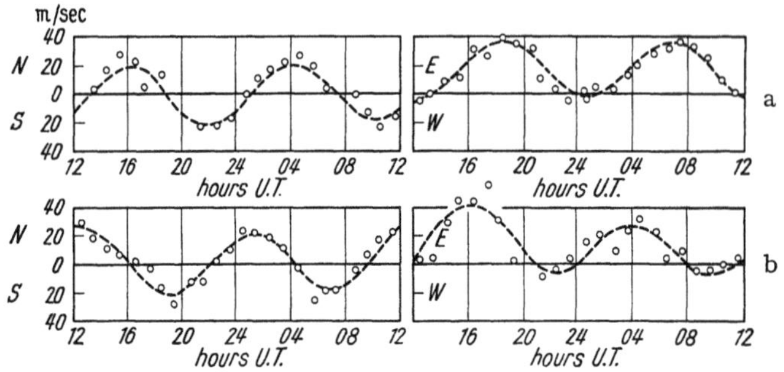

Fig. 16 a and b. The diurnal variation of the N-S, E-W wind components as determined by the radio echo technique (a) for 1954 September 16-17, and (b) for 1954 September 22-23. The results refer to a mean height between 90 and 95 km.

clockwise direction with a period of 12 hours. Similar results throughout the year show that the wind vector is towards the North at 06 hours and 18 hours except during the autumn months when the phase of the periodic wind component

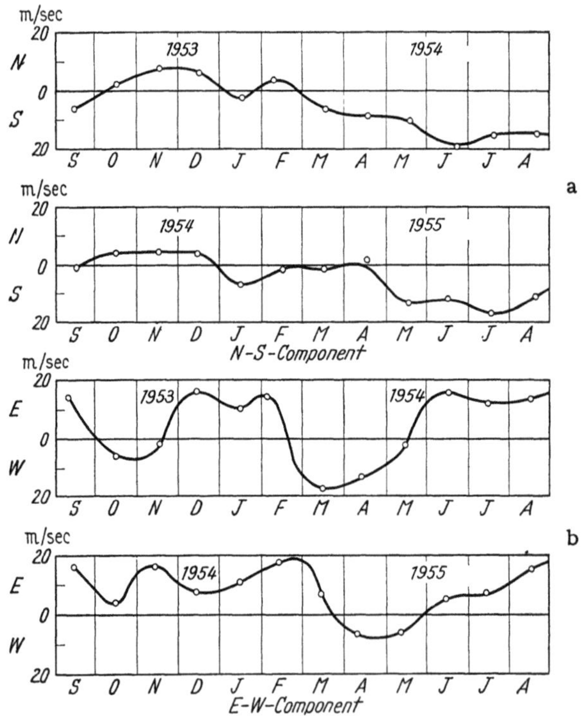

Fig 17 a and b. The variation of the prevailing component of the wind throughout the year as deduced from the radio echo meteor observations. (a) The N-S component, (b) the E-W component. The results refer to a mean height between 90 and 95 km.

changes by 360°. Fig. 17 shows the prevailing component which is towards the North in winter and towards the South in summer. The prevailing East West components are stronger and are towards the East in winter and summer and towards the West in spring and autumn.

These results refer to a mean height between 90 and 95 km. In recent work GREENHOW and NEUFELD[45] have investigated the variations of the wind with altitude over the range 80 km. to 100 km. Marked changes in phase and amplitude of the semidiurnal component are found over this height range. As an example, the results for a complete day are given in Fig. 18, which shows the changes in this component for September 16–17, 1954 for four separate mean heights of 85, 90, 95 and 100 km. Twelve hour harmonic dials for the periodic component at these heights for a complete year are given in Fig. 19. The broken

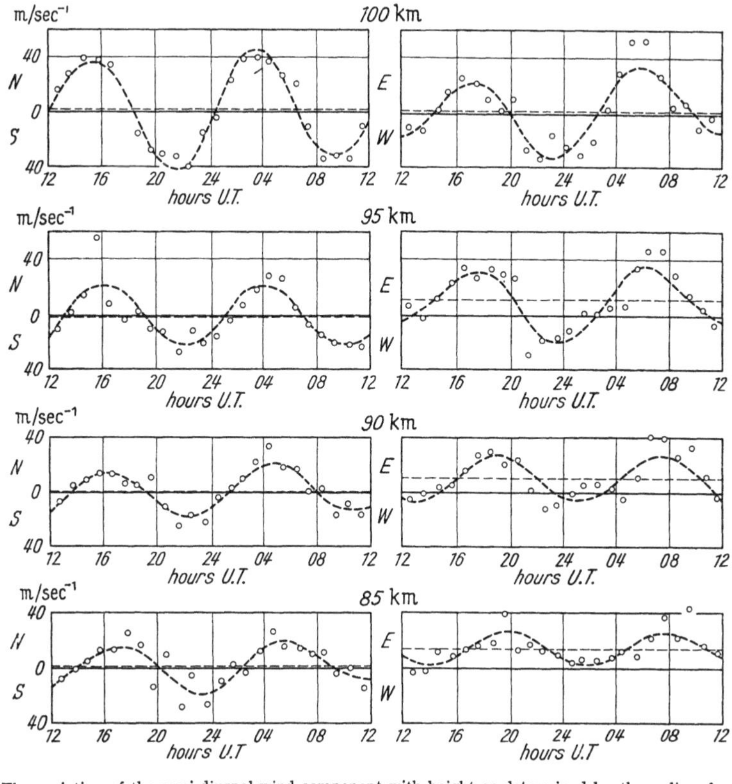

Fig. 18. The variation of the semi-diurnal wind component with height as determined by the radio echo technique for 1954 September 16-17.

section of these curves refers to the months of September and November when the phase undergoes a complete reversal. This diagram shows clearly both the increase in amplitude of the component wind with height and the phase change of about 90°. The average phase change is about 5 degrees km.$^{-1}$ and the increase in amplitude with height about 1 m. sec.$^{-1}$ km.$^{-1}$.

The prevailing wind also shows an increase in strength with height of about 0.5 m. sec.$^{-1}$ km.$^{-1}$. The variations in the E–W and N–S components of this prevailing wind for two heights are shown in Fig. 20. It is particularly interesting to note that the E–W component at 100 km. blows towards the W during the summer and autumn, but at 85 km. the component remains towards the E during the autumn.

The cause of this semidiurnal component is believed to be the result of thermal and tidal effects of the sun, increased by resonance at the heights under

[45] J. S. GREENHOW and E. L. NEUFELD: Phil. Mag. **1** (Series 8), 1157 (1956).

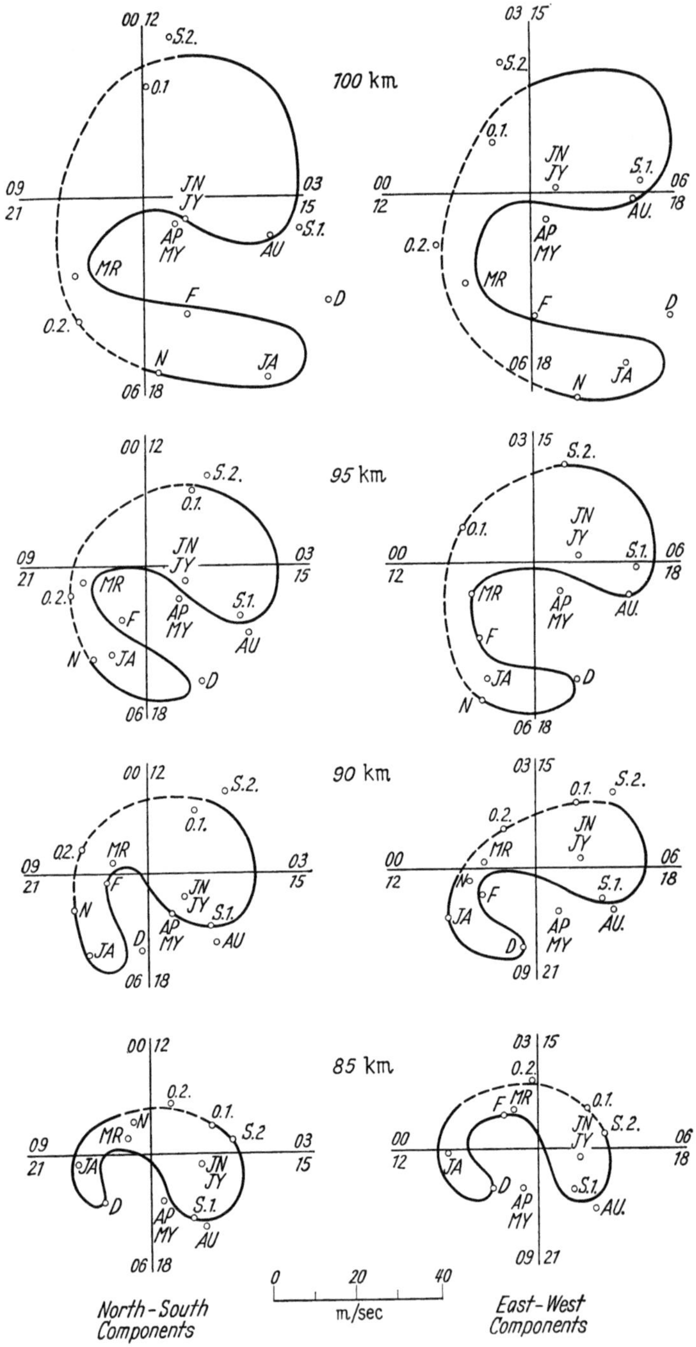

Fig. 19. Twelve hour harmonic dials for the N-S, E-W components of the upper atmospheric wind between 85 and 100 km. as determined by the radio echo technique.

discussion[46]. GREENHOW and NEUFELD show that the observed variations in amplitude and phase of the semidiurnal component are compatible with the predictions of this theory.

[46] See for example, M. V. WILKES: Oscillations of the Earth's Atmosphere. Cambridge 1949.

In earlier work, using a pulse technique, Greenhow[47] made a preliminary study of the degree of turbulence in the 80 to 100 km. region. He found the r.m.s. turbulent velocity to be of the order of 30 m. sec.$^{-1}$. These turbulent velocities are now being investigated in more detail with the coherent pulse techniques.

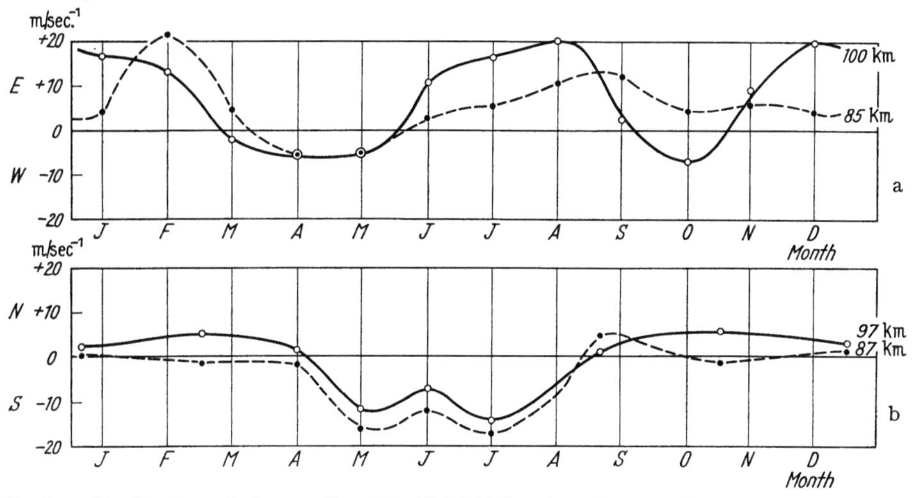

Fig. 20a and b. The change in the prevailing wind with height throughout the year as determined by the radio echo meteor technique. (a) The E-W component at 85 and 100 km. (b) The N-S component at 87 and 97 km.

10. Recombination and diffusion of the ionized column. The radio echoes obtained from the ionized meteor trails are generally very short lived. In the case of a radio wave scattered from meteors below the visible limit for which Eq. (4.3) applies the echo lasts only for a fraction of a second. For bright meteors [Eq. (4.4)] the echoes may last for several seconds. In deriving Eqs. (4.5) and (4.6) for the lifetime of the radio echo it has been assumed that diffusion is the predominant influence in the dispersal of the ionized matter and the consequences of this will be discussed in this section. There has, however, been considerable discussion as to the probable effects of other processes such as recombination or attachment and these will first be discussed.

α) The effects of recombination and attachment. In an early discussion of the problem of echo durations Lovell[48] compared the relative influences of diffusion and recombination on a meteor trail containing 10^{10} electrons/cm. Assuming that the recombination coefficient $\bar{\alpha}$ is the same as for the E region (10^{-8} c.c./sec.), the density in such a trail would be reduced to 5×10^{7} electrons/c.c. after 1 second. (This assumes that the diameter of the ionized column is of the order of the mean free path.) On the other hand if the diffusion rate is of the order of 1 to 4 m./sec. then, on account of diffusion, the volume density will decrease to about 10^{5} electrons/c.c. after 1 second, suggesting that recombination is unimportant compared with the effects of diffusion.

The discussion was reopened by Greenhow[20], because of difficulties in the interpretation of the durations of echoes from dense trails for which Eqs. (4.4) and (4.6) were appropriate. He investigated the variation of duration ΔT with λ for echoes with durations greater than 0.5 seconds, and his results are shown

[47] J. S. Greenhow: J. Atmos. a. Terres. Phys. **2**, 282 (1952). See also the chapter by Kertz in this volume.

[48] A. C. B. Lovell: Sci. Progr. **38**, 22 (1950).

in Fig. 21. The diffusion theory predicts a variation as λ^n with $n=2$ [Eq. (4.6)], which was found to be the case for echoes with $\Delta T < 2$ seconds on $\lambda = 4.2$ m. On the other hand for greater values of ΔT, Fig. 21 shows that the prediction of the diffusion theory is not obeyed; hence for these very dense trails some processes other than diffusion must be influencing the echo duration.

Since these trails have line densities of 10^{13} electrons/cm. or more recombination would be expected to be important if $\bar{\alpha} \sim 10^{-8}$ c.c./sec., but GREENHOW pointed out that this value of $\bar{\alpha}$ originally assumed by LOVELL was probably orders of magnitude too high and that the appropriate value would be that for the recombination of electrons with positive N_2 and O_2 ions for which $\bar{\alpha} = 10^{-12}$ c.c./sec.[49]. Attachment of the electrons to neutral particles n_0 must also be considered. Thus the rate of fall of electron density in the trail will be

$$\frac{d N_e}{dt} = -\bar{\alpha} N_e N^+ - \bar{\beta} N_e n_0 \tag{10.1}$$

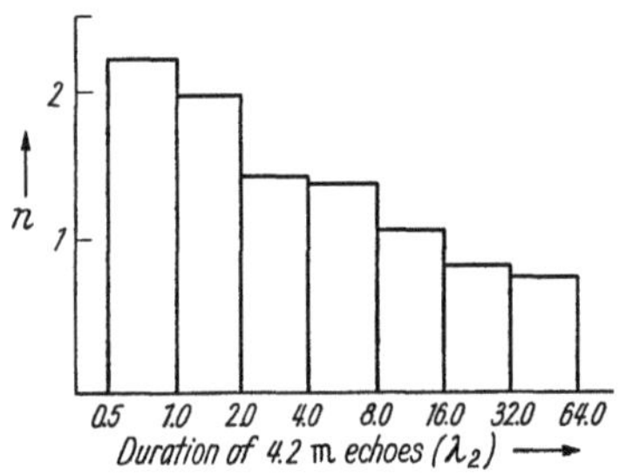

Fig. 21. Variation of the duration of radio echoes from meteor trails with change in wavelength λ. The histogram shows the exponent n from $T_1/T_2 = (\lambda_1/\lambda_2)^n$ as a function of T_2, where T_1 and T_2 are the durations of a meteor echo at wavelengths λ_1 and λ_2.

where $\bar{\beta}$ is the attachment coefficient.

Taking $N_e = N^+$, the first term in (10.1) gives, after t seconds

$$\frac{N_e}{N_0} = 1 + \bar{\alpha} N_0 t \tag{10.2}$$

where N_0 is the initial electron density. As an example, consider a trail of 10^{15} electrons/cm. If the initial radius is 10 cm. the volume density near the centre is 3×10^{12} electrons/c.c. According to (4.6) diffusion alone will reduce this to 1.3×10^7 electrons/c.c. (the critical density for $\lambda = 8.4$ m.) in 100 seconds. Eq. (10.2) shows that in this time recombination alone would reduce the density to 10^{10} electrons/c.c. and hence can be neglected in comparison with diffusion.

With regard to the second term in (10.1) the value of $\bar{\beta}$ will be that appropriate to the attachment of electrons to neutral O_2 molecules for which $\bar{\beta} = 10^{-16}$ c. c./sec.[50]. Unless $n_0 > 10^4 N_e$, this term can therefore be disregarded. Since at 95 km., $n_0 \simeq 10^{14}$ particles/c.c. it can be seen from the above discussion that for the trail in question attachment would only become comparable with recombination after 100 seconds. The additional decrease in electron density due to attachment is therefore small*.

The conclusion from these considerations, that recombination effects in meteor trails are small compared with diffusion in the dispersal of the ionized trail, depends on the assumption that the appropriate value for $\bar{\alpha}$ is close to the theoretical value for radiative recombination (10^{-12} c.c./sec.) which is 10^4 times less than the value of $\bar{\alpha}$ found experimentally in the E region. This assumption has been strongly criticised, notably by FEINSTEIN[51], who considers that the E region value of $\bar{\alpha}$ should also apply to meteor trails. In this case the duration of an echo on $\lambda = 8$ m., from trails with line densities in excess of 10^{13} electrons/cm. would be limited to about 8 seconds. It is inappropriate here to enter into a discussion of the possible reasons for different values of $\bar{\alpha}$ in the E region and in

---

[49] D. R. BATES, R. A. BUCKINGHAM, H. S. W. MASSEY and J. J. UNWIN: Proc. Roy. Soc. Lond., Ser. A **170**, 322 (1939).

[50] D. R. BATES and H. S. W. MASSEY: Phil. Trans. A **239**, 269 (1943).

* The departure of n from $n=2$ in Fig. 21 for long durations has been shown by GREENHOW to be due to the turbulent effects discussed in Sect. 9, which cause the trail to break up.

[51] J. FEINSTEIN: Proc. Phys. Soc. Lond. B **65**, 741 (1952).

meteor trails. The questions involved have been considered at some length by KAISER and GREENHOW[52] and by KAISER[53], who conclude finally that meteor atoms and not air molecules are primarily involved and hence there is no difficulty in setting an upper limit of about 10^{-11} or 10^{-12} c.c./sec. for the value of $\bar{\alpha}$ in meteor trails. At the present time it is generally believed that for meteors in the height range 80 to 110 km. considered here, the effects of recombination or attachment processes are very small compared with the influence of diffusion.

β) The measurement of diffusion coefficients. In the light of the above it is reasonable to assume that Eqs. (4.5) and (4.6) can be applied to the behaviour of the meteor trails under discussion here. In principle, either expression can be used for the determination of the diffusion coefficient D. However, the duration of the high density trails ($\alpha > 10^{12}$ electrons/cm.), depends on α, as well as on λ and D and it is preferable to base measurements of the diffusion coefficients on the low density trails ($\alpha < 10^{12}$ electrons/cm.) for which the duration varies only as λ and D. From Eq. (4.5) the duration T, to $1/e$ of the initial amplitude, is given by

$$T = \lambda^2/16\pi^2 D. \tag{10.3}$$

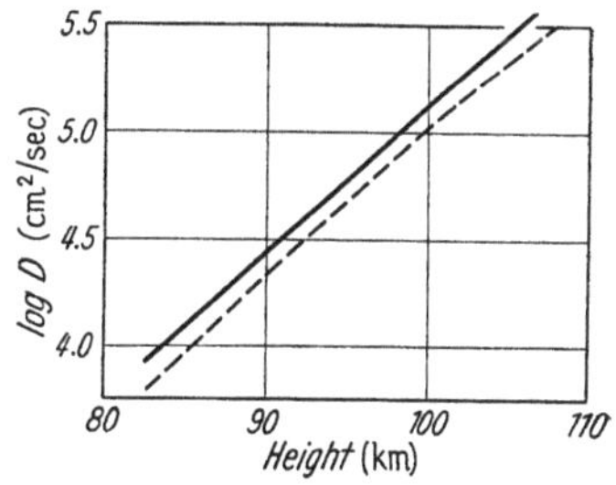

Fig. 22. The diffusion coefficient D measured from 80 to 110 km. height in the radio echo meteor investigations. The full line is the adopted experimental result, and the broken line the theoretical variation of D with height.

It is important to notice that for given λ and D this decay time constant is the same for all trails. Examples of echoes from such trails are shown in Fig. 2b. They are characterised by a rapid initial rise in amplitude, followed by the amplitude variations due to diffraction effects superimposed on the general decay in amplitude according to Eq. (10.3). These echoes have been used by GREENHOW and NEUFELD[54] and by WEISS[55] to determine the variation of D with height in the 80 to 110 km. region. Both sets of results are similar although the techniques differed, a pulse method being used in the first case and a continuous wave method in the latter. GREENHOW and NEUFELD selected only echoes showing exponential decays and measured the decay times to one half the maximum amplitude (i.e. to $\log_e 2T$) by drawing smooth curves through the diffraction fluctuations. The heights were derived from data obtained in the radio echo height measurements mentioned in Sect. 7. The final results for the variation of D in the 80 to 110 km. region are shown in Fig. 22, from which it is seen that D increases linearly from 10^4 cm.²/sec. at 80 km. to 10^5 cm.²/sec. at 100 km.

A simple ambipolar theory of the diffusion process given by KAISER[53] suggests that the diffusion coefficient D is related to the classical kinetic theory value of the ionic diffusion coefficient D_i by

$$D = D_i\left(1 + \frac{\Theta_e}{\Theta_i}\right) \tag{10.4}$$

where Θ_e and Θ_i are the electron and ion temperatures respectively. Assuming $\Theta_e = \Theta_i = \Theta$ and that the ionic mass μ equals the mass of neutral air molecules

[52] T. R. KAISER and J. S. GREENHOW: Proc. Phys. Soc. Lond. B **66**, 150 (1953).
[53] T. R. KAISER: Phil. Mag. Suppl. **2**, 495 (1953).
[54] J. S. GREENHOW and E. L. NEUFELD: J. Atmos. a. Terres. Phys. **6**, 133 (1955).
[55] A. A. WEISS: Austr. J. Res. **8**, 279 (1955).
[56] ROCKET PANEL: Phys. Rev. **88**, 1027 (1952).

(10.4) becomes

$$D = \frac{8}{3}\lambda_i\left(\frac{2k\Theta}{\pi\mu}\right)^{\frac{1}{2}} \text{cm.}^2/\text{sec.} \tag{10.5}$$

where λ_i is the ion mean free path, and k is BOLTZMANN's constant.

Since $\lambda_i = \frac{1}{\sqrt{2}\,\pi\, n_a\, \sigma_i^2}$ where n_a is the atmospheric particle density and σ_i the mean collision diameter for the meteor atoms and air molecules we obtain

$$D = \frac{3.5}{\pi\, n_a\, \sigma_i^2}\left(\frac{k\Theta}{\pi\mu}\right)^{\frac{1}{2}} \text{cm.}^2/\text{sec.} \tag{10.6}$$

for the theoretical relation between D and n_a. For comparison with the experimental results Eq. (10.6) is plotted in Fig. 22, by using the rocket panel air densities[56] and by assuming $\sigma_i = 9.2 \times 10^{-8}$ cm.*. The agreement is very reasonable.

11. The total meteor influx and contribution to the ionization in the ionosphere. *α) The influx of meteors.* A full discussion of the total meteor influx into the atmosphere has been given by LOVELL[1]. In the case of sporadic meteors in the magnitude range from about 0 to +9 there is fairly good visual and radio echo data on the frequency and mass distribution. The conclusion is that down to this magnitude limit there are approximately 10^9 meteors incident on the whole earth's atmosphere per day. The distribution (see Sect. 6) is such that each magnitude contains about the same mass of approximately 20 kilograms per day over the whole earth's surface. Hence in this range of 10 magnitudes there is believed to be about 200 kg of meteors entering the entire atmosphere per day. Beyond these limits there is less certainty. For meteors brighter than magnitude 0, there is reasonable visual and photographic evidence that the same mass distribution applies to magnitude −3. There is very little evidence of the distribution between −3 and −10, after which the meteors probably fall as meteorites**. At the other extreme the experimental evidence is negligible for meteors fainter than +10 or +11, and at present one can only assume that the same distribution continues until magnitudes between +20 or +30 at which limit the particles will be dispersed by radiation pressure from the sun before entering the atmosphere. On this basis the evidence on the mass and space distribution of the sporadic meteors assembled by LOVELL is summarised in Table 3. In calculating the space distribution it has been assumed as an approximation that the meteors are moving at the parabolic velocity limit.

Excluding meteorites, these figures indicate that there is probably between 700 and 800 kg. of sporadic meteors entering the whole earth's atmosphere per day. The total energy brought into the atmosphere is approximately 10^{19} ergs per day and the space density in the region of 10^{-24} to 10^{-25} gm./c.c.

In the case of the major showers the calculations are rather less certain, even in the 0 to +10 magnitude range, since it is clear from Sect. 6 that many uncertainties remain in the mass distribution. Even so, useful data can be obtained by assuming, as an approximation, that the mass distributions are the same as for the sporadics. The data in Table 4 are summarised from the information given by LOVELL[1], for the magnitude range of Table 3, and taking into account the number of days for which the showers are active.

Although the major showers are the spectacular occurrences in meteor phenomena they bring a total mass into the atmosphere per annum which is only about one third or one quarter of that brought in by the sporadic meteors. In addition

* The value obtained by RAMSAUER and BEECK for collisions between ionized alkali and oxygen molecules.

** The total daily mass of meteorites falling to the earth's surface is about 550 kg.

Table 3. *The daily influx of sporadic meteors over the entire earth's surface.*

	Meteor magnitudes	Mass range of individual particles	Number per day	Mass per day kg	Number per c.c.	Density gm./c.c.
Extra-polation	−10 to −8	500 to 100 g	~300	~60	4×10^{-28}	7.5×10^{-26}
	−7 to −6	100 to 10 g	~2500	~40	3×10^{-27}	5.0
	−5 to −4	10 to 1 g	$\sim1.8\times10^{4}$	~40	2×10^{-26}	5.0
	−3 to −1	1 g. to 100 mg.	$\sim4\times10^{5}$	~60	5×10^{-25}	7.5
Observational data	0 to +1	100 to 10 mg.	10^{6}	40	1.5×10^{-24}	5.0
	2 to 4	10 to 1 mg.	3.6×10^{7}	60	4.5×10^{-23}	7.5
	5 to 6	1 to 0.1 mg.	1.9×10^{8}	40	2.4×10^{-22}	5.0
	7 to 9	0.1 to 0.01 mg.	3.3×10^{9}	60	4.1×10^{-21}	7.5
	10 to 12	0.01 to 0.001 mg.	6.5×10^{10}	60	8.1×10^{-20}	7.5
Extra-polation	13 to 20	$10^{-3}-10^{-7}$	$\sim10^{19}$	160	$\sim10^{-11}$	~20
	21 to 30	$10^{-7}-10^{-11}$	$\sim10^{29}$	160	$\sim10^{-1}$	~20

to the showers of regular occurrence listed in Table 4, the occasional great meteoric storms such as those of the Giacobinids of October 1946, or of the Leonids in 1799, 1833 and 1866 bring very great additional meteoric material into the atmosphere. For example, although the Giacobinid shower of October 10, 1946 lasted only for a few hours, LOVELL[1] has estimated that the meteor influx amounted to about 70000 kg., a figure comparable with the total mass of all recurrent showers for a whole year. The total mass and densities in the orbits of these shower meteors have also been estimated[1]. Although not of immediate concern to this geophysical discussion it is interesting to mention that the calculated total mass of meteoric material in the orbits of the major showers is about 10^{12} kg. and the density is in the region of 10^{-23} to 10^{-24} gm./c.c.

β) The total meteor contribution to the ionization of the E region. From time to time there has been considerable discussion about the contribution of meteors to the ionization in the *E* region. Some of the earlier workers, notably SKELLETT[57] and PIERCE[58] assumed that the major part of the energy of the meteor was spent in ionization and they concluded that ionization due to meteors made a prominent contribution to the normal level of ionization in the E region. In a brief review of the subject in 1950, LOVELL[48] assumed values of 0.01 for the ionization probability (β in Sect. 2), and estimated the rate of ion production to be about $q=3\times10^{-5}$ electrons/c.c./sec. for sporadic meteors, rising to 10^{-3} electrons/c.c./sec. during major showers and to 10^{-1}/c.c./sec. during meteoric storms like the 1946 Giacobinid shower. If the recombination coefficient is that appropriate to the E region ($\bar{\alpha}=10^{-8}$ c.c./sec.) these would give equilibrium electron concentrations $\left(N=\sqrt{q/\bar{\alpha}}\right)$ of about 17 electrons/c.c. for sporadics, rising to 300 electrons/c.c. during major showers to about 3×10^{3} electrons/c.c. during the 1946 Giacobinid shower. These figures have to be compared with the average

Table 4. *Meteoric mass per year entering the entire atmosphere from the major meteor showers.*

	Sporadic meteors (as Table 3)	Daytime Arietids + ζ-Perseids	Daytime β-Taurids	Quadrantids	Perseids	Geminids	Other showers
No. of days of activity . . .	365	16	11	2	9	6	23
Total mass per year kg. . .	485×10^{3}	41.5×10^{3}	5.1×10^{3}	13.3×10^{3}	11.9×10^{3}	28.7×10^{3}	29.9×10^{3}

[57] A. M. SKELLETT: Proc. Inst. Radio Engrs. **23**, 132 (1935).
[58] J. A. PIERCE: Phys. Rev. **71**, 88 (1947).

values of E region ionization, amounting to the order of 10^5 electrons/c.c. during a midsummer day to 10^4 electrons/c.c. at night. Thus the contribution of meteors to E region ionization is small, although at night intense showers might have an appreciable influence. In this connection it is of interest that Pierce[58] observed a reflection layer which persisted for several hours on a frequency of 3.5 Mc/s during the 1946 Giacobinid shower.

The question has recently been investigated again by Kaiser[59] who has deduced the ionization contribution from the radio echo measurements of the electron line densities without introducing the rather uncertain factor of the ionizing efficiency β. He finds that sporadic meteors should produce 2.3×10^{-4} electrons/c.c./sec. independent of height [at least over the range of heights for which the sporadic meteors obey the mass law with $s=2$ (Sect. 6)]. This is ten times greater than the rate deduced in the earlier calculations by Lovell but is compatible with the revised value for the ionization efficiency of ~ 0.1 compared with the value of 0.01 assumed in his calculations. The question of the significance of this in the E region ionization depends on the value to be assumed for the recombination coefficient $\bar{\alpha}$. If the normal E region value of 10^{-8} applies, then the equilibrium electron density is about 150 electrons/c.c. On the other hand, as discussed in Sect. 10, the recombination rate in the meteor trails is not greater than about 10^{-11} or 10^{-12} c.c./sec., and if this value applies then the equilibrium electron density would be about 10^4 electrons/c.c.

Kaiser speculates on the nature of the ionospheric layer set up by this sporadic meteor influx. The condition that the production of electrons is continuous over a large volume is that a meteor must pass through the distribution of ionization from a previous meteor before it is significantly reduced below its initial value. Kaiser shows that on this basis the ionization should be continuous from about 130 km. to 115 km., but below this height the individual trails will not join up to form a continuum and the ionization will be patchy. The normal value of the night time E region electron density is 10^4 electrons/c.c. The density of the continuum produced by the sporadic meteors is greater than 10^2 electrons/c.c. If the recombination coefficient is significantly less than the E region value of 10^{-8} c.c./sec. then it is possible that sporadic meteors could be a significant factor in the maintenance of the nocturnal E region.

Kaiser also studied the effects of the shower meteors, taking into account the recent evidence about the mass distribution (Sect. 6). In the case of the intense day time streams, the equilibrium electron density is unlikely to exceed 10^4 electrons/c.c. even if the most favourable values of $\bar{\alpha}$ are assumed. This is not significant compared with the normal summer daytime E region ionization. For the major night time showers such as the Perseids and Geminids, the conclusion is that the equilibrium electron density is comparable to that produced by the sporadic meteor ionization at 100 km. for the Perseids and at 93 km. for the Geminids. The probability that meteoric storms produce a temporary reflecting layer has already been mentioned.

The above discussion refers to the contributions which meteor ionization might make to the normal E region ionization. With regard to the abnormalities in the E region, there is little doubt that at least some forms of sporadic E are associated with meteor activity. The existing evidence about this was assembled by Lovell[60] in 1948, and since that time there has been no significant clarification of a very complex problem.

[59] T. R. Kaiser: Phil. Mag. Suppl. **2**, 495 (1953). — J. Atmos. a. Terres. Phys. Suppl. **2**, 119 (1955).

[60] A. C. B. Lovell: Phys. Soc. Rep. Progr. Phys. **11**, 415 (1948).

Sound Propagation in Air.

By

EVERETT F. COX.

With 15 Figures.

1. Definitions and units. *Sound*, as used in this chapter, means [*1*] an alteration in pressure, stress, particle displacement, particle velocity, etc., propagated in an elastic material, or the superposition of such propagated alterations. It need not create an auditory sensation.

To be audible, pressure alterations impinging on a human ear must lie roughly within the range of 15 to 20000 cycles per second, and even at one kilocycle per second must be stronger than 2×10^{-4} *microbars*, or *dynes per square centimeter*. Alterations too rapid for the human ear to perceive are called *ultrasonic*; those occurring at too slow a rate are termed *infrasonic* or *microbarometric*.

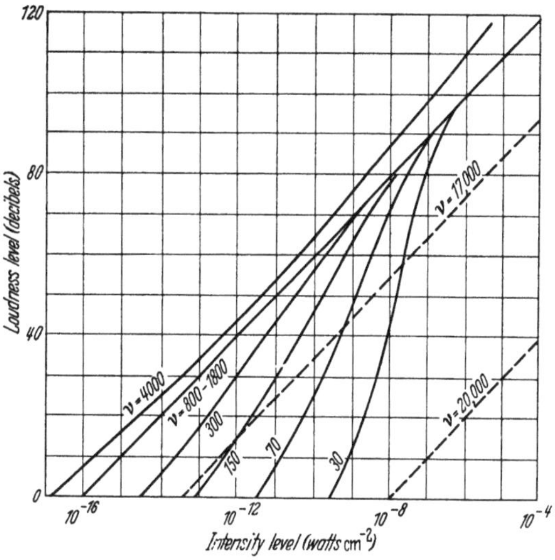

Fig. 1. Relation between loudness and intensity for pure tones of various frequencies (FLETCHER [*2*]). The straight line marked $\nu=800$ to 1800 cycles per second *defines* loudness level. Threshold of audibility at 1000 cycles per second, for an average human ear, occurs near 10^{-16} watts cm.$^{-2}$.

Intensity of sound at a point in space is the amount of energy passing through a unit area in unit time, the area being taken normal to the direction of sound propagation. It is proportional to the square of frequency, and to the square of the fluctuation amplitude of some function such as pressure. LOUDNESS of sound in its psychological sense has been shown [*2*] to increase as the logarithm of the intensity for frequencies in the range 800 to 1800 vibrations per second. For frequencies both above and below this range psychological loudness is not solely a function of intensity but also involves frequency and wave form (Fig. 1).

Through still dry air at normal temperature and pressure (i.e., 0° C, 760 mm. Hg), pressure undulations which are tiny compared with ambient pressure travel with essentially the same speed, 331.45 m. per second, regardless of frequency. Dispersion can be ignored in most instances when frequencies are less than about 10^8 cycles per second, i.e., phase velocity and group velocity of sound in still dry air at NTP are equal. Infrasonic waves in air thus have wave lengths exceeding 20 m., ultrasonic waves are shorter than 2 cm., and wave lengths between these limits constitute the audible range.

The physicist's definition of sound sets no bounds on magnitudes of alterations. Undulations of pressure, for example, may be "insignificant" or "significant" compared with the ambient pressure of the gas transmitting the sound. Air at one atmosphere ambient pressure ($\approx$ one *bar* $= 10^6$ dynes per square centimeter) can transmit a sound signal with a peak overpressure ripple of a millimicrobar, or, equally well a peak overpressure amounting to megabars.

When pressure or density disturbances are truly *infinitesimal* compared with ambient conditions, wave forms do not change rapidly during transmission. But the "phase" velocity of propagation does change with amplitude of disturbance. Marked changes in wave form take place during transmission of large perturbations (Fig. 2). High-amplitude parts of a perturbation travel faster than low-amplitude parts, and a gradual increase in pressure becomes more and more abrupt [3] until eventually the transition in pressure level from ambient to ambient-plus-peak-overpressure occurs in approximately one mean free molecular path length. The fundamental frequency of a significant pressure-perturbation wave form, including both a compression and rarefaction phase, may be infrasonic, yet the sound will be audible. FOURIER analysis of a sudden transition contains terms of many frequencies, including audio frequencies.

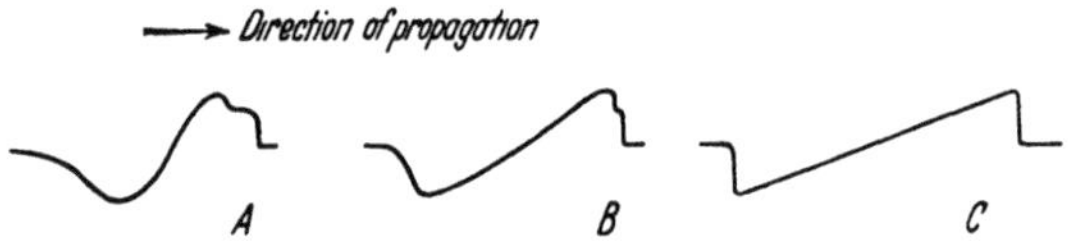

Fig. 2. Evolution of finite-overpressure wave forms into N-waves (*from* [3]). The small added pressure hump near the front of the wave form in diagram A moves progessively nearer to the front, merging with it in diagram C. Simultaneously the entire wave form grows longer.

The steep-fronted perturbation, which continues its travel at a speed greater than that of a low-amplitude sound, is termed a *shock wave*. If p represents the perturbation overpressure, P the ambient pressure, γ the ratio of specific heat of the transmitting gas at constant pressure to its specific heat at constant volume, the speed of the shock wave is

$$v = c\sqrt{1 + \frac{p}{P} \cdot \frac{\gamma + 1}{2\gamma}}, \tag{1.1}$$

where c is the speed of sound of truly infinitesimal amplitude, determinable from the LAPLACE equation (2.1).

2. LAPLACE equation for speed of sound. In a gas, pressure fluctuations constituting sound cause temperature fluctuations. High-pressure points are hot points; low-pressure places have depressed temperatures. If temperature gradients were steep, and if the gas were a sufficiently good conductor of heat, the process of sound transmission through the gas would be isothermal. But except near shock fronts, and for ultrasonic waves having wave lengths comparable to the mean free path of gas molecules, temperature gradients in all ordinary sound waves are too small to permit much heat flow between condensations and rarefactions. Sound transmission through a gas is therefore strictly an adiabatic process [4] when wave lengths are sufficiently long and pressure alterations small.

In his *Principia* NEWTON derived correctly an expression for the propagation speed, c, of a small disturbance through an elastic fluid having ambient density ϱ and volume modulus of elasticity k. The relation is $c = \sqrt{k/\varrho}$. Next, NEWTON incorrectly assumed the transmission process to be isothermal and accepted BOYLE'S law in his determination of k, obtaining $k = P$, the ambient pressure. for an ideal gas. NEWTON'S theoretical value for speed of sound, $c = \sqrt{P/\varrho}$, fell considerably below the measured value in air.

More than a century had passed when LAPLACE, recognizing the temperature changes associated with sound propagation, introduced the adiabatic modulus into NEWTON'S correct relation and obtained

$$c = \sqrt{\frac{\gamma P}{\varrho}} \tag{2.1}$$

for an ideal gas. This theoretical value of c agrees excellently with experiment.

Too-hasty conclusions based on partial reading and understanding of eq. (2.1) ofttimes lead to scientific blunders. It is common knowledge that the upper atmosphere is less dense than air near the earth's surface; and by noting only the denominator term ϱ while ignoring the numerator term P, one might infer that because air density decreases with altitude, sound speed would increase with altitude. This is incomplete reasoning. An even worse blunder results when, by noting that water is much more dense than air, it is concluded from eq. (2.1) that sound would travel more slowly through water than air. LAPLACE'S equation refers specifically to a gas, which water is not.

Direct substitution from the equation of state for an ideal gas,

$$\frac{P}{\varrho} = \frac{RT}{M}, \tag{2.2}$$

(T is temperature in degrees absolute, or Kelvin) permits a somewhat less confusing expression of LAPLACE'S equation:

$$c = \sqrt{\frac{\gamma R T}{M}}. \tag{2.3}$$

The gas constant per mole, $R(=8.314 \times 10^7$ erg deg^{-1} mole$^{-1})$, is a universal constant, not depending on the kind of gas. The molecular weight of the gas, M, depends on weights of the atoms and the polyatomicity of the molecules. The ratio of specific heats, γ, depends mainly on polyatomicity, being approximately 5/3 for monatomic gases, 7/5 for diatomic, 8/6 for triatomic. Equation (2.3) shows clearly that, within the range of validity of the ideal gas laws, sonic velocity for small disturbances depends only on *kind* and *temperature* of the gas, and that it is not directly dependent on gas pressure or density.

Dry air, a mixture of nitrogen, oxygen, argon, carbon dioxide, and traces of other gases and impurities, has an effective weight of $M = 28.97$ g. per mole, and its γ is 1.403. Water vapor in the air can be more readily handled through the temperature term of eq. (2.3) than through the γ and M terms.

Under some conditions effects of water vapor cannot be neglected in acoustical measurements, and relative humidity can be a rapidly changing function in the atmosphere. It thus becomes efficient to use *virtual temperature*, T^*, instead of measured temperature, T, to determine speed of sound in still damp air. Virtual temperature of a parcel of moist air is that temperature necessary to make the density of a parcel of dry air equal to that of the moist air under the same pressure. If ε is water-vapor pressure and P is total pressure, then

$$T^* = \frac{T}{1 - \frac{0.377\,\varepsilon}{P}}. \tag{2.4}$$

Introduction of numerical values for air into eq. (2.3) results in the formula

$$c = 20.07 \sqrt{T^*} \text{ meters per second}, \tag{2.5}$$

where virtual temperature, T^*, is expressed in degrees Kelvin.

3. Macro- and micrometeorology. Man's senses affected directly by weather are relatively sluggish compared with sight or hearing. Probably this is why instruments commonly used for measuring and recording meteorological functions are similarly slow in response. Liquid-in-glass or bimetallic-expansion thermometers generally have large heat capacities; wind-speed and -direction indicators have large inertias; and popular hygrometers are slow in absorbing and releasing water. Even when fast-response sensing elements such as small thermocouples are employed, they are ofttimes connected to long-time-constant recorders, enabling the observer to ignore with ease the marked and rapid fluctuations in meteorological functions at the point of measurement.

Moreover, "official" weather observation stations, which report "surface" temperature, humidity, wind, and pressure, do not position their sensing elements at the earth's surface but place them at "prescribed" heights over a "prescribed" type of surface and at least a "prescribed" distance from trees or buildings. The necessity for prescription and deliberate use of sluggish instruments admits steep and variable gradients with space, rapid changes with time. Meteorological data in common usage refer to *macro*-phenomena [5] and weather services generally disregard intentionally those changes taking place in times shorter than one hour or distances less than one hundred meters.

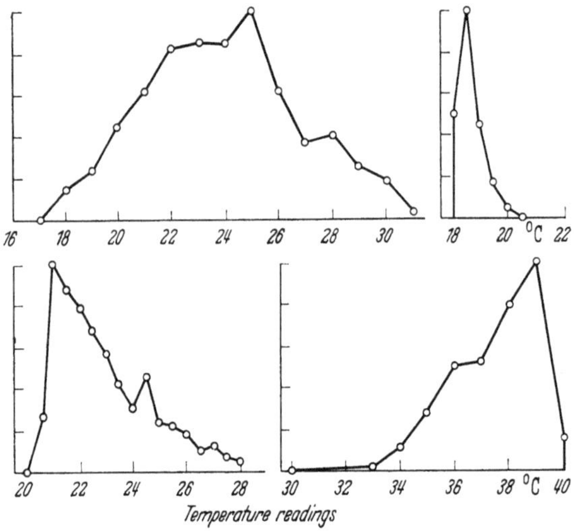

Fig. 3. Four frequency distributions of temperature, observed under different circumstances using sensitive, fast-acting resistance thermometers; measurements were made at a point, at intervals of six seconds, and for half-hour periods [6].

But sound, as a rapidly propagated phenomenon, notices inhomogeneities which man's senses overlook. *Micrometeorology* affects sound velocity [6], [7]. As a direct consequence finesse in computation of acoustical functions in atmospheric transmission problems is essentially meaningless beyond 70 or 80% accuracy when the only available data are those commonly called "weather data". Except for experiments conducted under carefully controlled laboratory conditions, the *practical* form of eq. (2.5) is

$$c = 20.0\sqrt{T} \approx (331 + 0.60\, C^\circ) \text{ meters per second} \tag{3.1}$$

(C° is measured temperature in degrees Centigrade) since micrometeorological fluctuations of real temperature in the atmosphere between source and receiver are generally more significant than speed changes producible by humidity (Fig. 3).

4. The law of sound refraction. The LAPLACE equation for speed of sound, in any of its forms, deals with propagation of a small disturbance through a gas at rest. But large volumes of the normal atmosphere are seldom, if ever, at rest. Instead the atmosphere appears to consist of a multitude of "cells" of various shapes and sizes, having different temperatures and different motions.

Quite exact equations may be written [8] to describe sound propagation through a nonisothermal and moving gaseous medium. In these derivations the scientist can carefully distinguish between vector velocity of sound at a point and the normal to the wave front at the point, and may symbolize exact virtual temperature, T^*, and vector wind, $\boldsymbol{w}$, at each point in space and each instant of time. But after all this care is taken, there are never sufficient micrometeorological data available to permit exact application in resultant transmission equations. When transmission path lengths under consideration are hecto- or kilometers and meteorological data are obtained from instruments in common use, acoustical experiments should not be expected to check theory better than 60 to 90%. By admitting these imperfections in advance of derivations, a very simple theory, agreeing with measurements as well as more sophisticated theory, may be formulated.

In the remainder of this chapter this simple theory will be employed.

Relative to a fixed point in terrestrial space, sound *velocity*, $\boldsymbol{V}$, as distinguished from sound speed, is a *vector* quantity made up of two components: the vector wind, $\boldsymbol{w}$, at the point; and a vector, $\boldsymbol{c}$, whose magnitude is given by eq. (3.1) and whose direction is normal to the wave front at the point; hence

$$\boldsymbol{V} = \boldsymbol{c} + \boldsymbol{w}. \tag{4.1}$$

While updrafts and downdrafts certainly exist, and wind in a turbulent "cell" may be circulatory in any plane, the atmosphere is customarily considered as stratified, with its layers parallel to the earth's (smooth) surface. Vertical winds are neglected with respect to horizontal winds. Sometimes a layer is treated as if it were isothermal and as if all the air in that layer traveled with a common velocity, $\boldsymbol{w}$, in which case thermal and wind *shears* exist only at discrete levels between layers. More commonly, temperatures and (horizontal) vector winds are observed at selected altitudes, and by straight-line interpolation between observation altitudes one assumes constant *gradients* or shears in wind direction, wind speed, and temperature, all scalar quantities.

Wind speed seldom exceeds ten or fifteen per cent of sound speed, at least within the troposphere. So except for studies of the noise from aeroplanes, meteors, or airburst bombs, or the noise *above* a ground-level explosion, *wind velocity in the direction of sound propagation may be set equal to the horizontal wind at each altitude.* Furthermore, *horizontal gradients of temperature and wind are assumed negligible over the area of sonic experimentation.*

Following MILNE [8], a *sound ray* is now defined as a curve such that the tangent at each point is in the direction of resultant sound velocity. Energy transport is mainly along these rays, and divergence or convergence of rays indicates decreasing or increasing energy concentration.

A vector form [9] of SNELL's law, consonant with the approximations and assumptions already made, supplies a means of computing the elevation angle, Θ, measured from the horizontal, of each sound ray at each point:

$$\boldsymbol{V} = \boldsymbol{A} \cos \Theta, \tag{4.2}$$

where $\boldsymbol{A}$, invariant for any selected sound ray, is the velocity of the intersection of the wave front with the horizontal. Thus toward any chosen azimuth (subscript φ) from a sound source, the elevation angle of the sound ray at altitude h is related to the starting elevation of the ray, Θ_0, according to the formula

$$(V_h)_\varphi \sec \Theta_h = (V_0)_\varphi \sec \Theta_0. \tag{4.3}$$

On the basis that horizontal winds, $\boldsymbol{w}$, are usually much smaller than sound velocity, $\boldsymbol{c}$, changes in azimuth during propagation have been neglected here.

When sound velocity, V_φ, decreases with height, sound rays in that azimuthal plane curve upward; if sound velocity increases with altitude, the rays curve downward. Ray paths are segments of circles, all having the same curvature $dV_\varphi/V_\varphi\, dh$, over altitude spans wherein the velocity gradient with altitude is constant[1].

5. Absorption and Dispersion. Clean dry air at normal temperature and pressure is a notably poor absorber of sound. And for persons who appreciate music, yet cannot afford close-in seats at symphony concerts, it is fortunate that air causes trivial dispersion: simultaneous notes sounded by the piccolo and the bass viol reach the distant listener at the same time.

But the earth's atmosphere contains air which is not all clean and dry, not all at normal temperature and pressure, and sound covers frequencies far above and below audible. Brief mention of both absorption and dispersion in the real atmosphere is therefore needed.

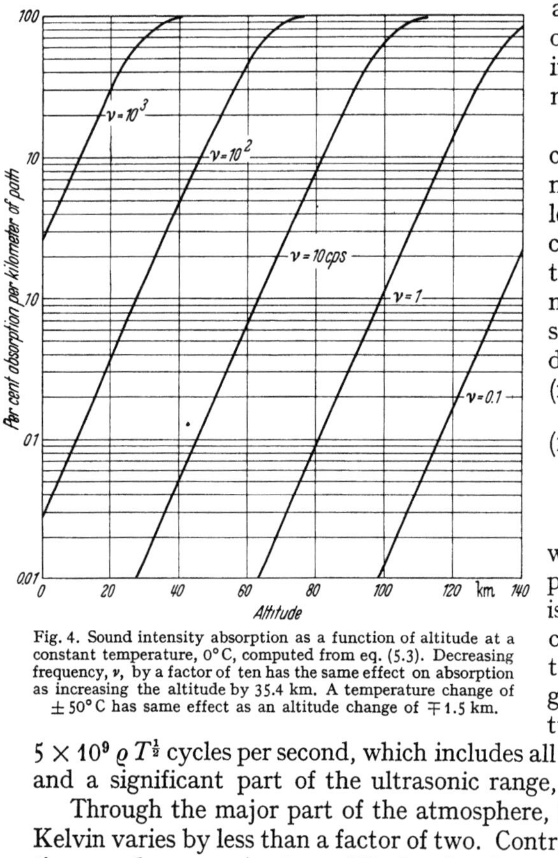

Fig. 4. Sound intensity absorption as a function of altitude at a constant temperature, 0° C, computed from eq. (5.3). Decreasing frequency, ν, by a factor of ten has the same effect on absorption as increasing the altitude by 35.4 km. A temperature change of $\pm 50°$ C has same effect as an altitude change of ∓ 1.5 km.

Dry air is a viscous, thermal-conducting medium containing molecules whose mean free path lengths depend on density. The classical-theory intensity absorption coefficient, (2α), defined as the negative logarithmic derivative of sound intensity with respect to distance progressed by a plane (nondivergent) wave, is [*10*], [*11*]

$$(2\alpha) = 9.5 \times 10^{-9} \times \frac{\nu^2}{\varrho\, T} \text{ per kilometer}, \qquad (5.1)$$

where ν is sound frequency expressed in cycles per second, ϱ is air density in grams per cubic centimeter, and T is absolute temperature. This expression is good within one order of magnitude for frequencies less than $5 \times 10^9\, \varrho\, T^{\frac{1}{2}}$ cycles per second, which includes all infrasonic and audible frequencies and a significant part of the ultrasonic range, at least at sea level.

Through the major part of the atmosphere, temperature expressed in degrees Kelvin varies by less than a factor of two. Contrarily, air at sea level is a thousand times as dense as air at an altitude of 50 km. and a million times as dense as air at 100 km. Density decreases nearly exponentially with height h, measured in kilometers, according to the formula

$$\varrho_h = \varrho_0 \exp(-h/7.7), \qquad (5.2)$$

where the number 7.7 km. has been determined as the "e-height" (S. CHAPMAN's *scale-height*) of the atmosphere by averaging values [*12*] at 20 km. levels from

[1] See also the corresponding discussions on seismic ray-paths in the chapter by BULLEN (Sect. 17, 23 to 25) in Volume XLVII of this Encyclopedia. For earlier work on sound propagation in air, see H. BENNDORF: Phys. Z. **30**, 97 (1929) and the article by O. MEISSER: Luftseismik, in WIEN-HARMS, Handbuch der Experimentalphysik, Bd. 25, Teil 3, S. 211. Leipzig: Akademische Verlagsgesellschaft 1930. Further references in Sect. 11.

sea level to 100 km. Consequently for frequencies less than $8 \times 10^6 \exp(-h/7.7)$ cycles per second, the sound intensity absorption coefficient is

$$(2\alpha) = 7.7 \times 10^{-6} \nu^2 T^{-1} \exp(+h/7.7) \text{ per kilometer.} \tag{5.3}$$

Forty per cent of the energy in a one-hundred-cycle sound would be absorbed (Fig. 4) during passage over a 10-km. path at an altitude of 40 km.

Dispersion, always associated with absorption, is shown by the same classical theory [*10*] to be trivial for all ordinary situations. Phase velocity exceeds group velocity, but for horizontally traveling waves, the difference is only $8.4 \times 10^{-22} \nu^2/\varrho^2 T^{\frac{1}{2}}$ m. per second. Water vapor and contaminants change [*13*] the absorption coefficient by factors of 10 to 20 for certain frequencies, and no doubt play a proportionate role in dispersion.

6. Reflection [*14*]. Sound propagation through various media depends on their elasticities and densities. When sound waves traveling in one medium impinge on another medium, some energy goes in, some is bounced back. How much enters and how much is reflected is determined by their relative compressibilities, relative densities, incident angle, and abruptness of the transition boundary.

At a horizontal boundary, undisturbed pressures are the same on both sides, and sound perturbation pressures and normal components of particle velocities must be the same in the two media. These conditions determine the fraction of incident energy reflected in medium 1,

$$f = \left[\frac{\frac{\varrho_2}{\varrho_1} - \frac{\tan\Theta_2}{\tan\Theta_1}}{\frac{\varrho_2}{\varrho_1} + \frac{\tan\Theta_2}{\tan\Theta_1}}\right]^2 \tag{6.1}$$

and the fraction of incident energy admitted through the boundary into medium 2,

$$a = 1 - f. \tag{6.2}$$

By substitution from eq. (4.3), the reflection factor becomes

$$f = \left\{\frac{\frac{\varrho_2}{\varrho_1} - \frac{\sqrt{V_1^2 - V_2^2\cos^2\Theta_1}}{V_2\sin\Theta_1}}{\frac{\varrho_2}{\varrho_1} + \frac{\sqrt{V_1^2 - V_2^2\cos^2\Theta_1}}{V_2\sin\Theta_1}}\right\}^2. \tag{6.3}$$

This equation, and in fact eq. (4.3), shows existence of a *critical angle* in the "slower" of two media. The angle Θ_2 in the "faster" medium becomes zero when the incident angle becomes

$$(\Theta_1)_{\text{crit}} = \operatorname{arc\,cos}\frac{V_1}{V_2}. \tag{6.4}$$

For all incident angles Θ_1 less than $(\Theta_1)_{\text{crit}}$, sound energy undergoes total internal reflection in the "slower" medium 1.

Eq. (6.3) also shows the possibility of another interesting angle, $(\Theta_1)_{\text{trans}}$, at which all energy is transmitted:

$$(\Theta_1)_{\text{trans}} = \operatorname{arc\,sin}\sqrt{\frac{\left(\frac{V_1}{V_2}\right)^2 - 1}{\left(\frac{\varrho_2}{\varrho_1}\right)^2 - 1}}, \tag{6.5}$$

which is real provided either

$$\frac{V_1}{V_2} > 1 \quad \text{and} \quad \frac{\varrho_2}{\varrho_1} > 1, \tag{6.6}$$

or

$$\frac{V_1}{V_2} < 1 \quad \text{and} \quad \frac{\varrho_2}{\varrho_1} < 1. \tag{6.7}$$

At normal incidence the reflection factor is

$$f = \left[\frac{\frac{\varrho_2}{\varrho_1} - \frac{V_2}{V_1}}{\frac{\varrho_2}{\varrho_1} + \frac{V_2}{V_1}}\right]^2, \tag{6.8}$$

and change in the factor near normal incidence is second order in change of incident angle. Near the critical angle

$$f = 1 - \frac{4\varrho_1 V_1 [\Theta_1 - (\Theta_1)_{\text{crit}}]}{\varrho_2 \sqrt{V_2^2 - V_1^2}}, \tag{6.9}$$

where angles are measured in radians.

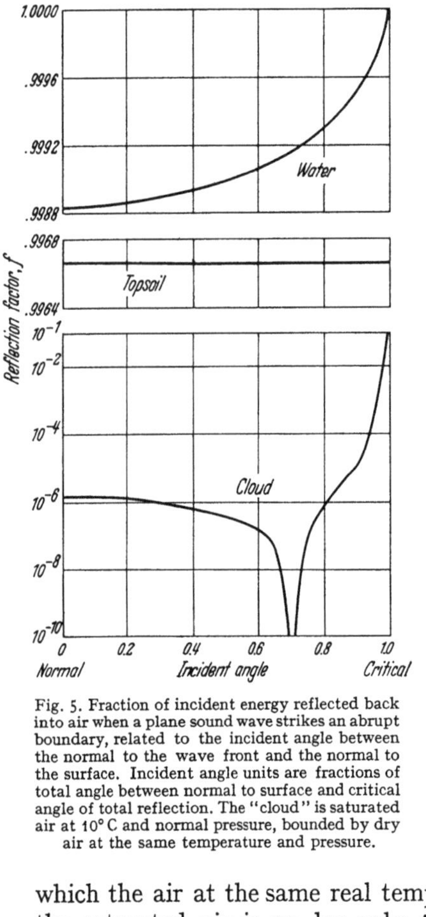

Fig. 5. Fraction of incident energy reflected back into air when a plane sound wave strikes an abrupt boundary, related to the incident angle between the normal to the wave front and the normal to the surface. Incident angle units are fractions of total angle between normal to surface and critical angle of total reflection. The "cloud" is saturated air at 10° C and normal pressure, bounded by dry air at the same temperature and pressure.

Topsoil [*15*] has a density about 1.2×10^3 times that of air, and sonic speed through topsoil is essentially the same as through air. When sound, traveling through air, strikes smooth earth, reflected energy is therefore at least 300 times as great as admitted energy. Water, which is eight-hundred times as dense as air and transmits sound 4.3 times as fast, reflects back into air at least 860 times as much energy as it admits. In many instances, then, the earth's surface may be regarded as a perfect reflector of sound even for normal incidence (Fig. 5).

Clouds and other meteorological "discontinuities" reflect insignificant amounts of sound energy except for incident angles near, at, or below critical. Suppose sound traveling through dry air at 10° C and normal pressure reaches an abrupt boundary beyond which the air at the same real temperature is saturated. Virtual temperature of the saturated air is up by only 1.3° C; hence sound speed is up by a mere 0.23%. Density of the saturated air, or cloud, is but 0.45% less than that of the dry air. Near normal incidence only one-millionth of the incident sound energy is reflected. Condition (6.7) is fulfilled, and the boundary becomes completely transparent at an incident angle of 45°. Toward criticality at 4° the reflection factor changes rapidly, for there eq. (6.9) becomes

$$f = 1 - 60\,[\Theta_1 - (\Theta_1)_{\text{crit}}]. \tag{6.10}$$

"Abrupt boundary" or "gradual transition" is an almost meaningless expression unless there is some unit of measure. Such a measure is the wave length of sound. Whereas a farm plat may look flat and homogeneous to an audible wave length of 5 m., it is a rugged, heterogeneous aggregate to ultrasonic waves having only 1 mm. between compressions. The transition phase between clear air and dense cloud can be an abrupt boundary to waves of 1 km. in length, yet appear most gradual to waves 30 cm. in length. And although the equations of this section dealing with reflection have contained no wave length or frequency terms, a boundary having a thickness small compared with these wave lengths has been implied. Shaped obstacles, such as an echelon, can of course cause wave-length-selective reflection; but natural terrain, forests, vegetation-covered fields, building-spotted villages, and even clouds usually reflect long waves more efficiently simply because these surfaces cause more destructive interference among short waves. A practical value for the long-distance reflection factor of topsoil, as determined from explosions (Sect. 10), is substantially less than shown in Fig. 5.

7. Transmission channels through the atmosphere. Any region of the atmosphere between two levels of equal sound velocity, provided velocity is depressed between the two levels, may be properly called a transmission *channel*. Sound velocity at any altitude is really a vector quantity because wind is involved. The vertical region constituting a transmission channel in one azimuthal sector may not be a transmission channel for any other sector.

Sound rays emanate in all directions from a point source. If such source is placed anywhere in the region between two air levels having greater sound velocity than that at the source, some rays emitted with an upward component will be refracted downward, some of those emitted with a downward component will be bent upward.

In a practical sense, at least for long wave lengths, the earth can be regarded as a possible boundary of a channel for transmission of sound through the atmosphere. Suppose that, for some azimuthal sector, the maximum value of sound velocity through the (known) atmosphere occurs at an elevation above the earth boundary, and a nondirectional sound source is placed between the earth's surface and the altitude of maximum velocity. Rays starting downward will bump the ground and be reflected. Some of those rays emitted upward and some of those once reflected from earth will be refracted downward. Following the first set of escapes through the upper layers, ray paths remaining in the transmission channel consist of those alternately refracted downward and reflected upward[1].

As in the study of optics, treatment of propagation by ray construction ignores certain phenomena associated with the wave nature of sound. Optical ray paths define sharp shadows and areas of luminous intensification. Sound ray paths through the atmosphere define zones of intensification and sharp zones of silence.

Actually some sound energy crosses the borderline from zones of noise into zones of silence. "Shadow" lines are not sharp, and diffraction of energy, being a wave phenomenon, causes greater amplitudes [*16*] of low-frequency sound than of high-frequency sound at any selected point within a zone of "silence"; intensity of monochromatic sound falls essentially exponentially with distance from the shadow boundary and exponentially as $\nu^{\frac{1}{3}}$. Thus in any practical problem dealing with sound propagation through the atmosphere, lack of precise micrometeorological knowledge invariably masks noticeable diffraction for all except exceedingly infrasonic frequencies.

8. Sources and receivers. Anything that makes a noise can of course be used as a source to study sound transmission through air. But the remaining four sections of this chapter really deal with open-air propagation to distances exceeding ten kilometers, and only a few sources are capable of supplying measurable

[1] On low velocity channels in the ocean see the chapter on surface waves and guided waves in Volume XLVII, Sect. 6, p. 136.

perturbation pressures or intensities at these distances. *Explosions*, either planned or accidental, have been the sources most fruitful in supplying data.

TNT releases 4.6×10^6 joules of energy per kilogram weight! It and other forms of explosives are effective and relatively inexpensive sources for long-distance sound-propagation studies.

Naturally, too, anything able to detect a sound and distinguish it from background noise may be a *receiver* in simple acoustical studies. Human ears meet these requirements over a remarkable frequency range—ten octaves—and in association with good timepieces have supplied much fundamental data in the subject of sound.

But a scientist is generally interested in obtaining quantitative measurements, preferably recordings, of intensity or some related alteration such as pressure; and when possible, he desires to express results in standard, not purely relative, units. The human ear is not readily calibrated; it fails to register ultrasonic and infrasonic frequencies, and it can provide no records for later and repeated study.

Calibration of receiving and recording systems for absolute measurements in the audible or ultrasonic frequency ranges is indeed intricate work [*17*]. But again in view of the fact that the following four sections deal with long-distance transmission, much of it through the rarefied atmosphere above the tropopause, the remainder of this section will discuss *low-frequency equipment.*

Receivers placed on flat level land at distances between 10 and 100 km. from a point source on the ground, such as a TNT explosion, may or may not find a sound signal. If toward the azimuth of the receivers sound velocity decreases with altitude and never exceeds that at ground level, sound rays from the source curve upward and overpass receivers. On the other hand, a positive gradient in sound velocity extending up from earth level forms a transmission channel (Fig. 7), and the ground toward this azimuth is completely sound-covered. By the time sound waves reach distances large compared with channel depth, most rays have struck ground many times. Ultrasonic and most audible frequencies higher than low-bass rumbles have disappeared through destructive interference on reflection. Energy remaining in the sound signal is essentially all infrasonic.

Should it occur that, toward the receivers, a negative gradient in sound velocity connects with the earth, and this negative gradient is topped by a positive gradient extending to a velocity exceeding the earth level value (Fig. 10), a transmission channel exists, but the ground is not completely sound-covered. *Zones of silence* and *zones of sound* are separated by certain select rays. Except for the diffracted energy discussed in the previous section, receivers in "silent" zones observe no signals. However, in the zone of noise on its earth strike nearest the source, sounds from an explosion contain much high-frequency as well as low-frequency energy. Zones of sound resulting from successive reflections off the earth contain less and less high-frequency energy. Good measurements of acoustic intensity or perturbation pressures near an explosive source then require receivers having broad frequency bands spanning the infrasonic, audible, and even part of the ultrasonic ranges. The author doubts whether such instruments now exist. Infrasonic instruments are sufficient for far-away measurements. "Nearby" and "far-away" derive their meanings from the meteorological situation at the time of measurement.

With but few exceptions only infrasonic frequencies and on rare occasions low-bass rumbles are observed at 100 km. or more from even gigantic explosions. Cylindrical divergence reduces all intensities, and atmospheric absorption and scattering on reflection take out the ultrasonic frequencies even for allowed paths through the troposphere. Sound signals observed at distances beyond 100 km.

have, however, usually traveled from source to receiver by paths reaching to altitudes greater than 30 km., where absorption takes heavy toll of all but infrasonic energy. Truly long-distance acoustic measurements may thus be made exclusively with infrasonic instruments.

While an instrument for measuring and recording infrasonic *intensity* would be desirable for acoustic experimentation, the author has not heard of its invention. Instead, experimenters have been (or have had to be) satisfied with pressure-perturbation recording devices, or those which record rate of change of pressure. Both types are usually named *microbarographs*, although several so-named should be called *milli*barographs if *bar* is to mean 10^6 dynes cm.$^{-2}$.

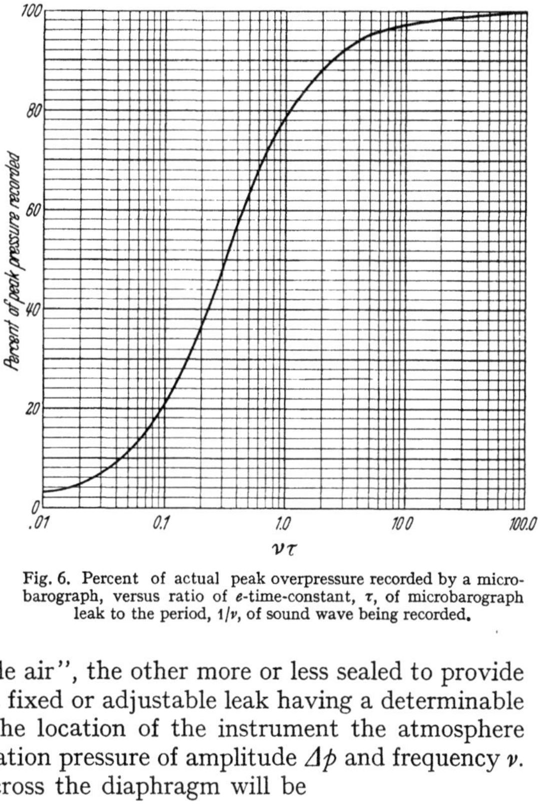

Fig. 6. Percent of actual peak overpressure recorded by a microbarograph, versus ratio of e-time-constant, τ, of microbarograph leak to the period, $1/\nu$, of sound wave being recorded.

Just as a barometer or barograph really measures pressure *difference* between (Torricellian) vacuum and (unperturbed) air pressure, so a microbarograph attempts to record small variations of air pressure from "ambient" [*18*]. Immediate problems arise, frequency- or period-wise, in attempting to define just what is meant by "ambient". Natural meteorology and wind turbulence cause pressure variations which may be random or periodic, slow or rapid, weak or strong. What, then, does "ambient" pressure mean?

By construction, each microbarograph instrument gives its own graduated definition. A diaphragm or BOURDON tube separates two "chambers", one exposed rather directly to "outside air", the other more or less sealed to provide a reference pressure. The seal is a fixed or adjustable leak having a determinable e-time-constant, τ. Suppose at the location of the instrument the atmosphere contains a periodic sound perturbation pressure of amplitude Δp and frequency ν. The difference in pressure, p_d, across the diaphragm will be

$$p_d = (\Delta p)\,[\exp(-t/\tau)]\,[\sin 2\pi\nu t], \tag{8.1}$$

and the ratio of the maximum pressure difference to the true sound pressure amplitude is

$$\frac{(p_d)_{\max}}{\Delta p} = \left[\exp\left(-\frac{\arctan 2\pi\nu\tau}{2\pi\nu\tau}\right)\right][\sin(\arctan 2\pi\nu\tau)]. \tag{8.2}$$

When the e-time-constant of the leak is equal to the period of perturbation, maximum pressure difference between the two "chambers" is 79% of the sound pressure (Fig. 6). The diaphragm essentially ignores sound pressures having periods greater than 25 times the e-time-constant of the leak since the maximum pressure difference across the diaphragm is less than 10% of the true variation.

Reference to the first sentence of this chapter, the definition of sound, shows the impossibility of distinguishing completely between *noise* produced by natural causes and *sounds* we desire o tmeasure. Both are really sound. The scientist must simply decide in advance of experimentation which frequencies he chooses to record and which he will ignore and adjust the microbarograph leak accordingly.

Small turbulent cells, which produce considerable noise background on a microbarograph, travel with velocities quite different from sound velocity. Advantage may be taken of this fact to reduce background noise. Instead of a single opening to the "exposed" chamber of a microbarograph, several pipes or hoses [*19*] may be run from that chamber to afford a random distribution of parallel openings. With openings spaced over distances small compared with desired wave lengths, yet large compared with turbulent cell dimensions, considerable improvement in signal-to-noise ratio is possible.

Some type of transducer is attached to the diaphragm or Bourdon tube in a microbarograph. This may be a mirror between light source and moving photographic paper, or the coil of an electromagnetic pickup, or one side of a capacitance, or the magnetic shunt of an inductance. Each system [*18*] has some advantages. Several types of microbarograph are capable of being calibrated in standard units by means of manometers.

9. Blast damage. Use of explosives as a sound source for acoustical propagation studies has some inherent dangers. Handling of explosive material with all respect due it reduces hazards to personnel. Careful advance study of weather factors will enable the experimenter to avoid or reduce property damage and accompanying lawsuits.

The infinite number of objects classed as property, and the associated multitude of natural oscillation frequencies, makes impossible any prediction of damage resulting from resonance—wolf notes—induced in the objects by incident sounds. Similarly damage to wall and ceiling plaster in residences is essentially unpredictable since plaster may be adhering well or hardly at all to the laths.

Window-glass breakage is better behaved. Large plate-glass picture or display windows crack or break when *peak* pressure undulations of the outdoor air exceed about 1.5 millibars. Eight-millibar peak pressures definitely break small ($\sim$25 cm. on a side) windows. Open doors and windows allow buildings to breathe, indoor pressures follow outdoor variations, pressure differences are lessened, and damage is reduced.

10. Simple cases of refraction in the troposphere [*20*]. While bombs, artillery shells, and aeroplanes emit noise at points off the earth's surface, many more experiments are conducted with the source adjacent to terra firma. This and the two following sections of this chapter therefore treat propagation of sound from sources at the air-earth boundary only. Because of the great differences in densities of soil and air, essentially all energy from the source enters the upper hemisphere.

Basic equations for computing ray paths and associated sound travel times, t, are combinations of eqs. (4.2) or (4.3) with geometric definitions

$$dR_\varphi = \cot\Theta_h \cdot dh \tag{10.1}$$

and

$$dt = \frac{dh}{(V_h)_\varphi \sin\Theta_h}, \tag{10.2}$$

where R_φ is horizontal distance and h is altitude measured from the source. The sound velocity component, $(V_h)_\varphi$, and ray elevation angle, Θ_h (Sect. 4), are functions of altitude h.

With the understanding that the basic formula for refraction, eq. (4.3), applies to a particular azimuth, φ, the subscript (φ) will henceforth be omitted.

Case 1.—Sound velocity generally decreasing with altitude, never exceeding its value at the earth's surface.—Sound rays emanating from the source curve upward; none return to earth. Except for diffracted energy (Sect. 7) a zone of silence extends from very near the source out to about 100 km., where ozonosphere-returned signals (Sec. 11) may be observed.

Case 2.—Sound velocity increasing linearly with altitude from earth to top of the inversion at height h_i, decreasing thereafter (Fig. 7).—Rays leaving the source at angles between grazing and arc cos V_0/V_i curve downward enough to strike earth. Rays starting at steeper angles escape from the inversion channel.

The limiting ray, that starting at $\Theta_0 = \text{arc cos } V_0/V_i$ and becoming horizontal at height h_i, strikes earth at a distance

$$R_{\max} = 2h_i \sqrt{\frac{V_i + V_0}{V_i - V_0}} \approx 2.8\, h_i \sqrt{\frac{V_0}{V_i - V_0}}\,. \tag{10.3}$$

Rays starting at lesser angles strike earth between the source and $R_{\max}$ at distances

$$R = \frac{2h_i V_0}{V_i - V_0} \tan\Theta_0, \tag{10.4}$$

and all rays duplicate their path forms over and over again by reflection and refraction.

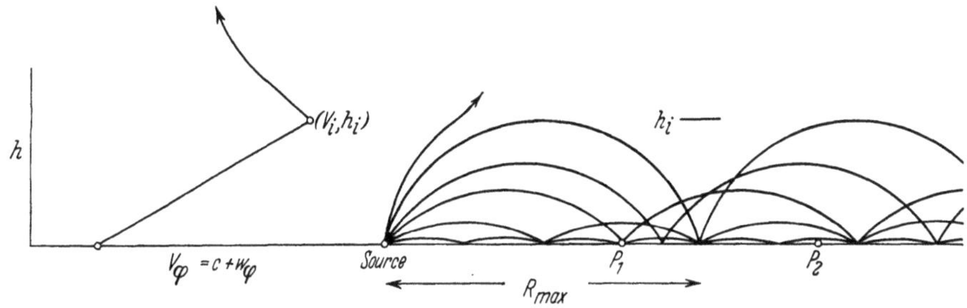

Fig. 7. When sound velocity toward some azimuth increases uniformly with altitude, sound rays strike ground at all points in the azimuth.

Every spot of earth in this direction from the source has an infinite number of rays striking it. Sound travel time, t, from source back to earth depends on sound-velocity gradient and starting angle:

$$\left.\begin{aligned} t &= \frac{2h_i}{V_i - V_0} \log\tan\left(\frac{\pi}{4} + \frac{\Theta_0}{2}\right) \\ &= \frac{h_i}{V_i - V_0} \log\left\{\frac{\sqrt{R^2 + \left(\frac{2h_i V_0}{V_i - V_0}\right)^2} + R}{\sqrt{R^2 + \left(\frac{2h_i V_0}{V_i - V_0}\right)^2} - R}\right\}. \end{aligned}\right\} \tag{10.5}$$ [1]

Higher-traveling sound arrives first at any selected point, the bouncing sound taking slightly longer. As a result of the spread in travel times, an explosion under these meteorological conditions is heard as a rumble instead of a sharp crack.

Let us now proceed for this particular meteorological case to derive [*21*] a prediction of intensity or perturbation pressure at any point on the ground when a point source emitting power E is at the ground-air interface.

The increment of power emitted through an elemental solid angle at starting elevation angle Θ_0 will be

$$dE = E \cdot \cos\Theta_0 \cdot d\Theta_0 \cdot \frac{d\varphi_0}{2\pi}\,. \tag{10.6}$$

Provided Θ_0 is less than arc cos V_0/V_i, this increment of power will impinge on the earth between radii R and $R + dR$ and cover an elemental area

$$dA = dR \cdot R \cdot d\varphi_0. \tag{10.7}$$

[1] Some writers use different mathematical terminology: $(V_i - V_0)\, t/2h_i$ is called the *anti-gudermannian* of Θ_0; Θ_0 is the *gudermannian* of $(V_i - V_0)\, t/2h_i$.

The area normal to the direction of sound propagation will be $dA \cdot \sin \Theta_0$, so the *direct* sound intensity at the surface will be

$$I_{\text{direct}} = \frac{E \cot \Theta_0}{2\pi R} \cdot \frac{d\Theta_0}{dR}, \tag{10.8}$$

which, according to eq. (10.4), is for *this meteorological case only* equivalent to

$$I_{\text{direct}} = \frac{E \cos^2 \Theta_0}{2\pi R^2}. \tag{10.9}$$

In a practical sense the $\cos^2 \Theta_0$ term may be taken as unity since arc cos V_0/V_i is never very large.

Total incident intensity at a point on the ground consists of that coming directly from the source, plus that received after one reflection from an area one-fourth as large at half the distance, plus that received after two reflections from an area one-ninth as large at one-third the distance, etc. Some energy is lost on each impact (Sec. 6), the reflected intensity being f times the incident intensity. The point of intensity measurement may lie within the distance $R_{\max}$ computed from eq. (10.3), like point P_1 of Fig. 7, or beyond it at some point like P_2. At the point of measurement reflection raises the intensity by a factor $(1+f)$. Total intensity at a distance R from the source will therefore be

$$\left.\begin{aligned} I &= \frac{E(1+f)}{2\pi R^2} \sum_{n=N}^{n=\infty} f^n \\ &= \frac{E}{2\pi R^2} \left(\frac{1+f}{1-f}\right) f^N, \end{aligned}\right\} \tag{10.10}$$

where N is the whole number of times $R_{\max}$ from eq. (10.3) divides into R. The root-mean-square perturbation pressure, measured side-on to the intensity front, is

$$p_{\text{RMS}} = \sqrt{\varrho_{\text{air}} V_0 I}. \tag{10.11}$$

If instead of a continuous source having power E one employs an explosive releasing energy W, the acoustical energy passing through a unit area normal to the direction of propagation will be

$$\xi = \frac{(\varepsilon W)}{2\pi R^2} \frac{(1+f)}{(1-f)} f^N, \tag{10.12}$$

where ε is the fraction of released energy which remains acoustical energy after the blast wave has traversed the distance R. Near the explosion, nonadiabatic processes [*3*], [*22*] of shock wave propagation leave behind in the form of heat much of the originally released energy. For an exploding spherical TNT charge of radius r, an experimentally determined value [*23*] of ε, satisfactory for ratios of R/r greater than 40, is

$$\varepsilon = (4.7 + 1.1 \times 10^3\, r/R) \text{ per cent}. \tag{10.13}$$

Thus except for places close by large explosions, the efficiency factor is 4.7%.

Perturbation pressure is generally of more interest than normal energy density. A relation between the two is

$$\xi = \int_0^\tau \frac{p^2\, dt}{\varrho_{\text{air}} V_0}, \tag{10.14}$$

where integration extends for the effective time duration, τ, of the signal. Furthermore, one usually prefers to learn the *peak* value of overpressure instead of the root-mean-square value. If the pressure-versus-time wave form were sinusoidal, the p^2_{peak} would equal $2p^2_{\text{RMS}}$; if the wave form resembled the letter N (see ref. [*3*]), then p^2_{peak} would equal $3\,p^2_{\text{RMS}}$. Fig. 8 shows that both types, and others, are ordinarily present; practical predictions of peak overpressure are satisfactory if one assumes $p^2_{\text{peak}} = 2.5\, p^2_{\text{RMS}}$.

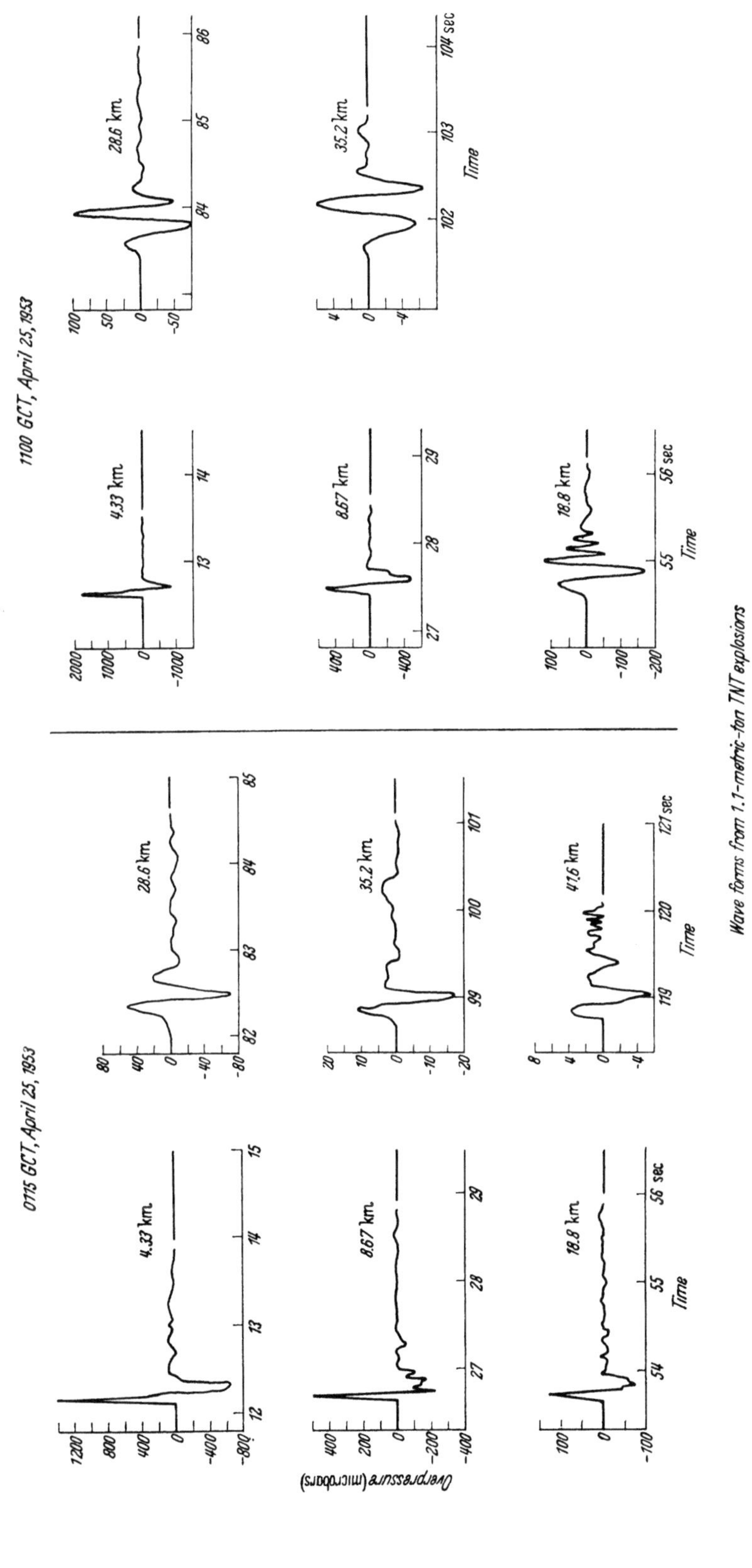

Fig. 8. Microbarograph records at increasing distances from two 1.1-metric-ton TNT explosions under conditions similar to those of Fig. 7. The wave form changes radically with distance, and signal duration increases slightly.

Little experimental and no known theoretical effort has been directed toward determining effective signal duration. A reasonable value of τ as observed at overpressures less than five per cent of a standard atmosphere (Fig. 8) appears for the linear inversion situation, Fig. 7, to be

$$\tau = 1.14\,(r R)^{\frac{1}{2}}/V_0\,(1-f)\times(1+R/R_{\max})^{\frac{1}{4}}\ \text{sec}. \tag{10.15}$$

For more complex situations, such as that shown by Fig. 10, a satisfactory expression for τ has not been determined.

Similarly, little effort has been devoted toward determining the energy reflection factor, f, over real terrain. Destructive interference reduces f significantly below the ideal values shown in Fig. 5 to about 0.7.

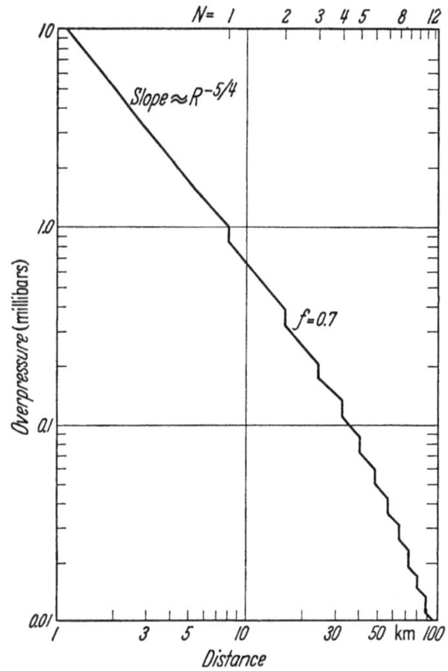

Fig. 9. Plot of peak overpressure versus distance from a surface explosion at sea level of one metric ton of TNT when the sound velocity increases linearly with altitude and gives a maximum direct range of 8 km.

By joining these empirical values together and simplifying by means of the LAPLACE equation (2.1), an expression is obtained for predicting relative peak overpressure for the linear inversion situation, Fig. 7:

$$\left.\begin{aligned}(p_{\text{peak}}/P) = 1.4\times 10^2\,(0.7)^{N/2}\times\\ \times\,(r/R)^{\frac{5}{4}}\,(1+R/R_{\max})^{\frac{1}{4}}\end{aligned}\right\}\tag{10.16}$$

A graph of this equation for a typical case is shown as Fig. 9.

Both scaling laws for explosive phenomena [24] are contained in and limited by eqs. (10.15) and (10.16). *Under similar weather conditions* ($R_{\max}$ the same), distances to equal overpressures, and time values at radii of equal overpressures, are proportional to the cube roots of charge weights.

Case 3.—Sound velocity decreasing linearly with altitude from earth to height h_d, then increasing linearly to height h_i, decreasing thereafter (Fig. 10).—To be of interest by differing from Case 1, the sound velocity, V_i, at height h_i must be larger than V_0. All rays curve upward until reaching height h_d after which they curve downward. Rays leaving the source at elevations steeper than $\arccos V_0/V_i$ will escape the troposphere; starting rays having lesser elevations will be trapped in the sound channel.

Sound will overpass a section of earth bordering on the source, then strike intensely in the manner of a focus. Whether the highest-traveling ray strikes near the inner or outer edge of the noise zone (Fig. 10) depends on the steepnesses and extents of the meteorological gradients.

Once they hit earth, sounds will be alternately reflected and refracted. Radial and tangential extents of successive noise zones increase as the number of the zone; intervening silent zones grow radially smaller with distance from the source until they disappear when adjacent noise zones overlap. In noise zones where overlap has not occurred, sound intensity at corresponding points in successive zones decreases at least as fast as the reciprocal square of the zone number.

Since incident intensity on the first strike is all direct, eq. (10.8) should be applicable. Fig. 10b shows that as the elevation angle, Θ_0, increases, R first decreases, then increases; hence eq. (10.8), derived from ray interpretation of sound phenomena, predicts infinite intensity at the inner boundary of the noise zone. Wave theory [16] significantly reduces intensity at this boundary, but agrees with ray sonics in proclaiming the boundary the place of maximum intensity.

Fig. 10 has been drawn with equal angular increments between emitted rays. Adjacent ray pairs thus bound equal increments of sound energy. The spread of these rays on striking earth gives a pictorial representation of the dispersal of energy. When the sound source is an explosion, damage to property may well be expected near the inner boundary of any close-in noise zone. Locations of these foci are determinable mathematically [9]. As stated in the preceding paragraph, computation of first-zone intensity may require more complex mathematics than ray construction, but practical answers do come from ray theory almost up to the point where $dR/d\Theta_0$ becomes zero.

More complicated meteorological situations may be treated mathematically, and computational procedures have been derived [9] for the general case. By applying such treatment to weather *forecasts* prior to explosive experimentation, claims for damage can usually be avoided.

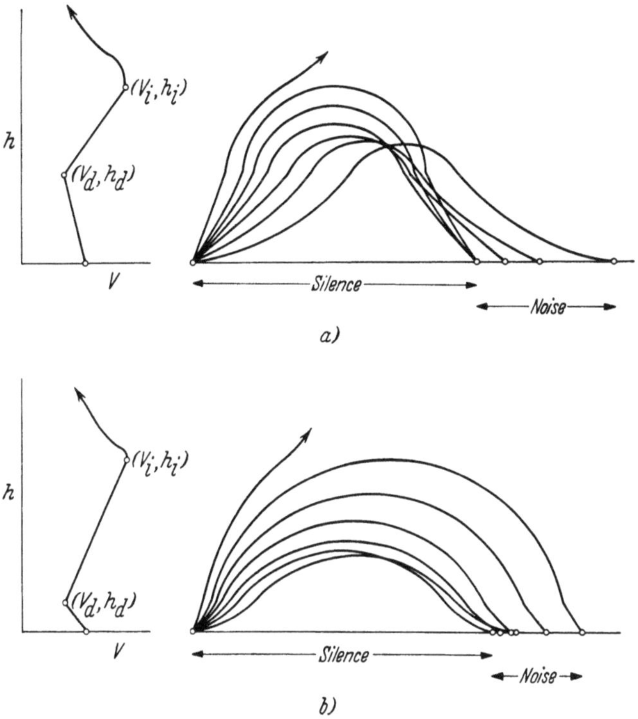

Fig. 10. a and b. When sound velocity toward some direction decreases, then increases, with altitude, a focus of energy is beyond a zone of silence.

11. Abnormal or anomalous sound propagation. Lengths of time required for sounds to travel from source to any receiver may be calculated using equations similar to (10.5), and the times will be found to differ by only tiny amounts for different allowable paths through the *troposphere*. Imagine, then, the surprise (Fig. 11) of an observer at a distance of 100 to 250 km from a single large explosion *when two distinct "booms" shock him*, the "booms" being separated in time by one to two minutes! First-arriving sound at such distances may be that coming through some troposphere transmission channel with a velocity readily computed from eqs. (4.1) and (3.1). This we call *normal* sound. The second signal, greatly delayed, has been called *abnormal* or *anomalous*.

Significantly retarded signals are never observed closer to the source than about 80 to 100 km., and the minimum distance, at least in the north temperate zone, varies [19], [24a] with season. When the troposphere contains no allowed transmission channel in the direction of observation, abnormal sound will be the first signal reaching earth in that direction after skipping over a large zone of silence.

In explanation of the skip phenomenon, first recognized [24b] during Queen Victoria's funeral in 1901, scientists first noted its similarity with Case 3 of the preceding section. Temperature, and hence sound speed, c, was generally believed in the early twentieth century to decrease with altitude. High-velocity winds at high elevations could create the skip zone and the far-distant noise zone in one direction from the source.

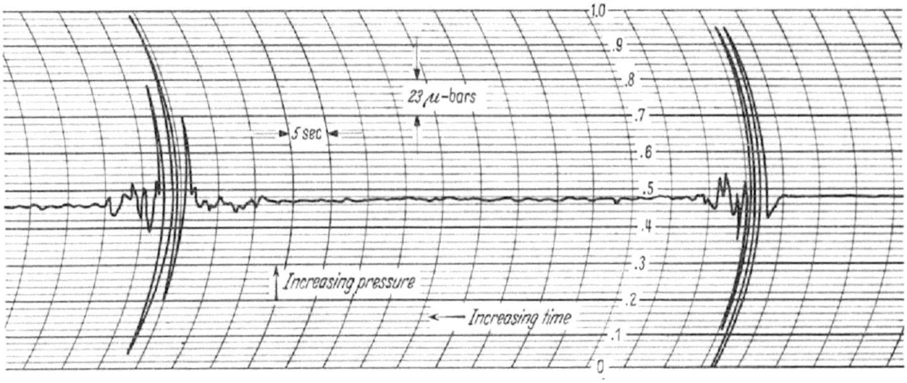

Fig. 11. Microbarograph records of two sounds, normal and abnormal, from a *single* explosion 182 km distant. The anomalous signal follows the normal signal by 77 sec.

But on December 14, 1903, the area of abnormal audibility from an explosion in Förde (Westfalen, Germany) subtended a horizontal angle of 100°, and an explosion at Witten-Annen, November 28, 1906, gave 150°. This lack of directionality, further substantiated when the abnormal zone *encircled* an explosion in Moscow on May 9, 1920, could not be explained by vector winds. Percentage increase, with altitude, of low-molecular-weight gases such as helium and hydrogen [eq. (2.3)] was considered [25] a possibility, but this hypothesis found little following.

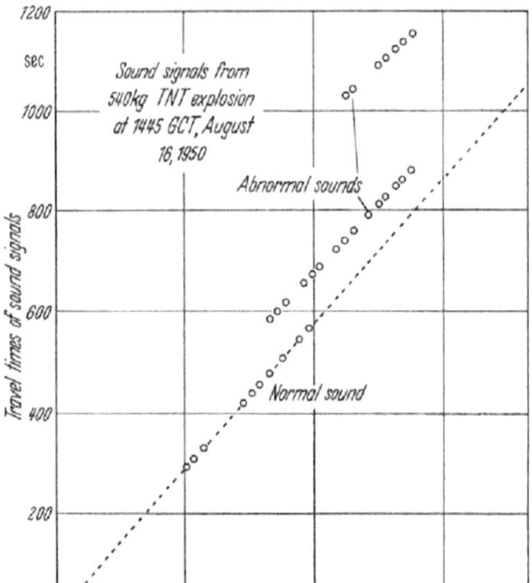

Fig. 12. Sound signal travel times versus distance between an explosion and a radial line of spaced microbarographs (JOHNSON and HALE [19]). Except during spring and autumn reception of both ozonosphere and ionosphere signals in the same direction from the shot point is rare.

Belief in cold upper atmosphere temperatures was shaken in 1922 when astrophysicists [26] found it impossible to account for the brightness of shooting stars—meteors—without hypothesizing at least earth-level temperatures at altitudes of around 60 to 80 km. On reading this revolutionary paper, another Briton, F. J. W. WHIPPLE, recognized there a potential explanation of anomalous sound and showed [27] that skip distances derived from such a theory could fit facts.

Data on observed sound travel times from scheduled explosions (Fig. 12) could now be employed to explore meteorology at altitudes greater than those reached by weather balloons. The problem was similar to that of seismology, inverted, of course, wherein man attempts to learn the structure of the earth at depths far below mines and wells. It is not surprising then to find seismologists ([*28*] to [*30*]) taking the lead in "measuring" temperatures above the isothermal layer, although F. J. W. WHIPPLE also remained active [*31*] in this field for a number of years. Recent publications, [*32*] to [*35*], give interesting historical accounts of progress, and of regress and digress, in these acoustical investigations at altitudes greater than 30 km.

Modern theory accredits raised temperatures in the 35 to 65 km. shell to ozone absorption of solar radiation even though the height of maximum temperature is considerably higher than the greatest ozone concentration [*36*]. Sound velocities in this shell appear to decrease trivially, if at all, at night [*33*]; and while this observation may be explainable for short nights, anomalous sound signals have also been recorded in the Arctic winter shadow [*37*]. This could not be a result of sunlight absorption there. Warm air must have moved into the Arctic shadow at those altitudes, or winds sufficient to counterbalance lower temperatures must have been present.

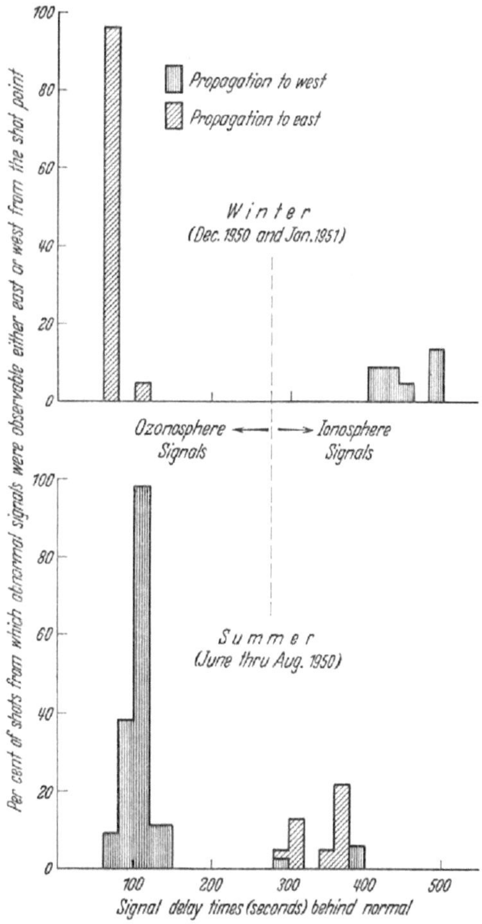

Fig. 13. Because of upper-atmosphere winds, anomalous sounds strike earth in preferred directions. Over southwest USA during summer and winter principal directions of ozonosphere and ionosphere winds are evidently opposing. During spring and autumn, preferred directions are less clear-cut. (From JOHNSON and HALE [*19*], corrected.)

Troposphere meteorology naturally affects the striking points of even those sound waves returning to earth from the ozonosphere. Yet in places where *measured* troposphere and isothermal-layer winds are consistently from the west, anomalous sounds are found to be inconsistent in a statistically predictable manner. In the north temperate zone, when the earth-level temperature is constant over the entire region of experimental observation[1], ozonosphere-returned sound signals are more prevalent (Fig. 13) west from the source in summer, east from the source in winter. Spring and autumn months produce a mixture. Lack of symmetry must result from winds.

When ozonosphere signals can be obtained in opposing directions from the source, the mean of computed ray-vertex velocities can be related to air temperature at the height of the vertex, and the difference gives some information about ozonosphere winds. The most comprehensive studies to date of both temperatures and winds at 30 to 60 km. above sea level are contributions [*33*], [*38*] of the United States Air Force Cambridge Research Laboratories, who have conducted

[1] Generally false when the area of observation spans the English Channel.

or arranged experiments in the Panama Canal Zone, Bermuda, Alaska, Southwest USA, East Coast (USA), and the Territory of Hawaii. They have found, in general, that temperatures reported by earlier experimenters (e.g., refs. [*28*] to [*31*]) are too high, the result of having assumed winds to be insignificant. Table 1 and Fig. 14 are extracted from their latest publications.

Thirty kilometers or so above the warmest level in the ozonosphere the temperature evidently reaches a low comparable to that at the tropopause. In this cold region at altitudes around 80 km., where noctilucent clouds sometimes

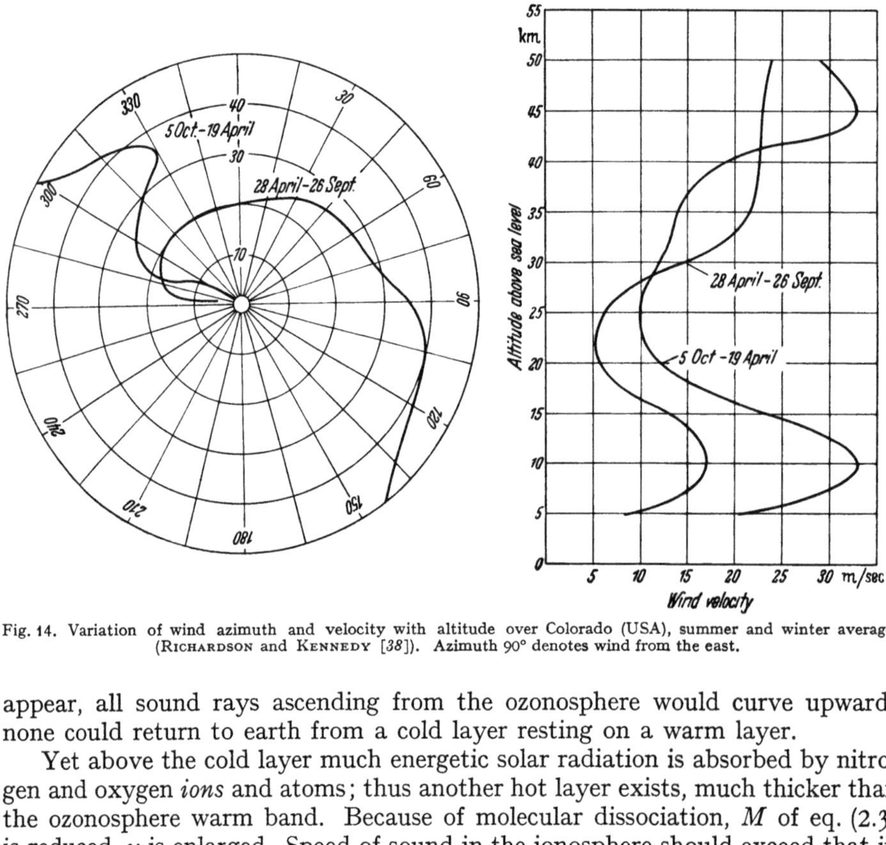

Fig. 14. Variation of wind azimuth and velocity with altitude over Colorado (USA), summer and winter averages (RICHARDSON and KENNEDY [*38*]). Azimuth 90° denotes wind from the east.

appear, all sound rays ascending from the ozonosphere would curve upward; none could return to earth from a cold layer resting on a warm layer.

Yet above the cold layer much energetic solar radiation is absorbed by nitrogen and oxygen *ions* and atoms; thus another hot layer exists, much thicker than the ozonosphere warm band. Because of molecular dissociation, M of eq. (2.3) is reduced, γ is enlarged. Speed of sound in the ionosphere should exceed that in equally high temperature bands nearer the earth. Consequently one might expect the hot ionosphere to send back to earth some of the sound rays which left the source at elevation angles too great for them to be captured by the ozonosphere-to-earth sound channel.

Discovery and identification of ionosphere-returned sound signals [*39*] had to await development of sensitive infrasonic recording equipment, a scheduled large explosion, and most important of all, a low background level of infrasonic noise at explosion time. The Helgoland blast on April 18, 1947, provided the necessary combination.

Fig. 4 shows clearly why earlier experimenters had not found or recognized ionosphere signals. Their minimum striking distance appears to be 200 to 250 km., of which perhaps 100 km. has been traversed at heights greater than 40 km. Absorption takes a high toll of all but deep [*40*] infrasonic-frequency energy. Human ears are ineffective for its observation.

Table 1. *Calculated temperatures and winds* (from CRARY [33], [38]).

Altitude (km.)	Temperature (°C)	Wind Speed (m. sec.$^{-1}$)	Wind azimuth (degrees[1] from north)
Alaska, January-February 1948			
20	−64	21	202[2]
25	−67	29	202[2]
30	−68	38	202[2]
35	−68	48	202[2]
40	−65	56	202[2]
45	−58	60	202[2]
50	−45	63	202[2]
55	−28	64	202[2]
60	−8	63	202[2]
Alaska, August 1949			
20	−57	9	189
25	−55	8	186
30	−52	13	140
35	−44	18	126
40	−27	20	122
45	−2	17	124
50	+19	7	171
Bermuda, July 1949			
20	−59	9	313
30	−35	3	107
40	−27	34	89
50	−9	49	82
Bermuda, December 1950			
20	−54	15	260
30	−40	25	283
40	−28	47	271
50	−13	56	240
Canal zone, July—August 1948			
30	−65	28	87
35	−56	38	97
40	−38	44	97
45	−17	42	105
50	+9	52	105
55	+35	40	107
Canal zone, January 1950			
20	−64	1	298
25	−56	4	223
30	−47	11	211
35	−33	21	207
40	−13	22	199
45	+10	26	186
Canal zone, November 1950			
25	−58	10	216
30	−30	31	216
35	−39	10	143
40	−14	27	124
45	+5	21	82
Florida, January 1951			
20	−73	32	266
25	−66	27	262
30	−56	22	261
35	−37	18	258
40	−8	11	252

In southwestern United States during a one-year period in which 214 explosions were made of 540 kg. of TNT each, ozonosphere-returned signals were observed sixty per cent of the time but ionosphere signals on only 24% of the shots (Fig. 13). In winter when ozonosphere signals struck earth only east from the source, ionosphere returns were generally west from the source; in summer the reverse was true, seeming to show statistically that ionosphere winds are opposite to those of the ozonosphere at that location.

A composite picture of anomalous sound-ray paths is presented as Fig. 15. As in Case 3 (Sect. 10) foci or caustics appear near inner boundaries of abnormal sound zones, both those produced by the ozonosphere and those produced by the ionosphere. Also as in Case 3, successive zones of sound increase in area, successive silent zones grow smaller; zones of sound eventually overlap. In a region such as at 380 to 500 km. in Fig. 15, the experimentalist must be careful to distinguish between arrival times of first-strike ionosphere signals and second-strike ozonosphere sounds. Identification is usually aided by the absorption curve (Fig. 4); although the main energy content of both signals will be infrasonic, ozonosphere signals will contain a higher percentage of high-frequency components because of having traveled less distance at great altitudes.

[1] Wind from the north and from the east have azimuths 0° and 90°, respectively.
[2] Approximate

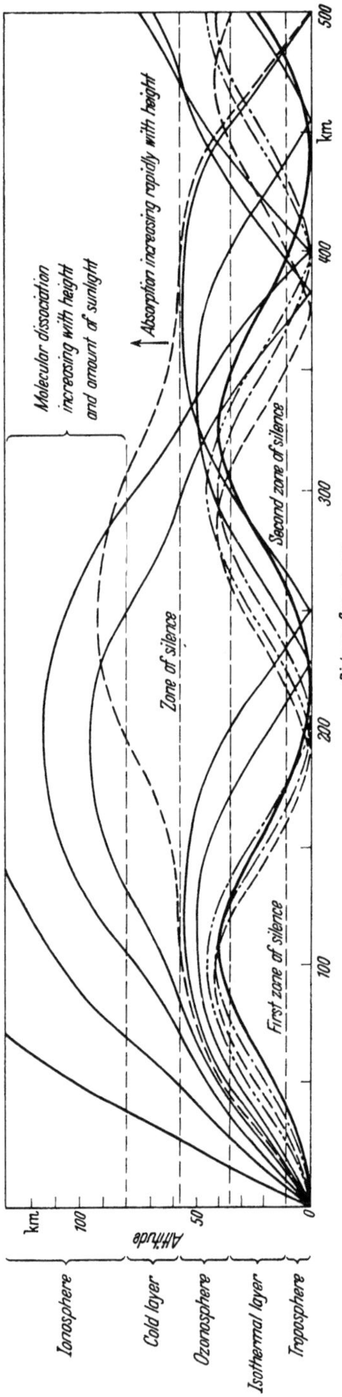

Fig. 15. Paths of abnormal sound signals (GUTENBERG [35]).

Too little is known today of wind and temperature structures and their variation at altitudes greater than 30 km. to allow more than educated guesses about the perturbation pressures that will strike earth in anomalous sound zones when explosives are fired. Minimum striking distances of ozonosphere returns, [*29*], [*35*] when plotted versus month of the year, give statistically a sinusoidal curve having a least value, 100 km., in February, and a maximum value, 200 km., in August. Distance to the caustic or point of maximum intensity is also sinusoidal and in phase with minimum striking distance but extends from about 125 to 240 km. Fortunately, though abnormal sound intensities cannot today be well predicted, even those from atomic bombs [*41*] seldom cause damage.

Rockets have confirmed [*12*] the 1922 hypothesis [*26*] of upper-atmosphere temperatures, and consequently F. J. W. WHIPPLE's explanation [*27*] of abnormal sound. But whereas sound measurements were able to contribute significant *temperature* information for the ozonosphere, it is doubtful because of their inability to distinguish among the M, γ, and T terms of eq. (2.3), whether they can help in this particular manner for the ionosphere. Infrasonic measurements should, however, continue to supply wind information for 30 to 60 km. heights until weather balloons take over the job—probably soon; they should also supply wind information for the ionosphere and sound-speed data for the ionosphere to be used in computing Mach numbers of rockets.

12. Natural oscillations of the atmosphere[1]. Gigantic explosions—Krakatoa in 1883, the Great Siberian Meteor in 1908, and possibly hydrogen bombs—are also sources of "sound" in its general definition. Both normal and abnormal sound zones would be expected from them.

But in addition these tremendous releases of energy excite the atmosphere to oscillate in its normal modes, and oscillating pressure waves may encircle the earth [*42*] at a speed of about 310 m. sec.$^{-1}$. Significance of these measurements is treated elsewhere in this volume; they are mentioned here largely as a note

[1] See Chapter by W. KERTZ on Atmospheric Tides in this Volume.

to experimentalists. Abnormal sounds were so named in part because of their abnormally slow apparent speeds from source to observers. An apparent speed of 310 m. sec.$^{-1}$ is certainly in the range of possibility, even probability, for ozonosphere and ionosphere signals. Abnormal signals might therefore be confused with natural-period signals; however, the ozonosphere or ionosphere would be expected to produce a variation in *interval* velocities between successive receivers, whereas one would predict that interval velocities would remain constant for natural-period signals.

References.

[*1*] American Standard Acoustical Terminology, American Standards Association, Inc., New York 1951.

[*2*] FLETCHER, H.: Bell. Labor. Rec. **13**, 130 (1935).

[*3*] DUMOND, J. W. M., E. R. COHEN, W. K. H. PANOFSKY and E. DEEDS: J. Acoust. Soc. Amer. **18**, 97 (1946).

[*4*] CONDON, E. U.: Amer. J. Phys. **1**, 18 (1933).

[*5*] GEIGER, RUDOLF: Mikroklima und Pflanzenklima. In Handbuch der Klimatologie, Vol. 1, Part D. Berlin: Gebrüder Borntraeger 1930. — See also the chapter on Dynamic Meteorology in this Volume.

[*6*] SCHILLING, H. K., C. E. DRUMHELLER, W. L. NYBORG and H. A. THORPE: Amer. J. Phys. **14**, 343 (1946).

[*7*] SCHILLING, H. K., M. P. GIVENS, W. L. NYBORG, W. A. PIELEMEIER and H. A. THORPE: J. Acoust. Soc. Amer. **19**, 222 (1947).

[*8*] FUJIWHARA, S.: Bull. Cent. Meteor. Obs. Japan **2** (1) (1912) and **2** (4) (1916). — MILNE, E. A.: Phil. Mag. **42**, 96 (1921).

[*9*] COX, E. F., H. J. PLAGGE and J. W. REED: Bull. Amer. Meteor. Soc. **35**, 95 (1954).

[*10*] LINDSAY, R. B.: Amer. J. Phys. **16**, 371 (1948).

[*11*] SCHRÖDINGER, E.: Phys. Z. **18**, 445 (1917).

[*12*] The Rocket Panel: Phys Rev. **88**, 1027 (1952).

[*13*] KNUDSON, V. O.: J. Acoust. Soc. Amer. **5**, 112 (1933). — KNESER, H. O.: J. Acoust. Soc. Amer. **5**, 122 (1933). — PIELEMEIER, W. H.: J. Acoust. Soc. Amer. **16**, 273 (1945); see also ref. [*7*].

[*14*] For a more complete treatment, see LORD RAYLEIGH, Theory of Sound, Dover Publication (1945), Sect. 270.

[*15*] HEILAND, C. A.: Geophysical Exploration. Prentice-Hall 1940.

[*16*] PEKERIS, C. L.: J. Acoust. Soc. Amer. **18**, 295 (1946). — HASKELL, N. A.: J. Appl. Phys. **22**, 157 (1951).

[*17*] BERANEK, L. L.: Acoustic Measurements. New York: John Wiley & Sons 1949.

[*18*] SHAW, W. N., and W. H. DINES: Quart. J. Roy. Meteor. Soc. **31**, 39 (1905). — MACELWANE, J. B., and J. E. RAMIREZ: Trans. Amer. Geophys. Un. **19**, 125 (1938). — BENIOFF, H., and B. GUTENBERG: Bull. Amer. Meteor. Soc. **20**, 421 (1939). — SAXER, L.: Helv. Phys. Acta **18**, 527 (1945). — COX, E. F., J. V. ATANASOFF, B. L. SNAVELY, D. W. BEECHER and J. BROWN: J. Meteorology **6**, 300 (1949). See also ref. [*19*].

[*19*] JOHNSON, C. T., and F. E. HALE: J. Acoust. Soc. Amer. **25**, 642 (1953).

[*20*] See ref. [*9*] for more complete presentation.

[*21*] This derivation starts like that given by R. EMDEN, Meteor. Z. **35**, 114 (1918).

[*22*] KIRKWOOD, J. G., and S. R. BRINKLEY jr.: OSRD 4814, 1945.

[*23*] CURTIS, W.: Ballistics Research Laboratory Memorandum Report 544, 1951.

[*24*] COLE, R. H.: Underwater Explosions, Princeton Univ. Press 1948. Chs. 4 and 7: Los Alamos Scientific Laboratory, The Effects of Atomic Weapons, McGraw-Hill (1950) eqs. (3.16.1) and (3.16.2).

[*24a*] WEGENER, A.: Meteor. Z. **42**, 261 (1925).

[*24b*] DAVISON, CHARLES: Quart. Rev. **452**, 51 (1917).

[*25*] VON DEM BORNE, G.: Phys. Z. **11**, 483 (1910).

[*26*] LINDEMANN, F. A., and G. M. B. DOBSON: Proc. Roy. Soc. Lond., Ser. A **102**, 411 (1922).

[*27*] WHIPPLE, F. J. W.: Nature, Lond. **111**, 187 (1923).

[*28*] WIECHERT, E.: Nachr. Ges. Wiss. Göttingen **1**, 49 (1925).

[*29*] WEGENER, A.: Z. Geophys. **1**, 297 (1925).

[30] GUTENBERG, B.: Z. Geophys. **2**, 101 (1926). — Handbuch der Geophysik, Bd. 9, S. 1. 1932.
[31] WHIPPLE, F. J. W.: Gerlands Beitr. Geophys. **31**, 158 (1931). — Quart. J. Roy. Meteor. Soc. **60**, 80 (1934); **61**, 285 (1935).
[32] COX, E. F.: Amer. J. Phys. **16**, 465 (1948).
[33] CRARY, A. P.: J. Meteorology **7**, 233 (1950).
[34] MATHUR, L. S.: Indian J. Meteorol. a. Geophys. **1**, 24 (1950).
[35] GUTENBERG, B.: Compendium of Meteorology. Amer. Meteor. Soc., Waverly Press, Baltimore, Md., 1951, pp. 366 to 375.
[36] PENNDORF, R.: Meteor. Z. **58**, 1 (1941). — GOWAN, E. H.: Proc. Roy. Soc. Lond., Ser. A **190**, 227 (1947). — See chapter on Atmospheric Ozone in this Volume.
[37] WÖLCKEN, K.: Z. Geophys. **10**, 222 (1934); see also ref. [33].
[38] CRARY, A. P.: J. Meteorology **9**, 93 (1952); **11**, 257 (1954). — J. M. RICHARDSON, and W. B. KENNEDY: J. Acoust. Soc. Amer. **24**, 731 (1952).
[39] COX, E. F., J. V. ATANASOFF, B. L. SNAVELY, D. W. BEECHER and J. BROWN: J. Meteorology **6**, 300 (1949).
[40] VELDKAMP, J.: J. Atmos. a. Terres. Phys. **1**, 147 (1951).
[41] GUTENBERG, B.: Bull. Seism. Soc. Amer. **36**, 327 (1946). — COX, E. F.: Sci. Amer. **188**, 94 (1953).
[42] Report of the Krakatoa Committee of the Royal Society, London, p. 70, 1888. — WHIPPLE, F. J. W.: Quart. J. Meteor. Soc. **56**, 287 (1930). — TAYLOR, G. I.: Proc. Roy. Soc. Lond., Ser. A **156**, 318 (1936). — WILKES, M. V., and K. WEEKES: Proc. Roy. Soc. Lond., Ser. A **192**, 80 (1947). — PEKERIS, C. L.: Phys. Rev. **73**, 145 (1948).

MIX
Papier aus verantwortungsvollen Quellen
Paper from responsible sources
FSC® C105338

If you have any concerns about our products,
you can contact us on
ProductSafety@springernature.com

In case Publisher is established outside the EU,
the EU authorized representative is:
Springer Nature Customer Service Center GmbH
Europaplatz 3, 69115 Heidelberg, Germany

Printed by Libri Plureos GmbH
in Hamburg, Germany

ENCYCLOPEDIA OF PHYSICS

EDITED BY

S. FLÜGGE

VOLUME XLVIII

GEOPHYSICS II

GROUP EDITOR

J. BARTELS

WITH 497 FIGURES

Springer-Verlag Berlin Heidelberg GmbH
1957

HANDBUCH DER PHYSIK

HERAUSGEGEBEN VON

S. FLÜGGE

BAND XLVIII

GEOPHYSIK II

REDAKTION

J. BARTELS

MIT 497 FIGUREN

Springer-Verlag Berlin Heidelberg GmbH
1957

ISBN 978-3-642-45883-5 ISBN 978-3-642-45881-1 (eBook)
DOI 10.1007/978-3-642-45881-1

Ursprünglich erschienen bei Springer-Verlag oHG. Berlin · Göttingen · Heidelberg 1957
Softcover reprint of the hardcover 1st edition 1957

Inhaltsverzeichnis.

The Physics of Clouds.

By

F. H. Ludlam and B. J. Mason.

With 25 Figures.

I. General considerations.

1. Introduction. Over most of the earth clouds and precipitation are the dominant elements of the weather, and their study includes, directly or indirectly, a large part of the science of meteorology. In the last fifteen years there has been a surge of activity in this science under the stimulus of development in civil and military aviation. The growth of cloud physics during this period has been fostered not only by this general invigoration, but also by recognition of the practicability of exerting some influence upon the behaviour of clouds and their capacity for producing rain, hail, lightning and other meteorological phenomena.

The energies associated with the production of these phenomena so greatly exceed those hitherto at man's disposal that any artificial modification seems at first thought quite impracticable. The possibility arises, however, from our ability to interfere, using comparatively little effort, with the microphysical processes which affect the formation of cloud particles and their growth or aggregation into precipitation elements. Following the demonstration of this ability the study of clouds has tended to divide into a *microphysics*, dealing with these purely physical processes, and a large-scale or *macrophysics* which is concerned with the meteorological processes of cloud formation and growth, and which provide a framework of environmental conditions confining the rates and duration of the microphysical events. This macrophysics might more properly be called a *dynamics*, since it is principally a study of atmospheric movements which are manifest in clouds.

It is unprofitable to emphasise this convenient partition of the subject, because all the various processes are intimately linked. For example, the growth or solidification of cloud droplets is accompanied by the realisation of great quantities of latent heat, profoundly influencing the motion of cloudy air masses, while that part of the motion which ultimately causes the dissolution of a cloud determines its duration and may set a definite limit to the size which its particles can attain. Progress in cloud physics has been hindered by a poor appreciation of these relations. The discovery of methods of exploiting the microphysical processes was made and pursued by physicists who were not always sufficiently aware of the limitations which are imposed by meteorological conditions. On the other hand, this can be attributed to the backward state of the macrophysical or meteorological studies, due mainly to the intrinsic difficulty of obtaining observational information about air motions on a scale too large for laboratory simulation yet too small to be defined by the observation net used in weather forecasting. The theme of this review is the development of precipitation in clouds; in it we shall wherever possible draw attention to the important relations between small- and large-scale processes.

2. Definitions. The *mixing ratio* is the ratio of the mass of water vapour to the mass of dry air with which it is associated. Moist air which can exist in equilibrium over a plane surface of pure liquid water is said to be *saturated*. The vapour pressure is then not exactly the saturation vapour pressure corresponding to equilibrium at the same temperature between the pure vapour and the liquid, but the difference amounts to less than ½% over the range of meteorological conditions and is usually ignored. The *relative humidity* of moist air is the percentage fraction which the actual mixing ratio forms of the mixing ratio of air saturated at the same temperature and pressure. If the relative humidity exceeds 100%, the excess is referred to as a *supersaturation*. At temperatures below 0° C water may exist in all three phases, and a distinction is made between saturation, and relative humidity, with respect to a liquid (supercooled) surface, and with respect to an ice surface. If damp air is cooled it eventually reaches a temperature, the *dew-point*, at which it is saturated with respect to the liquid, or the *frost-point* at which it is saturated with respect to ice. The saturation mixing ratio over ice is less than that over liquid water at the same temperature; thus when the dew-point lies below 0°C the frost-point is higher. Consequently ice crystals can exist in air which is not saturated with respect to liquid water, i.e. at relative humidities (with respect to liquid water) of less than 100%. Water droplets frequently occur in the atmosphere at temperatures below 0° C; they can persist only if the air is considerably supersaturated with respect to ice: by about 22% at −20° C and 47% at −40°C.

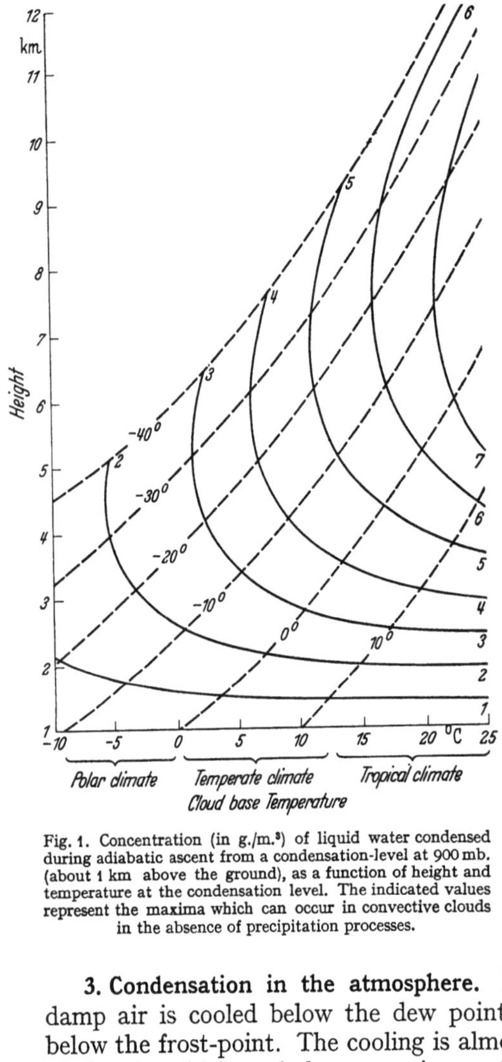

Fig. 1. Concentration (in g./m.³) of liquid water condensed during adiabatic ascent from a condensation-level at 900 mb. (about 1 km above the ground), as a function of height and temperature at the condensation level. The indicated values represent the maxima which can occur in convective clouds in the absence of precipitation processes.

3. Condensation in the atmosphere. In the atmosphere clouds form when damp air is cooled below the dew point or, with reservations discussed later, below the frost-point. The cooling is almost always brought about by the ascent of the air, which cools by expansion as the pressure decreases. Usually the air rises in such volumes and at such speeds that the expansion is practically adiabatic. The concentration of water condensed in an adiabatic ascent increases with the temperature at the level where condensation begins and with the distance risen from this level. Consequently clouds occur with a wide range of water contents; the maxima occur inside the upper parts of tropical clouds with low bases, and may reach 30 g./kg. of air (Fig. 1). Although cloud water thus occurs in rather small concentrations, the heat released during its condensation can warm the cloud air substantially except at low temperatures, when there is little condensate;

the rate at which rising cloud air cools is only about half the rate of 1° C/100 m. of ascent to which unsaturated air is subject. Clouds disappear during descending motion or by mixing with unsaturated air; often these processes occur together, and they may be helped by the precipitation of some of the cloud water.

II. Forms of clouds.

a) Classification.

4. Usual nomenclature. The present generally-recognised cloud nomenclature (a revised International Cloud Atlas containing definitions, descriptions and illustrations is in the press) rests on a classification proposed by Howard in 1803. Although based primarily upon the *appearance* of clouds to a ground observer, it is nearly a satisfactory *physical* classification, implying information on the constitution (liquid or solid particles) and dynamical structure of clouds. The principal forms are *stratus*, the clouds disposed in extensive horizontal layers; *cumulus*, the isolated heap clouds; *cirrus*, the high fibrous (ice) clouds, and *nimbus*, the precipitating clouds. In addition there are *cirrostratus* and *cirrocumulus*, high clouds in fibrous and dappled sheets respectively; *altostratus* and *altocumulus*, similar clouds of greater thickness in the middle troposphere; *stratocumulus*, a low-level layer cloud with a coarse dappled or waved appearance (*stratus* is now reserved for a low, *amorphous* layer); and *cumulonimbus*, the precipitating cumulus, usually of great vertical extent.

5. Physical classification. A physical, or dynamical classification need recognise only two fundamental classes. The first contains the *stratiform*, or layer clouds, which are horizontally extensive, often being caused by ascending motion with a vertical component of a few cm./sec. extending over regions hundreds of km. across. In the other class are the *cumuliform*, or convection clouds, formed by local and rapid ascending motions whose vertical component

Table 1. *Principal kinds of clouds.*

Kind of air motion	Typical vertical speeds cm. sec.$^{-1}$	Kind of cloud	Name in international cloud classification	Characteristic dimensions hori-zontal km.	Characteristic dimensions ver-tical km.	Charac-teristic life of particles min.	Characteristic precipitation
			Stratiform clouds:				
Simple orographic disturbances	10^2	wave clouds	species lenticularis	10	1	10	none
Widespread irregular stirring	10–10^2	low-level layer clouds, (fogs)	stratus stratocumulus	10^2 – 10^3	1	10^2	drizzle or light snow
Widespread regular ascent	10	multi-layer clouds	cirrostratus altostratus altocumulus nimbostratus	10^3	10	10^3	prolonged rain or snow
			Cumuliform clouds:				
Penetrative convection	10^2	small heap clouds	cumulus	1	1	10	none
	10^3	shower and thunder-clouds	cumulonimbus	10	10	30	showers of rain or hail

is measured in m./sec. It is convenient to sub-divide the layer clouds according to whether the ascent is a regular motion throughout much of the troposphere, or part of an irregular, stirring motion usually confined to a relatively shallow layer. It is also necessary to distinguish the *orographic* clouds which are formed where air streams flow over hills and mountains. In strong winds there can be updraughts in these clouds comparable in strength to those of the cumuliform clouds, and although the clouds are not always extensive, their persistence over hilly areas often produces striking effects on the distribution and intensity of precipitation.

The characteristics of these important kinds of clouds and the type of precipitation associated with them are summarised in Table 1 and described in the following sections. There are great variations in the properties of clouds, and the numerical values included in the table are only intended to show typical magnitudes.

b) Characteristics of cloud forms.

6. Orographic clouds. If the air is sufficiently moist orographic clouds form where it rises during flow over hills; where the air descends in the lee of the hills the clouds evaporate, and their horizontal extent is about equal to that of the hills. In generally fair weather they are usually thin and lens-shaped, clearly portraying the wave-like disturbances in the air stream, and they are conspicuous on account of their fixed position, even in strong winds. Their formation is favoured when the wind increases greatly with height, without much change in direction. In favourable conditions they may occur at any height in the troposphere, even over hills only a few hundred metres high, and over the mountains of Norway, Scotland, Iceland and Alaska examples are occasionally observed at heights of 20 to 30 km., well above the tropopause. On these occasions they often glow with bright iridescent colours produced by the diffraction of sunlight, so that they are called *nacreous* or *mother-of-pearl* clouds.

In bad weather, when the air generally is damp, the orographic clouds may be thick, more extensive and less easily recognisable, or the orographic disturbances may only cause local intensification of cloud-water concentration and updraught speed.

A comprehensive review of knowledge and theory concerning the dynamics of orographic disturbances has been given by CORBY [*1*]. LUDLAM [*2*] has described the characteristics of high-level wave clouds and BERGERON [*3*] has discussed the influence of various orographic clouds on the distribution and intensity of precipitation.

7. Layer clouds formed by widespread irregular stirring. Air is often cooled near the surface when the ground loses heat by radiation, or when the winds blow towards colder regions. In moderate or strong winds the cooling is spread upwards by turbulent, stirring motions, and during the cooling period the greatest temperature fall tends to occur at a level some several hundred metres above the surface (Fig. 2). The first condensation therefore often occurs at this level, and leads to the formation of a shallow layer cloud (stratus). In light winds, on the other hand, the maximum cooling occurs close to the surface and may result in a fog. When a layer of fog or low stratus cloud has formed beneath otherwise mainly clear skies there is usually a net loss of heat by radiation at the cloud tops, which intensifies the condensation and promotes a convective stirring throughout the cloud layer. The convection often imposes a pattern on the appearance of the cloud, producing a dappling or a waviness, even with cracks or clear spaces when the layer is thin, so that the cloud is described as stratocumulus rather than stratus. SCORER [4] has discussed the effect of this "slow convection" in shallow layers.

Inside the thicker layers, which can be a km. or more in depth, the convective motions may cause local up- and down-draughts of speeds up to, perhaps, 1 m./sec. Individual droplets are continually diffused to the upper and lower surfaces of the layer, and there evaporated, but inside clouds which are persistent, perhaps for several days, the great majority of the droplets probably have lives exceeding an hour.

It is observed that *drizzle*, a precipitation consisting of drops (radius less than 250 μ) smaller but more numerous than raindrops, usually falls from clouds

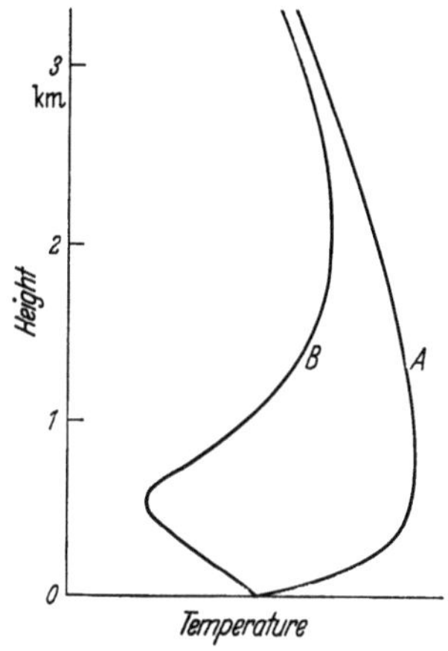

Fig. 2.

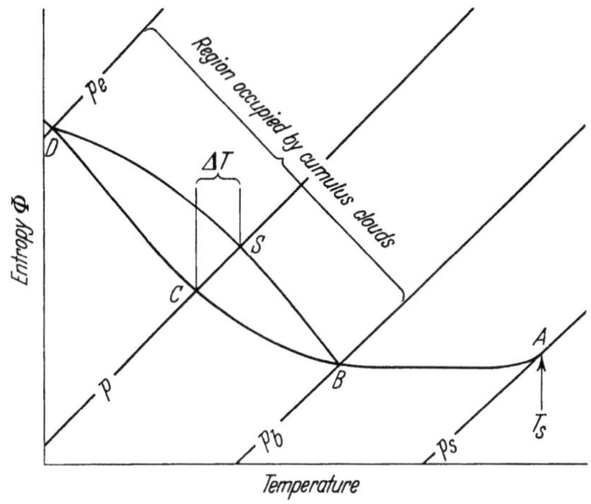

Fig. 3.

Fig. 2. Distribution of temperature with height in air masses cooled at the surface, A, in calm conditions, and, B, in a fresh wind. The distribution A is associated with the formation of fog, while B is associated with the formation of a shallow layer cloud (stratus) a few hundred metres above the surface.

Fig. 3. Characteristic distribution of temperature with height in clear air around scattered convection clouds, plotted upon a ***tephigram***. The *clear-air sounding* is represented by the line $ABCD$. Immediately above the point A, representing the temperature T_s and pressure p_s near the ground, the lapse-rate exceeds the dry adiabatic, but subsequently becomes practically equal to the ***dry adiabatic*** up to the ***condensation level*** B, where cloud bases are found. Above this level the lapse rate at first exceeds, and subsequently is less than the *saturated adiabatic* represented by the curve BSD, which describes the temperature changes in cloud air rising without mixing. The cloud tops are usually found near the equilibrium level D where this curve again meets the sounding-curve. At any level in the layer occupied by the clouds (at a pressure p) the buoyancy of the cloud air has a maximum corresponding to the maximum excess of temperature ΔT.

whose depth exceeds about 600 m., and is likely to reach the ground if the height of the cloud base above the surface is not greater than the cloud thickness.

In cold weather, when the clouds occur at air temperatures below 0° C, fallstreaks of precipitation consisting of very slight snow are often seen beneath even shallower layers and sometimes reach the ground. Thicker layers may cause a persistent slight precipitation of snow crystals, which may be aggregated into small snow-flakes. Individual crystals are some times so heavily coated with frozen droplets that they are described as *granular snow* (radius usually less than 500 μ).

8. Layer clouds formed by widespread regular ascent. In cyclones, pressure troughs, frontal regions and other bad weather systems a slow ascending motion occurs throughout much of the troposphere over regions hundreds of km. across. The maximum speed of the ascending motion amounts to 10 to 30 cm./sec. in the middle troposphere; the horizontal component of the wind is about 100 times greater. Clouds form in extensive layers at several levels; in the weaker systems and at the edges of more intense systems they remain shallow and separate, but within bad weather regions where there is widespread snow or rain the layers become thicker and fuse, or the clear spaces become filled with precipitation. At most levels air flows through the system in a period of several hours, and there is an overall convergence in the lower layers which feeds the ascending motion and the more or less steady precipitation.

Above the 0° C level the precipitation is in the form of snow; occasionally aircraft icing occurs in the snow region, indicating the presence of supercooled droplets, but more commonly, apparently, only the ice phase exists at temperatures more than a few degrees below 0° C. Sometimes strong convective motions develop inside the layer clouds, so that they may be regarded as having embedded cumuliform clouds, while wave disturbances over the hills are always in some degree superimposed upon the regular ascending motion. These irregularities ensure that the precipitation overland, although continuous for several hours or more as the storm system passes, is rarely "steady" or uniformly distributed.

In spite of the evident importance of these storm clouds there have been few systematic attempts to explore them. Several flights by CUNNINGHAM [5] are models of the detailed and thorough exploration which is needed.

9. Cumuliform clouds. These clouds represent the *penetrative convection* of large masses of buoyant air through an extensive relatively undisturbed environment, usually of clear air. The state of the atmosphere during such a convection caused by heating of the surface layers (in sunshine overland or during travel of air over the sea towards warmer regions), is shown in Fig. 3. Volumes of warmed air rise away from the surface, cooling at about the dry adiabatic rate (1° C/100 m.) until the dew-point is reached at the level of the cloud base, usually a kilometre or more above the ground. At this level the updraughts have a typical width of about a kilometre, and speeds of 1 to 5 m./sec., associated with a buoyancy determined by an excess of temperature over that in the environment amounting to a large fraction of a degree Centigrade (strictly, the buoyancy is related to the excess of *virtual temperature*—the temperature of completely dry air of the same density and at the same pressure as a sample of moist air).

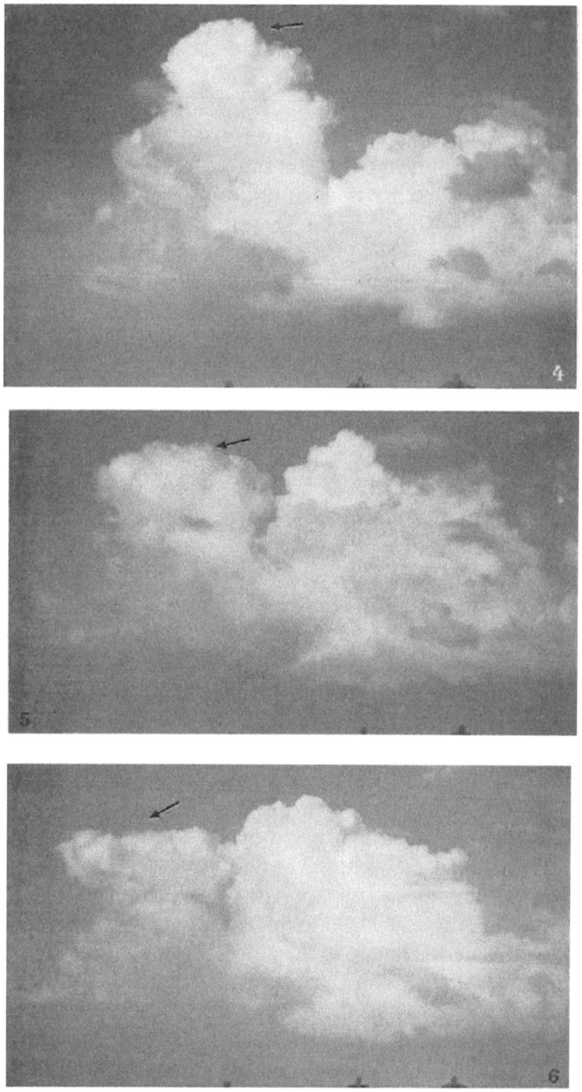

Fig. 4 (*1*—*6*). Photographs of cumulus taken at intervals of 4 minutes showing bubble structure. Cloud base is at 1300 m; the bubble diameter is about 1.5 km. The clouds move away from and slightly to the left of the observer; the wind shear is directed farther towards the left. The bubbles (one of which is indicated by the black arrows) emerge from the up-wind side of the cloud mass, rise to about 5300 m and subside and evaporate on the down-wind side.

Cumuliform clouds also develop when for any reason condensation occurs in a layer of air in which the decrease of temperature with height *(lapse rate)* exceeds that which occurs in rising cloud air (the *saturated adiabatic lapse rate*), as in the lower part of the layer shown occupied by clouds in Fig. 3. In such an unstable layer the temperature excess of a rising cloud can increase. The buoyancy is reduced by the weight of the condensed liquid, but in general also increases, and the cloud air accelerates upward. In small clouds (1 to 3 km. thick) the updraught rarely exceeds 5 m./sec., but in the deep clouds up to 10 km. thick which produce heavy showers, hail and thunderstorms, speeds of up to 30 m./sec. have been encountered.

Early theories of cumulus growth were purely thermodynamical, attempting to derive the height of the cloud tops and the speed of the updraughts within them from the temperature distribution observed to exist simultaneously in the clear air, assuming the rise of the cloud air to begin at the cloud base and proceed adiabatically. In their simplest form these theories greatly overestimate these features, and it has become necessary to take account of complications arising from compensatory motions in the clear air and the mixing and evaporation which occurs near the cloud edges. Moreover, it is recognised that the observed state of the clear air is itself a part of the convection process, and that more basic information is needed to *predict* the convection.

SCORER and LUDLAM [6] envisage the cloud air rising in the form of great "bubbles" whose caps contain a shrinking volume of unmixed air which is eventually completely eroded away by mixing at the sides and rear into a turbulent "wake". The bubbles are supposed to have rising speeds which approximate to limiting velocities representing a balance between the buoyancy and a resistance or drag due to the reaction from the environment. The mixing of cloud with clear air is accompanied by the evaporation of cloud particles and a chilling whose magnitude depends upon the quantity of water evaporated, the humidity of the clear air and the initial temperature excess of the cloud air. With ordinarily encountered values the chilling amounts to between 1 and 2° C, and is usually sufficient to produce a negative temperature excess in the mixed air; consequently there is a sinking motion on the outer surface of the rising air masses which promotes their erosion.

A large cumulus cloud seems to be made up of several rising bubbles, which emerge as prominences at the cloud top (Fig. 4), but is mainly composed of bubble wakes. Only in very localised regions do the temperature excess and liquid-water concentration approximate to the values corresponding to adiabatic ascent from the cloud base. The mixing processes are a powerful brake on the development of clouds, and in general a cloud can only become tall if its lower parts are sufficiently extensive to provide an adequate shield of expendable bubbles for the protection of the interior bubbles against mixing with clear air. The factors which determine the size of individual clouds are not known. In any given population of cumulus there are usually many more small than large clouds, and only a very few of these ever grow or become assembled into large clouds. Parts of the topography which become warmer than their surroundings, hills with slopes more directly facing the sun, and so on, are places over which the convection tends to become concentrated, and so are important generators of large clouds. Over featureless terrain or the oceans the factors which determine the size of clouds are quite unknown.

In clouds of a given size the large-scale properties important in the development of precipitation appear to be the vertical thickness and the updraught speeds. Although large clouds may persist as recognisable entities for an hour or even several hours, it may be observed that their individual parts have a much briefer existence. The prominences or towers which rise in quick succession from the central parts of the clouds are usually observed to suffer a rapid shrinking and subsidence, so that they are recognisable for only some fifteen minutes or less; it is inferred that the whole life of a bubble which rises from somewhere near the cloud base to form such a tower is usually less than half an hour. When precipitation forms in the towers it must evidently develop very rapidly.

Showers of heavy rain, hail or snow fall from large cumuliform clouds (cumulonimbus). Great variations are observed in the size of clouds which produce showers. Particularly in the tropics and in maritime air rather heavy rain sometimes falls from clouds about $1\frac{1}{2}$ km. thick, whereas inland, especially over hot, arid regions, there are occasions when no showers form in clouds several km. thick. In extra-tropical regions the tops of shower clouds almost always acquire a fibrous texture which is interpreted as due to *glaciation*, or the extensive formation of ice crystals.

III. The initial stages of condensation — condensation nuclei.

10. The physics of condensation on nuclei. In air freed from aerosol and from ions, condensation of water vapour occurs only if the supersaturation reaches several hundred per cent. In the atmosphere abundant foreign particles serve as ready-prepared aggregates (nuclei) for condensation so that supersaturations exceeding 1% are rarely achieved.

The equilibrium vapour pressure p_r over the surface of a droplet of pure water of radius r exceeds that over a plane water surface at the same temperature, p_∞, as expressed by the equation

$$\ln p_r/p_\infty = 2\sigma M/\varrho R T r \tag{10.1}$$

where σ, ϱ, M are respectively the surface tension, density and molecular weight of the liquid, R is the universal gas constant, and T the absolute temperature. Thus a drop cannot persist, much less grow, unless the environment is supersaturated by an amount required by Eq. (10.1). Solid, insoluble particles whose surfaces are wettable may serve as nuclei for condensation at supersaturations given very nearly by Eq. (10.1) which is plotted as the upper curve of Fig. 5 and evaluated in Table 2, from which we see that particles of $r > 10^{-5}$ cm. require supersaturations of less than 1%.

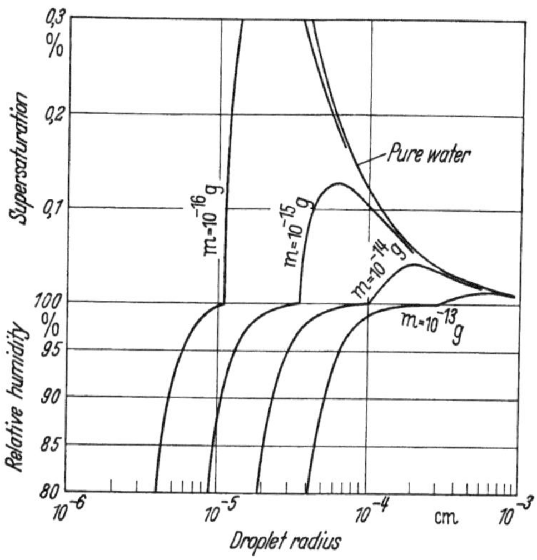

Fig. 5. The equilibrium relative humidity (or supersaturation) as a function of droplet radius for solution droplets containing the indicated masses of sodium chloride (from MASON [7]).

If a droplet is formed on a wholly or partially soluble nucleus, the equilibrium vapour pressure at its surface is reduced by an amount depending upon the nature and concentration of the solute, which means that condensation will be able to occur at a lower supersaturation than on an insoluble particle of the same size. The equilibrium vapour pressure p'_r over a droplet of solution may be written (MASON [7])

$$\frac{p'_r}{p_\infty} = \left[\exp \frac{2\sigma' M}{\varrho' R T r}\right] \left[1 + \frac{i\, m\, M}{W(\frac{4}{3}\pi r^3 \varrho' - m)}\right]^{-\varrho/\varrho'}, \tag{10.2}$$

where the primed symbols refer to the solution, m is the mass of the solute in grammes, W the molecular weight of the solute, and i VAN'T HOFF'S factor which depends upon the nature and concentration of the solute. When the solution droplet is in equilibrium with the surrounding atmosphere, $p'_r/p_\infty = H/100$, where H is the relative humidity. Eq. (10.2) can now be used to calculate the radius of droplets in equilibrium with atmospheres of given relative humidities as shown in Fig. 5. With increasing relative humidity, the equilibrium radius of the nucleus droplet increases until a critical value of the supersaturation is reached corresponding to the maximum of the relevant curve of Fig. 5. At this stage we may write to a high degree of accuracy:

$$\frac{H}{100} \simeq 1 + \frac{P}{r} - \frac{Q}{r^3}, \quad \text{where} \quad P = 3.2 \times 10^{-5}/T,$$

$Q = 8.6m/W$; the critical values of the radius and relative humidity being

$$r_c^2 = 3Q/P; \quad H_c/100 = 1 + (4P^3/27Q)^{\frac{1}{2}}.$$

If a nucleus droplet exceeds the critical radius and the supersaturation is maintained, it will continue to grow with a decrease of free energy to form a cloud droplet. In theory, it would grow without limit; but in practice, a stage is reached when the large numbers of rapidly growing droplets in a cloud extract the water vapour faster than it is released by the cooling air, the supersaturation is forced to retreat, and the growth rate of the droplet consequently decreases.

Table 2. *Critical radii and supersaturations for nuclei of various sizes.*

$\log_{10} m$ (g.)	−16	−15	−14	−13	−12	−11	−10	−9	−8
a) Hygroscopic nuclei of NaCl									
r (μ) at $H = 78\%$	0.039	0.084	0.185	0.39	0.88	1.85	4.1	8.8	18.5
r_c (μ)[1]	0.20	0.62	2.0	6.2	20	62	200	620	2000
$H_c - 100$ (= % supersat)[2]	0.42	0.13	4.2×10^{-2}	1.3×10^{-2}	4.2×10^{-3}	1.3×10^{-3}	4.2×10^{-4}	1.3×10^{-4}	4.2×10^{-5}
Radius of crystal (μ)	0.022	0.048	0.103	0.22	0.48	1.03	2.2	4.8	10.3

Droplet radius at $H = 100\%$ is approx. $r_c/\sqrt{3}$.

b) Insoluble wettable nuclei

$\log r$ (cm.)	−7	−6	−5	−4	−3
$100 \cdot p_r/p_\infty$	323	112.5	101.2	100.12	100.01

It will be seen from Fig. 5 and Table 2 that all sodium chloride nuclei of $m > 10^{-15}$ g. (crystal radius $> 5 \times 10^{-6}$ cm.) and all insoluble, wettable nuclei of $r > 10^{-4}$ cm require supersaturations of less than 0.1% to act as centres of continued condensation.

The general shapes of the curves in Fig. 5 have been verified experimentally. Although a crystal of NaCl becomes a droplet of saturated solution at 78% relative humidity, it is often possible to follow the equilibrium curve down to relative humidities as low as 40%, the solution remaining in the supersaturated state for several minutes before crystallising with explosive violence to produce one or more crystals.

Of course natural aerosols will not necessarily be either completely soluble or completely insoluble. Indeed, JUNGE [*8*], [*9*] finds that continental aerosol consists largely of insoluble particles coated with a thin layer of hygroscopic substance. At high humidities these *mixed nuclei* react to humidity changes like wholly soluble particles of equivalent size, but at humidities below about 70%, the solution coat shrinks until the particle becomes almost completely solid. The efficiency of hygroscopic nuclei, whether of uniform or mixed composition, will depend upon their size and chemical nature. The extent to which completely non-hygroscopic particles are involved in cloud-droplet formation is not known, but, in any case, they may be a rather small constituent of atmospheric aerosol.

[1] For other solutes of molecular weight W, multiply by $(58.5/W)^{\frac{1}{2}}$.

[2] For other solutes of molecular weight W, multiply by $(W/58.5)^{\frac{1}{2}}$.

11. The concentration, size and size-distribution of condensation nuclei. Atmospheric aerosols cover a wide range of particle sizes from about 10^{-7} cm. radius for small ions to more than 10 μ for the largest salt and dust particles, and an enormous range of total concentrations. The small, ubiquitous ions almost certainly play no role in atmospheric condensation because of the very high supersaturations (ca. 300%) required for their activation, while the largest particles may usually be rare at cloud levels because of their large fall speeds.

The size distribution of atmospheric aerosol is represented in Fig. 6. It is convenient to divide the spectrum into three size groups:

a) those small nuclei which are detected in an AITKEN counter, i.e. AITKEN *nuclei* having radii between 5×10^{-7} cm. and 2×10^{-5} cm.;

b) *large nuclei* with 2×10^{-5} cm. $< r < 10^{-4}$ cm.; and

c) *giant nuclei* with $r > 1$ μ.

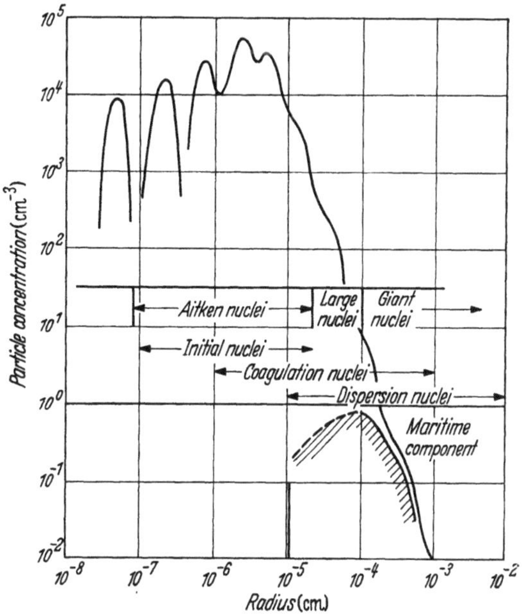

Fig. 6. A generalised representation of the size distribution of natural aerosols in heavily polluted air over land (after JUNGE [9]).

The size range of nuclei detected in an AITKEN counter is not well established but electron-micrographs of the deposit suggest the range 5×10^{-7} cm. to 2×10^{-5} cm., and there is other evidence to suggest that the very small ions ($r \simeq 10^{-7}$ cm.) are not detected. The sizes of charged nuclei of radius up to 10^{-5} cm. may be determined from measurements of ionic mobility; these indicate that the mean radius of AITKEN nuclei is about 5×10^{-6} cm. Positive identification of the nature of AITKEN nuclei is very difficult even under the electron microscope unless they have a very distinctive crystalline habit, not only because of their small size but because they may evaporate in the high vacuum or be affected by the electron bombardment. A very large number of determinations of the nucleus content of the air has been made with the AITKEN counter in many different localities; a comprehensive review was given by LANDSBERG [*10*]. They reveal an enormous range of particle concentrations from only a few per cm.^{3} over the oceans and in the upper air to values of, perhaps, 10^6/cm.^{3} in industrial cities. Measurements made over the land are difficult to interpret; not only do they appear to be correlated with a number of interrelated meteorological factors but they are often dominated by local influences. Over the oceans the situation is less complex but unfortunately the observations are few and sporadic. The concentrations are usually a few hundreds per cm.^{3} but occasionally are as low as 2/cm.^{3}, and there appears to be little correlation with windspeed or the height of the waves.

A number of methods have been devised for collecting the large nuclei for their subsequent examination under either the optical or electron microscope. One may employ an air jet of high velocity, thermal or electrostatic precipitation to ensure a high collection efficiency with a relatively large collector, or alternatively, use very narrow collectors (e.g. spiders' threads) and correspondingly low air speeds. Although the large nuclei can be resolved by the optical microscope it is difficult to measure and identify the smallest of them. Here the electron microscope proves useful though it can deal with only very small samples and suffers from the disadvantages mentioned above.

Measurements made by DESSENS [11] to [13] in the open country revealed that the concentrations of large hygroscopic nuclei varied from day to day but, on average, were of order 100/cm.³. At 78% relative humidity the majority of these formed droplets of radius less than 0.2 μ with only occasional ones of $r > 1$ μ. The largest nuclei were identified as consisting of sea salt. JUNGE [14] has measured the size distribution of aerosol at a number of places in Germany and finds that particles of $r > 0.1$ μ obey a law of the form $N = A/r^3$ where N is the number of nuclei of radius greater than r and A is a constant. In cloud-free air over the land JUNGE found the concentration of large nuclei to vary between 45 and 360/cm.³ with an average value of 130/cm.³, these forming a rather constant component of the aerosol in contradistinction to the large fluctuations in the population of AITKEN nuclei. There is very little information on the concentrations of large nuclei over the oceans. MOORE and MASON [15] found hygroscopic nuclei with 2×10^{-14} g. $< m < 10^{-11}$ g. in concentrations of up to 12/cm.³ which did not appear to vary greatly with the windspeed, while WOODCOCK [16], collecting only particles of $m > 10^{-13}$ g., found concentrations of 4/cm.³ at 500 m. over Hawaii and 12/cm.³ when flying at 15 m. near Cape Cod, Mass.

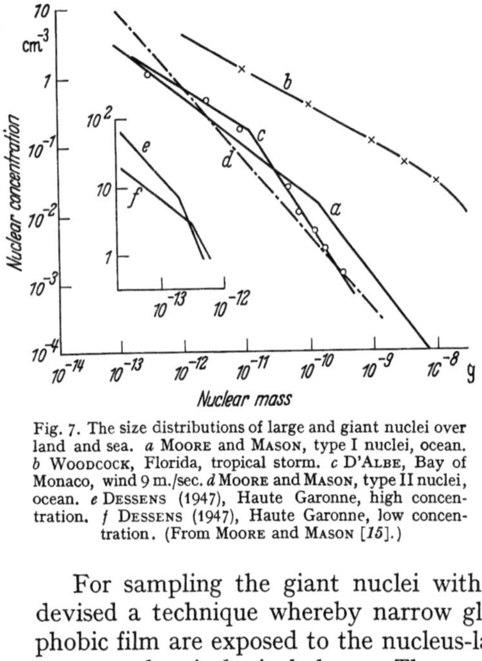

Fig. 7. The size distributions of large and giant nuclei over land and sea. *a* MOORE and MASON, type I nuclei, ocean. *b* WOODCOCK, Florida, tropical storm. *c* D'ALBE, Bay of Monaco, wind 9 m./sec. *d* MOORE and MASON, type II nuclei, ocean. *e* DESSENS (1947), Haute Garonne, high concentration. *f* DESSENS (1947), Haute Garonne, low concentration. (From MOORE and MASON [15].)

For sampling the giant nuclei with $r > 1$ μ, WOODCOCK and GIFFORD [17] devised a technique whereby narrow glass or metal rods coated with a hydrophobic film are exposed to the nucleus-laden air and on which collected droplets assume a hemispherical shape. The masses of the nuclei can be determined by exposing them to an atmosphere of known relative humidity and measuring the diameters of the hemispherical droplets formed around them. Using this method, WOODCOCK has found that with surface windspeeds up to 10 m./sec., salt nuclei of $m > 10^{-12}$ g. are present in concentrations of up to 500/litre, while nuclei as large as 5×10^{-9} g. may be present to the extent of 1/litre. Increasing winds were associated with a rather consistent pattern of increases in both the numbers and sizes of particles; in a tropical storm with winds of 30 m./sec., WOODCOCK found nuclei of $m > 10^{-11}$ g. in concentrations of about 1/cm.³, and some very large ones ($m = 4\times10^{-8}$ g.) in concentrations of 2/litre. In strongly stable conditions the concentration of giant nuclei may fall off rapidly with height, but if the lower layers are well mixed, they may be fairly uniformly distributed up to heights of at least 2 km.

From measurements made on board a Weather Ship in the N. Atlantic 250 miles from land, MOORE and MASON [15] found that their size distributions of hygroscopic nuclei fell into two classes (see Fig. 7). One of these closely resembled the curves obtained by WOODCOCK and showed a rapid fall off in the concentration of nuclei above a certain critical size which was a function of the windspeed, and such a distribution is believed to be representative of the suspended salt nuclei. There was a marked increase in the numbers of these larger nuclei when white-caps appeared, and with windspeeds of 15 m./sec. the maximum concentration of nuclei with $m > 2\times10^{-14}$ g. was 10/cm.^{3}. The second type of distribution occurred only with windspeeds of less than 7 m./sec. and was of the same form as that obtained by JUNGE for continental aerosol. The implication was that when the current rate of production of nuclei at the sea surface was small, there was a preponderance of nuclei of land origin. An interchange of nuclei between land and sea is also indicated by the fact that sea-salt nuclei have been found more than 100 km. inland.

JUNGE's measurements on the giant-nucleus content of continental aerosol reveal total concentrations of between 1 and 10 particles/cm.^{3}, those of the largest particles ($r > 15\,\mu$) being only about 10/m.^{3}. Particles of $r > 5\,\mu$ consisted almost entirely of solid matter but smaller particles generally contained a proportion of hygroscopic substance.

12. The nature and origin of condensation nuclei. Atmospheric aerosol originates in three main ways:

a) by condensation and sublimation of vapours during the formation of smokes and in gaseous reactions;

b) by the mechanical disruption and dispersal of matter, e.g. the formation of dust and spray;

c) by coagulation of nuclei which may lead to larger particles of mixed constitution.

The first mechanism is probably responsible for most of the AITKEN nuclei with 10^{-7} cm. $< r < 10^{-5}$ cm., while larger particles are produced by disruption and coagulation. Smoke particles mostly lie in the size range 10^{-6} to 10^{-4} cm. and may consist of soot particles (carbon), oils, tars, and hygroscopic substances such as H_2SO_4. Chemical reactions between the nitrogen, oxygen and water vapour of the air with various trace gases (e.g. SO_2, Cl_2, NH_3, O_3 and the oxides of nitrogen) may be important sources of nuclei; here solid particles may play an important role by absorbing the gases and water vapour and thereby concentrating the substance—perhaps in solution. The mechanical disintegration of rocks and soil produces particles of $r > 10^{-5}$ cm., while droplets as small as $r = 2\times10^{-5}$ cm. containing only 10^{-15} g. of salt may be produced by the bursting of air bubbles in sea water. The most important feature of the coagulation process is probably the sweeping up of small hygroscopic nuclei by larger solid particles to form mixed nuclei.

It seems likely that during cloud formation all the large and giant nuclei present will be activated together with a proportion of the AITKEN nuclei which will vary with the particular circumstances. Hygroscopic nuclei, which are most efficient, are thought to originate mainly from the sea and as combustion products.

Whether or not the sea is the main source of cloud-droplet nuclei has long been a matter of argument. The matter cannot yet be regarded as finally settled but the evidence is beginning to point *against* the sea being an adequate source. Thus, while cloud droplet populations are usually of the order a few hundreds per cm.^{3}, the concentrations of nuclei large enough to be positively identified as sea salt rarely exceed 10/cm.^{3} even in very strong winds. Smaller salt nuclei may, of course, exist in appreciably high concentrations, but these should be detected in an AITKEN counter and according to MOORE [*18*], AITKEN counts over the sea do not increase appreciably with windspeed.

It is also questionable whether nuclei can be produced at the sea surface at a rate sufficient to replace those lost by the atmosphere during precipitation. SIMPSON [19] estimated that the necessary rate of production during periods of spray formation would be 50000 nuclei/cm.^{2} sec. There are reasons for believing that this figure may be too high by a factor of, perhaps, five, but even so, MASON [7] has estimated that even in a moderate gale, the production rate of salt nuclei with $m > 2\times 10^{-14}$ g. would be only 100/cm.^{2} sec. and that of all salt nuclei ($m > 10^{-15}$ g.), only 1000/cm.^{2} sec. This all suggests that *sea-salt nuclei* contribute perhaps only *one-fifth* to the average total numbers of nuclei involved in cloud formation and that the rest largely consist of *combustion nuclei* which are produced in vast numbers by natural and man-made fires.

Some direct support for this view comes from some preliminary experiments by KUROIWA [20], [21], OGIWARA and OKITA [22] and by YAMAMOTO and OHTAKE [23] in Japan in which the nuclei of small numbers of cloud- and fog-droplet nuclei collected a few kilometres from the coast have been examined under the electron microscope. All these authors find that combustion products are the most abundant source of nuclei in the droplet residues, with soil particles being comparable in importance to salt nuclei. It is dangerous to generalise on the basis of such a few observations, but they are suggestive and point to the need for further work. Considerable interest also attaches to some recent work by JUNGE [14], [24] in which he finds that most of the large nuclei ($0.1\ \mu < r < 1\ \mu$) he has collected in Germany and on the Atlantic coast of the United States consist mainly of ammonium sulphate while the giant particles contain variable, though sometimes considerable amounts of sodium chloride and nitrate.

IV. The growth of droplets in cloud and fog.

13. The processes involved. The growth rate of individual cloud droplets depends not only on the hygroscopic and surface tension forces and the ambient humidity as discussed in the last section, but also on the processes which determine the rate of transfer of water vapour to and heat from the droplet. With a population of droplets the problem becomes more complicated since all the droplets compete for the available water vapour; their growth rate then depends upon the concentration, size distribution and nature of the nuclei, the rate of cooling of the air, its temperature and the scale and intensity of the turbulence in the cloud. The relative importance of these factors can only be estimated by theoretical computations whose crucial test is the accurate prediction of observed droplet-size spectra.

a) Observational results.

14. The size distribution of cloud droplets. Much effort has been devoted to developing experimental techniques for measuring the sizes of cloud droplets. Direct methods which involve the capture of the droplets on suitably prepared and exposed surfaces and the later counting under the microscope of the droplets or their impressions with reference to their diameters, although tedious, appear to be the most reliable. The accuracy of these methods depends upon obtaining a true, representative sample from the cloud or fog, preventing evaporation between sampling and counting, and avoiding shattering or coalescence of the captured droplets. The droplets are usually caught on slides coated either with oil into which the droplets penetrate and become completely submerged, or with magnesium oxide in which they produce craters, the diameters of which can be related to those of the droplets by prior calibration. The mean size of the droplets and some indication of their size distribution have been obtained in supercooled clouds, from measurements of the differing catches of rime per unit length on rotating cylinders of different diameters, the collection efficiency of which depends upon the droplet size, the cylinder diameter and the wind speed.

Optical methods based upon a determination of intensity distribution within a diffraction pattern or upon the absorption of infra-red radiation at different wavelengths by the cloud promise to be less tedious, but are relatively complicated and will need to be checked against the direct methods.

Typical values for some of the parameters which characterise measured droplet concentrations and size distributions in different types of cloud are summarised in Table 3. The very limited data which are available indicate the tendency for

Table 3. *Characteristics of cloud-droplet spectra.*

r_m = Mean radius, i.e. the sum of all the droplet radii divided by the total number of droplets.
r_d = Mode radius, i.e. the radius corresponding to the maximum number of droplets.
r_p = Predominant radius, i.e. the radius corresponding to the maximum volume of water.

Cloud type	Concentration (droplets/cm.³)	r_m μ	r_d μ	r_p μ	Range μ	Authority
Fair-weather cu.	300	9	6	13	3–33	WEICKMANN and AUFM KAMPE [25]
			3.5	6.5	1–10	DIEM [26]
Cu. congestus	64	24	6	51	3–83	WEICKMANN and AUFM KAMPE
	150–600		5.5		2–40	ZAITSEV [27] at 500 m. above base
Cumulonimbus	72	20	5		2–100	WEICKMANN et al.
Stratocumulus	350	4	3.5	6.5	1–12	DIEM
	500			10–20	3–25	FRITH [28]
	310	8				BRICARD [29]
Altostratus	450	5	4.5	7	1–13	DIEM
Nimbostratus	330	6	4	12	1–20	DIEM
	175	11				BRICARD
Stratus	260	6	4	9	1–22	DIEM
			7		2–40	NEIBURGER [30]
	664	5.3				BRICARD

clouds which are unlikely to precipitate (fair-weather Cu, thin layers of Sc and As) to consist of relatively larger concentrations of small droplets, and for the thicker, denser clouds (Cu-congestus, Cb, Ns) to contain considerably larger droplets. Low stratus and fog also generally consist of rather low concentrations of relatively large droplets, but their constitution largely depends on their life history.

Fig. 8 reproduced from ZAITSEV's paper, shows the variation of droplet-size distribution, droplet concentration and liquid-water content with height in a particular cumulus congestus. Near the cloud base the spectrum tends to be narrow and the droplets small, but about 2 km. higher up the droplets are larger and the spectrum broader; this trend becomes reversed again towards the top of the cloud. Very similar results were obtained by WEICKMANN and AUFM KAMPE. Some typical droplet spectra for different clouds are shown in Fig. 9.

15. The liquid-water content of clouds. A detailed knowledge of the magnitude and distribution of liquid water in a cloud assumes importance in cloud dynamics since they indicate the degree of mixing which takes place between the rising cloudy air and its clear environment and because changes in the water content are accompanied by large energy changes. Also, the water content largely determines the rate of growth of precipitation elements by accretion.

Unfortunately there is, as yet, no entirely satisfactory method of measuring liquid-water content which gives a continuous easily-interpreted record over the

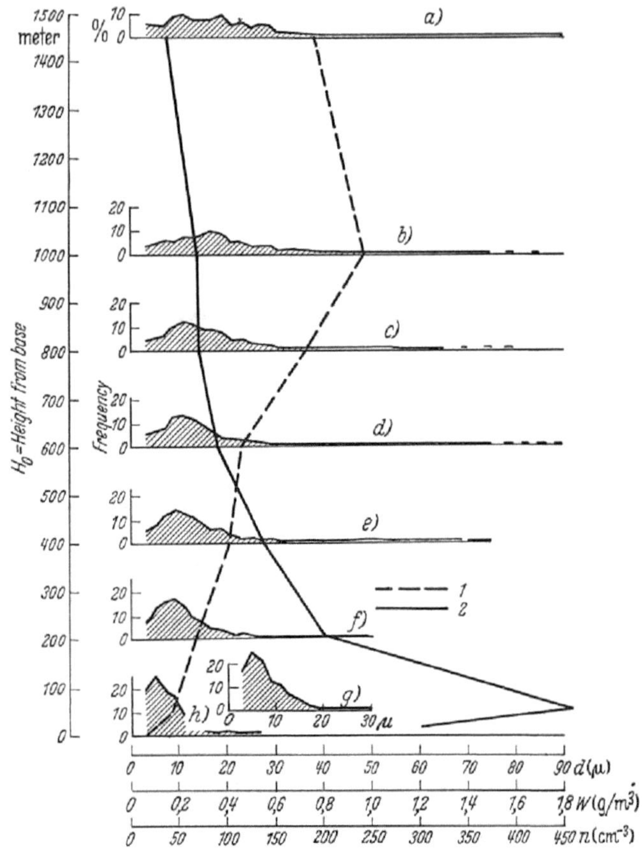

Fig. 8. The droplet-size distribution, liquid-water content and droplet concentration as functions of height in a cumulus congestus. *1* = liquid water content W = in g./m.³. *2* = number of droplets N = per cm.³. n = number of measured droplets. H_0 = the height of the middle of the layer in m. from the base. (*a*) n = 4321, H_0 = 1450; (*b*) n = 4062, H_0 = 1000; (*c*) n = 5010, H_0 = 800; (*d*) n = 11883, H_0 = 600; (*e*) n = 21277, H_0 = 400; (*f*) n = 17267, H_0 = 225; (*g*) n = 8236, H_0 = 100; (*h*) n = 4489, H_0 = 25. (From ZAITSEV [27].)

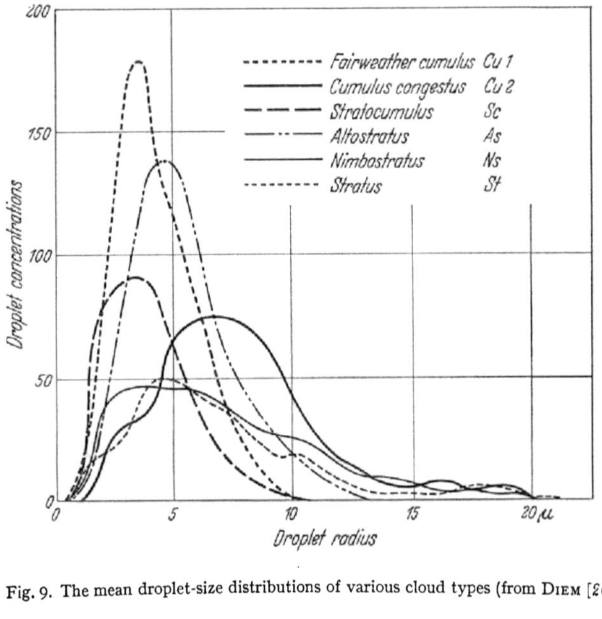

Fig. 9. The mean droplet-size distributions of various cloud types (from DIEM [26]).

required range of values, has a rapid response, is robust, reliable, and capable of accurate calibration. Instruments available at present for use in non-supercooled clouds are the capillary collector (VONNEGUT [31]), and an instrument designed by WARNER and NEWNHAM [32] in which the droplets are absorbed by a paper tape, the electrical resistance of which gives a measure of the water content. A fairly satisfactory instrument for use in supercooled clouds is the rotating disc icing meter which records the thickness of a rime deposit and can be refrigerated for use in warm clouds. Thermal methods in which the droplets impinge on an electrically-heated cylinder and one measures either the power required to keep the latter dry, or its temperature and therefore resistance change produced by the cooling effect of the evaporating droplets, also appear promising. It is also possible to determine the water content from simultaneous measurements of visibility in the cloud and the droplet-size distribution; the former has been measured by AUFM KAMPE [33] using a searchlight and a photocell mounted on the wing of an aircraft.

Ideally, it would be desirable to measure the horizontal distribution of liquid water in a cumulus at several different levels simultaneously in order to obtain an "instantaneous" picture which could be correlated with the thermal and dynamical structure, but such comprehensive information has not yet been obtained. The most detailed picture has been provided by ZAITSEV [27] (see Fig. 10). He finds that, in general, the water content first increases with height above cloud base and later decreases again. It is lower near the periphery of the cloud than in the centre—a result, no doubt, of the cloudy air mixing with the drier environment at the edges of the cloud. In smaller cumulus the water content rarely exceeds 1 g./m.³, but in large clouds much higher values are found, and over small localised regions, may approach the maximum theoretical value calculated on the assumption of adiabatic ascent of the air from cloud base. A long series of measurements with rotating riming cylinders in stratocumulus layers reveal values ranging from 0.06 to 1.3 g./m.³ with average values of 0.30 g./m.³ (KLINE and WALKER [34]).

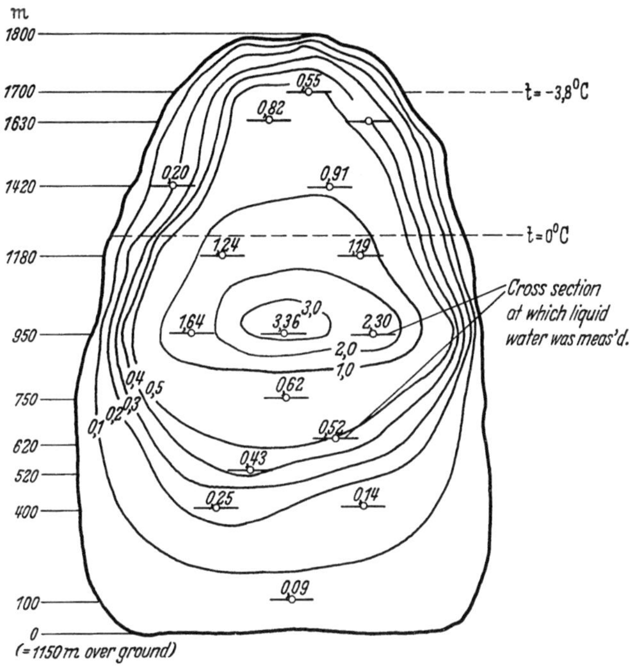

Fig. 10. The distribution of liquid water in a cumulus congestus. Heights in metres above cloud base (which was at 1150 m.) (from ZAITSEV [27]).

b) Theoretical studies of cloud-droplet growth.

There are two conceivable processes by which nucleus droplets may attain radii of several microns and so form cloud or fog: the diffusion of water to, and its condensation upon their surfaces, and growth by the coalescence of droplets moving relative to each other by virtue of BROWNian motion, small-scale turbulence and differential rates of fall under gravity.

16. Growth of a droplet by condensation. The growth rate of an isolated droplet at rest, and in equilibrium with an infinite atmosphere maintained at constant temperature and supersaturation can be expressed by the equation

$$r\frac{dr}{dt} = \frac{D\,M}{\varrho\,R\,T}\,[p - p'(r)] = \frac{D\,M}{\varrho\,R\,T}\,[S\,p_s(T) - p'(r)] \tag{16.1}$$

where p is the vapour pressure in the undisturbed environment (remote from the droplet), $p'(r)$ the corresponding quantity at the droplet surface, $p_s(T)$ the saturation vapour pressure of the environment, $S = p/p_s(T)$ the saturation ratio and the other symbols are as already defined.

Table 4. *Rate of growth of a droplet by condensation on salt nuclei (after* BEST *[35]).*
Temperature $T = 273°$ K Pressure = 900 mb.

Nuclear Mass (g.)	10^{-14}	10^{-13}	10^{-12}
% Supersaturation = 100 (S−1)	0.05	0,05	0.05
Radius (μ)	Time (sec.) to grow from initial radius of 0.75 μ		
1	2.4	0.15	0.013
2	130	7.0	0.16
5	1000	320	62
10	2700	1800	870
15	5200	4200	2900
20	8500	7400	5900
30	17500	16000	14500
40	44500	43500	41500

The growth rate is limited not only by the rate of diffusion of vapour but also by the rate at which the latent heat of condensation can be dispersed to the environment. Thus the surface temperature T_r of the droplet is higher than that of its surroundings by an amount given by the equation

$$r\frac{dr}{dt} = \frac{K}{L\,\varrho}\,(T_r - T)$$

or

$$T - T_r = \frac{L\,D}{K\,R\,T}\,[S\,p_s(T) - p'(r)]\,, \tag{16.2}$$

where K is the thermal conductivity of the air and L the latent heat of condensation.

The equilibrium vapour pressure $p'(r)$ over the surface of the very dilute solution droplet is given quite accurately by the following approximate form of Eq. (10.2)

$$\frac{p'(r)}{p_s(T_r)} = \exp\frac{2\sigma\,M}{\varrho\,R\,T_r}\left[1 - \frac{8.6\,m}{W\,r^3}\right]. \tag{16.3}$$

Eqs. (16.1) to (16.3) allow the growth rate of a droplet to be calculated for specified values of m, T and S, some specimen results being given in Table 4.

When the drop has grown to such a radius that the dissolved salt and the curvature of the drop no longer exert an appreciable influence on the vapour pressure in the immediate vicinity of the surface the expression for the growth rate reduces to

$$r\frac{dr}{dt} = (S-1)\Big/\left[\frac{J\,L^2\,M\,\varrho}{K\,R\,T^2} + \frac{\varrho\,R\,T}{p_s(T)\,D\,M}\right], \tag{16.4}$$

where J is the mechanical equivalent of heat.

17. Growth of a droplet by accretion. A drop which is larger than, and consequently moves with a velocity relative to its smaller neighbours, will overtake and collide with some of them lying in its fall path. If a fraction E_2 of the droplets which collide with the larger drop coalesce with it, the latter will grow at a rate given by

$$\frac{dm}{dt} = 4\pi R^2 \varrho \frac{dR}{dt} = E_1 E_2 \pi R^2 w (V - v), \tag{17.1}$$

where E_1 is the *collision efficiency* defined as the ratio of the actual collision cross-section to the geometrical cross-section of the drop, E_2 is the *coalescence efficiency*, w the concentration of liquid water in the form of smaller droplets, and V, v are the fall velocities of the large and smaller drops respectively. The product $E_1 E_2$ will be termed the *collection efficiency* and denoted by the symbol E.

Since both E and V are increasing functions of R, it is clear that growth by accretion, once started, will accelerate, and that the growth will be represented by some such curve as Fig. 11 b. Whether the condensation process or the accretion process is the more important in the growth of cloud droplets will depend upon the values of the radius r^* at which the two curves of Fig. 11 intersect and the time t^* taken for this radius to be reached under the prevailing conditions. Values of the collision efficiencies E_1 of drops (radius R) for smaller droplets (radius r) have been computed by LANGMUIR [*36*] treating the smaller droplets as points. This approximate treatment leads to error in excess of 10% for values of $r/R > 0.05$, but a rough correction for the finite size of the smaller drop can be made by replacing LANGMUIR's value E_L by $(E_L^{\frac{1}{2}} + r/R)^2$. For values of r/R greater than about 0.25, however, the whole procedure breaks down because of mutual interference between the flow patterns set up by the two drops; the problem then becomes mathematically intractable. While there is some experimental evidence to indicate that collisions between large drops ($R \simeq 1$ mm.) and small droplets ($r \simeq 10\ \mu$) usually result in coalescence, the coalescence efficiency following collisions between droplets of more nearly the same size has not yet been established experimentally, but for the purposes of our calculations will be taken as unity. Integration of Eq. (17.1) shows that a droplet of radius $R = 25\ \mu$ falling through a cloud at temperature 0° C and 900 mb. pressure containing 1 g. of liquid water per m.³ in the form of droplets of $r = 6\ \mu$, would grow to a radius of 30 μ in 12 minutes and to 50 μ in 38 minutes. These times are very short compared with those given in Table 4 for growth by condensation. If the supersaturation in the cloud were 0.05%, the rates of growth by condensation and coalescence would become equal at $R = 15\ \mu$; above this size growth by coalescence would predominate. However, in the strong updraughts which may exist in vigorous cumulus clouds, considerably larger supersaturations (approaching 1%) may be achieved, so that droplets of radius 30 μ may grow by condensation on large salt nuclei in periods of 10 to 15 minutes.

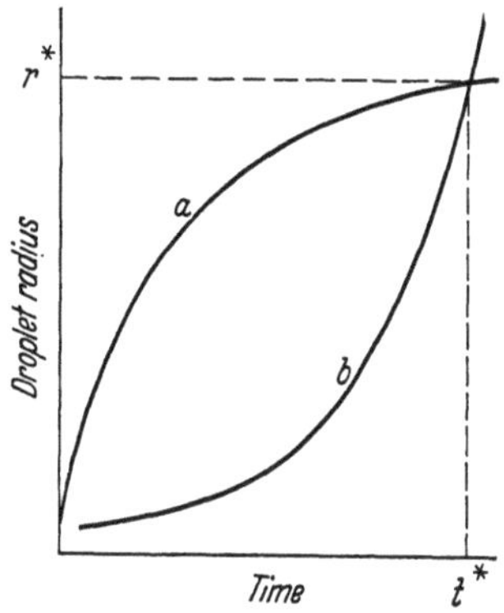

Fig. 11. The growth of a cloud droplet by condensation (curve a) and by coalescence (curve b).

18. The growth of a population of droplets. The growth rate and the ultimate size of a particular group of droplets will depend upon the rate and duration of the cooling process, the initial temperature of the air, the concentration and size-distribution of the nuclei. Once droplets of different sizes appear, coalescences become possible, giving rise to even greater inhomogeneity; but as we have seen, this process is not likely to exert a major influence on the droplet-size spectrum until the droplets have already grown to a radius of several microns by condensation. This same conclusion may be drawn from the calculations made by SCHUMANN [37] who estimated that, under typical conditions, 140 minutes would be required for droplets to grow to radius of 8 μ by coalescence.

A detailed study of the growth, by condensation alone, of cloud droplets in air cooled at a uniform rate, with the object of predicting changes in the droplet-size distribution with time, has been undertaken by HOWELL [38]. Starting from an assumed concentration and size distribution of nuclei, equations analogous to (16.1) to (16.3) set up in differential form suitable for numerical integration are used to compute the subsequent march of supersaturation and droplet-size distribution for different rates of ascent of the air. His initial conditions and a summary of his results are shown in Table 5 from which it appears that a droplet may grow to $r = 5\ \mu$ in times varying from about 100 to 2500 sec., depending upon the size of the nucleus and the rate of ascent. These times are much shorter than the 110 minutes required by SCHUMANN's coalescence theory.

The following further conclusions may be drawn from HOWELL's calculations. Higher rates of cooling produce larger peak supersaturations and cause a larger proportion of the smaller nuclei to be activated. Condensation occurs first on the larger nuclei; if they are present in only small numbers, and so are unable to assimilate the vapour as fast as it is released by the cooling air, the supersaturation rises and allows smaller nuclei to become effective. The maximum concentration of drops coincides with the peak supersaturation. Thereafter, the supersaturation falls because vapour diffuses to the many condensation centres more quickly than it can be replaced by cooling of the air and now the smallest droplets begin to evaporate. It appears that in clouds containing a few hundred nuclei per cm.^{3} the supersaturation may attain a peak value of only about 0.1% and will surpass 1% only under extreme circumstances.

The most uniform droplet spectrum was produced with a relatively few nuclei in rapidly expanding air, but even with slow rates of cooling associated with the activation of a large range of nuclear sizes the final droplet-size spectrum was much narrower than that usually observed in natural clouds. This suggests that the rather broad spectra which are observed in all except newly-formed clouds cannot be explained solely in terms of the growth of even a broad size distribution

Table 5. *The growth of non-uniform groups of nuclei under various atmospheric conditions.*

Rate of ascent (cm. sec.$^{-1}$)	400	30	1.5
Total nucleus concentration (per cm.^{3})	500	2000	667
Range of nuclear masses (moles)	10^{-19} to 10^{-15}	10^{-16} to $10^{-13.8}$	10^{-16} to $10^{-13.5}$
Peak supersaturation (%)	0.36	0.072	0.056
Time to attain peak supersaturation (sec.)	25	70	400
Mass range of activated nuclei (moles)	10^{-17} to 10^{-15}	$10^{-15.65}$ to $10^{-13.8}$	$10^{-15.5}$ to $10^{-13.5}$
Concentration of activated nuclei (per cm.^{3})	250	410	40
Radius at peak supersaturation (μ)	1.5 to 3	1.5 to 4.5	2 to 8
Time taken to grow to $r = 5\ \mu$ (sec.)	150–100	2200–100	2450–30
Modulus of size distribution (m)* at 100 m. above cloud base	0.06	0.33	0.23
Volume-median radius at 100 m. above cloud base	~3 μ	~5.5 μ	~14 μ

* $m = \sqrt{2\sigma / r_{vm}}$ where r_{vm} = volume-median radius and σ = standard deviation of drop volumes. A small value of m denotes a more nearly uniform spectrum.

of nuclei in uniformly cooled air. Spatial variations in vertical velocity and associated variations in supersaturation followed by mixing may be invoked to account for a greater degree of inhomogeneity, but it is difficult to set up a realistic cloud model which will produce the type of droplet spectra which is observed.

This has led MASON [*39*] to suggest that the size of the droplets may be largely determined by the time which they spend in the interior of the cloud before being transferred by turbulent diffusion to the drier air outside. By treating the turbulent motion of the droplets in the cloud as a random walk, it is possible to calculate what fraction of drops will remain inside the cloud for a time Δt longer than average and attain a radius Δr greater than average. Such calculations indicate that even if the supersaturation, temperature and nucleus size were uniform throughout the cloud, droplet spectra approaching the observed shape could be produced just as a result of this transfer process.

Thus the size distribution of cloud droplets is controlled not only by the microphysical processes of condensation on nuclei, but also by the larger-scale processes of cloud dynamics which will determine the cloud dimensions, the degree of mixing with the environment, the distribution of vertical velocity, the scale and intensity of the turbulence, and the period for which individual droplets will remain in the cloud.

V. The initial stages of ice formation—ice-forming nuclei.

19. Introduction. In view of the generally accepted idea that the co-existence of ice crystals and supercooled water droplets is *often* essential to the release of precipitation from clouds, and the fact that the high cirrus clouds are composed almost entirely of ice, a study of the initiation and development of the ice phase in the atmosphere assumes fundamental importance.

Although appreciable supercooling of large quantities of water, e.g. lakes, ponds is not observed, cloud droplets commonly exist in the liquid state at temperatures well below 0° C. Water-droplet clouds are a common occurrence at $-10°$ C and have on occasion been observed at temperatures as low as $-35°$ C, while minute droplets ($r \simeq 5\ \mu$) of highly purified water have been supercooled to $-40°$ C in the laboratory.

The supercooled state is metastable and its existence is attributed to the absence of suitable nuclei to initiate freezing. The stability of supercooled clouds at temperatures above about $-15°$ C speaks for the rarity of efficient ice-forming nuclei in the atmosphere. Besides the freezing of liquid droplets there is also the possibility that ice crystals may form by the direct *sublimation* of vapour upon solid foreign particles. The relative importance of "*freezing nuclei*" and "*sublimation nuclei*" has often been debated, but experiments have shown that the great majority of nuclei fail to produce ice crystals until the air is saturated relative to liquid water, and no substance has been found to act *initially* at or near ice saturation. This strongly suggests that ice crystals are rarely formed without the intervention of the liquid phase. It may be sufficient for the liquid to exist as a thin film, or to form in the pores of a solid particle by capillary condensation, in which case it may do so before the air achieves 100% relative humidity.

20. The spontaneous nucleation of supercooled water. A number of experimenters have reported on the possibility of supercooling small droplets ($r < 5\ \mu$) to about $-40°$ C before freezing occurs. CWILONG [*58*] and FOURNIER D'ALBE [*61*], by expanding atmospheric air in small cloud chambers, found that while at temperatures above $-40°$ C a few ice crystals appeared among a much larger number of water droplets, below $-41°$ C there was a sudden increase in the number of crystals, the fog being composed almost entirely of ice. SCHAEFER (1948), however, quotes a critical temperature of $-38.9°$ C at which ice crystals are produced in a supercooled cloud simply by passing a cold body through it. Thus a pellet of mercury (melting point $-38.9°$ C) leaves ice crystals in its wake as long as it remains solid, but, at the instant of melting, this property disappears. In such an experiment a thin sheath of air in immediate contact with the pellet is suddenly cooled from $-10°$ C to about $-39°$ C, producing a very dense cloud of tiny droplets ($r \simeq 1\ \mu$) by *spontaneous condensation*. When these droplets freeze they grow rapidly and are observed as glittering ice crystals; if they do not freeze they quickly evaporate behind the pellet.

As foreign nuclei are excluded in these tiny droplets nucleation of the ice phase must occur by chance association of the molecules in the supercooled water (i.e. homogeneous or spontaneous nucleation). The rate of production of nuclei of critical size in the supercooled liquid, J cm.$^{-3}$ sec.$^{-1}$, may be written (MASON [*40*])

$$\log J = 32.84 + \log T - \frac{U}{2.303\, k T} - \frac{1.11 \times 10^{17}\, \sigma_{SL}^3}{L^2\, T \ln^2 T_0/T}, \tag{20.1}$$

where U is the activation energy for self diffusion of a molecule in the liquid, σ_{SL} the specific surface energy of the crystal-liquid interface, T the absolute temperature, L the latent heat of fusion per gramme, and $T_0 = 273°$ K.

Using the best estimates for $U = 4.8 \times 10^{-13}$ erg, $L = 58$ cal/g., and $\sigma_{SL} =$ 17 ergs/cm.^{2}, Eq. (20.1) predicts that at $-40°$ C, in a droplet of radius $5\ \mu$, a nucleus of critical size will, on average, be formed after 0.003 sec. and at $-41°$ C after 5×10^{-4} sec. The probability of a droplet freezing at a given temperature will be proportional to $t v$, where t is the time for which the temperature persists and v is the droplet volume. Although these times are in good agreement with observations made on the freezing of small droplets in cloud chambers, this is largely fortuitous since a 10% error in the value used for σ_{SL} can change these times by a factor of about 10^6. However, if we differentiate Eq. (20.1) with respect to temperature we obtain

$$\frac{d}{dT}(\log J) = \frac{1}{T} + \frac{U}{2.303\, k T^2} - \frac{1.11 \times 10^{17}\, \sigma_{SL}^3}{L^2\, T^2} \cdot \frac{(2 - \ln T_0/T)}{\ln^3 T_0/T}. \tag{20.2}$$

At $-41°$ C, $d(\log J)/dT = -0.75$, showing that the nucleation rate should increase about six-fold for each degree fall in temperature around $-40°$ C. This result is much more reliable than the calculated value of the absolute nucleation rate J since an error of $\pm 5\%$ in the value of σ_{SL} would cause this factor to vary between 5 and 7 only. MASON confirmed this prediction experimentally by observing the crystallisation of tiny water droplets supercooled in air. When a cold body of surface temperature $-39°$ C was passed through a supercooled cloud (temperature $-10°$ C) the tiny droplets produced by *spontaneous condensation* in its wake quickly evaporated and only a few ice crystals appeared. These probably resulted from the freezing of the larger droplets ($r \simeq 5\ \mu$) of the original cloud. At $-40°$ C the number of crystals increased about ten-fold, at $-41°$ C about ten-fold again, while at $-42°$ C the whole cloud was rapidly transformed into myriads of ice crystals, indicating that the spontaneously formed droplets were now freezing.

It appears that the spontaneous crystallisation of a supercooled droplet is a statistical phenomenon, the probability of nucleation depending upon the droplet volume, its temperature and the time for which it is cooled. MASON points out that the difference between the so-called "critical" temperatures, -38.9 and $-41°$ C, which have been quoted for the spontaneous appearance of ice crystals arises from differences in experimental design and the criteria adopted by different workers for the onset of ice crystal formation.

These experiments suggest that droplets will not *persist* in the supercooled state at temperatures below $-40°$ C in the atmosphere. There is evidence, however, that some cirrus clouds, and especially mother-of-pearl clouds, *form* at considerably lower temperatures. At temperatures below about $-80°$ C, which may prevail in mother-of-pearl clouds formed in the stratosphere, the condensate may be a "glassy ice" which crystallises only very slowly. Although optical phenomena indicate the cloud particles to be spherical, they may not necessarily be liquid.

21. The supercooling and freezing of water containing foreign nuclei. In cloud physics we are concerned with the temperatures at which airborne water drops, varying in diameter from a few microns to about 5 mm., will freeze and how these freezing temperatures may depend upon the drop size, the rate of cooling and the presence of foreign particles. Because of the experimental difficulties there have been very few detailed investigations on the freezing of airborne drops and we have therefore to examine the extensive literature which describes the freezing of water either held in glass containers or supported on metal surfaces, in the hope of finding well established relationships which may be applied to the freezing of drops in clouds. But an examination of the literature soon reveals serious discrepancies between the results of different workers and only very recently has some semblance of law and order appeared.

Since the time of FAHRENHEIT a very large number of determinations have been made of the freezing points of water contained in glass vessels, tubes or capillaries, the samples ranging in volume from about 1 litre to 1/100 $cm.^3$ and being of greatly different degrees of purity. A careful examination of the results does indicate that, in general, the freezing point increases with the volume of the sample and that it can be substantially lowered when impurities are removed by distillation and filtration. The freezing point may also be profoundly influenced by the walls of the container or the supporting surface; for this reason alone one must be careful in deducing the behaviour of airborne drops from the results of experiments in which solid surfaces are present. The fact that the freezing temperature of the same water sample is not generally reproducible to within 1° C or more suggests that, to some extent, freezing is a random process and that a given sample has only a statistical probability of freezing at a particular temperature. Thus the results of a particular experiment should be weighted according to the number of trials upon which it is based and one must not overestimate the significance of a rare occasion when an extraordinary high degree of supercooling may be obtained.

Of particular importance are the results of some recent experiments in which the freezing points of *large numbers* of samples of *widely different* volumes have been determined and which undoubtedly show a marked volume dependence. These results are plotted in Fig. 12.

HEVERLEY [*41*] studied the freezing of drops of water and dilute aqueous solutions which were supported on a small thermocouple junction or on waxed paper in a cryostat. The drop diameter varied from 50 μ to 1.1 mm. and the cooling rates from 1 to 20° C/min. The freezing points of the drops fell roughly on the same diameter-temperature curve (see Fig. 12a) irrespective of the origin and contamination of the water. The freezing points of drops of 400 μ to 1.1 mm.

diameter were roughly constant at $-16°$ C, but fell rapidly with decreasing diameter below 400 μ. The freezing points appeared to be independent of the cooling rate.

The freezing of about 5000 drops with diameters between 8.75 μ and 1 mm. on a metal plate was studied photomicrographically by DORSCH and HACKER [*42*]. The rate of cooling varied between 6 and 15° C/min. There was a considerable spread in the freezing points of a group of drops having the same diameter, the width of the probability—temperature curve at the 50% value varying between 2 and 6° C—the spread being generally greater for larger drops. The *average* freezing temperature for a group of drops is plotted against the corresponding diameter in Fig. 12b, which shows a linear relationship between the logarithm

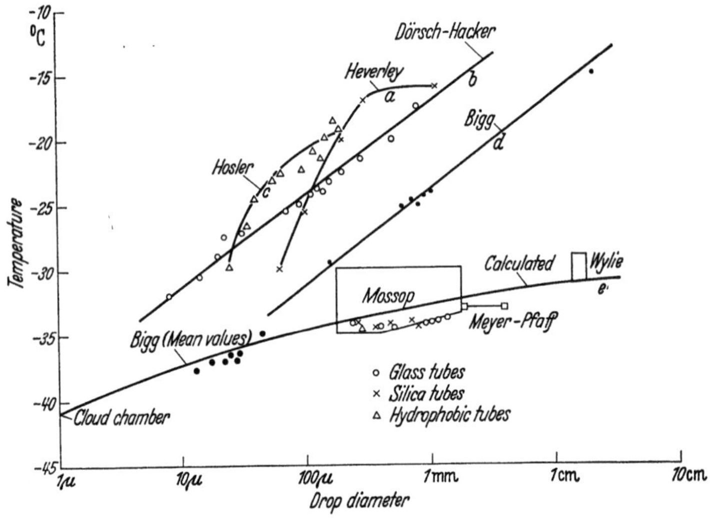

Fig. 12. The freezing points of water drops as a function of their diameter, as determined by various authors.

of the drop diameter and the freezing point. The results of a similar investigation by HOSLER [*43*], involving the freezing of 490 drops of distilled water on a platinum, surface are shown in Fig. 12c.

A very careful and extensive investigation has been carried out by BIGG [*44*] who eliminated the influence of solid surfaces by suspending drops of purified water at the interface of two immiscible liquids having different densities, where they were protected from infection by atmospheric impurities. His experiments covered a very wide range of droplet diameters from 15 μ to 2 cm. The cooling rate was usually 0.5° C/minute. The use of five different pairs of supporting liquids suggested that the observed freezing temperature was a property of the water and not of the surrounding medium. A large number of observations of the freezing temperature of drops of various sizes was made and divided into groups according to drop diameter, each group containing not less than 100 observations. The mean temperature of freezing for each group was taken as that below which half of the drops froze and is plotted against drop diameter in Fig. 12d. These results closely resemble those of DORSCH and HACKER in that both curves are of the form log diameter (or volume) $= K(T - T_0)$, but BIGG's temperatures are about 7° C lower.

From observations on 1089 drops BIGG also determined the frequency with which drops of a particular size froze as a function of temperature. The distribution curve is shown in Fig. 13; the integral of this curve gives the probability that a drop will freeze at a given temperature departure from the mean and is shown by the dotted line. BIGG was able to interpret his results in terms of a simple probability theory and deduced from Fig. 12d, which may be represented by the relation $\ln V = 16.2 - T_s$ where $T_s = T_0 - T$ is the degree of supercooling, that for a drop of volume V cm.³ cooled to T_s for t seconds, the probability P of freezing would be given by $\ln(1-P) = -A\,V\,t\,(e^{T_s} - 1)$. The experimental relationships obtained by BIGG and by DORSCH and HACKER can be interpreted in terms of the drops containing a statistical population of freezing nuclei having a distribution of the form $n = n_0\,e^{T_s}$ where n is the number of nuclei per unit volume active at T_s.

The generally accepted view that the freezing point of a water sample is dependent upon its *volume* has been challenged by HOSLER [43], who reported that the freezing points of samples contained in pyrex capillary tubes depended only upon the *diameter* of the tube and not upon the volume or surface area. However, much more careful experiments performed by MOSSOP [45] indicate that the HOSLERS' results were due to impurities on the inner walls of the tubes and that if these are coated with a hydrophobic film, volume-dependent freezing points are found.

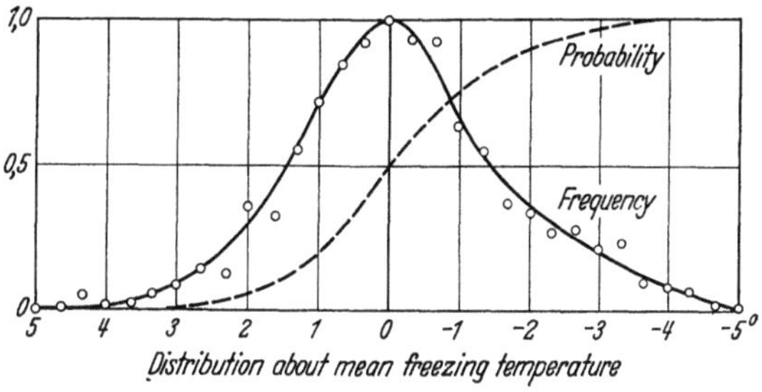

Fig. 13. The relative frequency of freezing of water drops of 1 mm. diameter as a function of temperature (after BIGG [44]).

Although BIGG removed the larger particles from his water samples they may well have contained small colloidal particles serving as freezing nuclei at relatively low temperatures; it is unlikely that his curve (Fig. 12d) represents spontaneous nucleation. This is also suggested by the fact that the freezing points of his very small droplets (diameter 15 to 50 μ) lay below this curve. Further evidence comes from the work of MOSSOP who studied the freezing of water in glass, silica and hydrophobically-coated tubes when extreme care was taken to clean the tubes and to exclude airborne impurities. The freezing points of his samples lay in the rectangle shown in Fig. 12. His *lowest* freezing points are marked separately and lie well below BIGG's curve, as do those obtained by MEYER and PFAFF [46] and WYLIE [47], who used glass containers. Finally, we plot the results of the cloud chamber experiments, which indicate that 1 μ droplets freeze within about ½ sec. at −41° C (MOSSOP [45]). It may well be that these *lowest* freezing points represent spontaneous nucleation—a suggestion which receives some support from the fact that Fig. 12e, calculated from Eq. (20.1) using the value of $\sigma_{SL} = 17.2$ erg cm.⁻² suggested by the cloud chamber experiments, passes rather close to the experimental points. Since the times for which the various samples were held at their indicated temperatures before freezing are unknown, but were probably different in the different cases, an exact comparison between theory and observation cannot be made, but it does seem likely that these lowest freezing points may have been the result of spontaneous nucleation while those of BIGG were not.

RAU [48] claims to have supercooled water to temperatures well below −41° C; he stated that when all nuclei were removed it was possible to supercool drops supported on a highly-polished metal plate to −72° C and to maintain them at that temperature for several hours. Below −72° C the drops froze uniformly without the selection of special sites but melted again on being heated above $-72°$ C. Attempts to repeat these observations have been made by CWILONG [*49*] and by BREWER and PALMER [*50*] who reported that they could not reproduce RAU's results unless they deliberately contaminated their water drops with organic vapours such as acetone or alcohol, and criticised RAU's experimental arrangement on the grounds that contamination could have arisen from his cooling system.

The freezing point of water may, of course, be influenced by the presence of salts in solution. Some salts (e.g. NaCl) depress the freezing point at all concentrations, while others (e.g. alkali iodides) (see MASON [*51*]) may raise the freezing point provided their concentration is not too high. However, the concentration of salt in cloud droplets is generally so low that their freezing points will not be appreciably affected.

22. The abundance and behaviour of atmospheric ice nuclei. Experimental techniques and results. The ice-nucleus content of the atmosphere has been studied by two main experimental techniques: observation of the freezing points of the condensation products formed on a chilled metal plate exposed to airborne contamination, and of the appearance of ice crystals at different temperatures in a supercooled cloud produced in expansion- and mixing-cloud chambers.

Using the first method WEICKMANN [*52*] found that when a metal plate was cooled in a chamber containing room air saturated with water vapour, no ice crystals appeared on the plate at relative humidities corresponding to ice-saturation, even at temperatures as low as $-40°$ C, while at ice-supersaturations of 30% and temperatures of $-39°$ C only a few crystals were formed. When water saturation was reached, however, the number of ice crystals increased very considerably, in consequence of which WEICKMANN was convinced that they were formed by the freezing of liquid water. He also obtained some evidence that the freezing was initiated by solid, insoluble particles.

RAU [*48*], [*53*] used a very similar technique to study the freezing of droplets with diameters ranging from about 300 μ to a few millimetres on highly polished metal surfaces. After admitting room air into the chamber freezing was often observed to start repeatedly at quite definite places in the drop, which RAU assumed to correspond to the locations of foreign nuclei. Plots of the fraction of drops freezing at different temperatures were assumed by RAU to represent the temperature spectra of the atmospheric freezing nuclei. These showed distinct maxima at $-12°$ C and about $-20°$ C, the former being more prominent for the larger drops, and the latter dominant for the smaller drops. Very few drops froze above $-4°$ C and the maximum supercooling obtained in the presence of nuclei was $-33°$ C.

In order to avoid the presence of solid surfaces which themselves may have nucleating properties, it is desirable to examine the condensate when suspended in air which can be cooled at a controlled rate to predetermined temperatures. The expansion chamber provides an obvious technique and has been employed by a number of workers whose results will now be summarised.

REGENER [*54*], having passed air through a bacteria filter, subjected it to very rapid expansions and examined the resultant condensate through a microscope. With the initial temperature below 0° C and with expansion ratios between 1.3 and 1.5 only supercooled water droplets appeared; at a ratio of 1.5 a few crystals appeared and with ratios between 1.6 and 1.8 (depending upon the initial temperature, which is not stated) a pure ice mist was formed. REGENER reports that water drops existed at temperatures down to $-50°$ C but this temperature appears to have been calculated on the basis of a perfectly adiabatic expansion and so may have been several degrees too low.

FINDEISEN and SCHULZ [55] studied the behaviour of ice-forming nuclei in outdoor surface air by using a large expansion chamber (capacity 2 m.³) and employing rates of expansion corresponding more nearly to those prevailing in the atmosphere. They investigated particularly the effect of varying the rate of expansion on the temperature at which ice crystals first appeared. On the average, the number of ice particles increased steadily from about 1/m.³ at −12° C to about 100/m.³ at about −20° C and 1/litre at −30° C. In this temperature range the threshold temperature for the appearance of ice crystals decreased with increasing rates of expansion (corresponding to vertical air velocities of 5 to 22 m./sec.) and was correlated with the rate prevailing at the time of cloud formation rather than with that which obtained later. The ice nuclei active in this regime were called "Type 1" nuclei. That some ice particles could be formed before water saturation was attained, was established by interrupting an expansion before the condensation point was reached; the formation of visible drops was thus prevented but thin water films may still have formed on the surfaces of any foreign insoluble particles by virtue either of their surface hygroscopicity (mixed nuclei) or their suitability for capillary condensation. The number of crystals appearing under such conditions was said to be not noticeably less than when crystals were produced in mixed clouds (FINDEISEN [56]).

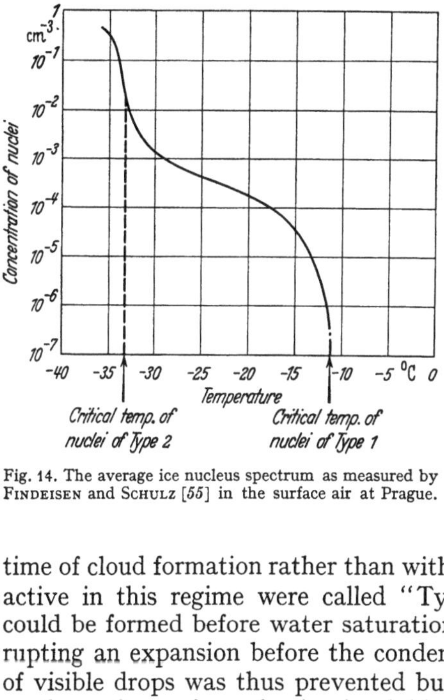

Fig. 14. The average ice nucleus spectrum as measured by FINDEISEN and SCHULZ [55] in the surface air at Prague.

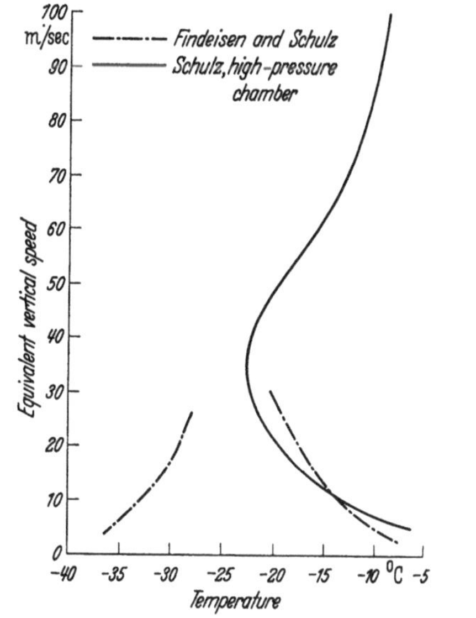

Fig. 15. The variation with rate of expansion of the air of the threshold temperature for the appearance of ice crystals in the large expansion chambers of FINDEISEN and SCHULZ.

When the temperature fell below about −32° C, the concentration of ice particles increased almost one thousand-fold to about 1/cm.³. The temperature at which this sudden increase occurred was higher for increased rates of cooling, the responsible nuclei now being termed "Type 2" nuclei. At still lower temperatures the numbers of crystals increased still further, but even at −40° C they were considerably less than those of the water droplets. The results of FINDEISEN and SCHULZ are summarized in Figs. 14 to 16.

SCHULZ [57] using a high pressure chamber of volume 0.65 m.³ expanded air initially at 10 atmospheres to atmospheric pressure, obtaining equivalent vertical speeds of from 5 to 100 m./sec. He determined the highest temperature at which Type 1 nuclei became active for different rates of cooling. Fig. 15

shows that for vertical speeds up to about 30 m./sec. the results agreed fairly well with those obtained with the low pressure chamber, but at higher speeds a reverse trend occurred; the threshold temperature now became *lower* during more rapid expansions.

CWILONG [*58*], [*59*] examined outdoor air using an expansion chamber of only a few hundred cm.^{3} capacity and high rates of expansion. He observed no ice crystals above −32° C, probably because of their rarity and the small volume of his chamber. At −33° C a few ice crystals appeared; at lower temperatures their concentration increased but became comparable to that of the water drops only when the temperature fell below −41° C. Measurements on the Jungfraujoch and over mid-ocean indicated that the −32° C nuclei were absent or extremely rare, few ice crystals appearing in the chamber until a temperature of −41.2° C was reached. Later, CWILONG [*60*] observed that the concentration of the −32° C nuclei relative to that of the −41° C nuclei at Wellington (N. Z) was lowered by the onset of a sea breeze.

FOURNIER D'ALBE [*61*] used a 2-litre chamber and illuminated a small volume (0.2 cm.^{3}) to facilitate the microscopic observation of individual particles. The minimum detectable concentration of ice particles was about 1/cm.^{3}. With outdoor air he obtained results in agreement with CWILONG. Between 0 and −32° C only water droplets appeared. At about −33° C a few ice crystals appeared simultaneously with the water drops (i.e. at the dew point); the critical temperature was not definite, but varied between −32 and −36° C. When the temperature fell below −41° C, a sudden increase in the number of ice crystals was observed, the fog being composed almost entirely of ice. Using carefully cleaned air FOURNIER D'ALBE found that with relative humidities of 112% relative to water (188% relative to ice) neither droplets nor crystals were formed even at temperatures below −42° C.

PALMER [*62*] used CWILONG's chamber on an aircraft and reported that ice nuclei which became active at −32° C in surface air were absent above inversions, stratus cloud or haze-tops, but that the −41° C nuclei existed at all heights up to 6000 ft., the ceiling of the observations. He also used a 5-litre chamber on the ground to detect the FINDEISEN Type 1 nuclei active above −32° C. Using expansion rates corresponding to vertical speeds of 100 to 200 m./sec. he found nuclei which produced ice crystals between −18 and −31° C, their numbers increasing about 500-fold over this temperature range (see Fig. 16). In common with FINDEISEN, PALMER observed occasional ice crystals at high ice supersaturations, but without a water cloud being formed.

A very simple apparatus consisting of a commercial refrigerating unit of about 70 litres capacity has been used by SCHAEFER [*63*] to study the ice nucleus content of surface air down to temperatures of about −20° C. The bottom of the chamber is cooled to −20° C and the top is open, the temperature inversion making the air in the chamber very stable. A supercooled cloud is produced by evaporating water into the chamber. Atmospheric air is drawn into the chamber by a pump and the number of glittering ice crystals observed in a powerful light beam is estimated visually. SCHAEFER reports on the 11800 ice nucleus counts made with such a chamber at 3-hourly intervals on Mt. Washington during 1948 to 1952. Crystal counts varied between 1 and 10^7/m.^{3} but usually lay between 10^2 and 10^5/m.^{3}, being less than 500/m.^{3} on nearly half of the occasions, and rarely exceeded 10^6/m.^{3}. There was some evidence that very high counts were associated with the arrival of air which had passed over arid regions of the United States during dust-storm activity.

A similar experimental arrangement was used by WORKMAN and REYNOLDS [64] to measure the numbers of ice nuclei which were effective in the temperature range -14 to $-30°$ C in New Mexico. The observations which were made over a two-month period, showed considerable spread and are represented by the shaded zone in Fig. 16. AUFM KAMPE and WEICKMANN [65] used a room-size cold chamber with an illuminated volume of about 1 m.^{3}. They estimated visually the concentrations of ice crystals which appeared at different temperatures. The trend of their results is roughly parallel to those of FINDEISEN and SCHULZ and of PALMER, but displaced to higher concentrations by a factor of 10^2 to 10^3.

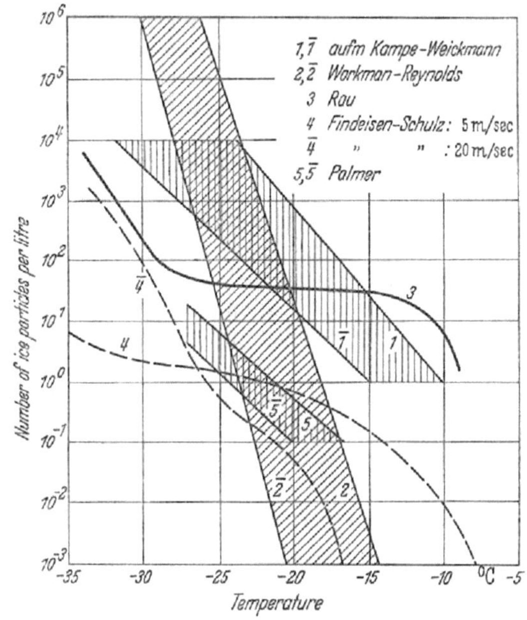

Fig. 16. The concentrations of ice particles appearing in expansion- and mixing-type cloud chambers as a function of the minimum temperatures attained (from RAU [67]).

The use of a cold chamber to measure ice nucleus concentrations from aircraft has been described by SMITH and HEFFERNAN [66]. A chamber of 76 litres capacity was cooled down at a steady rate and the number of crystals appearing in an illuminated volume counted visually at different temperatures. Precautions were taken against the results being vitiated by frost splinters leaving the walls of the apparatus. Cognisance was also taken of the fact that there is a time lag in the activation of ice nuclei and that those crystals which appear first may settle out before others have appeared, and that the loss by sedimentation will depend on the rate at which the chamber is cooled. The observed crystal concentrations were therefore corrected by factors which were experimentally determined by using silver-iodide smokes as test substances. The lower limit of nucleus detection varied a little with temperature, but was of the order 1 particle/litre. The authors report on having observed three types of ice nucleus. One of these appeared only below $-30°$ C but in reproducible concentrations of about 1/litre at about $-32°$ C, increasing to 10^6/litre (complete glaciation) at $-37°$ C, and apparently correspond to FINDEISEN'S Type 2 nuclei. The second type was usually present in variable quantities and had rather variable characteristics, detectable numbers appearing at threshold temperatures between -10 and $-20°$ C. Their concentration increased rapidly (with falling temperature) to between 10/litre and 10^4/litre and then the crystal numbers remained fairly steady until the temperature fell below $-32°$ C. The third type which was observed only occasionally was characterised by the appearance of a low concentration (about 1 per 3 litres) of comparatively efficient nuclei active at about $-10°$ C.

Crystals in concentrations of 1/litre were observed at temperatures below $-15°$ C on 7/30 days, below $-20°$ C on 22/30 days and above $-15°$ C on only 1/30 days. Apart from the fact that efficient nuclei appeared to be rather more abundant above inversions than below, there was no correlation between ice nucleus concentrations and altitude.

RAU [67] points out some of the disadvantages of the open cold chamber and describes some measurements he has made in Bavaria with a closed chamber in which the temperature distribution in the vertical was nearly uniform except at the very top. His average spectrum is shown in Fig. 16. This again shows a sharp rise below about −30° C, but the numbers of nuclei active above −20° C and especially above −10° C are considerably higher than those reported by other workers. It appears, however, that RAU may not have taken adequate precautions to prevent frost splinters from being dislodged from the chamber walls and these may have been responsible for these high counts.

23. Ice nuclei: Discussion of experimental data. To infer the ice nucleus and potential ice crystal content of clouds as a function of temperature from the freezing spectra of water samples and measurements made with expansion and mixing cloud chambers is a difficult and uncertain procedure. There can be no guarantee that the freezing-nucleus content of bulk water will correspond closely to that of the atmosphere and it is difficult to establish that the supporting medium is entirely without influence on the freezing process. In cloud chambers, apart from the inherent difficulties of obtaining a representative atmospheric sample and of detecting the very small concentrations of highly efficient nuclei, serious errors may arise from the nuclei having unknown time-lags and from the losses of nuclei by sedimentation during slow cooling. It appears that these methods are not capable of accurately determining the absolute concentrations of nuclei as a function of temperature, though relative values may be more reliable. Furthermore, the observed concentration of ice nuclei in clear air may bear very little relation to the numbers of ice crystals appearing in clouds at the same temperature since the latter may be greatly multiplied by fragmentation of delicate crystals and the bursting of drops during freezing.

So far only a very few measurements have been made at cloud level and since these indicate that the highly efficient nuclei may be very sporadic, it will be a long, laborious process to obtain adequate data on the ice-nucleus content of the free atmosphere and how this is influenced by geographical, topographical, meteorological and other factors. It is all the more important therefore, to examine the results of the various cloud chamber investigations for clues as to the properties, nature and possible origin of the nuclei.

It has been shown by MASON [68] that the background population of ice crystals appearing below −31° C in SMITH and HEFFERNAN'S cold chamber and the sudden large increase in the numbers appearing below −32° C in the large cloud chamber used by FINDEISEN and SCHULZ may be attributed to spontaneous nucleation. However, in the much smaller chambers of CWILONG and FOURNIER D'ALBE where the sensitive times and droplets are small ($d \simeq 1\ \mu$), detectable concentrations of crystals formed spontaneously would not appear unless the temperature fell below −36 and −38° C respectively. Crystals appearing at higher temperatures, and above −32° C in the FINDEISEN-SCHULZ chamber must therefore have been initiated by foreign ice nuclei. MASON [69], [68] suggests that these consist of either solid insoluble particles or of mixed nuclei with soluble and insoluble components. These insoluble nuclei probably consist largely of mineral dusts (silicates) and soil particles which are found to produce some ice crystals in a cold chamber at temperatures below −12° C, and particularly below −30° C; some evidence for this comes also from electron microscope examination of the nuclei found in the centres of snow crystals (KUMAI [70], AUFM KAMPE et al. [71]). These would be relatively rare over the sea, in the upper air and above inversions and haze-tops. The rare, efficient FINDEISEN

Type 1 nuclei are assumed to be *large* insoluble particles with a hygroscopic coat and will therefore be preferred centres of condensation. With slow expansions they may assimilate the vapour almost as fast as it is released by cooling, become very dilute and so act as ice nuclei at relatively high temperatures. With more rapid expansions, more nuclei will become centres of condensation so that the available water will be shared among a larger population of drops leaving the large nuclei *less* dilute and so less efficient as freezing nuclei than in slow expansions. In very rapid expansions, the much greater numbers of particles forming condensation and potential freezing nuclei (together with the reduced loss of droplets by sedimentation) may account for the reverse trend found by SCHULZ.

Similarly a rise in the threshold temperature for the FINDEISEN Type 2 nuclei may be a result of the large numbers of solution nuclei activated by the higher supersaturations associated with more rapid cooling and consequently of larger numbers of crystals produced by spontaneous freezing.

24. Artificial ice nuclei. *a) Silver iodide.* The fact that supercooled clouds are a common occurrence at temperatures above $-20°$ C with the implication that efficient ice-forming nuclei are often deficient in the atmosphere, and the possibility of inducing rain by supplying artificial nuclei, have stimulated a search for substances which may be dispersed into the atmosphere in finely-divided form to produce vast quantities of efficient, durable nuclei.

The most important discovery was made by VONNEGUT in November 1946, who found that silver iodide was highly effective in this respect. A dilute solution of the salt may be vaporised in a hot flame to produce about 10^{16} tiny crystals per gramme of silver iodide. When silver iodide smoke is introduced into a supercooled cloud, some ice crystals appear when the temperature falls below about $-4°$ C but the numbers rapidly increase with decreasing temperature until, at about $-15°$ C, most of the silver iodide particles appear to serve as ice nuclei. Silver iodide, which was chosen originally because of the similarity between the lattice constants of the hexagonal form and those of ice, is certainly the most effective ice-nucleating substance so far discovered. Its efficiency depends very much, however, upon the method of preparation and the presence of impurities. In fact, it occurs in two crystal forms at room temperature—a hexagonal (wurtzite) arrangement and a cubic (zincblende) arrangement, the cubic form being regarded as the stable form at temperatures below 135° C. The hexagonal form is the thermodynamically stable one between 135 and 146° C. Above 146° C both types may transform into a new high-temperature cubic form. The unit-cell dimensions of the three forms are as follows:

Substance	Crystal form	Unit dimensions (Å)
AgI	Hexagonal (wurtzite)	$a=4.58$, $c=7.49$
	Cubic (zincblende)	$a=6.47$
	"High temp." cubic	$a=5.03$
PbI_2	Hexagonal	$a=4.54$, $c=6.86$
Ice	Hexagonal	$a=4.52$, $c=7.37$

When silver iodide is vaporised at several hundred degrees centigrade, X-ray and electron-diffraction patterns of the smoke show that it consists of a mixture of the hexagonal and the cold cubic forms, the hexagonal form being the more dominant the higher the temperature of the source. Whether the hexagonal or cubic form provides the more efficient ice nuclei has not yet been established, but it is perhaps worth noting that the atomic array in the (111) planes of the cold cubic structure and the (001) planes of the hexagonal form are geometrically identical, the lattice spacing being 4.58 Å in each case.

Silver iodide generators used in cloud seeding make use of silver iodide dissolved with either sodium or potassum iodide in acetone. The solution is sometimes atomised through a nozzle with either compressed hydrogen or propane and the spray ignited; alternatively, the solution may be sprayed into a kerosene flame. Another method is to soak coke nuts in the solution and burn these in a furnace under forced ventilation.

Electron-diffraction patterns, taken at Imperial College by LISGARTEN (unpublished), of smokes produced by the vaporisation of AgI—NaI and AgI—KI mixtures suggest that those techniques may not be the most efficient. The patterns produced by AgI—KI mixtures show little or no trace of either silver iodide or of potassium iodide but suggest that mixed crystals are being formed. The AgI—NaI mixtures reveal no trace of hexagonal silver iodide but produce patterns consistent with those of either the cubic form of AgI or of NaI, which have an almost identical lattice spacing; alternatively, mixed crystals of both substances with unaltered lattice spacing may have been formed. X-ray and electron-diffraction patterns of smokes produced by the vaporisation of pure silver iodide crystals include broad, diffuse rings indicative of an amorphous rather than a crystalline structure. It seems that the silver iodide, having become partly decomposed by heating, fails to form good crystals during quenching; certainly electron micrographs show the particles to be malformed with few well-defined crystalline shapes.

If the silver iodide nuclei are produced in a reducing atmosphere such as will be provided if acetone and particularly hydrogen are used in the burning process, their surface layers may be reduced to metallic silver and their ice-nucleating ability largely inhibited. This effect appears to be greatly enhanced if the crystals are irradiated with ultra-violet light (cf. the reduction of silver bromide to metallic silver in a photographic emulsion) which explains why silver iodide may quickly lose its ice-nucleating properties when dispersed into the atmosphere during strong sunshine. According to tests made by SMITH, SEELY and HEFFERNAN [72] the activity of nuclei from a hydrogen burner decays to about 1/100-th of its initial value after an exposure of only 15 minutes.

The discovery of a method for producing large quantities of highly-efficient and durable silver-iodide particles is of considerable practical importance. Lead iodide, which is not so susceptible to photolytic decomposition, appears to merit further investigation as a possible substitute.

b) Ice-nucleating ability of other chemicals. Papers have appeared from several different laboratories reporting on the ice-nucleating ability of a wide variety of chemical compounds introduced either as smokes, sprays or as fine dusts into a supercooled cloud formed in an expansion- or mixing-cloud chamber. Except in the case of silver iodide, the results obtained by the different workers are very conflicting; for example substances reported as highly effective by one laboratory are found to be completely inactive by another.

Believing that, in many of these experiments, spurious results may have resulted from the presence of trace contaminants (particularly silver and iodine) in the air, and of trace impurities in the various chemicals, MASON and HALLETT (unpublished) have recently carried out very careful tests on substances reported as highly-effective nucleating agents by at least two independent workers. These include:

AgI, PbI_2, $HgCl_2$, $AgNO_3$, Ag_2SO_4, Ag_2S, $PbCl_2$, $NaCl$

I_2, NH_4I, NaI, KI, CdI_2, CoI_2, HgI_2, TlI.

They find that if spectroscopically pure materials are used and great care is taken to exclude all traces of iodine and silver from the air in the experimental chamber, the great majority of these substances fail to produce ice crystals in a supercooled cloud at temperatures above $-40°$ C.

The nucleation tests were carried out in a continuously-sensitive diffusion cloud chamber and in a 400-litre mixing cloud chamber. Before each experiment, the diffusion chamber was thoroughly cleaned, sealed and left for some hours for any particles present to fall out and leave very clean air. Artificial nuclei were introduced into the top of the chamber and their nucleating efficiency assessed by noting the temperature level below which more than one glittering ice crystal appeared for about every 10^4 supercooled droplets. The nuclei were produced either by spraying a solution or by vaporising the substance on an electrically-heated wire in an atmosphere of clean nitrogen to prevent oxidation effects. Blank tests were always carried out with the heated wire and with distilled water to check that they produced no effective nuclei.

In early tests, NH_4I, KI, NaI, and I_2 vaporised on an electrically-heated nickel wire all produced ice crystals in the diffusion chamber at temperatures above $-20°$ C as reported by other workers, but it was found later that these were initiated by nuclei of AgI or PbI_2 produced by the iodine compounds reacting with silver and lead impurities in the wire. When spectroscopically-pure nickel wire was used the four above-mentioned substances and also CoI_2, HgI_2, TlI and SnI_2 did not produce significant numbers of ice crystals at temperatures above $-40°$ C. Sprayed solutions of NH_4I, KI, NaI, I_2, CoI_2, SnI_2, MgI_2, M_1I_2, and NiI_2 were also found to be inactive.

When smokes and sprayed solutions of $AgNO_3$, Ag_2SO_4, $Pb(NO_3)_2$, and $PbCl_2$ were introduced into the mixing-cloud chamber held at $-20°$ C, the numbers of ice crystals were not significantly increased; but on introducing a small quantity of iodine vapour, the whole cloud was immediately transformed into ice. These and other tests led to the conclusion that these four salts are not capable of nucleating supercooled clouds and that the apparent high efficiency of the silver salts may be attributed to the formation of AgI in the presence of small quantities of iodine.

The only chemical compounds which were found to be definitely active as ice-forming nuclei at temperatures above $-40°$ C were AgI ($-5°$ C), PbI_2 ($-7°$ C) and $HgCl_2$ ($-25°$ C), the temperatures in brackets indicating the threshold limits of activity.

VI. The growth of ice crystals.

25. Introduction. We are now concerned with the growth of a crystal from the formation of an ice nucleus and to examine the factors which may determine its subsequent development. Ice crystals grow in an ice-supersaturated atmosphere by the diffusion of water vapour to, and its sublimation upon their growing surfaces. They may exist as individual units of simple geometrical shape, e.g. hexagonal prismatic columns (prisms) and thin hexagonal plates, or under suitable conditions, may develop as richly-branched (dendritic) forms. Several of these snow crystals may coagulate to form a snowflake (see Fig. 17).

26. The occurrence of ice crystals in natural clouds. The most comprehensive collections of ice crystals from natural clouds have been obtained by WEICKMANN [*52*] at different heights in the troposphere up to cirrus levels, the temperature of the sampling region being measured. His observations are analysed in

Table 6. Weickmann's *observations of predominant crystal forms in different cloud types.*

Level of observation	Temperature range	Cloud types	Crystal forms	Crystal sizes (approx.)
Lower troposphere	0 to −15° C	(nimbostratus) (stratocumulus) (stratus)	thin hexagonal plates	50 μ to ½ mm. diameter, 10 to 20 μ thickness
			star-shaped crystals showing dendritic structure	½ to 5 mm. diameter
Middle troposphere	−15 to −30° C	(altostratus) (altocumulus)	thick hexagonal plates	∼200 μ diameter
			prismatic columns single prisms and twins	∼200 μ long
Upper troposphere	< −30° C	isolated cirrus	clusters of prismatic columns containing funnel-shaped cavities	∼1 mm. diameter
			single hollow prisms	∼½ mm. long
		cirrostratus	single, complete prisms	∼100μ long, length to diameter = 1 to 5

Table 6. They indicate that in proceeding from high- through medium- to low-level clouds, and therefore to higher temperatures, there is a gradual transition from prisms (representing preferential growth along the principal axis), through thick plates to thin hexagonal plates, the latter being the result of preferential

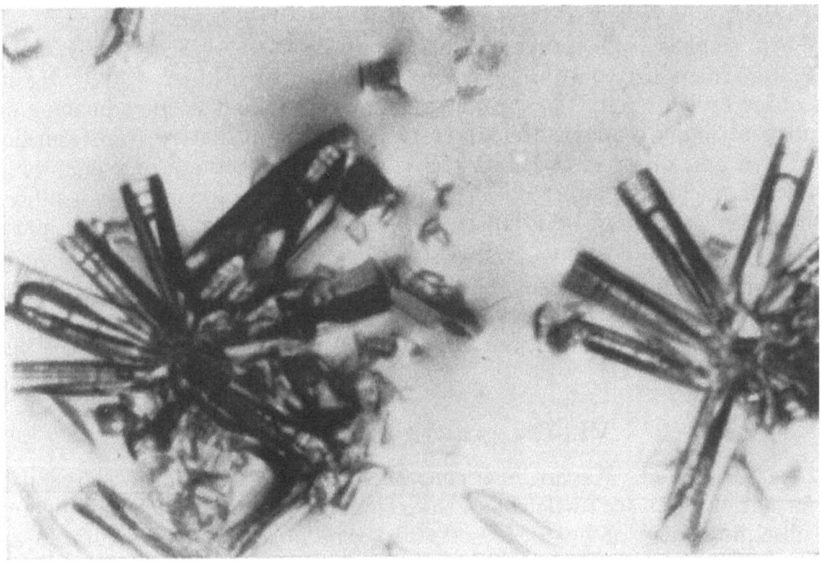

Fig. 17a–f. a) Clusters of hollow prismatic columns from isolated cirrus at −44° C (from Weickmann [52]). (Scale: The longer side of the photo is about 1.1 mm.)

growth in the direction of the secondary axes. At temperatures between −10 and −20° C the striking star-shaped crystals are dominant. They usually possess six arms which may grow rapidly and develop side branches to assume the fine dendritic form of Fig. 17d.

Needle-shaped crystals such as are shown in Fig. 17e are not represented in Weickmann's observations, but are often observed at the ground when the

Fig. 17 b. Complete prisms from cirrostratus at $-26°$ C (from WEICKMANN [52]) Magnification 90×.

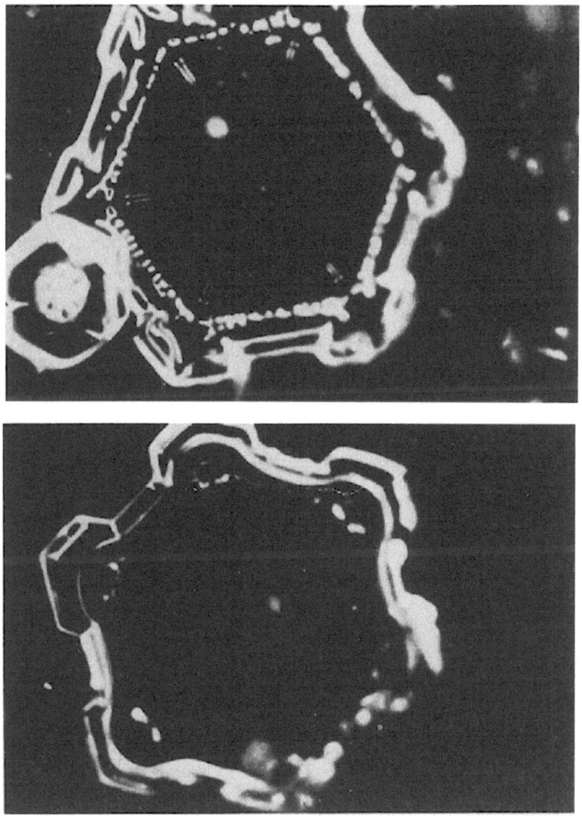

Fig. 17 c. Thin hexagonal plates. Magnification 50×.

Fig. 17d. A hexagonal plate which has developed into a richly-branched (dendritic) stellar crystal. Magnification 10×.

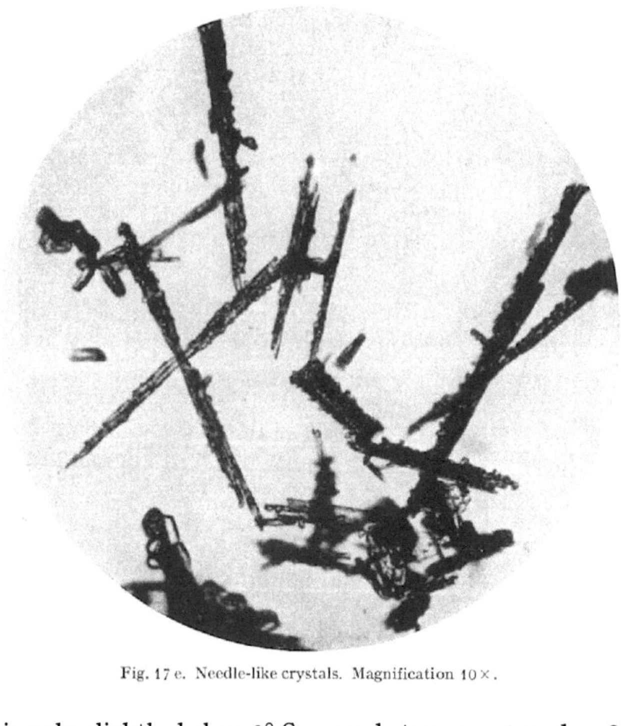

Fig. 17e. Needle-like crystals. Magnification 10×.

temperature is only slightly below 0° C—say, between −3 and −8° C. Besides the basic types of ice crystal, combination forms consisting of two or more of the basic structures are quite common, e.g. a prismatic column with end plates.

Snowflakes are agglomerates of individual crystals in which the stellar dendrites are generally dominant (Fig. 17f). Agglomeration of crystals appears to take place readily when the temperature is only just below 0° C (flakes containing a hundred or more crystals are quite common), but probably does not often occur at temperatures much below −10° C.

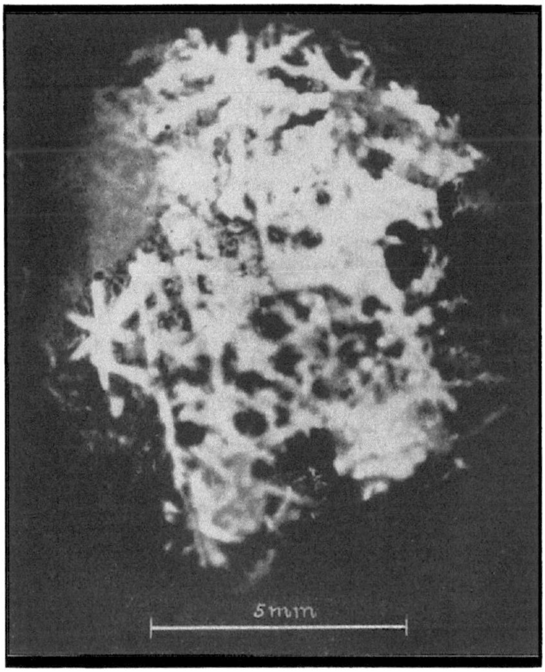

Fig. 17f. A snowflake composed of a number of stellar crystals.

27. Studies of ice crystal growth in the laboratory: Variation of crystal habit with temperature and supersaturation. A number of experiments have been designed to study the growth of ice crystals in the laboratory where the temperature and supersaturation of the environment can be controlled.

The growth of crystals in supercooled water clouds produced in room-size cold chambers whose temperature can be controlled, has been studied by AUFM KAMPE, WEICKMANN and KELLY [*73*] and by MASON [*74*]. MASON seeded the cloud near the top of the chamber with a pellet of dry ice, collected the crystals near the floor and photographed them under the microscope. The observed changes of crystal habit with temperature were very similar in both sets of experiments and are summarised in Table 7.

The classification of these laboratory-grown crystals according to temperature bears a marked similarity to that of natural snow crystals. The most striking feature of Table 7 is the remarkable sequence of habit, plates-prisms-plates-prisms, which occurs as the temperature is lowered from 0 to about −25° C.

NAKAYA [*75*] has grown crystals on a fine rabbit hair stretched on a frame and suspended in an air stream whose temperature and water vapour content could be varied. The development of an individual crystal was followed over a period of 30 minutes or so by time-lapse photography. The crystals were classified into 8 types and the correlation between habit and temperature was in broad agreement with the data of Tables 6 and 7. NAKAYA's observations are plotted in Fig. 18 where the ordinate is the saturation ratio of the environment, defined as the ratio of the measured concentration of water (liquid and vapour) in the air to the saturation vapour density of the air at its temperature T_a. NAKAYA interprets Fig. 18 as showing that the chief factor controlling the form of the crystal is the air temperature, since a given type of crystal can be obtained over a wide range of supersaturation. It appears, however, that the supersaturation is important in determining the transition from plates to stellar dendritic forms, since the latter appear only if the saturation ratio exceeds 1.1 and if T_a lies between −14 and −17° C.

Table 7. *Changes of crystal habit with temperature in artificially-produced water clouds (after* AUFM KAMPE *et al., and* MASON*).*

Temperature range	Crystal habit
0 to − 5° C	Simple, clear plates with no surface markings, some trigonal shapes
− 4 to − 9° C	Prisms, some showing marked cavities and similarity to needles
− 10 to − 25° C	Hexagonal plates showing ribs, surface markings and tendency to sprout at corners. Sector stars Dendritic stars, most prominent between − 14 and − 20° C
− 25 to − 40° C	Single prisms, twins, and hollow prisms. Aggregates of prisms and irregular crystals (AUFM KAMPE et al.)

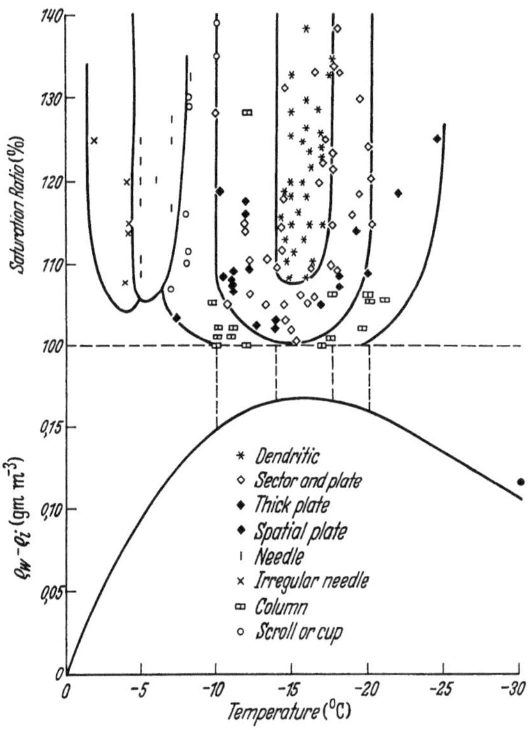

Fig. 18. (Upper part.) Relation between crystal type, "supersaturation" and air temperature as observed by NAKAYA [75]. (Lower curve.) The vapour-density excess against temperature.

A more detailed investigation of the factors which control the growth form of ice crystals has been undertaken by SHAW and MASON [76] who have measured the growth rates of the different faces of crystals growing on a metal surface in an atmosphere the temperature and supersaturation of which could be varied independently. For crystals grown at water saturation, the habit was found to alternate between prisms and hexagonal plates as the temperature was lowered according to the scheme of Table 6. Under conditions of constant temperature and supersaturation, the square of the linear dimensions of a crystal increase linearly with time. However, the growth rates of both prism and basal faces were often different on different crystals, while crystallographically similar faces of the same crystal sometimes grew at different rates. Furthermore, the growth rate of a particular face sometimes changed abruptly, even though the external conditions remained constant. These results indicate that the growth of crystal faces is not determined solely by the environmental conditions, but partly by certain individual characteristics of the faces themselves which control the rate at which molecules can be built into the crystal structure. It was also established that whether a crystal developed as a prism or a plate was determined very largely by the temperature, the supersaturation having no *systematic* effect. A minimum critical value of the supersaturation was required to start growth on any crystal face, but this varied in an apparently random manner from face to face and from crystal to crystal; there was no systematic difference between the critical supersaturations required for growth on the basal and prism faces.

28. The growth rates of ice crystals in supercooled water clouds. The following equation has been derived (MASON [74]) for the growth rate of an ice crystal growing from the vapour when at rest relative to the air and in equilibrium with the diffusion field:

$$\left.\begin{aligned}\frac{dm}{dt} &= 4\pi C\,\sigma/[J\,L^2 M/K\,R\,T^2 + \\ &\quad + R\,T/D\,M\,p_s(T)] = \frac{4\pi C\sigma}{f(T)},\end{aligned}\right\} \quad (28.1)$$

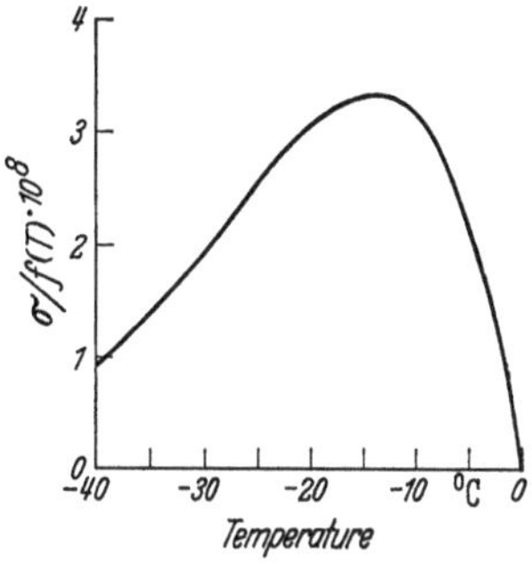

Fig. 19. The function $\sigma/f(T)$ as a function of temperature.

where m is the crystal mass, C the electrostatic capacity of the crystal regarded as a conductor of the same shape, $\sigma = p/p_s(T) - 1$ the supersaturation of the environment relative to ice and the other symbols are as already defined in Sect. 16. At constant air pressure, the term in the square brackets is a function oft emperature only. If the environment is saturated with respect to liquid water, σ is also a unique function of temperature; values of $\sigma/f(T)$ are plotted as a function of temperature in Fig. 19, the air pressure being 1000 mb. Eq. (28.1) can be used to calculate the growth rates of ice crystals whose shapes approximate to those of conductors of known capacity, e.g. for a sphere of radius r, $C=r$; for a thin circular disc (an approximation to the thin hexagonal plates), $C=2r/\pi$; for a prolate spheroid (approximation to a prism), $C=2ae\cdot\ln[(1+e)/(1-e)]$ where a is the major semi-axis and e the eccentricity.

Experiments designed to measure the rate of mass increase of crystals growing in a supercooled water cloud have been described by REYNOLDS [77] and MASON [74]. In both cases a supercooled cloud was produced in a large cold chamber and seeded with dry ice, the crystals being collected near the floor at successive time intervals and measured under the microscope—the masses being calculated from their linear dimensions. Making measurements on simple, regular hexagonal plates grown at $-2.5°$ C and on prisms grown at $-5°$ C, MASON found that for crystals of diameter less than 100 μ, the calculated and measured growth rates agreed within 20%. For appreciably larger crystals, when $vr/D \gg 1$, v being the fall velocity of the crystal and D the coefficient of diffusion, the diffusion field is no longer in a steady state, the simple theory breaks down, and dendritic growth may occur.

VII. The formation of precipitation.

29. General. Inside all clouds, presumably, the processes of condensation and aggregation produce particles of larger and larger size. Precipitation occurs when some of these particles reach such a size that they fall out of the clouds, and their sustaining updraughts. If the particles are able to survive the evaporation which occurs in the unsaturated air below the clouds, the precipitation reaches the ground; otherwise it may be noticed as fallstreaks or virga, which hang for some distance below the cloud base.

The occurrence of precipitation is strongly controlled by the motion of the cloud air. The air motion during the formation or renewal of the cloud, in combination with the properties of those aerosols it contains which act as condensation and freezing nuclei, determines the concentration, initial size-distribution and nature of the cloud particles. As soon as these have formed the microphysical processes of condensation and aggregation begin to deform the spectrum, setting about the production of precipitation elements. Because the air motion governs the dimensions, density and duration of the cloud it controls the rates of these processes; moreover, it determines the period during which they operate and thus the maximum size which the largest of the cloud particles can attain. If this should ensure the formation of precipitation, it is again the air motion which determines the maximum rate of precipitation, and its duration. Although the microphysical processes can be said to have an important influence on the air motion, for example in promoting the release of the latent heat of condensation or freezing and thus affecting the buoyancy of cloud air, for the most part their effect is not dependent upon the character of the cloud particles (concentration and size distribution). The factors of prime importance therefore appear to be the air motion and its aerosol content, and from specifications of them it should be possible to calculate quantitatively the course of cloud and precipitation development.

However, the laws governing the motion of cloud air, and the formation, growth and aggregation of cloud particles are not firmly established. For example, there is uncertainty about the efficiency of collision and coalescence of cloud droplets in the stage when they are nearly all the same size, about the processes of formation of ice crystals, about splintering processes and the manner in which snow crystals are aggregated into snowflakes. Moreover, the great variations in the concentration and properties of atmospheric aerosol, and the great complexity of atmospheric motions add to the difficulty of constructing a general quantitative theory of precipitation development. Some progress has been made with calculations, based on simple models of the air motion, of the growth of individual cloud particles; these give a reasonable qualitative explanation of the formation of precipitation in different kinds of cloud.

30. The role of condensation. In droplet clouds the condensation process alone is usually ineffective in producing precipitation. The size which droplets or crystals attain by condensation is mainly determined by their concentration and the amount of water vapour made available by the ascent of the cloud air. In the lower atmosphere, for example, droplet concentrations established during the formation of clouds are believed always to be of the order 100/cm.^{3}. During the continued ascent of air about 0.4 g./m.^{3} of water vapour is condensed during rise to a level 200 m. above the condensation level, and so the average radius of cloud droplets having the above concentration must there be rather less than 10 μ. If the rise continues the amount of vapour condensed may under the most favorable circumstances be about 30 times greater, corresponding to an average droplet radius of about 30 μ. Droplet concentrations arising during condensation in very clean air at higher levels may be one or two orders of magnitude less, but the amount of vapour to be condensed is also greatly reduced, and so the conclusion that condensation alone is unlikely to produce droplets of average radius exceeding 30 μ is not affected. Droplets of this size evaporate completely in unsaturated air after falling a distance of the order of a metre, and hence are incapable of reaching the ground as precipitation. The distance which can be fallen before evaporation is complete can be derived from Eq. (16.4), and is approximately proportional to the initial radius raised to a power between 3 and 4. It first becomes of the order of several hundred metres for an initial radius of about 100 μ, and consequently this can be regarded as the minimum size of a cloud particle which can be regarded as a precipitation element.

That the *average* radius of a droplet grown by condensation is unlikely to exceed 30 μ does not preclude the growth of *some* droplets to a considerably larger size. If all the droplets of a population are of pure water, or contain only dissolved substances from the Aitken nuclei and are therefore very dilute solutions, then as condensation proceeds the radii tend to become equal. On the other hand a disproportionate condensation may occur on the giant hygroscopic nuclei. It may be shown, however, that even in this case the condensation must be prolonged for at least several hours if the droplet radii are to approach 100 μ, at which the droplet fall-speeds are about 1 m./sec. In the layer-type clouds which have this duration the vertical air motion has a speed considerably less than this, and the growing droplets can be expected to fall out of the clouds before attaining even this size. On the other hand the orographic and cumuliform clouds which contain updraughts capable of sustaining droplets do not provide more than about one hour for droplet growth. Consequently condensation by itself is not regarded as a process capable of causing precipitation from droplet clouds.

In ice clouds the circumstances are changed by the relatively very low concentrations in which ice crystals form. The growth of these crystals is considered in more detail below.

31. Precipitation from layer clouds composed of droplets. In wholly liquid clouds the production of drizzle and rain drops is effected by the coalescence process. In the preliminary calculations which have been made it is usually assumed that inside the cloud a small concentration of droplets of radius 20 to 30 μ is present among the great majority whose average radius is several microns, and the growth of one of the large droplets is considered. The large droplets are considered to have arisen on giant nuclei soon after the cloud formation, or to be the result of prolonged condensation on a small concentration of droplets which by chance have been left for a long time within the cloud without having been brought to the cloud edges by eddying motions and there exposed to evaporation (Mason [*39*]).

If it is assumed that there are only feeble ascending motions (of less than 10 cm./sec.) within a well-stirred low-level layer cloud it can be shown by application of Eq. (17.1) (assuming the collection efficiency $E_1 E_2 \approx 1$) that such a large droplet can reach the size of a drizzle-drop ($r \approx 150$ μ) in falling from the top of a cloud about 1 km. thick, and raindrop size ($r \approx 500$ μ) in a cloud about 2 km. thick. At low temperatures or in clouds whose liquid water content is less than the maximum or predominantly held in droplets of radius less than about 7 μ (for which the initial collection efficiency is very small), these thicknesses must be increased considerably. If more substantial ascending motions (10 to 100 cm./sec.) are assumed in the cloud, the thickness required for drizzle-drop growth is considerably reduced, and that for raindrop growth slightly reduced.

These results accord with the observation that drizzle readily falls from low stratus clouds several hundred metres thick, particularly near hills and coasts (where the thickness may locally increase and where a more pronounced vertical motion may be introduced by orography), and that rain may fall when the clouds are more than about 1500 m. thick (see, e.g., Mason and Howorth [*78*]).

Apparently similar clouds occur frequently without causing any precipitation; for the time being it may be supposed that these are clouds which have formed only very recently, or which contain unusually little condensed water or unusually small droplets, or in which the required large droplets did not form.

Continuous precipitation can be maintained only if the cloud water is replenished by a persistent vertical motion (fed, perhaps over a large region, by a low-level convergence in the wind field) or a persistent direct cooling. The latter cooling may be due to net loss of heat by radiation at the upper surface of the cloud, or to the travel of the cloud air over a progressively colder surface. The typical rates of cooling due to these processes, however, can only just support a slight drizzle. On the other hand, a persistent vertical motion of 10 cm./sec. throughout a cloud 1 km. thick can support moderate drizzle, while one of 40 cm./sec. is needed to maintain moderate rain from a cloud 2 km. thick. Vertical speeds near the ground of more than 10 cm./sec. probably occur only locally in orographic disturbances, but must be extensive enough horizontally to allow time for the precipitation to be developed as the cloud air moves through them. This period, as indicated by approximate integration of Eq. (17.1), appears to be about an hour, which is longer than that available in most simple orographic clouds, demanding a horizontal extent of 30 or 40 km. in a fresh wind. Consequently precipitation forms in orographic clouds only when they occur over extensive hill slopes, or near coasts and long ridges lying nearly parallel to the wind; in these circumstances, however, the rate and amount of precipitation may be considerable.

Layer clouds of more than the minimum thickness of about 1 km. rarely occur well above the ground except in storm regions, when the great thickness and extent of the layers, and the persistent vertical motion of 10 cm./sec. or more, make the production of widespread rains practically inevitable.

32. Precipitation from layer clouds containing ice crystals. At temperatures only a little below 0° C the concentrations of ice crystals which form in clouds are extremely small or zero, and the difference in the saturated vapour pressures over ice and liquid water small, so that the growth of a crystal appearing in a supercooled cloud is little different from that of the droplets. At temperatures below about −10° C, however, crystals may appear during supercooled cloud formation in concentrations (of order 10^4/m.^{3}) which are significant for the production of appreciable precipitation, and yet low enough to allow a moderate vertical air motion (>10 cm./sec.) to maintain a state of saturation or near-saturation with respect to liquid water while they grow rapidly by sublimation. The growth therefore proceeds under supersaturations with respect to ice of up to 20 to 30%, and is much more rapid than the condensation which occurs on droplets under supersaturations which rarely reach 1%. Irrespective of the shape of the growing crystals (Table 6), fall speeds of 40 cm./sec. are attained after about 5 minutes, and they are therefore precipitated from shallow layer clouds soon after their formation. They may become visible below the cloud as fall-streaks (or virga), particularly at low temperatures, when their growth may continue in a layer several hundred metres thick in which there is a state of ice-supersaturation. The distinctive forms of such trails of crystals, to which most scattered high clouds (cirrus) bear strong resemblance, have been discussed by LUDLAM [*79*].

Application of Eq. (28.1) indicates that when the crystals form in deep layers of cloud they may attain a dimension (plate-diameter or prism-length) of 1 to 2 mm. after falling up to 2 km. relative to the air during the course of an hour.

At this stage the radius of the drop formed by their melting would be about 250 μ. Subsequently the growth by condensation becomes slower: even under the most favourable conditions it would need to be prolonged for several hours, and over a fall-path of several km., before a dimension twice as great could be reached. In general, therefore, the growth of individual crystals by condensation is halted by their arrival at the melting level before their melted size has exceeded that of drizzle drops. However, in the neighbourhood of this level it appears that individual crystals are often aggregated into snowflakes, 10 to 100 composing a single flake. This is presumably brought about by collisions due to differing settling speeds or horizontal relative motions during a fluttering fall, but the process has not been studied sufficiently to allow any quantitative discussion. After the snowflakes have melted the resulting drops may grow considerably by collision with cloud droplets at temperatures above 0° C before they finally reach the ground as rain. Indeed, if clouds over 2 km. thick occur below the 0° C level they may provide a substantial part of the rain water, and may even contribute raindrops formed wholly by coalescence (see, e.g. CUNNINGHAM [*5*]).

Above the melting level collisions with supercooled droplets may sometimes contribute to the growth of the ice crystals. Probably, however, such droplets occur in the snow region only locally where stronger vertical motions are produced by orographic or convectional motions. Certainly the riming of exposed bodies, due to the catch and freezing of liquid droplets, occurs even in moderate or heavy snowfall on mountain tops, where strong orographic updraughts are always present. In general, however, it appears that there are only crystals above the melting level; if so, the freezing of droplets cannot be the source of crystals which is needed to replace those precipitated in the snowfall, and it is likely that the splintering of the larger dendritic crystals makes the provision.

33. The structure of precipitating layer clouds as revealed by radar. The radar echoes received from the widespread, steady precipitation normally associated with warm fronts and depressions are characterised by the appearance of a particularly strong echo in the vicinity of the 0° C level, generally termed the *melting band*. This separates a relatively steady echo obtained from the snow above the melting region and a rapidly fluctuating echo from the rain below. The variation of radar echo intensity with height under these conditions can be explained both qualitatively and quantitatively in the following terms. Above the 0° C level the scattering particles are snow crystals and small snowflakes, growth and aggregation of which cause the echo intensity to increase with decreasing height. As the snowflakes fall through the 0° C level they start to melt, and since the reflectivity of a water particle is 5 times greater (at centimetre wavelengths) than an ice particle of equal volume, this melting will be accompanied by an increase in the echo intensity, which will be enhanced further by any aggregation of the wet flakes which may occur. If the precipitation rate is assumed to remain constant at all levels, the spatial concentration of the raindrops will be reduced relative to that of the parent snowflakes in the ratio of their respective terminal velocities. Thus just below the level where complete melting occurs, the echo intensity will show a sharp decrease, with the result that a narrow band of enhanced echo delineates the melting zone which has a depth of a few hundred feet. This interpretation of the radar echo has been confirmed by combined visual and radar observations from aircraft (e.g. BOWEN [*80*]).

The growth of ice crystals above the 0° C level can, in principle, be computed in terms of the temperature, pressure, humidity and liquid-water content of the cloud, and the echo intensity $I \propto \sum n\, m^2$, where n is the concentration of crystals of mass m, can be calculated as a function of height and compared with measured values. Such calculations have been made by WEXLER [81] who assumed the crystals to originate 3 km. above the 0° C level and to grow in a uniform up-draught of 10 cm./sec. as hexagonal plates of constant thickness. If they were assumed to grow solely by diffusion in a water-saturated atmosphere their computed masses on arriving at the 0° C level were much larger than those normally observed for single snow crystals, and also the computed radar-echo intensity increased with decreasing height much more rapidly than is observed. WEXLER therefore concluded that, in layer clouds with small updraught speeds, water droplets are not abundant and that the ice crystals do not grow at water saturation but at a humidity intermediate between water and ice saturation. By assuming that all the water created by the updraught within each height interval is deposited on the growing crystals within that region, so that there is no storage of water in the form of droplets, he was able to obtain much better agreement between computed and observed values of the radar-echo intensity as a function of height. But the results which emerge from such calculations depend a good deal on the assumptions made concerning the geometry of the growing crystals, the magnitude of the vertical air currents and whether or not a steady state prevails. That nimbostratus giving continuous precipitation probably does not usually contain appreciable concentrations of supercooled droplets is strongly indicated by PEPPLER'S [82] observations that icing was only rarely encountered by air craft in layer clouds at temperatures below −8° C.

There is some visual and radar evidence that aggregation of snow crystals occurs *above* the 0° C level, but this process is almost certainly accelerated in the melting zone. MASON [83] has recently shown that, in most cases, the intensity of the radar melting-band relative to the echoes from above and below can be interpreted only if one assumes a snowflake to aggregate, on average, with two or more others in the melting zone.

34. Precipitation from cumuliform clouds composed of droplets. If *showers* are to form in cumuliform clouds the processes of particle growth must be accelerated, on account of the relatively short duration (about half an hour) of the individual parts of the clouds. This acceleration is favoured by the large vertical displacement and correspondingly high cloud-water content of air which forms the upper parts of tall clouds. Moreover, the characteristically strong updraughts tend to keep growing precipitation elements in these regions of high liquid water content. On the other hand, the strength of the updraughts is such that the individual cloud masses may reach the summit-level and be exposed to mixing with clear air and evaporation before the microphysical accretion processes have the opportunity to produce precipitation elements. Evidently it is these features —the rate of the accretion processes, the thickness of the clouds, and the strength of the updraughts—which principally govern the formation of shower rain. The great variation in the size of clouds which produce showers suggests that in many clouds the favourable combination of conditions is only just realised, and n many others only narrowly missed.

LUDLAM [84] has investigated the formation of showers by the droplet coalescence process on the basis of a simplified model of cumulus cloud structure. It is assumed that near the base of the clouds a small proportion of droplets, formed upon giant hygroscopic nuclei, attain radii of 20 to 40 μ amongst a population of uniformly smaller droplets, whose radius is determined by their concentration and the cloud-water content. The latter is assumed to be that corresponding to adiabatic rise from the cloud base, so that the assumed conditions approximate to those inside the cores of the cloud bubbles. Since these are the densest parts of the cloud and contain the strongest updraughts, it is supposed that cloud particles are more likely to grow into precipitation elements within them than within the bubble wakes which compose the bulk of a cloud, although possibly droplets in these slowly dissipating regions may exist for longer periods than those in bubble cores, which are quickly lifted to the cloud tops and there exposed to a rapid evaporation.

It is then shown that once the radius of the majority of the cloud droplets reaches about 8 μ, at a level which with ordinary droplet concentrations lies several hundred metres above the cloud base, the collection efficiency of the large droplets, throughout the further course of their growth, is not likely to depart materially from a mean value of 0.85. If, moreover, the fall-speed of the cloud droplets (about 1 cm./sec.) is ignored, Eq. (17.1) may be reduced to

$$\frac{dR}{dt} = \frac{0.85}{4} w\,V = \frac{w\,V}{4.7} \qquad (34.1)$$

while the vertical motion of the growing large droplet is described by

$$dz = (U - V)\,dt \qquad (34.2)$$

U being the speed of the updraught at some height z above a reference level. If these equations are combined we obtain

$$4.7\frac{(U-V)}{V}\,dR = w\,dz.$$

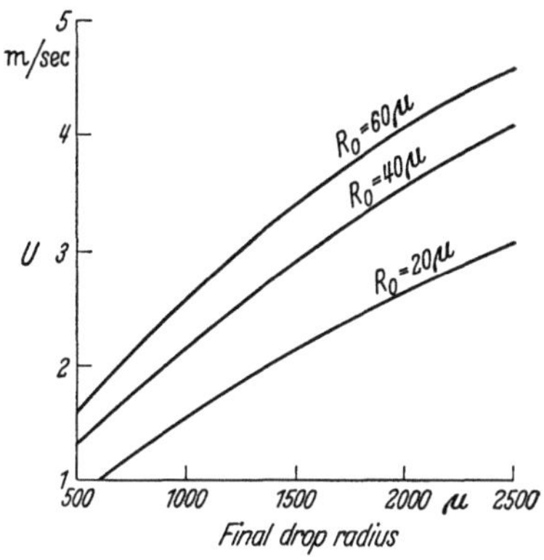

Fig. 20. Diameters of raindrops grown from droplets of initial radius 20 μ, 40 μ, 60 μ as a function of the steady updraught U.

It is now considered that w is a function only of z (for a particular cloud-base temperature), V is a function only of R (strictly, there is a small variation with z), and U is supposed to have a constant value (although usually it has a maximum value somewhere in the middle levels of the cloud). Then the growth of a droplet between two levels z_0 and z_1 from radius R_0 to R_1 may be evaluated by solving the equation

$$U\int_{R_0}^{R_1}\frac{dR}{V} - (R_1 - R_0) = \frac{1}{4.7}\int_{z_0}^{z_1} w\,dz. \qquad (34.3)$$

The equation can thus be used to determine the height to which a cloud containing a steady updraught U must extend in order that the largest of the droplets near the base shall reach the size of raindrops. Somewhat before this stage, the drops may acquire a fall-speed equal to, and subsequently greater than U. The drops then descend through the updraught and reach their maximum size at the level of the cloud base. The integral on the right of Eq. (34.3) is then zero, and it is seen that the maximum size of the drops is solely a function of the updraught speed U and their radii at the beginning of their growth by accretion. The relationship is shown in Fig. 20. The figure indicates that if the mean updraught in cumulus exceeds 3 or 4 m./sec., then raindrops may be formed having the maximum stable size of about 5 mm. diameter. The distortion produced by air resistance in drops of this size is such that small disturbances lead to their disruption into about ten drops of radius perhaps 1 mm. and a greater number of very small droplets. The larger of these drops may grow rapidly by coalescence with cloud droplets, and in dense clouds may themselves reach the unstable size within

3 to 4 minutes. The possibility of a chain reaction of raindrop multiplication therefore arises, as first pointed out by LANGMUIR [36]. The process is unlikely to be important unless the cloud updraught in the region of drop-breaking has a sustained speed about equal to the terminal velocity of the large drops (some 10 m./sec.).

It may not be necessary for a cloud to extend to a level above the peak of the drop-growth trajectory in a uniform updraught in order that shower formation shall occur. LUDLAM considers it sufficient for the droplets to attain radii

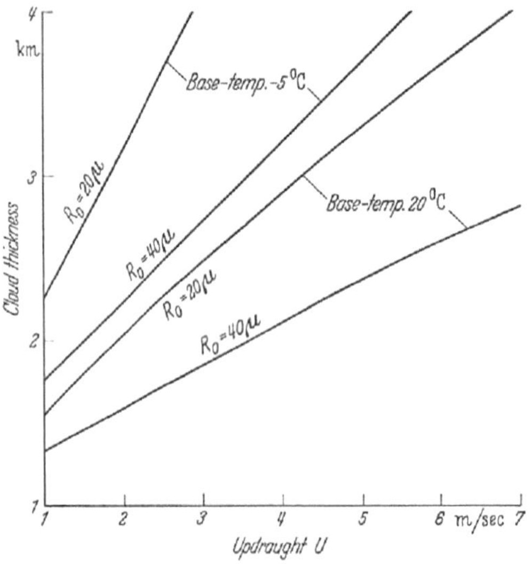

Fig. 21.

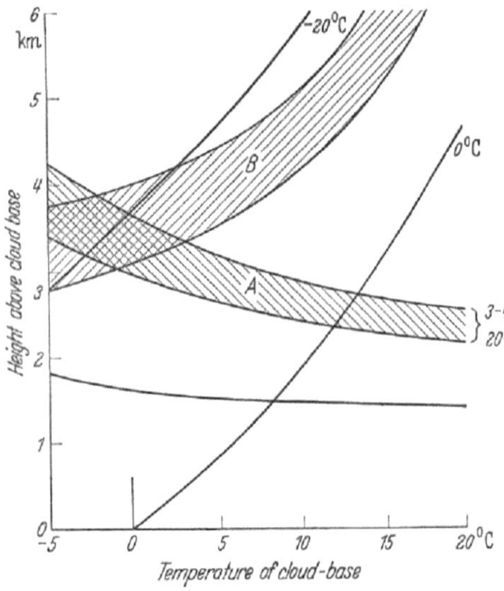

Fig. 22.

Fig. 21. The variation of the minimum cloud thickness for shower development as a function of the updraught speed U, for droplets of initial radii of 20 μ, 40 μ growing in clouds with base temperatures − 5° C and + 20° C.

Fig. 22. The thickness of isolated cumulus likely to be necessary before shower formation can begin, as a function of base-temperature. The cloud bases are assumed to be at the 900 mb. level, about 1 km. above the ground. Cumulus in maritime air masses, containing droplets of 20 to 40 μ radius near the base and average updraught speeds of 3 to 4 m./sec., are likely to produce showers by the coalescence process if the tops reach into the zone A. Cumulus reaching the lower, almost horizontal line, have about the smallest size likely ever to be associated with showers. Cumulus whose tops reach into the zone B are likely to produce showers by the ice-crystal growth process.

of about 150 μ (fall speed over 1 m./sec.), since in nature they will then settle rapidly out of the weaker updraughts within the cloud summits and escape the evaporation which is the fate of the smaller droplets. With updraughts of a few m./sec. the distinction is not important, however, on account of the rapidity of the growth once this size is reached. Fig. 21 shows the thickness of clouds, calculated in this way, in which showers may form, as a function of initial droplet size and updraught speed. The results are shown to vary substantially, when the updraughts are moderately strong, according to the temperature of the cloud base, which controls the amount of water condensed in the cloud. By extending such calculations, Fig. 22 is obtained. This figure probably portrays some essential features of shower formation by the coalescence process, in spite of the gross simplifications used in its construction. Among these features the following are important:

1. Rather small concentrations, of order perhaps 100/m.³, of droplets attaining radii of 20 to 40 μ within a few hundred metres of cumulus bases may readily lead to shower formation. It is known that in maritime air masses

there are usually sufficient giant condensation nuclei to ensure the introduction of the required large droplets. In continental air masses they are likely to be absent, so that shower formation by the coalescence process may there be greatly hindered.

2. There is a minimum thickness of cloud necessary for shower formation by the coalescence process. This thickness increases with the mean speed of the cloud updraughts, and decreases with rise in the temperature of the cloud base, and with increase in the size of the raindrop nuclei which enter the cloud base.

3. In clouds with the minimum updraught speed (≈ 1 m./sec.) characteristic of convection clouds, showers may form in clouds only about $1\frac{1}{2}$ km. thick. If the cloud base temperature exceeds about 8° C, showers may form in clouds whose tops fail to reach the 0° C level.

4. In clouds containing stronger updraughts the growing raindrops are lifted to levels where they may become frozen. The shower is then likely to consist at least partly of hail, and will not obviously be due to the coalescence process. When updraughts are strong, particularly also when the cloud base temperature is low, showers can form only in clouds several km. thick, whose summit level temperatures are well below 0° C.

The minimum shower cloud thickness of about $1\frac{1}{2}$ km. (in clouds containing feeble updraughts) predicted by the theory seems confirmed by observation. This may however be a coincidence. A number of factors hitherto neglected or unrecognised are likely to influence shower formation. For example, when the wind increases with height cumulus may lean so that the tops project beyond the lower parts. Large droplets growing in the summits then settle out into clear air and evaporate, whereas had they fallen through the body of the cloud they may have completed their growth into raindrops capable of reaching the ground. On the other hand the occurrence of layer clouds at the cumulus summit levels may retard the evaporation of the cloud towers and favour shower formation. Among the microphysical processes there is great uncertainty concerning the rate at which coalescence may begin in the absence of large droplets introduced by giant condensation nuclei, and concerning the effect of unusually high or unusually low concentrations of condensation nuclei, which may be expected respectively to retard and favour the coalescence process. The importance of these and other factors has not yet been estimated.

35. Precipitation from cumuliform clouds containing ice crystals. Although hardly any direct observational evidence is available, it is believed that crystals in concentrations significant for shower formation (of order 100/m.^{3}) form in the supercooled parts of cumulus when the temperature has fallen to a value which usually lies between -10 and -15° C, but that this threshold temperature varies considerably from day to day and place to place. These variations are attributed to changes in the concentration and properties of atmospheric ice nuclei. Very large variations are observed in the concentrations of crystals which form when samples of outdoor air are introduced into a supercooled cloud in a mixing chamber, but it is not known exactly how these may be related to the concentrations which would occur when the air is involved in natural cloud formation.

It has sometimes been assumed that shower formation occurs when cumulus summits reach to a level at which the threshold temperature is attained, but clearly some additional development is required if the ice crystals which then form are to reach a size which will ensure their continued growth inside the cloud into precipitation elements. By analogy with his derivation of the critical size to be attained by droplets, LUDLAM [85] considers that the ice particles must at the least reach a size such that

1. during their growth they have settled relative to the air a distance of a hundred metres or more,

2. they have acquired fall-speeds exceeding 1 m./sec.,

3. they can fall through several hundred metres of unsaturated air with insignificant shrinking (and thus settle into the main part of the cloud even if caught in the rapid evaporation of an entire cloud tower).

The size at which these conditions are fulfilled depends upon the shape and structure of the ice particle. If the average droplet radius in the cloud summits is less than some 7 μ, a growing crystal is unable to collect by collision a significant number, but rather must grow predominantly by condensation, perhaps eventually into a snow-crystal. However, in the upper parts of most cumulus the average droplet radius is likely to be considerably greater, and then it appears that accretion with the supercooled droplets provides the main contribution to the growth of a crystal once its diameter considerably exceeds 100 μ. As the accretion continues the condensation of vapour becomes steadily less important, and the ice particle becomes a roughly spherical aggregate of frozen droplets, containing many air enclosures. Its mean density may be small; observations of the density of soft hail and indirect evidence suggest values of 0.1 to 0.3 g cm^{-3}. For such densities the particles must attain radii of 300 to 500 μ for the conditions listed above to be fulfilled, involving a growth-period of 5 to 15 minutes following the formation of the crystals. The shorter period corresponds to the high droplet-concentrations to be expected in the upper parts of tall cumulus with high base-temperatures, and the longer to the small liquid content of cumulus with low base-temperatures (such as occur in cold weather, or in warm dry climates when the cloud base is very high). These periods imply that if a shower is to form in this way the cloud summits are likely to be several hundred metres to even two or three km. above the level at which crystals first form in significant concentrations, the extension being most marked in clouds with cold bases. In such clouds summit-level temperatures as low as -20 to $-25°$ C may *usually* need to be attained before shower formation is to be expected, whereas warm-base clouds may need to rise only a few hundred metres above the $-10°$ C level.

If now reference is again made to Fig. 22, it is possible to make a rough comparison of the efficiencies of the coalescence and ice particle mechanisms of shower formation. It is seen that the former is likely to be the more efficient in warm-base clouds and the latter in cold-base clouds, but that over a large intermediate range either the one or the other may be the more important, according to the particular updraught-strength or aerosol content, for example. Indeed, in many clouds both processes may contribute precipitation elements to a shower formed inside a cloud of more than the minimum size. In such clouds the complete description of shower formation and development must be very complicated.

More generally, it can be said that even the *qualitative* theory of shower production is complicated. The formation of a shower requires a favourable combination of large-scale and small-scale properties, to ensure that the processes of cloud-particle growth are completed within the useful life of the cloud. The size, speed and distribution of cloud updraughts, the cloud-base temperature, wind-shear, the concentration and nature of atmospheric aerosol, are all factors which are important and which are subject to great natural variation. A *quantitative* theory of shower rains has not yet been attempted.

36. The growth of hail. Inside the supercooled upper parts of large cumulus clouds very small concentrations of ice particles are likely to form at levels where the temperature is about $-10°$ C. At first these grow mainly by condensation of vapour; when the diameter of the particle exceeds about $100\,\mu$ then the growth continues predominantly by collision and accretion with unfrozen droplets. The particle rapidly becomes a roughly spherical aggregate containing many air enclosures; it has a white, opaque appearance characteristic of *soft hail (graupel)*. The density of soft hail may be less than $0.3\,\mathrm{g\,cm^{-3}}$, especially during the early stages of its growth or when the growth occurs mainly at low temperatures (-20 to $-30°$ C). For this reason, and because the hail develops in the upper parts of

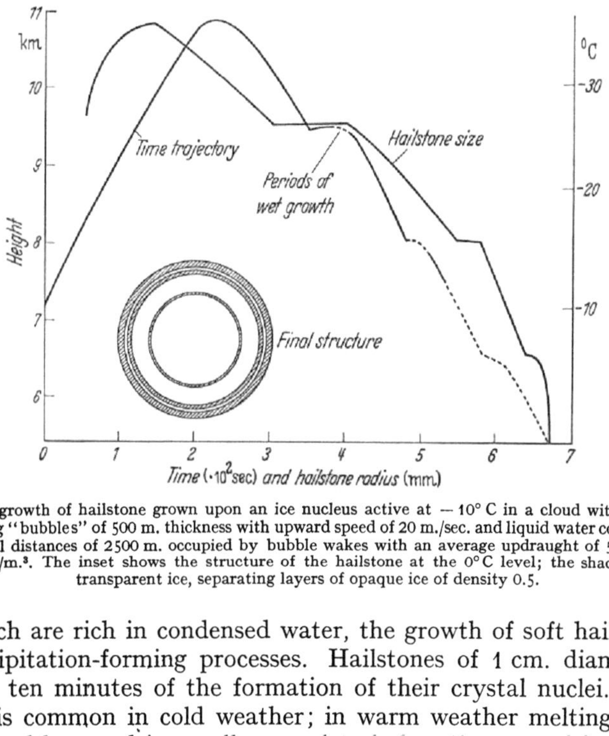

Fig. 23. Calculated growth of hailstone grown upon an ice nucleus active at $-10°$ C in a cloud with base at 900 mb., 20° C, and containing "bubbles" of 500 m. thickness with upward speed of 20 m./sec. and liquid water content of 6.5 g./m.³, separated by vertical distances of 2500 m. occupied by bubble wakes with an average updraught of 5 m./sec. and liquid water content of 2 g./m.³. The inset shows the structure of the hailstone at the 0° C level; the shaded rings represent transparent ice, separating layers of opaque ice of density 0.5.

cumulus, which are rich in condensed water, the growth of soft hail is the most rapid of precipitation-forming processes. Hailstones of 1 cm. diameter can be grown within ten minutes of the formation of their crystal nuclei.

Soft hail is common in cold weather; in warm weather melting begins at a height of several km. and is usually complete before the ground is reached. On the other hand the hailstorms of warm weather, in which true hail falls, are much heavier. True hailstones are partly or even wholly composed of glassy, transparent ice containing hardly any air enclosures. Often there is a core of soft ice, and sometimes the large stones have an onion-like structure, with several successive shells alternately of clear and opaque ice. The clear ice is deposited when the rate of collection of supercooled droplets is such that, even though the surface temperature of the hailstones rises to 0° C, the rate at which the latent heat of fusion is transferred to the air is insufficient to allow the freezing of all the liquid. This stage is reached when

$$E^2 R V > \frac{32\,\varrho\,(180\,D\,\Delta p - 0.3\,K T)^2}{\eta\, w^2\,(80 + T)^2} \cdots, \tag{36.1}$$

where ϱ is the density, K the coefficient of thermal conductivity, D the diffusion coefficient for vapour in air, η the coefficient of viscosity and $T°$ C the temperature of the air; Δp is the difference between the saturated vapour densities at 0° C and $T°$ C, and the other symbols have their usual meanings.

Subsequently the growth is controlled by the rate at which heat is transferred to the environment, and is described approximately by the equation

$$2\eta^{\frac{1}{2}}(80+T)\,R^{\frac{1}{2}}\,dR = V^{\frac{1}{2}}\,\varrho^{\frac{1}{2}}\,(600\,D\,\Delta p - K\,T)\,dt. \tag{36.2}$$

It appears that the concentration of liquid cloud water is the factor of prime importance in determining whether "wet growth" of a hailstone occurs. In particular, alternate wet and dry growth is likely to be due to passage through regions of relatively high and low liquid water concentration respectively, and which may be identified with the bubble cores and bubble wakes of which large cumulus are composed. Fig. 23 illustrates the growth of a hailstone calculated by numerical integration of Eqs. (17.1) and (36.2) on the basis of assumptions described in the figure legend.

Hailstones which are completely transparent arise by the freezing of drops in the supercooled regions of the clouds. These drops are carried into these regions during their growth by coalescence, or are shed from the surplus liquid accumulated on the surfaces of hailstones in the stage of wet growth. On account of their high density and fall-speed their growth remains at all times in the wet stage. Naturally large hailstones are grown only in the presence of very strong updraughts, of some 20 m./sec., which delay their fall out of the supercooled regions of the clouds.

37. The structure of showers as revealed by radar. The first evidence of shower formation inside a cumulus cloud is provided by a small radar echo from a region in its interior. This echo quickly becomes stronger and assumes the form of a more or less vertical column, some km. or more across, which spreads both upwards and downwards. While it grows upwards the top of the radar echo is usually several hundred metres below the top of the visible cloud; remarkably good relations have been observed between the rate of rise of the echo and the updraught-speeds observed inside the cloud, which are usually 1 to 2 m./sec. greater (cf. BYERS and BRAHAM [*86*]). The echo tops are often observed to rise in a number of surges separated by pauses or even a subsidence; these surges have been related to individual cloud towers.

In New Mexico the initial echo usually appears at a level where the temperature is about $-10°$ C, and the clouds show the "glaciated" appearance when the echo top reaches the $-20°$ C level. It has therefore been supposed that the precipitation forms by the growth of ice crystals. On the other hand, BATTAN [*87*] found that in Ohio thunderstorms over half of the initial echoes occur at temperatures above 0° C. This is strong evidence of the importance of the coalescence process in large clouds in which previously it had never been suspected that the ice-particle mechanism might not be entirely responsible for the precipitation.

In Australia conclusive radar evidence of the formation of showers solely by the coalescence mechanism has been obtained by STYLES and CAMPBELL [*88*] and DAY [*89*]: the echoes and the cloud tops have been below the 0° C level at the time of their maximum development.

Studies have been made (e.g. LANGILLE and GUNN [*90*]) of the distribution and changes of echo intensity in shower rains. These have served mainly to emphasise the rapidity with which the distribution of rain-water content may change, and the characteristic structure of sharply-defined columnar echoes, often surrounded by more persistent and less intense echo regions, perhaps containing a melting-level bright band, which are associated with slowly-decaying shower cloud residues.

VIII. The artificial stimulation of precipitation.

38. Introduction. The demonstration in recent years that suitable clouds may be made to release their precipitation by introducing artificial nuclei and the possibility of extending this technique to producing economically important increases in rainfall has created the greatest interest and a good deal of controversy.

Modern rainmaking experiments are based on three main assumptions:

1. that either the presence of ice crystals in a supercooled cloud is necessary to release precipitation by the BERGERON process (see Sect. 32), or the presence of comparatively large water droplets is essential to initiate the coalescence mechanism;

2. that some clouds precipitate inefficiently or not at all, because these agents are naturally deficient;

3. that this deficiency can be remedied by seeding the clouds artificially with either solid carbon dioxide (dry ice) or silver iodide to produce ice crystals, or by introducing water droplets or large hygroscopic nuclei.

The possibility of producing rain from supercooled clouds by the introduction of artificial sublimation nuclei was foreseen by FINDEISEN [*90*], but it was not until SCHAEFER [*92*] discovered that a tiny fragment of dry ice when dropped into a supercooled cloud resulted in the production of many millions of ice crystals, that it became possible to supply the nuclei in the vast numbers required. The first field trial was carried out by LANGMUIR and SCHAEFER on 13th November 1946 when three pounds of crushed dry ice were dropped along a 3 mile track into a supercooled altocumulus deck. Snow was seen to fall from the seeded cloud for a distance of about 2000 feet before evaporating in the dry air beneath.

A further important step in the history of cloud seeding was the discovery by VONNEGUT in November 1946 that minute crystals of silver iodide produced in the form of a smoke, acted as ice forming nuclei at temperatures below $-5°$ C. These discoveries led to experiments in cloud modification being undertaken in the United States, Australia, Canada, South Africa and a number of other countries, and to unprecedented activity in the field of cloud physics. The results of these trials will now be briefly summarised.

39. The experimental seeding of cumulus clouds: General remark. Experiments in which individual cumulus clouds are seeded from an aircraft are most suited to scientific study and analysis because one can control the quantity of seeding agent introduced into a particular cloud which will remain as an identifiable entity in which subsequent changes can be followed and compared with those occurring in neighbouring unseeded clouds.

40. Experiments on cumulus clouds with dry ice. Experiments in which a few pounds of dry ice crushed into pellets of about $\frac{1}{2}$ inch diameter are dropped into the tops of supercooled cumulus have been carried out mainly in South Africa [*93*], the United States (COONS et al. [*94*], [*95*]), Australia (SQUIRES and SMITH [*96*], BOWEN [*97*]) and Canada (ORR et al. [*98*]). In operations carried out in Ohio in 1948 and the Gulf States in 1949 by COONS et al., and during the Australian trials, the aircraft were fully equipped to make meteorological soundings, carried radar and were controlled by radar from the ground. It was therefore possible to detect whether the clouds were in the precipitation stage before seeding, to follow their development after seeding, and to compare them with unseeded clouds occurring within a radius of about 30 miles. The results of the five sets of trials are summarised in Tables 8 and 9.

The South African, early Australian and Canadian experiments all indicated that there is a high probability of inducing precipitation with dry ice if the cloud summit is colder than $-7°$ C and the depth of the supercooled region exceeds

Table 8. *Results of seeding cumulus with dry ice.*
S = supercooled clouds; NS = non-supercooled clouds.

Location		South Africa	Ohio		Gulf States	Australia	Canada	
Total number of experiments . .		36 (S)	46 (S)	12 (NS)	39 (S)	20 (S)	35 (S)	6 (NS)
Number in which precipitation observed after seeding but not before	Total . . .	24[1]	15	1	12	15	23	0
	Radar echo.	24[1]	15		8	4		
	Precipitation at ground .		4		2	10	12	
	Virga . . .			1	10	4	11	
Number of occasions on which natural rain observed within 30 miles.			37	9	39	1	15	
Number of clouds building at time of seeding			38	7	29			
Number of clouds building after seeding			4	2	0		2	0
Number of clouds dissipating after seeding			41	10	39		4	4

Table 9. *Variation of success in seeding cumulus with dry ice with cloud summit temperature and depth of supercooled region.*

Temperature range	South Africa			Australia			Canada			Combined successes
	Suc-cesses	fail-ures	percent success	suc-cesses	fail-ures	percent success	suc-cesses	fail-ures	percent success	
0 to $-$ 3° C	0	1	0	2	2	50	3	7	30	$\frac{5}{15}$ = 33%
$-$ 4 to $-$ 7° C	6	1	86	3	2	60	4	2	67	$\frac{13}{18}$ = 61%
$-$ 8 to $-$ 11° C	9	1	90	5	0	100	8	2	80	$\frac{22}{25}$ = 88%
$-$ 12 to $-$ 15° C	2	1	67	5	0	100	5	1	83	$\frac{12}{14}$ = 86%
$-$ 16 to $-$ 19° C	—	—	—	0	0	—	3	0	100	$\frac{3}{3}$ = 100%
$-$ 20 to $-$ 23° C	—	—	—	0	1	0	—	—	—	
$-$ 24 to $-$ 27° C	—	—	—	—	—	—	—	—	—	
Depth of supercooled region										
0 to 1500 ft.	—	—	—	3	1	75	1	5	17	$\frac{4}{10}$ = 40%
2000 to 3500 ft.	0	1	0	6	1	86	8	6	57	$\frac{14}{22}$ = 64%
4000 to 5500 ft.	8	2	80	2	3	40	9	0	100	$\frac{19}{24}$ = 79%
6000 to 7500 ft.	3	1	75	3	1	75	4	0	100	$\frac{10}{12}$ = 83%
8000 to 9500 ft.	2	0	100	1	0	100	1	0	100	$\frac{4}{4}$ = 100%
$<$10000 ft.	3	0	100							$\frac{3}{3}$ = 100%

about 4000 feet. This is confirmed by BOWEN [97] who, on the basis of about one hundred experiments performed in Australia during 1947 to 1951, draws the following conclusions:

1. With cloud-summit temperatures of $-7°$ C and colder there is a 100% chance of producing precipitation. At temperatures between -7 and

[1] Only 17 in which radar echo appeared within 1 hour of seeding.

0° C the chances of success fall off progressively, tending to zero at 0° C. At temperatures of about −15° C and below the results lose their significance because of the high probability of clouds producing rain naturally.

2. The time at which precipitation will appear at cloud base depends mainly on the thickness of the cloud; there is a gestation period of about 10 minutes, with an additional time of about 1 minute for every 800 feet of cloud.

3. The intensity of the precipitation from the cloud base increases with increasing cloud thickness.

4. A considerable fraction of the precipitation will reach the ground if the height of the cloud base is equal to, or less than, the cloud thickness.

But the summit temperature and the cloud depth cannot be the only parameters which will determine the chances of successful seeding; for example, the lifetime of the cloud, controlled by the dynamics, must be sufficient for particles to grow to precipitation size.

Occasionally, the seeding of an active cumulus is followed by spectacular growth of cloud (Kraus and Squires [*99*], Vonnegut [*100*], Orr et al. [*98*]) which may be a consequence of the latent heat liberated by the transformation of the supercooled water into ice (see Figs. 24a, b).

Approximate calculations indicate that, according to the criteria for shower production adopted in Sect. 35, the introduction of dry ice into cumulus having base temperatures above 10° C and relatively high water contents, might release a shower if the cloud extends a few hundred metres above the 0° C level. In colder clouds the minimum critical depth for shower production will be considerably greater with the cloud summit reaching levels at which natural ice nuclei may be abundant. In this case, the chances of successful seeding will be greatly reduced. Thus there is broad agreement between the theoretical predictions and the experimental results of Tables 8 and 9.

41. Seeding of individual cumulus with silver iodide. There are very few detailed reports on the results obtained by seeding individual cumulus clouds with silver iodide. Langmuir [*101*] describes two trials in New Mexico: one in which pellets of burning charcoal impregnated with silver iodide were dropped into a cloud from an aeroplane produced no marked effects; in the other, when a ground generator was used, the cloud grew rapidly after seeding, and within six minutes produced a very large radar echo and heavy rain at the ground.

An interesting technique has been used by Davies [*102*] in East Africa: small charges of gunpowder impregnated with 0.2 g. of silver iodide were carried aloft by balloons, a fuse being timed to detonate the charge near the 0° C level. The authors report that on 8 of the 15 days in which seeding was carried out in this manner the trials were apparently unsuccessful, on 3 days results were inconclusive, but on the remaining 4 days, rain fell downwind of the point of release of the balloons shortly after detonation of the charges. However, the experimental design and the quality of the observations were inadequate to determine with confidence whether the rain was probably due to the seeding.

42. The seeding of cumulus with water droplets and hygroscopic nuclei. Experiments carried out in South Africa (Anon [*93*]) and in Ohio (Coons et al. 1948) in which water drops (of perhaps $\frac{1}{2}$ mm. diameter) were introduced into the tops of growing cumulus only very rarely resulted in the appearance of precipitation—the general tendency was for the clouds to dissipate after seeding. Indeed it,

a

b

Fig. 24 a u. b. a) Area of New Mexico at 0945 a. m. before cloud seeding operation. The cumulus clouds are small with bases at about 15000 feet, their tops being supercooled. b) Towering cumulonimbus rising from region of cloud seeded 20 minutes earlier. (Photos copyright—Project Cirrus, Signals Corps Engineering Laboratory.)

can readily be shown that the spraying of a few gallons of water into the top of a cloud cannot produce a shower unless a process of raindrop multiplication occurs.

A much more efficient method is to introduce droplets of radius 20 to 30 μ into the *base* of a growing cloud and to capitalise on their subsequent growth during both their upward and downward journeys through the cloud. A droplet of initial radius 20 μ growing by coalescence in a cloud with base temperature 20° C and a uniform updraught of 3 m./sec. will after a total journey of 5 km. emerge with the maximum stable radius of 2.5 mm., the mass growth factor being nearly 2×10^6.

Eleven experiments along these lines have been undertaken in Australia (BOWEN [*103*]) in which water drops of median diameter 50 μ were sprayed at the rate of about 30 gallons/min. during traverses made at about 1000 feet above cloud base. On six of the seven occasions when the cloud thickness was less than 5000 feet, virga appeared within 10 to 25 minutes of seeding, after which the clouds dissipated. Neighbouring untreated clouds did not precipitate. In the few cases where the cloud thickness exceeded 5000 feet, there was a considerable fall of rain or hail within a short time of seeding and on three of these occasions neighbouring unseeded clouds did not rain. Calculations indicate that the introduction of drops of diameter 50 μ into the base of a cloud having base temperature 10° C and a steady updraught of 1 m./sec. could release a shower from a cloud of depth greater than about 1.5 km.—a result which agrees well with BOWEN's experiments.

Recent tests have been carried out by the University of Chicago group in which water spray in quantities of about 450 gallons per mile was introduced into the tops of tropical cumuli in the Caribbean area. Clouds were selected in pairs and one of each pair was treated. The results indicated that about one half of the treated clouds developed radar echoes while only about one quarter of the untreated clouds developed echoes.

In view of the important role which giant hygroscopic nuclei are believed to play in initiating the coalescence process, it would appear more economical to use such nuclei rather than water droplets as a seeding agent. Since dry salt crystals of diameter 10 μ will more than double their size while being carried through the first few hundred metres of cloud, the dispersal of about 100 g. of salt would be equivalent to that of about 1 gallon of water in 50 μ diameter droplets. Use of this method was first reported by DESSENS [*104*] with apparently some success, but it has been most extensively tried by DAVIES [*105*] in East Africa using their balloon-bomb technique. The bombs containing gunpowder and sodium chloride were arranged to explode near cloud base dispersing salt particles of diameter 5 to 100 μ, with a peak at about 50 μ. For the 38 days when salt was released, the total rainfall in an area 6 to 12 miles downwind of the release point was up to 6 inches in excess of that for intermediate unseeded days, but the rainfall over an area extending 5 miles upwind was also greater by 2 to 3 inches. Although these results cannot be regarded as conclusive evidence of the efficiency of salt-seeding, they are suggestive and indicate the desirability of further work.

43. The experimental seeding of layer clouds. The effects which are produced by the localised seeding of supercooled stratiform clouds from aircraft have been the subject of a number of investigations the more important of which have been carried out by *Project Cirrus* (SCHAEFER [*106*], COONS et al. [*107*], [*108*]), AUFM KAMPE et al. [*109*]. The usual procedure is to drop granulated dry ice at the rate of a few pounds per mile of flight along a well-defined track into the top of a supercooled stratus deck, and to photograph subsequent developments from above. Most of the clouds treated in these operations were too thin to produce precipitation at the ground, but spectacular changes in the appearance of the cloud often occurred. Along the seeded track the cloud becomes transformed completely into ice crystals. Gradually the turbulent motions in the cloud layer diffuse these crystals laterally, so that after a few minutes a long lane a few hundred yards broad becomes affected. After a while, the crystals at the edge of the lane become so diffused that each has a large share of the available water and grows large enough to fall out of the cloud layer into the drier air beneath. The crystals then evaporate, and the cloud disappears in a broad strip around the seeded track to leave a long hole in the cloud (Fig. 25).

Whether or not a hole will develop will depend upon the rate of seeding, the temperature, depth and water content of the cloud, the magnitude of the horizontal divergence or convergence and the intensity of lateral mixing, the

Fig. 25. A γ-pattern of ice crystals produced in a thin supercooled layer cloud by seeding with dry ice. The photograph was taken 36 minutes after the loop in the upper part of the picture was seeded. (Photo by courtesy of the General Electric Company, Schenectady, U.S.A.)

area affected depending largely on the last two factors. AUFM KAMPE et al report that clouds forming in regions of horizontal divergence could be cleared fairly readily with small amounts of dry ice, the seeded lanes generally spreading to widths of two miles within half an hour.

Thus cloud decks less than 3000 feet thick and colder than $-5°$ C could be cleared equally well with seeding rates between 1 and 20 pounds per mile, but thick, convective layers were sometimes underseeded with rates as high as 10 pounds per mile. When clouds only 500 feet in depth were seeded only about 10 minutes elapsed before the ground could be seen, but for depths of 2000 feet this time was increased to about 25 minutes. These experiments have obvious application to the landing of aircraft in conditions of low cloud, and in very cold weather it may be possible to disperse entire layers of supercooled fog and low cloud by using a silver iodide generator on the ground. But as far as the authors are aware, no systematic tests of this method of fog dispersal have yet been carried out.

44. Large-scale cloud seeding operations. The experiments just described have clearly demonstrated that the introduction of seeding agents into suitable clouds is often followed by observable, and sometimes, spectacular effects. They are, however, unlikely to tell us very much about the possibilities of modifying the rainfall from widespread cloud systems extending over thousands of square miles. For operations on this scale, it is the general practice to release silver iodide in the form of a smoke from ground generators producing about 10^{13} particles/sec., relying on the air currents to carry it up into the supercooled regions of the cloud. Trials of this type have been carried out on a large scale during the past eight years, mainly in the United States, Canada, and South America by commercial operators.

The generators are switched on when the arrival of suitable storms in the target area is imminent and kept burning during their passage, the working hypothesis being that seeding may cause an increase in the precipitation ove that which would fall naturally. The method of evaluation usually adopted is to form a regression for the target area rainfall on that in a control area and the assessment of the effects due to seeding takes the form of determining whether there are statistically significant departures from the normal historica relationship.

From the published accounts of large-scale seeding operations discussed in more detail by MASON [*110*], one can find no convincing evidence that large increases in rainfall can be produced consistently over large areas. Although it is feasible that, under suitable conditions, marginal effects of, say, 10 or 20% may be produced, in general operations in any one area have not been carried on for a sufficiently long period and designed for use with evaluation techniques sufficiently reliable and sensitive to separate effects of this magnitude from the ordinary variations in natural rainfall.

Apart from doubts about the validity of the evaluation procedures which have been used to assess massive cloud-seeding operations, they are conducted on certain assumptions which are insecurely based as may be illustrated by our present inability to answer the following leading questions.

a) How often in a given locality, and under what conditions are natural ice nuclei deficient, and when and in what quantities should they be supplied artificially?

b) When silver iodide smoke is released from the ground, in what concentrations does it reach the supercooled regions of the cloud and how will this vary with distance from the source?

c) For how long does silver iodide remain active as an ice-forming nucleus under specified atmospheric conditions?

Because of the technical difficulties involved in measuring the ice-nucleus content of air at cloud level and correlating this with meteorological factors, and also because the numbers of ice nuclei may bear very little relation to the numbers of ice crystals appearing in a cloud at the same temperature (because of splintering etc.), there appears to be little prospect of predicting, in a given situation, the requirement for artificial nuclei. One can only make an intelligent guess at the minimum concentration of nuclei necessary for rain production (probably of order 1/litre) and try to ensure that this concentration of active silver iodide nuclei is achieved at levels where they may forestall the natural ice nuclei, in general, between the -5 and $-15°$ C levels.

This brings us to the second problem; our present knowledge of the turbulent diffusion of particles over distances of ten or hundreds of miles is inadequate and its study must form an essential part of any serious investigation of the potentialities of large-scale cloud seeding. This problem is also closely linked with the fact that the ice nucleating properties of silver iodide may deteriorate when it is exposed to the atmosphere. There is some evidence that this deterioration is linked with photolysis of the surfaces layers by the ultra-violet component of solar radiation, but the decay characteristics appear to depend very much on the method and temperature of smoke production, the size of the particles, and the nature of impurities as well as on the atmospheric conditions.

Even if these fundamental problems were solved, the evaluation of a large-scale rainmaking experiment would still be very difficult. Since we are still far from being able to compute, on a purely physical basis, how much rain will fall naturally in given regions, the effects of seeding must necessarily be assessed statistically. In order to detect with confidence artificial variations of order 10% from the natural fluctuations, a carefully designed experiment will have to be continued for a period of several years. It would appear advisable to seed on only half of the potentially suitable occasions *selected at random*, the unseeded occasions being used as controls.

In speculating on the potentialities of cloud seeding, it is probably more realistic to regard this as a possible method of effecting a re-distribution of rainfall rather than of increasing the total amount falling on the whole globe. By inducing premature release of precipitation in certain regions, it might be possible to either augment or reduce the rainfall over land relative to that which normally occurs over the oceans. It is not obvious that seeding of the deep layer-cloud system of middle latitudes whose tops generally reach to levels where natural ice nuclei are abundant and where the natural processes have plenty of time to operate would produce a significant increase in rainfall, but it is conceivable that judicious seeding could forestall the natural processes and thereby effect some redistribution of the rainfall. Perhaps more promising as sources of additional rain or snow are the persistent orographic clouds produced by the ascent of damp air over large mountain barriers. Here the continuous generation of the appropriate concentrations of silver iodide nuclei near the windward edge may be expected to produce slight, persistent snow to the leeward, wich could reach the mountain surface before evaporating and accumulate in appreciable amount if maintained for many hours. Aside from average conditions, there may be a number of individual occasions in particular regions where natural nuclei are deficient and when the occurrence or non-occurrence of precipitation is decided by very marginal conditions. By exploiting such marginal conditions, cloud seeding may bring considerable benefit to those regions where even a small increase in rainfall would be of enormous economic value.

It is conceivable that the introduction of an excessive concentration of artificial ice nuclei into a cloud such that the resulting crystals would not be able to grow sufficiently large to fall out and reach the ground, could retard the production of precipitation, but the practical difficulties of "overseeding" over a large area are so great that it has not been seriously attempted. The prevention of large hail and of lightning by the overseeding of incipient thunderstorms are also intriguing possibilities which again have yet to be put seriously to the test.

Although it is unfortunate that greatly exaggerated claims have been made for the results achieved by cloud seeding, these experiences certainly represent an exciting development in meteorology and have given a new impetus to researches on the fundamental physics of clouds and precipitation. Their full potentialities have almost certainly not yet been realised and the next decade should witness many interesting new developments. Let us hope that the problems will be tackled with the imagination and enthusiasm they deserve.

References.

[1] Corby, G. A.: The airflow over mountains; a review of the state of current knowledge. Quart. J. Roy. Met. Soc. **80**, 491 (1954).
[2] Ludlam, F. H.: Orographic cirrus clouds. Quart. J. Roy. Met. Soc. **78**, 554 (1952).
[3] Bergeron, T.: The coastal orographic maximum of precipitation in autumn and winter. Tellus **1**, 15 (1949).
[4] Scorer, R. S.: Billow clouds. Quart. J. Roy. Met. Soc. **77**, 235 (1951).
[5] Cunningham, R. M.: The distribution and growth of hydrometeors around a deep cyclone. Techn. Rep. No. 18, 1952, M.I.T., Cambridge, Mass.
[6] Scorer, R. S., and F. H. Ludlam: Bubble theory of penetrative convection. Quart. J. Roy. Met. Soc. **79**, 94 (1953).
[7] Mason, B. J.: The physics of clouds. Oxford: Clarendon Press 1956 (in the press).
[8] Junge, C.: Das Wachstum der Kondensationskerne mit der relativen Feuchtigkeit. Ann. Met., Hamburg **3**, 128 (1950).
[9] Junge, C.: Die Konstitution des atmosphärischen Aerosols. Ann. Met., Hamburg Beih. **1952**.
[10] Landsberg, H.: Atmospheric condensation nuclei. Ergebn. kosm. Phys. **3**, 207 (1938).
[11] Dessens, H.: La brume et le brouillard étudiés á l'aide des fils d'araignées. Ann. Géophys. **2**, 276, 343 (1946).
[12] Dessens, H.: Les noyaux de condensation de l'atmosphère. Météorologie **1947**, 321.
[13] Dessens, H.: The use of spider's threads in the study of condensation nuclei. Quart. J. Roy. Soc. **75**, 23 (1949).
[14] Junge, C.: Die Rolle der Aerosole und der gasförmigen Beimengungen der Luft im Spurenstoffhaushalt der Troposphäre. Tellus **5**, 1 (1953).
[15] Moore, D. J., and B. J. Mason: The concentration, size distribution and production rate of large salt nuclei over the oceans. Quart. J. Roy. Met. Soc. **80**, 583 (1954).
[16] Woodcock, A. H.: Atmospheric salt particles and raindrops. J. Meteorology **9**, 200 (1952).
[17] Woodcock, A. H., and M. Gifford: Sampling atmospheric sea-salt nuclei over the ocean. J. Marine Res. **8**, 177 (1949).
[18] Moore, D. J.: Measurements of condensation nuclei over the North Atlantic. Quart. J. Roy. Met. Soc. **78**, 596 (1952).
[19] Simpson, G. C.: Sea salt and condensation nuclei. Quart. J. Roy. Met. Soc. **67**, 163 (1941).
[20] Kuroiwa, D.: Electron microscope study of fog nuclei. J. Meteorology **8**, 157 (1951).
[21] Kuroiwa, D.: Electron microscope study of atmospheric condensation nuclei. Sapporo, Hokkaido Univ. Studies on Fogs, 1953, p. 349.
[22] Ogiwara, S., and T. Okita: Electron microscope study of cloud and fog nuclei. Tellus **4**, 233 (1952).
[23] Yamamoto, G., and T. Ohtake: Electron microscope study of cloud and fog nuclei. Sci. Rep. Tôhoku Univ., Ser. V, Geophys. **5**, 141 (1953).
[24] Junge, C.: The chemical composition of atmospheric aerosols, I: measurements at Round Hill Field Station, June to July 1953. J. Meteorology **11**, 323 (1954).
[25] Weickmann, H. K., and H. J. aufm Kampe: Physical properties of cumulus clouds. J. Meteorology **10**, 204 (1953).
[26] Diem, M.: Messungen der Größe von Wolkenelementen, II. Meteor. Rdsch. **1**, 261 (1948).
[27] Zaitsev, V. A.: Liquid water content and distribution of drops in cumulus clouds. Nat. Res. Council of Canada, Techn. Translation No. 395, Ottawa, 1953, from Glav. geofiz. Observ. Trudy **19**, 122 (1950).
[28] Frith, R.: The size of cloud particles in stratocumulus cloud. Quart. J. Roy. Met. Soc. **77**, 441 (1951).
[29] Bricard, J.: La constitution des nuages au Pic du Midi. C. R. Acad. Sci., Paris **216**, 306 (1943).
[30] Neiburger, M.: Reflection, absorption and transmission of insolation by stratus cloud. J. Meteorology **6**, 98 (1949).
[31] Vonnegut, B.: A capillary collector for measuring the deposition of water drops on a surface moving through clouds. Rev. Sci. Instrum. **20**, 110 (1949).
[32] Warner, J., and T. D. Newnham: A new method of measurement of cloud-water content. Quart. J. Roy. Met. Soc. **78**, 46 (1952).
[33] aufm Kampe, H. J.: Visibility and liquid water content in clouds in the free atmosphere. J. Meteorology **7**, 54 (1950).
[34] Kline, D. B., and J. A. Walker: Meteorological analysis of icing conditions encountered in low-altitude stratiform clouds. N.A.C.A., Techn. Note 2306, 1951.

[35] BEST, A. C.: The size of cloud droplets in layer-type clouds. Quart. J. Roy. Met. Soc. **77**, 241 (1951).
[36] LANGMUIR, I.: The production of rain by a chain-reaction in cumulus clouds at temperatures above freezing. J. Meteorology **5**, 75 (1948).
[37] SCHUMANN, T. E. W.: Theoretical aspects of the size distribution of fog particles. Quart. J. Roy. Met. Soc. **66**, 195 (1940).
[38] HOWELL, W. E.: The growth of cloud drops in uniformly cooled air. J. Meteorology **6**, 134 (1949).
[39] MASON, B. J.: The production of rain and drizzle by coalescence in stratiform clouds. Quart. J. Roy. Met. Soc. **78**, 377 (1952).
[40] MASON, B. J.: The spontaneous crystallisation of supercooled water. Quart. J. Roy. Met. Soc. **78**, 22 (1952).
[41] HEVERLY, J. R.: Supercooling and crystallisation. Trans. Amer. Geophys. Un. **30**, 205 (1949).
[42] DORSCH, R. G., and P. T. HACKER: Photomicrographic investigation of spontaneous freezing temperatures of supercooled water droplets. N.A.C.A. Techn. Note 2142, 1940.
[43] HOSLER, C. L.: Ice crystal formation. Proc. Toronto Met. Conf. 1953. Roy. Met. Soc. **1954**, 253.
[44] BIGG, E. K.: The supercooling of water. Proc. Phys. Soc. Lond. B **66**, 688 (1953).
[45] MOSSOP, S. C.: The freezing of supercooled water. Proc. Phys. Soc., B **68**, 193 (1955).
[46] MEYER, J., and W. PFAFF: Zur Kenntnis der Kristallisation von Schmelzen. Z. anorg. allg. Chem. **224**, 305 (1935).
[47] WYLIE, R. G.: The freezing of supercooled water in glass. Proc. Phys. Soc. Lond. B **66**, 241 (1953).
[48] RAU, W.: Gefriervorgänge des Wassers bei tiefen Temperaturen. Schr. dtsch. Akad. Luftfahrtforsch. **8**, 65 (1944).
[49] CWILONG, B. M.: Observations on the incidence of supercooled water in expansion chambers and on cooled solid surfaces. J. Glac. **2**, 53 (1947).
[50] BREWER, A. W., and H. P. PALMER: Freezing of supercooled water. Proc. Phys. Soc. Lond. B **64**, 765 (1951).
[51] MASON, B. J.: Progress in cloud physics research. Arch. Met., Geophys. u. Biokl., Ser. A **6**, 1 (1953).
[52] WEICKMANN, H.: Die Eisphase in der Atmosphäre. Ber. dtsch. Wetterdienst, US-Zone **1949**, Nr. 6.
[53] RAU, W.: Nachweis der allgemeinen Gültigkeit des Gefrierkernspectrums. Geofisica Pura e Applicata **26**, 75 (1953).
[54] REGENER, E.: Versuche über die Kondensation und Sublimation des Wasserdampfes bei tiefen Temperaturen. Schr. dtsch. Akad. Luftfahrtforsch. **37**, 15 (1941).
[55] FINDEISEN, W., u. G. SCHULZ: Experimentelle Untersuchungen über die atmosphärische Eisteilchenbildung. I. Forsch. u. Erfahrb. Reichs-Wetterdienst, Berlin A **1944**, Nr. 27.
[56] FINDEISEN, W.: Experimentelle Untersuchungen über die atmosphärische Eisteilchenbildung. Meteor. Z. **59**, 349 (1942).
[57] SCHULZ, G.: Die Arbeiten und Forschungsergebnisse der Wolkenforschungsstelle des Reichsamtes für Wetterdienst in Prag. Ber. dtsch. Wetterdienst, US-Zone **1**, 3 (1947).
[58] CWILONG, B. M.: Sublimation in a WILSON Chamber. Proc. Phys. Soc. Lond. A. **190**, 137 (1947).
[59] CWILONG, B. M.: Sublimation in the atmosphere over the oceans. Nature, Lond. **161**, 62 (1948).
[60] CWILONG, B. M.: Sublimation in outdoor air and seeded sublimation. Nature, Lond. **163**, 727 (1949).
[61] FOURNIER D'ALBE, E. M.: Some experiments on the condensation of water vapour at temperatures below 0° C. Quart. J. Roy. Met. Soc. **75**, 1 (1949).
[62] PALMER, H. P.: Natural ice-particle nuclei. Quart. J. Roy. Met. Soc. **75**, 15 (1949).
[63] SCHAEFER, V. J.: The occurrence of ice crystal nuclei in the free atmosphere. Occ. Rep. 20, Project Cirrus, G.E.C., Schenectady, New York, 1950. — The concentration of ice nuclei at the summit of Mt. Washington. Occ. Rep. 35, Project Cirrus, G.E.C. Schenectady, New York, 1952.
[64] WORKMAN, E. J., and S. E. REYNOLDS: Thunderstorm Electricity. New Mexico School of Mines Progr. Rep. 6, New Mexico, 1950.
[65] AUFM KAMPE, H. J., and H. K. WEICKMANN: The effectiveness of natural and artificial aerosols as freezing nuclei. J. Meteorology **8**, 283 (1951).
[66] SMITH, E. J., and K. J. HEFFERNAN: Airborne measurements of the concentration of natural and artificial freezing nuclei. Quart. J. Roy. Met. Soc. **80**, 182 (1954).

[67] Rau, W.: Gefrierkerngehalt und Gefrierkernspektrum des natürlichen Aerosols. Meteor. Rdsch. **7**, 59 (1954).
[68] Mason, B. J.: The appearance of ice crystals in expansion- and mixing-cloud chambers. Bull. Obs. Puy de Dôme, **3**, 65 (1955).
[69] Mason, B. J.: The nature of ice-forming nuclei in the atmosphere. Quart. J. Roy. Met. Soc. **76**, 59 (1950).
[70] Kumai, M.: Electron-microscope study of snow-crystal nuclei. J. Meteorology **8**, 151 (1951). See also remarks by supplement.
[71] aufm Kampe, H. J., H. K. Weickmann and H. H. Kedesdy: Meteorology **9**, 374 (1952).
[72] Smith, E. J., B. K. Seely and K. J. Heffernan: The decay of ice-nucleating properties of silver iodide in the atmosphere. J. Meteorology **12**, 379 (1955).
[73] aufm Kampe, H. J., H. K. Weickmann and J. J. Kelly: The influence of temperature on the shape of ice crystals growing at water saturation. J. Meteorology **8**, 168 (1951).
[74] Mason, B. J.: The growth of ice crystals in a supercooled water cloud. Quart. J. Roy. Met. Soc. **79**, 104 (1953).
[75] Nakaya, U.: The formation of ice crystals. Compendium of Meteorology. Amer. Met. Soc. **1951**.
[76] Shaw, D., and B. J. Mason: The growth of ice crystals from the vapour. Phil. Mag. **46**, 249 (1955).
[77] Reynolds, S. E.: Ice crystal growth. J. Meteorology **9**, 36 (1952).
[78] Mason, B. J., and B. P. Howorth: Some characteristics of stratiform clouds over N. Ireland in relation to their precipitation. Quart. J. Roy. Met. Soc. **78**, 226 (1952).
[79] Ludlam, F. H.: The forms of ice clouds. Quart. J. Roy. Met. Soc. **74**, 39 (1948).
[80] Bowen, E. G.: Radar observations of rain and their relation to mechanisms of rain formation. J. Atmos. a. Terres. Phys. **1**, 125 (1951).
[81] Wexler, R.: Precipitation growth in stratiform clouds. Quart. J. Roy. Met. Soc. **78**, 363 (1952).
[82] Peppler, W.: Unterkühlte Wasserwolken und Eiswolken. Forsch. u. Erfahrb. Reichs-Wetterdienst, Berlin B **1** (1940).
[83] Mason, B. J.: Radar evidence for aggregation and orientation of melting snowflakes. Quart. J. Roy. Met. Soc. **81**, 262 (1955).
[84] Ludlam, F. H.: The production of showers by the coalescence of cloud droplets. Quart. J. Roy. Met. Soc. **77**, 402 (1951).
[85] Ludlam, F. H.: The production of showers by the growth of ice particles. Quart. J. Roy. Met. Soc. **78**, 543 (1952).
[86] Byers, H. R., and R. R. Braham: The thunderstorm, U.S. Weather Bureau, Washington D.C., 1949.
[87] Battan, L. J.: Observations on the formation and spread of precipitation in convective clouds. J. Meteorology **10**, 311 (1953).
[88] Styles, R. S., and F. W. Campbell: Radar observations of rain from non-freezing clouds. Austral. J. Phys. **6**, 73 (1953).
[89] Day, G. A.: Radar observations of rain at Sydney, N.S.W. Austral. J. Phys. **6**, 229 (1953).
[90] Langille, R. C., and K. L. S. Gunn: Quantitative analysis of vertical structure in precipitation. J. Meteorology **5**, 301 (1948).
[91] Findeisen, W.: Die kolloidmeteorologischen Vorgänge bei der Niederschlagsbildung. Meteor. Z. **55**, 121 (1938).
[92] Schaefer, V. J.: The production of ice crystals in a cloud of supercooled droplets. Science, Lancaster, Pa. **104**, 457 (1946).
[93] S. African Council for Scientific and Industrial Research, Pretoria. Artificial stimulation of precipitation, 1948.
[94] Coons, R. D., E. L. Jones and R. Gunn: Second Partial Report on the artificial production of precipitation; cumuliform clouds, Ohio, 1948. Res. Pap. 31, U.S. Weather Bureau, Washington, D.C., 1949.
[95] Coons, R. D., E. L. Jones and R. Gunn: Fourth Partial Report on artificial production of precipitation; cumuliform clouds, Gulf States, 1949. Res. Pap. 33, U.S. Weather Bureau, Washington, D.C., 1949.
[96] Squires, P., and E. J. Smith: The artificial stimulation of precipitation by means of dry ice. Austral. J. Sci. Res. A **2**, 232 (1949).
[97] Bowen, E. G.: Australian experiments on artificial stimulation of rainfall. Weather **7**, 204 (1952).
[98] Orr, J. L., D. Fraser and K. G. Pettit: Canadian experiments on artificially inducing precipitation. Bull. Amer. Met. Soc. **31**, 56 (1950).

[99] KRAUS, E. B., and P. SQUIRES: Experiments on the stimulation of clouds to produce rain. Nature, London **159**, 489 (1947).
[100] VONNEGUT, B., and K. MAYNARD: Spray-nozzle type silver iodide generator for airplane use. Bull. Amer. Met. Soc. **33**, 420 (1952).
[101] LANGMUIR, I.: Control of precipitation from cumulus clouds by various seeding techniques. Science, Lancaster, Pa. **112**, 35 (1950).
[102] DAVIES, D. A.: Artificial stimulation of rain at Kongwa. Nature, Lond. **167**, 102 (1951).
[103] BOWEN, E. G.: A new method of stimulating convective clouds to produce rain and hail. Quart. J. Roy. Met. Soc. **78**, 37 (1952).
[104] DESSENS, H.: Recherches sur la pluie artificielle à l'Observatoire du Puy-de-Dôme. Ann. Géophys. **6**, 331 (1950).
[105] DAVIES, D. A.: Artificial stimulation of rain at Kongwa. Nature, Lond. **169**, 1001 (1952).
[106] SCHAEFER, V. J.: Project Cirrus, Final Report, Part I, G.E.C., Schenectady, New York 1953.
[107] COONS, R. D., R. C. GENTRY and R. GUNN: First Partial Report on the artificial production of precipitation; stratiform clouds, Ohio 1948. Res. Pap. 30, U.S. Weather Bureau, Washington, D.C., 1948.
[108] COONS, R. D., R. C. GENTRY and R. GUNN: Third Partial Report on the artificial production of precipitation; orographic stratiform clouds, California 1949. Res. Pap. 33, U.S. Weather Bureau, Washington, D.C., 1949.
[109] AUFM KAMPE, H. J., H. K. WEICKMANN and J. J. KELLY: Seeding experiments in supercooled stratus clouds (to be published).
[110] MASON, B. J.: Design and evaluation of large-scale rainmaking experiments. Nature, Lond. **175**, 448 (1955).

Atmosphärische Elektrizität.

Von

R. Mühleisen.

Mit 44 Figuren.

A. Überblick.

1. Historisches. Schon bald, nachdem man erkannt hatte, daß gewisse Erscheinungen in der Natur auf Wirkungen der Elektrizität zurückgehen, wurde dies auch von gewissen Vorgängen in der Atmosphäre vermutet. Wall[1] soll 1708 als erster den *Blitz* mit elektrischen Funkenentladungen verglichen haben. Diese Vermutung von der Mitwirkung elektrischer Ladungen beim Gewitter wurde von anderen Forschern wie Gray[2], Kleist, Winkler und Franklin[3] (1750) weiter verfolgt und bestätigt. 1752 entdeckte dann Lemonnier[4], daß auch bei Schönwetter ein *elektrisches Feld* an der Erdoberfläche besteht, und untersuchte dessen Schwankungen.

In der folgenden Zeit haben hauptsächlich französische und italienische Forscher wie Volta, Saussure und Peltier[5] die Meßverfahren verbessert und das Verhalten des luftelektrischen Feldes genauer untersucht. So wurde gefunden, daß das elektrische Feld bei Schönwetter immer von oben nach unten gerichtet, der Betrag im Winter meist höher als im Sommer ist, und auch der tägliche Gang wurde bereits erkannt. Weitere Verbesserungen der Elektrometer und die Einführung des Tropfkollektors stammen von Thomson, tragbare Meßgeräte entwickelte Exner[6] und verwendete sie zu Messungen an verschiedenen Stellen auf Land und auf See.

Ganz neue Probleme traten auf, als entdeckt wurde, daß die Luft kein vollkommener Isolator ist, sondern in geringem Maße die Elektrizität leitet. Linss[7] hat wohl als erster 1887 erkannt, daß diese *Luftleitfähigkeit* ausreichen müßte, um das beobachtete luftelektrische Feld in wenigen Minuten zum Verschwinden zu bringen (vgl. Ziff. 2ε). Elster und Geitel[8] haben 1899 die Beobachtung veröffentlicht, daß atmosphärische Ionen die Träger der Leitfähigkeit sind. Die bis dahin geltende Anschauung, daß das luftelektrische Feld die Folge einer negativen Eigenladung der Erde sei, wurde damit hinfällig. Hess[9] vermutete die Ursache für diese Ionenbildung in einer *kosmischen Strahlung*. Diese konnte er durch Messungen im Hochgebirge nachweisen.

Verschiedene Hypothesen sind zur Erklärung der elektrischen Strömung in der Atmosphäre entwickelt worden; nach einer von ihnen sollten beispielsweise negativ geladene Korpuskeln auf die Erde treffen. Erst 1920 bis 1925 gewann

[1] W. Wall: Phil. Trans. Roy. Soc. Lond. **26**, 69 (1708).
[2] S. Gray: Phil. Trans. Roy. Soc. Lond. **39**, 400 (1735).
[3] B. Franklin: Phil. Trans. Roy. Soc. Lond. **47**, 289 (1751).
[4] L. G. Lemonnier: Mém. Acad. Sci. **2**, 233 (1752).
[5] A. Peltier: Ann. Chem. u. Phys. **3**, 129 (1842).
[6] F. Exner: Acad. Wien. 1886—1890.
[7] F. Linss: Meteor. Z. **4**, 345 (1887).
[8] J. Elster u. H. Geitel: Phys. Z. **1**, 245 (1899).
[9] V. F. Hess: Phys. Z. **12**, 998 (1911).

langsam die „Gegenstrom"-Theorie von WILSON [14] die Oberhand. Seine Behauptung, daß die *Gewitter* die Erzeuger der Luftelektrizität sind, wurde durch Untersuchungen der Wolkenpolarität mit Ballonen von SIMPSON und SCRASE [12] bestätigt. Diese konnten zeigen, daß die Gewitterwolken an der Oberseite stets positiv und an der Unterseite überwiegend negativ geladen sind. Der Transport der negativen Ladung zur Oberfläche der Erde über Spitzenentladungen war leicht nachweisbar und ebenso der entsprechende Transport positiver Ladungen zu dem gut leitenden Luftraum über 50 km Höhe, nachdem mit Ballonaufstiegen und zuletzt mit Raketen die elektrische Leitfähigkeit der Luft bis in größere Höhen gemessen worden war.

Einen wesentlichen Beitrag erfuhr die luftelektrische Forschung durch die Messungen auf dem Expeditionsschiff Carnegie [2]. Auf allen Weltmeeren wurde die elektrische Feldstärke gemessen und gefunden, daß diese, bei nahezu konstanter elektrischer Leitfähigkeit der Luft, einen überall gleichen *Tagesgang nach Weltzeit* hat. Daß dieser Tagesgang synchron mit der Weltgewittertätigkeit verläuft, bedeutet eine Bestätigung der WILSONschen Theorie.

Heute ist die Frage nach den elektrischen Vorgängen in den *Gewitterwolken* selbst noch ungeklärt. Man kennt zwar eine Menge von elektrischen Erscheinungen, welche für den Prozeß in Frage kommen, aber man weiß noch nicht, welche Art der Elektrizitätserzeugung die Hauptrolle spielt. Andere luftelektrische Probleme sind die Feinstruktur von Feld und Vertikalstrom, die Änderung der Eigenschaften der Ionen im Verlauf ihres Bestehens in der Atmosphäre und in den verschiedenen Höhen und ihre Gesetzmäßigkeiten, meteorologisch-luftelektrische Zusammenhänge, ferner biologische Wirkungen der luftelektrischen Erscheinungen und anderes.

2. Die luftelektrischen Elemente. α) Die verschiedenen atmosphärischen Erscheinungen sind so vielfältig miteinander gekoppelt, daß eine Einzelerscheinung nicht ohne Kenntnis der anderen Faktoren verstanden wird. Deshalb sollen zunächst in Kürze die wichtigsten Größen und ihre gegenseitigen Einwirkungen betrachtet werden.

β) In der überwiegenden Zeit ist das in der Atmosphäre vorhandene elektrische Feld nach unten gerichtet. Man spricht meist nicht von der *elektrischen Feldstärke* $\mathfrak{E}$, sondern von der Größe des *Potentialgradienten* oder des *Potentialgefälles* in der Richtung der Vertikalen z:

$$E = \partial V/\partial z = -\mathfrak{E}_{\perp}. \tag{2.1}$$

Der Potentialgradient ist meist positiv (d.h. die Erdoberfläche erscheint negativ geladen) und hat am Erdboden in ebenem Gelände und bei Schönwetter Werte von 1 bis 3 V/cm oder 100 bis 300 V/m. In bewohnten Gegenden kann seine Größe bis über 500 V/m ansteigen und ebenso in unebenem Gelände bei konvexer Krümmung, also auf Bergspitzen oder Bergrücken oder auch an der Oberseite von hohen Bäumen, Häusern usw. Bei Niederschlägen oder bei künstlichen Raumladungen schwankt der Potentialgradient stark und nimmt höhere positive und negative Werte an. In der freien Atmosphäre nimmt der Potentialgradient mit zunehmender Höhe über dem Erdboden ab, in 3 km Höhe z.B. findet man nur noch etwa 25 V/m, in 10 km Höhe 4 V/m usw.

Die täglichen und jährlichen Schwankungen des luftelektrischen Potentialgradienten, welche an einigen Landstationen schon über 50 Jahre ununterbrochen gemessen werden, haben sehr unterschiedlichen Charakter. Die einfachsten Verhältnisse herrschen über den Ozeanen. Dort fand S. J. MAUCHLY [2b] auf den Meßfahrten mit der Carnegie Tagesgänge mit einer einfachen Periode, wobei das

Minimum etwa um 4 Uhr GMT (also *Weltzeit*!) und das Maximum etwa um 18 Uhr GMT liegt (Fig. 1). Ganz ähnlich ist der Verlauf des Potentialgradienten in der Arktis.

γ) Das zweite luftelektrische Element ist die *Leitfähigkeit* der Luft. Über offenem Land findet man einen Wert von etwa 1 bis $2 \cdot 10^{-16} \Omega^{-1}$ cm^{-1} (im elektrostatischen Maßsystem 1 bis $2 \cdot 10^{-4}$sec^{-1}), entsprechend einem spezifischen Widerstand von 0,5 bis $1 \cdot 10^{16}\,\Omega$cm. Zum Vergleich damit seien der spezifische Widerstand von Seewasser mit etwa 20 Ωcm, von trockenem Boden 10^5 bis $10^7\,\Omega$cm erwähnt[1]. Die Luft-Leitfähigkeit kann jedoch noch um mehr als eine Zehnerpotenz kleiner sein und zwar in Luft, welche sehr viele Kondensationskerne oder Nebel- oder Dunstteilchen enthält. In der freien Atmosphäre nimmt die Leitfähigkeit mit zunehmender Höhe rasch zu. In 3 km Höhe beträgt sie etwa $5 \cdot 10^{-16}\,\Omega^{-1}$ cm^{-1}, in 10 km Höhe rund $30 \cdot 10^{-16}\,\Omega^{-1}$cm^{-1}.

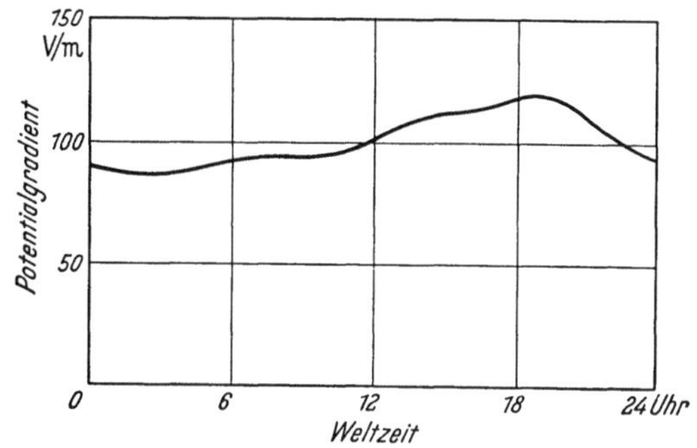

Fig. 1. Mittlerer taglicher Gang des luftelektrischen Potentialgradienten über den Ozeanen (gemessen auf den Fahrten mit der „Carnegie" 1909—1921).

δ) Die elektrische *Stromleitung* in der Luft der Troposphäre und unteren Stratosphäre erfolgt über die *atmosphärischen Ionen*. Diese bestehen aus Komplexen (clusters) von mehreren Molekülen, welche ihren Zusammenhalt einer elektrischen Elementarladung verdanken. Je nach ihrer Masse und der Zusammensetzung verändert sich ihre *Beweglichkeit*, d.h. Geschwindigkeit im elektrischen Einheitsfeld. Neben diesen sog. *Klein-Ionen* enthält die Luft Mittel- und Groß-Ionen, welche nur geringe Beweglichkeit haben, so daß sie auch bei großer Konzentration fast nicht zur Leitfähigkeit beitragen.

In der Nähe des Erdbodens finden wir im Kubikzentimeter Luft etwa 100 bis 1000 positive und negative Klein-Ionen, ferner 1000 bis 10000 Groß-Ionen und in stark verschmutzter Luft bis weit über 100000 Groß-Ionen. Die Groß-Ionen tragen im allgemeinen ebenfalls nur eine, die größeren teilweise auch zwei oder mehr Elementarladungen. Von größeren Höhen an haben auch die *Elektronen* eine ausreichend lange Lebensdauer, um zur elektrischen Leitfähigkeit beitragen zu können. Da die von außen auf die Erde fallende ionisierende Sonnenstrahlung in der Atmosphäre an Intensität verliert, bilden sich, wegen der mit zunehmender Luftdichte zunehmenden Absorption, Schichten mit Leitfähigkeitsmaxima aus, die *ionosphärischen* Schichten[2] in Höhen von 100 bis 400 km.

ε) Aus elektrischer Feldstärke und elektrischer Leitfähigkeit der Luft am Erdboden ergibt sich ein elektrischer *Leitungsstrom*, der sog. *Vertikalstrom* (air-earth current), der normalerweise von oben zum Erdboden gerichtet ist. Er hat die Größenordnung von $2 \cdot 10^{-16}$ Amp/cm^2 und ist leicht meßbar. Etwa die gleiche Größe des Vertikalstromes, errechnet aus Potentialgradient und Leitfähigkeit finden wir auch in der ganzen Troposphäre und unteren Stratosphäre. Eine einfache Rechnung ergibt, daß dieser Vertikalstrom das gemessene luftelektrische Feld in weniger als einer Stunde vollständig abgebaut hätte, wenn nicht eine Elektrizitätsquelle diesen Strom aufrechterhalten würde. Der Gedanke von Wilson [*14*], welcher diese Elektrizitätsquelle in den Gewitterwolken vermutete,

[1] Vgl. Bd. XLVII dieses Handbuches, Beitrag Cagniard, Ziff. 4, S. 409.
[2] Vgl. Beiträge Nicolet und Rawer, Bd. XLIX dieses Handbuches.

konnte inzwischen bestätigt werden durch verschiedene Untersuchungsergebnisse. Einmal tragen die Gewitterwolken, wie SIMPSON und SCRASE [*12*] durch Ballonaufstiege gezeigt haben, an der Oberseite positive und an der Unterseite überwiegend negative Ladungen. Die unterhalb und oberhalb aller Gewitterwolken gemessenen Ausgleichsströme (vgl. Ziff. 38) zusammen sind von solcher Größe, daß sie den Vertikalstrom in den Schönwettergebieten decken können. Die von den Gewittern erzeugten Ladungen verteilen sich in wenigen Minuten über die gesamte Erde, einerseits in der gut leitenden Ausgleichsschicht über 50 km Höhe, andererseits auf der als Leiter anzusehenden Erdoberfläche.

ζ) Innerhalb der Atmosphäre beobachtet man meist positive *Raumladungen*. Die Ursache hierfür ist zu suchen einerseits in der nach oben zunehmenden Leitfähigkeit der Luft, andererseits in dem in der Höhe nahezu konstanten Vertikalstrom. Vom Standpunkt der Elektrizitätsströmung durch die Atmosphäre sind die Raumladungen also die Folge der veränderlichen Leitfähigkeit. Zu diesen „*natürlichen*“ *Raumladungen* kommen noch vom Boden ausgehende künstliche oder „*anthropogene*“ *Raumladungen* [*11*] hinzu, welche mit dem Rauch, den Abgasen u. a. in die Atmosphäre gelangen. Diese sind insbesondere in dicht besiedelten Gebieten und deren Umgebung von sehr großem Einfluß auf das luftelektrische Feld und den Vertikalstrom am Erdboden und in niedrigen Höhen.

η) Die *Gewitter* als die hauptsächlichen Generatoren des luftelektrischen Feldes und des Vertikalstromes wurden wegen ihrer Fülle von Erscheinungen auf die verschiedenste Weise und teilweise mit großem Aufwand untersucht. Außer durch die Ballonaufstiege von SIMPSON und SCRASE hat man vom Boden und von Flugzeugen aus die elektrischen Felder um die Gewitterwolken gemessen. Insbesondere beim Rhön-Segelflugwettbewerb 1938 wurden wertvolle und vielseitige Ergebnisse gewonnen, hauptsächlich über die meteorologischen Verhältnisse innerhalb der Gewitterwolken. Dem gleichen Ziel diente ein großangelegtes Unternehmen, das „*Thunderstorm-Project*“ [*13*], in dem amerikanische Forscher beim Durchfliegen von Gewitterwolken meteorologische und andere Daten ermittelten. Ebenso sind die Begleiterscheinungen der Gewitter, nämlich die Blitze, ihr Spektrum und ihre langwellige elektro-magnetische Strahlung, die Spitzenströme und ähnliches auf ihre Entstehungsursache, auf ihren Ablauf und auf ihre Wirkungen hin untersucht worden[1]. Besonderes Interesse finden die seltenen aber vielleicht echten Phänomene des Kugelblitzes [*1*], für den es noch keine physikalische Erklärung gibt, und des Perlschnurblitzes.

B. Elektrische Ladungen in der Atmosphäre, Erzeugung und Vernichtung.

a) Ionisatoren der Atmosphäre.

3. Ionisierung durch kurzwelliges Licht. Die Ionisierungsspannungen der in der unteren Atmosphäre vorkommenden Atome und Moleküle N_2, O_2, Ar, CO_2 usw. betragen etwas mehr als 10 eV (für N_2 z.B. 15,8 eV, für O_2 12,5 eV). Nur ultraviolettes Licht mit Wellenlängen von weniger als 0,1 μ vermag demnach in Luft Ionen zu bilden. Solche Wellenlängen liefern in der Nähe des Erdbodens nur künstliche UV-Quellen, deren ionisierende Wirkung neuerdings von NORINDER und SIKSNA[2] und anderen untersucht worden ist. Da Licht mit Wellen-

[1] Vgl. Beitrag SCHONLAND, dieses Handbuch Bd. XXII, S. 576 (1956).
[2] H. NORINDER u. R. SIKSNA: Arch. Physik **5**, 471—491 (1952).

längen von 0,29 μ und darunter durch die Ozonschicht in 20 bis 30 km Höhe absorbiert wird, werden erst über 30 km Höhe Elektrizitätsträger durch Sonnenstrahlung (Licht und weiche Röntgenstrahlung) gebildet.

Ein zweiter Effekt kurzwelligen Lichtes, der *licht-elektrische* oder HALLWACHS-*Effekt,* ist ebenfalls auf Wellenlängen unterhalb 0,2 μ beschränkt. Deshalb wird die Emission von Elektronen durch Bestrahlung von Oberflächen mit kurzwelligem Licht ebenfalls nur bei künstlichen UV-Quellen eintreten und dabei überwiegend negative Ionen erzeugen. In der oberen Stratosphäre wäre es möglich, daß diese Art der Ionenbildung an sehr hoch geschleuderten Teilchen oder Eisnadeln zur Ionenbildung durch natürliches Sonnenlicht führt. Der Effekt ist aber im allgemeinen ohne Bedeutung.

4. Ionisierung durch kosmische Strahlung. In den mittleren Höhen der Troposphäre und in der Stratosphäre werden die atmosphärischen Ionen hauptsächlich durch die *kosmische Ultrastrahlung* gebildet. Unmittelbar am Erdboden erzeugt sie nur ein Fünftel der Gesamtionisation; in großen Höhen über 25 km wird ihre Wirkung durch diejenige der UV- und Röntgenstrahlung der Sonne überlagert. Die kosmische Ultrastrahlung hat ihren Ursprung im Weltenraum und fällt nahezu kontinuierlich aus allen Richtungen ein, abgesehen von gelegentlichen solaren Ausbrüchen. Diese primäre Komponente besteht wahrscheinlich hauptsächlich aus Protonen mit Energien von 10^9 bis 10^{17} eV. Beim Zusammenstoß dieser Protonen mit den Molekülen der Atmosphäre entstehen nun durch Kern-Prozesse weitere Teilchen, die zusammengenommen die sekundäre Komponente bilden und hauptsächlich aus Mesonen, Positronen und Elektronen, Lichtquanten und leichten und schwereren Nukleonen bestehen. Durch diese Vermehrung der primären Strahlung hat die Ionisierung in der Atmosphäre ein Maximum in etwa 13 km Höhe über dem Erdboden; sie nimmt nach unten infolge der zunehmenden Absorption, insbesondere der energieärmeren, weicheren Strahlung, ab.

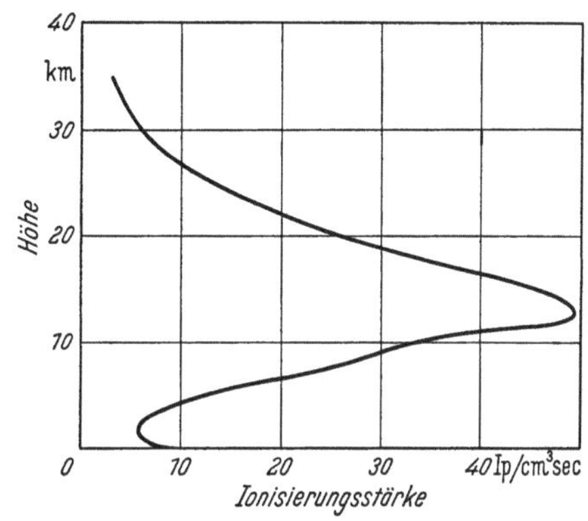

Fig. 2. Die Ionisierungsstärke in der freien Atmosphäre als Funktion der Höhe.

In unseren Breiten erzeugt die kosmische Ultrastrahlung am Erdboden etwa 2 Ionenpaare pro Kubikzentimeter und Sekunde. Die Zunahme der Ionisierung mit der Höhe ist mit bemannten und unbemannten Ballonen mit Hilfe von offenen oder geschlossenen Ionisationskammern gemessen worden. Die Ergebnisse sind von HESS [*18*] zusammengestellt worden. Für die gesamte Ionisierung in der Atmosphäre, also auch durch die radioaktive Strahlung, ist der mittlere Verlauf in Fig. 2 wiedergegeben. Die Ionisierung zeigt ein Maximum in 13 km Höhe; sie wird mit zunehmender Höhe dann kleiner, infolge der Verminderung der Luftdichte.

Da die anfliegenden geladenen Teilchen entsprechend ihrer Energie im Magnetfeld der Erde abgelenkt werden, hängt die Intensität der Ultrastrahlung von der geomagnetischen Breite ab. Von den Polen bis zu etwa 50° nördlicher und südlicher geomagnetischer Breite ist die Wirkung des Magnetfeldes klein: dagegen werden die Primärteilchen bei Annäherung zum geomagnetischen Äquator teilweise so abgelenkt, daß die Intensität der Ultrastrahlung besonders in größeren Höhen abnimmt. In Erdbodennähe ist das Defizit am Äquator nur etwa 14%, dagegen in 17 km Höhe etwa 80%. Die Ionisierung der Gesamtatmosphäre durch die Ultrastrahlung ist deshalb über dem Äquator rund 40% geringer als in nördlichen Breiten. BOWEN, MILLIKAN und NEHER[1] haben die Ergebnisse von Messungen der Ionenbildung in drei verschiedenen geomagnetischen Breiten gemäß Fig. 3 zusammengestellt. Die Folgen für die Leitfähigkeit der Atmosphäre werden in Ziff. 22 behandelt. Weitere Einzelheiten findet man in den Bänden XLVI und IL dieses Handbuches.

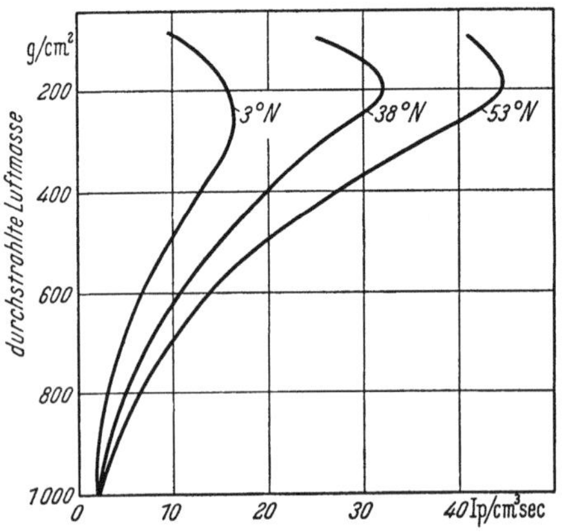

Fig. 3. Die Ionisierungsstärke der Ultrastrahlung als Funktion der durchstrahlten Luftmasse (Höhe) und der geomagnetischen Breite (Breiteneffekt), nach Ergebnissen von BOWEN, MILLIKAN und NEHER umgerechnet auf die freie Atmosphäre. Abszisse: Ionisierungsstärke in Ionenpaaren pro cm³ und sec.

5. Ionisierung durch radioaktive Substanzen im Erdboden und in der Atmosphäre. In der Nähe der Erdbodenoberfläche erfolgt die Ionenbildung zum größeren Teil nicht durch die kosmische Strahlung, sondern durch *radioaktive Substanzen*, welche sich teils im Erdboden, teils an der Erdoberfläche und teils in der Atmosphäre befinden. Dabei ionisieren die α-, β- und γ-Strahlen entsprechend ihrer Reichweite und ihrem Ionisierungsvermögen ganz verschieden.

α) Die radioaktiven Substanzen im Erdboden. In den verschiedenen Böden und Gesteinen kommen vor allem die Elemente der Uran- und Thoriumreihen und außerdem das radioaktive Kalium vor. Die radioaktiven Bestandteile der Gesteine sind in Bd. XLVII dieses Handbuches (Beitrag J. T. WILSON, R. D. RUSSELL, R. M. FARQUHAR) ausführlich diskutiert.

Die α-Strahlung dieser radioaktiven Stoffe wird im Erdboden zum größten Teil absorbiert, jedoch können gebildete Luft-Ionen durch Erdspalten oder -poren bis über die Erdoberfläche gelangen, insbesondere bei abnehmendem Luftdruck.

Die β-Strahlung ist in der Lage, dünne Bodenschichten zu durchdringen und die Luft über dem Erdboden zu ionisieren. In stark vermehrtem Maße vermag dies die γ-Strahlung.

β) Die radioaktiven Substanzen in der Atmosphäre. Der Gehalt der Luft an Radium-Emanation und Thorium-Emanation verändert sich örtlich und zeitlich. Da die Emanation aus Ritzen und Poren des Erdbodens nachströmt, wird diese *Exhalation* beeinflußt vom Fallen und Steigen des Luftdruckes, von der Erdbodenbedeckung und deren Änderungen durch Austrocknung oder Niederschläge, Bewachsungen usw. Nach vielen Messungen findet man über festem Boden zwischen 20 und $400 \cdot 10^{-18}$ Curie/cm³ Radium-Emanation [*18*]. Der Gehalt an Thorium-Emanation ist etwa 10000mal größer, die Ionisierung jedoch wegen des langsameren Zerfalls nur etwa 60% von derjenigen der Radium-Emanation.

Die Folgeprodukte der Emanation bleiben in der Luft, da sich diese Atome an Aerosolteilchen anlagern und damit auch bei ihrem weiteren Zerfall zur

[1] I. S. BOWEN, R. A. MILLIKAN u. H. V. NEHER: Phys. Rev. **52**, 80 (1937); **53**, 217, 856 (1938).

Ionisierung der Luft beitragen. Auch wenn sie sich an der Erdoberfläche ablagern oder durch Niederschläge dorthin gelangen, trägt ihre Strahlung, hauptsächlich ihre α-Strahlung, noch zur Ionisation der Luft bei, besonders im Abstand von wenigen Zentimetern von der Erdoberfläche.

Die Abnahme der Emanation mit der Höhe ist stark verschieden; der je nach der Jahreszeit und Bewölkung verschieden starke *Austausch* und die unterschiedliche Nachlieferung aus dem Erdboden sind die Ursachen dafür. Ganz rohe Anhaltspunkte erhält man aus einer von PRIEBSCH[1] stammenden Tabelle 1.

Tabelle 1. *Abnahme der Radium-Emanation in der freien Atmosphäre mit der Höhe über dem Kontinent.*

Höhe über dem Boden m	Gehalt an Radium-Emanation je Masseneinheit Luft (in relativen Einheiten)
0,01	100
1	95
10	87
100	69
1 000	38
7 000	12
12 000	7,6

Über dem *Meer* ist der Gehalt an radioaktiven Stoffen infolge fehlender Nachlieferung sehr gering und beträgt etwa $1 \cdot 10^{-18}$ Curie/cm³ oder etwa $^1/_{100}$ des Mittelwertes über Land; in der Nähe der Küsten hängt er von der Windrichtung ab.

In letzter Zeit ist der Gehalt an künstlich radioaktiven Substanzen durch *atomtechnische Versuche* verschiedentlich meßbar angestiegen. HAXEL und SCHUMANN[2] u. a. haben Aktivitäten mit verschiedenen Zerfallszeiten gefunden; die langlebigen Spaltprodukte können sich wochenlang in der Atmosphäre halten.

γ) Ionisierung durch die radioaktiven Substanzen. HESS [*18*] hat zusammen mit anderen die durch die α-, β- und γ-Strahlen der radioaktiven Stoffe erzeugte Ionisierung untersucht und zusammengestellt. Bezeichnet man mit J die Zahl der je Sekunde im Kubikzentimeter gebildeten Ionenpaare, dann erzeugt die γ-Strahlung aus dem Erdboden Werte der Ionisierung wie in Tabelle 2 gegeben:

Tabelle 2. *Ionisation durch γ-Strahlung der im Erdboden enthaltenen radioaktiven Elemente* (nach HESS [*18*]).

	Uran-Radiumreihe	Thoriumreihe	Kalium	Summe
1. Über Eruptivgestein:				
Mittelwert	1,5 J	1,5 J	1,4 J	4,4 J
Höchstwert	3,0 J	3,0 J	2,2 J	8,2 J
Mindestwert	0,7 J	0,7 J	0,6 J	2,0 J
2. Über Böden aus Eruptivgestein:				
Mittelwert	1,2 J	1,2 J	1,1 J	3,5 J
Höchstwert	2,4 J	2,4 J	1,8 J	6,6 J
Mindestwert	0,5 J	0,5 J	0,5 J	1,5 J
3. Über Sedimentärgestein:				
Mittelwert	0,9 J	0,9 J	0,7 J	2,5 J
Höchstwert	1,8 J	1,8 J	1,4 J	5,0 J
Mindestwert	0,4 J	0,4 J	0,2 J	1,0 J
4. Über Böden aus Sedimentärgestein:				
Mittelwert	0,7 J	0,7 J	0,6 J	2,0 J
Höchstwert	1,4 J	1,4 J	1,2 J	4,0 J
Mindestwert	0,4 J	0,4 J	0,15 J	1,0 J

[1] J. PRIEBSCH: Phys. Z. **32**, 622 (1931).
[2] O. HAXEL u. G. SCHUMANN: Naturwiss. **40**, 458 (1953). Vgl. auch S. 550, Fußnote 3.

Die Ionisation durch die γ-Strahlen schwankt demnach zwischen ungefähr 1 und 8 J am Erdboden. Entsprechend ihrer Energie nimmt die Ionisierung mit der Höhe stark ab, wie Fig. 4 zeigt und zwar für die γ-Strahlung des Radiums

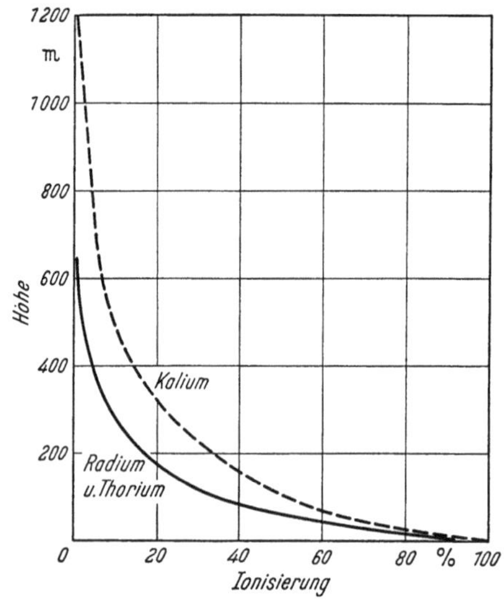

Fig. 4. Abnahme der Ionisierung durch γ-Strahlen aus dem Erdboden mit der Höhe.

und Thoriums bzw. des Kaliums. Die β-Strahlen dieser Stoffe ionisieren zusätzlich mit etwa $^1/_8$ der Wirkung der γ-Strahlen. Aus dem Meerwasser tritt trotz geringer Spuren von radioaktiven Salzen praktisch keine ionisierende Strahlung aus.

Die Emanationen in der Atmosphäre und ihre Folgeprodukte ionisieren vor allem durch ihre α-Strahlung. Da ihr Gehalt nicht nur örtlich, sondern auch zeitlich stark schwankt, ist der Anteil der durch sie gebildeten Ionen in der Atmosphäre ebenfalls sehr variabel. Nach Hess kann man mit Ionisierungsstärken rechnen, welche in Tabelle 3 aufgeführt sind:

Tabelle 3. *Ionisierung durch die in der Atmosphäre enthaltenen Substanzen.*

	Mittelwert	Maximum (geschatzt)	Minimum (geschatzt)
a) In Bodennähe (0 bis 10 m über dem Boden):			
Ionisation durch α-Strahlung der in der Luft befindlichen Radium- und Thorium-Emanation und ihrer Zerfallsprodukte . . .	4,4 J	13 J	1,4 J
Ionisation durch β-Strahlung	0,02 J	—	—
Ionisation durch γ-Strahlung	0,1 J	0,5 J	—
Totale Ionisation	4,5 J	13,5 J	1,4 J
b) In der freien Atmosphäre (Höhen von einigen 100 bis über etwa 1000 m):			
α-Strahlung	2 bis 3 J	—	—
β-Strahlung	0,04 J	0,1 J	—
γ-Strahlung	0,2 J	0,6 J	—
Totale Ionisation: etwa	2 bis 3 J		

6. Die Gesamt-Ionisierungsstärke und ihre Messung. *α) Zusammenfassung der Ionisatoren.* In den Ziff. 4 und 5 wurden die einzelnen Ionisatoren besprochen. Es wird klar, daß die Gesamt-Ionisierungsstärke noch mehr als die Einzelbeträge mit Ort und Zeit schwankt. Die folgende Übersicht gibt also nur rohe Anhaltswerte.

Die Gesamt-Ionisierungsstärke q über dem Festland beträgt nach Hess [*18*] etwa 10 Ionenpaare/cm³ sec über Eruptivgestein, etwa 8,7 Ionenpaare/cm³ sec über Sedimentärgestein. Die Extremwerte können dabei bis zu 4 und 22 J betragen. Im einzelnen darf man im Mittel mit folgenden Werten rechnen:

Tabelle 4. *Übersicht über die Beiträge zur Ionisierungsstärke in Bodennähe nach* Hess.

1. Ionisierung durch radioaktive Substanzen in der Luft:	α-Strahlung	4,4 J	
	β-Strahlung	0,03 J	
	γ-Strahlung	0,15 J	
	etwa	4,6 J	(1,4—13,5 J)
2. Ionisierung durch radioaktive Substanzen im Erdboden:	β-Strahlung	0,3 J	
	γ-Strahlung	3,2 J	
	etwa	3,5 J	(2—6 J)
3. Ionisierung durch kosmische Ultra-Strahlung in 50° geomagnetischer Breite:		1,9 J	
Gesamt-Ionisierungsstärke:	etwa	10 J	(4—22 J)

β) Meßmethoden. Die Messung der Ionisierungsstärke und erst recht der einzelnen Komponenten durch die verschiedenen Strahlenarten ist sehr schwierig und nur durch Differenzmessungen möglich. Die gesamte Ionisierung durch γ-Strahlen und Ultrastrahlung erhält man mit *Ionisationskammern.* Wählt man die angelegte Spannung so hoch, daß Sättigung herrscht, also alle Ionen aus der Kammer durch das elektrische Feld herausgeschafft werden, bevor sie rekombinieren können, dann liefert der Strom direkt die Zahl der gebildeten Ionen pro Sekunde und Kammerinhalt. Durch Messung mit dieser Ionisationskammer an dem gewünschten Ort und Vergleichsmessungen entweder in großer Höhe oder innerhalb dicker Bleiabschirmungen oder in Wasser von mehreren Metern Tiefe erhält man dann Ströme, die entweder die radioaktiven Bestandteile der Luft oder die Strahlung aus dem Erdboden oder die weiche Komponente der Ultrastrahlung nicht mehr enthalten, während die von den Gefäßwänden ausgehende Strahlung immer enthalten ist. Aus entsprechenden Differenzen ergeben sich dann einzelne Anteile. Um auch die β-Strahlung zu erfassen, muß man mit sehr dünnwandigen Ionisationskammern arbeiten, welche die β-Strahlung fast ungeschwächt hindurchgehen lassen, und die Versuche in gleicher Weise anstellen.

Am schwierigsten ist die Erfassung der Ionisierung durch α-Strahlen. Diese werden bereits von den dünnsten Wänden von Ionisationskammern absorbiert. Da die α-Strahlung größtenteils von den in der Luft schwebenden radioaktiven Teilchen stammt, muß man die Luft selbst in Ionisationskammern, sog. *Emanometer,* leiten und die Entladungsströme zu verschiedenen Zeiten nach dem Verschluß messen, zunächst mit emanationshaltiger und nachher mit emanationsfreier Luft. Es ist dabei zu beachten, daß der aktive Niederschlag der Emanation ebenfalls ionisiert und die Zerfallszeiten der Thorium-Emanation und deren Folgeprodukte andere sind als die der Radium-Emanation. Durch die Aufnahme der Abklingkurven läßt sich dann die Gesamt-Ionisation durch Radium- und

Thoriumgehalt aufteilen. Nähere Anleitungen für den Gebrauch solcher Emanometer hat ISRAEL[1] gegeben.

Bei einer älteren Methode von ELSTER und GEITEL[2] wird die Tatsache der elektrischen Ladung der in der Luft schwebenden radioaktiven Substanzen ausgenutzt. Spannt man in freier Luft einen Draht (Länge etwa 10 m) aus und legt an ihn eine negative Spannung von einigen 1000 V, dann sammelt dieser aus der vorbeistreichenden Luft die (meist positiv geladenen) radioaktiven Kondensationskerne. Durch Messung der Ionisierung in einer üblichen Ionisationskammer, in die der Draht gelegt wird, läßt sich — allerdings nur in grober Näherung — der Gehalt an radioaktiven Stoffen und damit deren Ionisierungsstärke errechnen.

Man kann auch die Emanation bei der Temperatur flüssiger Luft kondensieren und erhält damit zur Messung größere Mengen. Nach einer Methode von HAXEL und SCHUMANN[3] wird Luft durch Asbest-Filtergewebe hindurchgesaugt, wodurch quantitativ alle radioaktiven Bestandteile, nicht allerdings die Emanation selbst, absorbiert werden. Durch nachträgliche Messung der γ-Strahlung dieser Filter mittels GEIGER-MÜLLERschen Zählrohren kann man den radioaktiven Gehalt der Luft bestimmen und außerdem zusätzliche künstliche Verseuchungen durch Kernspaltprodukte identifizieren. Ein Warngerät für radioaktive Verseuchung hat unter anderen WILKENING[4] entwickelt.

Die Exhalation des Bodens und die dabei aus dem Boden entweichende Emanation erhält man nach JOLY und SMYTH[5], indem man unten offene Gefäße auf den Erdboden aufsetzt; über diesen wird Luft geleitet, die sich mit der ausströmenden Emanation vermischt; deren Ionisierung kann wie oben gemessen werden.

7. Die atmosphärischen Ionen. Durch die Wirkung eines hochenergetischen Korpuskular- oder γ-Strahls werden primär ein oder mehrere Elektronen aus der Elektronenhülle eines Atoms oder Moleküls entfernt. Bei der Dichte der Luft der Troposphäre und unteren Stratosphäre lagert sich dieses Elektron in kürzester Zeit, etwa $^1/_{1000}$ bis $^1/_{100}$ sec, an ein neutrales Atom oder Molekül an, so daß zunächst ein positives und ein negatives Atom- oder Molekül-Ion entsteht. Die elektrische Ladung dieser Ionen polarisiert aber nunmehr Nachbaratome oder -moleküle und ist imstande, einige von ihnen in Bruchteilen von Sekunden an sich zu ziehen und auf diese Weise *Molekülkomplexe (clusters)*[6] zu bilden. Diese sind nunmehr für längere Zeit, in der Größenordnung von Minuten, beständig; man bezeichnet diese Ionen als *Klein-Ionen*.

Die Klein-Ionen entstehen nicht nur durch ionisierende Strahlen, sondern können auch durch andere thermische[7] oder elektrische Effekte erzeugt werden, so z.B. durch Flammen, durch Spitzenentladungen[8] oder durch Reibungseffekte. Die Zahl der positiven bzw. negativen Klein-Ionen beträgt in Erdbodennähe 100 bis 1000 je Kubikzentimeter und nimmt mit zunehmender Höhe zu. Für ihre Konzentration in der Atmosphäre sind die Ionisierungs- oder Bildungsstärke einerseits und andererseits die ionenvermindernden Effekte maßgebend: zwischen diesen bildet sich ein Gleichgewichtszustand aus, welcher in Ziff. 21 behandelt wird.

Durch die Anlagerung dieser Klein-Ionen an Kondensationskerne, aber auch bei anderen thermischen oder chemischen Vorgängen entstehen die *Groß-Ionen*,

[1] H. ISRAEL: Gerlands Beitr. Geophys. **51**, 35 (1937).
[2] J. ELSTER u. H. GEITEL: Phys. Z. **2**, 590 (1901); **3**, 305 (1902); **4**, 96, 522 (1903).
[3] G. SCHUMANN: Arch. Met., Geophys. u. Biokl. A **9**, 204—223 (1956).
[4] M. H. WILKENING: Nucleonics **10**, 36 (1952).
[5] J. JOLY u. L. B. SMYTH: Phil. Mag. **24**, 632 (1912).
[6] A. W. OVERHAUSER: Phys. Rev. **76**, 250—254 (1949).
[7] R. SIKSNA: Ark. Fysik **5**, 531—543 (1952).
[8] J. J. NOLAN u. P. J. NOLAN: Proc. Roy. Ir. Acad. **40**, 11—59 (1931); **41**, 111—128 (1933).

die von sehr vielen Autoren[1,2,3] [*20*], [*37*] nach Masse, Konzentration und Zusammensetzung untersucht worden sind. Auch sie tragen zum großen Teil eine Elementarladung, haben aber im Vergleich zu den Klein-Ionen eine um einige Zehnerpotenzen größere Masse. Für ihre Zusammensetzung gilt das gleiche wie für die atmosphärischen Kondensationskerne (Ziff. 17). Sie können maritimen oder anthropogenen Ursprungs sein. Die *maritimen* Kerne gelangen, soviel wir heute wissen, beim Zerspritzen von Meerwasser, bei Brandung und Gischt in die Atmosphäre und können infolge ihrer verschwindend kleinen Sinkgeschwindigkeit monatelang in der Atmosphäre verbleiben. Über dem Festland sind aber die *anthropogenen* Kerne in der überwiegenden Mehrzahl. Diese entstehen hauptsächlich bei den Verbrennungen aller Art, bei chemischen Prozessen und werden vor allem in den besiedelten Gegenden in großer Zahl produziert. Ihr Durchmesser bewegt sich zwischen 10^{-7} und 10^{-4} cm, ihre chemische Zusammensetzung ist außerordentlich vielfältig. Größtenteils handelt es sich um Mischkerne, die durch Koagulation entstanden sind, und aus Chloriden, Sulfaten, Ammoniak, Magnesiumsalzen, Nitriten, Wasser und ähnlichem[4] bestehen. Diese Groß-Ionen kommen in der untersten Atmosphäre in Konzentrationen von 10^3 bis 10^5 je Kubikzentimeter vor.

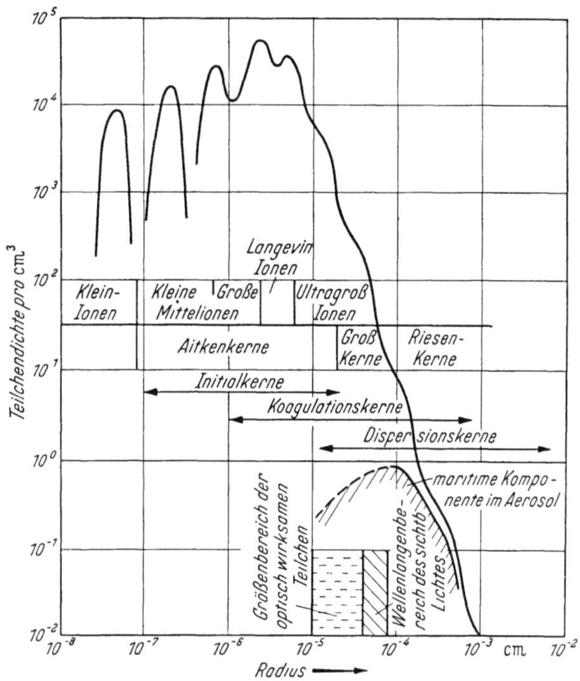

Fig. 5. Charakter der Größenverteilung der Ionen und Kondensationskerne im atmosphärischen Aerosol nach Junge.

Eine Gruppe zwischen Klein-Ionen und Groß-Ionen bilden noch die von Pollok[5] entdeckten und von Hogg [*19*] näher untersuchten *Mittel-Ionen*. Diese haben Radien von 10^{-7} bis 10^{-6} cm und scheinen aus Molekülansammlungen von etwa 2000 Molekülen und Vielfachen davon zu bestehen. Sie sollen vornehmlich nur in Luft vorkommen, welche durch großstädtische und industrielle Abgase verseucht ist und Schwefelsäure enthält.

Diese Mittel- und Groß-Ionen haben, ebenso wie die Kondensationskerne, eine eigene Entwicklungsgeschichte. Nach ihrer Entstehung koagulieren sie mit verschiedensten Halbwertszeiten mit anderen geladenen oder ungeladenen Teilchen so lange, bis sie infolge ihres Gewichtes sedimentieren. Das auf diese Weise entstehende Gleichgewicht hat eine Größenverteilung der Ionen und Kondensationskerne der verschiedenen Massen zur Folge, welche in einer von Junge [*23*] zusammengestellten Übersicht in Fig. 5 wiedergegeben ist. Weitere Einzelheiten sind in Ziff. 17 beschrieben.

[1] H. Israel: Gerlands Beitr. Geophys. **31**, 173 (1931); **36**, 24 (1932).
[2] J. J. Nolan u. J. G. O'Keeffe: Proc. Roy. Ir. Acad. **41**, 26—40 (1933).
[3] H. Norinder u. R. Siksna: Tellus **3**, 234 (1951).
[4] H. Cauer: Arch. Met., Geophys. u. Biokl. B **1**, 221 (1949).
[5] J. A. Pollok: Phil. Mag. **29**, 636—646 (1915).

8. Ionenzählung. Zur Ionenzählung sowie zur Messung ihrer Beweglichkeit (Ziff. 9) und der elektrischen Leitfähigkeit der Luft (Ziff. 10) verwendet man heute fast ausschließlich die Methode der „senkrechten Geschwindigkeiten". Durch einen Zylinderkondensator nach Fig. 6 wird die zu untersuchende Luft aspiriert *(Aspirationsmethode)*. Ein elektrisches Feld im Innern des Zylinderkondensators erteilt den im Luftstrom mit der Geschwindigkeit v laminar bewegten Ionen eine dazu senkrechte Komponente

$$u = k \cdot \mathfrak{E}, \tag{8.1}$$

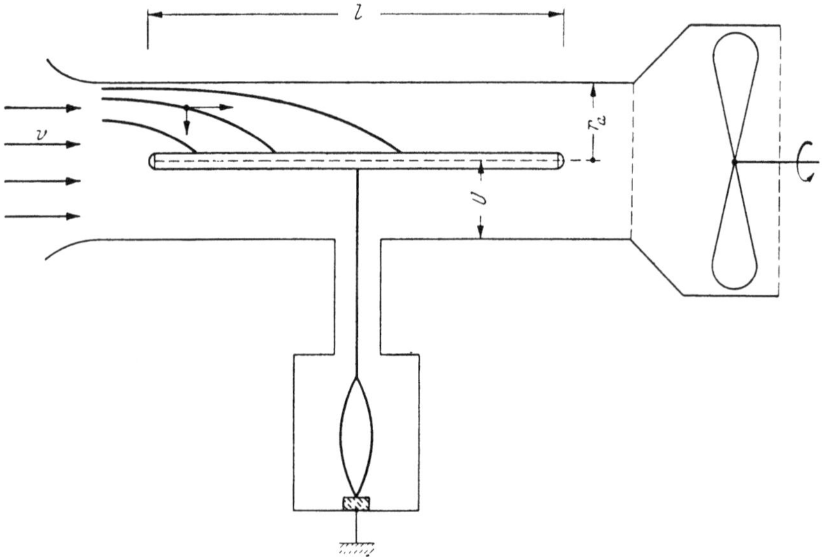

Fig. 6. Schematische Darstellung eines Zylinderkondensators zur Aspirationsmethode für Ionenzahlung, Messung der Beweglichkeit von Ionen und der elektrischen Leitfahigkeit der Luft.

wobei k die *Beweglichkeit* der Ionen und $\mathfrak{E}$ die elektrische Feldstärke bedeutet. Die Bahnen der Ionen im Zylinderkondensator sind Parabeln, als Folge des Verlaufs der Feldstärke

$$\mathfrak{E} = \frac{1}{r} \frac{U}{\ln \frac{r_a}{r_i}} \tag{8.2}$$

U = Spannung am Zylinderkondensator; r = Abstand eines Ions von der Zylinderachse; r_a, r_i = Radien des Außen- und Innenzylinders.

Bei genügend großer Spannung U werden alle Ionen, deren Beweglichkeit größer oder gleich einer bestimmten *Grenzbeweglichkeit* k_g ist, auf der Mittelelektrode vollständig abgefangen.

Diese Grenzbeweglichkeit k_g erhält man aus der Bedingung, daß die am äußeren Rande eintretenden Ionen gerade noch zur Mittelelektrode gelangen:

$$k_g = \frac{M}{4\pi C_1 U}. \tag{8.3}$$

Es bedeuten: M aspirierte Luftmenge in Kubikzentimeter je Sekunde, C_1 wirksame Kapazität im Zylinderkondensator.

Für die Zählung der *Klein-Ionen*, welche durch eine Lücke im Beweglichkeitsspektrum der Ionen von den Mittel- und Groß-Ionen getrennt sind, wählt man nunmehr k_g kleiner oder gleich der Beweglichkeit der kleinsten zu erwartenden Beweglichkeit. Dann tragen alle Klein-Ionen eines Vorzeichens zusammen eine elektrische Ladung pro Sekunde dQ/dt auf die Mittelelektrode, welche sich ergibt zu

$$dQ/dt = n \cdot e \cdot M, \tag{8.4}$$

Fig. 7. Ionenzahler nach EBERT.

woraus man n, die Zahl der positiven oder negativen Klein-Ionen je Kubikzentimeter mit Beweglichkeiten $\geq k_g$ erhält; e bedeutet die elektrische Elementarladung $= 1{,}6 \cdot 10^{-19}$ Coulomb.

Bei der Zählung der Groß-Ionen lenkt man zunächst durch ein vorgeschaltetes schwächeres elektrisches Feld alle Klein-Ionen und, wenn gewünscht, auch Mittel- und Groß-Ionen bestimmter Beweglichkeit aus dem aspirierten Luftstrom ab. Durch das folgende stärkere Feld in einem Zylinderkondensator mit passend gewählter Grenzbeweglichkeit erfaßt man die restlichen Groß-Ionen bis zu der gewählten Grenzbeweglichkeit k_g und einen weiteren Bruchteil langsamerer Ionen. Durch Variation der Grenzbeweglichkeiten verschiedener hintereinandergeschalteter Zylinderkondensatoren ergibt sich ein Bild des vorhandenen Ionenspektrums.

Für die praktische Messung ist früher vor allem der Ionenzähler nach EBERT[1] (s. Fig. 7) verwendet worden, bei welchem der Zylinderkondensator direkt auf ein empfindliches Einfaden-Elektrometer aufgesteckt ist und die zuvor aufgeladene Mittelelektrode durch die Ionen entladen wird *(Entlademethode)*.

Neuerdings leitet man die Ladung der Ionen über Höchstohm-Widerstände zur Erde ab und mißt mit empfindlichen Röhren-Elektrometern den Spannungsabfall am Widerstand. Mit einem solchen Gerät, z.B. nach Fig. 8 ist es möglich, Dauerregistrierungen der Zahl der positiven und negativen Ionen bequem durchzuführen[2]. Da bei der zweiten Methode, der *Aufladungsmethode*, der Außenzylinder eine elektrische Spannung gegen Erde hat, kann ein Meßfehler durch

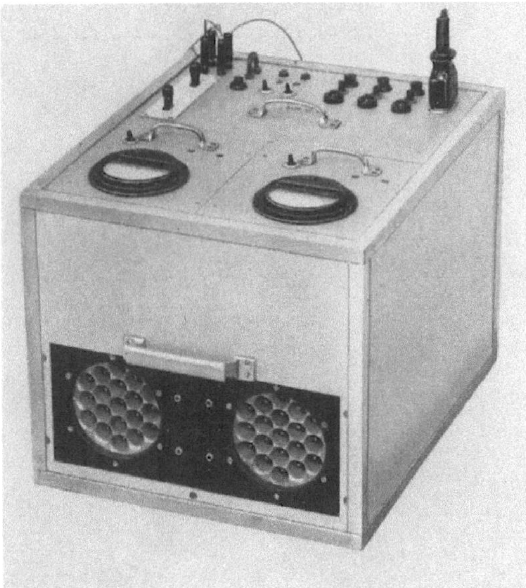

Fig. 8. Klein-Ionenzahler nach MÜHLEISEN und CREUTZBURG.

das Randfeld entstehen, durch welches die zu messenden Ionen vorzeitig abgelenkt werden und der Messung entzogen werden können. Auch müssen die Ionen-Eintrittsöffnungen vom Erdfeld abgeschirmt werden, um eine Ablenkung durch dieses zu vermeiden.

9. Die Beweglichkeit der Ionen und ihre Messung. Ionen bewegen sich in einem elektrischen Felde mit einer Geschwindigkeit u, welche proportional zur Feldstärke $\mathfrak{E}$ ist. Als Beweglichkeit k ist daher definiert:

$$k = \frac{u}{|\mathfrak{E}|} \left[\frac{\mathrm{cm}^2}{\mathrm{V\,sec}}\right]. \tag{9.1}$$

Sie hängt ab von der Größe des Ions und vom Luftdruck p und der Temperatur T der umgebenden Luft:

$$k_{p\,T} = k_{p_0 T_0} \cdot \frac{p_0}{p} \cdot \frac{T}{T_0} \,. \tag{9.2}$$

Bei den Klein-Ionen findet man meist zwei Gruppen von diskreten Beweglichkeiten in atmosphärischer Luft. Ihre Beweglichkeiten hängen offenbar von

[1] H. EBERT: Phys. Z. **2**, 662 (1901). — Verh. dtsch. phys. Ges. **7**, 2, 34 (1905).

[2] R. MÜHLEISEN u. U. CREUTZBURG: Klein-Ionenzähler. Noch unveröffentlicht.

anwesenden Spurengasen und dem Feuchtigkeitsgehalt der Luft[1] ab. Deshalb variieren die Meßergebnisse teilweise stark mit Ort und Zeit. Einige Werte für die Beweglichkeiten positiver und negativer Ionen k_+, k_- sind in Tabelle 5 enthalten, ergänzt durch Werte bei künstlicher Ionisierung in künstlicher Atmosphäre.

Tabelle 5. *Klein-Ionenbeweglichkeit in atmosphärischer Luft.*

Autor	Ort/Luftart	k_+	k_- $\left[\frac{cm^2}{V\,sec}\right]$
Mittelwert	Festland	1,4	1,9
MAUCHLY	Ozeane	1,58	1,66
J. ZELENY[2]		1,36	1,87
L. B. LOEB u. M. F. ASHLEY[3]		1,60	1,85
J. ZELENY[4]		1,04 u. 1,50	1,45 u. 2,40
R. SIKSNA[5]	Upsala	0,45 u. 2,00	0,7 u. 2,4
R. MÜHLEISEN, U. CREUTZBURG u. W. BLOSS	Stuttgart 1952—54	0,8	0,8
		1,40	1,60
		1,80	1,80
		2,50	2,80
H. SCHILLING[6]	Gereinigte Luft	2,0	2,45

Die Mittel- und Groß-Ionen zeigen ein mehr oder weniger kontinuierliches Spektrum der Beweglichkeiten. Ihre Werte sind in Tabelle 6 zusammengestellt.

Tabelle 6. *Beweglichkeit der Mittel- und Groß-Ionen.*

	Durchmesser (cm)	k $\left[\frac{cm^2}{V\,sec}\right]$
Mittel-Ionen	10^{-7} bis 10^{-6}	10^{-1} bis 10^{-2}
Groß- oder LANGEVIN-Ionen	2 bis $5 \cdot 10^{-6}$	10^{-3} bis $3 \cdot 10^{-4}$
Ultra-Groß-Ionen	$> 5 \cdot 10^{-6}$	$< 2{,}5 \cdot 10^{-4}$

Die Größe der Beweglichkeit der Ionen erhält man

1. aus dem Zusammenhang (vgl. Ziff. 22, S. 572) zwischen Leitfähigkeit und Ionen-Zahl

$$\Lambda = \lambda_+ + \lambda = \sum^{p} n_p^+ \cdot e \cdot k_p^+ + \sum^{q} n_q^- \cdot e \cdot k_q^- . \tag{9.1}$$

Mißt man also die Leitfähigkeit Λ (s. Ziff. 10) und die Klein-Ionenzahl n, so erhält man eine mittlere Beweglichkeit[7]

$$\bar{k} = \frac{\Lambda}{n \cdot e} . \tag{9.2}$$

2. Die Beweglichkeit der Klein-Ionengruppen kann man aus einer Strom-Spannungskurve an einem Ionenzähler erhalten. Nimmt man an, daß die angesaugte Luft Ionen einer bestimmten Beweglichkeit enthält, dann laufen die einzelnen Ionen je nach dem Ort ihres Eintritts in den Zylinderkondensator auf

[1] J. J. NOLAN: Proc. Roy. Ir. Acad. **39**, 82—99 (1930).
[2] J. ZELENY: Phil. Trans. Roy. Soc. Lond., Ser. A **195**, 193 (1900).
[3] L. B. LOEB u. M. F. ASHLEY: Proc. Nat. Acad. Sci., Wash. **10**, 351 (1924)
[4] J. ZELENY: Phys. Rev. **36**, 35 (1930).
[5] R. SIKSNA: J. Atmos. a. Terres. Phys. **4**, 106—111 (1953).
[6] H. SCHILLING: Ann. Phys. **83**, 23 (1927); **86**, 447 (1928).
[7] H. GERDIEN: Phys. Z. **4**, 632 (1903).

kongruenten Parabeln, und der Auftreffpunkt auf der Innenelektrode wird (nach Fig. 6) nur vom Eintrittsort abhängen. Bei niedrigen Saugspannungen unterhalb der Sättigungsspannung werden nur diejenigen Ionen auf der Mittelelektrode von der Länge l abgefangen, welche an der Öffnung des Zylinderkondensators innerhalb eines Kreises vom Radius r' eintreten; für diesen Radius gilt die Beziehung:

$$\frac{v\,(r'^2 - r_i^2)\ln\frac{r_a}{r_i}}{2 l k U} = \text{const}. \tag{9.3}$$

Wenn der Radius der Innenelektrode r_i gegen r' vernachlässigt wird, dann ist die Fläche, durch welche die abgefangenen Ionen eintreten, $F = \pi r'^2$ proportional der angelegten Spannung U. Da die Zahl der abgefangenen Ionen proportional

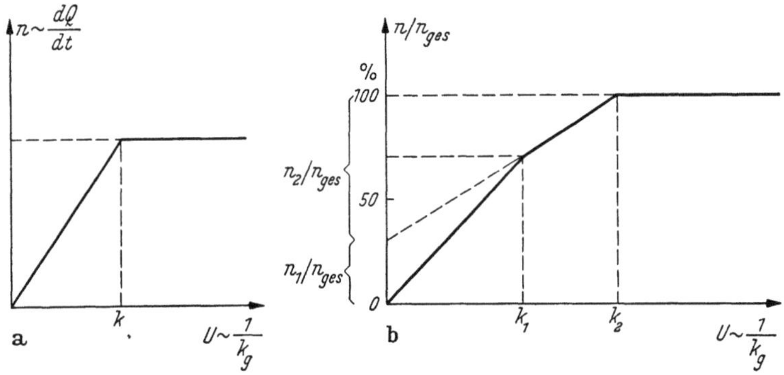

Fig. 9a u. b. Stromspannungscharakteristik im Zylinderkondensator zur Messung der Beweglichkeit von Ionen, a bei einheitlicher Beweglichkeit, b bei zwei Ionengruppen verschiedener Beweglichkeit.

dieser Fläche F und der Luftgeschwindigkeit im Kondensator v ist, bleibt somit die Menge der abgefangenen Ionen so lange proportional U, bis alle in den Zylinderkondensator eintretenden Ionen abgeschieden werden, also

$$F = F_{\text{ges}} \cong \pi r_a^2 \tag{9.4}$$

ist. Die pro Sekunde abgeschiedenen Ionen n erzeugen einen Strom dQ/dt, welcher im ersten Teil der Strom-Spannungskurve linear von U abhängt, im zweiten Teil konstant ist, wie Fig. 9a zeigt.

Sind mehrere Ionengruppen mit unterschiedlicher Beweglichkeit vorhanden, so setzt sich der Kurvenverlauf aus so vielen geraden Stücken zusammen, als es Beweglichkeitsgruppen gibt (Fig. 9b). Jede Knickstelle tritt bei einer Spannung U auf, für welche die zugehörige Grenzbeweglichkeit k_g des Zählers gleich der Beweglichkeit k der nunmehr vollständig abgeschiedenen Ionengruppe ist. Da

$$k_g \sim 1/U \tag{9.5}$$

ist, kann man die Abszisse in Fig. 9 direkt mit k_g beziffern und erhält sofort einerseits die Beweglichkeit der Ionengruppen und andererseits den Anteil der Ionen dieser Gruppe an der gesamten Ionenzahl aus dem rückwärtigen Schnittpunkt der Geraden mit der Ordinate. Die Ableitung dieses Zusammenhanges findet man unter anderem bei ISRAEL[1].

Beweglichkeitsmessungen im Bereich der Groß-Ionen[2] kann man ebenso ausführen, oder durch Zylinderkondensatoren mit unterteilter Innenelektrode, oder

[1] H. ISRAEL: Gerlands Beitr. Geophys. **31**, 171 (1931).
[2] P. LANGEVIN: C. R. Acad. Sci. Paris **140**, 232 (1905).

mit hintereinander geschalteten Zylinderkondensatoren verschiedener Daten. Mit den beiden letzten Methoden sind kontinuierliche Messungen ohne Spannungsänderungen möglich. Eine Anzahl anderer Methoden kommt für spezielle Beweglichkeitsuntersuchungen in Betracht[1,2,3].

10. Messung der Leitfähigkeit der Luft. Die elektrische Leitfähigkeit der Luft innerhalb der unteren Luftschichten ist die Folge der Beweglichkeit der Klein-Ionen im elektrischen Feld. Es gilt Formel (9.1). Eine indirekte Methode zur Leitfähigkeitsbestimmung beruht auf der Messung des Vertikalstromes (s. Ziff. 33) und der elektrischen Feldstärke im luftelektrischen Feld. Liegen die Meßstellen genügend nahe beisammen, ohne sich zu stören, dann erhält man die Leitfähigkeit aus

$$\Lambda = j/|\mathfrak{E}|, \tag{10.1}$$

wobei j die Vertikalstromdichte ist, also Vertikalstrom je Quadratzentimeter horizontale Fläche am Erdboden.

Bei der direkten Methode nach GERDIEN[4] legt man an einen Zylinderkondensator der oben beschriebenen Art (nach Fig. 6), durch welchen die zu untersuchende Luft aspiriert wird, eine so geringe Spannung, daß die Grenzbeweglichkeit größer als die Beweglichkeit der schnellsten Ionen ist; dann ist die auf einer Elektrode in der Zeiteinheit abgeschiedene Elektrizitätsmenge für jede Beweglichkeitsgruppe der Ionen gleich der Ladung der Ionen, welche innerhalb eines Kreises vom Radius r' eintreten (s. Ziff. 9).

$$d Q_{\pm}/d t = \pi r'^2 \cdot v \cdot n \cdot e. \tag{10.2}$$

Nach Gl. (9.3) ist angenähert

$$r'^2 = \frac{2 l k U}{v \cdot \ln \frac{r_a}{r_i}} \sim F' \tag{10.3}$$

also ist

$$\frac{d Q_{\pm}}{d t} = \frac{\pi \cdot 2 l k \cdot U \cdot n \cdot e}{\ln \frac{r_a}{r_i}} = 2\pi l \cdot \frac{U}{\ln \frac{r_a}{r_i}} \cdot \lambda_{\pm} = C \cdot U \cdot \lambda_{\pm}, \tag{10.4}$$

mit C = Gesamtkapazität des Systems.

Die Luftleitfähigkeit $\lambda_{\pm}$ ist also

$$\lambda_{\pm} = \frac{d Q_{\pm}/d t}{C \cdot U},$$

und ihre Messung wird unabhängig von der Luftgeschwindigkeit v. Sind mehrere Ionengruppen verschiedener Beweglichkeit vorhanden, dann gilt dieselbe Gleichung für die gesamte polare Leitfähigkeit, solange für keine Gruppe Sättigung erreicht ist.

b) Andere ladungserzeugende Effekte.

Neben der Ionisierung durch radioaktive und kosmische Strahlung führen noch andere Effekte zu elektrischen Ladungen in der Atmosphäre oder auf in der Luft vorhandenen Teilchen. Diese Erscheinungen spielen ihre größte Rolle bei der Entstehung der elektrischen Raumladung in Niederschlägen, hauptsächlich aber in den Gewitterwolken. Diese und ihre mutmaßliche Deutung werden

---

[1] E. RUTHERFORD: Proc. Cambridge Phil. Soc. **9**, 401 (1898).
[2] H. A. ERIKSON: Phys. Rev. **24**, 502 (1924).
[3] K. PRZIBRAM, im Handbuch der Physik, Bd. XXII/1, S. 344ff. Berlin: Springer 1933.
[4] H. GERDIEN: Phys. Z. **6**, 800 (1905). — Terr. Magn. **10**, 69 (1905).

im folgenden einzeln besprochen und ihre Auswirkungen auf die elektrische Ladung von Wolkenelementen in einer Übersicht im Anschluß an Ziff. 14 zusammengestellt. Ihre Wirkungen auf die elektrischen Felder in der Atmosphäre werden in Ziff. 27ff. behandelt.

11. Der Influenzeffekt. Schon ELSTER und GEITEL[1] haben 1913 eine elektrische Aufladung von Wolkentröpfchen unter der Wirkung der Influenz eines schon bestehenden elektrischen Feldes für möglich gehalten. Sie dachten an einen Zusammenstoß von Wolkentröpfchen verschiedener Größe und folglich verschiedener Fallgeschwindigkeit, von denen die größeren bei der normalen Richtung

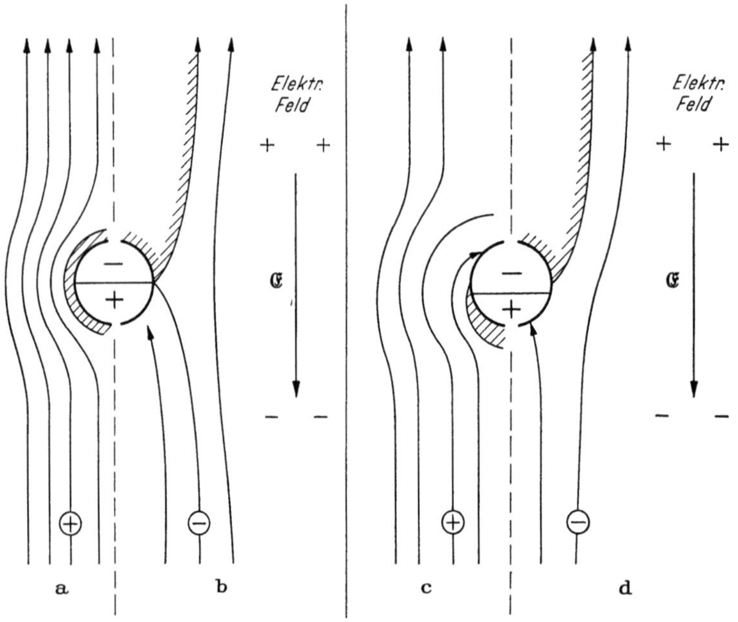

Fig. 10a—d. Bahnen der positiven und negativen Ionen im elektrischen Feld bei Umströmung eines a, b ungeladenen, c, d negativ geladenen, fallenden Wassertropfens, bezogen auf den Tropfen (Fallgeschwindigkeit > Ionengeschwindigkeit im Feld, Toträume schraffiert), nach WHIPPLE und CHALMERS und E. WALL.

des luftelektrischen Feldes oben negative und unten positive Ladung tragen. Kleinere Teilchen, welche von diesen polarisierten Tropfen eingeholt, aber wieder von ihnen abgestoßen werden, sind dann positiv geladen und werden durch die Schwerkraft von den negativ zurückbleibenden größeren Tropfen getrennt. Durch diese erste Erwähnung von Influenzwirkungen an Wolkenelementen angeregt, hat WILSON [38] die Wirkung fallender polarisierter Tropfen auf die Ionen der Luft einer Betrachtung unterzogen. Er konnte zeigen, daß geladene und ungeladene Teilchen, welche in einer Ionen enthaltenden Atmosphäre fallen, sich durch Einfangen von Ionen umladen. Zu diesem Vorgang haben WHIPPLE und CHALMERS[2] eine Theorie entwickelt, welche quantitative Angaben über die Umladungen von Tropfen erlaubt. Mit den experimentellen Untersuchungen von GOTT[3] und MÜHLEISEN und HOLL[4] ist diese Theorie in guter Übereinstimmung.

[1] ELSTER u. GEITEL: Phys. Z. **14**, 1287 (1913).
[2] F. J. W. WHIPPLE u. J. A. CHALMERS: Quart. J. Roy. Met. Soc. **70** (1944).
[3] I. P. GOTT: Proc. Soc. Lond., Ser. A **142**, 248 (1933); A **151**, 665 (1935).
[4] R. MÜHLEISEN u. W. HOLL: Geofisica Pura e Applicata **25**, 61—70 (1953).

Fällt ein Tropfen in einem elektrischen Feld, wo er polarisiert wird, so ist bei positivem Potentialgradienten die Oberseite negativ, die Unterseite positiv. Die elektrische Leitfähigkeit der Tropfen darf man dabei als genügend groß voraussetzen. In Fallrichtung bewegen sich dann bei dieser Feldrichtung die positiven Ionen, dem fallenden Tropfen entgegen aber die negativen Ionen. Die positive Unterseite wird dabei nach Fig. 10 negative Ionen einfangen, die negative Oberseite positive Ionen, aber weniger und nur dann, wenn die Fallgeschwindigkeit des Tropfens kleiner ist als die Ionengeschwindigkeit im Feld. Unter

Tabelle 7. *Umladung fallender Wassertropfen im elektrischen Feld, errechnet von* WHIPPLE *und* CHALMERS.

Q = elektrische Ladung der Wassertropfen; v = Fallgeschwindigkeit der Wassertropfen; a = Radius der Tropfen; λ = elektrische Leitfähigkeit der Luft; E = elektrische Feldstärke; k = Beweglichkeit der Klein-Ionen; 1 = positiv, 2 = negativ.

Ionen	Anfangsladung Q_0	Aufladegeschwindigkeit dQ/dt	Entladung Q_e	Gesetz
	$Q_0 > 3Ea^2$	$-4\pi\lambda_2 Q$	$-3Ea^2$	exponentiell
nur −-Ionen	$3Ea^2 > Q_0 > -3Ea^2$	$-\frac{\pi\lambda_2}{3Ea^2}(3Ea^2+Q)^2$	$-3Ea^2$	hyperbolisch
	$Q_0 < 3Ea^2$	0	Q_0	—
nur +-Ionen	$Q_0 > 3Ea^2$	0	Q_0	—
	$3Ea^2 > Q_0 > -3Ea^2$	$\frac{\pi\lambda_1}{3Ea^2}(3Ea^2-Q)^2$	$+3Ea^2$	hyperbolisch
$k_1E > v$	$Q_0 < -3Ea^2$	$-4\pi\lambda_1 Q$	$+3Ea^2$	exponentiell
nur +-Ionen	$Q_0 > 0$	0	Q_0	—
$k_1E < v$	$Q_0 < 0$	$-4\pi\lambda_1 Q$	0	exponentiell
+- und −-Ionen	$Q_0 > 3Ea^2$	$-4\pi\lambda_2 Q$	0	exponentiell
$k_1E > v$	$3Ea^2 > Q_0 > -3Ea^2$	$-4\pi\lambda_2 Q$	0	exponentiell
$\lambda_1 = \lambda_2$	$Q_0 < -3Ea^2$	$-4\pi\lambda_1 Q$	0	exponentiell
+- und −-Ionen	$Q_0 > 3Ea^2$	$-4\pi\lambda_2 Q$	$-0{,}515\,Ea^2$	exponentiell
	$3Ea^2 > Q_0 > 0$	$-\frac{\pi\lambda_2}{3Ea^2}(3Ea^2+Q)^2$	$-0{,}515\,Ea^2$	hyperbolisch
$k_1E < v$	$0 > Q_0 > -3Ea^2$	$-\frac{\pi\lambda_2}{3Ea^2}[(3Ea^2)^2 + 18Ea^2Q + Q^2]$	$-0{,}515\,Ea^2$	Tan, Cot
$\lambda_1 = \lambda_2$	$Q_0 < -3Ea^2$	$-4\pi\lambda_1 Q$	$-0{,}515\,Ea^2$	exponentiell

diesen Annahmen und für die verschiedenen Fälle kamen WHIPPLE und CHALMERS zu Ergebnissen für die sich einstellende End-Ladung und für die Umladegeschwindigkeit, welche in der Tabelle 7 auszugsweise zusammengestellt sind.

Nach der Theorie und den Meßergebnissen entstehen durch den WILSON-Effekt keine großen Ladungen auf Tropfen. Sie sind in der Größenordnung von 10^{-13} Coulomb bei einer Feldstärke von 1000 V/m. Der WILSONsche Mechanismus bestimmt vor allem die Ladung der am Erdboden ankommenden Niederschlagsteilchen. Eine ähnliche Annahme über die Influenzwirkung machte LUDLAM[1] für das Wegspritzen von Teilchen an wachsenden Hagelkörnern. Seine Vermutung, daß die elektrische Ladung durch die Influenz bedingt sei, wird aber nach den in Ziff. 13 und 14 behandelten Effekten zweifelhaft. Alle diese Effekte beschrieb im Zusammenhang WALL[2] unter der Bezeichnung „Asymmetrie-Effekt".

[1] F. H. LUDLAM: Quart. J. Roy. Met. Soc. **76**, 52 (1950).
[2] E. WALL: Wetter u. Klima **1**, 65 (1948).

12. Wasserfalleffekt. Bei dem 1892 von LENARD[1] näher untersuchten Wasserfalleffekt entstehen aus einem ungeladenen Wassertropfen beim Zerplatzen positive und negative Teile. Die Tropfen können zerplatzen, wenn sie am Erdboden aufschlagen, wenn kleinere von größeren Tropfen eingeholt werden, oder wenn sie bei Fallgeschwindigkeiten von > 8 m/sec oder in einem starken Luftstrom zerblasen werden. LENARD beobachtete, daß dabei die kleineren Tropfen und die Luft negativ, die größeren positiv geladen zurückbleiben. Die beobachteten Ladungen auf den Wassertropfen sind von der Größenordnung $\sim 10^{-10}$ Coulomb pro Gramm H_2O. Die Ursache für die Ladungstrennung ist in der elektrischen Doppelschicht in der Oberfläche des Tropfens zu suchen. Die negativen Ladungen sitzen bei reinem Wasser außen, die positiven weiter innen. Da diese Doppelschicht bereits durch geringe Beimengungen gelöster Substanzen in ihrem Vorzeichen umgekehrt wird, kann man beim Zerspritzen von schwachen Lösungen für die größeren Tropfen negative Ladungen und für die kleinen positive Ladungen erhalten. Der LENARD-Effekt kann außer in den Niederschlägen auch bei Brandung und Gischt, Wasserfällen u.a. auftreten. Auch beim künstlichen Zerspritzen und Zerstäuben von Wasser und wäßrigen Lösungen („Sprüh")[2] sind die Tröpfchen elektrisch geladen. Durch den LENARD-Effekt können auch Kleinionen entstehen, wenn nämlich die kleinen Tröpfchen verdampfen oder so hoch geladen sind, daß Ladungen absprühen. Auch die von Niederschlagsteilchen getroffenen metallischen Oberflächen, z. B. von Flugzeugen, werden elektrisch aufgeladen.

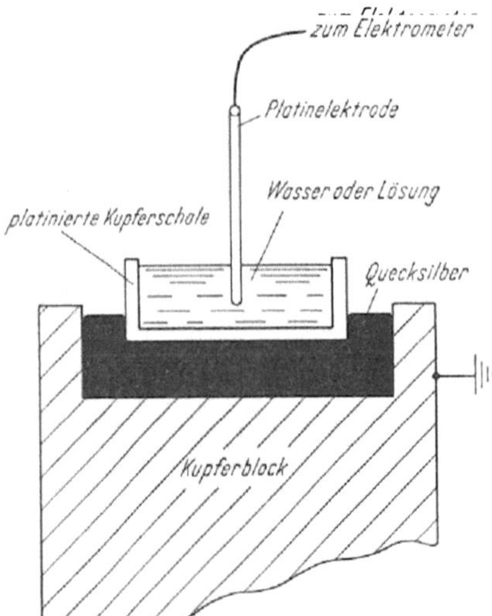

Fig. 11. Eine Versuchsanordnung von WORKMAN und REYNOLDS zur Messung der Potentiale zwischen fester und flüssiger Phase des Wassers oder wäßriger Lösungen.

13. Elektrische Effekte an Phasengrenzen des Wassers. Die Erscheinung, daß an einer Oberfläche flüssigen Wassers negative Ionen stärker absorbiert werden als positive Ionen, soll nach FRENKEL[3] zu der beobachteten vermehrt negativen Ladung von Nebeltröpfchen und Wassertropfen führen. Die Tropfen sollen sich dabei bis zu einem Potential von $-0{,}25$ V aufladen.

An der Phasengrenze zwischen Wasser und Eis treten unter bestimmten Voraussetzungen größere Potentialunterschiede auf. Schon FARADAY uud SOHNCKE[4] haben darauf aufmerksam gemacht, daß von Eis zurückspritzendes Wasser negativ geladen ist. Im Laboratorium haben WORKMAN und REYNOLDS [39] die Spannungen zwischen flüssiger und fester Phase eingehend untersucht und dabei folgende Anordnung benutzt: In einer platinierten Kupferschale nach Fig. 11, die durch umlaufendes Quecksilber gekühlt werden konnte, befand sich die Wasserprobe. Bei der Abkühlung des Wassers von unten durch die Kupferschale wurde die Spannung zwischen dieser Schale und einer in das Wasser eingetauchten Platinelektrode gemessen. Bei sehr reinem Wasser lud sich bei der Bildung eines Eis-Ansatzes die Platinelektrode im Wasser negativ auf, und zwar

[1] P. LENARD: Ann. Physik **47**, 463—524 (1915).
[2] E. E. DODD: Phys. Rev. **81**, 648 (1951).
[3] J. FRENKEL: J. Phys. Moskau **10**, 151, 158 (1946).
[4] L. SOHNCKE: Ann. Physik **28**, 550 (1886).

bis zu -90 V. Bei geringerem spezifischem Widerstand des Wassers, z.B. durch Auflösen von NaCl, wurden auch positive Spannungen bis zu $+25$ V gefunden, bei neueren Versuchen [*40*] an über 40 verschiedenen Lösungen Spannungen zwischen -232 V und $+75$ V. Diese Spannungen sind also viel größer als die normalen Kontaktspannungen zwischen zwei verschiedenen Stoffen.

Die Ursache für die Spannungen, welche mit Ladungsverschiebungen von etwa 10^{+13} bis 10^{+14} Elementarladungen/cm^3 verbunden sind, hängt offenbar nicht nur mit den Vorgängen an der Oberfläche, sondern auch mit Vorgängen im Kristallgitter selbst zusammen. Einige Autoren[1,2] vermuten eine Ladungsverschiebung an der Phasengrenze, an welcher die mit einem Wassermantel umgebenen Ionen dem vordringenden Eis ausweichen sollen, weil sie bei der regulären Eiskristallbildung nicht eingebaut — bestenfalls eingeschlossen — werden können. Je nach der Beweglichkeit der Ionen soll dann das Wasser das Vorzeichen der beweglicheren Ionen annehmen, während das Eis das umgekehrte Vorzeichen erhält.

Untersuchungen an natürlichen und künstlichen Eisteilchen wurden an vielen Orten angestellt. Alle verschiedenen Möglichkeiten bei der Ausbildung von Phasengrenzen sind erwogen worden; der Wasserdampf kann an Kondensationskernen kondensieren und bei fortgesetzter Kondensation zu größeren Wolkenteilchen anwachsen. Bei Temperaturen unter 0° C kann Sublimation an Sublimationskernen oder an schon vorhandenen Eisteilchen eintreten (Anreifung). Ferner können unterkühlte Wassertropfen gefrieren oder beim Zusammenstoß mit Eisteilchen zu Graupeln werden. Umgekehrt wird Wasser beim Schmelzen der Graupel von diesen beim Fallen weggerissen werden. Auch können die fallenden Graupel und Wassertropfen wieder verdampfen.

Vielerorts sind die elektrischen Effekte bei diesen Phasenumwandlungen des Wassers etwa unter den Bedingungen, wie sie in den Wolken herrschen, experimentell oder theoretisch untersucht worden; sie sind in die Übersicht (Ziff. 14) aufgenommen worden. Sehr große Ladungsbildungen fand LUEDER[3] bei natürlichem, unterkühltem Nebel. Er beobachtete die elektrische Aufladung eines im Nebel rotierenden Stabes bei Vereisung und leitet daraus die elektrische Aufladung von Graupelkörnern und die Elektrisierungsgeschwindigkeit ab. Die Grenzladung beträgt nach seinen Überlegungen (mit R in cm)

$$Q \approx -6 \cdot 10^{-9} \cdot R^2 \text{ Coulomb} \tag{13.1}$$

für Eiskörner mit einem Radius $R = 0{,}2$ cm also

$$Q = 2{,}4 \cdot 10^{-10} \text{ Coulomb}.$$

Die Geschwindigkeit der Elektrisierung ist nach LUEDER so groß, das Graupelkörner nach Entladung durch einen Blitz in einigen Sekunden ihre Grenzladung wieder angenommen haben können. Dieser Vorgang könnte bei der Erstarrung unterkühlten Wassers, bei der Vergraupelung und beim Abschmelzen von Eisteilchen zu hoch geladenen Teilchen führen. Fraglich ist aber, ob sich das Experiment von LUEDER auf Wolkenteilchen übertragen läßt. Immerhin fand auch WOLF[4] bei Untersuchungen im Laboratorium Aufladungen von Eiskörnern bis zu $+10^{-10}$ Coulomb.

Die elektrische Aufladung von Flugzeugen bei Vereisung beschreibt MEINHOLD[5].

[1] A. KUMM: Arch. Met., Geophys. u. Biokl. A **1951**, H. 5, 382—401.
[2] G. OPLATKA: Helv. phys. Acta **6**, 198—209, 244—246 (1933).
[3] H. LUEDER: Z. angew. Phys. **3**, 288—295 (1951).
[4] F. WOLF: Arch. Met., Geophys. u. Biokl. **9**, 242—263 (1956).
[5] H. MEINHOLD: Geofisica Pura e Applicata **19**, 1 (1951).

Übersicht. Elektrische Ladungs-

Autor	Vorgang	Voraussetzungen
ELSTER und GEITEL 1913	Kleine Wolkentröpfchen prallen von polarisierten Tropfen oder Hagelkörnern nach Zusammenstoß ab	elektrisches Feld
LUDLAM 1950	Von wachsenden Hagelkörnern spritzen Teile der anfrierenden Tröpfchen weg	elektrisches Feld $t < 0°$ C
WILSON 1929, WHIPPLE und CHALMERS 1944—1947, WALL 1948, WORMELL 1953	Selektiver Ioneneinfang durch polarisierte fallende Niederschlagsteilchen	elektrisches Feld
WALL 1948	Selektiver Ioneneinfang durch polare mit vertikaler Achse fallende Eiskristalle	$t < 0°$ C
FRENKEL 1944—1947	Bevorzugte Absorption negativer Ionen durch Wassertröpfchen	$t > 0°$ C
LENARD 1892, SIMPSON 1909, CHAPMAN 1950—1952	Verspritzen von Wassertropfen im Luftstrom oder bei Zusammenstoß	$t > 0°$ C starke Aufwinde
SIMPSON und SCRASE 1937, KUNKEL 1950	Zusammenstoß von Eiskristallen	$t < 0°$ C
LANGE 1943	Berührung von Eiskristallen und Graupeln	$t < -10°$ C
FARADAY und SOHNCKE 1886, FINDEISEN 1943	Kleine Tropfen zerspritzen auf Eis	unterkühltes Wasser
REIFFERSCHEID und FINDEISEN 1940—1943, KRAMER 1947/48, WEICKMANN und AUFM KAMPE 1950, LUEDER 1951, MEINHOLD 1951, MASON 1953	Sublimation bzw. Verdampfen von Reif unter Ablösung von kleinen geladenen Eissplittern, Reifbildung aus unterkühlten Tröpfchen, Eisbildung an Metallstäben	$t < 0°$ C
WORKMAN und REYNOLDS 1948 bis 1950	Langsames künstliches Gefrieren von schwachen Lösungen	$t < 0°$ C

14. Spezielle Effekte an Eisteilchen. Da die meisten Niederschläge über die Eisphase gebildet werden, sind elektrische Effekte an Eis- und Schneeteilchen für die Elektrizitätserzeugung in Wolken von Bedeutung. FINDEISEN[1] hat nach Untersuchungen und Beobachtungen eine Hypothese über die Elektrisierung von Eissplittern aufgestellt. Wegen des stark polaren Charakters des Eises wachsen die Eiskristalle hauptsächlich in Form von langen Nadeln, Dendriten, weiter. Brechen nun von größeren Schnee- oder Eisgebilden die äußersten Splitter ab, dann zeigen sich diese Splitter größtenteils hoch geladen. Da Dipole an sich neutral sein müßten, muß man annehmen, daß durch die Ionen der Luft die Dipolladungen am freien Ende neutralisiert werden derart, daß dann beim Abbrechen diese Oberflächenladungen als Teilchenladungen auftreten. Über das Vorzeichen besteht noch keine Übereinstimmung bei den verschiedenen experimentellen

[1] W. FINDEISEN: Meteor. Z. **60**, 145 (1943).

erzeugung über die Wasserphasen.

Elektrischer Effekt	Experimenteller Beweis	Elektrische Ladung der leichten/schweren Teilchen (bzw. der Luft)		
		ohne Feld	bei pos. Pot. Grad.	bei neg. Pot. Grad.
Trennung der Teilchen nach elektrischem Kontakt führt zur Feldverstärkung	0	0	±	∓
Wegspritzen von der unteren/oberen Hageloberfläche führt zur Feldverstärkung/Feldschwächung	0	0	+/− −/+	−/+ +/−
Aufladung der Niederschlagsteilchen. Spiegelbildeffekt. Unter bestimmten Bedingungen Feldverstärkung	+	0	(+) −	(−) +
Wenn positiver Pol der Kristalle unten, dann negative Aufladung	0	(±)		
Negative Tröpfchen lassen nach Absinken positiv geladene Luft zurück: Feldaufbau	?	(±)		
Wegen elektrischer Doppelschicht werden die kleinen wegspritzenden Tröpfchen negativ, der größere Tropfen positiv geladen	+	∓		
Eiskristalle negativ, Luft positiv Starke Ladungen beider Vorzeichen	0 +	(±) +/−		
Graupel positiv, Eiskristalle negativ	?	∓		
Eis wird positiv, zerspritztes Wasser negativ	+	∓		
Bei Sublimation: Reif ±, Eissplitter − Bei Verdampfen: Reif −, Eissplitter + Metallstäbe: −	+	+/−		
Hohe positive oder negative Spannungen je nach Stoffart und Konzentration (Größenordnung 100 V) zwischen flüssiger und fester Phase	+	+/−		

Untersuchungen. Nach KUMM[1] verhält sich die Zahl der positiven Splitter zu der der negativen wie 7:1. Die Größe der Ladungen beträgt etwa $7 \cdot 10^{-16}$ Coulomb im Mittel; einzelne Splitter tragen Ladungen bis zu $2 \cdot 10^{-14}$ Coulomb.

Auch Berührungen zwischen Eiskristallen und Graupeln sollen nach LANGE[2], Zusammenstöße von Eiskristallen untereinander nach SIMPSON und SCRASE[3] zu stark geladenen Teilchen führen.

WALL[4] versucht durch die Annahme, daß Eiskristalle in ganz bestimmter Orientierung polarisiert sind und gleichartig orientiert fallen, elektrische Felder zu erklären. Die polare Achse soll dabei vertikal ausgerichtet sein und die positive Ladung unten sitzen, durch welche negative Ionen eingefangen werden.

[1] Siehe Fußnote S. 561.
[2] E. LANGE: Meteor. Z. **60**, 154—161, 303—314 (1943).
[3] G. C. SIMPSON u. F. I. SCRASE: Geophys. Mem. Met. Off. London **1937**, No. 161, 309.
[4] Siehe Fußnote S. 559.

15. Reibung und andere Elektrisierungseffekte. In der Atmosphäre und am Erdboden können sich durch den Wind bewegte oder im Schwerefeld fallende Teilchen aneinander oder an festen Oberflächen reiben. Sind diese Stoffe gute Isolatoren, dann sind sie nach der Berührung jeweils mit entgegengesetztem Vorzeichen geladen, wobei erhebliche Ladungsdichten entstehen können. Das Vorzeichen der elektrischen Ladungen ergibt sich dabei meistens aus der COEHNschen Regel[1]; diese sagt aus, daß der Stoff mit der höheren Dielektrizitätskonstanten sich bei Berührung positiv, derjenige mit der niedrigeren Dielektrizitätskonstanten negativ auflädt. In die Atmosphäre kommen deshalb die meisten Teilchen in geladenem Zustand. Dies gilt für den Staub[2], für Sandkörner bei Sandstürmen, Rauchteilchen, außerdem für Eis- und Schneeteilchen bei Reibung aneinander[3] oder auf festen Eis- und Schneeschichten, also z.B. bei Schneewehen und anderem. Die Aufladungen können zu Entladungen bei Sand- oder Staubstürmen führen. Bekannt sind ebenso die gefährlichen Aufladungen in Papier- und Textilfabriken, Druckereien, in Mühlen, Bergwerken und anderswo. Bei der kurzzeitigen Berührung von *leitenden* festen Körpern treten als Folge der Kontaktpotentiale Ladungen auf, welche aber gering sind.

Es soll noch abschließend die Bildung von elektrischen Ladungsträgern in Flammen und bei Verbrennungen aller Art, bei Funken-, Bogen- und Spitzenentladungen erwähnt werden, bei welchen die Ladungsträger verschiedene Vorzeichen und Größe haben können.

c) Ladungsvernichtende Effekte.

Verschiedene Vorgänge, teils elektrischer, teils mechanischer Art, verursachen eine Verminderung der Ionenzahl in einem bestimmten Volumen. Man unterscheidet die *eigentliche Ionenvernichtung*, unter der man die Vereinigung von Ionen verschiedenen Vorzeichens versteht, und die *uneigentliche Ionenvernichtung*, d.h. den Abtransport von Ionen und die Anlagerung an größere Teilchen, Kondensationskerne, Nebel- und Niederschlagsteilchen.

16. Vereinigung der Ionen. Durch Vereinigung eines positiven Klein-Ions mit einem negativen verschwindet deren Ladung, und damit auch der Zusammenhalt des Molekülkomplexes. In der Nähe des Erdbodens ist dieser Vorgang von untergeordneter Bedeutung, da bei einer mittleren Klein-Ionenzahl von 500 pro Kubikzentimeter die Zusammenstöße derselben mit Kondensationskernen und Groß-Ionen, deren Konzentration rund 10 bis 100mal so groß ist, viel häufiger sind. Die Ionenvereinigung überwiegt aber bei sehr hohen Klein-Ionenkonzentrationen und in sehr reiner Luft, also in größeren Höhen über dem Erdboden und über den Ozeanen, die übrigen Effekte. Für das Verschwinden der Klein-Ionen gilt das Vereinigungsgesetz:

$$\partial n_1/\partial t = \partial n_2/\partial t = -\alpha n_1 n_2, \tag{16.1}$$

wobei α der Vereinigungskoeffizient (Rekombinations-Koeffizient) genannt wird. Sind gleich viel positive und negative Klein-Ionen vorhanden, ist also $n_1 = n_2$, dann gilt:

$$\partial n/\partial t = -\alpha n^2. \tag{16.2}$$

Die Geschwindigkeit der Vereinigung, also der Koeffizient α, hängt von der Diffusion bzw. der Beweglichkeit k der Klein-Ionen ab:

$$\alpha = f(k).$$

[1] A. COEHN: Ann. Phys. (4) **30**, 777 (1909).
[2] E. BODENSTEDT: Z. angew. Phys. **6**, 297—303 (1954).
[3] J. A. CHALMERS: J. Atmos. a. Terres. Phys. **2**, 337—339 (1952).

Den Einfluß von Temperatur T und Druck p auf den Vereinigungskoeffizienten beschreibt die Formel

$$\alpha_{p,T} = \alpha_{p_0 T_0} \left(\frac{p}{p_0}\right)^x \left(\frac{T_0}{T}\right)^y, \tag{16.3}$$

wobei, nach Angaben verschiedener Autoren, x zwischen 0,3 und 1,0, und y zwischen 1,5 und 3,0 schwankt. Für die atmosphärischen Ionen sind die Werte $x=0{,}3$ und $y=2{,}3$ nach O. H. GISH und K. L. SHERMAN[1] am wahrscheinlichsten, und für α_0 findet man ziemlich einheitlich den Wert

$$\alpha_0 = 1{,}6 \cdot 10^{-6}\,\mathrm{cm^3\,sec^{-1}}.$$

17. Kondensationskerne, Groß-Ionen und ihre Zählung. In der unteren Troposphäre wird die Verminderung der Konzentration der Klein-Ionen überwiegend durch deren Anlagerung an ungeladene und geladene Kondensationskerne verursacht, wobei die geladenen meist als Groß-Ionen bezeichnet werden. Deshalb sei einiges über deren Natur vorausgeschickt: Bei geringer Wasserdampfübersättigung in der Atmosphäre kommt es nur dann zur Kondensation, wenn eine Oberfläche mit nicht zu kleinem Krümmungsradius zur Verfügung steht, also an der Oberfläche fester hygroskopischer oder flüssiger Körper. Wegen der Zunahme des Dampfdruckes mit abnehmendem Krümmungsradius muß die Übersättigung, welche zur Kondensation führt, mit abnehmendem Krümmungsradius der vorhandenen Teilchen zunehmen. In völlig reiner Luft kondensiert der Wasserdampf erst bei einer Übersättigung von 420%.

Kleinste Teilchen von 10^{-7} cm Radius an sind als Ausgangspunkte der Kondensation bei geringer Übersättigung geeignet und werden deshalb als *Kondensationskerne* [*17*] bezeichnet. Die Zahl dieser Kerne im Kubikzentimeter Luft schwankt zwischen etwa 10^3 in sehr reiner Luft bis 10^5 und 10^6 in stark verunreinigter Atmosphäre; gelegentliche Spitzenwerte von 10^7 je Kubikzentimeter kommen kurzzeitig in Großstädten, Industrieanlagen und Räumen vor. Mit der Höhe nimmt die Kernzahl rasch ab, in 5000 m sind es nur noch 10 bis 1000 Kerne/cm³. Die örtlichen und zeitlichen Schwankungen über den Kontinenten erklären sich durch die Lage und Ergiebigkeit der Kernquellen, durch die horizontalen und vertikalen Luftbewegungen und durch die Verminderung der Konzentration als Folge von Koagulation und Sedimentation.

Die chemische Zusammensetzung dieser Kondensationskerne wechselt stark. CAUER[2] fand in der Atmosphäre Chloride mit einem Gehalt von im Mittel 30 γ/m³, Sulfate mit 12 γ/m³, Ammoniak 8 γ/m³, Magnesium, Nitrite und anderes mit geringeren Konzentrationen ($1\,\gamma = 10^{-6}$ g). Diese Verbindungen kommen aber nicht in reiner Form in einem Kondensationskern vor, sondern es handelt sich bei den Kernen im atmosphärischen Aerosol größtenteils um sog. *Mischkerne*, welche nach der Entstehung durch Koagulation der *Initialkerne* entstanden sind. In Abhängigkeit von den meteorologischen Bedingungen untersuchte JUNGE[3] die Zusammensetzung des Aerosols.

Die Initialkerne bestehen zum größten Teil aus den Produkten von Verbrennungsvorgängen, von Auspuffgasen, Schornsteinrauch usw. Diese koagulieren ohne Rücksicht auf die chemische Zusammensetzung und wachsen daher im Laufe ihres Verbleibens in der Atmosphäre. In einem alternden Aerosol nimmt deshalb die Zahl der Kondensationskerne durch Koagulation ab, während ihre Zusammensetzung sich zu Kernen größeren Radius verschiebt. Im Gleichgewicht zwischen Entstehung, Wachstum durch Koagulation und Ausfall durch Sedimentation

[1] O. H. GISH u. K. L. SHERMAN: Terr. Magn. **49**, 159 (1944).
[2] H. CAUER: Arch. Met., Geophys. u. Biokl. B **1**, 221 (1949).
[3] CHR. JUNGE: J. of Meteorology **11**, 323—333 (1954).

kommt es dann zu einer Größenverteilung der Kondensationskerne, wie sie in Fig. 5 bereits wiedergegeben ist. Das Abfallgesetz[1] bei den größeren Kernen lautet:

$$\frac{\partial N}{\partial \log r} = \frac{C}{r^3}, \tag{17.1}$$

wobei ∂N die Anzahl der Kerne pro Kubikzentimeter in einem Radiusbereich der logarithmischen Skala, r der Kernradius und C eine Konstante ist. Man kann dieses Gesetz als eine Art Kontinuitätsgleichung verstehen, welche im Bereich von 10^{-6} cm bis 10^{-3} cm gilt: die der Atmosphäre neu zugeführten Kerne haben meist kleinere, die ausfallenden Kerne größere Radien; im Zwischenbereich nimmt die Kernkonzentration N hauptsächlich durch Koagulation ab, wobei die Kernmasse proportional r^3 zunimmt, die gesamte Masse je Raumeinheit aber konstant bleibt.

Während das bisher betrachtete Größenwachstum irreversibel verläuft, gibt es auch ein reversibles Wachstum bei zunehmender Feuchtigkeit. Von einer relativen Feuchtigkeit von etwa 60% an umgeben sich die Kondensationskerne mit Wassermolekülen, die bei abnehmender Feuchtigkeit wieder verdampfen[2].

Eine besondere Rolle spielt die maritime Komponente der Kondensationskerne. Diese besteht hauptsächlich aus *Kochsalzteilchen*, die wegen ihres längeren Aufenthaltes in der Atmosphäre bereits zu größeren Kernen koaguliert sind. Diese Komponente liefert vielleicht vorwiegend die Gefrierkerne für die Sublimation und Eisteilchenbildung[3].

Für die Kondensation und ebenso für die Wirkung auf die Klein-Ionen spielt die Ladung der Kondensationskerne nur eine untergeordnete Rolle. Nach zahlreichen Messungen sind etwa zwei Drittel der Kondensationskerne größtenteils mit *einem* elektrischen Elementarquantum geladen. Wie die Ladungen auf Kerne von verschiedenen Radien verteilt sind, zeigt Fig. 12.

Zur Messung des Gehalts der Luft an Kondensationskernen diente früher ausschließlich der SCHOLZsche Kernzähler[4], eine Weiterentwicklung des AITKENschen Zählers: In einem zylindrischen Innenraum wird wasserdampfgesättigte Luft adiabatisch expandiert, durch die Abkühlung entsteht Übersättigung. Bei genügendem H_2O-Angebot und einem Expansionsverhältnis von 1,2 bilden sich um die meisten Kondensationskerne Wassertröpfchen, welche infolge ihrer Schwere auf eine Glasscheibe niederfallen und dort ausgezählt werden können. Die relative Feuchtigkeit ist dabei 300 bis 400% und reicht aus, um alle Kondensationskerne zu erfassen, noch nicht aber die Klein-Ionen, welche eine relative Feuchtigkeit von mehr als 420% benötigen, um als Kerne wirksam zu werden.

Der SCHOLZsche Kernzähler hat eine Reihe von Mängeln[5]. Einmal ist ein großer relativer Fehler damit verbunden, daß jeweils nur soviel Luft eingelassen werden kann, daß auf ein Zählquadrat einer Glasscheibe nur rund 7 Teilchen niederfallen, welche mit dem Auge in der kurzen Zeit bis zu ihrer Wiederverdampfung erfaßt werden müssen. Strömungen im Innenrohr und Wirbelbildung[6] können zu einer unregelmäßigen Ablagerung der Nebeltröpfchen auf der Glasscheibe und damit zu Fehlern führen. Weitere persönliche Fehler sind bei der Handhabung des Gerätes möglich. Wegen dieser und noch weiterer Fehler sind in letzter Zeit neue Kondensationskernzähler entwickelt worden. Erwähnt

[1] CHR. JUNGE: Ber. dtsch. Wetterdienst. US-Zone **1952**, Nr. 35, 201.
[2] H. DESSENS: Météorologie **1947**, 321—327.
[3] W. RAU: Arch. Met., Geophys. u. Biokl. A **8**, 185—203 (1955).
[4] J. SCHOLZ: Z. Instrumentenkde. **51**, 505—522 (1931). — Meteor. Z. **49**, 384—388 (1932).
[5] P. COURVOISIER: Höhenklima-Forschungen des Basler Physiologischen Instituts II. 89—97 (1948).
[6] J. C. THAMS u. W. WIELAND: Geofisica Pura e Applicata **20**, 138—156 (1951).

seien der Kondensationskernzähler von Pollak[1] mit photographischer Aufnahme der Tröpfchen, die automatischen Kernzähler von Bradbury und Meuron[2] und Vonnegut[3], der Zähler für Dauerbetrieb von Verzár[4] und ein Kondensationskernzähler mit kontinuierlicher Übersättigung von Holl und Mühleisen[5].

Mit allen Kernzählern lassen sich auch Aussagen über die Groß-Ionenkonzentration machen: Durch ein vor den Kondensationskernzähler eingeschaltetes elektrisches Feld können die Groß-Ionen abgefangen werden, so daß dann nur noch die ungeladenen Kondensationskerne zur Zählung gelangen. Differenzmessungen ergeben dann die Zahl der Groß-Ionen.

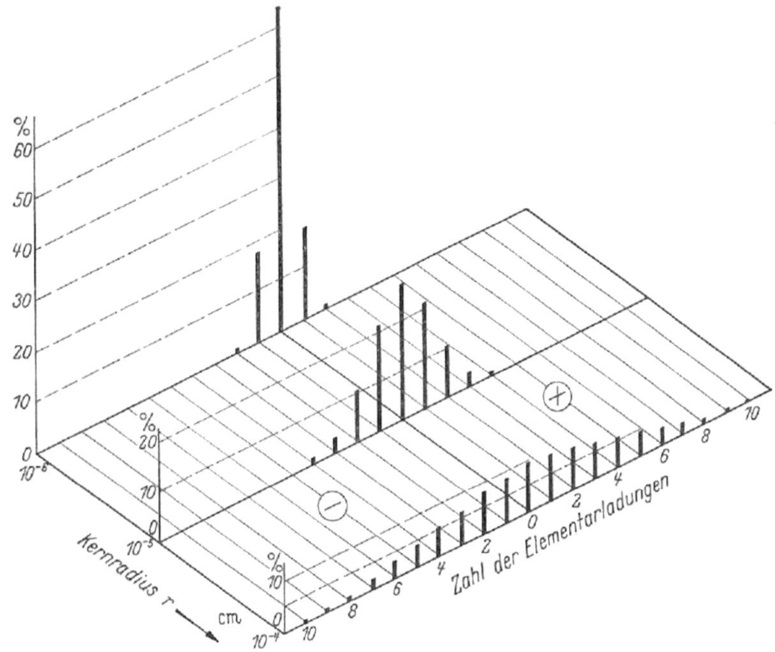

Fig. 12. Verteilung der Ladungen (in %) auf Kondensationskernen (Groß-Ionen) als Funktion des Kernradius r.

18. Anlagerung der Klein-Ionen an geladene und ungeladene Kerne. Klein-Ionen können sich an geladene oder ungeladene Kondensationskerne anlagern, wenn sie durch ihre Eigenbewegung sich diesen so weit nähern, daß elektrostatische Kräfte wirksam werden. In erster Näherung kann die Anlagerung durch eine dem Massenwirkungsgesetz entsprechende Formel ausgedrückt werden. Sind n_1 (n_2) die Zahlen der positiven (negativen) Klein-Ionen pro Kubikzentimeter und entsprechend N_0, N_1, N_2 die Zahlen der ungeladenen Kerne und der positiven und negativen Groß-Ionen pro Kubikzentimeter, dann ergibt sich die Abnahme der Klein-Ionenzahl durch Anlagerung zu

$$\frac{\partial n_1}{\partial t} = -\eta_{10} \cdot n_1 N_0 - \eta_{12} n_1 N_2 - \eta_{11} n_1 N_1 \tag{18.1}$$

[1] L. W. Pollak: Geofisica Pura e Applicata **22**, 75—83 (1952).
[2] T. E. Bradbury u. H. J. Meuron: Terr. Magn. **43**, 231 (1938).
[3] B. Vonnegut: Occ. Report Nr. 19 Project Cirrus, Report Nr. RL 300, 1. 1. 1950.
[4] F. Verzár: Arch. Met., Geophys. u. Biokl. A **5**, 372—376 (1953).
[5] W. Holl u. R. Mühleisen: Met. Rdsch. **8**, 165—168 (1955).

und
$$\frac{\partial n_2}{\partial t} = -\eta_{20} \cdot n_2 N_0 - \eta_{21} n_2 N_1 - \eta_{22} n_2 N_2. \quad (18.2)$$

Wenn man noch berücksichtigen will, daß ein geringer Teil der Groß-Ionen nicht eine, sondern mehrere Elementarladungen trägt, dann sind noch weitere Glieder in den Formeln (18.1) und (18.2) einzufügen. Die Koeffizienten η_{10} usw. nennt man die *Anlagerungskoeffizienten* zwischen Klein- und Groß-Ionen oder Kernen. Die von verschiedenen Autoren gemessenen Werte für die Koeffizienten streuen erheblich; einige Beispiele seien in Tabelle 8a aufgeführt:

Tabelle 8a. *Anlagerungskoeffizienten.*

	η_{10} $\times 10^{-6}$	η_{20} $\times 10^{-6}$	η_{12} $\times 10^{-6}$	η_{21} $\times 10^{-6}$
SCRASE[1]	0,58	1,07	2,35	2,96
PARKINSON[2]	0,6	1,1	2,4	4,5
NOLAN und DE SACHY[3]	6,8	7,6	8,7	9,7

Die Streuung der Werte ist nur zum Teil auf die Ungenauigkeit der schwierigen Messungen zurückzuführen.

Die Anlagerungskoeffizienten lassen sich auf die physikalischen Eigenschaften der beteiligten Teilchen zurückführen. Die älteren Theorien von LANGEVIN[4], WHIPPLE[5] und anderen, welche dies versuchten, sind verbessert worden durch eine neue Theorie von BRICARD[6]; dieser berücksichtigt die Diffusion der Klein-Ionen und die elektrostatischen Anziehungskräfte zwischen ihnen und den größeren Teilchen durch ihre Eigenladung und die Bildladung auf ihnen. Zur Vereinfachung werden Hypothesen eingeführt, welche plausibel sind: die Kondensationskerne und Groß-Ionen werden als Kugeln vom Radius a und mit leitender Oberfläche betrachtet; jede Berührung eines Klein-Ions mit einem in der Luft ruhend angenommenen Kern oder Groß-Ion soll zu einer Anlagerung führen.

Mit diesen Annahmen kommt BRICARD zu dem Gleichungssystem

$$\eta_{10} = \frac{4\pi a \cdot D_1}{I(0)}, \qquad \eta_{20} = \frac{4\pi a \cdot D_2}{I(0)}, \quad (18.3)$$

$$\eta_{11}^{p} = \frac{4\pi a \cdot D_1}{I(+p)}, \qquad \eta_{12}^{p} = \frac{4\pi a \cdot D_1}{I(-p)}, \quad (18.4)$$

$$\eta_{21}^{p} = \frac{4\pi a \cdot D_2}{I(+p)}, \qquad \eta_{22}^{p} = \frac{4\pi a \cdot D_2}{I(-p)}. \quad (18.5)$$

Hierin bedeuten D_1, D_2 die Diffusionskoeffizienten der positiven bzw. negativen Klein-Ionen, a der mittlere Radius der Kerne bzw. Groß-Ionen; $I(p)$ ist ein Integral, welches von BRICARD graphisch ausgewertet wurde aus den Beweglichkeits- und Diffusionskonstanten der Klein-Ionen, dem Radius a und der Ladungszahl p der Großionen; es nimmt für größere Ionenradien den Wert 1 an.

Da Groß-Ionen und Kerne nach ihrer Entstehung koagulieren, und zwar mit Halbwertszeiten, welche mit steigendem Radius der Teilchen stark zunehmen, ändern sich damit auch die Anlagerungskoeffizienten mit dem Alter der Kerne

[1] F. J. SCRASE: Geophys. Mem. Met. Off. London **1935**, No. 64.
[2] W. D. PARKINSON: Trans. of Oslo Meeting 1948 [*31*].
[3] J. J. NOLAN u. G. P. DE SACHY: Proc. Roy. Ir. Acad. A **37**, 71 (1927).
[4] P. LANGEVIN: Ann. de chém. et phys. **28**, 438 (1903).
[5] F. J. W. WHIPPLE: Proc. Phys. Soc. Lond. **45**, 367 (1933).
[6] J. BRICARD: J. Geophys. Res. **54**, 39 (1949).

und Groß-Ionen. Entsprechende Messungen haben NOLAN und FAHY[1] und NOLAN und KENNAN[2] ausgeführt. Aus den gemessenen Werten (Tabelle 8b u. c) erkennt man die Zunahme des Anlagerungs-Koeffizienten sowohl mit dem Alter der Kerne wie mit deren mittlerem Radius.

Tabelle 8b.
Resultierende *Anlagerungskoeffizienten*:

$$b_+ = \frac{\eta_{10} N_0 + \eta_{11} N_1 + \eta_{12} N_2}{Z}; \quad (Z = N_0 + N_1 + N_2).$$

NOLAN und FAHY (1945)[1].

Alter der Kerne Stunden	b_+	b_-
0—0,5	$2{,}4 \cdot 10^{-6}$	$2{,}9 \cdot 10^{-6}$
0,5—3	3,4	4,3
18—30	5,8	6,5
>48	8,4	9,2

Tabelle 8c. *Anlagerungskoeffizienten*:
NOLAN und KENNAN (1949)[2].

Radius (cm)	η_{10}	η_{12}
$1 \cdot 10^{-6}$	$0{,}28 \cdot 10^{-6}$	$2{,}7 \cdot 10^{-6}$
3	2,0	4,4
6	5,9	8,3
10	11,7	14,1
14	17,3	19,7

Diese Tabellen 8b und c zeigen den gemessenen und nach der Theorie zu erwartenden Zusammenhang zwischen Anlagerungskoeffizienten, dem Radius der Kondensationskerne und Groß-Ionen und der Verweilzeit des Aerosols; es wird so verständlich, daß Einzelmessungen nach Ort und Zeit unterschiedliche Werte ergeben.

19. Ausfall von Ionen mit Niederschlägen (Konvektionsstrom). Eine Anlagerung von Ionen an Nebel- und Niederschlagsteilchen findet teils aus denselben Ursachen wie an die Groß-Ionen und Kondensationskerne (Ziff. 18) statt, teils zusätzlich als Folge des in Ziff. 11 behandelten Influenzeffektes [*38*]. Durch diesen Effekt werden die Ionen eines Vorzeichens bevorzugt eingefangen. Die Ionen fallen dann mit den Teilchen aus und treten als Konvektionsstrom (Ziff. 34) am Erdboden auf. Auch auf Änderungen der Ionenzahl z.B. durch Spitzenentladungen reagiert der Konvektionsstrom merklich. So werden nach SIMPSON [*45*] die Ionen von Spitzenentladungen bei kräftigem Niederschlag wieder fast vollständig zur Erde zurückgeführt. Die Verminderung der Ionenzahl hängt also von deren Beweglichkeit, Ladung und Vorzeichen, vom vorhandenen elektrischen Feld und von der Stärke des Niederschlags ab.

Außer den Klein-Ionen werden auch Groß-Ionen und Kondensationskerne durch die Niederschläge teilweise aus der Atmosphäre entfernt. Mit den Kondensationskernen, an denen sich die Folgeprodukte der radioaktiven Emanationen abgelagert haben, verschwinden diese Ionisatoren zum Teil, was ebenfalls zu einer geringen Verminderung der Ionenzahl innerhalb der Atmosphäre führen kann.

20. Änderung der Ionenkonzentration durch den Leitungsstrom. In einem homogenen elektrischen Feld fließt, bei räumlich konstanter Ionenproduktion, ein stationärer *Leitungsstrom*, durch welchen an jeder Stelle die Zu- und Abfuhr der Ionen sich aufhebt, also die Ionenkonzentration konstant bleibt. Dagegen werden örtliche Unterschiede der Ionenkonzentration entstehen, wenn eine Divergenz des elektrischen Feldes besteht oder der Gradient der Ionenproduktion nicht verschwindet, also die Ionenproduktion sich stetig oder sprungweise ändert.

[1] P. J. NOLAN u. E. F. FAHY: Proc. Roy. Ir. Soc. A **50**, 233—260 (1945).
[2] P. J. NOLAN u. E. L. KENNAN: Proc. Roy. Ir. Soc. A **52**, 171—190 (1949).

Am deutlichsten wird dies an der Erdoberfläche, wo bei konstanter Ionenproduktion pro Volumeinheit zwar in der Luft direkt über dem Erdboden gleich viele positive und negative Ionen gebildet werden, die negativen bei positivem Potentialgradienten aber nach oben weglaufen ohne von unten ersetzt zu werden, während die positiven einerseits ihre Ladung am Erdboden abgeben, andererseits aber von oben her ersetzt werden. Das entstehende Defizit an negativen (positiven) Ionen bei positivem (negativem) Potentialgradienten wird *Elektrodeneffekt* genannt. Er ist nur bei Windstille durch Messung eindeutig nachweisbar (Ziff. 33β).

d) Das Ionisationsgleichgewicht und die Leitfähigkeit der Luft.

21. Theorie des Ionisationsgleichgewichtes. In atmosphärischer Luft verlaufen gleichzeitig die Vorgänge der Ionisierung (= Ionenproduktion) und der Ionenvernichtung. Für die zeitliche Änderung der Konzentration der Ionen einer Gruppe gelten Gleichungen von der Form

$$\frac{dn_1}{dt} = q_1 - \alpha n_1 n_2 - \eta_{12} n_1 N_2 - \eta_{10} n_1 N_0 - \eta_{11} n_1 N_1 + k_1 \frac{\partial (n_1 E)}{\partial h} \quad \text{usw.} \tag{21.1}$$

Ionisierung ↗ ; Kombination ↑ ; Anlagerung ↖ ↑ ↗ ; Elektrodeneffekt ↑

Da sich bei konstanter Ionisierung und Ionenvernichtung in Einstellzeiten von Minuten bis Viertelstunden[1] ein Gleichgewichtszustand ausbildet, genügt im allgemeinen die Kenntnis der Werte im *Gleichgewichtsfall*; diesen kann man ableiten, wenn man in den Gleichungen von der Form (21.1) die linken Seiten $dn/dt = 0$ setzt. Vernachlässigt wird meist der Einfluß des Leitungsstromes im Felde E, also der Elektrodeneffekt und die Vereinigung von Klein- und Groß-Ionen gleichen Vorzeichens; dann erhält man das Gleichungssystem:

$$q_1 = \alpha n_1 n_2 + \eta_{12} n_1 N_2 + \eta_{10} n_1 N_0, \tag{21.2}$$

$$q_2 = \alpha n_1 n_2 + \eta_{21} n_2 N_1 + \eta_{20} n_2 N_0, \tag{21.3}$$

$$Q_1 = \eta_{21} n_2 N_1 - \eta_{10} n_1 N_0, \tag{21.4}$$

$$Q_2 = \eta_{12} n_1 N_2 - \eta_{20} n_2 N_0, \tag{21.5}$$

$$Q_0 = \eta_{10} n_1 N_0 + \eta_{20} n_2 N_0 - \eta_{21} n_2 N_1 - \eta_{12} n_1 N_2. \tag{21.6}$$

Es bedeuten dabei

q_1, q_2: die Bildung von positiven (n_1) und negativen Klein-Ionen (n_2) je Kubikzentimeter und Sekunde.

Q_1, Q_2, Q_0: die Erzeugung oder Nachlieferung von Groß-Ionen N_1, N_2 und ungeladenen Kernen N_0 durch Verbrennungen, Rauch, Brandung und durch Lufttransport und anderes.

α, η: die entsprechenden Vereinigungs- und Anlagerungskoeffizienten.

Eine Wiedervereinigung zwischen Groß-Ionen entgegengesetzten Vorzeichens findet ebenfalls statt, aber so selten, daß diese Vereinigung bei den elektrischen Vorgängen keine Rolle spielt und nur bei der Koagulation von Kernen und

[1] C. G. STERGIS: J. Geophys. Res. **59**, 63—66 (1954).

Groß-Ionen von Interesse ist. Die entsprechenden Glieder lauten:

$$\frac{\partial N_1}{\partial t} = -\gamma N_1 N_2 = \frac{\partial N_2}{\partial t}, \tag{21.7}$$

wobei $\gamma \simeq 1{,}5 \cdot 10^{-9}\ \mathrm{cm^3\,sec^{-1}}$ angegeben wird[1,2].

Aus den theoretischen Ansätzen (s. Ziff. 18) lassen sich eine Anzahl Folgerungen ableiten, welche mit Beobachtungsergebnissen gut übereinstimmen:

1. In der Nähe des Erdbodens überwiegt die Anlagerung der Klein-Ionen an Kerne und Groß-Ionen. Damit erhalten die Gleichungen für das Ionisationsgleichgewicht die einfache Form:

$$q_1 = \eta_{12} n_1 N_2 + \eta_{10} n_1 N_0, \tag{21.8}$$

$$q_2 = \eta_{21} n_2 N_1 + \eta_{20} n_2 N_0. \tag{21.9}$$

Daraus ergibt sich dann ein linearer Zusammenhang zwischen Klein-Ionenzahl, Ionisierung und den Anlagerungskoeffizienten:

$$n_1 = \frac{q_1}{\eta_{12} N_2 + \eta_{10} N_0} \quad \text{usw.} \tag{21.10}$$

Diese Formel entspricht nahezu dem vereinfachten Wiedervereinigungsgesetz von SCHWEIDLER[3]

$$n = q/\beta, \tag{21.11}$$

in welchem β allerdings nicht konstant war.

2. Die Beziehung (21.10) läßt sich mit Hilfe der BRICARDschen Theorie noch weiter zurückführen auf Diffusionskonstanten und Kernradius. Man erhält so die Gleichungen:

$$n_1 \sim \frac{q_1}{4\pi \bar{a} \cdot D_1 \cdot Z}, \tag{21.12}$$

$$n_2 \sim \frac{q_2}{4\pi \bar{a} \cdot D_2 \cdot Z} \tag{21.13}$$

Z ist die Summe der Kerne und Groß-Ionen, also $= N_0 + N_1 + N_2$, $\bar{a}$ der effektive Radius der Kerne und Groß-Ionen, D_1, D_2 sind die Diffusionskoeffizienten der positiven bzw. negativen Klein-Ionen.

Experimentelle Untersuchungen des Zusammenhanges von Klein-Ionenkonzentration n und Kernkonzentration Z führten zu einer Beziehung[4]

$$n \sim 1/\sqrt{Z}.$$

Diese läßt sich deuten, wenn man einen Zusammenhang zwischen effektivem Kernradius $\bar{a}$ und Kernzahl Z annimmt[5]. Es müßte bei konstantem q in Gl. (21.12) und (21.13)

$$\bar{a} \sim 1/\sqrt{Z}$$

sein. Da hohe Kernzahlen nur in der Nähe der Kernquellen vorkommen, dort aber die Kerne noch nicht lange durch Koagulation anwachsen konnten, ist ein solcher Zusammenhang naheliegend.

---

[1] A. R. HOGG: Gerlands Beitr. Geophys. **41**, 32 (1934).
[2] H. KENNEDY: Proc. Roy. Ir. Acad. A **33**, 58 (1916).
[3] E. v. SCHWEIDLER: Wien. Ber. **127**, 953 (1918); **128**, 947 (1919); **133**, 23 (1924).
[4] J. J. NOLAN u. P. J. NOLAN: Proc. Roy. Ir. Acad. A **40**, 11—59 (1931).
[5] J. J. NOLAN u. V. H. GUERRINI: Proc. Roy. Ir. Acad. **43**, 5—24 (1953).

3. In *großer Höhe* über dem Erdboden oder über den Ozeanen wird das Ionisationsgleichgewicht wesentlich durch die Kombination der Klein-Ionen mitbestimmt. Aus der Ionisierung q und dem Kombinationskoeffizienten α läßt sich daher die Zahl der im Gleichgewicht vorhandenen Klein-Ionen n_1, n_2 errechnen.

Da $n_1 \approx n_2 = n$ ist, gilt $q = \alpha n^2$. Mit $\alpha = 1{,}6 \cdot 10^{-6}$ und q (Ozeane) $= 1{,}9$ erhält man

$$n = \sqrt{q/\alpha} = 1100 \text{ Ionenpaare/cm}^3. \tag{21.14}$$

Nahe der Meeresoberfläche muß, bei einem mittleren Gehalt von 500 bis 1000 Kernen und Groß-Ionen je Kubikzentimeter, auch die Anlagerung an diese berücksichtigt werden. So kommt man dann zu Klein-Ionenzahlen von etwa 500 bis 600/cm³ im Einklang mit Beobachtungen. Für 5 km Höhe ist der Vereinigungskoeffizient nur etwa

$$\alpha = 1 \cdot 10^{-6}; \tag{21.15}$$

für $q = 12$ Ionenpaare/cm³ sec erhält man dann eine Klein-Ionenzahl im Gleichgewicht von $n \approx 3500$ bei Abwesenheit von Kernen.

4. Weitere Folgerungen aus der Theorie von BRICARD sind: Das Verhältnis der Anlagerungskoeffizienten ist gleich dem Verhältnis der Diffusionskoeffizienten, nämlich

$$\frac{\eta_{10}}{\eta_{20}} = \frac{\eta_{12}}{\eta_{21}} = \frac{D_1}{D_2} = \text{const}. \tag{21.16}$$

Daraus ergibt sich unter Verwendung von Gl. (21.10) für das Verhältnis der positiven Klein-Ionen zu den negativen in kernhaltiger Luft, in welcher keine Raumladung vorhanden ist:

$$\frac{n_1}{n_2} = \frac{D_2}{D_1} = 1{,}1. \tag{21.17}$$

5. Aus der Diffusion der Klein-Ionen folgt ihre *mittlere Lebensdauer* τ. Sie kann am einfachsten errechnet werden aus der Ionisierung und der im Gleichgewicht vorhandenen Ionenzahl. Es ist:

$$\tau = n/q. \tag{21.18}$$

Für atmosphärische Luft in der Nähe des Erdbodens ergibt sich somit die mittlere Lebensdauer der positiven Ionen

$$\tau_1 = \frac{1}{\eta_{12} N_2 + \eta_{10} N_0} \tag{21.19}$$

und Entsprechendes für die negativen Klein-Ionen. Die mittlere Lebensdauer beträgt demzufolge zwischen etwa 30 bis 1000 Sekunden.

22. Elektrische Leitfähigkeit der Luft, Höhenabhängigkeit und Säulenwiderstand. Zur elektrischen Leitfähigkeit der Luft tragen in den unteren Luftschichten ausschließlich die vorhandenen *Ionen* bei. Die Leitfähigkeit hängt von der Zahl und der Beweglichkeit der Ionen ab. Bedeutet

Λ die Gesamtleitfähigkeit der Luft,

λ_+, λ_- die polaren Leitfähigkeiten,

n_{1p}, n_{2q} die Anzahl der positiven (1) und negativen (2) Ionen pro Kubikzentimeter, unterteilt in Gruppen (p, q) konstanter Beweglichkeit,

k_{1p}, k_{2q} die entsprechenden Beweglichkeiten [cm²/Volt sec],

e die elektrische Elementarladung,

dann ist:

$$\Lambda = \lambda_+ + \lambda_- = \sum^{p} n_{1p} \cdot e \cdot k_{1p} + \sum^{q} n_{2q} \cdot e \cdot k_{2q}. \tag{22.1}$$

Die Beweglichkeit der Klein-Ionen ist etwa 10^2 bis 10^5mal größer als die der Mittel- und Groß-Ionen. Deshalb wird die *elektrische Leitfähigkeit fast ausschließlich durch die Konzentration der Klein-Ionen* bestimmt. Die Meßmethoden sind in Ziff. 10 behandelt.

Da die für das Ionengleichgewicht entscheidenden Größen in Erdbodennähe, nämlich Ionisierungsstärke und Zahl der Kondensationskerne und Groß-Ionen,

Tabelle 9. *Mittlere elektrische Leitfähigkeit und Ionenzahl am Erdboden.*

Ort	Kernzahl Z cm^{-3}	Leitfähigkeit $\Omega^{-1}\,cm^{-1}$	Zahl der Klein-Ionenpaare $n_{1,2}$ cm^{-3}	tägliche Variation
Großstädte	10^5 $(5\cdot 10^3 - 5\cdot 10^6)$	$0{,}5\cdot 10^{-16}$	150 (30—1000)	groß
Kleinstädte	$3\cdot 10^4$ $(10^3 - 5\cdot 10^5)$	$1\cdot 10^{-16}$	250 (50—1000)	mittel
Offenes Land	10^4 $(10^2 - 10^5)$	$2{,}0\cdot 10^{-16}$	500 (100—1000)	schwach
Ozeane	10^3	$2{,}4\cdot 10^{-16}$	600 (400—1000)	fast Null

stark mit Zeit, Ort und Höhe variieren, ändert sich damit ebenso die Leitfähigkeit in den unteren Schichten stark. Einige Mittelwerte für die Klein-Ionenzahl und die elektrische Leitfähigkeit der Luft sind in Tabelle 9 enthalten.

Durch die vom Menschen beeinflußte Kern- und Groß-Ionen-Produktion im Verlauf des Tages erhält die elektrische Luftleitfähigkeit einen typischen Tagesgang mit einem Maximum in der späten Nacht und ein oder zwei Minima am Morgen und am Nachmittag, zwischen welchen sich noch ein schwächeres Maximum zur Zeit des größten Vertikalaustausches am Mittag ausbildet (Fig. 13).

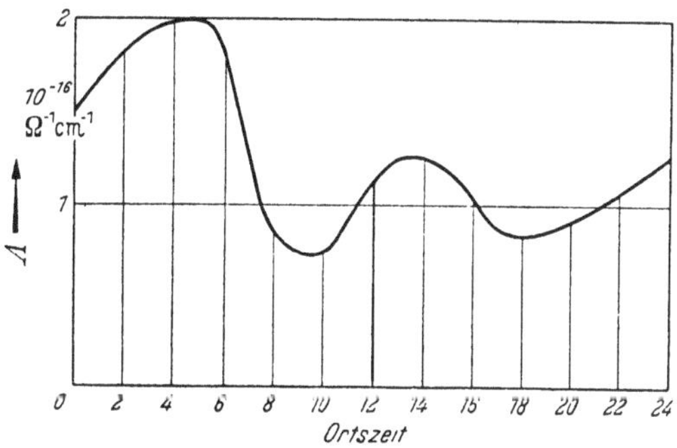

Fig. 13. Mittlerer Tagesgang der elektrischen Leitfähigkeit Λ der Luft in besiedelten Gegenden.

Mit der *Höhe* nimmt die elektrische Leitfähigkeit der Luft zuerst langsam, dann stark zu. Die Werte auf hohen Bergen in Bodennähe bzw. in der freien Atmosphäre in gleicher Höhe sind verschieden; am Boden sind die Werte höher, da außer der kosmischen Strahlung auch die radioaktive Strahlung vom Boden aus Ionen erzeugt.

Die elektrische Leitfähigkeit in der freien Atmosphäre ist gelegentlich von Freiballonen[1], von Flugzeugen aus oder mit selbstregistrierenden Geräten gemessen worden. Die Ergebnisse streuen, da Ionenerzeugung und -vernichtung durch den vertikalen Austausch stark beeinflußt werden. In Fig. 14 ist die Höhenabhängigkeit der elektrischen Leitfähigkeit nach einer empirischen Formel von Gish und Wait[2] graphisch dargestellt:

$$\Lambda = \Lambda_0 + A\cdot h^2 \tag{22.2}$$

in welcher Λ die elektrische Gesamtleitfähigkeit, Λ_0 ihr Wert am Boden, $A = 0{,}24$ und h die Höhe in der freien Atmosphäre in Kilometer bedeuten.

[1] A. Wigand: Ann. Phys. **66**, 81 (1921).
[2] O. H. Gish u. G. R. Wait: J. Geophys. Res. **55**, 473 (1950).

Die außerdem wiedergegebenen Meßresultate von SAGALYN und FAUCHER[1] von Flugzeugen aus zeigen die großen Schwankungen in den unteren Luftschichten, verursacht durch die Luftbeimengungen; der Kern- und Großionen-Gehalt ist bis zu Höhen, zu welchen der vertikale Luftaustausch reicht, groß, so daß die Zahl der Klein-Ionen klein ist und nur von Z [nach Gl. (21.10)] abhängt. Erst über dem Austauschniveau stimmen die Meßergebnisse mit der theoretisch aus der Kombination von Klein-Ionen sich ergebenden Konzentration $n=\sqrt{q/\alpha}$ weitgehend überein.

Von der Stratosphäre liegen nur einige wenige Ergebnisse von Messungen der elektrischen Leitfähigkeit der Luft vor. MACDONALD[2] weist auf einige

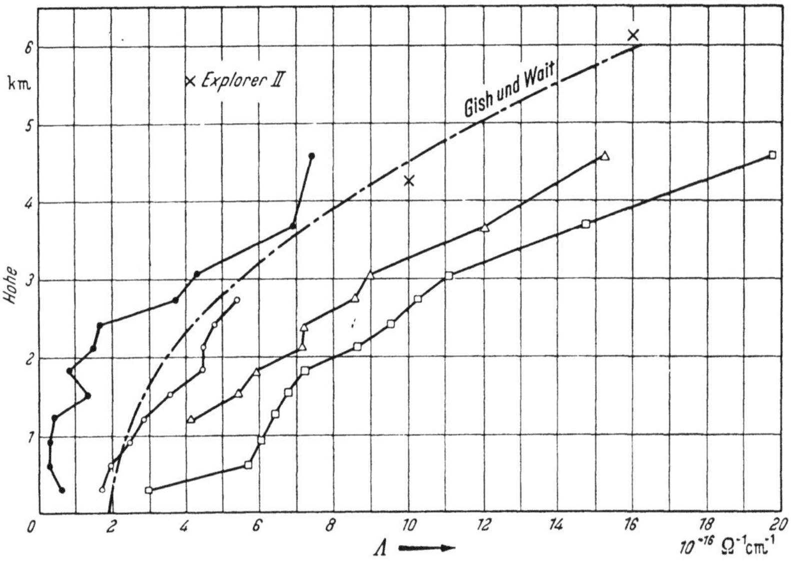

Fig. 14. Verlauf der elektrischen Gesamtleitfähigkeit in der freien Atmosphäre nach einer empirischen Formel von GISH und WAIT, nach Messungen im Flugzeug (CALLAHAN-SAGALYN und FAUCHER), und mit dem Ballon Explorer II.

Abweichungen und Unstetigkeiten des erwarteten Verlaufs mit der Höhe hin, welche noch ungeklärt sind. In neuester Zeit wurden diese Meß-Ergebnisse durch Radiosonden-Aufstiege bestätigt[3].

Rechnerische Ansätze für die Höhenabhängigkeit auf Grund der Ionisierungsstärke der Ultra-Strahlung und des Wiedervereinigungskoeffizienten sind mehrfach gemacht worden[4,5] und geben Anhaltspunkte für die Leitfähigkeit. Eine Unsicherheit ist aber die mit der Höhe und der Reinheit der Luft sich verändernde Beweglichkeit der Ladungsträger. Dazu kommt, daß die effektive Beweglichkeit der negativen Träger im ganzen stark mit abnehmendem Druck zunimmt; dies erklärt sich durch die mit abnehmendem Druck wachsende Lebensdauer freier Elektronen. Berücksichtigt man dieses, dann kommt man nach ISRAEL zu Werten der Leitfähigkeit, wie sie die Kurve 2 in Fig. 15 wiedergibt. Mit Hilfe von Echos von kurzwelliger Radiostrahlung haben GARDNER und PAWSEY[6]

[1] R. C. CALLAHAN-SAGALYN u. G. A. FAUCHER: J. Atmos. a. Terres. Phys. **5**, 253—272 (1954).

[2] J. P. MACDONALD: Trans. Amer. Geophys. Un. **34**, 529—533 (1953).

[3] L. KOENIGSFELD: Inst. Roy. Met. Belg., Publ. Ser. B **1955**, Nr. 18.

[4] E. LENZ: Hochfrequenztechn. **43**, 47—51 (1934).

[5] H. ISRAEL u. H. W. KASEMIR: Ann. Géophys. **5**, 313—324 (1949).

[6] F. F. GARDNER and J. L. PAWSEY: J. Atmos. a. Terres. Phys. **3**, 321—344 (1953).

in Höhen unter 100 km Anhaltspunkte gewonnen für die Konzentration freier Elektronen. Man darf wohl erst ab 65 km Höhe mit einem je nach Tageszeit verschiedenen Verlauf und einem raschen Anstieg ihrer Dichte rechnen. Nach diesen Meßresultaten ist ein Verlauf der Leitfähigkeit nach Kurve 3 wahrscheinlicher.

Für das luftelektrische Gesamtproblem, nämlich für den luftelektrischen Haushalt, interessiert der *Gesamtwiderstand* zwischen Erdoberfläche und luftelektrischer Ausgleichsschicht. Man nennt den Widerstand einer senkrechten

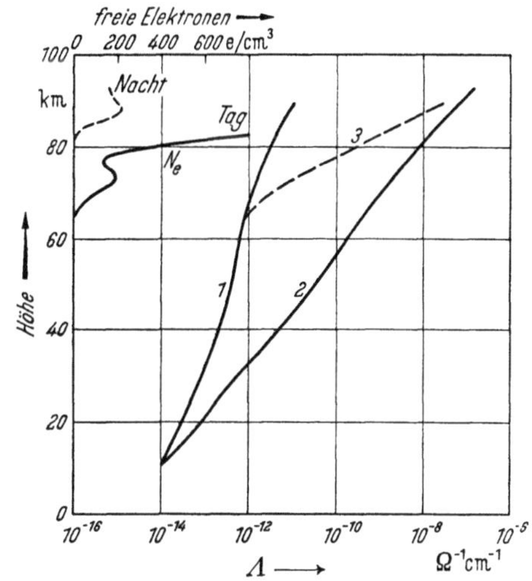

Fig. 15.

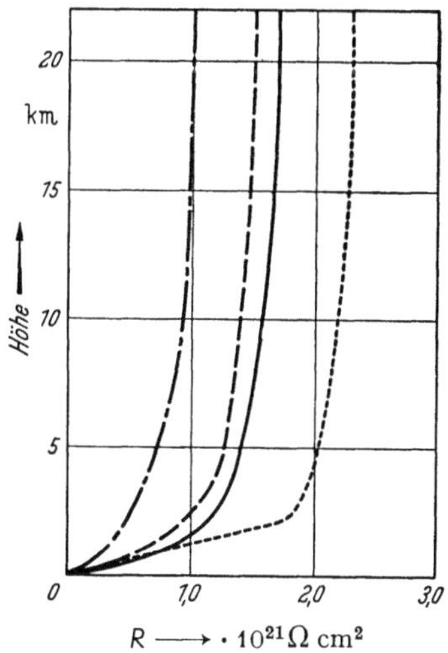

Fig. 16.

Fig. 15. Elektrische Leitfähigkeit der Luft in der Stratosphäre. Kurve 1 und 2 nach Rechnungen von ISRAEL und KASEMIR auf Grund der Ionisierung durch Ultrastrahlung bei Berücksichtigung von Druck und Temperatur in der Atmosphäre bei einem angenommenen Zusammenhang zwischen Vereinigungskoeffizient α und Druck p: $\alpha \sim p^{0,3}$. Kurve 1 ohne Berücksichtigung freier Elektronen, Kurve 2 bei Berucksichtigung freier Elektronen; Kurve 3 berücksichtigt eine gefundene Elektronenverteilung durch Radarechos.

Fig. 16. Säulenwiderstände errechnet aus Flugzeugmessungen der elektrischen Leitfähigkeit der Luft von R. CALLAHAN-SAGALYN und G. A. FAUCHER.

Luftsäule von 1 cm² Fläche vom Erdboden ausgehend den *Säulenwiderstand* der Luft[1,2]. Es ist also:

$$R = \int_0^h r\,dh = \int_0^h \frac{1}{\Lambda}\,dh. \tag{22.3}$$

Hierin bedeutet $r = 1/\Lambda$ den spezifischen Widerstand der Luft.

Dieser Säulenwiderstand ist von der Größenordnung $10^{21}\,\Omega\,cm^2$; der Hauptanteil desselben liegt aber in den unteren Luftschichten der Troposphäre. Eine Zusammenstellung von CALLAHAN-SAGALYN und FAUCHER[3] zeigt den Verlauf des Säulenwiderstandes bei einigen Flügen und bei dem Ballonaufstieg Explorer II, errechnet aus Messungen der Leitfähigkeit (Fig. 16). Man erkennt daraus die wichtige Tatsache, daß der Säulenwiderstand je nach der Luftreinheit in den unteren Luftschichten mehr als um den Faktor 2 variieren kann.

23. Die Ionosphäre [*22*] und [*33*]. In der Atmosphäre über 20 km Höhe ionisieren außer der kosmischen Ultrastrahlung auch die kurzwelligen Sonnenstrahlen.

[1] G. R. WAIT: Terr. Magn. **47**, No. 3 (1942).

[2] O. H. GISH: Terr. Magn. **49**, 159—168 (1944).

[3] R. C. CALLAHAN-SAGALYN u. G. A. FAUCHER: J. Atmos. a. Terres. Phys. **5**, 253—272 (1954).

Weil die Absorption dieser Strahlung mit zunehmender Luftdichte stärker wird, die Intensität aber dadurch abnimmt, entstehen Schichten, normalerweise die E- und F-Schicht, mit einem Maximum der Ladungsträger je Volumeneinheit in bestimmten Höhen, welche noch von der Art des Absorptionsprozesses abhängen. Einfallswinkel der Strahlen, Kombination der Ionen und Elektronen, Vertikalaustausch in der Atmosphäre und anderes spielen dabei eine Rolle. Die Maxima und Minima dieser Schichten variieren sehr stark mit der Tageszeit und außerdem beim Eintritt außergewöhnlicher Strahlung in die Erdatmosphäre, also bei Meteorfall und nach chromosphärischen Eruptionen auf der Sonne. Dann entsteht vorübergehend noch eine tieferliegende D-Schicht, und die anderen Schichten können aufspalten. Rohe Anhaltswerte über die Konzentration der Elektronen als Funktion der Höhe vermittelt die Fig. 17.

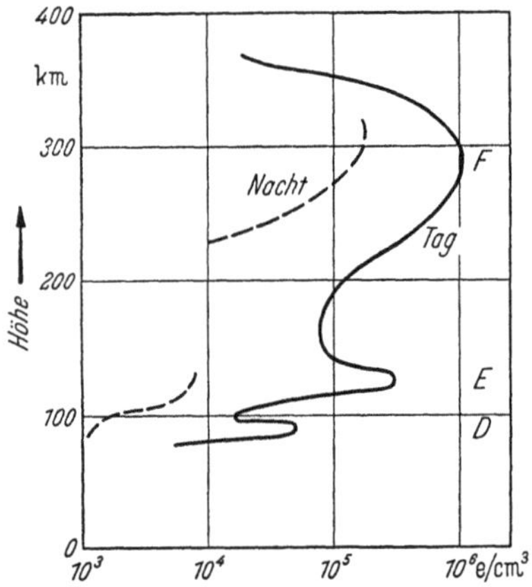

Fig. 17. Mittlere Konzentration der freien Elektronen in der Ionosphäre.

Diese Schichten werden heute durch Raketenaufstiege [25] erforscht, daneben aber laufend untersucht und überwacht durch Echolotungen[1] mittels kurzer Radiowellen, welche je nach Wellenlänge, Einfallswinkel und Trägerkonzentration reflektiert werden. Diese Methoden liefern heute schon viele Einzelheiten, von denen einerseits die Zusammensetzung der hohen Atmosphäre und die photochemischen Prozesse, andererseits die Horizontal- und Vertikalbewegungen in der Ionosphäre besonders interessieren. *Ein Einfluß der Vorgänge in ihr auf die luftelektrischen Erscheinungen am Erdboden ist bis jetzt nicht bekannt geworden.* Für die Ausbreitung von Radiowellen im Bereich der Kurz-, Mittel- und Langwellen dagegen ist der Zustand der Ionosphäre von großer Bedeutung, und deshalb auch für die Ausbreitung der Funkstörgeräusche von fernen Wolkenentladungen (Blitzen), der Atmospherics[2,3] (vgl. S. 587 oben).

C. Das luftelektrische Feld.

24. Messungen des luftelektrischen Feldes. Der luftelektrische Potentialgradient

$$E = \partial V/\partial h$$

oder die (nach unten gerichtete) Feldstärke

$$-|\mathfrak{E}| = \partial V/\partial h \qquad (24.1)$$

ist im freien Gelände unter Schönwetterbedingungen von der Größenordnung $+1$ V/cm $= +100$ V/m.

α) Zur Feldstärkenmessung an der Oberfläche der Erde oder ausgedehnter Körper kann man nach einer von WILSON[4] eingeführten Methode die elektrische Flächenladungsdichte σ auf einer dem Feld ausgesetzten, an der Erdoberfläche isoliert angebrachten Platte messen und erhält daraus den Potentialgradienten

$$E = -\sigma/\varepsilon_0 \qquad (\varepsilon_0 = 8{,}86 \cdot 10^{-12}\ \text{Coulomb/Volt} \cdot \text{m}) \qquad (24.2)$$

[1] Kapitel Ionosphäre in Bd. IL dieses Handbuches.

[2] W. O. SCHUMANN: Z. angew. Phys. 6, 346—352 (1954).

[3] K. G. BUDDEN: Phil. Mag. (7) 43, 1179—1200 (1952).

[4] C. T. R. WILSON: Proc. Cambridge Phil. Soc. 13, 363 (1906). — Phil. Trans. Roy. Soc., Ser. A 221, 13—115 (1921).

Die Fig. 18 zeigt die in Höhe der Erdoberfläche angebrachte „WILSON-*Platte*", welche durch eine bewegliche größere Platte abgedeckt und freigegeben werden kann. Zu Beginn einer Messung wird die dem elektrischen Felde ausgesetzte Platte geerdet, dann die Erdung getrennt und die Platte abgedeckt. Die am Elektrometer abgelesene Spannung U ist danach

$$\left.\begin{aligned} U &= \frac{Q}{C_g} = \frac{\sigma\cdot F}{C_g} \\ &= \frac{-\varepsilon_0\cdot E\cdot F}{C_g}. \end{aligned}\right\} \tag{24.3}$$

F ist dabei die Fläche der WILSON-Platte, C_g ihre Gesamtkapazität.

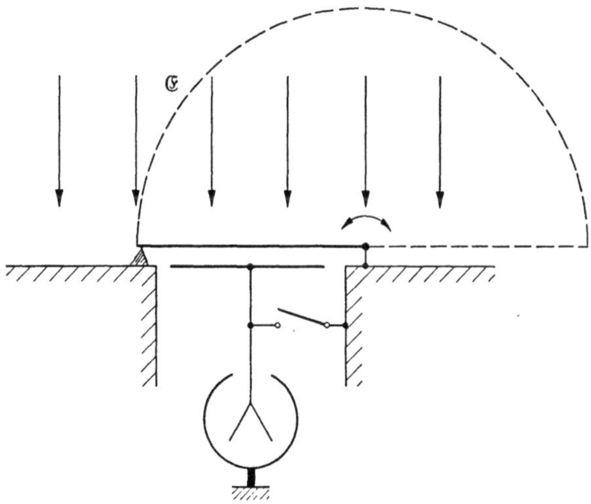

Fig. 18. WILSON-Platte zur Messung der luftelektrischen Feldstärke.

β) Eine Weiterentwicklung dieser WILSONschen Methode sind die *Feldmühlen*[1,2,3] in den verschiedensten Ausführungen. Bei ihnen wird z. B. eine feststehende isolierte Platte durch eine drehbare Scheibe mit Öffnungen periodisch abgedeckt und freigegeben. Es können aber auch ein oder zwei, auf der Achse eines Elektromotors befindliche, isolierte Zylindersegmente umlaufen, wobei laufend das eine oder andere Segment dem elektrischen Feld ausgesetzt wird. Das Prinzip geht aus Fig. 19 hervor. Leitet man die Oberflächenladung über einen hohen Widerstand R ab, dann entsteht an ihm eine Wechselspannung mit der Frequenz der Abdeckungen. Ist Q_0 die Oberflächenladung bei freier Platte, F_0 ihre Fläche, dann variiert die Oberflächenladung Q mit der Umdrehungsfrequenz,

$$Q = Q_0\cdot\sin\omega t, \tag{24.4}$$

wobei

$$Q_0 = -\varepsilon_0\cdot E\cdot F_0 \tag{24.5}$$

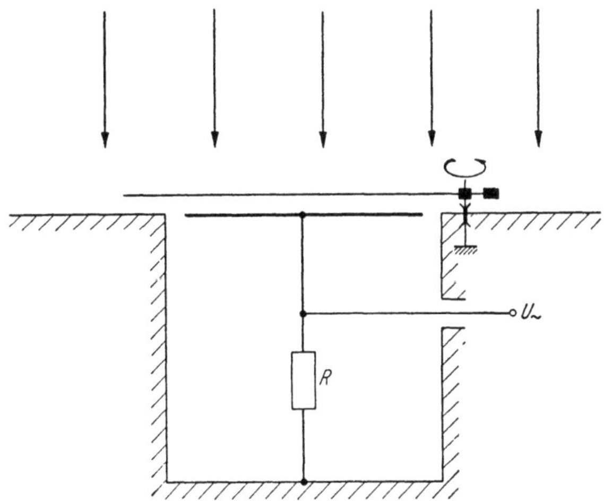

Fig. 19. Prinzipanordnung zur Feldmühle.

ist. Die Wechselspannung ergibt sich aus

$$U = R\frac{dQ}{dt} = -\varepsilon_0\cdot\omega\cdot E\cdot F_0\cdot R\cos\omega t. \tag{24.6}$$

[1] H. LUEDER: Meteor. Z. **60**. 340–351 (1943).
[2] H. W. KASEMIR: Tellus **3**, 240 (1951).
[3] B. F. J. SCHONLAND: Proc. Phys. Soc. **63**, 402 (1950).

Für die Verwendbarkeit dieser Formel, aus welcher E ermittelt werden kann, ist Voraussetzung, daß die Zeitkonstante

$$\tau = R \cdot C \quad \text{kleiner als } 1/\omega \text{ ist;} \tag{24.7}$$

C ist dabei die Kapazität der mit R verbundenen Platte und der Anschlußleitungen. Im allgemeinen wird man eine Feldmühle empirisch eichen.

γ) Am häufigsten wird die luftelektrische Feldstärke aus *Potentialdifferenzen* bestimmt. Für ihre Messung am Erdboden genügt dabei die Messung des Potentials V in *einer* bestimmten Höhe h über dem Boden, wobei dann

$$E = V/h \tag{24.8}$$

gleich dem Mittelwert im Höhenbereich 0 bis h ist. In der freien Atmosphäre, z.B. von Ballonen aus, muß sinngemäß das Potential in zwei verschiedenen Höhen gemessen werden.

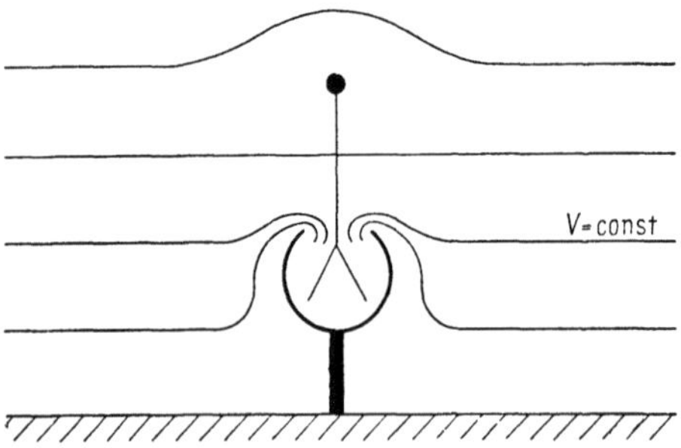

Fig. 20. Zur Erklärung der Wirkungsweise radioaktiver Potentialsonden.

Aus den Gesetzen der Elektrostatik leiten sich die Grundlagen für die üblichen Methoden der Messung von Potentialen in einem Medium sehr geringer Leitfähigkeit wie in Luft ab.

1. Bringt man einen ungeladenen Körper freischwebend in ein elektrisches Feld, dann nimmt er sofort das vorher ungestörte Potential seines Ortes an. Durch Influenzwirkung des Feldes entsteht auf seiner Oberfläche die elektrische Ladungsverteilung eines Dipols.

2. Verbindet man denselben Körper, beispielsweise eine Kugel, mit Erde, dann lädt sich dieser Körper, entsprechend der nunmehr überhöhten Feldstärke an seiner Oberfläche auf:

$$Q = \int_F \sigma \, dF, \quad \text{mit} \quad \sigma = -\varepsilon_0 \cdot E. \tag{24.9}$$

3. Ist der Probekörper, z.B. eine Kugel, mit einem statischen Voltmeter verbunden (Fig. 20), dann wird dieses zunächst einen Wert des Potentials anzeigen, der zwischen dem des vorher ungestörten Ortes der Kugel und dem des Gehäuses des Voltmeters liegt. Auch jetzt besteht eine überhöhte Feldstärke an der Kugeloberfläche und deshalb eine vermehrte Oberflächenladungsdichte.

4. Vergrößert man die Oberfläche des Probekörpers um ein Vielfaches in Richtung der ungestörten Äquipotentialfläche, so wird die auf ihm befindliche Oberflächenladung allmählich durch die Ionen der Luft abgeleitet. Das Voltmeter zeigt dann ein gegenüber dem Fall 3 höheres Potential an.

5. Man erhält das Potential der ungestörten Äquipotentialfläche, in welcher sich der Probekörper befindet, wenn man durch eine Erhöhung der Luftleitfähigkeit für schnelle und vollständige Ableitung seiner Oberflächenladung sorgt. Diese erreicht man mit radioaktiven *Kollektoren*[1,2], welche von EXNER eingeführt wurden. Am besten verwendet man reine α-Strahler kurzer Reichweite, also z.B. Polonium. Früher waren Flammenkollektoren, Zerstäuberkollektoren oder Wassertropfkollektoren im Gebrauch.

[1] A. WIGAND u. H. KIRCHNER: Gerlands Beitr. Geophys. **17**, 379 (1927).
[2] H. ISRAEL: Arch. techn. Messen **5**, 656—3.

δ) Bei der Forderung nach einer kleinen Einstellzeit wird man starke radioaktive Präparate verwenden, welche eine Aktivität von mindestens 0,01—0,1 mC haben. Da diese Kollektoren selbst Ionen erzeugen, welche bei Trennung im elektrischen Erdfeld zu Raumladungen führen, können nach Untersuchungen von MÜHLEISEN[1] durch sie erhebliche Fehler bei der Potentialmessung entstehen, besonders bei starken Präparaten und bei Windstille. Deshalb sollte ein Kollektor von der geerdeten Umgebung möglichst weit entfernt sein, mindestens 0,5 bis 1 m. Dann liegen an ihm aber normalerweise Potentiale um 100 V, bei Niederschlägen ein Vielfaches davon. Dieses Potential des Kollektors kann mittels Elektrometer oder statischer Voltmeter gemessen werden.

Zur *Registrierung* des luftelektrischen Potentials werden aber vielfach Elektrometerröhren benutzt; der geringe aussteuerbare Bereich der Eingangsspannung erfordert dann eine Unterteilung der Spannung des Kollektors durch Höchstohm-Widerstände[2]. Diese Widerstände in der Größenordnung bis etwa $10^{12}\,\Omega$ belasten den Kollektor, was zu Meßfehlern führen kann. Die genannten zwei Fehler lassen sich stark vermindern, wenn man ein Spezial-Röhrenvoltmeter nach Fig. 21 verwendet, welches den Kollektor praktisch nicht belastet[3]. Dieses erlaubt die Verwendung eines schwachen Kollektors und bringt trotzdem eine Verbesserung der Einstellzeit und der wirksamen Isolation dadurch, daß man Schutzringe um die Isolatoren automatisch nahezu auf dem Potential des Kollektors mitführen kann.

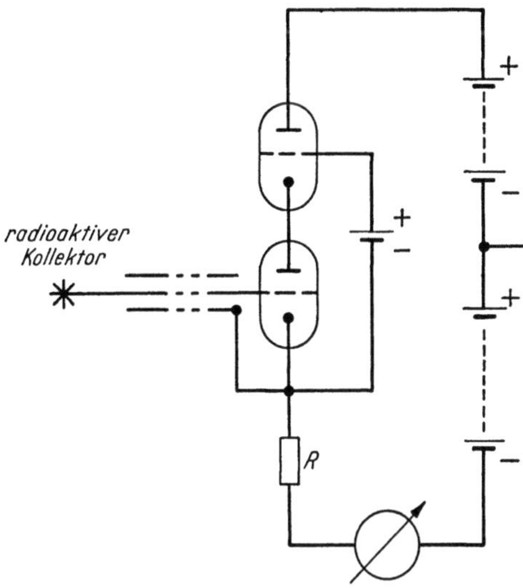

Fig. 21. Prinzipanordnung zu luftelektrischen Potentialmessungen mit radioaktivem Kollektor und Röhrenvoltmeter nach EHMERT und MÜHLEISEN.

In gewissen Fällen kann eine Schaltung nach KOENIGSFELD und PIRAUX[4] günstig sein, bei der das zu messende Potential an die Anode einer Triode gelegt wird und damit der auf das schwach positive Gitter fließende Elektronenstrom gesteuert wird.

Dauerregistrierungen des luftelektrischen Potentialgradienten wurden früher mit Registrier-Elektrometern ausgeführt. Heute erlauben die Elektrometerröhren oder Röhrenvoltmeter-Schaltungen den direkten Anschluß handelsüblicher Schreiber.

ε) Im allgemeinen wird eine *Reduktion* der Potentialanzeigen auf das *ebene Feld* erwünscht sein, da die Potentialkollektoren vielfach an Gebäuden oder in unebenem, teilweise bewachsenem Gelände aufgestellt werden müssen. Einzelne Bäume, Häuser oder Masten beeinflussen das elektrische Potential in der Ebene in einer Entfernung, welche größer als ihre Höhe ist, praktisch nicht mehr. Der Einfluß durch Bergrücken oder Täler wurde von BENNDORF[5] rechnerisch erfaßt.

Zur Messung des *Reduktionsfaktors* wird auf einer kurzgeschnittenen Wiese oder auf einem anderen ebenen Gelände, in möglichst geringer Entfernung vom

[1] R. MÜHLEISEN: Z. Naturforsch. **6**a, 667—671 (1951).
[2] K. BURKHART: Z. Meteor. **1**, 212 (1947).
[3] A. EHMERT u. R. MÜHLEISEN: Z. angew. Phys. **5**, 43—47 (1953).
[4] L. KOENIGSFELD u. P. L. PIRAUX: Mémoires Inst. Roy. Mét. Belg. **45**, 1—25 (1951).
[5] H. BENNDORF: Sitzgsber. Akad. Wiss. Wien, Abt. IIa **109**, 923 (1900).

Ort der zu reduzierenden Potentialmessung eine transportable Potential-Meßeinrichtung aufgebaut. Früher wurde dazu ein in konstanter Höhe von z. B. 2 m ausgespannter langer Draht mit einem Elektrometer oder statischen Voltmeter verbunden. In der Mitte des Drahtes war ein radioaktives Präparat befestigt. Es genügt aber, wenn an einer Stange über dem statischen Voltmeter der radioaktive Kollektor aufgesteckt ist, womit das ungestörte Potential in der Höhe des Kollektors erhalten wird. Man prüft die Genauigkeit des so erhaltenen Reduktionsfaktors durch Vergleich über längere Zeiten bei nicht zu schwachem Wind.

ζ) Eine besondere Schwierigkeit bei der Dauerregistrierung der luftelektrischen Größen ist die Erhaltung guter *Isolation*. Durch Heizung und Abschirmung der Isolatoren und Verwendung guten Isoliermaterials, z. B. Plexiglas, kann man zwar den Einfluß hoher Feuchtigkeit auch bei Niederschlägen beseitigen. Darüber hinaus können die Isolatoren noch durch Staub und Rauch allmählich verschmutzen oder durch Spinnweben usw. überbrückt werden.

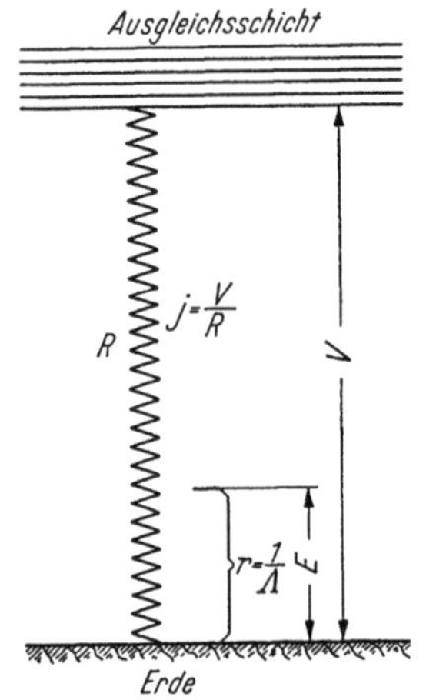

Fig. 22. Zur Veranschaulichung der Messung des Potentialgradienten.

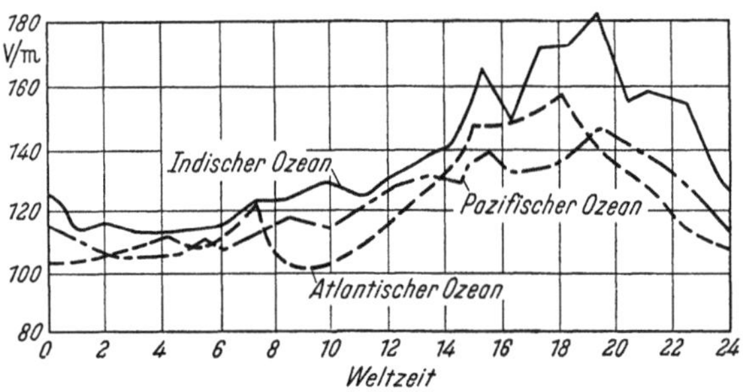

Fig. 23. Täglicher Gang des Potentialgefälles nach Weltzeit. CARNEGIE 1915 bis 1921.

25. Das Schönwetterfeld über See und Land. Die Ursache für das Schönwetterfeld ist die positive Aufladung der Ausgleichsschicht in über 50 km Höhe durch die Gewitter und die negative Ladung an der Erdoberfläche. Als Folge der Spannung V zwischen dieser Schicht und Erde, in der Größenordnung von 400000 V, fließt ein vertikaler Leitungsstrom j durch die Atmosphäre, und an jedem Teilstück des Säulenwiderstandes R entsteht dabei nach Fig. 22 eine Potentialdifferenz. In erster Näherung ist somit die Feldstärke gleich dem Spannungsabfall an der Einheitslänge des Säulenwiderstandes 1 cm (oder 1 m) und damit

$$E = -j \cdot r = \frac{-j}{\Lambda} = \frac{V}{R\Lambda}. \qquad (25.1)$$

α) Über den Ozeanen ändert sich im Laufe eines Tages sowie im jährlichen Verlauf die Leitfähigkeit der Luft praktisch nicht, und deshalb ist die Feldstärke proportional zur Spannung der Ausgleichsschicht V. Nach den Untersuchungen mit der Carnegie von BAUER und SWANN [2] und anderen hat der Potentialgradient über dem Meere einen täglichen Gang mit einem Minimum um 4 Uhr GMT und einem Maximum um 18 Uhr GMT, wie in Fig. 23 für die Ozeane dargestellt (vgl. auch Fig. 1, S. 543).

Einen um etwa 20% verminderten Wert des Potentialgradienten am Äquator gegenüber größeren Breiten über 40° N und S glaubt GISH[1] gefunden zu haben;

[1] O. H. GISH: Terr. Magn. 47, 323 (1942).

die Ursache, der Breiteneffekt der Ultrastrahlung (= Abnahme zum Äquator hin), führt zu einem größeren Säulenwiderstand R über den äquatorialen Breiten und entsprechend zu einem verminderten Vertikalstrom und Potentialgradienten.

β) Über wenig besiedeltem Land hat die Luftleitfähigkeit Λ einen täglichen Gang mit einem Maximum in den späten Nachtstunden. Mit Sonnenaufgang und mit Beginn menschlicher Tätigkeit nimmt, zuerst in niedrigen Höhen, die Leitfähigkeit der Luft ab, was durch die Vermehrung der Kondensationskerne und den Austausch nach Erwärmung der Bodenschichten hervorgerufen wird. Im ganzen steigt damit gemäß Gl. (25.1), der Potentialgradient E im Laufe des Tages an und erreicht ein Maximum, je nach der Verdichtung der Luftverunreinigungen. Ein Beispiel zeigt die Fig. 24a. Normale Werte des Potentialgradienten über Land sind 60 bis 100 V/m bei Nacht und 80 bis 150 V/m bei Tage.

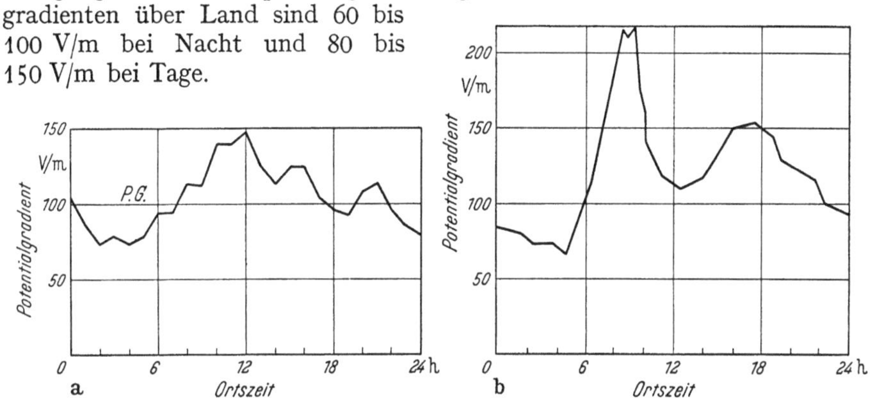

Fig. 24a u. b. Typische Tagesgänge des Potentialgradienten, a in offener Landschaft, b in dicht besiedelter Gegend.

Versuche, einen Tagesgang, der ähnlich wie auf den Ozeanen verläuft, auf hohen Berggipfeln zu finden, unternahmen unter anderen HOLZER[1] und ISRAEL[2]. Die Messungen zeigen aber, daß auch dort der Vertikalaustausch Änderungen der Luftleitfähigkeit bewirkt und örtliche Einflüsse auftreten können.

γ) In stärker besiedeltem Gebiet und in Städten sind die Schwankungen der luftelektrischen Feldstärke größer. Man beobachtet ein- und zweifach-periodische Tagesgänge mit Maxima am Vormittag und eventuell am Nachmittag und Minima in der späten Nacht und am Mittag. Dies entspricht teilweise dem Verlauf der Feldstärke über offenem Land, unterscheidet sich aber durch größere Amplituden und durch die mittägliche Einsattelung. Die größeren Amplituden entstehen durch die stärkere Leitfähigkeitsverminderung und durch künstliche positive Raumladungen in den untersten Luftschichten [*11*]; die Einsattelung am Mittag, welche besonders an heiteren Sommertagen gefunden wird, hat ihre Ursache in dem kräftigen Vertikalaustausch durch Sonneneinstrahlung. Dieser Austausch verteilt die leitfähigkeits-vermindernden Kondensationskerne und die positiven Raumladungen in einem größeren Höhenbereich und führt gleichzeitig dem Boden reinere Luft zu. So kommt es zur Vergrößerung des Säulenwiderstandes, zugleich aber auch der Leitfähigkeit am Boden, somit also zu einer Abnahme der Feldstärke um die Mittagszeit nach Fig. 24b.

---

[1] R. E. HOLZER: University of California. 1955.

[2] H. ISRAEL, H. W. KASEMIR u. K. WIENERT: Arch. Met., Geophys. u. Biokl., Ser. A 3, 357—381 (1951); A 8, 72—94 (1955).

Schönwetterwerte in Städten sind 60 bis 200 V/m bei Nacht und 100 bis 700 V/m bei Tage. Die Fig. 25 vermittelt einen Eindruck von dem Tagesverlauf der Potentialgradienten in einer Stadt. Die verschiedenen Verläufe von Tagesgängen des luftelektrischen Potentialgradienten sind zum Teil korreliert mit meteorologischen Faktoren, wie dem Austausch[1], dem Wasserdampfgehalt[2], der Sichtweite und anderem[3].

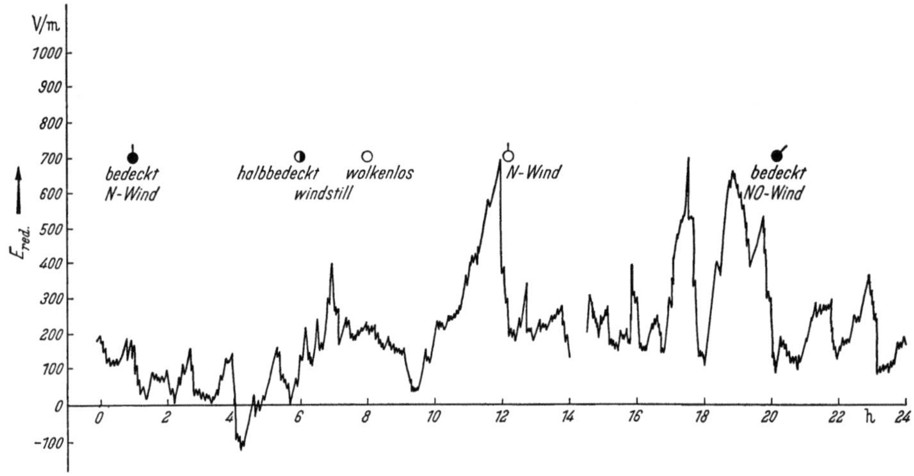

Fig. 25. Typischer Tagesverlauf des Potentialgradienten an einem Februartag in einer Großstadt.

26. Elektrische Felder bei Bewölkung und Nebel. Auch die luftelektrischen Verhältnisse bei Nebel und Bewölkung sind leicht zu verstehen mit Hilfe der am Eingang von Ziff. 25 gegebenen Übersicht und der Gl. (25.1). Verschiedene Messungen haben ergeben, daß die Luftleitfähigkeit in Wolken und im Nebel um den Faktor 2 bis 5,5, im Mittel etwa 3, herabgesetzt ist. Ist die Nebelschicht also von geringer Höhenausdehnung, dann ändert sich damit der Vertikalstrom j kaum; bei $j=\text{const}$ muß dann der Potentialgradient $E\sim 1/\Lambda$ sein, also um den Faktor 3 im Nebel ansteigen. Dasselbe findet man in einer nicht zu dicken, aber seitlich weit ausgedehnten Wolkenschicht.

Anders sind die Verhältnisse, wenn die Wolken- oder Nebelschicht sich über größere Höhen erstreckt. Dann wird der Säulenwiderstand R erheblich vergrößert, damit sinkt die Vertikalstromdichte; der Potentialgradient nimmt dann ungefähr normale Werte an. Messungen von ISRAEL und KASEMIR[4] bestätigen die zu erwartenden Feldverhältnisse in Nebeln; KASEMIR[5] hat die Felder für spezielle Wolkenformen berechnet.

Darüber hinaus gibt es aber noch Ausnahmefälle, bei denen die luftelektrische Feldstärke und auch der Vertikalstrom das *negative* Vorzeichen annehmen. CHALMERS[6] und MÜHLEISEN[7] haben solche Verhältnisse gefunden und studiert. Die Ursache sind negative Raumladungen, welche von sprühenden Hochspannungsleitungen bei hoher relativer Feuchtigkeit, von chemischen Betrieben oder ähnlichem ausgehen.

---

[1] H. ISRAEL: Arch. Met., Geophys. u. Biokl. **2**, 367—373 (1950).
[2] H. ISRAEL: J. Atmos. a. Terres. Phys. **1**, 26—31 (1950).
[3] U. KÜHN: Z. Meteor. **8**, 236—243 (1954).
[4] H. ISRAEL u. H. W. KASEMIR: Arch. Met., Geophys. u. Biokl., Ser. A **5**, 71—85 (1952).
[5] H. W. KASEMIR: Arch. Met., Geophys. u. Biokl. **5**, 56 (1952).
[6] J. A. CHALMERS: J. Atmos. a. Terres. Phys. **2**, 155 (1952).
[7] R. MÜHLEISEN: Z. Geophys. Sonderbd. 1953, 154—156.

27. Elektrische Felder bei Niederschlägen. Eine sehr ausführliche und umfassende Arbeit über die luftelektrischen Verhältnisse bei Niederschlägen aller Art stammt von SIMPSON [45]. Bei Bewölkung *ohne* Niederschlag herrschen am Erdboden fast immer annähernd normale Feldverhältnisse mit etwas verkleinerten Amplituden. Dagegen bilden sich bei Wolken *mit* Niederschlag Raumladungen, so daß die elektrischen Felder am Erdboden „gestört" sind und fast nur noch von den Wolken- und Raumladungen in der Troposphäre abhängen.

Durch Ladungstrennung in den Wolken, deren mögliche Einzelursachen in den Ziff. 11 bis 15 beschrieben wurden, bilden sich *Raumladungs-Gebiete* aus. Die negativen Ladungen befinden sich größtenteils an der Unterseite der Wolken, die positiven an der Oberseite derselben[1,2]. Die Folge davon sind bei nassem Niederschlag meist *negative* Felder unter den Wolken bis 20 V/cm. Bei ruhigem *Landregen* kann die Feldstärke dabei ziemlich konstant über längere Zeit sein; auch kommen periodische Schwankungen zwischen positiven und negativen Werten vor, von SIMPSON "wave pattern" genannt. Die Ursache für diesen in Fig. 26 gezeigten Verlauf ist noch nicht bekannt.

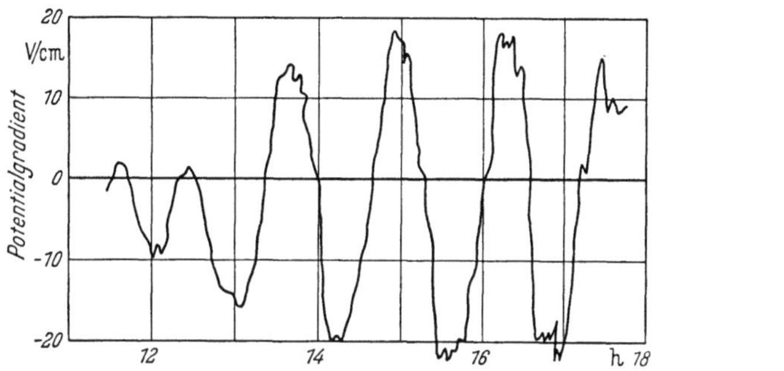

Fig. 26. Periodische Schwankungen des Potentialgradienten bei Landregen, von SIMPSON "wave pattern" genannt.

Bei *Schneefall* herrscht das positive Vorzeichen für den Potentialgradienten am Erdboden vor.

Bei lang andauernden Regen- und Schneefällen fällt allgemein auf, daß die luftelektrischen Felder sich allmählich dem Schönwetterwert nähern, obwohl dabei die Niederschlagsintensität gleich geblieben sein kann.

Bei kräftigen *schauer*artigen Niederschlägen in Form von Regen oder Schnee können die Feldstärken sehr hohe Werte bis zu etwa 100 V/cm annehmen, wobei es, wie beim Gewitter, zu Corona-Entladungen an Spitzen kommt, ohne daß wir aber (wegen der fehlenden Blitze) von einem Gewitter sprechen können. Nach Messungen der elektrischen Ladung der einzelnen Regentropfen[3] muß man annehmen, daß der fallende Regen nur wenig zum elektrischen Feld am Erdboden beiträgt.

28. Elektrische Felder bei Gewittern. Die *meteorologischen* Bedingungen und Vorgänge in den Gewitterwolken, auf welche zuerst kurz eingegangen werden soll[4], sind die Voraussetzung für die elektrischen Vorgänge. Die Bildung von Gewittern [42], [46] hängt im wesentlichen von der vertikalen Schichtung der Atmosphäre, also vom Verlauf von Temperatur und Feuchtigkeit in Abhängigkeit vom Luftdruck ab. Steigt ein Luftquantum auf, — sei es ausgelöst durch eine gegenüber der Umgebung höhere Temperatur, also geringere Dichte, oder durch das Einfließen von Kaltluft, — dann hängt die Größe des Auftriebs von der jeweiligen Temperaturdifferenz ΔT in den erreichten Höhen ab. Diese bleibt konstant, falls die Temperaturabnahme in der ruhenden Luft den adiabatischen Wert von 1° C pro 100 m Höhenunterschied hat. Die Temperaturdifferenz ΔT

[1] R. GUNN: Phys. Rev. (2) **71**, 181—186 (1947).
[2] G. C. SIMPSON: Quart. J. Roy. Met. Soc. **68**, 1—34 (1947).
[3] R. GUNN: Phys. Rev. (2) **71**, 181—186 (1947).
[4] Vgl. auch das Kapitel MASON und LUDLAM, Cloud Physics, in diesem Bande.

der aufsteigenden Luft gegenüber der Umgebung wird sich, bei einem mittleren Gradienten der ruhenden Luft von 0,7°/100 m also verkleinern; bei starker Erwärmung der erdbodennahen Schicht durch Sonneneinstrahlung und deshalb hohem Gradienten kann sie sich aber auch beim Aufsteigen vergrößern. Sobald aber beim Aufsteigen der *Taupunkt* erreicht wird und der Wasserdampf beim weiteren Aufsteigen kondensiert, führt die *Kondensationswärme* zu einer kleineren Temperaturabnahme des aufsteigenden Luftquantums, also zu einer Vergrößerung von ΔT und damit der Labilität, sog. *Feuchtlabilität,* und damit des Auftriebs. Ähnliches tritt dann beim *Gefrieren* der Wassertropfen ein, wo noch die latente *Gefrierwärme* frei wird.

Bei hohen Lufttemperaturen und -feuchtigkeiten am Erdboden, oder beim Hochheben feuchtwarmer Luft durch kalte, werden die Energien frei zur Bildung von Wärme- oder Frontgewittern. Es kommt in schmalen Zonen zu heftigen Aufwinden mit Vertikalgeschwindigkeiten bis zu 30 m/sec und daneben zu größeren Abwindgebieten mit geringen Vertikalgeschwindigkeiten. Viele Einzelheiten sind früher bei Segelflügen und vor einigen Jahren beim amerikanischen Thunderstorm-Projekt [*41*] untersucht und gefunden worden. Man kann demnach die Lebensgeschichte einer Gewitterwolke in drei Stadien einteilen: 1. das Cumulusstadium, in welchem Aufwind und zur Mitte starke Temperaturzunahme herrschen, 2. das Reifestadium mit starken Niederschlägen und Auf- und Abwinden, und 3. das Stadium des Absterbens ohne Aufwind mit schwächer werdendem Niederschlag.

In der Gewitterwolke wandelt sich die Phase des Wassers mehrmals um. Nach der Kondensation des Wasserdampfes an den Kondensationskernen, welche im allgemeinen in genügender Zahl vorhanden sind, wird das flüssige Wasser im Aufwind das Niveau mit der Temperatur 0° C überschreiten; die Wassertröpfchen können in unterkühltem Zustand höher gehoben werden. Die Bildung von Eisteilchen ist an Gefrierkerne gebunden und erfolgt nach atmosphärischen und Laboratoriums-Untersuchungen von RAU[1] und anderen bei tieferen Temperaturen als 0° C, vorwiegend nahe bei —4° C und —10° C. Dann laufen — des großen Wasserangebots wegen — Eisteilchenbildung und Vergrößerung zu Graupeln nebeneinander her. Die größeren Teilchen werden allmählich nicht mehr vom Aufwindstrom getragen und fallen deshalb nach unten, mindestens in Gebieten geringeren Aufwindes am Rande des Schlauches. Die leichteren Teilchen, vornehmlich Eiskristalle, werden höher getragen und erreichen zum Teil die Obergrenze der Cumulo-Nimben.

Im Abwindgebiet oder beim Fall der Graupel- und Hagelteilchen findet ein umgekehrter Effekt statt. Wenn diese Teilchen die 0°-Grenze durchfallen, dann schmelzen sie langsam ab und kommen als großtropfiger Regen am Erdboden an. Beim Abschmelzvorgang kann die äußere Wasserhaut zerrissen werden und der schwerere Hagel- oder Graupelkern getrennt von den leichteren Wassertröpfchen schneller fallen.

Die elektrischen Verhältnisse in einer Gewitterwolke sind noch nicht genau bekannt; sie lassen sich nach den heutigen Kenntnissen etwa folgendermaßen darstellen: nach den in den Ziff. 11 bis 14 beschriebenen Elektrifizierungs-Effekten können in der Gewitterwolke durch die Vorgänge bei der Eisteilchen-Bildung, bei der Vergraupelung, beim Abbrechen der äußersten Spitzen der Eisnadeln, beim Abschmelzen der Graupeln und vielleicht noch anderen Vorgängen elektrisch geladene Teilchen verschiedenen Vorzeichens entstehen. Wenn sich diese Teilchen durch ihre Masse und ihr Volumen unterscheiden, haben sie

[1] W. RAU: Geofisica Pura e Applicata **26**, 75 (1953).

sowohl in ruhender wie in bewegter Luft verschiedene Vertikalgeschwindigkeiten. Durch die Schwerkraft und durch die Aufwinde werden sie räumlich getrennt. Daraus folgt, daß dann auch die Ladungen dem Vorzeichen nach getrennt werden.

So kommt es zu *Raumladungs-Gebieten* innerhalb der Gewitterwolke, welche durch Untersuchungen von SIMPSON und SCRASE [*12*] räumlich festgelegt werden konnten. Diese ließen Ballone in die Gewitterwolken steigen und erhielten durch Registrierung der Spitzenströme die Polarität und ungefähre Stärke der Felder in Funktion der Höhe. Fig. 27 zeigt das Gesamtergebnis; an der Oberseite der Gewitterwolken befindet sich ein großes positives, darunter ein breites negatives Raumladungsgebiet, welches sich größtenteils bis an die Unterseite der Wolke erstreckt. Dort ist in der Nähe des Aufwindschlauches meist noch ein kleineres positives Raumladungsnest eingelagert, wo die Temperatur über 0° C ist. Dieses positive Raumladungsgebiet fehlt bei nichtgewittrigen Schauern im allgemeinen.

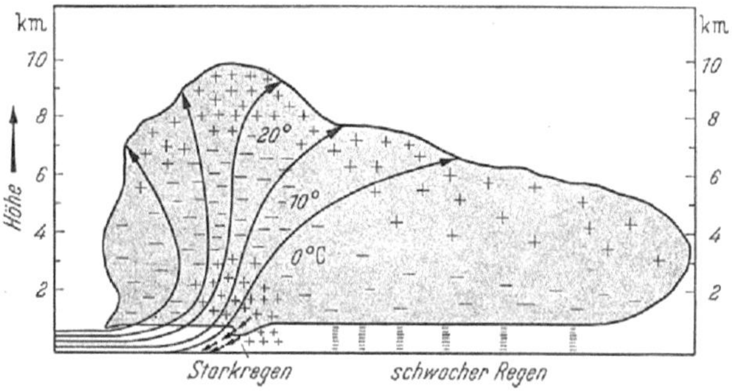

Fig. 27. Ladungsverteilung und Luftströmung in einer Gewitterwolke (SIMPSON und SCRASE).

Zur Erklärung dieser Ladungsverteilung in einer Gewitterwolke kann man heute nur Hypothesen machen. Folgende Grundsätze, von WALL[1] aufgestellt, treffen das Wesentliche:

1. Grundsatz: In den riesigen Abmessungen eines Gewitters können sonst unbeachtete Elementarvorgänge zur Auswirkung kommen, die in den kleinen Verhältnissen eines Laboratoriums keine Rolle spielen. Deshalb muß jeder Effekt in seiner *quantitativen* Auswirkung über die in Frage kommenden Räume abgeschätzt werden.

2. Grundsatz: In einem so riesigen Gebilde sind die physikalischen Bedingungen keineswegs homogen. Es werden also aller Erwartung nach neben- und übereinander *mehrere Mechanismen* der Elektrizitätsproduktion in Tätigkeit treten können.

Ferner kann gesagt werden:

3. Die Trennung großer Ladungsmengen in Cb-Wolken, welche aus den gemessenen elektrischen Feldern zu 10 bis 100 Clb abgeschätzt werden können, setzt voraus, daß starke Aufwinde die Niederschlagsteilchen hochheben und diese durch die Schwerkraft nach ihrer Größe und Masse getrennt werden. Leichtere Teilchen werden also höher getragen werden als schwerere. Nur durch starke Aufwinde läßt sich ein stationärer Betrieb des „Gewittergenerators" erklären.

4. Die Elektrisierungsprozesse müssen so wirken, daß Teilchen verschiedener Fallgeschwindigkeit Ladungen verschiedenen Vorzeichens tragen. Dies führt dann zu Schichten mit ausreichend großer Überschußladung eines Vorzeichens.

5. Welche von den Elektrisierungsvorgängen (Ziff. 11 bis 14) die Hauptrolle spielen, ob die Influenz durch ein schon vorhandenes Initialfeld dieses verstärkt (Fremderregung, „Asymmetrieeffekt", Ziff. 11) oder Vorgänge mit „Eigenpolung" (Ziff. 12 bis 14) die Raumladungen hervorbringen, ist noch unentschieden[2-4]. Wahrscheinlich ist, daß je nach der herrschenden Temperatur und Luftbewegung verschiedene Effekte nebeneinander mehr oder weniger wirksam sind. Dabei können auch abschwächende Effekte nebenhergehen. Die Übersicht nach Ziff. 14 läßt die große Zahl von in Betracht gezogenen Möglichkeiten erkennen.

---

[1] E. WALL: Wetter u. Klima **1**, 8—9 (1948).
[2] E. WALL: Wetter u. Klima **1**, 194 (1948).
[3] F. WOLF: Arch. Met., Geophys. u. Biokl. **9**, 242—263 (1956).
[4] D. MÜLLER-HILLEBRAND: Arch. för Geophys. **2**, 395—416 (1956).

Die elektrischen Feldstärken in Gewitterwolken wurden außer von SIMPSON und SCRASE[1] später auch von BYERS und BRAHAM[2] beim Thunderstorm-Projekt und von GUNN und anderen[3] untersucht. Während erstere mit ihrer einfachen Meßmethode nur geringe Feldstärken unter 100 V/cm gefunden hatten, sind von den anderen Autoren Feldstärken bis 1000 V/cm gemessen worden. Es scheint, als ob die für die Blitzentstehung notwendige Durchschlagsfeldstärke nur an wenigen Punkten kurzzeitig herrscht. Als Ausgangsorte für Blitze kommen hauptsächlich die Grenzen zwischen Ladungsgebieten verschiedenen Vorzeichens in Betracht, also in der Nähe des positiven Ladungsgebietes an der Unterseite und die Grenzfläche zwischen negativer und positiver Elektrizität im oberen Teil der Wolke, beide in der Nachbarschaft des Aufwindschlauches.

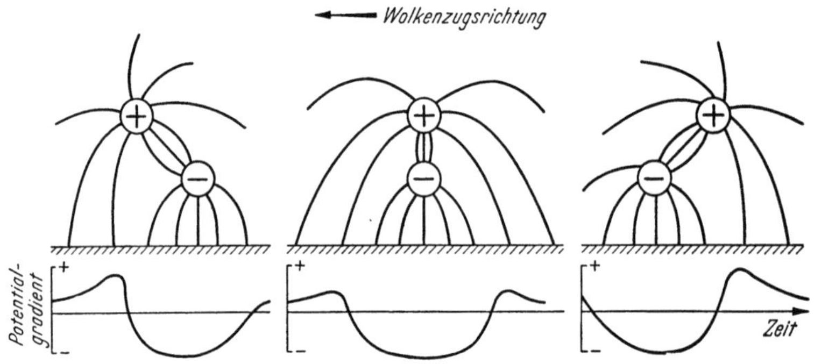

Fig. 28. Feldlinien und Verlauf des Potentialgradienten am Erdboden bei einem Wolkendipol, links beim Voreilen und rechts beim Nachhinken der oberen Wolkenpartie.

Die Feldstärken in der Umgebung von Gewitterwolken und unter diesen lassen nur in einfachen Fällen einen Schluß auf die Ladungsverteilung in der Wolke zu, nämlich nur dann, wenn man die Wolke durch einen Dipol oder einen Tripol ersetzen kann. WICHMANN [*47*] hat die Feldverläufe am Erdboden für verschiedene einfache Ladungsverteilungen gezeichnet. Alle Einzelheiten sind in der Fig. 28 enthalten. Unter ausgedehnten Gewitterfronten sind die Feldverläufe verwickelter. Die Werte der Feldstärke am Boden und entsprechend auch an anderen Orten in der Nachbarschaft einer Gewitterwolke sind von der Größenordnung 10 bis 25 *V*/cm und steigen nur gelegentlich bis über 100 V/cm an.

Die Beziehung zwischen dem Verlauf der Vertikalkomponente des luftelektrischen Feldes am Erdboden und einer Wolkenladung Q in der Höhe h und im Abstand r vom Meßpunkt wird im einfachsten Falle dargestellt durch die Formel:

$$E_{\perp} = \frac{2hQ}{r^3} + \frac{2}{c\,r^2}\cdot\frac{d}{dt}(hQ) + \frac{1}{c^2 r}\,\frac{d^2}{dt^2}(hQ) \tag{28.1}$$

(c = Lichtgeschwindigkeit).

Man sieht leicht, daß das rein statische Feld der Gewitterladungen Q [erster Term in (28.1)] in geringer Entfernung bereits auf kleine Werte absinkt; in einer Entfernung von 5 km vom Rande der Gewitterwolken weicht der Potentialgradient kaum mehr vom normalen Werte ab. Die Verschiebungsströme (zweiter Term) bei den Feldänderungen durch Blitzentladungen dagegen, welche bei ungenügender Anpassung der Meßgeräte mit erfaßt werden, reichen noch in größere Entfernungen, da sie nur mit $1/r^2$ abfallen. Die elektromagnetischen

[1] G. C. SIMPSON u. J. SCRASE: Proc. Roy. Soc. Lond., Ser. A **161**, 309 (1937).
[2] G. R. WAIT: Thunderstorm-Electricity-Report, Univers. of Chicago, [*13*], 1950.
[3] R. GUNN: Phys. Rev. **71**, 181 (1947).

Wellen, welche von Blitzentladungen ausgehen (dritter Term), den Radioempfang stören und *Atmospherics* genannt werden, sind im Frequenzbereich von 5 bis 30 kHz, wo sie ihre Maximalintensität haben, noch in Entfernungen von 10000 km wahrnehmbar: ihre Feldstärke nimmt nur mit $1/r$ ab; sie werden an ionosphärischen Schichten und am Erdboden reflektiert.

Die elektrischen Potentiale der Gewitterwolken-Partien sind schwer anzugeben. Während das Potential der luftelektrischen Ausgleichsschicht etwa 400000 V beträgt, sind die Potentiale in den Cumulonimben von der Größenordnung 100 bis 1000 Millionen Volt. Die statischen Feldsprünge am Erdboden bei Blitzentladungen nehmen Werte an, wie sie nach Messungen von WILSON[1], WORMELL[2], NORINDER[3], ISRAEL[4] und anderen in Tabelle 10 zusammengestellt sind.

Tabelle 10. *Gemessene Feldänderungen bei Blitzentladungen.*

Blitzentfernung km	Gesamt-feldschwankung V/m	Feldanstieg (V/m)/sec
1	100000	$5 \cdot 10^6$
10	1000	$5 \cdot 10^5$
30	50	$5 \cdot 10^3$
100	12	$12 \cdot 10^2$

29. Luftelektrische Felder bei speziellen Naturereignissen. Es gibt noch eine Anzahl von Naturereignissen, bei welchen ebenfalls erhebliche Raumladungen und entsprechend hohe Felder am Erdboden entstehen können. Dazu gehören: Sandstürme in Wüstengebieten, Staubstürme, Vulkanausbrüche und Wintergewitter. Über einen Sandsturm an der Cheops-Pyramide berichtet W. v. SIEMENS[5]. Bei Vulkanausbrüchen haben verschiedene Autoren große Raumladungen in der abziehenden Rauchwolke festgestellt, so z. B. HATAKEYAMA[6] beim Ausbruch des japanischen Vulkans Yakeyama. Auch Blitze wurden gelegentlich über einem aktiven Vulkan beobachtet. Eine Abart der Sommergewitter stellen die Wintergewitter dar. Da die Temperaturen unter 0° C liegen, kann die Wasserphase dabei keine Rolle spielen. Als Ursache für die hohen elektrischen Ladungen können also nur Vorgänge an Schnee- und Eiskristallen angesehen werden.

Andere Naturerscheinungen haben offenbar nur indirekten Einfluß auf die luftelektrischen Verhältnisse am Erdboden. Sonnenfinsternisse haben schon vielerorts anormale Änderungen des Verlaufs des luftelektrischen Potentials hervorgebracht. Nach sehr zahlreichen Beobachtungen ist der Schluß von GISH[7] naheliegend, daß lediglich die verminderte Wärmestrahlung während einer Sonnenfinsternis über die Änderung von Wind, Temperatur und Feuchtigkeit Einfluß auf die luftelektrischen Elemente gewinnt.

30. Elektrische Raumladungen in der Atmosphäre und ihre Messung. Außer in den Wolken, in welchen sehr hohe elektrische Raumladungsdichten vorkommen können, ist auch überall sonst in der Atmosphäre eine zwar geringe, aber nachweisbare Raumladungs-Dichte ϱ vorhanden. Die Ursachen sind einmal Schwankungen der elektrischen Leitfähigkeit der Luft Λ, die bei stationärem Vertikalstrom j mit einem Gradienten der elektrischen Feldstärke $\partial E/\partial h$ einhergehen. Es ist

$$\varrho = -\varepsilon_0 \frac{\partial E}{\partial h}; \tag{30.1}$$

[1] C. T. R. WILSON: Phil. Trans. Roy. Soc. Lond., Ser. A **221**, 73—115 (1921).
[2] T. W. WORMELL: Phil. Trans. Roy. Soc. Lond., Ser. A **238**, 249—303 (1939).
[3] H. NORINDER: Nature, Lond. **224**, 303 (1939).
[4] H. ISRAEL: Gerlands Beitr. Geophys. **55**, 314—333 (1939). — Meteor. Z. **61**, 1—8 (1944).
[5] W. v. SIEMENS: Phys. Bl. **10**, 167—169 (1944).
[6] H. HATAKEYAMA: J. Geomagn. a. Geoelectr. **1**, No. 2 (1948).
[7] O. H. GISH: Terr. Magn. **49**, 123 (1944).

für $E = j/\Lambda$ erhält man dann bei konstantem j

$$\varrho = -\varepsilon_0 \cdot \frac{j}{\Lambda^2} \cdot \frac{\partial \Lambda}{\partial h} \cdot \tag{30.2}$$

Aus der normalen Leitfähigkeits-Zunahme mit der Höhe errechnet sich nur eine geringe Raumladungsdichte von etwa $10\,e/\text{cm}^3$. Direkt an der Erdoberfläche, wo durch den Elektrodeneffekt (Ziff. 33 β) eine Verarmung an negativen Ionen eintritt, kann die Raumladungsdichte bei Windstille höhere Werte annehmen, ebenso an der Schichtgrenze von Dunst, Nebel oder Wolken.

Die unregelmäßigen Schwankungen des luftelektrischen Potentials über dem Erdboden deuten auf weitere, aber räumlich begrenzte Raumladungen hin, welche in geringer Höhe treiben. Vermutungen über diese Zusammenhänge wurden früher schon häufig ausgesprochen[1,2,3], genauere Untersuchungen führte MÜHLEISEN [11] mit leicht transportablen Geräten zur Messung des Potentialgradienten in einer Großstadt und in verschiedenen Entfernungen von dieser aus. Durch weitere Messungen von Vertikalstrom, Klein-Ionenzahl und elektrischer Raumladungsdichte konnte gezeigt werden, daß die anthropogenen Quellen von Kondensationskernen, Rauch und Staub, mit diesen auch elektrische Raumladungen produzieren. Durch diese Messungen im Freien und durch ergänzende Laboratoriums-Versuche wurden die jederzeit reproduzierbaren Vorzeichen der künstlichen Raumladungen gefunden, von denen ein Teil in der Tabelle 11 aufgeführt sind. Man erkennt, daß in der Mehrzahl positive Raumladungen vorkommen, welche das normale luftelektrische Feld vergrößern.

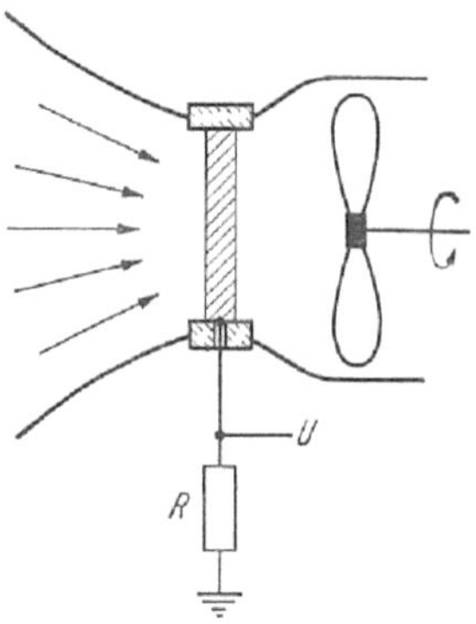

Fig. 29. Filtermethode zur Messung der elektrischen Raumladungsdichte der Luft nach OBOLENSKI.

Tabelle 11. *Vorzeichen künstlicher Raumladungen.*

Positiv	Negativ
Auspuff (Benzinmotor)	Benzinfeuer, offen
Auspuff (Dieselmotor)	Petroleumfeuer, offen
Schornsteinrauch	Alkoholfeuer, offen
Kohlenpulverfeuer	Holzwollefeuer, offen
Lokomotivdampf	Wasserdampf, rein (aus Metalldüse)
Dampfwolke (Gaswerk)	

In wenigen 100 m Entfernung von den Raumladungsquellen sind Änderungen vom Normalwert des luftelektrischen Potentialgradienten in der Größenordnung von 100 bis 500 V/m häufig gefunden worden. Je nach den meteorologischen Bedingungen werden diese Raumladungen vom Wind und durch den Vertikalaustausch fortgetragen und konnten, trotz des Abbaus der Raumladungen durch die Leitfähigkeit der Luft, noch nach mehr als einer Stunde in Windrichtung nachgewiesen werden; diese künstlichen Raumladungen können also noch nach Stunden wirksam sein. Das mitunter z.B. in ländlichen Gegenden gefundene Maximum des Potentialgradienten gegen Mittag läßt sich erklären, wenn man annimmt, daß die am Morgen in Siedlungen erzeugten Raumladungen aus großen Entfernungen mit dem Wind hergetragen werden.

Die elektrische Raumladungsdichte der Luft kann mit der von W. N. OBOLENSKI[4] beschriebenen Filtermethode gemessen werden; ein bestimmtes Luft-

[1] J. BROWN: Terr. Magn. **38**, 161 (1933); **40**, 413–424 (1935).
[2] H. HATAKEYAMA u. S. HUDIMATO: Geophys. Mag. Tokyo **23**, 15 (1951).
[3] F. J. SCRASE: Geophys. Mem. Met. Off. London **7**, Nr. 67 (1935).
[4] W. N. OBOLENSKI: Ann. Phys. **77**, 644 (1925).

quantum wird dabei durch ein isoliertes Filter gesaugt und dessen Aufladung gemessen, wie Fig. 29 zeigt. Bei einer zweiten Methode von MÜHLEISEN und

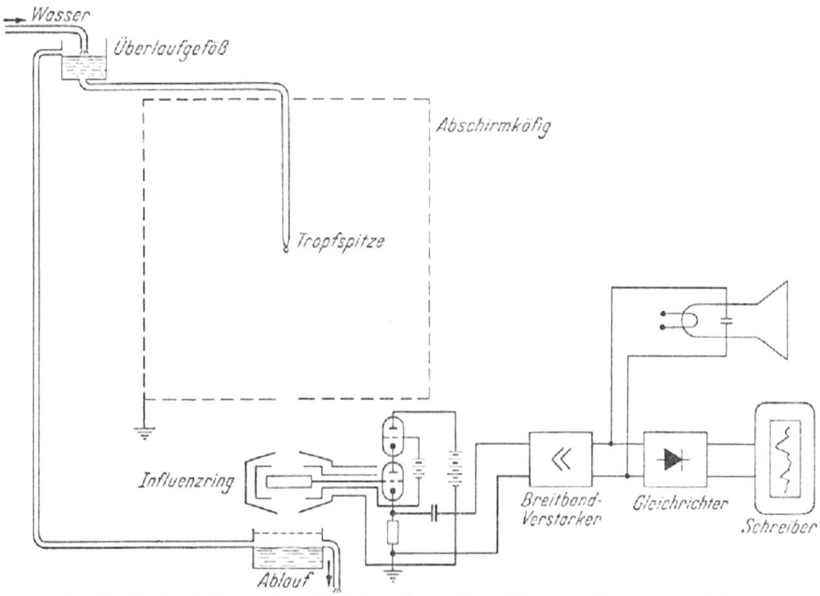

Fig. 30. Zur Registrierung der elektrischen Raumladungsdichte nach MÜHLEISEN und HOLL.

HOLL[1] werden auf fallenden Wassertropfen Ladungen influenziert durch die Raumladung der umgebenden Luft. Durch Messung der Tropfenladungen gewinnt man Werte für die Raumladungsdichte bis herab zu fünf Elementarladungen je Kubikzentimeter mit einer zeitlichen Auflösung von $^1/_3$ sec. Die Anordnung im Prinzip zeigt Fig. 30.

31. Feldverlauf mit der Höhe. Wenn im einfachsten Fall das luftelektrische Feld E_h am Erdboden ($h = 0$) und in der Atmosphäre nur von der Spannung zwischen der hohen Ausgleichsschicht und dem Erdboden herrührt, dann gilt für den Potentialgradienten bei quasistationären Verhältnissen:

$$E_h \approx 1/\Lambda_h \qquad (31.1)$$

d.h. also, Potentialgradient und elektrische Leitfähigkeit Λ_h der Luft sind in jeder Höhe h umgekehrt proportional. Aus dem normalen Verlauf von Λ_h mit der Höhe für reine, wolken- und dunstfreie und gut durchmischte Atmosphäre (Ziff. 22) ergibt sich der normale Verlauf der elektrischen Feldstärke, wie in Fig. 31 dargestellt.

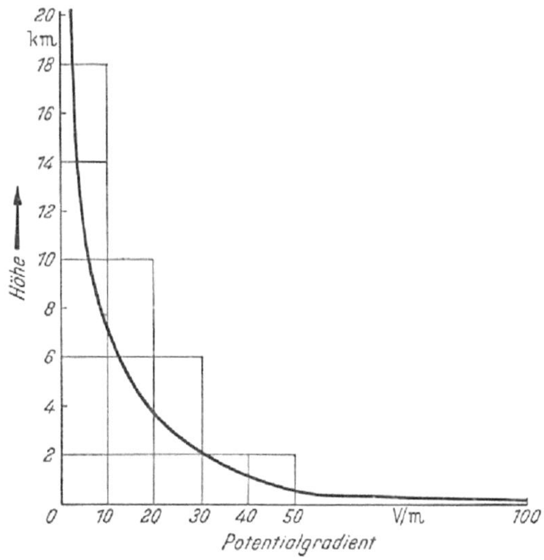

Fig. 31. Normaler Feldverlauf in der freien Atmosphäre bei reiner gut durchmischter Luft.

[1] R. MÜHLEISEN u. W. HOLL: Geofisica Pura e Applicata 22, 189—194 (1952).

Abweichungen von diesem theoretischen Verlauf findet man zunächst in der Nähe des Erdbodens, weil die dort treibenden elektrischen Raumladungen (s. Ziff. 31) ihr eigenes Feld erzeugen. Beispiel: Eine Schicht von großer Horizontalausdehnung, welche eine angenommene konstante Raumladungsdichte von $\varrho_1 = 100\,e/\mathrm{cm}^3$ bis zur Höhe von 100 m haben soll, vergrößert den Potentialgradienten am Erdboden um den Wert

$$\Delta E = \frac{1}{\varepsilon_0} \cdot \int_0^h \varrho\, dh = \frac{1}{\varepsilon_0} \varrho_1 \cdot h. \tag{31.2}$$

Die Rechnung ergibt für ΔE: ≈ 200 V/m. Mit zunehmender Höhe nimmt also dann der Potentialgradient von 300 V/m am Erdboden auf rund 100 V/m in 100 m Höhe ab; von da ab ist er normal. In wirklichen Fällen wird natürlich der Abfall, entsprechend der Raumladungsverteilung, unregelmäßiger sein.

Andere Abweichungen vom normalen Verlauf des Potentialgradienten werden dort auftreten, wo durch Kondensationskerne, Groß-Ionen und anormale Ionisation die elektrische Leitfähigkeit Unstetigkeiten aufweist, insbesondere also auch in Nebel- und Wolkenschichten (s. Ziff. 26). Solche Schwankungen des Potentialgradienten in verschiedenen Höhen wurden durch Aufstiege mit Fessel- und Registrierballonen bestätigt. Auch in der Stratosphäre haben IDRAC[1] und KOENIGSFELD[2] sehr hohe Werte von 10 bis 20 V/m in Höhen um 20 km gefunden, welche aber noch wegen der Schwierigkeiten der Feldstärkemessung in der freien Atmosphäre unsicher sind.

D. Die elektrischen Ströme in der Atmosphäre.

Unter dem Einfluß des elektrischen Feldes in der Atmosphäre und der Schwerkraft entstehen verschiedene Arten von Ladungstransporten: Ein *Leitungsstrom* durch Ionen und Elektronen, ein *Konvektionsstrom* über die Ladungen der Aerosolteilchen und Niederschlagsteilchen, *Spitzenstrom* bei hohen Feldern und *Erdblitze.*

32. Methoden und Ergebnisse der Messung des vertikalen Leitungsstromes. Setzt man eine hochisolierte Platte an der Erdoberfläche, nach Art der in Ziff. 24 erwähnten WILSON-Platte (Fig. 18), dem elektrischen Feld aus, dann fließt einmal ein *Leitungsstrom* I_L auf sie:

$$I_L = \Lambda \cdot |\mathfrak{E}| \cdot F, \tag{32.1}$$

wobei Λ die elektrische Gesamtleitfähigkeit der Luft, $\mathfrak{E}$ die an der Platte herrschende Vertikalkomponente der elektrischen Feldstärke und F die Fläche der Platte bedeuten. Zu diesem Leitungsstrom kommt aber noch ein *Verschiebungsstrom* bei Änderungen des elektrischen Feldes:

$$I_V = \frac{d\mathfrak{D}}{dt} \cdot F = \varepsilon_0 \frac{d\mathfrak{E}}{dt} \cdot F. \tag{32.2}$$

Darin bedeuten ε_0 die Verschiebungskonstante $8{,}86 \cdot 10^{-12}$ Coulomb/V·m, $\mathfrak{D}$ die dielektrische Verschiebung. Da dieser Verschiebungsstrom praktisch die gleiche Größenordnung haben oder, bei rasch wechselnden Feldern, 10 bis 100mal größer als der Leitungsstrom sein kann, ist die besondere Schwierigkeit bei der Vertikalstrommessung die Unterbindung des Verschiebungsstromes.

---

[1] P. IDRAC: C. r. Acad. Paris **182**, 1634 (1926).

[2] L. KOENIGSFELD: Aus Thunderstorm-Electricity, [*13*] Bd. 24—45, Univers. of Chicago 1953.

C. T. R. WILSON war der erste, der einwandfreie Messungen durchgeführt hat. Er schirmte die Vertikalstromplatte periodisch durch eine Abdeckplatte ab und maß elektrometrisch die in einer bestimmten Zeit auf die Platte geflossene Ladung. Durch die Abschirmung des Feldes vor und nach der Exposition ist der Feldeinfluß ausgeschaltet.

SCRASE [*56*] hat in Kew die Methode für Dauerregistrierungen verbessert, indem er gemäß Fig. 32 auf das eine Quadrantenpaar eines Elektrometers I die Ladung der Platte, auf das andere die durch die Schwankungen des luftelektrischen Potentials über einen Kondensator C influenzierte Ladungen geleitet hat. Durch die gleichzeitige periodische Erdung von Platte und Influenzierungs-Kondensator sollte man einen Ausschlag des Elektrometers erhalten, welcher dem wahren Leitungsstrom, also Gesamtstrom minus Verschiebungsstrom, und

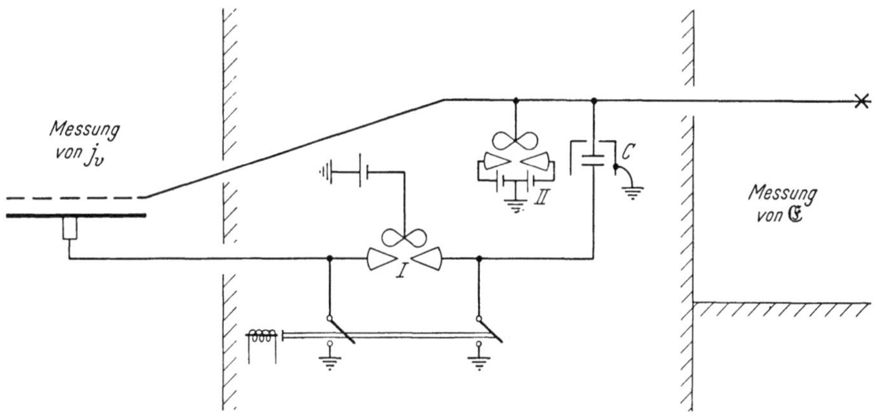

Fig. 32. Vertikalstromregistrierung nach SCRASE. Methode zur Kompensation der Verschiebungsströme bei Feldschwankungen.

der Expositionszeit proportional ist. Diese Methode führte aber nicht zu befriedigenden Resultaten, da das Potential eines radioaktiven Kollektors nicht exakt den wirklichen Feldschwankungen folgt und deshalb die Verschiebungsströme auch nicht exakt kompensiert werden können. SCRASE hat deshalb folgenden Ausweg eingeschlagen: er legte über die Vertikalstromplatte ein engmaschiges, hochisoliertes Drahtnetz, welches das natürliche elektrische Feld abschirmte, aber mit dem Potentialkollektor verbunden war. Der Strom auf die Platte war zwar dann nicht durch das natürliche Feld, aber doch durch ein dem natürlichen nahezu entsprechendes Feld verursacht; dies hat aber zur Folge, daß nunmehr der Verschiebungsstrom an der Platte exakt über den Kondensator C kompensiert werden kann.

Neuerdings ist es üblich, den auf die Platte fließenden Strom nach Fig. 33 über einen hohen Widerstand R zur Erde abzuleiten und die Spannung an diesem Widerstand $U = I \cdot R$ mittels Röhrenvoltmeter zu messen. Durch einen Kunstgriff hat KASEMIR[1] dabei die Verschiebungsströme eliminiert. Wird der parallel zum Widerstand geschaltete Kondensator C passend gewählt, dann kann man erreichen, daß der Verschiebungsstrom gerade zur Aufladung bzw. Entladung des Kondensators C dient. Bei einer Feldänderung dE soll die Spannungsänderung an der Platte dU lediglich proportional der Leitungsstromänderung sein, also:

$$dU = R \cdot dI_L = R \cdot \Lambda \cdot F\, d|\mathfrak{E}|. \tag{32.3}$$

[1] H. W. KASEMIR: Diss. Aachen 1954.

Dies wird erreicht, wenn die Influenzladung durch den Verschiebungsstrom

$$dQ = \varepsilon_0 \cdot F \cdot d|\mathfrak{E}| \tag{32.4}$$

zur Aufladung des Kondensators verwendet wird, also gleich $C \cdot dU$ ist. Es muß deshalb sein

$$\varepsilon_0 \cdot F \cdot d|\mathfrak{E}| = R \cdot \Lambda \cdot F \cdot d|\mathfrak{E}| \cdot C. \tag{32.5}$$

Daraus

$$\varepsilon_0/\Lambda_0 = R \cdot C. \tag{32.6}$$

Diese einfache Beziehung besagt, daß die Verschiebungsströme keinen Einfluß auf die angezeigte Spannung U haben, U also direkt proportional dem Leitungsstrom ist, wenn die Zeitkonstante des Systems Platte + Kondensator C und

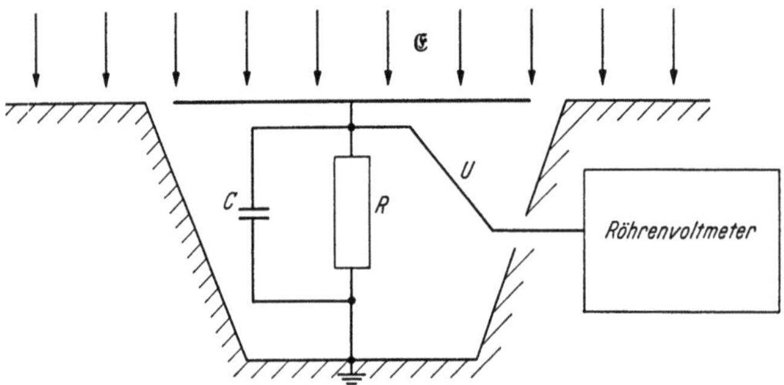

Fig. 33. Zur Messung des Vertikalstroms (Kapazität C zur Kompensation des Verschiebungsstromes nach KASEMIR).

Ableitwiderstand R gleich der „Zeit-Konstanten" der Luft ε_0/Λ_0 ist. Dieser letzte Ausdruck wird in der nächsten Ziffer erklärt; er ist bei einer Leitfähigkeit $\Lambda_0 = 2 \cdot 10^{-16}\,\Omega^{-1}\,\mathrm{cm}^{-1}$:

$$\varepsilon_0/\Lambda_0 = 450\,\mathrm{sec} = 7{,}5\,\mathrm{min}. \tag{32.7}$$

Registrierungen mit dieser Anordnung und großer zeitlicher Auflösung zeigen einen völlig befriedigenden Parallelgang von Potential und Vertikalstrom bis zu kurzzeitigen Einzelheiten. Lediglich bei größeren Abweichungen der Leitfähigkeit vom angesetzten Normalwert machen sich geringfügige Phasenverschiebungen und Kurvendeformationen bemerkbar. Für praktische Zwecke aber können sie bei Dauerregistrierungen in Kauf genommen werden.

Weitere Schwierigkeiten entstehen durch den VOLTA-*Effekt* zwischen der Vertikalstromplatte und der Umgebung, auf welche bereits SCRASE[1] aufmerksam gemacht hat. Die Aufrechterhaltung guter Isolation ist bei der ausgedehnten Platte oder dem Netz wiederum von besonderer Schwierigkeit. Geheizte Isolatoren und Wasserbecken oder Klebringe, um Spinnen und andere kleine Lebewesen abzuhalten, reichen aber im allgemeinen aus.

Zwei bewährte Schaltungen zur Vertikalstrommessung nach diesem Prinzip seien angegeben. Fig. 34 zeigt ein Gerät mit Elektrometerröhre nach KASEMIR[2], ein anderes Gerät mit normalen Verstärkerröhren und Netzanschluß die nächste Fig. 35.

Die Ergebnisse von Messungen des vertikalen Leitungsstromes weichen stark voneinander ab. Eine Zusammenstellung der Mittelwerte über längere Meßreihen gibt ISRAEL [*54*]. Während niedrige Werte für Kew bei London, München und Potsdam von $1{,}1 \cdot 10^{-16}$ Amp/cm² aufgeführt sind, melden andere Stationen,

[1] F. J. SCRASE: Geophys. Mem. Met. Off. London **7**, No. 58, 1—22 (1933).
[2] H. W. KASEMIR: J. Atmos. a. Terres. Phys. **2**, 32—37 (1951).

wie Durham und Peebles in England, Mittelwerte von $2{,}2 \cdot 10^{-16}$ Amp/cm². Größer sind die Ströme auf Bergstationen, weil dort bei gleicher Spannung zwischen Erde und Ausgleichsschicht der Säulenwiderstand der Luft kleiner ist.

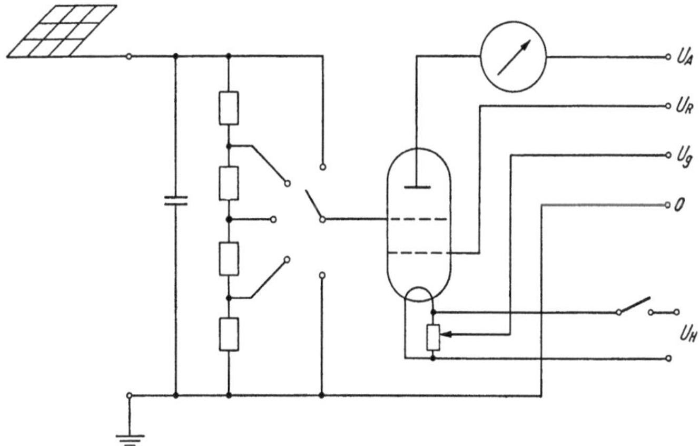

Fig. 34. Zur Messung des Vertikalstromes mit Elektrometerröhre nach KASEMIR.

Für die Unterschiede der Werte an Landstationen werden drei Ursachen maßgebend sein. Einmal sind verschiedenartige Geräte in Gebrauch, und die Reduktion der Werte auf ebenes Feld (Ziff. 24 ε) erfolgt in unterschiedlicher Weise. Zweitens birgt die *Auswahl* der für die Mittelwertbildung geeigneten

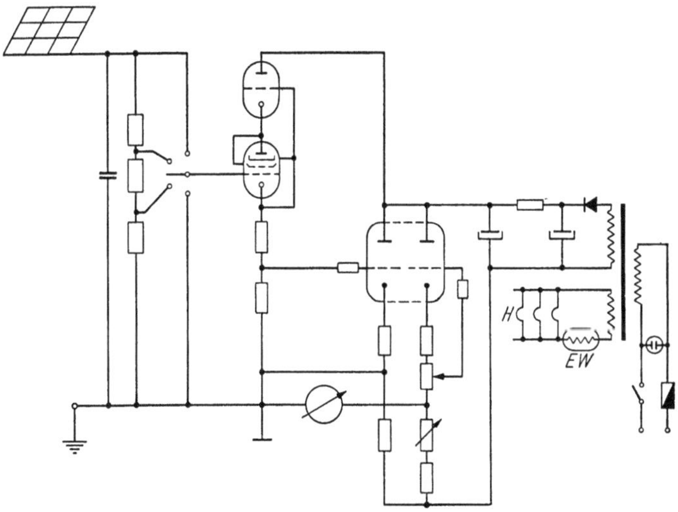

Fig. 35. Röhrenvoltmeter für Vertikalstrommessung nach MÜHLEISEN und CREUTZBURG.

„*ungestörten Tage*" und Stunden eine große Willkür in sich, die auch dadurch nicht ganz behoben wird, daß man als ungestörte Zeit alle Stunden ohne Niederschlag festsetzt. Drittens sind künstliche Raumladungen eine bisher viel zu wenig beachtete Störungsursache, welche vollkommen ortsabhängig auftreten und daher jede Vergleichsmöglichkeit ausschließen. Fig. 36, Potentialgradient (*PG*) und Vertikalstrom j_v in Stuttgart, gibt ein Beispiel für den morgendlichen Anstieg des Vertikalstromes in einer Großstadt: während die zunehmende Verunreinigung der Luft am Morgen ein Absinken des Vertikalstromes bewirken müßte, wird diese Erscheinung vollkommen überdeckt durch die starke Zunahme künstlicher positiver Raumladungen in den untersten Luftschichten. Außerhalb von Städten ist der Effekt geringer, aber nicht zu vernachlässigen. Die aus elektrischer Feldstärke und elektrischer Luftleitfähigkeit *berechneten* Werte des Vertikalstromes liegen vielerorts etwas höher als die direkt gemessenen.

Die mittlere Vertikalstromdichte über den Ozeanen ist etwa $3{,}2\cdot 10^{-16}$ Amp/cm², berechnet aus Feldstärke und Leitfähigkeit[1]. Auf dieselbe Weise sind Werte für die freie Atmosphäre von WIGAND[2] und ROSSMANN[3] ermittelt worden — mit dem Ergebnis, daß der Vertikalstrom bis 7000 m Höhe etwa konstant $1\cdot 10^{-16}$ Amp je Quadratzentimeter ist — und von GISH und WAIT [*51*] bis 15 km Höhe.

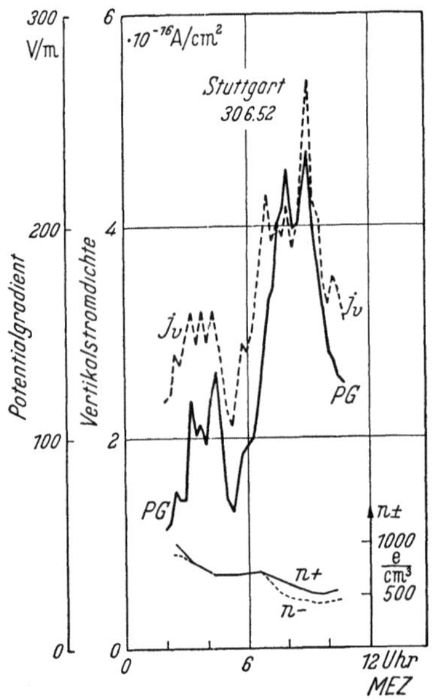

Fig. 36. Beispiel des Verlaufs von Potentialgradient, Vertikalstrom und Klein-Ionenzahl in einer Großstadt.

Eine theoretische Möglichkeit zur Messung des luftelektrischen Vertikalstroms ist dadurch gegeben, daß man aus dem Linien-Integral der magnetischen Feldstärke längs eines geschlossenen Weges auf den gesamten Strom (A Ampere) durch die umschlossene Fläche schließen kann:

$$\oint \mathfrak{H}\cdot d\mathfrak{s} = 4\pi A/10 .$$

Die numerische Berechnung solcher Linien-Integrale aus den Ergebnissen erdmagnetischer Vermessungen ergab für Flächen der Ausdehnung 10° geographischer Breite und 30° geographischer Länge hypothetische Strom-Dichten der Größenordnung 1000 Amp über 10000 km², während die luftelektrischen Messungen nur etwa 1500 Amp für die gesamte Erde (510 Mill km²) ergeben (Ziff. 38). Die erdmagnetisch errechneten Vertikalströme erscheinen also mindestens 10^4mal stärker als die luftelektrisch gemessenen Ströme. BENNDORF[4] hatte in der Aufklärung dieser Diskrepanz ein Grundproblem der luftelektrischen Forschung gesehen. Tatsächlich ist diese Aufklärung trivial und von den Erdmagnetikern wiederholt gegeben worden[5,6]: die allgemeinen luftelektrischen Vertikalströme sind viel zu schwach, um einen erdmagnetisch nachweisbaren Effekt zu geben, mit Ausnahme von nahen Blitzschlägen. Wenn man also aus erdmagnetischen Vermessungen die Linien-Integrale berechnet, so erhält man, wegen der unvermeidlichen Fehler der erdmagnetischen Karten, Schlußfehler, die keine physikalische Bedeutung haben. Heute berechnet man solche Linien-Integrale nur noch in der Absicht, daraus Verbesserungen des erdmagnetischen Kartenmaterials abzuleiten, wobei die Korrektheit der erdmagnetischen Werte nach dem genügend genauen Verschwinden der Integrale beurteilt wird.

Über den fragwürdigen „*vertikalen Erdstrom*" vgl. Sect. 23 im Beitrag CAGNIARD, Band XLVII dieses Handbuches, S. 434.

33. Strömungstheorie des Leitungsstromes und Elektrodeneffekt. Für quasistationäre Verhältnisse, also bei nur langsam veränderlicher Spannung U zwischen luftelektrischer Ausgleichsschicht und Erde und bei nur langsam veränderlichem Säulenwiderstand R ist die Vertikalstromdichte eine reine Funktion dieser Größen

$$j_v = U/R. \tag{25.1}$$

Anders verhält sich der Vertikalstrom bei schnelleren Änderungen.

---

[1] S. J. MAUCHLY: Res. Dept. Terr. Magn. Carnegie Inst. Wash. Publ. 175 **5**, 385—424 (1926).
[2] A. WIGAND: Ann. Phys. (4) **66**, 81 ff. (1921).
[3] F. ROSSMANN: Ber. dtsch. Wetterdienst. US-Zone **1950**, Nr. 15, 1—54.
[4] H. BENNDORF: Phys. Z. **26**, 81 (1925). — Z. Geophys. **1**, 147 (1924/25).
[5] F. DYSON u. H. FURNER: Mon. Not. Roy. Astronom. Soc., Geophys. Suppl. **1**, 76(1923).
[6] AD. SCHMIDT u. J. BARTELS: Gerlands Beitr. Geophys. **55**, 292 (1939).

α) Leitungsstrom. Eine Theorie des Leitungsstromes entwickelte KASEMIR[1]. Aus den MAXWELLschen Gleichungen lassen sich die folgenden Beziehungen zur Ableitung verwenden:

$$j = \Lambda \cdot \mathfrak{E}$$

$$\mathfrak{D} = \varepsilon_0 \cdot \mathfrak{E}$$

$$\operatorname{div}\left(\frac{\partial \mathfrak{D}}{\partial t} + j\right) = 0 .$$

Mit

$$\frac{\partial \mathfrak{D}}{\partial t} = \frac{\varepsilon_0}{\Lambda} \cdot \frac{\partial j}{\partial t}$$

wird

$$\operatorname{div}\left(\frac{\varepsilon_0}{\Lambda} \cdot \frac{\partial j}{\partial t} + j\right) = 0 . \quad (33.1)$$

Diese Gleichung läßt sich integrieren, und man erhält

$$\frac{\varepsilon_0}{\Lambda} \cdot \frac{\partial j}{\partial t} + j = j_s \quad (33.2)$$

mit der Randbedingung

$$\operatorname{div} j_s = 0 . \quad (33.3)$$

Die Differentialgleichung (33.2) ergibt integriert:

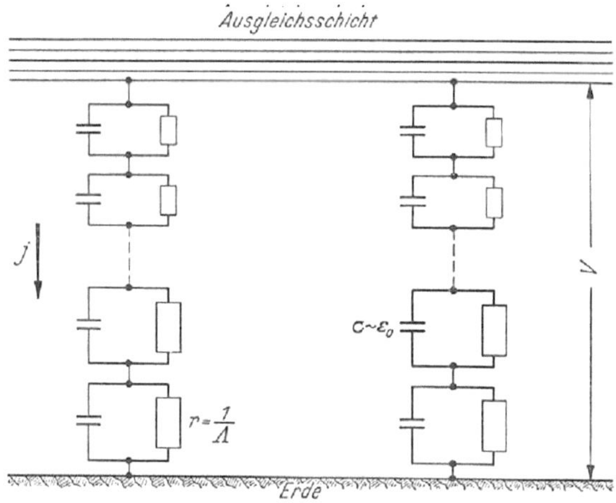

Fig. 37. Ersatzschaltbild für den komplexen Säulenwiderstand in der Atmosphäre.

$$j = e^{-\frac{\Lambda}{\varepsilon_0} t} \left(j_a + \frac{\Lambda}{\varepsilon_0} \int_0^t j_s e^{-\frac{\Lambda}{\varepsilon_0} t} \, dt \right) . \quad (33.4)$$

Versteht man in dieser Gleichung j_s als eine nur langsam mit der Zeit sich ändernde Größe der Vertikalstromdichte, also als quasistationär, dann wird die endgültige Form für j:

$$j = j_a e^{-\frac{\Lambda}{\varepsilon_0} t} + j_s \left(1 - e^{-\frac{\Lambda}{\varepsilon_0} t} \right) . \quad (33.5)$$

Diese Gleichung gibt uns den Verlauf des Vertikalstromes von einem Anfangswert j_a zum quasistationären Endwert j_s. Man sieht, daß der Anfangswert nach einer e-Funktion mit einer Zeitkonstanten $\tau = \varepsilon_0/\Lambda$ verschwindet, während der quasistationäre Endzustand sich mit derselben Zeitkonstanten einstellt. Und ferner erkennen wir den bereits in der vorhergehenden Ziffer abgeleiteten Ausdruck ε_0/Λ als *Zeitkonstante im „luftelektrischen Stromkreis"*. Eine entsprechende Gleichung läßt sich für den Potentialgradienten E anschreiben.

Von der Elektrotechnik her kennt man *Einschalt-Vorgänge* dieser Art bei einer Kette von hintereinandergeschalteten Widerstand-Kondensatorengliedern. Denken wir uns also den Säulenwiderstand in der Atmosphäre nach Fig. 37 unterteilt in z.B. 1 cm hohe Teile und ersetzen jedes dieser Teilstücke durch einen Widerstand $r = 1/\Lambda$ und eine parallel geschaltete Kapazität, welche gleich ε_0 sein soll, dann hat das einzelne Glied dieser Kette gerade diese Zeitkonstante

$$\tau = \varepsilon_0 \cdot r = \varepsilon_0/\Lambda . \quad (33.6)$$

[1] H. W. KASEMIR: Arch. Met., Geophys. u. Biokl. 3, 84 (1950).

Die einzelnen Glieder zwischen Erdboden und luftelektrischer Ausgleichsschicht haben dementsprechend nach oben abnehmende Zeitkonstanten. In den verschiedenen Höhen ergeben sich somit Einstellzeiten τ, deren Mittelwerte in Tabelle 12 zusammengestellt sind.

Ein Einschaltvorgang im luftelektrischen Stromkreis spielt sich also folgendermaßen ab, wie an Hand des Ersatzschaltbildes Fig. 37 leicht zu übersehen ist: Angenommen die Spannung der Ausgleichsschicht steige momentan vom Wert 0 auf den Wert V an, dann würde sich im ersten Augenblick eine Spannungsverteilung am Säulenwiderstand einstellen, welche sich nach den Kapazitäten je Zentimeter richtet. Diese entsprechen aber dem konstanten Wert ε_0. Entsprechend der großen Leitfähigkeit in der Höhe werden sich dort die Raumelemente gemäß der kleinen Zeitkonstanten schnell entladen, in der Nähe des Erdbodens dagegen wegen des größeren Widerstandes der Raumelemente mit größeren Zeitkonstanten aufladen. Es werden also hohe Ausgleichsströme in großer Höhe sehr schnell den stationären Endzustand herstellen, in der Nähe des Erdbodens aber ein niedriger Anfangswert sich allmählich mit der Zeitkonstanten von etwa 500 sec dem quasistationären Endwert nähern. Ähnliches erfolgt bei jeder Änderung der Spannung der Ausgleichsschicht. Bei künstlichen elektrischen Raumladungen, welche in niedriger Höhe und mit beschränkter räumlicher Ausdehnung über den Meßort ziehen, wird der Endzustand sofort erreicht, weil in dem niedrigen Höhenbereich ein nur geringer Unterschied zwischen dem kapazitiven und dem OHMschen Spannungsteiler herrscht.

Tabelle 12. *Die „Zeitkonstanten" oder Einstellzeiten in der Atmosphäre.*

Höhe km	$\tau = \frac{8{,}86 \cdot 10^{-14}}{\Lambda\,(\Omega^{-1}\,\mathrm{cm}^{-1})}$ sec
0	450 sec
5	70 sec
10	18 sec
20	5 sec
40	0,5 sec

Wählt man wie KASEMIR[1] bei der Vertikalstrommessung nach Ziff. 32 für die Größe der Zeitkonstanten des Ableitwiderstandes an einer Vertikalstromplatte denselben Wert wie den der Zeitkonstanten des untersten Teilstückes des Säulenwiderstandes, dann bedeutet dies im Sinne der Elektrotechnik eine „*Anpassung*" des Meßgerätes an den Stromkreis.

β) Der Elektrodeneffekt. Unter Elektrodeneffekt versteht man die Abweichung der Konzentration der Ionen eines Vorzeichens in der unmittelbaren Nähe einer Elektrode eines stromdurchflossenen Kondensators. Der luftelektrische Vertikalstrom besteht bei vorliegender Volumionisation und positivem Potentialgradienten aus dem Strom der positiven Ionen von oben nach unten und dem Strom der negativen Ionen von unten nach oben. Da in nächster Nähe des Erdbodens die negativen Ionen normal nach oben strömen, aber keine Nachlieferung vom Erdboden her erfolgt, so stellt sich ein Mangel in der Nähe des Erdbodens ein. Dieser Mangel verschwindet theoretisch erst in bestimmter Höhe über dem Erdboden, nämlich am Ende der Bahn, welche ein Ion im Laufe seiner Lebensdauer zurücklegt[2]. In Wirklichkeit wird dieser Elektrodeneffekt meist durch die Turbulenz der Luftbewegung am Erdboden und außerdem durch die am Erdboden verstärkte Ionisierung — auf diese Tatsache hat CHALMERS[3] hingewiesen — praktisch zunichte gemacht. Lediglich bei Windstille kann ein eindeutiger Unterschied zwischen der Zahl der negativen und der Zahl der positiven Ionen, wie erwartet, gemessen werden, woraus eine Raumladung von ungefähr 500 Elementarladungen pro Kubikzentimeter resultieren kann.

[1] H. KASEMIR: Diss. Aachen 1954.
[2] H. LETTAU: Gerlands Beitr. Geophys. **52**, 365 (1941).
[3] J. A. CHALMERS: Quart. J. Roy. Met. Soc. **72**, 199—205 (1946).

34. Konvektionsstrom. Eine horizontal angebrachte Platte zur Messung des Vertikalstromes nimmt auch Ladung niederfallender Teilchen auf, also von Niederschlags- und sinkenden Nebelteilchen und von sedimentierenden Aerosol- und Staubteilchen. Letztere haben wegen ihrer geringen Zahl elektrisch keine Bedeutung, dagegen führen die Niederschlagsteilchen erhebliche Elektrizitätsmengen mit sich.

Es können gemessen werden:

1. die elektrischen Ladungen auf den einzelnen Teilchen;
2. die Gesamtniederschlagsladung je Zeiteinheit, also der Niederschlagsstrom;
3. die Gesamtniederschlagsladung je Raum- oder Masseneinheit des Niederschlages.

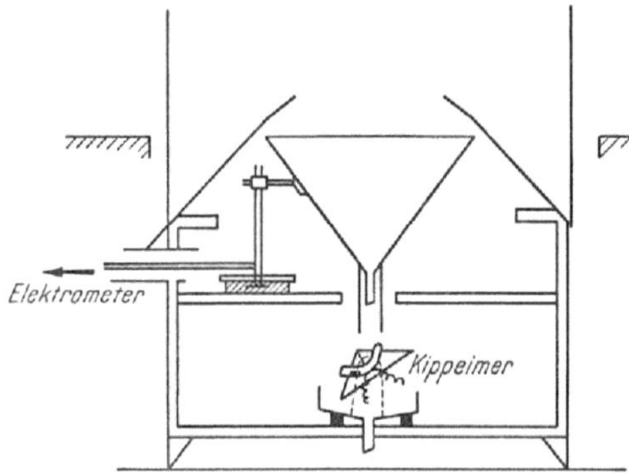

Fig. 38. Gerät zur Messung der Regenladung und Regenmenge nach SCRASE.

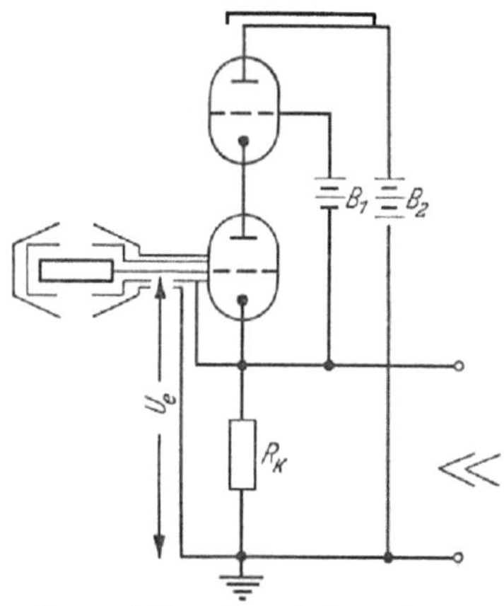

Fig. 39. Influenzringmethode zur Messung von Tropfenladungen.

Gesamtladung und Niederschlagsmenge erhält man durch eine Anordnung von SCRASE [57] gemäß Fig. 38. Dabei dienen die seitlich angebrachten Stäbe zur Abschirmung des elektrischen Feldes, die Abdeckbleche unten zur Verhinderung des Verlustes von Ladungen abspritzender leichter Teilchen und die Kippvorrichtung am Wasserauslauf zur Messung der Regenmenge.

Der *Regenstrom* i_R, also die vom Regen mitgeführte Ladung je Sekunde, und die *Regenladung* q, also die Ladung pro Kubikzentimeter Regenmenge, zeigt nach Stärke und Vorzeichen einen Zusammenhang mit dem luftelektrischen Feld. Werte von i_R sind in Tabelle 13 aufgeführt. Das Vorzeichen ist im allgemeinen dem Potentialgradienten entgegengesetzt („Mirror image effect"), was durch den Influenzeffekt (Ziff. 11) erklärt werden kann. SIMPSON hat diese Beobachtung in der Formel ausgedrückt, welche für kleine Feldstärken gilt:

$$q \approx -5 \cdot 10^{-12} (E - 4) \text{ (Clb/cm}^3\text{)},$$

wobei E der Potentialgradient in Volt/cm ist. Eine Begründung des Befundes, daß die Regenladung q gerade für den Schönwetterwert des Feldes am Meßort Kew ($P = +4$V/cm) $= 0$ wird, steht aus. Für große Feldstärken ist kein klarer Zusammenhang mehr zu erkennen. Lediglich bei kräftigem Niederschlag und gleichzeitigem Spitzenstrom scheint es so, als ob der letztere insgesamt wieder mit dem Niederschlag zur Erde zurückkehrt.

Zur Messung der Einzeltropfenladungen eignet sich die Methode von GUNN[1]; beim Fall geladener Teilchen durch einen isolierten und abgeschirmten Ring nach Fig. 39 wird durch Influenz ein Spannungsstoß erzeugt, welcher verstärkt

[1] R. GUNN: Phys. Rev. **71**, 181 (1947).

und registriert werden kann. Die unterste Grenze der meßbaren Ladung war bei GUNN etwa $3 \cdot 10^{-12}$ Clb, bei einer neuen Ausführung von MÜHLEISEN und HOLL[1] 10^{-15} Clb.

Die Ergebnisse einiger Messungen von Einzeltropfenladungen sind in Tabelle 13 aufgeführt.

Tabelle 13. *Elektrische Ladung einzelner Niedersschlagsteilchen und Größe von Niederschlagsströmen.*

	Q_{mittel} Coulomb	Q_{ma} Coulomb	$i_{R\,\text{mittel}}$ Amp/cm²	$i_{R\,\text{max}}$ Amp/cm²
Landregen	10^{-15} — 10^{-14}	$5 \cdot 10^{-13}$	$5 \cdot 10^{-15}$	$5 \cdot 10^{-14}$
Schauer-Gewitter .	10^{-13} — 10^{-12}	$5 \cdot 10^{-12}$	$1 \cdot 10^{-14}$	$5 \cdot 10^{-12}$
Graupelschauer . .	10^{-11}	10^{-10}		$1 \cdot 10^{-11}$
Schnee	10^{-13} — 10^{-11}	$5 \cdot 10^{-11}$	$1 \cdot 10^{-14}$	$5 \cdot 10^{-13}$

Bemerkenswert ist, daß in allen Regen gleichzeitig Ladungen beider Vorzeichen vorkommen. Von vielen Beobachtern wird für das Verhältnis der Zahl der positiv geladenen zu derjenigen der negativ geladenen Tropfen etwa 1,5:1 angegeben[2-5] [*53*]; die Werte schwanken aber im einzelnen von 1,1:1 bis 30:1. Die Beziehung zwischen Tropfenladung und Tropfengröße hat SMITH[6] ausführlich untersucht.

Aus allen gefundenen Beziehungen darf man folgern, daß für die Regenladungen am Erdboden vor allem der WILSONsche Influenzeffekt (Ziff. 11) maßgebend ist, erst in zweiter Linie — und vor allem bei den größeren Tropfen — die Anfangsladung in der Wolke.

35. Spitzenstrom. Der Ausgleich der elektrischen Ladungen von den Gewitterwolken zur Erde erfolgt nur zum geringen Teil durch den Leitungsstrom. Bei hohen Feldstärken treten Corona-Entladungen von allen hochgelegenen Punkten an der Erdoberfläche, also von Masten, Häusern, Bäumen, Gräsern usw. auf; diese Spitzenströme fließen in der Größenordnung von Mikroampere und sind demgemäß in der Lage, wesentlich größere Ladungen zu transportieren als der normale Leitungsstrom.

Den Zusammenhang zwischen Spitzenstrom und elektrischer Feldstärke und Höhe der Spitze haben unter anderen WHIPPLE und SCRASE [*58*], SIMPSON und SCRASE [*12*] und CHALMERS[7] untersucht. Von den ersteren stammt die Formel

$$i_S = C_1 (E^2 - E_0^2). \qquad (35.1)$$

Es ist i_S der Strom einer Spitze, E das elektrische Feld in der Ebene, E_0 die Einsatzfeldstärke, unterhalb welcher kein Spitzenstrom fließt. Die Konstante C_1 hängt von den Dimensionen der Spitze ab, E_0 noch vom Vorzeichen des Feldes. Ein der Wirklichkeit gut entsprechendes Experiment stammt von SCHONLAND[8], welcher einen 4 m hohen Baum unten abgesägt und auf Isolatoren gestellt hat und damit den gesamten Spitzenstrom messen konnte. Er erhielt die in der

[1] R. MÜHLEISEN u. W. HOLL: Geofisica Pura e Applicata **25**, 61—70 (1953).
[2] J. A. CHALMERS u. F. PASQUILL: Proc. Phys. Soc. Lond. **50**, 1 (1938).
[3] S. K. BANERJII: Quart. J. Roy. Met. Soc. **56**, 305 (1930).
[4] F. J. SCRASE: Met. Off. Geophys. Mem. 1938, Nr. 75.
[5] B. F. J. SCHONLAND: Proc. Roy. Soc. Lond., Ser. A **118**, 233 (1928).
[6] L. G. SMITH: Quart. J. Roy. Met. Soc. **81**, 23—47 (1955).
[7] J. A. CHALMERS: J. Atmos. a. Terres. Phys. **3**, 346—347 (1953); **6**, 149—159 (1955).
[8] B. F. J. SCHONLAND: Atm. Electr., [*11 a*], S. 39. London: Methuen & Co. 1953.

Fig. 40 aufgezeichneten Werte. Andere Untersuchungen führten zu dem Ergebnis, daß die natürlichen Spitzen in einem mittleren effektiven Abstand von 25 m stehen[1].

Für manche Untersuchungen bei Gewittern und Schauern sind die Spitzenströme ein bequemes Meßmittel für die Feldstärke, da sie wegen ihrer Größe ohne Verstärkung registriert werden können. Leider ist aber die Beziehung (35.1) nur in grober Näherung zutreffend und noch keine bessere, allgemein gültige Formel bekannt.

Der Spitzenstrom besteht aus Ionen eines Vorzeichens, bei positivem Potentialgradienten also negativen Ionen. Diese Ionen können nicht rekombinieren; sie können sich lediglich an Kondensationskerne und Groß-Ionen anlagern. Sie bilden oberhalb der Spitzen Raumladungen, bei positivem Feld also negative Raumladungen. Diese bewirken zunächst direkt an der Spitze, daß der Strom nicht in Form eines Gleichstromes, sondern in Form von Stromstößen, sog. TRICHEL-Stößen[2], fließt. Ebenso ist die Raumladungsdichte auf der Wegstrecke der Klein-Ionen sehr hoch, wo sich diese an Kondensationskerne und Groß-Ionen anlagern und damit ihre Beweglichkeit fast ganz einbüßen. Durch diese Raumladungen sind während des Auftretens von Spitzenströmen die am Erdboden beobachteten elektrischen Feldstärken viel geringer als in größeren Höhen[3], [*58*]. Wegen der gleichen Ursache ist die Beziehung zwischen Spitzenstrom und Feldstärke (35.1) abhängig von der jeweiligen Umgebung der Spitze und der Luftbewegung, welche die Raumladungen wegschafft.

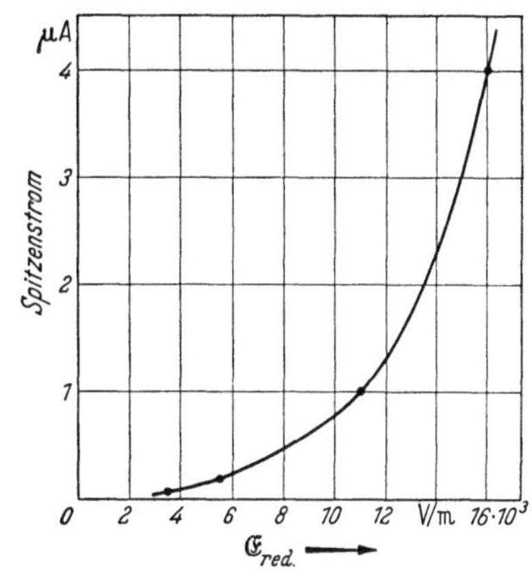

Fig. 40. Spitzenstrom gemessen von SCHONLAND an einem Baum von 4 m Höhe.

36. Ladungstransport durch Blitze. Die Entstehung und der Verlauf eines Blitzes wird durch die TOEPLERsche Theorie[4] des elektrodenlosen Funkens erklärt. Danach genügt als Voraussetzung für einen Blitz eine Feldstärke von 5000 bis 10000 V/cm in einem Punkt. Von diesem aus bildet sich durch Stoß-Ionisation ein Blitzkanal, der infolge der hohen Beweglichkeit der entstehenden Elektronen gegenüber der geringen Leitfähigkeit der Wolkenluft als Leiter angesehen werden kann. An den beiden Enden des Blitzkanals bilden sich laufend zur weiteren Stoßionisation ausreichende Feldstärken aus, solange der Blitz in Raumladungsgebiete entgegengesetzten Vorzeichens hineinstößt[5,6]. Die Fig. 41a und b machen dies für den Fall einer einfachen bipolaren Ladungsverteilung deutlich. Am Kopf des Blitzkanals ist die Feldstärke größer als an der betreffenden Stelle vor der Entladung. Dies erfolgt solange, bis ein Punkt mit der Feldstärke 0 im vorher ungestörten Raumladungsgebiet erreicht ist; von da an vermindert sich die Feldstärke am Kopf des Blitzkanals, und der Blitz findet sein Ende, wenn die Feldstärke am Blitzkopf die Durchbruchs-Feldstärke unterschreitet. In diesem Fall schlägt der Blitz nicht bis zur Erde durch, ist also ein *Wolkenblitz*.

[1] Siehe Fußnote 7, S. 598.
[2] G. W. TRICHEL: Phys. Rev. **54**, 1078 (1938).
[3] C. W. LUTZ: Gerlands Beitr. Geophys. **57**, 317 (1941).
[4] M. TOEPLER: Phys. Z. **20**, 299 (1919). — Mitt. der Hermsdorf-Schomburg Isolatoren G.m.b.H., Heft 25, 1926.
[5] B. F. J. SCHONLAND, D. J. MALAN u. H. COLLENS: Proc. Roy. Soc. Lond. **168**, 455 (1938).
[6] H. W. KASEMIR: Probl. kosm. Phys. **25** (1950).

Nach der Blitzentladung stellt sich sehr rasch die alte Feldverteilung wieder her, da die Raumladungen infolge ihres Sitzes auf den im elektrischen Feld praktisch unbeweglichen Wolkenteilchen nur zum kleinsten Teil mit dem Blitzkanal in Berührung kommen und weggeführt werden. Die Zeit zur Wiederherstellung der alten Ladungsverteilung (recovery time) kann ≥ 5 sec betragen[1]; sie ist ein Anzeichen für die Wirksamkeit des Vorgangs der Ladungserzeugung in der Wolke.

Wenn die positive Ladung im Wolkenkopf eine größere Elektrizitätsmenge enthält als die darunterliegende negative Partie, dann ist es möglich, daß der Blitz ein genügend großes Feld an seinem unteren Ende bis zur Berührung des Erdbodens vorfindet. Während er im Stadium des Wolkenblitzes noch die Gesamtladung 0 hat, aber gegenüber der Erde auf einem positiven oder negativen Potential liegen kann, nimmt der

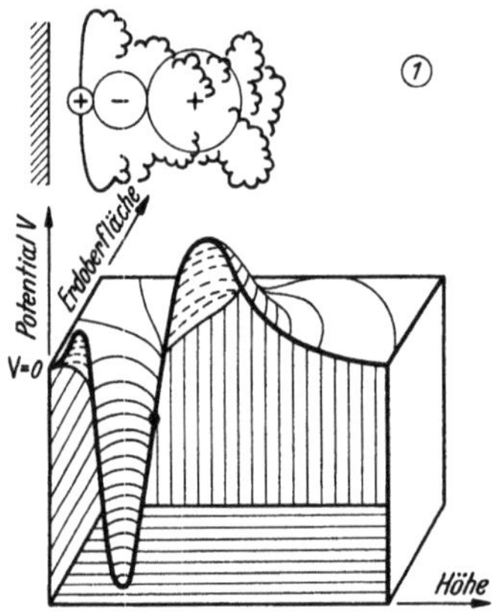

a

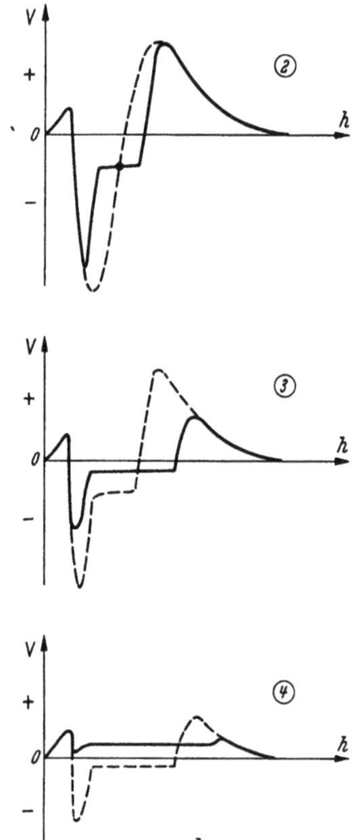

b

Fig. 41a u. b. a Potentialrelief eines Wolkentripols und b Potentialverlauf bei einer Entladung vom Punkt ● aus

Blitzkanal infolge seiner hohen Leitfähigkeit im Augenblick der Erdberührung das Potential 0 an und gleicht damit die am Erdboden angesammelte, dem ursprünglichen Potential des Blitzkanals und seiner Kapazität entsprechende Influenzladung aus. Diese zweite Phase, die nur beim *Erdblitz* auftritt, bezeichnet man als *Hauptentladung*, häufig auch als Rückentladung *(return stroke)*, bei welcher der größte Strom in derselben Richtung wie beim vorhergehenden *leader stroke* fließt. Im Augenblick der Erdberührung entsteht ein verstärktes elektrisches Feld am oberen Ende des Blitzkanals, und damit die Möglichkeit zum Weiterwachsen. Dadurch kann es zu weiteren Teilentladungen kommen.

Die Einzelheiten der Blitzentladungen werden mit optischen und elektrischen Methoden untersucht: Boys-Kamera[2], Blitzoszillographie[3] und magnetische Messungen[4]. Man erhält für die Gesamtdauer eines Blitzes im Mittel einige

[1] T. W. Wormell: Phil. Trans. Roy. Soc. Lond., Ser. A **238**, 289 (1939).
[2] C. V. Boys: Nature, Lond. **118**, 749 (1926).
[3] McEachron K. B.: Ionon. Franklin Inst. **227**, 149 (1939).
[4] C. M. Fonso u. H. P. Kuehni: Gen. Electr. Rev. **35**, 644 (1932).

Zehntelsekunden, gelegentlich aber auch bis zu 1 sec; die kürzesten *Teilentladungen* dauern 1 bis 20 μsec. Bei Erdblitzen kommen bis zu 25 Teilentladungen in einem Blitz vor mit einem mittleren Abstand von rund $^1/_{10}$ sec.

Beim Wolkenblitz werden nur geringe Ladungen ausgeglichen; dagegen gehen beim Erdblitz größere Ladungen von der Wolke zur Erde. Die Größe der transportierten Elektrizitätsmengen beträgt in einem Erdblitz etwa 10 bis 20 Clb im Mittel, maximal 160 Clb, wobei die Vorentladung etwa 0,5 bis 1 Clb enthält. Ihre Polarität ist meist negativ: es gehen etwa 3 (bis 15)mal so viel negative Blitze wie positive zur Erde. Die in Erdblitzen vorkommenden *Stromstärken* liegen zwischen 2000 und 200000 Amp. Sie wurden zumeist mit Magnetstäbchen gemessen, welche durch den Blitzstrom magnetisiert werden.

Über die Blitzforschung berichtet ausführlich SCHONLAND in Band XXII dieses Handbuches.

E. Die Elektrizitätsbilanz in der Atmosphäre.

37. Ersatzschaltbild des luftelektrischen Stromkreises. Das elektrische Feld in der Atmosphäre wird, wie wir gesehen haben, trotz der elektrischen Leitfähigkeit der Luft dauernd aufrechterhalten. Aus dem Säulenwiderstand R erhält man den Widerstand W der Gesamtatmosphäre zwischen Erde und luftelektrischer Ausgleichsschicht zu:

$$W = R/F_E, \qquad (37.1)$$

wobei F_E die Oberfläche der Erde von rund $500\cdot 10^6\ \text{km}^2$ ist; damit wird $W \approx 200\,\Omega$. Den Strom durch diesen Widerstand liefern vor allem die Gewitter, daneben (mit untergeordneter Bedeutung, und nur für die tiefen Schichten der Atmosphäre), sonstige Niederschläge und ortsbegrenzte Raumladungen.

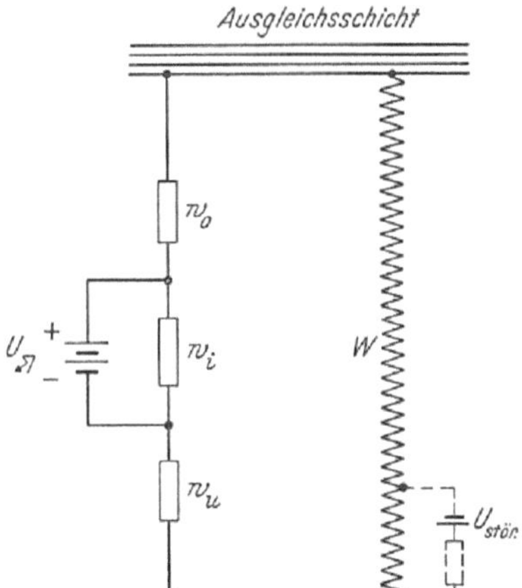

Fig. 42. Ersatzschaltbild des luftelektrischen Stromkreises: für ein Gewitter von 50 km²: $w_0 \approx 4\cdot 10^8\,\Omega$ oberhalb des Gewitters, $w \approx 2\cdot 10^8\,\Omega$ in der Gewitterwolke, $w_u \approx 8\cdot 10^8\,\Omega$ unterhalb des Gewitters; für die Gesamtatmosphäre: $W \approx 200\,\Omega$. $U_{\text{stör}}$ = Störspannungen als Folge niedriger Raumladungen.

Den Stromfluß im „*luftelektrischen Stromkreis*" kann man in einem Ersatzschaltbild der in der Elektrotechnik üblichen Form darstellen. Ein von ISRAEL und LAHMEYER[1] gegebenes Schaltbild soll noch zu der in Fig. 42 gewählten Form erweitert werden durch die Einführung der nur in den niedrigen Luftschichten wirkenden Störquellen. Darin spielen die Leitungswiderstände von Gewitterober- und -unterseite zur Ausgleichsschicht w_0 und Erde w_u und der „Innenwiderstand" der Gewitter w_i als Belastung der „Spannungsquelle" eine bedeutende Rolle; man sieht leicht, daß von der elektrischen Leistung der Gewitter nur ein kleiner Bruchteil für den Vertikalstrom in den Schönwettergebieten übrig bleibt.

Da der Strom durch den Innenwiderstand w_i sich zum Strom über den Außenwiderstand $w_0 + w_u + W$ nach der Rechnung von ISREAL und LAHMEYER wie 6:1 verhält, fließt demnach nur $^1/_7$ der produzierten Strommenge nach außen. Das Spannungsverhältnis ist noch ungünstiger für den Außenkreis und beträgt

[1] H. ISRAEL u. G. LAHMEYER: Terr. Magn. **53**, 373–386 (1948).

etwa $\frac{w_0 + w_u}{Z} : W = 1500:1$, wobei Z die geschätzte Anzahl (etwa 4000) von gleichzeitig wirksamen Gewitterzellen auf der Erde bedeutet.

Diese Zahlen sind rohe Näherungen, denn für die Widerstände im Gewitterbereich ist der normale Leitfähigkeitsverlauf mit der Höhe angenommen worden; dies scheint nach Messungen unter und über den Cumulonimben berechtigt zu sein, ist wahrscheinlich aber für den Wolkenbereich selbst fraglich.

Die kleineren künstlichen Raumladungen beeinflussen nur die untersten Luftschichten, da ihre Oberseiten noch nicht an Räume mit hoher Leitfähigkeit grenzen. Deshalb haben sie nur örtlichen Charakter, beeinflussen aber stark die lokalen Bodenwerte von Potentialgradient und Vertikalstrom.

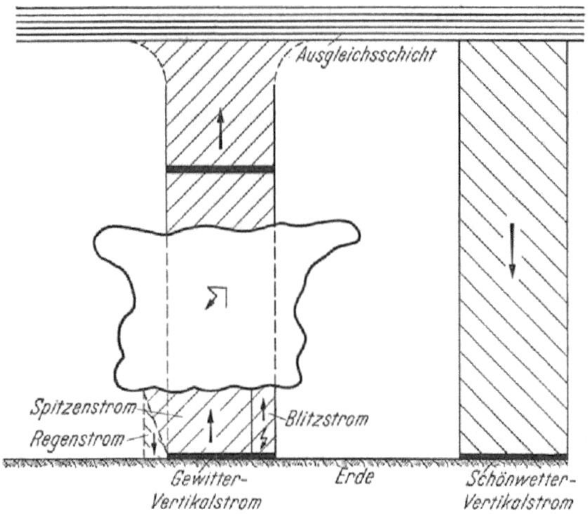

Fig. 43. Elektrizitätsbilanz der Gesamtatmosphäre.

38. Die Leistung der Gewitter und Elektrizitätsbilanz. Die Hypothese von WILSON [14] über die Aufrechterhaltung des Erdfeldes durch den „Gegenstrom" der Gewitter findet ihre Bestätigung durch eine Bilanz der Ströme an Hand des Ersatzschaltbildes von Fig. 43. Auf der Seite der Spannungsquellen stehen die Gewitter mit ihren oben positiven und unten hauptsächlich negativen Ladungen. Ihre Gesamtleistung ergibt sich aus folgenden Daten.

Die mittlere Stromstärke über einer Gewitterzelle haben GISH und WAIT [51] aus gemessener Feldstärke und Leitfähigkeit ermittelt; bei einer angenommenen mittleren Gesamtfläche einer Gewitterwolke von 50 km² ergibt sich etwa 0,5 Amp im Mittel. Auf einen entsprechenden, aber weniger sicheren Wert kommt man über Spitzenstrommessungen am Boden, weil man nur sehr schwer den effektiven Spitzenabstand ermitteln kann. Eine Möglichkeit ergibt sich aus der Beobachtung, daß nahezu die gesamten über Spitzen nach oben fließenden Ladungen bei sehr starkem Niederschlag wieder mit diesem zur Erde zurückkehren; die Spitzenstromstärke dividiert durch die gleichzeitig gemessene Regenstromdichte (Amp cm^{-2}) gibt nach SIMPSON [45] annähernd die Fläche, die der gewählten Spitze äquivalent ist. So gewinnt man ebenfalls einen Stromwert von etwa 0,5 Amp je 50 km² Fläche.

Der Außenwiderstand, über den dieser Strom fließen muß, ist nach ISRAEL[1] und gemäß Fig. 42: $w_0 + w_u + W = 12 \cdot 10^8\ \Omega$. Daraus errechnet man eine mittlere „Klemmenspannung" an einer Gewitterzelle von etwa $6 \cdot 10^8$ V und, beim Strom von 0,5 Amp, eine nach außen abgegebene Leistung von

$$3 \cdot 10^8 \text{ Watt.}$$

Nimmt man nach ISRAEL an, daß der „Innenwiderstand" einer Gewitterzelle w_i aus der Leitfähigkeit in der ungestörten Luftschicht zwischen 3 und 7 km Höhe errechnet werden darf — die wegen der Wolkenelemente verringerte Ionenleitfähigkeit kann durch die Elektronenleitung bei Wolkenblitzen aufgewogen

[1] H. ISRAEL: Geofisica Pura e Applicata **24**, 3—11 (1953).

werden — dann fließt über $w_i \approx 2 \cdot 10^8\,\Omega$ 6mal soviel Strom als über den Außenwiderstand. Dann wäre die elektrische Gesamtleistung einer Gewitterzelle

$$\text{rund } 2 \cdot 10^9 \text{ Watt.}$$

BRAHAM[1] kommt unter anderen Annahmen nur auf ein Neuntel dieses Wertes: $2{,}25 \cdot 10^8$ Watt. Der Versuch, eine Beziehung dieser elektrischen Leistung zu dem mechanisch-thermodynamischen Energieumsatz in einem Gewitter aufzustellen, führt bei beiden Verfassern zu vorläufig stark divergierenden Resultaten.

Nun ist zur Berechnung der Leistung aller Gewitter auf der Erde die Kenntnis ihrer Häufigkeit notwendig. Von BROOKS [*59*] stammt eine Gewitterstatistik für die Gesamterde. Danach sollen im Durchschnitt immer 1800 Gewitter gleichzeitig ablaufen, deren mittlere Dauer 1 Stunde beträgt. Da ein Gewitter durchschnittlich aus 2 bis 3 Zellen besteht, sind also rund 4000 Gewitterzellen je Stunde aktiv. Aus dem Vertikalstrom je Gewitterzelle von 0,5 Amp ergibt sich eine Lieferung von allen Gewittern von

$$\underline{I_{\text{Gew}} \approx -2000 \text{ Amp.}}$$

Dieser Gewittervertikalstrom kann auch aus dem integralen Spitzenstrom, dem Blitzstrom und dem Niederschlagsstrom unterhalb der Cumulonimben ermittelt werden, wie es Fig. 43 veranschaulicht. Je nach Verfasser[2,3] weichen diese Werte erheblich voneinander ab. Für den Spitzenstrom, welcher bei Gewittern und Schauern positive und negative Werte annimmt, wird gezeigt, daß er die Größenordnung von 0,5 Amp je Gewitterzelle haben kann und mehr negative als positive Ladungen zur Erde bringt. Die durch die Erdblitze zur gesamten Erde transportierte Ladung pro Zeiteinheit wird mit

$$I_{\text{Bl}} = -320 \text{ bis } 700 \text{ Amp}$$

angegeben.

Für den Niederschlagsstrom wird im allgemeinen der Strom auf die gesamte Erdoberfläche für alle Arten von Niederschlägen genannt:

$$I_N \approx +320 \text{ Amp.}$$

Da die Masse der Niederschlagsladungen bei Gewittern auftritt, steht also dem Gewitter-Vertikalstrom hauptsächlich der Vertikalstrom in den Gebieten der Erde ohne Niederschlag entgegen; hierfür sind heute folgende Werte am wahrscheinlichsten: der Vertikalstrom über den Ozeanen bei einer mittleren Dichte von $3{,}2 \cdot 10^{-16}$ Amp/cm^2 ergibt

$$I_{V_{\text{Ozean}}} \approx +1200 \text{ Amp;}$$

über den Kontinenten in reiner Luft ist die mittlere Vertikalstromdichte rund $2 \cdot 10^{-16}$ Amp/cm^2, daraus der Gesamtstrom:

$$I_{V_{\text{Land}}} \approx +300 \text{ Amp;}$$

der gesamte Vertikalstrom im niederschlagsfreien Gebieten beträgt demnach

$$\underline{I_V \approx +1500 \text{ Amp.}}$$

Durch die Unterschiede der luftelektrischen Verhältnisse an den verschiedenen Orten, von welchen die Meßwerte zu dieser Bilanz verwendet werden, ist eine bessere Übereinstimmung im allgemeinen gar nicht zu erwarten. Ohne Zweifel ist es deshalb berechtigt, die Gewitter als die hauptsächlichen Urheber

---

[1] R. R. BRAHAM jr.: J. Meteorology **9**, 227—242 (1952).
[2] T. W. WORMELL: Proc. Roy. Soc. Lond. **127**, 567 (1930).
[3] F. J. WHIPPLE: Quart. J. Roy. Met. Soc. **55**, 351 (1929).

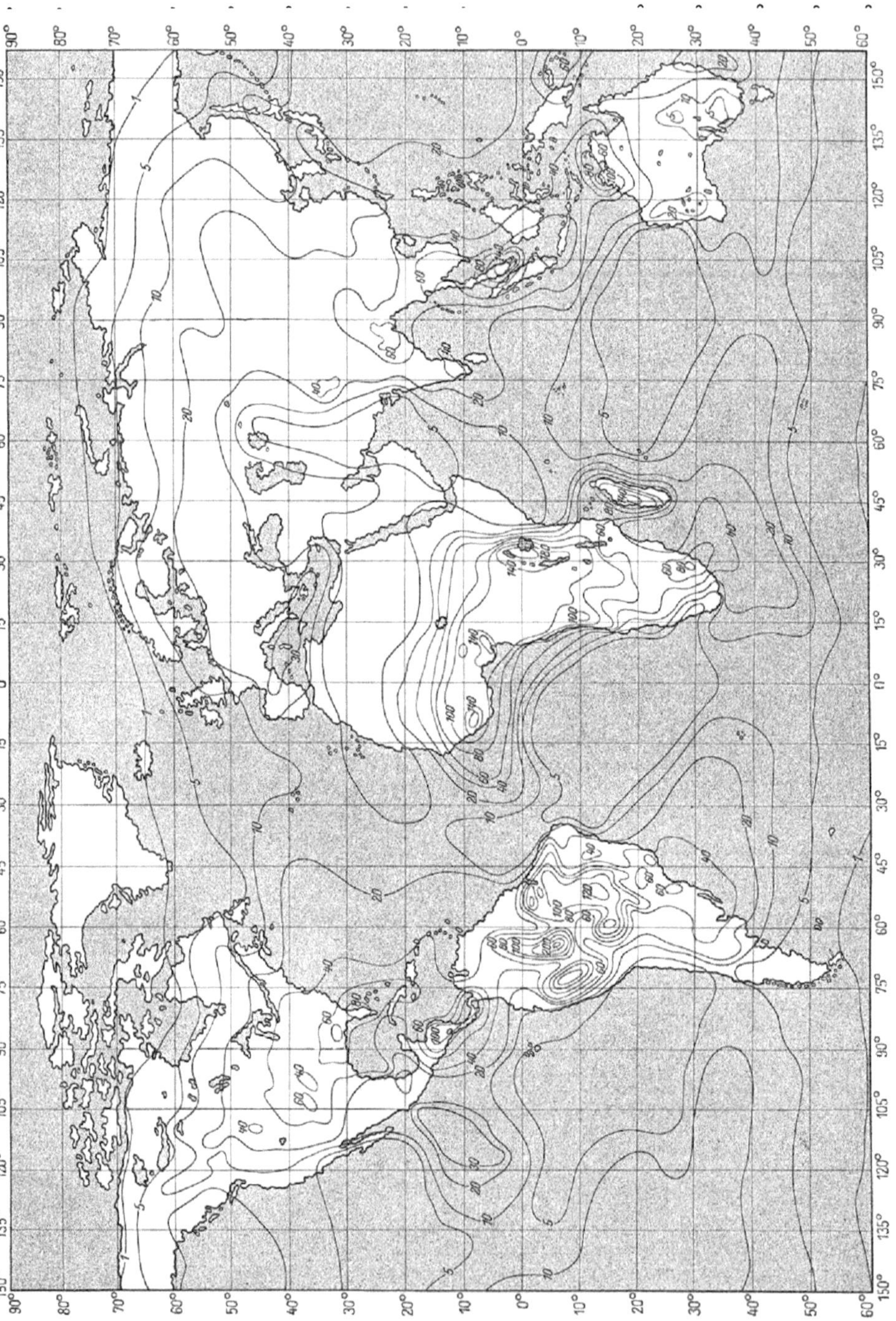

Fig. 44. Mittlere jährliche Häufigkeit der Tage mit Gewitter auf der Erde.

des luftelektrischen Schönwetterfeldes anzusehen, wie es WILSON als erster vermutet hat. Diese Anschauung wurde außerdem durch die gleichlaufende Tages variation der Weltgewittertätigkeit und des Potentialgradienten über den Weltmeeren von MAUCHLY bestätigt. Wie auch eine neueste Statistik der World

Meteorological Organisation[1] über die jährliche Häufigkeit der Tage mit Gewitter veranschaulicht (s. Fig. 44), treten die meisten Gewitter als Wärmegewitter in den frühen Nachmittagsstunden nach Ortszeit am häufigsten in der Nähe des Äquators auf. Wegen der vorgegebenen Verteilung der Landmassen nach der geographischen Länge ergibt sich so das Maximum der Gewittertätigkeit gegen 19 Uhr GMT (Weltzeit) und das Minimum um 4 Uhr GMT.

Bibliographie.

Zu A. Überblick.

[*1*] ANIOL, R.: Der Kugelblitz. (Eine Bibliographie.) Met. Rdsch. **1**, 220—222 (1954). (Kurze Zusammenstellung von Beobachtungen und Hypothesen.)

[*2a*] BAUER, L. A., u. W. F. G. SWANN: Results of atmospheric-electric observations made aboard the "Galilee" (1907—1908) and the "Carnegie" (1909—1916). Res. Dept. Magn., Carnegie Inst. Wash. Publ. No. 175 **3**, 361—422 (1917).

[*2b*] MAUCHLY, S. J.: Studies 1915—1921, l. c. Bd. **5**, 388—402 (1926).

[*3*] CHALMERS, J. A.: Atmospheric electricity. Oxford: Clarendon-Press 1949. (Anschauliche, ausführliche Darstellung des Gesamtgebiets mit Bibliographie bis 1947.)

[*4*] CHALMERS, J. A.: Atmospheric electricity. Rep. Progr. Phys. **17**, 101—134 (1954). (Übersicht und Diskussion neuerer Arbeiten bis 1953 auf dem Gesamtgebiet.)

[*5*] DRUMMOND, A. J.: A century of progress in atmospheric electricity at Kew Observatory. Ref. Met. Rdsch. **1**, 60 (1947).

[*6*] FRENKEL, J.: A theory of the fundamental phenomena of atmospheric electricity. J. of Phys. **8**, 285—304 (1944). (Versuch einer Feldtheorie auf Grund des Voltapotentials von Wolkenelementen.)

[*7*] GISH, O. H.: Universal aspects of atmospheric electricity. Comp. of Met., Amer. Met. Soc., S. 101—119, 1951. (Allgemeine Übersicht über Tatsachen und offene Probleme.)

[*8*] ISRAEL, H.: Das Gewitter. Leipzig: Akademische Verlagsgesellschaft 1950. (Einzelbeiträge zur Meteorologie des Gewitters, zu den Gewittertheorien, den Entladungen; ausführliches Schrifttum 1939—1949.)

[*9*] ISRAEL, H.: Atmosphärische Elektrizität. Zusammenfassung der Arbeiten von 1936 bis 1946 und Literaturangaben in Fiat Rev (= Naturforschung und Medizin in Deutschland 1939—1946) Bd. 17, Teil 1, S. 165—183.

[*10*] ISRAEL, H.: Luftelektrizität. In LANDOLT-BÖRNSTEIN, Bd. 3 Astronomie und Geophysik, Kap. Meteorologie, S. 704—717. 1952.

[*11*] MÜHLEISEN, R.: Die luftelektrischen Elemente im Großstadtbereich. Z. Geophys., Sonderbd. **1953**, 142—159. (Untersuchung und Beschreibung der Ursachen für verschiedene Arten luftelektrischer Störungen, hauptsächlich durch künstliche Raumladungen.)

[*11a*] SCHONLAND, B. F. J.: Atmospheric electricity. London: Methuen & Co. 1953. Gute Einführung.

[*12*] SIMPSON, G. C., and F. J. SCRASE: The Distribution of Electricity in Thunderclouds. Proc. Roy. Soc. Lond., Ser. A **161**, 309 (1937). (Ergebnisse von Messungen der Polarität der Wolkenladungen mit Hilfe der Spitzenströme mittels Ballonaufstiegsgeräten.)

[*13*] *Thunderstorm Electricity*: Report of Conference on Thunderstorm Electricity, held at Chicago April 10—14, 1950. Publ. by Department of Met. Univ. of Chicago, Oct. 1950. (Aufbau und Dynamik der Gewitter, elektrische Ladungserzeugung, Geophysikalische Aspekte, Blitze und Atmospherics.)

[*14*] WILSON, C. T. R.: Gegenstromtheorie zur Aufrechterhaltung des luftelektrischen Feldes. Phil. Trans. Roy. Soc. Lond., Ser. A **221**, 73ff. (1920).

[*15*] WORMELL, T. W.: Atmospheric electricity; some recent trends and problems Quart. J. Roy. Met. Soc. **79**, 3—38 (1953).

Zu B. Elektrische Ladungen in der Atmosphäre, Erzeugung und Vernichtung.

Vergleiche auch dieses Handbuch, Bd. XXI, Elektronen-Emission, Gasentladungen I. 1956.

[*16*] BRICARD, P.: L'équilibre ionique de la basse atmosphère. J. Geophys. Res. **54**, 39 (1949). (Erweiterte Theorie des Ionisationsgleichgewichts in atmosphärischer Luft.)

[*17*] BURKHARDT, H., u. H. FLOHN: Die atmosphärischen Kondensationskerne. Berlin 1939. (Zusammenfassende, ausführliche Darstellung.)

[1] W. M. O. BULLETIN **5**, 1 (1956).

[18] HESS, V. F.: Die Ionisierungsbilanz der Atmosphäre, Ergebnisse der Kosmischen Physik, Bd. 2, S. 95—152. Leipzig 1933. (Ausführliche Darstellung der Ionisierungsbilanz der Atmosphäre, zahlreiche Meßergebnisse.)
[19] HOGG, A. R.: The intermediate Ions of the Atmosphere. Proc. Phys. Soc. **51**, 6 (1939). (Es werden Gruppen von Mittelionen mit diskreten Beweglichkeiten in schwefelsäurehaltiger Luft gefunden, Masse 2200 Moleküle und ein Vielfaches davon.)
[20] ISRAEL, H.: Schwere Ionen in der Atmosphäre. Z. Geophys. **7**, 123 (1931).
[21] ISRAEL, H.: Ionen und Kerne; eine kritische Studie. Gerlands Beitr. Geophys. **57**, 261—282 (1942).
[22] JOUAUST, R.: L'Ionosphère. Paris 1946.
[23] JUNGE, CHR.: Die Konstitution des atmosphärischen Aerosols. Ann. Met. Hamburg, Beih. **1952**. 1. Bildung der Kondensationskerne und ihr Wachstum. 2. u. 3. Bisherige Staubzählmethoden und Doppel-Konimeter. 4. Kernwachstumsmessungen nach der Spinnfadenmethode von DESSENS. 5. Elektronen-Mikroskopaufnahmen. 6. u. 7. Zusammenfassung und Literaturverzeichnis.
[24] LUEDER, H.: Vergraupelungselektrisierung als Ursache der Gewitterelektrizität. Z. angew. Phys. **3**, 288 (1951). (Experimentelle und theoretische Untersuchung.)
[25] NEWELL jr., H. E.: High Altitude Rocket Research. New York: Acad. Press Inc. 1953. (Enthält Daten über gemessene Ionenkonzentrationen in der Hochatmosphäre.)
[26] NOLAN, J. J.: The effect of water vapour on the mobilities of negative ions in air. Proc. Roy. Ir. Acad. **39**, 82—99 (1930).
[27] NOLAN, P. J.: Recombination of ions in atmospheric air. II. The law of recombination of ions and nuclei. Proc. Roy. Ir. Acad. **38—41**, 49—59 (1927—1933).
[28] NOLAN, P. J.: The Influence of Condensation Nuclei and Dust particles on atm. Ionisation. Proc. Roy. Ir. Acad. **41**, 61—69 (1933).
[29] NOLAN, P. J., and C. O'BROLCHAIN: Recombination of ions in atmospheric air. I. Investigation of the Decay-coefficient by SCHWEIDLER's Method. Proc. Roy. Ir. Acad. **38**, 40—48 (1929).
[30] NOLAN, J. J., and J. G. O'KEEFFE: Multiply charged Large Ions. Proc. Roy. Ir. Acad. **41**, 26—40 (1933).
[31] PARKINSON, W. D.: Review of the literature on the ion balance of the lower atmosphere since 1939. Int. Assoc. Terr. Magn. Electr. Bull. Nr. 13 Trans. Oslo 1948 Meeting, S. 307, 1950.
[32] PLUVINAGE, PH.: Etude Théorique et expérimentelle de la Conductibilité électrique dans les nuages non orageux. Ann. Géophys. **2**, 31—54 (1946).
[33] RAWER, K.: Die Ionosphäre, ihre Bedeutung für Geophysik und Radioverkehr. Groningen, Holland: P. Noordhoff N. V. 1953.
[34] REYNOLDS, S. E.: Thunderstorm-precipitation growth and electrical charge generation. Bull. Amer. Met. Soc. **34**, 117—123 (1953).
[35] ROSSMANN, F.: Luftelektrische Messungen mittels Segelflugzeugen. Ber. dtsch. Wetterdienst. US-Zone **1950**, Nr. 15. (Feld- und Ionen- und Konimetermessungen bis zur Höhe von 7000 m.)
[36] SCHWEIDLER, E. v.: Zur Berechnung des Ionisationsgleichgewichts in kernhaltiger Luft. Gerlands Beitr. Geophys. **57**, 283 (1941). (Theoretische Untersuchungen.)
[37] WAIT, G. R., and W. D. PARKINSON: Ions in the Atmosphere. Comp. of Met., Amer. Met. Soc., S. 120—127, 1951. (Über Lebensdauer, Beweglichkeit der Klein-Ionen, Ionisationsgleichgewicht u. a.)
[38] WILSON, C. T. R.: Some thundercloud problems. J. Franklin Inst. **208**, 1 (1929). (Elektrische Ladungsaufnahme fallender Wassertropfen durch den Influenzeffekt.)
[39] WORKMAN, E. J., and S. E. REYNOLDS: A suggested Mechanism for the Generation of Thunderstorm-Electr. Phys. Rev. **74**, 709 (1948). (Ladungstrennung an Phasengrenzen des Wassers erzeugt hohe elektrische Spannungen.)
[40] WORKMAN, E. J., and S. E. REYNOLDS: El. Phenom. occuring during the freezing of dilute aqueous Solution. Phys. Rev. **78**, 254 (1950). (Fortsetzung der Versuche von 1948.)

Zu C. Das luftelektrische Feld.

[41] BYERS, H. R., and R. R. BRAHAM: Thunderstorm structure and circulation. J. Meteorology **5**, 71—86 (1948).
[42] FAUST, H.: Kaltfronten und Gewitter. Ber. dtsch. Wetterdienst. US-Zone **1951**, Nr. 24.
[43] GUNN, ROSS: Electric Field Intensity Inside of Natural Clouds. J. Appl. Phys. **19**, 481—484 (1948). (Feldmessungen bei Wolkendurchflügen.)
[44] HERATH, F.: Versuchsergebnisse der luftelektrischen Höhenforschung am Aeronaut. Observat. Lindenberg. Ber. dtsch. Wetterdienst. US-Zone **1951**, Nr. 22. (Messungen mittels Drachen und Fesselballon.)

[45] SIMPSON, G. C.: Atmospheric electricity during disturbed weather. Geophys. Mem. Met. Off. London **1949**, No 84, 1—68. (Ergebnisse umfangreicher Untersuchungen des Potentialgradienten und der Regenelektrizität und ihrer Beziehungen.)

[46] WALL, E.: Das Gewitter. I—III. Wetter u. Klima **1**, 7, 65, 193, 321 (1948). I. Die Elementarvorgänge der Elektrizitätsproduktion. II. Der Asymmetrie-Effekt. III. Das Gewitter als Ganzes.

[47] WICHMANN, H.: Grundprobleme der Physik des Gewitters. Wolfenbüttel u. Hannover 1948. — Zur Theorie des Gewitters. Arch. Met., Geophys. u. Biokl., Ser. A **5**, 187—230 (1953). (Behandlung verschiedener Gewitterprobleme: Meteorologie, elektrische Felder, Ladungstrennung, Blitzentstehung und -wirkungen. Theorien.)

Zu D. Die elektrischen Ströme in der Atmosphäre.

[48] BANERJI, S. K., and S. R. LELE: Electric charges of raindrops. Proc. Nat. Inst. Sci. India **18**, 93 (1952). Ref. Phys. Abstr. **55**, 9248 (1952). (Über Messungen der Ladung einzelner Tropfen und theoretische Überlegungen zur Bildung der Tropfenladungen.)

[49] BOWEN, E. G.: Radar observations of rain and their relation to mechanisme of rain formation. J. Atmos. Terres. Phys. **1**, 125 (1950). Phys. Ber. **31**, 692 (1952). (Radar-Echos erlauben Aussagen über Kondensations- und Gefrierniveaus und über die Phasen des Wassers.)

[50] CHALMERS, J. A., and W. W. MAPLESON: Point discharge currents from a captive Balloon. J. Atmos. a. Terres. Phys. **6**, 149—159 (1955).

[51] GISH, O. H., and G. R. WAIT: Thunderstorms and the Earth's General Electrification. J. Geophys. Res. **55**, 473 (1950). (Ergebnisse von Feld- und Leitfähigkeitsmessungen beim Flug über Gewitterwolken.)

[52] GUNN, ROSS: Electronic apparatus for the determination of the physical properties of freely falling raindrops. Rev. Sci. Instrum. **20**, 291—296 (1949). (Ladung, Masse und Geschwindigkeit einzelner Tropfen wird gleichzeitig erfaßt.)

[53] GUNN, ROSS: The free electric charge on Precipitation inside an active Thunderstorm. J. Geophys. Res. **55**, 171 (1950). (Mit der Influenzringmethode werden beim Durchflug durch Wolken Einzelladungen von Wolkenteilchen und Tropfen bestimmt: max 10^{-10} Clb.)

[54] ISRAEL, H.: Der luftelektrische Vertikalstrom. Ann. Géophys. **10**, 2 (1954). (Zusammenfassung der Meßergebnisse zahlreicher Stationen und Versuch einer Deutung der Tagesgänge.)

[55] SCHONLAND, B. F. J., H. COLLENS and D. J. MALAN: Progressive Lightnings. Proc. Roy. Soc. Lond., Ser. A **143**, 654 (1934); **152**, 595 (1935). (Untersuchungen der Blitzentwicklung an Hand der photographischen Aufnahmen mit einer rotierenden Kamera.)

[56] SCRASE, F. J.: The Air-Earth-Current at Kew-Observatory. Observations of atmospheric Electricity at Kew 1843—1931. Geophys. Mem. Met. Off. London **58**, 1—21 (1933). (Methode und Ergebnisse der Vertikalstrommessungen.)

[57] SCRASE, F. J.: Electricity on Rain. A discussion of Records obtained at Kew Observatory 1935/36. Geophys. Mem. Met. Off. London **1938**, No. 75.

[58] WHIPPLE, F. J. W., and F. J. SCRASE: Point discharge in the electric Field of the Earth. Geophys. Mem. Met. Off. London **1936**, No. 68.

Zu E. Die Elektrizitätsbilanz in der Atmosphäre.

[59] BROOKS, C. E. P.: The distribution of thunderstorms over the globe. Met. Off. Geophys. Mem. **1925**, No. 24. (Gewitterstatistik an Hand der Meldungen meteorologischer Beobachter.)

[60] ISRAEL, H.: Die Höhe der luftelektrischen Ausgleichsschicht. Ann. Géophys. **2**, 253 bis 257 (1952). (Rechnung an Hand von Annahmen über den Leitfähigkeitsanstieg mit der Höhe.)

[61] ISRAEL, H.: Der Elektrizitätshaushalt der Erdatmosphäre. Naturwiss. **29**, 700 (1941). (Betrachtungen zum Elektrodeneffekt, Austausch und Leitfähigkeit in der bodennahen Schicht und zu den Gewittertheorien.)

[62] ISRAEL, H.: Gewitterbeobachtungen in Südwestdeutschland. Ber. dtsch. Wetterdienst. US-Zone **1952**, Nr. 42, 63. (Statistik über den Wasserumsatz, Häufigkeit und Dauer der Gewitter.)

[63] WAIT, G. R.: The relative frequency of thunderstorm throughout the day over the oceans and in the tropics. Terr. Magn. **34**, 237—239 (1929). (Bemerkungen zur Gewitterstatistik.)

Oceanography.

By

H. U. SVERDRUP.

With 26 Figures.

A. Introduction.

The following presentation deals primarily with the general physics of the oceans. Tides and waves will be discussed in separate chapters of this volume, except that a brief treatment of internal waves is given here, because in the presence of internal waves caution has to be excercised when discussing other phenomena. The chemistry of the sea water is touched upon, because in some cases knowledge of the distribution of chemical constituents helps to throw light on the character of the circulation.

When dealing with instruments only the principles of the different types are indicated, but no detailed descriptions are given.

Detailed geographical distributions of temperatures, salinities, currents and so on are not described but an attempt is made to present the physical processes which govern these distributions.

The list of recent literature is necessarily restricted, but numerous further references may be found in many of the quoted papers.

Text-Books.

[1] ISAACS, JOHN D., and C. O. D. ISELIN: Oceanographic Instrumentation. Nat. Acad. of Sci. Nat. Res. Council, Div. of Physical Sci. Publ. 309, pp. 233.

[2] PROUDMAN, J.: Dynamical Oceanography, pp. 409. London: Methuen & Co., and New York: John Wiley & Sons 1953.

[3] SVERDRUP, H. U., M. W. JOHNSON and R. H. FLEMING: The Oceans, pp. 1087. New York: Prentice-Hall 1942.

[4] LANDOLT-BÖRNSTEIN: Zahlenwerte und Funktionen usw., 6. Aufl., Bd. 3, S. 426—541. (Contributions by G. DIETRICH, J. JOSEPH, K. KALLE, W. HANSEN, W. HORN, F. NUSSER.) Berlin: Springer 1952.

B. Oceanographic instruments and methods.

1. Vessels and their equipment. Vessels to be used for oceanographic work should be sturdy, have a large cruising radius and satisfactory accommodations for laboratories and storage. They should have low freeboard in order to ensure safe handling of delicate instruments and ample deck space for winches and gear.

The *winches* vary a great deal in construction from relatively light hydrographic winches (used for collecting water samples, making vertical hauls with small nets, or lowering light instruments) to very heavy winches (needed for obtaining long cores of bottom deposits or undertaking dredging of the bottom at very great depths). Winches for trawling at moderate depths are of intermediate size. The winches are operated electrically or hydraulically, or, in some instances by steam. It is always necessary to be able to control the speed of the winch accurately and quickly.

The *wire ropes* to be used must be flexible, of high tensile strength and, if feasible, corrosion resistant. Stainless steel is superior but expensive, and in practice tinned or galvanised steel ropes are more in use. A *deck laboratory* where water samples can be drawn and thermometers read is essential. Otherwise the needed laboratory space on board depends upon the length of the cruise, the character of the studies to be undertaken and other factors.

2. Soundings and bottom samples. Wire soundings for determining the depth to the bottom have practically been discontinued, and soundings are now made acoustically. The principle is that a signal is sent out from a diaphragm, the echo from the sea bottom is picked up by a hydrophone, and the time interval between signal and echo is recorded. Ultrasonic signals may be used, and the depth may be read off from a continuous graphical record. The instrument is usually adjusted to a sound velocity between 1450 and 1500 m. sec.$^{-1}$, but the average velocity in any given region will generally be somewhat different (see Sect. 15). A correction of a few per cent may have to be applied in order to obtain the exact depth, but in many cases the uncorrected values are reported.

Knowledge of the exact locality at which a sounding is made is equally important as the sounding itself. On the open ocean the *position* of the vessel still has to be determined by dead reckoning and astronomical fixes and cannot be ascertained with the accuracy which in many cases is desirable. Near shore recent development of such aids to navigation as DECCA and SONAR makes it, however, possible to determine the location at sea with very great precision, the principle being that specific radio signals are sent from fixed stations on land and are received on board ship by special instruments, the records of which give the exact position of the vessel.

Samples of the bottom deposits can be obtained by various devices which fall in three categories: dredges, snappers and coring devices. In recent years great emphasis has been placed upon obtaining long undistorted cores of deep sea sediments. This problem has been successfully solved by KULLENBERG[1] who designed the ingenious *piston corer* by means of which up to 20 m. long cores have been taken from great depths (see Sect. 10).

3. Observations of temperature. The sea surface temperature is measured by taking a surface water sample with a bucket and measuring the temperature with an ordinary thermometer. Certain simple precautions are necessary. At shore stations and on board ship thermographs are used for obtaining a continuous record of the temperature near the surface. On board ship the thermometer bulb is generally mounted in one of the intake pipes for the cooling water to the engine and is connected to the recording instrument by a capillary. The temperature is actually recorded at a depth a few metres below the surface, but only in very calm and clear weather will this value deviate significantly from the temperature of the very surface.

A number of countries have collected numerous observations of sea surface temperatures made on board merchant vessels sailing all oceans, and mean monthly charts from all ocean areas have been prepared. For the North Sea and the waters surrounding the British Isles so many observations are made regularly that charts of the surface temperature in individual months are being published under the auspices of the *International Council for the Exploration of the Sea*, Copenhagen (Fig. 1).

[1] B. KULLENBERG: Svensk. Hydr. Biol. Komm. Skr., Ser. Hydr. 1, Nr. 2 (1947).

The *protected reversing thermometer* is used for accurate determination of subsurface temperatures between the surface and the greatest depths. The thermometer is sent down in *set* position with the main bulb at its lower end. At the upper end is a small bulb which is connected with the main bulb by a capillary. At a short distance above the main bulb the capillary tube is bent in a loop below which the capillary is severely constricted and branched. In set position the mercury fills the loop, the capillary and part of the upper small

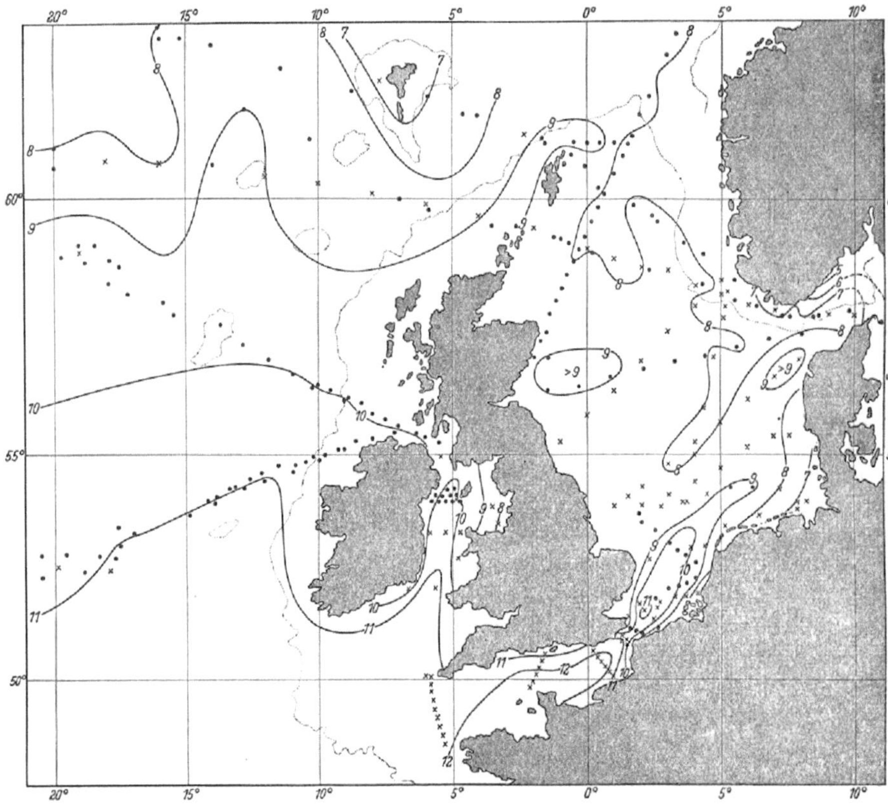

Fig. 1. Surface temperature, December 1954.

bulb, and the amount of mercury above the constriction depends on the temperature. The entire thermometer is enclosed in a glass tube that can withstand the pressure at the greatest oceanic depths. The space between the protecting tube and the main bulb of the thermometer is filled by mercury in order to facilitate the conduction of heat (Fig. 2).

When the thermometer has been lowered to the desired depth and has been left there for a sufficiently long time to attain the temperature of the surrounding water, the thermometer is reversed by some special device. At reversal the column of mercury breaks off at the constriction, the mercury fills the smaller bulb and part of the capillary, thus indicating the temperature when reversal took place.

Generally the thermometer is read on board at a temperature that differs from that at reversal, therefore since the broken off amount of mercury may have expanded or contracted, the reading has to be corrected in order to find the true temperature. A thermometer to determine the temperature at reading is also enclosed in the protecting tube; it has about the same thermal capacity as the broken off part of the main thermometer. The correction can easily be determined with an accuracy of 0.01° (e.g. BÖHNECKE[1]).

For determining the depth of reversal an *unprotected* reversing thermometer is frequently used in conjunction with the protected. The construction of the thermometer itself is similar to that of the protected, but the protecting glass tube is open at one end. The main bulb of the thermometer is, therefore, subjected to pressure and after reversal the reading depends on both the temperature and the pressure when reversal took place. This reading also has to be corrected because of difference in temperature at reversal and reading. After correction the pressure at reversal is obtained from the difference between the corrected reading of the unprotected thermometer and the true water temperature as determined by the protected thermometer. Expressed as depth, the accuracy is about ±5 m. at depths less than 1000 m. and about 0.5% at greater depths.

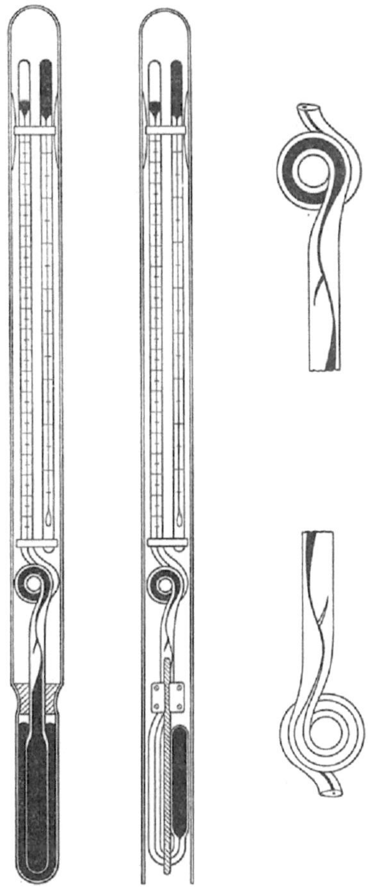

Fig. 2. Deep sea reversing thermometer, protected and unprotected, in set position, that is, before reversal. To the right the constricted part of the capillary is shown.

The reversing thermometers have been in use in their present form since about 1900 and are still the standard instruments. The accuracy which can be obtained by the best constructions is very high and is often increased by using two thermometers in conjunction, in which case the probable error is about ±0.01°. Their disadvantage is that they give only the temperature at the specific depths at which they are reversed, wherefore temperatures between these depths must be obtained by interpolation. For this reason several types of instruments have been constructed which give a continuous record of the temperature when lowered.

The *bathythermograph* (SPILHAUS[2]) is the most widely used recording instrument. The temperature sensitive element operates a stylus that moves over a blackened glass plate which in turn is moved by a pressure element. In this manner a record of temperature against pressure is obtained between the surface and a depth of 150 to 300 m., depending upon the design of the instrument. The bathythermograph has turned out to be especially valuable for the study of details of the temperature structure of the surface layer where sharp changes of the temperature gradient often are found.

[1] G. BÖHNECKE: Dtsch. Atlant. Exped. „Meteor“ 1925—1927, Wiss. Erg. 4, Teil 1, 208. (1932).

[2] A. F. SPILHAUS: J. Marine Res. 1, 95 (1938).

MOSBY[1] has constructed a "*thermosound*" which can be used to any depth. The temperature sensitive element is a specially treated brass wire that is mounted on an invar frame. The length of the brass wire is recorded by a stylus on a disk that is slowly rotated by a propeller when the instrument is being lowered or hauled up.

For special purposes *resistance thermometers* or *thermocouples* are used.

4. Water sampling devices. A variety of water sampling devices has been developed depending on the purpose for which the water sample is to be collected. In principle all water samplers must be designed in such a manner that a sample which has been enclosed at a known depth can be brought to the surface without being contaminated by water from other levels.

One of the most commonly used devices is the NANSEN *reversing water bottle.* This water bottle consists of a cylinder, the ends of which can be closed by stop cocks. When the water bottle is sent down the stop cocks are both open, and the lower end of the bottle is fastened securely to the wire, whereas the upper end is only hooked on the wire. When the desired depth has been reached a small weight, a "messenger", is sent down along the wire. As this strikes the upper end of the bottle, the upper fastening is released and the bottle turns over by its own weight, thereby closing both stop cocks and trapping a water sample.

Generally, reversing thermometers are attached to the water bottle such that the temperature is measured at the depth at which the water sample is taken.

5. Determination of salinity and density. The most common method of determining the salinity of a water sample is to determine the chlorinity by titration with silver nitrate and compute the salinity from the empirically established relation between chlorinity and salinity (Sect. 11).

The salinity can also be obtained from measurements of the electrical conductivity or the refractive index of the sample. Methods based on the use of these properties require strict temperature control because both the conductivity and the refractive index depend also on temperature. Measurements of the electrical conductivity are, however, being used successfully. Instruments have also been designed, which record the electrical conductivity continuously when the instrument is lowered in the sea. From this record and a simultaneous record of the temperature the salinity can be derived.

The density can be *computed* from the chlorinity, using an empirical relation (Sect. 13). For the computation of currents (Sect. 37) the density must be known with an accuracy of about 1×10^{-5} and the salinity, expressed approximately as grams of solid matter in grams of water, must be known with about the same accuracy. The density can also be determined directly by pycnometer or hydrometer, but if these methods are to yield the desired accuracy, they are laborious and time consuming.

Samples of surface water have been collected from all oceans by merchant vessels or other vessels, and have been sent to various national institutes for analysis. On the basis of these, mean monthly charts of surface salinity have been prepared for all oceans. For the North Sea and the waters around the British Isles similar charts are being prepared for individual months by the International Council for the Exploration of the Sea.

6. Observations of currents. Knowledge of the surface currents of the oceans is derived mainly from ship's observations. When a ship's position has been ascertained by an astronomical fix, the next position is computed by means

[1] H. MOSBY: Assoc. Internat. d'Océanogr. Phys. Procès-Verb. **1940**, Nr. 3.

of the course that has been steered and the distance which has been covered according to the log (dead reckoning). A new fix generally shows a slightly different position, and the departure from the position by dead reckoning is taken as resulting from the action of a surface current. Numerous ships report these currents to national institutions which on the basis of the reports have prepared mean monthly charts of surface currents for all oceans as well as other representations. In nearshore regions the ship's observations are supplemented by results from experiments with drift bottles, and close to shore accurate observations of a drifting body may be made by sighting the direction to the body from two or more fixed stations on land, or by other devices.

In recent years the accurate determination of position at sea by such aids to navigation as DECCA or SONAR has made possible detailed studies of surface currents within areas where positions can be ascertained by one of the radio methods. The principle employed is the same as that by means of which currents are derived from the discrepancy between positions by dead reckoning and astronomical fixes, the difference being that instead of using astronomical fixes at intervals of 24 hours or more, exact positions can be obtained every hour or even more frequently by means of the radio signals. The use of this method has revealed numerous details in the structure of the surface currents (Sect. 44).

Still another new method has rendered new details. This is based on the fact that an electrical field is induced as sea water, being a conductor, moves in the magnetic field of the earth[1]. Instruments have been designed for the measurement of the field induced both by surface currents and subsurface currents, but the interpretation of the measured values presents difficulties. The problem is dealt with by LONGUET-HIGGINS et al.[2].

For the measurement of *subsurface currents* so many different types of *current meters* have been designed that it is impossible to give details. A review of the present instruments with a comprehensive list of literature is given by BÖHNECKE[3]. Generally speaking the instruments fall in two categories: the instruments that measure the flow by the *pressure* which is exerted on a given surface and the instruments that measure the flow by counting the number of revolutions of a *propeller* or a cup in a known time interval. As examples of the former type may be mentioned the NANSEN *Pendulum Current Meter* which was designed to be mounted on a tripod to be lowered to the bottom for measuring very weak currents. By means of a clockwork the pendulum is lowered at regular intervals, and a stylus makes a mark on a waxed plate that is oriented by means of a compass needle. In this manner both velocity and direction may be recorded.

EKMAN'S *Current Meter* is the classical example of a propeller instrument by means of which the direction is also recorded, using a magnetic needle and an ingenious arrangement. Other instruments are designed to give a continuous record of the velocity and direction of the current, and each design has its advantages and disadvantages.

Observations of subsurface currents from *anchored vessels* have rendered a number of interesting results, but have also demonstrated the great complexity of the ocean currents. Superimposed upon the large scale *permanent currents* and the oscillating *tidal currents* are numerous other types of motion associated with the presence of series of *internal waves* and *irregular eddies* of varying dimensions. Determination of the large scale currents of the ocean would be a

---

[1] See article by CAGNIARD in this Encyclopedia, Vol. XLVII, p. 435f.

[2] M. S. LONGUET-HIGGINS, M. E. STERN and H. STOMMEL: Papers Phys. Oceanogr. a. Meteorol. **13**, No. 1 (1954).

[3] G. BÖHNECKE: Assoc. Internat. d'Océanogr. Phys. Publ. Sci. **1955**, Nr. 14.

hopeless task if it were to be based on direct measurements, because observations at each locality would have to be carried out continuously at many depths for more than a month in order to eliminate the effects of tidal currents and the many types of irregular flow. Our knowledge of the large-scale currents is, therefore, based on the *computation of currents* from the observed distribution of mass (Sect. 44).

7. Other types of observations. The collection of water samples for determination of salinity and for a number of standard chemical analyses, and the measurements of temperatures represent *routine observations* that are carried out on all oceanographic cruises. Current measurements are made in connection with more special investigations, and besides other types of observations may be undertaken occasionally. Among such may be mentioned measurements of the *penetration of light* by means of photoelectric or photronic cells, often in connection with the use of filters to determine the penetration of various parts of the spectrum and measurements of transparency. The *velocity of sound* may be measured directly instead of being derived from observed values of temperature, salinity and pressure. Recently the *geothermal gradient in bottom deposits* has been measured directly by means of thermometers attached to the barrel of a coring apparatus that may penetrate up to 20 m. into the bottom mud. *Sonic instruments* have also been used for other purposes than obtaining the depth to the bottom. Such echo-sounding equipments have demonstrated the existence of a "*deep scattering layer*", the nature of which has not yet been satisfactorily explained.

Several oceanographers now hold the opinion that for many ocean regions the methods introduced by the turn of the century for determining temperature and salinity and for computing currents have been used to the limit of what they can render. There still exist large areas, such as the South Pacific which must be explored in the classic manner, but in many cases new advances will depend upon more detailed studies which require new approaches both as far as instruments and theoretical considerations are concerned.

C. The ocean basins.

8. Area, volume and depths of the oceans. Of the surface of the earth 361×10^6 km.^{2} (70.8%) are covered by water. From Table 1 in which are entered the ocean and land areas between parallels of latitude it is seen that by far the greater ocean areas are found in the southern hemisphere, particularly between 30 and 60° S. If a finer subdivision had been used, it would be seen that this latitude range extends to about 70° S.

Table 1. *Distribution of water and land between parallels of latitude* (*after* KOSSINA[1], 1921).

Latitude	Northern hemisphere				Southern hemisphere			
°	Water 10^6 km.^{2}	Land 10^6 km.^{2}	Water %	Land %	Water 10^6 km.^{2}	Land 10^6 km.^{2}	Water %	Land %
90—75	7.266	1.496	82.9	17.1	0.522	8.239	6.0	94.0
75—60	9.993	15.652	38.9	61.1	19.721	5.924	76.9	23.1
60—45	17.540	23.137	43.0	57.0	40.087	0.590	98.3	1.7
45—30	29.246	23.586	55.4	44.6	48.098	4.734	91.0	9.0
30—15	40.082	21.261	65.4	34.6	47.035	14.308	76.8	23.2
15—0	50.568	15.149	77.0	23.0	50.901	14.816	77.4	22.6
90—0	154.695	100.281	60.7	39.3	206.364	48.611	80.9	19.1

[1] E. KOSSINA: Veröff. Berl. Inst. Meeresk., N. F. A **1921**, Nr. 9.

Table 2. *Area, volume and mean depth of oceans and seas (after* KOSSINA[1] *and* STOCKS[2]).

Region	Area 10^6 km.^{2}	Volume 10^6 km.^{3}	Mean depth m.
Atlantic Ocean } excluding adjacent seas	82.216	318.078	3868
Pacific Ocean } excluding adjacent seas	165.246	707.555	4282
Indian Ocean } excluding adjacent seas	73.443	291.030	3963
Large mediterranean seas	29.479	45.017	1512
Small mediterranean seas	2.331	0.402	172
Marginal seas	8.079	7.059	874
All oceans and adjacent seas	360.794	1368.471	3793

For further details see LANDOLT-BÖRNSTEIN [4], p. 460 (article DIETRICH).

Geographically the seas of the earth comprise the three *oceans*, the Atlantic, the Pacific and the Indian Ocean, and a number of *adjacent seas* including the *large mediterranean seas*, such as The Mediterranean and the Black Sea, the Arctic Sea and Gulf of Mexico, the *small mediterranean seas* such as the Baltic Sea and the Hudson Bay, and the *marginal* seas such as the North Sea, the Bering Sea and others. Table 2 shows the area, volume and mean depths of oceans and seas.

9. **Bottom topography.** Until about the year 1920 our knowledge of the topography of the ocean bottom was very inadequate because *soundings* had to be carried out by means of *lead and wire*, requiring many hours where the depth to the bottom was great. This time consuming method has been replaced by the use of *echo-sounding* equipment by means of which a continuous record of the depth to the bottom can be obtained from a ship under way. The result is that soundings are being accumulated so fast that it is impossible to keep the *bathymetric charts* up to date.

When only a few deep soundings were yet available, it was generally believed that the deep basins of the oceans were characterised by a flat and contourless bottom, but the echo-soundings have revealed that generally the bottom of the sea is as irregular as the surface of the land. Mountain ranges or isolated mountains rise from great depths without reaching the sea surface, and in several areas deep and narrow trenches are present. The *greatest depth* which so far has been reported, has been found in the Mariana Trench in the Pacific, in about lat. 11° N., long. 143° E. Gr. According to a critical analysis by CARRUTHERS and LAWFORD[3] the depth is 5940 fathoms or 10863 m. Table 3 shows some of the other greatest depths which have been found in the different oceans.

Table 3. *Some of the greatest depths of the oceans.*

Locality		Depth m.
North Atlantic Ocean,	Puerto Rico Trough . .	8750
South Atlantic Ocean,	South Sandwich Trench	8264
North Pacific Ocean,	Japan Trench	10374
	Philippines Trench . .	10497
	Mariana Trench . . .	10863
South Pacific Ocean,	Tonga-Kermadec Trench	9427
	Peru Basin	7635
Indian Ocean,	Sunda Trench	7455

[1] E. KOSSINA: Veröff. Berl. Inst. Meeresk., N. F. A. **1921**, Nr. 9.
[2] TH. STOCKS: Dtsch. Atlant. Exped. „Meteor“ 1925—1927, Wiss. Erg. **3**, Teil 1. (1938).
[3] J. N. CARRUTHERS and A. L. LAWFORD: Nature, Lond. **169**, 601 (1952).

10. Sediments. In general the ocean bed is covered by sediments, but ridges and isolated peaks may be free from sediments because of the action of currents. All sediments in shallow water and most of the sediments in near shore areas are derived from the adjacent land and are called *terrigenous* in contrast to the sediments of the deep ocean basins which are called *pelagic*. The latter may be mainly of organic origin, in which case they are called *oozes*, or they may be mainly of inorganic origin and are called *red clays*. The oozes are calcareous or silicious. Of the former the *globigerina* ooze dominates, formed by the shells of the pelagic *foraminifera*. Of the latter *diatom* ooze, formed by the frustules of diatoms (microscopic plants) covers the larger areas. All told 268 million km.^{2} of the ocean bottom are covered by pelagic deposits; of these, 126 million km.^{2} by globigerina ooze, 31 by diatom ooze and 102 by red clay, and the remaining 9 by other oozes. All these sediments are very fine, the red clay being by far the finest. The median diameters of the particles in the three types are 7.8, 7.0 and 1.1 microns, respectively.

At present the rate of growth of the deep sea deposits is very slow. According to the more recent estimates the average rate of deposition of globigerina ooze is of the order of 2 mm. in 1000 years and that of red clay is about 1 mm. in 1000 years. For red clay values as low as 0.04 mm. in 1000 years have been reported (PETTERSSON and KOTSCHI[1]).

Recently it has become possible to obtain undistorted cores of a length of up to 20 metres, using the KULLENBERG piston corer. In red clay such a core may represent material which has been laid down during the last 20 million years or more, that is during a period of time which is significant even from a geological point of view. The study of these cores, and of others which have been obtained previously by other methods, has revealed that in globigerina ooze *warm and cold water foraminifera* alternate, indicating that changes in the temperature of the ocean waters have taken place in the time interval during which the sediments were deposited. These matters are being further investigated.

In many deep basins layers of *coarser sediments* are found between the fine deposits which dominate at great depths. It has been suggested that the coarser sediments have been carried to considerable distances from the continental shelf by so-called density currents, or turbidity currents (HEEZEN and EWING[2]). Terrigenous sediments may be deposited on a slope until the slope becomes so steep that the limit of stability is exceeded. In this case "slumping" may take place, that is, a considerable bulk of the sediments may start to slide down the slope, and in doing so the sediments may be whirled up from the bottom and become suspended in the water, whereby the density of the water is increased and a current is created which may carry substantial amounts of coarser material into deep water. The action of suspension currents has been demonstrated in laboratories, but so far the existence in the sea of such currents has not been directly observed.

The total thickness of the sediment carpet covering the ocean bottom has been determined by means of *seismic methods*[3] developed by W. MAURICE EWING and used by himself in the North Atlantic, by RUSSELL RAITT in the North Pacific, and by HANS PETTERSSON on board the Swedish research vessel "Albatross" in the tropical part of all three oceans. In the Atlantic the total thickness of the sediments was found to be up to 4000 m., but in the Pacific and the Indian

[1] H. PETTERSSON and H. KOTSCHI: Geochim. et Cosmochim. Acta **2**, 81 (1952).

[2] B. C. HEEZEN and M. EWING: Amer. J. Sci. **250**, 849 (1952).

[3] See the articles by M. EWING and F. PRESS in this Encyclopedia, Vol. XLVII, p. 153 and p. 251.

oceans the recorded thickness nowhere exceeds 300 to 400 m. It has been suggested that in the latter oceans the deeper sediments may have been covered by *lava flows*. In several places where soft sediments were expected, the bottom was found to be formed by lava of recent origin.

D. The waters of the oceans.

11. Salinity. Of the known elements at least 49 have been found in sea water or in marine organisms, but several are present in so small amounts that it has been impossible to determine their concentrations (see Table 4).

Table 4. *Elements present in solution in oceanic sea water of chlorinity = 19.00 ‰, given by weight in parts per million. Exclusive of dissolved gases.*

Element	Parts per million	Element	Parts per million	Element	Parts per million
Chlorine . . .	18980	Lithium . .	0.1	Cerium . . .	0.0004
Sodium . . .	10561	Phosphorus .	0.001−0.10	Silver . . .	0.0003
Magnesium .	1272	Barium. . .	0.05	Vanadium .	0.0003
Sulphur . . .	884	Iodine . . .	0.05	Lanthanum .	0.0003
Calcium . . .	400	Arsenic . .	0.01−0.02	Yttrium . .	0.0003
Potassium . .	380	Iron	0.002−0.02	Nickel . . .	0.0001
Bromine . .	65	Manganese .	0.001−0.01	Scandium .	0.00004
Carbon . . .	28	Copper . . .	0.001−0.01	Mercury . .	0.00003
Strontium . .	13	Zinc	0.005	Gold	0.000006
Boron	4.6	Lead	0.004	Radium . .	$0.2-3\times 10^{-10}$
Silicon . . .	0.02−4.0	Selenium . .	0.004	Cadmium . .	present
Fluorine . . .	1.4	Cesium . . .	0.002	Cobalt . . .	present
Nitrogen[1] . .	0.006−0.7	Uranium . .	0.0015	Tin	present
Aluminium .	0.5	Molybdenum	0.0005		
Rubidium . .	0.2	Thorium . .	0.0005		

Because of the many different ions, sea water is such a complex solution that its total contents of dissolved solids cannot easily be determined by direct chemical analysis. By international conventions, the *salinity* of the sea water has, therefore, been defined as

"the total amount of solid material in grams contained in one kilogram of sea water when all carbonate has been converted to oxide, the bromine and iodine replaced by chlorine, and all organic matter completely oxydised".

The numerical value of the salinity thus defined is slightly lower than the actual concentration of dissolved solid matter.

Analysis of a large number of samples from different parts of the oceans has revealed that, regardless of the absolute concentrations, the *relative amounts of all the major constituents are virtually constant.* Therefore, the salinity of a water sample can be determined by determining the amount of one single of the major components. Chlorine is present in relatively large quantities and can be determined easily and accurately, for which reason the determination of the *chlorinity* of the sea water has been standardised and has been taken in general use. The chlorinity is also a conventionally defined quantity: The number giving the

chlorinity in grams per kilogram of a sea water sample is identical with the number giving the mass in grams of "atomic weight silver" just necessary to precipitate the halogens in 0.3285233 kilogram of the sea-water sample.

For the reason for this particular definition, see, for instance, SVERDRUP et al. [3] pp. 51 and 52.

[1] Nitrogen in combined forms. See also K. KALLE, in LANDOLT-BÖRNSTEIN [4], p. 476.

The empirical relation between salinity and chlorinity is

$$\text{Salinity} = 0.03 + 1.805 \times \text{Chlorinity}. \tag{11.1}$$

Both salinity and chlorinity are expressed in parts per thousand or per mille, for which the symbol "$^0/_{00}$" is used.

12. Variable constituents. Sea-water also contains the *atmospheric gases* in solution, but the ratios between these differ from the ratio of the same gases in the atmosphere because of differences in solubility and because the contents of CO_2 and O_2 in the water are altered by biological processes.

The contents of *dissolved oxygen*, which generally is reported as millilitres in one litre of sea-water of temperature 20° C (ml/L), is often discussed when dealing with problems related to the circulation of the waters (Sect. 45). The reason is that oxygen can be added to the sea-water only when the water is in contact with the *atmosphere*, or if oxygen is produced by the marine plants when *photosynthesis* takes place. Photosynthesis occurs, however, only in the surface layer because of the rapid extinction of the light with increasing depth. Below the photosynthetic zone the oxygen contents is reduced by the oxydation of organic matter. Although these processes are going on continuously, the oxygen contents of the water of the open ocean appears to remain constant in any given locality, meaning that a *steady state* has been reached at which the consumption of oxygen in a given locality is compensated for by replenishment of oxygen by processes of mixing and by currents. Thus, the general distribution of oxygen depends partly on biological activity and partly on physical processes, wherefore knowledge of the oxygen distribution may facilitate the discussion of the latter. This applies particularly when dealing with the deep water circulation.

Table 4 indicates that certain elements, such as silicon, nitrogen (in compounds) and phosphorus are present in *variable* amounts. These amounts are, however, in all instances so small that the variations are irrelevant to the salinity of the water, but they are of fundamental importance to the biological processes in the sea. In the photosynthetic zone nitrates, phosphates, silicates and other compounds are used by the marine plants for building up organic tissues or skeletons, and below the photosynthetic zone these compounds are returned in solution when the organic matter disintegrates and is oxydised. The concentrations in the deep water of these compounds depend, however, not only upon these processes, but also upon mixing and circulation, for which reason knowledge of the concentrations is useful when dealing with the flow of the deep-water.

13. Density. The density of sea-water depends on salinity, temperature and pressure and is indicated by the symbol $\varrho_{s,\vartheta,p}$. When dealing with numerical values, space is saved by introducing

$$\sigma_{s,\vartheta,p} = (\varrho_{s,\vartheta,p} - 1)\,1000, \tag{13.1}$$

for instance, $\varrho_{s,\vartheta,p} = 1.02674$ is represented by $\sigma_{s,\vartheta,p} = 26.74$.

The density of sea-water at the pressure at which the sample was collected is called the *density in situ* and is generally expressed as $\sigma_{s,\vartheta,p}$. At *atmospheric* pressure and at the temperature, ϑ, at which the sample was collected, the density is expressed by the symbol σ_t, where t now stands for temperature. At a temperature of 0° the corresponding symbol is σ_0. It should be observed that σ_0 is a function of salinity or chlorinity only, the empirically established relation being:

$$\sigma_0 = 0.069 + 1.4708\,\text{Cl} - 0.00157\,\text{Cl}^2 + 0.0000398\,\text{Cl}^3 \tag{13.2}$$

KNUDSEN's Hydrographical Tables[1] contain corresponding values of σ_0, chlorinity and salinity; a few values follow:

$S\,^0/_{00}$ . . .	10	20	30	35	40
$\mathrm{Cl}\,^0/_{00}$ . . .	5,53	11,07	16.61	19.37	22.14
σ_0	8.01	16.07	24.10	28.13	32.17

The density at atmospheric pressure and temperature t can be written as

$$\sigma_t = \sigma_0 - D \tag{13.3}$$

where the correction term D which has been determined experimentally, depends in a complicated manner upon σ_0 (salinity) and temperature. The term is entered in KNUDSEN's Hydrographical Tables. Since the density as expressed by σ_t is widely used in oceanography, special tables have been prepared (MATTHEWS[2]), from which σ_t can be obtained directly from temperature and salinity.

The *density in situ* depends also on pressure because sea-water is slightly compressible. The relationship is complicated because the compressibility is a function of all three variables, temperature, salinity and pressure. The relation ship has been established by experiments, and correction terms have been tabulated.

In many problems in dynamical oceanography the *specific volume*, α, is used instead of density. In this case it is practical to introduce the *anomaly* of the specific volume as defined by:

$$\delta = \alpha_{s,\vartheta,p} - \alpha_{35,0,p} \tag{13.4}$$

where $\alpha_{35,0,p}$ represents the specific volume of water of salinity $35\,^0/_{00}$, temperature 0°, and at pressure p. Omitting terms that are so small that they always can be neglected, the anomaly can be written as

$$\delta = \delta_s + \delta_\vartheta + \delta_{s,\vartheta} + \delta_{s,p} + \delta_{\vartheta,p} = \Delta_{s,\vartheta} + \delta_{s,p} + \delta_{\vartheta,p}. \tag{13.5}$$

The term $\Delta_{s,\vartheta}$ is readily obtained from σ_t because

$$\alpha_{s,\vartheta,0} = \frac{1}{\varrho_{s,\vartheta,0}} = 1 - \frac{10^{-3}\sigma_t}{1+10^{-3}\sigma_t} = \alpha_{35,0,0} + \Delta_{s,\vartheta} = 0.97264 + \Delta_{s,\vartheta} \tag{13.6}$$

and, therefore,

$$\Delta_{s,\vartheta} = 0.02736 - \frac{10^{-3}\sigma_t}{1+10^{-3}\sigma_t}. \tag{13.7}$$

The quantity $\Delta_{s,\vartheta}$ has been tabulated by LAFOND[3] who has also prepared tables for computing σ_t when $\Delta_{s,\vartheta}$ has been determined.

The field of mass in the ocean can be described by the field of the specific volume as expressed by the two terms, $\alpha_{35,0,p}$ and δ. The term $\alpha_{35,0,p}$ describes the distribution of mass in an ocean of salinity $35\,^0/_{00}$ and temperature 0° in which the specific volume depends on pressure only. Therefore the surfaces of $\alpha_{35,0,p} =$ const. coincide with *isobaric surfaces* and these deviate so little from surfaces of equal depth that for all practical purposes $\alpha_{35,0,p}$ is a function of depth only. The field of $\alpha_{35,0,p}$ is a constant one, wherefore the actual field of mass is always described adequately by the anomaly of specific volume, δ, as defined by (13.4).

[1] M. KNUDSEN: Hydrographical Tables. Copenhagen; G. E. C. Gad 1901.
[2] D. J. MATTHEWS: Conseil Perm. Internat. p. l'Explor. de la Mer, Copenhagen 1932.
[3] E. C. LAFOND: H. O. Publ. No. 614, U. S. Navy Hydr. Office, Washington D. C. 1951.

14. Thermal properties[1]. The salt contents of sea water lead to a *freezing point depression*, $\Delta\vartheta_f$, which has been determined experimentally as a function of chlorinity:

$$\Delta\vartheta_f\,(^\circ\mathrm{C}) = -\,0.0966\,\mathrm{Cl} - 0.0000052\,\mathrm{Cl}^2. \tag{14.1}$$

For the depression as a function of salinity the following values apply:

S, ‰ . .	5	10	15	20	25	30	35
$\Delta\vartheta$, °C . .	−0.27	−0.53	−0.80	−1.07	−1.35	−1.63	−1.91

The thermal expansion of sea water depends upon temperature, salinity and pressure. At atmospheric pressure pure water contracts with increasing temperature between 0 and 4° and expands at temperatures above 4°. Thus, pure water has its maximum density at 4°. For sea water the temperature of maximum density decreases with increasing salinity. At a salinity of 24.7‰ it coincides with freezing point of the water (−1.332°) and at lower salinity freezing begins before the maximum density has been attained.

The thermal conductivity of sea water is somewhat smaller than that for pure water, but its numerical value is of no practical importance because in the sea the transfer of heat by *"eddy" conductivity* dominates.

The specific heat of sea water decreases somewhat with increasing salinity. Since sea water is slightly compressible, it is necessary to distinguish between specific heat at constant pressure, c_p, and at constant volume, c_v. For water of a given salinity the ratio c_p/c_v increases with increasing temperature and increasing pressure. It can be computed from the formula

$$\frac{c_p}{c_v} = 1 + \frac{T\,e^2}{\varrho\,K\,J} \tag{14.2}$$

where T is the absolute temperature, e is the thermal expansion, ϱ the density, K the true compressibility and J the mechanical equivalent of heat. With sea water of salinity $S = 34{,}85$‰, c_p/c_v increases from 1.0004 at 0° C to 1.0207 at +30° C (after [4]).

The latent heat of vaporisation has not been examined for sea water, but it is generally assumed that it does not differ significantly from that of pure water. Between temperatures ϑ of 0° and 30° C the formula L (gcal) $= 596 - 0.52\vartheta$ is considered sufficiently accurate.

Since sea water is slightly compressible, the effects of *adiabatic* processes have to be considered. The temperature which a water sample would attain if brought adiabatically to the surface from a given depth (pressure) at which it has a temperature ϑ, is called the *potential temperature* Θ. The adiabatic change of temperature, $\Delta\vartheta$, is a complicated function of temperature, salinity and pressure, but convenient tables have been prepared (HELLAND-HANSEN[2]). From these it is, for instance, found that water of salinity 34.85‰, which at a depth of 4000 m., has a temperature of 2°, is cooled 0.35° when brought adiabatically to the surface, that is, the potential temperature is 1.65°. In several deep basins the temperature increases towards the bottom, but the potential temperature decreases or is constant.

15. Other properties[3]. *The vapour pressure* in air over sea water is slightly lower than over fresh water. Calling the saturation pressure over fresh water e

[1] For further details see LANDOLT-BÖRNSTEIN [*4*], p. 426ff.

[2] B. HELLAND-HANSEN: Michael Sars N. Atlantic Deep-Sea Exp. 1910, Sci. Res. **1**, 2 (1930).

[3] See the tables in LANDOLT-BÖRNSTEIN [*4*].

and that over sea water e_s, one has

$$e_s = e(1 - 0.00053\,S). \tag{15.1}$$

For the open oceans where the salinity varies between $33^0/_{00}$ and $38^0/_{00}$, one has with sufficient accuracy $e_s = 0.98\,e$.

The *viscosity* of sea water decreases rapidly with increasing temperature and increases slightly with increasing salinity. With a salinity of $35^0/_{00}$ the viscosity is 18.9×10^{-3} g. cm.$^{-1}$ sec.$^{-1}$ at 0° and 9.6×10^{-3} at 25°. The true viscosity of sea water is of importance to the rate of sinking of small marine organisms, but it is of no significance when dealing with problems in dynamics because turbulent motion always prevails, and an *eddy viscosity* has to be introduced.

Similarly, when dealing with *diffusion* the ordinary coefficient of diffusion, about 2×10^{-5} cm.^{2} sec.$^{-1}$, has to be replaced by a coefficient of *eddy* diffusion which has a large value that depends upon the turbulent state of the motion.

The surface tension, S.t., of sea water increases with decreasing temperature and with increasing chlorinity (salinity):

$$\text{S.t., (dyne/cm.)} = 75.64 - 0.144\,\vartheta + 0.0399\,\text{Cl}. \tag{15.2}$$

In the sea the value is generally lower because of impurities.

The refractive index increases with decreasing temperature and increasing chlorinity. It also varies with the wave length of the light and is generally stated for the D-line of sodium. At 25° the value is

$$n_{D,25^\circ} = 1.3325 + 0.00033\,\text{Cl}. \tag{15.3}$$

The electrical conductivity increases with increasing temperature and increasing salinity. Expressed in ohm^{-1} cm.$^{-1}$ it varies

at 0° from about 0.009 at chlorinity $5^0/_{00}$ (salinity $9.06^0/_{00}$) to about 0.03 at chlorinity $20^0/_{00}$ (salinity $36.13^0/_{00}$)

and at 25° from about 0.016 to about 0.055 for the same chlorinities.

Both the refractive index and the conductivity have been used for determining the chlorinity of samples of sea water, but the methods require very accurate temperature control.

The *velocity of sound* increases with increasing temperature, salinity and pressure. Under conditions found in the open ocean it varies between the limits of about 1440 m./sec. and 1600 m./sec., the latter value pertaining to the water at the greatest oceanic depths. The average velocity with which sound travels from the surface to the bottom of the sea, is called the "*mean sounding velocity*" and can be computed for any region for which the distributions of temperature and salinity are known. If true depths are to be obtained by sonic methods, the mean sounding velocities must be used, but sonic depth finders are generally adjusted to a constant velocity between 1460 and 1500 m./sec. and corrections are seldom applied.

Exact knowledge of the variation of the velocity of sound with depth is essential when dealing with *refraction* of sound "beams" that are sent out horizontally, for instance by *asdic*.

16. Absorption of radiation[1]. When dealing with the absorption of radiation of wave length λ a coefficient of absorption, $\varkappa_\lambda$, is defined by the equation

$$dI_\lambda = -\varkappa_\lambda I_\lambda\,dn. \tag{16.1}$$

[1] See the article on "Meeresoptik" by J. Joseph in Landolt-Börnstein [4], p. 441 ff.

Here I_λ is the energy at wave length λ and dI_λ is the decrease in energy in the direction of the beam in an infinitesimal layer of thickness dn. Integrating between the limits $n=N$ and $n=N+L$, one obtains

$$\varkappa_\lambda = \frac{1}{L} \ln \frac{I_{\lambda,N}}{I_{\lambda,N+L}}. \tag{16.2}$$

This equation also serves as a definition of the coefficient of absorption, the numerical value of which depends upon the unit of length that is used. In physics this unit is one cm., but in oceanography one metre is commonly used as the unit of length, for which reason the numerical values given here are 100 times larger than the corresponding values entered in textbooks of physics. At the same time the coefficient is defined in a somewhat different manner and is, therefore, not called the coefficient of absorption, but the *coefficient of extinction*. The reason for this different definition is that in oceanography the main interest is attached to the manner in which downwards travelling radiation decreases with depth. This rate can be expressed by means of a coefficient of extinction that is defined by the equation:

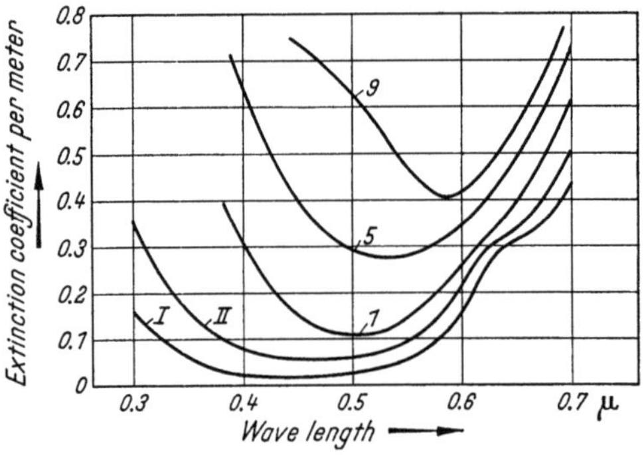

Fig. 3. Extinction coefficients of radiation in different types of sea water. Coefficients in clear and average oceanic water shown by curves marked I and II; in coastal water of different turbidity by curves marked 1, 5 and 9.

$$\varkappa_\lambda = \ln \frac{I_{\lambda,z}}{I_{\lambda,z+1}} \tag{16.3}$$

where z is the depth, taken positive downwards.

With a clear sky the extinction coefficient which is defined in this manner depends upon the altitude of the sun because the direct radiation from the sun penetrates obliquely into the sea. This dependency is, however, not very important, partly because the refraction is so great that even with the sun close to the horizon the sun's rays form an angle of 49° with the vertical after having passed the sea surface, and partly because of the great scattering which rapidly reduces the obliquity of the rays. Furthermore, the effect of the scattering on the rate of decrease of the incident radiation is so great and so variable that the effect of the sun's altitude becomes less important.

Laboratory measurements of the coefficient of absorption have not given quite consistent results, but they all show that for distilled water the coefficient is small for the visible part of the spectrum of the sun and very large for the infra-red part. They also show that for pure sea water the true absorption coefficient differs insignificantly from that for distilled water.

In the ocean, numerous measurements of the extinction coefficient have been carried out, using photoelectric or photronic cells and various coloured filters in order to determine the coefficient within narrow bands of wave length. According to these measurements the extinction coefficient for a given wave length varies between very wide limits. Even in the clearest oceanic water it is greater than the corresponding coefficient for pure water and in turbid coastal water it is very much greater.

The great variability of the extinction coefficient of sea water is illustrated in Fig. 3 which shows the coefficient as function of the wave-length for different types of sea water (JERLOV[1]). It is seen that with increasing turbidity of the water the region of minimum extinction coefficient *(maximum transparency)* is shifted towards the red end of the spectrum.

[1] N. G. JERLOV: Rep. Swedish Deep Sea Exp. 1947—1948. 3, No. 1, 1 (1951).

There are two reasons for the large values of the coefficient. In the first place the ocean waters always contain some very fine *suspended matter* which increases the scattering. In the second place there are generally present varying amounts of certain metabolic products derived from plants or animals, which with a common name are called "*yellow substance*". This yellow substance absorbs strongly at wave lengths shorter than 0.5 μ. It is present in appreciable quantities in turbid coastal waters where it causes the wave length of maximum transparency to be shifted from about 0.45 μ (for pure water) to 0.55 μ or even a little higher.

The *colour* of sea water varies from blue to green. The blue colour of the sea water is comparable to the blue colour of the sky and is explained as resulting from the scattering of the light by the water molecules or by very small suspended particles. The greenish colour results from the combined effect of this scattering and the absorption of wave lengths shorter than about 5 μ by the "yellow substance". The "yellow substance" is found in greater quantities near the coasts and in higher latitudes. Therefore, the greenish colour is more frequent in these regions, whereas the blue colour is typical of most open oceanic regions of middle and lower latitudes.

"Discoloured" water, brownish, reddish or milklike, is caused by the presence of numerous small plants or, near coasts, by suspended silt. Thus, the colour after which the Red Sea has been named is due to abundance of certain algae.

17. Eddy viscosity, conductivity and diffusivity. The coefficients of viscosity, heat conductivity and diffusion which can be determined in the laboratory are applicable to conditions in the sea only if the water is completely at rest or in *laminar motion*. These conditions are never fulfilled, for which reason it is necessary to replace the ordinary coefficients by "*eddy coefficients*" which depend upon the degree of turbulence.

In the presence of turbulence, momentum as well as heat or dissolved substances are carried from one layer to another by the random motion of smaller or larger masses of the fluid, whereby the transfer of properties is greatly increased. The random motion is superimposed upon a more or less simple average flow pattern.

The concept of *eddy viscosity* was introduced in oceanography already in 1902 when V. W. Ekman examined the effect of winds towards producing currents. At that time he dealt only with the effect of exchange of momentum in a *vertical* direction, but subsequently it has been shown that it is equally necessary to introduce a *horizontal* eddy viscosity when dealing with horizontal motion. The numerical values of the vertical and horizontal eddy viscosities differ greatly, in agreement with the fact that there exist fundamental differences between the random motions in a vertical and a horizontal plane. In the vertical direction the dimensions of eddies are limited by the depth to the bottom and, what is generally of greater importance, vertical motion is impeded because of the *stable stratification* of the waters. Where the stratification is stable (density increasing with depth) a small water mass which has been brought to another level will be subjected to Archimedean forces that are directed toward the level at which the mass has the same density as its surroundings. In the case of horizontal motion the boundaries exercise a far less restricting influence on the dimensions of the possible eddies and at the same time the motion is not impeded by Archimedean force.

Another effect of the stable stratification is that numerically the *coefficient of eddy diffusivity* becomes smaller than the corresponding *kinematic eddy viscosity*. G. I. Taylor has shown that such must be the case, and Jacobsen has pointed out that whereas elements in turbulent motion may rapidly give off their momentum to the surroundings, other properties may be exchanged slowly, and that before equalisation has been attained, the elements may be moved to new surroundings by gravitational forces.

Numerical values of *vertical* eddy coefficients have been computed, either directly from observed conditions or by making use of theoretical results that are based on laboratory study of turbulence. The wide range of the computed values and their magnitude as compared to the ordinary coefficients are evident from the following brief compilation.

Ordinary kinematic viscosity 0.019 to 0.009 $cm.^2 sec.^{-1}$.
Kinematic eddy viscosity 2 to 7500 $cm.^2 sec.^{-1}$.
Ordinary diffusivity 2×10^{-5} $cm.^2 sec.^{-1}$.
Eddy diffusivity 0.02 to 100 $cm.^2 sec.^{-1}$.

The range in the values of the eddy coefficients is generally attributed to the variable effects of stability and the speed of the currents because intense turbulence is associated with small stability and strong currents.

The *horizontal* coefficients vary within even wider limits, from values of the order of 10 up to 10^8 $cm.^2 sec.^{-1}$ or more. The smallest values are applicable to the diffusion as observed when dealing, for instance, with small spots of dye, and the largest to diffusion on an ocean-wide scale. The results clearly indicate that the coefficients to be used depend upon the scale of the phenomenon, a conclusion which is in agreement with recent theories of *isotropic turbulence* (STOMMEL[1]). These theories assume that turbulence can be characterised by a continuous series of eddies of all sizes. When dealing with horizontal motion the average flow pattern upon which the random fluctuations are superimposed, is obtained by averaging over a smaller or larger area. When averaging over a small area the corresponding random fluctuations will be small, but when averaging over a large area the random fluctuations will be large, and the same will apply to the eddy coefficients.

When a continuous series of eddies are present, energy must be supplied to the larger in order to maintain a steady state. In the case of the oceans the energy for maintenance of the larger eddies is supplied by the wind, and this energy is transferred to smaller and smaller eddies and is finally transferred to heat.

According to theory the eddy coefficients applicable to such a system are proportional to $L^{\frac{4}{3}}$ if the turbulence is isotropic. Here L represents the linear dimensions of the region under consideration. Small scale experiments in the sea have indicated that this so-called $\frac{4}{3}$ law holds when dealing with horizontal diffusion, but this result may be fortuitous, because the assumptions of the theory are not valid in that case. It must suffice to underline that the horizontal eddy coefficients vary with the dimensions of the area under examination, and that for ocean-wide areas the value is of the order of 10^8 $cm.^2 sec.^{-1}$. This value will be used later on when discussing how the large-scale ocean currents are maintained.

It is evident from the above that the present knowledge of the turbulence in the sea is very fragmentary. Only a few measurements have been made of the random fluctuations that characterise the turbulent motion, and all values of the eddy coefficients are derived not from a study of the random velocities, but from the observed distributions of properties and their variation. Applications to the problems of turbulence in the sea of experimental results from laboratories or theoretical studies have, therefore, been limited and a further discussion is not possible.

[1] H. STOMMEL: J. Marine Res. **8**, 199 (1949).

18. Properties of sea ice[1]. Formation of ice begins when the sea water has been cooled to freezing point corresponding to its salinity (Sect. 14). With a smooth sea, elongated ice crystals appear, and as freezing continues, these crystals form a matrix consisting of a network of pure ice within which brine is enclosed in numerous small spaces. With waves present, freezing may start simultaneously within a surface layer, and the sea may suddenly become covered by a lead-coloured mass of plastic sea ice which similarly consists of pure ice containing numerous cells in which brine is enclosed. In general any sample of sea ice has, therefore, a certain salinity which can be determined by analysing the melted sample. This salinity depends partly upon the rapidity of freezing and partly upon changes in the brine content that have taken place after freezing. It may vary between 10 to 15‰ and 0‰. Sea ice may also contain gas bubbles.

The brine content, and to a minor extent the gas bubbles, exercise a marked influence on the physical properties of the ice. *The specific heat of sea ice* depends upon the salinity and the temperature of the ice and is very great if the salinity is high and the temperature close to freezing point. Thus sea water of salinity 5‰ has a specific heat of 5.66 cal/gr. at $-2°$ and 0.56 at $-20°$. For pure ice the corresponding values are 0.48 and 0.47 cal/gr., respectively. The reason for this feature is that when the temperature of a sample of sea ice is altered, part of the enclosed brine freezes or part of the ice melts to dilute the brine because the concentration of the enclosed brine depends upon the temperature of the ice. The specific heat of sea ice is, therefore, not a specific heat in the ordinary sense because energy associated with freezing or melting is involved.

The latent heat of fusion cannot be designated in the ordinary manner because when dealing with sea ice, some melting takes place whenever the temperature of the sample is raised. The latent heat of fusion may, therefore, be given as the heat required for melting one gram of sea ice of stated temperature and salinity.

The coefficient of thermal expansion also varies with salinity and temperature. At temperatures close to freezing point sea ice *expands* with decreasing temperature because relatively large amounts of brine freeze, but at low temperatures it contracts.

The coefficient of thermal conductivity is generally smaller than that of pure ice because of the presence of gas bubbles.

During *summer*, ice in hummocks and pressure ridges is heated to zero degree, but before this stage has been reached, the enclosures containing brine have been enlarged so much that the ice has become porous and the brine has trickled down. When melted, this "*old ice*" renders practically pure water.

During *winter*, the salt contents of the upper layers of the ice is also decreased by the following process: The temperature of the ice increases with depth. Owing to this increase, the top of a drop of enclosed brine will be at a lower temperature than the bottom of the drop. The equilibrium concentration of the brine will, therefore, be different at the top and at the bottom; a circulation must take place resulting in a downwards displacement of the drop because the tendency towards establishing equilibrium must lead to melting of ice at the bottom and freezing at the top.

E. The sea-air boundary layer.

19. The stress of the wind. Energy is transferred from the atmosphere to the ocean by the stress that the wind exerts on the sea surface. This energy transfer leads to the development of *waves*, the development and maintenance of the pure *wind current* and the maintenance of the large-scale *permanent currents*

[1] For further details see LANDOLT-BÖRNSTEIN [4], pp. 438 and 534ff.

in the upper layers of the oceans. It also leads to the *piling-up* of water against coasts by on-shore winds, which under extreme conditions may lead to disastrous inundations of low-lying areas (Sects. 38, 39 and 40).

For these reasons knowledge of the relation between the observed wind over the ocean and the stress is highly desirable and considerable effort has been directed towards the establishment of such a relationship. Direct measurements of the *stress*, τ, have not been undertaken, and it is questionable if such measurements ever will be possible. Many indirect approaches have, however, been made in order to determine the resistance coefficient, γ^2, which G. I. TAYLOR has defined by the equation:

$$\tau = \gamma_a^2 \varrho' v_a^2 \tag{19.1}$$

where ϱ' is the density of the air. The numerical value of γ^2 will depend upon the height, a, at which the wind velocity v_a, is measured.

The first conclusions as to the stress of the wind have been drawn from observations of wind profiles in the lowest 10 to 15 m. above the sea surface, using relations between velocity profiles and stress, that have been established by means of laboratory studies of turbulent motion and theoretical considerations. Laboratory experiments show that if the density of the moving fluid does not vary with height (that means in the case of the atmosphere, if the stratification of the air in the boundary layer is indifferent), the velocity increase in the turbulent layer is proportional to the logarithm of the distance from the surface.

Over a surface that is hydrodynamically *smooth* there exists next to the surface a very thin layer *(boundary layer)* in which the flow is laminar and in which the velocity increases *linearly* with the distance from the fixed surface. Above the laminar layer the flow is turbulent and the velocity increases with the logarithm of the distance. The stress is independent of distance. In this case the relation between the velocity in the turbulent region at a distance a from the boundary surface and the stress, τ, can be written in the form

$$\frac{v_a}{\sqrt{\tau/\varrho'}} = \frac{1}{\gamma_a} = 5.5 + 5.75 \log \frac{a\,\varrho'}{\mu} \sqrt{\tau/\varrho'} \tag{19.2}$$

where μ is the viscosity of the air.

When dealing with a *rough* surface it is necessary to introduce a "*roughness length*", z_0, that characterises the surface. In this case the flow is turbulent throughout, and the relation between the velocity and the stress is given by

$$\tau = \varrho' \frac{k_0^2}{\left(\log \frac{a + z_0}{z_0}\right)^2} v_a^2 \tag{19.3}$$

where $k_0 = 0.4$ is a universal constant (v. KÁRMÁN'S constant).

Observations of wind profiles over the sea have shown that with an indifferent stratification, which generally prevails, the logarithmic law is valid. Some evidence indicates that at low wind velocities the sea surface may be smooth, in which case one has approximately

$$\tau \approx 0.9 \times 10^{-3} \varrho' v_{10}^2, \tag{19.4}$$

where v_{10} is the wind velocity at 10 m.

At moderate and higher wind velocities the sea surface is rough, and ROSSBY[1] has compiled evidence to show that the roughness length is independent of the wind velocity and equal to 0.6 cm. With this value:

$$\tau = 2.9 \times 10^{-3} \varrho' v_{10}^2. \tag{19.5}$$

[1] C.-G. ROSSBY: Papers Phys. Oceanogr. a. Meteorol. 4, No. 3 (1936).

A second method makes use of the fact that the stress can be found if the geostrophic departure of the wind is known at the sea surface and up to a height of several hundred metres (SHEPPARD and OMAR[1]). The method assumes that the *geostrophic wind*[2] can be accurately determined and that, in the layer under consideration, the departure of the *observed wind* from it is due to frictional influence only. This method has rendered values of the resistance coefficient of about $\gamma_{10}^2 = 1 \times 10^{-3}$.

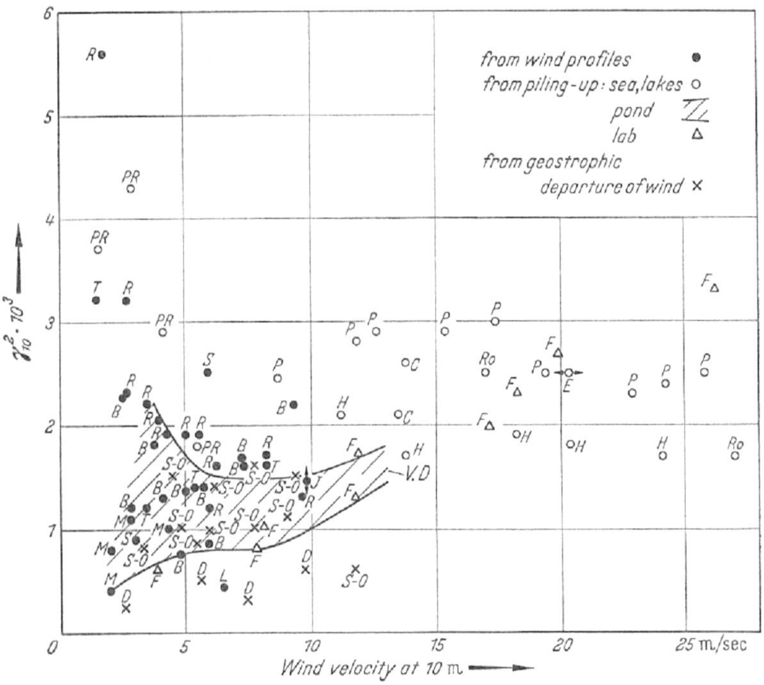

Fig. 4. The resistance coefficient, $\gamma_{10}^2 = \tau/\varrho' v_{10}^2$, as function of the wind velocity at 10 m., according to different authors. *B* H. BRUCH: Veröff. Berl. Inst. Meeresk. **1940**, H. 38, 1. — *C* R. H. CORKAN: Phil. Trans. Roy. Soc., Ser. A **242**, 493 (1950). — *D* C. S. DURST: Mon. Not. Roy. Astronom. Soc., Geophys. Suppl. **5**, 369 (1949). — *E* V. W. EKMAN: Ark. Mat., Astronom. Fys. **2**, 11 (1905). — *F* J. R. D. FRANCIS: Proc. Roy. Soc. Lond., Ser. A **206**, 356 (1951). — *H* I. HELA: Geophysica, Helsinki **3**, 148 (1948). — *J* JOHNSON (unpubl.). — *M* MONTGOMERY (C.-G. ROSSBY and R. B. MONTGOMERY: Papers Phys. Oceanogr. a. Meteorol. **31**, No. 3 (1955).] — *P* E. PALMÉN and E. LAURILA: Soc. Sci. fenn., Comment. phys.-math. **10**, Nr. 1 (1938). — *PR* PALMÉN-ROLL [H. U. ROLL: Geofisica Pure e Applicata **21** (1952)]. — *R* H. U. ROLL: Ann. Meteorol. **1**, 139 (1948). — *Ro* J. R. ROSSITER: Phil. Trans. Roy. Soc. Lond., Ser. A **246**, 371 (1953). — *S-O* P. A. SHEPPARD and M. H. OMAR: Quart. J. Roy. Meteorol. Soc. **78**, 583 (1952). — *S* W. SHOULEIKIN: Beitr. Geophys. **20**, 99 (1928). — *L* R. C. SUTCLIFFE: Quart. J. Roy. Met. Soc. **62**, 3 (1936). — *T* C. W. THORNTHWAITE and M. HALSTEAD: Trans. Amer. Geophys. Un. **22**, 249 (1942). — *V.D* W. H. VAN DORN: J. Marine Phys. **12**, 249 (1953).

The third method is based on a study of the relation between the wind stress and the piling-up of water against a boundary [Eq. (39.2)]. When a steady state is reached and bottom friction is negligible the stress acting on a unit area of the surface is balanced by the component of gravity acting on a water column of unit area cross section. Values of the stress at different wind velocities have been obtained from observations of the slope of the sea surface during high winds or storms in the North Sea or in the Baltic, from similar observations in shallow lakes, in a small pond, and in laboratory tanks.

In Fig. 4 all available values of the resistance coefficient, γ_{10}^2, have been plotted against the wind velocity at 10 metres. The scatter of the values is very great,

[1] P. A. SHEPPARD, and M. H. OMAR: Quart. J. Roy. Met. Soc. **78**, 583 (1952).

[2] See chapter by ELIASSEN and KLEINSCHMIDT in this volume, Sect. 12, p. 29.

particularly at wind velocities below about 10 m. sec.$^{-1}$. It is, therefore, not surprising that different authors have suggested different relationships between γ_{10}^2 and v_{10}, or, what amounts to the same, have suggested that the stress should be expressed by an equation of the type $\tau = \gamma_{10}^2 \varrho' v_{10}^n$ where n differs from 2. This form assumes that γ^2 is not a pure number, but a function of v.

On the basis of wind profile results and other observations MUNK[1] has suggested that there exists a *critical wind speed* between 6 and 8 m.sec.$^{-1}$ below which the sea surface behaves as a smooth surface and above which it behaves as a rough surface of constant "roughness length" At wind velocities above 10 m. PALMÉN and LAURILA[2] found $\gamma_{10}^2 = 2.4 \times 10^{-3}$, whereas NEUMANN[3] using PALMÉN's and other observations derived the expression $\gamma_{10}^2 = 9 \times 10^{-2} v_{10}^{-\frac{1}{2}}$ and FRANCIS[4] on the basis of laboratory experiment obtained $\gamma_{10}^2 = 0.13 \times 10^{-3} v_{10}$. From the compilation in Fig. 4 it appears that none of these expressions has particular merits, although MUNK's concept of a critical wind speed in the vicinity of 8 m.sec.$^{-1}$ appears to be slightly supported. Furthermore, it is impossible to decide whether the great scatter of the γ^2-values is due to experimental errors or if it may be related to circumstances which have not been given due consideration, such as the stability of the air or deviations from a steady state.

The conclusion to be drawn is that no definite relationship has been established between the stress and the wind, but if rough values of the stress are to be derived from wind observations the simple relation

$$\tau = 2.4 \times 10^{-3} \varrho' v_{10}^2 \qquad (19.6)$$

may be used, particularly if the wind velocities under consideration vary over a wide range because then the stress values for low velocities enter with a small weight only, because the stress is proportional to the square of the velocities.

20. The flux of water vapour. A study of the flux of water vapour through the boundary layer is of particular interest in connection with the problem of evaporation from the sea surface. The flux, which can be assumed to be independent of height a in the lowest 10 to 20 m., can be written in the form

$$F = \varrho' A \frac{dq}{da} = \text{const}, \qquad (20.1)$$

when A is the *eddy diffusivity* in the air and q is the *specific humidity*, that is, the mass of water vapour per unit mass of moist air. If the flux is directed upwards, it equals the *evaporation* E, and in this case dq/da is negative. Therefore

$$E = -\varrho' A \frac{dq}{da}. \qquad (20.2)$$

From laboratory studies, theoretical considerations and observations of wind and humidity profiles in the boundary layer it is concluded that in the turbulent region, at a given height a,

$$A \approx k_0 \gamma_a v_a a. \qquad (20.3)$$

Since $\varrho' A\, dq/da = \text{const}$, it follows that q must vary with the logarithm of height, wherefore MONTGOMERY[5] has introduced an *evaporation coefficient*, defined by

$$\Gamma_a \equiv -\frac{1}{q_s - q_a} \frac{dq}{d \log a} \qquad (20.4)$$

where q_s represents the specific humidity at the sea surface and q_a the specific humidity in the air at a given height, a. From the above relations we obtain:

$$E = \varrho' k_0 \gamma_a v_a \Gamma_a (q_s - q_a). \qquad (20.5)$$

[1] W. H. MUNK: J. Marine Res. 6, 203 (1947).
[2] E. PALMÉN and E. LAURILA: Soc. Sci. fenn. Comment. phys.-math. 10, Nr. 1 (1938).
[3] G. NEUMANN: Z. Meteor. 2, 193 (1948).
[4] J. R. D. FRANCIS: Proc. Roy. Soc. Lond., Ser. A 206, 356 (1951).
[5] R. B. MONTGOMERY: Papers Phys. Oceanogr. a. Meteorol. 7, No. 4 (1940).

Various theoretical expressions can be derived for Γ, depending upon the assumptions as to the character of the flux of water vapour. The most probable assumption appears to be that next to the sea surface there exists a very thin layer, the thickness of which is inversely proportional to the wind velocity, through which water vapour is transported upwards by *molecular diffusion*, and that above this layer the transport takes place by *eddy diffusion* (SVERDRUP[1]). This process may take place even over a rough surface where the turbulent motion starts at the very surface, because a small air mass that strikes the surface may immediately give off its momentum, but may not immediately attain the specific humidity that prevails at the surface. If it remains at rest for a short while in one of the hollows of the rough surface, its vapour content will increase by the effect of molecular diffusion, meaning that statistically a thin diffusion layer exists. However, different assumptions can be made as to the thickness of the diffusion layer and as to the transition from molecular to eddy diffusion. With two different assumptions Γ_a turns out to be nearly independent of the wind velocity. Therefore

$$E \approx K'_a (e_s - e_a)\, v_a, \tag{20.6}$$

where

$$K'_a = \varrho' k_0 \gamma_a \Gamma_a \frac{0.621}{p}. \tag{20.7}$$

Here e represents the vapour pressure and p the atmospheric pressure (both in the unit millibar), and $q = 0.621\, e/p$ has been introduced. According to the two assumptions and with $a = 10$ m. the value of K'_{10} lies between 1.0×10^{-9} and 1.5×10^{-9}, using c.g.s.-units, provided e is expressed in mbar and p is put equal to 1000 mbar. It is more practical to express the evaporation in mm./24 hours and the wind velocity in m.sec.$^{-1}$. In that case

$$E = K_{10} (e_s - e_{10})\, v_{10} \text{ (mm./24 hours)} \tag{20.8}$$

and K_{10} lies between 0.12 and 0.19.

No great significance can be attributed to the numerical value of K_{10}, but the fact that the above considerations lead to a simple relationship between evaporation, water vapour difference and wind velocity is of considerable value.

In several cases the interest is attached not to the evaporation but to the *energy* that the surface loses because of the evaporation. Expressing the energy, Q_e, in units of heat we have

$$Q_e = L E = L \varrho' k_0 \gamma_a v_a \Gamma_a (q_0 - q_a) \tag{20.9}$$

where L is the latent heat of vaporisation.

If warm, moist air flows over a cold sea surface, the direction of the flux will be reversed, condensation will take place on the sea surface and energy is transferred from the air to the sea. These conditions may prevail in spring and summer in higher latitudes.

21. The flux of heat. Since the sea surface generally is slightly warmer than the air at, say, 10 m., heat is transported away from the sea surface. The loss of heat by the sea surface can be written in the form

$$Q_h = - c_p \varrho' A \frac{d\vartheta}{da} \tag{21.1}$$

where c_p is the specific heat of the air (0.240 cal/gr.)[2] and ϑ is the air temperature. Strictly speaking the gradient of the *potential* air temperature should

[1] H. U. SVERDRUP: Comp. of Meteorology, p. 1071, Amer. Meteorol. Soc. Boston 1951.
[2] The unit cal ≡ gram calorie is used throughout.

have been introduced, but in the lowest few metres this gradient differs only slightly from that of the ordinary temperature.

By a reasoning similar to that applied when dealing with the flux of water vapour [Eq. (20.5)] we obtain:

$$Q_h = c_p \varrho' k_0 \gamma_a v_a \Gamma_a (\vartheta_s - \vartheta_a) \tag{21.2}$$

where the value Γ_a is the same as in the preceding case, provided that the mechanism of heat transport is the same as that of the transport of water vapour. The latter condition can be expected to be nearly fulfilled.

Instead of computing the heat flux directly, it is often more advantageous to compute the ratio, $R = Q_h/Q_e$. From (20.9) and (21.2) we obtain:

$$R = \frac{Q_h}{Q_e} = \frac{c_p}{L} \frac{\vartheta_s - \vartheta_a}{q_s - q_a}. \tag{21.3}$$

Introducing $q \approx 0.621\, e/p$, $p = 1000$ mbar, $c_p = 0.240$ cal/gr., $L = 585$ cal/gr., and measuring e in mbar:

$$R = \frac{Q_h}{Q_e} \approx 0.66 \frac{\vartheta_s - \vartheta_a}{l_s - l_a}. \tag{21.4}$$

This ratio is called the "BOWEN *ratio*" because it was first derived by I. S. BOWEN, but in a different manner.

The heat flux will be directed downwards if the air is warmer than the sea surface. This situation is more frequent than the situation in which the water vapour flux is directed downwards (see end of Sect. 20).

F. The heat budget of the oceans.

22. The general heat budget of the atmosphere and the oceans. The exchange of energy between the earth as a whole and the space around it is determined by processes of *radiation* only. The solar energy which is received at the limit of our atmosphere is lost partly by reflection and partly by radiation to space from the surface of the earth, the clouds, and the atmosphere.

The distribution of temperature over the surface of the earth is, however, not determined by processes of radiation only. While for the earth as a whole the amounts of energy received and lost by radiation are in balance, this is not true at any given latitude. Both the effective insolation (the incoming radiation from the sun minus the reflected part) and the outgoing terrestrial radiation[1] decrease from the equator to the poles; but at the equator the effective insolation is greater than the terrestrial radiation, whereas at the poles it is less. In computing these quantities, different authors have reached somewhat different results, but they all agree as to the main features. Fig. 5 shows the average annual values of the incoming effective insolation and the outgoing terrestrial radiation between parallels of latitude in the Northern Hemisphere, according to HOUGHTON[2]. Between the equator and about 35° N. the earth receives more energy than it loses, whereas north of latitude 35° N. the loss exceeds the gain.

It follows from the curves in Fig. 5 that if only radiative processes were active, heating would take place in lower latitudes and cooling in higher. Since there is no evidence of any persistent change in the temperature of the atmosphere or the oceans, it must be concluded that, on an average for the whole year, heat is transported horizontally from lower to higher latitudes. This transport reaches its maximum at 38° N., where it attains a value of about 6.3×10^{16} cal min.$^{-1}$.

[1] See the article by F. MÖLLER in this volume.

[2] H. G. HOUGHTON: J. Amer. Meteorol. Soc. **11**, 1 (1954).

It is mainly effected by the atmosphere but, as will be shown later on, also to some extent by the ocean currents. This is true when dealing with conditions in the Northern Hemisphere, to which the curves in Fig. 5 apply. For the Southern Hemisphere it appears probable that there the ocean currents are less important in the transport of heat into polar latitudes.

The oceans play another and even more important rôle in regulating the *temperature distribution*. Where surplus energy is received, it is used mainly for *evaporation*; and later this surplus energy becomes available to the atmosphere when condensation takes place, often in a region far removed from where the evaporation occurred. Knowledge of the *heat budget* of the ocean and of ocean currents and the interaction between ocean and atmosphere is, therefore, indispensable to an understanding of the temperature distribution at the surface of the earth and in the atmosphere.

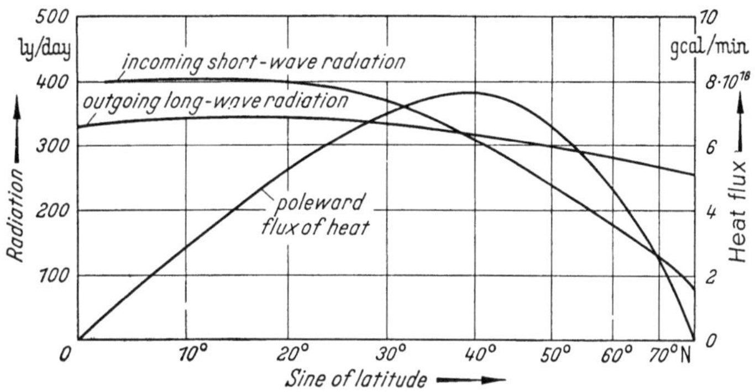

Fig. 5. Incoming and outgoing radiation and required polewards flux of heat (northern hemisphere) according to HOUGHTON (1954). — Radiation given in ly/day = gcal cm^{-2} day^{-1}.

For all oceans and for the whole year the average amounts of energy received and lost must balance, because there is no evidence that the temperature of the waters changes appreciably over periods as short as a century or so. The oceans receive or lose energy mainly by processes of radiation, by exchange of sensible heat with the atmosphere and by evaporation from the sea surface or condensation on the surface. The effects of other processes, such as the conduction of heat from the interior of the earth, the transformation of kinetic energy into heat, and energy derived from chemical processes in the sea, are of very small importance to the heat budget; but the energy that is transmitted to the sea by the stress of the wind on the sea surface is, as will be shown, of prime importance in the maintenance of the currents.

The *heat flux from the interior of the earth* has been found to be about 40 gcal $cm.^{-2}$ $year^{-1}$, that is, 10^{-4} gcal $cm.^{-2}$ $min.^{-1}$ (REVELLE and MAXWELL[1], BULLARD[2]). The *kinetic energy* that is transmitted to the sea by the stress of the wind and part of the energy of the tidal currents are dissipated by friction and transformed into heat; the average amounts are of the same order of magnitude as the heat flux from the interior of the earth. Both quantities are very small compared to the radiation values. Finally, in areas with abundant *plant life*, up to 0.8% of the incoming radiation is used by the plants for photosynthesis; but the average for all oceans and the whole year is very much smaller.

23. Processes of radiation. The average annual amount of energy which the sea receives from sun and sky varies with latitude; and in a given latitude it varies with the cloudiness. Measurements have been carried out over some

[1] R. REVELLE and A. E. MAXWELL: Nature, Lond. **170**, 199 (1954).
[2] E. C. BULLARD: Proc. Roy. Soc. Lond., Ser. A **222**, 408 (1954).

ocean areas; but where no direct observations are available, the average amount may be computed from the empirical relation

$$Q = k(1 - 0.071\,\overline{C})\,\overline{h}\,(\text{gcal cm.}^{-2}\,\text{min.}^{-1}) \tag{23.1}$$

where $\overline{C}$ is the average cloudiness on the scale 0 to 10, $\overline{h}$ is the average altitude of the sun in degrees, and k is a factor which depends upon the transparency of the atmosphere and which appears to vary between 0.023 at the equator and 0.027 in latitude 70°.

Part of the short-wave radiation (wave lengths in and near the visible spectrum) that reaches the sea surface is reflected from the surface. The reflection of the direct radiation from the sun depends upon the altitude of the sun. With the sun 60, 30, and 10° above the horizon, the reflected amounts are 2.1, 6.0, and 34.8%, respectively. For diffuse radiation a reflection of 6.6% has been determined theoretically (BURT[1]). On clear days, measurements have given values in agreement with the foregoing; but on overcast days, when one might expect the radiation to be diffuse, values between 8 and 15% have been found (NEIBURGER 1948).

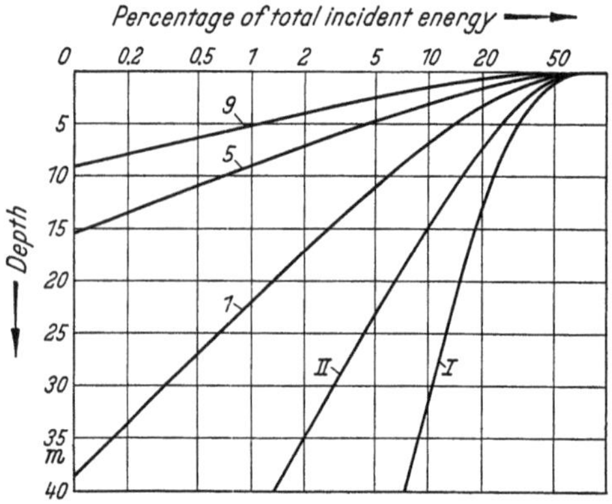

Fig. 6. Percentages of total energy reaching different depths in clear and average oceanic water (curves I and II) and in coastal water of different turbidity (curves 1,5 and 9).

Even on a clear day, part of the radiation received at the sea surface is diffuse, and this circumstance must be taken into account when computing the reflection loss on a clear day for different altitudes of the sun. One obtains the following approximate values:

Altitude of sun . . .	5°	10°	20°	30°	40°	50°	90°
Percentage reflected .	40	25	12	6	4	3	3

These values are applicable if the sea surface is smooth. In the presence of waves, the reflection is somewhat decreased.

The part of the radiation from sun and sky, Q_s, which is not reflected from the sea surface, penetrates into the sea, where it is absorbed. By far the greatest fraction of this energy is absorbed close to the sea surface, because water has an appreciable transparency only in the wave-length range 0.4 to 0.6 μ (Sect. 16). This feature is illustrated in Fig. 6, which shows the percentage amounts of energy that reach different depths under different conditions.

The great absorption of radiation near the sea surface does, in most cases, not lead to the development of a thin, warm surface layer. The reasons are that part of the energy is lost immediately by long-wave radiation (wave-lengths around 10 μ) from the sea surface, part is given off to the atmosphere as sensible heat, and the remaining part is rapidly distributed over a thicker layer because of the eddy conductivity. Only on an absolutely calm day with a smooth sea may the afternoon temperature at the surface be a few degrees higher than the temperature at a depth of a few metres.

[1] W. V. BURT: Trans. Amer. Geophys. Un. **34**, 199 (1953).

Table 5. *Energy available for evaporation in areas between parallels of latitude, and evaporation rates determined by different methods.*

Interval of latitude (degrees)	Area of North Atl. and North Pacific (10^{16} cm.^{2})	Q_r $\left(\frac{\text{gcal}}{\text{cm.}^2\,\text{day}}\right)$	R (JACOBS, adjusted)	Q_e $\left(\frac{\text{gcal}}{\text{cm.}^2\,\text{day}}\right)$	L (cal)	Evaporation (mm day^{-1}) From Q_e	From met. obs. (JACOBS)	From obs. at sea (WÜST)
60—50 N.	9.3	80	0.31	61	592	1.03	1.66	1.18
50—40	13.9	113	0.21	93	590	1.58	1.81	2.11
40—30	18.7	164	0.15	143	586	2.44	3.31	2.90
30—20	24.1	239	0.11	215	583	3.69	3.54	3.50
20—10	26.8	253	0.10	230	582	3.95	3.68	3.61
10—0	27.8	239	0.10	217	582	3.73	3.07	3.20

Because of the combined effect of the processes mentioned, the range of the *diurnal variation* of the surface temperature is nearly always small, generally less than 1°; and the range of the *annual variation* of surface temperature is small also, except in some coastal waters, where a layer of fresh water may effectively suppress the mixing processes.

Only during the hours of the day when the sun is above the horizon does the sea receive energy by radiation from sun and sky. During all hours it loses energy by its own *long-wave radiation*; but at the same time it also receives radiation from the water vapour and the carbon dioxide in the air and from the clouds. The net result is a radiation loss which will be called *back-radiation*, Q_b. With a clear sky, the back-radiation amounts to 0.15 to 0.20 cal cm.$^{-2}$ min.$^{-1}$. With a completely overcast sky, it is, on an average, reduced to only 17% of the value with a clear sky under corresponding conditions; but, depending upon the thickness of the clouds, the percentage may vary between 75 and 0.

The average amount of radiation that *penetrates* the sea surface, Q_s, is, in all latitudes, greater than the loss of energy from the sea surface by back-radiation, Q_b, which results in a *radiation surplus* $Q_r = Q_s - Q_b$. In Table 5 are entered average annual values of Q_r between parallels of latitudes (SVERDRUP[1]). Averaged for the whole year and all oceans, the radiation surplus Q_r must be given off to the atmosphere, but not necessarily when and where it was received. Part of the surplus may be carried by currents to other regions, and part may be stored during the warm season and given off during the cold season. Thus the oceans exercise a great *equalising influence on the climate* of the earth.

Let Q_c be the amount of energy that is carried by the ocean *currents* into, or out of, any given area; Q_e the energy used for *evaporation* within the area; and Q_h the energy that is given off to the atmosphere as sensible *heat*. The energy budget of the area may then be written

$$Q_r + Q_c = Q_e + Q_h. \tag{23.2}$$

For the oceans as a whole and for the whole year, Q_c must be zero, and

$$Q_r = Q_e + Q_h. \tag{23.3}$$

24. Evaporation. It has been shown (Sect. 21) that the ratio, R, between the amounts of energy given off from the sea surface as sensible heat and the amount that is used for evaporation can be computed from meteorological observations at sea [Eq. (21.4)]. Therefore, if the ratio R can be established, the average annual evaporation from the oceans as a whole ($Q_c = 0$), can be obtained from the relation:

$$\overline{E} = \frac{\overline{Q}_r}{L(1+R)}. \tag{24.1}$$

[1] H. U. SVERDRUP: Comp. of Meteorology, p. 1071, Amer. Meteorol. Soc. Boston, 1951.

This relation has been used by W. SCHMIDT[1] and revised by H. MOSBY[2] who found the average evaporation to be 106 cm./year. The still further revised values, some of which are entered in Table 5, give an average value of 101 cm./year between latitudes 70° S. and 70° N. On the assumption that the ocean currents do not carry any energy into or out of the total areas of regions between parallels of latitude, the evaporation can also be computed for these regions, and the values that are computed from Q_e in this manner are also entered in Table 5.

In order to obtain more detailed information it is necessary to use other approaches. Measurements of evaporation have been carried out on a number of expeditions and have been carefully analysed, particularly by WÜST[3] who found a value of 93 cm./year for the average annual evaporation between latitudes 70° S. and 70° N. The observations are not sufficiently numerous to permit a study of the evaporation from different ocean areas in different seasons, but WÜST has been able to derive average values for regions between parallels of latitudes.

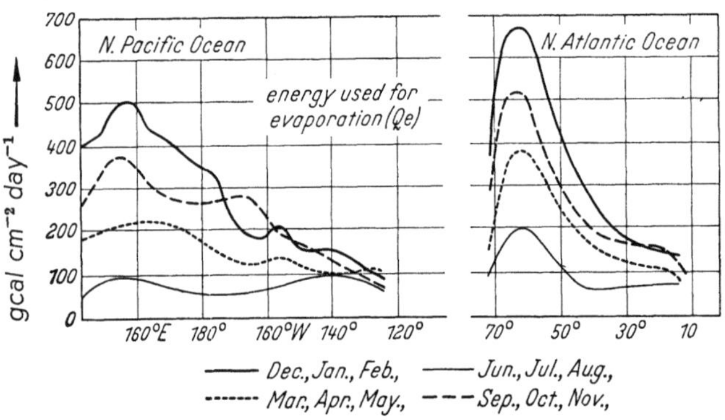

Fig. 7. East-west and seasonal variation in evaporation in latitude 35–40° N in the Pacific and Atlantic Oceans.

Still further information can be obtained by computing the evaporation from the flux of water vapour through the air-sea boundary layer on the basis of *meteorological* observations over the oceans. Such computations have been made by JACOBS[4] and ALBRECHT[5]. The computations are based on the information as to the differences $(e_s - e_a)$ and $(\vartheta_s - \vartheta_a)$, (see Sects. 20 and 21), which are derived from climatological charts of the oceans. WÜST[6] and DIETRICH[7] have pointed out that because of systematic errors in the usual observations on board ships, these differences may deviate greatly from the true values, the errors being in some cases up to 100%. The deviations are not systematic and vary from one region to another and from season to season; therefore, evaporation values that are based upon climatological data must be unreliable.

This objection is only partly valid when dealing with the results that JACOBS has presented: JACOBS made use of the approximate formula (20.8), but instead of using an uncertain theoretical value for K_{10} he adjusted his K_{10}-value such that, on an average for the northern oceans which he considers, he obtains the value of the evaporation which is found from the available energy. In this manner an average deviation of the climatological value of the differences $(e_s - e_{10})$ and $(\vartheta_s - \vartheta_{10})$ from the true values is taken into account, but the regional distributions and the annual variations may still be in error because, as already mentioned, the deviations may vary with region and season. Nevertheless, the main features that are found by JACOBS and ALBRECHT are so striking that they deserve confidence. The computation has been carried out for the northern hemisphere only, because climatological charts for the southern hemisphere are lacking.

The characteristic features of the North Pacific and the North Atlantic (see Fig. 7) are:

[1] W. SCHMIDT: Ann. d. Hydrogr., Berlin **43**, 111 (1915).
[2] H. MOSBY: Ann. Hydrogr., Berlin **64**, 281 (1936).
[3] G. WÜST: Festschr. Norbert Krebs (Länderk. Forsch.) Stuttgart 1936.
[4] W. C. JACOBS: J. Marine Res. **5**, 37 (1942).
[5] F. ALBRECHT: Ann. Meteorol. **2**, 129 (1949).
[6] G. WÜST: Dtsch. hydrogr. Z. **3**, 111 (1950).
[7] G. DIETRICH: Dtsch. hydrogr. Z. **3**, 314 (1950).

1. The evaporation is very great in the western parts of the oceans where the currents, the Kuroshio and the Gulf Stream, carry warm water into higher latitudes, and small over the eastern parts where cold currents carry water to the south.

2. In middle and higher latitudes the evaporation is great in winter and small in summer. This feature is particularly conspicuous in the western parts of the oceans where in winter cold air from the continents flows out over the warm ocean such that the difference $(e_s - e_{10})$ becomes great.

The total *water budget* of the oceans cannot be examined separately because part of the evaporation from the oceans is compensated for by run-off from land. According to the latest compilation (Reichel[1]) the best values for the water budgets of the ocean and land areas are:

Annual totals of		
evaporation from the oceans		345×10^3 km.^{3}
precipitation on the oceans	315×10^3 km.^{3}	
run-off from land	30×10^3 km.^{3}	345×10^3 km.^{3}
evaporation from land areas		70×10^3 km.^{3}
precipitation on land areas	100×10^3 km.^{3}	
run-off from land	-30×10^3 km.^{3}	70×10^3 km.^{3}

25. Heat and total energy given off to atmosphere. Since approximate values of $R = Q_h/Q_e$ [Eq. (21.4)] can be derived from climatological charts of the oceans, it is possible to find the amount of heat that is given off directly to the atmosphere in different regions and different seasons. Jacobs found that

1. The ratio Q_h/Q_e increases with latitude.

2. In middle latitudes the ratio is much greater over the western than over the eastern parts of the northern oceans. Over the eastern (cooler) parts, that is off the coasts of Spain-Morocco and California, the heat flux is reversed in summer, and heat is transferred from the air to the sea.

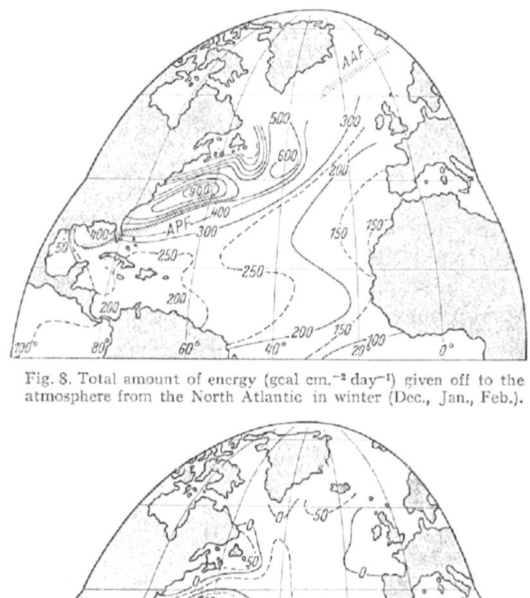

Fig. 8. Total amount of energy (gcal cm.$^{-2}$ day^{-1}) given off to the atmosphere from the North Atlantic in winter (Dec., Jan., Feb.).

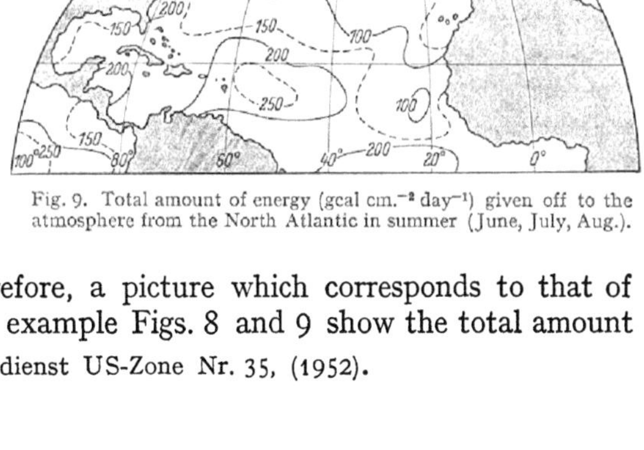

Fig. 9. Total amount of energy (gcal cm.$^{-2}$ day^{-1}) given off to the atmosphere from the North Atlantic in summer (June, July, Aug.).

In general the regional distribution of the energy given off from the sea as *heat* corresponds to the regional distribution of energy that is used for *evaporation* and the seasonal variations of the two quantities are similar. The *total energy exchange* shows, therefore, a picture which corresponds to that of the picture of *evaporation*. As an example Figs. 8 and 9 show the total amount

[1] E. Reichel: Ber. dtsch. Wetterdienst US-Zone Nr. 35, (1952).

of energy, expressed in gcal cm.$^{-2}$ day^{-1}, that is given off from the North Atlantic ocean in winter (Dec. to Febr.) and in summer (June to Aug.).

26. Transport of heat by ocean currents. Fig. 5 showed that on an average for the year a large amount of energy must be transported from lower to higher latitudes in order to equalise the effect of the gain of energy by processes of radiation in lower latitudes and the corresponding loss in higher latitudes. It is ordinarily assumed that this transport is accomplished by the atmosphere, but from the preceding discussion of the energy exchange between the ocean and the atmosphere it is evident that, at least in the Northern hemisphere which has been discussed, the ocean currents must also contribute.

The amount of energy which is carried into or out of any given region is, according to Eq. (23.2),

$$Q_c = (Q_e + Q_h) - Q_r. \tag{26.1}$$

Here a positive value of Q_c indicates that energy is carried into the region, a negative that it is carried out. Assuming that the average values for E and R between parallels of latitude are approximately correct, we find from the data in Table 5 that in the Northern hemisphere the ocean currents must (on an average) carry energy out of the regions to the south of about latitude 30° N., and into the regions to the north of that latitude. The amounts that on an average are carried to the north across the different latitudes are entered in the bottom line of Table 6.

Table 6. *Amounts of energy, in 10^{16} cal min.$^{-1}$, that are carried across parallels of latitude in the North Atlantic and the North Pacific Oceans.*

	North latitude						
	0°	10°	20°	30°	40°	50°	60°
North Atlantic[1]	0.4	0.8	1.1	1.3	0.8	0.6	0.3
North Pacific[1] .	0.0	0.7	0.6	0.4	0.0	0.0	0.0
Both oceans[1] .	0.4	1.5	1.7	1.7	0.8	0.6	0.3
Both oceans[2] .	0.4	1.2	1.5	1.7	0.8	0.6	0.3

It is assumed that a *transport of energy* takes place *across the equator*. This assumption as well as the numerical value of the transport is based exclusively on oceanographic observations which show that in the *Atlantic* warm surface water is carried across the equator off the coast of South America, whereas cold deeper water flows south. A corresponding transport does not take place in the Pacific.

On the basis of the charts by JACOBS and of charts representing the distribution over the oceans of the average radiation surplus, charts of Q_c for the North Atlantic and the North Pacific oceans can be prepared. Charts of the radiation surplus can be derived from KIMBALL'S[3] charts of the radiation that reaches the sea surface, correcting them for reflection and back radiation and adjusting the values such that the averages between parallels of latitude correspond approximately to the values on Table 5.

Figs. 10 and 11 show charts of Q_c for the North Atlantic and the North Pacific. Negative values indicate that energy is carried away from the region by ocean currents. The charts can also be interpreted as showing the amount of *radiation surplus* that is stored in the water and not given off locally by evaporation and heat conduction. Stream lines for the transport of the upper water masses by the prevailing currents are also entered. It is evident that energy is stored in the water where it flows south and is given off where it flows north,

[1] From charts of energy carried by currents.
[2] From average values in Table 5.
[3] H. H. KIMBALL: Month. Weath. Rev. **56**, 393 (1928).

the greatest amounts being released from the Gulf Stream and the Kuroshio waters between latitudes 30 and 40° N., off the east coasts of North America and Japan. The charts show the annual averages, but according to Figs. 8 and 9 the by far largest amounts of energy are given off in winter.

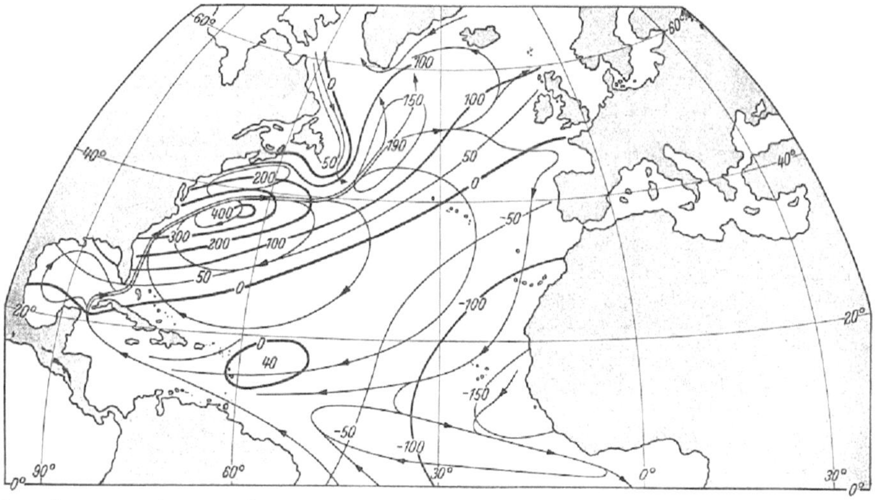

Fig. 10. Average annual amounts of energy (gcal cm.$^{-2}$ day^{-1}) that in the North Atlantic are carried into or out of any area by the ocean currents. Lines of water transport of the upper layer masses are entered.

From the charts in Figs. 10 and 11 it is possible to find the amounts of energy that the ocean currents in the two oceans transport in regions between parallels of latitude. The values are entered in Table 6. If the data had been consistent the sum of the transports in the two oceans should agree with the transport

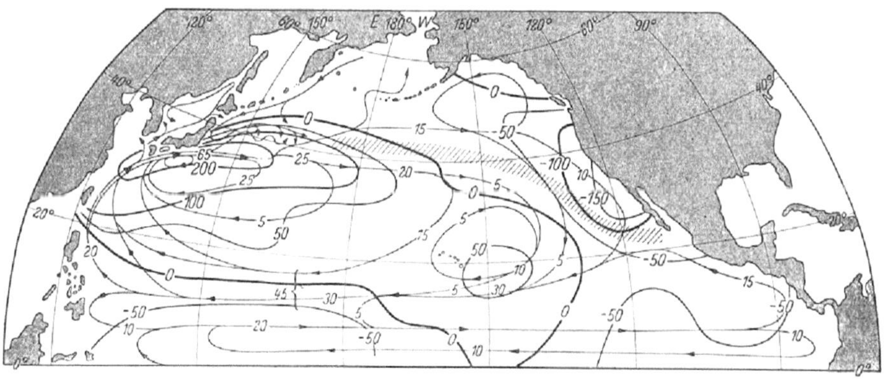

Fig. 11. Average annual amounts of energy (gcal cm.$^{-2}$ day^{-1}) that in the North Pacific are carried into or out of an area by the ocean currents. Lines of water transport of the upper layers are entered.

computed from the values of the average evaporation, but no complete adjustment has been made of the values of the radiation surplus that have been derived from KIMBALL'S charts. Still, it may be expected that the main features shown in Figs. 10 and 11 are correct. It is, for instance, noteworthy that in the Pacific practically no energy is carried across latitude 60° N., whereas in the Atlantic an appreciable amount of energy is carried to the north across that latitude. Oceanographic observations render a rough check on this value. The transports of warm water to the north and cold water to the south across 60° N. in the Atlantic amount to between 6×10^6 and 8×10^6 tons/sec. or between 3.6×10^{14} and 4.8×10^{14} g./min. and the temperature difference between the warm and the cold water is about 6°. According to these values the transport of heat across 60° N. is about between 0.2×10^{16} and 0.3×10^{16} gcal/min., in fair agreement with the values in Table 6.

G. General distribution of salinity, temperature and density.

27. Salinity of the surface layer. In all oceans the surface salinity reaches a maximum value in the central parts, in about latitude 30° N. or 30° S. The highest values, somewhat above 37‰, are found in the North and South Atlantic Oceans. In the North and South Pacific the maximum values are about 35.5‰ and 36.5‰, respectively, and in the Indian Ocean the corresponding value is about 36‰. The average surface salinity between parallels of latitude varies in a north-south direction; it is low near the equator, increases to a maximum between latitudes 20 and 30° north or south and decreases polewards from these latitudes. This variation depends upon the difference between the evaporation and the precipitation (WÜST[1]):

$$S = S_0 + k(E - P), \tag{27.1}$$

where S_0 represents a standard salinity and where E and P represent annual values of evaporation and precipitation, respectively. For all oceans combined the relation takes the form:

$$S = 34.60 + 0.0175\,(E - P) \tag{27.2}$$

where E and P are measured in centimeters. Somewhat different values of S_0 are found for the different oceans, but the factor k is nearly the same. In coastal waters run-off may lead to very low salinity values.

The *annual* variation of the surface salinity is small. For the central part of the North Atlantic the annual range is only about 0.05‰, but off New Foundland it is about 0.7‰. The maximum value occurs in early spring in agreement with the fact that evaporation is greatest in winter. In low latitudes a small *diurnal* variation has been ascertained.

In the surface layer the salinity is generally constant within an upper mixed layer. Below this the salinity decreases or increases, depending upon the relation of the surface salinity to that of the deep water. In low latitudes it generally decreases, in higher latitudes it increases, but there are several exceptions from this rule, depending upon the character of the currents. Only in the high northern and southern latitudes of the Atlantic are regions encountered in which the salinity in winter is nearly constant from the surface to the greatest depths. Details of the salinity distribution in the sub-surface layer will be discussed in the chapter dealing with water masses.

28. Temperature of the surface layer. In general the surface *temperature* decreases from the equator towards the poles, but the decrease is not uniform. In lower and middle latitudes, currents carrying warm water to the north or the south follow the eastern coasts of the continents, whereas currents carrying cold water towards lower latitudes follow the western coasts. For this reason we find that both in the Atlantic and the Pacific Oceans the extent of latitude with high ocean surface temperature is relatively narrow off the western coasts

[1] G. WÜST: Festschr. Norbert Krebs (Länderk. Forsch.) Stuttgart 1936.

and wide off the eastern. In high latitudes in the northern hemisphere we find the reverse condition, the temperature is high along the western coasts and low along the eastern. The explanation of these features will be given when dealing with the currents.

The *annual* range of the surface temperature reaches a maximum between latitudes 30 and 40° N. or S., in agreement with the fact that the range of annual variation of the radiation income is greatest in about that latitude. In the tropics the annual range of the surface temperature is about 2°, in the North Atlantic and North Pacific the average maximum range in about 40° is about 8.5 and 10°, respectively, but in the three southern oceans the corresponding ranges are all between 5° and 6°.

The *diurnal* range of the surface temperature is on an average only 0.2 to 0.3°, but on clear and calm days it may exceed 1°. The values apply to the open ocean; in closed coastal waters both the annual and the diurnal ranges may be much larger.

The *depth* to which the *annual variation* of temperature penetrates and its character at different depths depends upon several factors, such as the effects of heat conduction, lateral displacements of currents and vertical motion. In areas in which the two latter factors are of minor importance, it is possible to compute the *coefficient of eddy conductivity* (Sect. 17) from observed temperatures at different depths. Such computations have rendered values between 2 and 80 g./cm./sec. The larger values have been found in regions of strong currents where the turbulence may be expected to be intense. In these regions the annual variation may be observed to a depth of about 200 m., but in general it is negligible below 100 m.

In most ocean areas there exists in all seasons an *upper mixed layer* within which the temperature is constant, but the thickness of this layer varies with the season and with the wind conditions. In winter the layer is thick, particularly with a strong wind, but in summer it is shallow and in calm weather it may be lacking, that is, in calm weather a temperature decrease with depth may begin at the surface. The transition zone between the upper mixed layer and the colder deeper layers, the *thermocline,* may be so sharp that it approaches a discontinuity.

The character of the temperature distribution below the thermocline will be dealt with in the Chapt. H (Sect. 32ff.) on water masses.

29. Density and stability. In all oceans the density generally increases with depth, irrespectively of the effect of compressibility. Disregarding the effect of pressure, the density depends on temperature and salinity. In the surface layer the density may be altered because of heating or cooling, evaporation, precipitation, run-off from land or freezing of ice, but at sub-surface depths the density is changed by processes of mixing only. If the density of the surface water is increased, the water must sink and must be replaced by a mass from some sub-surface depth. In such case the water will sink to a depth where the density prevails which it has attained, and it will spread out at that depth. Similarly, if a convergence exists at which surface water must sink, the sinking water will spread out at the depth at which its density corresponds to the prevailing density.

In general the density of the surface layers decreases from the tropics towards higher latitudes. In all latitudes the surface density is greater in winter than in summer, and the very lowest surface values are found in certain regions in high latitudes in winter. Between about 40° S. and 40° N. the surface density is so low through the entire year that within that region sinking of surface water hardly contributes to the formation and maintenance of the deeper water masses of the oceans, but south and north of the mentioned latitudes where the surface density decreases towards the poles, sinking takes place because of convergences and winter cooling. *In the highest latitudes the water of the greatest density sinks and fills the deepest parts of the ocean basins.* Therefore, there exists a correspondence between the variation with latitude of the surface density in winter, south and north about 40° latitude, and the variation of density with depth in the oceans. It will be shown that conditions in the *Atlantic* ocean are of importance to the character of the deep and bottom water in *all* oceans.

Since the density of the water generally increases with depth, the stratification is, as a rule, stable. The *stability* in respect to vertical displacements can be expressed by means of the difference between the density which a mass of water attains when displaced vertically a distance Δz and the density of its new surroundings. The specific volume has previously (Sect. 13) been expressed by means of a standard specific volume and an anomaly, that is a sum of several terms. Similarly the density can be expressed as

$$\varrho_{s,\vartheta,z} = \varrho_{35,0,z} + \varepsilon_s + \varepsilon_\vartheta + \varepsilon_{s,\vartheta} + \varepsilon_{s,z} + \varepsilon_{\vartheta,z} = 1 + 10^{-3}\sigma_t + \varepsilon_{s,z} + \varepsilon_{\vartheta,z} \qquad (29.1)$$

where the depth z, is used as variable instead of the pressure p. A unit mass of water which is moved vertically a distance Δz will then attain a density that differs from its new surroundings by the amount

$$\Delta\varrho = 10^{-3}\Delta\sigma_t + \frac{\partial\varepsilon_{s,z}}{\partial S}\Delta S + \frac{\partial\varepsilon_{\vartheta,z}}{\partial\vartheta}\Delta\vartheta - \frac{\partial\varrho}{\partial\vartheta}\Delta\theta \qquad (29.2)$$

where $\Delta\sigma_t$, ΔS and $\Delta\vartheta$ represent the variations of σ_t, S and ϑ on the distance Δz and where $\Delta\theta$ represents the adiabatic change of temperature. The stability, E, can be taken as proportional to the acceleration to which the unit mass will be subjected if displaced a vertical distance of unit length:

$$E = \lim_{\Delta z\to 0}\frac{1}{\varrho}\frac{\Delta\varrho}{\Delta z} = \frac{1}{\varrho}\frac{d\varrho}{dz}. \qquad (29.3)$$

Omitting the factor $1/\varrho$, the practical expression becomes

$$E = 10^{-3}\frac{d\sigma_t}{dz} + \frac{\partial\varepsilon_{s,z}}{\partial S}\frac{dS}{dz} + \frac{\partial\varepsilon_{\vartheta,z}}{\partial\vartheta}\frac{d\vartheta}{dz} - \frac{\partial\varrho}{\partial\vartheta}\frac{d\theta}{dz}. \qquad (29.4)$$

HESSELBERG and SVERDRUP[1] have published tables giving the terms in the above equations. Use of these tables has demonstrated that down to a depth of about 2000 m. the stability is expressed with sufficient accuracy by means of the *first* term. POLLAK[2] has introduced another definition of the stability, which has certain advantages and which renders slightly different numerical values.

In the very deepest waters the salinity is often constant. In this case the expression for the stability is reduced to

$$E = \frac{\partial\varrho}{\partial\vartheta}\left(\frac{d\vartheta}{dz} - \frac{d\theta}{dz}\right) = \frac{\partial\varrho}{\partial\vartheta}\frac{d\Theta}{dz} \qquad (29.5)$$

meaning that it depends only upon the difference between the actual and the adiabatic temperature gradients, or upon the gradient of the potential temperature Θ.

[1] TH. HESSELBERG and H. U. SVERDRUP: Bergens Museums Aarbok 1914/15, Nr. 15.
[2] M. J. POLLAK: J. Marine Res. 13, 101 (1954).

The stratification is stable, indifferent or unstable depending upon whether the potential temperature decreases with depth, is constant or increases with depth. In many basins the temperature increases with depth, but in no case have increasing potential temperatures been found. Indifferent stratification is approached, but unstable conditions have not been encountered in the deep or bottom water.

30. Effect of vertical convection, convergences and divergences. Near the sea surface indifferent stability or slight instability is frequently found in the well mixed *top layer* which is nearly *homothermal* and *homohaline*. In middle and higher latitudes instability of the surface layer generally develops in winter because of cooling, and in these circumstances vertical convection starts, which in a few localities may reach to great depths and lead to renewal of the deep water.

The depth to which the winter convection reaches depends upon the large scale variation of density with depth. In the tropics the upper layers are so warm and the winter cooling so small that the winter convection is of no importance to the stratification. The same is true in lower and middle latitudes except in a few more or less enclosed adjacent seas where evaporation increases the surface salinity so much that during winter the reduction of the temperature raises the surface density to such values that water may sink from the surface to the greatest depths, meaning that convection currents may reach to the bottom.

These conditions are encountered particularly in the *Mediterranean Sea* where the surface salinity reaches values up to 39‰ because evaporation greatly exceeds the sum of precipitation and run-off. The lowest winter temperatures occur off the northern coasts of the Mediterranean, where the surface salinity is in the neighbourhood of 38.4‰, and in these regions the density may in winter become uniform from top to bottom, leading to renewal of the deep water.

Similar, but less pronounced conditions are encountered in the Red Sea where the salinity may reach a value of 41‰, and in the northern part of the Gulf of California.

In high latitudes the salinity of the upper layer is generally reduced because of excess of precipitation over evaporation, and winter cooling leads, therefore, in most regions only to the development of a moderately thick convection layer. This, for instance, is the case in the North Pacific where the winter convection in the region of the Aleutian Islands reaches to a depth of to 200 to 300 m. At some distance off the Antarctic Continent the corresponding depth is only about 100 m. in spite of the fact that there the salinity of the surface layer is increased in winter by freezing of ice.

On the other hand, on the continental shelves of the Antarctic Continent and particularly in the Weddell Sea area, the winter cooling and the increase in salinity because of freezing of ice may lead to the formation of water of such a great density ($\vartheta = -1.9°$, $S = 34.62‰$ and $\sigma_t = 27.9$) that it sinks from the shelves and mixes with water of slightly higher temperature and lower salinity, forming the *Antarctic Bottom Water* with a temperature of about $-0.4°$ and a salinity of 34.66‰.

In the northern part of the North Atlantic entirely different conditions are present. There the last branches of the Gulf Stream System carry water of high salinity to relatively high northern latitudes. The water of the branch that follows the south coast of Iceland towards Greenland and subsequently bends around Cape Farewell and continues along the west coast of Greenland, becomes diluted by admixture of less saline water from the East Greenland Current whereby its salinity is reduced to a little above 34.9‰. When this water in winter is cooled to between 3° and 2.5°, it attains a sufficiently high density (σ_t about 27.8) to sink to the greatest depth, thus renewing the deep and bottom water of the Atlantic.

In the Norwegian Sea a corresponding process takes place, but there the surface water has to be cooled to nearly freezing point before it sinks. The cold deep water that is formed does not flow into the North Atlantic, but fills the deep parts of the Polar Sea.

The processes which have been described, do not attain the same intensity every winter. The deep water which is formed is, therefore, not homogeneous, but within it the density increases somewhat towards bottom. This implies that during one winter when the processes at the surface lead only to a moderate increase of the surface density the winter convection will reach to a moderate depth only, whereas in a winter with a great increase of the surface density the convection may reach to the bottom. Thus, renewal of the very bottom water takes place intermittently.

Sinking of surface water also takes place in regions of *converging* currents. In this case the sinking water spreads at the depth at which otherwise water of such density is found. Conversely, water rises to the surface in regions of divergence, but except around the Antarctic Continent such rising takes place from depths that rarely exceeds a few hundred metres.

The density of the water varies not only in a vertical direction, but also horizontally. These horizontal differences in density can exist only in the presence of currents. Their relation to the currents will be discussed later on (Sect. 43).

31. Water masses. The water masses of the ocean can be described and classified by their temperature-salinity structure. For this purpose it is convenient to make use of the *temperature-salinity diagram*, the *T-S* diagram, which was introduced by HELLAND-HANSEN, who showed that when in any region the corresponding temperature and salinity values from different depths are plotted against each other, the values generally lie on a well defined curve or within a narrow band Observations in the surface layer are disregarded because in this layer local conditions and annual changes alter the relations. Below the surface layer a *water mass* is defined by a *T-S* curve or by a band. A single point in a *T-S* diagram defines a *water type*, but a water type is encountered only in such special localities as the deeper portions of basins where homogeneous water may be present.

The character of a *T-S* diagram is evident from Figs. 12 and 13, which also demonstrate the characteristic features of the water masses of the South Atlantic Ocean. In two diagrams are plotted observations between depths of 100 m. and 4400 m. at three stations in the South Atlantic that were occupied by the "Meteor" in 1925-27. The locations of the stations form a triangle of side length about 4100 km., but in spite of the great distances between the stations there is a great similarity between the water masses in the three localities.

In the depth interval 100 to 600 to 700 m. the temperature and salinity values all fall within a narrow and nearly straight band, although several characteristic differences are present. It may, for instance, be observed that at station 74 (off South Africa) water of a temperature of about 10.5° and a salinity of about 34.9‰ is found at a depth of 500 m. whereas at station 246 (close to the equator) water of these characteristics is encountered at about 250 m.

The water mass above a depth of 600—700 m. is called *Central Water*. It is probably formed by sinking in winter of surface water in the region between 30 and 40° S., because in this region where converging currents are known to be present the T-S relation of the surface water in a north-south direction is similar to the vertical T-S relation in the entire central part of the South Atlantic. In the diagram, curves for σ_t have been plotted, partly to illustrate the

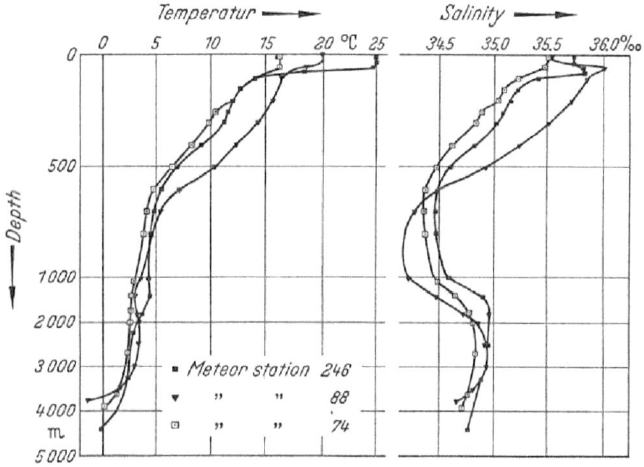

Fig. 12. Vertical distribution of temperature and salinity at "Meteor" stations 246 (2° 40′ S., 16° 37′ W.), 88 (33° 44′ S., 46° 01′ W.) and 74 (34° 00′ S., 13° 17′ E.). Note change of vertical scale at a depth of 1000 m.

stable stratification of the Central Water, partly to illustrate that water that sinks from the sea surface spreads out along a σ_t-surface.

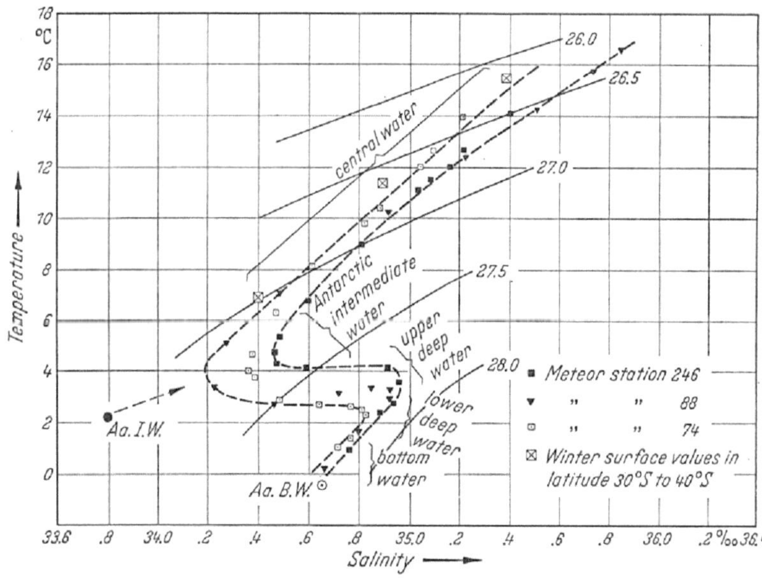

Fig. 13. Temperature and salinity at the same "Meteor" stations as in Fig. 12, plotted in a T-S diagram.

Below the Central Water a *salinity minimum* at depths between 800 and 1000m. is evident at all stations, but the salinity at the minimum decreases towards the north. This salinity minimum is related to the sinking of water at the Antarctic Convergence in about latitude 50° S. The water that sinks has a temperature of 2.2° and a salinity of about 33.8‰, as indicated in Fig. 13. As this water spreads to the north it becomes mixed with the over- and underlying waters, and a water mass, the *Antarctic Intermediate Water*, is formed, the limiting salinity values of which must be selected somewhat arbitrarily.

The deep water below the Intermediate Water is considered as composed of two parts, the *upper and lower deep water*. Both these water masses originate in the North Atlantic and move south, whereby the salinity decreases because of mixing with the over- and underlying waters. The bottom water is very cold and represents water that flows north from the depths around the Antarctic Continent. The characteristics of the *Antarctic Bottom Water* is also indicated in Fig. 13.

From this discussion it is evident that when examining the water masses of a region the study of minima and maxima of the T-S curve plays an important part. For the Atlantic a discussion of these features was undertaken by WÜST[1] who introduced the *"core method"* (*„Kernschicht-Methode"*) which in combination with the T-S diagram has proved to be a very powerful tool when analysing the structure of the waters of the oceans.

H. The water masses of the oceans.

32. The water masses of the Atlantic Ocean. The character of the water of the North Atlantic is influenced by the inflow of relatively warm and saline water from the Mediterranean, and by the transport of saline water into high northern latitudes by the last branches of the Gulf Stream System.

A great exchange of water takes place through the *Strait of Gibraltar*, where Atlantic Water flows in near the surface and Mediterranean deep water flows out along the bottom. In the Mediterranean the evaporation greatly exceeds the sum of precipitation and run-off from land, for which reason the salinity of the inflowing water is increased to values above 39‰. Winter convection, particularly in the Thyrrenian Sea, leads to the formation of a deep- and bottom-water which has a temperature of a little above 13° and a salinity of about 38.4‰. Slightly different values are found in the different basins, both temperature and salinity of deep water increasing towards the east.

The water that flows from the Mediterranean into the North Atlantic along the bottom of the Strait of Gibraltar has at the eastern entrance to the strait a density of 28.97, expressed as σ_t. In the strait, intensive mixing takes place such that at the western end of the strait the water has a salinity of about 37.2‰ and a σ_t of about 28.12. The temperature remains at 13°. The density of this water is so great that outside the strait it sinks from a depth of about 300 m. to a depth of about 1000 m., but as it sinks, its salinity is reduced to about 36.45‰, its temperature to about 12°, and its σ_t-value to 27.74. This density corresponds roughly to the characteristic density at depths between 1000 and 2000 m. in the Atlantic, and, therefore, the Mediterranean Water spreads between these depths, but loses more and more of its character because of mixing with adjacent layers.

In the North Atlantic, Mediterranean Water can be traced off the European west coast, occasionally as far north as to Shetland, and it can be recognised over the greater part of the central part of the ocean. In a T-S diagram it appears as a bulge on the otherwise smooth curve that characterises the central water of the North Atlantic.

Part of the Mediterranean Water turns to the south where it spreads below the Antarctic Intermediate Water and appears as a salinity maximum in the

[1] G. WÜST: Deutsch. Atlant. Exped. „Meteor" 1925—1927, Wiss. Erg. **6**, Teil 1, 2 (1935).

upper deep water (Fig. 14). WÜST has been able to show that traces of the Mediterranean Water can be found past South Africa, and an analysis of the Antarctic waters indicates that its influence is present even in the southern part of the Indian Ocean, south of 50° S.

The influence of the Mediterranean Water is, however, not limited to the depths at which it can be clearly recognised. Down to a depth of about 1500 m. the salinity of the water of the North Atlantic is nearly 0.5 $^0/_{00}$ higher than that of the South Atlantic, and this striking difference must be ascribed to the effect of the Mediterranean Water, or, in the last instance, to the effect of the excessive evaporation in the Mediterranean Sea.

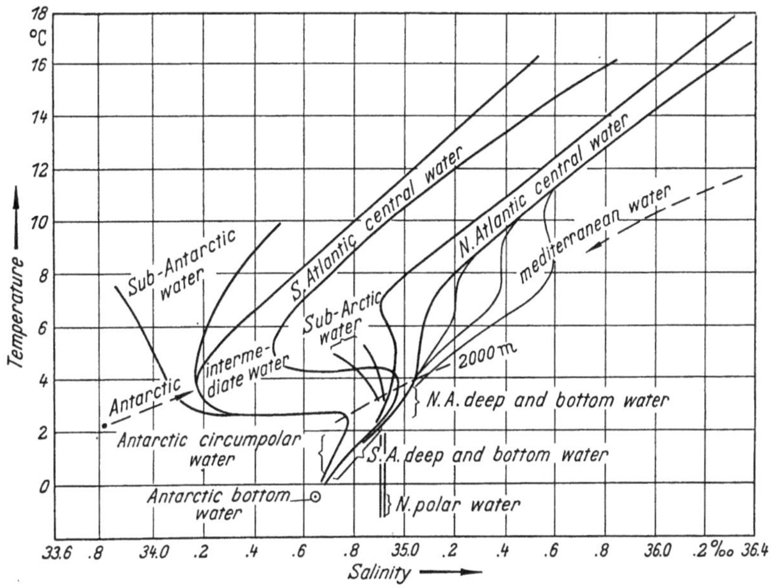

Fig. 14. Water masses of the Atlantic Ocean.

The character of the water masses of the Atlantic Ocean is shown schematically in Fig. 14 in which the data which were presented in Fig. 13 have been incorporated, and in which the character of the Sub-Arctic and Sub-Antarctic waters also has been indicated. The figure demonstrates the higher salinity of the waters of the North Atlantic. In the deeper layers we find a continuous change in the water masses as we pass from north to south,—in contrast to conditions in the Pacific where the northern and southern water masses are separated by an equatorial mass. In the upper layers the central water masses are separated by a narrow transition zone in about 10° N. Thus, the central water of the South Atlantic crosses the equator.

Characteristic differences exist between the sub-polar waters of the north and the south. The Sub-Arctic Water, which is found around the southern part of Greenland, has a salinity between 34.7 and 34.8 whereas the salinity of the Sub-Antarctic Water lies between 33.8 and 34.5. This difference is due to the circumstance that in the North Atlantic branches of the Gulf Stream System carry Central water of high salinity, above 35.2 $^0/_{00}$, into high latitudes where it mixes with Arctic Water that is carried south by the East Greenland Current. By this mixing, water is formed of a salinity that varies between 34.7 and 34.9, and when this water in winter is cooled to about 3°, the density of the most saline water ($\sigma_t = 27.8$) is high enough for the water to sink to the greatest depths and form the bottom water of the North Atlantic and the deep water of the entire Atlantic Ocean.

On the other hand, the salinity of the Sub-Antarctic Water is so low that winter cooling cannot increase its density so much that it can sink to any appreciable depth and contribute to the deep and bottom water. However, formation of bottom water takes place in the extreme southern part of the Atlantic Ocean, where the waters of the continental shelf around the Weddell Sea attain an especially high density because of freezing of ice. Under extreme conditions the salinity of the water may reach a value of 34.62‰, at which freezing point is at $-1.9°$ and $\sigma_t = 27.89$ at freezing. When this water sinks along the slope of the shelf it mixes with the Antarctic Circumpolar Water which near the shelf has a salinity of 34.62‰, a temperature of 0.5° and a σ_t of 27.84. The result of the mixing is the formation of a bottom water of salinity 34.66, a temperature of $-0.4°$ and a σ_t of 27.86. This dense water spreads to the north in the Atlantic Ocean where it can be traced across the equator to about 10° north latitude. It also spreads to the east, along the Antarctic Continent, where to the south of the Indian Ocean some sinking of cold water from the shelf may occur.

Between about latitude 50° S. and the Antarctic Continent a very uniform water mass, the *Antarctic Circumpolar Water*, is present, with salinities between 34.70 and 34.76 and temperatures between 0.5 and 2.5°. Both salinity and temperature decrease from north to south. This water mass is formed by mixing of the North Atlantic Deep Water with Antarctic Intermediate and Bottom Water. The highest salinities and temperatures indicate a slight admixture of Mediterranean Water.

A special type of deep and bottom water is found in northern adjacent seas, the *Norwegian Sea* and the *Polar Sea*. In the Norwegian Sea mixing and winter cooling lead to the formation of a deep and bottom water which has a salinity of 34.92, a minimum temperature of about $-1.2°$ and a maximum σ_t of 28.11. This water has the highest density of all water masses except the Mediterranean bottom water, but is prevented by the Shetland-Faroe-Iceland-Greenland ridge from flowing into the North Atlantic. The deep water of the Norwegian Sea flows, however, into the Polar Sea, crossing the ridge that runs in about 80° N. between Svalbard and Greenland. According to the analysis by WÜST[1], the greatest depression of the ridge is between 1750 and 2000 m. deep and the inflowing water has a salinity of about 34.9‰ and a temperature between -1.0 and $-1.2°$. In the Polar Sea the temperature increases adiabatically towards the bottom and at a depth of 5000 m. it is, therefore, about $-0.65°$. This value applies to the basin that lies north of the Norwegian Sea. Observations to the north of Alaska (WORTHINGTON[2]) have shown that there the temperature of the water below a depth of 2000 m. is 0.3 to 0.5° higher than the temperature at the same depths to the north of Svalbard-Franz Josef Land. It was therefore suggested that a submarine ridge must separate the deeper portions of the Beaufort Sea from the rest of the Polar Sea.

From the analysis of tidal observations FJELDSTAD[3] concluded that: "A possible explanation (of certain discrepancies) would be that north of the Beaufort Sea there might be a ridge with comparatively small depths. The almost synchronity of the tide along the coast of Alaska demands that this conjectured

[1] G. WÜST: Veröff. dtsch. Wiss. Inst., Kopenhagen, **1**, Arktis, Nr. 6 (1942).

[2] L. V. WORTHINGTON: Trans. Amer. Geophys. Un. **34**, 543 (1953).

[3] J. E. FJELDSTAD: The Norw. N. Polar Exped. "Maud" 1918—1925. Sci. Results **4**, Nr. 4 (1936).

ridge or plateau extends unbrokenly from the Siberian Shelf to the Arctic Archipelago." According to reports in USSR. periodicals which have been translated into English (WEBSTER[1]) and German (Petermanns Mitt.[2]) made by aviators of the Soviet Union who have landed on the ice have demonstrated the existence of such a ridge, the *Lomonosov ridge,* that runs from the New Siberian Islands to the northern part of Greenland. The ridge rises to depths less than 1000 m. with saddle depths of 1500 to 1600 m.

33. The water masses of the Indian Ocean. The structure of the waters of the Indian Ocean is less complicated than that of the Atlantic (Fig. 15). The Sub-Antarctic Water to the south of 40° S. is similar to that of the Atlantic. Between latitudes 10° S. and 40° S. and to a depth of 500 to 800 m. a central water mass dominates, which is practically identical with the central water mass of the South Atlantic. Within the same region the Antarctic Intermediate Water is responsible for the salinity minimum on the *T-S* curve at a depth of 700 to 900 m. Below a depth of about 1000 m. there exists a gradual transition from the deep water in the central part of the ocean with a salinity of about 34.8‰ and a temperature between 1.5 and 4° to the Antarctic Circumpolar Water.

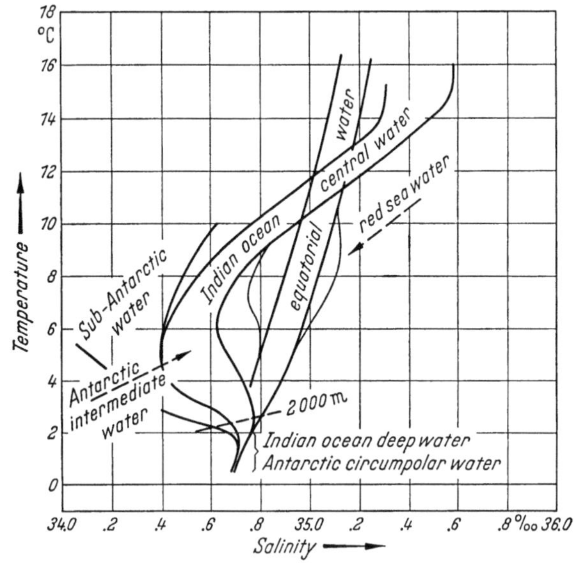

Fig. 15. Water masses of the Indian Ocean.

To the north of 10° S. we find an entirely different water mass, the *Equatorial Water,* which except near the top is characterised by higher salinities. These higher salinities probably reflect the influence of *Red Sea Water,* which, however, plays a minor part in the Indian Ocean than does the Mediterranean Water in the Atlantic. The exchange of water between the Red Sea and the adjacent part of the Indian Ocean is not so well known as the exchange of water through the Strait of Gibraltar, but it is probable that the volume of Red Sea Water that flows into the Indian Ocean is only about one sixth of the volume of Mediterranean Water that flows into the Atlantic. Still, the Red Sea Water can be recognised at a number of stations to the north of 10° S. and it can also be traced to about 25° S. along the African east coast.

The origin of the equatorial water of the Indian Ocean is not clear, but it is probable that in the upper layers its character is influenced by conditions in the Pacific Ocean with which communication exists, and that the salinity of the lower layers is increased by addition of Red Sea Water.

34. The water masses of the Pacific Ocean. In a vertical direction the structure of the water masses of the Pacific is simple, probably because of the absence of high salinity water from adjacent seas, but in a horizontal direction conditions

[1] C. J. WEBSTER: Arctic 7, 59 (1954).
[2] Petermanns geogr. Mitt. 9, 70 (1955).

are complicated because of the great width of the ocean and great differences between eastern and western regions (Fig. 16).

Proceeding from south to north we find first in the Antarctic region the *Circumpolar Water* which is still more uniform than farther to the west. The *Sub-Antarctic Water* is similar to that of the other southern oceans, and is characterised by a relatively high salinity.

To the north of the Sub-Antarctic Water we encounter two types of central water, a western and an eastern. The *Western Central Water* which extends to long. 160° W. is identical with the central water masses of the Indian Ocean and the South Atlantic. This supports the assumption that these water masses are

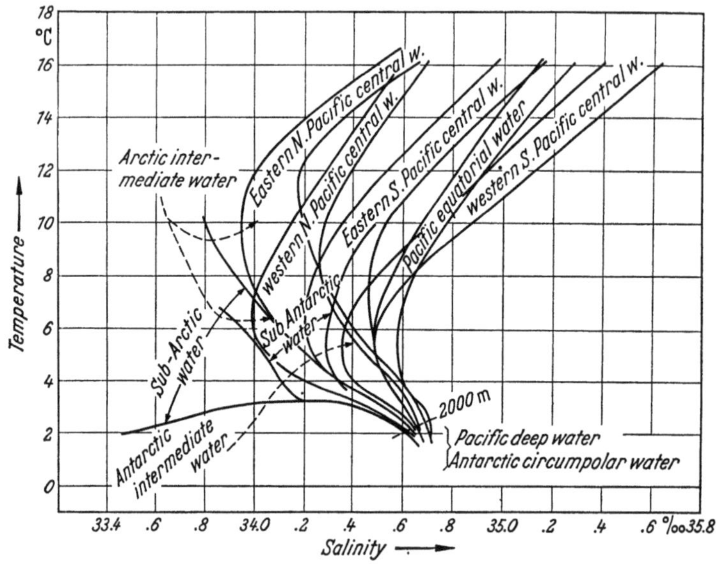

Fig. 16. Water masses of the Pacific Ocean.

formed by the sinking of water in the subtropical convergence region where, because of similar climatological conditions, we find similar horizontal T-S relations in winter. The *Eastern Central Water* between 160 and 90° W. has a considerable lower salinity and appears to be influenced by admixture of Sub-Antarctic Water. This is in agreement with the character of the currents because the southward extension of South America forces part of the Sub-Antarctic Water to flow north along the coast of Chile.

Below the central water masses the *Antarctic Intermediate Water* is found. This water is probably also responsible for the salinity minimum in the equatorial region, but no trace of it appears to the north of the equatorial region.

In the tropics of the Pacific, in a belt which off America extends from 10° S. to 20° N., but which north of New Guinea is narrowed to a few degrees of latitude, we find an *Equatorial Water Mass* which in its upper part is intermediate in character between the central water masses of the South Pacific, and in its lower part appears to be influenced by the Antarctic Intermediate Water.

In the North Pacific we find again a western and an eastern type of central water, which both are characterised by low salinities and which are more shallow than other central water masses. The *North Pacific Western Central Water* occupies by far the greater part of the ocean, the *Eastern Central Water* being confined to a small region to the northeast of the Hawaiian Islands. Below the central water masses *Arctic Intermediate Water* is present at depths that vary between 300 and 800 m.

To the north of about 45° N. we find the *Sub-Arctic Water* which covers a large part of the ocean. A mixture of this water and equatorial water is found off the North American west coast to about 25° S. The Sub-Arctic Water of the North Pacific is characterised by a very low salinity which is caused by the excess of precipitation over evaporation in the northern part of the ocean. It seems probable that the low salinity of this water is also responsible for the low salinity of the central water masses of the North Pacific. The contrast be-ween these conditions and the conditions in the North Atlantic is striking. In the latter ocean the high salinities must be ascribed in part to the influence of the Mediterranean Water and in part to the shape and the depths of the ocean which make it possible for currents to carry high salinity water into high latitudes, thereby suppressing the effect of the excess of precipitation.

The discussed differences in the character of the waters exist only in the upper layers. Below a depth of about 1500 m. or better below 2000 m. the waters are remarkably uniform, as evident from Fig. 16. A few systematic differences exist between the deep and bottom water of the South and North Pacific, but these are too small to be seen from the curves in Fig. 16. The bottom water of the South Pacific has a salinity slightly above $34.70\,{}^0/_{00}$ and a temperature that may be as low as 1.0°, whereas the bottom water of the North Pacific has a salinity a little below 34.7 and a temperature around 1.5°. The *oxygen contents* show greater differences, varying between 2 and 3 ml./L in the deep water of the North Pacific and 3.5 and 4.5 ml./L in the South Pacific. The contrast to conditions in the Atlantic is again striking and is explained by the fact that in the Pacific no active formation of bottom water takes place, whereas in the Atlantic bottom water is formed both in the north (salinity $34.90\,{}^0/_{00}$, temperature 3°) and in the south (salinity $34.66\,{}^0/_{00}$, temperature $-0.4°$).

The character of the deep and bottom water in the Atlantic and Pacific Oceans is illustrated by the meridional sections in Figs. 25 and 26 (Sect. 45).

I. Applications of hydrodynamics.

35. The equations of motion. In oceanography a *lefthanded coordinate* system is generally used with the *z-axis positive downwards.* The equations of motion, in an approximation sufficient for most oceanographic applications, then take the form:

$$\left.\begin{aligned}
\frac{dv_x}{dt} &= \frac{\partial v_x}{\partial t} + \frac{\partial v_x}{\partial x} v_x + \frac{\partial v_x}{\partial y} v_y + \frac{\partial v_x}{\partial z} v_z = -\alpha \frac{\partial p}{\partial x} + 2\omega \sin\varphi\, v_y + \alpha R_x,\\
\frac{dv_y}{dt} &= \frac{\partial v_y}{\partial t} + \frac{\partial v_y}{\partial x} v_x + \frac{\partial v_y}{\partial y} v_y + \frac{\partial v_y}{\partial z} v_z = -\alpha \frac{\partial p}{\partial y} - 2\omega \sin\varphi\, v_x + \alpha R_y,\\
\frac{dv_z}{dt} &= \frac{\partial v_z}{\partial t} + \frac{\partial v_z}{\partial x} v_x + \frac{\partial v_z}{\partial y} v_y + \frac{\partial v_z}{\partial z} v_z = -\alpha \frac{\partial p}{\partial z} + g + \alpha R_z.
\end{aligned}\right\} \quad (35.1)$$

They express that the acceleration equals the sum of the acting forces, the pressure force (α, specific volume, p, pressure), the gravitational force (g, acceleration of gravity), the Coriolis force (ω, angular velocity of earth, φ, latitude) and the frictional force. To these equations must be added the equation of continuity,

$$\frac{\partial v_x}{\partial x} + \frac{\partial v_y}{\partial y} + \frac{\partial v_z}{\partial z} = \frac{1}{\alpha}\frac{d\alpha}{dt} \quad (35.2)$$

and boundary conditions.

In the above form the equations of motion, with latitude φ positive, apply to the northern hemisphere (the Coriolis force acts to the right). When dealing with conditions in the southern hemisphere the signs of the Coriolis terms in the first two equations are reversed (with φ).

The above equations are too complicated to be solved completely, but simplifications can be introduced, depending upon the character of the problem.

36. The field of pressure. It is obvious that in many problems dealing with currents knowledge of the pressure gradients is needed, that is, knowledge of the field of pressure. The pressure at different depths in the ocean cannot be determined *directly* with sufficient accuracy, but distances between isobaric surfaces can be computed, meaning that it is possible to determine an *internal* field of pressure, from which a field of *relative* pressure gradients can be derived.

For this purpose we turn to the last of the three equations of motion (35.1). In this equation the last (friction) term on the right hand side can always be neglected, and vertical accelerations are always negligible except when dealing with short surface waves. Disregarding problems related to wind waves and swell we may, therefore, use the *hydrostatic equation:*

$$g\,dz = \alpha\,dp \tag{36.1}$$

from which distances between isobaric surfaces can be found by integration, provided that the specific volume is known as a function of pressure. The latter condition is fulfilled if temperature and salinity, from which the specific volume can be computed (Sect. 13), have been determined at a sufficient number of depths, because in the first approximation the pressure increases by one decibar (one bar $= 10^6$ dynes cm.$^{-2}$) when the depth increases by one metre. This implies that observations which are made at known depths can be interpreted as made at known pressures, the latter being expressed by means of the same numbers as the depths if decibar is used as unit of pressure and metre as unit of depth.

Instead of computing the *geometric* distance between isobaric surface it is practical to compute the *geopotential differences* dD, that is the work performed when moving a unit mass vertically from one isobaric surface to another. As a practical unit for dD, V. BJERKNES introduced the "*dynamic metre*", with which $dD = g\,dz/10$, if g and dz are measured in the metre-ton-second system (MTS). In the same system the unit of pressure is one centibar which equals 1/10 of a decibar. With D in *dynamic metres* and p in *decibar* the hydrostatic equation therefore takes the form

$$dD = \alpha\,dp$$

Introducing

$$\alpha = \alpha_{35,0,p} + \delta, \tag{36.2}$$

one obtains

$$D_2 - D_1 = \int_{p_1}^{p_2} \alpha_{35,0,p} + \int_{p_1}^{p_2} \delta\,dp = (D_2 - D_1)_s + \Delta D, \tag{36.3}$$

meaning that the dynamic or geopotential distance can be expressed as a sum of a standard distance and an *anomaly*. The standard pressure field is represented by a series of parallel isobaric surfaces, and this field lacks therefore a relative field of force. The latter field is completely determined by the field of anomalies of dynamic depths as obtained from the equation:

$$\Delta D = \int_{p_1}^{p_2} \delta\,dp. \tag{36.4}$$

When a sufficient number of stations are available, the relative field of pressure can be fully represented by a series of charts showing the *dynamic topography* of isobaric surface *relative* to one selected *isobaric surface*. *Absolute topographies* can be determined only if the shape of the reference surface is known.

37. Computation of currents. For computing currents from the distribution of pressure it is assumed that accelerations and frictional forces are negligible, whereby the first two equations of (35.1) are reduced to:

$$\lambda v_y = \alpha \frac{\partial p}{\partial x}, \qquad \lambda v_x = -\alpha \frac{\partial p}{\partial y}, \tag{37.1}$$

where

$$\lambda = 2\omega \sin\varphi.$$

In some instances it is practical to replace the horizontal pressure gradient by the slope of the isobaric surface. The x-profile of the isobaric surface is defined by:

$$\frac{\partial p}{\partial x} dx + \frac{\partial p}{\partial z} dz = 0 \tag{37.2}$$

or, using the hydrostatic Eq. (36.1)

$$\frac{\partial p}{\partial x} dx + \frac{g}{\alpha} dz = 0. \tag{37.3}$$

Therefore, the slope in the x-direction is

$$i_{p,x} = \frac{dz}{dx} = -\frac{\alpha}{g}\frac{\partial p}{\partial x} \tag{37.4}$$

meaning that

$$\alpha \frac{\partial p}{\partial x} = -g\, i_{p,x} \tag{37.5}$$

and, similarly

$$\alpha \frac{\partial p}{\partial y} = -g\, i_{p,y}. \tag{37.6}$$

Consequently the currents are obtained from the equations:

$$v_x = \frac{g}{\lambda} i_{p,y}, \qquad v_y = -\frac{g}{\lambda} i_{p,x}. \tag{37.6}$$

The computed currents represent true currents if the slopes of the relative isobaric surface refer to a level isobaric surface at which the horizontal motion vanishes. In this case Eq. (37.1) corresponds to the equation of the *geostrophic wind*, which is used extensively in meteorology. In oceanography a great deal of effort has been directed towards the establishment of a "*surface of no motion*" to which the computed currents can be referred. Here it must suffice to state that in the Pacific Ocean good agreement between observed and computed currents is obtained by selecting a surface of no motion at a constant depth of about 1500 m., but in the Atlantic Ocean a variable depth must be introduced in agreement with the much more complex character of the circulation in the Atlantic.

The components of the total *mass transport* by currents are:

$$M_x = \int_0^d \varrho v_x dz, \qquad M_y = \int_0^d \varrho v_y dz \tag{37.8}$$

where d represents "the depth of no motion". Similarly, the components of the *volume transport* equal

$$T_x = \int_0^d v_x dz, \quad T_y = \int_0^d v_y dz. \tag{37.9}$$

The mass and volume transports can be computed directly from the observed distribution of mass without first finding the velocities. We define a function

$$P = \int_0^d p\, dz \tag{37.10}$$

which can be computed by making use of the hydrostatic equation. Integration of (37.1) then gives

$$M_x = -\frac{1}{\lambda}\frac{\partial P}{\partial y}, \quad M_y = \frac{1}{\lambda}\frac{\partial P}{\partial x}. \tag{37.11}$$

Eq. (37.1) on which the computation of currents is based states only that, neglecting acceleration and friction, there exists a mutual *relationship* between the field of *pressure* and the field of *motion*. If the field of pressure is known, as has been assumed, the field of motion can be computed, but vice versa we may also state that if the field of motion were known we could have computed the corresponding field of pressure. Therefore, we cannot claim that the currents are *caused* by the distribution of pressure, which in turn is determined by the distribution of mass. We can only state that, neglecting acceleration and friction, a mutual relationship exists.

The computation of currents from the pressure distribution implies that the motion is directed along the isobars or along the contour lines of the isobaric surfaces. Furthermore, if the distribution of pressure is stationary the isopycnals must coincide with the isobars and must be parallel at all levels. These requirements were expressed by HELLAND-HANSEN and EKMAN who introduced them under the name "*the law of parallel solenoids*". Observations show that within the regions of the major currents the requirements are nearly fulfilled when dealing with *average* conditions, meaning that on an average a *stationary state* exists, and that the computation of currents from the observed distribution of mass can be expected to render reliable results.

However, *friction* cannot be completely neglected, although a good approximation is obtained by disregarding it. Actually, kinetic energy is always being dissipated, and therefore the existing circulation cannot prevail unless energy is added. In the ocean only negligible amounts of heat are transformed into kinetic energy because heating and cooling take place at the same pressure. The remaining possibility is that the *ocean currents are maintained by the prevailing winds*, and this appears probable because in general there is a striking agreement between the direction of the prevailing winds and the prevailing currents. In the regions of the trade winds the currents run in the direction of the wind, towards the west, and, similarly, between 35 and 50° N where winds from the west dominate, the currents run to the east. Between latitudes 20 and 35° N. anticyclonic circulation prevails both in the atmosphere and in the ocean whereas north of about 50° N. cyclonic circulations are present (Gulf of Alaska, Bering Sea, Labrador Sea, Norwegian Sea).

There exist two *striking discrepancies between the winds and the currents:* To the north of the Equator a *counter current* towards the east is embedded between the equatorial currents to the west, and in both oceans *the western parts of the anticyclonic gyrals are compressed* such that the transport to the north takes place by narrow and swift currents (the Gulf Stream and the Kuroshio). A theory that shall account for the relationship between winds and current must explain the reason for these discrepancies. It will be shown that this can be done, but first we will consider the more immediate effects of the wind towards producing currents or piling-up water.

38. The pure wind currents. In order to deal with the "pure wind current" we return to the first two equations of motion and omit all terms except the Coriolis terms and the frictional terms. When dealing with the latter we assume that only the effect of vertical turbulence enters. The frictional stress exerted on a horizontal surface has then the components:

$$\tau_x = \mu_e \frac{d v_x}{d z} \quad \text{and} \quad \tau_y = \mu_e \frac{d v_y}{d z} \tag{38.1}$$

where μ_e is the eddy viscosity. The frictional force per unit mass has the components:

$$\alpha R_x = \alpha \frac{d}{dz}\left(\mu_e \frac{d v_x}{d z}\right) \quad \text{and} \quad \alpha R_y = \alpha \frac{d}{dz}\left(\mu_e \frac{d v_y}{d z}\right). \tag{38.2}$$

The equations of horizontal motion then take the form:

$$\left.\begin{aligned} \lambda \varrho v_y + \frac{d}{dz}\left(\mu_e \frac{d v_x}{d z}\right) &= 0, \\ \lambda \varrho v_x - \frac{d}{dz}\left(\mu_e \frac{d v_y}{d z}\right) &= 0. \end{aligned}\right\} \tag{38.3}$$

The boundary conditions are:

$$\tau_{a,x} = -\left(\mu_e \frac{d v_x}{d z}\right)_0, \quad \tau_{a,y} = -\left(\mu_e \frac{d v_y}{d z}\right)_0, \tag{38.4}$$

where τ_a is the stress that the wind exerts on the sea surface, and

$$v \to 0 \quad \text{when} \quad z \to \infty. \tag{38.5}$$

The first condition expresses that at the sea surface the stress that the wind exerts on the water must balance the stress that the water exerts on the air.

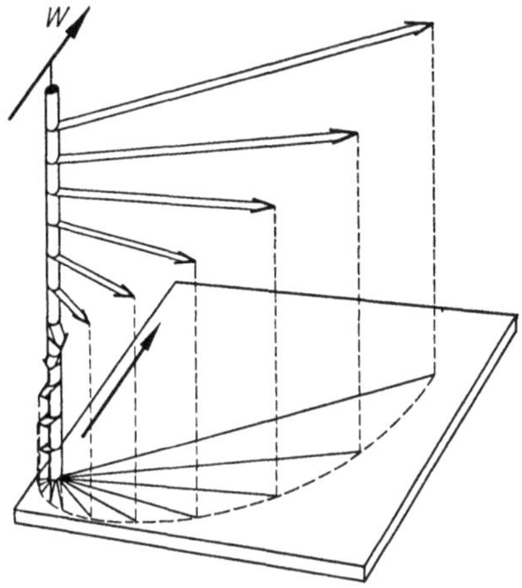

Fig. 17. Schematic representation of the pure wind current in deep water (northern hemisphere) showing the decrease of the velocity and the turning to the right of the direction as the depth increases.

Using these boundary conditions we obtain by integration

$$M_x = \frac{1}{\lambda} \tau_{a,y}, \qquad M_y = -\frac{1}{\lambda} \tau_{a,x}. \tag{38.6}$$

This means that the total transport by the wind current is directed 90° to the right of the stress of the wind (in the southern hemisphere to the left). This result is independent of the manner in which the eddy viscosity varies with depth.

In order to learn how the velocity of the wind current varies with depth it is necessary to introduce certain assumptions as to the character of the eddy viscosity. The simplest assumption, $\mu_e =$ constant, was introduced by V. W. Ekman in 1902 when he first discussed the problem.

Integration renders:

$$\left.\begin{aligned} v_x &= v_0 e^{-\frac{\pi}{D} z} \cos\left(45^\circ - \frac{\pi}{D} z\right), \\ v_y &= v_0 e^{-\frac{\pi}{D} z} \sin\left(45^\circ - \frac{\pi}{D} z\right), \end{aligned}\right\} \tag{38.7}$$

where

$$v_0 = \frac{\tau_a}{\sqrt{\mu_e \varrho\, 2\omega \sin\varphi}}, \qquad D = \pi \sqrt{\frac{\mu_e}{\varrho\, \omega \sin\varphi}}. \tag{38.8}$$

The character of the current is very simple: At the surface the current is deflected 45° to the right of the direction of the stress (the wind), the angle of deflection increases linearly with depth and the velocity decreases logarithmically with depth (Fig. 17). Other assumptions have been made regarding the eddy viscosity, but in all cases similar major features have been found. When dealing with the pure wind current it is obvious that the energy that is dissipated by friction must be supplied by the wind. Multiplying the first of Eq. (38.3) by v_x and the second by v_y and subtracting we obtain

$$v_x \frac{d}{dz}\left(\mu_e \frac{dv_x}{dz}\right) + v_y \frac{d}{dz}\left(\mu_e \frac{dv_y}{dz}\right) = 0. \tag{38.9}$$

Now

$$v_x \frac{d}{dz}\left(\mu_e \frac{dv_x}{dz}\right) = \frac{d}{dz}\left(v_x \mu_e \frac{dv_x}{dz}\right) - \mu_e \left(\frac{dv_x}{dz}\right)^2. \tag{38.10}$$

Integrating between the limits 0 and d where $d > D$ and introducing the boundary conditions

$$\left(\mu_e \frac{dv_x}{dz}\right)_0 = -\tau_{a,x}, \tag{38.11}$$

and

$$z = d, \quad v_x = 0, \tag{38.12}$$

we obtain

$$\int_0^d v_x \frac{d}{dz}\left(\mu_e \frac{dv_x}{dz}\right) dz = v_{x,0}\, \tau_{a,x} - \int_0^d \mu_e \left(\frac{dv_x}{dz}\right)^2 dz. \tag{38.13}$$

Treating the second term in Eq. (38.9) in a similar manner we find:

$$v_{x,0}\, \tau_{a,x} + v_{y,0}\, \tau_{a,y} = \boldsymbol{v}_0 \cdot \boldsymbol{\tau}_a = \int_0^d \mu_e \left[\left(\frac{dv_x}{dz}\right)^2 + \left(\frac{dv_y}{dz}\right)^2\right] dz \tag{38.14}$$

meaning that the power per unit area which is exerted by the stress of the wind equals the total dissipation of energy by friction in a water column of unit area cross section.

The transport at right angles to the wind accounts for the *upwelling* of cold sub-surface water that occurs off the coasts of California, Morocco, Peru and Southwest Africa. Where the wind blows more or less permanently parallel to the coast and with the coast to the left (right in the southern hemisphere) of an observer with the back to the wind, the warm surface layer is carried away from the coast. Near the coast the water must be replaced by water that rises towards the surface. Observations show that this water comes only from a depth of 200 to 300 m. but has a temperature that is many degrees below the average surface temperature of the latitude in question. This process is responsible for the low temperature off the coasts that were mentioned.

39. Piling-up of water by wind. The above results apply to the open ocean where the stress of the wind may be balanced by the total effect of the Coriolis force. When dealing with a narrow body of water the stress has to be balanced in a different manner. Consider a long, shallow bay with the wind blowing into the bay. In this case the slope of the sea surface cannot be neglected. Omitting acceleration but retaining the friction term in the previous form, integration of the equations of horizontal motion gives:

$$\left.\begin{aligned} \tau_{a,x} + \tau_{d,x} &= -\lambda M_y - \int_0^d g \varrho\, i_{p,x}\, dz, \\ \tau_{a,y} + \tau_{d,y} &= \lambda M_x - \int_0^d g \varrho\, i_{p,y}\, dz \end{aligned}\right\} \tag{39.1}$$

where d is the depth to the bottom and τ_d is the stress at the bottom. Placing the x-axis in the direction of the bay, considering that when a steady state has been reached the transport must be zero, and assuming the water to be homogeneous so that the inclination of the isobaric surfaces is independent of depth, we obtain:

$$\tau_a - \tau_d = -g\varrho i_p d. \tag{39.2}$$

In this case the sum of the stresses exerted at the upper and lower surfaces of the body of water is balanced by the component of gravity acting on the entire body of water. Eq. (39.2) has been used for computing the stress as a function of wind velocity (Sect. 19).

Piling-up of water also takes place in the open ocean. In the tropics where the trade winds blow steadily from the east, the light surface water is accumulated in the western part of the oceans. At a depth of a few hundred metres the temperature and salinity do not vary with longitude, indicating that there the isobaric surfaces do not slope in an east-west direction. The sea surface, on the other hand, must rise towards the west because of the piling-up of the warm and light surface water. The oceanographic observations indicate a slope of 3.8×10^{-8} in the Atlantic and of 4.5×10^{-8} in the Pacific.

40. General relation between winds and currents. Having dealt with these special cases of the effects of the wind stress towards transporting water or maintaining slopes of isobaric surface we can revert to the general problem of the relation between the prevailing winds and the prevailing currents. For this purpose we again make use of the equations of motion, taking the effects of both vertical and horizontal eddy viscosities into account. We write the frictional components of the frictional term in the form

$$\left.\begin{aligned} \alpha R_x &= \alpha \frac{\partial}{\partial z}\left(\mu_e \frac{\partial v_x}{\partial z}\right) + h_x, \\ \alpha R_y &= \alpha \frac{\partial}{\partial z}\left(\mu_e \frac{\partial v_y}{\partial z}\right) + h_y \end{aligned}\right\} \tag{40.1}$$

and introduce

$$H_x = \int_0^d h_x\, dz, \qquad H_y = \int_0^d h_y\, dz. \tag{40.2}$$

Omitting the field accelerations which according to experience are small, and integrating the equations of motion we obtain

$$\left.\begin{aligned} \frac{\partial M_x}{\partial t} &= -\frac{\partial P}{\partial x} + \lambda M_y + \tau_{a,x} + H_x, \\ \frac{\partial M_y}{\partial t} &= -\frac{\partial P}{\partial y} - \lambda M_x + \tau_{a,y} + H_y. \end{aligned}\right\} \tag{40.3}$$

We recognise that previously we have used the first two terms on the right hand side of (40.3) for computing the mass transport from the distribution of pressure, the second and third terms for computing the mass transport by the pure wind current, and the first and third terms for computing the slopes of isobaric surface under specialised conditions.

We now eliminate P by cross differentiation, observing that for an incompressible fluid continuity requires

$$\frac{\partial M_x}{\partial x} + \frac{\partial M_y}{\partial y} = 0 \tag{40.4}$$

and that, with the x-axis east-west, $\lambda = 2\omega \sin\varphi$ is independent of x. Using the designation $\beta = d\lambda/dy$ we obtain:

$$\frac{\partial}{\partial t}\left(\frac{\partial M_x}{\partial y} - \frac{\partial M_y}{\partial x}\right) = \beta M_y + \left(\frac{\partial \tau_{a,x}}{\partial y} - \frac{\partial \tau_{a,y}}{\partial x}\right) + \left(\frac{\partial H_x}{\partial y} - \frac{\partial H_y}{\partial x}\right). \tag{40.5}$$

This is a *vorticity equation* which states that the local change of vorticity of the mass transport equals the sum of three terms which all have their specific physical significance. Assuming that on an average a *steady state* exists, the condition which in that case has to be fulfilled can be written:

$$\left(\frac{\partial \tau_{a,x}}{\partial y} - \frac{\partial \tau_{a,y}}{\partial x}\right) + \left(\frac{\partial H_x}{\partial y} - \frac{\partial H_y}{\partial x}\right) + \beta M_y = 0 \tag{40.6}$$

meaning that in every locality the wind-stress vorticity must be balanced by vorticity related to lateral friction and by the term βM_y which EKMAN calls the *planetary vorticity*.

Using this equation STOMMEL[1] has given a qualitative explanation of the westward intensification of the ocean currents. Assume a symmetrical anticyclonic wind-circulation over a closed ocean area in the northern hemisphere, meaning that the wind-stress vorticity is negative over the entire area. We may off-hand expect that this stress will produce an anticyclonic circulation in the sea. The effect of horizontal friction at the boundaries must be to counteract the effect of the wind-stress, and the second term of (40.6) must be positive. The third term is negative for a water mass that moves to the north and positive for a mass that moves to the south. This is a consequence of the conservation of angular momentum or, what means the same, of the variation of the Coriolis parameter with latitude.

The vorticity of the wind-stress is the same at both boundaries because the wind system is assumed to be symmetrical. A symmetrical oceanic circulation would imply that the vorticity of the frictional term would be the same at the western and eastern boundaries and that the planetary vorticity would have the same numerical value, but would be of opposite sign.

Calling the numerical values of the three terms in Eq. (40.6) a, b, and c, respectively, it follows that with symmetrical currents we must have

$$\text{off western boundary:} \quad -a + b - c = 0,$$
$$\text{off eastern boundary:} \quad -a + b + c = 0$$

but these conditions require $c = 0$, that is, no north-south transport off the boundaries, contrary to our assumptions.

An anticyclonic circulation is possible only if the transport to the north off the western boundary of the ocean is greatly increased whereby the lateral shearing stresses are increased correspondingly. Over the remaining part of the ocean the transports may be small and the lateral stresses correspondingly small. STOMMEL illustrates this situation by tabulating some arbitrary values of the vorticity terms in an asymmetric circulation:

	Strong north flowing current off western boundary	South flowing currents over rest of ocean
Wind stress (a)	− 1.0	− 1.0
Frictional (b)	10.0	0.1
Planetary (c)	− 9.0	0.9
Totals	0.0	0.0

STOMMEL concludes that the interesting consequences of this reasoning are "(1) the fact that although energy is added to the oceans by the work of the wind over the entire surface, it is dissipated chiefly in the strong western current;

[1] H. STOMMEL: Bull. Amer. Met. Scc. **32**, 21 (1951).

(2) that a good representation of the circulation outside of the western currents can be obtained from a knowledge of the wind stress alone, independent of lateral friction".

Before the general theory had been developed SVERDRUP[1] had shown, in agreement with STOMMEL's conclusion, that the equatorial counter current could be computed from a knowledge of the stress of the wind only, and that it developed in the region of minimum stress.

Neglecting lateral friction when dealing with equatorial currents and assuming that the stress is directed mainly east-west, Eq. (40.5) is reduced to

$$\beta M_y = -\frac{\partial \tau_{a,x}}{\partial y}. \tag{40.7}$$

From (40.4) and (40.7) follows, when R is the radius of the earth:

$$\frac{\partial M_x}{\partial x} = \frac{1}{2\omega \cos\varphi}\left(\frac{\partial \tau_{a,x}}{\partial y}\tan\varphi + R\frac{\partial^2 \tau_{a,x}}{\partial y^2}\right). \tag{40.8}$$

Integrating from 0 to Δx and assuming a north-south vertical boundary at $x=0$ where continuity requires $M_x = 0$ we obtain:

$$M_x = \frac{\Delta x}{2\omega\cos\varphi}\left(\overline{\frac{\partial \tau_{a\,x}}{\partial y}}\tan\varphi + R\overline{\frac{\partial^2 \tau_{a\,x}}{\partial y^2}}\right) \tag{40.9}$$

where average values of the derivatives of the stress are introduced. The second term in the parenthesis is much larger than the first and therefore

$$M_x \approx \frac{\Delta x}{2\omega\cos\varphi} R\overline{\frac{\partial^2 \tau_{a\,x}}{\partial y^2}}. \tag{40.10}$$

Thus, the mass transport can be computed from knowledge of the wind stress only. The result can be compared to the value that can be found from oceanographic observation, using the relation

$$M_x = -\frac{1}{\lambda}\left(\frac{\partial P}{\partial y} - \tau_{a,y}\right) \tag{40.11}$$

provided that $|\varphi| > 0$. The minus sign applies to $\varphi > 0$ (northern hemisphere), the plus sign to $\varphi < 0$. It will be shown that in the case of the Pacific Ocean the two methods lead to results that are in remarkable agreement with each other, demonstrating that there the equatorial currents are maintained by the prevailing winds.

In order to tackle the general problem it is necessary to introduce an expression for the *lateral frictional force*. This is done by writing

$$\tau'_x = A\frac{\partial v_x}{\partial x}, \quad \tau'_y = A\frac{\partial v_y}{\partial y} \tag{40.12}$$

where τ' is the stress acting on a vertical surface and A is the *lateral eddy viscosity* which will be assumed to be constant. Doing this, the final equations can be simplified by defining a mass transport stream function, ψ, such that $M_x = -\partial\psi/\partial y$ and $M_y = \partial\psi/\partial x$. Assuming no accelerations the differential equation of mass transport then takes the form (MUNK[2])

$$(A\nabla^4 - \beta\,\partial/\partial x)\,\psi = -\operatorname{curl}_z \boldsymbol{\tau}_a \tag{40.13}$$

where

$$\nabla^4 = \partial^4/\partial x^4 + 2\,\partial^4/\partial x^2\,\partial y^2 + \partial^4/\partial y^4$$

[1] H. U. SVERDRUP: Proc. Nat. Acad. Sci. U.S.A. **33**, 318 (1947).
[2] W. H. MUNK: J. Meteorol **7**, 79 (1950).

is the biharmonic operator and $\mathrm{curl}_z\, \boldsymbol{\tau}_a$ is the vertical vorticity of the wind stress. This equation as well as Eq. (40.7) bring out the interesting feature that the mass transport does not depend upon the magnitude of the wind stress, but upon its vorticity. By adding other assumptions it is possible to compute not only the mass transport, but also the distribution of velocities.

The statement that the vorticity of the wind stress determines the mass transport does not apply if one deals with a *zonal canal* around the earth and currents maintained by zonal winds. MUNK and PALMÉN[1] have discussed this problem. Instead of using spherical coordinates they consider a canal on a rotating cone, limited by the distance r_0 and r_1 from the top of the cone. In this case:

$$M_x = \frac{\tau_a}{18A}\left(r_1^3 - r_0^3 - \frac{6r_1^2 r_0^2}{r_1 + r_0}\log\frac{r_1}{r_0}\right) \tag{40.14}$$

that is the transport is proportional to the stress and inversely proportional to A. The application of these equations to oceanographic problems is dealt with in Sect. 44.

After the above presentation had been prepared an important contribution to the dynamics of ocean currents has been published by NEUMANN[2] who has re-examined the horizontal wind-driven ocean currents, taking into account the spherical shape of the earth, the average vertical density stratification, and the variable depth of the lower boundary of the circulation system, that is the depth of the layer of no motion (see Sect. 44). NEUMANN concludes that:

"The analysis shows much evidence that the ocean tends to eliminate certain asymmetries of the flow with respect to the driving wind field by adjusting its level of no motion. East-west asymmetries, resulting from the variation of the Coriolis parameter with latitude, for example, are eliminated by adjusting the meridional slope of the level of no motion so that the slope is proportional to the sine of the latitude. Characteristic zonal slopes are another consequence of the adjustment, resulting from different factors, including the convergence of the meridians."

41. Internal waves. When dealing with *wave motion*, local accelerations cannot be neglected, but in most problems the effects of field accelerations, friction and the Coriolis acceleration are not considered. The equations of motion then take the form

$$\frac{\partial v_x}{\partial t} = -\alpha\frac{\partial p}{\partial x}, \quad \frac{\partial v_y}{\partial t} = -\alpha\frac{\partial p}{\partial y} \quad \text{and} \quad \frac{\partial v_z}{\partial t} = -\alpha\frac{\partial p}{\partial z} + g. \tag{41.1}$$

To these must be added the equation of continuity and boundary conditions.

When dealing with a homogeneous liquid with a free surface, only one type of waves with maximum vertical amplitude at the free surface is possible, but if *two layers* of different density are present, boundary conditions must also be satisfied at the interface between the two layers, whereby new solutions are obtained. The theory shows that in such a system, in addition to the ordinary waves, internal waves may be present with maximum vertical amplitude at the interface. In a liquid within which the density varies continuously with depth the theory leads, as might be expected, to the conclusion that an infinite series of internal waves is possible, first order waves with one maximum vertical displacement between the free surface and the bottom, second order waves with

[1] W. H. MUNK and E. PALMÉN: Tellus 3, 53 (1951).
[2] G. NEUMANN: Met. Papers New York Univ. 2, No. 4 (1955).

two maxima and so on (FJELDSTAD[1]). The theory renders the velocity of progress of such waves and the relative magnitudes of the amplitudes, but gives no information as to the phases or the periods[2].

Observations at sea have shown that *internal waves of tidal period* occur very frequently in the ocean, and that internal waves also may be present of the period characteristic of the *inertia oscillation* related to the effect of the Coriolis acceleration[3]. The manners in which these waves are produced and maintained are not known. The waves are mentioned here because their presence introduces uncertainties as to the reality of some of the conclusions as to currents, that are based on the observed distribution of mass (Sect. 44).

J. Currents of the oceans.

42. General considerations. The following description will deal primarily with the more or less *permanent currents*. No reference will be made to the tidal currents (discussed in DEFANT'S contribution in this volume) or to the currents related to internal waves or to irregular eddies, although such currents are always present and complicate the actual conditions.

Our knowledge of the *surface currents* is mainly derived from ships' observations (Sect. 6), supplemented by results from drift bottle experiments and special investigations. Knowledge of the *sub-surface currents* of the upper layers is obtained by computing the currents from the observed distribution of mass (Sect. 37). In the deeper layers the differences in density are so small that this method fails, and conclusions as to the flow of the deep water are, therefore, based on distributions of temperature and salinity taking differences in oxygen contents and other variable constituents into account.

The permanent currents of the oceans can be divided in three groups:

the currents that are maintained by excess of *evaporation* or *precipitation* in certain regions,

the currents that are maintained by the energy that is transferred to the sea by the *wind*, and

the currents that are maintained because of *cooling* in high latitudes.

Instead of describing the currents of the different ocean regions these three groups will be dealt with.

43. Currents maintained by excess of evaporation or precipitation and run-off. In the open ocean the excess of evaporation or of precipitation and run-off is never sufficient for the maintenance of measurable currents (EKMAN[3]), but in many instances the exchange of water between a more or less closed basin and the adjacent open ocean is determined by the difference $(E-P)$ in the basin, where E as previously means evaporation, but where P now stands for the sum of precipitation and run-off (from the surrounding land). The exchange is determined by the conditions that the net flow of salt into the basin must be zero, and that the amount of water in the basin must remain constant. Calling the mass transport into the basin M_i and the mass transport out of the basin M_u and calling the corresponding average salinities S_i and S_u the two conditions can be written:

$$M_i S_i = M_u S_u, \tag{43.1}$$

and

$$M_i = M_u + (E - P). \tag{43.2}$$

From these relations follows:

$$M_i = (E - P) \frac{S_u}{S_u - S_i}. \tag{43.3}$$

[1] J. E. FJELDSTAD: Geof. Publ. **10**, No. 6 (1933).

[2] See DEFANT'S article on tidal phenomena in water in this volume, Sect. 5.

[3] V. W. EKMAN: Ann. Hydrogr., Berlin **54**, 261 (1926).

If evaporation exceeds precipitation and run-off $(E-P)$ is positive and S_u is greater than S_i, but if evaporation is less than precipitation and run-off, $(E-P)$ is negative and S_u is smaller than S_i.

The exchange of water through the Strait of Gibraltar represents the classical example for the type of currents that develop when the evaporation in the basin, in this case in the Mediterranean Sea and the Black Sea combined, exceeds the sum of precipitation and run-off. According to measurements of the current that flows in through the Strait of Gibraltar the inflow amounts to about 1.75×10^6 tons per sec. The average salinity of the inflowing water is about $36.25\,^0/_{00}$ and that of the outflowing $37.75\,^0/_{00}$. With these values one obtains $(E-P)=0.07\times10^6$ tons per sec., in good agreement with other estimates.

Similar conditions exist in the Red Sea, but there they are complicated by the effect of winds which change with the season. A small outflow of saline water also takes place from the inner part of the Gulf of California. The saline water that flows out from the Mediterranean Sea and from the Red Sea exercises a remarkable influence upon the distribution of salinity in the Atlantic and Indian Oceans, but the outflow from the inner part of the Gulf of California is far too small to be of importance.

Extreme conditions of a similar nature are found in some lagoons in the Gulf of Mexico. The communications between these lagoons and the sea are so narrow and shallow that they permit only a small exchange of water, and correspondingly the salinity in the lagoons may increase to very high values.

Basins or lagoons in which evaporation is in excess, are found in lower latitudes only. In higher latitudes precipitation and run-off are in excess in more or less closed seas, e.g. the Baltic or the northern seas comprising the Norwegian Sea and the Polar Sea. When communication with the open sea is restricted, low salinity water flows out near the surface, and high salinity water flows in along the bottom of the connecting straits or sounds. This type of flow is encountered in the sounds and belts that separate the Baltic from the Kattegat. With more than one opening, low salinity water may flow out through one and high salinity water in through another. This situation applies to the northern seas where less saline Arctic water flows out between Iceland and Greenland, whereas the saline Atlantic water flows in through the Faroe-Shetland channel. It should, however, be added that the currents in the northern seas are not maintained only by the excess of precipitation and run-off, but also by the prevailing winds.

Extreme conditions are found in many of the *Norwegian fjords* into which rivers carry large quantities of fresh water. An inner deep portion of the fjord is often separated from the outer fjord by a shallow threshold which effectively reduces the inflow of the deeper, saline water. Under these conditions the outflowing water has a very low salinity and the inflow approaches zero [Eq. (43.3)]. When inflow takes place only intermittantly the deeper water in the inner basin may become stagnant during long periods and its oxygen contents completely consumed.

The conditions in the *Black Sea* correspond to those in nearly closed basins in higher latitudes. Because of run-off from the great rivers that empty into the Black Sea, the sum of precipitation and run-off exceeds evaporation. Communication between the Black Sea and the inner part of the Mediterranean takes place through the narrow and shallow straits of Bosporus and the Dardanelles where the low salinity water of the Black Sea flows out near the surface and the saline Mediterranean water flows in along the bottom. Considerable mixing between these waters takes place in the shallow straits. The circulation is so weak that the deeper waters in the Black Sea are stagnant below a depth of about 200 m. where no oxygen, but hydrogen sulphide is present.

44. The currents of the upper layers of the oceans. Knowledge of the surface currents is, as already stated, mainly based on results of ships' observations, whereas knowledge of the sub-surface currents is derived by computing the currents from the distribution of mass. The latter computation requires the establishment of a *"layer of no motion"* because the direct computation renders only information as to the currents at one level relative to the currents at another. When searching for a reference level several approaches have been attempted.

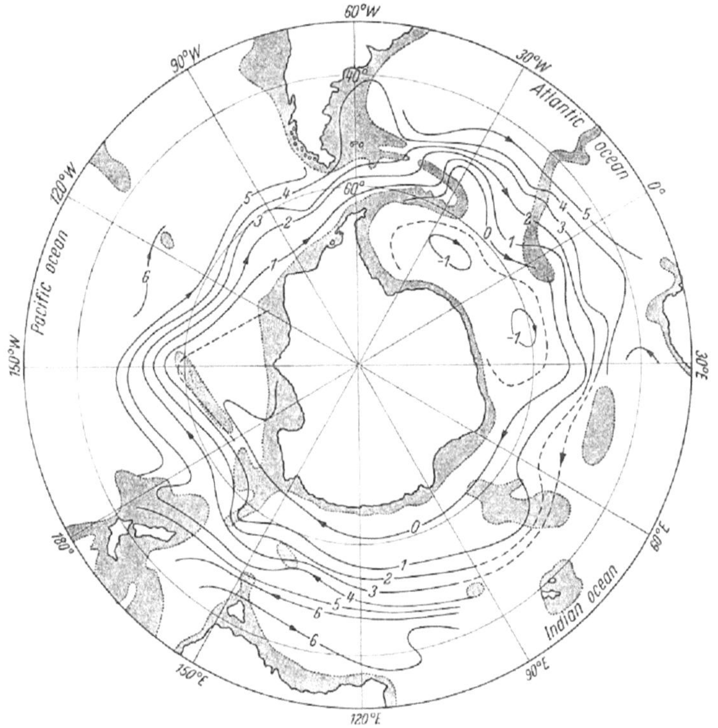

Fig. 18. Transport around the Antarctic Continent referred to the 3000-decibar surface. Between two lines the transport is about 20 million cubic metres per second.

In some cases the structure and character of the deep water suggest that the deep water is practically motionless, in which case a reference level can be placed at the top of the deep water. Such conditions are encountered in the North Pacific, where in particular the low oxygen contents of the deep water indicates that no appreciable circulation can be present. When dealing with the North Pacific it may, therefore, be expected that reliable results are obtained if the reference level is placed at a depth of 1500 m., or to be more exact, at a pressure of 1500 decibar.

In other instances the reference level must be placed as deep as possible because measurable differences in density are present at the greatest depths and because there may be no reason for assuming that the direction of the current changes with depth. Such conditions are met with around the Antarctic Continent, where the Circumpolar Current runs from west to east. There, the validity of selecting a very deep reference level is supported by the fact that the current evidently is deflected by submarine ridges, the tops of which may be 2000 m. below the surface although they rise about 2000 m. from the general ocean bed (Fig. 18). Transports above the 3000 decibar surface have been computed, but it is possible that the currents reach to even greater depths, perhaps to the bottom of the sea, and that, therefore, the computed transports are too small.

Another method is based on the application of the equation of continuity in some form or another. When dealing with the South Atlantic Ocean the mass structure indicates that in the upper layers water is transported from the South to the North Atlantic. It follows that deep water must flow in the opposite direction, and continuity requires that equal masses are carried in both directions. The limitation imposed upon the character of the currents by this requirement makes it possible to ascertain the level at which the currents must be zero.

A third method was introduced by DEFANT[1] and applied to the Atlantic Ocean. It was found that at certain intervals of depth the differences in the dynamic depths of given isobaric surfaces as obtained from observations at

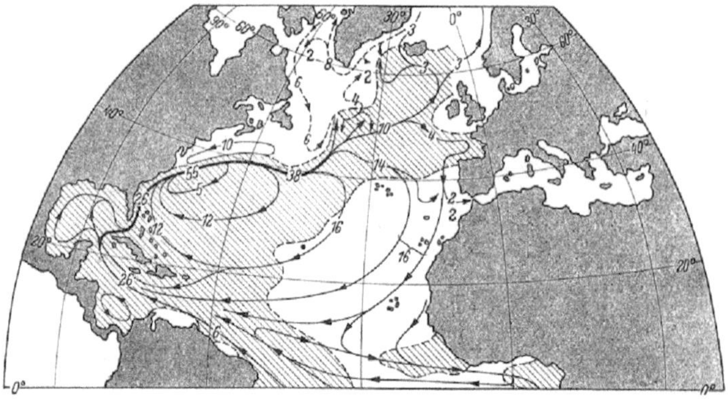

Fig. 19. Transport of Central and Subarctic water in the Atlantic Ocean. Lines with arrows indicate direction of transport and inserted numbers indicate the transported volumes in millions of cubic metres per second.

adjacent stations were nearly constant. DEFANT showed that it is plausible to assume that this layer represents a layer of no motion and found that in the South Atlantic the depth of this layer increases fairly regularly from about 500 m. near the equator to about 2500 m. in latitude 50° S. In the North Atlantic the depth also increases with latitude, but great irregularities occur in the region of the Gulf Stream off the east coast of North America.

When dealing with the most complicated of the well explored oceans, the North Atlantic, the requirement as to continuity in the transport of salt and heat may render added assistance in selecting a level of no motion, but in this case several uncertainties are still present. The transports in the North Atlantic and the North Pacific Oceans are shown in Figs. 19 and 20.

In the preceding discussion emphasis has been placed on the computation of mass transports and not on the representation of velocities at stated selected levels. The reason is that when dealing with transports a single chart gives considerable information, whereas, if velocities were to be represented, numerous charts are needed. For this purpose charts of the *dynamic topography of isobaric surfaces* are prepared. If these can be referred to a level isobaric surface, and if accelerations and friction can be neglected, the current is directed along the isolines and the velocity is inversely proportional to the distance between the lines.

Charts of dynamic topography and transport charts ought to be based on simultaneous observations, but the collection of such requires the simultaneous

[1] A. DEFANT: Dtsch. Atlant. Exped. „Meteor" 1925—1927, Wiss. Erg. 6, Teil 2, 5 (1941).

operation of a number of ships. Even in instances in which special cooperation has been arranged, the number of ships has been too small for obtaining simultaneous data from a large area. The error that may arise for this reason is not too serious because repeated observations in the same locality show that local changes are not very important if only the major features of the currents are considered. The presence of internal waves (Sect. 41) introduces irregularities in the pattern, but these are of minor importance if major features are examined. However, if details are to be studied, internal waves of tidal period may make the interpretation difficult (DEFANT[1]).

The general pattern of currents in the upper layers can be briefly described as follows: In the *Polar Sea* a small anticyclonic circulation is present in the region between Alaska, Greenland and the North Pole. In the *northern parts of the oceans* and particularly in the adjacent seas in higher northerly latitudes cyclonic

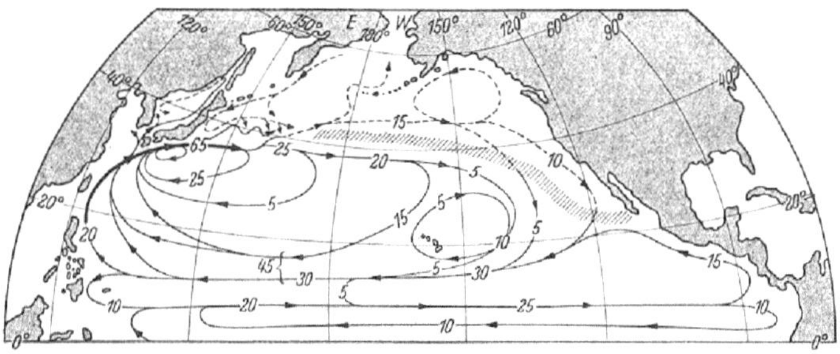

Fig. 20. Transport of Central and Subarctic water in the North Pacific Ocean. (Meaning of lines and numbers as in Fig. 19.)

circulations prevail, for instance in the Norwegian Sea, the Labrador Sea, the Bering Sea and the Sea of Japan. In the *central parts of the oceans* of both hemispheres anticyclonic circulations are present, with by far the strongest currents off the eastern coasts of the continents. Between the anticyclonic circulations of the Atlantic and Pacific Oceans the *equatorial currents* flow from east to west with a *counter current* towards the east on the northern side of the equator. In the northern part of the *Indian Ocean* the currents vary with the monsoon, and an equatorial current is present only in one season of the year. The cyclonic circulations are lacking in higher southerly latitudes, but instead the *Antarctic Circumpolar Current* flows from west to east around the Antarctic Continent. Very near the continent currents towards the west prevail.

This pattern of currents corresponds to the pattern of prevailing winds, for which reason it appears probable that the energy for maintaining the currents is derived from the stress that the wind exerts in the sea surface. Theoretical considerations support this view, particularly because these account both for the westward intensification of the anticyclonic currents in mid latitudes and for the presence of the equatorial counter currents (Sect. 40).

The quantitative relation between wind stress and currents has been examined by several authors. When dealing with the eastern portion of the equatorial counter current of the Pacific Ocean, SVERDRUP[2] and REID[3] have obtained a satisfactory agreement between the mass transport as computed from oceanographic observations and from the wind stress (Fig. 21). The problem of the transport by the currents of the North Atlantic and the North Pacific Oceans

[1] A. DEFANT: J. Marine Res. **9**, 120 (1950).
[2] H. U. SVERDRUP: Proc. Nat. Acad. Sci. U.S.A. **33**, 318 (1947).
[3] R. O. REID: J. Marine Res. **7**, 74 (1948).

has been dealt with by Munk and collaborators (Munk[1]; Munk and Carrier[2]; Munk, Groves and Carrier[3]); Hidaka[4]; Hansen[5] and others. Munk and collaborators and Hidaka have mainly used the approach discussed in Sect. 40, but have also taken the higher order terms into consideration, whereas Hansen has dealt with the relationship as a boundary value (Eigen value) problem.

The *qualitative* agreement between the transport as computed from the wind stress and from the oceanographic observations is very good as evident from Figs. 22 and 23 that show the results obtained by Munk and Carrier for the North Pacific. Similar agreement is also obtained for the North Atlantic where the computations based on the wind stress reveal the existence of an eddy on the right hand side of the Gulf Stream, the presence of which is also demonstrated by the oceanographic observations when these are carefully analysed.

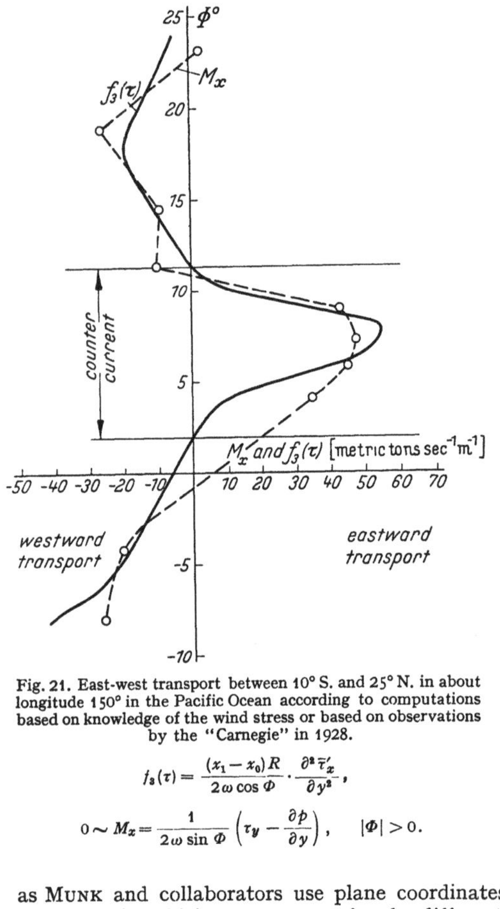

Fig. 21. East-west transport between 10° S. and 25° N. in about longitude 150° in the Pacific Ocean according to computations based on knowledge of the wind stress or based on observations by the "Carnegie" in 1928.

$$f_3(\tau) = \frac{(x_1 - x_0)R}{2\omega\cos\Phi}\cdot\frac{\partial^2 \tau'_x}{\partial y^2},$$

$$0 \sim M_x = \frac{1}{2\omega\sin\Phi}\left(\tau_y - \frac{\partial p}{\partial y}\right), \qquad |\Phi| > 0.$$

The *quantitative* agreement is not so satisfactory. From the wind stress Munk (1950) obtained transport values which are only about half of those derived from the observations at sea, 36 and 39×10^6 m.³/sec. for the maximum transport by the Gulf Stream and the Kuroshio, respectively, against "observed" values of 55 to 74 and 65×10^6 m.³/sec. Munk points out that the discrepancy cannot be due to an erroneous value of the horizontal eddy diffusivity (5×10^7 cm.² sec.⁻¹) because in this case the transports are independent of the numerical value of A, and also that it is not probable that the wind stress has been grossly underestimated. More refined computations by Munk and collaborators have not altered the results. Hidaka, on the other hand, obtains very good numerical agreement for the transport of the Kuroshio. One difference in the treatment is that Hidaka uses spherical coordinates whereas Munk and collaborators use plane coordinates. It seems uncertain if this difference in treatment is sufficient to account for the different results.

The above conclusions may be considerably modified in view of more recent theoretical results obtained by Neumann (Sect. 40). On the basis of these results and Defant's chart of the depth of the layer of no motion in the South Atlantic Neumann concludes that there the effect of the change with latitude of the Coriolis parameter is nearly cancelled by the increase with latitude of the depth of layer of no motion. This would imply that, in agreement with observations, the circulation of the South Atlantic should not show any conspicuous westward

[1] W. H. Munk: J. Meteorol. **7**, 79 (1950).
[2] W. H. Munk and G. F. Carrier: Tellus, **2**, 158 (1950).
[3] W. H. Munk, G. W. Groves and G. F. Carrier: J. Marine Res. **9**, 218 (1950).
[4] K. Hidaka: Geophys. Notes **3**, No. 23 and No. 38 (1950); **4**, No. 3 (1951).
[5] W. Hansen: Dtsch. hydrogr. Z. 4, 161 (1951).

intensification. For the North Atlantic NEUMANN's preliminary computation has rendered values of the transport in good agreement with those shown in Fig. 19.

When dealing with the Gulf Stream the computations on the basis of the wind stress do not explain the presence of the counter current on the left hand side (the western side) of the Gulf Stream, nor do they account for the increase

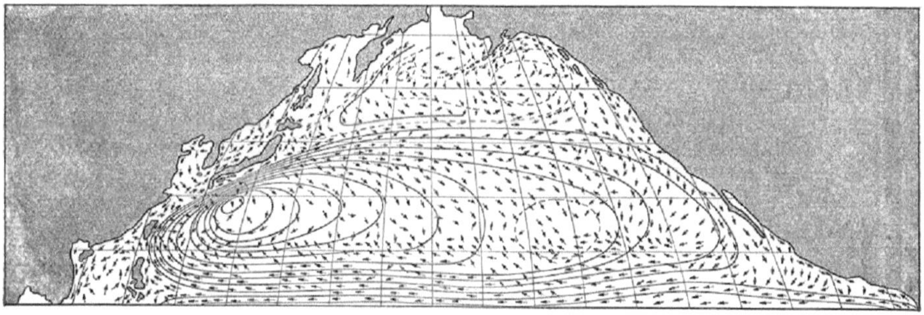

Fig. 22. Transport chart of the North Pacific, based on oceanographie observations. Transport between stream lines is six million metric tons per second.

of the transport in the direction of flow which is indicated by the oceanographic observations. ROSSBY[1] has treated the Gulf Stream from the Strait of Florida and towards the northeast as a *jet stream* and has developed a theory which does account for the two last named features. It is possible that in order to reach a satisfactory agreement between the theoretical conclusions and the observed conditions, part of the jet stream theory must be considered, or that the effect of cooling in the direction of flow has to be taken into account.

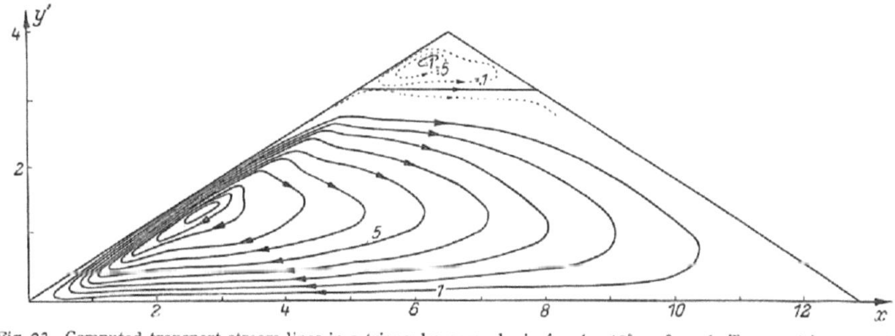

Fig. 23. Computed transport stream lines in a triangular ocean basin for $A = 10^8$ cm.^{2} sec^{-1}. Transport between stream lines is about three million metric tons per second.

CHARNEY[2] has dealt with the Gulf Stream as an internal boundary problem. He obtains a satisfactory agreement between computed and observed increase in the transport of the current. His theory leads also to the interesting conclusion that at a given distance from its origin, which is taken as the Strait of Florida, the current must leave the coast and that this separation does not require the presence of a protrusion of the coast line.

In the case of the Antarctic Circumpolar Current the theory pertaining to flow in a canal around the earth has been applied by MUNK and PALMÉN[3]. They find that the transport computed from the wind stress is ten to one hundred times greater than the transport as

[1] C.-G. ROSSBY: Papers Phys. Oceanogr. a. Meteorol. **5**, No. 1 (1936).
[2] J. G. CHARNEY: Proc. Nat. Acad. Sci. U.S.A. **41**, 731 (1955).
[3] W. H. MUNK and E. PALMÉN: Tellus **3**, 53 (1951).

computed from oceanographic observations. The discrepancy may be due to the selection of a too small value for the eddy viscosity, but this is not likely because a value of 10^8 cm.^{2} per sec., which they have used, has been derived from other observations within the Antarctic region. It seems more probable, either, that the transports computed from the oceanographic observations are too small because they assume no motion at a depth of 3000 m., whereas the currents may reach to the bottom, or, that the stress of the wind is partly balanced by the effects of the submarine ridges which are evident on the transport chart (Fig. 19) and by stress at the bottom, or that both these features enter.

These conclusions may also be modified in the view of the results by NEUMANN, who on the basis of his theory obtains perfect agreement between the computed and observed transport by the Antarctic Circumpolar Current.

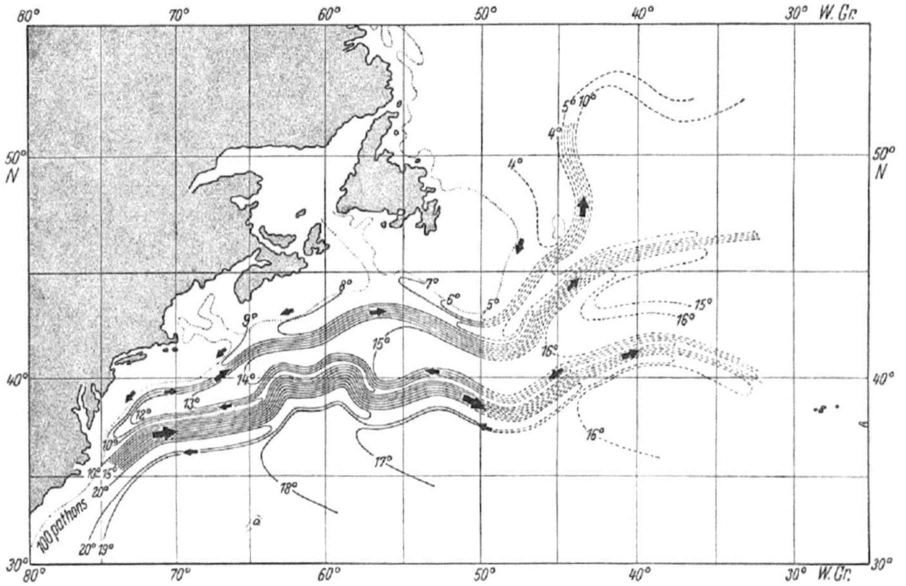

Fig. 24. Schematic chart of the temperature at a depth of 200 m. in the Gulf Stream area, showing the streakiness of the current.

Summarising the preceding discussion it may be stated that considerable evidence has been accumulated to show that the currents of the upper layers are maintained by the prevailing winds, but quantitative agreement has not been obtained between transports computed from wind stress and from oceanographic observations.

So far we have dealt only with the major features of the currents. Detailed examinations, partly of the temperature structure where swift currents are present and partly of the currents using radio location of the ship at frequent intervals or the geomagnetic method (Sect. 6), have revealed the existence of numerous minor features. Within the *Gulf Stream* the presence of well defined *eddies* has been demonstrated and, in addition, a peculiar *streakiness* of the current has been found to exist (Fig. 24). The study of these details opens up a number of new problems.

45. The deep water circulation of the oceans. When dealing with the deep water circulation we will consider only the *meridional circulation*. This is best done in connection with an examination of the vertical meridional sections, showing the distribution of temperature, salinity and oxygen in the Atlantic and Pacific Oceans (Figs. 25 and 26).

The temperature and salinity sections of the *Atlantic* show that deep (and bottom) water that is formed in the North Atlantic flows south together with Mediterranean Water which is responsible for the high salinities between 1000 and 1500 m. in the North Atlantic. To compensate for the flow to the south

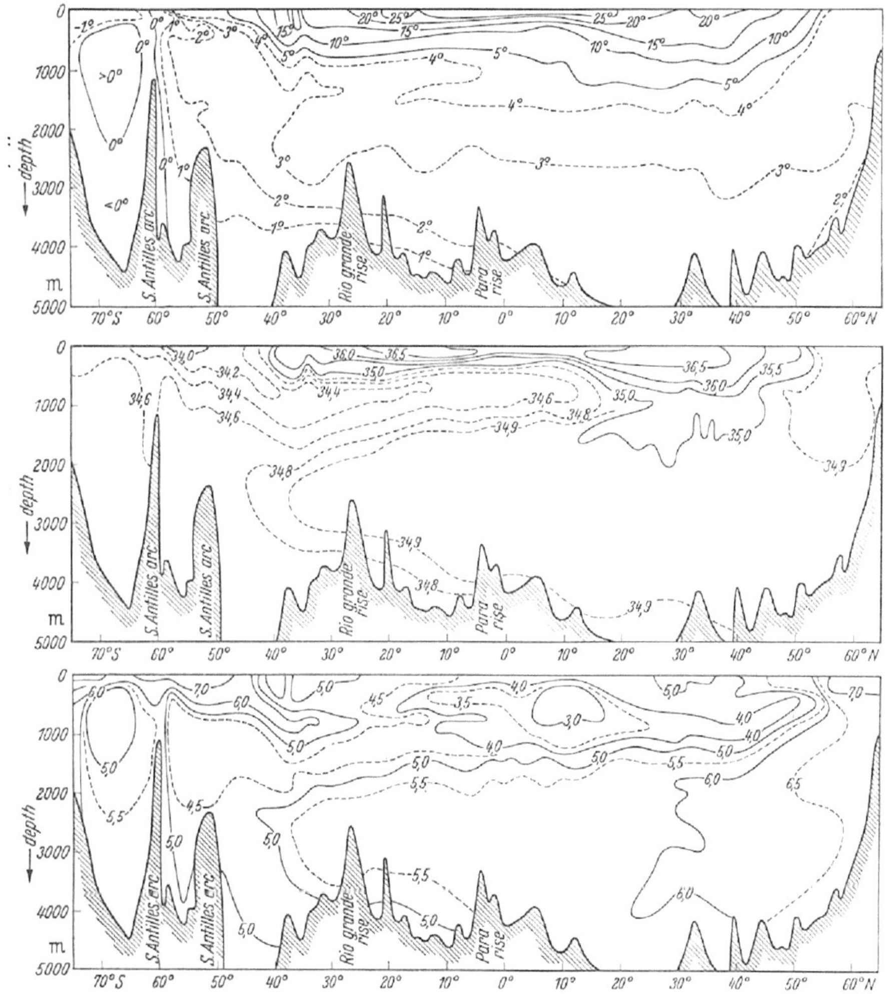

Fig. 25. Vertical sections showing distributions of temperature, salinity and oxygen in the Western Atlantic Ocean (after WÜST).

Antarctic Intermediate Water and Antarctic Bottom Water flow north and, even larger, quantities of central water cross the equator, flowing north along the coast of South America. The estimated values are:

To the south, North Atlantic deep water,		9×10^6 m.³/sec.
To the north, Antarctic Intermediate Water	2×10^6	
Antartic Bottom Water	1×10^6	
Upper Water	$\underline{6 \times 10^6}$	9×10^6 m.³/sec.

An independent circulation is present in the South Atlantic where parts of the northflowing intermediate and bottom water mix into the deep water and return to the south with this water, which finally rises towards the surface to the south of the Antarctic Convergence. Computations show that between latitudes

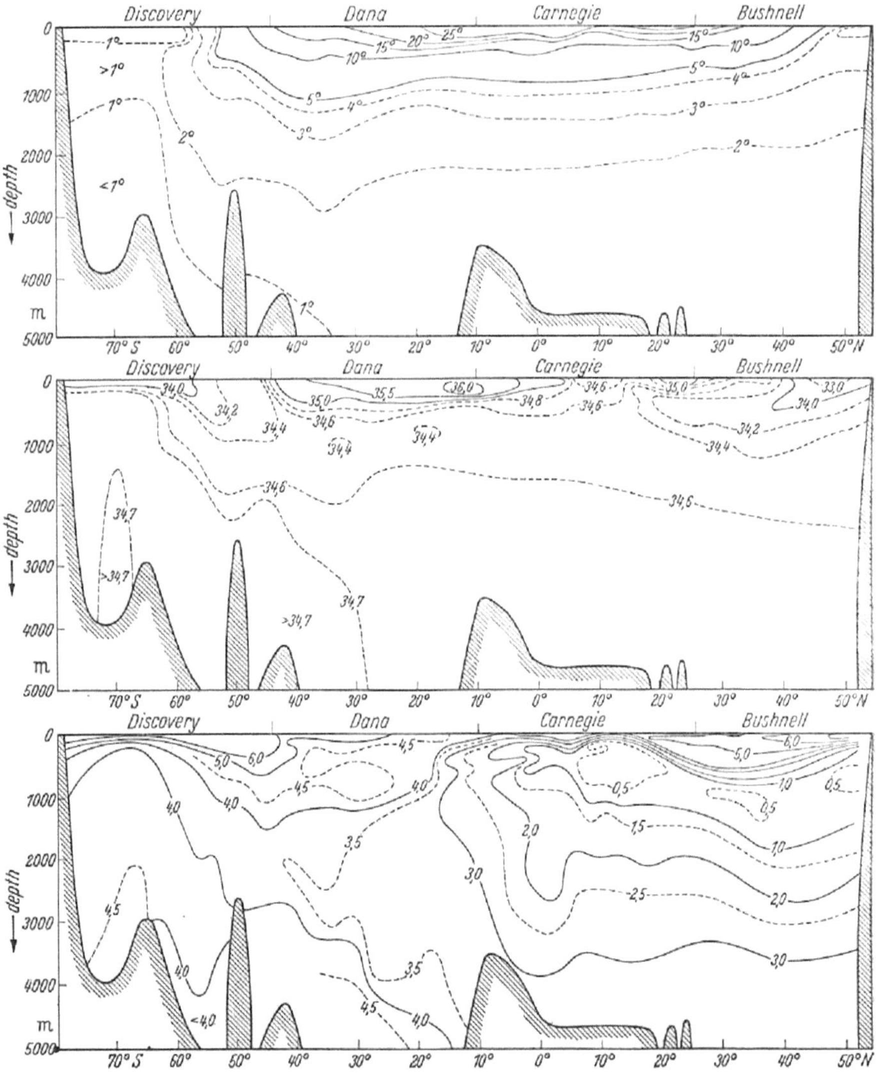

Fig. 26. Vertical sections showing distributions of temperature, salinity and oxygen in the Pacific Ocean, approximately along the meridian of 170° W. (after SVERDRUP et al.).

20 and 30° S. an amount of deep water of 18×10^6 m.³/sec. flows to the south, meaning that in these latitudes about 9×10^6 m.³/sec. of intermediate and deep water is returning to the south.

The values of the *oxygen contents* confirm these conclusions. The high oxygen values of the deep water in the northern part show that the water must recently have been in contact with the atmosphere, and the decreasing values towards the south indicate flow to the south. The rapid decrease to the south of 30° S.

indicates admixture from above and below of water that is poorer in oxygen. Finally the high oxygen content near the Antarctic Continent indicates that there formation of bottom water takes place.

The fact that about two thirds of the north-flowing water is upper water is also reflected in the distribution of phosphates, nitrates and silicates in the Atlantic Ocean. The upper water is depleted of these compounds because of the production of organic matter by marine plants, but the deeper layers show higher concentrations because of decomposition of organic matter. The net transport of phosphate, nitrate and silicates across the equator must, however, be zero, meaning that since the north-flowing water is relatively poor in these compounds, the deep water of the North Atlantic must also be relatively poor. This is borne out by observations. Stating the concentrations in μ g-atoms per 20° litre of phosphorus, nitrogen and silicon, respectively, we find:

	P	N	Si
North Atlantic deep water	0.5—1.0	10—15	15—25
South Atlantic deep water	1.0—2.0	15—40	30—80

In the North Atlantic the values are nearly constant in a north-south direction, but in the South Atlantic they increase a great deal from north to south, also in agreement with the character of the circulation.

In the *Pacific Ocean* entirely different conditions are encountered. In the South Pacific there probably exists a weak meridional circulation related to the transport to the north of Antarctic Intermediate Water and Bottom Water and a return flow to the south of deep water, but no water crosses the equator and no perceptible meridional circulation is present in the North Pacific. These conclusions are also supported by the distribution of oxygen, and the other variable constituents. The oxygen contents of deep water of the South Pacific is low, between 3.5 and 4 ml/L, indicating that no formation of deep water takes place in that ocean, and the values in the North Pacific are even lower, between 1.0 and 3 ml/L. The phosphate contents of the South Pacific deep water is about the same as that of the extreme South Atlantic, 2 μ g-atoms of phosphorus per 20° litre, but in the North Pacific the corresponding value lies between 2 and 3.5 μ g-atoms.

The meridional circulation of the southern part of the Indian Ocean is similar to that of the South Pacific and the South Atlantic. A circulation across the equator is lacking or is very weak.

A discussion of the deep water circulation is not complete without considering the *zonal circulation around the Antarctic Continent*, where the Antarctic Circumpolar Current carries deep and bottom water from west to east. These water masses are very uniform, but some systematic changes appear to take place during one circulation of the water. To the south of South Africa an admixture of traces of Mediterranean water at depths between 1000 and 2000 m. and the presence of newly formed bottom water lead to relatively large differences in temperature and salinity within a vertical water column, but during the flow to the east these differences are smoothed out. Thus, in 50° S. and 30° W. and between the bottom and 1000 m. the salinity and temperature vary between 34.66 and 34.76 ‰, and between 0° and 2.5°, respectively, whereas at the western entrance to Drake Passage the corresponding values are 34.70 and 34.72 ‰, and 0.5 and 2.0°.

The deep water of the Indian Ocean and the Pacific Oceans must be considered as originating from the Antarctic Circumpolar and Bottom Water and is, therefore, much more uniform than the deep water of the Atlantic. The deep water of the North Pacific is nearly homogeneous and has a slightly lower salinity than that of the South Pacific, probably because of admixture of low salinity upper water. It is probably very slowly renewed by an inflow across the equator in the western part of the ocean and an outflow in the eastern part. Since the character of the Antarctic Circumpolar water is mainly determined by the processes of sinking in high latitudes in the Atlantic Ocean it follows that these processes together with the character of the circulation are responsible for the character of the deep water of the other oceans.

The deep water circulation is generally ascribed to the effect of *cooling in high latitudes*. It must, however, be pointed out that vertical convection due to cooling does not necessarily lead to the development of deep sea currents, and that there exists no reason why such vertical circulation shall lead to transport of water from one hemisphere to the other. The intense meridional circulation of the Atlantic can, therefore, not be the result only of the sinking of water of high density in high latitudes, but must be considered as maintained because the character of the prevailing winds together with the character of the ocean boundaries force upper water masses to flow north across the equator off the coast of South America. Thus, *the winds are in part responsible for the deep water circulation as well.*

The final conclusion is, therefore, that except for the exchange of water between adjacent, more or less isolated seas and the oceans, the currents of the oceans are mainly maintained by the effect of the prevailing winds, but the character of the currents depends on differences in density which are maintained by thermal processes.

Oberflächen-Wellen des Meeres.

Von

HANS ULRICH ROLL.

Mit 33 Figuren.

Vorbemerkung. Die Erforschung der Oberflächen-Wellen des Meeres befindet sich zur Zeit in einer entscheidenden Phase. Nachdem die Grundlagen für die mathematische Behandlung der reinperiodischen Wasser-Oberflächenwellen — größtenteils bereits im vorigen Jahrhundert — gelegt wurden, ist man seit etwa einem Jahrzehnt (angeregt durch praktische Erfordernisse) bemüht, den wirklichen Seegang mit all seinen Unregelmäßigkeiten zu messen und die Ergebnisse der Messung sowie die zugrundeliegenden Gesetzmäßigkeiten mathematisch zu formulieren, um eine rationelle Basis für eine Vorausberechnung des Seeganges und seiner vielfachen Wirkungen zu schaffen.

Diese Entwicklung ist in vollem Gange. Daher ist es hier nicht möglich, ein abgeschlossenes Gebäude der Erkenntnis zu errichten. Das Hauptgewicht der vorliegenden Darstellung (Teil I) liegt auf der Beschreibung derjenigen Ergebnisse, die als grundsätzlich und als verhältnismäßig gesichert anzusehen sind. Hierunter fallen die allgemeinen Untersuchungen über die Form und Bewegung der Meeres-Wellen sowie die Bemühungen um eine möglichst genaue Erfassung und realistische mathematische Darstellung des unregelmäßigen Seegangs-Gefüges. Dagegen sind die Vorgänge bei Entstehung, Wachstum, Zerfall und Vernichtung der Meeres-Wellen zum großen Teil sehr kompliziert und noch ungeklärt. Deshalb werden diese Umwandlungs-Prozesse hier kürzer behandelt (Teile II und III).

I. Form und Bewegung.

1. Einteilung der Meeres-Wellen. Unter dem Begriff *Oberflächen-Wellen* des Meeres könnte man *im weiteren Sinne* zunächst alle Meeres-Wellen zusammenfassen, bei denen — im Gegensatz zu *internen Wellen* — die Bewegung an der Meeres-Oberfläche *sichtbar* wird.

Es ist jedoch üblich, die Bezeichnung Oberflächen-Wellen *im engeren Sinne* zu verwenden und darunter nur diejenigen Wellen zu verstehen, bei denen die Bewegung — im wesentlichen — auf die Meeres-Oberfläche *beschränkt* bleibt. Hierdurch ergibt sich bereits eine Unterteilung der Oberflächen-Wellen des Meeres und zwar sind zunächst *Art und Umfang der Bewegung* maßgebend für die Wahl der Bezeichnung.

Bei den Oberflächen-Wellen im engeren Sinne ist die senkrechte Amplitude der Bewegung eines Teilchens nicht sehr verschieden von der waagerechten: Amplitude und Länge dieser Wellen sind klein gegenüber der Wassertiefe. Man pflegt deshalb diese Oberflächen-Wellen auch als *Tiefwasser-Wellen* zu bezeichnen, obwohl sie, sofern die Wellen nur kurz genug sind, auch auf flachem Wasser vorkommen können.

Das Gegenstück zu den Oberflächen-Wellen des Meeres im engeren Sinne wären Wellen, bei denen die an der Oberfläche sichtbare Bewegung *bis zum Grunde* reicht. Man könnte diese Form *Grund-Wellen* nennen. Hier überwiegt der waagerechte Anteil der Bewegung den senkrechten, und die Wellenlänge ist größer als die doppelte Wassertiefe. Für solche Wellenformen findet man gelegentlich die Bezeichnung *Seichtwasser-Wellen*, — die allerdings den zugeordneten Begriff ebenso unvollständig charakterisiert wie die Bezeichnung Tiefwasser-Wellen, da auch hier nicht der Absolutwert der Wassertiefe, sondern nur ihr Verhältnis zur Wellenlänge entscheidend ist. Wird im Grenzfall die Wellenlänge sehr groß gegenüber der Wassertiefe, sprechen wir von *Flut-Wellen.* Hier herrscht die horizontale Komponente der Bewegung vor. Einen Spezialfall der Grundwellen stellen die *Brandungs-Wellen* dar. Sie entstehen, wenn die Wassertiefe in Richtung der Wellen-Wanderung kontinuierlich bis auf Null abnimmt.

Wollen wir zum Ausdruck bringen, welche *Kräfte* bei der Wellen-Bewegung im Wasser in erster Linie maßgebend sind, so gebrauchen wir die Bezeichnung *Kapillar-Wellen* für Wellen, die hauptsächlich unter der Wirkung der Oberflächenspannung stehen (Länge unter 1,72 cm), während wie die längeren Wellen *Schwere-Wellen* nennen, da bei ihnen die Schwerkraft vorherrscht.

Die Oberflächen-Wellen des Meeres lassen sich ferner nach der *Art ihrer Erzeugung* einteilen. Die *Wind-Wellen* verdanken ihre Entstehung der Wirkung des

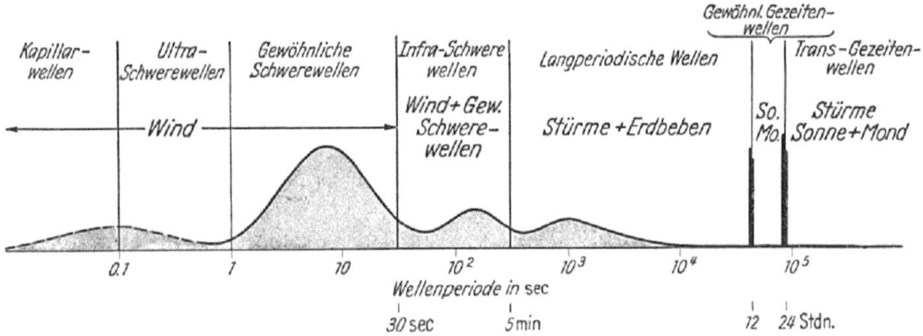

Fig. 1. Versuch einer Klassifikation der Meereswellen nach der Wellenperiode (nach Munk). Bezeichnungen, anregende Kräfte und relative Amplituden (Kurve).

Windes. Ihre Perioden liegen zwischen 0 und etwa 30 sec. Wir nennen sie insbesondere *Windsee,* wenn sie noch unter der anfachenden Kraft des erzeugenden Windes stehen. Wind-Wellen, die nicht mehr dieser Windeinwirkung unterliegen — sei es daß sie aus ihrem Entstehungsgebiet herausgewandert sind, oder sei es, daß der Wind an Stärke abgenommen hat — nennen wir *Dünung.*

Mittelbar durch den Wind erzeugt sind die sog. *Brandungs-Schwebungen* (surf-beats)[1]. Wir verstehen hierunter Wasserstands-Schwankungen an der Küste von wenigen Zentimetern Amplitude und einer Periode von einigen Minuten. Sie werden dadurch hervorgerufen, daß die Höhe der Brandungs-Wellen und damit der Wassertransport in Richtung auf die Küste regelmäßigen Schwankungen unterliegt, die durch Struktur-Eigentümlichkeiten der Wind-Wellen bedingt sind.

Plötzliche Veränderungen des Meeresbodens (mit Erdbeben verbunden) sind die Ursache von sehr langen Meereswellen, die Perioden von etwa 15 min aufweisen und unter dem Namen *Tsunami* vor allem im Pazifischen Ozean bekannt sind.

Von einem fahrenden Schiff geht ein eigentümliches System von Keil- und Querwellen aus, wofür sich der Name *Schiffs-Wellen* eingebürgert hat.

Weniger auf dem Meere als in Versuchsbecken und -kanälen der Laboratorien treten Oberflächen-Wellen auf, die durch eine punktförmige oder eine streckenförmige Störung des Wasserspiegels erzeugt werden. Im ersten Falle erhalten wir *Ring-Wellen* mit kreisförmigen Wellenfronten, während wir im zweiten Falle von *Kanal-Wellen* oder Wellen mit gerader Front sprechen.

Verwenden wir schließlich die Wellenperiode als Ordnungsmaß, so läßt sich ein *Spektrum aller Meeres-Wellen* entwerfen, wie es etwa Munk [7] versuchsweise getan hat (Fig. 1). Die angegebene Kurve soll die relativen Amplituden andeuten.

[1] W. H. Munk: Trans. Amer. Geophys. Un. **30**, 849 (1949).

Danach ist die Energie bei den gewöhnlichen Schwere-Wellen und den gewöhnlichen Gezeiten konzentriert.

Diese Übersicht über die verschiedenen Bezeichnungsweisen erhebt keinen Anspruch auf Vollständigkeit. Sie soll nur dazu dienen, die Grenzen abzustecken, zwischen denen sich der folgende Abschnitt bewegen wird. Wir beschränken uns darin auf *wind-erzeugte Oberflächen-Wellen des Meeres im weiteren Sinne*, d.h. unter Einschluß der wind-erzeugten Seichtwasser- und Brandungs-Wellen.

Die Maßzahlen, durch die wir Größe und Bewegung dieser Wellen beschreiben, sind die bei Wellen üblichen:

$$\textit{Amplitude } a,\ \textit{Länge } L,\ \textit{Periode } T \text{ und } \textit{Geschwindigkeit } C. \tag{1.1}$$

Bei den Meeres-Wellen wird überdies häufig der Begriff

$$\textit{Wellenhöhe } H \tag{1.2}$$

verwendet. Wir verstehen darunter den Höhenunterschied zwischen Wellenberg und Wellental. Die *Wellen-Steilheit* ist definiert als das Verhältnis Wellen-Höhe/Wellen-Länge. Die grundsätzliche Unregelmäßigkeit der Periodizitäten im Seegang wird uns später nötigen, diese nur auf periodische Erscheinungen anwendbaren Begriffe zweckentsprechend zu präzisieren.

a) Ergebnisse der „klassischen" Wellentheorie.

2. Nicht-lineare Gleichungen der wirbelfreien Wellen-Bewegung. Unter der Voraussetzung von Wirbelfreiheit, Reibungslosigkeit und Inkompressibilität lassen sich Schwere-Wellen an einer ebenen Flüssigkeits-Oberfläche durch folgende Gleichungen und Bedingungen charakterisieren [*1*], [*9*]:

Potential-Gleichung

$$\varphi_{xx} + \varphi_{yy} + \varphi_{zz} = 0. \tag{2.1}$$

BERNOULLI-Gleichung

$$\frac{p}{\varrho} = \varphi_t - g z - \frac{1}{2} (\varphi_x^2 + \varphi_y^2 + \varphi_z^2). \tag{2.2}$$

Kinematische Grenzbedingung für den Meeresboden

$$\varphi_n = 0 \quad \text{für} \quad z = -h(x, y). \tag{2.3}$$

Grenzbedingung für Druckkontinuität an der Meeresoberfläche

$$\eta = \frac{\varphi_t}{g} - \frac{1}{2g} (\varphi_x^2 + \varphi_y^2 + \varphi_z^2) \quad \text{für} \quad z = \eta. \tag{2.4}$$

Kinematische Grenzbedingung für die Meeresoberfläche

$$\eta_t = -\varphi_z \quad \text{für} \quad z = \eta. \tag{2.5}$$

Hierin bedeuten:

x, y, z rechtwinklige Raumkoordinaten; $z = 0$ an der ungestörten Oberfläche; z positiv nach oben; t Zeit; φ Geschwindigkeit-Potential; p Druck; ϱ Dichte; g Schwerebeschleunigung; $z = \eta(x,y,t)$ Gleichung der Meeresoberfläche; $z = -h(x,y)$ Gleichung des Meeresbodens; Indices x, y, z, t partielle Ableitungen nach x, y, z, t; Index n Ableitung nach der Normalen einer Fläche.

Wegen der nicht-linearen Terme in (2.4) (2.2) ist es bis jetzt nicht gelungen, dieses Gleichungssystem explizit und vollständig zu lösen. Es existieren lediglich Näherungs-Lösungen. Die bekannteste stammt von STOKES[1] und liefert die

[1] G. G. STOKES: Camb. Trans. **8**, 441 (1847).

Gleichung des Wellen-Profils für den Fall unendlich großer Wassertiefe in Form einer Reihe, deren Konvergenz durch Levi-Civita[1] bewiesen wurde. Die ersten drei Glieder dieser Reihenentwicklung des Wellenprofils stimmen mit der Gleichung einer *Trochoide* überein, und für den gleichen Grad der Annäherung gewinnt die Wellengeschwindigkeit die Form [vgl. (1.1) und (1.2)]

$$C = \sqrt{\frac{g\,L}{2\pi}\left(1 + \pi^2 \frac{H^2}{L^2}\right)}. \tag{2.6}$$

Bei der Ausbreitung solcher Wasser-Oberflächen-Wellen tritt somit eine *Dispersion* auf, da die Wellen-Geschwindigkeit von der Wellenlänge abhängt. Stokes konnte ferner zeigen, daß die Grenzform der *steilsten Wellen* durch scharfe Kämme mit einem Winkel von 120° zwischen Luv- und Leehang gekennzeichnet ist. Nach Michell[2] ist die Höhe dieser steilsten Wellen gleich 0,142 L, also etwa gleich dem siebenten Teil ihrer Länge. Die Wasserteilchen an den Kämmen dieser steilsten Wellen bewegen sich mit der gleichen Geschwindigkeit vorwärts wie die Welle selbst.

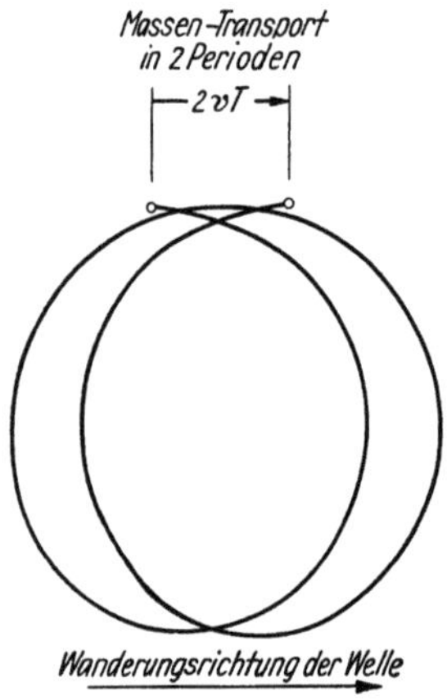

Fig. 2. Orbitalbahnen der Wellen endlicher Amplitude in tiefem Wasser (nach Stokes).

Die Oberflächen-Wellen nach Stokes haben ferner die bemerkenswerte Eigenschaft, daß die *Orbitalbahnen* der Wasserteilchen nicht geschlossen sind. Die Wasserteilchen beschreiben beim Durchlaufen einer Welle schleifenförmige Bahnen (Fig. 2), wobei die der Wellenwanderung gleichgerichtete Bewegung im Wellenkamm schneller ist als die entgegengesetzte im Wellental. Aus diesem Grunde bewegen sich die Wasserteilchen in der Wanderungsrichtung der Wellen. Jede solche Wellenbewegung ist also mit einer *Strömung* verbunden, deren Geschwindigkeit V in der Tiefe z angenähert durch die folgende Formel gegeben wird

$$V = \frac{\pi^2 H^2}{L^2}\, C\, e^{\frac{4\pi z}{L}} \tag{2.7}$$

und mit wachsender Tiefe rasch abnimmt (Tiefe z nach Verabredung negativ gerechnet). Laboratoriums-Versuche bestätigen im wesentlichen die Ergebnisse der Theorie von Stokes bis herab zu Wassertiefen von 0,2 L (Morison and Crooke[3]). Bei wind-erzeugten Wellen werden die Bahnen in unmittelbarer Nähe der Wasser-Oberfläche durch eine beträchtliche Translation in Windrichtung verformt (Johnson und Rice[4]).

Während die vorstehende Theorie der Oberflächen-Wellen endlicher Amplitude einige charakteristische Züge des Seeganges — insbesondere die Eigenschaften steiler Wellen — gut wiedergibt, versagt sie in dem folgenden wichtigen Punkt:

Wegen der Nicht-Linearität einiger Ausgangsgleichungen ist es nicht möglich, durch lineare Superposition von Lösungen neue Lösungen zu finden. Daher erhalten wir als Ergebnis stets rein-periodische Wellen, die der unregelmäßigen Seegangs-Struktur nicht gerecht werden können. Eine weitere Annäherung an die Wirklichkeit ist also zur Zeit nur durch Linearisierung der Ausgangsgleichungen möglich.

---

[1] T. Levi-Civita: Math. Ann. **93**, 264 (1925).

[2] Michell: Phil. Mag. (5) **36**, 430 (1893).

[3] J. R. Morison u. R. C. Crooke: Un. of Calif., Inst. Eng. Res., Berkeley, Calif., Series No. 3 Issue No. 344, 1953.

[4] J. W. Johnson u. E. K. Rice: Trans. Amer. Geophys. Un. **33**, 845 (1952).

3. Lineare Gleichungen der wirbelfreien Wellen-Bewegung. Setzen wir das Geschwindigkeits-Potential φ als so klein voraus, daß die Quadrate von φ und seinen Ableitungen vernachlässigt werden können, reduzieren sich die Gleichungen (2.1) bis (2.5) zu

Potential-Gleichung $$\varphi_{xx} + \varphi_{yy} + \varphi_{zz} = 0, \tag{3.1}$$

BERNOULLI-Gleichung $$\frac{p}{\varrho} = \varphi_t - g z, \tag{3.2}$$

Grenzbedingungen
$$\varphi_n = 0 \quad \text{für} \quad z = -h(x, y), \tag{3.3}$$
$$\eta = \frac{\varphi_t}{g} \quad \text{für} \quad z = 0, \tag{3.4}$$
$$\eta_t = -\varphi_z \quad \text{für} \quad z = 0. \tag{3.5}$$

Hierin ist mit genügender Genauigkeit zur Erfüllung der Grenzbedingungen an der Meeresoberfläche $z=0$ an Stelle von bisher $z=\eta$ gesetzt worden. Diese Gleichungen sind linear, gestatten also die Superposition von Lösungen. Sie gelten aber streng genommen nur für Oberflächen-Wellen mit infinitesimalen Amplituden. Wir können zwei Typen von Lösungen unterscheiden, nämlich *periodische* und *nicht-periodische* Zeitfunktionen.

α) Periodische Lösungen. Besitzt die Potentialfunktion φ die Periode T, so gilt

$$\varphi(x, y, z, t) = \varphi(x, y, z, t + T). \tag{3.6}$$

Wir können die Funktionen φ und η als einfach-harmonisch in der Zeit ansehen und diese Zeitabhängigkeit wie folgt abtrennen:

$$\varphi(x, y, z, t) = \varphi^*(x, y, z)\, e^{-\frac{2\pi i t}{T}}, \tag{3.7}$$

$$\eta(x, y, t) = \eta^*(x, y)\, e^{-\frac{2\pi i t}{T}}. \tag{3.8}$$

Mit diesen Ansätzen reduzieren sich die Gln. (3.1) und (3.3.) bis (3.5) wie folgt:

$$\varphi^*_{xx} + \varphi^*_{yy} + \varphi^*_{zz} = 0, \tag{3.9}$$

$$\varphi^*_n = 0 \quad \text{für} \quad z = -h(x, y), \tag{3.10}$$

$$\eta^* = -\frac{2\pi i}{g T}\varphi^* \quad \text{für} \quad z = 0, \tag{3.11}$$

$$\eta^* = \frac{T}{2\pi i}\varphi^*_z \quad \text{für} \quad z = 0. \tag{3.12}$$

Exakte Lösungen dieser reduzierten Gleichungen liegen bisher nur für konstante oder linear veränderliche Wassertiefe h vor, jedoch nicht für willkürliche Funktionen $z = -h(x, y)$.

Die einfachste und am häufigsten benutzte Lösung beschränkt sich auf *konstante Wassertiefe* und führt auf einen unendlich langen Zug kongruenter Sinus-Wellen mit der Amplitude a und *unendlich langen geraden Wellenkämmen*, die in der Richtung Θ wandern. Diese Lösung läßt sich in folgende Form bringen:

$$\varphi^* = a g T i\, e^{i\left[\frac{2\pi}{L}(x\cos\Theta + y\sin\Theta) + \delta\right]} \frac{\operatorname{Cos}\frac{2\pi}{L}(z + h)}{\operatorname{Cos}\frac{2\pi}{L} h}, \tag{3.13}$$

$$\eta = a\cos\left[\frac{2\pi}{L}(x\cos\Theta + y\sin\Theta) - \frac{2\pi t}{T} + \delta\right]. \tag{3.14}$$

Hierin ist δ eine willkürliche Phasenkonstante. Für die Wellengeschwindigkeit C ergibt sich aus (3.13), (3.12) und (3.11)

$$C = \frac{L}{T} = \sqrt{\frac{g L}{2\pi} \operatorname{Tan} \frac{2\pi h}{L}}. \tag{3.15}$$

Ist die Wellenlänge L kleiner als die doppelte Wassertiefe (dies trifft auf den Ozeanen praktisch immer zu), so gilt mit sehr guter Näherung

$$\operatorname{Tan} \frac{2\pi h}{L} \cong 1$$

und wir erhalten als Wellen-Geschwindigkeit der *Tiefwasser-Wellen*

$$C = \frac{L}{T} = \sqrt{\frac{g L}{2\pi}} = \frac{g}{2\pi} T. \tag{3.16}$$

Über tiefem Wasser hängt die Wellen-Geschwindigkeit somit nur von der Wellenlänge ab. Ist dagegen die Wellenlänge L groß gegen die Wassertiefe h, gilt angenähert

$$\operatorname{Tan} \frac{2\pi h}{L} \cong \frac{2\pi h}{L},$$

so daß sich für die Geschwindigkeit der *Flutwellen* ergibt:

$$C = \sqrt{g h}. \tag{3.17}$$

Die Geschwindigkeit der Flut-Wellen wird also nur durch die Wassertiefe bestimmt. Eine Dispersion ist hier nicht mehr vorhanden.

Es läßt sich zeigen, daß jedes Wasserteilchen beim Durchgang einer Welle (3.14) eine elliptische Schwingung ausführt, deren Periode gleich der Wellenperiode T ist. Lassen wir die x-Richtung mit der Wanderungsrichtung der Wellen zusammenfallen, wählen wir also $\Theta = 0$, so sind die horizontalen und vertikalen Halbachsen der *elliptischen Orbitalbahnen* gegeben durch:

$$a \frac{\operatorname{Cos} \frac{2\pi}{L} (z + h)}{\operatorname{Sin} \frac{2\pi}{L} h} \tag{3.18}$$

und

$$a \frac{\operatorname{Sin} \frac{2\pi}{L} (z + h)}{\operatorname{Sin} \frac{2\pi}{L} h}. \tag{3.19}$$

Die Orbitalbewegung verläuft im gleichen Sinne wie die Wellenwanderung, wenn das Teilchen sich auf einem Wellenkamm befindet, jedoch entgegengesetzt in einem Wellental. Beide Halbachsen der elliptischen Orbitalbahnen nehmen bei Annäherung an den Meeresboden ab. Unmittelbar am Meeresboden $z = -h$ verschwindet die vertikale Halbachse, während die horizontale einen endlichen Wert behält. Die Orbitalbahn entartet dort zu einer Strecke (Fig. 3). Diese Ergebnisse wurden durch experimentelle Untersuchungen im wesentlichen bestätigt, bis auf einen gewissen Massentransport, den die lineare Theorie nicht liefert (Wiegel und Johnson [7]).

Im Falle von Tiefwasser-Wellen ($L < 2h$) sind $\exp\left[-\frac{2\pi}{L}(z+h)\right]$ und $\exp\left(-\frac{2\pi}{L} h\right)$ kleine Größen, so daß die durch (3.18) und (3.19) gegebenen Halbachsen der Orbitalellipsen gleich werden. Die Wasserteilchen bewegen sich dann in *Kreisen*, deren Radien durch

$$a\,e^{\frac{2\pi z}{L}} \tag{3.20}$$

gegeben sind und mit wachsender Tiefe rasch abnehmen (Fig. 3).

Die *Energie* wandernder Oberflächen-Wellen mit unendlich langen geraden Kämmen setzt sich in jedem Zeitpunkt je zur Hälfte aus potentieller und kinetischer Energie zusammen. Auf die Flächeneinheit bezogen ist ihr Gesamtbetrag

$$E = \frac{g\varrho}{2}a^2, \tag{3.21}$$

also gleich der Arbeit, die erforderlich ist, um eine Flüssigkeitsschicht von der Dicke a um die Höhe $a/2$ zu heben.

In ähnlicher Weise wie die Wellen mit geraden Kämmen läßt sich ein System *kreisförmiger Wellen*, das von einem Punkt ausgeht, für den Fall konstanter Wassertiefe behandeln [*9*]. Diese Wellenform kann Bedeutung gewinnen beim Studium der Wellenentstehung in eng begrenzten rotationssymmetrischen Sturmgebieten wie den tropischen Wirbelstürmen.

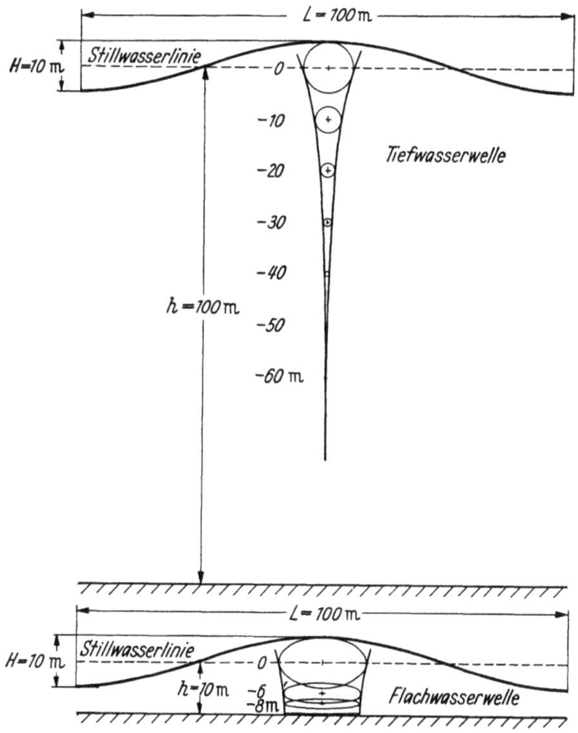

Fig. 3. Orbitalbahnen der Wellen infinitesimaler Amplitude in tiefem und flachem Wasser.

β) Nicht-periodische Lösungen. Die bisherigen Lösungen der linearen Gln. (3.1) bis (3.5) waren rein-periodische Zeitfunktionen, d.h. sie bestanden aus unendlich langen Folgen kongruenter Wellen. In Wirklichkeit jedoch besitzen die Wellenzüge auf dem Meere stets endliche Länge, und überdies treten innerhalb eines solchen Wellenzuges starke Unterschiede in Form und Bewegung der Einzel-Wellen auf. Die nächstliegende Möglichkeit, dem wahren Bilde des Seeganges näher zu kommen, ist bei einem linearen Gleichungssystem die *Superposition von periodischen Lösungen* unter Verwendung des Fourier-Prinzips. Im allgemeinsten Falle würde dieses Verfahren eine Integration der durch (3.7), (3.13) und (3.14) gegebenen Funktionen des Geschwindigkeitspotential φ und der Gestalt der Meeresoberfläche η über ein kontinuierliches Spektrum der Wellenperioden und alle Wellenrichtungen erfordern. Beschränken wir uns auf Tiefwasserwellen ($L < 2h$) und benutzen statt der Wellenperiode T die durch

$$\omega = 2\pi/T \tag{3.22}$$

definierte Wellen-Kreisfrequenz ω, dann würde z.B. das Fourier-(Doppel)-Integral für das Geschwindigkeitspotential unter Benutzung von (3.7) und (3.13) folgende Gestalt erhalten

$$\varphi = \int_0^\infty \int_\pi^{+\pi} -\frac{g\,a}{\omega}\, e^{\frac{\omega^2 z}{g}} \sin\left[\frac{\omega^2}{g}(x\cos\Theta + y\sin\Theta) - \omega t + \delta\right] d\omega\, d\Theta. \qquad (3.23)$$

Hierin sind Amplitude a und Phase δ Funktionen der Wellenkreisfrequenz ω und der Wellenrichtung Θ, für die allerdings geeignete Beziehungen vorliegen müssen. Aus diesem Geschwindigkeitspotential läßt sich auf Grund der Grenzbedingung (3.4) die Gestalt der Meeresoberfläche ermitteln.

Mit Hilfe des FOURIERschen Integraltheorems haben bereits CAUCHY und POISSON Wellenprobleme untersucht und unter anderem eine Darstellung der Wellensysteme gegeben, die auf tiefem Wasser durch eine lokale Oberflächenstörung erzeugt werden [*1*]. Diese Untersuchungen haben indessen geringe Beziehungen zu den winderzeugten Meereswellen.

4. Gruppen-Geschwindigkeit und Energietransport. Bei der Behandlung der nicht-periodischen Lösungen verdient folgender Umstand besondere Beachtung: Betrachten wir eine isolierte *Wellengruppe*, die wir uns durch Superposition von einfach-harmonischen Wellenzügen nach dem FOURIER-Prinzip entstanden denken können, so ist nicht nur die Wellen-(Phasen-)geschwindigkeit C der Einzelwellen von Bedeutung, sondern auch die Geschwindigkeit C_g der Wellengruppe. Sie ist gegeben durch

$$C_g = C - L\frac{dC}{dL} \qquad (4.1)$$

und somit abhängig von der dispersiven Natur der Wellen-Ausbreitung. Aus (4.1) folgt mit Hilfe von (3.15) für die Wasseroberflächenwellen

$$C_g = \frac{C}{2}\left(1 + \frac{4\pi h}{L \operatorname{Sin}\frac{4\pi h}{L}}\right). \qquad (4.2)$$

Das Verhältnis C_g/C wächst somit, wenn $4\pi h/L$ abnimmt. Für Tiefwasserwellen ($L < 2h$) ist $C_g/C = \frac{1}{2}$, die Gruppengeschwindigkeit ist also gleich der halben Wellengeschwindigkeit, während bei Flutwellen ($L \gg h$) die Gruppen-Geschwindigkeit gleich der Wellengeschwindigkeit wird.

Daß die Gruppen-Geschwindigkeit von Tiefwasserwellen geringer ist als deren Wellengeschwindigkeit, läßt sich leicht durch Beobachtungen verifizieren: Verfolgt man eine einzelne Welle am Ende einer Gruppe mit den Augen, so sieht man, daß sie innerhalb der Gruppe langsam nach vorne rückt, um, wenn sie die Spitze der Wellengruppe erreicht hat, zu verschwinden.

Die Gruppengeschwindigkeit hängt eng mit dem *Energietransport* zusammen. Berechnen wir nämlich die Energie E', die durch eine senkrecht zur Wanderungsrichtung der Wellen stehende Ebene in der Breiten- und Zeiteinheit hindurchgeht, so erhalten wir als Mittelwert nach LAMB [*1*]

$$E' = \frac{g\varrho}{4} a^2 C\left(1 + \frac{4\pi h}{L \operatorname{Sin}\frac{4\pi h}{L}}\right) \qquad (4.3)$$

und weiter unter Berücksichtigung von (4.2) und (3.21)

$$E' = \frac{E\,C}{2}\left(1 + \frac{4\pi h}{L \operatorname{Sin}\frac{4\pi h}{L}}\right) = E\,C_g. \qquad (4.4)$$

Hieraus läßt sich entnehmen, daß der Transport der Gesamtenergie E mit Gruppengeschwindigkeit stattfindet, und daß

$$\frac{E}{2}\left(1+\frac{4\pi h}{L\,\mathrm{Sin}\,\frac{4\pi h}{L}}\right) \tag{4.5}$$

derjenige Teil der Gesamtenergie ist, der sich mit der Wellengeschwindigkeit C ausbreitet. Nach (4.4) können wir also das Verhältnis Gruppen-Geschwindigkeit/Wellen-Geschwindigkeit auffassen als den Bruchteil der Gesamtenergie, der mit Wellengeschwindigkeit wandert und im Falle der Tiefwasserwellen ($L<2h$) gleich $\frac{1}{2}$ ist.

Das Verständnis für diesen Vorgang wird vertieft, wenn wir uns daran erinnern, daß die Gesamtenergie einer Welle je zur Hälfte aus potentieller (E_p) und kinetischer Energie (E_k) besteht. Dabei ist die *potentielle Energie* eine periodische Funktion, die in Phase mit der Schwingung der Meeres-Oberfläche, also mit der Wellen-Geschwindigkeit fortschreitet, während die *kinetische Energie* gleichmäßig auf die ganze Wellen-Länge verteilt und unabhängig von der Wellen-Geschwindigkeit ist (Fig. 4). Für Tiefwasser-Wellen gilt

$$\overline{E}_p = E_k = \frac{E}{2} = \frac{g\,\varrho}{4}\,a^2. \tag{4.6}$$

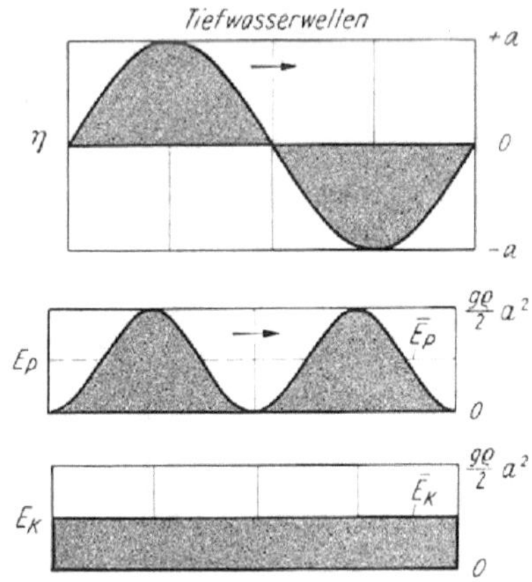

Fig. 4. Änderung der Meeresoberfläche (η), der potentiellen (E_P) und kinetischen (E_K) Energie entlang einer Wellenlänge (nach SVERDRUP und MUNK).

Wir betrachten das Fortschreiten einer aus einer endlichen Anzahl von kongruenten Einzelwellen bestehenden Wellen-Gruppe in ruhiges Wasser. Wenn die vorderste Welle um eine Wellenlänge in das ruhige Wasser vorrückt, muß sie darin Orbital-Bewegungen erzeugen und dabei die Hälfte ($E/4$) der von ihr mitgeführten potentiellen Energie $E/2$ in kinetische Energie umwandeln. Sie wird also an Höhe verlieren. Bei jedem weiteren Vorrücken um eine Wellenlänge wird sich dieser Prozeß wiederholen. Das Ergebnis läßt sich aus der Tabelle 1 ablesen, wobei angenommen ist, daß jeweils im Zeitabstand einer Wellenperiode der Energiebetrag $E/2$ von links zugeführt wird. In allgemeinen Zahlen ausgedrückt ist also die Energie der dem Wellen-Erzeuger unmittelbar benachbarten

Tabelle 1. *Energieverteilung bei der Ausbreitung einer Gruppe von Tiefwasserwellen* (nach SVERDRUP und MUNK [*4*]).

Anzahl der vorhandenen Wellenperioden	Anzahl der vorhandenen Wellen							Gesamtenergie der Gruppe
	1	2	3	4	5	6	7	
1	$\frac{1}{2}E$							$\frac{1}{2}E$
2	$\frac{3}{4}E$	$\frac{1}{4}E$						$\frac{2}{2}E$
3	$\frac{7}{8}E$	$\frac{4}{8}E$	$\frac{1}{8}E$					$\frac{3}{2}E$
4	$\frac{15}{16}E$	$\frac{11}{16}E$	$\frac{5}{16}E$	$\frac{1}{16}E$				$\frac{4}{2}E$
5	$\frac{31}{32}E$	$\frac{26}{32}E$	$\frac{16}{32}E$	$\frac{6}{32}E$	$\frac{1}{32}E$			$\frac{5}{2}E$
6	$\frac{63}{64}E$	$\frac{57}{64}E$	$\frac{42}{64}E$	$\frac{22}{64}E$	$\frac{7}{64}E$	$\frac{1}{64}E$		$\frac{6}{2}E$
7	$\frac{127}{128}E$	$\frac{120}{128}E$	$\frac{99}{128}E$	$\frac{64}{128}E$	$\frac{29}{128}E$	$\frac{8}{128}E$	$\frac{1}{128}E$	$\frac{7}{2}E$
	Richtung der Ausbreitung →							

Welle am größten. Sie beträgt nach n Perioden

$$\frac{2^n - 1}{2^n} E$$

und nähert sich bald der Gesamtenergie E. Die mit der Wellen-Geschwindigkeit C vorrückende Welle besitzt, der fortgesetzten Energie-Verluste wegen, nur noch $^1/_2{}^n$ der Gesamt-Energie, während die mit der Gruppengeschwindigkeit $C/2$ fortschreitende Zentral-Welle gerade die Hälfte $E/2$ der Gesamt-Energie enthält.

Besonders deutlich wird der *frontartige Charakter* der mit Gruppengeschwindigkeit wandernden *Zentralwelle*, wenn wir zu größeren Wellenzahlen n übergehen. Für $n = 900$ zeigt Fig. 5 die Energie- und Höhenverteilung einer in ruhiges Wasser fortschreitenden Wellen-Gruppe nach einer (näherungsweisen) Berechnung von SVERDRUP und MUNK [4]. In einem sehr schmalen Bereich von nur 50 m nimmt hier die Wellenhöhe von 10 auf 90% ihres endgültigen Wertes zu, so daß wir abschließend sagen können: Zwar wird bei jeder Welle die halbe Energie mit der Wellen-Geschwindigkeit transportiert. Wegen der beim Vorrücken in ruhiges Wasser kontinuierlich auftretenden Energie-Verluste sind die Wellen *vor* der mit der Gruppen-Geschwindigkeit $C_g = C/2$ wandernden Zentralwelle jedoch so niedrig, daß sie praktisch bedeutungslos sind. Erst kurz vor dem Eintreffen der Zentral-Welle treten merkbare Wellenhöhen auf, so daß wir — mit einer für alle praktischen Zwecke ausreichenden Genauigkeit — die Gruppen-Geschwindigkeit als maßgebend für die Ausbreitung eines begrenzten Wellenzuges ansehen können.

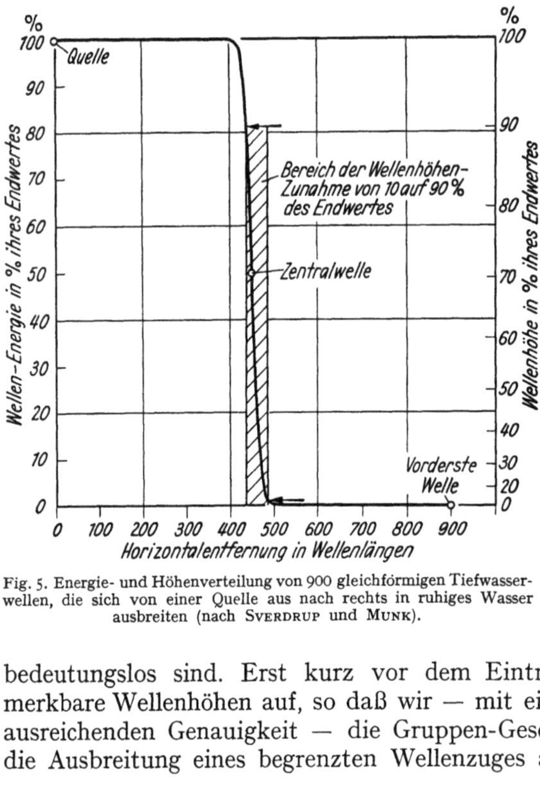

Fig. 5. Energie- und Höhenverteilung von 900 gleichförmigen Tiefwasserwellen, die sich von einer Quelle aus nach rechts in ruhiges Wasser ausbreiten (nach SVERDRUP und MUNK).

5. GERSTNER's Wirbelwellen. Weniger um der praktischen Bedeutung als des historischen Interesses willen sei eine mögliche Form der Tiefwasser-Wellen kurz erwähnt, für die GERSTNER[1] und später unabhängig davon RANKINE[2] exakte Gleichungen angaben. Beide setzten für die Wasserteilchen kreisförmige Orbital-Bahnen an und erhielten nach Anwendung der dynamischen Gleichungen für die Koordinaten eines Teilchens in der x, z-Ebene zur Zeit t

$$\left.\begin{aligned} x &= x_0 - a\, e^{\frac{2\pi z_0}{L}} \sin\frac{2\pi}{L}(x_0 - C t), \\ z &= z_0 + a\, e^{\frac{2\pi z_0}{L}} \cos\frac{2\pi}{L}(x_0 - C t), \end{aligned}\right\} \tag{5.1}$$

[1] F. GERSTNER: Abh. Kgl. Böhm. Ges. Wiss. 1802 [Gilberts Ann. Physik **32**, 412 (1809)].
[2] RANKINE: Phil. Trans. Roy. Soc. Lond. **108**, 127 (1863).

wobei x_0, z_0 die Koordinaten des Mittelpunktes des Orbitalkreises bedeuten. Die Wellenlinien sind Trochoiden. Aus der Forderung der Druckkonstanz an der Wasseroberfläche ergibt sich die Bedingung

$$C = \sqrt{\frac{g L}{2\pi}} = \frac{g T}{2\pi}, \tag{5.2}$$

die mit der entsprechenden Formel für harmonische Tiefwasser-Wellen (3.16) übereinstimmt. Für die Halbmesser r der Orbitalkreise findet man in Übereinstimmung mit (3.20)

$$r = a\, e^{\frac{2\pi z_0}{L}}. \tag{5.3}$$

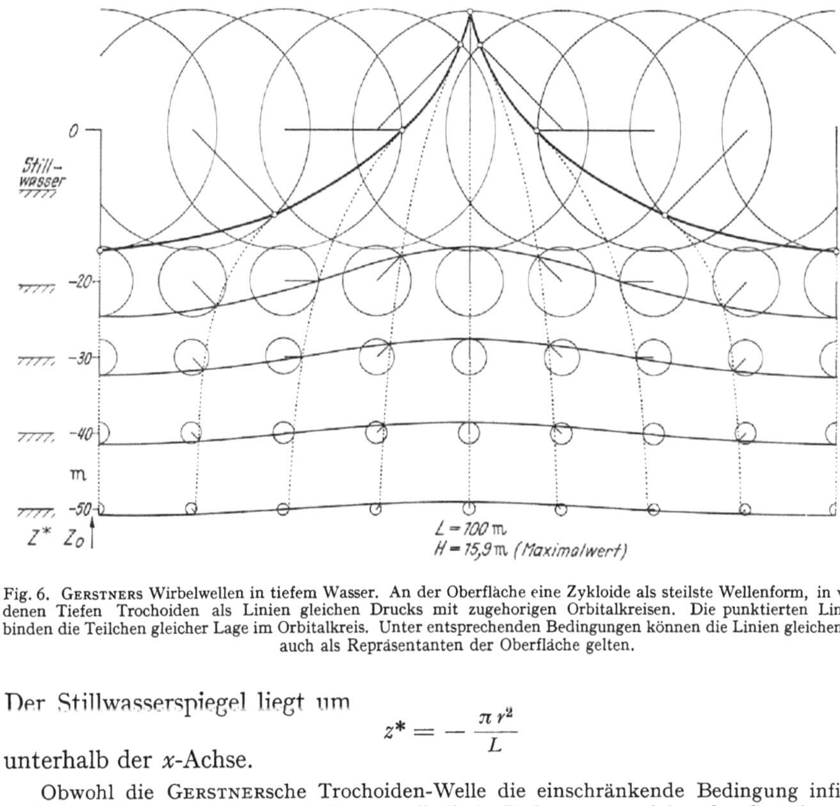

Fig. 6. GERSTNERS Wirbelwellen in tiefem Wasser. An der Oberfläche eine Zykloide als steilste Wellenform, in verschiedenen Tiefen Trochoiden als Linien gleichen Drucks mit zugehorigen Orbitalkreisen. Die punktierten Linien verbinden die Teilchen gleicher Lage im Orbitalkreis. Unter entsprechenden Bedingungen können die Linien gleichen Drucks auch als Repräsentanten der Oberfläche gelten.

Der Stillwasserspiegel liegt um

$$z^* = -\frac{\pi r^2}{L} \tag{5.4}$$

unterhalb der x-Achse.

Obwohl die GERSTNERsche Trochoiden-Welle die einschränkende Bedingung infinitesimaler Amplituden nicht kennt, ist ihre physikalische Bedeutung aus folgenden Gründen gering:

a) Die *steilste Wellenform* für eine vorgegebene Wellen-Länge ist eine *Zykloide* und besitzt einen fast rasiermesserähnlichen Wellenkamm (Fig. 6). Dies widerspricht den Beobachtungen.

b) Die mit diesen Trochoiden-Wellen verbundene Flüssigkeits-Bewegung ist *nicht wirbelfrei* und zwar ist der Wirbel dem Umlaufsinn der Wasserteilchen in den Orbitalbahnen entgegengesetzt gerichtet.

Eine Erweiterung der GERSTNERschen Theorie der Trochoidenwellen auf den Fall endlicher Wassertiefe wurde durch GAILLARD (s. WIEGEL u. JOHNSON [7]) gegeben.

b) Die wirkliche Struktur der Meereswellen.

6. Beobachtungs- und Meßmethoden. Die Luftaufnahme Fig. 7 vermittelt einen Eindruck von dem Gefüge des Seeganges und läßt zugleich deutlich werden, wie wenig realistisch es ist, diese Struktur durch einen konservativen Wellenzug konstanter Periode und Amplitude annähern zu wollen.

Tabelle 2. *Methoden der Seegangsmessung.*

D Richtung, *H* Höhe, *L* Länge, *T* Periode, *C* Geschwindigkeit der Wellen.

Nr.	Methode der Messung[1]	Meßprinzip	Erhaltene Wellengrößen					Registrierung möglich?	Einsatzfähig	
			D	*H*	*L*	*T*	*C*		im tiefen Wasser	im flachen Wasser
1	visuell	Höhenschätzung (Augenhöhe = Wellenhöhe)	*D*	*H*				nein	×	
2		Zeitdifferenzen an einem „festen" Punkt (Pfahl, Boje Schaumfleck)	*D*			*T*		nein	×	×
3		Vergleich mit bekannten Längen	*D*		*L*			nein	×	×
4		System von Peilstangen oder Bojenreihen	*D*		*L*	*T*	*C*	nein		×
5	optisch	Stereophotogrammetrische Filmaufnahmen (vom Schiff, Flugzeug oder von der Küste aus)	*D*	*H*	*L*		*C*	ja	×	×
6		Filmaufnahmen eines fest auf dem Meeresboden stehenden oder verankerten Maßstabes	*D*	*H*		*T*		ja		×
7		Filmaufnahmen gegen eine parallel zur Wellenrichtung stehende Wand		*H*	*L*	*T*	*C*	ja		×
8		Filmaufnahmen vom Flugzeug aus	*D*		*L*	(*T*)	(*C*)	ja	×	×
9		Anvisierung der Wellenkämme mit Theodolit	*D*					nein		×
10		Theodolit mit eingebautem perspektivischem Visiernetz	*D*	(*H*)	*L*	*T*	*C*	nein		×
11		Photographie und Photometrie von Sonnen- und Mondreflexen der Meeresoberfläche		*H* /	*L*			nein	×	×
12		Photographie der sonnenlicht-erzeugten Wellenbilder auf dem Meeresboden	*D*		*L*	*T*	*C*	nein		×
13	akustisch	Echolot über Wasser		*H*		*T*		ja		×
14		Umgekehrtes Echolot auf dem Meeresboden		*H*		*T*		ja		×
15		Umgekehrtes Echolot auf einem Unterseeboot		*H*				ja	×	
16	hydrodynamisch	Schwimmer in vertikaler Führung		*H*		*T*		ja		×
17		Schwimmer mit Registriergerät für vertikale Beschleunigungen		*H*		*T*		ja	×	×
18		Schwimmer durch Filmaufnahmen von der Küste aus verfolgt	*D*	*H*		*T*		ja		×

19		Serie von Schwimmern, die Kontakte betätigen		*H*		*T*		ja		×
20		Schwimmer mit daranhängendem Wasserdruck-Schreibgerät		*H*		*T*		ja	×	
21		Schwimmer mit daranhängendem Vertikal-Strommesser		*H*		*T*		ja	×	
22		Spieren-Boje mit Dämpfungsscheibe } Filmaufnahme der Wasserstandsschwankungen	*D*	*H*		*T*		ja	×	
23		Spieren-Boje mit Dämpfungsscheibe } Wasserdruckregistrierung in geringer Tiefe		*H*		*T*		ja	×	
24		Wasserdruck-Registrierung am Meeresboden		*H*		*T*		ja		×
25		Wasserdruck-Registrierung am Schiff, dazu Messung der vertikalen Beschleunigungen		*H*		*T*		ja	×	×
26		Wasserdruck-Registrierung in geringer Tiefe mit einer senkrechten, unten offenen Röhre		*H*		*T*		ja		×
27	aerodynamisch	Messung der Schwankungen des (statischen) Luftdrucks auf einem Schwimmer		*H*		*T*		ja	×	×
28	thermodynamisch	Adiabatische Temperaturänderungen in einer gasgefüllten Gummikammer am Meeresboden		*H*		*T*		ja		×
29	elektrisch	Serie von Widerständen mit Kontakten, die eintauchend kurzgeschlossen werden		*H*		*T*		ja		×
30		zwei parallele Drähte als Elektroden senkrecht im Wasser		*H*	(*L*)	*T*	(*C*)	ja		×
31		Kapazitätsänderung eines Kondensators gebildet aus einem isolierten Draht und dem Seewasser		*H*		*T*		ja		×
32		Aufzeichnung der Wellenkontur am Schiff durch zahlreiche über die Schiffsaußerhaut verteilte Sonden		*H*	*L*	*T*	*C*	ja	×	
33	elektronisch	Höhenmessung durch Radar vom Flugzeug aus		*H*	*L*			ja	×	×
34		Photographie des Seegangsechos auf dem Radar-Bildschirm	*D*		*L*		*C*	ja	×	×
35		Messung der DOPPLER-Verschiebungen bei Radiowellen, die von der Meeresoberfläche reflektiert werden	*D*		*L*		*C*	ja	×	×
									18	28

[1] *Nicht* Methode der *Übertragung* der Meßwerte!

Die *grundsätzliche Unregelmäßigkeit* der Meereswellen hat zur Folge, daß wir an die Meßinstrumente des Seeganges die Forderung stellen müssen, alle Einzelheiten dieser komplizierten Bewegung zu erfassen und festzuhalten. Dabei würde eine kontinuierliche Aufzeichnung aller Bewegungskomponenten des Seeganges an einem einzigen Punkt der Meeresoberfläche keinesfalls genügen, vielmehr müßte eine solche für einen genügend großen Bereich und streng genommen auch in allen von der Wellenbewegung erfaßten Tiefen erfolgen. Diese an ein ideales Seegangsmeßgerät zu richtenden Forderungen sind so hoch, daß sie heute von

Fig. 7. Luftaufnahme einer Windsee (aus [*13*]).

keinem Instrument praktisch erfüllt werden können. Wir verfügen zur Zeit nur über Geräte, die einzelne Teile des Gesamtvorganges mehr oder minder genau messen.

Neben der regellosen Seegangsstruktur bedeutet das *Fehlen fester Bezugspunkte* auf dem freien Ozean eine weitere Erschwerung der Wellenmessung, die dadurch hauptsächlich auf das flache Wasser des Küstenvorfeldes verwiesen wird.

Tabelle 2 gibt einen Überblick über die derzeit bekannten Methoden der Meereswellenmessung unter Verzicht auf Einzelheiten wie Arbeitsweise, Genauigkeit, Vor- und Nachteile der einzelnen Verfahren (s. hierzu [*17*]).

Von den 35 erwähnten Meßmethoden können 18 auf hoher See, 28 im flachen Wasser des Küstenvorfeldes eingesetzt werden. Es überwiegen also eindeutig — insbesondere wenn zusätzlich die Qualität der Methoden gewertet wird — die küstengebundenen Instrumente und es besteht ein schwerwiegender Mangel an brauchbaren Wellenmeßmethoden für die freie See. Demzufolge sind wir zur Zeit auch hauptsächlich über die Meereswellen des Küstenvorfeldes informiert, wo Bodeneinflüsse, wie Refraktion und Dissipation die ursprüngliche Gestalt und Bewegung der winderzeugten Tiefwasserwellen bereits verändert haben. Dabei wirkt sich hier insbesondere ungünstig aus, daß die überwiegend verwendeten Druckmeßgeräte am Meeresboden erstens eine *selektive Wirkung* haben, d. h. vor allem die langperiodischen und weniger die kurzperiodischen Bestandteile des Seeganges anzeigen und zweitens die *Umwandlung der gemessenen Druckschwankungen* am Meeresboden in Wasserstandsänderungen an der Meeresoberfläche problematisch ist, da die Reduktionsformeln der klassischen Theorie entstammen, also nur für reinperiodische Wellen und einen starren Meeresboden gelten (Ewing und Press [*5*], Snodgrass [*17*]).

Es muß ferner darauf hingewiesen werden, daß die meisten der angegebenen Verfahren nur eine Aufzeichnung der Wasserstandsänderung als Funktion *einer* Veränderlichen liefern, die im allgemeinen die Zeit, gelegentlich auch die Horizontalentfernung ist. Im Falle der zeitlichen Amplitudenregistrierung an einem festen Punkt werden Wellenhöhe und Wellenperiode, im zweiten Falle (bei der Erfassung der räumlichen Amplitudenunterschiede entlang einer horizontalen Geraden) Wellenhöhe und Wellenlänge gemessen, sofern diese Gerade parallel zur Wellenrichtung verläuft. Beide Verfahren beschränken sich also auf eine zweidimensionale Betrachtungsweise, die naturgemäß unvollständig ist. Eine kontinuierliche zeitliche Aufzeichnung der Wasserstandsänderung an allen Punkten einer horizontalen Geraden (etwa durch Filmaufnahme) würde bereits drei Dimensionen erfassen, aber erst eine zeitliche Registrierung der Bewegung eines größeren Teilgebietes der Meeresoberfläche würde die Seegangsbewegung im vierdimensionalen Raumzeitkontinuum vollständig abbilden und die bisher nicht erfaßten Bestimmungsstücke des Seeganges (Wellenrichtung und Kammlänge) festhalten.

Das einzige Verfahren, das der idealen Forderung nahekommt und genaue Werte für die Höhe, Länge und Geschwindigkeit der Meereswellen sowie deren Richtung und seitliche Anordnung erbringt, ist die stereophotogrammetrische Seegangsaufnahme mit schneller Bildfolge. Leider ist seine Anwendung zeitraubend, kostspielig und daher verhältnismäßig selten. Aus diesem Grunde bedeutet die kürzlich von Cartwright[1] angegebene Methode, die es erlaubt, außer der Höhe und Periode auch die Richtung der Wellenzüge mit einem Schiffs-Wellenschreiber bei nicht zu komplizierten Seegangserscheinungen zu messen, eine willkommene Verbesserung.

7. Form und Größe der Wellen. Wir beschränken uns auf Ergebnisse von Messungen an natürlichen Meereswellen und verzichten auf die Wiedergabe von Laboratoriumsresultaten.

Über *topographische Einzelheiten* der Meeresoberflächenwellen unterrichten am besten Auswertungen stereophotogrammetrischer Seegangsaufnahmen (Schumacher [*17*], [*6*], Weinblum und Block[2]). Fig. 8 gibt als Beispiel einen solchen Wellenplan. Die Darstellung durch Höhenschichtlinien läßt die komplizierte Gestalt der Meeresoberfläche gut erkennen. Als Ergänzung hierzu dienen Profilschnitte; sie liefern Wellenform, Neigung der Luv- und Leeabhänge, Scheitelwinkel usw.

Es zeigte sich[2], daß von *Wellenprofilen*, die sich durch einfache mathematische Funktionen wiedergeben lassen, nicht gesprochen werden kann. Gelegentlich finden sich trochoidenähnliche Formen, oft aber werden gebrochene Geraden mit Abrundungen am Berg und im Tal den Beobachtungen am ehesten gerecht.

Die *Wellensteilheit* — ausgedrückt durch das Verhältnis Wellenhöhe/Wellenlänge — schwankte bei Schumacher zwischen 1:17,6 und 1:92,2, maximale Häufigkeit bei 1:40.

[1] D. E. Cartwright: Proc. Roy. Soc. Lond., Ser. A **234**, 382 (1956).
[2] G. Weinblum u. W. Block: Jb. Schiffbautechn. Ges. **37**, 214 (1936).

Die *Neigung* der vorderen und rückwärtigen Wellenabhänge wird überschläglich durch das Verhältnis der senkrechten zur waagerechten Projektion des Abhanges gegeben. Bei den von Schumacher vermessenen Wellen (63% Windsee-, 37% Dünungswellen) wiesen die vorderen Abhänge eine Neigung von durchschnittlich 0,06, im Maximum von 0,14 (~8°) auf, während für die rückwärtigen

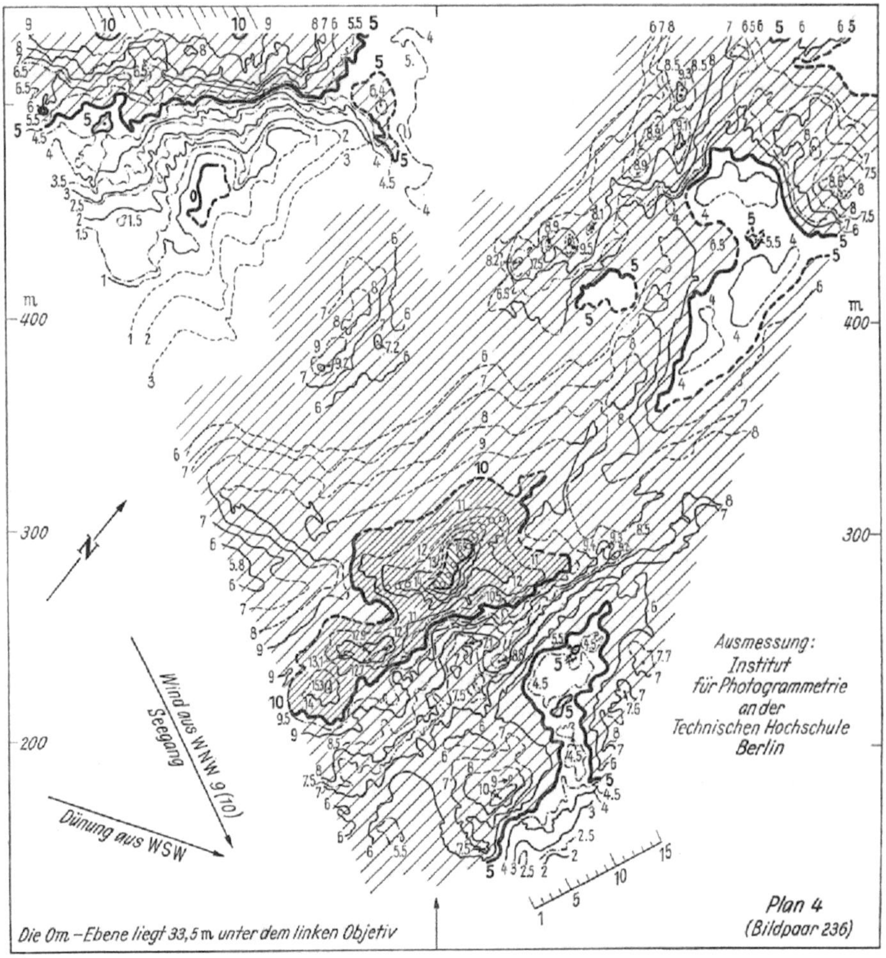

Fig. 8. Wellenplan nach stereophotogrammetrischen Aufnahmen auf Schnelldampfer „Europa" (nach Schumacher).

Abhänge entsprechend 0,05 und 0,10 (~6°) gefunden wurde. Die Neigung des vorderen Abhanges war also im Mittel 1,2mal so groß wie die des rückwärtigen. Diese angenäherten Neigungswerte wurden jedoch von den größten wahren Neigungen weit übertroffen, die beim vorderen Abhang im Mittel das 2,3fache des obigen Verhältnisses der Hangprojektionen, beim rückwärtigen Abhang das 1,8fache erreichten. In Einzelfällen können die größten wahren Neigungen beim vorderen Abhang das vier- bis siebenfache des Verhältnisses der Projektionen betragen. Es wurden dort Neigungen von 0,44 (~24°) gemessen.

Die mittleren *Scheitelwinkel* der von Schumacher untersuchten Wellen fielen größtenteils (80%) in den Bereich 149° bis 165°, etwa 10% lagen zwischen 122° und 145°, 10% über 165°.

Als *Maximalwert* der Wellenhöhe fanden WEINBLUM und BLOCK 16 m, die maximale Höhendifferenz auf diesem Wellenplan betrug sogar 18,5 m. Dabei ist anzunehmen, daß noch höhere Wellen der Aufnahme entgingen. Erstaunlich war dabei die rasche Zunahme der maximalen Wellenhöhe von 7 auf 18,5 m in nur 2 Std. Als Höchstwerte der Wellenlänge wurden bei WEINBLUM und BLOCK etwa 200 m, bei SCHUMACHER 350 m, bei ROLL[1] (Radarmessung) 410 m erreicht.

Wellenbeobachtungen der Ozeanwetterschiffe (ROLL[2], BROWN[3]), Wellenmessungen auf einem See (BURLING[4]) sowie Laboratoriumsversuche (FRANCIS[5]) zeigten ferner einen deutlichen Effekt der *thermischen Schichtung* auf die Größe der Meereswellen. Ist die Lufttemperatur (in einigen Metern Höhe gemessen) um 6° C kälter als die Temperatur des Oberflächenwassers, so wurden für Wellenhöhe und Wellenlänge im Mittel um 22% bzw. 15% größere Werte erhalten als bei Temperaturgleichheit zwischen Luft und Wasser, obwohl die Windgeschwindigkeit (in etwa 20—30 m Höhe gemessen) die gleiche war. Diese Unterschiede lassen sich darauf zurückführen, daß in Kaltluft (infolge starker Turbulenz und geringer vertikaler Windscherung) größere Windkräfte auf die Meeresoberfläche einwirken als bei ausgeglichener Temperaturverteilung.

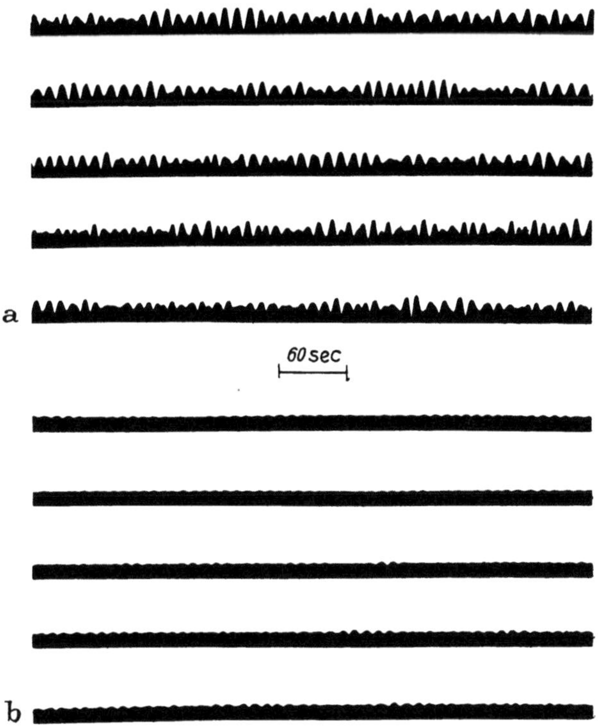

Fig. 9a u. b. Registrierungen von a Windsee- und b Dunungswellen (aus [13]).

8. Charakteristische Eigenschaften von Windsee und Dünung. Ergebnisse visueller Beobachtungen der Meereswellen lassen sich folgendermaßen zusammenfassen:

Die einzelnen Wellen der *Windsee* gleichen Gebirgszügen mit scharfen, teilweise überbrechenden Kämmen, die nur etwa zwei- oder dreimal so lang sind wie die Wellenlänge. Die großen Wellen sind von zahlreichen kürzeren Wellen bis hinab zu den Kapillarwellen überlagert. Die Höhe aufeinanderfolgender Wellenkämme ist verschieden. Auch kann nicht davon gesprochen werden, daß die Wanderungsrichtung der Wellen streng konstant sei, vielmehr erscheinen zuweilen Wellenzüge, deren Richtung um 30 oder mehr Grad von der Hauptrichtung abweicht, ohne daß eine entsprechende Schwankung der Windrichtung

[1] H. U. ROLL: Bücherei der Funkortung **4** II, 56 (1954).
[2] H. U. ROLL: Dtsch. Hydrogr. Z. **5**, 111 (1952).
[3] P. R. BROWN: Marine-Observer **23**, 94 (1953).
[4] R. W. BURLING: Proc. V. Conf. on Coastal Engineering, U.S. Council on Wave Research, The Eng. Foundation **1**, 1955.
[5] J. R. D. FRANCIS: Weather **9**, 163 (1954).

auftritt. Bemüht man sich, einer besonders markanten Welle mit den Augen zu folgen, so muß man feststellen, daß sie meist schon nach kurzer Zeit im Wellengemisch des Seeganges untertaucht.

Im Gegensatz zu der soeben geschilderten turbulenten Windsee zeichnet sich die *Dünung* durch eine größere Regelmäßigkeit oder (besser gesagt) einen geringeren Grad von Unregelmäßigkeit aus. Die Dünungswellen sind flacher als die Windseewellen und besitzen runde Kämme, die etwa 6—8mal so lang sind wie die Wellenlänge. Die Höhen aufeinanderfolgender Dünungswellen weichen nicht sehr voneinander ab. Die Wellenrichtung ist einheitlich. Es ist möglich, eine markante Welle verhältnismäßig lange mit den Augen zu verfolgen, ehe sie wieder im Wellengemisch untertaucht.

Ein Teil der genannten Kennzeichen von Windsee und Dünung läßt sich auch aus den in Fig. 9 gegenübergestellten Registrierungen von Windsee- und Dünungswellen ablesen.

9. Wahre und scheinbare Wellen. Die üblichen Bestimmungsstücke einer wandernden Welle, wie Amplitude, Länge, Periode und Geschwindigkeit beziehen sich — strenggenommen — nur auf ideale Wellenzüge, d. h. auf eine unendlich lange Folge kongruenter Wellen, und sind für einen solchen Wellenzug konstante Größen. Da die Meereswellen von diesem einfachen Bilde erheblich abweichen, müssen wir die in Ziff. 1 eingeführten Begriffe und Bezeichnungen in geeigneter Weise modifizieren [*9*], [*12*], [*13*].

Die mit Auge und Meßgerät feststellbaren, unregelmäßigen Erhebungen und Senkungen der Meeresoberfläche wollen wir *scheinbare Wellen* nennen. Die *scheinbare Wellenhöhe* $\tilde{H}$ bedeutet sodann den Höhenunterschied zwischen einem Wellenberg und dem vorhergehenden Wellental. Unter der *scheinbaren Wellenperiode* $\tilde{T}$ sei die Zeitdifferenz, und unter der *scheinbaren Wellenlänge* $\tilde{L}$ der Abstand zwischen zwei aufeinanderfolgenden Wellenbergen verstanden.

Diese so definierten Wellengrößen charakterisieren das äußere Erscheinungsbild, die beobachtbaren Eigenschaften des Seeganges. Ihre Werte schwanken zeitlich und räumlich in unregelmäßiger Weise und es ist insbesondere nicht anzunehmen, daß sie die für harmonische Wellen geltenden Gesetzmäßigkeiten wie z.B. (3.15) oder (3.16) befolgen werden.

Die *wahren Wellen* des Seeganges sind Auge und Meßgerät zunächst verborgen. Sie werden erst dann sichtbar, wenn es durch eine geeignete Analyse gelingt, die Amplituden und Perioden der harmonischen Wellenkomponenten zu finden, deren Superposition das unregelmäßige Erscheinungsbild der Meereswellen erzeugt. Diese harmonischen Teilwellen des Seegangsspektrums wollen wir als wahre Wellen ansehen und ihre Bestimmungsstücke H, L, T usw. als wahre Wellengrößen, auf die die Ergebnisse der klassischen Wellentheorie anwendbar sind.

10. Statistische Eigenschaften der scheinbaren Wellengrößen. Als Grundmaterial dienen hauptsächlich die — in geringen Tiefen vor der Küste, unter annähernd stationären äußeren Bedingungen gewonnenen — Registrierungen der wellenbedingten Druckschwankungen am Meeresboden (Fig. 9). Eine Umrechnung (die zwar nur für wahre Wellen gilt, aber meist auch auf scheinbare Wellen angewendet wird) liefert sodann ein Bild des Seeganges in der Form *zeitlicher Profilschnitte an einem festen Punkt.* Durch genaue Ausmessung solcher Registrierungen wird versucht, statistische Gesetzmäßigkeiten zwischen den scheinbaren Wellengrößen zu gewinnen. Hierbei wurden folgende Resultate erzielt:

Die aus einer Wellenregistrierung an zahlreichen (zufällig ausgewählten oder äquidistanten) Zeitpunkten entnommenen Ordinaten gruppieren sich um ihren Mittelwert im Sinne einer GAUSS*schen Normalverteilung* (RUDNICK[1], PIERSON und MARKS[2]) mit einer Streuung, die der mittleren Gesamtenergie pro Flächeneinheit proportional ist.

Ordnen wir die N Wellen einer Registrierung nach ihrer Wellenhöhe $\tilde{H}$ an, so läßt sich die Höhenverteilung durch folgende *Mittelwerte der scheinbaren Wellenhöhe* charakterisieren:

$\bar{\tilde{H}}$: Mittel über alle N Wellen.

$\tilde{H}_{\frac{1}{3}}$: Mittel über die $N/3$ höchsten Wellen *(significant wave height = charakteristische Wellenhöhe)*.

$\tilde{H}_{\frac{1}{10}}$: Mittel über die $N/10$ höchsten Wellen.

Zwischen diesen Mittelwerten, sowie zwischen ihnen und dem Maximalwert $\tilde{H}_{\max}$ bestehen folgende empirische Beziehungen:

SEIWELL [5][3]	1948,	USA-Ostküste	$\tilde{H}_{\frac{1}{3}}/\bar{\tilde{H}} = 1{,}57$	(10.1)
PUTZ[4]	1952,	USA-Westküste	$= 1{,}60$	
WATTERS[5]	1953,	Neuseeland	$= 1{,}58$	
DARLINGTON[6]	1954,	Nordatlantik	$= 1{,}603$	
WATTERS[5]	1953,	Neuseeland	$\tilde{H}_{\frac{1}{10}}/\bar{\tilde{H}} = 1{,}94$	(10.2)
DARLINGTON[6]	1954,	Nordatlantik	$\tilde{H}_{\max}/\bar{\tilde{H}} = 2{,}398$	(10.3)

Die *mittlere Abweichung* $\sigma_{\tilde{H}}$ der *Wellenhöhenverteilung* ist proportional zur mittleren Wellenhöhe $\bar{\tilde{H}}$:

PUTZ[4]	1952,	USA-Westküste	$\sigma_{\tilde{H}} = 0{,}494\,\bar{\tilde{H}} + 0{,}037$ [m]	(10.4)
DARLINGTON[6]	1954,	Nord-Atlantik	$\sigma_{\tilde{H}} = 0{,}536\,\bar{\tilde{H}}$	

Aus der Existenz solcher empirischen Beziehungen zwischen den charakteristischen Mittelwerten der scheinbaren Wellenhöhe, sowie aus der linearen Abhängigkeit der Streuung vom Mittelwert läßt sich schließen, daß die statistischen Verteilungen der scheinbaren Wellenhöhe bei den verschiedenen Seegangszuständen sehr ähnlich sein müssen und durch eine *einparametrige Schar von (unsymmetrischen) Verteilungsfunktionen* angenähert werden können, deren Parameter nur vom Mittelwert $\bar{\tilde{H}}$ abhängt. PUTZ[4] und WATTERS[5] haben geeignete Funktionen angegeben.

Die Verteilung der *scheinbaren Wellenperioden* $\tilde{T}$ ist durch ähnliche Zusammenhänge gekennzeichnet. Die mittlere scheinbare Periode $\tilde{T}_{\frac{1}{3}}$ der $N/3$ höchsten Wellen, die sog. *charakteristische Wellenperiode*, und die mittlere Abweichung $\sigma_{\tilde{T}}$ ändern sich linear mit dem Mittelwert $\bar{\tilde{T}}$ über alle N scheinbaren Wellen (PUTZ[4], DARLINGTON[6]). Allerdings weichen die von beiden Autoren

[1] P. RUDNICK: Proc. II. Berkeley Symp. on Mathem. Statistics, Univ. Calif. Press 627, 1951.
[2] W. J. PIERSON u. W. MARKS: Trans. Amer. Geophys. Union **33**, 834 (1952).
[3] H. R. SEIWELL: Mass. Papers Phys. Oceanogr. Met. **10**, No. 4 (1948).
[4] R. R. PUTZ: Trans. Amer. Geophys. Union **33**, 685 (1952).
[5] J. K. A. WATTERS: New Zealand J. Sci. Technol. B **34**, 408 (1953).
[6] C. R. DARLINGTON: Quart. J. Roy. Met. Soc. Lond. **80**, 619 (1954).

gegebenen Konstanten voneinander ab. Als Verteilungsfunktion schlägt PUTZ – im Gegensatz zu der unsymmetrischen Verteilungsfunktion für die Wellen*höhe* – eine GAUSS-Verteilung vor.

Die Höhen und Perioden der *einzelnen* scheinbaren Wellen scheinen nahezu voneinander *unabhängig*[1] zu sein, mit Ausnahme einer gewissen Tendenz der höchsten Wellen, überwiegend bei mittleren Perioden aufzutreten. (Dies gilt wohl vor allem für Dünungswellen; bei jungen Windsee-Wellen im Wind-Wellenkanal fanden JOHNSON und RICE[2] gewisse Zusammenhänge.)

Auch bestehen nur geringe Beziehungen zwischen den Höhen *aufeinanderfolgender* Wellen sowie deren Perioden.

Neben diesen aus Registrierungen herausdestillierten Eigenschaften der scheinbaren Wellengrößen stehen die Ergebnisse anderer Methoden. Wir nennen

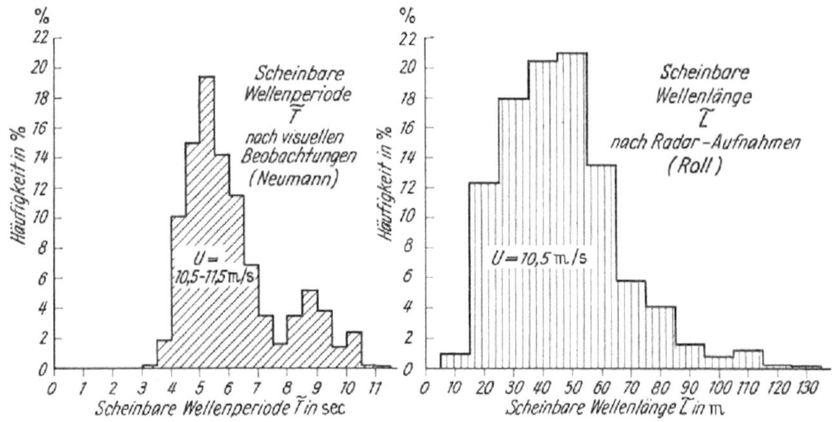

Fig. 10. Häufigkeitsverteilungen scheinbarer Wellenperioden und -längen in einer Windsee.

z. B. die von NEUMANN [*10*] auf Grund von visuellen Beobachtungen bei bestimmten Windgeschwindigkeiten gewonnenen *Häufigkeits-Verteilungen der scheinbaren Wellenperiode in der vollentwickelten Windsee* (Fig. 10). Diese „ozeanischen" Verteilungen werden sich zweifellos von den im Küstenvorfeld mit Druckschreibern auf dem Meeresboden erhaltenen Resultaten unterscheiden. Im Gegensatz zu PUTZ[1] sind hier die Verteilungen auch nicht symmetrisch. Die Breite des Periodenspektrums ist erheblich, es erstreckt sich im Mittel von 0,5 $\tilde{\tilde{T}}$ bis 2 $\tilde{\tilde{T}}$. Als Ergänzung hierzu zeigt Fig. 10 eine bei annähernd gleichen äußeren Verhältnissen durch Ausmessung von Radaraufnahmen der Seegangstörung (ROLL[3]) erhaltene *Häufigkeitsverteilung der scheinbaren Wellenlänge*, die der Periodenverteilung sehr ähnelt.

Nicht unerwähnt bleiben sollen schließlich Wellen-Statistiken, die mehr *klimatologischen Charakter* tragen (SEIWELL[4]: 2 Stationen an der Ostküste Nordamerikas, DARBYSHIRE[5]: 2 Wetterschiffs-Stationen im östlichen Nordatlantik, 1 Station in SW-England, ROLL[6]: 10 Wetterschiffs-Stationen im Nordatlantik,

[1] S. 689, Fußnote 4.
[2] S. 674, Fußnote 4.
[3] S. 687, Fußnote 1.
[4] S. 689, Fußnote 3.
[5] J. DARBYSHIRE: Nature, Lond. **174**, 827 (1954). — Marine-Observer **25**, 115 (1955).
[6] H. U. ROLL: Seewetteramt Hamburg Einzelveröff. Nr. 1 u. 6, 1953 u. 1954.

VENKATARMAN[1]: Schiffsbeobachtungen in der Arabischen See). Sie unterrichten über die mittleren Seegangsverhältnisse, ihre lokalen Besonderheiten und jahreszeitlichen Unterschiede.

11. Spektrale Analyse von Seegangsregistrierungen. Grundlage für diese Form der Analyse ist die Vorstellung, daß die Wellenbewegung der Meeresoberfläche als Resultat einer linearen Superposition von harmonischen Wellen nach dem FOURIER-Prinzip aufgefaßt werden kann. Die Gesamtheit dieser wahren Wellen liefert das *Spektrum* des Seeganges und gibt somit die Verteilung der Wellenenergie auf die verschiedenen Periodenbereiche an. Während die scheinbaren Wellen das äußere Bild des Seeganges mit statistischen Mitteln beschreiben, können wir das Energiespektrum als den physikalischen Hintergrund dieser Erscheinungen ansehen.

Das starke zahlenmäßige Anwachsen der Seegangs-Registrierungen sowie der große Zeitaufwand, den rechnerische Methoden erfordern, ließ die Konstruktion von geeigneten Maschinen zur Wellen-Analyse in England[2,3] und den USA (KLEEBA [*5*]) zweckmäßig erscheinen. Bei diesen *Wellen-Analysatoren* wird ein Streifen mit einer zeitlich begrenzten Wellenregistrierung von meist etwa 20 min Dauer nach einer Vorbehandlung, die den Bereich unterhalb der Wellenaufzeichnung schwärzt, auf dem gleich langen Umfang eines Rades befestigt. Durch Drehung dieses Rades wird die Registrierung an einem optischen System vorbeigeführt, das die Aufzeichnung beleuchtet und das von dieser gestreute Licht mittels einer Photozelle in elektrische Stromschwankungen umwandelt, die anschließend einem Vibrations-Galvanometer mit geeigneter Resonanzfrequenz zugeführt werden. Durch kontinuierliche Änderung der Drehgeschwindigkeit des Rades wird erreicht, daß nacheinander die in der Wellenregistrierung vorhandenen Periodizitäten mit dem Oszillations-Galvanometer in Resonanz kommen. Diese Resonanzen werden als Funktion der Periode auf einem Diagramm aufgezeichnet, das entsprechend der Drehgeschwindigkeit des Rades bewegt wird. Auf diese Weise wurden die in Fig. 11 dargestellten „*Wellenspektren*" erhalten, die die typische Seegangsstruktur auf die Interferenz zahlreicher Perioden zurückführen.

Auffällig ist die gezackte Form dieser Spektren. Sie kann einerseits dadurch erklärt werden, daß nicht alle im Seegang vorhandenen Perioden zur Resonanz mit dem Galvanometer kommen können, sondern nur diejenigen, bei denen die Länge der Registrierung ein ganzzahliges Vielfaches ist. Eine Verlängerung der Registrierung würde also die Dichte der resonanzfähigen und daher anzeigbaren Perioden erhöhen. Zum anderen können die Zacken offenbar auch dadurch bedingt sein, daß der Resonanzbereich der verwendeten Galvanometer zu eng und die Geschwindigkeitsabnahme des Rades zu groß war. Zur Erzielung der bestmöglichen Analyse kommt es offenbar darauf an, die Länge der Registrierung, die Bandbreite des Galvanometers und die Drehgeschwindigkeit des Rades günstig aufeinander abzustimmen (PIERSON [*9*]).

SEIWELL und WADSWORTH[4,5] äußerten Bedenken gegen diese Form der Analyse, vor allem deswegen, weil hier eine zeitlich begrenzte Aufzeichnung der Wellen — durch Wiederholungen beliebig verlängert — verwendet wird und daher die physikalische Bedeutung der erhaltenen Perioden zweifelhaft sei. Sie glaubten, den geophysikalischen Gehalt der Registrierungen durch die Methode der *verallgemeinerten harmonischen Analyse* besser zu erfassen. Hierzu werden der

[1] K. S. VENKATARMAN: Indian J. Met. Geophys. **7**, 179 (1956).

[2] N. F. BARBER, F. URSELL, J. DARBYSHIRE u. M. J. TUCKER: Nature, Lond. **158**, 328 (1946).

[3] G. E. R. DEACON: Quart, J. Roy. Met. Soc. Lond. **75**, 227 (1949).

[4] H. R. SEIWELL u. G. P. WADSWORTH: Science, Lancaster, Pa. **109**, 271 (1949).

[5] H. R. SEIWELL: Science, Lancaster, Pa. **112**, 243 (1950).

Wellenregistrierung in konstanten Zeitabständen von 1 oder 2 sec Werte entnommen. Diese zeitliche Wertreihe wird mit sich selbst korreliert, wobei jeweils Phasenverschiebungen um konstante Beträge vorgenommen werden. Das Ergebnis ist ein *Autokorrelogramm* (Fig. 11), das in vielen Fällen einfach periodisch erscheint. SEIWELL und WADSWORTH folgerten daraus, daß der ozeanische Seegang keine komplizierte Interferenzstruktur durch Kombination vieler Perioden besitzt, sondern vielmehr häufig aus einer *einzigen Sinuswelle* besteht, die

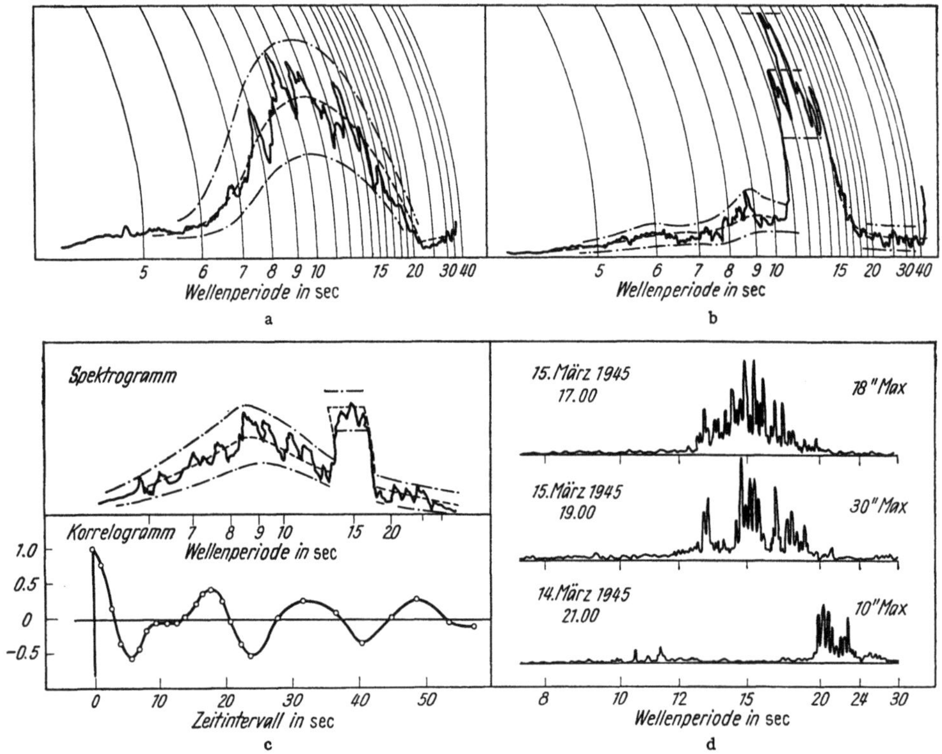

Fig. 11a–d. Wellenspektren und Autokorrelationsfunktion, erhalten durch maschinelle Analyse. a Wellenspektrum nach einer Druckregistrierung in 37 m Wassertiefe bei Bermuda. Typisches Windsee-Spektrum, durch Druck-Filterung Perioden unter 7 sec unterdrückt (nach SEIWELL[1]). b Wellenspektrum nach einer Druckregistrierung in 24 m Wassertiefe bei Cuttyhunk (Ostküste der USA). Typisches Dunungs-Spektrum aus großer Entfernung mit einem schmalen Periodenbereich von 11–15 sec (nach KLEBBA [5]). c Wellenspektrum und Autokorrelationsfunktion nach einer Druckregistrierung bei Guam. Lokale Windsee und Dünung aus großer Entfernung (nach RUDNICK[2]). d Wellenspektren nach Druckregistrierungen in Pendeen, England. Oben zwei Windsee-Spektren, unten Spektrum der vorauslaufenden Dünungswellen (nach BARBER und URSELL[3]).

von *zufälligen Schwingungen überlagert* ist. Sie ordneten die zyklische Komponente der vorherrschenden meteorologischen Situation zu, während die oszillatorische Komponente auf zufällige lokale Einflüsse zurückgeführt wurde.

Mag auch in einigen Fällen von Dünungswellen mit sehr engem Frequenzband die Deutung von SEIWELL und WADSWORTH angemessen sein, aus Analysen anderer Registrierungen ist jedenfalls ersichtlich, daß von der allgemeinen Existenz *einer* vorherrschenden zyklischen Komponente im Seegang nicht gesprochen werden kann. Wird die Autokorrelation nur über eine genügende Anzahl von Phasenverschiebungen fortgesetzt, ergibt sich meistens ein Absinken

[1] H. R. SEIWELL: Proc. Nat. Acad. Sci. U. S. A. **35**, 518 (1949).
[2] S. 689, Fußnote 1.
[3] N. F. BARBER u. F. URSELL: Phil. Trans. Roy. Soc. Lond. **240**, 527 (1948).

des Korrelationskoeffizienten in unregelmäßigen Schwingungen auf Beträge nahe Null. Die weitere Auswertung dieser Korrelogramme wäre dann wiederum nur durch FOURIER-Analyse, also durch die Rückkehr zum ersten Verfahren möglich.

Demnach wäre in erster Linie die Entwicklung verbesserter Wellen-Analysatoren anzustreben, die die erwähnten Fehler vermeiden und ein ausgeglichenes Wellenspektrum mit zuverlässig definierten Ordinatenwerten liefern (PIERSON [*9*]). Ein geeignetes Gerät, das elektronische Methoden benutzt, wurde kürzlich von PIERSON[1] beschrieben.

PIERSON und MARKS[2] haben ferner auf der Grundlage der in Ziff. 15 erläuterten Seegangsdarstellung eine Analysenmethode angegeben, die auf das *Spektrum der Autokorrelationsfunktion der Druckregistrierung* führt, aus dem sodann durch (erlaubte) Anwendung der klassischen Beziehungen das Energiespektrum der wellenbewegten Meeresoberfläche entsteht.

c) Theorie des Seeganges auf statistischer Basis.

Die durch neue Meßmethoden erweiterte Kenntnis von der wirklichen Struktur der Meereswellen hat Versuche im Gefolge gehabt, die Erfahrungstatsachen durch eine Theorie zu verbinden und zu erklären. Wegen der bekannten Unregelmäßigkeit der Meereswellen wird es nicht wundernehmen, daß diese theoretischen Ansätze sich von der klassischen Wellentheorie durch die Einbeziehung statistischer Überlegungen unterscheiden.

12. Die statistische Verteilung der scheinbaren Wellenhöhen. LONGUET-HIGGINS[3] berechnete theoretisch die Verteilung der scheinbaren Wellenhöhen unter folgenden 2 Voraussetzungen:

α) Das Wellenspektrum besteht aus einem einzigen schmalen Frequenzband.

β) Der Seegang entsteht durch Superposition einer großen Anzahl harmonischer Wellen, deren Amplituden klein und deren Phasen zufällig verteilt sind.

Der Voraussetzung β) liegt die Vorstellung zugrunde, daß die sich überlagernden Wellen aus verschiedenen Teilen eines Wellenentstehungsgebietes stammen, dessen Abmessungen groß gegenüber der Wellenlänge sind. Demnach können die Phasen der einzelnen Teilwellen als voneinander unabhängig betrachtet werden.

N konsekutive Wellen mit den Amplituden $\tilde{a}_1, \tilde{a}_2 \ldots \tilde{a}_N$ werden nach fallenden Werten dieser Amplitude geordnet. $\tilde{a}^{(p)}$ sei der Mittelwert der ersten pN Wellenamplituden ($0 \leq p \leq 1$), also $\tilde{a}^{(1)}$ die mittlere Amplitude aller N Wellen, $\tilde{a}^{(\frac{1}{3})}$ ihr „charakteristischer" Wert. Ferner wird durch

$$\bar{a}^2 = \frac{1}{N} (\tilde{a}_1^2 + \tilde{a}_2^2 \cdots + \tilde{a}_N^2) \tag{12.1}$$

eine der mittleren Energie der Wellen proportionale Maßzahl eingeführt. Unter den genannten Voraussetzungen gelangte LONGUET-HIGGINS zu folgenden Ergebnissen:

a) Verteilung der scheinbaren Wellenamplituden. p ist der Anteil derjenigen Amplituden, deren Werte die Größe r übersteigen. (Darstellung $r/\bar{a}$ als Funktion von p in Fig. 12.)

Beispiel: 10% aller N Amplituden sind höher als 1,52 $\bar{a}$
20% aller N Amplituden sind höher als 1,27 $\bar{a}$ usw.

[1] W. J. PIERSON, jr.: U.S. Beach Erosion Board, Techn. Mem. No. 56, 1954.
[2] S. 689, Fußnote 2.
[3] M. S. LONGUET-HIGGINS: J. Marine Res. 9, 245 (1952).

b) Verhältnis der Mittelwerte scheinbarer Amplituden. $\tilde{a}^{(p)}/\tilde{a}^{(1)}$ wird als Funktion von p gewonnen (Fig. 12).

Beispiel: $\tilde{a}^{(\frac{1}{3})} = 1{,}598\,\tilde{a}^{(1)}$

$\tilde{a}^{(\frac{1}{10})} = 2{,}032\,\tilde{a}^{(1)}$.

c) Maximale Wellenamplitude. Die wahrscheinlichsten Werte $\mu\,(\tilde{a}_{\max})$ der maximalen Wellenamplitude, bezogen auf die mittlere scheinbare Wellenamplitude $\tilde{a}^{(1)}$, wachsen mit der Anzahl N der Wellen (Fig. 13).

Beispiel: $N = 10$ $\mu\,(\tilde{a}_{\max}) = 1{,}786\,\tilde{a}^{(1)}$

$N = 100$ $= 2{,}452\,\tilde{a}^{(1)}$

$N = 1000$ $= 2{,}982\,\tilde{a}^{(1)}$.

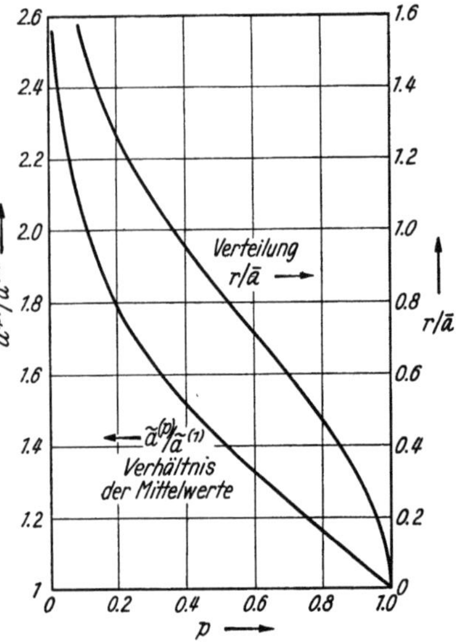

Fig. 12. Charakteristik der Wellenamplituden im Falle eines schmalen Wellenspektrums (nach Longuet-Higgins).

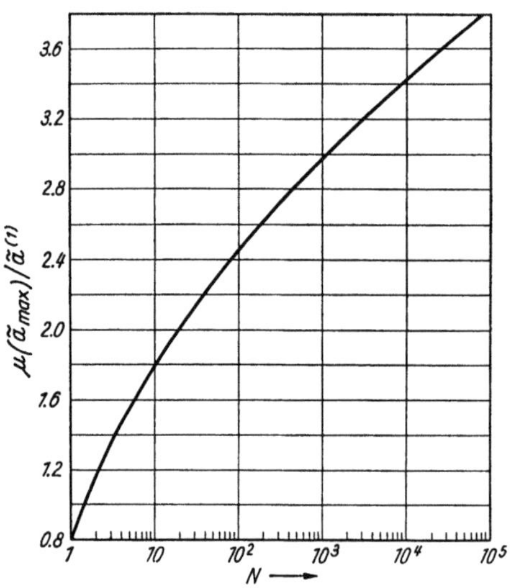

Fig. 13. Wahrscheinlichste Werte der maximalen Wellenamplitude μ ($a_{\max}$) bezogen auf die mittlere Wellenamplitude $a^{(1)}$ als Funktion der Anzahl N der Wellen (nach Longuet-Higgins).

d) Charakteristik der scheinbaren Wellenamplituden und -höhen. Die maßgebenden Verhältniswerte werden auf das Energiemaß $\bar{a}$ bezogen:

	Amplitude	*Höhe* (näherungsweise)	
Häufigster Wert	$\tilde{a}_f = 0{,}707\,\bar{a}$	$\tilde{H}_f = 1{,}414\,\bar{a}$	(12.
Mittlerer Wert	$\tilde{a}^{(1)} = 0{,}886\,\bar{a}$	$\tilde{\bar{H}} = 1{,}772\,\bar{a}$	
Charakteristischer Wert	$\tilde{a}^{(\frac{1}{3})} = 1{,}416\,\bar{a}$	$\tilde{H}_{\frac{1}{3}} = 2{,}832\,\bar{a}$	
Mittlerer Wert d. $N/10$ höchsten Werte	$\tilde{a}^{(\frac{1}{10})} = 1{,}800\,\bar{a}$	$\tilde{H}_{\frac{1}{10}} = 3{,}600\,\bar{a}$.	

Ein Vergleich mit (10.1) und (10.2) zeigt, daß die theoretischen Ergebnisse unter (b) mit den empirischen Beziehungen gut übereinstimmen, obwohl die Voraussetzung eines schmalen Frequenzbandes bei den Messungen sicher nicht immer erfüllt war. Aus dieser Tatsache könnte man schließen, daß die statistische Verteilung der scheinbaren Wellenhöhen nicht wesentlich von der Breite des Frequenzbandes abhängt. Das Entscheidende scheint vielmehr die Zufallsverteilung der Phasen zu sein.

Obwohl die in (c) theoretisch erhaltene Abhängigkeit der maximalen Wellenamplitude von der Anzahl der beobachteten scheinbaren Wellen plausibel erscheint und die von Zeit zu Zeit im Seegang auftretenden maximalen Wellenhöhen jedem Beobachter wohl vertraut sind, wurde dieses Ergebnis durch DARLINGTON[1] — möglicherweise wegen zu geringer Variation von N — nicht bestätigt gefunden. Die für sehr große N theoretisch erhaltenen maximalen Amplituden dürften allerdings kaum auftreten, da die Seegangsstruktur wegen der Veränderlichkeit des Windes selten über die hierzu erforderlichen langen Zeiten hinweg konstant bleibt.

13. Das Energiespektrum der Windsee. Als ein weiterer wichtiger Schritt auf dem Wege zu einer rationellen Seegangsdarstellung ist NEUMANNS [*12*] Versuch anzusehen, das Energiespektrum der Windsee zu berechnen. Er setzt dieses Spektrum als kontinuierlich voraus und definiert in Anlehnung an (3.21) die mittlere Gesamtenergie E pro Flächeneinheit durch

$$E = \frac{g\varrho}{2}\int\limits_0^\infty [a(\omega)]^2\, d\omega \quad [\text{erg.}\, \text{cm}^{-2}]. \tag{13.1}$$

Hierin bedeutet $[a(\omega)]^2$ den Beitrag der spektralen Frequenzen $\omega = 2\pi/T$ im Bereich $\omega \pm d\omega$ zur Gesamtenergie. $[a(\omega)]^2$ ist daher proportional zur spektralen Energiedichte und gilt als Repräsentant des *Energiespektrums*. Es besitzt die Dimension $[\text{cm}^2\, \text{sec}]$. Zuweilen wird das Energiespektrum auch als Funktion der Periode betrachtet. Die Dimension von $[a(T)]^2$ ist dann $[\text{cm}^2\, \text{sec}^{-1}]$. Zur Vereinfachung wird ferner die zur Gesamtenergie proportionale Größe

$$E^* = \int\limits_0^\infty [a(\omega)]^2\, d\omega \quad [\text{cm}^2] \tag{13.2}$$

eingeführt. E^* ist gleich der doppelten Streuung der GAUSS-Verteilung, durch die sich die Abweichungen der Meeresoberfläche vom Mittelwert darstellen lassen, und stimmt außerdem mit dem durch (12.1) definierten Energiemaß $\bar{a}^2$ überein, so daß sich aus E^* sofort die Charakteristik der scheinbaren Wellenhöhen $\widetilde{H}$ ableiten läßt.

Es gilt nun einen formelmäßigen Ausdruck für das Energiespektrum $[a(\omega)]^2$ zu finden. Hierfür benutzt NEUMANN visuelle Beobachtungen ausgereifter Wellen und Ergebnisse von Wellenregistrierungen, sowie die Vorstellung, daß die bei bestimmten Werten des Verhältnisses Wellenperiode/Windgeschwindigkeit beobachteten maximalen Wellenamplituden dann erreicht werden, wenn die zugehörigen harmonischen Teilwellen in Phase sind. Somit können diese maximalen Wellenamplituden als angenähertes Maß für die spektrale Energiedichte angesehen werden. Als Ergebnis dieser Betrachtungen findet NEUMANN das *Energiespektrum der vollentwickelten Windsee* in folgender Form

$$[a(\omega)]^2 = \frac{C}{\omega^6}\, e^{\frac{-2g^2}{\omega^2 U^2}}. \tag{13.3}$$

Hierin ist U die Windgeschwindigkeit in cm/sec, g die Schwerebeschleunigung und die Konstante $C = 4{,}80 \cdot 10^4\ [\text{cm}^2\, \text{sec}^{-5}]$. Die charakteristische Form dieser Energiespektren ist aus Fig. 14 ersichtlich. Wir erkennen zunächst, daß die Energie der ausgereiften Windsee sich auf einen mehr oder weniger breiten Spektralbereich konzentriert, obwohl rein theoretisch alle Perioden möglich sind. Die Maxima der Wellenenergie verschieben sich dabei mit wachsender Windgeschwindigkeit in Richtung größerer Perioden T also geringerer Frequenzen ω. Für die Wellenperiode

[1] S. 689, Fußnote 6.

maximaler Energie T_{max} oder für die zugehörige Frequenz ω_{max} läßt sich aus (13.3)

$$\omega_{max} U = \sqrt{\tfrac{2}{3}}\, g \qquad \text{oder} \qquad T_{max} = 0{,}785\, U \begin{bmatrix} T \text{ in sec,} \\ U \text{ in m/sec} \end{bmatrix} \tag{13.4}$$

errechnen, woraus wir erkennen, daß das *Produkt aus der Frequenz maximaler Wellenenergie und der Windgeschwindigkeit konstant* ist. Diese Beziehung wird — verhältnismäßig gut — durch DARBYSHIRE [*11*] bestätigt, der aus Windfeld- und Wellenanalysen für die ausgereifte Windsee die empirische Gleichung $T_{max} = 0{,}74\, U$ herleitete.

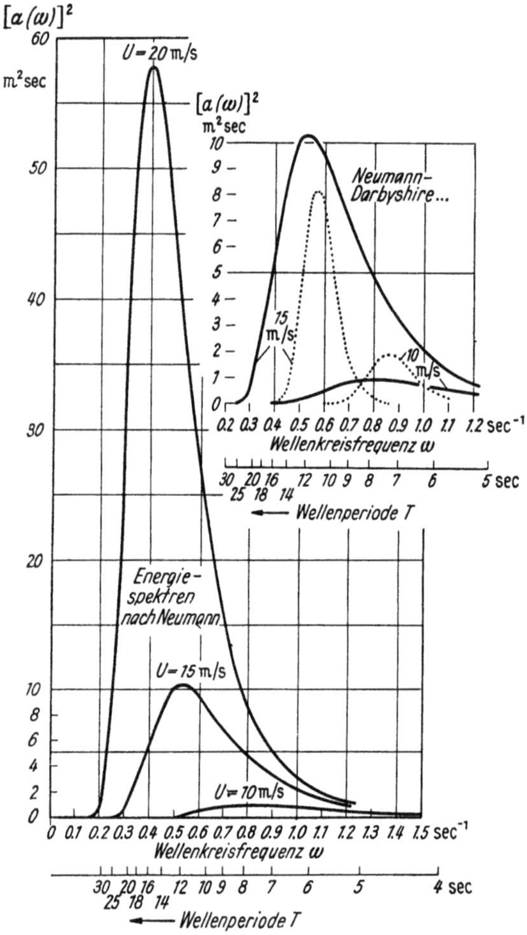

Fig. 14. Energiespektren der ausgereiften Windsee nach NEUMANN und DARBYSHIRE.

Das NEUMANN-Spektrum zeigt noch weitere bemerkenswerte Eigenschaften. Es sinkt z.B. für Wellenfrequenzen, deren zugehörige Gruppengeschwindigkeiten größer als die Windgeschwindigkeit sind, praktisch auf Null ab. Wenn daher die vordere Begrenzung eines Wellenerzeugungsgebietes mit der Windgeschwindigkeit wandert — dies ist bei Kaltfronten meist der Fall —, ist es unmöglich, daß Wellen ihrem Entstehungsgebiet vorauseilen.

Aus (13.1) und (13.3) läßt sich durch Integration über alle Wellenfrequenzen ω eines bestimmten Bereiches $\Delta\omega$ die *Wellenenergie* E pro Flächeneinheit oder die Größe E^* berechnen, die in diesem *Spektralbereich* vorhanden ist. Für die praktische Anwendung ist insbesondere die in Fig. 15 wiedergegebene Form gedacht, wobei die Integration von $T = 0$ bis zur oberen Grenze T oder von ω bis $\omega = \infty$ erstreckt wurde. Die den Abszissenwerten T oder ω zugehörigen Ordinaten geben jeweils die entsprechenden E^*-Werte an, oder die gemäß (12.2) berechneten scheinbaren Wellenhöhen.

Wir können schließlich die Integration über alle Perioden von 0 bis ∞ (oder alle Frequenzen von ∞ bis 0) ausdehnen und gelangen sodann zu der Beziehung

$$E = 3{,}125 \cdot 10^{-9}\, U^5 \; [\text{erg cm}^{-2}], \tag{13.5}$$

aus der wir ersehen, daß die auf die Flächeneinheit entfallende *Gesamtenergie E der ausgereiften Windsee mit der fünften Potenz der Windgeschwindigkeit anwächst.* Abweichend hiervon gelangte DARBYSHIRE [*11*] zu dem Ergebnis, daß die Gesamtenergie mit der dritten Potenz der Windgeschwindigkeit zunimmt. In diesem

Zusammenhang haben COX und MUNK[1] darauf hingewiesen, daß dieser Unterschied offenbar auf die Aufstellung des von DARBYSHIRE benutzten Wellenschreibers (Druckmessung in 17 m Wassertiefe) zurückzuführen sei, wodurch Wellen mit Perioden unter 6 sec unterdrückt wurden. Da diese kurzen Wellen erheblich zur Gesamtenergie beitragen, sei das DARBYSHIRE-Spektrum als zu schmal und demgemäß die Gesamtenergie als zu gering anzusehen (vgl. Fig. 14, Nebenfigur). Es zeigte sich ferner bei der von COX und MUNK auf Grund von Aufnahmen der Sonnenreflexe durchgeführten Statistik der Böschungswinkelverteilung der Meeresoberfläche (die übrigens ebenfalls den Charakter einer GAUSS-Verteilung aufwiesen), daß das NEUMANN-Spektrum den Beobachtungstatsachen wesentlich besser gerecht wurde als das von DARBYSHIRE vorgeschlagene.

Die unter Verwendung empirischen Ausgangsmaterials, jedoch im wesentlichen auf Grund theoretischer Überlegungen durchgeführte Berechnung des Energiespektrums der voll entwickelten Windsee durch NEUMANN hat also eine Reihe von bemerkenswerten Resultaten erbracht, von denen einige sich leicht verifizieren lassen, während andere noch der Bestätigung durch geeignete Messungen bedürfen. Eine Zusammenstellung der aus dem NEUMANN-Spektrum ableitbaren *Maßzahlen der ausgereiften Windsee* enthält Tabelle 3. Hierbei wurden die „unteren“ und „oberen“ Grenzperioden dadurch bestimmt, daß 3% der totalen Energie vom kurzwelligen Ende und 5% vom langwelligen abgeschnitten gedacht wurden. Diese willkürliche Maßnahme findet eine gewisse Rechtfertigung darin, daß die so ermittelten Grenzperioden durch visuelle Beobachtungen (Fig. 10) bestätigt werden. Eingedenk seiner empirischen Grundlagen können vom NEUMANN-Spektrum auch kaum Aussagen über die sehr kurzen und sehr langen Wellen erwartet werden.

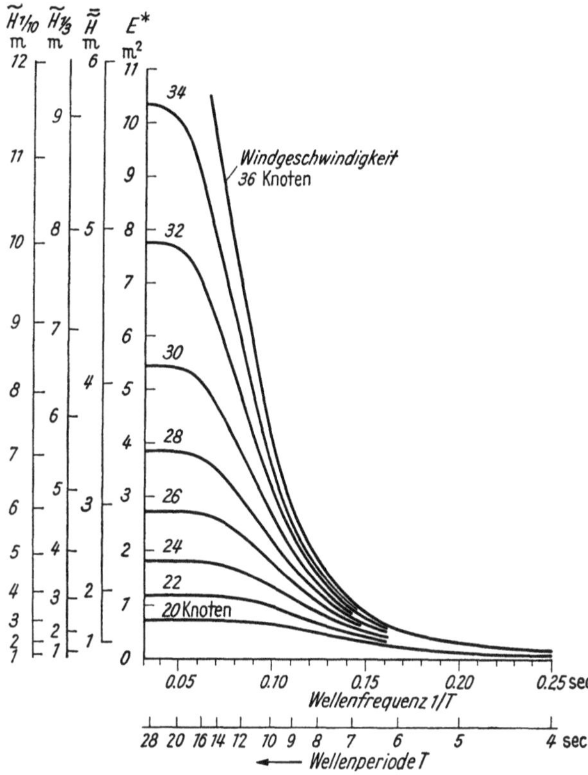

Fig. 15. Energiesummenspektrum und Höhencharakteristik der scheinbaren Wellen in der ausgereiften Windsee für Windgeschwindigkeiten von 20—36 Knoten (nach NEUMANN).

Kritische Bemerkungen zum NEUMANN-*Spektrum* haben kürzlich ROLL und FISCHER[2] veröffentlicht. Danach ist die von NEUMANN gegebene Darstellung insofern unbefriedigend, als das Maximum der Wellenenergie bei verschiedenen Perioden liegt, je nachdem das Spektrum als Funktion der Wellenperiode oder der Wellenfrequenz betrachtet wird. Man kann diese Schwierigkeit vermeiden, wenn man in beiden Darstellungsformen die unabhängige Variable logarithmisch unterteilt. Dies hat eine Änderung der Beziehungen (13.4) und (13.5) zur Folge. Insbesondere ergibt sich die Gesamtenergie des ausgereiften Seeganges nunmehr

[1] C. COX and W. MUNK: J. Mar. Res. **13**, 198 (1954).
[2] H. U. ROLL u. G. FISCHER: Dtsch. Hydrogr. Z. **9**, 9 (1956).

Tabelle 3. *Maßzahlen der ausgereiften*

Wind			Wind-					
			Hohencharakteristik m			Periodencharakteristik sec		
Beaufort	Knoten	mittlere Windgeschwindigkeit m/sec	$\overline{\widetilde{H}}$	$\widetilde{H}_{\frac{1}{3}}$	$\widetilde{H}_{\frac{1}{10}}$	$T_u - T_0$	T_{max}	$\overline{\widetilde{T}}$
1	1— 3	1,2	0,01	0,02	0,02	bis 1	0,47	0,33
2	4— 6	2,6	0,06	0,09	0,11	0,4— 2,8	2,0	1,4
3	7—10	4,4	0,18	0,30	0,36	0,8— 4,9	3,4	2,4
4	11—16	6,9	0,55	0,88	1,12	1,6— 7,6	5,4	3,9
5	17—21	9,8	1,3	2,1	2,7	2,8—10,6	7,7	5,4
6	22—27	12,6	2,5	4,0	5,2	3,8—13,6	9,9	7,0
7	28—33	15,7	4,5	7,0	8,8	4,8—17,0	12,4	8,7
8	34—40	19,0	7,0	11,3	14,2	6,0—20,5	14,9	10,5
9	41—47	22,6	11,0	17,4	22,2	7,0—24,2	17,7	12,5
10	48—55	26,5	15,8	25,2	32	8,0—28,2	20,8	14,7
11	56—63	30,6	22,2	35,0	45	10 —32	24	17

als proportional zur 4. Potenz der Windgeschwindigkeit. Gleichzeitig öffnet sich ein Weg zur Neubestimmung der Konstanten C, auf dem WALDEN[1] zu einem Wert für C gelangt, der den Beobachtungen besser gerecht wird als der von NEUMANN angegebene (vgl. Ziff. 23).

Die *theoretische* Ableitung einer Häufigkeitsverteilung der scheinbaren Wellenperioden — etwa entsprechend den Ergebnissen von LONGUET-HIGGINS für die scheinbaren Wellenhöhen — ist noch nicht gelungen.

14. Statistische Interpretation der beobachtbaren Wellengrößen. Das in Ziff. 13 abgeleitete Energiespektrum der Windsee enthält keine unmittelbar beobachtbaren Wellengrößen. Für seine Prüfung und Anwendung in der Praxis muß es daher mit solchen scheinbaren Größen des Seeganges in Verbindung gebracht werden. Wir folgen dabei Gedankengängen von PIERSON[2], der die von RICE[3] erhaltenen Ergebnisse einer mathematischen Analyse verwandter elektrischer Vorgänge auf die Probleme des Seeganges anwandte.

Die *mittlere scheinbare Wellenperiode* $\overline{\widetilde{T}}$ wird gemeinhin definiert als der mittlere Zeitabstand zwischen aufeinanderfolgenden Wellenkämmen gemessen an einem festen Ort. PIERSON gelangt zunächst im Anschluß an RICE und unter Verwendung des NEUMANN-Spektrums zu der folgenden Beziehung für den mittleren Zeitabstand $\overline{T}^*$ der *zero up-crosses*, d.h. der Nullstellen mit Übergang von negativen zu positiven Amplitudenwerten, innerhalb einer Wellenregistrierung

$$\overline{T}^* = \sqrt{\frac{\int\limits_0^\infty T^4 e^{\frac{-2g^2T^2}{4\pi^2U^2}}\,dT}{\int\limits_0^\infty T^2 e^{\frac{-2g^2T^2}{4\pi^2U^2}}\,dT}} \tag{14.1}$$

und zeigt sodann, daß bei einer ausreichend langen Wellenregistrierung der aus den Nullstellen erhaltene Periodenmittelwert $\overline{T}^*$ mit dem aus den Wellenkämmen ermittelten $\overline{\widetilde{T}}$ genügend genau übereinstimmt,

$$\overline{T}^* = \overline{\widetilde{T}}. \tag{14.2}$$

[1] H. WALDEN: Dtsch. Hydrogr. Z. **9**, 14 (1956).
[2] W. J. PIERSON jr.: Trans. Amer. Geophys. Un. **35**, 747 (1954).
[3] S. O. RICE: Bell Syst. Techn. J. **23**, 282 (1944).

Windsee (nach NEUMANN).

see

	Wellenlängencharakteristik m			Mindest-Windwirklänge km	Mindest-Winddauer Std
Beaufort	$\tilde{L}$	L_{max}	$\tilde{\tilde{L}}$	F_m	t_m
1	bis 0,9	0,34	0,12	0,1	0,07
2	0,16— 8,1	6	2	1,2	0,7
3	0,70—25	18	6	11	2,3
4	2—60	47	16	45	4,8
5	8—116	93	31	120	9,2
6	15—193	153	51	260	15
7	24—300	240	80	540	24
8	37—440	345	115	980	37
9	51—610	490	163	1780	52
10	66—830	675	225	2900	73
11	104—1060	900	301	4600	101

Es sei übrigens vermerkt, daß wegen der in einer Wellenregistrierung häufig auftretenden Nebenmaxima die Gesamtzahl der Maxima im Seegang wesentlich größer ist (PIERSON fand um 73%) als die Anzahl der „zero up-crosses", d.h. der Nullstellen mit Übergang von negativen zu positiven Amplitudenwerten, die ja mit der Anzahl der Wellen oder Wellenkämme gleichzusetzen ist.

Durch Integration von (14.1) finden wir

$$\frac{\tilde{\tilde{T}}}{U} = \frac{\pi\sqrt{3}}{g}, \tag{14.3}$$

$$\tilde{\tilde{T}} = 0{,}555\, U. \tag{14.4}$$

Durch Anwendung von (13.4) erhalten wir hieraus die bemerkenswerte Beziehung

$$T_{max}/\tilde{\tilde{T}} = \sqrt{2} = 1{,}414, \tag{14.5}$$

aus der wir entnehmen können, daß die *mittlere scheinbare Wellenperiode der Windsee wesentlich kleiner ist als die Periode der energiereichsten Wellen.* Dies ist bis zu einem gewissen Grade durch das Beobachtungsverfahren bedingt. Bei einer bestimmten Beobachtungsdauer haben nämlich die Wellen kürzerer Periode häufiger Gelegenheit, aufzutreten und gewertet zu werden, als die längeren. Dieser Effekt muß zu einer Überbetonung der kurzperiodischen Wellen und damit zu einer Erniedrigung des Mittelwertes der scheinbaren Wellenperioden führen.

Da T_{max} als die Periode der energiereichsten Wellen in der Windsee gleich sein muß der scheinbaren Periode der höchsten aus dem Wellenentstehungsgebiet herauslaufenden Dünungswellen, läßt sich (14.5) zugleich als Beziehung zwischen der scheinbaren Periode $\tilde{\tilde{T}}$ der voll angefachten Windsee und der scheinbaren Periode $\tilde{\tilde{T}}_D$ der aus diesem Windfeld herauslaufenden Dünung auffassen:

$$\tilde{\tilde{T}}_D/\tilde{\tilde{T}} \leq \sqrt{2}. \tag{14.6}$$

Es erhebt sich ferner die wichtige Frage, welche *Beziehung zwischen den Mittelwerten der scheinbaren Wellenperiode* $\tilde{\tilde{T}}$ *und der scheinbaren Wellenlänge* $\tilde{\tilde{L}}$ besteht. Bisher war es allgemein üblich, auch auf diese scheinbaren Wellengrößen die klassische Beziehung (3.16) anzuwenden (sofern es sich um Tiefwasserwellen handelte), obwohl diese Gleichung streng genommen nur für konservative Wellen gilt. PIERSON hat dieses Problem ebenfalls behandelt, und zwar auf der Grundlage eines auf zwei Dimensionen erweiterten Energiespektrums der voll angefachten Windsee.

Es ist bekannt, daß die Meereswellen keine unendlich langen Kämme besitzen, sondern mehr oder weniger kurzkämmig sind. Um dieser Beobachtungstatsache Rechnung zu tragen, setzt PIERSON im Anschluß an eine Arbeit von ARTHUR [5] an, daß die (langkämmigen) Windseewellen nicht nur in der Windrichtung ($\Theta = 0$) wandern, sondern beiderseits um $\Theta = \pm \frac{\pi}{2}$ von dieser abweichen können, wobei die Abnahme der Amplituden dieser schräglaufenden Wellenzüge ein cos-Gesetz befolgt. Das charakteristische Gefüge der kurzkämmigen Windsee ergibt sich somit durch Interferenz dieser schräglaufenden langkämmigen Wellenzüge. Das auf diese Weise erweiterte Energiespektrum hängt somit von ω und Θ ab und läßt sich in folgender Form schreiben

$$[a(\omega, \Theta)]^2 = \frac{2C}{\pi \omega^6} \cdot e^{\frac{-2g^2}{\omega^2 U^2}} \cdot \cos^2 \Theta. \tag{14.7}$$

Dieser Ansatz gilt für den Wellenrichtungsbereich $-\frac{\pi}{2} < \Theta < \frac{\pi}{2}$, für andere Θ-Werte soll die Energie Null sein. Die scheinbare Wellenlänge $\tilde{L}_0$ wird jeweils von Kamm zu Kamm in der Windrichtung $\Theta = 0$ gemessen. Auch die Länge der schräglaufenden Wellen wird auf diese Richtung bezogen. Durch Integration über alle Wellenlängen und -richtungen ergibt sich in einem analogen Berechnungsverfahren wie bei der mittleren scheinbaren Wellenperiode die folgende Gleichung für den mittleren Abstand der Kämme in einer kurzkämmigen Windsee gemessen in Windrichtung zu einer bestimmten Zeit

$$\overline{\tilde{L}}_0 = \frac{\pi U^2}{g}. \tag{14.8}$$

Durch Kombination von (14.8) und (14.3) erhalten wir schließlich die gesuchte Beziehung zwischen den Mittelwerten der scheinbaren Wellenlänge und scheinbaren Wellenperiode in folgender Form

$$\overline{\tilde{L}}_0 = \frac{2}{3} \frac{g \overline{\tilde{T}}^2}{2\pi}. \tag{14.9}$$

Unter der Voraussetzung, daß das Energiespektrum von NEUMANN gilt und der Ansatz von ARTHUR über die Veränderlichkeit der Wellenrichtung zutrifft, ergibt sich also, daß die *mittlere scheinbare Wellenlänge der kurzkämmigen Windsee nur $\frac{2}{3}$ des Wertes erreicht, der sich aus der mittleren scheinbaren Wellenperiode mit Hilfe der „klassischen" Formel* (3.16) *berechnen läßt.* Dieses theoretische Ergebnis ist von großer Bedeutung, zählte doch die viel benutzte „klassische" Formel (3.16) bisher zu den am besten durch Beobachtungen gestützten Beziehungen der Wellentheorie. Eine exakte Bestätigung der neuen Formel steht indessen noch aus, wenn auch neuere Beobachtungen darauf hindeuten, daß die nach der „klassischen" Gleichung aus der scheinbaren Wellenperiode berechneten scheinbaren Wellenlängen im allgemeinen zu groß sind.

15. Seegangsdarstellung durch Energiespektrum und Zufallsphase. Eine rationelle Theorie der Meereswellen wird sich vor allem darum bemühen müssen, ein zutreffendes und zugleich handliches mathematisches Modell für den wirklichen Seegang zu finden, dessen Hauptmerkmal — wie wir gesehen haben — die Unregelmäßigkeit seiner Periodizitäten ist. Erst auf Grund eines solchen mathematischen Modells ist es möglich, die Aussagen von Seegangsregistrierungen in allgemeiner Form zu fassen und für die Vorausberechnung der Meereswellen zu benutzen.

Zunächst ist klar, daß die Darstellung des Seeganges durch *eine harmonische Welle*, bei der das ganze Energiespektrum auf eine einzige spektrale Frequenz zusammenschrumpft, wegen der Variabilität der natürlichen Seegangsgrößen unbefriedigend bleiben muß. Dies gilt auch für den Fall, daß hierbei — wie es in den Arbeiten von SVERDRUP und MUNK [*4*] u. a. geschehen ist — die in Ziff. 10 erwähnten „charakteristischen" (significant) Werte der scheinbaren Wellenhöhe und Wellenperiode zur Seegangsdarstellung benutzt werden.

Man könnte sodann versuchen, den Seegang durch eine FOURIER-*Reihe*, also durch ein periodisches Wellensystem darzustellen, dessen Energiespektrum aus einer Anzahl diskreter Frequenzen besteht. Ganz abgesehen davon, daß dieses Verfahren wegen der hohen Ordnungszahl der bedeutungsvollen FOURIER-Komponenten (bei einer Beobachtungsdauer von 20 min kommt es auf die Terme von 50 bis etwa 400 an) sehr mühsam ist, würde es immer nur eine wahre Abbildung des analysierten Ausschnittes einer Wellenregistrierung liefern. Über die Wellen davor und danach würde dieses Modell — abgesehen von der trivialen Aussage durch beliebige Wiederholung des analysierten Ausschnittes im Falle rationaler Frequenzverhältnisse — nichts erkennen lassen (DENIS and PIERSON[1]).

Eine Darstellung des natürlichen Seeganges mit Hilfe des FOURIER-*Integraltheorems* böte zwar den Vorteil, daß ein kontinuierliches Amplitudenspektrum, d.h. die Wellenamplitude als kontinuierliche Funktion der Wellenfrequenz erscheinen würde. Aber auch hier beschränkt sich die so gefundene mathematische Form auf die zwar exakte aber ausschließliche Wiedergabe des analysierten Ausschnittes, den eine Registrierung liefert, und gestattet keine Schlüsse auf die Gestalt des Seeganges außerhalb dieser Zeitspanne.

Im Anschluß an Verfahren der theoretischen Statistik hat nun PIERSON[2] [*9*] eine *vierte mathematische Darstellungsform des natürlichen Seeganges* vorgeschlagen, die hinsichtlich Wirklichkeitsnähe und Handlichkeit vor den drei oben erwähnten Methoden gewisse Vorzüge besitzt. Er stellt die *Bewegung der Meeresoberfläche an einem bestimmten Punkt* — der ohne Beschränkung der Allgemeinheit zum Nullpunkt gemacht werden kann — durch das folgende „Integral" als reine Zeitfunktion dar:

$$\eta(t) = \int_0^\infty \cos[\omega t + \delta(\omega)] \sqrt{[a(\omega)]^2 d\omega}. \tag{15.1}$$

Dieses auf den ersten Blick befremdende „Integral" ist kein Integral im RIEMANNschen Sinn. Es kann nicht nach den üblichen Verfahren berechnet werden. Es soll vielmehr nur eine mathematische Abstraktion darstellen, die durch eine Partialsumme beliebig genau angenähert werden kann.

In (15.1) bedeutet $[a(\omega)]^2$ das Energiespektrum des Seeganges in der durch (13.1) definierten Form. Die Phase $\delta(\omega)$ trägt den Charakter einer Zufallsphase. Alle Werte aus dem Bereich 0 bis 2π sind gleich wahrscheinlich und voneinander unabhängig. $\delta(\omega)$ ist dabei für jedes ω in Übereinstimmung mit

$$W[0 < \delta(\omega) < 2\pi\alpha] = \alpha \tag{15.2}$$

zu wählen, das heißt die Wahrscheinlichkeit, daß $\delta(\omega)$ in den Bereich 0 bis $2\pi\alpha$ fällt, ist gleich α.

Zur Bestimmung des „Integrals" (15.1) unterteilen wir die ω-Achse durch eine Reihe willkürlicher Punkte $0, \omega_1, \omega_2, \ldots, \omega_n, \ldots \omega_{2q}$. Wir berechnen zunächst $[a(\omega)]^2$ für alle ungeraden Punkte $2n+1$, wählen beliebige Phasen in Übereinstimmung mit (15.2) und erhalten das gesuchte „Integral" als Grenzwert der

[1] M. ST. DENIS and W. J. PIERSON jr.: On the motion of ships in confused seas, Trans. Soc. Naval Arch. and Mar. Eng., New York, **61**, 280 (1953).

[2] S. 698, Fußnote 2.

folgenden Partialsumme

$$\eta(t) = \lim_{\substack{\omega_{2n+2}-\omega_{2n}\to 0\\ \omega_{2q}\to\infty}} \sum_{n=0}^{q} \cos\left[\omega_{2n+1} t + \delta(\omega_{2n+1})\right] \sqrt{[a(\omega_{2n+1})]^2 (\omega_{2n+2} - \omega_{2n})}. \tag{15.3}$$

In der Praxis wird es nicht nötig sein, die Unterteilung „unendlich fein" zu machen und den Frequenzausschnitt „unendlich groß" zu wählen. Vielmehr wird die gesuchte „Integral"-Darstellung des Seeganges (15.1) bereits bei genügend dichter, aber endlicher Unterteilung und bei einem genügend großen, aber endlichen Frequenzbereich nicht mehr wesentlich von der Näherungssumme (15.3) abweichen.

Diese neue Darstellungsform des natürlichen Seeganges zeichnet sich dadurch aus, daß zu einem einzigen Energiespektrum $[a(\omega)]^2$ nicht mehr nur eine einzige Wellenform, sondern wegen des Zufallscharakters der Phasen eine ganze *Klasse von Funktionen* $\eta(t)$ gehört, die alle als mögliche Bewegungsformen der Meeresoberfläche aufgefaßt werden müssen. Diese Tatsache bewirkt zunächst, daß willkürlich herausgegriffene Punkte jeder dieser Seegangsdarstellungen eine GAUSS*sche Normalverteilung* bilden. Dieses Ergebnis konnte an Hand von Wellenregistrierungen verifiziert werden (RUDNICK[1], PIERSON and MARKS[2]). Abweichungen von der Normalverteilung sind indessen bei steilen Wellen d.h. im nicht-linearen Fall zu erwarten. Es kann ferner festgestellt werden, daß eine durch (15.1) gegebene Darstellung einer Wellenregistrierung sich *bei Verlängerung nicht wiederholt*. Eine solche mathematische Seegangsdarstellung gilt also beliebig lange. Ihre statistischen Eigenschaften ändern sich nicht, solange die äußeren Wirkungen, wie z.B. die Windgeschwindigkeit, gleich bleiben. Es ist somit auf diese Weise ein *Modell des Seeganges* gefunden worden, das sich auf die mathematische Wiedergabe dessen beschränkt, was *allen Wellenregistrierungen* bei gleichen äußeren Bedingungen *gemeinsam* ist. Durch Anwendung dieser Analyse auf eine spezielle Seegangsregistrierung sind wir daher in der Lage, die unveränderlichen statistischen Eigenschaften dieses Bruchstückes des gesamten Seegangsablaufes festzustellen und damit die Natur aller möglichen Seegangsformen anzugeben. Die Bedeutung einer einzelnen Wellenregistrierung wird hierdurch erheblich erweitert. Voraussetzung für ein solches Vorgehen ist allerdings eine einwandfreie Ableitung des Energiespektrums $[a(\omega)]^2$ entweder auf Grund von Seegangsregistrierungen oder auf theoretischem Wege. Für die Analyse der Wellenregistrierung haben PIERSON und MARKS[2] entsprechende Verfahren angegeben. Bei Anwendung des von NEUMANN berechneten theoretischen Energiespektrums (13.3) bedarf es nur der Angabe der Windgeschwindigkeit, allerdings steht dessen vollständige experimentelle Überprüfung noch aus.

Die obigen Entwicklungen galten der Darstellung der Seegangsbewegung an einem bestimmten Punkt, sowie wir sie mit den üblichen Seegangsmessern feststellen. Wenn wir uns nun der *Erweiterung dieses Modells auf den Seegang eines gewissen Teilgebietes der Meeresoberfläche* zuwenden, müssen wir zwischen Wellen mit langen Kämmen und solchen mit kurzen Kämmen unterscheiden. Wir wollen hier nur PIERSONS Ergebnis für *kurzkämmige Wellen* skizzieren, da diese Form weitaus häufiger vorkommt als die Wellen mit langen Kämmen. Wie bekannt wird dieses charakteristische Seegangsgefüge durch die Interferenz langkämmiger Wellenzüge erzeugt, die beiderseits bis zu $\Theta = \pm\frac{\pi}{2}$ von der Windrichtung $\Theta = 0$ abweichen. Für diesen recht allgemeinen Fall erhalten wir an Stelle von (15.1)

[1] S. 689, Fußnote 1.
[2] S. 689, Fußnote 2.

die folgende Darstellung des kurzkämmigen Seeganges

$$\eta(x, y, t) = \int_0^\infty \int_{-\frac{\pi}{2}}^{+\frac{\pi}{2}} \cos\left[\frac{\omega^2}{g}(x\cos\Theta + y\sin\Theta) - \omega t + \delta(\omega,\Theta)\right] \sqrt{[a(\omega,\Theta)]^2\, d\omega\, d\Theta}. \quad (15.4)$$

Dieses „Integral" ist für die praktische Auswertung wie (15.1) durch eine entsprechende Partialsumme anzunähern. Wir bemerken, daß nunmehr sowohl das Energiespektrum $[a(\omega, \Theta)]^2$ als auch die Zufallsphase $\delta(\omega, \Theta)$ von zwei Variablen, der Wellenfrequenz ω und der Wellenrichtung Θ, abhängen. Wiederum kommt es darauf an, das (nunmehr zweidimensionale) Energiespektrum des Seeganges zu bestimmen. Ein Verfahren, diese Aufgabe mit Hilfe von stereophotogrammetrischen Seegangsaufnahmen vom Flugzeug aus zu lösen, hat MARKS[1] angegeben. Ergebnisse solcher Messungen sind indessen noch nicht bekannt geworden. Zur Zeit steht daher für die Berechnung des zweidimensionalen Energiespektrums des kurzkämmigen Seeganges nur das — vorwiegend auf theoretischem Wege erstellte — erweiterte NEUMANN-Spektrum (14.7) zur Verfügung, dessen Gültigkeit noch nachzuweisen bleibt.

Somit sind wir zu einer mathematischen Darstellung des natürlichen Seeganges gelangt, die auf der Vorstellung basiert, daß die Gestalt der Meeresoberfläche beschrieben werden kann durch die *Interferenz einer „unendlichen" Anzahl von Sinus-Wellen infinitesimaler Höhe, die in verschiedenen Richtungen wandern, verschiedene Frequenzen besitzen und sich in zufälligen Phasen überlagern.* Zur Zeit scheint dies die allgemeinste und zugleich wirklichkeitsnächste Darstellung des Seeganges zu sein. Sie trägt der grundsätzlichen Unregelmäßigkeit der Meereswellen Rechnung, indem sie sich auf Aussagen über die unveränderlichen statistischen Eigenschaften des Seeganges beschränkt, und bemüht sich andererseits, eine Brücke zu schlagen zu den wirklich beobachtbaren Bestimmungsstücken des Seeganges, den scheinbaren Wellenhöhen, Wellenperioden und Wellenlängen sowie ihren Mittelwerten.

Unter Verwendung dieser Seegangsdarstellung sind Voraussagen von Mittelwerten und statistischen Verteilungen der Wellengrößen möglich. Die wesentlich weitergehende Frage nach der Höhe der Meeresoberfläche an einem bestimmten Punkt zu einer bestimmten späteren Zeit läßt sich hiermit jedoch nicht beantworten, auch dann nicht, wenn man die Vorausberechnung nur für eine kurze Zeit verlangt. Die Lösung dieses Problems erfordert vielmehr die Kenntnis der „Vorgeschichte" der Meeresoberfläche an diesem Punkt sowie an einigen Punkten seiner Umgebung. Mit diesen Fragen hat sich COTE[2] beschäftigt.

II. Entstehung und Wachstum.

a) Entstehung von Wasseroberflächenwellen durch Wind.

Es besteht kein Zweifel darüber, daß der Wind den Seegang erzeugt, nur das „Wie" ist problematisch. Wenn ein Windhauch über eine bislang glatte Wasseroberfläche streicht, entstehen schlagartig kleine Wellen Sie werden Rippel- oder Kräuselwellen genannt, oft auch bezeichnet man sie als Initialwellen, da sich aus diesem Anfangsstadium die großen Wellen des Seeganges entwickeln.

16. Einfluß der Oberflächenspannung. Bei diesen Initialwellen von wenigen Zentimetern Länge sind nur geringe Massen beteiligt, dafür aber treten starke Krümmungen der Wasseroberfläche auf. Somit ist plausibel, daß beim Übergang

[1] W. MARKS: Trans. Amer. Geophys. Un. **35**, 758 (1954).
[2] L. J. COTE: U.S.Publ. "Ship and Waves", Chap. 6, p. 73. 1955.

von den großen zu den Initialwellen der Einfluß der Schwere abnimmt, die Bedeutung der Oberflächenspannung jedoch wächst. Berücksichtigen wir diese Tatsache durch die Einführung der Kapillaritätskonstanten γ [dyn cm^{-1}], so müssen die Gln. (3.15) und (3.16) für die Wellengeschwindigkeit C folgende Ergänzung erfahren:

$$C = \sqrt{\left(\frac{g L}{2\pi} + \frac{2\pi\gamma}{\varrho_w L}\right) \operatorname{Tan} \frac{2\pi h}{L}} \quad \text{für Flachwasserwellen,} \tag{16.1}$$

$$C = \sqrt{\frac{g L}{2\pi} + \frac{2\pi\gamma}{\varrho_w L}} \quad \text{für Tiefwasserwellen.} \tag{16.2}$$

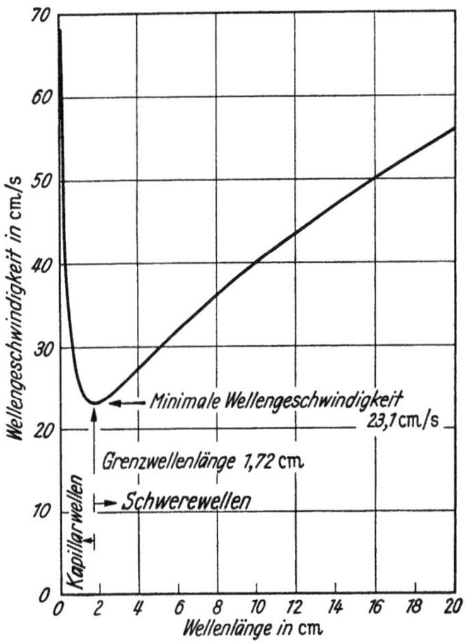

Fig. 16. Zusammenhang zwischen Wellenlänge und Wellengeschwindigkeit unter der Wirkung von Schwere und Oberflächenspannung.

Hierin bedeutet ϱ_w die Dichte des Seewassers, gegenüber der die Luftdichte vernachlässigt wurde. Der Zusammenhang (16.2) ist in Fig. 16 dargestellt. Mit abnehmender Wellenlänge wird die Wellengeschwindigkeit ebenfalls geringer, jedoch endet diese Abnahme bei der *Grenzwellenlänge* 1,72 cm. Hier halten Schwerkraft und Oberflächenspannung sich das Gleichgewicht und die Wellengeschwindigkeit erreicht ihren kleinsten Wert von 23,15 cm/sec. Bei weiter abnehmender Wellenlänge treten wir in den Bereich der reinen *Kapillarwellen* ein, der durch eine rasche Zunahme der Wellengeschwindigkeit gekennzeichnet ist. Diese theoretischen Ergebnisse sind durch Messungen bestätigt worden (Matthiesen[1] [2], Chinn[2], Kaplan[3]). Sie sind für die Wellenentstehung von großer Bedeutung.

17. Instabilität der Wasseroberfläche. Beim Studium der Wellenentstehung ist besonders die Frage der Existenz und Größe einer *Mindest-Windgeschwindigkeit zur Wellenanregung* interessant. Bereits Lord Kelvin[4] und v. Helmholtz[5] untersuchten dieses Problem unter der Voraussetzung reibungsfreier Potentialströmung bei Berücksichtigung der Oberflächenspannung. Sie fanden, daß Wellen allein infolge *dynamischer Druckänderungen* bestehen können (Unterdruck und Sogwirkung über den Wellenbergen, Überdruck über den Wellentälern). Der ebene Wasserspiegel erweist sich danach als instabil, kleine zufällige Störungen verstärken sich zu Wellen, sobald die Geschwindigkeit des darüberstreichenden Windes 671 cm/sec überschreitet. Beobachtungen führten jedoch auf wesentlich geringere Werte für die Mindest-Windgeschwindigkeit [2].

Neuerdings hat Munk[6] diese hohe „kritische Windgeschwindigkeit" von Lord Kelvin und v. Helmholtz zur Erklärung von Erscheinungen in der Grenzschicht Wasser—Luft (Impulsaustausch, Verdunstung) herangezogen, die für eine plötzliche Änderung des hydro-

[1] L. Matthiesen: Ann. Phys. Chem. **38**, 118 (1889).

[2] A. J. Chinn: Inst. Eng. Res., Univ. of Calif., Berkeley, Calif., Techn. Rep. HE 116—302, 1949 (unveröffentlicht).

[3] K. Kaplan: Inst. Eng. Res., Univ. of Calif., Berkeley, Calif., Techn. Rep. HE 116—316, 1950 (beide zitiert nach R. L. Wiegel u. J. W. Johnson [7]).

[4] Lord Kelvin (W. Thomson): Phil. Mag. (4) **42**, 368 (1871).

[5] Helmholtz, v.: Sitzgsber. Akad. Wiss. Berlin **1889**, 761; **1890**, 853.

[6] W. H. Munk: J. Mar. Res. **6**, 203 (1947).

dynamischen Zustandes der Meeresoberfläche (glatt—rauh) beim Überschreiten dieses Wertes sprechen. Laboratoriumsversuche an Flüssigkeitsoberflächen verschiedener Viskosität führten FRANCIS[1] ebenfalls dazu, die KELVIN-HELMHOLTZ-Instabilität in einem anderen Lichte zu sehen. Danach scheint diese kritische Windgeschwindigkeit reell zu sein. Unter natürlichen Bedingungen wird diese Instabilität allerdings durch andere Effekte verdeckt.

Ferner hat WÜST[2] den Ansatz von Lord KELVIN und v. HELMHOLTZ durch Berücksichtigung der *Flüssigkeitsreibung und der Grenzschichten in Luft und Wasser* verbessert. Er erhielt bei Reibung — jedoch Vernachlässigung der Grenzschichten — etwa 160 cm/sec und bei zusätzlicher Einführung laminarer Grenzschichten an der Trennfläche beider Medien etwa 70 cm/sec für die Mindestwindgeschwindigkeit. Diese Ergebnisse entsprechen den Beobachtungen wesentlich besser als der frühere hohe Wert.

18. Einseitiger Winddruck. Von einer anderen Seite faßte JEFFREYS[3] dieses Problem an. Er führte die Wellenanfachung auf die durch *Leewirbelbildung* verursachten Verschiedenheiten des Winddruckes auf Lee- und Luvhänge der Wellenberge zurück. Er setzte diesen Winddruck gleich

$$s\,\varrho_L (U-C)^2 \frac{\partial\eta}{\partial x}, \tag{18.1}$$

worin s den Widerstandsbeiwert der Wellen, ϱ_L die Luftdichte, $(U-C)$ die relative Windgeschwindigkeit Wind-Wellen und $\partial\eta/\partial x$ die Neigung der Wasseroberfläche in der Windrichtung bedeutet. Durch Energiebetrachtungen gelangte JEFFREYS zu einer Ungleichung für die Mindest-Windgeschwindigkeit. Eine zahlenmäßige Auswertung war jedoch nicht möglich, da der Widerstandsbeiwert s der Wellen unbekannt war.

MOTZFELD[4] bestimmte sodann solche *Widerstandsbeiwerte* durch Messungen der Druck- und Geschwindigkeitsverteilung *an feststehenden Wellenprofilen im Windkanal.* Bei seinen anschließenden Rechnungen erhielt er jedoch keine Mindest-Windgeschwindigkeit zur Wellenanregung, vielmehr erwies sich die Wasseroberfläche als instabil für jede, auch die kleinste Windgeschwindigkeit, ein Ergebnis, das in erster Linie auf die Vernachlässigung der Oberflächenspannung durch MOTZFELD zurückzuführen ist.

NEUMANN[5] bemühte sich zunächst um eine zahlenmäßige Erfassung der *Widerstandsbeiwerte für wirkliche Wellen.* Aus Messungen der Schiefstellung des Meeresspiegels unter der Wirkung des Windes erhielt er den „effektiven Widerstand" der Meeresoberfläche, den er aus dem Druckwiderstand der Wellen und dem reinen Reibungswiderstand der Wasseroberfläche zusammensetzte und insgesamt wie einen Druckwiderstand behandelte. Die als Funktion der Windgeschwindigkeit erhaltenen Widerstandszahlen wurden mit entsprechenden empirischen Wellendaten verglichen und ergaben — abweichend von JEFFREYS und MOTZFELD — eine einfache *Proportionalität zwischen Widerstandsbeiwert und Wellensteilheit.* Ausgehend von der Vorstellung, daß die vom Winde der Welle zugeführte Energie mindestens gleich der durch Reibung im Wasser vernichteten Energie sein muß, wenn die Welle angefacht oder erhalten werden soll, gelangte NEUMANN schließlich zu einer Ungleichung, die — bei tiefem Wasser — als untere Grenze der zum Anfachen der Initialwellen nötigen Windgeschwindigkeit $U_m =$ 69,6 cm/sec liefert, ein Wert, der mit dem auf anderem Wege gefundenen Ergebnis von WÜST gut übereinstimmt. Für die Wellenlänge L_0 dieser Initialwellen gibt

[1] J. R. D. FRANCIS: Proc. V. Conf. on Coastal Engineering, U.S.Council on Wave Research, The Eng. Foundation, 12, 1955.
[2] W. WÜST: Z. angew. Math. Mech. **29**, 239 (1949).
[3] H. JEFFREYS: Proc. Roy. Soc. Lond., Ser. A **107**, 189 (1925); **110**, 241 (1926).
[4] H. MOTZFELD: Z. angew. Math. Mech. **17**, 193 (1937).
[5] G. NEUMANN: Z. Met. **2**, 193 (1948). — Dtsch. Hydrogr. Z. **2**, 187 (1949).

Neumann folgende auf (16.2) basierende Beziehung an:

$$L_0 = \frac{\pi}{9g} U_m^2 \pm \sqrt{\left(\frac{\pi}{9g} U_m^2\right)^2 - \frac{4\pi^2 \gamma}{g \varrho_w}}. \qquad (18.2)$$

Setzen wir $U_m = 69{,}5$ cm/sec und $\gamma/\varrho_w = 73$ cm^3 sec^{-2}, so verschwindet die Wurzel und wir erhalten $L_0 = 1{,}72$ cm. Es handelt sich also bei den Initialwellen um die in Ziff. 16 erwähnten Grenzwellen, die durch die minimale Geschwindigkeit $C_0 = 23{,}15$ cm/sec gekennzeichnet sind.

Dieses Ergebnis hängt damit zusammen, daß die den Wellen durch den Winddruck zugeführte Energie ein Maximum erreicht, wenn die Windgeschwindigkeit gleich der dreifachen Wellengeschwindigkeit ist. Für das Anfachen der langsamsten Wellen mit der Geschwindigkeit von 23,15 cm/sec ist also gerade der Einfluß eines mit $3 \times 23{,}15 = 69{,}5$ cm/sec Geschwindigkeit wehenden Windes am günstigsten.

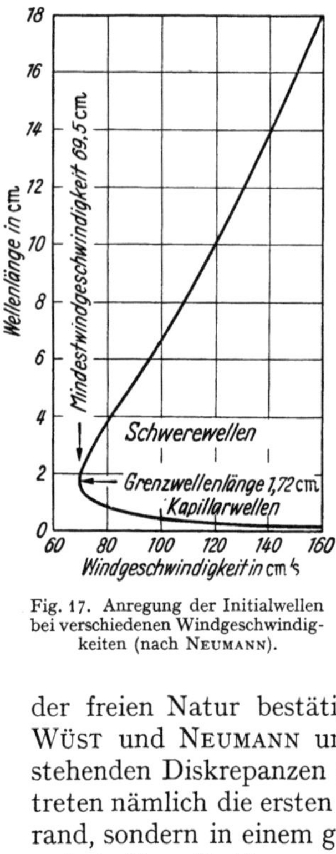

Fig. 17. Anregung der Initialwellen bei verschiedenen Windgeschwindigkeiten (nach Neumann).

Bei jedem Wind, der stärker ist als 69,5 cm/sec, tritt nach Neumann durch das Zusammenwirken von Schwere und Kapillarität eine *Aufspaltung der Initialwellen* in zwei Systeme ein, wobei das eine aus *Schwerewellen* besteht, die also länger sind als 1,72 cm, und das andere aus *Kapillarwellen* mit Längen unterhalb dieses Grenzwertes (Fig. 17).

19. Beobachtungen und Messungen. Man hat sowohl in der freien Natur als auch in den Wind-Wellen-Kanälen der Laboratorien versucht, die Mindest-Windgeschwindigkeit und die Größe der Initialwellen zu bestimmen. Beide Verfahren führten auf verschiedene Resultate. Lagen doch die von Scott-Russell[1] und Jeffreys[2] *auf freien Wasserflächen* gewonnenen Minimalwerte zwischen 85 und 104 cm/sec, während Stanton[3], Weinblum[4] und Wüst[5] sowie Schooley[6] *aus Windkanalversuchen* 200—300 cm/sec als Mindestgeschwindigkeit ableiteten. Neue Beobachtungen und Messungen von Roll[7] über die Wellenentstehung in der freien Natur bestätigen im wesentlichen die theoretischen Ergebnisse von Wüst und Neumann und erscheinen insbesondere geeignet, zwischen den bestehenden Diskrepanzen zu vermitteln. In einem abgeschlossenen Wasserbecken treten nämlich die ersten deutlich erkennbaren Wellen nicht unmittelbar am Luvrand, sondern in einem gewissen *leewärtigen Abstand* von diesem hervor (Fig. 18), wobei diese Entfernung mit abnehmender Windgeschwindigkeit rasch wächst und bei 100 cm/sec Windgeschwindigkeit etwa 40 m betragen dürfte. Wenn also überhaupt der Mechanismus der Wellenanregung in Natur und Laboratorium als gleich angesehen werden kann, dann könnte die Tatsache, daß im Windkanal meist kürzere Wellenrinnen benutzt wurden, als Erklärung dafür gelten, daß das Auftreten der ersten Wellen dort durchweg höhere Windgeschwindigkeiten erforderte als in der freien Natur.

[1] J. Scott-Russell: Rep. 14. Meeting Brit. Assoc. London 311, 1845.
[2] S. 705, Fußnote 3.
[3] Th. Stanton, D. Marshall u. R. Houghton: Proc. Roy. Soc. Lond., Ser. A **137**, 283 (1932).
[4] G. Weinblum: Naturwiss. **26**, 193 (1938).
[5] S. 705, Fußnote 2.
[6] A. H. Schooley: Trans. Amer. Geophys. Un. **36**, 273 (1955).
[7] H. U. Roll: Ann. Met. **4**, 269 (1951).

ROLL versuchte die Existenz dieses *glatten Gebietes* zwischen Luvrand und Wellenbeginn mit dem Zustand der Grenzschichten in Verbindung zu bringen, die sich in Luft und Wasser vom Luvrand aus leewärts ausbilden werden und wies ferner auf die merkwürdige Beobachtungstatsache hin, daß dieses glatte Gebiet am Luvrand nur scheinbar wellenfrei ist, vielmehr dort Wellen zu erkennen sind, die schräg und zum Teil beinahe senkrecht zur Windrichtung laufen (nach beiden Seiten abgelenkt), wobei der Ablenkungswinkel bei großen Windgeschwindigkeiten größer zu sein scheint als bei geringen. Eine Deutung dieser *schräglaufenden Wellenzüge* wurde auf Grund der Vorstellung versucht, daß dort

Fig. 18. Wellenentstehung auf Teichen. Ausbildung eines glatten Streifens (G) zwischen dem Luvrand (L) und dem Wellenbeginn (B); W = Windrichtung.

turbulente Luftquanten streifend die Wasseroberfläche berühren, sie deformieren (etwa wie eine wandernde Druckstörung, vgl. LAMB [*1*]) und dadurch diese Wellen erzeugen, die sich in Anlehnung an die Theorie der Schiffswellen als *Bugwellen wandernder Druckstörungen* auffassen lassen und die letzten Endes als die Ursache für die (durch kurzkämmige und schräglaufende Wellen gekennzeichnete) Struktur der Windsee angesehen werden müssen.

20. Böentheorie der Wellenentstehung. Ähnliche Beobachtungen erwähnte ECKART[1]. Er wies ferner auf eine unbefriedigende Tatsache hin, die allen vorstehend skizzierten Theorien der Wellenentstehung gemeinsam ist: Es wird stets das *Vorhandensein* kleiner Störungen der Meeresoberfläche *vorausgesetzt* und lediglich untersucht, welche Verhältnisse vorliegen müssen, damit sich solche infinitesimalen Störungen *verstärken*. Das Grundproblem, nämlich die Aufdeckung der *Ursache solcher Störungen*, wurde aber bisher nicht behandelt. ECKART betrachtete daher zunächst die Wirkung einer einzelnen *Böe* auf die Wasseroberfläche, wobei unter einer solchen ein Zentrum hohen oder tiefen Druckes verstanden wird, das sich mit der mittleren Windgeschwindigkeit fortbewegt. Der Winddruck soll stets senkrecht zur (gestörten) Wasseroberfläche wirken. Es zeigt

[1] C. ECKART: J. Appl. Phys. **24**, 1485 (1953).

F m	$\overline{H}$ cm	$\overline{L}$ cm	$\overline{H/L}$
5 – 6	0,9	7,8	1 : 9
9 – 10	1,5	15,0	1 : 10
15 – 16	2,0	21,2	1 : 11
35 – 36	2,8	35,5	1 : 13

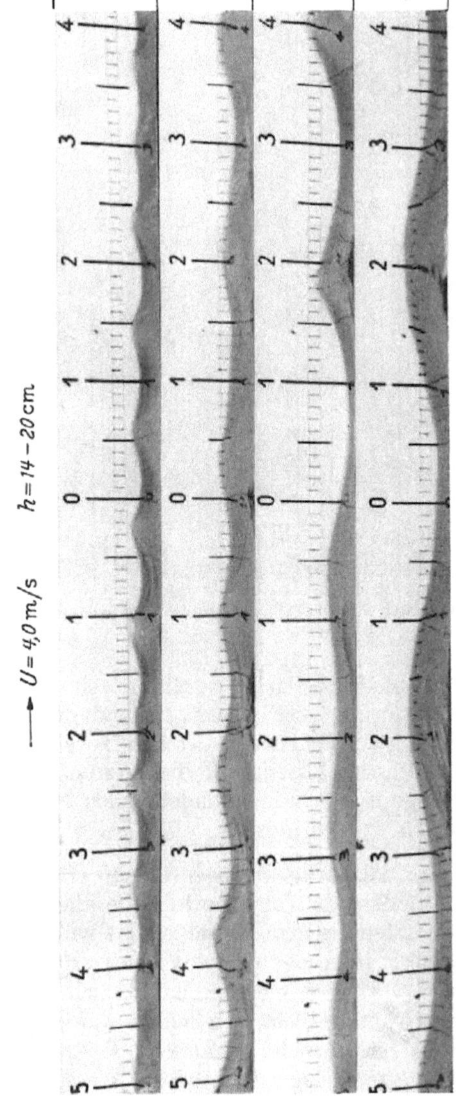

Fig. 19. Einfluß der Windwirklänge F auf das Anwachsen der Windsee (nach Wellenaufnahmen).

sich, daß eine solche Böe ein *V-ähnliches „Kielwasser"* erzeugt, das nach Abklingen der Böe als eine Gruppe freier Schwerewellen weiterwandert. Die statistische Behandlung einer Vielzahl solcher Böen führte allerdings zu keinem befriedigenden Ergebnis hinsichtlich der Wellenhöhe, da die zur Erzeugung der beobachteten Wellen nötigen Luftdruckunterschiede etwa 10 mal so groß sein müßten, wie sie wirklich sind.

Fassen wir zusammen, so läßt sich zur Zeit sagen, daß die verschiedenen Theorien der Wellenentstehung offenbar deshalb nicht zum Ziele geführt haben, weil mehrere Faktoren gleichzeitig wirksam sind, bei den theoretischen Ansätzen aber immer nur der eine oder der andere Faktor Berücksichtigung fand. Man wird also erst dann eine befriedigende Theorie erwarten können, wenn es gelingt alle bedeutungsvollen Komponenten wie Instabilität der Meeresoberfläche, normaler Winddruck, tangentiale Windschubspannung und die Wirkung vertikaler Windböen in geeigneter Weise zu kombinieren.

b) Anfachung und Ausbreitung der Meereswellen unter der Wirkung des Windes (Windsee).

21. Windwirklänge und Winddauer. Die Existenz hoher Windwellen läßt sich nicht durch kurzzeitige örtliche Einwirkung des Windes erklären. Man muß vielmehr zu dem Schluß gelangen, daß die damit verbundene hohe Energiedichte im wesentlichen durch den Energietransport „von weiter her" zustande kommt. Wir müssen also erwarten, daß die Dimensionen der Windsee über tiefem Wasser außer von der Windgeschwindigkeit von einer weiteren Größe abhängen werden, die die Länge dieses durch einheitliche Windeinwirkung ausgezeichneten Transportweges angibt. In der angelsächsischen Literatur hat sich hierfür der Ausdruck *fetch* eingebürgert, im Deutschen findet man *wirksame Windbahn, Streichlänge des Windes* oder auch *Windwirklänge*. Die Bedeutung dieses Begriffs für das Anwachsen der Windsee läßt sich unmittelbar aus der Fig. 19 ersehen, die vier nahezu gleichzeitig bei gleicher Windgeschwindigkeit, aber verschiedenen Windwirklängen in der freien Natur gewonnene Wellenaufnahmen zusammenfaßt.

Die Erfahrung lehrt ferner, daß die Windsee an einem bestimmten Ort außer von der Windgeschwindigkeit und der Windwirklänge noch von einer dritten Größe, der *Winddauer*, abhängt, wobei zwischen Windwirklänge und Winddauer gewisse Beziehungen bestehen.

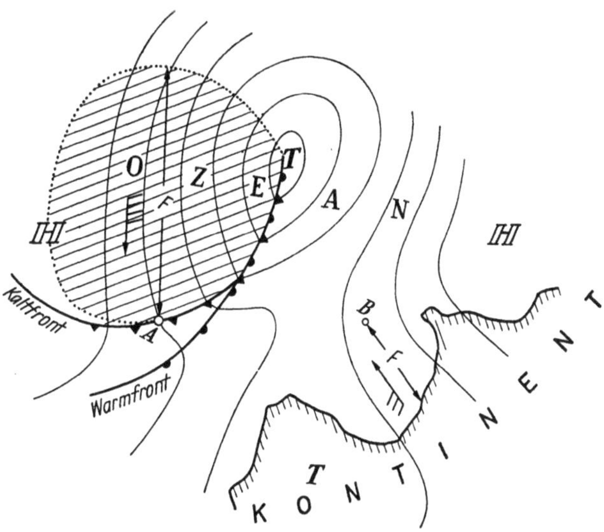

Fig. 20. Bestimmung der Windwirklänge aus der Wetterkarte. Beobachtungsort *A*: Windwirklänge *F* ist gegeben durch Länge des Windfeldes einheitlicher Richtung und Geschwindigkeit. Beobachtungsort *B*: Windwirklänge *F* ist begrenzt durch Küstenlinie.

Die Zusammenhänge werden am deutlichsten sichtbar, wenn wir uns zwei Idealfälle vorstellen:

Wir nehmen zunächst an, daß die *Winddauer unbegrenzt* sei. Die Windsee soll also nur von der Windgeschwindigkeit und der Windwirklänge abhängen. Dieser Fall ist bei kurzen Windwirklängen, wie sie z.B. der Fig. 19 zugrunde liegen, mit guter Annäherung verwirklicht. An jedem dieser vier durch eine bestimmte Windwirklänge gekennzeichneten Punkte bleiben die Wellengrößen konstant. Die Windsee ist dort stationär. Wandern wir jedoch in Richtung zunehmender Windwirklängen, so stellen wir ein Anwachsen der Wellendimensionen fest. Dieses setzt sich so lange fort, bis bei einer gewissen Windwirklänge der zu der betreffenden Windgeschwindigkeit gehörende stationäre Endzustand der „ausgereiften Windsee" erreicht wird, bei dem die vom Winde der Welle gelieferte Energie und die Energiedissipation in der Welle sich das Gleichgewicht halten. Dieser Idealfall ist also gekennzeichnet durch *Zeitunabhängigkeit an allen Orten bei gleichzeitigen örtlichen Unterschieden*. Wir können solche Verhältnisse am ehesten in kleinen Seegebieten bei langer Winddauer erwarten.

Beim zweiten idealisierten Beispiel setzen wir das von einem einheitlichen Winde überwehte *Seegebiet* als *unbegrenzt* voraus. Die Windsee wird bestimmt durch Windgeschwindigkeit und Winddauer. Nach dem plötzlichen Einsetzen eines bestimmten Windes beobachten wir ein gleichmäßiges Anwachsen der Windsee an allen Orten. Dies ist verständlich, da — wegen des als unendlich vorausgesetzten Ozeans — die Windwirklängen überall gleich sind und gleich bleiben. Schließlich wird nach einer bestimmten Winddauer gleichfalls der stationäre Endzustand der „ausgereiften Windsee" eintreten. Die Kennzeichen dieses zweiten Idealfalles sind: *Keine örtlichen Unterschiede, aber gleichmäßige zeitliche Änderung an allen Orten*. Solche Verhältnisse werden wir am ehesten auf den Ozeanen bei kurzer Winddauer antreffen.

Die Wirklichkeit weicht im allgemeinen von diesen Idealfällen insofern ab, als beide Größen Windwirklänge und Winddauer gleichzeitig begrenzt sind. Es muß daher jeweils festgestellt werden, durch *welche* dieser beiden Größen die Windsee im Einzelfall kontrolliert wird. Dabei ist diejenige Größe als bestimmend für die Windsee anzusehen, die die Wellenentwicklung in einem früheren Stadium begrenzt als die andere. Nur für den stationären Endzustand der ausgereiften Windsee entfällt diese Entscheidung, da in diesem Falle Windwirklänge und Winddauer auf die gleichen Wellendimensionen führen.

Die *Größe der Windwirklänge* bestimmen wir mit Hilfe synoptischer Wetterkarten aus der Länge der atmosphärischen Windfelder einheitlicher Richtung und Geschwindigkeit (Fig. 20 A). Bei ablandigen Winden kann die Windwirklänge durch die Küstenlinie begrenzt sein (Fig. 20 B).

Die *Größe der Winddauer* wird ermittelt aus geeigneten Windregistrierungen sowie angenähert aus einer Folge synoptischer Wetterkarten.

Beiden Bestimmungsverfahren haften Schwierigkeiten an, insbesondere wenn die Windfelder wandern und sich verändern.

22. Das Anwachsen der charakteristischen Wellen. Es ist klar, daß die in Ziff. 21 geschilderten Beobachtungstatsachen nicht zu verstehen sind, wenn wir die Windsee als aus „konservativen" Wellen bestehend ansehen. So blieben auch alle älteren Versuche unbefriedigend, das Anwachsen der Windsee auf Grund der klassischen Wellentheorie und empirischer Werte darzustellen. Die zeitliche Entwicklung und die örtlichen Unterschiede der Windsee, also die Einflüsse von Winddauer und Windwirklänge lassen sich nur erfassen, wenn wir die klassische Vorstellung aufgeben, daß die Meereswellen „konservativ" sind und bei ihrer Wanderung ihre Identität behalten. Wir können vielmehr diesen Beobachtungstatsachen nur dann gerecht werden, wenn wir die Windsee als eine Überlagerung zahlreicher Wellenzüge verschiedener Höhe, Periode und Richtung auffassen.

Als erster Schritt auf diesem neuen Wege ist die von Sverdrup und Munk [*4*] veröffentlichte halbempirische Theorie anzusehen. Hier wird die unregelmäßige Windsee durch die in Ziff. 10 erläuterten Begriffe der ***charakteristischen (significant) Wellenhöhe und Wellenperiode*** beschrieben. Die Energieübertragung vom Winde auf die Wellen soll sowohl durch *normalen Winddruck* als auch durch *tangentiale Windschubspannung* erfolgen. Durch den Winddruck kann die Welle nur dann Energie gewinnen, wenn sie langsamer ist als der Wind ($C < U$). Sie muß durch den Luftwiderstand Energie verlieren, falls ihre Geschwindigkeit die des Windes übersteigt ($C > U$). Die tangentiale Windschubspannung sorgt jedoch — nach dem Ansatz von Sverdrup und Munk — dafür, daß auch im zweiten Falle — Welle schneller als der Wind — vor allem durch die Erhöhung der Partikelgeschwindigkeit im Wellenkamm die Welle insgesamt gesehen Energie gewinnt. Durch Kombination dieser Vorstellungen über die Energieübertragung Wind—Wellen mit den in Ziff. 21 geschilderten Beobachtungstatsachen und den in Ziff. 4 erwähnten Ergebnissen über den Energietransport durch Wellen gelangen Sverdrup und Munk, unter der Voraussetzung tiefen Wassers und konstanter Windgeschwindigkeit, zu folgenden Differentialgleichungen der *Energie-Übertragung*

stationärer Fall $$\frac{C}{2}\frac{dE}{dF} + \frac{E}{2}\frac{dC}{dF} = R_T \pm R_N, \tag{22.1}$$

nicht-stationärer Fall $$\frac{dE}{dt} + \frac{E}{C}\frac{dC}{dt} = R_T \pm R_N. \tag{22.2}$$

Hierin bedeutet E die mittlere Energie pro Flächeneinheit, C die Geschwindigkeit der charakteristischen Wellen, F die Windwirklänge, t die Winddauer, R_T oder R_N die durch tangentiale Windschubspannung oder normalen Winddruck bewirkte Energieübertragung in erg cm^{-2} sec^{-1}. Auf den rechten Seiten der beiden Gleichungen gelten die positiven Vorzeichen, wenn $C < U$, die negativen, wenn $C > U$ ist. Die Energiedissipation durch innere Reibung und Turbulenz im Wasser wird bei dieser Energiebilanz nicht berücksichtigt.

Um die unbekannten Wellengrößen E (gleichbedeutend mit dem Quadrat der Wellenhöhe) und C zu erhalten, bedienen sich SVERDRUP und MUNK einer *empirischen Beziehung zwischen der Wellensteilheit* H/L und dem Verhältnis

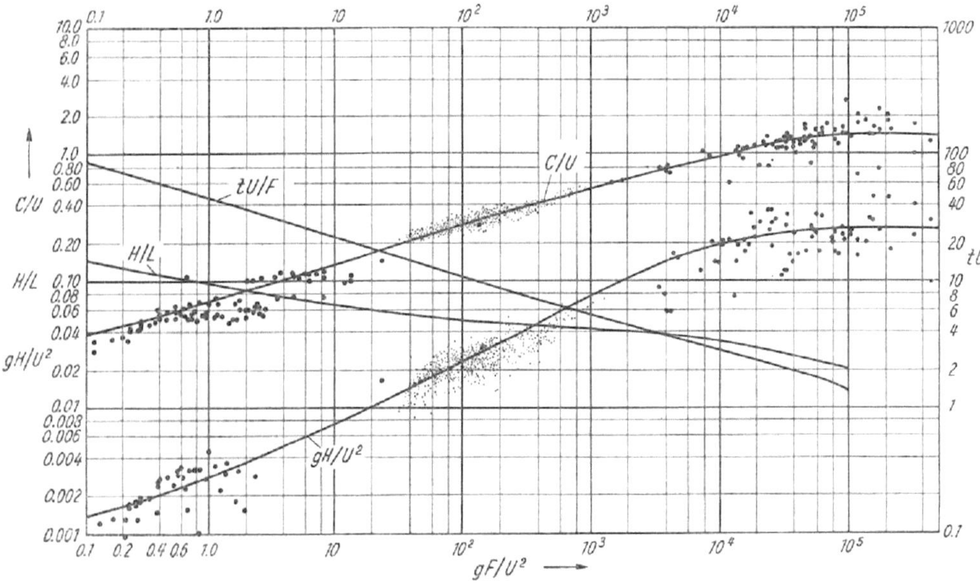

Fig. 21. Höhe H und Geschwindigkeit C der charakteristischen Wellen als Funktion von Windgeschwindigkeit U, Windwirklänge F und Winddauer t (nach BRETSCHNEIDER).

Wellengeschwindigkeit/Windgeschwindigkeit C/U, welches sie als *Wellenalter* bezeichnen. Es sei vermerkt, daß die Wellengeschwindigkeit C und die Wellenlänge L aus der charakteristischen Wellenperiode T mit Hilfe der klassischen Beziehung (3.16) berechnet werden. Die Lösungen werden nach gewissen, zum Teil empirischen Ansätzen für die normalen und tangentialen Windschubspannungen erhalten und in der Form von Diagrammen der dreidimensionslosen Parameter

$$g H/U^2, \quad H/L, \quad C/U, \quad g F/U^2, \quad t U/F$$

dargeboten. Wir geben in Fig. 21 die von BRETSCHNEIDER[1] [8] auf Grund zahlreicher neuer Messungen verbesserte Fassung eines dieser Diagramme wieder, die sich von der ursprünglichen nicht grundsätzlich, aber doch zahlenmäßig unterscheidet.

Es sei allerdings darauf hingewiesen, daß BRETSCHNEIDER eine etwas abweichende Definition der charakteristischen Wellen benutzt. Die mittlere Periode der ausgeprägten hohen Wellen wird hier als charakteristische Periode angesehen und zur Bestimmung einer fiktiven Wellenzahl der Registrierung herangezogen. Ein Drittel dieser so bestimmten Wellenzahl gibt sodann die Anzahl der Wellen, über die zur Errechnung der charakteristischen Wellenhöhe zu mitteln ist.

[1] C. L. BRETSCHNEIDER: Trans. Amer. Geophys. Un. **33**, 381 (1952).

Hieraus lassen sich die von einer bestimmten Windgeschwindigkeit U erzeugten charakteristischen Wellengrößen H, C und L, die am Ende einer bestimmten Windwirklänge F erwartet werden müssen, leicht entnehmen. Allerdings muß zunächst geprüft werden, ob die beobachtete Winddauer t dazu ausreicht, um die gesamte Windwirklänge F wirksam werden zu lassen. Hierzu dient die Kurve tU/F als Funktion von gF/U^2, die zu jedem Wertepaar U und F diejenige (minimale) Winddauer angibt, nach der keine zeitliche Zunahme der Wellengrößen am Ende von F mehr auftritt. Ist die beobachtete Winddauer gleich der minimalen Winddauer oder größer als diese, werden die Wellengrößen durch die Windwirklänge F bestimmt und das Diagramm ist anwendbar. Erweist sich die beobachtete Winddauer als kleiner, dann sind die Wellengrößen von der Winddauer abhängig und es müssen andere Darstellungen zu ihrer Bestimmung benutzt werden. Für die praktische Anwendung hat Bretschneider geeignete Diagramme entwickelt.

Die Theorie von Sverdrup und Munk ist wiederholt durch *Vergleich mit Messungen* geprüft worden. Die Resultate sind nicht einheitlich.

So berichteten Isaacs und Saville [*5*] über das Ergebnis ausführlicher Untersuchungen vor der nordamerikanischen Westküste. Die Theorie erwies sich dabei als zuverlässig hinsichtlich der Wellenhöhe (80% wichen höchstens um 0,6 m von der Messung ab) sowie der Ankunftszeiten besonderer Wellenhöhen-Zunahmen. Die berechneten Wellenperioden befriedigten weniger, sie waren hauptsächlich zu niedrig. Bei den jungen Wellen des Wattenmeeres fand Roll[1] im großen und ganzen die theoretischen Werte bestätigt, während Johnson[2] bei Wellenmessungen auf einer kalifornischen Lagune gute Übereinstimmung nur hinsichtlich der Wellenhöhe fand, bei der Wellenperiode jedoch um 20% höhere Werte als die theoretischen feststellte. Zu ähnlichen Resultaten gelangte Pierson[3] durch Vergleich von Wellenmessungen vor der Ostküste Nordamerikas mit berechneten Werten. Donn[4] dagegen fand bei einer Untersuchung der Wellen im westlichen Nordatlantik mit Hilfe von Seegangsschreibern Abweichungen zwischen Messung und Rechnung vor allem bei der Wellenhöhe und wies dabei auf die dem Berechnungsverfahren anhaftenden subjektiven Einflüsse hin (Festlegung der Windwirklänge und Winddauer aus Wetterkarten). Inzwischen wurden von Kaplan[5], Walden[6] und Wilson[7] Analysenmethoden entwickelt, die (insbesondere bei wandernden Wellen-Entstehungsgebieten) eine praktische Hilfe bieten und zur Vereinheitlichung der Methoden beitragen können. MacDougal[8] schließlich verglich Wellenhöhenschätzungen aus Nordatlantik und Mittelmeer mit berechneten Werten und gelangte zu einer guten Übereinstimmung, die jedoch angesichts der Ungenauigkeit der verwendeten Beobachtungen nur bedingt gewertet werden kann.

Das Ergebnis dieser sowie weiterer Vergleiche — insbesondere seien die sorgfältigen *Laboratoriumsversuche* von Johnson und Rice[9] erwähnt, die bessere Vergleichsmöglichkeiten als die bisherigen Feldbeobachtungen bieten — fand seinen Niederschlag in den durch Bretschneider[10] [*8*] verbesserten Diagrammen zur Berechnung der Wellengrößen, deren Brauchbarkeit Datz[11] durch die Untersuchung eines besonderen Falles an der nordamerikanischen Westküste bewies. Die berechneten Wellenhöhen und Wellenperioden stimmten mit den gemessenen überein, die Ankunftszeit der höchsten Wellen lag um nur 2 Std zu früh. (Neuere Messungen von Burling[12] auf einem künstlichen Wasserbecken wurden durch

[1] H. U. Roll: Dtsch. Hydrogr. Z. **2**, 268 (1949).
[2] J. W. Johnson: Trans. Amer. Geophys. Un. **31**, 386 (1950).
[3] W. J. Pierson jr.: U. S. Beach Erosion Board, Techn. Mem. No. 24, 1951.
[4] W. L. Donn: Trans. Amer. Geophys. Un. **30**, 507 (1949).
[5] K. Kaplan: U. S. Beach Erosion Board, Techn. Mem. No. 35, 1953.
[6] H. Walden: Dtsch. Hydrogr. Z. **7**, 129 (1954).
[7] B. W. Wilson: U.S.Beach Erosion Board, Techn. Mem. No. 73, 1955.
[8] F. MacDougal: Trans. Amer. Geophys. Un. **32**, 878 (1951).
[9] S. 674, Fußnote 4.
[10] S. 711, Fußnote 1.
[11] M. Datz: Bull. Beach Erosion Board **7**, (4) 1 (1953).
[12] S. 687, Fußnote 4.

BRETSCHNEIDERs Kurven allerdings nur bedingt wiedergegeben.) Diese theoretischen Verfahren zur Berechnung der Windsee haben in den USA verbreitet praktische Anwendung gefunden, insbesondere bei der Erstellung von Statistiken der Wellenhöhen und -perioden auf Grund eines langjährigen Materials historischer Wetterkarten *(hindcast-Verfahren)* für solche Orte, für die ausreichende Wellenmessungen nicht existieren.

Neben diesen Überprüfungen und Anwendungen der theoretischen Ergebnisse von SVERDRUP und MUNK sowie von BRETSCHNEIDER seien einige *Ergänzungs- und Verbesserungsvorschläge* erwähnt. So bemühte sich ARTHUR [*5*] darum, der bekannten *Richtungsvariabilität* der kurzkämmigen Windsee durch entsprechende Erweiterung der Diagramme von SVERDRUP und MUNK Rechnung zu tragen, während THIJSSE [*6*] eine Ergänzung für den Bereich der *Seichtwasserwellen* auf Grund von Laboratoriumsversuchen vorschlug. WALDEN[1] untersuchte die Windsee für den Fall eines *gleichmäßig zunehmenden Windes*. SAVILLE[2] diskutierte den Einfluß der *Breite des Windfeldes* auf die Ausbildung der Windsee. Dieser durch die Richtungs-Variabilität der Wellen bedingte Effekt kann besonders bei langen und schmalen Windfeldern merkbar werden.

SCHAAF und SAUER[3] übten Kritik an der von SVERDRUP und MUNK verwendeten Beziehung für die *Energieübertragung durch tangentiale Windschubspannungen* R_T und bemerkten, daß dabei Terme fortgefallen seien, die bei dem erstrebten Näherungsgrad nicht hätten vernachlässigt werden dürfen. Dies ist insofern nicht unwichtig, als bei der genauen Analyse sich nur für $C < U$ positive R_T-Werte ergeben, so daß dann auch die Energieübertragung durch tangentiale Windreibungskräfte auf *die* Wellen beschränkt bliebe, die langsamer sind als der Wind. Es steht dahin, ob dieses an Sinus-Wellen abgeleitete Ergebnis für die unregelmäßige Windsee zutrifft. Einstweilen muß es daher offen bleiben, wie die langen Wellen mit $C > U$, die erwiesenermaßen in der Windsee existieren, ihre Energie beziehen.

Schließlich hat sich YOSHIDA[4] mit dem von SVERDRUP und MUNK verwendeten Ansatz beschäftigt, bei dem die vom Winde auf die Wellen durch normale und tangentiale Kräfte übertragene Energie in bestimmter Weise zugunsten der Erhöhung von Wellenhöhe und Wellengeschwindigkeit aufgeteilt wird. YOSHIDA gelangte — ausgehend von den hydrodynamischen Grundgleichungen — zu einer abweichenden Form dieser *Energieaufteilung*, wobei sich ergab, daß die Zunahme der Wellengeschwindigkeit ausschließlich durch die Wirkung der tangentialen Windreibung zustandekommt.

23. Das Wachstum der spektralen Komponenten der Windsee. Die Darstellung der Windsee-Entwicklung mit Hilfe der charakteristischen Wellen bedeutet zwar schon einen erheblichen Fortschritt gegenüber früheren Versuchen. Angesichts der komplexen Natur des Seeganges läßt sie aber — insbesondere was die Verteilung der Wellenperioden angeht — noch viele Wünsche offen, kommt es doch bei manchen Problemen weniger auf die charakteristischen als auf die kurzen und langen Wellen des Spektrums an.

Es hat daher nicht an Bemühungen gefehlt, die Entwicklung des gesamten Windseespektrums zu erfassen. Auf Grund von Wellenmessungen vor den britischen Westküsten und auf Seen sowie Analysen der jeweils zugehörigen stationsnahen Windfelder hat DARBYSHIRE [*11*] zunächst empirische Zusammenhänge

[1] H. WALDEN: Dtsch. Hydrogr. Z. **8**, 236 (1955).
[2] T. SAVILLE jr.: U.S.Beach Erosion Board. Techn. Mem. No. 70, 1954.
[3] S. A. SCHAAF u. F. M. SAUER: Trans. Amer. Geophys. Un. **31**, 867 (1950).
[4] K. YOSHIDA: Geophys. Notes Geophys. Inst. Tokyo **4**, (6) 1 (1951).

zwischen Wind und Wellen sowohl für die ausgereifte Windsee als auch für die Anfachungsstadien herausgearbeitet und sie sodann durch physikalische Vorstellungen zu stützen versucht. Er ging dabei von der Grundannahme aus, daß *kein Energieaustausch zwischen den verschiedenen Wellen-Komponenten des Spektrums* stattfindet, daß diese vielmehr unabhängig voneinander mit der ihrer Periode entsprechenden Gruppengeschwindigkeit $C_g = C/2$ wandern und unter der Wirkung des Windes anwachsen. Zu dieser Annahme gelangte man durch den Vergleich der gemessenen Anfangs- und Endzeiten bestimmter Wellenperioden an der Küste mit den aus Wetterkarten abgeleiteten Zeiten, zu denen diese Wellen im entfernten Sturmfeld zu entstehen begannen und aufhörten.

DARBYSHIRE fand empirisch, daß das *Wellenspektrum bereits nach einer Windwirklänge von 200—300 Seemeilen voll ausgebildet* ist. Unter diesen Gleichgewichtsbedingungen ist die längste Wellenperiode des Spektrums ausgedrückt in Sekunden gleich $\frac{1}{3} U_g$, mit U_g = maximale Gradientwind-Geschwindigkeit in Knoten. Die Periode $T_{\max}$ der höchsten Wellen des Spektrums ergab sich — ausgedrückt in Sekunden — gleich $U_g/4$, eine Beziehung, die übrigens der von NEUMANN angegebenen Gl. (13.4) weitgehend entspricht. Die Steilheit der höchsten Wellen scheint proportional $1/\sqrt{U_g}$ zu sein.

Diese und weitere empirische Beziehungen fanden schließlich einen zusammenfassenden formelmäßigen Ausdruck in Gleichungen für das *Amplituden-Spektrum*, die sowohl für das ausgereifte Stadium als auch — in Abhängigkeit von der Windwirklänge — für die Anfachungszustände gegeben werden. In der Nomenklatur der Ziff. 13 nimmt das DARBYSHIRE-Spektrum im allgemeinen Falle die folgende Form an

$$[a(\omega)]^2 = \frac{8\pi^3 C'}{\omega^4} e^{-2{,}0\left[\frac{2\pi g}{\omega U_g} - 4{,}6\,(1+1{,}25\cdot 10^{-4} F)\left(1-e^{-0{,}23\sqrt{F}}\right)\right]^2}. \tag{23.1}$$

Hierin ist U_g die Gradientwind-Geschwindigkeit, die DARBYSHIRE aus praktischen Gründen (Wetterkarten-Auswertung) allen seinen Betrachtungen zugrunde legte. Für die Windgeschwindigkeit U in Nähe der Meeresoberfläche nahm DARBYSHIRE 0,667 U_g an. Die Windwirklänge F ist im Hinblick auf die praktischen Anwendungen in Seemeilen einzusetzen. Für große F ergibt sich aus (23.1) das einfachere Energie-Spektrum der ausgereiften Windsee. Die Konstante C' bestimmte sich zu 22 cm^2 sec^{-3}.

Mit Hilfe der obigen Gleichung kann die Höhe jeder spektralen Komponente berechnet werden. Dabei ist für jede Periode T diejenige Länge als Windwirklänge anzusehen, über der die Windgeschwindigkeit in Knoten größer als $3\,T$ ist. Die von DARBYSHIRE mitgeteilten Ergebnisse von Vergleichen zwischen Messung und Berechnung befriedigen. Demgegenüber stehen die in Ziff. 13 erhobenen Bedenken gegen die schmale Form des DARBYSHIRE-Spektrums sowie das Ergebnis eines Vergleiches von DATZ[1], bei dem die Ankunftszeit einer starken Wellenhöhenzunahme durch das DARBYSHIRE-Verfahren um 16 Std zu spät angegeben wurde. Es müssen daher weitere Erfahrungen abgewartet werden.

Beiträge zum Problem der Anfachung spektraler Windsee-Komponenten wurden ferner von NEUMANN [*10*], [*12*] vorgelegt. In diesen Arbeiten bemühte sich NEUMANN der komplexen Natur der Windsee zunächst durch die Annahme von *drei charakteristischen Hauptwellen* gerecht zu werden. Später hat er diese Konzeption — zumindest für den Fall der vollentwickelten Windsee — zugunsten eines kontinuierlichen Energie-Spektrums verlassen und will die drei charakteristischen Wellen als *Durchgangsstadien* aufgefaßt wissen, als gewisse Diskon-

[1] S. 712, Fußnote 11.

tinuitäten des Wachstums, die die Wellen bei ihrer Entwicklung zu durchlaufen haben. Nach NEUMANNs Auffassung entwickeln sich nach dem Einsetzen des Windes aus den Initialwellen zunächst kurzperiodische steile Wellen *(See)*, die bald instabil werden und überbrechen, aber ständig durch Energiezufuhr seitens des Windes regeneriert werden, so daß sie einen quasi-stationären Zustand darstellen. Wenn dieses Stadium voll ausgebildet ist, beginnen sich längere Wellen zu entwickeln, die etwa mit der Windgeschwindigkeit wandern *(Zwischenwelle)*. Bei genügend großen Werten von Windwirklänge und Winddauer werden schließlich *lange Wellen* auftreten. Dabei erscheint NEUMANN eine Energieübertragung von den kurzen auf die langen Wellen durch folgenden Mechanismus möglich: Eine steile, kurz vor dem Instabilwerden befindliche „See" wird von einer „langen Welle" unterlaufen, bricht auf dem Kamm der „langen Welle" über — eine häufig beobachtete Erscheinung — und gibt dabei einen gewissen Teil ihrer Bewegungs- und Lageenergie an diese „lange Welle" ab. (Diese Prozesse der Energieübertragung zwischen den einzelnen Komponenten des Spektrums hat YOSHIDA[1] theoretisch behandelt.)

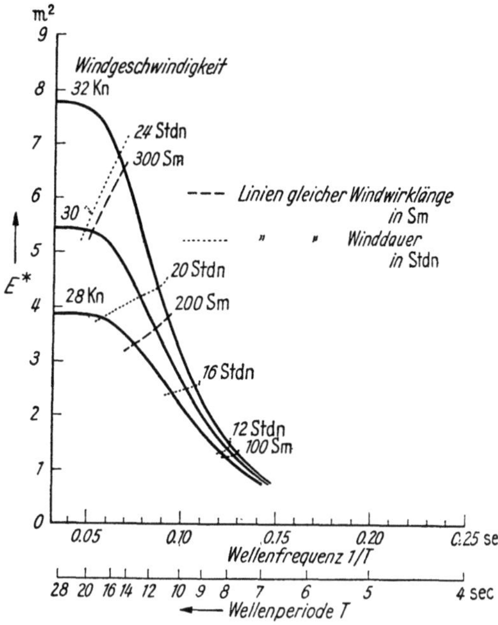

Fig. 22. Begrenzung des Energiesummenspektrums der sich entwickelnden Windsee durch Windwirklänge und Winddauer für Windgeschwindigkeiten von 28—32 Knoten (nach NEUMANN).

Die Bestimmung der Wellengrößen für die *Anfachungszustände* in Abhängigkeit von Windgeschwindigkeit, Windwirklänge und Winddauer erfolgt durch Energiegleichungen, da der Seegang nur so lange wächst, wie die Energiezufuhr durch den Wind die Energieverluste durch innere Reibung und Turbulenz im Wasser übertrifft, die NEUMANN hier (im Gegensatz zu SVERDRUP und MUNK) berücksichtigte. Auch bei der mathematischen Formulierung der Energieübertragung vom Winde auf die Wellen wich NEUMANN von den erwähnten Autoren dadurch ab, daß er die periodischen Änderungen der tangentialen Windschubspannung entlang des Wellenprofils, sowie die Windabhängigkeit des „effektiven" Reibungsfaktors in Rechnung stellte. Die erhaltenen Ergebnisse ließen sich für die drei erwähnten charakteristischen Wellen mit Hilfe der gleichen dimensionslosen Wellengrößen, wie sie SVERDRUP und MUNK benutzten, in Diagrammen darstellen.

Für die praktische Anwendung eignet sich besonders die Form der *Energiesummenspektren*, wie sie bereits für die ausgereifte Windsee in Fig. 15 gegeben wurde. Fig. 22 zeigt das gleiche Energiesummenspektrum ausgestattet mit den Begrenzungslinien für die Winddauer und die Windwirklänge, die auf Grund der obigen Ergebnisse berechnet wurden. Der Schnittpunkt der Windgeschwindigkeitskurve mit der in Frage kommenden Linie für Winddauer oder Windwirklänge gibt die durch diese Größen bewirkte *Begrenzung des Windsee-Spektrums* an, wobei wie früher der geringere Wert von beiden Möglichkeiten zu wählen

[1] K. YOSHIDA: Geophys. Notes Japan **3**, No. 24, 1 (1950). — J. Oceanogr. Soc. Japan **7**, 49 (1951).

ist. Wir erhalten damit den Teilbetrag der Gesamtenergie $E = \frac{g\varrho}{2} E^*$, den die Wellenbewegung unter den vorliegenden Bedingungen aus der Windströmung entnehmen kann. Aus ihm läßt sich wie bekannt die Verteilung der scheinbaren Wellenhöhen berechnen. Die obere Grenze der Wellenperiode ist ebenfalls annähernd durch diesen Schnittpunkt gegeben. Allerdings muß man durch eine kleine Korrektur der Tatsache Rechnung tragen, daß das Spektrum nicht abrupt abgeschnitten werden kann, vielmehr Wellen mit etwas längeren Perioden im Anfangsstadium vorhanden sein werden.

Eine von JAMES[1] veröffentlichte *Prüfung* des vorliegenden Verfahrens durch Vergleich mit Seegangsregistrierungen ergab eine gute Übereinstimmung auch in den Einzelheiten der Seegangscharakteristik. Von anderen Methoden der Windseeberechnung (z.B. SVERDRUP und MUNK) unterscheidet sich dieses Verfahren vor allem durch die kurzen Anfachungszeiten der ausgereiften Windsee, ein Umstand, der noch sorgfältig untersucht werden muß.

Es ist klar, daß ein Verfahren, das sich bemüht, der komplizierten Seegangsstruktur gerecht zu werden, nicht einfach sein kann. Die Hauptschwierigkeiten liegen jedoch nicht auf diesem mehr rechnerischen Gebiet, sie treten vielmehr bei der Bestimmung der Windwirklänge und der Winddauer auf, die zur Zeit am stärksten subjektiven Einflüssen Spielraum läßt, und sind daher bei allen Verfahren mehr oder weniger die gleichen. In der Praxis können unzählig viele verschiedene Kombinationen erscheinen und Bedingungen zusammenwirken, so daß sich kaum Richtlinien aufstellen lassen, die alle Lagen erfassen. Es wird daher noch viel von den Kenntnissen, den Erfahrungen und dem Urteilsvermögen des Betreffenden abhängen, der diese Verfahren benutzt.

Eine ausführliche Erklärung der vorliegenden Berechnungsmethode mit einer Anleitung für ihre Anwendung haben PIERSON, NEUMANN und JAMES [*13*] veröffentlicht.

Trotz seiner unbestreitbaren Vorzüge, die vor allem in der Berücksichtigung des Seegangs-Spektrums liegen, scheint das von PIERSON, NEUMANN und JAMES entwickelte Verfahren nach neueren Untersuchungen jedoch noch einige *Schwächen* zu besitzen. DEARDUFF[2] wendete diese Methode auf den gleichen Fall wie DATZ[3] and und mußte feststellen, daß die nach PIERSON-NEUMANN-JAMES berechneten Wellen wesentlich höher als die gemessenen ausfielen und verspätet eintrafen. Zu einem ähnlichen Resultat gelangte WALDEN[4] beim Vergleich von zahlreichen Windseebeobachtungen im Nordatlantik mit theoretischen Werten.

Er stellte darüber hinaus fest, daß die beobachteten Wellenhöhen am besten durch das von BRACELIN[5] angegebene Berechnungsverfahren wiedergegeben werden, während die gemessenen Wellenperioden mit den nach PIERSON-NEUMANN-JAMES berechneten Werten übereinstimmen. (WALDEN[6] wertete dieses Ergebnis praktisch aus, indem er auf dieser Grundlage ein neues Diagramm zur Berechnung der Windsee entwarf.)

Auch ein von RATTRAY und BURT[7] an einer Sturmwetterlage angestellter Vergleich zwischen den Verfahren von SVERDRUP-MUNK, BRETSCHNEIDER, PIERSON-NEUMANN-JAMES und Wellenbeobachtungen ergab bei den Wellenhöhen gute Übereinstimmung für die beiden erstgenannten Methoden, weniger gute bei dem dritten Verfahren. Doch waren die Unterschiede bei diesem Test nicht so auffallend wie vorher erwähnt. Somit läßt sich aus diesen zum Teil widersprechenden Mitteilungen zur Zeit nur entnehmen, daß die Theorie von PIERSON-NEUMANN-JAMES hinsichtlich der Wellenhöhen noch verbesserungsbedürftig ist (vgl. Ziff. 13).

---

[1] R. W. JAMES: Trans. Amer. Geophys. Un. **35**, 153 (1954).
[2] R. F. DEARDUFF: Bull. Beach Erosion Board **9** (1), 5 (1955).
[3] S. 712, Fußnote 11.
[4] H. WALDEN: Ann. Meteor. Hamburg **6**, 296 (1953/54).
[5] P. BRACELIN: Brit Naval Weather Service, Mem. No. 147/52, 1952.
[6] H. WALDEN: Ann. Meteor. Hamburg **7**, 213 (1955/56).
[7] M. RATTRAY jr. u. W. V. BURT: Deep-Sea Research **3**, 140 (1956).

Abschließend sei auf den *grundsätzlichen Unterschied* der Auffassungen von DARBYSHIRE und NEUMANN über das Anwachsen der Windsee hingewiesen. Während DARBYSHIRE annimmt, daß die einzelnen Wellenkomponenten *unabhängig voneinander* wachsen, können bei NEUMANN die *längeren Wellen erst dann* erscheinen und weiterwachsen, wenn die *kürzeren voll ausgebildet* sind, d.h. die Energie erreicht haben, die für sie im Spektrum der ausgereiften Windsee zur Verfügung steht. GELCI[1] und Mitarbeiter haben versucht, zwischen diesen Standpunkten zu vermitteln, und auf Grund von empirischen Befunden einen Wachstums-Mechanismus vorgeschlagen, bei dem einerseits alle Perioden gemäß DARBYSHIRE gleichzeitig anwachsen, andererseits die Maxima der Wellenenergie, die nach bestimmten Andauerzeiten erreicht werden, bei den gleichen Perioden wie bei NEUMANN auftreten und sich im Endstadium NEUMANNs Spektrum der ausgereiften Windsee ergibt.

Die bisher erwähnten theoretischen Ansätze trugen der Tatsache keine Rechnung, daß der Wind nicht nur die Wellen, sondern gleichzeitig auch den *Triftstrom* erzeugt. Eine vollständige Theorie müßte daher die Energieübertragung Wind→Wellen und Wind→Strom sowie den Energieaustausch Wellen↔Strom berücksichtigen. YOSHIDA[2] hat als erster dieses Problem angegriffen, nachdem ABDULLAH [5] und THOMPSON [5] Oberflächenwellen in strömendem Wasser theoretisch behandelt hatten. Das Fehlen der erforderlichen empirischen Daten verhinderte die Formulierung endgültiger Ergebnisse.

III. Zerfall und Vernichtung.

a) Umgestaltung der Wellen nach Aufhören des Windes (Dünung).

24. Maßzahlen der Dünung. Sobald die anfachende Kraft des Windes aufhört oder nachläßt, bildet sich die Windsee zur Dünung um, deren Eigenschaften in Ziff. 8 beschrieben wurden. Es handelt sich also um Wellen, die der vorderen Begrenzung eines Windfeldes vorauseilen, oder um solche, die nach plötzlicher Auflösung eines Sturmes innerhalb seines Bereiches aus den Windseewellen gebildet werden.

Fig. 23 orientiert über die zu untersuchende Situation sowie über die verwendeten Maßzahlen. H_F und T_F seien Höhe und Periode der charakteristischen Wellen des Windseespektrums an der Luvkante (Punkt A) eines Wellenentstehungsgebietes von der Länge F und der Breite B. Die Winddauer t_F sei groß genug, um die gesamte Windwirklänge F zum Tragen kommen zu lassen. Das Problem der Dünung besteht nun darin, die Gesetzmäßigkeiten zu entdecken, nach denen sich die Umwandlung der Wellen außerhalb des Entstehungsgebietes vollzieht, d.h. wir suchen nach Beziehungen zwischen

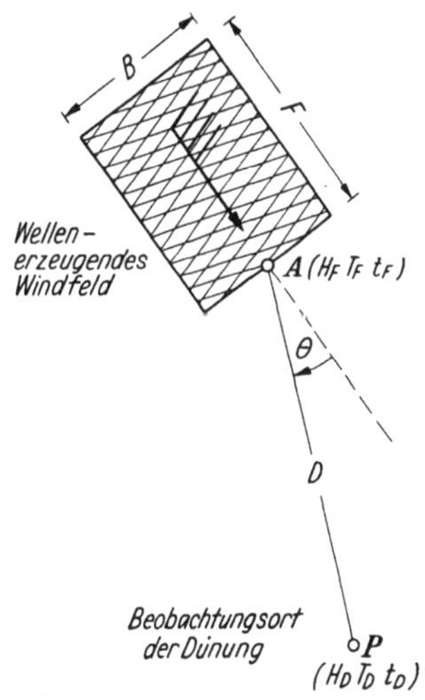

Fig. 23. Maßzahlen der Dünung.

den charakteristischen Größen H_D und T_D der Dünung im Punkte P und der Zeit t_D ihres dortigen Auftretens einerseits und

den Maßzahlen der Windsee im Punkte A zur Zeit $t=0$ sowie der Dünungslaufstrecke D und dem Winkel Θ der seitlichen Ausbreitung andererseits.

[1] R. GELCI, H. CAZALÉ u. J. VASSAL: Bull. d'Inform. C. O. É. C. **8**, (4) 169 (1956).
[2] K. YOSHIDA: J. Oceanogr. Soc. Japan **7**, 97 (1951).

Aus Beobachtungen ist, mehr qualitativ als quantitativ, bekannt, daß die Wellenhöhen mit wachsender Dünungslaufstrecke abnehmen, die Wellenperioden jedoch wachsen. Es handelt sich darum, diese Tatsachen physikalisch zu interpretieren und Verfahren für eine exakte Berechnung der Dünung zu entwickeln.

25. Wirkung von Luftwiderstand, ozeanischer Turbulenz und Gezeiten. Der erste Versuch, die Dünungswellen quantitativ zu behandeln, erfolgte durch SVERDRUP und MUNK [4], die zunächst nur das Hineinlaufen der Dünung in eine windstille Zone betrachteten. Sie berechneten den durch den *Luftwiderstand* eintretenden Energieverlust mit dem gleichen Ansatz wie für R_N in Ziff. 22 und auch dem gleichen Widerstandsbeiwert (was fraglich erscheint) und vernachlässigten Energieverluste durch tangentiale Schubspannungen. Sie erhielten die Periodenänderung T_D/T_F als Funktion von D und T_F, sowie anschließend die Dünungslaufzeit t_D/T_F und die Höhenabnahme H_D/H_F als Funktionen von T_D/T_F. Um die Wirkung von Gegen- und Rückenwinden auf die Dünungsentwicklung zu beschreiben, wurde eine *„effektive“ Dünungslaufstrecke* eingeführt, die sich zur *wirklichen* verhält wie die Wellengeschwindigkeit relativ zur Luft $(C \pm U)$ zur Wellengeschwindigkeit C. Die Ergebnisse der Rechnung schienen den (damals noch recht spärlichen und unzuverlässigen) Messungen zu entsprechen.

SVERDRUP[1] wandte sodann das gleiche Prinzip auf ein Periodenspektrum an und konnte nachweisen, daß durch den Luftwiderstand eine *relative Auslöschung* in der Weise stattfindet, daß die kurzen Wellen schneller an Höhe verlieren als die langen. Dadurch verschiebt sich das Maximum der Wellenhöhe in Richtung der größeren Perioden.

Dieses Ergebnis wurde von DARBYSHIRE [11] empirisch bestätigt, der außerdem feststellte, daß diese Höhenabnahme für alle Perioden *bei Rückenwinden wesentlich schwächer* ausgeprägt ist. Er fand ferner, daß hohe Wellen proportional stärker ausgelöscht werden als niedrige Wellen gleicher Periode. Dieser Befund deutet darauf hin, daß der *Widerstandsbeiwert* sich *proportional zur Wellensteilheit* ändert, wie es ja auch von NEUMANN im Gegensatz zu JEFFREYS und MOTZFELD erhalten wurde (vgl. Ziff. 18). Unter Anwendung dieser Ergebnisse führte auch DARBYSHIRE die Dünungsumwandlung zurück auf die verschiedenartige Auslöschung der einzelnen, unabhängig voneinander wandernden spektralen Komponenten durch den Luftwiderstand, und zwar mit befriedigendem Erfolg.

Mit der bereits in Ziff. 23 erwähnten Annahme, daß Windseewellen der Periode T sich dann ausbilden, wenn die Gradientwindgeschwindigkeit $3\,T$ (in Knoten) erreicht oder überschreitet, hat DARBYSHIRE ferner die Wellenentstehungsgebiete und die zugehörigen Dünungslaufstrecken D zeitlich sowie örtlich festgelegt und diejenigen Fälle ausgesondert, bei denen entlang der Dünungslaufstrecke Rückenwinde herrschten. Es zeigte sich, daß die hierbei ermittelte empirische *Gruppengeschwindigkeit* der Dünungswellen bei einer Gradientwindgeschwindigkeit von etwa 20 Knoten (entlang D gemittelt) mit der klassischen, aus der Periode errechneten Gruppengeschwindigkeit übereinstimmte, bei Zunahme dieser Gradientwindgeschwindigkeit jedoch *abnahm* ,und zwar etwa proportional dem Quadrat dieser Windgeschwindigkeit, so daß bei einem mittleren Gradientwind von 40 Knoten die Gruppengeschwindigkeit nur noch 80% des klassischen Wertes betrug. Die Tatsache, daß dieser Effekt bei Rückenwinden auftrat, legt die Vermutung nahe, hierin eine Wirkung der *ozeanischen Turbulenz* zu erblicken.

GROEN und DORRESTEIN[2] haben sich mit dieser Frage beschäftigt und kamen zu dem Ergebnis, daß die für den Ozean bekannten Turbulenzkoeffizienten

[1] H. U. SVERDRUP: Trans. Amer. Geophys. Un. **28**, 407 (1947).

[2] P. GROEN u. R. DORRESTEIN: Nature, Lond. **165**, 445 (1950).

ausreichen, um die *Höhenabnahme in der Dünung* und wahrscheinlich auch die *Periodenzunahme* zu erklären, wenn man diesen Koeffizienten nach v. WEIZSÄCKER[1] als proportional zur $\frac{4}{3}$-Potenz der Wellenlänge ansetzt. Nach ihrer Auffassung ist es sehr unwahrscheinlich, daß der für die Energieübertragung Wind→Wellen durch normalen Winddruck abgeleitete Widerstandskoeffizient unverändert auf Dünungswellen angewendet werden könne, weil bei diesen meist nur eine Bewegung der Luft relativ zur Wellenform, aber nicht relativ zum Wasser stattfindet. BOWDEN[2] gelangte — allerdings mit einem von GROEN und DORRESTEIN abweichenden Ansatz für den Turbulenzkoeffizienten (proport. zur Geschwindigkeit und Amplitude der Wellen) — ebenfalls zu dem Resultat, daß damit die Dünungsumwandlung vollständig erklärt werden könne. Eine Entscheidung darüber, ob Luftwiderstand oder Turbulenz oder beide Faktoren gemeinsam wirksam sind, war indessen nicht möglich.

Auch eine erneute Untersuchung durch GROEN[3] brachte, trotz Anwendung verfeinerter Methoden (Unterscheidung zwischen „welleneigener" und „wellenfremder" Turbulenz, Wellenspektrum) keine endgültige Beantwortung der Frage, in welchem Umfange Luftwiderstand und ozeanische Turbulenz an der Dünungsumwandlung wirklich beteiligt sind. Es ließ sich lediglich Aufschluß darüber gewinnen, wie groß die Turbulenzkoeffizienten sein müssen, um bei Abwesenheit von Luftwiderstand und alleiniger Wirkung äußerer („wellenfremder") Turbulenz die beobachteten Veränderungen der Dünung zu erzielen.

Schließlich sei die Beeinflussung der Dünungswellen durch *Gezeitenströme* erwähnt, über die BARBER und URSELL[4], BARBER[5], DEACON [*5*] und DARBYSHIRE [*11*] berichtet haben. Wellenregistrierungen bei Perranporth vor der Südwestküste Englands zeigten auffallende *Schwankungen der Wellenperiode* im Rhythmus von 12 Std, wobei die Periode während des Stromhöchstwertes in Nordostrichtung vermindert und während des Strommaximums in Südwestrichtung erhöht war. Diese Periodenoszillationen lagen in der Größenordnung von etwa 1 sec und ließen sich rein kinematisch erklären und korrigieren. Es bestanden auch Anzeichen für eine *Beeinflussung der Wellenhöhen durch die Gezeitenströme*, wobei die größte Verminderung der Wellenhöhe jeweils mit dem Maximum des Gezeitenstromes (unabhängig von seiner Richtung) zu koinzidieren schien. Diese Erscheinung ließe sich als Turbulenzeffekt deuten.

26. Dispersion und Winkel-Ausbreitung. Während die Bedeutung von Luftwiderstand und Turbulenz für die Umgestaltung der Dünungswellen noch umstritten ist, können wir die Wirkung der *Dispersion* als gesichert ansehen. Wegen der Beziehungen (3.15) und (3.16) ist das Wasser für die Oberflächenwellen ein dispersives Medium, *lange Wellen wandern schneller als kurze*. Die langperiodischen Komponenten eines Wellenspektrums werden daher den kurzperiodischen kraft ihrer größeren Gruppengeschwindigkeit „davonlaufen", wenn wir wie bisher eine voneinander unabhängige Wanderung der einzelnen Wellenkomponenten annehmen. Die Folge hiervon ist, daß — zu einem festen Zeitpunkt betrachtet — die *scheinbare Wellenperiode mit wachsender Dünungslaufstrecke zunimmt* und umgekehrt an einem festen Ort eine zeitliche Abnahme dieser Periode auftritt. Da die (ursprünglich auf das Wellenentstehungsgebiet konzentrierte) Energie sich hierbei auf einen weit größeren Bereich ausbreitet, ist mit diesem Vorgang der

[1] C. F. v. WEIZSÄCKER: Z. Physik **124**, 614 (1948).
[2] K. F. BOWDEN: Phil. Mag., Ser. 7 **41**, 907 (1950).
[3] P. GROEN: Kon. Nederlands Met. Inst., Mededelingen en Verhandelingen No. 63, 1954.
[4] N. F. BARBER u. F. URSELL: Phil. Trans. Roy. Soc. Lond. **240**, 527 (1948).
[5] N. F. BARBER: Proc. Roy. Soc. Lond., Ser. A **198**, 81 (1949).

Dispersion notwendig eine Abnahme der Wellenenergie pro Einheitsfläche, also eine *Reduktion der charakteristischen Wellenhöhenwerte* verbunden.

Auf die Bedeutung dieses Dispersionseffektes hat bereits Darbyshire [*11*] hingewiesen. Er hat dementsprechend, bei seiner Untersuchung der selektiven Auslöschung, die Wirkung der Dispersion dadurch ausgeschaltet, daß er die einzelnen Komponenten getrennt betrachtete.

Bretschneider[1] führte erstmalig die Umgestaltung der Dünungswellen ausschließlich auf den Dispersionseffekt zurück und behandelte diesen quantitativ auf der Basis der *charakteristischen Wellen*. Seine Ausgangswerte sind H_F, T_F, F und D. Schon hierin unterscheidet sich sein Ansatz von dem Sverdrup und Munks (Ziff. 25), die hierfür lediglich T_F und D benutzten. Durch diese Erweiterung wird das maßgebende Windseespektrum am Ende der Windwirklänge F vollständiger als bisher charakterisiert. Bretschneider brachte seine Ergebnisse in die Form dimensionsloser Beziehungen, wobei ein Diagramm (Fig. 24)

$$D/H_D \text{ als Funktion von } D/F \text{ und } D/H_F$$

und somit die *Höhenabnahme* darstellt, während ein zweites Diagramm (Fig. 25)

$$D/T_D^2 \text{ als Funktion von } D/F \text{ und } D/T_F^2$$

für die *Periodenzunahme* entworfen wurde. (Hier wurde im Nenner der „dimensionslosen" Größen D/T_D^2 und D/T_F^2 die Schwerebeschleunigung aus praktischen Gründen fortgelassen.) Für $D\to 0$ gilt naturgemäß $D/H_D\to D/H_F$ und $D/T_F^2\to D/T_D^2$. Alle verfügbaren Meßwerte wurden eingetragen. Sie scheinen die Rechnung in ausreichender Weise zu bestätigen. Die (meist geringen) Abweichungen könnten auf den Einfluß von Rücken- oder Gegenwinden zurückgeführt werden. Die Dünungslaufzeit t_D wird aus der Gruppengeschwindigkeit C_{Dg} der Dünungswellen mit der Periode T_D und der Dünungslaufstrecke D gemäß $t_D = D/C_{Dg}$ errechnet.

Aus den Diagrammen kann folgender Einfluß der Windwirklänge im Wellen-Entstehungsgebiet auf die spätere Dünungs-Umgestaltung entnommen werden: Für eine vorgegebene Wellenperiode T_F der Windsee und eine vorgegebene Dünungslaufstrecke D sind Perioden*zu*nahme und Höhen*ab*nahme in der Dünung um so *größer*, je *kürzer* die *Windwirklänge* ist. Erklärung: Eine Verringerung der Windwirklänge F bei gleichbleibender charakteristischer Windsee-Periode T_F bedingt eine Erhöhung der Windstärke und damit der Höhe und Steilheit der Windsee, aber auch eine Verkürzung der in den Dispersionsbereich eintretenden „Wellenfelder" und damit eine Beschleunigung ihres Zerfalles.

Pierson, Neumann und James [*13*], [*9*] legten ihrer Untersuchung der Dünungsumwandlung das *Energiespektrum der kurzkämmigen Windsee* zugrunde. Als wirkende Faktoren sahen sie — unter Vernachlässigung anderer Einflüsse (Luftwiderstand, Turbulenz, Rücken- und Gegenwinde) — die *Dispersion* der spektralen Komponenten und außerdem die *Winkelausbreitung* an. Dieser Ansatz ergibt sich zwangsläufig aus der in Ziff. 15 beschriebenen Darstellung der kurzkämmigen Windsee. Wenn die Komponenten der Windsee im Entstehungsgebiet verschiedene Richtungen aufweisen, dann ist es folgerichtig, auch bei der Ausbreitung der Dünungswellen verschiedene Richtungen bis maximal $\pm\pi/2$ beiderseits der Windrichtung zuzulassen.

Fig. 26 erläutert die notwendigen Überlegungen für den einfachen Fall eines *linienförmigen Wellenentstehungsgebietes* der Breite B. Diese Strecke ist als Quelle

[1] S. 711, Fußnote 1.

von Dünungswellen verschiedener Periode und Richtung anzusehen, die vom Zeitpunkt $t=0$, ab jeweils bestimmte Zeiten lang, ausgesendet werden. Wir

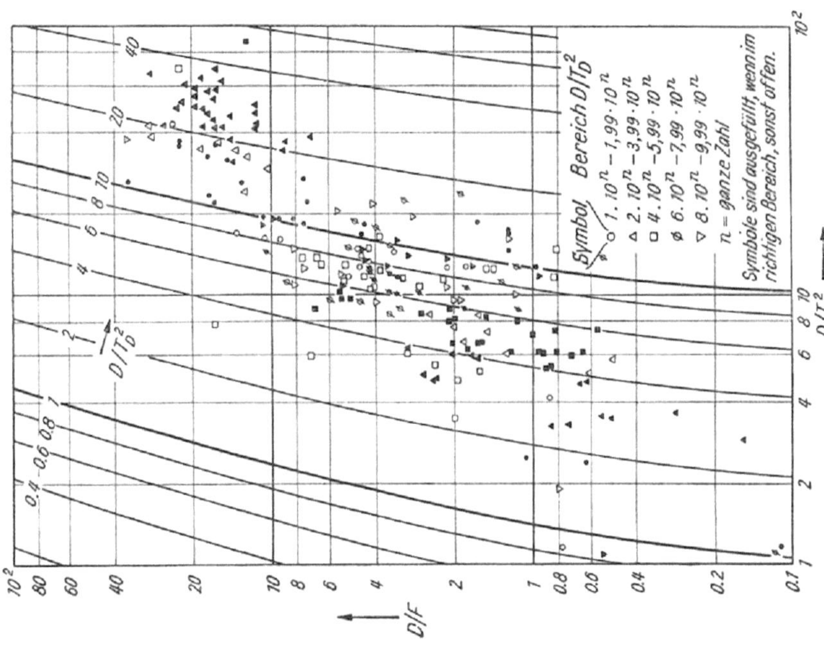

Fig. 25. Periode T_D der charakteristischen Wellen einer Dünung als Funktion von Windwirklange F (Seemeilen), Periode der Windsee T_F (Sekunden), Dünungslaufstrecke D (Seemeilen) (nach BRETSCHNEIDER).

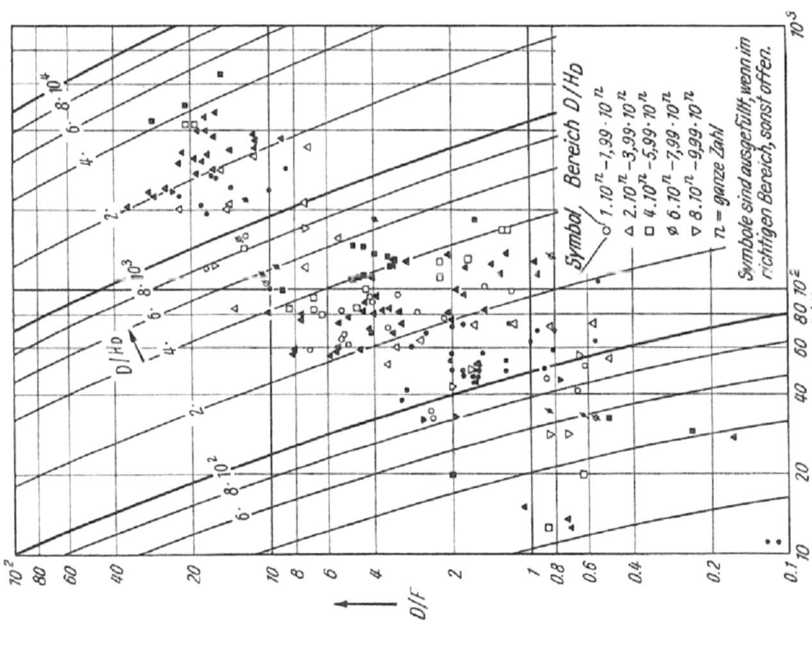

Fig. 24. Höhe H_D der charakteristischen Wellen einer Dünung als Funktion von Windwirklänge F (Seemeilen), Höhe der Windsee H_F (Fuß), Dünungslaufstrecke D (Seemeilen) (nach BRETSCHNEIDER).

fragen nach den Bereichen der Wellenperiode T und der Wellenrichtung Θ (Windrichtung $\Theta=0$), die zu einem bestimmten Zeitpunkt t_D am Beobachtungsort P in der Entfernung D vorhanden sein können. Durch einfache kinematische

Betrachtungen unter Verwendung der zu jeder Periode gehörenden Gruppengeschwindigkeit ergibt sich:

T_1 als Perioden der Wellen, die P zur Zeit t_D gerade passiert haben;

T_2 als Periode der Wellen, die P zur Zeit t_D gerade erreichen;

Θ_1 als Richtungswinkel der Wellen, die gerade „oberhalb" von P vorbeilaufen;

Θ_2 als Richtungswinkel der Wellen, die gerade „unterhalb" von P vorbeilaufen.

Zur Berechnung von T_1 ist die Kenntnis der Zeitdauer der Aussendung von Wellen der Periode T_1 an der Quelle, zur Bestimmung von Θ_1 und Θ_2 die Kenntnis der Breite B des Wellenentstehungsgebietes erforderlich.

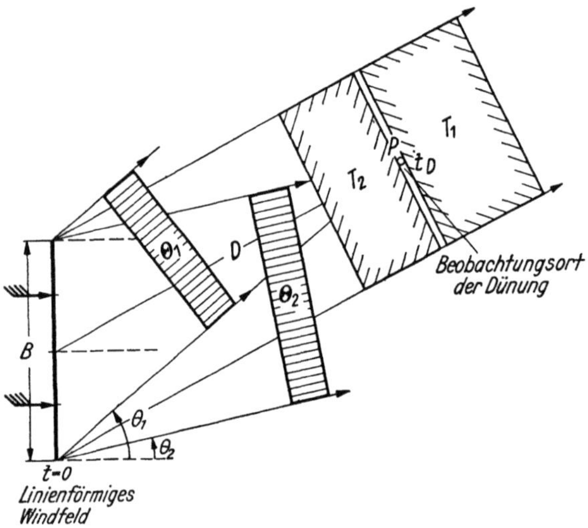

Fig. 26. Maßzahlen eines Dünungsfilters bei einem linienförmigen Wellenentstehungsgebiet.

Die zur Zeit t_D bei P vorhandenen Wellenperioden bzw. Wellenrichtungen fallen also in den Bereich T_1 bis T_2 bzw. Θ_1 bis Θ_2. Durch sie wird aus dem ursprünglich breiten Energiespektrum der Windsee ein mehr oder weniger schmales Band herausgeschnitten, das die Eigenschaften der Dünung am Ort P zur Zeit t_D beschreibt. Dieses Berechnungsverfahren gleicht also der Anwendung eines *Filters*, dessen Maßzahlen bestimmt werden durch die Eigenschaften des Wellenentstehungsgebietes sowie Ort und Zeit der Dünungsbeobachtung. Hier wurde nur der einfache Fall einer linienförmigen Wellenquelle behandelt. Länge (F), Bewegung und Veränderung des Wellenentstehungsgebietes komplizieren die Bestimmung der Maßzahlen des Filters.

Mit den durch Anwendung des Filters erhaltenen Grenzperioden T_1 und T_2 wird aus dem Energiespektrum der Windsee im Wellenentstehungsgebiet (Fig. 22) derjenige *Energiebetrag* bestimmt, der in diesem *Periodenbereich* vorhanden ist (Fig. 27). Dieser Energiebetrag wird sodann mit Hilfe einer Verteilungskurve der Wellenenergie auf die verschiedenen Wellenrichtungen, der die in (14.7) verwendete Richtungsabhängigkeit zugrundeliegt, auf den *Restbetrag* reduziert, der auf den *Richtungssektor* zwischen Θ_1 und Θ_2 entfällt (Fig. 28). Aus dieser Restenergie kann die Charakteristik der scheinbaren Wellenhöhen in bekannter Weise abgeleitet werden.

Aus (14.6) ergibt sich übrigens, daß die mittlere scheinbare Periode $\overline{\overline{T}}_D$ der höchsten Dünungswellen, die ja der Periode $T_{\max}$ der energiereichsten Wellen der Windsee gleicht, maximal gleich $\sqrt{2}\,\overline{\overline{T}}$ werden kann, wenn $\overline{\overline{T}}$ die mittlere scheinbare Wellenperiode im Wellenentstehungsgebiet ist. Die *Zunahme der Dünungswellenperiode* mit der Dünungslaufstrecke ist also *begrenzt*.

Diese von PIERSON, NEUMANN und JAMES vorgeschlagene Behandlung der Dünungsausbreitung hat etwas Bestechendes an sich. Mit einfachsten Vorstellungen und Methoden wird hier ein realistisches Bild von der Umwandlung der Windsee in Dünung erzielt, das eine Reihe von Beobachtungstatsachen mühelos deutet, wie z.B. die bekannte Regelmäßigkeit „alter" Dünungen, die sich hier einfach dadurch erklärt, das Frequenzband und Richtungsbereich der Dünung um so schmäler, die Kammlängen daher um so größer und die Perioden um so regelmäßiger werden müssen, je weiter sich der Beobachtungsort vom Wellenentstehungsgebiet entfernt. Hierbei bewährt sich die von den bisherigen Auffassungen abweichende Annahme der Winkelausbreitung, die allerdings auch zur Folge hat, daß Dünungswellen aus Richtungen auftreten können, in deren rückwärtiger Verlängerung (Großkreis) kein durch gleiche Windrichtung gekennzeichnetes Wellenentstehungsgebiet existiert. Allerdings scheint die Energieverteilung auf die Wellenrichtungen noch mehr oder weniger hypothetisch zu sein und daher noch der Bestätigung zu bedürfen. Soweit bekannt, zeitigte diese Methode bisher gute Erfolge (JAMES[1]).

Die zusätzliche Wirkung von Luftwiderstand, Rücken- und Gegenwinden sowie

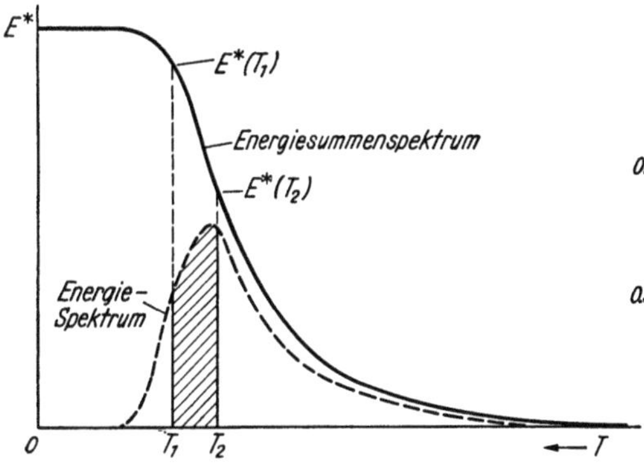

Fig. 27. Wirkung der Grenzperioden T_1 und T_2 auf die Energie der Dünungswellen (nach PIERSON, NEUMANN und JAMES).

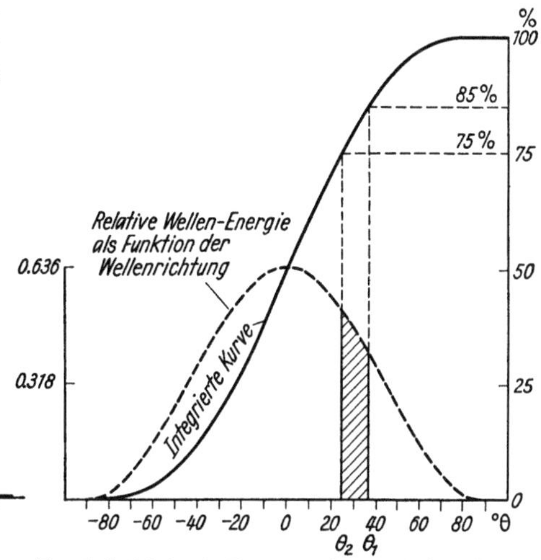

Fig. 28. Reduktion der Dünungswellenenergie durch Beschränkung auf den Wellenrichtungssektor zwischen Θ_1 und Θ_2 (nach PIERSON, NEUMANN und JAMES).

ozeanischer Turbulenz bliebe noch zu klären. Es ist möglich, daß diese Effekte von minderer Bedeutung sind, da sie hauptsächlich die kurzperiodischen Dünungsanteile beeinflussen dürften, die ohnehin von der lokalen Windsee überdeckt werden.

Die praktische Anwendung der Verfahren zur Berechnung der Dünung wird naturgemäß dadurch erschwert, daß die erzeugenden Windfelder sich zeitlich und räumlich verändern. Mit diesem Problem hat sich WALDEN[2] auf der Basis der Theorie von PIERSON-NEUMANN-JAMES beschäftigt.

Ausgehend von ihren in Ziff. 23 erwähnten Gedankengängen haben ferner GELCI[3] und Mitarbeiter ein Verfahren zur Berechnung der Dünung entwickelt, bei dem der Transport an Wellenenergie in Richtung auf den Beobachtungsort fortlaufend berechnet wird. Hierbei wird auch die Einwirkung des Windes auf die Dünung berücksichtigt. Dieses Verfahren hat sich bei der Dünungsvorhersage für die atlantische Küste Nordafrikas bewährt.

b) Transformation der Wellen vor der Küste (Brandung).

Soweit die Energie der winderzeugten Meereswellen nicht auf dem Meere selbst durch turbulente Reibung, Luftwiderstand, dämpfende Windeinwirkung oder andere Effekte aufgezehrt wird — wir haben Grund, dies nur zu einem

[1] S. 716, Fußnote 1.

[2] H. WALDEN: Dtsch. Hydrogr. Z. 7, 59, 174, 183 (1954).

[3] R. GELCI: Bull. Sc. C. O. E. C. du Maroc, Ser. 2 No. 2, 25, 1954. — R. GELCI u. H. CAZALÉ: Ann. Serv. du Globe et de Mét. de l'Inst. Sc. Chérifien, Année 1947, 59. 1954.

kleinen Teil anzunehmen — werden diese Energien laufend an die Küsten der Kontinente und Inseln transportiert und dort in andere Formen (Strömungen, Sandtransport) umgewandelt oder in der Brandung durch Turbulenz vernichtet.

27. Die Wellen in der Übergangszone. Unsere bisherigen Betrachtungen galten fast ausschließlich Tiefwasserwellen, d.h. wir sahen für die harmonischen Teilwellen des Seeganges (und meist auch für die charakteristischen Wellen) die Beziehung (3.16) als gültig an. Dieser Gültigkeitsbereich endet nach den klassischen Vorstellungen praktisch bei der Wassertiefe $h = L/2$. Dort beginnen die Wellen *den Meeresboden zu fühlen* und damit die Gl. (3.15) zu befolgen. Diese Beziehung gründet sich allerdings auf zwei wesentliche Voraussetzungen, nämlich die Kleinheit der Amplituden (wegen der Linearisierung) und die Konstanz der Wassertiefe.

Beide sind im Küstenvorfeld nicht oder zumindest nicht streng erfüllt. Welche anderen Darstellungsmöglichkeiten existieren?

Der *Veränderung der Wassertiefe* vor der Küste wurde bisher nur für die linearisierten Gleichungen, also für infinitesimale Amplituden, Rechnung getragen. STOKER[1] und FRIEDRICHS[2] behandelten Wellen mit unendlich langen Kämmen, die parallel, PETERS[3] [6] untersuchte solche, die schräg zu einer geraden Strandlinie verlaufen, wobei der nach See zu abfallende Meeresboden durch eine Ebene gebildet wird. ECKART [6] gelang eine angenäherte Lösung für den allgemeinen Fall. Die Unterschiede gegenüber der für konstante Wassertiefe abgeleiteten Beziehung (3.15) ergaben sich dabei jedoch für alle h/L-Werte zu höchstens 4%, so daß wir weiter (3.15) anwenden können, sofern die Neigung des Meeresbodens gering ist.

Die *Einbeziehung endlicher Amplituden* würde den Verzicht auf Linearität der Grundgleichungen erfordern und damit auf die Möglichkeit, den unregelmäßigen Seegang durch Superposition von harmonischen Teilwellen anzunähern. Zwar ist es FUCHS [6] gelungen, kurzkämmige Wellen endlicher Amplitude nach STOKES darzustellen. Dies würde der Superposition zweier langkämmiger STOKES-Wellen entsprechen. Die Überlagerung einer beliebigen endlichen oder unendlichen Anzahl solcher Wellen ist indessen bisher nicht erreicht worden. Aus diesem Grunde (und vielleicht auch der Einfachheit wegen) wird heute vielfach der linearen Theorie der Vorzug gegeben vor den approximativen Lösungen der nicht-linearen Gleichungen (STOKES[4], STRUIK[5], LEVI-CIVITA[6]), die naturgemäß nur auf konservative Wellen führen. Wir müssen dann allerdings unsere Betrachtungen auf eine *Übergangszone* beschränken, deren seewärtige Begrenzung etwa bei $h/L = 2$ liegt und die im Flachwasser vor der Küste dort endet, wo die Zunahme der Wellenhöhen die Brecherentwicklung einleitet. Die eigentliche *Brandungszone* fällt also nicht in diesen Bereich.

Die Umwandlung der Wellengrößen vor der Küste besitzt ein grundsätzliches Merkmal: Die *Konstanz der Periode.* Hierbei sind im strengen Sinne nur Perioden der spektralen Teilwellen gemeint. (Jede Änderung der Wellenperiode würde eine ständige Zu- oder Abnahme der Wellenzahl zwischen zwei festen Punkten ihres Weges bedingen, was nicht angängig ist.)

Wir betrachten die Umgestaltung von Wellen, die *senkrecht zu den Tiefenlinien* auf die Küste zu wandern. Bezeichnen wir jetzt mit C_0 und L_0 die durch

[1] J. J. STOKER: Quart. Appl. Math. **5**, 1 (1947).
[2] K. O. FRIEDRICHS: Comm. Pure Appl. Math. **1**, 109 (1948).
[3] A. S. PETERS: Comm. Pure Appl. Math. **5**, 87 (1952).
[4] S. 673, Fußnote 1.
[5] D. J. STRUIK: Math. Ann. **95**, 595 (1926).
[6] S. 674, Fußnote 1.

(3.16) gegebenen Tiefwasserwerte der Wellengeschwindigkeit und Wellenlänge und verstehen unter C und L nunmehr die Flachwasserwerte, so erhalten wir aus (3.15) für das Verhältnis dieser Größen bei konstanter Periode T

$$\frac{C}{C_0} = \frac{L}{L_0} = \operatorname{Tan} \frac{2\pi h}{L}. \tag{27.1}$$

Die Änderung der Wellenhöhe ergibt sich unter der Voraussetzung, daß der Energietransport in Richtung auf die Küste konstant ist, aus der Gl. (4.4) zu

$$\frac{H}{H_0} = \sqrt{\frac{E}{E_0}} = \sqrt{\frac{C_0}{C} \frac{1}{1 + \frac{4\pi h}{L \operatorname{Sin} \frac{4\pi h}{L}}}}. \tag{27.2}$$

Durch Kombination von (27.1) und (27.2) folgt die Änderung der Wellensteilheit. In der vorliegenden Form sind diese Beziehungen noch nicht befriedigend, da rechts die unbekannte Wellenlänge L über Flachwasser erscheint. Durch Iteration lassen sich jedoch beide Gleichungen als Funktion der *relativen Wassertiefe* h/L_0 ausdrücken. Das Ergebnis zeigt Fig. 29. Während Wellenlänge und Wellengeschwindigkeit mit abnehmender Wassertiefe ebenfalls geringer werden, steigt die Wellenhöhe — nach einer anfänglichen durch $1 + \frac{4\pi h}{L \operatorname{Sin} \frac{4\pi h}{L}}$ verursachten leichten Abnahme um 9% — infolge der Zunahme von C_0/C ständig an. Einen ähnlichen Verlauf können wir dementsprechend auch bei der Wellensteilheit feststellen. Diese Ergebnisse der linearen Wellentheorie sind gemäß SVERDRUP und MUNK[1] durch Feld- und Laboratoriumsmessungen im wesentlichen bestätigt worden, während spätere Untersuchungen von IVERSEN [6] im Wellenkanal ergaben, daß die lineare Theorie den Messungen nur für kleine Wellensteilheiten H_0/L_0 und große relative Wassertiefen h/L_0 annähernd entspricht. Bei größeren Wellensteilheiten tritt H_0/L_0 als zusätzlicher Parameter auf; dann muß die nichtlineare Theorie endlicher Amplituden zur Darstellung herangezogen werden.

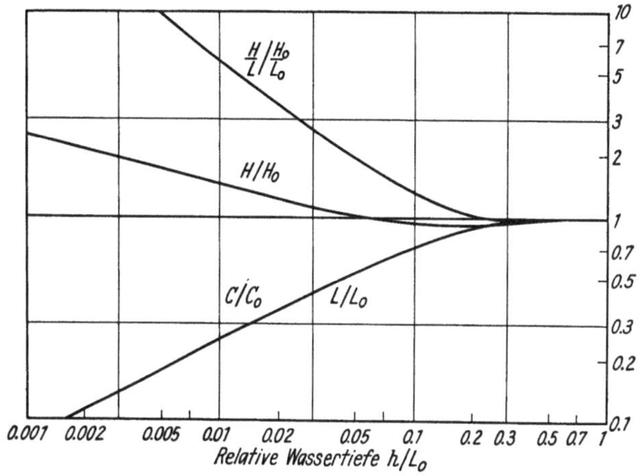

Fig. 29. Änderung der Wellengrößen mit abnehmender „relativer Wassertiefe" h/L_0 (nach der linearen Theorie).

Wenn wir nunmehr auch Wellenzüge einbeziehen, die schräg auf eine Küste treffen, so müssen wir berücksichtigen, daß solche Wellen eine *Refraktion* erleiden. Hierbei kommen die küstennahen Teile eines Wellenkammes wegen der dort geringeren Wassertiefe langsamer voran als die küstenfernen Partien. Wir beobachten daher bei solchen Wellen die Tendenz, allmählich küstenparallel zu werden (Fig. 30). Die Richtungsänderung der Wellenkämme folgt dem Brechungsgesetz von SNELLIUS, wobei die Wellengeschwindigkeiten gemäß (3.15) aus den Wassertiefen zu bestimmen sind. Wegen der unregelmäßigen Gestalt

[1] H. U. SVERDRUP u. W. H. MUNK: Trans. Amer. Geophys. Un. **27**, 828 (1946).

des Meeresbodens kommen theoretisch-rechnerische Verfahren kaum in Frage. Die Lage der Wellenkämme und deren orthogonale Trajektorien werden daher im allgemeinen mit graphischen Methoden ermittelt, die auf Grund von *Prinzipien der geometrischen Optik* entwickelt wurden (JOHNSON, O'BRIEN und ISAACS[1], MUNK und TRAYLOR[2], ARTHUR, MUNK und ISAACS[3]).

Mit der Refraktion ist außer der Änderung der Wellenrichtung auch eine der Wellenhöhe verbunden. Diese läßt sich unmittelbar aus dem Verlauf der orthogonalen Trajektorien, insbesondere deren *Konvergenz* und *Divergenz* erschließen. Unter der Voraussetzung, daß keine Energie senkrecht zu diesen Trajektorien transportiert wird, ist einfach die Änderung des Abstandes s zweier

Fig. 30. Refraktion einer langperiodischen, aus der Sudhemisphäre stammenden Dünung vor der kalifornischen Küste. [Luftaufnahme aus W. H. MUNK u. M. A. TRAYLOR: J. of Geology 55, 1 (1947).]

benachbarter Trajektorien ein Maß für die *Änderung der Wellenhöhe*. Quantitativ läßt sich dieser Zusammenhang durch eine entsprechende Ergänzung von (27.2) formulieren. Bedeutet s_0 den Abstand dieser orthogonalen Trajektorien in tiefem Wasser, so wird

$$\frac{H}{H_0} = \sqrt{\frac{C_0}{C} \, \frac{1}{1 + \dfrac{4\pi h}{L \operatorname{Sin} \dfrac{4\pi h}{L}}} \, \frac{s_0}{s}}. \tag{27.3}$$

Konvergenzen und damit Erhöhungen der Wellenhöhe finden sich häufig an Küstenvorsprüngen, die sich unter Wasser fortsetzen, Divergenzen und damit Wellenhöhenabnahmen dagegen über submarinen Rinnen (Canyons). Es ist durch Anwendung dieser einfachen Verfahren gelungen, örtliche Unterschiede des Wellenangriffs auf Küstenbauten zu erklären und gefährdete Stellen zu bestimmen (HORRER[4], Fig. 31).

[1] J. W. JOHNSON, M. P. O'BRIEN u. J. D. ISAACS: U. S. Hydr. Office Publ. No. 605, 1948.
[2] W. H. MUNK u. M. A. TRAYLOR: J. Geology **55**, 1 (1947).
[3] R. S. ARTHUR, W. H. MUNK u. J. D. ISAACS: Trans. Amer. Geophys. Un. **33**, 855 (1952).
[4] P. L. HORRER: Bull. Beach. Erosion Board **4**, (3) 1 (1950).

In einer Kritik dieser Methoden hat PIERSON [*9*] darauf hingewiesen, daß die benutzten Beziehungen streng genommen nur für reinperiodische Wellen gelten und auf die charakteristischen Wellen nicht angewendet werden dürfen. Für eine durch ein enges Frequenzband ausgezeichnete Dünung mögen die durch Benutzung charakteristischer Wellen entstandenen

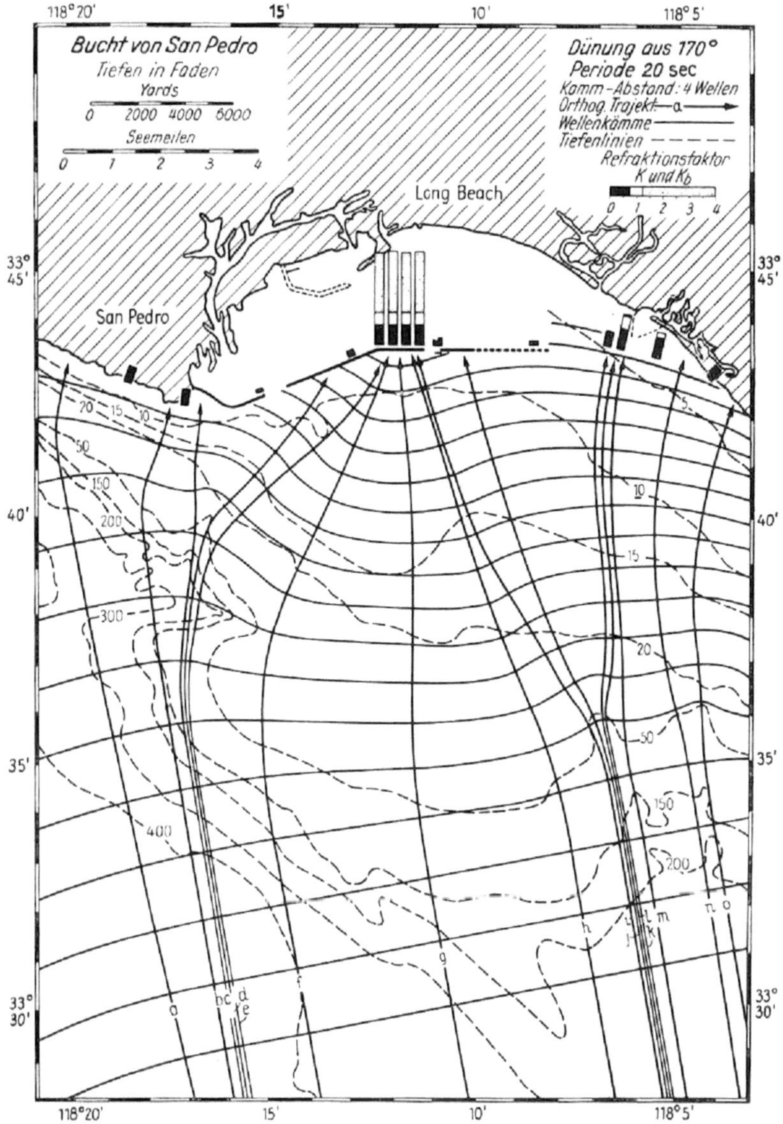

Fig. 31. Refraktionsdiagramm für die Bucht von San Pedro in Kalifornien (nach HORRER).

Fehler noch nicht stark ins Gewicht fallen, umsomehr jedoch bei dem breiten Spektrum der Windsee, das kurze und lange Perioden enthält, die bei der Refraktion getrennt behandelt werden müssen. Der Versuch, ein solches Spektrum mit Hilfe *einer* charakteristischen Wellenperiode der Refraktion zu unterwerfen, muß daher zu falschen Ergebnissen führen. PIERSON hat im Anschluß an (15.4) Seegangsdarstellungen für die Übergangszone geliefert, die genau wie im tiefen Wasser „GAUSSische" Eigenschaften aufweisen, und auch Verfahren [*13*] für die *einwandfreie Behandlung der Refraktion* angegeben. Allerdings läßt es sich dabei nicht vermeiden, das zweidimensionale Seegangsspektrum sowohl nach der Frequenz als auch nach der Wellenrichtung in eine endliche Anzahl von Elementen zu zerlegen und auf jedes dieser Elemente eine Refraktions- und eine Richtungsfunktion anzuwenden, die graphisch aus der Refraktion bei bestimmten harmonischen Frequenzen und Richtungen ermittelt wurden. Die so behandelten Elemente müssen dann wieder zu einem Spektrum zusammengefügt werden, das nun die Bewegung der Meeresoberfläche an einem bestimmten Punkt der Übergangszone repräsentiert. Es bleibt abzuwarten, ob dieses zwar sehr korrekte, aber auch sehr mühsame Verfahren sich bewähren wird.

Für manche praktischen Zwecke ist es erwünscht, die *Laufzeit* der Wellen zwischen zwei Punkten verschiedenen Küstenabstandes zu kennen. Hierfür hat Fuchs[1] unter der Annahme harmonischer Wellen und konstanter (geringer) Bodenneigung eine Beziehung angegeben, die durch Sibul[2] im Laboratorium und durch Wiegel[3] im Feldversuch geprüft und für ausreichend genau befunden wurde.

Bei der bisherigen Betrachtung der Wellenumwandlung in der Übergangszone wurde vorausgesetzt, daß die zwischen zwei benachbarten orthogonalen Trajektorien transportierte Wellenenergie konstant bleibt. Die erhaltenen Änderungen der Wellengrößen ergaben sich also allein durch die *Grenzbedingungen*, nämlich das Heranrücken des als fest und undurchlässig gedachten Meeresbodens. Will man das Problem vollständig behandeln, müssen folgende energieändernden Einflüsse berücksichtigt werden:

a) Energie-Dissipation durch Bodenreibung und Perkolation (oszillatorische Sickerbewegungen des Wassers in einem permeablen Meeresboden);

b) Energie-Zerstreuung durch Diffraktion;

c) Energie-Reflexion an steilen Abhängen;

d) Energie-Verluste durch Turbulenz (Brandung);

e) Energie-Zufuhr durch den Wind.

In den Bereich dieses Abschnittes fallen a), b) und e). Die Wirkungen der *Bodenreibung und Perkolation* werden die einzelnen Teile des Wellenspektrums verschiedenartig beeinflussen. Man kann also eine selektive Auslöschung erwarten. Wegen der Nicht-Linearität der Dissipationsfunktion ist jedoch eine spektrale Betrachtung dieser Vorgänge noch nicht gelungen. Man beschränkte sich bisher auf rein-periodische Wellen bzw. charakteristische Wellengrößen.

Nachdem Putnam und Johnson[4] sowie Putnam[5] die Energie-Dissipation berechnet hatten, die harmonische Wellen geringer Steilheit durch Bodenreibung und Perkolation erleiden, haben Bretschneider und Reid[6] diese Verluste entlang der orthogonalen Trajektorien für die Fälle konstanter Wassertiefe sowie konstanter und variabler Bodenneigung integriert. Diese Lösungsform eignet sich insbesondere für die Behandlung von *Dünungswellen*, die aus dem tiefen Wasser kommend in das Küstenvorfeld hineinlaufen.

Bretschneider[7] hat ferner die *Entstehung von Windsee-Wellen im flachen Wasser* unter Berücksichtigung von Windwirkung, Bodenreibung und Perkolation untersucht, wobei die entsprechenden Beziehungen bei Tiefwasserwellen[8] mit

[1] R. A. Fuchs: Univ. of Calif., Inst. Engineering Research, Wave Res. Lab., Techn. Rep. Ser. 29 Issue 55, 1954.

[2] O. Sibul: Univ. of Calif., Inst. Engineering Research, Wave Res. Lab., Techn. Rep. Ser. 29 Issue 57, 1954.

[3] R. L. Wiegel: Univ. of Calif., Inst. Engineering Research, Wave Res. Lab., Techn. Rep. Ser. 29 Issue 56, 1954.

[4] J. A. Putnam u. J. W. Johnson: Trans. Amer. Geophys. Un. **30**, 67 (1949).

[5] J. A. Putnam: Trans. Amer. Geophys. Un. **30**, 349 (1949).

[6] C. L. Bretschneider u. R. O. Reid: U.S. Beach Erosion Board, Techn. Mem. No. 45, 1954.

[7] C. L. Bretschneider: U.S. Beach Erosion Board, Techn. Mem. No. 51, 1954.

[8] S. 711, Fußnote 1.

den oben erwähnten Ergebnissen über die Energieverluste am Meeresboden kombiniert wurden. Es zeigte sich, daß — neben der Winddauer und der Windwirklänge — Wassertiefe und Bodengestalt das Wachstum der Windsee entscheidend begrenzen.

Messungen an Dünungs- und Windsee-Wellen im Golf von Mexiko[1] bestätigten die theoretischen Ableitungen in ihrer Form und lieferten zahlenmäßige Unterlagen über die Konstanten der Reibung. Sie ließen ferner vermuten, daß außer den berücksichtigten Effekten auch Energieverluste durch Mitschwingen eines nicht-starren Meeresbodens möglich sind.

Über den *Beginn des Einflusses der Wassertiefe* auf die Wellenausbreitung sagen Laboratoriumsmessungen von SIBUL[2] einiges aus. Danach werden die (charakteristischen) Wellenhöhen für $h/H_0 < 5$, die Wellenperioden für $h/L_0 < 0{,}2$ durch die Wassertiefe reduziert. Auch die Höhencharakteristik wird im Flachwasser geändert. So ergab sich hier z.B. $\tilde{H}_{\frac{1}{3}}/\bar{\bar{H}} = 1{,}44$ und $\tilde{H}_{\max}/\bar{\bar{H}} = 1{,}93$ im Gegensatz zu den für tiefes Wasser abgeleiteten Beziehungen (10.1) und (10.3)

Die *Energie-Zerstreuung durch Diffraktion* und die damit verbundenen Wellenvorgänge sollen hier nicht behandelt werden. Es werden dabei mit Erfolg Verfahren benutzt, die den entsprechenden Methoden der Optik ähnlich sind. Als praktische Anwendungen seien erwähnt: Ausbreitung von Wasseroberflächenwellen hinter dem Kopf oder hinter einer Lücke eines Wellenbrechers oder hinter einer Hafeneinfahrt[3].

28. Brandende Wellen. An die in Ziff. 27 behandelte Übergangszone schließt sich küstenwärts die Brandungszone an. Sie ist gekennzeichnet durch eine rasche Zunahme von Wellenhöhe und Wellensteilheit bis zu dem Punkt, an dem die Wellen instabil werden, überbrechen, ihre Form verlieren und ihre Bewegungsenergie in Turbulenz oder andere Energieformen umsetzen.

Der *Vorgang des Überbrechens der Wellen* läßt sich mit der wünschenswerten Vollständig- und Genauigkeit bisher nur im Laboratorium an künstlich erzeugten Wellen untersuchen (SUQUET[4], IVERSEN [*6*], MASON [*6*]).

Nach MASON geht das Überbrechen folgendermaßen vor sich: Bei Annäherung an die Brandungszone wird die Welle dadurch deformiert, daß die Geschwindigkeit des vorderen Hanges reduziert und dieser vom seewärtigen Abhang überholt wird. Hierbei scheinen außer der Abnahme der Wassertiefe auch Geschwindigkeit und Tiefe des Rückstromes von der vorhergegangenen Welle eine Rolle zu spielen. Unter dieser Wirkung wird die Frontseite der Welle laufend steiler und schließlich nahezu vertikal, um sodann instabil zu werden und überzubrechen. Es werden 3 Typen von Brechern unterschieden. Die charakteristischen Merkmale des *schäumenden* (spilling) und des *stürzenden* (plunging) Brechers lassen sich aus den Aufnahmen von MASON (Fig. 32) ablesen. Der dritte Typ, der *Schwall*-Brecher (surging breaker) ist dadurch gekennzeichnet, daß die Vorderseite der Welle weder wie beim schäumenden Brecher allmählich reduziert wird noch wie beim stürzenden Brecher in einem plötzlichen Überrollen zusammenbricht, sondern in größeren Teilen turbulent wird und sich als schäumender Wasserschwall den Strand hinaufbewegt. Die genannten Brechertypen sind nicht als Entweder-Oder-Kategorien aufzufassen, sondern mehr als Grenzfälle der möglichen Zwischenformen. Bei einer hohen Wellensteilheit im tiefen Wasser

[1] C. L. BRETSCHNEIDER: U.S. Beach Erosion Board, Techn. Mem. No. 46, 1954.

[2] O. SIBUL: U.S. Beach Erosion Board, Techn. Mem. No. 72, 1955.

[3] J. A. PUTNAM u. R. S. ARTHUR: Trans. Amer. Geophys. Un. **29**, 481 (1948). — F. L. BLUE jr. u. J. W. JOHNSON: Trans. Amer. Geophys. Un. **30**, 705 (1949). — J. H. CARR and M. E. STELZRIEDE [*6*]. — J. H. DUNHAM [*7*]. — J. W. JOHNSON [*8*].

[4] F. SUQUET: Houille blanche **5**, 343 (1950).

und wenig geneigtem Meeresboden scheinen die schäumenden Brecher bevorzugt aufzutreten (Windsee-Brandung), während bei geringen Anfangssteilheiten die stürzenden Brecher überwiegen, insbesondere bei einem starken Anstieg des Meeresbodens und kräftigem Wellen-Rückstrom (Dünungs-Brandung). Schwall-Brecher sind speziell bei extrem niedrigen Wellensteilheiten im tiefen Wasser und starker Neigung des Meeresbodens zu erwarten (Iversen [*6*]). Allgemein sind die aus gleichen Tiefwasserwellen sich entwickelnden Brecher bei steilem Anstieg des Meeresbodess höher als bei flachem.

Die Messungen der *Kammwinkel beim Überbrechen* erscheinen so wenig einheitlich wie die Verfahren zu ihrer Bestimmung. Mason fand einen weiten Bereich von 88° bis 145°, während die (allerdings in hiervon abweichender Weise bestimmten) Werte von Iversen für die Strandneigungen 1:10, 1:20 und 1:50

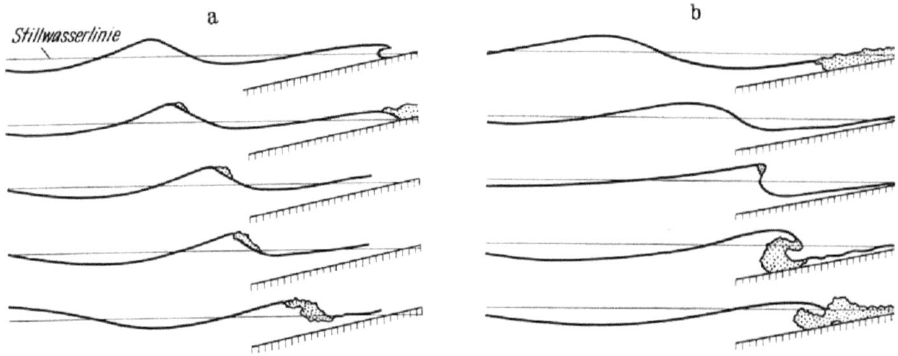

Fig. 32a u. b. Brecher-Entwicklung nach Laboratoriumsversuchen von Mason. a Schäumender Brecher; b Stürzender Brecher.

im Mittel bei 121°, 127° und 133° lagen. Hieraus würde sich — abgesehen von den Unterschieden in der Form (Unsymmetrie) — eine verhältnismäßig gute Übereinstimmung mit dem theoretischen Wert von 120° (Stokes[1]) zumindest für stärkere Neigungen des Meeresbodens ergeben.

Für die *mathematische Formulierung der geschilderten Vorgänge* liegen verschiedene Ansätze vor. Zunächst muß festgestellt werden, daß von linearisierten Theorien in diesem Bereich steiler Wellen kein Erfolg zu erwarten ist, nachdem sie bereits in der Übergangszone nur bedingt zu verwenden waren.

Es stände nunmehr die auf Stokes zurückgehende nicht-lineare Theorie für solche Fälle endlicher Amplituden zur Verfügung. Ihre Anwendung führt jedoch auf unendliche Reihen, die um so langsamer konvergieren, je mehr man sich dem Ort des Überbrechens nähert[2]. Dieses Verfahren erweist sich daher als unhandlich. Auch weicht es insofern von der Wirklichkeit ab, als die Asymmetrie der Brandungswellen nicht wiedergegeben wird.

Die Beobachtung lehrt nun, daß die in das flache Wasser der Brandungszone hineinlaufenden Wellen mehr und mehr ihre ursprüngliche, zur Stillwasserlinie annähernd symmetrische Form verlieren und als schmale Kämme erscheinen, die durch breite flache Täler voneinander getrennt sind. Solche Wellenkämme könnten nahezu als selbständige, isolierte Erscheinungen aufgefaßt werden. Ihre charakteristische Größe ist weniger die Wellenlänge als die Wellenhöhe.

Unter diesem Gesichtspunkt bietet sich die Theorie der *Einzelwelle* zur Anwendung auf die Brandungswellen an. Eine solche Welle besteht aus einem ein-

[1] S. 673, Fußnote 1.
[2] W. H. Munk [*5*].

zelnen, unendlich langen Wellenkamm, der vollständig über dem Stillwasserspiegel liegt und sich mit seinen beiden Abhängen allmählich diesem zuneigt, um im Unendlichen dieses Niveau zu erreichen (Fig. 33). Ihre Geschwindigkeit ist durch $C = \sqrt{g(h+H)}$ gegeben. Die Partikelbewegung in der Einzelwelle erfolgt in Orbitalbahnen, die unten abgeflachten Ellipsen ähneln. Insgesamt findet ein Massentransport in der Wellenrichtung statt. Er ist an der Oberfläche stärker als am Boden.

Diese spezielle Wellenform — von SCOTT-RUSSELL 1834 erstmalig beobachtet, von BOUSSINESQ, MCCOWAN u.a. theoretisch (vgl. LAMB [*1*], THORADE [*2*]), von DAILY und STEPHAN[1] sowie IPPEN und KULIN[2] experimentell untersucht — hat KELLER [*5*] kürzlich erneut diskutiert und MUNK [*5*] bei der mathematischen Behandlung der Wellen in der Brandungszone benutzt. Hierbei bedeutet die Anwesenheit einer Küste eine im mathematischen Modell nicht enthaltene Beschränkung und Veränderung (Rückstrom). Die Analogie hat also ihre Grenzen. Unter der Voraussetzung der Konstanz der Wellenperiode, der Erhaltung der Energie und der Anwendbarkeit der Theorie der Einzelwelle gelangte MUNK zu den folgenden, bei Abwesenheit von Refraktion geltenden Beziehungen für die *Brecherhöhe* H_B und die *Wassertiefe* h_B *beim Überbrechen*

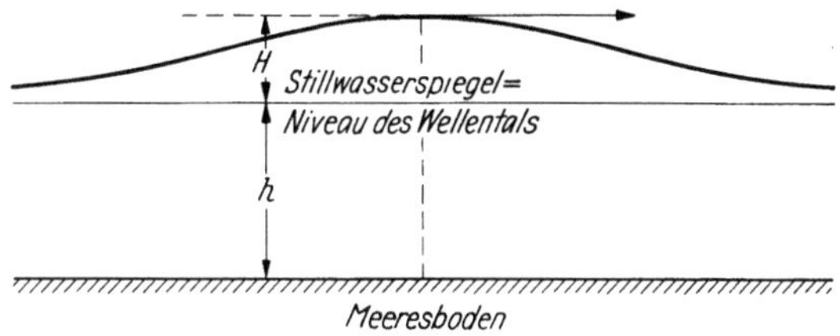

Fig. 33. Einzelwelle.

$$\frac{H_B}{H_0} = \frac{1}{3{,}3\sqrt[3]{H_0/L_0}}, \tag{28.1}$$

$$\frac{h_B}{H_B} = 1{,}28. \tag{28.2}$$

Als *Instabilitätsbedingung* wurde dabei gefordert, daß die Partikelgeschwindigkeit im Wellenkamm gleich C wird. Für den *Kammwinkel* ergab sich sodann 120° wie bei der Theorie von STOKES.

Ein *Vergleich der Ergebnisse* von Feld- und Laboratoriumsmessungen mit (28.1) zeigt zwar im Mittel gute Übereinstimmung. Die Streuung der Einzelwerte ist jedoch beträchtlich und wird durch die Theorie nicht erklärt. Ein geringer Reibungseinfluß ist angedeutet. Die Aussage der Beziehung (28.2) hat sich mit einer gewissen — möglicherweise durch die Ungenauigkeit der Beobachtungen bedingten — Streuung in der Natur bestätigt gefunden. Hierbei sei betont, daß auch bei diesen Ableitungen die Wassertiefe als konstant vorausgesetzt wurde und wir bei stärkeren Änderungen der Wassertiefe füglich keine Übereinstimmung erwarten dürfen. Die unsymmetrische Form der Brandungswellen läßt sich auch aus der Theorie der Einzelwelle nicht ableiten. Als seewärtige Grenze ihres Anwendbarkeitsbereiches gab MUNK die Wassertiefe 1,4 h_B an. In tieferem Wasser führte die lineare Theorie auf bessere Resultate.

Trotz Brauchbarkeit in mancher Hinsicht hat die Theorie der Einzelwelle noch keine vollständig befriedigende Lösung des Brandungsproblems gebracht. Daher seien den Bemühungen um eine bessere Darstellung der Brandung noch einige Worte gewidmet.

[1] J. W. DAILY u. S. C. STEPHAN jr.: Proc. III. Conf. on Coastal Engineering, US-Council on Wave Research, The Eng. Foundation, 13, 1953.

[2] A. T. IPPEN u. G. KULIN: Proc. V. Conf. on Coastal Engineering, U.S.Council on Wave Research, The Eng. Foundation, 27, 1955.

Erwähnt sei zunächst die von Miyazaki[1] vorgelegte, *nicht-lineare Theorie der Brandungswellen*, die die zweidimensionale Wellenbewegung über einem Meeresboden konstanter geringer Neigung unter Voraussetzung der Wirbelfreiheit bis zur Instabilität durch näherungsweise Integration der Grundgleichungen behandelt. Die Ergebnisse weichen von denen Munks nicht unerheblich ab, so ist z.B. das Verhältnis h_B/H_B nicht konstant, sondern weist einen (allerdings geringen) Gang mit H_0/L_0 auf. Wegen der erwähnten Streuung der Meßwerte ist zur Zeit keine Entscheidung darüber möglich, welche Theorie den Messungen besser gerecht wird. Eine Darstellung der Profile von Brandungswellen wurde von Miyazaki nicht gegeben.

Stoker [*5*] hat ferner versucht, die auf Lagrange zurückgehende *Theorie der Flachwasserwellen* in Anlehnung an Entwicklungen der Gasdynamik (Stoßwellen) auf die Brandungswellen anzuwenden. Hierbei ist die Wellenlänge groß gegenüber der Wassertiefe, die Amplitude jedoch nicht notwendig klein gegen diese. Die Wellengeschwindigkeit ist $C=\sqrt{g(\eta+h)}$. Als Ergebnis liefert diese nicht-lineare Flachwassertheorie Erklärungen dafür, warum und wann Wellen überbrechen, sowie Angaben über die dabei zu erwartenden Profile. Die Theorie scheint insbesondere für die Erfassung nicht-stationärer Brandungsvorgänge, wie z.B. den stürzenden Brecher, geeignet zu sein, während die Theorie der Einzelwelle eher auf solche quasi-stationären Wellenbewegungen wie den schäumenden Brecher angewendet werden kann.

Schließlich müssen wir in diesem Zusammenhang auf eine Studie von Biesel [*6*] hinweisen, in der die Grundgleichungen der wirbelfreien Wellenbewegung zunächst für den zweidimensionalen Fall unter der alleinigen Voraussetzung näherungsweise gelöst werden, daß die Quadrate sowie höheren Ableitungen der Neigung des Meeresbodens vernachlässigt werden dürfen. Es ergeben sich *unsymmetrische Wellenprofile* mit steilen Lee- und flacheren Luvhängen. Diese Unsymmetrie hat die Tendenz, sich rasch zu verstärken, wenn die relative Wassertiefe gering wird. Bei zunehmender Wassertiefe wird diese Asymmetrie dagegen weniger deutlich und verschwindet für $h/L=0{,}27$. Die von Biesel berechneten Profile von Brandungswellen beim Überbrechen gehören zu den mathematischen Darstellungen dieses Vorganges, die der Wirklichkeit am nächsten kommen.

Es wurde bereits angedeutet, daß durch den Vorgang der Brandung nicht alle ankommende Wellenenergie durch Turbulenz vernichtet wird, sondern daß dort oder in der Übergangszone eine Umwandlung der Wellenenergie in andere Energieformen stattfindet. Wir denken hierbei in erster Linie an die *küstennahe Wasserzirkulation*, bestehend aus küstenparallelen und Rip-Strömungen, dann aber auch an den von den Wellen bewirkten *Sandtransport*, an die *Wellenwirkungen auf Küstenbauten* und schließlich an mehr untergeordnete Erscheinungen wie die langperiodischen *Brandungsschwebungen* und gelegentliche *Reflektionseffekte*. Ein Teil dieser Probleme, wie z.B. der Sandtransport, besitzt große wirtschaftliche Bedeutung im Rahmen des Küstenschutzes und hat sich daher zu einem umfangreichen Forschungsgebiet ausgeweitet. Auf die zahlreichen damit zusammenhängenden Fragen kann an dieser Stelle weder ausführlich noch andeutungsweise eingegangen werden.

Literatur.

[*1*] Lamb, H.: Hydrodynamics, 6th ed. Cambridge Univ. Press 1932. Deutsche Ausgabe: Lehrbuch der Hydrodynamik. Leipzig u. Berlin 1931. (Abschnitt IX: Umfassende mathematische Behandlung der Oberflächenwellen auf der Grundlage der klassischen Theorie.)

[1] M. Miyazaki: Oceanogr. Mag. Japan **3**, 65 (1951); **4**, 13 (1952).

[2] THORADE, H.: Probleme der Wasserwellen. Hamburg: H. Grand 1931. (Messung und Theorie der Oberflächenwellen des Meeres nach dem Stande von 1931.)
[3] CORNISH, V.: Ocean waves and kindred geophysical phenomena. Cambridge 1934. (Ergebnisse langer und gründlicher Beobachtungen, mit einem mathematischen Anhang von H. JEFFREYS.)
[4] SVERDRUP, H. U., and W. H. MUNK: Wind, sea and swell: Theory of relations for forecasting. U. S. Hydrogr. Office Publ. No. 601. 1947. (Halbempirische Theorie basierend auf der Konzeption der charakteristischen Wellen.) Dazu: Techniques for forecasting wind waves and swell. U. S. Hydrogr. Office Publ. No. 604, 1951.
[5] HAURWITZ, B. u. a.: Ocean surface waves. Ann. New York Acad. Sci. **51**, 343—572 (1949). (Ergebnisse einer Tagung über Meereswellen; 15 Vorträge.)
[6] U. S. Nat. Bureau of Standards: Gravity Waves. Proc. NBS Semicentennial Symposium on Gravity Waves (33 Vorträge). U. S. National Bureau of Standards Circular 521, 1952.
[7] JOHNSON, J. W. (Editor): Proceed. I. Conf. on Coastal Engineering, U. S. Council on Wave Research, The Eng. Foundation 1951. (Darin zusammenfassende Aufsätze über Grundprinzipien der Wellenbewegung im Hinblick auf die praktische Anwendung im Rahmen des Küstenschutzes.)
[8] JOHNSON, J. W. (Editor): Proceed. II. Conf. on Coastal Engineering, U. S. Council on Wave Research, The Eng. Foundation 1952. (Wie [7].)
[9] PIERSON, W. J.: A unified mathematical theory for the analysis, propagation, and refraction of storm generated ocean surface waves, Part I and II. New York Univ., Coll. of Eng., Dept. of Meteorology 1952. (Mathematische Formulierung einer möglichst realistischen Darstellung von Windsee, Dünung und Refraktion.)
[10] NEUMANN, G.: On wind generated ocean waves with special reference to the problem of wave forecasting. New York Univ., Coll. of Eng., Dept. of Meteorology 1952. Sowie: Über die komplexe Natur des Seeganges, Teil 1 u. 2. Dtsch. Hydrogr. Z. **5**, 95, 252 (1952). (Theorie der Windsee-Entwicklung auf der Grundlage von Beobachtungen.)
[11] DARBYSHIRE, J.: The generation of waves by wind. Proc. Roy. Soc. Lond., Ser. A **215**, 299 (1952). (Zusammenfassung der Ergebnisse der britischen Seegangsmessungen und deren physikalische Deutung auf der Grundlage des Wellenspektrums.)
[12] NEUMANN, G.: On ocean wave spectra and a new method of forecasting wind—generated sea. U. S. Beach Erosion Board, Techn. Mem. No. 43, 1953. Sowie: Zur Charakteristik des Seeganges. Arch. Meteorol. Geophys. Bioklimat. A **7**, 352 (1954). (Energiespektrum der Windsee und Folgerungen daraus.)
[13] PIERSON, W. J. jr., G. NEUMANN and R. W. JAMES: Practical methods for observing and forecasting ocean waves by means of wave spectra and statistics. New York Univ., Coll. of Eng., Dept. of Meteorology and Oceanogr. Techn. Rept. No. 1, 1953. (Praktische Anwendung von [9], [10], [12].)
[14] BRUNS, E.: Handbuch der Wellen der Meere und Ozeane. Berlin 1953. (Ausführliche Darstellung von Wellenmeßmethoden und Sammlung von Meßergebnissen mit Blickrichtung auf die wasserbautechnischen Erfordernisse.)
[15] PIERSON, W. J.: Wind generated gravity waves, Advances in Geophysics **2**, 93, New York 1955. (Derzeitiger Stand der statistischen Seegangstheorie unter dem Blickwinkel: Die Meeresoberfläche als quasi-stationärer GAUSS-Prozeß.)
[16] MICHE, R.: Propriétés des trains d'ondes océaniques et de laboratoire. Paris 1954. (Eingehende, hauptsächlich theoretische Abhandlung über die Wasseroberflächenwellen auf dem Ozean und im Laboratorium unter Einbeziehung statistischer Methoden.)

Literatur über Meßmethoden.

[17] SCHUMACHER, A.: Stereophotogrammetrische Wellenaufnahmen. Wiss. Ergebnisse Dtsch. Atlant. Expedition „Meteor" VII (2), Berlin 1939.
CALDWELL, J. M.: An ocean wave measuring instrument. U. S. Beach Erosion Board, Techn. Mem. No. 6, 1948.
FOLSOM, R. G.: Measurement of ocean waves. Trans. Amer. Geophys. Un. **30**, 691 (1949).
SNODGRASS, F. E.: Ocean wave measurements. Univ. of Calif., Inst. Eng. Res. Berkeley Calif., Series No. 2 Issue No. 342, 1952. Oder: Oceanographic Instrumentation. U. S. Nat. Research Council, Conf. Rancho Santa Fe, Calif. 139, 1952.
TUCKER, M. J.: Sea wave recording. Dock Harb. Author. **34**, 207 (1953).
VALEMBOIS, J.: Les appareils réalisés à Chatou pour la mesure de la houle naturelle. Proc. V. Conf. on Coastal Engineering, U.S. Council on Wave Research, The Eng. Foundation, 170, 1955.
GERHARDT, J. R., K. H. JEHN and I. KATZ: A comparison of step-, pressure- and continuous-wire-gage wave recordings in the Golden Gate Channel. Trans. Amer. Geophys. Un. **36**, 235 (1955).

Gezeitenkräfte.

Von

JULIUS BARTELS.

Mit 17 Figuren.

A. Elementare Diskussion.

1. Allgemeines. „Gezeitenkräfte“ werden beschrieben durch ein System von Massenbeschleunigungen derselben Art wie die Fallbeschleunigung g, erzeugt durch NEWTONsche Gravitation und Zentrifugalbeschleunigung. Speziell für die Erde sind Mond und Sonne gezeiten-erzeugende Himmelskörper, deren Wirkungen sich addieren.

Das NEWTONsche Gesetz bestimmt die Bahnen der Himmelskörper; sie bewegen sich so, als ob jeweils die gesamte Masse im Schwerpunkt vereinigt wäre. Auch hinsichtlich der Erzeugung der lunaren und solaren Gezeitenkräfte auf der Erde wirken Mond und Sonne wie Punktmassen. Infolge der endlichen Ausdehnung der Erde sind jedoch die Anziehungen des Mondes und der Sonne nicht überall auf der Erde gleich; ihre Unterschiede sind die *Gezeitenbeschleunigungen*. Sie lassen sich durch ein Potential V beschreiben, mit $\partial V/\partial s$ als Beschleunigungskomponente in Richtung s.

Wenn die gesamte Erde völlig starr wäre, so würden nur diese *(„primären“)* Gezeitenkräfte auftreten. Da sich aber, unter der Wirkung der Gezeitenkräfte, Massen auf der Erde verschieben, ändert sich das Gravitationspotential des Erdkörpers um den Betrag, der den verlagerten Massen entspricht. Diese Wirkungen (man könnte sie *„sekundäre“* Gezeitenkräfte nennen) können dieselbe Größenordnung wie die primären Gezeitenkräfte erreichen (Ziff. 21 d); hier sollen zunächst nur die primären behandelt werden.

Besondere Kapitel dieses Handbuches behandeln die *Wirkungen* der Gezeitenkräfte auf den Erdkörper, auf die Ozeane, die Atmosphäre und Ionosphäre, einschließlich der erdmagnetischen Gezeiten. Diese sind vielfach nicht *allein* durch die eigentlichen (primären und sekundären) Gezeitenkräfte bedingt: Mit den solaren (Gravitations)-Gezeitenkräften konkurrieren verschiedene Teile der Sonnenstrahlung mit Tagesperioden, wie Wärmestrahlung in Bodennähe, dissoziierende und ionisierende Wirkungen in der Höhe. Die Diskussion im vorliegenden Kapitel ist so angelegt, daß sie die Möglichkeit solcher superponierter Wirkungen beachtet und die Trennung der verschiedenen Einflüsse in den beobachteten Erscheinungen vorbereitet. Vgl. auch Ziff. 23.

2. Berechnung der Gezeitenbeschleunigungen. Am einfachsten lassen sich die Gezeitenkräfte ausdrücken in einem Koordinatensystem, $\boldsymbol{K}$, dessen Ursprung O sich mit dem Erdschwerpunkt bewegt, und das keine Rotation ausführt *(„Revolution ohne Rotation“)*. Jeder Punkt von $\boldsymbol{K}$ beschreibt also, abgesehen von einer konstanten Verschiebung, dieselbe Bahn wie der Erdschwerpunkt O.

Die Gezeitenbeschleunigungen ergeben sich dann folgendermaßen[1]: Es sei f = Gravitationskonstante, M, S = Massen des Mondes und der Sonne, R, R_s = Abstände ihrer Schwerpunkte vom Erdschwerpunkt O; r, r_s diejenigen von einem beliebigen Punkte P; V_m, V_s die Gravitationspotentiale des Mondes und der

[1] Ziff. 24 (am Schluß) stellt alle verwendeten Symbole zusammen.

Sonne. Dann ist, in einem Inertialsystem $\boldsymbol{I}$, die anziehende Beschleunigung durch Mond und Sonne gegeben durch grad $(V_m + V_s)$, also

$$\boldsymbol{b}_0 = \operatorname{grad}\left(\frac{fM}{R} + \frac{fS}{R_s}\right) \text{ in } O, \qquad \boldsymbol{b} = \operatorname{grad}\left(\frac{fM}{r} + \frac{fS}{r_s}\right) \text{ in } P. \tag{2.1}$$

Unter der gemeinsamen Anziehung von Sonne und Mond beschreibt O eine Bahn, in der $\boldsymbol{b}_0$ durch eine Führungsbeschleunigung $\boldsymbol{t} = -\boldsymbol{b}_0$ kompensiert wird. Dieselbe Führungsbeschleunigung herrscht auch in allen Punkten (z.B. in P), die sich mit O im Koordinatensystem $\boldsymbol{K}$ bewegen. Auf eine Masse im Punkte P, der relativ zu $\boldsymbol{K}$ fest ist, wirkt deshalb die *Gezeitenbeschleunigung* $(\boldsymbol{b} - \boldsymbol{b}_0)$ (vgl. auch das anschauliche Modell in Ziff. 7g und h, S. 742).

Beim Ansatz (2.1) sind Erde, Sonne und Mond als kugel-symmetrische Massen angenommen, deren Gravitationsanziehungen aufeinander also ebenso wirken, wie wenn die jeweiligen Massen im Schwerpunkt vereinigt wären. JUNG[1] hat gezeigt, daß diese Vereinfachung gerechtfertigt ist: die Abplattung der Erde z.B. bewirkt Abweichungen, die nur das 10^{-6}-fache der Gezeitenkräfte erreichen, auf die wir uns hier beschränken.

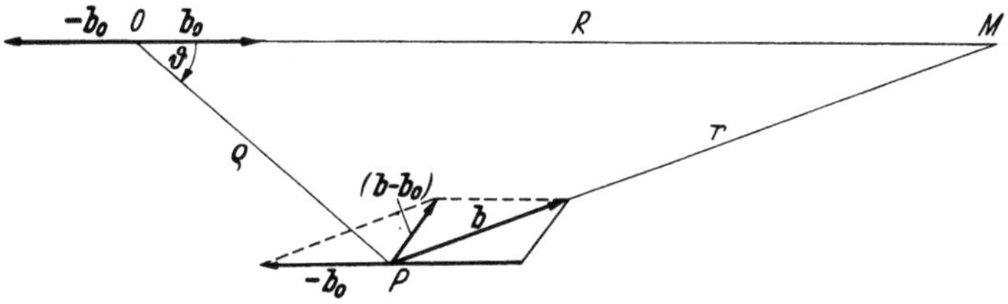

Fig. 1. Entstehung der lunaren Gezeitenbeschleunigung $\boldsymbol{b} - \boldsymbol{b}_0$.

3. Lunares Gezeitenpotential. $\boldsymbol{b}$ ist bereits durch (2.1) als Gradient eines Potentials dargestellt. Um für das Feld der überall gleichen Vektoren $-\boldsymbol{b}_0$ ein Potential zu finden, betrachten wir zunächst den Mondanteil von $-\boldsymbol{b}_0$: Er hat überall den Betrag fM/R^2 und die Richtung MO. Vorübergehend betrachten wir ein rechtwinkeliges Koordinatensystem um O, mit z in Richtung OM. Dann ist das Potential des homogenen Feldes $\boldsymbol{b}_0$ gleich $|\boldsymbol{b}_0| z$, oder, mit ϑ = geozentrische Zenitdistanz des Mondes (Fig. 1), gleich $fM\varrho\cos\vartheta/R^2$. Um im Erdschwerpunkt O $V = 0$ zu erhalten, ziehen wir noch fM/R ab, ein Ausdruck, der in jedem Augenblick überall gleich ist und deshalb keinen Anteil zu grad V liefert. Damit wird das *lunare Gezeitenpotential*

$$V_m = fM\left[\frac{1}{r} - \frac{1}{R} - \frac{\varrho\cos\vartheta}{R^2}\right]. \tag{3.1}$$

Analog ist das Potential der *solaren* Gezeitenkräfte

$$V_s = fS\left[\frac{1}{r_s} - \frac{1}{R_s} - \frac{\varrho\cos\vartheta_s}{R_s^2}\right]. \tag{3.2}$$

Die solaren und lunaren[2] Gezeitenkräfte addieren sich, entsprechend der Linearität der Gln. (2.1).

Im Dreick (Fig. 1) OPM ist $r^2 = R^2 + \varrho^2 - 2R\varrho\cos\vartheta$, also

$$\frac{1}{r} = \frac{1}{R}\left\{1 - 2\frac{\varrho}{R}\cos\vartheta + \left(\frac{\varrho}{R}\right)^2\right\}^{-\frac{1}{2}}. \tag{3.3}$$

[1] K. JUNG: Gerlands Beitr. Geophysik **64**, 278 (1955).

[2] Es ist üblich, bei analogen Größen wie r, R, c, ϑ nur diejenigen für die *Sonne* mit dem Index s zu markieren, diejenigen für den *Mond* dagegen nur im Notfall mit dem Index m zu kennzeichnen.

Für Punkte an der Erdoberfläche (ϱ = Erdradius) ist (ϱ/R) für den Mond 1/60, für die Sonne 1/23600. Man entwickelt (3.3) nach Potenzen von (ϱ/R):

$$\left.\begin{aligned}\frac{1}{r} = \frac{1}{R}\Big\{1 + \frac{\varrho}{R}\cos\vartheta - \frac{1}{2}\left(\frac{\varrho}{R}\right)^2 + \\ + \frac{3}{2}\left(\frac{\varrho}{R}\right)^2\cos^2\vartheta - \frac{3}{2}\left(\frac{\varrho}{R}\right)^3\cos\vartheta + \frac{5}{2}\left(\frac{\varrho}{R}\right)^3\cos^3\vartheta + \left(\frac{\varrho}{R}\right)^4(\cdots)\Big\}.\end{aligned}\right\} \quad (3.4)$$

Dies ist die bekannte Entwicklung nach zonalen Kugelfunktionen

$$\frac{1}{r} = \frac{1}{R}\sum_{n=0}^{\infty}\left(\frac{\varrho}{R}\right)^n P_n(\vartheta) \quad (3.5)$$

mit

$$\left.\begin{aligned} &P_0(\vartheta) = 1, \qquad P_1(\vartheta) = \cos\vartheta, \\ &P_2(\vartheta) = \tfrac{1}{2}(3\cos^2\vartheta - 1) = \tfrac{1}{4}(3\cos 2\vartheta + 1), \\ &P_3(\vartheta) = \tfrac{1}{2}(5\cos^3\vartheta - 3\cos\vartheta) = \tfrac{1}{8}(5\cos 3\vartheta + 3\cos\vartheta).\end{aligned}\right\} \quad (3.6)$$

(3.5) in (3.1) eingesetzt, gibt

$$V_m = \frac{fM\varrho^2}{R^3}\left(P_2(\vartheta) + \frac{\varrho}{R}P_3(\vartheta) + \left(\frac{\varrho}{R}\right)^2(\cdots)\right) = V_2 + V_3 + \cdots. \quad (3.7)$$

Doodson [1] führt als *Gezeitenkonstante* ein

$$G(\varrho) = \tfrac{3}{4} fM\,\varrho^2/c^3. \quad (3.8)$$

Hierin ist c die *„mittlere Mondentfernung"*, definiert als große Halbachse der „intermediären Mondbahn", die den konstanten Teil der Sonnenstörungen (Ziff. 12c) enthält.

Man kann $G(\varrho)$ auch anders ausdrücken: Es sei a = Äquatorradius der Erde, ψ = äquatoriale Horizontalparallaxe des Mondes entsprechend der Entfernung r, ψ_0 diejenige entsprechend der mittleren Entfernung c, also

$$\sin\psi = \frac{a}{r}, \qquad \sin\psi_0 = \frac{a}{c}. \quad (3.9)$$

An Stelle der Gravitationskonstanten f kann man die Fallbeschleunigung g_1 einführen, die auf der Oberfläche einer *nicht*-rotierenden Kugel herrschen würde, die dasselbe Volumen und dieselbe Masse hätte wie die Erde,

$$g_1 = \frac{fE}{\varrho_1^2}, \quad (3.10)$$

mit E = Erdmasse, ϱ_1 = *mittlerer Erdradius*. So wird aus (3.8)

$$G(\varrho) = \frac{3}{4}\,\frac{M}{E}\,\frac{g_1\varrho_1^2}{c^3}\,\varrho^2 = \frac{3}{4}\,\frac{M}{E}\,\frac{g_1\varrho_1^2\sin^3\psi_0}{a^3}\,\varrho^2 \quad (3.10+)$$

(3.6) und (3.8) in (3.7) eingesetzt gibt

$$\left.\begin{aligned} V &= V_2 + V_3 + \cdots \\ &= G(\varrho)\left[\left(\frac{c}{R}\right)^3\left(\cos 2\vartheta + \frac{1}{3}\right) + \frac{1}{6}\,\frac{\varrho}{c}\left(\frac{c}{R}\right)^4(5\cos 3\vartheta + 3\cos\vartheta) + \cdots\right].\end{aligned}\right\} \quad (3.11)$$

Zahlwerte. Horn verwendet im Landolt-Börnstein [5] $g_1 = 982{,}04$ cm/sec², $\varrho_1 = 6371{,}221$ km, $a = 6378{,}388$ km, $\sin\psi_0 = 0{,}016593$, oder, in Bogensekunden, 3422″,54, $E/M = 81{,}53$. Das gibt

$$G \equiv G(\varrho_1) = 26206\ \text{cm}^2/\text{sec}^2, \quad (3.12)$$

und allgemein

$$G(\varrho) = \left(\frac{\varrho}{\varrho_1}\right)^2 G. \tag{3.12+}$$

In V_3 kann man für ϱ/c einsetzen $(\varrho/a)\sin\psi_0$.

Neuere Bestimmungen für E/M weichen (nach HORN [5]) von 81,53 um $\pm 0{,}26$ ab. Dem entspricht die Unsicherheit in G von

$$\mp 84\,\text{cm}^2/\text{sec}^2, \tag{3.13}$$

rund 1/300 des Wertes.

4. Solares Gezeitenpotential. In V_s (3.2) erscheinen die Sonnenmasse S, die große Halbachse c_s der ungestörten Erdbahnellipse, und die mittlere äquatoriale Horizontalparallaxe $\psi_{0s} = 8\overset{''}{,}790$. An Stelle von $G(\varrho)$ tritt also

$$\left.\begin{aligned} G_s(\varrho) &= \frac{S}{M}\left(\frac{c}{c_s}\right)^3 G(\varrho) = \frac{3}{4}\,\frac{S}{E}\,\frac{g_1\,\varrho_1^2}{c_s^3}\,\varrho^2 \\ &= \frac{S}{E}\cdot\frac{E}{M}\left(\frac{c}{c_s}\right)^3 G(\varrho) = \frac{S}{E}\cdot\frac{E}{M}\left(\frac{\sin\psi_{0s}}{\sin\psi_0}\right)^3 G(\varrho). \end{aligned}\right\} \tag{4.1}$$

Für die Berechnung benutzt HORN [5] das dritte KEPLERsche Gesetzt in seiner exakten Form $f[S + (E + M)] = 4\pi^2 c_s^3/U^2$, mit U = Dauer des siderischen Jahrs = 365,256 mittlere Sonnentage. Mit $fE = \varrho_1^2 g_1$ und $\sin\psi_{0s} = a/c_s$ wird

$$\frac{S}{E}\sin^3\psi_{0s} = \frac{4\pi^2 a^3}{U^2\,\varrho_1^2\,g_1} - \left(1 + \frac{M}{E}\right)\sin^3\psi_{0s}, \tag{4.2}$$

und

$$\left.\begin{aligned} G_s(\varrho) &= 0{,}46051\,G(\varrho) \\ &\pm 0{,}00150. \end{aligned}\right\} \tag{4.3}$$

Die angegebene Unsicherheit stammt aus E/M.

5. Gleichgewichtsgezeiten. Es sei F das Potential der Fallbeschleunigung, zusammengesetzt aus der Wirkung der Gravitation der Erdmasse und der Zentrifugalbeschleunigung infolge der Erdrotation. Eine Niveaufläche $F = \text{const} = F_0$ wird durch die Gezeitenkräfte (des Mondes, z. B.) ein wenig deformiert zu

$$F + V = F_0. \tag{5.1}$$

Dabei werde die Niveaufläche um ζ gehoben. Wenn z die Höhe bedeutet, ist in genügender Näherung,

$$F = F_0 + \zeta\cdot\partial F/\partial z = F_0 - \zeta\cdot g, \tag{5.2}$$

mit g = Fallbeschleunigung. Einsetzen in (5.1) ergibt

$$\zeta = \frac{V}{g}. \tag{5.3}$$

Folgende Modellvorstellung für ζ ist nützlich: Die Erde sei völlig starr gedacht Sie sei ganz bedeckt von einer hypothetischen Flüssigkeit, deren Dichte so gering sei, daß das sekundäre Gezeitenpotential (Ziff. 1) gegen das primäre zu vernachlässigen wäre; mit anderen Worten, die Gravitationsanziehung der, durch Gezeitenströme verschobenen Massen soll klein sein gegen die Gezeitenbeschleunigungen. Zu irgendeiner Zeit habe die Oberfläche dieser Flüssigkeit genau die Form der Niveaufläche (5.3) von $(F + V)$, auf der die Resultante aus Fallbeschleunigung und Gezeitenbeschleunigung senkrecht stünde. Längs der Oberfläche würde also die horizontale Kraftkomponente verschwinden (diejenige der

Gezeitenkräfte würde durch die Komponente der Fallbeschleunigung kompensiert). Deshalb heißt ζ die Höhe der *Gleichgewichtsflut*. ζ könnte, in diesem Modell, an Pegeln abgelesen werden, die fest mit dem starren Erdkörper verbunden wären (Fig. 2).

Wenn die Flüssigkeit inkompressibel angenommen wird, so muß die gestörte Niveaufläche dasselbe Volumen einschließen wie die ungestörte, also

$$\int_0^\pi \zeta \cdot \sin\vartheta \, d\vartheta = 0. \tag{5.4}$$

Wenn g konstant wäre, so wäre dies für $\zeta = V/g$ erfüllt, da $P_n(\vartheta) \sin\vartheta$ im Mittel über die Kugel verschwindet. Auch für die tatsächliche Erde, wo vom Äquator zum Pol g um $1/189$ zunimmt und ϱ um etwa $1/300$ abnimmt, ist (5.4) genügend genau erfüllt.

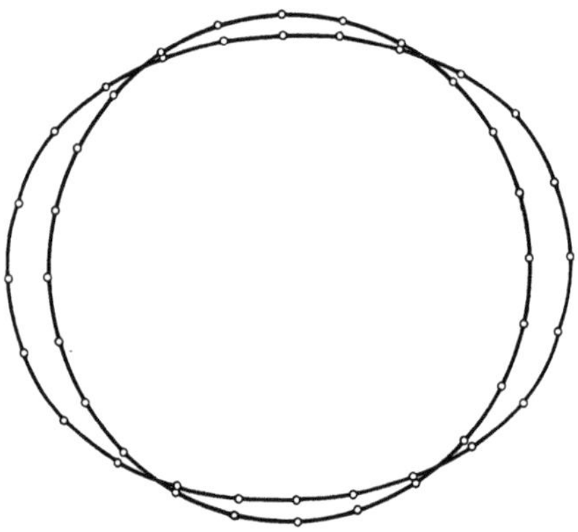

Fig. 2. Lunare Gleichgewichtsflut, etwa 3 Millionen mal vergrößert. Der Mond steht rechts oder links.

Da V und ζ von der Zeit abhängen, ist die Frage nach der tatsächlichen Wirkung der Gezeitenkräfte ein *dynamisches* Problem. Die Lösungen brauchen keinerlei Ähnlichkeit mit der Gleichgewichtsflut zu haben. ζ ist aber zur Veranschaulichung der Gezeitenkräfte geeignet, denn die horizontalen Komponenten der Gezeitenbeschleunigungen längs einer ungestörten Niveaufläche der Fallbeschleunigung sind

$$\partial V/\partial s = g \, \partial \zeta/\partial s, \tag{5.5}$$

also gleich den Komponenten der Fallbeschleunigung längs der Niveaufläche der Gleichgewichtsflut.

6. Das Verhältnis der solaren und lunaren Gezeitenkräfte. Nach (4.3) erreichen die solaren Gezeitenkräfte nur 46 Prozent der lunaren. Daß überhaupt Sonne und Mond Gezeiten derselben Größenordnung bewirken, begründet man qualitativ oft damit, daß die größere Nähe des Mondes durch die größere Masse der Sonne nahezu kompensiert wird. Man kann diese bemerkenswerte Kompensation über mehr als sieben Zehnerpotenzen hinweg anschaulich machen, wenn man von der alltäglichen (jedoch nicht weniger merkwürdigen) Beobachtung ausgeht, daß Sonne und Mond uns am Himmel nahezu *gleich groß erscheinen*, wie totale Sonnenfinsternisse demonstrieren. In der Tat sind, in mittleren Entfernungen, die scheinbaren Radien der Sonnenscheibe $\eta_s = 15'59\overset{''}{.}6$, der Mondscheibe $\eta_m = 15'32\overset{''}{.}6$. Die tatsächlichen Radien sind, genügend genau, gleich $\eta_s \cdot c_s$ und $\eta_m \cdot c$. Wenn d_s und d_m die mittleren Dichten sind, wird das Verhältnis der Massen

$$\frac{S}{M} = \frac{d_s}{d_m} \left(\frac{\eta_s}{\eta_m}\right)^3 \cdot \left(\frac{c_s}{c}\right)^3. \tag{6.1}$$

Nach der ersten Formel (4.1) wird also

$$\frac{G_s}{G} = \frac{d_s}{d_m} \cdot \left(\frac{\eta_s}{\eta_m}\right)^3 = \frac{d_s}{d_m} \cdot 1{,}087. \tag{6.2}$$

Das Verhältnis G_s/G wird also hauptsächlich durch das Verhältnis der mittleren *Dichten* bestimmt,

$$\frac{d_s}{d_m} = \frac{1{,}41}{3{,}33} = 0{,}423. \tag{6.3}$$

Die größere Nähe des Mondes macht sich jedoch darin bemerkbar, daß die *Unterschiede* der „Hochwasser" im Zenit und Nadir beim Mond noch merklich sind, bei der Sonne dagegen bedeutungslos; vgl. die Diskussion zu V_3, Ziff. 5d.

Wenn Erde, Mond und Sonne ziemlich genau in einer Geraden liegen (*Syzygien*, Neu- oder Vollmond), addieren sich die Wirkungen zu besonders großen kombinierten Gezeiten, *Spring*-Flut und -Ebbe. Etwa einen viertel Monat später (*Quadraturen*, erstes oder letztes Mondviertel), wenn Mond, Erde und Sonne im rechten Winkel zueinander stehen, sind die kombinierten Gezeiten besonders klein, *Nipp*-Flut und -Ebbe. Da V_2, und damit ζ_2, proportional $\cos 2\vartheta$ sind, verhalten sich die größten Deformationen der Gleichgewichtsfläche ζ zur Zeit der Syzygien und der Quadraturen, nach (4.3), ungefähr wie

$$(G + G_s):(G - G_s) = 1{,}46:0{,}54 = 2{,}7:1\,. \tag{6.4}$$

Das Verhältnis G_s/G der solaren und lunaren Gezeitenkräfte findet sich übrigens auch wieder im Verhältnis der solaren und lunaren Anteile in der *luni-solaren Präzession*, der Westwanderung des Frühlingspunktes. Dabei verhält sich die Erde wie ein Kreisel, mit einem Ring um den Äquator (wegen der Erdabplattung). Infolge der differentiellen Gravitationswirkung der Sonne und des Mondes auf den Äquatorring im Zenit und im Nadir (also wie bei den Gezeiten!) wirken Sonne und Mond so, daß sie den Erdäquator in die Ebene ihrer Bahn (Ekliptik) ziehen wollen; die Erde, als Kreisel, weicht senkrecht zum störenden Drehmoment aus. Man kann die Winkelgeschwindigkeit des Frühlingspunktes infolge der *lunaren* Präzessionswirkung schreiben

$$\frac{d\Psi}{dt} = -2\,\frac{G}{a^2\,\omega}\cdot\frac{C-A}{C}\cos J \tag{6.5}$$

mit a = Äquatorradius der Erde, ω = Winkelgeschwindigkeit der Erdrotation, C und A Trägheitsmomente der Erde um die Polachse und eine Äquatorachse, J = Winkel zwischen Mondbahn und Äquator. Für die *solare* Präzession muß man in (6.5) G durch G_s ersetzen, und J durch die Ekliptikschiefe ε.

7. Die Gezeitenkräfte als Funktionen der Zenitdistanz.

a) Der Hauptterm des Gezeitenpotentials des Mondes in mittlerer Mondentfernung, $c = R$, ist, nach (3.11) und (3.12),

$$V_2 = \left(\frac{G}{\varrho_1^2}\right)\varrho^2\left(\cos 2\vartheta + \frac{1}{3}\right). \tag{7.1}$$

Daraus ergibt sich die radiale Komponente Z der Gezeitenbeschleunigung (nach außen positiv gerechnet) — die man auch auffassen kann als negative Gezeitenstörung δg der Fallbeschleunigung g, weil die radiale Richtung nur wenig von der Lotrichtung abweicht —,

$$Z(= -\delta g) = \frac{\partial V_2}{\partial \varrho} = \frac{2G(\varrho)}{\varrho}\left(\cos 2\vartheta + \frac{1}{3}\right). \tag{7.2}$$

Die horizontale Komponente H, positiv in Richtung auf den Mond, ist

$$H = -\frac{\partial V_2}{\varrho\,\partial\vartheta} = \frac{2G(\varrho)}{\varrho}\sin 2\vartheta\,. \tag{7.3}$$

Die Störung der Lotrichtung durch die Gezeitenbeschleunigung ist

$$\frac{H}{g}, \tag{7.4}$$

mit g = Fallbeschleunigung am Orte der Beobachtung. Fig. 3 gibt die Gezeitenbeschleunigungen als Vektoren mit den Komponenten H und Z.

b) Für den mittleren Erdradius $\varrho = \varrho_1$ ist die Konstante in (7.2) und (7.3), nach (3.12),

$$2G(\varrho_1)/\varrho_1 = 82{,}26 \cdot 10^{-6}\ \text{cm/sec}^2 = 0{,}08226\ \text{milligal} \tag{7.5}$$

mit der Einheit milligal $= 10^{-3}$ cm/sec², die bei relativen Schweremessungen gebräuchlich ist.

Wir setzen

$$C = \frac{G(\varrho_1)}{\varrho_1} = 41{,}13 \cdot 10^{-6}\ \text{cm/sec}^2. \tag{7.5 +}$$

Für (7.4) erhalten wir einen repräsentativen Wert, wenn wir setzen $\varrho = \varrho_1$ und $g = g_0 = 979{,}8$ cm/sec² = Mittel über die Erdoberfläche. Dann wird

$$\left.\begin{aligned} \frac{g_0}{2C} &= 979{,}8/82{,}3 \cdot 10^{-6} \\ &= 11{,}91 \cdot 10^6, \end{aligned}\right\} \tag{7.6}$$

und die *lunare* Schwankung der Lotrichtung (7.4) wird

$$\left.\begin{aligned} \frac{H}{g_0} &= \frac{1}{11{,}91 \cdot 10^6} \sin 2\vartheta \\ &= 0{,}''0173 \sin 2\vartheta. \end{aligned}\right\} \tag{7.7}$$

Ein 119 km langes Pendel würde auf starrer Erde ± 1 cm ausschlagen, wenn der Mond in $\pm 45°$ Höhe über dem Horizont stünde.

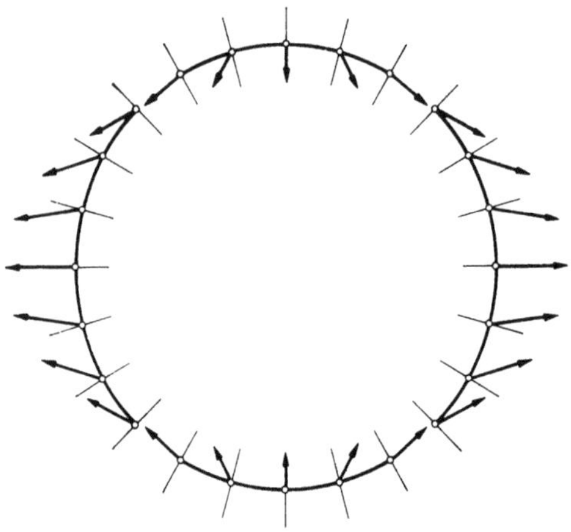

Fig. 3. Vektoren der lunaren Gezeitenbeschleunigung auf einem Meridianschnitt der Erde durch den Mond (dieser kann rechts oder links stehen.)

c) Nach (7.1), (3.12) und (5.3) wird die Höhe der *lunaren* Gleichgewichtsflut, für $\varrho = \varrho_1$ und $g = g_0$, mit $G(\varrho_1)/g_0 = (26206/979{,}8)$ cm,

$$\zeta = 26{,}7\ \text{cm}\ (\cos 2\vartheta + \tfrac{1}{3}). \tag{7.8}$$

In den beiden Punkten, in denen der Mond im Zenit oder Nadir steht, ist die Niveaufläche um 35,6 cm gehoben; auf dem größten Kreis $\vartheta = 90°$, wo der Mond auf- oder untergeht, ist sie um 17,8 cm gesenkt. Die Gesamtamplitude ist also 53,4 cm. — Für die *solaren* Gezeiten sind die Amplituden, nach (4.3), mit 0,4605 zu multiplizieren, also

$$\zeta_s = 12{,}3\ \text{cm}\ (\cos 2\vartheta_s + \tfrac{1}{3}). \tag{7.9}$$

Im Sinne von (6.4) verhalten sich die Gesamtamplituden (lunar + solar) zur Zeit der Syzygien und Quadraturen etwa wie 78 cm : 29 cm.

d) Das Glied V_3 im *lunaren* Gezeitenpotential (3.11) drückt folgendes aus: Die NEWTONsche Anziehung des Mondes nimmt wie $1/R^2$ ab. Dort, wo der Mond im Zenit steht, ist deshalb der Überschuß seiner Anziehung über den Wert im Erdschwerpunkt etwas größer als der Unterschuß im Nadirpunkt. V_3, als P_3 nach (3.6), verschwindet auf den drei Kreisen $\vartheta = 39{,}°2$, 90° und 140,°8. Bei mittlerer Mondentfernung ($R = c$) ist, nach (3.11),

$$V_3(\vartheta = 0°) = \frac{4}{3} G(\varrho) \cdot \frac{\varrho}{c} = -V_3(\vartheta = 180°). \tag{7.10}$$

Für $\varrho = \varrho_1$ und $g = g_0$ wird demnach das Zenit-„Hochwasser" der lunaren Gleichgewichtsflut um $\frac{8}{3}\frac{G\varrho_1}{c\,g_0} = 1{,}18$ cm höher als das Nadirhochwasser; dieser Unterschied ist 1/45 der gesamten Deformation (53,4 cm) der Niveaufläche durch das lunare V_2-Potential. Genauer

$$\zeta_3 = 0{,}0740\,\text{cm} \cdot \left(\frac{c}{R}\right)^4 (5\cos 3\vartheta + 3\cos\vartheta), \tag{7.11}$$

$\zeta_4 = V_4/g$ für den Mond ist überall auf der Erde kleiner als 0,02 cm.

Im Gezeitenpotential der *Sonne* ist bereits V_3 gegen V_2 bedeutungslos, weil in (3.11), im Term für V_3, ϱ/c_s an die Stelle von ϱ/c tritt; und $c_s/c = 149{,}5/0{,}384 = 390$.

e) Auf der Erdoberfläche ändert sich ϱ nur um die Größenordnung 1/300 (= Erdabplattung); bis auf Bruchteile von 1 Prozent sind also die Gezeitenbeschleunigungen dieselben wie für die Radien $\varrho = a$ (Äquator) oder $\varrho = \varrho_1$ (mittlerer Erdradius) berechnet.

Mit der Entfernung ϱ vom Erdschwerpunkt ändern sich die Gezeitenbeschleunigungen Z und H, nach (7.2) und (7.3), proportional ϱ. Die Deformation der Niveauflächen durch die Gleichgewichtsgezeiten $\zeta_2 = V_2/g$ nimmt viel stärker zu: Denn $V_2 \sim \varrho^2$ [nach (7.1)], und $g \sim 1/\varrho^2$, also $\zeta_2 \sim \varrho^4$. In der F_2-Schicht, in der Höhe von etwa $a/20 = 320$ km, ist ζ_2 bereits das 1,22fache des Wertes am Boden. Im Abstand $\varrho = 5a$ — wo mitunter der (erdmagnetisch erschlossene) Ringstrom angenommen wird — ist ζ_2 625mal so groß wie am Boden; dort ist die reine Gravitationsanziehung der Erde auf 40 cm/sec² gefallen, und die lunare Gezeitenbeschleunigung auf 0,0004 cm/sec² gestiegen.

f) Nach (3.11) hängt V_2 von der Entfernung R des fluterzeugenden Gestirns ab wie $1/R^3$. Es sei e die numerische Exzentrizität der Bahnellipse; dann verhalten sich die größten und kleinsten Abstände von der Erde wie $(1+e):(1-e)$. Für die Mondbahn und die Erdbahn sind (vgl. auch Ziff. 12)

$$e_m = 0{,}05490, \qquad e_s = 0{,}01673. \tag{7.12}$$

Die von V_2 herrührenden Gezeitenkräfte verhalten sich also beim Mond, zur Zeit des mittleren Perigäums und Apogäums, wie 1,391:1; bei der Sonne, zur Zeit des Perihels und Aphels, wie 1,106:1.

Da $V_3 \sim 1/R^4$, vermindert sich die Wirkung von V_3, vom Perigäum zum Apogäum, wie 1,55:1.

g) Nach Ziff. 2 ist das System der Gezeitenbeschleunigungen ein Vektorfeld, für dessen Beschreibung sich das nicht-rotierende Koordinatensystem $\boldsymbol{K}$ eignet (Ziff. 2). Für einen Punkt P' im üblichen Koordinatensystem, das mit der Erde rotiert (Länge, Breite, Höhe) erhält man die Gezeitenbeschleunigungen, indem man für jeden Augenblick die Lage von P' im System $\boldsymbol{K}$ betrachtet. Außer den Gezeitenkräften wirkt in P' die Gravitationsbeschleunigung der Erde und die Zentrifugalbeschleunigung infolge der Erdrotation, die im Potential der Fallbeschleunigung g zusammengefaßt sind. Auf Massen, die sich relativ zur Erde bewegen, wirken außerdem noch die Coriolisbeschleunigungen, wenn man ihre Bewegungen relativ zu einem mit der Erde rotierenden System betrachtet.

Das System $\boldsymbol{K}$ hat den Vorzug, daß für Punkte, die in ihm ruhen, die Führungsbeschleunigungen überall gleich groß sind. Gelegentlich hat man zur Berechnung der Gezeitenkräfte auch andere Koordinatensysteme benutzt, z.B. dasjenige, das im Monat einmal um den gemeinsamen Schwerpunkt der Erde und des Mondes rotiert. Das führt aber auf unübersichtliche Verhältnisse, wenn man die Einflüsse des Mondes und der Sonne zusammenfassen will. Irrtümliche Auffassungen des Einflusses der Rotation oder Revolution auf die Gezeitenkräfte scheinen nahe zu liegen[1]; sie liegen darin, daß dabei jede Masse sowohl wegen ihrer *Schwere* (durch NEWTONsche Gravitationskräfte) wie wegen ihrer *Trägheit* (Trägheitskräfte, Führungsbeschleunigungen) beschleunigt wird.

h) Es erscheint deshalb nicht überflüssig, das *Wesen der Gezeitenkräfte* noch einmal an einem einfachen *Modell* zu veranschaulichen: Fig. 4a zeigt ein starres Kreuz, das als Ganzes auf die Sonne S frei fällt. Das Kreuz selbst sei gewichtslos; in seinem Mittelpunkt und an seinen vier Enden befinden sich gleich große Punktmassen. Fig. 4b zeigt, für zwei Lagen des Kreuzes, maßstabsgetreu die Gravitationsbeschleunigungen, die von der Sonne auf die fünf Massen ausgeübt werden. Bei kugeliger Symmetrie wäre die resultierende Beschleunigung genau so groß, als ob alle fünf Massen in O vereinigt wären; bei unserer Konfiguration ist die Resultierende $\boldsymbol{b}_0'$ zwar nicht genau gleich der Beschleunigung $\boldsymbol{b}_0$ in O, aber der Unterschied ist bei der angenommenen Kleinheit des Kreuzes relativ zur Sonnenentfernung (AC/OS) nicht

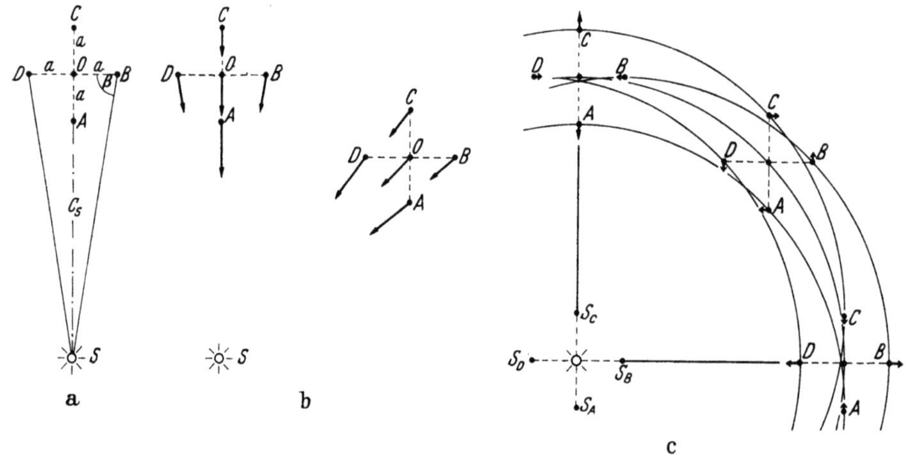

Fig. 4a—c. Modell zur Veranschaulichung des Wesens der Gezeitenkräfte.

von Belang, seine Vernachlässigung entspricht der Beschränkung des Gezeitenpotentials auf V_2. Für Wesen *(Astronauten)*, die sich während des Falles in A, B, C, D festhalten, und die sich alle mit der gleichen Beschleunigung $\boldsymbol{b}_0'$ *bewegen*, wären nur die vektoriellen Differenzen $(\boldsymbol{b}_A - \boldsymbol{b}_0')$ usw. *spürbar*. Diese sind in Fig. 4c, gemäß Fig. 1 konstruiert, dargestellt; es sind genau die Gezeitenbeschleunigungen.

Tatsächlich ist, in der V_2 entsprechenden Näherung,

$$|\boldsymbol{b}_A - \boldsymbol{b}_0'| = f S \left[\frac{1}{(c_s - a)^2} - \frac{1}{c_s^2}\right] \sim 2 f S\, a/c_s^3,$$

$\boldsymbol{b}_C - \boldsymbol{b}_0'$ ist ebenso groß, aber entgegengesetzt gerichtet. Ferner

$$|\boldsymbol{b}_B| = f S/(c_s^2 + a^2), \quad \text{und} \quad |\boldsymbol{b}_B - \boldsymbol{b}_0| = |\boldsymbol{b}_B| \cos\beta \sim |\boldsymbol{b}_B| \cdot a/c_s \sim f S\, a/c_s^3.$$

Die nach außen gerichteten Beschleunigungen in den Punkten A und C sind also doppelt so groß wie die nach innen gerichteten in den Punkten B und D. Auch in den anderen Lagen des Kreuzes (z.B. Sonne unter 45°) läßt sich die Konstruktion von Fig. 1 leicht durchführen.

Fig. 4c veranschaulicht noch die „*Revolution ohne Rotation*" unseres Koordinatensystems $\boldsymbol{K}$: Das Kreuz beschreibt, ohne Rotation, eine KEPLERsche Ellipse, wobei die Führungsbeschleunigung genau entgegengesetzt der resultierenden solaren Gravitationsbeschleunigung ist. Die fünf Punkte beschreiben fünf kongruente Ellipsen um Brennpunkte S (für O), S_A für A, ...

---

[1] Zusammenstellung bei ALOYS MÜLLER, Gerlands Beitr. Geophysik. **10**, 125—151 (1910). Zwei weitere Fälle in den Ann. Hydrogr., Hamburg: E. HOPF, **35**, 122—130 (1907), diskutiert mit W. SCHWEYDAR, l. c. 179 u. 375f.: R. SPITALER, **49**, 220—231 (1921), diskutiert mit A. DEFANT, E. KLEINSCHMIDT u. J. BARTELS **50**, 22—26, 322—330 (1922).

Wenn man also sagt, daß die Gezeitenbeschleunigungen gleich den vektoriellen Differenzen $(\boldsymbol{b}_A - \boldsymbol{b}_0')$ sind, so bedeuten darin $\boldsymbol{b}_A$ die Gravitationsbeschleunigung der Sonne, während $(-\boldsymbol{b}_0')$ als Führungsbeschleunigung $\boldsymbol{t}$ hinzutritt. Bei dieser Auffassung werden auch die komplizierteren Führungsbeschleunigungen richtig berücksichtigt, die bei der wirklichen Erdbahn auftreten, unter der gemeinsamen Wirkung der Sonne und des Mondes. Das ist in Ziff. 1 und 2 geschehen; die elementaren Betrachtungen an Hand der Fig. 4 lassen das Wesentliche hervortreten.

8. Der tägliche Gang der Gezeitenbeschleunigungen an einem festen Ort der rotierenden Erde.

a) Wir führen nunmehr die zu Anfang von Ziff. 7g angedeutete Umrechnung durch. Nach (7.1), (7.2), (7.3) braucht man nur die augenblicklichen geozentrischen Zenitdistanzen ϑ und ϑ_s des Mondes und der Sonne zu kennen (außer den Entfernungen R und R_s), um die Komponenten Z und H zu berechnen, sowie die Gleichgewichtsflut ζ nach (7.8) und (7.9).

In der üblichen Projektion vom Erdmittelpunkt auf die Himmelskugel seien (Fig. 5) C der Himmelspol, M der Mond (Deklination δ, Zenitdistanz ϑ), P der Beobachtungsort (geozentrische[1] Breite φ) und CP sein Meridian. Wahre Mondzeit τ_1 sei von der unteren Kulmination gerechnet, in Grad, so daß der Stundenwinkel des Mondes gleich $(\tau_1 - 180°)$ ist. Dann ist im Dreieck CMP

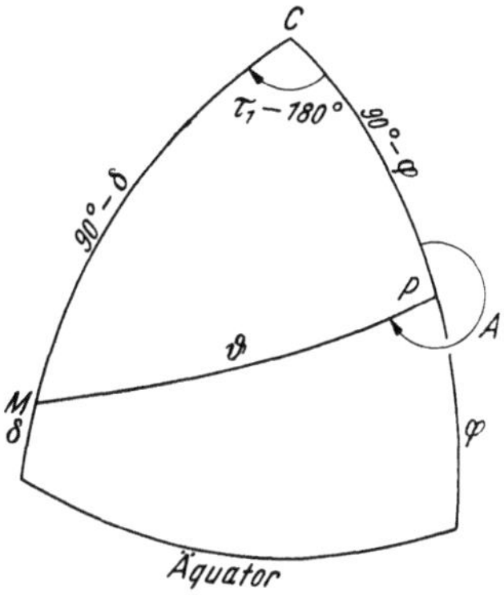

Fig. 5. Himmelsdreieck.

$$\cos\vartheta = \sin\varphi\sin\delta + \cos\varphi\cos\delta\cos(\tau_1 - 180°). \quad (8.1)$$

Man entnimmt δ und τ_1 einem Astronomischen Jahrbuch *(Almanac)* und kann dann ϑ berechnen; daraus findet man Z, H und ζ. Das Azimut des Mondes, also die Richtung von H, findet man aus

$$\sin A = \cos\delta\sin\tau_1/\sin\vartheta. \quad (8.2)$$

Wenn man nur *wenige* Werte der Gezeitenbeschleunigungen berechnen will, genügt dieses direkte Verfahren.

Übersichtlicher ist es, wenn man $\cos 2\vartheta = 2\cos^2\vartheta - 1$ durch (8.1) als Funktion von φ, δ und τ_1 ausdrückt. (3.11) wird dann

$$\left.\begin{aligned} V_2 = G(\varrho)\cdot\left(\frac{c}{R}\right)^3\cdot\Big[3\left(\tfrac{1}{3} - \sin^2\delta\right)\left(\tfrac{1}{3} - \sin^2\varphi\right) - \\ - \sin 2\varphi\sin 2\delta\cos\tau_1 + \cos^2\varphi\cos^2\delta\cos 2\tau_1\Big]. \end{aligned}\right\} \quad (8.3)$$

$$\left.\begin{aligned} V_3 = G(\varrho)\cdot\frac{\varrho}{c}\cdot\left(\frac{c}{R}\right)^4\Big[\frac{1}{3}(3\sin\varphi - 5\sin^3\varphi)(3\sin\delta - 5\sin^3\delta) - \\ - \frac{1}{2}\cos\varphi(1 - 5\sin^2\varphi)\cos\delta(1 - 5\sin^2\delta)\cos\tau_1 + \\ + 5\sin\varphi\cos^2\varphi\sin\delta\cos^2\delta\cos 2\tau_1 - \\ - \frac{5}{6}\cos^3\varphi\cos^3\delta\cos 3\tau_1\Big]. \end{aligned}\right\} \quad (8.4)$$

Für die solaren Gezeiten sind zu ersetzen $G(\varrho)$ durch $G_s(\varrho)$ nach (4.3), (c/R) durch (c_s/R_s), δ durch δ_s, und τ_1 durch die wahre Sonnenzeit t_1.

[1] Die *geozentrische* Breite φ eines Ortes A ist die Neigung seiner Verbindung mit dem Erdmittelpunkt gegen den Erdäquator; die *geographische* Breite B ist die Neigung des Lotes in A gegen den Äquator. Der Unterschied kann $11'\!,6$ erreichen, $\varphi = B - 11'\!,6\sin 2B$.

Die Berechnungen von (8.3) und (8.4) sind Spezialfälle der Transformation zonaler Kugelfunktionen $P_n(\vartheta)$, in bezug auf die Achse OM, auf die Achse OC, mittels (8.1), nämlich

$$P_n(\vartheta) = \sum_{m=0}^{n} P_n^m(90° - \varphi)\, P_n^m(90° - \delta) \cos m(\tau_1 - 180°). \tag{8.5}$$

Dabei sind die nach ADOLF SCHMIDT normierten Kugelfunktionen gemeint, nämlich

$$P_n^0 \equiv P_n \qquad \text{wie in (3.6)},$$

und

$$P_n^m = [2(n-m)!/(n+m)!]^{\frac{1}{2}} \sin^m\vartheta\, d^m P_n(\vartheta)/d(\cos\vartheta)^m, \tag{8.6}$$

speziell

$$\left.\begin{aligned} &P_1^1 = \sin\vartheta;\\ &P_2^1 = \tfrac{1}{2}\sqrt{3}\sin 2\vartheta, \qquad P_2^2 = \tfrac{1}{2}\sqrt{3}\sin^2\vartheta;\\ &P_3^1 = \tfrac{1}{4}\sqrt{6}\sin\vartheta(5\cos^2\vartheta - 1), \quad P_3^2 = \tfrac{1}{2}\sqrt{15}\sin^2\vartheta\cos\vartheta, \quad P_3^3 = \tfrac{1}{4}\sqrt{10}\sin^3\vartheta. \end{aligned}\right\} \tag{8.7}$$

(8.1) und (8.3) sind identisch mit (8.5) für $n=1$ und 2. Ebenso liefert (8.5), mit $n=3$, die Form (8.4) von V_3, aus (3.7) oder (3.11).

Die Symmetrie der Formel (8.1) in δ und φ bleibt natürlich in V_2 und V_3 erhalten.

b) Analog zur wahren Mondzeit τ_1 und wahren Sonnenzeit t_1, beide von den *unteren Kulminationen* aus gerechnet, wollen wir vorübergehend noch die (Orts)-*Sternzeit* θ^* einführen, gerechnet von der *unteren* Kulmination des Frühlingspunktes; die Astronomen rechnen Sternzeit $\theta = \theta^* - 180°$ von der *oberen* Kulmination aus.

Die Zeiten für den Greenwich-Meridian seien durch den Index o bezeichnet; sie sind mit den Zeiten auf dem Meridian L östlich Greenwich verbunden durch

$$\tau_1 = \tau_{10} + L, \qquad t_1 = t_{10} + L, \qquad \theta^* = \theta_0^* + L. \tag{8.8}$$

Mit den Rektaszensionen α und α_s des Mondes und der Sonne gilt

$$\left.\begin{aligned} \tau_1 &= \theta^* - \alpha, & t_1 &= \theta^* - \alpha_s\\ \tau_1 &= \theta_0^* + L - \alpha, & t_1 &= \theta_0^* + L - \alpha_s. \end{aligned}\right\} \tag{8.9}$$

In astronomischen Jahrbüchern *(Almanac)* findet man die Sternzeit θ, woraus sich unser θ^* durch Änderung um 12 Std. $= 180°$ ergibt; ferner δ und α, sowie δ_s und α_s. Die relativen reziproken Entfernungen (c/R) und (c_s/R_s) sind dort durch die *Parallaxen* ψ und ψ_s gegeben, auch durch die scheinbaren Halbmesser.

c) Folgende größten und kleinsten Werte treten auf: δ_s schwankt innerhalb eines Jahres zwischen $\pm\varepsilon = \pm 23° 27'$. (c_s/R_s) liegt zwischen 0,9833 (Perihel) und 1,0170 (Aphel). Die Mondbahn ist im Mittel um $i = 5°8'$ gegen die Ekliptik geneigt. Je nach der Länge N des aufsteigenden Knotens der Mondbahn (der in 18.6 Jahren einmal die Ekliptik umläuft, Ziff. 12c) schwankt δ innerhalb eines Monats zwischen den Grenzen $\pm J$, wobei J zwischen 28°,5 und 18°,5 liegt. Die Schwankung der Mondentfernung innerhalb eines Monats hängt davon ab, wie die Sonne zur Richtung der großen Mondbahnhalbachse steht: Steht die Sonne *in Richtung* der großen Achse der Mondbahn, schwankt ψ zwischen 54',0 und 61',5, entsprechend (c/R) zwischen 0,9466 und 1,0781; steht die Sonne *quer* zur Achse, sind diese Grenzen in ψ 54',3 und 59',2, in (c/R) 0,9589 und 1,0378 (vgl. Ziff. 12d; die hier gegebenen Werte nach HORN [5]).

d) Die *Beschleunigungskomponenten* X nach Norden, Y nach Osten und Z nach außen sind

$$X = \partial V/\varrho\,\partial\varphi, \qquad Y = \partial V/\varrho\cos\varphi\,\partial L, \qquad Z = \partial V/\partial\varrho. \tag{8.10}$$

Wir beschränken uns auf V_2 (für V_3 vgl. Ziff. 20). Für die Differentiation nach L muß man τ_1 und t_1 durch L ersetzen, gemäß (8.8). Die Werte gelten für mittleren Erdradius $\varrho=\varrho_1$, mit $C=G(\varrho_1)/\varrho_1=41{,}13$ microgal aus (7.5+).

$$\left.\begin{aligned} X = &- C\cdot\left(\frac{c}{R}\right)^3\cdot[\sin 2\varphi\,(1-3\sin^2\delta)+\\ &+2\cos 2\varphi\sin 2\delta\cos\tau_1+\sin 2\varphi\cos^2\delta\cos 2\tau_1],\end{aligned}\right\}\tag{8.11}$$

$$Y=+2C\cdot\left(\frac{c}{R}\right)^3[\sin\varphi\sin 2\delta\sin\tau_1-\cos\varphi\cos^2\delta\sin 2\tau_1],\tag{8.12}$$

$$\left.\begin{aligned} &Z(=-\delta g)=+2C\left(\frac{c}{R}\right)^3\times\\ &\times\left[\frac{1}{3}(1-3\sin^2\delta)(1-3\sin^2\varphi)-\sin 2\varphi\sin 2\delta\cos\tau_1+\cos^2\varphi\cos^2\delta\cos 2\tau_1\right].\end{aligned}\right\}\tag{8.13}$$

Für die *solaren* Gezeiten sind die Zahlwerte, nach (4.3), mit 0,4605 zu multiplizieren, ferner treten (c_s/R_s) an Stelle von (c/R), δ_s an Stelle von δ, und t_1 an Stelle von τ_1. Für andere Abstände ϱ vom Erdschwerpunkt sind die Komponenten mit (ϱ/ϱ_1) zu multiplizieren. Vergleiche auch die Darstellungen bei Tomaschek (S. 776—781) und Defant (S. 882—886).

9. Die geographische Verteilung der Gezeitenkräfte.

a) Die in Ziff. 8 abgeleiteten Ausdrücke für V_2 und V_3 sowie für die Beschleunigungskomponenten X, Y, Z waren zunächst gedeutet für einen festen Ort (Breite φ, Länge L) als *Funktionen der Zeit*, speziell der lokalen wahren Mondzeit τ_1 oder Sonnenzeit t_1. Wie schon bei der Berechnung von Y (8.12) geschehen, kann man diese Formeln, nach (8.9), auch deuten als *geographische Verteilungen*, für einen bestimmten Augenblick, den man z.B. fixieren kann durch eine bestimmte Sternzeit θ_0^* in Greenwich.

b) Am Pol ($\varphi=\pm 90°$) verschwinden alle Terme in (8.11) bis (8.13) außer dem konstanten Term in Z, sowie den ganztägigen Gliedern in X und Y. Es mag zunächst verwundern, daß X und Y in Polnähe sich so schnell von Ort zu Ort ändern sollen, jedoch liegt das nur am Polarkoordinatensystem: Die ganztägigen Glieder, mit $\tau_1=\tau_{10}+L$, setzen sich nahe dem Pol ($\varphi\sim 90°$) für einen bestimmten Augenblick (z.B. $\tau_{10}=0$), als $X=2C\cos L$, $Y=2C\sin L$, zu einem einheitlichen Beschleunigungsvektor zusammen, der auf den Mond weist und sich mit ihm dreht, wobei sich sein Betrag langsam proportional $\sin 2\delta$ ändert. Das entspricht der Formel (7.3).

c) Es genügt, die superponierten lunaren und solaren Gezeitenbeschleunigungen als Funktionen der Zeit nur für 2 oder 3 ausgewählte Breitengrade zu berechnen (zunächst unter Beschränkung auf V_2); daraus ergeben sich dann die Werte für alle anderen Breiten auf einfache Weise. In der Tat kann man z.B. die Schwereschwankung $\delta g=-Z$, nach (8.13), folgendermaßen schreiben (J. Goguel [*10*], hier leicht abgeändert)

$$-Z=\delta g=2A(1-3\sin^2\varphi)+B\sin 2\varphi+2C\cos^2\varphi,\tag{9.1}$$

wobei A, B und C Funktionen der Zeit sind, die von den Augenblickswerten der Größen R, R_s, δ, δ_s, τ_1 und t_1 abhängen. Mit

$$N=B+(C+3A),\qquad S=-B+(C+3A),\qquad P=-4PA\tag{9.2}$$

wird

$$\delta g=P+N\cos\varphi(\cos\varphi+\sin\varphi)+S\cos\varphi(\cos\varphi-\sin\varphi).\tag{9.3}$$

Daraus sieht man, daß P, $P+N$, $P+S$ gleich δg am Pol und in den Breiten $\varphi=+45°$ (Nord) und $\varphi=-45°$ (Süd) sind.

Dasselbe Schema gilt für V_2 und für $\zeta=V_2/g$.

Für X zeigt (8.11), daß

$$X=X_0\cos 2\varphi+X_{45}\sin 2\varphi,\tag{9.4}$$

und nach (8.12) ist

$$Y=Y_0\cos\varphi+Y_{90}\sin\varphi.\tag{9.5}$$

V_3 läßt sich ähnlich zerlegen (vgl. Ziff. 20).

10. Diskussion der Hauptteile der Potentiale V_2 und V_3. In den Formeln (8.3), (8.4) und (8.11) bis (8.13), betrachtet als Funktionen der Zeit an einem festen Ort (φ, L), sind mehrere Variable zeitabhängig. Am schnellsten ändern sich t_1 (Umlauf ein Sonnentag) und τ_1 (Umlauf ein Mondtag); δ und R schwanken zwischen ihren Extremwerten innerhalb eines Monats, δ_s und R_s innerhalb eines Jahres. Nach der Abhängigkeit von τ_1 und t_1 ergibt sich zwanglos eine Einteilung: Aber auch die Amplituden der Glieder mit $\cos\tau_1$, $\cos t_1$ usw. ändern sich mit der Zeit; ebenso wachsen die Argumente, τ_1 und t_1, ungleichförmig, weil die Rektaszensionen α und α_s sich ungleichmäßig ändern.

Immerhin geben die Formeln (8.3), (8.4), (8.11) bis (8.13) eine grobe Einteilung der Glieder, die sich zu V_2 und V_3 überlagern: Das jeweils erste Glied (*ohne* τ_1 oder t_1) heißt *langperiodisch*, das zweite (mit $\cos\tau_1$ oder $\cos t_1$) *ganztägig* (diurnal), das dritte (mit $\cos 2\tau_1$ oder $\cos 2t_1$) *halbtägig* (semidural); in V_3 tritt noch ein *dritteltägiges* (terdiurnal) Glied (mit $\cos 3\tau_1$) hinzu.

Bei der Diskussion geht man am besten vom Fall $\varphi=\delta=0$ aus, weil dann $\tau_1=180°\pm\vartheta$, $t_1=180°\pm\vartheta_s$, so daß die Formeln aus Ziff. 8, für V_2, sich zu den Formeln (3.11) und (7.2) und (7.3) vereinfachen: aus $\cos 2\vartheta$ wird dabei $\cos 2\tau_1$ oder $\cos 2t_1$. Hinsichtlich V_2 liest man aus den Formeln ab:

Die *langperiodischen* Glieder in $\zeta=V_2/g$ sind symmetrisch zum Äquator $\varphi=0$, in X antisymmetrisch (Vorzeichenwechsel mit φ). In Y treten sie nicht auf. Die Amplituden sind — innerhalb der Grenzen von δ — am größten, wenn Mond oder Sonne im Äquator stehn.

Die *ganztägigen* Glieder sind in X symmetrisch zum Äquator, in ζ, Y und Z antisymmetrisch. Die Amplituden verschwinden für $\delta=0$; ihr absoluter Betrag wächst mit δ.

Die *halbtägigen* Glieder sind in ζ, X und Z symmetrisch zum Äquator, in Y antisymmetrisch. Sie sind am größten, wenn $\delta=0$; aber auch für den größtmöglichen Betrag von δ (28°,5) ist das lunare Glied um nur 23 Prozent kleiner als für $\delta=0$.

V_3 wird vom *dritteltägigen* Glied mit $\cos 3\tau_1$ beherrscht, jedoch nicht ausschließlich, denn schon für $\delta=0$ und $c=R$ ist am Äquator ($\varphi=0$)

$$V_3=-\frac{1}{6}\,G(\varrho)\cdot\frac{\varrho}{c}\,(3\cos\tau_1+5\cos 3\tau_1)\,.$$

V_3 liefert also einen nicht unbeträchtlichen Anteil zum *ganztägigen* Glied im Gesamtpotential.

B. Harmonische Analyse.

11. Vorbereitungen zur harmonischen Entwicklung. Um die komplizierte Zeitabhängigkeit von V leichter zu übersehen, schreibt man V als Summe von Sinus- oder Cosinusgliedern („*Tiden*") mit konstanten Frequenzen und Amplituden; die Amplituden sind nur abhängig vom *Ort*, nämlich von der Entfernung ϱ vom Erdschwerpunkt und von der geozentrischen Breite φ. Dabei sind einige astronomische Begriffe nützlich, insbesondere sog. „*mittlere Längen*" (mean longitudes), gemessen in der Ekliptik vom gleichzeitigen mittleren Frühlingspunkt aus, für den Mond (s), die Sonne (h), das Perigäum der Mondbahn (p), den aufsteigenden Knoten der Mondbahn (N) und das Perigäum der Sonnenbahn (p_s). Bis auf sehr kleine Reste ändern sich diese mittleren Längen *gleichförmig* mit der Zeit, und zwar wachsen sie alle (ostwärts), mit Ausnahme von N (der Schnittpunkt der Mondbahn mit der Ekliptik wandert *west*wärts), weshalb man in der Gezeitentheorie vielfach $N'=-N$ verwendet.

Die folgenden Formeln (aus HORN [5]) verwenden die Zeit T gemessen in Julianischen Jahrhunderten zu je 36525 mittleren Sonnentagen, $T=0$ am mittleren Greenwich Mittag, 1899 Dezember 31. Es ist

$$\left.\begin{aligned} s &= 270{,}^\circ 43659 + 481\,267{,}^\circ 89057\,T + 0{,}00198\,T^2 + 0{,}000002\,T^3, \\ h &= 279{,}^\circ 69668 + 36000{,}^\circ 76892\,T + 0{,}00030\,T^2, \\ p &= 334{,}^\circ 32956 + 4069{,}^\circ 03403\,T - 0{,}01032\,T^2 - 0{,}00001\,T^3, \\ N &= 259{,}^\circ 18328 - 1934{,}^\circ 14201\,T + 0{,}00208\,T^2 + 0{,}000002\,T^3, \\ p_s &= 281{,}^\circ 22083 + 1{,}^\circ 71902\,T + 0{,}00045\,T^2 + 0{,}000003\,T^3. \end{aligned}\right\} \quad (11.1)$$

An Stelle der *wahren* Zeiten τ_1 und t_1 werden eingeführt die *mittlere* Mondzeit τ und die *mittlere* Sonnenzeit t. Sie stehen mit der Sternzeit θ^* (vgl. Ziff. 8b) in den Beziehungen

$$\theta^* = t + h = \tau + s; \qquad \tau = t + h - s. \quad (11.2)$$

Aus den Zeiten θ_0^*, τ_0 und t_0, die im Greenwich-Meridian gelten, erhält man durch Addition von L die gleichzeitigen Werte auf einem Meridian in der Länge L östlich von Greenwich.

In *Tabelle 1* sind die „Frequenzen" durch die Änderung pro mittlere Sonnenstunde für die wichtigsten Variablen gegeben, ferner die Perioden (Umlauf um 360°).

Tabelle 1. *Mittlere Zeiten und Längen: Änderungen σ um 1900, und Perioden.*

	σ in °/Stunde	Periode = 360°/σ
θ	15,0410686	0,997270 Tage = Sterntag
t	15,0000000	1,000000 Tage = mittlerer Sonnentag
τ	14,4920521	1,035050 Tage = 24 Std 50,47 min = mittlerer Mondtag
s	0,5490165	27,321582 Tage = tropischer Monat
h	0,0410686	365,242199 Tage = tropisches Jahr
p	0,0046418	8,847 Jahre = Umlaufzeit des mittleren Perigäums
N'	0,0022064	18,613 Jahre = Umlaufzeit des mittleren Knotens
p_s	0,0000020	20940 Jahre = Umlaufzeit des Perihels
$s-N$	0,5512229	27,21222 Tage = mittlerer drakonitischer Monat
$s-p$	0,5443747	27,55455 Tage = mittlerer anomalistischer Monat
$s-h$	0,5079479	29,53059 Tage = mittlerer synodischer Monat
$h-p_s$	0,0410667	365,25964 Tage = mittleres anomalistisches Jahr
$h-p$	0,0364268	411,78471 Tage = vgl. Ziff. 12d
$s-2h+p$	0,4715211	31,812 Tage = Evektionsperiode, Ziff. 12d

Wichtige *gemeinsame Vielfache* dieser Perioden:

1653,5 Tage ~ 60,5 tropische oder 60 anomalistische oder 56 synodische Monate;

Saros-Periode = 6585,33 Tage = 14 Normaljahre + 4 Schaltjahre + $11^1/_3$ Tage = etwa 233 synodische oder 239 anomalistische Monate oder 19 Perioden von $(h-N)$, daher Wiederholung der Sonnen- und Mondfinsternisse;

Metonischer Cyclus = 19 Jahre = 235 synodische = 254 tropische Monate; Wiederholung des Kalenders der Mondphasen.

Werte für h, s, p und N am 1. Januar jedes Jahres, 1911—1970, sowie Hilfstafeln zur Berechnung der Werte für jeden Tag des Jahres im LANDOLT-BÖRNSTEIN ([5], p. 278) oder (für 1850—1975) im Geophysikalischen Mondalmanach [6].

12. Die Bahnbewegungen der Sonne und des Mondes.

a) Die augenblicklichen Stellungen der Sonne und des Mondes relativ zur Erde sind gegeben durch die Koordinaten *relativ zur Ekliptik* (wahre Länge h_1 der Sonne; wahre Länge s_1 und Breite β des Mondes) und die reziproken Entfernungen c_s/R_s und c/R. Alle diese Größen lassen sich ausdrücken durch unendliche Reihen folgender Art:

$$\sum_{BCDEF} k_{BCDEF} \cdot \begin{Bmatrix} \cos \\ \sin \end{Bmatrix} (Bs + Ch + Dp + EN' + Fp_s), \tag{12.1}$$

wobei die k Konstante sind, und B, C, D, E, F ganze Zahlen, positiv, negativ oder 0. Wenn man $s = \sigma_s T + s_0$, $h = \sigma_h T + h_0, \ldots$ schreibt, kann man für das Argument schreiben

$$(B\sigma_s + C\sigma_h + \cdots + \varepsilon_{ABCDE})\, T. \tag{12.1a}$$

In der Sprache der Hochfrequenzphysik (vgl. auch Ziff. 19) kann man sagen, daß diese Formeln durch wiederholte *Modulation* der Amplituden und Phasen entstehen: Dabei werden TAYLOR-Entwicklungen des sin und cos verwendet, ferner werden alle Potenzen und Produkte von Termen verschiedener Frequenzen in Terme mit konstanten Amplituden und festen Frequenzen verwandelt, mittels Formeln wie

$$2 \sin\alpha \cos\beta = \sin(\alpha+\beta) + \sin(\alpha-\beta).$$

b) Wir wollen die ersten Glieder der Terme vom Typ (12.1) mitteilen und ihre Bedeutung veranschaulichen:

Für die Bewegung der *Sonne* geben die allgemeinen Formeln für eine elliptische Bahn (HORN [4])

$$\left.\begin{aligned} h_1 &= h + \left(2e_s - \tfrac{1}{4} e_s^3\right) \sin(h - p_s) + \tfrac{5}{4} e_s^2 \sin 2(h - p_s) + \\ &\quad + \tfrac{13}{12} e_s^3 \sin 3(h - p_s) + \cdots, \end{aligned}\right\} \tag{12.2h}$$

$$\left.\begin{aligned} \frac{c_s}{R_s} &= 1 + \left(e_s - \tfrac{1}{8} e_s^3\right) \cos(h - p_s) + e_s^2 \cos 2(h - p_s) + \\ &\quad + \tfrac{9}{8} e_s^3 \cos 3(h - p_s) + \cdots, \end{aligned}\right\} \tag{12.2c}$$

bis zu den Gliedern mit e_s^3. Mit der numerischen Exzentrizität der Erdbahn $e_s = 0{,}016750$ ($= 57{,}6$ Bogenminuten) wird

$$h_1 = h + 115' \sin(h - p_s) + 1{,}'2 \sin 2(h - p_s) + \cdots, \tag{12.3h}$$

$$\frac{c_s}{R_s} = 1 + 0{,}01675 \cos(h - p_s) + 0{,}00028 \cos 2(h - p_s) + \cdots. \tag{12.3c}$$

Die Bedeutung der Formel für c_s/R_s ist klar; über die genaue Definition von c_s vgl. HORN [4]. Die Formel für h_1 drückt aus, daß die wahre Sonne sich im Perigäum schneller bewegt als die mittlere Sonne. Deshalb eilt die wahre Sonne, vom Perigäum bis zum Apogäum, der mittleren Sonne voraus.

Die Differenz $(h_1 - h)$ gibt einen Anteil zur *Zeitgleichung* $(t_1 - t)$, die nach (8.8) durch die ungleichförmige Änderung der Rektaszension α_s bewirkt wird. Darüber lagert sich die Wirkung der Ekliptikschiefe, denn h_1 ist längs der Ekliptik gemessen, α_s längs des Äquators.

c) Die *Mond*-Bewegung ist komplizierter: Er läuft in einer Bahn, die um durchschnittlich $i = 5°8'$ gegen die Ekliptik geneigt ist; dieser Winkel schwankt zwischen $4°59'$ und $5°18'$. Der aufsteigende Knoten (= Schnittpunkt der Mondbahn mit der Ekliptik) verändert seine mittlere Länge N rückläufig (also westwärts), mit einem Umlauf in 18,6 Jahren. Wenn der *auf*steigende Knoten in den Frühlingspunkt fällt, addiert sich i zur Schiefe der Ekliptik ($\varepsilon = 23°27'$); die Neigung der Mondbahn gegen den Äquator ist dann $J = 28°35'$. Nach 9,3 Jahren fällt der *ab*steigende Knoten in den Frühlingspunkt, und $J = 18°19'$.

Wenn J groß ist, erreicht der Mond große Deklinationen; der Vollmond kulminiert dann im Sommer besonders tief, im Winter besonders hoch am Himmel, wie z. B. im Jahre 1950.

Das mittlere Perigäum p verschiebt sich ostwärts, Umlauf in 8,85 Jahren.

Die Bahnellipse des Mondes wird durch die Anziehung der Sonne auf den Mond gestört. Diese *solaren Störungskräfte*, die auf den Mond wirken, kann man auffassen als solare Gezeitenkräfte im Abstand $\varrho = 60{,}3\ a$ vom Erdmittelpunkt. Da die Gezeitenbeschleunigungen proportional ϱ zunehmen, die Erdanziehung aber $\sim 1/\varrho^2$ abnimmt, ist in der mittleren Mondentfernung das Verhältnis Erdanziehung zur Sonnengezeitenkraft nach [(4.3) und (7.6)] maximal gleich

$$(1{,}19 \cdot 10^7/0{,}46)/60{,}3^3 = \text{rund } 120:1\,. \tag{12.4}$$

Die *Richtungen* der solaren Störungskräfte entsprechen Fig. 3, wobei jetzt die Anfangspunkte der Pfeile den Mond bei seiner Bahn um die Erde bedeuten. Bei Voll- und Neumond wird also der Mond relativ von der Erde weggezogen, beim ersten und letzten Viertel dagegen auf die Erde zu beschleunigt. Dies gibt die „*Variation*“ mit dem Argument $2(s-h)$, vgl. Abschnit d.

E. Brown hat in seiner Mondtheorie die Reihenentwicklungen für s_1, c/R (oder Parallaxe) und β soweit getrieben, daß man diese Größen bis auf 0,″1 genau berechnen kann. Seine Ausdrücke für die Länge s_1 und die reziproke Entfernung c/R enthalten je etwa 155 periodische Terme, deren Koeffizienten größer als 0,″1 sind, und über 500 kleinere Terme, die noch für die angestrebte Genauigkeit zu berücksichtigen sind. Die Breite β (Abstand von der Ekliptik), die nur wenig über 5° hinausgeht, kann man mit etwa der halben Zahl der Terme ausdrücken.

d) In den folgenden Ausdrücken sind, wie in (12.3 h), die Koeffizienten der Winkelgrößen in Bogenminuten angegeben:

$$\left.\begin{aligned} s_1 = s + 377'\sin(s-p) + 13'\sin 2(s-p) + 76'\sin[(s-p) - 2(h-p)] + \\ + 40'\sin 2(s-h) - 11'\sin(h-p_s) + \cdots, \end{aligned}\right\} \tag{12.5}$$

$$\left.\begin{aligned} \beta = 308'\sin(s-N) + 17'\sin(2s-p-N) - 17'\sin(p-N) + \\ + 10'\sin[2(s-h) - (s-N)] + \cdots, \end{aligned}\right\} \tag{12.6}$$

$$\left.\begin{aligned} \frac{c}{R} = 1 + 0{,}05450\cos(s-p) + 0{,}00297\cos 2(s-p) + \\ + 0{,}01002\cos[(s-p) - 2(h-p)] + 0{,}00825\cos 2(s-h) + \cdots. \end{aligned}\right\} \tag{12.7}$$

In den Formeln für s_1 und c/R treten zunächst die elliptischen Glieder auf, entsprechend (12.2), mit $e = 0{,}05450$. Die weiteren Glieder sind solare Störungen der Mondbahn, die wie am Schluß des Absatzes c) qualitativ zu veranschaulichen sind: Zunächst die Glieder mit

$$2(s-h) = \text{Variationsargument}, \tag{12.8}$$

die als *Variation* bezeichnet werden. Die Glieder mit dem Argument

$$(s-p) - 2(h-p) = s - 2h + p = \text{Evektionsargument} \tag{12.9}$$

stellen die *Evektion* dar. Sie interferieren mit den elliptischen Hauptgliedern [mit $(s-p)$]: Sie verstärken sie für $(h-p)=0°$ oder 180°, d.h. etwa alle halbe Jahre (genauer alle 206 Tage), wenn die Sonne in Richtung der großen Achse der Mondbahn steht. Aus dem Verhältnis der Koeffizienten in s_1 und c/R kann man entnehmen, daß zu diesen Zeiten die Exzentrizität der Mondbahn um fast 20% größer ist als durchschnittlich; 103 Tage vor- oder nachher, wenn die Sonne in Richtung der kleinen Achse der Mondbahn steht, ist die Exzentrizität um ebenso viel kleiner. Die Evektion bedeutet also eine Verlängerung der Mondbahnellipse in Richtung der Verbindungslinie Erde—Sonne.

Der Term in s_1 mit $(h-p_s)$ heißt *jährliche Variation.*

e) In der Breite β drückt das erste Glied die Schiefe der mittleren Mondbahn gegen die Ekliptik aus. Zum Verständnis des zweiten und dritten Terms faßt man sie zusammen:

$$\left.\begin{aligned}17'\{\sin[(s-N)+(s-p)]-\sin[(s-N)-(s-p)]\}\\=34'\cos(s-N)\sin(s-p).\end{aligned}\right\}\qquad(12.10)$$

Die ersten drei Glieder in β sind also zusammen

$$308'[\sin(s-N)+0{,}11\cos(s-N)\sin(s-p)],\qquad(12.11)$$

was aber nichts anders ist als

$$\sim 308'\sin(s_1-N).\qquad(12.12)$$

Denn, nach (12.5), mit $377'=0{,}110$, ist

$$\sin(s_1-N)=\sin[(s-N)+(s_1-s)]\sim\sin[(s-N)+0{,}110\sin(s-p)],$$

oder, mit $\sin 0{,}11\sim 0{,}11$, $\cos 0{,}11\sim 1$, der Klammerausdruck in (12.11).

Das vierte Glied in β (Evektion in Breite) kann man auch schreiben

$$10'\sin[(s-N-2(h-N)].$$

Es überlagert sich dem ersten Glied: Die Summe des ersten und letzten Gliedes beträgt, für $h=N$, $318'\sin(s-N)$, für $h=N+90°$ dagegen $298'\sin(s-N)$. Die Mondbahn wird also im zweiten Falle, durch die Sonnenanziehung, mehr in die Ekliptikebene gezogen.

13. Die Argumentzahl und die geodätischen Funktionen. Aus den in Ziff. 12 diskutierten Größen berechnen sich die Deklination δ und, über die Rektaszensionen α, die wahren Stundenwinkel. Dadurch werden die Argumente in (12.1) um mindestens eine gleichförmige Zeitvariable vermehrt. Für diese wählt DOODSON [*1*] die *mittlere lokale Mondzeit* τ. Diese ist, für einen Ort in der östlichen Länge L, verbunden mit der mittleren Sonnenzeit t_0 und der mittleren Mondzeit τ_0 sowie der Sternzeit θ_0^*, die gleichzeitig im Greenwich-Meridian gelten, durch

$$\tau=\tau_0+L=t_0-s+h+L,\qquad t=t_0+L\qquad(13.1)$$

$$\theta^*=\tau+s=t+h,\qquad \theta^*=\theta_0^*+L,\qquad(13.2)$$

wie in (11.2).

Für die Argumente

$$A\tau+Bs+Ch+Dp+EN'+Fp_s\qquad(13.3)$$

führt DOODSON [*5*] eine „*Argumentzahl*" ein,

$$A(B+5)(C+5)\cdot(D+5)(E+5)(F+5),\qquad(13.4)$$

z.B. $2\tau - 3s + 4h + p - 2N' + 2p_s \equiv \{229{,}637\}$; an Stelle der seltenen 10 wird X geschrieben, und E für 11.

Um die Abhängigkeit von der geographischen Breite φ auszudrücken, multipliziert DOODSON [1] die Kugelfunktionen (3.6) und (8.7) mit solchen Faktoren, daß sie sämtlich, für $\varrho = \varrho_1$, den numerischen Extremwert $\pm G \equiv \pm G(\varrho_1)$ annehmen (3.12); diese nennt er *Geodätische Funktionen.* So werden alle Tiden, auch die solaren (!), als Vielfache von G ausgedrückt; G_s, nach (4.3), wird nicht mehr verwendet.

Aus V_2 stammen die Geodätischen Funktionen (für $\varrho = \varrho_1$)

$$G_0 = \tfrac{1}{2} G (1 - 3 \sin^2 \varphi), \quad G_1 = G \sin 2\varphi, \quad G_2 = G \cos^2 \varphi \tag{13.5}$$

und aus V_3 stammen, ebenfalls[1] für $\varrho = \varrho_1$,

$$\left.\begin{aligned} G_0^* &= \tfrac{1}{2} G \sin\varphi\, (3 - 5 \sin^2 \varphi), & G_1^* &= 0{,}7262\, G \cos\varphi\, (1 - 5 \sin^2 \varphi), \\ G_2^* &= 2{,}5981\, G \sin\varphi \cos^2 \varphi, & G_3^* &= G \cos^3 \varphi. \end{aligned}\right\} \tag{13.6}$$

Für $\varrho \neq \varrho_1$ ist zu setzen

$$G_k(\varrho, \varphi) = \left(\frac{\varrho}{\varrho_1}\right)^2 G_k; \quad G_k^*(\varrho, \varphi) = \left(\frac{\varrho}{\varrho_1}\right)^3 G_k^*. \tag{13.7}$$

So wird das Gesamtpotential V_2 zu einer Summe von Gliedern der Form (die *Koeffizienten* $K_{ABC.DEF}$ sind reine Zahlen; die physikalische Dimension steckt in G, mit der Einheit $\mathrm{cm^2/sec^2}$):

$$K_{ABC.DEF}\, G_A(\varrho, \varphi) \begin{Bmatrix} \cos \\ \sin \end{Bmatrix} (A\tau + Bs + Ch + Dp + EN' + Fp_s) \tag{13.8}$$

mit cos für $A = 0$ oder 2, und sin für $A = 1$.

Aus V_3 kommt entsprechend

$$K^*_{ABC.DEF}\, G^*_A(\varrho, \varphi) \begin{Bmatrix} \cos \\ \sin \end{Bmatrix} (A\tau + Bs + Ch + Dp + EN' + Fp_s) \tag{13.9}$$

mit sin für $A = 0$ oder 2, cos für $A = 1$ oder 3. (Der Grund dafür, weshalb bei festem A entweder nur cos oder nur sin vorkommt, wird in Ziff. 15—18 ersichtlich.)

14. Verzeichnis der wichtigsten Tiden. Die DOODSONsche Entwicklung des Gezeitenpotentials (1) ist auch im LANDOLT BÖRNSTEIN [5] vollständig wiedergegeben. Tabelle 2 gibt eine Auswahl, mit folgenden Spalten:

1. Herkunft: M oder S, lunar oder solar, aus V_2 stammend (3.11); M^*, lunar, aus V_3.

2. Symbol: $\boldsymbol{M}$ oder $\boldsymbol{S}$, meistens getrennt; $\boldsymbol{MS}$ kennzeichnet lunare Tiden, die von der Störung der Mondbahn durch die Sonne herrühren (Ziff. 11d). Die Indices 0, 1, 2, 3 kennzeichnen den Faktor A von τ oder t im Argument; diese Indices sind jedoch nicht überall verwendet. $\boldsymbol{a}$ = annual = jährlich, $\boldsymbol{sa}$ = semiannual = halbjährlich; $\boldsymbol{m}$ = monthly, $\boldsymbol{f}$ = fortnightly = halbmonatlich, $\boldsymbol{t}$ = termonthly = drittelmonatlich, $\boldsymbol{q}$ = quarter-monthly. Die übrigen Bezeichnungen sind mehr oder weniger konventionell.

3. Koeffizient: Vgl. Bemerkung vor (13.8). In dieser Spalte ist der 10^5fache Wert angegeben; den größten Koeffizienten, 0,90812, hat $\boldsymbol{M}_2$.

4. Argument: Wie in (13.4) erklärt.

[1] Bei G_0^* und den dazutretenden Koeffizienten sind die Zahlenfaktoren nach HORN ([5], S. 273) berichtigt.

5. Geodätische Funktion: G_A oder G_A^*, vgl. (13.5) bis (13.7) und (3.12). Man erkennt leicht, daß sie die in (8.3) bis (8.4) ausgedrückte Abhängigkeit von φ richtig wiedergeben.

6. Zusammensetzung des Arguments: Die konventionelle Schreibweise aus der dritten Spalte der Tabelle ist hier durchsichtiger geschrieben, um die Herkunft deutlich zu machen, durch die Zusammensetzung aus folgender Art gleichförmig wachsender Argumente:

θ^* = Sternzeit;
τ = mittlere Mondzeit = $\theta^* - s$;
t = mittlere Sonnenzeit = $\theta^* - h$;
s = mittlere Länge des Mondes; wächst um 360° während eines tropischen Monats;
h = mittlere Länge der Sonne; wächst um 360° während eines tropischen Jahres;
$(s - p)$ = „mittlere Anomalie“ = Abstand des mittleren Mondes vom mittleren Perigäum; wächst um 360° in einem anomalistischen Monat = Argument der elliptischen Ungleichheit des Mondes 1. Ordnung; $2(s-p)$, $3(s-p)\ldots$ = elliptische Ungleichheiten 2., 3. Ordnung;
$(s-2h+p)$ = Argument der Evektion (12.9);
$2(s - h)$ = Argument der Variation, (12.8);
$(h - p_s)$ = solares Gegenstück zu $(s-p)$ = Argument der elliptischen Ungleichheit 1. Ordnung in der Bewegung der Sonne (12.2) sowie der jährlichen Ungleichheit in der Bewegung des Mondes (12.5);
$N' = -N$, mit N = mittlere Länge des aufsteigenden Knotens der Mondbahn.

W: Diese Größe soll die praktische Berechnung des Arguments erleichtern. W drückt das Argument durch t (an Stelle von τ) aus, ersetzt also $\theta^* = t+h$, $\tau = t-s+h$. An Stelle des Arguments (13.3) tritt also

$$W \equiv At + u(t) \equiv At + bs + ch + Dp - EN + Fp_s \pm n\cdot 90^\circ; \qquad (14.1)$$

mit $b = B - A$, $c = C + A$. Ferner wird ein passend gewähltes Vielfaches ($n =$ 0, ± 1, oder 2) von 90° hinzugefügt, um zu erreichen, daß die Zeitabhängigkeit, anstatt durch $\pm$ sin oder $\pm$ cos, stets ausgedrückt wird durch

$$+\cos W = +\cos[At + u(t)]; \qquad (14.2)$$

man beachte jedoch, daß die Geodätischen Funktionen negative Vorzeichen haben können!

W_0 sei das Argument auf dem Meridian von Greenwich ($L = 0$) zur Weltzeit t_0; dann ist es *gleichzeitig*, auf dem Meridian $L°$ östlicher Länge, gleich $(AL + W_0)$. *Frequenz* σ ist die Änderung des Tidenarguments pro Stunde, also während t sich um 15° ändert. Sie setzt sich additiv, mit den Koeffizienten A, b, c, D, E, F, aus den Frequenzen der Grundveränderlichen t, s, h, p, N und p_s zusammen, die in Tabelle 1 gegeben sind, in der Einheit °/Stunde. Entsprechend der dekadischen Anordnung der Tiden nach dem Argument wächst σ in der Tabelle von Anfang bis Ende.

Werte des Arguments für bestimmte Zeiten: Auf dem Meridian von Greenwich (mittlere Ortszeit t_0, wie immer, in Grad, 15°/Stunde, gerechnet) ist das Argument des cos gleich

$$At_0 + u(t_0). \qquad (14.3)$$

Für ein bestimmtes Datum, Weltzeit Mitternacht, $t_0 = 0°$, sei $u(0) = u_0$, also $= bs + ch + Dp - EN + Fp_s \pm n\cdot 90°$, wobei für s, h, p, N, p_s die für diesen Augenblick geltenden Werte einzusetzen sind. Für andere Zeiten t_0 (in Grad gerechnet, also gleich $t_0/15$ Std) am gleichen Tage (am gleichen Datum) ist dann das Argument gleich

$$\sigma t_0/15 + u_0. \qquad (14.4)$$

Auf dem Meridian in der östlichen Länge $L°$, mit der mittleren Ortszeit

$$t = t_0 + L,$$

ist das Argument *im gleichen Augenblick* um AL größer, also gleich

$$\sigma t_0/15 + AL + u_0 = \sigma t/15 + AL - \sigma L/15 + u_0. \tag{14.5}$$

Wenn man mit *Zonenzeit* z des Standardmeridians S rechnet, so ist $z = t_0 + S = t + (S - L)$; für Orte auf dem Meridian L ist also $t = z + (L - S)$ zu setzen, und das Argument (14.5) wird

$$\sigma z/15 + AL - \sigma S/15 + u_0. \tag{14.6}$$

Über die praktische Rechnung vgl. Ziff. 21.

Entstehung: Die Namen deuten folgendes an: *Elliptische* Tiden beruhen auf der Elliptizität der Bahnen, enthalten $(s - p)$ oder $(h - p_s)$ oder Vielfache davon (Glieder 1., 2., 3. Ordnung).

Deklinations-Tiden beruhen auf der Schiefe der Ekliptik als Sonnenbahn und (genäherte) Mondbahn; sie enthalten s oder h.

Evektions-Tiden enthalten $(s - 2h + p)$, nach (12.9); *Variations*-Tiden enthalten $2(s - h)$, nach (12.8).

Einteilung der Tiden: Nach der ersten Ziffer A des Arguments teilt man die Tiden in *Klassen*, die während eines *Tages* etwa 0, 1, 2 oder 3 Perioden durchlaufen. Als *Gruppen* sind in der Tabelle solche Tiden abgeteilt, die gleiche Ziffern A und B haben; sie unterscheiden sich durch die Zahl der Umläufe pro *Monat*. Unter den Tiden mit gleichen A, B und C (deren Argumente noch innerhalb eines *Jahres* nur wenig auseinander laufen), heißt *Stammtide* diejenige mit dem größten Koeffizienten K.

Auswahl: Aus der ausführlichen Tabelle DOODSONs ([*1*], abgedruckt in [*5*]) die wichtigsten Stammtiden, ferner alle Tiden mit Koeffizienten K über 0,001, darüber hinaus noch einige, die trotz kleinerer Koeffizienten von Interesse sind.

15. Erläuterung der wichtigsten Tiden: Langperiodische.

a) Die detaillierte Entwicklung des Gezeitenpotentials kann hier nicht vorgeführt werden; das Resultat ist in Tabelle 2 übernommen. Es ist jedoch nützlich, hier zu veranschaulichen, wie die verschiedenen Elemente der Bahnbewegungen zusammenwirken, um die einzelnen Tiden zu erzeugen; dabei soll auf die letzte Dezimale und auf höhere Glieder verzichtet werden.

Wir gehen aus von V_2 in der Form (8.3) und seinem solaren Äquivalent. Hinsichtlich der Geodätischen Funktionen ist schon in Ziff. 14 festgestellt worden, daß sie die Abhängigkeit von der Breite φ richtig wiedergeben. Wir können uns also auf die Betrachtung der Abhängigkeit von der *Zeit* beschränken; wir setzen immer $\varrho = \varrho_1$, unter Hinweis auf (13.7).

b) Die Geodätische Funktion ist

$$G_0 = \tfrac{3}{2} G \left(\tfrac{1}{3} - \sin^2\varphi\right),$$

positiv am Äquator, negativ jenseits $\pm\varphi = 35°15',9$.

Aus (8.3) und (4.3), folgt also der langperiodische Teil von V_2, für $\varrho = \varrho_1$,

für den Mond $$V_m = G_0 \cdot \left(\frac{c}{R}\right)^3 \cdot 2 \quad \left(\tfrac{1}{3} - \sin^2\delta_m\right), \tag{15.1m}$$

für die Sonne $$V_s = G_0 \cdot \left(\frac{c_s}{R_s}\right)^3 \cdot 0{,}921\left(\tfrac{1}{3} - \sin^2\delta_s\right). \tag{15.1s}$$

Tabelle 2. *Die wichtigsten Tiden*. Allgemeine Erläuterungen in Ziff. 13 und 14, spezielle Bemerkungen in Ziff. 15—18.

1	2	3	4	5	6	7	8	9
Herkunft	Symbol	Koeffizient $K \cdot 10^5$	G_k	Argument	Zusammensetzung des Arguments	Argument von $+\mathrm{ccs}\,[At+u(t)]$	Frequenz σ in °/Stunde	Entstehung
					Langperiodische Glieder (Ziff. 15)			
M	$\boldsymbol{M_0}$	$+50458$	G_0	cos {055,555}	0	0	0,000000	Konstante Mondtide
S	$\boldsymbol{S_0}$	$+23411$	G_0	cos {055,555}	0	0	0,000000	Konstante Sonnentide
M	...	$-\ 6552$	G_0	cos {055,565}	N'	$-N+180°$	0,002206	Zu $\boldsymbol{M_0}$
S	**Sa**	$+\ 1176$	G_0	cos {056,554}	$(h-p_s)$	$+h\ -p_s$	0,041067	Ellipt. Tide 1. Ord. zu $\boldsymbol{S_0}$
S	**Ssa**	$+\ 7287$	G_0	cos {057,555}	$2h$	$+2h$	0,082137	Deklinationstide zu $\boldsymbol{S_0}$
S	**Sta**	$+\ 427$	G_0	cos {058,554}	$2h+(h-p_s)$	$+3h\ -p_s$	0,123204	Elliptische Tide zu **Ssa**
M	**MSm**	$+\ 1587$	G_0	cos {063,655}	$(s-2h+p)$	$+s\ -2h\ +p$	0,471521	Evektionstide zu $\boldsymbol{M_0}$
M	**Mm**	$+\ 8254$	G_0	cos {065,455}	$(s-p)$	$+s\ -p$	0,544375	Ellipt. Tide 1. Ord. zu $\boldsymbol{M_0}$
M^*	...	$+\ 1052$	G_0^*	sin {065,555}	s	$+s\ -90°$	0,549017	Aus V_3
M	**MSf**	$+\ 1370$	G_0	cos {073,555}	$2(s-h)$	$+2s\ -2h$	1,015896	Variationstide zu $\boldsymbol{M_0}$
M	**Mf**	$+15642$	G_0	cos {075,555}	$2s$	$+2s$	1,098033	Deklinationstide zu $\boldsymbol{M_0}$
M	...	$+\ 6481$	G_0	cos {075,565}	$2s+N'$	$+2s\ -N$	1,100240	Zu **Mf**
M	**MStm**	$+\ 569$	G_0	cos {083,655}	$2s+(s-2h+p)$	$+3s\ -2h\ +p$	1,569554	Evektionstide zu **Mf**
M	**Mtm**	$+\ 2995$	G_0	cos {085,455}	$2s+(s-p)$	$+3s\ -p$	1,642408	Ellipt. Tide 1. Ord. zu **Mf**
M	...	$+\ 1241$	G_0	cos {085,465}	$2s+(s-p)+N'$	$+3s\ -p\ -N$	1,644614	Zu **Mtm**
M	**MSqm**	$+\ 478$	G_0	cos {093,555}	$2s+2(s-h)$	$+4s\ -2h$	2,113929	Variationstide zu **Mf**
					Ganztägige Glieder (Ziff. 16)			
M	$\boldsymbol{2Q_1}$	$+\ 955$	G_1	sin {125,755}	$(\tau-s)-2(s-p)$	$t\ -4s\ +h\ +2p\ -90°$	12,854286	Ellipt. Tide 2. Ord. zu $\boldsymbol{O_1}$
M	σ_1	$+\ 1153$	G_1	sin {127,555}	$(\tau-s)-2(s-h)$	$t\ -4s\ +3h\ -90°$	12,927140	Variationstide zu $\boldsymbol{O_1}$
M	...	$+\ 1360$	G_1	sin {135,645}	$(\tau-s)-(s-p)-N'$	$t\ -3s\ +h\ +p\ +N\ -90°$	13,396455	Zu $\boldsymbol{Q_1}$
M	$\boldsymbol{Q_1}$	$+\ 7216$	G_1	sin {135,655}	$(\tau-s)-(s-p)$	$t\ -3s\ +h\ +p\ -90°$	13,398661	Ellipt. Tide 1. Ord. zu $\boldsymbol{O_1}$
M	ϱ_1	$+\ 1371$	G_1	sin {137,455}	$(\tau-s)-(s-2h+p)$	$t\ -3s\ +3h\ -p\ -90°$	13,471514	Evektionstide zu $\boldsymbol{O_1}$

M	...	$+\ 7105$	G_1	$\sin\{145{,}545\}$	$(\tau - s) - N'$	$t\ -2s\ +h$	$+N$	$-90°$	13,940830	Zu $\boldsymbol{O}_1$
M	$\boldsymbol{O}_1$	$+37689$	G_1	$\sin\{145{,}555\}$	$(\tau - s)$	$t\ -2s\ +h$		$-90°$	13,943036	Eintäg. Haupt-Mondtide
M	$\boldsymbol{\tau}_1$	$-\ 491$	G_1	$\sin\{147{,}555\}$	$(\tau + s) - 2(s - h)$	$t\ -2s\ +3h$		$+90°$	14,025173	Variationstide zu $\boldsymbol{K}_{1m}$
M	...	$-\ 1065$	G_1	$\sin\{155{,}455\}$	$(\tau - s) + (s - p)$	$t\ -s\ +h\ -p$		$+90°$	14,487410	Zu $\boldsymbol{O}_1$
M	$\boldsymbol{M}_1$	$-\ 2964$	G_1	$\sin\{155{,}655\}$	$(\tau + s) - (s - p)$	$t\ -s\ +h\ +p$		$+90°$	14,496694	Ellipt. Tide 1. Ord. zu $\boldsymbol{K}_{1m}$
M	χ_1	$-\ 566$	G_1	$\sin\{157{,}455\}$	$(\tau + s) - (s - 2h + p)$	$t\ -s\ +3h\ -p$		$+90°$	14,569548	Evektionstide zu $\boldsymbol{K}_{1m}$
S	$\boldsymbol{\pi}_1$	$+\ 1029$	G_1	$\sin\{162{,}556\}$	$(t - h) - (h - p_s)$	$t\ -2h\ +p_s$		$-90°$	14,917865	Ellipt. Tide 1. Ord. zu $\boldsymbol{P}_1$
S	$\boldsymbol{P}_1$	$+17554$	G_1	$\sin\{163{,}555\}$	$(t - h)$	$t\ -h$		$-90°$	14,958931	Eintäg. Haupt-Sonnentide
S	$\boldsymbol{S}_1$	$-\ 423$	G_1	$\sin\{164{,}556\}$	$(t + h) - (h - p_s)$	$t\ -p_s$		$+90°$	15,000002	Ellipt. Tide 1. Ord. zu $\boldsymbol{K}_{1s}$
M	$\boldsymbol{K}_{1m}$	-36233	G_1	$\sin\{165{,}555\}$	$(\tau + s) = \theta^*$	$t\ +h$		$+90°$	15,041069	Eintäg. Deklin.-Haupttide
S	$\boldsymbol{K}_{1s}$	-16817	G_1	$\sin\{165{,}555\}$	$(t + h) = \theta^*$	$t\ +h$		$+90°$	15,041069	Eintäg. Deklin.-Haupttide
M	...	$-\ 7182$	G_1	$\sin\{165{,}565\}$	$(\tau + s) + N'$	$t\ +h$	$-N$	$+90°$	15,043275	Zu $\boldsymbol{K}_{1m}$
S	$\boldsymbol{\psi}_1$	$-\ 423$	G_1	$\sin\{166{,}554\}$	$(t + h) + (h - p_s)$	$t\ +2h\ -p_s$		$+90°$	15,082135	Ellipt. Tide 1. Ord. zu $\boldsymbol{K}_{1s}$
S	φ_1	$-\ 756$	G_1	$\sin\{167{,}555\}$	$(t + 3h)$	$t\ +3h$		$+90°$	15,123206	Eintäg. Deklin.-Tide 2. Ord.
M	$\boldsymbol{\vartheta}_1$	$-\ 566$	G_1	$\sin\{173{,}655\}$	$(\tau + s) + (s - 2h + p)$	$t\ +s\ -h\ +p$		$+90°$	15,512590	Evektionstide zu $\boldsymbol{K}_{1m}$
M	$\boldsymbol{J}_1$	$-\ 2964$	G_1	$\sin\{175{,}455\}$	$(\tau + s) + (s - p)$	$t\ +s\ +h\ -p$		$+90°$	15,585443	Ellipt. Tide 1. Ord. zu $\boldsymbol{K}_{1m}$
M	$\boldsymbol{SO}_1$	$-\ 492$	G_1	$\sin\{183{,}555\}$	$(\tau + s) + 2(s - h)$	$t\ +2s\ -h$		$+90°$	16,056964	Variationstide zu $\boldsymbol{K}_{1m}$
M	$\boldsymbol{OO}_1$	$-\ 1623$	G_1	$\sin\{185{,}555\}$	$\tau - 3s$	$t\ +2s\ +h$		$+90°$	16,139102	Eintäg. Deklin.-Tide 2. Ord.
M	...	$-\ 1039$	G_1	$\sin\{185{,}565\}$	$\tau - 3s + N'$	$t\ +2s\ +h$	$-N$	$+90°$	16,141308	Zu $\boldsymbol{OO}_1$
M	$\boldsymbol{\upsilon}_1$	$-\ 311$	G_1	$\sin\{195{,}455\}$	$\tau - 3s + (s - p)$	$t\ +3s\ +h\ -p$		$+90°$	16,683476	Ellipt. Tide 1. Ord. zu $\boldsymbol{OO}_1$
					Halbtägige Glieder (Ziff. 17)					
M	$\boldsymbol{3N}_2$	$+\ 259$	G_2	$\cos\{225{,}855\}$	$2\tau - 3(s - p)$	$2t\ -5s\ +2h\ +3p$			27,350980	Ellipt. Tide 3. Ord. zu $\boldsymbol{M}_2$
M	ε_2	$+\ 671$	G_2	$\cos\{227{,}655\}$	$2\tau - 2(s - p) - (s - 2h + p)$	$2t\ -5s\ +4h\ +p$			27,423834	Evektionstide zu $\boldsymbol{2N}_2$
M	$\boldsymbol{2N}_2$	$+\ 2301$	G_2	$\cos\{235{,}755\}$	$2\tau - 2(s - p)$	$2t\ -4s\ +2h\ +2p$			27,895355	Ellipt. Tide 2. Ord. zu $\boldsymbol{M}_2$
M	μ_2	$+\ 2777$	G_2	$\cos\{237{,}555\}$	$2\tau - 2(s - h)$	$2t\ -4s\ +4h$			27,968208	Gr. Variationstide zu $\boldsymbol{M}_2$
M^*	...	$-\ 569$	G_2^*	$\sin\{245{,}555\}$	$2\tau - s$	$2t\ -3s\ +2h$		$+90°$	28,435088	Aus V_3
M	$\boldsymbol{N}_2$	$+17387$	G_2	$\cos\{245{,}655\}$	$2\tau - (s - p)$	$2t\ -3s\ +2h\ +p$			28,439730	Gr. ellipt. Tide 1. Ord. zu $\boldsymbol{M}_2$
M	$\boldsymbol{\nu}_2$	$+\ 3303$	G_2	$\cos\{247{,}455\}$	$2\tau - (s - 2h + p)$	$2t\ -3s\ +4h\ -p$			28,512583	Gr. Evektionstide zu $\boldsymbol{M}_2$

Tabelle 2. (Fortsetzung.)

1	2	3	4	5	6	7	8	9
Herkunft	Symbol	Koeffizient $K \cdot 10^5$	G_k	Argument	Zusammensetzung des Arguments	Argument von $+\cos[At + u(t)]$	Frequenz σ in °/Stunde	Entstehung
M	γ_2	$-$ 273	G_2	cos {253,755}	$2\tau-(s-p)+(s-2h+p)$	$2t\ -2s\ \ +2p\ \ +180°$	28,911251	Evektionstide zu $\boldsymbol{N_2}$
M	α_2	$-$ 314	G_2	cos {254,556}	$2\tau-(h-p_s)$	$2t\ -2s\ +h\ +p_s\ \ +180°$	28,943038	Gr. Tide d. jährl. Ungl. z. $\boldsymbol{M_2}$
M	…	$-$ 3386	G_2	cos {255,545}	$2\tau-N'$	$2t\ -2s\ +2h\ \ +N\ +180°$	28,981898	Zu $\boldsymbol{M_2}$
M	$\boldsymbol{M_2}$	$+$90812	G_2	cos {255,555}	2τ	$2t\ -2s\ +2h$	28,984104	Halbtäg. Haupt-Mondtide
M	β_2	$+$ 276	G_2	cos {256,554}	$2\tau+(h-p_s)$	$2t\ -2s\ +3h\ -p_s$	29,025171	Kl. Tide d. jährl. Ungl. z. $\boldsymbol{M_2}$
M	δ_2	$+$ 107	G_2	cos {257,555}	$2(\tau+s)-2(s-h)$	$2t\ -2s\ +4h$	29,066242	Variationstide zu $\boldsymbol{K_{2m}}$
M	λ_2	$-$ 670	G_2	cos {263,655}	$2\tau+(s-2h+p)$	$2t\ -s\ \ +p\ \ +180°$	29,455625	Kl. Evektionstide zu $\boldsymbol{M_2}$
M	$\boldsymbol{L_2}$	$-$ 2567	G_2	cos {265,455}	$2\tau+(s-p)$	$2t\ -s\ +2h\ -p\ \ +180°$	29,528479	Kl. ellipt. Tide 1. Ord. zu $\boldsymbol{M_2}$
M^*	…	$+$ 525	G_2^*	sin {265,555}	$2\tau+s$	$2t\ -s\ +2h\ \ -90°$	29,533121	Aus V_3
M	…	$+$ 643	G_2	cos {265,655}	$2(\tau+s)-(s-p)$	$2t\ -s\ +2h\ +p$	29,537763	Ellipt. Tide zu $\boldsymbol{K_{2m}}$
S	$\boldsymbol{2T_2}$	$+$ 101	G_2	cos {271,557}	$2t-2(h-p_s)$	$2t\ \ -2h\ +2p_s$	29,917866	Ellipt. Tide 2. Ord. zu $\boldsymbol{S_2}$
S	$\boldsymbol{T_2}$	$+$ 2479	G_2	cos {272,556}	$2t-(h-p_s)$	$2t\ \ -h\ +p_s$	29,958933	Gr. ellipt. Tide 1. Ord. zu $\boldsymbol{S_2}$
S	$\boldsymbol{S_2}$	$+$42286	G_2	cos {273,555}	$2t$	$2t$	30,000000	Halbtägige Haupt-S.-Tide
M	$\boldsymbol{S_2}$	$+$ 72	G_2	cos {273,555}	$2\tau+2(s-h)$	$2t$	30,000000	Kl. Variationstide zu $\boldsymbol{M_2}$
S	$\boldsymbol{R_2}$	$-$ 354	G_2	cos {274,554}	$2t+(h-p_s)$	$2t\ \ +h\ -p_s$	30,041067	Kl. ellipt. Tide 1. Ord. zu $\boldsymbol{S_2}$
M	$\boldsymbol{K_{2m}}$	$+$ 7858	G_2	cos {275,555}	$2\theta^*=2(\tau+s)$	$2t\ \ +2h$	30,082137	Halbtäg. Dekl.-Tide zu $\boldsymbol{M_2}$
S	$\boldsymbol{K_{2s}}$	$+$ 3648	G_2	cos {275,555}	$2\theta^*=2(t+h)$	$2t\ \ +2h$	30,082137	Halbtäg. Dekl.-Tide zu $\boldsymbol{S_2}$
M	…	$+$ 3423	G_2	cos {275,565}	$2(\tau+s)+N'$	$2t\ \ +2h\ \ -N$	30,084343	Zu $\boldsymbol{K_{2m}}$
M	ζ_2	$+$ 123	G_2	cos {283,655}	$2(\tau+s)+(s-2h+p)$	$2t\ +s\ \ +p$	30,553656	Evektionstide zu $\boldsymbol{K_{2m}}$
M	η_2	$+$ 643	G_2	cos {285,455}	$2(\tau+s)+(s-p)$	$2t\ +s\ +2h\ -p$	30,626512	Ellipt. Tide 1. Ord. zu $\boldsymbol{K_{2m}}$
M	…	$+$ 168	G_2	cos {295,555}	$2(\tau+s)+2s$	$2t\ +2s\ +2h$	31,180170	Zu $\boldsymbol{K_{2m}}$
					Dritteltägige Glieder (Ziff. 18)			
M^*	…	$-$ 326	G_3^*	cos {345,655}	$3\tau-(s-p)$	$3t\ -4s\ +3h\ +p\ \ +180°$	42,931781	Ellipt. Tide 1. Ord. zu $\boldsymbol{M_3}$
M^*	$\boldsymbol{M_3}$	$-$ 1188	G_3^*	cos {355,555}	3τ	$3t\ -3s\ +3h\ \ +180°$	43,476156	Dritteltäg. Haupt-M.-Tide

c) Wahre Länge h_1, Deklination δ_s und Ekliptikschiefe ε (Fig. 6) sind verbunden durch

$$\sin\delta_s = \sin\varepsilon \sin h_1. \tag{15.2s}$$

Für den *Mond* kommt durch die Neigung der Mondbahn gegen die Ekliptik eine kompliziertere Beziehung heraus (vgl. Abschnitt e); aber näherungsweise, und im großen Durchschnitt, gilt

$$\sin\delta_m \sim \sin\varepsilon \sin s_1. \tag{15.2m}$$

In erster Näherung wird h_1 durch h ersetzt. Dann wird die Klammer mit δ_s in (15.1s)

$$\left.\begin{aligned} \tfrac{1}{3} - \sin^2\delta_s &= \tfrac{1}{3} - \sin^2\varepsilon\sin^2 h = \tfrac{1}{3} - \tfrac{1}{2}\sin^2\varepsilon + \tfrac{1}{2}\sin^2\varepsilon\cos 2h \\ &= 0{,}254 + 0{,}079\cos 2h. \end{aligned}\right\} \tag{15.3}$$

Wir erkennen hierin [immer mit $G = 26206\ \mathrm{cm^2/sec^2}$, nach (3.12)] das konstante Sonnenglied

$$\boldsymbol{S_0} \equiv K\cos\{055{,}555\} = +0{,}2341\, G_0, \tag{15.4}$$

und das halbjährige (semi-annual) Glied

$$\left.\begin{aligned} \boldsymbol{Ssa} &\equiv K\cos\{057{,}555\} \\ &= 0{,}0729\, G_0 \cos 2h. \end{aligned}\right\} \tag{15.5}$$

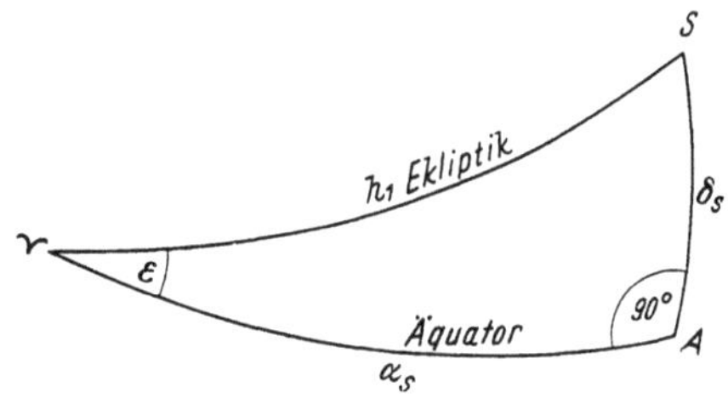

Fig. 6. Dreieck am Frühlingspunkt.

Die analogen Glieder für den Mond enthalten s statt h. So ergibt sich das konstante Mondglied

$$\boldsymbol{M_0} \equiv K\cos\{055{,}555\} = +0{,}5046\, G_0 \tag{15.6}$$

und das halbmonatliche (fortnightly) Glied

$$\boldsymbol{Mf} = K\cos\{075{,}555\} = +0{,}1564\, G_0 \cos 2s \tag{15.7}$$

d) Die Summe $\boldsymbol{S_0} + \boldsymbol{M_0}$ entspricht einer Gleichgewichtsflut (nach Ziff. 7c)

$$\zeta = +0{,}739 \cdot \left(\frac{G}{g_0}\right) \cdot \frac{1}{2}(1 - 3\sin^2\varphi) = +9{,}9\ \mathrm{cm}\,(1 - 3\sin^2\varphi). \tag{15.8}$$

Die Niveaufläche (das Geoid) wird also *dauernd* an den Polen um 20 cm gesenkt, am Äquator um 10 cm gehoben. Diese zusätzliche Abplattung der Erde kann man entstanden denken durch die Wirkung von zwei Massenringen, auf denen die Massen der Sonne und des Mondes gleichmäßig verteilt sind, und die die Erde in der Äquatorebene mit den Radien c und c_s umgeben.

e) $\boldsymbol{M_0}$ und $\boldsymbol{Mf}$ entsprechen einem fiktiven Mond, der sich gleichförmig in der *Ekliptik* bewegt. Die Abweichung der Mondbahn von der Ekliptik muß sich in Gliedern äußern, die N enthalten. In grober Näherung wollen wir die Deklination des Mondes, δ_m, ersetzen durch $(\delta_0 + \beta)$, mit δ_0 = Deklination eines Punktes in der Länge s auf der Ekliptik, und β = Breite = Abstand des Mondes von der Ekliptik. Nach (12.6), mit arc $308' = 0{,}09$, und mit $\cos\beta \sim 1$, $\sin\beta \sim \beta$, $\cos\delta_0 \sim 1$, wird genähert

$$\left.\begin{aligned} \tfrac{1}{3} - \sin^2[\delta_0 + 0{,}09\sin(s - N)] &= \tfrac{1}{3} - [\sin\delta_0 + 0{,}09\sin(s - N)]^2 \\ &\sim \tfrac{1}{3} - \sin^2\delta_0 - 0{,}18\sin\delta_0\sin(s - N). \end{aligned}\right\} \tag{15.9}$$

Die ersten beiden Glieder wurden im Abschnitt c behandelt. Das Zusatzglied [nach (15.2s)]

$$-0{,}18\sin\varepsilon\sin s\sin(s - N) \tag{15.10}$$

wird, mit

$$2\sin x\sin y = \cos(x - y) - \cos(x + y), \tag{15.11}$$

und mit $\sin\varepsilon = 0{,}398$

$$- 0{,}036 \cos N + 0{,}036 \cos(2s - N). \qquad (15.12)$$

So erklären sich [mit (15.1*m*)] die Glieder mit (nahezu) entgegengesetzt gleichen Koeffizienten

$$K \cos\{055{,}565\} = -0{,}0655\, G_0 \cos N, \quad \text{mit 18,6jähriger Periode}, \qquad (15.13)$$

$$K \cos\{075{,}565\} = +0{,}0648\, G_0 \cos(2s - N), \quad \text{mit halbmonatlicher Periode}. \qquad (15.14)$$

Sie interferieren mit $\boldsymbol{M_0}$ und $\boldsymbol{Mf}$ in leicht verständlicher Weise.

f) Die Exzentrizität der Mondbahn wirkt sich folgendermaßen aus: Nach (12.7) ist $(c/R)^3 \sim 1 + 0{,}164 \cos(s - p)$. In (15.3) setzen wir, an Stelle von h oder s, nach (12.5), mit arc $377' = 0{,}110$,

$$s_1 \sim s + 0{,}110 \sin(s - p).$$

Dann wird aus (15.1*m*) und (15.3)

$$[1 + 0{,}164 \cos(s - p)] \cdot [0{,}254 + 0{,}079 \cos 2s_1]. \qquad (15.15)$$

Die üblichen Entwicklungen und Vernachlässigungen geben, mit (15.11), zunächst

$$\left.\begin{aligned} \cos 2s_1 &\sim \cos[2s + 0{,}22 \sin(s - p)] \sim \cos 2s - 0{,}22 \sin 2s \sin(s - p) \\ &= \cos 2s + 0{,}11 \cos(3s - p) - 0{,}11 \cos(s + p). \end{aligned}\right\} \qquad (15.16)$$

In (15.15) eingesetzt, kommen, außer den schon behandelten Gliedern $\boldsymbol{M_0}$ und $\boldsymbol{Mf}$, zwei weitere Glieder, die sich so näherungsweise ergeben als

$$0{,}042 \cos(s - p) + 0{,}015 \cos(3s - p).$$

Diese finden sich, nach Multiplikation mit $2G_0$ gemäß (15.1*m*), in Tabelle 2 als

$$\boldsymbol{Mm} \equiv K \cos\{065{,}455\} = +0{,}0825\, G_0 \cos(s - p) \quad \text{(monatlich)}, \qquad (15.17)$$

$$\boldsymbol{Mtm} \equiv K \cos\{085{,}455\} = +0{,}0300\, G_0 \cos(3s - p) \quad \text{(drittelmonatlich)}. \qquad (15.18)$$

Im Perigäum $(s = p)$ verstärken sich $\boldsymbol{M_0}$ und $\boldsymbol{Mm}$, sowie $\boldsymbol{Mf}$ und $\boldsymbol{Mtm}$; im Apogäum $(s = p + 180°)$ wirken die Glieder gegeneinander.

g) Die Exzentrizität der Sonnenbahn bewirkt die entsprechenden, aber wesentlich kleineren Glieder

$$\boldsymbol{Sa} \equiv K \cos\{056{,}554\} = +0{,}0118\, G_0 \cos(h - p_s) \quad \text{(jährlich)}, \qquad (15.19)$$

$$\boldsymbol{Sta} \equiv K \cos\{059{,}554\} = +0{,}0043\, G_0 \cos(3h - p_s) \quad \text{(dritteljährlich)}. \qquad (15.20)$$

h) Wie man aus (12.5) und (12.7) ersieht, tritt in s_1 und (c/R) das Evektionsargument $(s - 2h + p)$ in derselben Form auf wie das elliptische Argument $(s - p)$, jedoch mit Koeffizienten, die nur etwa $^1/_5$ so groß sind. Man kann deshalb die Rechnungen aus dem Abschnitt f ohne weiteres übertragen und erhält die Evektionstiden.

$$\boldsymbol{MSm} \equiv K \cos\{063{,}655\} = +0{,}0158\, G_0 \cos(s - 2h + p) \quad \text{(monatlich)}. \qquad (15.21)$$

$$\left.\begin{aligned} \boldsymbol{MStm} &\equiv K \cos\{083{,}655\} \\ &= +0{,}0057\, G_0 \cos[2s + (s - 2h + p)] \quad \text{(drittelmonatlich)}. \end{aligned}\right\} \qquad (15.22)$$

$\boldsymbol{MSm}$ verstärkt $\boldsymbol{Mm}$, und $\boldsymbol{MStm}$ verstärkt $\boldsymbol{Mtm}$ für $h = p$ oder $h = p + 180°$, wie es dem Wesen der Evektion entspricht.

i) Genau so liefern die Glieder mit $2(s-h)$ die Variationstiden

$$\boldsymbol{MSf} \equiv K\cos\{073{,}555\} = +0{,}0137\, G_0 \cos 2(s-h) \qquad \text{(halbmonatlich).} \quad (15.23)$$

$$\boldsymbol{MSqm} \equiv K\cos\{093{,}555\} = +0{,}0048\, G_0 \cos[2s+2(s-h)] \quad \text{(viertelmonatl.).} \quad (15.24)$$

16. Die ganztägigen Tiden.

a) Entsprechend dem zweiten Term in (8.3) ist die Geodätische Funktion

$$G_1 = G\sin 2\varphi,$$

mit Vorzeichenwechsel am Äquator; und der ganztägige Teil von V_2 wird, für $\varrho=\varrho_1$,

für den Mond
$$V_m = -\,G_1\cdot\left(\frac{c}{R}\right)^3 \sin 2\delta_m \cos\tau_1. \qquad (16.1\,m)$$

für die Sonne
$$V_s = -\,0{,}4605\, G_1\cdot\left(\frac{c_s}{R_s}\right)^3 \sin 2\delta_s \cos t_1. \qquad (16.1\,s)$$

b) In erster Näherung vernachlässigen wir den Unterschied zwischen Rektaszension und Länge, setzen also die wahren Stundenwinkel gleich den mittleren, $\tau_1\sim\tau$, $t_1\sim t$. Aus (15.2m) folgt, mit $\cos\delta_m\sim 1$, und $s_1\sim s$, für den Mond

$$\sin 2\delta_m \sim 2\sin\varepsilon\sin s, \qquad (16.2)$$

$$-\cos\tau\sin 2\delta_m \sim -\,2\sin\varepsilon\cos\tau\sin s = -\sin\varepsilon\,[\sin(\tau+s)-\sin(\tau-s)] \qquad (16.3)$$

entsprechend für die Sonne

$$-\sin\varepsilon\,[\sin(t+h)-\sin(t-h)]. \qquad (16.4)$$

$(\tau+s)=(t+h)=$ Sternzeit θ^*; $\tau-s=\theta^*-2s$ und $t-h=\theta^*-2h$ sind die Stundenwinkel fiktiver Himmelskörper, die sich doppelt so schnell in ihrer Bahn bewegen wie Mond und Sonne.

Die Darstellung des ganztägigen Gezeitengliedes als Summe von Sinuswellen konstanter Amplitude führt also auf je zwei Glieder mit entgegengesetzt gleichen Amplituden. Das eine Paar ist

$$\left.\begin{aligned} \boldsymbol{K}_{1m} &= K\sin\{165{,}555\} = -\,0{,}3623\, G_1 \sin\theta^*,\\ \boldsymbol{K}_{1s} &= K\sin\{165{,}555\} = -\,0{,}1682\, G_1 \sin\theta^*. \end{aligned}\right\} \qquad (16.5)$$

Diese führen also beide auf Sternzeitglieder; ihre Summe ist $\boldsymbol{K}_1$, die *luni-solare* Tide.

Das andere Paar ist

$$\boldsymbol{O}_1 \equiv K\sin\{145{,}555\} = +\,0{,}3769\, G_1\sin(\tau-s) \quad \text{(lunar)}, \qquad (16.6)$$

$$\boldsymbol{P}_1 \equiv K\sin\{163{,}555\} = +\,0{,}1755\, G_1\sin(t-h) \quad \text{(solar)}. \qquad (16.7)$$

c) Den Einfluß der wechselnden Schiefe J der Mondbahn, also der Knotenbewegung (N), findet man wie in Ziff. 15e. Zunächst wird

$$\sin 2\delta_m \sim 2\,[0{,}398\sin s + 0{,}09\sin(s-N)]. \qquad (16.8)$$

Multiplikation mit $\cos\tau$ führt hauptsächlich auf die in Ziff. 16b diskutierten Glieder; dazu kommen zwei Glieder mit entgegengesetzt gleichen Koeffizienten, genauer

$$K\sin\{145{,}545\} = +\,0{,}0710\, G_1\sin[\tau-(s-N)], \qquad (16.9)$$

$$K\sin\{165{,}565\} = -\,0{,}0718\, G_1\sin[\tau+(s-N)]. \qquad (16.10)$$

Das erste Glied interferiert mit $\boldsymbol{O}_1$, das zweite Glied mit $\boldsymbol{K}_{1m}$, im Sinne einer Verstärkung, wenn $N = 0°$ (große Neigung J der Mondbahn), einer Abschwächung, wenn $N = 180°$ (kleine Neigung J). Die Amplituden von $\boldsymbol{K}_{1m}$ und $\boldsymbol{O}_1$ schwanken, infolge dieser Superposition, sozusagen zwischen etwa $+0{,}448$ und $0{,}306$, innerhalb 18,6 Jahren.

d) Für die Sonne wollen wir die Rechnung in Abschnitt c verbessern, indem wir den Unterschied zwischen h_1 und Rektaszension α_s berücksichtigen. Der Stundenwinkel (Sternzeit minus Rektaszension) der wahren Sonne ist

$$t_1 = \theta^* - \alpha_s. \tag{16.11}$$

Aus Fig. 6 folgt

$$\cos\delta_s \cos\alpha_s = \cos h_1; \qquad \cos\delta_s \sin\alpha_s = \cos\varepsilon \sin h_1; \qquad \sin\delta_s = \sin\varepsilon \sin h_1. \tag{16.12}$$

Durch Verwendung von Formeln wie

$$2\sin^2 h_1 = 1 - \cos 2h_1, \qquad 2\sin x \cos y = \sin(x+y) + \sin(x-y) \qquad \text{usw.}$$

folgt

$$\begin{aligned}
-\sin 2\delta_s \cos t_1 &= -2\sin\delta_s \cos\delta_s \cos(\theta^* - \alpha_s)\\
&= -2\sin\varepsilon \sin h_1 \cos h_1 \cos\theta^* - 2\sin\varepsilon\cos\varepsilon\sin^2 h_1 \sin\theta^*\\
&= -\sin\varepsilon\sin 2h_1\cos\theta^* - \sin\varepsilon\cos\varepsilon\sin\theta^* + \sin\varepsilon\cos\varepsilon\cos 2h_1 \sin\theta^*\\
&= \frac{1}{2}\sin\varepsilon\sin(\theta^* - 2h_1) - \frac{1}{2}\sin\varepsilon\sin(\theta^* + 2h_1) - \sin\varepsilon\cos\varepsilon\sin\theta^* +\\
&\quad + \frac{1}{2}\sin\varepsilon\cos\varepsilon\sin(\theta^* + 2h_1) + \frac{1}{2}\sin\varepsilon\cos\varepsilon\sin(\theta^* - 2h_1)\\
&= \left.\begin{aligned}&-\sin\varepsilon\cos\varepsilon\sin\theta^* + \sin\varepsilon\cos^2\left(\frac{\varepsilon}{2}\right)\sin(\theta^* - 2h_1) -\\ &-\sin\varepsilon\sin^2\left(\frac{\varepsilon}{2}\right)\sin(\theta^* + 2h_1)\end{aligned}\right\} \qquad (16.13)\\
&= -0{,}365\sin\theta^* + 0{,}382\sin(\theta^* - 2h_1) - 0{,}016\sin(\theta^* + 2h_1). \qquad (16.14)
\end{aligned}$$

Wenn wir hierin h_1 durch h ersetzen, also den Einfluß der Elliptizität außer acht lassen, gibt diese Formel mit guter Näherung die solaren Glieder $\boldsymbol{K}_{1s}$ und $\boldsymbol{P}_1$, ferner ein Zusatzglied

$$\varphi_1 \equiv K\sin\{167{,}555\} = -0{,}0076\, G_1 \sin(t + 3h). \tag{16.15}$$

e) Die entsprechende Rechnung für den Mond liefert, außer $\boldsymbol{K}_{1m}$ und $\boldsymbol{O}_1$, noch

$$\boldsymbol{OO}_1 \equiv K\sin\{185{,}555\} = -0{,}0162\, G_1 \sin(\tau + 3s). \tag{16.16}$$

f) Den Einfluß der Elliptizität berechnen wir ebenso wie in Ziff. 15f, ausgehend von den dortigen Formeln für $(c/R)^3$ und s_1. Für den Mond wird (16.14), wenn wir im letzten kleinen Glied τ durch τ_1 ersetzen

$$-\sin 2\delta_m \cos\tau_1 = -0{,}365\sin\theta^* + 0{,}382\sin(\theta^* - 2s_1) - 0{,}016\sin(\tau + 3s). \tag{16.17}$$

Darin wird, ähnlich wie in (15.16),

$$\sin(\theta^* - 2s_1) \sim \sin(\theta^* - 2s) - 0{,}110\sin(\theta^* - s - p) + 0{,}110\sin(\theta^* - 3s + p)$$

und schließlich, wenn wir durchweg τ statt $(\theta^* - s)$ schreiben und für s_1 einsetzen, wie in (15.16),

$$\begin{aligned}
-\left(\frac{c}{R}\right)^3 \sin 2\delta_m \cos\tau_1 = {}&[1 + 0{,}164\cos(s - p)]\,[-0{,}365\sin(\tau + s) + 0{,}382\sin(\tau - s) -\\
&- 0{,}042\sin(\tau - p) + 0{,}042\sin(\tau - 2s + p) - 0{,}016\sin(\tau + 3s)].
\end{aligned}$$

Die wesentlichen Zusatzglieder, die p enthalten, ergeben sich beim Ausmultiplizieren zu

$$\left.\begin{aligned} &+0{,}073\sin(\tau-2s+p)-0{,}030\sin(\tau+2s-p)-\\ &\qquad -0{,}030\sin(\tau+p)-0{,}011\sin(\tau-p)\,. \end{aligned}\right\} \tag{16.19}$$

Dies sind die Tiden, die bei exakter Rechnung werden

$$\boldsymbol{Q}_1\equiv K\sin\{135{,}655\}=+0{,}0722\,G_1\sin[(\tau-s)-(s-p)]\,, \tag{16.20}$$

$$\boldsymbol{J}_1\equiv K\sin\{175{,}455\}=-0{,}0296\,G_1\sin[(\tau+s)+(s-p)]\,, \tag{16.21}$$

$$\boldsymbol{M}_1\equiv K\sin\{155{,}655\}=-0{,}0296\,G_1\sin[(\tau+s)-(s-p)]\,, \tag{16.22}$$

$$K\sin\{155{,}455\}=-0{,}0106\,G_1\sin[(\tau-s)+(s-p)]\,. \tag{16.23}$$

Die Schreibweise der Argumente läßt den Zusammenhang der Tiden mit den Stammtiden hervortreten, nämlich $\boldsymbol{Q}_1$ und (16.23) mit $\boldsymbol{O}_1$, $\boldsymbol{J}_1$ und $\boldsymbol{M}_1$ mit $\boldsymbol{K}_{1m}$. $\boldsymbol{Q}_1$, $\boldsymbol{J}_1$ und $\boldsymbol{M}_1$ verstärken ihre Stammtiden im Perigäum.

g) Die elliptischen Glieder zweiter Ordnung in s_1 und (c/R), mit $2(s-p)$ führen, nach demselben Schema, auf Tiden, deren größte analog $\boldsymbol{Q}_1$ ist, nämlich

$$2\boldsymbol{Q}_1\equiv K\sin\{125{,}755\}=+0{,}0096\,G_1\sin[(\tau-s)-2(s-p)]\,. \tag{16.24}$$

h) Die *solaren* elliptischen Glieder ergeben sich ebenso wie im Abschnitt f, sind aber wesentlich kleiner:

$$\pi_1\equiv K\sin\{162{,}556\}=+0{,}0103\,G_1\sin[(t-h)-(h-p_s)]\,, \tag{16.25}$$

$$\psi_1\equiv K\sin\{166{,}554\}=-0{,}0042\,G_1\sin[(t+h)+(h-p_s)]\,, \tag{16.26}$$

$$\boldsymbol{S}_1\equiv K\sin\{164{,}556\}=-0{,}0042\,G_1\sin[(t+h)-(h-p_s)]\,. \tag{16.27}$$

Ihre Stammtiden sind $\boldsymbol{P}_1$ und $\boldsymbol{K}_{1s}$.

i) Die (lunaren) Evektionsglieder, mit $(s-2h+p)$, und die Variationsglieder mit $2(s-h)$, ergeben sich nach demselben Schema wie im Abschnitt f zu

$$\varrho_1\equiv K\sin\{137{,}455\}=+0{,}0137\,G_1\sin[(\tau-s)-(s-2h+p)]\,, \tag{16.28}$$

$$\vartheta_1\equiv K\sin\{173{,}655\}=-0{,}0057\,G_1\sin[(\tau+s)+(s-2h+p)]\,, \tag{16.29}$$

$$\chi_1\equiv K\sin\{157{,}455\}=-0{,}0057\,G_1\sin[(\tau+s)-(s-2h+p)]\,, \tag{16.30}$$

$$\sigma_1\equiv K\sin\{125{,}755\}=+0{,}0115\,G_1\sin[(\tau-s)-2(s-h)]\,, \tag{16.31}$$

$$\boldsymbol{SO}_1\equiv K\sin\{183{,}555\}=-0{,}0049\,G_1\sin[(\tau+s)+2(s-h)]\,, \tag{16.32}$$

$$\tau_1\equiv K\sin\{147{,}555\}=-0{,}0049\,G_1\sin[(\tau+s)-2(s-h)]\,. \tag{16.33}$$

Der Zusammenhang mit den Stammtiden $\boldsymbol{O}_1$ und $\boldsymbol{K}_{1m}$ ist in allen Fällen deutlich.

17. Die halbtägigen Tiden.

a) Sie sind der wesentliche Teil von V_2, der dritte Term in (8.3), mit der Geodätischen Funktion

$$G_2=G\cos^2\varphi\,,$$

$$\left.\begin{aligned} &\text{für den Mond} && V_m=G_2\left(\frac{c}{R}\right)^3\cos^2\delta_m\cos2\tau_1\,,\\ &\text{für die Sonne} && V_s=0{,}4605\,G_2\left(\frac{c_s}{R_s}\right)^3\cos^2\delta_s\cos2t_1\,. \end{aligned}\right\} \tag{17.1}$$

b) Die Hauptglieder entsprechen fiktiven Gestirnen, die sich gleichförmig im Himmelsäquator bewegen:

$$\boldsymbol{M}_2 \equiv K \cos\{255{,}555\} = +\,0{,}9081\, G_2 \cos 2\tau\,, \tag{17.2}$$

$$\boldsymbol{S}_2 \equiv K \cos\{273{,}555\} = +\,0{,}4229\, G_2 \cos 2t\,. \tag{17.3}$$

Wenn die fiktiven Gestirne mit den realen identisch wären, so wären diese Koeffizienten für den Mond 1,0000, für die Sonne 0,4605. Die Schiefe der tatsächlichen Bahnen und die ungleichförmige Bewegung in diesen Bahnen — beide beim Mond stärker als bei der Sonne — reduzieren diese idealen Koeffizienten auf die Bruchteile 0,9081 (Mond) und 0,9184 (Sonne).

c) Für die Sonne führen wir eine Näherungsrechnung analog zu Ziff. 16d. Es ist

$$\cos^2\delta_s \cos 2t_1 = \cos^2\delta_s \cos(2\theta^* - 2\alpha_s) = \cos^2\delta_s(\cos 2\theta^* \cos 2\alpha_s + \sin 2\theta^* \sin 2\alpha_s)\,. \tag{17.4}$$

Aus (16.12) folgt

$$\cos^2\delta_s \sin 2\alpha_s = \cos\varepsilon \sin 2h_1 \tag{17.5}$$

$$\cos^2\delta_s \cos 2\alpha_s = (\cos^2 h_1 - \cos^2\varepsilon \sin^2 h_1) = \frac{1}{2}\sin^2\varepsilon + \frac{1}{2}(1+\cos^2\varepsilon)\cos 2h_1\,. \tag{17.6}$$

Aus $\cos^2\left(\frac{\varepsilon}{2}\right) = \frac{1}{2} + \frac{1}{2}\cos\varepsilon$ und $\sin^2\left(\frac{\varepsilon}{2}\right) = \frac{1}{2} - \frac{1}{2}\cos\varepsilon$ folgen

$$\cos^4\left(\frac{\varepsilon}{2}\right) + \sin^4\left(\frac{\varepsilon}{2}\right) = \frac{1}{2} + \frac{1}{2}\cos^2\varepsilon\,, \qquad \cos^4\left(\frac{\varepsilon}{2}\right) - \sin^4\left(\frac{\varepsilon}{2}\right) = \cos\varepsilon\,. \tag{17.7}$$

Einsetzen in (17.5) und (17.6) liefert für (17.4)

$$\left.\begin{aligned}\cos^2\delta_s \cos 2t_1 &= \frac{1}{2}\sin^2\varepsilon\cos 2\theta^* + \cos^4\left(\frac{\varepsilon}{2}\right)\cos(2\theta^* - 2h_1) + \sin^4\left(\frac{\varepsilon}{2}\right)\cos(2\theta^* + 2h_1)\\ &= 0{,}079\cos 2\theta^* + 0{,}919\cos 2t_1 + 0{,}002\cos(2t_1 + 4h_1)\,.\end{aligned}\right\} \tag{17.8}$$

Wenn man hierin t_1 durch t, h_1 durch h annähert, wird das zweite Glied zu $\boldsymbol{S}_2$; das erste wird

$$\boldsymbol{K}_{2s} \equiv K \cos\{275{,}555\} = 0{,}0365\, G_2 \cos 2\theta^*\,. \tag{17.9}$$

d) Das lunare Analogon zu (17.8) liefert im mittleren Term $\boldsymbol{M}_2$, im ersten und dritten

$$\boldsymbol{K}_{2m} \equiv K \cos\{275{,}555\} = +\,0{,}0786\, G_2 \cos 2\theta^*\,, \tag{17.10}$$

$$K \cos\{295{,}555\} = +\,0{,}0017\, G_2 \cos(2\tau + 4s)\,. \tag{17.11}$$

e) Die elliptischen halbtägigen Tiden werden nach demselben Schema berechnet wie die ganztägigen (Ziff. 16f). Im wesentlichen ergeben sich ein Hauptglied erster Ordnung, $\boldsymbol{N}_2$, ferner ein kleineres Glied erster Ordnung, $\boldsymbol{L}_2$, sowie ein Glied zweiter Ordnung, $2\boldsymbol{N}_2$, mit $2(s-p)$,

$$\boldsymbol{N}_2 \equiv K \cos\{245{,}655\} = +\,0{,}1739\, G_2 \cos[2\tau - (s-p)]\,, \tag{17.12}$$

$$\boldsymbol{L}_2 \equiv K \cos\{265{,}455\} = -\,0{,}0257\, G_2 \cos[2\tau + (s-p)]\,, \tag{17.13}$$

$$2\boldsymbol{N}_2 \equiv K \cos\{235{,}755\} = +\,0{,}0230\, G_2 \cos[2\tau - 2(s-p)]\,. \tag{17.14}$$

f) Die elliptischen Sonnentiden sind wesentlich kleiner:

$$\boldsymbol{T}_2 \equiv K \cos\{272{,}556\} = +\,0{,}0248\, G_2 \cos[2t - (h - p_s)]\,, \tag{17.15}$$

$$\boldsymbol{R}_2 \equiv K \cos\{274{,}554\} = -\,0{,}0035\, G_2 \cos[2t + (h - p_s)]\,, \tag{17.16}$$

$$2\boldsymbol{T}_2 \equiv K \cos\{271{,}557\} = +\,0{,}0010\, G_2 \cos[2t - 2(h - p_s)]\,. \tag{17.17}$$

g) Der Einfluß der Knotenbewegung führt auf zwei lunare Glieder mit entgegengesetzt gleichen Amplituden, die zu den Stammtiden $\boldsymbol{M}_2$ und $\boldsymbol{K}_{2m}$ gehören, nämlich

$$K\cos\{255{,}545\} = -0{,}0339\,G_2\cos(2\tau + N)\,, \tag{17.18}$$

$$K\cos\{275{,}565\} = +0{,}0342\,G_2\cos(2\theta^* - N)\,. \tag{17.19}$$

h) Analog zu $\boldsymbol{N}_2$ entstehen das Hauptevektionsglied

$$\nu_2 \equiv K\cos\{247{,}455\} = +0{,}0330\,G_2\cos[2\tau - (s - 2h + p)] \tag{17.20}$$

und das Hauptvariationsglied

$$\mu_2 \equiv K\cos\{237{,}555\} = +0{,}0278\,G_2\cos[2\tau - 2(s - h)]\,. \tag{17.21}$$

18. Die aus V_3 hervorgehenden Glieder. Tabelle 2, in der die Herkunft dieser Tiden durch M^* markiert ist, beschränkt sich auf

eine langperiodische Tide mit $\sin s$,

zwei halbtägige Tiden, etwa gleicher Größe, mit $-\sin(2\tau - s)$ und $+\sin(2\tau + s)$, deren Summe proportional $\sin s\cos 2\tau$ ist,

das dritteltägige Hauptglied

$$\boldsymbol{M}_3 \equiv K\cos\{355{,}555\} = -0{,}0119\,G_3^*\cos 3\tau \tag{18.1}$$

mit dem elliptischen Glied

$$K\cos\{345{,}655\} = -0{,}0033\,G_3^*\cos[3\tau - (s - p)]\,. \tag{18.2}$$

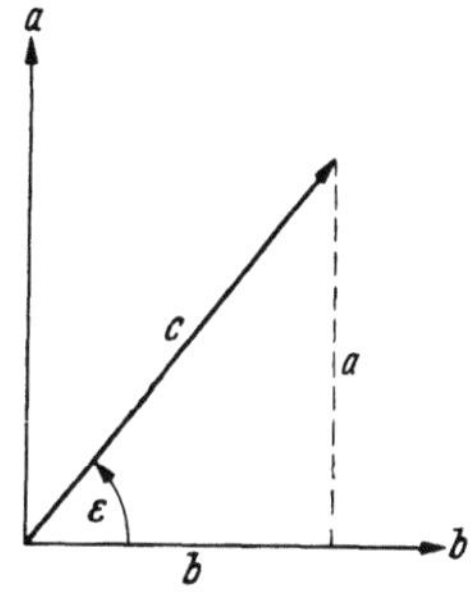

Fig. 7. Schema für eine Periodenuhr (harmonic dial).

Dieses elliptische Glied aus V_3 erreicht das 0,274fache des Koeffizienten von M_3, ist also relativ stärker als in V_2, mit $\boldsymbol{N}_2 : \boldsymbol{M}_2 = 0{,}191$, entsprechend der Bemerkung am Schluß von Ziff. 7f.

19. Veranschaulichung durch Periodenuhren.

a) Superposition und Modulation von Sinus-Wellen lassen sich in *Periodenuhren (harmonic dials)* veranschaulichen.

Eine Welle mit der Periode T und der Frequenz $\omega = 2\pi/T$,

$$\left.\begin{aligned} a\cos\omega t + b\sin\omega t &= c\sin(\omega t + \varepsilon) \\ &= \text{Imaginärteil von } \left[c\cdot e^{i(\omega t+\varepsilon)} = c\cdot e^{i\varepsilon}\cdot e^{i\omega t}\right] \end{aligned}\right\} \tag{19.1}$$

läßt sich durch die komplexe Amplitude $c\cdot e^{i\varepsilon}$ als Vektor in der komplexen Zahlenebene darstellen. In der Geophysik ist es üblich geworden[1], die rechtwinkligen Komponenten a nach oben, b nach rechts aufzutragen (Fig. 7); dadurch erreicht man, daß bei einer Welle von $T = 12$ Std. Periode der Vektor auf die Uhrzeit des Maximums der Welle zeigt, abzulesen an einer Randskala wie auf dem Zifferblatt (dial) einer gewöhnlichen Uhr.

b) In einer Periodenuhr für die Periode T sind alle Sinus-Wellen dieser Periode durch feste Vektoren gegeben; es genügt, die *Endpunkte* der Vektoren zu markieren. Da (19.1) in a und b linear ist, lassen sich Wellen derselben Periode T durch einfache Vektorenaddition superponieren.

[1] J. Bartels: Z. Geophysik **3**, 389–397 (1927); Terr. Magn. **40**, 1–60 (1935); Chapter XV in S. Chapman and J. Bartels, Geomagnetism, Vol. 2. Oxford Univ.-Press 1940. — Vgl. auch das Kapitel „Analysis of geophysical functions..." in diesem Handbuch, Band XLIX.

c) Fig. 8 zeigt eine Periodenuhr für die halbtägige Hauptmondwelle $\boldsymbol{M}_2 = K\cos 2\tau$, mit der Periode $T = 12$ Mondstunden $= 12{,}42$ Sonnenstunden, Frequenz $\omega = 28{,}^\circ 98$/mittlere Sonnenstunde. Für den Anfang der Kurve (rechts) ist angenommen, daß genau um Mitternacht mittlerer Neumond eingetreten ist

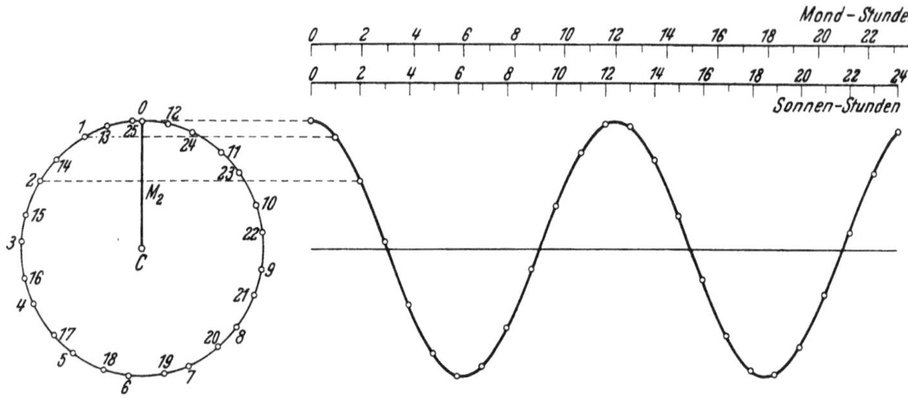

Fig. 8. Periodenuhr für die halbtägige Hauptmondwelle $\boldsymbol{M}_2$, und Konstruktion der üblichen Kurvendarstellung daraus.

$(\tau = t = 0)$. Die Kurve für $\cos 2\tau$ (rechts) kann so konstruiert werden, daß man den Vektor in der Periodenuhr, entsprechend dem Zeitfaktor $e^{i\omega t}$, gleichförmig *links herum* rotieren läßt; die senkrechte Komponente ist jeweils die Kurvenordinate, wie das für die Sonnenstunden 0, 1, 2, ... gezeigt ist.

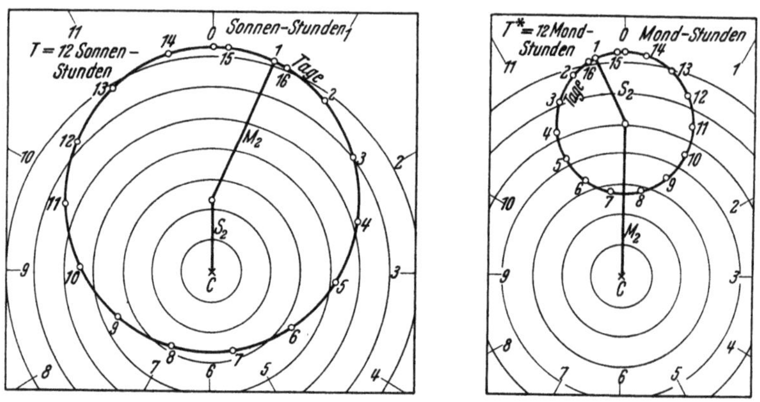

Fig. 9. Überlagerung der Tiden $\boldsymbol{S}_2$ und $\boldsymbol{M}_2$, in Periodenuhren veranschaulicht: links für die Periode von $\boldsymbol{S}_2$, rechts für die Periode von $\boldsymbol{M}_2$. Zahlen am Kreis bedeuten Tage nach dem Neumond, der anfänglich mit Mitternacht zusammenfallend angenommen ist.

d) In einer Periodenuhr für die Frequenz ω lassen sich auch Sinuswellen anderer Frequenzen ω^* darstellen, denn

$$c^* \sin(\omega^* t + \varepsilon^*) = c^* \sin[\omega t + (\omega^* - \omega)\, t + \varepsilon^*] = c \sin(\omega t + \varepsilon),$$

mit

$$c = c^*, \qquad \varepsilon = (\omega^* - \omega)\, t + \varepsilon^*. \tag{19.2}$$

Dies gibt einen Kreis mit dem Radius c um das Zentrum C, der im Uhrzeigersinne (für $\omega^* < \omega$) oder entgegengesetzt durchlaufen wird. Auch die Kurvenkonstruktion (Abschnitt c) bleibt möglich: Während das Blatt mit der Periodenuhr sich mit der Frequenz ω links herum dreht, dreht sich der Vektor-Endpunkt, dessen vertikale Koordinate die Kurve liefert, auf diesem Blatt, indem er den Kreis mit der Frequenz $(\omega^* - \omega)$ durchläuft.

Als Beispiel zeigt Fig. 9 die Überlagerung der sonnen- und mondtägigen Haupttiden: *links* in einer Periodenuhr für $T = 12$ *Sonnen*stunden, also mit $\boldsymbol{S}_2$ als festem Vektor, rechts in einer Periodenuhr für $T = 12$ *Mond*stunden $= 12{,}42$ Sonnenstunden, mit $\boldsymbol{M}_2$ als festem Vektor. Auf dem Kreis, der $\boldsymbol{S}_2 + \boldsymbol{M}_2 = \boldsymbol{M}_2 + \boldsymbol{S}_2$ darstellt, ist der Anfang dei Zeitzählung so gewählt, daß er mit einem genau auf Mitternacht fallenden mittleren Neumond zusammenfällt; anschließend sind die Zeiten jeweils 1, 2, 3, ... Sonnentage später markiert.

Fig. 10 gibt die *übliche* Darstellung der Überlagerung $S_2 + M_2$, ebenfalls von einem auf Mitternacht fallenden mittleren Neumond ausgehend. Alle Einzelheiten der Interferenz der beiden Wellen lassen sich auch aus den beiden Periodenuhren (Fig. 9) ablesen: sowohl das An- und Abschwellen der Amplituden, wie die Verfrühung oder Verzögerung des Eintritts der Maxima und Minima.

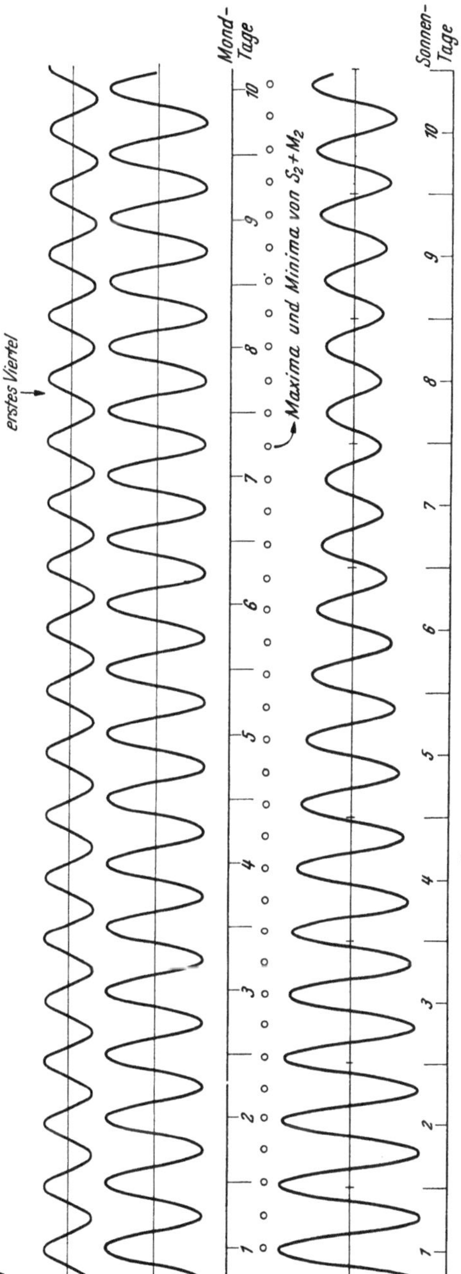

Fig. 10. Überlagerung von $\boldsymbol{S}_2$ und $\boldsymbol{M}_2$ in üblicher Darstellung.

e) Die Entstehung der *ganztägigen* Tiden (Ziff. 16b) läßt sich ebenfalls in der Periodenuhr veranschaulichen, z. B. nach (16.4)

$$\left.\begin{aligned} 2\sin h\cos t &= \sin(t+h) - \\ &\quad - \sin(t-h) \\ &= \sin(t+h) + \\ &\quad + \sin(t-h+180^\circ) \\ &\sim \boldsymbol{K}_{1s} + \boldsymbol{P}_1 . \end{aligned}\right\} \quad (19.3)$$

Links steht eine Welle $\cos t$ mit Amplituden $2\sin h$, die sich im Laufe eines Jahres periodisch ändern, $=0$ zur Zeit der Äquinoktien ($h = 0^\circ$ oder 180°), $+2$ im Sommersolstitium ($h = 90^\circ$, auf der Nordhalbkugel), -2 im Wintersolstitium ($h = 270^\circ$). *Rechts* stehen zwei Wellen mit konstanten Amplituden und konstanten Frequenzen; sie durchlaufen in einem Jahr (365,25 Tage) 366,25 Umläufe (Sterntage) oder 364,25 Umläufe.

Fig. 11 veranschaulicht dies in Periodenuhren für die Periode eines Sonnentages (365,25 Umläufe pro Jahr); sie bedarf keines weiteren Kommentars.

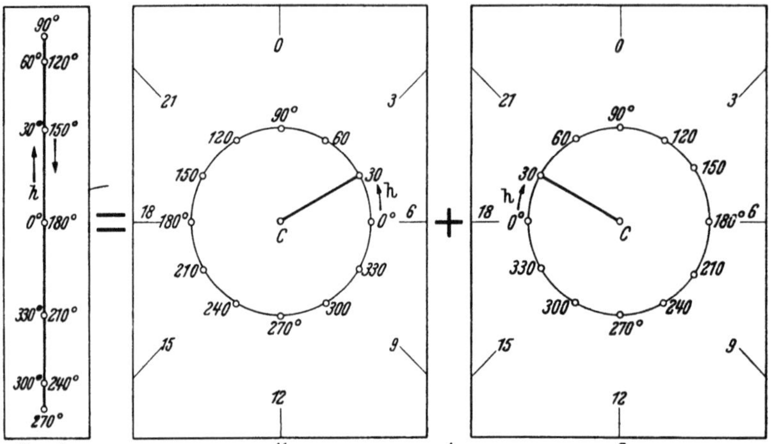

Fig. 11. Veranschaulichung der solaren ganztägigen Tiden in Periodenuhren für die Periode 1 Sonnentag = 24 Sonnenstunden.

Nach demselben Schema — mit s statt h, Mondtag statt Sonnentag, Monat statt Jahr — läßt sich $\boldsymbol{K}_{1m}+\boldsymbol{O}_1$ veranschaulichen (16.3), nach Art einer Amplituden-Modulation.

f) Der Einfluß der *Elliptizität* der Mondbahn ist dagegen keine reine Amplitudenmodulation, da im Perigäum zur Verstärkung der Amplituden auch eine Verlängerung des wahren Mondtages tritt.

$$\left.\begin{array}{l}\boldsymbol{M}_2+\boldsymbol{N}_2\sim\cos 2\tau+\\ \qquad +0{,}19\cos\left[2\tau-(s-p)\right]\end{array}\right\}\qquad(19.4)$$

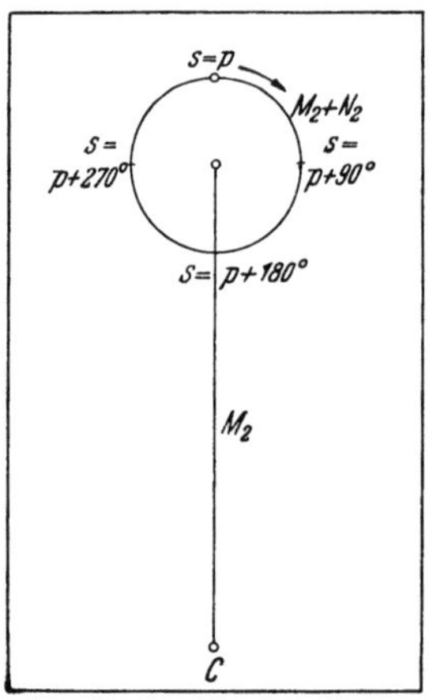

Fig. 12. Überlagerung $\boldsymbol{M}_2+\boldsymbol{N}_2$.

stellt sich also wie in Fig. 12 dar, in einer Periodenuhr für die Periode eines halben Mondtages. Gegen den mittleren Mond läuft der wahre Mond in seiner scheinbaren Bahn von Osten nach Westen am Himmel, in dem halben Monat um das Perigäum ($s=p$) *langsamer*, weil seine wahre Länge (nach Osten) dann schneller wächst als die mittlere (12.5); etwa sieben Tage nach dem Perigäum kulminiert der wahre Mond deshalb über 26 Zeitminuten später als der mittlere.

g) In einer Periodenuhr für 2τ (Periode = halber Mondtag) läßt sich der Einfluß der Elliptizität genauer darstellen[1] durch epicyclische Überlagerung von

$$\begin{array}{lll}\boldsymbol{M}_2\sim+100{,}00\cos 2\tau & & \text{Phase } \varepsilon=90^\circ;\\ \boldsymbol{N}_2\sim+19{,}15\cos\left[2\tau-(s-p)\right], & & \varepsilon=90^\circ-(s-p);\\ \boldsymbol{L}_2\sim-2{,}83\cos\left[2\tau+(s-p)\right], & & \varepsilon=270^\circ+(s-p);\\ 2\boldsymbol{N}_2\sim+2{,}53\cos\left[2\tau-2(s-p)\right], & & \varepsilon=90^\circ-2(s-p).\end{array}$$

[1] J. BARTELS u. H. F. JOHNSTON: Terr. Magn. **45**, 496 (1940).

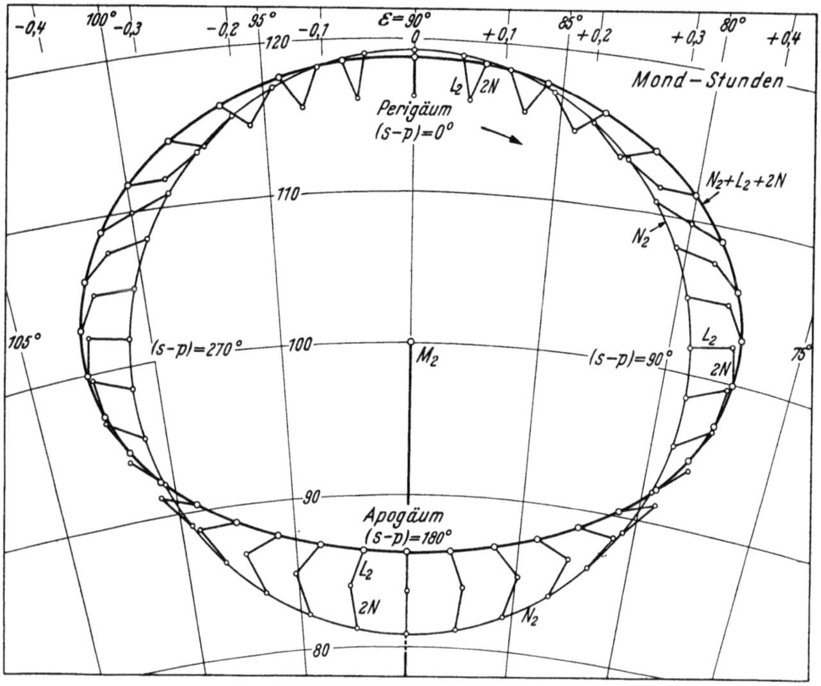

Fig. 13. Überlagerung $\boldsymbol{M}_2+\boldsymbol{N}_2+\boldsymbol{L}_2+2\boldsymbol{N}$, Periodenuhr mit der Periode eines halben Mondtages, oberer Teil.

Fig. 13 zeigt den oberen Teil der Periodenuhr: $\boldsymbol{N}_2$ beschreibt in einem anomalistischen Monat [Periode $(s-p)$] einen vollen Kreis im Uhrzeigersinn, $2\boldsymbol{N}$ beschreibt zwei solche Umläufe, und $\boldsymbol{L}_2$ beschreibt einen Umlauf im entgegengesetzten Sinn. Fig. 14 zeigt noch einmal das Ergebnis in der vollen Uhr, im Vergleich zu $\boldsymbol{M}_2+\boldsymbol{N}_2$ allein. — Für die Sonne siehe S. 934.

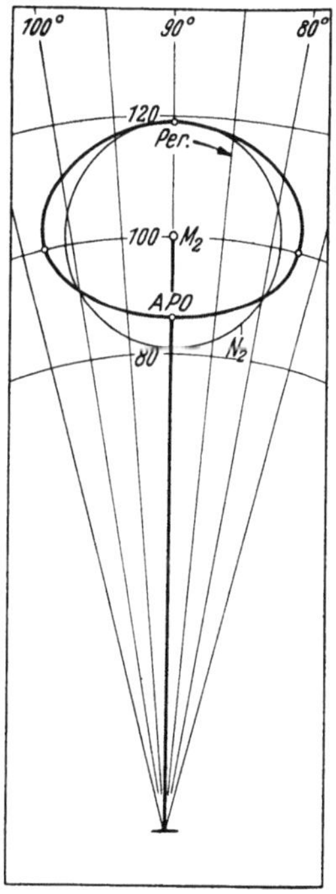

Fig. 14. Überblick für das Ergebnis von Fig. 13, mit $\boldsymbol{M}_2+\boldsymbol{N}_2$ zum Vergleich.

20. Harmonische Analyse der Beschleunigungskomponenten. Wir differenzieren (13.8) und (13.9) gemäß (8.10). Mit (13.1) wird V_2 eine Summe von Tiden der Form

$$K \cdot \left(\frac{\varrho}{\varrho_1}\right) \cdot G_A(\varphi) \cdot \begin{Bmatrix}\cos\\ \sin\end{Bmatrix} (AL + W_0), \tag{20.1}$$

wobei wir dei Deutlichkeit wegen schreiben $G_A(\varphi)$ für die Geodätischen Funktionen (13.5), und, nach (14.1),

$$\left.\begin{aligned} W_0 = A t_0 + (B-A)\,s + (C+A)\,h - {}\\ - EN + Fp_s \pm n \cdot 90^\circ. \end{aligned}\right\} \tag{20.2}$$

$(AL + W_0)$ ist identisch mit dem Argument der Tide (in V), wie es in Spalte 7 der Tabelle 2, gemäß (14.2), ausgedrückt ist; W_0 ist, nach (13.1), das gleichzeitige Argument der Tide auf dem Greenwich-Meridian. Wir setzen ferner vorübergehend (7.5)

$$2C = 2G(\varrho_1)/\varrho_1, \quad \text{also} \quad C = 41{,}13 \cdot 10^{-6}\,\text{cm/sec}^2; \tag{20.3}$$

K sind die Koeffizienten aus Spalte 3 der Tabelle 2. Die typischen Terme (langperiodisch, ganz- und halbtägig durch Indices 0, 1, 2 bezeichnet) sind dann:

$$X_0 = C \cdot \left(\frac{\varrho}{\varrho_1}\right) \cdot \frac{3}{2} \sin 2\varphi \cdot K \cos (W_0 + 180^\circ),$$

$$X_1 = C \cdot \left(\frac{\varrho}{\varrho_1}\right) \cdot 2 \cos 2\varphi \cdot K \cos (L + W_0 - 90^\circ),$$

$$X_2 = C \cdot \left(\frac{\varrho}{\varrho_1}\right) \cdot \sin 2\varphi \cdot K \cos (2L + W_0 + 180^\circ),$$

$$Y_0 = 0,$$

$$Y_1 = C \cdot \left(\frac{\varrho}{\varrho_1}\right) \cdot 2 \sin\varphi \cdot K \cos (L + W_0),$$

$$Y_2 = C \cdot \left(\frac{\varrho}{\varrho_1}\right) \cdot 2 \cos\varphi \cdot K \cos (2L + W_0 + 90^\circ),$$

$$Z_0 = C \cdot \left(\frac{\varrho}{\varrho_1}\right) \cdot (1 - 3 \sin^2\varphi) \cdot K \cos W_0,$$

$$Z_1 = C \cdot \left(\frac{\varrho}{\varrho_1}\right) \cdot 2 \sin 2\varphi \cdot K \cos (L + W_0),$$

$$Z_2 = C \cdot \left(\frac{\varrho}{\varrho_1}\right) \cdot 2 \cos^2\varphi \cdot K \cos (2L + W_0).$$

Bei V_3 ist, in (20.1), $\left(\frac{\varrho}{\varrho_1}\right)^3$ und $G_A^*(\varphi)$ nach (13.6) zu schreiben:

$$X_0^* = C \cdot \left(\frac{\varrho}{\varrho_1}\right)^2 \cdot \frac{3}{2} \cos\varphi\, (1 - 5 \sin^2\varphi) \cdot K \cos (W_0 - 90^\circ),$$

$$X_1^* = C \cdot \left(\frac{\varrho}{\varrho_1}\right)^2 \cdot 0{,}726 \sin\varphi\, (15 \cos^2\varphi - 4) \cdot K \cos (L + W_0 + 180^\circ),$$

$$X_2^* = C \cdot \left(\frac{\varrho}{\varrho_1}\right)^2 \cdot 2{,}598 \cos\varphi\, (1 - 3 \sin^2\varphi) \cdot K \cos (2L + W_0 - 90^\circ),$$

$$X_3^* = C \cdot \left(\frac{\varrho}{\varrho_1}\right)^2 \cdot 3 \sin\varphi \cos^2\varphi \cdot K \cos (3L + W_0 + 180^\circ).$$

$$Y_0^* = 0,$$

$$Y_1^* = C \cdot \left(\frac{\varrho}{\varrho_1}\right)^2 \cdot 0{,}726\, (1 - 5 \sin^2\varphi) \cdot K \cos (L + W_0 + 90^\circ),$$

$$Y_2^* = C \cdot \left(\frac{\varrho}{\varrho_1}\right)^2 \cdot 2{,}598 \sin 2\varphi \cdot K \cos (2L + W_0),$$

$$Y_3^* = C \cdot \left(\frac{\varrho}{\varrho_1}\right)^2 \cdot 3 \cos^2\varphi \cdot K \cdot \cos (3L + W_0 + 90^\circ).$$

$$Z_0^* = C \cdot \left(\frac{\varrho}{\varrho_1}\right)^2 \cdot \frac{3}{2} (1 - 3 \sin^2\varphi) \cdot K \cos W_0,$$

$$Z_1^* = C \cdot \left(\frac{\varrho}{\varrho_1}\right)^2 \cdot 0{,}726 \cdot 3 \cos\varphi\, (1 - 5 \sin^2\varphi) \cdot K \cos (L + W_0),$$

$$Z_2^* = C \cdot \left(\frac{\varrho}{\varrho_1}\right)^2 \cdot 2{,}598 \cdot 3 \sin\varphi \cos^2\varphi \cdot K \cos (2L + W_0),$$

$$Z_3^* = C \cdot \left(\frac{\varrho}{\varrho_1}\right)^2 \cdot 3 \cos^3\varphi \cdot K \cos (3L + W_0).$$

Aus Tabelle 2, Spalten 3 und 7, kann man nunmehr die Tiden in X, Y, Z anschreiben. Für einige der wichtigsten Tiden folgen diese Ausdrücke, für $\varrho = \varrho_1$, sämtlich in der Einheit 10^{-6} cm/sec^2 = micro-gal:

$$\boldsymbol{Mf} \equiv X_0 = 9{,}65 \sin 2\varphi \cos(2s + 180°);$$
$$Y_0 = 0;$$
$$Z_0 = 6{,}44\,(1 - 3\sin^2\varphi)\cos 2s.$$
$$\boldsymbol{K}_{1m} \equiv X_1 = 29{,}8 \cos 2\varphi \cos\theta^*;$$
$$Y_1 = 29{,}8 \sin\varphi \cos(\theta^* + 90°);$$
$$Z_1 = 29{,}8 \sin 2\varphi \cos(\theta^* + 90°).$$
$$\boldsymbol{M}_2 \equiv X_2 = 37{,}4 \sin 2\varphi \cos(2\tau + 180°);$$
$$Y_2 = 74{,}7 \cos\varphi \cos(2\tau + 90°);$$
$$Z_2 = 74{,}7 \cos^2\varphi \cos 2\tau.$$
$$\boldsymbol{M}_3 \equiv X_3^* = 1{,}47 \sin\varphi \cos^2\varphi \cos 3\tau;$$
$$Y_3^* = 1{,}47 \cos^2\varphi \cos(3\tau - 90°);$$
$$Z_3^* = 1{,}47 \cos^3\varphi \cos(3\tau + 180°).$$

Um X und Y als Komponenten der *Lotschwankung* auszudrücken, in der Einheit 0,001 Bogensekunden, muß man diese Zahlenwerte mit 206,265/979,8 = 0,210 multiplizieren; für M_2 wird der Faktor bei Y_2 also 0,″0157.

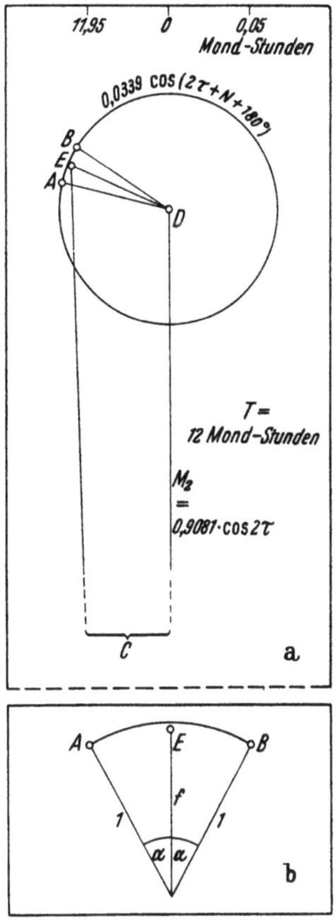

Fig. 15a u. b. Die Summe der Stammtide $\boldsymbol{M}_2$ und einer verwandten Tide im Jahre 1956.

21. Zusammenfassung von verwandten Tiden, praktische Rechnung.

a) Bei der Analyse von Meeresgezeiten sind meistens stündliche Pegelablesungen für ein Jahr gegeben. Innerhalb dieses Jahres lassen sich dann alle Tiden zusammenfassen, bei denen die drei ersten Ziffern der Argumentzahl (13.4) dieselben sind, zu einer Tide mit der Frequenz der Stammtide, also der Tide mit dem größten Koeffizienten K unter ihnen.

Typisches Beispiel: Stammtide $M_2 = \{255{,}555\} \sim 0{,}9081 \cos 2\tau$, und die verwandte Tide $\{255{,}545\} \sim 0{,}0339 \cos(2\tau + N + 180°)$. Im Laufe des Jahres 1956 z. B. ändert sich N von 256,°3 auf 236,°9, also um 19,°4. In einer Periodenuhr (Fig. 15) mit der Frequenz der Stammtide (Periode 12 Mondstunden) beschreibt die Tide {255,545} im Jahre 1956 nur ein 19,°4 langes Bogenstück ihres Kreises, vom Radius $K = 0{,}0339$. Der Schwerpunkt E eines Bogenstücks AB vom Radius K und Zentriwinkel 2α liegt auf der Winkelhalbierenden in der Entfernung fK vom Mittelpunkt; dabei ist f der Mittelwert von $\cos x$ von $x = 0$ bis $x = \alpha$, also[1]

$$f = (\sin\alpha)/\alpha. \tag{21.1}$$

Im Beispiel ist $2\alpha = 19{,}°4$, also $f = \sin 9{,}°7/\text{arc } 9{,}°7 = 0{,}9962$. Folglich ist für {255,545}, im Durchschnitt des Jahres 1956 näherungsweise zu setzen

$$0{,}0338 \cos(2\tau + 66{,}°6) = 0{,}0134 \cos 2\tau - 0{,}0310 \sin 2\tau,$$

Hierzu kommt $\boldsymbol{M}_2$ $\quad = 0{,}9081 \cos 2\tau,$

und die Summe der beiden Tiden wird $\quad 0{,}9220 \cos(2\tau + 1{,}°9).$

b) Vor der Zusammenfassung von Tiden, die aus V_2 und V_3 stammen, müssen die Koeffizienten K mit den Werten der Geodätischen Funktionen für die Breite des Ortes multipliziert werden.

[1] Tafel für f bei K. STUMPFF, Tafeln und Aufgaben zur Harmonischen Analyse usw. Berlin: Springer 1939.

Sowohl für das Potential wie für die Beschleunigungskomponenten läßt sich diese Zusammenfassung allgemein formulieren (für einen bestimmten Ort): Mit den Koeffizienten $H = KG_A$ und $H_j = K_j G_A$ oder $K_j G_A^*$, und mit den konstanten Anfangsphasen η und η_j für Stammtide und verwandte Tiden, soll die Zusammenfassung für ein bestimmtes Zeitintervall $t = t_2$ bis t_3 ausgedrückt werden. Die Stammtide ist

$$H \cos(\sigma t + \eta), \tag{21.1}$$

eine verwandte Tide sei

$$\left.\begin{aligned} H_j \cos(\sigma_j t + \eta_j) &= H_j \cdot \cos[\sigma t + \eta + (\sigma_j - \sigma)t + (\eta_j - \eta)] = H_j \cos(\sigma t + \eta) \times \\ &\times \cos[(\sigma_j - \sigma)t + (\eta_j - \eta)] - H_j \sin(\sigma t + \eta) \sin[(\sigma_j - \sigma)t + (\eta_j - \eta)]. \end{aligned}\right\} \tag{21.2}$$

Die Summe der Stammtide und aller verwandten Tiden ist

$$\begin{aligned} \cos(\sigma t + \eta)\{C + \textstyle\sum C_j \cos[(\sigma_j - \sigma)t + (\eta_j - \eta)]\} - \sin(\sigma t + \eta) \cdot \textstyle\sum C_j \times \\ \times \sin[(\sigma_j - \sigma)t + (\eta_j - \eta)] \equiv \cos(\sigma t + \eta) \cdot A(t) - \sin(\sigma t + \eta) \cdot B(t). \end{aligned}$$

In der Periodenuhr für die Frequenz σ der Stammtide (Ziff. 19) wird diese Summe durch die epicyclische Bewegung des Endpunktes eines Vektorenzuges dargestellt, mit den Beträgen H und H_j, und den relativen Frequenzen $(\eta_j - \eta)$. Den Schwerpunkt dieser epicyclischen Bahn erhalten wir durch Mittelbildung der Koeffizienten $A(t)$ und $B(t)$ von t_2 bis t_3. Wir schreiben kurz c_j und s_j für die Mittelwerte von {cos oder sin} $[(\sigma_j - \sigma)t + (\eta_j - \eta)]$ und erhalten für die gesuchte Zusammenfassung

$$\cos(\sigma t + \eta) \cdot (C + \sum_j C_j \cdot c_j) - \sin(\sigma t + \eta) \sum_j C_j \cdot s_j = C \cdot f \cdot \cos(\sigma t + \eta + u) \tag{21.3}$$

mit

$$f \cos u = 1 + \sum_j (C_j/C) \cdot c_j, \qquad f \sin u = \sum_j (C_j/C) \cdot s_j. \tag{21.4}$$

Durch die graphische Darstellung in Periodenuhren lassen sich solche Rechnungen leicht kontrollieren.

c) Die im Abschnitt b beschriebene Zusammenfassung von verwandten Tiden ist bei der harmonischen Analyse von Pegelablesungen (Meeresgezeiten) üblich. Bei der Untersuchung von Gezeiten in anderen Erscheinungen muß man die Rechnung oft anders anlegen, weil das Medium selbst sich ändert. Am stärksten ist dies bei den erdmagnetischen und den ionosphärischen Gezeiten der Fall, weil der Ionisationszustand schon am Tage und in der Nacht verschieden ist, und weil er außerdem in komplizierter Weise von wechselnder solarer Strahlung (Wellen und Korpuskeln) abhängt. Aber auch bei den Meeresgezeiten werden die idealen Verhältnisse gestört durch meteorologische Einflüsse (Sturmfluten) und die Verzerrung der reinen Gezeitenwellen in seichtem Küstenwasser. Bei den Gezeiten der festen Erde stören thermische (meteorologische) Einflüsse, und die atmosphärischen Gezeiten sind in höheren Breiten fast völlig von Wettererscheinungen verdeckt.

Wie man in diesen Fällen die Gezeitenkräfte berechnet, die für die vorliegende Beobachtungsreihe gelten, ist in den einzelnen Abschnitten über Gezeiten behandelt. Wenn man sich auf $\boldsymbol{M_2}$ beschränkt, liegt die Mondphasenzahl $\nu = (s - h)/15$ fertig tabuliert vor ($\nu = 0$ ist Neumond, 6 Erstes Viertel, 12 Vollmond, 18 Letztes Viertel). Für $\boldsymbol{N_2}$ kann man die Daten der mittleren Perigäen benutzen [*5*], [*6*], [*7*]. Über eine zweckmäßige Einteilung des Jahres vgl. Ziff. 23.

Für die Reduktion von Beobachtungen an Gravimetern und Drehwaagen wegen der Gezeitenkräfte liegen *Nomogramme* vor für die Beschleunigungskomponenten (8.11) bis (8.13), von Elkins [*8*] für Z, von Malone [*9*] für X und Y; allgemeine Tafeln für Z von J. T. Pettit [*15*]. Für die Vertikalkomponente Z

sind für 1954 ausführliche Tafeln von GOGUEL [12], [13], vgl. auch Ziff. 9c, und Kurven von MORELLI [10], [11] veröffentlicht; ein Beispiel in Fig. 16.

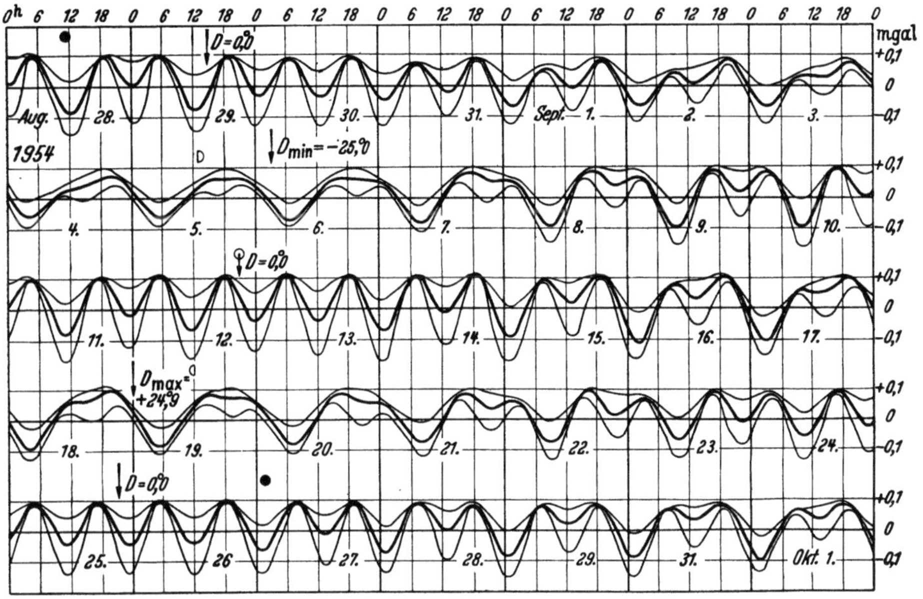

Fig. 16. Gezeitenschwankungen der Schwerkraft auf dem Meridian 15° E in den Breiten 65° N (obere Kurve), 45 N (stärker gezogen) und 25° N (untere Kurve), vom 28. August bis zum 1. Oktober 1954. Mitteleuropäische Zeit. Nach C. MORELLI [11]. Nur wenn die Deklination des Mondes ebenso verschwindet wie diejenige der Sonne — um den 12. und 26. September 1954 — ist der Charakter der Gezeiten überall halbtägig, sonst nur am Äquator.

d) Über die *sekundären* Gezeitenkräfte, die durch die Massenverlagerungen bewirkt werden (Ziff. 1), berichtet das Kapitel TOMASCHEK, S. 785 ff. und S. 811. GOGUEL [12], [13] multipliziert die primäre Gezeitenkomponente Z mit dem allgemeinen Faktor 1,2.

22. Die DARWINsche Entwicklung.

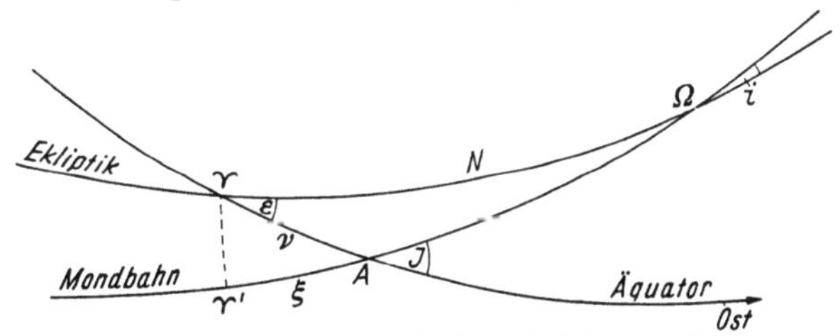

Fig. 17. Definition der Mondbahnabschnitte ν und ξ.

a) In der Ozeanographie ist die DOODSONsche Entwicklung [1], die auch hier behandelt worden ist, international angenommen. In der Literatur finden sich oft noch die älteren DARWINschen [2], [3] Ausdrücke. Diese unterscheiden sich von den DOODSONschen dadurch, daß die Länge N des Mondbahnknotens nicht explizit im Argument auftritt. Der Einfluß von N auf die Phasen und Amplituden der Tiden wird als Abhängigkeit von der wechselnden Neigung J der Mondbahn gegen die Ekliptik ausgedrückt. Außerdem erscheinen, als Phasenkorrekturen, der Abschnitt ν auf dem Äquator, zwischen Frühlingspunkt ϒ und Schnittpunkt A mit der Mondbahn, und seine Projektion ξ auf die Mondbahn, wie in Fig. 17 gezeigt. Man behandelt also J bei den lunaren Gezeiten ebenso wie die Ekliptikschiefe ε bei den solaren Gezeiten.

Hinsichtlich der Bearbeitung von Meeresgezeiten, bei denen man sich im allgemeinen auf Beobachtungen innerhalb eines Jahres beschränkt, war die DARWINsche Methode gerechtfertigt, weil sich der Einfluß von N in so kurzer

Zeit nur wenig ändert und deshalb wie in Ziff. 21 behandelt werden kann. Bei anderen Gezeitenwirkungen behandelt man aber Beobachtungen, die viele Jahre auseinander liegen können, wobei die konsequente Doodsonsche Methode vorteilhafter ist.

J, ν und ξ sind als Funktionen von N tabuliert [14]; einige typische Werte sind (die Symmetrie oder Antisymmetrie der Werte relativ zu $N = 180°$ ist ersichtlich):

$N=$ 0°	30°	60°	90°	120°	150°	180°	210°	240°	...
$J=$ 28°,60	28°,02	26°,38	23°,98	21°,32	19°,16	18°,31	19°,16	21°,32	...
$\nu=$ 0,00	+5,48	+10,07	+12,75	+12,33	+7,85	0,00	−7,85	−12,33	...
$\xi=$ 0,00	+4,94	+9,14	+11,68	+11,41	+7,32	0,00	−7,32	−11,41	...

b) Bei den wichtigsten Tiden treten bei Darwin folgende Ausdrücke auf (hier mit A_0, A_1, B_1 und A_2 bezeichnet), die aus den Neigungen ε, i und J zusammengesetzt sind:

$$\left.\begin{aligned} A_0 &= \frac{\sin^2 J}{\sin^2\varepsilon\cos^4\left(\frac{i}{2}\right)}, & A_1 &= \frac{\sin J\cos^2\left(\frac{J}{2}\right)}{\sin\varepsilon\cos^4\left(\frac{\varepsilon}{2}\right)\cos^4\left(\frac{i}{2}\right)}, \\ B_1 &= \frac{\sin 2J}{\sin 2\varepsilon\left(1-\frac{3}{2}\sin^2 i\right)}, & A_2 &= \left(\frac{\cos\left(\frac{J}{2}\right)}{\cos\left(\frac{\varepsilon}{2}\right)\cos\left(\frac{i}{2}\right)}\right)^4. \end{aligned}\right\} \tag{22.1}$$

Wenn wir die Doodsonschen Geodätischen Funktionen (13.5) verwenden, lauten einige Darwinsche lunare Tiden

$$\begin{aligned} \boldsymbol{Mf} &\equiv 0{,}156\cdot G_0\cdot A_0\cdot\cos 2(s-\xi), \\ \boldsymbol{O}_1 &\equiv 0{,}377\cdot G_1\cdot A_1\cdot\cos(\tau-s+2\xi-\nu+90°), \\ \boldsymbol{Q}_1 &\equiv 0{,}073\cdot G_1\cdot A_1\cdot\cos[\tau-s-(s-p)+2\xi-\nu+90°], \\ \boldsymbol{K}_{1m} &\equiv 0{,}362\cdot G_1\cdot B_1\cdot\cos(\theta^*-\nu-90°), \\ \boldsymbol{M}_2 &\equiv 0{,}908\cdot G_2\cdot A_2\cdot\cos 2(\tau+\xi-\nu), \\ \boldsymbol{N}_2 &\equiv 0{,}176\cdot G_2 A_2\cdot\cos[2(\tau+\xi-\nu)-(s-p)]. \end{aligned}$$

Ausführliche Entwicklungen und Rechentafeln zur Darwinschen Ausdrucksweise bei Schureman [14].

23. Einteilung des Jahres für geophysikalische Gezeitenrechnungen. Für die Untersuchung erdmagnetischer und ionosphärischer Gezeiteneffekte hat sich eine von Bartels [16] vorgeschlagene neue Einteilung des Jahres bewährt, die mit dem mittleren Winter-Solstitium beginnt, und die 24 Gruppen von je 15 oder 16 Kalender-Tagen umfaßt. Die Zahl *(Season-number)*

$$\sigma = 48(h° + 90°)/360°$$

läuft vom Winter-Solstitium zum nächsten von 0 bis 48. Die 24 Tagesgruppen werden mit der ungeraden σ-Zahl in der Mitte gekennzeichnet. Die ersten Tage in jeder Gruppe sind:

Gruppe	01	03	05	07	09	11
Erster Tag	Dez. 22	Jan. 6	Jan. 21	Febr. 5	Febr. 20	März 7
Gruppe	13	15	17	19	21	23
Erster Tag	März 23	April 7	April 22	Mai 7	Mai 22	Juni 6
Gruppe	25	27	29	31	33	35
Erster Tag	Juni 21	Juli 6	Juli 21	Aug. 6	Aug. 21	Sept. 6
Gruppe	37	39	41	43	45	47
Erster Tag	Sept. 21	Okt. 6	Okt. 22	Nov. 7	Nov. 22	Dez. 7

Man kann die Gruppen zusammenfassen, z.B. in 12 Paare, beginnend mit dem Paar 47, 1 [oder (−1, +1)], ein Satz von 30 Tagen, der zu Beginn des Tages Dez. 22 zentriert ist. Dieser Satz könnte durch $\sigma = 0$ bezeichnet werden, und die späteren Sätze mit 4, 8, 12, ..., 44. Diese Jahreseinteilung ist analog zu der gewöhnliche Monatseinteilung, aber besser.

Auch andere Gruppeneinteilungen sind leicht möglich, z.B. in 8 Sätze, entweder

$$1+3+5, \quad 7+9+11, \ldots,$$

oder, wenn man Sätze haben möchte, die genau zu den Solstitien und Äquinoktien und mitten dazwischen zentriert sind, in symbolischer Schreibweise

$$\text{halb } 3+5+7+\text{halb } 9 \text{ (Zentrum 6)},$$

$$\text{halb } 9+11+13+\text{halb } 15 \text{ (Zentrum 12)}, \ldots$$

24. Durchlaufend gebrauchte Bezeichnungen.

a Äquatorradius der Erde.
a annual, Ziff. 14.2.
a Koeffizient von cos, Ziff. 19a.
A Azimut (8.2).
A Faktor von τ oder t im Argument (13.3).
b Faktor von s in (14.1).
b Koeffizient von sin, Ziff. 19a.
B geographische Breite, Ziff. 8a, Fußnote.
B Faktor von s (13.3).
c Amplitude, Ziff. 19a.
c mittlere Mondentfernung (12.7).
c_s mittlere Sonnenentfernung (12.2c).
c Faktor von h in (14.1).
C $= 41{,}13 \cdot 10^{-6}$ cm/sec², (7.5+), Ziff. 20.
C Himmelspol, Fig. 5.
C Faktor von h, (13.3).
D Faktor von p, (13.3).
e, e_s numerische Exzentrizitäten der Mondbahn und der Erdbahn, Ziff. 12.
E Masse der Erde.
E Faktor von N' (13.3).
f Gravitationskonstante.
f = fortnightly, Ziff. 14.2.
f $= \sin\alpha/\alpha$, (21.1).
F Potential der Fallbeschleunigung (5.1).
F Faktor von p_s (13.3).
g Fallbeschleunigung.
g_0 (7.6), g_1 (3.10).
G (ϱ) Gezeitenkonstante (3.8), (3.10+), (3.12)
G $= G(\varrho_1)$ (3.12).
G_s (ϱ) (4.1).
G_k Geodätische Funktionen (13.5) bis (13.7).
h mittlere Länge der Sonne.
h_1 wahre Länge der Sonne (12.3h).
H horizontale Gezeitenbeschleunigung (7.3)
i Winkel Mondbahnekliptik, Ziff. 12c.
J Winkel Mondbahnäquator, Ziff. 12c.
$\boldsymbol{J_1}$ (16.21).
K Koeffizienten (13.8), Ziff. 14.3.
$\boldsymbol{K_{1m}}$, $\boldsymbol{K_{1s}}$ (16.5), Ziff. 19c, Ziff. 20.
$\boldsymbol{K_{2m}}$, $\boldsymbol{K_{2s}}$ (17.9) und (17.10).
L geographische Länge östlich von Greenwich.
$\boldsymbol{L_2}$ (17.13), Ziff. 19g.
m als Index: lunar, oft weggelassen.
m monthly, Ziff. 14.2.
M Masse des Mondes.
$\boldsymbol{M_0}$ (15.6), $\boldsymbol{M_1}$ (16.22), $\boldsymbol{M_2}$ (17.2), Ziff. 19, $\boldsymbol{M_3}$ (18.1), Ziff. 20, ***Mf*** (15.7), Ziff. 20, ***Mm*** (15.17), ***Mtm*** (15.18), ***MSm*** (15.21), ***MStm*** (15.22), ***MSf*** (15.23), ***MSqm*** (15.24).
N mittlere Länge des aufsteigenden Knotens der Mondbahn, Ziff. 11, Fig. 17.
$N' = -N$, Ziff. 11.
$\boldsymbol{N_2}$ (17.12), Ziff. 19g, $\boldsymbol{2N_2}$ (17.14), Ziff. 19g.
$\boldsymbol{O_1}$ (16.6), Ziff. 19e, $\boldsymbol{OO_1}$ (16.16).
p mittlere Länge des Perigäums der Mondbahn, Ziff. 11.
p_s mittlere Länge des Perihels, Ziff. 11.
P_n zonale Kugelfunktion (3.6).
P_n^m zugeordnete Kugelfunktion (8.6).
$\boldsymbol{P_1}$ (16.7).
q quarter-monthly, Ziff. 14.2.
$\boldsymbol{Q_1}$ (16.20), $\boldsymbol{2\,Q_1}$ (16.24).
R, R_s Abstände des Erdmittelpunktes vom Mond oder von der Sonne.
r, r_s Abstände beliebiger Punkte der Erde.
$\boldsymbol{R_2}$ (17.16).
s als Index: solar.
s mittlere Länge des Mondes.
s_1 wahre Länge des Mondes.
sa semi-annual, Ziff. 14.2.
S Masse der Sonne.
S Länge des Standardmeridians für Zonenzeit (14.6).
$\boldsymbol{S_0}$ (15.4), $\boldsymbol{S_1}$ (16.27), $\boldsymbol{S_2}$ (17.3), ***Sa*** (15.19), $\boldsymbol{SO_1}$ (16.32), ***Ssa*** (15.5), ***Sta*** (15.20).
t mittlere Sonnenzeit, t_0 in Greenwich.
t_1 wahre Sonnenzeit, t_{10} in Greenwich (8.8).
t third-monthly, Ziff. 14.2.
T Zeit in Jahrhunderten, Ziff. 11.
T Periodendauer, Ziff. 19.
$\boldsymbol{T_2}$ (17.15), $\boldsymbol{2T_2}$ (17.17).
V Potential, $\partial V/\partial s$ Beschleunigung in Richtung s.
V_2, V_3 ... (3.7); V_3 (7.10).
W Argument (14.1), W_0 (20.2).
X, Y, Z Gezeitenbeschleunigungen nach Norden, Osten, oben, (8.10) bis (8.13), (7.2).

α, α_s Rektaszensionen (8.9).
β Breite = Abstand von der Ekliptik, Ziff. 12a, (12.6).
δ, δ_s astronomische Deklinationen des Mondes und der Sonne.
ε Ekliptikschiefe.
ε Phase, Ziff. 19.
ζ, ζ_s Gleichgewichtsflut (5.3), (7.8), (7.9).
η, η_s scheinbare Radien des Mondes und der Sonne (6.1).
ϑ, ϑ_s Zenitdistanzen des Mondes und der Sonne (3.1), (3.2).
ϑ_1 (16.29).
θ astronomische Sternzeit = Stundenwinkel des Frühlingspunktes.
θ^* Sternzeit in diesem Kapitel, analog zu t und τ von der *unteren* Kulmination des Frühlingspunktes aus gerechnet, Ziff. 8b.
θ_0^* für Greenwich (8.9).
μ_2 (17.21).
ν Mondphasenzahl, Ziff. 21c.
ν abhängig von N, Ziff. 22.
ν_2 (17.20).
ψ, ψ_0, ψ_{0s} lunare und solare Horizontalparallaxen (3.9), Ziff. 8c.
ψ_1 (16.26).
ξ abhängig von N, Ziff. 22.
π_1 (16.25).
ϱ Entfernung eines Punktes der Erde vom Schwerpunkt (3.1).
ϱ_1 mittlerer Erdradius (3.10).
ϱ_1 (16.28).
σ Frequenzen (12.1a); Seasonnumber. (Ziff. 23)
σ_1 (16.21).
τ mittlere Mondzeit (13.1).
τ_1 wahre Mondzeit (8.1), τ_{10} für Greenwich.
τ_1 (16.33).
φ geozentrische Breite, Ziff. 8a.
φ_1 (16.15).
χ_1 (16.30).
ω Frequenz, Ziff. 19.

Literatur.

[1] DOODSON, A. T.: The harmonic development of the tide-generating potential. Proc. Roy Soc. Lond., Ser. A **100**, 305—329 (1922).
[2] DARWIN, G. H.: Brit. Assoc. Advancem. Sci. Report for 1883, 49—118; abgedruckt in Scientific Papers Vol. 1. Cambridge 1907.
[3] BÖRGEN, C.: Ann. Hydrogr., Hamburg **12** (1884).
[4] HORN, W.: Die astronomischen Grundlagen des harmonischen Verfahrens zur Berechnung der Gezeiten. Arch. dtsch. Seewarte u. Marine-Observatoriums **61**, Nr. 8, 28 (1941).
[5] BARTELS, J., u. W. HORN: Gezeiten-Kräfte. In LANDOLT-BÖRNSTEIN, Zahlenwerte und Funktionen aus Physik usw., 6. Aufl. Bd. 3, Astronomie und Geophysik, S. 271—283. Berlin: Springer 1952.
[6] BARTELS, J., u. G. FANSELAU: Geophysikalischer Mond-Almanach. Z. Geophys. **13**, 311—328 (1937). English version of the description in Terr. Magn. **43**, 155—158 (1938).
[7] BARTELS, J., u. G. FANSELAU: Geophysikalische Mondtafeln 1850—1975. Geophys. Inst. Potsdam Abh. Nr. 2. Berlin: Springer 1938.
[8] ELKINS, THOMAS A.: Nomograms for computing tidal gravity. Geophysics **8**, 134—145 (1943).
[9] MALONE, A. B.: A nomogram for the computation of tidal deviations of the vertical. Geophysics **17**, 615—619 (1952).
[10] MORELLI, C.: Variazione diurna della gravità in Europa. Nota 2ª-Fondamenti teórici. Ann. Geofisica, Roma **6**, 295—307 (1953).
[11] MORELLI, C.: Tidal gravity corrections for 1954. Geophysical Prospecting **2**, Suppl., 32—42 (1954). — Ann. Geofisica, Roma **6**, 579—583 (1953).
[12] GOGUEL, JEAN: A universal table for the prediction of the luni-solar tidal gravity correction. Geophysical Prospecting **2**, Suppl., 2—5 (1954).
[13] GOGUEL, JEAN: Tidal gravity corrections for 1954. Geophysical Prospecting **2**, Suppl., 6—31 (1954); for 1956. Geophysical Prospecting **3**, Suppl. No. 4 (1955).
[14] SCHUREMAN, P.: Manual of harmonic analysis and prediction of tides. Washington, U.S. Coast & Geodetic Survey, Special Publ. No. 98, 1941.
[15] PETTIT, JOHN T.: Trans. Amer. Geophys. Union **35**, 193—202 (1954).
[16] BARTELS, J.: A table of daily integers, 1902—1952, seasonal, solar, lunar and geomagnetic. Geophys. Inst. Univ. Alaska Sci. Report No. 2 AF 19 (604)—503, 1954.

Tides of the Solid Earth.

By

RUDOLF TOMASCHEK.

With 37 Figures.

A. Introduction.

1. Tidal field. General remarks. a) The tidal forces (as described in the preceding article "Gezeitenkräfte") act on every particle of the Earth's globe. On or near the surface of the Earth the observations show not only the immediate action of these forces on suitable instruments, but also, as the Earth is an elastic yielding body, that the numerical values and their variation with time will be changed by this yielding. It is one of the main problems discussed in this article to establish the quantitative features of this effect and to derive by it new information on the interior of the Earth. However, there exist additional influences which bring about elastic or other deformations of the Earth's crust. One of these is the indirect effect of the maritime tides, which is of a similar type to the primary tidal force. Other influences have time-variations comparable with the periods of the tidal forces and often attain considerable amplitudes. They are mainly of meteorological origin and combine with geological or tectonic conditions.

As the tidal fields are only about 10^{-7} of the normal gravitational field on the Earth's surface, it is a major experimental task to disentangle the different influences and to dissect the complicated observed movements into their different components. Apart from the direct geophysical interest, investigations of this kind are of immediate practical importance as they establish a knowledge of the incessantly varying background of small gravity variations. This can be used to *reduce* accurate measurements of gravity (as they are currently performed in prospecting or for geodetic purposes) to a constant reference system[1].

b) Because of the experimental arrangements the observer generally measures either the vertical or the horizontal components of the gravity variations. The direction of the vertical component Z is taken as positive in the direction opposite to the normal gravitational field. The maximum of the tidal action therefore corresponds to a measured minimum of gravity. The horizontal component H is positive in the direction of the azimuth of the tide producing body.

The vertical variations are usually given in milligals ($=10^{-3}$ cm/sec²). The horizontal components are generally transformed into the deviation of the vertical H/g and expressed in milliseconds of arc ($=10^{-3}$ seconds of arc) and involve therefore the local value of gravity. Taking the mean acceleration $g_0=979.8$ cm per sec², the actual acceleration at the observing station being g, it is

1 milligal $=1.02061\times10^{-6}\,g_0$ cm/sec² corresponding to deviation of 210.517 (g_0/g) millisec,

$10^{-6}\,g_0=0.9798$ milligals corresponding to 206.266 (g_0/g) millisec,

1 millisecond corresponds to $4.7502\times10^{-3}\,(g/g_0)$ milligal or $4.8481\times10^{-9}\,(g/g_0)$.

[1] R. NEUMANN: Geophysical Prospecting **2**, Nr. 4, 290 (1954). — H. HAALCK: Gerlands Beitr. Geophys. **64**, No. 1, 1–15 (1954). See also H. JENSEN, Bull. Géod., N. S. **1950**, No. 18, 267 (1950).

c) From the theory (see "Gezeitenkräfte", p. 735 and 739) the vertical component Z and the horizontal component H of the tide producing forces of the Moon can be obtained as

$$Z = g_1 \cdot \frac{M}{E}\left(\frac{\varrho_1}{R}\right)^2 \left[\frac{\cos\vartheta - \varrho/R}{\{1 - 2(\varrho/R)\cos\vartheta + (\varrho/R)^2\}^{\frac{3}{2}}} - \cos\vartheta\right], \tag{1.1}$$

$$H = g_1 \cdot \frac{M}{E}\left(\frac{\varrho_1}{R}\right)^2 \left[\frac{\sin\vartheta}{\{1 - 2(\varrho/R)\cos\vartheta + (\varrho/R)^2\}^{\frac{3}{2}}} - \sin\vartheta\right]. \tag{1.2}$$

The symbols used correspond to those given in the article *Gezeitenkräfte*, Sect. 24, p. 773, namely

$$g_1 = f\,E/\varrho_1^2 = 982.04 \text{ cm/sec}^2, \tag{1.3}$$

M and E the masses of the Moon and the Earth,
ϱ_1 the Earth's mean radius,
R the distance between the centre of the Earth and that of the Moon,
ϱ the distance of the point under consideration from the centre of the Earth; at the surface $\varrho = \varrho_1$,
ϑ geocentric zenith distance of the Moon,
f gravitation constant.

Eqs. (1.1) and (1.2) represent completely the vertical and horizontal components of the lunar tide-producing force at any point of the Earth. Similar formulas describe the tidal forces originated by the Sun.

The maximal values are for the Moon

$$Z☽_{\max} = 0.144 \times 10^{-6}\, g_1 = 0.1414 \text{ milligal},$$

$$H☽_{\max} = 0.107 \times 10^{-6}\, g_1 = 0.105 \text{ milligal corresp. to } 22.1 \text{ millisec},$$

and for the Sun

$$Z\odot_{\max} = 0.054 \times 10^{-6}\, g_1 = 0.053 \text{ milligal},$$

$$H\odot_{\max} = 0.041 \times 10^{-6}\, g_1 = 0.040 \text{ milligal corresp. to } 8.5 \text{ millisec}.$$

The greatest possible tidal forces produced by the combination of both influences occur approximately every 1600 years. Data of this condition are 3500 B.C., 1900 B.C., 250 B.C., 1433 A.D. and 3300 A.D. They have been credited with an influence on climate and earthquakes.

Eqs. (1.1) and (1.2) can be simplified by developing them into a series according to ascending powers of ϱ/R, this being a small fraction with an approximate maximum value of 0.018. Thus in sufficient approximation

$$\left.\begin{aligned} Z = {} & \frac{3}{2} g_1 \frac{M}{E}\left(\frac{\varrho_1}{R}\right)^2\left(\frac{\varrho}{R}\right)\left(\cos 2\vartheta + \frac{1}{3}\right) + \\ & + \frac{3}{2} g_1 \frac{M}{E}\left(\frac{\varrho_1}{R}\right)^2\left(\frac{\varrho}{R}\right)^2 (5\cos^3\vartheta - 3\cos\vartheta), \end{aligned}\right\} \tag{1.4}$$

$$\left.\begin{aligned} H = {} & \frac{3}{2} g_1 \frac{M}{E}\left(\frac{\varrho_1}{R}\right)^2\left(\frac{\varrho}{R}\right)\sin 2\vartheta + \\ & + \frac{3}{2} g_1 \frac{M}{E}\left(\frac{\varrho_1}{R}\right)^2\left(\frac{\varrho}{R}\right)^2 \sin\vartheta\,(5\cos^2\vartheta - 1). \end{aligned}\right\} \tag{1.5}$$

Eqs. (1.4) and (1.5) express the tidal forces of the Moon to within 0.04% and those of the sun to a far higher degree of accuracy. Usually only the first term of these equations is used and the second term, called the "lesser forces" is neglected.

Most tables and nomograms (see Gezeitenkräfte, p. 770 and 774, Bibliogr. No. 8 to 13 and 15) in use for tidal calculations are based on the first term only. It represents about 98% of the tidal field of the Moon and about 99.998% of that of the Sun. This is sufficient for today's observational accuracy, even for the effect of the Moon, but it has to be borne in mind that if only the first term is used in calculating the theoretically expected values for comparison with observation, then from the very beginning a possible error of up to 2 to 3% may be introduced in the values for the Moon's effect.

Introducing the mean distance c and putting

$$\frac{3}{2} g_1 \frac{M}{E} \left(\frac{\varrho_1}{c}\right)^3 = C$$

then for the Earth's surface $(\varrho = \varrho_1)$

$$Z = C \cdot \left(\frac{c}{R}\right)^3 \left(\cos 2\vartheta + \frac{1}{3}\right) = C'(p) \left(\cos 2\vartheta + \frac{1}{3}\right), \tag{1.6a}$$

$$H = C \cdot \left(\frac{c}{R}\right)^3 \sin 2\vartheta = C'(p) \sin 2\vartheta, \tag{1.6b}$$

where $C'(p)$ is a function of the parallax.

The value for the Moon's mean parallax $= 0.016593$ gives for the Earth's surface $(\varrho = \varrho_1)$

$$C = 0.08226 \text{ milligal} = 0.83956 \times 10^{-7}\, g_0,$$

corresponding for horizontal deflection to $17.317\, g_0/g$ milliseconds.

The values for the Sun are obtained by multiplying these figures by 0.46051.

The forces given by Eq. (1.6) taken for the surface derive from a potential

$$W_2 = \frac{3}{4} \frac{M}{E} g_1 \cdot \frac{\varrho^2 \varrho_1^2}{R^3} \left(\cos 2\vartheta + \frac{1}{3}\right) = 26206 \left(\frac{c}{R}\right)^3 \left(\frac{\varrho}{\varrho_1}\right)^2 \left(\cos 2\vartheta + \frac{1}{3}\right) \text{cm}^2/\text{sec}^2. \tag{1.7}$$

W_2 will be frequently used in the later part of this article.

d) The tidal influence of other heavenly bodies may be discussed here, as todays instruments are so sensitive and constant that amplitudes of regular variations amounting to 10^{-3} milligals or 0.6 milliseconds of arc could be detected, if they occurred strictly periodically, and a sufficient long period of observation is taken.

The gravitational field of the sun has a strength of about 600 milligals at a point in space coinciding with the surface of the Earth, this being about 180 times that of the Moon. The corresponding maximal gravitational fields of the planets are: Venus 1.9×10^{-2} milligal, Mars 1.4×10^{-3} milligal, Jupiter 3.6×10^{-2} milligal and that of the Galaxy may be of the order of 3×10^{-5} milligal. However, these gravitational fields are not measurable by an observer on the Earth, although some of them would be within the reach of the instruments. Since the inertial and gravitational mass are equivalent, then these gravitational fields are compensated for by the centrifugal field of the Earths orbital movement and by the accelerations due to the deviations of the orbit from circularity. But whereas the gravitational mass is subject to the central symmetrical field of the gravitating celestial body, the inertial mass is conditioned by the homogeneous field of the reference system of the fixed stars. This difference gives rise to the tidal forces, wherever the extension of the Earth is not too small compared with the distances of the celestial bodies. They amount for the different planets to the following maximal values: Venus 10^{-5} milligal. Jupiter 10^{-6} milligal and Mars 10^{-7} milligal. The effects are far beyond the limits of experimental verification. But it is remarkable that the tidal forces of the planets on the surface of the Sun, where they are of the order of magnitude of several 10^{-5} milligals and in their sum approach 10^{-4} milligals seem to be sufficiently effective in producing significant effects in the Sun's corona[1].

Regarding the tidal influence of the Galaxy on the Earth due to the asymmetrical position of the Sun in the Galaxy, it is of the order of 10^{-19} milligals and even the annual variation due to the orbital movement of the Earth around the Sun is only about 10^{-15} milligals.

---

[1] K. G. MELDAHL: Tidal Forces in the Sun's Corona due to Planets. Copenhagen: Berlinske Forlag 1938, 1940 and 1952. See also Nature, Lond. **1950**, No. 4191. — C. N. ANDERSON: J. Geophys. Res. **59**, 455 (1954). — The *apparent size* of the tide-producing bodies is essential, see p. 738.

e) In order to calculate the theoretical values of the tidal forces for a given point and time of observation, tables and nomograms have been developed (see Gezeitenkräfte, Sect. 21 c, p. 770). The direction of the horizontal component of the tidal forces can be easily shown graphically by the diagram Fig. 1. It is obtained by placing a tracing of a stereographic projection of meridians and parallels over another one showing the direction and magnitude of the horizontal tidal forces[1]. The angle between the equator lines of both tracings is made to correspond with the declination of the Moon. In Fig. 1 the case of a declination of 15° N is shown. Considering the case of latitude 30° N, at 0^h the force is 1.0

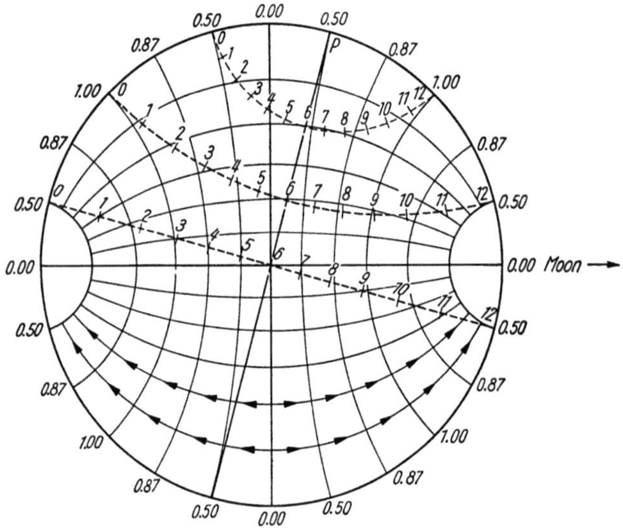

Fig. 1. Diagram of lunar horizontal tidal forces in certain latitudes hour by hour with Moon 15° N of the equator. *P* is the position of the north pole. (From Admiralty Manual of Tides.)

and is directed due south, at 2^h it has a magnitude of about 0.9 and is directed approximately SW. At about 5.5^h the force is zero and there is a change in direction thereafter from west to east. At 12^h it is 0.5 due south.

2. Tidal field on the rotating Earth. a) An observer on the Earth's surface will rotate in the slowly varying field of the tidal forces. The coordinate system most conveniently to be used is the equatorial system, where the place of the observer is given by longitude λ and latitude φ and the position of the tide-producing body by its declination δ and the hour angle Θ for $\lambda = 0$ and $t = \Theta - \lambda$ for the observer. The $\cos\vartheta$ of the zenith distance can be expressed by

$$\cos\vartheta = \sin\varphi\sin\delta + \cos\varphi\cos\delta\cos t. \tag{2.1}$$

Neglecting terms of the fourth order of the parallax the components of the tidal forces are obtained as the sum of three terms of different periodicity (see also Gezeitenkräfte, Sect. 8).

Vertical component. With the abbreviation introduced in (1.6) it is

$$\left.\begin{aligned} Z = {} & C'(p)\cdot\left(\tfrac{1}{2}-\tfrac{3}{2}\sin^2\varphi\right)\left(\tfrac{2}{3}-2\sin^2\delta\right) && \text{[longperiod]},\\ & + C'(p)\cdot\sin 2\varphi\sin 2\delta\cos t && \text{[diurnal]},\\ & + C'(p)\cdot\cos^2\varphi\cos^2\delta\cos 2t && \text{[semidiurnal]}.\end{aligned}\right\} \tag{2.2}$$

[1] A. T. DOODSON and H. D. WARBURG: Admiralty Manual of Tides, Chap. III. London 1941.

The first term is independent of the rotation, but varies with the longperiod changes of the declination and the distance of the tide-producing body, the second term includes variations multiplied by $\cos t$ that is with the period between two culminations. Its periodicity is diurnal. The third term is of double this periodicity. These three terms are called *classes* or *species* and afford an important classification in the further development of the tidal forces. The amplitudes of the different species depend differently on the latitude of the observer and the declination of the tide-producing body.

Horizontal components. For reasons of instrumental practice it is convenient to resolve the horizontal component (which has the direction towards the azimuth A of the tide-producing body) into two perpendicular directions. The South and the West are regarded as positive directions and the angles are reckoned from the south through the west. Neglecting terms of the 4th order of parallax it is from (1.6)

$$\left.\begin{aligned} H_s &= C'(p)\sin 2\vartheta\cos A, \\ H_w &= C'(p)\sin 2\vartheta\sin A. \end{aligned}\right\} \tag{2.3}$$

Transformation into the equatorial system can be performed by using the relations

$$\sin\vartheta\cos A = -\cos\varphi\sin\delta + \sin\varphi\cos\delta\cos t, \tag{2.4}$$

$$\sin\vartheta\sin A = \cos\delta\sin t. \tag{2.5}$$

The result is

$$\left.\begin{aligned} H_s = {} & \tfrac{3}{4}C'(p)\sin 2\varphi\,(\tfrac{2}{3} - 2\sin^2\delta) && \text{[longperiod]}, \\ & -C'(p)\cos 2\varphi\sin 2\delta\cos t && \text{[diurnal]}, \\ & +\tfrac{1}{2}C'(p)\sin 2\varphi\cos^2\delta\cos 2t && \text{[semidiurnal]}, \end{aligned}\right\} \tag{2.6}$$

$$\left.\begin{aligned} H_w = {} & C'(p)\sin\varphi\sin 2\delta\sin t && \text{[diurnal]}, \\ & +C'(p)\cos\varphi\cos^2\delta\sin 2t && \text{[semidiurnal]}. \end{aligned}\right\} \tag{2.7}$$

The west component has no longperiod term.

The horizontal component in any given direction with the azimuth B (measured from south through west) is

$$\left.\begin{aligned} H_B = {} & H_s\,(\text{longperiod})\cdot\cos B \\ & + H_s\,(\text{diurnal})\cdot\cos B + H_w\,(\text{diurnal})\cdot\sin B \\ & + H_s\,(\text{semidiurnal})\cdot\cos B + H_w\,(\text{semidiurnal})\cdot\sin B. \end{aligned}\right\} \tag{2.8}$$

The longperiod term varies with a monthly and fortnightly period for the Moon and with an annual and semiannual period for the Sun.

b) *The lesser field.* It has already been shown that the terms containing the 4th power of the parallax constitute only about 2% of the total tidal field of the Moon and are neglegible for that of the Sun. Introducing a factor

$$D'(p) = C'(p)\cdot(\varrho_1/c) = 1.3649\times 10^{-3}\ \text{milligal}$$

for mean parallax, the following formulas can be derived for the vertical Z^0 and horizontal H^0 components of the lesser field in a similar way:

$$\left.\begin{aligned} Z^0 = {} & \tfrac{5}{2}D'(p)\sin\varphi\,(\cos^2\varphi - \tfrac{2}{5})\sin\delta\,(5\cos^2\delta - 2) && \text{[longperiod]}, \\ & +\tfrac{15}{4}D'(p)\cos\varphi\,(\cos^2\varphi - \tfrac{4}{5})\cos\delta\,(5\cos^2\delta - 4)\cos t && \text{[diurnal]}, \\ & +\tfrac{15}{2}D'(p)\sin\varphi\cos^2\varphi\sin\delta\cos^2\delta\cos 2t && \text{[semidiurnal]}, \\ & +\tfrac{5}{4}D'(p)\cos^3\varphi\cos^3\delta\cos 3t && \text{[terdiurnal]}, \end{aligned}\right\} \tag{2.9}$$

$$\left.\begin{aligned} H_s^0 = & -\tfrac{5}{2}\, D'(p) \cos\varphi\, (\cos^2\varphi - \tfrac{4}{5}) \sin\delta\, (5\cos^2\delta - 2) && \text{[longperiod]},\\ & + \tfrac{15}{4}\, D'(p) \sin\varphi\, (\cos^2\varphi - \tfrac{4}{15}) \cos\delta\, (5\cos^2\delta - 4) \cos t && \text{[diurnal]},\\ & - \tfrac{15}{2}\, D'(p) \cos\varphi\, (\cos^2\varphi - \tfrac{2}{3}) \sin\delta \cos^2\delta \cos 2t && \text{[semidiurnal]},\\ & + \tfrac{5}{4}\, D'(p) \sin\varphi \cos^2\varphi \cos^3\delta \cos 3t && \text{[terdiurnal]}, \end{aligned}\right\} \quad (2.10)$$

$$\left.\begin{aligned} H_w^0 = & \ \tfrac{5}{4}\, D'(p) \cdot (\cos^2\varphi - \tfrac{4}{5}) \cos\delta\, (5\cos^2\delta - 4) \sin t && \text{[diurnal]},\\ & + \tfrac{5}{2}\, D'(p) \sin 2\varphi \sin\delta \cos^2\delta \sin 2t && \text{[semidiurnal]},\\ & + \tfrac{4}{5}\, D'(p) \cos^2\varphi \cos^3\delta \sin 3t && \text{[terdiurnal]}. \end{aligned}\right\} \quad (2.11)$$

For the west component there is no longperiod term.

3. Harmonic analysis of observations. As shown in "Gezeitenkräfte" part B (p. 746ff.), the variations of the tidal forces due to the alterations of the astronomical elements of the Moon and the Sun can be expressed by the sums of cosine functions with constant amplitudes and frequencies, the amplitudes being only functions of the geocentric latitude of the observer (and his distance ϱ from the centre of the Earth). Each constituent, characterized by a letter, represents the product of a solid harmonic of degree two (Fig. 2) and a harmonic function of time, each class (except the longperiod terms) having periods which are nearly diurnal or semidiurnal. The class is added to the letter of each constituent or partial tide as an index, for example M_2 for the principal semidiurnal partial tide of the Moon or O_1 for the strongest lunar diurnal tidal constituent. The 18.61 years movement of the Moon's node is generally accounted for by using a slowly varying node factor f for the amplitude and a slowly varying small angle u for adapting the phases. Observations at different times are similarly reduced to mean values (see pp. 771 f.).

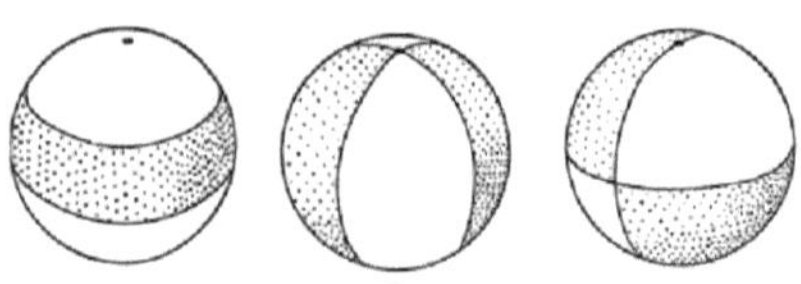

Fig. 2a—c. Types of long period (a), semidiurnal (b) and diurnal tides (c).

Then the Eqs. (2.2) can be written for example for the vertical component

$$\left.\begin{aligned} Z = & \ C'(p)\cdot(\tfrac{1}{2} - \tfrac{3}{2}\sin^2\varphi)\cdot \textstyle\sum f F \cos(V+u) + C'(p)\cdot \sin 2\varphi \sum f F \cos(V+u)\\ & + C'(p)\cdot\cos^2\varphi \textstyle\sum f F \cos(V+u). \end{aligned}\right\} \quad (3.1)$$

Similar equations describe the horizontal components (2.6) and (2.7)

Thus all the partial tides can be written as

$$T_0 = f R_0 \cos(V+u) \qquad (3.2)$$

where f is a factor varying in a period of 18.61 years (node factor),

u is an angle varying in a period of 18.61 years,

V is an angle varying steadily at the mean speed of the constituent,

R_0 is the amplitude of the constituent.

A list of the constituents generally used in the calculation of Earth tides is given in Table 1. The changes of amplitude of a semidiurnal and a diurnal constituent with latitude are shown in Fig. 3. It can be seen that for a diurnal constituent the vertical component of the tidal vector is high in medium latitudes, whereas the south component of the horizontal vector is extremely small.

The theoretical values as mentioned above, are called the *Theoretical Equilibrium Tide*; they are used as a standard of reference for the Earth tides, and an observed tidal constituent, obtained by suitable harmonic analysis of the

Table 1. *List of main harmonic constituents of equilibrium tide.*

Class	Symbol	Origin	Mean amplitude		Period hours	Relative coefficient
			Z microgal	H/g_0 millisec [1])		
Semidiurnal	M_2	lun. princ.	$76.0 \cos^2 \varphi$	$16.0\, p_2$	12.42	0.908
	S_2	sol. princ.	$35.5 \cos^2 \varphi$	$7.48\, p_2$	12.00	0.423
	N_2	lun. ellipt.	$14.6 \cos^2 \varphi$	$3.10\, p_2$	12.66	0.174
	K_2	lunisolar	$9.62 \cos^2 \varphi$	$2.03\, p_2$	11.97[2]	0.115
Diurnal	K_1	lunisolar	$44.5 \sin 2\varphi$	$9.38\, p_1$	23.93[3]	0.531
	O_1	lun. decl.	$31.6 \sin 2\varphi$	$6.69\, p_1$	25.82	0.377
	P_1	sol. decl.	$14.8 \sin 2\varphi$	$3.12\, p_1$	24.07	0.176

measurements is related to the corresponding equilibrium constituent by expressing the observed partial tide in the form

$$T = f R \cos(V + u - \varkappa) \tag{3.3}$$

where R is the amplitude observed and $\varkappa$ is the phase lag of the observed partial tide behind the phase of the equilibrium partial tide. R and $\varkappa$ are said to be

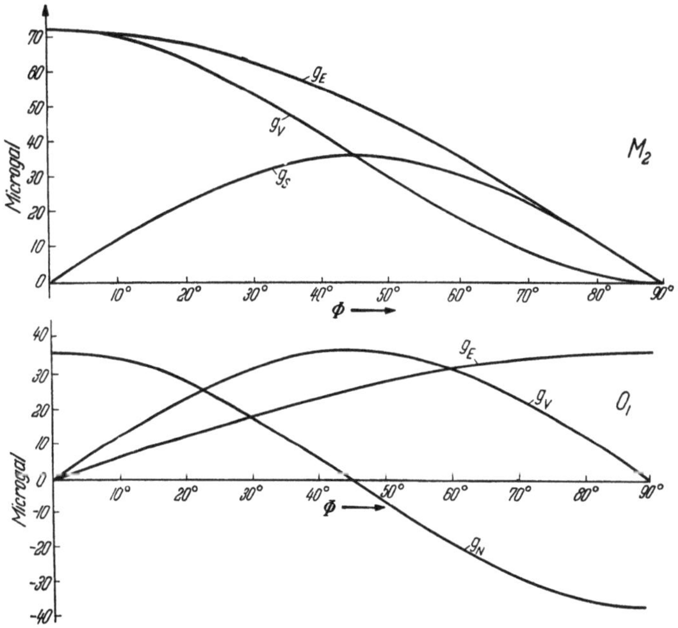

Fig. 3. The components of the equilibrium tidal gravity vector for different latitudes. For inclination 28° 36′ of the Moon's orbit (after BERROTH).

the *harmonic constants* of the observed partial tides and they are characteristic for the elastic behaviour of the Earth and the disturbing influences.

The advantage of treating the experimental results by harmonic analysis lies in the possibility of eliminating by this treatment the systematic influences of meteorological and other disturbing factors. These may decisively disturb the validity of a comparison of the total tidal equilibrium forces as represented by

[1] $p_1 = -\cos 2\varphi$ for H_s and $= \sin\varphi$ for H_w; $p_2 = \frac{1}{2}\sin 2\varphi$ for H_s and $= \cos\varphi$ for H_w.
[2] Half a sidereal day.
[3] Sidereal day.

the formulas in Sect. 2 with the experimental results. As these disturbances often contain periodicities of a solar day of varying amplitudes, the partial tide to be determined experimentally with highest accuracy will be that of the highest amplitude which has a period most strongly deviating from the solar day or its fractions. This is M_2 (Table 1) and practically most of the observations are able to establish only the harmonic constants of this constituent with some reliability. If sufficient long periods of observation are possible several other constituents may be determined with some accuracy. They are shown in Table 1.

Of these S_2 is often suspect because of possible meteorological influences which also give rise to an indirect solar diurnal S_1 tide, although a constituent of this type is not contained in the direct tidal effects. Its period is very near to K_1, which has a period of a sidereal day and it makes a determination of this constituent unreliable. The amplitude of the disturbances of S_1 and S_2 period are surprisingly high even at considerable depth (see Sect. 20). The north-south component of S_2 has been found disturbed to about 60% in the mean of several years' observations at a depth of more than 600 ft.

Examples of the variations in the harmonic constants encountered by measurements of the Earth tides are given in Tables 2 and 3. Table 2 shows various monthly values obtained by MICHELSON and GALE [2] using their pipe method and those measured by DOODSON and CORKAN [3] with horizontal pendulums as examples for conditions near the surface of the Earth. Table 3 contains yearly mean values of SCHWEYDAR'S [4] horizontal pendulum measurements at a depth of 189 m.

In order to give an impression of the experimental accuracy and the limiting conditions for harmonic analysis these tables may be discussed briefly. As for the pipe measurements which were fairly undisturbed by sea tides and can be regarded as typical for a continental station, the ratio of the extreme values for M_2 is $0.75/0.62=1.20$ occurring even in succeeding months, the maximum range for $\Delta\varkappa$ is 4°5. The figures for O_1 are $2.04/0.40=5.1$ and $\Delta\varkappa=157°$. This shows quite clearly the disastrous influence of meteorological conditions on diurnal constituents.

From the horizontal pendulum measurements which are typical for a coastal station the following results are obtained for M_2: Ratio $63.0/42.4=1.48$ and $\Delta\varkappa=12°$ and for O_1: ratio $34.0/1.5=22.6$ (!) and $\Delta\varkappa=292°$. This shows that

Table 2. *Monthly means of harmonic constants of Earth tide measurements (tilt) by pipes at a continental station* (MICHELSON-GALE [2]) *and at a coastal station* (DOODSON-CORKAN [3]) *by horizontal pendulums.*

Tide	M_2				O_1			
Month	pipes		hor. pend.		pipes		hor. pend.	
	rel. ampl.	$\varkappa$		$\varkappa$	rel. ampl.	$\varkappa$	millisec	$\varkappa$
1	0.68	1°4	57.0	−44°	1.22	−7°	7.8	6°
2	0.62	3°8	63.0	−50°	0.97	2°	9.7	157°
3	0.75	0°3	59.5	−46°	1.66	118°	7.8	113°
4	0.70	2°1	55.2	−49°	2.04	87°	4.4	55°
5	0.68	−0°3	61.5	−49°	0.34	−15°	8.8	42°
6	0.69	0°6	58.2	−44°	0.52	22°	1.5	78°
7	0.70	2°7	42.4	−41°	0.48	−4°	2.4	−135°
8	0.67	0°8	43.8	−39°	0.40	0°	10.0	103°
9	0.67	−0°7	62.5	−49°	0.45	18°	6.3	−22°
10	0.66	1°9	62.0	−42°	0.80	−39°	15.1	8°
11	0.67	0°0	51.2	−40°	0.43	−29°	34.0	58°
12	0.67	1°5	53.1	−51°	0.54	+26°	14.6	142°

Table 3. *Reduced*[1] *yearly means of harmonic constants of Earth tide measurements by horizontal pendulums at a depth of 189 m* (SCHWEYDAR [4]).

Tide	M_2		S_2		N_2		O_1	
Year	millisec	$\varkappa$	millisec	$\varkappa$	millisec	$\varkappa$	millisec	$\varkappa$
1	4.94	+4°3	3.84	−6°9	1.21	16°1	3.03	−45°3
2	4.30	+3°7	4.16	−8°3	0.51	6°4	3.43	−56°6
3	4.81	−0°3	5.43	−5°4	0.95	−9°2	2.48	−23°1
4	4.10	−0°9	4.00	−7°9	0.76	11°6	1.93	−52°1
5	3.95	+4°3	3.99	−10°1	0.77	−4°7	3.05	−51°0

very long periods of observation are necessary in order to eliminate those variations. Even the use of long series of observations and the entire elimination of actual temperature influences as in the case of Table 3 yield for different *years* the following maximum ranges: M_2: 4.94/3.95 = 1.25, $\Delta\varkappa = 5°2$; S_2: 5.43/3.84 = 1.42, $\Delta\varkappa = 4°7$; N_2: 1.21/0.51 = 2.4, $\Delta\varkappa = 25°3$ and O_1: 3.43/1.93 = 1.8, $\Delta\varkappa = 33°5$.

This discussion may serve as a valuable example for assessing the necessary criteria, when studying the experimental results, and demonstrates clearly the variability of geophysical data of this kind. Furthermore it has to be borne in mind that there exist indirect influences which have the same periodicity as the direct tidal effects and can therefore not be eliminated by harmonic analysis. This creates an additional serious difficulty in obtaining the undisturbed effects.

Because of these disturbing influences the harmonic analysis using different methods sometimes yields slightly different results. In the main constituents differences of several percent may occur as has been shown by a special study[2]. For this reason the analysis of longer periods of observations should be done preferably in monthly batches, as then the variations due to the disturbances can be followed up. A convenient method of harmonic analysis, using 29 days of observations has been developed by DOODSON[3]. Recently a different method of similar practical simplicity has been developed by LECOLAZET[4].

B. The yielding of the solid Earth without oceans.

4. Resonance problems. Primary, secondary and indirect effects. Whereas the oceans respond to the tractive forces, the horizontal components, the Earth is subject as a whole to the tidal stresses, and yields to them because of its elasticity. This deformation—considered separately from the water on its surface—is called an *Earth tide*. As the shallowness of the oceans makes their natural periods comparable with the periods of the diurnal and semidiurnal tidal forces and, therefore, profoundly influences their amplitude and phases, the *oceanic tides* are of a *dynamic type* (see the article by DEFANT in this volume). In contradistinction, the tides of the "solid" Earth can be regarded as being of the *equilibrium type*, that is *the relative amplitudes and phases of the constituents (partial tides) should correspond to the theoretical ratios*. The expression "solid" Earth shall be used for distinguishing these tides from those of the oceans. In fact the core of the

[1] Reduced for varying f (node factor).

[2] J. T. PETIT, L. B. SLICHTER and L. LA COSTE: Trans. Amer. Geophys. Un. **34**, 174 (1953). On the influence of drifts see O. MEISSNER: Astronom. Nachr. **184**, 398 (1910).

[3] A. T. DOODSON: Internat. Hydrogr. Rev., May **1954**. *Remark:* The figures in the first line of Table 3 of that paper should read 1.1, and −1.1; furthermore $C_{p_0} = X_{p_0} + Y_{p_0}$. In Table 4: XO_a is −1175 and +145; XO_b is 85 and −1127. A simplified method (in most cases quite sufficient for work on earth tides) is given in Admiralty Tide Tables, Part III, Sect. III. London 1936.

[4] P. LECOLAZET: Ann. Géophys. **12**, Nr. 1, 59 (1956).

Earth has to be regarded as fluid (and not only imperfectly elastic solid). The influence of this fact will be considered later (Sect. 6c).

The Earth has time to adjust itself to the rapidly changing forces even of the diurnal and semidiurnal tides. The natural frequencies of oscillations of the tidal type (Fig. 2) of the solid Earth are fairly short. A liquid sphere of the size and mean density of the Earth would have a natural period of 94 min. An incompressible elastic sphere of the size and mean density of the Earth and the rigidity of steel would have a natural period of 55 min. Recently the free period of the Earth (by taking the mean observed density and elastic values) has been calculated to be $44^{\mathrm{m}}17^{\mathrm{s}}$[1]. Furthermore, the seismic observations[2] have shown that an elastic impulse travels the length of one diameter in 22 min (PKP wave).

The experimental arrangements do not measure exclusively the "*direct*" or "*primary*" Earth tides. There exist obvious effects which shall be called "*indirect*" *effects*. These are of tidal periodicity (although with different relative amplitudes and phases) but they are caused by indirect action of the tidal forces on the solid Earth, mainly by the oceanic tides. Their varying load will give rise to elastic deformations which are transmitted downwards from the surface, but it will also give rise to effects which are not of elastic type, as will be shown later (Sect. 22, 23). Moreover, there exist other effects which are not of tidal periodicity; they will be called "*secondary*" *effects*. So do the meteorologically caused maritime and barometric pressure-changes seem to act not only in a purely elastic way, but also movements of a seesaw type or even lateral undulations seem to exist. It has been calculated that natural periods of this type are in the order of an hour, but the possibility cannot be excluded of very much longer periods which may possibly give rise to resonance phenomena in certain regions of the Earth[3].

The measurements therefore contain a complex sum of effects which can be characterized by vectors and the results can be expressed as a vectorial sum

$$\boldsymbol{M} = K \cdot \boldsymbol{E} + \boldsymbol{I} + \boldsymbol{S}. \tag{4.1}$$

Here $K \cdot \boldsymbol{E}$ is the direct effect which will be proportional to the equilibrium theoretical tide by a factor K characterized by the type of measurement and the elasticity of the Earth. The indirect effects are contained in $\boldsymbol{I}$ which is of tidal periodicity and is mainly of oceanic origin, but may also include tectonic influences. $\boldsymbol{S}$ are the secondary effects, which can be written as a vectorial sum

$$\boldsymbol{S} = \boldsymbol{S}_G + \boldsymbol{S}_M + \boldsymbol{S}_T, \tag{4.2}$$

where $\boldsymbol{S}_G$ is the effect of geological and tectonic structure, $\boldsymbol{S}_M$ is the meteorological pressure effect (including also maritime surges) and $\boldsymbol{S}_T$ is the effect of variations of temperature, that is, either direct effects on the instruments, or secondary effects such as those due to strain influences in the Earth's crust.

The secondary effects can be eliminated by harmonic analysis although this is possible sometimes only with difficulty and at the cost of very long series of observations; the indirect effects are not affected by harmonic analysis directly. But as their phases and amplitudes are not proportional to the equilibrium tide values, they may be eliminated to a certain degree by comparing their influence on different constituents, if e.g. the ratio of indirect effects of different constituents is known from other sources (see Sect. 19).

The following sections of this chapter deal with the direct effect $K \cdot \boldsymbol{E}$ only, disregarding the other complications.

[1] C. L. PEKERIS, J. Geophys. Res. **61**. 405 (1956).
[2] See this Encyclopedia, Vol. XLVII, p. 96.
[3] R. SCHUMANN: Z. Geophys. **11**, 1–9 (1935).

5. Deformation and potential. An entirely solid Earth which adapts itself to the tidal forces will be considered first. As shown in the article „Gezeitenkräfte", Sect. 5 and Fig. 2 p. 737 there, the deformed equipotential surface F_{rg} for a perfectly rigid Earth has the form of the equilibrium tide, as it would be expressed by a very thin perfectly fluid layer which has the property of adapting itself momentarily to the acting forces (Fig. 4a). In the direction S of the celestial tide-producing body the mean elevation of the equipotential surface by the Moon is 35.6 cm and the mean depression on the great circle perpendicular to this direction is 17.8 cm. The analogous data for the action of the Sun are 16.4 and 8.2 cm. The maximum possible deformations are therefore 52.0 and 26.0 cm, the amplitude thus being 78 cm.

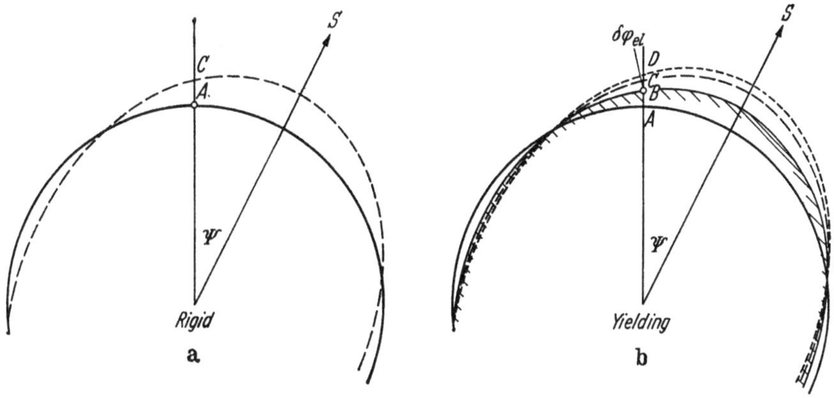

Fig. 4a and b. Deformation of equipotential and actual surface due to the tidal potential (a) rigid, (b) yielding Earth.

An elastic solid Earth will yield to the tidal forces and will be deformed in the manner shown by Fig. 4b. Because of its rigidity it will not quite reach the equipotential surface C, as an ideal liquid of density zero would do, but it will assume some intermediate height at B which will depend on the elastic properties and the density of the Earth. Buth this deformation has simultaneously another effect. The different distribution of the yielding masses deforms the equipotential surface C to another shape D. Furthermore, because of the bodily deformation of the Earth the observer has been shifted from A to B (Fig. 5).

a) The potential W_{el} at a station on the surface of the yielding Earth can therefore be expressed by

$$W_{\text{el}} = W_0 + W_2 + V_2 - g \cdot u_r. \tag{5.1}$$

Here W_0 is the potential due to gravity, W_2 the tidal potential [Eq. (1.7)]. For the surface ($\varrho = \varrho_1$),

$$W_2^{☽} = 26206 \left(\frac{c}{R}\right)^3 \left(\cos 2\vartheta + \frac{1}{3}\right) \text{cm}^2/\text{sec}^2, \tag{5.2a}$$

$$W_2^{\odot} = 0.46051\, W_2^{☽}. \tag{5.2b}$$

V_2 is the potential due to the changed mass distribution. The last term takes account of the displacement of the observer, as, with sufficient approximation,

$$g = -\left(\frac{d W_0}{d r}\right)_{r=\varrho_1}. \tag{5.3}$$

Love [5] has shown that u_r can be assumed to be proportional to W_2, so that

$$u_r = h \cdot W_2 / g. \tag{5.4}$$

Furthermore, V_2 can be calculated as being the potential of an additional layer of masses having the thickness u_r, and it is proportional to W_2

$$V_2 = k \cdot W_2. \tag{5.5}$$

V_2 can be calculated in a similar way to that employed when calculating the eccentricity of the Earth due to its rotation, using CLAIRAUT's theorem. It is

$$V_2 = \frac{3}{4}\,\frac{k^2}{\varrho_1^3}\,(A - C)\left(\cos 2\vartheta + \frac{1}{3}\right). \tag{5.6}$$

Here C is the moment of inertia of the layer about an axis from the Earth's centre to the celestial body and A the value of the two other principal moments of inertia. It can be shown that independently of the rate of change of density and elasticity within the Earth, k is proportional to r^{-5} for outer points, as W is proportional to r^{-2}. The number k can therefore be considered as a constant on the Earth's surface.

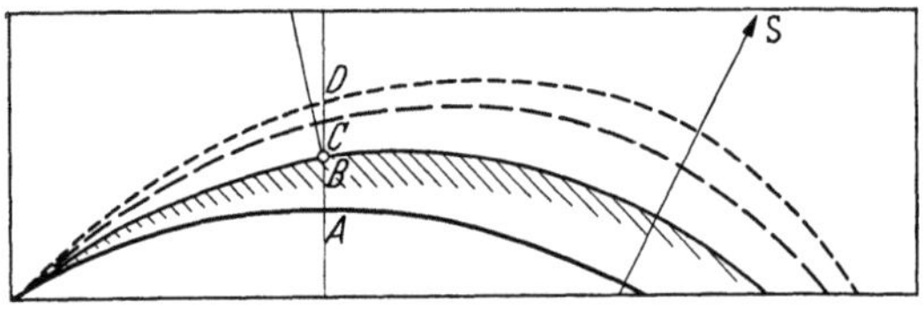

Fig. 5. Observer on yielding Earth. A undisturbed position, B tidal position, C equilibrium tide, D potential surface for yielding Earth.

$$W_{\mathrm{el}} = W_0 + W_2 + V_2 - g\cdot\delta r = W_0 + W_2(1 + k - h),$$
$$V_2 = k\cdot W_2; \quad g\cdot\delta r = h\cdot W_2$$

$\delta H_{\mathrm{el}} = \delta H_{rg}(1 + k)$	Deformation of equipotential surface.
$\delta H_{\mathrm{el}} - \delta r = \delta H_{rg}(1 + k - h)$	Dislocation of equipotential surface relative to observer. Observable equilibrium tide.
$\delta g_{\mathrm{el}} = \delta g_{rg}(1 + h - \frac{3}{2}k)$	Change of gravity for observer.
$\delta\theta_{\mathrm{el}} = \delta\theta_{rg}(1 + k - h)$	Deflection of pendulum or level relative to observer.
$h\cdot\delta\theta_{tg}$	Tilt of Earth's surface.
$(1 + k)\cdot\delta\theta_{rg}$	Tilt of equipotential surface.
$u_\varphi = \frac{l}{g}\cdot\frac{\delta W_2}{\delta\varphi}; \quad u_\lambda = \frac{l}{g\cos\varphi}\cdot\frac{\delta W_2}{\delta\lambda}$	Horizontal displacements.
$E_\varphi = \frac{l}{g\,a}\cdot\frac{\delta^2 W_2}{\delta\varphi^2}; \quad E_\lambda = \frac{l}{g\,a\cos^2\varphi}\cdot\frac{\delta^2 W_2}{\delta\lambda^2}$	Strain.
$\delta\varphi_{\mathrm{el}} = \delta\varphi_{rg}(1 + k - l); \; \delta\lambda_{\mathrm{el}} = \delta\lambda_{rg}(1 + k - l)$	Variations in longitude and latitude.

From these equations we have for a point on the Earth's surface

$$W_{\mathrm{el}} = W_0 + W_2(1 + k - h). \tag{5.7}$$

For a perfectly rigid Earth, $h = k = 0$.

b) The deformation of the equipotential surface is

$$\frac{1}{g}(W_2 + V_2) = F_{\mathrm{el}} = F_{rg}(1 + k). \tag{5.8}$$

The deformation of the equipotential surface is increased by the elasticity of the Earth as k is a positive number. The deformation relative to the observer is

$$F_{\mathrm{el}} - u_r = F_{rg}(1 + k - h). \tag{5.9}$$

This will be also the apparent height of tides for a liquid which is able to assume the shape of the equipotential surface (that is actually for long-period tides), measured by an observer connected with the yielding Earth.

c) The *change of gravity at the surface* can be found as follows: The variational part of the tidal deforming potential on the surface is according to (5.1) and (5.3) (writing now r for ϱ)

$$u_r \frac{d W_0}{d r} + V_2 + W_2. \tag{5.10}$$

The variation in gravity is

$$\Delta g = u_r \frac{d^2 W_0}{d r^2} + \frac{\partial V_2}{\partial r} + \frac{\partial W_2}{\partial r}. \tag{5.11}$$

From LAPLACE's equation

$$\nabla^2 W_0 = 0 \tag{5.12}$$

follows

$$\frac{d^2 W_0}{d r^2} = -\frac{2}{r} \frac{d W_0}{d r} = \frac{2g}{r}. \tag{5.13}$$

For V_2, taking into account (5.6), it follows that

$$\frac{\partial V_2}{\partial r} = -\frac{3}{r} V_2. \tag{5.14}$$

Furthermore regarding Eq. (5.2)

$$\frac{\partial W_2}{\partial r} = \frac{2 W_2}{r}. \tag{5.15}$$

Inserting the last three equations into (5.11)

$$\Delta g = (2h - 3k + 2) \frac{W_2}{r}. \tag{5.16}$$

Because of (5.15) this can be also written

$$\Delta g = \left(1 + h - \frac{3}{2} k\right) \left(\frac{\partial W_2}{\partial r}\right)_{r=\varrho_1}, \tag{5.17}$$

where

$$\left(\frac{\partial W_2}{\partial r}\right)_{r=\varrho_1} = \Delta g_{rg} \tag{5.18}$$

is the change of gravity on a perfectly rigid Earth with $h = k = 0$, corresponding to the vertical component of the theoretical equilibrium tide Z. On the yielding Earth therefore the gravitational variation will be modified by a factor G, called the *gravimetric factor*

$$G = 1 + h - \tfrac{3}{2} k \tag{5.19}$$

compared with the theoretical equilibrium tide.

d) The *deflection of the vertical* relative to the observer, that is relative to the Earth's surface is found by differentiating $F - u_r$ for a horizontal direction. It is

$$\delta \varphi_{\mathrm{el}} = \delta \varphi_{rg} (1 + k - h); \tag{5.20}$$

here $h \cdot \delta \varphi_{rg}$ is the angle between the original and the deformed surface, and $(1 + k) \cdot \delta \varphi_{rg}$ the angle between the original and the deformed equipotential surfaces. The deformation of the surface is, as Fig. 5 shows, in the same direction as the tidal component of the horizontal force and therefore compensates part of its effect. The factor

$$D = 1 + k - h \tag{5.21}$$

is for this reason called the *diminishing factor* and is smaller than 1.

e) There exist also *lateral displacements*. They can be assumed to be proportional to the tidal potential by a factor l which was first introduced by SHIDA (1912). It is sometimes also called LAMBERT'S constant. The surface values of the displacements in colatitude θ and longitude λ are

$$u_\theta = \frac{l}{g} \frac{\partial W_2}{\partial \theta}, \qquad u_\lambda = \frac{l}{g \sin\theta} \frac{\partial W_2}{\partial \lambda}. \tag{5.22}$$

whereas the radial displacement, according to (5.4) is given by

$$u_r = \frac{h}{g} \cdot W_2.$$

6. The characteristic numbers. The three numbers h, k and l characterize the elastic behaviour of a spherical entirely solid yielding Earth on its surface free from oceans and neglecting the lesser forces.

a) In order to give an idea how these numbers depend on the elastic constants, the results of calculations of KELVIN for the ideal case of a homogeneous and incompressible Earth may be mentioned. With d being uniform density, μ the rigidity and ϱ_1 being the mean radius, the surface values are

$$h = \frac{5f}{2f+1}, \tag{6.1}$$

$$k = \frac{3f}{2f+1}, \tag{6.2}$$

$$l = \frac{\frac{3}{2}f}{2f+1}, \tag{6.3}$$

where

$$f = \frac{g d \varrho_1}{19\mu}. \tag{6.4}$$

There exist the relations

$$k = 0.6\,h = 2l \tag{6.5}$$

and

$$l = 0.3\,h = 0.5\,k. \tag{6.6}$$

For a value of $k = 0.288$, which seems to be fairly reliable, as shown later (Sect. 9), μ becomes 1.5×10^{12} dyn/cm². This gives

$$k = 0.29; \quad h = 0.48; \quad l = 0.14$$

for this rather drastic simplification. See further Sect. 12.

The meanings of the three numbers from their definitions as given in Sect. 5 is as follows:

h represents the ratio between the height of the tide of the solid Earth to the theoretical equilibrium heigth;

k represents the ratio of the additional potential due to the tidal deformation of the yielding Earth to the tidal potential;

l represents the ratio between the horizontal displacement of the yielding Earth to the theoretical equilibrium tide.

b) It is possible to determine experimentally the values of $D = 1 + k - h$; $G = 1 - \frac{3}{2}k + h$; $L = 1 + k - l$ and of k and l directly. The observational methods are as follows:

α) The expression $D = 1 + k - h$ can be obtained in different ways.

1. From the observation of the equilibrium tides as shown in Eq. (5.9). The semidiurnal and diurnal tides are too near to the resonance frequencies of the

different oceans and sea basins to be usable for this purpose. But the fortnightly tide should already be sufficiently remote from these frequencies to give reasonable results.

2. Tilting of the surface of lakes or of fluids in pipes. A body of water whose natural period is short compared with the tidal periods will set itself in such a way that its surface is an equipotential surface. If u is the elevation of the surface, then

$$-g u + (1 + k) W_2$$

will be constant. It is not possible to measure directly the difference in level. But as the bottom of the shallow water body moves itself with the Earth its elevation being $h W_2/g$, the elevation u' of the free surface relative to the bottom is given by

$$u' - (1 + k - h) W_2/g = \text{constant} \tag{6.7}$$

and can be determined, as the constant is given by the fact that the volume of water remains constant with respect to time (or can be assumed as varying linearily with time within tidal periods).

3. It has ben already shown (Sect. 5d) that the deflection of the vertical relative to the adjacent ground is measured by the same diminishing factor D. Special instruments can be constructed to perform such measurements. They are called *horizontal pendulums*.

β) Two methods exist for determining the gravimetric factor

$$G = 1 + h - \tfrac{3}{2} k.$$

The first uses very careful observations of *pendulum clocks* and their comparison with *quartz clocks*, the second observes the minute changes of the weight of a mass in special instruments called *gravimeters*.

γ) The value of

$$L = 1 + k - l$$

can be obtained by astronomical observations of the deflections of the vertical referred to the Earth's axis. As the local vertical is defined by the direction of gravity, the effect would be shown by variations of latitude and longitude in tidal periods, if the axis of rotation was fixed in space. Certain corrections seem necessary as the tidal potentials also produce couples affecting the Earth as a whole and perturbing the axis of rotation.

δ) The number l can be obtained by measuring the horizontal strain developed in the Earth's surface with the aid of special strainmeters or deformographs.

Taking the X-axis to the south and the Y axis to the west along the Earth's surface the components e_{xx}, e_{yy} and e_{xy} of the horizontal linear strain can be expressed, using Eqs. (5.22) and (5.23) as

$$e_{xx} = \frac{1}{\varrho_1} \cdot \frac{\partial u_\theta}{\partial \theta} = \frac{l}{\varrho_1 g} \frac{\partial^2 W}{\partial \theta^2}, \tag{6.8}$$

$$e_{yy} = \frac{1}{\varrho_1 \sin\theta} \cdot \frac{\partial u_\lambda}{\partial \lambda} = \frac{l}{\varrho_1 g \sin\theta} \cdot \frac{\partial^2 W}{\partial \lambda^2}, \tag{6.9}$$

$$e_{xy} = \frac{1}{\varrho_1} \frac{\partial u_\lambda}{\partial \theta} + \frac{1}{\varrho_1 \sin\theta} \cdot \frac{\partial u_\theta}{\partial \lambda} = \frac{l}{\varrho_1 g \sin\theta} \cdot \frac{\partial^2 W}{\partial \theta \, \partial \lambda}. \tag{6.10}$$

For example, putting the amplitude of the M_2-tide

$$A_{M2} = C'(p) F,$$

the components of strain become

$$\left.\begin{aligned} e_{xx} &= 2l A_{M2} \cos 2\theta \cos 2\lambda, \\ e_{yy} &= -4l A_{M2} \cos 2\lambda, \\ e_{xy} &= -6l A_{M2} \cos\theta \sin 2\lambda. \end{aligned}\right\} \tag{6.11}$$

Observing in a direction with the cosines of direction being σ and τ, the tidal strain can be expressed by

$$e_t = e_{xx}\sigma^2 + e_{yy}\tau^2 + e_{xy}\sigma\tau. \tag{6.12}$$

ε) The number k can be determined by analyzing the movement of the pole[1] regarding the period of the free oscillation in latitude. EULER showed that a small disturbance from rotation about the axis of greatest moment of inertia makes the axis of rotation move in a cone about the axis of greatest moment. The period of revolution is $A/(C-A)$ times the period of rotation, the moments being (A, A, C).

In the case of the Earth this period should be 305 days. This period however could not be found, but in addition to a yearly period another period of about 14 months has been observed. It has been shown that this longer period, the so called CHANDLER's period, corresponds to EULER's period which is lengthened to this amount by the elastic yielding of the Earth. This period gives a measure of the elastic strain which is always symmetrical about the instantaneous axis.

If EULER's period is

$$T_0 = \frac{A}{C-A} \tag{6.13}$$

CHANDLER's period T can be expressed by

$$T = \frac{A + k\omega^2 \cdot \varrho_1^5/3f}{C - A - k\omega^2 \cdot \varrho_1^5/3f} \tag{6.14}$$

where f is the gravitational constant, ω the angular velocity of Earth's rotation, ϱ_1 the surface radius and k is LOVE's number, characterizing the external potential due to the deformation.

As a first approximation we obtain, introducing

$$\mu' = \frac{\omega^2 \varrho_1}{g} = \frac{\omega^2 \varrho_1^3}{f M} \tag{6.15}$$

where $\mu' = 1/288$ is the ratio of centrifugal force to gravity on the equator, M being the mass of the Earth

$$\frac{k\omega^2 \varrho_1^5}{3fA} = \frac{k}{3} \cdot \frac{\mu' \varrho_1^2 M}{A}. \tag{6.16}$$

Furthermore

$$\frac{3}{2}\,\frac{C-A}{M\varrho_1^2} = e - \frac{1}{2}\mu' \tag{6.17}$$

where $e = 1/298$ is the eccentricity of the Earth. This gives

$$k = \left(\frac{2e}{\mu'} - 1\right)\left(1 - \frac{T_0}{T}\right) = 0.9330\left(1 - \frac{T_0}{T}\right). \tag{6.18}$$

c) It has to be borne in mind that the characteristic numbers as derived here depend on a statical theory, any intertial terms being neglected in comparison with elastic terms. This restriction is not quite satisfactory as there exists evidence that the core of the Earth behaves like a fluid. Putting the rigidity of the

[1] See this Encyclopedia, Vol. XLVII, p. 610.

core to zero gives $k = 0.37$ which is too high and indicates that the assumptions are yet too simplified. Another complication arises from the fact that a liquid core would show possible motions of its own. JEFFREYS [1] has shown that the effect of a liquid core, namely a shortening of the free period and a reduction of the amplitude of the 19 years nutation, are represented satisfactorily by suitably adapting the elasticity of the shell. He could show that the shell bends in such a way as to produce comparatively little motion in the core. Nonetheless, the fluid core should exercise some influence on the diurnal tides, whereas the semi-diurnal, fortnightly and monthly tides are not affected. Quite apart from other influences such as the oceanic tides or those of meteorological origin, the values of the characteristic numbers derived from diurnal terms could differ significantly from those of the semidiurnal terms for this reason. See end of Sect. 13, p. 822.

C. Experimental methods and results.

7. Diminishing factor $D = 1 + k - h$. a) The first method consists in using the equilibrium tides. The fortnightly and monthly ocean tides should be sufficiently uninfluenced by resonance effects. DARWIN (1881) obtained from the registrations of 14 ports

$D = 0.675$ from the fortnightly tides

and $D = 0.680$ for the monthly tides.

SCHWEYDAR[1] using a larger material obtained

$D = 0.626$

and $D = 0.605$ respectively.

Here the effect of the attraction of the tidal masses of the oceans and that of the forces due to the rotation of the Earth (CORIOLIS forces) on the water have to be applied as corrections. Neither effect is well known, so that the values obtained by those determinations are subject to some doubt.

b) The second method uses, according to a suggestion by PROUDMAN[2], the semidiurnal and diurnal tides in lakes, as the resonance period in this case is short compared with the tidal periods. Observations in the Red Sea were not quite satisfactory, but Lake Baikal yielded fairly good values. The results were

diurnal tides:	STERNECK (1928)	$D = 0.73$
semidiurnal tides:	STERNECK (1928)	$D = 0.55$
	GRACE (1931)	$D = 0.54$
	AKSENTCHEVA (1948)	$D = 0.72$
		$D = 0.55$

the last date referring to another station. The tides are in phase with the local tidal equilibrium forces.

MELCHIOR[3] has analyzed a fairly undisturbed period of 29 days of observations made by J. CAPART at Albertville on Lake Tanganika. He obtains

$$D = 0.56\,(\pm 0.06)$$

and a phase difference of $+9°5'$, the factor of reduction of the readings being known to about $\pm 10\%$.

[1] W. SCHWEYDAR: Gerlands Beitr. Geophys. **9**, 41–77 (1908).

[2] J. PROUDMAN: Month. Not. Roy. Astronom. Soc. Geophys. Suppl. **2**, 96 (1928). — Bull. Sect. Océan. Conseil. Int. Recherches **11**, 1–6 (1928).

[3] P. J. MELCHIOR: Bull. Acad. Roy. Belg., Cl. Sci. **1956**, 368.

c) The third method is that of observing the tidal variations of the heights of the water at the ends of long pipes or basins.

α) The classic determination of this kind was performed by MICHELSON and GALE in 1914 and 1916/17 [2]. They used two tubes of 15 cm diameter filled with water. These tubes were situated N—S and E—W respectively and had a length of 153 m each. They were buried at a depth of about 2 m. The changes in height of the water surface were observed by an interference method. If N is the dislocation of the interference picture, measured in numbers of fringes, then

$$\Delta h = \frac{N \cdot \lambda}{2(n-1)} \tag{7.1}$$

where λ is the wavelength and n is the refractive index of the water. Using mercury light ($\lambda = 4358$ Å), the dislocation of the image by one fringe corresponds to a change of the height of the water surface relative to that of the Earth of 6.38×10^{-5} cm. The measurements were performed at both ends of each tube. Some of the results have been already given in Table 2. The mean values of the harmonic constants for M_2 and O_1 are:

$$\left.\begin{aligned} D_{\mathrm{NS}} &= 0.68; \quad \varkappa = +1^\circ\,15' \\ D_{\mathrm{EW}} &= 0.69; \quad \varkappa = +6^\circ\,04' \end{aligned}\right\} M_2$$

and

$$\left.\begin{aligned} D_{\mathrm{NS}} &= 0.78; \quad \varkappa = +12^\circ \\ D_{\mathrm{EW}} &= 0.70; \quad \varkappa = +0^\circ\,30' \end{aligned}\right\} O_1 .$$

The point of observation was Yerkes Observatory ($\varphi = 42^\circ\,34'$ N, $\lambda = 88^\circ\,33'$ W) which is a typical continental station.

β) Similar observations have been made by EGEDAL and FJELDSTADT[1] (1934) using a water pipe of 2.7 cm diameter and 108 m length in a tunnel at Bergen (Norway). The readings were taken at both ends by niveau variometers, mainly consisting of a cylindrical glass float whose vertical movements are transmitted to a mirror by a vertical glass rod. The azimuth was N 18° E to S 18° W and the period of observations was 15 days. The observed tidal variations of diurnal type seem to be fairly undisturbed by oceanic influences, but the semidiurnal tides were too high. They have therefore been corrected for an influence of the maritime tides, as the tidal range increases rapidly along the coast of Norway when passing from south to north (nearly in the direction of the pipe). The observed amplitude of M_2 of 2.47 μ and a phase angle of 253°.8 is found to consist of two components namely a tide assumed as static with amplitude 1.93 μ phase angle 200°.9 and a load tide of amplitude 2.02 μ, phase angle 303°.4, so that the load influence would have about the same amplitude as the static Earth tide. The corrected results are

semidiurnal	M_2	$D = 0.58$	diurnal	O_1	$D = 0.75$
	N_2	0.66		K_1	0.60
	S_2	0.63			

γ) NÖRLUND[2] performing a precise levelling between Danish isles by means of a pipe of 18 km length has observed variations of M_2 period of the level of about 2 to 3 mm. He obtains $D = 0.8$.

[1] J. EGEDAL and J. E. FJELDSTADT: Appdx. Naut.-Meteor. Annual 1936; Geofysiske Publ. **11**, No. 14. Norsk. Videnskaps-Akad. Oslo 1937.

[2] N. E. NÖRLUND: Danish Geod. Inst. Skrifter 3 Raekke **6** and **8** (1945, 1946). See also E. GIGAS: Dtsch. Geod. Komm. Reihe A **1952**, No. 3, 15.

δ) ZERBE[1] has made observations of a similar kind in the DAVID TAYLOR Model Basin (in a suburb of Washington), using a distance of 810 m at an azimuth W 16° N to E 16° S. The gauges were plunger and dial type indicating instruments. The smallest division was 0.0001 inch, one revolution of the pointer corresponded to 0.02 inch. The average range of the observed tides was about 0.002 inch. The period of observations was seven days. The ratio of the observed to the equilibrium ranges was 0.75. As the station is not far from the coast the influence of the oceanic tides has to be determined. The correction is difficult, but it seems that the loading effect is greater than the Earth tides and both differ in phase by about 80°. The most probable corrected value is $D = 0.74$ with no significant phase lag against the theoretical equilibrium tide.

ε) Discussion of the results obtained by these methods shows that MICHELSON-GALE's are the most extended ones, but they have not yet been corrected for possible maritime influences. Probably these corrections are not very large as observations on other continental stations show. The observation periods of the other measurements are fairly short and as they are situated near the coasts the corrections are difficult and not certain. The value of

$$D = 0.7_2$$

can be regarded as a general result of the observations of this type, with no indication of a significant phase difference.

A repetition of experiments of this type in a suitable mine or a subterranean lake over a period of several years is highly desirable, especially because by the use of this method, which integrates over a fairly large distance, peculiarities of a strictly local character are avoided.

d) α) The most frequently used instrument for determining D is the horizontal pendulum[2], consisting of a mass which is supported in such a way that it can swing about a nearly vertical axis. If the axis of the pendulum coincides with the plumb line, then it is in indifferent equilibrium (apart from the torsional forces of a suspension, if any, which shall be neglected at first). If the axis is inclined by the angle i the pendulum mass moves till its zero position is situated in the plane which contains the plumb line and the axis (Fig. 6a). If the pendulum is tilted by the angle φ in a direction perpendicular to this plane the pendulum moves and its deflection α increases as the angle i decreases (Fig. 6b; in fact the angles are very small). As a pendulum set up with a given inclination i is sensitive only to horizontal components of the tidal force or tilts perpendicular to the plane of its zero position, two pendulums are necessary, set up in perpendicular directions, in order to determine the actual deflection of the vertical.

The angle i is being determined by comparing the period T_1 of the horizontal pendulum with the period T_0 of the same pendulum when its axis is horizontal. As in the first case the effective force of gravity is only $g \cdot \sin i$, the period is given by

$$T_1 = 2\pi \sqrt{\frac{l}{g \sin i}} \tag{7.2}$$

and it is

$$\sin i = T_0^2 / T_1^2 . \tag{7.3}$$

The static deflection is

$$\delta\alpha = \frac{\delta\varphi}{\sin i} = \delta\varphi \left(\frac{T_1}{T_0}\right)^2 \tag{7.4}$$

[1] W. B. ZERBE: Trans. Amer. Geophys. Un. **30**, 357 (1949).
[2] See also this Encyclopedia, Vol. XLVII, p. 42.

where $\delta\varphi$ is the component of the variation of the vertical, perpendicular to the plane of the pendulum. For small angles i a correction has to be added, taking into account the influence of the torsional forces of the suspension[1].

The horizontal pendulum has been first devised by HENGLER (1832), later by PERROT (1862) and was mainly developed by ZÖLLNER (1869)[2]. There exist also forms using pointed pivots by VON REBEUR and a combination of pivots and suspension by MILNE-SHAW. For the purposes of Earth tide measurements only the *suspension type* is sufficiently reliable.

The weights of the pendulum mass vary between 70 and 15 gr, the lengths of the suspensions between 20 to 10 cm. Generally a period of the order of magnitude of 20 to 30 sec, compared with T_0 of about 0.7 to 0.9 sec is sufficient for

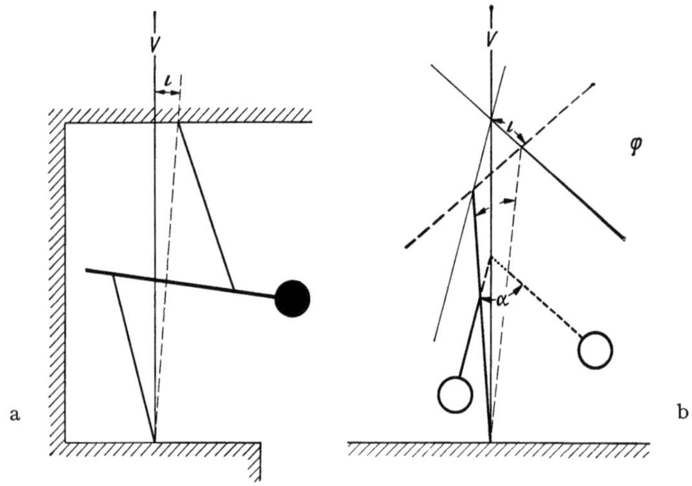

Fig. 6a and b. Principle of horizontal pendulum.
Fig. 6b shows, from top to bottom, the angles i, φ and α. Please note: that the letter φ has not been inserted.

the registration of the Earth tides. When using higher sensitivity difficulties may arise at certain stations or at certain times due to drift movements. By using optical paths of about 5 to 3 m for recording the movements of the pendulum mirrors, amplifications of 1 mm of the curve corresponding to tilts of 0.015 to 0.05 sec of arc can be easily obtained. In some cases amplifications of 0.001 sec of arc per mm have been used.

An instrument capable of very high sensitivity is the horizontal double pendulum by LETTAU[3] where two pendulums are coupled together, the second being suspended from the pendulum bar of the first pendulum. Sensitivities better than 0.001 sec of arc per mm can be obtained in the recordings.

Fig. 7 shows a picture of an instrument of the ZÖLLNER type. When in use it is covered by an airtight cylinder. In Fig. 8 records of simultaneous measurements at two different stations taken with instruments of similar type are reproduced. The sensitivity on the original curves is in this case about 0.001 sec/mm.

[1] See A. ROMBERG: Bull. Seism. Soc. Amer. **21**, 224 (1931). — P. CALOI: Ann. di geofis. **3**, No. 4, 451 (1950). — W. F. BONCHKOVSKIJ: Akad. Nauk USSR, Trudy Geophys. Inst. **1949**, Nr. 5 (132), 49 (1949).

[2] For historical information see G. H. DARWIN [Sci. Papers **1**, No. 13, 414ff. (1881)] and W. F. BONCHKOVSKIJ [Akad. Nauk USSR. Trudy Seismolog. Inst. **1940**, No. 99]. The mathematics of the instrument are given by H. BOUASSE: Séismes et séismographes. Paris 1927. See also K. JUNG: Handbuch der Experimentalphysik, Bd. 25, 2, S. 324. (1928).

[3] H. LETTAU: Gerlands Beitr. Geophys. **51**, 250 (1937).

A modification of the usual ZÖLLNER type pendulum[1] has been constructed by ELLENBERGER[2]. The pendulum itself has the shape of a disk with an eccentric mass. The suspension wires are attached to the centre of the disk and are connected with the frame by short springs. Measurements with this instrument have not yet been published.

Fig. 7. ZOELLNER horizontal pendulum (type: SCHAFFERNICHT-TOMASCHEK).

β) The difficulties of these observations do not consist in insufficient sensitivity of the instruments but rather in uncontrollable, probably meteorologically or tectonically caused drifts and variations. If the wires or ribbons of the instruments are properly treated by long slight annealing under load and oscillations, any noticeable instrumental drift can be avoided. The often observed drifts are due either to insufficient treatment of the suspension or are real tilts of the stations. The influence of temperature is of course very disturbing, as elastic strains due to temperature variations are transferred to considerable depths in the Earth's crust. Modern stations have therefore been mostly chosen in deep mines, a depth below surface between 100 to 200 m being most suitable. Not every mine can be used, since rock movements caused by geological pressures often occur. Salt mines seem to have special advantages.

γ) *Results.* Due to the unavoidable disturbances harmonic analysis of the results is necessary. Those for the M_2 tide obtained in Europe and Asia between latitudes 30° to 56° and longitudes +10° W to 150° E are summarized in Fig. 9. There exist no published measurements yet of Earth tides with horizontal pendulums in America[3]. A Table of all the measurements with commenting data is given in the report of MELCHIOR [*1*]. The values show strong local variations. It is difficult to take a mean value as the observations have not the same weight. But the more extended measurements performed in mines suggest values of D between 0.55 to 0.8. It is obvious that the observed values contain disturbing influences, but there is no immediate evidence of a relation to oceanic disturbances. The problem of correcting the data will be treated in Chap. D and E.

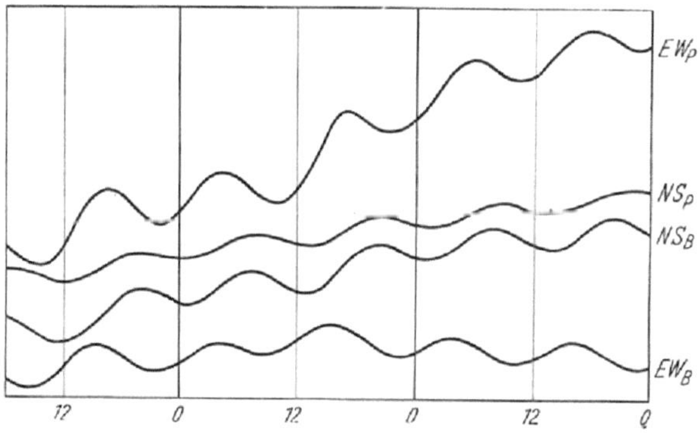

Fig. 8. Simultaneous records of horizontal Earth tides by horizontal pendulums at two stations (*P* Pillnitz, *B* Beuthen) (after GNASS). Size about 1/5 of the originals.

[1] W. SCHAFFERNICHT: Ann. d. Phys. **29**, 349 (1937). — G. GNASS: Z. Geophys. **16**, 1 (1940).

[2] H. ELLENBERGER: Z. Vermessungswes. **1954**, Sonderh. 2, 14.

[3] See F. NAPIER DENISON: Bull. Seism. Soc. Am. **28**, 39 (1938).

Examples of the best available results have already been given in Table 3, where the NS components of SCHWEYDAR'S measurements have been reproduced. Table 4 contains his 5 years' means of the harmonic constants for a number of constituents. The amplitudes are given as the ratios of the observed to the theoretical amplitudes, except for the meteorological term S_1. This table is an example of perhaps the best data at present available. An inspection of it shows

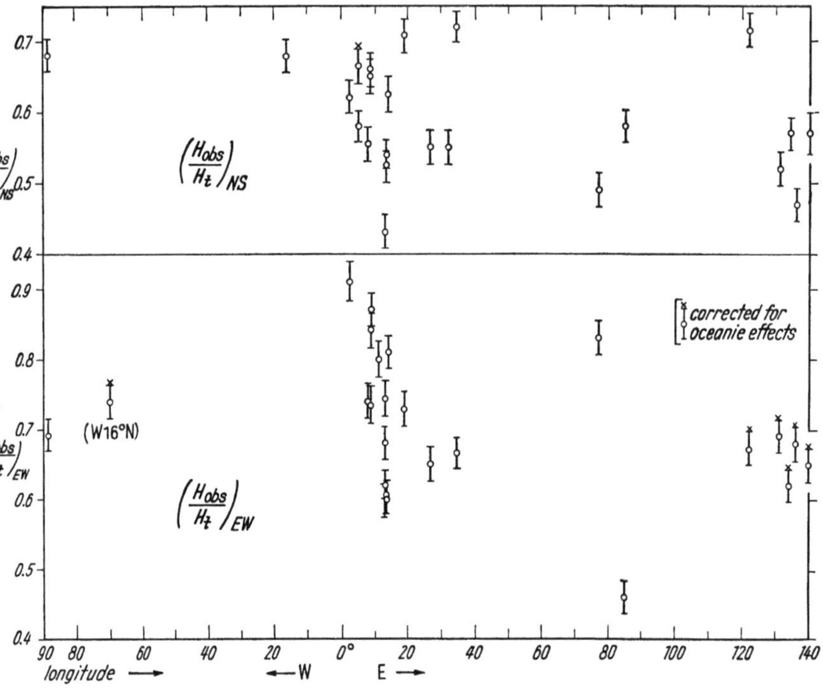

Fig. 9. Measurements of the Diminution Factor of M_2 by means of horizontal pendulums and pipes.

that D is by no means identical for the different constituents and also not for the different directions. It reveals the enormous difficulties in disentangling the complicated observations. Especially the European observations of D for M_2 show a systematic difference between the values in the direction of the meridian and the prime vertical, the latter values being higher than the meridional values.

Table 4. *5 yearly means of harmonic constants determined by horizontal pendulum observations at Freiberg at 189 m depth by* SCHWEYDAR [*4*].

Tide	West		North	
	H_{ob}/H_{th}	$\varkappa$	H_{ob}/H_{th}	$\varkappa$
M_2	0.607 ± 0.018	− 0°.8 ± 1°.6	0.540 ± 0.019	+ 12°.5 ± 1°.0
S_2	0.667 ± 0.048	− 4°.6 ± 3°.2	1.086 ± 0.050	− 31°.1 ± 3°.1
N_2	0.684 ± 0.052	− 7°.8 ± 4°.1	0.640 ± 0.060	+ 16°.9 ± 5°.8
K_2	0.675 ± 0.096	+ 2°.7 ± 7°.6	0.867 ± 0.122	− 8°.8 ± 8°.6
K_1	0.810 ± 0.018	+ 1°.6 ± 0°.3	0.502 ± 0.116	− 138°.9 ± 13°.1
O_1	0.788 ± 0.040	+ 0°.6 ± 4°.4	1.321 ± 0.291	− 186°.9 ± 5°.1
P_1	0.966 ± 0.068	+ 9°.5 ± 5°.0	1.630 ± 0.263	− 109°.0 ± 10°.0
S_1	0.57 millisec	+ 97°.0	2.07 millisec	− 91°.4

See Sect. 17b. But the observations of ORLOV[1] show the opposite behaviour for Poltava ($\lambda = 34°$ E) with $D_N = 0.80$ and $D_E = 0.69$ and for Tomsk ($\lambda = 84°$ E) with $D_N = 0.60$ and $D_E = 0.48$.

There exist only few observations of the *diurnal* tides, where O_1 is of special interest because, thanks to the large deviation of its period from a solar day, it should be least disturbed by temperature effects. The observed values are given in Table 5. The European values are definitely higher than the values obtained from the semidiurnal observations.

Table 5. *Ratios of observed to theoretical amplitudes of O_1 constituent by horizontal pendulums.*

Station	EW	NS	Observer
Freiberg (Sax.) . . .	0.788	1.321	SCHWEYDAR 1910—1915
Freiburg (Bad.) . . .	1.294	0.575	HAID 1907/1908
Durlach	1.439	0.602	HAID 1907/1908
Potsdam	0.64	0.91	HECKER 1902—1909
Barim	0.615	0.411	NISHIMURA 1941
Kamigamo	0.78 (corr)		NISHIMURA 1939

The long-period variations have been studied by EBLÉ[2] at Paris. Although he had a unique material of 72 lunar months at his disposal, the amplitudes obtained were about 3 times larger than theoretically expected. Even the East-West component [see Eq. (2.7)] showed amplitudes of the same magnitude. It is not known whether this result is caused by oceanic tides.

Summarizing the results of observations covering about half a century of enormous labour and skill, the results show that the ratio of observed to theoretical values is far from giving a definite value of D as expected. According to these observations D is somewhere near 0.7, a value which accords with the figures obtained by other methods, but numerous experiments have revealed the existence of strong local or regional influences which have to be taken into account and have to be investigated. They have a significance of their own as will be shown in Chap. E, F and G. It will be shown after their discussion that the most probable value derived from observations with horizontal pendulums is

$$D = 0.72.$$

8. Gravimetric factor $G = 1 + h - \frac{3}{2}k$, and a survey on gravimeters. This factor characterizes the influence of the yielding of the Earth on the variations of the vertical component of gravity. They can be observed by the variations in the indications of pendulum clocks and in variations of the weight of an elastically suspended mass in gravimeters.

a) Observations with pendulum clocks. A pendulum clock having a half period corresponding to

$$T = \pi \sqrt{\frac{l}{g}} \tag{8.1}$$

will change its period to T' by a change of gravity of dg

$$T' = \pi \sqrt{\frac{l}{g \mp dg}} = T \pm \frac{T}{2} \frac{dg}{g}. \tag{8.2}$$

From this

$$\frac{T' - T}{T} = \left| \frac{1}{2} \frac{dg}{g} \right| = \frac{\Delta t}{T} = \frac{1}{2g} \left(\frac{\partial W_2}{\partial r} \right)_{r=\varrho_1} \tag{8.3}$$

[1] A. ORLOV: Izv. Akad. Nauk USSR., Ser. geogr. i geofiz. **1939**, Nr. 1, 3—29.
[2] L. EBLÉ: Ann. Géophys. **6**, Nr. 2, 128 (1950).

and

$$\Delta t = \frac{1}{2}\int \frac{1}{g}\left(\frac{\partial W_2}{\partial r}\right)_{r=\varrho_1} dt = \int \frac{1}{g}\cdot\frac{W_2}{\varrho_1}\,dt, \tag{8.4}$$

where Δt is the correction of the pendulum clock. Expressing Δt in seconds gives

$$\Delta t^{(s)} = \frac{86400}{2\pi}\int \frac{1}{g}\,\frac{W_2}{\varrho_1}\,dt. \tag{8.5}$$

For the M_2 tide an oscillation of about $\pm 1.5\times 10^{-4}$ sec with a period of $12^{\mathrm{h}}\,25^{\mathrm{m}}$ is to be expected.

The first investigation of this kind was performed by BROWN and BROUWER[1] at Washington by comparing three SHORTT clocks with a quartz clock over a period of 147 days be means of a LOOMIS chronograph. They found a mainly diurnal period, probably due to temperature effects and a M_2 oscillation of a mean amplitude of $1.37(\pm 0.07)\times 10^{-4}$ sec compared with a *theoretical* value of 1.53×10^{-4} sec. It is

$$G' = Z_{\mathrm{obs}}/Z_{\mathrm{th}} = 0.90 \text{ (Washington)}.$$

As the observations by ZERBE show (Sect. 7c δ), an oceanic load tilt of about the same amplitude as the tidal tilt exists at Washington. It is not known how far this may influence the value of G'.

The same ratio, namely 0.92 has been obtained for Washington by using as comparison with the clocks the latitude observations performed by means of a photographic zenith tube by SOLLENBERGER and CLEMENCE[2]. This ratio involves the tidal latitude influence (Sect. 10) and it is not possible to separate those two effects in their observations.

HOFFROGE[3] has been able to measure the tidal variations of gravity by comparing two SCHULER pendulums with a quartz clock from two series of observations of only three days each, reaching an accuracy of about ± 0.05 millisec. The series of observation are too short for determining a reliable value of G.

The most extended comparison of pendulum and quartz clocks has been made by STOYKO[4] using 5 pendulum clocks and 4 years of observations. His mean results using two different methods of analysis I and II are given in Table 6a.

Table 6 a. *Gravimetric factors from comparison of pendulum clocks with quartz clocks at Paris* (STOYKO).

Year		1940	1941	1942	1943	Total mean
Mean of five clocks	I $G=$	1.25	1.14	1.38	1.19	1.21
	II $G=$	1.24	1.18	1.13	1.26	1.18
	$\varkappa=$	$-0^{\mathrm{h}}.84$	$+0^{\mathrm{h}}.93$	$-1^{\mathrm{h}}.13$	$-0^{\mathrm{h}}.38$	$-0^{\mathrm{h}}.41$
Type of clock		44 *Sh*	1185 *L*	1228 *L*	1229 *L*	1372 *L*
Mean of four years	I $G=$	1.24	1.13	1.17	1.27	1.27
	II $G=$	1.00	0.97	1.34	1.19	1.40
	$\varkappa=$	$-0^{\mathrm{h}}.12$	$-1^{\mathrm{h}}.10$	$-0^{\mathrm{h}}.11$	$-0^{\mathrm{h}}.16$	$-0^{\mathrm{h}}.55$

The total mean of these observations is

$$G' = 1.195 \pm 0.043 \text{ (Paris)};\quad \varkappa = -6^{\circ}.1.$$

[1] E. W. BROWN and D. BROUWER: Month. Not. Roy. Astronom. Soc. **91**, 575 (1931).
[2] P. SOLLENBERGER and G. M. CLEMENCE: Astronom. J. **48**, 78 (1939).
[3] CH. HOFFROGGE: Geofisica Pura e Applicata **13**, 118 (1948).
[4] M. N. STOYKO: Bull. Astronomique **14**, 1 (1949).

At the same place in Paris observations of the deflections of the vertical have been performed by EBLÉ from 1912 to 1917. They gave $D = 0.79$, which indicates a fairly normal behaviour of this station although the value is fairly high. It is not known how far the measurements are influenced by the Atlantic tidal loads.

b) *Observations with gravimeters.* The most extended measurements of G have been performed by gravimeters, especially since they have become a commercial instrument frequently used for prospecting, especially for oil, by determining the local gravity differences produced by the varying density of underground rocks.

α) The principle of all practically used gravimeters is to measure the variation in weight of a mass by means of the elastic deformation of a spring and measuring the corresponding changes of lengths or angles. As it is necessary to detect changes in weight of 10^{-8} or even 10^{-9} in tidal measurements, very delicate instruments have to be used. There exist more than a hundred ingenious suggested methods for the construction of gravimeters, but only a small number has been really developed for practical use[1] and only those which are most sensitive and constant are usable for tidal observations. The main problems involved are as follows: Using the lengthening of a helical spring by the mass involves the measuring of variations of 10^{-8} of the extension of the spring, if changes of gravity of 10^{-8} are to be determined. For practical reasons, especially in order to avoid temperature influences, the total extension will not exceed about 10 cm. Changes of about a millionth of a millimeter have therefore to be observed. Even then temperature has to be kept as constant as possible, that is within about 0.01° C or even less. Pressure variations influence the buoyancy and have either to be excluded or compensated. The invention of isoelastic alloys which show a change of YOUNG's modulus with temperature as small as $\pm 10 \times 10^{-6}$ per degree C made instruments of this kind practically possible.

β) Besides the linear systems which use the extension or rotation of a simple helical or spiral spring and have to rely on the measurement of extremely small changes of lengths or angles, there exists another arrangement for the detection of small changes of weight. It consists of using devices which have a theoretically unlimited sensitivity. This is achieved by nearly compensating the elastic force by another labilizing force so that conditions nearing an unstable equilibrium are approached.

The sensitivity of a static system can be expressed by its natural period of oscillation. The longer the period, the more sensitive it is. If f is the elastic force and m the mass, a linear system will have a period of

$$T = 2\pi \sqrt{m/f}\,. \tag{8.6}$$

It is possible to develop systems where the elastic force or moment, characterized by a constant f, is counteracted by an adjustable "labilizing" force or moment, characterized by a constant a. Then the natural period of this "*astatized*" *system* will be

$$T_a = 2\pi \sqrt{m/(f - a)}\,. \tag{8.7}$$

If $f = a$, the system is in unstable equilibrium and shortly before this state is reached, it becomes extremely sensitive. The periods of commercial astatized gravimeters having a reading accuracy of about 0.01 milligals are in the order of

[1] See for example: O. MEISSER: Z. Geophys. **11**, 221 (1935). — A. GRAF: Gravimeter. Arch. techn. Messen V. 651–1 to 651–6 (1943); V. 651–7, Liefg. 202 (1952) or L. NETTLETON: Geophysical Prospecting for Oil. New York 1940.

4 to 12 sec, those of linear systems are of about 1 sec. For purposes of tidal observations periods of 30 to 40 sec have been used.

The principle of astatization may be explained as shown in Fig. 10a and b. A mass m which is affixed to a lever is held by torsion in a horizontal position as in Fig. 10a. The equation for equilibrium is

$$f(\gamma+\alpha)_{\alpha=90^\circ} = m\,l\,g \tag{8.8}$$

and from this

$$d\alpha = \frac{f}{m\,l}\,dg. \tag{8.9}$$

For small changes of gravity the effect is proportional to the variation of gravity, the system is linear.

If, on the other hand the mass and the torsional angles are adapted as shown in Fig. 10b then

$$f(\gamma+\alpha) = m\,g\,l\sin\alpha. \tag{8.10}$$

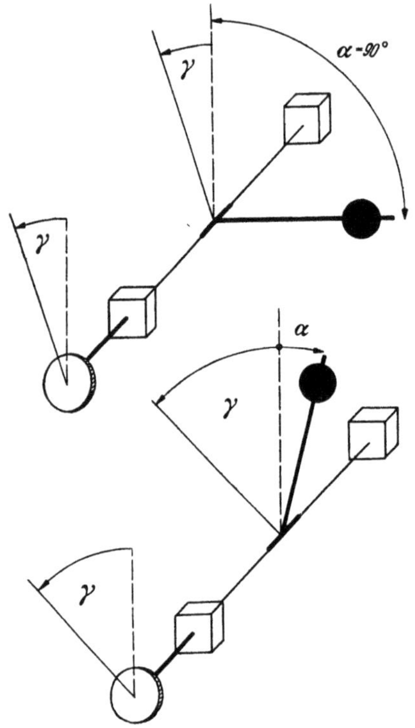

Fig. 10. Principle of linear (a) and astatic (b) gravimeter.

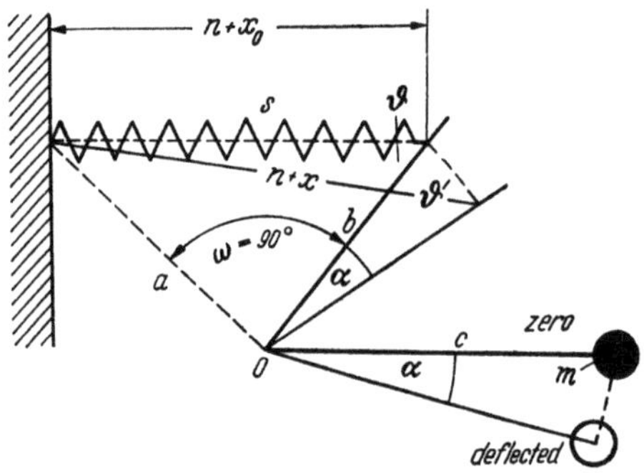

Fig. 11. Astatization of GALITZIN-type gravimeters.

A decrease in gravity now decreases not only α, but also the momentum $ml\sin\alpha$. There will exist a limiting angle at which the pendulum topples over. As from (8.10)

$$\frac{dg}{d\alpha} = \frac{f}{m\,l}\left(\frac{\sin\alpha-(\gamma+\alpha)\cos\alpha}{\sin^2\alpha}\right) \tag{8.11}$$

the astatic position is given by

$$\sin\alpha = (\gamma+\alpha)\cos\alpha \tag{8.12}$$

or

$$\gamma = \operatorname{tg}\alpha - \alpha \tag{8.13}$$

when $d\alpha/dg$ becomes infinite.

Most of the systems in practical use derive from the principle of GALITZIN'S seismograph. It is shown in Fig. 11. The mass m is affixed to a lever of length c, which has its axis at 0. This lever is rigidly connected to another lever of length b, which is held by a spring of length $n+x_0$, where n is the length of the unloaded spring. The length of $n+x_0$ may be chosen in such a way that $\omega=90^\circ$ in the zero position of the instrument. An increase in gravity moves the mass to its lower position, rotating the rigid lever system through the angle α and extending the spring to the length $n+x$. The angles ϑ and ϑ' are those between the lever b and the direction of the axis of the spring in the zero position and its changed place. The elastic constant of the spring is f. It is

$$m\,g\,c \cdot \cos\alpha = f\,x\,b \cdot \sin\vartheta' \tag{8.14}$$

further

$$\frac{a}{n+x} = \frac{\sin\vartheta'}{\sin(\omega+\alpha)} = \frac{\sin\vartheta'}{\cos\alpha} \tag{8.15}$$

then

$$m\,g\,c = \frac{f\,a\,b\,x}{n+x}. \tag{8.16}$$

As

$$\frac{dg}{g} = \frac{n}{x\,(n+x)}\,dx \tag{8.17}$$

and

$$(n+x)^2 = a^2 + b^2 + 2\,a\,b\sin\alpha \tag{8.18}$$

it is

$$dx = \frac{a\,b\cos\alpha}{n+x} \cdot d\alpha \tag{8.19}$$

and

$$d\alpha = \frac{x\,(n+x)^2}{n\,a\,b\cos\alpha} \cdot \frac{dg}{g} \tag{8.20}$$

or for the case $\alpha \to 0$, where also $x \to x_0$

$$d\alpha = \frac{x_0(x_0+n)^2}{n\,a\,b} \cdot \frac{dg}{g}. \tag{8.21}$$

Therefore the sensitivity can be made infinite if $n=0$ and ω is exactly 90°. The case $n=0$ can be obtained by using "zero" springs, which are produced by feeding the wire under tension at an angle when coiling the simple helical spring. By this the windings are compressed and a certain minimum weight is necessary to extend the spring.

Generally ω will deviate from 90° by a small angle ε. Then from (8.16)

$$m\,g\,c \cdot \cos\alpha = \frac{f\,x\,b\,a \cdot \cos(\alpha+\varepsilon)}{n+x} \tag{8.22}$$

and Eq. (8.20) is now

$$d\alpha = \frac{x\,(x+n)^2}{n\,a\,b - \varepsilon\,x\,(x+n)^2} \cdot \frac{dg}{g}. \tag{8.23}$$

Then infinite sensitivity is achieved for

$$n\,a\,b = \varepsilon\,x\,(x+n)^2. \tag{8.24}$$

The instruments are constructed in such a way as to make possible a change of ω by a screw and to adapt the natural period of the instrument to high values empirically.

As the angle between the levers c and b does not enter into the calculation, the spring must not necessarily be horizontal. It may have any direction, provided always that the angle covered by it, when viewed from the axis, is about 90°. But the axis and the mass must be in a horizontal plane in order to make the instrument as little sensitive to inclinations as possible.

Practically all the astatic instruments have to be used with a *reading spring*. This is a very fine spring, much weaker than the main spring, which brings the mass back to its zero position at each reading. The gravity values are then determined by reading the movements of this spring. If this is not done the nonlinearity of these instruments has to be determined by suitable calibration.

γ) Types of instruments. 1. *Bifilar gravimeter.* This instrument invented by SCHMIDT[1] was the first to be used in detecting tidal variations of gravity (SCHWEYDAR 1912[1]). In an improved form it has been extensively used for such

[1] AUG. SCHMIDT: Gerlands Beitr. Geophys. **4**, 109 (1900). — W. SCHWEYDAR: Sitzgsber. Akad. Berlin I, **1914**, 454 (1914). — R. TOMASCHEK and W. SCHAFFERNICHT: Ann. d. Phys. (5) **15**, 787 (1932).

measurements with photographic registration (TOMASCHEK-SCHAFFERNICHT 1932[1]). It is of the astatic type working as follows:

The mass is suspended both by means of a helical spring (of *WT* 10 alloy) and, as Fig. 12 shows, by a bifilar suspension. If the spring is rotated by the angle α_1 the bifilar suspension will revolve through the smaller angle φ. This can be continued till ϱ becomes larger than 90°. As the torsional momentum of the spring steadily increases, the opposite torsional momentum of the bifilar suspension eventually reaches a maximum and begins to decrease when φ exceeds a certain value which in practical cases is about 145°. It is larger than 90° as the part p of the total weight P which is supported by the bifilar suspension also increases with increasing φ. Each variation of the weight can be

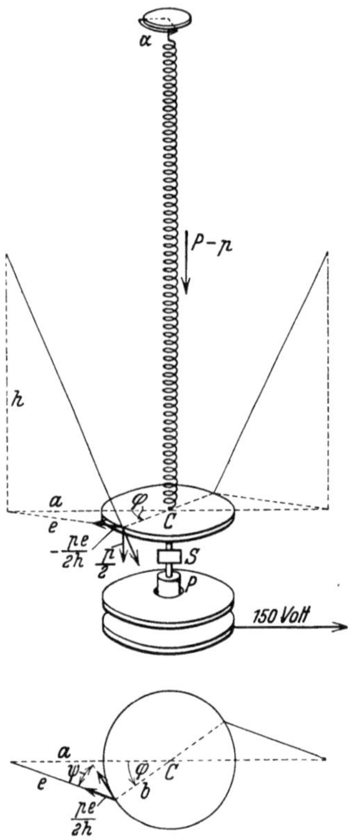

Fig. 12.

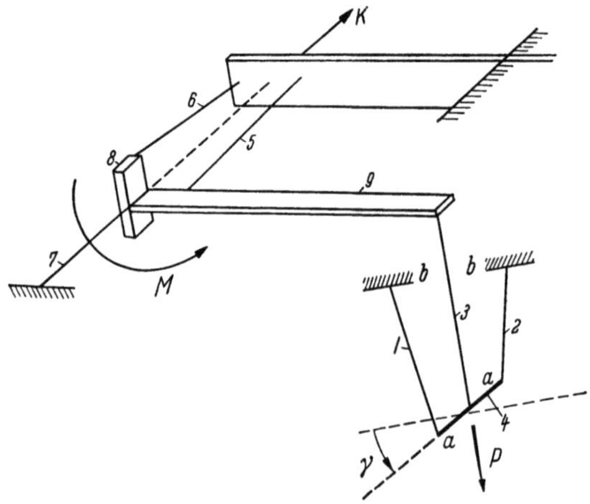

Fig. 13.

Fig. 12. SCHMIDT Bifilar Gravimeter, with electrostatic calibration after TOMASCHEK-SCHAFFERNICHT. *C* disk, *P* mass, *S* mirror. p weight suspended by bifilar wires.

Fig. 13. Principle of double bifilar gravimeter by ELLENBERGER.

measured by the rotation of the system which becomes very sensitive near the point of instability. With the symbols given in Fig. 12 and f being the elastic constant of the spring and

$$s = \frac{a\,b\,p}{h} \cdot \cos\varphi + f \tag{8.25}$$

the instrument acts according to

$$\frac{dg}{g} = -\frac{h\,s}{a\,b\,P \sin\varphi} \cdot d\varphi . \tag{8.26}$$

Natural periods of 30 to 40 sec have been used with this instrument which is capable of photographic recording. An electric condenser is attached to the instrument for calibration purposes. The theory of this instrument has been further developed especially in respect of high sensitivity and linearity by VOIT[2].

2. A Double Bifilar Gravimeter has been constructed by ELLENBERGER[3] especially for tidal measurements, having a small size and a wide range of linearity.

[1] See footnote 1, p. 801.
[2] H. VOIT: Geofisica Pura e Applicata **15**, 1 (1949).
[3] H. ELLENBERGER: Veröff. dtsch. geodät. Komm., Reihe C **1952**, Nr. 2.

Fig. 13 shows the principal arrangement. The mass 4 is partly suspended by two wires 1 and 2 of a bifilar arrangement. The greater part of the weight is taken up by a second horizontal bifilar system instead of a helical spring. The wires 5, 6, and 7 of this system are stretched by the force k of a springy element and transfer a momentum M on the part 8 to 9 which stretches the suspension 3.

A double bifilar gravimeter of a different type has also been constructed by ICHINOHE[1], using two very fine wires for connecting the spring with the disk C (Fig. 12).

3. Another gravimeter which has been extensively used for tidal measurements[2] is the FROST Gravimeter (Fig. 14). As the figure shows it is of the GALITZIN type with a main spring inclined by about 45° to the horizontal. It has automatic buoyancy compensation and a double thermostat. It has to be read visually which limits its application to Earth tide measurements. The main difficulty, as in all instruments which are not completely airtight, is the compensation of pressure variations. Furthermore, in practically all instruments the fluctuations of temperature caused by the thermostats give rise to apparent oscillations of several hours period.

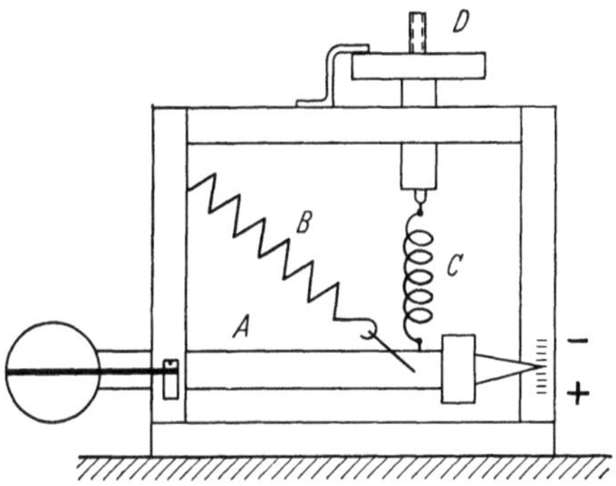

Fig. 14. Schematic diagram of FROST Gravimeter.

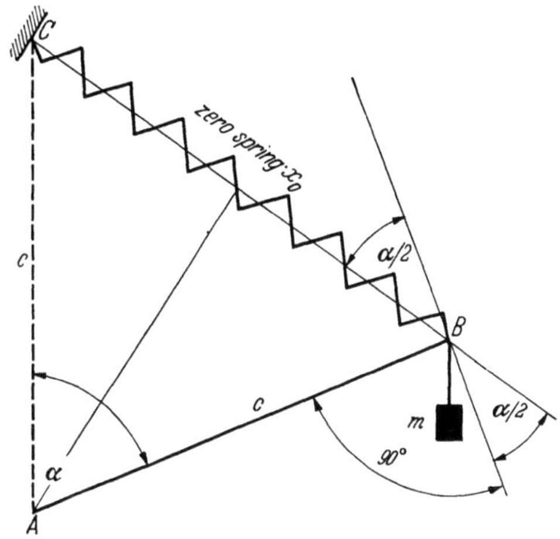

Fig. 15. Principle of LACOSTE-ROMBERG Gravimeter.

4. The gravimeter by LACOSTE-ROMBERG is very well developed and has a large range of linearity. Modern models print the readings of the instruments automatically. The principle which has been given by REINGANUM (1907) is shown in Fig. 15. In this case the length of the lever c which supports the mass is equal to the distance of the fixed end of the spring from the axis. Then, using a zero spring ($n = 0$) having an elastic constant f, it is

$$m g c \cdot \sin\alpha = c f x_0 \cdot \cos\frac{\alpha}{2}. \tag{8.27}$$

As

$$\sin\frac{\alpha}{2} = \frac{x_0}{2} \cdot c \quad \text{and} \quad \sin\alpha = 2\sin\frac{\alpha}{2}\cos\frac{\alpha}{2},$$

this becomes

$$m g = c f \quad (\text{independent of } \alpha). \tag{8.28}$$

Near to this position, the instrument can be kept very sensitive with little variation of its natural period. See also paragraph 9, p. 806.

[1] T. ICHINOHE, Mem. Coll. Sc. Univ. Kyoto, Ser. A **27**, No. 3, 289 (1955).

[2] L. H. TARRANT: Month. Not. Roy. Astronom. Soc., Geophys. Suppl. **6**, 278 (1952). — R. TOMASCHEK: Month. Not. Roy. Astronom. Soc., Geophys. Suppl. **6**, 286, 372 (1952). — Nature, Lond. **175**, 937 (1955).

5. Other well developed and reliable commercial gravimeters have occasionally been used for tidal gravity work; examples are the North American, Western, HUMBLE and HEYLANDT Gravimeters which are of similar type to those described above. Two other types use quartz springs, the MOTT-SMITH Gravimeter and the WORDEN Gravimeter. Particularly the latter represents a different type as it has a system built entirely from quartz (Fig. 16a). This has many advantages as all connections can be fused together. It has no thermostat, but uses a vacuum flask for insulation purposes and the system itself is enclosed in an airtight container. The disadvantage for tidal measurements[1] lies in its strong steady *drift* which is not sufficiently constant for tidal work although it is quite satisfactory for the usual practical purposes[2]. Generally it can be said that modern metal springs seem to be far superior to quartz springs for tidal work.

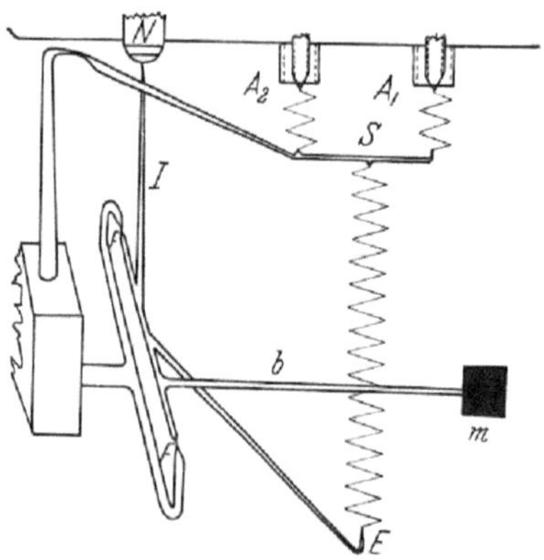

Fig. 16a. Schematic diagram of the WORDEN Gravimeter.

6. Of the linear gravimeters the *Interference Gravimeter* (TOMASCHEK-SCHAFFERNICHT[3]) has been used for tidal measurements. (Fig. 16b). The mass of a mirror is employed as loading element. This is suspended by a simple helical spring and moves in a direction perpendicular to its surface and opposite to a total reflecting prism whose surface is parallel to the mirror. The interference fringes of a mercury lamp are photographed continuously.

The absolute linearity of instruments of this type is of great advantage when the changes of force are so small as in tidal measurements and photographic records over long periods with stronger drifts are intended.

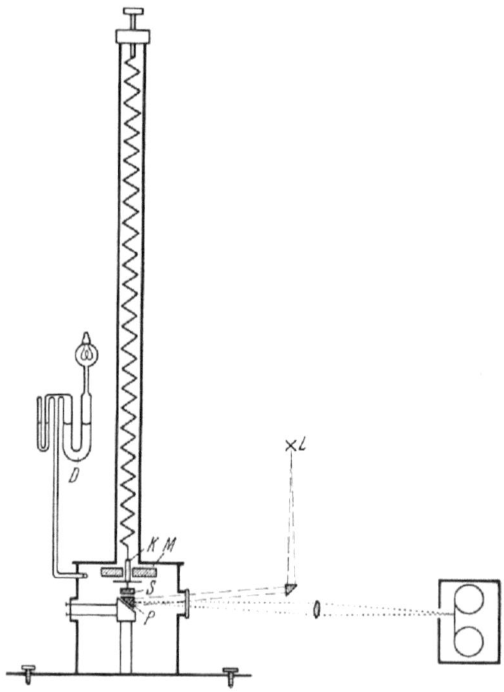

Fig. 16b. Interference Gravimeter

The Multiple-Reflection Gravimeter (TOMASCHEK-SORBER[4], HOYT-Gulf[5]) is of the same linear type and works by the rotation of an extended helical spring using multiple reflexion between the mirror of the instrument and a fixed mirror or that of a second spring which moves in the opposite direction. In order to

[1] R. TOMASCHEK: Month. Not. Roy. Astronom. Soc., Geophys. Suppl. **1957**.
[2] G. P. WOOLARD: Geophysics **15**, No. 1, 1 (1950).
[3] R. TOMASCHEK and W. SCHAFFERNICHT: Astronom. Nachr. **248**, No. 5929 (1933).
[4] H. SORBER: Phys. Z. **37**, 599 (1936). — R. TOMASCHEK: Naturwiss. **25**, 179 (1937).
[5] A. HOYT: USA Patent No. 2 131 737. — R. D. WYCKOFF: Geophysics **1**, No. 1 (1941).

obtain stronger rotation for a certain lengthening a helical ribbon is used for making the spring.

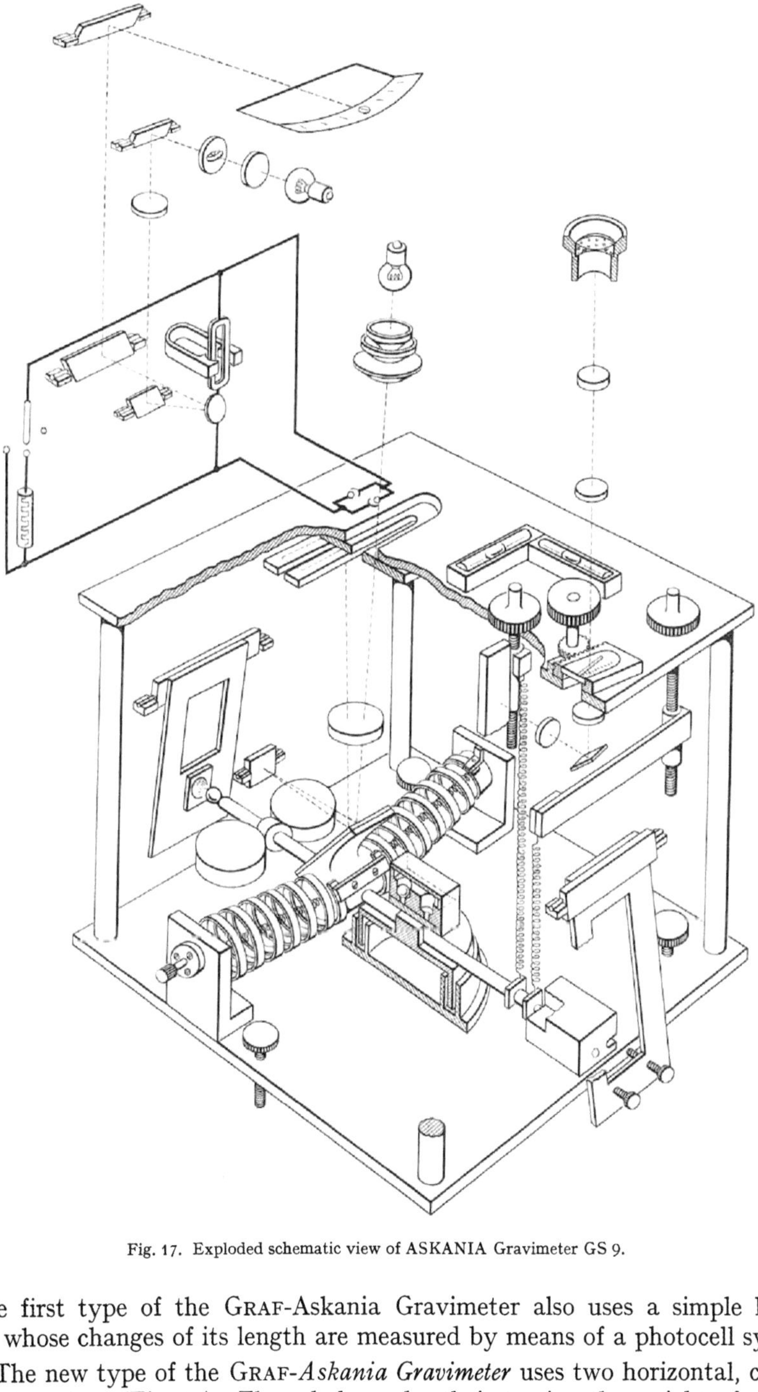

Fig. 17. Exploded schematic view of ASKANIA Gravimeter GS 9.

The first type of the GRAF-Askania Gravimeter also uses a simple helical spring whose changes of its length are measured by means of a photocell system.

7. The new type of the GRAF-*Askania Gravimeter* uses two horizontal, conical torsional springs (Fig. 17). These balance by their torsion the weight of a mass

in a way similar to the system shown in Fig. 10. The mass is rigidly connected with a mirror. The very small optical deflections caused by the changes in gravity are taken up by a differential photocell system, where they are transformed into electrical variations which can be easily registered.

8. As many of the instruments have been developed for underwater measurements with distant indication, such devices, given a sufficiently sensitive gravimeter, can be easily adapted for recording tidal measurements[1]. Fig. 18 shows a record of Earth tides by an Askania Gravimeter[2] where the wax recorder is driven by a motor which itself is steered by a photoelectric device, which magnifies the actual movement of the pointer.

9. An improved instrument for measurement of tidal variations of gravity has been quite recently developed by CLARKSON and LACOSTE[3]. It has the necessary provisions for making continuous measurements at least ten times as accurate as regular gravity meters. This has been achieved by increasing the amplification of the measuring system of the former LACOSTE Gravimeter by a factor of 16 and using a servomechanism and selsyn to operate it as a selfrecording system. A special thermostat holds the temperature of the heater box to within approximately 0.001° C. The box is sealed to remove the influence of barometric pressure variations. An outside heaterbox has been added to maintain the ambient temperatures to within a few tenth of a degree Centigrade. As a reading system comparison photocells are used. One photocell receives its illumination by means of an optical path including a concave mirror on the end of the beam of the gravimeter. This light passes through a small slit so that any motion of the beam results in a change in this illumination. The drift of the meter is approximately 30 microgals per month.

δ) Limitations. The gravimeters are less affected by local conditions than tiltmeters are, but they are seriously limited by pressure and temperature effects in tidal work, as this needs about twenty times higher accuracy than is usual in exploration work. If the gravimetric system is not perfectly sealed off, pressure effects are difficult to eliminate or to determine. The calibration in pressure chambers may give quite erroneous values, even regarding the sign of the correction, as adiabatic effects and thermal airstreams inside the gravimeter are quite troublesome. As to temperature influences, outer temperature variations reflect considerably inside the instrument, even if double thermostats are used which themselves often show oscillations characteristic of cybernetic systems. There always exist elastic hysteresis drifts which may last for several months, especially if the instrument has to be set up at a point whose absolute gravity value differs greatly from that of a previous station. These drifts are also influenced by outer temperature variations and therefore often variable. Measurements in mines with constant temperature are therefore always preferable. On the other hand older instruments are often surprisingly free from elastic drift and it may be less than 0.1 milligals per month.

ε) Results. 1. The observations may be divided into two classes, those which have been harmonically analyzed and those where an overall comparison of the observed and theoretical equilibrium values has been made. The first class is generally more reliable, as harmonic analysis, especially of the M_2 constituent, removes some of the pressure and temperature influences and generally the

[1] B. BREIN: Dtsch. Geodät. Komm., Reihe B **1954**, Nr. 21. See also P. LECOLAZET: Ann. Géophys. **12**, Nr. 1, 59 (1956).

[2] R. SCHULZE: Askania Warte No. 49, 16 (1956).

[3] H. N. CLARKSON and L. J. B. LACOSTE: Trans. Amer. Geophys. Un. **37**, No. 3, 266 (1956).

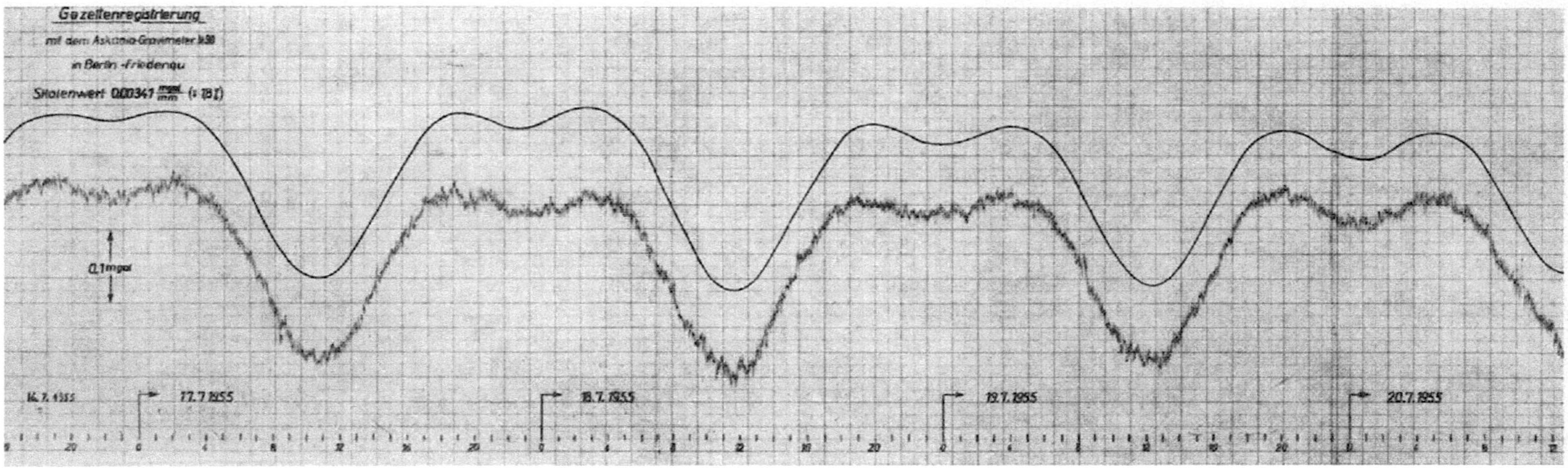

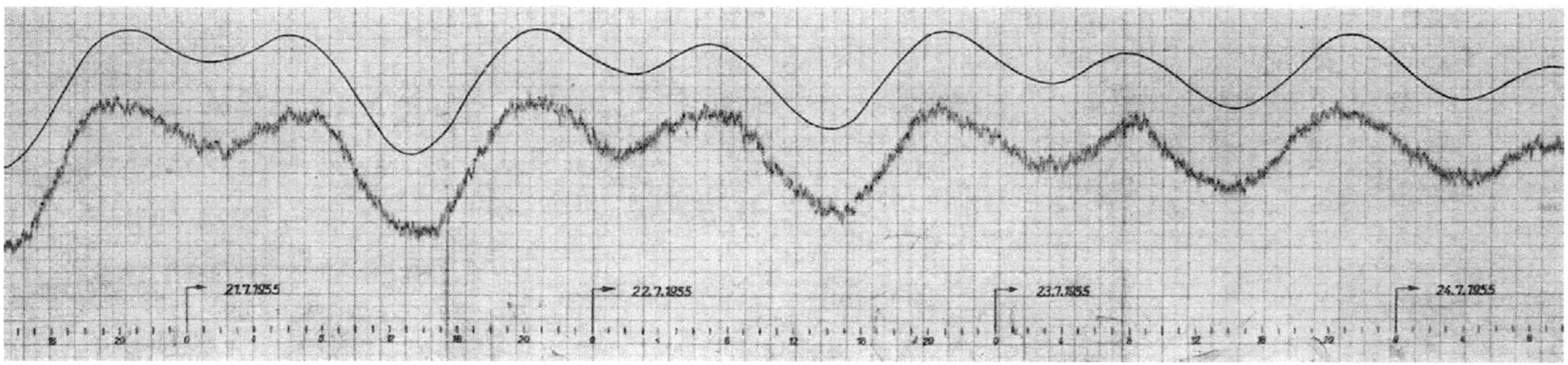

b

Fig. 18a and b. Record of Earth tides by the ASKANIA Gravimeter GS 11. Smooth curve = calculated tide.

periods of observation are longer. The second class makes no distinction between semidiurnal and diurnal constituents which theoretically could give different ratios of observed to theoretical amplitudes (Sect. 6c). In this case observed and theoretical ranges are compared, also theoretical and observed times of maximum and minimum. It has to be taken into consideration that the observed range derived from a study of maxima and minima is larger than one deduced from harmonic analysis (LAMBERT [I] 1955, Appx. II):

If the gravity values are given by

$$Z = A\cos(at+\alpha) + B\cos(bt+\beta) + C\cos(ct+\gamma) + \dots \tag{8.29}$$

and A is larger than the other amplitudes, then the mean semirange is given by

$$Z_m = A + \frac{1}{4Aa^2}\left[Bb^2 + Cc^2 + Dd^2 + \cdots\right]. \tag{8.30}$$

If non-tidal influences are present the right side of this equation has further to be multiplied by a factor > 1 in order to take account of accidental irregularities (say 1.02 in the case of sea tides). The two methods may therefore differ by several percent.

2. The observations made by gravimeters are quite numerous although unequal in accuracy. The reader may refer to the lists given by LAMBERT [*1*] and MELCHIOR [*1*].

The only values available regarding extended records by gravimeters are by LECOLAZET (1956) whose provisional values[1] are partly reproduced in Table 6b. His results for the whole period of observation are:

M_2:	$G' = 1.212 \pm 0.005$;	$\varkappa' = +0.°11 \pm 0°25$
S_2:	1.231 ± 0.014;	$-2.°9 \pm 0.°8$
N_2:	1.224 ± 0.025	$+1.°5 \pm 1.°2$
K_1:	1.201 ± 0.01	$-2.°17 \pm 0.°5$
O_1:	1.176 ± 0.01	$-2.°04 \pm 0.°5$

It has to be taken into account that as far as the absolute values are concerned there exists an uncertainty of the calibration of about $\pm 1\%$.

As Table 6b shows the variations of gravimeter measurements are very much less than those of the horizontal pendulums as shown in Table 2 which are very much more affected by local conditions.

Table 6b. *Monthly means of harmonic constants of gravimeter observations at Strassbourg* (LECOLAZET).

Mean date	M_2		S_2		N_2		K_1		O_1	
	G'	$\varkappa'$	G'	$\varkappa'$	G'	$\varkappa'$	G'	$\varkappa'$	G'	$\varkappa'$
3. 11. 1954	1.23	+0.°17	1.22	−2.°56	1.22	+0.°56	1.20	−1.°88	1.20	−2.°77
24. 11.	1.21	−0.°21	1.20	−2.°41	1.24	−0.°26	1.18	−1.°45	1.18	+0.°02
15. 12.	1.22	−0.°36	1.21	−1.°82	1.17	+1.°14	1.18	−1.°81	1.17	+0.°31
5. 1. 1955	1.19	+0.°57	1.21	−2.°23	1.18	−2.°43	1.17	−1.°41	1.17	−1.°56
26. 1.	1.21	+0.°20	1.22	−3.°61	1.15	−2.°31	1.22	−2.°04	1.15	−3.°25
16. 2.	1.20	+0.°73	1.23	−3.°43	1.22	+1.°01	1.22	−1.°59	1.19	−2.°66
9. 3.	1.19	+0.°19	1.27	−3.°64	1.39	+2.°81	1.26	−3.°22	1.16	+0.°31

[1] Information by Prof. LECOLAZET to the author. Extended and definite values will be published in 1957.

A simultaneous series of worldwide observations has been undertaken from May 9 to 23, 1949 with 26 stations participating[1]. Fig. 19 gives the values of the ratios of observed to theoretical values for the M_2 component of those stations whose results have been subjected to harmonic analysis. It includes also some other stations of the same type. A standard error of $\pm 4\%$ has been assumed in this drawing, as experience shows that careful measurements of this type due to calibration, pressure and temperature influences may attain this accuracy.

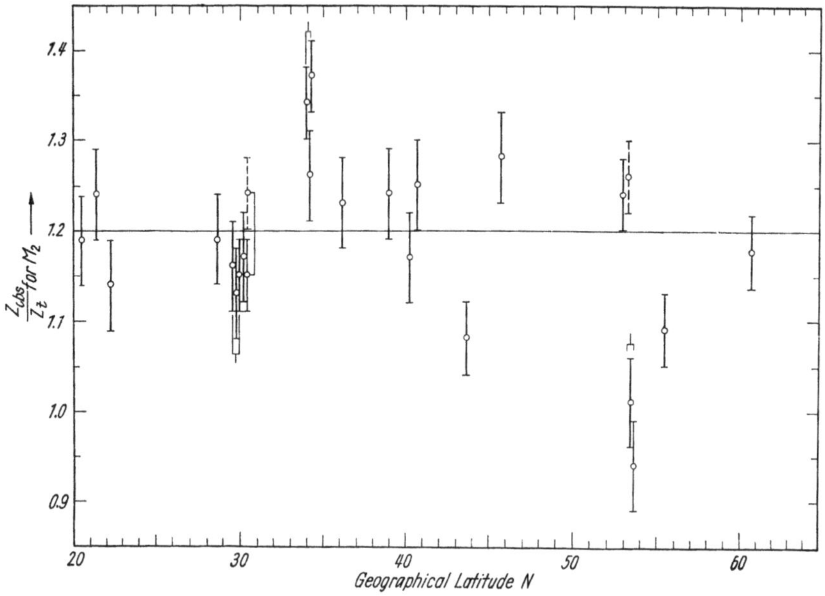

Fig. 19. Gravimetric factor for M_2 according to observations after 1948.

Table 7 includes also the values for the constituents S_2, O_1 and K_1.

The mean values derived by LAMBERT from a discussion of the results are:

	$(Z_{obs}/Z_{th})_m$	$\varkappa_m$
M_2:	1.191 ± 0.012	$1^\circ.99 \pm 0^\circ.95$
S_2:	1.213 ± 0.022	$4^\circ.92 \pm 1^\circ.41$
K_1:	1.192 ± 0.026	$-0^\circ.84 \pm 0^\circ.86$
O_1:	1.178 ± 0.016	$+0^\circ.70 \pm 0^\circ.77$.

The final result is given as

$$\text{semidiurnal tides: } G = 1 + h - \tfrac{3}{2}k = 1.198;\ +2^\circ.97$$

$$\text{diurnal tides: } 1.186;\ -0^\circ.22.$$

It is questionable whether the difference between the semidiurnal and diurnal values is significant. Furthermore the phase difference which gives information on possible friction effects (Sect. 15), is very small for both classes.

[1] B. BAARS: Geophysical Prospecting 1, No. 2, 82 (1953).

Table 7. *Ratio of observed to theoretical amplitudes G' and phase lags ϰ' of different constituents of gravimetric earth tide observations.*

No.	Stations		M_2		S_2		O_1		K_1	
	Lat. N	long. W	G'	$ϰ'$	G'	$ϰ'$	G'	$ϰ'$	G'	$ϰ'$
1	20°28′	89°43′	1.19	− 5°.3	1.29	+10°.9	1.28	− 2°.9	1.78	− 0°.9
2	21°18′	157°49′	1.24	+11°.2	1.36	+17°.0	1.56	− 0°.8	1.30	−10°.4
3	22°10′	98°10′	1.14	+ 1°.9	1.17	+13°.9	1.16	+ 2°.9	1.35	− 2°.4
4	29°42′	95°26′	1.16	+ 1°.5	1.17	− 7°.6	1.15	− 1°.6	0.89	− 3°.8
5	29°42′ II	95°26′ II	1.15	− 4°.0	1.16	− 8°.7	1.16	− 6°.6	1.21	− 4°.5
6	30°04′	98°12′	1.17	− 0°.6	1 32	− 2°.8	1.26	+ 0°.3	1.24	− 7°.2
7	30°17′	97°44′	1.15	+ 2°.1	1.10	+ 6°.3	1.05	− 0°.3	1.16	+ 1°.1
8	34°04′	118°26′	1.34	+10°.3	1.37	+12°.4	1.37	+ 4°.8	1.44	− 0°.5
9	34°04′ II	118°26′ II	1.37	+10°.8	1.40	+ 7°.9	1.36	+ 2°.2	1.51	− 0°.5
10	34°09′	118°10′	1.26	+ 7°.4	1.30	+ 5°.1	1.24	+ 0°.6	1.38	+ 3°.8
11	36°09′	95°54′	1.23	+ 2°.9	1.54	−11°.1	1.11	− 3°.1	1.36	−19°.3
12	38°54′	77°02′	1.24	− 2°.3	1.07	+ 1°.7	1.17	+ 0°.8	1.18	− 6°.2
13	40°15′	103°49′	1.17	+ 3°.9	1.01	− 4°.3	1.17	− 5°.1	1.13	− 2°.6
14	40°46′	111°51′	1.25	0°.0	1.07	+ 8°.9	1.06	−21°.2	1.39	+ 8°.4
15	43°40′	79°24′	1.08	+ 2°.6	1.11	+ 8°.0	1.04	+ 0°.9	0.97	− 2°.0
16	45°44′	75°43′	1.28	+ 4°.8	1.42	+ 5°.2	1.26	+ 2°.9	1.17	+ 2°.6
17	53°07′	00°59′	1.22	−11°.0	1.29	− 7°.0	1.11	+ 6°.0	1.07	+ 3°.0
18	53°33′	113°29′	1.01	+ 5°.5	1.37	− 6°.4	1.15	− 7°.6	1.13	− 0°.3
19	53°33′ II	113°29′ II	0.94	− 2°.4	1.47	+ 4°.7	1.32	− 1°.5	1.06	− 7°.1
20	55°39′	03°11′	1.08	− 8°.0	1.07	−12°.0	1.13	+12°.0	1.06	− 1°.0
21	60°46′	00°51′	1.18	+14°.2	1.20	− 5°.8	1.06	+ 3°.5	1.10	− 6°.1

BAARS finds the following mean values from the 1949 campaign by comparing the observed and theoretical *ranges*

$$G = 1.24; \quad ϰ = -0°.48.$$

LAMBERT discusses 10 tropical stations of this campaign and obtains from using the larger maxima and minima

$$G = 1.23; \quad ϰ = 0.$$

The somewhat larger values of these range calculations are in good agreement with the remarks made above [Eq. (8.30)] on the expected differences in both methods, so that the final values as given above seem to be quite reliable, as far as it is possible to judge from 14 days series of observations.

As no phase difference for the bodily Earth tides is to be expected (Sect. 15), from this point of view it could be assumed that the smaller the observed phase difference is the greater is the weight that can be attached to an observation. Doing this yields as a mean for M_2

$$G_{M_2} = 1.163$$

a value which comes fairly near to JEFFREY'S expectation for semidiurnal tides, if the movement of the core is taken into account (Sect. 13).

A detailed discussion of the observations shows that there exist individual variations which greatly exceed the purely experimental limits. There are definite indications on several stations (see for example the high Californian values No. 8, 9, 10 in Table 7) that these differences are not instrumental. The situation is similar to that discussed in Sect. 7 about the variations of the values of D and the problem will be treated in detail in Chap. D, E and F.

The value of G obtained by gravimetric measurements agrees very well with that obtained by Stoyko from clock observations, so that the value of

$$G = 1.20$$

seems to be well founded.

η) Determination of the total vector. Simultaneous measurements of the vertical and horizontal components of gravity offer the possibility of determining not only the deviations of the vertical and the tilt of the Earth's surface connected with them, but also the variations in height of the point of observation. Such a combination of measurements with gravimeters and horizontal pendulums makes the reconstruction of the actual movements of the Earth's crust possible. Their main value is that they make possible an analysis of the tilt movements of secondary or indirect character in respect to their origin, as the absolute vertical up and down movements are given by the gravimetric results. But in order to extract these effects the primary tidal effects have to be determined first.

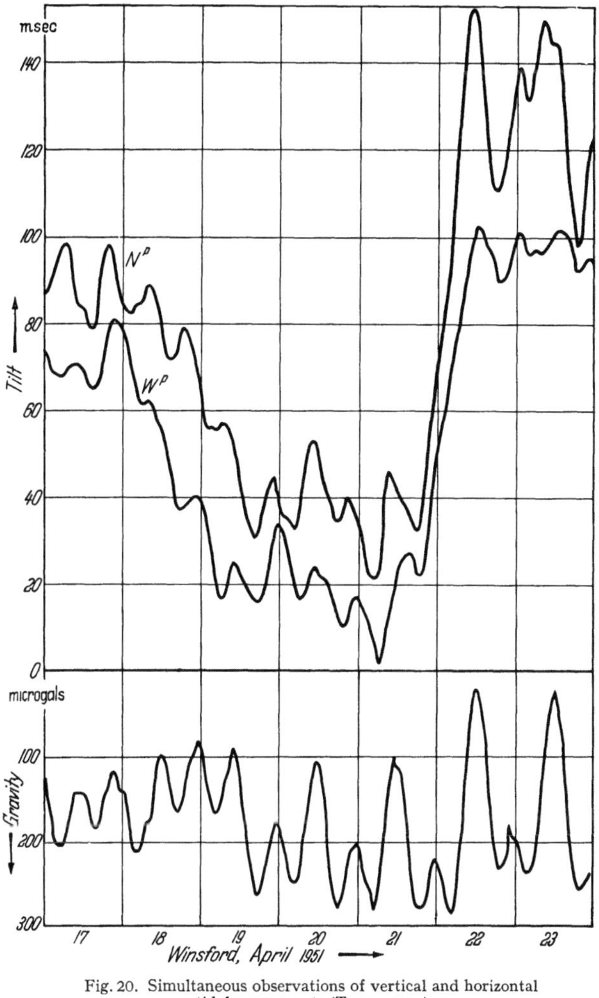

Fig. 20. Simultaneous observations of vertical and horizontal tidal components (Tomaschek).

Fig. 20 shows the records of such a simultaneous investigation. In addition to the tidal variations (which are included in Fig. 20) a strong drift displacement is shown which occurs not only in tilt but also simultaneously in the gravimetric curve. Fig. 21 shows the theoretical variations of the total gravity vector in the upper part and the observed tidal portion in the lower part, whereas the residuals are reproduced in Fig. 35 (Sect. 23). In this case the values of D and G were respectively 0.72 and 1.20. This yields $h = 0.44$. According to (5.4) h describes the ratio between the height of the tide of the solid Earth and the theoretical equilibrium height. The latter is given according to „Gezeitenkräfte", p. 740 (7.8) and (7.9) to

$$\zeta_m = 26.7 \text{ cm} (\cos\vartheta_m + \tfrac{1}{3})$$

and

$$\zeta_s = 12.3 \text{ cm} (\cos\vartheta_s + \tfrac{1}{3})$$

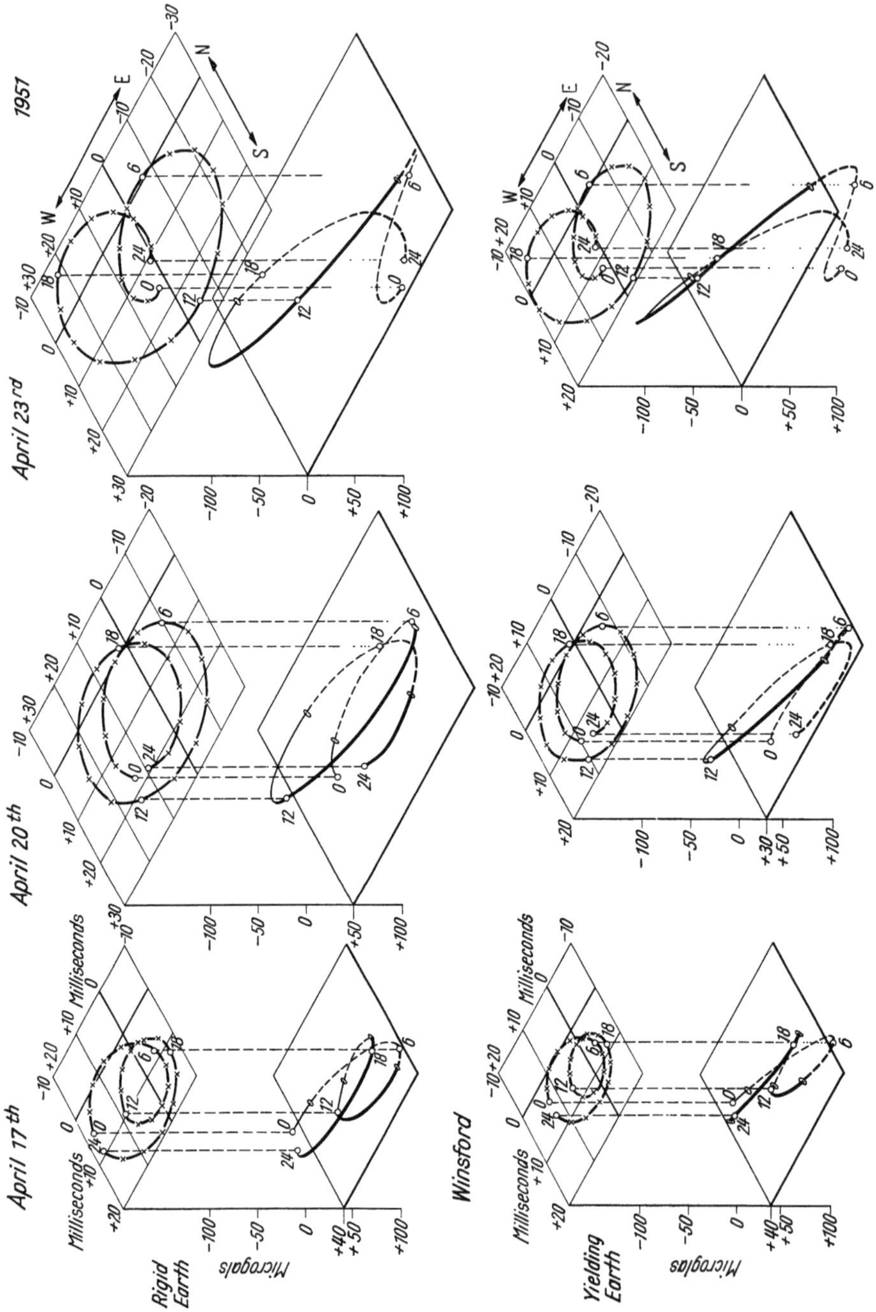

Fig. 21. Tidal variations of the total gravity vector for rigid and yielding Earth ($D = 0.72$; $G = 1.20$). Drift and load effects are eliminated (TOMASCHEK).

from this the height of the Earth tides can be calculated thus

$$z_m = 11.7 \text{ cm} \left(\cos\vartheta_m + \tfrac{1}{3}\right), \tag{8.31}$$

$$z_s = 5.4 \text{ cm} \left(\cos\vartheta_s + \tfrac{1}{3}\right). \tag{8.32}$$

For the magnitude of the residual movements see Sect. 23.

9. Determination of k from the pole movement. The formula (6.18) for k needs the determination of CHANDLER'S period of the pole movement. The difficulties arise from two factors. Firstly there is superimposed a yearly period which has its origin in meteorological disturbances, especially in the regular displacement of air masses by the great seasonal pressure variations in Central Asia. Secondly the free vibration is not continuous, but actuated by random perturbations, so that it may have a larger amplitude at certain times and, besides dying down by damping, it may even reverse its sign at other times . It is therefore necessary to select periods of time when a high amplitude of the free vibration is dying down in rather undisturbed conditions. This seems to have been the case from 1908 to 1921. An analysis of this period by JEFFREYS[1] gives

$$T = 439.0 \pm 5.8 \text{ solar days}$$

and a time of relaxation of 15.1 ± 1.8 years.

Using this determination gives

$$k = 0.288 \pm 0.008.$$

Another determination by POLLAK (1927)[1] yields $k = 0.287$ and LABROUSTE[1] using observations from 1900 to 1940 obtained

$$k = 0.272.$$

A value of

$$k = 0.28$$

should therefore be regarded as reliable, especially as this determination of k has the great advantage of being free from all local influences as it is obtained from the behaviour of the integrated Earth's globe as a whole. Some investigators consider it for this reason as the most reliably known of the three characteristic numbers. But the discussion concerning the proper value of T cannot yet be regarded as closed[2]. This question will be further discussed in Sect. 12.

10. Tidal latitude variations. Determination of L. The long period latitude variations of the magnitude of about 0.3 sec of arc due to the pole movement are overlaid by semidiurnal and diurnal tidal variations, although the amplitudes of these terms are very minute (about 0.″01). As the astronomer refers his measurements to fluid levels, he observes the change of the level surface relative to the Earth's axis and this is proportional to $(1 + k)$, but as the strain of the Earth's globe also displaces his own position in horizontal direction and by this, changes his true longitude and latitude, he actually observes effects proportional to

$$L = 1 + k - l.$$

In order to extract the tidal effect from the observational data of latitude the original large change of latitude of annual and 14 months' periods has to be eliminated by subtracting a general smoothed value of their change from the individual values of the star observations and the minute residuals have to be analyzed harmonically.

---

[1] Sir H. JEFFREYS: Month. Not. Roy. Astronom. Soc. **100**, 139 (1940). — L. W. POLLAK: Gerlands Beitr. Geophys. **16**, 108 (1927). — H. and Y. LABROUSTE: Ann. Géophys. **2**, Nr. 2, 115 (1946).

[2] A. H. WALKER and A. YOUNG: Observatory **75**, No. 884, 1—4 (1955). — Month. Not. Roy. Astronom. Soc. **115**, No. 4, 443 (1955).

Such an analysis has been undertaken at Greenwich by JONES[1] using the period 1911 to 1936. L was obtained as approximately 1, but fairly large corrections had to be applied for the tidal load influence of the nearby Thames. MARKOWITZ and BESTUL[2] obtained for Washington a value of $L = 1.14$, analyzing the 24-year period 1915 to 1939. There exist six international stations at 39° latitude, with longitudes 141°08′ E (Mizusawa), 66°53′ E (Tschardjui), 8°19′ E (Carloforte), 77°12′ W (Gaithersburg), 84°25′ W (Cincinnati), 123°13′ W (Ukiah), which specialize in latitude determinations. Their results have been analyzed by various observers (see MELCHIOR [1]), the most comprehensive investigation being that by NISHIMURA[3]. He used in some cases 35 years of observations and Fig. 22 shows his final curves. The influence of tidal oceanic load deformations can be shown to be very small when compared with its influence in tilt measurements. Applying a correction for this, the values shown in Table 8 have been obtained. The weighted mean given by NISHIMURA is

$$L = 1 + k - l = 1.20 \pm 0.10 .$$

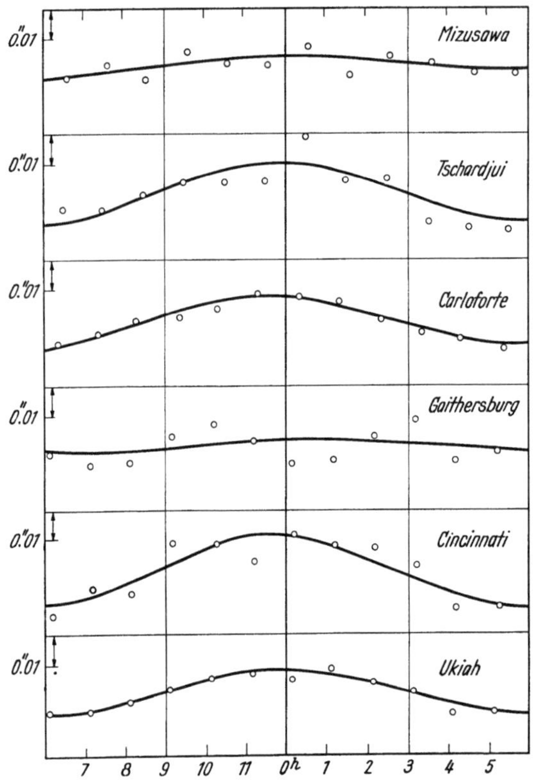

Fig. 22. Tidal variation of latitude at six International Latitude Observatories, 1900—1935 (NISHIMURA).

There exists an unexplained disparity between continental and coastal stations although the values had been corrected for the oceanic effect. It is

$L = 1.49$ for continental stations

and

$L = 1.07$ for coastal stations.

MELCHIOR [1] derives as an unweighted mean of all analyses of the different observers

$$L = 1.11 .$$

The problem of determining a definite value of L by latitude variations cannot yet be regarded as solved.

Table 8. *M_2 constituent of tidal latitude variations* (NISHIMURA).

Station	M_2	L	$\varkappa$	Analyzed period years
Carloforte . . .	0.″0083	1.08 ± 0.06	$-6°$	35
Ukiah	0.″0082	1.06 ± 0.06	$+9°$	35
Tschardjui . . .	0.″0101	1.31 ± 0.19	$-6°$	19
Cincinnati . . .	0.″0128	1.66 ± 0.18	$-6°$	16

[1] H. SPENCER JONES: London, H. M. Stat. Off. 1939; Month. Not. Roy. Astronom. Soc. **99**, 196 (1939).
[2] W. M. MARKOWITZ and S. M. BESTUL: Astronom. J. **49**, No. 10, 81 (1941).
[3] E. NISHIMURA: Trans. Amer. Geophys. Un. **31**, 357 (1950).

11. Strain measurements. Determination of *l*. By measuring tidal changes of the strain of the Earth's crust the number l can be obtained directly by means of Eq. (6.12). Such observations have been made by SASSA, OZAWA and YOSHIKAWA[1]. They used an extensometer constructed in 1941 by SASSA (Figure see MELCHIOR [*1*]; see also this Encyclopedia vol. XLVII, p. 65). It consists of a superinvar wire, 1.6 mm in diameter, which is stretched practically horizontally between two fixed bases about 20 m apart. A weight of about 350 gr is suspended from the midpoint. The sag of the stretched wire is approximately 1/70 of its length. The suspended weight moves up and down in accordance with the change in distance between the two fixed bases. This movement of the weight is converted into the revolution of a mirror by means of a bifilar suspension. Optical registration gives a sensitivity of a change of strain of about 10^{-8} per mm on the record.

Table 9. *Tidal strain observations* (OZAWA).

Station	Direction	l	Period months
Osakayama	S 38° W	0.035 ± 0.011	18
Osakayama	S 76° W	0.091	1
Osakayama	N 2° E	0.021	0.5
Makimine	N 57° W	0.025 ± 0.012	12

A year's obervations in the Osaka-Yama tunnel yielded the value of

$$l = 0.05$$

by applying a correction for elastic deformation by the load of the oceanic tides having an amplitude of about 50% of the one observed.

A second series of observations analyzed by OZAWA yielded the values given in Table 9, if no oceanic correction is applied. The attempts to assess the influence of the sea tides have shown that it is not possible to explain the discrepancies as caused by elastic load deformations. It is therefore impossible to correct the values so far obtained for l.

For this reason the question of the best experimental value of l must be left open. The observations indicate that probably

$$l = \text{about } 0.05 .$$

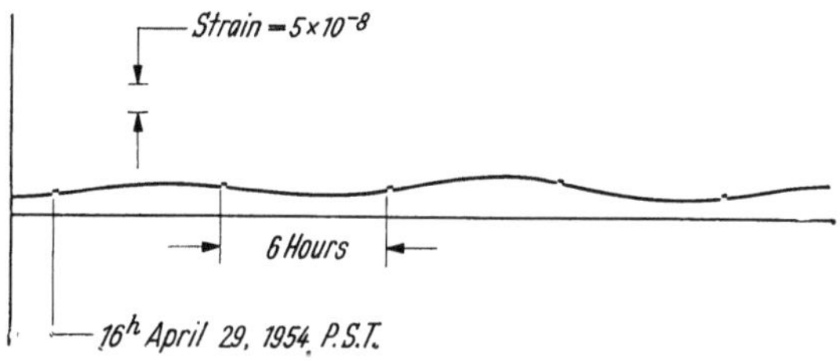

Fig. 23. Record of tidal strain variation by H. BENIOFF's strainmeter. Size about 1/3 of original record.

Remarkable tidal strain curves as shown in Fig. 23 have been obtained by BENIOFF[2] with his strainmeter. It consists of two steel piers set into the rock at points 20 m apart. An indicator pipe of about 20 m length made of fused quartz extends to within a short distance of the other pier. The pipe is suspended by 12 wire supports which are longitudinally compliant and relatively rigid in the transverse direction (Fig. 24). When the strain varies in the Earth's crust the two piers either approach or recede from each other. The free end of the pipe is thus displaced relative to the adjacent pier and as it bears a magnetic coil at its end this relative motion serves to actuate a variable-reluctance transducer, which feeds into a galvanometer. The instrument has been set up in the mountains

[1] K. SASSA, I. OZAWA and S. YOSHIKAWA: Disaster Prevent. Res. Inst. Kyoto, Bull. No. 3, 1952.

[2] H. BENIOFF: Bull. **171**, State of California, Dep. of Nat. resourc. Div. of Mines, p. 149, 1955. See also this Encyclopedia, Vol. XLVII, p. 64.

North of Glendora in California. Extended results on tidal measurements have not yet been published. Another arrangement, the *rolling deformograph* or *extensometer* (Fig. 25) has also been extensively used by Japanese observers[1]. It uses

Fig. 24. Linear strain seismometer by BENIOFF. End of the quartz pipe with transducer.

the rotation of a very thin cylinder which is actuated by the longitudinal displacement of a long rod (Fig. 25a). As Fig. 25b shows the extension of the rod A to A_2 (corresponding to the deformation D of the ground) is connected with a rotation of the cylinder of radius r. It is

$$A A_1 + A_1 A_2 = B_2 K_1 + A_1 B_1 = 2 r \alpha = D. \quad (11.1)$$

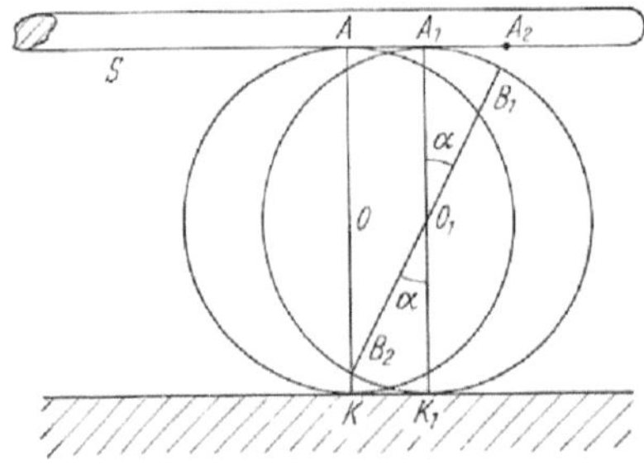

Fig. 25. Deformograph by W. BONCHKOVSKII and L. LATYININA. (a) Schematic diagram, (b) geometry of action.

A mirror is rigidly connected with the cylinder and optical recording is used. If L is the length of the path of light and the deflection on the record is d then

$$D = 2 r \alpha = \frac{r d}{L}. \quad (11.2)$$

Several silica tubes, 7 mm in diameter and of 1 m length each, are joined together and each is supported by metal rollers on smooth glass plates. The last indicating roller has, in the Japanese version, a diameter of 6 mm. Thus it is easy to record the relative movement between the end of the tube and the ground with a sensitivity of 10^{-7} of the length of the rods; this corresponds to the magnitude of the deformations caused by the tides ot the ocean at near coastal

[1] T. HAGIWARA, T. RIKITAKE and J. YAMADA: Bull. Earthq. Res. Inst. Tokyo **26**, 23 (1948); **27**, 35 (1949). See also W. F. BONCHKOVSKIJ and L. A. LATJINIA: Izv. Akad. Nauk USSR., Ser. Geofiz. **1955**, Nr. 3, 275.

stations. In order to determine l realiably on a continental station an accuracy about 20 times higher than this is needed. This can be achieved by using a very thin indicating cylinder of about 0.3 to 0.1 mm in diameter.

D. Discussion of results.

12. The numerical results. *a) Consistency of the experimental data.* Scanning the evidence of the incredible amount of more than half a century's patient labour, ingenious experimentation and laborious calculations spent on the problems of determining the three characteristic numbers, the result is, that although there exists a general conformity of the numerical results, the knowledge of the exact values remains yet far from satisfactory.

There is no doubt that the influence of the oceans is of great importance, firstly because of the potential of the dislocated water masses and secondly because of their deformative action on the Earth's crust. These effects are very difficult to eliminate as their periodicities are of tidal origin. An even more disturbing fact is that on the Earth's surface these corrections seem to express the actual behaviour to a small degree only, as there exist other effects of meteorological and tectonic origin which are of the same or even greater magnitude and which by their very nature can only be assessed more or less regionally or locally.

Except for the method of using pole movements which is the least individually influenced, all other methods have to be treated on a fairly equal footing and it has to be considered which values can be regarded as least influenced by local or regional conditions. In some cases this can be done by using the means of a great number of independent determinations, but even in the most favourable case, that of the gravimetric factor, the material is limited. For example Table 7 does not contain a single station on the European or Asiatic continent! And what is known of measurements in continental Europe deviates widely from the average. Strong efforts are expected to be made during the International Geophysical Year 1957/58 to extend the measurements over a worldwide area and intensify them by long series of observations[1] so that it is hoped that in several years time the situation may have radically changed.

Table 10 gives a list of what are probably the best values from the experimental standpoint. The data in brackets give values which would be yet compatible with experimental results. The table also contains the values according to different theoretical expectations as treated later on.

Table 10. *Experimental and theoretical values for tidal elastic factors.*

Experimental	Theoretical			
	incompressible homogeneous Earth		TAKEUCHI	
	$\mu=8\times10^{-11}$	$\mu=1.6\times10^{12}$	Model 1	Model 2
$h=$ [0.46 (0.60)]	0.78	0.46	0.587	0.610
$k=$ 0.29 (0.27)	0.47	0.28	0.290	0.281
$l=$ 0.08 (0.05)	0.23	0.14	0.068	0.082
$D=$ 0.72 (0.68)	0.69	0.82	0.703	0.671
$G=$ 1.20 (1.15)	1.08	1.03	1.152	1.198
$L=$ 1.22 (1.20)	1.23	1.14	1.222	1.199

[1] Commission pour l'étude des marées terrestres. Commun. Obs. Roy. Belg. Nr. 100, Sér. Géophys. Nr. 36. Bruxelles 1956. See also W. D. LAMBERT, J. D. BOULANGER and P. J. MELCHIOR: Programme marées terrestres AGI. Dtsch. Geod. Komm., München, Wiss. Übersetzungsdienst H. 8, 1956.

When considering the experimental values as given in Table 10 the strange fact emerges that it is not easily possible to bring them into a consistent system from which conformable values for k, h and l can be derived.

If $k=0.29$ is regarded as the best known value, as it is for example by JEFFREYS, then $D=0.72$ and $G=1.20$ yield different values of h, namely 0.57 and 0.635 respectively. On the other hand k determined by these two equations yields $k=0.16$ which seems to be much too low. MELCHIOR [1] from slightly different values also obtains $k=0.19$. On the contrary, recent calculations by WALKER and YOUNG lead to a CHANDLER period of 462 days, corresponding to $k=0.32$, although this may be too high. MELCHIOR[1] has also shown that there exists some doubt about the relaxation time of CHANDLERS motion. On the other hand calculations by LARMOR (1915) suggest a lengthening of the period by the influence of the ocean tides by about 10 days. This would diminish the k values but the actual oceanic tides are not sufficiently well known.

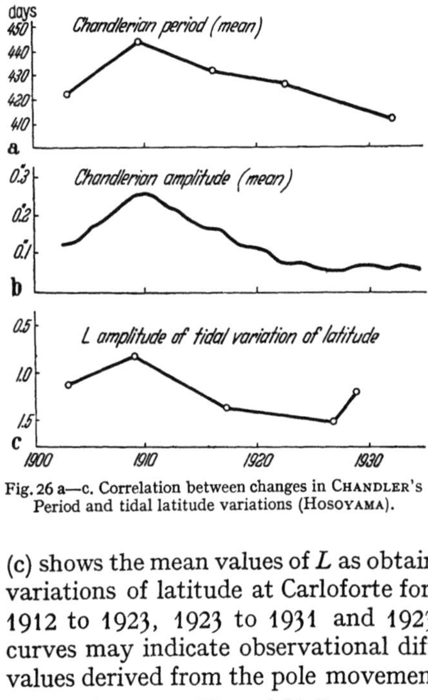

Fig. 26 a—c. Correlation between changes in CHANDLER's Period and tidal latitude variations (HOSOYAMA).

There may exist real variations of the observed values of k and L, as determined by astronomical observations. This has been shown by HATTORI and HOSOYAMA[2]. Fig. 26 shows in (a) the mean values of the CHANDLERian periods for the three stations Mizusawa, Ukiah and Carloforte for the epochs 1900 to 1906, 1907 to 1912, 1913 to 1919, 1920 to 1925 and 1926 to 1938. The CHANDLERian amplitudes are shown in (b); they are the running means of six successive years of the mean values at the three stations. (c) shows the mean values of L as obtained by NISHIMURA, calculated from the tidal variations of latitude at Carloforte for the subepochs 1900 to 1906, 1906 to 1912, 1912 to 1923, 1923 to 1931 and 1923 to 1935. The strong correlation of these curves may indicate observational differences in k and make the accuracy of the values derived from the pole movements seem less than it is generally assumed.

Furthermore, H. and Y. LABROUSTE[3] have found that three main components of the pole movement can be distinguished, namely of 365.2; 427.5 and 441 mean solar days period, they are of the same importance and have fairly constant amplitudes. Similarly LEDERSTEGER[4] regards the CHANDLER movement as being essentially a beat oscillation having a beat period of a double revolution of the Moon's nodes. MELCHIOR[5] suggests as an explanation the interference of two components attributed, respectively, to the Earth's core and mantle. There seems therefore to exist some ambiguity as to the true value of CHANDLER's period.

If the value of $k=0.28$ is assumed to be nearly correct, either D or G must be regarded as systematically disturbed. On the whole it seems more likely that D is less well known than G, because the tiltmeters are much more liable to local

[1] J. P. MELCHIOR: Rend. Acad. naz. Lincei, Cl. Sz. Fis., Ser. VIII **19**, Nr. 3/4, 137 (1955).
[2] K. HOSOYAMA: Trans. Amer. Geophys. Un. **33**, No. 3, 345 (1952).
[3] H. and Y. LABROUSTE: Ann. Géophys. **2**, Nr. 2, 115 (1946).
[4] K. LEDERSTEGER: UGGI. Bull. Géod. **1952**, Nr. 23, 67.
[5] P.-J. MELCHIOR: Bull. Géod. **1952**, Nr. 23, 59. See also the comprehensive report on this subject Obs. Roy. Belg. Monographie Nr. 3, 1954.

disturbances. Taking $G = 1.20$ as shown by the analysis of LAMBERT, D would become $D = 0.66$. This is in accordance with NISHIMURA's observations at the continental station at Barim. But the European stations, especially those far from the coast and after correction for sea tide influences, suggest a higher value. The same is shown by the influence of tectonic structure (Sect. 23). The question here emerges of possible *regional differences* in the values of h, k and l. LAMBERT has further directed attention to the circumstance that no theory yet exists, taking into account the flattening of the Earth.

We see from this that although a general agreement on the experimental values has already been reached, more exact measurements, specially directed towards the determination of the secondary and indirect effects, are necessary. A better knowledge of the tides of the high sea also remains as an important problem to be solved.

b) Comparison of the experimental data with theoretical expectations. The characteristic numbers are functions of the density and elasticity distributions in the interior of the Earth. A comparison between the experimental results with the values to be expected for a homogeneous, incompressible Earth as they are given for two different rigidities μ in Table 10 shows at once that these are not compatible with experience. This was soon realized and for half a century strenuous efforts have been made to deduce the inner structure of the Earth from tidal experimental data. A number of formal solutions has been given by HERGLOTZ, DARWIN, PRESCOTT, SCHWEYDAR, HOSKINS, JEFFREYS, ROSENHEAD, PREY, BOAGA and others. A comprehensive summary of these papers up to 1914 has been given by JANNE[1].

The assumptions made are that the Earth can be regarded as perfectly elastic for deformations of this type and that its matter is incompressible ($\lambda = \infty$); that the density and rigidity are functions of r only and that the effects of inertia can be neglected. ROCHE's density distribution

$$d = d_0\left[1 - \beta\left(\frac{r}{\varrho_1}\right)^2\right] \tag{12.1}$$

and homogeneous rigidity have first been used. PREY also discussed the problem, taking $\lambda = \infty$ and ROCHE's law for density, but assuming the rigidity distribution

$$\mu = \mu_0\left(\frac{r}{\varrho_1}\right)^2\left[1 - \alpha\left(\frac{r}{\varrho_1}\right)^2\right]. \tag{12.2}$$

LOVE (1907) and HOSKINS (1920) allowed for compressibility, the first assuming $d = \text{const}$, $\mu = \text{const}$ and $\lambda = \text{const}$. HOSKINS used ROCHE's distribution and

$$\lambda = \mu = \mu_0\left[1 - \alpha\left(\frac{r}{\varrho_1}\right)^2\right]. \tag{12.3}$$

In several cases one discontinuity of density and rigidity has been taken into account (SCHWEYDAR 1907, JEFFREYS 1915, 1926). ROSENHEAD (1929) took also account of oceans covering the Earth to different, but uniform depth, the core being supposed fluid. No agreement for semidiurnal tides with the observational values can be obtained. All the methods imply the existence somewhere of a much lower rigidity than is compatible with seismological data. The assumption of a liquid core gives too high a value for k, which indicates that a model closer to the observed heterogeneity of the layers has to be used.

[1] M. H. JANNE: Mem. Soc. Roy. Sci. Liège, III, s. **10** (1914).

13. Comparison of the tidal results with seismological knowledge. The first method adapted to a detailed application of seismological knowledge was given by STONELEY (1926)[1]. He used a numerical method which could be applied to any distribution of density and rigidity, besides taking account of compressibility. With the values available to him a yielding of about two thirds of that observed was obtained. A great advance has been made by TAKEUCHI (1950) [6]. He solved the fundamental equations by numerical integration, using 6 different Earth models based on data as provided by modern seismology. Three assumptions are made, in addition to the basic starting point that the tidal equilibrium can be dealt with as static. They are:

1. That the Earth is in a state of initial stress. This may be regarded as a hydrostatic pressure balancing the self gravitation of the Earth in the initial state (= without tides).

2. That the stress in the Earth may be regarded as consisting of the initial stress combined with an additional stress.

3. That the additional stress is taken to be connected with the strain, measured from the initial state as unstressed state, by HOOKE's law.

After introducing 4 boundary conditions the equations have been numerically solved for the case of equilibrium tidal deformation of the Earth. Numerical data used in the computation are:

1. Velocity distributions of P and S seismic waves within the Earth as obtained by GUTENBERG and RICHTER (1935) and WADATI (1933).

2. Density distribution by BULLEN.

3. The surface density has been assumed to be either

$$d_0 = 3.0 \quad \text{or} \quad d_0 = 2.7\,\text{gr/cm}^3.$$

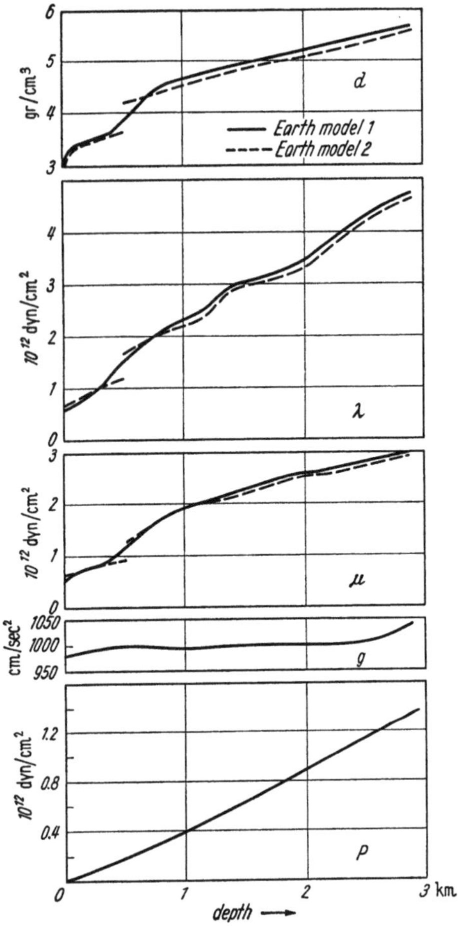

Fig. 27. Distribution values for TAKEUCHI's Earth models.

4. At 500 and 2900 km depth, discontinuous surfaces were assumed.

5. Several cases of rigidity in the so-called liquid core beneath 2900 km depth were assumed, including zero.

Fig. 27 shows the assumed distributions of d, λ, μ, g and pressure. Two slightly different models for the shell have been used. In the Earth model I the value of d_0 of BULLEN (1940) is applied and the values of λ and μ are calculated by the formulas

$$V_p = \sqrt{(\lambda + 2\mu)/d_0} \quad \text{and} \quad V_s = \sqrt{\mu/d_0} \qquad (13.1)$$

where V_s and V_p are the longitudinal and transversal seismic velocities. In the Earth model II the value obtained by BULLEN (1936) is used for d_0 and the values of λ and μ are calculated from (13.1) using this value of d_0 and the values of

[1] R. STONELEY: Month. Not. Roy. Astronom. Soc., Geophys. Suppl. 1, No. 7, 356 (1926).

JEFFREYS (1939) from the depth from zero to 500 km, and those obtained by GUTENBERG and RICHTER and WADATI for 500 to 2900 km. The so-called JEFFREYS discontinuity is assumed to be at a depth of 500 km. The values of g are calculated by

$$g = f M/r^2 \tag{13.2}$$

where f is the gravitation constant and M the mass enclosed by a sphere of radius r as calculated from the density distribution. The value of the initial hydrostatic pressure P_0 is calculated from formula

$$\frac{d P_0}{d r} = - d_0 \cdot g \tag{13.3}$$

using the values of g and d_0 obtained above. The values of g and P_0 differ but little for both models. Table 11 shows the assumed boundary conditions in CGS-units. For both models the density distribution within the Earth's core as obtained by BULLEN 1936 is used. In Earth model I the density thus adopted will give a little too low a value for the density within the core.

Table 11. *Assumed boundary conditions in CGS units* (TAKEUCHI).

Depth km	Earth model I			Earth model II		
	d_0	λ	μ	d_0	λ	μ
0	3.0	0.6×10^{12}	0.5×10^{12}	3.0	0.68×10^{12}	0.6×10^{12}
Above 500	3.89	1.520	1.120	3.69	1.197	0.926
Below 500				4.22	1.644	1.217
Above 2900	5.68	4.734	3.010	5.53	4.608	2.931

Under these assumptions TAKEUCHI obtains

Model I $k = 0.290$; $h = 0.587$; $l = 0.068$
Model II 0.281 0.610 0.082.

These values and the factors derived from them have been included in Table 10.

There remains to be discussed the problem of determining the largest value of rigidity of the Earth's core compatible with the observations of Earth tides. Four different models have been calculated by TAKEUCHI (III to VI). In these models the internal constitution above the 2900 km discontinuity is assumed to be the same as that of model II, but below this discontinuity, the density and elastic constants are assumed to be homogeneous. The values for d_0, λ and μ are taken as follows: d_0 is chosen so as to make $g_0 = 982$ cm/sec² for $b = 6370 - 2900 = 3470$ km. The value of d_0 thus determined is 10.13 gr/cm³. The value of λ within the Earth's core is about $\lambda = 6.0 \times 10^{12}$ at a depth of 2900 km and $\lambda = 15 \times 10^{12}$ at the centre of the Earth. This gives a mean value of 11×10^{12} for λ. For the rigidity within the core the following succession of values has been taken. $\mu = 10^7$, 10^9, 10^{11} and 10^{13} CGS respectively for models III, IV, V and VI. The results are given in Table 12. They show no significant change

Table 12. *Values of the characteristic numbers corresponding to different values of the rigidity of the core* (TAKEUCHI [*6*]).

	μCGS =				
	0	10^7	10^9	10^{11}	10^{13}
k	0.281	0.275	0.275	0.243	0.055
h	0.606	0.601	0.600	0.530	0.109
l	0.082	0.081	0.081	0.083	0.092

for a rigidity between zero and 10^9, whereas higher values of μ lead to figures which are not compatible with experience. The largest value of the rigidity of the core conforming with the observations is about 10^9 to 10^{10} dyn/cm². Zero rigidity, as for a liquid, is also consistent with the data.

TAKEUCHI'S results agree satisfactorily with the experimental data as shown in Table 10. Taking the theoretical values as a guide the experimental results between the following limits seem the most reliable:

$$k = 0.27 \text{ to } 0.29; \quad h = 0.59 \text{ to } 0.61 \text{ and } l = 0.07 \text{ to } 0.08.$$

With the exception of l which is not yet known with sufficient accuracy, these figures are well within the experimental range. Although they are not identical with the values which seem experimentally the most reliable according to the present state of affairs, a fairly clear idea has been established of the values to be expected for static tides. In this way a close agreement has been reached in the pictures of the interior of the Earth obtained from both seismic and tidal information.

The problem has yet to be solved as to how far deviations from the assumption of the Earth tides being static influence the results. Motions that change the direction of the axis of rotation are not included in a static theory. They would influence the variations of latitude, the nutations and the diurnal tides. JEFFREYS, using TAKEUCHI'S theory of the shell and assuming a homogeneous incompressible core predicts a period of about 460 days for the free motion and a significant correction to the amplitude of the 19-yearly nutation. Because of the influence of the dynamic effects of the core the numerical values of the factors D and G are changed. The values expected are given by JEFFREYS (1956) as

	$l+k-h$	$l+h-\frac{3}{2}k$
semidiurnal tides	0.704	1.152
diurnal tides K_1	0.712	1.185
P_1	0.676	1.209
O_1	0.658	1.221

Therefore effects amounting up to 7% could result. Although these theoretical values for the semidiurnal tides are definitely lower (especially for G) then today's assumed means, they are quite within the range of the experimentally possible values (p. 810). The experimental values of the diurnal tides are not yet sufficiently known to give an indication of significant differences.

14. Distribution of tidal values in the interior of the Earth. TAKEUCHI [1] has also determined the distribution of the tidal values within the shell of the Earth, using his previously mentioned models. Fig. 28 shows these distributions as a function of depth. The radial displacement u_r caused by the tidal attraction of the Moon and the additional potential K of the Earth are almost constant. The lateral displacement of the Earth u_θ increases monotonously from the surface towards the 2900 km discontinuity, whereas the radial and lateral stresses R_r, θ_r and Λ_r have maxima and minima values. For the tidal action of the Moon the units become

$$u_r \sim 66; \quad u_\theta \sim 99; \quad R_r \sim 10^5; \quad \theta_r \sim 1.5 \times 10^5; \quad K \sim 3.6 \times 10^4.$$

This gives for the surface

$$\left.\begin{aligned} &u_r = 22\,\text{cm}; \quad u_\theta = 4.4\,\text{cm}; \quad u_\lambda = R_r = \theta_r = \Lambda_r = 0;\\ &K = 10^4\,\text{CGS units}. \end{aligned}\right\} \tag{14.1}$$

The maximum value 2.6×10^4 CGS of the radial stress R_r occurs at a depth of 1400 km, that of the lateral stress $\theta_r = 6.3\times10^4$ CGS at 2200 km. The volume divergence at the surface of the Earth is found to be

$$\Delta = 2.6\times10^{-8}. \tag{14.2}$$

As the elastic constants of the rocks are of the order of magnitude 10^{12}, the stress caused by the tidal attraction of the Moon is about 10^4 CGS.

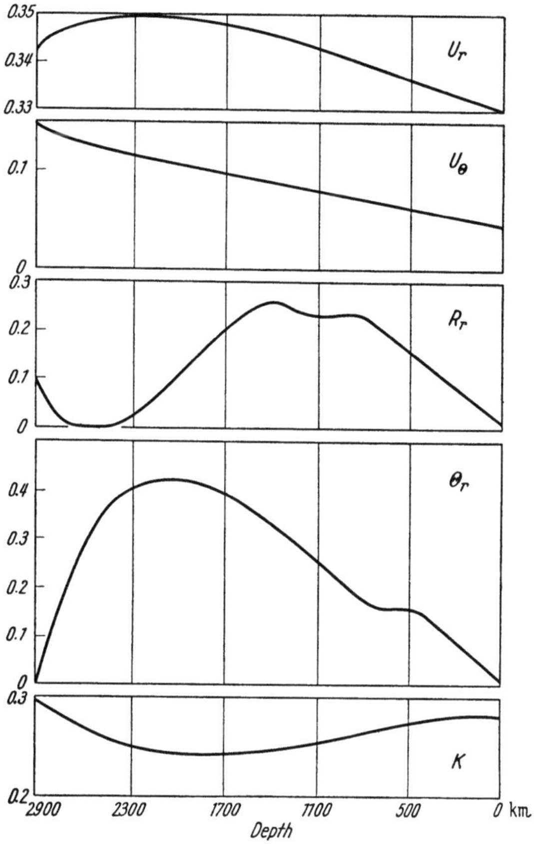

Fig. 28. Distribution of tidal displacements and stresses in the interior of the Earth (TAKEUCHI).

(1) Radial displacement u_r	Unit: $a\,S_2/4\pi f$.
(2) Lateral displacement u_θ	$(a/4\pi f)\cdot\partial S_2/\partial\theta$.
(3) Radial stress R_r	$(10^{12}/4\pi f)\cdot S_2$.
(4) Lateral stress Θ_r	$(10^{12}/4\pi f)\cdot\partial S_2/\partial\theta$.
(5) Deformation potential K	$a^2 S_2$.
$S_2 = f M c^{-3} P_2(\cos\theta)$	$a = \varrho_1$.

15. Tidal friction. If the Earth's substance departs from perfect elasticity a part of its tidal energy must be converted into heat; tidal friction comes into play. Its effect is shown in Fig. 29. Regarding the semidiurnal tide due to the Moon, in the absence of friction (Fig. 29a) the maximum elevation will be at a point where the Moon is in zenith or nadir. If however friction is present, it will delay the times of maximum elevation at P. Thus the highest tide at P will not occur till some time after it has passed the zenith or nadir position of the Moon, the high tides being at B or C. The observed maximum of the effect will therefore *precede* the Moon and the phase lag $\varkappa$ should be negative, if the tidal deformation is lagging in time as a result of internal friction. It is evident from Fig. 29b that in case of friction the Moon will tend to retard the rotation of the Earth and that the reaction to this force will tend to accelerate the Moon in its orbit. Astronomically this effect is part of a complex phenomenon which is called the *secular acceleration* (see Vol. XLVII, p. 2).

The height of the tidal bulge at latitude φ and longitude λ, assuming the Moon being in the plane of the equator, is given by

$$z = u_r \cos^2\varphi \sin 2\lambda. \tag{15.1}$$

If ε is the angle between the meridian of the Moon and the meridian of maximum tide, the couple N, tending to retard the rotation is given by

$$-N = \frac{8}{5}\pi f M \frac{\varrho_1^4}{c^3} u_r \cdot d \cdot \sin 2\varepsilon \tag{15.2}$$

where d is the density of matter composing the bulge. In order to obtain the rate at which energy E is being expended N has to be multiplied by $\omega - \Omega$, the angular velocity of the Moon relative to a point on the Earth. It is

$$-\frac{dE}{dt} = \frac{8}{5}\pi f M \frac{\varrho_1^4}{c^3} u_r \cdot d \cdot (\omega - \Omega) \sin 2\varepsilon . \tag{15.3}$$

For this tidal energy dissipation JEFFREYS has calculated from the observed secular acceleration a value of $-dE/dt$ of about 1.5×10^{19} erg/sec. If this amount is assumed to be due entirely to bodily friction within the solid Earth the dependence of ε on the mean viscosity of the Earth will make possible the calculation of the value of viscosity causing this dissipation.

It can be assumed that this viscosity ν is of an elasticoviscous plastic type as suggested by MAXWELL, which means that it can be expressed by a relation

$$\nu = \mu \tau \tag{15.4}$$

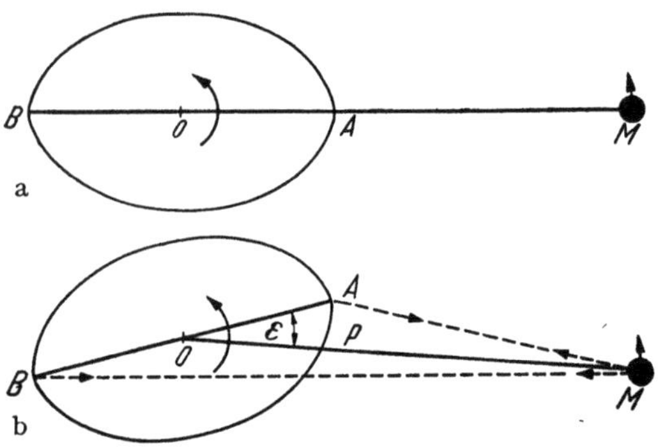

Fig. 29 a and b. Influence of tidal friction on phase lag.

where τ is a relaxation time having the following meaning. If, for example, the substance be strained by a certain shearing stress and maintained in that position, the stress decreases exponentially with time and at the time τ it will have decreased to $e^{-1} = 0.368$ of its initial value. It is possible, following a procedure given by Sir G. H. DARWIN, to calculate the mean relaxation time of the material of which the Earth must be built, in order to dissipate the rate of energy stated above. Assuming a mean rigidity of about 1.5×10^{12} CGS it is found to be of the order of magnitude of two to three days and ε would be about one degree, corresponding to $\varkappa = -2°$.

This relaxation time is in sharp disagreement with the experiences derived from the pole movement and that of the secular readaption of the Earth's crust after the disturbances caused by the load of the ice masses during the ice age. Recent investigations have shown that the damping of the pole movement may be of the order of magnitude of 15 years or even considerably more (Sects. 9 and 12). The values obtained for the slow retrogression of the ice-age deformations indicate a value of $\nu = 3 \cdot 10^{22}$ CGS corresponding to a relaxation time of about 800 years[1]. This shows that even a value of $\varkappa = -2°$ would be widely in excess of that allowed for by the physical constitution of the Earth. A mean relaxation time of the Earth's substance of 50 years would produce a phase lag of less than 1″, which in its order of magnitude would be compatible with the observations of the pole movement.

There exists a second reason why the bodily friction can be only very small. JEFFREYS has shown that "a large fraction of the energy dissipation of 1.5×10^{19} erg/sec can be accounted for by the friction due to the oceanic tides in shallow seas and that there is nothing to prove that it is incapable of accounting for the whole of it. In particular, any portion of the secular acceleration can be attributed to bodily tidal friction only as a result of independent proof that bodily friction must produce such an amount".

[1] E. NISKANEN: Ann. Acad. Sci. fenn., Ser. A, **3**, No. 21 (1949). — B. GUTENBERG: Nature, Lond. **177**, 887 (1956).

This proof has not yet been forthcoming and it is therefore a justifiable assumption that there exists no phase lag of any measurable amount in the tides of the solid Earth. One can concur with TAKEUCHI's statement that from this point of view the smaller the observed phase lag of the Earth tide is the greater is the weight that can be attached to an observation.

E. The influence of the oceans.

16. Introductory remarks. As $\frac{7}{11}$ of the Earth's surface are covered by oceans then oceanic tides must exert an influence even on continental stations. Their effect consists of three components:

1. The gravitational attraction of the shifted water masses on the point of observation (component A).
2. The deformation of the solid Earth due to the shift of the water masses. This acts in a general way deforming the whole globe of the Earth and in a more special way by tilting or moving the observing station (component B).
3. The change of potential due to the changed distribution of masses (component C).

In three directions the investigations of these oceanic influences have been carried out in order to correct the direct observations for the influence of oceanic tides. It is the formal, the numerical and the empirical method which will be discussed in this chapter. A short glimpse at our knowledge of the oceanic tides seems to be necessary here[1]. The actual sea tides are in no way static, but widely and irregularly disturbed by resonance phenomena. But there exist certain constituents which may be truly regarded as static, such as the annual variations or those having CHANDLER's period of about 14 months. It is questionable whether the monthly and fortnightly tides can be treated as static, as they are affected by the forces of the rotation of the Earth, the CORIOLIS forces. A theory treating the tides as static may, therefore, be able to obtain the correction for the pole movement and perhaps, at least partly, that of the monthly and fortnightly tides.

For the diurnal and semidiurnal tides there exists a dynamic theory by LAPLACE and HOUGH. It makes possible the treatment of the problem of oceans of uniform depth.

These two possibilities have been used in the first direction described above. The second type of investigations relies on tidal charts. But these are not yet in a satisfactory state, as the tides in the open seas are as yet largely unknown and have to be inferred from coastal stations. On the contrary there exists the hope of obtaining information about their behaviour from the observations of the Earth tides. The third way depends on the knowledge of the ratios of the amplitudes of different partial tides in the oceans and this can be given with some better evidence than the values themselves.

These problems have to be treated here in some detail especially as the correction of the observed values of the tides of the solid Earth is of utmost importance, as the figures in the previous chapters have shown. But this discussion is important also because of the possibility, mentioned above, of obtaining by the necessary adaptions some information on the tides of the open sea themselves.

17. The formal solutions. *a) Static theory.* It is assumed that the different mass distribution of the oceanic tides on the surface of the Earth can be treated as a thin layer of different surface density which can be expressed by a spherical

[1] See also the following article by DEFANT, especially pp. 919ff.

harmonic
$$T = T_n \varrho_1 S_n = T_n (W_n)_{r=\varrho_1}, \tag{17.1}$$

where T_n is a constant, W_n is a solid spherical harmonic of degree n and S_n is a spherical surface harmonic of degree n. This surface density gives rise to a potential A within the Earth. It is

$$A = A_n W_n = \frac{4\pi f \varrho_1}{2n+1} T_n W_n. \tag{17.2}$$

The Earth is deformed by the potential A and the surface traction

$$R_r = -T_n g (W_n)_{r=\varrho_1}. \tag{17.3}$$

By the elastic deformation of the Earth there arises an additional potential. The sum of this potential and A shall be denoted by

$$K = K_n(r) W_n. \tag{17.4}$$

In this way the equations for the deformation of the Earth by surface loads can be reduced to the same type of equation as in the theory of the Earth tides.

Assuming a distribution of the surface density according to a spherical harmonic of the degree $n=2$ and using the numerical data of his Earth model TAKEUCHI[1] could show that

$$K_2(\varrho_1) = 0.7347\, A_2. \tag{17.5}$$

Assuming an ocean of uniform depth having static tides of the type S_2 the depth will assume the form

$$H = \frac{H_2}{\varrho_1^2} [W_2]_{r=\varrho_1}. \tag{17.6}$$

Condensing the load of density $d_w = 1.03$ gr/cm³ to the surface of the Earth the density distribution becomes

$$T = d_w \frac{H_2}{\varrho_1^2} [W_2]_{r=\varrho_1} \tag{17.7}$$

which is of the same type as that assumed above. If the Earth were perfectly rigid K_2 would be equal to A_2. The additional density caused by the deformation of the Earth is -0.265 times the primary surface density and adding this negative density to that produced by the oceans A_2 a total density is obtained of 0.7347 the primary surface density.

The numerical integration of the Earth model also yields the result that the vertical displacement is

$$u_r = -0.9665 \frac{A_2}{g} W_2. \tag{17.8}$$

Applying this for example to the measurements by horizontal pendulums, the pendulum will be attracted directly by the effects $A+C$, whereas the effect B affects it by the load tilt of the surface (Sect. 16). The effects $A+C$ on the horizontal pendulum are in the direction s

$$\Theta_{A+C} = 0.7347 \frac{A_2}{g} \frac{\partial W_2}{\partial s}. \tag{17.9}$$

The load tilt is

$$\Theta_B = \frac{\partial u_r}{\partial s} = -0.9665 \frac{A_2}{g} \frac{\partial W_2}{\partial s}. \tag{17.10}$$

This tilt shifts the pendulum in the same direction as does the attraction of the tidal waters[2]. The total effect is therefore

$$\Delta_2 = 1.7012 \frac{A_2}{g} \frac{\partial W_2}{\partial s}. \tag{17.11}$$

[1] TAKEUCHI: J. Fac. Sci. Univ. Tokyo, Sect. II 7, Part II, 1 (1951).

[2] The components $A+C$ do not affect the ground but tilt the horizontal plane in the opposite direction.

Since $\frac{A_2}{g}\frac{\partial W_2}{\partial s}$ is the action of the tidal water masses, if the Earth were perfectly rigid, the effect on the pendulum due to the yielding of the Earth to an oceanic tide of the type of a surface harmonic S_2 is 1.701 times that of a perfectly rigid Earth.

The Earth models give by numerical integration the results that for a tide of the spherical surface harmonic type of S_2 and of 1 m height the radial u_r and the lateral u_Θ displacements, the radial stress R_r and the total potential A_2 have the values on the surface

$$u_r = -11\text{ cm};\quad u_\Theta = 5.5\text{ cm};\quad R_r = -10^5\text{ CGS};\quad K = 0.79\times 10^4\text{ CGS}. \quad (17.12)$$

TAKEUCHI applies these results to the calculations of the *annual* variation of latitude which is caused mainly by the annual variation of meteorological elements such as the worldwide distribution of air and snowfall and which can

Table 13. *Influence of higher harmonics of loads.*

Degree of harmonic	$n=2$	4	8
additional density / primary density	-0.265	-0.113	-0.0635
total density / primary density	0.735	0.887	0.9365
actual effect / attractive effect	1.701	1.918	2.1779

Table 14.

Distance	$\Delta'(r)$
(0)	3.805
200	3.097
400	2.716
600	2.606
800	2.564
1000	2.543

be assumed to act in a static manner and in first approximation according to a spheric harmonic of type S_2. Calculations by JEFFREYS (1916) and ROSENHEAD (1929) have shown that a factor of about 0.7 has to be applied to the observed contributions. The factor 0.73 following TAKEUCHI's calculations is in good agreement with it and gives account of the masses of the Earth's surface depressed by the loads.

Interesting results have been obtained for the distribution of loads corresponding to spherical harmonics of higher degrees. The ratios of the additional densities of the condensed surface layers caused by the deformation of the Earth, further that of the total density and that of the actual effects on a pendulum compared with the purely attractive effect of the masses on a rigid Earth are given in Table 13.

As will be shown in Sect. 18b it is of interest to know how deep the load influence penetrates into the interior of the Earth to a considerable degree. TAKEUCHI's calculations show that the depths where the ratio of the radial displacement to that at the surface becomes $\frac{1}{2}$ diminishes with the higher degrees of harmonics, giving the following values: $n=4$: 1800 km; $n=8$: about 500 km.

For higher harmonics the deformation by a surface load can be regarded as superficial. For loads where the first zero is less than 1800 km apart from its pole [as it is for $P_8(\cos\Theta)$] the effects can be calculated by a model of a non-gravitating Earth with a *plane* surface. Using a simplified distribution of λ and μ the following values for $\Delta'(r)$, the ratio of the total effect to the attraction of the masses as a function of distance of the masses, are given by Table 14.

The secondary effects can then be calculated in the following way: The whole area of the ocean is divided into many concentric circular areas D_n ($n=1, 2, 3, \ldots$) with the observing station in the centre. The deflection by the tidal masses m_n is calculated by

$$\Theta(n) = \frac{f\,m_n}{g\,r_n^2}. \tag{17.13}$$

Multiplying Θ_n with the $\Delta'(r)$ interpolated from Table 14 gives the total tilting (for example in NS and EW direction) and adding the values $\Delta'(r)\,\Theta_n$, bringing into consideration the phase relations, gives the secondary effect of the oceanic tide.

This method has not yet been applied to actual cases, but it gives a good theoretical foundation to the numerical calculations reported in Sect. 18.

b) Calculation for dynamic tides. TAKEUCHI's results on the harmonic deformations cannot immediately be applied to the diurnal and semidiurnal tides as they are not static; on the contrary they are mainly opposite in phase to the static tides. But there exists a dynamic theory of the oceanic tides by HOUGH[1] to which it may be applied. This has not yet been done, but simplified Earth models have been used in connection with HOUGH's theory in order to estimate the oceanic effects on continental stations. It shows for example that tides of a period of a sidereal day cannot develop, especially K_1. (This is actually true for the Atlantic, but not for the Pacific and the Indian Ocean, a consequence of the distribution of the continents.) Furthermore it is shown that there exist natural periods coinciding for M_2 or S_2 with depths of uniform oceans of 26044 feet or 29180 feet respectively. There does not yet exist a more general theory taking account of the variations of the depths within the oceans and the land distribution, but HOUGH's theory seems to render the general features quite satisfactorily to a first approximation.

SCHWEYDAR[2] calculated the influence of the oceanic tides on horizontal pendulum measurements using the simplified assumption of an ocean covering the whole Earth with uniform depth of 4800 m, which may be given here as an example. If W_2/g is the static equilibrium height of the ocean and θ the pole distance, the dynamic height for an ocean of 4800 m depth can be expressed by

$$\left.\begin{aligned} T &= \frac{W_2}{g}\,[-1.0647 + 0.60095\,(7\cos^2\theta - 1) - \\ &\qquad - 0.12136\,(33\cos^4\theta - 18\cos^2\theta + 1)] \\ &= H_2 + H_4 + H_6 \ldots. \end{aligned}\right\} \tag{17.14}$$

It can be seen that the main term is opposite in sign to the static tide. Assuming as in Sect. 17a (p. 825) that the water tides can be regarded as a surface layer of varying surface density the radial component u_0 of the elastic displacement can be written:

$$u_0 = h\,\frac{W_2}{g} - h_2 H_2 - h_4 H_4 - h_6 H_6 \ldots. \tag{17.15}$$

Assuming a homogeneous Earth (e.g. $\mu = 2\times 10^{12}$ CGS; $d = 5.5$ gr/cm³; $d_w = 1.03$ gr/cm³) the coefficients can be written (compare Sect. 6a)

$$\left.\begin{aligned} h &= \frac{5}{19\dfrac{\mu}{\varrho_1 g d} + 2\dfrac{d - d_w}{d_w}}, & h_2 &= \frac{2}{5}\,\frac{d_w}{d}, \\ h_4 &= \frac{1}{\dfrac{51}{4}\,\dfrac{\mu}{\varrho_1 g d} + \dfrac{d - d_w}{d_w}}, & h_6 &= \frac{1}{\dfrac{33}{2}\,\dfrac{\mu}{\varrho_1 g d} + \dfrac{d - d_w}{d_w}}. \end{aligned}\right\} \tag{17.16}$$

---

[1] S. S. HOUGH: Phil. Trans. London **191**, 139 (1898). See the article by KERTZ in this volume, p. 950.

[2] W. SCHWEYDAR: Veröff. preuß. geodät. Inst., N. F. **1916**, Nr. 66.

The disturbance of the potential on the Earth's surface is $W_2 + V - u_0 g$ and the deflection of the pendulum due to this is in the meridian

$$\eta_N = \frac{1}{\varrho_1 g} \frac{\partial}{\partial \theta} (W_2 + V - u_0 g) \tag{17.17}$$

and in the prime vertical it is

$$\eta_E = \frac{1}{\varrho_1 g} \frac{\partial}{\partial \lambda} (W_2 + V - u_0 g)\,. \tag{17.18}$$

Putting

$$\left.\begin{aligned}
\alpha &= 1 + \frac{3}{5} \frac{d - d_w}{d} h - h\,,\\
\alpha_2 &= 1.0647 \left(-\frac{3}{5} \frac{d - d_w}{d} h_2 + h_2 + \frac{3}{5} \frac{d_w}{d}\right),\\
\alpha_4 &= 0.60095 \left(-\frac{1}{3} \frac{d - d_w}{d} h_4 + h_4 + \frac{1}{3} \frac{d_w}{d}\right),\\
\alpha_6 &= 0.12136 \left(-\frac{3}{13} \frac{d - d_w}{d} h_6 + h_6 + \frac{3}{13} \frac{d_w}{d}\right)
\end{aligned}\right\} \tag{17.19}$$

it can be written

$$\frac{1}{g}(W_2 + V - u_0 g) = \frac{W_2}{g} [\alpha - \alpha_2 + \alpha_4 (7 \cos^2\theta - 1) - \alpha_6 (33 \cos^4\theta - 18 \cos^2\theta + 1)]. \tag{17.20}$$

As on a rigid Earth

$$\eta_N^{rg} = \frac{1}{\varrho_1 g} \frac{\partial W_2}{\partial \theta} \tag{17.21}$$

and

$$\eta_E^{rg} = \frac{1}{\varrho_1 g \sin\theta} \frac{\partial W_2}{\partial \lambda} \tag{17.22}$$

and as

$$\frac{W_2}{g \varrho_1} \cos\theta \sin\theta = \frac{1}{2} \sin^2\theta\, \eta_N^{rg} \tag{17.23}$$

the deflection of the vertical becomes

$$\left.\begin{aligned}
\eta_N = \eta_N^{rg} [\alpha - \alpha_2 + \alpha_4 (7 \cos^2\theta - 7 \sin^2\theta - 1)\\
- \alpha_6 (33 \cos^4\theta - 66 \sin^2\theta \cos^2\theta + 18 \sin^2\theta - 18 \cos^2\theta + 1)],
\end{aligned}\right\} \tag{17.24}$$

$$\eta_E = \eta_E^{rg} [\alpha - \alpha_2 + \alpha_4 (7 \cos^2\theta - 1) - \alpha_6 (33 \cos^4\theta - 18 \cos^2\theta + 1)]. \tag{17.25}$$

As the diminution factor is

$$D = \eta / \eta_{rg}$$

it is

$$D_E - D_N = 7 \alpha_4 \sin^2\theta - 6 \alpha_6 \sin^2\theta (11 \cos^2\theta - 3)\,. \tag{17.26}$$

The result is that

1. The value of D is diminished by the influence of the oceanic tides.

2. The oceanic tides even of a uniform ocean covering the whole Earth introduce an asymmetry between the values of D in the meridian and in the prime vertical.

This result can be used to explain the asymmetry observed in the European results of horizontal pendulum measurements (Sect. 7d γ, p. 796 and Fig. 9).

With the numerical values for μ and d as given above it is

$$\alpha_4 = 0.0475 \quad \text{and} \quad \alpha_6 = 0.0070\,.$$

Therefore

M_2: $D_E - D_N$ = calc. + 0.091; obs. + 0.151 at Freiburg (Baden)
+ 0.072 + 0.067 Freiberg (Saxony).

This is a good agreement, especially as at Freiburg (Baden) a certain EW influence by coastal tides has to be suspected.

There is also to be expected an oceanic influence on the phases and it can be calculated by the values given above that a phase difference of maximal about 4° in a latitude of 45° in the azimuths ±35° and ±145° is introduced in

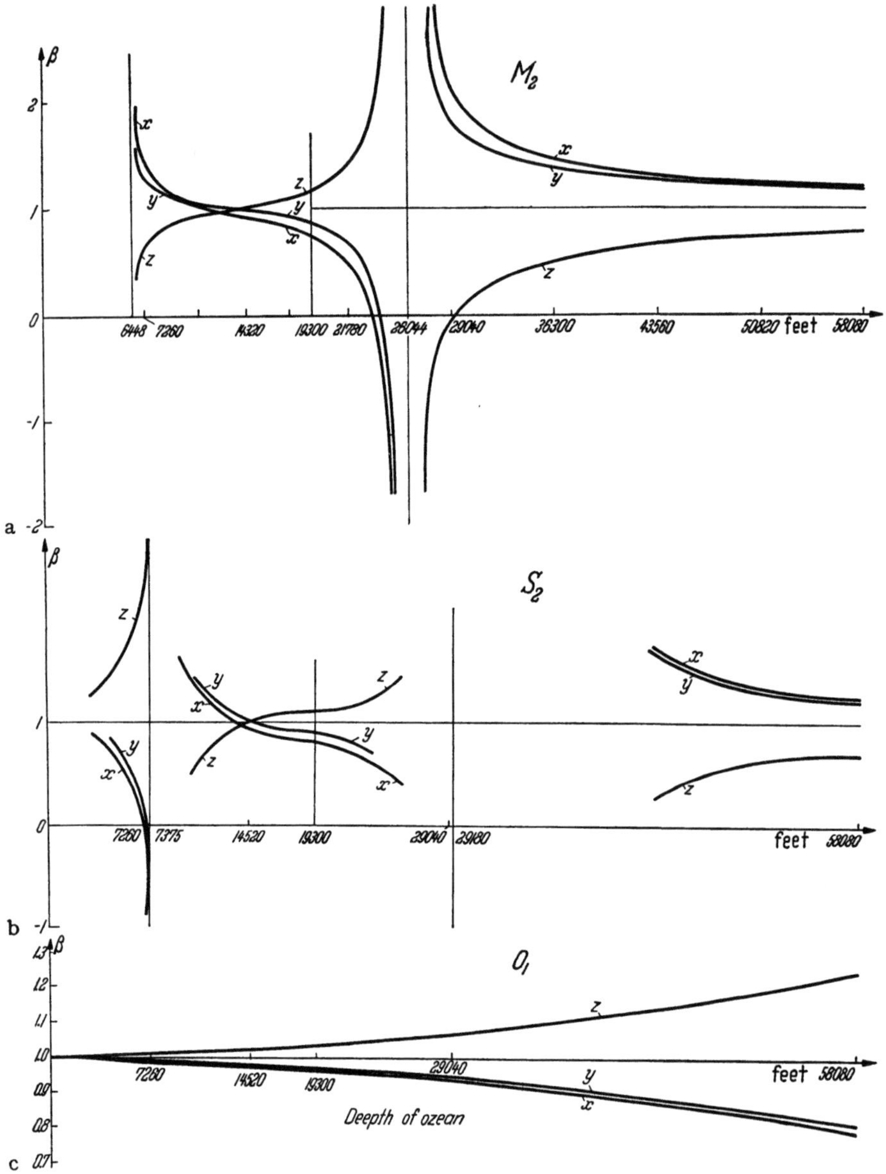

Fig. 30 a—c. Influence on the characteristic ratios of Earth tides by oceans of different uniform depth (BERROTH). $\beta = 1$ no influence of oceanic tides.

the direction opposite to that which would exist by solid body friction. In the case of the oceanic tides $\varkappa$ is positive, the pendulums lag behind the moon.

These investigations have been taken up in a more general way by BERROTH[1], by extending them to different depths and to the vertical component. Fig. 30a, b and c show the influence of different uniform depths of oceans extending over

[1] A. BERROTH: Sitzgsber. Ges. Wiss. Göttingen, math.-physik. Kl. **1932**, Nr. 5, 449.

the whole globe for the horizontal and the vertical components for the partial tides M_2, S_2 and O_1 at a pole distance of $\theta = 39° 06'$ (Freiberg, Saxony) and using the same general constants as SCHWEYDAR. In Fig. 30 the ordinates denote a number β which has to supplant the number 1 in the expression for

$$D = 1 + k - h \quad \text{and} \quad G = 1 - \tfrac{3}{2} k + h$$

in order to obtain the Earth tides in the presence of oceans as

$$D' = \beta + k - h \quad \text{and} \quad G' = \beta - \tfrac{3}{2} k + h. \tag{17.27}$$

As the oceanic resonance depths are different for M_2 and S_2 the corrections for both are also different. A comparison of D or G deduced by both partial tides could therefore give a further independent check of the actual depth to be assumed, if the constituent S_2, as observed, can be regarded as sufficiently free from temperature influences.

It can be further seen that the influence of diurnal tides is nearly negligible for any actual depth. For this reason the diurnal constituents should be nearly free from oceanic influences. Unfortunately there exist strong meteorological disturbances of diurnal period which add considerable uncertainty to these constituents, as also the observed phase difference show. The influence of the oceanic tides in middle latitudes on the factor G will be to increase it. Taking the normal mean depths of the oceans, the corrected values will be lower than the observed values.

Applying the values for a depth of 4800 m from Fig. 30 to the values of SCHWEYDAR it is found that the inequality in different directions is removed, but the corrected value of $D = 0.62$ is fairly low. BERROTH has also tried to calculate the influence of the distribution of the continents on the oceanic tides extending DARWIN'S similar calculations and he obtained the result that the vertical displacement on a continent for the M_2 constituent is only $\frac{1}{2}$ to $\frac{1}{3}$ of that calculated for a uniform ocean.

The general impression of these important investigations is that they give at least a first approximation and show that the influence of oceanic tides on continental stations is fairly small (confirming the results presented in Sect. 18) if only elastic and gravitaional effects are taken into account. The corrections for the sea tides do not remove the greater part of the observed discrepancies. For this reason it has to be assumed that significant disturbances of a different kind exist. See Chap. F, p. 835, and G, p. 839.

18. The numerical method. This method has been mainly developed and used by the Japanese observers[1] as there the experiments have to be performed fairly or very near to the shore. A careful correction of the results for tidal influences has therefore to be made and the deformation of the Earth produced by the tidal loads are of paramount importance. The results are also of great consequence for tidal gravity measurements as they throw some light on their discrepancies. The three effects A, B and C, as mentioned in para 16, have again to be taken into account, effect B being the largest for coastal stations.

a) In calculating the effect A, the attraction of the oceanic tides, the following numerical method has been developed: Let h be the surface of the sea above mean sea level. A segmental area may be determined by the part of a ring, limited by two segments of radii r_1 and r_2, having the observing station as a

[1] See for example R. TAKAHASHI [Bull. Earthq. Res. Inst. Tokyo **10**, 145, 531 (1932)] where also prior literature is given and NISHIMURA [7].

centre and extending to an angle between the azimuths A_1 and A_2, measured from south to west. Then the deflection of the plumb line due to the attraction of the seawater in this direction and distance will be

$$\delta_N^a = \frac{f\, d_w h}{g \sin 1''} \cdot \ln \frac{r_2}{r_1} \cdot (\sin A_2 - \sin A_1)\,, \tag{18.1}$$

$$\delta_W^a = \frac{f\, d_w h}{g \sin 1''} \cdot \ln \frac{r_2}{r_1} \cdot (\cos A_2 - \cos A_1)\,. \tag{18.2}$$

h is a function of time and place, for example for the M_2 tide

$$h = H_m \cos(2t - \gamma) = H_m \cos\gamma \cos 2t + H_m \sin\gamma \sin 2t \tag{18.3}$$

where H_m and γ are functions of r and A.

The actual values have to be taken from cotidal charts and the single segments have to be extended over areas where the amplitude and phase can be taken to vary only within a certain small interval. The numerical integration has to be taken over n such steps. Putting

$$\ln \frac{r_{n+1}}{r_n} = p; \qquad \cos A_{n+1} - \cos A_n = q; \qquad \sin A_{n+1} - \sin A_n = s \tag{18.4}$$

then for all values of n, where p, q and s are constants, the deflection of the vertical is given by

$$\varDelta_N = \frac{f \cdot d_w p\, s}{g \cdot \sin 1''} \left[(\Sigma H_n \cos\gamma_n) \cos 2t + (\Sigma H_n \sin\gamma_n) \sin 2t \right], \tag{18.5}$$

$$\varDelta_W = \frac{f \cdot d_w p\, q}{g \cdot \sin 1''} \left[(\Sigma H_n \cos\gamma_n) \cos 2t + (\Sigma H_n \sin\gamma_n) \sin 2t \right]. \tag{18.6}$$

Another correction has to be applied, if areas appear above sea level at low tides as it is the case in coastal waters.

b) Effect B, the influence of the tidal load on the Earth's surface is generally calculated following a method given by BOUSSINESQ (1878). According to this the tilt of the plane surface of an elastic body extending over an infinite half space with LAME's constants λ and μ due to a load of density d is given by

$$\delta^l = \frac{\lambda + 2\mu}{4\pi\mu(\lambda + \mu)}\, g\, d \frac{\partial}{\partial x} \int \frac{h\, d\sigma}{r}\,. \tag{18.7}$$

On the other hand, the deflection of the vertical due to the attraction of the same load will be

$$\delta^a = \frac{f\, d}{g} \frac{\partial}{\partial x} \int \frac{h\, d\sigma}{r}\,. \tag{18.8}$$

The tilt of the surface due to the load is equal in amount to the deflection of the vertical by the same load multiplied by

$$m = \frac{\lambda + 2\mu}{4\pi\mu(\lambda + \mu)} \cdot \frac{g^2}{f}\,, \tag{18.9}$$

$$B = m \cdot A\,. \tag{18.10}$$

Taking $\lambda = \mu$ and expressing μ in units of 10^{11} CGS this becomes

$$m = \frac{17.2}{\mu}\,. \tag{18.11}$$

At more exact way would be to use TAKEUCHI's method, but it has not yet been applied to actual cases. As his calculations show (Sect. 17a) the influence of the load becomes the more superficial the smaller its extension. The different parts of the oceans used in the numerical integration will therefore affect different depths with different elastic properties. Table 14 has already taken account of this.

By simultaneous observations on two stations NISHIMURA [7] could eliminate the effect of the Earth tides and evaluate the action of the oceanic tides. He could show that the main part of the oceanic effect is caused by the tides within a distance of 1000 km. He found empirically that m can be expressed as function of r by

$$m(r) = \frac{12.6}{r+3.0}, \tag{18.12}$$

where r is the mean distance measured in degrees between the sea segment and the point of observation.

c) Effect C has already been discussed in Sect. 17a. As TAKEUCHI's values have not yet been applied in detail to actual measurements the empirical value of NISHIMURA is usually taken. The effect is negative and proportional to A, so that

$$C = -nA, \quad \text{with} \quad n = 0.5. \tag{18.13}$$

A similar value has been obtained from the annual pole movement, as derived in Sect. 17a where the figure given there corresponds to $1-n$. TAKEUCHI's values are lower, expecially for higher harmonics (Table 13).

The total oceanic effect can therefore be expressed by

$$\Delta^0 = (1 + m - n)A \tag{18.14}$$

with the values given above.

NISHIMURA found an effective value of about

$$m = 2.71$$

for the upper mantle to about 1000 km depth and from this the effective rigidity becomes $\mu = 6.5 \times 10^{11}$ dyn/cm², assuming POISSON's ratio to be 0.27. This value is fairly low comparing it with the values assumed by TAKEUCHI (Fig. 27), but the upper layers may have a prevailing influence in this type of determination as TAKAHASHI has suggested.

It is yet an open question whether these corrections for oceanic tides are sufficient to remove the discrepancies in the observations at different stations. In the actual Earth's crust there exist many topographical irregularities and geological heterogeneities which will influence the results, and the elastic properties of the rocks deviate considerably from HOOKE's law for slow movements. There were few instances for example where a solution of the BOUSSINESQ type could be applied satisfactorily to the tidal stress measurements[1].

19. The empirical method. For the reasons stated above, a purely empirical method has been developed by CORKAN and DOODSON using the ratios of different tidal constituents of the sea and Earth tides. They assume that for partial tides of the same class, the ratio of the oceanic effects on the Earth tides is in the same proportion as the ratio of the oceanic tides themselves. Under the assumption that the diminution factor D is the same for partial tides of the same class, the ratio of the undisturbed Earth tides is given by their theoretical values for example

$$\frac{S_2}{M_2} = 0.465; \quad \frac{N_2}{M_2} = 0.193 \text{ (see Table 1)}. \tag{19.1}$$

[1] I. OZAWA: Disaster Prev. Res. Inst. Bull. No. 3, p. 4 1952.

The ratios of the partial sea tides for a given area can be deduced from observations and harmonic analysis. For the Northern Atlantic the values are

$$\frac{S_2'}{M_2'} = 0.320 \quad \text{and} \quad \frac{N_2'}{M_2'} = 0.190. \tag{19.2}$$

The observed deflection or tilt T due to a certain partial tide can be expressed as the sum of the direct tidal tilt DT_d and the superposed tilt T_s due to the sea tides. If it is assumed that there are no other disturbing influences it is

$$T = D\,T_d + T_s. \tag{19.3}$$

The expressions for the partial tilts are of the form

$$T = N\cos(A - \varkappa); \quad T_d = N_d\cos(A - b); \quad T_s = N_s\cos(A - s). \tag{19.4}$$

From this

$$\left.\begin{aligned} D\,N_d\cos b + N_s\cos s &= N\cos\varkappa,\\ D\,N_d\sin b + N_s\sin s &= N\sin\varkappa. \end{aligned}\right\} \tag{19.5}$$

It can be assumed that there is no phase lag in the undisturbed Earth tides, that is $b = 0$.

Then

$$\operatorname{tg} s = \frac{N\sin\varkappa}{N\cos\varkappa - D\,N_d} \tag{19.6}$$

$$N_s = \frac{N\sin\varkappa}{\sin s}. \tag{19.7}$$

To each value of D, special values of s and N_s correspond. In order to reduce them for other partial tides to the scale of M_2, the N_d are reduced by the theoretical ratios and the N_s by the empirical ratios as given above under (19.1) and (19.2). The results for a certain set of s values are graphically plotted with $N_s(M_2)$ as abscissa and $D\,N_d$ as ordinates. The (nearly straight) lines meet within a small area, forming a triangle whose centre of gravity (taking the proper weights for the different partial tides) determines the best values of s for each partial tide and from this, the actual values of N_s and $D\,N_d$ can be obtained.

Furthermore it is possible to calculate the influence of b on D for different values of b and so to get rid of the assumption of its being zero. Then the comparison of different stations, yielding different D after the simplified method, can be compared and the value of b can be determined which yields the same D for both stations. The method has been applied to observations at Bidston by CORKAN[1] yielding $D = 0.77$. Its application to a combination of the Bidston results with those at Bergen by CORKAN[1] gives 0.65 and $b = +8°$. This result is not very probable and indicates that the assumptions are not applicable in this case. The difficulty is that S_2 cannot be determined with sufficient accuracy from observations which are made near the surface. Only observations in deep mines and of sufficiently long periods of observations can be reliable. MELCHIOR[2] has applied CORKAN's method to SCHWEYDAR's measurements at Freiberg (Saxony) (see Table 4). Assuming $b = 0$ he obtains as the weighted mean, using the partial tides M_2, S_2 and N_2 in the East-West direction:

$$D = 0.725.$$

[1] R. H. CORKAN: Month. Not. Roy. Astronom. Soc., Geophys. Suppl. **4**, 481 (1939); **6**, 431 (1953).

[2] P.-J. MELCHIOR: Acad. Roy. Belg. Bull., Cl. Sci. **40**, 382 (1954).

This seems to be the most reliable value of D, but it is in contradiction to the values obtained in Eastern Asia. This indicates that the oceanic effect is not the only one disturbing the measurements, but that there must exist other indirect effects.

F. Meteorological influences.

These influences originate mainly in the temperature and pressure variations. Both deform the Earth's surface and thanks to elastic stresses caused by their variations influence not only the region where these variations occur, but even remoter parts of the interior or the surface of the Earth. Their amplitude may sometimes considerably exceed the direct tidal amplitudes of the solid Earth. The knowledge of these factors and their correction is for this reason of great importance. But this part of the movements of the Earth's surface has not yet obtained the attention which should be due to it, mainly in consequence of the complex character of the phenomena. But some general features have already been discovered, so that general corrections seem possible.

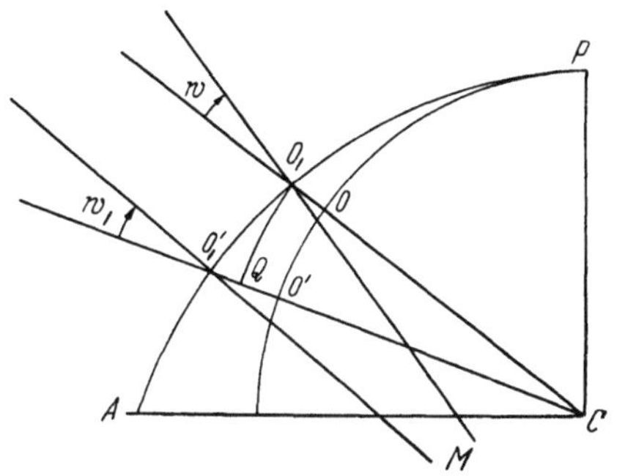

Fig. 31. Diagram of the thermal deformation of the Earth's globe.

20. Periodic influences. *a) Temperature.* The main periodic influences are a diurnal and, to a lesser degree, a semidiurnal period; higher harmonics are also present. These periodic disturbances penetrate surprisingly deep into the ground in the form of elastic stresses. The tilt components are quite considerable and therefore mainly disturbing in horizontal pendulum observations. Mean diurnal amplitudes of 100 milliseconds of arc have been observed at a depth of 10 feet, 10 millisec at about 75 feet and 1.5 millisec at 800 feet. Even at this depth the influence is about 25% of the M_2 amplitude. It can only be eliminated by harmonic analysis if sufficiently long periods (about a year) are available (Table 3).

Ehlert[1] and Schweydar[1] have shown that near the surface, these periodic deformations occur even on rainy days without any significant local temperature variation and it has been concluded that there must exist a more general phenomenon consisting of a deformation of the Earth's globe by the thermal radiation of the sun.

In Fig. 31 the arc POO' denotes the surface of the undeformed Earth PO_1O_1' the surface of the Earth heated by the sun, which is assumed to be in the direction CA. By the deformation the station at O will be moved to O_1 and a point immediately connected with it O' to O_1'. The radii of curvature intersect in the direction of M. The angles w and w_1 are the deflections due to the temperature tilt. If r is the radius of the ellipsoid at O_1 then $O_1'Q$ equals dr, with the angle $PCO_1 = 90° - \varphi'$ and $O_1'CO_1 = -d\varphi'$

$$w = -\frac{dr}{r\,d\varphi'}. \tag{20.1}$$

Introducing $CA = a$, $CP = b$ and

$$\frac{x^2}{a^2} + \frac{y^2 + z^2}{b^2} = 1$$

and putting $b = \varrho_1$ it follows

$$r = \frac{\varrho_1}{\sqrt{1 - e^2 \cos^2 \varphi'}} \tag{20.2}$$

[1] R. Ehlert: Gerlands Beitr. Geophys. **3**, Nr. 1, 147 (1896). — Schweydar: Gerlands Beitr. Geophys. **7**, 33 (1905).

which gives

$$w = \frac{e^2 \sin\varphi' \cos\varphi'}{1 - e^2 \cos^2\varphi'} \tag{20.3}$$

where φ' is identical with the zenithal distance of the sun ϑ_s. Therefore in sufficient approximation

$$w = e^2 \sin\vartheta_s \cos\vartheta_s . \tag{20.4}$$

If h is the maximum deformation

$$a = \varrho_1 + h \tag{20.5}$$

then, neglecting h^2

$$e^2 = \frac{2h}{\varrho_1 + 2h} \tag{20.6}$$

and with the same approximation

$$W = \frac{h}{\varrho_1} \sin 2\vartheta_s = w_0 \sin 2\vartheta_s . \tag{20.7}$$

This can be developed into the meridional component w_s, the declination of the sun being δ and θ the pole distance of the station. It is

$$\left.\begin{aligned} w_s &= [m \sin 2\delta \cos(t - M) + n \cos^2\delta \cos(2t - N)]\, w_0 \\ m \cos M &= -\cos 2\theta, \qquad n \cos N = \sin\theta \cos\theta, \\ m \sin M &= \sin\theta, \qquad n \sin N = \cos\theta . \end{aligned}\right\} \tag{20.8}$$

Consequently, w_s consists mainly of a semidiurnal period, the diurnal part being comparatively small.

The greater part of the observed strong diurnal amplitude cannot be explained by a *general* thermic deformation. There must exist *local* factors which are responsible for it, such as the asymmetrical heating of mountain ranges or hill ridges.

The semidiurnal period should show a semiannual variation of its amplitude; this is in accordance with observation.

The semidiurnal general temperature effect seems to be fairly well documented. From one year's observations at Heidelberg SCHWEYDAR finds as mean semidiurnal effect $w_0 = 0''.0039$, corresponding to $h = 2.2$ cm. This value should be corrected for the potential of the deformed masses, but the experimental data are lacking which would be needed to asses a mean value of w_0 for larger areas. As the meteorological and geological elements are different at other stations, the deformation h will vary from station to station. It is assumed that the maximum of the solar influence will take place at 2^h p.m. A phase lag against this maximum of 6 hours has been observed for the semidiurnal period and of 15 hours for the diurnal period. As the latter is empirically dominant the total maximum of temperature tilt occurs at about 6^h in the morning. Similar observations[1] have been made concerning gravimetric measurements (ΔS_1-effect), where the amplitudes were 6 microgal for ΔS_1 with a positive maximum at about 8^h in the morning and nearly of the same amplitude for ΔS_2.

The influence of thermo-elastic deformations in tunnels and galleries, and also the general case of the thermo-elastic deformation of the Earth's crust near oceans has been treated by HOMMA[2].

b) Barometric influences. There exist periodic barometric variations due to the diurnal temperature oscillations and the tides of the atmosphere[3]. In atmospheric tides the resonance effects are very pronounced. Nevertheless their effect

[1] R. TOMASCHEK: Month. Not. Roy. Astronom. Soc., Geophys. Suppl. **6**, No. 5, 286 and No. 6, 372 (1952).

[2] S. HOMMA: The Geophys. Magaz. Tokyo **23**, 125 (1952).

[3] See article by KERTZ in this volume, p. 928.

on Earth tide measurements is fairly small. The largest influence is semidiurnal, quite rapidly diminishing towards higher latitudes. These periods may influence the harmonic constituent S_2 in some measurements as e.g. with gravimeters which are not entirely compensated for buoyancy.

21. Non-periodic influences. a) Atmospheric pressure variations will influence the Earth's surface by changing its load. DARWIN[1] has shown that this influence cannot be neglected. He assumes an incompressible semi-infinite plane solid body of rigidity μ and on its surface a sinusoidal mass distribution of parallel ridges and valleys

$$p = d_m h_0 \cos\frac{x}{b} \tag{21.1}$$

where d_m is the density, h_0 the amplitude, $d_m \cdot h_0 = p_0$ the maximum pressure, and x the horizontal direction perpendicular to the ridges, the distance between maximum and minimum pressure being $L/2$, and $b = L/2\pi$. The potential of this mass distribution is at a depth z

$$V = \frac{3g\, d_m h_0}{2\varrho_1 d_e} b\, e^{-\frac{z}{b}} \cos\frac{x}{b} \tag{21.2}$$

where d_e is the mean density of the Earth.

On the surface ($z=0$) the forces are in horizontal direction

$$\frac{\partial V}{\partial x} = -\frac{3}{2}\,\frac{g\, d_m h_0}{\varrho_1 d_e} \sin\frac{x}{b} \tag{21.3}$$

in vertical direction

$$\frac{\partial V}{\partial r} = -\frac{3}{2}\,\frac{g\, d_m h_0}{\varrho_1^2 d_e} b \cos\frac{x}{b}. \tag{21.4}$$

The vertical elastic displacement due to the periodic pressure variation is given by

$$u_r = \frac{g\, d_m h_0}{2\mu} b\left(1 + \frac{z}{b}\right) e^{-\frac{z}{b}} \cos\frac{x}{b} \tag{21.5}$$

and on the surface

$$u_0 = \frac{g\, d_m h_0}{2\mu} b \cos\frac{x}{b}. \tag{21.6}$$

The gravitational components are approximately (for not too extended areas) in horizontal direction

$$\frac{\partial(-u g)}{\partial x} = \frac{g^2 d_m h}{2\mu} \cdot \sin\frac{x}{b} \tag{21.7}$$

in vertical direction

$$\frac{\partial(-u g)}{\partial r} = -u \frac{dg}{dr}. \tag{21.8}$$

If $L/2 = 2500$ km is assumed ($b = 8\times 10^7$ cm), with a difference of 25 mm Hg in pressure ($h = 1.25$ cm; $d_m = 13.6$ gr/cm³; $h \cdot d_m = 17.0$ gr/cm²) and NISHIMURA's empirical value of $\mu = 6\times 10^{11}$ dyn/cm² is taken, then

$$u_0 = 1.1 \text{ cm}.$$

The ground is 2.2 cm higher under the barometric depression than under the barometric high.. The horizontal deflection comes to

$$\partial V/\partial x = 7.1 \text{ microgal} = 1.5 \text{ millisec},$$

[1] Sir G. H. DARWIN: Coll. Pap. I, 444. Cambridge, Univ. Press 1907.

whereas slope (21.2) bears to deflection the same ratio as $\frac{1}{3}\varrho_1 d_e$ to μ/g and is therefore twice the amount of attraction. The total influence on a horizontal pendulum will therefore be about 4.5 millisec; that is of the order of the larger tidal constituents of the solid Earth tides. As $dg/dr = -3.086$ microgal/cm, the

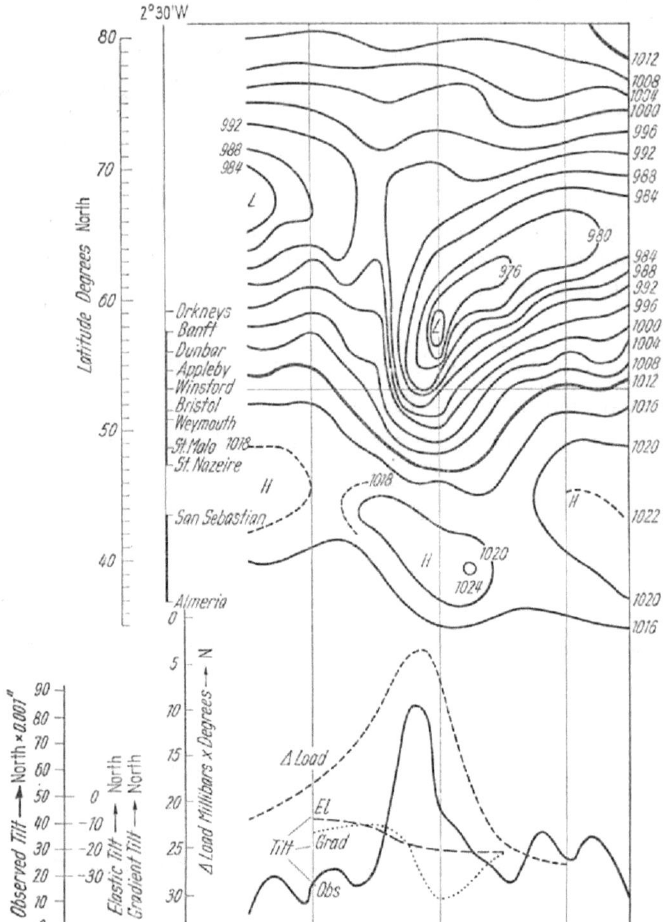

Fig. 32. Meteorological influence on Earth tilt at Winsford, probably due to tilting of Caledonian bloc (TOMASCHEK).

influence on a gravimeter will be about 7 microgal between maximum and minimum pressure due to the elasticity of the ground. This is a smaller influence than that of tilt, but in no way negligible.

Observations on the load influence of barometric pressure distributions with the horizontal pendulum have been discussed by ISHIMOTO[1], LETTAU[1] and TOMASCHEK[2]. See also[3].

Fig. 32 shows such a tilt movement accompanying the passing of a meteorological depression. In this figure the meridional pressure distribution as a function of time is shown

[1] M. ISHIMOTO: Bull. Earthq. Res. Inst. Tokyo **4** (1928). — H. LETTAU: Meteor. Z. **54**, 453 (1937). — Gerlands Beitr. Geophys. **51**, 250 (1937).

[2] R. TOMASCHEK: Geofisica Pura e Applicata **25**, 17 (1953).

[3] T. HAGIWARA and others: Bull. Earthq. Res. Inst. Tokyo **24**, 159 (1946).

In the lower part the observed meridional tilt component (including the tidal movement) is shown by the full curve. The tilts at Winsford calculated for purely elastic deformation due to the pressure distribution and also the local gradient tilt are shown by dotted lines. It is obvious that the observed tilt cannot be explained in this way. Only the curve load which is calculated under the assumption of the tilting of a coherent bloc with an axis at 58° N is able to offer an explanation. This indicates that the treatment of the deformations due to the pressure influences as purely elastic is, at least in many cases, quite remote from the actual behaviour of the Earth crust.

In order to study the actual movements the additional knowledge of the variation of the vertical component of gravity is necessary (see Fig. 35).

b) Although the inner accuracy of tiltmeter measurements over a period of one month is about two to four percent, there exist far larger differences between

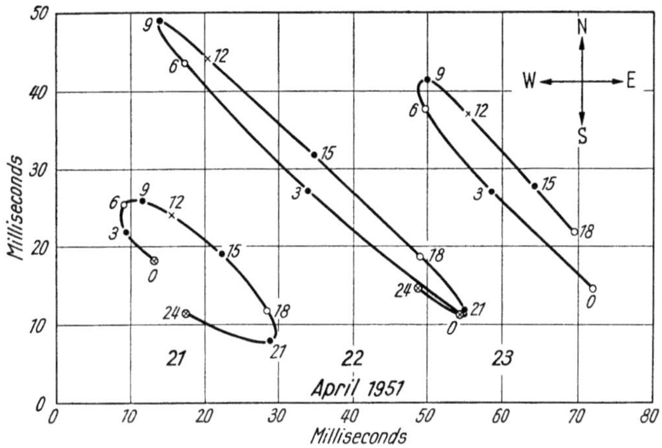

Fig. 33. Residuals of meteorological origin in tilt measurements at Winsford.

means of different months (see Table 2). NISHIMURA [7] has studied such variations in detail. He finds for two horizontal pendulums, when comparing the monthly means for one year, fluctuations of ± 13 and ± 21 % respectively. These fluctuations seem to originate from meteorologically caused tilts. Fig. 33 shows an example of such tilt residuals persisting for several days; such tilts repeat quite frequently during certain seasons. In this figure the Earth tides and oceanic influences have already been removed. The constancy of the direction of the tilts of this kind indicates the manifestation of special geological features as discussed in more detail in the next chapter.

c) *Influence of precipitations.* The elastic deformation of the Earth's crust caused by the local loading by great masses of snow has been discussed by STEINHAUSER[1] and the effect of its thawing has been measured with the horizontal pendulum by LETTAU[1].

G. Influences of the geological and tectonic structure.

22. Geological conditions. These influence the Earth tide measurements in a double way. Firstly the stresses in the rocks, especially at greater depth, give rise to strong drift movements which make it impossible at times to obtain any reasonable results at all. For this reason a moderate depth of about 200 m seems the optimum for placing the instruments to record earth tides. Secondly, asymmetric geological features in the neighbourhood such as faults, induce periodical residual movements. NISHIMURA [7] has studied a special case at Beppu using

[1] F. STEINHAUSER: Gerlands Beitr. Geophys. **41**, 466 (1934). — H. LETTAU: Gerlands Beitr. Geophys. **54**, Nr. 3, 179 (1939).

six stations. Fig. 34 shows the residual movements after elimination of Earth tides and oceanic influences. Two of the six stations (D and E) are especially

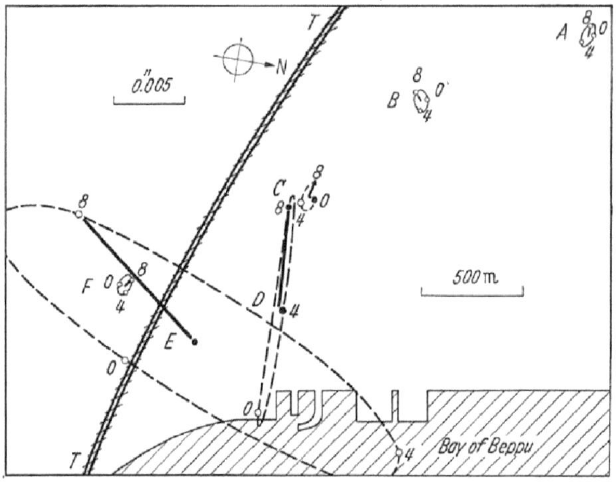

Fig. 34. Influence of a fault on tilt measurements (NISHIMURA).

influenced by a great active fault TT which divides the area into two parts which are known to have a vertical movement relatively to each other. The thick arrows

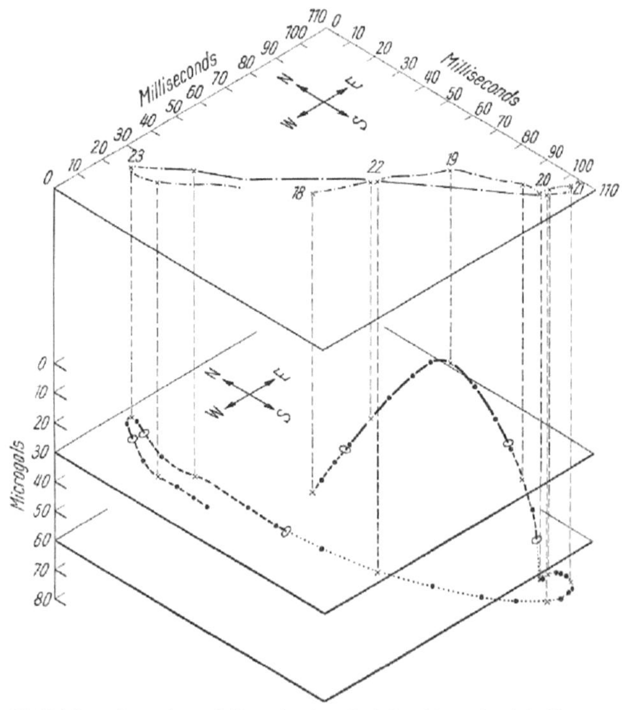

Fig. 35. Total gravity vector variations of meteorological and tectonic origin (TOMASCHEK).

indicate the tilt at the hour when Beppu Bay is at high tide. It seems, when the alluvial plain at the north of the fault (left of TT) is bent by the oceanic tide, that the fault line behaves like a free boundary, so that points like D and E

tilt towards the fault line rather than towards the ocean itself. The point F on firm ground of volcanic rock is only very little affected. This shows that the mosaic structure of the surface layer of the Earth can be investigated by a dense net of tilt meter stations, especially in coastal regions.

23. Tectonic conditions. For a given station the meteorologically induced tilts are remarkably independent in their direction from the situation of the inducing pressure distribution. The constant direction of these residual tilts shows that they are caused by structural elements. In this case the tilt is practically perpendicular to the CALEDONian structure which stretches from Norway to the southern part of the British Isles. Simultaneous observations with gravimeters and horizontal pendulums make it possible to determine not only the tilt, but also the elevation of these movements as shown in Fig. 35 which depicts the three-dimensional representation of Fig. 20 and contains the residuals after elimination of the tidal influences. The vertical movement which was connected with special pressure distributions amounts up to about 15 cm. The tilt has again not the direction of an elastic tilt. It indicates an oscillation of the CALEDONian structure perpendicular to its axis. The tilts induced by barometric pressure changes can therefore give information on the tectonic structure of certain regions.

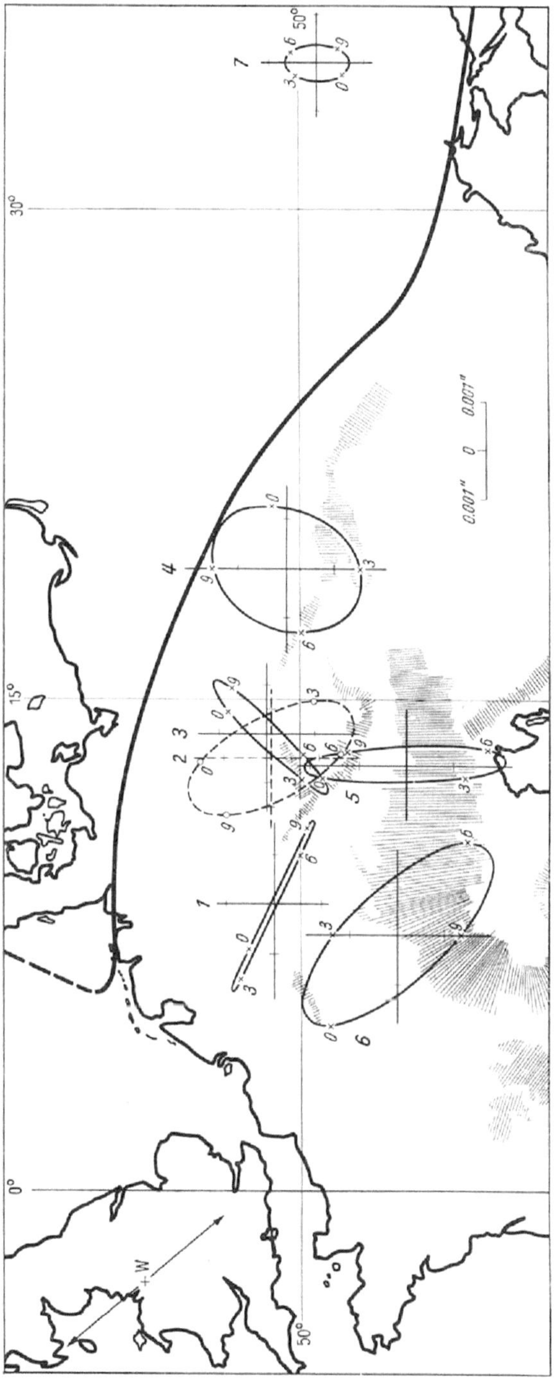

Fig. 36. Distribution of residuals of uniformly reduced M_2-tilts reflecting the tectonic structure. (Thick line: Approximate direction of border of the Baltic Shield.)

Another method leading to similar results is given by the analysis of the residuals of the regular tidal tilt movements. An example is shown by Fig. 36 where the observations of the residual terms due to M_2 on the European Continent are reproduced, after having been reduced in a uniform way[1]. A pattern of the residual tilt emerges which compares remarkably well with the tectonic structure of the continent, even in special features, such as the different axis of inclination in stations 2 and 3. In the western part of Europe a certain Atlantic influence is superposed. It seems that such measurements may help in revealing tectonic structures and their connections, especially when tilt observations can be connected with gravity measurements, so that local tilts can be distinguished from extended ones.

H. Miscellaneous topics.

24. Tidal changes in wells and volcanos. There exist several observations on tidal variations of the levels of wells or the water in mines (see Melchior [*1*]

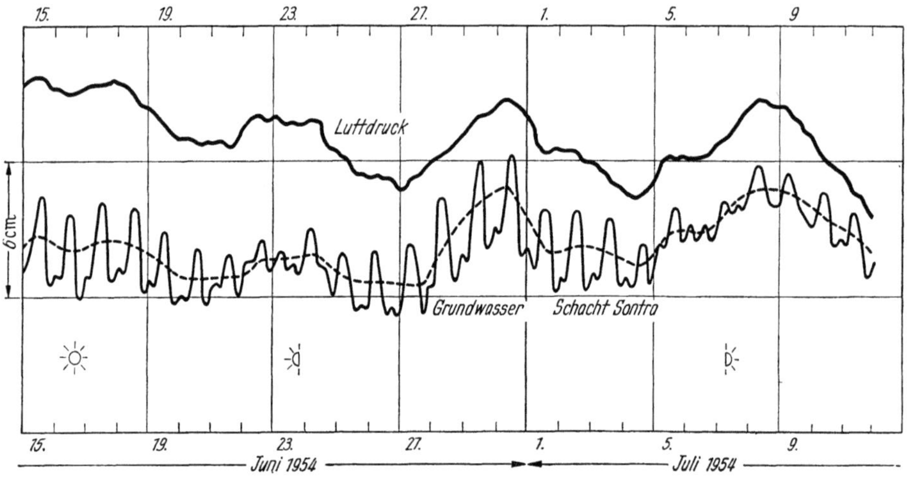

Fig. 37. Tidal variations of water level in a mine (Meissner).
Full Moon June 16, New Moon June 30; Quarters June 23 and July 8.

Chap. IX). The greater part of such variations is due to the barometric pressure variations which act with a time difference on the free water surface on the one side and on the water included in the capillaries of the surrounding rock on the other side[2], but genuine tidal effects are clearly effective. Fig. 37 shows the tidal variations as observed in a mine with a very well expressed difference between the spring and neap tides. Lambert[3] has given an explanation by showing that the tidal stresses in the Earth and the consequent increase in volume of the cavities must cause a sinking of the level. There exists for this reason an opposition in phase between the height of the water in the wells and the height of the tides of the solid Earth. This is in agreement with the observations. Table 15 shows data of the harmonic analysis of the variations of the water level for two subsequent months observed at Turnhout (Belgium) in a bore hole of 2000 m depth[4]. The opposition in phase is clearly seen in the values for M_2 which is the least

[1] R. Tomaschek: Nature, Lond. **173**, 143 (1954).
[2] R. Mügge: Z. Geophys. **20**, Nr. 2, 65 (1954). — R. Meissner: Z. Geophys. **21**, Nr. 2. 104 (1955).
[3] W. Lambert: U.S. Coast and Geod. Surv. Spec. Publ. No. 223, 1940.
[4] P. Melchior: Obs. Roy. Belg. 1956, in print.

disturbed partial tide. PEKERIS[1] has given a quantitative deduction and shows that by the lunar semidiurnal tide, a change of level of 1 cm in a cylindrical well of 1 m radius can be brought about by the tidal variations of the volume of a water-bearing hemisphere of only 100 m radius. He suggests observing these Earth tides by means of a dilatometer; a cavity of 10^6 cm^3 filled with a liquid and sealed by a capillary would undergo changes of the order of magnitude of 10 mm^3. This is in good agreement with TAKEUCHI's value of the volume divergence [Eq. (14.2)] which would give about 26 mm^3. No experiments seem yet to have been done on this line.

The effect of similar strains in oilwells has been observed by SPERLING[2]. An influence on volcanic processes is possible as has been shown for Vesuvius by IMBO[3] where semidiurnal lunar and solar influences seem to exist. It may be mentioned that perfect seismograms can also be obtained by the variation of the level of water in the mines[5].

Table 15[4]. *Harmonic constants of tides in a bore hole at Turnhout. January and February 1956.* (MELCHIOR 1956.)

	H mm		$g°$	
M_2	14.78	14.92	186.6	183.6
S_2	6.94	6.58	212.6	205.8
K_1	13.61	14.68	164.1	159.2
O_1	11.81	11.52	153.1	150.2

25. Earthquakes and strainrelease. An effect of the stresses which are connected with Earth tides on the occurence of earthquakes has been claimed[6]. It has been shown in Sect. 14 that the tidal stresses near the Earth's surface are about 10^4 CGS. This small value[7] makes it rather somewhat unlikely that the Earth tides themselves may have an influence on earthquakes, although trigger effects in very unstable situations cannot be excluded. BRAZEE[8] discussed from this point of view the earthquakes in Kern County, California in 1952. It was noted that the horizontal component of the tidal force might perhaps have a greater effect than the vertical component. The maximum frequency of earthquakes occurred when the horizontal vector was nearly parallel to the San Andreas Fault. The tidal load of the oceans should have a larger influence than the Earth tides themselves, as the local stresses in this case may be considerably higher. Furthermore, the influence of deformations by barometric pressure, as calculated in Sect. 21 may not be negligible[9].

The same may be true for the trigger effect in mine disasters[10] where the meteorological load influences may be more effective than the direct tides.

The records of horizontal pendulums have shown that in some cases a worldwide stress release by earthquakes can be observed[11]. The study by horizontal

[1] C. L. PEKERIS: U.S. Coast and Geod. Surv. Spec. Publ. No. 223, p. 23, 1940.

[2] K. SPERLING: Erdöl u. Kohle 6, 446 (1953). — R. TOMASCHEK: J. Inst. Petrol. **38**, 591 (1952).

[3] G. IMBO: Ann. Oss. Vesuv, Ser. VI, **1** (1954).

[4] g is the phase angle adjusted to the standard time kept at the place.

[5] See MÜGGE and MEISSNER above. R. C. VORKIS: Water Resourc. Bull. May 1953, p. 25. Further obs. and lit. in G. C. PARKER and V. T. STRINGFIELD: Economic Geology **45**, 441 (1950).

[6] L. A. COTTON: Bull. Seism. Soc. Amer. **12**, 47 (1922); Proc. 3rd Pan-Pacific Sci. Congr. (Tokyo) **2**, 1511, 1516 (1928); but see D. M. Y. SOMMERVILLE: Month. Not. Roy. Astronom. Soc., Geophys. Suppl. **3** (1932). Examples R. N. GHOSH: Current Sci. **3**, 61 (1934).

[7] See also CH. TSUBOI: Bull. Earthq. Res. Inst. **11**, 275 (1933).

[8] R. J. BRAZEE: Trans. Amer. Geophys. Un. **35**, 365 (1954).

[9] S. K. BANERJI: Current Sci. **3**, 409 (1935).

[10] See also K. MAINKA: Beitr. angew. Geophys. **6**, 241 (1937).

[11] R. TOMASCHEK: Nature, Lond. **176**, 24 (1955). See also H. BENIOFF: Bull. Geol. Soc. Amer. **62**, 331 (1951).

pendulums in seismic active regions[1] of the developing tilts and their change in time is a very important tool and these investigations are undertaken with the aim of making possible the prediction of earthquakes. Complementary measurements with registering gravimeters (Fig. 35) could show how far these tilts are local or regional.

26. Influence of Earthtides on the length of the day. The tides will influence the angular velocity of the Earth by changing the moments of inertia relative to the axis of rotation[2]. This effect has been calculated by ANDERSSON[3] under the same assumptions concerning the constitution of the interior of the Earth as made by SCHWEYDAR (Sect. 12b). These assumptions are incompressibility, ROCHE's law for density and rigidity and a value of $k=0.27$. The result is that the moon causes a time difference of $0\overset{s}{.}001$ within 7 days, (half of the fortnightly period) and the sun $0\overset{s}{.}008$ within a quarter of a year (half of the semiannual period). The variation due to the movement of the node of the moon's orbit is $0\overset{s}{.}278$ within 9 years. The maximum variation within 9 monts is $0\overset{s}{.}042$ and within 3 months $0\overset{s}{.}021$. It is mainly due to the change of the longitude of the moon's node.

MARKOWITZ[4] has analyzed the variable rotation of the Earth for the period 1951 to 1954 comparing the astronomical observations made with the photographic zenith tubes at Washington and Richmond with two quartz clocks. Although the main amplitude of the one year term of 30 milliseconds is apparently caused by meteorological effects, the observed semiannual term Ssa of 10 milliseconds and the fortnightly Mf and monthly Mm terms of 1 millisecond each are in good agreement with the values calculated by ANDERSSON.

27. Influence of the Earth's cosmic motion on the Earth tides. An influence of the cosmic motion of the Earth should show in the terms K_1 and K_2 which have periods of one sidereal day and half a sidereal day respectively, whereas influences of the revolution of the Earth in its orbit would show in S_1 and S_2.

COURVOISIER[5] using several different methods claims to have detected a cosmic effect. This effect was regarded as a contraction of the Earth's globe brought about by the LORENTZ-contraction which in turn is caused by its cosmic movement of several hundred km/sec together with the sun towards an apex at about $AR=84°$; $\delta=40°$. However measurements with the Bifilargravimeter and the principally simpler Interference Gravimeter[6] showed that the claimed effect on gravity does not exist. This result has been confirmed by the later pendulum and gravity measurements. There is no indication of an effect of the cosmic movement of the Earth beyond the limits of today's accuracy of observation.

Another effect has been deduced theoretically by SZOLNOKI[7]. It is supposed that for an observer on the rotating Earth there exists an EÖTVÖS effect which is due to the change in velocity caused by the combination of the rotation of the Earth round its axis and its orbital

---

[1] K. SASSA and E. NISHIMURA: Trans. Amer. Geophys. Un. **32**, 1 (1951). — E. NISHIMURA and K. HOSOYAMA: Trans. Amer. Geophys. Un. **34**, 597 (1953). — E. NISHIMURA: Disaster Prev. Res. Inst. Bull. No. 6, 1953. — W. F. BONCHKOVSKIJ: Akad. Nauk USSR., Publ. Inst. Seism. **1940**, Nr. 99. — Akad. Nauk USSR., Trudi Geofiz. Inst. **1955**, Nr. 25 (152), 134. — W. F. BONCHKOVSKIJ and S. NAMSARAI: Akad. Nauk USSR., Trudi Geofiz. Inst. **1954**, Nr. 22 (149), 3. — P. CALOI and M. C. SPADEA: Comm. Sci. X. Assembl. U. G. G. I. Roma 1954, p. 18. An excellent example is given by K. HOSOYAMA in Fig. 3 of J. Phys. of the Earth **1**, No. 2, 78 (1952).

[2] See also this Encyclopedia, Vol. XLVII, p. 21.

[3] F. ANDERSSON: Ark. Mat., Astronom. Fys., Ser. A **26**, Nr. 8 (1938). See also N. STOYKO: C. R. Acad. Sci., Paris **230**, Nr. 7, 620 (1950).

[4] W. MARKOWITZ: Trans. Amer. Geophys. Un. **36**, No. 3, 519 (1955).

[5] L. COURVOISIER: Astronom. Nachr. **226**, Nr. 5416 (1926); **230**, Nr. 5519 (1927); **234**, Nr. 5599 (1928); **237**, Nr. 5684 (1930). — Ann. Guébhard-Severine **8**, 264 (1932).

[6] R. TOMASCHEK and W. SCHAFFERNICHT: Astronom. Nachr. **244**, Nr. 5844 (1932).

[7] I. SZOLNOKI: Ann. d. Phys. (4) **67**, 73 (1922).

movement round the sum. This effect is similar to the change in gravitational acceleration of a body moving east or west on the surface of the rotating Earth. It has been calculated that for a latitude of 47° the effect on the vertical component should be mainly diurnal with an amplitude of about 11 milligals at the solstices and about 7 milligals at the equinoxes. This effect does not exist. The measurements with best temperature constancy show that, taking into account the elastic deformation of the Earth, no deviations from the expectations exist which would exceed about one thousandth of the expected effect.

28. The problem of an absorption of gravity. The effect on the Earth tides of a possible absorption of the gravitational force of the Sun or the Moon by the mass of the Earth has been discussed by RUSSELL[1]. Today's measurements are not sufficiently undisturbed and the amplitudes of the different harmonic constituents are not yet sufficiently well known as to exclude the possibility of effects of the order of several percents of the tidal action. Observations during an eclipse by TOMASCHEK[2] have shown that, caused by the interposed mass of the Moon, no effect exists which may exceed 2 to 3 thousandths of a milligal. The interposition of the Moon between the observer and the Sun could certainly not have altered the intensity of the field due to the Sun by more than one part in a hundred thousand. This yields an upper limit for an absorption coefficient of gravity of less than 10^{-14} C. G. S. units, when an exponential law of absorption is assumed.

Bibliography.

[1] Comprehensive reports on Earth tides:
MELCHIOR, PAUL-J.: Les Marées terrestres. Observ. Roy. Belg. Monograph **1954**, Nr. 4, p. 134. — Excellent review with nearly complete bibliography and a special discussion of the astronomical observations[3].
LAMBERT, W. D.: Rapports Généraux sur les Marées de l'Ecorce Terrestre. Travaux de l'Association de Géodésie de U.G.G.I. Reports to the meetings of the U.G.G.I. in 1927, 1930, 1933, 1936, 1939, 1948, 1951, 1953, 1955. — These reports give an excellent picture of the actual development of the problem and contain several important appendices.
TAKEUCHI, H.: J. Faculty Sci. Univ. Tokyo, Sect. II, **7**, Part II, 1—153 (1951). — More from the theoretical and mathematical point of view. Contains many original contributions.
Shorter reports:
TOMASCHEK, R.: Ergebn. exakt. Naturw. **12**, 36—81 (1933).
GOUGENHEIM, A.: Bull. d'Inform. du Comité central de Océanographie **3**, Nr. 8 (1951).
ELLENBERGER, H.: Sonderh. Z. Vermessungswes. **1954**, H. 2, 32p.
See also JEFFREYS, Sir H.: The Earth, 3rd edit. Chapt. VII and VIII. Cambridge: Univ. Press 1952.
[2] MICHELSON, A. A., and H. G. GALE: Astrophys. J. **39**, 105 (1914); **50**, 330 (1919).
[3] DOODSON, A. T., and R. H. CORKAN. Month. Not. Roy. Astronom. Soc., Geophys. Suppl. **3**, No. 6, 203 (1934).
CORKAN, R. H.: Month. Not. Roy. Astronom. Soc., Geophys. Suppl. **4**, No. 7, 481 (1939).
[4] SCHWEYDAR, W.: Zentralbur. Internat. Erdmessung, N. F. **38** (1921), 114 p.
[5] LOVE, A. E. H.: Some Problems of Geodynamics, Chapt. IV. 1911.
[6] TAKEUCHI, H.: Trans. Amer. Geophys. Un. **31**, No. 5, 651 (1950).
[7] NISHIMURA, E.: Trans. Amer. Geophys. Un. **31**, No. 3, 357 (1950).

---

[1] H. N. RUSSELL: Astrophys. J. **54**, 334 (1921).
[2] R. TOMASCHEK: Nature, Lond. **175**, 937 (1955).
[3] A supplement to this bibliography has been given by P. MELCHIOR, Marées terrestres, Bull. d'Inform CSAGI, Uccle, No. 1, 1956.

Flutwellen und Gezeiten des Wassers.

Von

A. DEFANT.

Mit 27 Figuren.

I. Flutwellen (Tidal waves).

1. Flutwellen in Kanälen konstanten Querschnittes. Als *lange Wellen* oder *Flutwellen* bezeichnet man solche Wellen an der Oberfläche einer Wassermasse, bei denen die Wellenlänge im Verhältnis zur Wassertiefe groß ist. Der Charakter der Wellenbewegung ist dann anders als bei den *Oberflächenwellen*, die in diesem Handbuch von H. ROLL behandelt werden (S. 671).

Die für „lange" Wellen geltenden Grundbeziehungen kann man leicht aus den allgemeinen Bewegungsgleichungen ableiten, wenn man die für diese Wellen charakteristischen Merkmale gleich anfangs in Rechnung stellt Diese sind in erster Linie folgende: Bei der in bezug auf die Wellenlänge geringen Wassertiefe treten die vertikalen Beschleunigungen gegenüber den horizontalen so sehr zurück, daß sie unberücksichtigt bleiben können. Der Druck an einem beliebigen Punkt der Wassermasse ist dann stets gleich dem statischen Druck, der durch die Wasserhöhe über dem Punkt gegeben ist. Damit wird aber die horizontale Bewegung für alle Wasserteilchen in Ebenen senkrecht zur Fortpflanzungsrichtung der Welle stets dieselbe. Die Horizontalgeschwindigkeit u der Wasserteilchen ist dann nur eine Funktion dieser Richtung x und der Zeit t.

Nimmt man einen geraden Kanal mit konstantem *rechteckigem* Querschnitt (x Längsrichtung des Kanals, z positiv aufwärts) und einer Wasserhöhe h im Ruhezustand, dann wird, bei einer zweidimensionalen (x, z) Wellenstörung, die Wasserhöhe $h+\zeta$ und der Druck an einem Punkt z

$$p = p_0 + g\varrho(h + \zeta - z),$$

wenn p_0 der als konstant angesehene äußere Druck ist, g = Fallbeschleunigung, ϱ = Wasserdichte. Die Bewegungsgleichung hat, wenn das als klein angesehene Advektionsglied $u\frac{\partial u}{\partial x}$ vernachlässigt wird, die Form

$$\frac{\partial u}{\partial t} = -g\frac{\partial \zeta}{\partial x}. \tag{1.1}$$

Führt man an Stelle der Horizontalgeschwindigkeit u die horizontale Wasserversetzung ξ bis zur Zeit t ein $(u = d\xi/dt)$, so erhält man die für Flutwellen charakteristische *Bewegungsgleichung*

$$\frac{\partial^2 \xi}{\partial t^2} = -g\frac{\partial \zeta}{\partial x}. \tag{1.2}$$

Die *Kontinuitätsgleichung* für Wasserbewegungen im Kanal hat die Form

$$\zeta = -h\frac{\partial \xi}{\partial x}, \tag{1.3}$$

und die vertikale Geschwindigkeit v der Wasserteilchen ist

$$v = -z\frac{\partial u}{\partial x}; \tag{1.4}$$

sie ist proportional der Höhe z des Teilchens über dem Boden. Eliminiert man aus diesen Gleichungen entweder ζ, ξ oder u, so erhält man die Differentialgleichung der Wellenbewegung in einem solchen Kanal

$$\frac{\partial^2 \xi}{\partial t^2} = g\,h\,\frac{\partial^2 \xi}{\partial x^2}; \tag{1.5}$$

Sie gilt in derselben Form auch für ζ und u.

Eine vollständige Lösung derselben ist

$$\xi, \zeta \quad \text{oder} \quad u = F(x - c\,t) + f(x + c\,t), \tag{1.6}$$

wobei $c = \sqrt{g\,h}$ ist. Sie stellt zwei Flutwellen dar, von denen sich die erste in der $+x$-Richtung, die zweite in der $-x$-Richtung mit der Geschwindigkeit c fortpflanzt. Diese LAGRANGEsche *Fortpflanzungsgeschwindigkeit langer Wellen* ist unabhängig von der Wellenlänge und proportional der Wurzel aus der Wassertiefe.

Auf lange Wellen, die sich in einer Richtung *ohne* Formveränderung fortpflanzen, läßt sich (wie bei den Oberflächenwellen) der RAYLEIGHsche *Kunstgriff der stationären Bewegung* anwenden. Man erteilt der ganzen Wassermasse eine der Fortpflanzungsgeschwindigkeit c entgegengesetzte Geschwindigkeit $-c$; dann wird die Bewegung stationär und die Anwendung des BERNOULLIschen Theorems ergibt für die Oberfläche

$$\frac{p}{\varrho} + g\,\zeta + \frac{c^2}{2}\left(1 - \frac{\zeta}{h}\right)^2 = \text{const.} \tag{1.7}$$

Die Bedingung $p = \text{const}$ für die freie Oberfläche liefert durch Reihenentwicklung $g\,\zeta + \frac{3}{2}\frac{\zeta^2}{h^2} + \cdots = c^2\frac{\zeta}{h}$. Wenn ζ^2 gegen h^2 klein ist, bekommt man daraus für c obige LAGRANGEsche Formel; eine weitere Annäherung ist

$$c = \sqrt{g\,h\left(1 + \frac{3}{2}\frac{\zeta}{h}\right)}. \tag{1.8}$$

Die Wellengeschwindigkeit ist zwar auch hier unabhängig von der Wellenlänge, aber die einzelnen Teile der Welle wandern nicht gleich schnell. Dadurch kann sich die Welle nicht ohne Formveränderung fortpflanzen. AIRY[1] hat durch sukzessive Annäherung zeigen können, daß die vorderen Wellenabhänge immer steiler, die hinteren sanfter werden, bis schließlich die Wellenkämme sich überstürzen und brechen. FJELDSTAD[2] hat für diesen Fall eine geschlossene Lösung gegeben, auf die hier verwiesen sei.

Ist der Kanalquerschnitt zwar in der x-Richtung konstant, aber beliebiger Form[3], so bleibt die Bewegungsgleichung (1.2) ungeändert, aber in die Kontinuitätsgleichung geht natürlich die Form des Kanalquerschnitts ein. Bei ungestörtem Wasser sei der Querschnitt S, seine Breite in der Wasserlinie b; dann erhält man, wenn keine Querströmungen im Kanal vorhanden sind, als Kontinuitätsgleichung

$$\frac{\partial (S\,\xi)}{\partial x} + b\,\zeta = 0 \tag{1.9}$$

[1] G. B. AIRY: Tides and waves. Encycl. metrop., § 208. London 1842.
[2] J. E. FJELDSTAD: Astrophys. Norv. **3**, 10 (1941).
[3] P. KELLAND: Trans. Roy. Soc. Edinburgh **14**, 497 (1893).

Wenn S und b konstant sind, ist $\bar{h} = S/b$ (die „*mittlere*“ *Tiefe* des Kanalquerschnitts) und (1.9) nimmt die Form (1.3) an, d.h. die Wellengeschwindigkeit folgt der LAGRANGEschen Formel, wenn man für h die mittlere Tiefe $\bar{h}$ nimmt. Ist die Breite b im Bereich der Änderungen des Wasserspiegels nicht konstant (flache Neigung der Uferböschungen), dann kann nicht mit dieser einfachen Formel gerechnet werden. MCCOWAN[1] hat den Fall eines Kanals mit schrägen Wänden behandelt.

AIRY[2] hat die Ansicht vertreten, daß in einem breiten Kanal, in dem Querströme wahrscheinlich sind, der Umstand, daß an jeder Stelle die Welle mit einer der Tiefe entsprechenden Geschwindigkeit fortschreitet, die Verzögerung der Welle an den seichteren Ufern eine *Krümmung* der Wellenkämme bedingt — in ähnlicher Weise, wie an einem flachen Strand die Dünnung stets der Küste zueilt, auch dann, wenn weit draußen bei größeren Tiefen die Welle der Küste parallel laufen sollte. Diese Ansicht ist schon bald nach ihrer Äußerung als nicht stichhaltig erkannt worden, aber erst PROUDMAN[3] hat gezeigt, daß wenigstens für einen Kanal parabolischen Querschnitts der Teil der AIRYschen Vermutung betreffs der Krümmung der Wellenkämme nicht zurecht besteht. Die Kammlinie ist stets eine Gerade quer zum Kanal, der Einfluß der verschiedenen Geschwindigkeiten der einzelnen Wellenteile gleicht sich aus.

Hingegen bleibt die *Wellenhöhe* quer zum Kanal nicht konstant und nimmt gegen die Ufer hin zu. Der Strom fließt bei Hochwasser in der Kanalrichtung, der Querstrom ist Null; bei fallendem Wasser wird die gegen die Kanalmitte gerichtete Strömung stärker und erreicht den Höchstbetrag, wenn der Wasserspiegel durch die Ruhelage geht, dann wird sie wieder schwächer und wird Null bei Niedrigwasser, wo der Strom nun wieder in der Kanalrichtung, aber jetzt entgegengesetzt wie früher setzt. Dann geht bei steigendem Wasser die Querströmung von der Mitte zu den Ufern. Die Stromfigur ist demnach eine Ellipse mit verschiedenem Drehsinn an beiden Ufern; in der Kanalmitte ist nur alternierender Strom vorhanden. Die Bewegungsgleichungen und die Kontinuitätsgleichung haben in diesem Fall die Form:

$$\frac{\partial u}{\partial t} = -g\frac{\partial \zeta}{\partial x}, \quad \frac{\partial v}{\partial t} = -g\frac{\partial \zeta}{\partial y} \quad \text{und} \quad \frac{\partial}{\partial x}(h u) + \frac{\partial}{\partial y}(h v) + \frac{\partial \zeta}{\partial t} = 0. \tag{1.10}$$

Außerdem ist die Bedingung zu erfüllen:

$$h v = 0 \quad \text{für} \quad y = \pm a, \tag{1.11}$$

wenn a die halbe Kanalbreite ist (Koordinatenursprung für y in der Kanalmitte). Setzt man eine Lösung in der Form

$$\zeta = Z\cos(\sigma t - \varkappa x), \quad u = U\cos(\sigma t - \varkappa x) \quad \text{und} \quad v = V\sin(\sigma t - \varkappa x) \tag{1.12}$$

worin $\sigma/\varkappa = c$ ist, dann müssen die Z, U, V als Funktionen von y allein die Beziehungen erfüllen

$$U = \frac{g}{\sigma} Z, \quad V = -\frac{g}{\sigma}\frac{dZ}{dy} \quad \text{und} \quad \frac{d}{dy}\left(h\frac{dZ}{dy}\right) + \left(\frac{\sigma^2}{g} - \varkappa^2 h\right) Z = 0. \tag{1.13}$$

Die Randbedingung (1.11) wird

$$h\frac{dZ}{dy} = 0 \quad \text{für} \quad y = \pm a. \tag{1.14}$$

[1] MCCOWAN: Phil. Mag. (5) **33**, 250—265 (1892).
[2] G. B. AIRY: Tides and waves, § 358—363. London 1842.
[3] J. PROUDMAN: Phil. Mag. (6) **49**, 466 (1925).

Ist der Querschnitt des Kanals parabolisch, dann ist

$$h = h_0\left(1 - \frac{y^2}{a^2}\right), \tag{1.15}$$

worin h_0 die maximale Tiefe in der Kanalmitte ist. Die mittlere Wassertiefe ist $\bar{h} = \frac{2}{3} h_0$. Die Lösung der Gleichungen für eine in ζ zur Kanalmitte symmetrischen Wellenform liefert für die Wellengeschwindigkeit

$$c = \sqrt{g\bar{h}}\left[1 - \frac{4\pi^2\alpha^2}{90}\right], \tag{1.16}$$

worin $\alpha = a/\lambda$ mit λ als Wellenlänge bedeutet. In erster Annäherung gilt somit die LAGRANGEsche Formel, wenn als Wassertiefe die mittlere Wassertiefe im Kanalquerschnitt genommen wird.

Ein Zahlenbeispiel mit einer Kanalbreite von 200 km, $h_0 = 100$ m, einer Wellenperiode $T = 12^h\,25^m$ (halbtägige Gezeitenperiode) ergibt folgende Verteilung der Wellenhöhe und des Verhältnisses v/u quer zum Kanal (Tabelle 1). Die Wellengeschwindigkeit wird $c = 0{,}9966\sqrt{g\bar{h}}$ und man erkennt, daß die Annäherung an die LAGRANGEsche Formel groß ist. Bei fast konstanter Wellenhöhe wandert

Tabelle 1.

	Kanal-mitte					Ufer
$y/a =$	0	0.1	0.2	0.3	0.4	0.5
Wellenhöhe =	100	100	101	102	103	105
$v/u =$	0.000	0.037	0.073	0.109	0.144	0.178

in einem parabolischen Kanal die Welle mit geradlinigen Kämmen und konstanter Geschwindigkeit. Es ist nicht zu bezweifeln, daß auch bei komplizierteren Querprofilen des Kanals der Einfluß der Bodenkonfiguration auf die Ausbildung der Wellen, was die Wellenkämme und die Wellengeschwindigkeit betrifft, von ähnlich kleinem Einfluß ist, wie im betrachteten Fall.

2. Flutwellen in Kanälen veränderlichen Querschnitts. Wird in der Kontinuitätsgleichung (1.9) die Querschnittsfläche $S = b\bar{h}$ gesetzt, dann wird sie:

$$\zeta = -\frac{1}{b}\frac{\partial}{\partial x}(b\,\bar{h}\,\xi).$$

Die mittlere Tiefe $\bar{h}$ und die Breite b sind nun Funktionen von x und die Bewegungsgleichung nimmt die Form an:

$$\frac{\partial^2\zeta}{\partial t^2} = \frac{g}{b}\frac{\partial}{\partial x}\left(\bar{h}\,b\,\frac{\partial\zeta}{\partial x}\right). \tag{2.1}$$

Das Verhalten von langen Wellen in Kanälen mit stetig veränderlichem Querschnitt rechteckiger Form hat zuerst GREEN untersucht. Er konnte zeigen, daß die Querschnittsänderungen zur Folge haben, daß die Amplitude der Welle in einem solchen Kanal sich umgekehrt proportional zur Quadratwurzel der Breite und zur vierten Wurzel der Wassertiefe verhält. Dies folgt, wie RAYLEIGH[1] zeigen konnte, in einfacher Weise aus dem Satz von der Erhaltung der Energie: Die Gesamtenergie einer Welle der Länge λ und der Amplitude a in einem Kanal der Breite b ist $E = 2Ca^2 b\lambda$, worin C eine Konstante ist. Wenn keine Reflexion

[1] Lord RAYLEIGH: Phil. Mag. (5) **1**, 260 (1876).

der Welle im Kanal stattfindet und die Periode dieselbe bleibt, muß, da $\lambda = cT = T\sqrt{gh}$ ist, die Größe $a^2 b \sqrt{h} = \text{const}$ sein. Dies bedeutet, daß a im Fortschreiten der Welle im Kanal sich wie $b^{-\frac{1}{2}} h^{-\frac{1}{4}}$ ändern muß. Dieses GREENsche Gesetz gilt aber nur unter der Voraussetzung, daß die Änderung von b und h im Bereich einer Wellenlänge λ klein sind.

Es ist nicht zu erwarten, daß bei den großen Wellenlängen der Gezeiten und der Seebebenwellen (Tsunami) das GREENsche Gesetz gilt. Die Frage, die sich THORADE[1] gestellt hat, war deshalb berechtigt: Wie verhält sich eine lange Welle, wenn die Wassertiefe sich *rasch* ändert? Er berechnet Fälle, in denen eine fortschreitende lange Welle von einem tiefen Ozean in ein seichtes Meer übergeht, und findet, daß die Amplitudenänderung nicht mehr der GREENschen Formel entspricht. Zugleich ändert sich aber auch die Laufzeit (Periode) der Welle. Die Untersuchungen von THORADE haben große Bedeutung im Falle der Gezeitenwellen, die von der Tiefsee auf den Schelf übertreten. Leider gibt es fast keine Beobachtungen, um die theoretischen Ergebnisse an der Erfahrung zu prüfen.

3. Stehende Wellen in abgeschlossenen Wassermassen (Seiches in Seen und Meeresbuchten). *α) Konstanter rechteckiger Querschnitt. Einfluß der Reibung auf die Eigenperiode.* Die Bewegungsgleichung (1.5) hat auch eine Lösung in der Form einer *stehenden Welle,* da die Superposition zweier fortschreitender Wellen mit entsprechender Phasenverschiebung immer zu einer solchen führt. Diese stehende Welle kann man sich auch entstanden denken durch *totale Reflexion* einer einfallenden an einer senkrechten Wand mit der reflektierten. Erfolgt die Reflexion bei $x = 0$ an der senkrechten Wand und ist ζ_1 die einfallende ζ_2 die reflektierte Welle, dann ist

$$\zeta_1 = \frac{a}{2}\cos(\sigma t - \varkappa x), \quad \zeta_2 = \frac{a}{2}\cos(\sigma t + \varkappa x) \quad \text{und} \quad \zeta = \zeta_1 + \zeta_2 = a\cos\varkappa x \cos\sigma t. \tag{3.1}$$

Die Erfüllung der Bewegungsgleichung liefert wieder $c = \sigma/\varkappa = \sqrt{gh}$ und die Periode der Schwingung $T = \lambda/c$ oder

$$T_n = \frac{2l}{n\sqrt{gh}} \qquad \text{(MERIANsche Formel)}, \tag{3.2}$$

wenn n eine beliebige ganze Zahl und l die Länge des *rechteckigen Troges* bedeuten. Für $n = 1$ erhält man die *längste Eigenperiode* der schwingenden Wassermasse $T_1 = \frac{2l}{\sqrt{gh}}$; in der Mitte des Troges $\left(x = \frac{l}{2}\right)$ ist ein Knotenpunkt vorhanden ($\zeta = 0$) mit maximalen Werten von ξ, was auch aus der Lösung für ξ

$$\xi = -\frac{a}{h\varkappa}\sin\varkappa x \cos\sigma t \quad \text{und} \quad u = \frac{ca}{h}\sin\varkappa x \sin\sigma t \tag{3.3}$$

ohne weiteres folgt. $n = 2, 3, \ldots$ gibt die zwei-, drei- usw. knotige Schwingung, wobei an den Enden des Troges stets Schwingungsbäuche vorhanden sind. Fig. 1 gibt die Stromlinien einer zweiknotigen Schwingung. In den *Schwingungsbäuchen* ist, mit einem Maximum der Vertikalbewegung, keine Horizontalverschiebung der Teilchen vorhanden ($u = 0$); in den *Knotenpunkten* verhält es sich umgekehrt.

Die Wasserbewegungen bei Schwingungen in einem Becken unterliegen Reibungseinflüssen einerseits durch die *innere Reibung,* andererseits durch *Grenzflächenreibung* am Boden und an den Wänden des Beckens. Letzterer

[1] H. THORADE: Mitt. math. Ges. Hamburg **6**, 217 (1926).

Einfluß dürfte mehr ausgeben als ersterer, da beim Hin- und Herrollen der gesamten Wassermasse im Becken die Bewegungen der einzelnen Teilchen recht geordnet vor sich gehen dürften. Die Reibungskraft kann in einfacher Weise proportional der Horizontalgeschwindigkeit gesetzt werden, also $-\beta\,\partial\xi/\partial t$, wobei β die Dimension [$\sec^{-1}$] hat. Die neue Bewegungsgleichung

$$\frac{\partial^2\xi}{\partial t^2}+\beta\frac{\partial\xi}{\partial t}-c^2\frac{\partial^2\xi}{\partial x^2}=0, \tag{3.4}$$

mit den Randbedingungen $\xi=0$ für $x=0$ und l, hat dann eine geschlossene Lösung in der Form:

$$\xi=A\,e^{-\frac{\beta}{2}t}\sin\varkappa x\cos\left[t\sqrt{c^2\varkappa^2-\frac{\beta^2}{4}}-\varepsilon\right]. \tag{3.5}$$

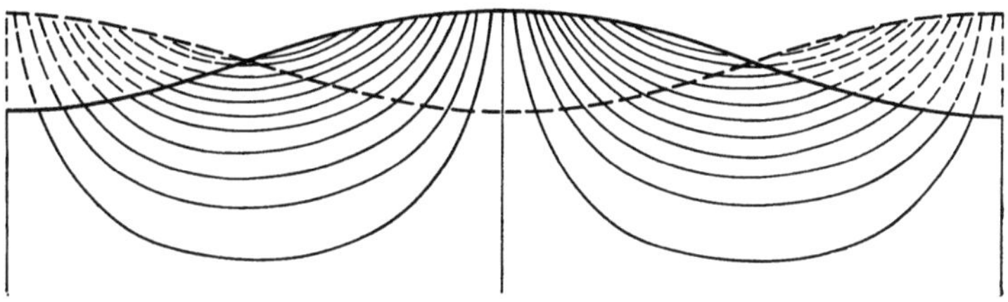

Fig. 1. Stromlinien einer zweiknotigen stehenden Schwingung.

Der Dämpfungsfaktor ist durch $e^{-\frac{\beta}{2}t}$ gegeben und aus dem Abklingen der Amplitude mit der Zeit läßt sich das *logarithmische Dekrement* Λ der gedämpften Schwingung ermitteln. Man findet $\Lambda=\beta\,T/4$. Bezeichnet man mit T_n die Periode der Schwingung ohne Reibungseinfluß, so wird jene der gedämpften Schwingung

$$T_R=T_n\left(1+\frac{\beta^2\,T_n^2}{32\pi^2}+\cdots\right). \tag{3.6}$$

β ist bei Wassertiefen von 50 bis 100 m von der Größenordnung $10^{-5}\,\sec^{-1}$. Der Faktor von T_n^2 in der Klammer wird dann größenordnungsgemäß 3×10^{-13}, so daß der Einfluß der Reibung auf die Periode im allgemeinen recht klein bleiben wird.

Will man eine genauere Bestimmung des Einflusses der *inneren* Reibung haben, dann muß man von den vollständigen Bewegungsgleichungen mit den üblichen Gliedern der turbulenten Reibung und den entsprechenden Randbedingungen ausgehen. Das Problem wird dann zweidimensional (x horizontal, z vertikal). Ist ν der kinematische Reibungskoeffizient [$\text{cm}^2\,\sec^{-1}$], der Koordinatenursprung in der ungestörten Wasseroberfläche, die Wassertiefe $-h$, so lauten Bewegungs- und Kontinuitätsgleichung:

$$\frac{\partial u}{\partial t}=-g\frac{\partial\zeta}{\partial x}+\nu\frac{\partial^2 u}{\partial z^2}\quad\text{und}\quad\frac{\partial}{\partial x}\int\limits_{-h}^{0}u\,dz+\frac{\partial\zeta}{\partial t}=0. \tag{3.7}$$

Dazu kommen die Randbedingungen $u=0$ für $x=0$ und l, sowie für $z=-h$. An der Oberfläche ($z=0$) muß ferner $\partial u/\partial z=0$ sein. Eine Lösung derselben hat Defant[1] gegeben. Man findet für die Eigenperiode des schwingenden Systems eine komplizierte Formel, aus der man nur aus Zahlenbeispielen entnehmen kann, wie stark sich die Wirkung der turbulenten Reibung auf die Verlängerung der Periode auswirkt. Beispiel: Für ein abgeschlossenes Meer von

[1] A. Defant: Beitr. Phys. fr. Atmosph. **19** (1932).

$l = 100$ km und $h = 20$ m Tiefe ergibt sich mit einem Turbulenz-Reibungskoeffizienten $\nu = 100\ \mathrm{cm^2\,sec^{-1}}$ eine Eigenperiode von $T_R = 4{,}35$ Std, während die MERIANsche Formel 3,96 Std liefert. Die Verlängerung der Periode beträgt somit 9,9%. Im allgemeinen ist aber die Verlängerung kleiner.

β) Variabler Querschnitt. Seiches-Theorien. Längliche Seen können als Becken von veränderlicher Breite und Tiefe angesehen werden, und die in ihnen enthaltenen Wassermassen können zu periodischen Schwingungen angeregt werden, wenn sie einmal aus dem Gleichgewicht gebracht worden sind. In systematischer Weise hat zuerst 1869 F. A. FOREL[1] die Seespiegelschwankungen des Genfer Sees untersucht. Diese Schwingungen werden dort „*Seiches*" genannt und seit dieser Zeit hat diese Bezeichnung für solche regelmäßige Seespiegelschwankungen, die an jedem See feststellbar sind, in die wissenschaftliche Literatur Eingang gefunden. FOREL verglich die aus den Beobachtungen abgeleiteten Perioden der Eigenschwingungen des Sees mit den aus der MERIANschen Formel $T = 2l/\sqrt{gh}$ berechneten, in der für h die mittlere Tiefe des Sees genommen wurde. Der Vergleich fiel nicht besonders gut aus und er blieb auch unbefriedigend, als man nach DU BOYS an Stelle der MERIANschen Formel $T = 2\int_0^l \frac{dx}{\sqrt{gh}}$ setzte, worin man für h nun die Tiefe längs des „Talweges" des Sees nahm.

Eine volle Erklärung der Probleme der Seiches länglicher Seen ergab erst die hydrodynamische Theorie solcher Schwingungen; sie gibt wohl über alle hier auftretenden Fragen eine erschöpfende Antwort. Im allgemeinen gilt diese Theorie nur für längliche Seen, das sind solche, bei denen im wesentlichen nur eine Horizontalbewegung der Wasserteilchen in der Richtung des Talweges auftritt. Die Bewegungsgleichung hat die frühere Form (1.2), die Kontinuitätsgleichung wird entsprechend (1.9)

$$\zeta = -\frac{1}{b(x)}\frac{\partial}{\partial x}[S(x)\xi], \tag{3.8}$$

wenn $b(x)$ und $S(x)$ die Breite an der Oberfläche und die Querschnittsfläche des Sees an der Stelle x des Talweges sind. Die zu erfüllenden Randbedingungen sind $\xi = 0$ für $x = 0$ und l, wenn l die Länge des Talweges ist.

Verschiedene Methoden sind zur Ermittlung der Schwingungsformen der Wassermassen länglicher Seen ausgearbeitet worden Die wichtigsten sollen hier kurz dargelegt werden.

CHRYSTAL*sche Methode*[2]. Führt man in (1.2) und (3.8) zwei neue Variable

$$u = S(x)\cdot\xi \quad \text{und} \quad v = \int_0^x b(x)\,dx \tag{3.9}$$

ein, so nehmen sie, mit

$$S(x)\cdot b(x) = \sigma(v), \tag{3.10}$$

die Form an:

$$\frac{\partial^2 u}{\partial t^2} = g\,\sigma(v)\frac{\partial^2 u}{\partial v^2} \quad \text{und} \quad \zeta = -\frac{\partial u}{\partial v}. \tag{3.11}$$

Hat ein See konstante Breite b, jedoch veränderliche Tiefe $h(x)$, dann ist mit $u = h\cdot\xi$

$$\frac{\partial^2 u}{\partial t^2} = g\,h(x)\frac{\partial^2 u}{\partial x^2} \quad \text{und} \quad \zeta = -\frac{\partial u}{\partial x}. \tag{3.12}$$

---

[1] F. A. FOREL: Le Léman, Bd. 2. Lausanne 1895.

[2] G. CHRYSTAL: Proc. Roy. Soc. Edinburgh **25**, 328—337, 637—647 (1904/05). — Trans. Roy. Soc. Edinburgh **41** (1905).

Die Gln. (3.11) und (3.12) sind identisch, d.h. die Schwingungen eines beliebig geformten Sees $[b(x), S(x)]$ sind geometrisch gesehen dieselben wie die eines Sees rechteckigen, aber in der Längsrichtung veränderlichen Querschnitts, wenn man nur $\sigma(v)$ als Talweg des Sees auffaßt Die Kurve mit den zwei Veränderlichen v [das ist die Oberfläche des Sees von dem einen Ende $(x=0)$ bis zum Querschnitt $S(x)$] als Abszisse und $\sigma(v)$ [das ist das Produkt aus $S(x)$ und $b(x)$ an der Stelle x] als Ordinate, wird als „*Normalkurve*" des Sees bezeichnet und ist für die Schwingungsformen des Sees maßgebend. Wählt man in (3.11) $u = A(x)\cos(\omega t + \varepsilon)$ mit $\omega = 2\pi/T$ als Frequenz der Schwingung, so ergibt die erste Gleichung

$$\frac{d^2 A}{d x^2} + \frac{\omega^2}{g \hbar(x)} A = 0 \tag{3.13}$$

in ihrer Lösung die Schwingungsformen des Sees konstanter Breite, aber veränderlichen rechteckigen Querschnitts. Läßt sich die Normalkurve eines Sees durch Teilstücke mehrerer einfacher Kurven $h(x)$ genügend genau approximieren, dann sind auch die Schwingungskurven dieses Sees bestimmt. CHRYSTAL hat für einige einfache Längsprofile $h(x)$ die Lösungen von (3.13) gegeben, so für horizontalen Boden, für Längsprofile aus zwei geneigten Geraden, für konkav und konvex parabolische Längsprofile usw. und damit die Unterlagen zur Ermittlung der Schwingungsformen von Seen mit recht komplizierten Normalkurven gegeben. Zur Festlegung der Eigenperiode der verschieden-knotigen Seiches führt die Lösung einer transzendenten Gleichung, die im Näherungsverfahren gelingt. Die Ermittlung der Schwingungsformen des Sees selbst (der vertikalen und horizontalen Verschiebungen längs des Sees) erfordert weitere, ziemlich umständliche Rechenarbeit. Deshalb ist die CHRYSTALsche Methode nicht sehr oft praktisch benützt worden; wo sie angewandt wurde, hat sie zu sehr befriedigenden Ergebnissen geführt Die eingehendste Anwendung auf einen größeren See, den Vettersee in Schweden hat F. BERGSTEN[1] gegeben. Siehe auch für den Gardasee DEFANT[2], für mehrere oberitalienische Seen CALOI[3].

DEFANT*sche Methode*[4]. Setzt man Ausdrücke der Form

$$\xi = \xi_0(x)\cos\left(\frac{2\pi}{T}t - \varepsilon\right) \quad \text{und} \quad \zeta = \zeta_0(x)\cos\left(\frac{2\pi}{T}t - \varepsilon\right),$$

wo ξ_0 und ζ_0 nur Funktionen von x sind, in die Bewegungs- und Kontinuitätsgleichung (1.2) und (3.8) ein, so erhält man für die endlichen Änderungen $\Delta\xi$ und $\Delta\zeta$ längs einer kleinen Strecke Δx des Talweges die Beziehungen:

$$\Delta\zeta = \frac{4\pi^2}{g T^2}\xi \cdot \Delta x \quad \text{und} \quad \xi = \frac{1}{S}\int_0^x \zeta b\, dx. \tag{3.14}$$

Dazu kommt noch die Randbedingung $\xi = 0$ für $x = 0$ und l. Wenn $\alpha = \frac{4\pi^2}{g T^2}\Delta x$ gegeben ist, gestatten diese Gleichungen eine schrittweise Berechnung von ξ und ζ für alle Querschnitte längs des Talweges Die Eigenperiode T des Sees ist aber nicht bekannt und soll erst ermittelt werden Es kann aber hierfür in erster Annäherung der Wert aus der MERIANschen Formel genommen werden (h die mittlere Tiefe des Sees). Dann ist α gegeben. Legt man die Querschnitte sehr eng, dann verlaufen in erster Annäherung die Änderungen von ξ und ζ zwischen zwei aufeinanderfolgenden Querschnitten linear und den Gleichungen

[1] F. BERGSTEN: Geogr. Annaler. Stockholm 1926.

[2] A. DEFANT: Sitzgsber. Wien. Akad. **1908**.

[3] P. CALOI: Verschiedene Arbeiten in Ann. di Geofisica, Roma ab 1949.

[4] A. DEFANT: Ann. Hydrogr.. Hamburg **1918**.

kann man für die praktische schrittweise Berechnung von ξ und ζ die bequeme Form

$$\left.\begin{aligned} &\zeta_2 = \zeta_1 + \alpha \frac{\xi_1 + \xi_2}{2}, \\ &\xi_2 = -\frac{1}{S_2\left(1 + \frac{\alpha v_2}{4 S_2}\right)} \left[q_1 + \left(\zeta_1 + \alpha \frac{\xi_1}{4}\right) v_2\right] \quad \text{und} \quad q_2 = q_1 + \frac{\zeta_1 + \zeta_2}{2} v_2 \end{aligned}\right\} \tag{3.15}$$

gegeben werden. Die Größen mit dem Index 1 bzw. 2 gelten für zwei aufeinander folgende Querschnitte; v_i ist die Oberfläche des Sees zwischen dem $(i-1)$-ten und dem i-ten Querschnitt; q_i ist eine Hilfsgröße und bedeutet die in der Zeit $T/2$ durch den i-ten Querschnitt hindurchgeschobene Wassermasse (q_0 ist Null). Man beginnt die Rechnung an einem Ende des Sees und setzt dort $\xi_1 = 0$ und ζ_1 willkürlich 100 cm. Dann geben die Formeln (3.15) die Werte ξ_2 und ζ_2 des zweiten Querschnittes und mit diesen berechnet man die des dritten usw. Kommt man zum anderen Ende des Sees, so muß, falls α (d.h. die Periode T der Eigenschwingung des Sees) richtig gewählt wurde, q und ξ dort Null sein. Im allgemeinen ist dies nicht der Fall, aber bei wiederholter Rechnung mit abgeändertem T bzw. α zeigt sich, daß T bald zwischen kleinen Werten von q bzw. ξ eingeschlossen und so beliebig genau ermittelt werden kann.

Die Methode, die sich auch bei komplizierter Gestaltung des Sees anwenden läßt, ergibt neben der Periode zugleich auch die Verteilung von ξ und ζ längs des Sees, also auch die Lage der Knotenlinien und die ganze Schwingungsform der Seiches. Mit ξ sind aber auch die mit den Seiches verbundenen Strömungen innerhalb des Sees gegeben; denn es ist

$$u = \frac{d\xi}{dt} = \frac{2\pi}{T} \xi_0(x) \cos\left[\frac{2\pi}{T}\left(t + \frac{T}{4}\right) - \varepsilon\right]. \tag{3.16}$$

Durch diese Umstände ist diese Methode den anderen weit überlegen.

Methode von HIDAKA[1]. Sie beruht auf eine Anwendung des RITZschen Theorems der Variationsrechnung bei der Integration der CHRYSTALschen Differentialgleichungen (3.11). Wählt man als neue Variable $z = v/a$, wo a die ganze Oberfläche des Sees ist, dann ist $\sigma(z)$ die Normalkurve des Sees und mit $u = A(x) \cos\left(\frac{2\pi}{T} t + \varepsilon\right)$ erhält man:

$$\frac{d^2 A}{dt^2} + \frac{\lambda}{\sigma(z)} A = 0, \quad \text{worin} \quad \lambda = \frac{4\pi^2 a^2}{g T^2} \tag{3.17}$$

ist. Diese Differentialgleichung ist mit den Randbedingungen $A(0) = A(1) = 0$ zu lösen. Sie ist nach dem RITZschen Theorem äquivalent der Beziehung:

$$\int_0^1 \left\{\left(\frac{dA}{dz}\right)^2 - \frac{\lambda}{\sigma(z)} A^2\right\} dz = \text{Minimum}. \tag{3.18}$$

Setzt man für $A(z)$ die Reihe

$$A(z) = \sum_{i=0}^{m} z(1-z) z^i A_i \tag{3.19}$$

dann lassen sich die unbekannten A_i aus der Minimumbedingung (3.18) in langwieriger Rechnung ermitteln. Meistens begnügt man sich in (3.19) mit den Gliedern bis $m = 2$ und man erhält dann zur Ermittlung von A_0, A_1, A_2 drei simultane

[1] K. HIDAKA: Mem. Imp. mar. Obs. **6** (1936).

homogene Gleichungen, die nur bestehen können, wenn ihre Determinante verschwindet. Diese liefert eine kubische Gleichung zur Bestimmung von λ und damit der Eigenperiode der ein-, zwei-, und dreiknotigen Seiche des Sees. HIDAKA hat seine Methode auf den Yamanakasee, der eine schöne regelmäßige Form hat, angewandt und Perioden erzielt, die mit den beobachteten ausgezeichnet übereinstimmen. Eine Anwendung auf die große, fast völlig abgeschlossene Wassermasse der Ostsee hat G. NEUMANN[1] gegeben.

Methode von ERTEL[2]. Diese Methode schließt sich an die wellenmechanische Störungsrechnung an. Geht man von der Gl. (3.17) aus, führt in ihr an Stelle von λ die Frequenz der Eigenperiode $\omega = 2\pi/T$ ein, so bekommt man:

$$\frac{d^2 A}{d z^2} + \frac{\omega^2}{\varphi(z)} A = 0, \quad \text{worin} \quad \varphi(z) = \frac{g}{a}\,\sigma(z) \tag{3.20}$$

ist.

Im Falle eines Beckens konstanter Breite und konstantem Querschnitts wird φ konstant, $\varphi_0 = g\bar{h}/l^2$, wenn l die Länge und $\bar{h} = S/b$ die Tiefe des Sees sind. Die Lösung für diesen Fall unter den üblichen Randbedingungen ist:

$$A_0 = \sqrt{2}\sin\pi z, \tag{3.21}$$

wobei A_0 entsprechend $\int\limits_0^1 A_0^2\,dz = 1$ durch den Faktor $\sqrt{2}$ normiert ist. Als Periode der ersten Eigenschwingung ergibt sich die MERIANsche Formel. Zur Lösung der allgemeineren Gl. (3.20) macht ERTEL den in der Störungstheorie üblichen Ansatz:

$$\left.\begin{aligned} A &= A_0 + \Delta A \\ \varphi &= \varphi_0 + \Delta\varphi \end{aligned}\right\} \quad \text{und} \quad \frac{\omega^2}{\varphi} = \frac{\omega_0^2}{\varphi_0} + \Delta\left(\frac{\omega^2}{\varphi}\right). \tag{3.22}$$

Die Gl. (3.20) reduziert sich damit auf

$$\frac{d^2(\Delta A)}{d z^2} + \frac{\omega_0^2}{\varphi_0}\Delta A = -\Delta\left(\frac{\omega^2}{\varphi}\right)\cdot A_0. \tag{3.23}$$

Der Störungsterm $-\Delta\left(\frac{\omega^2}{\varphi}\right)$ transformiert ERTEL unter Vernachlässigung von Größen zweiter Ordnung in $\frac{\Delta\omega^2}{\varphi_0} - \omega_0^2\frac{\Delta\varphi}{\varphi_0^2}$, was, wie FR. DEFANT[3] zeigte, nur erlaubt ist, wenn $\frac{\Delta\varphi}{\varphi_0} \ll 1$ bleibt. Dies ist aber nicht immer der Fall, weshalb man bei der Form (3.23) bleiben muß. Diese Differentialgleichung hat nur dann eine Lösung, wenn der Störungsterm auf der rechten Seite orthogonal auf die Lösung des homogenen Falls ist. Dies fordert

$$\int\limits_0^1 -\Delta\left(\frac{\omega^2}{\varphi}\right) A_0^2\,dz = 0. \tag{3.24}$$

Man bekommt dann mit (3.21) für die Frequenz der einknotigen Seiche

$$\omega^2 = \frac{\omega_0^2}{2}\,\frac{1}{\int\limits_0^1 \frac{\sin^2\pi z}{\left(1+\frac{\Delta\varphi}{\varphi_0}\right)}\,dz}. \tag{3.25}$$

---

[1] G. NEUMANN: Arch. dtsch. Seewarte **61** (1941).

[2] H. ERTEL: Sitzgsber. preuß. Akad. Berlin **1933**.

[3] F. DEFANT: Arch. Meteor., Geophys. u. Bioklimat. **6**, 218–241 (1953).

Für $\Delta\varphi = 0$ (rechteckiges Becken $\varphi = \varphi_0$) wird das Integral gleich $\frac{1}{2}$ und $\omega = \omega_0$. Für die zwei- oder mehrknotigen Seiches ist in (3.25) statt πz zu setzen $2\pi z$ oder $n\pi z$. FR. DEFANT hat diese Methode erstmalig zur Bestimmung der Eigenschwingungen des Michigansees angewandt und hierfür Perioden bekommen, die mit den aus anderen Verfahren berechneten um weniger als 1% abweichen.

Es liegt noch eine Methode von PROUDMAN[1] vor, die ebenfalls an die Differentialgleichung (3.17) anschließt. DOODSON, CAREY und BALDWIN[2] haben sie zur Berechnung der Eigenperioden des Genfer Sees erprobt. Sie hat gegenüber dem CHRYSTALschen Verfahren den Vorteil, daß eine Anpassung der Normalkurve an einfache analytische Kurvenstücke nicht erforderlich ist. Eine „japanische" Methode (HONDA, TERADA, JOSHIDA und ISITANI[3]) ist ein Spezialfall der ERTELschen Methode und gibt wegen der auch von ERTEL zuerst benützten Vernachlässigung von Glieder höherer Ordnung in $\Delta S/S_0$ und $\Delta b/b_0$ nicht besonders gute Resultate gegenüber den anderen Verfahren, die alle mehr oder minder gleichwertig sind.

γ) Stehende Schwingungen in teilweise offenen Wassermassen. Wenn eine Wassermasse in einer Beckenform an einer Stelle mit einer wesentlich ausgedehnteren in Verbindung steht, kann sie ebenfalls freie Schwingungen ausführen. Diese sind dann Eigenschwingungen von nur *einseitig* geschlossenen Kanälen oder Meeres- bzw. Seebuchten. Bei solchen Schwingungen können die erforderlichen Wassermassen durch die Mündung der Bucht aus der vor ihr liegenden größeren, ausgedehnteren Wassermasse ohne Schwierigkeit in horizontaler Richtung bezogen werden, so daß an der Mündung der Bucht stets eine Knotenlinie der stehenden Welle vorhanden sein wird. Die längste Eigenschwingung der Bucht muß deshalb identisch mit jener eines Beckens sein, das aus zwei, zur Mündung spiegelbildlich gleichen Hälften besteht. Für einen einseitig geschlossenen Kanal der Länge l und konstanten rechteckigen Querschnitts der Tiefe h ist die längste Eigenperiode

$$T = \frac{4l}{gh}. \tag{3.26}$$

Wenn die Breite des Kanals an der Mündung b ein merklicher Bruchteil der Länge l der Bucht ist, dann werden bei den Schwingungsvorgängen an der Mündung nicht allein Bewegungen in der Längsrichtung der Bucht auftreten; es kann dort Wasser auch von der Seite herangesaugt oder seitwärts abgestoßen werden. In Analogie zu ähnlichen Vorgängen in einseitig offenen Pfeifen kann man erwarten, daß dadurch die Eigenperiode etwas verlängert wird. Die verbesserte Periode ist dann $T(1+\varepsilon)$ und die *Mündungskorrektion* ε läßt sich durch Übertragung diesbezüglicher Untersuchungen von RAYLEIGH[4] für Pfeifen zu

$$\varepsilon = \frac{b}{\pi l}\left[\frac{3}{2} - \ln\frac{\pi b}{4l} - \gamma\right] \tag{3.27}$$

ableiten, worin γ die EULERsche Konstante 0,577215 ist. Für verschiedene Verhältnisse $b:l$ wird:

Breite/Länge	= 1	$\frac{1}{2}$	$\frac{1}{3}$	$\frac{1}{4}$	$\frac{1}{5}$	$\frac{1}{10}$	$\frac{1}{20}$
Korrektionsfaktor $(1+\varepsilon)$	= 1,371	1,297	1,240	1,203	1,176	1,110	1,1066

J. PROUDMAN[5] hat gezeigt, daß diese Mündungskorrektion nur für Kanäle gilt, bei denen an der Mündung die Küste rechtwinklig abzweigt und eine gleichförmige

[1] J. PROUDMAN: Proc. Math. Soc., Lond. **14** (1941).
[2] A. T. DOODSON, R. M. CAREY u. R. BALDWIN: Trans. Roy. Soc. **52** Edinburgh **1920**.
[3] HONDA, TERADA, JOSHIDA, ISITANI: J. Coll. Sci. **24** (1908).
[4] Lord RAYLEIGH: Theorie des Schalles. Übers. von F. NEESEN, 2. Teil, § 265.
[5] J. PROUDMAN: Mon. Not. Roy. Astronom. Soc., Geophys. Suppl. 1, 247–270 (1925).

Stromverteilung quer zur Mündung vorhanden ist. Ist der Küstenverlauf trichterförmig, indem er mit der Mittelachse des Kanals einen Winkel $n\pi$ bildet, dann tritt an Stelle von (3.27) die Formel

$$\varepsilon = \frac{1}{2n}\frac{\pi}{bl}\left[\beta_n - \ln\frac{1}{2n}\frac{b}{4l} - \gamma\right], \quad \text{mit} \quad \beta_n = n\int\limits_0^1 \frac{1-(1-t)^n}{t}\,dt. \tag{3.28}$$

Der Einfluß der Gestalt der Mündung erweist sich als recht groß. Man findet für

				nach Gl. (3.27)
$n\pi =$	45°	90°	135°	90°
Mündungsfaktor $(1+\varepsilon) =$	1,285	1,259	1,233	1,265

Die Ermittlung der Eigenperiode von Meeresbuchten und Meeresstraßen unregelmäßiger Gestalt läßt sich nach den früher dargelegten Methoden durchführen, nur muß man die Randbedingungen entsprechend ändern.

δ) Schwingungen gekoppelter Systeme. Die theoretische Behandlung von Seebecken- oder Meeresbuchten-Kombinationen, wie sie in der Natur häufig vorkommen, nach den üblichen hydrodynamischen Methoden trifft auf große Schwierigkeiten, die sich zwar überwinden lassen, aber nicht sehr übersichtliche Resultate ergeben. Einfachere Fälle hat ZEILON[1] (Schwingungen des Gulmarfjords) und G. NEUMANN[2] untersucht. Es handelt sich dabei zumeist um komplizierte Beckengestaltungen, bei denen durch starke Querschnittsverengungen der ganze Schwingungsbereich nicht mehr durch eine stetige Schwingungsfunktion allein erfaßt werden kann. Natürlich kann dabei jeder Teil für sich schwingen, aber zumeist gibt es Schwingungen des ganzen Beckenkomplexes, die eine ganz andere Größenordnung in der Periode haben. G. NEUMANN[2] hat den Gedanken gehabt, den Begriff der „*Impedanz*" schwingender Systeme, der sich in der Theorie elektrischer und akustischer Schwingungen sehr fruchtbar erweisen hat, auch auf die Schwingungen begrenzter Wassermassen anzuwenden. Die in dieser Theorie abgeleiteten Formeln gelten zwar nur für Becken konstanten rechteckigen Querschnitts, lassen sich aber durch kleine formale Änderungen auch auf Seen und Buchten unregelmäßiger Gestalt übertragen.

Nimmt man an, daß die periodischen Wasserbewegungen $\xi \sim e^{i\varkappa x}$ durch eine auf das Wasser in horizontaler Richtung wirkende periodische Kraft X erzwungen werden, dann lautet die Bewegungsgleichung

$$\frac{\partial^2\xi}{\partial t^2} + \sigma^2\xi = X, \tag{3.29}$$

worin $\varkappa c = \sigma$ die Eigenfrequenz des Systems ist. Ist ω die Frequenz der Kraft, also $X = A\,e^{i\omega t}$, und setzt man $u = \frac{\partial\xi}{\partial t} = U e^{i\omega t}$ und $\xi = \frac{1}{i\omega} U e^{i\omega t}$, dann wird aus (3.29)

$$U\left(i\omega + \frac{\sigma^2}{i\omega}\right) = A. \tag{3.30}$$

Die Größe

$$Z = \frac{A}{U} = i\omega + \frac{\sigma^2}{i\omega} \tag{3.31}$$

kann man in Analogie zur Theorie elektrischer und akustischer Schwingungen die *Impedanz* des hydrodynamischen Schwingungssystems nennen. Man kann aus $Z=0$ die Eigenperiode des Systems berechnen; denn für $Z=0$ gibt (3.31) $\omega=\sigma$.

[1] N. ZEILON: Svensk. Hydrogr.-Biol. Komm. Skr. **5** (1913).
[2] G. NEUMANN: Ann. Meteor. Hamburg **1944**, 65, 193.

NEUMANN hat für verschiedene Glieder eines gekoppelten Schwingungssystems die ihnen zukommenden Impedanzen Z berechnet. Die wichtigsten Fälle sind folgende ($q = bh$ ist die Querschnittsfläche eines Verbindungskanals, $S = bh$ die eines Sees, die Dichte des Wassers $= 1$, λ die Wellenlänge):

Einseitig geschlossenes Becken:

$$Z = -\frac{ic}{S} \operatorname{ctg} \frac{\sigma l}{c} \quad \text{mit} \quad l = \frac{\lambda}{4}; \tag{3.32}$$

für ein beiderseits geschlossenes Kanalstück gilt dieselbe Formel mit $l = \lambda/2$.

Beiderseits offener Kanal

$$Z = \frac{i\sigma}{S} \operatorname{tg} \frac{\sigma l}{c} \quad \text{mit} \quad l = \frac{\lambda}{2}. \tag{3.33}$$

Enge Abflußöffnung oder enger Kanal der Länge l

$$Z = \frac{i\sigma l}{2}. \tag{3.34}$$

Bei der Behandlung von Eigenschwingungen kombinierter Beckensysteme ist zu beachten, daß sich die Impedanzen der einzelnen Teilstücke in derselben Weise addieren wie die Widerstände elektrischer Stromkreise oder akustischer Schwingungskreise. Dabei ist zu unterscheiden, ob die Teilbecken „parallel" oder in Serie „geschaltet" sind. Sind z.B. drei Kanäle vorhanden, so kann sich der Kanal I in die Kanäle II und III gabeln. Hier liegen diese zwei Kanäle mit den Impedanzen Z_2 und Z_3 „parallel" hinter I. Die Gesamtimpedanz P von II und III ist dann entsprechend den Regeln $\frac{1}{P} = \frac{1}{Z_2} + \frac{1}{Z_3}$ oder $P = \frac{Z_2 Z_3}{Z_2 + Z_3}$. Der Kanal I ist hingegen mit (II + III) in „Serie geschaltet", so daß die Gesamtimpedanz des ganzen Systems $Z = Z_1 + P = Z_1 + \frac{Z_2 Z_3}{Z_2 + Z_3}$ ist. Die Eigenfrequenz ergibt sich dann aus $Z = 0 = \frac{1}{Z_1} + \frac{1}{Z_2} + \frac{1}{Z_3}$. Setzt man für Z_i (3.32) ein, so erhält man als Periodengleichung

$$b_1 c_1 \operatorname{tg} \frac{\sigma l_1}{c_1} + b_2 c_2 \operatorname{tg} \frac{\sigma l_2}{c_2} + b_3 c_3 \operatorname{tg} \frac{\sigma l_3}{c_3} = 0.$$

In ähnlicher Weise erhält man:

für den Fall: Kanal I offen gegen das freie Meer, II und III geschlossen

$$-b_1 c_1 \operatorname{ctg} \frac{\sigma l_1}{c_1} + b_2 c_2 \operatorname{tg} \frac{\sigma l_2}{c_2} + b_3 c_3 \operatorname{tg} \frac{\sigma l_3}{c_3} = 0;$$

für den Fall: Kanal I und II offen, III geschlossen:

$$-b_1 c_1 \operatorname{ctg} \frac{\sigma l_1}{c_1} - b_2 c_2 \operatorname{ctg} \frac{\sigma l_2}{c_2} + b_3 c_3 \operatorname{tg} \frac{\sigma l_3}{c_3} = 0;$$

für den Fall: alle drei Kanäle offen:

$$b_1 c_1 \operatorname{ctg} \frac{\sigma l_1}{c_1} + b_2 c_2 \operatorname{ctg} \frac{\sigma l_2}{c_2} + b_3 c_3 \operatorname{ctg} \frac{\sigma l_3}{c_3} = 0.$$

NEUMANN hat auch für weitere Beckenkonfigurationen die Periodengleichungen abgeleitet, auf die hier für Anwendungen besonders verwiesen sei. Die Theorie der Impedanz gekoppelter Schwingungssysteme hat sich in manchen Fällen der Natur bewährt.

4. Fortschreitende Wellen in Kanälen rechteckigen Querschnitts bei Berücksichtigung der Erdrotation. *α) Allgemeine Beziehungen.* Die mit fortschreitenden oder stehenden Wellen verknüpften Wasserversetzungen stehen auf der rotierenden Erde wie jeder bewegte Körper unter der Wirkung der Corioliskraft. Ihre Ignorierung ist in den meisten Fällen nicht gestattet, da ihre Größenordnung gleich der bei der Wellenbewegung auftretenden horizontalen Druckkräften ist. Ist die periodische Horizontalgeschwindigkeit U, dann ist die Corioliskraft $2\Omega \sin\varphi \cdot U$, wenn Ω die Winkelgeschwindigkeit der Erdrotation und φ die geographische Breite ist. Diese Kraft steht senkrecht auf U und wechselt mit U ihr Vorzeichen. Mit der auf die Wasserteilchen in lotrechter Richtung wirkenden Schwerkraft g hat sie eine Resultierende, die um den kleinen Winkel γ gegen die Vertikale geneigt ist: $\operatorname{tg}\gamma = \frac{2\Omega\sin\varphi}{g}\,U$. Herrscht in jedem Augenblick in der Wassermasse statisches Gleichgewicht, so muß die Wasseroberfläche mit der Horizontalen ebenfalls den Winkel γ bilden. Bei $\varphi = 45°$ wird $\operatorname{tg}\gamma = 1{,}027 \times 10^{-7} \cdot U$ und bei $U = 20\ \text{cm sec}^{-1}$ wird $\gamma = 0{,}4''$. Die Neigung der Oberfläche ist somit sehr klein. In einem Kanal der Breite b ist die Erhebung bzw. Senkung der Oberfläche an den Rändern durch die Wirkung der Corioliskraft $\frac{b}{2}\operatorname{tg}\gamma$, kann somit bei genügend großem b ziemlich viel ausgeben. Aus (3.16) kann man sogleich folgern, daß die durch die periodischen Schwankungen von U bedingten *Querschwingungen* gegenüber den Längsschwingungen eine *Phasenverschiebung* um ein Viertel der Periode aufweisen.

Die Grundgleichungen zur genauen Erfassung der Wirkung der Erdrotation auf eine in einem rechteckigen Kanal (Wassertiefe h konstant) fortschreitenden Welle sind folgende (x Fortpflanzungsrichtung der Welle, y quer hierzu, $2\Omega \cdot \sin\varphi = f$, Coriolisparameter):

$$\frac{\partial u}{\partial t} - f v = -g\frac{\partial\zeta}{\partial x}, \quad \frac{\partial v}{\partial t} + f u = -g\frac{\partial\zeta}{\partial y} \quad \text{und} \quad \frac{\partial\zeta}{\partial t} = -h\left(\frac{\partial u}{\partial x} + \frac{\partial v}{\partial y}\right). \tag{4.1}$$

Bei einer periodischen Störung in ζ, u und v in der Form $e^{i\sigma t}$ nehmen sie die Form an:

$$i\sigma u - f v = -g\frac{\partial\zeta}{\partial x}, \quad i\sigma v + f u = -g\frac{\partial\zeta}{\partial y} \quad \text{und} \quad i\sigma\zeta = -h\left(\frac{\partial u}{\partial x} + \frac{\partial v}{\partial y}\right). \tag{4.2}$$

Die Elimination von u und v oder ζ führt zur Schwingungsgleichung in einem rotierenden System:

$$(\nabla^2 + k^2)\,\zeta \quad \text{oder} \quad u \quad \text{oder} \quad v = 0, \tag{4.3}$$

worin $\nabla^2 = \frac{\partial^2}{\partial x^2} + \frac{\partial^2}{\partial y^2}$ der LAPLACEsche Operator und $k^2 = \frac{\sigma^2 - f^2}{g h}$ ist.

β) KELVINsche Wellen. Für einen Kanal der Breite b, an dessen Wänden stets die Querbewegung $v = 0$ sein muß, läßt sich eine Lösung der Wellengleichungen (4.3) leicht geben:

$$u = \frac{g}{c}\zeta, \quad v = 0 \quad \text{und} \quad \zeta = A\,e^{-\frac{f}{c}y}\cos(\sigma t - \varkappa x). \tag{4.4}$$

Die Wellengeschwindigkeit $c = \sqrt{g h}$ wird durch die Erdrotation nicht beeinflußt. Dagegen nimmt die Wellenamplitude im Kanalquerschnitt von der einen Seite des Kanals, die der Rotation vorangeht, nach der anderen Seite entsprechend einer Exponentialfunktion zu. Tabelle 2 gibt für verschiedene Kanalbreiten eine Übersicht über die vorkommenden Änderungen. Man erkennt, daß bei einiger Breite des Kanals sich die ganze Wellenbewegung vorzugsweise auf die Seite des Kanals beschränkt, die der Rotation nachfolgt. Nach dem Entdecker dieses Wellentypus nennt man diese Wellen „KELVIN"-Wellen.

Tabelle 2. *Wellenamplitude quer zum Kanal bei* KELVIN-*Wellen* (in Prozent der Amplitude auf der Kanalseite, die der Rotation nachfolgt).

	Breite des Kanals in km						
	10	25	50	100	250	500	1000
$\varphi = 45°$, $h = 90$ m, $c = 30$ m sec^{-1}. . . .	96.6	91.8	84.2	70.9	42.4	16.2	3.2
$\varphi = 90°$ (Pol), $h = 50$ m, $c = 22.1$ m sec^{-1}.	93.6	87.9	72.0	51.9	19.4	3.9	0.1_5

Bei dieser Wellengattung besteht in jedem Augenblick ein Gleichgewicht zwischen dem Druckgefälle quer zum Kanal und der Corioliskraft. Diese Eigenschaft ist aber, wie PROUDMAN[1] zeigte, an rechteckigen Kanalquerschnitt gebunden. Bei anders geformten Kanalquerschnitten kommt es zu periodischen Querströmen, die aber nicht symmetrisch zur Kanalmitte entwickelt sind. PROUDMAN hat speziell einen Kanal parabolischen Querschnitts in numerischem Beispiel berechnet, wobei neben den erwähnten Querbewegungen auch festgestellt wurde, daß die Rotation die Wellengeschwindigkeit c herabgesetzt, und zwar um so mehr, je breiter der Kanal ist.

POINCARÉ*sche Wellen*. Außer diesen KELVINschen Wellen sind in einem Kanal noch andere Wellen möglich, die nach ihrem Entdecker POINCARÉsche Wellen genannt werden. Nimmt man eine Lösung der Bewegungsgleichungen in der Form (1.12) an und die Tiefe des rechteckigen Kanals konstant, dann müssen U, V und Z, die nur Funktionen von y sind, die Beziehungen

$$\sigma U - f V = \varkappa g Z, \quad \sigma V + f U = -g \frac{dZ}{dy} \quad \text{und} \quad \frac{d^2 Z}{dy^2} + \left(\frac{\sigma^2 - f^2}{c^2} - \varkappa^2\right) Z = 0 \tag{4.5}$$

erfüllen. Die Grenzbedingungen an den Kanalwänden $y = \pm a$ sind wieder $v = 0$. Läßt man n eine ganze *ungerade* Zahl ($n = 1, 3, 5 \ldots$) sein und setzt man

$$\frac{\sigma^2 - f^2}{c^2} - \varkappa^2 = \left(\frac{n\pi}{2a}\right)^2 \tag{4.6}$$

dann erhält man in einfacher Weise eine vollständige Lösung in der Form:

$$\left.\begin{aligned} u &= A\left(\frac{\sigma}{\varkappa h}\cos\frac{n\pi}{2a}y + \frac{\pi g}{2af}\sin\frac{n\pi}{2a}y\right)\cos(\sigma t - \varkappa x), \\ v &= -A\frac{f}{\varkappa h}\left(1 + \frac{\pi c^2}{4a^2 f^2}\right)\cos\frac{n\pi}{2a}y\sin(\sigma t - \varkappa x), \\ \zeta &= A\left(\cos\frac{n\pi}{2a}y + \frac{\pi\sigma}{2af\varkappa}\sin\frac{n\pi}{2a}y\right)\cos(\sigma t - \varkappa x). \end{aligned}\right\} \tag{4.7}$$

Die Eigenperiode der n-knotigen Querschwingung des Kanals ist $T_q = 4a/nc$, die Periode der Trägheitsschwingungen $T_t = \frac{2\pi}{f} = \frac{12 \text{ Sternstunden}}{\sin\varphi}$ und jene der fortschreitenden Längswelle im Kanal $T = 2\pi/\sigma$. Die Beziehung (4.6) wird damit

$$1 - \left(\frac{T}{T_t}\right)^2 - \left(\frac{T}{T_q}\right)^2 = \frac{c^2\varkappa^2}{\sigma^2}. \tag{4.8}$$

Dies bedeutet, daß solche POINCARÉsche Wellen in einem Kanal nur möglich sind, wenn sowohl $T < T_t$, wie auch $T < T_q$ ist, d.h. die Welle muß sowohl kürzer als ein halber Pendeltag, als auch kürzer als die Eigenperiode der Querschwingungen des Kanals sein. Diese Bedingungen schränken die Möglichkeit des Auftretens solcher POINCARÉscher Wellen wesentlich ein. In schmalen Kanälen

[1] J. PROUDMAN: Phil. Mag. (6) **49**, 465—475 (1925).

sind so bei Gezeitenperioden nur KELVINsche Wellen möglich; aber POINCARÉsche Wellen werden in breiten kanalartigen Ozeanteilen bei der Ausbildung der Gezeiten vielleicht eine erhebliche Rolle spielen.

Ist die Tiefe des Kanals nicht im Querschnitt konstant, so werden die mathematischen Schwierigkeiten erheblich. Für einen Kanal parabolischen Querschnitts siehe die früher erwähnten Untersuchungen von PROUDMAN; bei diesen Untersuchungen ist noch keine Dissipation der Wellenenergie angenommen. In einer weiteren Arbeit PROUDMANs[1] wurde zwar für den Mittelteil des Kanals ebenfalls reibungslose Bewegung der Wassermassen, aber an den beiden Seitenrändern bei geringen Wassertiefen eine Dissipation der Energie angenommen. Dies entspricht besser den Verhältnissen in der Natur und läßt passendere Anwendungen zu.

Noch eine Art von Schwingungen ist in einem Kanal möglich. Man erhält sie am bequemsten, wenn man von den Gl. (4.2) und (4.3) ausgeht. Eine Lösung der letzteren Gleichung für v ist:

$$v = e^{-s_n x + i \frac{n\pi}{2a} y}, \quad \text{wenn} \quad s_2^n = \left(\frac{n\pi}{2a}\right)^2 - k^2 \tag{4.9}$$

ist, wobei $\frac{n\pi}{2a} > k$ sein muß. Da v für $\pm a$ verschwinden soll, kann v nur die Formen haben:

$$\left.\begin{aligned} v &= A_n\, e^{-s_n x} \sin \frac{n\pi}{2a}\, y \cdot e^{i\sigma t} \quad \text{für ganze } \textit{gerade} \text{ Zahlen } n = 2, 4, \ldots, \\ v &= A_n\, e^{-s_n x} \cos \frac{n\pi}{2a}\, y \cdot e^{i\sigma t} \quad \text{für ganze } \textit{ungerade} \text{ Zahlen } n = 1, 3, \ldots. \end{aligned}\right\} \tag{4.10}$$

Mit diesen Beziehungen lassen sich mittels (4.2) auch die dazugehörigen Werte von ζ und u leicht berechnen. Diese Schwingungen haben die Form *stehender* Querwellen im Kanal mit Amplituden, die längs des Kanals von der Stelle $x = 0$ nach einer e-Potenz abnehmen. Sie beschränken sich somit praktisch auf einen relativ kleinen Kanalteil. Solche Schwankungen hat TAYLOR eingeführt, um die Bedingung der Totalreflexion einer KELVINschen Welle in einem einseitig geschlossenen Kanal zu erfüllen (siehe Abschn. γ). Will man hingegen die Schwingungen eines beiderseitig geschlossenen rechteckigen Kanals ermitteln, dann müssen, um die Randbedingungen zu erfüllen, zu den KELVINschen Wellen POINCARÉsche Wellen hinzugenommen werden (siehe den Schluß von Abschnitt γ).

γ) Reflexion einer KELVINschen Welle. Eigenschwingungen eines rotierenden rechteckigen Beckens. Die totale Reflexion einer KELVIN-Welle an einer senkrechten Querwand eines Kanals kann — wie gezeigt werden wird — nicht in der einfachen Weise durch Überlagerung einer einlaufenden und einer an der Wand reflektierten Welle erfolgen, da es dabei keinen Querschnitt im Kanal gibt, an dem die Horizontalbewegung der Superposition zu allen Zeiten verschwindet.

Die Überlagerung einer in der $+x$-Richtung laufenden KELVIN-Welle und einer in entgegengesetzter Richtung laufenden ist

$$\zeta = A\, e^{-\frac{f}{c} y} \cos(\sigma t - \varkappa x) + A\, e^{+\frac{f}{c} y} \cos(\sigma t + \varkappa x). \tag{4.11}$$

Hochwasser $\partial\zeta/\partial t = 0$ verlangt, daß

$$\frac{\operatorname{tgh} \frac{f}{c} y}{\operatorname{tg} \varkappa x} = \operatorname{tg} \frac{2\pi}{T}\left(t + \frac{T}{4}\right) \tag{4.12}$$

[1] J. PROUDMAN: Proc. Math. Soc. Lond., Ser. II **49**, 211 (1946).

ist. Dies ist die Gleichung der Linien gleichen Hochwassers; diese fallen für $t=0$, $\frac{1}{4}T$, $\frac{1}{2}T$ und $\frac{3}{4}T$ mit den Koordinatenachsen zusammen, während in der Nähe des Ursprungs für jede Hochwasserlinie $\frac{y}{x}=\frac{\varkappa c}{f}\operatorname{tg}\frac{2\pi}{T}\left(t+\frac{T}{4}\right)$ ist. Fig. 2 zeigt die Amplituden- und Phasenverteilung der Überlagerung der beiden KELVIN-Wellen für $A=\frac{1}{2}$, $k=1$ und $\alpha=\frac{f}{c}=0.7$. Die angenommenen Werte entsprechen etwa einer Kanalbreite von 400 km bei einer Wassertiefe von 40 m in $44\frac{1}{2}°$ Breite; die Wellenperiode ist zu 12 Std angenommen. Auf nicht-rotierender Erde würde die Überlagerung den Charakter einer stehenden Welle haben. Diesen hat sie bei Rotation völlig verloren: Die Welle zerfällt in aufeinder folgende *Zellen,* und in jeder ist eine *Drehwelle (Amphidromie)* mit einer sternförmigen Anordnung der Linien gleicher Phase vorhanden. Die Amplituden sind an den Rändern des Kanals am größten und im Zentrum der Amphidromie Null. Die Aufeinanderfolge solcher Amphidromien erweckt den Eindruck, daß auf der linken Seite des Kanals eine Welle mit schwankenden Amplituden in der einen Richtung, auf der rechten Seite in der entgegengesetzten Richtung fortschreitet. Im Kanal gibt es nur alternierenden Strom in der Längsrichtung, und es bewegt sich das Wasser auf beiden Kanalseiten jeweils in entgegengesetzten Richtungen; Querbewegungen gibt es nicht.

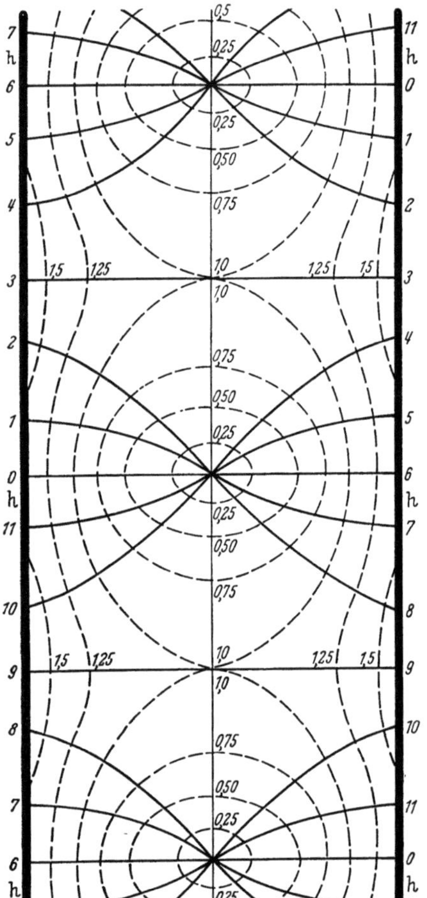

Fig. 2. Überlagerung zweier gegeneinander laufender KELVIN-Wellen in einem unendlich langen Kanal konstanten rechteckigen Querschnitts. Periode = 12 Std, Amplitude jeder Welle = 1, Parameter $\varkappa = 1$, $\alpha = 0{,}7$, entsprechend einer Kanalbreite etwa 400 km, Kanaltiefe 40 m, geographischen Breite $\varphi = 44\frac{1}{2}°$.

Man erkennt — wie eingangs angedeutet — daß diese Superposition zweier KELVINschen Wellen nirgends quer zum Kanal verschwindende Amplituden der Horizontalbewegung aufweist, so daß man hier ohne Störung der Wellenbewegung eine senkrechte Wand aufrichten könnte. Dann hätte man an dieser Stelle eine totale Reflexion einer KELVIN-Welle einer in einem einseitig geschlossenen Kanal einlaufenden KELVIN-Welle. Dieses Problem der totalen Reflexion einer KELVIN-Welle an einer Querwand hat G. J. TAYLOR[1] gelöst. Er konnte zeigen, daß eine solche Reflexion an einem geschlossenem Ende eines Kanals nur stattfindet, wenn im innersten Kanalteil sich eine Störbewegung ausbilden kann, deren Horizontalbewegung jene der Überlagerung der ein- und auslaufenden KELVIN-Wellen gerade aufhebt. Diese Störbewegung besteht aus Querschwingungen im innersten Kanalteil mit in der Kanalrichtung abklingenden Amplituden gemäß der Wellenart der Beziehungen (4.9). Diese Querwellen lassen sich stets in ihren

[1] G. J. TAYLOR: Proc. Math. Soc. Lond., Ser. II **20** (1920).

Amplituden und Phasen so bestimmen, daß ihre Horizontalbewegung in der Kanalrichtung am geschlossenen Ende des Kanals gleich und entgegengesetzt der Horizontalbewegung der Superposition der beiden KELVIN-Wellen ist. Ihre Zusammenfassung ergibt dann die Erfüllung der für das geschlossene Ende des Kanals notwendigen Randbedingung $u=0$. Die totale Reflexion ist aber an die Bedingung geknüpft

$$\sigma^2 - f^2 < g h \frac{\pi^2}{b^2}, \tag{4.13}$$

worin b die Breite des Kanals bedeutet. Da $\sigma = \frac{2\pi}{T}$ und $f = \frac{2\pi \sin\varphi}{12\,\mathrm{Std}}$ sind und, da $2b/\sqrt{gh} = T_q$ die Eigenperiode des Kanals in der Querrichtung ist, läßt sich dieser Bedingung auch die Form

$$\left(\frac{T_q}{T}\right)^2 < 1 + \left(\frac{T_q \sin\varphi}{12\,\mathrm{Std}}\right)^2 \tag{4.14}$$

geben, worin die Perioden in Stunden zu nehmen sind. Nimmt man für die Welle die halbtägige Gezeitenperiode (12 Std), dann erhält man aus (4.14)

$$\frac{T_q}{T} < \frac{1}{\cos\varphi}, \tag{4.15}$$

für 45° Breite $=1{,}41$, d.h. die Eigenperiode der Querschwingungen darf bei Gezeitenwellen in mittleren Breiten etwa das $1\frac{1}{2}$fache der Gezeitenperiode nicht überschreiten, wenn Totalreflexion am geschlossenen Ende des Kanals (Bucht) eintreten soll.

Wenn die TAYLORschen Ergebnisse streng nur für Kanäle rechteckigen Querschnitts gelten, so lassen sie doch auch bei komplizierteren Becken-Konfigurationen die Möglichkeit der Reflexion von eindringenden KELVIN-Wellen richtig abschätzen Will man die Schwingungsverhältnisse in einem solchen einseitig geschlossenen Kanal überblicken, dann muß man einzelne spezielle Fälle numerisch ausrechnen. TAYLOR hat für einen Kanal, der in den Dimensionen ungefähr der Nordsee entspricht ($\alpha=0{,}7$, Periode der Wellen 12 Std), die Verteilung sowohl der Amplituden und Phasen, als auch die der Ströme numerisch berechnet und in den Fig. 3 und 4 veranschaulicht. Der Vergleich mit der Fig. 2 zeigt, welche Abweichungen der Abschluß des Kanals bewirkt. Die innere Amphidromie vor diesem ist auf allen Seiten regelmäßig ausgebildet und anderer Gestaltung als die in Fig. 2, die ja die reine Superposition der ein- und auslaufenden KELVIN-Welle ist. Auch hier sind die Amplituden an den Wänden des Kanals am größten und erreichen in den beiden Ecken der Bucht Maxima.

Die Ströme verlaufen in den äußeren Teilen der Bucht stets alternierend und parallel dem Kanalrand; sie sind hier identisch mit jenen der Superposition der beiden KELVIN-Wellen. Im inneren Kanalteil gehen die Ströme stets parallel der Wand, aber sie sind hier Drehströme, die Stromfiguren sind Ellipsen, deren Exzentrizität um so kleiner ist, je mehr man sich von der Wand entfernt. Diese Drehströme verlaufen im Sinne der Erdrotation, und die Achse der Stromfigur ist knapp vor dem Kanalabschluß parallel der Querwand, weiter draußen aber parallel den Längswänden. Gleichzeitig nimmt die Quergeschwindigkeit nach außen zu an Stärke ab und verschwindet völlig im äußeren Kanalteil. Die Stromverteilung für vier Phasenzeiten findet man bei DEFANT[1].

[1] A. DEFANT: Ann. Hydrogr. Hamburg **1923**.

Die Eigenschwingungen eines rotierenden, allseits geschlossenen Beckens kreisförmiger (LAMB, Hydrodynamik[1]) oder elliptischer Form (JEFFREYS[2]) lassen sich relativ einfach ermitteln. Für rechteckige Becken konstanter Tiefe hat erst TAYLOR[3] eine Lösung gegeben. Die Grundwelle ist wieder die Superposition zweier gegeneinander in der einen Richtung des Kanals laufender KELVINscher Wellen, bei denen an den Längswänden die Bedingung $v=0$ ja von vornherein

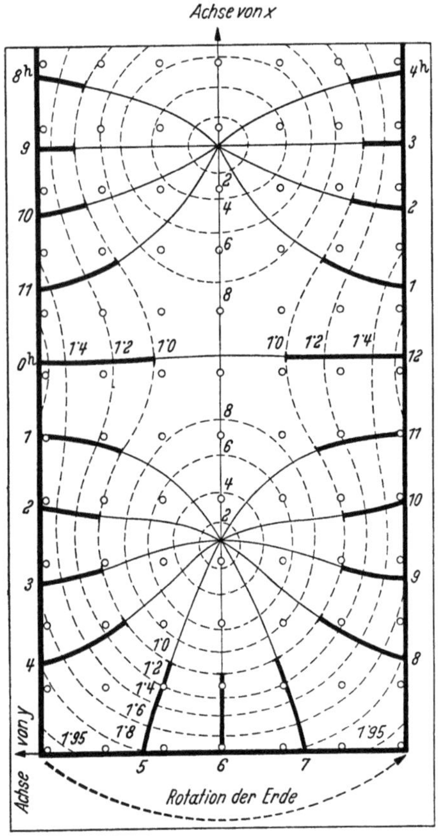

Fig. 3.

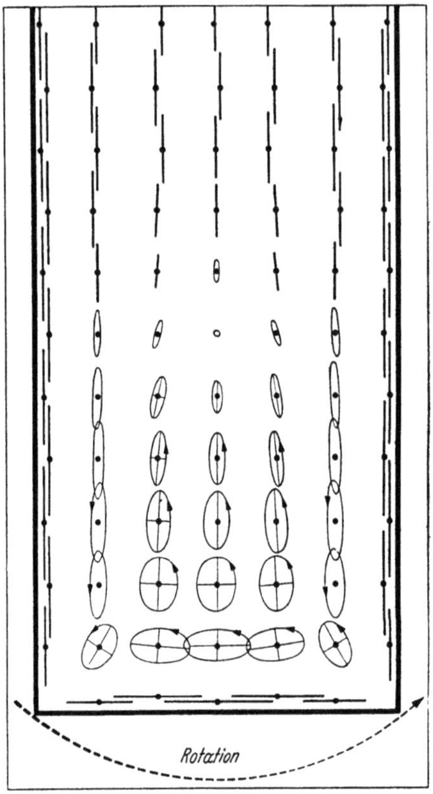

Fig. 4.

Fig. 3. Totale Reflexion einer KELVIN-Welle an einem geschlossenen Ende einer rechteckigen Bucht auf der rotierenden Erde: Phasen- und Amplitudenlinien (nach G. J. TAYLOR).

Fig. 4. Stromfiguren zum Gezeitenbild der Fig. 3.

erfüllt ist. Zur Erfüllung der Randbedingung $u=0$ an den beiden Querwänden müssen in diesem Fall POINCARÉsche Wellen hinzugenommen werden. Die genaue Ermittlung der Periode erfordert einige Rechenarbeit, ist aber an sich nicht schwierig. TAYLOR hat für ein doppelt so langes wie breites Becken, dessen Rotationsperiode gleich der längsten Eigenperiode des *nicht* rotierenden Systems ist ($f/c=\alpha=1$) gefunden, daß $\sigma/c=0{,}429$ ist. Da ohne Rotation $\sigma/c=0{,}5$ ist, sieht man, daß die Periode bei Rotation um 14% verlängert wird. Für lange schmale Becken nähert sich σ/c rasch 0,5, so daß für solche Becken der Einfluß der Rotation klein ist. Hingegen macht sich die Verlängerung der Periode um so

[1] H. LAMB: Lehrbuch der Hydrodynamik. Deutsche 2. Aufl., Leipzig u. Berlin 1931.
[2] H. JEFFREYS: Proc. Math. Soc. Lond., Ser. II **23**, 449—476 (1923).
[3] G. J. TAYLOR: Proc. Math. Soc. Lond., Ser. II **20** (1920).

bemerkbarer, je breiter das Becken im Verhältnis zu seiner Länge ist. Die stehende Welle ohne Rotation verwandelt sich in eine in der Richtung der Rotation umlaufenden Amphidromie.

5. Interne Wellen. In geschichtetem Wasser oder in Wassermassen, in denen sich die Dichte mit der Tiefe ändert, sind Wellen anderer Art möglich als jene, die an der freien Oberfläche einer Wassermasse auftreten können und die in den vorhergehenden Abschnitten besprochen wurden. Da sie im Innern der Wassermassen auftreten, bezeichnet man sie als *Grenzflächenwellen* bzw. *interne Wellen.* Ihr charakteristisches Merkmal ist, daß in ihnen die größten vertikalen Verlagerungen der Wasserteilchen an der Grenzfläche zwischen den übereinander gelagerten Medien (d.h. in einiger Tiefe unter der freien Oberfläche) auftreten und die Amplitude dieser Verlagerungen im allgemeinen wesentlich größer ist, als die der gewöhnlichen Wellen an der freien Oberfläche. Das Auftreten solcher Wellen an der Grenzfläche zweier Schichten ist lange Zeit der Aufmerksamkeit der Beobachter entgangen, da bei ihnen (auch bei größerer vertikaler Schwankung der Grenzfläche) die freie Oberfläche des oberen Mediums nur in ganz unwesentlicher Weise gestört wird und praktisch in Ruhe bleibt. Erst in neuer Zeit brachte mit Steigerung der Häufigkeit und Genauigkeit der Beobachtungen auch im Inneren der Wassermassen die Möglichkeit, solche Wellen eingehender zu erforschen.

α) Interne Wellen an Grenzflächen. Die Theorie von Wellen an der gemeinsamen Grenzfläche zweier übereinander gelagerter Flüssigkeiten ist zuerst von G. STOKES[1] entwickelt worden. Für eine Wellenbewegung, die mit der Entfernung von der Grenzfläche rasch abnimmt und praktisch in größerer Entfernung von ihr verschwindet, ergibt sich als Gleichung der durch die interne Welle gestörten Grenzfläche

$$\zeta = a \cos \varkappa x \cdot e^{i\sigma t}. \tag{5.1}$$

Aus der Erfüllung der kinematischen und dynamischen Randbedingungen an der Grenzfläche[2] folgt

$$\sigma^2 = g\varkappa \frac{\varrho_2 - \varrho_1}{\varrho_2 + \varrho_1}. \tag{5.2}$$

Mit $\sigma/\varkappa = c$ für die Fortpflanzungsgeschwindigkeit einer fortschreitenden internen Welle der Wellenlänge $\lambda = 2\pi/\varkappa$ erhält man

$$c^2 = \frac{g\lambda}{2\pi} \frac{\varrho_2 - \varrho_1}{\varrho_2 + \varrho_1}. \tag{5.3}$$

Das Vorhandensein des oberen Mediums setzt die Wellengeschwindigkeit der internen Welle im Verhältnis $\sqrt{\frac{\varrho_2 - \varrho_1}{\varrho_2 + \varrho_1}}$ herab. Bei den in der Natur vorkommenden Fällen kann $\varrho_2 - \varrho_1$ von der Größenordnung 10^{-3} g/cm³ sein, so daß c gegenüber den Oberflächenwellen auf rund $\frac{1}{45}$ vermindert ist, was recht beträchtlich ist.

Fig. 5 gibt nach V. BJERKNES[3] die Stromlinien und Orbitalbahnen in den beiden übereinander gelagerten Wasserschichten bei einer nach rechts fortschreitenden Welle. Dabei ist besonders zu beachten, daß an der Grenzfläche auch

[1] G. G. STOKES: Cambridge, Trans. **8** (1847). — Papers **1**, 212.

[2] Hier und im folgenden sind stets folgende Bezeichnungen benutzt worden: Oberschicht: Dichte ϱ_1, Mächtigkeit h_1; Unterschicht: Dichte ϱ_2, Mächtigkeit $h_2 - h_1$; ganze Wassertiefe h_2. Außerdem $\Delta\varrho = \frac{\varrho_2 - \varrho_1}{\varrho_2}$.

[3] V. BJERKNES: Geofys. Publ. **2**, Nr. 4 (1921).

eine Unstetigkeit der Bewegung vorhanden ist. Die Normalkomponente der Geschwindigkeit ändert sich beim Überschreiten der Grenzfläche zwar stetig, aber die Horizontalkomponente ändert an der Grenzfläche ihr Vorzeichen, so daß hier eine *Wirbelschicht* vorhanden ist.

In der Natur sind nur Fälle realisiert, in denen die Oberfläche der oberen Wassermasse als freie Oberfläche anzusehen ist. In diesem Falle führen die kinematischen und dynamischen Randbedingungen für die Grenzfläche, sowie für

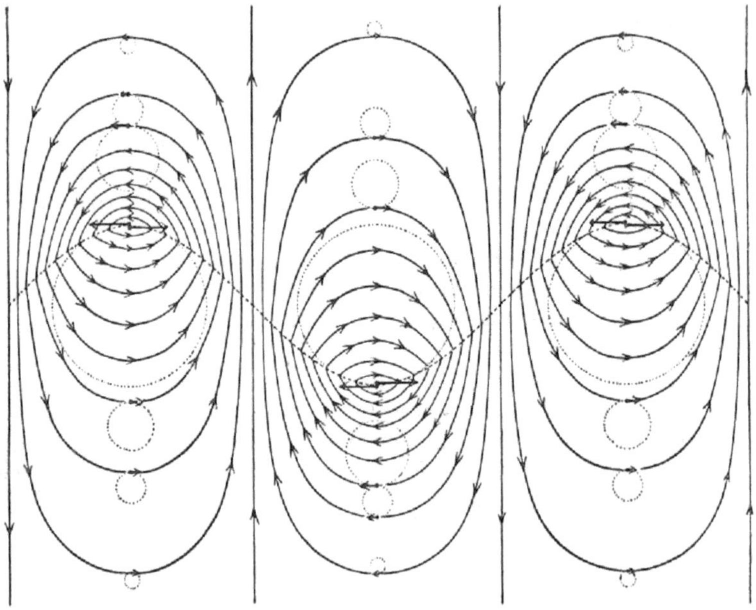

Fig. 5. Stromlinien (ausgezogen) und Orbitalbahnen (feinpunktiert) in einer von links nach rechts fortschreitenden inneren Welle an der Grenzflache zweier verschieden dichter Medien (nach V. Bjerknes).

die freie Oberfläche für die Fortpflanzungsgeschwindigkeit c von Wellen der Wellenlänge $\lambda = 2\pi/\varkappa$ zur quadratischen Gleichung

$$\left.\begin{aligned} &c^4\left(\varrho_2 \operatorname{ctgh}\varkappa(h_2-h_1)\operatorname{ctgh}\varkappa h_1+\varrho_1\right)- \\ &\quad -c^2\varrho_2\left(\operatorname{ctgh}\varkappa h_1+\operatorname{ctgh}\varkappa(h_2-h_1)\right)\frac{g}{\varkappa}+(\varrho_2-\varrho_1)\frac{g^2}{\varkappa^2}=0.\end{aligned}\right\} \quad (5.4)$$

Sie zeigt, daß bei jeder Wellenlänge *zwei voneinander verschiedene Wellensysteme* möglich sind. Ist der Dichtesprung $\varrho_2-\varrho_1$ an der Diskontinuitätsfläche klein, dann lassen sich in erster Annäherung folgende zwei Lösungen finden:

$$c_1^2=\frac{g}{\varkappa}\operatorname{tgh}\varkappa h_2, \qquad c_2^2=\frac{g}{\varkappa}\,\frac{\varrho_2-\varrho_1}{\varrho_2\operatorname{ctgh}\varkappa h_1+\varrho_1\operatorname{ctgh}\varkappa(h_2-h_1)}. \quad (5.5)$$

Die erste Wellenart ist identisch mit den Wellen auf einer einzigen Wasserschicht der Tiefe h_2. Bei dieser Wellenart nimmt auch die Vertikalkomponente der Verlagerung der Wasserteilchen von der Oberfläche an stetig mit der Tiefe ab. Die Oberfläche hat die größte Amplitude, die Grenzfläche eine kleinere, und überdies ist die Phase der Wellenbewegung an der Oberfläche wie an der Grenzfläche dieselbe. Diese *„äußere" Welle* unterscheidet sich in keiner Weise von den gewöhnlichen Wellen an der Oberfläche einer homogenen Wassermasse.

Die zweite mögliche Wellenart hat die größte Amplitude an der Grenzfläche; diese nimmt von hier nach oben und unten rasch ab. Dies ist somit eine *interne* Welle. In der oberen Schichte ist zwischen der Grenzfläche und der freien Oberfläche stets ein Tiefenhorizont vorhanden, in dem die Vertikalbewegungen verschwinden. Diese Tiefe liegt immer recht nahe der freien Oberfläche. Die innere Grenzfläche und die freie Oberfläche schwingen somit mit entgegengesetzter Phase, aber die Amplitude an der freien Oberfläche ist so klein, daß sie kaum wahrgenommen werden kann.

Im allgemeinen wird in der Natur die Tiefe $(h_2 - h_1)$ der unteren Wassermasse als groß gegenüber der Wellenlänge anzusehen sein. Dann kann man $\operatorname{tgh} \varkappa (h_2 - h_1) = 1$ setzen und die Werte der Gl. (5.5) werden:

$$c_1^2 = \frac{g}{\varkappa} \quad \text{und} \quad c_2^2 = \frac{\varrho_2 - \varrho_1}{\varrho_2 \operatorname{ctgh} h_1 + \varrho_1} \cdot \frac{g}{\varkappa}; \tag{5.6}$$

als Verhältnis der Amplituden an der freien Oberfläche und an der Grenzfläche erhält man:

$$-\frac{\varrho_2 - \varrho_1}{\varrho_1} e^{-\varkappa h_1}. \tag{5.7}$$

Wenn ϱ_2 von ϱ_1 sich nur wenig unterscheidet, ist dies stets eine kleine Zahl.

Wird hingegen die Dicke h_1 der oberen Schichte klein gegenüber der Wellenlänge, dann kann man an Stelle von $\operatorname{ctgh} \varkappa h_1 = \frac{1}{\varkappa h_1}$ setzen und er wird in erster Annäherung:

$$c_1^2 = \frac{g}{\varkappa} \quad \text{und} \quad c_2^2 = \frac{\varrho_2 - \varrho_1}{\varrho_2} g h_1; \tag{5.8}$$

das obige Verhältnis der Amplituden wird

$$-\frac{\varrho_2 - \varrho_1}{\varrho_2}. \tag{5.9}$$

Den Fall von mehreren übereinander gelagerter Schichten verschiedener Dichte hat GREENHILL[1] untersucht. Von besonderem Interesse ist vielleicht, daß bei drei Schichten die Diskontinuitätsflächen mit entgegengesetzter Phase schwingen können, so daß den Wellentälern (-bergen) der einen Grenzfläche Wellenberge (-täler) der anderen entsprechen. Betreffs weiterer theoretischer Fragen siehe auch H. SOLBERG[2], der auch die Verhältnisse bei keilförmigen Schichten behandelt hat.

Interne Wellen lassen sich natürlich auch im Laboratorium in der Wellenrinne, in der zwei Wasserarten verschiedener Dichte übereinander gelagert sind, erzeugen und genauer verfolgen, wenn die eine Wasserart anders gefärbt ist. Derartige Versuche in geschichtetem Wasser hat vor allem J. SANDSTRÖM[3] ausgeführt.

β) Interne Wellen von Oberflächen- und Flutwellencharakter. Die Beziehungen (5.5) bis (5.9) gelten zunächst nur für „*kurze*" Wellen, bei denen die *Wellenlänge klein gegenüber der Wassertiefe* ist. Da die Energie, die erforderlich ist, um solche innere Wellen an Grenzflächen hervorzurufen, sehr gering ist, werden irgendwelche geringe Druckstörungen oder Tangentialspannungen an der Oberfläche einer geschichteten Wassermasse imstande sein, an inneren Grenzflächen interne

[1] GREENHILL: Amer. J. Math. **9** (1887).
[2] H. SOLBERG: Geofys. Publ. **5**, Nr. 9 (1928).
[3] J. SANDSTRÖM: Ann. Hydrogr. Hamburg **1908**, 6—23.

Wellen zu erzeugen oder zu erhalten. In Verbindung mit diesen kurzen internen Wellen steht die Erscheinung des „*Totwassers*": In der norwegischen Seemannssprache bedeutet „im Totwasser liegen" nicht vorwärts kommen. Vor Flußmündungen in norwegischen Fjorden und anderswo oder in der Nähe von schmelzenden Eismassen (Treibeis in Polargebieten) breitet sich oft eine relativ dünne Wasserschichte süßen oder schwach salzigen Wassers über die Unterschicht salzreicheren (schwereren) Wassers aus. Bei ruhigem Wetter und Zurücktreten der Gezeiten (Erhaltung der Schichtung) zeigt sich, daß Schiffe, wie Segler bei flauem Wind oder Ruderbote bei langsamer Fahrt, auf das empfindlichste in ihrem Vorwärtskommen und Kurshalten gehemmt werden und der Eindruck erweckt wird, als ob das Schiff von einer mystischen Kraft nach rückwärts gezogen würde. Die geringsten äußeren Umstände z.B. das Aufkommen von Wind, Vorbeifahrt eines anderen Schiffes in voller Fahrt sind oft imstande die Totwasserwirkung zum Verschwinden zu bringen. Die Erscheinung wurde besonders von FR. NANSEN auf seiner Polarfahrt auf der „Fram" näher beobachtet und untersucht.

Nach den theoretischen und experimentellen Untersuchungen von W. EKMAN[1] liegt die Ursache des Totwassers in der Erregung von internen Wellen durch das fahrende Schiff, wodurch der Wellenwiderstand des Schiffes wesentlich gesteigert wird. Durch die Fahrt des Schiffes werden im geschichtetem System zwei Wellenarten, eine äußere und eine innere Welle erzwungen. Ihre Geschwindigkeiten sind durch die Gl. (5.6) gegeben. Die äußeren Wellen sind sehr kurz und bedingen den normalen Wellenwiderstand der Schiffsbewegung. Bei entsprechender Schiffsgeschwindigkeit ist aber auch die Möglichkeit der Erregung interner Wellen gegeben. Aus der zweiten der Gl. (5.6) läßt sich ableiten, daß der Wert $\varkappa = 2\pi/\lambda$ nur dann reell ist, wenn

$$c^2 < \frac{\varrho_2 - \varrho_1}{\varrho_2} g h_1 \tag{5.10}$$

ist. Dies bedeutet, daß das Totwasser sich nur dann einstellt, wenn die Fahrtgeschwindigkeit des Schiffes unter diesem kritischen Wert c bleibt. Übersteigt sie ihn, dann werden nur die äußeren Wellen erregt und der Wellenwiderstand ist praktisch derselbe, wie wenn nur eine einzige Wassermasse vorhanden wäre. Diese Ergebnisse werden durch die Beobachtungen in der Natur ebenso bestätigt, wie durch die Modellversuche EKMANs.

Ist die Wellenlänge der internen Wellen lang gegenüber den Dicken der beiden Schichten, dann kann man in der Gl. (5.4) an Stelle der Cotangenten $1/\varkappa(h_2 - h_1)$ bzw. $1/\varkappa h_1$ setzen und mit genügender Genauigkeit wird aus ihr:

$$c^4 - c^2 h_2 g + \left(1 - \frac{\varrho_1}{\varrho_2}\right) g^2 (h_2 - h_1) h_1 = 0.$$

Ihre beiden Wurzeln sind nun

$$c_1 = \sqrt{g h_2} \quad \text{und} \quad c_2 = \sqrt{\frac{\varrho_2 - \varrho_1}{\varrho_2} g \frac{h_1 (h_2 - h_1)}{h_2}}. \tag{5.11}$$

Bei der ersten dieser Wellen ist die Wellengeschwindigkeit identisch mit jener von Flutwellen an der Oberfläche der ganzen Wassermasse bis zum Boden. Dies ist die äußere Welle, die hier nicht weiter behandelt wird. Der Wert c_2 ist die Wellengeschwindigkeit der *inneren* Welle. Ist die Dicke der unteren Wasserschichte $h_2 - h_1$ groß gegenüber h_1, dann reduziert sich diese Gleichung auf die zweite der Gl. (5.8). Die Amplitude der internen Welle ist an der Grenzfläche

[1] W. EKMAN: The Norw. North Polar Exp. 1893—1896. Christiana 1906. — Siehe auch Ann. Hydrogr. Hamburg **1904**, 562.

am größten und nimmt nach oben und unten ab. An der Wasseroberfläche ist sie entsprechend der Gl. (5.9)

$$\zeta_0 = -Z\,\frac{\varrho_2 - \varrho_1}{\varrho_2}\,, \tag{5.12}$$

wenn Z die Amplitude an der Grenzfläche ist. Das negative Vorzeichen deutet an, daß hier die Phase der Welle gegen jene der Grenzfläche entgegengesetzt ist. Nimmt man z.B. $\varrho_2 - \varrho_1 = 1{,}0245 - 1{,}0235 = 0.001$ g/cm³, was schon ein großer Wert ist, so wird, bei $Z = 10$ m, ζ_0 von der Größenordnung 1 cm. Diese Amplitude ist im allgemeinen so klein, daß sie unbeachtet bleibt. Am Boden ist keine vertikale Verschiebung der Wasserteilchen vorhanden, so daß hier die interne Welle ebenfalls verschwindet. Nimmt man an (was in erster Annäherung erlaubt ist), daß die Amplituden der internen Welle sich sowohl in der oberen wie in der unteren Wasserschichte linear mit Entfernung von der Grenzfläche ändern, dann ist die Änderung pro Längeneinheit in der oberen Schichte Z/h_1, in der unteren $Z/(h_2 - h_1)$. Aus der Amplitude der vertikalen Verlagerung kann aber in jeden Fall auf die Amplitude der horizontalen Verlagerung der Wasserteilchen geschlossen werden, da

$$u = c_2\,\partial\zeta/\partial z \tag{5.13}$$

sein muß. Mit den obigen Werten erhält man so für die beiden Schichten

$$U_1 = c_2\,\frac{Z}{h_1} \quad \text{und} \quad U_2 = -\,c_2\,\frac{Z}{h_2 - h_1}\,. \tag{5.14}$$

Die Amplituden der horizontalen Geschwindigkeiten sind somit konstant in jeder Schichte; sie ändern aber an der Grenzfläche unstetig ihr Vorzeichen, sind auf beiden Seiten der Grenzfläche verschiedener Richtung. Da aus Kontinuitätsgründen $U_1 h_1 = U_2 (h_2 - h_1)$ sein muß, sind diese Geschwindigkeitsamplituden verkehrt proportional der Dicke der beiden Schichten und man erhält:

$$U_1 = Z\sqrt{\Delta\varrho g\,\frac{h_2 - h_1}{h_2 h_1}} \quad \text{und} \quad U_2 = -\,Z\sqrt{\Delta\varrho g\,\frac{h_1}{(h_2 - h_1)\,h_2}}\,. \tag{5.15}$$

Die Größenordnung dieser Geschwindigkeiten ersieht man am besten an einem Zahlenbeispiel: $\varrho_2 = 1{,}025$, $\varrho_2 - \varrho_1 = 10^{-3}$, $g = 981$ cm sec^{-2}, $h_1 = 50$ m, $h_2 - h_1 = 200$ m und $Z = 15$ m. Man erhält: $U_1 = 18.6$ und $U_2 = 4.7$ cm sec^{-1}. Dies zeigt, daß schon bei internen Wellen mäßiger Amplitude die mit ihnen verbundenen *horizontalen* Verlagerungen der Wasserteilchen relativ große Werte annehmen, die bei Strommessungen ohne besondere Schwierigkeiten festgestellt werden können. Die entsprechende Wellengeschwindigkeit der internen Welle ist hier 63 cm sec^{-1}, jene der gewöhnlichen Flutwelle an der Oberfläche hingegen 4953 cm sec^{-1}. Beispiele für solche interne Flutwellen werden später gegeben.

γ) Interne Wellen bei stetiger Dichtezunahme mit der Tiefe. Eine umfassende Theorie der internen Wellen in Wassermassen, in denen die Dichte stetig mit der Tiefe variiert, hat J. F. FJELDSTAD[1] entwickelt und auch eine Methode angegeben, alle internen Wellen, die bei einem gegebenen Dichteaufbau möglich sind, zu berechnen. Im Falle einer fortschreitenden harmonischen Welle mit der Periode $T = 2\pi/\sigma$ und einer Wellenlänge $\lambda = 2\pi/\varkappa$, also der Fortpflanzungsgeschwindigkeit $c = \sigma/\varkappa$, kommt es auf die Integration der Differentialgleichung

$$\frac{d^2\zeta}{dz^2} - \frac{\varkappa^2}{\sigma^2}\,g\,\varphi\,\zeta = 0 \tag{5.16}$$

[1] J. E. FJELDSTAD: Geofys. Publ. **10** Nr. 8 (1935).

an, worin ζ die vertikale Verlagerung der Wasserteilchen bedeutet. Da $\partial\zeta/\partial t = w$ die vertikale Geschwindigkeit der Wellenstörung ist, kann man an Stelle von ζ auch w einführen. Die Randbedingungen sind, daß am Boden $\zeta = 0$ und an der Oberfläche konstanter Druck herrscht. Schließt man die gewöhnliche „äußere" Welle aus, dann ist die Randbedingung für die Oberfläche in erster Annäherung ebenfalls $\zeta = 0$. Die Größe $\varphi = -\frac{1}{\varrho_0}\frac{d\varrho_0}{dz}$ ist durch den *mittleren* Dichteaufbau der Wassermassen als Funktion von z gegeben.

Die nähere Prüfung dieser Differentialgleichung zeigt, daß sie unendlich viele Lösungen hat. Sie entsprechen unendlich vielen internen Wellen derselben Periode T, aber verschiedener Wellengeschwindigkeit und verschiedener Verteilung von ζ. Man bezeichnet sie als *interne Wellen nullter, erster, zweiter usw. Ordnung*, wobei die Welle nullter Ordnung die gewöhnliche Oberflächenwelle ist, die wir durch die vereinfachte Randbedingung ausgeschlossen haben. Bei der Welle erster Ordnung ist die vertikale Verlagerung in der ganzen Wassermasse von der Oberfläche bis zum Boden von derselben Richtung, Null an diesen Grenzen und mit einem Maximum in einer bestimmten Tiefe. Bei einer Welle zweiter Ordnung hat die obere Schicht bei ihren vertikalen Verlagerungen eine andere Richtung als die untere Schicht und es gibt zwei maximale Amplituden; bei Wellen dritter Ordnung gibt es drei Maxima, und so fort. Die Wellengeschwindigkeiten verhalten sich angenähert wie $1:\frac{1}{2}:\frac{1}{3}$ usw.

Die Theorie FJELDSTADS gilt nur für *lange Wellen* und zunächst ohne Berücksichtigung der Erdrotation. Will man auch diese in Betracht ziehen, dann werden die Schwierigkeiten mathematischer Art größer. Aber es gibt auch hier wieder Wellen von KELVINschen Typus, bei denen keine Querbewegungen zur Fortpflanzungsrichtung auftreten, dafür ändert sich die Wellenamplitude in dieser Querrichtung nach einer e-Potenz. Diese Änderung kann bei internen Wellen z. B. in einem Kanal so stark sein, daß sich diese Wellen praktisch nur auf einer Seite des Kanals auswirken. FJELDSTAD hat eine Methode der schrittweisen Lösung der Differentialgleichung nach der vereinfachten Integrationsmethode von STÖRMER gegeben, mit der bei gegebenem Parameter $\frac{\varkappa^2}{\sigma^2} g\varphi$ die Verteilung von ζ und w leicht bestimmt werden kann. Auf diese Methode sei hier besonders verwiesen.

Eine Erweiterung dieser Theorie interner Wellen hat P. GROEN[1] gegeben; er erweitert die Gültigkeit der Theorie auch auf *kurze Wellen*. Für eine spezielle Dichtezunahme mit der Tiefe (zwei im wesentlichen homogene Schichten mit einer Übergangsschicht dazwischen) werden die Rechnungen analytisch bis zum Schluß durchgeführt. Die Dicke der Übergangsschicht geht in die Rechnung als freier Parameter ein und kann beliebig gewählt werden. Dadurch dürften seine Ergebnisse ziemlich allgemein gelten und angewandt werden können. Die Übereinstimmung mit den Folgerungen aus der FJELDSTADschen Theorie sind bemerkenswert, nur zeigt sich, daß es bei einer gegebenen Dichteverteilung eine Untergrenze für die Periode interner Wellen gibt. Diese Tatsache wurde von GROEN auch für jede beliebige Dichtezunahme mit der Tiefe bewiesen. Diese Minimalperiode dürfte die Periode der *Eigenschwingungen* der gestörten stabilen Dichteverteilung um ihren Gleichgewichtszustand sein und entspricht der Periode, die der Schicht maximaler Stabilität angehört. Diese Eigenperiode ist recht klein, z. B. bei der Dichtezunahme $\Delta\varrho = 10^{-4}$ pro Meter von der Größenordnung einiger Minuten. Es dürften dies jene Stabilitätsschwingungen sein, die man auf ganz anderem Wege theoretisch schon früher ermittelt hat, und die auch durch die

[1] P. GROEN: Meded. Verh. Kon. Nederl. Meteor. Inst., Ser. B., Deel II **1948**, Nr. 11.

Beobachtungen einigermaßen belegt sind. K. KALLE[1] und G. NEUMANN[2] haben auf solche kurzperiodische Schwingungen innerhalb der Wassermassen hingewiesen.

Auch bei der theoretischen Behandlung *zellulärer* Wellen, die auf A. E. H. LOVE[3], BRUNSIDE[4], V. BJERKNES[5] und L. PRANDTL[6] zurückgeht, ergibt sich, daß eine Wassermasse mit nach unten zunehmender Dichte ein den homogenen Medien völlig fremdes Verhalten zeigt. Die Theorie geht aus von den gewöhnlichen linearisierten Bewegungsgleichungen. Meistens wird im Gleichgewichtszustand eine Dichtezunahme mit der Tiefe nach einer e-Potenz angenommen derart, daß $\varrho_0 = \varrho_B\, e^{-\frac{z}{H}}$ wird, wenn z vom Boden aus gezählt wird; H ist jene Höhe über dem Boden, in der die Dichte auf den e-Teil gesunken ist (LAMB[7]). Wenn es sich um „kürzere" Wellen handelt, dann ergibt die Lösung eine Wellenbewegung in stabil geschichteten Medien in *Zellenform*, d.h. der ganze von den Schwingungen erfaßte Raum ist in „Zellen" von bestimmten Ausmaßen unterteilt; in jeder Zelle geht die Schwingung wie zwischen festen Wänden vor sich, und zwar in jeder Zelle mit einer gemeinsamen Periode.

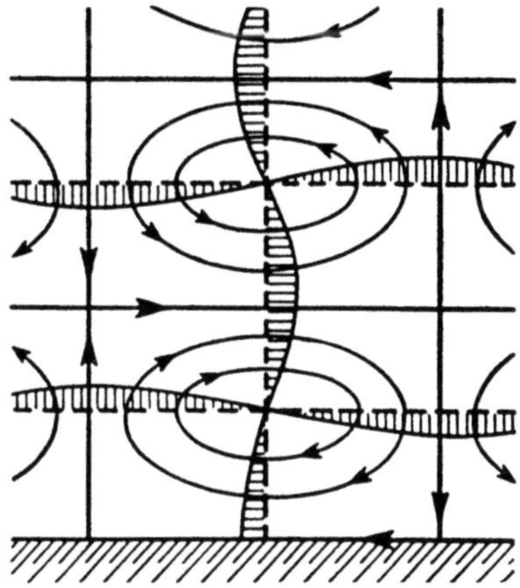

Fig. 6. Zellulare Schwingung in einem Medium, dessen Dichte mit der Höhe über dem Boden abnimmt (schematisch).

Einfacher sind *stehende zellulare Schwingungen* zu behandeln. Ein möglicher Typus ist folgender:

$$\left.\begin{aligned} \xi &= A\, e^{z/2H} \cos\varkappa x \left[\frac{1}{2H\varkappa}\cos\varepsilon z - \frac{\varepsilon}{\varkappa}\sin\varepsilon z\right]\cos\sigma t, \\ \zeta &= A\, e^{z/2H} \sin\varkappa x \cos\varepsilon z \cos\sigma t. \end{aligned}\right\} \tag{5.17}$$

ξ und ζ sind die horizontalen und vertikalen Verlagerungen der Teilchen; $\lambda_x = 2\pi/\varkappa$ und $\lambda_z = 2\pi/\varepsilon$ die horizontale und vertikale Wellenlänge der Schwingung. Obige Fig. 6 gibt eine schematische Darstellung einer solchen inneren Schwingung. Man erkennt, die Aufteilung des ganzen Raumes in „Zellen" gleicher Form. Die Schwingungsdauer hängt von der Stabilität der Wasserschichtung und von den linearen Dimensionen der Zellen ab. Für kleinere horizontale Wellenlängen erhält man eine praktisch konstante Schwingungsdauer

$$T = 2\pi\sqrt{\varrho_0/g\Gamma}, \tag{5.18}$$

wenn

$$\Gamma = -\,d\varrho_0/dz = \varrho_0/H$$

ist und mit der Stabilität der Schichtung zusammenhängt. Die Periode ist unabhängig von der Wellenlänge, während bei den gewöhnlichen Oberflächenwellen $T = \sqrt{2\pi\lambda_x/g}$, also abhängig von der Wellenlänge ist. Es scheint, daß die schon öfters registrierten rasch verlaufenden Temperaturschwankungen solche zellulare

[1] K. KALLE: Ann. Hydrogr. Hamburg **70**, 383 (1942).
[2] G. NEUMANN: Naturwiss. **33**, 282 (1946).
[3] A. E. H. LOVE: Proc. London Math. Soc. **22**, 307 (1891).
[4] BRUNSIDE: Proc. London Math. Soc. **20**, 392 (1889).
[5] V. BJERKNES: Geofys. Publ. 2, Nr. 4 (1921).
[6] L. PRANDTL: Strömungslehre, 2. Aufl., S. 331. Braunschweig 1942.
[7] H. LAMB: Lehrbuch der Hydrodynamik, 2. Aufl., S. 423 u. f., Leipzig u. Berlin 1931.

Schwingungen oder Stabilitätsschwingungen sind. Wenigstens macht G. NEUMANNS[1] Analyse solcher Beobachtungen in der Ostsee die Annahme, daß es sich um derartige Schwingungen handelt, recht wahrscheinlich.

δ) Interne Wellen in abgeschlossenen Wassermassen (Seen und Buchten) und im Ozean (besonders mit Gezeitencharakter). Durch Reflexion fortschreitender interner Wellen an den Rändern abgeschlossener Wassermassen werden interne Wellen in diesen die Form *stehender interner Wellen* annehmen. In Becken konstanter Tiefe und rechteckigen Querschnittes sind die Perioden der freien stehenden internen Schwingungen bei zweifach geschichtetem Wasser (l die Länge des Beckens, $n = 1, 2, 3 \ldots$) gegeben durch:

$$T_n = \frac{2l}{n} \sqrt{\frac{1}{g\Delta\varrho} \frac{h_2}{(h_2 - h_1)\, h_1}}\,. \tag{5.19}$$

Ist das Becken nicht allseitig geschlossen, sondern einseitig offen (Bucht), dann tritt in dieser Formel $4l$ and Stelle von $2l$.

Da in Seen und in Buchten eine lange Zeit des Jahres hindurch gut entwickelte Sprungschichten der Dichte vorhanden sind, müssen interne stehende Schwingungen eine häufige Erscheinung sein, da sie leicht erregbar sind. E. WEDDERBURN[2] hat für schottische Seen, F. M. EXNER[3] für den Wolfgangsee das häufige Vorhandensein solcher stehender interner Wellen nachgewiesen und zeigen können, daß ihre Perioden mit den theoretisch ermittelten befriedigend übereinstimmen.

Ein besonders schönes Beispiel für Temperaturseiches in einem See sind jene des Madü-Sees in Hinterpommern in der Zeit vom 25. Juli bis 14. August 1909 (W. HALBFASS[4] und W. WEDDERBURN[5]). Fig. 7 gibt den Verlauf der Isothermen in der Schicht zwischen 10 und 20 m Tiefe während dieser Zeit. Die vollausgezogenen Kurven sind die Isothermen für eine Station am Nordende, die gestrichelten jene für eine Station am Südende des langgestreckten Sees. Der völlig inverse Verlauf dieser beiden Kurvenscharen zeigt eindeutig, daß man es mit einer einknotigen internen Welle *(Seiche)* von rund 6 m Schwingungsweite zu tun hat. Ihre Periode ist fast genau 25 Std. Die für einen rechteckigen Trog umgerechneten Dimensionen der schwingenden Wassermasse des Madü-Sees ergab: $l = 13{,}75$ km, $h_1 = 23$ m, Temperatur der oberen Wasserschicht $t_1 = 17°$; $h_2 - h_1 = 15$ m, Temperatur der unteren Schicht $t_2 = 8°$. Damit erhält man die Periode der einknotigen internen Seiche $T = 91\,000$ sec = rund 25 Std, in guter Übereinstimmung mit dem beobachteten Wert.

Außerordentlich lange Perioden interner Wellen hat O. PETTERSSON[6] im Gulmarfjord an der Station Bornö aus täglichen Feststellungen der Schichtung der Wassermassen des Fjordes abgeleitet. Die Schwingungen haben eine Hauptperiode von rund 14 Tagen und Amplituden bis zu 25 m. E. M. WEDDERBURN[7] hat zeigen können, daß es sich bei diesen internen Wellen um stehende Schwingungen des ganzen Skagerrak handeln dürfte. Eine eingehende Untersuchung der internen Wellen des kalifornischen Golfes hat H. U. SVERDRUP[8] gegeben.

[1] G. NEUMANN: Dtsch. hydrogr. Z. **2**, 52 (1949).
[2] E. M. WEDDERBURN: Trans. Roy. Soc. Edinburgh **29** 420 (1905). — Proc. Roy. Soc. Edinburgh **28** (1907); **29** (1909).
[3] F. M. EXNER: Sitzgsber. Wien. Akad. II a, **117**, 9ff., 1295ff. (1908).
[4] W. HALBFASS: Int. Revue ges. Hydrobiol. u. Hydrogr. **3**, Suppl. 1, 1 (1911).
[5] E. M. WEDDEBURN: Int. Revue ges. Hydrobiol. u. Hydrogr. **3**, Suppl. 1, 1 (1911).
[6] O. PETTERSSON: Publ. Circonst. Nr. 47, Kopenhagen 1909.
[7] E. M. WEDDERBURN: Proc. Roy. Soc. Edinburgh **29**, 602 (1909).
[8] H. U. SVERDRUP: Assoc. Oceanogra. Phys. Procès Verb. Nr. 3, S. 170, 1904 und Proc. Sixth Pac. Sci. Congr. Berkeley 1939.

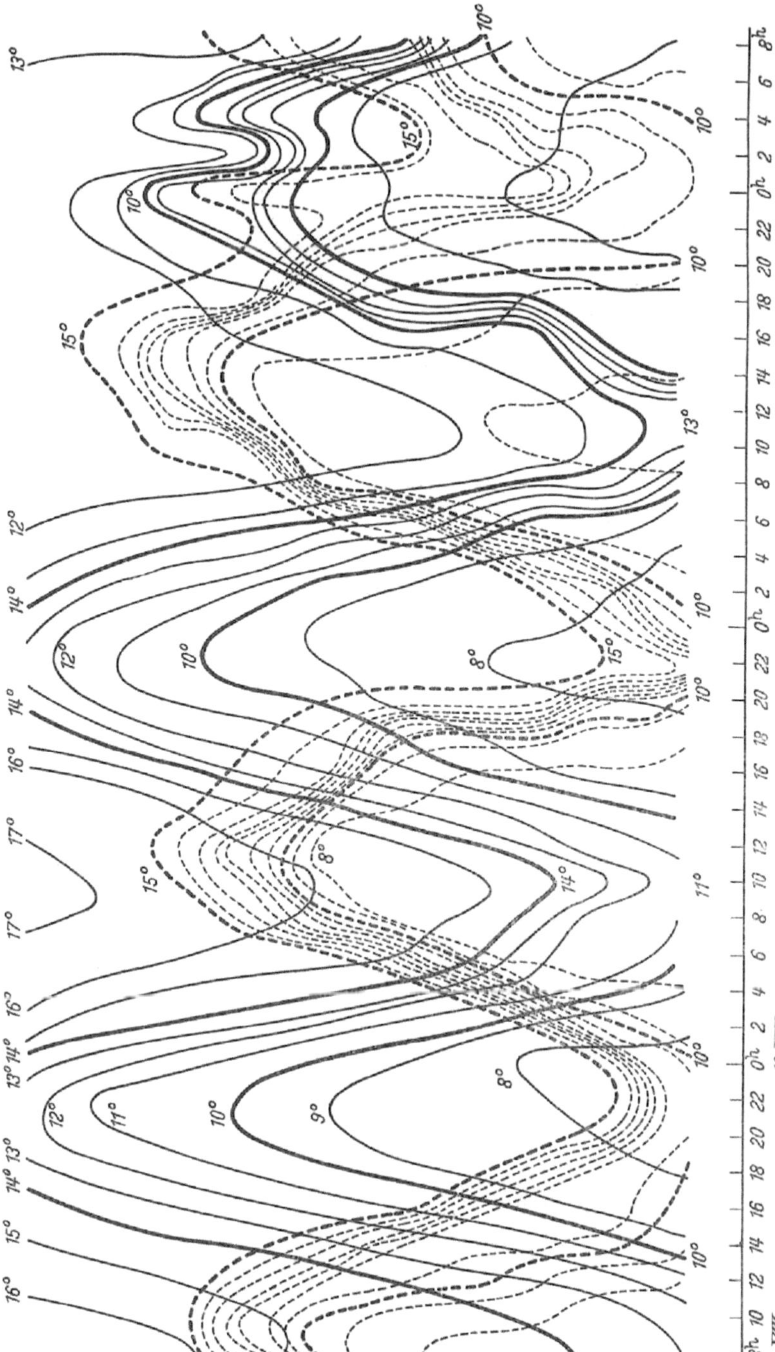

Fig. 7. Interne Seiches im Madü-See in Hinterpommern für die Zeit vom 9. bis 12. August 1909. Die vollausgezogenen Kurven: Isothermen zwischen 10 und 20 m Tiefe für eine Station am Nordende des Sees. Die gestrichelten Kurven dasselbe für eine Station am Südende des Sees.

Ihre Periode ergibt sich aus Beobachtungen zu rund sieben Tagen und W. H. MUNK[1] hat auf theoretischem Wege erwiesen, daß der orographischen Konfiguration des Golfes gemäß die interne Welle erster Ordnung eine Periode von 7 Tagen, jene der zweiten Ordnung eine solche von 14,8 Tagen haben muß. Es ist nicht unwahrscheinlich, daß die Erregung der 7tägigen Periode durch die 13,6tägige Mondtide erfolgt, da $2 \times 7 = 14$ näher an 13,6 steht als die Periode der stehenden Welle zweiter Ordnung.

Die übliche Methode, aus laufenden Serienmessungen der ozeanographischen Faktoren von einem verankerten Schiff aus die vertikale Amplitude der internen Wellen abzuleiten, ist folgende: Aus den beobachteten Änderungen in fortlaufenden Beobachtungen einer konservativen Eigenschaft des Wassers (z.B. seiner Temperatur, des Salzgehaltes oder des Sauerstoffgehaltes) läßt sich ohne Mühe — unter der Voraussetzung, daß das Gefälle dieser Eigenschaften in horizontaler Richtung klein ist — die Größe der vertikalen Verlagerung der Wassermassen, die diese Änderungen bedingte, berechnen. Ist S diese Eigenschaft, so lautet die Bedingung, daß das Wasser S unverändert behält[2]:

$$\frac{dS}{dt} = \frac{\partial S}{\partial t} + u\frac{\partial S}{\partial x} + w\frac{\partial S}{\partial z} = 0.$$

Setzt man $S = S_0(z) + s(x, z, t)$, so wird, bei kleinem $\partial S/\partial x$, und da $w = \partial\zeta/\partial t$ ist, $s = -\zeta\frac{dS_0}{dz}$ in erster Annäherung, oder $\zeta = -\frac{s}{\frac{dS_0}{dz}}$.

Die vertikale Verlagerung ζ läßt sich demnach dort am besten ermitteln, wo sich das vertikale Gefälle dS_0/dz der Eigenschaft leicht und genau bestimmen läßt.

Daß es echte interne Wellen im Ozean gibt, ergaben schon die ersten längeren Serienmessungen, die ausgeführt wurden (B. HELLAND-HANSEN und F. NANSEN[3]). Die Hauptwellen folgen neben meteorologischen Einflüssen in erster Linie den Gezeitenperioden und scheinen sich dort überall einzustellen, wo mehr oder minder deutliche Sprungschichten der Dichte vorhanden sind. Beim Studium des Wasseraustausches zwischen Ostsee und Kattegat durch den großen Belt hat OTTO PETTERSSON[4] zuerst solche interne Wellen mit halbtägiger Gezeitenperiode gefunden. Dann ist das Studium interner Wellen der Sprungschichten der Dichte im freien Ozean vor allem im Sommer 1910 von der „Micheal Sars"-Expedition im Nordatlantischen Ozean betrieben worden, später in mehr systematischer Weise auf den Ankerstationen der „Meteor"-Expedition 1925 bis 1927 in den tropischen und subtropischen Teilen des Atlantischen Ozeans, auf den Ankerstationen der „Snellius"-Expedition 1929/30 in den ostindischen Gewässern und auf den Ankerstationen der späteren Fahrten des „Meteor" 1937/38. Neben der Feststellung des thermo-halinen Zustandes in kurzen Intervallen wurden auf diesen Ankerstationen auch Strommessungen in verschiedenen Tiefen durchgeführt, so daß schon von einigen Punkten im freien Ozean ein zusammenhängendes Beobachtungsmaterial über interne Wellen vorhanden ist. Die Analyse dieses Beobachtungsmaterials hat ergeben, daß interne Wellen überall existieren und nicht etwa nur an scharf ausgebildete Grenzflächen (Sprungschichten)

---

[1] W. H. MUNK: J. Mar. Res. **4** Nr. 1 (1941)

[2] u und w sind die horizontale (Richtung x) und die vertikale (Richtung z) Geschwindigkeit des Stromes. ζ ist die vertikale Verlagerung eines Wasserteilchens und t die Zeit.

[3] B. HELLAND-HANSEN u. F. NANSEN: Norwegian Sea. Norw. Fish. u. Mar. Invest. Rep. **2** Nr. 2 (1909).

[4] O. PETTERSSON: Siehe A. DEFANT C. R. Congr. int. Geogr. Amsterdam, Sect. IIb **12**, 9 (1938).

gebunden sind, sondern auch da auftreten, wo eine mehr oder minder gleichmäßige Zunahme der Dichte den Aufbau der oberen Wassermassen charakterisiert. Sie sind also Erscheinungen weitester Verbreitung. Daß dabei eine enge Beziehung zwischen der vertikalen Verlagerung der Wassermassen in der Welle und den Schwankungen der Stromgeschwindigkeit bestehen muß, war vorauszusehen und wurde auch in den meisten Fällen durch die Strommessungen belegt. Gerade diese gemeinsame Analyse der Strommessungen und der Serienbeobachtungen gibt die Unterlagen zur Prüfung der Theorie der internen Wellen und gibt zugleich ein Bild des inneren Ablaufes der ganzen Erscheinung.

Die Theorie von FJELDSTAD hat sich bei den auf Ankerstationen zur Beobachtung gelangten internen Wellen längerer Periode sehr gut bewährt. Nur ein Fall soll hier herausgegriffen werden, und zwar die von FJELDSTAD[1] selbst vorgenommenen Messungen im Herdalefjord (Norwegen). Die Schmalheit des Fjords läßt erwarten, daß die Wellen, die hier gefunden wurden, wahrscheinlich vom einfachen KELVINschen Typus waren. Aus 88stündigen Temperatur- und Salzgehaltsmessungen, sowie aus gleichzeitigen Strommessungen wurden durch harmonische Analyse folgende Amplituden und Phasen der Vertikalbewegung, sowie jene des Stromes ermittelt.

Tabelle 3. *Interne Wellen im Herdalefjord, Sommer 1934. Station 1: 88 Std. Messungen.*

Tiefe m	Vertikale Verlagerung				Tiefe m	Strom				
	Beobachtungen		theor. Berechnung			in Wellenrichtung	quer dazu	Phase τ in Grad	theor. Berechnung	
	Ampl. ζ cm	Phase ε in Grad	Ampl. ζ cm	Phase ε in Grad		u cm/sec	v cm/sec		u cm/sec	τ in Grad
0	63	22	64,3	22,7	5	8,5	0,2	244	12,5	231
10	170	50	167,4	47,8	10	10,5	0,1	239	9,4	233
15	233	49	229,0	50,0	15	5,9	0,5	246	6,7	230
20	273	49	273,8	48,9	35	3,4	0,8	249	2,6	226
30	364	51	362,6	46,8	50	1,0	0,1	213	1,1	227
100	577	27	517,5	46,8	100	1,3	0,9	113	1,4	46

Es handelt sich dabei um die halbtägige Gezeitenwelle; u ist die Stromgeschwindigkeit in der Fortpflanzungsrichtung der Welle, die mit der Fjordrichtung übereinstimmt, v die Stromkomponente quer hierzu. Die Kleinheit von v bezeugt, daß die interne Welle tatsächlich angenähert dem KELVIN-Typus entsprach. Die Theorie ergab aus dem mittleren vertikalen Dichteaufbau für die ersten vier internen Wellen folgende Fortpflanzungsgeschwindigkeiten: 63, 33, 23, 17 cm/sec. Nimmt man an, daß die beobachteten vertikalen Verlagerungen im wesentlichen durch diese vier Wellen festgelegt sind, dann läßt sich die resultierende theoretische Verlagerung für jede Tiefe berechnen und mit den Beobachtungen vergleichen. Die Tabelle zeigt, daß die Übereinstimmung gut ist. Bei den Strommessungen ist zu beachten, daß der gewöhnliche Gezeitenstrom, der zur Welle nullter Ordnung gehört, in den theoretischen Werten nicht enthalten ist, wohl aber in den beobachteten. Den geringen Abweichungen nach muß der Strom im Fjord im wesentlichen nur von den inneren Gezeitenwellen herrühren.

Von besonderem Interesse sind die internen Wellen mit *Gezeitenperiode* im freien Ozean. Auf fast allen Ankerstationen gelangen solche innere Gezeitenwellen zur Beobachtung, und es scheint deshalb, daß sie von weiter Verbreitung sein müssen. Ein Beispiel soll zeigen, welch große Amplituden solche innere

[1] J. E. FJELDSTAD: C. R. Congr. int. Geogr. Amsterdam, Ser. IIb **12**, 49.

Gezeitenwellen haben können und wie regelmäßig die ganze Erscheinung abläuft. Es sind die internen Wellen der Ankerstation „Meteor“ 385 der Deutschen Nordatlantischen Expedition, 12. bis 14. Februar 1938 (O. v. SCHUBERT[1], A. DEFANT[2]). Sie liegt 700 sm vcm Festland entfernt, mitten im Nordatlantischen Ozean über großen Tiefen von 2900 m auf dem mittelatlantischen Rücken. Der Dichteaufbau zeigt bis 60 m Tiefe die homogene Deckschicht, darunter die subtropische Sprungschicht mit dem maximalen Dichtegefälle in etwa 100 m Tiefe; darunter nimmt die Dichte rasch von 1,0247 auf 1,0271 zu. Die Sprungschichte war der Sitz ausgesprochener, besonders gut ausgebildeter innerer Gezeitenwellen von 12 Mondstunden Dauer. Fig. 8 gibt eine Darstellung der internen Gezeitenwellen auf dieser Ankerstation. Alle drei dargestellten Kurven weisen einen streng periodischen Verlauf mit der Mondperiode von 12 Mondstunden auf. Die harmonische

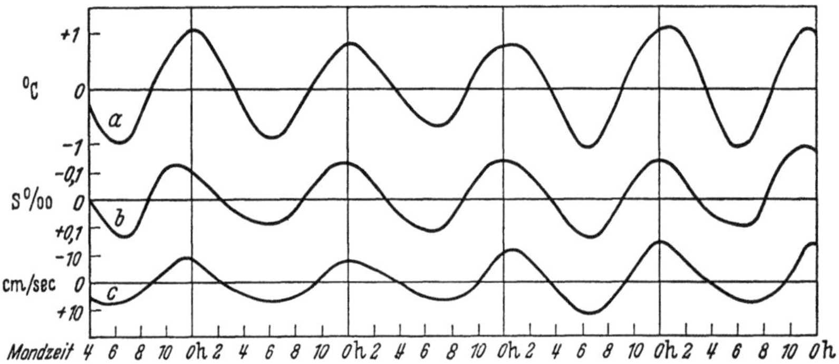

Fig. 8. Interne Gezeitenwelle auf der „Meteor“-Ankerstation 385; 16° 48,3′ N, 46° 17,1′ W. 12. bis 14. Februar 1935. *a* Abweichungen der Mitteltemperatur der Schicht 70 bis 120 m Tiefe vom mittleren Wert. *b* Abweichungen des Salzgehaltes in 80 m Tiefe vom mittleren Wert (inverse Kurve!). *c* Abweichungen der Stromstärke der *N*-Komponente in 50 m Tiefe (inverse Kurve!). Zeit in Mondstunden: 0^h Zeit der oberen Kulmination des Mondes in Greenwich.

Analyse der Beobachtungen ergab als Amplitude der vertikalen Verlagerung der 24°-Isotherme 12,1 m, als Phase 0.24^h Mondzeit in Grw. Die Strommessungen ergaben für die Oberschicht eine Amplitude 8,4 cm/sec, für die Unterschicht 9,0 cm/sec, während die Phase 6,6 bzw. 10.6^h waren. Wäre *nur* eine halbtägige interne Gezeitenwelle vorhanden, müßte der Phasensprung an der Sprungschichte 6 Std sein und nicht 4, wie die harmonische Analyse ergab. Es müssen noch innere Wellen höherer Ordnung schwach vertreten gewesen sein, aber vor allem wird der Gezeitenstrom der normalen Gezeit Störungen bedingen, die deutlich wahrnehmbar sein müssen.

ε) Die Eerregung interner Wellen, insbesonders solcher mit Gezeitencharakter. Trägheitswellen. Bei der Frage nach den *Ursachen* für die Ausbildung interner Wellen muß man sich klar sein, daß schon kleine Energiebeträge ausreichen, um solche Schwankungen von Grenzflächen zwischen verschieden dichten Flüssigkeiten oder auch vertikale Verlagerungen der Schichten in einer stetig geschichteten Wassermasse hervorzurufen. Kleine Druckstörungen an der Oberfläche reichen schon aus, um große vertikale Verlagerungen im Innern eines geschichteten Mediums einzuleiten. Solche Störungen des inneren Gleichgewichtes können bei günstigen Bedingungen durch *Resonanz* zu gewaltiger Amplitude anwachsen. Es läßt sich so erwarten, daß beim Auftreten von Störungen des inneren Gleichgewichtes die Rückkehr zu einem neuen Gleichgewichtszustand von internen Wellen begleitet sein wird, die allmählich durch Reibung abklingen.

[1] O. v. SCHUBERT: Ann. Hydrogr. Hamburg **72**, Beih. Januar (1944).
[2] A. DEFANT: Dtsch. hydrogr. Z. **5**, 231—245 (1952).

Daß *meteorologische Einflüsse* in hervorragendem Maße bei der Ausbildung interner Wellen beteiligt sein werden, läßt sich schon aus den leicht zu überblickenden Vorgängen bei der Entstehung interner Wellen in Seen erkennen. WEDDERBURN[1] hat solche Untersuchungen an schottischen Seen durchgeführt. In ähnlicher Weise hat H. PETTERSSON[2] die täglichen Schwankungen des Aufbaues der Wassermassen an der ozeanographischen Station Bornö (Gulmarfjord) benützt, um Beziehungen zu gleichzeitigen meteorologischen Änderungen festzustellen. Untersuchungen der Korrelationsbeziehungen zwischen den vertikalen Verlagerungen im Innern der Wassermassen und der Windstärke hat diesen meteorologischen Einfluß zahlenmäßig bestätigt. Der Windstau in den Oberflächenschichten im Inneren des Fjordes bei Wind von See her ist mit einer Senkung der Grenzfläche zwischen Ober- und Unterwasser (zumeist mit einer Verzögerung von 1 bis 2 Tagen) begleitet, und in entgegengesetzter Weise verlaufen die Änderungen bei Wind seewärts. Natürlich können nicht alle Schwankungen auf diese einfachen Vorgänge zurückgeführt werden, aber der Windstau bildet, wenigstens bei mehr oder minder abgeschlossenen Wassermassen, die Hauptursache für die Ausbildung interner Wellen. ZEILON[3] hat durch sehr sorgfältige Untersuchungen der internen Seiches des Gulmarfjordes festgestellt, daß die Stromvorgänge nahe der Mündung des Fjordes in den Kattegat die äußere Veranlassung für die Erregung interner Wellen im Fjord sind. Die Bodenkonfiguration an der Mündung spielt hierbei eine große Rolle. In Versuchen in der Wellenrinne hat später ZEILON[4] diese Ansicht über die Erregung interner Wellen in seichteren Meeresgebieten gut zu stützen vermocht.

Im freien Ozean sind topographische Einflüsse vom Meeresboden her kaum als Erregungsquelle der häufigen internen Wellen dortselbst zu erwarten. Eher sind hier äußere Einflüsse von Seite der Atmosphäre her anzunehmen, aber diese werden kaum Gezeitenperioden aufweisen, und gerade solche Periode treten besonders häufig in Erscheinung. O. PETTERSSON[5] hat sich in einer Reihe von Arbeiten mit diesem Problem befaßt und die Ansicht vertreten, daß in erster Linie die *vertikale* Komponente der Gezeitenkräfte für die Erregung interner Gezeitenwellen verantwortlich sei. Es läßt sich aber auf theoretischem Wege zeigen (A. DEFANT[6]), daß sich durch solche Kräfte Gezeitenwellen an inneren Grenzflächen *nicht* einstellen können. Die hierfür notwendigen Bedingungen sind bei den gegebenen Verhältnissen innerhalb des Meeres nirgends erfüllt. In den Amplitudenausdrücken für die erregten internen Gezeitenwellen erscheint im Nenner stets ein Ausdruck der Form

$$c^4 - h_2 g c^2 + \Delta\varrho\, h_1 (h_2 - h_1)\, g^2 = D. \qquad (5.20)$$

c ist die durch die fluterzeugende Kraft gegebene Fortpflanzungsgeschwindigkeit der erregten Gezeitenwellen der Wellenzahl $\varkappa$. Der Fall der Resonanz tritt ein, wenn $D = 0$ ist und die Wurzeln der Gl. (5.20) sind dann angenähert

$$c_1^2 = g h_2 \quad \text{und} \quad c_2^2 = g \Delta\varrho \frac{h_1 (h_2 - h_1)}{h_2}. \qquad (5.21)$$

[1] E. M. WEDDERBURN: Trans. Roy. Soc. Edinburgh **29**, 420 (1905). — Proc. Roy. Soc. Edinburgh **28** (1907); **29** (1909).

[2] H. PETTERSSON: Geogr. Ann. **1920**, 32.

[3] N. ZEILON: Kungl. Vetens. Akad. Handl. Stockholm **1912**.

[4] N. ZEILON: Medd. Göteborgs Högskolas Oceanogr. Inst. **1934**, Nr. 8.

[5] O. PETTERSSON: Siehe besonders Tidvattnetts Problem I bis IV. Arkiv Mat., Astronom. Fys. A **23** (1933); **24**, Nr. 16 u. 17 (1934); **25** Nr. 1 (1935).

[6] A. DEFANT: „Meteor"-Werk 7/1, S. 295. Berlin 1932. — Siehe auch J. JOHNSTONE's Mem. Vol. 310. Liverpool 1934.

Der erste Wert c_1 gehört der normalen freien Flutwelle der Oberfläche an, der zweite, c_2 der freien inneren Welle. Der Fall der Resonanz für innere Gezeitenwellen tritt somit nur ein, wenn c gleich der Fortpflanzungsgeschwindigkeit c_2 *freier* innerer Wellen der Grenzfläche ist. Dies ist aber in der Natur nie der Fall; denn c ist von der Größenordnung 150 bis 200 m/sec, c_2 jedoch von der Größenordnung von 1 bis 2 cm/sec.

Das sehr häufige Auftreten interner Gezeitenwellen im freien Ozean muß daher einen anderen Grund haben. Bei der theoretischen Ableitung obiger Beziehung wurde stets die Erdrotation vernachlässigt. Aber im freien Ozean ist diese Vernachlässigung wohl kaum erlaubt. Führt man die Corioliskraft ein (DEFANT[1], HAURWITZ[2]), dann tritt im Nenner der Ausdrücke für die Amplitude der vertikalen Verlagerung der Wasserteilchen der Ausdruck:

$$\varepsilon^4 - h_2 g \varepsilon^2 + \Delta \varrho g^2 h_1 (h_2 - h_1) = D_e \tag{5.22}$$

auf. Er hat dieselbe Form wie (5.20), nur steht statt c^2 nun $\varepsilon^2 = \frac{\sigma^2 - f^2}{\varkappa^2}$; $\frac{\sigma^2}{\varkappa^2} = c^2$ ist die durch die fluterzeugende Kraft gegebene Fortpflanzungsgeschwindigkeit der Gezeitenwelle, und in dem Coriolisparameter $f = 2\omega \sin\varphi$ steckt der Einfluß der Erdrotation. Während c von der Größenordnung 200 m/sec ist, und D stets eine große Zahl ist, kann ε klein sein und nahe an die Wellengeschwindigkeit freier innerer Wellen kommen, wodurch D_e eine kleine Größe wird und der Fall der Resonanz erreicht wird. Ist $T_i = 12$ Std/sin φ die Trägheitsperiode $= 2\pi/f$, so wird $\varepsilon^2 = c^2\left[1 - \left(\frac{T}{T_i}\right)^2\right]$. Für 30° Breite wird z. B. $\varepsilon = 0$ für die ganztägige Gezeit, für 74° Breite wird $\varepsilon = 0$ für die halbtägige Gezeit. Ist aber ε^2 sehr klein, reduziert sich D_e auf $D_e' = \Delta \varrho g^2 h_1 (h_2 - h_1) - g h_2 \varepsilon^2$ und dies wird Null, wenn

$$\varepsilon^2 = \Delta \varrho\, g (h_2 - h_1)\, h_1/h_2 = c_2^2;$$

das ist das Quadrat der Geschwindigkeit freier interner Wellen. Die Erdrotation ist demnach ein Faktor, der die Erregung großer interner Wellen ermöglicht und der erreicht, daß die internen Gezeitenwellen eine stärkere periodische Wellenbewegung werden als die Gezeit der Oberfläche.

Auf eine weitere Erscheinung, die mit den inneren Gezeitenwellen in Verbindung steht, möge hier hingewiesen werden. Mit den inneren Gezeitenwellen treten oft, ja fast immer auch interne *Trägheitsschwingungen* auf. Da die Periode der Trägheitswellen $T = 12$ Std/sin φ von der Größenordnung der Gezeitenwellen ist, vermag ihr gleichzeitiges Auftreten das äußere Bild der Schwankungen stark zu beeinflussen. Es erscheint, daß, wie die von den fluterzeugenden Kräften erregten inneren Gezeiten, auch die reinen inneren Trägheitswellen von weltweiter Ausbildung in den Ozeanen sind.

Ein Fall soll auch hier das Zusammenspiel von inneren Gezeitenwellen und Trägheitswellen illustrieren. Die 90stündige Ankerstation des „Altair" am 16. bis 20. Juni 1938 (A. DEFANT[3]) zeigte zunächst bis 800 m Tiefe — so weit reichten die Strommessungen — die normale Gezeit. In den oberen Schichten war die Amplitude der Stromstärke 8,3 cm/sec, die Phase 0,5 Mondstunden nach Grw., das Verhältnis der großen zur kleinen Achse der Stromfigur betrug 0,43, ihr Drehsinn war cum sole, alles Eigenschaften, die einer normalen Gezeit im freien Ozean eigen sind. Gleichzeitig ergab die Analyse der Strommessungen eine 17 Std-Welle, die sich, mit dem Gezeitenstrom gekoppelt, in den Stromkurven

[1] A. DEFANT: J. Mar. Res. **2**, 111 (1950).

[2] B. HAURWITZ: Trans. Amer. Geophys. Un. **31**, Nr. 1 (1950). — Arch. Meteor., Geophys. u. Bioklimat., Ser. A, **7** 406—424 (1954).

[3] A. DEFANT: Ann. Hydrogr. Hamburg **68** (1940).

durch deutliche Schwebungen kundtat. Diese Welle ist eine reine Trägheitsschwingung: Die theoretische Periode einer solchen Welle beträgt in 44° 33′ Breite 17,1 Std; die Amplituden der beiden Komponenten waren einander gleich, der Strom war rechtsdrehend und die Ostkomponente folgte der Nordkomponente um $\frac{1}{4}$ der Periodendauer nach. Dies sind wieder Eigenschaften, die nur *reinen* Trägheitswellen zukommen. Diese Trägheitsschwingung zeigte sich aber auch in allen Schwankungen der ozeanographischen Faktoren, so daß kein Zweifel besteht, daß neben der Gezeitenwelle auch eine Trägheitswelle im gesamten Beobachtungsmaterial vorhanden war. Die Vermutung ist gerechtfertigt, daß das ganze Meeresgebiet um die Altairkuppe durch eine atmosphärische Störung vor der Verankerungszeit aus dem Gleichgewicht gebracht wurde und dann das System durch Eigenschwingungen allmählich einer neuen Gleichgewichtslage zustrebte. Im rotierenden System der Erde haben aber bei den großen Dimensionen des schwingenden Systems diese Eigenschwingungen die Periode der Trägheitswellen. — Auch im Kattegat und in der Ostsee hat man ähnliche Erscheinungen angetroffen, in der Ostsee allerdings nur reine Trägheitsschwingungen, da dort die Gezeiten völlig zurücktreten (T. GUSTAFSON and B. KULLENBERG[1]).

Setzt man in den Bewegungsgleichungen und in der Kontinuitätsgleichung eines Systems eines zweifach geschichteten Meeres alle variablen Größen proportional $e^{i\sigma t}$, also periodisch der Frequenz σ, so erhält man nach einigen Umbildungen als Ausdrücke für die vertikale Verlagerung ζ_1 der Oberfläche und ζ_2 der inneren Grenzfläche folgende Beziehungen[2]:

$$\left.\begin{aligned} \left[h_1 + (h_2 - h_1)\frac{\varrho_1}{\varrho_2}\right]\nabla^2\zeta_1 + (h_2 - h_1)\,\Delta\varrho\,\nabla^2\zeta_2 + \frac{\sigma^2 - f^2}{g}\zeta_1 = 0 \\ \nabla^2\zeta_1 + \frac{\sigma^2 - f^2}{g\,h_1}(\zeta_1 - \zeta_2) = 0. \end{aligned}\right\} \qquad (5.23)$$

Läßt man nun ζ_1 und ζ_2 einer Schwingung folgen, deren Wellenzahl $\varkappa$ ist, wenn also $\zeta_n = A_n e^{i\varkappa x}$ ist $(n = 1, 2)$, dann erhält man zwei homogene Gleichungen für A_1 und A_2, deren Determinante Null sein muß. Sie führt zu einer quadratischen Gleichung für die Größe $\frac{\sigma^2 - f^2}{g}$, die die Frequenz der Welle mit der Wellenzahl $\varkappa$ im *rotierenden* System festlegt. Die zwei Lösungen dieser Gleichung sind:

$$\sigma_o^2 = f^2 + \varkappa_1^2 g h_2 \quad \text{und} \quad \sigma_i^2 = f^2 + \varkappa_2^2 g\,\Delta\varrho\,\frac{h_2 - h_1}{h_2}\,h_1. \qquad (5.24)$$

Die erste gehört zu den gewöhnlichen Oberflächenwellen, die zweite zu den internen Wellen. Die Eigenfrequenz einer Schwingung an der Oberfläche des schwingenden Systems bei *nicht rotierender Erde* ist aber $\sigma_1^2 = \varkappa_1^2 g h_2$; andererseits ist die Eigenfrequenz der inneren Welle $\sigma_2^2 = \varkappa_2^2 g\,\Delta\varrho\,\frac{h_2 - h_1}{h_2}\,h_1$.

Es wird damit aus den Beziehungen (5.24)

$$\sigma_o^2 = f^2 + \sigma_1^2 \quad \text{und} \quad \sigma_i^2 = f^2 + \sigma_2^2. \qquad (5.25)$$

Bei großer Ausdehnung des schwingenden Systems ist aber namentlich $\sigma_2 \ll f$, so daß in großer Annäherung $\sigma_i = f$ wird, d.h. *bei rotierender Erde* ist die Periode der freien inneren Schwingungen stets gleich der Periode der Trägheitsschwingungen. Dieses wichtige Ergebnis macht vieles, was man auf Ankerstationen im freien Ozean beobachten kann, verständlich.

---

[1] T. GUSTAFSON u. B. KULLENBERG: Svenk. Hydrogr.-Biol. Komm. Skr. Ny Ser. **13** (1936). Siehe auch B. KULLENBERG u. J. HELA: Svenk. Hydrogr.-Biol. Komm. Skr., Ny Ser. **16** (1942).

[2] ∇^2 ist der LAPLACEsche Operator $\frac{\partial^2}{\partial x^2} + \frac{\partial^2}{\partial y^2}$.

ζ) Stabilität interner Gezeitenwellen, interne Brandung. Die gezeitenperiodischen Verlagerungen der Sprungschichte der Dichte erfolgen im freien Ozean in den allermeisten Fällen in völlig symmetrischer Form, das Wellenprofil ist durchaus symmetrisch. Aber bei geringer Wassertiefe, bei inneren Gezeitenwellen in Schelfmeeren oder seichten Meeresstraßen, ist dies oft nicht der Fall. Die innere Welle zeigt Degenerationserscheinungen. Ein typisches Beispiel hierfür bieten die inneren Gezeitenwellen in der Straße von Gibraltar (A. DEFANT[1]). Bei Nipptide verlaufen diese Wellen bei gut entwickelter Sprungschichte zwischen dem nach Osten gerichteten atlantischen Oberstrom und dem nach Westen gerichteten mediterranen Unterstrom normal, das Wellenprofil ist symmetrisch. Aber bei Springtide und vielleicht ungünstigen Wetterverhältnissen, also bei verstärkten Gezeitenströmen oder Grundströmen degeneriert die innere Welle so stark, daß der eine Wellenast nahezu senkrecht steht. Die Welle steht knapp vor dem Umkippen. Das Wellenprofil ist sehr unsymmetrisch und die innere Welle nimmt die Form einer *internen Bore* an. In dieser Form durchwandert sie wahrscheinlich den ganzen Bereich der Meeresstraße. Weitere Fälle instabiler innerer Wellen haben die Beobachtungen auf Ankerstationen auf dem westafrikanischen Schelf zwischen den Kanarischen und Kap Verdeschen Inseln ergeben.

Es scheint demnach, daß es auch bei den internen Gezeitenwellen zu *inneren Brandungserscheinungen* kommen kann. Man darf sich natürlich nicht vorstellen, daß diese internen Brandungen mit jener Verhemenz und Schnelligkeit sich abspielen, wie jene bei den gewöhnlichen Oberflächenwellen. Entsprechend der geringen Wellengeschwindigkeit der internen Wellen geht der Vorgang äußerst langsam vor sich.

Bei der Frage nach den Ursachen für die Steigerung der Asymmetrie des Wellensprofils bis zur Ausbildung der inneren Brandung kann man zunächst an dieselben Ursachen denken, die die Profiländerung bei den gewöhnlichen Oberflächenwellen bedingen. Wenn die Amplitude der Welle groß wird gegenüber der mittleren Wassertiefe, besteht immer die Neigung die Wellenvorderseite steiler zu stellen, bis schließlich die Welle brandet. Auch bei den internen Wellen wird so (bei geringer Wassertiefe am Schelf) die geringe Wassertiefe für eine Steigerung der Asymmetrie des Wellenprofils verantwortlich zu machen sein.

In den Fällen interner Gezeitenwellen können aber auch Fälle *dynamischer Instabilität* vorliegen. Wenn zwei Wasserschichten großer Mächtigkeit übereinander lagern und sich mit den Geschwindigkeiten U_1 und U_2 bewegen, dann ist die Wellengeschwindigkeit an der Grenzfläche gegeben durch

$$c = \frac{\varrho_1 U_1 + \varrho_2 U_2}{\varrho_1 + \varrho_2} \pm \sqrt{\frac{g}{\varkappa}\,\frac{\varrho_2 - \varrho_1}{\varrho_2 + \varrho_1} - \varrho_1\,\varrho_2\left(\frac{U_1 - U_2}{\varrho_1 + \varrho_2}\right)^2}\,. \tag{5.26}$$

Das erste Glied rechts stellt die konvektive Geschwindigkeit dar; sie ist ein Mittelwert der beiden Grundströme. Gegenüber dieser mittleren Geschwindigkeit ist die Wellengeschwindigkeit von Wellen der Länge $\lambda = 2\pi/\varkappa$ durch den Wurzelausdruck gegeben. Das erste Glied davon ist die Wellengeschwindigkeit im ruhenden System. Durch das zweite Glied kann diese Geschwindigkeit imaginär werden, d.h. es kommt zur *Instabilisierung* der Wellen. Dies tritt ein, wenn

$$(U_1 - U_2)^2 > \frac{g}{\varkappa}\,\frac{\varrho_2^2 - \varrho_1^2}{\varrho_1 \varrho_2} \tag{5.27}$$

ist, d.h., wenn Wellenlänge und Dichtesprung gegeben sind, bei genügend großem Geschwindigkeitssprung an der Grenzfläche.

[1] A. DEFANT: Arch. Meteor., Geophys. u. Bioklimat., Ser. A **1** (1948)

In den Fällen in der Natur sind die Mächtigkeiten der zwei übereinandergelagerten Wasserschichten klein und die obere hat überdies eine freie Oberfläche. Man erhält dann für lange Wellen, wie es die inneren Gezeitenwellen sind, die Ungleichung

$$(U_1 - U_2) > \sqrt{\Delta\varrho g(h_2 + \Delta\varrho h_1)} > \sqrt{\Delta\varrho g h_2}. \tag{5.28}$$

Für die Werte, die größer sind als der Ausdruck rechts, werden die inneren Gezeitenwellen *instabil* und neigen zur inneren Brandung. Aus Zahlenwerten ersieht man, daß schon relativ kleine Geschwindigkeitsunterschiede in den Grundströmen ausreichen, um eine Instabilität zu erzwingen. So muß unter gegebenen Umständen die Tendenz interner Gezeitenwellen zu einer Degeneration des Wellenprofils bis zur Ausbildung von inneren Bore-Erscheinungen groß sein.

Besonders auffallend sind die inneren Gezeitenwellen in der Straße von Messina, die sich zwischen dem unteren jonischen und dem oberen tyrrhenischen Wasser ausbilden. Hier kommt es wegen der geringen Mächtigkeit der Oberschicht bei günstigen Verhältnissen zu sehr turbulenten Stromwirbeln, die sich aus den inneren Gezeitenwellen entwickeln und als *Scylla*- und *Charibdis*-Wirbel bekannt sind. Die Beobachtungen, die auf der Ankerstation des Forschungsschiffes „Marsigli" in der Straße von Messina gesammelt wurden (A. DEFANT[1]), haben eindeutig gezeigt, daß die inneren Gezeitenwellen gelegentlich instabil werden und dann durch Ungleicheit der Ströme im Straßenquerschnitt zu großen zyklonalen und antizyklonalen Wirbeln vertikaler Achse führen.

II. Theorie der Gezeiten.

6. Überblick über die Beobachtungstatsachen. Man versteht unter *Gezeiten (Ebbe und Flut, Tiden)* jene periodischen vertikalen Änderungen des Wasserspiegels des Meeres, die sich im wesentlichen zweimal des Tages einstellen. Mit diesen vertikalen Verlagerungen des Meeresniveaus sind auch horizontale Verschiebungen der Wassermassen verknüpft *(Gezeitenströme)*. Das rhythmische Auf- und Abwallen des Meeres geht völlig gesetzmäßig vor sich und wird nur durch Wind und Wetter gestört, gelegentlich sogar sehr stark *(Sturmfluten)*. Man erkennt bald, daß die Gezeiten, die in den Ozeanen ein erdumfassendes Phänomen sind, eine Erscheinung darstellen, die durch ein Kräftesystem verursacht sein muß, dem alle Wassermassen der Erde in gleicher Weise unterworfen sind. Ein solches Kräftesystem kann nur kosmischer Natur sein, d.h. mit der Stellung der Erde im Weltraum in Beziehung stehen.

Den höchsten Wasserstand einer Gezeitenwelle nennt man *Hochwasser*, den niedrigsten *Niedrigwasser*, das Steigen des Wasserspiegels *Flut*, das Fallen *Ebbe*. Der Wasserstandunterschied zwischen Hoch- und Niedrigwasser ist der *Tidenhub (doppelte Amplitude der Gezeitenwelle)*. Hoch- und Niedrigwasser folgen sich im Mittel in einem Intervall von 12 Std und 25 min. Dies weist auf einen Zusammenhang mit der Mondbewegung hin; denn auch der Mond verspätet sich bei seinem Durchgang durch den Meridian täglich um 50 min. Eine Tidenperiode entspricht somit im Mittel genau einem halben mittleren Mondtag. Man kann so im allgemeinen die Eintrittszeit des Hochwassers auf jene des Meridiandurchganges des Mondes beziehen und dieser (im Mittel konstante) Zeitunterschied wird für einen Ort (Hafen) mittlere *Hafenzeit* oder mittleres *Mondflutintervall* genannt. Da bei Voll- und Neumond (Syzygien) Mond und Sonne zugleich durch den Ortsmeridian gehen, die Uhrzeit vom Durchgang der Sonne durch den Meridian

[1] A. DEFANT: Ann. Hydrogr. **68**, 145 (1940). Siehe auch Naturwiss. **28**, 729 (1940)

aus gezählt wird, gibt die Hafenzeit auch an, um wieviel Uhr bei Voll- und Neumond Hochwasser an einem Ort eintritt. Die einzelnen Hochwasserintervalle schwanken um etwa ± 1 Std um das mittlere Flutintervall; diese Änderungen sind aber nicht regellos verteilt und kehren in etwa halbmonatlichen Zwischenzeiten wieder. Das ist die *halbmonatliche Ungleichheit* der Hochwasserzeit. Sie steht in Beziehung zu den Phasen des Mondes und ihre Periode beträgt die Hälfte eines synodischen Monats, d.i. 14,77 Tage. Auch der Tidenhub schwankt mit dieser Periode.

Die höchsten Hochwasser und die niedrigsten Niedrigwasser stellen sich stets mit gewissen Verspätungen nach Voll- und Neumond ein. Man nennt sie *Springtiden*, ihre Eintrittszeit ist die *Springzeit*. Die kleinsten Tidenhübe *(Nipptiden)* treten mit ähnlichen Verspätungen nach dem ersten und letzten Viertel des Mondes (Quadraturen) zur Nippzeit ein. Die Tatsache der halbmonatlichen Ungleichheit in Höhe und Zeit zeigt, daß nicht der Mond allein, sondern auch die Sonne als Ursache der Gezeiten in Betracht kommt.

Neben diesen wichtigeren Ungleichheiten gibt es nocht weitere, die sich erst durch sorgfältige Analyse der Beobachtungen eines längeren Zeitraumes feststellen lassen. Es sind dies folgende:

1. Die *parallaktische* oder monatliche *Ungleichheit* mit der Periode des anomalistischen Monats (27,55 Tage). Sie hängt von der wechselnden Entfernung des Mondes von der Erde ab. Mit gewissen Verspätungen ist nahe der Erdnähe des Mondes (Perigäum) der Tidenhub groß, nahe der Erdferne (Apogäum) hingegen klein.

2. Die *Deklinationsungleichheit*. Es kommt dabei auf den absoluten Betrag der Deklination des Mondes an, so daß ihre Periode die eines halben tropischen Monats, d.i. 13,66 Tage ist. Die größten Tidenhübe folgen mit gewissen Verspätungen den Durchgängen des Mondes durch den Äquator nach. Außer der lunaren Deklinationsungleichheit gibt es auch eine solare, die von den Deklinationsänderungen der Sonne abhängt; ihre Periode ist ein halbes Jahr.

3. Die *tägliche Ungleichheit* zeigt sich in einer abwechselnden Erhöhung und Erniedrigung der beiden täglichen Hoch- bzw. Niedrigwasserhöhen (unsymmetrische Gezeitenkurven). Die größten Höhenunterschiede treten mit gewissen Verspätungen nach der größten nördlichen oder südlichen Deklination des Mondes ein und verschwinden nach Äquatordurchgang des Mondes. Ihre Periode ist deshalb ein ganzer tropischer Monat. Für die Sonne gilt Analoges.

Schon diese kurze Übersicht zeigt, daß die Gezeiten zwar eine sehr verwickelte, aber völlig gesetzmäßig ablaufende Erscheinung sind, die sich nur durch systematische Sammlung von Beobachtungen und planmäßiges Studium der Zusammenhänge aufhellen läßt.

7. Gezeitenkräfte und ihr Potential. Gleichgewichtstheorie der Gezeiten. Die Grundlage für die Erklärung der Gezeitenerscheinungen bildet das System der fluterzeugenden Kräfte, das von den Gestirnen auf die Massen der Erde ausgeübt wird. Von solchen Gestirnen kommen für die Meeresgezeiten nur Mond und Sonne in Frage. Denn die maximale fluterzeugende Kraft eines Gestirns ist proportional seiner Masse und verkehrt proportional der dritten Potenz seiner Entfernung von der Erde. Diese Abhängigkeit zeigt sofort, daß gegenüber den fluterzeugenden Kräften von Mond und Sonne jene der anderen Planeten infolge ihrer geringen Masse und ihrer großen Entfernung völlig zurücktreten. Sie läßt auch das Verhältnis der fluterzeugenden Kräfte von Mond und Sonne leicht abschätzen. Ist S bzw. M die Sonnen- bzw. Mondmasse und s bzw. a die Entfernung Sonne—Erde bzw. Mond—Erde und ist E die Erdmasse, R der Erdradius, so

hat man:
$$\frac{S}{M}\cdot\frac{a^3}{s^3}=\frac{333\,200\,E}{\frac{1}{81,5}E}\cdot\left(\frac{60\,R}{23\,360\,R}\right)^3=0,460\sim\frac{4}{9}<\frac{1}{2}. \tag{7.1}$$

Die fluterzeugende Kraft der Sonne erreicht somit nicht die Hälfte jener des Mondes. Diese letztere kann man mit einer uns geläufigen Kraft z.B. mit der Schwerkraft (Fall-Beschleunigung g) vergleichen. Man hat dann in guter Annäherung:

$$\frac{\text{max. Flutkraft des Mondes}}{\text{Schwerkraft}}=2\frac{M}{E}\cdot\left(\frac{R}{a}\right)^3=2\frac{1}{81,5}\cdot\frac{1}{60^3}\sim\frac{1}{9}\cdot 10^{-6}. \tag{7.2}$$

Die fluterzeugende Kraft des Mondes ist rund der neunmillionste Teil der Schwerkraft, ist somit außerordentlich klein.

Für die Entwicklung des fluterzeugenden Potentials eines störenden Körpers (Mondes) ist in diesem Handbuch der Abschnitt „Gezeitenkräfte“ vorgesehen. Hier soll nur das wiederholt werden, was für das Verständnis des späteren notwendig ist.

Ist ϑ die Zenitdistanz des Mondes, so hat sein fluterzeugendes Potential in seinem Hauptglied die Form

$$V=m\,R\left(\frac{1}{3}-\cos^2\vartheta\right),\quad\text{mit}\quad\frac{m}{g}=\frac{3}{2}\frac{M}{E}\left(\frac{R}{a}\right)^3. \tag{7.3}$$

Für den Mond wird:

$$\frac{m}{g}=8,57\cdot 10^{-8},\quad\text{für die Sonne}\quad\frac{m}{g}=3,78\cdot 10^{-8}. \tag{7.4}$$

Wählt man die *horizontale Komponente* der fluterzeugenden Kraft K_h positiv in der Richtung zu jenem Punkt der Erdoberfläche, der senkrecht unter dem Monde liegt, dann ist:

$$K_h=\frac{1}{R}\frac{\partial V}{\partial\vartheta}=\frac{m}{2}\sin 2\vartheta. \tag{7.5}$$

Die *vertikale Komponente* bedeutet eine Gezeitenstörung der Schwerkraft und ist positiv nach unten gezählt:

$$\delta g=-\frac{m}{2}\left(\cos 2\delta+\frac{1}{3}\right). \tag{7.6}$$

Gegenüber der Fallbeschleunigung g ist diese Störung δg außerordentlich klein und trägt zur Bewegung der Wassermassen so wenig bei, daß sie im Allgemeinen ganz außer Betracht bleiben kann.

Infolge der Eigenbewegung des störenden Körpers (Mondes) und infolge der Rotation der Erde verschiebt sich das System der fluterzeugenden Kraft ständig auf der Erde. Ist δ die Deklination des Mondes, φ und λ die geographischen Koordinaten des betrachteten Punktes auf der Erdoberfläche, so ist für diesen:

$$\left.\begin{aligned}&\cos\vartheta=\cos\chi\cos\Delta+\sin\chi\sin\Delta\cos(t+\lambda-180^\circ),\\ \text{mit}\quad&\Delta=90^\circ-\delta,\quad\chi=90^\circ-\varphi\end{aligned}\right\} \tag{7.7}$$

und t = wahrer Stundenwinkel des Mondes in Greenwich, gezählt von der *unteren* Kulmination. In (7.3) eingesetzt erhält man, mit $-mR/2=A$,

$$\left.\begin{aligned}V=&A\,(\cos^2\Delta-\tfrac{1}{3})(3\cos^2\chi-1)\\&-A\sin 2\delta\sin 2\varphi\cos(t+\lambda)\\&+A\cos^2\delta\cos^2\varphi\cos 2(t+\lambda).\end{aligned}\right\} \tag{7.8}$$

Man kann nach dieser Entwicklung von V drei Gattungen von Glieder unterscheiden:

1. Das *erste Glied* enthält den Stundenwinkel nicht. Die Deklination kann für kurze Zeit als konstant angesehen werden. In bezug auf χ oder die geographische Breite φ stellt das Glied eine einfache zonale Kugelfunktion P_2 dar: Symmetrie zur Erdachse, Knotenlinien in den Parallelkreisen $\varphi = \pm 35° 16'$. Eine Schwankung dieses Anteils des Gezeitenpotentials rührt vom Glied $(\cos^2 \varDelta - \frac{1}{3}) = \frac{1}{2}(\cos 2\varDelta + \frac{1}{3})$ her, bedeutet also eine *Deklinationsungleichheit* und hat die halbe Periode der Deklinationsschwankung; beim Mond die Dauer eines halben tropischen Monats (rund 14 Tage), bei der Sonne die Dauer eines halben Jahres.

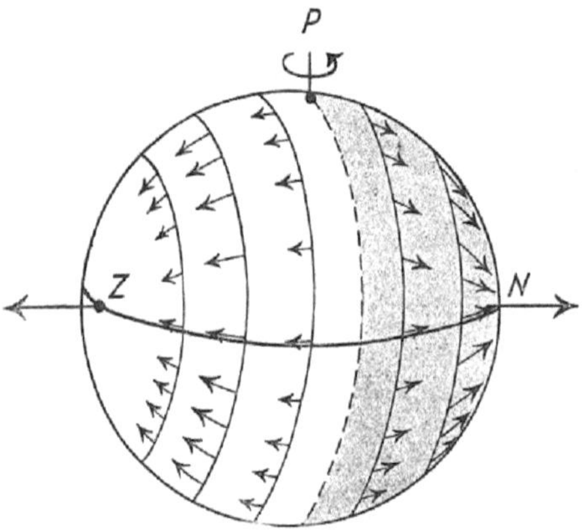

Fig. 9. Verteilung der horizontalen Komponente der fluterzeugenden Kraft auf der Erdoberflache. In Z steht der störende Mond im Zenith, in N im Nadir. (Vgl. auch Artikel TOMASCHEK in diesem Band, Fig. 1, S. 778.)

2. Das *zweite Glied* hat die Periode der Umlaufzeit des störenden Körpers, also eines Mond- bzw. Sonnentages. Knotenlinien sind am Äquator ($\varphi = 0°$) und in jenem Meridian, der vom störenden Körper um 90° entfernt ist. Die maximalen Werte finden sich im Meridian des störenden Körpers, und zwar 45° nördlich und südlich des Äquators (tesserale Kugelfunktion P_2^1). Das Glied stellt somit in der Hauptsache den *ganztägigen Anteil* am Gezeitenpotential dar.

3. Das *dritte Glied* hat die Periode der *halben* Umlaufzeit des störenden Körpers, also eines halben Mond- bzw. Sonnentages. Knotenlinien finden sich an den Meridianen 45° westlich und östlich von jenem des störenden Körpers (sektorielle Kugelfunktion P_2^2).

In dieser Entwicklung von V ist sowohl δ wie auch A konstant angenommen worden. Berücksichtigt man auch diese Veränderungen, dann kommt man zur *harmonischen Analyse des Gezeitenpotentials*, in der dieses in eine Reihe von cos-Gliedern mit konstanter Amplitude, konstanten Perioden und festen Phasen entwickelt wird. Betreffs dieser Entwicklung siehe den Beitrag Gezeitenkräfte.

Will man die Einwirkung der fluterzeugenden Kraft des Mondes auf einen die ganze Erde gleichmäßig bedeckenden Ozean untersuchen, dann ist es am besten, die Verteilung (7.5) der horizontalen Komponente dieser Kraft auf der Erdoberfläche zu betrachten: Unter ihrer Einwirkung strebt das Wasser gegen den Zenitpunkt Z und den Nadirpunkt N zu fließen, und es bilden sich an diesen Konvergenzpunkten Wasseransammlungen, während in einem Großkreis 90° davon der Wasserspiegel sinkt (siehe Fig. 9). Durch die Ausbildung dieser Flutberge entstehen Druckgradienten im Meer, und die durch die fluterzeugenden Kräfte in Gang gekommenen Strömungen erlöschen, wenn Druckgradient und Flutkraft an jeden Ort sich im Gleichgewicht befinden. Die Gezeiten sind (als statisches Problem aufgefaßt) nur eine Folge dieses Gleichgewichtszustandes. Ist $\bar{\zeta}$ die Erhebung der Gezeit über der Ruhelage, so ist $g\bar{\zeta}$ das Schwerepotential, und die Gleichgewichtsbedingung erfordert, daß die Summe der Potentiale der Schwerkraft und Flutkraft konstant ist *(Niveaufläche)*. Das ergibt

$$\bar{\zeta} = -\frac{V}{g} + C\,, \tag{7.9}$$

worin C eine Konstante ist. Mit (7.3) und $mR/g = H$, wird

$$\bar{\zeta} = H(\cos^2\vartheta - \tfrac{1}{3}) + C. \tag{7.10}$$

Die Gleichgewichtsform der Meeresoberfläche *(Flutsphäroid)* entspricht einer zonalen Kugelfunktion zweiter Ordnung, deren Achse durch den Mond geht. Hochwasser tritt für $\vartheta = 0°$ und 180° auf, d.h. wenn der Mond im Zenit oder im Nadir steht, während Niedrigwasser sich bei auf- bzw. untergehendem Mond einstellt. Man findet $\bar{\zeta}_{\max} = \frac{2}{3}H$ und $\bar{\zeta}_{\min} = -\frac{1}{3}H$. Der Tidenhub ist somit H und zwar ist H für die Mondgezeit 0,55 m, für die Sonnengezeit 0,24 m, während die vereinigte Wirkung beider 0,79 m ergibt.

Wählt man für V die Entwicklung (7.8), dann gibt es im wesentlichen drei Arten von *Partialfluten*: langperiodische, tägliche und halbtägige Fluten. Alles, was über das Gezeitenpotential erörtert wurde, läßt sich auf diese Partialtiden der Gleichgewichtstheorie übertragen. Die Beträge der einzelnen Glieder zum Gezeitenpotential können in solchen der Größe $\bar{\zeta}$ umgerechnet werden, und dies ist auch die Form, in der sie zumeist gegeben werden. Man gewinnt dadurch eine anschaulichere Darstellung der Verteilung des Gezeitenpotentials auf der Erdoberfläche.

Die vereinte Wirkung von Mond und Sonne läßt sich für die halbtägigen Gezeiten leicht überblicken. Führt man in (7.8) statt des Stundenwinkels $(t+\lambda)$ des störenden Körpers in der geographischen Länge λ die Sternzeit Θ und die Rektaszension α ein, so ist die Gleichgewichtsflut der vereinten halbtägigen Mond- und Sonnengezeit

$$\bar{\zeta} = A\cos 2(\Theta - \alpha) + A'\cos 2(\Theta - \alpha'), \tag{7.11}$$

worin die Größen mit dem Strich sich auf die Sonnengezeit beziehen. Setzt man $\Theta - \alpha' = \Theta - \alpha + (\alpha - \alpha')$, so läßt sich (7.11) die Form geben:

$$\left.\begin{aligned} &\bar{\zeta} = B\cos 2(\Theta - \alpha + \beta) \\ &\text{mit} \\ &B = \sqrt{A^2 + A'^2 + 2AA'\cos 2(\alpha - \alpha')} \quad \text{und} \quad \operatorname{tg} 2\beta = \frac{(\sin 2(\alpha - \alpha')}{A + A'\cos 2(\alpha - \alpha')} \end{aligned}\right\}. \tag{7.12}$$

Wenn $(\alpha - \alpha') = 0°$ oder 180° wird, also bei Konjunktion und Opposition von Mond und Sonne, oder bei Voll- und Neumond, so wird B ein Maximum; das sind die *Springfluten*. Bei $(\alpha - \alpha') = 90°$ oder 270°, also zur Zeit der Quadraturen, wird B ein Minimum; das sind die *Nippfluten*.

Auch die tägliche Ungleichheit in Höhe und Zeit wird durch die Gleichgewichtstheorie verständlich. Wenn der störende Körper nicht im Äquator steht, fällt die Drehachse des Flutsphäroids nicht mehr mit der Erdachse zusammen und diese Asymmetrie hat eine tägliche Ungleichheit in der Flut zur Folge. Die Gleichgewichtstheorie verlangt, daß für alle Orte der Erde das Mondflutintervall (Hafenzeit) Null ist. Da dieses aber erfahrungsgemäß alle Werte zwischen 0 und 6 Std annehmen kann, besteht hier ein unüberbrückbarer Widerspruch zwischen Theorie und Beobachtung. Auch die Tatsache, daß im allgemeinen überall die Gezeitenströme auftreten, während es solche in der Gleichgewichtstheorie nicht gibt, zeigt, daß diese Theorie kaum als Grundlage für die Erklärung der Gezeitenerscheinungen zu benützen ist. Dieses Mangels war sich NEWTON, auf den diese Theorie zurückgeht, schon voll bewußt.

Diese Gezeitentheorie setzt voraus, daß die Himmelskörper, die die Gezeiten hervorrufen, ihre Stellung gegenüber der Erde so langsam ändern, daß der geforderte Gleichgewichtszustand stets erhalten bleibt. Tatsächlich sind aber die zeitlichen Änderungen nicht so langsam, und vor allem stört die Begrenzung der Wassermassen durch die Kontinente. Man kann somit kaum erwarten, daß diese Theorie den Verhältnissen in der Natur gerecht werden kann. Höchstens für die langperiodischen Partialtiden (Periode 14 Tage und mehr) mag sie in erster Annäherung Geltung haben.

Man hat versucht (THOMSON und TAIT[1]), später G. H. DARWIN und H. H. TURNER[2], Verbesserungen an die Gleichgewichtstheorie anzubringen, indem man den Einfluß der Land- und Meerverteilung auf der Erde in Rechnung stellte. In dieser verbesserten Theorie tritt das Hochwasser nicht mehr synchron mit dem Maximum des Gezeitenpotentials ein, es gibt also in diesem Fall eine „Hafenzeit", die von Ort zu Ort verschieden ist, aber dieser Einfluß ist nicht sehr bedeutend.

Eine weitere Verbesserung betrifft die Berücksichtigung der gegenseitigen Anziehung der Wasserteilchen im deformierten Wasservolumen des Ozeans. Zum Gezeitenpotential des störenden Körpers tritt dann noch das Gravitationspotential der im Gezeitensphäroid verschobenen Wassermassen. Bei einem Meer das die ganze Erde bedeckt, erhöht sich V um $-\frac{3}{5}\frac{\varrho}{\varrho_m}g\bar{\zeta}$, worin ϱ die Dichte des Wassers, ϱ_m die mittlere Dichte der Erde ist ($\varrho/\varrho_m = 0{,}18$). Die Berücksichtigung der Anziehung der durch die Flut gestörten Wassermassen erhöht die Gleichgewichtsflut um 12%. H. POINCARÉ[3] hat dabei die Land- und Meerverteilung mitberücksichtigt. Die neue Verbesserung erweist sich aber als nicht sehr bedeutend.

8. Die dynamische Theorie der Gezeiten. An Stelle der Gleichgewichtstheorie hat ein Jahrhundert nach NEWTON (1774) LAPLACE eine dynamische Theorie gesetzt, nach der die fluterzeugenden Kräfte statt der feststehenden Flutberge *Flutwellen* hervorrufen, deren Periode durch die fluterzeugenden Kräfte gegeben sind. Das Gezeitenproblem wird dadurch zu einem Problem der *Flüssigkeitsbewegung*. Bei der Ausbildung dieser erzwungenen Flutwellen spielen aber außer den periodischen Kräften auch noch verschiedene andere Faktoren eine Rolle, in erster Linie die Tiefe und Gestaltung der Meeresbecken, die Corioliskraft und die Reibungseinflüsse der verschiedenen Art. Den Ausgangspunkt bilden die hydrodynamischen Bewegungsgleichungen, in denen die fluterzeugenden Kräfte in der früher angegebenen Form auftreten. In ihrer Allgemeinheit lassen sich diese Gleichungen nur in den einfachsten Fällen lösen und auch hier nur unter vereinfachenden Annahmen, die nur rohe Annäherungen an die in der Natur vorkommenden Verhältnissen darstellen. Durch die periodischen Kräfte werden *erzwungene Flutwellen* erregt, aber es ist klar, daß (namentlich zur Erfüllung der Randbedingungen) auch *freie Wellen* auftreten werden, die die Eigenperioden des schwingenden Systems festlegen und deshalb besonders wichtig sind, weil sie bei Resonanz mit der Periode der fluterzeugenden Kraft einen bedeutenden Einfluß auf Amplitude und Phase der erzwungenen Wellen haben.

Die dynamische Theorie der Gezeiten ist in ihren Details sehr verwickelt und erfordert auch bei einfacher morphologischer Gestaltung der Ozeane einen erheblichen mathematischen Aufwand. Schon LAPLACE[4] ist es gelungen, die theoretischen Gezeiten eines homogenen, die ganze Erde gleichförmig bedecken-

---

[1] THOMSON u. TAIT: Natural Philosophy, § 808. Cambridge 1883.

[2] G. H. DARWIN u. H. H. TURNER: Proc. Roy. Soc. Lond. **1** (1886). Auch DARWIN: Pap. I, 328.

[3] H. POINCARÉ: Théorie des mareés. Leçons de mécanique céleste 3. 60. Paris 1910.

[4] LAPLACE: Mém. Acad. Roy. Sci. **1775** u. **1776**. Auch Mécanique céleste, 4. Buch Kap. 1. 1779.

den Weltmeeres zu berechnen, aber die Ergebnisse sind recht dürftig und haben keine Bedeutung für das Verständnis des Bildes der irdischen Gezeiten der Ozeane erlangt[1]. Erst in neuerer Zeit scheinen sich Methoden zu entwickeln, die den komplizierten Umrissen der Ozeane einigermaßen gerecht zu werden versuchen und deshalb mehr erfolgversprechend sind.

Nimmt man für die Dichte des Wassers $\varrho = 1$, die Poldistanz ϑ, die geographische Länge λ und für den Erdradius R, dann lauten die LAPLACEschen hydrodynamischen Bewegungsgleichungen [bei Vernachlässigung der Vertikalbeschleunigung (7.6) gegenüber der Schwerebeschleunigung und der vertikalen Komponente der Corioliskraft] analog zu (4.1):

$$\frac{\partial u}{\partial t} - 2\Omega v \cos\vartheta = -\frac{g}{R}\frac{\partial}{\partial\vartheta}(\zeta - \bar{\zeta}), \quad \frac{\partial v}{\partial t} + 2\Omega u \cos\vartheta = -\frac{g}{R}\frac{\partial}{\sin\vartheta\,\partial\lambda}(\zeta - \bar{\zeta}). \tag{8.1}$$

Die dritte Gleichung reduziert sich auf die statische Grundgleichung. Zu diesen Gleichungen kommt noch die Kontinuitätsgleichung, die bei variabler Wassertiefe h lautet:

$$\frac{\partial\zeta}{\partial t} + \frac{1}{R\sin\vartheta}\left\{\frac{\partial h\, u \sin\vartheta}{\partial\vartheta} + \frac{\partial h\, v}{\partial\lambda}\right\} = 0. \tag{8.2}$$

Wenn $\bar{\zeta}$ periodischer Natur ist, werden es auch die Größen u, v, und ζ sein und alle haben den Zeitfaktor $e^{i(\,t + s\lambda + \varepsilon)}$ gemeinsam (s ganzzahlig). Ist die Tiefe h nur eine Funktion von ϑ, so reduzieren sich die Gleichungen auf:

$$\left.\begin{aligned}
i\sigma u - 2\Omega v \cos\vartheta &= -\frac{g}{R}\frac{\partial}{\partial\vartheta}(\zeta - \bar{\zeta}),\\
i\sigma v - 2\Omega u \cos\vartheta &= -\frac{i g s}{R\sin\vartheta}(\zeta - \bar{\zeta}),\\
i\sigma\zeta &= -\frac{1}{R\sin\vartheta}\left\{\frac{\partial h\, u\sin\vartheta}{\partial\vartheta} + i s h v\right\}.
\end{aligned}\right\} \tag{8.3}$$

Aus den beiden ersten Gleichungen lassen sich u und v als Funktionen von ζ berechnen und mit der dritten erhält man eine Differentialgleichung für ζ allein, deren Lösungen die erzwungenen Flutwellen für einen die ganze Erde bedeckenden Ozean liefern. Je nach s sind verschiedene Gattungen von Wellen möglich:

1. $s = 0$ gibt die langperiodischen Glieder in V, was nach der Bezeichnung von LAPLACE zu den „*Schwingungen erster Gattung*" führt. Zu diesen gehören besonders die 14tägige Mondflut und die halbjährige Sonnenflut.

2. $s = 1$ ergibt die „*Schwingungen zweiter Gattung*" davon als wichtigste die ganztägige Mond- und Sonnengezeiten.

3. $s = 2$ sind die „*Schwingungen dritter Gattung*": halbtägige Mond- und Sonnengezeiten.

Die Lösungen für diese drei Wellengattungen können hier nur in den Grundgedanken dargelegt werden, und wir folgen hierin den vereinfachten Darstellungen von KELVIN[2], AIRY[3] und DARWIN[4]. Setzt man

$$\zeta - \bar{\zeta} = \zeta', \quad \frac{\sigma}{2\Omega} = f, \quad \frac{\Omega^2 R}{g} = m, \quad \beta = \frac{4mR}{h} = \frac{4\Omega^2 R^2}{g h} \quad \text{und} \quad \cos\vartheta = \mu, \tag{8.4}$$

[1] Für die Atmosphäre, die dem Bilde eines zusammenhängenden Ozeans entspricht, behalten die LAPLACEschen Entwicklungen ihre Bedeutung. Vgl. den Beitrag von W. KERTZ „Atmosphärische Gezeiten" in diesem Band, S. 947.

[2] W. THOMSON (Lord KELVIN): Phil. Mag. (4) **50**, 279 (1875), (Pap. IV, 248).

[3] AIRY: Tides and waves. Encycl. Metrop. § 95 1845.

[4] G. H. DARWIN: Encycl. Brit., 9. Aufl., Bd. 23, S. 359.

so ergibt eine einfache Rechnung für den Fall $s=0$ die Differentialgleichung für ζ:

$$\frac{\partial}{\partial\mu}\left(\frac{1-\mu^2}{f^2-\mu^2}\frac{\partial\zeta'}{\partial\mu}\right)+\beta\zeta'=0. \tag{8.5}$$

Mit $\zeta'=\zeta$ erhält man die *Eigenschwingungen* des Weltmeeres.

Ohne Rotation ($f=\infty$ und $\beta f^2=\sigma^2R^2/gh$) sind die Eigenfrequenzen durch $\sigma^2=n(n+1)gh/R^2$ gegeben (n eine ganze Zahl); die einfachsten Schwingungen haben die Form von Kugelfunktionen.

Wenn Rotation vorhanden ist, setzt LAPLACE in (8.5) eine Lösung von ζ in Potenzreihen nach steigenden Potenzen von μ an und bestimmt die Koeffizienten derselben aus den Grenzbedingungen. Dies führt zu ihrer Bestimmung zu unendlichen Kettenbrüchen. Diese berühmte Lösung LAPLACEs hat zu Auseinandersetzungen zwischen AIRY, FERREL[1] und Lord KELVIN[2] geführt; letzterer hat die Lösung als richtig erwiesen und weiter ausgebaut. Bei den in der Natur gegebenen Verhältnissen konvergieren die Kettenbrüche und auch die Potenzreihen für ζ recht schnell. Für die kürzeste Periode jener Schwingung, bei der sich die Oberflächenerhebung wie eine zonale Kugelfunktion zweiter Ordnung P_2 ($\cos\vartheta$) verhält, erhält man bei Wassertiefen von 17700 m oder 2210 m die Perioden der freien Schwingungen $11^h\,35^m$ oder $32^h\,49^m$.

Die *erzwungenen Schwingungen erster Gattung* ($s=0$) gehen vom Potential wert $\bar\zeta=A_1(\frac{1}{3}-\cos^2\vartheta)\cos(\sigma t+\varepsilon)$ aus. Die Bestimmung der Koeffizienten der Lösung ζ geschieht auf ähnliche Weise wie bei den Eigenschwingungen. Im Falle der 14tägigen Mondflut ist f das Verhältnis eines Sterntages zu einem Mondmonat und beträgt 0,0365. Damit wird $f^2=0{,}00133$, so daß in erster Annäherung $f=0$ gesetzt werden kann. Ähnliches gilt für die anderen Gezeiten

Tabelle 4. *Verhältnisse der Tidenhübe zu den Werten der Gleichgewichtsgezeiten.*

			β: 40	20	10	5
			Wassertiefe in m: 2210	4430	8850	17700
Langperiodische Gezeiten	Äquator	LAPLACE	0,455	—	0,708	0,817
		HOUGH	0,426	0,551	0,681	0,796
	Pol	LAPLACE	0,154	—	0,470	0,651
		HOUGH	0,140	0,266	0,443	0,628
Halbtägige Gezeit S_2	Äquator	LAPLACE	+7,43	−1,82	+ 11,26	+1,92
		HOUGH	+7,95	−1,50	−234,87	+2,14
$\sigma=2\Omega=0{,}9350$: M_2	Äquator	HOUGH	−2,42	−1,80	+ 11,07	+1,92

langer Periode. Die Tabelle 4 gibt in der Zeile „LAPLACE" das Verhältnis der Hubhöhen dieser Partialtiden zu den Hubhöhen der entsprechenden Gleichgewichtsflut am Äquator und am Pol für drei Tiefen des Weltmeeres. Das positive Vorzeichen bedeutet, daß diese Gezeiten langer Periode überall direkt sind, d.h. Hochwasser tritt stets synchron mit jenem der Gleichgewichtstide auf. Bei den Tiefen, die der tatsächlichen mittleren Tiefe des Meeres entsprechen, ist aber diese Gezeit weniger als halb so groß als bei der Gleichgewichtstide.

[1] FERREL: Tidal researches. U.S. coast surv. Rep. 1874, 154.

[2] W. THOMSON (Lord KELVIN): Phil. Mag. **50** (4) 227 (1875).

Bei den *erzwungenen Schwingungen zweiter Gattung* ($s=1$) hat das Potential die Form $\bar{\zeta}=A_2\sin\vartheta\cos\vartheta\cos(\sigma t+\lambda+\varepsilon)$, wobei σ von Ω nicht sehr verschieden ist. Setzt man einfachheitshalber $\sigma=\Omega$ oder $f=\frac{1}{2}$, so haben die Gleichungen eine geschlossene Lösung, wenn die Meerestiefe dem Gesetz $h=h_0(1-c\cos^2\vartheta)$ folgt. Am Äquator ist dann die Wassertiefe h_0, am Pol $h_0(1-c)$. Man erhält in einfacher Rechnung $\zeta=-\frac{8c}{\beta-8c}\bar{\zeta}$. Das Verhältnis der Tidenhübe zu jenen der statischen Gezeit ist nun konstant. Wenn $c=0$ ist, wird $\zeta=0$, für konstante Wassertiefe gibt es also *keine* ganztägigen Gezeiten im Ozean, wohl aber ganztägige Gezeitenströme. Dieses auffallende Ergebnis hielt LAPLACE in Anbetracht des Fehlens der ganztägigen Partialtiden in den europäischen Gewässern für sehr wichtig. Aber dieser Schluß ist unhaltbar, nachdem die Beobachtungen an vielen Stellen des Weltmeeres das Vorhandensein großer ganztägiger Gezeiten gezeigt haben.

Bei den *Schwingungen dritter Gattung* ($s=2$) ist das Potential durch $\bar{\zeta}=A_3\sin^2\vartheta\cos(\sigma t+2\lambda+\varepsilon)$ gegeben, wobei hier σ nahezu gleich $2\,\Omega$ wird. Die Annahme $\sigma=2\Omega$ und $f=1$ vereinfacht die Rechnung wesentlich und wieder gibt es eine geschlossene Lösung, wenn man $h=h_0\sin^2\vartheta$ wählt (Wassertiefe am Äquator h_0, am Pol aber Null). Diese Lösung ist $\zeta=-\frac{8}{\beta-8}\bar{\zeta}$. Bei Tiefen, die im Meer vorkommen ($\beta>10$) ist die Gezeit überall *invers*, d.h. dort, wo die statische Gezeit Hochwasser hätte, tritt Niedrigwasser auf. Am Pol ergeben sich bei diesem Tiefengesetz unendlich große Geschwindigkeiten, was bei der Tiefe Null zu erwarten war.

Auch für ein Meer *gleichförmiger Tiefe* hat LAPLACE die halbtägige Gezeit ($s=2$, $f=1$) berechnet. Das Ergebnis steht in der Tabelle 4 in der Zeile „Halbtägige Tiden, LAPLACE". Bei allen Tiefen sind in Polnähe die Gezeiten direkt, auch am Äquator sind sie es, wenn die Tiefe größer als 8000 m ist; aber die Hubhöhen werden dann praktisch gleich jenen der Gleichgewichtsflut. Zwischen 8850 und 4430 m muß es eine kritische Tiefe geben, bei der die Gezeit am Äquator verschwindet und dann invers wird. Auch nicht weit unter 2210 m wird bald ein zweiter kritischer Wert auftreten. Bei inverser Flut am Äquator und direkter in polnahen Gebieten muß es ein oder mehrere Paare kreisförmiger Knotenlinien $\zeta=0$ geben, die symmetrisch auf beiden Seiten des Äquators liegen.

Einen Ausbau und eine Verbesserung der LAPLACEschen Theorie hat später S. S. HOUGH[1] gegeben. Statt nach Potenzreihen in $\cos\vartheta$ bzw. $\sin\vartheta$ führt er Reihenentwicklungen nach Kugelfunktionen durch; diese Reihen haben den Vorzug rascher zu konvergieren. Es gelingt ihm auch, die gegenseitige Anziehung der Wasserteilchen der Flutwellen zu berücksichtigen. Die Änderungen, die dadurch eintreten, kann man der Tabelle 4 entnehmen. Der Einfluß besteht in einer Verkleinerung der Amplitude, aber er ist nicht besonders groß. Stärker zeigt er sich in den Perioden der *freien* Schwingungen, indem für Meerestiefen von 17700 m oder 2210 m dieselben auf $9^h\,52^m$ oder $18^h\,4^m$ herabgesetzt werden.

Die *freien* Schwingungen der zweiten und dritten Gattung ($s=1$ und $s=2$) sind von HOUGH besonders untersucht worden. Die in verschiedener Richtung (mit und gegen die Rotation) um die Erde laufenden Wellen — die im Falle keiner Rotation, da sie genau gleich sind, in ihrer Überlagerung *stehende* Normalschwingungen von Kugelfunktionform ergeben — erhalten nun verschiedene Fortpflanzungsgeschwindigkeiten, so daß die stehenden Normalschwingungen in west- und ostwärts wandernde Wellen mit dem Charakter von tesseralen

[1] S. S. HOUGH: Phil. Trans. Roy. Soc. **189** 201 (1897); **191**, 139 (1898).

($s = 1$) und sektoriellen ($s = 2$) Kugelfunktionen zerfallen. Die westwärts wandernden, die schneller als die ostwärts wandernden sind, sind die wichtigeren, da sie dieselbe Richtung wie die Gezeitenkräfte haben. Tabelle 5 gibt die Perioden für jene Wellen „erster Klasse", die für die ganz- und halbtägigen Gezeiten in Frage kommen.

Bei den *erzwungenen* Wellen zweiter Gattung ($s = 1$) verschwindet die eintägige Gezeit entsprechend dem Ergebnis von LAPLACE, wenn $\sigma = \Omega$ ist. Bei der ganztägigen Mondflut $\sigma/\Omega = 0{,}9270$ sind die Amplituden für die tatsächlichen Meerestiefen sehr klein gegenüber jenen der Gleichgewichtsgezeit und invers zu diesen. Bei der halbtägigen Sonnentide S_2 ($s = 2$) stehen die Verhältniszahlen zur Gleichgewichtstide in Tabelle 4, und der Vergleich mit den Werten von

Tabelle 5. *Eigenperioden der westwärts wandernden Wellen auf der rotierenden Erde.*

	β			
	40	20	10	5
	Wasertiefe in m			
	2210	4430	8850	17700
Ohne Rotation ($\Omega = 0$)	$32^h\,49^m$	$23^h\,12^m$	$16^h\,25^m$	$11^h\,35^m$
Mit Rotation				
Zweiter Gattung ($s = 1$) westwärts	$14^h\,41^m$	$12^h\,51^m$	$11^h\,5^m$	$9^h\,8^m$
Zweiter Gattung ($s = 1$) ostwärts	$24^h\,24^m$	$19^h\,16^m$	$14^h\,50^m$	$11^h\,6^m$
Dritter Gattung ($s = 2$) westwärts	$17^h\,59^m$	$14^h\,52^m$	$12^h\,1^m$	$9^h\,24^m$
Dritter Gattung ($s = 2$) ostwärts	$38^h\,34^m$	$26^h\,54^m$	$18^h\,40^m$	$12^h\,55^m$

LAPLACE darüber zeigt den Einfluß der Anziehungskräfte der Flutdeformation. Die große Verhältniszahl bei $\beta = 10$ weist darauf hin, daß hier nahezu *Resonanz* mit einer freien Welle vorhanden sein muß. Tatsächlich zeigt in Tabelle 5 bei $\beta = 10$ die entsprechende freie Welle eine Periode von $12^h\,1^m$. Der Unterschied von 1 min gegenüber der Periode der Kraft erzwingt eine 250fache Vergrößerung der Amplitude gegenüber der Gleichgewichtstide. Eine zweite Resonanzstelle scheint nicht unweit von $\beta = 40$ zu liegen. Die Oberschwingung der freien Welle von $17^h\,59^m$ hat eine Periode von $12^h\,5^m$ Dauer. Die Differenz von 5 min kann aber nur eine Vergrößerung auf das 10fache hervorrufen. Die richtigen Resonanztiefen für Wellen mit der Periode von genau 12 Std, also für die halbtägige Sonnentide, sind 8894 und 2248 m. Zwischen diesen Werten ist die Gezeit invers. HOUGH hat auch für die M_2-Tide, wo $\sigma/2\Omega = 0{,}96350$ ist, die Verhältniszahlen berechnet; sie stehen in Tabelle 4. Die Resonanztiefen liegen nun bei 7938 und 1965 m, und für die tatsächlichen Meerestiefen auf der Erde sind so die halbtägigen Partialtiden auf alle Fälle invers — wie immer, für einen Ozean, der die ganze Erde gleichmäßig bedeckt.

9. Gezeiten in begrenzten Meeresbecken. Das Ziel, das sich die Gezeitenforschung stellt, ist die Aufklärung der tatsächlichen ozeanischen Gezeiten. Die Gezeitentheorien führen von der einfachen Gleichgewichtstheorie ausgehend über die LAPLACEsche dynamische Theorie eines die ganze Erde gleichförmig bedeckenden Ozeans zur theoretischen Untersuchung der Gezeiten begrenzter Meeresbecken auf der rotierenden Erde. Erst diese Untersuchungen ergeben Gezeitenbilder, die sich mit den Gezeiten der von Kontinenten ringsumschlossenen wirklichen Meere vergleichen lassen. Erst wenn eine größere Zahl von solchen theoretischen Gezeitenbilder für verschiedene Meerestiefen und bei den verschiedensten

Begrenzungen der Meere durchgerechnet vorliegen, ist die Möglichkeit eines tieferen Verständnisses der durch die Beobachtungen gegebenen Gezeitenbilder vorhanden.

Polarbecken gleichförmiger Tiefe, die von einem Breitenkreis in 60° bzw. 75° 30′ Breite begrenzt sind, hat G. R. GOLDSBROUGH[1] näher untersucht. Er fand, daß die langperiodischen Tiden bedeutungslos, die halbtägigen praktisch gleich jenen der Gleichgewichtstheorie sind. Hingegen werden die ganztägigen Tiden, die bekanntlich bei einem die ganze Erde bedeckenden Ozean fast verschwinden, die Haupterscheinung und dies ergeben auch tatsächlich die Gezeitenbeobachtungen polarer Meeresbecken. Auch den Fall eines von zwei Breitenkreisen begrenzten Ozeans, sowie eines Ozeans, der sich von Pol zu Pol erstreckt und von zwei Meridianen begrenzt ist (Zweieck auf der Kugel), hat GOLDSBROUGH[2] behandelt. Die mathematischen Schwierigkeiten werden aber dabei immer größer.

Einen bedeutsameren Fortschritt haben bei der Erfassung der Gezeiten begrenzter Meere J. PROUDMAN und A. T. DOODSON erzielt. Der erstere hat zuerst die TAYLORsche Methode für rechteckige Becken (siehe Abschnitt 4γ) auch auf anders geformte Ozeanbecken ausgedehnt und das Problem in allgemeiner Form auf ein unendliches System linearer Gleichungen zurückgeführt, das sich nach der Methode der unendlichen Determinanten lösen läßt. Die Ableitung der Gezeiten eines halbkreisförmigen Ozeans (PROUDMAN[3]) ergab die Möglichkeit, das Ergebnis mit den Gezeiten des Schwarzen Meeres zu vergleichen.

Rechnerisch schwieriger gestalten sich die Berechnungen der Gezeiten eines Ozeans, der von einem vollständigen Meridian begrenzt ist (Halbkugel). J. PROUDMANs[4] allgemeine Darlegungen über die Dynamik der Gezeiten haben hierfür die theoretischen Grundlagen gegeben; A. T. DOODSON[5] hat für die numerische Berechnung eine eigene Integrationsmethode entwickelt. Er hat die Gezeitenbilder solcher Ozeane für verschiedene Tiefen und zwar für die Partialtide $K_1(\sigma=\Omega)$ und $K_2(\sigma=2\Omega)$ berechnet. In den Fig. 10 und 11 ist jedesmal nur ein *Quadrant* des ganzen Ozeans (der eine ganze Halbkugel bedeckt) zur Darstellung gebracht, da stets Symmetrie zum Äquator und zum Zentralmeridian vorhanden ist. Die Gezeit ist durch Flutstundenlinien (cotidal lines) und durch Linien gleicher Amplitude (corange lines) gegeben. Die gewählte Meerestiefe ist 4420 m. Bei der Eintagstide K_1 erkennt man eine stetige Änderung der Phase vom Äquator gegen den Pol, wobei die größten Amplituden im Bereich des Pols auftreten. Im äquatorialen Teil sind die Phasenänderungen bei nur kleiner Amplitude gering. Die Erscheinung der ganztägigen Gezeit konzentriert sich somit auf die Polargebiete. Bei der Halbtagsgezeit K_2 treten Amphidromien auf. Bei der Tiefe 4420 m sind davon in jedem Quadranten *zwei* mit positiven Drehsinn vorhanden; dabei ist der Tidenhub an der Begrenzung stets am größten, mit maximalen Werten im West- und Ostpunkt des Äquators.

W. HANSEN[6] hat zur Ermittlung der Gezeiten eines beliebig begrenzten Meeresgebietes $\mathfrak{G}$ ein anderes Rechenverfahren benützt. Wenn in einem Meeresgebiet $\mathfrak{G}$ die Gezeitenwerte (ζ oder die Normalkomponente des Gezeitenstromes) auf der das Gebiet $\mathfrak{G}$ begrenzenden Randkurve $\mathfrak{C}$ vorgegeben sind, dann lehrt die Theorie, daß diese für das ganze Gebiet $\mathfrak{G}$ eindeutig festgelegt sind

[1] G. R. GOLDSBROUGH: Proc. Math. Soc. Lond. **14**, 31 (1913). Korrektur hierzu: A. T. DOODSON: Mon. Not. Roy. Astronom. Soc. Geophys. Suppl. **1** 541 (1928).

[2] G. R. GOLDSBROUGH: Proc. Roy. Soc. Lond. Ser. A **117**, 692—718 (1928).

[3] J. PROUDMAN: Mon. Not. Roy. Astronom. Soc., Geophys. Suppl. **2**, 32 (1928).

[4] J. PROUDMAN: Phil. Trans. Roy. Soc. Lond., Ser. A **235**, 237 (1936).

[5] A. T. DOODSON: Phil. Trans. Roy. Soc. Lond., Ser. A **237**, 311 (1938).

[6] W. HANSEN: Ann. Hydrogr. Hamburg **66**, 429—443 (1938).

(Randwertverfahren). Für ebene Koordinaten (u und v ost- und nordwärts gerichtete Komponente des Gezeitenstromes) und mit Berücksichtigung eines

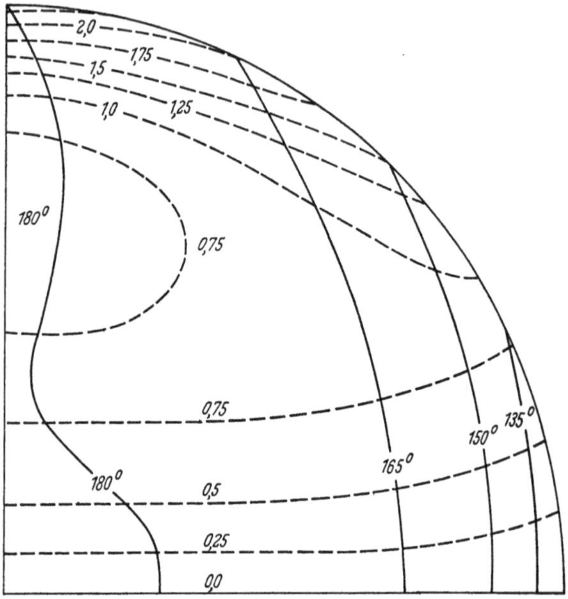

Fig. 10. Ganztägige Partialtide K_1 in einem von einem kompletten Meridian begrenzten Meer (nur ein Viertel dieses Meeres, also ein Oktant der Erde ist dargestellt): Flutstundenlinien ausgezogen, Linien gleicher Amplitude gestrichelt Meerestiefe 4420 m (nach A. T. DOODSON).

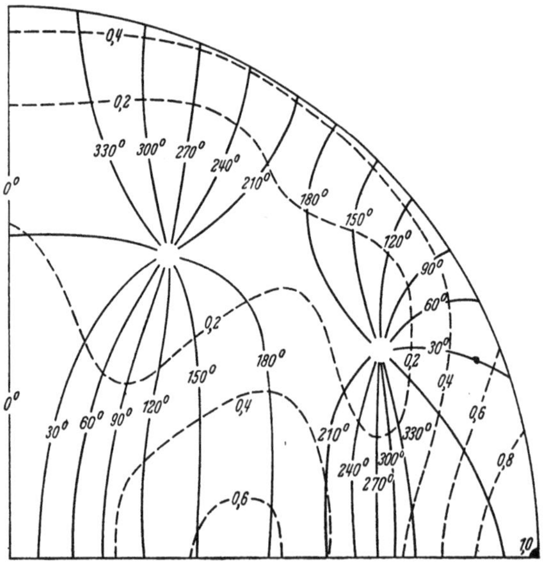

Fig. 11. Halbtägige Partialtide K_2 in einem von einem kompletten Meridian begrenzten Meer: Flutstundenlinien ausgezogen, Linien gleicher Amplitude gestrichelt. Meerestiefe 4420 m (nach A. T. DOODSON).

Reibungseinflusses proportional der Geschwindigkeit mit dem Reibungsparameter r lassen sich die Gln. (3.13) und (3.14) in die Differentialgleichungen

$$\left.\begin{aligned}\lambda u - f v + g\zeta_x &= g\bar{\zeta}_x\\ \lambda v + f u + g\zeta_y &= g\bar{\zeta}_y\end{aligned}\right\} \qquad -i\sigma\zeta + (h u)_x + (h v)_y = 0 \tag{9.1}$$

transformieren. Dabei ist $\lambda = r - i\sigma$ und $f = 2\Omega \sin\varphi$ und u, v und ζ sind proportional $e^{i\sigma t}$ gesetzt worden. Die als Index angehängten x und y bedeuten partielle Differentiationen nach diesen Achsen. Entsprechende Gleichungen lassen sich für größere Meeresgebiete auch in Polarkoordinaten ableiten. Die Abschätzung der für die tatsächlichen Verhältnisse in Meeresbecken und Ozeanen vorkommenden Größenordnungen der in den Gl. (9.1) auftretenden Beschleunigungen zeigt, daß diese Glieder alle derselben Ordnung sind (mit Ausnahme der aus dem fluterzeugenden Potential folgenden) und deshalb im Gleichungssystem beibehalten werden müssen.

Aus (9.1) läßt sich eine Gleichung für ζ allein ableiten:

$$\left.\begin{aligned} &\Delta(\zeta - \bar{\zeta}) + \frac{h_x}{h}(\zeta_x - \bar{\zeta}_x) + \frac{h_y}{h}(\zeta_y - \bar{\zeta}_y) + \\ &\quad + \frac{f}{h}\left[\frac{h_x}{h}(\zeta_y - \bar{\zeta}_y) - \frac{h_y}{h}(\zeta_x - \bar{\zeta}_x)\right] + \frac{i\sigma(f^2 + \lambda^2)}{g h \lambda}\,\zeta = 0. \end{aligned}\right\} \quad (9.2)^{1}$$

Zur Ermittlung der Verteilung von ζ im Meeresgebiet $\mathfrak{G}$ überdeckt man dieses Gebiet mit einem Gitternetz konstanter Maschenweite l. Für jeden Gitterpunkt innerhalb der Randkurve $\mathfrak{C}$ ersetzt man die Gl. (9.2) durch eine entsprechende Differenzengleichung und man erhält zur Bestimmung der ζ an den einzelnen Gitterpunkten ein System linearer Gleichungen, das lösbar ist, wenn ζ für die Randkurve $\mathfrak{C}$, also für die Küste, gegeben ist. Diese HANSENsche Methode ist praktisch auf die Nordsee, den Englischen Kanal und auf den Atlantischen Ozean angewandt worden (siehe Abschnitt 14b) und hat befriedigende Ergebnisse geliefert.

10. Die Kanaltheorie von AIRY. Während die LAPLACEsche Theorie von der Annahme eines die ganze Erde bedeckenden Meeres ausging und erst in neuerer Entwicklung auf die durch die Kontinente gegebenen Begrenzungen der Meere Rücksicht genommen wurde, hat AIRY[2] dem Phänomen der Gezeiten dadurch näher zu kommen gesucht, daß er die Schwingungsvorgänge in schmalen Kanälen, die in verschiedener Richtung die ganze Erde oder Teile derselben umspannen, näher studiert. Er ist dadurch der Begründer der „*Kanaltheorie*" der Gezeiten geworden. Sie bietet auch heutzutage in manchen Einzelheiten besonders für das Verständnis der Gezeiten in engen Meeresteilen, in Meeresstraßen und in Flußmündungen viel Beachtenswertes.

α) Erdumfassender Kanal gleichförmiger Tiefe. Die dynamischen Gleichungen für die Wasserbewegung in einem schmalen Kanal konstanten rechteckigen Querschnittes (Tiefe h) sind durch (1.2) und (1.3) gegeben. Nimmt man einen solchen Kanal, der den Erdäquator umspannt und wählt man statt der x-Koordinate die geographische Länge λ, dann nehmen sie (mit $c^2 = gh$, und R Erdradius) die Form an:

$$\frac{\partial^2 \xi}{\partial t^2} = c^2 \frac{\partial^2 \xi}{R^2 \partial \lambda^2} \quad \text{und} \quad \zeta = -h\frac{\partial \xi}{\partial \lambda}. \quad (10.1)$$

Die *Eigenschwingungen* eines solchen Kanals haben die Frequenz $\sigma_0 = sc/R$, worin s eine ganze Zahl ist; ihre Periode ist $T_0 = 2\pi R/sc$. Diese freien Wellen laufen alle mit der Geschwindigkeit c und benötigen zum Durchlaufen des Erdumfanges die Zeit T_0. Bei einer Geschwindigkeit von 1670 km/Std, was einer Tiefe von 22 km entspricht, hält die freie Welle mit der Sonne Schritt (Umlaufzeit 24 Std). Bei kleineren Meerestiefen laufen die Wellen langsamer, die Umlaufzeit ist größer (z.B. $h = 5000$ m, $c = 800$ km/Std und $T_0 = 50$ Std).

[1] $\Delta = \frac{\partial^2}{\partial x^2} + \frac{\partial^2}{\partial y^2}$ der LAPLACEsche Operator.

[2] AIRY: Tides and waves. Encycl. Metrop., Abschn. VI. 1845.

Bei den durch eine periodische horizontale Gezeitenkraft *erzwungenen Schwingungen* tritt zu den Gl. (10.1) rechts noch die Kraft K_h hinzu, die nach (7.5) durch $K_h = \frac{m}{2}\sin 2\vartheta$ gegeben ist; ϑ ist die Zenitdistanz des fluterzeugenden Körpers. Nimmt man als störenden Körper einen Mond, der in der Äquatorebene der Erde mit der scheinbaren Winkelgeschwindigkeit n umkreist, dann ist $\vartheta = nt + \lambda + \varepsilon$, und die erzwungenen Gezeiten im Kanal sind mit $nR/c = p$:

$$\zeta = \frac{H}{2}\,\frac{1}{1-p^2}\cos 2(nt+\lambda+\varepsilon) = \frac{H}{2}\,\frac{1}{1-\frac{\sigma^2}{\sigma_0^2}}\cos 2(nt+\lambda+\varepsilon). \tag{10.2}$$

Dabei ist $H = mR/g$ (für den Mond 55 cm, für die Sonne 24 cm) und $\sigma = 2n$ die Frequenz der erzwungenen Gezeit, $\sigma_0 = 2c/R$ die Frequenz der freien Welle ($s = 2$). $(H/2)\cdot\cos(\sigma + 2\lambda + \varepsilon)$ ist die Gezeit der Gleichgewichtstheorie. Die halbtägige Gezeit ist invers oder direkt, je nachdem $p \gtrless 1$ oder $c \lessgtr nR$ oder $\sigma \gtrless \sigma_0$ ist. Resonanz tritt auf, wenn die Periode der Kraft mit jener der freien Welle übereinstimmt. Für die sonnen- bzw. mondtägigen Halbtagszeit ist dies der Fall, wenn ein Kanal 22 bzw. 20,5 km tief ist. Für die Größenverhältnisse der Erde wird $\frac{\sigma_0^2}{\sigma^2} = \frac{g}{n^2 R}\cdot\frac{h}{R} = 311\cdot\frac{h}{R}$. Da für die Tiefen der Meere $\frac{h}{R}$ im Maximum etwa 1/600 ist, wird stets $\sigma > \sigma_0$ sein, so daß am Äquator die Gezeiten *invers* ausfallen werden, aber ihre Amplitude ist klein, bei $h = 4$ km nur 6 cm.

Für einen Kanal in der geographischen Breite φ und einer Bahn des Mondes in der Äquatorebene wird

$$\zeta = \frac{H\cos^2\varphi}{2}\,\frac{1}{1-p^2\cos^2\varphi}\cos 2(nt+\lambda+\varepsilon). \tag{10.3}$$

Wenn $c < nR$, werden nun die Gezeiten direkt oder invers, je nachdem $\cos\varphi \gtrless \frac{c}{nR}$ ist. Es ist dann möglich, daß in höheren Breiten die Gezeiten anders sich verhalten als in niedrigen.

Ist die Bahn des Mondes nicht in der Äquatorebene, dann muß für $\cos\vartheta$ die volle Beziehung (7.7) genommen werden. Zur halbtägigen Gezeit kommt nun noch eine ganztägige hinzu; sie verschwindet, wenn die Deklination des störenden Körpers Null ist. Die Amplitude der halbtägigen Gezeit ist aber im Verhältnis $\sin^2\Delta : 1$ verkleinert.

β) Kanäle beschränkter Ausdehnung. Umfaßt der (am Äquator gedachte) Kanal nur die graphische Länge $\lambda = \pm\alpha$, dann kommt zu den Gln. (10.1) noch die Randbedingung $\xi = 0$ für die Enden des Kanals $\lambda = \pm\alpha$ hinzu. Zählt man die Zeit vom Meridiandurchgang des Mondes in der Mitte des Kanals ($\lambda = 0$), so erhält man bei *Vernachlässigung der Trägheit* ($\partial^2\xi/\partial t^2 = 0$):

$$\zeta = \frac{H}{2}\left\{\cos 2(nt+\lambda) - \frac{\sin 2\alpha}{2\alpha}\cos 2nt\right\}. \tag{10.4}$$

Man nennt diese Lösung die „*verbesserte statische Theorie*". In der Kanalmitte $\lambda = 0$ wird:

$$\zeta = \frac{H}{2}\left(1 - \frac{\sin 2\alpha}{2\alpha}\right)\cos 2nt. \tag{10.5}$$

Es gibt keinen eigentlichen Knotenpunkt der Schwingung; auf einer kleinen Strecke des Kanals ändert sich die Hochwasserzeit sehr rasch. Für die Kanalenden $\lambda = \pm\alpha$ ist:

$$\zeta = \frac{HR_0}{2}\cos 2(nt \pm \alpha \mp \varepsilon_0), \tag{10.6}$$

wobei

$$R_0 \cos 2\varepsilon_0 = 1 - \frac{\sin 4\alpha}{4\alpha} \quad \text{und} \quad R_0 \sin \varepsilon_0 = \frac{\cos 4\alpha - 1}{4\alpha} \tag{10.7}$$

sind.

Ist α klein, dann wird in erster Annäherung $R_0 = 2\alpha$ und $\varepsilon_0 = -\frac{\pi}{4} + \frac{2}{3}\alpha$. Hochwasser tritt bei kleinem α am Ostende zur Zeit $9^h - \frac{\alpha}{45^0}$ ein, am Westende zur Zeit $3^h + \frac{\alpha}{45^0}$; bei sehr beschränkter Kanallänge ($\alpha \to 0$) reduzieren sich diese Zeiten auf 9^h und 3^h, und die Gezeit hat den Charakter einer einfachen stehenden Welle.

Bei Berücksichtigung der Trägheit erhält man dieselbe Beziehung (10.6), nur bedeuten jetzt

$$R_0 \cos 2\varepsilon_0 = \frac{p \sin 4\alpha - \sin 4p\alpha}{(p^2 - 1)\sin 4p\alpha} \quad \text{und} \quad R_0 \sin 2\varepsilon_0 = \frac{p(\cos 4p\alpha - \cos 4\alpha)}{(p^2 - 1)\sin 4p\alpha}. \tag{10.8}$$

Für kleine α gelten wieder die Beziehungen (10.7); Resonanz tritt bei $R_0 = \infty$ auf. Dies fixiert die Längen des Kanals, für die die freie Periode desselben gleich π/n wird, d.h. 12 Mondstunden oder einen halben Mondtag. Denn die Eigenperiode eines Kanals der Ausdehnung 2α ist $4\alpha R/c$ und, wenn dies π/n sein soll, wird $4p\alpha = \pi$.

Bei einem Kanal in der geographischen Breite φ tritt in (10.6) an Stelle von H $H\cos^2\varphi$ und für $p = \frac{nR}{c}\cos\varphi$.

Was für Verhältnisse in zonalen Kanälen herrschen, die man sich in den Ozeanen zwischen den Kontinenten abgegrenzt denken kann, läßt sich durch zahlenmäßige Auswertung obiger Formel ersehen. Solche Berechnungen haben G. PRÜFER[1] und G. DIETRICH[2] vorgenommen. Ein Beispiel gibt die folgende Tabelle

Erzwungene Querschwingungen im Atlantischen Ozean in 10° südlicher Breite, 35° W bis 13° E (zwischen Südamerika und Afrika). (Phasen bezogen auf Greenwich.)

$\lambda°$		35° W	26° W	21° W	16° W	11° W	6° W	1° W	4° E	13° E
M_2	Phase in °	75	72	65	58	338	258	251	244	241
	Amplitude in cm	63	55	41	22	5*	22	41	55	63
K_1	Phase in °	235	246	253	269	337	45	61	68	79
	Amplitude in cm	30	23	15	8	4*	8	15	23	30

für die M_2- und K_1-Tide in einem zonalen Kanal in 10° südlicher Breite. In einer uneigentlichen Knotenlinie geht in der Mitte des Kanals die Amplitude auf ein Minimum herab und die Phase ändert sich auf kurzer Strecke rasch um fast 180°. Im Atlantischen Ozean ist immer nur *eine* solche uneigentliche Knotenlinie vorhanden. Im Pazifischen treten mit Verbreiterung von Norden gegen den Äquator eine, zwei bis drei auf; im Südpazifik bleibt es bei zwei. Im Indischen Ozean sind zwei vorhanden. Solche Berechnungen sollen natürlich nur Andeutungen geben, wie sich ungefähr die Ausdehnung der Ozeane in ihrer west-östlichen Erstreckung zur Erregung von Schwingungen in dieser Richtung verhalten. Durch Querbewegungen wird aber in erster Linie weniger die Zahl der uneigentlichen Knotenlinien als ihre Form (durch Ausbildung von Amphidromien) beeinflußt.

[1] G. PRÜFER: Veröff. Inst. Meereskunde, N. F., A **1939**, H. 27.

[2] G. DIETRICH: Veröff. Inst. Meereskunde, N., F. A **1944**, H. 47.

11. Die harmonische Analyse der Gezeiten. Charakterisierung der Gezeiten durch die harmonischen Konstanten. Die dynamische Theorie lehrt, daß die durch die fluterzeugenden Kräfte erzwungene Flutwellen die Periode dieser Kräfte haben müssen, daß es aber — infolge der Kompliziertheit der Bodengestaltung der Ozeane und der Umrisse der Kontinente — nicht möglich ist, mit hinreichender Genauigkeit die Amplitude und Phase dieser Wellen für die verschiedenen Punkte der Meere theoretisch abzuleiten. Es besteht aber kein Zweifel, daß die Gezeiten überall völlig gesetzmäßig ablaufen, und die Beobachtungen zeigen auch, daß dies der Fall ist. Während so die *Perioden* der Flutwellen theoretisch durch die harmonische Analyse des Gezeitenpotentials ein für allemal festgelegt sind, bleiben ihre *Amplituden* und *Phasen* zunächst unbekannt. Sie können aber aus genügend langen Beobachtungen des Wasserstandes an einem Ort rechnerisch ermittelt werden. Da das Gezeitenphänomen sich immer wieder innerhalb eines bestimmten Zeitraumes gesetzmäßig wiederholt, werden die — aus den Beobachtungen abgeleiteten — Amplituden und Phasen der Flutwellen für einen Ort *konstante* Größen sein Den zu ihrer Ermittlung benützten Rechenvorgang bezeichnet man als *harmonische Analyse der Gezeiten*. Seit seiner Einführung durch W. THOMSON und G. H. DARWIN ist er zu der für praktische Zwecke sehr wichtigsten Methode der Auswertung von Gezeitenbeobachtungen geworden. Die neueste Darstellung des Verfahrens stammt von A. T. DOODSON[1]. Der Grundgedanke ist folgender:

Jedes harmonische Glied des Gezeitenpotentials in der Form $V = C\cos(\sigma t + V_0)$ erzwingt, nach der dynamischen Theorie, eine Partialtide der Form

$$\zeta = H\cos(\sigma t + V_0 - \varkappa). \tag{11.1}$$

σ ist die Frequenz (Winkelgeschwindigkeit) des Potentialanteils und auch die Frequenz der erzwungenen Partialtide, V_0 ist das Argument der Tide für 0^h des ersten Tages des Beobachtungsmaterials; es läßt sich aus den Bahnelementen der fluterzeugenden Körper, die sich proportional der mittleren Zeit ändern, berechnen. $\varkappa$ wird als die *Phase der Partialtide* am betreffenden Ort oder als *Kappazahl* bezeichnet. $\varkappa/\sigma$ gibt die Verspätung der Tide gegenüber dem Flutpotential, also die Größe des Hochwasserintervalls. H und $\varkappa$ sind Ortsfunktionen auf der Erde und von Hafen zu Hafen verschieden. Sie müssen aus den Wasserstandsbeobachtungen ermittelt werden; man nennt sie *die harmonischen Gezeitenkonstanten des Ortes*. Kennt man alle diese Größen, dann läßt sich, durch *harmonische Synthese*, für eine beliebige Zeit die Abweichung ζ des Wasserstandes vom Mittelwasser und (bei Kenntnis der Höhe des letzteren über Kartennull) die Wassertiefe zu dieser Zeit berechnen

Wenn laufende Registrierungen des Wasserstandes vorliegen, ist diese Aufgabe der harmonischen Analyse leicht zu lösen, wenn sie auch in dieser einfachen Form sehr viel Rechenarbeit erfordert. Ordnet man die Wasserstandswerte, die man jeweils den Registrierungen für jeden Zeitpunkt entnehmen kann, nach den Tidestunden einer Partialtide, deren H- und $\varkappa$-Werte man ermitteln will, in Reihen von je einer Periode (z.B. im Falle der M_2-Tide immer 12 bzw. 24 Mondstunden nach Meridiandurchgang des Mondes am Ort), so wird in jeder Reihe für diese Tide Hoch- und Niedrigwasser zur selben Stunde eintreten, während für alle anderen Tiden, die ja andere Frequenzen haben, die Werte sich in jeder Reihe verschieben. Im Mittel einer großen Zahl solcher Reihen wird die hervorgehobene Partialtide (im obigen Fall die M_2-Tide) sich aufsummieren, während die anderen um so mehr eliminiert erscheinen, je größer die Zahl der benützten Perioden ist.

[1] A. T. DOODSON: Phil. Trans. Roy. Soc. Lond., Ser. A **227**, 223 (1928).

Mathematisch läßt sich dies folgendermaßen darstellen: Die Wasserstandsbeobachtungen ζ lassen sich durch Überlagerung einer großen Zahl von Wellen verschiedenen Periode und Phase darstellen, derart daß

$$\zeta = H_0 \cos(\sigma_0 t - \varkappa_0) + \sum_{n=1}^{r} H_n \cos(\sigma_n t - \varkappa_n) = B_0 \cos\sigma_0 t + C_0 \sin\sigma_0 t + \sum_{n=1}^{r} H_n \cos(\sigma_n t - \varkappa_n) \quad (11.2)$$

ist. Die Partialtide mit dem Index 0 (Periode T_0) ist hervorgehoben, während die übrigen r in die Summe zusammengezogen sind. Multipliziert man mit $\cos\sigma_0 t$ bzw. $\sin\sigma_0 t$ und integriert man über t von 0 bis $p\,T_0$ (p eine ganze große Zahl), dann wird

$$\int_0^{pT_0} \zeta \begin{Bmatrix}\cos\\ \sin\end{Bmatrix} \sigma_0 t\, dt = \begin{Bmatrix}B_0\\ C_0\end{Bmatrix} \frac{p}{2} T_0 \pm \sum_{n=1}^{r} H_n \frac{\sigma_n}{\sigma_n^2 - \sigma_0^2} \left[\begin{Bmatrix}\sin\\ \cos\end{Bmatrix} \left(\frac{\sigma_n}{\sigma_0} 2p\pi - \varkappa_n\right) \pm \begin{Bmatrix}\sin\\ \cos\end{Bmatrix} \varkappa\right]. \quad (11.3)$$

Aus diesen Beziehungen lassen sich B_0 und C_0 und damit H_0 und $\varkappa$ ermitteln, wenn der Summenausdruck rechts verschwindet oder sehr klein ist. Das ist der Fall, wenn $\frac{\sigma_n}{\sigma_0} 2p\pi$ mit großer Annäherung $2s\pi$ wird (s wie p eine große ganze Zahl). Bei der S_2-Tide ist z.B. $\sigma_n = 15°/\text{Std}$, während für die M_2-Tide $\sigma_0 = 14{,}^\circ 4921/\text{Std}$ ist. In diesem Fall wird $\sigma_n/\sigma_0 = 1{,}03505$. Stehen Beobachtungen von einem Jahr zur Verfügung, dann ist $p\,T_0$ etwa 360 Tage, und mit $T_0 = \frac{1}{2}$ Tag wird p etwa 720. Das dieser Zahl am nächsten kommende Verhältnis von s und p, das obiger Zahl entspricht, ergibt sich zu 738/713. Als günstigste Länge des für die harmonische Ermittlung der M_2-Tide zu wählenden Zeitraumes findet man so $p\,T_0 =$ 369 Tage. Für eine große Zahl kurzperiodischer Partialtiden ergibt sich ebenfalls die Zeitlänge von 369 Tagen als günstigster Zeitraum, so daß man im allgemeinen diesen Zeitraum zur Trennung der einzelnen Partialtiden von einander gewählt hat. Ist das Beobachtungsmaterial kürzerer Dauer, dann muß man das Verfahren etwas abändern; dies hängt aber wesentlich ab von der Form, in der dieses Material gegeben ist.

Die numerische Berechnung einer Partialtide erfordert die Aufteilung des Beobachtungsmaterials in Zeiträume einer Periodenlänge dieser Tide, und innerhalb eines solchen, die Ablesung des Wasserstandes an einer Anzahl äquidistanter Abstände, dann die Mittelbildung aller Periodenreihen, um die *mittlere* Schwankung während eines Periodentages zu bekommen. Durch Wiederholung dieses Verfahrens mit anderen Periodenlängen lassen sich alle Tideneinflüsse voneinander trennen. Aber eine solche harmonische Analyse würde umfangreiche und mühevolle Rechenarbeit erfordern, und nur die Benützung von Registrierkurven des Wasserstandes wäre möglich, nicht aber ein Beobachtungsmaterial aus stündlichen Wasserstandsbeobachtungen. E. ROBERTS und G. H. DARWIN haben deshalb ein Verfahren entwickelt, das weniger Arbeit macht und auch Stundenwerte des Wasserstandes zu benützen erlaubt. Es erfolgt nur *eine* Reduktion der Flutkurve nach Sonnenstunden (24 Werte pro Tag). Im allgemeinen ist der Abstand einer vollen Periodenstunde ($= \frac{1}{24}$ der Periodenlänge) irgend einer betrachteten Partialtide von der vollen Stunde gewöhnlicher Sonnenzeit nicht größer als $\frac{1}{2}$ Std; deshalb kann, mit einem kleinen Fehler, der Wasserstand der zunächst liegenden vollen Sonnenstunde als jener der vollen Periodenstunde genommen werden. Sobald der Zeitunterschied mehr als $\frac{1}{2}$ Std ausmacht, muß man einen Wasserstandswert weglassen oder zweimal nehmen, je nachdem die Zeitzählung der Periodenuhr zur mittleren Sonnenzeit zurückbleibt oder voreilt. Man kann für jede Tide vorgedruckte Formulare benützen, in denen an bestimmten Stellen durch besondere Zeichen ausgeführt ist, wo *ein* Wasserstandswert in zwei aufeinanderfolgende Spalten zweimal oder in einer Spalte *zwei* aufeinanderfolgende Werte der gewöhnlichen Reduktion einzusetzen sind. Die Schreibarbeit bleibt immer noch recht groß, kann aber von Rechenkräften allein ausgeführt werden. Jede mittlere Wasserstandsschwankung, die durch Mittelbildung über alle Periodenlängen einer Partialtide gewonnen wurde, wird dann nach dem üblichen Verfahren der FOURIERschen Analyse in eine Reihe der einfachen, doppelten, usw. Frequenz der betrachteten Tide entwickelt. Man erhält so für die

Partialtide und für alle *Obertiden* die H- und $\varkappa$-Werte, also die zur Partialtide gehörigen Gezeitenkonstanten.

C. BÖRGEN[1] hat eine Erweiterung des DARWINschen Verfahrens gegeben, das eine weitere Ersparnis an Rechenarbeit bedeutet und auch mit einem weniger umfangreichen Material auszukommen vermag (s. auch K. HESSEN[2]). An die harmonischen Konstanten aus diesen Verfahren sind noch einige Korrekturen anzubringen, die sich leicht ermitteln lassen. Auch aus kürzeren Beobachtungsreihen lassen sich die Konstanten der wichtigsten Partialtiden mit einiger Genauigkeit ableiten. Man benötigt mindestens 15 oder 29 Beobachtungstage. Anweisungen für solche verkürzte Analysen hat A. T. DOODSON[3] gegeben.

Die Ermittlung der harmonischen Konstanten der *langperiodischen* Partialtiden geht von den Tagessummen der stündlichen Wasserstände aus, die durch einfache Reduktionsfaktoren vom Einfluß der kurzperiodischen Tiden noch genauer befreit werden. Im übrigen verfährt man nach ähnlichen Prinzipien wie früher. Betreffs Details siehe die Arbeiten von DARWIN und BÖRGEN, siehe auch K. HESSEN[2].

Zur Charakteristik der Gezeiten durch die harmonischen Konstanten. Die H- und $\varkappa$-Werte der harmonischen Analyse eignen sich gut, um eine Reihe von Eigenschaften der ozeanischen Gezeiten schärfer festzulegen. Die einzelnen Partialtiden haben eine sehr verschiedene Bedeutung. Schon wenige bestimmen bereits die wesentlichen Züge des Gezeitenverlaufes. Unter den wichtigsten sieben Partialtiden M_2, S_2, N_2, K_2 und K_1, O_1, P_1 treten besonders die zwei größten halb- und eintägigen Tiden M_2, S_2 und K_1, O_1 als die ausschlaggebenden hervor. Die $\varkappa$-Zahl von M_2 gibt angenähert bei Halbtagsgezeiten (vgl. weiter unten) die Eintrittszeit des Hochwassers nach Meridiandurchgang des Mondes (Mondflutintervall, Hafenzeit). Bei Voll- und Neumond gehen Sonne und Mond mittags gleichzeitig durch den Meridian, die beiden Partialtiden aber mit Verspätungen, die durch die $\varkappa$-Zahlen $\varkappa_M$ und $\varkappa_S$ gegeben sind. Der Unterschied der Winkelgeschwindigkeiten $30° - 28{,}984° = 1{,}016°$ pro Stunde oder $24{,}384°$ im Tag besagt, daß die M_2-Tide täglich um 0,84 Std = 50 min hinter der S_2-Tide zurückbleibt. An den Syzygien haben beide Tiden gleiche Phase, ihre Amplituden addieren sich zu $M_2 + S_2$; die Uhrzeit dieses Springhochwassers richtet sich nach der $\varkappa$-Zahl der S_2-Tide; sie ist $\varkappa_S/30°$ in Stunden. Die Verspätung der Springzeit gegen Voll- und Neumond ist durch $\varkappa_S - \varkappa_M$ gegeben. $\frac{\varkappa_S - \varkappa_M}{24{,}384} = 0{,}04\,(\varkappa_S - \varkappa_M)$ ist jene Größe, die als das „*Alter der Gezeit*" bezeichnet wird. Die Amplitudendifferenz $M_2 - S_2$ gibt die Amplitude der Nipptiden. Nach der Amplitude von S_2 richtet sich demnach die „tägliche Ungleichheit".

Bei eintägigen Gezeiten (vgl. weiter unten) bestimmt die Phase der K_1-Tide zusammen mit jener von O_1 im wesentlichen die Hochwasserzeit. Diese hat nun keinen Zusammenhang mehr mit dem Meridiandurchgang des Mondes. Die Phase von K_1 läuft nach Sternzeit, und in einem konstanten Zeitabstand vom Meridiandurchgang eines bestimmten Fixsternes tritt Hochwasser ein. Ohne Rücksicht auf O_1 gibt die $\varkappa$-Zahl von K_1 die Uhrzeit (Ortszeit) des Eintritts des Hochwassers am 21. Juni an; an jedem folgenden Tag tritt es 4 min früher ein; für jeden Monat 2 Std früher, so daß bei solchen Gezeiten das Hochwasser im Laufe des Jahres nacheinander sich an jeder Stunde des Tages einstellt. Durch die O_1-Tide erfolgt eine periodische Verstärkung und Abschwächung von K_1, ähnlich wie es die S_2-Tide bei M_2 macht. Auch bei eintägigen Gezeiten gibt es so den Gegensatz

[1] C. BÖRGEN: Ann. Hydrogr. **1894**.
[2] K. HESSEN: Ann. Hydrogr. **1920** 1, 441.
[3] A. T. DOODSON: The Admiralty Tide Tables, Bd. 3, London 1938.

von Spring- und Nipptiden, aber sie wiederholen sich nicht wie bei den halbtägigen Gezeiten alle 14,765 Tage, sondern alle 13,66 Tage. Die relative Geschwindigkeit ist hier $15{,}041° - 13{,}943° = 1{,}098°$ pro Stunde oder $26{,}3528°$ pro Tag. Die Springtiden treten alle $360°/26{,}353° = 13{,}66$ Tage ein, somit 26,74mal im Jahr. Bei den Halbtagsgezeiten gibt es aber nur 24,74mal solche Tiden im Jahr. Die Differenz $\varkappa_K - \varkappa_O$ gibt auch hier die Springverspätung an, und zwar ist $(\varkappa_K - \varkappa_O) \times 0{,}038$ die Verspätung in Tagen nach der größten Deklination des Mondes.

Zur Charakterisierung der Gezeiten eines Ortes kann man mit P. VAN DER STOK[1] das Verhältnis der Summe der Amplituden der eintägigen Partialtiden $K_1 + O_1$ zur Summe der Amplituden der halbtägigen $M_2 + S_2$ nehmen. Dieses Verhältnis wird um so größer, je größer die tägliche Ungleichheit ist. Ein Maximum erreicht es, wenn nur ein Hochwasser am Tage eintritt. Man bezeichnet

$$F = \frac{K_1 + O_1}{M_2 + S_2}$$

als die „*Formzahl der Gezeiten*". Eine Einteilung der Gezeitenformen nach der Formzahl hat vor allem A. COURTIER[2] gegeben:

$F = 0{,}0$ bis $0{,}25$: halbtägige Gezeitenform. Täglich zwei Hoch- und zwei Niedrigwasser. Die Hochwasserzeit folgt dem Meridiandurchgang des Mondes in nahezu gleichbleibendem Abstand. Mittlerer Hub zur Springzeit $2(M_2 + S_2)$.

$F = 0{,}25$ bis $1{,}5$: gemischte, überwiegend halbtägige Gezeitenform. Täglich zwei Hoch- und zwei Niedrigwasser, jedoch ist eine starke tägliche Ungleichheit vorhanden. Ihre größten Werte hat sie, wenn die Monddeklination ihr Maximum überschritten hat.

$F = 1{,}5$ bis $3{,}0$: gemischte, überwiegend eintägige Gezeitenform. Zeitweise nur ein Hochwasser am Tage, und zwar nach den Extremen der Monddeklination. Zu anderen Zeiten zwei Hochwasser am Tage mit starken Ungleichheiten in Höhe und Zeit. Der mittlere Springtidenhub is $2(K_1 + O_1)$.

$F = 3{,}0$ bis ∞: eintägige Gezeitenform. Täglich nur ein Hochwasser, zur Nippzeit; wenn der Mond durch den Äquator gegangen ist, können auch vorübergehend zwei Hochwasser vorkommen. Der mittlere Springtidenhub ist $2(K_1 + O_1)$.

Zur Veranschaulichung des Einflusses der Formzahl auf das Gezeitenbild eines Hafens sind in der Fig. 12 für den Monat März 1936 die Gezeitenkurven von vier Häfen mit verschiedener Formzahl zur Darstellung gebracht. Daß durch die vier verwendeten Haupttiden das Gezeitenbild in der Hauptsache erfaßt wird, erkennt man daraus, daß in allen vier Häfen durch diese rund 70% der Gesamtamplitude gegeben sind; nimmt man die weiteren wichtigen Partialtiden N_2, K_2 und P_1 hinzu, dann erhöht sich diese Prozentzahl nur auf 83%.

Der gesetzmäßige Ablauf der Gezeiten ermöglicht es, diesen vorauszurechnen und damit eine *Gezeitenvorhersage* zu gewinnen. Diese Vorhersage ist in den jährlich von den hydrographischen Ämtern herausgegebenen *Gezeitentafeln* enthalten; sie besitzen eine große praktische Bedeutung. Die Vorhersage beruht auf die für einen bestimmten Hafen charakteristischen Grundwerte der Gezeiten und kann nach zwei verschiedenen Verfahren durchgeführt werden: entweder nach dem sog. *non-harmonischen* oder nach dem *harmonischen Verfahren.*

Das *non-harmonische Verfahren* ist zuerst von J. W. LUBBOCK[3] entwickelt worden und verwendet in der Hauptsache die Angaben betreffend Zeit und Höhe der Hoch -und Niedrigwasser. Vor allem wird mit Mondflut-Intervallen gerechnet, und es ist deshalb hauptsächlich für halbtägige Gezeiten bestimmt, also für Häfen, in denen die tägliche Ungleichheit klein ist. Man zerlegt die Ungleichheiten in eine halbmonatliche, eine parallaktische Ungleichheit, in eine

[1] J. P. VAN DER STOK: Tides and tidal Streams in the Indian Archipelago. Batavia 1887.
[2] A. COURTIER: Serv. hydr. de la marine. Paris 1948.
[3] J. LUBBOCK: An elementary treatise on the tides. London 1839.

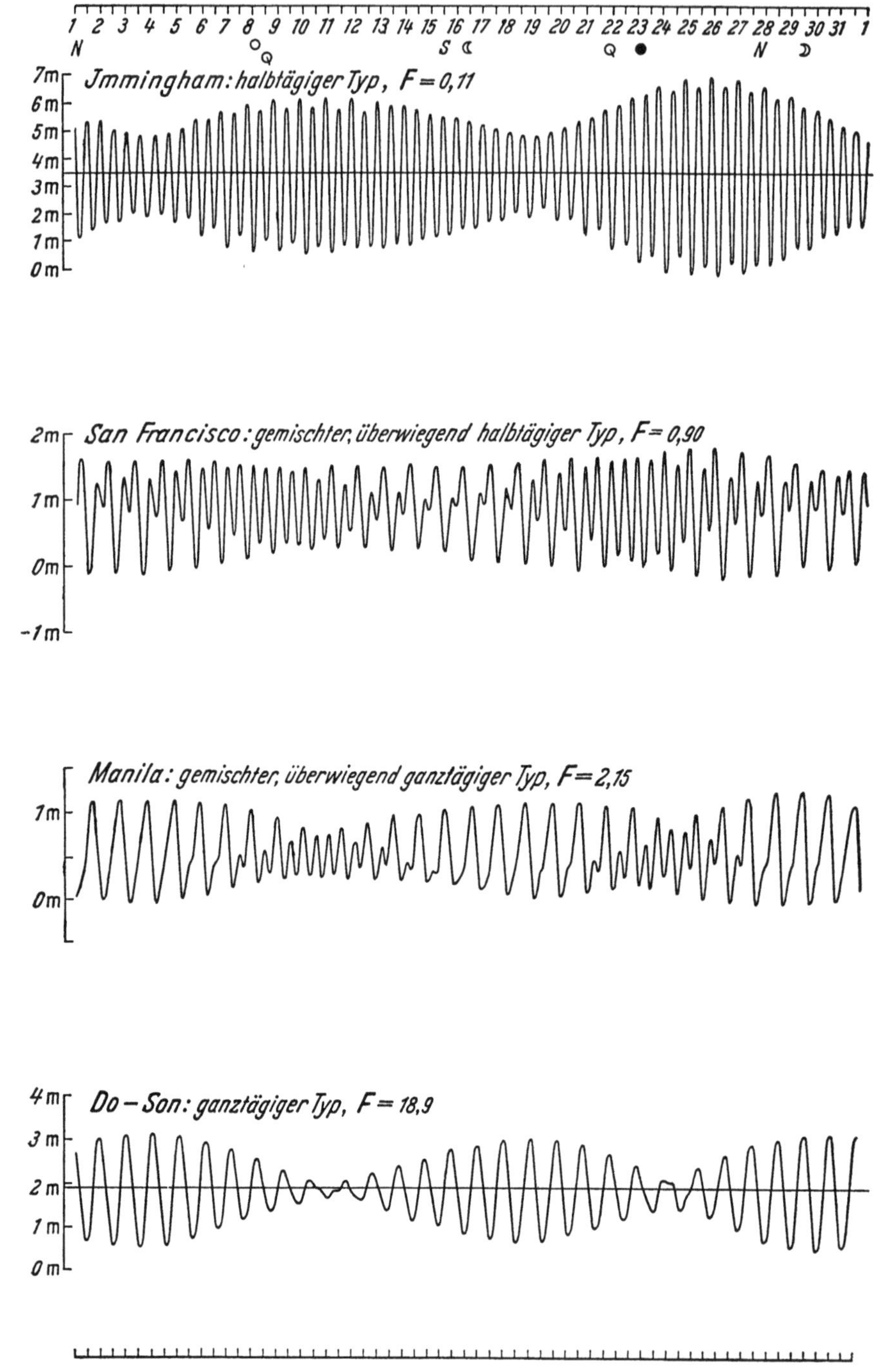

Fig. 12. Gezeitenkurven für den Monat März 1936 (nach den Deutschen Gezeitentafeln 1940, Bd. II). ●, ☽, ○, ☾ Mondphasen, *N* und *S* größte nördliche und südliche Deklination des Mondes, *Q* Äquatorialer Durchgang des Mondes.

Deklinationsungleichheit, in eine tägliche und eine jahreszeitliche Verbesserung. Alle diese werden völlig empirisch aus den Beobachtungen ermittelt (mindestens aus 9jährigen Beobachtungen).

Die gewöhnlichen Gezeitentafeln geben im allgemeinen nur die Höhe und Eintrittszeit von Hoch- und Niedrigwasser und gestatten die Wasserhöhe für eine beliebige Zeit nur nach umständlichen Rechnungen zu bestimmen. Die steigenden praktischen Erfordernisse nach genauer Kenntnis der Wassertiefe zu jeder Zeit kann man nur durch die Benützung des *harmonischen Verfahrens* gerecht werden. Bei Kenntnis der harmonischen Konstanten der wichtigsten Partialtiden für einen Küstenort läßt sich der Gezeitenverlauf eines beliebigen Tages, d.h. für jede Stunde die zu erwartende Wassertiefe über Kartennull berechnen. Für solche Berechnungen ist ein praktisches Verfahren entwickelt worden, das sich allgemein bewährt hat[1]. Berücksichtigt werden dabei zehn Partialtiden: die üblichen sieben Haupttiden (vier halbtägige und drei eintägige) und drei Seichtwassertiden, denen in Häfen in Gebieten seichten Wassers eine gewisse Bedeutung zukommt. Die Vernachlässigung der übrigen Partialtiden bedingt einige Fehler, die aber selten Beträge erreichen, die für die Schiffahrt von Bedeutung sind.

Die Berechnungen für die jährlichen Gezeitentafeln werden heutzutage durch besondere *Gezeitenrechenmaschinen* ausgeführt. Eine solche Maschine wurde zuerst von Lord KELVIN (1872) ersonnen und von E. ROBERTS verbessert. Die Standorte der nunmehr sehr großen Gezeitenrechenmaschinen sind (mit Angabe der Zahl der in ihnen berücksichtigten Partialtiden): Deutsches Hydrographisches Institut, Hamburg [*62*][2], Liverpool Tidal Institute [*42*][3], US Coast and Geodetic Survey, Washington [*37*][4]. Eine Übersicht über die bis 1926 konstruierten Maschinen gibt eine Spezialpublikation des Intern. Hydrogr. Bureau von Monaco[5].

Die größte, fortlaufend ergänzte und berichtigte Veröffentlichung aller bekannt gewordenen harmonischen Gezeiten- und Gezeitenstromkonstanten gibt das Internationale Hydrographische Bureau von Monaco[6] heraus; sie umfaßt 1950 etwa 2700 Orte. Die Deutschen Gezeitentafeln 1939 enthalten die harmonischen Konstanten für etwa 1800 Orte (bis zu zehn Tiden), die Englischen Gezeitentafeln 1938 für 2650 Orte und außerdem für rund 3500 weitere; die Konstanten für M_2, S_2 und K_1, O_1 sind aber hier durch Differenzen aus Standardstationen abgeleitet worden.

III. Gezeiten und Gezeitenströme in Landnähe und im freien Ozean.

12. Beziehungen zwischen Gezeit und Gezeitenstrom. Kennt man die Hubhöhe ζ, dann lassen sich die Stromkomponenten u und v aus den Beziehungen berechnen. Man kann diesen mit (4.2) die praktische Form

$$+i\sigma\zeta + (hu)_x + (hv)_y = 0 \quad \text{und} \quad i\sigma(u_y - v_x) - f(u_x + v_y) = 0 \qquad (12.1)$$

geben. Im allgemeinen beschreibt die Spitze des Gezeitenstrom-Vektors als Funktion der Zeit in einer Gezeitenperiode eine Ellipse *(Stromfigur)*. Setzt man $\zeta = \zeta_1 + i\zeta_2$, $u = u_1 + iu_2$ und $v = v_1 + iv_2$, so ergeben sich die Elemente derselben aus folgenden Formeln:

[1] Anleitung zur Berechnung der Gezeiten nach dem harmonischen Verfahren. Deutsche Gezeitentafeln 1940. Berlin 1939.

[2] W. HORN: Gezeiten des Meeres. Naturforschung und Medizin in Deutschland 1939—1946, Bd. 18. Wiesbaden 1948.

[3] A. T. DOODSON: Int. Hydrogr. Rev. **24** 65—72 (1947).

[4] A. T. DOODSON u. H. D. WARBURG: Admiralty Manual of Tides, London (1941). — P. SCHUREMAN: Manual of harmonic analysis and prediction of tides. Washington 1941.

[5] Int. Hydrogr. Bur. Monaco: Spec. Publ. No. 13, 1926.

[6] Int. Hydrogr. Bur. Monaco: Spec. Publ. No. 26, 1933ff.

Große und kleine Achse der Ellipse:

$$\frac{1}{\sqrt{2}}\sqrt{u_1^2+u_2^2+v_1^2+v_2^2 \pm \sqrt{(u_1^2+u_2^2+v_1^2+v_2^2)^2-4(u_1v_2-u_2v_1)^2}}\,. \qquad (12.2)$$

Achsenrichtung δ und Eintrittszeiten τ_0 der Stromextreme:

$$\operatorname{tg} 2\delta = \frac{2(u_1v_1+u_2v_2)}{u_1^2+u_2^2-v_1^2-v_2^2} \quad \text{und} \quad \operatorname{tg} 2\sigma\tau_0 = \frac{2(u_1u_2+v_1v_2)}{u_1^2+v_1^2-u_2^2-v_2^2}\,.$$

Der Gezeitenstrom ändert sich im allgemeinen sowohl in Richtung und in der Stärke etwas mit der Tiefe. Hierin zeigen sich in erster Linie die *Reibungseinflüsse* des Meeresbodens. Die Stromgeschwindigkeit als Funktion der Tiefe

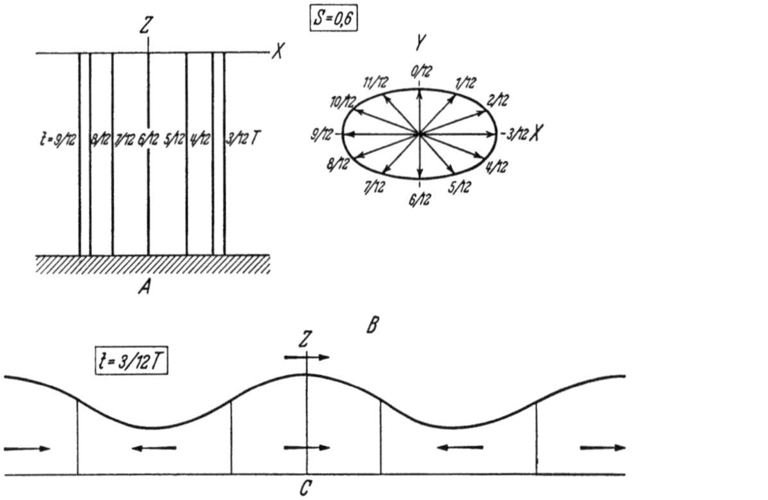

Fig. 13 A–C. Gezeitenwelle in einer rotierenden, weit ausgedehnten Wsaserschicht (*ohne* Reibung): A Lage eines Stromfadens zu verschiedenen Phasen. B Stromfigur (in allen Tiefen gleicher Art). C Wellenform und Stromverteilung in der Kanalrichtung = Fortpflanzungsrichtung der Welle (nach H. U. SVERDRUP).

läßt sich, bei nicht zu großen Wassertiefen, empirisch gut durch Potenzgesetze $v = v_0(z/h)^n$ darstellen, worin v_0 die Geschwindigkeit an der Oberfläche und $z=0$ der Boden sind. In Flußmündungen hat n die Größenordnung 1/5. Will man den Reibungseinfluß genauer in Rechnung stellen, muß er in die allgemeinen Bewegungsgleichungen (4.1) Aufnahme finden; sie nehmen dann die Form an:

$$\left.\begin{aligned} &\frac{\partial u}{\partial t} - f v = -g\frac{\partial \zeta}{\partial x} + \nu\frac{\partial^2 u}{\partial z^2}, \\ &\frac{\partial v}{\partial t} + f u = -g\frac{\partial \zeta}{\partial y} + \nu\frac{\partial^2 v}{\partial z^2} \quad \text{und} \quad \frac{\partial \zeta}{\partial t} = -h\int_0^h\left(\frac{\partial u}{\partial x}+\frac{\partial v}{\partial y}\right). \end{aligned}\right\} \qquad (12.3)$$

Hierin ist ν der Reibungskoeffizient (turbulente).

Lösungen dieser Differentialgleichungen findet man bei H. SVERDRUP[1], J. E. FJELDSTAD[2] und H. THORADE[3]. Die wichtigsten Ergebnisse lassen sich den beiden Fig. 13 und 14 entnehmen: Fig. 13 gibt die Verhältnisse in einer fortschreitenden Gezeitenwelle auf einer rotierenden, unendlich ausgedehnten Wasser-

[1] H. SVERDRUP: Geofys. Publ. **4**, No. 5 (1926).
[2] J. E. FJELDSTAD: Scien. Res. Norweg. North Polar Exp. „Maud", Bd. 4, No. 3. Bergen 1929.
[3] H. THORADE: Arch. dtsch. Seewarte **46** (1928). — Gerlands Beitr. Geophys. **21**, 290 (1929).

schicht konstanter Tiefe *ohne* Berücksichtigung eines Reibungseinflusses wieder. Solche Wellen sind nur möglich, wenn die Periode der Welle kürzer als die Hälfte der Rotationsdauer ist ($s < 1$). Die auftretenden Querströme bedingen elliptische Stromfiguren (Drehströme), die für die ganze Schichte dieselben sind. Das Verhältnis der maximalen zur minimalen Geschwindigkeit ist s. Die Wellengeschwindigkeit ist $c = \sqrt{gh \cdot \frac{1}{1-s^2}}$, also größer als bei nicht rotierender Wasserschicht.

Wenn Reibung am Boden angenommen wird, ändern sich die Verhältnisse wie auf Fig. 14. Die Wellengeschwindigkeit wird durch die Reibung verlangsamt, wobei die Wellenamplitude allmählich abnimmt. Die Stromfigur ändert sich nun mit der Tiefe, wobei die Ellipse mit Annäherung an den Boden schmäler wird. Die maximale Stromstärke tritt *vor* dem Hochwasser auf; dieser Zeitunterschied nimmt mit der Tiefe zu. — In Wirklichkeit sind die Verhältnisse nicht so einfach wie in Fig. 14; die Fig. 15 gibt als Beispiel die Stromfiguren bei Helgoland.

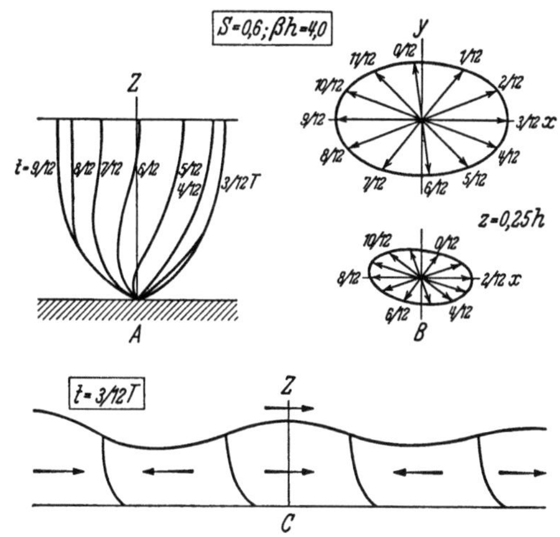

Fig. 14 A—C. Gezeitenwelle in einem rotierenden ($s = 0.6$), sehr weiten Kanal bei *Reibung*. In B oben: Stromfigur an der Wasseroberfläche, darunter in $\frac{1}{4}$ der Wassertiefe über dem Boden, sonst wie in Fig. 9 (nach H. U. SVERDRUP).

Ohne irgendwelche nähere Angaben über die Ursache der Gezeiten geben schon die Bewegungsgleichungen allein einige wichtige Beziehungen zwischen den Gezeiten und den Gezeitenströmen, sowie auch allgemeine Schlüsse über die Beeinflussung des Gezeitenbildes durch den Küstenverlauf und durch Inseln. Untersuchungen dieser Art hat vor allem J. PROUDMAN[1] vorgenommen. Die Gezeit an einem Ort (eine einzelne Partialtide) läßt sich darstellen durch

$$\zeta = H\cos(\sigma t - \varepsilon) = \zeta_1 \cos\sigma t + \zeta_2 \sin\sigma t. \tag{12.4}$$

ζ_1 und ζ_2 sind unabhängig von der Zeit; Hochwasser tritt bei $t = \varepsilon/\sigma$ auf. $\varepsilon = 0$ oder $\zeta_2 = 0$ bedeutet eine fortschreitende Gezeitenwelle (maximaler Flutstrom bei Hochwasser); $\varepsilon = 90°$ oder $\zeta_1 = 0$ hingegen eine stehende Welle (maximaler Flutstrom bei Mittelwasser). Sind ψ und ψ' die Winkel, die die Flutstundenlinien ($\operatorname{tg}\varepsilon = \text{const}$) und die Hubhöhenlinien ($H = \text{const}$) mit der x-Achse bilden, dann findet man:

$$\operatorname{tg}\psi = \frac{\zeta_1\zeta_{2x} - \zeta_2\zeta_{1x}}{\zeta_1\zeta_{2y} - \zeta_2\zeta_{1y}} \quad \text{und} \quad \operatorname{tg}\psi' = \frac{\zeta_1\zeta_{1x} + \zeta_2\zeta_{2x}}{\zeta_1\zeta_{1y} + \zeta_2\zeta_{2y}}. \tag{12.5}$$

Geht man von den Bewegungsgleichungen (4.1) aus und legt man an einem Ort die x-Achse in die Richtung des maximalen Flutstromes und zählt die Zeit von seinem Eintritt aus und setzt $u = U\cos\sigma t$, $v = V\sin\sigma t$, so reduzieren sich diese Gleichungen auf:

$$\left.\begin{aligned} gHH_x &= \zeta_2(\sigma U + fV), & gHH_y &= -\zeta_1(\sigma V + fU), & \operatorname{tg}\psi &= -\frac{\sigma U + fV}{\sigma V + fU}\operatorname{ctg}\varepsilon, \\ gH^2\varepsilon_x &= \zeta_1(\sigma U + fV), & gH^2\varepsilon_y &= \zeta_2(\sigma V + fU), & \operatorname{tg}\psi' &= \frac{\sigma U + fV}{\sigma V + fU}\operatorname{tg}\varepsilon. \end{aligned}\right\} \tag{12.6}$$

[1] J. PROUDMAN: Mon. Not. Roy. Astronom. Soc., Geophys. Suppl. **1925**, 243.

Aus den letzten Formeln folgt allgemein, daß $\psi' = \psi + \frac{\pi}{2}$ ist, d.h. bei Reibungslosigkeit stehen Flutstundenlinien und Hubhöhenlinien stets aufeinander senkrecht.

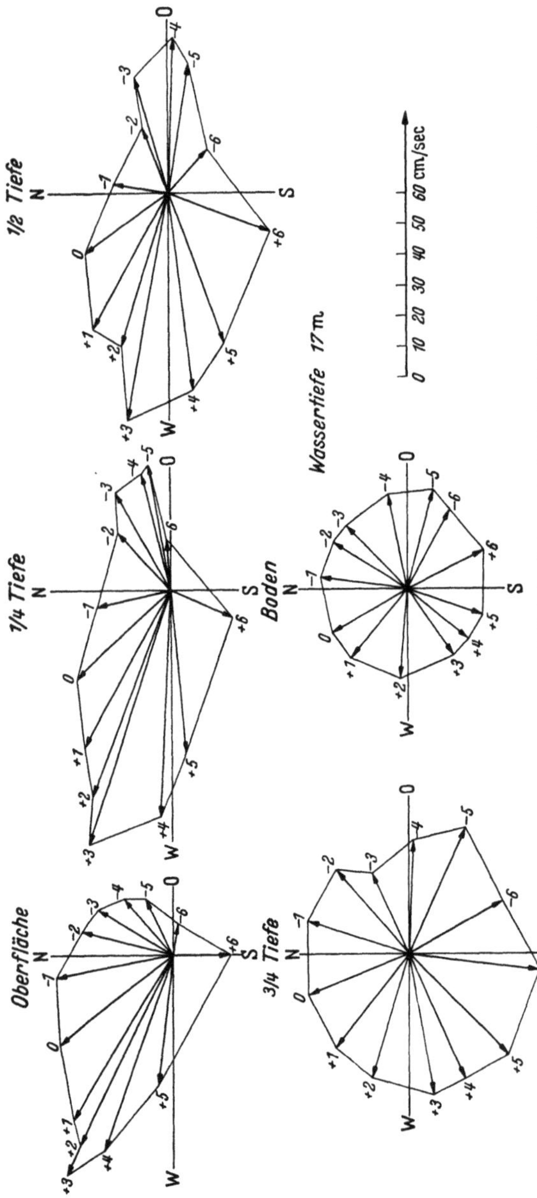

Fig. 15. Stromfiguren bei Helgoland ($\varphi = 54° 16' 1''$; $\lambda = 7° 46'' 3'$) in verschiedenen Tiefen. 0^h: Monddurchgang in Greenwich.

Nimmt man *eine fortschreitende Gezeitenwelle*, dann ist $\varepsilon = \zeta_2 = 0$, d.h. $H_x = 0$ und $\varepsilon_y = 0$. Diese Verhältnisse sind in der Fig. 16a zur Darstellung gebracht. Die Flutstundenlinien haben die Richtung des Stromes bei Mittelwasser, die Hubhöhenlinien die Richtung des Stromes bei Hoch- bzw. Niedrigwasser. Bei einer *stehenden Welle* hat man $\varepsilon = 90°$ und $\zeta_1 = 0$ d.h. $H_y = 0$ und $\varepsilon_x = 0$ und die Verhältnisse der Fig. 16b. Auch für diesen Fall gilt dieselbe Feststellung: die Flutstundenlinien folgen der Richtung des Stromes bei Mittelwasser, die Hubhöhenlinien jener bei Hoch- bzw. Niedrigwasser, so daß dieser Satz allgemeine Geltung hat.

Man kann auch den Einfluß der *Reibung* in Rechnung stellen. Dann stehen die Kurvenscharen nicht mehr senkrecht zueinander, und es kommt auf die Größe der Reibungsglieder an, wie groß diese Winkel wird. Im allgemeinen dürfte er aber nicht stark von 90° abweichen.

PROUDMAN hat in ähnlicher Weise auch die Störungen der Flutstundenlinien und der Hubhöhenlinien durch den orographischen Küstenverlauf (Meeresbuchten und Inseln) untersucht. Die Ergebnisse haben eine praktische Bedeutung, wenn man auf Grund von Gezeitenstromkarten das getreue Gezeitenbild eines Nebenmeeres konstruieren will.

Bestimmung der Gezeiten aus Gezeitenstrommessungen.

Da die Bewegungs- und die Kontinuitätsgleichung eine Verknüpfung von vertikaler Gezeit und Gezeitenströmen darstellen, muß es möglich sein, aus Strommessungen auf die vertikale Gezeit zu schließen. Da die Neben- und Randmeere sich zumeist auf den Schelfen erstrecken, ihre Tiefe deshalb selten 200 m übersteigt, sind Strommessungen in ihnen ohne zu große Schwierigkeiten auszuführen, und heutzutage besitzt man von vielen Meeren Darstellungen der Stromverteilung von Stunde zu Stunde der Gezeitenperiode (*Stromatlanten*). Wenn auch diese Stromkarten im allgemeinen nur praktischen Zwecken dienen, lassen sie sich doch verwerten, um Schlüsse auf die vertikale Gezeit auf der ganzen Fläche des Meeres zu ziehen. Hierfür sind zwei Methoden entwickelt worden. Die erste stammt von A. DEFANT[1] und stützt sich auf die Kontinuitätsgleichung und stellt im wesentlichen ein Differentiationsverfahren dar. Die zweite, von J. PROUDMAN und A. T. DOODSON[2] verwendet die Bewegungsgleichungen allein; sie ist im wesentlichen ein Integrationsverfahren längs Linien quer über das Nebenmeer von einer Küste zur anderen. Beide Methoden sind auf die Gezeiten der Nordsee angewandt worden.

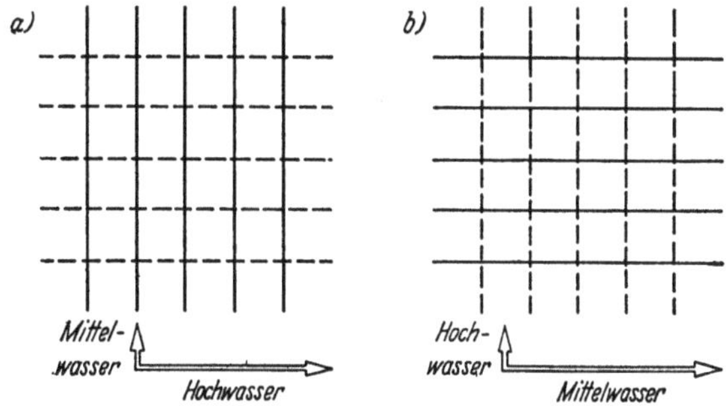

Fig. 16a u. b. Flutstundenlinien (ausgezogen) und Hubhöhenlinien (gestrichelt). a bei einer fortschreitenden Gezeitenwelle ($\varepsilon = 0°$), b bei einer stehenden Gezeitenwelle ($\varepsilon = 90°$).

Setzt man

$$\zeta = \zeta_1 \cos\sigma t + \zeta_2 \sin\sigma t,$$
$$u = u_1 \cos\sigma t + u_2 \sin\sigma t,$$
$$v = v_1 \cos\sigma t + v_2 \sin\sigma t,$$

so zerfällt die einfache Kontinuitätsgleichung (4.1) in die beiden Beziehungen:

$$\zeta_1 = \frac{1}{\sigma}\left[(h u_2)_x + (h v_2)_y\right] \quad \text{und} \quad \zeta_2 = -\frac{1}{\sigma}\left[(h u_1)_x + (h v_1)_y\right]. \tag{12.7}$$

Zerteilt man die Fläche eines Nebenmeeres durch ein Gitternetz in eine große Zahl flächentreuer Quadrate, dann lassen sich in guter Annäherung bei Kenntnis der Stromwerte u_1, u_2 und v_1, v_2 an allen Eckpunkten die beiden Werte ζ_1 und ζ_2 für die Mitte eines jeden Quadrates berechnen. Die Zusammenfassung gibt dann die Amplitude und Phase der Gezeit an diesen Punkten; sie ist unabhängig von den Werten der Nachbarschaft und auch unabhängig von den Küstenwerten.

Die Methode von PROUDMAN-DOODSON berücksichtigt in den Bewegungsgleichungen (4.1) auch die Reibungskräfte, und zwar nach dem TAYLORschen Ansatz, indem rechts in den Gleichungen diese zu $-\frac{F}{h}$ und $-\frac{G}{h}$ angesetzt werden. Nimmt man $F = F_1 \cos\sigma t + F_2 \sin\sigma t$ und $G = G_1 \cos\sigma t + G_2 \sin\sigma t$ und wählt für die M_2-Tide die Zahlenwerte

$$a = 10^7 \cdot \frac{\sigma}{g} = 1{,}432, \qquad b = 10^7 \cdot \frac{f}{g} = 1{,}487 \sin\varphi \quad \text{und} \quad c = \frac{10^7}{g h} = \frac{1{,}019}{h:10^4},$$

dann nehmen die Gleichungen die Form an:

$$\left.\begin{aligned} 10^7 \zeta_{1x} &= -a u_2 + b v_1 - c F_1, & 10^7 \zeta_{2x} &= a u_1 + b v_2 - c F_2, \\ 10^7 \zeta_{1y} &= -a v_2 - b u_1 - c G_1, & 10^7 \zeta_{2y} &= a v_1 - b u_2 - c G_2. \end{aligned}\right\} \tag{12.8}$$

Wenn längs einer Linie Gezeitenstrommessungen gegeben sind, lassen diese Gleichungen die Gradienten der Werte ζ_1 und ζ_2 längs der Linie berechnen. Schließt man diese Werte an einem Küstenpunkt an, in dem ζ_1 und ζ_2 bekannt

[1] A. DEFANT: Ann. Hydrogr. **1923**, 177.
[2] J. PROUDMAN u. A. T. DOODSON: Phil. Trans. Roy. Soc. Lond., Ser. A **224**, 185 (1924).

sind, so bekommt man die vertikale Gezeit längs der ganzen Linie. Die Genauigkeit des Verfahrens hängt von der Zahl und Güte der benützten Stromwerte ab.

Beide Methoden haben in ihrer Anwendung auf die Gezeiten der Nordsee gute, übereinstimmende Ergebnisse geliefert und bilden beim Vorhandensein guter Stromatlanten für Nebenmeere gute Verfahren, sich über die Gezeit auf der ganzen Fläche derselben zu orientieren.

13. Die Gezeiten der Neben- und Randmeere. Einflüsse von Reibung und Erddrehung. In folgenden Ausführungen werden vor allem Einflüsse der Konfiguration der Meeresbecken in den Vordergrund gestellt. Sie betreffen in erster Linie die Gezeiten von Meeresteilen, die in sich abgeschlossen sind und nur an einzelnen Stellen durch mehr oder minder breite Öffnungen mit dem freien Ozean in Verbindung stehen. Der Fall eines rechteckigen Meeresbusens konstanter Tiefe, zunächst ohne Berücksichtigung von Erdrotation und Reibung, gibt das Grundsätzliche. Die Gezeiten eines solchen Beckens haben zweifachen Ursprung: Die *fluterzeugenden Kräfte* werden periodische horizontale Wasserverlagerungen erzwingen, bei denen der Umstand, daß durch den Mündungsquerschnitt zum freien Ozean stets ein ungehinderter Bezug von Wasser aus dem freien Ozean möglich ist, an der Mündung eine Knotenlinie erzwungen wird. Andererseits verursachen die Gezeiten des freien Ozeans an der Mündung ein *Mitschwingen* der mehr oder minder abgeschlossenen Wassermassen des Beckens mit der äußeren Gezeit. Die tatsächlichen Gezeiten des Meeresbusens sind so die Überlagerung dieser beiden Komponenten: *der selbständigen Gezeit und der Mitschwingungsgezeit.*

Ein rechteckiges Becken der Länge l und der Wassertiefe h habe bei $x=0$ das geschlossene Ende, bei $x=l$ die Mündung in den freien Ozean. Die selbständige Gezeit unter der Einwirkung der horizontalen Flutkraft $X=m\cos(\sigma t+\varepsilon)$ hat die Form (Randbedingungen: $\xi=0$ für $x=0$ und $\zeta=0$ für $x=l$):

$$\left.\begin{aligned}\xi &= \frac{2m}{\sigma^2\cos\nu\pi}\sin\nu\pi\frac{y}{2}\sin\nu\pi\left(1-\frac{y}{2}\right)\cos(\sigma t+\varepsilon)\,,\\ \zeta &= \frac{m T_K\sqrt{gh}}{2\pi g\cos\nu\pi}\sin\nu\pi(y-1)\cos(\sigma t+\varepsilon)\,.\end{aligned}\right\}\qquad(13.1)$$

Dabei ist $y=x/l$, Mündung $y=0$, geschlossenes Ende $y=1$; die Eigenperiode des geschlossen gedachten Kanals ist $T_f=2l/c$, die Periode der Kraft $T_K=2\pi/\sigma$. Die Größe $\nu=\frac{T_f}{T_K}=\frac{\sigma l}{c\pi}$ fixiert die Form der selbständigen Gezeit im Meeresbusen. Sie hat den Charakter einer *stehenden* Welle. Die Lösung wird unbrauchbar, wenn ν die Werte $\frac{1}{2}, \frac{3}{2}, \frac{5}{2}$ usw. annimmt. In diesem Falle tritt *Resonanz* mit der erregenden Kraft auf: die Gezeiten werden dabei wesentlich verstärkt, theoretisch werden die Amplituden unendlich groß. Die Fig. 17 gibt oben die Verteilung der Amplitude im Meeresbecken bei verschiedenen ν-Werten; die Amplitude ist hierbei um so größer, je tiefer das Becken ist. Für die vereinigte Mond- und Sonnengezeit wird der Faktor $\frac{m T_K c}{2\pi g}=0{,}0273\sqrt{h}\cos\varphi$ und man erkennt daraus, daß die selbstständigen Gezeiten im allgemeinen nur kleine Amplituden haben werden (für $h=200$ m wird der Amplitudenfaktor rund 4 cm). Nur bei Resonanz gelangen sie zu einiger Bedeutung.

Bei den *Mitschwingungsgezeiten* sind die Randbedingungen: am geschlossenen Ende $(y=0)$ $\xi=0$ und an der Mündung $(y=1)$ $\zeta=Z\cos(\sigma t+\varepsilon)$, was die Gezeit im offenen Meer vor der Mündung sein soll. Die Mitschwingungsgezeiten haben die Form:

$$\xi=Z\frac{l}{\nu\pi h}\frac{\sin\nu\pi y}{\cos\nu\pi}\cos(\sigma t+\varepsilon)\,,\qquad \zeta=Z\frac{\cos\nu\pi y}{\cos\nu\pi}\cos(\sigma t+\varepsilon)\,.\qquad(13.2)$$

Auch dieser Gezeitenanteil hat den Charakter einer *stehenden* Welle; bei den ν-Werten $\frac{1}{2}, \frac{3}{2}, \frac{5}{2} \ldots$ tritt, wie bei der selbständigen Gezeit, Resonanz auf. Fig. 17 unten gibt wieder die Verteilung der Amplituden längs des einseitig geschlossenen Kanals. Da die Hubhöhen der Gezeit vor der Mündung zumeist recht groß sind, können die Mitschwingungsgezeiten in Nebenmeeren von ausschlaggebender Bedeutung werden.

Hat das Meeresbecken eine kompliziertere Konfiguration, so lassen sich zur Ermittlung beider Gezeitenanteile dieselben Methoden anwenden, die zur Ermittlung der Eigenperioden des Beckens benützt wurden (siehe Ziff. 3). Am bequemsten gestalten sich die Berechnungen bei Benützung der schrittweise numerischen Integration der vom Zeitfaktor befreiten Bewegungsgleichungen [siehe Gl. (3.15)]. Bei der Mitschwingungsgezeit beginnt man die Rechnung am geschlossenen Ende

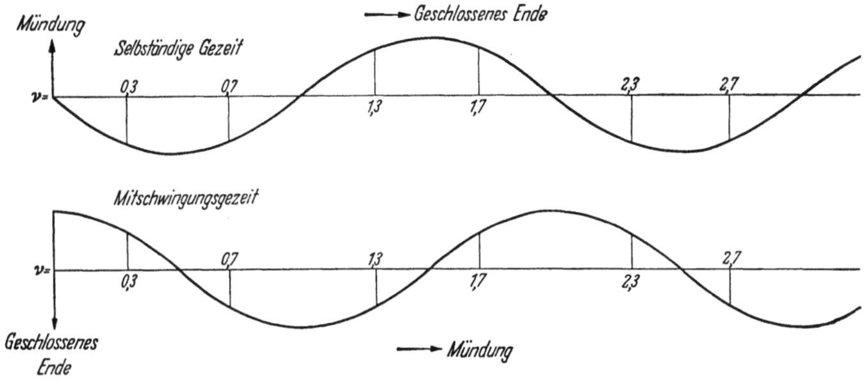

Fig. 17. Oben: Amplitudenverteilung der selbständigen Gezeit. Unten: Amplitudenverteilung der Mitschwingungsgezeit in einem rechteckigen Meeresbecken konstanter Tiefe in Abhängigkeit von $\nu = T_f/T_k$. (Es gilt immer nur der Teil links vom Strich, der den ν-Wert angibt.)

mit $\xi = 0$ und willkürlichem ζ. Die Größe α ist nun durch die Gezeitenperiode fixiert. Man erhält am offenen Ende ($x = l$) die Amplitude ζ_l. Ist Z die Amplitude im freien Ozean vor der Mündung, dann multipliziert man alle Größen der numerischen Berechnung mit Z/ζ_l. Die Berechnungen bleiben richtig, da sich ein Proportionalitätsfaktor in ξ und ζ stets heraushebt.

Bei der selbständigen Gezeit lassen sich dieselben Gleichungen benützen, doch muß man in ihnen bei der ersten noch die Größe ΔH und bei der zweiten nach $\alpha \frac{\xi_1}{4}$ die Größe $\frac{\Delta H}{2}$ einfügen. Dabei ist $\Delta H = \frac{m}{g} \Delta x$. Die Randbedingungen sind am geschlossenen Ende ($x = 0$) $\xi_0 = 0$, an der Mündung ($x = l$) hingegen $\zeta_l = 0$. ζ_0 am geschlossenen Ende (Querschnitt 0) muß so gewählt werden, daß die zweite Bedingung erfüllt ist. Das erfordert vielleicht eine wiederholte Berechnung, bis der richtige Wert ζ_0 gefunden wird. Man kann die Rechnung aber abkürzen durch den Kunstgriff, daß die Hinzunahme einer Mitschwingungsgezeit, die die Forderung $\zeta_l = 0$ erzwingt, stets möglich ist ohne die Gültigkeit der ganzen Berechnung zu gefährden.

Einfluß der Erdrotation. Betreffs des Einflusses der Corioliskraft auf die stehenden Wellen der selbständigen und der Mitschwingungsgezeit in Nebenmeeren siehe die Ausführungen in Ziff. 4. Die durch die Corioliskraft auftretenden Querschwingungen geben mit den Längsschwingungen *Amphidromien* an Stelle der Knotenlinien, die *contra solem* verlaufen. Aus den alternierenden Gezeitenströmen werden Drehströme mit elliptischer Stromfigur; nur bei den KELVIN-Wellen ist das nicht der Fall, da hier stets Gleichgewicht zwischen Quergefälle und Corioliskraft besteht. Eine genaue Lösung, wie der Einfluß der Corioliskraft sich bei dem Mitschwingen eines rechteckigen Meeresbecken konstanter Tiefe mit der äußeren Gezeit gestaltet, hat G. J. TAYLOR gegeben (siehe hierzu Ziff. 4, d). Dort, wo der Einfluß der Corioliskraft klein ist (schmale Becken, geringe Geschwindigkeiten und niedrige Breiten) haben beide Gezeitenanteile mehr den Charakter stehender Wellen; aber dort, wo die Corioliskraft sich voll entfalten kann, hat man an Stelle der Knotenlinien Amphidromien, mit Drehströmen im Inneren der Bucht.

Einfluß der Reibung. Auch bei den periodischen horizontalen Bewegungen der selbständigen und der Mitschwingungsgezeit wird die Wellenenergie in erster Linie durch die *Turbulenzreibung* herabgesetzt. Die einfachste Art, die Reibung quantitativ in Rechnung zu stellen, ist, in den Bewegungsgleichungen für die

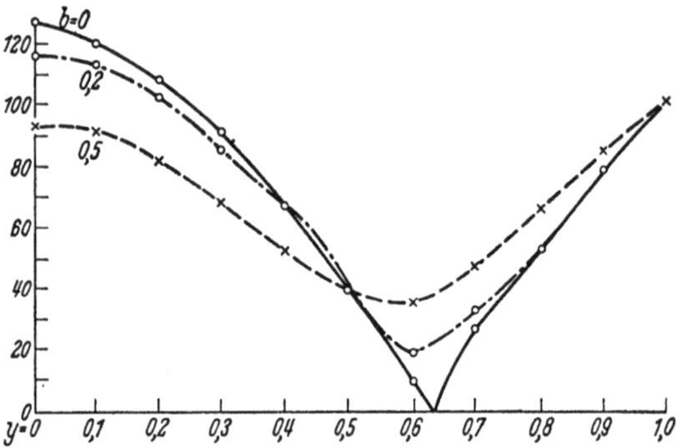

Fig. 18. Amplitudenverteilung der halbtägigen Gezeit in einer rechteckigen Meeresbucht konstanter Tiefe ($\nu = 0{,}8$) bei verschiedenen Reibungswerten b. (Mündung bei $y = 1$: dort Amplitude 100 cm, Phase 0^h.)

horizontale Wasserverlagerung ein Reibungsglied der Form $\beta\, \partial\xi/\partial t$ aufzunehmen [wie in Gl. (3.4)]. Die Reibung wäre dann proportional der horizontalen Geschwindigkeit. Diese Annahme entspricht nicht ganz dem tatsächlichen Verhalten einer turbulenten Bewegung, hat aber den Vorzug, daß man die Rechnungen bis zum Ende durchführen und so den Reibungseinfluß voll überblicken kann. Die Größe β ist keine Konstante, sie hängt von der Wassertiefe und der Beschaffenheit des Bodens ab. In der Theorie tritt sie stets in der Verbindung $b = \beta/\sigma = \beta T/2\pi$ auf. b ist eine reine Zahl. Als Größenordnung von b hat sich ergeben:

$h =$	100	50	30	10 m und darunter
$b =$	0,1 — 0,2	0,5	0,5 — 1,0	1,0 — 2,0.

Der Reibungseinfluß läßt sich sowohl bei der selbständigen, wie auch bei der Mitschwingungsgezeit durch die Integration der erweiterten Bewegungsgleichungen zahlenmäßig erfassen (DEFANT[1]). Die Änderungen, die die Amplituden- und Phasenverteilung längs der Meeresbucht erfahren, sind um so kräftiger, je größer die Reibung ist. In der Nähe der Knotenlinien mit ihren größten horizontalen Geschwindigkeiten ist die Reibung am stärksten fühlbar. Der Phasensprung um eine halbe Periode, der hier sonst vorhanden ist, wird bei Reibung aufgehoben und in einem allmählichen Übergang der Phase verwandelt. Dadurch nimmt die Gezeit den Charakter einer fortschreitenden Welle an, aber die Lage der ursprünglichen Knotenlinie verrät sich durch eine Zusammendrängung der Flutstundenlinien. Auch in der Amplitudenverteilung verschwindet die Nullstelle

[1] A. DEFANT: Denkschr. Akad. Wiss. Wien, Math.-Naturwiss. Kl., I. Teil **1919**. — Probleme des Meeres in Landnähe. Probleme der Kosmischen Physik, Bd. 6. Hamburg 1925.

an der Knotenlinie und statt dieser tritt ein Minimum der Hubhöhe auf, das um so schwächer ist, je stärker die Reibung ist. Fig. 18 und 19 gibt ein Beispiel für die Mitschwingungsgezeit in einer Meeresbucht bei $\nu = 0{,}8$ und verschiedenen Werten von b. Auch bei Nebenmeeren unregelmäßiger Konfiguration läßt sich bei der schrittweisen Berechnung der beiden Gezeitenanteile die Reibung berücksichtigen; ein Verfahren hierzu hat DEFANT[1] gegeben.

Wenn bei der Ausbildung der Mitschwingungsgezeit die von der Mündung einlaufende, fortschreitende Welle durch Reibung geschwächt wird, wird auch die am geschlossenen Ende der Bucht reflektierte Welle nicht mehr die Energie der einlaufenden haben. Ihre Superposition ergibt dann nicht mehr eine stehende, sondern eine fortschreitende Welle. Viele Buchten und Nebenmeere laufen auf große Strecken in ihren inneren Teilen in seichte Randgebiete aus, auf denen ein beträchtlicher Teil der Gezeitenenergie aufgezehrt wird und für die Reflexion verloren geht. In solchen Nebenmeeren hat, auch wenn sonst ihre Tiefe recht

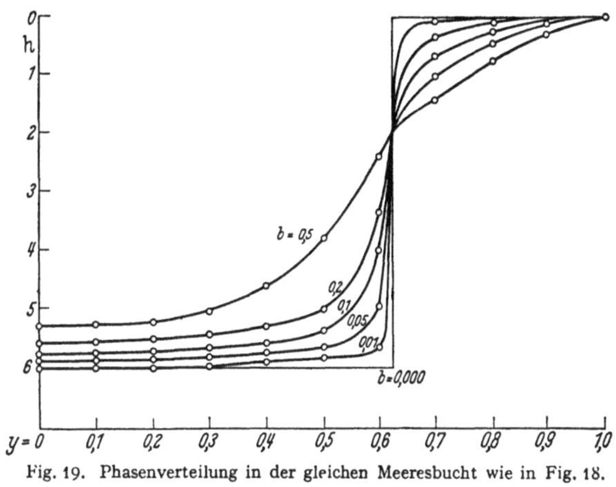

Fig. 19. Phasenverteilung in der gleichen Meeresbucht wie in Fig. 18.

groß ist, die Mitschwingungsgezeit den Charakter einer stehenden Welle verloren (DEFANT[2]). Ein einfacher Fall ist folgender: Eine Bucht münde bei $y = 1$ in ein gezeitenführendes Meer mit der Gezeit $Z \cos \sigma t$. Die Zeit wird demnach von der Hochwasserzeit an der Mündung gezählt. Die Amplitude der eindringenden Welle (in der x-Richtung) sei a Die rücklaufende Welle habe aber die Amplitude b, so daß der Energieanteil $(a^2 - b^2)/a^2$ bei der Reflexion am inneren Ende verloren gegangen ist. $b = a$ bedeutet totale Reflexion. a/b ist durch den Reibungsverlust gegeben. Dann ist die Verteilung der Hubhöhe längs der ganzen Bucht mit Berücksichtigung der Randbedingungen durch folgende Gleichung gegeben:

$$\left.\begin{aligned} \zeta &= Z\left[\frac{\cos\varepsilon}{\cos\nu\pi}\cos\nu\pi y\cos(\sigma t-\varepsilon) - \frac{\sin\varepsilon}{\sin\nu\pi}\sin\nu\pi y\sin(\sigma t-\varepsilon)\right] \\ \text{mit}\qquad & \frac{a-b}{a+b}\operatorname{tg}\nu\pi = \operatorname{tg}\varepsilon. \end{aligned}\right\} \tag{13.3}$$

Für $a = b$ geht diese Gleichung in die Form der stehenden Welle der einfachen Mitschwingungsgezeit über. Mit Reibung hat sie nun der Charakter einer *Überlagerung zweier orthogonaler Wellen* (Phasenunterschied $\frac{1}{4}$ der Periode) angenommen. Längs der Bucht verspätet sich die Phase allmählich immer mehr und erreicht am inneren Ende den vollen Betrag ε.

Auch wenn ein Nebenmeer an sich eine solche Konfiguration hat, daß in ihm die Gezeiten durch Reibung nur sehr wenig gestört sind, wird seine Mitschwingungsgezeit mit dem äußeren Meer nicht mehr die Form einer einfachen stehenden Welle haben, wenn in den inneren Endteilen ein Teil der Energie nicht mehr ins Nebenmeer zurückkehrt, sondern verloren geht. Dies ist auch der Fall, wenn z. B. eine große Meeresbucht am inneren Ende in ein sehr seichtes Randmeer übergeht, in dem die Dissipation der Gezeitenenergie groß ist. Auch in der tiefen Meeresbucht wird dann die Mitschwingungsgezeit als eine Superposition zweier orthogonaler stehender Wellen erscheinen. Dieser Fall ist wichtig für die Erklärung natürlicher Fälle in den Ozeanen.

[1] Siehe Fußnote 1, S. 908.
[2] A. DEFANT: Ann. Hydrogr. **1928**, 274.

Eine bessere Annahme über die *Reibung* erhält man, wenn man sie proportional dem *Quadrate* der mittleren Geschwindigkeit setzt. Dieser TAYLORsche Reibungsansatz ergibt als Reibungskraft

$$F = k \varrho u^2 , \tag{13.4}$$

wenn u die mittlere Geschwindigkeit, ϱ die Dichte des Wassers und k eine Konstante zwischen 0,002 und 0,0016 cgs-Einheiten ist, falls man von den beobachteten Werten in natürlichen Gerinnen ausgeht. Bei stärker gestörter Bodengestaltung kann sie aber weit höhere Werte annehmen. Die von der Reibungskraft geleistete Arbeit ist dann $k \varrho u^3 = k \varrho U^3 \cos^3 \frac{2\pi}{T} t$. Der Mittelwert von $\cos^3 \frac{2\pi}{T} t$ für eine volle Periode ist $4/3\pi$, so daß sich als Betrag der Dissipation der Energie $\frac{4 k \varrho}{3\pi} U^3$ ergibt.

TAYLOR[1] hat zuerst nach dieser Beziehung den Energieverlust der Gezeitenbewegung im Irischen Kanal ermittelt. Im Mittel ergeben die Beobachtungen $U = 2\frac{1}{2}$ Knoten $= 114$ cm/sec; man erhält dann als mittleren Energieverlust 1300 erg pro cm² Oberfläche und sec. Der früher von R. O. STREET[2] ermittelte Wert von nur 7 erg cm^{-2} sec^{-1} beruht auf irrige Annahme. TAYLOR hat auch eine andere Methode angewandt, um für dasselbe Meer den Energieverlust zu bestimmen. Er ermittelt die Energiemenge, die dieser Meeresteil von den beiden Einfallstoren im Norden und Süden erhält, außerdem die direkt durch die Gezeitenkräfte dem Meere zugeführte Energie. Diese beiden Energien E_a und E_i müssen durch den einzigen Energieverlust durch Reibung kompensiert sein, da im Mittel über eine volle Gezeitenperiode die Energie weder zu- noch abnehmen kann: $R = E_a + E_i$. Die numerische Berechnung ergab einen Energieverlust durch Reibung von 1530 erg cm^{-2} sec^{-1}, was mit dem obigen Wert der Größenordnung nach übereinstimmt.

S. F. GRACE[3] hat einen anderen Weg beschritten, um die Reibung bei Gezeiten in Nebenmeeren abzuschätzen. Die Kontinuitätsgleichung läßt bei langgestreckten Meeresbecken mit fast vollständiger longitudinaler Gezeitenbewegung für jeden Querschnitt die mittlere Geschwindigkeit u der M_2-Tide berechnen; die Bewegungsgleichung ergibt dann als einzige Unbekannte die Reibungskraft, aus der man die Reibungskonstante k ermitteln kann. GRACE hat diese Methode auf die Gezeiten des Golfes von Bristol und eines Teiles des Englischen Kanals, sowie des seichten Golfes von Suez angewandt. Es ergibt sich, daß die Phase der Reibung in erster Annäherung dieselbe ist wie die des Gezeitenstromes, daß aber die Reibungskonstante sehr unregelmäßig schwankt. Der Mittelwert für den Bristolkanal ist $2{,}6 \times 10^{-3}$, in guter Übereinstimmung mit $2{,}4 \times 10^{-3}$, den TAYLOR gefunden hatte. Dieser Wert entspricht einer Reibungskonstante $b = 1{,}3$, ebenfalls in guter Übereinstimmung mit einem Wert, den DEFANT gefunden hatte. Im Englischen Kanal findet man im Mittel $9{,}3 \times 10^{-3}$, aber die Schwankungen sind hier wesentlich größer. Für den Golf von Suez findet man $k = 3{,}6 \times 10^{-3}$ oder $b = 0{,}11$; dieser Wert ist kleiner als einer mittleren Tiefe von nur 44 m entspricht. Diese größere Reibung im mittleren Teil des Golfes dürfte von den ungewöhnlich starken Unregelmäßigkeiten der Bodenkonfiguration herrühren. Alle diese Untersuchungen zeigen, daß der TAYLORsche Reibungsansatz die Beeinflussung der Gezeiten durch Reibung im richtigen Sinne wiederzugeben vermag, daß man aber doch nur ein rohes Bild dieser Beeinflussung erhält, solange

[1] G. J. TAYLOR: Phil. Trans. Roy. Soc. Lond., Ser. A **220** (1919).
[2] R. O. STREET: Proc. Roy. Soc. Lond., Ser. A **93** (1917).
[3] S. F. GRACE: Mon. Not. Roy. Soc. Astronom. Geophys., Suppl. **1931** und **1937**.

man nur die Reibung im Mittel und nicht auch im vertikalen Stromaufbau in Rechnung stellt, was aber wesentlich schwieriger ist (siehe hierzu auch BOWDEN and FAIRBAIRN[1]).

14. Die Gezeiten in Neben- und Randmeeren und im freien Ozean. *α) Einleitung.* Es ist nicht möglich in diesem Rahmen eine erschöpfende Darstellung der Gezeitenerscheinungen in allen Neben- und Randmeeren zu geben, auch nicht soweit diese eine eingehende Untersuchung erfahren haben. Es soll nur in einer Übersicht darüber, ohne in Einzelheiten zu gehen, gesagt werden, welche dieser Meere auf Grund der vorliegenden Beobachtungen eine spezielle Behandlung erfahren haben und bei welchen wenigstens von einer einigermaßen gesicherten Grundlage einer Deutung gesprochen werden kann. Ein besonders einfaches Nebenmeer — das Adriatische Meer — soll im Anschluß an die allgemeine Übersicht eine speziellere Darstellung erfahren.

β) Übersicht über die bisher vorliegenden Bearbeitungen. Auf Grund der in den vorhergehenden Abschnitten dargelegten Methoden sind die Gezeitenerscheinungen einer ganzen Reihe von Neben- und Randmeeren bearbeitet worden. Dabei sind diese Untersuchungen jedesmal verschieden weit getrieben worden, von der erschöpfenden theoretischen Berechnung des ihnen zukommenden Gezeitenbildes und eines eingehenden Vergleiches mit den Beobachtungstatsachen auf Grund der harmonischen Konstanten der Küstenorte und der Verteilung der Gezeitenströme in ihnen, bis zur schematischen, mehr qualitativen Abschätzung, wie sich auf Grund der morphologischen Konfiguration des Becken und seiner Verbindung mit dem offenen Ozean die Gezeiten abspielen sollten. Die folgende Übersicht ist vielleicht nicht ganz vollständig, aber die wichtigste Literatur darüber dürfte komplett angeführt sein.

Nordsee. Einen geschichtlichen Überblick über die Entwicklung der Kenntnisse ihrer Gezeiten hat H. THORADE[2] gegeben. Die ersten auf hydrodynamischer Grundlage beruhenden Untersuchungen stammen nach den von ihnen dargelegten Methoden (s. Ziff. 8) von A. DEFANT[3] und J. PROUDMAN and A. T. DOODSON[4]. Die neueste Bearbeitung auf Grund der harmonischen Konstanten der M_2-Tide und Analyse der entsprechenden Gezeitenströme hat W. HANSEN[5] gegeben. Die Nordsee ist wahrscheinlich dasjenige Nebenmeer, dessen Gezeiten am genauesten bekannt sind. Im wesentlichen sind es Mitschwingungsgezeiten mit dem Atlantischen Ozean im Norden, die durch die Erdrotation und die Reibung stark beeinflußt sind.

Ostsee. Eine eingehende Synthese auf Grund der harmonischen Konstanten einer größeren Anzahl von Orten stammt von R. WITTING[6]. Die halbtägigen Gezeiten sind eine Überlagerung der direkt in der Ostsee erzeugten und der Mitschwingungsgezeiten mit jenen im Kattegat, letztere scheinen die bedeuteren zu sein. Die Eintagtiden, die etwa gleich stark wie die halbtägigen sind, sind selbständige Gezeiten der Ostsee.

Englischer Kanal. Nach einer langen Periode mehr qualitativer Betrachtungen lieferten ausführliche Untersuchungen auf hydrodynamischer Grundlage A. DEFANT[7], der die Gezeiten im Kanal im wesentlichen als Mitschwingungsgezeiten mit denen vor den beiden Kanalmündungen ansah, und A. T. DOODSON and R. H. CORKAN[8], die das auf S. 905 dargelegte Verfahren zur Fixierung des Gezeitenbildes benützten. Beide Arbeiten führten zu durchaus befriedigenden Ergebnissen. Eine genaue Analyse der Gezeitenströme in diesem Kanal verdankt man W. HANSEN[9].

---

[1] K. F. BOWDEN u. L. A. FAIRBAIRN: Proc. Roy. Soc. Lond., Ser. A **214**, 371 (1952).
[2] H. THORADE: Peterm. geogr. Mitt. Erg.-Heft **209**, 195 (1930).
[3] A. DEFANT: Ann. Hydrogr. **1923**, 57—64 177—186.
[4] J. PROUDMAN u. A. T. DOODSON: Phil. Trans. Roy. Soc. Lond., Ser. A **224**, 185—219 (1924).
[5] W. HANSEN: Dtsch. hydrogr. Z., Erg.-H. **1** (1952)
[6] R. WITTING: Tidvattnen i Oestersjön och Finska viken. Helsingfors 1911.
[7] A. DEFANT: Denkschr. Akad. Wiss. Wien, Math.-naturwiss. Kl., IV.—V. Teil **1919**.
[8] A. T. DOODSON u. R. H. CORKAN: Phil. Trans. Roy. Soc. Lond., Ser. A **234** (1932).
[9] W. HANSEN: Dtsch. hydrogr. Z. **3**, H. 3/4 (1950).

Irischer Kanal. In ähnlicher Weise wurde auch dieser Kanal untersucht. Es liegen vor: A. DEFANT[1] und A. T. DOODSON and R. H. CORKAN[2], sowie J. PROUDMAN and A. T. DOODSON[3]. Auch die irischen Gezeiten sind in der Hauptsache auf Impulse zurückzuführen, die die Wassermassen vom Atlantischen Ozean an beiden Mündungen erhalten.

Europäisches Mittelmeer. Die Gezeiten dieses fast in sich geschlossenen Nebenmeeres haben eingehende Untersuchung vor allem durch R. STERNECK[4] erfahren. Sie sind auch hier namentlich im westlichen Becken eine Superposition selbständiger Gezeiten und Mitschwingungsgezeiten mit den aus der Straße von Gibraltar kommenden Impulsen. Auch die eintägigen Gezeiten, sowie die auffallende Verteilung der Gezeitenformzahl lassen sich auf diese Weise erklären.

Adriatisches Meer. Siehe Abschn. γ.

Schwarzes Meer. Es muß hinsichtlich der Gezeiten als völlig abgeschlossene Wassermasse angesehen werden. Deshalb bieten seine Gezeiten theoretisch einiges Interesse. Arbeiten darüber liegen von R. STERNECK[5] vor; siehe hierzu auch die theoretischen Erwägungen von S. F. GRACE[6].

Rotes Meer. Die einfache Gestalt dieses langgestreckten, tiefen Nebenmeeres macht es für eine Prüfung der dynamischen Gezeitentheorien besonders geeignet. Die ersten Untersuchungen dieser Art stammen von A. DEFANT[7]. Auch hier erweisen sich die Gezeiten als die Superposition der Mitschwingungsgezeiten mit jenen des Golfes von Aden und selbständiger Gezeiten. Das Verhältnis der beiden Komponenten ist bei der M_2-Tide 3:4. Eine neuere Untersuchung hat S. F. GRACE[8] gegeben, die im wesentlichen dasselbe ergibt. Betreffs der Erdgezeiten im Roten Meer siehe Ziff. 17.

Persischer Golf. Die charakteristischen Gezeiten dieses Nebenmeeres, für die nur wenige harmonischen Konstante vorliegen, haben bisher nur eine einzige Darstellung durch A. DEFANT[9] erfahren, in der eine mögliche Deutung derselben durch selbständige und Mitschwingungsgezeiten gegeben wird.

Ostindisches Archipel. Die Gezeiten dieses verworrenen Meeresstraßensystemes haben eine ausgezeichnete Darstellung durch P. VAN DER STOK[10] erfahren, durch die das komplizierte Gezeitenbild eindeutig festgelegt wurde. Hier bemerkt man besonders deutlich, daß die Gezeiten dieses Archipels in erster Linie von den Eigenperioden der einzelnen Teilbecken und Kanäle abhängen, aber eine eingehende dynamische Analyse der ganzen Erscheinungen steht noch aus.

Ostchinesische Randmeere. Eine Zusammenstellung der harmonischen Konstanten dieser Nebenmeere (Ostchinesische See, Gelbes Meer mit Pwok Hai und den Golf von Liautung, Japanisches Meer und Ochotskisches Randmeer) hat S OGURA[11] gegeben, auch Karten der ungefähren Verteilung der Phasen- und Hubhöhenlinien sind veröffentlicht worden. Nur für die innersten Teile des Gelben Meeres hat (auf Grund von Gezeitenstrommessungen) OGURA[12] genauere Gezeitenbilder dieser Gebiete geben können.

Golf von Mexiko. Die Gezeiten dieses ausgedehnten Mittelmeeres sind nur schwach entwickelt. Nach einigen Arbeiten darüber mehr qualitativer Art hat S. F. GRACE[13] eine hydrodynamische Erklärung derselben versucht, die auch in methodischer Hinsicht interessant ist. Es scheint, daß die geringe Entwicklung der Gezeiten hier dadurch zustande kommt, daß die Amphidromie der selbständigen Gezeit und jene der Mitschwingungsgezeit entgegengesetzt verlaufen und sich dadurch gegenseitig größtenteils aufheben.

---

[1] A. DEFANT: Sitzgsber. Wien. Akad., Math.-naturwiss. Kl., Teil IIa **129** (1920).

[2] Siehe Fußnote 8, S. 911.

[3] J. PROUDMAN u. A. T. DOODSON: Merseyside. Handbook of Liverpool. **1923**.

[4] R. STERNECK: Sitzgsber. Wien. Akad., Math.-naturwiss. Kl., Teil IIa **122** (1913); **124** (1915); **131** (1922). — Siehe auch Ann. Hydrogr. **1923**, 215.

[5] R. STERNECK: Ann. Hydrogr. **1926**, 289. — Siehe auch Sitzgsber. Wien., Akad. Math.-naturwiss. Kl., Teil IIa **131**, 81 (1922).

[6] S. F. GRACE: Mon. Not. Roy. Astronom. Soc.. Geophys. Suppl. **1931**.

[7] A. DEFANT: Denkschr. Akad. Wiss. Wien, Math.-naturwiss. Kl., II. Teil **96**, 110 (1919). — Ann. Hydrogr. Hamburg **1926**, 185.

[8] S. F. GRACE: Mont. Not. Astronom. Soc., Geophys. Suppl. **1930**, 274.

[9] A. DEFANT: Denkschr. Akad. Wiss. Wien, Math.-naturwiss. Kl., III. Teil **96** (1919)

[10] P. VAN DER STOK: Mededeel. Verh. kgl. Nederl. Meteor. Inst. **1910**, Nr. 102. Siehe auch Ann. Hydrogr. **1911** ,354.

[11] S. OGURA: Bull. Hydrogr. Dep. Imp. Jap. Navy **7** (1933).

[12] S. OGURA: Bull. Hydrogr. Dep. Imp. Jap. Navy **12** 47, 89 137 (1933); **13**, 91, 133, 179, 431, 471 (1934); **15**, 147, 181 (1936).

[13] S. F. GRACE: Mon. Not. Astronom. Soc., Geophys. Suppl. **1932** 70; **1933**, 156.

Golf von St. Lorenz. Das Gezeitenbild dieses Nebenmeeres und die darin vorkommenden Gezeitenströme hat vor allem W. BELL DAWSON[1] in mehreren Arbeiten klargelegt. Eine dynamische Erklärung hat H. SPRINGSTUBBE[2] versucht, doch es bleibt in ihr noch manches genauer zu erfassen übrig.

Betreffs der Gezeiten des *Nordpolarbeckens*, das in der Hauptsache aber als ein Teil des Nordatlantischen Ozeans aufzufassen ist, siehe vor allem die Arbeiten von A. DEFANT[3], R. STERNEK[4], sowie vor allem jene von H. U. SVERDRUP[5].

γ) Die Gezeiten der Adria. Das Adriatische Meer hat die Form eines weit ins Festland eindringenden Kanals von etwa 820 km Länge, der ins Jonische Meer in der Straße von Otranto mündet. Die mittlere Breite ist 166,5 km, die Mündung ist nur halb so breit und hat eine Querschnittsfläche von rund 37 km². Die Tiefenkarte läßt zwei verschiedenartige Becken erkennen; im nördlichen senkt sich der Meeresboden gleichmäßig nach Süden bis zu einer Querbarriere, die es vom Südbecken trennt. Das letztere hat die Form eines Kessels mit Maximaltiefen von etwas über 1200 m in der Mitte. Die harmonische Analyse der Pegelregistrierungen einer großen Zahl von Orten dankt man W. v. KESSLITZ[6], der mit Hinzunahme einiger italienischer Stationen (M. TENANI[7]) eine ausgezeichnete Beschreibung der Gezeitenerscheinungen diese Nebenmeeres gegeben hat. Die Verteilung der harmonischen Konstanten der Halbtagstiden längs der Küste deutet auf eine gut ausgebildete Amphidromie hin, die wohl eine Wirkung der Corioliskraft durch Ausbildung von Querschwingungen ist. Da die Phasen der Halbtagstiden in den einzelnen Teilen des Meeres recht verschieden sind, jene der Eintagstiden hingegen untereinander nur wenig differieren, muß innerhalb der Adria eine große Mannigfaltigkeit der Gezeitenform vorhanden sein.

Zur Erklärung der Gezeiten wurde die Adria zunächst als einseitig offener Kanal betrachtet, der mit der Gezeitenbewegung des Jonischen Meeres mitschwingt (A. DEFANT[8]). Die Theorie dieser *Mitschwingungsgezeiten*, sowohl der halbtägigen wie der ganztägigen Tiden, zeigte bereits, daß diese Annahme im wesentlichen das Richtige trifft, wobei das Gezeitenbild schließlich durch die Erddrehung noch stark modifiziert wird. Eine Zusammenfassung der Ergebnisse und Überprüfung durch erweiterte Beobachtungen hat dann R. STERNECK[9] vorgenommen, wobei er sich auch die Frage vorlegte, ob sich eine bessere Übereinstimmung erzielen liese, wenn auch der *direkte* Einfluß der Gezeitenkräfte in Rechnung gestellt wird. Zur Berechnung der Mitschwingungsgezeiten wurden 40 Querschnitte benützt und nach der üblichen Methode für *alle* Partialtiden die Amplitudenverteilung der Längsschwingung berechnet. Fig. 20 gibt als Beispiel für die M_2- und K_1-Tide die Längsverteilung längs der Adria. Die vollausgezogene Linie zeigt die beobachteten Werte, die gestrichelte den theoretischen Verlauf bei bloßem Mitschwingen, die strichpunktierte den theoretischen Verlauf bei Mitberücksichtigung der direkten Gezeitenkräfte. Diese Verteilung schmiegt sich etwas besser an die beobachtete an, aber man erkennt, daß die Mitschwingungsgezeit die Hauptrolle spielt. Die Theorie ergibt die Knotenlinie genau an der durch die Beobachtungen festgelegten Stelle. Die gute Übereinstimmung auch in bezug auf die Hubhöhen läßt erwarten, daß Reibungseinflüsse weit zurücktreten.

[1] W. BELL DAWSON: Depart. naval serv. Ottawa **1907**, **1913**; **1915**; **1920**. — Auch Bull Geogr. Soc. Philadelphia **1920**, 35.

[2] O. SPRINGSTUBBE: Arch. dtsch. Seewarte **53** (1934).

[3] A. DEFANT: Ann. Hydrogr. **1924**, 153, 177.

[4] R. STERNECK: Ann. Hydrogr. **1928**, 81.

[5] H. SVERDRUP: Geofys. Publ. **4**, 5 (1926).

[6] W. v. KESSLITZ: Denkschr. Akad. Wiss. Wien, Math.-naturwiss. Kl. (1919).

[7] M. TENANI: Rend. R. Accad. naz. Lincei, Ser. VI **10** (1929).

[8] A. DEFANT: Ann. Hydrogr. **1914**, 270, 556; **1915**, 147.

[9] R. STERNECK: Denkschr. Akad. Wiss. Wien, Math.-naturwiss. Kl. **96** (1919). — Ergänzungen hierzu Akad. Anzeiger, Wiener Akad. **1919**, Nr. 19.

Die ganztägigen Tiden zeigen keine Knotenlinie, sondern nur eine Hebung und Senkung des ganzen Wasserspiegels mit der äußeren Gezeit, wobei die Amplitude gegen das innere Ende des Kanals hauptsächlich wegen der Verengung des Querschnittes etwas zunimmt.

Auf die horizontalen Wasserverschiebungen der Längsschwingung wirkt die Corioliskraft und bedingt Querschwingungen. Sie lassen sich bei Kenntnis der Phase und Amplitude der theoretischen Längswelle genau berechnen, aber auch aus den Beobachtungen ableiten. Der Vergleich (Fig. 21) zeigt eine völlig befriedigende Übereinstimmung. Die Zusammenfassung der Längsschwingung und der Querschwingungen ergibt das Gezeitenbild der Adria, das in Fig. 22 wiedergegeben ist. Die Wiedergabe der Gezeitenkonstanten der Küstenorte durch die Theorie ist sowohl bei den Halbtags- wie Eintagstiden eine vollständige, so daß wohl kein Punkt unerklärt bleibt.

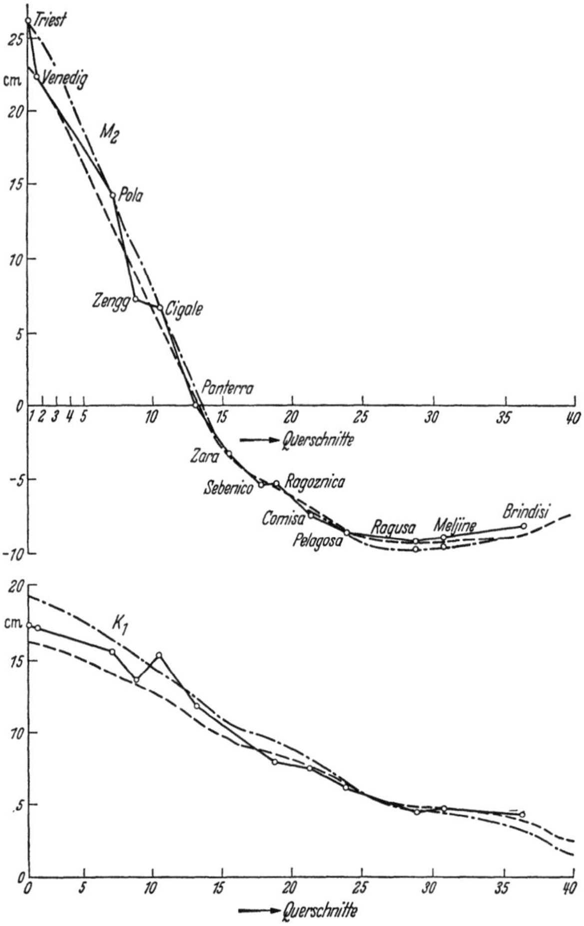

Fig. 20. Verteilung der Amplituden der M_2- und der K_1-Tide in der Längsschwingung der Adria. ——— beobachtete Werte, ------ theoretische Werte bei bloßem Mitschwingen mit dem Jonischen Meer, —·—·— bei Mitberücksichtigung der direkten, selbständigen Gezeit.

δ) Die Gezeiten der offenen Ozeane. Ein genaues Bild der Weltmeergezeiten läßt sich nur schwer mit einem größeren Grad von Wahrscheinlichkeit für seine Richtigkeit entwerfen, da als Beobachtungen nur die harmonischen Konstanten der Küstenorte und weniger landferner Inseln zur Verfügung stehen und das Zeichnen von Flutstunden- und Hubhöhenlinien über die weiten Wasserflächen deshalb stets subjektiv bleibt. Natürlich gewinnt ein solches Bild an Wahrscheinlichkeit, wenn man bei solchen Konstruktionen die Ergebnisse der hydrodynamischen Theorien soweit als möglich heranzuziehen versucht. Die ersten Karten, die solche Darstellungen geben (W. WHEWELL 1833 und 1836) haben derzeit nur mehr historisches Interesse. Auch die Karten von R. A. HARRIS[1] befriedigen heutzutage nicht mehr, da sie auf zu hypothetischen Grundlagen beruhen. R. STERNECK[2] hat wohl zuerst einigermaßen vertrauenswürdige Darstellungen

[1] R. A. HARRIS: Manual of Tides IV/B. U.S. Coast and Geod. Surv. Rep. App. **5**, 315 (1904).

[2] R. STERNECK: Sitzgsber. Wien. Akad., Math.-naturwiss. Kl., Teil IIa **1920**, 131; **1921**, 363. Siehe auch Ann. Hydrogr. **1922**, 145.

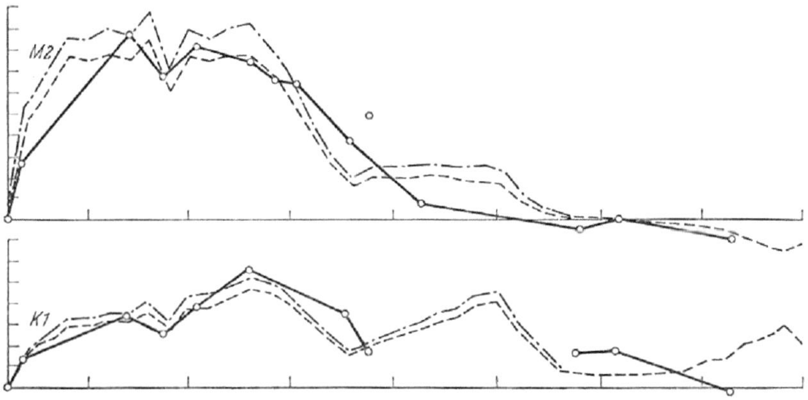

Fig. 21. Die durch die Corioliskraft erzwungenen Querschwingungen der M_2- und der K_1-Tide in der Adria. (Kurvenerklärung wie in Fig. 20.)

(Kurvenerklärung wie in Fig. 20.)

Fig. 22. Gezeitenbild der M_2-Tide in der Adria. (Die Zahlen an den Stationen sind die Phasenwerte nach STERNECK, die Querlinien sind die Querschnitte, die zur theoretischen Berechnung gewählt wurden.)

gegeben. Er zerlegt die beobachtete Gezeitenbewegung überall in zwei orthogonale Systeme (siehe S. 909) und zeichnet auf Grund dieser Zerlegung zwei Systeme synchroner Schwingungen für den ganzen Ozean, die zusammen dann das Gezeitenbild für die ganze Fläche desselben ergeben. Eine neuere, auf das ganze derzeit zur Verfügung stehende Material an Gezeitenkonstanten (siehe S. 901) fußende Darstellung der Flutstundenlinien aller drei Ozeane hat G. DIETRICH[1] gegeben, die wohl derzeit die beste sein dürfte. Trotzdem bleibt auch ihr noch sehr viel subjektives haften und die Gezeitenbilder stellen nicht mehr als ein mit den Randwerten zusammenpassendes, wahrscheinliches Schwingungssystem der betrachteten Partialtide dar.

Es wäre von besonderem Interesse, die nur auf Grund der Beobachtungen entworfenen Flutstundenlinien der Ozeane theoretischen Gezeitenkarten gegenüber zu stellen und mit diesen zu vergleichen. Aber es ist bisher nicht möglich gewesen, für die Ozeane Gezeitenbilder unter Berücksichtigung aller in Betracht kommenden Faktoren *nur* auf Grund theoretischer Erwägungen abzuleiten. Die komplizierte Konfiguration der Ozeane, das Mitschwingen der Wassermassen mit den Gezeiten der angrenzenden Meere, die Corioliskraft und die Reibung stellen zu viele Umstände dar, die von Einfluß sind und ihre gegenseitige Einwirkung erschwert die Lösung des Problems. Am ehesten für den Atlantischen Ozean läßt sie sich vielleicht etwas weiter führen. Das Gezeitenbild dieses Ozeans kann man einigermaßen als gesichert ansehen. A. DEFANT[2] hat schon 1922 versucht, dieses Bild durch Längs- und Querschwingungen verständlich zu machen, wobei er die für Nebenmeere ausgearbeiteten Methoden auf den Atlantischen Ozean (aufgefaßt als riesiger, einseitig geschlossener Kanal, der im antarktischen Wasserring mündet) übertrug. Es scheint, daß eine Mitschwingungsgezeit mit den Gezeiten im Wasserring im Süden und eine selbständige Gezeit zusammen mit den durch die Erddrehung bedingten Querschwingungen die Haupterscheinungen der atlantischen Tiden zu erklären vermögen. Eine Erweiterung dieser Theorie hat er später[3] im „Meteor-Werk" gegeben. In ihr wird (entsprechend den neueren Erfahrungen über die Ausbildung des Gezeitenbildes in Nebenmeeren) angenommen, daß ein Teil der von Süden in den Atlantischen Ozean in Form einer fortschreitenden Welle eindringenden Gezeitenenergie im Norden auf den ausgedehnten Schelfen des Arktischen Meeres und durch die Eisbedeckung desselben vernichtet wird und nicht mehr in der reflektierten Welle in den

Tabelle 6. *Halbtägige Gezeit auf ozeanischen Inseln im Atlantischen Ozean.*

Station	Geographische Koordinaten		Halbtägige Gezeit		Welle	
	φ°	λ°	Ampl. ζ cm	Phase Mondst. Gr. Z.	1^h bis ζ_{1h} cm	4^h bis ζ_{4h} cm
Westmannaeyar, Heimaey. .	63,4° N	20,3° W	137	$6{,}5^h$	−132	+35
Westl. Azoren (Mittelwert) .	38,7° N	28,5°	60	$2{,}1^h$	+ 50	+ 33
Bermuda-Inseln	32,3° N	64,8°	55	$11{,}2^h$	+ 32	− 44
2 westl. Kap Verde-Inseln . .	17,0° N	25,2°	63	$8{,}0^h_2$	− 54	− 32
Fernando Noronha und Rocas	3,8° S	33,1°	105	$6{,}6^h_2$	− 103	+ 2,1
Ascension	7,9° S	14,4°	45	$6{,}3^h_3$	− 42	+ 16
St. Helena	15,9° S	5,7°	45	$3{,}4^h_8$	+ 12	+ 43
Trinidad und Martin Vaz . .	20,5° S	29,1°	42	$5{,}5^h$	− 30	+ 30
Tristan da Cunha.	37,0° S	12,3°	75	$0{,}8^h_2$	+ 75	− 7

[1] G. DIETRICH: Veröff. Inst. Meereskunde, N.F. A **1944**, Nr. 47. — Siehe auch Z. Ges. Erdkunde **1944**, 69.

[2] A. DEFANT: Ann. Hydrogr. **1924**, 153, 177; **1926**, 133.

[3] A. DEFANT: Wissenschaftliche Ergebnisse der Deutschen Atlantik Expedition „Meteor", Bd. VII, 1. Teil, S. 256. Berlin 1932.

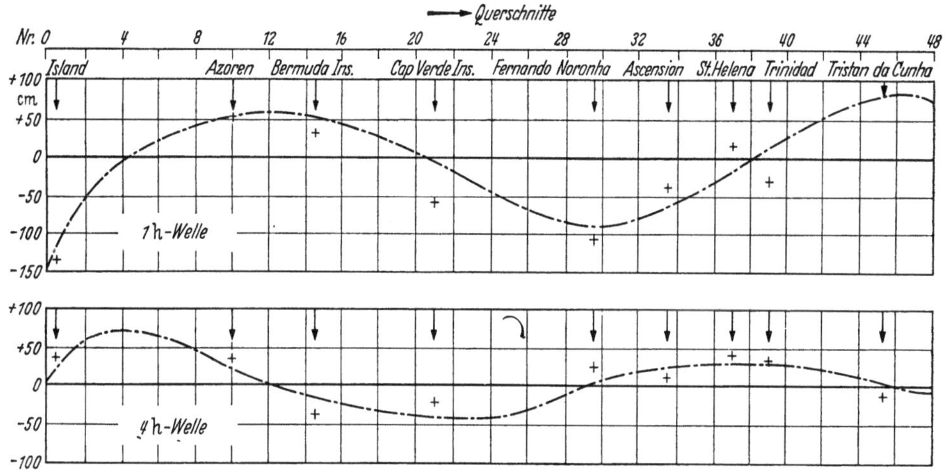

Fig. 23. Verteilung der Amplituden der 1^h- und der 4^h-Welle auf der Mittelachse des Atlantischen Ozeans und die Gezeitenbeobachtungen auf den ozeanischen Inseln.

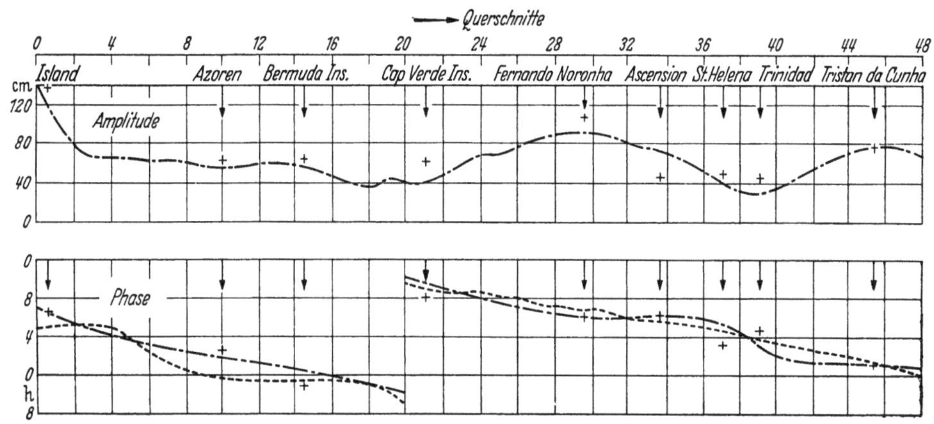

Fig. 24. Theoretische Halbtagsgezeit auf der Mittelachse des Atlantischen Ozeans (+ beobachtete Werte auf den ozeanischen Inseln; bei der Phase: ------ Verlauf nach der STERNECKschen Darstellung der Flutstundenlinien).

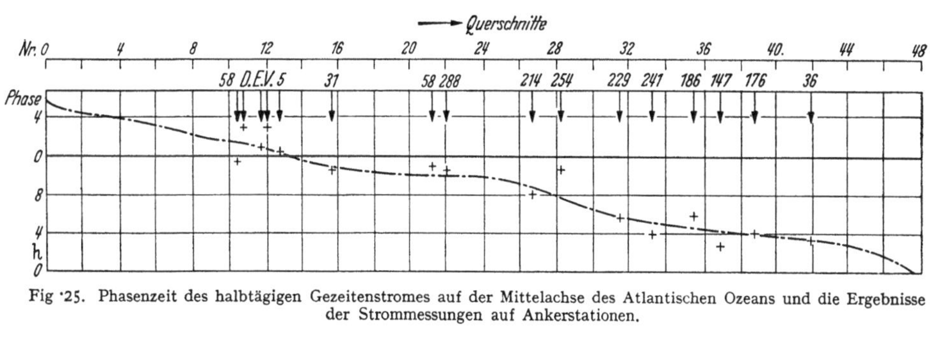

Fig. 25. Phasenzeit des halbtägigen Gezeitenstromes auf der Mittelachse des Atlantischen Ozeans und die Ergebnisse der Strommessungen auf Ankerstationen.

Atlantischen Ozean zurückkehrt (siehe S. 909). Die Gezeiten müssen dann im Atlantik als Überlagerung eines gegeneinander um $\frac{1}{4}$ der Periode verschobenen Wellengefüges erscheinen. Die reinen Mitschwingungsgezeiten haben die Phase 1^h G.Z. im Süden, die dazu orthogonale Schwingung die Phase 4^h und beide lassen sich numerisch unter Berücksichtigung der Gezeitenkraft *für die Mittelachse des Ozeans* berechnen. Dabei müssen natürlich die Randbedingungen berücksichtigt

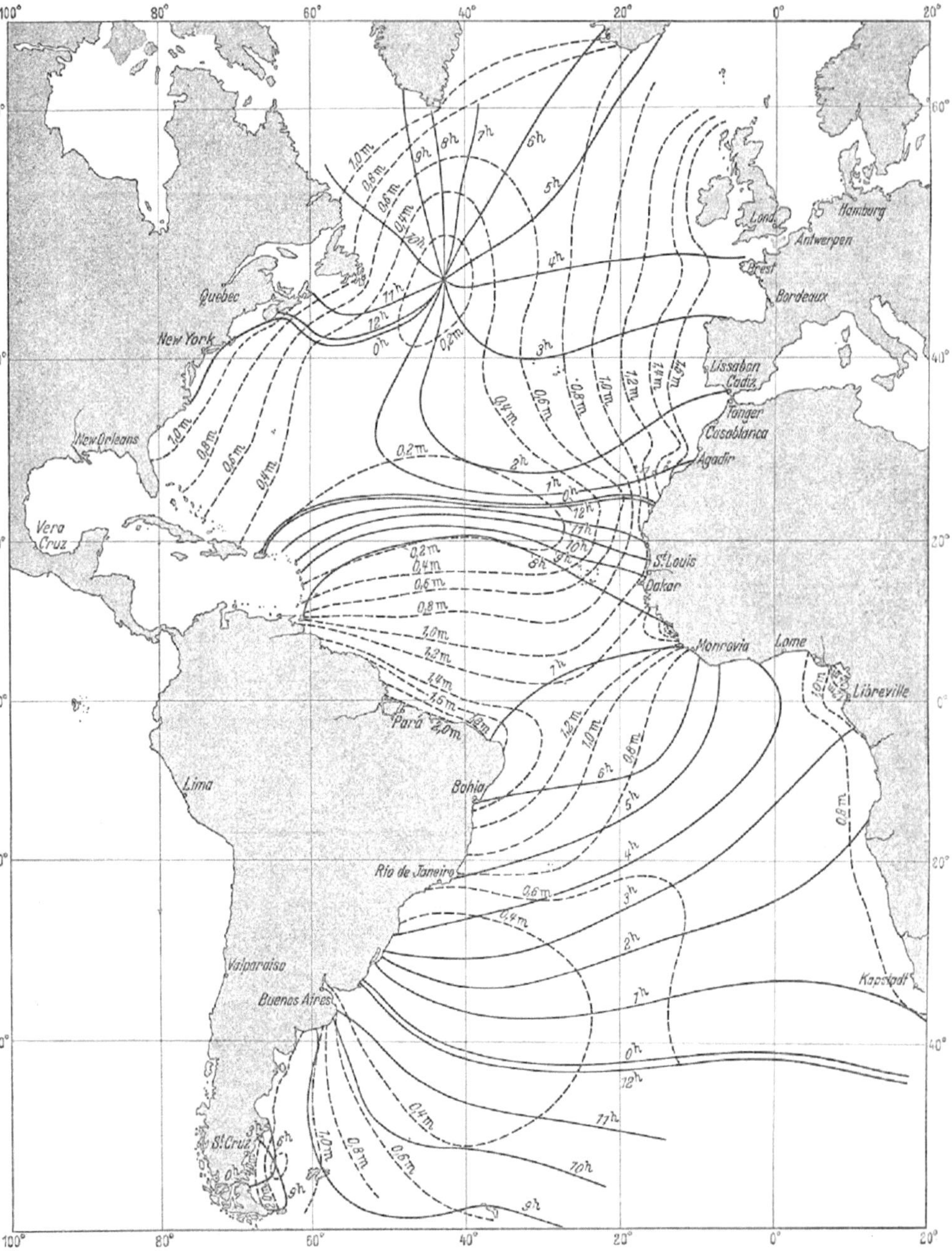

Fig. 26. Theoretische Gezeiten im Atlantischen Ozean, —— Linien gleichen Hochwasserunterschiedes gegen den Durchgang des Mondes durch den Greenwicher Meridian; ------ Linien gleichen Tidenhubes der halbtägigen Hauptmondtide M_2 in Metern (nach W. HANSEN).

werden. Die Übereinstimmung der berechneten Wellenpaare in Amplitude und Phase mit den Beobachtungen auf Inseln nahe der Mittelachse des Atlantik (Tabelle 6) ist gut, wie es die Fig. 23 zeigt. Der Atlantische Ozean verhält sich wenigstens auf seiner Mittelachse ähnlich einem schmalen Kanal (siehe Fig. 24). Auch die Phasen der Gezeitenströme fügen sich gut in diese Theorie (Fig. 25) ein. Weniger gut steht es mit den Amplituden derselben, da sie in Wirklichkeit stärker sind, als die Theorie sie ergibt. Dieser Widerspruch klärt sich aber auf, wenn man bedenkt, daß die Erdrotation nach der Theorie der Flutwellen eine solche Verstärkung erzwingt.

J. PROUDMAN[1] hat einen anderen Weg gewählt, um die Atlantischen Gezeiten theoretisch zu erfassen. Die Schwingungen, die die Gezeiten zusammensetzen, müssen jedenfalls den Flutwellenarten entsprechen, die in Kapitel I näher besprochen wurden. PROUDMAN berücksichtigt 1. die selbständige Gezeit, die bei Kenntnis der Kräfte berechnet werden kann; 2. zwei KELVINsche Wellen, die je von Norden und Süden in den Ozean eindringen und 3. die zwei einfachsten POINCARÉschen Wellen, die ebenfalls je von Norden und Süden eindringen. Für diese vier Wellen lassen sich mit Berücksichtigung der morphologischen Konfiguration des Ozeans die *relativen* Hubhöhen und Phasen, sowie die entsprechenden Gezeitenströme ermitteln. Die Superposition der letzteren vier Wellen muß mit den Beobachtungen von vier frei gewählten Küstenorten übereinstimmen, was zur Erfüllung der Randbedingungen und zur Überführung der relativen Werte in absolute genügt. Die Übereinstimmung des berechneten Gezeitenbildes mit dem aus den Beobachtungen abgeleiteten ist nur teilweise befriedigend; es ist möglich, daß die Vermehrung der POINCARÈschen Wellen ein etwas besseres Ergebnis erzielen wird lassen.

W. HANSEN[2] hat seine auf S. 891 dargelegte Methode (Randwertverfahren) zur Bestimmung der atlantischen Gezeiten benützt. Dies erfordert eine sehr große Rechenarbeit, da die Gezeitenwerte für die Punkte eines engen Gitternetzes, das den ganzen Ozean überdeckt, berechnet werden müssen. Das Ergebnis der Berechnungen steht in Fig. 26 und deckt sich gut mit dem beobachteten Gezeitenbild. Es ist zu bemerken, daß hier im theoretischen Bild auch Hubhöhenlinien für die ganze Fläche des Ozeans mitgeteilt werden.

IV. Die Gezeiten des Meeres in Beziehung zu jenen der Erdfeste.

15. Theoretische Grundlagen. Eine der wichtigsten, aber meistens stillschweigend gemachten Annahmen, die sowohl bei der Diskussion der Gezeitenbeobachtungen, wie auch bei den theoretischen Überlegungen gemacht wird, ist, daß die Erde vollkommen starr ist. Nach den Ergebnissen der geophysikalischen Forschung ist diese Annahme nicht mehr haltbar und die Gezeiten der Erdkruste vermögen in gewissen Fällen in merklicher Weise auf die ozeanischen Gezeiten zurückzuwirken und sie in ihrer Ausbildung zu beeinflussen. Wäre der Meeresboden völlig starr und unbewegt, dann soll bei Hochwasser an einem Punkt, an dem ein Pegel P steht, der Wasserspiegel ζ über Mittelwasser stehen. Dabei ist angenommen, daß die Gezeiten der Gleichgewichtstheorie folgen. Der Wasserspiegel wird zu dieser Zeit horizontal stehen und entsprechend der Richtung der fluterzeugenden Kraft wird auf beiden Seiten des Punktes ein Niveaugefälle $\bar{\alpha}$ nach beiden Seiten vorhanden sein. Da Gleichgewicht zwischen der horizontalen Komponente der Flutkraft und der Druckkraft bestehen muß, wird diese Druckkraft $g \operatorname{tg} \bar{\alpha}$ sein, oder $g\bar{\alpha}$, da $\bar{\alpha}$ sehr klein ist.

Diese einfache Vorstellung ändert sich: 1. wenn die Änderung der Fluthöhe durch die Anziehung des von den fluterzeugenden Kräften hervorgerufenen Flutberges berücksichtigt wird und 2. wenn ein den fluterzeugenden Kräften nachgiebiger Erdkörper angenommen wird. Durch das Hinzutreten des Potentials des Flutberges (siehe S. 886) tritt eine Erhöhung des Wasserstandes ein, und

[1] J. PROUDMAN: Mon. Not. Roy. Astronom. Soc. **105**, 244 (1944).
[2] W. HANSEN: Dtsch. hydrogr. Z. **2**, 44 (1949).

am Pegel wird eine Hubhöhenvermehrung um $\bar{\zeta}_0$ beobachtet. Die Größenordnung dieser Zunahme beträgt für ozeanische Verhältnisse etwa $\frac{1}{8}$ der ursprünglichen Hubhöhe.

Durch die Wirkung der fluterzeugenden Kräfte deformiere sich weiter der Erdkörper und am Punkte, an dem P steht, erhebe sich nun der Boden um ζ_0. Um denselben Betrag wird auch der Pegel in P gehoben. Nicht aber hebt sich der Wasserspiegel; denn seine Erhebung über den Erdmittelpunkt wird nur vom Gleichgewicht zwischen der Flutkraft und den Druckkräften geregelt. Höchstens könnte der Flutberg des Erdkörpers insofern eine kleine Änderung bedingen, als im Wasserflutberg der untere Teil nun statt von Wasser von fester Erde eingenommen wird und dadurch das Potential der Anziehung des Flutberges etwas geändert wird.

Man findet auf diese Weise die Erhebung des Hochwassers zu

$$\zeta = \bar{\zeta} + \bar{\zeta}_0 - \zeta_0 \tag{15.1}$$

Ebenso wird, wenn $\bar{\alpha}_0$ und α_0 die entsprechenden Gefälle des Wasserspiegels sind, die horizontale Beschleunigung zu

$$g\alpha = g(\bar{\alpha} + \bar{\alpha}_0 - \alpha_0). \tag{15.2}$$

Bei den Untersuchungen der Gezeiten der Erdfeste pflegt man $\bar{\zeta}_0$ und ζ_0 proportional $\bar{\zeta}$ zu setzen, wie es ja bei den Gleichgewichtstiden der Fall sein muß. Dann ist $\zeta_0 = h\bar{\zeta}$ und $\bar{\zeta}_0 = k\bar{\zeta}$ und es wird

$$\zeta = (1 - h + k)\bar{\zeta} = \gamma\bar{\zeta}. \tag{15.3}$$

$\gamma = 1 - h + k$ bedeutet dann das Verhältnis der beobachteten, von den fluterzeugenden Kräften erzwungenen Hubhöhe zu jener, die bei vollkommen starrer Erde zur Beobachtung gelangen würde.

16. Bestimmung der Erdgezeiten aus den Meeresgezeiten langer Periode. Eine Ermittlung der Größe γ für einen Ort aus dem Vergleich der *beobachteten* (aus der harmonischen Analyse abgeleiteten) Amplitudenwerte einer Partialtide und der *theoretisch* aus der Gleichgewichtstheorie unter der Annahme einer absolut starren Erde abgeleiteten läßt sich bei halb- und eintägigen Gezeitenwellen nicht durchführen, da bei diesen nicht angenähert eine der Gleichgewichtstheorie entsprechende Ausbildung der Gezeiten vorliegt. Bei den *Gezeiten mit langer Periode* (in Betracht kommen vor allem die 14tägige und die monatlichen Partialtide) ist hingegen mit einiger Wahrscheinlichkeit anzunehmen, daß sie ihre vollen Gleichgewichtswerte aufweisen würden, wenn die Erde absolut starr wäre. Die Berechnung der letzteren ist eine leichte Aufgabe, die Schwierigkeiten liegen vielmehr an der Gewinnung eines zuverlässigen Beobachtungsmaterials für die halb- und ganzmonatlichen Partialtiden. Die ersten Vergleiche dieser Art, die den Nachweis der Erdgezeiten wirklich erbrachten, rühren von W. THOMSON (Lord KELVIN[1]) und G. H. DARWIN[2] her. Letzterer benützte die harmonischen Konstanten der von 33 Beobachtungsstationen abgeleiteten Werte der halbmonatlichen Mondflut und fand $\gamma = 0{,}675$, d.h. die beobachtete Hubhöhe ist ungefähr

---

[1] W. THOMSON u. P. G. TAIT: Treatise on natural philosophy, Bd. 1, Teil 2. Oxford 1874. Oder Deutsche Ausgabe von Helmholtz-Wertheim, Bd. 1, Teil 2, S. 419. Braunschweig 1874.

[2] G. H. DARWIN: Phil. Trans. Roy. Soc. Lond., Part I **1879**. Siehe auch Ebbe und Flut, 2. Aufl., S. 241ff. Leipzig u. Berlin 1911.

ein Drittel kleiner als der nach der statischen Theorie berechnete Wert. W. SCHWEYDAR[1] hat nach derselben Methode ein umfangreicheres Material von 194 Beobachtungsjahren verarbeitet und erhielt im wesentlichen dasselbe Ergebnis: Aus der halbmonatlichen Mondflut $\gamma = 0{,}6614$, aus der monatlichen $\gamma = 0{,}6422$. Die Fluten hielten dabei die theoretischen Zeiten des Hochwassers ein, so daß die Annahme von Gleichgewichtsfluten gerechtfertigt erscheint.

Später ist es gelungen, durch Horizontalpendel einen direkten Nachweis der Gezeiten der Erdfeste zu erbringen. In den abgeleiteten Werten für γ zeigten sich aber große Verschiedenheiten, die wahrscheinlich von sekundären Erscheinungen herrühren und hauptsächlich durch die von Meeresgezeiten auf die Festländer ausgeübte Druckbelastung und durch die Anziehung der bewegten Wassermassen hervorgerufen werden. Nach SCHWEYDAR[2] scheint 0,841 der plausibelste Wert für γ zu sein, während A. PREY[3] 0,74 als den wahrscheinlichsten hält. H. JEFFREYS[4] gibt als zuverlässigsten Wert 0,66 an. Man sieht daraus, daß über den wahren Wert von γ noch keine genügende Übereinstimmung besteht.

17. Bestimmung der Erdgezeiten aus den Meeresgezeiten halbtägiger Periode. Einen Fortschritt in der Benützung der Meeresgezeiten zur Ermittlung der Gezeiten der festen Erde erbrachte der Nachweis von J. PROUDMAN[5], daß die hydrodynamische Theorie der Meeresgezeiten, wenigstens die für kanalartige schmale Gewässer weit genug entwickelt ist, um auch die Erdgezeiten in Rechnung zu stellen. Die Beschränkung auf schmale Gewässer, deren Bodenform genau bekannt sein muß, bedeutet, daß es erlaubt sein soll, von Querströmen abzusehen. Hingegen lassen sich auch die kurzperiodischen Partialtiden heranziehen, was wegen der größeren Genauigkeit ihrer harmonischen Konstanten einen großen Vorteil bedeutet. PROUDMAN hat für die Berechnung der Größe γ zwei Methoden entwickelt, die hier kurz erörtert werden sollen.

Wird x in der Richtung der Längsachse des schmalen, kanalartigen Meeres genommen und bedeutet u der Mittelwert des Gezeitenstromes auf einen Querschnitt senkrecht zu x, während keine Querströme vorhanden sind, so wird u nur eine Funktion von x und t sein. Die Bewegungsgleichungen haben dann die Form:

$$\frac{\partial u}{\partial t} = -g \frac{\partial}{\partial x} (\zeta + \zeta_0 - \bar{\zeta} - \bar{\zeta}_0) \quad \text{und} \quad f u = -g \frac{\partial}{\partial y} (\zeta + \zeta_0 - \bar{\zeta} - \bar{\zeta}_0), \tag{17.1}$$

während die Kontinuitätsgleichung die Form:

$$\frac{\partial}{\partial x} (S u) + b \frac{\partial \zeta}{\partial t} = 0 \tag{17.2}$$

annimmt, wo S die Querschnittsfläche und b die Breite des Querschnitts an der Stelle x bedeuten. Die zweite der Gln. (17.1) dient dazu, nach der Ermittlung von u die längs der Küsten beobachteten Hubhöhen auf die Kanalmitte zu reduzieren.

Bei der *ersten* Methode wird der Unterschied zwischen der ohne Annahme einer Erdgezeit berechneten Gezeit und der beobachteten festgestellt. Man

---

[1] W. SCHWEYDAR: Gerlands Beitr. Geophys. **9** (1907). Siehe auch Veröff. kgl. preuß. Geod. Inst., N.F. **1912**, Nr. 54, 57.

[2] W. SCHWEYDAR: Zentralbureau der Int. Erdmessung, N.F. Nr. 38, Berlin 1921.

[3] A. PREY: Gerlands Beitr. Geophys. **23** (1929).

[4] H. JEFFREYS: The Earth, 2. Aufl. Cambridge: University Press 1924.

[5] J. PROUDMAN: Bull. Nr. 11, Sect. d'Océanographie du Conseil de Recherches, Venedig 1928.

setzt, wie üblich,

und
$$\left.\begin{aligned} \zeta &= H\cos(\sigma t - \varkappa) = \zeta_1\cos\sigma t + \zeta_2\sin\sigma t \\ u &= U(\cos\sigma t - \varepsilon) = u_1\cos\sigma t + u_2\sin\sigma t \end{aligned}\right\} \tag{17.3}$$

und erhält so aus (17.2)

$$\Delta\, S_1 u_1 = -\,s\,\sigma H\sin\varkappa_1 b\,\Delta x \quad \text{und} \quad \Delta\, S_1 u_2 = +\,\sigma H\cos\varkappa_1 b\,\Delta x. \tag{17.4}$$

Bei Kenntnis des Tidenhubes längs des ganzen Kanals läßt sich aus diesen Beziehungen u_1 und u_2, d.h. die Stromgeschwindigkeit für jeden Querschnitt berechnen. Mit diesen Werten ist nach der ersten Gl. (17.1) auch die Änderung dieser Stromgeschwindigkeit während einer Zeiteinheit für alle Kanalabschnitte gegeben, so daß in der ersten Gl. (17.1), da $\frac{\partial}{\partial x}(\zeta - \bar{\zeta})$, d.h. $\frac{\partial}{\partial x}(\zeta_1 - \bar{\zeta}_1)$ und $\frac{\partial}{\partial x}(\zeta_2 - \bar{\zeta}_2)$ aus den Beobachtungen und aus der Theorie der fluterzeugenden Kräfte bekannt sind, nur noch $\frac{\partial}{\partial x}(\zeta_0 - \bar{\zeta}_0)_1$ und $\frac{\partial}{\partial x}(\zeta_0 - \zeta_0)_2$ unbekannt sind. Sie können auf diese Weise berechnet werden. Die Methode liefert also durch numerische Integration die Hubhöhen $\zeta_0 - \bar{\zeta}_0$ bis auf eine unbekannte Konstante, sie gibt demnach nur die *relative* Erdtide, d.i. den Unterschied der Hebung des Meeresbodens ζ_0 gegen jene, bedingt durch die Anziehung des Wasserflutberges.

Die *zweite* Methode ist weniger von der Genauigkeit der Beobachtungen abhängig. Ersetzt man u und $\bar{\zeta}$ durch $u e^{i\sigma t}$ und $\zeta e^{i\sigma t}$, so wird aus den Beziehungen (17.1) und (17.2), wenn man $\zeta = \gamma\zeta_1 + l\zeta_2$ und $u = \gamma u_1 + l u_2$ setzt

$$\left.\begin{aligned} \gamma\left[i\sigma u_1 + g\frac{\partial\zeta_1}{\partial x} - F\right] + l\left(i\sigma u_2 + g\frac{\partial\zeta_2}{\partial x}\right) &= 0, \\ \gamma\left[\frac{\partial S u_1}{\partial x} + i\sigma b\zeta_1\right] + l\left(\frac{\partial S u_2}{\partial x} + i\sigma b\zeta_2\right) &= 0. \end{aligned}\right\} \tag{17.5}$$

Die beiden Ausdrücke in den eckigen Klammern, für sich gleich Null gesetzt, geben mit den Randbedingungen (für $x=0$, $\zeta_1=0$ und $Su_1=0$) eine durch die Kraft $g\bar{\zeta}$ erzwungene Schwingung, hingegen die beiden Ausdrücke in den runden Klammern mit der Randbedingung: für $x=0$, $\zeta_2 = 1$ cm und $Su_2 = 0$ eine freie Schwingung mit einstweilen willkürlichen Randbedingungen. Eine allgemeine Lösung dieser Gleichungen ist:

$$\left.\begin{aligned} \zeta &= \{(1-h+k)\,\zeta_1 + l\,\zeta_2\}\,e^{i\sigma t}, \\ u &= \{(1-h+k)\,u_1 + l\,u_2\}\,e^{i\sigma t}. \end{aligned}\right\} \tag{17.6}$$

Ein Vergleich mit den Beobachtungen an mindestens einer Station (Querschnitt) gestattet sowohl die Unbekannte $\gamma = 1 - h + k$, als auch l zu ermitteln, wobei noch die Möglichkeit besteht, Phasenunterschiede festzustellen (insgesamt vier Unbekannte).

Eine Anwendung dieser Methoden für den Baikalsee und das Rote Meer hat S. F. GRACE[1,2] gegeben. Er benützte beim Baikalsee die zweite Methode und wiederholte (zunächst die Erdgezeiten ignorierend) die Rechnungen von R. STERNECK[3]; er gelangt zu denselben Ergebnissen wie dieser (an den Seeenden findet er die Amplituden 15,7 und 12,3 mm). Bei Berücksichtigung der

[1] S. F. GRACE: Mon. Not. Roy. Astronom. Soc., Geophys. Suppl. **2**, 273–296 (1930).
[2] S. F. GRACE: Mon. Not. Roy. Astronom. Soc., Geophys. Suppl. **2**, 301 (1931).
[3] R. STERNECK: Ann. Hydrogr. **1928**, 221–225.

Erdgezeiten erhält er bei Zugrundelegung der Beobachtungen in der Petschannaja-Bucht für die M_2-Tide $\gamma = 1 - h + k = 0{,}54$ und für die K_1-Tide $\gamma = 0{,}73$; dabei werden gleichzeitig die Amplitudenwerte auf die Größe der beobachteten herabgesetzt. Das Mittel der beiden γ-Werte stimmt zufällig mit dem JEFFREYSschen Wert überein. Es kann als gesichert angesehen werden, daß die Erdgezeiten den Widerspruch, der sich beim Vergleich zwischen der Gezeitentheorie, die wegen der Kleinheit, Schmalheit und großen Tiefe des Baikalsees der einfachen Gleichgewichtstheorie folgen müßte und den Beobachtungen ergeben hatte, vollständig beseitigen; es ist also nicht notwendig, hierfür, wie R. STERNECK es getan hatte, Reibungseinflüsse in besonderer Weise verantwortlich zu machen.

Nachdem schon PROUDMAN das Rote Meer als besonders geeignet für die Ermittlung der Erdgezeiten vorgeschlagen hatte, hat GRACE eine derartige Untersuchung für die M_2-Tide auch für dieses Nebenmeer sowohl nach der ersten, wie auch nach der zweiten Methode vorgenommen. Bei der zweiten Methode bleiben nach Fortlassung von wenig gesicherten harmonischen Konstanten einiger Orte noch genügend Werte übrig, um die Bestimmung der vier unbekannten Größen zu ermöglichen Diese Bestimmung ergab das Resultat, daß in auffallender Weise $\gamma = 1 - h + k = -0{,}42$ ist. Dieser Wert fällt vollkommen aus der Reihe der sonstigen Werte heraus und für ihn läßt sich auch keine Erklärung geben. Sicher ist nur, daß durch ihn die Beobachtungen besser wiedergegeben werden können, als ohne Berücksichtigung der Erdgezeiten.

Eine Anwendung der ersten Methode läßt vielleicht die Gründe für dieses abweichende Verhalten erkennen. Falls die Annahme, daß die Erdgezeit der Gleichgewichtstide proportional anzunehmen ist — das ist die Annahme in der zweiten Methode — zurecht besteht, müßte die *relative* Erdtide (der Überschuß der Erdtide über die durch die gegenseitige Anziehung bedingte Anschwellung der Wassertide) dieselbe Gestalt haben, wie die Gleichgewichtstide selbst (keine Phasenverschiebung). In Wirklichkeit ist aber dies, wie die erste Methode zeigen kann, nicht der Fall; die entsprechenden Kurven sind recht unruhig, was wohl auf Beobachtungsungenauigkeiten zurückzuführen sein dürfte (H. THORADE[1]). Vergleicht man ganz roh den Unterschied beider Kurven, dann kommt man nach GRACE zu einem Verhältnis der beiden Gezeitenanteile 1:4, woraus $h - k = \frac{1}{4}$ oder $\gamma = 1 - h + k = 0{,}75$ folgt. Die Übereinstimmung mit den früheren Werten ist nun ganz gut, aber diese Schätzung ist recht ungenau und mit Sicherheit läßt sich wohl nur sagen, daß die Gezeiten des Roten Meeres sich ohne Berücksichtigung der Erdtiden sicher weniger genau überblicken lassen als mit ihrer Berücksichtigung, und daß dies vor allem deshalb der Fall ist, weil die von den Erdgezeiten hervorgerufenen Beschleunigungen ungefähr von derselben Größenordnung sind, wie die der fluterzeugenden Kräfte Man ist aber im Roten Meer auch nicht ganz sicher, ob nicht die Reibungseinflüsse, die vom Golf von Suez stammen, die Verhältnisse doch etwas ändern, so daß ohne nähere Berücksichtigung dieser eine genaue Berechnung nach diesen Methoden nicht mehr angängig ist.

V. Die Gezeitenreibung in Beziehung zu kosmischen Problemen.

18. Die Gezeitenreibung und die Kanaltheorie der Gezeiten. Für eine Anzahl interessanter kosmischer Fragen, die für den Ablauf mancher Erscheinungen im Planetensystem und bei der Planetenentwicklung von größter Wichtigkeit sind, spielt die Gezeitenreibung eine ausschlaggebende Rolle. Diese Rolle war wohl zur Zeit, als die Massen der Erde und des Mondes — um bei diesen beiden

[1] H. THORADE: Ann. Hydrogr. **1933**, 49—56.

Gestirnen zu bleiben — noch flüssig waren, eine bedeutende und beim Monde war, als er noch näher an der Erde stand, die durch die Anziehungskräfte der Erde hervorgerufene Gezeit sehr groß. Jeder Flutberg übt eine reibende Wirkung auf den rotierenden Himmelskörper aus, und beide Flutberge suchen die Rotation des Himmelskörpers zu verkleinern, wirken auf ihn wie Hemmschuhe bei einem sich drehenden Rad. Es verhält sich so, daß die Reibung durch die Rotation der Erde einerseits die Flutberge beständig mitzuschleppen trachtet, so daß die Verbindungslinie Erde—fluterzeugender Himmelskörper vorauseilen wird, andererseits aber die beiden Flutberge als Gegenwirkung hemmend auf die Rotation der Erde zurückwirken. Wie sich dies genauer verhält, läßt sich am bequemsten überblicken, wenn man die von einer fluterzeugenden Kraft erzwungenen Gezeiten in einem äquatorialen Kanal bei Vorhandensein eines Reibungseinflusses berechnet. Der Fall ohne Reibung wurde in Abschnitt 10 näher behandelt. Zur dortigen Bewegungsgleichung (10.1) tritt bei Reibung auf der rechten Seite noch das Reibungsglied $-\beta \frac{\partial \xi}{\partial t}$ hinzu, in dem die Reibungswirkung proportional der Stromgeschwindigkeit gesetzt wird (s. Ziff 13). Für die horizontale Störungskraft $X = -m \sin 2(nt + \lambda + \varepsilon)$ lautet die Lösung für die Oberflächenerhebung im Kanal (s. H. LAMB)[1]:

$$\zeta = \frac{H}{2} \frac{1}{\left[(1-p^2)^2 + \frac{\beta^2 n^2 R^4}{4c^4}\right]^{\frac{1}{2}}} \cos 2(nt + \lambda + \varepsilon - \chi), \tag{18.1}$$

worin χ durch

$$\operatorname{tg} 2\chi = \frac{1}{2} \frac{\beta n R^2}{c^2} \frac{1}{1-p^2}; \quad 0 < \chi < \frac{\pi}{2} \quad \text{und} \quad p = \frac{nR}{c} \tag{18.2}$$

gegeben sind. Wenn $\beta = 0$, geht die Lösung in die Gl. (10.2) des Falles ohne Reibung über. Da $nt + \lambda + \varepsilon$ der Stundenwinkel des Mondes für den Meridian eines beliebigen Punktes λ ist, ergibt (18.1), daß das Hochwasser dem Meridiandurchgang des Mondes in einem Intervall t_1 folgt, das durch $nt_1 = \chi$ gegeben ist. Wenn $p > 1$ ist, wird bei sehr kleiner Reibung nach (18.2) $\chi = 90°$ und die Gezeiten sind *invers*. Wird der Reibungseinfluß merklich, dann liegt der Wert von χ zwischen 90° und 45° und Hochwasser wird um das Zeitintervall $90° - \chi$ *beschleunigt*. Bei $p < 1$ wären die Gezeiten *direkt* und χ liegt zwischen 0° und 45°; das Hochwasser wird um die diesen Winkel entsprechende Zeit *verzögert*. Diese beiden Fälle sind in der Fig. 27 veranschaulicht. In der Richtung M liegt der fluterzeugende Körper, der sich in der Ebene des Äquators bewegt.

Man sieht, wie in beiden Fällen die Anziehung des fluterzeugenden Körpers auf die beiden Flutberge einem Kräftepaar gleichwertig ist, das den Drehimpuls des rotierenden Körpers mit der Zeit zu verkleinern sucht. Die auf diese Weise auftretende verzögernde Wirkung des Mondes auf die Rotation der Erde ist natürlich auch jetzt vorhanden, wenn sie auch in der Lebensgeschichte der Erde und des Mondes früher viel stärker gewesen sein muß als jetzt. Der Tag wird immer länger. Um welchen Betrag es sich dabei handelt, ist noch eine offene Frage; 1 sec in 100000 Jahren wird als Maximum angesehen. In derselben Weise könnte die Gezeitenreibung auch zur Erklärung der säkularen Beschleunigung der Länge des Mondes, die etwa 5″ bis 12″ im Jahrhundert beträgt, herangezogen werden. Denn diese wäre dann eine direkte Folge der Verlängerung der Umdrehungszeit der Erde. Nimmt man für die säkulare Beschleunigung des Mondes den Wert 9″ im Jahrhundert an, so erfordert dies nach JEFFREYS[2] einen

[1] H. LAMB: Lehrbuch der Hydrodynamik, Deutsche Ausgabe, 2. Aufl. § 319, S. 637. 1931.

[2] H. JEFFREYS: The Earth, Kap. 14/3, S. 216. Cambridge: University Press 1924. — Phil. Trans. Roy. Soc. Lond. **201** (1920).

Energieverbrauch von $1{,}39 \times 10^{19}$ erg/sec, der von der Flutreibung geliefert werden muß. Nach ADAMS und DELAUNAY ist der Wert von 12″ an eine Verlängerung der Rotationsdauer der Erde um rund 1 sec gebunden.

Man hat früher die Tatsache, daß das Hochwasser dem Meridiandurchgang des Mondes um mehrere Stunden vorausgehen oder auch folgen kann (Hafenzeit oder Mondflutintervall) dem Einfluß der Gezeitenreibung zugeschrieben (AIRY[1]). Die neueren Theorien haben gezeigt, daß diese Erscheinung ein Ausfluß dynamischer Prinzipien darstellt und auch bei Abwesenheit jeglicher Reibung auftreten muß. Die Phasendifferenzen, die durch die Formel (18.1) erklärt werden können, bleiben auch im Falle eines äquatorialen Kanals von etwa 3510 m Tiefe (etwa die mittlere Tiefe der Meere) sehr klein. Aus (18.2) erhält man mit den Zahlenwerten auf S. 894 (Ziff. 10α)

$$\operatorname{tg} 2\chi = -\frac{1}{1 - 311^{\mathrm{h}}/R}\,\frac{1}{n\tau} = -0{,}191\,\frac{2\pi}{n\tau},$$

worin $\tau = 2/\mu$ der Dämpfungsfaktor der freien Schwingungen ist. Dieser ist im vorliegenden Falle ein beträchtliches Vielfaches von $2\pi/n$ (Mondtag). Die durch die Reibung bedingte Änderung der Hochwasserzeit wäre mit $\frac{2\pi}{n\tau} \cdot 22$ min vergleichbar, könnte also höchsten wenige Minuten betragen. Die beobachteten Phasenunterschiede lassen sich also damit nicht erklären.

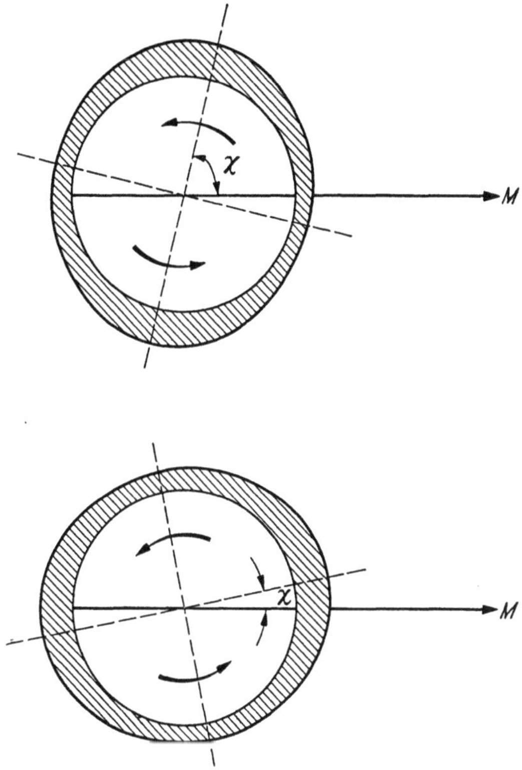

Fig. 27. Wirkung der Gezeitenreibung auf die Gezeitenwelle in einem äquatorialen Kanal (oben bei $p > 1$, unten bei $p < 1$; die gekrümmten Pfeile geben die Richtung der Erdrotation; in der Richtung M liegt der fluterzeugende Mond).

19. Der Energieverbrauch im Meer und die Turbulenzreibung der Gezeiten. Da der geforderte Energieverbrauch unmöglich von einer Flutreibung im festen Erdkörper geliefert werden kann, müßten im wesentlichen die Gezeiten der Ozeane hierfür aufkommen. Aber es scheint festzustehen, daß, wie schon dargelegt, die gewöhnliche Flutreibung nicht hinreicht, die in der Natur auftretenden Phasenverschiebungen zu erklären, so daß diese Flutreibung auch nicht ausreichen würde, den für die säkulare Beschleunigung von Sonne und Mond notwendigen Energieverbrauch verständlich zu machen. Eine Abschätzung, welche Energiemengen da zur Verfügung stehen, läßt sich leicht geben. Die Komponenten des Gezeitenstromes u und v (die Bodenschichten ausgeschlossen) müssen den Bewegungsgleichungen (12.3) genügen. Da Resonanzerscheinungen bei den großen Räumen der freien Ozeane kaum in Frage kommen, wird die Amplitude der Gezeit von derselben Größenordnung wie jene der Gleichgewichtsgezeit sein: $\bar{\zeta}$. Im offenen Ozean werden dann u und v die Größenordnung von 1 cm/sec haben. Die Reibungskraft ist $k\varrho(u^2 + v^2)$ und wirkt entgegen der Resultierenden des Gezeitenstromes. Der Energieverbrauch pro Flächeneinheit ist dann

$$k\varrho\,(u^2 + v^2)^{\frac{3}{2}}. \tag{19.1}$$

[1] AIRY: Tides and waves. In Encycl. Metrop., § 459. 1845.

Die Größe k hat einen Wert zwischen 0,002 und 0,0016. Der Energieverbrauch pro Quadratzentimeter wird somit von der Größenordnung 0,004 erg/sec sein. Die Fläche des Weltmeeres ist fast $3{,}62 \times 10^{18}$ cm², so daß der Gesamtverbrauch an Gezeitenenergie im ganzen Weltmeer von der Größenordnung 10^{16} erg/sec wird. Das ist nur ein kleiner Bruchteil, der benötigt wird. Die gewöhnliche Gezeitenreibung kommt deshalb für die Erklärung der säkularen Beschleunigungen kaum in Frage.

Nachdem G. J. TAYLOR[1] nachgewiesen hatte, daß der Energieverlust bei den Gezeiten der Irischen See (s. Ziff. 13) den großen Betrag von rund 6×10^{17} erg/sec liefert, der allein schon 60mal größer ist, als jener, der für den freien Ozean gefunden worden ist, ist die Vermutung von H. JEFFREYS[2] nicht von der Hand zu weisen, daß die Gezeitenreibung in den Neben- und Randmeeren und auf den ausgedehnten Schelfen für die von den Astronomen geforderte Dissipation an Energie verantwortlich zu machen ist. In diesen seichten Meeren tritt meistens eine wesentliche Vergrößerung der Hubhöhen auf, derart daß die Gezeitenströme weit über 1 cm/sec hinausgehen. Zwar sind die Flächen der so in Betracht kommenden Meeresteile klein gegenüber jener des Weltmeeres, aber, da der Energieverbrauch pro Flächeneinheit proportional der dritten Potenz der Geschwindigkeit ist, kann dieser Umstand doch den Ausschlag geben. Daß diese Überlegungen an Wahrscheinlichkeit sehr gewinnen, erkennt man, daß die kleine Irische See schon ein $1/_{20}$ des geforderten Betrages liefert.

JEFFREYS hat die TAYLORsche Methode der Abschätzung des Energieverbrauches auf eine große Zahl von Nebenmeeren und Meeresstraßen, soweit die vorhandenen Beobachtungen es zuließen, angewandt, um eine Schätzung des Gesamtwertes zu erhalten. Seine Berechnungen (s. Tabelle 7) beziehen sich auf

Tabelle 7. *Mittlerer Verbrauch an Reibungsenergie bei den Gezeiten* (Einheit: 10^{18} erg/sec)

Europäische Meere		Asiatische Meere		Nordamerikanische Meere	
Irische See	0,6	Südliche Chinasee. .	klein	Hudson-Straße . .	0,2
Englischer Kanal . .	1,1	Gelbes Meer	1,1	Hudson-Bai . . .	klein
Nordsee	0,7	Ochotskisches Meer .	0,4	Fundy-Bai . . .	0,4
		Beringmeer	15,0	Nordwest-Pssage .	1,6
		Malakka-Straße . .	1,1		

Zusammen: $2{,}2 \times 10^{19}$ erg/sec

Springtiden und geben so einen maximalen Wert. Der totale Energieverbrauch für alle in Betracht gezogenen Meeresteile ergibt sich zu $2{,}2 \times 10^{19}$ erg/sec. Das ist mehr als der geforderte Betrag verlangt. Es ist aber zu bedenken, daß eine Reduktion auf mittlere Verhältnisse notwendig ist.

Die Überlagerung des Gezeitenstromes der Mond- und Sonnengezeit ergibt, wenn ersterer durch $A \cos \sigma t$, letzterer durch $A \nu \cos (1-\nu)\, \sigma t$ gegeben ist, in ähnliche Weise wie in Ziff. 7, S. 885 durchgeführt ist, den totalen Strom zu

$$A\,[\cos \sigma t + \nu \cos (1-\nu)\, \sigma t] = A\,(1 + \nu^2 + 2\nu \cos \nu \sigma t)^{\frac{1}{2}} \cos\left(\sigma t - \operatorname{arc\,tg} \frac{\sin \nu \sigma t}{1 + \nu \cos \nu \sigma t}\right).$$

Die Amplitude des Stromes bei Springtide ist $A(1+\nu)$. Die Dissipation an Energie ist proportional der dritten Potenz der Amplitude, so daß das Verhältnis des mittleren Energieverbrauches zu jenem bei Springtide das Mittel von $(1+\nu^2+2\nu\cos\nu\sigma t)^{\frac{3}{2}} : (1+\nu)^3$ über eine ganze Gezeitenperiode sein wird. Vernachlässigt man ν^6, so ist der Zähler davon $(1+\frac{9}{4}\nu^2+\frac{9}{64}\nu^4)$. Das mittlere Verhältnis der Hubhöhen der Mond- und Sonnentiden ist für Küstenorte etwa $\nu = 1:2{,}73$ und, wenn man annimmt, daß das mittlere Verhältnis der Stromstärken etwa gleich groß ist, ergibt sich für das obige Verhältnis 0,51.

[1] G. J. TAYLOR: Phil. Trans. Roy. Soc. Lond., Ser. A **220**, 1 (1919).
[2] H. JEFFREYS: Phil. Trans. Roy. Soc Lond., Ser. A **221**, 239 (1920).

Diese Reduktion liefert als mittleren Energieverbrauch durch die Gezeiten $1,1 \times 10^{19}$ erg/sec, das sind 80% des geforderten Betrages. Man erkennt, daß dieser Energieverbrauch ohne weiteres ausreicht, um die beobachtete säkulare Beschleunigung des Mondes zu erklären. Dies gilt um so mehr, als gewiß eine ganze Anzahl von Nebenmeeren und Schelfgebiete noch unberücksichtigt geblieben sind. die zu dem oben berechneten Betrag weitere Beiträge liefern können. Der Patagonische Schelf, das Nordamerikanische Archipel, die Barentsee, der ausgedehnte Nordsibirische Schelf, die zahlreichen Fjorde Norwegens, Grönlands und anderer Küstengebiete und andere liefern gewiß Beträge, die beachtenswert sind und für die noch fehlenden 20% völlig ausreichend sein werden.

Eine rohe Schätzung läßt sich auch auf folgendem Wege finden. Die gesamte Schelffläche des Weltmeeres (Tiefen bis 200 m beläuft sich auf $27,50 \times 10^6$ km² $= 27,5 \times 10^{16}$ cm². Für die mittlere Stromstärke während einer Gezeitenperiode kann man für den Schelf als Ganzes etwa $\frac{1}{2}$ Knoten, das ist rund 30 cm/sec veranschlagen. Die Formel (19.1) liefert dann als Energieverbrauch am Schelf $1,53 \times 10^{19}$ erg/sec, was dem geforderten Betrag recht nahe kommt.

Als Reibungsverluste der Gezeitenbewegung kommen aber noch jene in Frage, die die Gezeitenenergie nach den Untersuchungen von H. U. SVERDRUP an den ausgedehnten Eisflächen der Arktis und Antarktis erleidet. So ist es kaum zu bezweifeln, daß die Gezeitenreibung reichlich für die von den Astronomen geforderte Dissipation an Energie zur Erklärung der säkularen Beschleunigung der Gestirne aufkommen kann, wenn die Turbulenzreibung der Gezeiten auf den großen Schelfflächen der Meere in Rechnung gezogen wird; jene der freien Ozeane reicht aber hierfür nicht aus.

Atmosphärische Gezeiten.

Von

W. Kertz.

Mit 19 Figuren.

A. Einleitung.

1. Begriffsbestimmung. Bedeutung für die Kenntnis der Atmosphäre. Literaturhinweise. Abweichend vom üblichen Sprachgebrauch bezeichnet man auch solche atmosphärischen Schwingungen als gezeitenartig, die nicht durch die Gravitationswirkung von Sonne oder Mond erregt werden, sondern durch Strahlung. Weil sie, wie die gewöhnlichen Gezeiten, von einem Himmelskörper außerhalb der Erde angeregt werden, haben sie planetarischen Charakter, d.h. sie erstrecken sich über die ganze Erde. Wegen der Erdrotation werden sie periodisch erregt und besitzen Perioden von $\frac{1}{1}, \frac{1}{2}, \frac{1}{3}, \frac{1}{4}, \ldots$ Tag. Es sollen aber nur solche Vorgänge als Gezeiten angesprochen werden, bei denen es sich um Schwingungen mit echten Luftbewegungen handelt, so daß z.B. die Ionisationserscheinungen ausgeschlossen sind, obwohl auch sie durch Strahlung erregt werden.

Die Gezeitenschwingungen der Atmosphäre äußern sich in den Geschwindigkeitskomponenten der Luftteilchen, in ihrer Temperatur, Dichte und im Druck. Am Grunde der Atmosphäre beobachten wir die Gezeiten vorwiegend als *Druckschwankungen*. Fig. 1 zeigt den zeitlichen Verlauf des Luftdrucks an 8 Stationen, die über die ganze Erde verteilt sind, an acht aufeinanderfolgenden Tagen. Für Stationen verschiedener Breite zeigen sich morphologisch große Unterschiede: In den Tropen herrscht eine regelmäßige halbtägige Welle vor, deren Phase fest an die Ortszeit gebunden ist. In höheren Breiten wird sie durch unperiodische, witterungsabhängige Schwankungen überdeckt.

Die halbtägige, *mit Ortszeit wandernde Welle* stellt die größte aller beobachteten Gezeitenschwingungen dar. Sie wird üblicherweise durch das Symbol W_2^2 mit den Indices des Hauptgliedes P_2^2 ihrer Kugelfunktionsentwicklung bezeichnet. Sie ist symmetrisch zum Äquator, und die Amplitude[1] der Bodendruckvariation hat dort ein Maximum mit etwa 1,5 mbar. Außer ihr hat man noch eine ganze Reihe andere Wellen gefunden, die mit der Ortszeit wandern:

W_4^3	$\frac{1}{3}$tägig	antisymmetrisch.	Max. Ampl. 0,2 mbar in 30° Breite
W_3^3	$\frac{1}{3}$tägig	symmetrisch.	Max. Ampl. 0,06 mbar am Äquator
W_5^3	$\frac{1}{3}$tägig	symmetrisch.	Max. Ampl. 0,02 mbar am Äquator
W_4^4	$\frac{1}{4}$ tägig	symmetrisch.	Max. Ampl. 0,02 mbar am Äquator
W_5^4	$\frac{1}{4}$ tägig	antisymmetrisch.	Max. Ampl. 0,02 mbar in 25° Breite

Für W_4^3 ist charakteristisch, daß die Welle vom Sommer zum Winter ihre Phase um 180° dreht und dabei während der Äquinoktien verschwindet.

Auffälligerweise gibt es nur eine sehr kleine planetarische *ganztägige Welle*. Zwar werden an den meisten Stationen große ganztägige Luftdruckschwingungen

[1] Amplitude hier immer als Faktor der Sinuswelle: Gesamtschwankung von W_2^2 am Äquator also 3 mbar. Bezeichnungen und Symbole sind in Ziff. 29 zusammengestellt.

beobachtet; die geographische Verteilung ihrer Amplituden und Phasen ist aber ganz unregelmäßig, und sie sind stark witterungsabhängig. Eine einigermaßen regelmäßige, witterungsunabhängige Welle erhält man nur aus Beobachtungen auf tropischen Inseln oder Schiffen auf äquatornahen Ozeanen (vgl. Ziff. 22).

Neben den mit Ortszeit wandernden Wellen hat man noch eine halbtägige stehende Schwingung Z_2 gefunden, die nach *Weltzeit* abläuft. Das Hauptglied ihrer Kugelfunktionsentwicklung ist P_2^0. Ihre Amplitude erreicht einen Maximalwert von etwa 0,1 mbar an den Polen. Sie übertrifft dort sogar W_2^2, deren Amplitude zu den Polen hin stark abfällt.

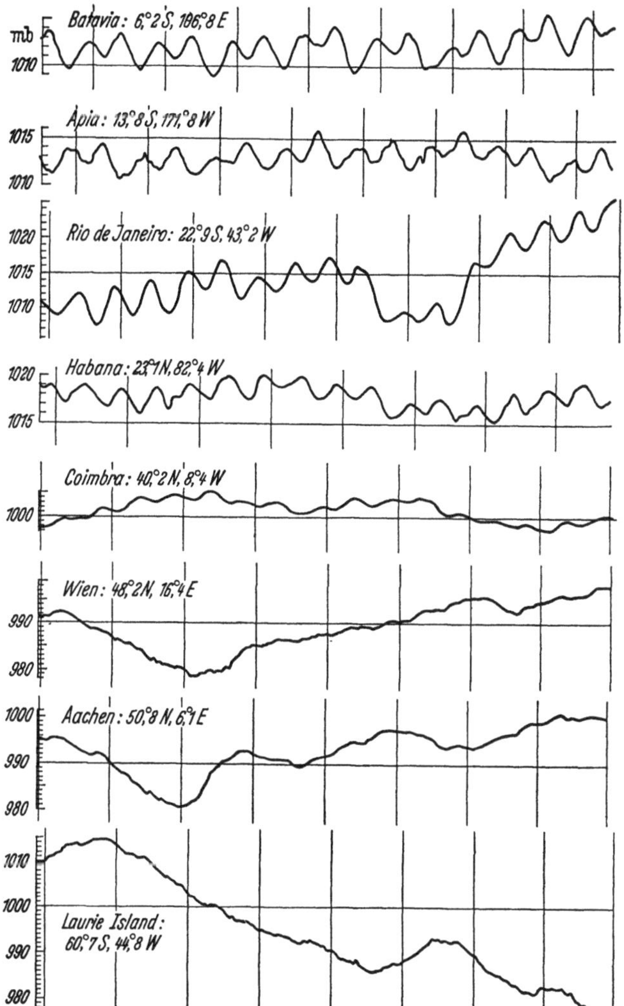

Fig. 1. Gang des Luftdruckes 21. bis 28. August 1904 an acht Stationen, die über die ganze Erde verteilt sind. Ortszeitmitternacht durch senkrechte Striche gekennzeichnet. — Man erkennt den typischen Unterschied im Druckverlauf innerhalb und außerhalb der Tropen.

Außer diesen durch die Sonne angeregten Schwingungen gibt es auch eine lunare Welle L_2. Sie ist wie W_2^2 symmetrisch zum Äquator und hat am Äquator eine Maximalamplitude von 0,1 mbar.

Diese gezeitenartigen Druckschwankungen sind im allgemeinen klein gegen die unperiodischen Änderungen. Sie konnten nur mit Hilfe statistischer Methoden aus den beobachteten Luftdruckwerten herauspräpariert werden. Durch Überlagerung sehr vieler Perioden kann man die unperiodischen Bestandteile eliminieren. Einzelheiten über die statistischen Verfahren siehe Bd. IL Bartels und v. d. Bijl, Analyse geophysikalischer Funktionen der Zeit und des Ortes.

Auf eine Fehlerquelle bei der statistischen Bearbeitung muß aber hier noch besonders hingewiesen werden. Die thermische Erregung durch die Sonnenstrahlung ist abhängig von der Beschaffenheit der Unterlage (Land, Meer, Gebirge) und vom Wetter. Daher dürften viele der berechneten solaren Amplituden und

Phasen außer den planetarischen auch noch lokale Anteile enthalten. BARTELS hat Verfahren zur Trennung von lokalem, wetterabhängigem und planetarischem Anteil angegeben[1].

Weil die periodischen Luftdruckschwankungen zu klein sind, um auf das Wetter einen merklichen Einfluß auszuüben, ist das Studium der atmosphärischen Gezeiten für die Meteorologie nur in theoretischer Hinsicht von Interesse. Für die *Theorie* sind sie aber in doppelter Hinsicht von Bedeutung: Die Beobachtung der Experimente, die Sonne und Mond täglich mit der Erdatmosphäre anstellen, stellt eine der wenigen Möglichkeiten dar, die Grundlagen der Theorie atmosphärischer Vorgänge zu überprüfen, weil man ja sonst kaum die Möglichkeit hat, geophysikalische Versuche in ähnlich großem Stil anzustellen. Zum andern gehen in die Theorie Größen ein, die der direkten Beobachtung bisher nur schwer zugänglich sind, wie die Höhenabhängigkeit der Temperatur der Atmosphäre im Ruhezustand und die durch Strahlung bewirkte Temperaturvariation in der Höhe. Durch Berechnung mehrerer Modelle mit verschiedenen Parameterwerten dieser Größen und Vergleich mit den Beobachtungen hofft man Aussagen über sie zu gewinnen.

Weil für die Physik der Atmosphäre der Vergleich der aus verschiedenen Modellrechnungen gewonnenen Ergebnisse mit den Beobachtungen so wichtig ist, ist in der vorliegenden Darstellung die Theorie an die erste Stelle gesetzt. Die Beobachtungsergebnisse werden erst unter Ziff. 15 und 16 und Ziff. 18 bis 22 referiert. Dort sind die aus den Beobachtungen abgeleiteten, ausgeglichenen geographischen Verteilungen der Amplituden und Phasen und ihre jahreszeitlichen Schwankungen angegeben. Detailliertere Darstellungen und Diskussionen der Beobachtungsergebnisse findet man in [*4*], [*3*] and [*2*], letzteres besonders für L_2. Numerische Werte und Quellenzitate in [*5*]. Besonders hingewiesen werde auf die umfangreiche Materialsammlung über die solaren Wellen in [*8*].

Aus der Phase von W_2^2 erkennt man, daß diese Variation nicht allein durch Gravitationskräfte angeregt sein kann. Es wäre dann auch verwunderlich, daß L_2 kleiner ist als W_2^2, während das lunare Glied M_2 des Gezeitenpotentials 2,2mal so groß ist wie das solare S_2. Nimmt man dagegen an, daß W_2^2 durch eine thermische Erregung zustande kommt, die stärker ist als die Gravitationserregung, so ist schwer verständlich, warum dann nicht eine noch wesentlich größere *ganz*tägige Druckwelle beobachtet wird. Lord KELVIN gab 1882 hierfür eine Erklärung, indem er die Vermutung aussprach, daß W_2^2 durch Resonanzverstärkung auf den beobachteten großen Wert käme. Wie das Meer, so stellt auch die Atmosphäre ein schwingungsfähiges System mit vielen verschiedenen Eigenschwingungen dar. Die Berechnung dieser Eigenschwingungen ist für die Atmosphäre einfacher als für das Meer, weil die Lufthülle die ganze Erde lückenlos bedeckt, während das Meer durch die Kontinente zerrissen wird. Erste quantitative Untersuchungen [*9*] ergaben, daß für W_2^2 durchaus eine starke Resonanzvergrößerung möglich sei, während für eine ganztägige Welle eine solche nicht in Frage kommt. CHAPMAN gab 1924 [*10*] eine quantitative Erklärung der Entstehung von W_2^2 aus einem thermischen und einem Gravitationsanteil, die beide etwa gleich groß sein sollten. Die Resonanzvergrößerung muß dabei etwa 100fach sein (vgl. Ziff. 18).

Beim weiteren Ausbau der Resonanztheorie wurde die Stellung von W_2^2 als Eigenschwingung der Atmosphäre wieder fraglich. Während zu W_2^2 ein h-Wert (vgl. Ziff. 11) von etwa 8 km gehört, liegt der Eigenwert $\hat{h}$ der Atmosphäre bei 10 km (vgl. Ziff. 12). Aus dieser Schwierigkeit fanden aber TAYLOR und

[1] J. BARTELS: Ann. Hydrogr., Berlin **51**, 153–160 (1923). — Beitr. Physik frei. Atmosph. **11**, 51–60 (1923) (durchgeführt am Beispiel Potsdam). — J. SPAR: Bull. Amer. Meteor. Soc. **33**, 339–343 (1952). — B. HAURWITZ: Bull. Amer. Meteor. Soc. **36**, 311–317 (1955) für HAMILTON, Bermuda.

PEKERIS einen Ausweg [*15*], [*16*]: Sie stellten fest, daß es Atmosphärenmodelle gibt, die mehrere Eigenwerte besitzen. Insbesondere besitzt eine Atmosphäre mit einer warmen Zone zwischen 35 und 70 km außer dem Eigenwert bei 10 km einen zweiten bei 8 km. Da man einen Temperaturanstieg bei 35 km auch aus anderen Beobachtungen erschlossen hatte, erschien diese Theorie sehr brauchbar. Man zog daher aus dem Vergleich von Theorie und Beobachtung der atmosphärischen Gezeiten den Schluß, daß dem Temperaturanstieg in 35 km ein Abfall in größerer Höhe folgen müsse. Spätere Temperaturmessungen bei Raketenaufstiegen bestätigten diese Annahmen qualitativ, aber nicht quantitativ: Die Temperaturerhöhung in der Ozonosphäre[1] ist nicht groß genug, um den zweiten Eigenwert bei 8 km zu liefern.

Aber auch innerhalb der Theorie hatten sich Widersprüche gezeigt: Die von TAYLOR und PEKERIS aufgestellte, von WILKES [*6*] ausgebaute neuere Resonanztheorie war ganz auf die Erklärung von W_2^2 zugeschnitten. Die anderen Wellentypen, insbesondere Z_2, erklärte sie nicht (vgl. Ziff. 19 bis 21).

SIEBERT [*21*] konnte mit einem einfachen Atmosphärenmodell, das nur *einen* Eigenwert bei 10 km besitzt, alle halb- und dritteltägigen Druckwellen befriedigend erklären, indem er eine *thermische Erregung* durch eine höhenunabhängige Temperaturvariation annahm (Strahlungsglied). Es zeigt sich (Ziff. 21), daß man auf diese Weise auch die $\frac{1}{4}$tägigen Variationen richtig erhält. Es erhebt sich nun natürlich wieder die Frage, warum dann die ganztägige Druckwelle nicht größer ausfällt, weil das ganztägige Strahlungsglied der Temperatur doch sicher größer ist als das beobachtete halbtägige. Eine gewisse Rolle spielt die Resonanzvergrößerung, die ja auch in der SIEBERTschen Theorie enthalten ist. Trotzdem müßte die ganztägige Welle aber noch etwa doppelt so groß sein wie die auf Inseln oder Schiffen in tropischen Ozeanen beobachtete (Ziff. 22). Dies Problem ist noch nicht befriedigend gelöst. Es gibt aber einige Gesichtspunkte, die daraufhindeuten, daß die ganztägige Welle in der Atmosphäre unterdrückt wird.

Die vorliegende Darstellung lehnt sich eng an die SIEBERTschen Untersuchungen an und weicht daher in der Deutung der Beobachtungsergebnisse von der herkömmlichen Resonanztheorie ab, die z.B. in [*1*] bis [*7*] vertreten wird. Es werden aber in Ziff. 12 die Grundzüge der neueren Resonanztheorie geschildert und in Ziff. 15 bis 16 und 18 bis 22 die Ergebnisse beider Theorien den Beobachtungen gegenübergestellt.

2. Zusammenhang mit anderen Erscheinungen. Während die Gezeitenschwingungen der Atmosphäre für den Wetterablauf am Grunde der Atmosphäre ohne Bedeutung sind, wirken sie sich stark aus auf die Vorgänge in der hohen Atmosphäre. Auch dort sind im allgemeinen die Einflüsse der solaren Gezeiten (S) stärker als die der lunaren (L). S läßt sich aber sehr viel schlechter isolieren, weil die Gezeitenbewegungen durch nicht gezeitenartige, solare Erscheinungen (Ionisation, Temperaturvariationen) verdeckt sind. Deshalb sind die Wirkungen von L besser bekannt als die von S.

a) Von diesen Folgeerscheinungen der Gezeitenbewegungen in der Höhe sind am längsten und genauesten bekannt die L_2 Variationen des äußeren Anteils des *Erdmagnetfeldes*[2]. Die „*Dynamotheorie*" verbindet die erdmagnetischen Variationen mit den Gezeitenbewegungen der ionisierten Luftteilchen. Durch die Bewegung der leitfähigen Luft im permanenten Magnetfeld der Erde werden in der Ionosphäre Ströme erzeugt. Das zu diesen Strömen gehörige Magnetfeld liefert die beobachteten lunaren Schwankungen des erdmagnetischen Feldes.

---

[1] Vgl. das Kapitel über Ozon von PAETZOLD in diesem Bande.

[2] Dieses Handbuch, Bd. IL. S. CHAPMAN u. E. H. VESTINE: Geomagnetism: Solar and lunar diurnal variations.

Bei der Durchrechnung der Dynamotheorie und einem Vergleich der Ergebnisse mit den Beobachtungen erkennt man, daß die Gezeitenbewegung in der Höhe des zu L_2 gehörigen Stromsystems die entgegengesetzte Phase haben muß wie das zu L_2 gehörige Windsystem am Boden. Diese Tatsache kann durch die neuere Resonanztheorie zwanglos erklärt werden (Ziff. 12).

b) Bei den direkten *Ionosphären*beobachtungen[1] findet man L_2 Variationen sowohl in den Änderungen δh[2] der wahren (und scheinbaren) Höhe als auch im Verlauf der Grenzfrequenzen f[2]. Den Zusammenhang mit den Druckschwankungen δp und Dichteschwankungen $\delta \varrho$ für eine wahre Höhe h kann man überschlagsmäßig nach folgender Formel berechnen:

$$\frac{\delta p}{p} = \frac{\delta h}{H} = \frac{\delta \varrho}{\varrho} = 2\frac{\delta f}{f} \quad \text{mit } H = \text{Skalenhöhe [vgl. (8.3)]}. \tag{2.1}$$

Für das Niveau der E-Schicht erhielt man aus Beobachtungen äquatornaher Stationen ebenfalls Phasenumkehr gegen die Schwingung am Boden, wie man sie auch aus den erdmagnetischen Beobachtungen erschlossen hat. Dagegen ergaben die Beobachtungen der englischen Station Slough einen Verlauf, der mit der Bewegung am Boden gleichphasig ist. Für L_2 im Niveau der F_2-Schicht ergeben die Beobachtungen *aller* Stationen die Phasenumkehr. Die geographische Verteilung von $\delta h(F_2)$ läßt sich annähernd beschreiben durch

$$\left.\begin{aligned} L_2\colon\ \delta h(F_2) &= 4\sin^2\vartheta \sin(2\tau + 252°)\ \text{km}, \\ \vartheta &= \text{Poldistanz},\ \tau = \text{mittlere Mondzeit.} \end{aligned}\right\} \tag{2.2}$$

Die Amplitude beträgt das 18500fache des Gleichgewichtswertes [vgl. (2.2) und (3.3)]. In der E-Schicht ergibt sich ein Faktor 2000. Die Beobachtungen von L_2 in $f(F_2)$ an einzelnen Tagen streuen stark[3]; jedoch ist L_2 in $f(F_2)$ wesentlich deutlicher als im Luftdruck am Boden.

c) L_2 ist auch in der am Boden beobachteten *kosmischen Strahlung* zu erkennen[4]. Bei den Gezeitenbewegungen der Atmosphäre wird die Isobarenfläche, in der die Primärstrahlung die Mesonen erzeugt, periodisch auf und ab bewegt. Die erzeugten Mesonen haben deshalb verschieden lange Wege bis zum Erdboden zurückzulegen. Wegen des natürlichen Zerfalls und der Zusammenstöße auf diesem wechselnden Wege schwankt auch die Zahl der überlebenden Mesonen periodisch. Aus den Beobachtungen folgt, daß die Atmosphäre in 18 bis 20 km Höhe eine L_2-Gezeitenbewegung ausführt, die etwa das 10fache der Gleichgewichtsflut beträgt. Die Phase ist zwar etwas gegen den Bodenwert gedreht, aber nicht entgegengesetzt.

d) Der Einfluß der *solaren* Gezeiten läßt sich verhältnismäßig gut aus den *erdmagnetischen* Beobachtungen herausschälen. Es zeigt sich wieder die Phasenumkehr in der Höhe des S_q-Stromsystems. In diesem Niveau beträgt S_2 etwa das 12fache von L_2 (im Luftdruck am Boden 19fach).

[1] Dieses Handbuch, Bd. IL. J. KAPLAN: Ionospheric winds. — K. RAWER: Ionosphären-Lotungen und ihre Ergebnisse.

Den Zusammenhang mit den atmosphärischen L_2 Gezeiten untersucht S. MATSUSHITA, J. Geomag. a. Geoelect. **1**, 17—21 (1949) (dort weitere Literatur). Eine Zusammenstellung vieler Beobachtungswerte von L_2 in der F_2-Schicht in Special Report on Tidal Phenomena in the Ionosphere, Union Radio-Scientif. Internat. 1950, Tabelle IV. Zusatz: S. MATSUSHITA, Rep. Ionosph. Res. Japan **4**, 57—58 (1950). E_s-Schicht: S. MATSUSHITA, Rep. Ionosph. Res. Japan **7**, 45—52 (1953); J. Geomag. a. Geoelectr. **8**, 91 (1955); J. A. THOMAS and A. C. SVENSON, Austral. J. Physics **8**, 554 (1955).

[2] Wegen der Bezeichnung vgl. Ziff. 29a.

[3] J. BARTELS: Ber. dtsch. Wetterd. US-Zone **12**, 30—33 (1950).

[4] A. DUPERIER: A lunar effect on cosmic rays. Nature, Lond. **157**, 296 (1946).

S zeigt in den *erdmagnetischen* Variationen von Huancayo nur eine kleine Variabilität mit der Jahreszeit (Maxima der Amplituden zur Zeit der Äquinoktien), während L im Juni viel schwächer ist als im Dezember[1].

e) Schwieriger ist es, S in den direkten *Ionosphären*beobachtungen nachzuweisen[2]. Das Verhältnis von S zu L scheint in der F_2-Schicht viel kleiner zu sein.

Die großen Unterschiede zwischen den Gezeiten in der Ionosphäre und am Boden legen den Schluß nahe, daß die ionosphärischen Gezeiten von den Gezeiten der unteren Atmosphäre unabhängig sind und ganz selbständig erzeugt werden[1].

f) Aus den Beobachtungen der Variationen der *kosmischen Strahlung* folgt für das Niveau 18 bis 20 km Höhe $S_2/L_2 = 6$. Also zeigt sich auch hier schon, daß S_2/L_2 mit der Höhe abnimmt.

g) Von der Höhe der 100 mbar-Fläche (16 km) liegen auch schon *Wind*beobachtungen für S_1 und S_2 vor, die aus Radio- und Radar-Sondenaufstiegen gewonnen wurden[3]. Für S_2 ergeben sich etwa die gleichen Windgeschwindigkeiten, wie man sie für den Boden aus den Druckbeobachtungen errechnet (vgl. Ziff. 18).

h) Stärkere Abweichungen zeigt die aus den Beobachtungen an *Meteoren* erschlossene Richtung und Stärke der Gezeitenwinde[4].

B. Quellen der Gezeitenerregung.

I. Gravitation (Gezeitenpotential).

Zur Berechnung der Anregung durch Gravitationskräfte können die Ergebnisse aus dem Artikel von J. Bartels, Gezeitenkräfte in diesem Band (im folgenden abgekürzt B. G.) übernommen werden. In den beiden folgenden Ziffern werden sie nur noch so umgeformt, daß sie sich unmittelbar auf die Gezeiten der Atmosphäre anwenden lassen.

Als Bezugsgröße für die Untersuchung der gravitationsbedingten Gezeiten benutzt man zweckmäßig die Druckvariation $\overline{\delta p}$, die bei den in B.G. Ziff. 5 beschriebenen „*Gleichgewichtsgezeiten*" auftreten würde. In der Literatur schreibt man gewöhnlich $\bar{\zeta}$ statt ζ für die Höhe der Gleichgewichtsflut und $-\Omega$ statt V für das Gezeitenpotential. Ist ϱ_0 die Dichte und p_0 der Druck der ruhenden Luft und die Skalenhöhe $H = p_0/g\varrho_0$ [vgl. (8.3)], so gilt

$$\overline{\delta p} = \frac{\bar{\zeta}}{H} p_0 = g\,\varrho_0\,\bar{\zeta} = -\,\varrho_0 \Omega. \tag{3.1}$$

3. Lunare Glieder. Das größte Glied des Gezeitenpotentials ist das *halbmondentägige* Glied M_2. Es bewirkt die Variation L_2. Am Erdboden beträgt diese

$$L_2\colon\ \overline{\delta p(0)} = 3{,}363\, P_2^2(\vartheta) \sin(2\tau + 90°) \cdot 10^{-2}\,\text{mbar}, \tag{3.2}$$

$$\left.\begin{aligned} \bar{\zeta} &= 28{,}0\, P_2^2(\vartheta) \sin(2\tau + 90°)\ \text{cm}, \\ &= 24{,}2 \sin^2\vartheta \sin(2\tau + 90°)\ \text{cm}. \end{aligned}\right\} \tag{3.3}$$

---

[1] J. Bartels: Indian J. Meteor. a. Geophys. **5**, 69—74 (1954).

[2] Theoretische Untersuchungen über S in F_2: T. Shimazaki: J. Geomagn. a. Geoelectr. **6**, 68—82 (1954). Zusammenstellung der Harmonischen von S_1 und S_2 für $\delta h(F_2)$ in verschiedenen geographischen Breiten, J. Radio Res. Lab. Tokyo **1**, 15 (1954); vgl. auch J. Geomagn. a. Geoelectr. **6**, 157—159 (1954). — S. Matsushita: J. Geomagn. a. Geoelectr. **1**, 41 (1949); **2**, 9 (1950).

[3] D. H. Johnson: Nature, Lond. **174**, 363—364 (1954). — Quart. J. Roy. Meteor. Soc. **81**, 1—8 (1955).

[4] Vgl. in diesem Band A. C. B. Lovell: Geophysical aspects of meteors.

Für die monatliche Änderung dieser halbtägigen Welle infolge der *Elliptizität* der Mondbahn siehe B. G. Fig. 13 und 14. Die große halbtägige elliptische Tide 1. Ordnung N_2 beträgt 19% von M_2. Die durch K_{2m} erfaßten Schwankungen der *Deklination* bewirken eine monatliche Amplitudenmodulation von $\pm 8{,}7\%$. Diese Glieder werden aber gewöhnlich bei den Analysen zusammen mit der Amplitudenmodulation der halbsonnentägigen Glieder herausgemittelt. (Aus B. G. Ziff. 17 ist zu ersehen, daß K_{2s} und K_{2m} formal identisch sind. Da aus einem Beobachtungsmaterial gewöhnlich zuerst die solaren Glieder berechnet werden, wird K_2 insgesamt zu diesen geschlagen vgl. Ziff. 4.) Daher zeigt sich ein Einfluß der Deklinationsänderungen des Mondes nur in dem halb so großen Glied {255,545}: Dieses stellt eine halbmondentägige Welle dar, deren Phase $-N'$ innerhalb von 18,6 Jahren die Werte von 0 bis 360° durchläuft (vgl. Ziff. 15).

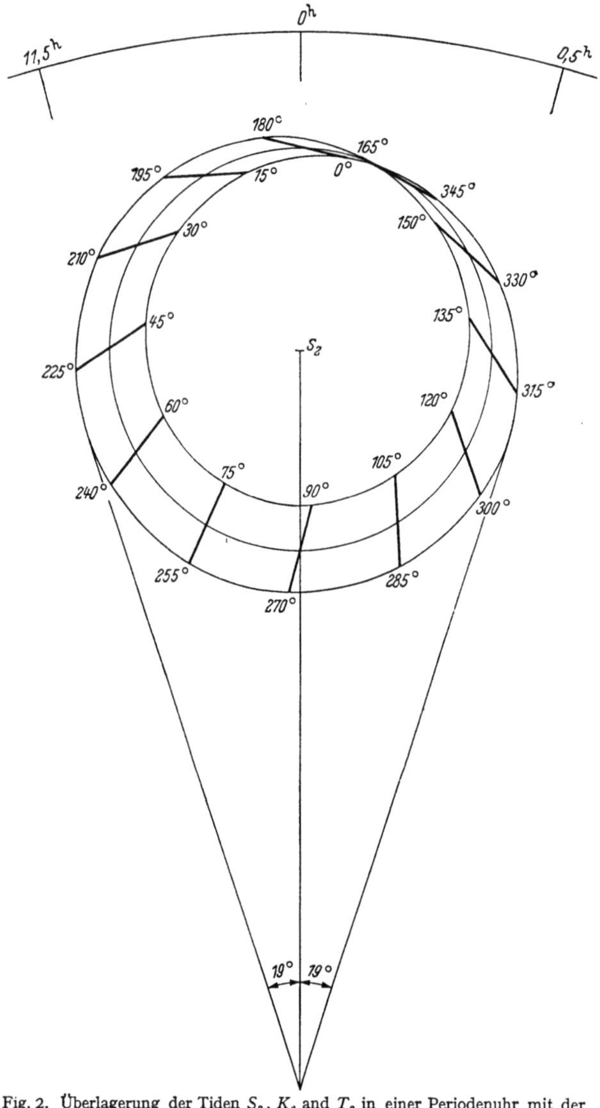

Fig. 2. Überlagerung der Tiden S_2, K_2 and T_2 in einer Periodenuhr mit der Periode eines halben Sonnentages. Die eingetragenen Zahlen bedeuten die mittlere Länge h der Sonne. Äquinoktien $h=0$ und 180°, Juni-Solstitium $h=90°$, Dezember-Solstitium $h=270°$.

Die ganztägigen lunaren Glieder $O_1 + K_{1m}$ ergeben eine L_1-Schwingung mit einer Schwebungsperiode von einem tropischen Monat. Da aber bei der gewöhnlichen Analyse wieder K_{1m} zusammen mit K_{1s} eliminiert wird, wäre nur O_1 aus den Beobachtungen zu entnehmen

$$L_1:\quad \overline{\delta p(0)} = 1{,}40 P_2^1(\vartheta)\sin(\tau - s)\cdot 10^{-2}\,\text{mbar} \tag{3.4}$$

mit s = mittlere Länge des Mondes in der Ekliptik. L_1 konnte aber noch für keine Station nachgewiesen werden.

4. Solare Glieder. Das Hauptglied ist das halbtägige S_2. Es beträgt am Boden:

$$S_2:\quad \overline{\delta p(0)} = 1{,}566 P_2^2(\vartheta)\cos 2t'\cdot 10^{-2}\,\text{mbar}. \tag{4.1}$$

Durch Überlagerung der elliptischen und Deklinationstiden erfahren Amplituden und Phasen Änderungen im Laufe eines Jahres. Da in den Beobachtungen eine

Trennung der lunisolaren Glieder K_{1m} und K_{1s} nicht möglich ist, werden beide bei der Analyse der halbsonnentägigen Luftdruckschwankung zusammen erhalten. Fig. 2 zeigt die jährliche Änderung von S_2. Die Amplitude durchläuft 2 Maxima während der Äquinoktien und 2 Minima zur Zeit der Solstitien. Das Winterminimum ist etwas tiefer als das Sommerminimum.

Die ganztägige Druckvariation wird gebildet durch K_{1s} und P_1 und beträgt am Boden

$$S_1: \quad \overline{\delta p(0)} = 1{,}303\, P_2^1(\vartheta) \sin h \cos(t' + 180°) \cdot 10^{-2}\,\text{mbar} \tag{4.2}$$

h = mittlere Länge der Sonne. In den Beobachtungen läßt sich aber die Deklinations-Mondtide K_{1m} davon nicht trennen. Sie muß eine zusätzliche jahreszeitliche Schwankung bringen mit

$$S_1^*: \quad \overline{\delta p(0)} = 1{,}270\, P_2^1(\vartheta) \sin(t + h + 180°) \cdot 10^{-2}\,\text{mbar}. \tag{4.3}$$

Die Gravitationserregung der drittel- und vierteltägigen Wellen ist verschwindend klein.

II. Strahlung von der Sonne.

Das thermische Gegenstück zum Gezeitenpotential ist bisher erst sehr unvollständig untersucht worden [*24*], [*20b*]. Primäre Ursache der thermischen Erregung ist die von der Sonne kommende Strahlungsenergie. Diese wird zum Teil in der Atmosphäre (beim Einfall sowie bei der Reflexion vom Erdboden) absorbiert (Strahlungsglied der Temperaturvariation Ziff. 5). Der andere Teil wird erst an der Erdoberfläche in Wärme umgewandelt und dann der Atmosphäre von unten her durch turbulenten Wärmeaustausch zugeführt (Austauschglied der Temperatur Ziff. 6).

5. Harmonische Analyse des Tagesganges der Strahlung. Bezeichnet man mit ζ die Zenitdistanz der Sonne und mit S_0 die Solarkonstante, so erhält man für die auf ein horizontales Flächenelement treffende Strahlungsleistung S

$$S = \begin{cases} S_0 \cos\zeta & \text{tags}, \\ 0 & \text{nachts}. \end{cases} \tag{5.1}$$

Für die Zenitdistanz ζ gilt näherungsweise, wenn man die wahre Ortszeit durch die mittlere ersetzt:

$$\cos\zeta = \cos\vartheta \sin\delta - \sin\vartheta \cos\delta \cos t'. \tag{5.2}$$

Um die thermische Erregung für die $\frac{1}{1}$-, $\frac{1}{2}$-, $\frac{1}{3}$-, $\frac{1}{4}$tägige Gezeitenwelle kennenzulernen, muß man den durch (1) und (2) gegebenen „Tagbogen“ der Strahlung harmonisch analysieren:

$$S = S_0 \sum_{\sigma=0}^{\infty} c_\sigma(\vartheta) \sin(\sigma t' + \varepsilon_\sigma). \tag{5.3}$$

Die Amplituden $c_\sigma(\vartheta)$ und die Phasen ε_σ sind noch von der Deklination δ abhängig. Die Phasen ε_σ werden so bestimmt, daß $c_\sigma(\vartheta)$ auf dem größeren Teil der Nordhalbkugel positiv wird. Die berechneten Amplituden und Phasen für $\sigma = 1, 2, 3, 4$ sind in Fig. 3 für die Äquinoktien ($\delta = 0°$) und Solstitien ($\delta = \pm 23{,}°45$) angegeben.

Während der *Äquinoktien* ($\delta = 0°$) sind alle Koeffizientenfunktionen (ausgezogene Kurven) symmetrisch zum Äquator. Die $\frac{1}{3}$-, $\frac{1}{5}$-, ...tägigen Variationen verschwinden. Die $\frac{1}{1}$tägige Amplitude ist etwa 2mal so groß wie die $\frac{1}{2}$tägige und 12mal so groß wie die $\frac{1}{4}$tägige. Zur Zeit der *Solstitien* ($\delta = \pm 23{,}°45$) (gestrichelte Kurven) verschiebt sich in der Amplitudenfunktion $c_1(\vartheta)$ der ganztägigen Schwingung das Maximum auf die Sommerhalbkugel. (Die Amplitudenfunktion für $\delta = -23{,}°45$ erhält man aus der für $\delta = +23{,}°45$ gezeichneten, indem man Nord- und Südhalbkugel vertauscht.)

$c_1(\vartheta)$ läßt sich auffassen als Summe einer zum Äquator symmetrischen Funktion, deren Amplitude von den Äquinoktien zu den Solstitien etwas kleiner wird, und einer kleineren antisymmetrischen Funktion, die ihr Vorzeichen vom Nordsommer ($\delta = +23{,}^\circ45$) zum Nordwinter ($\delta = -23{,}^\circ45$) vertauscht und die während der Äquinoktien ($\delta = 0^\circ$) verschwindet.

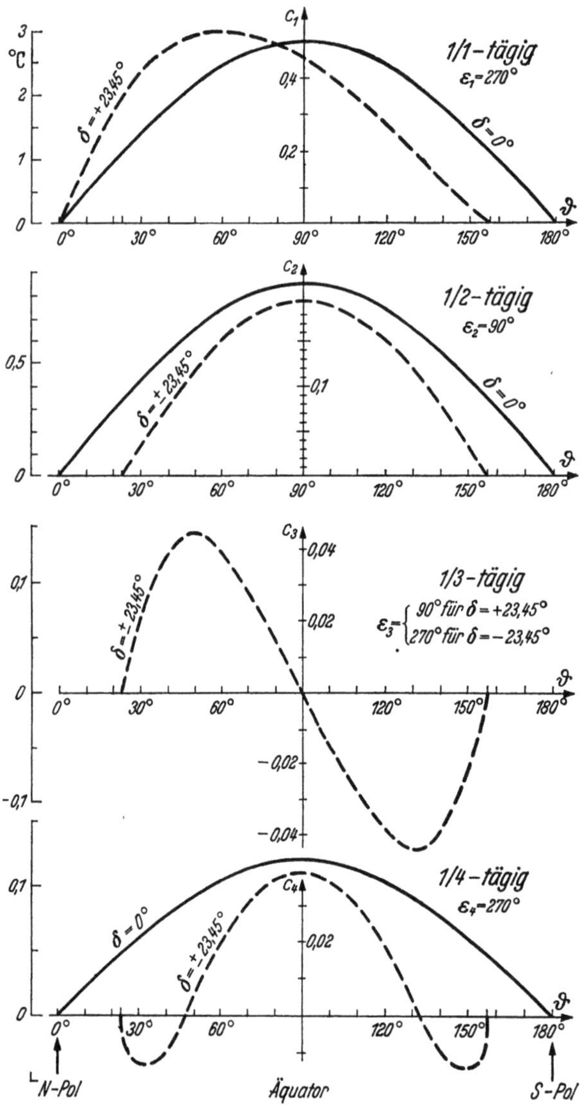

Fig. 3. FOURIER-Koeffizienten der Entwicklung des Tagesganges der Sonnenstrahlung S/S_0 [Gl. (5.3)]. — Am linken Rand ist eine Temperaturskala angebracht zur Deutung der Koeffizienten als Amplituden der Temperaturvariationen gemäß Gl. (6.5) [*20b*].

Die $\frac{1}{3}$tägige Oszillation verschwindet während der Äquinoktien. Ihre Amplitude ist kaum größer als die der $\frac{1}{4}$tägigen Schwingung. Ihre Amplitudenfunktion ist antisymmetrisch zum Äquator, mit Maximalwerten in 40° Breite. Sie wechselt ihr Vorzeichen vom Nordsommer zum Nordwinter.

Die Amplitudenfunktion der $\frac{1}{4}$tägigen Variation zeigt eine stärkere Veränderung während der Solstitien: in mittleren Breiten ändert sie ihr Vorzeichen und bildet Nebenmaxima aus.

Die Phasen ε_σ ergeben sich so, daß alle Wellen zu Mittag Extremwerte annehmen.

Die in der Atmosphäre absorbierte Strahlung wird aber durch ein gegen (5.1) etwas *abgeändertes* Gesetz beschrieben. Nach MÜGGE und MÖLLER[1] ist die vom Wasserdampf absorbierte Strahlung näherungsweise proportional zu $(\cos\zeta)^{0,7}$. Der hierzu gehörige Tagbogen ist „eckiger" als der normale. Die höheren Koeffizienten in einer FOURIER-Entwicklung sind also relativ größer als die in (5.3). Insbesondere verschwindet die dritteltägige Variation auch während der Äquinoktien nicht. Sie ist nahezu proportional zu $\sin\vartheta$. Der zugehörige Phasenwinkel beträgt 90°.

6. Planetarische Temperaturwellen. Die geographische und jahreszeitliche Variation der vom Erdboden ausgehenden $\frac{1}{1}$-, $\frac{1}{2}$-, $\frac{1}{3}$- und $\frac{1}{4}$tägigen thermischen Erregung könnte man aus der harmonischen Analyse langjähriger Beobachtungen eines dichten Netzes von Beobachtungsstationen ableiten. Das vorhandene Netz

[1] R. MÜGGE u. F. MÖLLER: Z. Geophys. **8**, 53—64 (1932).

ist leider viel zu weitmaschig, und es ist schwer abzuschätzen, welchen Einfluß der Abstand der Klimastationen vom Meer auf die Berechnung der planetarischen Temperaturwellen hat. SIEBERT hat in [*21c*] die jahreszeitlichen Variationen für europäische Stationen in 50 bis 60° Nord untersucht. HAURWITZ und MÖLLER haben in [*24*] die geographische Verteilung des Jahresmittels der halbtägigen Temperaturwellen nach einem geeigneten Mittelungsverfahren berechnet unter Berücksichtigung der Land-Wasserverteilung und einer Karte der Jahresmittel der Tagesschwankung der Temperatur. Um zu Näherungswerten für die geographische und jahreszeitliche Variation aller interessierender planetarischer Temperaturwellen zu kommen, wurde in [*20b*] ein noch schematischerer Weg eingeschlagen, und es wurden „*synthetische Klimawerte*“[1] berechnet. Diese synthetischen Werte wurden berechnet aus den geometrischen Verhältnissen bei der Sonneneinstrahlung (Ziff. 5) unter Berücksichtigung des Wärmetransportes durch Leitung und Austausch an der Land- und Meeresoberfläche.

a) Zusammenhang zwischen Strahlung und Temperatur an der Erdoberfläche. Wenn man in erster Näherung mit einem konstanten Koeffizienten der turbulenten Diffusion K und Austauschkoeffizienten A rechnet, gilt für den *Wärmetransport in der Luft* die „Wärmeleitungsgleichung“:

$$\frac{\partial \tau}{\partial t'} = K \frac{\partial^2 \tau}{\partial z^2}, \tag{6.1}$$

wobei τ die durch die einfallende Strahlung bewirkte Temperaturänderung bedeutet. Ist diese proportional zu $e^{i\sigma t'}$, so erhält man für $z \geqq 0$ die Lösung

$$\tau_\sigma(z, t') = \tau_\sigma(0)\, e^{-\frac{1}{2}(1+i)\sqrt{\frac{\sigma}{K}}z} \cdot e^{i\sigma t'}, \quad \sigma > 0. \tag{6.2}$$

Die zugehörige Stromdichte $S_\sigma^*(z, t')$ des nach oben fließenden Wärmestromes erhält man aus

$$S_\sigma^*(z, t') = -c_p A \cdot \frac{\partial \tau_\sigma}{\partial z}. \tag{6.3}$$

Dann gilt an der Erdoberfläche ($z=0$):

$$S_\sigma^*(0, t') = c_p A\, \tau_\sigma(0) \frac{1}{2}(1+i) \sqrt{\frac{\sigma}{K}}\, e^{i\sigma t'} = c_p A \sqrt{\frac{\sigma}{K}}\, \tau_\sigma(0)\, e^{i\left(\sigma t' + \frac{\pi}{4}\right)}. \tag{6.4}$$

Eine formal mit (6.1) übereinstimmende Gleichung gilt auch für die *Wärmeleitung im Erdboden* nach unten, wenn man K durch die Temperaturleitfähigkeit a^2 ersetzt. Für $z \leqq 0$ muß man dann aber in der Lösung (6.2) den ersten Exponenten negativ ansetzen. Für die Dichte des Wärmestromes nach unten wird in (6.3) $-c_p A$ durch $+\lambda$, die Wärmeleitfähigkeit, ersetzt. Die doppelte Vorzeichenänderung hebt sich in (6.4) auf, und man kann den Wärmeaustausch nach oben und die Wärmeleitung nach unten zusammenfassen, wenn man für K und A *Effektivwerte* verwendet, die sich aus den Werten für den Austausch nach oben und die Wärmeleitung nach unten zusammensetzen. K und A sollen im folgenden die Bedeutung solcher Effektivwerte haben.

Die von der Erdoberfläche nach oben und unten abgehende Wärmestromdichte $S^*(0, t')$ muß gleich der Bestrahlungsstärke $S(t')$ sein, so daß man aus (6.4) und (5.3) den Zusammenhang zwischen der Einstrahlung und der Temperatur änderung $\tau(0, t')$ an der Erdoberfläche erhält:

$$\tau(0, t') = S_0 \sum_{\sigma=1}^{\infty} \sqrt{\frac{K}{\sigma}} \frac{1}{c_p A} c_\sigma \sin(\sigma t' + \gamma_\sigma) \quad \text{mit} \quad \gamma_\sigma = \varepsilon_\sigma - 45^\circ. \tag{6.5}$$

---

[1] H. LETTAU: Synthetische Klimatologie. Ber. dtsch. Wetterd. US-Zone **38**, Weickmann-Heft, 127–136 (1952).

Die Amplituden der Harmonischen der Temperatur entstehen also aus denen der Einstrahlung durch Multiplikation mit $\sqrt{\frac{K}{\sigma}}\frac{1}{c_p A}$. Jede Temperaturschwingung läuft um 45° hinter der anregenden Strahlungsvariation her.

Eine gute Bestätigung der Beziehung (6.5) findet man beim Vergleich der für Potsdam abgeleiteten Temperaturschwingungen (Fig. 4, rechte Seite) mit den aus Fig. 3 entnommenen Werten für $\vartheta = 38°$. Alle gezeichneten Wellen zeigen im mittleren Verlauf eine Phasendifferenz von 45° zwischen Einstrahlung und Temperatur. Aus einem Vergleich der Amplituden sieht man, daß auch die Amplitudenverkleinerung gemäß $1/\sqrt{\sigma}$ im großen und ganzen gut erfüllt ist. Die Abweichungen dürften auf einer jahreszeitlichen Änderung des Austauschkoeffizienten beruhen. — Im jahreszeitlichen Verlauf erkennt man bei der *ganztägigen* Welle die für die Strahlung berechnete Amplitudenverkleinerung vom Sommer zum Winter auf $\frac{1}{5}$ des Sommerwertes. Man sieht auch, daß die Amplitudenänderung von den Äquinoktien zum Sommer nur etwa $\frac{3}{4}$ der Änderung von den Äquinoktien zum Winter ist. — Die *halbtägige* Temperaturwelle zeigt wie die Strahlung im Laufe des Jahres 2 Maxima zur Zeit der Äquinoktien und Minima während der Solstitien. Die Amplituden verhalten sich dabei wie 2:1. — Die *dritteltägige* Welle zeigt die charakteristische Phasenumkehr vom Sommer zum Winter. Dadurch werden im Tagesgang der Strahlung und der Temperatur die Verbreiterung oder Verschmälerung des Mittagsmaximums im Sommer oder im Winter bewirkt. Zur Zeit der Äquinoktien verschwindet die Amplitude in der Temperatur nicht ganz, ist aber klein. — Die *vierteltägige* Welle zeigt in mittleren Breiten einen doppelten Phasenwechsel. Die Amplitude erreicht 2 Hauptmaxima zur Zeit der Äquinoktien und 2 Nebenmaxima während der Solstitien. Dabei verhalten sich in Potsdam die Amplituden der Haupt- zu denen der Nebenmaxima wie 9:4. Bei Temperaturanalysen von weiter südlich gelegenen Stationen verschwinden die Nebenmaxima, wie es der Theorie entspricht.

b) Einfluß der Land-Wasserverteilung. Im Meer ist die effektive Wärmeleitfähigkeit wegen der turbulenten Durchmischung viel größer als im Erdboden. Daher sind die Amplitudenfaktoren $\frac{\sqrt{K}}{A} = \frac{1}{\varrho\sqrt{\sigma}}$ über dem Meere viel kleiner als auf dem Land. Näherungsweise kann man setzen

$$\frac{\sqrt{K}}{A} = \left[\frac{\sqrt{K}}{A}\right]_{\text{Land}} \cdot k(\vartheta, \varphi), \qquad k(\vartheta, \varphi) = \left\{\begin{matrix} 1 \text{ über Land,} \\ 0 \text{ über Meer.} \end{matrix}\right\} \tag{6.6}$$

Diese Näherung entspricht konstanter Temperatur im Tagesgang über dem Meer. Für die Rechnung ist es bequem, $k(\vartheta, \varphi)$ in einer FOURIER-Entwicklung nach der geographischen Länge φ zu verwenden

$$k(\vartheta, \varphi) = \sum_{m=0}^{\infty} a_m(\vartheta) \cos m\varphi + b_m(\vartheta) \sin m\varphi. \tag{6.7}$$

Dann wird

$$\tau(0, \vartheta, \varphi; t') = \sum_{\sigma} \sum_{m} U_{\sigma,m} \tag{6.8a}$$

mit

$$U_{\sigma,m} = S_0 \left[\frac{\sqrt{K}}{c_p A}\right]_{\text{Land}} \cdot \frac{c_\sigma}{\sqrt{\sigma}} (a_m \cos m\varphi + b_m \sin m\varphi) \sin(\sigma t' + \gamma_\sigma). \tag{6.8b}$$

Beim Ausmultiplizieren in (6.8b) behalten nur die Glieder $U_{\sigma,0}$ ihren Charakter als Wellen bei, die mit Ortszeit t' die Erde umkreisen. Ihre Amplitudenfunktion wird durch Multiplikation mit $a_0(\vartheta)$ modifiziert.

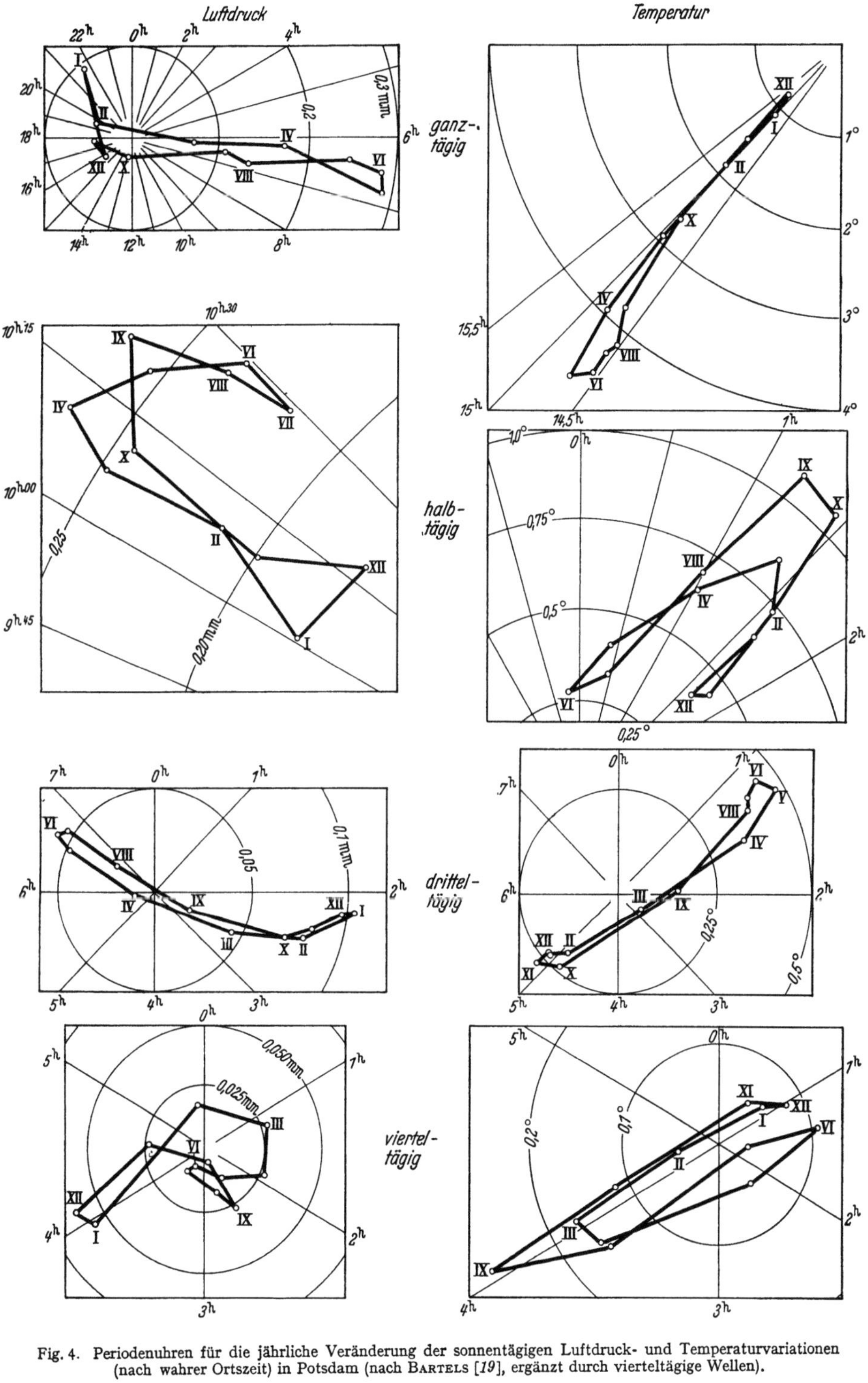

Fig. 4. Periodenuhren für die jährliche Veränderung der sonnentägigen Luftdruck- und Temperaturvariationen (nach wahrer Ortszeit) in Potsdam (nach BARTELS [19], ergänzt durch vierteltägige Wellen).

Um die Bedeutung der anderen Teilwellen mit $m > 0$ zu erkennen, empfiehlt es sich, in (6.8b) die Weltzeit

$$t = t' - \varphi \tag{6.9}$$

einzuführen. Durch Anwendung einfacher Formeln für die Produkte trigonometrischer Funktionen erhält man

$$U_{\sigma,m} = \frac{1}{2}\cdot\left[S_0 \frac{\sqrt{K}}{c_p A}\right]_{\text{Land}} \cdot \frac{c_\sigma}{\sqrt{\sigma}}\,[a_m \sin\Phi - b_m \cos\Phi + a_m \sin\Psi + b_m \cos\Psi] \tag{6.10a}$$

mit den Argumentfunktionen

$$\Phi = \sigma t + \gamma_\sigma + (\sigma + m)\,\varphi, \qquad \Psi = \sigma t + \gamma_\sigma + (\sigma - m)\,\varphi. \tag{6.10b}$$

Zu Φ gehören Wellen, die mit der Geschwindigkeit $\sigma/(\sigma + m)$ nach Westen wandern, während die zu Ψ gehörigen Wellen mit der Geschwindigkeit $\sigma/(\sigma - m)$ nach Westen

Tabelle 1. *Amplituden* C_n^s *und Phasen* $\left\{\begin{matrix}\varepsilon_n^s\\ \gamma_n^s\end{matrix}\right\}$ *in den Kugelfunktionsentwicklungen* $\sum_n C_n^s P_n^s(\vartheta)\sin\left(\sigma t' + \left\{\begin{matrix}\varepsilon_n^s\\ \gamma_n^s\end{matrix}\right\}\right)$ *der mit Ortszeit* t' *wandernden* $\left\{\begin{matrix}\text{Strahlungs-}\\ \text{Temperatur-}\end{matrix}\right\}$ *wellen.*

	Strahlung S/S_0				Berechnete Temperatur				Beobachtete Temperatur Jahresmittel	
	Äquinoktien $\delta=0°$		Solstitien $\delta=\pm 23{,}45°$		Äquinoktien $\delta=0°$		Solstitien $\delta=\pm 23{,}45°$			
	C_n^s	ε_n^s in Grad	C_n^s	ε_n^s in Grad	C_n^s °C	γ_n^s in Grad	C_n^s °C	γ_n^s in Grad	C_n^s °C	γ_n^s in Grad
					ganztägig ($\sigma=1$)*					
P_1^1	0,500	270	0,456	270	0,748	225	0,685	225		
P_3^1	—	—	0,000		0,083	225	0,076	225		
P_5^1	—	—	0,001	90	0,072	45	0,070	45		
P_7^1	—	—	0,001	270	0,057	225	0,054	45		
P_2^1	—	—	0,199	∓90	0,439	225	0,402	225		
P_4^1	—	—	0,050	∓90	0,158	225	0,145	225		
P_6^1	—	—	0,011	∓90	0,239	45	0,126	45		
P_8^1	—	—	0,005	±90	0,019	45	0,029	225		
					halbtägig ($\sigma=2$)					
P_2^2	0,270	90	0,226	90	0,283	45	0,235	45	0,301	54
P_4^2	0,052	90	0,005	270	0,087	45	0,018	45	0,113	68
P_6^2	0,023	90	0,019	270	0,009	45	0,042	225		
P_8^2	0,013	90	0,013	270	0,036	45	0,004	45		
P_3^2	—	—	—	—	0,144	45	0,094	45	0,368	65
P_5^2	—	—	—	—	0,100	225	0,031	45		
P_7^2	—	—	—	—	0,000		0,042	225		
					dritteltägig ($\sigma=3$)					
P_4^3	—	—	0,060	±90	—	—	0,055	45, 225		
P_6^3	—	—	0,017	±90	—	—	0,017	45, 225		
P_8^3	—	—	0,001	∓90	—	—	0,001	225, 45		
P_3^3	—	—	—	—	—	—	0,014	45, 225		
P_5^3	—	—	—	—	—	—	0,035	45, 225		
P_7^3	—	—	—	—	—	—	0,007	45, 225		
					vierteltägig ($\sigma=4$)					
P_4^4	0,069	270	0,049	270	0,072	225	0,035	225		
P_6^4	0,050	270	0,013	90	0,036	225	0,007	45		
P_8^4	0,014	270	0,015	90	0,011	225	0,015	45		
P_5^4	—	—	—	—	0,020	225	0,004	225		
P_7^4	—	—	—	—	0,023	225	0,006	45		

* Bei den ganztägigen berechneten Temperaturwerten ist für die Solstitien das Mittel aus Sommer und Winterwerten angegeben.

oder Osten ziehen, je nachdem $(\sigma - m)$ positiv oder negativ ist. Die sich mit Ortszeit westwärts bewegende Sonne regt also auf dem Umweg über die Land-Wasserverteilung Temperaturwellen an, die mit ganz anderen Geschwindigkeiten nach Westen oder sogar nach Osten laufen.

Für die Temperaturvariation ist diese Zerlegung in Teilwellen $U_{\sigma,m}$ nur eine Rechenoperation. Für die Gezeitenschwingungen der Atmosphäre aber gewinnen die einzelnen Partialwellen selbständige Bedeutung, da sie, unabhängig voneinander, ihnen entsprechende Druckwellen anregen. Da diese mit verschiedenen Resonanzfaktoren verstärkt werden und ihre Phasenwinkel gedreht werden, hat ihre Überlagerung eine ganz andere geographische Verteilung, als sie die ursprünliche Temperaturverteilung (6.8) besitzt.

c) Kugelfunktionsentwicklungen. Für die Anwendung auf die atmosphärischen Gezeiten werden die Amplituden nach Kugelfunktionen entwickelt. Man setzt

$$\sigma \pm m = s \tag{6.11}$$

und entwickelt die Koeffizienten in $U_{\sigma,m}$ nach den Kugelfunktionen P_n^s, damit die Ordnung der Kugelfunktionen mit der Wellenzahl zu φ übereinstimmt. [Zur Normierung der Kugelfunktionen vgl. Beitrag BARTELS, Formel (8.6), S. 744].

In den Tabellen 1 und 2 sind die Koeffizienten der Kugelfunktionsentwicklungen für die bei den atmosphärischen Gezeiten wichtigsten Wellen angegeben.

In Tabelle 1 stehen die Entwicklungen für die mit Ortszeit laufenden Wellen. In den beiden ersten Spalten wurden zum Vergleich die Entwicklungen für die in Fig. 3 dargestellten Amplitudenfunktionen der Strahlung S/S_0 angegeben.

Tabelle 2. *Amplituden C_n und Phasen γ_n in den Kugelfunktionsentwicklungen $\sum_n C_n P_n(\vartheta) \sin(\sigma t + \gamma_n)$ der stehenden (Nord-Süd-) Schwingungen nach Weltzeit t.*

	Berechnete Temperatur		„Beobachtete" Temperatur		Berechnete Temperatur	
	C_n °C	γ_n in Grad	C_n °C	γ_n in Grad	C_n °C	γ_n in Grad
	ganztägig $(\sigma = 1)$ $\delta = 0°$				dritteltägig $(\sigma = 3)$ $\delta = +23^{\circ}{,}45$*	
P_0	0,200	251			0,0025	49
P_2	0,303	47			0,0028	123
P_4	0,134	203			0,0054	240
P_6	0,154	310			0,0066	309
P_8	0,174	126			0,0060	97
P_1	0,168	285			0,0073	104
P_3	0,255	90			0,0096	283
P_5	0,093	246			0,0008	92
P_7	0,062	168			0,0007	144
	halbtägig $(\sigma = 2)$ $\delta = 0°$				vierteltägig $(\sigma = 4)$ $\delta = 0°$	
P_0	0,028	224	0,024	219	0,0059	323
P_2	0,034	195	0,040	214	0,0114	155
P_4	0,099	32	0,104	56	0,0100	347
P_6	0,015	241			0,0052	169
P_8	0,047	239			0,0027	256
P_1	0,076	180	0,076	194	0,0064	204
P_3	0,098	338	0,112	359	0,0154	24
P_5	0,080	79			0,0116	203
P_7	0,068	235			0,0010	33

* Für $\delta = -23^{\circ}{,}45$ ändern sich alle angegebenen Phasen um 180°.

Die „beobachteten" Temperaturwerte (Spalte 5) stammen aus der Analyse von HAURWITZ und MÖLLER [24]. Um die berechneten Temperaturwerte daran (und an die in Tabelle 2 aufgeführten) möglichst gut anzupassen, wurde

$$S_0 \left[\frac{\sqrt{K}}{c_p A}\right]_{\text{Land}} = 4{,}8 \cdot 10^{-2}\ \text{Grad}\ \text{sec}^{-\frac{1}{2}}$$

gesetzt. Bei den Phasenwinkeln für die Solstitien sind öfters 2 Vorzeichen oder 2 Werte angegeben. Dann gilt das obere Vorzeichen oder der erste Wert für das Nord-Sommer-Solstitium (Juni) und das andere für das Nord-Winter-Solstitium (Dezember).

In Tabelle 2 stehen die Kugelfunktionsentwicklungen für die Weltzeitglieder. Ihre Phasen sind viel unregelmäßiger als die in Tabelle 1.

C. Bewegungsgleichungen und ihre Lösung für Modellatmosphären.

Bei der Aufstellung der Theorie der gezeitenartigen Schwingungen der Atmosphäre muß man die wirklichen Verhältnisse und Vorgänge in der Atmosphäre durch ein vereinfachtes physikalisches Modell ersetzen, damit das Problem lösbar wird. Das Modell muß aber der Natur soweit angepaßt sein, daß die Abweichungen keinen wesentlichen Einfluß mehr auf die Lösungen haben.

Zum Teil sind die natürlichen Gegebenheiten nicht genau genug bekannt, z.B. die Temperaturschichtung der hohen Atmosphäre. Dann versucht man durch die Berechnung verschiedener Modelle den Einfluß dieser unbekannten Größen abzuschätzen. In den folgenden Ziffern wird besonderer Wert darauf gelegt, auf die bekannten und möglichen Unterschiede zwischen Modell und Wirklichkeit hinzuweisen (vgl. auch [29a]).

7. Bewegungsgleichungen. Ihrer physikalischen Natur nach sind die gezeitenartigen Schwingungen der Atmosphäre *Schwerewellen*. Zu ihrer Berechnung geht man gewöhnlich von den linearisierten und auf Kugelkoordinaten transformierten EULER-Gleichungen aus:

$$\frac{\partial u}{\partial t} - 2\omega v \cos\vartheta = -\frac{1}{a\varrho_0}\frac{\partial p}{\partial \vartheta} - \frac{1}{a}\frac{\partial \Omega}{\partial \vartheta}, \tag{7.1}$$

$$\frac{\partial v}{\partial t} + 2\omega u \cos\vartheta = -\frac{1}{a\varrho_0 \sin\vartheta}\frac{\partial p}{\partial \varphi} - \frac{1}{a\sin\vartheta}\frac{\partial \Omega}{\partial \varphi}, \tag{7.2}$$

$$0 = -\frac{\partial p}{\partial z} - \varrho_0 \frac{\partial \Omega}{\partial z} - g\varrho. \tag{7.3}$$

Im Koordinatensystem ϑ (= Poldistanz), φ (= geographische Länge), z (= Höhe über der Erdoberfläche) bedeuten u, v, w die Süd-, Ost- und Vertikalkomponente des Geschwindigkeitsvektors $\boldsymbol{v}$; t = Weltzeit, ω = Winkelgeschwindigkeit der Erdrotation, p = Druck, ϱ = Dichte der Luft [s. auch (8.1)], Ω = Gezeitenpotential, g = mittlere Schwerebeschleunigung der Erde, a = Erdradius. Weiter muß die *Kontinuitätsgleichung* erfüllt sein

$$\frac{D\varrho}{Dt} + \varrho_0 \chi = 0, \tag{7.4}$$

wobei D/Dt die substantielle Änderung bedeutet, und

$$\chi = \operatorname{div} \boldsymbol{v} = \frac{1}{a\sin\vartheta}\frac{\partial}{\partial\vartheta}(u\sin\vartheta) + \frac{1}{a\sin\vartheta}\frac{\partial v}{\partial\varphi} + \frac{\partial w}{\partial z}. \tag{7.5}$$

Diese Gleichungen gelten unter folgenden vereinfachenden Voraussetzungen:

a) Es wird von dem Einfluß der Bodenreibung und der inneren Reibung abgesehen[1].

b) Die Luftgeschwindigkeit und ihre Änderung werden als so klein angesehen, daß quadratische Glieder vernachlässigt werden können. Diese Voraussetzung ist sicher nicht erfüllt: in großen Höhen (vgl. Ziff. 13) und in der (sehr kleinen) Umgebung der Pole, in der die Ableitungen nach $\sin\vartheta\,\partial\varphi$ große Werte annehmen[2].

c) Vertikalgeschwindigkeit w und Vertikalbeschleunigung $\partial w/\partial t$ seien so klein, daß man sie aus den Bewegungsgleichungen streichen kann ($w \ll u, v$; $\partial w/\partial t \ll g$). Letzteres bewirkt, daß in jeder Vertikalen nicht nur im Ruhezustand, sondern auch während der Bewegung Gleichgewicht herrscht. H. Solberg, der diese Behandlungsweise abwertend die "*quasistatische Theorie*" nennt, hat selbst eine Berechnung ohne diese Vernachlässigung („dynamische Theorie") für das Meer in Angriff genommen [*26*]. Es zeigt sich dabei, daß bei Schwingungen, deren Periode länger ist als 12 Sternstunden, Abweichungen auftreten, während für die schnelleren Schwingungen die "quasistatische" Näherung brauchbar ist.

d) Die wirkliche Erde wird durch eine Kugel ersetzt. Dazu muß man sich zugleich das Potential der Erdschwere und der Zentrifugalkraft so abgeändert denken, daß die Äquipotentialflächen konzentrische Kugeln sind, damit das Modell in sich widerspruchsfrei bleibt [*29a*]. Es wird aber weiter vorausgesetzt, daß sich alle Vorgänge dicht an der Erdoberfläche abspielen, so daß man den Radiusvektor r durch den Erdradius a ersetzen kann, wenn r als Faktor einer anderen Variablen auftritt und g und $\partial\Omega/\partial z$ als konstant angenommen werden dürfen.

8. Die ruhende Atmosphäre. Die Zustandsgrößen Druck p, Temperatur T und Dichte ϱ der Luft im Ruhezustand ($\Omega \equiv 0$, $\boldsymbol{v} \equiv 0$) sollen durch einen Index 0 gekennzeichnet werden und die Änderungen bei der Schwingung durch δ, so daß allgemein gilt

$$p = p_0 + \delta p, \qquad T = T_0 + \delta T, \quad \varrho = \varrho_0 + \delta\varrho. \tag{8.1}$$

Zusätzlich zu den Modellannahmen in Ziff. 7 wird nun noch vorausgesetzt:

e) Die Zustandsgrößen der ruhenden Atmosphäre sollen nur von der Höhe z über der Erdoberfläche abhängen (vgl. Ziff. 25a).

f) Für die ruhende und die bewegte Luft soll die Gleichung für ideale Gase gelten

$$p = \frac{R}{M}\varrho T, \tag{8.2}$$

wobei dasselbe M für die ganze Atmosphäre gelten soll.

Zur Bestimmung der Zustandsgrößen der Atmosphäre hat man weiter die statische Grundgleichung

$$\frac{\partial p_0(z)}{\partial z} = -g\,\varrho_0(z) = -\frac{p_0(z)}{H(z)}, \quad \text{mit} \quad H(z) = \frac{R\,T_0(z)}{M g}. \tag{8.3}$$

$H(z)$ wird (nach S. Chapman), die *Skalenhöhe* genannt. $H(0)$ ist die Höhe einer homogenen, inkompressiblen Ersatzatmosphäre, die überall die Bodendichte $\varrho_0(0)$ hätte und die am Boden den gleichen Druck $p_0(0)$ liefern würde wie die wirkliche Atmosphäre.

[1] In älteren Rechnungen, z.B. [*9*], wurde gelegentlich eine geschwindigkeitsproportionale Reibung eingeführt. Die Abweichungen von der Wirklichkeit sind aber so groß, daß sie nicht einmal eine brauchbare quantitative Abschätzung über Art und Größe eines möglichen Reibungseinflusses erlauben.

[2] Vgl. H. Koschmieder: Dynamische Meteorologie, S. 297. 2. Aufl. Leipzig 1941.

Es bleibt dann aber immer noch eine Größe, etwa $T_0(z)$, unbestimmt. Unsere Kenntnisse über den Verlauf von $T_0(z)$ in größeren Höhen sind noch gering. Um eine Anschauung vom Verhalten der Lösungen zu bekommen, erweist sich die Diskussion dreier stark vereinfachter Modelle als zweckmäßig:

1. Isotherme Atmosphäre

$$T_0(z) = T_0(0) = \text{const}; \qquad H = H(0). \tag{8.4}$$

Aus (8.2) und (8.3) folgt

$$p_0(z) = p_0(0)\, e^{-\frac{z}{H}}, \qquad \varrho_0(z) = \varrho_0(0)\, e^{-\frac{z}{H}}. \tag{8.5}$$

Die Atmosphäre ist nach oben unbegrenzt, da der Druck erst im Unendlichen zu Null wird. Diese unnatürliche Eigenschaft des Modells führt zu Schwierigkeiten in der Theorie. Sie tritt auch bei allen Atmosphärenmodellen auf, die in der Höhe mit einer isothermen Schicht enden.

2. Atmosphäre mit linearem Temperaturgradienten

$$T_0(z) = T_0(0) - \alpha z, \qquad H(z) = H(0) - \varepsilon z, \qquad \text{mit} \qquad \varepsilon = R\alpha/M g. \tag{8.6}$$

Für den Druck ergibt sich

$$p_0(z) = p_0(0)\left(1 - \varepsilon \frac{z}{H(0)}\right)^{1/\varepsilon}. \tag{8.7}$$

Daraus ergibt sich als obere Grenze der Atmosphäre

$$z^\circ = H(0)/\varepsilon. \tag{8.8}$$

Die Verhältnisse werden besonders einfach, wenn α gleich dem adiabatischen Temperaturgradienten ist:

$$\alpha = \frac{\varkappa - 1}{\varkappa} \frac{M g}{R} = \frac{\varkappa - 1}{\varkappa} \frac{T(0)}{H(0)}; \qquad \varepsilon = \frac{\varkappa - 1}{\varkappa} \tag{8.9}$$

(vgl. Ziff. 12).

3. Zweischichtiges Modell. Von BARTELS [*19*] und TAYLOR [*15*] wurde zur besseren Annäherung an die wirkliche Atmosphäre ein zweischichtiges Modell in die Theorie eingeführt. Dieses zeigt einen linearen Temperaturabfall in der unteren und konstante Temperatur in der oberen Schicht. Neuerdings hat SIEBERT [*21*] wieder auf dieses Modell zurückgegriffen. Er benutzt folgende Zahlenwerte:

$$\left.\begin{aligned} T_0 &= \begin{cases} 288^\circ - 5{,}5^\circ z & \text{für } 0 \leqq z \leqq 12\,\text{km (Troposphäre)} \\ 220^\circ & \text{für } z > 12\,\text{km (Stratosphäre)}, \end{cases} \\ H &= \begin{cases} 8{,}435\,\text{km} - 0{,}1611\, z & \text{für } 0 \leqq z \leqq 12\,\text{km} \\ 6{,}502\,\text{km} & \text{für } z > 12\,\text{km}. \end{cases} \end{aligned}\right\} \tag{8.10}$$

Auch dieses Atmosphärenmodell besitzt keine obere Grenze.

An Hand dieser drei Modelle kann man viele Erscheinungen der atmosphärischen Gezeiten studieren. Die wirkliche Temperaturschichtung ist wesentlich komplizierter (vgl. Fig. 11). Man hat auch in der neueren Resonanztheorie mehrere solcher komplizierten Modelle in Rechnung gestellt (vgl. Ziff. 12). Man tut aber gut daran, die Theorie zunächst an einfacheren Modellen zu überprüfen, weil dabei die Vorgänge überschaubar bleiben. SIEBERT ist sogar der Meinung, daß man alle wesentlichen Züge der am Boden beobachteten atmosphärischen Gezeiten mit dem 3. Modell erfaßt.

9. Die Zustandsänderungen. Zu den bisherigen Voraussetzungen in Ziff. 7 und 8 fügen wir hinzu:

g) Die Zustandsänderungen sollen bei Gravitationsanregung adiabatisch erfolgen. Diese Annahme ist gerechtfertigt, da ein Wärmeaustausch in der Luft durch Wärmeleitung oder Konvektion wegen der Großräumigkeit der Wellen nicht in Frage kommt (vgl. aber Ziff. 25c).

Den 1. Hauptsatz der Thermodynamik schreibt man in einer für das vorliegende Problem günstigen Form:

$$J\,dt = c_v\,dT - \frac{p}{\varrho^2}\,d\varrho, \tag{9.1}$$

mit $J = J(z, \vartheta, \varphi; t) =$ auf die Masseneinheit bezogene Wärmemenge, die der Atmosphäre durch Austausch oder Strahlung pro Zeiteinheit zugeführt wird. J entspricht dem Gezeitenpotential Ω.

Benutzt man die Beziehungen

$$R = M(c_p - c_v), \qquad \varkappa = c_p/c_v \tag{9.2}$$

und vernachlässigt quadratische Terme, so erhält man aus (9.1)

$$\frac{R}{M(\varkappa-1)}\,\frac{DT}{Dt} = \frac{p_0}{\varrho_0^2}\,\frac{D\varrho}{Dt} + J = \frac{gH}{\varrho_0}\,\frac{D\varrho}{Dt} + J. \tag{9.3}$$

Aus der Gasgleichung (8.2) bekommt man

$$\frac{dp}{p} + \frac{d\varrho}{\varrho} = \frac{dT}{T}. \tag{9.4}$$

Damit kann man auf die gleiche Weise aus (9.1) eine Gleichung für die Druckänderungen gewinnen:

$$\frac{Dp}{Dt} = g\varkappa H\,\frac{D\varrho}{Dt} + (\varkappa - 1)\,\varrho_0 J. \tag{9.5}$$

Wegen der benutzten Vereinfachungen kann man für die substantiellen Änderungen schreiben

$$\frac{Dp}{Dt} = \frac{\partial\,\delta p}{\partial t} + w\,\frac{\partial p_0}{\partial z}, \tag{9.6}$$

und ebenso für ϱ und T. Mit Hilfe von (7.4) und (8.3) gewinnt man aus (9.5)

$$\frac{1}{g\varrho_0}\,\frac{\partial\,\delta p}{\partial t} - w = -\varkappa H\chi + \frac{\varkappa - 1}{g}\,J. \tag{9.7}$$

Die Variation der Wärmezufuhr wird als Temperaturvariation gemessen. Bezeichnet man mit τ (wie in Ziff. 6) die Temperaturvariation, die ausschließlich als Folge der die Erregung bewirkenden Wärmezufuhr auftritt, so gilt bei geringer Vertikalbewegung

$$J = c_p\,\frac{\partial\tau}{\partial t}. \tag{9.8}$$

Die gesamte Temperaturänderung δT setzt sich aus zwei Komponenten zusammen, der oben definierten primären, erregenden Variation τ [wie in (9.8)] und einer sekundären Variation $\tilde{\tau}$, die durch die thermodynamischen Vorgänge bei der Wellenbewegung (Kompression und Expansion) hervorgerufen wird.

$$\delta T(z) = \tau(z) + \tilde{\tau}(z). \tag{9.9}$$

Am Boden ($z = 0$) mit $w(0) = 0$ läßt sich $\tilde{\tau}$ aus den vorstehenden Formeln leicht berechnen:

$$\tilde{\tau}(0) = \frac{\varkappa - 1}{\varkappa}\,\frac{T_0(0)}{p_0(0)}\,\delta p(0). \tag{9.10}$$

10. Separation der Variablen. Zunächst müssen die fünf Gln. (7.1) bis (7.4) und (9.5) für die fünf Unbekannten $u, v, w, \delta p, \delta\varrho$ zu einer Gleichung für eine Unbekannte zusammengefaßt werden. Man wählt dazu üblicherweise die in (7.5) eingeführte Variable $\chi = \operatorname{div} \boldsymbol{v}$. Da nur zeitlich periodische Lösungen gesucht werden, wird angenommen, daß alle Veränderlichen durch den Faktor $e^{i\sigma t}$ von der Zeit abhängen. Dann erhält man aus (7.1) und (7.2), mit $f = \sigma/2\omega$,

$$u = \frac{\sigma}{4a\,\omega^2 (f^2 - \cos^2\vartheta)} \left\{ i\frac{\partial}{\partial\vartheta} + \frac{\operatorname{ctg}\vartheta}{f}\frac{\partial}{\partial\varphi} \right\} \left(\frac{\delta p}{\varrho_0} + \Omega \right), \tag{10.1}$$

$$v = \frac{i\sigma}{4a\,\omega^2 (f^2 - \cos^2\vartheta)} \left\{ \frac{i\cos\vartheta}{f}\frac{\partial}{\partial\vartheta} + \frac{1}{\sin\vartheta}\frac{\partial}{\partial\varphi} \right\} \left(\frac{\delta p}{\varrho_0} + \Omega \right). \tag{10.2}$$

Setzt man diese Größen in (7.5) ein, so erhält man

$$\chi - \frac{\partial w}{\partial z} = \frac{i\sigma}{4a^2\omega^2} F\left(\frac{\delta p}{\varrho_0} + \Omega \right), \tag{10.3}$$

mit dem Differentialoperator F:

$$F = \frac{1}{\sin\vartheta}\frac{\partial}{\partial\vartheta}\left(\frac{\sin\vartheta}{f^2 - \cos^2\vartheta}\frac{\partial}{\partial\vartheta} \right) + \frac{1}{f^2 - \cos^2\vartheta}\left[\frac{1}{\sin^2\vartheta}\frac{\partial^2}{\partial\varphi^2} + \frac{i}{f}\cdot\frac{f^2 + \cos^2\vartheta}{f^2 - \cos^2\vartheta}\frac{\partial}{\partial\varphi} \right]. \tag{10.4}$$

Aus der statischen Grundgleichung (7.3) und der Kontinuitätsgleichung (7.4) erhält man

$$i\sigma\left(\frac{1}{\varrho_0}\frac{\partial\,\delta p}{\partial z} + \frac{\partial\Omega}{\partial z} \right) = \frac{g}{\varrho_0}\frac{\partial\varrho_0}{\partial z} w + g\chi. \tag{10.5}$$

Aus der thermodynamischen Gl. (9.7) wird schließlich

$$\left(\chi - \frac{\varkappa - 1}{\varkappa}\frac{J}{gH} \right) = \frac{w}{\varkappa H} - \frac{i\sigma}{\varkappa}\frac{\delta p}{p_0}. \tag{10.6}$$

Aus diesen drei Gleichungen [(10.3), (10.5), (10,6)] kann man δp und w eliminieren und erhält unter Vernachlässigung eines Gliedes mit $\partial^2\Omega/\partial z^2$ (entsprechend Voraussetzung *d* in Ziff. 7) die Differentialgleichung

$$\left.\begin{aligned} &\frac{\partial}{\partial z}\left(H\frac{\partial}{\partial z} - 1 \right)\left\{ \chi - \frac{\varkappa - 1}{\varkappa}\frac{1}{gH} J \right\} \\ &\quad = \frac{g}{4a^2\omega^2} F\left\{ \frac{dH}{dz}\left(\chi - \frac{\varkappa - 1}{\varkappa}\frac{1}{gH} J \right) + \frac{\varkappa - 1}{\varkappa}\left(\chi - \frac{J}{gH} \right) \right\}. \end{aligned}\right\} \tag{10.7}$$

Diese Gleichung läßt sich vereinfachen, wenn man die neue Variable

$$x = \int_0^z \frac{d\zeta}{H(\zeta)} \tag{10.8}$$

einführt (WILKES [*17*]). Durch diese Transformation rückt die obere Grenze aller Atmosphärenmodelle $\to \infty$. Weiterhin setzt man zweckmäßig (SIEBERT [*21*] und ähnlich SAWADA [*25c*])

$$\chi - \frac{\varkappa - 1}{\varkappa}\frac{J}{gH} = y\,e^{\frac{1}{2}x}. \tag{10.9}$$

Dann erhält man für die so eingeführte Variable y

$$y'' - \frac{1}{4}y = \frac{g}{4a^2\omega^2} F\left\{ \left(\frac{dH}{dx} + \frac{\varkappa - 1}{\varkappa} H \right) y - \frac{\varkappa - 1}{\varkappa^2}\frac{J}{g} e^{-\frac{1}{2}x} \right\}. \tag{10.10}$$

Um eine Separation der Variablen durchführen zu können, setzt man Entwicklungen an für y, Ω und J nach den Eigenfunktionen $\Psi_n(\vartheta, \varphi)$ des Operators F. Diese Eigenfunktionen genügen der Eigenwertgleichung

$$\frac{g}{4a^2\omega^2} F \Psi_n = -\frac{1}{h_n} \Psi_n \tag{10.11}$$

mit den Eigenwerten $(-1/h_n)$. Die Entwicklungen lauten

$$\left.\begin{aligned} y &= \sum y_n(x)\, \Psi_n(\vartheta, \varphi)\, e^{i\sigma t}, \\ \Omega &= \sum \Omega_n(x)\, \Psi_n(\vartheta, \varphi)\, e^{i\sigma t}, \\ J &= \sum J_n(x)\, \Psi_n(\vartheta, \varphi)\, e^{i\sigma t}. \end{aligned}\right\} \tag{10.12}$$

Setzt man dies in (10.10) ein, so erhält man

$$\sum_n \left(y_n'' - \frac{1}{4} y_n\right) \Psi_n = -\sum_n \frac{1}{h_n} \left\{\left(\frac{dH}{dx} + \frac{\varkappa - 1}{\varkappa} H\right) y_n - \frac{\varkappa - 1}{\varkappa^2} \frac{J_n}{g} e^{-\frac{1}{2}x}\right\} \Psi_n. \tag{10.13}$$

Da die Ψ_n ein orthogonales Funktionensystem bilden (vgl. Ziff. 11), muß die Gleichung koeffizientenweise gelten:

$$y_n'' - \frac{1}{4} y_n + \frac{1}{h_n}\left(\frac{dH}{dx} + \frac{\varkappa - 1}{\varkappa} H\right) y_n = \frac{1}{h_n} \frac{\varkappa - 1}{\varkappa^2} \frac{J_n}{g} e^{-\frac{1}{2}x}. \tag{10.14}$$

Die Gln. (10.11) und (10.14) stellen die Bewegungsgleichungen des Systems nach der Separation der Variablen dar. Die Lösungen von (10.11) werden in Ziff. 11 und die von (10.14) in Ziff. 12 diskutiert.

Setzt man für u, v, w, δp und δT gleichartige Entwicklungen an wie (10.12), so erhält man für die Koeffizienten in symbolischer Operatorenschreibweise, wobei die Operatoren auf die Funktionen $\Psi_n(\vartheta, \varphi)$ anzuwenden sind:

$$u_n(x) = \frac{\varkappa g h_n e^{\frac{1}{2}x}}{4a\omega^2(f^2 - \cos^2\vartheta)} \left(\frac{dy_n}{dx} - \frac{1}{2} y_n\right)\left(\frac{\partial}{\partial\vartheta} - \frac{i}{f} \operatorname{ctg}\vartheta \frac{\partial}{\partial\varphi}\right) \tag{10.15}$$

$$v_n(x) = \frac{i\varkappa g h_n e^{\frac{1}{2}x}}{4a\omega^2(f^2 - \cos^2\vartheta)} \left(\frac{dy_n}{dx} - \frac{1}{2} y_n\right)\left(\frac{\cos\vartheta}{f} \frac{\partial}{\partial\vartheta} - \frac{i}{\sin\vartheta} \frac{\partial}{\partial\varphi}\right), \tag{10.16}$$

$$w_n(x) = -\frac{i\sigma}{g} \Omega_n(x) + \varkappa h_n e^{\frac{1}{2}x}\left[\frac{dy_n}{dx} + \left(\frac{H(x)}{h_n} - \frac{1}{2}\right) y_n\right], \tag{10.17}$$

$$\delta p_n(x) = \frac{p_0(0)}{H(x)}\left[-\frac{e^{-x}}{g} \Omega_n(x) + \frac{\varkappa h_n}{i\sigma} e^{-\frac{1}{2}x}\left(\frac{dy_n}{dx} - \frac{1}{2} y_n\right)\right], \tag{10.18}$$

$$\left.\begin{aligned} \delta T_n(x) = \frac{M}{R}\Bigg\{&\frac{\Omega_n(x)}{H(x)} \frac{dH}{dx} - \frac{\varkappa g h_n}{i\sigma} \times \\ &\times e^{\frac{1}{2}x}\left[\frac{\varkappa - 1}{\varkappa} \frac{H(x)}{h_n} + \frac{1}{H(x)} \frac{dH}{dx}\left(\frac{d}{dx} + \frac{H(x)}{h_n} - \frac{1}{2}\right)\right] y_n + \frac{\varkappa - 1}{\varkappa} \frac{J_n(x)}{i\sigma}\Bigg\}. \end{aligned}\right\} \tag{10.19}$$

Für $f < 1$, d.h. Schwingungen mit Perioden, die größer sind als 12 Sternstunden, gibt es ϑ-Werte, für die der Nenner in (10.15) und (10.16) verschwindet („*kritische Breiten*"). BRILLOUIN[1] zeigte aber, daß dort notwendigerweise gleichzeitig der Zähler verschwindet, so daß keine Singularität auftritt.

11. Die LAPLACEsche Gezeitengleichung. Die Gl. (10.11) ist schon von LAPLACE behandelt worden, als er die Gezeitenbewegungen eines Meeres berechnete, das die ganze Erdkugel mit konstanter Tiefe h bedeckt.

[1] M. BRILLOUIN: C. R. Acad. Sci., Paris **194**, 801—804 (1932).

Wenn man in (7.3) $\varrho = \text{const}$ setzt, so folgt

$$\delta p/\varrho = g(\bar{\zeta} - \zeta), \tag{11.1}$$

wobei $\zeta = \zeta(\vartheta, \varphi, t)$ die Erhebung der gestörten Oberfläche über der ungestörten und $\bar{\zeta} = -\Omega/g$ die Erhebung der Gleichgewichtsflut bedeutet. Dann folgt aus (7.1) und (7.2), daß u und v unabhängig von z sind. Wenn man in (7.4) und (7.5) vom Meeresboden $z=0$ bis zur Oberfläche $z=h$ über z integriert, erhält man wegen des Verschwindens der Vertikalkomponente der Geschwindigkeit am Boden und $D\varrho/Dt \equiv 0$

$$\frac{h}{a\sin\vartheta}\frac{\partial}{\partial\vartheta}(u\sin\vartheta) + \frac{h}{a\sin\vartheta}\frac{\partial v}{\partial\varphi} + \frac{\partial\zeta}{\partial t} = 0. \tag{11.2}$$

Sind $u, v, \zeta \sim e^{i\sigma t}$, so erhält man nach Elimination von u und v

$$\frac{g}{4a^2\omega^2}F(\zeta - \bar{\zeta}) - \frac{1}{h}\zeta = 0 \tag{11.3}$$

mit dem Operator F aus (10.4). Für den Fall der freien Schwingungen ($\bar{\zeta} \equiv 0$) ist (11.3) identisch mit (10.11). Dies ist der Grund für die Bezeichnung der Eigenwerte des Operators F durch $-1/h_n$, sowie für den häufig verwendeten Ausdruck „*äquivalente Tiefe*“ für h_n selbst.

Einen klaren Überblick über die Lösungen von (10.11) oder (11.3) mit $\bar{\zeta} \equiv 0$ erhält man, wenn man zunächst die möglichen Schwingungsformen bei nichtrotierender Erde ($\omega = 0$) diskutiert.

$$\lim_{\omega\to 0}\frac{\sigma^2}{4\omega^2}F(\zeta) = \frac{1}{\sin\vartheta}\frac{\partial}{\partial\vartheta}\left(\sin\vartheta\frac{\partial\zeta}{\partial\vartheta}\right) + \frac{1}{\sin^2\vartheta}\frac{\partial^2\zeta}{\partial\varphi^2} = -\frac{\sigma^2 a^2}{g h}\zeta. \tag{11.4}$$

Das ist die Differentialgleichung der Kugelflächenfunktionen. Es gibt nur dann Lösungen, die auf der ganzen Kugel endlich sind, wenn

$$\sigma^2 a^2/g h = n(n+1) \text{ mit ganzzahligem } n. \tag{11.5}$$

Durch (11.5) werden bei gegebenem h die möglichen Eigenfrequenzen σ_n festgelegt. Häufig dreht man aber auch den Rechnungsgang um und geht von einem (durch die Anregung) gegebenen σ aus und berechnet dann aus (11.5) die zugehörigen *Resonanztiefen* h_n. Für die Gezeiten der Atmosphäre stellt h einen geeigneten Parameter zur Berechnung der Resonanzvergrößerung eines Wellentypes dar. In dem oberen Teil der Fig. 5 ist der durch Gl. (11.5) gegebene Zusammenhang zwischen h, σ und n bei nichtrotierender Erde dargestellt. In dem unteren Teil sind die entsprechenden Beziehungen für die rotierende Erde angegeben. h ist dabei außer von σ und n noch von der Wellenzahl s abhängig. Vgl. auch die Unterschrift zu Fig. 5[1].

Zu diesen Eigenwerten gehören die $(2n+1)$ Eigenfunktionen

$$\zeta_n = P_n^s(\vartheta)\begin{Bmatrix}\cos\\ \sin\end{Bmatrix} s\varphi, \qquad s = 0, 1, 2, \ldots, n. \tag{11.6}$$

Da das Vorzeichen von σ durch (11.5) nicht bestimmt wird, gibt es zu jeder Eigenfrequenz west- und ostwärts wandernde Wellen der gleichen Form. Durch Superposition können stehende Wellen entstehen.

---

[1] Die Gl. (11.3) ist in h linear, in σ dagegen nicht. Deshalb eignen sich die Resonanztiefen besser zur Rechnung als die Resonanzfrequenzen. Eine Eigenwerttheorie für die Frequenzen kann man nicht mit der Gl. (11.3) durchführen, sondern muß das ganze System der EULERschen Gleichungen verwenden [*20a*], 1. Teil.

Der Wellentyp wird bestimmt durch die Parameter n und s der Kugelflächenfunktionen (11.6). Diese besitzen als Nullstellen (Knotenlinien der Wellen)

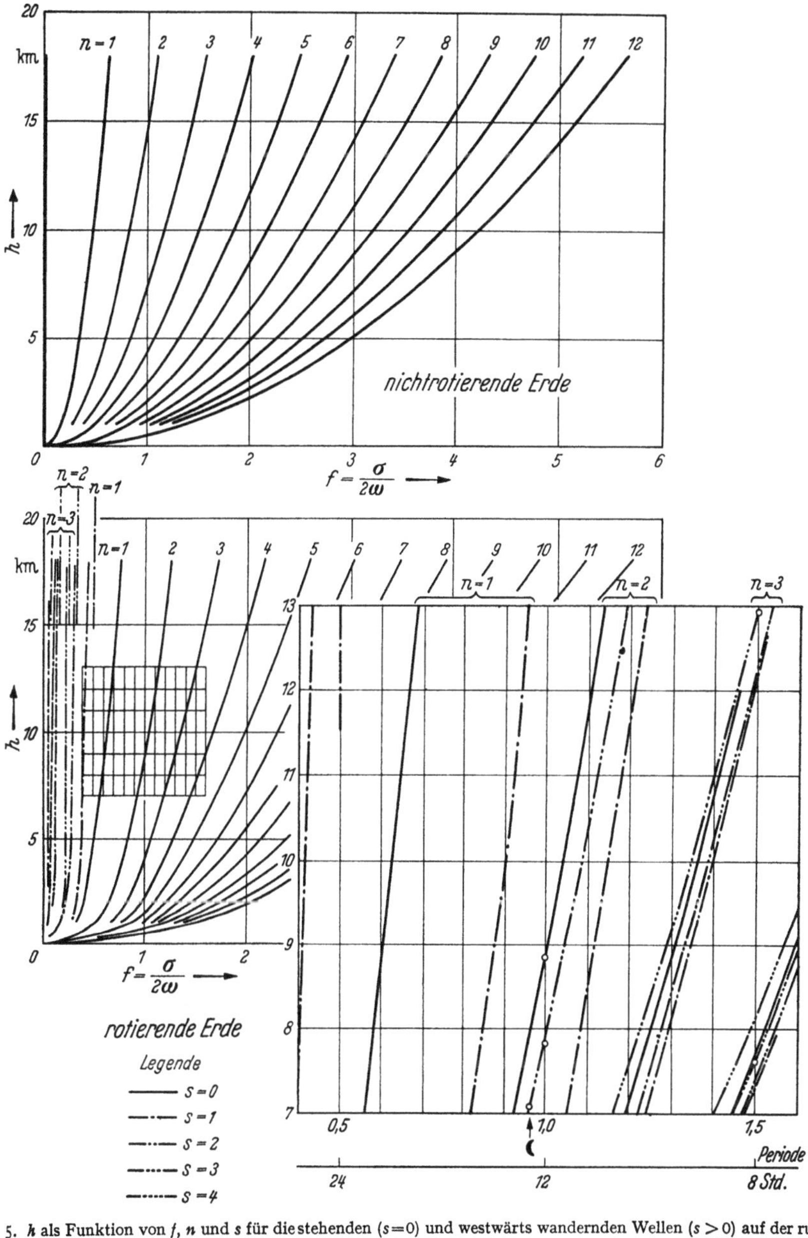

Fig. 5. h als Funktion von f, n und s für die stehenden ($s=0$) und westwärts wandernden Wellen ($s>0$) auf der ruhenden (oben) und rotierenden Erdkugel (unten). Der interessierende Bereich $0{,}5 \leq f \leq 1{,}5$ ist unten noch einmal vergrößert dargestellt. Bei der nichtrotierenden Erde fallen alle Kurven für $s=0, 1, \ldots, n$ zusammen. Durch die Rotation spalten sie auf (nur in dem vergrößerten Ausschnitt eingezeichnet). $h \to \infty$ ist auf der nichtrotierenden Erde unendlichfacher Eigenwert zu $f=0$. Durch die Erdrotation spaltet dieser auf. $h \to \infty$ tritt aber auch weiterhin als Eigenwert für diskrete f-Werte auf [Gl. (11.7)]. Der größte von ihnen ist $f=\frac{1}{2}$.

$2s$ Meridiane ($\varphi = \text{const}$) und $(n-s)$ Breitenkreise ($\vartheta = \text{const}$). Dadurch wird die Kugel in Gitterrechtecke geteilt, zwischen denen das Vorzeichen wechselt. Im Fall $s=0$ entarten diese zu Breitenzonen. Daher nennt man diesen Fall den *zonalen* Fall. Im anderen Grenzfall $n = s$ werden die Flächen gleichen Vorzeichens zu *Sektoren*, die durch Meridiane begrenzt werden.

Außer den Wellen gibt es auf der ruhenden Erde noch ein ganzes System stationärer Meeresströmungen ($\sigma = 0$). Zu ihrer Berechnung muß man wieder auf die Bewegungsgleichungen in Ziff. 7 zurückgreifen. Für $\sigma = 0$ folgt zwar $\zeta \equiv 0$, aber u und v brauchen nicht zu verschwinden. Es gibt vielmehr unendlich viele linear unabhängige Lösungen u_n, v_n.

Durch die Erdrotation[1] werden die ostwärts wandernden Wellen verzögert und die westwärts wandernden beschleunigt. Dadurch wird die $(2n+1)$fache Entartung der Eigenwerte weitgehend aufgehoben. Es bleiben nur noch doppelte Eigenwerte, die von n und s ($s > 0$) abhängen. Die Phase in φ bleibt unbestimmt. Auch der unendlichfache Eigenwert $\sigma = 0$ spaltet auf: aus den Meeresströmungen werden bei einsetzender Rotation langsam westwärts wandernde Wellen. Nur im zonalen Fall ($s = 0$) gibt es auch auf rotierender Erde noch ein System von unendlich vielen stationären Meeresströmungen, durch die aber im Gegensatz zur ruhenden Erde die Meeresoberfläche deformiert wird.

Den für unsere Theorie wichtigen Zusammenhang $h = h(\sigma, n, s)$ zeigt die Fig. 5 im oberen Teil für die ruhende und im unteren für die rotierende Erde. Bei der Aufspaltung des unendlichfachen Eigenwertes $h \to \infty$ für $f = 0$ ergeben sich Kurven, die den asymptotischen Werten

$$f_\infty = \frac{s}{n(n+1)} \tag{11.7}$$

zustreben. Die zugehörigen asymptotischen Eigenfunktionen sind

$$\zeta_{n\infty}^s = \left((n+1)^2 (n+s) P_{n-1}^s - n^2 \sqrt{(n+1)^2 - s^2}\, P_{n+1}^s\right) \begin{Bmatrix}\cos\\ \sin\end{Bmatrix} s\varphi\ ^2. \tag{11.8}$$

Für die Gezeiten wichtig ist der Fall $n = 1$ $s = 1$, weil dann $f_\infty = \frac{1}{2}$ und $\zeta_{1\infty}^1 \sim P_2^1$ (da $P_0^1 \equiv 0$). Für das Hauptglied P_2^1 einer nach Sternzeit ganztägigen Gravitationserregung ist also $h = \infty$. Deshalb kann eine solche Tide in einem Meer, das die ganze Erde bedeckt, nicht auftreten. Die in f nur wenig verschiedene ganztägige lunare Welle (3.4) müßte sehr klein werden.

Tabelle 3.

HOUGHsche Funktionen der halbtägigen wandernden solaren Wellentypen:

$s = 2$, $f = 0{,}99727$, $\sigma = 1{,}4544 \cdot 10^{-4}\ \text{sec}^{-1}$;

$n = 2$: $h_2 = 7{,}85$ km

$\Theta_2(\vartheta) = P_2^2 - 0{,}339\, P_4^2 + 0{,}041\, P_6^2 - 2 \cdot 10^{-3}\, P_8^2 + - \cdots$

$n = 4$: $h_4 = 2{,}11$ km

$\Theta_4(\vartheta) = 0{,}202\, P_2^2 + P_4^2 - 0{,}819\, P_6^2 + 0{,}24\, P_8^2 - 0{,}04\, P_{10}^2 + - \cdots$

$n = 6$: $h_6 = 0{,}957$ km

$\Theta_6(\vartheta) = 0{,}13\, P_2^2 + 0{,}755\, P_4^2 + P_6^2 - 1{,}72\, P_8^2 + 0{,}88\, P_{10}^2 - 0{,}2\, P_{12}^2 + - \cdots$

---

[1] Von der Lösung bei ruhender Erde ausgehend, verbesserte S. S. HOUGH die ersten Rechnungen von LAPLACE zu diesem Thema, indem er Entwicklungen nach Kugelfunktionen ansetzte. Er führte geeignete Rechenverfahren ein und gab Lösungen für die meisten interessierenden Fälle an. Phil. Trans. Roy. Soc. Lond., Ser. A **189**, 201—257 (1897); **191**, 139—185 (1898); vgl. auch Enzyklopädie der Mathematischen Wissenschaften, Bd. 6, I B, S. 3—83. Leipzig 1925. Neue Berechnungsmethoden für ein durch Meridiane begrenztes Meer bei E. A. HYLLERAAS, Astrophys. Norv. **3**, 139—164 (1939), 247—271 (1941). — Geofys. Publ. **13**, Nr. 10 (1943).

[2] Diese asymptotischen Eigenfunktionen sind eng verknüpft mit ROSSBY's „Waves in the westerlies", J. Marine Res. **2**, 38 (1939). ROSSBY behandelt den ebenen Fall. Übertragung auf die Kugel von B. HAURWITZ, J. Marine Res. **3**, 254—267 (1940). — Allerdings sollten nach dem Obenstehenden diese Wellen gerade nicht auftreten können (wegen $h = \infty$).

Tabelle 3. (Fortsetzung.)

Entwicklung von P_m^2 nach den HOUGHschen Funktionen der vorstehenden Art:

$$P_2^2(\vartheta) = 0{,}939\,\Theta_2 + 0{,}231\,\Theta_4 + 0{,}0703\,\Theta_6 + \cdots$$
$$P_4^2(\vartheta) = -0{,}177\,\Theta_2 + 0{,}638\,\Theta_4 + 0{,}235\,\Theta_6 + \cdots$$

HOUGHsche Funktionen der halbtägigen stehenden solaren Wellentypen[1]:

$s = 0$, $f = 0{,}99727$, $\sigma = 1{,}4544 \cdot 10^{-4}\,\mathrm{sec}^{-1}$

$n = 2$: $h_2 = 8{,}85$ km
$$\Theta_2(\vartheta) = P_2 - 0{,}386\,P_4 + 0{,}044\,P_6 - 2\cdot 10^{-3}\,P_8 + - \cdots$$
$n = 4$: $h_4 = 2{,}21$ km
$$\Theta_4(\vartheta) = 0{,}229\,P_2 + P_4 - 0{,}846\,P_6 + 0{,}24\,P_8 - 0{,}04\,P_{10} + - \cdots$$
$n = 6$: $h_6 = 0{,}981$ km
$$\Theta_6(\vartheta) = 0{,}15\,P_2 + 0{,}767\,P_4 + P_6 - 1{,}75\,P_8 + 0{,}89\,P_{10} - 0{,}2\,P_{12} + - \cdots$$

Entwicklung von P_m^0 nach HOUGHschen Funktionen der vorstehenden Art:

$$P_2(\vartheta) = 0{,}923\,\Theta_2 + 0{,}254\,\Theta_4 + 0{,}0797\,\Theta_6 + \cdots$$
$$P_4(\vartheta) = -0{,}198\,\Theta_2 + 0{,}617\,\Theta_4 + 0{,}232\,\Theta_6 + \cdots$$

HOUGHsche Funktionen der dritteltägigen wandernden solaren Wellentypen:

$s = 3$, $f = 1{,}4959$, $\sigma = 2{,}1817 \cdot 10^{-4}\,\mathrm{sec}^{-1}$

$n = 3$: $h_3 = 12{,}89$ km
$$\Theta_3(\vartheta) = P_3^3 - 0{,}105\,P_5^3 + 5{,}2\cdot 10^{-3}\,P_7^3 - 10^{-4}\,P_9^3 + - \cdots$$
$n = 4$: $h_4 = 7{,}66$ km
$$\Theta_4(\vartheta) = P_4^3 - 0{,}164\,P_6^3 + 0{,}012\,P_8^3 - 5\cdot 10^{-4}\,P_{10}^3 + - \cdots$$
$n = 5$: $h_5 = 5{,}09$ km
$$\Theta_5(\vartheta) = 0{,}0675\,P_3^3 + P_5^3 - 0{,}216\,P_7^3 + 0{,}020\,P_9^3 - 10^{-3}\,P_{11}^3 + - \cdots$$
$n = 6$: $h_6 = 3{,}62$ km
$$\Theta_6(\vartheta) = 0{,}115\,P_4^3 + P_6^3 - 0{,}265\,P_8^3 + 0{,}030\,P_{10}^3 - 2\cdot 10^{-3}\,P_{12}^3 + - \cdots$$
$n = 7$: $h_7 = 2{,}71$ km
$$\Theta_7(\vartheta) = 8{,}3\cdot 10^{-3}\,P_3^3 + 0{,}161\,P_5^3 + P_7^3 - 0{,}312\,P_9^3 + 0{,}042\,P_{11}^3 - + \cdots$$

Entwicklung von P_m^3 nach HOUGHschen Funktionen der vorstehenden Art:

$$P_3^3(\vartheta) = 0{,}993\,\Theta_3 + 0{,}102\,\Theta_5 + 0{,}0160\,\Theta_7 + \cdots$$
$$P_4^3(\vartheta) = 0{,}982\,\Theta_4 + 0{,}155\,\Theta_6 + \cdots$$
$$P_5^3(\vartheta) = -0{,}0665\,\Theta_3 + 0{,}960\,\Theta_5 + 0{,}197\,\Theta_7 + \cdots$$

HOUGHsche Funktionen der halbtägigen wandernden lunaren Wellentypen:

$s = 2$, $f = 0{,}96350$, $\sigma = 1{,}4052 \cdot 10^{-4}\,\mathrm{sec}^{-1}$

$n = 2$: $h_2 = 7{,}07$ km
$$\Theta_2(\vartheta) = P_2^2 - 0{,}375\,P_4^2 + 0{,}050\,P_6^2 - 3\cdot 10^{-3}\,P_8^2 + - \cdots$$
$n = 4$: $h_4 = 1{,}85$ km
$$\Theta_4(\vartheta) = 0{,}227\,P_2^2 + P_4^2 - 0{,}951\,P_6^2 + 0{,}32\,P_8^2 - 0{,}06\,P_{10}^2 + - \cdots$$
$n = 6$: $h_6 = 0{,}825$ km
$$\Theta_6(\vartheta) = 0{,}19\,P_2^2 + 1{,}03\,P_4^2 + P_6^2 - 2{,}42\,P_8^2 + 1{,}5\,P_{10}^2 - 0{,}5\,P_{12}^2 + - \cdots$$

Entwicklung von P_m^2 nach HOUGHschen Funktionen der vorstehenden Art:

$$P_2^2(\vartheta) = 0{,}927\,\Theta_2 + 0{,}230\,\Theta_4 + 0{,}0584\,\Theta_6 + \cdots$$
$$P_4^2(\vartheta) = -0{,}193\,\Theta_2 + 0{,}564\,\Theta_4 + 0{,}174\,\Theta_6 + \cdots$$

---

[1] HAURWITZ [*24*], [*35*], [*36*] benutzt in den Entwicklungen für den halbtägigen zonalen Fall an Stelle der HOUGHschen Funktionen die einfachen Ausdrücke (11.9b), da der wahre Wert von f sehr nahe an 1 liegt. — Die Kenntnis der Funktionen (11.9b) ermöglicht es zu übersehen, in welcher Weise die Hauptglieder $P_n(\vartheta)$ durch die zusätzlichen Terme in den Reihenentwicklungen verändert werden. Zum Beispiel wird die Amplitude von ϑ_2 am Äquator ebenso groß wie an den Polen, während sie bei P_2 nur halb so groß ist.

Für den Spezialfall $f=1$ und $s=0$ lassen sich nach SOLBERG [*26*a] Eigenwerte und Eigenfunktionen in geschlossener Form angeben:

$$h_n = \frac{1}{g}\left(\frac{4a\omega}{n\pi}\right)^2, \qquad n = 1, 2, \ldots \tag{11.9a}$$

und

$$\zeta_n = \cos\frac{n\pi}{2}\mu \text{ für } n \text{ gerade}, \zeta_n = \sin\frac{n\pi}{2}\mu \text{ für } n \text{ ungerade mit } \mu = \cos\vartheta. \tag{11.9b}$$

Im allgemeinen Fall lassen sich die Eigenfunktionen für die rotierende glatte Erde darstellen durch

$$\zeta_n = \Theta_n(\vartheta)\begin{Bmatrix}\cos\\ \sin\end{Bmatrix} s\varphi. \tag{11.10}$$

Für Θ_n liefert das HOUGHsche Lösungsverfahren Reihenentwicklungen nach den Kugelfunktionen P_ν^s. SIEBERT hat für sie den Namen „HOUGH*sche Funktionen*" vorgeschlagen [*21b*]. Tabelle 3 gibt die für die Theorie der atmosphärischen Gezeiten wichtigsten von ihnen mit den zugehörigen h-Werten, nach SIEBERT [*21d*]. $\Theta_n(\vartheta)$ hängt noch von s und h ab.

Die Wasserteilchen bewegen sich in allen Fällen auf Ellipsenbahnen, die am Äquator zu Geraden entarten. Die Hauptachsen der Ellipsen liegen in Richtung der Längen- und Breitenkreise. Bei nichtrotierender Erde hat man im zonalen Fall nur Bewegungen in Nord-Südrichtung. Durch die Rotation entstehen daraus Ellipsen, bei denen sich die Nord-Süd- zur Ost-Westachse verhält wie $\sigma/2\omega\cos\vartheta$. Im sektoriellen Fall $(n=s)$ verhalten sich diese Achsen bei nichtrotierender Erde wie $1/\mathrm{ctg}\,\vartheta$.

12. Gleichung für die Höhenabhängigkeit. Für die höhenabhängigen Amplituden ist die Gl. (10.14) zu lösen, die in etwas veränderter Schreibweise lautet:

$$y_n'' + \mu^2(x)\, y_n = \frac{1}{h_n}\frac{\varkappa-1}{g\varkappa^2} J_n e^{-\frac{1}{2}x}, \quad \text{mit} \quad \mu^2(x) = \frac{1}{h_n}\left(\frac{dH}{dx} + \frac{\varkappa-1}{\varkappa} H\right) - \frac{1}{4}. \tag{12.1}$$

Die Größe $\mu^2(x)$ wird von WEEKES und WILKES [*17*] als „*äquivalenter Brechungsindex*" bezeichnet, weil man (12.1) formal auffassen kann als die Differentialgleichung der Wellenausbreitung mit variablem Brechungsindex $\mu^2(x)$. Das Vorzeichen von μ^2 bestimmt dann die Ausbreitung der Energie.

Am Boden $x=0$ muß die Vertikalkomponente der Geschwindigkeit verschwinden (vgl. aber Ziff. 23)

$$w_n(0) = -i\sigma\frac{\Omega_n(0)}{g} + \varkappa h_n\left\{\left[\frac{dy_n}{dx}\right]_{x=0} + \left[\left(\frac{H}{h_n} - \frac{1}{2}\right) y_n\right]_{x=0}\right\} = 0. \tag{12.2}$$

Für die Berechnung von Atmosphärenmodellen ist wichtig der Fall eines *konstanten Temperaturgradienten*. Dann gilt

$$H(z) = H(0) - \varepsilon z; \qquad H(x) = H(0)\, e^{-\varepsilon x}; \qquad \frac{dH}{dx} = -\varepsilon H(x). \tag{12.3}$$

Damit lautet die Differentialgleichung

$$y_n'' + \left\{\left(\frac{\varkappa-1}{\varkappa} - \varepsilon\right)\frac{H(0)}{h_n} e^{-\varepsilon x} - \frac{1}{4}\right\} y_n = \frac{1}{h_n}\frac{\varkappa-1}{g\varkappa^2} J_n e^{-\frac{1}{2}x}. \tag{12.4}$$

Die zweckmäßigste Form der mathematischen Behandlung des vorliegenden Problems besteht darin, daß man zunächst das homogene Problem mit $J_n \equiv \Omega_n \equiv 0$ löst. Physikalisch bedeutet das, daß man zuerst die freien Schwingungen und dann erst die erzwungenen Schwingungen berechnet. In diesem Fall gibt es mit $w_n(0) = 0$ nur bestimmte Werte von h, für die das Problem lösbar ist. Dies sind die Eigenwerte der Atmosphäre oder des betrachteten Atmosphärenmodells. Sie sollen mit $\hat{h}$ bezeichnet werden.

Die Gl. (12.4) mit $J_n = 0$ ist am leichtesten zu lösen für eine Atmosphäre mit $\varepsilon = \frac{\varkappa - 1}{\varkappa}$, die sich in *konvektivem Gleichgewicht*[1] befindet. Aus der Differentialgleichung wird

$$y'' - \tfrac{1}{4} y = 0 \quad \text{mit der Lösung} \quad y = A\, e^{-\frac{1}{2}x} + B\, e^{+\frac{1}{2}x}. \tag{12.5}$$

Die Randbedingung am Boden ergibt:

$$\frac{1}{2}(-A + B) + \left(\frac{H(0)}{\hat{h}} - \frac{1}{2}\right)(A + B) = 0, \quad \text{also} \quad B = \left(\frac{\hat{h}}{H(0)} - 1\right) A. \tag{12.6}$$

Bei der zu B gehörigen Lösung nehmen die Amplituden nach oben zu, während die Amplituden bei der zu A gehörenden Lösung abnehmen. Um das Problem bestimmt zu machen, ist noch eine zweite Randbedingung an der oberen Grenze der Atmosphäre für $x \to \infty$ erforderlich. Diese erhält man durch die Berechnung der kinetischen Energie in einer vertikalen Säule mit dem Querschnitt 1 und durch die Forderung, daß die kinetische Energie endlich bleiben soll:

$$\int_0^\infty \varrho_0(x)\, \mathfrak{v}^2(x)\, H(x)\, dx < \infty. \tag{12.7}$$

Diese Forderung läßt sich umformen [*21d*] in

$$y = o\left(\frac{1}{\sqrt{x}}\right) \quad \text{d.h.} \quad \lim_{x \to \infty} \left(\sqrt{x} \cdot y(x)\right) = 0. \tag{12.8}$$

Um dieser Bedingung zu genügen, muß $B = 0$ sein. Damit die Lösung dann nicht identisch verschwindet, muß sein

$$\boxed{\hat{h} = H(0)\,.} \tag{12.9}$$

$H(0)$ ist also der Eigenwert der Atmosphäre im konvektiven Gleichgewicht.

Ganz entsprechend findet man für eine *isotherme Atmosphäre* ($H = \text{const}$) die Lösung

$$y = A\, e^{-\frac{1}{2}\beta x} + B\, e^{+\frac{1}{2}\beta x}, \quad \beta = \sqrt{1 - 4\frac{\varkappa - 1}{\varkappa}\frac{H}{\hat{h}}}\,. \tag{12.10}$$

Für die Befriedigung der Randbedingungen

$$B = 0, \quad \boxed{\hat{h} = \varkappa H} = 1{,}4 H = \tfrac{7}{5} H. \tag{12.11}$$

Damit wird

$$\beta = \sqrt{1 - \tfrac{8}{7} \cdot \tfrac{5}{7}} = \tfrac{3}{7} = 0{,}2857. \tag{12.12}$$

Für eine Atmosphäre mit einem *beliebigen*, konstanten Temperaturgradienten erhält man die Lösungen in Form von Bessel-Funktionen [*16a*]. Die Lösung, die den beiden Randbedingungen (12.2) und (12.8) genügt, hat die Form

$$y = A\, J_{1/\varepsilon}(\xi) \quad \text{mit} \quad \xi = \frac{2}{\varepsilon}\sqrt{\left(\frac{\varkappa - 1}{\varkappa} - \varepsilon\right)\frac{H(0)}{\hat{h}}}\; e^{-\frac{1}{2}\varepsilon x}. \tag{12.13}$$

---

[1] Setzt man voraus, daß auch andere als adiabatische Zustandsänderungen auftreten, so muß ε einen anderen Wert haben, z. B. $\varepsilon = 0$ bei isothermen Zustandsänderungen. V. Bjerknes nennt Atmosphären bei denen Zustandsänderungen und Temperaturschichtung so aufeinander abgestimmt sind, daß immer konvektives Gleichgewicht herrscht: *autobarotrop*. Die Schwingungen autobarotroper Atmosphären oder auch nur Atmosphärenschichten lassen sich einfach berechnen. Deshalb konnten auch kompliziertere Atmosphärenmodelle mit mehreren in sich autobarotropen Schichten auf diese Weise berechnet werden: A. Defant: Beitr. Physik frei. Atmosph. **12**, 112—137 (1926). — B. Haurwitz: [*22*], [*23*].

Es gibt dann im allgemeinen nicht nur einen, sondern unendlich viele Eigenwerte. Sie bestimmen sich aus

$$\hat{h}_\nu = 4\,\frac{\frac{\varkappa-1}{\varkappa}-\varepsilon}{\varepsilon^2}\,\frac{H(0)}{\xi_\nu^2} \quad \text{mit} \quad \xi_\nu J_{1/\varepsilon}(\xi_\nu) = 2\left(\frac{\frac{\varkappa-1}{\varkappa}}{\varepsilon}-1\right) J_{(1/\varepsilon)-1}(\xi_\nu). \qquad (12.14)$$

Für ein Atmosphärenmodell, bei dem der lineare Temperaturabfall der Troposphäre beliebig verlängert wäre, liegt der erste Eigenwert bei $1{,}1\,H(0)$ und der zweite bei $\frac{1}{6}H(0)$ (Fig. 9, Modell I), die folgenden werden immer kleiner. Für die atmosphärischen Gezeiten ist nur der größte von Bedeutung. Für ihn hat SIEBERT eine brauchbare Näherungsformel[1] aus folgender Überlegung gewonnen:

Er setzt in der Differentialgleichung (12.3) (mit $J_n \equiv 0$) $e^{-\varepsilon x} = 1$. Diese Näherung ist richtig für den größten Wert von ε $\left(\varepsilon = \frac{\varkappa-1}{\varkappa}\right)$ und den kleinsten Wert $\varepsilon = 0$. Dann läßt sich die Gleichung elementar integrieren, und man erhält den Eigenwert

$$\boxed{\hat{h} = \frac{H(0)}{1-\frac{\varkappa-1}{\varkappa}+\varepsilon} = \frac{\varkappa H(0)}{1-\varkappa\,\varepsilon}.} \qquad (12.15)$$

Man sieht sofort, daß die beiden Grenzfälle $\varepsilon = \frac{\varkappa-1}{\varkappa}$ und $\varepsilon = 0$ richtig dargestellt werden. Ferner gilt (12.15) exakt für eine Temperaturverteilung gemäß

$$H = H(0)\left[\frac{\varkappa\,\varepsilon}{\varkappa-1}\,e^{-\frac{\varkappa-1}{\varkappa}x} + 1 - \frac{\varkappa\,\varepsilon}{\varkappa-1}\right].$$

Bei der Aufstellung der neueren Resonanztheorie (vgl. Ziff. 1) wurden von PEKERIS [*16a*], WILKES [*6a*], JACCHIA und KOPAL [*30*] die Eigenwerte und Eigenfunktionen für viele, zum Teil recht komplizierte Atmosphärenmodelle berechnet. Der Rechenaufwand war dabei sehr groß und nur mit dem Einsatz von Differentialanalysatoren zu bewältigen.

Es soll hier versucht werden, die wesentlichen Züge dieser Rechnungen an einfachen, übersehbaren Modellen aufzuzeigen. Dazu wählen wir Atmosphärenmodelle, die sich aus isothermen Schichten zusammensetzen lassen. Als Maß für die Temperaturen werden durchweg die Skalenhöhen $H = \frac{R\,T_0}{M\,g}$ benutzt. Die Lösungen in den einzelnen Abschnitten erhält man aus (12.10). Damit die Vertikalgeschwindigkeit w an den Schichtgrenzen stetig ist, müssen

$$y \text{ und } \frac{dy}{dx} + \left(\frac{H}{h} - \frac{1}{2}\right) y \qquad (12.16)$$

stetig sein. Dabei soll eine möglicherweise vorhandene Vertikalbewegung der Grenzfläche nicht mit berücksichtigt werden.

Das einfachste Modell dieser Art besteht aus einer zweischichtigen Atmosphäre (Fig. 6). Im unteren Abschnitt soll die Temperatur doppelt so groß sein wie im oberen Abschnitt: $H_{\mathrm{I}} = 2$, $H_{\mathrm{II}} = 1$. Die Höhe der Grenzfläche sei $x = 6{,}5$. Diese Atmosphäre besitzt zwei Eigenwerte. Der größere $\hat{h}_1 = 2{,}76$ hat nahezu den Wert $\varkappa H_{\mathrm{I}} = 2{,}80$, den man erhielte, wenn die ganze Atmosphäre isotherm wäre und die Temperatur des unteren Abschnittes besäße. Der andere ist $\hat{h}_2 = 1{,}217$ und liegt in der Nähe von $\frac{8}{7}H_{\mathrm{II}}$. Für die zu $\hat{h}_1$ gehörige Lösung gilt im unteren Abschnitt

$$y_{\mathrm{I}} = 1{,}042\,(e^{-0{,}207\,x} - 0{,}0417\,e^{+0{,}207\,x}), \qquad 0 \leq x \leq 6{,}5.$$

[1] Andere Näherungsformeln findet man bei SOLBERG [*27b*].

Daran schließt sich im oberen Ende an:

$$y_{\mathrm{II}} = 1{,}395\, e^{-0{,}382\,x}, \quad 6{,}5 \leq x$$

y nimmt ständig ab. Die Lösung unterscheidet sich nur wenig von der Lösung (12.11) für die isotherme Atmosphäre.

Ganz anders aber die Lösung, die zu dem kleineren Eigenwert $\hat{h}_2$ gehört: Im unteren Abschnitt wird sie durch trigonometrische Funktionen dargestellt,

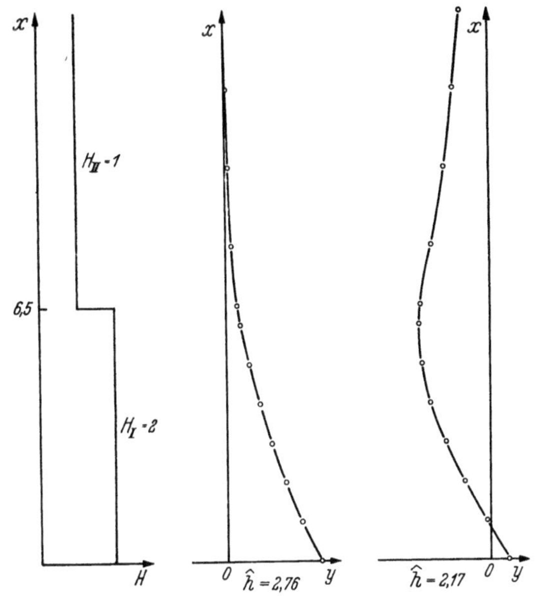

Fig. 6.

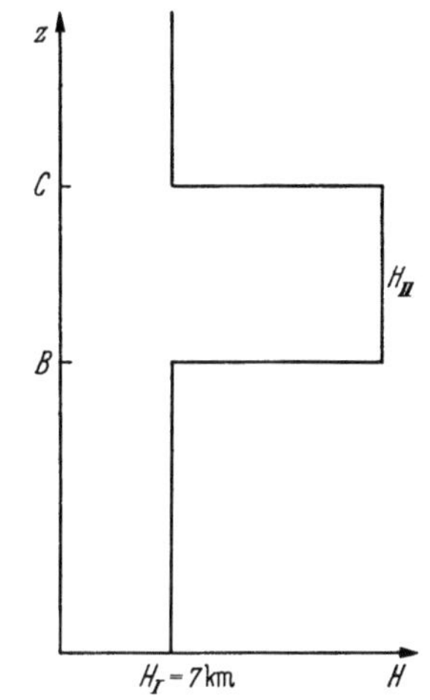

Fig. 7.

Fig. 6. Temperaturverlauf $H(z)=RT_{\mathfrak{v}}(z)/Mg$ und Eigenfunktionen im Zweischichtmodell zur Veranschaulichung des unterschiedlichen Verhaltens der in der neueren Resonanztheorie diskutierten Lösungen.

Fig. 7. Dreischichtmodell. Schemabild.

die aus (12.10) hervorgehen, wenn $\hat{h} < 4\,\frac{\varkappa - 1}{\varkappa}\,H = \frac{8}{7}\,H$ und damit β imaginär wird. Die Lösung für den unteren Abschnitt lautet daher:

$$y_{\mathrm{I}} = \cos(0{,}320x) - 3{,}585 \sin(0{,}320x), \quad 0 \leq x < 6{,}5.$$

Sie wechselt bereits in der Höhe $x = 0{,}85$ ihr Vorzeichen. Von da ab bleibt ihre Phase umgekehrt. Das Maximum erreicht sie dicht unterhalb der Grenzfläche. Im oberen Abschnitt dagegen muß wieder eine reine Exponentiallösung gelten, damit (12.8) befriedigt werden kann. Daher gibt es Lösungen dieser Art bei unserem zweischichtigen Modell nur innerhalb der Grenzen

$$\tfrac{8}{7} H_{\mathrm{II}} < \hat{h}_2 < \tfrac{8}{7} H_{\mathrm{I}}. \qquad (12.17)$$

Die Lösung für den oberen Abschnitt ist

$$y_{\mathrm{II}} = 8{,}00\, e^{-0{,}122\,x}, \quad 6{,}5 \leq x.$$

Sie bewirkt ein (langsames) Abklingen der Amplitude.

Mit einem dreischichtigen Modell gemäß Fig. 7 kommt man den Verhältnissen in der wirklichen Atmosphäre schon etwas näher: Der isotherme Abschnitt von

0 bis B stelle die Troposphäre und den unteren Teil der Stratosphäre dar. $H_{\mathrm{I}} = 7$ km (entspricht 239° K) wird so gewählt, daß $\varkappa H_0 = 9{,}8$ km etwa den Haupteigenwert der wirklichen Atmosphäre ergibt. Darüber liegt eine warme Zwischenschicht mit der Skalenhöhe H_{II} und der Dicke $(C-B)$, die der Ozonosphäre

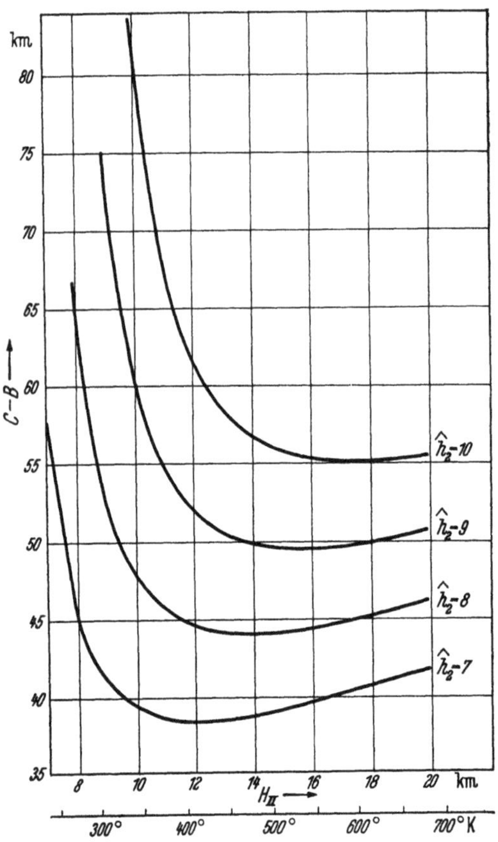

Fig. 8.

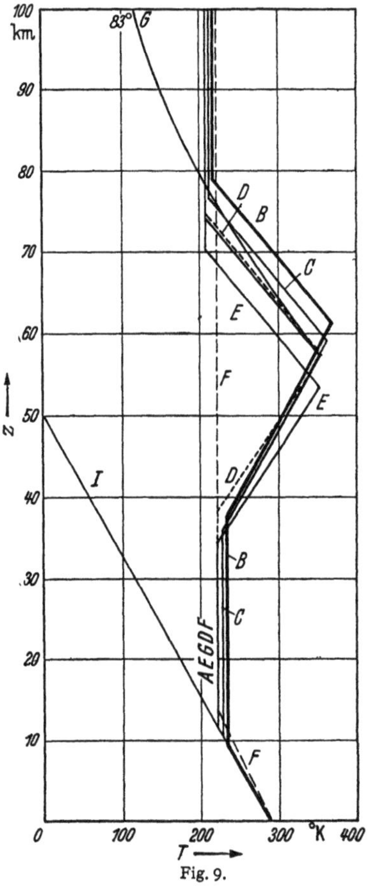

Fig. 9.

Fig. 8. Abhängigkeit des zweiten Eigenwertes $\hat{h}_2$ im Dreischichtmodell (Fig. 7) von $(C-B)$ und H_{II} nach der Näherungsformel (12.18) für $n=1$. Unter der Abszissenachse ist eine Leiste mit den zu H_{II} gehörigen Temperaturen angebracht.

Fig. 9. Von Pekeris benutzte Modellatmosphären. Alle Modelle außer I haben einen Eigenwert bei 10 km. Der größte Eigenwert von I liegt bei 9,36 km. Der zweite Eigenwert beträgt für A 7,815; B 8,192; C 8,001; D 7,815; E 7,654; G 7,815 und I 1,420 km. F hat keinen zweiten Eigenwert [16a].

entsprechen soll. Wenn B genügend groß ist ($> 3\,H_{\mathrm{I}}$), so werden die Eigenwerte zweiter Art durch den Näherungswert

$$\hat{h}_n \approx \frac{4\,\frac{\varkappa-1}{\varkappa}\,H_{\mathrm{II}}}{1+(2n-1)^2\left(\frac{\pi H_{\mathrm{II}}}{C-B}\right)^2}, \qquad n = 1, 2, \ldots \tag{12.18}$$

recht genau beschrieben. In diesem Fall ist der Abstand zwischen einem Maximum und einer Nullstelle von y gleich $(C-B)$. Von den formal unendlich vielen Nullstellen von (12.18) können nur diejenigen auftreten, die größer als $\frac{8}{7} H_{\mathrm{I}}$ sind. Für die Gezeiten der Atmosphäre kommt nur der Wert zu $n=1$ in Frage. Die Fig. 8 stellt die nach (12.18) berechneten Eigenwerte zweiter Art als Funktion von H_{II} und $(C-B)$ dar.

Die Entdeckung, daß es Atmosphärenmodelle gibt, die außer dem Haupteigenwert noch kleinere Eigenwerte besitzen, stammt von TAYLOR [*15*]. PEKERIS war der erste, der Atmosphärenmodelle angab, die außer dem Eigenwert bei 10 km noch einen bei 8 km haben, und zwar die in Fig. 9 dargestellten Modelle mit konstanten Temperaturgradienten.

Wenn bei einem schichtweise isothermen Atmosphärenmodell in einer Schicht $h < \frac{8}{7} H$, so treten dort trigonometrische Lösungen an die Stelle der Exponentiallösungen. Dabei können Knotenlinien und Phasenwechsel auftreten. Da in der hohen Atmosphäre Phasenumkehrungen beobachtet werden (vgl. Ziff. 2), ist es wahrscheinlich, daß dort dieser Fall eintritt. Deshalb braucht es sich aber nicht um eine Resonanzschwingung zu handeln.

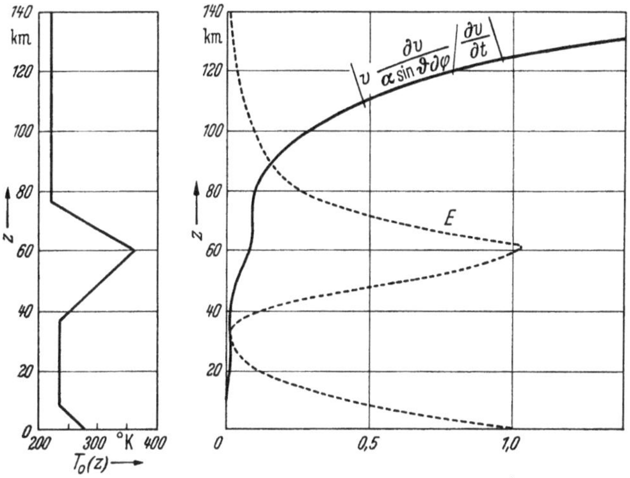

Fig. 10. Verhältnis des nach Ziff. 7b vernachlässigten quadratischen Gliedes $v \frac{\partial v}{a \sin\vartheta\, \partial\varphi}$ zu dem mitgeführten Term $\frac{\partial v}{\partial t}$ als Funktion der Höhe. Bei der Schwingung im Modell *B* von Fig. 9 mit $\hat{h} = 8{,}192$ km. — In die gleiche Darstellung ist die kinetische Energie *E* der Welle in relativen Einheiten eingetragen. Man erkennt, daß bei der zu diesem Eigenwert gehörigen Schwingung ein großer Teil der Energie in höheren Schichten lokalisiert ist, trotz der dort herrschenden geringen Dichte (nach PEKERIS [*16e*]).

13. Verhalten der Lösungen in großen Höhen. In den Darstellungen (10.15) bis (10.17) haben alle Geschwindigkeitskomponenten den Faktor $e^{\frac{1}{2}x}$. Die Geschwindigkeiten wachsen deshalb nach oben an. In Ziff. 7 war aber unter b vorausgesetzt, daß die Geschwindigkeiten so klein seien, daß man quadratische Glieder vernachlässigen kann. Diese Voraussetzung kann von einer bestimmten Höhe ab nicht mehr zutreffen. PEKERIS hat in [*16d*] das Verhältnis der vernachlässigten quadratischen Größen $v \frac{\partial v}{a \sin\vartheta\, \partial\varphi}$ zu der in der Rechnung mitgeführten Größe $\partial v/\partial t$ berechnet. Für das Modell *B* aus Fig. 9 mit $h = 8{,}19$ km ist dieses Verhältnis in Fig. 10 dargestellt. Wenn man verlangt, daß die vernachlässigte Größe kleiner als $\frac{1}{10}$ der mitgeführten sein muß, so sieht man, daß das Formelsystem bis zu einer Höhe von etwa 80 km richtig ist, darüber hinaus aber falsch wird. Man müßte für den oberen Bereich eine nichtlineare Theorie verwenden. Das ist bis jetzt noch nicht geschehen und würde auch einen sehr großen mathematischen Aufwand erfordern. Anfänge in [*16e*].

Mit dieser Unvollkommenheit der Theorie ist eine andere aufs engste verknüpft: Es ist allgemein üblich anzunehmen, daß die Atmosphäre ganz oben isotherm ist. Dort gelte die Skalenhöhe H_E. Dann folgt aus der Randbedingung

(12.7), daß

$$h > 4\frac{\varkappa - 1}{\varkappa} H_E = \frac{8}{7} H_E. \tag{13.1}$$

Schwingungen mit kleinerem h sollen also nicht auftreten können. Das sind aber nach Fig. 5 die längerperiodischen. Setzt man für H_E Werte ein, die unseren Vorstellungen über den Bau der hohen Atmosphäre entsprechen, so dürften nach (13.1) einige der beobachteten Schwingungen nicht auftreten. Man darf deshalb in (13.1) wohl keine physikalische Randbedingung sehen, sondern nur eine Grenze für das Versagen des benutzten Formelsystems. Es kann gut möglich sein, daß die Energie in einer Säule mit Einheitsquerschnitt endlich bleibt, auch wenn $h < \frac{8}{7} H_E$ ist und damit das Integral in (12.8) formal divergiert. Dies würde nur besagen, daß die berechneten Geschwindigkeiten in der Höhe wegen der Vernachlässigung der quadratischen Glieder dann so falsch wären, daß das Modell zu Widersprüchen führt. Auch physikalisch wäre es nur schwer verständlich, daß die Temperatur der allerhöchsten Atmosphärenschichten entscheiden soll, ob Schwingungen am Boden auftreten können oder nicht.

Da eine nichtlineare Theorie für die höchsten Atmosphärenschichten nicht vorhanden ist und (12.8) zu Widersprüchen führt, hat man versucht, (12.8) durch eine andere Randbedingung zu ersetzen. Der erste Versuch in dieser Richtung stammt von WEEKES und WILKES [*17*][1].

Sie betrachten den vertikalen Fluß der mit der Schwingung verbundenen Energie: Sei S die Energiestromdichte, so gilt bei zeitlicher Periodizität von S für den zeitlchen Mittelwert über eine Periode

$$S = \overline{\operatorname{Re}(\delta p) \cdot \operatorname{Re}(w)} = \tfrac{1}{2} \operatorname{Re}(\delta p \cdot w^*). \tag{13.2}$$

(Re = Realteil, * = konjugiert komplexe Größe). Für einen bestimmten Wellentyp folgt dann aus (10.17) und (10.18)

$$S_n = \frac{\varkappa^2 h_n p_0(0)}{2\sigma} [\Theta_n(\vartheta)]^2 \operatorname{Im}\left(y_n^* \frac{d y_n}{d x}\right) \tag{13.3}$$

(Im = Imaginärteil). WEEKES und WILKES fordern nun, daß S_n stets positiv ist. Sie gehen dabei von der Vorstellung aus, daß sowohl bei Gravitationserregung als auch bei thermischer Erregung die Energie vorwiegend der bodennahen Atmosphäre zugeführt wird, so daß in großen Höhen die Energie stets aufwärts fließen muß. Dann folgt

$$\operatorname{Im}\left(y_n^* \frac{d y_n}{d x}\right) > 0. \tag{13.4}$$

SIEBERT [*21d*] weist darauf hin, daß der benutzte Formelapparat an keiner Stelle Glieder besitzt, die einer Energievernichtung oder Erzeugung entsprechen, und daß man deshalb konsequenterweise auch in der Randbedingung keine solchen Glieder einführen sollte. Er fordert daher, daß

$$\operatorname{Im}\left(y_n^* \frac{d y_n}{d x}\right) = 0. \tag{13.5}$$

Bei der Betrachtung der erzwungenen Schwingungen (Ziff. 14 und 15) sieht man, daß seine Forderung einen glatten Übergang in das durch (13.1) gekennzeichnete Intervall gewährt, so daß die Randbedingung (13.5) als zweckmäßig erscheint.

[1] Vgl. auch M. V. WILKES: Nature, Lond. **164**, 281 (1949).

D. Erzwungene Schwingungen.

I. Gravitation.

14. Resonanzkurven bei Gravitationserregung. Bei Gravitationsanregung gilt die Differentialgleichung

$$y_n'' - \frac{1}{4} y_n = -\frac{1}{h_n}\left(\frac{dH}{dx} + \frac{\varkappa - 1}{\varkappa} H\right) y_n, \tag{14.1}$$

mit der Randbedingung am Boden

$$\left(y_n' - \frac{1}{2} y_n\right) + \frac{H}{h_n} y_n = \frac{i\sigma}{\varkappa g h_n} \Omega_n \quad \text{für} \quad x = 0, \tag{14.2}$$

und mit (12.7) als oberer Randbedingung. Ist $y = \Psi_{h_n}$ eine Lösung der Differentialgleichung (14.1) mit fest vorgegebenem h_n, die außerdem noch der Randbedingung (12.7) am oberen Ende der Atmosphäre genügt, so erhält man eine Lösung von (14.2) mit dem Ansatz $y_n = Q\,\Psi_{h_n}$ und bestimmt die Konstante Q zu

$$Q = \left[\frac{\frac{i\sigma}{\varkappa h_n g}\Omega_n}{\Psi_{h_n}' + (H/h_n - \frac{1}{2})\,\Psi_{h_n}}\right]_{x=0}. \tag{14.3}$$

Wenn h_n gleich einem Eigenwert $\hat{h}$ der Atmosphäre wird, so verschwindet nach (12.4) der Nenner und $Q \to \infty$, wie man es für eine ungedämpfte Schwingung im Resonanzfall erwartet.

Es interessiert vor allem die Druckschwankung an der Erdoberfläche $x = 0$. Nach (10.18) erhält man mit $w_n(0) = 0$

$$\delta p_n(0) = -p_0(0)\frac{\varkappa}{i\sigma} y_n = \left[\frac{-\frac{H}{h_n}\Psi_{h_n}}{\Psi_{h_n}' + (H/h_n - \frac{1}{2})\,\Psi_{h_n}}\right]_{x=0} \cdot \overline{\delta p_n(0)} = V_{h_n}\,\overline{\delta p_n(0)}. \tag{14.4}$$

Der Ausdruck in den eckigen Klammern gibt die Resonanzvergrößerung eines Wellentypes mit dem Parameter h_n.

Für die Atmosphäre im konvektiven Gleichgewicht ist

$$\boxed{V_{h_n} = \frac{H(0)}{H(0) - h_n}.} \tag{14.5}$$

Es treten die Phasenwinkel 0 und 180° auf, je nachdem $h_n < H(0)$ oder $h_n > H(0)$ ist. Für eine isotherme Atmosphäre ist

$$\boxed{V_{h_n} = \frac{H(0)}{H(0) - \frac{1}{2} h_n (1 + b_n)},} \tag{14.6}$$

$$b_n = \sqrt{1 - 4\frac{\varkappa - 1}{\varkappa}\frac{H(0)}{h_n}} \quad \text{für} \quad h_n > 4\frac{\varkappa - 1}{\varkappa} H(0). \tag{14.7a}$$

Für $h_n < 4\frac{\varkappa - 1}{\varkappa} H(0)$ hat (14.1) keine Lösung, die die Randbedingung (12.7) und (12.8) erfüllt. Die von SIEBERT benutzte Randbedingung (13.5) liefert die gleiche Resonanzfunktion (14.6) mit

$$b_n = -\sqrt{4\frac{\varkappa - 1}{\varkappa}\frac{H(0)}{h_n} - 1} \quad \text{für} \quad h_n < 4\frac{\varkappa - 1}{\varkappa} H(0). \tag{14.7b}$$

V_{h_n} und seine Ableitung nach h_n sind dann stetig. Die WILKEssche Randbedingung (13.4) liefert dagegen

$$V_{h_n}^* = \frac{H(0) - \frac{1}{2}h_n(1 - i\,b_n)}{H(0) - \frac{1}{2}h_n/\varkappa}. \tag{14.8}$$

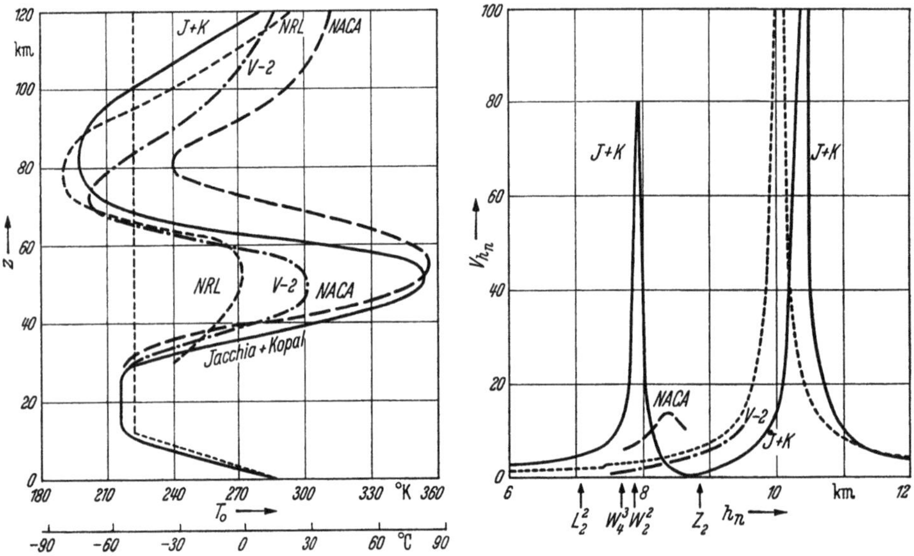

Fig. 11. Links Temperaturprofile für die Erdatmosphäre und Modellatmosphären. Das NRL-Profil ist eine Zusammenfassung von Raketenmessungen. Für die drei anderen Modelle, die das Temperaturmaximum in 50 km Höhe zeigen, wurden von JACCHIA und KOPAL [30] die rechts stehenden Resonanzkurven bei Gravitationserregung gerechnet. Außerdem ist das von SIEBERT benutzte zweischichtige Atmosphärenmodell und die von ihm berechnete Resonanzkurve eingetragen (die gleiche wie in Fig. 12) (nach [21b]).

Auch diese Lösung geht stetig in (14.6) über, aber ihre Ableitung springt. Die Resonanzfunktion ist jetzt komplex wie bei einem Oszillator mit Dämpfungsglied. Dadurch treten auch Phasenwinkel auf, die von 0 und 180° verschieden sind.

In Fig. 11 sind einige Resonanzfunktionen für verschiedene Atmosphärenmodelle aufgezeichnet, und in Fig. 12 ist noch einmal vergrößert die SIEBERTsche Resonanzfunktion für das zweischichtige Modell dargestellt.

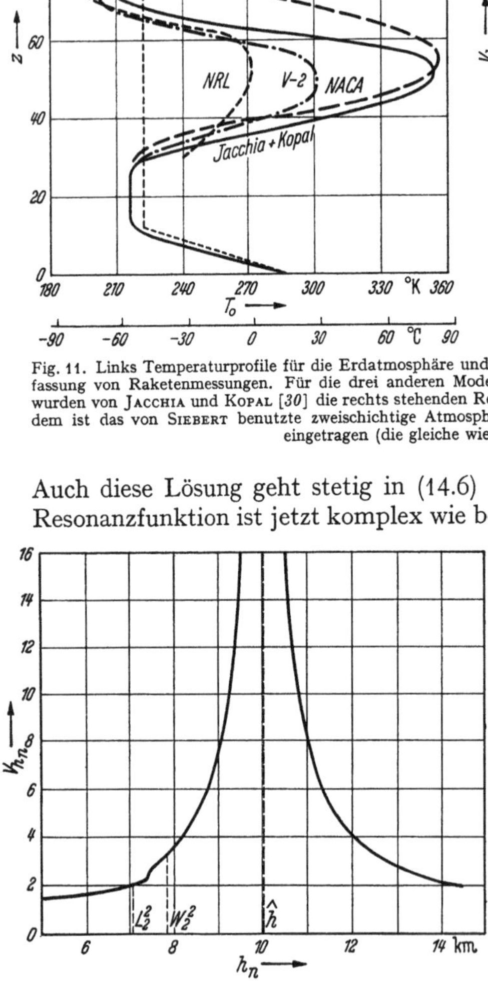

Fig. 12. Resonanzkurve des zweischichtigen Atmosphärenmodells bei gravitationsbedingter Erregung mit Angabe der h-Werte der wichtigsten Hauptwellentypen (nach SIEBERT [21b]).

15. Lunare Wellen. Das Jahresmittel der betrachteten lunaren halbtägigen Druckwelle L_2 läßt sich nach SIEBERT [21d] darstellen durch die Kugelfunktionsentwicklung

$$\left.\begin{aligned} L_2\colon \delta p(0) &= 0{,}071\,[P_2^2 - 0{,}30\,P_4^2 + \\ &+ 0{,}1\,P_6^2]\sin(2\tau + 75°)\ \text{mbar}. \end{aligned}\right\} \quad (1$$

Die Abhängigkeit von der Poldistanz ist in Fig. 13 dargestellt. Die Phase beträgt 75°, deshalb läuft die Welle um 15°, das entspricht $\frac{1}{2}$ Std, hinter der Erregung her [vgl. (3.2)].

Zum Vergleich mit der Theorie benötigt man Entwicklungen von Erregung und Beobachtung nach HOUGHschen Funktionen. Gemäß Tabelle 3 erhält man aus (3.2) für die Erregung

$$\overline{\delta p(0)} = 0{,}0312\,\Theta_2(\vartheta)\sin(2\tau + 90°)\ \text{mbar} \tag{15.2}$$

und für die beobachtete Bodendruckvariation

$$\delta p(0) = [0{,}070\,\Theta_2 + 0{,}004\,\Theta_4 + \cdots]\sin(2\tau + 75°)\ \text{mbar}. \tag{15.3}$$

Die Entwicklung (15.1) wird also schon weitgehend durch das Θ_2-Glied in (15.3) dargestellt. Es genügt, dieses Hauptglied zu betrachten. Die Resonanzvergrößerung ist dann 2,2fach und stimmt befriedigend mit der von SIEBERT berechneten Resonanzvergrößerung überein (Fig. 12).

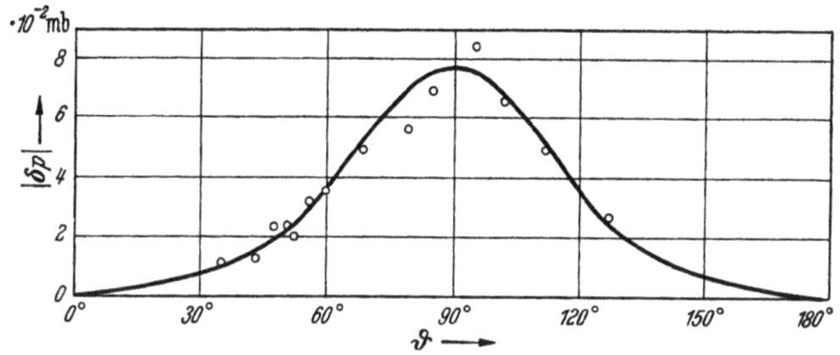

Fig. 13. Jahresmittel der Amplitude von L_2 als Funktion der Poldistanz (nach SIEBERT [*21 d*]).

Die Phasendifferenz von 15° zwischen Druckwelle und Erregung könnte durch die (in der Theorie vernachlässigte) Bodenreibung hervorgerufen sein.

Außer den Mittelwerten (15.1) sind von CHAPMAN, BARTELS und Mitarbeitern durch sorgfältige statistische Analysen eine Fülle von Einzelheiten über die lunare Gezeitenwelle der Atmosphäre bekannt geworden, die nur zum ganz kleinen Teil theoretisch verarbeitet wurden. Zusammenstellung in [*2*], [*4*], [*5*], [*13*].

a) Die Amplituden von L_2 hängen nicht nur von ϑ ab, sondern sind auch längenabhängig. Es gibt große Werte (20% größer als das Mittel) am südlichen indischen Ozean und kleine Werte (30% kleiner als das Mittel) an der Westküste von Nord- und Südamerika.

b) Amplituden und Phasen zeigen einen deutlichen *Jahresgang*, der an allen Stationen auf der Nord- und Südhalbkugel gleichartig verläuft: Im Nordwinter ist auf beiden Erdhalbkugeln die Amplitude am kleinsten, und das Maximum tritt etwa 1 Std später ein als im Nordsommer und zur Zeit der Äquinoktien. Die Ursache könnte in einer jahreszeitlichen Änderung der Temperaturverteilung in der Atmosphäre zu suchen sein. Um hierfür eine Theorie aufzustellen, hat SAWADA [*25a*], [*25b*], [*25d*] die lunaren Schwingungen für 20 verschiedene Atmosphärenmodelle mit der Randbedingung (13.4) berechnet. Dann bekommt er wie in (14.8) auch eine Variation der Phasen. Auf Grund dieser Annahme kann er die beobachteten Änderungen durch eine Erhöhung der Temperatur der Ozonosphäre während des Nordsommers und eine Erniedrigung während des Nordwinters erklären.

c) Die Wirkung anderer Gezeitenterme als M_2 auf die halbtägige Welle ist nur wenig untersucht. Die Wirkung von N_2 ist für eine Reihe von Stationen sicher nachgewiesen [*19*]. In den Einzelheiten zeigen sich jedoch Unstimmigkeiten. Wie in Ziff. 3 ausgeführt, können wegen der üblichen statistischen Bearbeitungsweise die Deklinationsänderungen nur an dem kleinen Glied {255,545} studiert werden. Untersuchungen dieser 18,6jährigen Variation für Batavia ([*19*], S. 37) zeigen eine befriedigende Übereinstimmung.

d) Zu den größten Erfolgen der statistischen Methoden gehört die CHAPMANsche Berechnung der zu L_2 gehörigen kleinen sekundären Temperaturvariation $\tilde{\tau}(0)$ in Batavia. Nach (9.10) errechnet man für Batavia aus den Druckbeobachtungen $\delta p(0)$ eine halbmondentägige Temperaturwelle von

$$\tilde{\tau}(0) = 0{,}0072^\circ \text{ C} \sin(2\tau + 65^\circ).$$

(Der Phasenwinkel von $L_2(\delta p)$ beträgt für Batavia 65°, nach neueren Rechnungen 68°, abweichend von dem Mittelwert 75° für die ganze Erde.) Mit den Daten aus 63 Jahren berechnete CHAPMAN tatsächlich

$$\tilde{\tau}(0) = 0{,}0086^\circ \text{ C} \sin(2\tau + 67^\circ).$$

Im Anschluß an diese Rechnung [*12a*] hat CHAPMAN noch untersucht, wie sich der Wärmetransport in der Unterlage (Land oder Meer) auswirken würde [*12b*]. Über Land würden die Ergebnisse nicht wesentlich abgeändert, aber an der Meeresoberfläche würde der adiabatische Temperatureffekt durch turbulenten Austausch fast völlig verschwinden. Erst in einigen 100 m Höhe über dem Wasser wäre wieder der volle Effekt zu erwarten.

e) Statistische Analysen für die zu L_2 gehörigen Winde sind bisher nur sehr spärlich vorhanden. Nur für Mauritius hat CHAPMAN aus 16jährigen Beobachtungsdaten Windvektoren abgeleitet. Die Werte sind aber noch nicht genügend gesichert [*2*].

f) Lunare ganztägige Wellen (zu 0_1 gehörig) haben sich bisher noch für keine einzige Station mit Sicherheit nachweisen lassen. Im Luftdruck treten sie offenbar, wie auch die solaren ganztägigen Variationen, stark zurück. Aber auch statistisch sind sie benachteiligt. Man würde viel längere Beobachtungsreihen benötigen, um sie vor dem statistischen Hintergrund mit Sicherheit zu erfassen, als bei den halbtägigen.

16. Solare Wellen. Die halbtägige wandernde Welle W_2^2 ist die größte aller planetarischen Wellen. Die zugehörige Schwankung des Bodendruckes wird im Jahresmittel dargestellt durch

$$W_2^2:\quad \delta p(0) = 1{,}23\,[P_2^2 - 0{,}182\,P_4^2]\sin(2t' + 158^\circ)\text{ mbar} \qquad (16.1)$$

(HAURWITZ [*36*]). Das Kugelfunktionsaggregat läßt sich näherungsweise durch $\sin^3\vartheta$ darstellen[1]. Doch ist dies wohl mehr Zufall, und die Darstellung (16.1) ist für den Vergleich mit der Theorie bequemer. Zu einem solchen Vergleich benötigt man eine Entwicklung nach HOUGHschen Funktionen. Gemäß Tabelle 3 ist

$$W_2^2:\quad \delta p(0) = [1{,}186\,\Theta_2 + 0{,}168\,\Theta_4 + \cdots]\sin(2t' + 158^\circ)\text{ mbar}. \qquad (16.2)$$

Dagegen erhält man aus (4.1) für die Erregung durch Gravitationswirkung:

$$\overline{\delta p(0)} = 0{,}0147\,\Theta_2 \sin(2t' + 90^\circ)\text{ mbar}. \qquad (16.3)$$

Die beobachtete Welle ist also 87mal größer als die Erregung. Diese große Verstärkung war einer der Gründe für die Aufstellung der Resonanztheorie. Aber auch die Phasen stimmen in (16.2) und (16.3) nicht überein. Dies erklärte CHAPMAN [*10*] (vgl. auch Ziff. 18) dadurch, daß W_2^2 nur teilweise durch Gravitationswirkung erzeugt wird.

[1] Zuerst abgeleitet von P. JAERISCH, Meteor. Z. **24**, 481–498 (1907). Bekannt geworden durch G. C. SIMPSON: Quart. J. Roy. Meteor. Soc. **44**, 1–19 (1918).

Die jahreszeitliche Veränderung von W_2^2 (Fig. 4) zeigt Maxima der Amplitude zur Zeit der Äquinoktien. Dies ist nach Ziff. 4 und 5 sowohl für die gravitationsbedingte als auch für die thermische Erregung zu erwarten.

Eine gravitationserregte ganztägige Welle wird nicht beobachtet (vgl. Ziff. 22)

II. Thermische Erregung.

17. Verschiedene Arten der thermischen Erregung und Resonanzkurven. Für die thermische Erregung gilt die Differentialgleichung:

$$y_n'' - \frac{1}{4} y_n + \frac{1}{h_n}\left(\frac{dH}{dx} + \frac{\varkappa - 1}{\varkappa} H\right) y_n = \frac{1}{h_n} \frac{\varkappa - 1}{g \varkappa^2} J_n e^{-\frac{1}{2}x} \tag{17.1}$$

mit der Randbedingung am Boden

$$y_n' - \frac{1}{2} y_n + \frac{H}{h_n} y_n = 0 \quad \text{für } x = 0. \tag{17.2}$$

Man sieht, daß y_n wieder unendlich groß werden muß, wenn h_n gleich einem der Eigenwerte $\hat{h}$ wird, da die linke Seite von (17.1) sonst verschwinden würde und nicht den durch die rechte Seite geforderten Wert annehmen kann.

Es gibt nun verschiedene Arten der thermischen Anregung je nach der Art der primären Temperaturvariation $\tau(z)$. Wichtig sind die beiden folgenden Fälle:

a) Die eingestrahlte Wärmeenergie breite sich durch Turbulenz (Austausch A) von der Erdoberfläche nach oben aus. Dann läßt sich $\tau_n^A(z)$ wie in (6.2) darstellen durch

$$\tau_n^A(z) = \tau_n^A(0)\, e^{-\frac{1}{2}(1+i)\sqrt{\sigma/K}\, z}. \tag{17.3}$$

Nach der Transformation von z auf x gilt für eine isotherme Atmosphäre exakt (und für andere Atmosphären näherungsweise), wenn die für K gemessenen Werte verwendet werden

$$J_n(x) = i\sigma c_p \tau_n^A(0)\, e^{-kx} \quad \text{mit} \quad k = \tfrac{1}{2}(1+i)\, H(0) \sqrt{\sigma/K}. \tag{17.4}$$

Als Lösungen von (17.1) und (17.2) erhält man für die Atmosphäre im konvektiven Gleichgewicht

$$\delta p_n(0) = -\frac{H(0)}{(H(0) - h_n)(1+k)} \frac{p_0(0)}{T_0(0)} \tau_n^A(0) \tag{17.5}$$

da $|k| \gg 1$, gilt näherungsweise

$$\boxed{\delta p_n(0) = \frac{\sqrt{K/\sigma}\, e^{\frac{3}{4}\pi i}}{H(0) - h_n} \frac{p_0(0)}{T_0(0)} \tau_n^A(0).} \tag{17.6}$$

Für eine isotherme Atmosphäre ergibt sich

$$\delta p_n(0) = -\frac{H h_n (1 - b_n + 2k)}{[2H - h_n(1+b_n)]\left[(1+k)\, k\, h_n + \frac{\varkappa - 1}{\varkappa} H\right]} \frac{p_0(0)}{T_0(0)} \tau_n^A(0) \tag{17.7}$$

mit b_n nach (14.7). Näherungsweise folgt daraus für große $|k|$:

$$\boxed{\delta p_n(0) = \frac{\sqrt{K/\sigma}\, e^{\frac{3}{4}\pi i}}{H - \frac{1}{2} h_n (1 + b_n)} \frac{p_0(0)}{T_0(0)} \tau_n^A(0).} \tag{17.8}$$

Charakteristisch sind die Phasendifferenzen von 135° und 315° für h_n-Werte, die kleiner bzw. größer als der Eigenwert der Atmosphäre sind.

b) Der zweite wichtige Fall ist der, daß die erregende Temperaturwelle in jeder Höhe die gleiche Phase und Amplitude hat:

$$\tau_n^S(z) = \text{const.} \tag{17.9}$$

Ein solches Glied könnte etwa durch eine gleichmäßige Absorption von Strahlung in der Atmosphäre entstehen. W. SCHMIDT[1] fand ein höhenunabhängiges „Strahlungsglied" bei der Diskussion von Temperaturbeobachtungen in verschiedenen Höhen auf dem Eiffelturm. LETTAU[2] verwirft jedoch seine Deutung als Strahlungsterm und sieht in (17.9) nur eine Korrektur des Austauschgliedes. Er findet, daß die Annahme eines zeit- und höhenunabhängigen Austauschkoeffizienten eine zu große Vereinfachung darstellt. Deshalb könne man die Temperaturvariation in der Höhe nicht allein durch einen Term der Art (17.3) darstellen. Vgl. Ziff. 28 und die dort angegebene Literatur.

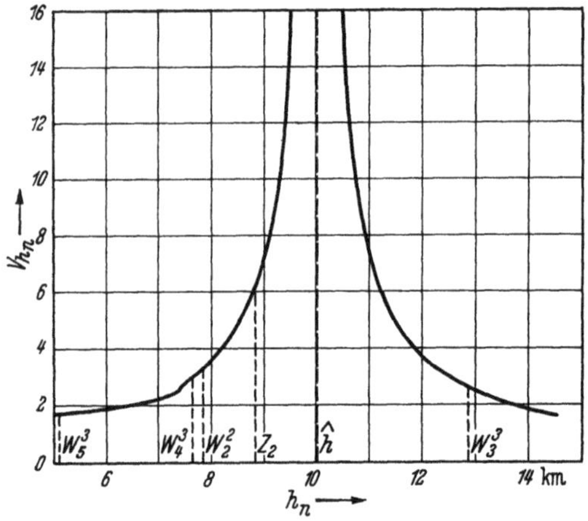

Fig. 14. Resonanzkurve des zweischichtigen Atmosphärenmodells bei thermischer Erregung durch ein höhenunabhängiges Strahlungsglied mit Angabe der h-Werte der wichtigsten Hauptwellentypen (nach SIEBERT [*21b*]).

Unabhängig von speziellen Annahmen über das Zustandekommen eines Gliedes der Gestalt (17.9) soll hier (nach [*21d*]) die Theorie für eine höhenunabhängige primäre Temperaturwelle aufgestellt werden. Der Kürze halber sprechen wir bei einer Erregung der Art (17.3) von einer Erregung durch *Austausch* (A) und bei einer Erregung durch ein Glied (17.9) von einer Erregung durch *Strahlung* (S).

Es folgt für eine Atmosphäre im konvektiven Gleichgewicht die Resonanzfunktion

$$\boxed{\delta p_n(0) = -\frac{H(0)}{H(0)-h_n}\,\frac{p_0(0)}{T_0(0)}\,\tau_n^S} \tag{17.10}$$

und für eine isotherme Atmosphäre

$$\boxed{\delta p_n(0) = -\frac{h_n(1-b_n)}{\frac{\varkappa-1}{\varkappa}\,[2H-h_n(1+b_n)]}\,\frac{p_0(0)}{T_0(0)}\,\tau_n^S.} \tag{17.11}$$

Ein Vergleich von (17.5) und (17.10) ergibt das Verhältnis der Druckvariationen, die durch ein Austauschglied $\tau_n^A(0)$ und Strahlungsglied τ_n^S erregt sind, zu

$$\left|\frac{[\delta p_n(0)]_S}{[\delta p_n(0)]_A}\right| = |k|\left|\frac{\tau_n^S}{\tau_n^A(0)}\right|. \tag{17.12}$$

Ein Strahlungsglied τ^S in der Variation der Bodentemperatur ist also um den Faktor $|k|\approx 100$ wirksamer als ein gleichgroßes Austauschglied. Etwa das gleiche Verhältnis erhält man auch für die isotherme Atmosphäre, wenn $h_n \approx H$.

SIEBERT [*21d*] hat die Resonanzfunktion auch für das durch (3.10) beschriebene Atmosphärenmodell berechnet (Fig. 14), wobei er annahm, daß das Strahlungsglied nur in der Troposphäre wirksam ist. Die in Fig. 14 dargestellte

[1] W. SCHMIDT: Sitzgsber. Akad. Wiss. Wien IIa **127**, 1889—1985 (1918). — Meteor. Z. **37**, 49—59 (1920). — Der Massenaustausch in freier Luft und verwandte Erscheinungen. Problem kosmischer Physik, Bd. 7, 118 S. Hamburg 1925.

[2] H. LETTAU: Atmosphärische Turbulenz. Leipzig 1939. § 34, 35.

Funktion ist so normiert, daß für die halbtägige wandernde Welle ($n=2$, $s=2$) $h=7{,}85$ km in den Resonanzkurven für die Gravitation (Fig. 12) und für die Strahlungserregung (Fig. 14) die gleiche Vergrößerung auftritt. Die beiden Figuren stimmen dadurch gut überein. WILKES [*18*] und HAURWITZ [*24*] haben Analogien zwischen thermischer Erregung durch ein Austauschglied und Gravitationserregung angegeben. Für den praktischen Gebrauch sind das sehr brauchbare Näherungen. Die beiden Resonanzkurven müssen ja die durch das Atmosphärenmodell bestimmten gleichen Resonanzstellen besitzen. Man wird durch solche Analogien aber verleitet zu übersehen, daß die thermische Erregung im Gegensatz zur Gravitationserregung ja nicht nur durch die Bodenwerte von τ_n beschrieben werden kann, sondern nur durch die ganze Funktion $\tau_n(z)$. Je nach der Art der Erregung werden die Resonanzstellen mehr oder weniger scharf.

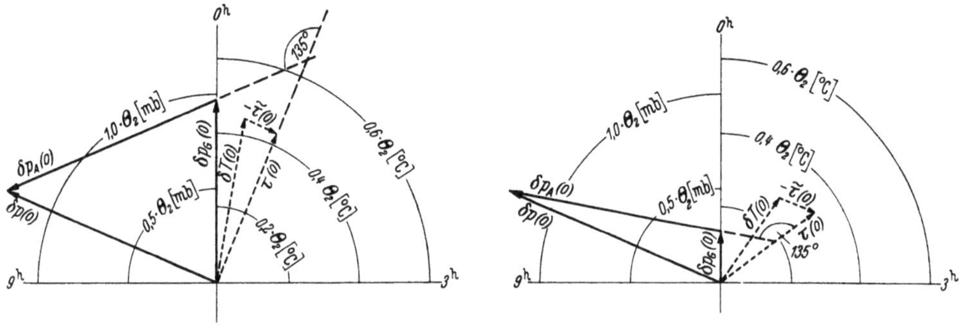

Fig. 15. Periodenuhr für die Erklärung von W_2^2 nach der Resonanztheorie. $\delta p(0)$ setzt sich zusammen aus einem durch Gravitation erregten Anteil $\delta p_G(0)$ und einem thermisch durch ein Austauschglied der Temperaturvariation erregten Anteil $\delta p_A(0)$. Links mit dem von CHAPMAN benutzten Werten fur $\delta T(0)$, rechts mit dem von HAURWITZ und MÖLLER gefundenen Wert.

18. Halbtägige wandernde Wellen. Die *Darstellung* der halbtägigen wandernden Welle ist bereits in (16.1) und (16.2) gegeben worden.

Die Entstehung dieser Welle kann auf verschiedene Weise erklärt werden; dabei beschränken wir uns auf das Hauptglied Θ_2:

a) *Resonanztheorie* (CHAPMAN [*10*][1]). Wegen ihrer Phasenlage kann W_2^2 keine reine Gravitationswelle sein, sonst müßte ihre Phase bei 90° [nach (4.1)] liegen. Darum nahm CHAPMAN an, daß sie zum Teil durch Gravitation und zum Teil thermisch, und zwar durch ein Austauschglied, erregt sei. Für die Anteile, die von beiden Erregungsarten herkommen, kann man die Phasenlage angeben. Für den Gravitationsanteil δp_G beträgt die Phase 90°. Für den thermischen Anteil findet man sie aus den Beobachtungen der Bodentemperatur. CHAPMAN fand aus dem Mittel äquatornaher Stationen

$$W_2^2\colon\ \delta T(0) \approx 0{,}43\,\Theta_2 \sin(2t' + 80°)\ °\mathrm{C}. \tag{18.1}$$

Die sekundäre Temperaturwelle beträgt nach (9.10)

$$W_2^2\colon\ \tilde{\tau}(0) = 0{,}096\,\Theta_2 \sin(2t' + 158°)\ °\mathrm{C}. \tag{18.2}$$

Wird diese von $\delta T(0)$ abgezogen, so bleibt

$$W_2^2\colon\ \tau(0) = 0{,}43\,\Theta_2 \sin(2t' + 67°)\ °\mathrm{C} \tag{18.3}$$

(Fig. 15a). Die dadurch erregte Druckwelle δp_A hat den Phasenwinkel $67° + 135° = 202°$. Die beobachtete Druckwelle mit dem Phasenwinkel von 158° kann

[1] Durch Berücksichtigung der neuen Darstellung (16.1) für den Bodendruck werden die zahlenwerte gegen die CHAPMANschen etwas geändert.

aus den beiden Anteilen mit Phasenwinkeln von 90° und 202° nur entstehen, wenn

$$\left|\frac{\delta p_G(0)}{\delta p_A(0)}\right| = 0{,}77. \tag{18.4}$$

Auch aus den Amplituden läßt sich dieses Verhältnis bestimmen. Sowohl für eine isotherme Atmosphäre als auch für eine Atmosphäre im konvektiven Gleichgewicht gilt nach (14.6) und (17.6), oder (14.5) und (17.8)

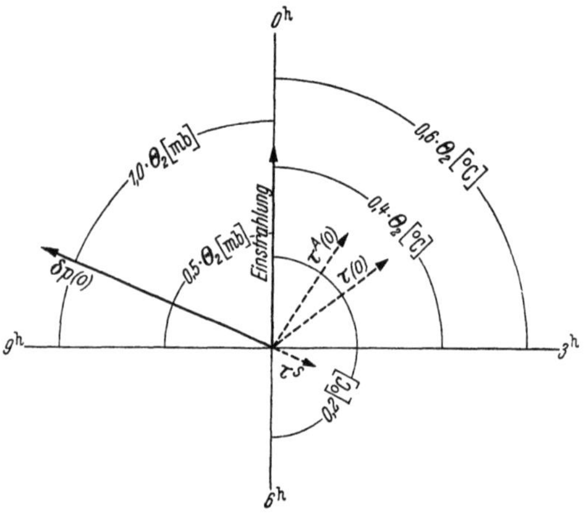

Fig. 16. Periodenuhr für die Erklärung von W_2^2 durch thermische Erregung mit einem Strahlungsglied τ^S der Temperaturvariation (SIEBERT) und Zerlegung der primären Temperaturvariation $\tau(0)$ am Boden in Strahlungs- (τ^S) und Austauschglied [$\tau^A(0)$]. [$\tau(0)$ hat den gleichen Wert wie in Fig. 15 rechts.]

$$\left|\frac{\delta p_G(0)}{\delta p_A(0)}\right| = \left|\frac{\Omega(0)/g}{\tau(0)/T_0(0)}\right| \sqrt{\frac{\sigma}{K}}\,. \tag{18.5}$$

Mit dem CHAPMANschen Wert (18.3) ergibt sich für das Verhältnis (18.5) der Wert 1,0 in befriedigender Übereinstimmung mit (18.4).

Man bekommt für die einzelnen Anteile:

$$\left.\begin{aligned} |\delta p_G(0)| &= 0{,}98\,\Theta_2\text{ mbar},\\ |\delta p_A(0)| &= 1{,}28\,\Theta_2\text{ mbar}.\end{aligned}\right\} \tag{18.6}$$

Nach (4.1) gilt aber

$$\left.\begin{aligned} &|\overline{\delta p_G(0)}|\\ &\quad=(0{,}0147\,\Theta_2+\cdots)\text{ mbar}.\end{aligned}\right\} \tag{18.7}$$

Die Resonanzvergrößerung müßte also 0,98/0,0147 = 67fach sein.

Benutzt man aber die Werte von HAURWITZ und MÖLLER [*24*], denen ein umfangreiches Beobachtungsmaterial zugrunde liegt, so ergibt sich ein etwas anderes Bild (Fig. 15b). Es ist

$$W_2^2\colon\ \delta T(0) = 0{,}282\,\Theta_2 \sin(2t' + 54°)\,, \tag{18.8}$$

$$\tau(0) = 0{,}320\,\Theta_2 \sin(2t' + 36°)\,. \tag{18.9}$$

Damit erhält man aus den Phasenbeziehungen

$$\left|\frac{\delta p_G(0)}{\delta p_A(0)}\right| = 0{,}23\,. \tag{18.10}$$

Dagegen liefern die Amplituden für dasselbe Verhältnis nach (18.5) den Wert 1,51. Die Amplitude der Temperaturwelle müßte also 6,6mal größer sein als HAURWITZ und MÖLLER fanden. Das ist sehr unwahrscheinlich.

b) Eine andere Erklärung gibt SIEBERT (Fig. 16). Er nimmt an, daß die Druckwelle durch ein *Strahlungsglied* der Temperatur τ^S erregt wird. Die Temperaturwelle muß entgegengesetzte Phase haben wie $\delta p(0)$. Aus der Resonanzkurve Fig. 14 findet er, daß

$$\tau^S = -\frac{1}{11{,}0}\,\delta p(0) = 0{,}108\,\Theta_2 \sin(2t' + 338°)\,. \tag{18.11}$$

Fig. 16 zeigt die zugehörige Zerlegung der primären Temperaturwelle τ am Boden (der gleichen wie in Fig. 15b) in das Austausch- τ^A und Strahlungsglied τ^S. Das Austauschglied spielt für die Druckwelle keine Rolle mehr, da ja keine Resonanzverstärkung stattfindet.

Über das Zustandekommen des Strahlungsgliedes kann man sich nach SIEBERT folgende Vorstellung machen: Die einfallende halbtägliche Strahlungsvariation hat nach Tabelle 1 einen Phasenwinkel von 90°. Die durch Absorption entstehende Temperaturwelle kommt um 90° später, muß also einen Phasenwinkel von 0° haben. Wahrscheinlich wäre dies die richtige Lage von τ^S. Dann müßte $\delta p(0)$ eine Phase von 180° besitzen. Der beobachtete Phasenwinkel von 158° deutet auf eine Verspätung durch Bodenreibung hin, wie wir sie auch bei der lunaren Gravitationswelle gefunden haben (vgl. Ziff. 15).

Im *Jahresgang* zeigt W_2^2 an nahezu allen Stationen auf der Erde Maxima zur Zeit der Äquinoktien und Minima zur Zeit der Solstitien. Diese Eigenschaft besitzen nach Ziff. 4 und 6 sowohl die Gravitations- als auch die thermische Erregung, so daß man hieraus nicht zwischen den beiden Theorien entscheiden kann.

Über eine Darstellung der zu W_2^2 gehörigen Winde vgl. STOLOV[1].

19. Halbtägige stehende Wellen. Die halbtägige stehende Welle nach Weltzeit Z_2 wurde von GREELY und AD. SCHMIDT[2] in den Druckbeobachtungen polnaher Stationen entdeckt, lange bevor man irgendeine Vorstellung über die Ursache ihrer Entstehung hatte. Sie läßt sich beschreiben durch die Darstellung

$$Z_2\colon\ \delta p(0) = 0{,}085\ P_2 \sin(2t + 118°)\ \text{mbar} \tag{19.1}$$

(HAURWITZ [36], vgl. auch [35]). Für die Entwicklung nach HOUGHschen Funktionen erhält man nach Tabelle 3

$$Z_2\colon\ \delta p(0) = (0{,}0785\,\Theta_2 + 0{,}0216\,\Theta_4 + \cdots) \sin(2t + 118°)\ \text{mbar}. \tag{19.2}$$

Die Überlagerung von W_2^2 und Z_2 ergibt in den Polgebieten eine eigentümliche geographische Verteilung der zusammengesetzten halbtägigen Druckwelle mit je zwei Amphidromiepunkten auf jeder Halbkugel[3].

Für Z_2 kommt nur eine thermische Erregung in Frage. Nach Ziff. 6 beträgt die entsprechende Temperaturwelle, die durch die Land-Wasserverteilung erzeugt wird, am Erdboden:

$$\left.\begin{aligned}\delta T(0) &= 0{,}040\,P_2 \sin(2t + 214°) + 0{,}104\,P_4 \sin(2t + 56°)\ °\text{C}\\ &= 0{,}056\,\Theta_2 \sin(2t + 220°) + 0{,}055\,\Theta_4 \sin(2t + 60°) + \cdots.\end{aligned}\right\} \tag{19.3}$$

Wir beschränken uns wieder auf das Hauptglied Θ_2. Die sekundäre Temperaturwelle ergibt sich aus (9.10) zu

$$\tilde{\tau}(0) = 0{,}006\,\Theta_2 \sin(2t + 118°)\ °\text{C}. \tag{19.4}$$

Somit bleibt für die primäre Erregung

$$\tau(0) = 0{,}057\,\Theta_2 \sin(2t + 230°)\ °\text{C}. \tag{19.5}$$

Vom Standpunkt der Resonanztheorie aus ist die beobachtete Phase schwer verständlich: Würde die Erregung durch ein Austauschglied erfolgen, so müßte nach (17.8) für einen h-Wert, der größer ist als der Eigenwert $\hat{h}$, zwischen $\delta p(0)$ und $\tau(0)$ ein Phasenwinkel von 45° liegen, nach den Beobachtungen beträgt er aber 112°.

Genau so ungünstig liegen die Verhältnisse bei der Amplitude: Die Erklärung der Amplitude von Z_2 stellt für die Resonanztheorie eine große Schwierigkeit

[1] H. L. STOLOV: J. Meteorology **12**, 117—140 (1955).

[2] A. W. GREELY: The meteorological results of the Lady Franklin Bay Expedition 1881 to 1883. New York 1894. — AD. SCHMIDT: Meteor. Z. **7**, 182—185 (1890).

[3] Eine Karte, die die Überlagerung von W_2^2 und Z_2 veranschaulicht, bei F. J. W. WHIPPLE, Quart. J. Roy. Meteor. Soc. **44**, 20—22 (1918) und HAURWITZ [36].

dar. Der zu dem Hauptglied Θ_2 gehörige h-Wert von 8,85 km liegt zu weit von dem zu dem Hauptglied von W_2^2 gelegenen Wert 7,85 km entfernt, als daß noch eine wirksame Resonanzverstärkung erwartet werden könnte. Die Resonanzkurve von JACCHIA und KOPAL (Fig. 11) geht bei 8,7 km sogar fast bis auf 0 herunter. BARTELS vermutete, daß die Unterschiede der zu Z_2, W_2^2 (und W_4^3) gehörigen h-Werte durch die Gestalt der Erdoberfläche ausgeglichen würden und daß ihre h-Werte alle mit dem Eigenwert der Erdatmosphäre zusammenfielen. Für die Resonanztheorie treten dann aber neue Schwierigkeiten auf (vgl. Ziff. 25).

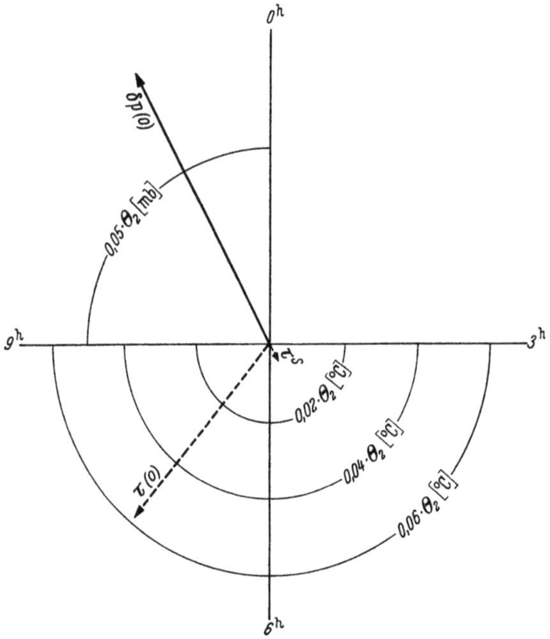

Fig. 17. Periodenuhr für die Erklärung von Z_2 durch thermische Erregung mit einem Strahlungsglied τ^S der Temperaturvariation. Dieses muß durch die zu der primären Temperaturerregung $\tau(0)$ gehörige Reflexion vom Erdboden entstanden ein.

Wesentlich einfacher kann man Z_2 durch ein Strahlungsglied erklären (Fig. 17). Zwar gibt es kein von außen kommendes Strahlungsglied, das auf Z_2 führt, die *Ausstrahlung* des Erdbodens könnte aber ein solches Weltzeitglied liefern. Um die beobachtete Bodendruckwelle zu erzeugen, müßte das zugehörige Strahlungsglied der Temperatur τ^S eine Phase von 298° haben. Zwischen $\tau(0)$ und τ^S liegt dann ein Phasenwinkel von 68°.

Aus der Resonanzkurve (Fig. 14) kann man entnehmen, daß

$$\left.\begin{aligned} \tau^S &= 0{,}049\,|\delta p(0)| \\ &= 0{,}0038\,\Theta_2 \end{aligned}\right\} \qquad (19.6)$$

sein muß. Also beträgt das durch die Ausstrahlung des Erdbodens verursachte Strahlungsglied der Temperatur

$$\tau^S = \tfrac{1}{15}\,\tau(0)\,^1. \qquad (19.7)$$

Es erscheint fraglich, ob die geographische Verteilung von $S_2(\delta p)$ und $S_2(\delta T)$ schon hinreichend bekannt sind, um die Resonanzeigenschaften der anderen Partialwellen in (19.2) oder allgemeinerer Partialwellen der Form

$$c(\vartheta)\sin(2t + 2\varphi + \varepsilon)$$

zu untersuchen. KERTZ[2] hat das von HAURWITZ in [*36*] zusammengestellte Material daraufhin untersucht und fand als stärkste Partialwelle außer W_2^2 und Z_2

$$10{,}70\,P_3^3(\vartheta)\sin(2t + 3\varphi + 88°)\cdot 10^{-2}\,\text{mbar}.$$

Ihre Amplitude ist so groß wie die von Z_2. Sie besteht aus drei Hoch- und drei Tiefdruckgebieten am Äquator, die die Erde in $\frac{3}{2}$ Tagen umkreisen.

[1] Bei der Erklärung von W_2^2 in Fig. 16 wurde das durch die Reflexion des Erdbodens bewirkte Strahlungsglied vernachlässigt. Setzt man ein solches Glied nach (19.7) von der Größe $\frac{1}{15}\tau(0)$ an, so erhält man ein kleines Korrekturglied zu τ^S, das keine wesentliche Änderung herbeiführt.

[2] W. KERTZ: Components of the semidiurnal pressure oscillation. Scientific Report No. 4, Project No. 299 New York University, College of Engineering, Dept. of Meteorlogy and Oceanography, 1956.

20. Dritteltägige Wellen. Fraglos werden die dritteltägigen Wellen thermisch erregt. Ihr Hauptglied W_4^3 spiegelt deutlich den Charakter der Erregungsfunktion wider. Es erreicht seine Extremwerte während der Solstitien, und zwar mit wechselndem Vorzeichen vom Sommer zum Winter. Während der Äquinoktien verschwindet es. Man darf daher die dritteltägige Welle nicht im Jahresmittel suchen, weil sich darin ihr größter Bestandteil heraushebt.

Die dritteltägige Welle zeigt eine große Regelmäßigkeit. Darauf hat HANN [8] wiederholt hingewiesen. Auch in der geographischen Verteilung folgt sie weitgehend der Erregungsfunktion: Ihr Hauptglied W_4^3 ist antisymmetrisch zum Äquator und besitzt Extremwerte in der Mitte jeder Halbkugel. Dieses Glied wird dargestellt durch (Zahlenwerte nach M. SIEBERT, bisher unveröffentlicht)

$$\left.\begin{aligned} W_4^3:\quad \delta p(0) &= \quad 0{,}269\, P_4^3 \cdot \sin 3t' + \begin{Bmatrix} 175^\circ \\ 355^\circ \end{Bmatrix} \text{mbar} \begin{Bmatrix} \text{Juli} \\ \text{Januar} \end{Bmatrix} \\ &= (0{,}254\, \Theta_4 + \cdots) \sin 3t' + \begin{Bmatrix} 175^\circ \\ 355^\circ \end{Bmatrix} \text{mbar} \begin{Bmatrix} \text{Juli} \\ \text{Januar} \end{Bmatrix} \end{aligned}\right\} \tag{20.1}$$

Außer dieser antisymmetrischen Welle gibt es aber noch zwei symmetrische dritteltägige Wellen, die das ganze Jahr über eine konstante Phase haben. Ihre Jahresmittel können beschrieben werden durch

$$W_3^3:\quad \delta p(0) = 0{,}075\, P_3^3 \sin(3t' + 10^\circ)\,\text{mbar}, \tag{20.2}$$

$$W_5^3:\quad \delta p(0) = 0{,}039\, P_5^3 \sin(3t' + 10^\circ)\,\text{mbar}. \tag{20.3}$$

W_4^3 läßt sich sowohl vom Standpunkt der Resonanztheorie (mittels des Austauschgliedes der Temperaturwelle) als auch durch Strahlungserregung erklären. Das entsprechende Glied der Bodentemperatur ist nach Tabelle 1 und 3

$$W_4^3:\quad \delta T(0) = 0{,}054\, \Theta_4 \sin\left(3t' + \begin{Bmatrix} 45^\circ \\ 225^\circ \end{Bmatrix}\right) {}^\circ\text{C} \begin{Bmatrix} \text{Nord-Sommer} \\ \text{Nord-Winter} \end{Bmatrix}. \tag{20.4}$$

Die sekundäre Temperaturwelle wird nach (9.10)

$$\tilde{\tau}(0) = 0{,}021\, \Theta_4 \sin\left(3t' + \begin{Bmatrix} 175^\circ \\ 355^\circ \end{Bmatrix}\right) {}^\circ\text{C}. \begin{Bmatrix} \text{Juli} \\ \text{Januar} \end{Bmatrix}. \tag{20.5}$$

Dann ergibt sich die primäre Temperaturwelle zu

$$\tau(0) = 0{,}070\, \Theta_4 \sin\left(3t' + \begin{Bmatrix} 31^\circ \\ 211^\circ \end{Bmatrix}\right) {}^\circ\text{C} \begin{Bmatrix} \text{Juli} \\ \text{Januar} \end{Bmatrix}. \tag{20.6}$$

Da die h-Werte von W_4^3 mit $h = 7{,}66$ km und W_2^2 mit $h = 7{,}85$ km so nahe zusammenliegen, würden beide Wellen durch Resonanz stark vergrößert werden. Die Phasendifferenz zwischen $\delta p(0)$ und $\tau(0)$ liegt mit 144° auch recht nahe bei dem Sollwert 135° für die thermische Erregung durch ein Austauschglied.

Genau so leicht läßt sich W_4^3 aber auch durch Erregung durch ein Strahlungsglied der Temperatur beschreiben. Dieses Glied müßte nach Fig. 14 sein:

$$\tau^S = 0{,}026\, \Theta_4 \sin\left(3t' + \begin{Bmatrix} 355^\circ \\ 175^\circ \end{Bmatrix}\right) {}^\circ\text{C} \begin{Bmatrix} \text{Juli} \\ \text{Januar} \end{Bmatrix}. \tag{20.7}$$

Da die Primäre Strahlung nach Tabelle 1 im Juli und Januar eine Phase von 90° bzw. 270° besitzt, so kommt τ^S wie zu erwarten um etwa 90° nach der Strahlung.

Die Erklärung der symmetrischen Wellen scheitert bisher an einem Fehlen entsprechender Temperaturwerte. Nach Ziff. 5 und 6 sollten überhaupt keine Glieder mit einer während des ganzen Jahres konstanten Phase auftreten. Lediglich in der absorbierten Strahlung (Ziff. 5 Schluß) tritt ein solcher Term auf. Diese könnte zu einer Strahlungserregung führen. Für die Resonanztheorie dürfte eine Erklärung recht schwer werden, weil die h-Werte (12,89 zu W_3^3 und 5,09 zu W_5^3) weit von den beiden Resonanzstellen entfernt sind. Quantitative Aussagen lassen sich aber vorläufig nicht machen.

21. Vierteltägige Wellen. Über die vierteltägige Druck- und Temperaturvariation gibt es bisher nur wenige Untersuchungen[1]. Am Zustandekommen der vierteltägigen Druckschwankungen sind eine Reihe von Partialwellen beteiligt. $S_4(\delta p)$ zeigt starke jahreszeitliche Variationen (Tabelle 4).

W_4^4 ändert sich verhältnismäßig wenig im Laufe des Jahres. Die ganzjährige Schwankung MASO$-{}^1/_2$(JFND+MJJA) ist etwas größer als die halbjährige JFND$-$MJJA. Der zugehörige h-Wert ist größer als der Eigenwert der Atmosphäre. Deshalb müssen für diese Welle Strahlungsglied der Temperatur und Bodendruckwelle in Phase sein. Da die vierteltägige Strahlungskomponente zu P_4^4 nach Tabelle 1 eine Phase von 270° hat, muß das Maximum der Strahlungstemperatur bei 180° liegen. In der Nähe dieses Wertes liegen auch die beobachteten Phasen der Druckwellen in Tabelle 4.

Tabelle 4. *Amplituden c_n und Phasen ε_n für die ${}^1/_4$-tägigen Partialwellen $W_n^4 = c_n P_n^4(\vartheta) \times \sin(4t'+\varepsilon_n)$ in den Äquinoktialmonaten (**M**ärz, **A**pril, **S**eptember, **O**ktober), dem Wintersolstitium (**J**anuar, **F**ebruar, **N**ovember, **D**ezember), dem Sommersolstitium (**M**ai, **J**uni, **J**uli, **A**ugust) und dem Jahresmittel.*

	h_n	MASO		JFND		MJJA		Jahresmittel	
	km	c_n 10^{-3} mb	ε_n	c_n 10^{-3} mb	ε_n	c_n 10^{-3} mb	ε_n	c_n 10^{-3} mb	ε_n
W_4^4	15,4	17,4	173°	7,9	244°	19,1	180°	13,4	187°
W_5^4	10,2	27,9	204°	98,1	207°	24,2	359°	35,1	213°
W_6^4	7,2	34,7	336°	24,4	215°	26,8	222°	15,5	260°
W_7^4	5,5	13,9	252°	28,2	191°	22,8	331°	10,8	248°

W_5^4 zeigt eine sehr starke ganzjährige Variation. Die halbjährige ist dagegen verhältnismäßig klein. Auffallend ist die große Amplitude in JFND. Da der zugehörige h-Wert in der Nähe der unbezweifelten Resonanzstelle $\hat{h} = 10$ km der Atmosphäre liegt, dürfte es sich hier um einen wirklichen Resonanzeffekt handeln. Leider kennt man die zugehörigen planetarischen Temperaturwellen nicht genau genug, um die Resonanzeigenschaften bis in die Details studieren zu können. Die Resonanzvergrößerung von W_5^4 scheint aber im Nordwinter wesentlich größer zu sein als zu den anderen Jahreszeiten.

W_6^4 besitzt eine große halbjährige und eine kleine ganzjährige Variation. Eine starke halbjährige Änderung ist auch für die entsprechende planetarische Temperaturwelle zu erwarten (Tabelle 1 Phasenwechsel zwischen Äquinoktien und Solstitien), da $W_6^4(\delta T)$ wesentlich durch die Ausbildung des Nebenmaximums der ${}^1/_4$-tägigen Welle zur Zeit der Solstitien in Fig. 3 beeinflußt wird.

Bei W_7^4 schließlich überwiegt wieder die ganzjährige Variation.

[1] S. K. PRAMANIK: The six-hourly variations of atmospheric pressure and temperature. Mem. Roy. Meteor. Soc. **1**, 35—57 (1926). — W. KERTZ: The seasonal variations of the six-hourly planetary pressure and temperature waves. Scientific Report No. 5, Project No. 299 New York University, College of Engeneering, Dept. of Meteorology and Oceanography, 1956.

22. Ganztägige Wellen. Für die ganztägige Welle ist die Anregung am stärksten (vgl. Ziff. 5 und 6). Sie zeigt aber sehr viele örtliche und wetterabhängige Unregelmäßigkeiten. Nur über den äquatornahen Ozeanen und auf tropischen Inseln ist sie einigermaßen regelmäßig. Nach Beobachtungen mit dem Forschungsschiff Meteor beträgt sie auf dem Atlantischen Ozean zwischen 20° N und 23° S[1]:

$$\delta p(0) = 0{,}38 \sin(t' + 9°) \text{ mbar}. \tag{22.1}$$

Über die lokalen Unterschiede: Einfluß der relativen Höhe, der Lage zum Meer, die Witterung vgl. [*3*], [*4*].

Für die Theorie bietet die ganztägige Welle große Schwierigkeiten, da ihr planetarischer Anteil wesentlich kleiner ist als W_2^2, obwohl die zugehörige thermische Erregung doch mindestens doppelt so groß ist wie bei der halbtägigen Welle. Dies war ein anderer Grund, der zur Aufstellung der Resonanztheorie führte (vgl. Ziff. 16). Diese erklärt die Kleinheit und Unregelmäßigkeit in W_1^1 durch die Tatsache, daß ihr im Gegensatz zu W_2^2 keine Eigenschwingung der Erdatmosphäre entspricht. Nimmt man an, daß sich das zu W_1^1 gehörige Strahlungsglied genau so zu dem entsprechenden Glied von S/S_0 in Tabelle 1 verhält, wie sich die Strahlungsglieder von W_2^2 und W_4^3 zu den entsprechenden Termen verhalten, so ergibt sich ein Strahlungsglied von

$$\tau^S = 0{,}210\, P_1^1 \sin(t' + 180°)\ °\text{C}. \tag{22.2}$$

Der zu Θ_1 gehörige h-Wert ist sehr klein. Er liegt bei $h = 1$ km, so daß keine Resonanzvergrößerung mehr wirksam ist. Dann berechnet[2] man die theoretische Druckvariation zu

$$W_1^1: \quad \delta p(0)_{\text{theor}} = 0{,}7\, P_1^1 \sin(t' + 0°) \text{ mbar}. \tag{22.3}$$

Für $\vartheta = 90°$ ist dieser berechnete Wert fast doppelt so groß wie der beobachtete (22.1).

Eine Erklärung dieses Widerspruches steht noch aus. Man kann aber schon einige Gesichtspunkte anführen, die zu seiner Erklärung führen könnten:

1. Bei der Ableitung von (22.3) haben wir stillschweigend angenommen, daß P_1^1 nahezu mit der Houghschen Funktion Θ_1 identisch ist. Das braucht aber für kleine h nicht mehr der Fall zu sein. Es könnte vielmehr so sein, daß in der Entwicklung von P_1^1 nach Houghschen Funktionen das Glied mit Θ_1 nur noch einen ganz kleinen Anteil liefert, während der Hauptanteil zu einem Wert mit sehr großem h gehört (Fig. 5, $f = 0{,}4986$) und deshalb nach (17.10) keinen Beitrag mehr liefert. Die Berechnung ist in diesem Bereich sehr schwierig, weil alle Näherungsverfahren für kleine h nur sehr schlecht konvergieren.

Das in der Strahlung zur Zeit der Solstitien auftretende Glied mit P_2^1 dürfte jedenfalls nach (Ziff. 11 Schluß) nur auf eine sehr kleine Druckwelle führen, da sie für $f = 0{,}5$ verschwinden muß.

2. Nach Solberg ist es für Perioden, die größer sind als 12 Sternstunden, überhaupt nicht statthaft, mit unserer „quasistatischen" Theorie zu rechnen (vgl. [*26*]). Bei der Berechnung nach der exakten „dynamischen" Theorie führt die Berechnung einer ganztägigen Welle zu Widersprüchen, die darauf schließen lassen, daß eine solche Welle überhaupt nicht existieren sollte [*29a*].

---

[1] J. Reger in Wiss. Erg. dtsch. Atlant. Exped. Meteor 1925 bis 1927. **14**, 357 (1938).

[2] Zur Berechnung wurde der Wert $\delta p_n(0) = -\dfrac{p_0(0)}{T_0(0)}\,\tau_n^S$ benutzt, der aus der Formel für eine Atmosphäre in konvektivem Gleichgewicht (17.10) mit $h = 0$ hervorgeht. In der Formel für eine isotherme Atmosphäre führt der Grenzübergang $h \to 0$ auf $\delta p_n(0) = 0$. Anscheinend ist dabei der Gültigkeitsbereich der Formel überschritten.

III. In den Voraussetzungen nicht berücksichtigte störende Einflüsse.

23. Bewegung der Unterlage. Bisher haben wir als Randbedingung am Erdboden immer benutzt $w(0)=0$. In Wirklichkeit liegt die Unterlage aber nicht fest, sondern vollführt selbst Gezeitenbewegungen (Meeresgezeiten und Gezeiten der festen Erde). Dieser Einfluß läßt sich leicht in die Theorie einbauen, wenn man die Bewegung der Oberfläche $\zeta(t)$ nach HOUGHschen Funktionen entwickelt

$$\zeta(t) = \sum \zeta_n \Theta_n e^{i(\sigma t + s\varphi)}. \tag{23.1}$$

Für jede einzelne Komponente gilt dann

$$w_n(0) = i\sigma\zeta_n\Theta_n e^{i(\sigma t + s\varphi)}. \tag{23.2}$$

Diese Glieder addieren sich einfach zu dem Graviationspotential. Die Formeln werden am einfachsten, wenn man wie in Ziff. 11 die Abkürzung benutzt $\bar{\zeta} = -\Omega/g$. Dann tritt an die Stelle der Randbedingung (14.2):

$$\left(y_n' - \frac{1}{2}y_n\right) + \frac{H}{h_n}y_n + i\sigma\bar{\zeta}_n^* = 0 \quad \text{für} \quad x = 0, \tag{23.3}$$

in der jetzt

$$\bar{\zeta}_n^* = \bar{\zeta}_n - \zeta_n \tag{23.4}$$

an die Stelle des Gravitationspotentials getreten ist. Bei den Gezeiten der festen Erde gilt in guter Näherung

$$\zeta = k\bar{\zeta}. \tag{23.5}$$

Dann ist

$$\bar{\zeta}^* = (1-k)\bar{\zeta}. \tag{23.6}$$

Diese Korrektur wurde von BARTELS ([*19*] p. 34—36 oder [*3*] 199—200) in die Theorie der atmosphärischen Gezeiten eingeführt. Er benutzte sie zur Erklärung einer Diskrepanz bei der lunaren Welle in der älteren Resonanztheorie: Wenn man annimmt, daß W_2^2 genau an der Resonanzstelle für eine autobarotrope Atmosphäre liegt, so ergibt sich für L_2 eine etwa 12fache Resonanzvergrößerung. Beobachtet wird aber eine nur 2fache. BARTELS erklärte diesen Unterschied durch ein Nachgeben der gesamten Wasser- und Landoberfläche gemäß (23.5) und erhielt nach (23.6) $1-k=\frac{3}{12}$, also $k=0{,}75$ als Mittelwert für die ganze Erde. Die neueren Resonanzkurven liefern für L_2 sofort den Wert 2. Vielleicht darf man daraus rückwärts schließen, daß im Mittel für Land und Meer keine Bewegung entsprechend (23.5) vorhanden ist.

Die mond- und sonnentägige Bewegung der ganzen Erdoberfläche ist noch zu wenig bekannt, als daß man schon daran denken könnte, die dadurch erzwungenen atmosphärischen Gezeiten zu berechnen, vgl. [*1*].

24. Gebirge als Strömungshindernisse. BARTELS hat auf den Einfluß der großen Gebirge, insbesondere der westamerikanischen Kettengebirge, als Strömungshindernisse für die Gezeiten der Atmosphäre hingewiesen [*19*]. KERTZ hat diesen Einfluß quantitativ nach der Methode der Störungsrechnung (für Eigenwertprobleme) untersucht [*20a*]. Auf der glatten Erde bleibt für die Eigenfunktionen die Phase in φ unbestimmt, oder anders ausgedrückt, die Eigenfunktionen mit dem Faktor $\sin s\varphi$ und $\cos s\varphi$ haben den gleichen h-Wert. Bei der Durchführung der Theorie der Gezeitenschwingungen über einer „rauhen" Erdoberfläche spaltet auch dieser doppelte Eigenwert des Problems (10.3) im allgemeinen noch auf. Benutzt man als stark vereinfachendes Modell einen schmalen Kanal, der längs eines Breitenkreises verläuft und dessen Bodenerhebung durch $q(\varphi)$ beschrieben wird, so erhält man näherungsweise die h-Werte

$$h = h^{(0)} \pm \frac{1}{\pi} \int_{-\pi}^{+\pi} q(\varphi) \cos 2s\varphi \, d\varphi, \tag{24.1}$$

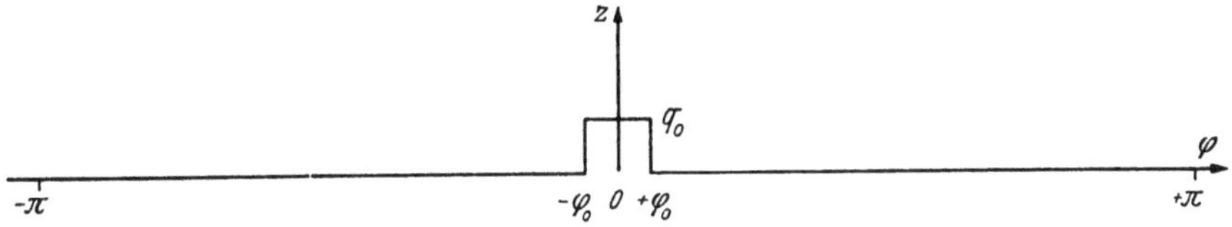

Fig. 18. Modell zur Berechnung des Einflusses eines Gebirges als Strömungshindernis für die wandernden Wellen [*20a*].

wobei $h^{(0)}$ der Eigenwert ist, der zu einem glatten Boden ($q \equiv 0$) gehört. Das $+$-Zeichen gehört zu einer Eigenfunktion, in deren FOURIER-Entwicklung das Hauptglied $\sin s\varphi$ ist, und das $-$-Zeichen zu einer Funktion mit dem Hauptglied $\cos s\varphi$. Nimmt man als Störung der Oberfläche ein „Gebirge" mit recht-

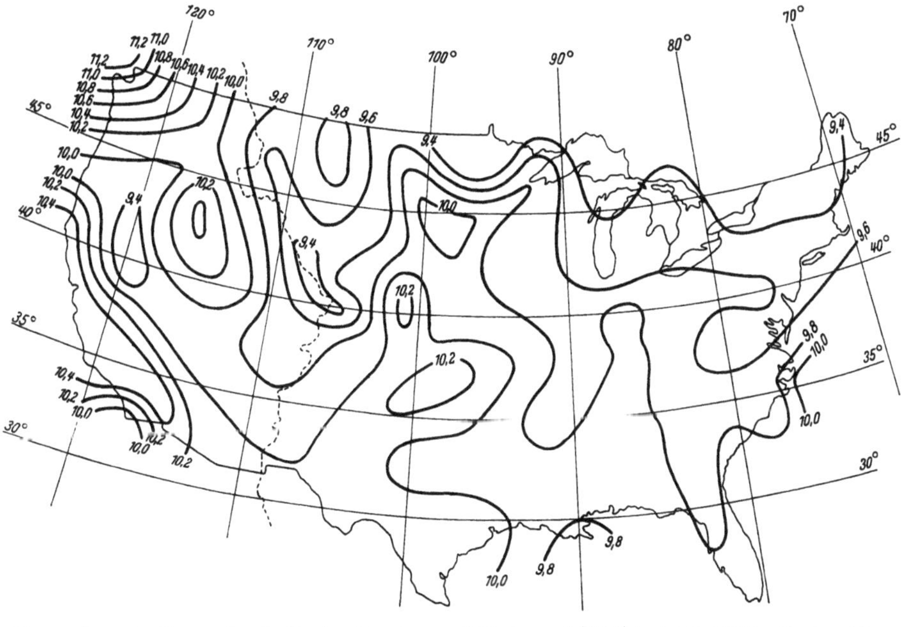

Fig. 19. Karte der Phasenzeiten in Stunden (eingetragene Zahlen) von $S_2(\delta p(0))$ in den Vereinigten Staaten. Man erkennt die Verfrühung des Eintritts des Maximums am Ostrand und die Verspatung am Westrand des Kettengebirges [aus J. SPAR, Bull. Amer. Meteor. Soc. 33, 438—441 (1952)].

eckigem Querschnitt nach Fig. 18 und beginnt die φ-Zählung in der Mitte dieses Gebirges, so bekommt man für das Integral in (24.1):

$$\frac{1}{\pi} \int_{-\pi}^{+\pi} q(\varphi) \cos(2s\varphi) \, d\varphi = \frac{q_0 \sin 2s\varphi_0}{\pi s}. \tag{24.2}$$

Wählt man, um die amerikanischen Kettengebirge anzunähern, $q_0 = 4$ km $\varphi_0 = 10°$, so erhält man für $s = 2$ (zu W_2^2 passend) die Änderung von h zu 0,41 km und für $s=3$ (zu W_4^3 passend) 0,37 km. Es gibt also bei den wandernden Wellen je einen erhöhten und einen erniedrigten h-Wert. Der erhöhte zu W_2^2 und W_4^3 rückt näher an den h-Wert von Z_2 heran. Es ist aber sehr unwahrscheinlich, daß alle drei durch den Einfluß der Gebirge auf der Erde bis auf den Haupteigenwert der Atmosphäre bei 10 km ansteigen. Würde auf diese Weise eine Resonanz entstehen, so würden die erzwungenen Schwingungen auch von den beobachteten sehr verschieden sein:

Die anregenden wandernden Wellen setzen sich aus zwei gleich großen Wellen mit $\sin s\varphi$ und $\cos s\varphi$ zusammen. Wegen der verschiedenen h-Werte bekommen die durch sie angeregten Schwingungen eine unterschiedliche Resonanzverstärkung. Dadurch entsteht eine laufende Welle mit an- und abschwellender Amplitude und Phasenänderungen über dem Gebirge, und zwar tritt das Maximum bei den vorliegenden Verhältnissen am Ostrand des Gebirges früher ein als am Westrand. Ein Effekt dieser Art zeigt sich auch in den sorgfältigen Analysen der Beobachtungswerte von SPAR[1] (Fig. 19). Läge der zu W_2^2 gehörige h-Wert in unmittelbarer Nähe einer Resonanzstelle, so müßte die Phasenänderung aber wesentlich stärker sein, als die Beobachtungen zeigen.

25. Örtliche und zeitliche Variationen der „Konstanten" der Theorie. In die Theorie gehen eine Reihe Größen ein, deren örtliche und zeitliche Variationen in der Theorie nicht berücksichtigt wurden. Zum Teil wurde aber in Einzeluntersuchungen der Einfluß dieser Veränderlichkeit abgeschätzt.

a) Die Temperaturverteilung $T_0(z)$ wurde als längen- und breitenunabhängig angenommen (vgl. Ziff. 8e). In [*1*], [*15a*], [*19*] wurde untersucht, welche Änderungen die Theorie erfährt, weil $T_0(z)$ breitenabhängig ist. Diese Änderungen sind nur gering.

b) Es ist angenommen worden, daß die Gezeitenschwingungen in einer relativ zur Erde ruhenden Atmosphäre stattfanden. CHIU [*33*] hat die Schwingung in einer Atmosphäre untersucht, die eine west-östliche Grundströmung mit konstanter Winkelgeschwindigkeit besitzt. Es ergab sich nur ein sehr geringer Einfluß auf die Eigenperiode.

c) Die Zustandsänderungen wurden immer als adiabatisch angenommen. Das ist nicht richtig in der Nähe der Meeresoberfläche, weil im Meer ein großer Wärmespeicher zur Verfügung steht, der bewirkt, daß die Zustandsänderungen isotherm erfolgen müssen. CHAPMAN [*10*] fand aber, daß der Einfluß des Meeres sich auf die untersten 100 m beschränkt und daher für die ganze Schwingung bedeutungslos ist.

d) Damit das Strahlungsglied wirklich in jeder Höhe eine konstante Amplitude haben soll, müßte die Absorption in der Atmosphäre völlig homogen sein. Das ist sicher nicht der Fall, aber vielleicht gibt es hinreichende konstante Mittelwerte. Hierzu fehlen Untersuchungen und Beobachtungen.

e) Die Gebirge dürften sich nicht nur als Strömungshindernisse für die Gezeitenschwingung bemerkbar machen, sondern bei der thermischen Erregung auch sehr stark durch das Anheben der Heizflächen. Auch hierüber liegen noch keine Untersuchungen vor.

f) Von den Variationen des Austauschkoeffizienten war schon in Ziff. 6 die Rede. Sein Einfluß auf die Bodentemperatur muß sich bei den durch Rückstrahlung erzeugten Gliedern Z_2, W_5^4 auch stark auf die Gezeitenschwingung der Atmosphäre auswirken.

[1] J. SPAR: Bull. Amer. Meteor. Soc. **33**, 438—441 (1952).

E. Freie Schwingungen.

26. Taylorsches Theorem. Taylor [*15*] zeigt, daß der Eigenwert $\hat{h}$ der Atmosphäre auch aus der Ausbreitungsgeschwindigkeit V „langer ebener" Wellen bestimmt werden kann. Für diese gilt

$$\boxed{V = \sqrt{g\hat{h}}}\,. \tag{26.1}$$

Diese Formel wird als „Taylor*sches Theorem*" bezeichnet. Es ist aus einem Grenzübergang hervorgegangen. Unendlich lange ebene Wellen können auf einer Kugeloberfläche nicht existieren. Daher kann (26.1) nur für einen bestimmten Wellenlängenbereich gelten, der nach unten durch die Forderung „langer" Wellen und nach oben durch die Geometrie der Kugel begrenzt wird.

Eine Ableitung der Formel unter Berücksichtigung der Kugelgeometrie wurde in [*21d*] durchgeführt. Dabei wird von der Erdrotation abgesehen, da die Wellen sich so schnell ausbreiten, daß die Corioliskraft nicht wirksam werden kann. Dagegen wird das Beschleunigungsglied dw/dt mitgeführt. Die Störung wird in der Form von zonalen Kugelfunktionen $P_n(\vartheta)$ um den Störungsherd herum angesetzt. Dann bedeutet die Forderung „langer" Wellen, daß

$$n(n+1) \ll (\varkappa - 1)\, a^2/\hat{h}^2. \tag{26.2}$$

Das bedeutet $n \leqq 38$, wenn man verlangt, daß die Fehler $< 1\,\%$ sind; läßt man Fehler bis zu 10% zu, so gilt $n \leqq 122$. Dann erhält man für die Frequenz die gleiche Formel wie (11.5):

$$\sigma_n^2 = \frac{g\hat{h}}{a^2}\, n(n+1)\,. \tag{26.3}$$

Man führt mit Hilfe der Wellenlänge λ_n die Phasengeschwindigkeit V ein:

$$V = \frac{\sigma_n}{2\pi}\,\lambda_n\,. \tag{26.4}$$

λ_n bestimmt man aus dem Abstand der Nullstellen $\vartheta_{\nu+2}$ und ϑ_ν von $P_n(\vartheta)$. Für hinreichend große n gilt näherungsweise

$$\lambda_n - a(\vartheta_{\nu+2} \quad \vartheta_\nu) = \frac{4\pi a}{2n+1}\,. \tag{26.5}$$

Dann folgt

$$V = \frac{\sqrt{4n(n+1)}}{2n+1}\,\sqrt{g\hat{h}}\,. \tag{26.6}$$

V ist also eine Funktion der Wellenlänge. Für $n \geqq 5$ geht die Formel aber hinreichend genau in (26.1) über. In diesen Bedingungen äußert sich die Kugelgeometrie. Also gilt (26.1) nach (26.2) und (26.6) für

$$\left.\begin{matrix}1000\ \text{km}\\ 330\ \text{km}\end{matrix}\right\} < \lambda < 7000\ \text{km},\quad \text{Fehler}\ \left.\begin{matrix}< 1\,\%\\ < 10\,\%\end{matrix}\right\}. \tag{26.7}$$

27. Beobachtungen. Beim Ausbruch des Krakatau am 26. August 1883[1] wurden in der Atmosphäre Druckwellen angeregt, die den Voraussetzungen der vorigen Ziffer entsprachen. Ihre Wellenlänge lag zwischen 1300 und 1400 km (Periode etwa 1,2 Std). Aus den Beobachtungen tropischer Stationen ergab sich eine Ausbreitungsgeschwindigkeit von 318,8 m/sec und auf der Strecke vom

[1] Strachey: Report of the Krakatoa-Comittee of the Roy. Soc. (1888); vgl. [*15a* und *b*], [*27b*]; vgl. auch M. Ewing u. F. Press: Trans. Amer. Geophys. Union **36**, 53—60 (1955).

Krakatau nach England 314 m/sec. Dem entspricht der Eigenwert

$$\hat{h} = 10{,}37\,\text{km} \quad \text{bzw.} \quad 10{,}06\,\text{km}.$$

PEKERIS [*16b*] fand bei einer erneuten Bearbeitung des Materials eine zweite Druckwelle mit einer Ausbreitungsgeschwindigkeit von 280 m/sec, die zu einem zweiten Eigenwert $\hat{h}_2 = 8{,}0$ km gehören würde. Die Existenz dieser Welle ist aber nicht sicher.

Solche langperiodischen Störungen wie beim Krakatauausbruch sind bisher nicht wieder beobachtet worden. Beim Fall des Sibirischen Meteors am 30. Juni 1908[1] wurden Druckwellen mit Perioden von etwa 2 min und Wellenlängen von etwa 40 km registriert. Die Fortpflanzungsgeschwindigkeit betrug 316 m/sec. Diese Beobachtungen lassen sich aber wegen (26.7) nicht nach dem TAYLORschen Theorem auswerten. SOLBERG [*27b*] und PEKERIS [*16c* und *d*] haben Untersuchungen über die Ausbreitung solcher „kurzperiodischen" Druckwellen angestellt, die diese Beobachtungen erklären. Man erhält daraus aber keinen Beitrag zur Theorie der atmosphärischen Gezeiten.

Auch die Druckstörungen, die durch Versuche mit Wasserstoffbomben im Jahr 1954 ausgelöst wurden, zeigen Perioden und Wellenlängen, die erheblich kleiner sind, als es der Bedingung (26.7) entspricht[2].

28. Schlußbemerkungen. Obwohl die Gezeiten der Atmosphäre zu den am besten durchforschten meteorologischen Erscheinungen gehören, bleiben noch viele ungelöste Probleme bestehen, herrschen doch über die wichtigsten Prinzipien zu ihrer Erklärung — Resonanztheorie oder nicht — verschiedene Ansichten. HOLMBERG[3] hat versucht, die Resonanzeigenschaften der Atmosphäre im Sinne der Resonanztheorie für die Stabilisierung der Erdrotation verantwortlich zu machen. Damit würde er zugleich einen Grund für das anderenfalls zufällige Zusammentreffen der Periode einer Eigenschwingung der Atmosphäre mit der Periode ihrer äußeren Erregung angeben. Doch sind seine Abschätzungen nicht recht überzeugend ([*21b*] p. 87—88).

Aber auch in den Details gibt es viele zu bearbeitende Fragen. Man sollte vor allem die Untersuchung der atmosphärischen Gezeiten nicht auf W_2^2 beschränken, sondern sollte alle Wellen in Betracht ziehen. Die beobachteten Unregelmäßigkeiten in der geographischen Verteilung [*13*] und die jahreszeitliche Variation der solaren [*21c*] und lunaren [*2*], [*25a*] Wellen bieten viele interessante Probleme. Die thermische Erregung ist bisher nur recht unvollkommen bekannt; so gibt es nur wenige Angaben über die planetarischen Temperaturwellen am Boden [*24*], [*20b*]. Für die Temperaturvariationen in der Höhe liegen überhaupt nur Einzelwerte vor[4]. In dieses Gebiet fallen auch die Fragen der beobachteten Phasenänderungen der Wellen innerhalb der Troposphäre [*28*], [*29*].

Schließlich ist das große Gebiet der Gezeiten in der hohen Atmosphäre sowohl nach der Seite der Beobachtungen (vgl. Ziff. 2) als auch nach der theoretischen Seite noch weitgehend offen. Zu einer theoretischen Erklärung und Berechnung wird man aber vermutlich nicht mit einer linearen Theorie auskommen [*16e*].

---

[1] F. J. W. WHIPPLE: Quart. J. Roy. Meteor. Soc. **56**, 287—304 (1930); **60**, 510 (1934). — R. SÜRING: Meteor. Z. **47**, 490—492 (1930). — S. SKREB: Meteor. Z. **48**, 148 (1931). — I. S. ASTAPOWITSCH: Quart. J. Roy. Meteor. Soc. **60**, 493 (1934).

[2] Geophys. Mag. Tokyo **26**, 215—223 (1955).

[3] E. R. R. HOLMBERG: Month. Not. Roy. Astr. Soc. Geophys. Suppl. London **6**, 325—330 (1952). Vgl. auch dieses Handbuch, Bd. XLVII (1956), pp. 6—7 und 14.

[4] A. WAGNER: Gerlands Beitr. Geophys. **37**, 315—344 (1932); **52**, 408—449 (1938). — H. STAPF: Der tägliche Luftdruck- und Temperaturgang in der freien Atmosphäre in seiner Abhängigkeit von den Jahreszeiten. Diss. Berlin 1934, 43 S. — R. HOLZAPFEL u. M. MANIG: Meteor. Rdsch. **8**, 85—88 (1955).

29. Durchlaufend gebrauchte Bezeichnungen und Zahlenwerte. *a)* Es ließ sich nicht vermeiden, gelegentlich die gleichen Symbole mit verschiedenen Bedeutungen zu benutzen: In Ziff. 2 wurden die in der Ionosphärenliteratur gebräuchlichen Symbole (f, h), in Ziff. 3 und 4 die bei der Behandlung des Gezeitenpotentials üblichen Bezeichnungen (τ, s, h) und in Ziff. 11 die in der Ozeanographie verwandte Größe (ζ) benutzt. S_σ bedeutet allgemein die $1/\sigma$-tägige solare Variation, in Ziff. 5 aber bedeutet S die einfallende Strahlungsleistung und S_0 die Solarkonstante. $k(\vartheta, \varphi)$ ist in Ziff. 6 die Landbedeckungsfunktion, in Ziff. 17 ist $k = H(0)\sqrt{i\,\sigma/K}$ u. dgl. m.

b) Variationen und Wellentypen.

S_σ und L_σ kennzeichnen die solare und die lunare $1/\sigma$-tägige Variation.

W_n^s mit Ortszeit $(s = \sigma)$ wandernde Gezeitenwelle mit dem Hauptglied P_n^s (eigentlich Θ_n siehe unter h).

Z_n zonale halbtägige Gezeitenschwingung nach Weltzeit mit Hauptglied P_n.

$U_{\sigma,m}$ allgemeine planetarische Wellen (6.10).

c) Konstanten.

a mittlerer Erdradius = 6371 km.

ω Winkelgeschwindigkeit der Erdrotation = $7{,}2921 \cdot 10^{-5}$ sec^{-1}.

g mittlere Schwerebeschleunigung der Erde = 979,8 cm/sec^2.

R universelle Gaskonstante = $8{,}314 \cdot 10^7$ erg/Mol · Grad.

M mittleres Molekulargewicht der Luft = 28,97.

c_p spezifische Wärme bei konstantem Druck.

c_v spezifische Wärme bei konstantem Volumen.

$\varkappa = c_p/c_v$ Adiabatenexponent = 1,400 für trockene Luft.

d) Zeitvariable.

t Weltzeit.

t' mittlere Ortszeit = $t + \varphi$.

τ mittlere Mondzeit.

σ Kreisfrequenz einer Schwingung.

Alle Zeitangaben werden gewöhnlich im Winkelmaß gemessen 1 Tag $\mathrel{\hat=} 2\pi \mathrel{\hat=} 360°$.

$f = \sigma/2\omega$ (S_1: 0,49864; S_2: 0,99727; S_3: 1,49590; S_4: 1,99454; L_1: 0,48175; L_2: 0,96350).

e) Koordinaten und Geschwindigkeiten.

ϑ Poldistanz, vom Nordpol nach Süden positiv gerechnet (= 90° − geograph. Breite).

φ geographische Länge, nach Osten positiv gerechnet.

z Höhe über der Erdoberfläche, vertikal nach oben positiv gerechnet.

r Abstand vom Erdmittelpunkt = $a + z$.

$\boldsymbol{v}$ (u, v, w) Geschwindigkeit eines Luftelementes an der Stelle (z, ϑ, φ) zur Zeit t.

u Südkomponente von $\boldsymbol{v}$ (Nordwind).

v Ostkomponente von $\boldsymbol{v}$ (Westwind).

w Vertikalkomponente von $\boldsymbol{v}$.

χ div $\boldsymbol{v}$.

V Phasengeschwindigkeit „langer Wellen" (26.4).

λ Wellenlänge der „langen Wellen" (26.5).

Bei der Bezeichnung von Funktionen von $(z, \vartheta, \varphi; t)$ sind gelegentlich, wenn es der Zusammenhang erlaubte, zur Abkürzung von hinten an Symbole fortgelassen.

f) Zustandsgrößen.

p_0 ungestörter (statischer) Druck in der Höhe z.

δp Druckvariation an der Stelle (z, ϑ, φ) zur Zeit t im Zusammenhang mit den Schwingungen.

$p = p_0(z) + \delta p(z, \vartheta, \varphi; t)$ = Gesamtdruck.

Ebenso:

$\varrho = \varrho_0(z) + \delta\varrho(z, \vartheta, \varphi; t)$ = Dichte.

$T = T_0(z) + \delta T(z, \vartheta, \varphi; t)$ = Temperatur in °Kelvin.

$$H(z) = \frac{p_0(z)}{g\,\varrho_0(z)} = \frac{R\,T_0(z)}{M\,g} = 2{,}929 \cdot 10^{-2} \cdot T_0(z)\ [\mathrm{km}].$$

Bodenwerte (INA):

$p_0(0) = 1013$ mbar; $\varrho_0(0) = 1{,}226$ kg/m^3; $T_0(0) = 288$ °K; $H_0(0) = 8{,}435$ km.

α linearer Temperaturgradient in Grad/100 m (8.6).

$= \frac{R\alpha}{Mg} = 0{,}2929\,\alpha$ (8.6).

z^0 obere Grenze einer Modellatmosphäre (8.8).

K Koeffizient der turbulenten Diffusion (6.1).

A Austauschkoeffizient $= K\varrho_0$ (6.3).

$k = H(0)\sqrt{i\sigma/K}$ (17.4).

g) Größen für die Anregung.

Ω $\Omega(z, \vartheta, \varphi; t)$ = auf die Masseneinheit bezogenes Potential der Gezeitenkraft.

$\bar{\zeta} = -\Omega/g$ Erhebung der Gleichgewichtsflut.

J $J(z, \vartheta, \varphi; t)$ auf die Masseneinheit bezogene und der Atmosphäre durch Austausch oder Strahlung pro Zeiteinheit zugeführte Wärmemenge.

τ Temperaturvariation durch die Wärmezufuhr J (9.8).

$\tilde{\tau}$ sekundäre Temperaturvariation (9.9).

δ Deklination der Sonne.

ζ Zenitdistanz der Sonne.

c_σ Amplitude der Strahlungsfunktion (5.3).

ε_σ Phase der Strahlungsfunktion (5.3).

γ_σ Phase der Temperaturvariation am Boden (6.5).

Die Indizes G, A, S beziehen sich auf die Erregung durch **G**ravitation, ein **A**ustauschglied und ein **S**trahlungsglied der Temperatur.

h) Lösungen.

F Operator in der LAPLACEschen Gleichung (10.4).

$P_n^s(\vartheta)$ Kugelfunktion in der Normierung nach AD. SCHMIDT [vgl. Formel (8.6), S. 144].

$\Theta_n(\vartheta)$ HOUGHsche Funktion (noch abhängig von s und h) (Tabelle 3).

h Parameter zur Kennzeichnung der Resonanzeigenschaften eines Wellentypes (10.11).

$\hat{h}$ Eigenwert der Atmosphäre (Ziff. 12).

x (10.8).

y (10.9).

Bibliographie.

A. Zusammenfassende Darstellungen.

[*1*] CHAPMAN, S., S. K. PRAMANIK u. J. TOPPING: The worldwide oscillations of the atmosphere. Gerlands Beitr. Geophys. **33**, 246—260 (1931). (Kurze Beschreibung aller beobachteten Wellentypen und ihrer Erregungsquellen. Beobachtungsmaterial zu $S_3(\delta T)$. Einige Bemerkungen zur Theorie.)

[*2*] CHAPMAN, S.: Atmospheric tides and oscillations. Compendium of Meteorology. Boston, Amer. Meteor. Soc. **1951**, 510—530. (Darstellung und Diskussion der Beobachtungsergebnisse, besonders für L_2. Die ungelösten Probleme und Widersprüche zwischen Theorie und Beobachtung werden deutlich herausgestellt.)

[*3*] BARTELS, J.: Gezeitenschwingungen der Atmosphäre. Wien-Harms, Handbuch der Experimentalphysik, Bd. 25, Teil 1, S. 161—210. 1928. (Beobachtungsergebnisse. Prägnante Darstellung der älteren Resonanztheorie. Zugleich Zusammenfassung von [*19*]. Eine Korrektur bei H. KOSCHMIEDER, Dynamische Meteorologie, 2. Aufl., S. 315. Leipzig 1941.)

[*4*] BARTELS, J.: Sonnen- und mondentägige Luftdruckschwankungen. Lehrbuch der Meteorologie, HANN-SÜRING, 5. Aufl., Bd. I, S. 276—306. Leipzig: W. Keller 1939. (Ausführliche Darstellung der Beobachtungen mit besonderer Betonung der meteorologisch wichtigen Gesichtspunkte. Überblick über die statistischen Methoden zu ihrer Ableitung.)

[*5*] BARTELS, J., S. CHAPMAN u. W. KERTZ: Gezeitenartige Schwingungen der Atmosphäre. Landolt-Börnstein, Physikalische, chemische und technische Zahlenwerte, Bd. III, S. 674—685. 1952. (Neuste Zusammenstellung von Zahlenwerten für alle beobachteten Luftdruckschwingungen.)

[*6a*] WILKES, M. V.: Oscillations of the earth's atmosphere. 76 S. Cambridge: University Press, 1949. (Zusammenfassende Darstellung der neueren Resonanztheorie.)

[*6b*] WILKES, M. V.: Worldwide oscillations in the earth's atmosphere. Quart. J. Roy. Meteor. Soc. **78**, 321—336 (1952). (Kurze Zusammenfassung von [*6a*] unter Berücksichtigung der inzwischen gewonnenen Ergebnisse aus [*18*].)

[*7*] MITRA, S. K.: The upper atmosphere. 2. Aufl., Kap. II, S. 31—50. Calcutta: Asiatic Soc. 1952. (Kurze Zusammenfassung der Beobachtungen und Theorie mit besonderer Berücksichtigung der Ergebnisse für die hohe Atmosphäre.)

B. Beobachtungsergebnisse.

Beobachtungsdaten von Einzelstationen und ausführliche Quellenzitate in [5]. Ein sehr umfangreiches Beobachtungsmaterial sammelte J. v. HANN. Dieses ist bis heute noch nicht vollständig ausgenutzt. Deshalb sollen hier noch einmal die wichtigsten seiner Arbeiten über dieses Gebiet zusammengestellt werden. Darüber hinaus findet man aber eine sehr große Anzahl kleinerer Arbeiten, meist mit Beobachtungsdaten in Meteor. Z. **1**—**38**.

[8a] Untersuchungen über die tägliche Oszillation des Barometers. Denkschr. Akad. Wiss. Wien **55**, 49—121 (1889); **59**, 297—356 (1892); **95**, 1—64 (1918).

[8b] Bemerkungen zur täglichen Oszillation des Barometers. Sitzgsber. Akad. Wiss. Wien, math.-nat. Kl. Abt. II **93**, 981—994 (1886).

[8c] Der tägliche Gang des Barometers an heiteren und trüben Tagen, namentlich auf Berggipfeln. Sitzgsber. Akad. Wiss. Wien, math.-nat. Kl. Abt. IIa **104**, 505—564 (1895).

[8d] Weitere Beiträge zu den Grundlagen für eine Theorie der täglichen Oszillation des Barometers. Sitzgsber. Akad. Wiss. Wien, math.-nat. Kl. Abt. IIa **107**, 63—159 (1898).

[8e] Über die tägliche Drehung der mittleren Windrichtung und über eine Oszillation der Luftmassen von halbtägiger Periode auf Berggipfeln von 2 bis 4 km Seehöhe. Sitzgsber. Akad. Wiss. Wien, math.-nat. Kl. Abt. IIa **111**, 1615—1711 (1902).

[8f] Die jährliche Periode der halbtägigen Luftdruckschwankung. Sitzgsber. Akad. Wiss. Wien, math.-nat. Kl. Abt. IIa **127**, 263—365 (1918).

[8g] Die ganztägige Luftdruckschwankung. Sitzgsber. Akad. Wiss. Wien, math.-nat. Kl. Abt. IIa **128**, 379—506 (1919). [Korrekturen: Tätigkeitsbericht Preuß. meteor. Inst. 1920—1923, S. 110 (1924).]

[8h] Täglicher Gang des Barometers auf Pikes Peak (4308 m) und zu Colorado Springs (1856 m). Meteor. Z. **16**, 87—91 (1899).

[8i] Über die Abnahme der Amplitude der halbtägigen Druckwelle mit der Höhe. Meteor. Z. **36**, 218—219 (1919).

[8k] Lehrbuch der Meteorologie. Leipzig, 3. Aufl. 1915; 4. Aufl. (zusammen mit R. SÜRING) 1926; 5. Aufl. siehe [4].

[8l] Über eine doppelte tägliche Periode der Windkomponenten auf den Berggipfeln. Meteor. Z. **20**, 501—512 (1903).

[8m] Die tägliche Variation der Windstärke auf den Berggipfeln in Südindien in ihrer Beziehung zu der täglichen Luftschwankung. Meteor. Z. **25**, 220—222 (1908).

[8n] Der tägliche Gang der Windgeschwindigkeit auf dem Gipfel des Misti in seiner Beziehung zu den Luftdruckänderungen. Meteor. Z. **27**, 319—321 (1910).

[8o] Die jährlichen und täglichen Änderungen in der Richtung und Stärke des Südost-Passats im Atlantischen Ozean. Meteor. Z. **32**, 56—61 (1915).

[8p] Der tägliche Gang der Westkomponente der Luftbewegung zu Johannesburg und die doppelte tägliche Luftdruckschwankung. Meteor. Z. **32**, 467—469 (1915).

[8q] Der tägliche Gang der Temperatur in der inneren Tropenzone. Denkschr. Akad. Wiss. Wien **78**, 249—366 (1906).

[8r] Der tägliche Gang der Temperatur in der äußeren Tropenzone. A. Das amerikanische und afrikanische Tropengebiet. Denkschr. Akad. Wiss. Wien **80**, 317—404 (1907); B. Das indische und australische Tropengebiet. Denkschr. Akad. Wiss. Wien **81**, 21—113 (1908).

C. Einzeluntersuchungen zur Theorie der atmosphärischen Gezeiten.

[9a] MARGULES, M.: Über die Schwingungen periodisch erwärmter Luft. Sitzgsber. Akad. Wiss. Wien, math.-nat. Kl. Abt. IIa **99**, 204—227 (1890).

[9b] MARGULES, M.: Luftbewegungen in einer rotierenden Sphäroidschale. Sitzgsber. Akad. Wiss. Wien, math.-nat. Kl. Abt. IIa **101**, 597—626 (1892); **102**, 11—56; 1369—1421 (1893).

[9c] TRABERT, W.: Die Theorie der täglichen Luftdruckschwankung von MARGULES und die tägliche Oszillation der Luftmassen. Meteor. Z. **20**, 481—501; 544—563 (1903). (In *a* und *b* erste vollständige Darstellung der älteren Resonanztheorie. Ausführliche Interpretation der Ergebnisse in c.)

[10] CHAPMAN, S.: The semidiurnal oscillation of the atmosphere. Quart. J. Roy. Meteor. Soc. **50**, 165—195 (1924). (Klassische Darstellung der Resonanztheorie von W_2^2 und ihre Zerlegung in Gravitations- und thermischen Anteil.)

[11a] CHAPMAN, S.: The lunar tide in the earth's atmosphere. Quart. J. Roy. Meteor. Soc. **45**, 113—139 (1919).

[11b] CHAPMAN, S.: The lunar tide in the earth's atmosphere. Proc. Roy. Soc. Lond., Ser. A **151**, 105—117 (1935).

[11c] CHAPMAN, S., and K. K. TSCHU: The lunar atmospheric tide at twenty-seven stations widely distributed over the globe. Proc Roy. Soc. Lond., Ser. A **195**, 310—323 (1948). (Zusammenfassung der Arbeiten über L_2.)

[12a] CHAPMAN, S.: The lunar diurnal variation of atmospheric temperature at Batavia, 1866—1928. Proc. Roy. Soc. Lond., Ser. A **137**, 1—24 (1932).
[12b] CHAPMAN, S.: On the theory of the lunar tidal variation of atmospheric temperature. Mem. Roy. Meteor. Soc. **4**, 35—40 (1932).
[13] CHAPMAN, S., and K. C. WESTFOLD: A comparison of the annual mean solar and lunar atmospheric tides in barometric pressure, as regards their worldwide distribution of amplitude and phase. J. Atmos. a. Terrest. Phys. **8**, 1—23 (1956).
[14] LAMB, H.: On atmospheric oscillations. Proc. Roy. Soc. Lond., Ser. A **84**, 551—572 (1911). (Auf dieser Arbeit baut die neuere Resonanztheorie auf, bei der Atmosphärenmodelle mit mehr als einem Eigenwert behandelt werden.)
[15a] TAYLOR, G. I.: Waves and tides in the atmosphere. Proc. Roy. Soc. Lond., Ser. A **126**, 169—183, 728 (1929).
[15b] TAYLOR, G. I.: The resonance theory of semidiurnal atmospheric oscillations. Mem. Roy. Meteor. Soc. **4**, 43—51 (1932).
[15c] TAYLOR, G. I.: The oscillations of the atmosphere. Proc. Roy. Soc. Lond., Ser. A **156**, 318—326 (1936). (Grundlagen der neueren Resonanztheorie.)
[16a] PEKERIS, C. L.: Atmospheric oscillations. Proc. Roy. Soc. Lond., Ser. A **158**, 650—671 (1937). (Erste Berechnung von Atmosphärenmodellen mit einem zweiten Eigenwert bei 8 km.)
[16b] PEKERIS, C. L.: The propagation of a pulse in the atmosphere. Proc. Roy. Soc. Lond., Ser. A **171**, 434—449 (1939).
[16c] PEKERIS, C. L.: The propagation of a pulse in the atmosphere, Part II. Phys. Rev. **73**, 145—154 (1948).
[16d] PEKERIS, C. L.: Free oscillations of an atmosphere in which temperature increases linearly with height. NACA, Technical Note 2209, 1—24. Washington (Oct. 1950). (*b*, *c* und *d*: Theoretische Untersuchungen über die Ausbreitung freier Wellen der Atmosphäre, die durch einen explosionsartigen Vorgang angeregt wurden.)
[16e] PEKERIS, C. L.: Effect of quadratic terms in differential equations of atmospheric oscillations. NACA, Technical Note 2314, 1—10. Washington (March 1951).
[16f] PEKERIS, C. L.: A method of solving the nonlinear differential equations of atmospheric tides with application to an atmosphere of constant temperature. (Final Rep., Contract NAw-6001). 25 pp. Princeton: Princeton Univ. 1951. (*e* und *f*: Anfänge einer nichtlinearen Theorie für die Gezeitenschwingungen in großen Höhen.)
[16g] PEKERIS, C. L.: The present status of the resonance theory of atmospheric tides. Washington: National Bureau of Standards 1952 (Circular 521), 201—208.
[17] WEEKES, K., and M. V. WILKES: Atmospheric oscillations and the resonance theory. Proc. Roy. Soc. Lond., Ser. A **192**, 80—99 (1947). (Berechnungen der Eigenwerte und Eigenfunktionen für kompliziertere Atmosphärenmodelle im Blick auf die Gezeitenschwingungen der hohen Atmosphäre.)
[18] WILKES, M. V.: The thermal excitation of atmospheric oscillations. Proc. Roy. Soc. Lond., Ser. A **207**, 358—370 (1951). (Erweiterung von [10]. In Gl. (18) muß man das Glied $+\frac{h}{H}\frac{dH}{dz}$ streichen vgl. [21] und [31].)
[19] BARTELS, J.: Über die atmosphärischen Gezeiten. Preuß. meteor. Inst. Abh. 8, Nr. 9 (Veröff. Nr. 346) 51 S. (1927). (Theorie, Statistische Methoden der Bearbeitung der Beobachtungen, Berechnung von L_2 in Potsdam und Hamburg aus insgesamt 66jährigen Beobachtungen. Grundlage zu [3].)
[20a] KERTZ, W.: Theorie der gezeitenartigen Luftschwingungen als Eigenwertproblem. Ann. Meteor. Hamburg **4**, 1. Beiheft (1951). [Anwendung der Störungsrechnung zur Gewinnung explizierter Formeln für die h-Werte auf rotierender glatter Erdkugel und Untersuchung der Änderung durch Erhebungen auf der Oberfläche (Amerikanisches Kettengebirge).]
[20b] KERTZ, W.: Die thermische Erregungsquelle der atmosphärischen Gezeiten. Nachr. Akad. Wiss. Göttingen IIa **1956**, Nr. 6, 145—166. (Grundlagen zu Ziff. 5 und 6 dieses Handbuchartikels.)
[21a] SIEBERT, M.: Zur Theorie der thermischen Erregung gezeitenartiger Schwingungen der Erdatmosphäre. Naturwiss. **41**, 446—447 (1954).
[21b] SIEBERT, M.: Über die gezeitenartigen Schwingungen der Erdatmosphäre. Ber. dtsch. Wetterd. Nr. 22, 65—71 u. 87—88, 1956.
[21c] SIEBERT, M.: Analyse des Jahresganges der $1/n$ tägigen Variationen des Luftdruckes und der Temperatur. Nachr. Akad. Wiss. Göttingen IIa **1956**, Nr. 6, 127—144.
[21d] SIEBERT, M.: Zur Theorie der freien und erzwungenen gezeitenartigen Schwingungen der Erdatmosphäre. Diss. Göttingen 1955. (Aufstellung und Durchführung der Theorie der thermischen Anregung durch Strahlungsglieder zur Ablösung der Resonanztheorie. Außerdem enthält die Arbeit *d* eine Fülle sorgfältiger Berechnungen

der HOUGHschen Funktionen und der geograph. Verteilung der Luftdruck- und Lufttemperaturwellen.)

[*22a*] HAURWITZ, B.: The oscillations of the atmosphere. Gerlands Beitr. Geophys. **51**, 195—233 (1937).

[*22b*] HAURWITZ, B.: Über die Eigenschwingungen einer zweifach geschichteten autobarotropen Atmosphäre und die atmosphärischen Gezeiten. Meteor. Z. **54**, 69—70 (1937). (Verallgemeinerung der MARGULESschen Rechnungen auf ein zweischichtiges Atmosphärenmodell, wobei die Zustandsänderungen in jeder Schicht autobarotrop erfolgen sollen.)

[*23*] HAURWITZ, B.: Zur Resonanztheorie der halbtägigen Luftdruckschwankung. Ber. dtsch. Wetterd. US-Zone, Kissingen Nr. 38, 12—16, 1952. (Diskussion versch. Einzelprobleme der Resonanztheorie an übersichtlichen Modellen.)

[*24*] HAURWITZ, B., u. F. MÖLLER: The semidiurnal air-temperature variation and the solar air tide. Arch. Meteor. Geophys. u. Biokl., Ser. A, Symposium-H., 170—182, 1955. (Erstmalige Ableitung der halbtägigen planetarischen Temperaturwellen aus den Beobachtungen.)

[*25a*] SAWADA, R.: The atmospheric lunar tides. Meteor. Papers **2**, No. 3 (1954).

[*25b*] SAWADA, R., and B. HAURWITZ: The lunar air tide. Ann. de Géophys. **11**, 145—147 (1955). (Modellrechnungen zur Erklärung der jahreszeitlichen Variation von L_2 durch Temperaturänderungen in der Ozonosphäre.)

[*25c*] SAWADA, R.: On the role of thermal excitation in the atmospheric tides. Geophys. Mag. Tokyo **26**, 267—281 (1955). (Überarbeitung von [*18*], übersichtlichere Formulierung. Der Erklärung der thermischen Erregung von W_2^2 ist aber wohl die in diesem Handbuchartikel in Ziff. 5 und 6 gegebene vorzuziehen.)

[*25d*] SAWADA, R.: The atmospheric lunar tides and the temperature profile in the upper atmosphere. Geophys. Mag. **27**, 213—236 (1956).

[*26a*] SOLBERG, H.: Über die freien Schwingungen einer homogenen Flüssigkeitsschicht auf der rotierenden Erde I. Astrophys. Norv. **1**, 237—340 (1936).

[*26b*] THORADE, H.: Die Gezeiten in neuer Beleuchtung [besonders nach H. SOLBERG]. Ann. Hydrogr. **64**, 381—386 (1936). (Anfang zu einer Theorie der Gezeitenschwingungen, bei der die Vertikalbeschleunigung nicht mehr vernachlässigt wird. Diskussion in [*b*].)

[*27a*] SOLBERG, H.: Wellenbewegungen einer isothermen Atmosphäre. Report of the 18. Scandinavian Naturalist Congress in Copenhagen 1929, 26.—31. August.

[*27b*] SOLBERG, H.: Schwingungen und Wellenbewegungen in einer Atmosphäre mit nach oben abnehmender Temperatur. Astrophys. Norv. **2**, 123—172 (1936). (Forführung und Erweiterng der Unterusuchungen von [*14*].)

[*28*] BJERKNES, J.: Atmospheric tides. J. Mar. Res. **7**, 154—162 (1948). (Eine neue Theorie, in der die Phasenverschiebung der Temperaturwelle in der Höhe für die Vergrößerung von W_2^2 verantwortlich gemacht wird.)

[*29a*] THRANE, P.: Some hydrodynamical properties of simple atmospheric oscillations with applications to the semidiurnal oscillation. Geofys. Publ. **18**, No. 1 (1951).

[*29b*] THRANE, P.: The tilt of a 12 hourly atmospheric tidal wave. Quart. J. Roy. Meteor. Soc. **81**, 9—17 (1955). (Einzelheiten zu der in [*28*] beschriebenen Theorie.)

[*30*] JACCHIA, L. G., and Z. KOPAL: Atmospheric oscillations and the temperature profile of the upper atmosphere. J. Meteorology **9**, 13—23 (1952). (Ergebnisse der Berechnung von Resonanzkurven nach der PEKERIS-WILKESschen Theorie für verschiedene Atmosphärenmodelle.)

[*31*] SEN, H. K., M. L. WHITE: Thermal and gravitational excitation of atmospheric oscillations. J. Geophys. Res. **60**, 483—495 (1955).

[*32*] WHITE, M. L.: Gravitational and thermal oscillations in the earth's upper atmosphere. J. Geophys. Res. **61**, 489—499 (1956).

[*33*] CHIU, W.: On the oscillations of the atmosphere. Arch. Meteor., Geophys. u. Biokl., Ser. A **5**, 280—303 (1953) (vgl. Ziff. 25b).

[*34*] HAURWITZ, B.: Atmospheric oscillations. Final Report, Project No. 122, Research Division, College of Engineering, New York University 1954. (Es wird über eine ganze Reihe neuerer New Yorker Arbeiten zu atmosphärischen Gezeiten berichtet. Literaturangaben.)

[*35*] SEPULVEDA, GLORIA M., and B. HAURWITZ: The geographical distribution and seasonal variation of the semidiurnal pressure oscillation in high latitudes. Scientific Report No. 299—2, Research Division, College of Engineering, New York University 1956.

[*36*] HAURWITZ, B.: The geographical distribution of the semidiurnal solar air tide. Meteor. Papers **2**, No. 5 (1956).

[*37*] HAURWITZ, B.: The standing semidiurnal oscillation in an atmosphere of variable depth. Scientific Report No. 3, Project No. 299 New York University, College of Engineering, Dept. of Meteorology and Oceanography, 1956.

Physical Volcanology.

By

S. SAKUMA and T. NAGATA.

With 27 Figures.

Introduction.

The scientific study of volcanoes started at the dawn of the geology and geography. The most part of the studies of volcanic phenomena have been done up to the present from the geological and geographical viewpoints, as the descriptions on earthquakes had been done by geologists till the middle of the last century. Geological, petrographical and chemical investigations have supplied very fruitful results. But volcanology also has not been allowed to remain indifferent to the current of the rapid development in the fields of physics and geophysics. Not a few efforts have been made to replace the past qualitative descriptions of volcanic phenomena by quantitative measurements after the methods of physics or of geophysics. However, it is still not easy to construct a vivid image concerning the mode of causation of volcanic activity based upon the results of geophysical methods only. At present, the results are interpreted usually on the basis of some older idea, not on the basis of advanced physics or geophysics. Physical and geophysical methods have been employed only to make the description of phenomena, as of old, more quantitative.

The principal purpose of this article is to review the results obtained up to date (by new as well as classical methods) from the standpoints of modern geophysics and physics, and to arouse interest in volcanic phenomena among geophysicists and physicists. For this purpose, after the characteristics of volcanic activities have been described briefly, the results of geophysical observations of volcanoes are summarized. Quantitative results are cited particularly. As observations of this kind have been made at a limited number of volcanoes, the sources of available data are greatly restricted. On the basis of those observations and experiments, several problems are brought forward. Possible ways to solve the problems and some approaches to interpretation are also described. But they are merely tentative ones meant to be taken as suggestions. The geological, petrographical and geochemical results are not described in detail here, in spite of their importance in the present volcanology.

A. Movement of ejected materials.

1. Outburst, flight and fall of ejecta. The forces which are met by a flying lava-fragment are that of gravity and that due to the resistance of surrounding gas or the air[1]. The movement can be described by the following differential equations, if it is assumed that every component of flight is affected only by the

[1] T. MINAKAMI: Bull. Earthq. Res. Inst. Tokyo **20**, 65–106 (1942). The reader is referred to the list of original papers there.

component of resistance in that direction.

$$\left.\begin{aligned} \frac{d^2 x}{dt^2} + \lambda\left(V_x - \frac{dx}{dt}\right)^2 &= 0,\\ \frac{d^2 y}{dt^2} + \lambda\left(V_y - \frac{dy}{dt}\right)^2 &= 0,\\ \frac{d^2 z}{dt^2} + \lambda\left(V_x - \frac{dz}{dt}\right)^2 + g &= 0,\\ \lambda = \frac{3k\varrho'}{4d\varrho}, & \end{aligned}\right\} \qquad (1.1)$$

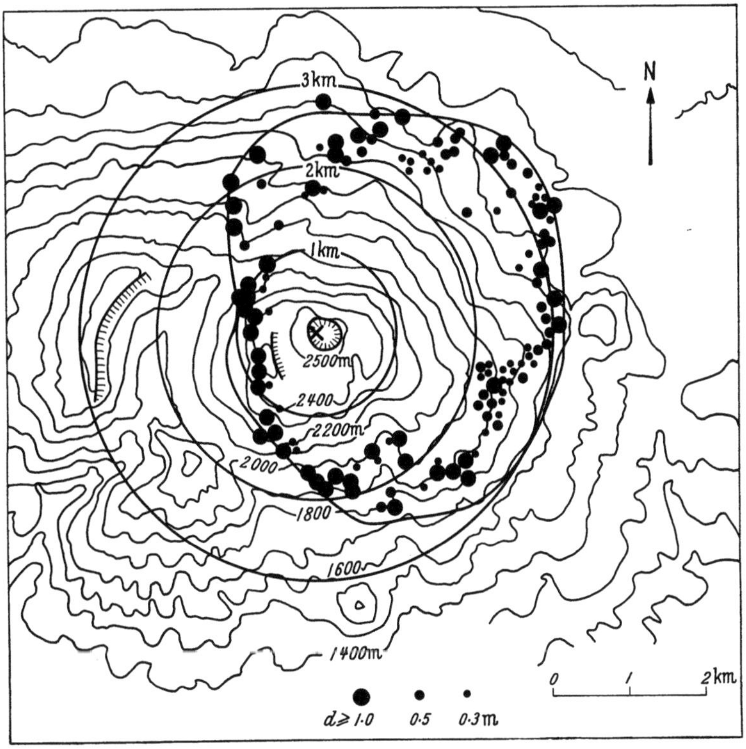

Fig. 1. Distance of the farthest flight of volcanic bombs, at the explosion of Asama (Japan) on April, 16, 1937. Initial velocity v_0 was estimated at 170 m/sec (after MINAKAMI).

where z is taken upwards, ϱ' and ϱ are the densities of gas and lava respectively, d is the diameter of the fragment, k is the coefficient of resistance, V is the velocity of the surrounding gas or the air. On various assumptions on V and λ, one can obtain x, y, z at any instant t. For example, when V is neglected,

$$\left.\begin{aligned} x &= \frac{1}{\lambda}\ln(\lambda t + v_0\cos\theta + 1),\\ y &= 0,\\ z_{\uparrow} &= \frac{1}{\lambda}\ln\sin\left\{\sqrt{\lambda g}\,t + \operatorname{arc\,cot}\left(\sqrt{\frac{\lambda}{g}}\,v_0\sin\theta\right)\right\} + \frac{1}{\lambda}\ln\sqrt{1 + v_0^2\frac{\lambda}{g}\sin^2\theta},\\ z_{\downarrow} &= -\frac{1}{\lambda}\ln\operatorname{Cos}\left\{\sqrt{\lambda g}\,t + \operatorname{arc\,tan}\left(\sqrt{\frac{\lambda}{g}}\,v_0\sin\theta\right)\right\} + \frac{1}{\lambda}\ln\left(\sqrt{\frac{\lambda}{g}}\,v_0\sin\theta\right), \end{aligned}\right\} \qquad (1.2)$$

where v_0 and θ are the initial velocity and the angle of ejection respectively. Because the effect of resistance of gas is negligibly small for such large bombs as $d > 50$ cm, v_0 can be estimated fairly precisely from the distance of flight or height of the large projectiles, the orbits of which are beautifully parabolic. The initial velocity of ejecta amounts to as much as 2×10^2 m/s. Fluctuation among the estimated velocities for many bombs from one explosion is rather small.

Rough estimation is possible concerning the kinetic energy which was necessitated to hurl out the mass (M) of lava, if the total volume of ejecta also is measurable.

$$\text{Kinetic energy} = \tfrac{1}{2} M v_0^2. \tag{1.3}$$

The estimation of the kinetic energy is indispensable to discuss energetics of volcanoes. Kinetic energy consumed in one explosion sometimes amounts to 10^{20} ergs (see [*25*]).

Ejections of solid or plastic lava are accompanied with rapid emission of gases. Most of the gas is water-vapour. Subsequently, estimation of the internal pressure of the gas just before the eruption may be possible by assuming a certain appropriate mechanism of acquirement of velocity for the lava fragments. Such great velocity of larger bombs as above 100 m/s cannot be explained by their drift due to the gas-stream, and accordingly the impulsive work done to the ejecta (of cross-section S) by gases of high pressure (P) seems to be a good explanation.

$$M v_0 = \int_0^\tau P(t)\, S\, dt. \tag{1.4}$$

In general,

$$P_{t=0} = \delta\, \varrho\, v_0^2, \tag{1.5}$$

where ϱ is the density of lava and δ is a numerical factor around 1. If P is constant through the time τ in which the work is done, $\delta = \frac{1}{2}$. If P decreases linearly with time to zero at $t = \tau$, $\delta = \frac{4}{3}$. There is no ground at present for stating which is nearer to reality[1].

The smaller and the lighter the fragments are, the more they are influenced by the movements of gas or the air. Fine ash and light pumice are transported by the rising smoke column high above the ground, are drifted by the upper wind, spread due to turbulence in the air and then fall by gravity back to the ground in the order of their size and weight.

2. Ascent of smoke columns. Volcanic activity is symbolized often by a violently rising smoke column taking the shape of a cauliflower. The gigantic column of turbulent gas and ash may be regarded as a *jet* that gushes out into the semi-infinite atmosphere. The diameter of the vortex-ring in the smoke is nearly proportional to the height, and this means that the average mixture length (after G. I. TAYLOR's definition) is proportional to height. Velocity distribution within such a jet is given by[2]

$$\frac{dz}{dt} \sim \frac{1}{z}, \quad \text{or} \quad z \sim \sqrt{t}. \tag{2.1}$$

Actually, some examples of rise of volcanic smoke give the empirical relation $z \sim t^{0.6}$ which agrees fairly well with the above considerations. In addition, if

[1] Pressure of gases at the time of violent explosion is estimated at more than 3×10^2 bars sometimes [*25*].

[2] W. TOLLMIEN: Z. angew. Math. Mech. **6**, 468–478 (1926).

the process of transfer of both momentum and heat is admitted and PRANDTL number is taken as 1, the temperatures of smoke and of surrounding air follow

$$T - T_{\text{air}} \sim \frac{1}{z}. \tag{2.2}$$

The temperature distribution has been proved to be correct at several solfataras.

If the height of smoke is defined as the height where the rising velocity becomes equal to the velocity of the wind there (u_0), the height h should be proportional to the upward velocity (w_0) of the smoke at a certain definite height z_0:

$$h = \left(\frac{z_0}{u_0}\right) w_0. \tag{2.3}$$

Actually, the smoke rises higher in proportion to the initial velocity of bombs for that eruption, as far as one certain volcano is concerned. The smoke columns of those volcanoes which eject red-hot materials rise higher in comparison with those from phreatic explosions. Aero-thermodynamical considerations, in which buoyancy is taken into account, should be introduced too[1].

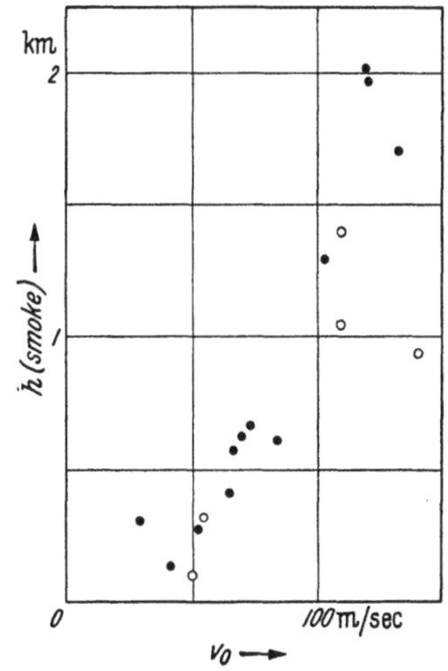

Fig. 2. Height of smoke columns and the initial velocity of ejection. Closed circles: eruptions associated with hot ejecta (Volcano Aso), open circles: phreatic explosions (Volcano Usu) (SAKUMA).

At any rate, a smoke column rises high in proportion to its initial velocity and transports fine fragments into the upper atmosphere. The well-known eruption of Krakatau in 1883 caused the anomalous optical phenomena (green

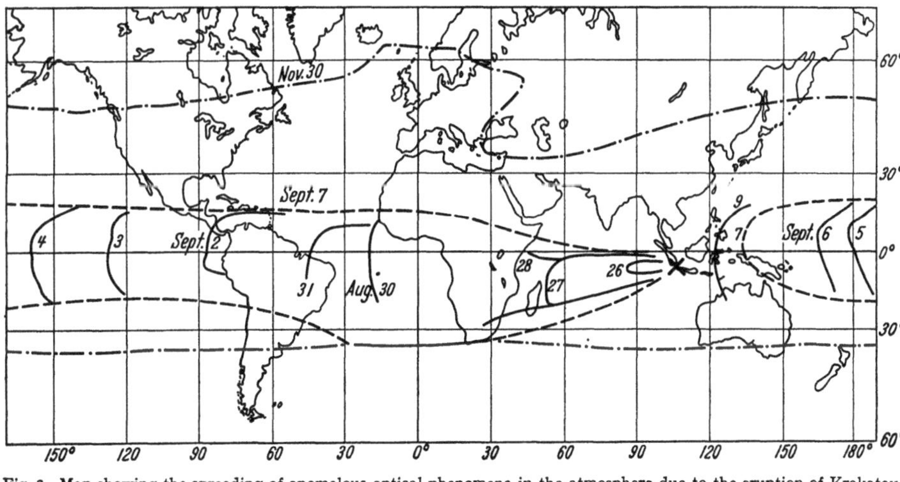

Fig. 3. Map showing the spreading of anomalous optical phenomena in the atmosphere due to the eruption of Krakatau on Aug. 26, 1883.

uns, red twilight glow, haze and corona), over the world by the fine dust which was blown up to the height of 36 to 18 km (see [*12*]).

Hot avalanche (*nuée ardente, Glutwolken*) is a phenomenon of volcanological importance. It may be explained also by taking the idea of turbulence into account, but no quantitative observation is available at present.

[1] B. R. MORTON, Sir G. TAYLOR and J. S. TURNER: Proc. Roy. Soc. Lond., Ser. A **234**, 1—23 (1956).

3. Motion of fluidal lavas and viscosity, its field studies and laboratory experiments. Emission of incandescent, most fluidal lava takes place in the form of fountains along fissures in a caldera or on the flank, around which swelling lava-pools are formed sometimes. In extreme instances the fountains may attain heights as great as 200 m or more. Fluidal lavas flow to the lower level naturally. The entire front of the most fluidal lava advances as a unit with a rolling motion. The front of the moderately fluidal lava with semi-solidified skin advances by successive protrusions of one small bulbous toe after another[1]. Less fluidal lava flows down between the banks of accumulated small fragments of solidified lava. It was shown that the flows are laminar in character under ordinary circumstances [*19*]. Much less fluidal lava moves down like a large heap of lava-blocks. However, the upper part of the heap moves forward more rapidly than the lower part, and at the front is rolled under and buried by the lava coming after it. That is to say, the large heap as a whole behaves like a liquid too (see [*22*]).

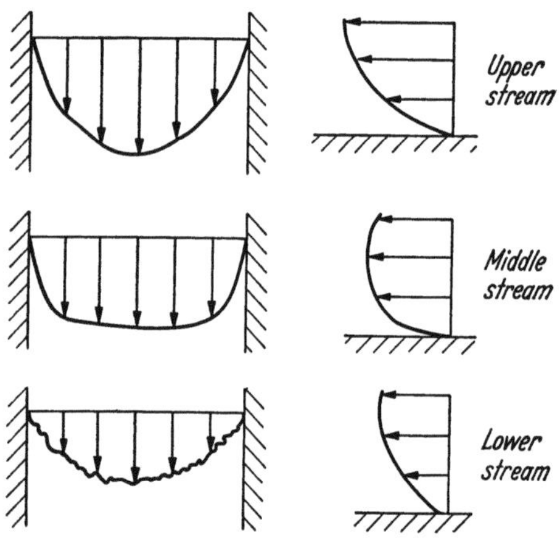

Fig. 4. Distribution of the flowing velocity within a lava stream in the 1951 eruption of O-shima in Japan (after MINAKAMI).

In order to estimate the coefficient of viscosity, the velocity (V), width, thickness (d) of the lava-flow which is flowing on a slope with inclination (i), are required to be measured together. Viscosity (η) is given as follows[2],

$$\eta = \frac{V \varrho g \sin i \, d^2}{\alpha} \tag{3.1}$$

where α is a numerical factor which is chosen according to the form of the cross-section of the flow. When V is taken as the maximum velocity of the surface, α should be 2 for infinitely wide flow, and 4 for the flow of semi-circular section. The viscosity is so much affected by temperature that simultaneous measurements of temperature is necessary. Viscosity of lavas in their natural molten state appears to be dependent on petrographical and chemical composition too. Besides, the viscosity in a certain phase of activity is often different from that of a similar lava in a different phase of the same activity.

The viscosity of the lavas of basaltic volcanoes is of the order of 10^3 to 10^5 poises at the time of ejection and that of acidic lava seems to exceed 10^6 poises even at the time of extrusion. The lavas which form domes or spines must be far more viscous than 10^{10} poises at the time of extrusion (Table 1).

Viscosity of silicate melts of various compositions has been measured by glass-engineers and petrologists[3]. At above the crystallization temperatures, viscosity is expressed in the following form as a function of temperature,

$$\eta = \eta_0 \exp(W/R\,T). \tag{3.2}$$

Increase of the fraction of SiO_2 or Al_2O_3 in the melt increases viscosity, and addition of Na_2O or K_2O decreases it. Remelted "dry" melts of various volcanic

[1] G. A. MACDONALD: Amer. J. Sci. **251**, 169–191 (1953). — C. K. WENTWORTH: J. Geology **62**, 425–438 (1954).

[2] R. L. NICHOLS: J. Geology **47**, 290–302 (1939).

[3] G. W. MOREY: The properties of glass. New York: Reinhold 1938. Extensive listing of original papers will be seen there.

Table 1. *Viscosity of lavas, with simultaneous temperature measurements, and the observed maximum temperature (T_m) [16], [19], [20], [22].*

Volcano [1]	Year	Viscosity (poise)	Temperature (°C)	T_m (°C)	Author
Kilauea	1912	—	—	1185	DAY and SHEPHERD
	1952	$2-4\times10^4$	1100—1050	1155	MACDONALD
Mauna Loa	1919	4×10^4	1200—1150	—	NICHOLS
	1950	4×10^3	1070	1120	MACDONALD
Parícutin	—	10^5-10^6	1070	—	KRAUSKOPF
	1944	—	—	1200	ZIES
Nyamuragira	1938	—	—	1145—1160	VERHOOGEN
O-shima	1937	5×10^4	1060	1060	NAGATA
	1950	2×10^5	1050	1050	MURAUCHI
	1951	6×10^3	1125	1125	MINAKAMI
		2×10^4	1108	1125	MINAKAMI
		7×10^4	1083	1125	MINAKAMI
		2×10^5	1038	1125	MINAKAMI
Vesuvius	1904	—	—	1100	BRUN
	1913	—	—	1200	PERRET
	1944	9×10^4	—	—	IMBÒ
Stromboli	1901	—	—	1150	BRUN
Etna	1951	—	—	1110—1120	TAZIEFF
Miyakejima	1940	7×10^5	1000	965	HAGIWARA and NAGATA
Asama	1936	—	—	1050	MINAKAMI
Sakurajima	1946	$3-6\times10^6$	1000	—	HAGIWARA
		10^9-10^{10}	900	—	NAGATA
Usu	1947	—	—	980	NAKAMURA

rocks were also examined in the laboratory at the range of 1450 to 1150° C[2]. Those melts from acidic rocks are more viscous than those from basic rocks. The results appear to be consistent qualitatively with the observed nature of lavas in the field. Abrupt increase in viscosity takes place below 1150° C, where crystallization of plagioclase and other minerals is observed.

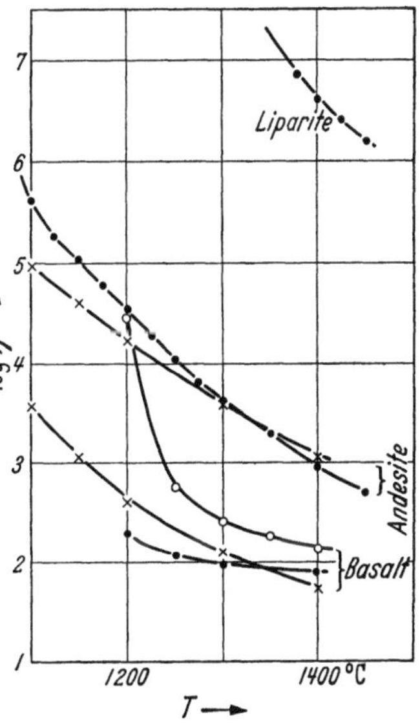

Fig. 5. Viscosity of dry melts from various volcanic rocks (after KANI and VOLAROVIČ).

At temperature below 1150° C, viscosity of the actual lava-specimens was measured also in the laboratory[3]. The coefficient of viscosity obtained in the laboratory is not only exceedingly higher than that observed in the natural state of the material, but increases more and more as time goes on, when the specimens are kept at high temperatures. This and other proved characters of the lava specimens below the crystallization temperatures imply nothing but that the lavas in the field as well as in the laboratory are not under the state of equilibrium corresponding to the temperature, and that the viscosity is not a function of temperature only. The factors by which the observed fluidity of lava is ruled must be principally the quantity and quality of glass in the groundmass. Although some differences or changes

[1] The volcanoes are arranged nearly in the order of acidity of lava (basic: upper).

[2] M. P. VOLARVIČ et al.: C. R. Acad. Sci. USSR **1**, 321—333 (1936). — K. KANI: Proc. Imp. Acad. Jap. **10**, 29, 79 (1934).

[3] S. SAKUMA: Bull. Earthq. Res. Inst. Tokyo **32**, 215—230 (1954).

in the groundmass are noticed by microscopic studies, it is not known yet what sort of weak bond within the glass is responsible for the viscous flow. At any rate, the viscosity of lava is influenced by its chemical composition at its glassy state, and it is affected by the thermal history which the lava has experienced below 1150° C.

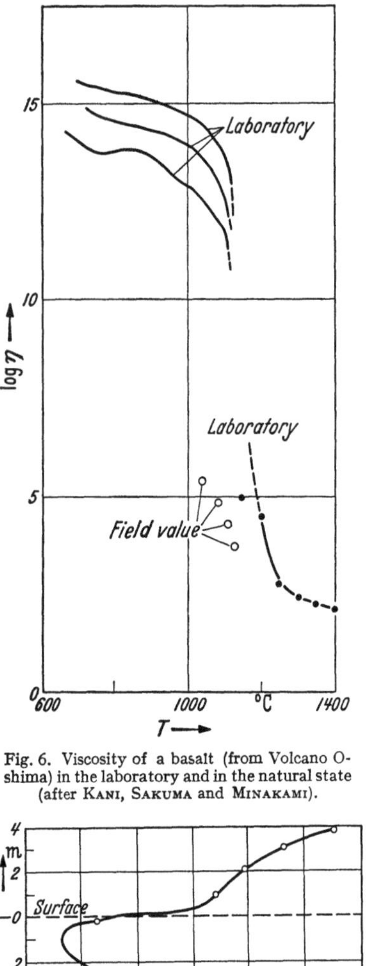

Fig. 6. Viscosity of a basalt (from Volcano Oshima) in the laboratory and in the natural state (after KANI, SAKUMA and MINAKAMI).

4. Temperature of lava. Measurement of temperature is usually done by *optical pyrometer* from a distance. Readings of pyrometers should be corrected for the effect of absorption due to volcanic fumes and ordinary mist [*17*]. Moreover, personal error is sometimes unexpectedly large in bad conditions. Accordingly, the results obtained by *thermocouples* are more reliable, but the use of this method is limited to the cases when the circumstances are favourable for a worker to have access to the very point where measurements are to be made.

The temperature of lavas at the time of ejection appears variable according to the type of rock and the phase of activity. Little is known concerning the temperature of lavas in the volcanoes of acidic rocks, because precise measurements of temperature at such activities as dome-building or hot avalanches have scarcely been made. But in general, it may be said that more basic lavas extrude at higher temperature, though the effect of combustion of gas (or effect of flare) raises the temperature to some extent. Basaltic lavas well out at the temperature range 1200 to 1100° C, andesitic lavas extrude below 1100° C and dacitic or liparitic rocks come to the surface at temperatures of about 1000° C. Yet this is a rough approximation. The difference in the temperature of extrusion would be one of the most important factors to control the characteristics of eruptive activities (Table 1, references group II all).

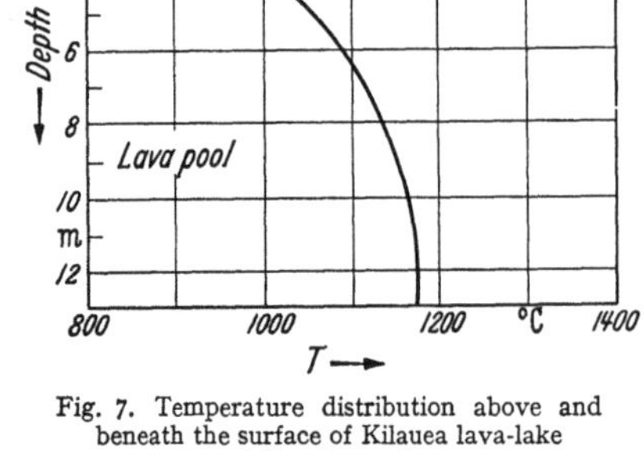

Fig. 7. Temperature distribution above and beneath the surface of Kilauea lava-lake (after JAGGAR).

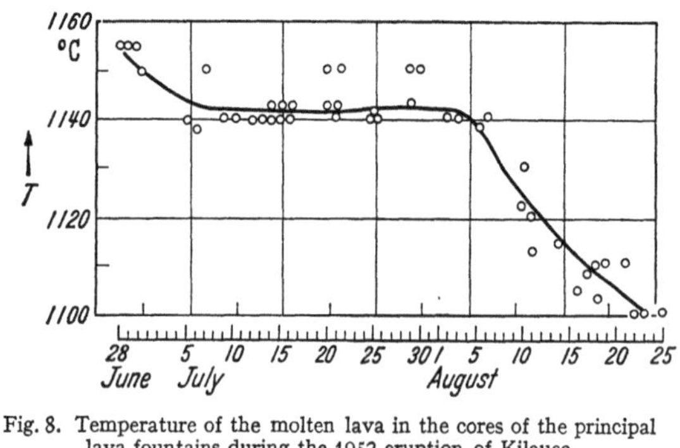

Fig. 8. Temperature of the molten lava in the cores of the principal lava fountains during the 1952 eruption of Kilauea (after MACDONALD).

B. Deformation of the earth's surface. (Geodetic investigations.)

5. Topographical changes. Changes in topography provide one of the common and remarkable features of volcanic activities. They are: opening of new craters or fissures; rise and fall of crater bottom; rapid growth of cones by accumulation of lava and cinders; destruction of volcanic bodies by explosions; outflow of enormous volume of fluidal lava; and development of dome or spine. The above mentioned change of topography is due to the addition of new materials which have been ejected from a depth to the surface of the earth, or the removal of materials which had occupied a part of the surface. Changes of this sort are noticed with naked eyes and can usually be described in detail. The total volume of the ejected substance is one of the measures of magnitude of activities and it appears to relate in a certain simple manner to the total energy emitted in the activity.

Table 2. *Total volume of ejecta of representative activities of various volcanoes.*

Name	Year	Volume of ejecta (km^3)
Tambora	1815	150
Krakatau	1883	18
Sakurajima	1914	2.2
Parícutin	1943–1952	1.4
Bandaisan	1888	1.2
Taal	1911	0.5
Mauna Loa	1950	0.46
Asama	1783	0.2
Komagatake	1929	0.1
Mauna Loa	1942	0.08
O-shima	1950/1951	0.03

6. Deformation of the earth's surface revealed by precise surveys. There is another kind of change of the earth surface which is not the product of accumulation or of removal of rocks at the surface. It is the deformation of the earth's

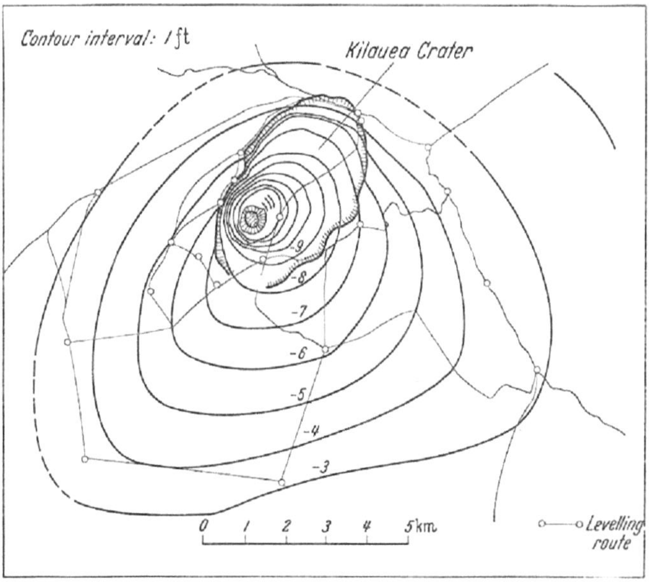

Fig. 9. Depression around the Kilauea Caldera during 1921 to 1927 (after WILSON).

surface caused by volcanic action over wide areas, which is the main object of precise measurements, because such deformations are too small to be noticed by other means. Movements in *vertical direction* are usually ascertained by repeated checking the heights of bench-marks along levelling routes which run through the volcanic district concerned.

For example, by virtue of the results of precise levellings done soon after the great eruption of Sakurajima, Japan, in 1914, depression of the ground on a large scale was brought to light. The total volume of the subsidence amounted to about 1.3 km³, which corresponded to about half of the lava flows and pumice ejected

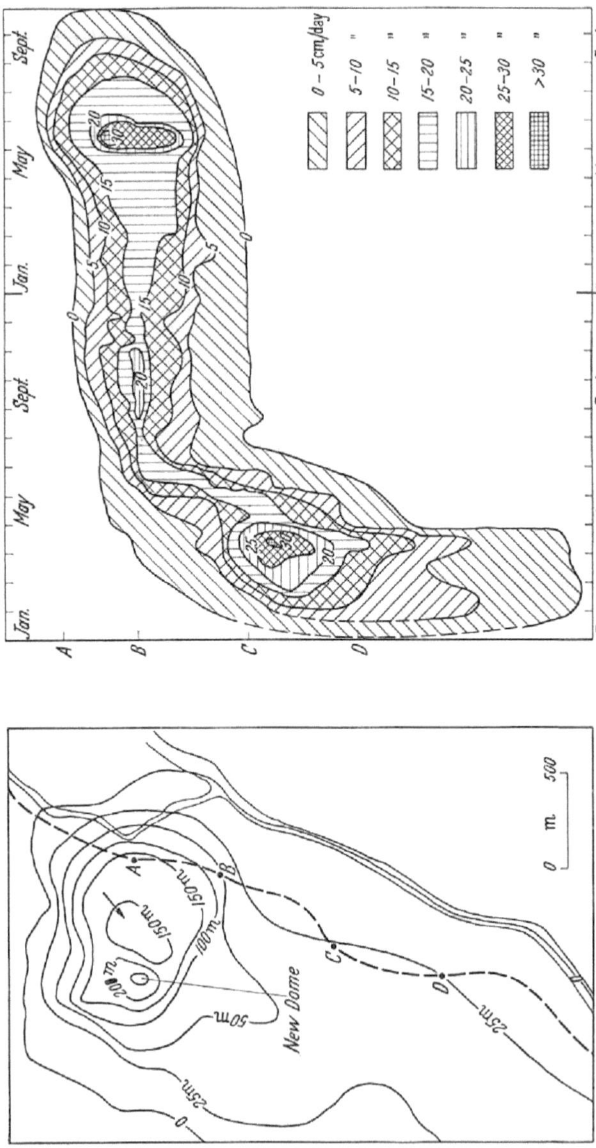

Fig. 10. Daily rising velocity of the ground (right) and the total amount of upheaval (along $A\,B\,C\,D$, left), caused by the 1943 to 1945 activity of Usu (after MINAKAMI and KANEKO)

during the eruption. The circular depression seems to have been caused by the eruption, not by the crustal deformation of any other origin, and it may be said that the eruptive activity affected the area within 50 km. The centre of the depression seems to coincide not with the active vent but with the centre of the caldera, near the rim of which the Sakurajima volcano stands. According to precise levellings which were carried out in 1915, 1919, 1932 and 1946 along the same line of levels, the depression progressed a little up to 1915 and recovery has taken place since then. The average rate of recovery in the recent 30 years is estimated at about 6×10^{12} cm³/year [*14*].

Similar depressions over wide areas at the time of eruptions and succeeding recoveries were noticed also at Kilauea[1], Hawaii, and in case of the activities of Usu (1910) and Komagatake in Japan (1929) [*13*] to [*15*].

The ground close to the site of activity does not subside at the time of eruption, but is usually remarkably upheaved. Extreme upheaval took place on the

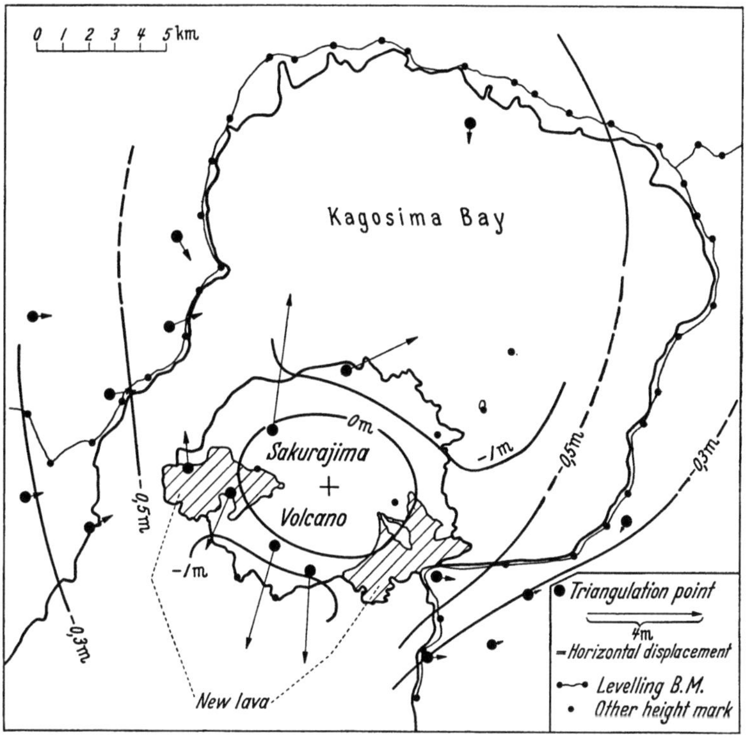

Fig. 11. Elevation, depression and horizontal displacement of the ground due to the 1914 eruption of Sakurajima (after OMORI).

eastern flank of Usu prior to and during the birth of the new parasitic dome in 1943 to 1945. The centre of upheaval shifted gradually to the north, having culminated in the outburst there and in the extrusion of a red-hot lava-dome at last [*18*]. Similar upheavals in volcanoes have been observed at the eruptions of Sakurajima (1914) and Miyakejima, Japan, (1940) by triangulations, at Usu (1910) and Asama (1935 to 1951) by levellings (see [*14*], [*16*], [*20*]).

Movements in the *horizontal direction* can be studied by triangulations, but only a few data are available now. At the time of the flank eruption of Miyakejima in 1940, the earth's surface deformed as though the ground was torn open by the fresh lava while it pushed its way ahead through the ground. Strain distribution at the surface could also be calculated on the basis of the horizontal movements. Dilatation and maximum shear were of the order of 1×10^{-3} at the maximum [*16*], [*18*], [*20*]. Similar deformation has been observed at Kilauea[1], Sakurajima (1914) and Usu (1943 to 1945) (see [*13*], [*14*], [*18*]).

[1] R. M. WILSON: Res. Publ. Hawaii Univ. **10**, 5–56 (1935).

The deformations which have been observed on other volcanoes in the world are not contradictory in general to the above described characters, though the observations are less accurate or they cover only limited areas.

7. Characteristics of crustal deformations. The above described deformations which occur in a volcanic region at the time of activity are classified into two sorts for convenience.

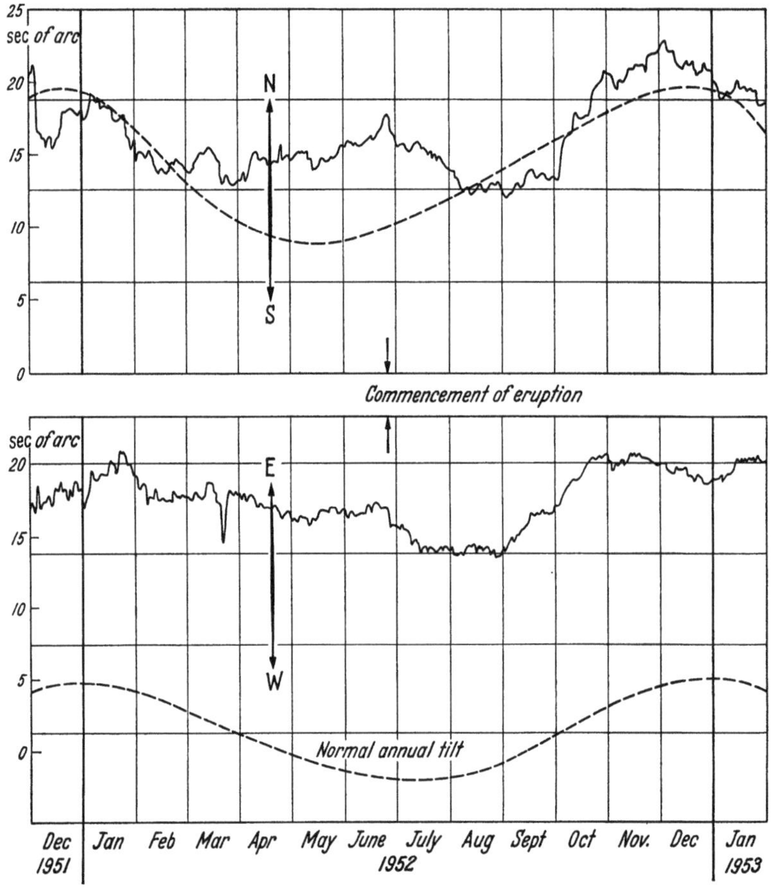

Fig. 12. Ground tilting on the NE rim of Kilauea Caldera during 1952. Arrow indicates commencement of eruption (after MACDONALD).

(a) One takes place *close to the site* of the eruptive activity, the deformation is great and the affected area is limited. Subsequently, the gradient of deformation (strain) is large enough to exceed the yield limit of soil or of rocks, and sometimes to cause fractures. The principal feature of a movement of this sort is upheaval, sometimes culminating in extrusion of fresh lavas.

(b) The other sort of deformation extends typically over a *wide surrounding area*, often including formations of non-volcanic origin. Its feature is the subsidence associated with ejection of an enormous volume of juvenile materials. The subsidences usually recover to some extent.

The difference in the affected area between these two kinds of deformations seems to be due probably to the difference in the depth of origin. The local and more remarkable deformation has probably been caused by a force of more shallow origin, though the exact depth is not known. It will be useful to apply the mathematical theory of elasticity and plasticity to the problems of deformation of the earth's surface caused by the force of internal origin.

8. Continuous observation of crustal movement. Besides the repeated precise geodetic measurements, continuous registrations of crustal deformation are being carried out at several volcanoes[1]. Observation of the tilt of the ground by means of pendulums of long period is the most common. Annual, diurnal, meteorological and other variations which have no connection with volcanic activity should be eliminated carefully by smoothing, by subtracting a normal curve and by other means. Tilt of the ground thus obtained often shows close correlation with volcanic conditions. Instruments near by the Kilauea caldera show the outward tilt responding upheaval of the interior of the caldera, prior to the eruptive activities [*13*], [*19*]. The approaching swarm of explosions of Asama can be forecast fairly reliably by a certain characteristic tilting which begins from a month or so before the activity [*25*]. But it should be kept in mind that the tilt observed by tiltmeters of pendulum type at a locality does not always agree with the actual large scale crustal movements, neither in direction nor in magnitude.

C. Elastic waves from volcanoes. (Seismological investigations.)

9. "Volcanic earthquakes" of ordinary appearance. Those earthquakes which take place near active volcanoes are considered customarily to be of volcanic origin. There is, however, no criterion at present to distinguish "volcanic earthquake" from the ordinary earthquakes, except for their occurrence in swarms in a rather limited part of the crust in volcanic districts and for the fact that some of the seismic activities are followed by eruptive activities of the volcanoes. The earthquakes which are discussed in this article are limited to those which take place significantly in swarms locally beneath the respective volcanic formations concerned, and especially to those which are followed, accompanied or preceded by volcanic activities. Those earthquakes are here classified tentatively into the following three kinds: earthquakes of *ordinary* appearance, *explosion-earthquakes*, and *volcanic tremors*.

In this section, the earthquakes of the first kind are treated the various phases of which can be identified on seismograms without much difficulty. Their focal depths are not extremely shallow, and the form of the seismogram is similar to that of an ordinary earthquake.

Not all of the seismic activities are associated with surface activities. But eruptive activity is always preceded and accompanied by seismic activity, which begins several days or several months prior to the actual eruptions. In most cases, the forerunning seismic activity tends to diminish shortly before the commencement of eruption. Seismological observation at a single station only may be able to detect such a restlessness of the volcano qualitatively. Really, *predictions* of forthcoming eruptions have been made successfully at several volcanoes, on the basis of statistical studies on the relation between the frequency of earthquakes and the eruptions (see [*25*], [*19*], and so on). In spite of lack of knowledge on the nature of the earthquakes, the statistical method is still of practical importance.

[1] G. Imbò: Bull. Volcanol., Ser. II **8**, 117–118 (1949). — T. A. Jaggar and R. H. Finch: Bull. Seism. Soc. Amer. **19**, 38–51 (1929).

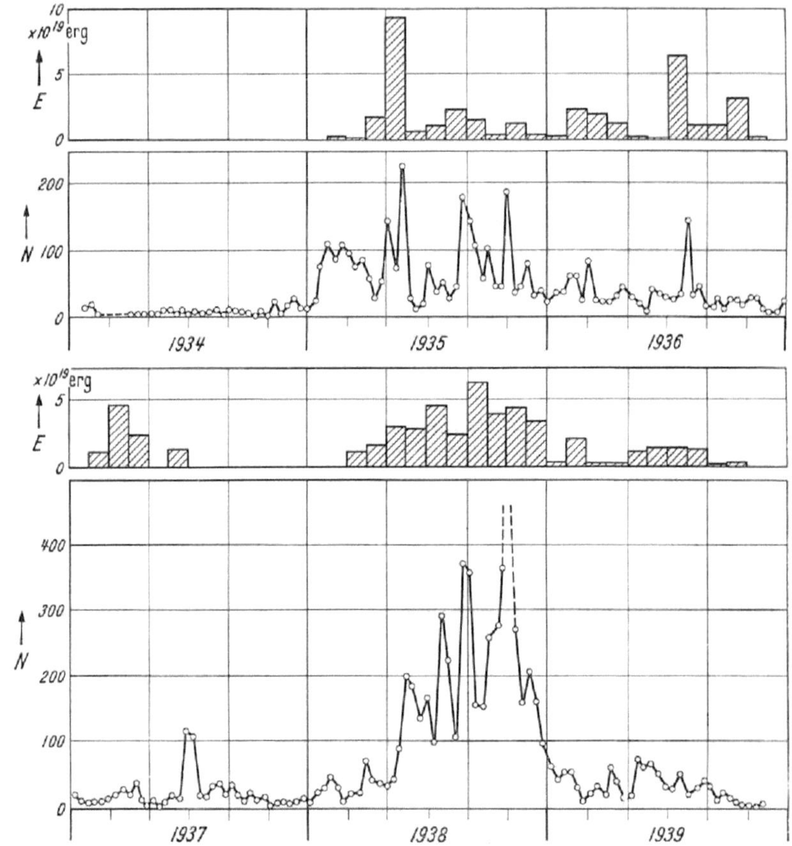

Fig. 13. 10-days' sum (N) of the earthquakes, registered by a 350 times seismograph at Asama Volcano Observatory, and the monthly sum (E) of the kinetic energy of ejecta (after MINAKAMI).

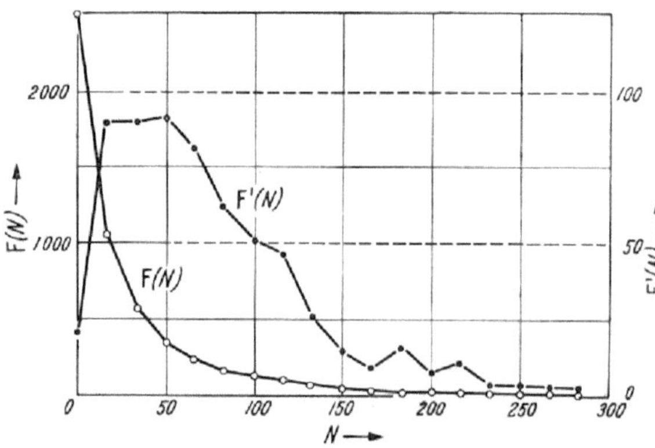

Fig. 14. Frequency (F) of 10-days' sum (N) of earthquakes at Asama, during 1934 to 1951. The frequency $F'(N)$ in the 10-days prior to eruptions (closed circles) is significantly different from the frequency distribution $F(N)$ through all the periods (open circles) (after MINAKAMI).

By the observations made by networks of seismographs, one can determine the seismic foci and trace the migration of them, if the phenomenon exists. For example, an eruption of either Kilauea or Mauna Loa is preceded by a series of

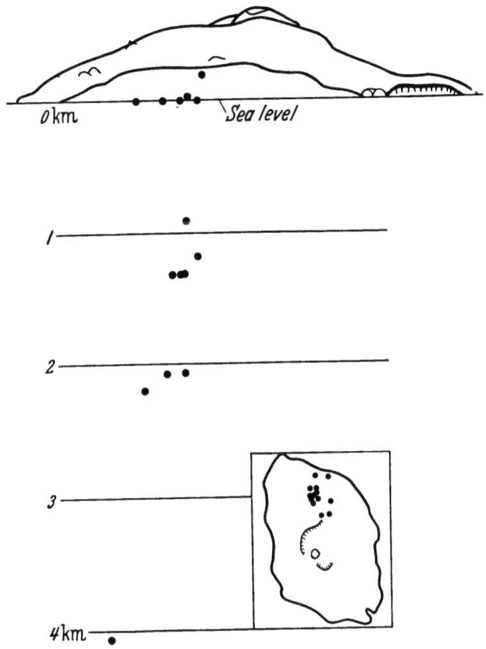

Fig. 15. Position of foci of the earthquakes which took place beneath O-shima during June, 1936 (TAKAHASI and NAGATA).

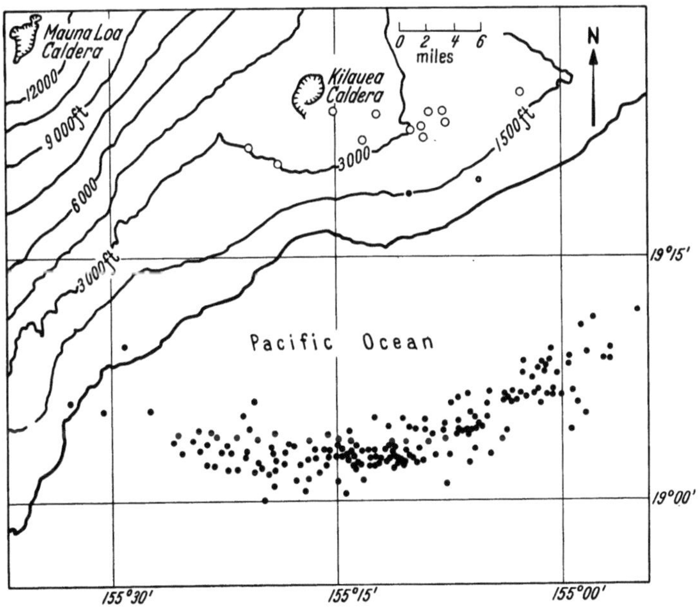

Fig. 16. Map of part of the southeastern coast of Hawaii, showing the location of foci during March to May, 1952. Trend of fault-system is nearly parallel to the coast. Open circles: Kilauea quakes preceding its July-eruption (after MACDONALD).

earthquakes[1]. These earthquakes occur initially at a depth of 30 to 50 km, but as the series progresses the places of origin become gradually closer to the surface and often shift laterally, finally centering at a place close to the site of the coming eruption. A similar tendency has been observed at Usu and other volcanoes

[1] R. H. FINCH: Bull. Seism. Soc. Amer. **33**, 237–241 (1943).

([*13*], [*18*] to [*20*]). But in volcanoes other than in Hawaii, earthquakes at such a depth as 30 to 50 km have not yet been identified as of volcanic origin. Most of the earthquakes originate within several kilometers beneath the craters. The foci assemble sometimes within the area of such a simple form, as a sheet or a

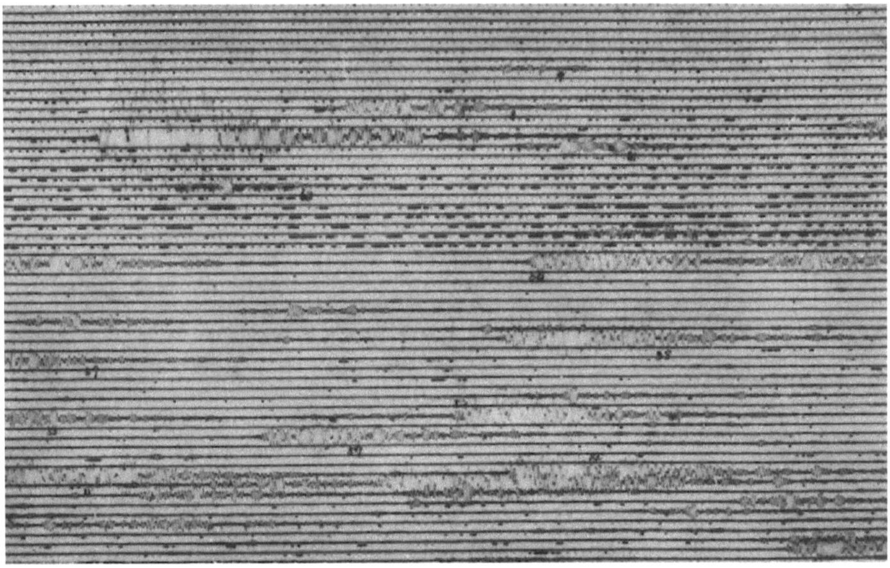

a

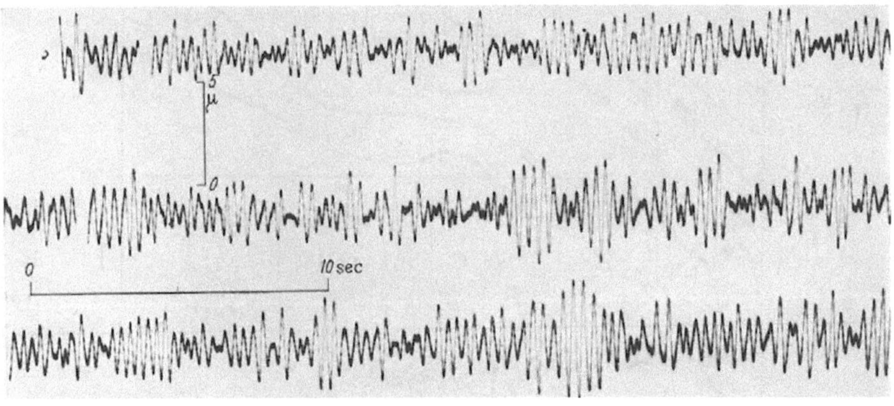

b

Fig. 17a and b. Two types of "volcanic earthquakes" (after MINAKAMI). (a) Volcanic earthquakes of ordinary type, recorded at Volcano Asama by a moving coil type seismograph for horizontal component ($T_0 = 1$ sec, $T_g = 0.2$ sec). Paper speed ca. 110 sec from left end to right end of the figure; the light dots in the figure are second marks. Magnification on the figure ca. 3000 fold. (b) Volcanic tremor of O-shima in 1950, registered by a mechano-optical seismograph at 4 km from the erupting vent.

cylinder. The trend of the dense part of the foci appears often to coincide with a pre-existing vent or to run in parallel to the established rift zones[1]. Precise studies by many seismographs of high magnification in network should be performed more commonly on and around active volcanoes, especially to learn the mechanism of the generation of such earthquakes.

[1] R. TAKAHASI and T. NAGATA: Zisin, Ser. I **11**, 161—167 (1939).

Seismic waves of the earthquakes of shallow origin, say within 1 km, are affected so much by superficial structures that identification of phases is rendered impossible even at short epicentral distances. Therefore, it is not an easy matter to study the small shallow earthquakes, which may be an important clue to know the subterranean state before eruptions.

10. Explosion-earthquakes. A volcanic explosion is always accompanied by an earth-shaking [*25*]. The *explosion-earthquake* occurs almost simultaneously with the outburst of gas and ejecta. Minute investigations of travel time suggest that the earthquake takes place a little earlier than the outburst, as though the focus is a few hundred meters deep. Initial motion of the earthquake of this sort is in question. At the time of the violent explosion of Asama in 1950, almost all of the initial motions of many seismograms were directed away from the volcano. (The range of the epicentral distance was 10 to 200 km.) Whereas, the explosion-earthquakes of Aso, Japan, seem to be caused by tension-cracking, judged from their initial motion observed near the crater[1]. The mechanism of an outburst itself would be an interesting problem.

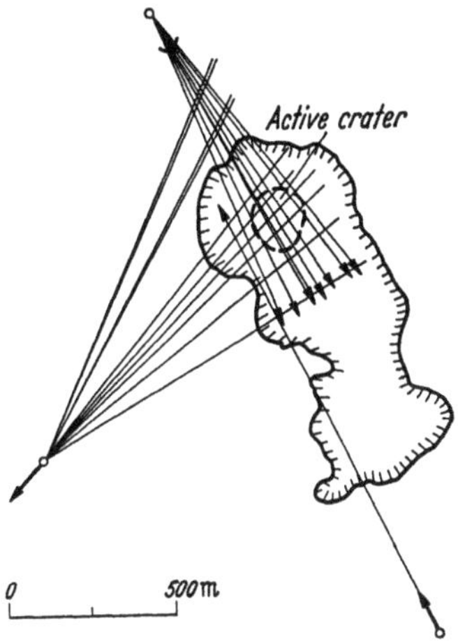

Fig. 18. Direction of initial motion of explosion-earthquakes at Aso (after SASSA).

The more violent is the explosion, the larger is the maximum amplitude of the explosion-earthquake. A relation can be obtained between the magnitude of explosions and the amplitudes of the ground vibrations at a definite station. The amplitude may be adopted as a magnitude scale of the eruption.

Energy released in explosion-earthquakes may be estimated by the same method as that used in seismology[2]. Seismic energy of volcanic explosion in its order of magnitude is 10^{-2} to 10^{-4} of the kinetic energy of ejecta (see Sect. 1).

11. Volcanic tremors. Sinusoidal wave trains which are observed at the time of volcanic activity are customarily called *volcanic tremors*. Some of them continue for a long time without intermission in spite of ebbs and flows in their amplitude. There is certainly a kind of tremor, the amplitude of which is in good accord with the intensity of the eruptive activity of the crater at the time. Many examples have been observed at the eruptions of volcanoes of various types, especially at those of Strombolian type in which small explosions take place every several seconds (see [*16*], [*19*], [*20*], [*22*], [*25*]). This sort of tremor may well be considered as nothing but a succession of explosion-earthquakes. Judging from the decrease of amplitude or of intensity of acceleration with distance, the depth of origin does not exceed a few hundred meters. But, a part of the observed tremors of this sort are caused by the air-shocks which strike the ground *near* the seismograph. Especially, when observations are made by seismographs of short period, the ground vibration is found to be in close correlation with the sound [*22*].

---

[1] K. SASSA: Mem. Coll. Sci. Kyoto Univ., Ser. A **18/19** (1935/1936).

[2] K. E. BULLEN: An introduction to the theory of seismology. New York: Cambridge Univ. Press 1953. See also this Encyclopedia, Vol. XLVII, p. 81.

There is another kind of sinusoidal oscillations which are not accompanied with simultaneous eruptive activities[1]. They are sometimes continuous through days and nights; sometimes they occur in separate trains. When observations are made by seismographs of high magnification and of various periods, more than two kinds of tremors are usually registered.

It is difficult to identify initial motion or other phases in the tremors. But some approaches to their nature are possible by studying the changes in amplitude and in period, in connection with the changes in the eruptive activities. For example, here will be introduced the changes of the character of the tremors of the 1951 activities of O-shima, Japan, when the stage of repose and that of

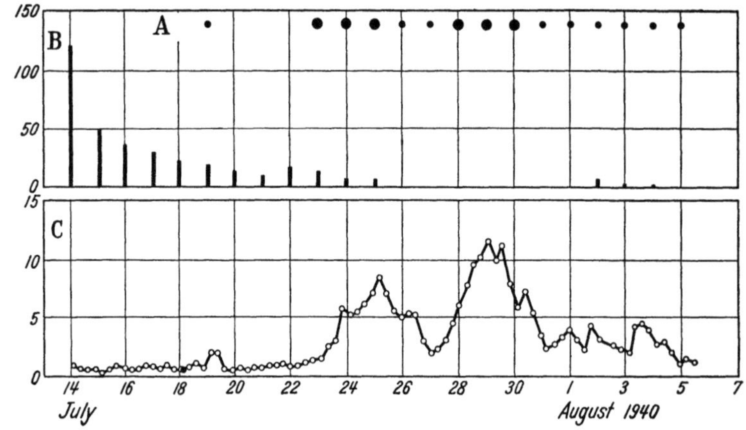

Fig. 19. Seismic activity associated with the volcanic activity of Miyakejima, 1940. A: intensity of detonation, arbitrary scale, B: daily frequency of earthquakes, C: amplitude of tremors (unit 3 μ) (after HAGIWARA and MINAKAMI).

eruption came alternately [20]. Tremors of comparatively short period and of small amplitude were recorded intermittently at 2 km from the crater, when the crater looked calm. The tremors increased both in frequency of occurrence and in amplitude as the eruptive activities approached. And at last, the tremors became continuous shortly before the commencement of active stages. The period of the tremors was predominantly 0.2 to 0.4 sec during the calm stages, and 0.2 to 0.9 sec in the most active stages. The tremors of intermediate character were also observed for several hours or several days before and after each eruptive stage (Fig. 20).

The phenomena will be explained consistently by assuming that the origin of the tremors of short period is always at a depth of several hundred meters, and that another origin of tremors moves upward or sinks downward gradually in accordance with the rise or fall of the head of fresh lava, causing the tremors of long period. General increase of their amplitude with approach of eruptive activity may be the result of increase of subterranean stress. Similar changes in amplitude and period of tremors have been observed at several other volcanoes (for example Vesuvius 1944, Aso 1933).

If it is assumed that the maximum amplitude (A) of tremors of period T declines during propagation with distance Δ in the following way:

$$A = A_0(T)\,\Delta^{-n} \exp\{-k(T)\,\Delta\}, \tag{11.1}$$

[1] R. H. FINCH: Bull. Seism. Soc. Amer. **39**, 73—78 (1949). — T. A. JAGGAR: Bull. Seism. Soc. Amer. **10**, 155—275 (1920). — IMBÒ, G.: Bull. Volcanol., Ser. II **16**, 161—170 (1955). — K. SASSA: Mem. Coll. Sci. Kyoto Univ., Ser. A **18/19** (1935)/1936).

n approaches $\frac{1}{2}$ for the waves of comparatively long period and 1 for the waves of shorter period. That is, the tremors of *short* period may be regarded as *body waves* principally and those of *longer* period are mainly the *surface waves*. Even though the fact does not always imply existence of essential difference in the mode of generation of tremors at the presumed two sources, this idea may be helpful in speculating about the cause of the tremors.

If the above described two-sources model is an acceptable one, new questions are aroused inevitably. They will be discussed in the next section.

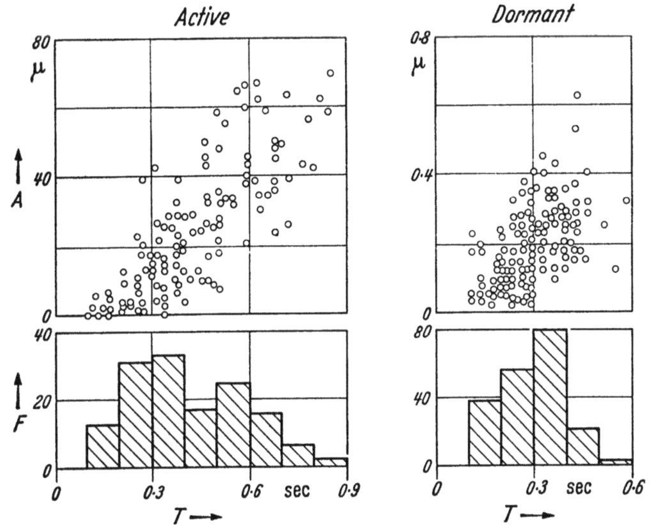

Fig. 20. Amplitude and period of tremors, observed at active stage (left) and at the dormant stage (right) of O-shima. A amplitude, T period, F frequency of occurrence (after MINAKAMI).

12. Magnitude-frequency relation of the earthquakes of volcanic origin, and further problems. The relation between magnitude M and frequency $N(M)$ of occurrence is expressed by the empirical formula

$$\log N(M) = a + b\,(8 - M), \tag{12.1}$$

for the earthquakes of non-volcanic origin, and b is commonly 1.0 to 0.8 from the destructive to the locally perceptible earthquakes[1]. On the other hand, another empirical relation[2]

$$N(a)\,da = k\,a^{-m}\,da \quad (a\text{: maximum trace amplitude}) \tag{12.2}$$

has been proved to be valid for aftershocks and other micro-earthquakes. The values of m were estimated at 1.9 to 1.8 for all cases. Because b is connected with m by

$$b = m - 1, \tag{12.3}$$

one and the same magnitude-frequency relation holds for the broad range of magnitude scale for earthquakes of non-volcanic origin.

Similar empirical relation between amplitude and frequency was proved to hold for the earthquake of volcanic origin too. The values of m are really 1.8 to

[1] B. GUTENBERG, and C. F. RICHTER: Seismicity of the earth. Princeton: Univ. Press (1954).

[2] M. ISHIMOTO et K. IIDA: Bull. Earthq. Res. Inst. Tokyo **17**, 443–478 (1939). — T. ASADA, Z. SUZUKI and Y. TOMODA: Bull. Earthq. Res. Inst. Tokyo **29**, 289–293 (1951).

1.9 for the earthquakes of ordinary appearance, which originated at a little deeper part within several kilometers beneath the crater. However, for the earthquakes which are accompanied by outbursts or which are of extremely shallow origin, the values of m were found to vary from 2.5 to 4 or more.

Summarizing the considerations offered in the four sections above, one may notice that the tentative classification of "volcanic earthquakes" into the three sorts is imperfect. The deeper earthquakes of volcanic origin seem to be *seriate* to the ordinary earthquakes, as far as the magnitude-frequency relation is concerned. It would be right to consider that the origin of stress causing such earthquakes of "volcanic" origin is a volcanic one, and that the mechanism of generation of elastic waves, perhaps fractures in the crust, is similar to that of ordinary quakes. On the other side, there exist explosion-earthquakes which occur at shallow depths and associate with emission of gas or other ejecta. In this case, the mechanism of generation of elastic waves may well differ from that of ordinary earthquakes.

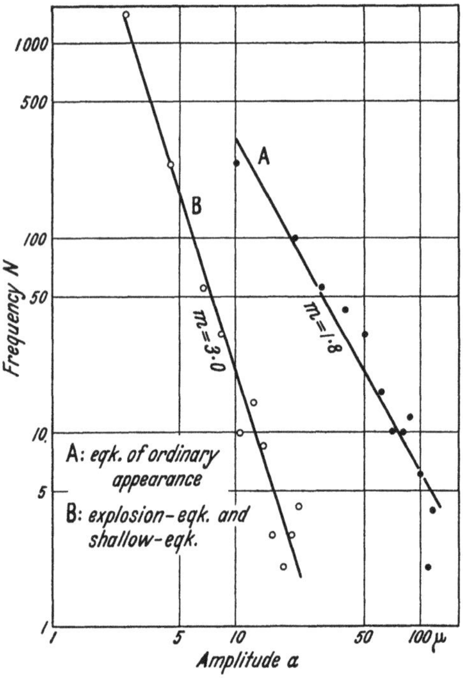

Fig. 21. Amplitude-frequency relation $N(a)\,da = k\,a^{-m}\,da$ of the earthquakes of volcanic origin, observed at Asama (1911—1913) (SAKUMA).

The nature of the non-eruptive shallow earthquakes, such as some of the tremors, remains unknown. Two possible explanations for that sort of quakes will be proposed here. One is: origin of stress is common to deep and shallow earthquakes, like the pressure owing to the intruding lava, while the mode of fracture as well as attenuation of waves is variable according to depth. The other is: one kind of origin, such as oscillation of lava reservoir or percussion of lava to the wall and so on, which is different from that of deeper quakes, exists corresponding to one kind of earthquake (or tremor). It should be not impossible to learn seismometrically or by other means which assumption is correct. For example, one way is: to ascertain whether the origin is within the hot lava or in the surrounding crust. At any rate, it would be an important subject of volcanology as well as of seismology to learn the mechanism of the generation of those "volcanic earthquakes".

13. Air- and water waves from volcanoes. Detonations of explosions have been studied by many authors (for examples see [*14*]); the mode of propagation will be referred to in another article of this Encyclopedia[1].

Sound waves from volcanoes include the oscillations of sufficiently long period to be registered by ordinary barographs and to cause damage to buildings. Sometimes the pressure amounts to the order of 2×10^{-2} bars at 10 km or so distant from the crater[2]. If the dimension of the origin were known, the pressure at the origin could be estimated.

[1] E. F. COX: Sound propagation in air. This Volume, p. 455.
[2] S. SAKUMA: Bull. Earthq. Res. Inst. Tokyo **29**, 605—615 (1951).

Sounds from submarine eruptions can travel over long distance through SOFAR channel in the ocean with intensity strong enough to be recorded by hydrophones[1].

Submarine volcanic activities are often accompanied by *tsunamis,* that is, great sea waves. Observations are done by tide-gauges or by the naked eye. The ejected gas of high pressure from the submarine vent, upheaval of a certain part of the sea bottom, and the sudden fall or inflow of ejecta of enormous volume are supposed to be responsible for the sea waves of such long period. The tsunami from Volcano Myojin-sho, off the coast of Japan, in 1952, which was estimated at about 2 m high near the site, was actually caused by the first kind of event[2]. The great tsunami from Krakatau in 1883 was supposed to have originated from the causes of the second and the third kinds [*12*].

D. Geomagnetic and other investigations to learn about the state of the interior.

14. Geomagnetic investigations. Physics of magnetism of volcanic rocks has been so developed in these twenty years that a special article is inserted in this Encyclopedia[3], and it is needless to state here the detail of the magnetic properties of rocks. The magnetization[4] of *volcanic rocks* was found to result mainly from the *thermo-remanent magnetism* and partly from the induced magnetism of the ferromagnetic minerals contained in the rocks themselves. The direction *in situ* of the thermo-remanent magnetization of lava coincides with that of the geomagnetic field at the time of its solidification, and it agrees with the present field within twenty or thirty degrees, as far as the lavas ejected in the Alluvial age are concerned. Although the total intensity of magnetization of volcanic rocks is not very strong (10^{-2} to 10^{-4} emu/gr), it is stronger than that of other kinds of rocks (10^{-3} to 10^{-5} emu/gr). Accordingly, the geomagnetic field in any volcanic region is expected to be much disturbed by the presence of any magnetized volcanic body.

Absolute measurements of the three elements of the geomagnetic field and surveys of relative values of the field intensity have been carried out on many volcanoes. As a result, the local anomaly of characteristic form was brought to light, which was explained well by the magnetization of the volcanic body in the direction of the present terrestrial field there[5]. The apparent intensity of magnetization of a volcanic body, estimated from the anomaly, is not contradictory to the experimental results on the magnetic properties of rocks.

Geomagnetic prospecting may also be possible if there exists distinct magnetic contrast among various volcanic formations in the given district. Actually, attempts to learn about the depth of the base of a volcano from the results of survey or to distinguish hidden dykes or craters from other surrounding sediments have been carried out with some success.

The possibility that anomalous change in the earth's crust at the time of volcanic eruptions might be accompanied by some anomalous change in geomagnetic field has been proposed frequently. Among many attempts made to

[1] R. S. DIETZ and M. J. SHEEHY: Bull. Geol. Soc. Amer. **65**, 941—956 (1954). See also this Encyclopedia Vol. XLVII, p. 135.

[2] R. MORIMOTO and J. OSSAKA: Bull. Earthq. Res. Inst. Tokyo **33**, 221—250 (1955).

[3] S. K. RUNCORN: Magnetism of rocks. This Encyclopedia, Vol. XLVII, p. 470.

[4] T. NAGATA: Rock magnetism. Tokyo: Maruzen 1953.

[5] T. MINAKAMI and S. SAKUMA: Proc. Seventh Pacific Sci. Congr. 2, 544—557 (1953). — Bull. Volcanol., Ser. II **18**, (1956). A listing of original papers will be seen there.

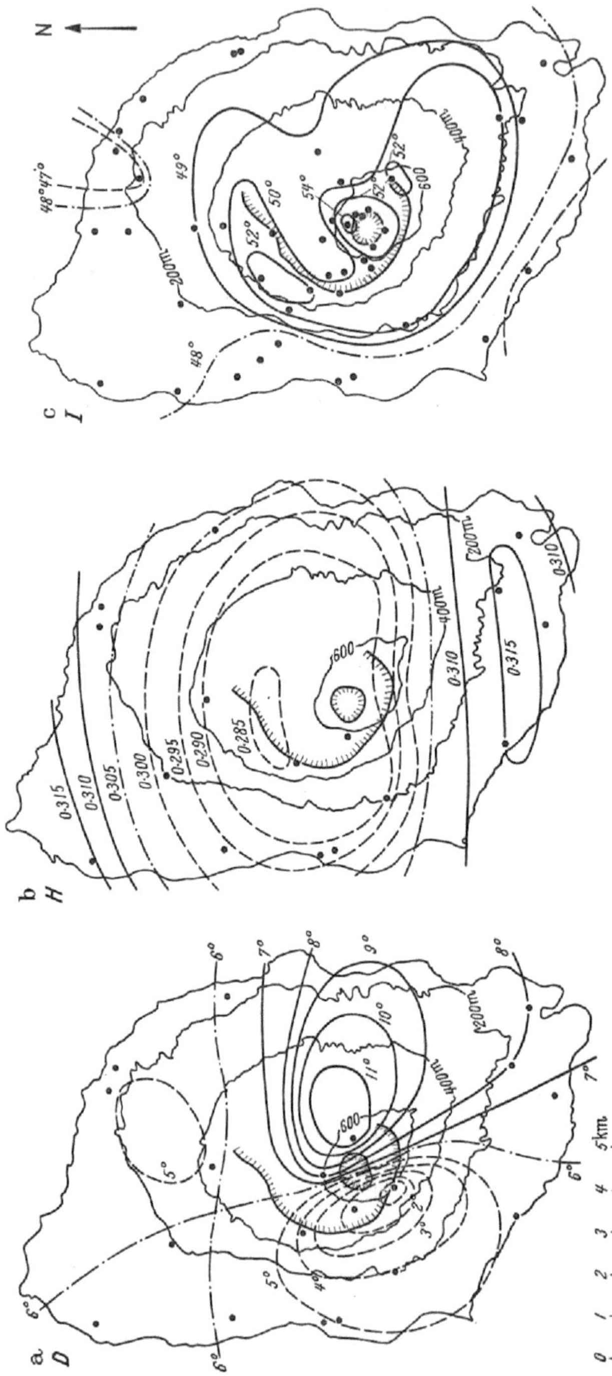

Fig. 22. Geomagnetic field on Volcano O-shima for 1951.4. (a) westerly declination, (b) horizontal intensity (in Gauss), (c) dip (inclination) (after RIKITAKE and YOKOYAMA).

solve this problem, the most reliable data[1] may be those obtained at Volcano O-shima (Mihara) on the occasion of its eruptive activity beginning in 1950 [*20*]. Magnetic dip-surveys were carried out repeatedly over the volcano at various stages of the eruption. By comparing the first survey with the second one, it

[1] T. RIKITAKE and I. YOKOYAMA: J. Geophys. Res. **60**, 165–172 (1955).

was found that the *geomagnetic dip-angle* decreased during about two months from the time of the first start of volcanic activity to its very active stage when plenty of lava flowed out. The decrease was systematically distributed, being larger in the central part of the volcano. The maximum decrease amounted to as much as 30′. The magnetic change corresponds to the demagnetization of the spherical volume of 2 km in radius which is situated at 5 km beneath the crater of the volcano. Gradual recovery of the geomagnetic change was also found by the repeated surveys up to October, 1953, when the volcano resumed activity again. Similar decrease in dip took place associated with the 1953 eruption. Decrease of dip-angle amounted to 4′ at the central part of the volcano.

Continuous recording of geomagnetic *declination* was carried out too. The base-line value of magnetogram was checked by absolute measurements. The change of declination is in good accord with the results of dip-surveys. Increase of westerly declination there, which means demagnetization of the volcanic body, took place at the time of volcanic activities.

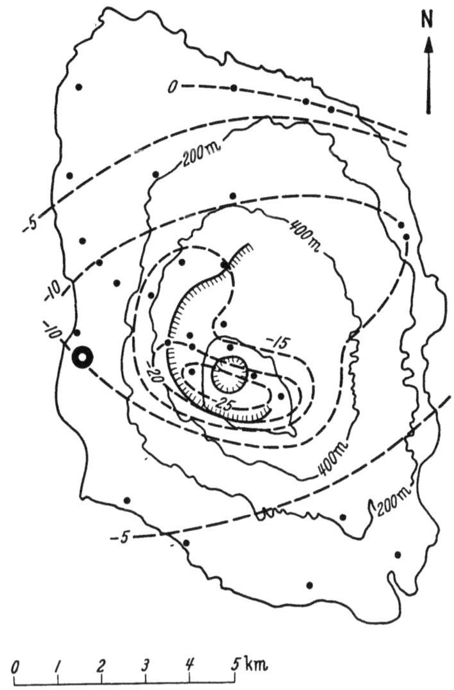

Fig. 23. Change in geomagnetic dip in minutes of arc associated with the 1950 eruption of O-shima. Open circle: the place where continuous recording of declination is done (after RIKITAKE).

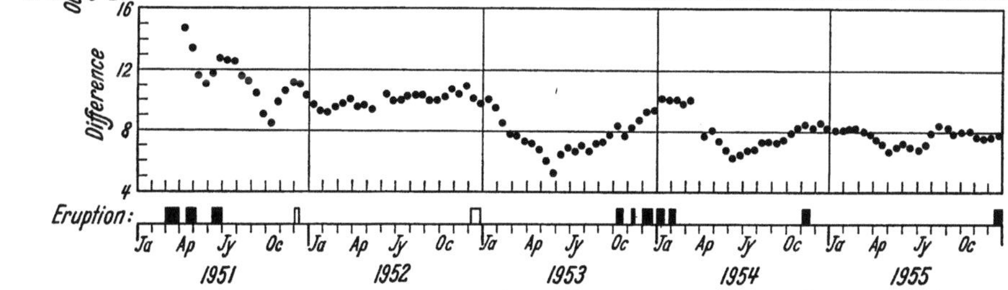

Fig. 24. The changes in 15-days' mean of geomagnetic declination observed at O-shima at O in Fig. 23. ●: absolute measurements. The periods of volcanic (black columns) and seismic (white columns) activities are also illustrated for reference. (After YOKOYAMA and RIKITAKE.)

Similar, but somewhat less reliable, results were obtained in Volcano Miyake-jima at its severe eruption in 1940 [*16*], [*20*].

As for the explanation of apparent demagnetization or magnetization, it seems most reasonable to suppose that heating and cooling within the volcano are responsible. If this view is accepted, the rough extent of the region in which the temperature exceeds the CURIE point (ca. 600° C) can be estimated. However, it is not an easy matter to understand such rapid changes in temperature. The heating may be explained by the injection of high temperature gas through crackes, but the cooling at such a rate would not be possible if heat conduction alone is taken into account. At any rate, subterranean thermal states may be inferred from geomagnetic changes. On the other hand, the change in geomagnetic field due to cooling of lavas at the earth's surface has been studied by many authors [16], [17], [22].

Field studies of magnetic and thermal behaviours of volcanoes and laboratory tests of magnetic and thermal properties of volcanic rocks would be interesting subjects in volcanology.

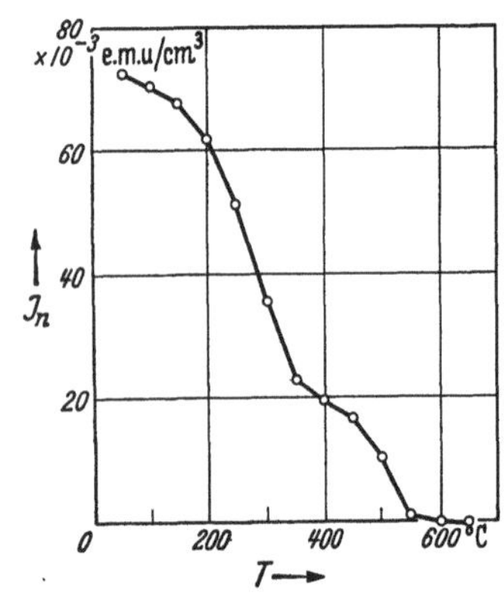

Fig. 25. Development of thermo-remanent magnetization of the O-shima lava in cooling (NAGATA).

15. Gravimetric investigations. The idea to use gravimetric methods in exploring[1] the subterranean structure of volcanoes is not a new one. Surveys by torsion balance or by gravimeter have been done on several volcanoes [26]. But it is not easy to infer the nature of the subterranean structures, because the corrections for altitude and for topographies are difficult, practically as well as theoretically. It is possible to learn about the average density of some protruded parts of volcanic bodies and the distribution of subterranean mass beneath the comparatively level ground. The density of Asama, for instance, was estimated at 1.88 from the torsion balance survey[2]. Change of gravity, if any, at the time of volcanic activity is an interesting subject to be studied. The change which amounted to 0.2 to 0.4 milligals was reported[3] at the 1950/1951 eruption of O-shima. The discussion of such small change calls for minute examinations on various sources of apparent gravity change not of subterranean origin.

16. Seismic prospecting of underground volcanic structures. Seismic refraction or reflection method is available to study subterranean structure of volcanoes, though actual examples are still rare. According to the seismic prospecting done in 1953 and 1954[4], the interior of the new dome of Usu, born in 1944, was found to be composed of rocks of relatively high velocity and the steep slope of the internal hard rock could be traced down to a certain depth. The velocity of the seismic waves of the new lava, which was measured in the laboratory from supersonic propagation, was not very different from the velocity of the surrounding rocks. However, YOUNG's modulus of the lava at high temperatures was found to be greater than that at room temperature. Though the physical interpretation for the increase of elasticity of the lavas and some other silica-rich glass is not yet made complete, the field results may be explained well by supposing that the intruded lava in the dome is still hot. This supposition is also supported by geothermal and geomagnetic surveys[5] made there at the same time.

17. Electrical phenomena. Atmospheric electric disturbances are usually observed at the time of volcanic explosions. Lightning discharge in the smoke

[1] F. SIGNORE: Bull. Volcanol., Ser. II **2**, 173—182 (1937).
[2] T. MINAKAMI: Bull. Earthq. Res. Inst. Tokyo **20**, 40—64 (1942).
[3] K. IIDA, M. HAYAKAWA and K. KATAYOSE: J. Geography, Tokyo **60**, 36—38 (1951).
[4] M. HAYAKAWA et al.: Geol. Surv. Japan Rep. **1957**.
[5] S. SAKUMA and T. MURASE: Geophys. Bull. Hokkaido Univ. **4**, 1—32 (1956).

and also radio communication disturbances are most common [12], [15], [22]. Electric charge within volcanic smoke is estimated from the observations of atmospheric electric potential gradient at the earth's surface. Principal change of the gradient under the smoke is to the upward direction. Negative charge seems to occupy the bottom part of the smoke probably. Change in the downward direction is observed rarely. The total charge (of one sign) in a smoke cloud from a mild eruption is estimated at 0.3 to 0.5 coulomb[1]. The magnitude of the charge may be variable with the intensity of the eruption. Though the origin of the charge is ascribed at present to frictional electricity on the basis of some experiments, reliable physical interpretation is not yet formulated.

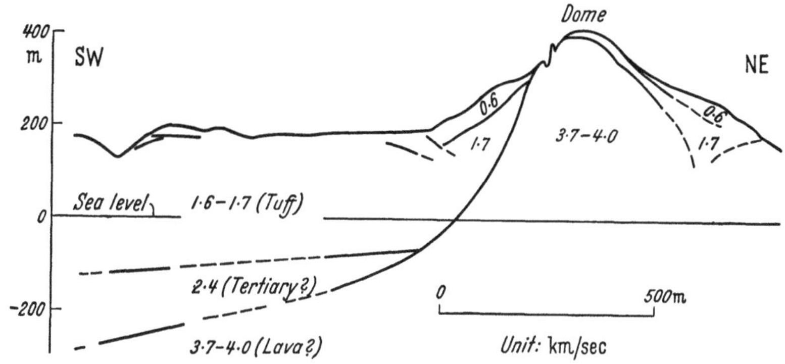

Fig. 26. Structure of the new dome of Usu determined by seismic prospecting (after HAYAKAWA).

Changes in earth-current are often observable at the time of volcanic activity, but the cause thereof is still unknown. Existence of mass of highly conductive lava (of high temperature) may affect the earth-current and other phenomena. Actually, diffraction of radio waves near hot lava mass could be observed[2] at the time of the activity of O-shima in 1950.

E. Volcanic energy and development of magma.

18. Gas content and heat of the erupted lava. Estimates of the rate of outflow of gas are made by measuring the dimension of fume cloud or by measuring the diameters of vents and the velocity of issuing gas. Having compared the rate of gas-issue with the amount of simultaneously ejected fresh lava, several authors obtained the proportion of gas to lava: 2.1 to 0.15% by weight for the Kilauea activity in 1950, 0.4% for the Mauna Loa activity in 1940, 0.7% for the 1938 eruption of Nyamuragira in Congo, 1.1% for the 1945 activity of Parícutin, Mexico, and 0.67% in the 1929 eruption of Vesuvius [17], [19]. Because of the many uncertainties in the data, the results represent only an order of magnitude. They do, however, serve to demonstrate that the *weight* proportion of volatiles to liquid or solid phase is small. Meanwhile, the *volume*-relationships are just the reverse, and accordingly, the important rôle played by the gas phase should be kept in mind.

The mass and heat of gases emitted by the explosions which do not accompany ejection of much lava are to be estimated by a method similar to the above, but the error in such an estimation will be large. The quantity of gases from the Valley of Ten Thousand Smokes, Katmai, was estimated at 2.6×10^7 l/sec (100° C,

[1] H. HATAKEYAMA: J. Meteor. Soc. Japan **21**, 49, 420, 426 (1943).
[2] T. RIKITAKE: Bull. Earthq. Res. Inst. Tokyo **29**, 150–159 (1951).

1 atm) and 2×10^6 ton/year for water-vapour and other gases respectively. A more precise method of estimation of the quantity of issuing gas should be developed in future. At those volcanic activities which are accompanied by ejection of lava-flows, rough estimation of heat escaping from the volcano may be done comparatively easily, from the observed mass of lava, temperature and assumed specific heat of the lava.

19. Energetics within volcanoes. It would be of some use to remark on the mode of transformation of energy within volcanoes. At first, the fact should be emphasized that transfer or transformation of energy is accompanied by transfer of a part of the working substances themselves from the working system to the worked system.

Let a closed surface be supposed which just envelopes in itself the upper part of the intruding lava mass. The surface can be imagined so adequately that all of the observed local crustal deformations and earthquakes may be caused by the mass within it. The mass within the surface is supplied with work W_1 and heat Q_1 from beneath, and it supplies work W_2 and heat Q_2 to the surrounding crust or to the air. The internal energy (in wide sense, including free energy) of the system itself may also well vary as much as ΔU. The acquired kinetic energy by the mass is written as ΔE. Of course, energy transformation within the system is out of question. Then

$$W_1 - W_2 + Q_1 - Q_2 = \Delta E + \Delta U. \tag{19.1}$$

The work exerted upon the surrounding earth's crust is classified into three sorts. The first among them is the elastic strain energy (Φ) stored in the surrounding rocks. The second is the seismic energy (S) caused probably by fractures due to intrusion, and the third is the energy (P_1) required to maintain the permanent deformation of the earth. These three are not mutually independent. The strain energy may be released by fracture, changing itself into seismic energy and potential energy. The total work done during a certain period (τ) is given by summing up the elastic potential energy in the crust at the last instant of the period, the total seismic energy emitted during the period, and the work done against the gravity by the permanently deformed crust.

$$W_2 = \int_0^\tau dW_2 = \Phi(\tau) - \Phi(0) + \int_0^\tau dS + P_1(\tau) - P_1(0). \tag{19.2}$$

The kinetic energy acquired by the lava mass itself to squeeze it out of a pre-existing vent or a new fissure is to be divided into the following two sorts; the potential energy required to bring the mass to the mouth (P_2) and its kinetic energy of ejection (K).

$$\Delta E = P_2 + K, \qquad K = \tfrac{1}{2} M v_0^2. \tag{19.3}$$

Next, the loss of heat (Q_2), by the issue of lavas or gases of high temperature and by conduction, should be taken into account too. Estimation of W_2, ΔE, and Q_2 may be thus possible theoretically. Practically, however, observations are needed on velocity and temperature of lava and gases, on seismicity, on crustal deformations over wide area, and on topographical changes. Estimations on some of the energies have been done by many authors, concerning heat, kinetic energy of bombs and so forth. In spite of that, estimation on the *whole* of the above mentioned energy dissipation throughout the course of a volcanic activity is rare at present. Further quantitative measurements available for the solution of the following problems would be of importance to volcanology; what kind of energy is principal among them and how much is the total energy required?

Simple estimations of volcanic energy lead one often to the conclusion that the heat energy (Q_2) is the principal item in the dissipating energy of various forms. This conclusion might not be denied, as far as only the first law of thermodynamics is concerned. To add the heat to the work may be acceptable, in order to obtain the total energy which is transferred from the interior of the earth. But, to compare simply the heat Q_2 with the work W_2 and ΔE is somewhat misleading, because the heat which is transferred to the surface by the ejected lava mass has very little potentiality actually to do work. The heat which is essential to volcanic activity is that which is lost to or added from beneath or surroundings (Q_1), resulting in a display of kinetic energy.

One will be able to estimate the total energy lost near the surface: $Q_2+W_2+\Delta E$ on the basis of various measurements. The total energy lost near the surface should have been derived from both the supplied work and heat from beneath and the decrease in the internal energy of the lava mass. The sign of Q_1-Q_2 or of W_1-W_2 would be variable according to the choice of the dimensions of the closed surface. But if it be assumed that the system can be so constructed as to make $W_1=0$ and $Q_1 \neq 0$, considerations become less confusing.

Is it possible to infer the loss or gain of heat Q_1 itself in the depths of the earth? One way has been opened by geomagnetic investigations (Sect. 14). The heat necessary to raise or lower the temperature of a certain subterranean volume through the CURIE point is not always beyond estimation[1]. Concerning the last term ΔU, knowledge is still poor. The origin of W_1 and Q_1 should be traced back to the changes in the internal energy of the subterranean mother magma. The changes in silicate melts will be referred to in the next section.

The mechanism of transformation of energy should be considered too. But as to the mechanism, the first law of thermodynamics gives no information. The mode of partition of energy can be discussed at least phenomenologically. For example, P_2/K for the lava-fountain activity equals nearly to h/h', where h and h' are the original depth of the lava and the attained height of the fountain respectively. The seismic energy of an explosion-earthquake is 10^{-2} to 10^{-4} of the kinetic energy of the ejecta, as far as the activity of explosive type is concerned. The mode of partition of energy at the instant of explosion is discussed here as an example. When one's eyes are fixed upon the site of explosion just before the event, the principal part of the energy is limited to the elastic energy stored in the surrounding medium. Just after the explosion, the elastic energy and decrease of the internal energy of gas become transformed to the kinetic energy of ejecta and gas, to the seismic and accoustic energy and to the work done to the air by expansion of the gas. Because the time taken in explosion is short enough for the process to be regarded as adiabatic, and because the seismic energy is far less than the kinetic energy, the decrease in the internal energy is equated to the sum of the acquired kinetic energy of ejecta and gas and the work done to the air by expansion. Then $W_1=Q_1=Q_2=0$. And

$$-\Delta U = \tfrac{1}{2}(M+M')\,v_0^2 + p_e \int dV. \tag{19.4}$$

For ideal gas,

$$J\{(c_{0v}T_0 - c_v T) + r\}\,M' = \frac{1}{2}(M+M')\,v_0^2 + \frac{M'}{A} R\,T. \tag{19.5}$$

where

M' = mass of gas,
r = latent heat,
c_v = specific heat of ideal gas,
A = molecular weight of the gas.

[1] I. YOKOYAMA: Bull. Earthq. Res. Inst. Tokyo **34**, 21, 185 (1956). Q_1 for the 1953 activity of O-shima was 10^{24} ergs, for example.

Then M' may be estimated approximately by substituting the constants from the observation of the kinetic energy of ejecta by taking T as a parameter. There is no available measurement of T just after the explosion and no ground to estimate T for the reason that its values should vary in accord with the path of gaseous work. Let it be assumed that $A = 18$, $c_v = 0.3$, $T_0 = 1300$ °K and $T = 500$ °K; the latent heat is not taken into account. For the explosion of Asama on April 20, 1935 ($M = 2.2 \times 10^{11}$ gr, $K.E. = 2.3 \times 10^{19}$ erg), M' is estimated at 3×10^9 gr, being 1.5% of the ejected lava. This discussion on such a great accumulation of assumptions is offered merely to illustrate the limit of application of the law of the conservation of energy.

20. Volcanism and development of magma. Characters and magnitude of volcanic phenomena are really influenced apparently by the *viscosity* of the intruding and extruding lavas. But when the phenomena of subterranean origin are discussed, the problem should not be treated from the viewpoint of viscosity only. Temperature and pressure at the time of intrusion and extrusion should be taken into account too. Intrusion or extrusion of lava is caused at least partly by the high internal pressure of the lava. The source of energy (including heat) which causes the various volcanic phenomena is, at present, to be traced back to phase transitions within the subterranean silicate melts of complicated composition, i.e., to the magma. The mode of partition of energy among the various forms also should be treated in relation with these materials. The apparent effects of viscosity on volcanic phenomena lead the students to the most important and complicated problem in volcanology.

The lava is the product of magma at its last stage of evolution. The magma is the parent source of the igneous rocks with various mineral associations. In order to understand and interpret the complex phenomena of crystallization, responding to changes in thermodynamical environment brought about by cooling or by changes in pressure, it is necessary to study the phase equilibrium relations in a large number of silicate systems. Items of fundamental information concerning the origin of minerals have been obtained mainly by the staff of the Geophysical Laboratory of the Carnegie Institution of Washington. (An extensive listing of papers will be seen in reference [*28*].)

Those physical behaviours which seem to be in the closest relation with volcanic activity are those due to concentration or separation of volatile substances in the melt. Some phase equilibrium relations in the silicate melts, in special connection with water-vapour phase, have been investigated to this date. The solubility of water in a granitic glass[1] at 900° C is 3.75, 5.75, 8.15, 8.90 and 9.35 weight per cent at 0.5, 1, 2, 3, and 4 kilobars respectively. The solubility of water in this glass decreases by 0.3% per 100° C rise in temperature in the 600 to 1200° C temperature range, i.e., at 800° C the solubility of water is 6.0%, at 1200° C it has decreased to 5.4%.

The ternary system H_2O—K_2SiO_3—SiO_2 has been studied[2] over the temperature range from 200° C to above 1000° C. The work comprised (a) a determination of the composition and properties of the various stable phases which can coexist with solution and vapour, (b) of the change in composition of these solutions with temperature, (c) and the approximate determination of the corresponding 3-phase pressure. The example concerning H_2O—$K_2Si_2O_5$ system shows that the existence of H_2O lowers the crystallization temperature of $K_2Si_2O_5$, that concentration of H_2O in remnant liquid phase increases with the development of crystallization, and that the pressure also increases as the crystallization develops with fall of temperature.

---

[1] R. W. GORANSON: Amer. J. Sci. **22**, 481—502 (1931).

[2] G. W. MOREY and C. N. FENNER: J. Amer. Chem. Soc. **39**, 1173—1229 (1917).

If the decrease in temperature takes place within the actual subterranean magma, its internal pressure will increase in a manner similar to that illustrated above and will result in "boiling"—a separation of gas phase. Decrease in external pressure may also be another cause of separation of gas phase. Though no experimental data are available at present concerning the actual magma, thermodynamical considerations are possible on the basis of the data on the fundamental silicate melts, and a study of this sort would be of much importance[1].

Computation of free energy of crystallization of minerals or of nucleation of gas vesicles would be not impossible. Really, some results in this direction have been reported concerning reaction relations among representative minerals.

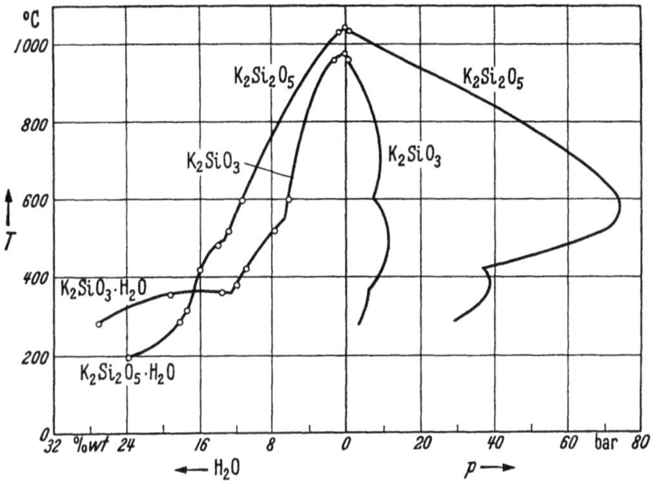

Fig. 27. The solubility or melting point curves of the binary systems $H_2O-K_2SiO_3$ and $H_2O-K_2Si_2O_5$ (left), and the $P-T$ curves for the monovariant systems (right) (after MOREY).

However, the thermodynamical relation within the magma seems to be so complicated that simple inference based on a monovariant system would be insufficient, even if one assumes the system to be in equilibrium.

In order to understand the volcanic phenomena in connection with development of crystallization in the silicate melt, the rate of nucleation and the velocity of growth should be investigated concerning minerals as well as volatile substances. This should be done because the actual volcanic phenomena do not take place under thermodynamical equilibrium, but they are transient essentially. The initial stage of eruptive activity would be the stage of nucleation and growth of bubbles in the silicate melt which is in non-equilibrium state.

When the number of vesicles is small and the melt is fluidal, large bubbles will escape from the surface of lava. In contrast to that, when the rate of nucleation of vesicles is extremely great, the bubbles will coalesce with each other and the result will be an explosion of ash-laden smoke. If the diffusion of gas molecules in the melt is also taken into account in connection with the growth of bubbles, the type of activity would be expected to be dependent more on nucleation rate of bubbles than on viscosity of the melt, because viscosity will affect both the growth and the rise of the bubbles[2]. Though the above supposition may be an audacious one, it would serve to emphasize the importance of nucleation.

---

[1] J. VERHOOGEN: Thermodynamics of a magmatic gas phase. Dept. Geol. Sci. Univ. Calif. Publ. **1949**, 91—135.

[2] J. VERHOOGEN: Mechanics of ash formation. Amer. J. Sci. **249**, 729—739 (1951).

Actually, the nucleation rate itself may have some correlation with viscosity of the melt. The rate (ν) of nucleation may be expressed as follows:

$$\nu \sim \exp(\Delta G/RT). \tag{20.1}$$

ΔG is a quantity which would be influenced by surface tension of the melt and by difference in chemical potential between the melt and the gas phase.

Accordingly, the studies of the physical properties of silicate melts should be developed in future, together with those of phase-equilibrium relations in the melts.

References.

Group I. Books and papers written principally from the viewpoint of classical volcanology, not of geophysics. They sometimes contain, however, detailed descriptions or original ideas on the phenomena of geophysical importance.

[1] WOLFF, F. v.: Der Vulkanismus, Bd. 1 u. 2., Stuttgart: Ferdinand Enke 1913—1929. — Various types of volcanic activities are described. Hypothetical interpretations are proposed on the basis of physico-chemical properties of rock melts, obtained to the date of publication. Supplementary volume contains geographical and petrographical descriptions on world volcanoes.

[2] WOLFF, F. v.: Plutonismus und Vulkanismus. In Handbuch der Geophysik, Bd. III, Liefg. 1. Berlin 1930.

[3] SAPPER, K.: Vulkankunde. Stuttgart: Engelhorn 1927.

[4] RITTMANN, A.: Vulkane und ihre Tätigkeit. Stuttgart: Ferdinand Enke 1936. — The classical standard book contains (i) classification of type of activities, (ii) description of volcanic products, (iii) form and structure of volcanoes, (iv) physico-chemical properties of magma. In the last chapter, is discussed the possible increase of gas pressure in the remnant liquid phase as a cause of volcanic activity, though not quantitatively.

[5] International Volcanological Association: Catalogue of the Active Volcanoes of the World. Napoli 1951. — Contains outline of geographical distribution, form and petrographical descriptions of lavas, and date of eruptive activities.

[6] SAPPER, K., T. A. JAGGAR and I. FRIEDLAENDER: Volcanology. In: Physics of the Earth series. Dover 1931. — Outline of classical volcanology is described well. Interpretations of geophysical observations are sometimes too hypothetical.

[7] LACROIX, A.: La Montagne Pelée et ses Eruptions. Paris: Mason 1904. — It contains detailed descriptions of the Pelée type activity. Seismic activity and crustal deformation are described too.

[8] FRIEDLAENDER, I. (editor): Z. Vulkanologie **1—17** (1914—1938). — This journal contains principally geological and geographical studies and descriptions of activities of many volcanoes in the world.

[9] PERRET, F. A.: Vesuvius Eruption of 1906. Carnegie Inst. Publ. No. **339** (1924).

[10] PERRET, F. A.: The Eruption of Mont Pelée, 1929—1932. Carnegie Inst. Publ. No. 4 **58** (1935).

[11] WILLIAMS, H.: Calderas and their Origin. Dept. Geol. Sci. Univ. Calif. Publ. **25**, 239—346 (1941). — This is familiar to volcano-geologists. Many geological and geophysical results up to the date of issue are referred to in this paper.

Group II. The books or comprehensive reports in which the results of expeditions or of continuous observations by geophysical methods are introduced principally. They are often referred to in this article.

[12] Royal Society of London: Eruption of Krakatoa and Subsequent Phenomena. London 1888. — The eruption of Krakatau in 1883, one of the most violent explosions in our history, is described by many physicists in this big volume. Those phenomena of worldwide influence, such as optical phenomena, tsunami and air-waves, are investigated especially well.

[13] JAGGAR, T. A.: Origin and Development of Craters. Geol. Soc. Amer. Mem. **21**, 1—508 (1947). — Contains detailed descriptions of activities of especially Hawaiian volcanoes and their geophysical studies, and a list of original papers.

[14] OMORI, F.: The Sakurajima Eruptions and Earthquakes. Bull. Earthq. Invest. Comm. Japan **8** (1914—1922).
OMORI, F.: The Usu-san Eruptions and Earthquakes and Elevation Phenomena. Bull. Earthq. Invest. Comm. Japan **5** and **9** (1911—1920).

OMORI, F.: The Eruption and Earthquakes of the Asama-yama. Bull. Earthq. Inv. Comm. Japan **6** and **8** (1912—1919). — The papers written by OMORI are not only of historical importance, but are also valuable at present. Especially the results of the precise surveys over wide areas around active volcanoes have been cited by many volcanologists.

[15] TSUYA, H., C. TSUBOI et. al.: The Eruption of Komagatake, Hokkaido, in 1929. Bull. Earthq. Res. Inst. Tokyo **8**, 239—319 (1930).
KOZU, S. et al.: The Eruption of Komagatake. Sci. Rep. Saito Gratitude Found. **15** (1932). — Topographical changes, crustal deformations and seismicity are investigated.

[16] TSUYA, H., R. TAKAHASI, T. HAGIWARA, T. MINAKAMI, T. NAGATA and S. OMOTE: The Eruption of Miyakejima in 1940. Bull. Earthq. Res. Inst. Tokyo **19**, 260—401 (1941). — The earthquakes, tremors, geomagnetic changes and crustal deformations are studied precisely, which associated with the flank- and summit eruption of the volcano.

[17] VERHOOGEN, J.: Les Eruptions 1938—1940 du Volcan Nyamuragira. Institut des Parc Nationaux du Congo Belge, Exploration du Parc National Albert. Fasc. 1, Brussels 1948. — Contains the physical studies on behaviour of fluidal lava, temperature, gas, geomagnetic change, spectrum of flare and other phenomena.

[18] MINAKAMI, T., T. ISHIKAWA and K. YAGI: The 1944 Eruption of Volcano Usu in Hokkaido, Japan. Bull. Volcanol., Ser. II **11**, 45—157 (1951). — Is a comprehensive report on geological and geophysical studies done prior to and during the birth of the new dome. The results of seismological and geodetic works are most interesting. Cites literature concerning the volcano.

[19] MACDONALD, G. A.: Activity of Hawaiian Volcanoes during the years 1940—1950. Bull. Volcanol., Ser. II **15**, 51—96 (1954).
MACDONALD, G. A.: Hawaiian Volcanoes during 1951; 1952. — U.S. Geol. Survey Bull. **996**-D, 141—215 (1954); **1021**-B, 15—108 (1955). — They describe the activities of Hawaiian volcanoes and the results of geophysical observations done by Hawaiian Volcano Observatory.

[20] MINAKAMI, T.: Report on the Volcanic Activities in Japan during 1939—1947. Bull. Volcanol., Ser. II **10**, 45—58 (1950).
MINAKAMI, T. and S. SAKUMA: Report on the Volcanic Activities and Volcanological studies concerning them in Japan during 1948—1951. Bull. Volcanol., Ser. II **14**, 79 to 130 (1953).
MINAKAMI, T.: Report on Volcanic Activities and Volcanological Studies in Japan for the period from 1951—1954. Bull. Volcanol., Ser. II **18** (1956). — In them, are compiled principally the geophysical investigations of the volcanoes which were active in the respective periods. The volcanoes are: Miyakejima, Usu, Sakurajima, Asama, Aso and so on. Listing of original papers will be seen in them.

[21] SIGNORE, F.: Le Ricerche Vulcanologiche in Italia nel Triennio 1948—1950. Bull. Volcanol., Ser. II **14**, 133—149 (1952). — An extensive listing of original papers concerning Italian volcanoes will be seen there.

[22] HAGIWARA, T., T. MINAKAMI, T. NAGATA, T. RIKITAKE, S. MURAUCHI, Y. HARADA, S. SAKUMA and others: The Eruption of Sakurajima in 1946. Bull. Earthq. Res. Inst., Tokyo **24**, 143—228 (1946).

[23] WILLIAMS, H., R. E. WILCOX et al.: Activity of Paricutin Volcano. Trans. Amer. Geophys. Un. **28** (1947). — Continuous observations of volcanic conditions are described.

[24] FOSTER, H. L. and A. C. MASON: 1950 and 1951 eruptions of Miharayama, O-shima Volcano, Japan. Bull. Geol. Soc. Amer. **66**, 730—762 (1955).

[25] MINAKAMI, T.: On Explosive Activities of Andesite Volcanoes and their Forerunning Phenomena. Bull. Volcanol., Ser. II **10**, 59—87 (1950).

[26] TAKAHASI, R. and T. NAGATA: Geophysical Studies of Volcano Mihara, Oosima Island, Part 1—5. Bull. Earthq. Res. Inst. Tokyo **15—17** (1937—1939).

[27] Netherlands East Indies Volcanological Survey, Bulletin. **1—34** (1927—1930). Monographs 1927—1940.
Annali del Osserv. Vesuv. 1924—1954. — They contain description of activities of the volcanoes, seismological and other observations.

[28] BIRCH, F. (editor): Handbook of Physical Constants. Geol. Soc. Special Papers **36**, 1—325 (1942). — Physical properties of minerals and rocks are tabulated with references.

Sachverzeichnis.

(Deutsch-Englisch.)

Bei gleicher Schreibweise in beiden Sprachen sind die Stichwörter nur einmal aufgeführt.

Subject Index.

(Englisch-German.)

Where English and German spelling of a word is identical the German version is omitted.

Table des matières

pour la contribution écrite en français:

J. Bricard: Diffusion des radiations par les gouttes d'eau en suspension dans l'atmosphère.

MIX
Papier aus verantwortungsvollen Quellen
Paper from responsible sources
FSC® C105338

If you have any concerns about our products,
you can contact us on
ProductSafety@springernature.com

In case Publisher is established outside the EU,
the EU authorized representative is:
Springer Nature Customer Service Center GmbH
Europaplatz 3, 69115 Heidelberg, Germany

Printed by Libri Plureos GmbH
in Hamburg, Germany